THE BIOLOGY OF INVERTEBRATES

McGraw-Hill Series in Organismic Biology

Consulting Editors

Gardiner: *The Biology of Invertebrates*
Kramer: *Plant and Soil Water Relationships: A Modern Synthesis*
Philips: *Introduction to the Biochemistry and Physiology of Plant Growth Hormones*
Price: *Molecular Approaches to Plant Physiology*

THE BIOLOGY OF INVERTEBRATES

Mary S. Gardiner

Professor Emeritus of Biology, Bryn Mawr College,
and
Research Associate, Department of Invertebrate Zoology,
Smithsonian Institution

McGraw-Hill Book Company

New York St. Louis San Francisco Düsseldorf Johannesburg
Kuala Lumpur London Mexico Montreal New Delhi
Panama Rio de Janeiro Singapore Sydney Toronto

THE BIOLOGY OF INVERTEBRATES

Library of Congress Catalog Card Number 70–37094

07–022725–x

4567890 HDHD 7987

The illustrators of this book were Leon Connelly, Carolyn B. Gast, and Elenore Schewe. This book was set in Baskerville, printed by Halliday Lithograph Corporation, and bound by The Book Press, Inc. The designer was Richard Paul Kluga; the illustrations were prepared by John Cordes, J & R Technical Services, Inc. The editors were James R. Young, Jr., and Susan Gamer. Sally Ellyson supervised production.

ὀλβιος ὅστις τῆς ἱστορίας
ἔσχε μάθησιν,
μήτε πολιτῶν ἐπὶ πημοσύνην
μήτ᾽ εἰς ἀδίκους πράξεις ὁρμῶν,
ἀλλ᾽ ἀθανάτου καθορῶν φύσεως
κόσμον ἀγήρων, πῇ τε συνέστη
καὶ ὅπῃ καὶ ὅπως.

Euripides fr. 910 (Nauck)

Happy is he who has knowledge from research and
does not turn to injury of his fellows or to
unjust deeds, but looks upon the ageless order
of eternal nature (to learn) in what way and
where and how it came to be.

CONTENTS

SECTION SEVEN: Mechanisms of Integration

SECTION EIGHT: Reproduction, Development, and Regeneration

PREFACE

This book has been a long time in the making. It was begun after many years of acquainting college students with the fascinations and challenges of invertebrate biology, years during which it became evident that what seems relevant biological information to most of them pertains particularly to vertebrates, especially man and the other mammals, or, now, to the molecular aspects of microorganisms and viruses. The aim of the book is to relate invertebrates to biology as a whole and to show some of the ways they deserve attention as members of Earth's population—for the most part highly successful members—as well as to point out the many areas of their biology that are still virtually unknown and that offer opportunities for research on basic questions applicable to biological systems of all kinds. It is intended for students who already have some background in science and general biology.

Obviously such a book cannot be all-inclusive. For one reason, more is known about some groups of invertebrates than others; the coverage of groups about which we know more is necessarily more thorough than that of phyla and species about which information is scanty. Whether the former groups may serve as adequate models for the latter is still to be shown. A second reason is that limitation of space precludes the possibility of presenting even all the available information extensively. Therefore, representative cases have been selected to illustrate general concepts; and specific cases have been selected to illustrate unusual, interesting, and (it is hoped) provocative facts. More complete information is to be found in the comprehensive treatises on invertebrate zoology listed in General and Special References.

Unlike other books that deal primarily with invertebrates, or with invertebrates and vertebrates together, this one does not treat them systematically. Rather, it treats them functionally, cutting across (but not disregarding) taxonomy to show how by multifarious adaptations they have adjusted to multifarious conditions on Earth. I have recognized, as any teacher must, that various aspects of biology have different appeals to students' interests; consequently, the invertebrates are considered here in respect to their structure, their ecology and ethology, and their biochemistry, physiology, and phylogeny, in the hope that the reader's attention and imagination will be captured by one or another of these facets. Little emphasis, however, has been put on gross anatomy, with the expectation that this will be learned where necessary by personal observation in the laboratory or elsewhere.

Some of this book was written at Bryn Mawr College during my final years of teaching there; some at the Mt. Desert Island Biological Laboratory, Salisbury Cove, Maine; and some at the Marine Biological Laboratory, Woods Hole, Massachusetts. It was completed in the Museum of Natural History of the Smithsonian Institution, Washington, D.C., where many courtesies have been extended to me as a Visiting Scientist and later as Research Associate in the Department of Invertebrate Zoology. I am especially indebted to Dr. Donald Squires, who was then Director of the Department and who originally made it possible for me to be associated with it; to Klaus Reutzler, David Pawson and Maureen Downey, who have continued to be sympathetic and generous hosts; and to the staffs of the Division of Echinoderms and Lower Invertebrates and the Library of the Museum, who have consistently shown me every consideration and helped me in many ways. I am grateful to the editors of the College and University Division of the McGraw-Hill Book Company for their patience and tolerance during the many years this book was in preparation, and for the constructive advice given me by their various reviewers and consultants.

Especial thanks are due to the artists who with endless pains interpreted my crude sketches and transformed them into works of art. Elenore Schewe, at Bryn Mawr, and Leon Connelly and Carolyn Gast, at the Smithsonian, bore with my efforts at representation and explanation, worried with me over techniques, procedures, and accuracy, and never complained about my impatience and demands. To their forebearance and skill in producing illustrations of such excellence is due much of the value of this book. The sources from which their drawings have been derived and adapted are listed at the end of the book (pages 883–886) and are also mentioned in the legends of the illustrations.

Others who helped materially in the preparation of the book are the typists, Mrs. G. Dawson Howell and Mrs. Edward Price, who worked long hours to turn my copy into clear and readable manuscript. I am indebted, too, to Mabel Lang, Professor of Greek at Bryn Mawr College, for identifying and translating the fragment from Euripides quoted just before the contents.

Information has been gathered from many books, reviews, and articles in scientific journals. The principal works of reference that have been used are listed at the end of the book (pages 881–883); at the end of each chapter there is a selected bibliography pertaining particularly to its contents.

I am grateful to the following for permission to use material from their publications:

Science, Washington, D.C.
Scientific American, W. H. Freeman and Company, San Francisco
The Macmillan Company, New York
G. and C. Merriam Company, Springfield, Massachusetts
The National Academy of Sciences, Washington, D.C.
The Rockefeller University Press, New York

I am indebted to the following for photographs:

The American Museum of Natural History, New York
The British Columbia Research Council, Vancouver, B.C.
Dr. Arthur L. and Dr. Laura H. Colwin, Queens College of the City University of New York
Dr. Thomas Eisner, Cornell University
The Field Museum of Natural History, Chicago
The General Biological Supply House, Chicago
H. Lou Gibson, Biological Photographic Association
The Institute of the History of Medicine, Johns Hopkins University
Dr. Alan Kohn, University of Washington, Seattle
Dr. Vivian T. Nachmias, Haverford College
The New York Zoological Society, New York
Dr. David Pawson, Smithsonian Institution, Washington, D.C.
Dr. H. G. Richards, Academy of Natural Sciences, Philadelphia
Dr. Waldo Schmitt, Smithsonian Institution, Washington, D.C.
U.S. Department of Agriculture, Bureau of Entomology and Plant Quarantine, Washington, D.C.
U.S. Department of the Interior, Fish and Wild Life Service, Washington, D.C.
Dr. Neal Weber, Swarthmore College

Mary S. Gardiner

section one

INTRODUCTION

chapter I THE DISTINGUISHING FEATURES, DISTRIBUTION, AND RELATIONSHIPS OF INVERTEBRATES

chapter I

THE DISTINGUISHING FEATURES, DISTRIBUTION, AND RELATIONSHIPS OF INVERTEBRATES

- A. DISTINGUISHING CHARACTERS OF INVERTEBRATES
- B. GEOLOGIC RECORD
- C. TYPES OF ENVIRONMENTS
  - 1. Aquatic environments: salt, fresh, and brackish water
    - a. AQUATIC HABITATS: BENTHOS, NEKTON AND PLANKTON
    - b. TRANSIENT PONDS AND POOLS
    - c. RIVERS AND STREAMS
    - d. SWAMPS AND BOGS
    - e. LITTORAL ZONE
  - 2. Terrestrial and aerial environments
    - a. TERRESTRIAL HABITATS: BURROWS, SURFACE, CAVES
    - b. AERIAL LIFE
- D. DISTRIBUTION OF INVERTEBRATES IN RELATION TO TEMPERATURE
- E. DISTRIBUTION OF INVERTEBRATES IN RELATION TO MINERALS
- F. EXPLOITATION OF NEW AREAS
  - 1. Invasions
  - 2. Methods of control of undesirable invaders
- G. BIOLOGICAL RELATIONSHIPS
  - 1. Predator-prey
  - 2. Temporary associations
  - 3. Habitual associations
    - a. COMMENSALISM
    - b. SYMBIOSIS
      - (1) Mutualism
      - (2) Parasitism

A. DISTINGUISHING CHARACTERS OF INVERTEBRATES

Division of the vast numbers of animals inhabiting earth into two major groups, Chordata and non-Chordata, is convenient and rational. Chordata are distinguished by the presence, at some time in their life cycles, of an internal supporting rod, the chorda (notochord), lying parallel and dorsal to the digestive tract. In most, this is replaced in adult life by a jointed and flexible column of cartilaginous or bony units (vertebrae), and the terms "vertebrate" and "invertebrate" have become practically synonymous with "chordate" and "nonchordate." Strictly speaking, however, they are not synonymous, for a vertebrate-invertebrate distinction excludes two small but highly significant groups of animals, Hemichordata and Urochordata (Tunicata). These two groups, often referred to as Protochordata, are not included among the types of animals considered in this book, which deals exclusively with invertebrates. These are distinguished from vertebrates by the following salient anatomical features:

1. Absence of a notochord or vertebral column and absence of an internal bony skeleton
2. Absence of a hollow dorsal nerve cord
3. Dorsal situation of the heart (if present)
4. Development of specialized respiratory areas (if present) from the body wall, not the digestive tract

Animals so characterized are extremely diverse in size, in body plan, and in devices for adaptation to their respective habitats. A brief classification of them is given on pages 150–171, to which, as well as to the index, the reader may refer, if necessary, for identification of the species mentioned in this chapter. Their body plans, particular histology, and adaptations are discussed in later chapters.

Not only are invertebrates more diverse in their anatomical features than vertebrates, but they are also far more numerous, in respect to both number of species and number of individuals. Over 90 percent of the animal species, both living and fossil, that have been identified to date are invertebrates. The disparity in number is probably even greater than this, for there are few, if any, fossil remains of soft-bodied invertebrates. The records of some may be preserved as imprints or casts, but estimates of the numbers of others that must have populated the earth at some time but have left no trace can only be intelligent guesses at best. Insects alone represent about 75 percent of all the known invertebrate species, with ants the most numerous of all land animals, both in regard to number of species and number of individuals. There are at present over 3,500 recognized species of ant, many of which live in social communities often including 100,000 to 200,000 individuals. And in a single colony of leaf-cutting ants of the tropics (*Atta*), there may be as many as several million members. Swarms of pasture mosquitoes (*Aedes nigromaculus*) in an acre of southern California may number 2,000,000.

Invertebrates are also much older inhabitants of Earth than vertebrates and more general in their distribution through the great variety of environments it provides. Indeed, there are few places in water, land, or air that have not proved suitable as permanent or temporary dwelling places for one or another species of invertebrate. A brief consideration of their geologic record, the types of environments available, and the relationships of invertebrates to them, and to each other, offer an introduction to this group of animals. Their biology, in terms of their adjustments and adaptations to the various environments in which they are presently found, is the central theme of this book and is presented in some detail in later chapters.

B. GEOLOGIC RECORD

The record of invertebrate life found in fossil remains is invaluable in giving biologists a picture of some of the events that must have taken place since Earth became a planet on which biological systems could originate, sustain themselves, and reproduce. One can read in it, even while recognizing its gaps, the story of the succession of plant and animal life, the evolutionary exploits, successes, and failures that have occurred, and the extent to which modern forms that can be collected and studied today have retained, modified, or abandoned ancestral characters (Table I–1).

There is little record of living things in the rocks that geologists identify as the oldest, although it is probable that soft-bodied, worm-like animals burrowed in the mud at the bottom of the warm, shallow seas that covered the greater part of the earth's surface over 500 million years ago. Rocks recently discovered in the Canadian Arctic have yielded fossils of clam-like molluscs and of worms, estimated by measurements of the extent of decay of rubidium^{87} and strontium^{87} to be at least 720 million years old. But the most abundant records are found in the Cambrian rocks, which show that marine representatives of all the major invertebrate phyla were established by that period, with sponges, brachiopods, and arthropods in greatest profusion (Fig. I–1). Indeed, the abundance of types is so great that it seems likely that more than half of the total evolution of invertebrates had occurred by Cambrian times. In Ordovician rocks there are traces of early coral formations, and these also give evidence that arachnids of sorts had evolved from the arthropod stock from which, not long after as geologic time is measured, the first air breathers also developed. These creatures, anatomically like scorpions and millipedes, were presumably the first animals to explore and inhabit the shore line and the drier atmosphere enveloping the earth, at about the same time that plants were making their earliest adventures as terrestrial systems. These land plants became established in the succeeding Devonian period, when spiders and wingless (apterous) insect types like modern springtails (Collembola) also lived freely on land. Then corals and brachiopods were particularly abundant in the warm seas that they shared with rapidly evolving fishes and small amphibia, the first recorded tetrapods. Corals and brachiopods seem to have changed but little throughout the centuries, for modern species are in all important features essentially like those that flourished in the Ordovician and Devonian periods. With the emergence of new land areas and their population by the rapidly growing "coal vegetation," there appears to have been rapid development among the arthropods. Spiders, scorpions, and centipedes were established as permanent types, as well as several hundred species of insects, all of them adapted to life on land. In the seas, gastropod molluscs were also in the process of rapid evolutionary development and, of other invertebrate types, echinoderms and foraminiferans were especially abundant. In the Permian rocks, representing the most recent period of the Paleozoic era, are found fossils corresponding to some modern insect species such as cockroaches, mayflies, and dragonflies. One of these, *Meganeura moneyi*, with a wing spread of 28 in., probably reached the greatest size ever to be attained by any land insect. This rapid evolution of

(*a*)

(*b*)

FIG. I–1 *Fossil invertebrates.* ***(a)*** *Bottom:* Lingula cincinnatus, *a fossil brachiopod from Ordovician rocks near Cincinnati, Ohio. Top:* Lingula unguis *Linne, a modern brachiopod from Australia.* (*Courtesy of Horace G. Richards.*) ***(b)*** Phacops rana, *a trilobite from Devonian rocks near Silica, Ohio. These early arthropods have a number of characters in common with modern isopod crustaceans, including the ability to roll up into a ball like the common pillbug,* Armadillidium. (*Photograph by D. F. Reath.*)

TABLE 1-1 Geological Time Scale

Era	*Period*	*Epoch*	*Beginning of interval*	*Principal events in invertebrate evolution*
CENOZOIC (recent life)	Quaternary	Pleistocene (most recent)	1 million years ago	Increase in cheilostomatous bryozoans; spread of gastropod and pelecypod molluscs; terrestrial annelids (earthworms); great spread of pterygote insects; first copepods
	Tertiary	Pliocene (very recent)	13 million years ago	Radiolaria and Foraminifera abundant
		Miocene (moderately recent)	25 million years ago	Pulmonate molluscs wide-spread
		Oligocene (slightly recent	36 million years ago	First centipedes (chilopods)
		Eocene (Dawn of recent)	58 million years ago	Testate amoebae (Apterygote insects) First moths and butterflies (Lepidoptera); extinction of belemnoids
		Paleocene (early dawn of recent)	63 million years ago	Decline of brachiopods; decline of cyclostomatous bryozoans
MESOZOIC (middle life)	Cretaceous (Latin *creta*, "chalk")		135 million years ago	Belemnoids abundant; great increase in cyclostomatous bryozoans; increase of eulamellibranch and prosobranch molluscs; increase of malacostracans, cirripedes, chilopods, insects and arachnids; decline of brachiopods; decline and extinction of ammonites; decline of dibranchiate cephalopods; marked decline of merostomatous arthropods; extinction of Stromatoporidae (Hydrozoa). First euglenoid protozoans; first gorgonians (Hydrozoa); first cheilostomatous bryozoans; first pteropod and octopod molluscs; first ants, bees and wasps (Hymenoptera).
	Jurassic (Jura Mts.)		181 million years ago	Radiolaria abundant; belemnoids abundant; great spread of ammonites; dibranchiate cephalopods abundant; echinoids abundant; Foraminifera with siliceous tests; invasion of freshwater by gastropod molluscs; rock and wood boring pelecypods; first dinoflagellates; first ciliates (Tintinnidae); first decapod cephalopods (squids).
	Triassic (from "tripartite" division in Germany)		230 million years ago	Belemnoids widespread; ammonoids (ammonites) flourishing; continued increase of diplopods, insects, and arachnids; first modern crinoids (Articulata); first modern reef building corals. First pennatulid Cnidaria; first cirripedes; first modern dragonflies (Odonata); first flies (Diptera).

TABLE I—1 Continued

Era	*Period*	*Epoch*	*Beginning of interval*	*Principal events in invertebrate evolution*
PALEOZOIC (ancient life)	Permian (Perm, a province in Russia)		280 million years ago	Brachiopods abundant; cockroaches (Orthoptera) widespread; increase of insects and arachnids. Extinction of rugose and tabulate corals; extinction of eurypterids, trilobites, and blastoids. First crickets (Orthoptera), mayflies (Ephemeroptera), beetles (Coleoptera), and bugs (Hemiptera).
	CARBONIFEROUS: Pennsylvanian		320 million years ago	Brachiopods abundant; millipedes abundant; Foraminifera with calcareous tests. First pulmonate gastropods (probable); first chilopods and winged insects (dragonflies and cockroaches).
	CARBONIFEROUS: Mississippian		345 million years ago	Gordiaceans ("horsehair worms"); dibranchiate cephalopods (belemnoids); holothurians. First opisthobranch gastropods; first ophiuroids.
	Devonian (Devonshire, England)		405 million years ago	Radiolaria abundant; great glass sponges abundant; eurypterids abundant; great development of brachiopods and increase of arachnids (spiders, mites, and pycnogonids). Decline of nautiloid molluscs and rise of ammonoids; decline of merostomata. Extinction of carpoids, cystoids, and somasteroids. First siphonophores; first branchiopods and malacostracans; first diplopods and springtails (Collembola). Thysanura(?).
	Silurian (the Silures, an ancient British tribe)		425 million years ago	Siliceous sponges abundant; growth of coral reefs widespread; great development of brachiopods; cystoids abundant; increase of ostracods. Decline of trilobites continuing. First scaphopod molluscs and first prosobranch gastropods; first air breathing arachnids (scorpions); myriapods; first mandibulate arthropods.
	Ordovician (the Ordovices, an ancient British tribe)		500 million years ago	Gymnolaematous bryozoans abundant; period of greatest differentiation of brachiopods; period of maximum development of trilobites (early) and of beginning decline (late). Nautiloids abundant, some with coiled shells; chitons (Amphineura); pelecypods (mussels, scallops, eulamellibranchs). Great spread of gastropods. Tubiculous annelids, with calcareous tubes (Spirorbis); ostracods, cirripedes, and malacostracans; blastoids, cystoids, crinoids, somasteroids, asteroids, ophiuroids, and echinoids (without Aristotle's lantern). Extinction of eocrinoids. First foraminiferans; first alcyonarians (Tabulata) and Zoantharians (Rugosa); first pyonogonids.
	Cambrian (Roman name for Wales)		530+ million years ago	Representatives of most invertebrate phyla already established. Radiolarians; sponges with siliceous spicules; hydrozoans (Stomatoporidae), scyphozoans (abundant), brachiopods with chitinous shells and articulate brachiopods with calcareous shells; gastropod molluscs with uncoiled, symmetrical shells; tetrabranchiate cephalopods (nautiloids) with straight or curved shells; annelids of several types, some resembling polychaetes; onychophorans; trilobites; xiphosurans and eurypterids (Merostomata); branchiopods; chaetognaths (possibly), eocrinoids, carpoids (stalked echinoderms), holothurians?
Proterozoic (earlier life)	Pre-Cambrian		?	Siliceous sponge spicules and trails of floating and crawling animals or plants found.
Ancheozoic (primitive life)	"Period of uncertainty"		?	Direct evidence of life lacking.

Source: Kulp, J. L. *Science*, 1961, v. 133, pp. 1105–1114—Geological Time Scale.
Moore, Lalicker, Fisher. 1952, *Invertebrate Fossils* (McGraw-Hill).
Kummel, B. 1961, *History of the Earth* (Freeman).
Shrock, R. R. and Twenhofel, W. H. 1953, *Principles of Invertebrate Paleontology* (McGraw-Hill).

insect types was paralleled by that of the cephalopod molluscs and, among the vertebrates, of the reptiles. Both of these became dominant types in the succeeding Triassic period, the reptiles populating the seas and streams and, as small dinosaurs, dominating the land fauna; among aquatic invertebrates, the now extinct ammonites were dominant, while crustaceans and echinoderms were evolving into modern types. Still later, in the Jurassic, when in the line of vertebrate evolution the great dinosaurs and flying reptiles were flourishing and reptile-like birds and primitive mammals first appeared, the ammonites reached their climax and soon became extinct. In this period and in the Cretaceous that followed it, the modern types of aquatic and terrestrial invertebrates were well established. For example, the species of ant that is most abundant in Baltic amber, the translucent, fossilized resin of Oligocene pine trees (*Pinetes succinifera*), is essentially identical with *Formica fusca*, the mound-building black ant that today is the commonest of its kind in Europe and North America (Fig. I–2). The discovery of two specimens of worker ants embedded in a lump of amber found at the base of a bluff in Cliffwood, New Jersey, and known to have been deposited in the Upper Cretaceous period, places the separation of ants from wasp-like insects as far back as approximately 100 million years ago. It also establishes the fact that social life had evolved among these insects at that time.

Evolution of the most successful vertebrates, the birds and mammals, apparently did not proceed to any great extent until the Cenozoic era when, in the Oligocene epoch, the extending areas of meadowland favored the evolutionary progress and specialization of avian as well as of insect species and grazing mammals. The mammals reached their greatest distribution and diversity in the Miocene epoch, but the human species may not have arisen until the Pleistocene, perhaps 2 million years ago, when many of the larger mammals were becoming extinct. The living world into which man emerged, whenever that event did take place, was therefore already populated with a multiplicity of living things. The oldest of these were the plants and invertebrate animals, which had, for a still uncalculable number of years, been testing their abilities to adapt to the many and changing environmental conditions on this planet. The success of the invertebrates is measured not only by their numerical superiority, but by their tremendous diversity in form and functional competence.

FIG. I–2 *Ant in amber. (Field Museum of Natural History.)*

C. TYPES OF ENVIRONMENTS

In order to support the life of any animal species, an environment must offer adequate food and, for most, a source of water, and the various chemical elements associated with the construction and biochemical operation of living systems. Other major factors determining the suitability of an environment are light, temperature, and pressure. These factors are quite different in different places on Earth and, to varying degrees, are inconstant in any given place. Light and temperature, for example, show both seasonal and diurnal differences in intensity, as well as local fluctuations of longer or shorter duration. Yet there are few places in water, on land, or in the air to which invertebrates of one kind or another have not become adapted during the long course of their evolution. Exceptional areas, where there is virtually no life, are some of the Norwegian fjords, the Black Sea, and a basin off the coast of Venezuela. Recently a "dead sea" area has been discovered behind the sand dunes of Cape Lookout, North Carolina. There is no circulation of water there and, owing to the decay of small organisms that drifted in through the only inlet and had no means of egress, there is so little oxygen that existence is impossible for almost every form of life.

1. Aquatic environments

These may be in salt, fresh, or brackish waters. In each case the inhabitants are subjected to different conditions, primarily in respect to mineral and oxygen content, pressure, light, and temperature.

SALT WATER (MARINE) ENVIRONMENTS Salt water covers over 72 percent of the earth's surface and in 50 percent of this area, equal to that of all the land masses combined, the water is over 2½ miles deep. The salinity of seawater is conventionally expressed as the total weight in grams of its salts per 1,000 grams (g) (liter) of "pure" water. At 20°C this (determined from samples taken from many areas of the oceans) is about 35 g/liter or 35‰. The composition of "average" seawater, in respect to its mineral salt content and its biologically most important ions, is given in Table I–2. Salinity is, in general, greater in equatorial than in polar regions, because of evaporation in the former and melting ice in the latter. Salinity in seas like the Mediterranean and the Baltic is different in different regions. It is higher in the eastern part of the Mediterranean than in the western, where the water mixes with that of the Atlantic Ocean. In the Baltic, salinity increases gradually from 5‰ at the head of the Gulf of Bothnia to 30‰ in the Kattegat where the Baltic opens into the North Sea. Ionic concentration in the great inland lakes such as the Dead Sea and the Great Salt Lake is as much as 200‰, more than six times that of average seawater. Salinity in all these bodies of water varies, too, with depth and is, in general, less at or near their surfaces than lower down.

In very general terms, the principal ions of seawater are represented in approximately the same proportions, but in different absolute amounts, in the cell and body fluids of animals. It is vital to living systems that these ions remain within a physiological range and be neither diluted nor concentrated, gained nor lost, to an extent that damages the effective operation of the system. Mechanisms for the regulation of ionic concentration and balance and the tolerance of different organisms to them are discussed in Chapter XIII. Ability for such regulation determines to a great extent the aquatic habitats in which an invertebrate species may live. This ability is so limited in some marine species of sponges, cnidarians, worms, and echinoderms that they are restricted to regions where "average" conditions prevail. Echinoderms, for instance, are practically excluded from the more dilute regions of the Baltic where, on the other hand, the polychaete worm *Arenicola* flourishes. The polychaete *Nereis diversicolor* can live in nearly freshwater in the Danish fjords, as well as in seawater with a salinity of 40‰. The mussel *Mytilus variabilis*, the crab *Neptunus pelagicus*, and the sea urchin *Centechinus*

TABLE 1-2 Composition of "Average" Seawater

Mineral salts
(Expressed in percentages by weight of total dissolved salt)

NaCl	$MgCl_2$	$MgSO_4$	$CaSO_4$	K_2SO_4	$CaCO_3$	$MgBr_2$	Fe, Sr, Silicates, Phosphates Nitrates
77.8	10.9	4.7	3.6	2.5	0.3	0.2	0.06

Most (biologically) important ions
(Expressed in percentages of total ionic concentration)

Na^+	Cl^-	Mg^{++}	Ca^{++}	K^+	CO_3^-	Br^-	SO_4^{--}
30.59	55.29	3.72	1.20	1.11	0.21	0.19	7.69

setosus live in the Red Sea where the salinity is never less than 40‰. Tremendous numbers of the brine fly *Ephydra* inhabit the shores of the Great Salt Lake, often in masses more than a foot wide that include about 370 million flies per mile. This fly has also been found in saltworks in water fully saturated with NaCl. The brine shrimp *Artemia* and the copepod *Tigriopsis* can also spend the major part—and indeed all—of their life cycles in pools saturated with salt.

FRESHWATER ENVIRONMENTS Freshwater environments include lakes and ponds, rivers and streams, and creeks and brooks, as well as smaller, often transient, bodies of water such as pools, puddles, and rivulets that tend to dry up when there is little rainfall. The concentration of inorganic salts in these varies considerably. The purest of freshwater is rainwater; the waters of ponds, lakes, rivers, and streams contain dissolved salts whose quantity and quality depends upon the nature of the underlying soil. Freshwaters are characterized as hard or soft according to their mineral content. This may be, on the average, 1.3 g/liter (1.3‰) in hard water and as little as 0.057 g/liter (0.057‰) in soft. Not only does the mineral content differ in different bodies of water but it alters from time to time in the same body with the seasons, local temperature conditions, and rainfall.

Invertebrates that live in freshwater have to adjust to a medium with a lower salt content than their cell and tissue fluids. Yet there are few invertebrate phyla in which some genera or species have not made the adjustment; notable exceptions are ctenophores, brachiopods, and echinoderms. But all bodies of freshwater support abundant populations of other kinds of invertebrates that spend all or part of their lives in or on the water or along the shore. A sample taken from any quiet pond or pool, or the backwater of a stream, will be found to be teeming with invertebrates of several kinds, and the microscopic examination of a drop of pond water can be a most exciting and illuminating experience.

BRACKISH WATER ENVIRONMENTS The mineral content of brackish waters lies between that of "average" salt and fresh. Brackish waters are found in estuaries and marshes, where there may be all degrees of mixing of salt and fresh, and where there may be areas, or times, when the water is as salty as that of the sea or as fresh as that of a river or lake. In marshes, which sometimes extend many miles inland as well as along the sea coast, the water inland may be fresh and that along the shore as salty as the adjacent ocean. In estuaries, the flowing water of the river may reach the ocean's edge at low tide, while at high tide the salt water may travel far up the river.

Animals that live in the brackish waters of marshes and estuaries must adjust to a saline or a freshwater medium at frequent intervals. Yet many accomplish this, like the flatworm *Gyratrix*, which makes the transition from a marine, brackish, and freshwater existence at least once a day. Sampling of the invertebrate population of a river mouth in the northern part of the Gulf of Mexico showed that bivalve molluscs (clams, oysters, and mussels), the gastropod molluscs *Neritina* and *Urosalpinx* (oyster drill), shrimps, and blue crabs, and even starfish lived there. Some of these were general in their distribution along the estuary, like the blue crabs that could tolerate the entire range of its salinity. Others were restricted to narrower limits and were found only in its upper or lower reaches.

a. AQUATIC HABITATS Aquatic animals are designated benthic, nektonic, planktonic, or littoral according to the areas they customarily inhabit. Animals that habitually live on the bottom, either at great depths or in the shallower regions along the continental slopes or shelves of the oceans or the sloping margins of lakes, constitute the benthos (Greek *benthos*, "depths of the sea"). These may be crawling or burrowing species, or ones that attach themselves to the bottom and lead a sessile or sedentary existence. Active and strong swimmers constitute the nekton (Greek *nektos*, "swimming"), and feeble swimmers and drifters, which depend for transportation largely upon currents and surface movements of the water, constitute the plankton (Greek, *planktos*, "wandering"). Littoral (Latin *litus*, "seashore") species inhabit the shore line of bodies of water. The conditions under which the benthos, the nekton, the plankton, and the littoral species live are quite different, but none has proved entirely restricting to one or another species of invertebrate. Nor are the distinctions hard-and-fast ones, and no limits can be set to the distribution of species within any one of these categories.

BENTHOS Conditions for the benthos are comparatively stable. Solar energy does not penetrate beyond the upper levels of oceans and freshwater lakes and ponds, so that in their depths the temperature is equable and the darkness continuous (Fig. I–3). In the Atlantic Ocean, for example, there are no seasonal variations in temperature below 600 ft, although the water becomes increasingly colder with greater depth, approaching or reaching 0°C in the abyssal regions. Similarly, the water at the bottom of lakes 200 ft deep or more remains near 4°C summer and winter (Fig. I–4). Benthic animals are subjected to high atmospheric pressures, for atmospheric pressure [at sea level equivalent to that of 760 millimeters (mm) of mercury] increases about 1 atmosphere (atm) for every 32 ft of depth. Therefore, an animal living 30 ft below the surface sustains a pressure of nearly 2 atm and one living a mile below, a pressure of about 166 atm.

By means of echo-sounding devices, very complete and accurate topographical maps have been made of the sea floor, and life in the abyssal regions is becoming better and better known through refinements of the older methods of dredging and trawling, the use of cameras equipped with electronic flash and triggering devices, and television. Yet the greater part of the

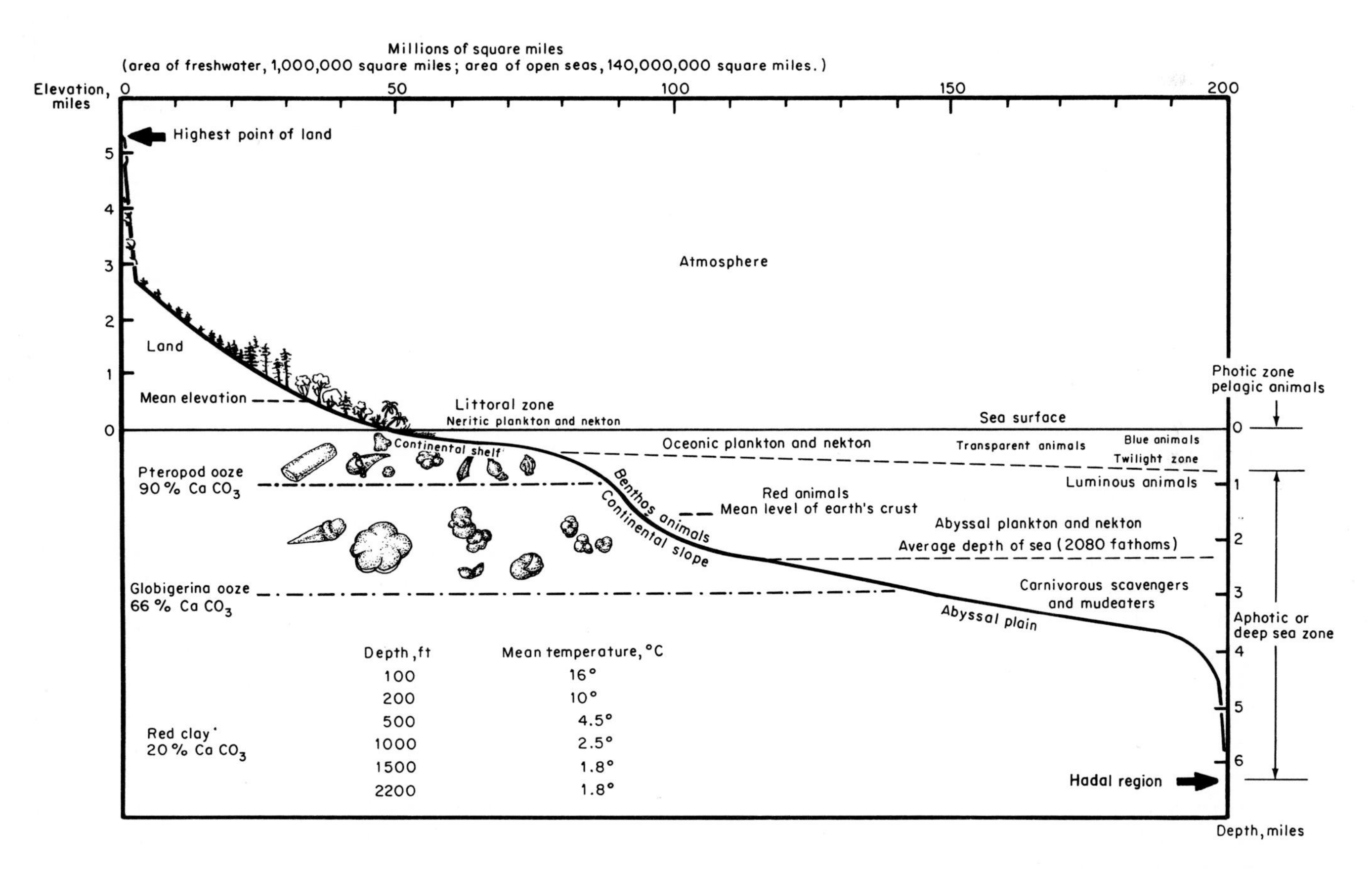

FIG. I–3 *The regions, depths, and bottom deposits of the ocean, and the altitude of the land. (After Pearse.) Animal life is distributed over the whole range of the ocean, while plant life is restricted to the upper 1,800 to 2,400 ft.*

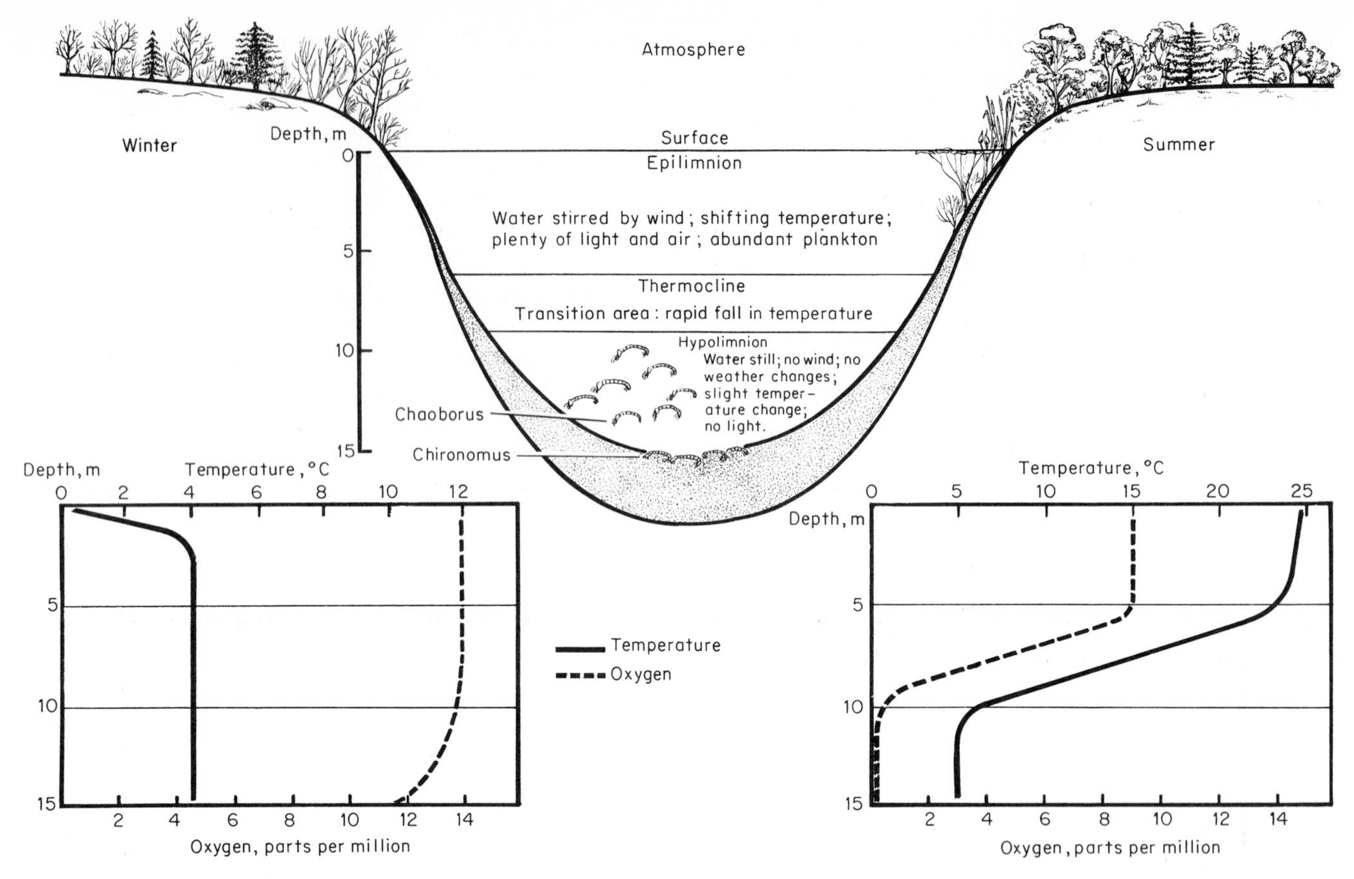

FIG. I–4 *Winter and summer conditions of oxygen supply and temperature in a pond. (After Deevey.) Lack of oxygen excludes most animals from the hypolimnion, yet in summer the transparent "phantom larvae" of the midge* Chaoborus *live there by day and migrate to the epilimnion by night, and the larvae of another midge,* Tendipes (Chironomus), *called "bloodworms" because of their hemoglobin, habitually live in the bottom mud.*

sea floor is still unsampled, and the depths of many lakes remain to be explored.

The deepest areas of the sea are those within the great trenches, cuts along the sea bottom that may extend for many miles. The maximum depth of the Philippine Trench is 6½ miles; that of the Mariana Trench, east of the Mariana Islands, 7 miles; and that of the Kermadec Trench, which stretches from the coast of New Zealand to the Tonga Islands, 5 miles. In samples taken from the Philippine Trench by biologists of the Danish vessel *Galathea* in 1951, eight distinct species were found, representatives of sea anemones, bivalve molluscs, polychaete annelids, amphipod crustaceans, and holothurian echinoderms. Eighteen different species were taken from the deepest part of the Kermadec Trench. These invertebrates live at a pressure of 1,000 atm, a temperature of near 0°C, and in stagnant, or at best very slowly moving, water. They are virtually cut off from contact with other marine species, and live in physiological and geographical isolation. About seven different species of invertebrates have been collected at depths of 2 to 5 miles in various oceans. These included hydroids and sea anemones, bivalve and gastropod molluscs, the scaphopod mollusc *Dentalium* (tooth shell), polychaete annelids, amphipod and decapod crustaceans, sea cucumbers, and sea urchins. Among the 21 species of polychaetes taken from depths of more than 3 miles were three not previously known. One species found in the Kermadec Trench corresponded to specimens collected in 1948 off the Canary Islands, on the other side of the world, by a Danish deep-sea expedition. Three other species corresponded to those found in shallow coastal regions, indicating that they must be able to stand tremendous differences in pressure and temperature.

Among benthic forms found along the continental shelf are sponges; alcyonarian corals and sea pens; polychaete and sipunculid worms; brachiopods; carnivorous gastropods; octopods; pogonophorans; crinoids; and ophiuroids. At the bottom of the shallower waters, along the shelf, are sponges; hydroids and sea anemones; and, in warm waters, madreporarian (reef-building) corals; flatworms; bryozoans; molluscs of all kinds; sipunculids and echiuroids; arthropods, represented by many kinds of crustaceans, by pycnogonids (sea spiders), and, on the Atlantic coast of North America and the southeastern coast of Asia, by the "horseshoe crab" *Limulus*; starfishes; sea urchins; brittle stars; and sea cucumbers (Fig. I–5).

Conditions at the bottom of a lake or deep pond are much like those in the depths of the sea. There is no light, and the temperature range at depths of 30 ft and below is 2 to 4°C, summer and winter. There is a great deficiency of oxygen, especially in summer when the warmer surface waters tend to stay on top without much mixing with those below; in winter, when the surface water cools and sinks, the oxygen concentration at these depths is greater but still rarely exceeds 0.012 to 0.014 milliliters/liter (ml/liter). Permanent benthic species include protozoans, nematodes, molluscs such as pill clams and snails, and oligochaete annelids such as *Nais* and *Tubifex*. Much of the benthos, however, is composed of larval and juvenile stages not only of these residents but of many insects that lay their eggs in water or on water plants. The newly hatched young crawl or sink down to the bottom and complete their development there. Prominent among these are the larvae of the midges *Tendipes* (*Chironomus*) and *Tanytarsus*. By day, larvae of the midge *Chaoborus* (*Corethra*) are also there, but these make nightly excursions to the surface and join the plankton. Nymphs and adult stone flies of the apterous genus *Capnia* have been dredged from the bottom of Lake Tahoe in the western United States. Presumably these plecopterans spend their entire lives crawling among the plants growing on the bottom, at depths of 200 to 264 ft, never emerging as others of their order do for even a brief terrestrial existence. The total population of lake or pond benthos may be very large, the weight of samples from a number of them averaging 10 grams/square meter (g/m^2). As much as 47 g/m^2 have been taken from the bottom of Lake Mendota near Madison, Wisconsin, and about 35 g/m^2 from Linsley Pond, near New Haven, Connecticut.

NEKTON AND PLANKTON Conditions for the nekton and plankton are very different from those for the benthos, and much less constant. There are variations, often wide, in light, temperature, pressure, turbulence, and turbidity of the water. The depth to which light penetrates into bodies of water has been measured by exposing sensitive photographic plates at different depths and by photometry, using an apparatus consisting of a photoelectric cell in a watertight case whose top is closed by a window that can be covered with appropriate filters. This instrument can be lowered into the water and the intensity of light at any selected level recorded on the deck of a boat by a meter connected with the photoelectric cell. Such measurements have shown that the transmission of light through water varies greatly with different conditions. Intensity of illumination at the surface and turbidity of the water are two of the most important variables. Transmission is greatest in the open ocean at or near the equator, where the sun's rays fall vertically on the water and there is little turbidity. It is progressively reduced toward the poles and during the ascent and descent of the sun as, in both cases, the rays strike the surface more and more obliquely. As light passes through water its intensity is reduced by reflection, absorption by the water and by particles in it, and by scattering. Loss by scattering is greatest in turbid water and in that in which there is an abundance of phyto- and zooplankton. Both photographic and photometric measurements have shown that the red rays of the spectrum are quickly absorbed and are lost 12 to 18 ft below the surface. Light in the green range is absorbed more slowly, and, in very pure waters, that in the blue-violet most slowly. In turbid water, how-

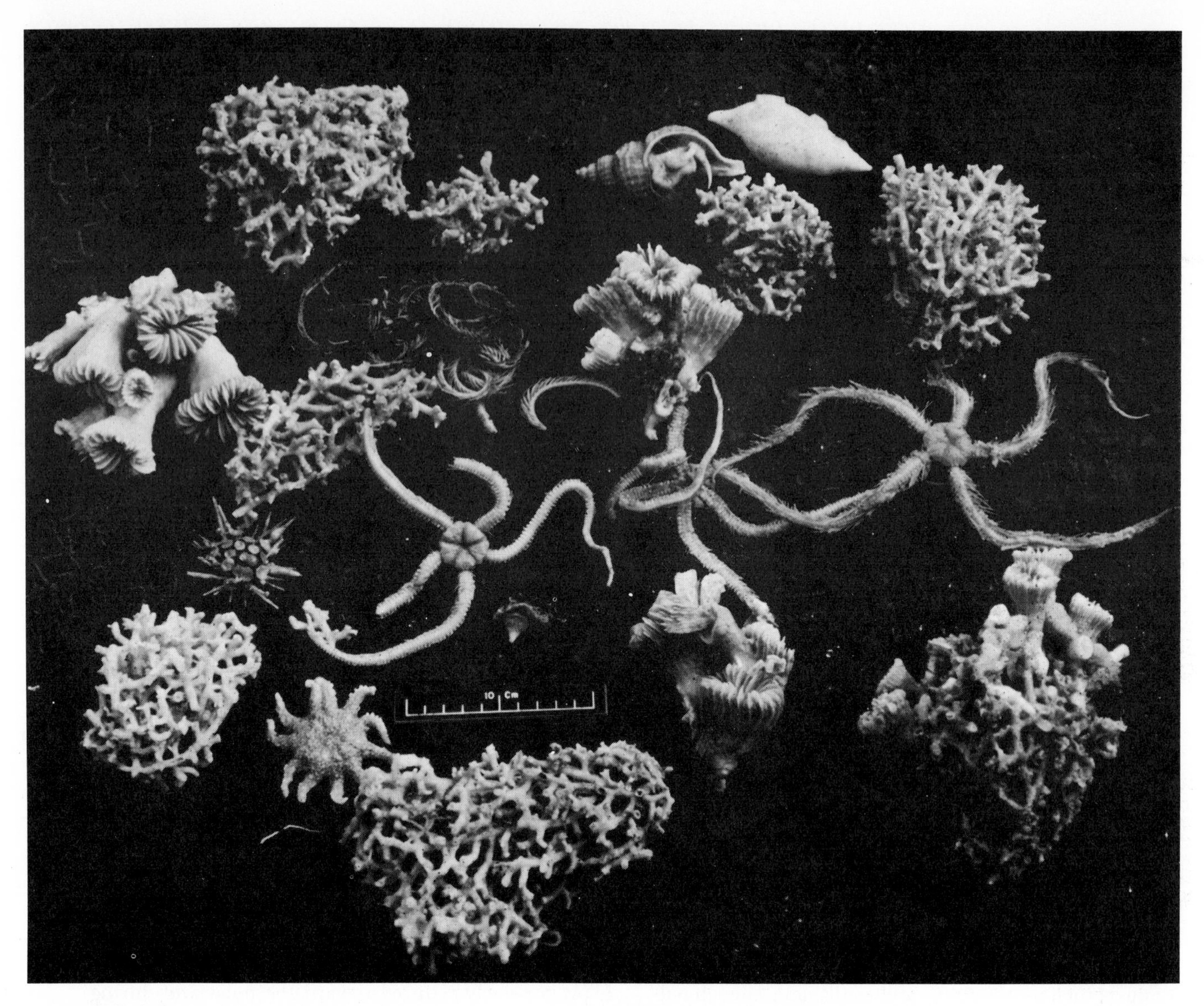

FIG. I–5 *Specimens, including corals, bryozoans, molluscs, and echinoderms, brought up in a net from a depth of 300 to 1,000 fathoms in the Campbell Plateau, sub-Antarctic waters off the coast of New Zealand. (Courtesy of Dr. David Pawson.)*

ever, green light penetrates further than blue-violet. But penetration of any light is never very great, and, in oceans, the photic zone is in general limited to the upper 1,800 to 2,400 ft. In freshwater lakes, where turbidity is greater than in the ocean, the photic zone is even less. This leads to a biological conditioning of aquatic environments, for chlorophyll-containing plants can only live where the light is adequate for photosynthesis. Animals therefore share the photic zone with plants, but not the waters below it.

Temperature of surface waters varies with that of the air above them, turbulence with the winds and air currents passing over them. These factors also affect the oxygen content. This depends upon:

1. Diffusion from the air with which the water is in contact

2. Turbulence
3. Solubility of the gas in waters of different mineral content
4. Local conditions

It has been calculated that if the distribution of oxygen depended upon diffusion alone it would take 42 years for a molecule that entered the water at the water-air interface to penetrate 812 ft below the surface. But winds, currents, and other agents that bring about displacement of surface waters mix them with those below, and oxygen is therefore distributed more quickly than by diffusion alone. The concentration of oxygen, however, even in the uppermost levels of water, is low compared with its concentration in air, where it represents about 20 percent of the gases in the mixture that composes earth's atmosphere. Salinity affects the solubility of oxygen in water, as does temperature. In freshwater, fully saturated with atmospheric gases, the oxygen concentration ranges from 10.3 ml/liter H_2O at 0°C to 5.6 ml/liter H_2O at 30°C and is 0 at 100°C. In the Atlantic Ocean, maximal values of 82 ml O_2/ liter H_2O have been found in the surface layers in the temperate zone and, in the intermediate and deep waters, values of 4.5 to 6 ml O_2/liter H_2O. Local conditions also affect the concentration of oxygen for nekton and plankton. Where plants are abundant in the photic zone, their contribution of oxygen as a by-product of photosynthesis is significant. On the other hand, especially in stagnant water, gaseous by-products of the activities of organisms of decay and decomposition tend to replace oxygen and so reduce its concentration. Respiratory adaptations to these conditions are discussed in Chapter XI.

Animals living in surface waters are subject to fluctuations in barometric pressure with changing meteorological conditions. They are minor in relation to the increase in pressure with increasing depth but still are evident some distance under water and give a certain degree of pressure inconstancy to these waters.

Invertebrates are but poorly represented in the nekton. Large shrimps and prawns are found in the nekton of tropical rivers, but there are few, if any, in the nekton of lakes and ponds anywhere. The predominant members of the marine nekton are cephalopod molluscs, such as squids and cuttle fish, and some crustaceans, such as the larger shrimps and swimming (portunid) crabs.

Marine and freshwater plankton, on the other hand, teem with invertebrates. There are two main areas of marine plankton:

1. Oceanic, in the open sea above the continental slope and abyssal plain
2. Neritic, above the continental shelf

The oceanic zone is by far the greater in extent and is largely free from land influences. The neritic zone, extending from the high tide level of the shore to the edge of the continental shelf, has depths of water ranging from inches to 600 to 700 ft, and is greatly influenced by the conditions of the land it borders and overlies. It is, in general, richer in zooplankton than the oceanic zone, but both support enormous numbers of animals representing a great number of species. In both, the population is fluctuating and shifting, subject to seasonal, monthly, daily, and even hourly changes. Samples of plankton, taken at different places and at different times in the same place, may be both qualitatively and quantitatively different, and probably no sample is a complete representation of all the animals that make up the plankton of a given region. Migrations are a factor in these shifts of planktonic populations; the vertical migrations of some plankters that move downward in bright light and upward in dim light are well-known, if not well-understood, phenomena (Fig. I–6).

Larvae and young of most invertebrates are transient members of the marine plankton, their presence and abundance depending upon the spawning habits of their parents and the respective lengths of their juvenile lives. At least 70 percent of marine invertebrates pass, in their life histories, through larval stages that have comparatively long planktonic lives. Temperature is an important factor in determining the larval composition of plankton; in northern seas, the planktonic life of a larva may last 2 to 3 months in the colder seasons but 2 to 3 weeks in summer.

Permanent or adult members of marine plankton include protozoans, cnidarians, ctenophores, nemerteans, chaetognaths (arrowworms), pteropod and heteropod molluscs, annelids, and crustaceans. Pro-

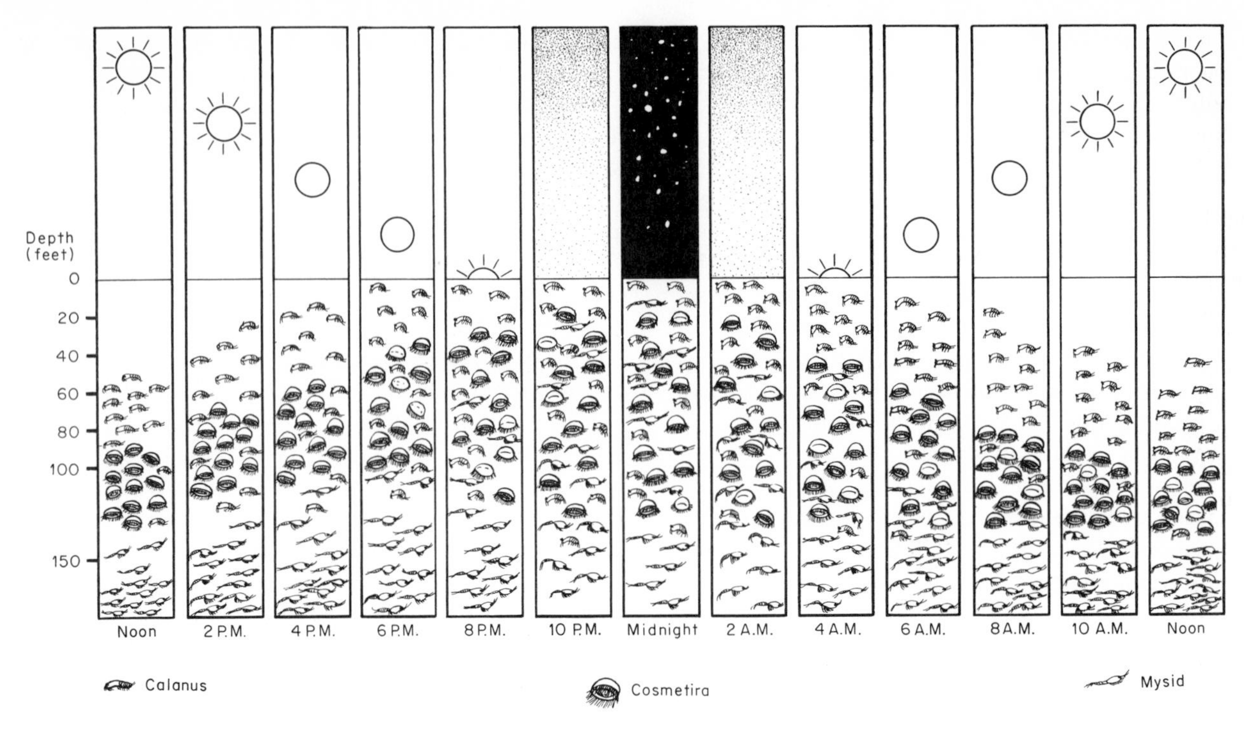

FIG. I–6 *The vertical migration of marine plankton in a 24-hr period. (After Russell and Yonge.) While the upper levels of the sea may be almost barren by day, they are teeming with life at night.*

tozoa are represented by foraminiferans, radiolarians, and dinoflagellates, often present in vast numbers. Samples of water dipped from the surface of the Gulf of California contained the large dinoflagellate *Noctiluca scintillans* (1 mm in diameter) in concentrations of 3,000,000 individuals per liter of sample. Jellyfish (scyphozoans) and ctenophores are abundant in most plankton, and pteropod molluscs, so numerous in certain areas that their delicate shells, sinking to the bottom when the animal dies, make conspicuous deposits there (pteropod ooze). In most areas, crustaceans, especially copepods and euphausiid shrimps, are dominant members of the plankton. The comparatively large copepod *Calanus finmarchicus* (up to 5 mm long) is particularly prominent in the upper waters of the northern seas. In southern seas, it is found at greater depths where the temperature corresponds to that of the surface waters off Greenland and Norway. An even larger species, *Calanus hyperboreus* (9 mm or more in length), is a surface plankter in arctic waters, where it is found at depths of 15 to 30 ft. But it is found half a mile down in more southerly waters, like the Norwegian Sea, the Atlantic Ocean off the Grand Banks of Newfoundland, and the Sargasso Sea, that great eddy in the North Atlantic between the coasts of the United States and Africa. This species of copepod must, therefore, be able to tolerate pressures from 1 to 100 atm. Shrimp-like euphausiids, the "krill" of fishermen, are an inch or two long and are often the only animals caught in plankton samples from both northern and southern waters. Indeed, these crustaceans may be so abundant that they have been held responsible for the "deep-scattering layer" (DSL), 150 to 500 ft in thickness, that has been detected by echo sounders over all the seas at depths of between 900 to 2,700 ft. Recent investigations, however, indicate that siphonophores, especially the Physonectidae and probably *Nanomia* (*Stephanomia*) *bijuga*, are more directly responsible. Observers in the U.S. Navy bathyscaphe *Trieste* in descents off San Diego, California, have seen concentrations of these cnidarians, estimated as great as 300 per 1,000 m^3, in zones corresponding to the upper and lower levels of the DSL. The float (pneumatophore) of *N. bijuga* is a small, gas-filled bubble, at the apex of the hydrozoan colony, which seems to have all the requisite properties of a perfect resonator. Expulsion of the gas, when the colony rises through the water, produces a column of bubbles that could also reflect sound.

Many members of the marine plankton are luminescent and their presence can be detected at night by the light they emit. The bodies of some, also, are brightly colored, so that large numbers of them make colored patches or streaks on the water. Dinoflagellates, particularly, are responsible for the "red water" often reported by fishermen, but the bodies of copepods and euphausiids may also make colored water if enough of them congregate together.

Freshwater plankton differs conspicuously from marine plankton in the absence of larval stages of bottom dwellers. The plankton of lakes and permanent ponds consists largely of flagellate and ciliate protozoans, rotifers, small crustaceans (copepods, ostracods, branchiopods, and, especially, cladocerans), and mites. A number of species also customarily attach themselves to the surface film of the water or glide along it. Hydras, for instance, fasten themselves to it by their pedal disks, with the column hanging down into the water, and mosquito larvae and pupae thrust their air tubes through it, keeping the rest of their bodies underwater. A small cladoceran, *Scapholoberis*, clings to it, ventral side uppermost. Planarians and snails glide along its under surface, and whirligig beetles and water striders skim or skate over its surface.

b. TRANSIENT PONDS AND POOLS Transient ponds and pools are smaller and shallower than lakes and permanent ponds and tend to dry up during periods of drought and at the end of summer. There is less variety of life in them than in permanent bodies of freshwater, for only those species that can successfully endure dehydration at some period of their life cycles can live there. Their zooplankton contains protozoans that can survive dehydration, often in an encysted state, and many species of rotifers and cladocerans whose eggs can withstand dehydration. Some water bugs, which can fly elsewhere when the water dries up, frequent transient bodies of water, on whose bottoms are frequently found larvae of those species of caddis flies (Trichoptera) that protect themselves from drought by digging deep into the mud and building around themselves cases of small pebbles, sticks, and leaves, cemented together with a secretion from their oral glands.

c. RIVERS AND STREAMS Rivers and streams (which may also be permanent or transient) offer an environment of moving freshwater. The current may be swift or slow, and the water may flow over sandy, muddy, or rocky bottoms and between steep or gently sloping banks. In general, there is little plankton in them, for the movement of the water, and often its turbidity, kills most of the small organisms that may enter from tributaries and backwaters. Rotifers are the only true river plankton, and whirligig beetles almost the only invertebrates found on the surface of moving water. But where the water flows gently, along the banks and in the backwaters and eddies, conditions approximate those of a lake or a pond, and a great variety of animals can survive there. Many invertebrates, too, live on the bottom of rivers and streams, especially along the shores of those with muddy or sandy bottoms where there is usually much vegetation. These inhabitants include protozoans, flatworms, bryozoans, bivalve molluscs, crustaceans (crayfish, shrimps, and prawns), and even some beetles.

Rocky bottoms, especially beneath rapids, present an environment in which only animals that can brace or fasten themselves in some way can withstand the rush of water. Among the stones and rocks of most freshwater rapids are found flatworms, snails, leeches, larvae of riffle beetles (Psephenidae) called "water pennies" because of their flat, copper colored bodies, larvae of midges and caddis flies, and nymphs of stone flies and mayflies. Even Niagara Falls abounds in mayfly nymphs.

d. SWAMPS AND BOGS Aquatic animals live also in swamps and bogs, especially when filled with water. Bogs, being shallower than swamps, tend to dry out sooner so that inhabitants vulnerable to dehydration often suffer destruction. In both, the water is stagnant and both are rich in vegetation and decomposing organic material. Conditions of life are exacting. The oxygen content of the water is practically nil, while that of hydrogen sulphide (H_2S) and other products of decomposition and reduction may be high, and there are wide fluctuations in temperature. Principally because of this, the animal populations vary with the seasons and may at times be very dense, although usually not very diverse. These populations include whirligig beetles, water striders, bugs, and spiders on the surface and, associated with submerged vegetation, protozoans, nematodes, rotifers, snails, ostracods, amphipods, isopods, water bugs, and beetles, as well as insect larvae and nymphs. Few animals are found on the bottom when the swamp or bog is water-filled, but, in times of drought, leeches, snails, insects, and other animals may bury themselves in the mud as protection against dehydration.

e. LITTORAL ZONE The littoral zone presents many hazards to invertebrate life. Animals living along the seashore are exposed to the pounding of the waves and the ebb and flow of the tides. On the sandy beaches, they live alternately in dry or damp sand and under water as the tide flows out and in. On rocky shores, they may be left on bare rocks or in temporary pools at low tide, exposed to light and air in the one case and to evaporating and warming waters in the other. Littoral species of lakes and ponds and streams are also in danger of being left high and dry when the water line for any reason becomes lower and, like those in temporary ponds and pools, are subject to the dangers of dehydration.

In spite of these hazards, many invertebrates make their homes on shores. Littoral inhabitants along the shores of oceans may include some of the neritic plankton left behind in pools or caves among the rocks when the tide goes out. Tide pools, especially those high on the shore, are very unstable environments, and their inhabitants are subjected to great changes in temperature, in availability of water, and in its salinity, when it is concentrated by evaporation or diluted by rainfall. Nevertheless, these pools are the homes of a great variety of invertebrate species—sponges, hydroids and sea anemones, mussels and snails, many kinds of crustaceans and starfishes, and sea urchins. Sandy shores, too, abound in invertebrate marine species that can burrow, like tubiculous polychaetes, clams and oysters, sandhoppers and similar crustaceans, and so, by either living permanently below the surface of the sand or retreating there when the tide goes out, can escape the more drastic effects of the changing tides. Along most coastlines the tidal zone also includes barnacles, mussels, limpets, and other snails. These animals can withstand the force of the water flowing in and flowing out by devices for attaching themselves firmly to some fixed object, and can withstand periods of low tide and exposure to air by withdrawing into their shells.

Along the shores of lakes and ponds, and even of rivers where the current is slow, many species may flourish, especially when there are rooted plants to which they may attach. Freshwater sponges, hydras, tubiculous rotifers, bryozoans, worms, snails, and mites all can live on the stems of plants or on their submerged leaves. Nymphs of dragonflies and damselflies and larvae of water bugs burrow among the roots of plants whose aerial portions may be well above the water's surface. Crayfishes live in crevices along the shore or in burrows that they make in clay or sandy banks.

Inhabitants of littoral zones may be able to lead a wholly aquatic existence or be forced by circumstance into an essentially terrestrial one. This may happen if they are washed up on the shore or if they are stranded there when the water for any reason recedes. Many doubtless perish through dehydration, asphyxiation, or chilling, but those in which protective devices have evolved can withstand terrestrial conditions for longer or shorter periods. Examples of these adaptive devices are given in later chapters.

2. Terrestrial and aerial environments

The land and the atmosphere above it offer a wide range of habitats and support a large and vigorous fauna. The most significant variables are light, temperature, and water. Oxygen, representing 20 to 21 percent of the mixture of gases that make up the atmosphere is sufficient at sea level at least, where its partial pressure (i.e., its share of total atmospheric pressure) is equivalent to 152 to 159 mm of mercury, with minor fluctuations with changing meteorological conditions. Pressure and oxygen, however, become gradually reduced with altitude, atmospheric pressure decreasing by about 0.03 atm for every 1,000 ft of elevation. At an altitude of 10,000 ft on a mountain slope or in the air, atmospheric pressure is equivalent to about 507 mm of mercury and the partial pressure of oxygen to about 102 mm. At an altitude of 53,000 ft, or rather more than 10 miles, both are zero. Oxygen deficiency is a major danger for air-breathing animals at high elevations, and lowered atmospheric pressure is also of considerable importance.

There are great variations in the intensity and duration of light in terrestrial and aerial environments. In every latitude, the length of day and night and the angle of the sun to the earth's surface change with the changing seasons. Inhabitants of open spaces and the air meet not only seasonal and diurnal changes in the intensity of light but also changes of longer or shorter duration according to the cloud cover. There is never much light in the depths of a forest or beneath any heavy vegetation, and none below the ground or in the recesses of deep caves. Similarly, there are seasonal and diurnal changes in temperature and local fluctuations of greater or lesser magnitude. Land dwellers may be exposed to periods of drought or reduced rainfall, with a consequent limitation of their water supply, or to torrential rains, with their ecological consequences of floods, washouts, and erosion.

Nevertheless, invertebrates are found at all altitudes up to at least 16,000 ft; they are found in open regions such as fields, meadows, pastures, and lawns; in shaded ones, such as forests and woodland; in caves; and below the ground. They are found in polar and tropical regions, where the temperature is almost uniformly cold or hot, and in temperate regions where the temperature rises and falls. And they are found in damp, humid regions and in periodically or permanently dry ones.

a. TERRESTRIAL HABITATS

BURROWERS Terrestrial burrowers, like aquatic plankters, are a shifting population, because many of them adjust their positions in relation to temperature and other climatic conditions. But any small sample of soil, depending upon its organic content and physical nature, may be expected to yield numbers of representative protozoans, flatworms, nematodes, annelids, collembolans, mites, and insects (especially coleopterans and hymenopterans), as well as the eggs and larvae of surface dwellers that use this area for nests and nurseries. Loamy soils, rich in organic matter, are more heavily populated than hard, clayey ones. Surveys of the top 10 in. of soil in English meadows have led to population estimates of several hundred million inhabitants per acre. Similar studies of the upper 5 in. of woodland in North Carolina, including the "litter" and humus levels of oak and pine forests, give an estimated population of 124 to 125 million per acre, with mites the predominant members. And samples of the soil population in a forest area of Denmark revealed that the macroscopic invertebrates alone numbered some 3,000 to 20,000 for areas 1 m square and 25 cm deep, with earthworms predominating in rich soil and arthropods in poor.

SURFACE DWELLERS Terrestrial surface dwellers are principally molluscs and arthropods, the former represented by snails and slugs and the latter by almost all the arachnid species (scorpions, mites, spiders, and daddy-longlegs), by isopod crustaceans (sowbugs and pillbugs), by centipedes and millipedes, and by nearly every order of insect. Spiders are especially abundant among surface dwellers; a census of them in an acre of rough grassland in Sussex, England, gave a population in late summer of something over 2½ million. The ubiquitous ant, dividing its time between the surface and its underground nest, also represents a large percentage of the surface population, as well as of that below the ground. Seventy-three nests of the Allegheny mountain ant alone were found in a 10-acre area in Maryland; the total population of ants in that locality was estimated at about 27 individuals per square foot, or some 1,200,000 per acre.

Seasonally, grasshoppers, crickets, and certain species of beetles may seem almost equally numerous and, in migrations, the estimated numbers of individuals temporarily in one place reach astronomical figures. A swarm of migrating butterflies, several hundred miles in width, may number millions at a time. During the invasion of Nebraska by "locusts" (actually the Rocky Mountain grasshopper) some years ago, the swarm was estimated to be 100 miles wide, 300 miles long, 0.5 to 1 mile high, and composed of more than 124 billion insects.

CAVE DWELLERS Cave dwellers live in an environment that is partly aquatic and partly terrestrial. The floor of a cave may be covered with water, enabling such truly aquatic species as amphipod crustaceans and crayfishes to live there. Or, in its inner recesses, the floor may be damp or even dry and the air humid or dry. Those that live deep within the cave are protected from winds and storms, are in constant darkness, and are in essentially equable, cool temperatures. In some caves food is abundant, even in the darkest regions, by the droppings (guano) of bats that hang in caves by day and during bad weather conditions. Little on the whole is known about the life histories of cave invertebrates or even of all the troglodytic (Greek *troglodytes*, "one who creeps into holes") species, but among regular inhabitants of caves are known to be (as well as amphipods and crayfishes) flatworms, snails, scorpions, spiders, daddy-longlegs, ticks and mites, millipedes, crickets, beetles, cockroaches, moths, and larvae of many kinds of insects. Some of these spend their entire lives there, but others, like the American cave crickets *Hadenoecus subterranus* and *Centhophilus stygius*, go outside at night to forage and return at dawn. Their droppings also provide food for the other inhabitants that never leave the cave.

b. AERIAL LIFE Insects are the only invertebrates that can be said to have adopted an aerial existence, and then only as adults. Aerial life is perilous, at least in the lower levels of the atmosphere, for winds and storms are hazards and even air currents present difficulties for small flying objects. Moreover, there is little food to be found there and those insects that spend most of their adult lives on the wing either, like mayflies, live on food stored during their juvenile aquatic stages or, like bees and many wasps, visit flowers to obtain nectar. Airplanes have made it possible for man to explore life in the air, especially in the upper levels, for by flying in different areas and at different heights with suitable traps, investigators have been able to collect and to identify much of the aerial population. This has proved to be large and distributed in two principal zones—the terrestrial (lower), up to about 1,000 ft, and the upper, 1,000 to 16,000 ft above the ground. Estimates of the number of invertebrates in a column of air extending 50 to 14,000 ft up over 1 square mile in Louisiana indicate that there is an average of 25 million individuals there throughout the year, with a range of some 12 million in January to 36 million in May. In the terrestrial zone are the larger and stronger insect fliers, such as dragonflies, mayflies, beetles, moths, butterflies, bees, and wasps. In the upper zone is the shifting population of the aerial plankton, composed of the weaker fliers and the drifters. It includes spiders and mites, as well as thrips, aphids, fleas, tiny flies, and hymenopterans. The cosmopolitan distribution of some species may be the result of the transportation of such airborne animals over considerable distances. The same species of ciliates, rhizopods, rotifers, tardigrades, and cladocerans are found on every continent; a possible explanation for this is that they have traveled from one place to another either independently in a dehydrated state or as passengers on the identified members of the aerial plankton. Either way they could have crossed geographical barriers otherwise impassable to them.

D. DISTRIBUTION OF INVERTEBRATES IN RELATION TO TEMPERATURE

It is well known that in a dehydrated state small invertebrates can endure extreme conditions, especially of temperature, that would be lethal to them in an active state. But there are some invertebrate species whose whole life cycles are spent at one or the other extreme of the physiological temperature range (approximately 0 to 50°C). An intensive survey of marine invertebrate life, conducted by members of the Arctic Research Laboratory in the vicinity of Point Barrow on the northern coast of Alaska, has revealed an abundance of life in the cold waters along the shores of the Arctic Ocean and the Beaufort Sea, where the annual temperature range is 1.8 to 4.0°C. Although the number of species represented here was less than in comparable regions of temperate and tropical waters, the number of individuals was far greater. The species identified included sponges, cnidarians, ctenophores, bryozoans, some pteropods and the clam *Hiatella arctica*, mysids, euphausiids, copepods and cirripedes, and echinoids and holothurians. On one small rock, dredged in late summer from a depth of 217 ft, 200 individuals and colonies, representing over 53 different species, were found to be occupying an area of about 11 in.^{2}. Equally well adapted to existence at the lower end of the temperature scale is the Alaskan stone fly, *Nemoura columbiana*, which lives continually at a temperature close to 0°C, mating on the frozen ceilings of ice caverns where its young also develop. At least 60 species of *Mesenchytraeus*, the glacier worm, live on the bare ice of the polar regions, sharing this bleak and apparently barren habitat with the "glacier fleas," which are really collembolans and not fleas at all. The orthopteran *Grylloblatta* lives habitually in the permafrost and dies when taken from it and held in even a chilly human hand. Live nematodes were thawed out from the ice by members of Sir Ernest Shackleton's expedition to Antarctica in 1909.

At the other end of the physiological thermal range, amebae are found in the surface scum of hot springs where the measured temperature is around 55°C; and larvae of the midge *Tendipes* live and mature at a temperature of almost 50°C in the pools of Yellowstone Park. The highly adaptable brine fly *Ephydra* can live and breed in these waters, as well as in the tide pools of the rocky Maine coast, and mites and tiger beetles, too, populate thermal springs and pools in a temperature range of 43 to 46°C. Polychaetes, molluscs, crustaceans, and echinoderms live in the Persian Gulf, where the temperature of the water may reach 35°C, the highest temperature recorded for any of the seas. An isopod, *Exosphaeroma thermophilium*, lives in hot springs and the amphipod *Gammarus limnaeus* is, like *Ephydra*, adaptable to a wide temperature range, for it can live both in warm springs and in cold arctic ponds. The tiny crustacean *Thermosbaena mirabilis* has been found crawling about on the shaded rocky walls of a hot spring near Tunis, North Africa, where the temperature reaches 44°C. Another member of this order (Thermosbaenaca), *Thermobathynella adami*, has been reported from a spring with a temperature in the vicinity of 55°C.

But the majority of invertebrates find their best living conditions in more temperate environments. Since none has a precise internal temperature-regulating mechanism like that of birds and mammals they meet the temporary rises and falls in external temperature by adjustments that may be considered either physiological or behavioral, a distinction that is one of degree rather than of kind. Among the physiological adjustments are the very general response to temperature drops of reduced metabolic activity and the equally general response to temperature elevations of increased metabolic activity. In most invertebrate species, increased metabolic activity is reflected in increased reproductive activity, so that perpetuation of the species is provided for, in addition is guaranteed, even if the existing adults are unable to survive their immediate environmental conditions. Encystment is also often a means of escape from too exacting temperatures; in this way an animal protects itself by secreting a firm case around its body in which it exists during the period of stress at a minimal level of metabolic activity. Others, with permanent shells or cases, may withdraw into them, shutting themselves off more or less completely from external conditions. Or they may show gross behavioral responses by moving from an unfavorable area to one that is better suited to them. Thus aquatic species move from one level to another, ground dwellers burrow deeper in extremes of cold and heat, and aerial species may migrate. Butterflies are best known and probably most dramatic in their migratory responses to temperature conditions. In North America the Monarch, or common milkweed butterfly, regularly flies south in the autumn and some, at least, make a return journey of 1,000 miles or more the following spring to keep in a favorable temperature zone.

Some invertebrates have means of regulating the temperature of their surroundings, and so of their bodies, without changing their locale. When the temperature around them falls they may increase their metabolism, especially of their muscles. In low temperatures, contractions of the thoracic (flight) muscles have been recorded in bees whose wings, nevertheless, remained stationary. Essentially they were shivering.

Bees and other social insects conserve metabolic heat by huddling together in the hive or nest. With a marked drop in temperature, bees take turns in flying around the cluster and especially around the brood cells to provide additional heat. This is one way in which they keep the hive between 34.5 and 35.5°C, the optimum for development of the brood. They cool the hive, when the outside temperature is above this, by fanning the air with their wings, and when the outside heat is considerable, by making use of the evaporation of water. Certain members of the colony act as water collectors and hunt and suck up water instead of nectar. On their return to the hive they are met by other workers who eagerly draw the water from their crops and deposit a drop of it in each brood cell. Evaporation is promoted by keeping the air in motion with their wings and, if the heat is extreme, the surface of the drop is enlarged by spreading it out as a thin film which is renewed as soon as it vaporizes. By this means the hive can be kept at an even 35°C, even when it is in full sunlight in an external temperature of 70°C.

Terrestrial arthropods also make use of evaporation as a means of lowering the temperature around their bodies when the external temperature is high. Those that live in moist places may take advantage of the external water for this, but others must supply it from their own bodies by transpiration (Latin *trans*, "across," and *spirare*, "to breathe"), which may occur over the general body surface if the cuticle is permeable; in insects the external openings of the respiratory tubes are the usual sites of water loss. The advantage of cooling the body by the evaporation of transpired water must be very delicately balanced against the disadvantage of excessive water loss. Laboratory studies upon isopods under controlled conditions of temperature and relative humidity have shown that their body temperatures can be lowered through evaporation by 1 to 7°C in 30 min. Larger species with greater water reserves, or slower rates of transpiration, can cool themselves in this way longer than small ones. It remains to be shown how effectively this means is used in natural conditions, but it seems quite evident that there is a mechanism available by which the body temperature can be lowered when the external temperature is high, at least for a short period of time.

E. DISTRIBUTION OF INVERTEBRATES IN RELATION TO MINERALS

The presence and concentration of minerals, directly reflecting the composition of the earth's crust, are also important factors in determining the extent to which a locality can support life and the kind of life it does support. The essential elements, designated macronutrients because they are needed in relatively large amounts, are carbon, hydrogen, oxygen, nitrogen, potassium, calcium, magnesium, sulfur, and phosphorus. Micronutrients (trace elements), needed in only very small amounts, include boron, chlorine, cobalt, copper, iron, manganese, molybdenum, sodium, vanadium, and zinc. Iodine is also an essential trace element for many organisms, and aluminum, lead, and nickel are at present suspected of being so.

The biological roles of these elements have only been partially elucidated. Many are known to be combined with organic molecules as structural components of the organism as a whole or as compounds that are vital to the operation of its cells. Sulfur is a component of many proteins and is found in mucopolysaccharides. There is iodine in the scleroproteins of polychaete annelids and in the integument of arthropods, as well as in the organic skeletons of sponges and some cnidarians. Phosphorus is a component of nucleotides and nucleic acids, and iron is included in the prosthetic group of many enzymes as well as in the heme moiety of cytochromes and hemoglobins.* Hemocyanins, probably the most important of invertebrate respiratory pigments, contain copper. The copper ion (Cu^{++}) also participates in some enzymatic reactions. It may be necessary for the synthesis of cytochrome and hemoglobin, and it is known to be important in the hardening or tanning of arthropod skeletons. Other ions are also implicated in enzymatic reactions which, in their absence, either do not occur at all or else do so with decreased efficiency. Molybdenum is needed for the effective utilization of copper and may also be important in some enzymatic reac-

* Cytochromes are enzymes that operate in the final stages of the transfer of electrons to molecular oxygen and so are components of most cells that derive energy from aerobic respiration. Hemoglobins are respiratory pigments in the blood and coelomic fluids of some invertebrates, and in the muscles of a number of species.

tions, as magnesium and manganese are known to be. Calcium acts as a regulator of membrane permeability and so to a great extent determines what may enter or leave a cell. It is also incorporated into the shells and exoskeletons of many invertebrate species, where magnesium is also frequently found. Cells contain potassium and sodium, and all tissue fluids contain their chloride salts.

All these elements are included in the 48 in sea salt, the residue of seawater after evaporation, and most of them are found, in smaller amounts, in freshwater. Aquatic animals may thus derive them from the water around them by direct absorption through their permeable surfaces or from their food; terrestrial animals can only obtain them from the water they drink and the foods they eat, whose content depends upon the nature of the soil in their environments.

In any area the distribution of these elements, largely as their mineral salts, may be a limitation to the kind and number of species inhabiting it, for there may be too little of a necessary element or so much that it is toxic rather than beneficial. A region deficient in calcium cannot support a population of animals that use it for their shells. Only about 5 percent of the molluscan species of temperate regions can obtain enough calcium for their vital processes and for shell deposition from very soft water with a calcium content less than 3 mg/liter. About 40 percent can live in moderately soft water with a calcium content of 3 to 10 mg/liter; about 55 percent can live in water with a calcium content of 10 to 25 mg/liter; and 3 to 5 percent can survive only in hard water with a calcium content greater than 25 mg/liter. But most of the species that can satisfy their calcium demands in soft water can also live in hard water, sometimes reflecting this difference in habitat in the thickness and mass of their shells and in their growth rates.

Direct correlation between the copper and iron content of seawater and the geographical distribution of oysters has also been conclusively shown. Fixation of motile oyster larvae as "spat," as well as their metamorphosis to adults, takes place only where the concentration of copper is above 0.05 to 0.06 mg/liter, so that oyster beds do not become established where the copper concentration is below this. Barnacle larvae also require a certain amount of copper for attachment, but concentrations above the optimal level prevent this and may even prove toxic to the larvae. However, there is a good deal of difference in tolerance to copper, which seems to depend upon the evolution of mechanisms for eliminating it or for storing it in inactive form when it is present in the environment above the animal's immediate needs. The mussel *Mytilus*, the abalone *Haliotis*, and the snail *Astraea*, inhabiting Pacific waters where the natural concentration of copper is about 0.001 mg/liter, have lived successfully in water in which the copper concentration has been experimentally raised to 0.10 to 0.20 mg/liter. The clam *Paphia* has survived as long as 60 days in water containing as much as 3.0 mg/liter. The sensitivity of barnacle larvae to copper has been exploited in the protection of ships' hulls and propellers, and underwater installations from encrustation by them and other marine organisms such as molluscs and bryozoans. Copper, or a mixture of mercury and copper, added to the paint used on them will prevent the animals' attachment and so their fouling. But barnacles, can recover from copper poisoning when restored to ordinary seawater, for it has been experimentally shown that they eliminate as much as two-thirds of the element that they have absorbed.

F. EXPLOITATION OF NEW AREAS

1. Invasions

Modern invertebrates, especially arthropods, have shown themselves extremely competent in the invasion of new areas, adaptation to them, and exploitation of them. Some of these invasions have occurred by chance through natural causes; others have come about directly or indirectly through the agency of man. Evidence for the dispersal of many species of small size, either as adults or juveniles, through winds and air currents is offered both by analyses of the terrestrial and planktonic zones of the atmosphere and by the discoveries of insects far out at sea. Spruce aphids, for example, found on the snows of Spitzbergen, must have been carried from the forests of the Kola Peninsula 800 miles away. Birds and mammals in whose feathers or fur small invertebrates may be entangled are natural vehicles for transporting species over distances they could not cover by themselves. Man, too, has provided means for the potential worldwide distribution of species of all kinds in the vehicles he has devised for his transport and communication needs. In some instances, his introduction of species into new areas has been deliberate, in order to increase, improve, or vary his food supply or else to combat a species destructive to his crops, his domestic animals, or himself.

Some of these introductions have proved fruitless but others have brought in species that have sooner or later become acclimated and widespread. Many have adopted new sources of food and wrought great changes in the ecological picture of the area. For example the Colorado potato beetle, *Leptinotarsa decemlineata*, native to the eastern part of the Rocky Mountain area from Colorado to Mexico, feeds upon the weed *Solanum rostratum* (sandbur or nightshade). A balanced relationship between insect and plant was preserved until about 1859, when the white potato, *S. tuberosum*, began to be cultivated there. *Leptinotarsa* speedily forsook the native species of *Solanum* for the newcomer, and throve upon it, doing great damage to an important food crop. By 1874, its feeding and breeding grounds had extended east to the Atlantic coast. Its invasion of Europe began in 1920, when some beetles were accidentally admitted at the port of Bordeaux. By 1935, their offspring had extended throughout France, and by 1956 had crossed the Pyrenees into Spain and Portugal and spread south through northern Italy and also westward through an extensive area.

Mischance first and then natural methods of distribution are responsible for the spread of the gypsy moth, *Lymantria* (*Porthetria*) *dispar*, whose larvae (caterpillars) feed voraciously upon trees of many kinds. In 1869, some eggs were brought to his home in Medford, Massachusetts, by a French astronomer working at the Harvard Observatory, who was interested also in silk-producing caterpillars. Adult females, although winged, are incapable of flight so that dispersion of the species seemed unlikely. But their larvae are so light and buoyant that they have been trapped by airplanes at heights of 2,000 ft. A few larvae accidentally escaped from the Medford collection and were the progenitors of a rapidly growing population of moths that ultimately spread all over Massachusetts and the rest of New England, as well as over New York, New Jersey, and parts of Canada. As each brood was hatched, the caterpillars swarmed over oak, willow, poplar, apple, and many other kinds of trees, defoliating them to an extent from which many could not recover. Invasion of Maryland, Virginia, and neighboring states is now threatened.

The introduction and acclimatization of the Japanese beetle, *Popillia japonica*, is another example of accidental arrival and rapid spread of an insect pest (Fig. I–7). *Popillia* is rarely a pest in its native Japan, where there are natural checks to its overpopulation, but it rapidly became one in the United States, feeding on over 250 species of plants, including many fruit and shade trees as well as ornamental plants and food crops, and defoliating them (Fig. I–8). A few adults, probably brought in with a shipment of irises or azaleas from Japan, were first noticed in 1916 in a New Jersey nursery; in the following year they occupied about an acre, in the next year about 3 miles, all around the original spot. Each year the spread was increased in concentric circles until, in 1923, the infested area included 2,442 square miles in New Jersey and Pennsylvania. In 1941 it extended over 20,000 miles in Pennsylvania, New Jersey, Delaware, and parts of Virginia, New York, and Connecticut; more recently, the beetles have reached North Carolina, West Virginia, and Ohio.

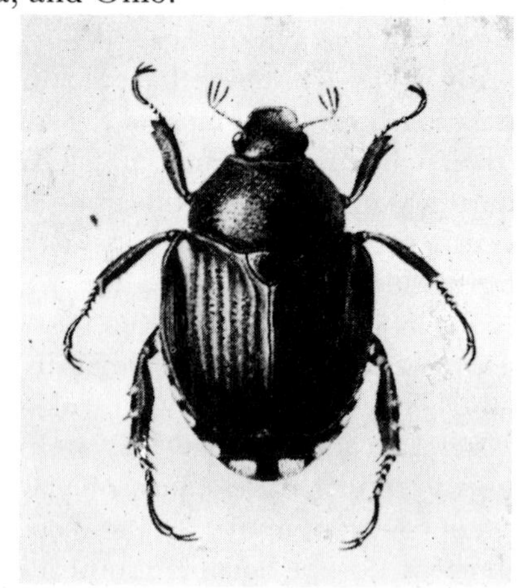

FIG. I–7 Popillia japonica (*adult*). (*U.S. Department of Agriculture, Bureau of Entomology and Plant Quarantine.*)

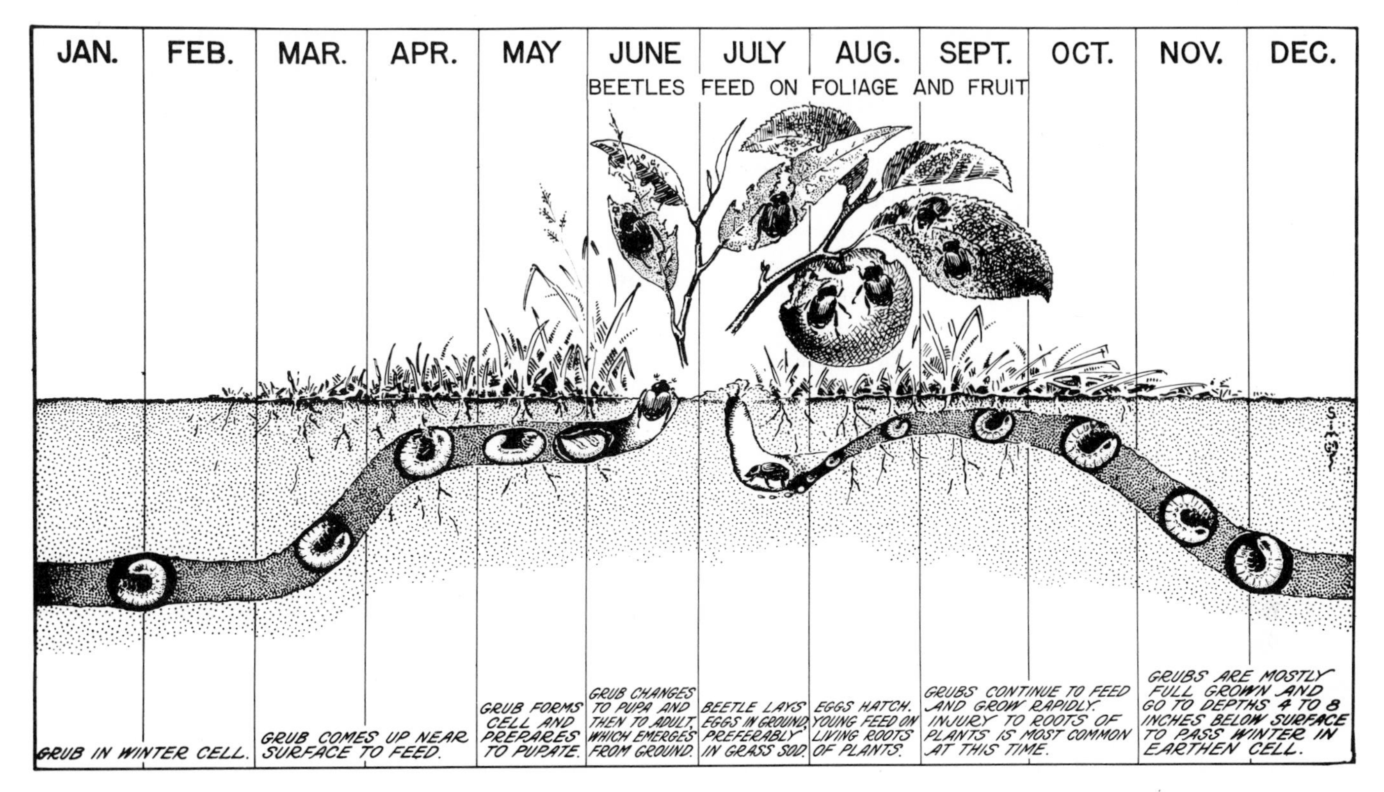

FIG. I–8 *The annual life cycle of* Popillia japonica. (*U.S. Department of Agriculture, Bureau of Entomology and Plant Quarantine.*)

A more recent invader of the United States is *Solenopsis saevissima*, popularly known as the "fire ant" because of the burning sensation accompanying its sting, which also causes allergic and febrile reactions in humans and, in some authenticated cases, fatality. *Solenopsis saevissima* is native to central and northern Argentina and part of Uruguay, where its natural foods are seeds, honeydew, and the flesh of other insects, and its natural mode of dispersal is the nuptial flights of winged queens and males, which may carry them as far as 5 miles from the parent colony. After mating, the queen excavates a mound in the earth and establishes a fast-growing colony that in 4 to 5 months may have a population of over 1,000 workers. Fire ants came into Alabama in 1918, probably on ships from Buenos Aires or Montevideo, and for 10 years lived within the city limits of Mobile without doing enough damage to attract attention. But in 1944 they underwent a population explosion, attributed to the arrival, probably also by ship, of a smaller, lighter colored variety of the species that seemed more adaptable to conditions of life in the United States. *Solenopsis saevissima* var. richteri is now a serious pest to farmers, for it has expanded its native diet to include the seeds of valuable food crops and the flesh of newborn poultry and livestock. Five years ago it was estimated that the crop damage from these ants in two counties in southern Alabama reached 50,000 dollars and that the total agricultural loss to the southern United States was several millions of dollars. Some of the damage they cause is indirect, resulting from the difficulty of ploughing infested land and of getting labor to do it. As many as 58 fire ant mounds have been found in a single acre of land, and workers have an understandable reluctance to try to put a plough through such territory as well as an understandable fear of the consequences of disturbing the inhabitants of a mound.

Insects are by no means the only successful

invaders. *Achatina fulica*, a snail native to East Africa, has slowly spread, largely by accidental dispersal, to Madagascar and its neighboring islands and then on to those in the Pacific Ocean. Known as the "giant snail" because of its large size, it is a nocturnal scavenger and eats almost everything—dead animals, living and dead vegetation, fruits, and flowers, among these especially orchids. It was brought into the island of Maui (Hawaiian group) by a resident who sent to Japan for a dozen specimens in answer to an advertisement encouraging their breeding for food and therapeutic values, especially for rheumatism and kidney troubles. The snails are hermaphroditic and extremely prolific, and the Hawaiian snail farm flourished. Shortly after its start the owner sent half a dozen specimens to a friend on the island of Oahu, and from these two focal points the snails have spread over both these islands, and also to Kauai and Hawaii. They have done enormous damage to vegetation and are now menacing the orchid crop as they come near to Hilo, the "orchid capital of the world." Because of their size they are hazards to traffic, for a 9-in. specimen can cause a vehicle to skid. They have now been brought into Florida, a gift from a small boy to his grandmother, where they are causing some havoc and much concern.

A well-known crustacean invader is the crab *Eriocheir sinensis*, popularly known as the "mitten" or "woolhanded crab" because the terminal segments of its first pair of thoracic appendages are so thickly covered with bristles that they seem fur-coated. Although native to China, this crab was found in the German river Weser; how it got there is still problematical. Perhaps some larvae were drawn into a steamer's water tanks in China, survived the voyage, and somehow got out again when the ship reached its German harbor, or perhaps some specimens were entangled in the seaweed of a fouled ship's bottom and so were carried from China to Germany. The first explanation is authenticated by the fact that two good-sized larvae of *Eriocheir* were actually found in 1932 in the ballast tanks of a German ship in Hamburg harbor. But however it got to Germany, the Chinese crab has become quite at home in European rivers and has populated those from the Vistula to the Seine. The adult crabs live and breed in the brackish water of these rivers, as they do in the Chinese rivers, but the larvae at a certain (megalops) stage of their development go upstream and grow to maturity in freshwater. They have been found 800 miles up the Yang Tse Kiang and 400 miles up the Elbe.

2. Methods of control of undesirable invaders

There are many instances of successful invasions, but less attention has been paid to those that have not upset the ecological balance than to those that have. Various means have been taken to prevent the entrance of invaders that might prove destructive, and to destroy or control those that have. Imposition of a rigid quarantine and other restrictions on the importation of plants and animals from one area to another seems to have been successful in keeping the potato beetle out of the British Isles, in spite of its successful invasion of Europe. But there is always a chance that, even with the utmost care, some individuals may escape the most searching inspection or survive the longest quarantine. Another means of control, after the invader has gained a hold in its new situation, is the introduction of a natural destroyer—a predator, a parasite, or a disease-producing organism specific to it. The praying mantis, *Stagmomantis carolina*, common in the southern United States, was brought north in the hope that it would establish itself there and prey upon Japanese beetles as it does on other insects, and the tachinid fly, *Centeter cinera*, was brought from Japan with the same hopeful expectations. These flies lay their eggs in the beetles and their hatched larvae become destructive parasites to their hosts. Another, and more successful, method is the use of "milky disease" which attacks the grubs in the ground (Figs. I–9 and I–10), but this entails treatment of extensive areas in infested regions.

Various predatory snails have been tried as counterpests against *Achatina*; *Gonaxias*, which attacks the body of the snail directly, has recently been introduced into Hawaii. In some cases, the counterpests have not adapted to the new area as successfully as the pests, and the introduction of a new species always carries the danger that if it does become adapted, it may turn out to be a pest itself.

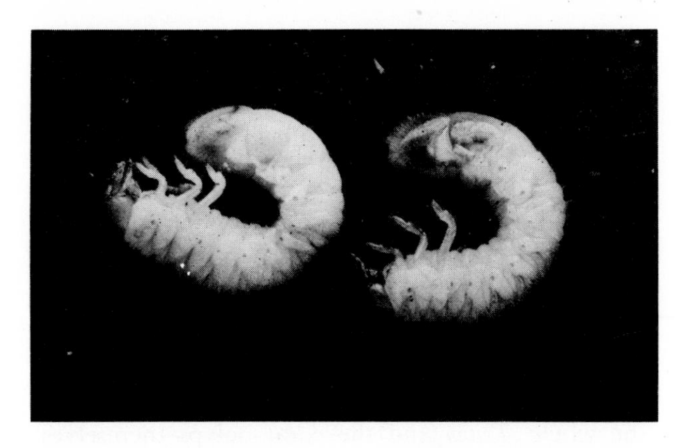

FIG. I–9 *"Milky disease" in* Popillia japonica. *Left: Healthy larva. Right: Larva infected with* Bacillus popilliae, *the agent of milky disease.* (*U.S. Department of Agriculture, Bureau of Entomology and Plant Quarantine.*)

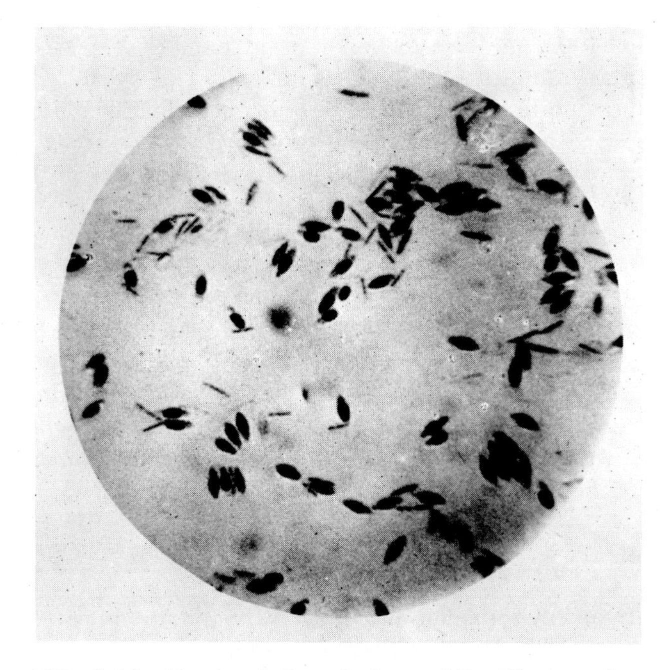

FIG. I–10 *Hemolymph from the larva of* Popillia japonica *infected with milky disease showing spore and rod forms of the organism* (*multiplied 1,200 times*). (*U.S. Department of Agriculture, Bureau of Entomology and Plant Quarantine.*)

The most direct method of control is by trapping, or in some way collecting, the undesirable newcomer and destroying it. Much of this has been done laboriously by hand, but a more effective means of trapping is to use some attractive substance to lure the pests to a suitable trap. Much attention is being given now to such substances, especially sex attractants, which are produced usually by the female and will bring males from a considerable distance. After many years of work, the sex attractant of the gypsy moth has been chemically identified and synthesized. Traps smeared with a sticky substance to hold the moths and baited with the synthetic substance to lure them are now used to detect gypsy moths in any suspected area and to capture them, so that they can be destroyed.

Another method of control is sterilization of one or both sexes. This method has been successful with the screw worm fly, *Chrysomia macellaria*, which first appeared in Texas in 1842 and has been responsible for the deaths of hundreds of thousands of cattle, sheep, goats, and hogs. Adult female flies are attracted to the open wounds of animals; newly born mammals with exposed navels are especially vulnerable. Each female lays about 250 eggs that hatch in 24 hr into larvae that feed on the flesh around a wound, and the broods of a number of females attracted to the same site produce enough feeding larvae to consume an entire animal. About 10 years ago, a planned program of eradication was initiated in Texas. Flies, raised in captivity to the pupal stage, were sterilized by exposure to radioactive cobalt and, as adults, distributed by airplane over 140,000 square miles of pasture, range, and forest in Texas and adjoining states. The released flies mated, but of course produced no offspring. Chemical sterilants are also being developed and experiments undertaken to see if new invasions can be prevented by spreading bands of sterilized flies as a screen between infected and uninfected areas.

Methods such as these, and the use of counterpests, provide natural checks to the reproductive potential of a species establishing itself in a new area where it has no natural enemies. But they are long-term methods and their results are not felt immediately. Much more immediate, but much more dangerous to the balance of nature, is the use of poisons and pesticides. Many of these, especially insecticides, have been developed and widely used in recent years, although some have been only temporarily useful in destroying the pest they were designed for. Most natural populations include some individuals that are resistant to poison, and these resistant types survive, breed, and give rise to new generations of resistant types. A well-known illustration of this is the establishment of strains of flies and mosquitoes resistant to DDT, which was at first regarded as an infallible eradicant. The great danger in the widespread use of chemical warfare against pests is that it may lead to the elimination of wanted species as well as the unwanted ones. This most definitely has happened, especially in areas where there has been mass spraying; destruction of one species in an ecological community may remove the natural check to the spread of another, possibly an even less desirable one. The number of springtails (Collembola), for example, increased enormously in a selected plot of ground sprayed with DDT, which effectively destroyed its population of mites. Mites are active predators on springtails, which feed on microfungi in the soil, so that even this short chain of events changed the character and quality of the soil and the productivity of the crops grown on it. Much longer and more complex chains are known that have led to drastic ecological changes whose effects have not yet been fully assessed. The value of chemical pesticides is a controversial matter, and the potential dangers resulting from them have recently been brought most vividly to the attention of the general public. Some states are proposing, or adopting, laws prohibiting the use of DDT which, being fat-soluble and relatively stable, can move up the food chain and become a menace to all levels of life.

G. BIOLOGICAL RELATIONSHIPS

1. Predator-prey

The predator-prey relationship is universal and so well known that it hardly warrants discussion. Undisturbed, it is a great factor in the preservation of balance in animal populations; disturbed, or pursued beyond its natural limits, it may have drastic effects, as has been pointed out in the discussion of invasions and pesticides. An illustration of such an effect, recently brought to the attention of the general public as well as of scientists, is the destruction of areas of the Great Barrier Reef and of islands in the Pacific Ocean by predations of the crown-of-thorns starfish, *Acanthaster planci*, upon the coral polyps that contribute to their formation. This large starfish, as much as 2 ft in diameter, feeds on living corals and has already eaten away about 100 square miles of the reef. Three or four decades ago, *Acanthaster* was a rarity on the reef, but it has multiplied so rapidly in recent years that it is now abundant. The reason, or reasons, for this multiplication have not yet been solved. It may have happened because the natural predators of *Acanthaster* (the whelk *Triton* and the coral polyps themselves) have been reduced in number, the former by visitors to the reef who collect the shells because of their beauty, and the latter by the predations of the starfish. Possibly there has been some other upset in the balance of the reef's population, such as a population explosion of *Acanthaster*, or a reduction in the number of species that ordinarily consume its larvae. This starfish is very prolific, one female spawning a million or more eggs. If all these hatched and the larvae grew into adults, the potential increase in population would be enormous. Reduction in the number of the larvae of species which feed on organic material of microscopic dimensions such as *Acanthaster* eggs may result from dredging and channel construction in this area or from poisons such as DDT moving along the food chain and destroying some important link in it. Whatever the cause, the predation of adult *Acanthasters* and the consequent diminution of coral reef formation, will have profound effects upon the geography of that area of the Pacific and so upon all its inhabitants.

2. Temporary associations

In all environments and under all conditions, invertebrates may live as solitary individuals—to the extent that any individual can be solitary on this planet—or in associations of differing degrees of intimacy. They may live in communities of the same or different species, such as the congregations of crustaceans, molluscs, and echinoderms found along the seashore, without any apparent organization (Fig. I–11). Or they may live in colonies or societies with a high degree of organization, like the social insects where community

FIG. I–11 *An association of the edible mussel,* Mytilus edulis, *and the common rock barnacle,* Balanus balanoides, *exposed on the seashore at low tide.* (*U.S. Fish and Wildlife Service.*)

living in a single species is developed to such a degree that there is a rigid division of labor and of privilege between different castes and a unified participation in various activities that promote the welfare of the colony.

3. Habitual associations

Habitual associations range from the comparatively loose relationship of commensalism (eating at the same table) to symbiosis (living together), in which dependence of one partner upon the other, or of both partners upon each other, may be so great that life for one is impossible in the absence of the other. The German biologist Anton de Bary (1831–1880), who coined the term "symbiosis" in 1879, recognized in it two aspects or conditions—mutualistic and antagonistic—meaning by the former, a relationship from which both partners derive some benefit and by the latter, a relationship that is advantageous to one but detrimental to the other. Since commensalism is essentially an animal-animal association, de Bary did not concern himself with it but considered only those habitual relations that exist most commonly, but not exclusively, between plants and animals and are recognized by twentieth-century biologists as mutualism and parasitism. There are many grades and degrees of these habitual associations, and hard and fast lines cannot be drawn between them. Some conditions of commensalism may well be considered mutualism, and extreme conditions of mutualism verge on parasitism. There are far too many instances of habitual associations for consideration here. All of them are interesting, some fantastic.*

a. COMMENSALISM Commensalism may require only the constant proximity of the two species that share food, or it may demand physiological responses from one or both of the partners and possibly structural modifications as well. An inquiline (an "indweller in a place not his own") is an animal that lives in the nest or abode of another, or in a cavity in its body that opens to the outside, finding in these places food without disturbance to, or from, the host. All inquilines are commensals but not all commensals are inquilines. A variety of terrestrial insects regularly live in the burrows of moles and rodents, and aquatic invertebrates find the tubes of worms and the holes that crabs and crayfish excavate along the shore secure habitations where food is plentiful (Fig. I–12). The canals and flagellate chambers of sponges are usually filled with a host of small animals, invertebrate and vertebrate, obtaining protection there and taking their

* A great many of them are described and graphically illustrated in the three volumes of "Symbiosis," edited by S. Mark Henry, Academic, New York, 1966.

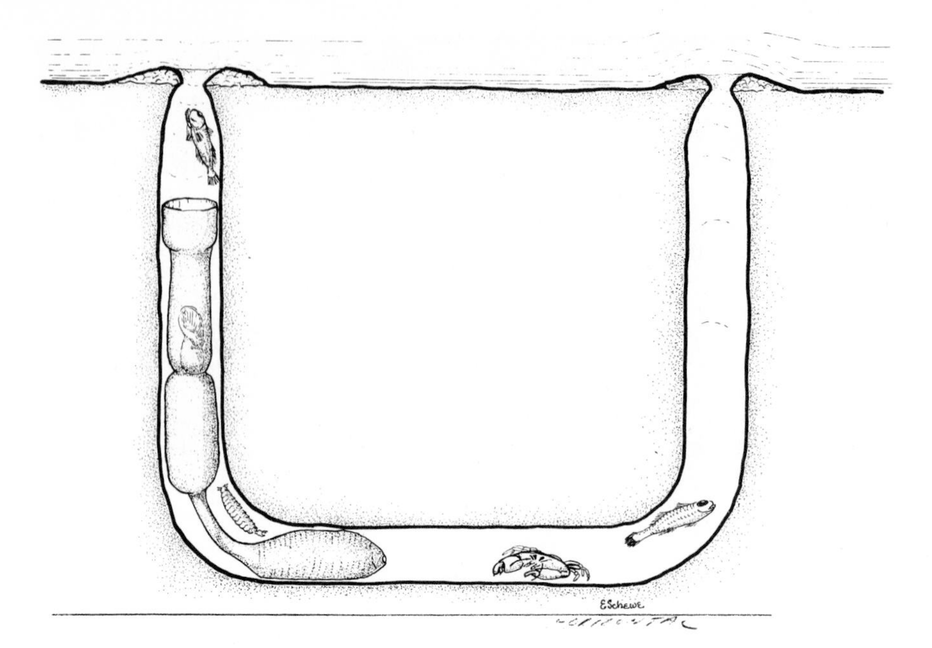

FIG. I–12 Urechis caupo, *a tubiculous echiuroid, with the inquilines that live with it in its mucus-lined burrow in the sand. By muscular movements of its body the worm keeps a current of water flowing through the tube, from which food particles are filtered off by means of the mucus funnel at the anterior end, which is secreted by epidermal glands. (After Fisher and MacGinitie.)*

nutrients from the food current swept along by the beating flagella of the sponge's choanocytes. As many as 17,128 individuals, representing 10 different animal species, have been taken from the cavities of a large logger head sponge (*Speciospongia vespara*). The little pearl fish, *Carapus* (*Fierasfer*), lives in the bodies of sea cucumbers and of starfish, and in the mantle cavities of bivalve molluscs, sheltering there from its predators, except for periodic excursions, only at night, to hunt for the small crustaceans on which it feeds. *Carapus* will live in aquaria without its habitual partner, but only if no other fish are there to eat the exposed inquiline.

Snails, worms, and other small animals frequently live in the shells that hermit crabs, in their turn, have taken over from gastropods when the molluscs no longer have use for them. *Nereilepas fucata* lives as such a cohabitant, resting quietly in the terminal whorls of the shell until the crab feeds, at which time it darts forward and literally snatches what it wants from the crab's mouth (Fig. I–13). At this moment the crab could easily make a meal of the worm, but it never seems to and tolerates the presence and habits of its inquiline in an arrangement that seems wholly to the worm's advantage and not at all to the crab's.

Some commensal relationships involve an even greater degree of tolerance than that of host and inquiline, and often physiological and structural modifications. A number of species find the tentacles of cnidarians an excellent refuge, depending upon the stinging capsules (nematocysts) of their host for protection and upon the food it captures for their nutriment. The amphipod *Hyperine medusarum*, along with a variety of little fishes, swims freely under the umbrella of the medusa *Rhizostoma cuveieri*; decapod crustaceans and small fish, too, are frequently found among the tentacles of sea anemones, and the Man-of-War fish, *Nomeus gonovii*, regularly associates with the Portuguese Man-of-War, the siphonophore *Physalia*. These commensals either do not stimulate discharge of the nematocysts or they are not affected by their toxins when they are discharged, although other animals may be poisoned by them. Such tolerance and immunity represent reciprocal adjustments of both members of the association, which are not yet understood.

One of the commonest examples of commensalism, and one verging on mutualism, is that between hydroids or sea anemones and hermit crabs. The hydroids *Hydractinia* and *Podocoryne* are often found attached to the shells that harbor the crabs, in positions near the aperture of the shell where they can reach for scraps as the crab feeds. They have also the benefit of transportation as the crab moves from place to place, and in turn they give the crab assistance and protection by discharging their nematocysts when food is to be caught or a predator attacked. More precise than this is the relationship between *Adamsia palliata* and *Eupagurus prideauxi*, because this particular species of sea anemone will live only in association with this particular species of crab. The anemone invariably attaches itself to the crab with its mouth opening behind the crab's, and directed ventrally, a position in which it can obtain without opposition a large share of the crab's prey. The association is so intimate that neither partner can survive alone. If the crab is removed from the shell, the anemone detaches itself and soon dies, even though it may settle on the bottom as if preparing for an independent life. Its death is not due to starvation, for even if fed it will not survive away from its habitual partner. The crab, in its turn, takes care always to have an anemone with it. The shell into which it inserts its abdomen is never large enough to cover its whole body, and serves more as a primary site of attachment for the anemone than as a shelter for the crab. When moving into a new shell, the crab either transfers its attendant anemone from the old one or quickly captures a new one. The anemone, once fixed on the shell, expands so as to cover the exposed surface of the crab's body and molds its shape more or less to conform to its host's contours. *Eupagurus prideauxi* thus is provided with a more flexible covering than are other hermit crabs, whose bodies are more completely inserted in the appropriated shells. It is, therefore, more agile in its movements. Its association with *Adamsia* can only have arisen through some means by which it attained immunity from the anemone's toxins, which have been shown to be fatal to crabs of other species when injected into them. One might suppose that for some genetic reason the tissues of *Eupagurus* contain natural antibodies lacking in other species, which made this association originally possible, or that during the long period of years in which it seems to have existed there has been a gradual adaptation of the partners to each other, and immunization of *Eupagurus*, with selection toward those individuals which could most successfully tolerate the

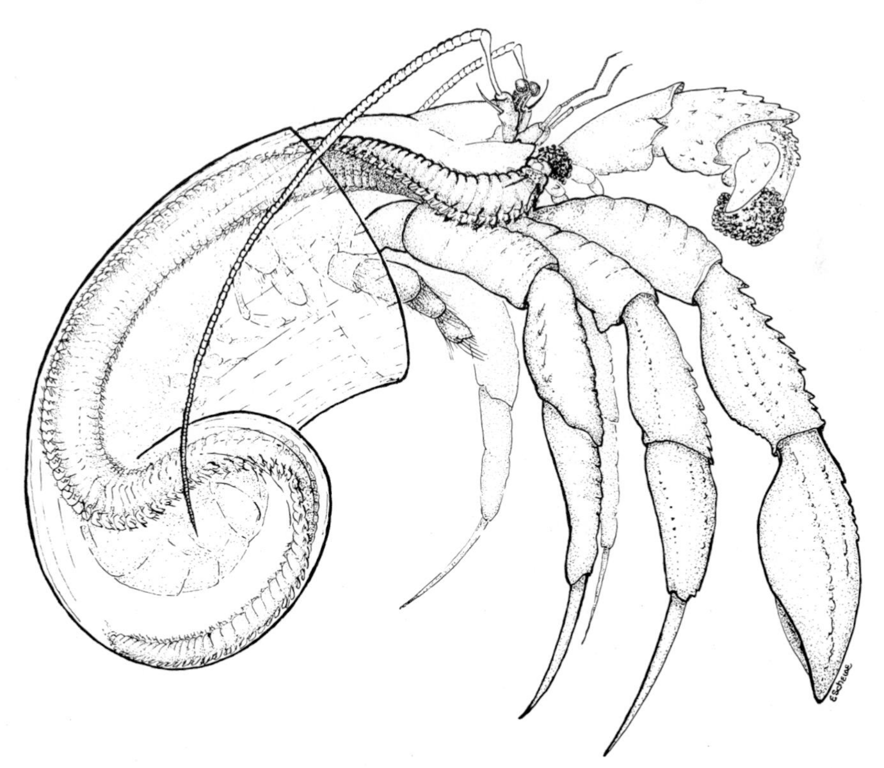

FIG. I–13 Nereilepas, *a polychaete worm, lives as a commensal in the abandoned gastropod shells inhabited by hermit crabs. In this drawing, the natural shell has been replaced by an artificial glass one, into which the crab has inserted itself, to show where the worm habitually lives, for it emerges only to share its host's meals, here a morsel of sea urchin ovary. (After Caullery.)*

commensal. The first alternative seems the more probable.

There are many other instances of tolerant commensalism to be found among terrestrial animals, especially in the associations between ant and termite colonies and the other animals that make their nests their homes. Among such myrmiciphilous (Greek *myrmex* "ant," and *philos*, "loving") forms are mites, springtails, flies, bugs, and a variety of beetles, living as larvae or adults in the ant hills or termitaria and permitted by their rightful inhabitants to take freely what food they can find and occupy what space they need. Some such commensals are actually sought and cultivated, sometimes with disastrous effects. One of the oldest and best known of these relationships is that between ants and aphids, which because of this were called "ant cows" by the Swedish natural historian and systematist Carolus Linnaeus (1707–1778). Aphids feed on plant juices, thrusting their mouthparts with great precision into the phloem and taking up the fluid from the phloem vessels. This is rich in sugar and in amino acids and amides. It is ingested in relatively large amounts, and a considerable quantity is eliminated from the hindgut of the aphid as "honeydew," a fluid with a high concentration of sugars and the hexahydric alcohol dulcitol. This is greatly prized by ants, and equally so by bees. An ant scout finding an assembly of aphids on a plant alerts the colony and many ants quickly collect at the site and feed on the extruded honeydew. An ant can also induce its emission by stroking an aphid's abdomen with its antennae. Some species of ant are so careful of their "cows" that they will rescue them when disturbed and transfer them as carefully to a safer place as they transfer their own larvae when in danger. In some cases there seems to be a high degree of mutual dependency between ant and aphid. The brown ant, *Lasius bruneus*, and aphids of the genus *Stomaphia*, for instance, both live in the bark of trees and one is never found there without the other.

The blood-red ant, *Formica sanguinea*, particularly prizes the ethereal secretions of the beetle *Lomachusa strumosa* (Fig. I–14). The ants seek the beetles, even stealing them from the nests of other colonies, and protect and feed the larvae. Adjustments to this kind of life must have evolved in the beetles, for their larvae

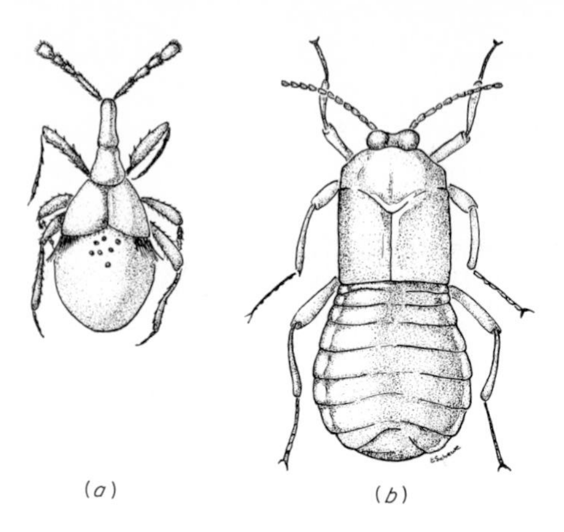

FIG. I–14 *Inquilines in ants' nests. (After Goetsch.)* ***(a)*** Claviger, *a small beetle tolerated in ants' nests because of its secretions. When the hairs at the junction of abdomen and thorax are touched, glandular secretions are discharged through pores on the dorsal surface of the abdomen and are eagerly licked up by the ants.* ***(b)*** Lomachusa (*adult*).

have very short labial palps and short, broad hypopharynges that are better adapted for taking the regurgitated food that the ants offer them than the more typical mouthparts of independently feeding beetles. They have also acquired the habit of moving their heads back and forth when they are hungry, characteristic of the ant larvae, and the adult ants respond to these signals by spitting out a drop of food for the beetle larvae just as readily as they do for their own. The larval secretions, although not nutritive like honeydew, are for some reason greatly valued by the ants, which even risk destruction of a colony to get them. Again, there is an interesting and peculiar relationship here. In spite of the careful feeding the ants give them, the beetle larvae prey upon the ant brood, but the very assiduity with which the ants care for the beetles protects the ants from complete annihilation. For when an ant larva is ready to pupate, the adults dig a small hollow in the ground in which the larva spins a thick cocoon and pupates. The ants treat the beetle larvae in just the same way, but the cocoons they spin are fine and delicate and easily damaged when the ants attempt to move them, as they frequently move their own pupae. When this happens the beetle pupa immediately begins to spin another cocoon to replace the damaged one but fails to accomplish pupation successfully, probably because of the increased strain put upon its resources by additional cocoon building. Thus, only the beetle larvae that the ants neglect grow into adults, and some sort of balance is maintained between the two groups of insects inhabiting the same anthill.

b. SYMBIOSIS (1) Mutualism Plant-animal associations offer the best illustrations of mutualistic symbiosis, and again ants and termites are well known for their external (ecto-) symbiosis with fungi. The leaf-cutting ants of Central and South America (genus *Atta*) pile up shredded leaves in their nests and inoculate them with fungus, carrying bundles of fungal mycelium from nest to nest in pockets in their hypopharynges (Fig. I–15). The workers tend these fungus gardens which provide food for the entire colony. Termites do the same kind of thing, carrying the fungus in their digestive tracts and spreading it in the termitarium with their feces. Both ants and termites tend the gardens so carefully and keep the mycelium so well cropped that fruiting bodies are rarely formed. This has made identification of the fungus difficult, but from samples taken from the nests and cultivated in vitro it has been established that all four of the main classes of fungi are represented in these gardens. The mutualistic aspect of this relationship is that through it the fungi are assured of perpetuation and the ants of a constant source of dietary protein, and probably also of essential vitamins.

FIG. I–15 *Queen of young* Atta sexdens *colony on her fungus garden. (Courtesy of Dr. Neal Weber, with permission from Science.)*

Internal (endo-)symbiosis is also practiced by termites, whose lives depend upon the continued presence in their digestive tracts of flagellate protozoans. Each species of termite has its particular species of flagellate which it harbors in a pouch of the hindgut. The flagellates engulf and, by means of a special enzyme, cellulase, that they and a small number of other animals can make, digest the particles of wood that the termites ingest as they cut their way through timber. The termites themselves are incapable of digesting wood. The termites feed on the flagellates, leaving enough to perform this essential function for them and taking care that there is always an adequate and mixed population of flagellates in their guts. For, deprived of this intestinal fauna, a termite dies unless fed by other members of the colony.

Many herbivorous (phytophagous) insects also maintain intestinal symbionts, usually bacteria, harboring them in enlargements of their guts called fermentation chambers, since in them the bacteria decompose the cellulose of the insects' food, and also provide them with vitamins. Microorganisms that live intracellularly in the abdominal cellular masses (mycetomes) also contribute to the metabolic processes of their hosts. These symbionts are transmitted from generation to generation in a rather complicated manner, either through the nurse cells that pass them on to the ovum or by migration into the future egg cells. Migration of microorganisms into future sperm cells has never been observed.

Endosymbiosis between algae and animals is a common phenomenon, found in about 150 genera of invertebrates. The algae are either green zoochlorellae, yellow-brown zooxanthellae, or blue-green cyanellae. These last are characteristic of freshwater protozoans, while zoochlorellae and zooxanthellae are regular inhabitants of certain species of sponges, cnidarians, ctenophores, turbellarian flatworms, rotifers, annelids, bryozoans, and molluscs. Zooxanthellae are limited to marine species, and zoochlorellae almost entirely to freshwater species, although there are some marine species that harbor them.

In some of the cases investigated, it has been found that each generation of individuals acquires the algae directly from its food; in others, although this may have been the original source, the algae, multiplying in the tissues of the host, are transmitted from generation to generation in its reproduction. The endosymbionts, showing various degrees of morphological alteration from their free-living states, are usually intracellular, but in the sea anemone *Phymactis clematis* and the jellyfish *Cassiopeia* they are lodged in the mesoglea. In the little flatworm *Convoluta roscoffennsis*, which lives along the sandy shores of the English Channel, its natural symbiont, *Platymonas convolutae*, occupies spaces between the epidermal cells, extending processes up to the surface of the flatworm's body.

The larvae of *Convoluta* are colorless when hatched and become green only after they have ingested flagellate algae. Experimental tests with strains maintained in the laboratory have shown that the worms turn green after the ingestion of a variety of chloroplast-containing species, but that they will eat these only in the absence of *P. convolutae* and establish true symbiotic relationship with this alga alone. Once ingested, the algal cell loses its theca, flagella, and eyespot and becomes stellate in form as its chloroplast lobes extend and push up toward the flatworm's surface. *Convoluta* commonly eats 5 to 30 flagellates per hour, and once the symbiotic relationship has been established with them, grows rapidly. If the relationship is not satisfactorily accomplished, the larva does not grow, in spite of voracious feeding on other organisms, and ultimately dies.

Mutualistic association with a photosynthesizing organism imposes certain conditions of life upon the host. Since light is required for at least one step in the complicated series of reactions that lead to the photosynthetic production of carbohydrate, animals that harbor algae must provide environments where light can penetrate. Most such hosts do live in photic environments, but there are exceptions, and in some cases precise behavior patterns have developed. *Convoluta*, for instance, regularly comes out from beneath the sand when the tide goes out and exposes itself to the sun. Minute as these flatworms are, they are present in such numbers along the coasts of Brittany and the Channel Islands that they make great green patches on the yellow sand that appear and disappear with startling suddenness at the ebb and flow of the tide. Yet such adjustments are not always necessary, for sea anemones with zooxanthellae have been collected off the coast of Florida from depths of 200 m and below. These live so well below the photic zone that it seems certain that the algae must be obtaining their food by some means other than photosynthesis. Indirect support of this assumption is provided by the experimental evidence that zooxanthellae in animals living in the photic zone do not survive more than 50 to 75 days of darkness.

Mutualism of course implies reciprocity between the two partners in the association, and in the case of such deep-sea partners it would seem that the algae must be living heterotrophically at the expense of their hosts. The extent of the independence of the members of these partnerships, or interdependence, can be tested by separating them and observing the effects on each of the absence of its natural associate. Separation can be achieved in a number of ways. Hosts can be deprived of their algae (made aposymbiotic) by keeping them in the dark and thus starving out the photosynthetic cells; by dissecting away the alga-containing cells; by treatment with 0.5 percent glycerine, which results in the complete elimination of the algae in about 8 days; or by preventing their ingestion, as has been done in the case of *Convoluta*. Algae can be isolated from the host by differential centrifugation, as well as by dissection, and cultured in appropriate media. Zoochlorellae isolated from *Paramecium bursaria* have been maintained in culture for many years.

Growth is known to be enhanced in *P. bursaria*, *Spongilla sp.*, and *Chlorohydra viridissima*, all harboring zoochlorellae. Tests of growth rates and survival times of symbiotic and aposymbiotic hydras showed that the growth of the former, as measured by production of buds, was about twice that of the latter, even when both were fed liberally with *Artemia salina* nauplii, a standard laboratory diet for hydras. Also, symbiotic hydras survived starvation as long as 4 weeks, while aposymbiotic specimens succumbed in 10 to 12 days.

Use of radioactive carbon, usually incorporated into the carbon dioxide used by the autotrophs in photosynthesis, has made it possible to trace the movement of their photosynthate. The form in which the carbon is released by zoochlorellae is usually one of the monosaccharides, maltose or glucose, and by zooxanthellae, glycerol. Radioautographs of host tissues show that in the animals tested these compounds have moved from the plant into the animal cells. They are essentially excretory products, for the sucrose manufactured by the algae is retained by them. As much as 45 to 50 percent of the carbon fixed in photosynthesis is transferred to the cells of *C. viridissima*, as is shown by the recovery of labeled proteins, nucleic acids, and other organic compounds from them. Lesser amounts have been found in the cells of other hosts, though still quantities sufficient to show that the algae contribute significantly to the nourishment of their partners. It is possible, too, that the algae provide vitamins and other nutrients. The host, too, makes some contribution to the photosynthesizing algae. The elimination of glycerol, for instance, from zooxanthellae isolated from the mollusc *Tridacna* and the reef coral *Pocillopora demicornis* has been found to increase sixteenfold when homogenized tissue from the host is added to the culture medium. This "host factor" is destroyed by boiling, indicating its protein, and possibly enzymatic, nature.

(2) Parasitism Parasitism is an interspecific relationship that is unilateral, one member of the partnership benefiting while the other suffers. A parasite derives its food entirely from its host and thus is wholly dependent upon it. Parasitism may be a temporary association, lasting for intervals of greater or lesser length in the lives of the partners, or it may be permanent, lasting the entire life span of one or the other. It may be an external association (ectoparasitism), the parasite fastening or holding itself on the surface of the host in such a way that it can withdraw food from it, or it may be an internal one (endoparasitism), the parasite living within the body spaces or cells of its host. The burden of establishing the relationship seems to rest with the parasite, and endoparasites at least must have had to solve the problems of how to invade their hosts, how to meet their respiratory and nutritive requirements, and how to withstand whatever defenses the host might muster to prevent their invasion or to weaken or destroy those the host might have. They had, moreover, the problem of satisfying their demands for existence and perpetuation without destroying their hosts as predators destroy their prey, for the survival and at least the minimal well-being of the host are necessary conditions for the survival of the parasite.

Parasitism has evolved in many invertebrate phyla. It is very old, and must have been instituted early in animal history, for parasitic gastropod molluscs (family Capulidae) have been found attached to crinoids of the Devonian and Triassic periods. Today there are more parasitic than free-living species, and more parasitic than free-living individuals. Moreover, there are few, if any, animals or plants that are not subject to invasion by parasites and the consequent depletion of their resources. About 10 percent of all the world's crop production is estimated to be lost annually because of parasites. Every class of Protozoa has parasitic species, and all orthonectids, dicyemids, and acanthocephalans are parasites. Some classes and orders of flatworms and nematodes are made up exclusively of parasitic species. Nematodes are known to parasitize nearly every known metazoan species, and are probably equally widespread as parasites of plants. There are parasitic molluscs, annelids, and arthropods, among which are species that are parasites only in their larval or juvenile stages and others only as adults.

Endoparasites may enter their hosts directly or indirectly. Direct entrance requires devices for biting or piercing the integument of a host; these are often assisted by carbohydrate or protein-splitting enzymes that soften or destroy the tissues through which the parasite must pass to reach its final destination in the host's body. Mechanical devices are illustrated by the hooks and cutting plates in the buccal cavity of hookworms, by means of which the nematode, in its infective stage, can rasp away the thick skin on the sole of the human foot, their usual point of entry, and by

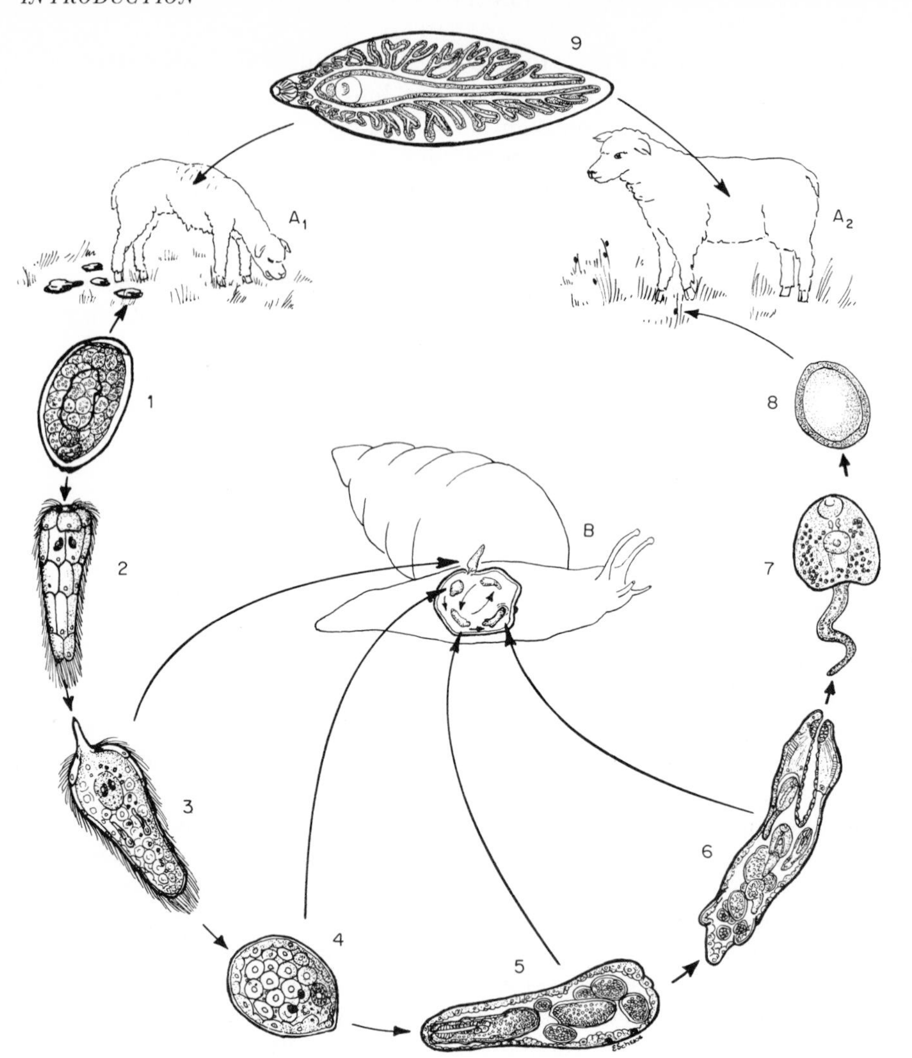

FIG. I–16 *Life cycle of the sheep liver fluke* Fasciola hepatica. *(From various sources.)* A_1*: The primary host, an infected sheep, grazing in a moist pasture.* (*1*) *Developing egg of the fluke, in which the outline of the embryo can be seen. This leaves the host's body with its feces.* (*2*) *Surface view of the free-swimming ciliated embryo* (*miracidium*) *which hatches from the egg. The paired eye spots later degenerate.* (*3*) *Embryo with everted head papilla with which it will bore into the tissues of a snail it may encounter.* (*4*) *A young sporocyst into which the miracidium metamorphoses in the tissues of the snail.* (*5*) *A mature sporocyst containing a redia nearly ready to emerge.* (*6*) *An adult redia with mouth, pharynx, and gut containing one daughter redia and some developing cercariae.* (*7*) *A free cercaria.* (*8*) *An encysted cercaria.* (*9*) *An adult fluke in the liver and bile passages of an infected sheep.*

A_2*: A healthy sheep that, feeding upon grass on which cercariae have encysted, becomes infected. While in the digestive tract, the cysts hatch into young flukes which bore through the intestinal wall into the peritoneal cavity and migrate to the liver where they mature and discharge masses of eggs to repeat the cycle.*

B: The intermediate or secondary host, Limnaea, *or a related pulmonate mollusc.*

the movable and protrusible papilla, often with sharp stylets, on the miracidia (first larval stage) of digenetic trematodes (Fig. I–16). Nematodes also employ chemical devices; an enzyme of the hyalinuridase complex has been demonstrated in flukes of the genus *Schistosoma* which decomposes hyaluronic acid, a highly viscous mucopolysaccharide that is the principal component of the ground substance of connective tissues. Collagenases, enzymes that destroy the collagen fibers of connective tissue, have been detected in *Strongyloides*, a nematode of wide distribution whose larvae penetrate the skin of man and many animals and, migrating through the vessels, finally reach the small intestine.

In indirect attack, a parasite may enter its host in active or encysted state in the foods it eats or the fluids it drinks, or it may depend upon another animal (vector) for introduction. Arthropods, especially insects, seem to have accepted the role of vector, for many of them carry parasites from host to host on some external surface, or harbor them within their bodies and infect the final host through a bite or sting. Among crustaceans, the copepod *Cyclops* is host to at least two human parasites, the guinea worm *Dracunculus medinensis* and the tapeworm *Dibrothriocephalus latum*. Crayfish and crabs are hosts to the lung fluke *Paragonimus*. Ticks and mites are the most frequent arachnid vectors; these are known to carry and transmit rickettsias, viruses, bacteria, and spirochaetes to vertebrate hosts. Certain families in at least five orders of insects have long been recognized as transmitters of protozoan and metazoan parasites. These include cockroaches (Orthoptera), *Triatoma*, the kissing bug (Hemiptera), sucking lice (Anopleura), fleas (Siphonaptera), and flies and mosquitoes (Diptera). Transmission of this kind demands very specific relationships between conveyor and conveyed and adjustments that involve mechanisms for the entrance of the parasite, for its lodging, and for its introduction into the final host without interference with the well-being and activity of the vector. In some cases, the parasite uses its vector as a place where juvenile individuals grow and change in structure before their introduction into the final host; in others, the parasites may multiply in the body of the vector without undergoing structural change; in still others, they may both multiply and develop.

Whatever the method of attack, chance plays a major role in the life of a parasite, for its existence depends upon successful meeting with and invasion of its host or hosts. On the other hand, most parasites are spared the necessity of foraging for food, and the degeneration of the organs of locomotion, ingestion and digestion, and special sense is common in parasitic species. Conversely, enlargement of the reproductive organs, prolific production of gametes, and, sometimes, elaborate devices for fertilization and development of the resulting embryos are characteristic of them.

The deleterious effects of a parasite upon its final host may be mechanical, chemical, or both, as well as nutritional. Mechanically, parasites may destroy tissues in their penetration into a body or, if they are sufficiently numerous, may block vessels or passages through which materials should move. Chemically, they may release toxic substances or cause disadvantageous or destructive alterations in the host's biochemistry. For example, the nematode *Pratylenchus penetrans* hydrolyses amygdalin, a glucoside characteristic of the plant genus *Amygdalus* to which peaches belong. In this hydrolysis, benzaldehyde and hydrogen cyanide are released which cause necrosis in the cells of the roots of peach trees, the sites of attack and operation of *Pratylenchus*. Two other nematodes, *Ditylenchus dipsaci* and *Meloidogyne hapla*, release a substance, most probably an enzyme, that inactivates auxin, a plant growth hormone, with the consequent stunting of infected plants. Many similar examples could be cited.

There is great specificity in many host-parasite relationships, although some parasites may complete their life cycles successfully in hosts as divergent as birds and mammals. In every case, the host must provide a suitable environment for the parasite, and the parasite, on its part, must be adaptable to conditions within its host, or hosts. Moreover, the parasite must not destroy the host and must preserve some kind of balance between its demands and the survival of its host. In some cases, this relationship is so precise that while there may be but little morphological distinction between two parasites, each can survive only in its particular host. Thus nematodes (roundworms) of the genus *Ascaris*, infecting man and pig, are structurally indistinguishable, but cross infections prove that they are different. Ascarids from a pig introduced into a human digestive tract do not develop normally, and the carbohydrates extracted from them have different antigenic properties, although the proteins have not. Also, conditions in the host may affect the morphological character of the parasite; thus, cercariae of *Fasciola hepatica* (sheep liver fluke) developing in the tissues of a guinea pig, a rabbit, or a cow are so different that they can easily be mistaken for those of three different species.

At least two aspects of host-parasite specificity are known. One results from the inability of a potential host to provide the necessary nutrients for the parasite and the other from the inherent possession of, or the ability to make on demand, substances or antibodies that are lethal to the invading parasite at some stage of its life cycle. In the first case, for example, vitamin deficiency in a presumptive host may confer partial or complete immunity upon the host. This is true for the malarial parasite of man (several species of the genus *Plasmodium* of the protozoan class Sporozoa), for it cannot survive in individuals deficient in pantothenic acid. Similarly, chickens which are subject to avian malaria through infection with *Plasmodium lophurae* are protected if they are deficient in riboflavin. The fact is also well established that human malarial parasites cannot multiply in individuals with the "sickle-cell trait," the effect of a recessive mutant gene(s) that is known to be responsible for the substitution of the amino acid valine in the position in the beta chain of the globin portion of the molecule occupied by glutamic acid in normal hemoglobin. In individuals homozygous for the mutant gene (genotype ss), the erythrocytes are crescentic or sickle-shaped and are deficient in oxygen-carrying capacity, so that a severe anemia results. In individuals heterozygous for the mutant gene (genotype +s), only a fraction of the erythrocytes sickle and do so only when exposed to low oxygen tension. These people, therefore, only suffer from anemia when under oxygen stress and enjoy immunity from malaria. Either the plasmodium cannot penetrate their red blood cells, or if it does penetrate, cannot derive adequate nutrition from them.

Immunity from a particular parasite may also result from the presence in the body of the potential host of substances that act as antibodies. Immunity of this kind is innate and is also an expression of the genetic constitution of the individual. Man has such immunity from infection by all but two of the many species of *Trypanosoma* (flagellate protozoan) that are disease producing and destructive to many kinds of African animals, especially ungulates. Trypanosomes are blood parasites, introduced into the victim by the bite of a vector, the tsetse fly *Glossina* (Fig. I–17). Various species of *Glossina* act as vectors for the various species of *Trypanosoma*. Because of a natural component of human blood serum, man is immune to all species but *T. gambiense* and *T. rhodesiense* which, when introduced into his body, produce the type of encephalitis known as sleeping sickness. The absence of this component in the blood of other mammals permits multiplication of all species of the parasite in their bodies and makes them victims of those against which man is naturally protected.

Acquisition of immunity after invasion by a parasite has been well demonstrated in the rat, which is subject to infection by a nematode, *Nippostrongylus muris*. The larvae of *Nippostrongylus* penetrate the skin of the rat, migrate through its tissues, and finally reach the small intestine where they pierce the mucosa and draw nutriment from the underlying lamina propria. They mature in about 2 weeks, and the females lay large numbers of eggs that leave the rat's body with its feces and develop into infective larvae. In the next week, most of the adults are also eliminated from the

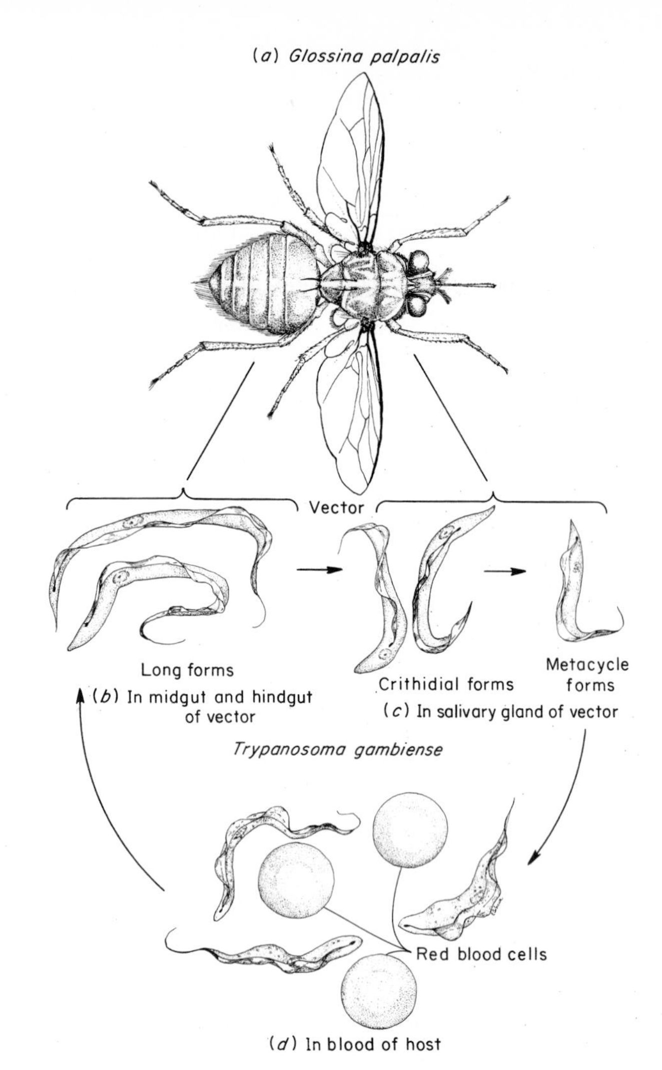

FIG. I–17 *Life cycle of a hemoflagellate. (After Mackie, Hunter, and Worth.)* Glossina palpalis ***(a)****, the African tsetse fly that lives always in the vicinity of water and feeds on crocodiles and other animals, including humans, and is the vector for* Trypanosoma gambiense, *the agent of sleeping sickness in man.* Trypanosoma gambiense: ***(b)*** *Long forms from the midgut and hindgut of* Glossina, *where they multiply after ingestion by the fly.* ***(c)*** *Crithidial and metacyclic forms from the salivary glands of* Glossina, *to which they migrate after multiplication in the gut.* ***(d)*** *From the blood of a human host after the introduction of the parasite through the bite of an infected fly. Man and certain domestic animals, notably the pig, are reservoirs for the parasite. Flies drawing blood from an infected individual become infective themselves in 18 to 34 days after the meal, during which time the parasites multiply in the gut and migrate to the salivary glands.*

intestine and few, if any, eggs are subsequently passed. This destruction has been found to be due to the production of antibodies by the host. These are precipitins whose formation is induced by the secretions and excretions of the parasite and that, in turn, cause the formation of precipitates, demonstrable in vitro as well as in vivo, in the gut and other organs, and at all the orifices of the feeding larvae. Consequently, the larvae become stunted and they, as well as the adults, are more or less immobilized so that they cannot resist the intestinal peristalsis to which they are constantly exposed and so are propelled out of the digestive tract. There are a number of other instances of the inhibition of growth and reproduction, and even of destruction, of invading parasites by this kind of reaction on the part of the host.

Relationships between vectors (secondary host) and parasites are also specific and show, in many instances, remarkable degrees of adaptation. Although a parasite may, and usually does, produce marked effects upon its primary host, it seems to cause little or no disturbance to its vector. There are, for instance, some 200 species of the mosquito genus *Anopheles* but only about 24 of them are recognized as transmitters of human malarial parasites. Inspection of the stomachs of mosquitoes carrying *Plasmodium* have shown that in a single one there may be as many as 50 oocysts in which rapid multiplication of the infective stage is in progress, yet the mosquitoes show little or no effects. Similar indifference to the presence of parasites in their tissues is found in the molluscs that act as secondary hosts to digenetic trematodes.

A good deal of attention has recently been directed to the relationships between viruses and their arthropod vectors. It is a remarkable fact that a virus may multiply in the cells of a vector's body without causing any pathological symptoms. Although some studies have indicated that there may be some alterations in the cells of the fat bodies of insects carrying a virus, most of them have shown the effects to be minimal. Electron-microscopic studies have shown crystalline arrays of particles in the cytoplasm of the cells of the fat body and in the muscles of the clover leafhopper, *Agallia constricta*, vector of a virus causing wound tumors in clover. Similar microcrystals could be seen in the cells of the infected plants, along with the virus particles. The insect showed no signs of viral infection, although the virus was multiplying in its cells with all the presumed metabolic derangements. Yet the same viruses that produce such derangements, and have drastic, even lethal, effects in the body of the animal or plant to which they are specific pathogens, seem harmless to the vector that introduces them.

The presence of one strain of a virus in the cells of a vector may or may not prevent the introduction of another. It has been shown that leafhoppers carrying the California strain of the "yellows" virus cannot be simultaneously infected with the Eastern strain. On the other hand, mosquitoes may harbor both the Eastern and the Western strain of the virus that causes equine encephalitis, although the two strains are quite clearly distinguishable clinically. In 94 percent of the cases tested, mosquitoes infected with the Eastern strain of the virus and a week later allowed to suck blood known to contain the Western strain, were found to have as much of each type of virus as could be recovered after only a single infection. Moreover, both the Eastern and Western forms of the disease are clearly transmitted by a single vector.

Little is known of the origin or evolution of any of these relationships. In a number of modern phyla, there are closely related free-living and symbiotic species, yet the transition from one to another has never been observed and the course of events that have led to them remain obscure. It may reasonably be supposed that symbiotic relations may have arisen through inquilinism, the cohabitant becoming more and more dependent upon its host and more and more adapted to life upon it, or within it, until its absolute dependence became established. The difficulty, and in most cases the impossibility, of cultivating either member of a well-established partnership in the absence of the other indicates the extent to which the existence of each has become integrated with that of the other. One might suppose that some intracellular symbionts are on the way to becoming so closely integrated with the cell's machinery that they may become part of its genetic endowment as extranuclear genes. A case of such close, and possibly final, integration is the influence of a spirochaete upon sex ratios in the fruit fly *Drosophila*. Females infected with this microorganism transmit it from generation to generation through their eggs and so to all the resulting zygotes. It is lethal to males early in their development, and broods from strains of flies carrying the infection are therefore almost exclusively female. Even more extreme conditions are seen in the inheritance of sensitivity (and resistance) to carbon dioxide in *Drosophila* and of the killer property in *Paramecium*. The greater sensitivity to carbon dioxide in some strains of *Drosophila* than in others has been genetically related to the presence in the egg cell of a body designated as sigma, which has many of the properties of a virus. Similarly, particles called kappa, visible in the cytoplasm of certain varieties of *Paramecium aurelia*, are believed to be, or originally to have been, an infective agent of some sort. This agent, although dependent for its survival upon a dominant gene in its host, is transmitted through the cytoplasm and makes the animal possessing it a "killer" through its emission of substances that are lethal to nonkiller (sensitive) individuals of the same or other strains.

section two

INVERTEBRATE ARCHITECTURE

BODY PLANS

The bodies of invertebrates show great diversity both in size and in architectural pattern. Some of them are microscopic, like those of many Protozoa and Rotifera that are measurable in microns and visible to the human eye only when greatly magnified. Others may reach a considerable size, although there are no modern species as big as the biggest of modern vertebrates. The largest invertebrate on record is a specimen of the squid *Architeuthis*, taken off the Grand Banks of Newfoundland, with an overall length, including the outstretched tentacles, of 55 ft. Weighing some 30 tons, this animal had about one-fifth the weight of the largest whale and was heavier than the giants among fishes, the whale sharks and basking sharks. Between these dimensional extremes, there is a wide range of size and of weight in invertebrates. Among macroscopic arthropods alone, this range extends from such delicate and barely visible insects as aphids and coccids to spider crabs, crustaceans that have a total body span of 12 ft and a weight of 6 to 8 lb, and lobsters weighing more than 8 lb.

Although the architectural patterns appear extremely diverse, they can be resolved into a few master plans, most of which have proved sufficiently flexible to allow great latitude in modification to different ecological circumstances and which, in the course of the long evolutionary development of invertebrate species, have been adapted to many different conditions of living and habits of life. These master plans are the unicellular (acellular) plan; the cell-aggregate plan; the blind-sac plan; and the tube-within-a-tube plan (Fig. II–1). In contrast to the first, the last three of these are multicellular, or pluricellular, in construction. The plans thus fall naturally into the major categories of taxonomic systems, Protozoa and Metazoa.

A. SYMMETRY

In general, bodies built on any of these plans are either radially or bilaterally symmetrical. Radial symmetry, if complete, implies that a cut, or a line drawn, through any two opposite radii divides an organism into two equal and similar halves, as it would a sphere or a cylinder. Bilateral symmetry implies that such a division can be made by a cut along only one plane. Since most bilaterally symmetrical animals are those with well-defined anterior and posterior poles, this plane is the one passing along the midline from head to tail. Typically, the organs are arranged in pairs on either side of this plane, although there are many exceptions to such precise symmetry. Perfect radial symmetry is also rare among animals, for the planes along which the animal may be cut into similar halves are limited by the number and position of organs and structures that may not lie equally along all radii.

Radial symmetry is often associated with a sedentary or sessile habit of life. To an animal that moves but slowly or that spends the greater part, if not all, of its life attached to an immovable object, there are obvious advantages in being equally receptive, responsive, and protected on all sides. But radial symmetry restricts the development of specialized structures, for it demands their repetition as many times as there are planes of symmetry, and such multiple specializations have apparently been difficult of achievement, since they are so seldom found. Bilateral symmetry is usually, but not invariably, associated with an active existence and the habit of moving freely from place to place. It has not been restrictive to evolutionary development, and the great majority of animals, both vertebrate and invertebrate, are bilaterally symmetrical ones.

B. UNICELLULAR (ACELLULAR) PLAN

Many animals and plants are constructed on the one-cell plan, and from the wide variety of types represented by the 30,000 or more species of protozoans at present identified, it is evident that little limitation has been imposed upon their adjustment to every known kind of environment. The great French physiologist Claude Bernard (1813–1878) was the first to point out how fallacious it is to consider Protozoa simple animals, arguing that instead they should be regarded as complex and highly organized ones inasmuch as within a single cell all the activities are conducted that in Metazoa are divided between many cells. Within the unicellular body, special areas, or organelles, may be adapted for one or another of these activities (Fig. II–2). Such are the buccal cavities, gastric vacuoles, and cytopyges that function in the intake of food, the diges-

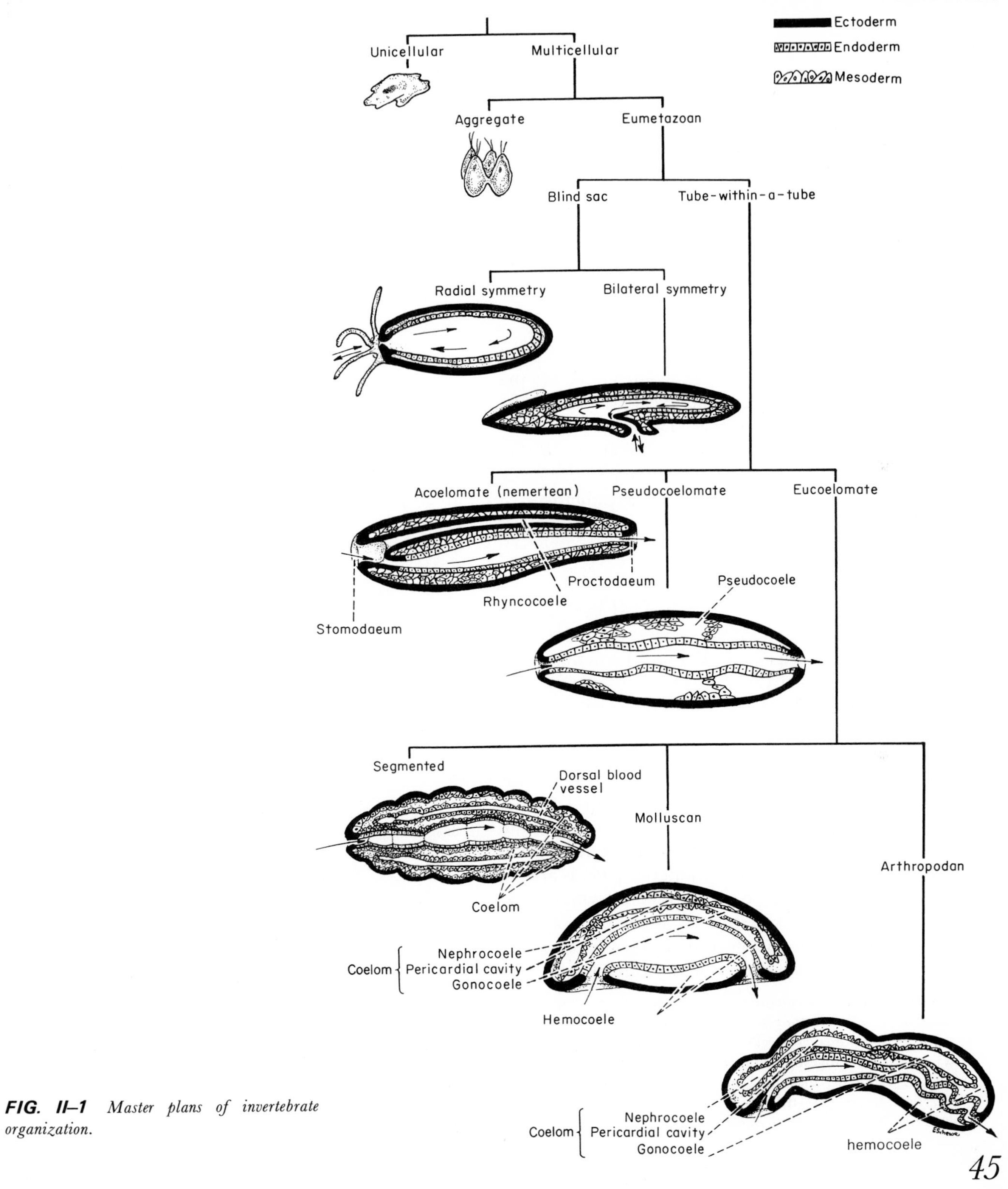

FIG. II–1 *Master plans of invertebrate organization.*

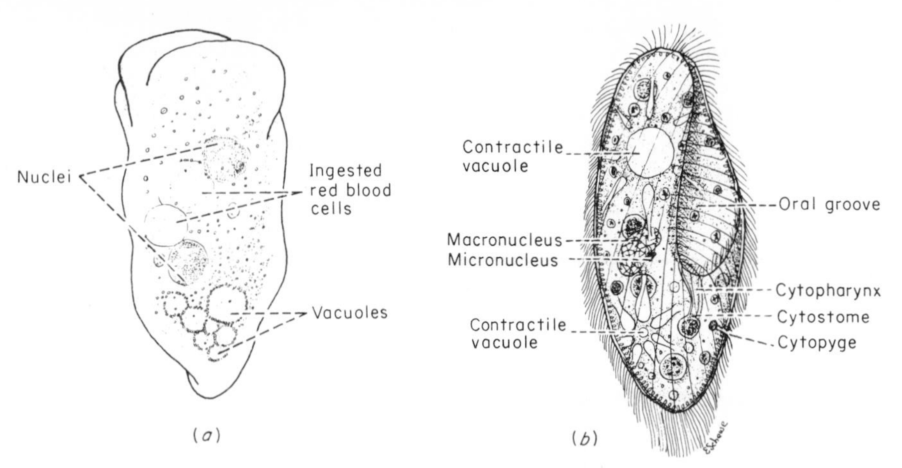

FIG. II–2 *Unicellular plan of organization.* ***(a)*** Entameba histolytica, *an intestinal parasite of man that has been successfully transferred to monkeys, rodents, rabbits, cats, dogs, and pigs. It has no localized organelles.* ***(b)*** Paramecium caudatum, *a fresh-water ciliate with localized organelles for the intake of food (oral groove, cytopharynx, cytostome) and for osmoregulation (contractile vacuoles), and a localized area for the elimination of undigested residue (cytopyge).*

tion, and the elimination of undigested residue; the vacuoles that store and eliminate excess water; and the flagella, cilia, and other specializations of the cell surface. The distribution of these organelles makes most protozoans asymmetrical individuals. Heliozoa and Radiolaria are, however, radially symmetrical.

C. MULTICELLULAR PLANS

1. Cell-aggregate plan

The cell-aggregate plan* is represented in Protozoa by Phytomonadina, and in Metazoa, by orthonectids, dicyemids, and sponges. In the Phytomonadina, the individual consists of a number of cells or zooids (four in some species of *Gonium*, 500 to 50,000 in *Volvox*), generally embedded in a common, gelatinous mass forming a flat, oval, or spherical colony (Fig. II–3). Each zooid is independent of its fellows in trophic and metabolic processes, but functional integration between the individuals is shown by the fact that the colony moves as a unit. In *Volvox* one region, which may legitimately be called the anterior pole, is always directed forward, and the body is radially symmetrical around this antero-postero axis. Some physiological division of labor is evident in these aggregates, for in many of them only certain cells are capable of reproduction. In *Volvox*, for instance, the reproductive cells, upon which the perpetuation of the species depends, are few in number, and invariably located opposite the anterior pole, where the somatic zooids are not only incapable of division but also smaller than the others in the colony.

In orthonectids and dicyemids, the distinction between reproductive and nonreproductive, or somatic cells, is also the extent of differentiation of cell types

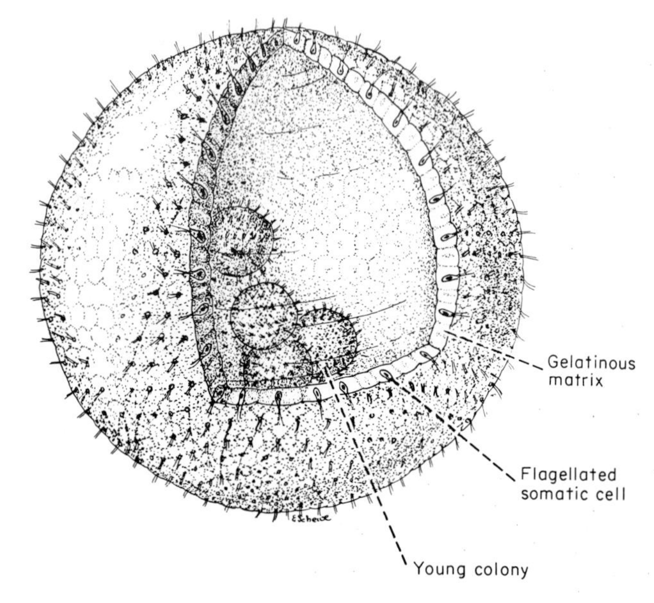

FIG. II–3 *Cell-aggregate plan.* Volvox*: The globular colony has been shown as if a segment were cut away, exposing the hollow cavity and the young colonies inside that have arisen by continued divisions of a reproductive cell derived from those forming the wall.*

* Possibly a group of minute animals belongs here. Originally found in a salt bed they were named *Salinella*. Their bodies consist of a single layer of cells surrounding a central cavity, open at both ends and digestive in function. Not related to any known group of animals, they have no systematic position. Further investigation of them—they can be maintained in saline aquaria—would be interesting and possibly rewarding.

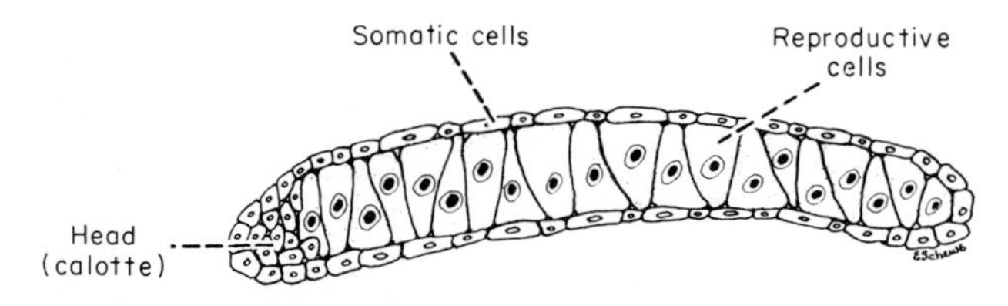

FIG. II–4 *Cell-aggregate plan. A female orthonectid,* Rhopaleura metschikovi. (*After McConnaughey.*)

(Fig. II–4). The body consists of a definite and limited number of cells, rarely more than 25 in all. These cells are assembled in two layers, an outer somatic layer and an inner reproductive one, sometimes represented by only a single cell. These organisms are, therefore, unique animals in that the mass of cells enclosed by the outer layer is partly reproductive, not digestive, in function. Head and trunk regions may be more or less distinguished in the body; the trunk cells are large, ciliated cells, with many vacuoles. The cells in the head region may be radially or bilaterally oriented, but the body as a whole conforms to no particular symmetrical pattern.

The bodies of sponges are composed primarily of an inner and an outer epithelium, surrounding a central cavity, the spongocoele, and separated from each other by a gelatinous layer into which cells from the outer epithelium migrate, giving it the characteristics of mesenchyme (Fig. II–5). The inner epithelium, composed of choanocytes, is nutritive in function and, in the more primitive members of the phylum, completely lines the central cavity. In the more highly evolved members, it is limited to certain areas directly or indirectly connected with the central cavity, which in these individuals is bounded by flattened, nonflagellate cells. Choanocytes are a variety of cyrtocyte, with a collar-like extension surrounding the basal part of a flagellum that extends into the spongocoele. They closely resemble individual choanoflagellates. Electron microscopy has confirmed light microscope observations made nearly 100 years ago that the collars are longitudinally striated and has revealed that they are made up of closely apposed microvilli. In representatives of freshwater Spongillidae, these fibrils (about 30 per collar) have diameters of 0.10 to 0.15 micron (μ). Two, less frequently several, fibrils may be held together by bridges, so that the collar serves as a filtering

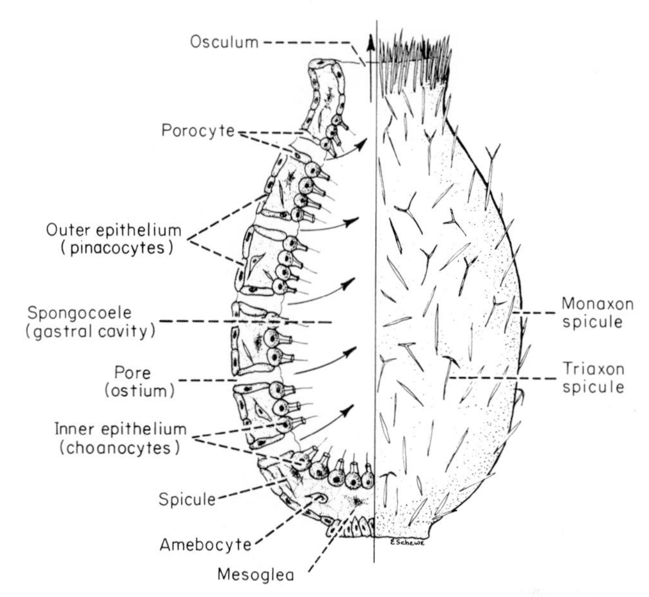

FIG. II–5 *Cell-aggregate plan. The general organization of the body of a sponge, with the right side in surface view and the left represented as if one wall were removed, exposing the spongocoele and the cells enclosing it.*

or trapping device. Early observations also showed that the collars were contractile. Sponges are sessile animals. Those which grow as cylindrical masses, attached by a base and with an opening, the osculum, at the free end, are radially symmetrical but the majority of species are asymmetrical and grow as irregular and branching masses over the substratum to which they are attached. The reproductive cells are derived from mesenchyme cells and may be lodged anywhere between the outer epithelium and the choanocyte layer; therefore they do not provide bodily landmarks as do those of *Volvox* or a dicyemid or orthonectid. Nor does the osculum distinguish an oral from an aboral pole, for the current of water created by the beating of the flagella in the choanocytes enters through numerous pores, or ostia, scattered throughout the body wall, and it leaves through the osculum, which therefore is not the homologue of a mouth (Fig. II–6).

2. Eumetazoan plans

Eumetazoa are animals with differentiated or specialized tissues, which in the more highly evolved species

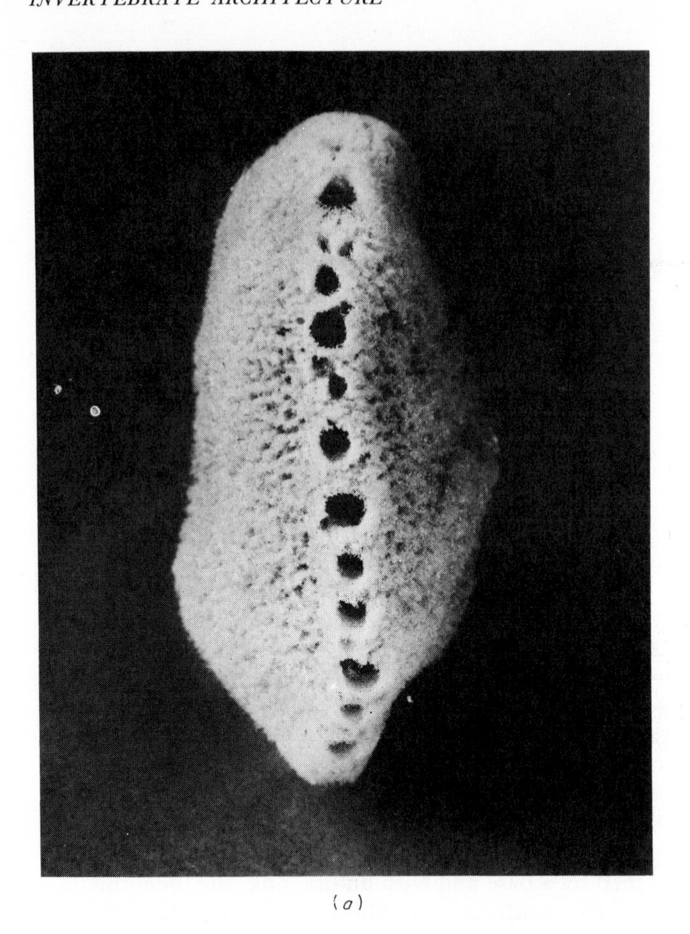

(a)

(b)

FIG. II–6 ***(a)*** *Reef sponge from the Bahama Islands.* ***(b)*** *Grass-wool sponge and sheepswool sponge attached to a common base.* (*U.S. Fish and Wildlife Service.*)

may be localized in organs that perform particular functions. On the basis of their symmetry, Eumetazoa may be distinguished as Radiata, with radial symmetry that is complete, or, far more frequently, limited to certain axes, and Bilateria, with complete or partial bilateral symmetry. On the basis of their body plans they may be separated into those with a blind-sac, radial plan; those with a blind-sac, bilateral plan; those with a tube-within-a-tube plan; with (eucoelomate) or without (acoelomate and pseudocoelomate) a true coelom; and those eucoelomates in which the typical pattern has been modified as either the arthropod or the molluscan plan.

a. BLIND-SAC, RADIAL PLAN Cnidaria and Ctenophora are eumetazoans of simple organization whose bodies show limited radial symmetry and are constructed on the blind-sac plan. In both of these phyla, the individual is composed of an outer epithelium, or epidermis, that covers it, and an inner one, or endodermis, that is primarily nutritive in function and is therefore a gastrodermis bounding the single internal cavity. Between these two tissues is mesoglea, invaded to a greater or lesser extent by migratory cells and so, like the corresponding layer in sponges, actually mesenchymatous. There is a permanent mouth opening which defines the oral pole, and hence also the aboral, through which food is passed to the gastral cavity. This is the only opening into the body, so that material that enters and leaves the gastral cavity,

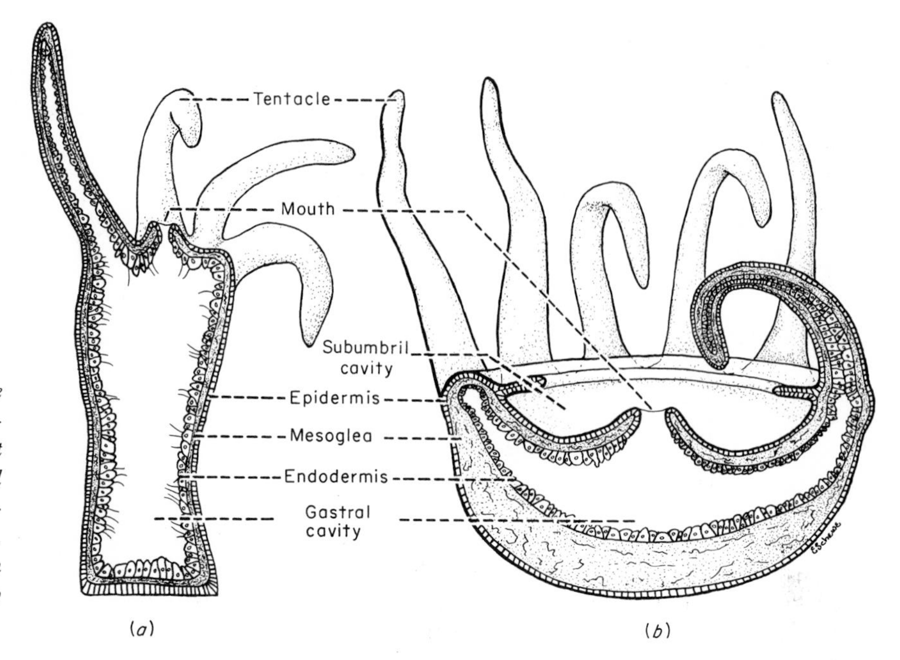

FIG. II–7 *Blind-sac, radial plan. The polyp and medusa plans of cnidarian organization.* ***(a)*** *Polyp plan, shown as if cut through the body so exposing the gastral cavity and the cells enclosing it.* ***(b)*** *Medusa plan, represented in the same way. The medusa is shown as if inverted from its natural position to indicate the homology between the medusa and the polyp plans.*

essentially a blind sac, must pass through the same orifice. The bodies of Cnidaria have in general the form of cylindrical polyps or disk-like medusae, either flattened or conical in shape; in both forms, the planes of symmetry along the oral-aboral axis are limited by the number and position of certain structures or organs (Fig. II–7). In polyps, these planes are determined by the number of tentacles that surround the mouth, the number of endodermal septa that project into the gastral cavity, or, as in most Anthozoa and Zoantharia, by the siphonoglyph, a ciliated groove running from the mouth opening inward along the stomodaeum, or so-called "pharynx" (Fig. II–8). This imposes a biradial symmetry upon the members of the subclass. Two similar halves can be obtained only by division in the sagittal or median plane along the long axis of the stomodaeum, bisecting the siphonoglyph. In medusoid individuals, the number of radial canals extending from the reduced gastral cavity to the circular canal at the periphery of the bell determines the possible planes of symmetry.

Ctenophora, whose bodies are typically conical, or, with great modification of the primary pattern, flattened in either the oral-aboral or the tentacular

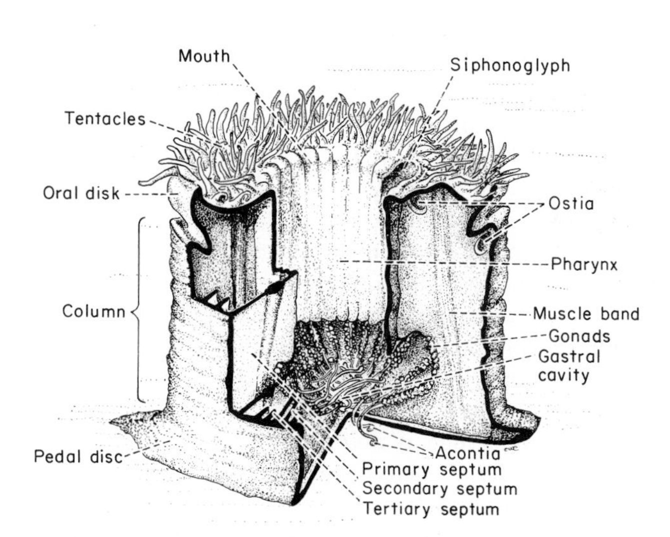

FIG. II–8 *Blind-sac, biradial plan. Dissected sea anemone,* Metridium *(Anthozoa), showing the biradial symmetry of the cylindrical body imposed by the presence of the siphonoglyph. The internal cavity is partially divided by pairs of septa. The primary pairs are attached to the wall of the pharynx; the secondary, tertiary, and subsequent pairs end free in the gastral cavity.*

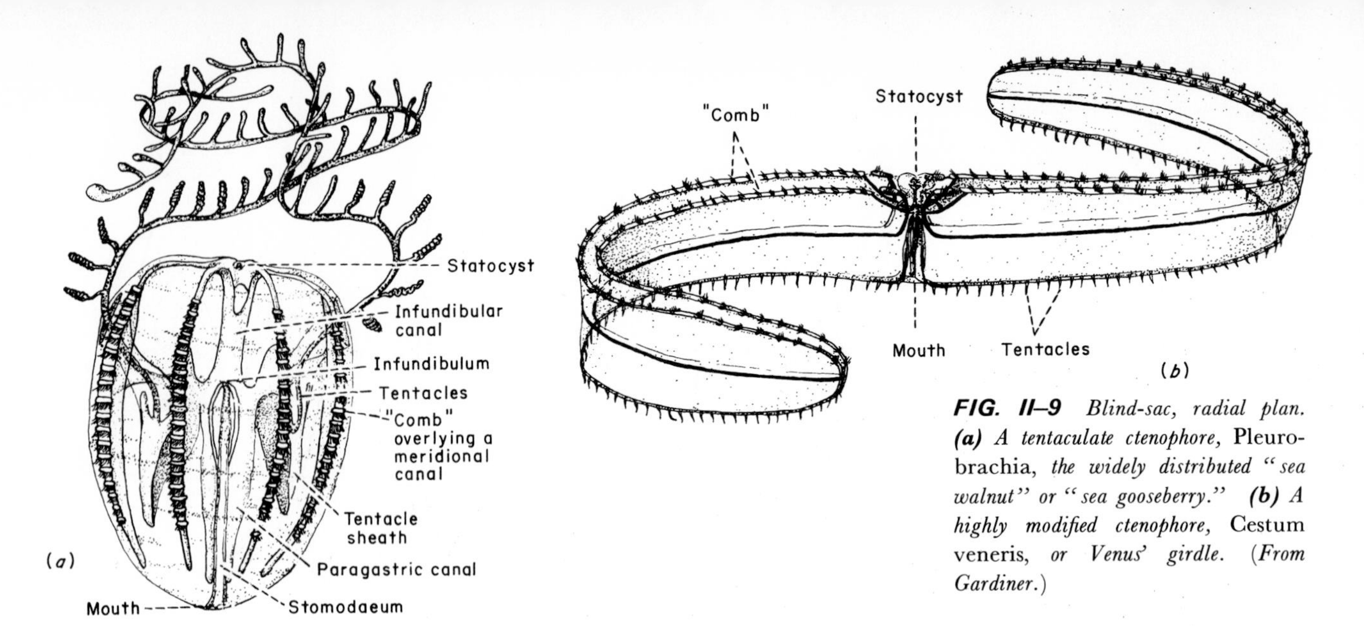

FIG. II–9 *Blind-sac, radial plan.* ***(a)*** *A tentaculate ctenophore,* Pleurobrachia, *the widely distributed "sea walnut" or "sea gooseberry."* ***(b)*** *A highly modified ctenophore,* Cestum veneris, *or Venus' girdle.* (*From Gardiner.*)

plane, are exclusively biradial forms, their symmetry determined by the two transverse canals originating from the stomach and, in Tentaculata, by the paired tentacles that lie on opposite sides of the body (Fig. II–9).

b. **BLIND-SAC, BILATERAL PLAN** In the platyhelminths, bilateral symmetry is combined with the blind-sac plan of bodily organization (Fig. II–10). In these animals, the gastral cavity, or enteron, generally represented by a pharynx and an intestine, has but a single external opening, the mouth. The intestine may be a simple endodermal pouch or may be extensively branched. In some of the most primitive forms, the acoelous turbellaria, there is no intestine but only a pharynx, or simply a mouth, communicating directly with the interior of the body. In the most highly specialized parasitic members of the phylum, the cestodes, both digestive tract and mouth are entirely wanting, and food is absorbed directly through the body wall. In all platyhelminths, the interior of the body, exclusive of the various organs, is filled with parenchyma, a mesenchymatous tissue. In polyclads, whose embryological development probably follows the primitive pattern most closely, this tissue is derived from one of the early blastomeres. The endoderm of the gut and the parenchyma in which it and the other organs are embedded, as well as some of the muscle fibers, originate by successive divisions and differentiations of this cell, the mesendoblast. The parenchyma is thus essentially mesoderm; some of its cells have the attributes of amoebocytes while others are syncytial, united into a loose network through whose interstices fluid circulates and amebocytes freely move. In some, these interstices may be of considerable size, simulating a perivisceral cavity but in no way homologous to the perivisceral space of typical eucoelomate animals.

c. **TUBE-WITHIN-A-TUBE PLAN** In animals constructed on a tube-within-a-tube plan, a second opening of the digestive tract, the anus, typically at or near the posterior end of the body, converts the blind-sac type of gastral cavity into a continuous tube enclosed within the outer tube of the body wall. The acquisition of this opening has distinct evolutionary advantages. Separation of the apertures for the intake of food and the output of undigested residue affords possibilities for the development of special devices for its capture, ingestion, mastication, and even preliminary digestion. Two openings also permit the specialization of areas along the tract for the storage of food, its progressive chemical disintegration, and its absorption, as well as for the concentration and ultimate defecation of the residue. In tube-within-a-tube animals the digestive tract may be embedded in parenchyma (acoelomate plan) or suspended in a cavity which may be either a pseudocoele or a true coelom (pseudocoelomate and eucoelomate plans).

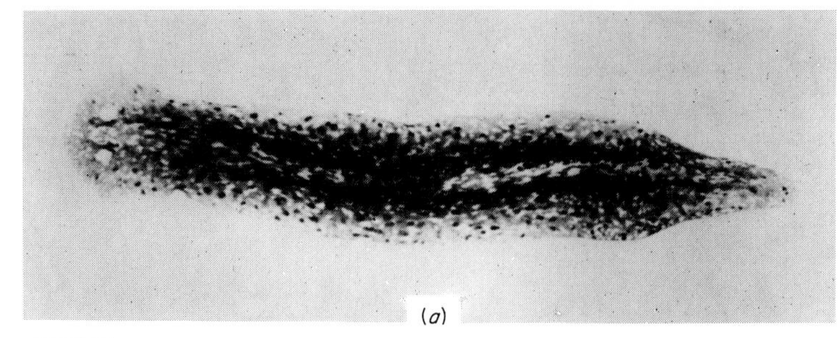

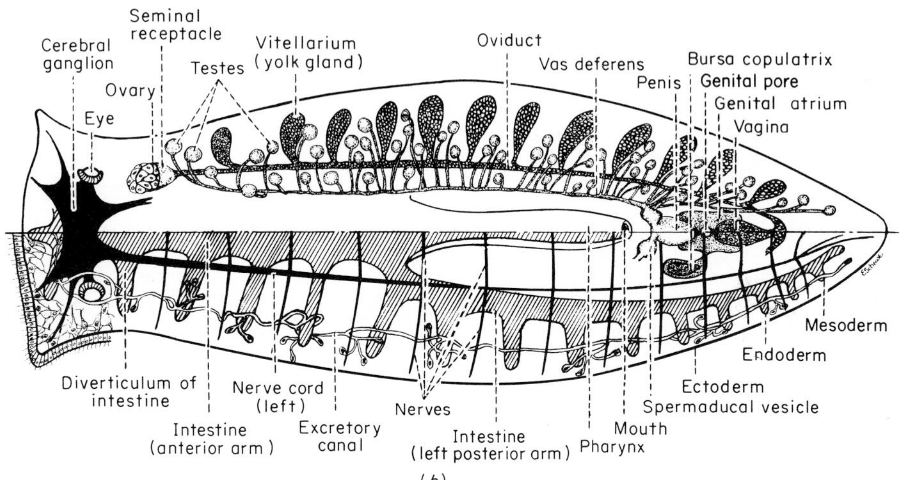

FIG. II–10 *Blind-sac, bilateral plan.* ***(a)*** Dugesia tigrina, *a turbellarian flatworm.* (*American Museum of Natural History.*) ***(b)*** *The internal anatomy of a triclad flatworm, such as* Dugesia. (*From M. S. Gardiner.*)

(1) Acoelomate plan One phylum of tube-within-a-tube animals, that of the Nemertinea (Rhynchocoela), is constructed on this plan, which is also the pattern of the blind-sac, bilaterally symmetrical flatworms. The nemerteans are the "ribbon worms," flat unsegmented animals that are particularly characterized by their unique possession of a long, eversible proboscis that lies, when unextended, in a tubular pouch or sheath that extends some distance along the length of the body beneath its dorsal wall. Except for this rhynchocoele, the body is solid in the sense that the digestive tract and other organs are surrounded by parenchyma cells, so that the only open spaces are the lumina of tubes and vessels (Fig. II–11).

(2) Pseudocoelomate plan In animals constructed on this plan, the perivisceral cavity represents the primary body cavity, the blastocoele of the embryo which is retained as a permanent internal space in the adult. The mesenchyme cells that migrate into it may become organized into connective tissue, muscles, or other specialized types of cells but not into mesothelium such as that bounding the perivisceral cavity of eucoelomate animals. This space is therefore homologous with the region that in Cnidaria and Ctenophora is filled with mesoglea and migrated cells, and in the flatworms and nemerteans with parenchyma. It is not homologous with the coelom, or secondary body cavity. The bodies of Acanthocephala, Rotifera, Gastrotricha, Kinorhyncha, Nematoda, and Nematomorpha are built according to this pattern (Fig. II–12).

(3) Eucoelomate plan A secondary body cavity, given in 1879 the name of coelom, from the Greek word *koilos*, meaning "hollow," by the German biologist Ernst Haeckel (1834–1919), is a perivisceral space, filled with fluid and bounded by mesothelium, that is, by an epithelium of mesoblastic origin, often called the peritoneum. Among existing animals, the coelom originates in two principal ways, either as an enterocoele from mesodermal pouches budded off the embryonic archenteron or as a schizocoele by separation, or split-

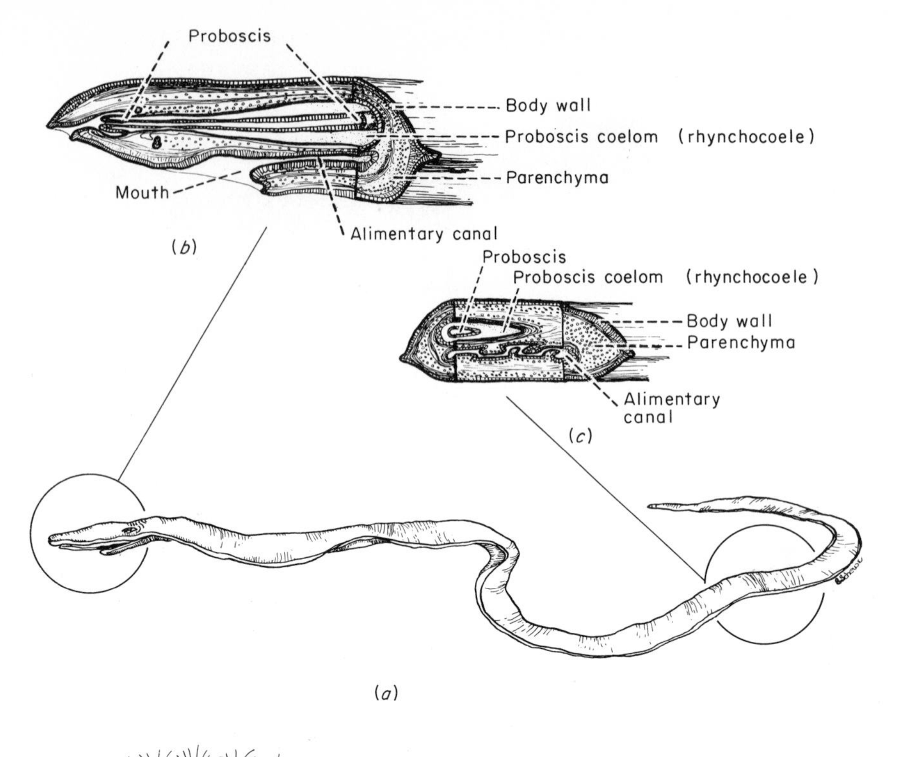

FIG II–11 *Acoelomate tube-within-a-tube plan.* ***(a)*** *External view of a nemertean.* ***(b)*** *The internal anatomy of the anterior region [enclosed in a circle in (a)].* ***(c)*** *The internal anatomy of the posterior region [enclosed in a circle in (a)].*

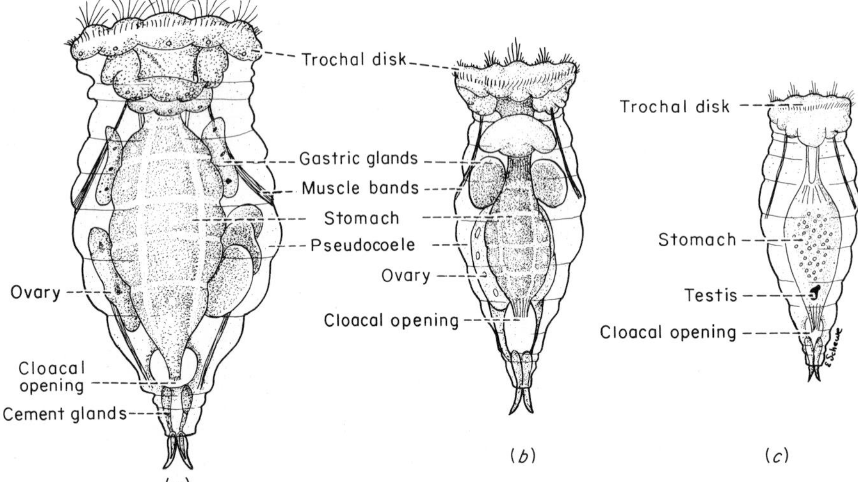

FIG. II–12 *Pseudocoelomate tube-within-a-tube plan. The anatomy of the rotifer* Epiphanes senta. ***(a)*** *Parthenogenetic female.* ***(b)*** *Sexual female.* ***(c)*** *Male.* (*From Gardiner.*)

ting, of masses of mesoderm derived from cells set apart at some early stage of embryonic development as mesoblasts or mesoderm-forming cells (Fig. II–13). In the first case, characteristic of Chaetognatha and Echinodermata, the coelomic pouches become detached from the wall of the archenteron, enlarge, and finally come to occlude the primary body cavity completely, or almost completely. In the second case, characteristic of chordates, as well as of all the remaining phyla of eucoelomate nonchordates with the exception of the Phoronida, the mesoderm first appears as bands or sheets of cells in the primary body cavity, later becoming separated into an outer (somatic) and an inner (splanchnic) layer by the migration or dissolution of the central cells. In the Phoronida, mesenchyme cells that migrate into the blastocoele become arranged as mesothelium bounding the coelomic cavity. The primary body cavity thus becomes occluded, or greatly reduced in extent, and replaced by the cavity secondarily formed. Also, the outer or peripheral layer of mesoderm becomes applied to the inner surface of the body wall as parietal or somatic mesoderm and the inner layer to the outer surface of the digestive tract and other viscera, as splanchnic or visceral mesoderm. In these locations, the cells of both may differentiate as muscles, connective tissue, and coelomic epithelium (mesothelium) that lines the coelomic cavity and covers all the organs suspended in it.

The acquisition of a coelom, like that of a second opening to the digestive tract, conferred certain advantages upon the animals in which it developed. It is in coelomate animals that the digestive tract, with its own musculature, can control the passage of food along its length independent of movements of the body as a whole and in which it has found space to expand, enlarging its area for the processing of food by the development of accessory glands and of a long intestine. Since this can be coiled within the coelomic cavity, proportionate increase in body length is not necessary. Similarly, the body muscles can move independently of the visceral ones, and other organs as well as those of the digestive system have room to grow and increase their functional capacities. But the coelom has also imposed certain conditions upon the animals possessing it, which had to be fulfilled if the coelomate pattern was to become the successful one that it has proved to be. These conditions arise primarily from the separation of the digestive area by an appreciable distance, and tissues of appreciable thickness, from other important areas of the body, making diffusion or even amebocyte transport an inefficient, or indeed ineffective, means of distribution of its products. Concomitant with the establishment of such a body plan must therefore have been the development of a transport system of such a kind that substances essential to the operation of the cells could be carried, in solution or suspension, from one part of the organism to another and the end products of cellular metabolism collected and ultimately disposed of. A circulatory system thus becomes obligatory to a coelomate animal—at least to one of any size—and such a system in which a fluid tissue, the blood, is propelled through a series of vessels, or of sinuses, is characteristic of coelomate animals. Echinoderms are an exception to such a generalization.

Possibly in the first coelomate animals, the coelom

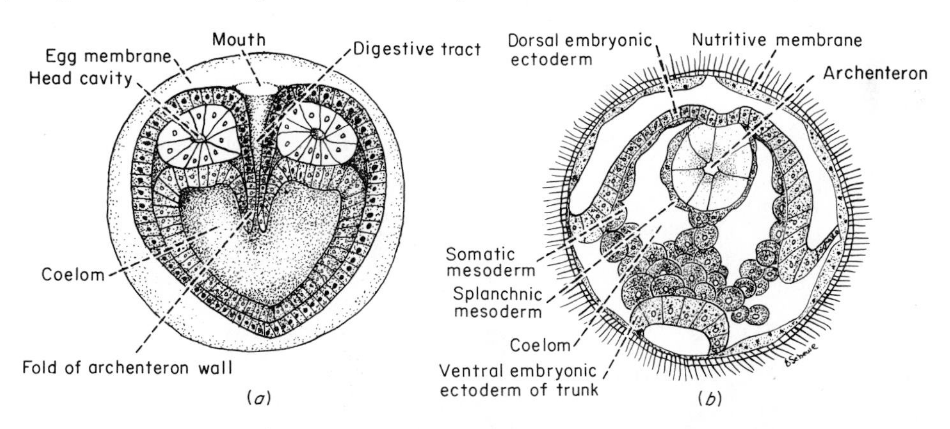

FIG. II–13 *Development of the coelom as enterocoele or schizocoele. (After Hyman.)* ***(a)*** *Section through an embryo of the arrow worm* Sagitta, *in which the coelom develops as an enterocoele by the backward growth of folds of the wall of the archenteron. These folds are directed inward, so that with their growth the archenteron becomes divided longitudinally into a median portion, the digestive tract, and two lateral ones, the coelomic sacs. The anterior portions of these sacs become separated from the rest and form a pair of head cavities, from whose walls the chaeta muscles are differentiated.* ***(b)*** *Section through a trochophore larva of* Sipunculus nudus *in which the coelom develops as a schizocoele by "splitting" of the mesodermal bands derived from repeated divisions of the mesoblast.*

furnished a place where metabolic wastes could be deposited until they could be eliminated from the body or stored in a biologically inactive state. Or possibly the coelom first functioned as a region into which reproductive cells were discharged and retained until set free from the body that produced them. The facts that the reproductive organs, if not the cells, of coelomate animals are mesodermal in origin, that they project into the coelom, and that the germ cells are shed into some part of it are supporting, but not conclusive, evidence for the latter concept of its original significance. Either function, or both together, necessitates some connection between the coelom and the external environment, unless the body is to disintegrate each time the germ cells are shed or the coelom relieved of its load of metabolic waste. Such connections, in the form of coelomoducts, seem to have developed along with the coelom itself. Among existing animals, these may be retained, with slight modifications, as genital ducts or gonoducts, or they may be wholly or partially obliterated, their surviving portions being united with the nephridia into a combined excretory organ (Fig. II–14).

(a) SEGMENTATION Many eucoelomate animals are segmented ones, with bodies that consist of a succession of segments or metameres in which the paired organs are serially repeated. A segment has been defined as "a region more or less distinctly marked off from the rest of the body by transverse grooves, surrounding the alimentary canal, containing a special coelomic cavity (more or less completely separated off

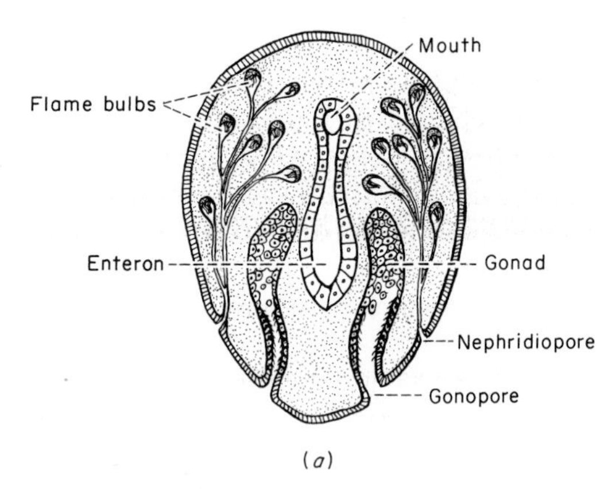

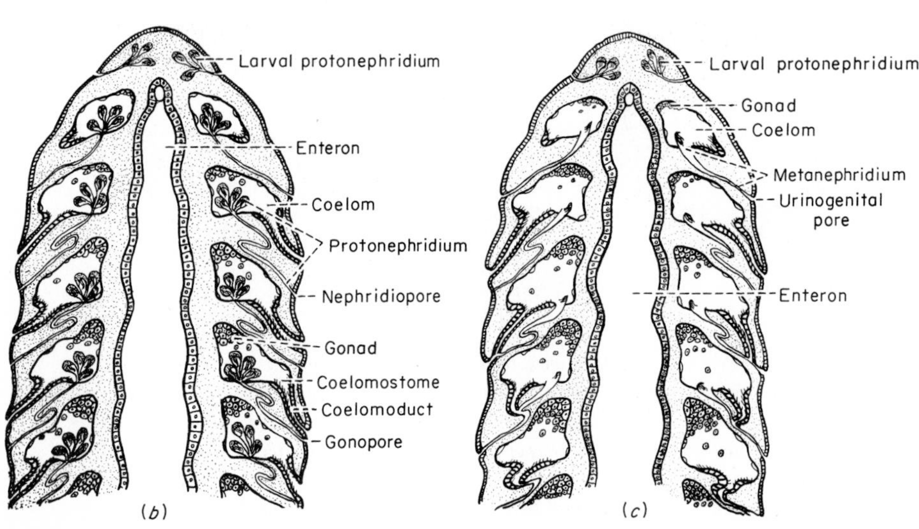

FIG. II–14 *The relationship of nephridia and coelomoducts.* ***(a)*** *Longitudinal section through a primitive flatworm, showing protonephridia ending internally in flame bulbs and externally in a pair of nephridiopores; and a pair of coelomoducts receiving the products of the gonads and discharging them to the outside through their external openings, the gonopores.* ***(b)*** *Longitudinal section through the anterior region of a primitive segmented worm showing, in the most anterior segment, a pair of protonephridia draining the mesenchyme (as in the larval stages of some modern annelids) and, in the posterior segments, protonephridia with flame bulbs lying in the coelomic cavity. In each segment, except the two most anterior, paired coelomoducts are shown, independent of the nephridia and opening to the outside through gonopores.* ***(c)*** *Longitudinal section through a primitive annelid showing the conversion of the protonephridia into metanephridia and their connection with the coelomoducts as mixonephridia. (From Gardiner.)*

from the coelom of adjoining segments by means of transverse septa), a pair of nephridia and of peritoneal funnels communicating with the exterior, a pair of ganglionic enlargements of the ventral longitudinal nerve cords, and (in polychaetes and arthropods) a pair of appendages."* Among modern animals, the segmented, coelomate tube-within-a-tube plan is most completely expressed by annelids, but even in the polychaetes, segmentation is not ideal. Their most anterior segments lack peritoneal funnels and nephridia and have the ganglia of several segments concentrated into a dorsal mass connected by commissures to the ventral chain, so constituting a more or less well-defined head, posterior to which the series of segments begins (Fig. II–15). In other segmented invertebrates there are further modifications. Certain segments may be wanting and others fused together with their organs united into common structures, or else reduced or entirely obliterated. Various theories have been advanced to explain the origin of segmentation in the animal kingdom and the survival value it obviously had. Possibly its value is due to more effective body movement, since a segmented body can provide a mechanism based on a series of consecutive motor units; possibly it offered a vital margin of safety, since in an organism in which there is a serial repetition of parts injury or damage to one area is not necessarily fatal. Whatever may have been its basis in the establishment of animal types, it must have been a distinctly significant factor for a modified segmental pattern is characteristic of the most successful animals, the arthropods and the vertebrates.

* E. S. Goodrich, *Q. J. Microsc. Sci.*, vol. 40, pp. 249–250, 1898.

FIG. II–15 *Segmented, coelomate tube-within-a-tube plan.* **(a)** *The polychaete annelid,* Neanthes virens *(the clam worm).* *(New York Zoological Society.)* **(b)** *The oligochaete annelid,* Lumbricus terrestris *(the earthworm).* *(U.S. Department of Agriculture, Bureau of Entomology and Plant Quarantine.)*

(b) THE MOLLUSCAN PLAN Molluscs are fundamentally bilaterally symmetrical, eucoelomate animals, but in the course of their evolution this basic pattern has been profoundly modified. In its most generalized form, the molluscan body is composed of a defined head, where photo-, chemo-, and tactoreceptors are localized, a visceral mass in which most of the organs lie, and a highly muscular region ventral to this, the foot. The central nervous system consists of five pairs of ganglia (cerebral, pleural, pedal, parietal, and visceral) located in the head, the visceral mass, and a foot; primitively, the respiratory apparatus consists of one or more pairs of ctenidia, or gills, which are outgrowths of the lateral body wall, each consisting of a vascular central shaft with pinnately arranged lateral projections, the gill filaments (Figs. II–16 and II–19). The coelom is much reduced and is represented in adult animals only by the cavities of the gonads and excretory organs and by the pericardial cavity around the heart. This reduction may have been effected by an unusual development of the elements of the primary body cavity into great blood spaces at the expense of the secondary body cavity, whose lumen consequently was occluded as its walls were pressed together in all but these three regions. Or it may have been the consequence of a tremendous distension of capillaries and their union into a common blood space, for in molluscs and arthropods with so-called "open" circulatory systems blood sinuses replace the capillary beds found in animals with "closed" circulations, where the blood passes along a continuous system of tubes with bores of varying diameter and walls of varying thickness. The perivisceral space in molluscs is thus not the coelom but a hemocoele. Another distinctive feature of molluscs is the mantle enclosing the greater part, or all, of the body. This is a double layer of epidermis, often vascular, that grows down from the middorsal body

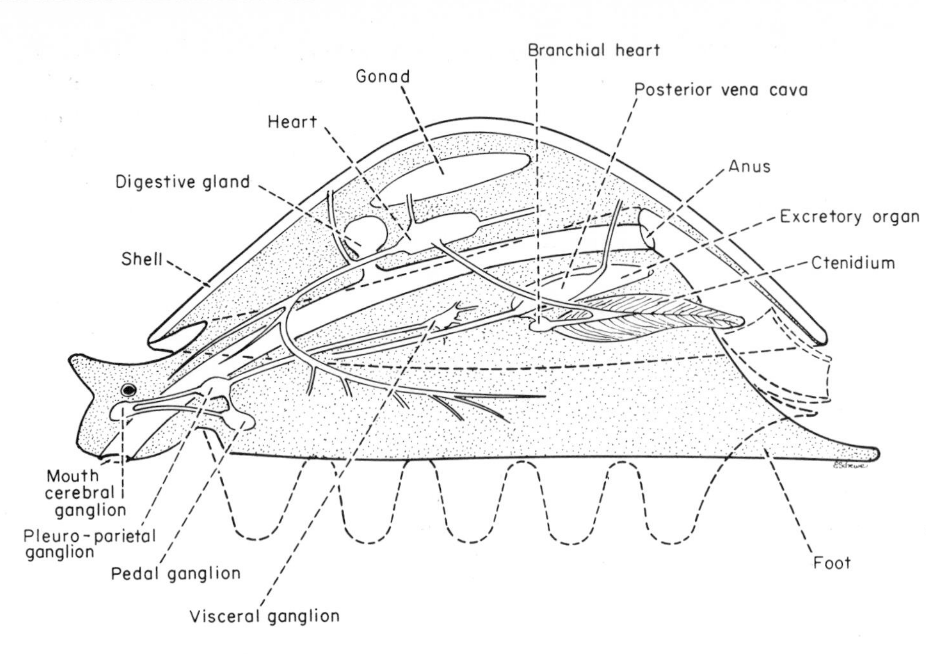

FIG. II–16 *The molluscan plan. The anatomy of a hypothetical unspecialized mollusc.* (*After Williams.*) *The dotted lines below the foot indicate the modification of this region into arms in the cephalopods. Those in the mantle cavity indicate the position of the funnel in cephalopods.*

wall around the visceral mass, separated from it by a space, the mantle cavity, in which the gills often lie and into which the anus and the reproductive and excretory ducts open. In the great majority of molluscan species, the cells of the mantle secrete materials for the shell that partially or entirely encloses the soft parts of the body.

The evolutionary origin of molluscs from a segmented ancestor and, thus, their phylogenetic relationship to annelids has long been postulated. This hypothesis is based primarily upon resemblances in cleavage patterns and larval forms of the two phyla. The recent discovery of a number of specimens dredged up from the depths of the sea off the west coast of Mexico and the Peru-Chile Trench off northern Peru has confirmed this relationship. Ten living specimens of *Neopilina galathea* and three additional shells of this species, previously supposed to have become extinct 280 million years ago, were brought up from a depth of 3,590 m off western Mexico by the Danish expedition that cruised from 1950 to 1952 on the ship *Galathea*. Within a comparatively thin and roughly circular shell, the organs of this mollusc are symmetrically arranged along the longitudinal axis, with five pairs of gills, five pairs of auricles receiving blood from them, five pairs of excretory organs, and five pairs of dorso-ventral foot retractor muscles. The gonads are also paired, their products passing out through the excretory organs. More recently, in 1958, an American expedition found four similar, but not identical, forms in material dredged from even greater depths in a locality some 1,300 miles south of the *Galathea*'s find. These specimens are considered to represent a new subgenus and species and have been given the name of *Neopilina* (*Vema*) *ewingi* after the oceanographic vessel the *Vema* and Dr. Maurice Ewing, the director of the Lamont Geological Observatory of Columbia University (Fig. II–17). *Neopilina (Vema) ewingi* has six pairs of gills, one more pair than *N. galathea*; it has a greater number of postoral tentacles than that species and an even more delicate and fragile shell. Both are included in the class Monoplacophora, and there are similarities between them and the chitons, in the class Polyplacophora (Fig. II–18).

The molluscan body plan has proved to be sufficiently flexible to permit the development of a great variety of types, for there are some 88,000 species of existing molluscs, and a large number of fossil ones. The modifications of the basic plan have resulted chiefly from alterations in the relations of the principal parts of the body, such as suppression of the head or

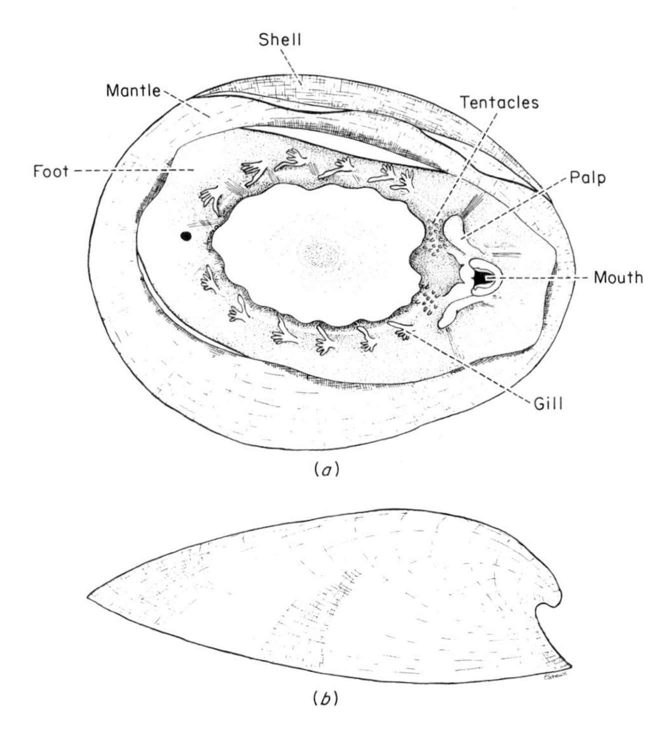

FIG. II–17 Neopilin (Vema) ewingi. (*After Menzies and Clarke.*) **(a)** *Ventral view.* **(b)** *Lateral view.*

decided accentuation of it with localization of the main nerve centers there; reduction of the foot or its adaptation as a blade-like structure in pelecypods (Fig. II–19*c* and *f*), a circlet of powerful arms, or tentacles, in the cephalopods (Fig. II–19*e*), or a bilobed, wing-like process in the pteropods (Fig. II–20); or enlargement of the visceral mass, giving additional space for the expansion of its contained organs and their associated structures. The development of a glandular outgrowth from the digestive tract, in which most of the digestion and absorption of the ingested food takes place, has produced, in gastropods especially, a marked dorsal enlargement of the visceral mass (Fig. II–19*a* and *b*). The coiling of this "dorsal hump" through the unequal growth of one or the other of its sides has made it possible for them to preserve this advantageous alimentary modification. Uncoiled, it would become an unwieldy, dome-shaped mass that could hardly have stayed erect without some kind of support, and its coiling in a flattened or ascending spiral is the solution to this difficulty with which the gastropods were provided. Partly in consequence of this, gastropods have lost the bilateral symmetry of the molluscan stock and are now asymmetrical animals.

This asymmetry is also a consequence of torsion, an event that occurs in embryonic or larval life. In

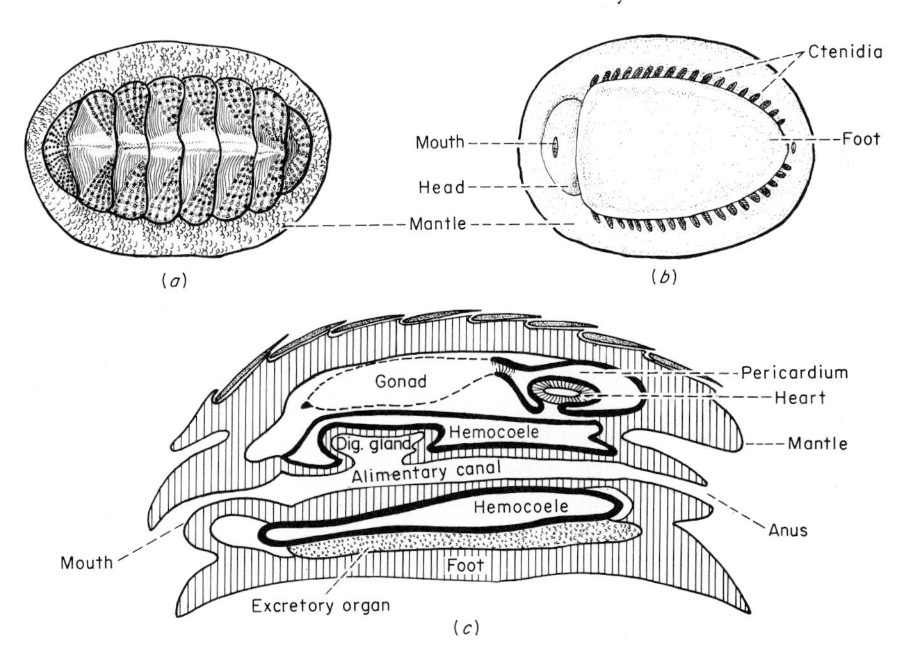

FIG. II–18 *The anatomy of a chiton.* **(a)** *The exterior from the dorsal surface.* **(b)** *The exterior from the ventral surface.* **(c)** *Section in the midsagittal plane.*

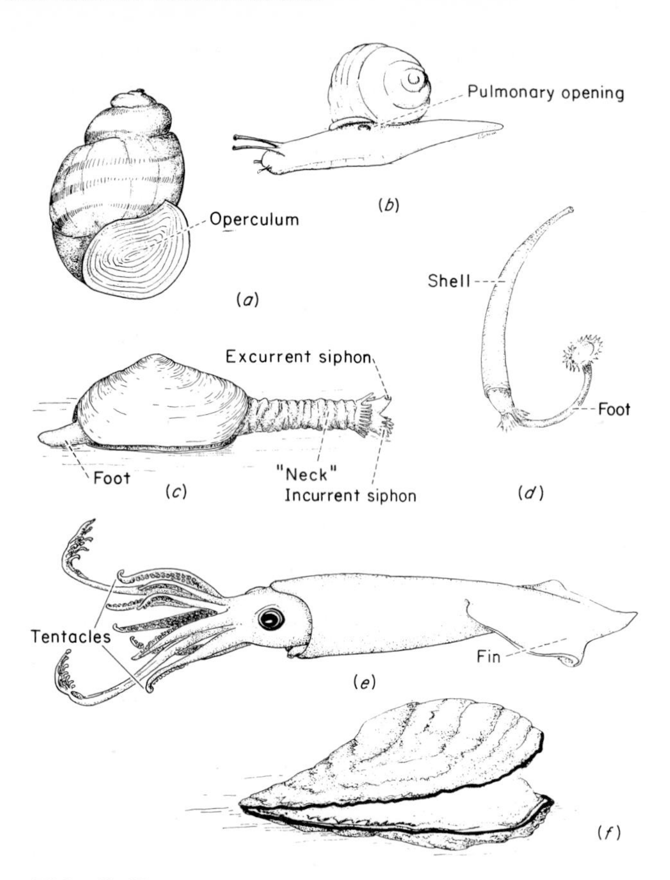

FIG. II–19 *Representative molluscs.* ***(a)*** Campeloma subsolida, *a freshwater snail, withdrawn into its shell.* ***(b)*** Helix, *a land snail, extended.* ***(c)*** Mya, *the soft-shell clam, with foot and siphons extended.* (*a, b, c after Cambridge Natural History.*) ***(d)*** Dentalium, *a scaphopod.* (*After Grassé.*) ***(e)*** Loligo pealii, *the squid.* ***(f)*** Ostrea, *the oyster, with open shell.*

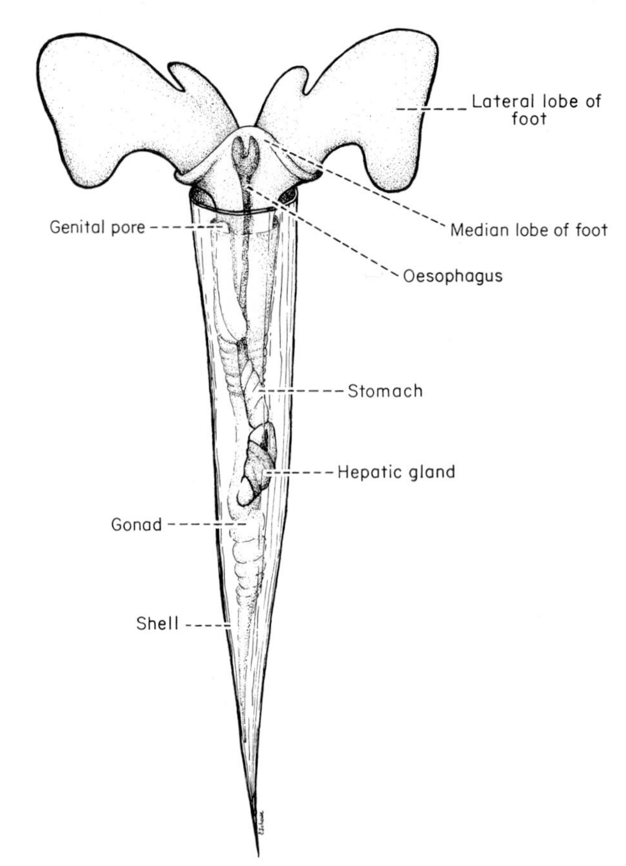

FIG. II–20 Clioacicula, *a pteropod, from the ventral surface.* (*After Lankester.*)

torsion, the dorsal hump is swung through an arc of 180°, so that the organs that lie in the neck of tissue connecting it with the rest of the body are reversed in their antero-postero positions, and therefore also in their positions relative to the right and left sides of the body. Thus, the anus comes to be directed anteriorly, the ventricle of the heart to lie posterior to the auricles, and the left auricle, ctenidium, and excretory organ to be shifted to the right side of the body, with their counterparts to the left (Fig. II–21). The right and left parietal and visceral ganglia are also reversed in position, and the commissures connecting the parietal with the cerebral ganglia are looped into a figure eight, one arm of the loop passing above the intestine, the other below it. This and the coiling of the visceral hump has led, in a number of species, to the suppression of organs on one side (usually the right), so that as adults they have but a single auricle, a single ctenidium, and a single gonad.

A number of theories have been advanced to explain the origin of torsion, and its evolutionary significance. One of these suggests that it occurred originally, by stages, in the adult; another, that it occurred in postlarval life and met the needs of adult life. Other theories connect it with embryonic life, meeting the needs of adult life in benthic regions or of embryonic life in pelagic. According to one theory, it is due to antagonism between the growth of the foot and the shell; according to another, it results from

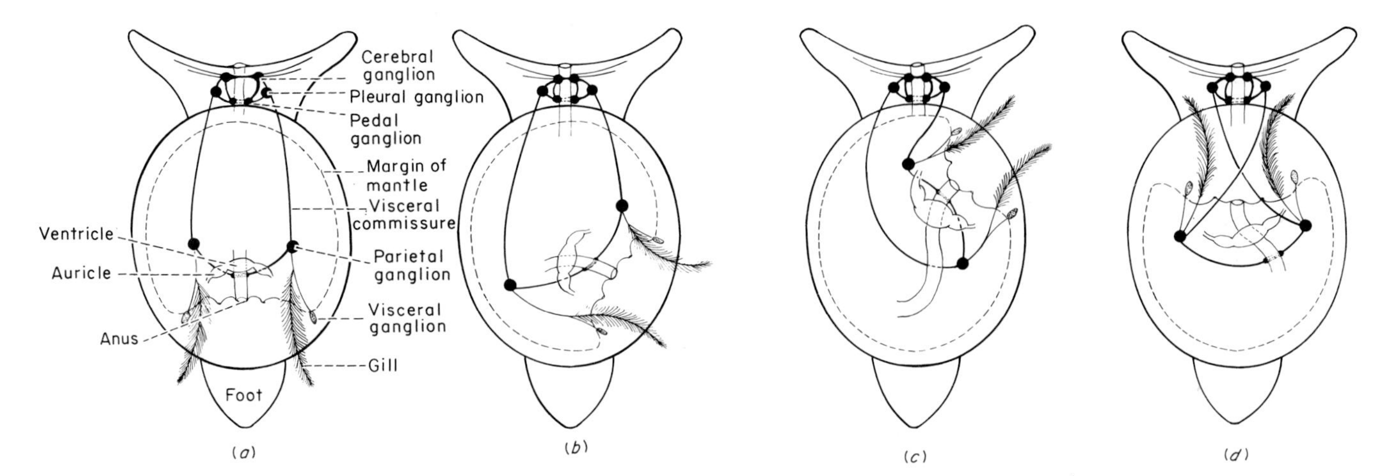

FIG. II–21 *Torsion in gastropods. (From Gardiner.)* ***(a)*** *Hypothetical primitive gastropod in dorsal view, showing the bilateral arrangement of the paired organs.* ***(b** and **c)*** *Hypothetical stages in the progress of torsion, whereby the organs on the left side of the visceral mass rotate through an arc of 180°.* ***(d)*** *Final position, characteristic of some modern gastropods. The gill, visceral, and parietal ganglia and auricle originally on the right side of the body now lie on the left, and the corresponding structures originally on the left side now lie on the right; the alimentary canal, originally a straight tube running antero-posteriorly, is bent into a loop with the anus opening anteriorly into the mantle cavity; the visceral commissure is thrown into a figure 8. In a number of modern branchiate gastropods, the gill and ganglion of one side or the other are reduced or absent.*

unequal growth on the two sides of the body so that one, enlarging faster than the other, coils around it. Careful analysis of the early development of certain prosobranch gastropods, particularly in relation to the larval musculature, has provided what is probably the most accurate explanation of torsion. In these, the mesodermal band on the right side is larger than that on the left, and in the abalone (ormer) *Haliotis tuberculata*, about 22 hr after fertilization, its cells begin to elongate and to differentiate as muscle cells. Because of this, and because the visceral hump is gradually displaced to the left, the young veliger larva is actually asymmetrical before torsion begins. From the muscle band is developed a powerful retractor muscle, consisting of six spindle-shaped cells, all attached at one end (the posterior) to the apex of the shell (Fig. II–22). Because of their connections, and because the corresponding muscle of the left side is not yet operative, the effect of the contraction of these cells is both to retract the mantle and the velum and also to rotate the cephalo-pedal mass in a counter-clockwise direction. In most of the genera investigated, torsion begins as soon as the muscle cells acquire retractor power and occurs in two phases. The duration of the first phase, in which the visceral hump is swung through 90°, depends upon the amount of yolk present in the egg; it is accomplished in 3 to 6 hr in *Haliotis* and in 10 to 15 hr in the limpet *Patella*. The second phase is not completed until the sixth day of larval life in *Haliotis*, when the animal has taken up a benthic existence, but in *Patella*, it takes only a few hours. In *Haliotis*, it starts when the larva is 29 to 35 hr old, but in *Calliostoma* there is no free larval stage, yet the embryo, while still within the egg membrane exhibits torsion similar to that of other gastropods. In very specialized prosobranches, such as *Viviparus*, torsion occurs before the development of any larval muscles, presumably by differential growth.

Torsion appears to be a reversible process, for in some genera the anus and the organs on either side of it lie posteriorly and the nerve commissures are untwisted. This is associated with, or accompanied by, uncoiling of the visceral hump and reduction in the size of the shell, or even with a complete absence of

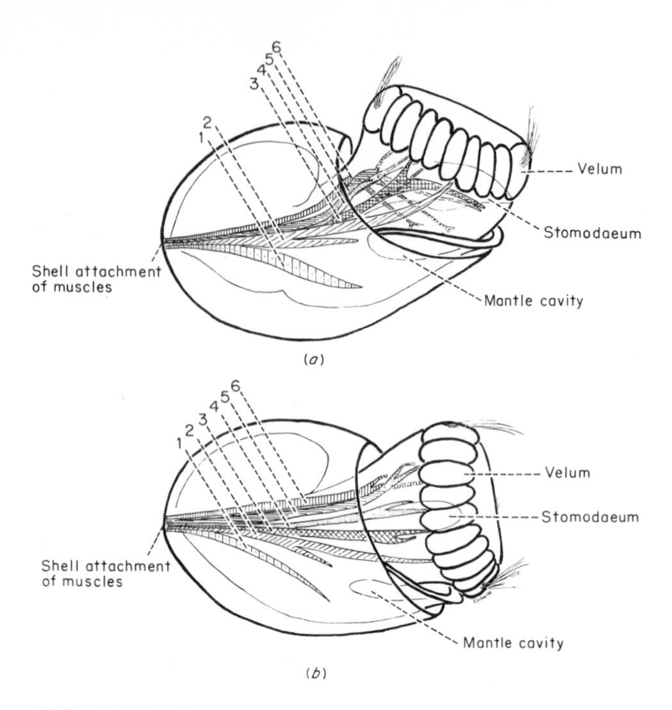

FIG. II–22 *Torsion in* Haliotis tuberculata. *(After Crofts.)* ***(a)*** *Veliger larva from the right side before torsion: 1, Muscle cell attached to outer region of the mantle on the right side. 2, Muscle cell, one of whose processes is attached to the mantle and the other to the underside of the velum, on the right. 3, Muscle cell, one of whose processes is attached to the epithelial cells under the velum on the dorsal surface of the right side and the other to the right wall of the stomodaeum. 4, Muscle cell that curves dorsally over the digestive tract and is attached as 3, but on the left side. 5, Muscle cell attached to the left side of the velum and the posterior rudiment of the foot, on the left side. 6, Muscle cell that curves over the digestive tract and is attached to the velum and to the foot on the left side.* ***(b)*** *Veliger after 90° torsion (from the dorsal view); (1–6) as in (a).*

shell. There appear also to be degrees to which detorsion may take place. For example, in the bubble shell, the opisthobranch *Bulla*, the shell and visceral hump are completely coiled but the internal organs show rotation through only 90°. The anus opens on one side of the body, and the other structures affected by torsion are directed laterally. But typically, detorsion is complete in opisthobranchs, and the nervous system is symmetrical and untwisted, and the ctenidium is directed posteriorly. Yet the facts of a single ctenidium and a single auricle, and an anus and genital aperture laterally and not posteriorly situated, show that this symmetry is secondarily acquired and that at some stage in ontogeny torsion and its reversal have taken place.

A larval mutation has been postulated to account for the origin of torsion. The effect of this, which must have occurred at an early developmental stage of the primitive mollusc from which the gastropods originated, was either to retard the development of one member of the pair of retractor muscles or to accelerate that of the other. Such a mutation could have survival value either for the larva or for the adult. Its advantage for the larva would be primarily to permit it to draw the head, its most vulnerable area, into the shell before the foot, and then to block the opening by the pedal mass; a secondary advantage would be the possibility of developing a hard or horny operculum on the foot that would fit closely in the aperture of the shell and so provide even greater protection for the body inside. For the adult, as well as for the larva, it would also be advantageous to have the opening of the mantle cavity anterior rather than posterior, as it seems to have been in the primitive molluscs, where the ctenidia lie posteriorly also. Since the mantle cavity is essentially a respiratory chamber, into which water with its dissolved oxygen is drawn by the action of cilia on the lateral cells of the gill filaments, water is drawn in from behind in an animal of this kind. When the creature moved forward, a current would be set up against the respiratory current; moreover, if, like most animals, it oriented itself upstream against a natural current, there would be two currents moving in opposition to the respiratory current. After torsion, all currents would flow in the same direction, thus reducing the work of the cilia and insuring the ventilation of the mantle cavity in almost any circumstances.

Possibly partly because of this ability to close off the mantle cavity, the gastropods have been able to adapt to a terrestrial existence and to invade the land. They are the only class of the phylum in which species have evolved feeding and other mechanisms enabling them to do this, for all the others are aquatic, living at various depths in the sea and in bodies of freshwater.

(c) THE ARTHROPOD PLAN The arthropod plan is that of a segmented, eucoelomate animal with, typically, a pair of jointed appendages on each segment. A distinctive feature of arthropod organization is tagmosis, the grouping of segments into functional units, or tagmata. This word, derived from the Greek *tagmas*, meaning "regiment," was first applied to the condition in arthropods by the English zoologist Sir Edwin Ray Lankester (1847–1929). Actually, it is evident to some degree in the tubiculous worms, where the body is divisible, on the basis of differences in the form of the parapodia, into an anterior thorax and a posterior abdomen. In arthropods, the number of tagmata are different in different classes, as are the number of segments included in them, so that they are not necessarily homologous regions. In the arachnids (Fig. II–23*a*), the anterior tagma is the prosoma, or cephalothorax, consisting probably of six segments, four of which bear locomotor appendages; the posterior tagma is the opisthosoma, or abdomen, of not more than thirteen segments, and a telson; in some species this tagma is subdivided into mesosoma and metasoma. In crustaceans, myriapods, and insects (Fig. II–23*b*, *c*, and *d*), the anterior tagma is the head, also of six segments, but never with locomotor appendages. The posterior tagma is the trunk of a variable number of segments, in which all may bear locomotor appendages, as they do in some crustaceans and in myriapods, or certain ones may lack them. This tagma may be subdivided into thorax and abdomen, as it is in the insects, where jointed locomotor appendages are developed only on the three segments of the

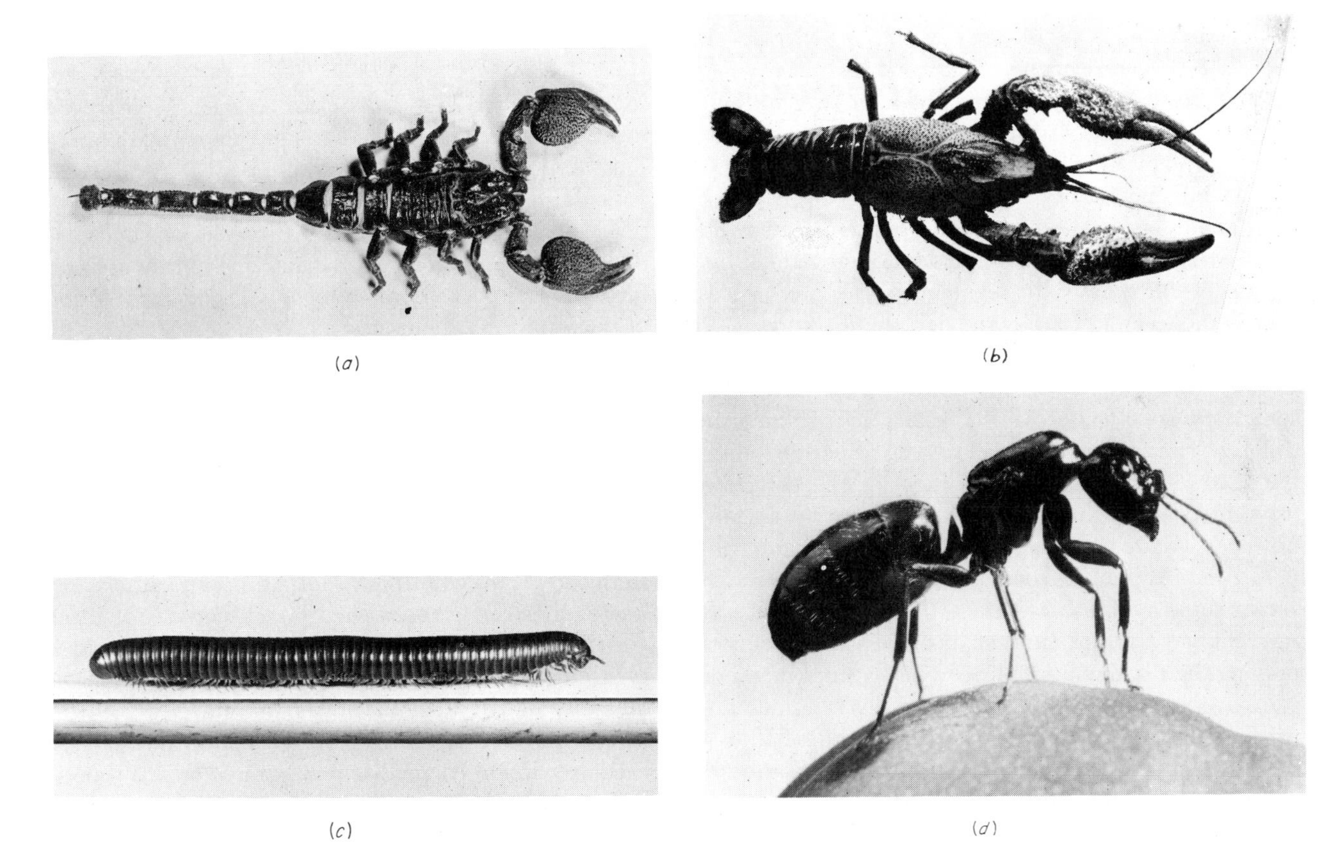

(a) (b) (c) (d)

FIG. II–23 *The arthropod plan.* ***(a)*** Pandinus dictator, *an African scorpion. (American Museum of Natural History.)* ***(b)*** Cambarus blandingi acutus, *a crayfish of the southern United States. (U.S. Fish and Wildlife Service.)* ***(c)*** *A millipede, showing metachronal movement of the legs. (American Museum of Natural History.)* ***(d)*** *The common black ant. (Courtesy of H. Lou Gibson.)*

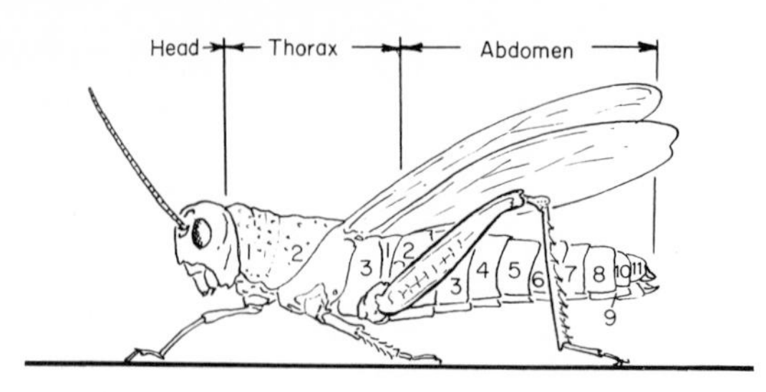

FIG. II–24 *External anatomy of an insect (grasshopper) showing the anterior and posterior tagmata. The posterior tagma is subdivided into a thorax of three segments (1, prothorax; 2, mesothorax; 3, metathorax) and an abdomen of 11 segments.*

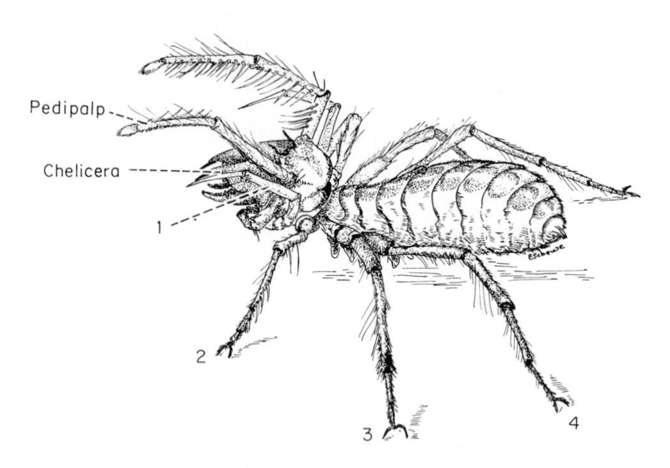

FIG. II–25 Galeodes arabs, *a solifuge (solpugid), the "tarantula" of Egypt and Arabia, with "head" raised in the attitude of defense. (After Grassé.) 1, 2, 3, 4: Walking legs.*

thorax (Fig. II–24). The stability of tagmata, once fixed, has been an important factor in arthropod evolution, imposing conditions and restrictions upon the possible extent of adaptive radiation. However, the Solifugae are evidence that tagmosis, even when established, can undergo radical change, for, in these tropical and subtropical spider-like arachnids, the first three segments of the prosoma are fused to form a head, while the last three are free (Fig. II–25). These are the only arachnids that can raise their heads, and there is also almost as much freedom of movement between the fifth and sixth segments of the prosoma as there is between its sixth segment and the abdomen.

Another characteristic feature of the arthropods is sclerotization of the epidermis. This chitinous, sometimes calcified, covering varies in thickness in different species, as well as in degree of permeability, but it always provides a great measure of protection from mechanical and physiological injury. It has, however, made certain anatomical adjustments imperative. Animals so enclosed must, for one thing, retain some permeable surfaces for gaseous and salt interchange and, for another, provide some means of moving their bodies from place to place, since methods of progression like those of soft-bodied animals are precluded. The arthropods have made these adjustments by the development of branchiae in aquatic species or of book lungs or tracheae in terrestrial species and of paired segmented and jointed appendages, or legs, which are perhaps the most distinctive feature of the phylum and that from which it derives its name. Arthropod appendages are ventro-lateral outgrowths from the main axis of the body, divided into a number of segments or articles. These are movably articulated together, and the entire appendage is similarly articulated to the body. At all these articulations, the integument of the articulating surface may be raised into one or two processes or condyles; elsewhere at the joint it is thin, membranous, and flexible, giving freedom of movement on contraction of the appropriate muscles.

The area of origin of the appendage makes a natural horizontal division of the body into a dorsum above and a venter below them. General sclerotization of the body, or of any given segment, would then produce a dorsal plate, or tergum, and a ventral plate, or sternum, with the legs articulated at the lower margin of the tergum and the lateral margin of the sternum. This condition is found in some arachnids, but in most arthropods the dorsum consists of three plates, a tergum (notum) covering the back, and a pleuron (epimeron) covering each side. Each leg is thus articulated with the lower margin of a pleuron and with the sternum. These plates have been variously modified in different groups of arthropods (Fig. II–26). What may be a primitive condition is found in the small shrimp-like crustacean *Anaspides*, native only to Tasmania where it lives in mountain pools of running water. In these animals, the tergum of each segment comes down to the bases of the legs on each side, but there is a small groove just above

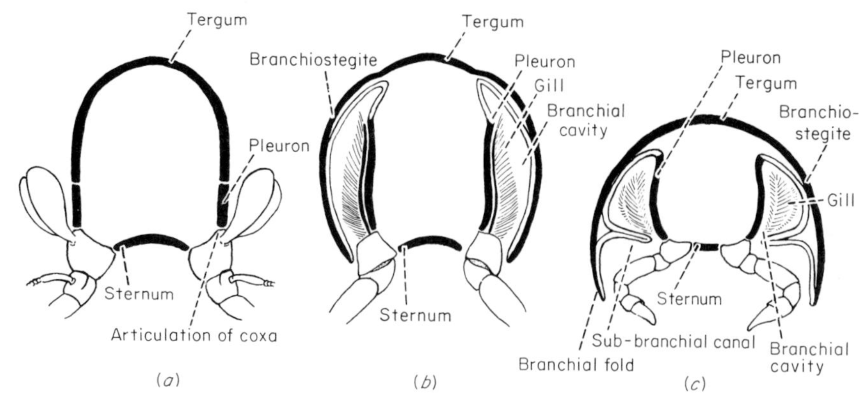

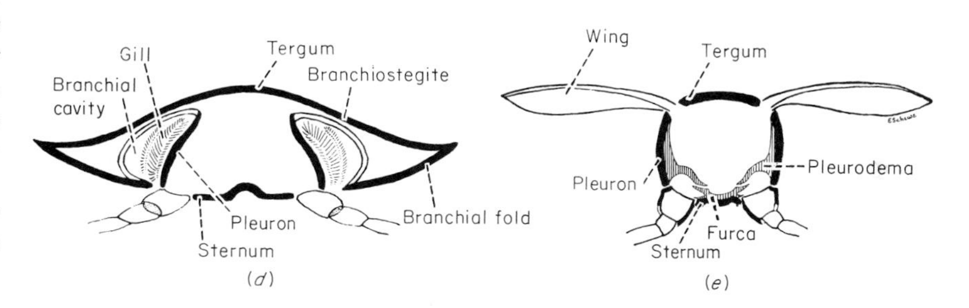

FIG. II–26 *Diagrammatic cross sections of arthropods, showing various modifications of the tergum. (After Snodgrass.)* **(a)** *Cross section through the thorax of a primitive crustacean,* Anaspides. **(b)** *Cross section through the thorax of a decapod crustacean (crayfish) with the tergal folds extending down as gill covers (branchiostegites).* **(c)** *Cross section through the thorax of the mole crab,* Emerita, *with the folds of the branchiostegites extending over the legs and enclosing subbranchial canals.* **(d)** *Cross section through the thorax of a crab* (Callinectes), *with the branchiostegites extending as lateral folds.* **(e)** *Cross section through a winged thoracic segment of an insect, with wings formed as lateral outgrowths of the tergum.*

each leg base that defines tergal and pleural regions. The branchiae are thin-walled outgrowths from the basal segment of each leg. In most decapod crustaceans, the tergum is extended as folds, the gill covers or branchiostegites, so that the branchiae of each side are enclosed in a chamber through which a current of water continually flows. These folds may extend downward, as they do in crayfish and lobsters, or laterally, as they do in crabs. In the mole crab, *Emerita*, they reach well down over the legs forming subbranchial canals below the branchial chambers. In insects, the terga (nota) of the last two thoracic segments have greatly extended lateral folds, which can be used as organs of flight.

The arthropod plan has proved to be even more flexible and productive of new types than the molluscan, for there are more than 800,000 existing species of arthropods, or about 10 times as many as there are molluscs. The arthropods have invaded every kind of habitat, adapting themselves to lives in waters of all kinds; in moist, dry, and apparently utterly arid areas on land; in the ground and in the air; and at low, high, and moderate temperatures. They are the most numerous and widely distributed of existing animals, and among them are some of the most highly specialized ones. To these extents, they may be considered the most successful. The principal evolutionary modifications of their basic plan have been (1) the condensation and concentration of somites and their contained organs, especially those of the anterior end, into a unified area; (2) the concentration of the primitively segmental ganglia and of primary sense organs in the cephalic region; and (3) the conversion of the generalized appendage, originally presumably a locomotor organ, into a wide variety of specialized structures adapted to the reception of external stimuli, to the capture, sifting, and mastication of food, to the cleaning of the body, and to the transmission of reproductive products and the incubation of fertilized eggs. In this latter modification, certain appendages may differ so markedly in the two sexes of a species that they serve as secondary sex characters, conspicuous external features distinguishing males from females. In the social insects, the modifications of the appendages for

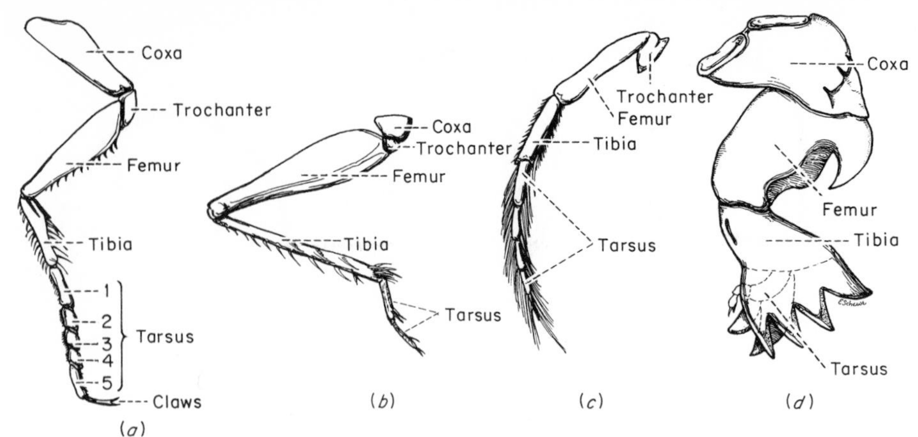

FIG. II–27 *Legs of insects, showing their adaptations to functions of various kinds.* ***(a)*** *Leg of cockroach* (Blatta) *adapted for running.* ***(b)*** *Leg of cricket* (Gryllus), *adapted for leaping.* ***(c)*** *Leg of water beetle* (Dytiscus), *adapted for swimming.* ***(d)*** *Foreleg of a mole cricket* (Gryllotalpa) *adapted for digging.* (*From Gardiner.*)

the role the individual is to play in the organized colony may be such as to make its caste easily recognized. The appendages, too, have been adapted to many kinds of locomotion—walking, running, hopping, leaping, and swimming—as well as to digging, grasping, holding, and carrying (Fig. II–27).

SELECTED BIBLIOGRAPHY

Clark, A. H., and R. J. Menzies: *Neopilina (Vema) ewingi,* A Second Living Species of the Paleozooic Class Monoplacophora, *Science,* vol. 129, pp. 1026–1027, 1959.

Crofts, D. R.: Muscle Morphogenesis in Primitive Gastropods and Its Relation to Torsion, *Proc. Zool. Soc. London,* vol. 125, pp. 711–750, 1955.

Kozloff, E. N.: Morphology of the Orthonectid *Rhopalura ophiocomae, J. Parasitol.,* vol. 55, pp. 171–195, 1969.

Rasmont, R., J. Bouillon, P. Castiaux, and G. Vandermeerssche: Ultra Structure of the Choanocyte Collar Cells in Fresh Water Sponges, *Nature,* vol. 181, pp. 58–59, 1958.

Tiegs, O. W., and S. M. Manton: The Evolution of the Arthropods, *Biol. Rev.,* vol. 33, pp. 255–337, 1958.

Williams, L.: "Anatomy of the Common Squid," pub. under auspices of the American Museum of Natural History, Leiden, E. J. Brill, 1909.

SKELETAL STRUCTURES

By definition, the skeleton of an organism consists of the more rigid tissues or structures that provide protection or support for softer and more delicate parts. This definition may properly be expanded to include, as another and perhaps the primary function of a skeleton, the provision of a means by which contractile elements may be antagonized and extended after a contraction. In vertebrates, the cartilaginous or bony internal skeleton to which most of the muscles are attached provides such antagonism as do also opposing sets of muscles. In bending a limb, for example, the flexor muscles do work not only to exert an effective pull upon the bone on which they are inserted but also to overcome the tension set up by the extensor muscles. Similarly, in straightening the limb, the extensor muscles act against the flexors, both to extend them after contraction and to overcome their tension when the limb is extended. Each set thus contributes to the relaxed state of the other, and this opposite, or antagonistic, action results from the location of their points of origin and insertion on a rigid tissue. Such a contribution is necessary to all effective muscular action, since contraction is the only active phase in the operation of muscle cells and extension is passive and brought about only by an external source of tension.

With the exception of molluscs and arthropods, the invertebrates have not provided their muscles with firm bases for attachment for antagonism of this kind nor selectively protected any especial tissue with skeletal structures, as vertebrates have protected their central nervous systems. Invertebrates have tended, on the other hand, to provide protection and support for their entire bodies by enclosing them in external skeletons, which are various in their composition, pattern, and origin. Muscle antagonism is effected by body fluids or plastic tissues, which provide a hydrostatic skeleton. Such a skeleton may be defined as a mass of fluid, or a plastic parenchyma, enclosed within a muscular wall. The viscosities of the fluids or the compactness of the tissues vary, but the general conditions that determine whether or not they may act as skeletons are that the fluids should be incompressible and essentially constant in volume and the tissues deformable. Thus, the contents of the gastral cavity, as well as the elastic mesoglea, of Cnidaria and Ctenophora provide for them a hydrostatic skeleton; the fluids in the perivisceral spaces provide it for pseudocoelomate and coelomate forms; and the hemolymph provides it for molluscs and arthropods. In acoelomate animals, the parenchyma acts in this capacity.

A. COMPOSITION

In respect to their chemical composition, exoskeletal structures may be broadly classified as inorganic or organic, depending upon the relative amounts of minerals or of carbohydrate and protein in them. If inorganic, the minerals, whose source is the animal's immediate environment, are usually deposited around or within an organic base, but their final accumulation may be so great that virtually all trace of organic substrate is lost. Calcium compounds, especially carbonates and phosphates, are those most commonly found in invertebrate skeletons. Calcium carbonate occurs in them in at least two crystalline forms—calcite, in which the system of crystallization is hexagonal, and aragonite, in which it is orthorhombic. Silicon is the mineral of next most frequent occurrence, although others may be used exclusively or in various mixtures. For example, strontium is used by the radiolarian *Acanthometra pellucidum* to make its skeleton of celestite, strontium sulphate in orthorhombic crystallization. Strontium may replace calcium in the construction of invertebrate skeletons, as it can in that of bone by vertebrates. This has been shown by experiments in which the mollusc *Physa*, raised in media with different concentrations of strontium, has made a shell containing more than 50 percent strontianite ($SrCO_3$) closely resembling its usual one containing aragonite ($CaCO_3$).

The organic materials of invertebrate skeletons are either carbohydrates, proteins, or carbohydrate-protein complexes. Some invertebrates, notably certain species of Protozoa, share with plants the capacity to make cellulose or hemicellulose, a complex and biologically very stable carbohydrate molecule. The protein products belong to the category of scleroproteins, characterized by their insolubility in water, in aqueous solutions of neutral salts, and in organic solvents. The two principal groups of scleroproteins are keratins, resistant to the action of proteolytic enzymes and of boiling water, and collagens, which can be enzymatically hydrolyzed and by prolonged boiling in water are converted to glue, or gelatin, a substance that is very viscous in solution and, in concentrations above 2 to 3 percent, forms stiff gels at temperatures up to about 20°C. Both keratins and collagens yield, on degradation, a variety of amino acids but they differ in the relative proportions of those that they have in common and in the absence of tryptophan and cysteine from the collagens. Spongin, the skeletal framework in some sponges; the gorgonins, which form the axial skeleton of sea feathers and sea fans; and conchiolin, the basis of the molluscan shell, are all collagenous scleroproteins. Spongin and gorgonin, both made by marine organisms, yield, in addition to the typical amino acids di-iodotyrosine and dibromotyrosine, iodine representing 1 to 2 percent of their substance. The "gelatinous" cases or tubes of many invertebrates belong to the category of collagens, although little is known about the exact composition of any of them.

The carbohydrate-protein complexes of invertebrate skeletal structures are conjugated proteins with glucose or other sugars or sugar derivatives in the molecule. In general, those with less than 4 percent carbohydrate are called glucoproteins, and those with more than 4 percent, mucoproteins, or mucoids. These molecules are long and thread-like. In solution they form a viscous mass that will, under appropriate conditions, solidify like gelatin. The ability to secrete mucin, a general term for such compounds in aqueous solution, is widespread throughout the animal kingdom. As moist, slimy substances, mucins may act as lubricants and serve also to entangle foreign particles; as hardened masses, they may provide a firm, protective covering for delicate underlying structures. Tectin is a glucoprotein elaborated by many species of Protozoa and used by them for a number of different purposes. Chitin is a similar compound made by many invertebrate species as well as by most lichens and some fungi (Zygomycetes, Phycomycetes, Ascomycetes, Basidiomycetes), but not by echinoderms or chordates. It is laid down, like cellulose, in definite fibrous form in intimate association with the protoplasm of the cell wall. The carbohydrate of chitin is a glucoseamine; this is linked, or very intimately mixed, with at least two proteins, one of which is soluble, the other insoluble, in water. The glucoseamine in this combination is highly polymerized, several hundred glucoseamine units being linked together in a long and very stable molecule, resembling the cellulose molecule but differing from it by the presence of an acetyl amine group ($\mathrm{O{-}\overset{\overset{NH}{|}}{C}{-}CH_3}$) on carbon 2 of each of the glucose units. Like the chain molecules of cellulose, those of chitin are arranged in a lattice configuration; in animals, the reorientation of the chitin molecules from the heterogeneous mass in which they are first deposited occurs not in immediate contact with, but at some distance from, the cells secreting them.

B. COLORATION

The skeletons of invertebrates, as well as their living tissues, are often colored, sometimes brilliantly, in a variety of hues and patterns. Skeletal colors may arise either from physical phenomena that cause the decomposition of white light or from biological ones, i.e., synthesis of pigments by the organism itself or incorporation of those it uses as food, or by processes of tanning.

Physical, or structural, coloration is the consequence of the distribution or orientation of materials within the skeletal substance so that light is modified in its passage through it. While the Tyndall effect, resulting from the scattering of light in the violet-blue range of the spectrum by bodies of minute dimensions (0.6 μ or less), is relatively rare in producing color in invertebrate skeletons, it does account, in some cases at least, for the blue and bluish colors in their integuments, eyes, and other superficial organs. Diffraction and interference are the two optical phenomena involved primarily in structural coloration of their inorganic or organic exoskeletons. When materials within the skeleton are arranged in parallel lines, or rows, some of the light rays are deflected as they are when passing through a fine grating. This results in diffraction and the consequent decomposition of white light into its component wavelengths, and produces fringes of parallel light or bands of dark and light colors. The fine lines on the cuticle of *Nereis* and some other worms, and in the wings of insects, and the linear arrangement of the crystals of aragonite and calcite in the prismatic layer of molluscan shells act as diffraction gratings and so give this effect. When the skeletal materials are arranged as thin laminae or plates, white light is also decomposed as the rays reflected back from the upper and lower surfaces of the laminae cancel each other out by interference. Bands or fringes of various colors result. This is the phenomenon responsible for the iridescence of many skeletons, or parts of them, as well as for the metallic coloration of insects like the Japanese beetle and the "gold bug." The nacreous layer of the molluscan shell is made up of more or less regularly spaced plates of calcium carbonate separated by thin layers of the protein conchiolin. When the plates are of unequal thickness, as they are in many bivalves, this layer displays the iridescent characteristic of mother-of-pearl. When the plates are of uniform thickness, corresponding to a particular wavelength of light, this region of the spectrum will be cut out and the substance will have the color, or range of colors, of the remaining light. Thus, in some species of *Haliotis* the inner layer of the shell is reddish or rosy, because the thickness of all the laminae (48 μ) corresponds to the wavelength of blue-green light, whose reflection is therefore eliminated by interference. If the laminae are thicker than the wavelengths of any light, there is no interference and all the rays are reflected back so that the material appears white. In many cases, structural coloration is the combined effect of diffraction and interference as it is, for example, in the molluscan shell at the junction of prismatic and nacreous layers. Moreover, the characteristic color of an animal's covering may be due to a combination of these schemochromes (colors due to physical configuration) and biochromes (pigments that result from biochemical reactions and that show color because of their absorption of light of different wavelengths).

Animal biochromes are many and various. They may be exogenous, derived from the environment or the food that the animal eats, or endogenous, products of metabolic reactions whose pathways and roles are but little understood. The dull red color of the calcareous case, or corallum, of the organ pipe coral, *Tubipora*, is for example due to the deposition in it of iron salts that have been accumulated by the living polyps inside. And in some cases the color of the skeleton may be due to the penetration of other organisms. The pink and greenish colors of the corallum of a number of stony corals is an illustration of this, for they are caused by the presence of red or green algae that have bored their way into the calcareous mass and made it their home.

Other exogeneous animal biochromes belong to the chemical category of carotenoids, highly unsaturated long-chain hydrocarbons that are synthesized by plants and are taken into the animal body in its food, either directly by herbivores or indirectly by carnivores that feed on herbivores. Carotenoids and their derivatives are soluble in lipids and are primarily yellow, orange, or red in color. When conjugated with proteins they are blue, green-violet, brown, or gray. They are stored in considerable quantities in the tissues of probably all animals, where they are known in many cases to be biologically active; those that are stored in the superficial tissues of invertebrates may diffuse into their exoskeletons and color them. It is the presence of such compounds in the carapace of many decapod crustaceans that gives them their characteristic colors. The green, brown, or blue colors of American lobsters and crayfish in life, for example, are due to conjugated carotenoids, at least two of which, astaxanthin and ovoverdin, have been extracted from the carapace. When heated, the carotenoid-protein combination is broken, and the "shell" turns red, the color of the carotenoid component of the complex. The Norway lobster, *Nephrops norvegicus*, is on the other hand red when alive, because the carotenoids that have been deposited in its carapace are unconjugated. Red spots on the wings of some Lepidoptera are usually due also to carotenoids. Perhaps most dramatic in their accumulation of these compounds are some of the beetles, especially the familiar ladybug, a member of the family Coccinellidae. These beetles feed on aphids, which in turn feed on plants, so that the ladybug derives its carotenoids second hand and deposits them, apparently unchanged from their state in the plant that made them, in its hardened forewings, or elytra.

Other biochromes that may be deposited in skeletal structures are nitrogenous compounds belonging to the chemical category of tetrapyrroles, in which there are many water-soluble pigments with a wide range of colors from red to violet. Porphyrins are cyclic tetrapyrroles which have in common a "mother molecule" of porphin, in which four pyrrole nuclei are joined by methane bridges in a ring configuration. Porphin seems to be of almost universal occurrence, in aerobic organisms at least, for it, or a porphyrin derived from it, is included in the molecule of the cytochromes. Porphyrins are also components of the molecules of the heme portions of hemoglobins and similar blood pigments, and of chlorophyll. Bilins are linear tetrapyrroles in which the ring is opened out so that the four pyrrole nuclei are united as an open chain. These include the bile pigments, such as biliverdin and bilirubin, that are formed in the vertebrate liver through degradation of hemoglobin. The structural formulae of porphin and bilin are shown in Fig. III–1.

Like carotenoids, both porphyrins and bilins may be accumulated in the tissues of invertebrates and may also be found in their skeletons, although their sources and significance in invertebrate metabolism are almost completely unknown. They may represent excretory products or metabolites derived from the food that the organism is unable to process further. For example,

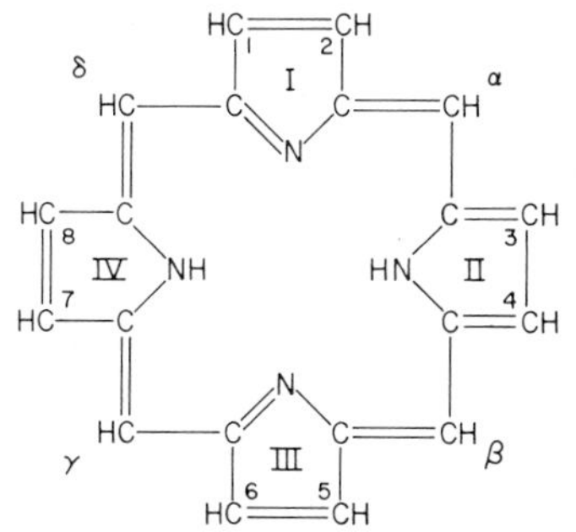

α, β, γ, δ = methane bridges uniting pyrrole rings I, II, III, and IV

(*a*) Porphin

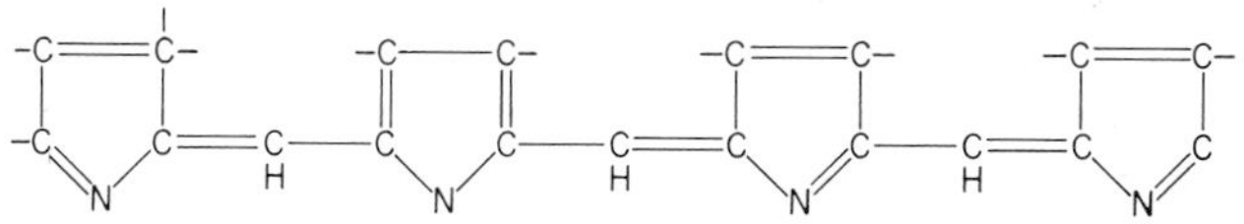

(*b*) Bilin

FIG. III–1 *Structural formulae of* **(a)** *porphin and* **(b)** *bilin.*

uroporphyrin, a common excretory product of many animals arising from the decomposition of hemoglobin, and conchoporphyrin, arising probably from the decomposition of chlorophyll, are found in the shells of many marine gastropod and lamellibranch molluscs, either free or bound with conchiolin, but are absent from those of terrestrial species. The corallum of *Heliopora* owes its distinctive blue color to the presence of a bilin called helioporobilin, which is chemically related to biliverdin. A similar compound, a water-soluble green pigment, has been found in the wings of the cabbage butterfly, *Pieris brassicae*, and a red pigment, rufescine, in the shells of some species of the gastropod molluscs *Turbo* and *Haliotis*.

Other nitrogenous compounds containing the pyrrole nucleus are the indigoid pigments, which are blue, green, or red, and the melanins, which are yellow, brown, or black. Indigoid pigments are known to be deposited in the shells of several species of gastropods. The brownish-purple pigment extracted from the shells of *Nucella lapillus*, a marine species, is an indigoid which the snail obtains from the mussels that it customarily eats and which is deposited unchanged in its shell. When experimentally restricted to a diet of barnacles, the shells of *Nucella* lack this pigment, so that its source from the natural food seems unquestionable. Melanins are of wide distribution, but are more frequent in the integument and the tissues of invertebrates than in their exoskeletons, although they may contribute directly to the colors of some molluscan shells and arthropod cuticles. Their very common location in epidermal cells, however, contributes to structural coloration by absorption of the light that penetrates the exoskeleton without diffraction or interference.

Pterins, possibly derived from purines that are in turn derivatives of nucleoproteins, may also be deposited in exoskeletons, particularly in the wings of Lepidoptera. Leucopterin, which is white, and erythropterin, which is deep red, have been isolated from the wings of several members of the Pieridae; in some of these, the erythropterin has been found in conjunction with guanopterin, a colorless derivative of the purine guanine. Xanthopterin, a yellow pigment, was isolated as long ago as 1899 from the wings of one of the sulfur butterflies, *Gonepteryx rhami* L. Because of their crystalline structure, these substances also may act as schemochromes, contributing to the structural coloration of the skeleton.

The roles that these nitrogenous compounds may play in the metabolism of the animals in whose skeletons or tissues they are found are almost entirely unknown. Many of them are known to be of physiological importance in vertebrates at least, or to be intermediate products of metabolism, or terminal products that are excreted. The origin and nature of these pigments in invertebrates, and the mechanisms of their deposition in exoskeletons offer a wide and fruitful field for investigation in comparative biochemistry and ecology.

Other cyclic organic compounds, the quinones, are also involved in skeletal characters. The echinoids are unique in making use of naphthoquinones in the coloration of their tests and spines. One of these, echinochrome, is a purple pigment and another, spinochrome, a deep red one. Both of these may exist in several different forms, and so in a variety of shades, in different species of sea urchins and even in different parts of the same animal. Different proportions of them, and also combinations of them with calcium to form dull green or violet salts, account for the range of colors that are found in the same individual or in different individuals of the same population. In *Paracentrotus lividus*, for example, a ratio of echinochrome to spinochrome of approximately 8:1 gives the test and spines a greenish color, while ratios of 1:4 and higher make them brownish violet and violet. Quinones are also, in many cases, involved in the stabilization of soluble proteins that may be included in the organic coverings of invertebrate bodies. This is essentially a process of tanning, in which the quinones probably unite with different amino groups of neighboring protein chains and form cross linkages between them, making the proteins insoluble and hardening the material. This also effects a color change, turning the previously colorless or translucent substance deep reddish-brown. There is evidence for quinone tanning in the central capsule membrane of the radiolarian *Thalassicola*; the external cortical layer of *Ascaris*; and the outer layer, or periostracum, of the shells of molluscs. It also takes place in the hardening and darkening of crustacean and insect cuticles, conspicuously so in those of bees and wasps. Compounds other than quinones are known also to be agents in tanning, and may be operative in the formation and coloration of other exoskeletons.

C. DISTRIBUTION AMONG SPECIES

1. Protozoa

Certain genera of flagellates, rhizopods, ciliates and sporozoans characteristically possess exoskeletons, and even those members of the phylum that are without them during their active phases may develop protective coverings during encystment, gamete formation, or production of infective stages (Fig. III–2). Among the dinoflagellates, which form a large proportion of the plankton of the sea but also inhabit fresh and brackish waters, members of the suborder Prorocentrinea typically have a bivalve shell made of cellulose, while in some of the peridinians, such as the genus *Ceratium* with many species common in salt- and freshwater, the body is enclosed in an armor of cellulose plates, sculptured in various ways and in some species prolonged into spines. Of the rhizopods, almost all species of Radiolaria, whose bodies are divisible into a central capsule, surrounded by a membrane of tectin and extracapsular cytoplasm, have a skeleton that is either radiate or concentric in pattern. With the exception of the Acantharia, this skeleton is siliceous in its composition; in the Acantharia, it is of strontium sulphate. In Heliozoa, the body may be enclosed in a latticework of tectin, as it is in the freshwater genus *Clathrulina*, or in a gelatinous envelope in which are deposited spicules or plates of silica. In *Acanthocystis*, both plates and scale-like spinules are formed; in *Clathrella*, the skeleton appears continuous but is actually made of discrete siliceous plates, cup-like in shape, between which delicate protoplasmic processes, the filopodia, radiate from the spherical central mass.

Most conspicuous among Rhizopoda for their skeletal structures are Foraminifera, comparatively large Protozoa of which the majority are marine. Some are pelagic, but most of them live at the bottom of the sea or among animal and plant growths. The bodies of these animals are encased in tests, primarily

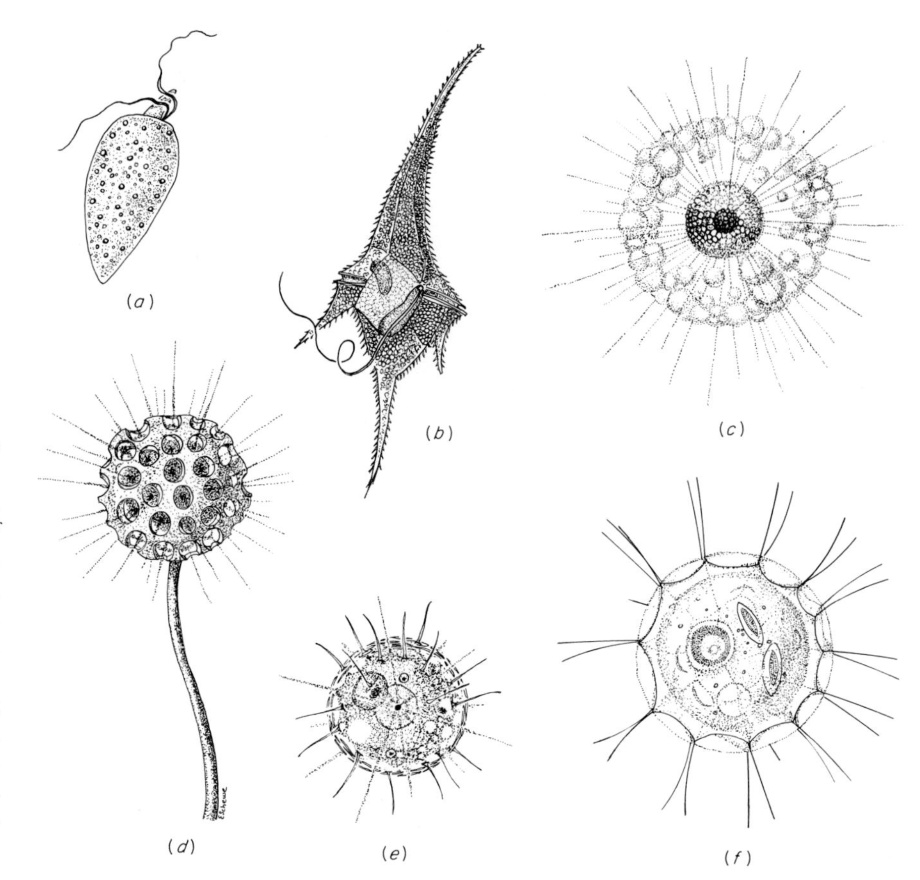

FIG. III–2 *Types of protozoan skeletons. (After Kudo.)* *Dinoflagellates:* ***(a)*** Prorocentrum micans *(36 to 52 μ long), one cause of "red water" in the sea.* ***(b)*** Ceratium hirundinella *(95 to 700 μ long).* *Radiolarians:* ***(c)*** Thalassicola nucleata *(2,000 to 3,000 μ in diameter). Marine, pelagic.* *Heliozoans:* ***(d)*** Clathrulina elegans *(diameter of capsule = 60 to 90 μ). Freshwater.* ***(e)*** Acanthocystis aculeata *(35 to 40 μ in diameter). Freshwater.* ***(f)*** Clathrella foreli *(40 to 55 μ in diameter). Freshwater.*

of chitin, to which may be added calcium salts, or more rarely, silicates, making a shell often of considerable thickness and comparatively large size. In some species, sand grains and other foreign particles are embedded in the secreted material, often in a precise and characteristic way (Fig. III–3). The patterns of these shells are peculiar to the individual genus and are frequently intricate in design; they may be composed of one or several chambers, arranged in linear or spiral order, and increase in size from the one first formed, with one, two, or many apertures through which the pseudopodia protrude. Those to whose shells foreign particles are added often show what seems to be a remarkable selectivity in the materials used, even when the environment provides a wide range of them. For example, one species of *Psammosphaera, P. fusca*, uses only sand grains and usually only those that are similar in color; another, *P. parva*, uses those of any color but of more or less uniform size and, in addition, adds a single perfect needle-shaped sponge spicule, rejecting those that are broken or too short to extend across its test, so that both tips will project from it. *Psammosphaera rustica*, on the other hand, uses broken as well as intact sponge spicules, fitting the broken pieces into appropriate areas of the framework of its shell. *Psammosphaera bowmanni* gathers only mica flakes, sparsely scattered on the sea bottom, and cements them together into an irregular and weak test.

In most of the ciliates a relatively thick but elastic pellicle closely invests the body. This may be sculptured in a variety of designs, elevated into longitudinal or spiral ridges or nodes of characteristic pattern, or prolonged into spines. The genus *Coleps*, with many salt- and freshwater species, is characterized by the presence of rectangular plates beneath the pellicle, which are arranged in longitudinal rows running from

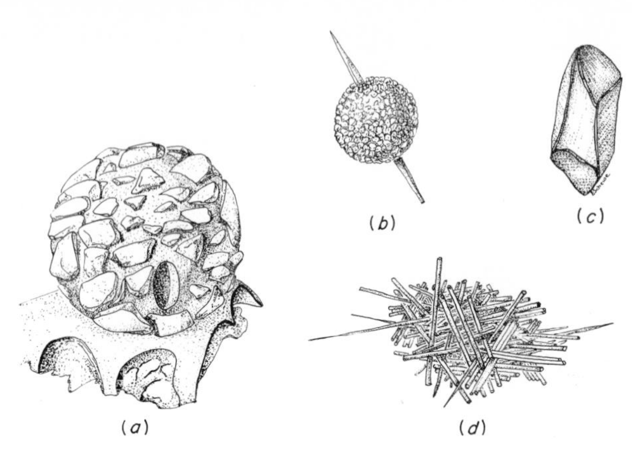

FIG. III–3 *Skeletons of* Psammosphaera. *(After Cushman.)* ***(a)*** Psammosphaera fusca, *resting upon substrate (diameter up to 4 μ).* ***(b)*** Psammosphaera parva *(0.3 to 0.75 μ in diameter).* ***(c)*** Psammosphaera rustica *(0.3 to 0.5 μ in diameter).* ***(d)*** Psammosphaera bowmanni *(0.4 to 0.6 × 0.25 to 0.35 mm).*

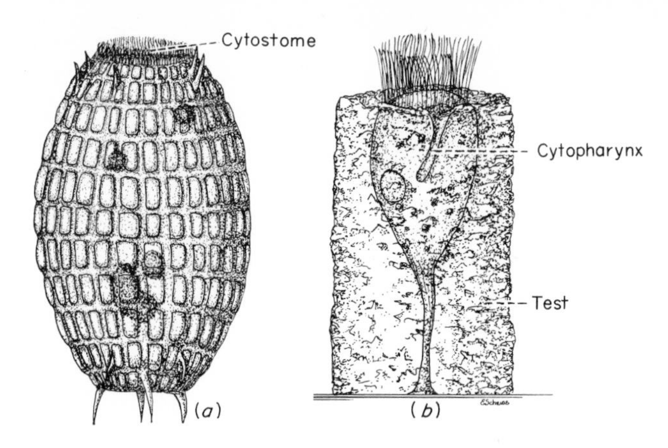

FIG. III–4 *Testaceous ciliates. (From Gardiner.)* ***(a)*** Coleps octospinous, *found in marshy waters. Discrete plates of firm consistency are embedded just below the ectoplasm, forming a kind of armor.* ***(b)*** Tintinnidium semiciliatum, *which surrounds its body with a chitinous test attached to freshwater plants.*

the oral to the aboral end of the cell and varying in number in the different species. Among the oligotrichous forms, the Tintinnidae, most of which are pelagic with a few species inhabiting brackish and freshwater, habitually secrete a case of tectin. This case is often cup- or vase-like, but sometimes irregular, in shape, with foreign particles adhering to it; the long cilia, which are limited to the area around the buccal cavity, extend from the opening of the cup in a feathery mass (Fig. III–4). The sessile genera of Peritricha, both marine and freshwater, make loricas of tectin to which their bodies may be attached at the basal end only or by the peristomial margin; in the stalked forms, the stalk of attachment may extend through the lorica to the object on which the organism is fastened. In those without a stalk, the lorica itself may be attached by the base or the side to a plant or some other kind of anchorage, the free end of the organism extending from the opening of the case.

The bodies of Suctoria are regularly covered by a pellicle and, in addition, those of some species, both salt- and freshwater inhabitants, are enclosed in a lorica which they more or less completely fill. While adult Sporozoa are without investment of any kind, the young are produced in cysts, which are often thick and relatively impenetrable and resistant (Fig. III–5). Similar protective coverings may be formed by adult Protozoa of all classes under certain conditions, such as, accumulation of metabolic wastes, oxygen deficiency, dehydration, and temperature and hydrogen ion concentrations outside their optimum range. Such encystment is very common among freshwater protozoa, though less usual in marine species. It may also be experimentally induced by abundant feeding, after which the animal embarks on a period of quiescence, surrounding itself with a protective wall until the contents of its food vacuoles are digested and assimilated. At the onset of encystment, the cell secretes a wall that is at first soft and later hardens to become the ectocyst, sometimes yellow or brown in color and with a lumpy or a spiny surface. In some species, this ectocyst has been found to be of carbohydrate, in others, of tectin or some other keratin; sometimes minerals, especially silicon, may be incorporated into the ectocyst or foreign particles attached to it. Within the ectocyst a delicate, transparent endocyst is secreted. Enclosed in such a cyst, the animal may endure extreme conditions for long periods of time and be carried by various agents from place to place, thus achieving geographic distribution of the species. Active individuals of *Colpoda*, for example, have been recovered from partially dried soils kept for 42 years and, under experimental conditions, from cysts exposed to a high vacuum (10^{-5} mg Hg), at 450°C, for 2 to 3 days, or

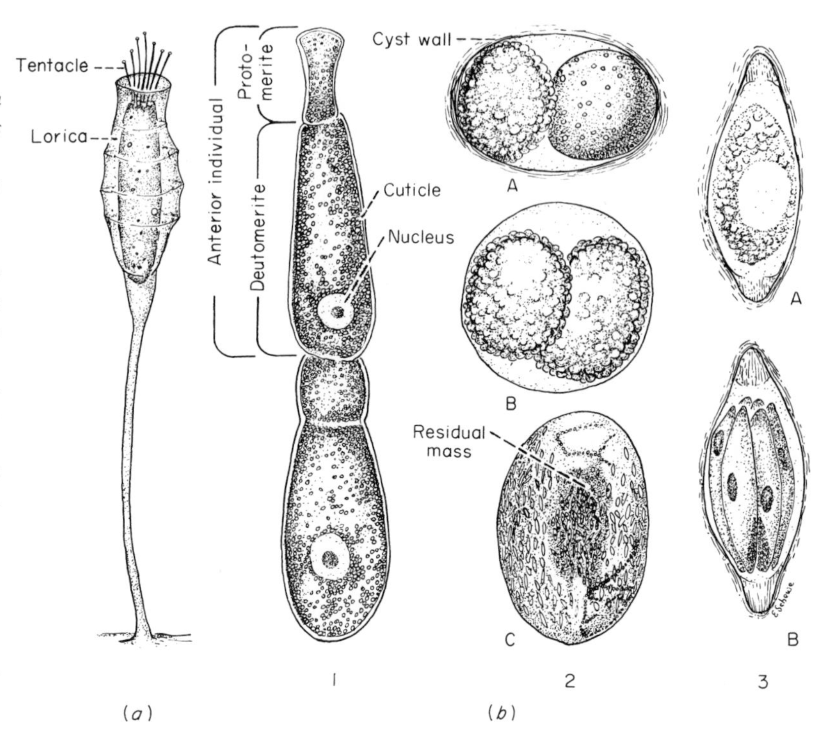

FIG. III–5 ***(a)*** Thecacineta gracilis, *a suctorian found in association with hydroids. The lorica measures 110 by 35 μ; the stalk, 200 by 4 μ. (After Kudo.)* ***(b)*** Gregarina blattarum, *a sporozoan parasitic in the digestive tract of cockroaches. (After Doflein.): (1) Two feeding individuals (gamonts) in the lumen of the gut. Each individual is composed of an anterior region, the protomerite, and a posterior region, the deutomerite, in which the nucleus lies. The two gamonts are joined together end to end in an association known as syzygy. (2) Stages in the transformation of gamonts into sporocysts. A, The anterior and posterior gamonts (primite and satellite) are enclosed in a common cyst (sporocyst). B, The nucleus of each undergoes many divisions and forms gametes. C, The gametes unite and the zygotes become oriented around a central residual mass of cytoplasm and disintegrating nuclei. (3) Sporocysts. A, Unripe sporocyst resulting from gametic union. B, Ripe sporocyst containing eight sporozoites. Within the sporocyst wall the zygote nucleus has divided three times to form eight haploid sporozoites which, when released, will become trophic individuals and eventually gamonts.*

kept 12½ days in liquid air at temperatures of −190°C and below.

2. Porifera

Isolated spicules of calcareous or siliceous composition or a felted mass of fibers of spongin, or both together, constitute the skeletons of the sponges. These spicules are secreted by cells that migrate from the outer epidermal layer to the gelatinous mass, or mesoglea, that lies between it and the inner epidermis. They become differentiated as scleroblasts or spongioblasts, the former concentrating the mineral salts that form the spicules, the latter elaborating the fibrous protein, spongin. The skeleton of these animals is thus essentially an endoskeleton, although the spicules usually project through the surface of the body. All sponges are aquatic in habitat and, with the exception of one cosmopolitan family, the Spongillidae, are marine, so that the only source of the minerals they use is the water in which they live, where the mineral concentration at the most is never very high. The concentration of silicon, for example, in seawater is in the neighborhood of 1.5 mg/liter H_2O; the condensation of this element into the long glassy spicules characteristic of Hexactinellidae represents the expenditure of relatively enormous amounts of energy on the part of the cells directly involved, as well as on the part of those concerned with maintaining a water current through the sponge and so providing the scleroblasts, in this case silicoblasts, with the materials they need.

The spicules, whether calcareous or siliceous, may be simple rods of varying length or very elaborate in design and, on the basis of size, are distinguished as megascleres and microscleres. On the basis of shape, they are classified as monaxon, when growth has occurred along a single axis in one or both directions; as tetraxon, when four arms radiate from a central point but not in the same plane; as triaxon, when three axes cross at right angles so that six rays extend from a central point; as polyaxon, when several arms radiate from a central point; and as spheres, when growth has occurred in concentric layers around a center. In the formation of a monaxon calcareous spicule, a single binucleate scleroblast produces, between its nuclei, a minute fiber or thread of organic material, around which calcium carbonate is gradually concentrated. The cell elongates, its two nuclei drawing apart, and ultimately separates into two cells, the founder and the thickener (Fig. III–6). The founder cell is the one chiefly concerned in the formation of the spicule, and moves slowly inward toward the gastral cavity leaving the free end of the growing spicule behind it. The thickener cell, which is the distal daughter cell of the original scleroblast, moves along the surface of the spicule depositing additional layers of calcium carbonate as it goes and gradually, after the spicule has attained its characteristic thickness and shape, moves into the inner region of the mesoglea, along with the founder cell. In the more elaborate triaxon and tetraxon calcareous spicules, three and four scleroblasts, respectively, are involved, each also dividing into founder and thickener cells which operate together to form one arm of the finished product. After the completion of the rays, the thickener cells continue to deposit calcium carbonate at the central point, where the rays unite, so that a mass of considerable thickness may be accumulated there. The spicules formed by the silicoblasts are often long and very complicated in design; the simpler ones appear to be products of the activity of individual cells, while the more elaborate arise in association with a syncytial mass, which may have its origin in repeated nuclear divisions of an original scleroblast, unaccompanied by cytoplasmic partition. The orientation of the spicules is usually quite precise. The basal rays of triradiate and quadriradiate types, for example, generally point in the direction opposite to that of the water current, and for this reason the flow of water was held responsible for their orientation. Recent experiments have shown, however, that this is independent of the current and must be related to other conditions. In these experiments, pieces below the osculum were cut off from the main body of the simple sponge *Leucosolenia* and mounted on the tips of fine glass tubes through which a current of water was passed. The spicules that were formed were all found to be oriented in the same direction, although the current was made to flow from the osculum toward the base, as well as in the natural direction from the base toward the osculum.

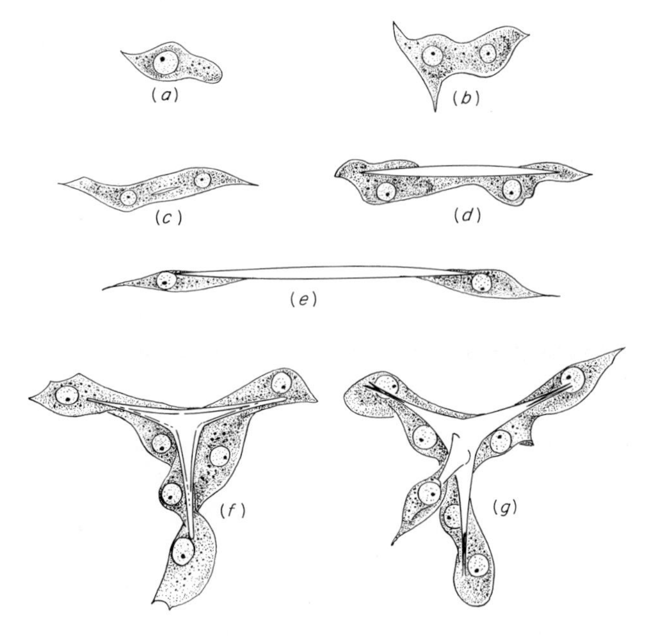

FIG. III–6 *Spicule formation in sponges. (After Woodland.)* ***(a)*** *Scleroblast.* ***(b)*** *Binucleate scleroblast, after division of nucleus.* ***(c)*** *Binucleate scleroblast with organic core of future spicule in the cytoplasm.* ***(d)*** *Elongation of scleroblast as spicule increases in size.* ***(e)*** *Division of scleroblast into founder and thickener cells and migration of cells along spicule.* ***(f)*** *Formation of triradiate spicule.* ***(g)*** *Formation of quadriradiate spicule.*

Spongin fibers are secreted by spongioblasts that become aligned in rows, the fiber secreted by each cell fusing with adjacent fibers to form a long, continuous strand; these strands become meshed and felted together as a flexible supporting framework. Spongin fibers are basically clear and homogeneous in appearance, but they may become impregnated with foreign material, such as fine particles of rock, and so lose this quality. In most of the Demospongia, siliceous spicules are intermingled with the fibers of spongin; only in Keratosa is the skeleton exclusively of spongin.

3. Cnidaria

The skeletons of Cnidaria are chitinous, horny, or calcareous in composition. They may be close or loosely fitting investments of chitin, as in the perisarc of at least 10 genera of colonial hydroids, or supporting frameworks of isolated calcareous spicules, as in the soft corals. Or they may be firm, axial skeletons, such as those of gorgonin, specific to each genus, that support the colonies of horny corals, or the compact and massive calcareous exoskeletons of the true, or stony, corals (Fig. III–7). The chitinous perisarc, or periderm, of hydroids is secreted by ectodermal cells; this either adheres closely to the outer surface as in *Pennaria* and similar members of this class or forms a loose case around the hydrocaulus, expanding into an open or closed cup around the zooids, as in *Obelia*.

In the soft corals the spicules are formed much as they are in sponges, by cells that migrate into the mesoglea and become differentiated as scleroblasts, secreting first an organic thread, then dividing as founder and thickener cells and depositing calcium salts around the organic core. The corals are inhabitants of the shallow waters of warm seas from which they, too, derive the minerals used in the construction of their spicules. These are characteristically rod-shaped, but in one family (Xeniidae) they have the form of thin oval disks, and in many species the contour of the rod may be lumpy with so many protuberances of various size on its surface that it looks like an irregular mass. Although they may be arranged in a definite pattern, often in double rows, the spicules are isolated and discrete elements, usually projecting from the animal's body. Gorgonin is laid down by an investing epithelium, whose source has not yet been satisfactorily determined, as the axial skeleton of the horny corals, found at nearly all depths in the warm seas. These skeletons may be colored, with orange, red, and purple the predominating shades, and may also be interspersed with calcareous spicules or impregnated to a greater or lesser extent with calcium salts. The most extreme instance of this latter condition is *Corallium*, the precious red coral of commerce, where gorgonin seems to be entirely lacking, the axial skeleton being made up of calcareous spicules embedded in a matrix of calcium carbonate. True, or stony, corals (Madreporaria) live in warm, shallow waters which provide

(a)

FIG. III–7 *Cnidarian skeletons.* ***(a)*** *A gorgonian (sea feather).* ***(b)*** *Live golden coral. (American Museum of Natural History.)*

(b)

them with the most favorable conditions for skeleton formation and reef building. Each polyp, about 10 mm in diameter, makes its own external skeleton wholly of aragonite. As the polyps grow and bud, large colonies are formed and the entire skeletal mass, or corallum, may reach enormous dimensions and a weight of several tons. As the colony expands into deeper waters, and the base becomes more and more massive, the living polyps are brought near the surface. This is of great importance to them because light is essential to photosynthesis in the zooxanthellae that all these corals harbor in special carrier cells of their gastrodermis. These symbionts may be indirect contributors to the process of skeleton formation. The association may therefore have been one that has had great effect in the evolution of reef-building corals. A number of hypotheses have been proposed to explain the mechanism of their skeleton formation. All of these accept the origin of the carbonate included in it as metabolic, the result of oxidative reactions going on primarily in the cells of the polyps themselves. Calcium is derived from the seawater, and experiments in which unfed polyps have been kept in seawater with labeled calcium, added as $Ca^{45}Cl$, indicate that Ca^{++} is directly absorbed and not taken in with food. One hypothesis for calcification proposes that the calcium so absorbed is transferred by some unknown mechanism to a site just outside the epidermis. Here it is adsorbed on a mucopolysaccharide, secreted by the epidermal cells, that acts as a template for the deposition of aragonite, which takes place according to this scheme:

$$Ca^{++} + 2HCO_3 \rightleftharpoons Ca(HCO_3)_2 \quad (1)$$

$$Ca(HCO_3)_2 \rightleftharpoons CaCO_3 + H_2CO_3 \quad (2)$$

carbonic anhydrase

$$H_2CO_3 \rightleftharpoons H^+ + HCO_3^- \quad (3)$$

$$H_2CO_3 \rightleftharpoons CO_2 + H_2O \quad (4)$$

Since the surface waters of tropical seas are supersaturated with $CaCO_3$, the velocities of reactions 1 and 2 depend upon the rate of removal of HCO_3 by enzyme systems (carbonic anhydrase) in the polyps and by their photosynthesizing symbionts. By day, both animal and plant operate together to effect the precipitation of calcium carbonate; by night, the polyp must act alone.

Other hypotheses assume that the calcium carbonate is formed within the epidermal cells and, in solution, is secreted to the exterior, where, on contact with the supersaturated seawater, it immediately crystallizes out as aragonite; or that it is stored as such in the epidermal cells which, when fully loaded, drop from the body and disintegrate, thus contributing their contents to the skeletal mass. However formed, the skeleton contains but one crystalline form of calcium carbonate, aragonite, and never calcite. Nor is magnesium included in it, although magnesium salts are freely available to the polyps and included in the skeletons of other organisms.

4. Platyhelminthes

Although the flatworms are generally thought of as "soft-bodied" invertebrates, members of the parasitic species at least provide their bodies with a cuticular covering secreted by gland cells of epidermal or mesodermal origin. In the trematodes this cuticle is a highly polymerized mucoprotein which owes its resistance to digestion by animal proteases to its intimate association with glycogen, the glycogen molecules so to speak enveloping the mucoprotein molecules and so protecting them from contact with the proteolytic enzymes. The biological properties of cestode cuticle are of particular interest. This may very likely also be a similar highly polymerized mucoprotein, though there is at present little precise information about its composition. For the survival of such intestinal parasites without mouth or alimentary canal, the cuticle must be permeable to the diffusible contents of the host's digestive tract, since the parasite depends upon these for its nutrition, and yet must preserve it from destruction by the digestive enzymes to which it is at the same time exposed. This resistance has been found to be maintained only in the living, intact animal, for if one is killed, or its cuticle torn or cut, the cuticle is destroyed as readily by proteolytic enzymes of animal origin as it is in the living state by enzymes of plant origin, such as ficin, obtained from the milky sap of the fig tree, or papain, from the latex of the pawpaw.

5. Rotifera, Nematoda, Nematomorpha

Many rotifers envelop their bodies in clear and transparent chitinous material, which may be thickened in longitudinal or circular ridges, or prolonged into several, or many, spines. Some construct mucous tubes, often studded with precisely arranged foreign particles (Fig. III–8). Two species of *Floscularia*, for example, *F. janus* and *F. pelula*, cement their own feces together into ovoid pellets, while two others, *F. ringens* and *F. conifera*, make, respectively, spheroidal or cylindricoconoid ones of excess food particles, swept into a ciliated glandular cup near the corona and there fastened together by the secreted material. When the pellets are made, the animal bends over and places each one in position, gradually building up course after course of the tube wall. The bodies of nematodes and gordiaceans are covered with a tough, resistant cuticle, secreted by the epidermis, often marked with delicate striations or extended as bristles; the cuticle does not contain chitin, although the eggshells of nematodes do, showing that at some stage of their life cycles, at least, the synthesis of chitin is possible. The cuticle of nematodes is, like that of trematodes and cestodes, a combination of mucoproteins that resist digestion by proteolytic enzymes other than those of plant origin. This protection allows them to live a parasitic existence in the digestive tracts of hosts of many species and also a saprozoic one in decomposing organic material, a situation in which many disintegrative enzymes must be operating.

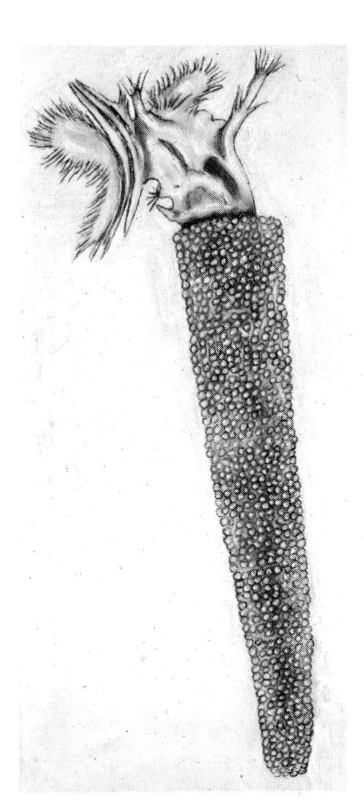

FIG. III–8 *The rotifer* Floscularia, *showing anterior end of animal extending from its case made of pellets of excess food particles.*

6. Bryozoa

More definite as exoskeletons are the cases, called zooecia, of the aquatic and, for the most part, marine Bryozoa whose colonies form masses of moss-like incrustations in the shallow waters of the seas and inland lakes and ponds (Fig. III–9). In *Plumatella*, one of the Phylactolemata, each zooecium consists of an ectocyst, primarily of polysaccharide but impregnated with calcium salts, which the living organism has extracted from its freshwater environment. This is sometimes studded with sand grains or other small bits of debris. Within the ectocyst is an endocyst, softer and more

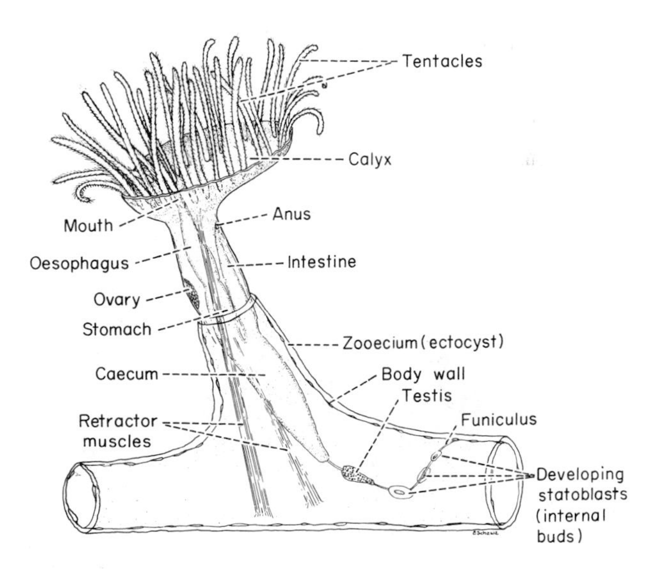

FIG. III–9 *A single individual (polypide) of the colonial freshwater bryozoan,* Plumatella. *(After Pennak.) The zooecium (ectocyst), continuous throughout the colony, is here shown cut off.*

delicate in character, which is continuous throughout all members of the colony. In the related genus, *Pectinatella*, the zooecium is without calcification. Among the Cheilostomata, *Bugula* builds a transparent yellowish zooecium of chitin, while the individuals of *Membranipora* are enclosed in boxes of calcified chitinous cuticle in which only the operculum, covering the opening like a lid, is predominantly chitin.

7. Brachiopoda

The hinged shell, with dorsal and ventral valves, peculiar to Brachiopoda is secreted by special cells in the mantle, two flaps of the body wall that surround the body proper; these cells are collected in papillae which penetrate into the finished shell (Fig. III–10). The shell is composed of an outer periostracum of organic material and an inner layer, relatively thick, in which the organic base is impregnated with calcium salts; between these is a third layer, which is thin and apparently made exclusively of calcium carbonate. In the Inarticulata, of which *Lingula* is perhaps the best known genus, the organic material is chitin and the inorganic material is in the form of more or less loosely assembled calcareous spicules; in the Articulata, such as *Terebratulina* and *Magellania*, chitin is absent and the calcium compounds are in the form of densely packed prisms so that the shell is firm and solid.

8. Mollusca

The skeletons of molluscs may be either external or internal. External skeletons are secreted by cells of the mantle and, while basically similar in construction and composition, vary greatly in pattern and design. Those of polyplacophorans, the chitons or coats-of-mail, consist of transverse plates articulated together; those of gastropods are single and often coiled to accommodate the visceral mass. In pelecypods, the shell has two valves oriented laterally to the animal's body and joined dorsally by a strong elastic ligament. The shells of scaphopods are tubular and open at both ends (Fig. II–19*d*). The evidence provided by fossil ammonites shows that the bodies of early cephalopods were enclosed in chambered shells coiled in a flat spiral, but among modern species only *Nautilus* makes such a shell, for the shells of cephalopods with 10 arms are not only reduced in size but essentially internal in that they are embedded in the fleshy mantle, and in those with eight arms they are absent.

In composition, molluscan shells consist of an organic base called conchiolin which is impregnated to greater or lesser extent with inorganic salts. Conchiolin is made of structural proteins, whose amino acid content shows specific variations and which have undergone quinone tanning. The inorganic component is largely calcium carbonate, although in marine species small amounts of magnesium carbonate may be in-

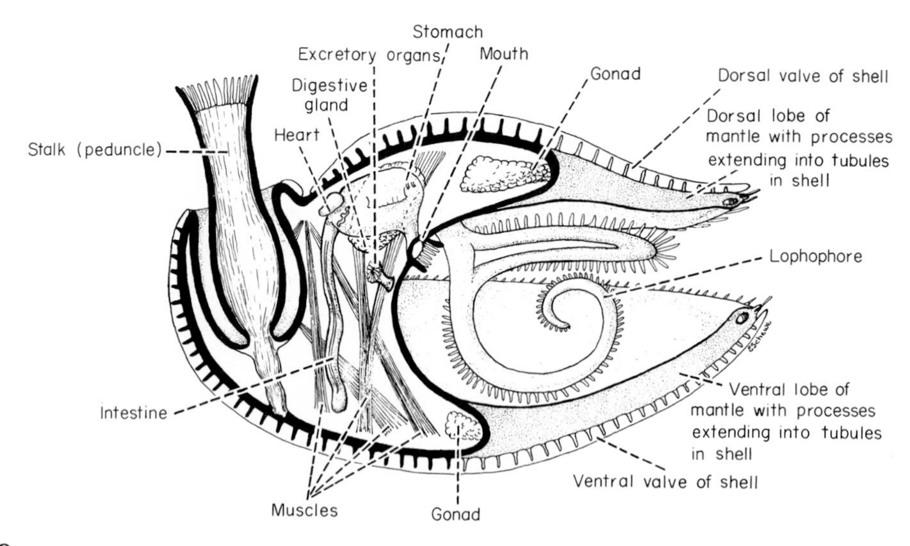

FIG. III–10 *A brachiopod. Diagram shows surfaces of mantle lobes at anterior end (right), with mantle cut open posteriorly to reveal the internal organs. (From various sources.)*

cluded, and, experimentally, strontium may be substituted for calcium. This suggests that these species might be useful as indicators of pollution of the atmosphere and water by atomic fission. In bivalves grown experimentally in calcium-free seawater, the organic component of the shell is laid down in its entirety, but calcification does not take place.

The construction of the shell has been most fully investigated in lamellibranchs and in gastropods, not only for its intrinsic interest but as a possible model for the mechanisms of calcification in general. In lamellibranchs, the valves of the shell consist of an inner layer of nacre, made of plates of calcium carbonate with conchiolin between them. External to this are one or two layers of crystalline calcium carbonate in which the crystals, either aragonite or calcite, are collected in prism-shaped bundles with a small amount of conchiolin, which represents only 1 to 10 percent of the dry weight of the shell. The outermost layer, or periostracum, is wholly conchiolin; this may be gradually worn away so that the prismatic layer is exposed (Fig. III–11). The calcium carbonate in the shells of most bivalves, both larval and adult, is in the form of aragonite. But in the adult oyster, *Crassostrea virginica*, it is mainly in the form of calcite, although the larval shell is aragonitic. This shift in crystalline pattern after attachment and metamorphosis of the larval oyster is not yet understood.

Examination of 27 species of bivalves and gastropods has revealed that the periostracum of their shells differs both in morphology and in composition, according to the species and the environment. In freshwater gastropods it is composed of two layers, in terrestrial species of one layer, and in marine species there is a single thin layer or an incomplete one. Fifteen amino acids have been identified in the conchiolins, differing in concentration in different species. In general, the periostracum of marine gastropods is lower in glycine content than that of the freshwater species, but that of marine bivalves is particularly rich in it. Experiments with the freshwater snail *Potamopyrgus* have shown that the salinity of the environment has an effect upon the composition of the periostracum. Thus, snails raised in freshwater and in 10, 40, and 50 percent seawater had increasing amounts of histidine, glutamic acid, isoleucine, and methionine, and decreasing amounts of glycine and aspartic acid with the increased salinity of the medium.

The construction of the shell occurs in two phases: first, the cellular elaboration and discharge of the fibrous proteins of conchiolin and, second, the deposition of the mineral salts. All the cells of the mantle epithelium contribute to the formation of the organic base of the nacreous layer, but only those at the edge of the mantle contribute to that of the middle layers and of the periostracum. In the restoration of a broken shell, however, all the cells may contribute conchiolin to the periostracum. The inorganic component appears first as a calcium phosphate, which is later converted to calcium carbonate by a mechanism that has not as yet been elucidated. The calcium carbonate is discharged in colloidal form and its crystallization takes place outside the cells that produce it. The source of the carbonate is primarily metabolic, while that of the calcium is external. Aquatic species obtain this from the water in which they live and the food that they eat, and terrestrial forms from the rocks and soils of their environments as well as from their food. Marine species have a constant and relatively abundant supply of Ca^{++}, which they absorb through the general surfaces of their bodies—the mantle, gills, and gut. This has been demonstrated by experiments using Ca^{45} and Sr^{90}, in which it has been shown that the labeled ions are adsorbed on the sheets of mucus discharged by the cells of the mantle and gills. This food sheet is swept along the gills, into which some

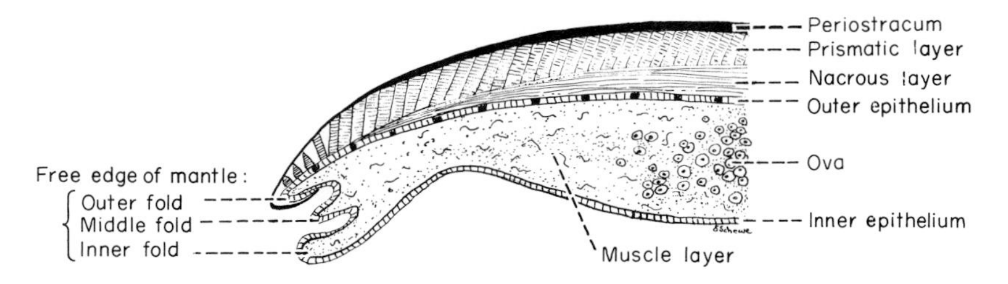

FIG. III–11 *Vertical section through the edge of the mantle and shell of the mussel* Mytilus. *The cells of the outer epithelium of the mantle secrete nacre; those of the inner fold of the free edge of the mantle, the periostracum and prismatic layer of the shell.* (*After Borradaile et al.*)

Ca^{++} passes, and is ultimately swallowed. Fresh-water forms have a less abundant supply, and their habitats are necessarily limited to areas where the waters and soil provide adequate amounts. Terrestrial species depend largely on their food and upon their ability to concentrate and to store the amount available to them. Conspicuous in the digestive glands of pulmonates are "lime cells" in which calcium is stored and used in shell formation as required. These cells are, however, not unique to terrestrial species, because calcium deposits are also found in the digestive glands of chitons and in the connective tissues of some marine gastropods, although pelecypods do not seem to maintain such reserves.

Shell deposition may be a continuous process, going on even when the animal is not feeding or growing. This has been shown by observations made on oysters during the winter, when they are quiescent, and on those kept in sterile seawater; under both these conditions, the dry weight of the shell increases slowly but steadily. Under natural conditions, periods of rapid and less rapid growth may be seen on pelecypod and gastropod shells as bands or rings, much like the growth rings of trees (Fig. III–12). The coloration of the shells may also show the cyclic deposition of pigments as bands or rings of different colors. This is perhaps particularly conspicuous in the shells of members of the genus *Liguus*, the tree snails of southern Florida and the islands of the Caribbean Sea, in which different populations of the same species may have shells of basically different colors with different colored bands of different widths.

The chambered shell of *Nautilus* has no periostracum and only two calcareous layers (Fig. III–13). The inner one is of nacre, as is also the siphuncle. The inorganic substance of the nacre is supported by fibers of chitin, visible with the electron microscope. These can be destroyed by treatment with the enzyme chitinase. The internal shell of the cuttlefish, *Sepia*, which accounts for about 9 percent of the total volume of the animal, is closely related structurally to that of *Nautilus* and to those of the fossils Nautiloidea, Ammonoidea, and Belemnoidea. In its construction, lamellae are successively laid down forming a series of chambers, the newer ones below the older. Each lamella has an organic base, which is calcified by the deposition of aragonite; successive lamellae are held apart by numerous pillars, irregularly placed. Conchiolin represents only 2 to 5 percent of the dry weight of a "cuttle bone," of which 95 to 98 percent is calcium carbonate. The chambers between the lamellae are divided by about six membranes running parallel to the lamellae, and each chamber is sealed off from the others as an independent unit. The upper surface of the shell, or cuttle bone, is covered by a thick calcified surface which extends downward laterally and anteriorly; the lower surface is closed by the curved distal ends of the lamellae. This surface, called the siphuncular surface because of its presumed homology with the siphuncle of the *Nautilus* shell, is separated from the viscera by a thick membrane, with numerous ampullae distributed along its free surface. The ampullae are connected by fine ducts which lead to open spaces, or sinuses, in the basement membrane, which are, in turn, in close connection with the underlying veins. In specimens of *Sepia officinalis*, whose shells have been carefully studied, the most recently formed chambers

FIG. III–12 *Shells of the pelecypod mollusc,* Lampsilis andontoides, *showing progressive enlargement and the periods of growth and cessation of growth marked as lines on the surface.* (*U.S. Fish and Wildlife Service.*)

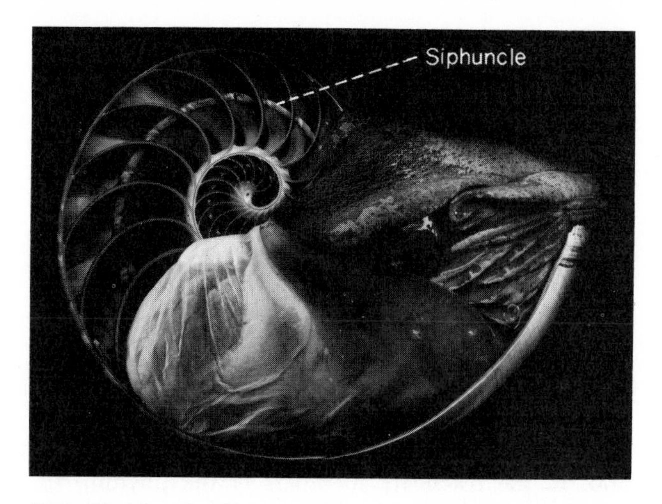

FIG. III–13 Nautilus pompilius, *the chambered nautilus, from Polynesian waters. The bisected shell shows the animal in the largest chamber, the last to be constructed, and the siphuncle, the continuous tube that passes through all the chambers and connects the one first formed with the last. The vacated chambers are filled with gas that differs from air only in its lower oxygen content. This gives the system buoyancy, so that the animal can support the increasingly heavy shell as it moves through the upper levels of the sea. (American Museum of Natural History.)*

had a volume of 1 to 2 ml and were filled with fluid, with concentrations of sodium and potassium essentially the same as those of seawater. The oldest chambers, with volumes less than 10^{-3} ml, were filled with gas, which had an average composition, in samples withdrawn from various chambers, of about 92 percent nitrogen, about 7 percent oxygen, and less than 0.5 percent carbon dioxide. The contents of the chambers between these were primarily gas, although in certain conditions there was some fluid in them. The distribution of liquid in the chambers is such that the animal can maintain a horizontal position in the sea, but can also change its posture by changes in the proportions of fluid and gas in the chambers of its shell and regulate its density in relation to that of the water in which it happens to be. The mechanism of the control is not yet clear, but it has been suggested that the fluid in the most recently formed chamber is pumped out by some active "osmotic" process and that gas slowly diffuses in from the surrounding tissues until equilibrium with them is reached, for the partial pressure of the gases in the chambers is never higher than in animal tissues in general. The relations of the veins to the ampullae in the membrane underlying the cuttle bone provide a means for connection between its chambers and the vascular fluid, so that anatomically a way seems open for exchange between them. The shell of *Sepia* seems to provide the animal not only with internal support but also with a means of adjusting its buoyancy.

In the squids the shell, also embedded in the tissues of the mantle, is entirely organic in composition and, in some species, bears close resemblance both histologically and histochemically to hyaline cartilage. *Lepidoteuthis grimaldi* from the Mediterranean has dermal scales similar in many ways to those of fish, and *Cranchia scabra*, found in Florida coastal waters, has rounded dermal projections made of spherical or ovoid cells embedded in a firm, relatively homogeneous matrix.

In the members of the phylum without exoskeletons, spicules may be deposited in the integument which give some support to the soft parts. In nudibranch gastropods (Fig. III–14), these are primarily calcareous but in some marine pulmonates they are siliceous. Internal tissues may also be supported by discrete spicules, and in cephalopods especially, by masses of chondroid tissue.

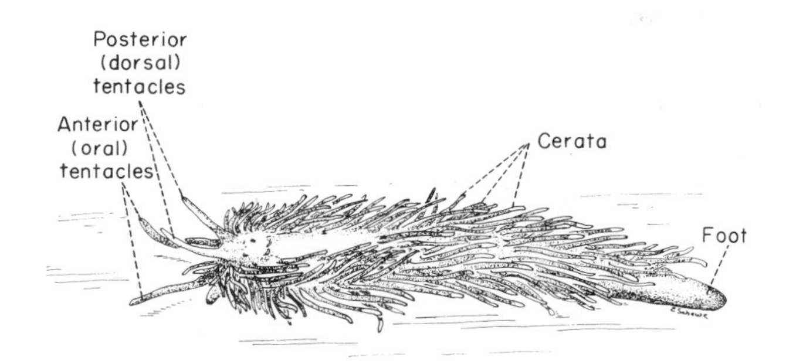

FIG. III–14 *A nudibranch mollusc,* Eolis, *the sea slug, lacking shell, mantle, and ctenidia. The cerata are projections of the body wall, containing extensions of the digestive gland and acting as respiratory areas. The posterior or dorsal tentacles have sense organs that appear to be sensitive to chemical stimuli. (After D. P. Wilson.)*

9. Annelida

Most annelids have nonchitinous cuticular coverings entirely of epidermal origin; in the earthworms this covering is a collagen. Although annelids do not use chitin as a body covering, their ability to make it is evident from the fact that it is found in the chaetae that project from the body wall, as well as in the jaws, denticles, and linings of some regions of the gut. Most members of the phylum secrete mucin in solution through the activity of epidermal glands. In earthworms, this secretion serves to keep the body moist, a condition that is essential to their gaseous interchange; in the tube-dwellers, it may serve to line the tube the animal makes in the sand or mud or to cement together sand grains and other fine particles to strengthen the walls against collapse. The tubiculous annelids, like other tube-dwellers, seem to show selective capacities in making their homes (Fig. III–15). Some species of *Sabella*, for example, choose only the finest particles of mud; *Terebella conchilega* picks out grains of sand and fragments of shells, groping with its tentacles to find them. It uses its tentacles also to put them in its mouth, where they are mixed with an oral secretion, and to insert them into the walls of the tube. The tubes of Serpulidae have an organic base of conchiolin to which is added calcium carbonate secreted by special tube glands in the peristomium.

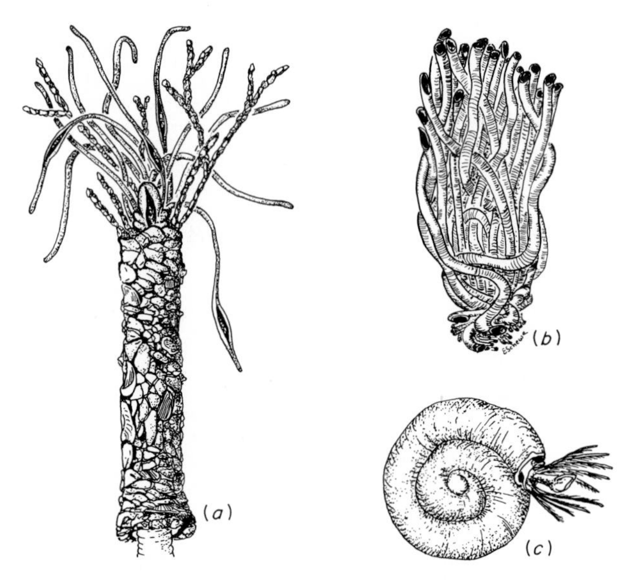

FIG. III–15 *Tubes of tube-dwelling annelids.* ***(a)*** *Anterior portion of* Terebella conchilega, *showing the worm in its case of sand grains and shell fragments embedded in mucus. (After Cambridge Natural History.)* ***(b)*** *Intertwined masses of the tubes of* Serpula. ***(c)*** Spirorbis *in its coiled tube. (After Carson.)*

10. Arthropoda

It is arthropods especially that have made use of chitin as a basis for the exoskeletons in which their bodies are typically enclosed and which may extend into some of their internal organs, lining the fore- and hindguts, and, in the tracheate members of the phylum, the tracheal tubes. The skeleton, which is secreted by cells of the epidermis, has two primary areas, an inner endocuticle and an outer epicuticle (Fig. III–16). The endocuticle is relatively thick and contains chitin, which in Crustacea represents 60 to 80 percent of the total organic material. The epicuticle is comparatively thin and is made of protein only, which becomes stiffened through quinone tanning. Both of these areas are generally resolvable into several sublayers. In the endocuticle, the chitin is bound with protein in a protein-chitin complex. In crustaceans and some diplopods, the endocuticle may become impregnated with calcium salts. There is evidence that in Crustacea the endocuticle also contains an acid mucopolysaccharide, which upon hydrolysis yields fucose, galactose, and glucose, and which may prove to have an important role in the processes of calcification. Calcification is particularly extensive in the decapods, giving lobsters and crabs the heavy shells which are characteristic of them.

In insects, as well as in crustaceans, ingrowths of the cuticle, or apodemes, provide an endoskeleton which serves to support and protect some organs and to provide a firm base for the attachment of muscles (Fig. III–17). In decapod crustaceans, apodemes in the thoracic region are united into arches over the nerve cord. This endophragmal skeleton protects the nerve cord and also gives greater efficiency to the performance of the thoracic muscles which are attached to it. In the thoracic segments of insects, the apodemes are particularly well developed and are

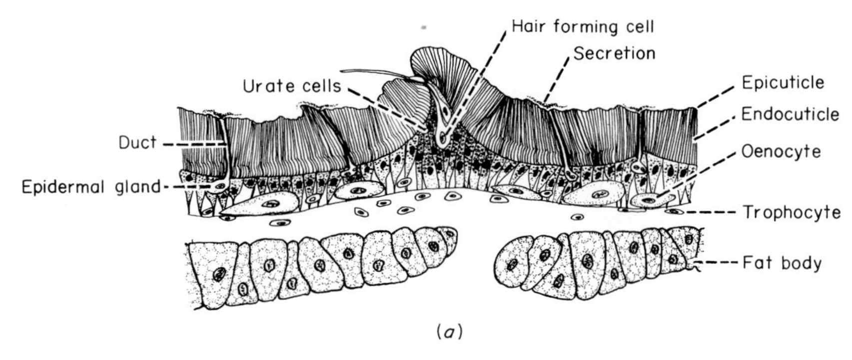

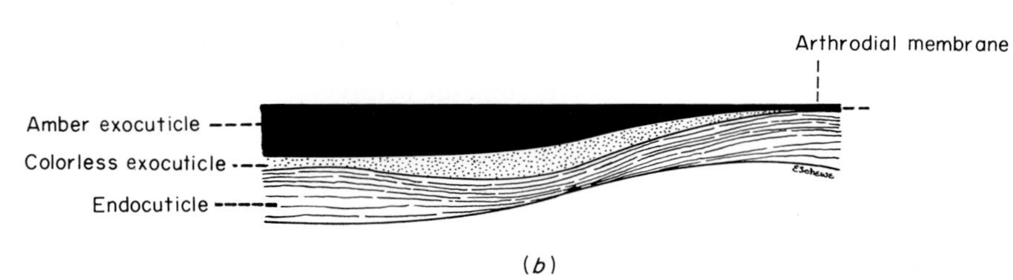

FIG. III–16 *Integument of insects.* ***(a)*** *Longitudinal section of tergites of the fourth nymphal instar of the blood-sucking bug,* Rhodnius prolixus. *(After Wigglesworth.)* ***(b)*** *Diagrams showing the changes in thickness of the cuticle of a cockroach,* Periplaneta americana, *in passing from a sclerite to an arthrodial membrane. (After Dennell and Malek.)*

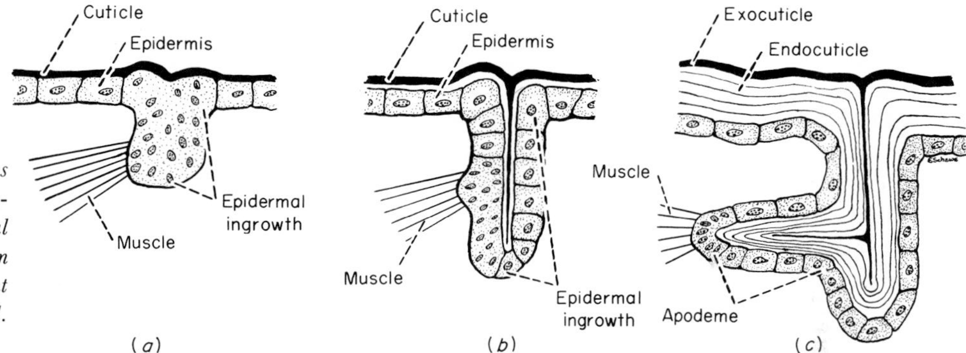

FIG. III–17 *Formation of apodemes in arthropods.* ***(a)*** *Initial epidermal ingrowth.* ***(b)*** *Deepening of epidermal ingrowth.* ***(c)*** *Lateral projection from ingrowth at site of muscular attachment forms apodeme. (After Grassé, vol. VI.)*

situated so as to provide broad areas for the insertion of muscles that move the legs and wings.

The cuticle of insects has been intensively studied from the point of view of its origin, composition, and chemical and physical properties. Its thickness, measurable in microns, varies in different species and even in different areas of the same individual. In the adult mealworm, *Tenebrio*, for example, it is about 4 μ thick in the dorsal plates, or tergites, of the abdomen and about 36 μ in the ventral plates, or sternites. The form and manner in which the endocuticle (procuticle) is secreted is not yet entirely clear, but electron micrographs show that when formed it has a laminar structure, visible as alternating light and dark bands, whose thicknesses may range from 0.2 to 10 μ. It is because of this laminar arrangement that many insects show structural coloration, as well as the chemical coloration resulting from sclerotization, which may be general or localized in particular areas or patches. The process of sclerotization, which results from the stepwise oxidation of tyrosine by the enzyme tyrosinase to the dark pigment melanin and from quinone tanning, usually begins at the interface between the endocuticle and the exocuticle, and spreads inward through the outer layer of the endocuticle. The degree of darkening of any given area depends upon the amount of substrate present. Where sclerotization does not take place, the cuticle remains membranous and flexible.

The epicuticle is deposited upon the surface of the endocuticle by unicellular or multicellular epidermal glands, whose ducts penetrate through the layers already laid down. In *Tenebrio*, the epicuticle can be resolved into four distinct layers: an inner one of cuticulin, then one of polyphenols, next one of wax, and finally an outer layer, or tectocuticle. The cuticulin layer is deposited first, most probably by oenocytes, and is believed to be of lipoprotein. Then droplets of fluid with a high content of polyphenols, presumably oxidative derivatives of tyrosine, are secreted onto the surface of the cuticulin and spread out over it forming a continuous sheet. The wax layer may be secreted by the epidermal cells in general, although the involvement of oenocytes also is indicated; this is first discharged in liquid form through the ducts of the glands and spreads, like the polyphenols, over the surface. The chemical nature of the wax is different in different species, but has the general properties of long-chain alcohols with long-chain fatty acids or long-chain paraffins. The melting points of the different waxes range from about room temperature to over 100°C. The molecules have been shown to have a definite orientation, forming up to 30 monolayers. After its discharge, the wax solidifies through evaporation of its more or less volatile solvent, and so provides a waterproof coating. The outer layer, or tectocuticle, is a mixture of chemical substances; it is a product of the epidermal glands and is very thin and delicate.

Such an exoskeleton has provided the insects with a light and flexible covering, which has allowed them to exploit the mechanism of flight and which has also given their bodies protection from what is probably their greatest physical danger, dehydration. Its extension within the body has afforded them a greater possibility of mechanical efficiency in muscular activity.

11. Echinodermata

The skeleton of echinoderms is primarily inorganic in composition and is the product of mesodermal rather than ectodermal tissues. In these respects, it resembles the skeleton of sponges and also of vertebrates, although it differs completely from the latter in its location and the absence of cells embedded in the matrix. The calcium salts of which it is made are obtained from the seawater and are concentrated by cells of mesodermal origin and deposited beneath the epidermis either as ossicles so small as to be legitimately called spicules or as large plates of characteristic size and shape. These are either articulated together so as to provide a flexible body, as in the asteroids and ophiuroids, or closely interlocked to form a rigid investment, as in the echinoids and crinoids (Fig. III–18). Typically these plates bear movable spines that may project through the surface epithelium giving the animals the bristling appearance and piercing protection for which their phylum was named. The plates may also be perforated by pores through which the tube feet of the unique water vascular system project. The pedicellariae of the asteroids and echinoids are shorter, broader, and blunter modifications of these spines, two or three being articulated together to form a pincer-like organ capable of seizing and grasping an object with great force and tenacity.

D. ADVANTAGES AND RESTRICTIONS

The development of skeletal coverings, whether thick or thin, delicate or heavy, primarily organic or inorganic, offers certain advantages to the organisms possessing them. They provide protection from mechanical or chemical injury, from the assaults of predators or competitors, from the invasion of harmful microorganisms or other parasites, and from the dangers of loss of water and other diffusible substances whose conservation is physiologically important. Such development has provided, too, potentialities for evolving powerful or highly specialized mechanisms for attack or defense, for the crushing or trituration of food, and for the transmission and deposition of sexual products. Yet it has also imposed certain limitations. These arise not only from the reduction of surface area through which gases and other solutes may diffuse in and out of the body but also from the weight of the structures themselves and, in the case of complete exoskeletons, from the restrictions upon growth. Animals with a massive encasement must be either slow moving or sedentary in their habits, like pelecypod and gastropod molluscs, decapod crustaceans, and echinoderms, or entirely fixed and sessile, like corals.

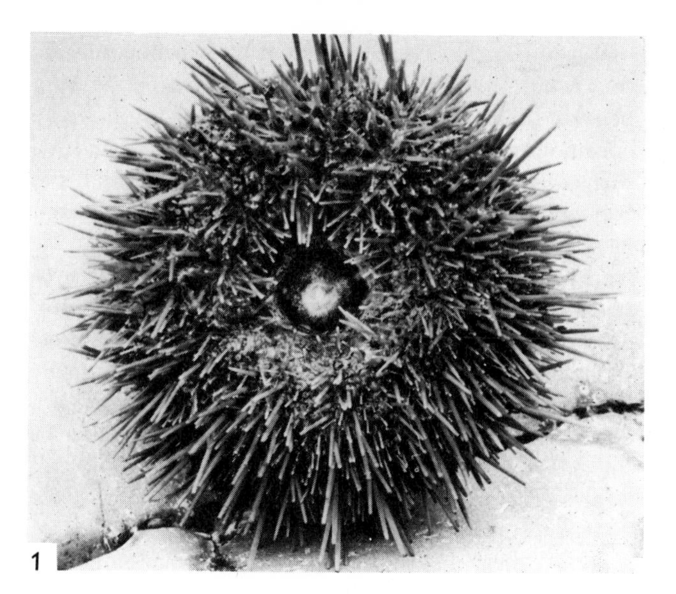

(a)

(b)

FIG. III–18 *Echinoderm skeletons.* ***(a)*** *The green sea urchin,* Strongylocentrotus droebachiensis, *from the northern seas, where it is eaten freely by gulls. It is also the principal food for sea otters and an important one for the aborigines: 1, Oral surface; 2, Aboral surface.* ***(b)*** *The starfish,* Asterias amurensis, *from the Aleutian Islands, whose body may reach 1 ft in diameter. (U.S. Fish and Wildlife Service.)*

The restrictions upon growth are perhaps the most serious ones an exoskeleton imposes. The animal enclosed in one must either curtail its growth and, having made a case to fit its size early in its life, remain within it without further growth, or the covering must be enlarged as the living animal grows or be periodically abandoned and a new one made to fit the increased dimensions of its occupant. The first is hardly a biological possibility, for while the bodies of all adult animals conform not only to a design but ultimately to a range of sizes that are characteristic of the species, growth is a continuous process and, in juvenile stages, is usually a very rapid one. Invertebrates have, however, adopted both of the other possibilities. In the tube-dwelling rotifers, for example, the tube is first made when the animal is young but is enlarged, as its occupant grows, by the continued addition of particles to its rim. As the body of the pelecypod and gastropod mollusc grows, new shell is laid down by the mantle cells. *Nautilus* and many of the Foraminifera have adopted the device of building a series of chambers coiled in a flat spiral. The animal moves, as it grows, into a newly constructed, larger chamber. In the corals, growth of the colony is effected by multiplication of its individual members; the new individuals secrete their own skeletons of appropriate size and the old ones leave theirs behind as they divide or die. The arthropods have elected the third possibility, so that as their bodies grow they undergo a series of molts, or ecdyses, during which they make a new cuticle or shell to conform to their new dimensions and crawl out of the old, abandoning it entirely. The phenomenon of molting is an elaborate and delicately controlled one, which involves the softening of the old skeleton as well as the formation and hardening of the new; in some insects, where the process has been carefully investigated, it is known to be regulated by neurosecretions that act as hormones.

E. ACCUMULATION AND DISPOSITION OF DISCARDED SKELETONS

The disposition of invertebrate skeletons, abandoned after a molt or left behind when the living organism dies, is a matter not only of biological but of economic importance. For, unless removed by some means, they would accumulate in masses that over the long periods of geologic time would become sufficiently enormous to alter radically the character of the earth's surface. It has been estimated, for example, that the copepods of the marine plankton alone produce each year several billion tons of chitin in the oceans of the world; other marine crustaceans and terrestrial arthropods are probably responsible for the production of several times this amount. Such chitinous remains, together with other organic and inorganic ones, represent a quantity of material that would soon become an appreciable layer, even when distributed over the whole surface of the earth, at the bottom of all bodies of water as well as on the surface of dry land. They represent moreover, the binding of considerable amounts of carbon and nitrogen, as well as of minerals, and therefore a reduction in the supply available to living biological systems. But investigations of marine sediments and of the soil have shown that neither contains significant amounts of chitin. Since the chitin molecule is known not to undergo spontaneous decomposition, some physical or biological agent must be concerned in its destruction and the release and restoration to the common supply of the elements bound in it. That bacteria are the primary agents in the removal of chitin residues has been shown by the isolation of microorganisms from marine sediments and sands and from decomposing crustacean shells. These bacteria attack the chitin molecule in one way or another, some deriving their complete carbon and nitrogen requirements from it. Chitinoclastic bacteria represent between 0.1 and 1 percent of marine mud dwelling species. They carry on their activities effectively, though very slowly, both aerobically and anaerobically, and at temperatures as low as 0°C. Probably bacteria are also responsible for the decomposition of much of the organic material incorporated into the shells of other marine animals, as well as those of freshwater and terrestrial species.

The inorganic material, exposed to such destructive physical agents as weathering and mechanical crushing may be reduced to fine particles, so contributing to the soil that covers the earth's surface, to sediments at the bottom of bodies of water, and to localized mineral deposits. Or the shells may remain virtually intact and, falling from the surface, accumulate on the sea or lake floor. Because of the consistency of these marine bottom deposits, the Scottish oceanographer Sir John Murray (1841–1914) called them oozes, distinguishing one from another on the basis of the type of skeleton present in greatest amount (Fig. I–3). In temperate and tropical seas, for example, the foraminiferan *Globigerina* is especially plentiful, and *Globigerina* ooze covers some 48 million square miles of the sea bottom, particularly that beneath the deeper waters of the Atlantic Ocean. In shallower waters, and nearer the equator, however, the character of the ooze is different, for here pteropods are the predominant members of the plankton; their delicate shells, sinking to the bottom when the animal dies, are the main constituent of the ooze. Pteropod ooze is common near coral islands and is found exclusively in equatorial waters.

Abandoned shells may pile up as mounds or ridges or, if they have been transported from the places where the animals lived, they may become incorporated into sedimentary rocks. In each event, they become factors determining the physical geography of the areas in which they accumulate and so influence the lives of other biological systems. Thus, the chalk cliffs that bound some of the land masses are made largely from the heaped up shells of foraminiferans, brachiopods, and molluscs. Coral reefs and islands, on the other hand, are formed primarily by madreporarians slowly but continuously depositing skeletal material, tier upon tier, until the mass reaches the surface of the water. Alcyonarians, gorgonians, and some of the foraminiferans are included, too, in the reef-building fauna, and coralline algae, whose cellulose walls are impregnated with calcium carbonate, also take part in the construction of these limy formations in the warm, shallow waters in which they thrive and multiply. The Great Barrier Reef, off the northeast coast of Australia, has been made by the activity of such animals and plants as these, steadily multiplying and individual by individual depositing its own kind of skeleton. Now an elevation 1,350 miles long, from the Tropic of Capricorn almost to the equator, and 10 to 120 miles offshore, it has been gradually built to such a height that at low tide it is exposed above the surface of the sea. This reef, the largest of all coral reefs, provides homes for enormous numbers of invertebrates and vertebrates.

SELECTED BIBLIOGRAPHY

Brown, C. H.: Quinone Tanning in the Animal Kingdom, *Nature*, vol. 165, p. 275, 1950.

Brown, C. H.: Some Structural Proteins of *Mytilus edulis*, *Q. J. Microsc. Sci.*, vol. 93, pp. 487–502, 1952.

Brown, H.: Protein Skeletal Materials in Invertebrates, *Exp. Cell Res.*, Supp. 1, pp. 351–355, 1949.

Chapman, G. B.: The Hydrostatic Skeleton in the Invertebrates, *Biol. Rev.*, vol. 33, pp. 338–371, 1958.

Comfort, A.: The Pigmentation of Molluscan Shells, *Biol. Rev.*, vol. 26, pp. 285–301, 1951.

Dennell, R., and S. R. A. Malek: The Cuticle of the Cockroach *Periplaneta americana*, *Proc. R. Soc. (London), B*, vol. 143, pp. 126–135, 1954.

Denton, E. J.: Buoyancy Mechanisms of Sea Creatures, *Endeavour*, vol. 22, pp. 3–8, 1963.

Denton, E. J., and J. B. Gilpin-Brown: The Buoyancy of the Cuttle Fish *Sepia officinalis* (L), *J. Mar. Biol. Assoc. U.K.*, vol. 41, pp. 319–342, 1961.

Denton, E. J. and J. B. Gilpin-Brown: Distribution of Gases and Liquid within Cuttlebone, *J. Mar. Biol. Assoc. U.K.*, vol. 41, pp. 365–381, 1961.

Denton, E. J. and D. W. Taylor: The Composition of Gas in the Chambers of the Cuttlebone of *Sepia officinalis*, *J. Mar. Biol. Assoc. U.K.*, vol. 44, pp. 203–207, 1964.

Fox, D. L.: Carotenoid and Indolic Biochromes of Animals, *Annu. Rev. Biochem.*, vol. 16, pp. 443–470, 1947.

Fox, D. L., and C. F. A. Pantin: The Colors of the Plumose Anemone *Metridium senile* (L), *Philos. Trans. R. Soc. (London), B*, vol. 230, pp. 415–450, 1940.

Fox, H. Munro: The Colors of Animals, *Endeavour*, vol. 14, pp. 40–47, 1955.

Glaser, R., and E. Lederer: Echinochrome et Spinochrome; dérivés methylés, distribution, pigments associés, *C. R. Acad. Sci., Paris*, vol. 208, pp. 1939–1942, 1939.

Goffinet, G.: Étude au microscope électronique des structures organisées des constituants de la conchioline du

Nautilus macrocephalus Sowerby, *Comp. Biochem. Physiol.*, vol. 29, pp. 835–839, 1969.

Goffinet, G., and Ch. Jeuniaux: Composition chimique de la fraction "nacroine" de la conchioline de nacre de *Nautilus pompilius Lamarck, Comp. Biochem. Physiol.*, vol. 29, pp. 277–282, 1969.

Goreau, T.: Problems of Growth and Calcium Deposition in Reef Corals, *Endeavour*, vol. 20, pp. 32–39, 1961.

Haskin, H. H.: Age Determination in Molluscs, *Trans. N.Y. Acad. Sci.*, vol. 16, pp. 300–394, 1954.

Hock, C. W.: Decomposition of Chitin by Marine Bacteria, *Biol. Bull.*, vol. 79, pp. 199–206, 1940.

Hyman, L.: The Occurrence of Chitin in the Lophophorate Phyla, *Biol. Bull.*, vol. 114, pp. 106–112, 1958.

Jones, W. C.: The Effect of Reversing the Internal Water Current on the Spicule Orientation in *Leucosolenia variabilis* and *L. complicata, Q. J. Microsc. Sci.*, vol. 99, pp. 263–278, 1958.

Kennedy, G. Y. and H. G. Vevers: The Biology of *Asterias rubens*. V. A Porphyrin Pigment in the Integument, *J. Mar. Biol. Assoc. U.K.*, vol. 32, pp. 235–247, 1953.

Kramer, S. and V. Wigglesworth: The Outer Layers of the Cuticle in the Cockroach *Periplaneta americana* and the Function of the Oenocytes, *Q. J. Microsc. Sci.*, vol. 91, pp. 63–72, 1950.

Meenakshi, V. R., P. E. Hare, N. Watabe, and K. M. Wilbur: The Chemical Composition of the Periostracum of the Molluscan Shell, *Comp. Biochem. Physiol.*, vol. 29, pp. 611–620, 1969.

Meenakshi, V. R., and B. Scheer: Acid Mucopolysaccharide of the Crustacean Cuticle, *Science*, vol. 130, pp. 1189–1190, 1959.

Odum, H. T.: Notes on the Strontium Content of Sea Water, Celestite Radiolaria and Strontianite Snail Shells, *Science*, vol. 114, pp. 211–213, 1951.

Person, P.: Cartilaginous Dermal Scales in Cephalopods, *Science*, vol. 164, pp. 1404–1405, 1969.

Richards, A. G.: "The Integument of Arthropods," University of Minnesota Press, Minneapolis, 1951.

Stenzel, H. B.: Oysters: Composition of the Larval Shell, *Science*, vol. 145, pp. 155–156, 1964.

Trueman, E. R.: Quinone Tanning in the Mollusca, *Nature*, vol. 165, pp. 397–398, 1950.

Tsujii, T., D. G. Shapp, and K. M. Wilbur: Studies on Shell Formation. VII. The Submicroscopic Structure of the Shell of the Oyster *Crassostrea virginica, J. Biophys. Biochem. Cytol.*, vol. 4, pp. 275–280, 1958.

Watabe, N., D. G. Sharp, and K. M. Wilbur: Studies on Shell Formation. VIII. Electron Microscopy of Crystal Growth of the Nacreous Layer of the Oyster *Crassostrea virginica, J. Biophys. Biochem. Cytol.*, vol. 4, pp. 281–286, 1958.

Webb, D.: Histology, Cytology and Embryology of Sponges, *Q. J. Microsc. Sci.*, vol. 78, pp. 51–70, 1935.

Wilbur, K. M., and L. H. Jodrey: Studies on Shell Formation. I. Measurement of the Rate of Shell Formation using Ca^{45}, *Biol. Bull.*, vol. 103, pp. 269–276, 1952.

Zobell, C. E., and S. C. Rittenberg: The Occurrence and Characteristics of Chitinoclastic Bacteria in the Sea, *J. Bacteriol.*, vol. 35, pp. 275–287, 1938.

chapter IV

HISTOLOGY

A. CELLS AND TISSUES

The Greek word for warp or web is *histos*, extended by early anatomists to mean "tissue," and so used as the root of the word meaning the study, and the construction, of the groups or masses of cells that are assembled in different parts of the body and perform special functions there. With the exception of some epithelia, no tissue is a wholly homogeneous one, composed of cells of one type only. In addition to cells of special types, tissues usually contain supporting or adjuvant cells of other kinds, but for convenience they are designated by the type of cell predominating in them as epithelial, supportive, transportive, muscular, nervous, or reproductive. Each of these represents a different kind, and often a different degree, of cell specialization, in which one or another of the general attributes of any cellular system may be more conspicuous or effective than the others. Knowledge of the morphological attributes of invertebrate tissues, and to some extent at least of their functional capacities, has been derived from the microscopic study of them in living animals sufficiently transparent to permit it; in maceration preparations or tissue cultures where isolated cells can be observed; and in fixed material, sectioned and stained by standard histological procedure. In recent years, electron microscopy has added enormously to our knowledge of the fine structure of cells, revealing with certainty the configuration of many cytoplasmic organelles already recognized by less refined and detailed means (Fig. IV–1).

From the study of cells in tissue culture, it has been suggested that there are only three fundamental types of cells—epitheliocytes, mechanocytes (supporting cells that may be primarily contractile, motile, or phagocytic), and amebocytes that are both motile and phagocytic—from which all distinctive and specialized ones have evolved. These cells—and indeed all active and functional cells—have certain features in common, and it is on the basis of differences in shape, size, and manifestation of primary functional activity that the histologist makes his distinction between tissues of different kinds.

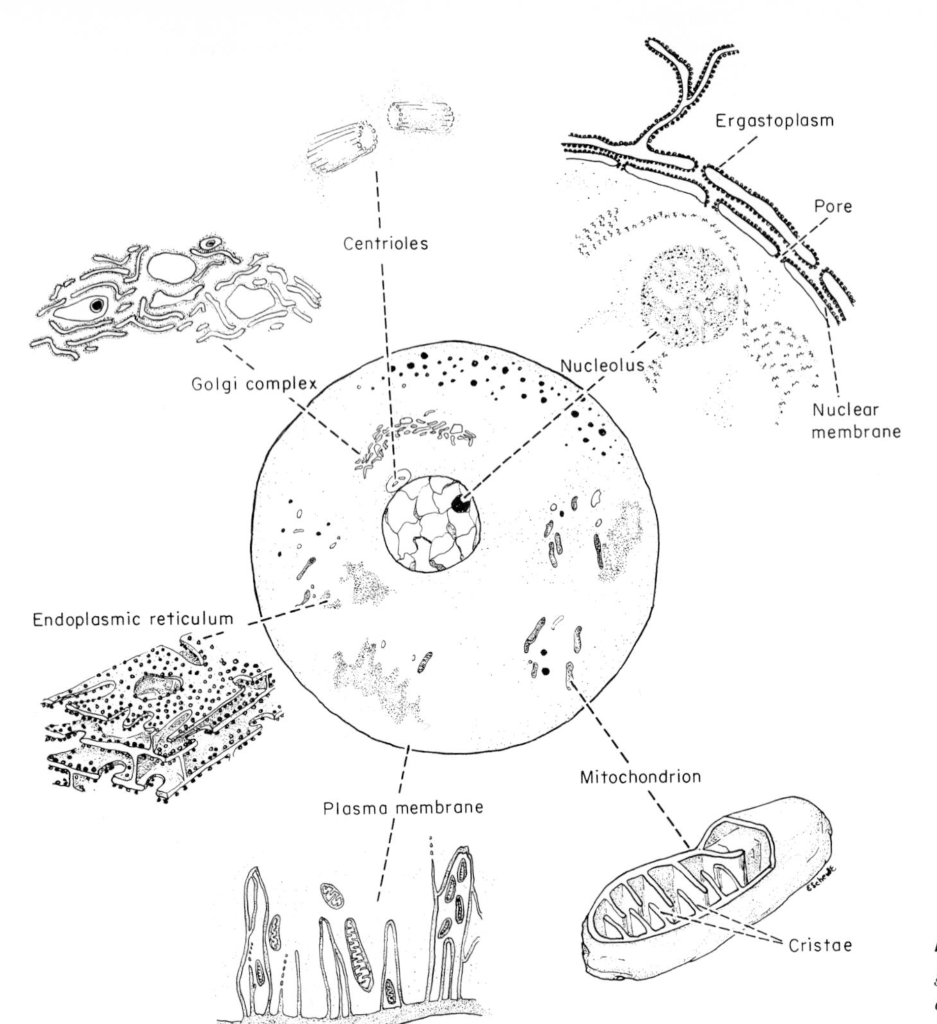

FIG. IV–1 *The components of a generalized cell. The enlargements of certain areas indicate details that have been revealed by electron microscopy.*

B. EPITHELIAL TISSUES

Epithelium is a tissue whose cells cover all surfaces of the animal body, both internal and external. Embryologically, it may be derived from ectoderm, endoderm, or mesoderm; anatomically, it may be located in any part of the body where there is an exposed surface. It may consist, as simple epithelium, of several superimposed layers. Its cells may be flattened so that they form a thin, or squamous, layer, or they may be cuboidal or columnar in shape. It may be primarily secretive or absorptive in function, sensory or in various ways protective of the underlying cells. This protection may be mechanical, reducing the chance of damage to delicate tissues by the rubbing or bumping of foreign objects against them, or it may be physiological, preventing loss or gain of water and salts or the invasion of microorganisms or other parasites.

In the invertebrates, epithelia are simple in character and not stratified as is common in the vertebrates. In some cases, as in the arrowworm, *Sagitta*, pseudostratified epithelium may be found; this is a type of epithelium which gives the appearance of stratification but is actually a simple epithelium in which some of the cells have been crowded inward so that they have almost entirely, or even entirely, lost contact with the surface.

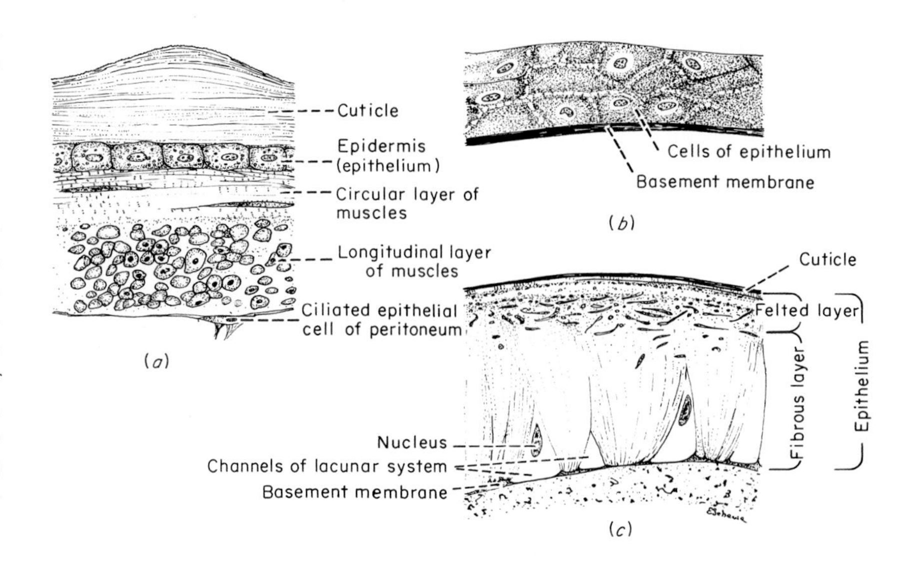

FIG. IV–2 *Types of epithelia.* ***(a)*** *Cuboidal epithelium from the trunk wall (in transverse section) of the sipunculid* Golfingia miniata. *(After Hyman, vol. V.)* ***(b)*** *Pseudostratified epithelium from the collarette of the arrowworm* Sagitta. *(After Hyman, vol. III.)* ***(c)*** *Syncytial epithelium from the body wall (in longitudinal section) of an acanthocephalan. (After Grassé, vol. IV.) Beneath the cuticle the epidermis has a felted structure of fine intercrossing fibers. Beneath this, the fibers are oriented perpendicularly to the surface of the body with their inner ends entangled in a basement membrane. Interspersed among the fibers are fluid-filled spaces, without walls. The nuclei are located in this layer of fibers, but there are no apparent cell walls.*

The external epithelium of invertebrates, or epidermis, is generally of cuboidal or low columnar type (Figs. IV–2 and IV–3). Sometimes it is syncytial, as in the five phyla often included in one (Aschelminthes), and often it is ciliated, as in nemerteans, echinoderms, and some flatworms and molluscs. In many species it is glandular or contains secretory cells that elaborate generalized materials, such as lubricants, adhesives, or cuticles, or more specific ones, such as odors, irritants, or poisons characteristic of a particular species. Certain cells of the external epithelium may also differentiate as sensory cells, receiving stimuli and either transmitting them directly to an effector cell, as is the case in many Cnidaria, or to connecting neurons through which they are relayed to the appropriate response centers. And in some invertebrates, notably the insects, epithelial cells contain pigments of various kinds.

The lubricating and adhesive substances are mucoproteins, secreted either by individual cells scattered among the nonsecretory ones of the integument or collected together in groups or clusters as glands. The mucous secretion may be a homogeneous one spread generally over the surface of the body, like the slime that covers the bodies of earthworms and provides a moist medium for the diffusion of gases across the body wall as well as a lubricant as the worms push through the soil. Or the secretion may be limited to some part of the surface such as that which exudes from the foot of a gastropod mollusc and makes a track along which the snail glides as the muscles of the foot undergo rhythmic contractions. Or the integumental glands may secrete mucoproteins of different kinds, some of which may have adhesive properties or, passing out through narrow apertures, emerge as threads of considerable tensile strength. Thus, in the tentacular epithelium of ctenophores there are groups of gland cells, three to seven in number, that secrete a sticky substance which may hold the animal temporarily to some object; similar

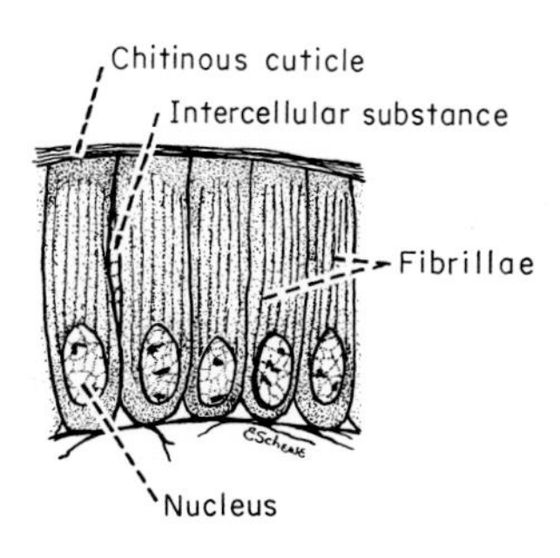

FIG. IV–3 *Columnar epithelial cells from the hindgut of the crayfish* Astacus fluviatilis, *showing the chitinous cuticle secreted by them. (After Janisch.)*

cells are found in the basal region of *Hydra*, and are localized in the adhesive disks on the tentacles of medusae and in the pedal glands of rotifers (Fig. IV–4). The byssus, or tuft of strong thread by which some of the pelecypod molluscs, like *Mytilus*, are held to stones or rocks, is secreted by integumental glands in the foot that are deeply invaginated and discharge their secretion through ducts. At least two sets of cells seem to be involved in the secretion, which, emerging from the foot as a very viscous fluid, is hardened by quinone tanning after it leaves the body to the characteristic thick, dark threads by which the animal attaches itself to a convenient substrate. In many species of insects, gland cells on the terminal segments of the legs produce a gummy secretion by which the animal anchors itself when at rest, walks up the straight, smooth wall of a pane of glass, or even, in an inverted position, walks along a ceiling (Fig. VI–28). The larvae of many insect species, too, elaborate a secretion in their labial glands, which are invaginations of the lining of the oral cavity and so are continuous with the integument. These secretions may be spun out as fine threads of considerable tensile strength. Weaver ants, of the genus *Cecophylla*, make use of this ability of their larvae in the construction of their nests. These nests are made of bundles of leaves bound together with threads. In making them, some workers pull the leaves together by grasping the edge of one in their jaws and hooking their legs into the edge of another. Other workers bring up the larvae, carrying them in their jaws and squeezing them slightly to induce secretion from the spinning glands, which the adults lack. These workers hold the heads of the larvae against the margin of a leaf and, as the thread emerges, move them across to the margin of the other leaf, and so back and forth until the leaves are bound together by a meshwork of criss-cross threads. The nest, when finally built, hangs down from the tree and provides a large and solid habitation for the warlike and predaceous colony. No other insects can live in trees inhabited by *Cecophylla*, so that they provide a natural means of protecting trees from parasitic insects. Recognizing this, the inhabitants of southern Asia, where weaver ants are indigenous, cut down the nests and hang them on citrus trees they wish to rid of aphids and other insects destructive to the crop.

In the Lepidoptera, particularly, these larval labial glands develop the substance known as silk. The larvae extrude the secretion as fine threads which form the cocoons enclosing the pupae. Because of the commercial value of its product, this secretion and the structure of the gland that makes it have been most carefully investigated in the silk moth, *Bombyx mori* (Fig. IV–5). In these larvae, the silk-forming glands

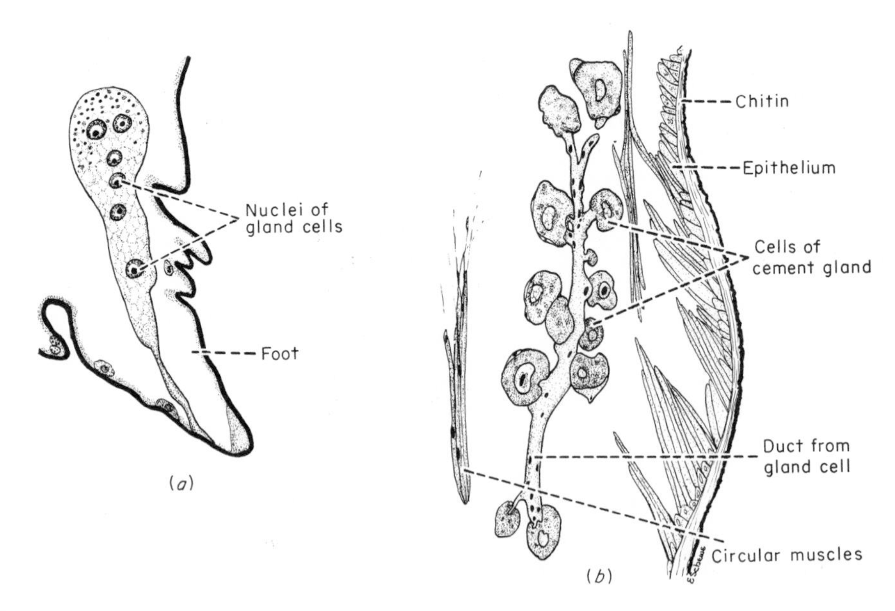

FIG. IV–4 *Glandular epithelium.* **(a)** *Pedal gland of a male* Epiphanes (*Hydatina*) senta. (*After Hermes.*) **(b)** *Longitudinal section of the stalk of the barnacle* Lepas anatifera, *showing part of the cement gland.* (*After Thomas.*)

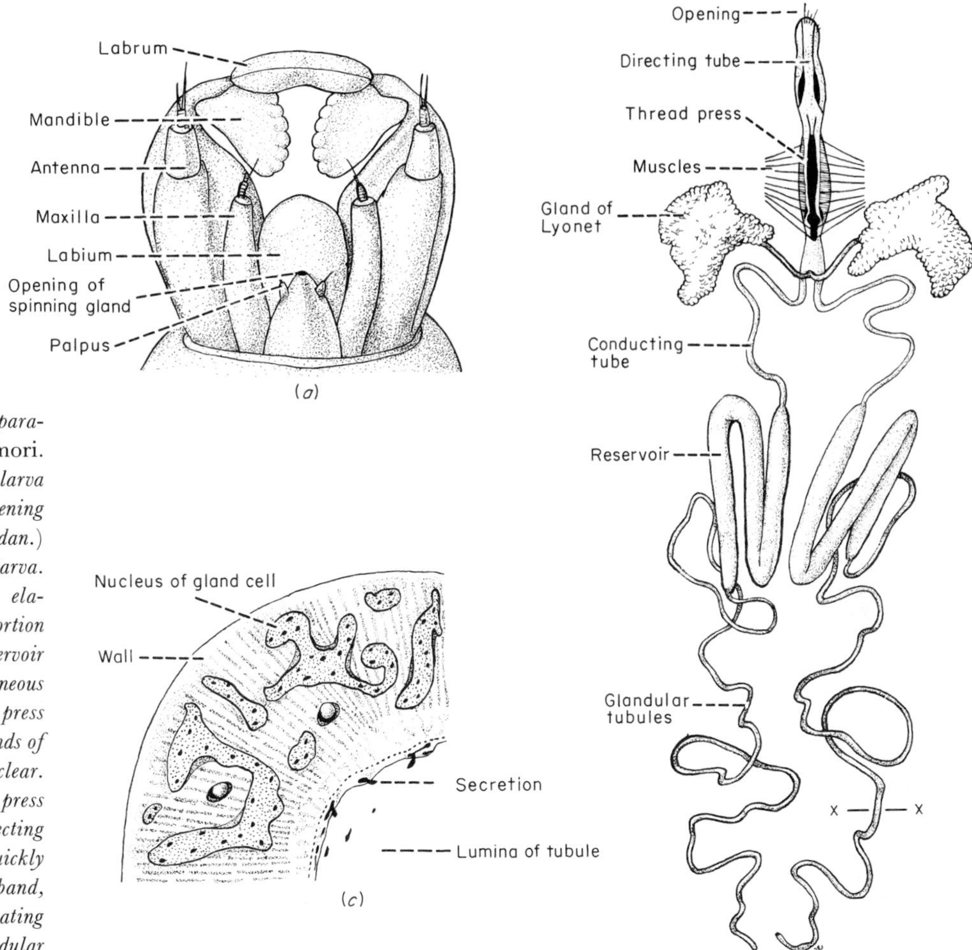

FIG. IV–5 *Silk-producing apparatus of the silkworm* Bombyx mori. ***(a)*** *Undersurface of the head of a larva showing the mouthparts and the opening of the spinning gland. (After Jordan.)* ***(b)*** *Spinning glands of the larva. (After Lesperon.) The secretion elaborated by the cells in the tubular portion of the gland is collected in the reservoir and then passes as a fairly homogeneous colloidal solution into the thread press into which open the ducts of the glands of Lyonet, whose function is not clear. The muscles attached to the thread press force the secretion through the directing tube from which it emerges as a quickly solidifying, double ribbon-like band, with a core of silk fibroin and a coating of sericene.* ***(c)*** *Section of glandular tubule in area indicated by x——x in b. (After Lesperon.)*

are long and tubular, with cells that are large and have extensively branching nuclei. In them, the components of silk are synthesized from amino acids, principally glycine, alanine, and tyrosine, supplied by the hemolymph; these components have been identified as silk fibroin, secreted by cells in the distal region of the gland, and sericine, a gelatinous substance secreted by cells nearer the midregion. After metamorphosis, these highly specialized cells function in the imago as salivary glands, producing a viscous secretion containing digestive enzymes. Dermal glands on the tarsi of the forelegs of the Embioptera (Fig. IV–6) make a similar silken secretion used to line the tunnels in which these fragile insects live, as do also cells of the Malpighian tubules in the larvae of some Coleoptera and Neuroptera from which enveloping cocoons are constructed.

Silk is also made by the spinning glands of adult spiders, developed as ectodermal invaginations of the buds originating on the fourth and fifth abdominal segments of the embryo (Fig. IV–7). These buds, essentially vestigial appendages, become the spinnerets through which the silk is passed when the web is spun. As many as seven different kinds of silk glands have been identified in spiders, each kind producing its own particular type of secretion. The orb-weaving spiders,

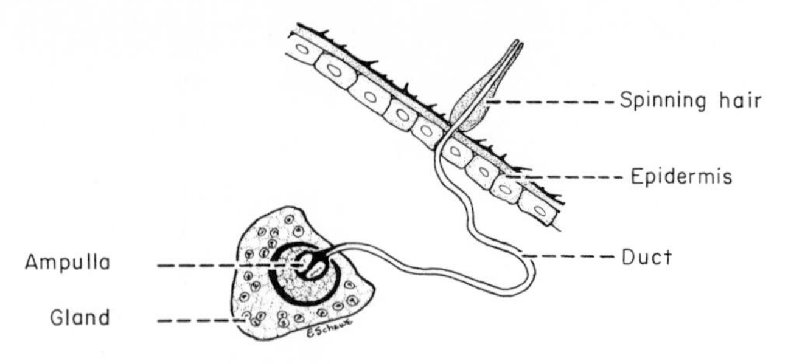

FIG. IV–6 *Silk gland from the enlarged metatarsus of* Embia, *opening at the tip of one of the short spine-like hairs on the under-surface of the leg. (After Lesperon.)*

the most highly evolved of the web builders, possess at least five of these seven glands and make five different kinds of silk which, forced out through the spinning tubes as fine threads, are used for specific purposes in the construction of the web (Fig. IV–8). Finer and even more delicate than that of the silkworm, spider silk can also be woven into fabric. In the seventeenth century, a vest of spider silk was made for Louis XIV, and some years later mittens and stockings made of it were presented to the Académie des Sciences; bed hangings woven from it were exhibited at the Paris Exposition in 1900. More recently, serious attempts to cultivate, on a large scale, species of *Nephila*, a spider that builds a web as much as 8 m in diameter with an exceptionally strong supporting line, have been made in the United States, in Madagascar, and in China. The silk is taken from the spider by fastening it in a miniature stanchion and stimulating it, by contact, to discharge the thread which is wrapped around a slender, slowly revolving cylinder as it is drawn out. In this way, from 50 to 300 m of thread can be extracted at a time from a single animal without injury to it. It can then be released and after a suitable interval used again. In quality this silk is superior to that of the silkworms. But because of the predatory habits of spiders, it is necessary to rear them in isolation and to feed them living food, making the cost of their silk production commercially impracticable. Spider silk is, however, the material of choice for making cross and guide lines in optical instruments and is widely used, though in small quantities, for this purpose.

Finer too than silkworm silk are the byssus threads spun by bivalve molluscs of the family Pinnidae. Those of *Pinna marina*, a Mediterranean species, are bronze in color and of a soft texture, though tough and elastic. Since the second century they have been collected, washed, carded, spun, and woven into Cloth of Gold, a gleaming fabric so soft that a good-sized piece of it can be drawn through a finger ring. Garments and trappings made of it were used by Roman emperors and European kings or given by them as costly gifts to others, for it was never available in amounts sufficient to put it into common use. It was because of the trappings of this material that the

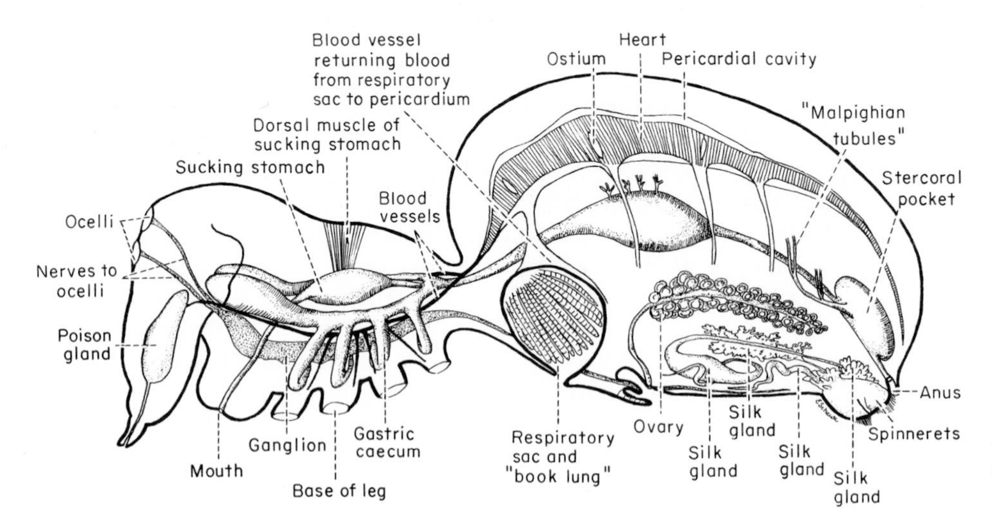

FIG. IV–7 *The internal anatomy of a spider,* Epeira diademata. *(From Gardiner.)*

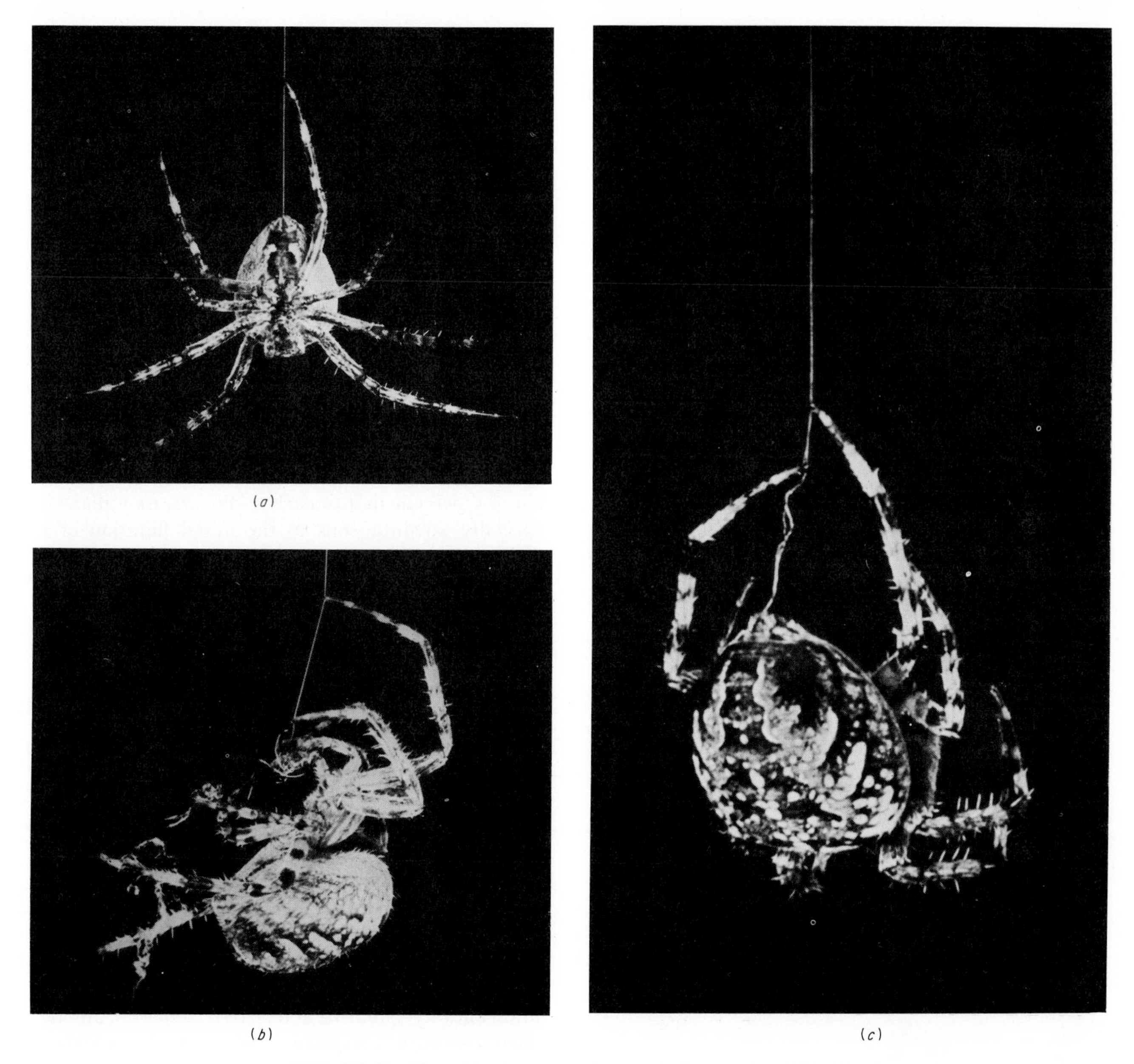

FIG. IV–8 *The spider* Aranea sericata *spinning a web.* (*American Museum of Natural History.*) **(a)** *Dropping the heavy dragline, the first one spun.* **(b)** *Climbing up the dragline.* **(c)** *Resting on the dragline.*

sumptuous picnic at which Henry VIII of England and Francis I of France met in 1520 became known as the Field of the Cloth of Gold. Production of the cloth in small amounts is still an industry in Taranto, Italy. The byssus threads of *Atrina rigida*, a species found along the southeastern coast of the United States, have also been made into cloth but never for commercial purposes, which can now be fulfilled by synthetic products.

Gland cells of the external epithelium may also produce gases. This is perhaps most dramatically illustrated by siphonophores, in many of which the polymorphic colony is kept afloat by a gas-filled bag, or pneumatophore. This is similar to the inverted bell of a medusa, but it is closed except for a small pore guarded by a sphincter muscle, through which gas may escape and so the specific gravity of the colony be regulated. The epithelial cells that form the inner lining of the bag constitute the gas gland, or pneumadena. In *Physalia* these are tall, columnar cells with branching basal processes embedded in the mesoglea (Fig. IV–9). Their secretion first appears as small granules near the nucleus, which swell and move distally as they become filled with gas. This is discharged into the pneumatophore, giving a mixture of gases which, on the average, contains 74.4 percent nitrogen, 14.4 percent oxygen, 1.1 percent argon, 0.4 percent carbon dioxide, and 8.9 percent carbon monoxide. Gas in the many tiny floats of the bathypelagic *Nanomia bijuga* has an amazingly high content of carbon monoxide, ranging from 80 to 92 percent. The secretion of this gas serves to inflate the pneumatophore, with the atmospheric gases diffusing into it and slowly replacing the carbon monoxide in the fully inflated float. But if it is artificially evacuated the cells of the gas gland become active again and secrete carbon monoxide at a rate, as measured in *Physalia*, of 7.5 to 120 µl/hr (microliters per hour). The high rate of secretion possible by cells of a gas gland has also been shown experimentally by emptying the pneumatophore of *Stephanomia* and finding that it is completely refilled in 30 min.

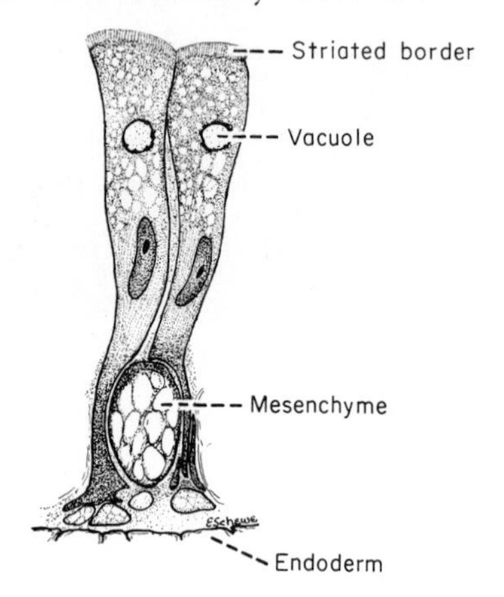

FIG. IV–9 *Gas-secreting cells from the epithelium of the float of the siphonophore* Physalia *(Portuguese Man-of-War). (After Dahlgren and Kepner.)*

Incubation of isolated pieces of pneumadenae with various substrates indicates that the source of the carbon monoxide is the amino acid serine. The vitamin folic acid (pteroylglutamic acid) is probably contributory to the reactions that lead to the evolution of carbon monoxide. Since this gas is a well-recognized inhibitor of the cytochrome system of enzymes, its production and use by siphonophores, and their apparent tolerance of it, is surprising. The low solubility of carbon monoxide in water, which is less than 0.3 percent that of carbon dioxide, may make it especially advantageous to the initial flotation of a colony. But though its maximum pressure in a float may be below that which would inhibit the cytochrome system, it is difficult to reconcile the secretion of a lethal gas by active cells, for those of the pneumadena have been shown, like those of the float, to have the usual pattern of aerobic respiration. It is possible, of course, that they are protected by some mechanism that is as yet unknown.

Many invertebrates, especially among the arthropods, have epidermal glands that discharge odorous or poisonous secretions. These may serve as a means of communication between different individuals or for marking trails to the nest or to a food source (see Chapter XIV). Also, light-producing gland cells are found in many species of invertebrates. This capacity for bioluminescence is shared by bacteria and a number of chordates but is apparently limited, among animals, almost entirely to marine and terrestrial species, among which it is erratic in its distribution (see Chapter VIII). Light-producing cells, as unicellular or multicellular glands, may be generally distributed over the body of the animal, as they are, for example, in luminescent cnidarian species, in the nemertean *Emplectonema kandai*, and in the ophiuroids. Or they may be localized in particular areas, as they are in Ctenophora, where they are located along the meridional canals, or in the luminous annelids, where they lie near the mouth or the anus. Further, they may be incorporated in elaborate and highly specialized photogenic organs, with lens and reflecting layer, as they are found in some of the crustaceans and coleopterans (Fig. IV–10).

In its modifications as neurosensory cells, epithelium exhibits one of the fundamental properties of protoplasm—the ability to accept and to respond to a stimulus external to it. Neurosensory cells may be selectively sensitive to light; to mechanical stimuli such as contact, pressure, and movement of air or water; to chemical stimuli; and possibly to thermal stimuli as well. Neurosensory cells are often slender, with one process extending to the surface of the epithelial layer and ending in a number of fine hair-like projections and another extending inward with branches connecting it to the effector cell or to some part of the nervous system. These cells may be individually wedged among other nonnervous ones of the epithelium, as they are, for example, in both the inner and outer epithelia of Cnidaria and in the external epithelium of earthworms, where they seem to act as photoreceptors (Fig. IV–11). Or they may be grouped together, along with cells of other types, to form sense organs of various kinds.

The external epithelium may also be modified for respiratory purposes, as it is in branchial or other surface areas where gaseous interchange is maximal. Here it is often very thin and delicate, composed of flattened cells whose long axes lie parallel to the surface forming a film-like membrane that facilitates the inward diffusion of oxygen and the outward diffusion of carbon dioxide along the respective concentration gradients (Fig. IV–12).

Internal epithelia, like external, may be cuboidal, columnar, or squamous; ciliated (flagellated), effecting the movement of materials across it; glandular or absorptive (Fig. IV–13). The epithelium, derived from endoderm that lines the digestive areas of the invertebrate enteron, is frequently ameboid, with pseudopodia-like processes extending from the free ends of the cells. It may also be glandular, synthesizing and discharging mucus into the lumen of the gut, but, unlike surface epithelium, forming zymogen granules, whose enzymes, liberated into the gastral cavity, bring about chemical disintegration of the food in the process of extracellular digestion. Evaginations

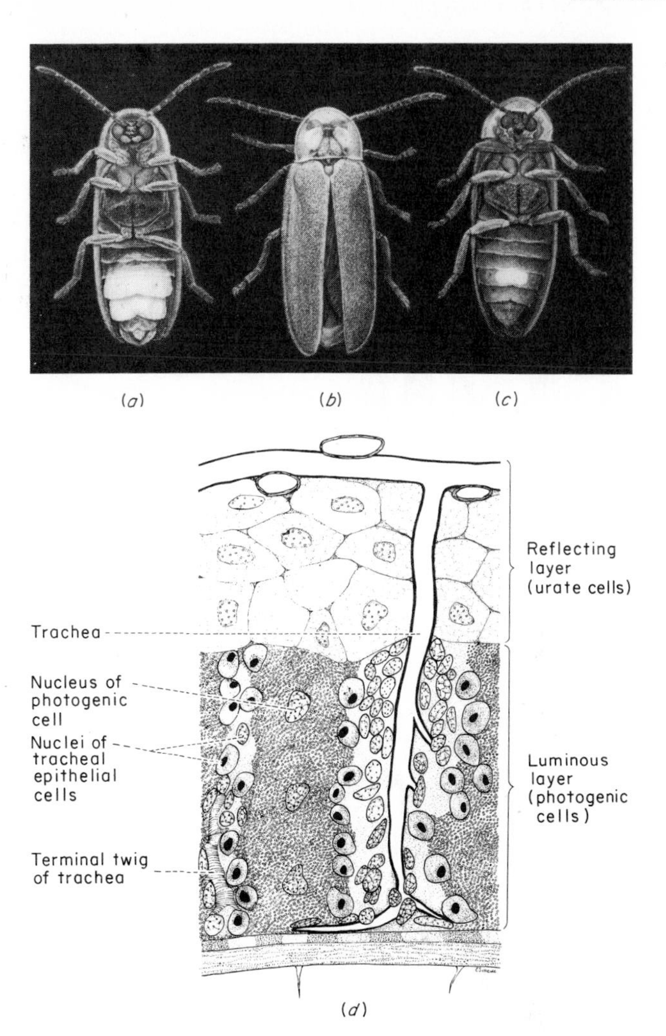

FIG. IV–10 *The firefly* Photinus marginellus. (*Courtesy of H. Lou Gibson.*) ***(a)*** *Male from the ventral side showing the large lantern.* ***(b)*** *Dorsal view of the insect.* ***(c)*** *Female from the ventral side.* ***(d)*** *Detail of photogenic organ.* (*From Gardiner.*)

of the gut may also become glandular areas, increasing its surface for generalized secretory purposes or developing regions for the production of specialized products. The endodermal epithelium may also be absorptive in nature, its cells actively engaged in taking up the end products of extracellular digestion.

In the excretory organs the internal epithelium (mesodermal or ectodermal in origin) may likewise be columnar and ciliated, or delicate and flattened as a covering membrane; functionally, it may be secretory

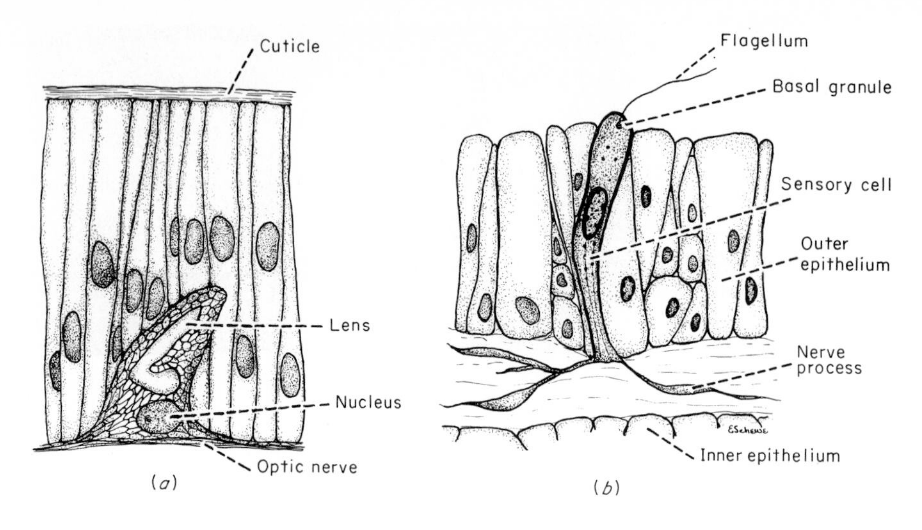

FIG. IV–11 *Neuroepithelium.* ***(a)*** *Epidermis from the prostomium of an earthworm showing one photoreceptor cell.* (*After Hess.*) ***(b)*** *Epidermis of* Hydra *showing a sensory cell.* (*After Burch.*)

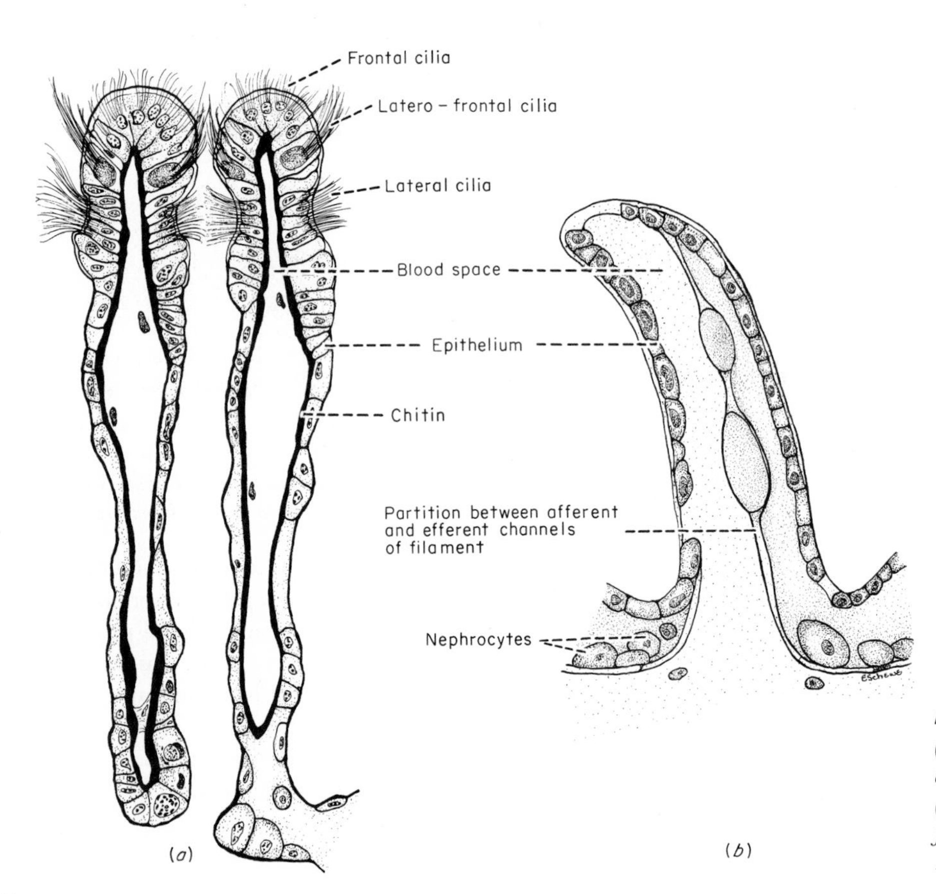

FIG. IV–12 *Respiratory epithelium.* ***(a)*** *Cross section of two gill filaments of a pelecypod mollusc.* (*After Atkins.*) ***(b)*** *Longitudinal section of a filament from an arthrobranch of the crayfish* Astacus fluviatilis. (*After Bock.*)

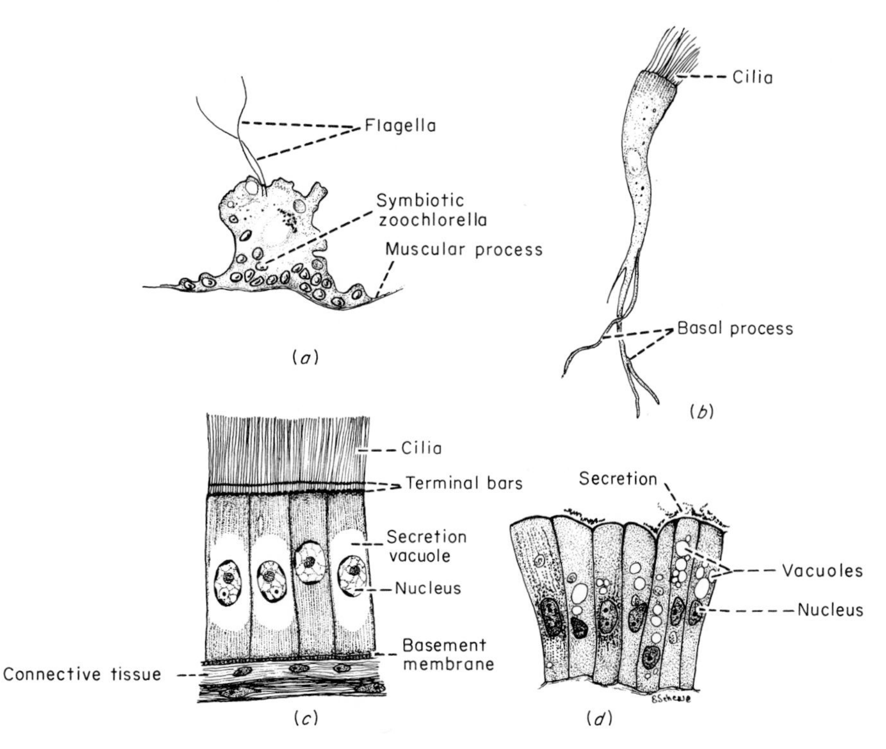

FIG. IV–13 *Endothelium.* ***(a)*** *Flagellate endodermal epitheliomuscular cell from* Chlorohydra viridis. *(After Goodrich.)* ***(b)*** *Ciliated endodermal cell from the stomach intestine of the earthworm* Lumbricus herculeus. *(After Goodrich.)* ***(c)*** *Cells from the wall of the small intestine of the nudibranch mollusc,* Melibe. *(After Agersborg.)* ***(d)*** *Cells from the wall of the canal of the midgut gland of the crayfish* Astacus leptodactylus.

or absorptive. In eucoelomate invertebrates, the peritoneum that lines the coelomic spaces is typically membranous, composed of a single layer of squamous or low columnar cells (Fig. IV–14).

C. SUPPORTIVE TISSUES

Supportive connective tissues are internal ones providing, as their name implies, a kind of framework for the body (Fig. IV–15). Most primitively, their cells are of irregular and inconstant shape, lying in a matrix that may be fluid, viscous, firm, or even rigid in consistency. The origins of these cells are different in different species of invertebrates and in some they are as yet unknown; the cells themselves are capable of movement, either active or passive, throughout the body and are, of all cell types, those most generally and widely distributed in the entire animal kingdom. The more differentiated cells are less mobile, sometimes fixed in position and often packed into compact masses of discrete cells or united together in syncytia. As such, they may constitute mesenchyme, in vertebrates typically an embryonic tissue retained in the adult in only a few localized areas but in most invertebrate species persisting throughout adult life as the primary supporting and connecting elements. Mesenchyme in which the cells are close together with relatively small intercellular spaces is often called parenchyma; this is the kind of tissue that is found between the body wall and the digestive tract in flatworms and nemerteans (Fig. IV–16). In rare cases among invertebrates, connective tissue cells may be aligned in bundles, forming short but stretchable ligaments such as those by which muscles in Crustacea are attached to the apodemes or the shell. In insects, in comparison to other inverte-

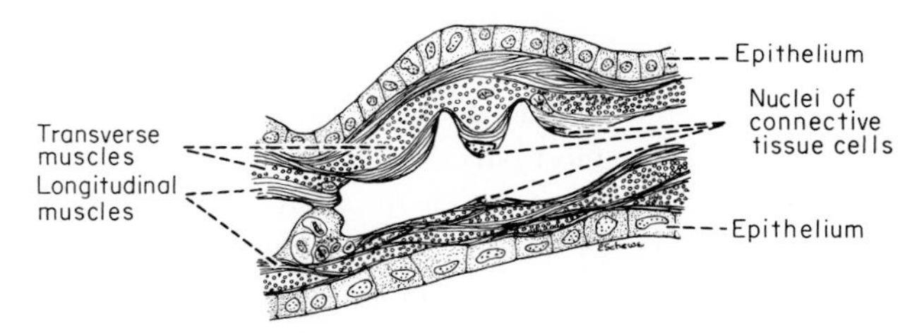

FIG. IV–14 *Cross section through the pericardium of the snail* Helix pomatia, *showing the low columnar epithelium that bounds the cavity.* *(After Nold.)*

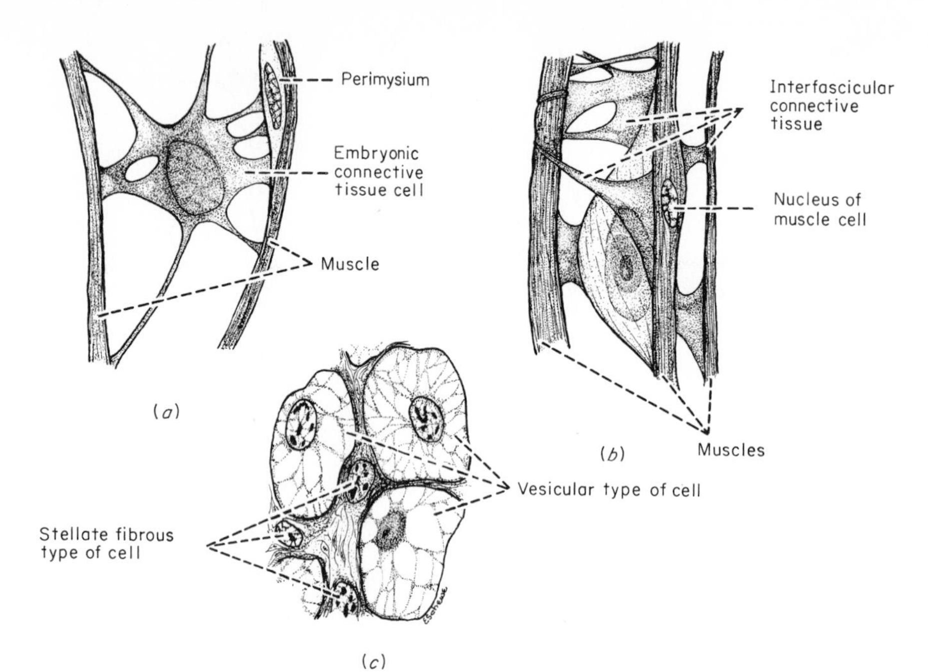

FIG. IV–15 *Connective tissue cells.* ***(a)*** *Embryonic connective tissue cell from the heart of the snail* Helix pomatia. (*After Nold.*) ***(b)*** *Interfascicular connective tissue from the heart of* Helix pomatia. (*After Nold.*) ***(c)*** *Connective tissue cells from the gut of the crayfish* Astacus fluviatilis. (*After Janisch.*)

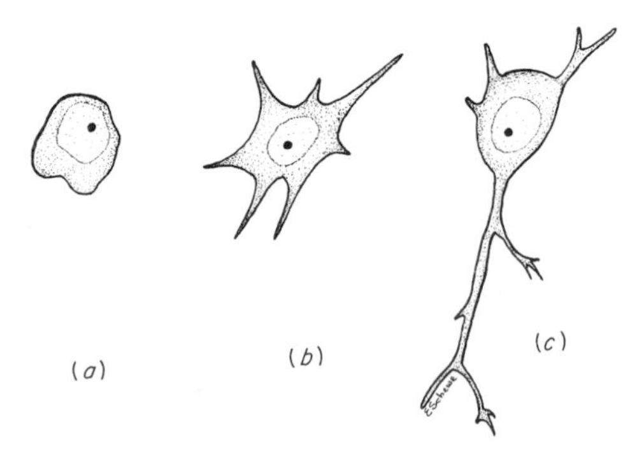

FIG. IV–16 *Planarian parenchyma cells cultured in vitro.* (*After Murray.*) ***(a and b)*** *Isolated parenchyma cells after 6 to 8 hr culture.* ***(c)*** *Isolated cell after 3 days culture.*

brates, connective tissue is sparse and their internal organs are for the most part supported by a fine meshwork of tracheoles into which the main tubes of their respiratory channels divide. But electron microscopy has revealed a delicate layer of connective tissue, like vertebrate connective tissue in that it contains collagen fibrils, covering all the organs and, in some species, forming strands between them.

The attributes of the mobile cells have been intensively studied in a number of invertebrate phyla. In general, they seem to be of two main types: a hyaline or nongranular cell and a granular one, in which the cell body is filled to a greater or lesser extent with discrete granules or in which they have apparently coalesced into one or more large masses (Fig. IV–17). The hyaline cells, because of their similarities to cells of the lymphoid series in vertebrates have been given the name of lymphoidocytes; their motion is an active, ameboid one and they are phagocytic, taking up particulate matter from their surroundings. The granular cells, which are neither ameboid nor phagocytic, with their closest vertebrate analogue the mast cell, have been given the name of trephocytes (Greek, *trephein*, "to nourish"), for they appear to produce and ultimately to liberate nutritive and growth substances.* Trephocytes are somewhat larger than lymphoidocytes, with a spherical or ovoid cell body and a nucleus that is small in relation to the cyto-

* These cells are frequently called trophocytes, a term originally used for the absorptive and nutritive cells of the insect fat body. Trephocyte has a broader connotation and will be used in this book for the granular, nonameboid supportive cells of invertebrates in general.

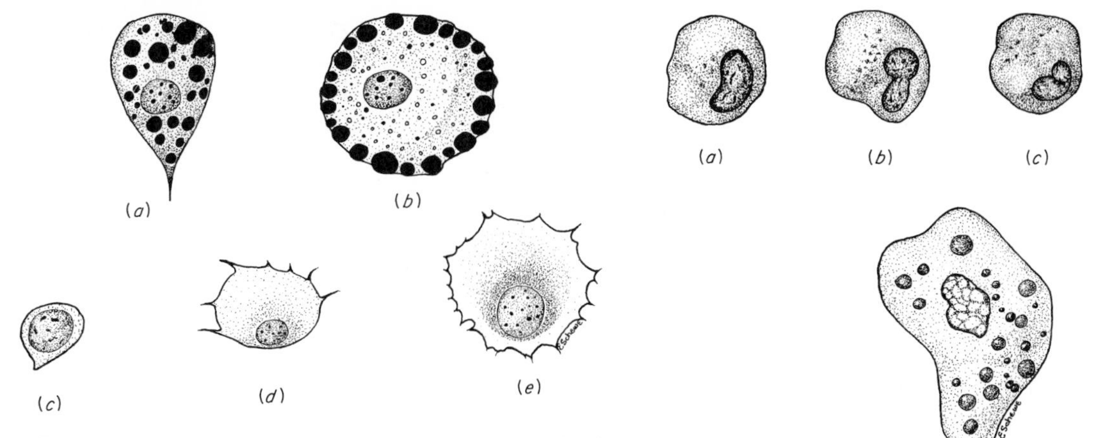

FIG. IV–17 *Trephocytes and phagocytes of the annelid* Eisenia foetida. (*After Liebman.*) ***(a)*** *Young trephocyte.* ***(b)*** *Mature trephocyte.* ***(c)*** *Lymphoidocyte.* ***(d)*** *Young macrophage* (*monocyte*). ***(e)*** *Macrophage.*

plasmic mass and smaller than that of the lymphoidocyte. In mature, or more highly differentiated trephocytes, the nucleus may be pycnotic or may show total disintegration so that the cell is enucleate. It is thus an "end cell," and has been regarded as the homologue, and even the prototype, of the vertebrate erythrocyte and erythroplastid. The cytoplasmic inclusions of trephocytes have been chemically identified as being most frequently of protein or lipid, or of protein and lipid in combination, but phospholipids, glycogen, and mucin are also of common occurrence. Sometimes the inclusions are colored. In echinoids, the characteristic naphthoquinone pigments are found in them; in some species in other phyla, the pigments have been considered to have a respiratory function.

The trephocyte appears to be a cell that is capable of taking up materials from its surroundings, of synthesizing them into a variety of compounds, and then of releasing them into the body fluids, tissues, or organs, or directly into another cell, either by total or partial disintegration or by some kind of transference across the adjacent membranes (Fig. IV–18). Removal of pieces from trephocytes by other cells has been reported, as has also the passive movement of a trephocyte toward another cell until it becomes closely applied to the surface where it seems to rest while giving up its material. The incorporation of trephocyte material

(a) (b) (c) (d)

(e)

FIG. IV–18 *Amoebocytes and trephocytes of the snail* Helix. (*After Nold.*) ***(a to d)*** *Amoebocytes with different forms of nuclei* (*magnified 1,840 times*). ***(e)*** *Trephocyte* (*magnified 1,730 times*).

by ovarian eggs may be very apparent; in annelids, penetration of trephocytes into the egg has been seen and, in echinoids, the echinochrome in the oocytes is described as being directly derived from trephocytes.

In a number of species, the trephocytes collect into conspicuous masses. In the earthworm *Eisenia foetida*, for example, there are two such masses, one at each end of the body, which function as regenerative organs because of their activity in the growth of new segments if the original ones have been lost or destroyed. Trephocytes (trophocytes) are similarly collected into masses in insects, where they form part of the cellular components of the so-called "fat bodies." Typically, these bodies are found in larvae, where they lie on either side of the digestive tract. In many insects, the fat body also contains urate cells, which accumulate the products of protein decomposition. During growth of the larva, the trephocytes become so enlarged as they increase their loads of fat, protein, and glycogen that their nuclei and even their boundaries become indistinguishable. These reserves are released during the period of pupation, the extent of the reduction of the storage products varying in different species (Fig. IV–19). Cells of the fat bodies may also contain carotenoid pigments, giving them characteristic colors.

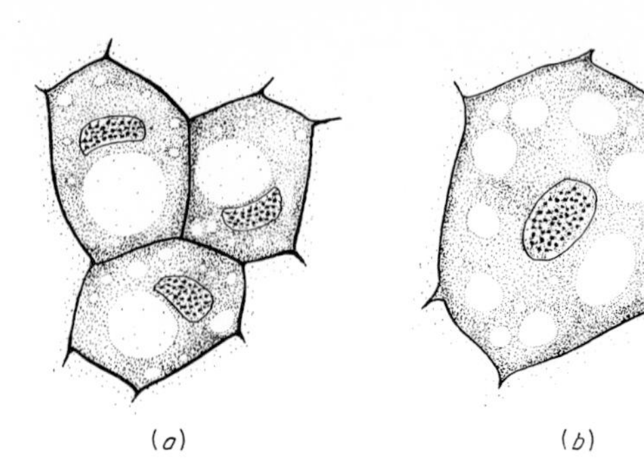

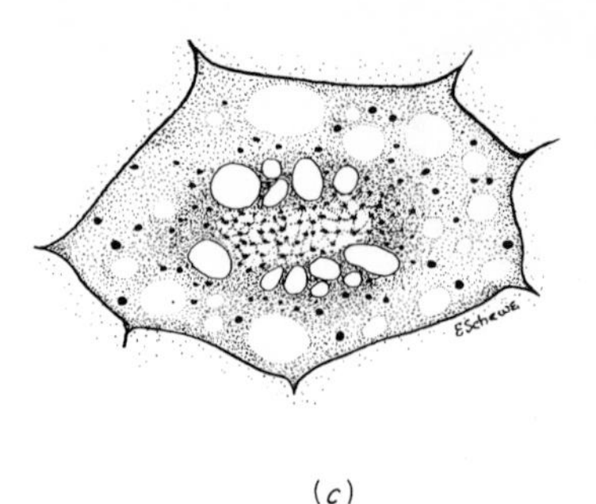

FIG. IV–19 *Fat cells from the bee* Apis. *(After Bishop.)* ***(a)*** *Larval fat tissue cell with a large oil vacuole.* ***(b)*** *Larval fat tissue cell with peripheral ring of oil vacuoles.* ***(c)*** *Late larval fat tissue cell, showing disruption of nucleus and accumulation of fat vacuoles.*

The principal functions of these generalized connective tissue cells are concerned with nutrition, excretion, growth, regeneration, and repair. To this extent they are protean, exhibiting many different morphological and physiological characters in accordance with the activity they are primarily engaged in. In some species they play a predominant role in digestion, capturing and digesting food particles and transmitting the end products to other cells. They may also take up and store waste products of metabolism or in some other way remove them from the body. They appear to be the cells that are mobilized at the sites of injury and that contribute to the processes of regeneration and repair, as well as to those of growth and reproduction in the intact animal.

Connective tissue cells also contribute to the matrix in which they lie. Recent chemical and structural studies of the mesoglea of some marine sponges have shown that in its organization and composition it resembles the connective tissue of vertebrates in being composed of collagen fibers and fibrils embedded in an amorphous carbohydrate-containing matrix. Similar studies of some Cnidaria have yielded similar results: the fibers laid down in the amorphous matrix by migratory cells are protein with an amino acid content like that of vertebrate collagen, giving the mesoglea elasticity as well as relatively high viscosity (Fig. IV–20).

The matrix that surrounds the supporting cells in localized regions of many molluscs is more rigid than this mesoglea and not fibrous. This kind of supporting tissue, known as chondroid tissue, is found in vertebrates in the course of cartilage development and resembles cartilage. The cells are quite large and are closely packed together with less intercellular substance than is characteristic of mature cartilage. Chondroid tissue forms the support of the dentary apparatus in many molluscs, is found in the head and tentacles of cephalopods, and may perhaps be even more widely distributed among invertebrates (Fig. IV–20). It is the firmest and most rigid supportive tissue found among them, for none of them has as yet achieved the ability to make bone.

Chromatophores, or pigment cells, are also com-

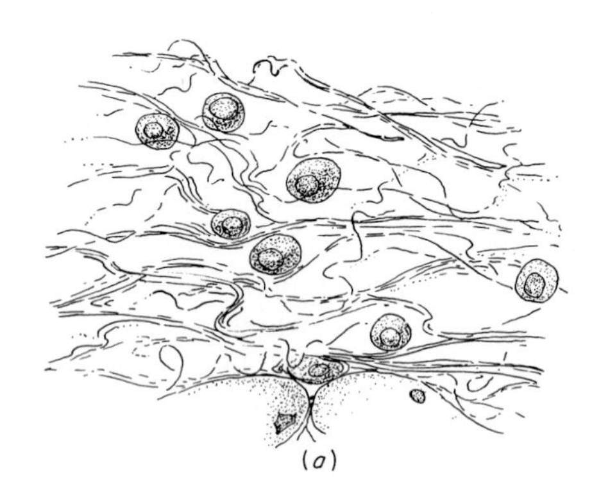

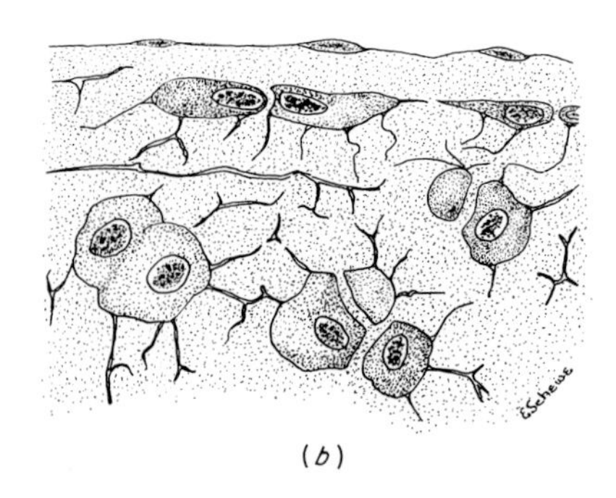

FIG. IV–20 *Sustentative tissues.* ***(a)*** *Mesoglea of the cnidarian* Aurelia aurita, *showing fibers. (After Chapman.)* ***(b)*** *Chondrioid tissue from a cephalopod mollusc,* Loligo *(squid). (After Dahlgren and Kepner.)*

ponents of connective tissue. These may elaborate pigments of various kinds which are either dissolved in their cytoplasm or, more frequently, distributed as granules or crystals throughout it. They may be monochromatic, synthesizing only one kind of pigment, or polychromatic, synthesizing two or more. They have been classified according to the predominating type of pigment in them as lipophores (erythrophores or xanthophores), melanophores, and leucophores (iridocytes or guanocytes). Lipophores are red or yellow because of their content of carotenoids; melanophores are dark because of their content of granules of yellow or black melanins; and leucophores are white or iridescent because of their content of fine crystals or plates of the purine, guanine. Some decapod crustaceans produce a blue pigment which is a carotoprotein. The chromatophores of cephalopods and crustaceans have been most widely studied, both for their pigment content and for the shifts in its distribution when the animals change color (Fig. IV–21). The cephalopod chromatophore is an elastic capsule, filled with pigment granules and surrounded by muscle cells, each apparently supplied with a nerve fiber, whose contraction draws the cell boundary out to such an extent that the diameter of an expanded cell may be 15 or 20 times as great as that in its contracted state. As the cell is extended, the pigment, or pigments, becomes more dispersed, changing the color pattern of the animal, which returns to its initial condition when the muscles relax and the elasticity of the chromatophore membrane causes the cell to collapse. Color change in these animals is thus the effect of a group of cells acting as a unit. In Crustacea, each chromatophore is of considerable size, possibly actually syncytial, with long, irregular and branching processes through which the pigment may be dispersed or from which it may be withdrawn and concentrated in the center of the cell as a dense mass. In the shifts of

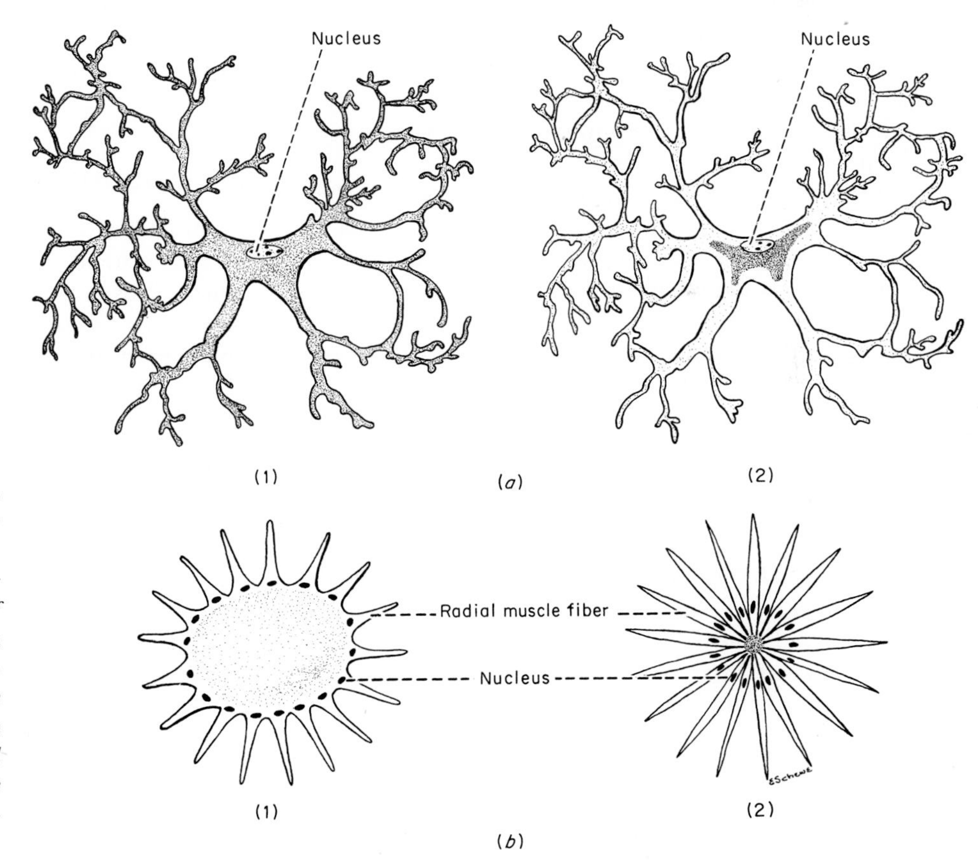

FIG. IV–21 *Diagrammatic representations of chromatophores, showing changes in dispersion of the pigment.* ***(a)*** *Crustacean chromatophores. (After Turner.) (1) Maximal dispersion of pigment. (2) Maximal concentration of pigment.* ***(b)*** *Cephalopod chromatophores. (After Parker.) (1) Muscle fibers, each one essentially a smooth muscle cell, in contraction with maximal dispersion of the pigment. (2) Muscle fibers in relaxation with maximal concentration of the pigment.*

pigment that bring about the color changes in these animals, each chromatophore acts independently and in response to hormonal stimulation; indeed, within one that is di- or polychromatic, each type of pigment may respond independently of the others, dispersing or concentrating without affecting the distribution of the others.

D. TRANSPORTIVE TISSUES

In the more simply organized invertebrates, the motile cells of the connective tissue and the fluid medium in which they lie serve as the only transport system, conveying materials from one part of the body to another and picking up, storing, or otherwise disposing of cellular waste and cell detritus. But in those with compact tissues especially and with more elaborate organs and organ systems, there is usually also a vascular transport system, a fluid tissue in many ways comparable to the blood of vertebrates and, like it, freely exchanging cells with the surrounding tissue fluids. Invertebrate blood may either be conveyed through a system of continuous tubes or vessels, as it is in the "closed" circulatory systems of nemerteans, phoronids, cephalopod molluscs, and most annelids, or it may flow through vessels and large blood spaces, or sinuses, as it does in the "open circulation" of brachiopods, of most molluscs, and of crustaceans. In either case, the cellular elements of the tissue are surrounded by a fluid matrix of low viscosity, whose movement through the various channels may be brought about by movements of the body as a whole, by those of the vessels themselves, or by those of localized areas along them acting as hearts. The cells of invertebrate bloods are, most generally, leucocytes, because any respiratory pigment the blood may contain is usually in solution in its fluid component, the plasma (Fig. IV–22). Exceptions to this are, among the annelids, the polychaetes *Glycera* and *Magelona*; the echiuroids *Urechis* and *Thalassema*; *Sipunculus*; and among the echinoderms, the holothurians *Sclerodactyla* (*Thyone*) and *Cucumaria*. Nucleated erythrocytes containing hemoglobin have been identified in the blood of *Glycera*, *Urechis*, *Cucumaria*, and *Sclerodactyla*; in *Sipunculus*, the nucleated cells contain hemerythrin

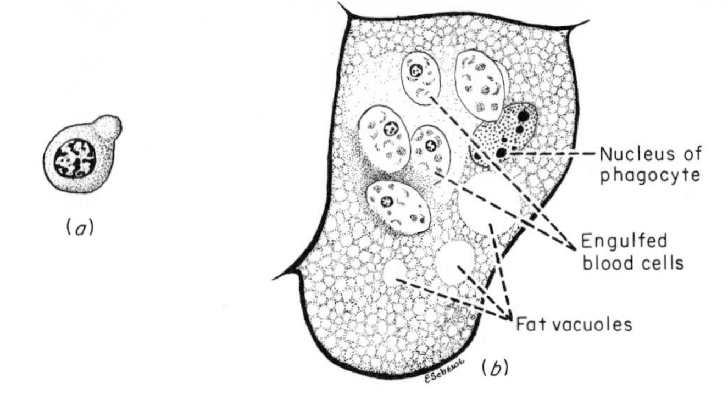

FIG. IV–22 *Transportive tissues. Cells from the hemolymph of the brine shrimp* Artemia. *(After Lochead and Lochead.)* ***(a)*** *Cell stained with hematoxylin and eosin.* ***(b)*** *Phagocytic storage cell, stained with Mallory's triple stain, showing five engulfed blood cells.*

rather than hemoglobin, while in *Thalassema* and *Magelona* hemoglobin is contained in nonnucleated corpuscles, considered specialized trephocytes and regarded as possible prototypes of vertebrate red blood elements (Fig. IV–23).

E. MUSCULAR TISSUES

Contractility is another general property of all protoplasmic systems, but it is the conspicuous property of muscle cells in which the contractile mechanism has reached its greatest degree of specialization and is not obscured by other cellular operations such as secretion, absorption, or division. This mechanism seems to reside principally, if not exclusively, in two proteins, actin and myosin, which make up a large proportion of the material of the microscopically visible bundles of long-chain molecules, the myofibrils (Fig. IV–24).

Histologically, muscle is usually classified as smooth, striated, or cardiac, the latter type being characteristic of rhythmically contracting hearts. Smooth muscle is, in general, associated with the maintenance of tension under extension, or muscle tonus by slow and sustained contraction, but it is also capable of effecting movement. The smooth muscle cell is usually spindle-shaped and tapering, with a single nucleus and, in the light microscope, optically homogeneous myofibrils that usually run parallel to

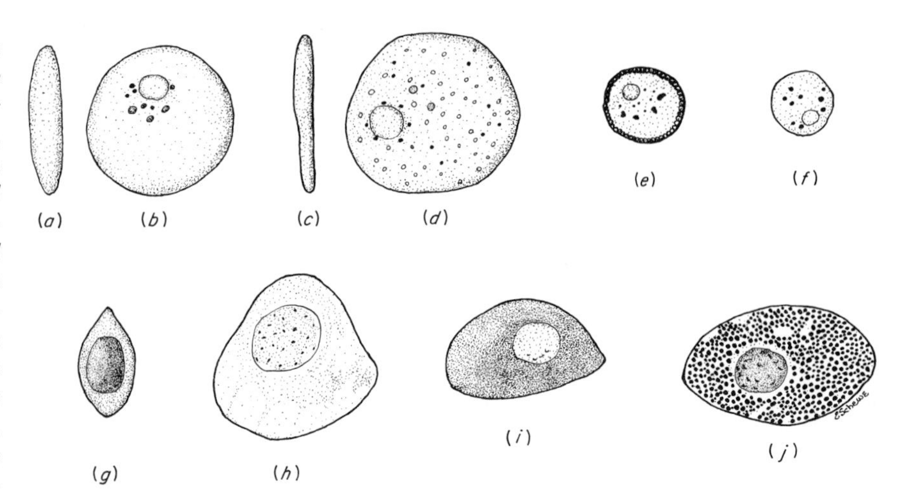

FIG. IV–23 *Transportive tissues.* ***(a and b)*** *Fresh corpuscle from the blood of the sipunculid* Golfingia gouldi. (*After Dawson.*) *a, side view; b, surface view, stained with neutral red.* ***(c and d)*** *Fresh corpuscles from the blood of the annelid* Glycera dibranchiata. (*After Dawson.*)*: c, side view; d, surface view, stained with neutral red.* ***(e)*** *Fresh corpuscle from the coelomic fluid of the holothurian* Sclerodactyla (Thyone) briareus, *stained with neutral red.* (*After Dawson.*) ***(f)*** *Fresh corpuscle from the hemolymph of the mollusc* Arca transversa, *stained with neutral red.* (*After Dawson.*) ***(g)*** *Lymphocyte from the hemolymph of the tarantula* Phormictopus canceroides, *stained with Wright's stain.* (*After Deevey.*) ***(h)*** *Hyaline leucocyte from the hemolymph of* Phormictopus, *stained with Wright's stain.* (*After G. B. Deevey.*) ***(i)*** *Medium-sized leucocyte with basophilic granules from the hemolymph of a tarantula.* (*After G. B. Deevey.*) ***(j)*** *Large leucocyte with eosinophil granules from the hemolymph of a tarantula, stained with Wright's stain.* (*After G. B. Deevey.*)

the long axis of the cell. Striated muscle is associated with quick contraction and relaxation. It is made up of syncytial (multinucleate) units, the muscle fibers, each containing many nuclei and a characteristic number of myofibrils enclosed in a delicate but visible membrane, the sarcolemma. In the light microscope, the myofibrils have a cross-banded appearance of alternating light and dark bands of different widths. In cardiac muscle, the fibers are united in a continuous meshwork of contractile elements. In all muscles, comparatively large particles, the sarcosomes or muscle mitochondria, are distributed in the sarcoplasm or muscle cytoplasm in close proximity to the myofibrils (Fig. IV–25).

The distinction between these types is not well defined. In invertebrate muscle, the myofibrils may be either slow or quick in contraction. In general, the slow fibrils lie in the central region of the contractile unit or are distributed generally through it; the fast fibrils are usually peripheral in position and are attached to the sarcolemma. Where both types of fibrils are present, the muscle may act to maintain tone as well as to contract rapidly.

Among the more simply organized invertebrates, myofibrils may be developed in limited areas of epithelial cells. In some Cnidaria, for example, the bases of the cells of both epidermis and gastrodermis are extended into two or more processes, with myofibrils running parallel to the long axes of these processes. Transitional stages between epitheliomuscular cells of this kind and wholly differentiated muscle cells are found in the deeper muscles of the body wall of actinarians and in the retractor muscles of their mesenteries. In these, the cell body has sunk below the surface and lies alongside the fibrillar region, although connection with the surface may still be retained by a slender protoplasmic thread. Distinct muscle fibers, with no connection to an epithelial cell, are, however, found in other Anthozoa and in Scyphozoa, where they either lie singly in the mesoglea or are collected into conspicuous muscle bands.

Similarly, in nematodes the myofibrils are concentrated in an expansion of an otherwise globular cell, the nucleus lying in a nonfibrillar, highly vacuolated portion. The fibrils are oriented parallel to the long axis of the body, and there is no circular or transverse musculature such as is found in most other animals of cylindrical shape (Fig. IV–26).

In general, however, invertebrate muscle consists of wholly specialized cells or of typical fibers (Fig. IV–27). The cells, which conform to the pattern of "smooth" muscle, may be distributed within the body

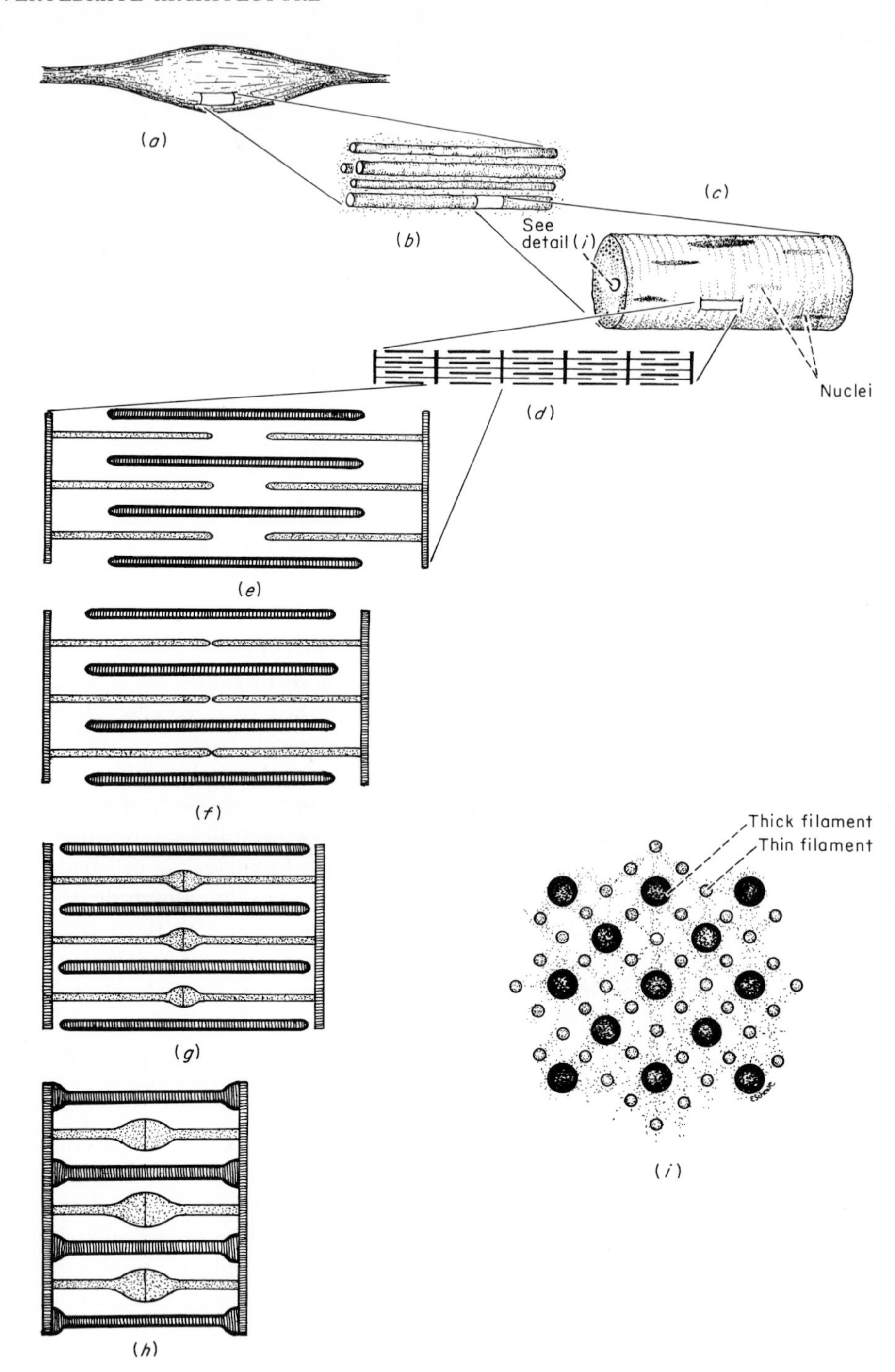

FIG. IV–24 *The construction of insect flight muscles and their contraction according to the sliding-filament theory.* (*After Huxley.*) ***(a)*** *An entire muscle.* ***(b)*** *Enlargement of one area, showing the fibers.* ***(c)*** *A single muscle fiber, enlarged.* ***(d)*** *Part of a myofibril, enlarged, showing the parallel arrays of thick and thin filaments and the repeating units of dark and light bands.* ***(e)*** *A single unit, enlarged.*
(f, g, and h) *Stages in contraction, during which the thin filaments move along the thick ones, meet, and crumple at the points of junction. The thick filaments also meet adjacent ones when the muscle is fully contracted.* ***(i)*** *Detail from transverse section of an insect flight muscle fiber, as seen at a magnification of 400,000 diameters with the electron microscope, showing the regular arrangement of thick and thin filaments.*

as isolated cells or in layers oriented in various planes relative to the main axis of the body, either alone or mingled with striated fibers. The fibers are for the most part arranged in bundles resembling those of vertebrate skeletal muscle.

The muscles of arthropods are typically cross-striated, like those of vertebrate skeletal muscles, but in many other invertebrate phyla the muscle fiber exhibits what has been called double oblique striation. That is, its surface, especially when in a state of contraction, shows a regular diamond-lattice pattern of fine lines. This type of fiber has been found in platyhelminths, nematodes, annelids, sipunculids, echinoderms, and molluscs, and has been most carefully examined in lamellibranchs and in the earthworm

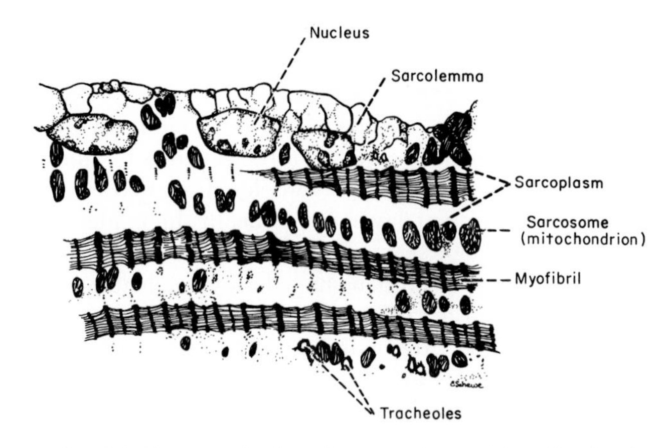

FIG. IV–25 *Muscle fiber of a wasp (magnified 4,680 times). (After Chapman.)*

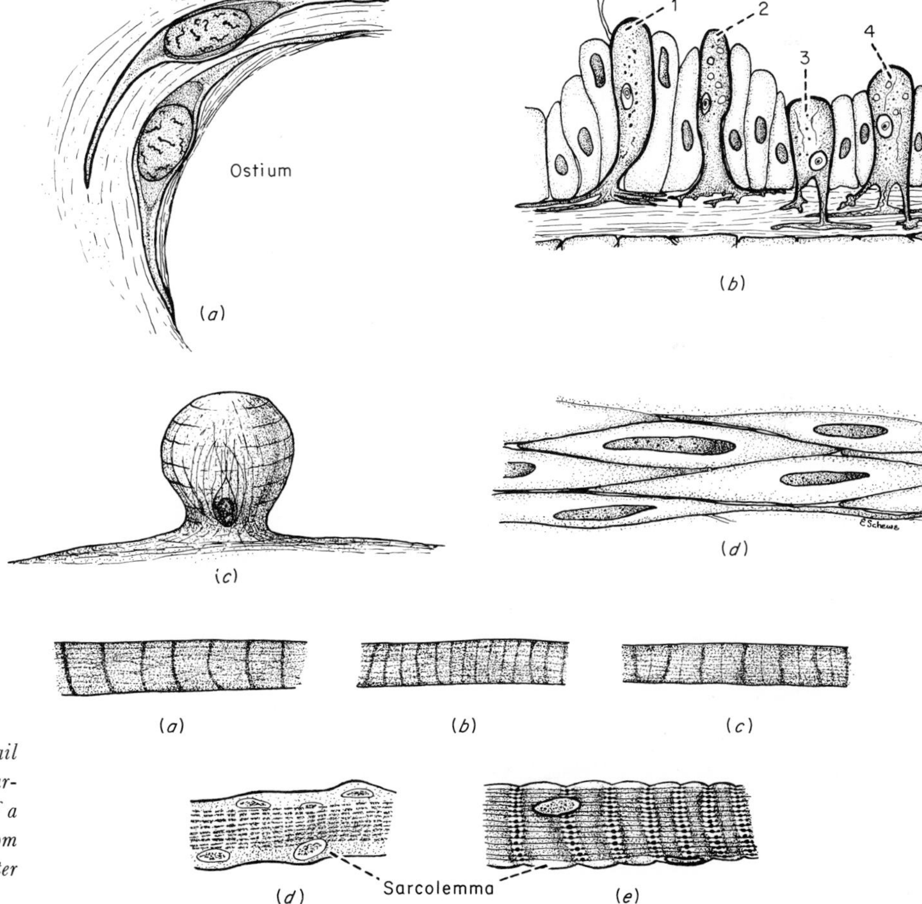

FIG. IV–26 *Contractile cells.* ***(a)*** *Myocytes around the pore of a sponge. (After Dendy.)* ***(b)*** *Epitheliomuscular cells (gastrodermis) of* Pelmatohydra oligactis. *(After Mueller.): 1, Cell with two flagella and five basal myonemes. 2, Cell with two basal myonemes and rootlets for attachment to mesoglea. 3, Cell with two rootlets, one attaching direct to the mesoglea, the other forming the sarc of a myoneme. 4, Cell with three rootlets, two attaching direct to the mesoglea, the other forming the sarc of a myoneme.* ***(c)*** *Muscle cell of the nematode* Ascaris. *(After Roskin.)* ***(d)*** *Smooth muscle cell from the body wall of the nudibranch mollusc* Melibe. *(After Agersborg.)*

FIG. IV–27 *Muscle cells.*
(a, b, and c) *Heart muscle cells from the snail* Helix pomatia. *(After Nold.)* ***(d)*** *Surface view of a muscle fiber from the larva of a honeybee. (After Snodgrass.)* ***(e)*** *Fiber from the leg muscle of an adult maybeetle. (After Snodgrass.)*

Lumbricus. In the lamellibranchs, obliquely striated muscle is far more general than cross-striated and in the majority of the genera studied contributes the "fast fibers" to the big adductor muscles that bring the valves of the shell together. In the earthworm, fibers of this kind are found in the muscles of the body wall, both in the circular layer just beneath the epidermis and in the much thicker inner longitudinal layer. This layer is made up of ribbon-shaped units, each attached by one edge to a radial septum of dense connective tissue, the other edge lying free in loose connective tissue. The units are attached at an angle, so that, in transverse section, septum and muscles are penniform. Electron microscopy has shown that there are two sets of myofibrils and a single nucleus in each unit (Fig. IV–28). Each set of fibrils is attached to one face of the flattened cell and is perpendicular to it when it is fully extended. All the fibrils in one set are parallel to each other, but those of the two sets are not, and the size of the angle they make with each other depends upon the degree of contraction. When fully contracted, the angle may be as great as 60°, giving the fiber a latticed or doubly striated appearance. Interpretation of the significance of these arrangements is still to be made.

The striated muscles of insects, especially those associated with movement of the wings in flight, show certain histological distinctions. Three main types have been described: tubular (lamellar), close-packed (microfibrillar), and fibrillar (dissociable). The tubular type seems to have appeared early in insect evolution, and is retained in the thoracic muscles (indirect flight muscles) of modern dragonflies and cockroaches. It is found also in the legs of most insects, and of spiders. As seen in the light microscope, it consists of a central core of fluid sarcoplasm in which lie many nuclei. Myofibrils, separated by sheets or lamellae of sarcoplasm, are radially arranged around this core. In the thoracic muscles, large mitochondria (sarcosomes) with diameters of 1 to 2 μ are regularly distributed between the myofibrils (Fig. IV–29).

Close-packed muscles are found in the flight muscles of grasshoppers and crickets and of some moths. In this type, the myofibrils, with diameters of 0.2 to

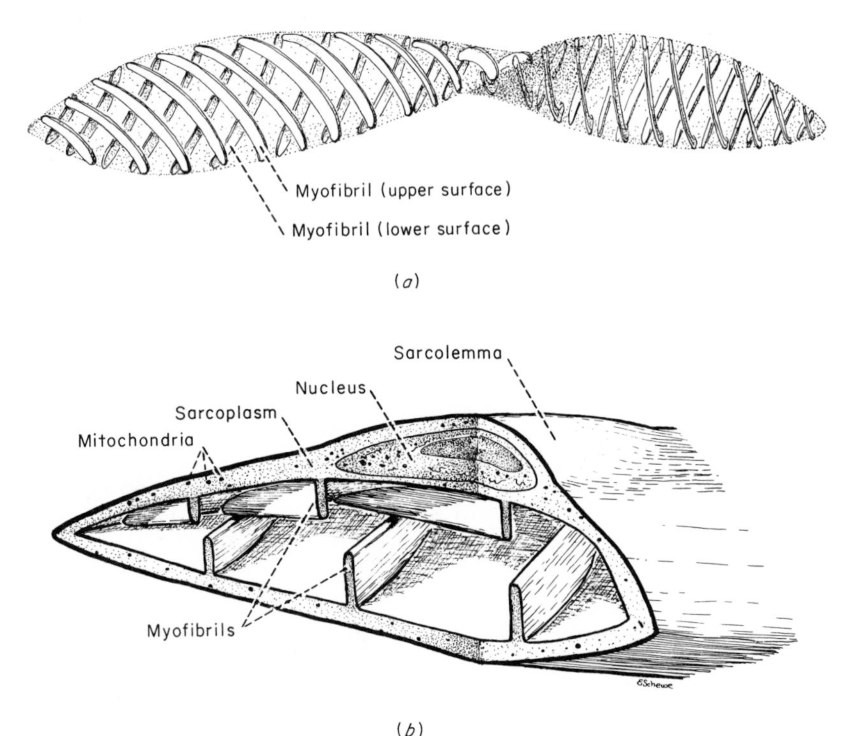

FIG. IV–28 *The arrangement of fibrils in the muscle fiber of the annelid* Lumbricus terrestris. (*After Hanson.*) ***(a)*** *Surface view of the ribbon-shaped fiber, showing the two rows of myofibrils, one attached to each surface.* ***(b)*** *Enlargement of a small area, as if in oblique section, showing the single nucleus of the fiber, the mitochondria distributed in the sarcoplasm, and the myofibrils.*

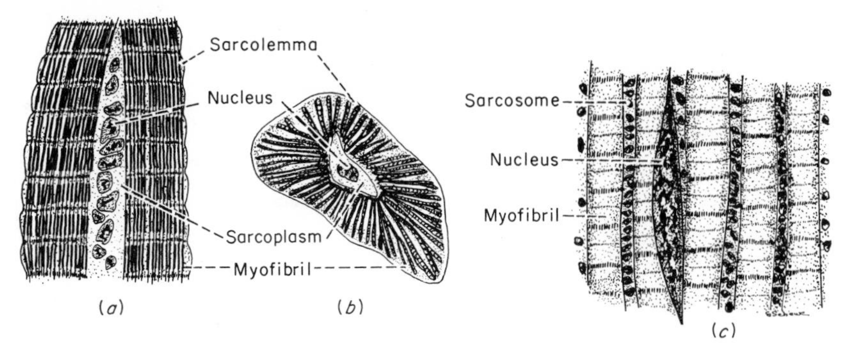

FIG. IV–29 *Tubular and fibrillar muscle of insects. (After Jordan.)* ***(a)*** *Longitudinal view of a fiber, in the relaxed condition, from the tubular muscle of the leg of the wasp* Vespa, *showing axial column of nuclei and myofibrils.* ***(b)*** *Cross section of a tubular fiber, showing radial arrangement of myofibrils.* ***(c)*** *Part of a longitudinal section of five adjacent myofibrils from a fibrillar muscle from the wing muscle of a wasp, showing columns of sarcosomes between the fibrils.*

0.6 μ, lie close together within the sarcoplasm, yet with large sarcosomes regularly distributed between them. The fiber is enclosed in a well-defined sarcolemma.

Fibrillar muscle is found in the thoracic muscles of insects with the highest capacities for flight, such as beetles, wasps, bees, and flies. The fibers are large, with diameters ranging from 70 μ in the aquatic beetle *Hydrous* to 1,800 μ in the fly *Rutilia.* Each fiber is enclosed in a delicate sarcolemma, beneath which lie many nuclei. In fresh muscle, these fibers can be easily separated from each other so that this type of muscle is known as dissociable. The myofibrils, which extend the entire length of the fiber, are also large; those in the fibers of *Hydrous*, for example, have diameters of 2.2 to 5.4 μ. They are separated by columns of conspicuously large and closely packed sarcosomes.

Another distinctive feature of insect muscle is its intimate association with the tracheal system. In tubular muscles, tracheoles are abundantly distributed over the surface of each fiber, but in close-packed and fibrillar muscles they form an intracellular network. Electron micrographs of thoracic muscles of the beetle *Tenebrio* show very intimate relationships between the tracheole and sarcolemma as the tracheole enters the fiber and passes between and among its myofibrils and sarcosomes (Fig. IV–30).

Invertebrate muscle differs also from vertebrate in its mechanical operations. This difference lies in the absence, in general, of its attachment to a rigid structure. Exceptions to this are found among molluscs, where muscles may be attached to shells or to chondroid tissue, and among arthropods, where they may be attached to endophragma or to apodemes (Fig. IV–31). Unattached muscles operate less efficiently than attached ones, and the contraction involved in their movement of any part demands a change in length greater than that required by attached muscles. Moreover, contraction of any one muscle affects the conditions under which all the others operate not only those of an individual antagonist, as is the case, for example, in the flexors and extensors of the vertebrate limb and in those of insects. In most of the invertebrates, this antagonism is supplied by hydrostatic skeletons. With such a system, pressure produced in any one area is transmitted equally and in all directions, so that contraction of any one muscle automatically affects the rest, by altering either their length or the degree of contraction they must preserve in order to maintain tonus. Under these conditions, localized action is not easily effected. Development of the

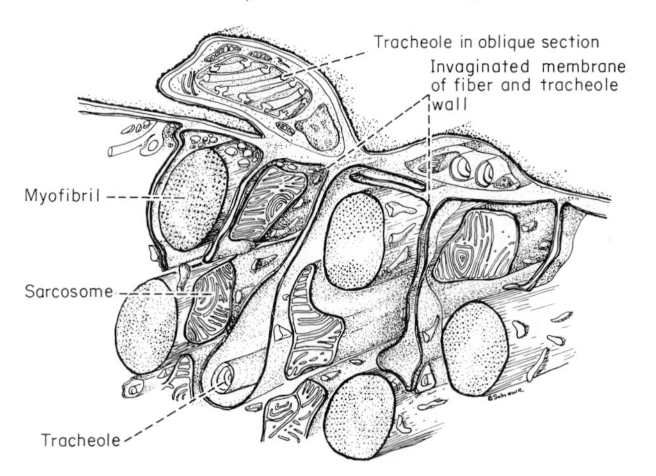

FIG. IV–30 *Structure of a flight muscle of the beetle* Tenebrio molitor. *(After D. E. Smith.) This detailed diagram shows the intimate relationship between the tracheoles, myofibrils, and sarcosomes.*

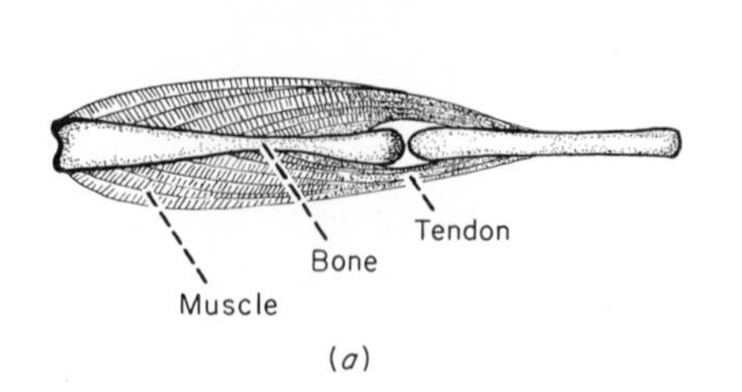

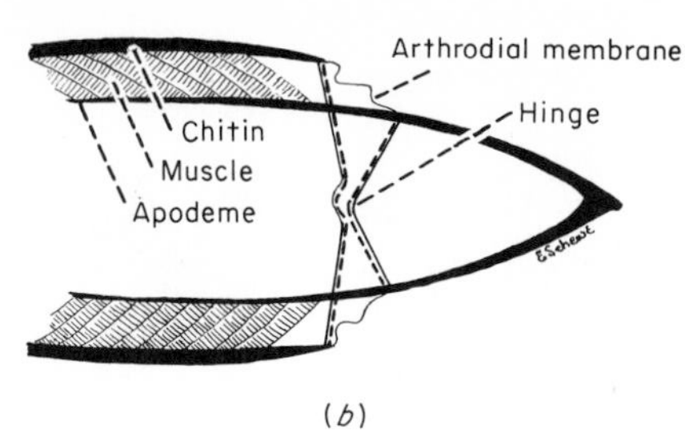

FIG. IV–31 *The relation of muscles to skeleton in vertebrates and in arthropods. (After Ramsay.)* ***(a)*** *Attachment of muscles between two bones in vertebrates.* ***(b)*** *Attachment of muscles to apodemes in arthropods.*

capacity for such action seems related to segmentation of the body, which would permit localized antagonistic action, and to the distribution of the units of contraction in sheets or bands in which all the myofibrils are similarly oriented, as in circular, transverse, oblique, or longitudinal muscles (Fig. IV–32). By this device, a limit is set to the complex and intricate nerve supply that would be demanded if a large number of units, oriented in all planes, were promiscuously distributed throughout the body. The thickness of the muscle layers, as well as the orientation of the units within them, determines the different degrees and directions of movements of animals with a hydrostatic skeleton. Where muscles have fixed attachments, however, the direction of movement is imposed by the nature of the articulation and is, therefore, limited to certain planes, as it is in the shells of bivalve molluscs and most appendages of arthropods.

Contraction, the active response of a muscle cell, is, as a rule, initiated by a nervous impulse. Cardiac muscle, however, may have the capacity to contract spontaneously in the absence of any nerve supply, and the contractile cells (myocytes) of sponges, that surround the pores, respond directly to environmental stimuli. The contraction of cardiac muscle is, however, regulated if not initiated by nerves, and, conversely, muscles that are ordinarily excited by a nervous impulse may continue to contract when nerve connections are severed or the passage of the impulse is chemically blocked. But muscular tissue is usually intimately associated with nervous tissue and muscular coordination with the distribution of appropriate nerves to appropriate units. This may be a condition secondarily imposed upon Metazoa, a means whereby the contractile mechanisms in individual cells were brought into control and harmonious responses ensured.

F. NERVOUS TISSUES

Nervous tissue consists of specialized epithelial cells in which two of the basic properties of all protoplasmic systems, sensitivity to stimuli and the ability to transmit the excitation so aroused to other cells, are the predominant ones. The stimuli to which these cells respond are various; they may be thermal, photic, chemical, mechanical, gravitational, or other manifestations of energy, but the response of the nerve cell to them is the same (Fig. IV–33).

In the lower Metazoa, nerve cells are usually peripherally located, forming a superficial network of irregularly shaped cells that transmit the excitation in all directions, either to other nerve cells or to muscle, gland, or other effector cells with which they may be in direct contact. In animals of more complicated organization, in which there are tissues derived from

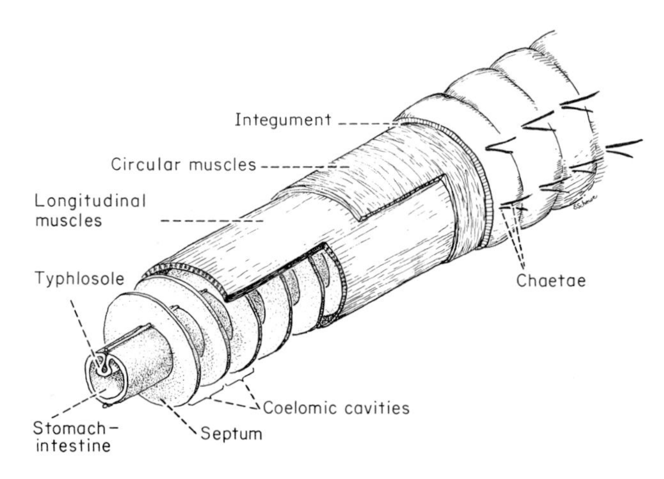

FIG. IV–32 *Part of an earthworm's body, showing the circular and longitudinal layers of muscles and the coelom divided into compartments by septa, so that a certain degree of localized action is possible. (After Parry.)*

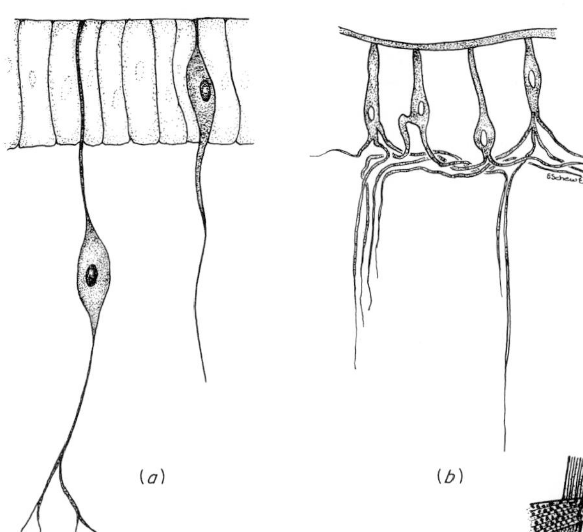

FIG. IV–33 *Primary sense cells.* (*After Hänstrom.*) ***(a)*** *Sense cell from the tentacle of the anthozoan* Cerianthus. ***(b)*** *Sense cell from the epithelium of the earthworm* Lumbricus.

the three primary cell layers, the majority of the nerve cells are more deeply situated and have become modified in certain ways that allow them not only to receive and transmit an excitation but also to conduct it over a considerable distance (Fig. IV–34).

The conducting unit of nervous tissue is called a neuron, and the neurons of invertebrate animals appear to differ in no essential respect from those of vertebrate animals. The individual cell consists of a cell body, in which the nucleus lies in a mass of cytoplasm, the neuroplasm, traversed by fine neurofibrils, with one or several processes extending from the cell body. The

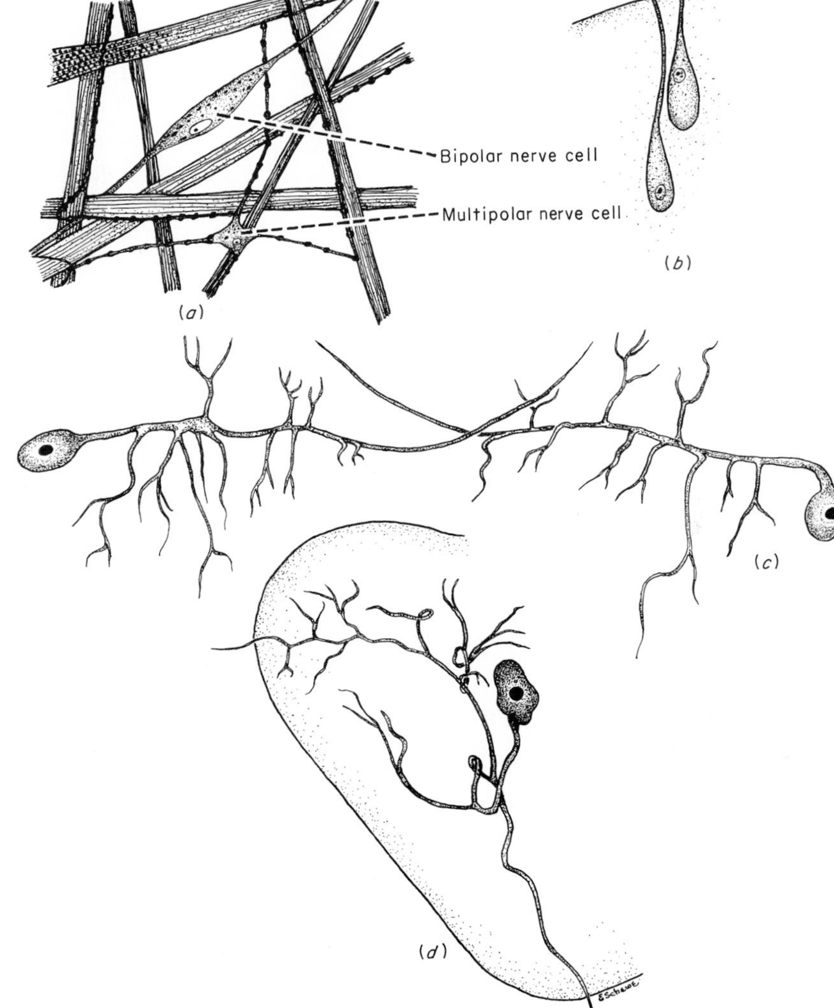

FIG. IV–34 *Nerve cells.* ***(a)*** *Two cells from the ectodermal nerve plexus of the cnidarian* Rhizostoma. (*After Bozler.*) ***(b)*** *Unipolar ganglion cells from the nematode* Ascaris. (*After Goldsmidt.*) ***(c)*** *A pair of ganglion cells from the ventral nerve cord of the polychaete* Arenicola. (*After Zawarzin.*)
(d) *Ganglion cells from the right commissure ganglion of a crayfish, stained with methylene blue.* (*After Orlov.*)

principal one of these processes, the axon, may be of considerable length and, together with the sheaths that invest it, is known as a nerve fiber. The central core of the fiber is the axis cylinder, whose cytoplasm, the axoplasm, flows out from the cell body and is likewise fibrillar in nature. The investing sheaths of some fibers are rich in the lipid myelin; in others, this myelin coat is absent. Axons terminate in fine branches intimately associated with the processes of the cell bodies of other neurons or with the surface of gland, muscle, or other responsive cells. Neurons with a single axon (unipolar) have apparently evolved from those in which several processes are equally efficient in conducting the excitation. In the most simply organized invertebrates there are many multipolar neurons; even in more complex types like annelids, motor neurons with two axons (bipolar) or three axons (tripolar) may be found in the cord along with unipolar ones (Fig. IV–35).

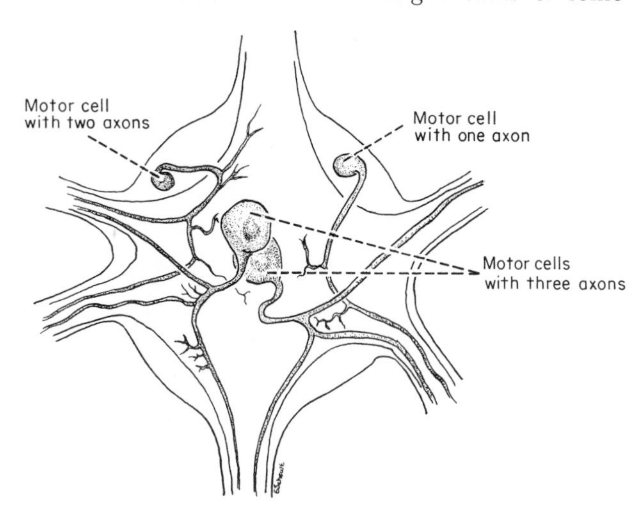

FIG. IV–35 *Four motor cells from a ganglion of the ventral nerve cord of the leech* Aulastomum. (*After Zawarzin.*)

In general, neurons are separate, discrete cells, numbers of which may be clustered in ganglia or bound together as nerves or cords in which there may also be nonconducting supportive cells (glia cells). These appear to have the general attributes of their homologues in the nervous system of vertebrates (Fig. IV–36). While some few exceptions to the general rule of neuron individuality have been found in vertebrates, fusion of the axons of several cell bodies is by no means uncommon in invertebrates. Such interneuronal fusion has been shown in crustaceans and in cephalopods. The giant axons of the stellar nerves in *Sepia* and *Loligo*, which run from the stellate ganglion to the muscles of the mantle, are formed by

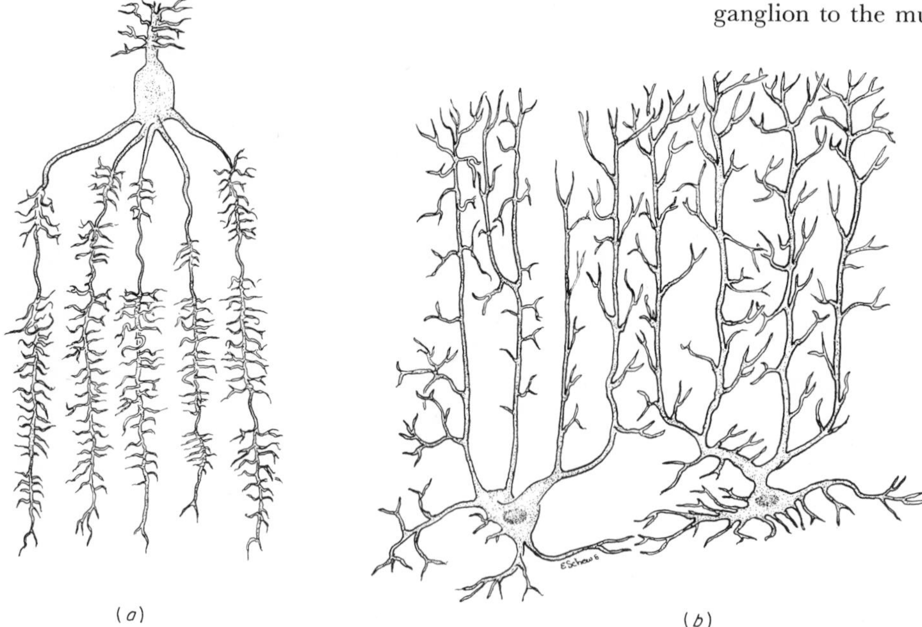

FIG. IV–36 *Glia cells from optic ganglia.* (*After Zawarzin.*)
(a) *From the horsefly* Tabanus.
(b) *From the cuttlefish* Sepia.

the fusion of the axons of some 300 to 1,500 nerve cells, resulting in a compound axon with a diameter of 200 to 1,000 μ. Functionally, such giant axons are valuable in the conduction of impulses for the performance of rapid actions involving the simultaneous stimulation and synchronous contraction of large groups of muscles. The rapid retraction response of an earthworm to touch at either anterior or posterior end is, for example, related to the giant axon system in the nerve cord. A syncytial arrangement of nerve cells, by union of these processes, has also been demonstrated in some Cnidaria.

G. REPRODUCTIVE TISSUES

Reproductive cells are those whose function it is, either singly or in combination, to produce new individuals and so to increase and propagate the species. Such cells may originate in various parts of the invertebrate body and either undergo the specific changes that transform them into mature sex cells (gametes) *in situ*, or they may migrate to particular areas of the body and there be localized in a sex organ, or gonad. They are liberated, in a few instances, by disintegration of the body or, more commonly, by discharge either directly or indirectly into ducts by which they are conveyed to the exterior. Most species of invertebrates show distinct sex differentiation as males or females, but there are some in which the single individual possesses both testes and ovaries, or a combined organ, the ovotestis, and functions either simultaneously or successively as male and female in the production of sperm and eggs.

Reproductive tissue is essentially epithelium from which the gametocytes or potential germ cells become differentiated. Other cells of the epithelium may become follicle cells, in which the maturing gametocyte or cluster of gametocytes is enclosed, or "nurse cells" that contribute nutritive material to the follicle cells (Fig. IV–37). The gametocytes, passing through the

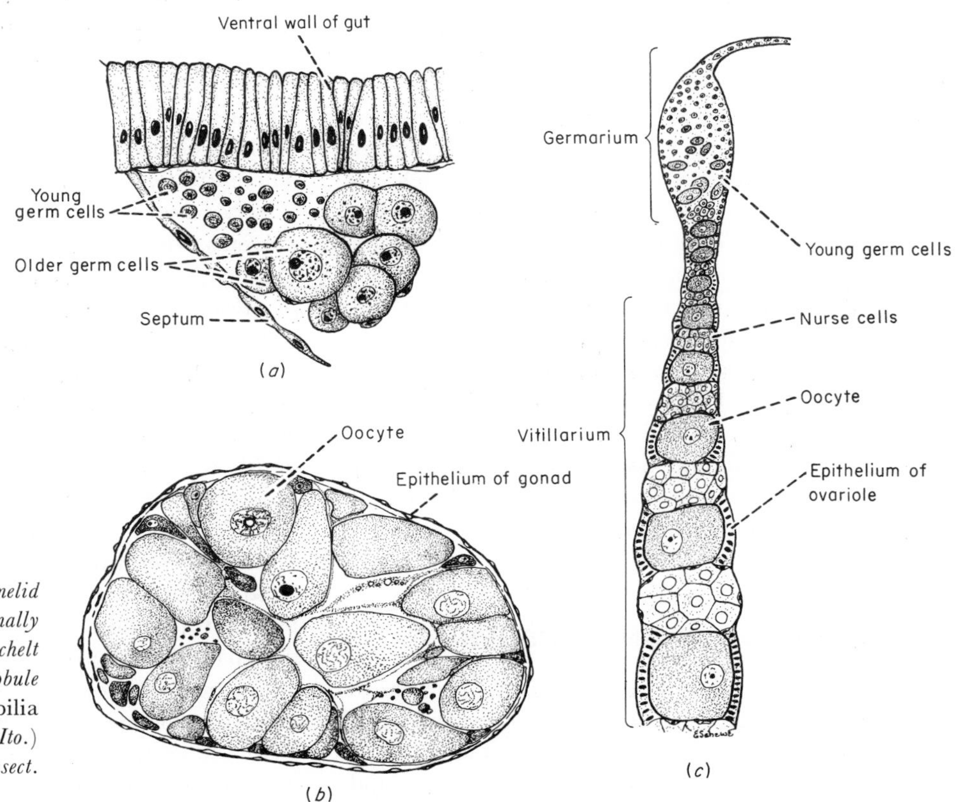

FIG. IV–37 *Reproductive cells.* ***(a)*** *Reproductive cells of the annelid* Ophryotrocha puerilis, *seasonally budded off the septa.* (*After Korschelt and Heider.*) ***(b)*** *Section of a lobule of an ovary of the sea urchin* Mespilia globulis. (*After Tennent and Ito.*) ***(c)*** *Egg tube of a coleopteran insect.* (*After Korschelt and Heider.*)

multiplication and growth phases and the meiotic divisions typical of sex cells, become definitive gametes which, in most species, are shed from the body and fertilized externally. Internal fertilization is characteristic of flatworms, pulmonate molluscs, insects, and some other invertebrates. Many species practising external fertilization have various devices for safeguarding the gametes once they are shed and for promoting their opportunities for union. But fertilization, or gametic union, is by no means an essential process in invertebrate reproduction, for natural parthenogenesis is not uncommon and in some species, notably among the rotifers, it is the only known means of reproduction. Parthenogenetic eggs, like those that require fertilization, are developed within the sex gland of a mature individual.

Most invertebrates are extremely prolific and produce enormous numbers of sex cells, many of which never fulfill their purpose. Depending on its size, a single oyster may lay 25 to 60 million eggs at a time. Queen bees lay 1,500 to 2,000 eggs per day, and termite queens 6,000 to 7,000. As these queens have been known to live 15 to 50 years, the number of their potential offspring reaches astronomical figures. It has been estimated that one pair of houseflies, for example, breeding from April to August, could produce 191,000,000,000,000,000,000 offspring. A single female aphid, also in one season and also if all her offspring lived, could be responsible for the addition of 1,560,000,000,000,000,000,000,000 new members to the population of plant lice. A single water flea (*Daphnia pulex*) could produce as many as 13,000,000,000,000 descendants. Many other species are almost as prolific, and many make very careful provision for the protection of their eggs and developing embryos. In spite of this, the natural hazards to which they are exposed prevent a large proportion of the potential offspring from reaching maturity.

H. UNIQUE CELLS AND TISSUES

There are certain types of cells and tissues peculiar to particular groups or species of invertebrates and without counterpart in any vertebrate species. These represent a very special kind of cell differentiation and offer what may prove to be a fruitful area of study for this aspect of cellular mechanisms. Among these unique types of cells are the cnidocytes of Cnidaria, the colloblasts of Ctenophora, the rhabdite-forming cells of Turbellaria (Platyhelminthes), the chlorogogue and botryoidal cells of Annelida, and the oenocytes and mycetocytes of Insecta.

Cnidocytes (nematocytes) originate from interstitial cells and are formed continuously during the active life of the animal. Each mature cnidocyte contains a nematocyst, or cnida (Greek *knide*, "nettle"), a product of the cell's secretory activity, but one with a high degree of structural detail. It consists of a capsule whose wall is most probably a collagen, within which a tubular thread is coiled in a characteristic manner. In some nematocysts there is a distinct operculum, or lid, closing the distal end of the capsule (Fig. IV–38*a*). At least 18 different kinds of nematocysts have been described, the distinctions being mainly in the character of the tubular thread. Certain types, or certain combinations of types, are characteristic of different species and have therefore been useful as taxonomic tools. Two major types are recognized: in one, the tube is closed at the tip; in the other, it is open. Nematocysts with closed threads are called astomocnidae (Greek *a*, "without," and *stoma*, "mouth"). There are two main categories of these: desmonemes (Greek *desmos*, "bond," and *nema*, "thread"), or volvents, which are of very general distribution among the members of the phylum; and rhopalonemes (Greek *rhopalon*, "club"), limited to Siphonophora and having a sac-like tube instead of a long coiled one. Nematocysts with threads open at the tip are called stomocnidae and are very varied in design. There are, for example, those with enlarged basal regions, or butts, of different shapes and those without them; there are those with stylets or barbs on the butt and those without them; and there are those in which the thread is of the same diameter throughout its length and those in which it tapers, or is swollen, near the base. This variety had led to elaborate systems of nomenclature to designate the different kinds.

Upon appropriate stimulation the thread is everted from the capsule and the nematocyst discharged from the cnidoblast. The threads of astomocnidae wrap around an object with which they come in con-

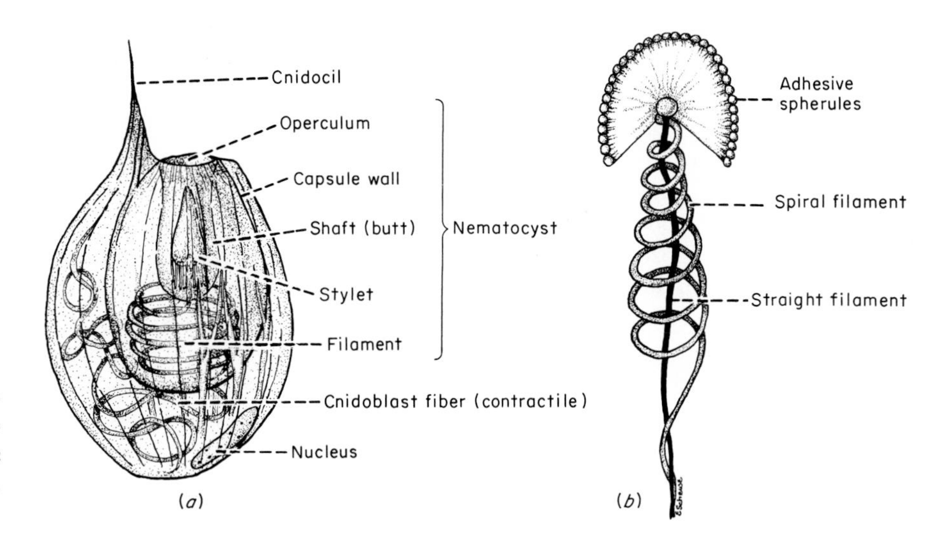

FIG. IV–38 *Cnidoblast and colloblast.* ***(a)*** *Cnidoblast with its contained penetrant nematocyst. (From various sources.)* ***(b)*** *Colloblast. (After Hyman, vol. I.)*

tact, and it is because of this action that desmonemes are more simply known as volvents (Latin *volvere*, "to turn" or "turn about"). A small crustacean such as *Daphnia*, on which freshwater hydras customarily feed, may be quickly immobilized by the discharge of a number of volvents when they are wound around its swimming appendages. Secretions are discharged from the open tips of stomocnidae that may be irritant, narcotic, or toxic to prey or predators, and the threads may penetrate into the tissues and so inject their contents into them. These are known as penetrants. Other stomocnidae, called in general glutinants, discharge adhesive substances in which organisms may be entangled or by which the cnidarian may temporarily anchor itself to some object. The source and chemical nature of the materials released by the stomocnidae have not yet been determined, but it seems likely that they, like the nematocyst itself, are products of the synthetic activity of the cnidoblast.

The sequence of events in the development of a nematocyst have been followed in great detail by electron microscopy, which has provided as complete a history of structural changes in a differentiating cell as any yet known. Preceding their transition to cnidocytes, the interstitial cells divide forming a cluster of 8 to 16 cells which remain attached to each other by cytoplasmic bridges. The subsequent events of differentiation take place simultaneously and similarly in all the cells of a cluster, so that they all produce nematocysts of the same type at the same time. The first morphological evidence of the conversion of an interstitial cell to a cnidoblast is the appearance of double membranes within the cytoplasm. In the comparatively undifferentiated interstitial cell there is little or no endoplasmic reticulum, but, especially during the period of division, ribosomes are very abundant throughout the cytoplasm. When division stops, and the endoplasmic reticulum appears in the differentiating cell, the ribosomes become localized on the membranes, which become very extensive and expanded into broad, flat cisternae. The cell then has the characteristic appearance of one in which the active synthesis of specific proteins is in progress. The products of synthesis appear to be collected in the endoplasmic reticulum and from it to be channelled into the Golgi substance, which plays a conspicuous role in nematocyst formation and remains closely applied to the apical pole of the organelle throughout its construction. The first indication of the nematocyst itself is the appearance of a distinct vesicle in which the capsular and tubular areas are distinguishable. As the vesicle grows by the accumulation of the material contributed to it, it displaces the nucleus and occupies the major portion of the cell. Growth at the apical end of the capsule, after it has reached its definitive size, results in the formation of the thread, which at first lies coiled in the cytoplasm but later, by some means, becomes coiled in the capsule. When the product is complete, Golgi substance and endoplasmic reticulum undergo regression and the internal differentiation of the organelle apparently proceeds independently of them. The cytoplasmic bridges disappear and the cnidocytes separate from each other and become distributed throughout the body, becoming especially concentrated in the tentacles in Hydrozoa and Scyphozoa. In Anthozoa, nematocysts are abundant also in the acontia where they have been described as directly embedded in the cytoplasm, the cnidoblast having degenerated after formation of the capsule. It is not certain whether this distribution is brought about by the active migration of the cells themselves or by their passive conveyance throughout the tissues in the gastral fluid. Their final location may, however, be some distance from that of their origin. In the tentacles of *Hydra* they are very definitely arranged in groups, as nettle batteries made up of a large central penetrant with a butt dilated at the base, around which there are three rings of cnidoblasts, each with a different type of nematocyst. The two inner rings contain stomocnidae without butts, and the outer ring is made up of desmonemes. These come into action before the penetrants, and trap and hold the prey until their toxins kill it.

Cnidocytes are independent effectors and respond directly to appropriate stimuli without any connection with the nerve net. What constitutes an appropriate stimulus and what is the mechanism of nematocyst discharge are still problems to be solved. The response is not the same in all cnidocytes, for each type appears to be more sensitive to certain stimuli than to others. In general, it is believed that the response is evoked by a combination of chemical and physical (mechanical) stimuli, the chemical action lowering the threshold of response to the stimulus of contact. The cnidocytes of *Hydra* have a small, hair-like projection, the cnidocil, on the exposed surface of the cell which has been thought to be the immediately responsive agent, acting as a trigger that sets off the changes that result in the discharge of the capsule and the eversion of its thread. Electron microscopy has shown that the cnidocil has the same pattern of fibrils as a typical cilium or flagellum (see page 186), and it is suggested that it is a modified cilium, also with relationship to the centriole. Although the cnidocil may be involved in nematocyst discharge in *Hydra*, it cannot be indispensable to the process, since it is absent from the acontia of Anthozoa.

Various theories have been proposed to explain the mechanism of discharge, none of which is satisfactory in all respects. It appears that conditions within the cnidocyte ordinarily determine the eversion of the thread, for this response has not been elicited from experimentally isolated nematocysts by any of the chemical or physical means yet used. It is probable that, on stimulation, water enters the capsule from the surrounding cytoplasm. One such osmotic theory suggests that this inflow of water increases the hydrostatic pressure in the capsule and so forces the thread out; another suggests that it leads to the swelling of colloidal material and so to eversion of the thread. There is also the possibility that the thread is everted as the result of contraction of the capsule, by contractile fibrils either within or surrounding it. Both mechanisms may, of course, be operative in nematocysts of different kinds. It has been suggested, from evidence of the effects of enzyme inhibitors such as sodium fluoride, that the nematocysts contain strategically placed enzymes which respond to chemical agents that act as effector substances. Once the nematocyst is discharged, the cnidocyte is spent and must be replaced by another if the offensive and defensive capacities of the animal are to be maintained. It has been calculated that 25 percent or more of the nematocysts in the tentacles of *Hydra* are lost in the capture of one nauplius of the brine shrimp *Artemia*, often used as food for laboratory cultures. Continuous renewal of these weapons of offense and defense is vital, and the transformation of interstitial cell to cnidoblast and cnidocyte offers one of the most dramatic and accessible demonstrations of cellular differentiation. For the cnidoblast is a highly specialized cell. It is biochemically capable of elaborating so complicated a structure as the nematocyst and its associated adherent and toxic substances; it is morphologically adapted to its position and function in the body; and it is physiologically capable of discriminating between stimuli and of accepting and transmitting selected stimuli to its contained capsule in such a way that the nematocyst is discharged.

Another mysterious and most interesting fact from the point of view of both physiology and ecology is the remarkable control of response shown by cnidocytes. Many Cnidaria harbor commensals which provoke no discharge of their nematocysts. The ciliates *Kerona* and *Trichodina*, for example, habitually live on the surface of hydras, gliding freely over the cnidocytes and even bending down their cnidocils, but not causing their discharge. Many Scyphozoa, and the particularly aggressive siphonophore *Physalia*, have small fish and crustaceans among their tentacles, but do not treat them as prey or as enemies. Observations of the behavior of such hosts indicate that in some way the commensals inhibit events that ordinarily lead to nematocyst discharge and not that they are immune to them, for free nematocysts are not found in them or in their vicinity.

Some animals that feed on Cnidaria do not digest the nematocysts but keep and make use of them instead. These animals must also have the ability to inhibit discharge, and they also show a high degree of selectivity in the type of nematocyst they retain. The turbellarian *Microstomum*, feeding on *Chlorohydra*, digests the cnidocytes, but their nematocysts are transferred to parenchyma cells. Volvents are digested here, but penetrants are transferred by migratory cells to the epidermis and very precisely oriented there. They may be discharged by the flatworm, a fact which implies that in the parenchyma or epidermal cell there is a mechanism equivalent to that in the cnidocyte itself. The sea slug *Eolis* feeds on the colonial hydroid, *Pennaria*, especially on its tentacles, and also conserves some nematocysts. It chooses only a particular type of penetrant with an elongated butt armed with several rows of spines (microbasic mastigophores). These are transferred from the stomach to the digestive gland, which extends out into the finger-like projections, or cerata, of the dorsal and ventral body wall (Fig. III–14). The nematocysts are ingested by the epithelial cells in the cnidosac, a small enlarged area at the tip of each ceras, and all but the microbasic mastigophores are digested. These are concentrated and oriented on the ceras and discharged when occasion arises.

The natural defensive and offensive organs of ctenophores are colloblasts, also called adhesive or lasso cells. They are special glandular epithelial cells of the tentacles which entirely, or almost entirely, cover them. From the nucleus of each potential colloblast is developed a straight filament that projects into the core of the tentacle and a contractile filament that is coiled around the straight one. Both these serve to anchor the head of the colloblast in the tentacle. The head of each colloblast is more or less crescentic and its convex distal surface is packed with granules that become a sticky and gummy secretion when discharged (Fig. IV–38*b*). Prey is entangled in this mass and so captured, for the colloblasts are primarily weapons of offense.

Among flatworms, Turbellaria have characteristic unicellular glands, situated either in the epidermis or in the parenchyma immediately below it, which secrete straight or curved rod-like bodies known as rhabdoids (Fig. IV–39*d*). These bodies are found in the gland cells or free in the parenchyma adjacent to them and can be discharged to the exterior of the animal, where they make a slimy or sticky mass that may have some toxic effects. There are various types of these inclusions in different species of turbellarians and two other kinds that are similar to them, pseudorhabdites and sagittocysts. Pseudorhabdites are intracellular amorphous masses of slimy material that are common in polyclads and are occasionally found in rhabdocoeles. Sagittocysts, found in only certain Acoela, are also intracellular, pointed vesicles with a central protrusible rod or needle. Histochemical tests have shown that the rhabdoids are made of protein, but electron microscopy has not disclosed any structural detail in them, as it has in the trichocysts of ciliate protozoans. These are similar rod-like bodies which likewise can be shot out of the body to form an entangling mass of very much extended threads, each as long as 40 μ. The thread is cross-banded, with a period of 600 to 650 angstroms (Å), and has a pointed body at its tip (Fig. IV–40). The trichocysts of some of the dinoflagellates have considerable resemblance to the nematocysts of cnidarians, and a relationship between them is a possible consideration. The polar capsules of the sporocysts of some sporozoans also have similarities with nematocysts (Fig. III–5).

In some of the annelids, mesoderm cells may be differentiated as specialized tissues peculiar to certain members of the phylum. In earthworms, for example, the mesothelium that covers the stomach intestine is

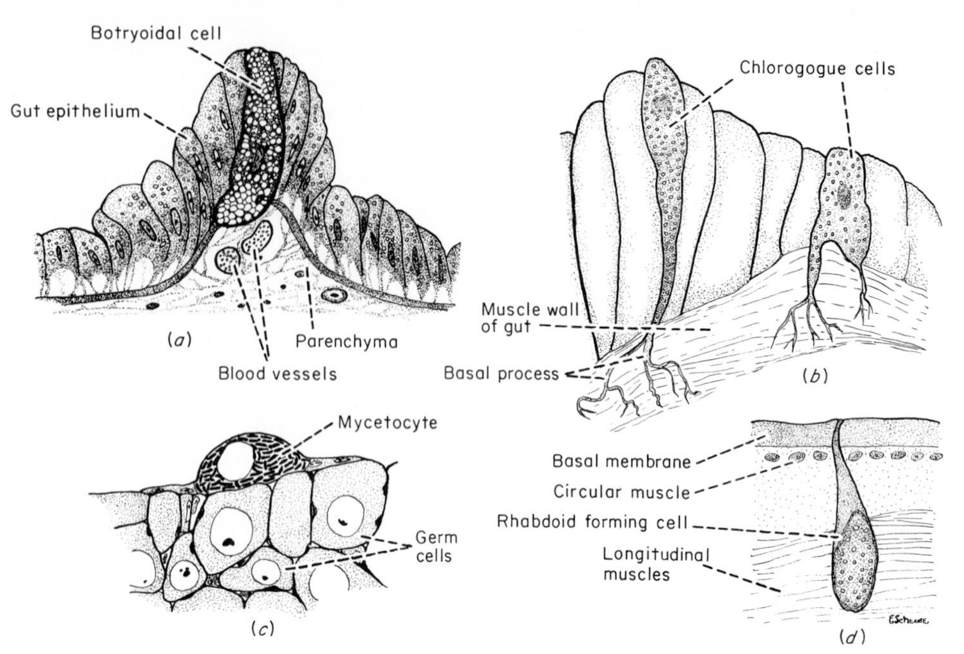

FIG. IV–39 *Special cells.* ***(a)*** *Epithelium from the gut of a leech, showing migratory botryoidal cells. (After van Emden.)* ***(b)*** *Chlorogocytes from the gut of the earthworm* Lumbricus herculeus. *(After Goodrich.)* ***(c)*** *Mycetocytes attached to the wall of an ovariole of a cockroach. (After Gier.)* ***(d)*** *Young rhabdoid-forming cell from a flatworm,* Phagocata gracilis. *(After Brönn, vol. IV.)*

modified as tall columnar cells, irregular in shape, and loaded with particulate matter, which is most likely coelomic waste that they have taken up. This is sufficiently yellow in color to give them their name of chlorogogen (Greek *chlor*, "pale green," and *agein*, "to lead") or chlorogogue cells (Fig. IV–39*b*). In leeches, there is a somewhat similar kind of tissue, called botryoidal tissue (Greek *botrys*, "bunch of grapes"). The botryoidal cells, forming grape-like masses between muscle cells and the mesenchyme cells that nearly fill the coelomic cavities of these annelids, are loaded with granules, most of them dark in color. These granules are chromolipids, oxidative derivatives of various lipids. Young botryoidal cells are ameboid and move about within the tissue spaces until they gather in the regions they occupy in the adult leech, where their distribution varies in different species, as their possible functions do also. These functions have been variously described as respiratory and excretory, as well as nutritive, in the sense that the botryoidal cells seem to accumulate and to store lipid substances that may be used as a food reserve (Fig. IV–39*a*).

In many arthropods, there are certain cells that act as functional accessories to the excretory system. These cells, known as nephrocytes, may be variously distributed throughout the body or collected in more or less definite sites. Whatever their location, they are morphologically similar and functionally so in their ability to take up substances from the tissue fluids and to store, or possibly metabolize, them. As storage areas they constitute "kidneys of accumulation," operating to remove from the circulating fluids undesirable or harmful products not dealt with by other tissues. In insects, cells of this type, which will take up experimentally injected dyes that are not excreted by the Malpighian tubules, are collected especially along the dorsal vessel as pericardial cells. In the grasshopper *Melanoplus differentialis*, for example, there are hundreds of these cells, surrounding the dorsal vessel in the thorax and abdomen and loaded with inclusion bodies. These bodies are rich in protein and lipoprotein, but do not contain glycogen, uric acid, or urate. Experiments of feeding and starving the grasshoppers have produced results that suggest that these

FIG. IV–40 *Discharged trichocyst of* Paramecium. *(After Jakus and Hall.)*

cells may take up peptides, proteins, and albuminoids from the hemolymph, change them in some way, and then release them back to the hemolymph. The widespread occurrence of pericardial cells in insects would imply that they have some important function or functions in insect metabolism.

Cyrtocytes are a distinct type of cell, identified so far in a limited number of invertebrates and with homologues in vertebrates. Basically a cyrtocyte is a cell with granular and vacuolated cytoplasm, a large nucleus, and, at one pole, a collar or tubular extension enclosing a flagellum or a tuft of cilia. The collar or tube is supported by rods of firm material, possibly chitinous in nature, that lie parallel to its long axis. The resemblance that such an arrangement of rods gives these cells accounts for their name, which is derived from the Greek word *churtos*, meaning "weir." Morphologically, flagellate protozoans with collars (choanoflagellates), the collar cells (choanocytes) of sponges, protonephridal terminal cells, and solenocytes of the excretory organs of some invertebrates fall into this category, as do the podocytes in the glomeruli of vertebrate kidneys and the terminal cells of the excretory organs of the cephalochordate *Branchiostoma lanceolatum*. All of these conform to the basic cyrtocyte type, although with variations of it. The podocyte, for example, so called because of the numerous interdigitating processes that extend from its basal pole into the basement membrane, is nonflagellate, and the terminal cells of the lancelet's excretory organs are a combination of podocyte and solenocyte. There are also differences in the numbers and arrangements of the rods and in the ways in which they are connected. The primary service of cyrtocytes is ultrafiltration, for the sieve-like nature of the tube, or collar, permits passage of fluids but prevents that of particles of colloidal dimensions. Thus, cells of this kind may serve as nutritive devices, as they do in choanoflagellates and sponges, or as excretory devices, as they do in the flatworms, gastrotrichs, nemerteans, and polychaetes in which they have been identified.

Oenocytes are cells peculiar to insects and are found in all pterygote orders. They are large cells of ectodermal origin to which numerous functions have been attributed. Originating in the epidermis, they may remain there between the epithelial cells and the basement membrane on which they rest, or they may project into the perivisceral cavity while still retaining connection with the epidermis. Or they may move away from the epidermis entirely and form clusters of cells in various positions throughout the body or embedded in the fat bodies (Fig. III–16). The fact that they undergo histological changes suggests that they may be actively concerned in metabolic processes, and cytochemical tests have demonstrated the presence of an oxidase in them that could function in intermediary metabolism; the periodicity of these changes may indicate that the oenocytes are concerned with the cyclic process of molting, although this connection has not yet been conclusively shown. Their role in the secretion of the wax that covers the cuticle has also been demonstrated, which may be their primary connection with the molting process.

Some insects also possess especially large cells, the mycetocytes, that are filled with microorganisms or bacteroids—fungi, yeasts, rickettsia-like organisms, and bacteria (Fig. IV–39*c*). These cells may be scattered along the mid- and the hindguts, either with connections or with no connections to its lumen, or near the reproductive tissues. The relation of these bacteroids to the host may be various, but in those cases where critical study has been made, their role seems to be an essential one, because the removal, destruction, or extirpation of the mycetome or mycetocyte-containing tissue is lethal to the insect that harbors them.

SELECTED BIBLIOGRAPHY

Epithelium

Florey, E.: Ultra-structure and Function of Cephalopod Chromatophores, *Am. Zool.*, vol. 9, pp. 429–442, 1969.

Hahn, W. E., and D. E. Copeland: Carbon Monoxide Concentration and the Effect of Aminopterin on Its Production in the Gas Bladder of *Physalia physalia*, *Comp. Biochem. Physiol.*, vol. 18, pp. 201–207, 1966.

Holmes, W.: The Color Changes and Color Patterns of *Sepia officinalis* (L), *Proc. Zool. Soc. London,* vol. 110, pp. 17–36, 1940.

Larimer, J., and E. A. Ashby: Float Gases, Gas Secretion and Tissue Respiration in the Portuguese Man-of-War, *Physalia, J. Comp. Cell. Physiol.*, vol. 60, pp. 41–47, 1962.

Osing, O.: Silk from the Sea, *Sea Front.,* vol. 14, pp. 18–21, 1968.

Pedersen, K. N.: Some Features of the Fine Structure and Histochemistry of Planarian Sub-epidermal Gland Cells, *Z. Zellforsch.,* vol. 50, pp. 121–142, 1959.

Stephenson, E. M., and C. Stewart: "Animal Camouflage," Adam and Charles Black, London, 1966.

Wittenberg, J. B.: The Source of Carbon Monoxide in the Float of the Portuguese Man-of-War, *Physalia physalia, J. Exp. Biol.,* vol. 37, pp. 698–705, 1960.

Supporting and transporting tissues

Ashurst, D. E.: The Connective Tissue of Insects, *Annu. Rev. Entomol.,* vol. 13, pp. 45–74, 1968.

Chapman, G.: Studies on the Mesogloea of Coelenterates, *Q. J. Microsc. Sci.,* vol. 94, pp. 155–176, 1953.

Chapman, G.: Studies on the Mesogloea of Coelenterates. II. Physical Properties, *J. Exp. Biol.,* vol. 30, pp. 440–451, 1953.

Gross, J., Z. Sokol, and M. Rougvie: Structural and Chemical Studies on the Connective Tissue of Marine Sponges, *J. Histochem. Cytochem.,* vol. 4, pp. 227–246, 1956.

Liebman, E.: The Trephocytes and Their Function, *Experientia,* vol. 3, pp. 442–451, 1947.

Wagge, L. E.: Amoebocytes, *Int. Rev. Cytol.,* vol. 4, pp. 31–78, 1955.

Muscular, nervous, and reproductive tissues

Bagby, R. W.: The Fine Structure of Myocytes in the Sponges *Microciona prolifera* and *Tedania ignis, J. Morphol.,* vol. 118, pp. 167–182, 1966.

Batham, E. J., and C. F. A. Pantin: The Organization of the Muscular System of *Metridium senile, Q. J. Microsc. Sci.,* vol. 92, pp. 27–54, 1951.

Bowden, J.: The Structure and Innervation of Lamellibranch Muscle, *Int. Rev. Cytol.,* vol. 7, pp. 295–333, 1957.

Chapman, G. B.: Electron Microscopy of Insect Muscle, *J. Morphol.,* vol. 95, pp. 257–262, 1954.

Hanson, J.: The Structure of the Smooth Muscle Fibers in the Body Wall of the Earthworm, *J. Biophys. Biochem. Cytol.,* vol. 3, pp. 111–121, 1957.

Rosenbluth, J.: Ultra-structure of Somatic Muscle Cells in *Ascaris lumbricoides. II., J. Cell Biol.,* vol. 26, pp. 579–592, 1965.

Sabrosky, C. W.: How Many Insects Are There?, *Syst. Zool.,* vol. 2, pp. 31–36, 1953.

Tiegs, O. W.: The Flight Muscles of Insects—Their Ancestry and Histology, with Some Observations on the Structure of Striated Muscle in General, *Philos. Trans. Roy. Soc. London, B,* vol. 238, pp. 221–348, 1954.

Unusual tissues

Blanquet, R., and H. Lenhof: A Disulfide-linked Collagenous Protein in Nematocyst Capsules, *Science,* vol. 154, pp. 152–153, 1966.

Chapman, G., and L. G. Tilney: Cytological Studies on the Nematocysts of *Hydra. I.* Desmonemes, Isorhizas, Cnidocils and Supporting Structures, *J. Biophys. Biochem. Cytol.,* vol. 5, pp. 69–78, 1959.

Chapman, G. B., and L. G. Tilney: Cytological Studies on the Nematocysts of *Hydra. II.* The Stenoteles, *J. Biophys. Biochem. Cytol.,* vol. 5, pp. 79–84, 1959.

Coe, W. R.: Unusual Types of Nephridia in Nemerteans, *Biol. Bull.,* vol. 58, pp. 203–216, 1930.

Glaser, O.: The Nematocysts of Eolids, *J. Exp. Zool.,* vol. 9, pp. 117–142, 1910.

Jakus, M. A.: The Structure and Properties of the Trichocysts of *Paramecium, J. Exp. Zool.,* vol. 100, pp. 457–485, 1945.

Jakus, M. A., and C. E. Hall: Electron Microscope Observations of the Trichocysts and Cilia of *Paramecium, Biol. Bull.,* vol. 91, pp. 142–144, 1946.

Kessel, R.: Electron Microscope and Cytochemical Studies on the Oenocytes of the Grasshopper *Melanoplus differentialis differentialis* Thomas, *Anat. Rec.,* vol. 137, p. 371, 1960.

Kessel, R.: Light and Electron Microscope Studies on Pericardial Cells of the Adult Grasshopper *Melanoplus differentialis differentialis* Thomas, *Anat. Rec.,* vol. 137, p. 371, 1960.

Kümmel, G., and J. Brandenburg: Die Reusengeisselzellen (Cyrtocyten), *Z. Naturforsch.,* vol. 166, pp. 692–697, 1961.

Lenhoff, H. M., E. S. Kline, and R. Hurley: A Hydroxyproline-rich Intracellular Collagen-like Protein of *Hydra* Nematocysts, *Biochem. Biophys. Acta,* vol. 26, pp. 204–205, 1957.

Lentz, T. L., and R. J. Barrnett: The Effect of Enzyme Substrates and Pharmacological Agents on Nematocyst Discharge, *J. Exp. Zool.*, vol. 149, pp. 33–38, 1962.
Phillips, J. H.: Isolation of Active Nematocysts of *Metridium senile* and Their Chemical Composition, *Nature*, vol. 178, p. 932, 1956.
Slautterback, D. B., and D. W. Fawcett: The Development of Cnidoblasts of *Hydra*: An Electron Microscope Study of Cell Differentiation, *J. Biophys. Biochem. Cytol.*, vol. 5, pp. 441–452, 1959.
Yanagita, T. M., and T. Wada: Physiological Mechanism of Nematocyst Response in Sea Anemone. VI. A Note on the Microscopical Structure of Acontium with Special Reference to the Situation of Cnidae within Its Surface, *Cytologia*, vol. 24, pp. 81–97, 1959.

SUGGESTED READING FOR SECTION TWO

Bayer, F. M., and H. B. Owre: "The Free Living Lower Invertebrates," Macmillan, New York, 1968.
Bourne, G. H.: "Division of Labor in Cells," Academic, New York, 1962.
Curtis, H.: "The Marvelous Animals: An Introduction to the Protozoa," Natural History Press for the American Museum of Natural History, New York, 1968.
Dobell, C.: "Antony v. Leeuwenhoek and his Little Animals," Russell and Russell, New York, 1932.
Goetsch, W.: "The Ants," The University of Michigan Press, Ann Arbor, 1957.
Hadzi, J.: "The Evolution of the Metazoa," Macmillan, New York, 1963.
Huxley, T. H.: "The Crayfish: An Introduction to the Study of Zoology," Kegan Paul, Trench and Trubner, London, 1880.
Millott, N. (ed.): "Echinoderm Biology," Symposium of the Zoological Society of London No. 20, Academic, New York, 1967.
Morley, D. W.: "The Ant World," Penguin Books, London, 1953.
Morton, J. E.: "Molluscs," Harper, New York, 1960.
Rees, W. J. (ed.): "The Cnidaria and Their Evolution," Symposium of the Zoological Society of London No. 16, Academic, New York, 1966.
Rounds, H. D.: "The Evolution of the Invertebrates," State University Press, Wichita, 1966.
Savory, T. H.: "The Spider's Web," Warne, London, 1952.
Thistleton, G. E.: "Crustacea and Arachnida," Evans, London, 1966.
Vandel, A.: "La Génèse du Vivant," Masson et Cie, Paris, 1968.
Wilmer, E. N.: "Cytology and Evolution," Academic, New York, 1960.

section three

CLASSIFICATION

chapter V THE CLASSIFICATION OF INVERTEBRATES

chapter V

THE CLASSIFICATION OF INVERTEBRATES

A. THE PURPOSE OF CLASSIFICATORY SYSTEMS

From the records that have been preserved it seems apparent that man, as soon as he began to be thoughtfully observant of the world in which he lived, felt the desire and the need for some systematic scheme in which to place the objects with which he found himself surrounded. In the history of early science and of biology, at least from the time of the ancient Greeks, there occur records of attempts to arrive at a satisfactory classification of living things. Some of these have been more successful than others, but all of them have been subjected to more or less drastic revisions in the light of expanding information and increasing understanding of the natural world. Current schemes will doubtless prove to be no exception to demands for constant revision.

Apparently one of the first distinctions to be made was that between animals and plants, with the institution of zoology and botany as different branches of biological science. With the more highly organized animals and plants such a distinction is not hard to make and seems a convenient, if not wholly valid, one, but it breaks down completely at the level of unicellular systems and cell aggregates, where neither structure and habits nor physiological processes and potentialities furnish satisfactory criteria for assigning organisms like euglenids or volvocales to one or the other of these categories.

But these attempts served at least to emphasize the fundamental unity of biological systems and, indeed, the fundamental unity of the biological and physical world. Even in the eighteenth century, minerals and stones—considered a part of the natural world although less perfect than those things that exhibited life—as well as diseases, were included in biological classificatory systems. Later ones excluded these, although all that has been learned about living and nonliving objects substantiates the belief that materially they have much in common, including in their composition the same atoms and elements, often in the same molecular configurations. We know, too, that these atoms are in a constant state of flux, passing from one object to another as physical and biological processes follow their inevitable course, and that an element that is now in a mineral or some other nonliving substance may later become part of a living organism, and vice versa. To this extent at least, the organism must be considered as continuous with its inorganic environment and in dynamic equilibrium with it.

Classificatory systems are to a great extent arbitrary, but they are essential, for it is expedient if not imperative, to reduce the multiplicity of living things to some kind of order, if only to make discussion and comparison of them possible. But the ideal goal of modern taxonomy is the expression of phylogenetic relationships, the establishment of a genealogical tree, or pedigree, for all living things. The alphabetical arrangements of earlier systematists contributed little to this desideratum, and it was only when comparative anatomy and comparative embryology were vigorously pursued that enough information was accumulated to furnish a basis for taxonomic systems that seemed to express the fundamental relationships between living things and to attribute some kind of natural order to them (Fig. V–1).

B. THE CONCEPT OF SPECIES

Underlying all systems of biological classification is the concept of species. It is species that are subdivided into the minor categories of subspecies, varieties, and races and that are collected into the major ones of genus, family, order, class, and phylum. Until almost the twentieth century, a species was considered as a group of organisms with one or more distinctive anatomical characters and by many of the early naturalists, including Carolus Linnaeus (1706–1778), to have been created as such, fixed and unchangeable (Fig. V–2). These concepts had to be revised, however, in the light of accumulating evidence provided by students of comparative embryology, comparative anatomy, and the mechanisms of inheritance. This evidence showed, on the one hand, that apparently close morphological

FIG. V–1 *A phylogenetic tree, showing the major groups of animals according to the approximate relative sizes and possible relationships of the groups.*

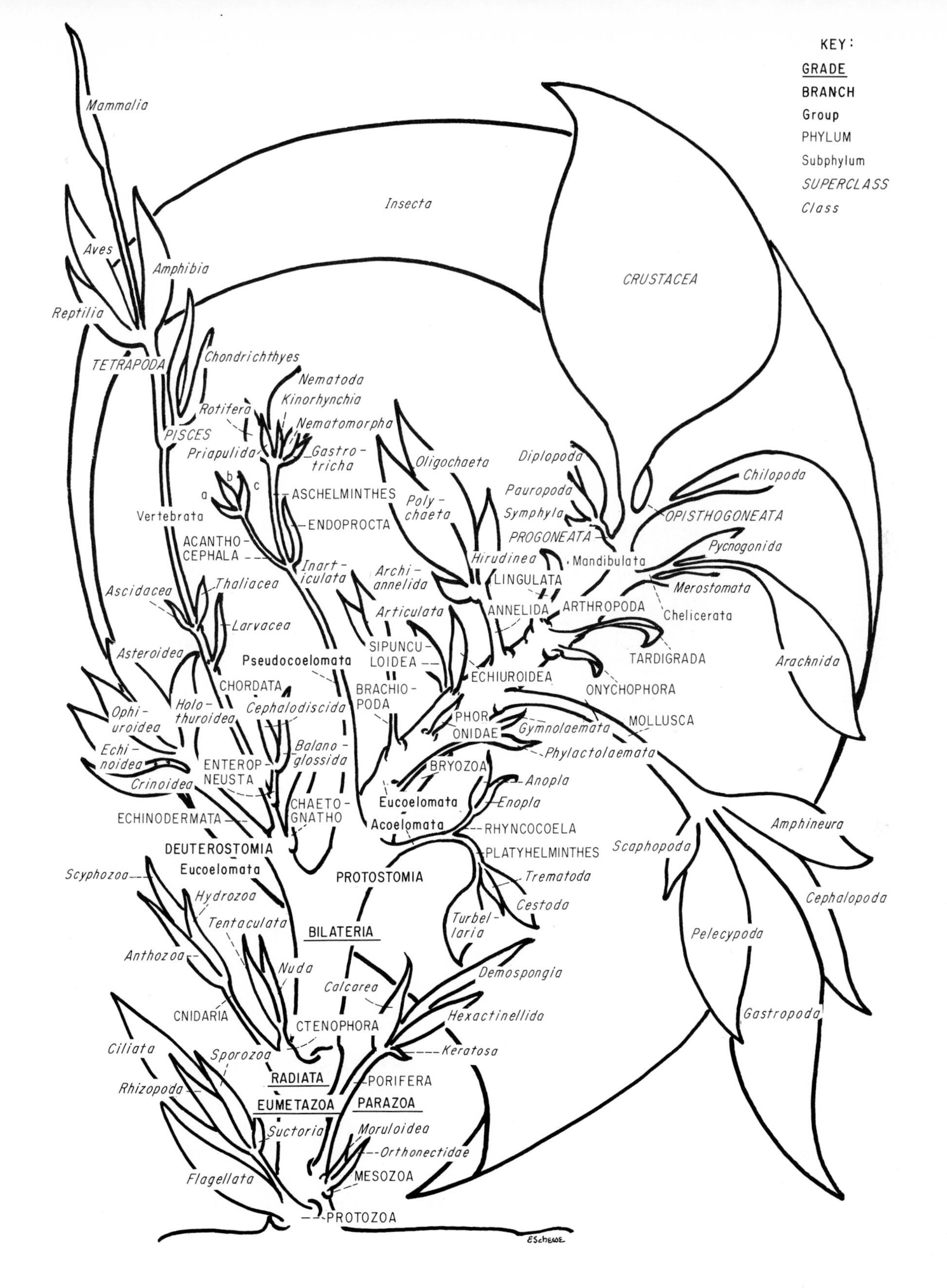
KEY:
GRADE
BRANCH
Group
PHYLUM
Subphylum
SUPERCLASS
Class
Mammalia
Insecta
Aves
Amphibia
CRUSTACEA
Reptilia
TETRAPODA
Chondrichthyes
Nematoda
Kinorhynchia
Rotifera
Nematomorpha
PISCES
Priapulida
Gastrotricha
Oligochaeta
Diplopoda
Chilopoda
a
b
c
ASCHELMINTHES
Polychaeta
Pauropoda
Symphyla
OPISTHOGONEATA
Vertebrata
ENDOPROCTA
PROGONEATA
Pycnogonida
ACANTHOCEPHALA
Inarticulata
Hirudinea
Mandibulata
Archiannelida
LINGULATA
Merostomata
Ascidacea
Thaliacea
ARTHROPODA
Chelicerata
Articulata
ANNELIDA
Larvacea
Asteroidea
SIPUNCULOIDEA
TARDIGRADA
Arachnida
Pseudocoelomata
ECHIUROIDEA
CHORDATA
BRACHIOPODA
ONYCHOPHORA
Ophiuroidea
Holothuroidea
Cephalodiscida
PHORONIDAE
MOLLUSCA
Gymnolaemata
Echinoidea
Balanoglossida
Phylactolaemata
ENTEROPNEUSTA
BRYOZOA
Crinoidea
Anopla
Eucoelomata
Enopla
CHAETOGNATHO
ECHINODERMATA
Acoelomata
RHYNCOCOELA
Amphineura
DEUTEROSTOMIA
Scaphopoda
Eucoelomata
PLATYHELMINTHES
Scyphozoa
PROTOSTOMIA
Trematoda
Hydrozoa
Cephalopoda
Cestoda
Tentaculata
BILATERIA
Turbellaria
Pelecypoda
Anthozoa
Nuda
Demospongia
Calcarea
Hexactinellida
CNIDARIA
Gastropoda
CTENOPHORA
Ciliata
Keratosa
Sporozoa
RADIATA
PORIFERA
Rhizopoda
EUMETAZOA
PARAZOA
Suctoria
Moruloidea
Orthonectidae
Flagellata
MESOZOA
PROTOZOA
ESchewe

FIG. V–2 *Carolus Linnaeus (1706–1778), Swedish botanist and natural historian, whose principle of identification by genus and species (binomial nomenclature) was immediately adopted and is in current use. (The Institute of the History of Medicine, Johns Hopkins University.)*

resemblances are often superficial ones to be referred to entirely different causes, and on the other, that an enormous variety of inheritable changes of type are possible. The modern concept of species is a genetic rather than a morphological one. It defines a species as a freely interbreeding population of possibly variant phenotypes which, unless prevented by geographic or physiological barriers, share in a common gene pool. Genes in this pool may mutate, so that within a species there may be individuals, stocks, races, or breeds that are unlike, even strikingly so, in appearance or in ways of life. All are, however, considered representatives of a single species if mating between them occurs naturally and results in the production of fertile offspring. Genic compatibility is thus implicit in this concept, for genic incompatibility would lead to the segregation of one stock from another and so, potentially, to the establishment of a new species. The degree of genic divergence between species, therefore, provides some measure of the closeness of their relationships. Divergence may be expressed in anatomical and physical characters, in habits and ways of life, or in subtle differences in biochemical processes and substances.

It was phenotypic variability within species and within even the offspring of a single individual that puzzled Charles Darwin (1809–1882) and yet formed the cornerstone of his theory of their origin through natural causes (Fig. V–3). He could offer no sound explanation for the fact of individual variability, although he showed in "The Origin of Species by Means of Natural Selection" (1859) that species could have arisen through preferential selection of the variants most able to meet the demands of their existing environments and to adapt to new ones, as conditions on Earth changed or they explored and settled in new areas. In this way, new species became established and, in their turn, produced variant types to become established as new species themselves. The crux lies not in variability per se but in its inheritability. Variations that are incidental in the life of the individual have no significance in species formation, and it is only those that can be transmitted from generation to generation that are selected for perpetuation or for extinction in the delicate balance and constant interplay between organism and environment. Their selection is determined by the sum total of environmental conditions and any variation may have positive survival value in one set of conditions, negative in another. Similarly, in some circumstances it may have neither and be, in effect, neutral.

C. THE BASES OF CLASSIFICATORY SYSTEMS

Taxonomy is based on recognition of homology, or correspondence in structure and in function. Homologies, originally sought in adult structures and established through study of the comparative mor-

FIG. V–3 *Charles Darwin (1809–1882), the English scholar whose theory of the origin of species through a process of natural selection provided the first acceptable explanation of evolution in the living world. (American Museum of Natural History.)*

phology of living and extinct forms, may also be found in juvenile stages and in developmental and physiological processes. Not only comparative anatomy, both gross and detailed, but comparative embryology, biochemistry, physiology, ethology, and ecology offer criteria for phylogenetic affinities and for taxonomic systems based on phylogenetic relationships. In such schemes, certain characters are recognized as fundamental ones, having general value and therefore common to all members of a group and, presumably, represented in the ancestral form from which they have all descended. From these may arise, through mutational changes, derived characters which have particular value to the more specialized descendants.

Since the earliest biological systems that have left recognizable traces in the Earth's crust are already well-defined ones—algae and representatives of all the major invertebrate groups—their origin and ancestry during the "period of uncertainty" can only be surmised. Two great gaps in the history of life on this planet must be filled, as well as a number of lesser ones, to make the story of life a continuous one. One of these large gaps is that between the nonliving matter of which the Earth and its atmosphere were composed and the living matter that must at some time have originated from it. The other is that between the organization of these first living systems and even the simplest unicellular or pluricellular systems that are known. The information needed to fill these is more a matter of intelligent conjecture than of fact, since there is as yet no direct evidence of intermediate types nor any way of experimentally repeating the processes that were actually involved. For duplication of the final products, even when this is possible, is not necessarily a repetition of the events as they may have taken place, however reasonable it may be to assume that the pathways were similar, if not identical.

D. THE ORIGIN OF BIOLOGICAL SYSTEMS

The most crucial event in the origin of biological systems must have been autocatalysis, or self-replication, and the assemblage of atoms into molecular configurations capable of this phenomenon. Theories of how this came about can be neither proved nor disproved; nor does it seem possible to know how the earliest biological systems were organized. Perhaps they conformed more or less to the accepted pattern of a generalized cell or perhaps to that of multicellular systems. If the former hypothesis is correct, then the origin of pluricellular systems becomes a problem for speculation and argument; if the latter, the derivation of single-celled systems from pluricellular ones is the central problem.

Over 50 years ago, the English biologist Clifford Dobell questioned the validity of the concept of a single-celled system, pointing out that the homology is between the protistan and the metazoan or metaphytan individual, and that the antithesis, therefore, is not between a body made up of one cell and one made up of many cells but between one of no cells and one of many cells. He therefore proposed the term "acellular" for the protista, including Protozoa and Protophyta, in contrast to multicellular, designating Metazoa and Metaphyta. This terminology and the concept behind it has been accepted by some biologists, critized and rejected by others. Those who reject it find it difficult, among other things, to reconcile the acceptance of the ovum as a cell and part of a multicellular body with refusal to recognize the fertilized egg as a cell. But, according to Dobell, while an egg that develops parthenogenetically is a fully accredited cell, a zygote is not, although after cleavage its products are cells.

However, whether they are to be regarded as acellular or as unicellular, the Protozoa represent a group of animals distinct from the Metazoa. It is undoubtedly false to consider them simple systems, for their bodies frequently show a high degree of localized differentiation in the various organelles associated with particular functions. Some biologists consider them even more complex than metazoans since they perform, within a single, often minute, protoplasmic mass, all the vital activities that in multicellular systems, large or small, are distributed between many cells. The extent to which they may be considered "primitive" systems is also questionable, since the fossil record does not show that they preceded other forms of animals in the population of Earth.

E. THE ORIGIN OF METAZOA

There are two main possibilities for the origin of Metazoa, if one assumes that the earliest biological systems were single cells. One of these possibilities is through the aggregation of such single and originally independent cells into groups, and their organization into patterns or plans through the division of labor between them. This may have come about through the failure of some of these cells, in one way or another, to continue to meet the conditions of independent existence. Assemblage into colonies, where one member could contribute what another lacked, would, therefore, become a necessity for survival. The other possibility is through the multiplication of organelles within the single unit and its subsequent cellularization, or partition, into a number of more or less discrete units, already to some extent organized and differentiated because of their location in the ancestral organism.

1. Cell aggregation

The first of these possibilities found its most vigorous support in the latter part of the nineteenth century, when much of scientific, and nonscientific, thought was directed toward evolution and phylogenetic relationships. It was substantiated by evidence produced largely by the German biologist Ernst Haeckel (1854–1919), who had made radiolarians, sponges, and medusae objects of intensive study. In following the embryological development of sponges, he recognized that the individual passed through a stage when it was essentially a hollow ball of cells, or blastula, and that later, by a process of invagination, it became converted to a two-layered system, or gastrula. This represented the simplest form of an organism possessing an enteron. These two stages he believed were passed through in the development of all animals and he postulated that the ancestors of all modern metazoan species were minute, diploblastic creatures, whose outer cells bore cilia that propelled them through the upper levels of the warm primeval seas and whose inner cells were concerned with the digestion and absorption of food particles. These were swept into the archenteron, or primitive gut, through a single opening, the blastopore. He called this hypothetical ancestor a gastraea, and his gastraea theory dominated biological thought for many years. Some animals, the coelenterates, a group that at this time included both cnidarians and ctenophores, remained in this two-layered, sac-like condition, but in others a third layer arose by divisions of cells in the inner and outer layers. This third layer was the mesoderm, situated between the outer ectoderm and the inner endoderm. Thus, there originated three primary layers, the germ layers, from which all the tissues and organs of a complex body were derived. They and their derivatives were, therefore, homologous throughout the entire animal kingdom. Like the gastraea theory, the germ layer theory had great influence upon Haeckel's contemporaries and equally upon scientists of later decades.

There is some evidence, both from adult forms and from developmental stages, to support the idea of metazoan evolution from a colony of cells in the form of a hollow ball like a blastula. The Volvocales are aggregations of cells that show some division of labor in that, in *Volvox* and certain species of *Eudorina*, a distinction can be made between somatic and reproductive cells. The mature *Volvox* colony is a fluid-filled ball of flagellated cells, enclosing a limited number of nonflagellated reproductive cells that have been differentiated from surface cells at what may be defined as the posterior pole of the sphere (Fig. II–3). The hypothetical ancestral cell aggregate may have had a similar organization, for it can be argued that the origin of Metazoa from Protophyta was more probable than their origin from Protozoa. The main ground for this argument is that since Protophyta absorb food materials equally on all surfaces, their assemblage into aggregates would present fewer problems than the assemblage of Protozoa with cytostomes and other localized organelles. Electron microscopy has also shown a similarity in photoreceptor structures in Protophyta and vertebrates, which offers at least a parallel between a fact of comparative cytology and an hypothesis of evolution. In many phytoflagellates, the stigma, or eyespot, is closely associated with a flagellum and is derived from a chromoplast either by its transformation *in toto* or by that of a part of it. In vertebrates, the retinal rod cell is divisible into an inner and an outer segment joined by a number of fibrils. These fibrils are identical in number and arrangement with the peripheral fibrils of a cilium and like them are enclosed in a membrane. In the one case, there appears to have been close adjustment of two organelles, chromoplast and flagellum, to one another to form the photosensitive unit; in the other, there appears to have been a differentiation of a primitive motile organelle into a similar structure. Both are manifestations of structural and functional adjustments between a ciliary organelle and a laminated system of vesicles containing carotene pigments.

Moreover, there is ample evidence, both natural and experimental, for the collection of isolated cells into multicellular aggregations. It is a regular event in the life cycle of Acrasiales, the cellular slime molds. When the spores develop into small amebae, they live independently as long as food is abundant. With its depletion, they stream together to form multicellular masses. One of the earliest experiments in which such association was induced was that performed on sponges in 1907 by the American biologist H. V. Wilson. By forcing pieces of *Microciona*, one of the Demospongia, through fine-meshed cloth, he was able to separate the cells from each other. These settled to the bottom of the vessel and began at once to fuse into small spherical conglomerates. When these masses were scattered over the surface of slides, they flattened and joined to form an encrusting mass in which the typical pattern of sponge organization soon became evident. More recently, vertebrate embryonic cells, dissociated by trypsin digestion, have been seen to reassemble on the floor of suitable containers filled with a nutrient medium. Whether their initial clumping is entirely the result of chance collisions or is influenced by their liberation of a mutually attractive substance, as has been found in the case of slime mold aggregations, still remains to be determined. An exudate, probably mucoprotein in nature, is known to form the substratum of aggregates that have reached macroscopic dimensions, and there is accumulating evidence that there is a secretion of sorts even from cells in suspension. Histogenesis may take place in the aggregates and their cells assume the histological features and patterns of organization characteristic of the organs from which they were derived.

There is some embryological evidence, too, for a possible relationship between a phytoflagellate colony and a metazoan at the level of poriferan organization. The development of the simplest modern sponges is peculiar in that inversion takes place at an early stage. The same phenomenon occurs in *Volvox*. In both, the initial cleavages are meridional so that in each case an open circlet of cells is derived from the egg. In *Volvox*, the successive cleavages are also meridional, but in the sponges an equatorial one follows the early meridional ones, separating an upper tier from a lower tier of cells (Fig. V–4). The upper tier is destined to become the epidermis, the lower, the gastrodermis, or choanocyte layer. In this layer, the flagella are directed inward until inversion occurs. At this time, the ball of cells is turned inside out, so that the free surfaces of the cells are reversed and the flagella of those of the future gastrodermis are directed outward at the anterior pole of the swimming larva, or amphiblastula. Later, in gastrulation, the flagellate cells move in so that they are enclosed by the epidermis and line the spongocoele, with their flagella directed toward

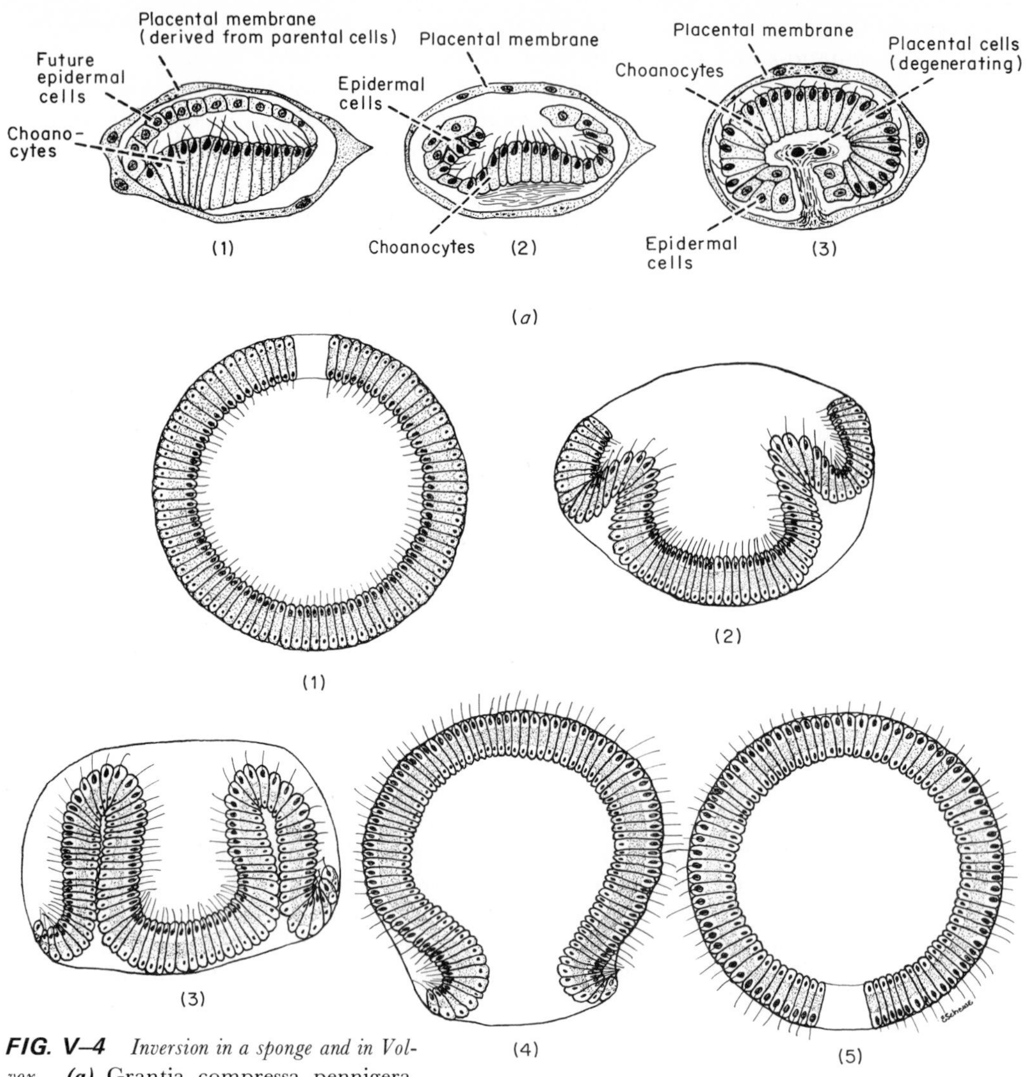

FIG. V–4 *Inversion in a sponge and in Volvox.* ***(a)*** Grantia compressa pennigera. *(From Dubosq and Tuzet.) (1) Blastula before inversion. (2) Beginning of inversion. (3) Inversion completed, with penetration of some of the "placental cells" into the central cavity; these degenerate and may provide nutriment for the embryo.* ***(b)*** Volvox aureus. *(After Zimmermann.) (1) 1,024-cell stage of a parthenogenetically developing egg. The flagella are directed inward. (2, 3, and 4) Stages in inversion. The ball of cells turns inside out through the opening, reversing the polar orientation of the cells. (5) Inversion completed. The flagella are directed outward.*

its cavity. Similarly, in *Volvox*, after a succession of meridional cleavages, the flagella are directed inward toward the fluid-filled cavity of the sphere, but also by inversion their orientation is reversed and they project from what then becomes the outer surface of each cell. Since inversion is not a general phenomenon in the development of sponges but is limited to a few species that are very simple in their organization, it may be argued that its occurrence in them and in *Volvox* is less indicative of phylogenetic relationships between them than of convergence, or the adoption of similar means to meet similar conditions. This principle is amply illustrated by widely divergent animals—as, for example, in the synthesis of hemoglobin and its utilization as a respiratory pigment by some invertebrates and almost all vertebrates, or in the similarity in body shape of earthworms and snakes, in the nephridial systems of annelids and the cephalochordate Branchiostoma, and in the visual organs of cephalopods and vertebrates. The occurrence of a unique embryological event in a phytoflagellate and in one or two species of sponges might equally well be a device to meet a unique situation arising in the early developmental stages of both as a general and fundamental process.

Cnidaria themselves present cogent embryological evidence against the origin of Metazoa from a gastraea as Haeckel conceived it. Gastrulation by invagination is, as Haeckel well knew, the exception rather than the rule among animals. In the majority of cnidarian species, the inner cells arise either from delamination, by tangential divisions of the surface cells of the blastula, or from ingression, the inward movement of some of them. Their typical gastrula is not a fluid-filled sac with a double wall but a stereogastrula, one whose outer wall encloses a continuous mass of cells that later adopt characteristic arrangements and become differentiated into particular types. The larva is an elongated, ciliated mouthless planula, with some concentration of nerve cells at one pole. The planula finally attaches itself to an object in the water; develops tentacles, mouth, and gastral cavity; and sooner or later completes the changes that will convert it to an adult of the class, order, and genus to which it is assigned.

However, some argument may be made for the planula representing an ancestral type, assuming that it changed its habit of life from a planktonic to a creeping one. Under these conditions, as it moved over the sea floor, there would be a positive advantage in developing a flattened body, greater sensitivity at the anterior end than elsewhere, and, ultimately, bilateral symmetry. Two possible lines have been suggested for the progression of such a creeping planuloid to a bilaterally symmetrical metazoan. One of these is through a ctenophore pattern to that of a flatworm; the other, directly to a flatworm pattern. The argument for the former avenue is based upon the existence of two modern genera of Ctenophora, *Coeloplana* and *Ctenoplana*, which, unlike the other members of this phylum, have flattened bodies and crawl upon the sea bottom. In a number of other respects, also, they can be said to resemble polyclad flatworms. Their symmetry is, however, actually biradial rather than bilateral, and their pattern of cleavage, which is biradial, differs from that of the polyclads, which is spiral (Figs. V–5 and V–9). Moreover, polyclads cannot be considered primitive types, for they most evidently have evolved from Acoela, minute, eyeless turbellarians without digestive or excretory systems. *Coeloplana* and *Ctenoplana*, as well as other members of the order Platyctenea, are more generally accepted as aberrant, even degenerate, ctenophores, so representing a terminal twig on the phylogenetic tree rather than the stem itself.

In support of the second hypothesis—that the acoelous flatworms were derived directly from an ancestor of planuloid organization—are homologies between the two both in habit of life and in bodily organization (Fig. V–6). Typical Acoela move along the bottom surfaces of the waters in which they live by means of cilia. There is no cavity within the body, which is filled with parenchymatous tissue. While they have a somewhat greater degree of structural

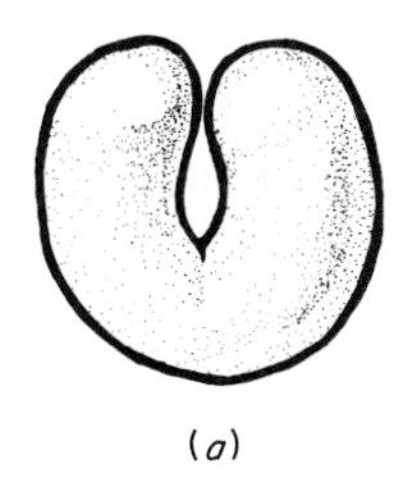

(a)

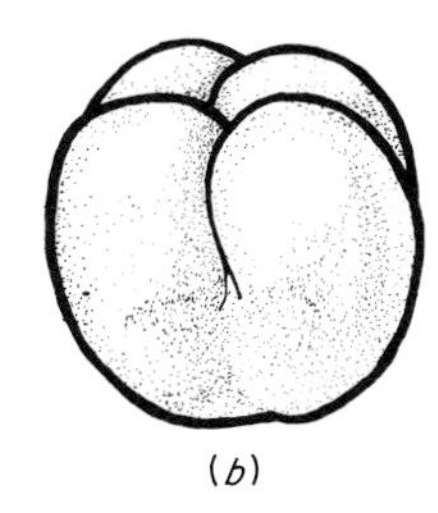

(b)

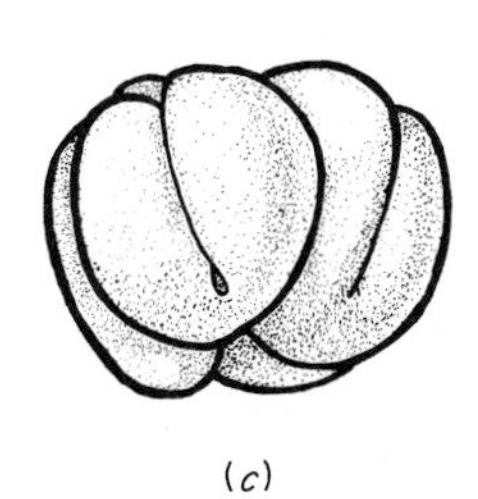

(c)

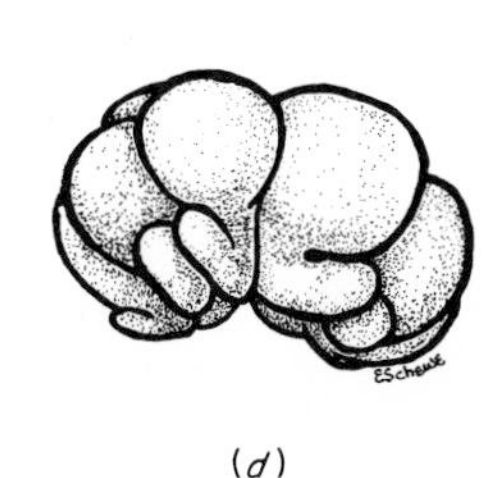

(d)

FIG. V–5 *Radial cleavage in a ctenophore,* Beroe ovata. *(After Zieglen.)* ***(a)*** *First cleavage.* ***(b)*** *Second cleavage.* ***(c)*** *Third cleavage.* ***(d)*** *Fourth cleavage, giving rise to micromeres.*

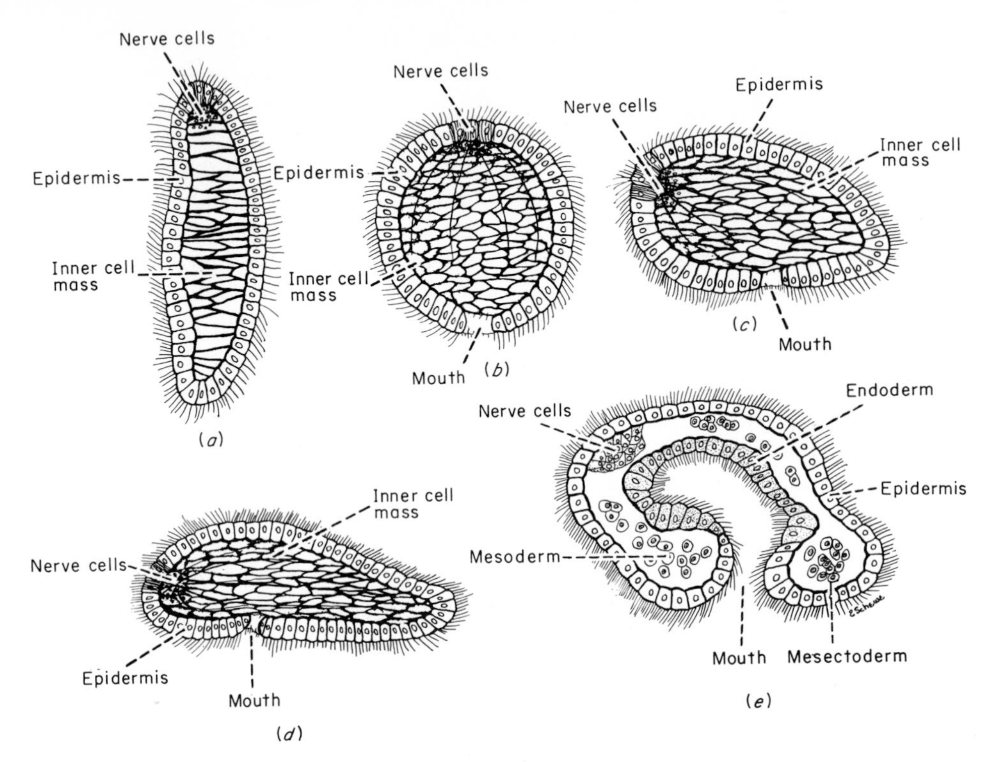

FIG. V–6 *The similarities between a planuloid larva and a flatworm.* ***(a)*** *Planula of* Aeginopsis mediterranea. *(After Korschelt and Heider.)* ***(b, c, and d)*** *Hypothetical stages in the conversion of a planuloid to an acoeloid type. (After Hyman.): b, mouth formed, shortening in oral-aboral axis, nerve cords originating from nerve center; c, body elongating, with anterior shifting of nerve center; d, stage comparable to an acoelous flatworm, with mouth opening directly into the cellular mass.* ***(e)*** *Longitudinal section of the larva of the polyclad* Planocera inquilina. *This might represent a rhabdocoeloid stage, with the mouth opening into a short, blind enteron. (After McBride.)*

organization than that of the planula's, the basic homologies between the two have been considered grounds for postulating their close phylogenetic relationship.

These hypotheses, then, assume the aggregation of cellular units, however they were initially formed, into multicellular aggregates; make radial symmetry the primitive pattern of organization; and place a cnidarian larval form at the base of the phylogenetic tree, as well as of the branch from which modern cnidarian species have arisen. They have, in the past quarter century, been carefully reexamined by a number of biologists, stimulated by the efforts of a Yugoslavian zoologist J. Hädzi to reconstruct the system of animal classification on more general grounds.

2. Cellularization

In this scheme, multicellularity is considered to have resulted as an alternative to cell aggregation from the polymerization of organelles in a single cell, leading to a multinucleate system, probably resembling a ciliate protozoan, and the partitioning of this system into units in which a single nucleus controlled or directed a limited area of cytoplasm. Rhabdocoele turbellarians are placed at the base of the metazoan line and Cnidaria are considered to have arisen from them by regressive evolution. Ctenophores are shown also to be derived from the flatworm stock, but at the polyclad rather than at the rhabdocoele level. The turbellaria, therefore, represent a type of primarily unsegmented eumetazoan, occupying a central position in the phylogenetic tree. Bilateral symmetry, rather than radial, is considered the primitive pattern of organization, and much importance is attached to segmentation. Indeed, it is argued that only four large animal groups should have the status of phyla—Ameria, or unsegmented animals; Polymeria, or animals with many segments; Oligomeria, or animals in which segments are reduced in number or entirely lacking; and Chordonia, or chordate animals in which segmentation has reappeared.

Proof of this interpretation of the origin of Eumetazoa rests upon the demonstration of a closer morphological similarity between the bilateral ciliate and the acoelous flatworm than between the volvocine colony and the planuloid larva. These homologies are found in adult forms rather than developmental stages, and their demonstration in modern protozoans and rhabdocoeles could be taken as evidence for a common ancestor of both.

It can be shown that gymnostomatous ciliates like *Remanella* and *Dileptus* have many things in common with rhabdocoeles (Fig. V–7). Their habitats and sources of food are similar; indeed there appear to be no important differences between them in nutrition, physiology, ethology, or ecology. The transition from ciliates of this type to a multicellular system is conceivable if one adopts the concept of "energids" first stated by the German botanist Julius Sachs (1832–1897). By energid, Sachs meant a metabolic unit, a nucleus, and the volume of surrounding cytoplasm that it could effectively control. Assuming that these energids in a multinucleate ciliate like *Dileptus* became wholly or partially separated from each other by a rearrangement of the membranes that seem to be part of every cell, then a syncytial or truly multicellular organism could appear. The nuclei of these energids, derived by mitosis, would necessarily be isogenic; in the undivided or acellular ciliate all the energids would be physiologically similar, but in the syncytial or multicellular organism they would become physiologically different because of their situation in different environments, such as epidermis, mesenchyme, or gastrodermis. The appearance of cell membranes, or possibly a rearrangement of the endoplasmic reticulum, would thus enforce energid, although not nuclear, heterogeneity and lead to the differentiation of characteristic cell types.

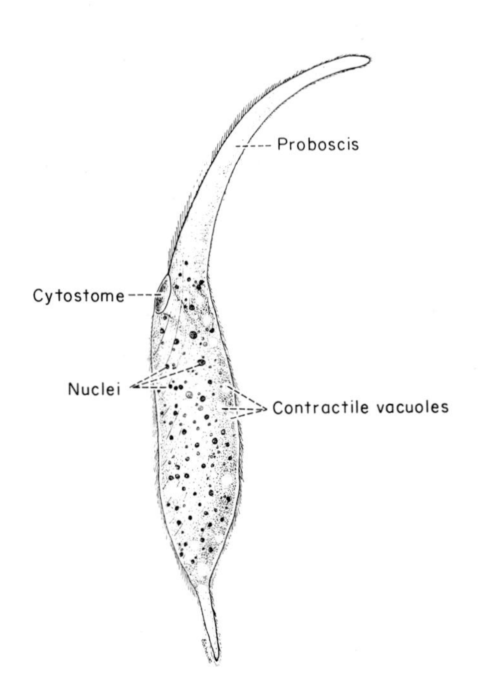

FIG. V–7 *Dileptus sp, a gymnostomatous ciliate.* (*After Kudo.*)

The origin of multiple cell systems through such a process of cellularization and individuation of physiological units has been criticized on evidence from insect embryology. Here the nucleus in the fertilized egg divides many times in a common cytoplasmic mass before cell walls become evident. This parallels the events presumed to have taken place in the protozoan or protistan ancestor of the Metazoa. Insects should then be regarded as among the most primitive of animals, but all evidence points to just the opposite—that they are among the most highly specialized. Cellularization is, however, not the most significant step in the transition from ciliate to rhabdocoele, for the tissues in some animals are syncytial.

Other postulates must also be made to reconcile the rhabdocoele organization with that of a cellularized gymnostome. According to Hädzi, mesoderm is not a new acquisition of Eumetazoa but was already localized as a potential tissue in the subcortical layer of the protozoan, along with the vacuoles and fibrillae within it. The vacuoles provided anlagen for the nephridial system and the fibrillae for the muscles, nerves, and connective tissue. Similarly, the gastric vacuoles served as anlagen for the enteron, their alignment making possible the construction of a digestive tube connected with the mouth. It may be argued against this that nuclei are not found in the cortical and subcortical regions of Protozoa, but appear to be confined to the more fluid endoplasm. Because of this, cellularization of the superficial areas is difficult to imagine. Moreover, the fibrillar neuromotor system of a ciliate like *Dileptus* does not resemble the nerve net of flatworms and, although homology can be shown between the trichocysts of ciliates and the rhabdoids and sagittocysts of many turbellarian and acoelous flatworms, secretory structures are more profuse and highly developed in the latter group of animals. The division of the nuclear apparatus into micronucleus and macronucleus, unique to the ciliates, and their agametic methods of reproduction present more serious problems. The first may be reconciled, perhaps, by the fact that macronuclei do not persist in the conjugation of such ciliates as *Paramecium* but are reconstructed from derivatives of a micronucleus resulting from the fusion of the gamete nuclei. Also, one of the possibly primitive features of *Remanella* is the inability of the macronuclei to divide and their enforced reconstruction, after each division, from micronuclei. But the presence, in the rhabdocoeles, of a fairly complicated reproductive system and a permanent gonopore, and especially, their production of cellular gametes present a real stumbling block to unqualified acceptance of this theory of the origin of Eumetazoa.

There is, of course, no real justification for assuming that Metazoa arose in only one way, any more than there is justification for assuming that the first biological systems were all identical. Since there is no record, and since the evidence that can be reconstructed seems to show that a variety of organic molecules may have been formed under the conditions that made possible the formation of any, it is as reasonable to suppose that several assemblages of these molecules, including self-replicating ones, may have arisen simultaneously, or nearly simultaneously. Each of these may have had a different way of achieving cellular status, or multicellular status. In the absence of definitive proof, arguments for the origin of living systems at any level must remain heuristic ones.

F. PHYLOGENESIS

Phylogenetic relationships between species that exist on Earth today, or that through their fossil remains are known to have existed on it at various periods of its history, are less speculative. There are, and there have been, radially symmetrical animals and bilaterally symmetrical ones, and the division of Eumetazoa into Radiata and Bilateria, as in the scheme represented in Fig. V–1, is a useful and convenient one, with no implication of which was the primitive and which the derived group.

1. Evidence from comparative embryology

On embryological grounds it is possible to divide the Bilateria into two groups, Protostomia and Deuterostomia, and to trace what seem to be two separate lines of ascent, or descent, from early bilateral types. Protostomia are characterized by a pattern of cleavage of the egg, in which the destiny of the cells first formed can be traced with certainty; by the formation of mesoderm from ingrowths of cells early separated from the other blastomeres as mesoblasts or mesoderm pole cells; by the persistence of the blastopore as the mouth opening, or by the formation of a mouth through a stomodaeum in the immediate vicinity of the closed blastopore; by the development of the coelom, in those forms that have one, as a schizocoele; and, if development is indirect, by a larval stage known as a trochophore (Fig. V–8*a*). Deuterostomia are characterized by a pattern of cleavage in which the early blastomeres exhibit a high degree of lability compared to those of Protostomia; by the formation of mesoderm and of the coelom as an enterocoele through outgrowths of the wall of the archenteron; by the persistence of the blastopore as the anus or the formation of a new anus through a proctodaeum in the immediate vicinity of the closed blastopore; and, if development is indirect, by a larva of the dipleurula type (Fig. V–8*b*). Such a distinction implies a dichotomy of the main trunk of the phylogenetic tree at the point where Bilateria were established as animal types, with evolutionary development in the protostomatous branch leading to the origin of those species now included in the division Schizocoela of the Eucoleomata (Bryozoa, Phoronidea, Brachiopoda, Mollusca, Sipunculoidea, Echiurodea, Annelida, Onychophora, Arthropoda, Pentastomida, and Tardigrada) and, in the deuterostomatous branch, to those in the division Enterocoela (Chaetognatha; Pogonophora, Echinodermata, Hemichordata, and Chordata).

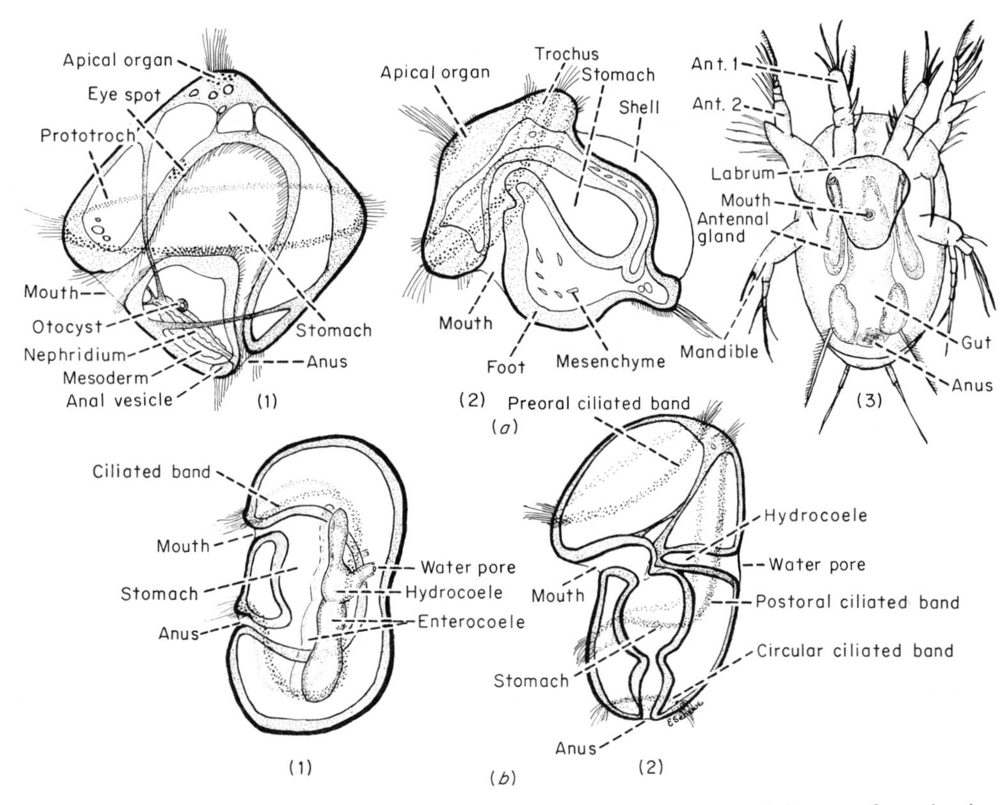

FIG. V–8 *Larval types.* ***(a)*** *Larvae of Protostomia (trochophore type). (1) Full grown, free swimming (4-day) larva of a marine annelid,* Eupomatus. *(After Shearer.) (2) Larva of a gastropod mollusc,* Patella. *(After Meisenheimer.) (3) Nauplius larva of a crustacean arthropod,* Cyclops. *(After Korschelt and Heider.)* ***(b)*** *Larvae of Deuterostomia (dipleurula type). (1) Echinoderm larva, based on the larval stage of a sea urchin. (2) Larva (Tornaria) of an enteropneustan,* Balanoglossus clavigerus. *(After Korschelt and Heider.)*

a. PROTOSTOMIA: CLEAVAGE AND LARVAL TYPES Spiral, or determinate, cleavage is seen typically in the eggs of polyclad flatworms, nemerteans, annelids, and most molluscs. But within each of the last two phyla at least, evolutionary changes have also occurred resulting in divergence in development as well as in adult structure. The superficial cleavage of the cephalopod egg is, for example, wholly unlike the spiral cleavage of the pelecypod or gastropod egg. There is no typical larval stage in cephalopod development; nor do oligochaete annelids pass through a larval period like that of polychaetes.

Typically in spiral cleavage, the first two divisions of the egg are meriodional, resulting in four cells, the macromeres, which are frequently of unequal size. The next cleavage is in the latitudinal plane and is markedly unequal, resulting in a group of four small cells, the first quartette of micromeres (Fig. V–9*b*). It is characteristic of spiral cleavage that the axis of the mitotic spindle of a dividing cell is oriented obliquely to the polar axis of the egg; these micromeres, therefore, do not lie directly over their corresponding macromeres but in the furrows between them. It has become customary to identify each of the macromeres by the capital letters A, B, C, and D, and the micromeres by the correlate small letters a, b, c, and d. Since the angle of the spindle may be either to the right or the left of the polar axis, a phenomenon that in molluscs is

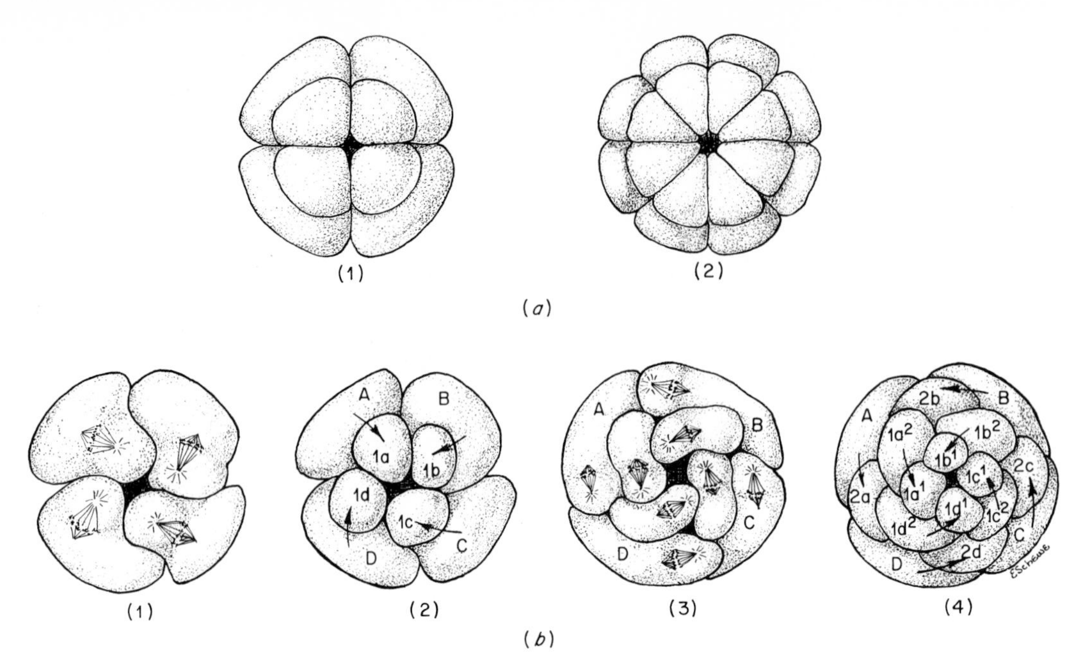

FIG. V–9 *Radial and spiral cleavage. (After Korschelt and Heider.)* ***(a)*** *Radial cleavage. (1) Eight-cell stage, resulting from two meridional and latitudinal cleavages. (2) Sixteen-cell stage.* ***(b)*** *Spiral cleavage. (1) Third cleavage, showing the transition from the four-cell to the eight-cell stage. (2) Eight-cell stage, showing the first quartette of micromeres. The blastomeres are conventionally labeled, with arrows indicating the relationships of micromeres and macromeres. (3) Fourth cleavage, showing the transition from eight-cell to sixteen-cell stage. (4) Sixteen-cell stage, after division of the first quartette of micromeres.*

known to be genetically determined, micromere a may lie between macromeres A and B, or A and D. In the first case, the cleavage is said to be dextrotropic, in the second, levotropic. A dextrotropic cleavage is followed by a levotropic one, and a levotropic by a dextrotropic, in continuously alternating sequence, as the macromeres give off successive groups of micromeres and the micromeres themselves divide. The large cells resulting from cleavage of the macromeres are labelled 1A, 1B, 1C, and 1D in the first division of this kind, 2A, 2B, 2C, and 2D in the second, and so on. The first quartette of micromeres are labeled 1a, 1b, 1c, and 1d, the second 2a, 2b, 2c, and 2d, and their derivatives are identified by the appropriate superscripts as $1a^1$ and $1a^2$; $1b^1$ and $1b^2$, etc., in each case the superscript ^{1} being used to designate the micromeres nearest the animal pole. This standardized identification has great value for it has been possible for embryologists to follow the fate of the blastomeres during the early phases of development and to determine the tissues or structures derived from each. Thus, derivatives of the first, second, and third quartette of micromeres give rise to the ectoderm, the set $1a^2$–$1d^2$, providing the cells that develop tufts of cilia and form the prototroch of the larva. Cells from the second and third quartettes give rise to the mesoderm that differentiates into the musculature of the larva. Derivatives of $3c^2$ and $3d^2$ become, in annelids, the larval nephridia. The macromeres 4A, 4B, 4C, and 4D, together with the derivatives of three of them, 4a, 4b, and 4c, become the endoderm, invaginating at gastrulation. The micromere 4d gives rise to two mesoblasts, which lie in the blastocoele and, by repeated divisions, form the two mesoderm bands that produce the definitive muscles, connective tissue, reproductive and excretory organs, and at least part of the blood vascular system.

Through this pattern of cleavage, the unsegmented egg is converted to a trochopore larva, a developmental stage characteristic of many modern species of marine annelids and molluscs. The absence of a larval stage in the development of freshwater and terrestrial species, which is usual but not invariable, may perhaps be explained by the more hazardous conditions of life in these environments which would reduce the chances of survival of so delicate a creature. In the majority of such species, a larval stage is either bypassed in the course of development or some means of protection evolved for it, often through a temporary parasitism.

The trochopore of the polychaete annelids is the most generalized found among modern species and conforms in many ways to an ideal architectural plan for an ancestor of Bilateria (Fig. V–10). The digestive tract is complete, and the space between it and the covering epithelium, a single layer of ectoderm cells, is the primary body cavity, filled to a considerable extent with mesenchyme and muscle cells of mesenchymatous origin. On either side of the gut there is a band of mesoderm derived from the mesoderm pole cells. The organ systems of the trochophore, other than the digestive, are represented by paired nephridia, each consisting of a tubule terminating internally in one or more solenocytes, and a ganglionic mass beneath a tuft of cilia at the apical pole with longitudinal nerve bands extending posteriorly and, in some species, connected to each other by circular bands. There are sometimes eyespots and statocysts and often a complicated muscular system consisting of bands along the subepidermal nerves and digestive tract and beneath the circular ciliary bands that effect its locomotion. The principal one of these ciliary bands, the prototroch, passes around the equator above the mouth; there may be another, the metatroch, also equatorial but below the mouth and a small third one, the paratroch, surrounding the anus. Feeding as well as locomotion is through ciliary action, for particulate matter is swept into the mouth and along the gut by the beat of cilia on the cells lining the oral cavity and the enteron.

This pattern of organization is essentially similar to that of a pseudocoelomate invertebrate and, indeed, its resemblance to that of a modern rotifer, *Trochosphaera*, has been suggested as evidence of close phylogenetic relationship between a hypothetical ancestor of the Protostomia and the rotifers and other pseudo-

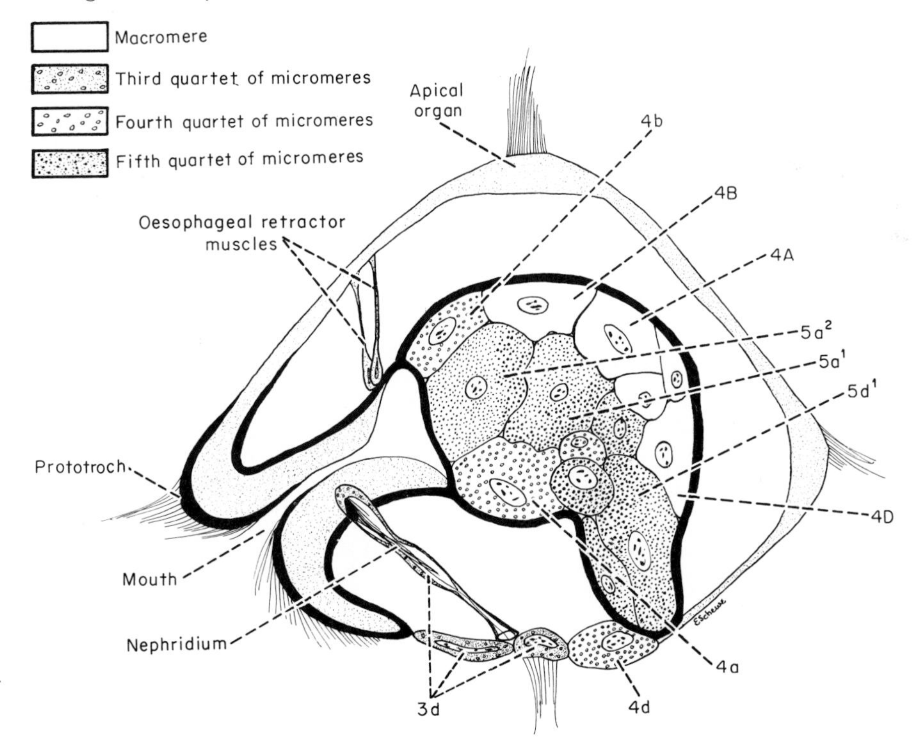

FIG. V–10 *Trochophore larva of* Polygordius. (*After Woltereck.*)

coelomate groups. The Austrian biologist Berthold Hatschek, toward the middle of the nineteenth century, advanced the idea that the trochophore represents the larval stage of a remote and vanished ancestor of all Protostomia. This he called a trochozoon, which, if it could be successfully related to acoelomate and pseudocoelomate groups, might be considered a common ancestor for all these Metazoa. But the rotifer *Trochosphaera* has been shown to be an aberrant rather than a primitive type of rotifer, so that the connection between it and the trochophore, or trochozoon, is by no means a direct, or even a clear, one. On Hatschek's assumption, too, it is difficult to account for existing acoelomate types, unless they are to be regarded as degenerate forms of higher Bilateria.

(1) Annelidan-arthropod relationships The arthropods are generally believed to have emerged from the annelid line as the most highly specialized group of Protostomia. There are anatomical similarities between the adults of Annelida and Arthropoda, such as a segmented body, in which the anterior segments are formed first and are therefore the oldest; a segmentally arranged nervous system, with one preoral, or dorsal, ganglion connected by circumoesophageal commissures to a ventral ganglionic chain; a complete digestive tract; and a coelom. These similarities offer the main bases for acceptance of the belief that the two groups have close phylogenetic relationships, at least to the extent of sharing a common ancestor. There is, however, little or no resemblance in cleavage patterns or larval stages between arthropods and annelids or molluscs. While cleavage in some Crustacea—branchiopods, copepods, and cirripedes—is determinate, the cell lineage in the forms in which it has been followed is quite different from that of any annelid or mollusc (Fig. V–11). The resemblance of the nauplius, the larval form of the most generalized Crustacea, to the trochophore is by no means striking, yet the nauplius is believed to be fundamentally of trochophore pattern to which arthropod features, such as chitinization of the epidermis, jointed appendages, and an overhanging upper lip, or labrum, have been precociously added to make it an "arthropodized trochophore." Internally, the digestive system of the nauplius does resemble that of the trochophore; however, its excretory organs are not nephridia but two pairs of coelomic sacs in which uric acid is deposited.

(2) Evolution of arthropods Attempts to establish phylogenetic relationships between the arthropods and other invertebrate phyla, and between the classes of the phylum itself, have met with great difficulties. The greatest of these arise from the absence, at present, of any discovered fossils that might represent connecting links or a common prototype and from the tremendous diversity of types within the phylum. This diversity has led to doubts as to whether it is really a natural group or an artificial assemblage of a number of dissimilar and unrelated species. Indeed, in most taxonomic schemes, the phylum has been subdivided into two subphyla: the Chelicerata, including those species whose anterior

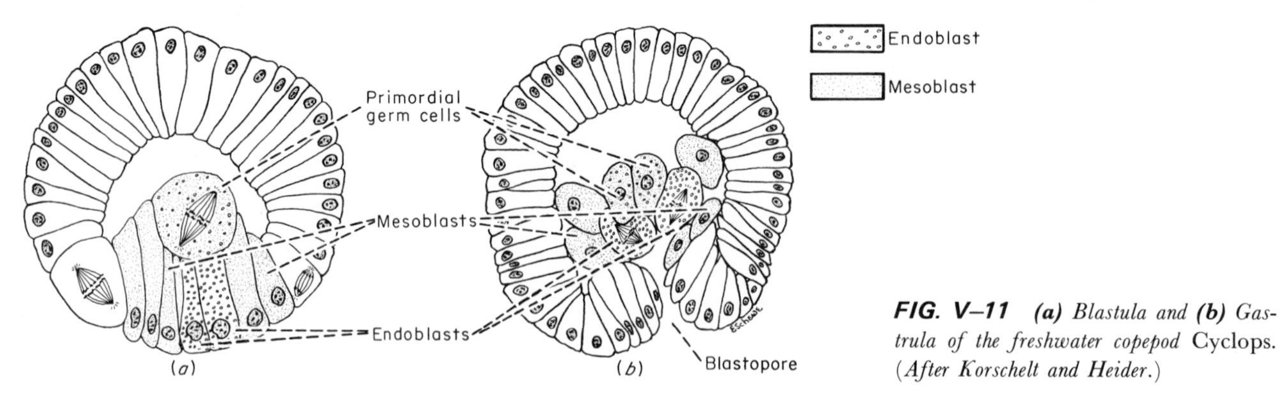

FIG. V–11 ***(a)*** *Blastula and* ***(b)*** *Gastrula of the freshwater copepod* Cyclops. (*After Korschelt and Heider.*)

appendages terminate in claws, or chelae; and the Mandibulata, including those species in which one pair of mouthparts is a jaw, or mandible, the two members of the pair moving laterally against each other to cut, chew, or grind food. This division seems a happier one than another: that into Branchiata and Tracheata, which makes the distinction on the basis of respiratory mechanisms and so, fundamentally, on the basis of habitats. Apparently, the Chelicerata, including the eurypterids, known now only as fossils, and the xiphosurans, pycnogonids, and arachnids began their emergence in the Ordovician period, as one of the great lines of arthropod evolution without evident relation to the myriapod-insect line of the Mandibulata, whose emergence, from evidence provided by fossils of centipede-like animals found in Upper Silurian and Devonian deposits, was somewhat later. The eurypterid-xiphosuran-arachnid assemblage does, however, show some affinity with the Crustacean line, which was already well differentiated in the Cambrian period. These resemblances are general ones, mainly branchiate respiration, compound eyes, and a large digestive gland in the cephalothorax. While phylogenetic relationships, within the crustacean and myriapod-insect lines at least, have been to some extent clarified, those between the major groups remain entirely speculative.

Speculation on the part of paleontologists and systematic zoologists has led to monophyletic, diphyletic, and polyphyletic theories of the origin of arthropods. The premise of the first is that all members of both Chelicerata and Mandibulata evolved from a single ancestral stock. One such theory presumes an ancestor that was essentially a lobopod annelid, in general resembling the onychophorans, which are believed also to be derived from this stock but to have separated from it before the definition of any truly arthropod characters. When these characters were developed, there is assumed a proto-arthropodan ancestor from which in pre-Cambrian times there arose Prototrilobita and Protomandibulata. The trilobite-chelicerate line developed from the former, the mandibulate from the latter. The Mandibulata then underwent two courses of evolutionary change: one, in the sea and in freshwater, to produce Crustacea; the other, on land, to produce Protomyriapods from which, later, myriapods and insects evolved. Another monophyletic theory considers trilobites the primordial arthropod group from which all classes have separately arisen. Trilobites, of which several thousand species have been described, were dominant arthropods of the early Palaeozoic seas, flourishing in the Cambrian and Silurian periods and becoming extinct in the Permian (Fig. I–1*b*). They are the oldest known animals with arthropod characters which are already well defined in even the earliest fossils. There is, therefore, a great gap between any known trilobite and an ancestor that might relate it to the annelids. The trilobite body, oval in form and dorso-ventrally flattened, is divided transversely into head, thorax, and pygidium, all of which bear paired and similar ambulatory appendages. This lack of differentiation of the appendages is the most generalized feature of the trilobites, although they can hardly be considered primitive arthropods. Rather, they are ones that had already reached a considerable degree of specialization for life in shallow or in deep waters. The head bears a pair of antennae and four pairs of appendages, which, like those of the thorax and pygidium, have been considered biramous and so prototypes of biramous crustacean appendages. But on more precise evidence, these have been shown to be actually uniramous (see Chapter VI). The pygidium, as well as the thorax, is clearly segmented in the earliest trilobites, but in the late ones the segments of the pygidium are united. The name trilobite is derived not from these three transverse divisions but from an apparent longitudinal division of the body into three lobes: a median axis, or rachis, and two lateral pleurae.

A major difficulty in accepting the trilobites as ancestral arthropods arises from the fact that they are already specialized forms, not generalized as the ideal ancestor should be if it is to give rise to a variety of species. Another more concrete difficulty concerns the evolution of winged insects. According to the trilobite theory, wings developed as expansions of the pleura, and it would therefore follow that winged insects preceded wingless ones. This is, however, contrary to the fossil record, for several specimens of what may be a primitive insect, the collembolan *Rhyniella praecursor*, have recently been found in mid-Devonian deposits. These animals, if they may legitimately be considered insects or in the insect line, antedate the earliest fossils of winged insects found in the Carboniferous period. Apterygote insects can hardly, then, be degraded pterygotes but rather their predecessors or precursors.

The obvious disparities between chelicerate and mandibulate arthropods have led to the postulation of their diphyletic origin and the assumption of a different ancestral type for each of these lines, separately evolved from the annelids. Moreover, the difficulties related to the inclusion of Crustacea in the same line of descent as Myriapoda and Insecta have seemed insurmountable to some systematists, so that a polyphyletic origin, from a number of ancestral types, has been proposed. These hypothetical ancestral forms are believed at some time to have branched off from a stock that also produced the annelids.

Whatever may have been its origin, the arthropod phylum in its evolution shows many instances of convergence. Probably the most clear-cut case of this is in the development of the Malpighian tubules which have arisen independently in the myriapod-insect line and in the arachnid line to meet the needs of nontoxic or dry renal excretion. Tracheae, as respiratory areas, must also have arisen at least twice in the same two lines, even more often if one accepts the trilobite theory of arthropod origin. On the other hand, on this theory, compound eyes become an ancestral character, but on polyphyletic theories, they must have evolved a number of times.

Phylogenetic relationships within the crustacean and the myriapod-insect assemblages have been traced to some extent, although all of them are by no means clear nor of general acceptance. It is generally agreed that the crustaceans are monophyletic in origin, although they show the greatest diversity of all the arthropod classes. The features that distinguish them are, primarily, the division of the body into head, thorax, and abdomen; the possession of a carapace, or fold of the integument, originating from the maxillary segment of the head and enveloping the body to greater or lesser extent; two pairs of antennae and other paired appendages that are primitively biramous but that have undergone great modification by specialization in the more evolved members of the class; and paired excretory organs that are coelomic in origin and found only in the antennal and maxillary segments of the head. Although all known crustaceans have undergone some degree of modification of these features, particularly in relation to feeding and locomotor mechanisms, fairly generalized forms are found among the Cephalocarida, Branchiopoda, Phyllocarida, Syncarida, and Mysidacea. All of these are marine, with more or less uniform segmentation of the trunk and biramous appendages that are nearly all alike. *Hutchinsoniella*, a recently discovered genus living in the soft mud of subtidal zones, appears to be a very primitive crustacean, in many respects resembling the fossil *Lepidocaris* found in Devonian deposits. It is generally accepted as being related to the forms that gave rise to modern branchiopods (Fig. V–12). *Hutchinsoniella* is about 3 mm in length, with a small carapace covering only the first thoracic segment. Its thoracic swimming legs are considered even closer to the primitive arthropod leg than the posterior biramous appendages of *Lepidocaris*. In other ways, it resembles branchiopods and copepods. Believed by some systematists to represent a new order of Branchiopoda, by others it is put into a new subclass, that of the Cephalocarida, and considered related to the ancestral stock from which Copepoda, Branchiura, and possibly Malacostraca originated. Another recent discovery is *Derocheilocaris*, assigned to the subclass Mystacocarida. *Derocheilocaris* is about 1 mm in length and swims like a nauplius by means of its antennae; it has only rudimentary appendages on the four free segments posterior to the head and none on the remaining six. It is questionable whether this is truly a primitive type, which may have arisen by neoteny from the nauplius of some crustacean ancestor, or whether it is a degenerate form adapted to life in the intertidal zone. With the discovery of more living specimens such as these, a more complete knowledge of the evolutionary history of Crustacea may be anticipated.

There is exceedingly scanty evidence of the early origin of myriapods and insects, and, again, the great differences between the species included in this assemblage make it difficult to confirm relationships between them. Indeed, the question may well be raised as to whether such relationships have actually existed. The general resemblance of the myriapod to an onychophoran like *Peripatus* suggests that they may have

FIG. V–12 Hutchinsoniella macracantha (*magnified 17 times*). (*Photograph from a model made by Thomas Bowman on exhibit in the U.S. National Museum; courtesy of the Smithsonian Institution and Dr. Waldo Schmidt.*)

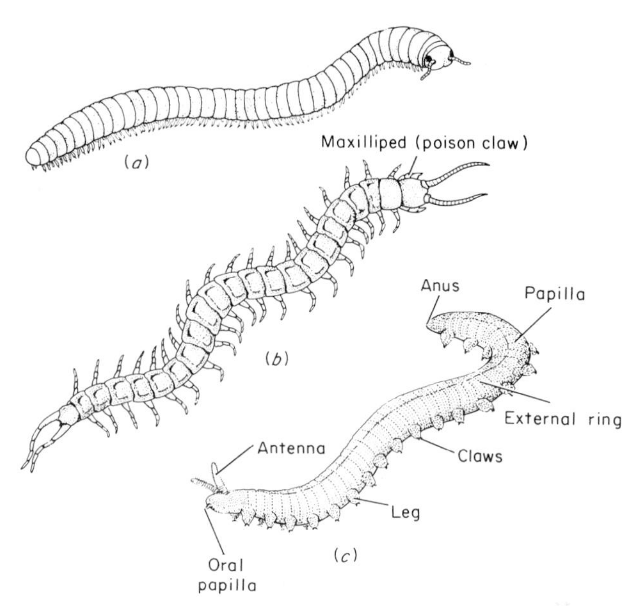

FIG. V–13 *Peripatus and diplopods.* ***(a)*** *The millipede* Julus. ***(b)*** *Centipede.* ***(c)*** Peripatus, *an onychophoran.* (*From Gardiner and Flemister.*)

arisen from an ancestor of this kind (Fig. V–13). The ancestry of *Peripatus* has been traced to mid-Cambrian forms, but the earliest fossil records of myriapods date only from the Upper Silurian period, and there seem at present to be no traces of antecedents. The fact that there is no tagmosis of the myriapod trunk, which is made up of a varying number of freely articulating segments bearing similar legs, is suggestive of a primitive organization, but the complexity of the head tagma denies this. The symphylans are considered the group most closely related to the insects and as representatives of the branch of myriapods from which insects may also have sprung. Their closest link is with the entognathous apterygotes, small wingless forms in which mandibles and maxillae are sunk in pockets formed by the union of the lower lip, or labium, with the lateral walls of the cranium. It is questioned whether these animals are really insects; the exclusion of one order of them, the Collembola, has been accepted by a number of entomologists, and the inclusion of another, the Protura, is considered doubtful. The Entotrophi (Diplura), therefore, remain as the accepted order of Insecta that bears the closest relationship to the myriapods as well as to the thysanurans. Although these are also apterygote, there is abundant evidence indicating that they are closer to the winged insects than to the entognathous forms.

The insect head tagma has been fixed since at least Devonian time. Symphyla have also an essentially insect head and the fact that embryos of thysanurans, orthopterans, and neuropterans have abdominal appendages gives further support to the symphylan-insectan relationship, implying the origin of hexopodous insects from polypodous ancestors. One of the still unresolved problems of insect evolution is the origin of their unique wings. This is discussed in Section Four, pages 205–207.

Phylogenetic relationships of arachnids are even less clear. Fossil scorpions have been found in Upper Silurian deposits, in association with eurypterids, yet marked differences between the two forms indicate that scorpions were already specialized by this time and make any direct connection between them and eurypterids doubtful. Similarity between the two may well be an expression of convergence. All the higher orders of arachnids seem to have appeared abruptly in the Devonian or Carboniferous periods. Nonetheless, the apparent absence of fossils in earlier deposits is not proof that arachnids of some kind, together with the evolving insects, did not populate the land. An unknown chelicerate ancestor has been postulated, from which the aquatic merostomes and pycnogonids, as well as the terrestrial arachnids, may have arisen. But any real evidence of the origin of Chelicerata, as well as of their evolution and of the relationships of merostomes, pycnogonids, and arachnids, is lacking. Since pycnogonids develop through a larval stage, called a protonymphon, which, like the nauplius of crustaceans, has three pairs of appendages, a common arthropod ancestor might be supposed. But such an assumption is not justifiable, for the appendages of a protonymphon are quite different from those of a nauplius. Difficulties arise also in attempts to establish relationships between pycnogonids and chelicerates in general, especially with respect to the number and location of the gonopores. Typically, both male and female chelicerates have but one pair of genital openings, but in pycnogonids, on the other hand, the females and, in most species, the males have gonopores on all, or at least several pairs of, appendages.

b. DEUTEROSTOMIA: CLEAVAGE AND LARVAL TYPES Cleavage in the eggs of deuterostomatous invertebrates is never spiral. In the echinoderms it results in the formation of a ciliated, hollow blastula, which moves freely through the water during its gastrulation and transition to a larva of the dipleurula type. The dipleurula is a small, bilaterally symmetrical animal with a complete digestive tract, two lateral coelomic pouches bounded by mesothelium that have arisen as evaginations of the archenteron, and one continuous band of strongly ciliated cells that passes ventrally above and below the mouth and laterally along the sides of the body [Fig. V–8*b* (1)]. In the subsequent development of members of the different classes, this generalized larval plan is converted to specialized ones, as seen in the bipinnaria and brachiolaria of asteroids, the plutei of ophiuroids and echinoids, and the auricularia of holothurians. The bilaterality of the echinoderm dipleurula is lost at different stages of embryology with the appearance of the hydrocoele on the left side of the body, and the future pentamerous plan of the adult is foreshadowed by the outgrowth of five blunt tubules from it to become the radial canals. While adult echinoderms exhibit a modified radial symmetry, restricted to five primary axes, study of their developmental stages shows that such symmetry has been secondarily acquired. Their adult symmetry has been superimposed upon an earlier bilaterality, indicating that the echinoderms belong with the Bilateria not the Radiata.

Similarities between young echinoderm larvae and the tornaria larva of indirectly developing Enteropneusta suggest a phylogenetic relationship between these two phyla, and an early separation of the echinoderm-chordate line from that of the molluscs, annelids, and arthropods. One similarity is in the location of the ciliated bands, the elevated ridges of cells whose beating cilia propel the larva through the water and sweep a current of food particles into its mouth. It has been suggested that these bands formed a point of departure for the embryological device by which the nerve cord in vertebrates is formed, the elevation and closing over of neural ridges. A second similarity between the two larval forms is in the origin and location of their coelomic cavities. A third is the fact that, in each, the hydrocoele is temporarily connected with the exterior by a pore on the dorsal surface.

Other evidence for placing the echinoderms in the chordate line is offered by the means of formation of their skeletal plates. These are products of mesoderm cells, as are the cartilage and bone of vertebrates. In this, the echinoderms differ from all other eumetazoan nonchordates, in which skeleton formation is accomplished by ectoderm cells. The chondroid tissue of molluscs may be an exception to this generalization.

c. VARIABILITY AND GENIC CHANGE All ideas of evolution since Darwin's accept adaptation as the primary factor in determining the survival of a species. But while Darwin recognized the fact of variability among individuals of the same species, and even of the same parentage, he did not know its cause; although he made it the cornerstone of his theory of the origin of species by natural selection. It was not until 1900 and the recognition of the significance of the "laws of inheritance" deduced by Gregor Mendel (1822–1884) from his studies on the hybridization of peas that variation could be given an acceptable scientific explanation. That it can result from changes, or mutations, in the nature of the material transmitted from generation to generation through the germ cells has been conclusively demonstrated again and again. In 1926 an American biologist and Nobel Laureate Thomas Hunt Morgan (1866–1945) formulated his concept of this material in the theory of the gene, which states that the characters of an individual are referable to paired elements (genes) in the germinal material; that these separate from each other at the maturation of the germ cells, so that only one member of each pair is represented in any one gamete; further, that the genes are bound together in orderly sequence in linkage groups; and that the frequency of crossing-over, or exchange of homologous segments between homologous chromatids, furnishes evidence for the linear order of the genes in each linkage group. The association of linkage groups, and hence of particular genes, with particular chromosomes is implicit in the theory as Morgan stated it. In fact, this condition had already been demonstrated when the theory was formulated.

Later revisions and expansions of the theory have had more to do with the nature of the gene itself and its operations than with the basic concept of paired, but not necessarily duplicate, sets of inheritable material present in the diploid individual, and with the mechanisms of their transmission from cell to cell and from generation to generation. Although it does not seem possible yet to define a gene in physico-chemical terms and to determine just what area and how much area it may occupy upon a chromosome, the term persists as a useful one, operationally at least. Genic change, or mutation, may occur at any time, or in any cell of the organism, but it is only those changes that occur in reproductive cells that are inheritable. These changes lead to the variability among individuals upon which natural selection may operate. They may affect conditions in the developmental or juvenile stages of an organism, or their effect may be apparent only in the adult phase. Mutations may arise through alterations in the gene itself, or through alterations in the genic system as a whole. Gene, or point, mutations occur through changes in the chemical composition of a base in the DNA molecule or in the sequence of its nucleotides. Since DNA carries the codes that determine the sequence of amino acids in proteins, changes in it will be reflected in protein synthesis and, since all enzymes are proteins, in all probability in the biochemical capacities of the cells of an organism. Changes in the genotype, or entire genic system, arise through chromosomal aberrations, such as loss or addition of genes or shifts in their position in relation to each other.

d. ONTOGENY AND PHYLOGENY It has been shown that interactions in an altered genotype may change the time relations in the operations of the genes so that manifestations of their activity appear earlier or later in the course of development than in an unaltered one. The term heterochrony has been proposed for these shifts, or displacements, along the time scale of developmental processes and its importance in the evolution of existing species stressed in the past 30 years by a number of biologists in different countries. This idea has arisen from analyses of the relations between phylogeny and ontogeny which Haeckel made the basis for his "biogenetical principle," or recapitulation theory. According to this, "the development of the individual is an abstract of the history of the genus." In its ontogeny every individual repeats the phylogeny of the race to which it belongs. This idea was most vigorously espoused by a contemporary of Haeckel's, a German doctor and biologist, Fritz Müller (1821–1897), who produced evidence from crustacean embryology to support it. He showed that the prawn *Penaeus* hatches as a nauplius, the larval form common to all the lower groups of Crustacea and found in isolated cases among higher ones. After its first molt, the rudiments of the forked tail and of four thoracic appendages are evident; this is the metanauplius stage. This is followed by the protozoea stage, when the four thoracic appendages become functional, and the remaining thoracic appendages and the anterior six abdominal segments, without appendages, appear. The single eye is still present, but there are also rudiments of the two compound eyes. The next larval stage is the zoea, in which the cephalothorax is clearly evident and all the thoracic appendages have the primitive biramous form. This name was given the little animal when it was first discovered in the plankton without knowledge of its position in a crustacean's ontogeny. The zoea molts into the mysis, with 13 pairs of cephalothoracic appendages, of which the maxillipeds and thoracic feet resemble those of the adult schizopod *Mysis*. A final, or critical, molt yields the adult, with 19 pairs of appendages of the typical crustacean form and pattern. According to Müller, all Crustacea are descended from a small, unsegmented species of animal with three pairs of legs, and the development of the prawn shows the stages through which this animal passed in evolving to the various crustacean groups. There is a corresponding stage in the development of every crustacean egg and a modification of the larval type in those that reproduce indirectly, which is imposed by, and appropriate to, its conditions of life. In the majority of decapod crustaceans, for example, the first larval stage is the zoea, with a large cephalothoracic shield, two compound eyes, and a median nauplius eye. The six posterior thoracic appendages are rudimentary or nonexistent. However, in the lobster, the embryo hatches as a mysis, and in the river crayfish, *Astacus*, essentially as an adult. Müller concluded, therefore, that the development of the individual is an historical document, an idea which he presented in 1864 in a paper "Für Darwin." Haeckel in particular received it with enthusiasm and incorporated it into his biogenetical principle, which included the gastraea theory as well as that of the primary germ layers.

Although the biogenetic, or recapitulation, theory has profoundly influenced biological thought, its inconsistencies with observed facts made it unacceptable to many biologists as a general law, and evoked confusion and controversy among them. It was given critical appraisal by Haeckel's contemporaries and immediate followers, and the relationships between ontogeny and phylogeny have recently been reexamined by American, European, and English biologists, for it is these relationships that have played a great part in the construction of systems of animal classification since Haeckel. It has been pointed out that ontogeny is the sequential expression of the zygotic powers of the cell. Each animal is the result of an ontogenetic process, so that any change in an adult form must be a result of change in this ontogenetic process. Developmental stages in closely related forms may follow the same pattern up to a certain point, but at some point divergence must occur to make them different as adults. Therefore, only comparable embryological stages of lower forms within any group can be represented in the ontogeny of higher ones, never an adult stage.

Changes that have arisen through mutation in ancestral ontogenies may be expressed in different ways and have different phylogenetic significance. They may, for example, be manifested as characters that appear only in juvenile stages, representing special adaptations to special conditions of embryonic or larval life. Disappearing later, without any marked effect upon the adult stage, they have little phylogenetic significance. For example, the larval forms of closely related species may be very different in appearance and even in habit, while the adults are very similar. The larvae of the mosquitoes *Culex* and *Corethra* (*Chaoborus*) and of the midge *Chironomus* (*Tendipes*) are quite dissimilar, but the imagos are barely distinguishable from each other. Another larval adaptation is the glochidium phase of the freshwater pelecypod *Anodonta*, which follows a typical veliger. The glochidium embeds itself in the gills of a fish and there leads a temporary parasitic existence, protected from predators and transported to new feeding and breeding grounds for its adult life. But as adults, these clams do not differ markedly from members of other genera that have no glochidial phase. On the other hand, mutational changes in an ontogeny may affect both young and adult stages and so have great phylogenetic significance. Torsion in gastropod molluscs is illustrative of this, for torsion is due to a change in the position of attachment, and so the direction of pull, of one of the six larval muscles of the veliger (Fig. II–21). This deviation from the general pattern of molluscan ontogeny separates the gastropods from the other members of the phylum, for it has no counterpart in any of them. Since the effects of torsion are evident in adult gastropods, the process cannot be regarded as entirely a larval adaptation, however useful it may have been to the preservation of the planktonic larva.

Heterochrony is more directly implicated in other kinds of change that may have occurred in ancestral ontogenies and be reflected in the characteristics of their descendants. Shifts in the usual time relations in embryogenesis may lead to acceleration or to retardation in the development of certain parts in respect to others. If, for example, the reproductive organs differentiate and become functional faster than other larval structures and organs, the animal becomes sexually mature, or adult, while still retaining its larval form and habits. This condition is known as neoteny, and is classically illustrated by the axolotl, the larva of a Mexican variety of the salamander, *Ambystoma tigrinum*, which, when sexually mature, still retains its gill slits and external gills. It is through neoteny that Hadži accounts for the origin of ctenophores, showing them to be not radially symmetrical animals but bilateral ones and believing that they have originated from the planktonic larvae of polyclads. The early larval stage of the polyclad *Planocera*, represented in Fig. V–6*e*, might equally well be an initial stage in the transition from flatworm to ctenophore or a more or less terminal one in the transition of planuloid to rhabdocoele through a creeping ctenophore-like ancestor. On this assumption, the ctenophore originated from bilaterally symmetrical flatworms that had reached the polyclad level of organization, and the resemblance between them and Cnidaria, with which they were grouped in the older taxonomic systems, is due to convergence caused by the planktonic mode of existence that both groups have adopted. Neoteny may also explain, at least partly, the evolution of the trilobites.

Retardation in the development of characters may lead to their reduction, so that in the adult stage they are vestigial, or to their complete suppression, so that in the adult they are missing. This might apply to the elimination of the foot in oysters, for up to the time of its attachment the oyster larva is like that of any other pelecypod, with a well-developed foot that is used for locomotion. The sessile adult has no foot. It might also be relevant in relation to the regressive evolution of Cnidaria from flatworms proposed by Hädzi, on the assumption that the original Cnidaria had a protonephridial and a reproductive system that disappeared; that their nervous systems became reduced with loss of special sense organs; and that likewise the mesoderm and muscular system were reduced and the digestive tract simplified. These changes led to the pattern of organization found in modern Anthozoa.

Homologies in early developmental stages have been particularly useful in determining the natural taxonomic position of aberrant species. It is, for example, entirely on embryological grounds that leeches have been recognized as annelids—their cleavage is a typical annelid one—and that cirripedes, passing through characteristic crustacean larval stages, are accepted as arthropods. In both cases, their phylum and class affinities are quite thoroughly obscured in the adults. This is especially true in the case of parasites, for in many cases an adult parasite is so highly modified structurally and functionally that it bears little or no resemblance to any recognized type. Thus, only study of its life history has revealed the crustacean and cirripede affinities of *Sacculina*, an internal parasite of crabs. The young of *Sacculina* hatch as free-living nauplii, yet already revealing their future parasitic habits by the absence of a digestive tract. During four molts, the nauplius becomes converted to a cypris larva, a stage also common to most Crustacea; settles down; and, like other barnacles, attaches by its antennules and metamorphoses to an adult state. But unlike free-living barnacles that find rocks, the piles of wharfs, and similar objects for their sites of attachment, the young *Sacculina* selects the body of a young, or recently molted, crab. Its attachment is accomplished only in darkness and only if the point of attachment happens to be a soft and vulnerable part of the crab's body. If these conditions are fulfilled, metamorphosis is initiated, and the cypris larva is converted to a small, undifferentiated cellular mass in the body of the crab. This mass then migrates to the abdomen, following the course of the intestine posteriorly from the original site of penetration and takes up a final position on the ventral surface of the intestine. There it enlarges, sending out long, ramifying processes through the tissues of the crab, much as a parasitic plant sends haustoria into the tissues of its host. Although reproductive and nervous tissue become differentiated, it appears little more than an amorphous lump between the crab's thorax and abdomen, so large sometimes that the abdomen is extended posteriorly rather than flexed under the thorax in the usual manner. All trace of crustacean affinities are lost in the adult *Sacculina*, and the only ground for considering it a crustacean and a cirripede is the parallelism in its development, up to the cypris stage, with other cirripedes.

On the other hand, embryology has thrown little light upon the systematic affinities of the Pentastomida, except to exclude them from the phylum Annelida. Adults of this group of animals are found only in the lungs and air passages of vertebrates, most frequently of reptiles. Their bodies are elongated and worm-like in form, with a chitinous covering; internally, there is a continuous body cavity, with segmentally distributed muscles and a complete reproductive system characteristic of the different sexes. But there are neither circulatory nor respiratory systems. In their development there are three larval stages. The first, or primary one, is passed within the egg and in the body of an intermediate host, apparently always a vertebrate and, for the genus *Linguatula*, a herbivorous mammal. In the second stage, after hatching, the larva has two pairs of movable appendages ending in chitinous hooks and a perforating apparatus at the anterior end; there is a mouth and a simple sac-like gut without a posterior opening. After two or three molts, the second stage larva reaches the tissues and encysts there, entering its third (quiescent) stage. It grows within the cyst, and molts several times before it emerges as an elongated adult, without appendages, and with rows of minute spines on its surface that give it an appearance of segmentation. It falls, or somehow makes its way, into the peritoneal cavity of the intermediate host. If this is eaten by the final host, the larva either migrates up the oesophagus and into the nasal passages or pierces the gut wall and from the gut enters directly into the lungs. There is nothing in its developmental history to link it with annelids and only slim evidence of an arthropod relationship furnished by the larval stages and successive molts. As yet, this group of parasitic animals remains in an anomalous position in relation to other invertebrates, and in modern classification is put into a phylum by itself, with no implication of relationships with any other group.

Data such as these from comparative embryology show that development in different species may follow parallel courses for longer and shorter periods and so may imply or reveal phylogenetic relationships. But it has also been shown that in no case is the adult form of an ancestral type repeated in the ontogeny of any member of a descendant group. The biogenetic law, as Haeckel conceived it, is therefore not a valid generalization.

2. Evidence from the comparative anatomy of adult forms

The earliest taxonomic schemes were based primarily upon homologies in adult forms. This is to a great extent a valid and reasonable basis. Few would deny that all invertebrates with jointed appendages have definite relationships and belong in one major group, although possibly not a phylum, or that among these only those with three pairs of legs belong in one subcategory while those with four belong in another. But the abundance of evidence that animals of undubitably diverse origin living under similar conditions tend to become superficially alike shows the inadequacy of such evidence alone in the determination of phylogenetic relationships. Thus, a long slender body is frequently associated with a burrowing habit of life, and to the ordinary observer the extended, flexible bodies of the Synaptidae (holothurians) with their crown of branched tentacles are as worm-like as those of the annelids and are justifications for considering them related. However, the internal anatomy and embryological development of *Synapta* shows clearly that it is an echinoderm and most closely allied to the holothurian stock. The cephalochordate *Amphioxus* has nephridia of the type characteristic of polychaete annelids; the muscles of some arthropods and of vertebrates are histologically similar; and hemoglobin as a respiratory pigment is found in some species of flatworms, nemerteans, nematodes, annelids, molluscs, and arthropods, as well as in all vertebrates. Such anatomical and physiological similarities must be considered as resulting from convergence in the evolution of these animals rather than as an indication of direct phylogenetic relationships between worms and cephalochordates or between insects and vertebrates. These characteristics reflect changes that have arisen independently in the different groups and have become fixed characters because of their survival value in each case.

3. Evidence from comparative physiology and biochemistry

Comparative physiology and comparative biochemistry have furnished less data for the taxonomist than have comparative embryology and anatomy. This is not because the data are of less potential value in determining systematic relationships but because there is less known of the physiology of invertebrates in general. Clearly, there are certain basic physiological mechanisms that must be employed by all animals eating and digesting the same kinds of foods and adjusting to the same kinds of habitats by adaptations of their respiratory, excretory, and osmoregulatory processes. All present evidence indicates that the biochemical events taking place in neural conduction and transmission and in muscular contraction are the same in all the invertebrates that have been investigated, and the same in them as in vertebrates. The most critical evidence for systematic relationships might be expected from studies of the proteins of different animals, since these are known to be highly specific. Study of the electrophoretic properties of the blood of a number of crustaceans, arachnids, and molluscs, for example, has shown that there are similarities in the mobility of the blood proteins in closely related forms but distinct differences in those belonging to different genera. Thus, three species of the fiddler crab (*Uca*) appear to have very similar, if not identical, serum proteins, as indicated by a similarity in position and size of the bands in the electrophoretic pattern. Likewise, two species of hermit crab (*Pagurus*) are much alike in their electrophoretic pattern, which is, however, unlike that obtained from *Uca*. Blood from a specimen of the green crab (*Carcinides*) collected off the coast of Scotland gives a pattern identical with that of specimens of the same species collected off the Massachusetts coast of the United States. On the other hand, the pattern produced by the blood of the American lobster (*Homarus americanus*) proved to be almost identical with that of the mole shrimp (*Emerita talpoida*). The fact that these two crustaceans are placed not only in different genera but in different families, Homaridea and Hippidae, respectively, suggests either caution in interpreting the results of this technique in the determination of systematic relationships or else a need for revising existing taxonomic schemes on the basis of information derived from it.

It was known before 1900 that animal proteins can act as antigens inducing the formation of antibodies, or substances capable of combining or reacting with the homologous antigen or a related compound. Such combination or reaction is recognized by the agglutination or precipitation of the homologous substance. In this way, chemical affinity or divergence between the two substances may be determined, and very slight differences in the configuration of protein molecules are detected by such a test. This method has been employed to test for the nature of proteins in the blood or body fluids of some invertebrates. In many cases, data from such systematic serological studies is conflicting, possibly because of lack of uniformity in technical procedures. At best, it is very incomplete. Yet in molluscs and in moths correlations have been shown between precipitin reactions and the systematic position of species within these groups that have already been established on morphological grounds. Such tests have also confirmed the phylogenetic relationships between molluscs and annelids, for which embryology furnished the first critical evidence. It seems likely that systematic serology, as one aspect of comparative biochemistry, an area of biology that is only beginning to be explored, may well clear up some of the puzzles in present taxonomic schemes, possibly revealing existing errors as well as confirming existing notions.

G. SCHEME OF CLASSIFICATION

The following abbreviated scheme of classification is intended to provide a frame of reference for the discussion of invertebrate animals. It includes only the nonchordates. The simpler animals usually included in the phylum Chordata, such as the acorn worms (subphylum Hemichordata), the tunicates (subphylum Urochordata), and the lancelets (subphylum Cephalochordata) are therefore omitted. Also, it follows the more or less traditional pattern in giving Protozoa the status of a nonchordate phylum. There are objections to this, principally because so heterogeneous a collection of single-celled organisms and cell aggregates can hardly be considered a natural group, or even a single one, and because, if such a group—the phylum Protozoa—is to include all the one-celled organisms and simple cell aggregates, it embraces certain ones whose characters are as plant-like as animal-like. Obvious examples of these are *Euglena, Chlamydomonas,* and *Volvox.* To overcome this difficulty, creation of two new kingdoms has been proposed: Protista for "protozoans" that are primarily animal-like, and Monera for those primarily plant-like. But this is not entirely satisfactory either, for at these levels of biological organization, particularly, divisions and distinctions become more arbitrary than logical. Where for instance do the slime molds (Acrasiales) belong, with a motile ameboid phase and reproduction typical of the true fungi? Such problems occur at all levels where fine distinctions are drawn. Moreover, as new species are discovered, genera become so large that they must be subdivided and new orders and new classes created. For taxonomy is far from being a static branch of biology. Taxonomic schemes are constantly under revision as the information upon which they are based expands, and what is accepted today may be discarded tomorrow. The following scheme has been compiled from a variety of sources and is intended as a help in placing the invertebrates mentioned in the text in reasonable relationship with others of their kind.

Taxonomy

A BRIEF SYNOPSIS OF CLASSIFICATION OF NONCHORDATE PHYLA (FROM VARIOUS SOURCES)

The figures given for the number of identified species (fossil and recent or recent only) in each phylum are approximate. Brackets indicate alternative names. Representative genera in each group are only those to which reference is made in the text. In some groups, the subdivisions have been given only as far as classes or subclasses; in others, they have been carried through orders. Groups of uncertain systematic position are listed at the end. For more detailed classification and descriptions, reference should be made to the following:

Grassé, P-P.:	"Traité de Zoologie,"	1952—.
Hyman, L. H.:	"The Invertebrates,"	1940—.
Kaestner, A.:	"Invertebrate Zoology,"	1964—.

PHYLUM PROTOZOA

Acellular, i.e., bodies not divided into cells; single, or in colonial aggregates. *c.* 30,000 species.

CLASS MASTIGOPHORA (Greek *mastix,* "whip," and *phorein,* "to bear"). [FLAGELLATA (Latin *flagellum,* "whip").] Locomotion by flagella (one or several at some stage of the life cycle).

Subclass Phytomastigina [Phytoflagellata] (Greek *phyton,* "plant"). Species with chromatophores and those closely resembling such species. *Volvox, Platymonas* [*Carteria*], *Euglena, Noctiluca, Gonyaulax* (both of the latter are in order Dinoflagellata (Greek *dinos,* "a whirling") [Peridineae (Greek *peridines,* "whirled")].

Subclass Zoomastigina [Zooflagellata] (Greek *zooion,* "animal"). No chromatophores; two or more flagella (usually). *Trypanosoma, Trichonympha.*

CLASS RHIZOPODA (Greek *rhiza,* "root," and *pous,* "foot"). [SARCODINA (Greek *sarcos,* "fleshy").] Locomotion by pseudopodia; often flagellate in certain phases; irregular or spherical in form; often with skeletons or shells, sometimes elaborate in design.

Order Rhizomastigina One long flagellum; ameboid in form; colorless; nutrition heterotrophic.

Order Amoebina No shell (naked), skeleton, or central capsule; pseudopodia usually lobose; ectoplasm nonvacuolated. *Ameba, Entameba.*

Order Testacea (Latin *testa*, "shell"). Simple shell, with single aperture; pseudopodia lobose or filose. *Arcella, Difflugia.*

Order Foraminifera (Latin *foramen*, "opening," and *ferre*, "to bear"). Shell or test (gelatinous, pseudochitinous, or calcareous) composed of one or many chambers, with one, few, or several openings through which the cytosome extends; pseudopodia slender, often branching and anastomosing; nutrition heterotrophic. *Elphidium* [*Polystomella*], *Globigerina.*

CLASS ACTINOPODA (Greek *aktinos*, "ray"). Pseudopodia numerous, slender, and supported by axial rod (axopodia).

Order Radiolaria (Latin *radiolus*, "little radius"). Body spherical, divided into central capsule composed of intracapsular cytoplasm, with from one to many nuclei enclosed in a capsular membrane of tectin, perforated by pores, and extracapsular cytoplasm; nutrition heterotrophic; exclusively marine. *Thalassicola, Acanthometra.*

Order Heliozoa (Greek *helios*, "sun"). Pseudopodia radiating from central spherical cytosome, enclosed in gelatinous envelope with embedded foreign particles or in a latticework of tectin; nutrition heterotrophic. *Actinophrys, Clathrulina.*

CLASS SPOROZOA (Greek *spora*, "seed"). [TELOSPORIDIA (Greek *telos*, "end").] Endoparasitic (inter- or intracellular) transmitted from host to host by encysted juvenile forms (sporocysts) or by naked ones (sporozooites).

Subclass Gregarinomorpha (Latin *grex*, "hard," and Greek *morphe*, "form"). Mature forms usually extracellular parasites. *Monocystis.*

Subclass Coccidiomorpha (Greek *kokkos*, "berry"). Mature forms intracellular, in gut or blood of host. *Plasmodium.*

CLASS CNIDOSPORIDIA (Greek *cnide*, "nettle"). Sporocysts with polar capsules.

CLASS CILIATA [CILIOPHORA] (Latin *cilium*, "eyelid"). Locomotion by cilia throughout life or in juvenile stages; usually with separation of nuclear apparatus into mega- and micronuclei.

Subclass Holotricha (Greek *holos*, "entire," and *thrix*, "hair"). Ciliation uniform over all of parts of body.

Order Gymnostomatida (Greek *gymnos*, "naked," and *stoma*, "mouth"). Cytopharynx [gullet], if present, nonciliated. *Dileptus.*

Order Suctorida (Latin *sugere*, "to suck"). [Acineta (Latin *akinetos*, "immovable").] Cilia lacking in mature phase; no cytostome; tentacles for food capture and intake; body round, oval, or branched, with stalk for attachment; mostly sessile. *Podophyra, Acineta.*

Order Chonotrichida (Greek *chone*, "crucible"). [Peritricha (Greek *peri*, "around").] Cilia limited to anterior end; peristome [oral groove] often spirally coiled and extending outward; body vase-shaped; ectocommensals on gills of amphipod crustaceans. Nuclei not separated into mega- and micronuclei.

Order Hymenostomatida (Greek *hymen*, "membrane"). Cytopharynx [gullet] with one or more undulating membranes acting as food scoops. *Tetrahymena, Paramecium.*

Order Trichostomatida (Greek *thrix*, "hair," and *stoma*, "mouth"). Peristome lined with rows of free cilia. *Colpoda, Balantidium.*

Order Astomatida No cytostome; endoparasitic in molluscs, annelids, crustaceans.

Order Apostomatida (Greek *apo*, "away from"). Rosette-like cytostome; parasitic, with complicated life history.

Order Thigmotrichida (Greek *thigma*, "touch"). Cilia responsive to contact; majority live in mantle cavity of mussels to which they attach by cilia.

Order Peritrichida (Greek *peri*, "around"). Cilia limited to adoral row passing from left to right around peristome; occasionally one or more posterior girdles; body usually bell- or vase-shaped, with stalk; sessile, often colonial. *Vorticella, Epistylis.*

Subclass Spirotricha (Greek *speira*, "coil"). Adoral zone of fused cilia (membranelles) beginning at right (anterior) margin of large peristome and passing along its left margin into cytopharynx.

Order Heterotricha (Greek *heteros*, "other," "different"). Cilia short, usually uniform in distribution; free-living or ecto- or endocommensals. *Stentor*.

Order Oligotrichida (Greek *oligos*, "few"). Cilia scarce or lacking; peristome circular, either elevated or depressed, and surrounded by adoral row of membranelles. Free-living, solitary, or colonial; some ectocommensals of hoofed mammals. *Halteria*.

Order Tintinnida (probably from Latin *tintinnare*, "to ring"). Conical or trumpet-shaped body, enclosed in gelatinous or pseudochitinous case; cilia in longitudinal rows. *Tintinnopsis*.

Order Entodiniomorpheda (Greek *ento*, "within," and *dinos*, "whirling"). Posterior end of body drawn out into three processes, one longer than the other two. Live in rumina and reticula of sheep and oxen.

Order Ctenostomatida (Greek *ctenos*, "comb"). Bodies compressed and enclosed in carapace. Free-living.

Order Hypotrichida (Greek *hypo*, "below"). Cilia fused into cirri and limited to undersurface of flattened body. *Euplotes*.

PHYLUM PORIFERA

(Latin *porus*, "a pore.") [PARAZOA (Greek *para*, "beside").] Sponges. *c.* 5,000 species. Body composed of two layers of cells; incipient tissue formation only; asymmetrical or radially symmetrical; body permeated by pores, canals, and chambers through which a current of water flows; one or more water exits or oscula; internal skeleton, if present, composed of spicules of calcium carbonate, silica or organic fibers, or both; sessile except in embryonic stages.

CLASS CALCAREA (Latin *calx*, "lime"). Skeleton of one-, three-, or four-rayed calcareous spicules; marine.

Order Asconosa (Greek *askon*, "a bladder"). [Homocoela (Greek *homos*, "one and the same," "common," and *koilos*, "hollow").] Body radially symmetrical; outer layer of flattened cells (epidermis) [pinacocytes]; inner layer of flagellate choanocytes; mesenchyme between; central cavity [spongocoele, gastral cavity]; single osculum at free end; numerous microscopic pores (ostia) perforating body wall through which current of water enters spongocoele; spicules project through body wall. *Leucosolenia*.

Order Syconosa (Greek *sykon*, "a fig"). [Heterocoela (Greek *heteros*, "other than," "different").] Choanocytes limited to flagellate chambers (radial canals) or outpocketings of the spongocoele. *Grantia*.

CLASS HEXACTINELLIDA (Greek *hex*, "six," and *aktis*, "ray"). [Triaxonida (Latin *tres*, "three," and Greek *axon*, "axis"); Hyalospongia (Greek *hyalos*, "glassy").] Glass sponges. Body radially symmetrical; skeleton of six-rayed (triaxon) siliceous spicules, or some modification of this pattern, discrete or intermeshed, and distinguishable as megascleres and microscleres; no surface epithelium; choanocytes limited to finger-like chambers; marine.

Order Hexasterophora (Greek *hex*, "six," *aster*, "star," and *phorein*, "to bear"). Microscleres hexasters. *Euplectella*.

Order Amphidiscophora (Greek *amphi*, "of both kinds," *disk*, "disk," and *phorein*, "to bear"). Microscleres with stellate disk at each end, never hexasters; sessile, with anchoring root tuft. *Hyalonema*.

CLASS DEMOSPONGIA (Greek *demos*, "populace"). Body irregular in shape or a rounded mass; skeleton of organic fibers (spongin), siliceous spicules (megascleres and microscleres), or both, or lacking entirely; leuconoid type; marine and freshwater.

Subclass Tetractinellida (Greek *tetra*, "four," and *aktis*, "ray"). Megascleres tetraxon; no spongin fibers; skeleton sometimes lacking.

Order Myxospongia (Greek *myxa*, "slime"). Spicules lacking. *Oscarella*.

Order Carnosa (Latin *carnosus*, "fleshy"). Megascleres and microscleres not sharply differentiated; skeleton may be lacking. *Haliclona*.

Order Choristida (Greek *chorisis*, "separation"). Megascleres and microscleres distinct.

Subclass Monaxonida (Greek *monos*, "one," and *axon*, "axis"). Megascleres monaxial; with, or without, spongin fibers.

Order Hadromerina (Greek *hadros*, "thick," and *meros*, "part"). Spongin lacking; megascleres predominantly with knob at end (tylostyles); microscleres, when present, astrose in form. *Cliona, Suberites.*

Order Halichondrina (Greek *hals,* "salt," and *chondros,* "cartilage"). Little spongin; megascleres mostly monaxon or diaxon, or both; microscleres rod-like, or lacking.

Order Poecilosclerina (Greek *poikilos,* "variegated," and *skleros,* "hard"). Megascleres often of two or more kinds, localized in distribution; intermeshed, united by spongin fibers. *Microciona.*

Order Haplosclerina (Greek *haploos,* "simple," and *skleros,* "hard"). Megascleres diaxon, without special localization; microscleres present or lacking; spongin generally present. *Spongilla.*

Subclass Keratosa (Greek *keras,* "horn"). Horny sponges. Skeleton exclusively of spongin fibers. *Spongia* [*Euspongia*].

The following two phyla include animals with radial, or modified radial, symmetry (Radiata).

PHYLUM CNIDARIA

(Greek *cnides,* "nettle.") [COELENTERATA (Greek *koilos,* "hollow").] *c.* 10,000 species. Animals with radial or biradial symmetry around an oral-aboral axis; two cell layers separated by a gelatinous mesoglea often containing cells; individuals either sessile, cylindrical polyps (often colonial) or free-swimming bell-shaped medusae; sac-like digestive cavity, with mouth, often surrounded by tentacles, the only opening; stinging capsules (nematocysts) distributed over the body surface; metagenesis and polymorphism frequent; aquatic only.

CLASS HYDROZOA (Greek *hydor,* "water"). Hydroids, medusae. *c.* 2,700 species. Radial symmetry tetramerous or polymerous; exclusively polypoid or medusoid [medusoid forms with velum, i.e., craspedote (Greek *kraspedousitae,* "to be bordered")], or polymorphic with both medusoid and polypoid forms; mesoglea noncellular; digestive cavity without stomodaeum or septa; sex cells mature in epidermis; solitary or colonial; chiefly marine.

Order Athecata (Greek *a,* "without," and *theca,* "case"). [Gymnoblastea (Greek *gymnos,* "naked," and *blastos,* "sprout"); Anthomedusae (Greek *anthos,* "flower").] Polyps (hydranths) without hydrotheca; gonophores naked; medusae free, bell-like, with ocelli but no statocysts; gonads on manubrium. *Tubularia, Velella, Hydractinia.*

Order Thecata [Calyptoblastea (Greek *calyptos,* "covered"); Leptomedusae (Greek *leptos,* "small," "fine").] Polyps (hydranths) with hydrotheca; gonophores enclosed in gonotheca; medusae flattened in oral-aboral axis; statocysts (usually); gonads on radial canals. *Obelia.*

Order Limnomedusae (Greek *limne,* "marsh," "pond"). Polyps minute, solitary, with and without tentacles. Medusae with velum and hollow tentacles; gonads on manubrium or radial canals. *Gonionemus.*

Order Trachymedusae (Greek *trachys,* "rough"). Medusae with statocysts and tentaculocysts with gastrodermal lithocytes; gonads on radial canals. *Geryonia, Liriope.*

Order Narcomedusae (Greek *narkos,* "naked"). Margin of bell scalloped; no manubrium; gonads on floor of digestive cavity. *Aegina.*

Order Actinulida (Greek *aktinos,* "ray"). Bell much reduced, consisting only of nerve ring, statocysts, and tentacles; ciliated.

Order Siphonophora (Greek *siphon,* "tube," and *phorein,* "to bear"). *c.* 150 species. Colonial; floating or free-swimming colonies composed of polypoid and medusoid individuals attached to a common stem or disk; polypoid members without tentacles; medusoid abortive in development and rarely free. *Physalia.*

CLASS SCYPHOZOA (Greek *skyphos,* "cup"). Jellyfish. *c.* 200 species. Medusae without velum (i.e., acraspedote); free-swimming or attached by aboral stalk; stomodaeum lacking; septa sometimes present dividing digestive cavity into four pockets (interradial); mesoglea cellular; gonads endodermal; marginal sense organs rhopalia (Greek *rhopalon,* "club") with endodermal statocysts; polypoid generation lacking, or represented by scyphistoma developing directly into adult or by strobilation (Greek *strobila,* "pine cone") giving rise to numerous medusae; marine.

Order Stauromedusae (Greek *stauros,* "cross"). [Lucernariida (Latin *lucerna,* "lamp").] Stalked jellyfish; *c.* 30 species. Sessile, attached by aboral stalk; marginal sense organs lacking or represented by modified tentacles; development direct from scyphistoma. *Lucernaria, Haliclystis.*

Order Cubomedusae (Latin *cubus,* "cube"). Four-sided jellyfish. *c.* 16 species. Free-swimming, cuboidal in form; margin of bell turned inward, resembling velum; tentacles solid, located singly or in groups at interradii. *Charybdea, Chironex.*

Order Coronatae (Latin *corona*, "crown"). Deep-sea jellyfish. *c.* 31 species. Free-swimming; flattened, dome-like, or conical in form; margin of bell scalloped and marked off by coronal groove. *Nausithoe.*

Order Semaeostomeae (Greek *semaia*, "military standard"). *c.* 50 species. Free-swimming; saucer- or bowl-like in form; margin of bell indented in eight or many lappets, with rhopalia between some or all of them; four corners of mouth drawn out into long, frilled lobes (oral arms). *Pelagia, Cyanea.*

Order Rhizostomeae (Greek *rhiza*, "root," and *stoma*, "mouth"). Many-mouthed jellyfish. *c.* 50 species. Free-swimming; flattened or bowl-like in form; margin of bell scalloped; aboral surface sometimes concave; tentacles lacking; oral arms fused. *Cassiopea, Rhizostoma.*

CLASS ANTHOZOA (Greek *anthos*, "flower"). Corals, sea anemones, sea pens, and sea pansies. *c.* 6,000 recent species. Exclusively polypoid; medusoid stage lacking; solitary or colonial; hexamerous, octomerous, or polymerous, biradial or radio-bilateral symmetry; oral end expanded radially into oral disk with hollow tentacles; digestive cavity with stomodaeum (usually with one or more sulci or siphonoglyphs) and endodermal septa fringed with filaments; mesoglea cellular, gonads endodermal; with or without endo- or exoskeleton.

Subclass Ceriantipatharia (Greek *keras*, "horn," *anti*, "against," and *pathein*, "to suffer"). Body elongated, enclosed in tube-like exoskeleton.

Order Antipatharia Black corals, thorny corals. *c.* 100 species. Colonial; horny skeleton thorny and branching; polyps with single and complete septa, six simple or eight branched tentacles; two siphonoglyphs.

Order Ceriantharia *c.* 50 species. Solitary, living inside cases made of a hardened slimy secretion in which sand grains or other foreign particles are embedded; bodies long, tentacles numerous, in two circlets; septa numerous, single, and complete; one siphonoglyph. *Cerianthus.*

Subclass Octocorallia (Greek *oktos*, "eight," and *korallion*, "coral"). Soft corals. *c.* 2,500 species. Colonial with either calcareous or horny endoskeleton; eight pinnate tentacles; eight endodermal septa attached to gullet (pharynx); one siphonoglyph.

Order Alcyonacea (Greek *alkyon*, "kingfisher"). Bases of polyps fused into a fleshy mass with oral ends only protruding; some dimorphic; skeleton of separate calcareous spicules. *Alcyonium* (dead men's fingers).

Order Gorgonacea (Greek *gorgos*, "terrible"). Horny corals. *c.* 1,200 species. Axial skeleton containing gorgonin (protein) usually with many lateral branches, separate or united by cross connections. Polyps short, rarely dimorphic, arranged on both sides of the skeletal axis. *Gorgonia.*

Order Pennatulacea (Greek *pennatus*, "feathered"). Sea pens. *c.* 300 species. One long axial polyp and many lateral polyps, always dimorphic (autozooids and siphonozooids), except at base. Skeleton of separate calcareous spicules. *Pennatula*, the sea pen.

Subclass Zooantharia [Hexacorallia (Greek *hex*, "six")]. Sea anemones and stony corals. Several thousand species. Solitary or colonial; skeleton, if present, always solid and not in form of spicules; always more, or less, than eight tentacles and septa (usually six, or multiples of six).

Order Zooanthiniaria Solitary or colonial; skeleton lacking; septal pairs usually consisting of one complete and one incomplete septum; mostly epizooic.

Order Corallimorpharia (Latin *corallum*, "coral"). Solitary or in small groups connected by coenenchyme. Aboral and flattened; nonretractile tentacles; siphonoglyphs reduced; septa irregularly arranged, many complete.

Order Actinaria Sea anemones. *c.* 1,000 species. Solitary; no skeleton; septa paired, either complete or incomplete; usually one or more siphonoglyphs. *Metridium, Adamsia, Calliactis.*

Order Ptychodactiaria (Greek *ptychos*, "fold"). Septal filaments without flagellated tracts; gonads stalked. Inhabit arctic waters.

Order Scleractinia (Greek *skleros*, "skeleton"). [Madreporaria (Latin *madre*, "mother," and Greek *poros*, "soft stone").] True or stony corals. *c.* 1,000 living species. Solitary or colonial (usually); compact calcareous skeleton; no siphonoglyph. *Fungia.*

PHYLUM CTENOPHORA

(Greek *ctenos*, "comb," and *phorein*, "to bear".) Comb jellies. *c.* 80 species. Animals with biradial symmetry; solitary; monomorphic; no stinging capsules (nematocysts); eight meridional rows of ciliary plates (combs) present throughout life or in larval stages (two genera only); aboral sense organ; aquatic.

CLASS TENTACULIFERA [Tentaculata] (Latin *tentare*, "to handle," "feel"). Tentacles present.

Order Cydippida (Greek *Kydippe*, "a Nereid"). Body round or ovoid in form; two branched tentacles, retractile into sheaths. *Pleurobrachia.*

Order Lobata (Greek *lobos*, "lobe"). Body laterally compressed, with oral end expanded into two contractile lobes; tentacles without sheaths, usually reduced. *Mnemiopsis.*

Order Cestida (Latin *cestus*, "girdle"). Body compressed in tentacular plane into a band or ribbon; tentacles reduced, with sheaths; four of the rows of ciliary plates rudimentary. *Cestum Veneris.*

Order Platyctenea (Greek *platys*, "broad," "flat," and *ctenos*, "comb"). Body compressed in oral-aboral plane to flattened form; creeping habit; two tentacles with sheaths; ciliary plates may be present in larvae only. *Ctenoplana.*

CLASS NUDA (Latin *nudus*, "naked"). [Atentaculata.] Tentacles lacking.

Order Beroida (Greek *Beroe*, "a nymph"). Body conical or thimble-shaped in form; meridional gastrovascular canals branched. *Beroe.*

The following phyla include bilaterally symmetrical animals (Bilateria); the first three are acoelomate.

PHYLUM PLATYHELMINTHES

(Greek *platys*, "flat," and *helmins*, "worm".) Flatworms. *c.* 12,700 species. Body dorso-ventrally flattened; unsegmented, enteron without anus.

CLASS TURBELLARIA (Latin *turba*, "disturbance"). Planarians. *c.* 3,000 species. Epidermis usually ciliated throughout life, with rhabdoids; simple life cycle, free-living (mostly).

Group I. Archeophora (Greek *arche*, "beginning," *oion*, "egg," and *phorein*, "to bear"). Female reproductive system primitive, with gonads not divided into yolk glands and ovary; ova endolecithal.

Order Macrostomida (Greek *makros*, "large"). [Opisthandropora (Greek *opisten*, "behind," *aner*, "man," and *poros*, "passage").] Simple pharynx; enteron unbranched; paired protonephridia; male gonopore ventral and posterior to female; one pair of nerve cords; no statocysts. *Macrostomum, Microstomum.*

Order Acoela (Greek *a*, "not," and *koilos*, "hollow"). Simple pharynx; enteron without lumen; no protonephridia or oviducts; three to six pairs of nerve cords; statocysts; exclusively marine. *Convoluta.*

Order Catenulida (Latin *catenula*, "little chain"). [Notandropora (Greek *noton*, "back," *aner*, "man," and *poros*, "passage").] Simple pharynx; enteron unbranched and without diverticula; gonads unpaired; single median protonephridium; male gonopore dorsal and anterior; no female gonopore; four nerve cords; statocysts; freshwater. *Stenostomum.*

Order Polycladida (Greek *poly*, "many," and *klados*, "branch"). Body broad and flattened; enteron with numerous radiating branches; nervous system with numerous radiating nerve cords; eyes numerous (a few eyeless species); ovaries and testes numerous and scattered; one or two gonopores; yolk glands lacking; marine (with a few exceptions). *Stylochus, Thysanozoon.*

Group II. Neoophora (Greek *neos*, "new"). Female genitalia divided into yolk gland and ovary; eggs ectolecithal.

Order Prolecithophora (Greek *pro*, "before," and *lekithos*, "yolk"). Pharynx plicate; enteron without diverticula; single gonopore, sometimes opening into buccal cavity.

Order Lecithoepitheliata (Greek *lekithos*, "yolk," *epi*, "upon," and *thele*, "nipple"). Enteron straight; no yolk glands; four pairs of nerve cords; marine, freshwater, and terrestrial.

Order Seriata. Pharynx plicate; enteron with lateral diverticula; paired yolk glands, arranged in row along yolk duct.

Suborder Proseriata Enteron not forked; four pairs of nerve cords (usually). *Monocelis.*

Suborder Tricladida (Greek *tria*, "three," and *klados*, "branch"). Elongate; pharynx usually directed posteriorly; enteron with three primary branches, one directed anteriorly and two posteriorly, with many diverticula; testes numerous; yolk glands; single gonopore; three to four pairs of nerve cords; marine, freshwater, and terrestrial.

Division Maricola (Latin *mare*, "sea," and *colere*, "to dwell"). [Retrobursalia (Latin *retro*, "backward," and *bursa*, "pocket").] Bursa copulatrix, when present, posterior to penis, or else with separate pore; three pairs of nerve cords; marine or brackish water. *Procerodes* [*Gunda*].

Division Paludicola (Latin *palus*, "marsh"). [Probursalia (Latin *pro*, "before").] Bursa copulatrix anterior to penis; three pairs of nerve cords; freshwater, sometimes brackish water. *Planaria, Dugesia* [*Euplanaria*], *Dendrocoelum*.

Division Terricola (Latin *terra*, "earth"). Bursa copulatrix, when present (rarely) posterior to penis; terrestrial, in damp locations. *Bipalium*.

Order Neorhabdocoela (Greek *neos*, "new," and *rhabdos*, "rod"). Pharynx bulbose (usually) protonephridia paired; yolk glands; one pair of nerve cords (usually); mostly free-living, marine, freshwater, and terrestrial.

Suborder Typhloplanoida (Greek *typhlon*, "caecum"). Mouth posterior; pharynx slightly protrusible; marine (few), freshwater (many), terrestrial (few). *Mesostoma*.

Suborder Dalyelloida (after Scotch naturalist John Dalyell —d. 1881). Mouth anterior; pharynx slightly protrusible; marine, freshwater; commensals, parasites. *Dalyella*.

Suborder Temnocephalida (Greek *temnein*, "to cut," and *kephale*, "head"). Cilia few or lacking; anterior end extended into 2 to 12 tentacles; posterior end with one or two adhesive disks; single gonopore; ectocommensal; freshwater. *Temnocephala*.

Suborder Kalyptorhynchia (Greek *kalyptein*, "to conceal," and *rhynchos*, "snout"). Anterior protrusible proboscis, independent of mouth. Three pairs of nerve cords (usually). Marine (mostly). *Gyratrix*.

CLASS TREMATODA (Greek *trema*, "hole," and *eidos*, "form"). Flukes. *c.* 6,250 species. No epidermis, rhabdoids, or external cilia; one or more suckers; enteron usually bifurcate, but sometimes simple with many branches; single ovary; life cycle simple or complex; ecto- or endoparasites.

Order Monogenea (Greek *monos*, "single," and *genesis*, "origin"). [Heterocotylea (Greek *heteros*, "different," and *kotyle*, "anything hollow").] *c.* 1,350 species. Oral sucker absent or small; anterior end usually with a pair of adhesive structures; posterior end with adhesive disk or hooks; paired anterior, dorsal nephridiopores; ecto- or endoparasites with simple life cycles. *Gyrodactylus, Polystoma*.

Order Aspidogastrea (Greek *aspis*, "shield"). [Aspidocotylea.] No oral sucker; no anterior adhesive structures; large ventral compound sucker, or row of ventral suckers without hooks; one posterior nephridiopore; endoparasitic, with simple life cycle. *Aspidogaster*.

Order Digenea (Greek *dyo*, "two," and *genesis*, "origin"). [Malacotylea (Greek *malakos*, "soft").] *c.* 4,900 species. Anterior sucker encircling mouth (usually), ventral sucker; no hooks; single nephridiopore; endoparasitic with complicated life cycles and several larval stages invading one or more hosts. *Fasciola, Clonorchis*.

CLASS CESTODA (Greek *kestos*, "girdle"). Tapeworms. *c.* 3,400 species. No epidermis, rhabdoids, or external cilia; body with cuticular covering undivided or made up of scolex (head) and a strobila, a series of segments (proglottides), of which the youngest are the most anterior, each containing a complete reproductive system of both sexes; no mouth or enteron; endoparasitic, principally in the intestines of vertebrates, with complex life cycles, usually involving two or more hosts.

Subclass Cestodaria. *c.* 15 species. Body undivided (monozooic); larvae with 10 hooks. *Gyrocotyle*.

Subclass Eucestoda (Greek *eu*, "true"). *c.* 3,400 species. Body elongated and ribbon-like, usually divided into segments, often very numerous (polyzooic); anterior end usually expanded into scolex with adhesive organs; larvae with six hooks. *Taenia*.

PHYLUM NEMERTINEA

(Greek *Nemertes*, one of the Nereids). [Rhyncocoela (Greek *rhynchos*, "snout," and *koilos*, "cavity").] Elongated, unsegmented worm-like animals, with eversible proboscis enclosed in a cavity (rhynchocoele) dorsal to the digestive tract; enteron complete (with anus); circulatory system present.

CLASS ANOPLA (Greek *enoplos*, "unarmed"). Proboscis without stylets; mouth posterior to ganglionic mass;

central nervous system beneath epidermis or embedded in musculature of body wall.

Order Palaeonemertini (Greek *palaios*, "ancient"). Musculature of body wall of two or three layers, third (innermost), if present, always circular; dermis gelatinous. *Cephalothrix.*

Order Heteronemertini (Greek *heteros*, "other," "different"). Musculature of body wall in three layers, innermost longitudinal; dermis fibrous. *Lineus, Cerebratulus.*

CLASS ENOPLA (Greek *enoplos*, "armed"). Proboscis usually with stylets; mouth anterior to ganglionic mass; central nervous system internal to body musculature.

Order Hoplonemertini (Greek *hoplon*, "weapon"). Proboscis with one or more stylets; enteron straight with one or more lateral diverticula. *Geonemertes.*

Order Bdellomorpha (Greek *bdella*, "leach," and *morphe*, "form"). [Bdellonermertini.] Proboscis without stylets; enteron sinuous; ectocommensals of bivalve molluscs. *Malacobdella.*

PHYLUM ENDOPROCTA

(Greek *endo*, "within," and *proktos*, "anus.") [Kamptozoa (Greek *kamptein*, "to bend").] *c.* 60 species. Body divided into stalk with basal attachment disks and calyx, and round or oval mass containing viscera with distal circlet of ciliated tentacles; paired protonephridia; digestive tract looped, both mouth and anus opening inside crown of tentacles; space between digestive tract and body wall filled with mesenchyme (regarded by some as pseudocoelom). Solitary or colonial; sessile. *Loxosoma, Pedicellina.*

The following six phyla include pseudocoelomate animals.

PHYLUM ACANTHOCEPHALA

(Greek *akantha*, "thorn," and *kephale*, "head.") Spiny-headed worms. *c.* 500 species. Body elongated, flattened, and rough; retractile and sheathed proboscis with rows of recurved spines; digestive tract lacking; endoparasitic, requiring two hosts; adults parasitic in digestive tracts of various vertebrates, larvae in arthropods.

CLASS ARCHIACANTHOCEPHALA (Greek *arche*, "beginning"). Spines on proboscis arranged concentrically; protonephridia present; endoparasitic in terrestrial vertebrates. *Monoliniformis.*

CLASS PALAEACANTHOCEPHALA (Greek *palaios*, "ancient"). Spines on proboscis arranged in alternating radial rows; protonephridia lacking; endoparasitic mostly in aquatic hosts (fish, crustaceans; some in ducks and geese).

CLASS EOACANTHOCEPHALA (Greek *eos*, "dawn"). Spines on proboscis arranged radially; protonephridia lacking; endoparasitic in aquatic hosts (fish, turtles, snails, as intermediate hosts).

The following five groups are often included in the single phylum Aschelminthes (Greek *askos*, "bladder," and *helmins*, "worm"). Characterized by unsegmented, or superficially segmented, bodies with a cuticular covering; straight or curved digestive tracts, usually with a highly differentiated pharyngeal region; usually with protonephridia; no respiratory or circulatory systems. In this scheme, however, the classes included in Aschelminthes have been given phylum status.

PHYLUM ROTIFERA

(Latin *rota*, "wheel," and *ferre*, "to bear.") *c.* 1,500 species. Microscopic in size; anterior ciliated disk (corona) and pharynx (mastax), with internal jaws, or trophi, consisting of seven cuticularized pieces (fulcrum, paired rami, paired unci, paired manubria); one pair of protonephridia; aquatic.

CLASS SEISONACEA (French *saiser*, "to seize"). Body elongated; corona represented by a few tufts of bristles only; mastax fulcrate, with elongated fulcrum and leaf-like manubria at anterior end; epizooic (ectoparasitic) on gills of *Nebalia*. *Seison.*

CLASS BDELLOIDA (Greek *bdella*, "leech"). Body elongated and jointed; corona typically with two trochal disks; anterior and retractile; mastax ramate, with fulcrum and manubrium reduced and unci large and plate-like with parallel ridges; females only; creeping or swimming habit. *Philodina.*

CLASS MONOGONATA (Greek *monos*, "one," and *gonia*, "reproduction"). Mastax never ramate; males more or less reduced structurally; swimming or sessile.

Order Ploima (Greek *ploimos*, "fit for sailing"). Corona typical; mastax virgate, cardate, forcipate, malleate, or incudate; lorica frequently present; swimming. *Epiphanes* [*Hydatina*], *Brachionus, Asplanchna.*

Order Floscularíacea (Latin *flosculus*, "floweret"). Corona circular or lobed; mastax malleo-ramate; males greatly reduced; sessile or swimming. *Floscularia, Trochosphaera.*

Order Collothecacea (Greek *kolla*, "glue," and *theke*, "case"). Anterior end expanded funnel, simple or lobed; ciliary circlets lacking; mastax mostly uncinate; males greatly reduced; sessile.

PHYLUM GASTROTRICHA

(Greek *gaster*, "stomach," and *thrix*, "hair.") *c.* 150 species. Microscopic, corona lacking, cilia on limited areas; cuticular covering often modified as scales, spines, or bristles; from two to many projecting cuticular tubes (adhesive tubes) with secretory cells; aquatic.

CLASS MACRODASYOIDEA (Greek *makros*, "large," and *dasys*, "thick"). Adhesive tubes numerous; protonephridia lacking; hermaphroditic; marine.

CLASS CHAETENIDEA (Greek *chaite*, "hair"). Two to four adhesive tubes located at posterior end only; one pair of protonephridia; parthenogenetic females only; freshwater (mostly). *Chaetonotus.*

PHYLUM KINORHYNCHA

(Greek *kinetos*, "moving," and *rhynchos*, "snout.") *c.* 60 species. Microscopic; cilia lacking; body composed of 13 to 14 rings, two of which constitute head, encircled by spines and containing short retractile proboscis with mouth; one pair of protonephridia; marine, in mud or sand.

PHYLUM NEMATODA

(Greek *nematos*, "thread.") Roundworms. *c.* 10,000 species. Body cylindrical, often with bristles but without cilia; subepidermal musculature of longitudinal fibers only; central nervous system in form of one or more circumenteric rings with ganglia and longitudinal cords; anterior, lateral sense organs (amphids), spiral, circular, or cyathiform (Greek *cyathus*, "cup"); consisting of paired cuticular depressions with gland cells and nerve endings, excretory system composed of one or two gland cells or of canals, or of both, but never protonephridial; parthenogenetic, hermaphroditic, and dioecious forms; aquatic, terrestrial, and parasitic.

CLASS ADENOPHOREA (Greek *adenos*, "gland"). [Aphasmida (Greek *a*, "without," and *phasmid*, "glandular structures in posterior region").]

Order Enoplida (Greek *enoplos*, "armed"). Anterior end with six labial papillae (sensory); 10 to 12 bristles in one or two circlets; amphids cyathiform. *Enoplus.*

Order Dorylaimida (Greek *dory*, "spear," and *eidos*, "form"). *c.* 200 species. Anterior end with two circles of sensory papillae of 6 and 10 each; buccal cavity with protrusible spear; amphids cyathiform. *Dorylaimus.*

Order Mermithida (Greek *mermis*, "cord," and *eidos*, "form"). Body very slender; papillae usually 16 in number; amphids cyathiform or reduced; intestine ends blindly (no anus) and functions as food reservoir; adults free-living in water or soil; juvenile stages parasitic in insects and some other invertebrates.

Order Trichurida (Greek *thrix*, "hair," and *oura*, "tail"). [Trichinelloidea (Greek *trichinos*, "hairy").] Mouth without lips; amphids reduced; pharynx embedded in two longitudinal rows of glandular cells (stichosome); adults parasitic in digestive tract principally of vertebrates, especially birds and mammals. *Trichiuris* (whip worm), *Trichinella.*

Order Dioctophymida. Mouth without lips; cephalic sense organs one or two circlets of papillae (6, 12, or 18); amphids reduced; adults parasitic in birds and mammals; intermediate hosts probably fish.

CLASS SECERNENTEA (Latin *secernere*, "to separate," "secrete"). [Phasmida.] Phasmids paired; many parasitic; free-living species, often saprophagous.

Order Rhabditida (Greek *rhabdos*, "rod"). [Anguillulida (Greek *anguilla*, "eel").] Cuticle ringed or smooth; amphids reduced to small pockets; cephalic sense organs all papillate; free-living (aquatic or terrestrial) or parasitic. *Rhabditis, Turbatrix aceti* (vinegar eel).

Order Tylenchida (Greek *tylos*, "knob," and *enchos*, "spear"). Stylet extensible. *Heterodera.*

Order Rhabdiasida (Greek *rhabdos*, "rod," and *dasy*, "thick"). Cuticle smooth; amphids reduced to small pockets; in life cycle, parasitic hermaphrodite or parthenogenetic females produce free-living young, which may be dioecious, giving rise to parasitic form, or may develop directly into parasitic form. *Strongyloides.*

Order Oxyurida (Greek *oxys*, "sharp," and *oura*, "tail"). *c.* 144 species. Cephalic sense organs papillate; amphids reduced to tubular pockets; adults parasitic in arthropods, some in man (pin worms).

Order Ascarida (Greek *askaris*, "intestinal worm"). Mouth with three prominent lips; amphids reduced; adults parasitic in intestines of vertebrates. *Ascaris*.

Order Strongylida (Greek *stronglos*, "round"). Mouth with conspicuous bursa copulatrix supported by muscular rays, typically 13 in number; adults parasitic in digestive tracts of vertebrates, especially mammals. *Anclystoma, Necator* (hookworms).

Order Spirurida (Greek *speira*, "coil"). Mouth usually with two, sometimes four or six, lateral lips; cephalic sense organs internal circlet of six reduced papillae and external circlet of four double or eight single papillae; amphids reduced; adults parasitic in digestive tract, eyes, nasal cavities, and sinuses of vertebrates.

Order Dracunculida (Latin *dracunculus*, "little dragon"). No definite lips; cephalic sense organs internal circlet of six and external circlet of four double or eight single papillae; amphids; reduced adults parasitic in connective tissue or coelom of vertebrates. *Dracunculus*.

Order Filarida (Latin *filium*, "thread"). Mouth without lips; amphids reduced; adults parasitic in blood vessels, coelomic cavities, vascular and connective tissue of vertebrates; young transmitted to final host by blood-sucking insect. *Wuchereria*.

PHYLUM NEMATOMORPHA

(Greek *nematos*, "thread.") [Gordiacea (Latin *Gordius*, "King of Phrygia cf. Gordian knot").] Horsehair worms. *c.* 230 species. Body long, slender; cuticle smooth; central nervous system an anterior ganglionic mass with midventral chain; excretory system lacking; enteron more or less degenerate at one or both ends; free-living as adults; juvenile stages parasitic in arthropods.

Order Gordioda Cuticle smooth; pseudocoele filled with mesenchyme; freshwater or terrestrial; juvenile stages parasitic in terrestrial or aquatic arthropods. *Gordius*.

Order Nectonematoida (Greek *nektos*, "swimming"). Cuticle with bristles; pseudocoele open, without mesenchyme filling; marine; juvenile stages parasitic in Crustacea.

The following phyla include eucoelomate animals, those with a secondary body cavity (true coelom) arising either as a space in the mesoderm (schizocoele) or as outgrowths from the enteron (enterocoele). The first 11 of these are schizocoelous.

PHYLUM BRYOZOA

(Greek *bryon*, "moss," and *zoion*, "animal.") [Polyzoa (Greek *polys*, "many"); Ectoprocta (Greek *ektos*, "outside," and *proktos*, "anus").] Body enclosed in gelatinous, calcareous, or chitinous case (zooecium); anterior end with circular or crescentic crown of tentacles, ciliated, and usually retractile; digestive tract looped, mouth opening within tentacular crown, anus outside it; excretory organs lacking; colonial; marine and freshwater.

CLASS GYMNOLAEMATA (Greek *gymnos*, "naked," and *laimos*, "throat"). Tentacular crown circular; mouth without epistome (lip).

Order Stenostomata (Greek *steno*, "narrow," and *stoma*, "mouth.") [Cyclostomata (Greek *kyklos*, "ring," "circle").] Opening of zooecium more or less flattened. *Crisea*.

Order Ctenostomata (Greek *ctenos*, "comb"). Opening of zooecium with tooth-like processes. *Bowerbankia*.

Order Cheilostomata (Greek *cheilos*, "lip"). Opening of zooecium with lid (operculum) which can be raised or lowered. *Bugula*.

CLASS PHYLACTOLAEMATA (Greek *phylassein*, "to guard"). *c.* 12 species. Tentacular crown horseshoe-shaped; mouth with epistome (lip). *Cristatella, Plumatella*.

PHYLUM PHORONIDA

(Greek *Phoronis*, surname of the river goddess Io.) Body slender, elongate, enclosed in leathery or membranous tube; anterior end with spiral or horseshoe-shaped crown of tentacles, ciliated, and retractile. Digestive tract looped; mouth inside tentacular crown, anus outside it; one pair of metanephridia; marine. *Phoronis*.

PHYLUM BRACHIOPODA

(Latin *brachium*, "arm," and *pous*, "foot.") *c.* 260 species. Body enclosed in bivalve calcareous shell, one valve dorsal, one ventral; anterior end with spiral crown of ciliated tentacles; anus sometimes lacking; one or two pairs of metanephridia; solitary; marine.

CLASS INARTICULATA (Latin *in*, "not," and *articulatus*, "jointed"). Two valves similar, without hinge or beak; anus present.

Order Atremata (Greek *a*, "not," and *trema*, "hole"). Ventral valve imperforate; peduncle emerges between valves. *Lingula.*

Order Neotremata (Greek *neos*, "new," and *trema*, "hole"). Ventral valve perforated; peduncle emerges through perforation.

CLASS ARTICULATA (Latin *articulatus,* "jointed"). Dorsal and ventral valves unlike, with hinge. Anus lacking. *Terebratulina.*

PHYLUM MOLLUSCA

(Latin *mollis*, "soft.") *c.* 128,000 species. Chitons, snails, bivalves, squids, and octopuses. Body covered by thin dorsal fold (mantle) generally secreting calcareous shell of one or more pieces; anterior region usually developed as head, ventral region as highly muscular foot used in burrowing, crawling, or swimming; digestive tract complete (mouth and anus), often U-shaped or coiled; mouth generally with radula bearing rows of small chitinous teeth; open circulatory system; coelom reduced to cavities of renal organs, gonads, and pericardium; respiration usually by gills (ctenidia); nervous system typically three pairs of ganglia (cerebral, pedal, visceral) connected by commissures; aquatic (marine or freshwater) and terrestrial.

CLASS POLYPLACOPHORA (Greek *polys*, "many," and *plax*, "flat"). [Loricata.] *c.* 1,000 species. Shell of eight overlapping calcareous valves; foot large and flat. *Chiton* (coat-of-mail); *Cryptochiton.*

CLASS APLACOPHORA. [Solenogastres (Greek *solen*, "pipe," and *gaster*, "stomach").] *c.* 150 species. Shell reduced to small calcareous spicules throughout mantle; foot rudimentary. *Chaetoderma.*

CLASS MONOPLACOPHORA (Greek *monos*, "one"). Three recent species. Shell of one piece; body almost bilaterally symmetrical, foot ventral; anus posterior. *Neopilina.*

CLASS GASTROPODA (Greek *gaster*, "stomach," and *pous*, "foot"). Snails, slugs. 105,000 recent species. Body, exclusive of head and foot (i.e., visceral mass), generally coiled within a spiral shell; head distinct, with one or two pairs of tentacles and a pair of eyes; foot large, flat, and muscular.

Subclass Prosobranchia (Greek *proso*, "in front," and *branchia*, "gill"). [Streptoneura (Greek *streptos*, "twisted," and *neuron*, "nerve").] *c.* 57,000 species. Shell often large, usually with calcareous or horny operculum; visceral mass looped, through torsion, with anus opening anteriorly; nerve chains twisted into figure eight and ctenidia anterior to heart.

Order Archaeogastropoda (Greek *archaios*, "ancient," "primitive"). Limpets. Ctenidia, when present, plume-like, with two rows of leaflets; nervous system little concentrated; marine. *Megathura* (keyhole limpet), *Haliotis* (abalone, ormer), *Patella* (limpet), *Trochus* (top shell).

Order Neritacea (Greek *nerita*, "seashell"). Right excretory organ degenerate; left functional. Freshwater and terrestrial. *Nerita.*

Order Mesogastropoda. Periwinkles, cowries. Organs of pallio-pericardial complex of right side lost. Ctenidia with one row of leaflets. Nervous system somewhat concentrated. *Littorina* (periwinkle), *Viviparus* (river snail), *Crepidula* (slipper limpet), *Janthina* (violet snail).

Order Neogastropoda. [Stenoglossa (Greek *stenos*, "narrowness," and *glossa*, "tongue").] Whelks, cone shells. Nervous system highly concentrated; reversible proboscis. *Buccinium*, *Busycon* (whelks), *Nassa* (dog whelk), *Conus* (cone shell), *Murex.*

Subclass Opisthobranchia (Greek *opisthen*, "behind," and *branchia*, "gill"). *c.* 1,300 species. Shell reduced or lacking; single ctenidium, when present, posterior to heart through detorsion of visceral mass. Marine, hermaphroditic.

Order Cephalaspidea (Greek *kephale*, "head," and *aspis*, "shield"). [Bullomorpha.] Head shield shaped for burrowing; single plicate ctenidium. *Bulla* (bubble shell).

Order Anaspidea. [Aplysiomorpha.] Shell reduced and internal; no head shield. *Aplysia* (sea hare).

Order Thecosomata (Greek *theke*, "case," and *soma*, "body"). Spirally coiled shell or modified nonspiral pseudoconch. Planktonic "pteropods" (Greek *pteron*, "wing," and *pous*, "foot") or "sea butterflies."

Order Gymnosomata (Greek *gymnos*, "naked"). No shell or mantle cavity; fast-swimming planktonic pteropods. *Clioacicula*.

Order Notaspidea (Greek *noton*, "back," and *aspis*, "shield"). Shell reduced and internal; body flattened, slug-like. *Pleurobranchus*.

Order Acochlidiacea (Greek *koklias*, "snail"). Small; visceral mass marked off as long hump from foot.

Order Sacoglossa (Greek *sakos*, "shield," and *glossa*, "tongue"). Sea slugs. Shell well developed and spirally coiled or reduced. Buccal cavity used for sucking. *Elysia*.

Order Nudibranchia (Latin *nudus*, "naked," "bare"). [Acoela (used also for order of Turbellaria).] Without shell; ganglia concentrated in head. *Doris* (sea lemon), *Eolis* (sea slug), *Archidoris*.

Subclass Pulmonata (Latin *pulmo*, "lung"). *c.* 35,000 species. Shell simple spiral, or lacking; ctenidia lacking, respiration through surface of mantle cavity opening by narrow contractile aperture; freshwater and terrestrial.

Order Basommatophora (Greek *basis*, "base," *omma*, "eye," and *phorein*, "to bear"). Eyes at base of nonretractile tentacles. *Lymnaea* (pond snail), *Planorbis* (ram's horn snail), *Physa*.

Order Stylommatophora (Latin stylus, "stake," Greek *omma*, "eye," and *phorein*, "to bear"). Eyes on tips of posterior pair of tentacles (retractile). *Helix* (land snail), *Limax, Arion* (land slugs).

CLASS SCAPHOPODA (Greek *skaphe*, "boat"). Tusk shells. 350 recent species. Body elongate, enclosed in tubular shell, slightly curved and open at both ends; mouth surrounded by delicate contractile tentacles; foot conical; marine. *Dentalium*.

CLASS BIVALVIA. [Pelecypoda (Greek *pelekys*, "hatchet"). Clams, oysters, mussels. *c.* 20,000 species. Shell with two hinged valves, lateral in position; drawn together by one or two large adductor muscles; margins of right and left lobes of mantle form siphons (posterior), directing flow of water in and out of mantle cavity; head and radula lacking; foot often hatchet-shaped, extending between valves of shell when moving; marine, some freshwater.

Subclass Protobranchia. Ctenidia with flat filaments; foot opens out to expose flat ventral surface. *Nucula* (nutshell), *Yoldia*.

Subclass Lamellibranchia (Latin *lamina*, "plate," "leaf"). Filaments of ctenidia elongated and reflected forming two-sided lamellae and used as feeding organs.

Order Taxodonta (Greek *taxis*, "division," and *odon*, "tooth"). Teeth of hinge equal in size and arranged in series. Gill filaments free. *Arca*.

Order Anisomyaria (Greek *a*, "not," *isos*, "equal," and *mys*, "muscle"). [Filibranchia.] Anterior adductor muscle larger than posterior. Gills with interlamellar junctions. Foot small, sometimes absent. *Pecten* (scallop), *Mytilus* (mussel), *Modiolus* (horse mussel), *Ostrea*, *Crassostrea*.

Order Heterodonta Hinge teeth unequal in size. Mantle edges usually united at one or more points and often produced into siphons. *Dreissensia*, *Venus*, *Mactra*, *Tridacna*.

Order Schizodonta (Greek *schizein*, "to split"). Gills lamellate; hinge teeth V-shaped (bifurcate), arranged inside each other. *Unio, Anodonta, Lampsilis* (all freshwater).

Order Adepedonta (*Adapa*, a mythical Babylonian demigod). Gills lamellate; mantle margins completely closed except for pedal gap. Shell gaping, with ligament weak or absent. No hinge teeth. Deep burrowers. *Solen, Mya, Pholas* (piddock), *Teredo* (shipworm).

Order Anomalodesmata (Greek *anomalos*, "irregular"). No hinge teeth; gills lamellate, mouth edges extensively fused; foot small; shell valves dissimilar. *Cuspidaria*, *Poromya*

Subclass Septibranchia. Gills transformed into muscular system, pumping water through the mantle cavity.

CLASS CEPHALOPODA (Greek *kephale*, "head," and *pous*, "foot"). Squids, cuttlefish, octopuses. 700 to 800 recent species. Shell external, internal, or lacking; head large; eyes well developed, complex; nervous system highly concentrated, mouth with radula and horny jaws, surrounded by arms or tentacles; marine.

Subclass Nautiloidea. Shell external, many-chambered, coiled, or straight; head with numerous retractile tentacles without suckers. *Nautilus* (pearly Nautilus).

Subclass Coleoida (Greek *koleos*, "sheath"). Shell internal and more or less rudimentary. Head with eight arms bearing suckers; some with two additional retractile tentaculate arms. Single gill and excretory organ.

Order Octopoda (Greek *okto*, "eight"). Eight uniform arms; suckers sessile without horny rims; no internal shell. *Octopus, Eledone* (lesser octopus), *Argonauta* (paper nautilus).

Order Sepiida. Cuttlefish. Calcareous or horny shell, if present, limited to dorsal portion of visceral sac (except in genus *Spirula*, in which it is coiled and located at the posterior end of the cylindrical body, i.e., opposite the head). *Spirula, Sepia, Heteroteuthis.*

Order Teuthidida (Greek *teuthis*, "squid"). Squids. Horny internal shell; body torpedo-shaped; pelagic swimmers.

Suborder Myopsida (Greek *myein*, "to shut," and *opis*, "vision"). Cornea of eye almost closed. *Loligo.*

Suborder Oegopsida (Greek *oigen*, "to open," and *opis*, "vision"). Cornea not closed. *Architeuthis.*

Order Vampyromorpha. Vampire squids. Eight long arms united by swimming web; two small retractile arms. *Vampyroteuthis.*

PHYLUM SIPUNCULA

(Greek *siphon*, "tube," and *eidos*, "form.") *c.* 275 species. Body elongated with crown of retractile lobes or tentacles; inversible proboscis, with chitinous papillae; intestine looped and twisted; anus anterior and middorsal; one pair of nephridia. *Golfingia* [*Phascolosoma*], *Sipunculus.*

PHYLUM ECHIUROIDA

(Greek *echis*, "adder," and *oura*, "tail.") *c.* 150 species. Body cylindrical or ovoid; mouth anterior; anus posterior; proboscis nonretractile; two or three pairs of nephridia; marine. *Bonellia, Echiurus, Urechis.*

PHYLUM ANNELIDA

(Latin *anellus*, "ring.") Worms. *c.* 8,700 species. Body elongated and usually segmented internally and externally; paired chitinous bristles (setae), few to many per segment; digestive tract tubular, complete; mouth anterior; anus posterior; closed circulatory system with longitudinal vessels with lateral branches in each segment; typically one pair of nephridia per segment; nervous system with paired anterior dorsal ganglia connected to midventral ganglionic chain; paired lateral nerves in each segment; marine, freshwater, and terrestrial.

CLASS POLYCHAETA (Greek *polys*, "many," and *chaite*, "hair"). 5,300 known species. Body with internal and external segmentation; numerous segments, with lateral parapodia bearing many setae; distinct head with tentacles. *Aphrodite, Nereis, Nephthys, Chaetopterus, Arenicola.*

CLASS MYZOSTOMARIA (Greek *myzein*, "to suck," and *stoma*, "mouth"). *c.* 130 species. Body disk-shaped; no external segmentation; five pairs of parapodia; ectoparasitic on echinoderms, principally crinoids. *Myzostoma.*

CLASS OLIGOCHAETA (Greek *oligos*, "small"). *c.* 3,100 species. Segmentation conspicuous externally and internally; head lacking; parapodia lacking; few setae per segment; freshwater and terrestrial. *Tubifex, Lumbricus, Pheretima, Allolobophora, Eisenia, Megascolides.*

CLASS HIRUDINEA (Latin *hirudo*, "leech"). *c.* 300 species. Body usually flattened dorso-ventrally, composed of 34 segments externally subdivided into many annuli; one large posterior sucker; often small anterior sucker; parapodia lacking; tentacles lacking; setae lacking (with one exception); coelom filled with connective tissue and muscles; marine, freshwater, and terrestrial. *Hirudo, Macrobdella.*

CLASS ARCHIANNELIDA (Greek *arche*, "beginning"). Body small; segmentation usually internal; parapodia and setae lacking. *Polygordius, Saccocirrus.*

PHYLUM ONYCHOPHORA

(Greek *onyx*, "claw," and *phorein*, "to bear"). *c.* 73 species. Body elongated, unsegmented; paired appendages and nephridia arranged in series; no distinct head. Anterior end with one pair of antennae and one pair of oral papillae; central nervous system of paired dorsal ganglia connected to a pair of separate nerve cords without ganglia but with transverse connections. *Peripatus, Peripatopsis.*

PHYLUM ARTHROPODA

(Greek *arthron*, "joint," and *pous*, "foot.") *c.* 815,000 species. Body segmented and jointed externally, with chitinous exoskeleton, divisible into head, thorax, and abdomen, variously distinct and fused; paired jointed appendages, primitively one per segment; straight, complete digestive tract; open circulatory system; central nervous system with paired dorsal (cerebral) ganglia connected to midventral ganglionic chain, primitively with one ganglion and paired nerves in each segment.

SUBPHYLUM CHELICERATA

(Greek *chele*, "claw," and *keras*, "horn"). Anterior pair of appendages jointed, with terminal claw.

CLASS MEROSTOMATA (Greek *meros*, "thigh," and *stoma*, "mouth"). *c.* four species. Mouth in cephalothorax between bases of walking legs.

Order Xiphosura (Greek *xiphos*, "sword," and *oura*, "tail"). "Horseshoe crabs." Cephalothorax covered by horseshoe-shaped carapace; abdomen broad; chelicerae with three joints; pedipalps and walking legs with six joints; marine. *Xiphosura* [*Limulus*].

CLASS PYCNOGONIDA (Greek *pyknos*, "compact," and *gone*, "knee"). Sea spiders. *c.* 440 species. Body short, slender; cephalothorax of five segments with seven pairs of appendages including proboscis and suctorial mouth, one pair of 10-jointed legs used to carry eggs by males in some species, and four or six pairs of eight- or nine-jointed walking legs; abdomen rudimentary; marine. *Nymphon.*

CLASS ARACHNIDA (Greek *arachne*, "spider"). Spiders, scorpions, daddy-longlegs. *c.* 30,000 species. Body usually divisible into cephalothorax and abdomen without appendages; no antennae or jaws; chelicerae, two pairs of pedipalps, and four pairs of walking legs; gonophores always on second abdominal segment; mostly terrestrial.

Order Scorpionida (Latin *scorpio*, "scorpion"). Scorpions. Body elongate, with compact cephalothorax broadly jointed to abdomen; abdomen long (12 segments); chelicerae small, three-jointed; pedipalpi six-jointed, with terminal heavy pincers; posterior six segments, narrow, terminating in poison claw. *Buthus, Pandinus.*

Order Pseudoscorpionida (Greek *pseudos*, "false"). Cephalothorax sometimes marked dorsally with two transverse grooves; abdomen of 11 similar segments without poison claw; chelicerae two-jointed; pedipalpi large, six-jointed, chelate. *Chelifer.*

Order Opiliones (Latin *opilio*, "shepherd"). Harvestmen, daddy-longlegs. Body short, ovoid; cephalothorax broadly joined to abdomen; chelicerae slender, three-jointed; pedipalps six-jointed, not chelate; legs long and slender, seven-jointed with many-jointed tarsi. *Phalangium.*

Order Acarina (Latin *acarus*, "mite"). Mites, ticks. Body small, compact, ovoid; no evident segmentation; cephalothorax and abdomen fused; chelicerae and pedipalpi various; legs six- or seven-jointed, widely separated; free-living or parasitic. *Acarus, Ixodes.*

Order Palpigradida (Latin *palpus*, "feeler," and *gradi*, "to step"). Micro whip scorpions. Body small; abdomen oval, of 11 segments with slender tail of 15 segments; chelicerae three-jointed, heavy; pedipalpi like walking legs. *Microthelyphonida.*

Order Uropygida (Greek *oura*, "tail," and *pyge*, "rump"). Whip scorpions. Chelicerae two-jointed, ending in hook or fang; pedipalpi jawed and prehensile; "tail" long and flexible. *Thelyphonida.*

Order Ricinuleida (Latin *ricinus*, "tick"). Cephalothorax with anterior movable projection (cucullus); chelicerae and pedipalpi chelate. *Ricinoides.*

Order Solifugae. [Solpugida (Latin *solpuga*, "venomous spider").] Sun spiders. Cephalothorax of six segments; chelicerae two-jointed, enlarged, chelate; pedipalpi six-jointed, leg-like; first pair of legs tactile, not locomotor; abdomen of 10 segments. *Galeodes.*

Order Araneae (Latin *aranea*, "spider"). Spiders. Body unsegmented; cephalothorax and abdomen narrowly joined; chelicerae two-jointed, small, with poison duct opening in claw; pedipalpi six-jointed, leg-like, with enlarged basal joint; legs seven-jointed. *Araneus* [*Epeira*], *Latrodectus.*

SUBPHYLUM MANDIBULATA

(Latin *mandibula*, "jaw.") Oral appendages modified as mandibles or jaws.

CLASS CRUSTACEA (Latin *crusta*, "shell"). *c.* 25,000 species. Head (five fused segments), with two pairs of antennae, a pair of jaws, and two pairs of maxillae; exoskeleton impregnated with calcium salts; appendages often biramous.

Subclass Branchiopoda (Greek *branchia*, "gill"). Segments various in number; appendages uniform, usually leaf-like; abdominal appendages lacking; antennules and maxillae reduced or absent.

Order Anostraca (Greek *a*, "not," and *ostrakon*, "shell"). Fairy shrimps. Carapace lacking; 11 to 19 pairs of thoracic appendages; eyes stalked. *Artemia, Chirocephalus.*

Order Notostraca (Greek *noton*, "back"). Tadpole shrimps. Dorsal, shield-shaped carapace; body of 40 to 60 segments; eyes sessile. *Triops* [*Apus*].

Order Conchostraca (Greek *konche*, "shell"). Clam shrimps. Carapace hinged, with adductor muscle, enclosing entire body. Ten to twenty pairs of trunk appendages.

Order Cladocera (Greek *klados*, "sprout," and *keras*, "horn"). Water fleas. Carapace without hinge or adductor muscle enclosing all the body but the head. Four to six pairs of thoracic appendages; single median compound eye. *Daphnia, Leptodora.*

Order Cephalocarida (Greek *kephalos*, "head," and *karis*, "shrimp"). Trunk divided into thorax (9 segments) and abdomen (10 segments). Appendages in thoracic segments only; two pairs of maxillae. No eyes. *Hutchinsoniella.*

Subclass Ostracoda (Greek *ostrakodes*, "shelly"). Carapace bivalve, compressed laterally, enclosing all the body; body indistinctly segmented; two pairs of thoracic appendages; first and second antennae locomotor.

Order Myodocopa (Greek *myo*, "close," "shut," and *kope*, "oar"). Carapace notched anteriorly; second antennae biramous, marine. *Cypridina.*

Order Podocopa (Greek *pous*, "foot"). Carapace unnotched; second antennae uniramous. *Cypris.*

Order Cladocopa (Greek *klados*, "sprout," "twig"). Second antennae biramous; exopodites and endopodites used in swimming.

Order Platycopa (Greek *platys*, "flat"). Second antennae biramous, flattened, and leaf-like. *Cytherella.*

Subclass Copepoda (Greek *kope*, "oar," and *pous*, "foot"). Body distinctly segmented; typically nine free thoracic segments without appendages on posterior four segments; three ocelli, often fused as median eye; free-living, commensal, and parasitic.

Order Calanoida (French *calandre*, "weevil"). First antennae long (23 to 25 segments). Second antennae biramous, not prehensile. Movable joint between fifth and sixth thoracic segments. *Calanus.*

Order Cyclopoida (Greek *kyklos*, "circle," and *ops*, "eye"). First antennae long (about 17 segments); second antennae only one joint. First four thoracic legs biramous. Movable joint between fourth and fifth thoracic segments. Thorax wider than abdomen. *Cyclops.*

Order Harpacticoida. First antennae short (eight segments or less). Movable joint between fourth and fifth thoracic segments. *Harpacticus.*

Order Monstrilloida (Latin *monstrum*, "something marvelous"). Second antennae and mouthparts missing; digestive tract greatly reduced; adults free-swimming, larvae parasitic on polychaetes. *Monstrilla.*

Order Notodelphyoida (Greek *noton*, "back"). Movable joint between fourth and fifth thoracic segments. Parasitic on tunicates. *Enterocola.*

Order Caligoida (Latin *caliga*, "shoe"). First antennae of two segments; second antennae prehensile; four pairs of swimming legs; fifth rudimentary; movable joint between third and fourth segments except in parasitic species.

Order Lernaeopododida (Latin *Lerna*, a marsh in Argolis). No movable articulation; male minute with one to two pairs of swimming legs; females without swimming legs; parasitic on marine and freshwater fishes. *Lernaeocera* [*Lernaea*].

Subclass Mystacocarida (Greek *mystax*, "upper lip," "moustache"). Body divisible into head with five pairs of appendages; thorax with maxillipeds and four pairs of appendages; and abdomen with appendages on the first four segments. Single median eye. Small, live in littoral zone. *Derocheilocaris.*

Subclass Branchiura (Greek *branchia*, "gill," and *oura*, "tail"). Fish lice. Body flat, with disk-like carapace; mouth suctorial; parasitic. *Argulus.*

Subclass Cirripedia (Latin *cirrus*, "curl," and *pous*, "foot"). Barnacles. Carapace a pair of folds (mantle) usually supported with calcareous plates enclosing body; body imperfectly segmented; abdomen rudimentary, usually with caudal styles; six pairs, or less, of thoracic appendages, slender, biramous and provided with fringe-like bristles; head appendages degenerate in adults; adults sessile, free-living, or parasitic.

Order Thoracica (Greek *thorax*, "chest"). Six pairs of thoracic appendages; free-living. *Balanus, Lepas.*

Order Acrothoracica (Greek, *akros*, "at the end"). Less than six pairs of thoracic appendages; borers in shells of molluscs.

Order Ascothoracica (Greek *askos*, "sac," and *thorax*, "chest"). Carapace a pair of soft folds enclosing body; body imperfectly segmented; abdomen rudimentary; six pairs of thoracic appendages; oral appendages modified for piercing and sucking.

Order Rhizocephala (Greek *rhiza*, "root," and *kephale*, "head"). Mantle without calcareous plates; appendages lacking; enteron lacking; parasitic. *Sacculina.*

Subclass Malacostraca (Greek *malakia*, "softness," and *ostrakon*, "shell"). Lobsters, crabs, shrimps, crayfish. Body distinctly segmented; typically 19 segments (head, 5; thorax, 8; abdomen, 6); exoskeleton of one or more of the thoracic segments as a cephalothoracic carapace; abdomen with appendages; caudal styles lacking.

Superorder Leptostraca (Greek *leptos*, "fine," "delicate"). [Phyllocarida (Greek *phyllon*, "leaf," and *karis*, "shrimp").] Carapace bivalve enclosing most of the body; abdomen of seven segments; thoracic appendages foliaceous; abdominal, biramous.

Order Nebaliacea (origin uncertain). Marine. *Nebalia.*

Superorder Syncarida (Greek *syn*, "with," and *karis*, "shrimp"). Carapace lacking; head united with first thoracic segment.

Order Anaspidacea (Greek *a*, "not," and *aspis*, "shield"). Thoracic appendages (except posterior or posterior two) biramous, with double series of gills (except last); abdominal appendages with reduced endopodites (except last, and first two in male); terminal abdominal appendages (uropods) expanded, with telson forming fan-like tail. *Anaspides.*

Superorder Peracarida (Greek *pera*, "pouch," and *karis*, "shrimp"). Carapace, when present, encloses first three or four thoracic segments; females with thoracic brood pouch.

Order Mysidacea (Greek *mysis*, "closing of lips or eyes"). Opossum shrimps. Carapace covering most of thorax; first pair of thoracic appendages modified as maxillipeds, others usually biramous; uropods and telson form fan-like tail fin. *Mysis, Hemimysis.*

Order Cumacea (Greek *kyma*, "wave"). Carapace with two anterior plates often joined over head and extended into rostrum; abdomen and uropods slender.

Order Tanaidacea (Greek *Tanais*, old name for the river Don). Carapace drawn out into anterior rostrum and extended laterally to enclose branchial cavity; first pair of thoracic appendages modified as maxillipeds. *Tanais.*

Order Isopoda (Greek *isos*, "equal," and *pous*, "foot"). Carapace lacking; body usually dorso-ventrally flattened; first pair of thoracic appendages modified as maxillipeds; aquatic and terrestrial; some parasitic. *Ligia, Armadillidium, Asellus, Idotea.*

Order Amphipoda (Greek *amphi*, "both," and *pous*, "foot"). Carapace lacking; body often laterally compressed; second and third pairs of thoracic appendages usually modified as gnathopods; three anterior pairs of abdominal appendages with many-jointed rami; three posterior with unjointed styliform rami. *Gammarus* (freshwater shrimp), *Caprella* (ghost shrimp).

Superorder Hoplocarida (Greek *hoplon*, "weapon," and *karis*, "shrimp"). Cephalothorax relatively short; abdominal appendages with branchiae.

Order Stomatopoda (Greek *stoma*, "mouth," and *pous*, "foot"). Mantis shrimps. Head with two movable anterior segments, bearing eyes and antennules; marine. *Squilla.*

Superorder Pancarida (Greek *pan*, "all").

Order Thermosbaenacea (Greek *therme*, "heat"). Occurring only in hot springs. *Thermosbaena.*

Superorder Eucarida (Greek *eu*, "true," and *karis*, "shrimp"). Carapace large, covering all of thorax; paired eyes on stalks; thoracic appendages with branchia.

Order Euphausiacea (Greek *eu*, "true," *phaneien*, "to make appear," and *ousia*, "substance"). Krill. Thoracic appendages biramous and similar; single series of branchia (podobranchs) on thoracic appendages. *Euphausia.*

Order Decapoda (Greek *deka*, "ten," and *pous*, "foot"). Prawns, crayfish, lobsters, crabs. Thoracic appendages mostly uniramous, with branchia in three series (podobranchs, arthrobranchs, and pleurobranchs); abdomen elongate or short.

Suborder Natantia (Latin *natare*, "to swim"). Swimming decapods. *Penaeus, Palaemon* [*Leander*], *Crangon* [*Crago*].

Suborder Reptantia (Latin *reptare*, "to creep"). Crawling decapods. *Cambarus, Astacus, Homarus, Nephrops, Cancer, Pagurus, Carcinus, Maia, Uca.*

SUPERCLASS PROGONEATA

(Greek *pro*, "before," and *gone*, "that which generates.") Head distinct, with one pair of many-jointed antennae, paired eyes, and two or three pairs of jaws; respiration by tracheae; gonopore on anterior segments.

CLASS PAUROPODA (Greek *pauros*, "small," and *pous*, "foot"). *c.* 60 species. Body small, cylindrical; of 12 segments; nine pairs of legs, antennae three-branched.

CLASS DIPLOPODA (Greek *diploos*, "double"). Millipedes. Body usually cylindrical; thorax of four segments, each with one pair of legs; abdomen of 20 to 100 segments (double), each with two pairs of legs; gonopore on third segment; terrestrial. *Spirobolus, Julus.*

SUPERCLASS OPISTHOGONEATA

(Greek *opisthen*, "behind.") Head distinct, with one pair of many-jointed antennae; paired eyes and two or three pairs of jaws; respiration by tracheae; gonopore on posterior segments.

CLASS CHILOPODA (Greek *cheilos*, "lip," and *pous*, "foot"). Centipedes. Body of 15 to 173 segments, flattened dorso-ventrally; posterior pairs of maxillae usually fused to form labium; one pair of legs on each body segment; first pair of legs four-jointed, hook-like, with poison duct opening in terminal claw; gonopore midventral on next to last segment; terrestrial. *Scolopendra, Lithobius, Scutigera.*

CLASS SYMPHYLA (Greek *syn*, "with," and *phyle*, "clan"). Twelve, or less, leg-bearing segments; feet with two claws. *Scutigerella.*

CLASS INSECTA (Latin *insecare*, "to cut"). [Hexopoda.] Body distinctly divided into head (5 segments), thorax (3 segments), and abdomen (typically 11 segments); head with one pair of antennae, mouthparts adapted for chewing, sucking, or lapping; thorax typically with three pairs of jointed legs and two pairs of wings (variously modified or lacking); excretory organs Malpighian tubules; development direct, without metamorphosis, or indirect, with gradual or abrupt metamorphosis.

Subclass Apterygota (Greek *a*, "without," and *pteron*, "wing"). [Ametabola, without metamorphosis.]

Order Collembola (Greek *kolla*, "glue," and *embolen*, "peg"). Springtails. Body small; abdomen of not more than six segments; antennae four-jointed; wings lacking; Malpighian tubules lacking; tracheae usually lacking; characteristic leaping movements, effected by ventral springing organ (furcula) on fourth abdominal segment released by hook on third abdominal segment; first abdominal segment with ventral tube receiving sticky secretion from gland in head region by which animal adheres to objects; direct development; terrestrial, in damp habitats. *Podura.*

Order Protura (Greek *protos*, "first," "early," and *oura*, "tail"). Body small; antennae lacking; eyes lacking; wings lacking; legs five-jointed; abdomen of twelve segments, first three with paired, minute appendages; direct development; terrestrial, in damp habitats.

Order Diplura (Greek *diploos*, "double"). Body small; antennae long; eyes lacking; abdomen of 11 segments, with conspicuous cerci; in damp and dark habitats. *Campodea.*

Order Thysanura (Greek *thysanos*, "tassel"). Bristletails. Body small; antennae long; chewing mouthparts; wings lacking; abdomen of 11 segments with two- or three-jointed cerci on terminal segment; direct development. *Lepisma* (silverfish), *Thermobia* (fire brat).

Subclass Pterygota. [Metabola, with metamorphosis either gradual (Hemimetabola) or abrupt (Holometabola).]

The following 13 orders are hemimetabolous.

Order Ephemeroptera (Greek *ephemeros*, "lasting but a day"). [Plectoptera (Greek *plektos*, "twisted").] Mayflies. Antennae with two large basal joints and a bristle-like apical region; mouthparts (chewing) vestigial; two pairs of wings usually, with hind wings smaller than fore wings; abdomen with two or three many-jointed terminal filaments; adults aerial, nymphs aquatic. *Ephemera, Baetis.*

Order Odonata (Greek *odon*, "tooth"). Damselflies and dragonflies. Body large, often brightly colored; compound eyes large, prominent; two pairs of wings with complex venation; abdomen slender; adults aerial, nymphs aquatic. *Agrion, Aeshna, Anax.*

Order Orthoptera (Greek *orthos*, "straight"). Grasshoppers, crickets, locusts, katydids, cockroaches, mantises, stick insects. Body medium or large; chewing mouthparts; fore wings narrow and parchment-like; hind wings broad and membranous, folding beneath fore wings; abdomen usually with cerci and ovipositor. *Acheta* [*Gryllus*] (cricket), *Schistocerca, Locusta, Melanoplus, Blatta* (cockroach).

Order Dermaptera (Greek *derma*, "skin," and *pteron*, "wing"). Earwigs. Body elongate; chewing mouthparts; fore wings short, leathery, unveined; hind wings large, membranous, veined, folding beneath fore wings; wings lacking occasionally. *Forficula.*

Order Plecoptera (Greek *plekein*, "to twine"). Stoneflies. Body medium to large; chewing mouthparts, often lacking in adults; antennae long, setose; two pairs of membranous wings; hind wings larger than fore wings; abdomen usually with two long, many-jointed cerci on terminal segments; nymphs aquatic, adults aerial. *Perla, Nemoura.*

Order Isoptera (Greek *isos*, "equal"). Termites, white ants. Thorax broadly joined to abdomen; chewing mouthparts; social; polymorphic (workers, soldiers, sexual males and females). *Kalotermes.*

Order Zoraptera (Greek *zoros*, "sheer"). Body small; antennae nine-jointed; wings, where present, membranous with anterior pair longer than posterior; biting mouthparts.

Order Embioptera (Greek *embios*, "living"). Web spinners. Body small, chewing mouthparts; males usually winged, females always wingless; first segment of tarsi of forelegs inflated, with spinning organs. *Embia.*

Order Corrodentia (Latin *corrodere*, "to gnaw"). Book lice. Body small, chewing mouthparts; two pairs of membranous wings, or none, roofed over abdomen at rest.

Order Mallophaga (Greek *mallos*, "lock of wool," and *phagein*, "to eat"). Biting lice. Body small or medium; dorso-ventrally flattened; head broad; chewing mouthparts; antennae three- to five-jointed; eyes reduced or lacking; wings lacking; legs short with one- to two-jointed tarsi with claws; ectoparasitic.

Order Anopleura (Greek *anoplos*, "unarmed"). Sucking lice. Body small, dorso-ventrally flattened; head narrow; piercing-sucking mouthparts; eyes reduced or lacking; antennae five-jointed; wings lacking; ectoparasitic on mammals. *Pediculus* (human louse).

Order Thysanoptera (Greek *thysanos*, "tassel"). Thrips. Body small, slender; rasping-sucking mouthparts; antennae six- to nine-jointed; two pairs of narrow wings, if present, fringed with long hairs; legs with one- to two-jointed tarsi, ending in protrusible bladder; parthenogenesis common, obligatory in some species in which males are unknown; terrestrial and aerial.

Order Hemiptera (Greek *hemi*, "half"). Bugs. Body usually large; piercing-sucking mouthparts; two pairs of wings (or none); fore wings thick and horny at bases, membranous distally; hind wings wholly membranous, folding under fore wings; aquatic and terrestrial.

Suborder Homoptera (Greek *homos*, "the same"). Wings membranous and alike. Cicadas, scale insects, leafhoppers. *Aphis, Phylloxera.*

Suborder Heteroptera (Greek *heteros*, "different"). Fore wings modified as hemielytrae, with leathery base. *Cimex, Dysdercus, Rhodnius, Corixa, Notanecta.*

The following 11 orders are holometabolous.

Order Neuroptera (Greek *neuron*, "nerve"). Lacewings. Body small, or large; antennae long; chewing mouthparts; two pairs of large, membranous wings, roofed over abdomen at rest; legs with five-jointed tarsi. *Chrysopa.*

Order Megaloptera (Greek *megas*, "large"). Dobson flies. Antennae long; chewing mouthparts; two pairs of large membranous wings, roofed over abdomen at rest; veins not forked near margins. *Corydalis, Sialis.*

Order Rhaphideidea (Greek *rhaphis*, "needle"). Snakeflies. Body large; two pairs of transparent wings; antennae long; chewing mouthparts; prothorax elongated; ovipositor of females long; larva terrestrial, adult aerial.

Order Mecoptera (Greek *mekos*, "length"). Scorpionflies. Body small; antennae long; chewing mouthparts; two pairs of membranous wings, roofed over abdomen at rest, legs long.

Order Hymenoptera (Greek *hymen*, "membrane"). Ants, bees, and wasps. Chewing or chewing-lapping mouthparts; two pairs of small membranous wings, those of each side interlocked in flight; ovipositor of female sawing, piercing, or stinging; larvae caterpillar-like and legless; pupae in cocoons; some species social. *Formica, Vespa, Bombus, Apis.*

Order Coleoptera (Greek *koleos*, "sheath"). Beetles. Body with heavy chitinous covering; antennae usually 11-jointed; chewing mouthparts; fore wings thick, leathery, unveined, meeting (at rest) in middorsal line and acting as wing cases (elytra); hind wings membranous, folding under fore wings at rest; wings sometimes lacking. *Dytiscus, Tenebrio, Tribolium, Leptinotarsa.*

Order Strepsiptera (Greek *strepsis*, "turning"). Body small; chewing mouthparts, when present; fore wings reduced to halteres in male; hind wings fan-shaped; antennae, eyes, wings, and legs lacking in female; females and larvae permanently endoparasitic in bees, wasps, and aphids. *Stylops.*

Order Trichoptera (Greek *thrix*, "hair"). Caddis flies. Body and wings covered with hairs; antennae long; mouthparts rudimentary; two pairs of membranous wings arched over abdomen at rest; larvae aquatic, living in tubes constructed of a silk-like secretion, often with foreign particles (sand, gravel, leaves, stems).

Order Lepidoptera (Greek *lepis*, "scale"). Moths and butterflies. Antennae long; eyes large; sucking mouthparts; maxillae fused as long coiled tube, other mouthparts rudimentary; two pairs of broad membranous wings covered with overlapping scales; larvae (caterpillar and cutworm) with chewing mouthparts and three pairs of true legs, prolegs on abdomen; aerial. *Galleria, Ephestia, Bombyx.*

Order Diptera (Greek *dyo*, "two"). Flies, mosquitoes. Body distinctly divided; abdomen with four to nine visible segments; fore wings transparent, veined; hind wings reduced to knob-like halteres; wings sometimes lacking; piercing-sucking and sponging mouthparts. *Culex, Anopheles, Drosophila, Tabanus, Calliphora.*

Order Siphonaptera (Greek *siphon*, "tube"). Fleas. Body small, laterally flattened; antennae short, in grooves; piercing-sucking mouthparts; wings lacking; legs long, with enlarged coxae and five jointed tarsi; larvae minute, legless; pupae in cocoons; adults periodically parasitic and blood-sucking on birds and mammals. *Pulex.*

PHYLUM PENTASTOMIDA

(Greek *pente*, "five.") [LINGUATULA (Latin *lingua*, "tongue").] Body elongated, unsegmented; cephalothorax short; abdomen long; two pairs of retractile hooks beside mouth; digestive tract complete (mouth and anus); circulatory, respiratory, and excretory organs lacking; endoparasitic in mammals.

PHYLUM TARDIGRADA

(Latin *tardus*, "slow," and *gradi*, "to step.") Water bearers. Body small, cylindrical, with rounded ends, unsegmented; four pairs of short legs with terminal claws; digestive tract complete (mouth and anus); circulatory, respiratory, and excretory organs lacking; terrestrial, in damp moss.

The following four phyla are enterocoelous.

PHYLUM CHAETOGNATHA

(Greek *chaite*, "hair," and *gnathos*, "jaw.") Arrowworms. 50 species. Body elongate, unsegmented, with lateral fins; two lobes with chitinous bristles lateral to mouth; digestive tract complete (mouth and anus) ventral; circulatory, respiratory, and excretory organs lacking; marine. *Sagitta.*

PHYLUM POGONOPHORA

(Greek *pogon*, "beard.") Beard worms. 43 species. Body slender, 10 to 35 mm long, divided into preannular and postannular regions by a pair of ridges (belts); single, several, or many long ciliated tentacles; no digestive tract; tubiculous; benthic.

CLASS ATHECANEPHRIA. Postannular adhesive papillae, few and irregularly scattered.

CLASS THECANEPHRIA. Postannular adhesive papillae, regularly arranged in transverse rows.

PHYLUM ECHINODERMATA

(Greek *echinos*, "spiny," and *derma*, "skin.") *c.* 5,700 species. Sea lilies, sea cucumbers, sea urchins, sand dollars, sea stars (starfish), serpent stars (brittle stars). Body secondarily radially symmetrical (larvae bilateral), usually pentamerous, without segmentation; mesodermal endoskeleton of scattered or compact (fixed or movable) calcareous plates and usually spines; coelom includes water-vascular system, with tube feet (podia) extending in rows along each radius; digestive system usually complete (mouth and anus); nervous system with circumoral ring and radial nerves; development generally indirect, through free-swimming larval stages; marine.

SUBPHYLUM PELMATOZOA

(Greek *pelma*, "sole.") Stalk for attachment on aboral surface either in juvenile stages or throughout life.

CLASS CRINOIDEA (Greek *krinoeidos*, "lily-like".) Feather stars, sea lilies. Body cup-shaped; branched arms; tube feet without suckers; spines and pedicellariae lacking.

Order Articulata. Tegmen flexible; arms articulated with calyx (theca). *Antedon* (feather star), *Metacrinus* (sea lily).

SUBPHYLUM ELEUTHEROZOA

(Greek *eleutheros*, "free.") Unstalked and free-moving.

CLASS HOLOTHURIODEA (Greek *holothuria*, "a kind of polyp"). Sea cucumbers. *c.* 500 species. Body elongated, cylindrical, with mouth and anus terminal; skeleton of microscopic plates, widely dispersed in body wall; arms, spines, and pedicellariae lacking; tube feet usually present; larva auricularia.

Order Aspidochirota (Greek *aspis*, "shield," and *cheir*, "hand"). Tentacles shield-shaped; numerous tube feet. *Holothuria, Stichopus.*

Order Elasipoda (Greek *elasos*, "bird"). Numerous tube feet; no respiratory tree; deep sea, benthic.

Order Dendrochirota (Greek *dendron*, "tree"). Tentacles around mouth irregularly branched; tube feet numerous. *Cucumaria, Thyone, Sclerodactyla.*

Order Molpadonia (probably after Molpadonia, minor Greek goddess). Tentacles small, hand-shaped; tube feet lacking. *Molpadia.*

Order Apoda. Tube feet lacking; respiratory tree lacking. *Synapta, Leptosynapta.*

CLASS ECHINOIDEA (Greek *echinus*, "spiny"). Sea urchins, sand dollars. Body hemispherical or disk-shaped; arms lacking, enclosed in rigid test, with many movable spines and three-jawed pedicellariae of firmly sutured plates; tube feet with suckers; larva pluteus.

Subclass Regularia. Sea urchins. Body globose; with special masticatory apparatus (Aristotle's lantern).

Order Cidaroidea. Rigid globular test; each ambulacral plate with a single large spine and circlet of small spines. *Cidaris.*

Order Lepidocentroida (Greek *lepis*, "scale"). Flexible test; spines hollow, often with poison glands. Few existing species; abyssal. *Phormosoma.*

Order Stirodonta (Greek *steros*, "solid"). Rigid test; spines solid; teeth of Aristotle's lantern keeled. *Arbacia.*

Order Aulodonta (Greek *aulos*, "hall"). Ambulacral plates compound, composed of three primary plates; spines solid; teeth unkeeled. *Diadema.*

Order Camorodonta (Greek *kamara*, "vaulted chamber"). Spines solid; pedicellariae numerous; teeth keeled. Includes most of the common sea urchins of shallow waters. *Temnopleurus, Toxopneustes, Lytechinus, Tripneustes [Hipponoe], Strongylocentrotus, Echinometra.*

Subclass Irregularia. Test flattened; periproct outside, and posterior to, system of apical plates.

Order Holectypoida (Greek *holos*, "whole," and *echtypos*, "worked in relief"). Anus separate from apical plates; peristome central. Mostly extinct.

Order Cassiduloida (Greek *cassis*, "helmet"). Test round or oval. Aristotle's lantern lacking in adults. Mostly extinct.

Order Clypeastroida (Latin *clypeus*, "shield," and Greek *aster*, "star"). Sand dollars, sea biscuits, cake urchins. Body flattened; anus marginal and on oral surface. *Clypeaster, Echinaruchnius, Mellita.*

Order Spatangoida (Greek *spatanges*, "kind of sea urchin"). Heart urchins. Body heart-shaped; anus eccentric in position; Aristotle's lantern lacking. *Echinocardium.*

CLASS ASTEROIDEA (Greek *aster*, "star," and *eidos*, "form"). Starfish, sea stars. Body star-shaped or pentagonal; arms not sharply set off from disk; pedicellariae and spines short; ambulacral groove open, with two to four rows of tube feet with suckers; larva bipinnaria or brachiolaria.

Order Platyasterida (Greek *platys*, "flat"). Five to many strap-shaped arms; ambulacral grooves wide flow. Mostly extinct. *Luidia.*

Order Paxillosida (Latin *paxillus*, "a peg"). Disk small; arms long and pointed; aboral surface with paxillae (special raised ossicles, each with a crown of spines that can be raised and lowered through a right angle to form a chamber; gland cells on spines secrete mucus). *Astropecten.*

Order Valvatida (Latin *valvatus*, "having folding doors"). Tube feet terminate in suckers; paxillae in some families; disk spherical or pentagonal. *Linckia.*

Order Spinulosa (Latin *spinulosus*, "spiny"). Papulae on both oral and aboral surfaces; pedicellariae rare; usually with paxillae on aboral surface. *Asterina, Patiria, Henricia, Acanthaster.*

Order Forcipulata (Latin *forceps*, "pincers"). Pedicellariae with cross jaws, like scissors; papulae on oral and aboral surfaces; skeleton reticulate; no paxillae. *Asterias, Pisaster.*

CLASS OPHIUROIDA (Greek *ophio*, "serpent," and *oura*, "tail"). Serpent stars, brittle stars. *c.* 2,000 species. Arms slender and sharply marked off from disk; tube feet in two rows, sensory, without suckers; pedicellariae lacking; madreporite on oral surface; larva ophiopluteus.

Order Phrynophiurida (Greek *phrynos*, "toad"). Disk and arms covered with skin overlying and obscuring plates and scales; in most, disk plates are absent or much reduced; arms straight or branched. *Ophioderma.*

Order Ophiuroida. Disk and arms usually scaly; arms straight and so articulated that only horizontal movement is possible. *Ophiura, Ophiactis, Ophiothrix.*

INVERTEBRATES OF UNCERTAIN (UNDETERMINED) SYSTEMATIC POSITIONS

ORTHONECTIDA (Greek *orthos*, "straight," and *nektos*, "swimming"). Endoparasitic with alternation of sexual and asexual generations; asexual phase (parasitic in marine turbellarians, nemerteans, bivalve and opisthobranch molluscs, annelids, and ophiuroids), a multinucleate plasmodium from which sexual individuals arise; sexual (free-swimming) phase, minute cellular organisms (not more than 7 mm long), with outer layer of ciliated cells (some cells contractile) enclosing sperm or ova. *Rhopalura.*

DICYEMIDA (Greek *dyo*, "two," and *kyema*, "embryo"). Endoparasitic with alternation of sexual and asexual generations. Asexual phase (parasitic in excretory organs of cephalopod molluscs), minute cellular organisms, with outer layer of 18 to 24 cells (somatoderm) enclosing a single elongated axial cell (axoblast), which in an immature host develops into a ciliated agamont (nematogen) capable of directly reinfecting the host, without escape to the outside, and of reproducing more nematogens by divisions of agametes arising from the axial cell. In sexually mature hosts, the nematogens give rise to rhombogens (Greek *rhombos*, "a spinning top"), similar to nematogens except that they produce coherent groups of cells (infusoriform larvae) that give rise to oocytes and tailless sperm and escape from the host through its excretory canals to a fate as yet unknown.

These two groups have been included in the Phylum Mesozoa (50 species) and variously regarded as intermediate between Protozoa and Metazoa or as retrograde Metazoa. While there is some resemblance between orthonectids and the nematogens and rhombogens of dicyemids, there seems to be no close relationship between these two groups, and it is suggested, at least until more is known about them, that they be given separate status and an undefined position.

PHYLUM (?) GNATHOSTOMULIDA (Greek *gnathos*, "jaw," and *stoma*, "mouth"). Nine species. Minute cylindrical, or slightly depressed, transparent body, divisible into head, collar, and tail regions; single layer of epidermal cells, each with a single flagellum; enteron straight, incomplete (no anus); pharynx complicated, with jaws; parenchyma poorly developed; no protonephridia; peripheral nervous system with concentration of nerve cells at "head" end; statocysts; worldwide distribution; marine, attached to substrate by secretion from adhesive tubes on surface of body.

Because their relationship seems closest to turbellarians these "lower worms" are sometimes considered a class in Platyhelminthes. They also show affinities to gastrotriches and rotifers and have been placed in the "phylum" Aschelminthes. Certain affinities with Achiannelida make their position all the more uncertain, and it has been suggested that they be assigned to a phylum of their own.

PHYLUM (?) PRIAPULIDA (Greek *priapus*, "phallus"). Eight species. Body divided into anterior inversible proboscis with small spines and trunk (ringed); digestive tract straight; mouth and anus terminal; protonephridia with solenocytes; marine.

The inclusion of the known species of the genus Priapulus in a class in the "phylum" Aschelminthes is based on certain recognized anatomical resemblances to the kinorhynches and nematodes especially, here given status as separate phyla. However, the presence of a spacious body cavity with epithelial covering and apparent mesenteries has led many systematists to include them among the coelomate animals, with possible relationships to the sipunculids and echiuroids. Incompleteness of embryological evidence as to the origin of this cavity and its homology with the coelom, and evidence that the membrane supporting the reproductive and excretory organs is not a true mesentery but an extension of the gonad epithelium leave these animals still in a doubtful systematic position.

SELECTED BIBLIOGRAPHY

Origins

Boyden, A.: Are There Any "Acellular Animals?", *Science*, vol. 125, pp. 155–156, 1957.

Corliss, J. O.: On the Evolution and Systematics of Ciliated Protozoa, *Syst. Zool.*, vol. 5, pp. 68–91; 121–140, 1956.

Curtis, A. S. G.: Cell Contact: Some Physical Considerations, *Am. Nat.*, vol. 94, pp. 37–56, 1960.

Dobell, C.: The Principles of Protistology, *Arch. Protistenk.*, vol. 23, pp. 269–310, 1911.

Fauré-Fremiet, E.: The Origin of the Metazoa and the Stigma of the Phytoflagellates, *Q. J. Microsc. Sci.*, vol. 99, pp. 123–129, 1958.

Fell, H. B.: Echinoderm Embryology and the Origin of Chordates, *Biol. Rev.*, vol. 23, pp. 81–107, 1948.

Hädzi, J.: "The Evolution of the Metazoa," Pergamon, New York, 1963.

Hanson, E. D.: On the Origin of the Eumetazoa, *Sys. Zool.*, vol. 7, pp. 16–47, 1958.

Hardy, A. C.: On the Origin of the Metazoa, *Q. J. Microsc. Sci.*, vol. 94, pp. 441–444, 1953.

Hyman, L.: The Transition from the Unicellular to the Multicellular Individual, in R. Redfield (ed.), "Biol. Symposia," vol. 8, "Levels of Integration in Biological and Social Systems," pp. 27–42, A.A.A.S., Washington D.C., 1942.

Rounds, H. D.: The Evolution of the Invertebrates, *Wichita State Univ. Bull.*, vol. 42, no. 2, 1966.

Sanders, H. L.: The Cephalocarida: A New Subclass of Crustacea from Long Island Sound, *Proc. Natl. Acad. Sci.*, vol. 41, pp. 61–66, 1955.

Zalokar, M.: "Degenerate" Coelenterates, *Syst. Zool.*, vol. 4, pp. 191–192, 1955.

Classification

Bigelow, A. S.: Classification and Phylogeny, *Syst. Zool.*, vol. 7, pp. 49–59, 1958.

Blackwelder, R. E., and A. Boyden: The Nature of Systematics, *Syst. Zool.*, vol. 1, pp. 26–33, 1952.

Gordon, I.: Importance of Larval Characters in Classification, *Nature*, vol. 176, p. 911, 1955.

Hädzi, J.: An Attempt to Reconstruct the System of Animal Classification, *Syst. Zool.*, vol. 2, pp. 145–154, 1953.

Levi, C.: Ontogeny and Systematics in Sponges, *Syst. Zool.*, vol. 6, pp. 174–184, 1957.

Myers, G. S.: The Nature of Systematic Biology and of a Species Description, *Syst. Zool.*, vol. 1, pp. 106–111, 1952.

Reidl, R. J.: Gnathostomulida: A New Animal Phylum, *Science*, vol. 163, pp. 445–452, 1969.

Phylogenesis

Boyden, A.: Precipitins and Phylogeny, *Am. Nat.*, vol. 68, pp. 516–536, 1934.

Boyden, A.: Review of Systematic Serology, *Physiol. Zool.*, vol. 15, pp. 109–145, 1942.

Boyden, A.: Serology in the Study of Animal Relationships, *Am. Nat.*, vol. 77, pp. 234–255, 1943.

Carter, G. S.: On Hädzi's Interpretations of Animal Phylogeny, *Syst. Zool.*, vol. 3, pp. 163–167, 1954.

de Laubenfels, M. W.: Are Coelenterates Degenerate or Primitive? *Syst. Zool.*, vol. 4, pp. 43–45, 1955.

Dodson, E. A.: A Note on the Systematic Position of the Mesozoa, *Syst. Zool.*, vol. 5, pp. 37–40, 1956.

Eaton, T. H. Jr.: Pedomorphosis: An Appraisal of the Chordate-Echinoderm Problem, *Syst. Zool.*, vol. 2, pp. 1–6, 1953.

Garstang, W.: The Theory of Recapitulation: A Critical Restatement of the Biogenetic Law, *J. Linnean Soc.*, vol. 35, pp. 81–101, 1922.

Grimstone, A. V.: Cytology, Homology and Phylogeny: A Note on "Organic Design," *Am. Nat.*, vol. 93, pp. 273–282, 1959.

Manton, S. M.: Locomotory Habits and the Evolution of the Larger Arthropodan Groups, *Symp. Soc. Exp. Biol.*, vol. 7, p. 339–376, 1953.

Sanders, H. L.: The Cephalocarida and Crustacean Phylogeny, *Syst. Zool.*, vol. 6, pp. 112–128, 1957.

Snodgrass, R. E.: Evolution of the Annelida, Onychophora and Arthropoda, *Smithsonian Misc. Publ.*, vol. 97, no. 6, 1938.

Thorson, G.: Reproduction and Larval Ecology of Marine Bottom Invertebrates, *Biol. Rev.*, vol. 25, pp. 1–45, 1950.

Tiegs, O. W., and S. M. Manton: The Evolution of the Arthropoda, *Biol. Rev.*, vol. 33, pp. 255–337, 1958.

Wilhelmi, R. W.: Serological Relations: Invertebrates, *Biol. Bull.*, vol. 87, pp. 96–105, 1944.

Woods, K., E. Paulson, R. J. Engle, and J. Pert: Starch Gel Electrophoresis of Some Invertebrate Sera, *Science*, vol. 127, pp. 519–520, 1958.

Yonge, C. M.: The Pallial Organs of the Aspidobranch Gastropoda and Their Evolution throughout the Mollusca, *Philos. Trans. R. Soc. London B.*, vol. 232, pp. 443–518, 1947.

SUGGESTED READING FOR SECTION THREE

Anfinson, C. B.: "The Molecular Basis of Evolution," Wiley, New York, 1959.

Blum, H. F.: "Time's Arrow and Evolution," 4th ed., Princeton, Princeton, N.J., 1968.

Darwin, C., and A. R. Wallace: "Evolution by Natural Selection," Cambridge, London, 1958.

deBeer, Sir G.: "Embryos and Ancestors," Oxford University Press, New York, 1940.

Mayr, E.: "Systematics and the Origin of Species," Dover, New York, 1942.

Simpson, G. G.: "Principles of Animal Taxonomy," Columbia, New York, 1961.

section four

LOCOMOTION

chapter VI LOCOMOTOR DEVICES AND MECHANISMS

LOCOMOTOR DEVICES AND MECHANISMS

A. LOCOMOTION BY CHANGES IN BODY SHAPE
- **1. By projections of the body**
  - a. PSEUDOPODIA
  - b. FORWARD PROTRUSION
- **2. By general contractions of the body**

B. LOCOMOTION BY MEANS OF SPECIAL APPENDAGES
- **1. Cell organelles: cilia and flagella**
- **2. Nonjointed appendages: parapodia, tube feet**
- **3. Jointed appendages: the arthropod limb**
  - a. DERIVATION FROM A UNIRAMOUS PROTOTYPE
  - b. DERIVATION FROM A BIRAMOUS PROTOTYPE
  - c. WALKING AND RUNNING
  - d. SWIMMING
  - e. PLOUGHING
  - f. BURROWING AND DIGGING
  - g. HOPPING AND LEAPING
- **4. Wings and the mechanism of flight in insects**

Motion is one of the most basic and vitally important functions of living matter, and it is exhibited in some form or other by all organisms in the active phase of their lives. It may be evident as streaming movements within the cells themselves or as the gross movement of entire bodies from place to place, but in every case it is a manifestation of energy changes within the system. Work is done in the translocation of a body through distance; the amount of work is proportional to the size of the body and the distance through which it is moved. Hence, the greater the motor activity of an organism, the greater are its energy expenditures and energy demands. These must be met, unless the organism is to consume its own substance, by the food it digests, assimilates, and metabolizes.

Locomotor mechanisms are those that effect movement of a body from place to place. In animals, they may be such as to cause movement of the body as a whole so that it progresses from point to point at varying rates of speed, or they may be such as to cause movements of parts of the organism only. These local movements may turn some area of the body in one direction or another, or they may effect the passage of other bodies along its surface or within its organs.

There are two basic mechanisms by which progressive movement of the body as a whole, or locomotion, is brought about. These are (1) changes in the shape of the body; and (2) the activity of special appendages, such as flagella, cilia, and undulating membranes, and of extensions of the body wall that are moved by muscles, such as the parapodia of polychaete annelids, the tube feet of echinoderms, the jointed appendages of arthropods and terrestrial vertebrates, and the wings of insects. Mobile invertebrates have adapted these mechanisms in a far greater number of ways than have vertebrates. But not all invertebrates are mobile. Some, at least in the adult phase of their lives, lead a sessile or sedentary existence, more or less permanently attached to some suitable substrate. For these, mobility of localized structures is essential, so that food may be swept or drawn to them.

Two primary factors determine which locomotor mechanism is appropriate for use, and so for modification and adaptation. These are the mechanical properties of the medium in which the animal lives and its own size. The most demanding and challenging of the various environmental media that Earth offers is the soil, where locomotion must be rectilinear and, while mechanically relatively simple, demands great efforts for a comparatively small return in absolute and relative velocity. There is greater friction to be overcome than in the air or water. Air, while presenting far more complex mechanical problems, permits easy change in direction and demands considerably less effort to attain high absolute and relative velocities. It is, however, a very unstable environment, and aerial organisms are subject to sudden and often drastic changes in conditions through wind, rain, and fluctuations in temperature and the intensity of light. Locomotion on the Earth's surface and in its waters presents a graded series of conditions between these two extremes of subterranean and aerial travel. In swimming, as in flying, progress is made by displacement of the surrounding medium, but a swimmer has greater advantage in the water's buoyancy than a flyer has in the buoyancy of the air. Crawlers, walkers, runners, and leapers on the surface of the Earth have the advantage of a firm substrate to push against, but the disadvantage of having to move over or around, fixed obstacles in their paths.

Locomotion by flagella, cilia, or undulating membranes is possible only in a fluid medium of relatively low viscosity and only for bodies of small size. For larger animals, a premium has been set upon the development of contractile cells and a strong and controlled musculature. In some species, ciliary action may be auxiliary to muscular, but in the majority of motile invertebrates, locomotion is effected primarily by muscular movement of the body or its parts. The two important requirements in locomotor systems using muscles are (1) that the force produced by the muscles be transferred so as to act between the body and the medium, and (2) that provision be made for reextending the contracted muscle, for the active phase of a muscle's functioning is shortening (contraction) and the passive one is extension (relaxation). In most of the invertebrates, this provision is met by the hydrostatic skeleton (see page 66).

A. LOCOMOTION BY CHANGES IN BODY SHAPE

1. By projections of the body

a. **PSEUDOPODIA** Locomotion by pseudopodia is characteristic of mesenchyme cells, of the migratory cells or amebocytes present in the bodies of all Metazoa, of the eggs of some animals—notably of sponges and some Cnidaria—and of Rhizopoda. Pseudopodia are temporary extensions of a cell that may be formed and retracted in different areas of its surface, so that its shape, within a characteristic range, is always changing. These extensions may be broad and blunt, like the lobopodia of *Amoeba*, *Pelomyxa*, and amebocytes in general; they may be slender and filamentous, like the filopodia of many Testacea and of fibroblasts; or they may be slender and pointed, supported by a fibrillar axial rod, like the more or less permanent axopodia of Heliozoa and Radiolaria (Figs. VI–1 and VI–2). In any of these forms, they may be branched or unbranched. The rhizopodia of Foraminifera are characteristically filamentous and branching, with anastomoses of the branches forming a network.

Protoplasmic streaming, a very general cellular phenomenon but one whose mechanism is not as yet understood, is involved in pseudopodium formation. Observations of an extending, or of a retracting, pseudopod reveal an axial stream of comparatively fluid endoplasm moving within a tube of firmer, more viscous ectoplasm. In rhizopods such as *Amoeba limax*, where a single, broad pseudopod is formed at the advancing, or anterior, end, the entire body may be taken to represent the tube, but in cells where a number of pseudopods may be formed simultaneously, each in

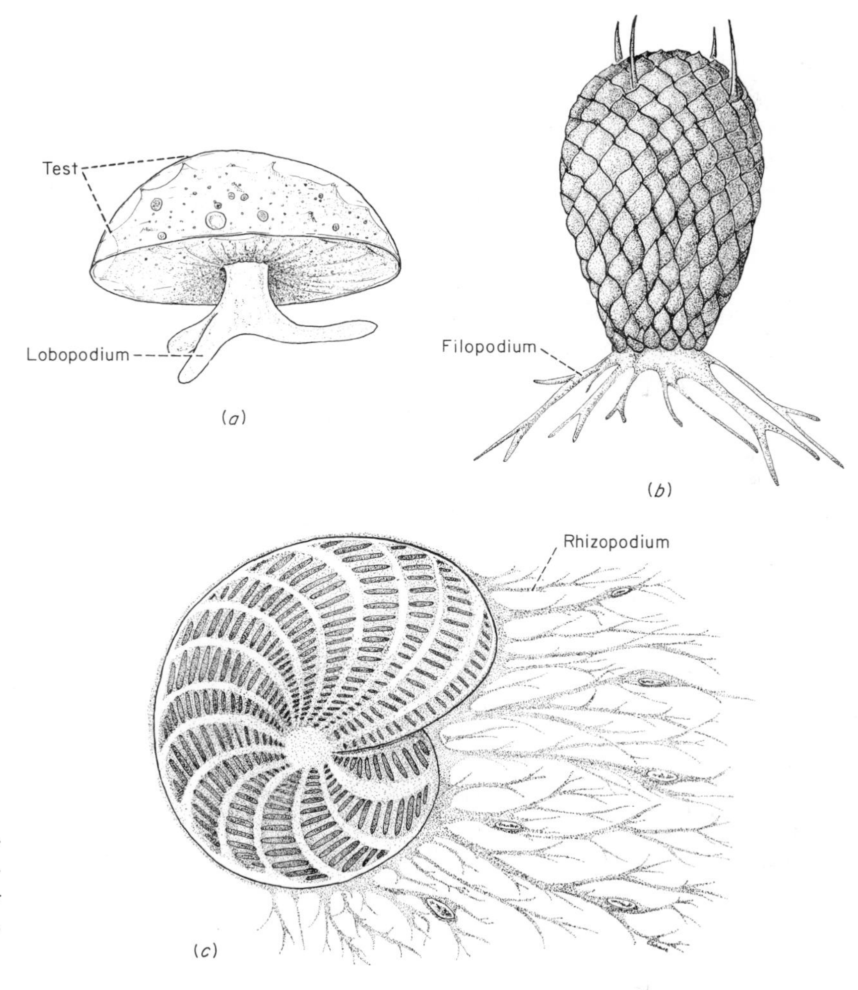

FIG. VI–1 *Types of pseudopodia.* ***(a)*** Arcella discoides *with lobopodia.* ***(b)*** Euglypha alveolata *with filopodia.* (*After Jahn.*) ***(c)*** Elphidium crispa, *a foraminiferan, with rhizopodia.* (*After Jahn.*)

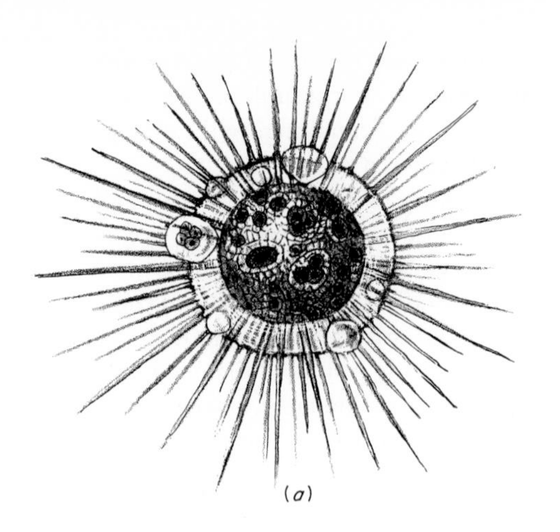

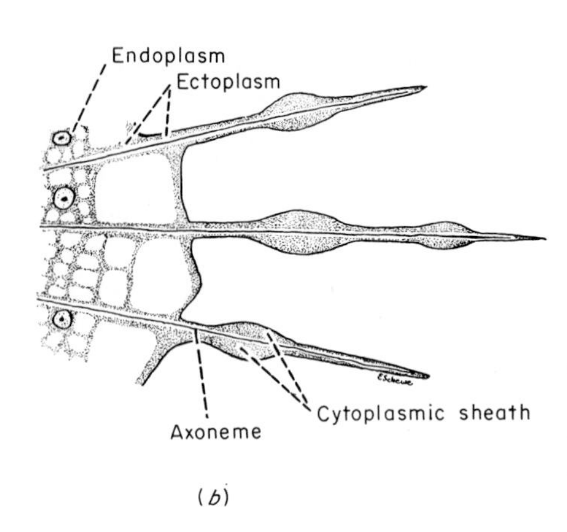

FIG. VI–2 *Axopodia.* ***(a)*** Actinosphaerium eichorni, *a heliozoan, with axopodia. (After Doflein.)* ***(b)*** *Detail of axopodium. (After Jahn.)*

itself may be regarded as the tube. The tube is enclosed in a pliable, relatively permeable sheath, the plasmalemma, whose thickness varies in different species; it may be several dozen microns in thickness in some species, while in others it is considerably thinner, and, in still others, entirely absent. Within the plasmalemma is a layer of hyaline, nongranular ectoplasm surrounding granular ectoplasm, in which particles may be seen in active Brownian movement, and which makes the inner lining of the tube. At the tip of the pseudopod, the zone of hyaline ectoplasm is wider than along the margins of the tube, forming a conspicuous hyaline cap (Fig. VI–3).

A necessary condition for effective pseudopodial locomotion, although not for pseudopodial formation, is the presence of a foundation or substratum to which the pseudopod may attach. During progressive movement, the plasmalemma is firmly fastened at a limited number of points both to the substratum and to the ectoplasmic tube, so that the tube is extended forward. Given adequate support, an ameboid cell may move in any direction at one time and may change direction. The speed at which it moves is influenced primarily by the viscosity of the medium and by temperature. Under similar conditions, the measured rates of *Amoeba*'s movement have been found to range from 0.5 to 4.5 μ/sec, with an average speed of about 1 μ/sec.

Pseudopodial formation has been most intensively studied in Rhizopoda, especially in *Amoeba*, and in the period since the first observations of ameboid movement by Rosel von Rosenhof in 1755 to the present day, a variety of theories have been offered in explanation of it, all adapted to the prevailing views of the composition and properties of cytoplasm. There is general agreement that the movement of the endoplasm is passive; the central problem becomes, therefore, the nature and the exact site of action of the force that moves it. Observation and experimental evidence have invalidated some of the older theories concerning this motive force, such as the surface tension theory which had very general acceptance for a number of years. Present theories have a common basis in the acceptance of a contractile mechanism, but differ in location of the contractile force and in interpretation of the contractile mechanism itself.

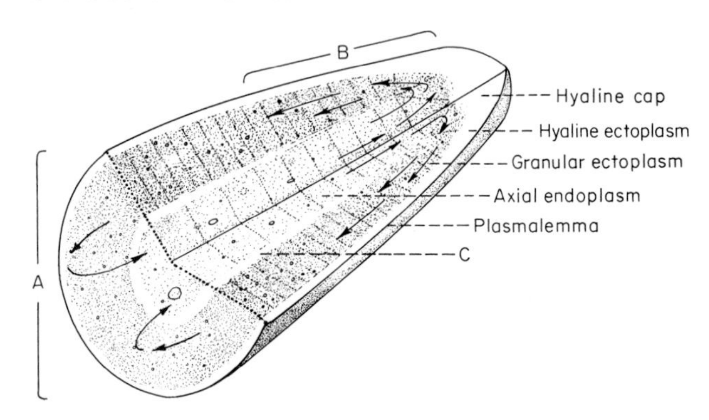

FIG. VI–3 *An advancing pseudopod. (After Allen.) The arrows indicate the direction of flow. A: site of the contractile mechanism according to the ectoplasmic-contraction hypothesis. B: site of the contractile mechanism according to the "fountain-zone" hypothesis. C: shear zone.*

The ectoplasmic-contraction hypothesis and the fountain-zone hypothesis both regard the ectoplasm (plasmagel) and the endoplasm (plasmasol) as contracted and uncontracted states of the same material, and locate the contractile mechanism in the ectoplasm. Although electron micrographs have not revealed any material in the ectoplasm comparable to the fibrils seen in muscle or other contractile systems, observation of its properties provides indirect evidence at least of its contractility. According to the ectoplasmic-contraction hypothesis, the contractile apparatus is a three-dimensional network of fibers held together by chemical as well as physical bonds with hyaline fluid trapped in the interstices of the mesh. It assumes that somewhere along the plasmagel tube a softening, or solation, occurs, into which endoplasm flows under pressure generated by contraction of the plasmagel in an opposite area. This pressure results from contraction, possibly folding, of the fibers and the consequent squeezing of fluid from the meshwork. Physical chemists have given the term syneresis to such a separation of a liquid from a gel. The pseudopod is thus pushed, or blown, out and may advance until it reaches a critical stage, usually a length of about 150 μ, when the pressure of the advancing stream is so reduced near its tip that it is just adequate to raise the membrane off the plasmagel. At this point, membrane and plasmagel are brought into contact, by some assumed process, and this contact then initiates contraction of the plasmagel at the tip, causing a backward flow running counter to the outward flow. Reduction in the diameter of the pseudopod by about 35 percent, when it has reached its maximum extension, ensures that this backward flow is now the principal one, and so the pseudopod is retracted. If the region of an advancing pseudopod is considered anterior, then, according to this hypothesis, the motive force is produced by contraction of the ectoplasm in the posterior region pushing the endoplasm forward.

The fountain-zone hypothesis, on the other hand, locates the motive force anteriorly and postulates that the endoplasm is pulled, rather than pushed, into a pseudopod. It visualizes the axial stream of endoplasm diverted, or everted, in regular streamlines, so creating a "fountain zone" where the lines run counter to the main axial flow. The cytoplasm then shortens and thickens as it is displaced through the fountain zone into the ectoplasm for here endoplasm is constantly being converted to ectoplasm. The conversion of ectoplasm into endoplasm takes place posteriorly, so that the axial endoplasm is continuously replenished. The fountain zone, rather than the posterior region of the ectoplasmic tube, is thus the site of the propagated contraction that draws, or drags, the endoplasmic stream into an advancing pseudopod. This theory has the advantage of showing how flow in two directions can occur simultaneously in a single pseudopod, a phenomenon, resembling cyclosis in plants, that is known to occur in the axopodia of Heliozoa and Radiolaria.

A third theory, derived particularly from study of movements in slime molds, attributes the motive force to shearing action arising in a zone between the axial endoplasm and the peripheral ectoplasm. If there is interaction between the fibrous components of the plasmasol and the plasmagel, as there is believed to be between the thick and thin interdigitating fibrils in certain types of muscles, contraction would result and the inner mass of endoplasm be drawn along the ectoplasmic tube. This is the mechanism that has been postulated to explain cyclosis in plants and streaming in slime molds and in the pseudopodia of Foraminifera. It is a mechanism that may have wider application than that of ectoplasmic or fountain-zone contraction, reducing all cytoplasmic movements to a common pattern.

There is as yet little decisive information as to the source of the energy that brings these changes about, but there is at least indirect evidence that ATP* is involved. Injection of ATP into amebae and slime molds causes local contraction, and viscosity changes in protein extracted from slime molds have been induced by the addition of ATP to them. It is a reasonable assumption that the biochemical events leading to the contraction of the ectoplasm, wherever this may take place, as well as the mechanism of its actual contraction, are essentially the same as in the shortening of muscle.

* ATP is an abbreviation conventionally used for the compound "adenosine triphosphate." Its importance as donor and acceptor of the phosphate radical in oxidation-reduction reactions is discussed in Chapter XII.

b. FORWARD PROTRUSION Movement of this kind, which bears some resemblance to pseudopodial locomotion, is practiced by some cnidarians and by some burrowing worms (Fig. VI–4). For example, one of the several methods of progression in *Hydra* is a gliding one, brought about by forward extension of the cells on the basal disk. Anthozoa with pedal disks can move at a rate of 8 to 10 centimeters per hour (cm/hr) by putting out a lobe of the disk in the direction of advance; the other side of the disk contracts simultaneously and so exerts a push on the extended lobe. The polychaete marine annelids, *Nephthys* and *Arenicola*, extend their proboscides to burrow through soft, damp sand. Each can progress rapidly through the sand or mud of the littoral zone by repeated eversions and retractions of the proboscis. In both species, there is little or no circular musculature and the coelom is only incompletely divided by septa. Contraction, primarily of the longitudinal muscles, drives the coelomic fluid anteriorly, setting up a hydrostatic pressure that leads to protrusion of the proboscis. By this means, *Nephthys* punches a hole in the sand while holding itself in place by means of dilated segments (from about segment XV to segments XL to L) and the extended parapodia attached to them. After the hole has been made, the proboscis is withdrawn through shortening of the proboscis retractor muscles and relaxation (extension) of the longitudinal body muscles, and the worm moves into the hole to begin the cycle again. *Nephthys* is unique among worms, and indeed among invertebrates, in having a system of ligaments. These are broad straps of tissue overlying the nerve cord and running to the body wall and parapodia. They are composed of bands of alternating elastic and inelastic proteins, which differ from each other in their chemical as well as in their physical properties. The chief antagonists of the longitudinal muscles are the dorso-ventral muscles of the body wall between the segments, and it is probable that these ligaments are the main factors in preventing lateral expansion of the body when the longitudinal muscles contract. They may also serve to resist the successive shocks to which the worm is subjected in its method of burrowing. This is essentially a series of convulsive jerks, through which its proboscis is thrust into the sand. Each impact of the proboscis against the sand produces a high, transient intracoelomic pressure, which the ligaments with their elastic properties seem well fitted to resist.

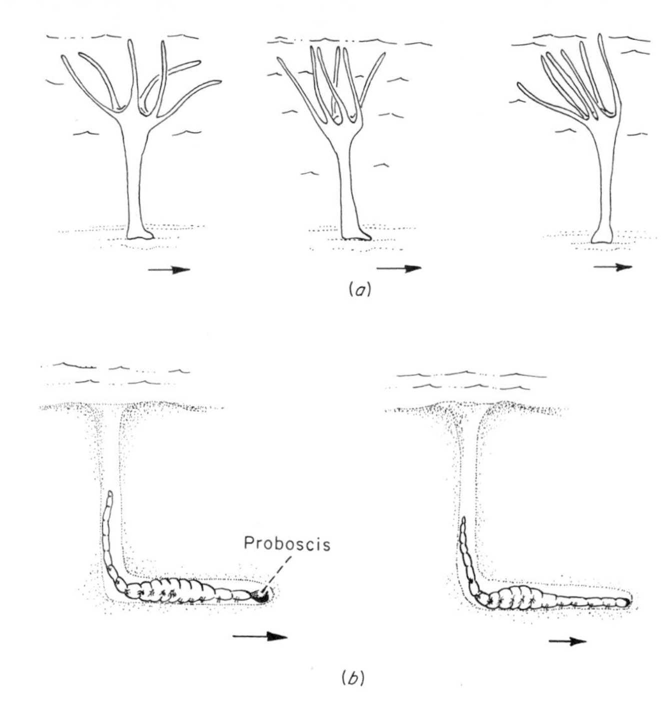

FIG. VI–4 *Locomotion by forward protrusion of the body.* ***(a)*** Hydra *moving by forward protrusion of the cells of the basal disk.* ***(b)*** Arenicola *making a burrow by successive eversions and retractions of its proboscis.*

Arenicola is not as active a burrower as *Nephthys* and pushes or scrapes away the sand grains with its everted proboscis to make a hole into which it can advance. Experiments have shown that the volume of its coelomic fluid is just adequate to provide maximum efficiency for this kind of progression, for if some is withdrawn the animal burrows more slowly than usual.

2. By general contractions of the body

Movement of this kind may be rhythmic or irregular, undulatory or peristaltic, depending upon the temporal relationships of neuromuscular stimulation. If impulses reach the appropriate muscles at regular intervals, the resulting movement is rhythmical; if they reach them sporadically, it is irregular. If the muscles in a given area on one side of the body are

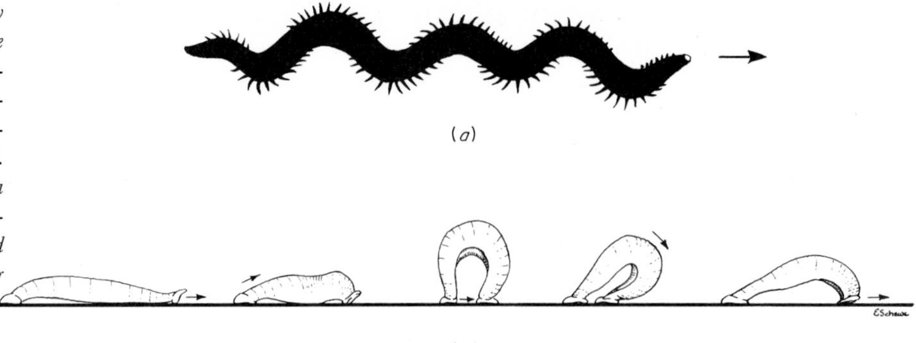

FIG. VI–5 *Locomotion by generalized body movements.* ***(a)*** *Undulation in the marine polychaete* Neanthes *brought about by contractions of the body musculature. The parapodia, moved in metachronous rhythm, help propel the animal through the water and wet sand. (After Gray.)* ***(b)*** *Looping movements of a leech, showing the animal holding by its posterior sucker while it extends its body and attaches to a new position with the anterior sucker, drawing the body along after it.*

stimulated to contract, while those of the corresponding area of the opposite side are extended, waves of contraction spread along the two sides in different phase, and the resulting movement is undulatory, writhing, or serpentine. If on the other hand, all those concerned with locomotion in a given area contract simultaneously, while those in the immediately adjacent posterior and anterior areas are extended, a wave of contraction spreads uniformly along the body in a peristaltic manner.

Movement by general contraction of the body results in progression by swimming, crawling, or looping (Fig. VI–5). Since the possession of a rigid covering limits the freedom of such general movements, it is employed particularly by those invertebrates whose bodies are not encased in exoskeletons or shells or which, like molluscs, can project a free muscular region from them. The planes of movement are determined by the orientation of the muscles within the body and the force of the movement by the amount of muscular tissue. Crawling and looping are often facilitated by the secretion of mucus which may form a track along which the animal can glide or else provide an adhesive area to hold part of the body in a fixed position while the rest is drawn up to it.

Medusoid Cnidaria swim by rhythmic contractions of their epitheliomuscular cells. Relatively rapid locomotion in a succession of jerky advances can be effected in this way, especially by hydroid medusae with dome-shaped bodies and wide vela. When the muscle processes are extended, the bell attains its maximum dimensions and water is drawn into the subumbril cavity through the opening in the velum. When the fibrils contract, water is forced out through this opening under a considerable amount of pressure, so that the animal is propelled forward at a rapid rate for some distance. A basically similar method of jet propulsion is practiced by some molluscs. Scallops (pelecypods) move through the water by rhythmic contractions of the stout adductor muscle that bring the valves of the shell together, forcing a stream of water out of the mantle cavity. On relaxation, water flows into the mantle cavity and is again forced out when the valves are clapped together (Fig. VI–6).

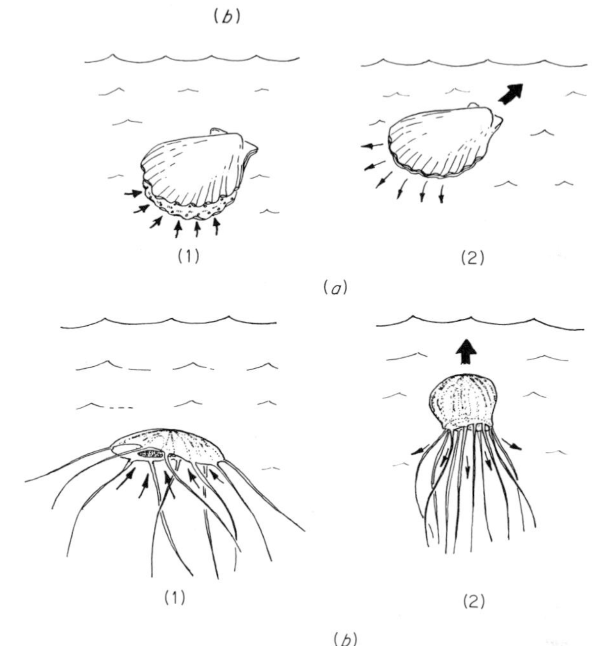

FIG. VI–6 *Locomotion by rhythmic contraction.* ***(a)*** *Locomotion of the scallop* Pecten. ***(b)*** *Locomotion of a medusa. The arrows indicate the direction of water flow (1) when the muscles are relaxed and (2) when the muscles are contracted.*

Squid and cuttlefish (cephalopods) dart backward or forward by the expulsion of water through the funnel, the direction depending upon which way the funnel is turned. The funnel connects with the mantle cavity, which fills with water flowing in through the collar when the mantle muscles are relaxed; when they contract and the collar is drawn tight around the neck, the water is forced out through the narrow opening of the funnel, and the animal is pushed either forward or backward. Octopuses, too, can swim by ejecting water through the funnel. In some species the arms are bound together by a membrane and contractions of the muscles in them displace a considerable amount of water and push the animal along (Fig. VI–7).

The bells of scyphozoan medusae also pulsate, but this action serves more to keep the animal afloat than for progressive movement. While some may swim slowly, the majority drift passively with the prevailing local currents, as do many of the siphonophores. Some of these, however, may also move actively by contraction of the muscles of those members of the colony that serve as nectocalyces, or swimming bells.

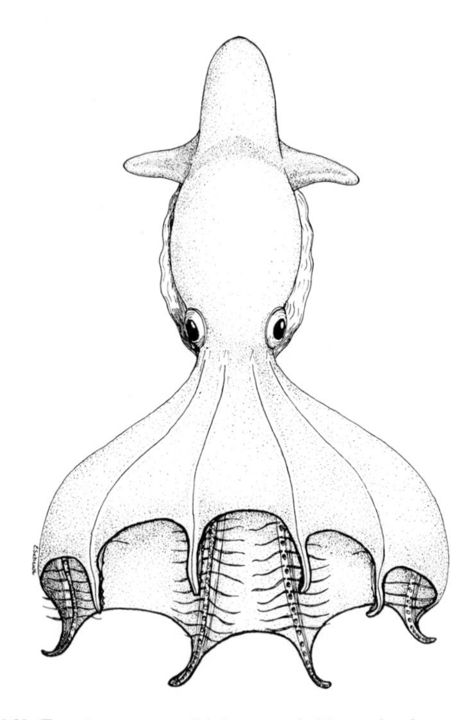

FIG. VI–7 *An octopus with its arms held together by a membrane.*

Undulatory movement is characteristic of many invertebrates with elongated bodies or with extended or flattened regions of their bodies. The crests and troughs of the waves may be in the dorso-ventral or in the lateral plane and in either case result in swimming or crawling. The waves may pass from anterior to posterior, or from posterior to anterior, to effect forward movement and be reversed to bring about movement in the opposite direction. Some few free-living nematodes, for example, swim with a whip-like motion by contractions of the muscles of their body walls, exclusively longitudinal in their orientation, which are so adjusted that the undulations are only in the dorso-ventral plane. Some nemerteans, and sometimes leeches, swim or crawl by similar undulations of their dorso-ventrally flattened bodies. In the polychaete worms, and sometimes in leeches, the waves pass along the sides of the body. In all the nereid polychaetes, these waves pass from tail to head as the animal moves forward, but in smooth-surfaced animals like leeches, and also among the vertebrates in eels and snakes, the waves pass from head to tail in forward movement.

The passage of waves of contraction along the flattened, highly muscular feet of aplacophoran, polyplacophoran, and gastropod molluscs results in their progression by a series of steps along surfaces that they lubricate with mucus secreted from pedal glands. As the wave passes along the length of the foot, the contracted areas are lifted off the substrate while the extended areas remain attached so that the length of each step is determined by the difference between the length of the musculature in the contraction phase of one wave and that in the relaxation phase. In pteropods, the foot is modified as two fins, or wings, which flap slowly up and down keeping the delicately shelled animal afloat and moving it slowly through the upper levels of the sea (Fig. II–20). The hatchet-shaped foot of most pelecypods can be extended or retracted by a similar series of wave-like contractions, allowing them to plough their way through mud or sand.

A number of annelids progress from place to place by peristaltic rather than undulatory movements of their bodies. Earthworms move in this way, but because the coelom is divided into segmental compartments by septa, its fluid is confined to each individual

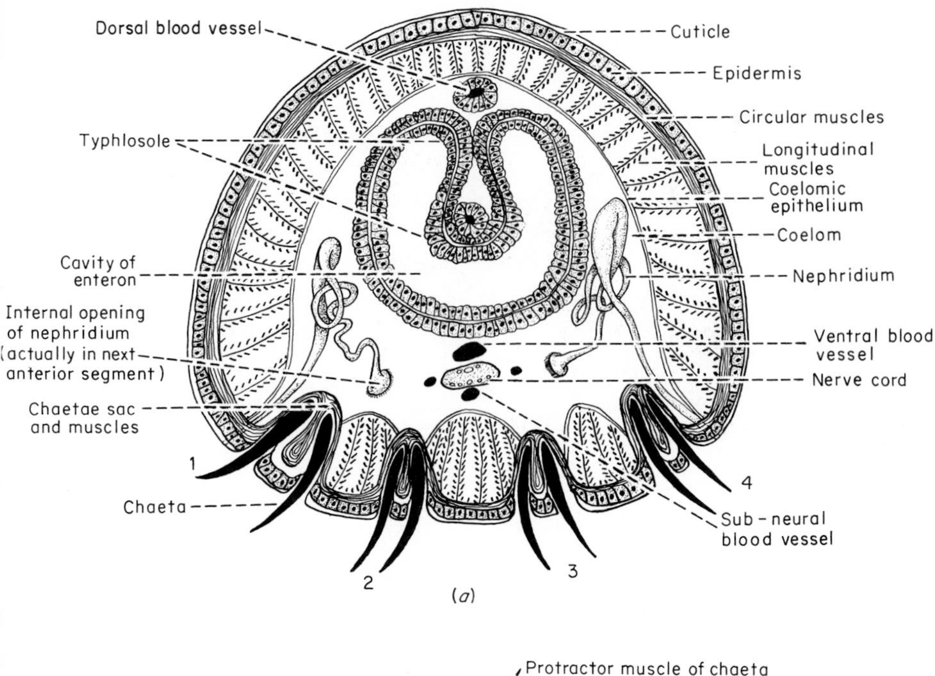

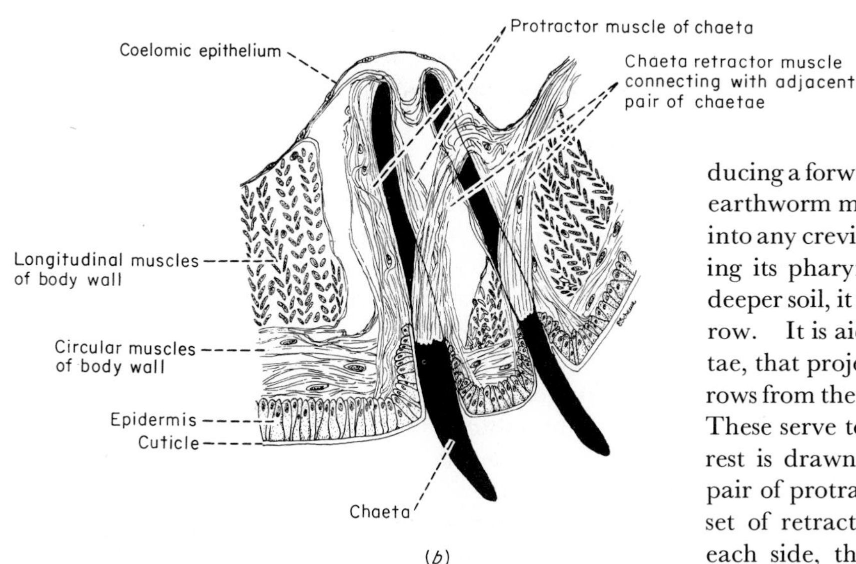

FIG. VI–8 *Mechanism of movement in an earthworm.* ***(a)*** *Cross section through a median segment of* Lumbricus terrestris, *showing the circular and longitudinal layers of muscles of the body wall and the four pairs of chaetae, numbered 1, 2, 3, and 4.* ***(b)*** *Detail of 3, to show chaetal muscles. Each chaeta can be extended and turned by contraction of the protractor muscles. The chaetae are withdrawn by contraction of the retractor muscles, which are attached to each chaeta of a pair and connect the two pairs on the same side of the animal.*

segment and the force produced by contraction of the muscles in any region is localized there rather than distributed generally along the length of the body, as it is in *Nephthys* and *Arenicola*. It has been calculated that the pressure of the coelomic fluid so developed in the earthworm is equivalent to forces between 1.5 and 8 grams which makes a very significant factor in producing a forward thrust of the body. In soft topsoil, the earthworm makes a burrow by pushing its anterior end into any crevice between soil particles and then expanding its pharynx to thrust them aside; in heavier and deeper soil, it literally eats its way along to make its burrow. It is aided in burrowing by the bristles, or chaetae, that project from its body in four double rows, two rows from the ventral and one from each lateral surface. These serve to anchor the body temporarily while the rest is drawn up. Moved by two sets of muscles, a pair of protractors from each individual chaeta and a set of retractors for the ventral and lateral pairs of each side, the chaetae can be extended, turned, or withdrawn, and are a great asset to the worm in crawling along a burrow (Fig. VI–8). *Neanthes* does not make use of its chaetae in crawling, but does use its parapodia which are moved by their own musculature, probably derived from the circular muscles of the body wall (Fig. VI–9).

Locomotion by general contraction of the body may also be aided by other devices for temporary

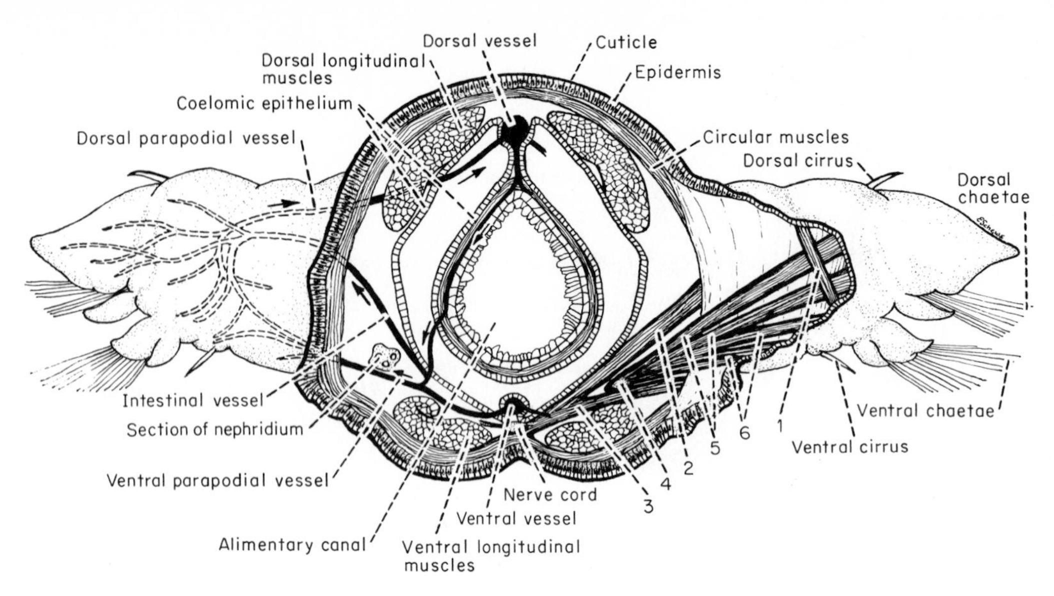

FIG. VI–9 *Cross section through a segment of* Neanthes, *showing the parapodial muscles, the muscles of the body wall, and the general course of the blood in the segmental vessels. The direction of blood flow is indicated by arrows. The musculature is shown on the right side of the diagram, the circulation on the left. 1, Intrinsic muscle of parapodium, between dorsal and ventral lobes. 2, Muscle drawing parapodium backward (posteriorly). 3, Muscle drawing parapodium forward (anteriorly). 4, Inner end of chaetal pouch. 5, Protractor muscles of dorsal acicula and chaetal pouch. 6, Protractor muscles of ventral acicula and chaetal pouch.*

anchorage, such as suckers and adhesive secretions, or for propulsion, such as cilia. Leeches, for example, loop their way along by attaching the anterior sucker to some surface, drawing the rest of the body up to it, attaching the posterior sucker, releasing the anterior one, and extending the free portion of the body to another point of attachment (Fig. VI–5*b*). The parasitic nemertean *Malacobdella*, which lives in the mantle cavities of bivalve molluscs, also has suckers that it uses in locomotion in the same fashion as the leech. *Pelmatohydra oligactis* may perform similar looping movements in travelling along a piece of pond weed, attaching its tentacles by the discharge of glutinant nematocysts which hold its anterior end in place while the basal disk is released from its attachment and, by contraction of the musculo-epithelial cells, brought up to the tentacles. Repetition of these movements results in progression by a series of somersaults. The nocturnal anthozoan, *Aiptasia carnea*, which habitually lives in a horizontal rather than a vertical position, moves backward by a sequence of events which involve attachment of the pedal disk, peristaltic contraction of the circular muscles of the column from the oral end to the midregion, followed by a strong contraction of the longitudinal septal muscles which results in retraction of the oral end. Attachment of the oral end in its new position, release of the pedal disk, and, finally, completion of the peristaltic wave in the circular muscles from the midregion to the base results in a "step" backward (Fig. VI–10). Other Anthozoa may "walk" on the tips of their tentacles, extending some of them to a point of attachment, bringing the rest, by muscular contraction, up to that point, attaching them there, and moving the others onward. Some can swim, although poorly, by coordinated lashings of their tentacles or by rapid alternate flexions of the column. Most octopuses crawl among rocks or over the sea bottom by fastening their tentacles to the substrate and dragging their bodies forward.

The holothurian *Astichopus*, abundant on the northwest coast of Puerto Rico at depths of 10 to 40 m, is a very active echinoderm and moves by peristaltic

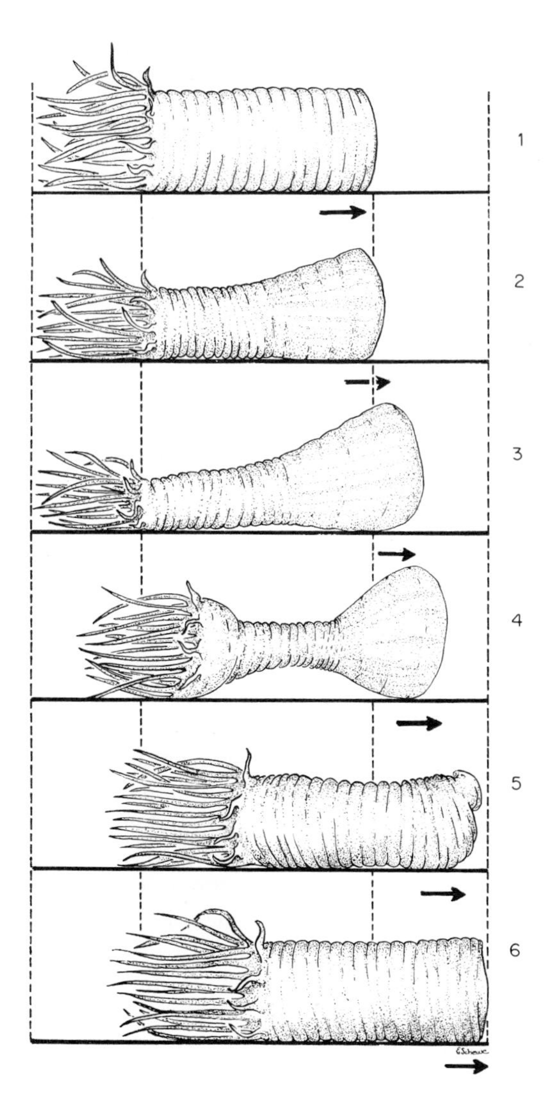

FIG. ***VI–10*** *Locomotion in* Aiptasia. *(After Hänstrom.) 1 to 6 show the muscular changes involved in a single backward "step."*

contraction of its body musculature. In the laboratory it has been observed to crawl at the rate of 0.15 to 0.25 m/min. At the beginning of movement, the posterior end is lifted off the substrate and the peristaltic wave begins there, proceeding anteriorly. Groups of tube feet are progressively detached from the substrate as the wave reaches them. Under certain conditions, notably sudden changes in the temperature of the water of the order of 3 to 4°C, contraction is far more rapid and, when the wave passes over the entire body in an interval as short as 3 to 5 sec, none of the tube feet is attached to the bottom and the animal moves forward in a leap or a bound. Somewhat less rapid contractions result in a fast walk.

In crawling, turbellarian flatworms attach the anterior end of the body by a mucoid secretion and draw the posterior end up to it by muscular action. This is manifested as waves of contraction involving both the circular and the longitudinal muscles, the waves either passing from anterior to posterior across the whole breadth of the body, as in triclads, or alternately along the two sides, as in polyclads. Differential action of local muscle groups leads to turning and twisting of the body.

In them, and in rotifers and some of the aquatic oligochaete annelids, muscular action may be supplemented or, in some circumstances, replaced by ciliary action (Fig. VI–11). The smallest Turbellaria probably move entirely by ciliary action, and larger ones

FIG. ***VI–11*** *Ciliary locomotion in Metazoa.* ***(a)*** *Flatworm with ventral ciliation, showing the mucous track along which the animal moves.* ***(b)*** *The freshwater annelid* Aelosoma *with ciliated anterior "head." This animal is in the course of transverse division, or fission, a means of reproduction usual in these oligochaetes.*

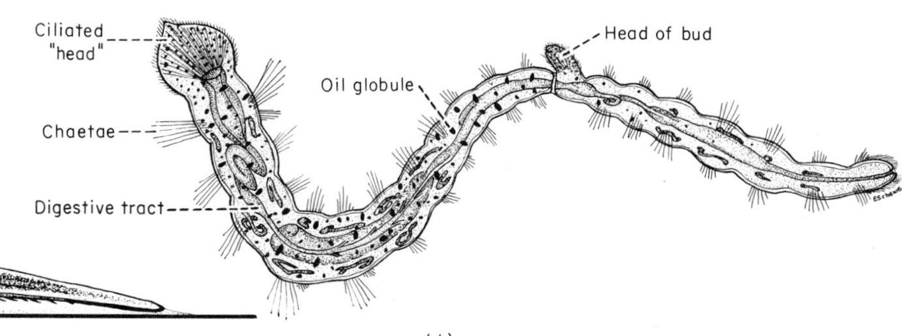

may glide by backward strokes of the cilia on their ventral surfaces along the slime track they secrete. The relative importance of these two mechanisms in ordinary locomotion has been tested by exposing animals to solutions of lithium chloride, which immobilizes cilia, and of magnesium chloride, which inactivates muscle. The elimination of ciliary action has been found to have no important effect upon the swimming of flatworms, while the elimination of muscular activity immobilizes them entirely.

Some rotifers can perform skipping or leaping movements by virtue of definitely oriented muscle bands, and some can also shorten and lengthen their bodies in a kind of telescoping motion. Cilia on the anterior segment of the freshwater oligochaete *Aelosoma* may be accessory to the undulation of its body, its primary means of locomotion.

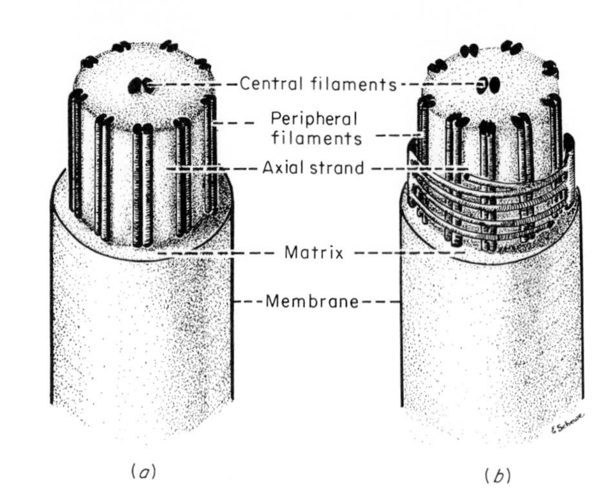

FIG. VI–12 *Fine structure of cilia and flagella. (After Fawcett.)* ***(a)*** *A cilium.* ***(b)*** *A flagellum from a spermatozoan.*

B. LOCOMOTION BY MEANS OF SPECIAL APPENDAGES

1. Cell organelles: cilia and flagella

Cilia and flagella are, like pseudopodia, projections of the cell surface but, unlike pseudopodia, they are ones that are more or less permanent and fixed in position. They may be generally distributed over the surface of a cell, or localized in special areas; they may be sparse or very numerous. Both are contractile vibratile organelles, and any distinction between them is one more of degree than of kind. Cilia are usually short, about 5 to 10 μ in length, while flagella are longer, reaching in some Protozoa lengths of 150 μ. As locomotor devices, cilia and flagella are effective only in a fluid medium and only for bodies of microscopic size, such as protozoans, rotifers, minute flatworms, sperm cells, embryos, and larvae. Ctenophores are the only animals of macroscopic dimensions that move exclusively by means of cilia. Ciliary locomotion is possible for ctenophores because of the extraordinary length of their cilia, which approximates 2 mm, and because of the association of many thousands of them into combs, or plates (Fig. II–9).

Electron microscopy has revealed many of the finer details of flagella and cilia and has shown a basic uniformity in their ultrastructure that implies basic uniformity in the mechanism of their operation (Fig. VI–12). The free portion of each, projecting beyond the surface of the cell, consists of a sheath enclosing a matrix within which are embedded two single and nine double filaments, or tubules, running parallel to the long axis. The sheath, which is actually the cell membrane, may be single or double and may have material adsorbed upon it. The two single filaments are centrally placed; the nine paired ones lie more or less symmetrically around the periphery. These peripheral filaments are made up of two subunits of unequal diameters with a common wall. In some cilia and flagella, there are tangential projections from the larger of the two subunits, forming arms that extend in the same direction (Fig. VI–13). The filaments converge at the tip and end independently there without making contact with each other or with the tip. In addition to these, there appear to be in flagella circular filaments that run around the central bundle, or axial strand. These seem to have their origin, or points of attachment, in two rod-like filaments on opposite faces of the strand and they branch and anastomose with each other.

While there is great uniformity in the structure of the free portions of these organelles, whatever their source, their relations with their basal bodies show some variation. The basal body is a spherical or cylindrical granule that lies beneath the surface of the

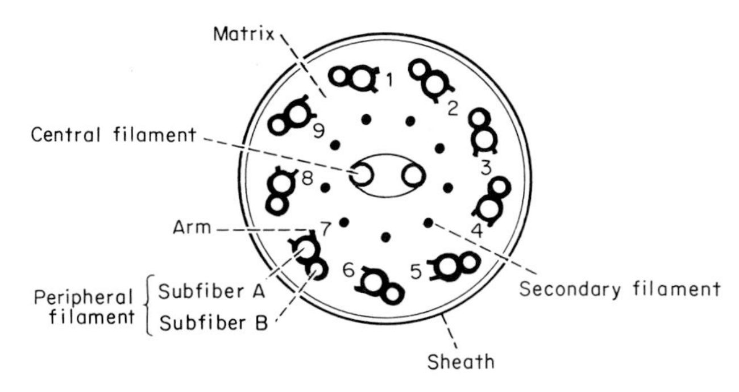

FIG. VI–13 *Diagrammatic cross section, looking toward the tip, of a flagellum from the protozoan* Pseudotrichonympha. (*After Gibbons and Grimstone.*) *The peripheral filaments are conventionally numbered 1 to 9. Similar patterns of filaments are found in the cilia on the corona of the rotifer* Philodina, *the flagella of sea urchin spermatozoa, and the cilia on the swimming plates of the ctenophore* Mnemiopsis.

cell and represents the proximal terminus of the cilium or flagellum. In ciliate Protozoa it has been designated as the kinetosome, in flagellates as the blepharoplast. It is a self-perpetuating body, capable of division or replication, whose homology with the centriole is well established. Electron microscopy has demonstrated structural homology for these two organelles, by showing that each has the same arrangement of nine peripheral fibrils. Studies of division in the gastrodermis of *Hydra* and of the metamorphosis of spermatid into flagellate sperm in a number of species have identified the basal body of the flagellum with the distal centriole of the dividing cell (Fig. VI–14).

FIG. VI–14 *Some of the forms of basal bodies of flagella and cilia.* (*After Fawcett.*) ***(a)*** *Base of a flagellum of a cell in the gastrodermis of* Hydra, *derived directly from the duplex centriole (diplosome), one centriole becoming the basal body proper and the other retaining its original orientation perpendicular to it.* ***(b)*** *Base of a cilium of* Paramecium. ***(c)*** *Base of a cilium of* Euplotes, *with the central filaments extending below the cell surface.* ***(d)*** *Base of a mammalian cilium, curved and closed.* ***(e)*** *Base of a cilium from the epithelium covering the typhlosole of the mollusc* Elliptio, *showing its fibrous rootlets.* ***(f)*** *Base of a cilium from the ciliated epithelium of a frog.*

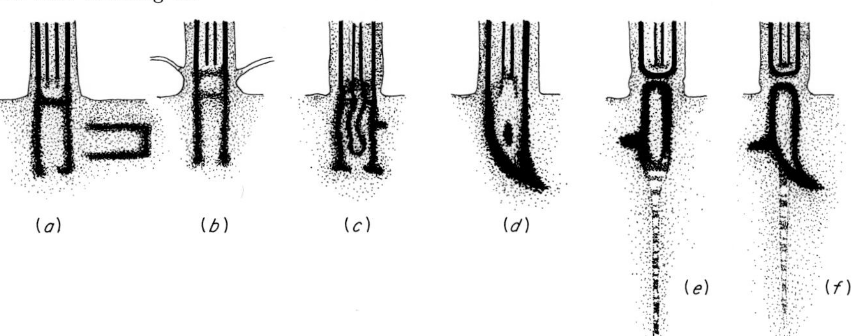

In many ciliated cells, rootlets, large cross-striated fibrils with a period of 550–700 Å, extend from the basal body into the cytoplasm. The number associated with each basal body varies from species to species but is constant for any given type of cell. They taper off and disappear as organized structures in the vicinity of the nucleus or may break up into many fine branches. Also, they may unite and, after union, branch, as has been shown for the cilia on the combs of *Pleurobrachia.* The cross striation of the fibrils, and its periodicity, suggests that they may be a collagen and that, as their name implies, they may serve to anchor the basal granules and keep them in position when the free portion of the organelle is in vigorous motion.

The typical movement of a flagellum is undulatory. Contraction begins at the base and progresses outward to the tip along the shaft as well as around it, with the waves often increasing in amplitude as they progress. This motion displaces the fluid medium and the flagellum, acting like the propellor of a ship, therefore pushes a cell that is free to move forwards, backwards, or sideways, depending upon the direction of the effective stroke (Fig. VI–15). Flagellate protozoans can not only move forward, slowly or rapidly, but they can change direction by swerving to right or to left; they can move backward and they can move laterally. Flagella may also lash in a pendular manner, in a straight backwards and forwards motion with only a slight flexure at the base. Beating in this manner, they move the organism in one direction only, and changes in its path are brought about by movement of the body.

Cilia may show undulatory motion, but they also move by flexion or by a combination of flexural and

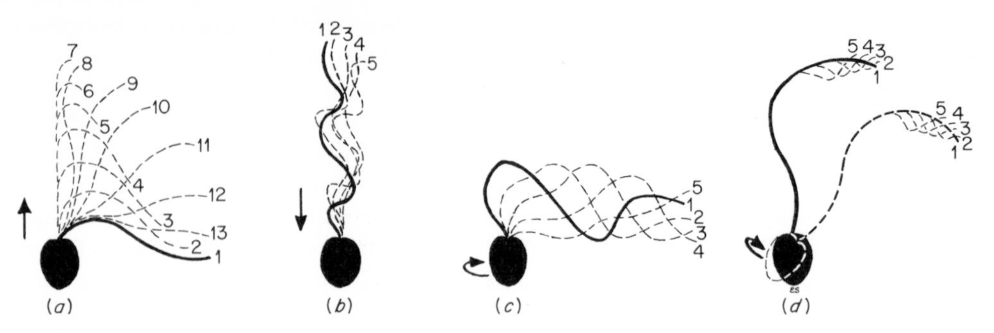

FIG. VI–15 *Flagellar movement. (After Krijsman.) The numbers, beginning with 1, show the successive positions of the flagellum in forward and backward beat.* ***(a)*** *Action of a flagellum in rapid forward movement of a motile cell. The numbers indicate the successive positions of the flagellum starting from 1, its position at the end of a beat.* ***(b)*** *Action of a flagellum in backward movement.* ***(c)*** *Action of a flagellum in turning a cell at right angles to its long axis.* ***(d)*** *Action of a flagellum in rotation of a cell.*

pendular action. If the effective stroke of a cilium is in an antero-postero direction along the animal's body, the fluid will be pushed behind it and the animal will move forward. If each cilium strikes the body directly, the motion will be only a forward one, but if it strikes it obliquely, there is rotation as well as forward movement. Thus, the bodies of paramecia and other asymmetrical ciliates rotate on their long axes as they move, and so they are able to pursue straight courses by compensating on each rotation for the change in direction that would otherwise result from the stronger ciliary action of the side with the oral groove. In combined flexural and pendular motion, as it has been observed and analyzed in the cilia on the gills of molluscs and on the trochi of their larvae, the effective stroke is a rapid one, with the cilium unbending and almost rigid. In the recovery stroke, the cilium appears limp and straightens out slowly from base to tip at about one-third the speed of the downward thrust (Fig. VI–16). In the slender cilia of *Opalina ranarum*, a protozoan parasitic in the gut of frogs, as well as in the stouter compound cilia of *Stentor polymorphus*, the recovery stroke cannot be separated from the effective stroke, the whole beat being a continuous process in which the base of the cilium is already bending toward recovery while the major part of its length is swinging downward in the effective stroke.

All the cilia on a cell may beat together or, more

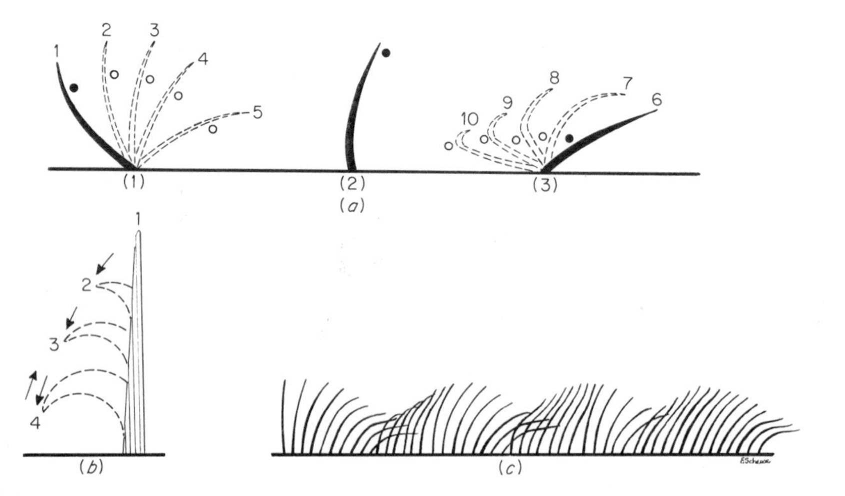

FIG. VI–16 *Ciliary movements.* ***(a)*** *Successive positions shown by numbers 1–5, of an active cilium on the latero-frontal cells of the gills of the mussel* Mytilus, *when stopped by a needle at different periods in the course of its beat. (After Carter.) (1) During the effective beat. (2) When the tip of the cilium strikes the needle. (3) During the recovery beat.* ***(b)*** *The composite structure and movement of a cilium on the latero-frontal cells of a gill of a lamellibranch mollusc. (After Atkins.) Numbers 1–4 indicate its successive positions.* ***(c)*** *Metachronal beat in a row of cilia.*

commonly, in sequence, with the motion beginning at one point on the cell surface and spreading like a wave over the remainder. In Metazoa, the cilia on adjacent cells may exhibit a similar coordination in metachronism. From observations on isolated cells, and from those on ciliated epithelia *in situ* in which the movement of cilia has been slowed down by anaesthetics or by increased viscosity of the medium, it has been suggested that at least three cellular mechanisms are involved in this coordination. One of these is responsible for the individual beat of each cilium, another regulates or determines the coordination of those in any one cell, and the third is active in regulating the coordination between cells.

Motion in both cilia and flagella seems to be autonomous, inherent in their physical and physiological constitution. Evidence for this has been provided by experiments in which pieces have been cut from the surfaces of ciliate protozoa; the cilia on these isolated pieces continue to beat for some time provided that their connections with their basal bodies remain intact. Moreover, the cilia on cells isolated from ciliated epithelia of metazoans will beat as long as the cell remains alive, in the absence of nervous connections. The beat is the result of localized contractions in the peripheral fibrils of the free portion, followed by relaxation of the contracted areas. The central pair of filaments does not appear to be involved in the contraction but does provide two main axes of symmetry, one bisecting the pair and the other passing between its members. A cilium in pendular motion bends at right angles to the first plane—that is, the one passing across the filaments—and if a number of cilia are beating together they are all oriented in the same way in respect to this plane. A line drawn between the central filaments passes across one of the peripheral filaments, which has conventionally been designated number 1, and the others 2, 3, 4, and so on in clockwise direction (Fig. VI–13).

A recently proposed theory of ciliary and flagellar action postulates that the impulse arises in the basal body at filament number 1, which initiates a propagated contraction up that filament, the excitation spreading with equal speed in both directions so that the adjacent filaments, numbers 2 and 9, then 3 and 8, contract, and so around the circumference of the cilium. It is assumed also, that the disturbance spreads to the central filaments, which are not contractile but conductile, and through them is conveyed to the distal ends of the peripheral filaments. The cilium would thus bend first along filaments 1, 2, and 9, whose contraction would be followed by that of the other opposite pairs in sequence and, later, by their relaxation. Such differences in timing of contraction could account for differences in the form of the beat of both cilia and flagella.

Flagella and cilia may also form parts of more complex systems or themselves be modified in various ways. For example, one of the two trailing flagella of trypanosomes is attached along the length of the body by a delicate film and thus forms part, and presumably the motile part, of an undulating membrane. Undulating membranes, like those of *Blepharisma*, may be formed also by the lateral fusion of cilia (Fig. VI–17). Lateral adhesion, if not actual fusion, is likewise evident in the combs of ctenophores, where the cilia in each row, but not those in different rows, are united

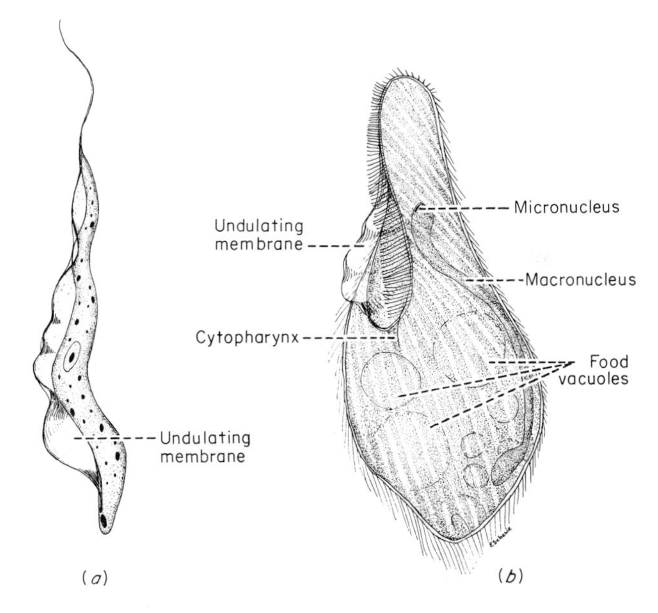

FIG. VI–17 *Undulating membranes.* **(a)** Trypanosoma, *a genus of parasitic flagellates, with one free flagellum and one attached to the body as an undulating membrane. (After Brown.)* **(b)** Blepharisma undulans, *a heterotrichous ciliate, with cilia fused along the gullet as an undulating membrane. (After Stolte.)*

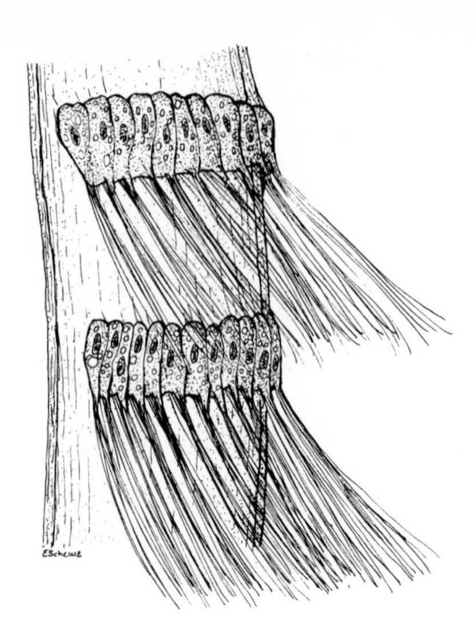

FIG. VI–18 *Cilia on the combs of a ctenophore. (After Hyman, vol. I.)*

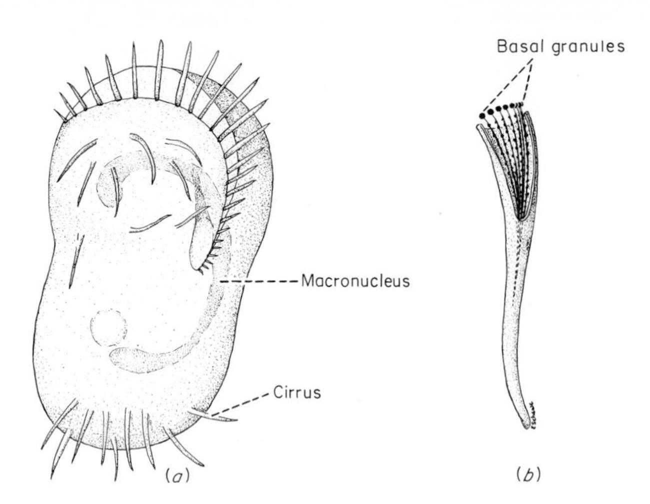

FIG. VI–19 *Cirri.* ***(a)*** Euplotes, *a hypotrichous ciliate common in fresh and brackish water. (After Pennak.)* ***(b)*** *Left lateral view of an anal cirrus of* Euplotes, *showing its composite structure. (After Taylor.)*

(Fig. VI–18). These cilia are unusual in that they have, in addition to the two tubular central filaments, a compact third filament, or midfilament. They, therefore, have a 9-plus-3 rather than the usual 9-plus-2 formula. They have also two lamellae which extend from the two opposite peripheral filaments to the sheath of the cilium and so more or less divide it into two compartments. These "comparting lamellae" are oriented in such a way that their distal regions lie at the points of contact between adjacent cilia in the same row. Fused cilia, without comparting lamellae, make up the stout cirri of protozoans like *Stylonichia* and *Euplotes* which can support their bodies upon their cirri and can also walk or creep upon them (Fig. VI–19). In *Euplotes*, each cirrus consists of four to six rows of cilia, with not more than six cilia in a row, which are united into a single thick and compact process. The cirri which are located in the anterior region (frontal) and in the posterior region (anal) have been shown to be indispensable for creeping movements; the anal cirri function particularly in backward movement and cooperate also in turning the body adorally and moving it backward in swimming, which is effected largely by the adoral membranelles. The variety of movements of which *Euplotes* is capable indicates a high degree of control and coordination of the cirri, which have a complex system of rootlets and fibrils associated with their basal granules.

Some flagella have delicate processes, the mastigonemes, extending from one or two surfaces of the sheath (Fig. VI–20). These processes may be modifications of the surface itself or may arise within the shaft and project from the surface. Such hispid flagella may increase the mechanical efficiency by increasing the surface of a flagellum and so make it more effective in propulsion. But motion, although intrinsic to flagella and cilia, is not an invariable attribute, for they may slow down or cease to beat entirely for longer or shorter intervals. Some cilia are permanently immobile, but are highly sensitive to stimuli of various kinds. Indeed, such sensitivity may be one of the primary functions of cilia and one that has been somewhat obscured by the attention that has been given to their mobility. Stationary cilia that are stiff and sensitive to tactile stimuli are found among the vibratory ones in *Stentor* and in *Coleps*, and the cnidocils that act as triggers for the discharge of nematocysts from the cnidoblasts of *Hydra* are modified cilia. Moreover, the photosensitive elements in the eyes of molluscs, amphibians, and mammals have been derived from cilia. Similarly, electron micrographs of the connection between the inner and outer segments of the

FIG. VI–20 *Hispid flagella.* ***(a)*** *Part of the flagellum of* Euglena gracilis *drawn from an electron micrograph, showing central core and lateral mastigonemes. (After Pitelka.)* ***(b)*** *Part of a flagellum of* Rhabdomonas incurvum *drawn from an electron micrograph, with mastigonemes distributed on both sides of the central axis. (After Pitelka.)*

rods and cones of the vertebrate eye show the nine peripheral double-filament pattern of the typical cilium or flagellum; in some cases, a central pair has also been evident. The critical evidence for the origin of these photosensitive elements from cilia comes from studies of morphogenesis in the outer segments of the rods in the retinas of young rodents, where it has been possible to trace the development of a primitive cilium into a fully differentiated rod.

In addition to the role that cilia play in sensory reception, they function also in respiration, in reproduction, in getting food, in the circulation of fluids within the body, and in locomotion.

2. Nonjointed appendages: parapodia, tube feet

In nereid polychaetes, locomotion through undulatory movements of the body as a whole may be augmented, especially in crawling, by forward and backward movement of the parapodia. These are segmentally distributed, fleshy projections of the lateral body walls, stiffened by acicula and moved by localized sets of muscles (Figs. VI–5*a* and VI–9). Some species of errant polychaetes can use them to raise their bodies off the substratum, keeping contact with it only by the ventral lobe of the parapodium (neuropodium). In this way, they may walk over a solid surface. Similarly, cephalopod molluscs may use the arms or tentacles into which their modified feet are divided to hold themselves on rocks or to crawl over a suitable substrate (Fig. VI–21). In swimming, squids hold their arms together and extended forward, but they may spread them apart and, like octopuses, crawl for short distances upon them.

The tube feet of echinoderms are derived locomotor organs, adapted from the feeding tentacles distributed along the ambulacral areas in modern crinoids and found in the fossilized predecessors of asteroids, echinoids, and holothurians. The significant step in the conversion of a feeding tentacle to a

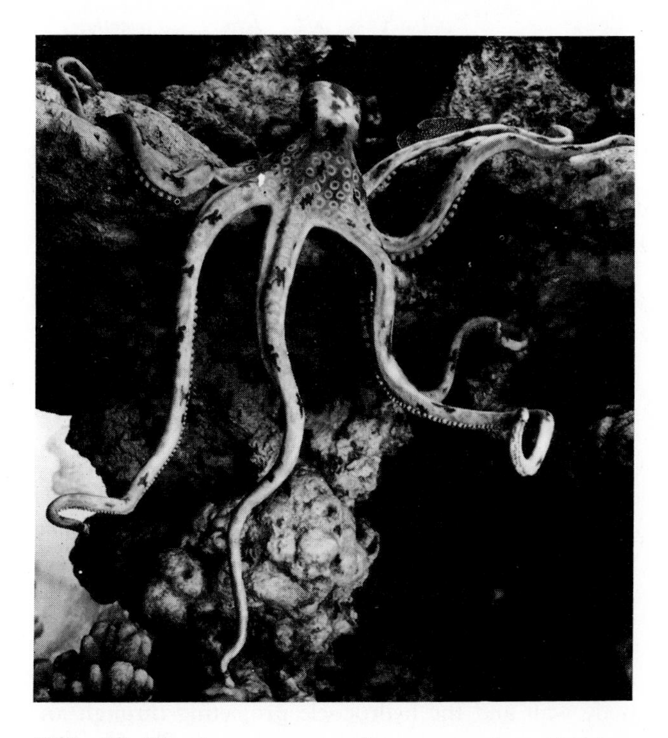

FIG. VI–21 *An octopus crawling over rocks. (American Museum of Natural History.)*

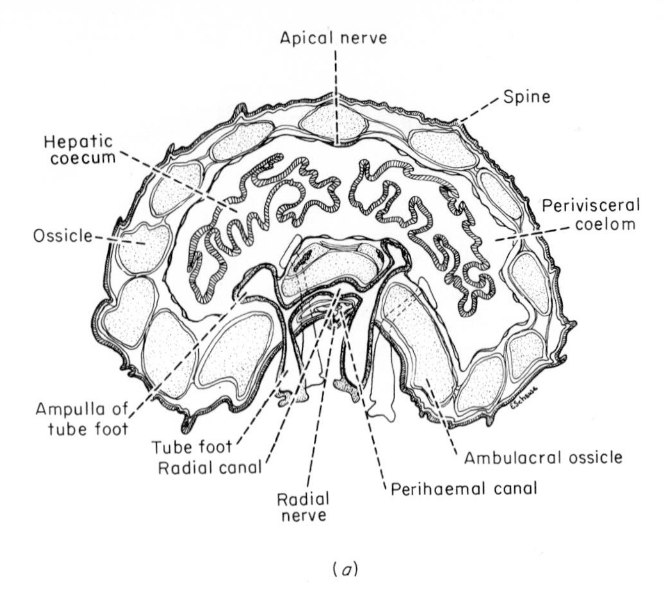

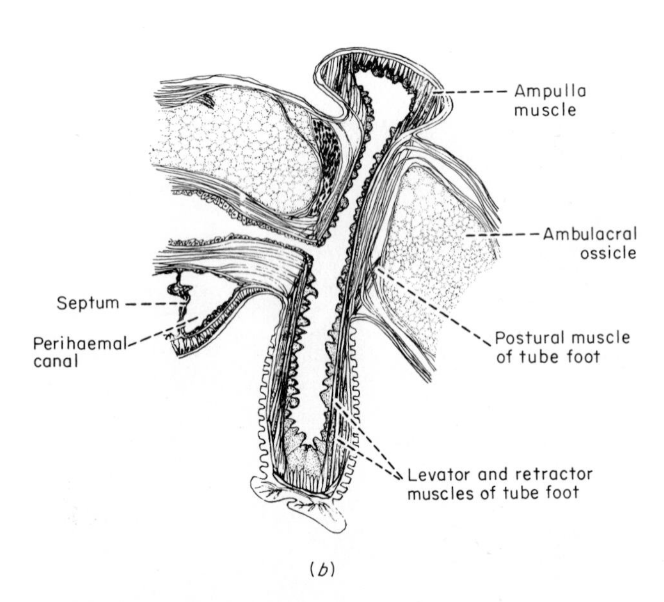

FIG. VI–22 *Mechanism of movement in a starfish.* ***(a)*** *Cross section of an arm, showing one pair of tube feet.* ***(b)*** *Detail of a tube foot, showing musculature. (After J. E. Smith.)*

podium, or tube foot, seems to have been the addition of an internal ampulla to the tubular extension of the body wall and the hydrocoele projecting through an aperture between the ambulacral plates (Fig. VI–22). In starfishes, in which the anatomy and mechanism of action of the podium are best known, the musculature and innervation of the ampulla is distinct and separate from that of the tubular extension, so that each acts independently of the other but reciprocally with it. The innervation, unlike that of the pedicellariae and spines, is supplied primarily from the central nervous system—that is, through the circumoral ring and radial cords—rather than through the diffuse peripheral system. The muscles of the tubular portion, or foot, are exclusively longitudinal; externally, they are enclosed in a thick layer of fibrous connective tissue that lies between them and the epidermis. The collagen fibers in the connective tissue at the base of the foot are circularly arranged, and muscle fibers, originating in the longitudinal muscles of the wall of the foot, are attached to this pad. In protraction, the muscles of the ampulla contract forcing the contained fluid into the foot, which is prevented from lateral expansion by the orientation of collagen fibers in the connective tissue sheath. Backflow of fluid into the radial vessel, with which each pair of tube feet is connected by fine channels, is prevented by valves in the walls of these channels. Hydrostatic pressure of the fluid driven into the foot brings about its extension by a mechanism similar to that producing the forward thrust of the proboscis in *Arenicola*. When the foot is fully extended, it is attached by mucus secreted in great quantities by the epidermal cells. Even firmer adhesion is afforded by the sucking action of the foot, whose central portion is drawn back by contraction of the muscle fibers attached to the connective tissue pad, making a vacuum cup. The pull that a single podium may exert through the combination of mucous adhesion and suction has been estimated to be equivalent to that of 25 to 30 grams. In retraction, the muscles of the foot contract, forcing the fluid back into the now relaxed ampulla. It has been shown that the ampulla has a fluid capacity just adequate to produce maximal extension of the foot, and there is no evidence that water enters or leaves the system during either protraction or retraction. When the foot is extended, the ampulla is empty, and when it is withdrawn, the ampulla is full.

Besides protraction and retraction, tube feet may bend, and they may also point in a given direction, making possible stepwise progression of the body. Pointing is effected through contraction of the postural muscle of the foot that is attached to the ambulacral ossicle lateral to it. Protraction, retraction, and bending are uncoordinated activities, for they may be carried on in each tube foot independently of others in the set, but postural activities, involving pointing and stepping, require the orientation of all the feet in an arm in a common direction. This orientation persists even if a part of the arm is held clear of the substrate, for the free tube feet continue to step in the same direction as those that remain attached. The dependence of this coordination upon the integrity of the central nervous system has been shown by transection experiments. If the nerve ring is cut through any interradius, the activity of the tube feet on the two sides of the cut becomes uncoordinated, and if it is cut through all the interradii, activity stops entirely.

3. Jointed appendages: the arthropod limb

Typically, arthropods have a pair of jointed appendages on each body segment, a characteristic which is one of the unifying features of this large and diverse phylum. These appendages are ventro-laterally placed and consist of a series of segments, or articles, differing both in number and in length (Fig. VI–23). They arise as extensions of the body wall into which muscles, derived from the body musculature, and nerves, originating in the segmental ganglia, extend. The exoskeleton of each segment of the appendage is sclerotized to varying degrees of thickness and may also be impregnated with calcium salts, but between the segments it always remains thin and flexible, making a movable joint. However, in the phyllopodia of branchiopods, usually considered the simplest and perhaps most primitive of living Crustacea, it is flexible throughout the length of the blade-like and unjointed appendage, whose shape is maintained by the pressure of the contained hemolymph (Fig. VI–24).

The typical, jointed limb may be divided into two distinct functional regions, a proximal one (coxite, protopodite, or sympodite) that connects with the body and a distal one (endopodite or telepodite) that is locomotor. It is essentially a system of hollow levers,

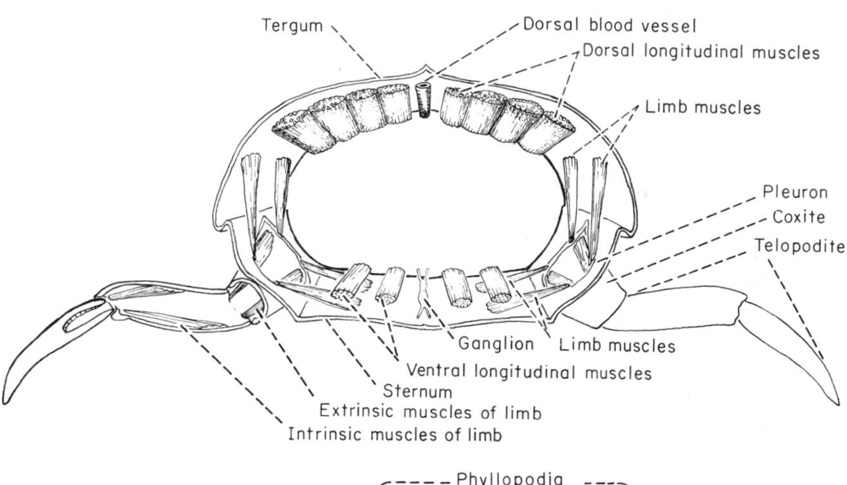

FIG. VI–23 *Transverse section through a leg-bearing segment of an arthropod. (After Grassé, vol. VI.)*

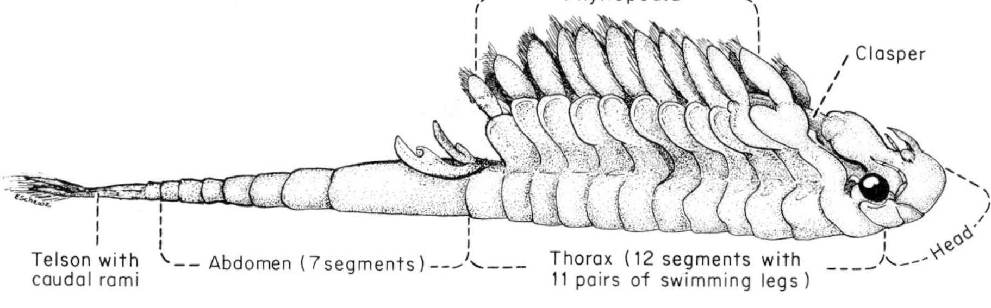

FIG. VI–24 Eubranchippus, *a fairy shrimp that is common in pools, where it customarily swims, ventral side uppermost, by oar-like movements of its phyllopodia. (After A. Morgan.)*

moved by internal muscles. Those muscles whose origins and insertions are within one segment are designated as intrinsic; those whose origin or insertions lie in different segments are designated as extrinsic. The joint is usually a simple hinge, permitting motion in only one plane, but in some cases rotation is possible. Usually, also, there are antagonistic sets of muscles, both flexors and extensors, but in spiders and scorpions some of the leg joints have flexor muscles only and their extension is effected through a fluid (hydrostatic) mechanism. The muscles are striated, capable of rapid and vigorous contraction, and may be attached to condyles at the articulation.

The derivation of the arthropod limb, like that of the arthropods themselves, is obscure and attempts to establish homologies between the various parts in different species or to discover a prototype from which all kinds of modern limbs may have evolved by transformation or reduction of parts have proved unsatisfactory for one reason or another. The difficulties in adopting a comprehensive scheme into which all actual observed types will fit arise from the great variety and diversity of these types and also from the failure, as yet, to uncover fossils that would provide prototypes or intermediate types. But a number of theories have been proposed, dating from the middle of the last century, all of which assume that originally all the appendages, except those of the head, in any one species were alike as they are in modern branchiopods, but that there has been a tendency for the anterior and posterior ones to become specialized for various functions, leaving only those in the midregion of the body for locomotion. These theories have fallen into two main categories: those that derive the appendage from a uniramous prototype, and those that derive it from a biramous one (Fig. VI–25).

a. DERIVATION FROM A UNIRAMOUS PROTOTYPE Such an appendage would consist of a single axis, with two parts, a basal coxite, and a distal telopodite. External extensions, or exites, may be inserted on this axis and also internal ones, or endites. One theory proposes that the lobopods of onycho-

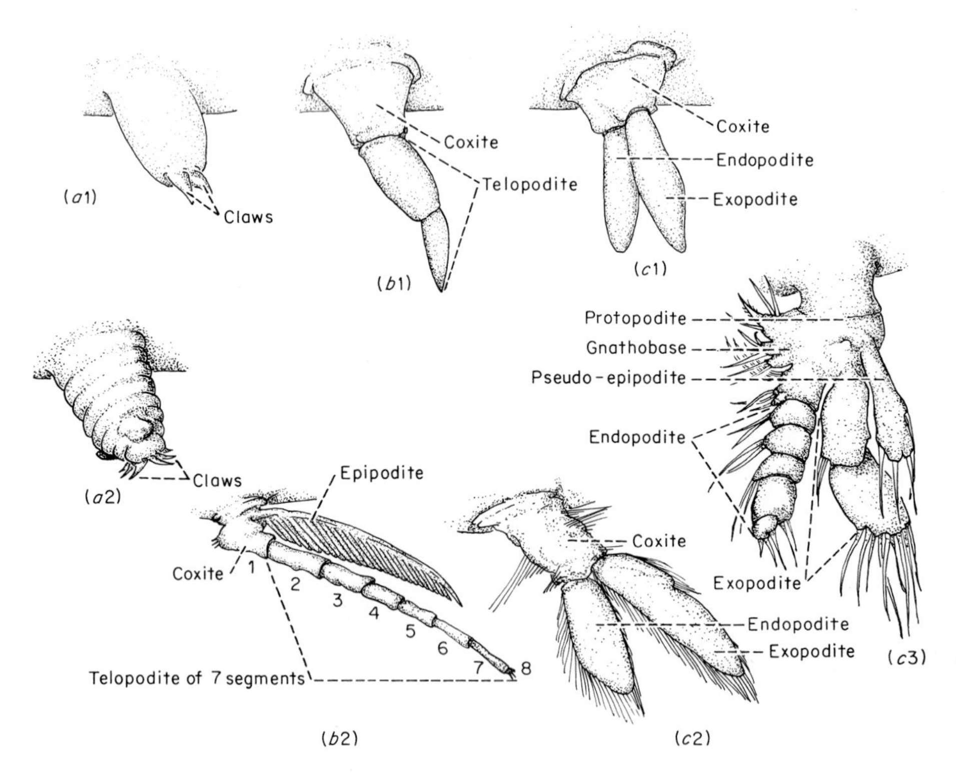

FIG. VI–25 *Derivation of the arthropod limb. (All figures represent the limb on the right side of the animal.)* ***(a1)*** *Hypothetical lobopod.* ***(a2)*** *Limb of* Aysheaia. *(After Hutchinson, from a conjectural restoration.)* ***(b1)*** *Hypothetical uniramous appendage.* ***(b2)*** *Limb of a trilobite. (After Snodgrass.)* ***(c1)*** *Hypothetical biramous appendage.* ***(c2)*** *Pleopod of lobster.* ***(c3)*** *First thoracic appendage of* Hutchinsoniella. *(After Sanders.)*

phorans are forerunners of such a limb. *Peripatus* has short, unjointed, but extensible appendages, which are paired and segmentally arranged and are moved by extrinsic muscles, as are all arthropod appendages in the extension of a limb. The most anterior of the appendages are modified as jaws and slime papillae, but in *Aysheaia*, discovered as a mid-Cambrian fossil, all the limbs are alike. The fauna associated with these fossils suggests that they lived in the sea, or in the littoral zone, so that they might equally well be ancestors of aquatic as of terrestrial arthropods. The development of the lobopod into an arthropodan appendage, by elongation and division into segments through differential hardening of the exoskeleton and readjustment of the musculature, would appear to be evolutionary steps that were not too difficult to achieve.

A second uniramous theory derives the jointed limb from phyllopodia. The argument for this rests upon the resemblance of the phyllopodium to the annelid parapodium and the general belief that the phyllopodous branchiopods are very simple and primitive Crustacea and that Crustacea include the least specialized arthropod types. There is, however, reason to think that branchiopod appendages are really specialized ones, derived from a generalized pattern rather than giving rise to one. Moreover, there is no embryological evidence for a phyllopodous origin of any arthropod appendage. The phyllopod theory has, therefore, not been widely accepted.

b. DERIVATION FROM A BIRAMOUS PROTOTYPE Such an appendage would consist of a base (protopodite or sympodite) on which are inserted an outer shaft, or blade, the exopodite, and an inner or medial one, the endopodite. The protopodite may also bear one or more lateral extensions with respiratory functions, the epipodites. The argument advanced for deriving all arthropod limbs from an appendage of this kind is that all the most primitive arthropods known have been thought to have biramous limbs and that they are found in the nauplius larva and also in the adults of a number of species of crustaceans. Moreover, most of the fossil arthropods found in the Cambrian rocks, and those of the diplopods found in the Carboniferous, have biramous limbs.

The oldest theory of the origin of the segmented appendage derives it from the trilobites. Except for the antennae, the limbs of trilobites are all alike and consist of an unsegmented base to which are attached outer and inner shafts, or blades. Such an appendage has long been considered biramous, but now, on good evidence, the outer shaft is recognized as an epipodite. The leg, therefore, is a uniramous one and not homologous with the biramous pleopods of Crustacea. It represents, rather, a generalized arthropod limb in that it contains all the primary segments found in the legs of any modern species. Although in the trilobites each of these segments has many fine, long plumose or pinnate processes, their basic structure resembles that of modern arachnids and their pattern is closely followed in the Pycnogonidae (Fig. VI–26) and in some of the legs of the Solifugae (Solpugidae). The tripartite coxa of pycnogonids, crustaceans, insects, and several orders of arachnids seems to be a secondary development and not a primitive feature, since the trilobite sympodite is unsegmented as is that of the nauplius larva and of the adult branchiopod.

The trilobite theory in its revised form thus accepts a uniramous prototype and so falls into the first category of theories. Its acceptance is open to qualification, since there is inevitable uncertainty in identifying limb segments where musculature is not known.

The recent discovery of living specimens of the apparently very primitive crustacean *Hutchinsoniella* may throw new light upon the origin of arthropod limbs. In this little animal, the first seven thoracic appendages are all alike in structure and in function and are believed to approximate very closely the ideal biramous pattern. Each consists of a two-jointed protopodite, with a flattened pseudopipodite borne on the lateral surface of the basipodite and a complex gnathobase with many lobes and spines on its medial surface. The exopodite has two segments and the endopodite presumably six. Five of these are distinct, and the sixth may be represented in the extension of the basipodite, which, however, lacks a definite suture. *Hutchinsoniella* is the only living crustacean yet known that has limbs that are apparently essentially triramous in design, with a well-developed gnathobase on the protopodite as well as a jointed exopodite and endopodite.

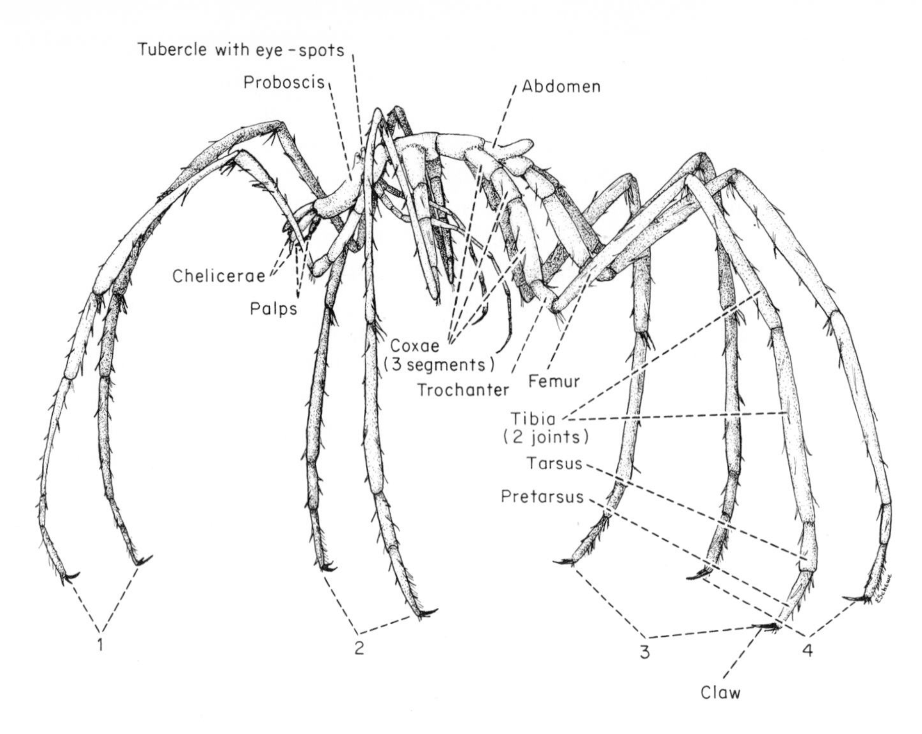

FIG. VI–26 Nymphon, *a pycnogonid. The body is composed of a cephalothorax and a short abdomen. 1, 2, 3, and 4 indicate the four pairs of ambulatory legs, each with eight joints. (After Miner.)*

However it may have evolved, the development of the pattern of the flexible limb has given the arthropods great advantages in locomotion, as well as a great range of possible modifications of the appendage for other purposes. Limbs of this kind enable the animal, for one thing, to hold its body up from the substrate, a great asset in walking and in running. But only a few of them really stand up upon their legs, for in most species the body is suspended from them so that the center of gravity is kept low and maximum stability preserved (Fig. VI–27). And, since the development of separate leg muscles has made it possible for the appendage to move independently of the body as a whole, movement is highly efficient in terms of energy expenditure. Energy need not be wasted in lateral undulations of the trunk, as it is in errant polychaetes whose walking by means of parapodia is always ac-

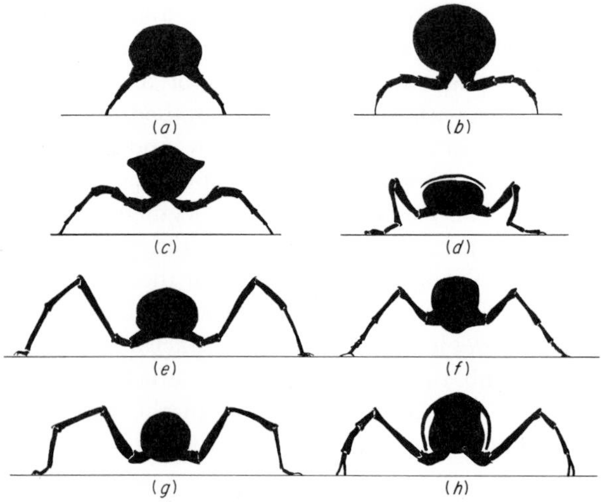

FIG. VI–27 *Diagrammatic cross sections through leg-bearing segments of representative arthropods, showing relation of limbs to the body and to the ground. (After Manton.)* ***(a)*** Pauropus, *a myriapod of the subclass Pauropoda. These are small animals, with bodies averaging 1 mm in length, with nine pairs of legs.* ***(b)*** Spirostreptus, *a myriapod of the subclass Diplopoda, with about 60 trunk segments.* ***(c)*** Polydesmus, *a myriapod of the subclass Diplopoda, with about 20 trunk segments.* ***(d)*** Forficula *(earwig), an insect of the order Dermaptera, showing body slung low between the legs and plantigrade position of their terminal segments.* ***(e)*** Buthus, *a scorpion (arachnid).* ***(f)*** Monomorium *(ant), an insect of the order Hymenoptera, with digitigrade position of the terminal segments of its legs.* ***(g)*** *A spider (arachnid).* ***(h)*** Astacus *(crayfish), a crustacean of the order Decapoda, showing unguligrade position of the terminal segments of its legs.*

companied by trunk movements. Further economy of effort is attained by ability to flex the limb in the middle of its propulsive stroke and so to eliminate the necessity of swinging it far out to the side in order to clear the substrate. Following this flexion, the limb can extend to its full length and so gain complete advantage of the propulsive force produced by contraction of the extrinsic muscles.

Such limbs have made every type of progressive movement, except flight, possible to all groups of arthropods. They can walk and run on land and in the water; they can swim, plough through mud or sand, burrow, dig, hop, leap, and climb. Indeed, a single species may employ one or more of these methods of locomotion at any stage of its life cycle. Furthermore, the limbs may function in various other ways—in feeding, in respiration, in defense and offense, and in the reception of external stimuli.

The number of appendages devoted primarily to locomotion varies in different groups of arthropods and frequently at different developmental stages in any one species. There are as many as 179 pairs of them in the adults of some of the geophilomorph chilopods, distributed along the entire lengths of their trunks. Among Crustacea, some branchiopods have between 40 and 60 pairs of phyllopodia, while decapods, as their name implies, have only five pairs of ambulatory appendages, or pereiopods, although in some, as for instance the lobster, the terminal segments of the most anterior pair are modified as claws that are more effective in offense and defense than in locomotion. Locomotion in lobsters is then actually effected by four pairs of legs. Typically, arachnids have four pairs of legs and insects, three; in these, as in the adult decapod crustaceans, they are limited to the thorax.

c. WALKING AND RUNNING Walking, with its correlatives, running and crawling, is the commonest means of progression among arthropods. Most terrestrial species are walkers, and the rhythms and gaits employed by different species, or by one species in different circumstances, can be analyzed by photography, or through records of tracks left on a suitable substrate. While the movement, or step, of each leg depends upon segmental mechanisms in the leg itself or in the segment to which it is attached, regular and orderly movement of all the appendages depends upon intersegmental coordinating mechanisms, as has been shown by experiments in which neural connections have been destroyed. Such coordination is essential to successful locomotion in any animal, but especially in one with many legs. Two main problems arise here: (1) to keep successive limbs from interfering with each other and so, in walking, to avoid stumbling, and (2) to keep enough limbs, and those close enough together, in contact with the substrate to maintain stability and to keep the body from sagging. In many-legged millipedes, the legs move metachronously, with the number involved in any one wave varying with the gait employed at any given time (Fig. II–23*c*). Each leg in a wave is slightly out of phase with the one immediately before and behind it, the phase difference, like that of the number of legs in each wave, depending upon the speed at which the animal is moving. Such a difference in timing is important for all animals with several sets of appendages, but it is especially important for those with short ones. With this adjustment, each successive foot can be put down in the same spot as the one immediately anterior to it. This is the situation in the many-legged millipedes, which never show high rates of speed and can escape predators more effectively by creeping into crevices than by running away. In the scutigeramorph chilopods, on the other hand, there are only 14 pairs of legs. These are of different lengths but the shortest are longer than those of geophilomorphs. Because of these differences in length, a leg can be put on the substrate lateral to the footprint of the leg anterior to it. Thus, a maximum stride can be obtained, and equal numbers of legs can be in the propulsive phase at all times. These adjustments and a number of other ones, including rapid contraction of the leg muscles so that a backward propulsive stroke can be accomplished in $\frac{9}{1,000}$ sec, enable these animals to move at very rapid rates. The house centipede, *Scutigera*, which has the added advantage of a plantigrade foot in which the several segments that are brought into contact with the substrate are covered with hairs that keep it from slipping, can run almost 2 ft/sec.

Insects are either plantigrade or digitigrade, which gives them a mechanical advantage over the

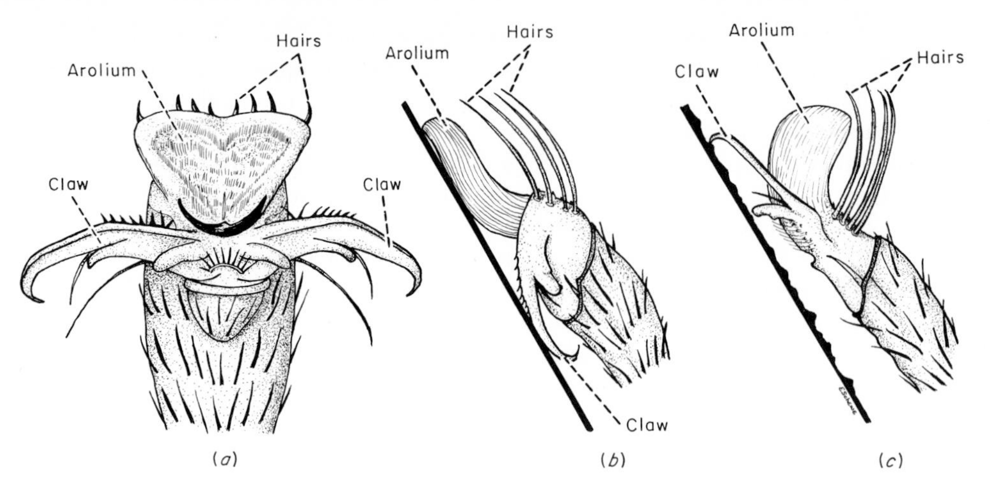

FIG. VI–28 *Tarsus of the honeybee.* **(a)** *Undersurface of the terminal segment.* **(b)** *Side view of the terminal segment, showing extension of arolium and retraction of claws when the insect is moving on a smooth surface.* **(c)** *Side view of a terminal segment, showing extension of claws and retraction of arolium when the insect is clinging to a rough surface.*

unguligrade Crustacea and other arthropod groups (Fig. VI–27). The reduction in the number of legs to three, and their length, has given them also a greater variety of gaits and of speeds. Experiments involving the isolation of thoracic segments have shown that each of these acts as a self-contained reflex unit for the stepping movements of the legs attached to it. These movements are of two principal kinds: (1) flexion of the tarsus which allows the insect to obtain a foothold on the substrate, with subsequent relaxation that releases the hold; and (2) flexion and extension of the leg as a whole, which supplies the propulsive force. By flexion of the tarsus, the pads, or arolia, on the undersurface of its joints are brought into contact with a smooth surface to which the insect may anchor itself by means of a sticky secretion exuded either from pores opening on the surface of the pad or from glandular chaetae, the tenent hairs. Or, on a rough surface, the flexed tarsus may act like a hook, bracing the insect between steps or holding it when at rest (Fig. VI–28). The legs of insects that walk on the surface of the water, like the hemipteran water strider *Gerris* and many gnats and flies, are covered on one or more surfaces of one or more articles with hairs that are coated with a nonwetting substance (Figs. VI–29 and VI–30). Because of these hydrofuge hairs, the surface film of the water is not broken but only slightly indented by the weight of the insect's body as it slides or skates freely over the surface.

Movement in insects may be forward, backward,

FIG. VI–29 Gerris, *a water strider, moving over water, whose surface film is slightly indented under its legs but not broken.* (*Courtesy of H. Lou Gibson.*)

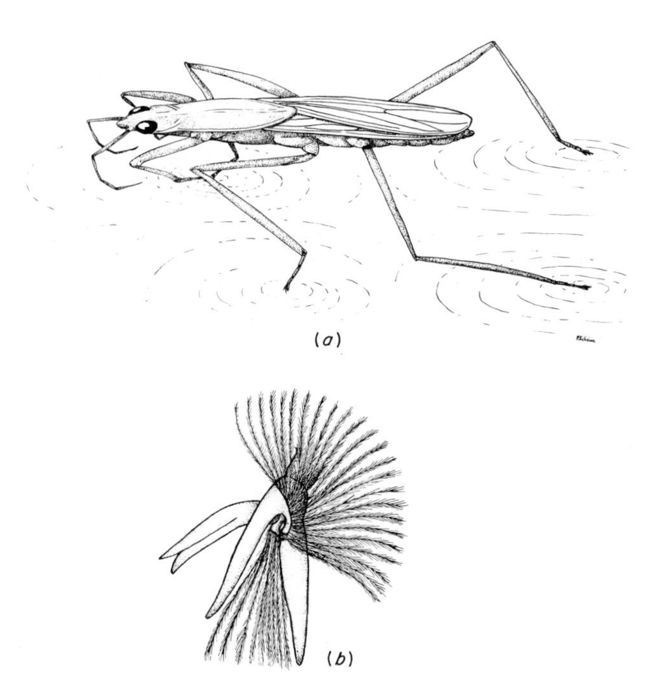

FIG. VI–30 Gerris. *(After Lutz.)* ***(a)*** *Insect on the surface of the water over which it moves by action of its meso- and metathoracic legs, holding the prothoracic pair up in a grasping position.* ***(b)*** *Tarsus of a mesothoracic leg of another water strider,* Rhagovelia, *showing hydrofuge hairs which enable it to maintain its position on the surface of the water. (After Pennak.)*

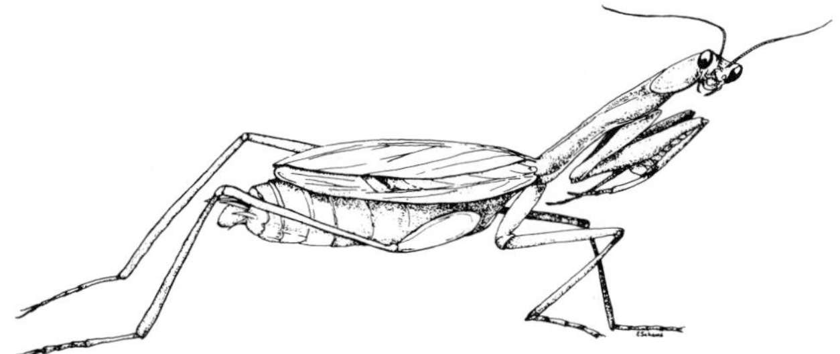

FIG. VI–31 Stagmomantis carolina, *a praying mantis, whose prothoracic legs are primarily prehensile rather than locomotor. (After Lutz.)*

or sideways and results from coordinated movements of the individual legs. In many walking and running insects, all six legs are used either simultaneously as in the true bugs (heteropteran Hemiptera) or, more frequently, in alternative movement, each member of a pair executing an opposite movement to that of its mate. In forward movement, for example, the first (prothoracic) leg and the last (metathoracic) leg of one side are elevated together with the second (mesothoracic) leg of the opposite side, while the body is supported on the other three legs in the extended position with their tarsi flexed. Mechanically, this is a position of maximum stability, the three legs in contact with the surface acting as a tripod to support the body. When the flexed legs have been extended again and contact made with new points ahead of the body, they form the tripod and the other three are released, flexed, and again extended to take up a position still further ahead. In these movements, the metathoracic legs and, to a lesser extent, the mesothoracic act as pushers, while the prothoracic legs act as tractors.

In actively predatory insects, the anterior pair of legs may be adapted for grasping rather than walking. This is shown, for example, in the Mantidae, which wave these prehensile appendages back and forth as they use their other four legs, separately and not alternatively, in their ordinary walking movements (Fig. VI–31). At a faster pace, however, the meso- and metathoracic legs of opposite sides are moved together, and, when the insect breaks into a run, the prothoracic legs are pressed into locomotor action. The sequence of movements is then like that of other running insects, the movement of each member of a pair of legs alternating with that of its mate on the opposite side. Similarly, *Gerris* uses only its meso- and metathoracic legs in skating, pushing itself over the surface of the water with the mesothoracic pair while steering with the metathoracic and holding the prothoracic pair up and out in a position to grasp the food it may be seeking.

d. SWIMMING The majority of truly aquatic arthropods are crustaceans, and for the most part the active swimmers among them are those of small size. They swim by movements of certain appendages, often wide and blade-like with surfaces further enlarged by fringes of strong hairs offering additional resistance to the water. In branchiopods, all the trunk appendages are generally used, moving in metachronal rhythm; in

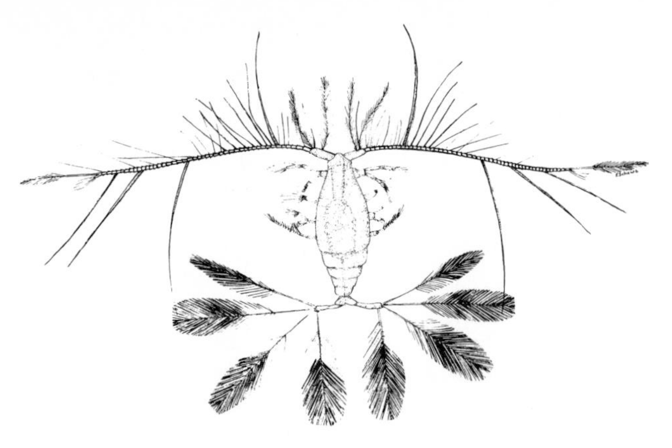

FIG. VI–32 Calanus, *a copepod and prominent member of the zooplankton that uses its four pairs of setose thoracic appendages in locomotion, with the plumose caudal rami helping to keep it afloat. (After MacGinitie and MacGinitie.)*

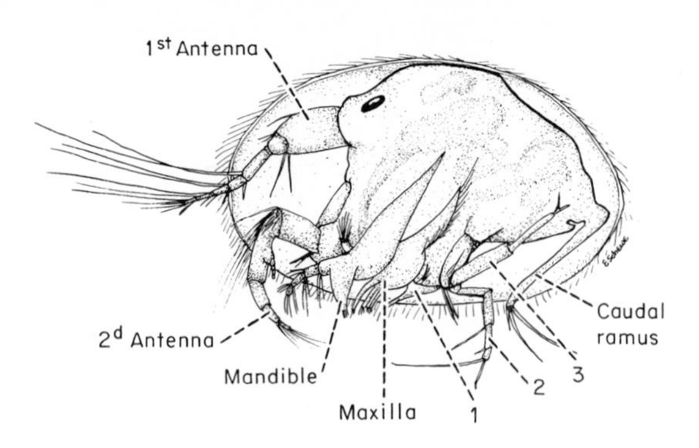

FIG. VI–33 Cypricercus reticulatus, *an ostracod that swims by beating of its first and second antennae and kicking thrusts of its caudal ramus. (After Pennak.) 1, 2, and 3 indicate first, second, and third pairs of legs.*

copepods, only the first four thoracic appendages are generally used, but the swimming ability of the animal is enhanced by modifications of other appendages in ways to keep it afloat (Fig. VI–32). Cladocerans and ostracods swim by means of their antennae; in both of these groups, the body is partially or entirely enclosed in a bivalved carapace from which, when opened, the antennae project. In ostracods, the appendages are so much reduced in size, as well as in number, that they are practically useless for any but subsidiary functions (Figs. VI–33 and VI–34). Larger Crustacea may swim slowly, mainly by paddle-like movements of their biramous pleopods. Speedier movement is attained by quick flexions of the tail under the abdomen, effecting a kind of jet propulsion.

There are some few insects whose mode of locomotion as adults is swimming. Of these the water boatman of the family Corixidae and the back swimmers of the family Notonectidae, both heteropteran Hemiptera, are truly aquatic with specialized swimming and respiratory organs (Fig. VI–35). While *Notonecta* lives largely on the surface of bodies of freshwater, it dives and can swim underwater, and *Corixa* can remain submerged for long periods, making use of its legs not only for swimming but also for clambering about on the underwater debris. At least two families of Coleoptera, the predaceous diving beetles (Dytiscidae), and the whirligig beetles (Gyrinidae) are

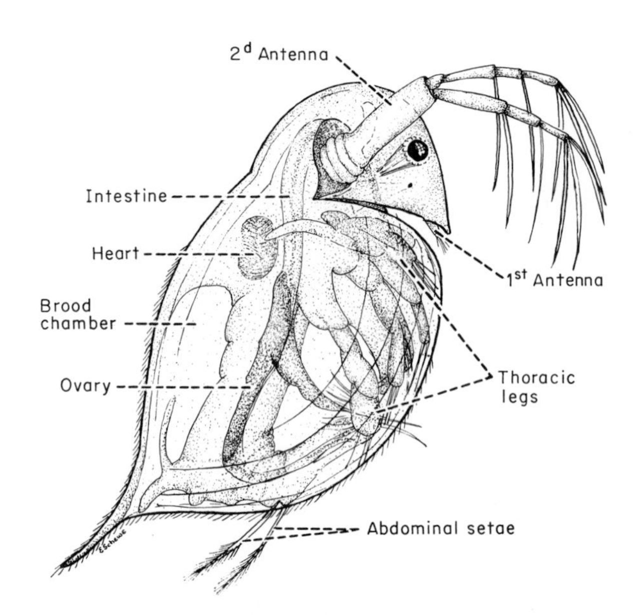

FIG. VI–34 Daphnia pulex (*water flea*), *a cladoceran that swims by the beating of its antennae and that can execute hopping movements by their more vigorous strokes. (After Pennak.)*

also swimmers and divers (Fig. VI–36). The whirligig beetles can lift themselves from the surface of the water and fly. In all these insects, swimming and diving are effected primarily by simultaneous movements of the metathoracic legs, of which the tibia and tarsal segments are usually flattened and equipped with fringes of chaetae.

Homework
teacher
time

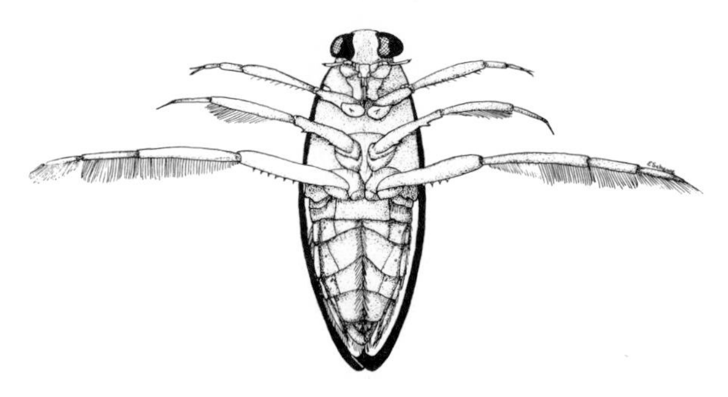

FIG. VI–35 Notonecta, *a back swimmer, whose swimming legs are fringed with hairs.* (*After Lutz.*)

e. PLOUGHING Ploughing through sand and mud on the part of aquatic arthropods, and through the loose humus and soft litter of a forest by terrestrial ones, is effected by much the same mechanisms as walking or swimming. These are, however, often supplemented by the action of some other part. *Limulus*, for example, can shove its heavy horseshoe-shaped prosoma through the sand by means of powerful backstrokes of its walking legs, followed by pushes by the long, pointed spine that is articulated to the opisthosoma (Fig. VI–37). This can be stuck into the sand at an angle and then straightened so that the entire body is forced forward. Similarly, to plough through the mud, caudal appendages of one kind or another are used by some crustaceans, such as certain branchiopods, notably some conchostracans and cladocerans, and some mysids.

f. BURROWING AND DIGGING Many arthropods make burrows for permanent or temporary habitation and make them in a number of different ways. Some, like termites (isopteran insects) and gribbles, isopod crustaceans of the genus *Limnoria*, literally eat their way through the wood, making long, interconnected tunnels in which they live. Others, like geophilomorph chilopods, insinuate their bodies into crevices already formed and then widen the space by shortening their bodies and so extending them laterally. They can do this because the dorsal and ventral sclerites of the integument slide over each other in a telescoping kind of action which is effected by powerful contractions of the leg musculature.

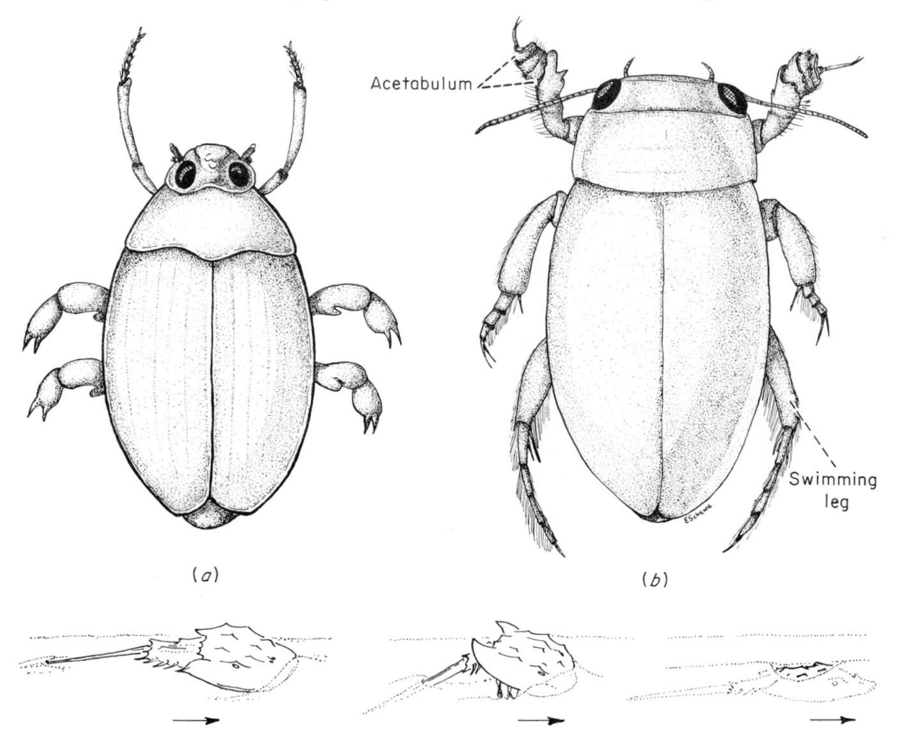

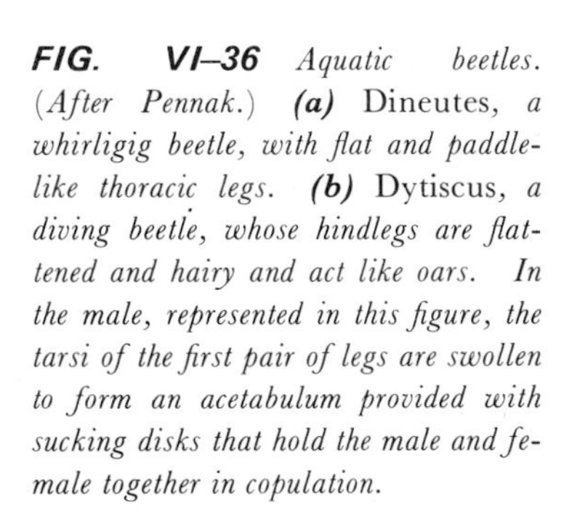

FIG. VI–36 *Aquatic beetles.* (*After Pennak.*) ***(a)*** Dineutes, *a whirligig beetle, with flat and paddle-like thoracic legs.* ***(b)*** Dytiscus, *a diving beetle, whose hindlegs are flattened and hairy and act like oars. In the male, represented in this figure, the tarsi of the first pair of legs are swollen to form an acetabulum provided with sucking disks that hold the male and female together in copulation.*

FIG. VI–37 *The horseshoe crab* Limulus *ploughing through sand, using its legs for excavation and its telson to give a powerful forward thrust.*

Their bodies can thus shorten and widen appreciably, while anchored in position by the legs, and later return to their ordinary dimensions. By burrowing in this way, the animal can escape from predators and also, in dry weather, from dangers of desiccation.

Other aquatic and terrestrial arthropods dig burrows and use a variety of appendages to do so. Some or all of the locomotor appendages may be especially modified for this, but usually a number are brought into action in a number of different ways.

There are two main problems for any burrowing animal. The first of these is the actual excavation of the hole, and the second is the removal of the excavated material from it. There is also the problem of keeping the appendages not actually concerned in the digging out of the way. The amphipod crustacean *Talorchestia*, the common beach flea, folds its antennae while digging head downwards with its second and third pairs of thoracic appendages (gnathopods) (Fig. VI–38). It sweeps the sand back to the telson and uropods with the first pair, or the first three pairs, of pereiopods, where it is thrown clear by sudden and quick extensions of the abdomen. While digging, *Talorchestia* braces its body in the hole with the second and third pairs of pereiopods, if they are not engaged in clearing the excavation, and then pushes itself downward into the hole with the fourth and fifth pairs.

Probably the crustacean best adapted for burrowing is the sand crab *Emerita*. The bodies of these animals are oval, and thus more streamlined than those

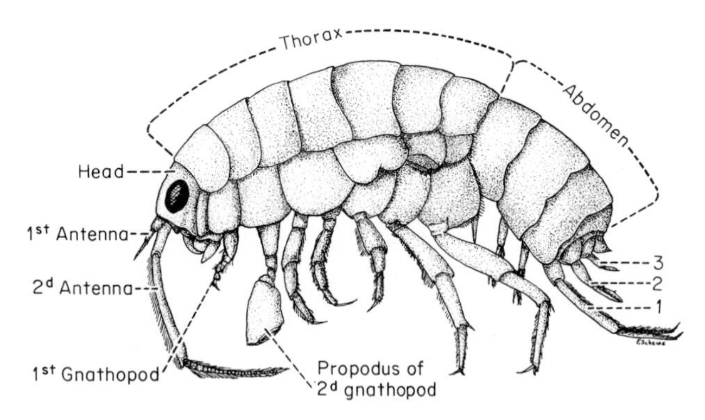

FIG. VI–38 Talorchestia longicornis, *a sandhopper, abundant in sandy beaches near the high-water mark, that burrows in the moist sand and uses its posterior abdominal appendages (1, 2, and 3) in jumping. (After Miner.)*

of other crabs, and all their appendages, except the short antennules, can be folded flat against the body. The first pair of pereiopods are longer than the others and are thick and strong. When preparing to burrow into the wet sand, the animal backs up, brings its uropods into contact with the sand, rotates them so that the sand is thrown up over its back and the body is moved backward into the enlarging hole. The second, third, and fourth pairs of pereiopods move the sand forward and help push the body backward, while the first pair, working alternately and in unison with the others, push the sand outwards and forwards. This can be done very rapidly, so that the crab disappears from view in a matter of seconds. Although burrowing is effected through the combined action of the pereiopods and the uropods, animals deprived of their uropods can still dig into the sand, although more slowly, but they cannot swim; deprived of their pereiopods, they can swim as well as ever, but they can burrow only very feebly.

Among the arachnids, a number of spiders dig burrows, usually making vertical tubes which are lined with silk and which are either open to the surface or closed in some way. Wolf spiders (Lycosidae) do this, but the most elaborate burrows are made by trapdoor spiders, the genus *Cteniza*. The burrow, several inches deep, is excavated by the chelicerae, which are furrowed, with one surface extended into a serrated comb, or rake, which loosens the earth and makes it into a ball. The balls of earth are then thrown out by the posterior pair of legs, which are particularly strong and provided with spines. Once the hole is made, silk is spun to line it and to make the door, or lid, which depending on its thickness may be either of the wafer or the cork type, forming in the one case a delicate hinged "door" or in the other a plug that fits well into the hole.

There are many burrowing insects, but those with the most conspicuous modifications of their legs are the mole crickets and the nymphs of cicadas. *Gryllotalpa*, the mole cricket, burrows in fairly light soil. Its forelegs are very broad and shaped like scoops, with the tibiae and tarsi adapted for cutting roots, on which, along with small underground insects, the crickets feed (Fig. II–27). Nymphs of the cicadan genus *Tibicen* (*Lyristes*) hatching out on the limbs of the trees, where the female has laid her eggs, fall to the ground and make burrows into it where they may live for a number of years before emerging as adults. The femurs of their forelegs are large and have strong spines; the tibiae are short and each has a stout, sharp process that acts like a pickaxe or mattock in dislodging the dirt; and the tarsi can be folded back out of the way while digging is in progress. The forelegs of digger wasps (Bembecidae) are curved, and the insect can dig rapidly with them in loose sand or soft earth.

g. **HOPPING AND LEAPING** Various devices are used for hopping and leaping, which may involve use of the body segments as well as the appendages and other specialized structures. Sand hoppers, when not digging, can walk on their five pairs of thoracic appendages, swim by simultaneous paddle-like strokes of the first three abdominal pairs, and hop by using the last three pairs of abdominal appendages, which are short and stiff. The tip of the abdomen is used to provide the propulsive force for a leap of considerable distance.

Springtails, the Collembola of somewhat anomalous systematic position, which live in damp soil, on the edges of ponds and streams, and in snowbanks, have highly specialized abdominal appendages that provide a mechanism that enables them to leap suddenly into the air (Fig. VI–39). The appendages of the fourth abdominal segment are partially fused together to form the furcula, which is turned under the body and held there, when the animal is at rest, by the tenaculum. This is derived from the appendages of the third abdominal segment, whose bases are fused while the distal parts are free. When the extensor muscles of the furcula contract, it is released from the tenaculum and pulled downwards and backwards so that it strikes the ground with enough force to propel the light body some distance through the air.

Saltatory insects use their legs for hopping and leaping. In some, like fleas (Siphonaptera), and crickets and grasshoppers (Orthoptera), the muscles of the femur of each metathoracic leg are greatly developed and are thick and strong, so that their contraction, and the consequent flexion and extension of the femural-tibial joint, is forcible enough to lift the entire body from the ground and to propel it for some distance through the air. Leafhoppers, tree hoppers, and jumping plant lice (homopteran Hemiptera) use a different mechanism, the force being supplied mainly at the coxa-trochanter articulation of the metathoracic pair of legs. The coxae of this pair are somewhat larger than those of the other pairs and form a passage for bands of thoracic muscles, which are attached to the coxa-trochanter joint and which are the ones effective in the flexion and extension of the leg. These insects, whose bodies are small and light, can leap several feet.

4. Wings and the mechanism of flight in insects

Among invertebrates, only the insects have achieved the capacity for flight, by the development of wings and muscular mechanisms to move them. They are the only flying animals that have gained this advantage without the sacrifice of a functional pair of legs. Insect wings are dorso-lateral outgrowths of, typically, the meso- and metathoracic segments and are extensions of the integument into which tracheae, nerves, and

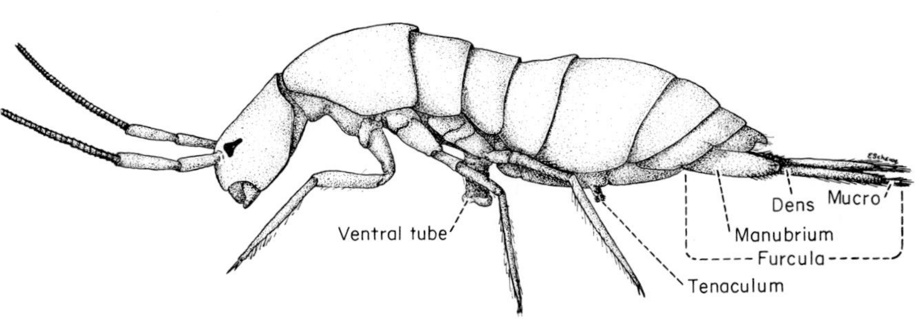

FIG. VI–39 Axelsonia, *a springtail (collembolan). (After Imms.) The first pair of embryonic appendages are fused to form the ventral tube which functions as an adhesive organ. The third pair are partially fused to form the tenaculum. The fourth pair are partially fused to form the furcula, consisting of the manubrium (completely fused bases) and the dentes, or free portions, each of which terminates in a claw-like process. In this drawing, the springtail is shown in a leaping position, with furcula extended.*

blood sinuses extend (Fig. II–26). The two layers of the integument may be only lightly chitinized, as they are in the membranous wings of many species, or more heavily so, as in the mesothoracic wings or leathery tegmina of Orthoptera and the heavy elytra of Coleoptera. They may also be covered with scales, as they are in the Lepidoptera, and with hairs. The course of the blood sinuses, through which an active circulation is maintained, is marked in all types of wings by more heavily sclerotized areas, the veins, which stand out conspicuously in the delicate membranous types. The pattern of venation differs in different species, and even in different varieties within a species, and has been used as a diagnostic character in some systems of insect classification. Typically, in adult insects, there are two pairs of wings, which may be of equal or nearly equal size, or one pair may be considerably larger than the other. In Diptera, the posterior (metathoracic) pair is much reduced and is represented by only two stumpy processes, the halteres (Fig. VI–40). There is a similar reduction of the fore wings in the males of Strepsiptera. There are, however, some insects that lack wings entirely throughout their whole life cycle, such as the silverfish (Thysanura), a primitive insect representing those in which wings presumably never evolved, and fleas (Siphonaptera) and lice (Mallophaga and Anopleura), representing degenerate forms descended from winged ancestors. In a number of species, whose life cycles represent a succession of different forms, there may be one or more winged phases; in some others, as for example, social insects like ants (Hymenoptera) and termites (Isoptera), newly hatched young may have wings but later regularly lose them. In some orders, such as the Embioptera and Strepsiptera, the males alone are winged and the females always without them (apterous).

At the base of the wing, where it joins the body, there is a membranous hinge reinforced by a number of small, chitinous plates, the axillary sclerites. At least two of these articulate with the dorsal thoracic sclerite, the notum. The axillary sclerites are particularly well developed in insects that fold their wings when at rest, flexing them over their backs at the end of a flight and extending them again in preparation for another; these sclerites provide attachment for the muscles that bring about these movements. Two more sclerites lie beneath each wing base, derived, presumably, from the lateral thoracic sclerite, or pleuron. One of these, the basalar, is anterior in position; the other, the subalar, lies posteriorly. The number and, to some extent, the positions of all these sclerites varies in different orders and genera of insects (Fig. VI–41).

FIG. VI–40 Atomosia puella, *a robber fly, showing membranous mesothoracic wings and metathoracic pair reduced to halteres.* (*U.S. Department of Agriculture, Bureau of Entomology and Plant Quarantine.*)

In all winged insects, the thoracic segments show a number of modifications both in muscular and in skeletal structures which represent adaptations to the combined mechanisms involved in movements of the legs in walking, running, or climbing and of the wings in flying. These modifications are less evident in the wingless prothoracic segment of modern insects than in the two posterior ones that bear the wings. In both the mesothoracic and the metathoracic segments, the notum is visibly marked off as an anterior alinotum and a posterior postnotum. The postnotum is connected to the posterior part of the pleuron, the epimeron. Each pleuron is enlarged, with a ventral coxal process, to which a leg is articulated, and a dorsal wing process, to which a wing is articulated. Alinotum, postnotum, and pleuron all have apodemes (anterior and posterior phragmata); the pleural apodeme, or pleurodema, extends some distance inward toward the midline. The pleuron is joined to the ventral exoskeletal plate, the sternum, by anterior and posterior apodemes (apophyses) which form a socket for the coxa of the leg.

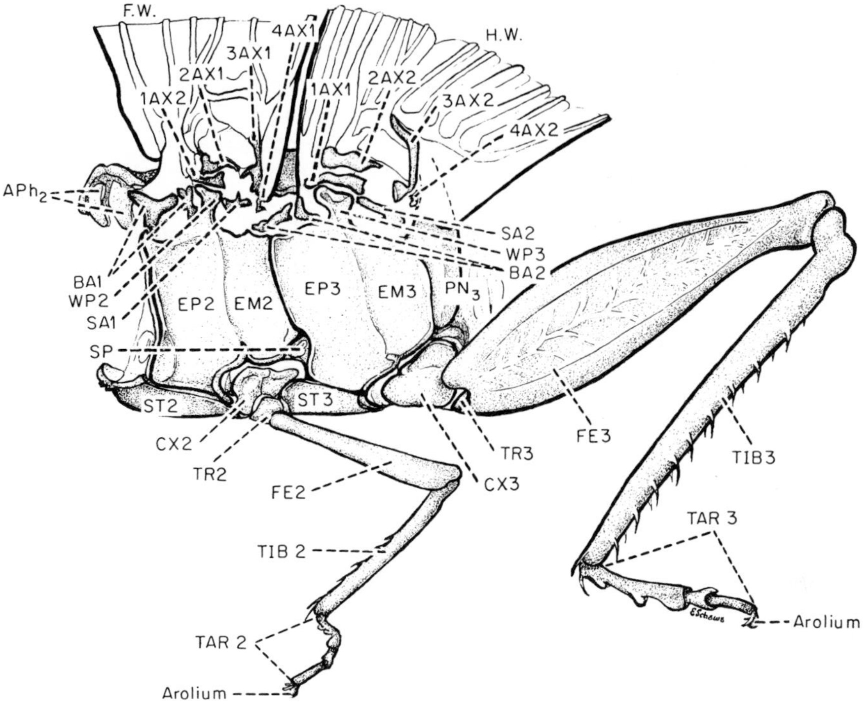

FIG. VI–41 *Wing-bearing segments (meso- and metathoracic) of the grasshopper* Dissosteira *from the lateral aspect, showing the proximal portions of the wings turned upward so as to expose the alar sclerites from the undersurface. (After Snodgrass.)* APh_2*: anterior phragma of mesothoracic notum. 1 AX1 to 4 AX1: axillary sclerites of fore wing. 1 AX2 to 4 AX2: axillary sclerites of hind wing. BA1 and BA2: basalar sclerites of fore and hind wings. CX2 and CX3: coxae of meso- and metathoracic legs. EM2 and EM3: epimerons of meso- and metathoracic pleura. EP2 and EP3: episterna of meso- and metathoracic pleura. FE2 and FE3: femora of meso- and metathoracic legs. F.W. and H.W.: fore and hind wings.* PN_3*: postnotum of metathorax. SA1 and SA2: subalar sclerites of fore and hind wings. SP: spiracle. ST2 and ST3: sterna of meso- and metathorax. TAR2 and TAR3: tarsi of meso- and metathoracic legs. TIB2 and TIB3: tibia of meso- and metathoracic legs. TR2 and TR3: trochanters of meso- and metathoracic legs. WP2 and WP3: wing processes of meso- and metathoracic pleura.*

Along the midventral line, the sternum is invaginated and supports a forked apodeme, the furca, whose branches are closely apposed to the pleurodema of each side. The muscles are usually arranged in two conspicuous sets—a longitudinal one, attached at the anterior and posterior borders of the notum, and a transverse one, connecting notum and sternum—and a number of less conspicuous sets more or less intricately arranged in relation to the wing sclerites (Figs. VI–42 and VI–43).

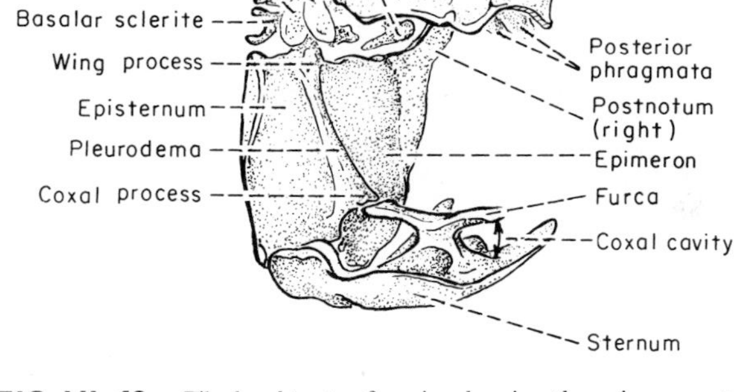

FIG. VI–42 *The hard parts of a wing-bearing thoracic segment of an insect, drawn to show the inner face of the right pleuron. (After Snodgrass.)*

The origin of insect wings is another problematical aspect of arthropod evolution, owing largely to the absence of any discovered fossils that might furnish clues about the progressive development of these new organs. Although it has been suggested that the earliest insects were winged ones and that modern apterygotes are degenerate types, paleontological evidence refutes this and indicates that the earliest insects were without wings and most probably aquatic, feeding on the prostrate plants in the shallow waters and swamps of the Devonian period. Winged insects appeared first in the Carboniferous, along with upright plants. These early pterygote forms were large, with two pairs of wings, each of equal size, one pair on the mesothoracic segment and one on the metathoracic. The completeness and apparent

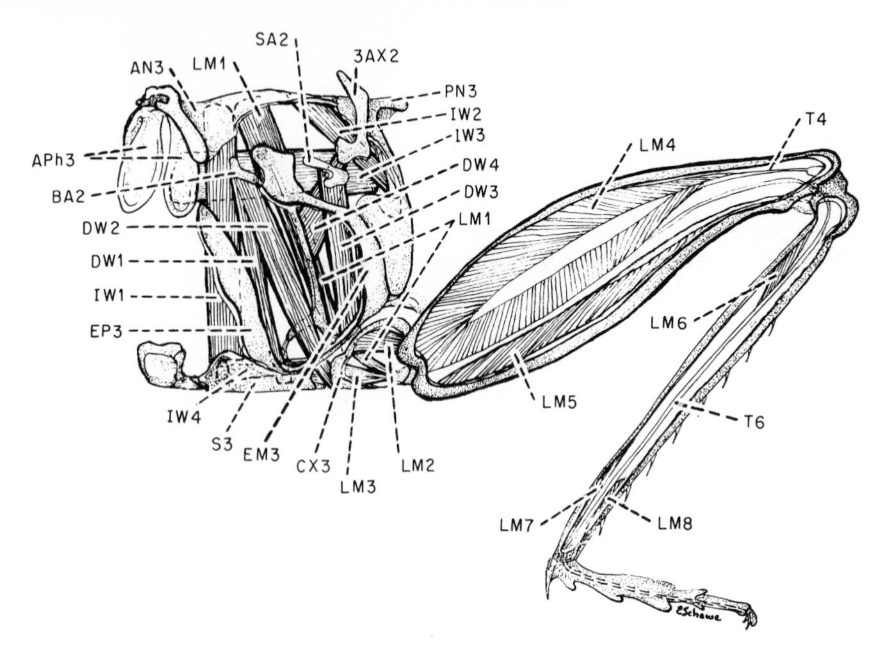

Wing Muscles, direct

DW1: first basalar muscle, attached to the basalar sclerite and the sternum anterior to the coxa, whose contraction draws the wing forward and tilts its anterior margin (costa) downward. Acts also to extend the wing after flexion. DW2: second basalar muscle, attached to the basalar sclerite and the coxa whose action on the wing is similar to that of DW1. DW3: subalar muscle, attached to the subalar sclerite and the coxa, whose contraction depresses the wing as a whole, draws it backward, and lowers its posterior margin. DW4: axillary muscle (pleuroalar), attached to the third axillary sclerite and the upper part of the pleurodema, whose contraction flexes the wing.

FIG. VI–43 *Metathoracic segment and leg of a male grasshopper,* Dissosteira. *The drawing is made with the sclerites of the left side partially cut away to expose the muscles. Only one set of thoracic muscles is shown. (After Snodgrass.)*

Sclerites

APh3: anterior phragma of metathoracic notum. AN3: alinotum of metathoracic notum. PN3: postnotum of metathoracic notum. BA2: basalar sclerite of hind wing. SA2: subalar sclerite of hind wing. 3AX2: third axillary sclerite of hind wing. EP3: Episternum of metathoracic pleuron. EM3: epimeron of metathoracic pleuron. S3: sternum of metathorax. CX3: coxa of metathoracic leg.

Wing Muscles, indirect

IW1: tergo-sternal muscle, attached to the notum and the sternum, whose contraction depresses the wing base by drawing the notum ventrally and thus elevates the blade. IW2: lateral oblique muscle, attached to the notum and its posterior phragma, whose contraction supplements that of the tergo-sternal muscle (IW1) in drawing the notum ventrally. IW3: dorso-longitudinal muscle, attached to the anterior and posterior phragmata of the notum, whose contraction arches the notum, thus raising the wing base and depressing the blade. IW4: longitudinal ventral muscle, attached to the sternal apophyses, whose contraction supplements that of the dorso-longitudinal muscle (IW3). In this drawing, the muscle is concealed by the sternum and is shown by dotted lines.

Leg Muscles

LM1: first tergo-coxal muscle, attached to the notum and anterior rim of the coxa, whose contraction draws the entire leg inward. It may act also as an elevator of the wing. LM2: anterior levator of the trochanter, attached dorsally to the anterior rim of the coxa and the anterior rim of the trochanter, whose contraction elevates the entire leg. LM3: coxal portion of depressor of trochanter, attached ventrally to the anterior rim of the coxa and the anterior rim of the trochanter, whose contraction lowers the entire leg. LM4: posterior levator to tibia, attached to the anterior and posterior walls of the femur and the sides of a thin, flat apodeme (T4) arising from the dorsal margin of the base of the tibia. Contraction of this powerful muscle elevates the tibia and is mainly responsible for the "hop." LM5: depressor of tibia, attached to the ventral surface of the femur and a long apodeme arising in the ventral membrane of the knee joint, whose contraction lowers the tibia. LM6: depressor of pretarsus and retractor of claws, attached to the ventral surface of the tibia and a long apodeme arising from a plate at the base of the claws (T6), whose contraction lowers the tarsus and retracts the claws. LM7: depressor of tarsus, attached to the distal part of the ventral wall of the tibia and the base of the tarsus, whose contraction lowers the tarsus. LM8: levator of tarsus, attached to the distal part of the dorsal wall of the tibia and the base of the tarsus, whose contraction elevates the tarsus.

Apodemes

T4: apodeme arising from the dorsal margin of the base of the tibia.
T6: apodeme arising from a plate at the base of the claws.

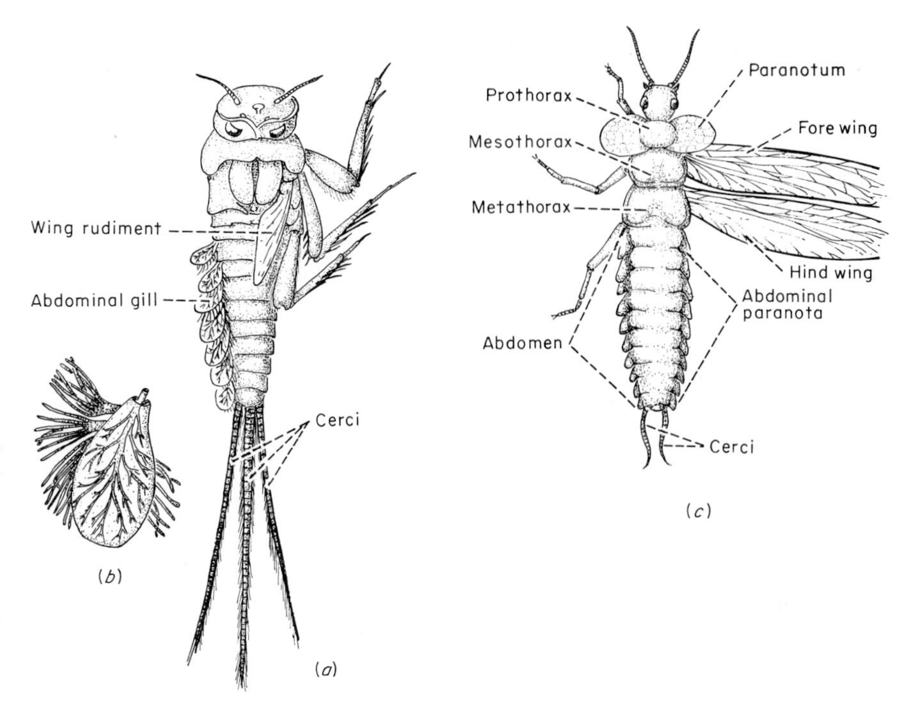

FIG. VI–44 *Tracheal gills and paranota. (After Imms.)* ***(a)*** *Nymph (eighth instar) of the mayfly* Heptagenia. *Wing rudiments and legs have been shown only on the right side, tracheal gills only on the left.* ***(b)*** *Enlargement of a tracheal gill, showing the distribution of the tracheal tubes within it.* ***(c)*** Stenodictya lobata. *The wings have been shown only on the right side, the legs only on the left.*

abruptness with which this change from a wingless body to one with wings came about leaves only to conjecture the transitional steps that may have occurred in the evolution of these organs that enabled the insects to exploit the air as a new habitat. Two principal theories have been advanced for the origin of wings, neither of which is satisfactory. One of these, the tracheal-gill theory, was proposed by the German comparative anatomist Karl Gegenbaur (1826–1903). It assumes that the early aquatic insects used paired outgrowths of the body wall both for locomotion and for respiration. Such organs are found along the sides of the abdomen in the nymphs of modern mayflies, at the sides of the thorax at the bases of the limbs in those of stone flies, and at the end of the abdomen in those of the smaller dragonflies (Fig. VI–44*a*). This theory would imply that insects made the transition directly from water to air by adaptations of such gills for flight rather than for swimming. The paranotal theory was proposed at about the same time by another German biologist, Fritz Muller (1821–1897). According to this, all segments of the insect body developed lateral expansions of the integument of the notum, but only those of the meso- and metathoracic segments became functional wings. Evidence for this is offered by the fossil *Stenodictya lobata* from Upper Carboniferous deposits, in which the prothoracic segment has very definite paranota and the abdominal segments have equally well defined, although smaller, ones (Fig. IV–44*c*). The prothoracic paranota are however, not as large as those of the meso- and metathoracic segments, and it has been clearly shown that their venation does not correspond to that of the true wings. They, therefore, cannot be regarded as homologous structures, and the theory is not substantiated by fact. It had, however, great appeal for the protagonists of the monophyletic theories of arthropod origin, since it provided the possibility of phylogenetic connection between insects and trilobites, whose pleurae might conceivably have expanded in this way.

However such integumental expansions originated, the animal possessing them was provided with an airfoil and the potentiality of free flight, once mechanisms for their movement and its control were established. Since each insect wing is hinged to the notum, and the hinge lies medial to the pleuron and can pivot upon it, all that is needed for its upward and downward motion is alternate depression and elevation of the notum. This can be, and in most insects is, brought about by contraction of the two main sets of thoracic muscles which presumably were present in each thoracic segment of the earliest hexapods. When the transverse (dorso-ventral, tergo-sternal) muscles of the segment contract, the margins of the notum are pulled downward and inward so that the bases of the wings are depressed and their surfaces raised horizontally. When these muscles are at full contraction, the longitudinal muscles are at full extension and so in the most favorable condition to give a maximum contraction. When these contract, the lateral margins of the notum are pulled upward and outward, the notum is arched, the bases of the wings are raised, and their surfaces are depressed. In both cases, the pull on the wing is indirect, and to be effective, the two wing-bearing nota and those of the metathoracic and the first abdominal segments must be firmly united; otherwise, the contractile force would be expended in drawing the nota together, not in arching each individually. Because of this arrangement of the indirect flight muscles, it follows that both wings in each segment must move together, each pair acting as a unit.

Such simple flapping movements are not adequate for flight, however, for flying, like swimming, requires the displacement of the medium in a direction opposite to that of the movement of the object. To act as propellers, the wings must turn on their long axes. This is achieved, in insect wings, by means of the "direct" muscles attached to the alar sclerites. In the downstroke, the wing blade is moved slightly forward (protracted) and its anterior margin (costa) tilted downward (pronated) by contraction of muscles attached to the basalar sclerite and the sternum (Fig. VI–43, DW1). In the upstroke, the blade is drawn backward (retracted) and the costa raised (supinated) by contraction of the muscles attached to the subalar sclerite and the coxa (Fig. VI–43, DW3). Minor adjustments of the wing path may be brought about by action of the muscles attached to the axillary sclerites (Fig. VI–43, DW4). The primary action of these muscles is, however, to retract and fold the wings in those insects in which the wings are carried over the dorsal surface of the body when the insect is not in flight. Pull on the axillary sclerites makes the wing crumple at its base and so swing posteriorly; it is uncrumpled and brought forward by contraction of the muscle attached to the subalar sclerite. In some insects, the wings are brought back flat to the "rest" position, but in others, especially in those with broad wings, they may be drawn into longitudinal folds by flexor muscles attached to the axillary sclerites and expanded by contraction of the muscle attached to the basalar sclerite.

All these sclerite and muscle adaptations for flight must have been established early in insect evolution, but how they evolved and how they developed into those that have given modern insects their remarkable flying powers is unknown. Perhaps the best guess is that as plants grew taller and insects climbed up on them in search of new feeding grounds, the animals' incipient wings allowed them to glide down to the ground again. For this, ability to tilt the wing would be most significant, and possibly the muscular adaptations for pronation and supination were the first to become fixed. The fossil remains of the earliest insects show that they were large animals; the wing spread, for example, of the fossil dragonfly *Meganeura* from the Upper Carboniferous deposits measures about 2 ft. Retraction and flexion of wings of this size must also have been advantageous, and though this ability seems to have been a secondary innovation, there is evidence that it also occurred early. Flapping movements may have been initiated by vibrations set up by the stream of air rushing past them as insects glided to the ground from the heights to which they had climbed and, then, a few changes in the distribution of sclerites and muscles could have permitted the vibration to be sustained by muscular action. Modern insects offer some evidence for the early establishment of these mechanisms. Mayflies and dragonflies, for example, representing orders of insects that were fully developed in the Carboniferous period, although they do not flex their wings, have all the muscles and sclerites associated with flexion as well as with flight, and dragonflies use all of these in flight movements alone. Cockroaches, whose ancestry dates from the same period, do flex their membranous metathoracic wings but have no transverse thoracic muscles and poorly developed longitudinal ones. They can fly, some with considerable range and speed, but the mechanism of the upward and downward movement of their wings, which must be due to a different arrangement of thoracic muscles than those in most modern insects, is not understood as yet. But since the insects, once they had acquired wings and a flight mechanism, had the air to themselves for some hundred million years, they had ample opportunity to evolve and to multiply with little selection pressure beyond that of environmental change. But when, later, this domain was invaded by vertebrates—first reptiles then birds and bats—this period of immunity was over, and while the number of insects must have decreased because of the predatory habits of their rivals, the number of species must have increased because of the rapid rate of evolution resulting from the exacting conditions imposed both by the climatic changes that took place in the Permian period and by the necessity of escaping from predators and of competing with them for food.

These evolutionary changes involve not only the actual machinery of flight but also metabolic mechanisms to sustain it and neuromuscular ones to initiate and maintain it. Many insects are capable of long, continuous flights. This is particularly true of migratory species. The Monarch butterfly (*Danaus plexippus*), for example, regularly leaves southern Canada and the northern United States in the autumn to fly to southern California, Florida, and the Gulf States, migrating northward again in the spring. It is doubtful if any individual makes the complete trip back to the summer habitat, and it is more likely that the females lay their eggs en route and then die, leaving their offspring to hatch out and complete the flight north. Monarchs have a cruising range of about 650 miles at a speed approximating 6 miles/hr; with a tail wind to help, one could at this speed fly across the Atlantic Ocean. The Red Admiral butterfly (*Vanessa atalanta*), which winters in northern Africa, flies across the Mediterranean each summer to visit the flowers in England, and the migratory phase of the desert locust, *Schistocerca gregaria*, which flies by day but rests in trees by night, has a range of over 200 miles when flying at a speed of 5½ to 6 miles/hr. A major problem for these insects is fuel supply; they must be able to ingest, possibly concentrate, and store an adequate amount of high-energy food, and also to mobilize and metabolize it so efficiently that as an energy source it is equal to the demands of the muscles that must lift the body off the ground, maintain it in the air against gravitational pull, and propel it forwards. The ways in which this problem has been met in different insect species are considered in the chapter on oxidative metabolism (Chapter XII).

Two types of neuromuscular flight mechanisms seem to have evolved. One of these has been termed synchronous, since as in ordinary neuromuscular mechanisms, the contraction of the flight muscles and, hence, the upward and downward movement of the wings follows immediately upon stimulation of the appropriate motor nerve, its conduction of the impulse, and the transmission of it to the muscle. The other mechanism is an asynchronous one, since the flight muscle contracts with greater frequency than the nerve to it conducts. The first mechanism is characteristic of insects with a low rate of wing beat, as has been established for the cockroach, *Periplaneta*, and the noctuid moth, *Agrotis*. The wings of *Periplaneta* can move at a rate of 28 beats per second, and measurements of the passage of the nerve impulse to the thoracic muscles show a 1:1 correspondence between impulse and beat. *Agrotis*, with a wing-beat frequency of 30 to 40 times per second operates on a similar neurogenic mechanism, each beat corresponding to the passage of an impulse along the nerve. The frequency of wing beat in some of the Diptera and Hymenoptera is, on the other hand, extremely rapid and, as measured by some modern methods, may reach more than 1,000 beats per second. Some of the earliest calculations of wing-beat frequency were made in 1869 by a French physiologist Etienne-Jules Marey (1830–1904) who made the tips of insect wings brush against a revolving smoked drum and so obtained a record of their movements with time. Similar but more precise kymograph methods have been used since Marey's original measurements. Other methods that are used are acoustic ones, by which the sound of the wings in motion is compared to the pitch of a tuning fork or other instruments vibrating at a known frequency or by which the sound, picked up by microphones, is amplified and recorded by an oscilloscope or similar instrument. Visual methods are also used by means of high-speed photography and of the stroboscope, whose intermittent flashes of light are timed to coincide with the wing-beat frequency. The wing is therefore illuminated at the same phase of its cycle and consequently appears stationary; the frequency of its beat can be read from an appropriate and calibrated timing device.

To make such records, the insect is fastened to a support by means of a fine thread attached to the dorsal surface of its body by wax or some other suitable material, and the initial stimulus to flight is given by the removal of a platform upon which it has been allowed to stand. Under these conditions, the wings will continue to beat, in stationary flight, until fatigue sets in or the supply of fuel is exhausted. The frequency of beat of representatives of a number of insect species has been determined by these means. The highest rate, that of more than 1,000 cycles per second (Hz) has been found in midges (Diptera), but other insects also have extremely fast rates. The fruit fly *Drosophila* can fly for two continuous hours, under these conditions, at an initial frequency of 300 beats per second and, even as fatigue sets in, maintains a rate of 100 beats per second. The blowflies *Calliphora* and *Lucilia* have measured frequencies of about 140 beats per second and the wasp *Vespa* (Hymenoptera) about 120 beats per second. These rates mean that, in the midges, the events of each complete contraction and relaxation of the flight muscles occupy $\frac{1}{1,000}$ sec; in *Drosophila*, $\frac{3}{100}$ sec at the maximum; and in the blowflies and wasps, something less than $\frac{1}{100}$ sec. This is 10 to 100 times faster than the time span for similar events in vertebrate muscle (0.1 sec). These conditions raise many questions concerning the muscular mechanism capable of excitation, contraction, and recovery in such short periods, as well as the metabolic processes involved.

The asynchronous nature of the operation of these insect flight muscles has been shown by experiments in which fine electrodes have been inserted in them and the differences in electrical potential that occur during flight recorded. The rise and fall in potential during a given period of time is shown as a series of peaks (spikes), on a suitable recording apparatus. When the frequency of the spikes recorded for intact specimens of *Calliphora* is compared to the frequency of the thoracic vibrations, and so of the wing beat, it is evident that there is but one large spike for approximately every 12 smaller ones. If the large spike registers, as it is believed to do, the arrival of a nervous impulse at the muscle and the small ones register muscular contractions, the wing is beating about 12 times more often than the muscle is being stimulated to contract through the orthodox nervous mechanism. Similar differences in timing, although of slightly different magnitude, have been shown for *Lucilia* and for *Vespa*. This divergence from the 1 : 1 ratio, which is characteristic of vertebrate somatic muscle and of the flight muscles of cockroaches and moths, suggests that the contractions that make such rapid wing beats possible are determined by conditions in the muscle itself rather than by those of nerve-muscle association. These conditions appear to be unique in the flight muscles of Diptera, Hymenoptera, Coleoptera, and some (the smaller) Homoptera, and in the sound-producing muscle (tymbal muscle) of cicadas, and to be correlated with their distinctive histology (Chapter IV, pages 108–109). For, while myogenic rhythms have appeared a number of times in animal evolution, notably in cardiac muscle of vertebrates, the flight muscles of insects showing a high frequency of wing beat seem to operate under biophysical principles of a different sort.

One theory of the operation of fibrillar muscle is that the arrival of a nervous impulse at a muscle fiber puts it in a state of excitation, altering the contractile fibers in such a way that they become sensitive to the stimulus of stretching, responding to it with a twitch whose duration is controlled by the load they must move. The contraction of one set of muscles stretches the antagonistic set, which then undergoes a twitch-like contraction. A rhythm of contractions is thus set up, which continues as long as the fibers remain in the excited state, but this state must be maintained by the arrival of nerve impulses above a critical frequency. When the impulses fall below this, the muscular contractions between them die out, and the wing beat slows or stops altogether. This is, therefore, quite a different mechanism than that by which rhythmic contractions are initiated and maintained in vertebrate heart muscle, which shows rhythmic contraction in the absence of any nervous connections whatsoever and concomitant changes in electrical potential of the external membrane of the muscle cell or fiber. Properties peculiar to membranes of these muscle fibers ensure their myogenic rhythms rather than, as in insect fibrillar muscle, properties of the fibrils within the fiber.

That the inertia of the wing is a primary factor in controlling the frequency of contraction in muscles of this kind has been shown by experiments in which a measured load was attached to the wings and those in which the wings have been clipped or amputated. While amputation of the wings of cockroaches and dragonflies has little effect upon the frequency of their thoracic movements, the rate of thoracic vibration increases markedly in wingless wasps and flies. In such insects, the ratio of thoracic vibration to spike potential, whose frequency shows no significant change, may be doubled or even trebled. It appears then that the synchronous motor mechanism is driven by a central nervous pacemaker located in each thoracic ganglion, whose activity is more less independent of the load, while the asynchronous mechanism is subject to peripheral control arising both from the load of the wings and from the pressure of the air through which they move.

Each of the wing-bearing segments is functionally an independent unit; the two members of each pair must move up and down in unison, but each pair can move independently of the other. Since wings, like any other aerial propulsive device, act most efficiently when the air is not turbulent but streamlined, independent action of the two pairs could present great mechanical disadvantages to insects, for the effect of a wing stroke is to set up air turbulence by creating a region of decreased air pressure before and above the insect's body and one of increased air pressure behind and below it. Insects have met, and avoided, this difficulty in a variety of ways. The Odonata, in which the wing movements of the two thoracic segments are alternate, have met it by timing the two separate beats in such a way that the wings move in antiphase, the posterior pair meeting the column of oncoming air before it is disturbed by the anterior pair. This can be seen in dragonflies, the best fliers among the Odonata, and indeed among most insects. The wings of locusts beat together, but because the beat of the hind wings has a greater amplitude than that of the fore wings, the hind wings are ahead of the fore wings during a great part of the stroke and thus avoid much of the air disturbance. Other insects have avoided the difficulty by the development of devices that hold the wings together, so that the two pairs act as a single functional unit, or by the reduction in size or functional capacity of one set of the pair. In Lepidoptera, for example, the wings are held together by a device known as the jugum or by one known as the frenulum (Fig. VI–45). In the jugate members

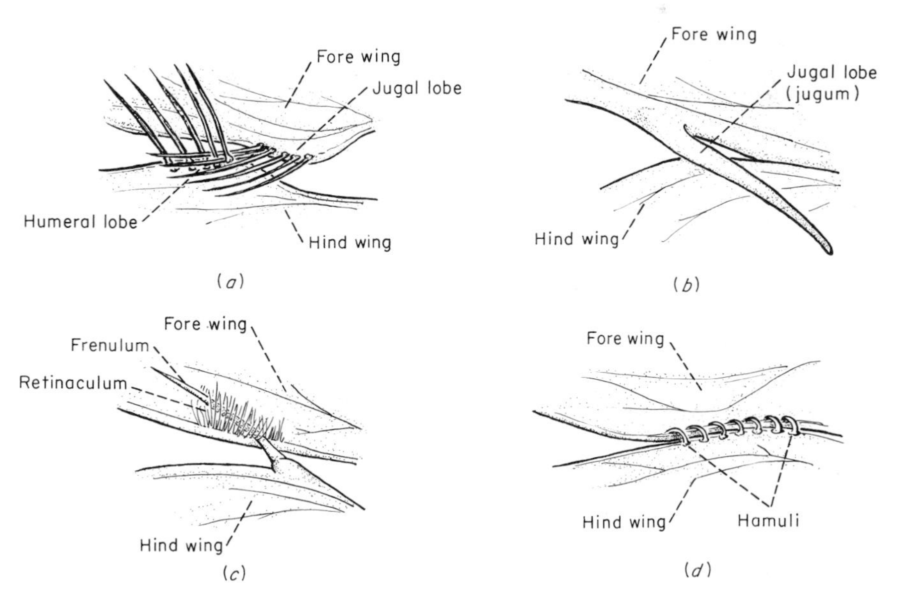

FIG. VI–45 *Devices for coupling fore and hind wings. Left wings only are shown, from the underside. (From various sources.)* ***(a)*** *The most primitive kind of jugate coupling as found in the scorpion flies (Mecoptera).* ***(b)*** *Jugate coupling in moths of the family Hepialidae (swifts). The jugal lobe is greatly extended and projects beneath the hind wing, which is also overlapped by the fore wing.* ***(c)*** *Frenate coupling in moths of the family Noctuidae (millers). The spine-like frenulum on the hind wing is held in place by the fringe of bristles (retinaculum) of the fore wing.* ***(d)*** *Hamate coupling in a wasp of the family Sphecidae (solitary wasps).*

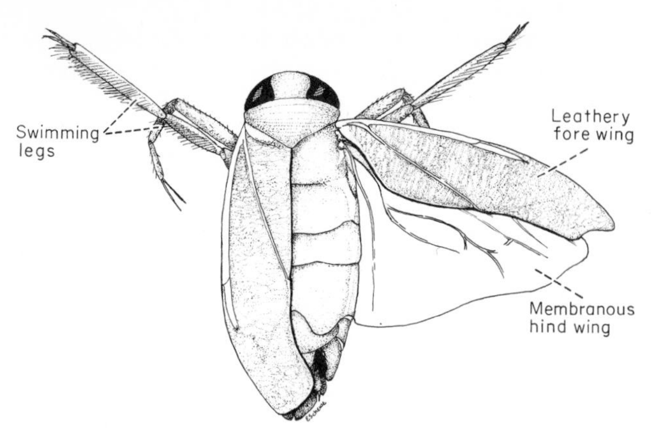

FIG. VI-46 Hespocorixa, *a water boatman, with leathery fore wings, membranous hind wings, and metathoracic legs fringed with swimming hairs.* (*After Hungerford.*)

of the order, the posterior margin of the fore wing bears a small bristled lobe, the jugal lobe, and the anterior margin of the hind wing bears a corresponding one, the humeral lobe. The wings are interlocked by a meshing of the bristles of the two lobes, so that both pairs move as one. In the frenate members, the fore wing has no lobe, but the humeral lobe of the hind wing, or else a long spine, the frenulum, serves to couple the wings together. In most of the Hymenoptera the anterior margins of the hind wings have a series of hooks (hamuli) that fit over a ridge running along the posterior margins of the fore wings and hold them together. In Orthoptera and Coleoptera, only the hind wings are membranous, and the fore wings are leathery or as heavily chitinized as the main sclerites of the body, so that they are either imperfectly functional or not functional at all and are held out straight from the sides, motionless, when the insect is in flight (Fig. VI-46). In Strepsiptera, the fore wings and in Diptera, the hind wings are reduced to halteres, very small stalks with a bulb at the distal end. In Diptera, each haltere is moved by a single muscle attached to the cuticle near its base and to the pleuron of the metathoracic segment. Contraction of this muscle elevates the haltere; when it relaxes, the elasticity of the hinge on which it moves causes its downward movement. The halteres move with the same frequency as the fore wings but in antiphase, so that it would seem that they are controlled by the same mechanism as the wings, although moving oppositely to them. Since they can sometimes be seen to oscillate when the insect is not in flight but walking around, it is apparent that they can move independently of the fore wings. They may act as stabilizers in flight and landing or in the takeoff. It has also been suggested, from studies of the effects of their removal or immobilization, that they are organs that stimulate the production of potential nervous energy for the wings and the legs, since without them some species are unable to fly with their ordinary competence, or even to fly at all; and some, indeed, cannot even walk without intact halteres.

While flight is the primary function of insect wings, they may be put to other uses. Temperature regulation by bees is one of these (Chapter I, pages 21-22), and wing motion may also be used for mating purposes. Male fruit flies, for example, perform a mating dance in which the wings are vibrated in a characteristic fashion, a device that may warn away other males or set up air currents by which distinctive odors are carried to the females. In many insects, vibration of the wings produces characteristic sounds that may or may not be appreciated by human ears without amplification but that are picked up as mating signals by sensitive receptors in the bodies of insects of the opposite sex.

SELECTED BIBLIOGRAPHY

Changes in body shape

Allen, R. D.: Ameboid Movement, in J. Brachet and A. E. Mirsky (eds.), "The Cell," vol. II, pp. 135-216, Academic, New York, 1959.

Allen, R. D.: A New Theory of Ameboid Movement and Protoplasmic Streaming, *Exp. Cell Res., Suppl.* No. 8, pp. 17-31, 1961.

Clark, M. E., and R. B. Clark: The Fine Structure and Histochemistry of the Ligaments of *Nephthys*, *Q. J. Microsc. Sci.*, vol. 101, pp. 138-148, 1960.

Clark, M. E., and R. B. Clark: The Ligamentary System and the Segmental Musculature of *Nephthys*, *Q. J. Microsc. Sci.*, vol. 101, pp. 149-176, 1960.

Eggers, F.: Zur Bewegungsphysiologie von *Malacobdella grossa* Mull., *Z. Wiss. Zool.*, vol. 147, pp. 101–131, 1936.

Glynn, P. W.: Active Movements and Other Aspects of the Biology of *Astichopus* and *Leptosynapta*, *Biol. Bull.*, vol. 129, pp. 106–127, 1965.

Goldacre, R. J.: The Role of the Cell Membrane in the Locomotion of *Ameba*, *Exp. Cell Res.*, Suppl. No. 8, pp. 1–16, 1961.

Goldacre, R. J., and J. J. Lorch: Folding and Unfolding of Protein Molecules in Relation to Cytoplasmic Streaming, Ameboid Movement, and Osmotic Work, *Nature*, vol. 166, p. 497, 1950.

Jahn, T. L.: Hydrodynamic Principles in the Locomotion of Micro-organisms, Excerpta Medica International Congress, series 91. *Second Int. Cong. Protozoology*, 1965.

Josephson, R. K., and S. C. March: The Swimming Performance of the Sea Anemone *Boloceroides*, *J. Exp. Biol.*, vol. 44, pp. 493–506, 1966.

Lissmann, H. W.: Mechanism of Locomotion in Gastropod Molluscs. I. Kinematics, *J. Exp. Biol.*, vol. 21, pp. 58–69, 1944.

Lissmann, H. W.: Mechanism of Locomotion in Gastropod Molluscs. II. Kinetics, *J. Exp. Biol.*, vol. 22, pp. 37–50, 1945.

Noland, L. E.: Protoplasmic Streaming: A Perennial Problem, *J. Protozool.*, vol. 4, pp. 1–6, 1957.

Porter, K. R.: "Submicroscopic Morphology of Protoplasm," Harvey Lectures, Academic, New York, 1956.

Trueman, E. R.: Bivalve Mollusks: Fluid Dynamics of Burrowing, *Science*, vol. 152, pp. 523–525, 1966.

Zenkevich, L. A.: The Evolution of Animal Locomotion, *J. Morphol.*, vol. 77, pp. 1–51, 1945.

Cilia and flagella

Afzelius, B.: Electron Microscopy of the Sperm Tail, Results Obtained with a New Fixative, *J. Biophys. Biochem. Cytol.*, vol. 5, pp. 269–278, 1959.

Afzelius, B.: The Fine Structure of Cilia from Ctenophore Swimming Plates, *J. Biophys. Biochem. Cytol.*, vol. 9, pp. 383–394, 1961.

Aiello, E. L.: Factors Affecting Ciliary Activity in the Gill of the Mussel *Mytilus edulis*, *Physiol. Zool.*, vol. 33, pp. 120–134, 1960.

Aiello, E. L.: Identification of the Cilio-excitatory Substance Present in the Gill of the Mussel *Mytilus edulis*, *J. Comp. Cell. Physiol.*, vol. 60, pp. 17–20, 1962.

Bradfield, J. R. G.: Fiber Patterns in Animal Flagella and Cilia, *Soc. Exp. Biol. Symp. IX*, pp. 306–334, 1955.

Chambers, R., and J. A. Dawson: Structure of Undulating Membranes, *Biol. Bull.*, vol. 48, pp. 240–242, 1925.

Child, F. M.: Some Aspects of the Chemistry of Cilia and Flagella, *Exp. Cell Res.*, Suppl. No. 8, pp. 47–53, 1961.

Fawcett, D.: Cilia and Flagella, in J. Brachet and A. E. Mirsky (eds.), "The Cell," vol. II, pp. 217–298, Academic, New York, 1959.

Gibbons, I. R., and A. V. Grimstone: On the Flagellar Structure in Certain Flagellates, *J. Biophys. Biochem. Cytol.*, vol. 7, pp. 697–716, 1960.

Gibbons, I. R., and A. J. Rowe: Dynein: A Protein with Adenine Triphosphatase Activity from Cilia, *Science*, vol. 149, pp. 424–425, 1965.

Lansing, A., and F. Laney: Fine Structure of the Cilia of Rotifers, *J. Biophys. Biochem. Cytol.*, vol. 9, pp. 799–812, 1961.

Roth, L. E.: Ciliary Coordination in the Protozoa, *Exp. Cell Res.*, Suppl. No. 5, pp. 573–585, 1958.

Schor, S.: Serotonin and Adenosine Triphosphate: Synergistic Effect on the Beat Frequency of Cilia in Mussel Gills, *Science*, vol. 148, pp. 500–501, 1965.

Seaman, G. R.: Localization of Acetyl Cholinesterase Activity in the Protozoan *Tetrahymena geleii*, *Proc. Soc. Exp. Biol. Med.*, vol. 76, pp. 169–170, 1951.

Sleigh, M. A.: The Form of Beat of Cilia in *Stentor* and *Opalina*, *J. Exp. Biol.*, vol. 37, pp. 1–10, 1960.

Sleigh, M. A.: "The Biology of Cilia and Flagella," Pergamon, New York, 1962.

Worley, L. J.: The Dual Nature of Metachronism in Ciliated Epithelium, *J. Exp. Zool.*, vol. 69, pp. 105–121, 1934.

Appendages

Boettiger, E. G.: The Machinery of Insect Flight, in B. T. Scheer (ed.), "Recent Advances in Invertebrate Physiology: A Symposium," pp. 117–142, University of Oregon Publications, Eugene, Oregon, 1957.

Hocking, B.: Aspects of Insect Flight, *Sci. Mon.*, vol. 85, pp. 237–244, 1957.

Parry, D. A.: Spider Leg-muscles and the Autotomy Mechanism, *Q. J. Microsc. Sci.*, vol. 98, pp. 331–340, 1957.

Parry, D. A.: Spider Hydraulics, *Endeavour*, vol. 19, pp. 156–162, 1960.

Parry, D. A., and R. H. J. Brown: The Hydraulic Mechanism of the Spider Leg, *J. Exp. Biol.*, vol. 36, pp. 423–434, 1959.

Parry, D. A., and R. H. J. Brown: The Jumping Mechanism of Salticid Spiders, *J. Exp. Biol.*, vol. 36, pp. 654–664, 1959.

Pringle, J. W. S.: Myogenic Rhythms, in B. T. Scheer (ed.), "Recent Advances in Invertebrate Physiology: A Symposium," pp. 99–116, University of Oregon Publications, Eugene, Oregon, 1957.

Pringle, J. W. S.: Excitation and Contraction of Flight Muscles of Insects, *J. Physiol.*, vol. 108, pp. 222–232, 1965.

Roeder, K. D.: Movements of the Thorax and Potential Changes in the Thoracic Muscles of Insects during Flight, *Biol. Bull.*, vol. 100, pp. 95–106, 1951.

Smith, J. E.: The Role of the Nervous System in Some Activities of Starfish, *Biol. Rev.*, vol. 20, pp. 29–43, 1945.

Smith, J. E.: The Mechanics and Innervation of the Starfish Tube-foot–Ampulla System, *Philos. Trans. R. Soc. London, B*, vol. 232, pp. 279–310, 1946.

Smith, J. E.: The Activities of the Tube-feet of *Asterias rubens L*: The Mechanics of Movement and Posture, *Q. J. Microsc. Sci.*, vol. 88, pp. 1–14, 1947.

Williams, C. M., and R. A. Galambos: Oscilloscopic and Stroboscopic Analysis of the Flight Sounds of *Drosophila*, *Biol. Bull.*, vol. 99, pp. 300–307, 1950.

SUGGESTED READING FOR SECTION FOUR

Clark, R. B.: "Dynamics in Metazoan Evolution: The Origin of the Coelom and Segments," Clarendon Press, Oxford, 1964.

Gray, J.: "Animal Locomotion," Norton, New York, 1968.

Pringle, J. W. S.: "Insect Flight," Cambridge University Press, Cambridge, 1957.

section five

NUTRITION

chapter VII

THE PROCUREMENT AND INGESTION OF FOOD

A. DEFINITION OF FEEDING TYPES

B. MACROPHAGY

- **1. Mechanisms for the ingestion of inactive food**
- **2. Mechanisms for scraping and boring**
- **3. Mechanisms for seizing and swallowing active prey and large particles**
  - a. MECHANISMS FOR SEIZING AND SWALLOWING FOOD WITHOUT PRELIMINARY PROCESSING
    - (1) Pseudopodial
    - (2) Tentacular
    - (3) Sucking or pumping
  - b. MECHANISMS FOR THE PHYSICAL OR CHEMICAL DISINTEGRATION OF FOOD BEFORE INGESTION

C. MICROPHAGY

- **1. Mechanisms for the intake of material in crystalline or colloidal solution**
  - a. MEMBRANE TRANSPORT
  - b. PINOCYTOSIS
- **2. Mechanisms for the generation and direction of food currents**
- **3. Filter feeding**
  - a. CILIARY FILTERS
  - b. MUCOID FILTERS
  - c. SETOSE FILTERS
  - d. EFFICIENCY OF FILTER FEEDING

D. FLUID FEEDING

- **1. Mechanisms for ingesting exposed (patent) fluids**
  - a. IN ENDOPARASITES
  - b. IN INSECTS
- **2. Mechanisms for ingesting internal (concealed) fluids**
  - a. IN NEMATODES
  - b. IN GASTROPOD MOLLUSCS
  - c. IN ANNELIDS
  - d. IN ARTHROPODS

Food is a prime necessity for all living things, and the means of obtaining and of processing it are the most important aspect of their adaptation to life on this planet. Every organism depends upon its environment for its food supply, and every animal depends, directly or indirectly upon plants for the satisfaction of its nutritive needs. These needs vary widely, according to individual capacities to synthesize substances essential to metabolism and growth. Some animals are able to do this from compounds that are simple in composition and general in distribution, but the majority are far more restricted in synthetic abilities and must derive a large measure, if not all, of their nutrients from complex substances. One limitation to full utilization of all that the environment might provide is therefore set by their own enzyme systems, because inability to digest, assimilate, or recombine into essential compounds the materials that may be offered, or even taken into the body, would result in starvation even in the midst of apparent plenty. Other limitations are set by the means of capturing and ingesting foods, and the greater the variety of food that an animal can utilize, the greater is its freedom. Development of mechanisms for obtaining nutriment is therefore of the greatest significance to the success of any species, for all of them are competing for the available food supply and those best adapted to obtain what they need from a variety of potential sources are the most likely to survive and to reproduce.

Some few animals are passive feeders able, as perhaps the earliest biological systems were, to nourish themselves on molecules that diffuse into their bodies directly from their environments. Intestinal parasites, such as some species of protozoans, and cestodes and acanthocephalans that have no digestive tracts, must support themselves in this way from products of their hosts' digestion. Tissue parasites, such as "mesozoans" and the cirripede *Sacculina*, live also on such a molecular diet. It is possible that some free-living protozoans also can satisfy their nutritional requirements from the molecular waste products of living organisms and the decomposition products of dead ones. Such products are present to a greater or lesser extent in all natural bodies of water, depending upon the abundance of life there and the extent of its decay. In the laboratory, certain species of protozoans are successfully cultured for generations in chemically defined solutions in which there is no particulate matter.

There are also fluid feeders among free-living and ectoparasitic invertebrates, whose diet consists exclusively of the body fluids, tissue juice, or secretions of other animals or of plants. Such fluids contain a variety of compounds in solution or suspension that are taken in through a localized orifice, the mouth, rather than by diffusion over the general surface of the body. Feeders of this kind include the blood-sucking trematodes, nematodes, annelids, insects, and arachnids, as well as the numerous species of insects that live on cell sap, the fluid in phloem vessels, or the nectar of flowers. Feeding of this kind involves, in addition to mechanisms for attachment, which are of particular importance to the ectoparasites, the development of devices for piercing the epidermis and other tissues of the donor, as well as for sucking or lapping up the nutrient fluid.

The great majority of invertebrates are, however, like the vertebrates, active feeders, eating more complex and solid food, particulate matter whose dimensions may be so small as to be measurable in fractions of a micron or so large as to be bigger than the animal ingesting it. This food may be of animal or of exclusively plant origin and either alive or dead. It may be found at the bottom of bodies of water, either on the surface or buried in the silt or mud; it may be found in the soil or on the surface of the ground; it may be suspended, floating, or swimming in the waters of oceans, lakes, ponds, and streams; or it may be found in the air. Wherever it is, and whatever it may be, it becomes available to the animal that can find it, bring it into its body, and make appropriate use of it. Adaptations for the exploitation of new feeding grounds and new sources of food are one of the most potent factors in animal evolution, in which, among other adjustments to new environments, high priority has been put upon locomotor devices and means of food capture and intake.

A. DEFINITION OF FEEDING TYPES

In many invertebrates the same organelles or organs are used for the procurement of food as for locomotion, but in many others, some of these have been diverted entirely to the purpose of food collection and its conveyance to the oral opening and have lost their locomotor function. Food brought to the oral opening may either be ingested—that is, taken into the body for presumptive digestion—or rejected. Some notion of the way in which an animal collects and ingests its food can be obtained by observation of it in its natural surroundings or in captivity when conditions are kept as close as possible to those of the natural habitat. Examination of the mouthparts and oral region will also provide some clues as to the nature of the food that is habitually eaten. Studies such as these have shown that, in one species or another, almost every conceivable device has evolved for the intake of foods that are extremely different both in chemical composition and in physical state and that come from many different sources. Some closely related species employ quite different devices and some widely divergent species—indeed, even those within different phyla—employ similar ones. These adaptations may have arisen in response to changing environmental conditions, as realization of the full genetic potential of the species concerned, or they may have arisen fortuitously, allowing the species to enjoy sources of food previously inaccessible to them or to move into new environments with new sources, different from those that they had previously encountered.

The actual food upon which an animal nourishes itself can be directly determined by observation of it in nature, when circumstances permit, or in the laboratory. The kind of food that is eaten under natural conditions can also be ascertained by examination of the contents of the digestive tracts of animals caught and sacrificed for this purpose. Laboratory observations may be misleading, for the habits of animals are known to change even when they are kept under the best experimental conditions. Temperature, light, humidity, and for aquatic animals, depth, salinity, oxygen supply, and pH are all factors that may induce alterations of natural habits. Carefully controlled laboratory studies have, however, provided a great deal of valuable information. Food preferences, if any, can be detected by offering the animals under observation a variety, or a selection, of natural foods or specially prepared synthetic diets of known composition. The food offered may be labeled with nontoxic dyes or radioactive compounds so that its ingestion and passage through the digestive tract can be followed. In the case of synthetic diets, these markers are directly incorporated into the food offered, but with natural ones it is necessary to introduce the label into the animal or plant upon which the species under observation ordinarily feeds.

Animals may be classified, according to the sources of their food, as omnivorous (polyphagous), carnivorous (zoophagous), herbivorous (phytophagous), or saprophagous. An omnivorous animal will eat organic matter of any kind and may act as a scavenger in cleaning up organic debris. A carnivorous animal subsists on the flesh or fluids of other animals; carnivorous invertebrates are frequently cannibals. A herbivorous animal uses plants exclusively as a source of food, and a saprophagous one feeds on decaying organic matter. Saprophagous invertebrates are detritus feeders and frequently scavengers. Animals may also be classified according to the physical characteristics of the food that they eat as macrophagous, microphagous, or fluid feeding. A macrophagous feeder ingests particles of comparatively large size, often indeed, whole organisms; a microphagous feeder ingests only small particles, usually measurable in microns, or minute organisms such as bacteria; and a fluid feeder, in the strictest sense, ingests only water containing organic material in solution. Blood-sucking invertebrates are usually considered fluid feeders, although the liquid they ingest contains whole cells. There is, therefore, little justification for making distinctions between predaceous animals that, by one means or another, expose the soft tissues of their prey and draw them through their mouths into their digestive tracts and those that suck up naturally exposed fluids or, by piercing the integument, gain access to blood, hemolymph, or plant sap and sponge or suck these fluids up. For purposes of the following discussion, however, animals with predatory habits are included among the species practicing macrophagy, and those that feed on surface fluids or, as ectoparasites, on those within another organism are considered fluid feeders, with recognition that this separation, like most biological classifications, is arbitrary and has little real meaning. Endoparasites that are passive fluid feeders are excluded from this consideration of devices for active feeding.

Many invertebrates can and do switch from one type of feeding to another in the course of development from egg to adult or even, from time to time, as adults. Molluscs and echinoderms are, for example, microphagous as larvae, yet as adults they may be predatory carnivores or practice some other form of macrophagy. Perhaps the best known and most dramatic of these changes is the metamorphosis of the biting-chewing phytophagous larvae of Lepidoptera into fluid-feeding adults with mouthparts and digestive tracts adapted to this very different kind of diet. There is obviously survival value in being able as an adult to change from one type of feeding to another, for a polyphagous animal is far more independent of the immediate conditions of its environment, in regard to food supply, than one limited either anatomically or physiologically to food of one particular kind. This lability has perhaps the greatest advantage for sessile or sedentary animals that cannot easily or quickly change their environments if their preferred food supply runs short. Such shifts are known to occur in a number of species under natural conditions and may occur in many more than are at present known.

B. MACROPHAGY

A macrophagous feeder may be an omnivore, a herbivore, or a saprophage. It may be a detritus feeder or scavenger, ingesting inactive food in the form of larger pieces of the remains of animals or plants, or it may be a predator, chasing, stalking, or otherwise capturing active prey. It may be a commensal, sharing the food caught by its partner, but for the most part macrophagous feeders are active animals, moving from place to place in quest of food and therefore very dependent upon their locomotor mechanisms.

1. Mechanisms for the ingestion of inactive food

It is a reasonable assumption that the most primitive metazoans had no feeding organs other than an intake opening into the digestive tract, comparable to that of modern rhabdocoeles, and that the aquatic ones at least fed by ingesting water and mud and digesting the organic material in it. Some modern species, primarily among the annelids and echinoderms, feed in this way. All of them are omnivorous, but while some species seem to feed indiscriminately, others appear to practice some selection from the great variety of material in the detritus. In the nonselective feeders, the contents of their digestive tracts are essentially identical with that of the substrates on which they are living; in the selective feeders, a preponderance of one or another component of the debris is found.

Among the annelids, a number of polychaetes are nonselective feeders, drawing in the substrate with their proboscides or protrusible pharynges. For example, *Arenicola* draws sand or mud into its oesophagus by the constant extension and retraction of its proboscis (see Fig. VI–4). This sequence of proboscis movements, beautifully adapted to the conditions of the worm's life, may occur once in every 5 sec; after about 10 or 12 "swallows" the worm rests briefly and then begins its feeding movements again. Similarly, most aquatic oligochaetes ingest the substrate, sometimes at levels 2 to 3 cm below its interface with the water. Earthworms literally eat their way through the soil, sucking it in with their muscular pharynges. Other polychaetes, largely the tube-dwellers, are more selective than *Arenicola* and search about with their tentacles to find what would seem to be the most suitable and richest area of the substrate before drawing it into their mouths.

Among the echinoderms, some echinoids and holothurians practice a similar method of feeding. The heart urchins (Spatangoidea), for example, push the substrate into their mouths as they burrow along, using their small oral spines for this purpose, or else scoop it up, exhibiting little if any selection. Some burrowing holothurians also shovel mud and sand into their mouths, in the course of their movement from place to place, or use their oral tentacles to push it in when they are stationary. Some species of nematodes are saprophagous, sucking into their mouths small dead animals and plants or the debris of larger ones in various stages of decay.

Most of these species are almost continuous feeders,

for the sand and earth which represents much of the material that enters their alimentary canals has no nutritive value at all and is eliminated unchanged in their feces. Far larger amounts of material must therefore be ingested, even from a rich substrate, than would be required by an animal that selected from such a mixture, by one means or another, only the particles that could be absorbed or digested and that would have the greatest nutritive value. Uneconomical as this method would seem to be as far as the individuals themselves are concerned, it has great influence on the ecological picture in its effect upon local topography. It has been calculated, for example, that if each member of a community of the polychaete *Euzonus*, consisting of as many as 8×10^{10} individuals spread over 1 square mile in the intertidal zone of the beaches it inhabits on the west coast of the United States, ingested 2,100 times its own weight of sand annually, the whole population might cycle as much as 7.3×10^{6} tons of sand a year. From this they would assimilate about 700 tons of organic matter and deposit the remainder in new areas or in new levels. Earthworms are responsible for a similar turnover in soil. Charles Darwin, one of the closest observers of the activity of earthworms, calculated that an average population will bring somewhat more than 0.2 in. of mold to the surface each year. Over a period of years, their castings, spread by wind and rain, could change the character of the upper level considerably and even account for the subsidence, or burial, of objects on the surface. Similar estimates of the activity of holothurians have been drawn from observations of *Stichopus* in waters off Bermuda. *Stichopus*, a large holothurian abundant in the littoral zone, fills and empties its alimentary canal at least twice a day. From observations of the rate of feeding and measurements of the capacities of the guts of representative specimens, it has been calculated that 500 to 1,000 tons of bottom deposit are eaten annually by a population of average size in an area of 1.7 square miles.

2. Mechanisms for scraping and boring

Some molluscs, some arthropods (crustaceans and insects), and some echinoderms feed on algae and small metazoans that form encrustations on rocks and other objects in the water. In these same major groups are species that obtain their food by boring through the protective coverings of other animals and drawing the soft parts into their mouths. Others bore through wood and ingest the resulting fine particles. All scrapers and borers are slow-moving or sedentary animals. Most of them are equipped in one way or another either to hold themselves in position in a favorable feeding ground or to hold their prey while they bore into it or break it up into small pieces. Feeding is usually a continuous process and a selective one to the extent that some species are herbivores and some carnivores. The carnivores, moreover, show some choice in the animals upon which they prey.

SCRAPERS Two typical molluscan structures, the foot and the radula (Fig. VII–1) have made scraping as a mode of feeding possible to amphineurans (polyplacophorans and aplacophorans) and the gastropods that practice it. The foot can hold the animal tightly to the object encrusted with the desired food and so give it a firm purchase for scraping, and the toothed radula can act as a very efficient rasper or file. Chitons are selective herbivores, feeding on algae and dis-

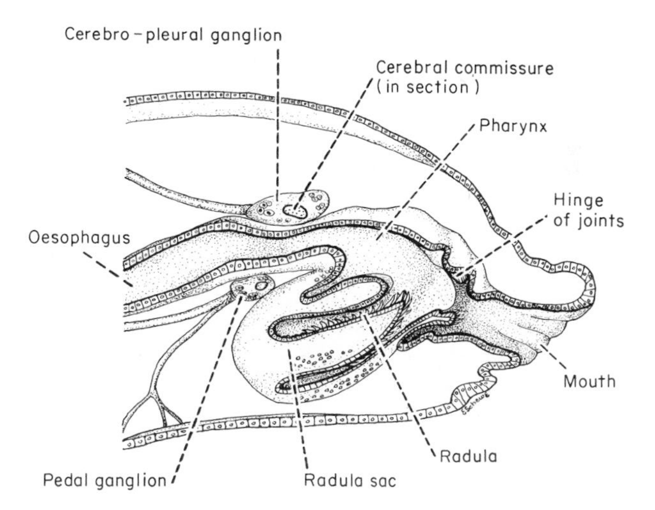

FIG. VII–1 *Radula of snail. Longitudinal section through the anterior end of the snail* Hancockia californica, *showing the mouth opening, the radula, and associated structures. Almost all gastropods have a radula; the teeth, usually arranged in rows, vary in number in different species from 16 to 750,000. (After Bronn, vol. III.)*

criminating between foods by means of the subradular organ. This is a sensory area on the roof of the subradular sac, a deep invagination that lies below the radular sac. The organ is periodically extended from the mouth, apparently to sample the environment. If suitable food is found, the organ is withdrawn and the radula is protruded from the oral cavity to scrape the food off. When the radula is retracted, the food, already reduced to small particles by the action of the radula teeth, is pressed against the roof of the mouth and enters the oesophagus entangled in mucus secreted by glands that lie at the back of the oral cavity.

Most gastropods have extensible proboscides whose cavity is continuous with that of the oesophagus. The radula lies in the proboscis and when this is extended it is brought into contact with the food and used as a scraper and rasper. In feeding from encrusted rocks, the rock itself may be rasped away and small particles from it, often containing organic material, are drawn into the alimentary canal along with the other food. The marine snail, *Littorina*, which is very common along the seashores, is considered to have been a major factor in the erosion of Eocene and Cretaceous silt stone. It has been calculated that a population of these snails on 1 square mile of shoreline at La Jolla, California, which may include as many as 8.6×10^8 individuals, could erode 2,260 tons of this stone in a year's time. This amount could contain as much as 57.2 tons of organic material, of which the snails might incorporate some 6.82 tons in their tissues, thus retrieving it from the geological and returning it to the biological world.

Although scraping is rare among Crustacea, it is practiced by a few copepods and decapods. The harpacticoids, largely benthic in habitat, may scrape food from the sea bottom, and *Cyclops* and *Heterope* (Cyclopoida) scrape off attached food, which is later broken up by their mouthparts. Some of the decapods that feed on algae use their third maxillipeds to scratch or scrape these plants from the rocks. The isopod *Idotea* is an omnivorous scavenger, with mouthparts adapted for scraping and biting off food. The largest of its appendages are the mandibles, which move sideways and act as the principal scrapers, but the first periopods also help in feeding by holding the food while scraping and biting is in progress.

The most highly specialized and remarkable scraping device has evolved in the echinoderms. This complex structure, characteristic of sea urchins, is known as Aristotle's lantern and is both masticatory and respiratory in function. It lies in a section of the coelom (lantern coelom) separated from the perivisceral coelom by the lantern membrane and consists of 40 calcareous ossicles and 60 muscles attached to them (Fig. VII–2). Movements of the teeth effect scraping and mastication. Changes in the volume of the lantern coelom are involved in respiration, for when these occur water flows in and out of the dermal branchiae—projections of the body wall whose cavities are continuous with the lantern coelom—with water carrying oxygen being pumped into it and water carrying respiratory wastes being pumped out. These movements and changes are brought about by contraction of the different sets of muscles in the following ways.

1. Ten protractor muscles (from aboral ends of alveoli to test): protrusion of teeth through mouth
2. Ten retractor muscles (from oral ends of alveoli to auricles): retraction and separation of teeth
3. Five interpyramidal (comminator) muscles (between adjacent pyramids): rocking teeth back and forth in sideways scraping motion
4. Rotula muscles (10 internal and 10 external, from rotulae to corresponding epiphyses): rocking teeth back and forth
5. Compass elevator muscle (five muscles united into a single pentagonal one attached to compasses): elevation of compasses and expansion of lantern coelom
6. Ten compass depressor muscles (from forked ends of compasses to test): depression of compasses and reduction of lantern coelom

As a scraping and masticatory device, Aristotle's lantern is exceedingly effective. Most sea urchins are omnivorous, but some show carnivorous or herbivorous preference. The carnivores usually subsist on such encrusting organisms as hydroids, bryozoans, barnacles, and tubiculous polychaetes, which are scraped from their locations and macerated, shell and all. They may, however, eat other echinoderms, molluscs,

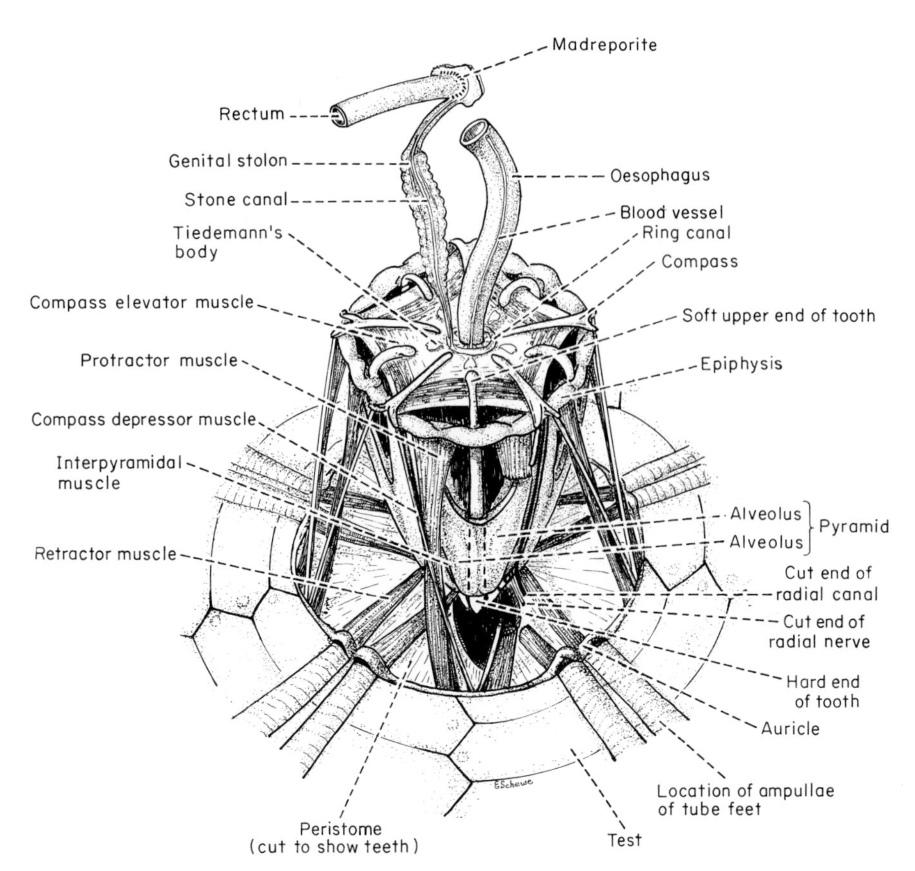

FIG. VII–2 *Aristotle's lantern. (From various sources.) This dentary apparatus, peculiar to sea urchins, acts both as a masticatory and respiratory mechanism. It consists of a number of calcareous plates, in sets of five, and a complicated set of muscles that move them. The five teeth, whose outer ends protrude through the mouth opening, are held in five jaws, each composed of two alveoli, fused at their outer ends, and ten epiphyses, immovably attached to the separate inner ends of each alveolus. Five flat plates, the rotulae (not shown in the figure) are attached in radial arrangement to the epiphyses and are overlaid by the compasses. The five jaws are held together by muscles; when these contract simultaneously, the teeth are brought together; when they contract successively, the teeth are rotated so that they act like an auger. Five pairs of protractor muscles are attached to the epiphyses and to the edge of the test; when these contract, the teeth are protruded. The teeth are retracted and the jaws opened by the action of five pairs of muscles attached to the surfaces of the alveoli and to the ossicles of the test. Five elevator muscles are attached to the compasses, holding them together and forming a kind of diaphragm around the oesophagus; when these contract the compasses are all raised and the lantern coelom (cavity of the lantern) is enlarged. Five pairs of depressor muscles, attached to the peripheral ends of the compasses, draw them down and so decrease the area of the lantern coelom, forcing fluid from it into the dermal branchiae.*

and crustaceans, which they first paralyze by toxic secretions from the pedicellariae and then hold with their tube feet while moving the teeth over the surface of the body. The herbivores prefer algae and diatoms but feed also on the larger seaweeds and, if necessary, on the bodies of dead animals. The destruction of kelp beds off the coast of southern California by sea urchins has become a matter of some concern, for besides providing habitats for a great variety of marine animals, kelp has commercial uses. So serious are the sea urchins' depredations that operations to keep their population in check are now being undertaken. The use of calcium oxide (quicklime) seems to be proving effective. This has deleterious effects upon the epidermis of the echinoderms, making them vulnerable to bacterial infection from which they die in about 1 week. Controlled liming is expected to save the kelp without total destruction of the urchins.

BORERS The habit of obtaining food by boring through the shells of their prey seems to have become established early in the evolutionary history of predatory gastropods. Collections made by members of the *Galathea* expedition included fossils of branchiopods and pelecypods from the middle Ordovician period, thus some 400 million years old, with holes drilled through them that in every respect resemble the holes made by modern boring gastropods. These modern species belong to six families: Cymantiidae, Muricidae, Thaididae, and Naticidae of the prosobranchs, and Helicidae and Oleancinidae of the pulmonates.

Recent investigations have made it clear that the

mechanical process of boring is supplemented by a chemical one. An accessory boring organ, consisting of tall columnar, ciliated, glandular cells located either on the sole of the foot or under the tip of the proboscis, is found in all boring predaceous snails but not in nonboring ones. When these glands are excised and put upon a smooth shell in a moist chamber, or on samples of aragonite or calcite of inorganic origin, their secretions make small depressions on the surface. This is not an ordinary process of etching, for the secretions are neutral not acid, and it is possible that they contain one or more chelating agents that extract the calcium from the minerals by combining with it as a water-soluble compound.* Close observations of boring operations in living oyster drills, *Urosalpinx*, show that at their beginning the snail, while holding the oyster firmly with its foot, extends its proboscis and radula and rasps away at the shell, frequently examining the site with the tip of the proboscis and forepart of the foot. When the operation is in full swing, mechanical and chemical procedures alternate. After a few minutes of rasping, the proboscis is withdrawn, and the accessory boring organ, which in *Urosalpinx* lies in the anterior midventral part of the foot, is brought over the site and held there. The foot clamps down upon the shell in such a way that the glandular organ is sealed off, preventing loss of its secretion or dilution by seawater. After an interval, which may last only a few minutes or as long as an hour, the foot is withdrawn, the site uncovered, and the proboscis is extended and rasping resumed. Ultimately, a neat round hole is made through all the layers of the shell, the proboscis is inserted in it and the radula tears away the soft tissues and draws them back into the oral cavity.

Prosobranch borers are selective, choosing only bivalve molluscs as their prey. Experiments have indicated that this selection results from their ability to receive chemical stimuli and to distinguish between them. These stimuli may not be directly from the prey itself but rather from the metabolites it releases. When, for example, a number of specimens of *Urosalpinx* were put into a chamber filled with clean seawater which was connected to two other chambers, one containing water that had flowed over oysters and the other water that had flowed over mussels, the snails moved into the one which had the greater concentration of end products of oxidative metabolism, regardless of which kind of bivalve had produced them. When the concentrations of these were equalized in the two side chambers, no preference was shown. It has been demonstrated experimentally, too, that *Urosalpinx* is attracted more strongly to young, actively growing oysters than to older, more slowly growing ones and gives a positive response to oxaloacetic acid, which is known to be an end product of the metabolic events involved in shell formation. At ordinary temperatures (10 to 25°C), the drills can consume oysters 10 to 33 mm long at the rate of 1½ per week. Laboratory observations have shown also that on an average, 20 drills can dispose of about 400 young oysters in 13 to 14 months.

Although chemoreception in gastropods seems to have been recognized since the pioneer observations of Jan Swammerdam (1637–1680), the chemosensitive site or sites are still open to question. There is evidence that in the prosobranchs chemoreception is the function of the osphradium, an area of columnar, presumably neuroepithelial cells located in the mantle cavity and well supplied with nerves from the central nervous system. Extirpation of this area, or organ, results in the absence of feeding responses in operated animals.

Some pelecypods are also effective borers, using their shells as augers, since molluscs in this class have no radula. Members of the family Pholadidae, commonly known as piddocks, bore into hard-packed clay and rocks but do so for protection rather than for getting food, because in their burrows they feed by the ciliary mechanisms characteristic of most bivalves. Their boring procedure is so effective, however, that young of the pholad *Martesia striata* are known to have drilled holes in the lead sheathing of power cables laid off the coast of Florida and so effectively riddled the 3-mm-thick wall that the wires inside short-circuited. The bivalves that bore for nutritive purposes are of

* Recent measurements of the pH of the secretion from the accessory boring organ of *Urosalpinx cinerea follyensis* by means of microelectrodes have shown it to be distinctly acid, with a minimum pH of 3.8. The nature of the acid produced is as yet unknown, but it effectively acidifies the seawater for a short distance around the bore-hole. This does not preclude the possibility that erosion of the shell is effected by a chelating agent, as well as by the acid component of the secretion.

the wood-boring family Teredinidae, especially the genera *Bankia* and *Teredo* (Fig. VII–3). These are worldwide in distribution and have been menaces to shipping ever since man constructed wooden rafts and boats and attempted to navigate in them. As early as 350 B.C., Theophrastus recorded the destructive effects of *Teredo*, and their influence upon early explorations was undoubtedly very great. Mariners feared their ravages upon their ships as much as they feared the ravages of scurvy upon their crews.

The bodies of these bivalves are superficially worm-like, for they are greatly extended in the antero-postero axis, with the mouth situated anteriorly and the anus posteriorly (Fig. VII–4). The body of an adult *Teredo navalis* may be 100 to 125 mm long and 5 mm in diameter. Because of their shape, they are popularly known as "shipworms." The visceral mass proper represents about one-quarter of the total body length; the remainder is primarily mantle and ctenidia. The shell is small and surrounds only the anterior tip of the body. Each valve is divisible into anterior, median, and posterior areas, or lobes; the anterior lobe of each is provided with sharp, pointed teeth and the median is also dentate, although its teeth are coarser.

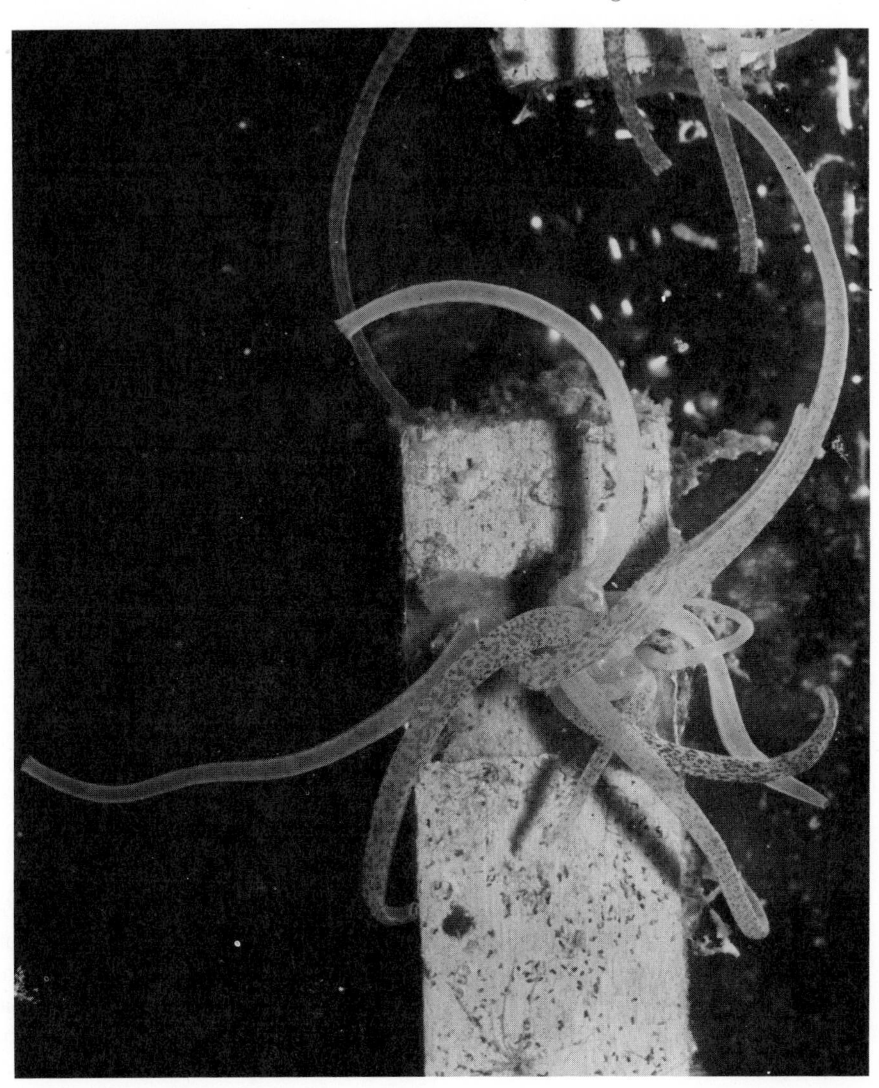

FIG. VII–3 Bankia setacea, *showing siphons extending from the wood into which the shell is boring. (British Columbia Research Council, Vancouver 8, B.C.)*

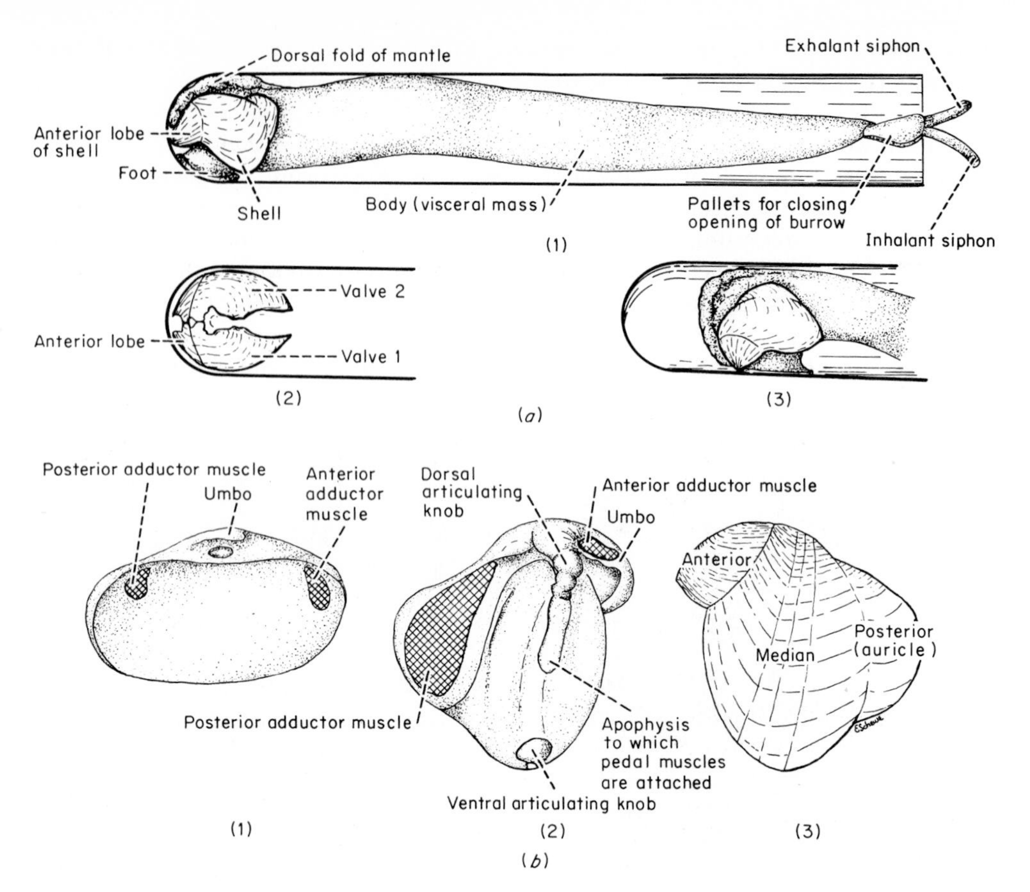

FIG. VII–4 *Boring mechanism of the shipworm* Teredo. *(After Russell and Yonge and Miller.)* ***(a)*** *(1) Position of the mollusc in a burrow, when boring in a forward direction. The bivalve shell, which by muscular movements effects the rasping of the wood, and the foot, which together with a flap of mantle that overlaps the shell, hold the animal in place while boring is in progress, are shown at the anterior end. At the posterior end are the two siphons which ordinarily extend from the burrow but can be withdrawn into it and the opening closed by the shelly pallets. The body, extending along the length of the burrow, contains the visceral organs suspended in a loose tissue through which the water entering through the inhalant siphon is filtered, any food particles in it passing forward to the mouth and the filtered water leaving through the exhalant siphon.* ***(a)*** *(2) Dorsal view of shell, showing the two valves. Each valve is extended into an anterior lobe, internally serrated. (3) Position of animal when boring a side passage at right angles to the former course of the burrow.* ***(b)*** *Homologous relations of the shell of* Teredo *to that of a nonboring bivalve. (1) Internal view of the left valve of the shell of Mya arenaria. (2) Internal view of the left valve of the shell of* Teredo navalis. *The apophysis to which the foot muscles are attached has no homologue in the typical bivalve shell and is found only in one other related group of boring molluscs, the Pholadidae. The ventral articulating knob is peculiar to the Teredinidae. (3) External view of the left valve of the shell of* Teredo, *showing the three primary lobes.*

Both valves are covered dorsally by a fold of the mantle [Fig. VII–4*a*(1)]. Pressure of this fold against one wall of the burrow and that of the foot against the other holds the valves in position for rasping action. This is effected by slow and continued contraction of the large posterior adductor muscle, which pulls the anterior margins of the valves apart, dragging them across the surface of the wood. The sharp teeth of the anterior lobes act as an advance boring tool while the teeth of the median lobes enlarge the burrow peripherally. After full extension, the two valves are drawn together again, without contact with the wood, by contraction of the small anterior adductor muscle. After each such backstroke, the foot muscles relax and the margins of the foot are spread out over the edges of the shell. Then contraction of the foot muscles brings the margins of the foot together again and the valves in position for a new stroke. In taking up the new position, the foot moves slightly to one side of its previous location. Since the posterior of the body is attached to the burrow and so is not free to move, this continued turning of the anterior end leads, after a series of strokes, to its rotation through 360°. Then the movement of the foot is reversed so that the body is unwound and the burrow continually enlarged as a smooth cylindrical tube. Changes in direction of the main course of the burrow are made by turning the anterior margins of the shell as illustrated in Fig. VII–4*a*(3).

The particles of wood scraped off in this way are swept into the mouth by currents set up by the combined action of cilia on the margin of the foot and in the oesophagus. This type of feeding begins as soon as the free-swimming larvae, which have a brief planktonic existence, settle down upon a piece of wood. They seem to exercise some selection in this, using the foot to explore the surface of the wood until a suitable place for penetration is found. This may be one already softened by decay. The larvae have chitinous, uncalcified shells, without teeth, and the means of their initial entrance into the wood is not yet clear. Possibly, enzymatic action reinforces the physical one of drilling. Metamorphosis begins when the larvae have bored only a little way; much of the foot is resorbed, the body elongates, and the shell becomes calcified and toothed. There is no evidence that boring in the adult is anything but a mechanical process, yet a very effective one. The life span of the animal, from the time of its penetration into wood until death is only 10 weeks; if prevented from penetration, degenerative processes begin in about 96 hr. Boring in the wood seems to be a continuous process and so effectively done by these small animals that a depression at least 1.2 mm deep can be rasped out in ½ hr, and a column of wood equal to the greatest dimensions of the adult body destroyed in the course of its natural life. *Bankia setacea* can in 1 year completely riddle a log 12 in. in diameter.

Octopuses drill holes in the shells of their molluscan prey. They use the radula to do this and have been observed, in aquaria, to cut small oval holes through the shells of a variety of bivalves and of *Ischnochiton conspicuous*. In about 3 hr an octopus weighing 48 grams drilled a hole in the shell of an abalone (*Haliotis fulgens*) weighing 19 grams. Venom elaborated in the posterior salivary glands of the octopus is discharged into the hole from ducts whose openings lie just below the radula. The effect of the venom is to relax the muscles of the prey, so that an abalone or a chiton loosens its grip on the substrate, and a bivalve opens its shell. The soft parts of the prey are thus exposed and made vulnerable to maceration by the beak or radula of the octopus.

Arthropodan borers are those like carpenter ants (*Camponotus*) and carpenter bees (*Xylocopa*) that tunnel deeply into wood and make nests there. There are also some true wood eaters. These are, among crustaceans, some marine harpaticoid copepods, marine isopods of the genus *Limnoria*, and at least one genus of amphipod, *Chelura*, and among insects, the termites. The habits of *Limnoria*, the gribble, and of the termites are best known. Gribbles, like shipworms, are worldwide in distribution and, like them, have done a great amount of damage to ships and wooden structures built in salt water. Their mandibles have sharp points and rigid surfaces that act as files and are used to chew into the wood. The wood scraps are then pushed into the mouth and pass into the digestive tract. Termites are also provided with strong chewing mouthparts and eat their way through trees and wooden buildings.

The presence of wood debris in the digestive tracts of these arthropods is evidence that it is actually ingested, but the extent to which it meets their nutritive needs is still an open question. Possibly associated fungi and protozoans provide some of the essential nutrients. It has been reported that gribbles attack only wood in which marine fungi have already begun the processes of decay and that they do not bore into fresh, sterile wood, but experiments have thrown doubt upon this conclusion. Pieces of clean fungus-free Douglas fir and yellow pine put into the waters of the Bay of Naples and into the Pacific Ocean at Friday Harbor, Washington, were in each case invaded by a native species of gribble. In some there was no evidence of mold; in others, mold was present but study of thin sections made parallel to the gribbles' tunnels showed no topographical relationship between the hyphae and the animals. In laboratory cultures, *Limnoria* was also observed to attack fresh wood, so that a preliminary softening seems unnecessary to the initiation of their boring. Bacteria, however, may be present and contribute to the utilization of the ingested wood and the general nutrition of the woodeater. Similarly, termites are most frequently found in wood that has already begun to rot, and it is well known that they harbor intestinal symbionts, flagellates, which digest the cellulose of the chewed-off wood and which are also consumed by their hosts (see page 33).

Many sea urchins use Aristotle's lantern as well as their spines to bore into rock or coral to make burrows in which they live. This is, like the boring of piddocks, for protection rather than nutrition. And, like piddocks, sea urchins may be destructive; reportedly they have bored into underwater steel girders and pilings and gnawed great gashes in wooden ones.

3. Mechanisms for seizing and swallowing active prey and large particles

Most of the invertebrates that obtain food by seizing it are provided with special organelles or organs for doing so. These may be pseudopodia, cytostomes, tentacles, suckers, protrusible pharynges, jaws or proboscides, chelae, and similar structures. Many such raptorial species are also equipped with poison glands or salivary glands, or both. The poison glands elaborate and discharge substances that paralyze or kill the prey, and the salivary glands secrete digestive enzymes that bring about at least preliminary chemical disintegration and softening of the food before it is actually taken into the alimentary canal. Some, too, have devices for its maceration or trituration before or after its ingestion. Some of them are hunters, moving about actively in search of prey, but others lie in wait for it, depending on chance to bring suitable food within reach. They are intermittent feeders, eating comparatively large masses of food at a time, and they are, to a great extent, selective ones.

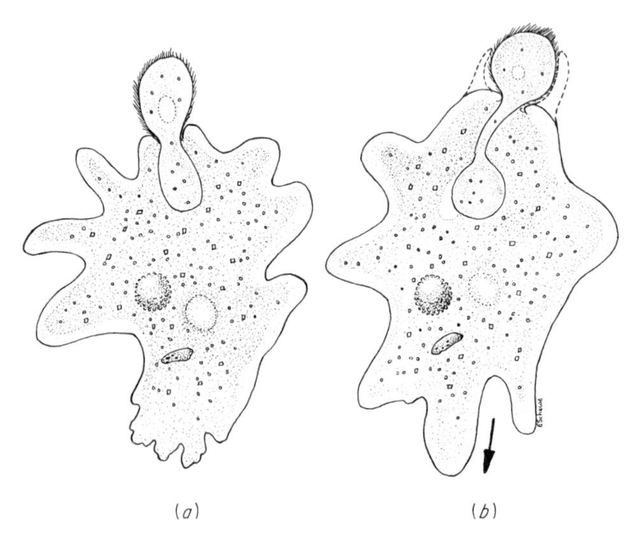

FIG. VII–5 *Ameba feeding on paramecium by circumfluence. (After Kepner and Whitlock.)* ***(a)*** *Contact of the prey with pseudopods of predator.* ***(b)*** *Ingestion of prey and formation of food cup.*

a. MECHANISMS FOR SEIZING AND SWALLOWING FOOD WITHOUT PRELIMINARY PROCESSING

(1) Pseudopodial The capture of food, followed by immediate ingestion, may be accomplished by pseudopodia, tentacles, and sucking mouths or pharynges. Pseudopodial methods are especially characteristic of free-living amebae, but food is ingested also by acoelous flatworms in essentially a pseudopodial manner (endocytosis). Amebae feed on protozoans and other small animals and on algae by circumvallation or circumfluence. In circumvallation the pseudopods surround the prey but do not make direct contact with it so that it is always enclosed in a film of droplets of water. In circumfluence, the pseudopods do make direct contact with the food and a vacuole is not formed until it is within the endoplasm (Fig. VII–5). In either case, pseudopodia are extended around the prey which the ameba encounters in its apparently random movement, so that it is trapped within the cell and ultimately included in a digestive vacuole in the endoplasm. The feeding habits of some of the soil amebae that eat flagellates and testate rhizopods have been carefully observed. *Thecameba sphaeronucleolus*, for example, appears able to identify prey within a distance of 20 to 30 μ and extends a single pseudopod toward it. This adheres to the prey—specifically to the flagellum of a flagellate—and then the plasmalemma disintegrates, and as the endoplasm flows back from the pseudopod, the prey is drawn down the pseudopodial tube into a food cup which closes around it.

Feeding in the giant freshwater ameba *Chaos chaos* has been followed under laboratory conditions and the course of events in the formation of food cups and subsequent digestion vacuoles analyzed by electron microscopy. When hungry amebae in culture are provided with *Paramecium aurelia*, the ciliates collect around each ameba, making random contact intermittently with its surface. As many as 10 to 20 may assemble around a single ameba. Some strike the surface, only indenting the plasmalemma, and then move directly away, but others can be seen to change their pattern of movement and to rotate or gyrate in the immediate vicinity of the ameba. This change is caused by the accumulation of clumps of granular material, derived from the mucopolysaccharide coat of the ameba, upon the cilia along one side or another of the paramecium. With this change in the pattern of movement, the chances of repeated contacts with the surface of the ameba increase, and with repeated contact, formation of a food cup is initiated. A single collision will not do this but contact or contacts lasting at least 2 sec will. Movement of the prey is not essential to evoke the response from the ameba, for paramecia immobilized by heat, or those very recently killed, will do it. The reaction may be the result of a combination between a labile substance on the cilia or pellicle of the paramecium and the mucopolysaccharide surface coat of the ameba. If so, feeding of *C. chaos* upon paramecium depends upon a definite interaction between the two protozoans and must be a highly specific one, although this is clearly not the case in all species of amebae (Fig. VII–6).

Formation of a food cup begins with a pseudopodial-like protrusion of the cytoplasm which lifts the plasmalemma outward around the point of contact with the paramecium. This point becomes the base of the cup and seems immediately to become a specialized region. The cytoplasm appears gelated, for there is no movement there, and electron micrographs show small vesicles, fibrils, and apparently paracrystalline material not evident anywhere else in the cell. The membrane of the forming food cup extends as deep folds into the subjacent cytoplasm, and as its margins converge they overlap and their membranes fuse. The prey is thus trapped in a pool enclosed by the plasmalemma of the ameba. Formation and closure of the food cup is complete in 2 to 12 sec and its conversion to a digestion vacuole in 1 to 10 min. The entrapped paramecium may swim vigorously for a time, making repeated contacts with the vacuolar membrane and so perhaps reinforcing the original stimulus, but after 5 to 15 min its movements slow down and finally cease. During the same interval, the volume of the vacuole, initially about 10 times that of the paramecium, decreases and the contours conform to the shape of the dead paramecium within it.

The events involve the transformation, at the point of contact with the paramecium, of a previously quiescent region of the plasmalemma into an active one, with the protrusion of the surface as the food cup and the alteration of the cytoplasmic components at its base. They involve also the transposition of about 10 percent of the plasmalemma from the surface to the interior of the ameba. Under controlled conditions of feeding, an individual *C. chaos* may form more than 100 food cups during a 24-hr cycle of growth and thus ingest, or remove from the surface, more than 10 times the amount of plasmalemma present at any one time. The organism must therefore be continuously engaged in replacing the plasmalemma as it is engaged in forming food cups and digestion vacuoles.

When feeding on organisms like diatoms, small crustaceans, and other platyhelminths, the acoelous flatworm *Convoluta paradoxa* moves along over the stones and rocks of the tide pools in which it lives, with one-third to one-half of its syncytial endodermal mass protruded from its mouth, and engulfs its food ameba-fashion. When larger prey is attacked, the flatworm attaches its posterior end to the substrate by means of secreted mucus and raises its anterior end. Any animal of suitable size that chances to swim by is grasped in the curled margins of the body and covered with mucus. The forepart of the body is then bent posteriorly so that the food is pressed into the mouth.

(2) Tentacular Tentacular feeding is characteristic of most cnidarians and of tentaculate ctenophores;

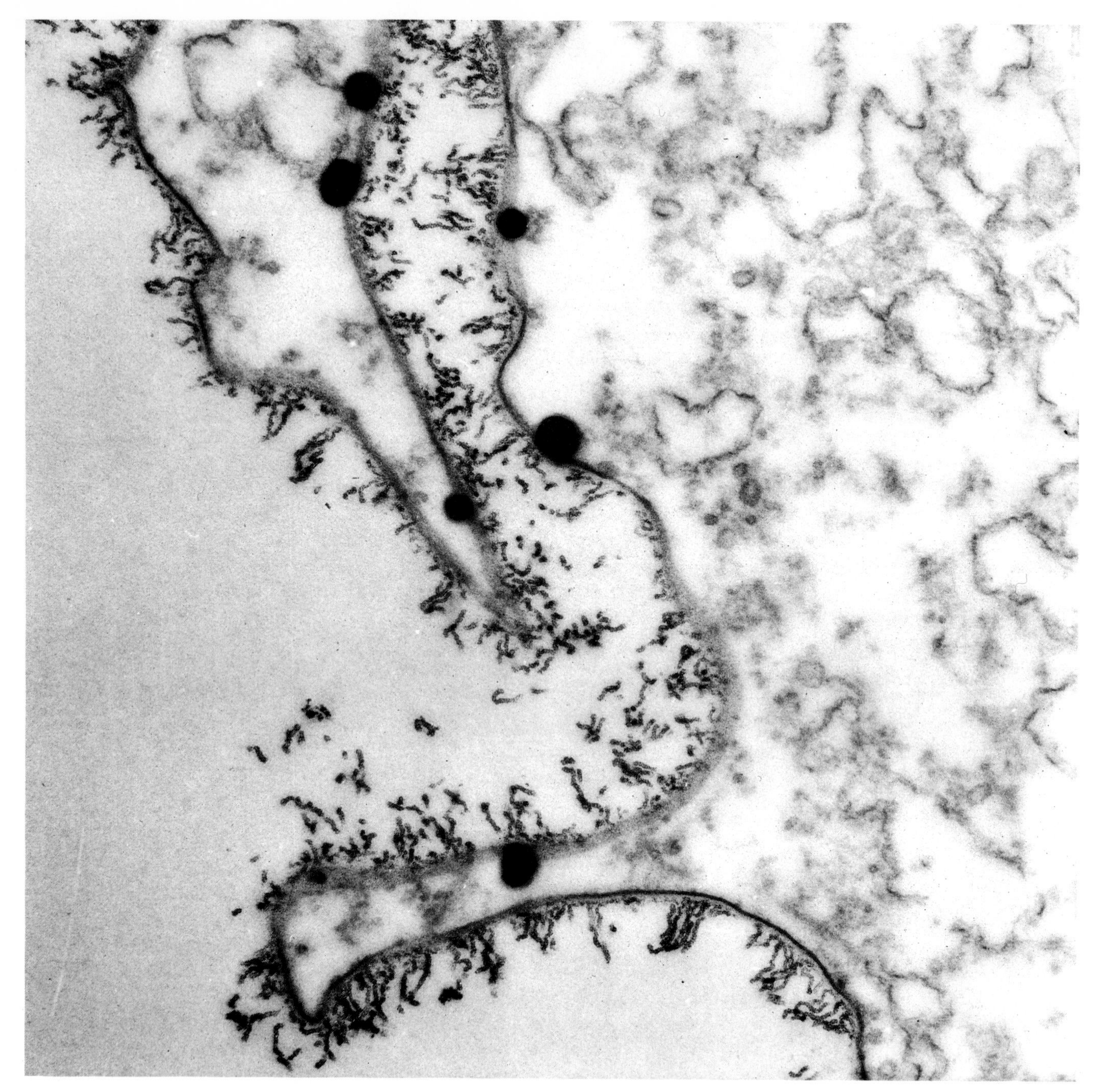

FIG. VII–6 *Surface coat of* Chaos chaos*; cross section through the surface and underlying cytoplasm of a pseudopod (multiplied 76,500 times). (Electronmicrograph by courtesy of Dr. Vivianne T. Nachmias.) The filaments of the outer layer of the surface coat are 150 to 200 mμ (1,500 to 2,000 Å) long, while the underlying continuous or amorphous layer is 150 to 200 Å thick. The very dense droplets lying just under the plasmalemma are characteristic of osmium-fixed* Chaos, *and are probably artifacts.*

all of these are carnivorous. Hydroids, siphonophores, medusae, sea anemones, and corals wave their tentacles about in the water and sting, paralyze, or entangle any animal that comes within reach of their nematocysts, excepting those with which a commensal or symbiotic relationship has been established. The tentacles then turn toward the mouth carrying the prey to it. Often animals of considerable size are captured and the mouth opening has to be widely stretched to accommodate them, or if this is impossible, the food may be rejected and the hunt begun again. Feeding is intermittent and one meal is usually disposed of before the search for another is begun. Yet sometimes a cnidarian will ingest more food at one time than it can handle and, if so, the excess is regurgitated. It has been shown, too, that the feeding reflex can be evoked in *Hydra* by glutathione. When this tripeptide, containing glycine, cystine, and glutamic acid, is introduced in concentrations of about 1×10^{-5} *M* into cultures of starved *Hydra* in buffered mineral solution, the animals respond by bending their tentacles toward the mouth, and in 0.5 to 1 min open their mouths, which stay open as long as 30 min or more, even in the absence of food. A number of other compounds, such as lactic, ascorbic, and acetic acids, and the proteolytic enzymes papain, ficin, and trypsin will also elicit the feeding response; indeed, any substance that stimulates the discharge of nematocysts seems also to evoke the feeding reflex. The colonial hydroid *Corymorpha* responds, when hungry, to glycine, *dl*-serine, and *l*-cysteine, at concentrations of 10^{-3} *M*, as it does to glutathione, and is most sensitive to glycine. The response in this case is not to the compounds themselves but to them in combination with an accompanying or immediately following tactile stimulus. Experimentally, contact with the tentacles may be made by a pipette or a fine glass rod, but under natural conditions probably the food particles provide the necessary secondary stimulus.

Most ctenophores feed on small members of the zooplankton which they catch with their colloblasts. The tentaculate species wipe the food from the tentacles into their mouths by drawing them across it. In some Nuda, the surface of the body is covered with small papillae, each of which has a sticky sucker at its apex. These hold on to any animal that makes contact with the ctenophore and the food so captured is passed from papilla to papilla until it reaches the mouth. In this way, *Beroe* may catch and ingest animals larger than its own body which, in large specimens, is as much as 8 in. long.

The members of at least one family of rotifers (Collothecidae or Floscularidae), which are sessile in habit, capture food by closing the fringed margins of the corona around small animals and so trapping them in the immediate vicinity of the mouth. In these species, the margins of the corona are extended into a number of tentacle-like lobes, which may be short or long and which are usually fringed with bristles, actually thick, nonmotile cilia. These lobes, while they cannot be elongated, can be spread apart as an open trap and brought together again as a closed one. The bristles help to prevent the escape of prey, which, after being caught, is swallowed whole.

(3) Sucking or pumping The majority of the invertebrates that catch their food and swallow it whole employ sucking or pumping mechanisms. Raptorial protozoa ingest their food at one or more localized areas. In ciliates, this is often at the tip of a permanent extension of the cell, a proboscis of sorts, and consists of cytostome and cytopharynx, a depression leading from the cytostome toward the interior of the cell. In Suctoria, sometimes considered degenerate ciliates, there are a number of such extensions, the so-called "tentacles," which are either knobbed or pointed. Raptorial ciliates feed on other protozoans, or rotifers and gastrotrichs, and seem to exercise some selection in their diet. The currents set up by their locomotion have little to do with their feeding, although their movement brings them into contact with suitable prey. *Didinium*, for example, feeds preferentially on *Paramecium* and uses its proboscis to impale a victim, which is then drawn in through the cytostome. This can be expanded to a diameter equal to that of the captor's body (Fig. VII–7). The structure of the proboscis has been observed in detail by electron microscopy, but the mechanism of its operation is still to be elucidated. In the laboratory, sufficient numbers of another predatory ciliate, *Dileptus*, have been known to decimate cultures of young planarians (*Dugesia dorotocephala*), apparently

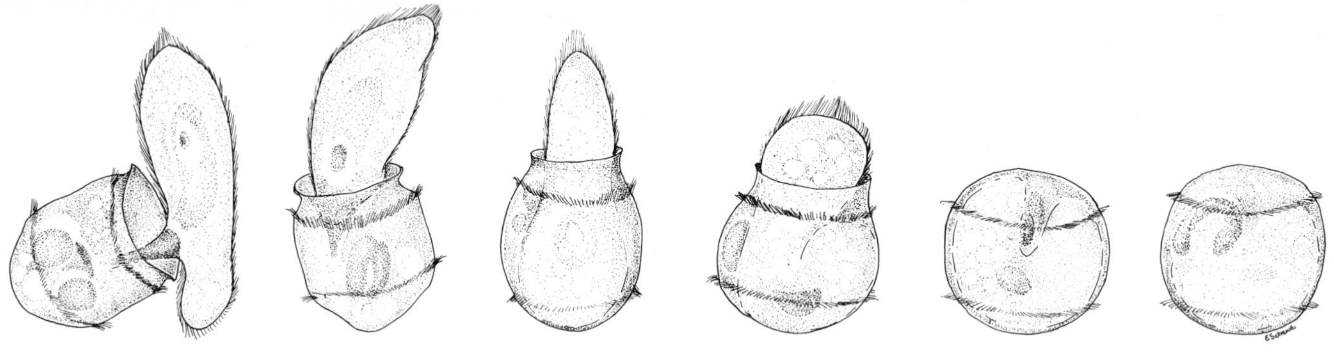

FIG. VII–7 Didinium *ingesting a paramecium.* (*After Calkins.*)

killing them by discharge of toxin-containing trichocysts which are especially abundant on the aboral surface of the proboscis. A single *Dileptus* has been seen to destroy a single *Stenostomum*; five can kill a small, recently metamorphosed pond snail (*Physa*), and a larger number can make away with a single hydra.

Suctoria feed mainly on ciliates and flagellates, and most will discriminate between real food and test substances such as carmine. Some are highly selective. *Podophrya fixa*, for example, feeds only on hypotrichous ciliates, while *Discophrya* (*Podophrya*) *collini* chooses holotrichs. The prey is held tightly on the tips of the tentacles, either impaled on the pointed ones or held by a sticky substance on the knobbed ones. Only a few of the knobbed tentacles act as cytostomes, and these enlarge to a diameter twice that of the others. Capture of food is accompanied by a wrinkling of the body surface, implying an increase in its area and therefore the formation of new plasma membrane. This expansion might be the important factor in the actual ingestion of food, creating a suction that would draw it into the body of the animal.

Feeding in *Tokophrya infusionum*, a small, sessile, freshwater suctorian has been closely followed by light microscopy and the details studied by electron microscopy. *Tokophrya infusionum* feeds exclusively on ciliates and seems unable to capture either flagellates or amebae. When, under laboratory conditions, *Tetrahymena*, a ciliate several times larger than *Tokophrya*, is supplied to them, the prey makes contact with the terminal knob of a tentacle entirely by chance, immediately adheres to it, and in a few seconds after contact is immobilized but not killed. These reactions suggest, first, that the knob of the tentacle may be covered with a sticky substance or that there may be some specific reaction between it and the cilia, which does not take place on contact with a flagellum or a pseudopod; and, second, that a toxin from the suctorian reaches the prey to cause its immobilization. This is not a lethal toxin, for the contractile vacuoles of the ciliate continue to operate and, if the animal is removed from the tentacle, it regains its mobility after a period of time whose length depends upon the length of time of its attachment to the knob. Directly after attachment, the tentacle broadens and a stream of granules flows along its periphery upward to the tip. Shortly after immobilization of the prey, its cytoplasm begins to flow down the center of the tentacle to the body of the predator. Mitochondria, basal bodies, and cilia of the victim remain recognizable and intact during this downward passage and do not undergo disintegration until taken into the body of the suctorian.

Tokophrya has 10 to 60 tentacles, depending on its age; each tentacle is a relatively long, stiff tube, consisting of a shaft about 1 μ in diameter, when not engaged in feeding operations, and a terminal knob about 2 μ in diameter. Each tentacle originates deep within the body of the suctorian and is therefore partly intracellular and partly extracellular. In electron micrographs, the shaft can be seen to be composed of two concentric tubes, with the lumen of the inner tube surrounded by tubular and possibly contractile fibrils. The cytoplasmic components of the prey travel downward in the inner tube, and the granules, which originate in the body of the suctorian and are most concentrated at the bases of the tentacles, travel upwards in the outer tube. These bodies have definite and consistent form, and apparent structure. They are pointed at one end, rounded at the other, and appear segmented. Because of their shape, they have been designated "missile-like" bodies and are believed to play an important role in the feeding process. Even in nonfeeding animals their pointed tips project through the plasma membrane of a tentacular knob. It is possible that they are responsible for the adhesion, as well as for the immobilization of the prey, and that they may contain and convey enzymes that initiate the events that precede the actual intake of food.

Rhabdocoele, triclad, and polyclad flatworms all have pharynges of different lengths and different degrees of complexity that can be protruded through the mouth and act as proboscides. The pharynx may be short and simple as it is in the rhabdocoeles *Macrostomum, Stenostomum,* and *Mesostomum,* or long and ridged on the inner surface as it is in triclads, in which it acts to disintegrate the food both mechanically and chemically, as well as to suck it in. The pharynx of *Macrostomum,* though short, is extremely distensible, as is also its mouth. This little flatworm, whose body is only 2 to 3 mm long and 0.5 mm wide, can feed on large ciliates, rotifers, nematodes, and small freshwater annelids and crustaceans. Although *Stenostomum,* also of minute dimensions, catches its prey by chance in a similar way, it exercises more selection, for it feeds mainly on large ciliates and rotifers and rejects other animals immediately after their capture. *Mesostomum* is a good deal larger and captures small oligochaetes, crustaceans, and insect larvae by wrapping its body around them. Movements of the body may crush the prey which is then sucked up by the short pharynx that is extended through the mouth.

The marine polyclad *Leptoplana* captures its food, consisting mainly of polychaetes, and isopod and amphipod crustaceans, by wrapping the anterior part of the body around one and then curling back until the food is brought to the short, extended pharynx which sucks it in. Small animals are ingested in a few minutes; for larger ones, an hour or more, as well as some mechanical disintegration, may be required before ingestion is complete. Some polyclads attach themselves to the surface of the prey, and thrust their pharynges into the soft parts of the victim's body, sucking them out. *Cycloporus,* for example, seems to feed exclusively on two genera of colonial tunicates, *Botryllus* and *Botrylloides,* attaching itself to the surface of the colony and sucking out the individual zooids one by one. *Stylochus* feeds on oysters in a similar manner.

Nemerteans also use their proboscides in the capture of food, but the nemertean proboscis has no direct connection with the digestive tract. It lies above the gut in a special elongated chamber, the proboscis cavity or rhynchocoele. It serves, however, to capture the food—mainly annelids but also molluscs and crustaceans—by coiling around them after eversion. When an animal is caught and firmly held, the proboscis is retracted and so the food is brought to the mouth into which it is sucked by pumping action of the muscles of the foregut.

Some anaspidean and nudibranch molluscs swallow their prey whole, using the radula as a holding rather than as a tearing or rasping organ. *Aplysia* (*Tethys*), the sea hare, for example, feeds on the eggs, embryos, and fry of shore fishes, enveloping them in two lobe-like tentacles. The nudibranch *Navanax inermis,* one of the sea slugs in bays and estuaries along the coasts of California and Mexico, is known to ingest whole bubble shell snails (*Bullaria*), as much as 1 in. long and nearly as wide. The intact shells are eliminated 24 hr or so later. *Navanax* is also a cannibal, feeding on others of its own species. Some land slugs swallow whole earthworms and other pulmonate snails, which they hold fast with the teeth of the radula while sucking out the body from its shell.

Many polychaete annelids feed on other small animals, including molluscs, crustaceans, and worms of their own as well as of other species. These are captured by the muscular, extensible pharynx, which in nereids and glycerids is provided with teeth or hooks. *Neanthes,* for example, has two large curved chitinous teeth that can be moved apart and then closed around the prey like pincers, as well as a number of smaller denticles which are on the outer surface of the extended pharynx, or proboscis (Fig. VII–8). All of these hold the prey firmly while the proboscis is being retracted and inverted in the act of swallowing. In some species

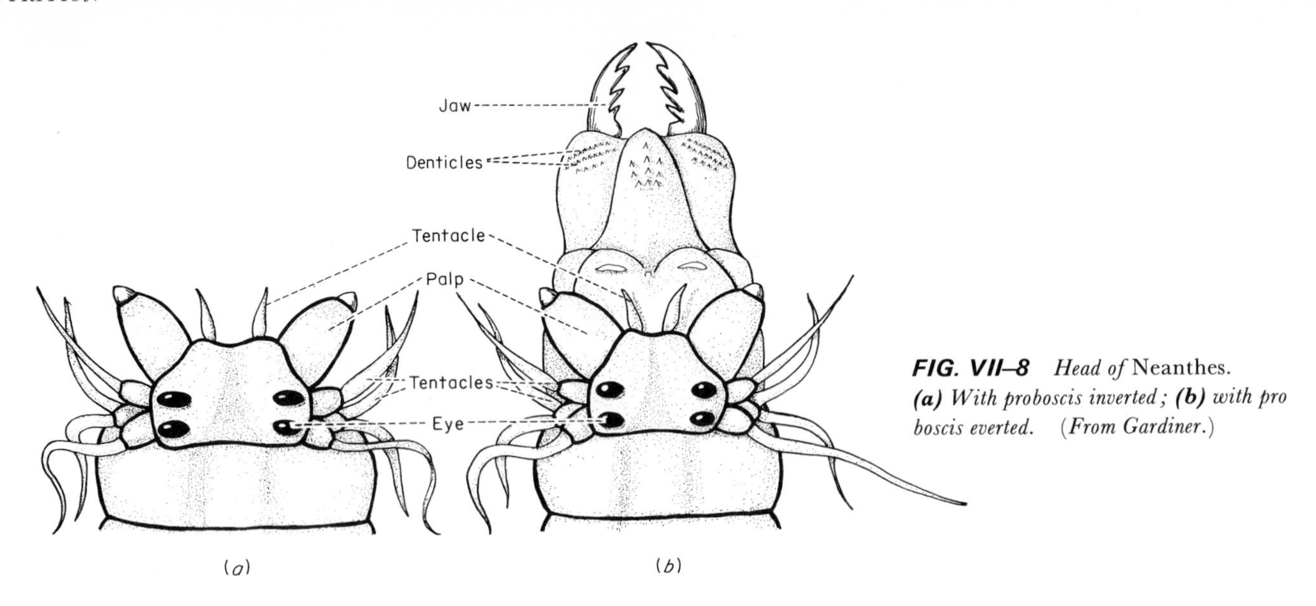

FIG. VII–8 *Head of* Neanthes. ***(a)*** *With proboscis inverted;* ***(b)*** *with proboscis everted. (From Gardiner.)*

the pharynx is long and is everted rapidly and with great force; its retraction is slower and may occupy several minutes if the captured animal is a large one.

Rhynchobdellid leeches have extensible pharynges and many of them are predaceous throughout their lives or in their juvenile stages. The proboscis is extended through the mouth and draws in worms, snails, insect larvae, and other small aquatic animals. Some of the terrestrial gnathobdellids are also predaceous, with strong sucking, but not eversible, pharynges. These feed on insect larvae in the soil and on slugs and earthworms. In feeding, the leech attaches itself to the prey by its anterior sucker and holds on firmly while the food is being drawn into the pharynx.

The starfish *Astropecten*, which lives on sandy bottoms, seems to be unique among echinoderms in swallowing its food whole. Its mouth is large, in comparison to its size, and like *Navanax*, it ingests living snails as well as scallops and other bivalve molluscs. Its tube feet are without suckers, so that it cannot hold its prey fast. This, as well as its habit of burrowing in the sand, may well be a primitive feature of echinoderms, retained in this genus. Ninety different species of animals, of which 73 were molluscan, were found in the stomach contents of 124 individuals examined. The ophiuroid *Ophiocomina* is both macrophagous and microphagous. Animals taken from the English Channel off Plymouth were seen to capture a living crustacean or a good-sized piece of food, by flexing the distal part of an arm laterally around the prey or piece of food offered them, and, by curving the entire arm orally, bring it to the disk, where it was pushed into the open mouth by the oral tube feet. Smaller pieces were caught on the tube feet and passed along the arm to the mouth.

b. MECHANISMS FOR THE PHYSICAL OR CHEMICAL DISINTEGRATION OF FOOD BEFORE INGESTION More efficient methods of feeding are practiced by invertebrates that reduce their food to small pieces before swallowing it, for these can be more easily ingested than large ones and present a greater surface area for the action of disintegrative enzymes. Efficiency is also increased if the food is lubricated and some digestion of it begun while it is still outside the body or in the oral cavity. Such preliminary processing allows a predaceous animal to feed on others as large as it can capture and hold or a herbivorous one to feed on plants too large to be swallowed whole.

Mechanical breakdown of the food mass may be accomplished by movements of the entire body or by movements of special parts, which may crush, grind, tear, or bite it into smaller pieces. Chemical breakdown is effected by the discharge of secretions, elaborated by epidermal glands, that contain at least some of the battery of enzymes required to reduce the complex carbohydrates, proteins, and fats of the food to their simpler, absorbable components. Secretion of mucus from epidermal glands often contributes to feeding in this manner by entanglement of the bits of food or by lubrication of the food mass. Some invertebrates are equipped only for mechanical disintegration, some only for chemical disintegration, and some for a combination of both.

The best known examples of protozoans that employ chemical means as a preliminary to the ingestion of their food are found in the family Vampyrellidae (Rhizopoda). Both marine and freshwater species of this genus feed on algae or small animals. The herbivorous species are able, presumably by virtue of a cellulase secreted from their bodies, to destroy the cellulose wall of an alga. They then insert their filopodia, or the entire body, through the hole and engulf the contents of the protoplast. Some vampyrellids are highly selective in their food. *Vampyrella spirogyra*, for example, feeds exclusively on *Spirogyra*, while *V. lateritia* will feed on other algae as well. In feeding, the animal travels along the surface of the alga, penetrating the wall at the intercellular junctions, or at some other point, and ingesting the protoplasts one by one.

Triclad flatworms feed on annelids, molluscs, crustaceans, and insect larvae by attaching their protruded pharynges to them. The pharynx of the triclad is longer than that of other Turbellaria and is of the plicate type—that is, it is folded back in a pharyngeal cavity whose opening is the mouth. When everted, it protrudes through the mouth, and the food is taken in through its external orifice and not through the mouth proper. Its walls are muscular, glandular, and well supplied with nerves. In feeding, the pharynx is moved over the surface of the prey and finally is thrust through some place in the integument weakened by the action of the pharyngeal muscles, and probably also by secretions from the pharyngeal glands. Body movements, strong contractions of the pharynx, and enzymatic action of its secretions all combine to disintegrate the flesh of the victim, often with great speed. The gregarious *Polycelis cornuta*, living in running streams, secretes mucus copiously as an aid to locomotion; this serves also as a food trap and one that is particularly effective in snaring arthropod prey. The polyclad *Leptoplana* feeds also on annelids and crustaceans, crushing the larger ones by movements of its body as well as of its pharynx. While small specimens may be ingested in no more than 5 min, larger ones may be sufficiently disintegrated only after at least 1 hr of vigorous muscular activity on the part of their captors, which use no trapping devices but catch their prey directly.

Chemosensitivity is an important asset to the turbellarian in its procurement of food. Planarians will move directly toward a dead animal, or a piece of meat, from which juices are diffusing. Recognition of live prey seems to depend upon stimulation of the chemoreceptors by direct contact or upon stimulation of the rheoreceptor cells that detect movements or currents in the water.

Nematodes and rotifers are equipped with cuticular devices for trituration. The mouth of a nematode is surrounded by projecting lobes, or lips, the number of lobes varying from species to species. In some, the cuticle of the lips is locally thickened as sharp, pointed teeth. The mouth leads into a buccal cavity, which, in turn, opens into the pharynx. Both buccal cavity and pharynx are lined by cuticle. The buccal cavity may be a narrow, uniformly chitinized tube, as it is in microphagous and fluid-feeding genera like *Rhabditis* and *Ascaris*, or an enlarged oval or cup-shaped space, in which the cuticle may be locally thickened and sclerotized as teeth (onchia) or as a stylet which can be protruded by the action of muscles (Fig. VII–9). In the family Talenchidae (order Rhabdasoidea), the stylet is formed by the apposition of two sclerotized areas and so forms a direct passage

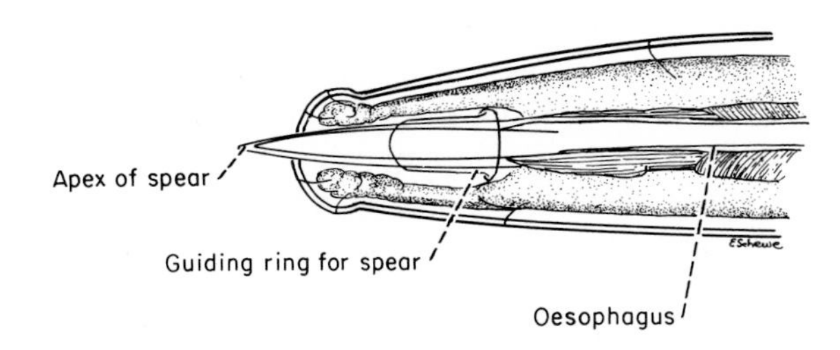

FIG. VII–9 *Anterior end of a female nematode* Dorylaimus fecundus, *showing the stylet.* (*After Ward and Whipple.*)

from the mouth to the pharynx. In the species that parasitize plants, the stylet is inserted in the plant tissues and their contents drawn up through it. In the Dorylaimoidea, some of which are phytophagous and some predatory, the stylet is sprear-shaped and, though contained in the buccal cavity, is actually formed in the pharynx. The teeth may be formed at the distal or the proximal end of the buccal cavity, or along its sides.

The pharynx is primarily a pumping organ to draw food through the mouth into the intestine. It is covered with radial muscles whose contraction, starting at the anterior end, dilates its lumen and opens valves that, when closed, prevent the regurgitation of food. The force of suction set up by these muscles is a considerable one, for it must counteract the pressure of the fluid in the pseudocoele which otherwise keeps the pharyngeal valves closed and the intestine collapsed. In *Ascaris*, this pressure has been measured and found to be equal to that of 70 mm Hg. A gland situated in the dorsal wall of the pharynx opens into the buccal cavity in some species, in others into the pharynx. Two ventral glands discharge their products at some point or another in the pharynx. The secretions from these glands contain hydrolytic enzymes and, in some carnivorous species, a toxin that paralyzes the prey.

Teeth and stylet are used by herbivorous, nonparasitic nematodes to bite or chew plant tissues. Predaceous carnivores, feeding on a variety of small metazoans—rotifers, oligochaetes, gastrotrichs, tardigrades, and also on other nematodes—seize their prey and tear it up with their lips and teeth; those with stylets may impale or puncture an animal and then, withdrawing the stylet, break it up with their teeth. In feeding, nematodes usually attach their lips firmly to the object and hold themselves to it by suction, in order to withstand the force of external resistance which otherwise would push them away from the food. Species that attack the roots of plants seem to hunt about for a weak spot through which the stylet can be most easily inserted. For example, *Heterodera*, which attacks the roots of beets, potatoes, and peas, selects the junction of a lateral with a primary root and arches its body so that the stylet is brought into direct line with the cell surface, which is struck repeatedly until penetration is achieved.

One of the distinguishing features of rotifers is their modified pharynx, or mastax, into which food may enter directly from the mouth or through a short buccal tube (Fig. VII–10). The mastax may be round, elongate, or trilobed; the wall is muscular and lined with cuticle which, in the basal region, is extended into seven firm, sclerotized pieces, the trophi, moved by small individual muscles. These pieces are a single, median fulcrum and three pairs of lateral ones, two rami, two unci, and two manubria. The two rami are attached to the fulcrum and these three pieces together constitute the malleus. Modification of these parts in rotifers of different feeding habits has resulted in eight different types' being found in modern species.

In many species two or more glands open into the mastax. These are called salivary glands because they are thought to secrete digestive enzymes, but conclusive evidence of this is lacking.

Most free-living rotifers are omnivorous, feeding on all members of the plankton which are brought to the vicinity of their mouths in the water currents set up by the cilia on the corona. The action of these cilia is the rotifer's primary means of locomotion. In the species with a mastax adapted for grinding, the small organisms swept into the mouth by the ciliary currents are broken up by the trophi, for all food must pass through the mastax. In raptorial species, the food is either grasped by the trophi or trapped in the lobed corona. Feeding is thus largely a matter of chance and is nonselective. There is some evidence that predatory carnivorous rotifers can detect living prey by tacto- or chemosensitivity, but there is no evidence that most make any selection of it. The creeping, freshwater genus, *Dicranophorus*, seems, however, to have well-developed discriminatory powers, for *D. isothes* picks only small cladocerans from the abundance of minute animals in the substrate, while *D. thysanus*, although rather more general in its tastes, confines itself to a diet of dead copepods, cladocerans, and oligochaetes.

The radula has provided molluscs with a masticatory device which the gastropods especially have put to various uses. Some of the herbivorous species, both marine and terrestrial, feed on plants of considerable size and break them up with the teeth of the radula.

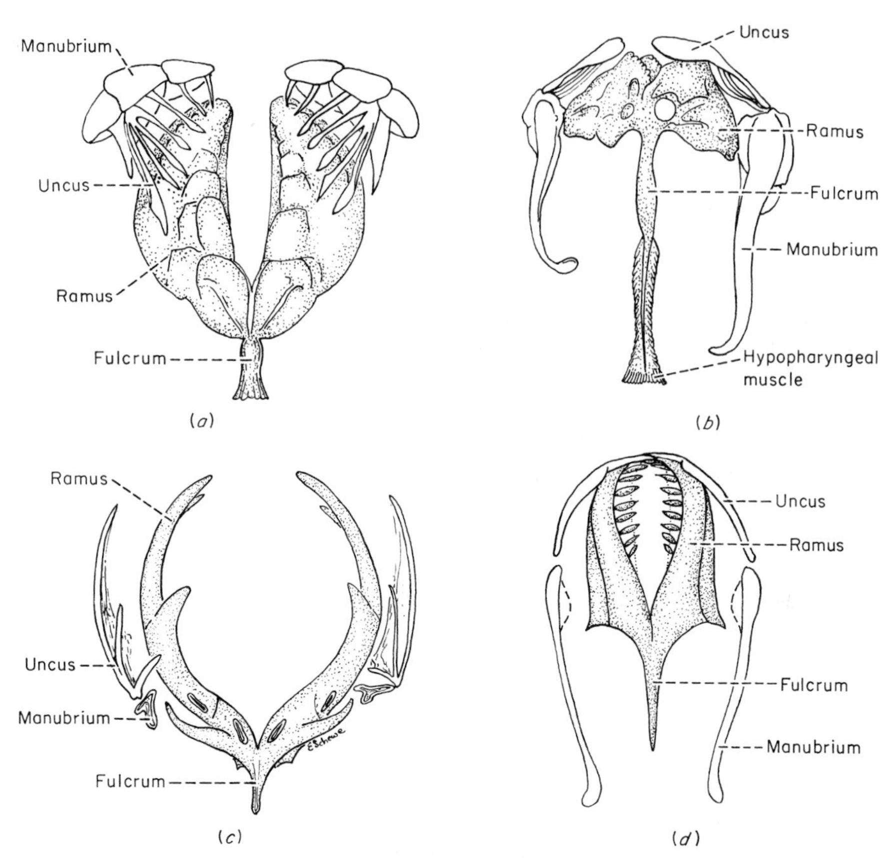

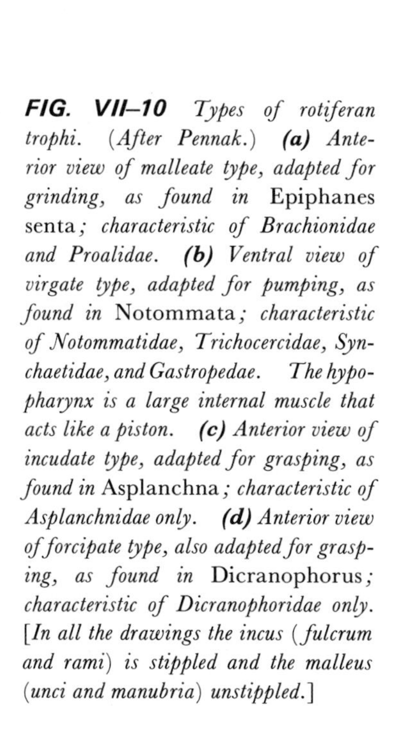

FIG. VII–10 *Types of rotiferan trophi.* (*After Pennak.*) ***(a)*** *Anterior view of malleate type, adapted for grinding, as found in* Epiphanes senta*; characteristic of Brachionidae and Proalidae.* ***(b)*** *Ventral view of virgate type, adapted for pumping, as found in* Notommata*; characteristic of Notommatidae, Trichocercidae, Synchaetidae, and Gastropedae. The hypopharynx is a large internal muscle that acts like a piston.* ***(c)*** *Anterior view of incudate type, adapted for grasping, as found in* Asplanchna*; characteristic of Asplanchnidae only.* ***(d)*** *Anterior view of forcipate type, also adapted for grasping, as found in* Dicranophorus*; characteristic of Dicranophoridae only.* [*In all the drawings the incus* (*fulcrum and rami*) *is stippled and the malleus* (*unci and manubria*) *unstippled.*]

Haliotis fulgens, for example, catches large pieces of seaweed with the anterior end of its foot, which is slender and prehensile, and holds them fast under its mouth while tearing bits off with the radula and swallowing them. The proboscides of some pteropods have various devices for seizing prey—hooks, suckers, or cephalocones, tentacle-like extensions that secrete a sticky fluid in which prey can become entangled. These are rapacious carnivores, often cannibals, which after seizing their prey reduce it to small pieces with their numerous radular teeth. Predaceous snails, other than the drills, have various ways of opening up a protected animal and making an entrance for the proboscis and radula. The shell of *Acanthina spirata*, for example, has a projection or spine on its outer lip. With this, the snail pries the valves of a small mussel or clam shell apart and inserts its proboscis into the gap so that the radula can reach the flesh and tear it off. Most of the pulmonates are herbivores, but there are some carnivorous species which live on earthworms and other snails and slugs. Worms are seized with the radula and macerated as this is drawn back over the odontophore. The shells of other gastropods may be rasped through until the soft parts are reached, or exposed tissues may be attacked directly.

Purple sea snails of the genus *Ianthina*, of which five species are known, are also carnivorous, and cannibalistic as well, feeding on anything they can catch, even members of their own species. These snails are dependent on chance for their food, for they drift with winds and currents along the surface of the sea, they are eyeless and respond to food only when it comes into contact with their tentacles. Each snail makes for itself a raft of air bubbles, which may be several inches long and from which its body, about $1\frac{1}{2}$ in. in diameter, hangs upside down in the water. Each bubble of the raft is built separately, the snail extending the cupped anterior end of its foot through the surface of the water, trapping air in the cup, and then secreting mucus from pedal glands around it. Each bubble as it is made is pressed into the proper position to make a float. *Ianthina* feeds readily on *Velella* and *Physalia*; a 4-in. *Physalia* has been observed to be almost completely devoured by two snails in less than a day. In feeding, the snail extends its proboscis, opens its mouth, and tears off the flesh of its victim with its radula. All species seem to be immune to the nematocysts of the cnidarians, possibly because of a purple secretion that they discharge when among the cnidarians or when feeding on them, which may in some way inactivate the toxins. The snails sometimes are found on the float of *Velella*, using it as a seagoing vessel. On the southeast coast of Florida, which they visit once nearly every year, they are often found beached with the cnidarians, but they are lost if cast a few feet above the watermark for they are such poor crawlers that they cannot cover even a short distance on land. Nor can they survive if submerged, for they must have access to air to make their rafts, for only when afloat do they come into contact with their food.

Cephalopod molluscs are all carnivorous, capturing living fish, crustaceans, and other molluscs with their long arms and tentacles. These are equipped with powerful suckers that grip the prey firmly. The radula is somewhat reduced, but at the mouth are two strong, curved chitinous teeth, heavily sclerotized. The lower tooth overlaps the upper and both together form a beak like that of a parrot but inverted. By movements of these teeth, flesh is bitten or torn off a captured animal, but the pieces are virtually unchanged as they pass over and along the radula. In the oral cavity they are mixed with mucus secreted by the anterior pair of salivary glands, which lie embedded in the muscles of the dorsal wall of the pharynx. A single median, posterior gland secretes tyramine, a derivative of tyrosine, which enters through the wounds made by the captor's teeth and poisons the prey.

In stalking prey, a squid moves slowly backward and forward through the water with its eight arms and two tentacles outstretched and pressed close together. When prey is sighted, the animal darts quickly toward it with arms spread apart and the two tentacles are driven against its body. If the captive is a fish, it is always grasped by the head and drawn back toward the mouth in an oblique position. It is held in this position by all the arms for 2 to 5 min while the head is bitten off. The body is then turned so that it is held horizontally in line with the body of the squid, with its tail sticking out between the arms. Pieces are then successively torn off in a series of transverse bites. The fish is eaten as a whole, except for its alimentary canal, the terminal vertebrae, and the caudal fin, which are all rejected. The rapidity with which squid can dispose of a meal—for digestion keeps pace with ingestion—makes it possible for them to take advantage of the abundance offered by an encounter with a school of fish and to feed continuously when in the midst of such plenty.

Octopuses, which live in crevices among rocks, lie in wait for their prey or hunt by moving over the rocks or sea bottom exploring each hole with a tentacle. Their food is mainly crabs and other crustaceans, or bivalve molluscs, for they do not move fast enough to prey successfully on fish. When eating clams or scallops, an octopus, if it does not bore through the shell, fastens its tentacles to the valves, pulls them apart, and bites out the body inside. A crab that passes by a lurking octopus is caught by the tentacles and enveloped in the interbrachial web between their bases. It is then pulled apart at the junction of cephalothorax and abdomen, killed by poison from the salivary glands, and within half an hour, its flesh, even to the tips of the appendages, is completely and neatly removed from the shell, which is left intact.

The pelagic *Tremoctopus* may make use of cnidarian nematocysts for defense and very likely also for offense. Pieces of the tentacles of *Physalia*, containing many undischarged nematocysts, were seen to be extended in very regular serial order between several suckers on the four dorsal arms of young females of *T. violaceus*, collected in the Pacific Ocean by investigators on a research vessel of the U.S. Bureau of Commercial Fisheries. Examination of the specimens caught in a dip net induced the discharge of the nematocysts, with consequent painful effects to the investigators, who were led to the belief that the stinging capsules might be used to quiet the small fish and members of the plankton on which this octopus ordinarily feeds. The captured octopods were very young females, not more than 40 to 72 mm long, or about one three-hundredth to one one-hundredth-and-seventieth the size of a full grown 4-ft adult, and though immature males have also been found holding *Physalia* tentacles in the same firm grip, there is as yet no information about their presence in the adults of either sex.

For the detection of food, squids, cuttlefish, and octopuses rely primarily on their eyes, well adapted for perception even in dim light, but *Nautilus* and *Argonauta* depend, respectively, on chemo- and tactoreceptors. This has been determined by observations of their behavior in nature and in aquaria when offered food of different kinds. Octopuses, squids, and cuttlefish will strike at moving objects but not at dead animals or pieces of their flesh unless they are moved about. *Sepia* at least seems to discriminate between shapes. Although shrimps, prawns, and small crabs may be attacked from any direction, full-grown crabs are struck only from behind, for the cuttlefish has apparently learned to recognize the menace of the pincer claws and keeps well away from them. *Nautilus* has been attracted to traps baited with pieces of boiled shellfish in even greater numbers than to those baited with live ones, and *Argonauta* draws to its mouth only those animals that come into contact with the interbrachial web of its first pair of arms. Chance thus plays a major part in the feeding of these cephalopods, whose vision is less acute than that of squids, cuttlefish, and octopuses, which see prey, recognize its value, and strike and grasp it with great accuracy.

Segmented appendages offer a wide range of possibilities for the development of feeding devices and have given those invertebrates possessing them a good deal of advantage in the exploitation of potential food sources. In onychophorans, one pair of appendages, with reduced claws, act as mandibles. These flank the mouth, which is encircled by lobes, or lips, and are shaped like sharp hooks with teeth. The upper surface of the oral cavity is chitinized, with a thick median ridge that bears a row of denticles and is moved by muscles. *Peripatus* feeds on a mixed diet, eating rotting wood, the excrement of wood-eating insects in whose galleries it often lives, small molluscs, worms, and caterpillars. Frequently it attacks termites, immobilizing them with a slimy secretion from the oral papillae and macerating them, like its other food, with its mandibles and oral denticles. Preliminary digestion takes place in the oral cavity through the activity of enzymes secreted by the salivary glands. These are modified nephrocoeles of the oral segment and produce both mucus and enzymes which are discharged through ducts that open on the floor of the mouth.

Limulus feeds on bottom-dwelling algae, worms, and molluscs. The food is picked up by the pincers formed by the two tarsal segments on the first four pairs of walking legs and macerated by their gnathobases as they pass it forward to the mouth. The gnathobases are movable spines, fringed with hairs, on the median side of the coxa of each of these appendages. The mouth, overhung by a labrum, lies just posterior to a pair of chelicerae inserted at each side of the labrum. Each chelicera is composed of three segments, the distal two forming strong pincers which grasp the food and put it into the mouth. By this means of feeding, *Limulus* can pick up food from a wide area as it moves along sandy and muddy bottoms, using its thoracic appendages simultaneously for locomotion and for feeding.

There is such great diversity in feeding habits among crustaceans that generalizations about the appendages employed in food capture and ingestion have little meaning. Indeed, an individual in almost any species may shift from one type of feeding to another depending upon its environment at any given time and the kind and abundance of food available to it there. For example, many of the marine isopods that usually feed on algae are known to devour barnacles, when living among them, and even to attack the lugworm *Arenicola*. And the hermit crab, *Coenobita*, which is adapted to life on land and habitually eats vegetation, will attack young terns in the breeding grounds and consume them.

No matter what the source or the nature of the food ingested, however, the same basic set of mouthparts seems to be used in fundamentally the same fashion. The mouthparts proper are the mandibles and, in order from anterior to posterior, the first pair of maxillae (maxillules), the second pair of maxillae (usually designated simply as maxillae), and the first, second, and third part of maxillipeds (Fig. VII–11). The actual ingestion of the food is usually performed by

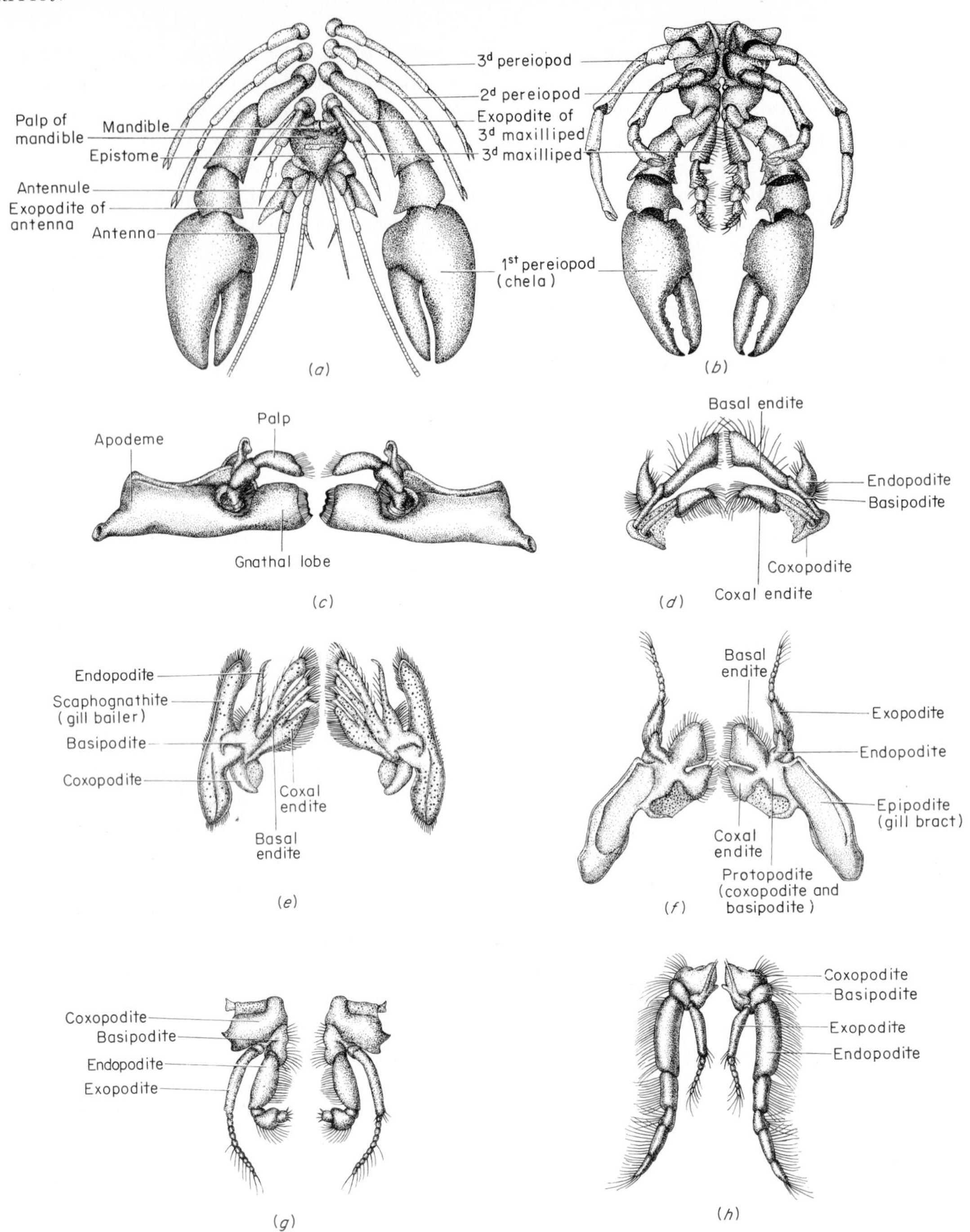

FIG. VII–11 *Crustacean mouthparts.* (*a, After Bullough; b to h, after Snodgrass.*) ***(a)*** *Cephalic and first thoracic appendages of the crayfish* Astacus, *dorsal view.* ***(b)*** *Third maxillipeds and first three pereiopods of the crayfish* Cambarus, *ventral view.* ***(c)*** *Mandibles of* Cambarus, *ventral view. Each consists of a broad basal plate attached to the mandibular segment of the body, a large gnathal lobe that projects below the mouth, and a palp of three segments.* ***(d)*** *First maxillae* (*maxillules*) *of* Cambarus, *ventral view.* ***(e)*** *Second*

the mandibles, maxillules, and maxillae, sometimes assisted by the labrum. The maxillipeds hold the food and bring it into position for mastication and ingestion, and sometimes to crush it. The pereiopods, particularly the first pair, are often used in the capture of food, especially by predators, and in some species the antennae contribute directly to feeding by sweeping the food toward the mouthparts. The unusually long antennae of the prawn *Sergestes*, a plankter of the deeper waters of the sea, have suggested to observers that they may be adapted for use in food capture. These appendages have a distinct kink, or bend, distal to which they are very flexible and move like a whiplash. Along the segments of the flexible parts are hook-like spines, which may fasten on prey encountered by the lashing antennae and bring it to the mouthparts proper.

Some macrophagous crustaceans are detritus feeders, picking up large pieces of plant or animal debris with their chelate appendages and bringing them to a position in front of the mouth for mastication and ingestion. Others, even small ones, are active predators. The little shrimp-like *Hemimysis*, for example, though usually microphagous, can seize arrowworms (*Sagitta*) as large as itself with its thoracic appendages; the animal is then held in front of the shrimp's mouth and torn up by the maxillules and mandibles before being taken into the oral cavity. Most lobsters and crabs catch worms, molluscs, other crustaceans, and echinoderms in the large crusher claw of their first pair of pereiopods or, in some, in the chelae of the second pair and hand the animal on to the third maxillipeds. These are so placed that they cover over the other mouthparts, but horizontal movement of their basal segments opens or closes a passageway to them. When the "door" is open, the distal segments of this pair of maxillipeds push the food through it. Sometimes the mandibles and the labrum hold the food while it is being macerated by the other mouthparts; sometimes these hold it while the mandibles chew it up.

In some genera, there are very special adaptations of the thoracic limbs for the capture and killing of prey. In *Lygiosquilla*, a stomatopod, the dactyl or movable joint of the second pereiopod can be drawn back into a groove and then suddenly released to pierce or cut an animal that passes by the burrow in which the crustacean habitually lives. Pistol shrimps of the genus *Alpheus* have developed probably the most remarkable mechanism of all. These shrimps, from 1 to 2 in. long, also live in burrows that they make, or find, and lie in wait for their prey. One chela is so much enlarged that it may be half the size of the whole animal. The dactyl extends well beyond the rest of the claw and can be raised upright and locked in position at an angle of 90° by the insertion of a protuberance on its base into a socket on the immovable jaw of the claw. This is equivalent to cocking the hammer of a gun, for when the dactyl is released it snaps down with great force, making a sound like the report of a pistol that is loud enough, from shrimps in an aquarium, to be heard throughout a good-sized laboratory. Populations of these shrimps around the Bermuda islands make an almost continuous "crackle," the noise they make reaching a measured frequency of 15 kilocycles per second (kHz). *Alpheus* lies in its burrow with its sensitive antennae extended. When prey is near, it crawls out and releases its pistol mechanism to stun the victim which it then picks up in its smaller claws. Sometimes it catches the victim first with these claws before stunning it with its pistol.

In myriapods and insects, whose bodies are clearly divided into head and trunk tagmata, the paired postoral appendages of the head—mandibles, maxillules, and maxillae—provide the main mechanism for feeding, although those of the trunk or of the thorax may be regularly or occasionally used for capture or collection of food. Chilopods are essentially carnivorous, feeding for the most part on larval or adult insects but also, in rare cases, on worms and slugs. *Scolopendromorpha*, the large centipede of the tropics, has been known to eat mice and small birds. Centipedes may turn, at times, to plant food and if their numbers are very great may

maxillae (maxillae) of Cambarus, *ventral view. The scaphognathites move back and forth causing water to flow through the gill chamber. The endites underlap the endites of the maxillules.* ***(f)*** *First maxillipeds of* Cambarus, *dorsal view. The posterior lobe of the scaphognathite of the maxilla fits into the concavity of the epipodite, so that a conduit from the gill chamber is formed.* ***(g)*** *Second maxillipeds of* Cambarus, *lateral view.* ***(h)*** *Third maxillipeds of* Cambarus, *lateral view. These project forward over the other mouthparts.*

be extremely destructive to vegetation. In general, their mandibles are elongated and not heavily sclerotized, and their maxillules are soft, triangular lobes that meet below the mouth and function as a lower lip. The two maxillae look more like legs than feeding organs, for they are long and slender, each with six segments. The first pair of trunk appendages are regularly used in feeding and so act as maxillipeds. Each consist of seven segments, the basal ones usually directed anteriorly while the distal ones hang downward with the terminal poison claws directed posteriorly. The duct of the poison gland, which lies within the appendage, opens on the pretarsus. These appendages are used to quiet or kill the prey and to convey it to the maxillules and mandibles, which crush or tear it and insert the pieces into the mouth.

Millipedes, on the other hand, are, with the exception of one family, primarily herbivores, feeding on a wide variety of dead or decaying leaves. The carnivorous species are said to attack earthworms and phalangids. The myriapod mandible is unique in that the basal plate is large and almost immovable; its functional part is a gnathal lobe which is freely movable by its own independent musculature. On the medial surface of each of these lobes there is distally a strong projecting tooth and proximally a serrated rasping surface. These lobes serve as the jaws. Below the mouth is a broad, flat appendage, the gnathochilarium (Greek *gnathes*, "jaw," *chilo*, "lip," and *arium*, "thing belonging to or connected with"), which acts as a lower lip. Opinions differ as to the composition of the gnathochilarium, some anatomists considering it a complex of maxillules and maxillae and others, the union of the two maxillules and the sternal plate of the segment to which they are attached. According to the latter interpretation, the diplopod head would be without a maxillary segment. In feeding, the first three pairs of legs may hold the leaves and even push them between the mandibles, thus serving as accessory feeding organs. The dry leaves are subjected to preliminary chemical as well as mechanical processing before being taken into the mouth. A secretion from the salivary glands that open just outside the mouth is first discharged upon the food. This softens the tissues of the leaf, at least that between the veins, so that it can be readily torn off or scraped off by the toothed lobes of the mandibles, and pushed into the mouth.

The paired appendages of the truly biting and chewing insects that are concerned directly with the ingestion of food are the mandibles, maxillules, and maxillae, but insects, like other arthropods, may bring their legs directly into action in the capture of food as well as in the pursuit of it. Some modern insects are omnivores, like cockroaches, some are herbivores, like grasshoppers, and some are carnivores, like Mantidae and predaceous beetles. And some of the holometabolous insects may be biters and chewers in their larval instars but change to a different method of feeding as imagos.

In most biting and chewing insects, the mandibles are strong, toothed plates, whose dentate margins can be brought together so that food can be bitten or torn and the pieces then further macerated (Fig. VII–12*c*). The maxillules hold the pieces of food and pass them forward toward the mouth. This pair of appendages is complex, although still retaining something of the primary leg-like structure (Fig. VII–12*d*). The basal segment, or coxa, is divided by a joint into a distal area, the stipes, and a proximal one, the cardo, that makes the articulation with the head. A jointed palpus originates from the stipes, which bears also two lateral endites, or gnathobases. The inner or medial one of these is called the lacinia and is toothed and has a fringe of long hairs. The outer, or lateral, one, called the galea, is a soft, thick lobe. Both lacinia and galea are independently movable. In feeding, the teeth of the lacinia hold the particles of food, which are also enclosed by the lobes of the galeae and passed over to the mandibles. When not engaged in feeding operations, the maxillules may be used for cleaning their palpi, the antennae, or the first pair of legs.

The combined maxillae form the labium, or underlip, of the insect (Fig. VII–12*f*). This is divisible into an immovable proximal or basal part, the postmentum, and a distal part, the prementum, which is movable on the base. In many insects, the postmentum is further subdivided into mentum and submentum. A jointed palpus originates from the prementum, at whose apex there are four lobes, an inner pair, the glossae, and an outer pair, the paraglossae. These four lobes are often called, collectively, the ligula and are the most active part of the labium in the feeding pro-

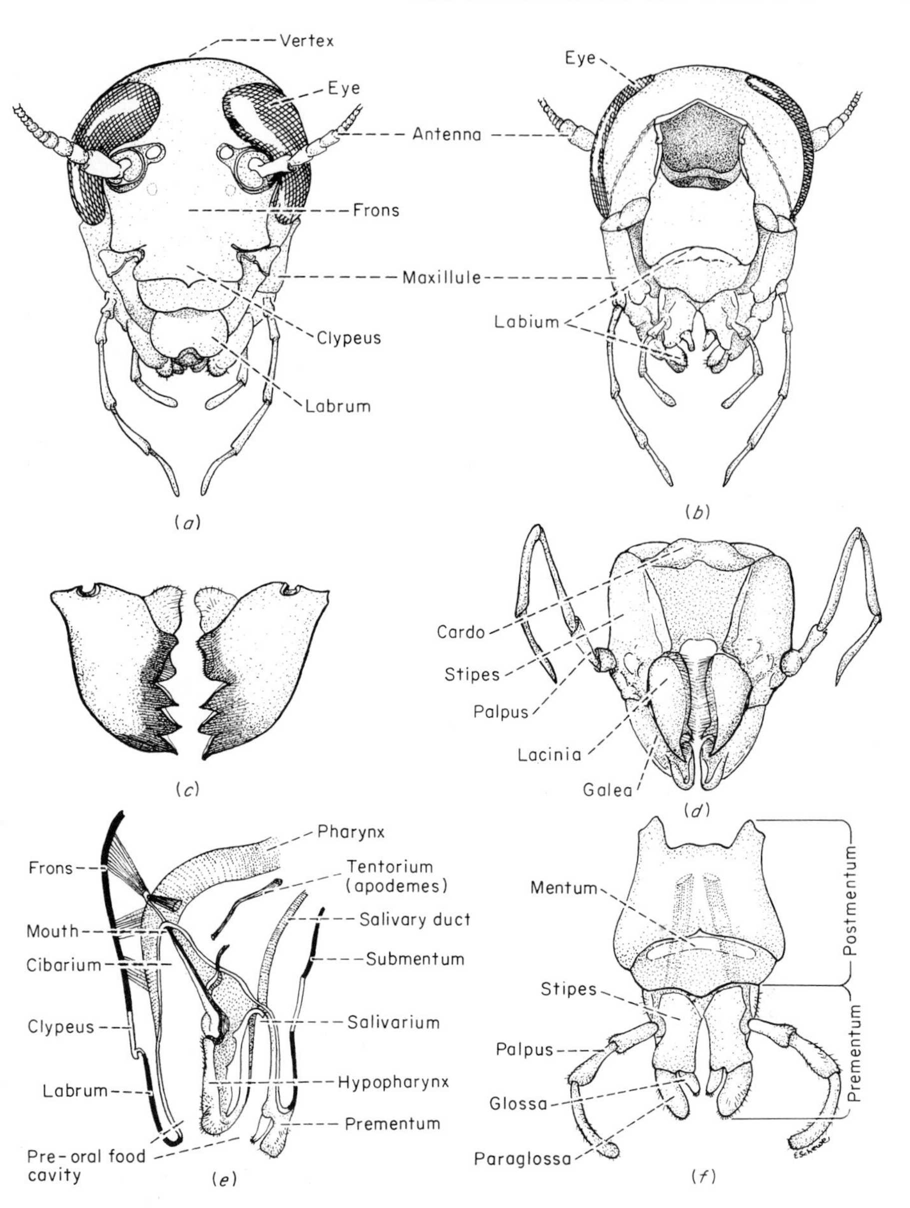

FIG. VII–12 *Insect mouthparts: mouthparts of a cockroach, illustrating a generalized type. (After Snodgrass.) If it is justifiable to assume that the first true insects were omnivorous as are modern cockroaches, mouthparts like these may well represent a primitive pattern from which the specialized ones of other insect genera have been derived.* ***(a)*** *Anterior view of head.* ***(b)*** *Posterior view of head.* ***(c)*** *Mandibles, anterior view.* ***(d)*** *Maxillules, posterior view.* ***(e)*** *Longitudinal section of lower part of the head, showing the position of the hypopharynx.* ***(f)*** *Labium, posterior view.*

cess, for they prevent the loss of bits of food from the mandibles. The labial palps are primarily sensory in function.

The entire set of mouthparts—labrum, mandibles, maxillules, and labium—enclose a space in which the food is first received and retained during its mastication, prior to its introduction through the mouth into the oral cavity and pharynx. This space is therefore a preoral food cavity, where food is both mechanically and chemically processed before its actual ingestion. The hypopharynx extending forward from the inner wall of the cavity, partially divides it into an upper and a lower chamber (Fig. VII–12*e*). The lower, or salivarium, receives secretions from the salivary glands whose ducts open at the base of the hypopharynx. The pieces of food are thus exposed to the action of the salivary enzymes while they are being chewed, much as they are in the mouth of a mammal. The upper chamber of the preoral cavity, the cibarium (Latin *cibus*, "food"), receives the processed food as it is passed on from the mandibles; the mouth opening is at the back of this chamber so that the food moves from cibarium to true oral cavity. The cibarium can be closed off from the rest of the preoral cavity when the base of the hypopharynx is pressed against the inner wall of the clypeus, that sclerite of the head to which the labrum is attached. The area of the cibarium closed off in this way can be expanded or diminished by the action of muscles on the walls of the clypeus. It can, therefore, act as a suction pump which is brought into play when the macrophagous insect drinks water. For all insects have need of water, and if the food does not supply an adequate amount, they must get it by drinking.

Echinoderms in general are not carnivorous, but starfishes are an exception for they feed freely on bivalve molluscs and are, accordingly, perpetual and particular enemies of oyster fishermen and the oyster industry. Most starfishes are scavengers as well and not at all selective in the debris that they eat. Some species feed on sponges, ascidians, sea urchins, sand dollars, and even other starfish. They make use of a remarkable and unique device for the chemical processing and ingestion of their food—eversion of the stomach through the mouth opening. In attacking an oyster or a clam, a starfish will attach the tube feet of its arms to the valves of the shell in such a way that the gape is directed toward the predator's mouth (Fig. VII–13). It then raises its disk so that its body is hunched over the shell and begins a long continuous pull to draw the valves apart. Ultimately, either by fatigue of the bivalve's adductor muscles, which hold the valves of its shell together, or by a greater counter force than they can exert—and assisted possibly by the liberation of a toxin that paralyzes the mollusc—the valves are slowly drawn apart. When the opening between them is wide enough, the starfish extrudes its stomach, turned inside out, from its mouth and wraps it around the soft tissues of the prey. These, therefore, are brought into direct contact with digestive enzymes and become gradually reduced to a rich nutrient fluid. When the meal is finished, the stomach is drawn back into the body by contraction of muscles attached to its five lobes and to skeletal plates in the arms.

Even scavenging starfish make use of this device.

FIG. VII–13 *Starfish eating a mussel.* (*U.S. Fish and Wildlife Service.*)

The bat star, *Patiria miniata*, for example, which is common along the Pacific coast of the United States, eats indiscriminately. In an aquarium it can be seen to extrude its stomach and flatten it against the glass wall from which the encrusted diatoms and algae soon disappear.

C. MICROPHAGY

Microphagy is always associated with an aquatic existence and frequently with a sessile or sedentary one. Since the ingested food particles are of minute dimensions, no advantage is placed on the development of masticating devices or means for preliminary digestion. Cutting, tearing, rasping, or other macerating mechanisms are conspicuously absent in strictly microphagous species, as are oral glands secreting digestive juices. Both may be present, however, in species which on occasion switch from microphagy to macrophagy. But for microphagy there is distinct advantage in a means of concentrating the food particles before their actual ingestion, thereby reducing the amount of water that would otherwise be concomitantly taken into the body. Viscid secretions from superficial glands often serve this purpose, either by entangling the particles or by acting as a filter through which they are strained. Various other devices for separating the water from the food have also evolved, especially among microphagous arthropods.

An effective filtering system must have, besides the filter itself, some means of maintaining a current of food-laden water through it, providing for the outflow of the filtrate and for the removal of the food particles from the filter and their introduction into the mouth, unless, as is the practice in some species, both filter and food are swallowed. There must also be provision for the removal of undesirable or unwanted material brought in with the incurrent stream and for the removal of feces and other waste material. Rejection currents and exhalant streams, preferably discharging some distance from the body, are accordingly valuable, if not essential, to filter feeding. All of these and coordination of action between them have evolved in the most efficient filter-feeding animals.

1. Mechanisms for the intake of material in crystalline or colloidal solution

a. MEMBRANE TRANSPORT In general, the food of microphagous invertebrates consists of the smaller members of the phyto- and zooplankton, and of bits of organic material that are suspended in the water or that have sunk to the bottom and lie on its surface or partially buried in it. These particles may range in diameter from several microns to fractions of a micron and so may include those of colloidal size, either in suspension or adsorbed on the surfaces of larger solid objects. Smaller particles, of less than 1 mμ (1 nm),* may be absorbed through the general body surface or through specialized areas of it. Intestinal parasites, without any other means of taking in food, must subsist in this way. The end products of their host's digestion, reduced to a physical state in which they can cross the cells of the host's gut, now traverse instead the epidermal cells of the parasite, either by diffusion or a biological process of active transport. Some protozoans can be maintained in the laboratory in sterile, chemically defined culture media in which there is no particulate matter, but there is no evidence that they do so in nature. It has been suggested that dissolved organic material, adsorbed on colloidal inorganic particles may be a dietary source for some microphagous organisms, but there is question about the intake and nutritive value of material in crystalline solution. Experiments with the solitary coral *Fungia scutaria* indicate that membrane transport can be operative. In these experiments, the mouths of the animals were plugged with paraffin and they were kept in seawater in which sugars labeled with C^{14} were dissolved in concentrations of 10 milligrams per liter (mg/liter), approximately equivalent to the total concentration of dissolved organic material in the coral's natural environment. Labeled glucose was subsequently found in their bodies, but labeled galactose did not enter. Similar experiments with the triclad planarian *Dugesia tigrina* indicate that protein molecules may be directly taken in by its cells.

Experiments on the usefulness of soluble carbohydrates to *Fasciola hepatica* cultured in vitro have shown that glycerol, glucose, and galactose are taken up from the medium and that glycerol and glucose are utilized in its metabolism. Maltose and sucrose, supplied in the medium, proved useless as energy sources, an implication of their failure to enter the body. The penetrability of organic molecules of various kinds has been tested on a number of marine invertebrates, in addition to *Fungia*. From these experiments it is apparent that amino acids such as glycine and phenylalanine are removed quite rapidly from dilute solutions (1×10^{-5} to 1×10^{-6} *M*) by most, and possibly by all, soft-bodied marine invertebrates but much more slowly by freshwater ones. Creatine is taken up from dilute solutions (2×10^{-5} to 2×10^{-3} *M*) by the annelids *Glycera* and *Nereis* and the bivalve mollusc (quahoq) *Mercenaria mercenaria*.

It has been shown in tracer experiments that various amino acids labeled with C^{14} are taken up by *Nereis virens*. Valine, glycine, and glutamic acid, presented in concentrations of 2×10^{-5} *M*, were taken up in greater amounts than the other compounds used, including proline, phenylalanine, and succinic acid. The question arises as to whether these are really

* In the Système International d'Unités (International System of Units), the basic units are length, meter (m); mass, kilogram (kg); and time, second (s). The abbreviation n (nano) represents a value of 1×10^{-9} of any one of these and is in general use at present. Thus 1 nm represents 1×10^{-9} m. It is equivalent to 1 millimicron (mμ), since 1 micron (μ) represents 1×10^{-3} millimeter (mm), or 1×10^{-6} m, and 1 mμ represents 1×10^{-3} μ.

metabolized by the organism, for their removal from the medium does not prove this. Indeed, evidence from experiments on the flatworm *Bdelloura candida* suggests that there is a net loss of amino acids to the medium rather than from it, a condition that may apply to other invertebrates as well. Yet experiments with echinoderms have shown that the tissues of these animals differ in their competence to assimilate organic molecules from the medium and to metabolize them after assimilation. Autoradiographs of various tissues of the holothurian *Cucumaria lactea* and the ophiuroid *Amphipholis squamata*, collected off the northeast coast of England and exposed to 6×10^{-6} *M* tritiated glycine or 1×10^{-5} *M* tritiated glucose, showed different degrees of labeling ranging from very intense to weak or absent. In *Cucumaria* the most conspicuous concentration of glycine was in the tube feet, particularly in the surface of the suckers. Labeling was also intense in the glands elaborating mucoprotein, with the implication that the assimilated glycine was incorporated into mucoprotein. These tissues showed no uptake of glucose, although the muscular layer and the coelomic epithelium of the tube feet were intensely labeled. Abundant glucose was also assimilated by the musculature of the buccal tube feet and jaw of *Amphipholis*, but was virtually absent from other muscles. The walls of the bursae of the ophiuroid assimilated both glycine and glucose intensely, as did the embryos in all stages of development brooded in the bursae. Possibly the uptake of organic molecules represents an important source of nutriment for the embryo after its yolk reserves have been exhausted, a possibility that may apply also to certain superficial tissues of the adult. But even though such uptake may have only slight general nutritional significance, it seems apparent from a study such as this that some of the tissues of echinoderms do absorb organic molecules from the medium surrounding them and do metabolize them.

This may also be how pogonophorans obtain food. These animals, living in chitinous tubes in the benthos, have no trace of a digestive tract at any stage of life, and their means of obtaining food is enigmatic. Possible organic particles, products of their own extracellular digestion or of bacterial decomposition, are absorbed through the delicate pinnules on their tentacles.

b. PINOCYTOSIS In experimental conditions such as these, and possibly in nature, where the surfaces of cells are exposed to solutions or suspensions of organic compounds, pinocytosis may be operative. This procedure was first observed in cells in tissue culture and suggested as a mechanism by which cells "drink." It is now recognized as one of wide distribution in cells of many kinds and one by which they take in macromolecules and colloidal particles through all exposed surfaces. Electron microscopy has revealed some of the details of the process and the possible fate of the pinocytotic vesicles formed in this way.

Pinocytosis involves active movements—that is, deformation—of the cell membrane and takes place either by elevation of the membrane as a thin fold or by its invagination. In the first case, which has been observed by electron microscopy in epithelial cells, the membrane is raised as a fold with a thickness of about 80 μ and of varying length. The cytoplasm within the fold is nongranular and devoid of inclusions. The single projection seems to bend back toward the surface of the cell; when contact is made the two membranes fuse. In this way a discrete vesicle of microscopic dimensions is formed, which sinks below the surface and travels through the cytoplasm (Fig. VII–14). In the second case, invaginations occur generally over the cell surface. These invaginations are so small as to be visible only with the electron microscope. As a result, minute flask-shaped vesicles about 80 Å in diameter are formed or slender channels that extend deep into the cytoplasm. Although connection may be retained with the surface and the surrounding medium for some time, ultimately vesicles are constricted off from the short necks of the flask-shaped superficial invaginations or from the inner ends of the deeper, canalicular ones. By these means the cell incorporates a minute droplet of water and solutes enclosed within a portion of its external membrane. The originally outer surface of the membrane becomes the inner surface of the vesicle, while the inner surface remains in contact with the cytoplasm.

Study of pinocytosis in *Chaos chaos* has shown that it is a two-step process, the first step involving selective binding and concentration of solutes in the medium and the second, their engulfment. The surface of *C. chaos* has been shown to be covered with radially

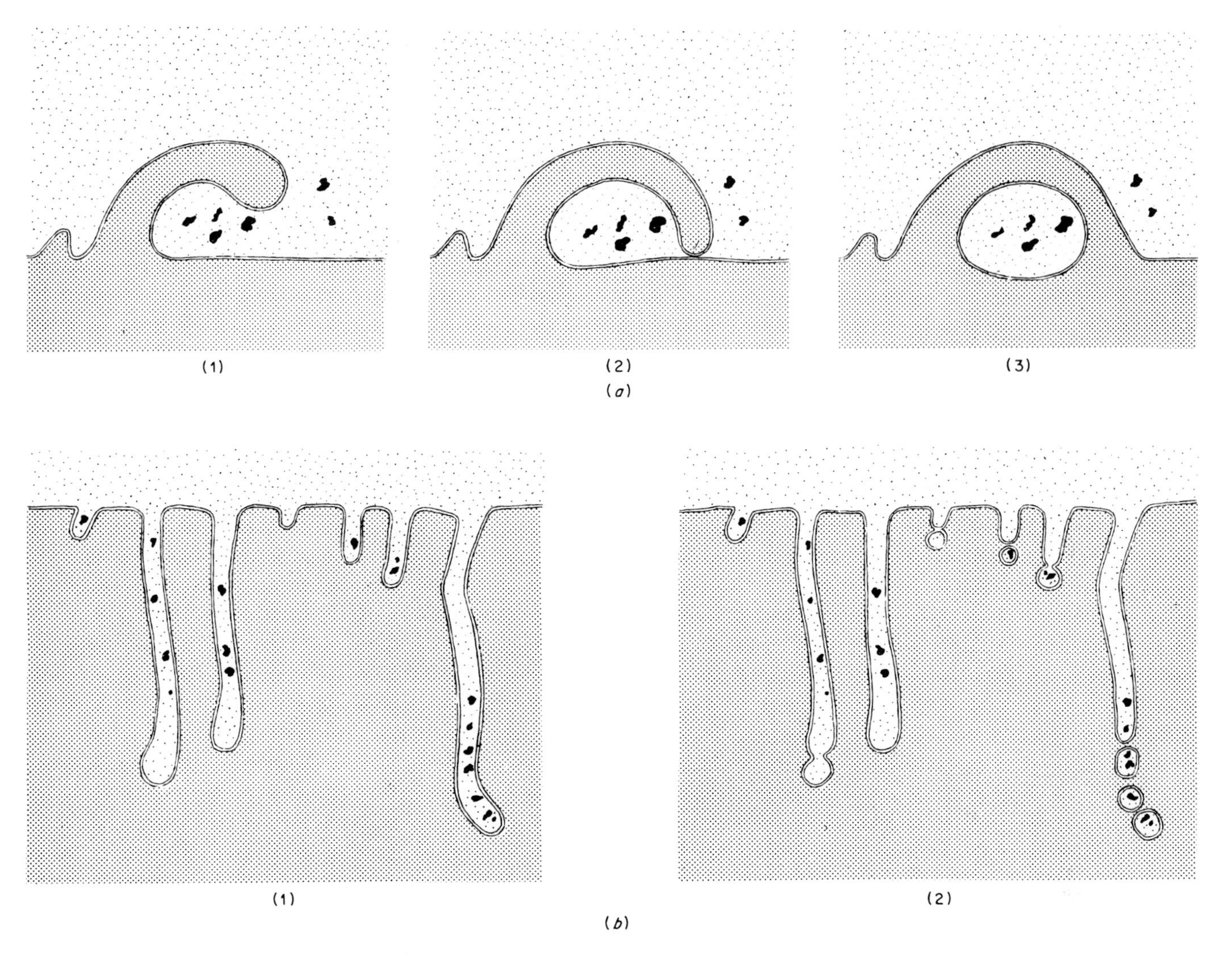

FIG. VII–14 *The probable course of events in pinocytosis.* ***(a)*** *Formation of a pinocytotic vesicle by elevation of the membrane.* ***(b)*** *Formation of a pinocytotic vesicle by invagination of the membrane.*

oriented filamentous processes that stain positively for polysaccharide. A similar polysaccharide coating has been demonstrated on the surfaces of cells of many types. Such a coat could provide sites for the selective binding of ions or molecules. That such binding is a necessary preliminary to pinocytosis has been shown by experiments with *C. chaos*, where the process can be induced by certain substances but not by others. Inducing substances are simple salts, amino acids, certain dyes (notably Alcian Blue), proteins, and nucleoproteins in cationic form. Anionic compounds are not inducers, nor are neutral ones, such as sugars, although the engulfment of these will take place in the presence of a suitable inducer. Pinocytosis seems therefore to be a selective response to certain compounds in the external medium, set off by a physicochemical reaction at anionic sites in the slime coat. When the pinocytotic vesicle is formed, the slime coat, now on its inner surface, is in some way destroyed. This event radically changes the permeability of the membrane, which is low while the membrane is on the external surface of the ameba. Once the pinocytotic vesicle is formed, however, water and some solutes appear to diffuse freely from it into the surrounding cytoplasm.

However widespread this phenomenon may be, it can apply as a feeding device only to organisms of small dimensions whose surfaces are freely exposed and whose nutritive demands are low. It may be the means by which food is taken in by the axopodia of radiolarians and heliozoans, whose function is nutritive rather than locomotor, and also by the rhizopodia of foraminiferans. It is known to occur in amebae, in addition to phagocytosis.

2. Mechanisms for the generation and direction of food currents

Mechanical selection of particles is inevitably practiced by microphagous animals, since the size of the aperture by which they enter the body, or of the pores of a filtering mechanism, determines which are ingested. Weight may also be a factor in selection, because heavier particles are not passed along by ciliary or flagellar action as readily as lighter ones. Evidence that has been obtained from a comparison of the stomach contents of certain microphagous feeders and of the variety of food available to them indicates that a certain degree of qualitative selection also may be made by biological means.

Some way of directing a flow of water toward the body and the oral aperture, if there is one, is a necessity for any microphagous animal, unless it is to depend only upon chance for suitable food to reach it. It is also advantageous that a flow be directed away from the body as an elimination or rejection current. Flagella and cilia are employed to create such water currents by many microphagous metazoans as well as by flagellate and ciliate protozoans. Movements of appendages serve the same purpose in some tubiculous annelids and in arthropods; the former may also pump water in and out of their tubes, or through them, by rhythmic movements of the entire body. In many cases, the flow of water so induced passes over a respiratory surface and thus provides the animal with oxygen as well as with food.

Microphagous species are almost always constant feeders but some have means of shutting off the current or of directing it away from the mouth when food is overabundant or, for other reasons, unwanted. They may do this by reducing the rate of movement of the parts that generate the food current, by stopping their movement entirely, or by shutting themselves in their cases or shells. This withdrawal cannot be for any great period of time if the food current is also a respiratory one, unless the animals can shift to anaerobiosis for their energy requirements.

3. Filter feeding

a. CILIARY FILTERS Ciliary mechanisms for creating such currents are widespread among invertebrates in their adult as well as in their juvenile stages. They are the general means of obtaining food in flagellate and ciliate protozoans, although photosynthesizing flagellates and raptorial ciliates are obvious exceptions to this generalization. But in heterotrophic flagellates, the currents set up by the moving flagella bring food particles into contact with the cell surface. The actual mechanism of their ingestion, even in the various species of *Euglena* where it has been most closely observed, is not clear. In choanoflagellates, food particles are swept by the action of the flagella into the area enclosed by the collar, where they are caught on its sticky surface, moved inward to the cell body, and there taken in, possibly by pinocytosis.

The cilia of ciliate protozoans provide food currents as well as locomotion for the animal. The food current is directed toward the cytostome, through a preoral chamber which may range from a slight depression of the cell surface to a deep groove. The cilia within this chamber may be longer than those on the rest of the body or may be set together or fused in such a way as to form one or several membranelles, or undulating membranes. The movement of these is so directed that the food particles are swept toward the oral aperture, or cytostome, where food vacuoles are successively formed. Figure VII–15 shows the courses of the food currents in the heterotrich *Stentor*.

In sponges, the lashing of the flagella of the choanocytes draws a current of water into the body and to the choanocytes, where it is filtered and the larger particles collected on the collar and ultimately, as in the choanoflagellates, ingested through the receptive area at its base. The water then passes out through the osculum, or oscula, so that under ordinary circumstances a continuous current is maintained through the body (Fig. VII–16). The food is extracted from it by the primary feeding cells and then taken up by amebocytes for digestion. The body of the sponge acts as a macrofilter because particles that are too large to pass through the ostia of simple sponges or the prosopyles of more complex ones are excluded from the current that enters the body. These larger particles may actually be available as a source of food, for they may be ingested by pinacocytes, the flattened cells that form the covering of the sponge's body. Since sponges are sessile animals, and often of large size, they are dependent upon an effective food current and so upon the abundance and variety of food within reach of it not only for their own nutrition but also for that of their numerous inhabitants. In order to furnish this a sponge must pass relatively enormous quantities of water through its body. This probably amounts, in larger specimens, to several hundreds of liters every 24 hr. The rate of flow in the body of the sponge is quite slow—something in the order of 0.01 mm/sec—and apparently can be regulated by the size of the

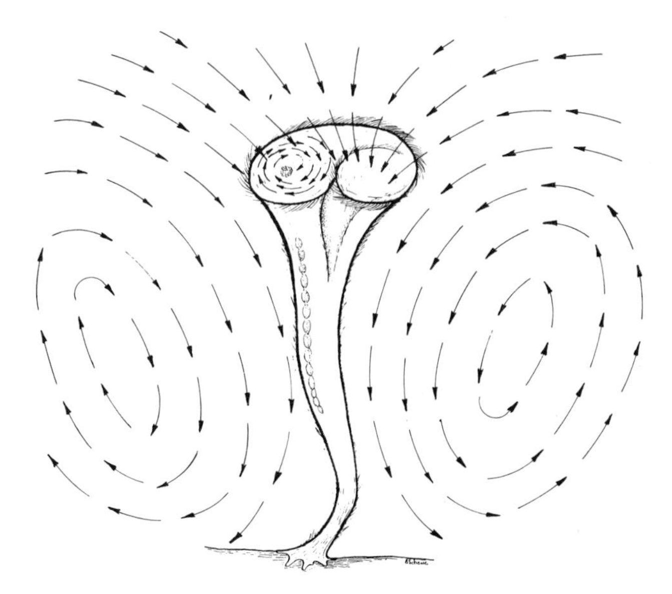

FIG. VII–15 *Food currents in* Stentor. (*After Shaeffer.*) *The arrows show the direction of currents set up in the water by beating of the cilia on the peristome, which sweep a food current into the mouth and roll rejected material into a ball-like mass which is ultimately dropped on the edge of the disk.*

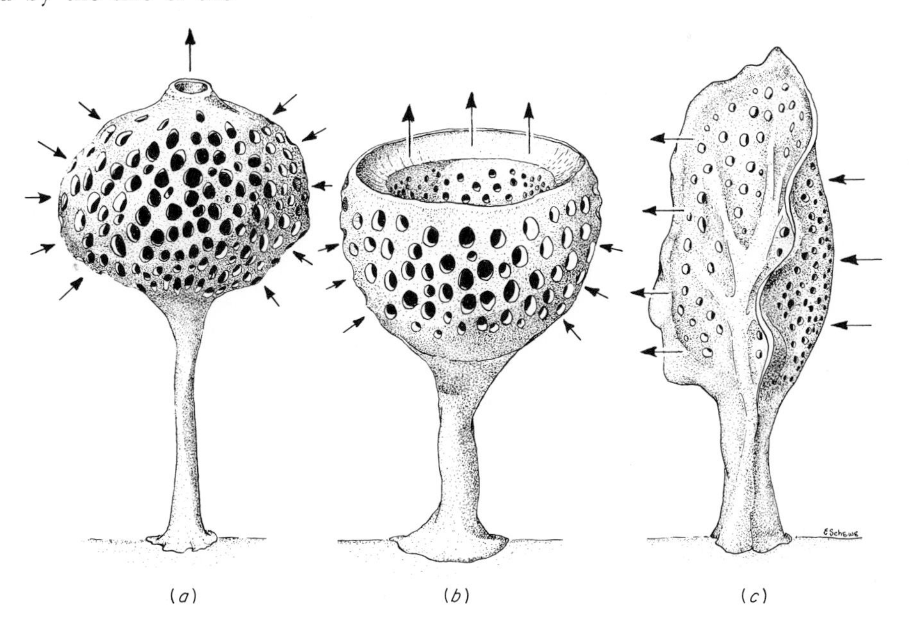

FIG. VII–16 *Food currents in sponges.* (*After Bidder.*) ***(a)*** Clathrina blanca, *in which the food current is drawn in from all surfaces of the body, the angle of supply being increased by elevation of the body on its slender stalk. The ejection stream passes out through the single osculum at the far end.* ***(b)*** Calyx lieberkuhnii (*Neptune's cup*), *in which the angle of supply is reduced owing to flattening of the free end of the body where there are numerous oscula.* ***(c)*** Phabellia conulosa, *in which because of the flattened fan-shaped body, the food current enters from one side only; the ejection stream leaves all over the surface of the other side.*

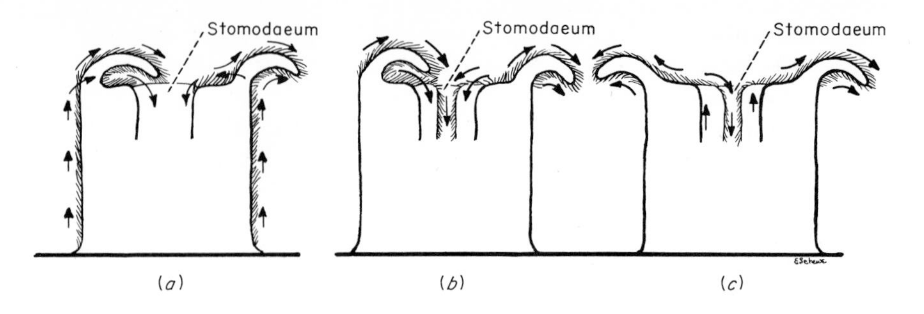

FIG. VII–17 *Feeding in Actinarians and Mareporarians.* (*After Carlgren.*) **(a)** Prothanthea, *a primitive actinarian from East Africa, with cilia distributed over the whole body. The arrows show the direction of the food currents, passing over the disk toward the mouth between the tentacles and in a similar direction when the tentacles are turned inward. The rejection current is shown in the tentacle turned outward.* **(b)** Sagartia, *in which external ciliation is limited to the disk and tentacles. The arrows show the direction of the incoming and outgoing food currents.* **(c)** Sagartia, *showing direction of currents in egestion.*

oscula. The action of the flagella in sponges is devoted entirely to physiological needs and not at all to locomotion, and their beat, although without coordination, sustains a current that is adequate for the nutritional and respiratory needs of a tremendously large number of cells.

Ciliary feeding may have been the mechanism of primitive cnidarians and is still practiced by some modern species. The epidermis of the body wall, tentacles, oral disk, and gullet of the East African actinarian *Prothanthea*, for example, is strongly ciliated, and a current continuously passes from aboral to oral surface of the column, from the periphery of the oral disk to the mouth, and from the bases of the tentacles to their tips (Fig. VII–17). Food particles in these currents are swept up the body and between the tentacles to the mouth and, when the tentacles are bent orally, from their tips to the mouth and hence inward through the gullet. In *Sagartia*, the ciliation is limited to the tentacles and disk, and incoming and outgoing currents are maintained in this area only. Similarly, *Fungia* and *Favia*, both native to southern Pacific waters, have been observed to feed by ciliary mechanisms (Fig. VII–18). The exposed areas of the body of *Fungia* are coated with mucus, whose discharge is apparently stimulated by the presence of suitable food. The food-containing mass is then drawn into the mouth by a current set up by the beat of the cilia in the stomodaeum alone. This behaviour was demonstrated by experiments in which particles were dropped on the oral disk; they remained stationary for a long period unless the mouth were open. The tentacles of *Fungia* are small and play little or no part in the feeding process. But they are a significant factor in the feeding of *Favia* because they can be seen to twist and turn toward the mouth when food is at hand. In this colonial genus,

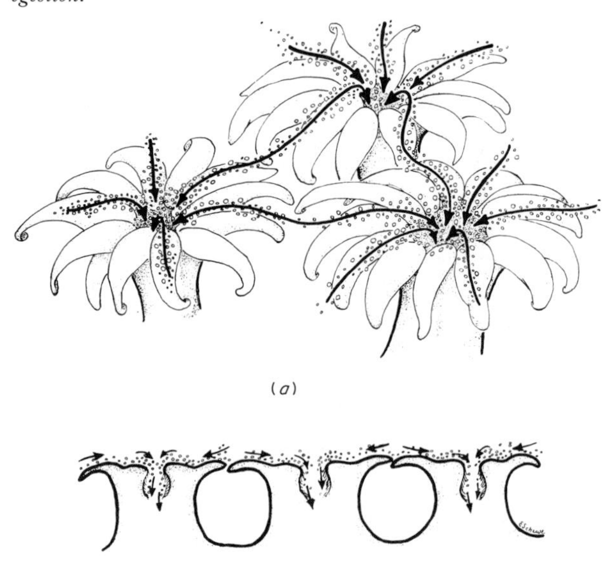

FIG. VII–18 *Feeding in corals.* (*After Duerden.*) **(a)** *Three polyps of* Favia, *a Hawaiian colonial coral, seen from above and showing the distribution of nutrient particles falling upon the colony. The particles are drawn into the mouths and down the stomodaea of the polyps entangled in mucous streams, which form first in the peristome and then extend over the entire disk and all the tentacles. The streams passing down the mouths of adjacent corals are often continuous, until they reach such tension that they break apart and pass to each of the polyps.* **(b)** *Cross section of three adjacent polyps, showing the passage of particles along the disks and down the stomodaea.*

the mucous secretion is continuous over adjacent polyps so that food particles are captured in it over a wide area and, as in *Fungia*, are carried into the gastral cavity by ciliary currents in the stomodaeum. Such primitive methods of feeding seem to have been supplanted in Anthozoa by macrophagous ones when, in the course of evolution, nematocytes became localized in nettle batteries on the tentacles. This localization can be correlated with reduction of ciliation on the body surface.

A similar method of ciliary feeding has been described for the moon jelly *Aurelia aurita*, which lives near the surface of the sea and feeds on plankton. The smaller members of the plankton, such as molluscan and annelid eggs and larvae, rotifers, copepods, diatoms, and so on, become entangled in mucus secreted by cells of the exumbrella and are carried to its rim partly by ciliary action and partly by contractions of the disk. Ciliated cells on the rim of the bell collect the food mass at eight points, corresponding to the positions of the ends of the adradial canals. When a sizable ball of this mass has accumulated at each of these points, the oral arms are moved in such a way that their separated limbs pass over each of the masses, which then are moved by ciliary action along the groove in each limb and toward the mouth.

All but raptorial rotifers depend upon the beat of cilia on their trochal disks to direct a food current into their mouths and so on to the mastax. Selection seems to be entirely quantitative, depending upon the size of the particles in relation to the opening into the mastax.

Endoprocts, ectoprocts, and brachiopods, all with a crown of ciliated tentacles, or lophophore, are exclusively ciliary feeders. Ingestion of food by various species of the marine endoproct *Loxosoma* has been carefully followed. The lophophore is oriented at approximately a right angle to the long axis of the body and, in feeding, its tentacles are usually spread wide apart so that the structure forms a shallow cup. They are ciliated on the inner surface only; the marginal cells of this surface bear long lateral cilia, while the three rows of central cells bear shorter, frontal cilia. The lateral cilia beat inward, perpendicular to the long axis of the tentacle, in metachronal rhythm, and draw a stream of water across the expanded lophophore. Particles carried inward in the stream are thrown onto the inner face of the tentacle, en-

FIG. VII–19 *Feeding in ectoprocts and endoprocts. (After Atkins.)* ***(a)*** *The lophophore of* Loxosoma crassicauda, *showing the ciliary currents and direction of beat of the lateral cilia of the tentacles. The ciliary tracts are shown on only four tentacles. The arrows on the others show the direction of the currents. The rejection current is set up by cilia (not shown) on the epistome.* ***(b)*** *Longitudinal section through the crown of a cyclostomatous ectoproct, showing the ciliation on one tentacle and the direction of the water currents.* ***(c)*** *Transverse section through the crown of a cyclostomatous ectoproct, showing the lateral cilia and direction of water currents across it.*

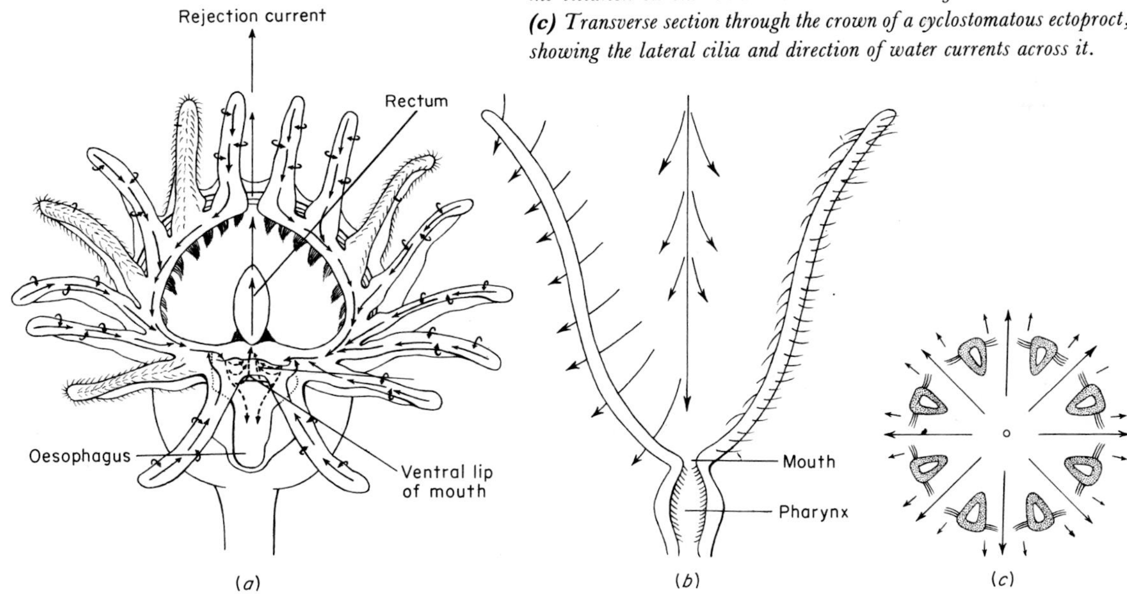

tangled in mucus, and moved orally by the downward beat of the frontal cilia to a ciliated groove, or vestibule, that leads into the mouth (Fig. VII–19*a*). When food is particularly abundant, the motion of varying numbers of cilia can be checked and the water current thus reduced. Also, food may be totally rejected by bending the tentacles inward and so closing the lophophoral cup, or by closing the mouth alone, or by regurgitation of particles that have already passed through it.

Ectoprocts employ a similar but, in some species at least, a rather less effective mechanism. The tentacles of each polypide are pushed out of its zooecium and spread apart in a funnel or bell-shaped configuration. The outer, nonciliated surfaces of the tentacles are provided with long hairs, or bristles, presumably tactile. They are usually two double rows of lateral cilia, whose movements are responsible for the main water current. Frontal cilia may be entirely absent, reduced in number along the tentacle, or limited to its base. In the ctenostomatous genus *Flustrella*, whose feeding operations have been closely watched, the frontal cilia are long and quite numerous; they cooperate in the production of the water current. The effective beat of the lateral cilia of ectoprocts is outward, not inward as it is in the endoprocts, so that water is drawn into the funnel or bell and is swept outward between the tentacles. This results in the loss of many food particles, but those that are retained in the inward and downward stream are drawn into the mouth by pumping action of the muscular pharynx, augmented by the strong beat of cilia on its epithelial cells (Fig. VII–19*b*). In species like *Flustrella hispida*, the motile frontal cilia on the tentacles help in directing the particles toward the mouth. Cessation of feeding and rejection of excess or undesirable food can be effected either by complete or by partial retraction of the tentacular crown (calyx) into the zooecium or, while it is still extended, by bringing the tips of the tentacles together and so closing off the incurrent stream, or by closing the mouth, or by reversal of the beat of the pharyngeal cilia.

The lophophore of brachiopods is very varied in pattern in different genera and species but always, like the rest of the animal's body, is enclosed in a bivalve shell. Therefore, as a preliminary to feeding, the dorsal and ventral valves of the shell must be separated to greater or lesser extent to permit the outside water to come into contact with the tentacles. These bear both lateral and frontal cilia. The lateral cilia beat, like those of the ectoprocts, outwards across the tentacles and create two incurrent streams of water, which enter the mantle cavity laterally. Small particles are thrown onto the face of the tentacle, caught in a film of mucus, and swept downward by the beat of the frontal cilia. The water leaves the shell in a single median stream. The arrangement of the lophophore determines the efficiency of the feeding process according to the extent to which it provides for separation of the filtered and unfiltered water. Although reversal of beat in the lateral cilia has not been observed, it has been seen in the frontal cilia so that particles too large to enter the mouth or those that may be rejected by other means can be moved to the tips of the tentacles, rather than to the base, and carried away in the median outgoing current.

Ciliary filter feeding has been developed to a high degree of perfection by bivalve molluscs. Some gastropods are also ciliary feeders as adults as well as larvae, but most of them have retained what was most probably the primitive molluscan mechanism: that of scraping or tearing their food with the toothed radula. The gastropods that are ciliary feeders are sessile, and, like the bivalves, have made use of their gills, primarily respiratory organs, as current producers and filtering devices. The limpet *Crepidula*, for example, is a ciliary feeder; the long filaments of the single gill stretch across the mantle cavity, dividing it into a left ventro-lateral inhalant chamber and a right dorso-lateral exhalant chamber. A current of water is drawn in by the action of rows of cilia on the anterior and posterior faces of the gill filaments. Particulate matter is filtered by the gill; the fine particles are deposited in a ciliated groove at the tips of the gill filaments, embedded in mucus, and rolled into a cylindrical mass which is carried toward the mouth. The limpet extends its radula with the marginal teeth spread apart. These close over when contact is made with the food mass which is caught between the rows. The radula is then retracted, and the mass passes into the pharynx. Larger particles carried in with the inhalant current are moved anteriorly and collected in

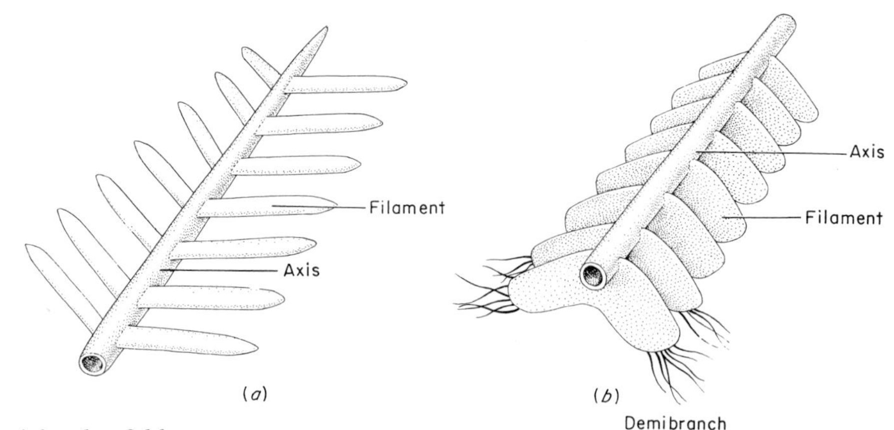

FIG. VII–20 *Molluscan gills.* ***(a)*** *Primitive ctenidium, or gill, showing the central shaft and paired lateral filaments extending from it.* ***(b)*** *Protobranch gill, with the filaments modified as triangular leaflets, ciliated at the ends. (After Atkins.)* ***(c)*** *Lamellibranch gill, showing its probable development through elongation and folding of the filaments. The fleshy interlamellar and interfilamentar junctions (A) are characteristic of eulamellibranchs; the ciliary junctions (B) of filibranchs.*

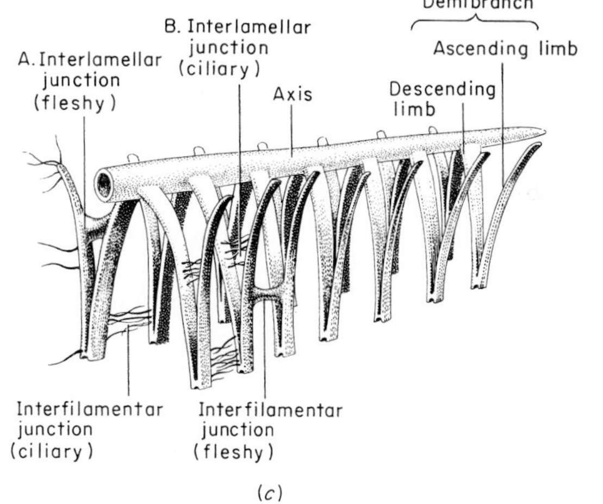

the food pouch, a deep groove in a semicircular fold of epidermis just in front of the mouth. There the particles are entrapped in a mucous pellet and either eaten or expelled.

The gill of *Crepidula*, and other ciliary feeding gastropods, functions as a food sieve, a water pump, and a respiratory organ. The radula, unlike that of other gastropods, is used for grasping not rasping.

Some pteropods are also ciliary feeders and some pulmonates seem to be at times. *Limnea*, for example, creeps along beneath the surface film of a pond or an aquarium and at intervals can be seen to lower the anterior end of its foot where minute particles are caught in mucus secreted by pedal glands. The mass is then moved posteriorly by ciliary action and when it is collected and compacted near the hind end of the foot, the animal bends its head backward and scrapes the food ball off.

It is, however, in the pelecypods, and especially in the eulamellibranchs, that ciliary feeding has evolved to the greatest degree of refinement. In eulamellibranchs, there is a pair of gills with one member of the pair on each side of the visceral mass; the central shaft of each pair runs antero-posteriorly parallel to the main axis of the body. The numerous filaments extending from this axis are relatively very long; presumably the gill, as it is found in common bivalves today, has evolved through bending, or folding, of each filament at a point, about midway along its length, marked by a small indentation. Each filament, therefore, seems folded back against itself, in a U or V shape, with one descending and one ascending arm of equal length (Fig. VII–20), giving the whole gill, in transverse section, the form of a W. All the descending arms constitute an inner lamella, continuous with the gill shaft, and all the ascending arms an outer lamella. The indentation at the angle between these arms makes a food groove. In each pair of gills, the tips or ends of the inner lamellae are fused together; the margin of the outer lamella of the outer (lateral) gill is in some species attached to the mantle and the corresponding lamella of the inner (median) gill to the wall of the foot. The mantle cavity is thus divided into an upper (dorsal) or suprabranchial chamber, and a lower (ventral) or infrabranchial one (Fig. VII–21). The only connection between the two is through spaces, or passages, in the gills. Since, in eulamellibranchs, adjacent lamellae are joined together by numerous fleshy interlamellar junctions and adjacent filaments by similar interfilamentar ones, these passages or

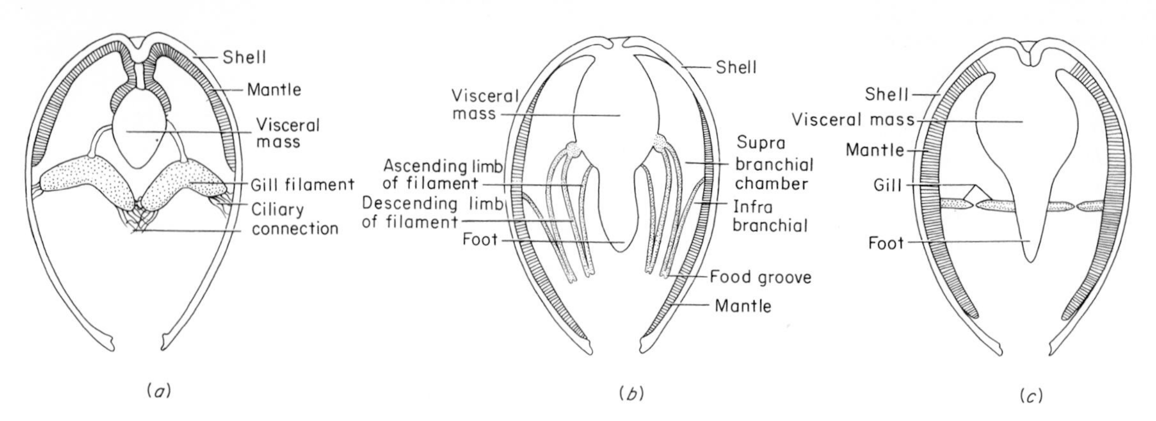

FIG. VII–21 *Diagrammatic representations of cross sections of molluscs, showing arrangement of their gills.* ***(a)*** *Protobranch* (Nucula). ***(b)*** *Lamellibranch.* ***(c)*** *Septibranch.*

water tubes are limited in number. The spaces between the interfilamentar junctions provide openings, the ostia, from the infrabranchial chamber of the mantle cavity to the water tubes, which, in turn, open dorsally into the suprabranchial chamber. The whole gill is thus a fleshy meshwork, supported by chitinous rods in the filaments and junctions, in which there are also blood vessels. A somewhat simpler and perhaps more primitive condition is exhibited by filibranch pelecypods such as the mussel *Mytilus* and the scallop *Pecten*, where there are fewer interlamellar junctions, and tufts of cilia, not fleshy bands, provide interfilamentar adhesion.

Success in diversion of the pelecypod gill from its primary function of respiration to that of nutrition required some modification of its ciliation and of the mantle's surface and edge. In addition to lateral and frontal tracts of cilia on the free edges of each gill filament, there are two small tracts, the latero-frontals; each of these lies between the frontal and a lateral tract (Fig. IV–12*a*). The mantle is richly vascularized and contributes to respiration to a much greater extent than it does in other molluscs. Two regions of its posterior margin, either close together or separated by some distance, are thick and muscular, often with fringed edges and deeply pigmented epithelium. These regions, the siphons, can be extended, sometimes a considerable distance, through a gap in the shell; elsewhere, margins of the two valves are held tightly together. One of these, the inhalant (ventral) siphon, through which water is drawn by action of cilia on the gills, opens into the infrabranchial chamber; the other, the exhalant (dorsal) siphon communicates with the suprabranchial chamber and provides an exit for the filtered water and other products that may be discharged into this part of the mantle cavity. This modification has made it possible for a bivalve to lie buried in the mud or sand and, with its siphons extended, to draw in the clearer waters above it and to expel the fouled ones. A further refinement of the feeding mechanism has been the development of labial palps, a pair of ciliated flaps overhanging the mouth which accomplish the final sorting of the food before its ingestion. All of these modifications have provided the eulamellibranchs with one of the most effective of filter-feeding systems.

The actual process of feeding can be observed by cutting a piece from one valve of the shell of a captured animal and inserting a glass plate in the opening. The course of the food current and the means of removal of particulate material from it can then be watched through a binocular microscope. This has been done with at least four different genera of pelecypods, each with a different habit of life, offered a suspension of diatoms or organic debris; the burrowing gaper clam, *Schizothaerus*, whose siphons are longer than its shell; the scallop *Pecten circularis*, that lives in the mud flats of the Pacific coast; the oyster *Ostrea* that attaches itself to some object so as to be above the surface of the sand or mud and thus keep clear of sedimentation; and the mussel *Mytilis californianus* that lives in the tidal zone. In all of these, the frontal cilia beat up and down, along the length of the ascending and descending limbs of the gill filaments; the latero-frontal cilia beat relatively slowly in a direction at right angles to that of the beat of the frontal rows; the lateral cilia, which are primarily responsible for the inflow of water, beat inward. At the beginning of feeding, a sheet of mucus is secreted by gland cells in the upper part of the gill and is spread downward over its surface by the beat of the frontal cilia. This sheet is the filter proper, because it traps and holds all the particles that have been admitted through the inhalant siphon; the water cleared of them passes into the ostia and on through the water tubes of the gill. As the sheet of mucus with its captured particles passes down over the gill, the heavier ones—such as sand grains and larger organisms or bits of debris—fall to the bottom of the infrabranchial chamber; the smaller ones are held in a mucus string in the food groove at the lower margin of the gill. This string is pushed forward along the margin of the gills to the labial palps, where the final selection and rejection of food takes place, and then into the mouth. In rejecting food, the palps are spread wide apart so that they are oriented at right angles to the mouth rather than around it. The mucous strands are moved to the bases of the lower, or ventral, palps and down their anterior margins to the edge of the mantle, where they, with the already discarded heavier particles, are cast out when the shell opens slightly.

Some microphagous annelids obtain their food by means similar to those practiced by ectoprocts, using cilia on their tentacles, or cirri, to create a food current as well as to filter the particles from it. Others make a filter of mucus, as the pelecypods do, but pump water in and out of their burrows, or through their tubes, by rhythmic contractions and expansions of their bodies. The ciliary feeders make their principal source of food either plankton or detritus covering the substrate. Most of these worms are marine polychaetes and either burrowers or makers of tubes. The free-living, freshwater oligochaete *Aeolosoma* is, however, also a ciliary feeder, sweeping microorganisms and particulate debris into its mouth in the currents generated by the beat of the cilia at the anterior end of the body. Terebellid (Latin *terebra*, "borer"), serpulid, and sabellid polychaetes are equipped with extensible and contractile tentacles, beautifully colored, which are extensions of the peristome and surround the head. This crown may be extended from the burrow, offering a large area for respiration and also for food collection. When the tentacles have many fine pinnules (as in serpulids and sabellids) and are fully expanded each looks like a feather and the whole set like an old-fashioned feather duster; hence, the animals are popularly known as "feather" or "feather duster" worms. The tentacles may be used to sweep over the substrate and select and collect food particles from it, or they may wave about in the water, trapping motile microorganisms or floating debris. In both cases the methods of food collection and ingestion appear to be similar, as they have been observed in a number of different species. A food current is produced by the action of the cilia on the tentacles, or their pinnules; the particles are trapped in mucus; and the string, or mass, is carried down to the mouth, where it is continuously swallowed. The process has been closely watched in the plankton feeder, *Sabella pavonina*, very common all around the British coast, which makes its tubes, mostly in estuarine mud, out of sand cemented with mucus. When feeding, the crown of tentacles, or radioles, and its pinnules are extended from the burrow and fully expanded, forming a broad and shallow funnel. Mucus-secreting cells are distributed all over the surface of the pinnules; the ciliated cells lie in four tracts: one frontal, on the inner surface of the pinnules; one abfrontal, on the outer surface; and two latero-frontal, between these. The cilia of these latero-frontal tracts are considerably longer than those in the frontal or abfrontal tracts. The abfrontal cilia, which beat strongly and continuously, are instrumental in drawing a food current into the funnel; the current is deflected toward the inner face of the pinnule by cilia in the latero-frontal tracts, whose beat is at right angles to that of the abfrontal, and directed down to the base of the pinnule by the frontal cilia, whose beat is parallel to the long axis of the pinnule. The particles are graded according to size in the basal folds, pairs of parallel ridges ciliated on all sides, that run along the oral surfaces of the tentacles. Particles too large to move between the folds are carried along their edges and cast off; smaller ones enter the folds and are moved toward the mouth, but only the smallest of them enter it. The others are discharged into ventral sacs that lie just below the mouth and are stored there until the sand grains among them are used for enlargement of the tube. The smallest particles are moved directly from the basal folds into the mouth. Examination of the contents of the gut shows that these are bits of debris, fine grains of sand, and, seasonally, diatoms, red and green algae, encysted flagellates, minute pieces of copepods, and the bristles of annelids and crustaceans.

Among the echinoids, the sand dollars (Clypeastridae) are also ciliary feeders, taking their food from the substrate on, or just below, which they live. The epithelium covering the club-shaped spines on the upper (aboral) surface of the flattened body is ciliated; the beat of these cilia creates a current in a direction opposite to that in which the animal is moving. The epithelium at the bases of the spines and covering the surface of the body is secretory; particles caught in the eddies arising at the posterior bases of the spines as the current flows past them are trapped in mucus and carried away in minute streams that flow away from these eddies. These little streams flow together into larger—but still very small—streams, and the confluence of these streams then forms larger ones that are moved over the sharp edges of the test and merge on the underside into five tracts that lead directly to the mouth.

At least one species of asteroid, the mud star *Ctenodiscus crispatus*, feeds in a similar fashion. Cilia on the epithelial cells in the grooves that run from aboral to oral surface between the marginal plates of the arms move mud and its contained organic debris to the podia, along which it is passed inward to the mouth. Transport in this manner has been demonstrated by mixing carmine particles with the substrate; these were later recovered from the stomach. This method of feeding does not provide a means of filtration and little or no selection, although it has been suggested that the cribriform organs (Latin *cribrum*, "sieve") may provide a discriminatory device of sorts. These organs are characteristic of the suborder Cribellosa and typically consist of vertical depressions between the marginal plates along the interradii on the sides of the arms. Within the depressions are a series of folds, like the leaves of a book, supported by calcareous plates and covered with ciliated epithelium.

b. MUCOID FILTERS Mucoid filterers are those that use mucous secretions to strain or sort the particles they finally ingest. The highly specialized tubiculous polychaete *Chaetopterus variopedatus* employs this method. The worm lives in a U-shaped tube constructed of mucus that is buried in the sand, except for the two tapered ends that open into the water just above the sand-water interface (Fig. VII–22). Most of the time the worm is fastened to the wall of the tube by suckers, which are modified neuropodia of the parapodia of certain segments. A stream of water is drawn through the tube by rhythmic movements of the enlarged notopodia of the fourteenth, fifteenth, and sixteenth segments that act as fans. The notopodia of the tenth segment are also enlarged, but they do not move; instead, they secrete mucus continuously and liberally from gland cells on their inner margins. This secretion extends in a sheet-like film between these two notopodia and is drawn out into a bag by cilia on cells along a groove in the dorsal body wall extending from these notopodia to the cupule, a small cup-like structure lined with ciliated cells that lies just anterior to the fans. All the water that goes through the tube must pass through this bag, which therefore constitutes the filter of the system. While the animal is feeding, a water current is constantly drawn through the tube and mucus is continuously secreted so that the bag is replenished at its anterior end while, at its posterior, it being rolled up into a ball by the cilia in the cupule. When the ball has reached a diameter of about 3 mm, or approximately that of a BB shot, the secretion of mucus stops, the fans stop their rhythmic movement, and the beat of the cilia in the dorsal groove reverses, so that the food pellet is propelled anteriorly and to the mouth. All but the largest particles in the food current, which as they enter are cast out in the region of the peristome, are caught in the filter and so are ingested.

The echiuroid *Urechis* feeds in a similar manner, by means of a mucus filter, although other echiuroids employ more conventional methods, extending their long proboscides from their burrows to forage for food and collecting it, when found, in strings of mucus that are carried to the mouth along ciliated tracts. The proboscis of *Urechis* is comparatively short; just posterior to it is a ring of mucous glands (Fig. VII–23).

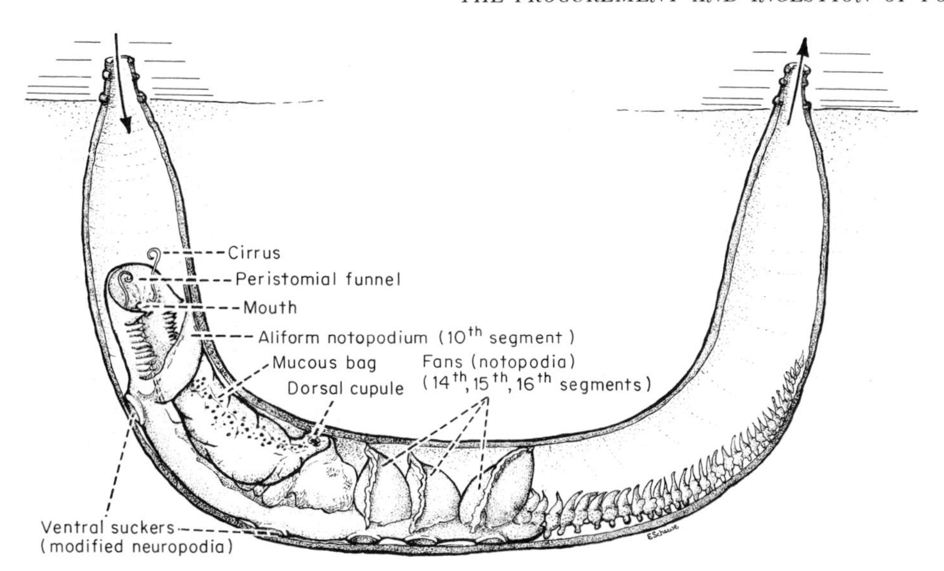

FIG. VII–22 Chaetopterus, *a tubiculous polychaete, in its tube whose ends extend above the sand.* (*After MacGinitie.*) *The worm lies ventral surface downward in its tube to which it is held by neuropodia modified as suckers. The notopodia of the tenth segment are much enlarged, with mucous glands on their inner surfaces. The notopodia of the fourteenth, fifteenth, and sixteenth segments are large and muscular, acting as fans to keep a current of water moving through the tube.*

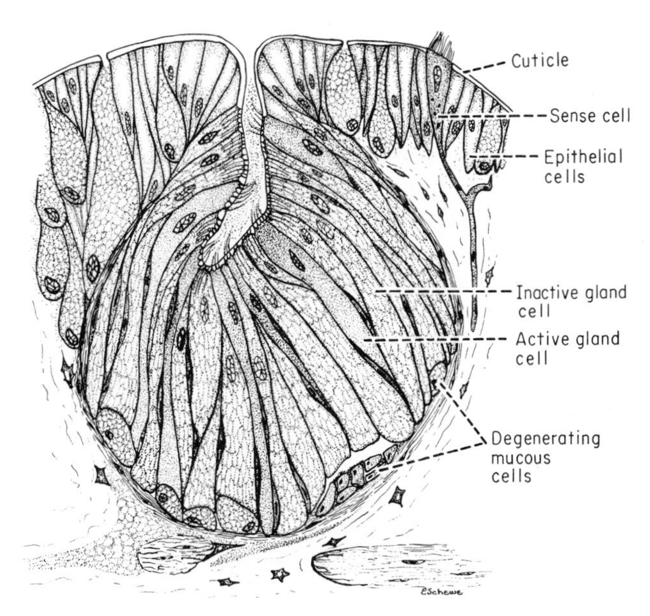

FIG. VII–23 *Slime net gland of* Urechis. (*After Newby.*)

A hungry *Urechis* moves toward one of the openings of its U-shaped burrow and swells out its body in the region of these glands until it makes contact with the walls of the burrow. A tube-shaped filter, into which the proboscis extends, is made by secretions from the gland cells while the worm is slowly backing away. A current of water is drawn through the tube by peristaltic contractions of the animal's body, which usually stop only when the filter is so clogged with particles that water can hardly pass through it. When filtration stops, *Urechis* contracts its body so as to detach itself from the loaded filter and, moving it forward over its head also by contraction of its body, grasps it with its proboscis and swallows it.

Some echinoderms are also microphagous, using mucoid secretions to trap or entangle particulate matter and, in some species, ciliary currents to move the mass to the mouth. Examination of the stomach contents of numerous species of crinoids has shown that their diet consists primarily of plankton—diatoms and algae, protozoans, small crustaceans, and larvae of various kinds. Events in the capture and ingestion of these plankters have been observed in *Antedon potasus* kept in an aquarium. In feeding, the animal spreads its arms apart with their pinnules wide open. Cells of the podial epithelium secrete mucus; food falling on the podia is thus caught and, when loaded, the podium bends quickly inward toward the ambulacral groove of the arm and the particles drop into it. The epithelium within the groove is both ciliated and glandular, and there are also ciliated cells in the interambulacral areas. Ciliary-induced currents in the ambulacral grooves move the mucous strings with their entrapped plankters toward the mouth and finally into it. Similar currents along the interambulacral surfaces are also orally directed but are prevented from entering the mouth by the ridges bordering the grooves. Material carried in these is, therefore, washed away and the surface of the crinoid kept clean, although some potential food may be lost in this way.

The burrowing echinoids, the spatangoids or heart urchins, and some of the holothurians are also mucoid feeders, trapping their food on certain appendages and then sticking them in their mouths to suck it off. Spatangoids have especially long podia, which, in feeding, are extended into the water above the burrow to pick up debris which is caught in a mucoid secretion from the epithelial cells. The podia are then drawn back to the mouth where the particles are removed from them and again extended to trap more. Holothurians such as *Sclerodactyla* (*Thyone*) and *Cucumaria* feed on plankton by extending their branched tentacles and spreading them wide in the water or by sweeping them through it or over the substrate. Plankters, entangled in the mucus that is spread over their surfaces, are cleaned from them when, one by one and sometimes in apparently definite rotation, they are bent back and thrust into the mouth. Other microphagous sea cucumbers use their tentacles as shovels to scrape up the mud and push it into their mouths.

Ophiocomina nigra, an ophiuroid frequently found where the alga *Laminaria* is abundant, constructs a mucus net from secretions of unicellular glands that are very generally distributed throughout its epidermis. Animals in aquaria, supplied only with minute particulate matter, have been seen to raise their disks slightly and spread their arms as widely as possible while swinging them from side to side. In this position, mucus is discharged over the general body surface, especially copiously over the spines on the arms from which it hangs in strands that extend between adjacent spines. Small animals and particles of detritus falling upon this net are trapped in it and, as the food accumulates, it is moved to the mouth either by ciliary currents flowing between the spines or by the tube feet. This may be a basic feeding device for these echinoderms to which they turn when material suitable for macrophagy is not available. It appears to be one wholly adequate for their nutrition, for animals were kept in the laboratory on a diet that demanded it for as long as 9 months, during which time their gonads matured and they successfully reproduced.

Among insects, larvae of the midge *Tendipes* (*Chironomus*) are mucoid feeders, employing a device similar to that of *Urechis*. These larvae are of wide distribution, commonly found among aquatic plants and at the bottom of ponds and pools and larger bodies of freshwater. The feeding of *T. plumosus* larvae has been closely followed in the laboratory, where they were kept in aquaria in glass U-tubes simulating the tubes that in their natural habitats they make for themselves out of sand or mud held together by secretions from the salivary glands. In the laboratory, the glass tubes were so placed that their ends opened into a layer of mud spread over a celluloid platform. It could be seen that a larva, held in a fixed position in the tube by the firm attachment of its posterior pair of legs to one wall, rotated the anterior part of its body, the head describing a full circle first in one direction and then in the other. While in rotation, a silk-like secretion poured from the salivary glands, was drawn out from the mouthparts by the prolegs, and was stretched across the lumen of the tube as a slightly concave sheet. When this was accomplished, the larva loosened its hold on the tube and backed a short way down it, holding with its mouthparts a thread of the salivary secretion attached to the middle of the sheet. In this way a funnel, open distally but closed proximally, was constructed, through which water for respiration as well as for feeding was drawn by rapid peristaltic undulations of the larva's body, proceeding in an antero-postero direction. After a very brief interval of filtration, determined apparently by factors other than the load in the net, the larva straightened out and ate the conical filter with all its attached particles, regardless of their nutrient value. The whole procedure of filter construction and consumption occupied $1\frac{1}{2}$ to 2 min and was repeated again and again while the larvae were under observation.

c. SETOSE FILTERS Microphagy in arthropods has evolved without benefit of either ciliated or mucus-secreting epithelia, for their chitinous cuticle has, in general, precluded such cellular differentiations. It is, nevertheless, practiced by a number of species of Crustacea and by the aquatic larvae of some insects. While in all of these the general mechanisms are similar, there are variations and no one method is characteristic of any single group. It seems apparent, therefore, that the habit has arisen independently in each of them and that the variant methods have evolved in relation to the conditions and demands of the environments common to members of these different groups.

Microphagous crustaceans are found among branchiopods, ostracods, copepods, cirripedes, leptostracans, mysidaceans, cumaceans, amphipods, euphausiids, and decapods. The appendages are used to create water currents; in some species, all of them are used for this purpose, as well as for respiration and locomotion; in others, the activity that generates a food current is confined to certain ones. The currents may be strained through rows of slender spines (setae), generally distributed over the appendages or on only a limited number and sometimes limited to certain segments of particular appendages. Such setose filters may be further refined by the presence of more delicate bristles, the setules or setulae, along the length of the seta. These are usually set at an angle to the long axis of the seta (chaeta), so that the whole system of apposed setae and interlocking setules forms a fine meshwork that acts as an effective sieve in retaining even the most minute particles. The food of filter-feeding crustaceans consists of plankton and detritus or very small organisms that live in the mud or sand, which may be stirred up by movements of the feeding organism.

Microphagous branchiopods, cirripedes, and leptostracans use all their limbs in feeding. The process has been followed in some detail in the freshwater branchiopods *Chirocephalus diaphanus* (Anostraca) and *Daphnia magna* (Cladocera), and in the cirripede *Balanus* and the marine leptostracan *Nebalia bipes*. While feeding devices in these species are in very general ways representative of the orders, there is much diversity in the refinements of the devices. These, like other mechanisms, are related to the natural habitats of the animal and the type of food available to the animal.

The branchiopod limb is a broad, biramous phyllopodium, on whose medial surface there is a series of lobes, or endites, each bearing a thick fringe of setae with many setules. The proximal endite is often enlarged as a gnathobase. *Chirocephalus*, like other fairy shrimps, swims on its back, propelled forward by the effective backward stroke of the 11 pairs of thoracic limbs. At the end of this stroke, the limbs lie back against the body, overlapping each other; at the end of the forward or recovery stroke, they are erect. They beat in metachronal rhythm, the movement of each limb beginning just a little after that of the one posterior to it. At each beat, the space between adjacent limbs is alternately enlarged, at the recovery stroke, and diminished, at the backstroke. The suction arising at each enlargement of this space, reinforced by the pressure of the water, draws the distal portions of each limb backward, so that each touches the one behind it. The space between each succeeding pair of limbs thus becomes a chamber, closed on all but the median side; here, it is separated from the channel along the ventral surface of the animal, between the bases of the limbs, by the setae on the medial endites. As each chamber in the series enlarges, at the upstrokes of the appendages, water is drawn into it from the ventral channel, and the particulate matter in it is held back by the filter of setae and setules. At the backstroke of the limbs, the pressure in these chambers rises as they become smaller, forcing apart the appressed parts of the limbs and so opening the chambers. The filtered water flows out and forms two ventro-lateral streams, which propel the animal forward. The particles held back by the setae are scraped off from those of one limb by those of another, and washed off by a current that flows from posterior to anterior in a groove along the ventral surface of the animal, between the bases of the limbs on the right and left sides. The origin of this current, which is counter to the primary or locomotor current, is in doubt; its motive force may be movements of the gnathobases or of other endites. However it is kept in motion, it carries the particles scraped or washed off the filtering setae forward, where they become entangled in a viscid secretion from the labrum and pushed into the mouth by the maxillules. The mechanism is neither selective nor very efficient, for much of the potential food is without doubt swept away in the main locomotor stream, even though it may be caught on the setae of the distal endites. To get enough to eat, a fairy shrimp must be an almost constant feeder and, hence, also a continuous mover.

The feeding process of the water flea *Daphnia*

follows much the same pattern. The particular differences are related to the presence of a bivalve shell and the reduced number of thoracic appendages. Moreover, movement of the antennae provides the principal means of locomotion, so that food collection is not necessarily dependent upon locomotion. In daphnids, the food current is drawn from an anteroventral direction into a single chamber, enclosed dorsally by the body wall, laterally by the shell and the five trunk limbs, and ventrally by the third and fourth pairs of these when they are brought together. These two pairs are the primary agents in the feeding process, for the chamber is enlarged, and water drawn into it, when they are erect, at the end of their forward stroke, and diminished in the backstroke, when the apposition of their tips precedes that of their bases. In this backstroke, water is forced either into the ventral food groove or into lateral streams between the appendages and so out of the chamber. Filtration is accomplished mainly by the setae of this third and fourth pair of appendages and seems to be somewhat selective, at least to the extent of the exclusion of large particles. It also seems to be very efficient, for *Daphnia* can be raised, and will breed, on a diet of bacteria alone. The filtered particles are carried forward to the mouth, as they are in *Chirocephalus*, very possibly by movement of the gnathobases.

The six pairs of thoracic appendages of sessile cirripedes are long, slender, many-jointed, plentifully supplied with setae, and are used exclusively for food collection. When *Balanus*, for example, is submerged and feeding, these cirri are extended between the separated valves of the shell and moved through the surrounding water like casting nets. When food particles are caught on the setae, the cirri bend orally and the food is scraped off them and pushed into the mouth by the maxillules.

Leptostracans of the genus *Nebalia* feed in a manner very similar to that of the branchiopods. *Nebalia bipes* is a littoral species, living in burrows or under rocks and stones along the shore line. The thoracic limbs are flattened and thickly fringed with setae. These are used, as in the branchiopods, to draw water in and out of chambers that, however, in *Nebalia* are closed by the forward strokes of the limbs and opened by the backward strokes.

Among amphipods and decapods there are species that make use only of the anterior pairs of limbs in their feeding maneuvers. In amphipods, the second and third pairs are modified as gnathopods. In some of the microphagous species, the terminal segments of these limbs are heavily fringed with setae which filter off food particles from water that is kept in motion by the pleopods. Some of these amphipods are active swimmers but feed only when clinging fast to a solid object; others live in burrows in the bottom mud or on its surface. These may use their gnathopods both to scrape up the mud and to strain particulate matter from it, but other species create currents and stir up the sediment with their pleopods, using the gnathopods only to filter the water that passes over them.

Most of the euphausiids are microphagous, using their thoracic limbs to create currents that are both locomotor and food collecting. They are pelagic in habit and themselves provide an important source of food for larger animals of the deep waters. *Euphausia superba*, the antarctic krill upon which many kinds of whales subsist, feeds mainly on diatoms which it draws into a ventral current by rotary movements of the exopodites of its thoracic limbs, none of which is specialized as a maxilliped. The food is filtered from this current, which is also a locomotor and respiratory one, by setae on the endopodites of these limbs. These form a sort of filtering basket. The particles are then scraped off the setae and passed forward to the mouth for ingestion.

Burrowing shrimps, such as *Calianassa* and *Upogebia*, are microphagous decapods that use certain thoracic appendages to strain their food from the mud and water. *Calianassa californiensis*, one of the ghost shrimps, lives along the Pacific coast of the United States in the mid and upper intertidal zone, where the sand or mud is soft enough to be filtered but firm enough to hold up the walls of their burrows. Its range is therefore somewhat limited, but the shrimps are abundant in favorable localities. Like earthworms, these shrimps eat as they burrow, scraping away the substrate with their second or third pairs of legs and sifting the coarse from the fine material through the setae on them. The retained particles, selected only according to size, are scraped off by setae on the third maxillipeds and sifted again by the other mouthparts as they are passed forward. It has been reported that in both *C. californiensis* and *C. major* this sifting is effective enough to retain only bacteria, on which both species can live as their exclusive diet.

The blue mud shrimp, *Upogebia pugettensis*, has a similar range of habitat. The first and second pairs of thoracic limbs of these decapods have long setae with numerous setules; these legs are used both for digging the burrow and for straining material fr[illegible] the food current that is circulated thro[illegible] [illegible]urrow by rhythmic movements of th[illegible] material retained on this fi[illegible] by the third ma[illegible] maxil-lip[illegible] sticky sec[illegible] les toge[illegible] w. [illegible] o-mur[illegible] *a*, "tai[illegible] l-ing, [illegible] "sho[illegible] mud t[illegible] endopod[illegible] thickly dis[illegible] particles filt[illegible] the filter by th[illegible] a tuft of setae, an[illegible] parts. In the co[illegible] further sorting of the [illegible]

One of the most hig[illegible] urans, both in habitat an[illegible] coral gall crab *Haplocarcin*[illegible] *supialis* are free-living, but th[illegible] the axils of branching coral [illegible] presence induces, in some way, [illegible] calcareous chamber around their bod[illegible] corallum. In feeding, the crab extends its [illegible] first and second maxillipeds from an opening [illegible] gall and filters plankton from the surrounding wa[illegible]

Some of the pea crabs, fiddler crabs, and herm[illegible] crabs are microphagous, also using particular thoracic appendages to obtain and sift their food. The pea crabs, *Polyonyx macrocheles* and *Pinnixia fransciascana*, are commensal with *Chaetopterus* and use the setose endopodites of their maxillipeds as filters; the tropical burrowing fiddler crab, *Uca signatus*, uses its first thoracic limbs (chelae) to bring mud to its mouthparts, which move apart to admit it, perform some sorting of it, and roll the rejected material into a little ball which is then dropped out; and the hermit crab, *Pagurus ochotensis*, stirs up the mud within reach with its chelae and sifts it with the second maxillipeds.

A good many crustaceans use their maxillae primarily as filters. Among these are certain copepods, mysidaceans, euphausiids, syncarids, cumaceans, tanaidaceans, and at least one genus of amphipod. The copepod *Calanus finmarchicus*, a plankton feeder and itself a member of the plankton that provides a source of food for fish and whales, has been watched in an aquarium stocked with a plentiful supply of food organisms, some of which were labeled with P^{32}. Locomotor and feeding activity are alike, for rotation of the distal parts of the biramous antennae and mandibular palps, as well of the maxillules, help to propel the animal forward and also draw a current of water into a chamber contained by the ventral body wall, the tips of the biramous thoracic limbs, used also in swimming, and the setae of the maxillae, which constitute the filter. This current flows in a posterior-antero direction, vibratory movements of the maxillae exerting suction that draws the current forward. This is further filtered by the setae of the maxillules, which [illegible]rovide a very fine screen. When food is particularly [illegible]bundant, *Calanus* slows down its feeding movements [illegible] stops them entirely.

Hemimysis, one of the oppossum shrimps, is micro[illegible]gous by habit but at times becomes predatory and [illegible]cks other animals. When feeding on small par[illegible], either in the plankton or stirred up from the [illegible]m sediments, currents are set in motion by rota[illegible] the distal segments of the thoracic limbs, which [illegible] a series of ellipses so that water is drawn toward [illegible] from all directions. These currents move the [illegible]rward, but water from them is also drawn in [illegible]he bases of the limbs to form a single respira[illegible]d current that moves anteriorly because of the [illegible]n of the maxillae. Particles filtered from the current by the setae on all segments of the maxillae are carried forward to the mandibles and drawn into the mouth by peristaltic contractions of the oesophagus. Some selection is exercised, for bits of material can be seen to be ejected at the sides of the mouth and to be carried away in the streams of filtered and oxygen-depleted water that pass out laterally between the maxillae and maxillules.

Syncarids, living in bodies of freshwater in other parts of the world than North America, likewise employ a maxillary filter to concentrate food particles. *Anaspides* and *Paranaspides*, both native to Tasmania, where the former is found in pools at elevations of 4,000 ft, use different means of drawing food to this filter. *Anaspides*, whose body may reach a length of 5 cm, captures quite large animals, such as worms and tadpoles, with its anterior thoracic limbs, passes them on to the mandibles for maceration, and then filters the pieces through the maxillary filter before ingesting them. *Paranaspides* obtains its food in a manner similar to that of the mysidaceans by creating water currents through movements of the exopodites of its thoracic limbs, which, however, move back and forth and not in an elliptical pattern. While the current they produce is forwardly directed and passes through the maxillary filter, it does not flow along a ventral channel as it is made to do in *Hemimysis*.

Feeding has been observed in the British genus of cumaceans, *Diastylis*, which, like others of its kind, lives in mud or sand along the shores of bodies of salt water, and in the tanaidacean *Apseudes*, which lives in similar habitats. Both of these filter food from respiratory water currents through setae on their maxillae. And the amphipod *Haustorius arenaurius*, which lives in burrows in sandy beaches, is unique among others of its order in that its filtration is performed, and its feeding current produced, by the maxillae rather than the thoracic limbs. These do initiate the respiratory current, but the maxillae alone pump water forward and strain food particles from it, which are then removed by the first pair of thoracic appendages (maxillipeds) and passed on to the mouth.

Maxillules and maxillae provide the setose filter for microphagous ostracods. Many of these are mud dwellers that use their antennae to agitate their surroundings and so to bring particles of debris and other material into the food current. This current is drawn in an antero-postero direction between the valves of the shell, usually by vibration of the epipodite of each maxilla, although this forward pull may be augmented by movements of other appendages as well. The current passes backward over the mouth and particles in it are trapped by setae on the maxillules or the palps of the mandibles, or by both together. In various species of Cypridina in which the events of feeding have been observed, the particles are entangled in a secretion from the labrum before transference to the mouth. In *Asterope* (order Myodocopa), there is no such secretion, and the material filtered by the interlocked setae on both maxillules and mandibles is scraped from them by stiff comb setae on the inner surface of the maxillae and pushed to the mouth by the maxillules.

Some crustaceans have even adapted their preoral appendages to the demands of microphagy. Among these are the gammarid amphipods and the mole crab, *Emerita*, an anomuran. These animals do not create their own food currents but depend upon movements in the water originating in other ways to bring food within reach of their antennules and antennae which furnish the filter. The gammarid *Haploops tubicula*, for example, makes a tube for itself whose opening is just above the surface of the sand. When feeding, the animal moves to the rim and hangs there with its plumose antennules extended in the direction of any current there may be. If this direction changes, the animal shifts its position accordingly, and if the water around it is motionless, it spreads its antennules out horizontally to catch any particles that may drift down from the upper levels. When a sufficiently ample collection of small particles is made, the antennules are drawn into the tube, the material is scraped off them by the gnathopods, and then inserted into the mouth. Other tubiculous species with shorter antennules often sweep them back and forth over the sand, collecting surface debris on them which is then conveyed to the mouth. Another gammarid, *Mephidipella macra*, is free-swimming but feeds only when at rest on the bottom, supported ventral side uppermost by its fifth, sixth, and seventh pereiopods. Its antennae and third and fourth pereiopods are extended as a trapping net into which, in addition to floating particles, bits of debris tossed up from the sand by one of the supporting pereiopods may also be caught. The food is then raked off the filter by the gnathopods and put into the mouth.

Mole or sand crabs of the genus *Emerita* live at tide

level, half buried in the sand, and depend upon the outgoing tide to bring their food to them. Their long, feathery antennae are extended above the sand and catch food particles as the waves move seaward over them. An entire population of *Emerita* will move up and down a beach with the changing tides, always embedding themselves at that level where the water will flow over them for the maximum length of time. The antennal filter is cleared of its catch by alternate passages of the antennae through the mouthparts. *Emerita talpoida*, common along the Atlantic coast of North America, seems wholly dependent upon movement of the waves for its food supply, but the Pacific coast species, *E. analoga*, can feed in quiet water by sweeping its antennae through it.

Some insects practice microphagy during their juvenile lives, using various filtering devices and, in most cases, depending upon natural water movements to bring the particles into contact with them. In consequence of this passivity, microphagous nymphs and larvae that live in running water are usually found oriented upstream, with their filters and mouths directed against the current.

The nymphs of most mayflies are macrophagous and use their mandibles to chew up large pieces of vegetation or to nibble submerged roots and stems of aquatic plants. The nymphs of *Isonychia* (*Chirotenetes*) are an exception, for they are microphagous and sift debris floating in the current through long setae on the anterior surfaces of their forelegs. These, when apposed and in conjunction with the setae of the mouthparts, make an effective sieve from which particles are removed and inserted in the mouth.

Among Diptera, the larvae of some midges and mosquitoes are filter feeders. Larvae of the dixa midges (Dixidae) live near the surface film of pools and puddles of freshwater and in the backwaters of swiftly running streams. Their mouthparts are setose and provide the filter through which microorganisms and small particles of debris are removed from the water. Movement of the mouthparts creates the current; the setae form the filter.

A highly specialized filtration device has developed in larvae of the mosquito *Chaoborus* (*Corethra*) (Fig. VII–24). Adults of this genus of Culicidae are commonly known as "phantom midges" and their larvae, because of their extreme transparency, as "phantom larvae." They are inhabitants of large ponds and lakes everywhere in the United States and feed primarily on the larger members of the zooplankton, such as *Daphnia* and *Cypris*. This food, which is captured by the prehensile antennae and the elongated labrum, is crushed by the mandibles and the pieces put into the mouth but not immediately swallowed. They are first sorted by the pharyngeal basket, a circular fringe of strong bristles at the back of the mouth, in which the material is held while in the process of partial digestion by enzymes discharged from the salivary glands. The softened material is then passed on to the stomach, while the undigestible parts are expelled from the basket when it is everted through the mouth.

Larvae of the blackfly *Simulium* have two conspicuous structures at the anterior end which are involved in feeding. Each of these consists of a stalk, or peduncle, which terminates distally in 30 to 60 long, curved setae. They are used as fans and strain plankton from the shallow waters of swift-running streams where the larvae live, attached to rocks or plants by hooks extending from a flat disk at the posterior tip of the body.

d. EFFICIENCY OF FILTER FEEDING For all microphagous animals the amount of energy that must

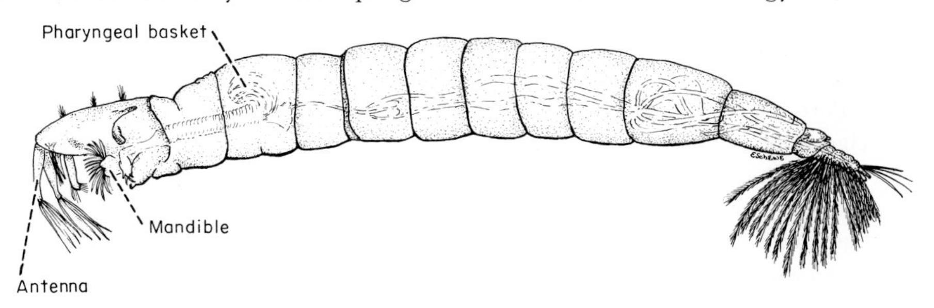

FIG. VII–24 *The "phantom larva" of* Corethra *about ⅔ in. long and almost completely transparent. The antennae are prehensile, capturing prey and passing it on to the mandibles. (After Miall.)*

be expended in obtaining food depends upon the quantity and quality of food available to them, both variable factors. Seasonal and diurnal variations in the distribution of plankton organisms are well-recognized conditions of their life; when and where plankton is plentiful, the sediment below is usually rich in organic material, drifting down from the upper levels. The floor of bodies of water may for other reasons be rich in nutrients, or it may be poor. The poorer it and the upper levels are, the harder must microphagous animals work in order to get enough to eat.

The energy demands are probably heaviest on the filter feeders, especially for those that must create their own food currents. The amount of energy required depends not only upon the abundance of suitable food but also on the efficiency of the filter and the sorting mechanisms in retaining the maximum amount of the nutrient material available and on the metabolic demands of the organism in respect to the operation of its filtration system over and above those for its basal maintenance, its growth, and its reproduction. When food is plentiful, less water has to be passed through the filter to screen an adequate amount than when it is scarce. It is important, too, for sessile filterers at least, to keep the outgoing stream of water well away from the incoming, in order to avoid mixing the depleted and fouled water with that which is bringing in nutrients and oxygen. Sponges do this by ejecting the water from the osculum at a speed considerably greater than that at which the food current enters or circulates within the canals and chambers, so that it is shot well away from the body. In most lamellibranchs, the exhalant siphon is smaller than the inhalant so that the water leaves under some pressure and flows away without diluting the incurrent stream appreciably. A number of tubiculous annelids have solved the problem by making their tubes U-shaped, so that the exit for the filtered water is some distance from the entrance for the food current.

Pore size is an important consideration for the efficiency of a filter. This has been measured, indirectly, in a number of species by testing the ability of their filters to hold back particles of known dimensions brought into contact with them experimentally. By adding proteins of different molecular weights, stained with Evans Blue so that their presence could be detected, to the water in aquaria containing *Chaetopterus* and *Urechis*, it has been shown that their mucous feeding nets will fully retain hemocyanin molecules, with a molecular weight of about 450,000, will partially retain serum globulin, with a molecular weight of about 176,000, and will freely transmit the smaller molecules of ovalbumin, with a molecular weight of about 44,000. It can, therefore, be calculated that the pore size of these nets is about 0.004 μ (40 Å). These filters are thus very efficient and may be even more so than these figures indicate, for the possibility cannot be discounted that strongly basic proteins may be held on them by the physicochemical phenomenon of adsorption, as polyvalent cations are known to be held on the mucous films of lamellibranchs. If this were the case, additional nutrients would be obtained, but there is no evidence that this does occur. Comparable tests of pore size in the mucous nets of *Tendipes* larvae made by passing test particles of known diameter across them, show that the filter will retain all particles with diameters of 17 μ or more, but hold back only a part of those with diameters of 12 μ. This size would, however, be adequate to filter off most members of the plankton and particles of organic debris of diverse origin. By eating the polysaccharide filter as well as its contents, all three of these animals get maximal returns for the work done in its construction and operation.

The diameters of the ostia and water tubes of the oyster *Crassostrea virginica* have been measured by introducing graphite particles of different dimensions into the water surrounding them. These experiments have shown that while particles 1 to 2 μ in diameter pass freely through the openings and tubes, those with diameters of 2 to 3 μ do not and are, therefore, filtered out of the water. The mussel *Mytilus edulis*, on the other hand, filters off the smaller particles as well; therefore, in natural conditions it should obtain a greater amount of particulate organic material for its food than the oyster.

In crustaceans, the distance between the setules when the setae of the filtering appendages are apposed or interlocked determines the porosity of the filter. This has been directly measured in a number of copepods and in *Euphausia superba*. In adult females of *Calanus finmarchicus*, where the final screening is done by the maxillules, the minimum distance between these setules is 5.7 μ; in juveniles at stage V it is 3.8 μ; and in those at an earlier stage (III), it is 3.2 μ. Setae on the thoracic limbs of *E. superba* serve as its filter plate, with a distance of 7 μ between the setules. Particles below these dimensions would, therefore, not be held by the filter and would escape in the outflowing stream and be lost as potential food. All those larger presumably would be retained, but shape as well as size and the chance orientation of long and slender particles on the filter are factors that also influence transmission or retention.

Calculations of the filtration rate have been made for some of the filter feeders. This rate is influenced not only by the quantity and, often, quality of food available but also by the physical factors of temperature, salinity, and pH of the water. The rate, under comparable conditions, and the constancy with which it is maintained are determined by the food require-

TABLE VII-1 Filtration Rates of some Representative Invertebrates

(Calculated as volume passed per hour)

Species	*Volume per mg body N, ml*	*Volume per ml O_2 consumed, liters*	*Volume per body weight*	*Volume per body length, ml*	*Temperature, °C*
Sponges[1]					
Grantia compressa	170	13			18 to 19
Sycon coronatum	170				18 to 19
Halichondria panica	65	14.3			18 to 19
Bivalve Molluscs					
Crassostrea virginica[2]	19 to 48	16			
Ostrea edulis (19 to 39 mm shell length)[3]				200 to 1000	
Venus striatula (18 to 21 mm shell length)[3]				50 to 200	
Mya arenaria (57 to 82 mm shell length)[3]				600 to 1,300	
Mytilus edulis[2]	120 to 100	14 to 15			
Worms					
Urechis[4]		20	30 to 900 ml/g fresh weight		17
Chaetopterus[4]	37				11 to 16
Crustaceans					
Calanus (V and VI juvenile stages)[5]		8	7 to 34 ml/mg dry weight		12.5

1. From clearance of graphite particles.
2. From clearance of graphite particles and of finely ground calcium carbonate and unicellular algae.
3. From clearance of cells of *Phaeodactylum* (alga) labeled with P^{32}.
4. From calculations of volume drawn out of glass tube of known dimensions at each peristaltic wave, and number of waves per minute.
5. From clearance rates of algae of different size (*Lauderia*, diameter 3 μ and *Ditallium*, diameter 60 to 90 μ).

ments of the animal, which can, in turn, be estimated from its metabolic and growth rates. During periods of optimal growth, the food requirements may be three to four times greater than those only for maintenance and survival.

Rates of flow in different animals have meaning only when they have some common denominator, such as body weight or, more precisely, the total nitrogen content of the body or its consumption of oxygen in a given time. Some of these rates, as determined by different means in representative invertebrates, are given in Table VII–1. It is of interest that the rates in animals using different kinds of filters, although divergent, are of the same order of magnitude. So many variables are involved in studies of this kind that consistent values could hardly be expected. The higher rate of filtration in *Ostrea edulis*, for example, may be correlated with the surface area of its gill, which is relatively larger than that of the other two species used in the same study (*Venus striatula* and *Mya arenaria*), or possibly with the pore size of its filter which allows small particles to escape. The wide range of values found for *Urechis* may be attributed to two different methods used in collecting the data. In one, in which lower values were found, the worms were inserted in glass U-tubes from which the water was siphoned out through a tube which led from the aquarium to a graduated cylinder; in the other, the glass tubes opened just above the surface of the sand in the aquaria, and the reduced frictional resistance in this arrangement may account for the higher values obtained.

Filter feeders provide for large turnovers in suspended material. It has, for instance, been estimated that a population of 1 million mussels (*Mytilus edulis*) spread over about $\frac{3}{4}$ acre of rocky shore, during the second year of their lives will pass some 22 million tons of seawater through their filtration systems, a weight comparable to that of the volume of an area of the sea 1 mile square and 25 ft deep. If the suspended material in this water has a concentration of 5 ppm, then the molluscs will remove some 121 tons of particulate material, some of which will shortly be returned as feces, and all of it, ultimately, at the death of the animal.

D. FLUID FEEDING

Ingestion of nutrients in aqueous solution is practiced by some endoparasitic flatworms and nematodes, by some molluscs and ectoparasitic annelids, and by many arthropods, free-living as well as ectoparasitic. All arachnids are fluid feeders, as are a number of insects of different kinds (Thysanoptera, Homoptera, Hymenoptera, Lepidoptera, and Diptera). In most, specialized mechanisms have evolved in connection with the habit of fluid feeding; these are primarily adaptations of the oral cavity, mouthparts, and digestive tract. Modifications of these structures, common to macrophagous species of the same groups, suggest that, although intake of nutrient fluids may have been a primitive means of obtaining food, the habit of fluid feeding in modern species with mouths and oral structures has been secondarily acquired in the course of their respective evolutions. Adaptations of the oral cavity and mouthparts have been directed toward their use in sucking or pumping fluid into the mouth and alimentary canal; in sponging or lapping up exposed fluids such as nectar, honey, and shed blood; and in penetrating the integument of animals and plants to get at internal, or concealed, fluids. Adaptations of the digestive tract have been directed primarily to elimination of excess water, for fluid feeders have a problem similar to that of filter feeders. While the latter have solved this by diverting much of the water in which their food is suspended before taking the food into the alimentary canal, the former have solved it by adaptations of the gut or excretory system that deals with the water after it has been ingested.

Fluid feeders are usually selective and often extremely specific in their choice of food and of its source. Their habits, in respect both to the type of food and to its source, may change, however, during any phase of their lives or from one phase to another. For example, thrips (Thysanoptera) ordinarily feed on the living tissues of plants but occasionally attack aphids and small mites and suck the juices from them. Nectar from flowers, or fluids from leaves or fruits, are the primary food of mosquitoes. Only the females have acquired the habit of piercing the integument of man and other animals to obtain a meal of blood which provides them with the necessary materials for the formation of their eggs. The fact that caged females of *Aedes aegypti* and *Culex tarsalis* can produce viable eggs after ingesting the body fluids of some moths, beetles, and spiders shows that invertebrate hemolymph may be an adequate substitute for vertebrate blood. There is no proof that mosquitoes do this in nature but the fact that in windswept areas they shelter in clumps of dense vegetation where there are many insect larvae implies that they may.

The most conspicuous examples of change in feeding habits from one phase of the life cycle to another are offered by Lepidoptera, whose larvae, with biting and chewing mouthparts, feed exclusively on vegetation while the adults suck nectar from flowers as well as the liquid products of the decomposition of fruits and animal remains.

1. Mechanisms for ingesting exposed (patent) fluids

a. IN ENDOPARASITES Among parasitic flatworms, trematodes (flukes) feed on shed blood and tissue exudates. Observations of the actual mechanisms of their ingestion are difficult to make as the animals cannot be maintained for any length of time outside the bodies of their hosts. It has been reported, however, that intestinal parasites compress the villi, to which they are firmly attached by means of their suckers, by wrapping their bodies around them. When a villus constricted in this way becomes congested with blood, the walls of the capillaries rupture and hemorrhage follows. The fluke swallows the blood as soon as it is shed. It is possible, too, that enzymes are discharged from its body onto the damaged cells of the villus and the products of such digestion thus contribute to the richness of the nutrient fluid.

Feeding by the hookworm *Anclystoma caninum* has been directly observed in loops of intestine drawn out of the abdominal cavity of an anaesthetized dog and slit open to expose the mucosa. In *Anclystoma* the buccal capsule is large and has, at its anterior end, a pair of cutting plates each of which has three hooks. Like other hookworms, *A. caninum* draws a bit of the mucosa into its buccal capsule and holds it firmly there with the hooks. Blood is squeezed out of the vessels and sucked into the gut of the parasite by contractions of the pharyngeal muscles that may occur as often as 120 to 250 times per minute. The blood passes rapidly through the digestive tract where the cells are digested and some fluid absorbed. Other species of intestinal flukes may not attach to the wall at all but suck in the contents of their host's digestive tract or tissue exudates arising from the inflammatory reaction caused by their presence in the gut. Similarly, evidence obtained from examination of the intestinal contents of certain lung worms indicates that they feed entirely on tissue exudates, for the composition of gut content and pulmonary exudate proved to be identical.

b. IN INSECTS The typical mouthparts of insects have been greatly modified in most of the fluid-feeding species and assembled into an elongated proboscis. In Diptera, all of which have, as adults, suctorial mouthparts, the labium is the most conspicuous part of the proboscis. The other trophi are contained, when at rest, in a groove along the anterior, or dorsal, surface of the labium. A corresponding groove on the posterior, or ventral, surface of the labrum forms a roof for this labial groove, or gutter. The labium is usually expanded distally into two fleshy lobes, or labella, believed, on evidence from comparative anatomy, to be derived from the palps. In species that sponge up liquids, many small tubules, open along one side, traverse the exposed surfaces of each labellum. These tubules are supported by chitinous rings which keep their lumina open and, because of their structural similarity to the tracheal tubes, are known as pseudotracheae, although they have nothing to do with respiration. Some of them unite into collecting tubes which converge at the distal end of the labial gutter, but the middle tubes run individually to this point. When the labella are flattened out over a liquid-covered surface, the fluid moves into and along these tubes by capillarity, as it does in the holes of a commercial sponge, and so collects at the tip of the labial

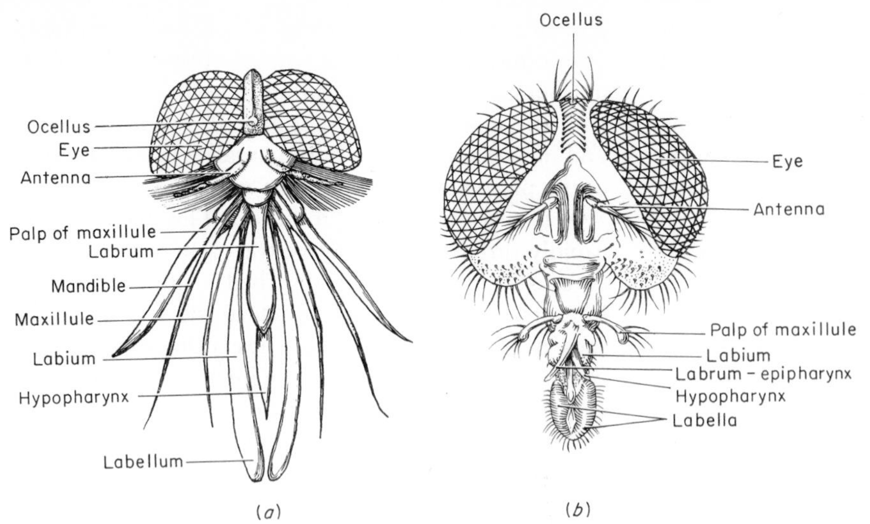

FIG. VII–25 *Insect mouthparts.*
(a) *Mouthparts of* Tabanus, *the horsefly, adapted for biting and sponging up the exposed fluid. (From Gardiner.)*
(b) *Mouthparts of* Musca domestica, *the housefly, adapted for sponging up exposed liquids. (After Ross.)*

gutter. In the horsefly *Tabanus* (Fig. VII–25*a*), equipped both to sponge up surface liquids and to expose the blood of its victims by biting into the integument, the mandibles are slender and tapering; the maxillules, whose lacinia are completely reduced, are represented only by palpi and galeae which, like the mandibles, are long, slender blades with fine teeth on their distal margins. The hypopharynx is also elongated as a sharp stylet with a central canal, through which secretions flow from the salivary glands. Female tabanids, like female mosquitoes, seem to have acquired the habit of feeding on blood secondarily, for the males feed on the honeydew extruded by aphids and coccids and on the juices of flowers. Females also will utilize these sources when blood is not available. In obtaining blood, the female tabanid uses her mandibles to scratch the skin of her victim and then thrusts them, together with the maxillules and hypopharynx, into the integument by a downward movement of her head. The sheathing labium buckles upward as the hard parts of the proboscis are inserted into the wound. Blood flows from the site of injury and does not clot because of an anticoagulant in the secretion of the salivary glands discharged through the hypopharynx. The labella are pressed down over the exposed blood which, as it is sponged up, is drawn into the food channel. The channel is made when the mandibles, now withdrawn from the puncture, are brought together in the midline beneath the labrum and so form a floor for its median groove. Suction is provided by movements of the muscles that control the size of the cibarium, which, therefore, acts as the pump. In the housefly *Musca domestica*, there is further reduction of mouthparts (Fig. VII–25*b*). Mandibles are lacking entirely and the maxillules are reduced to two small palpi and vestiges of the stipes. These flies are unable to cut or to pierce and so depend for their food entirely upon patent fluids.

Reduction and enlargement of certain mouthparts has occurred also in suctorial Hymenoptera and Lepidoptera, but different ones have participated in the construction of the sucking tube. All hymenopterans, whether solid or fluid feeders, have retained their mandibles, but bees and wasps put them to use in the construction of their hives and nests rather than in the maceration of food. The greatest modification of generalized mouthparts is apparent in honeybees, habitual feeders on honey, nectar, and honeydew but occasionally on sugar and other solids (Fig. VII–26). The mandibles of worker bees are larger than those of drones and smaller than those of queens and are used by the workers for handling the balls of pollen obtained from the flowers that they visit and for manipulating the wax in the comb. Each mandible is narrower in the midregion than at the base or the end, where it is expanded into a spoon shape, with the bowl directed inward. This concave face has one basal and two diagonal ridges across its surface. These spoons hold and handle the pollen and, in construction of the comb, grind up the wax which, though secreted in liquid form by the hypodermal wax glands, quickly hardens into little scales corresponding in shape and size to the wax plates through which it has been extruded. The scales are then passed forward by the legs to the mandibles, which cut off little pieces with the sharp edges of their "spoons," grind them into fine powder along the ridges on their inner surface, and apply the powder to the edges of the comb.

The proboscis of the honeybee is formed from the maxillules and the glossae of the labium. The laciniae and palpi of the maxillules are greatly reduced and the galeae are enlarged as long, flat, tapering blades. The palps of the labium, on the other hand, are long, but the paraglossae are only small lobes at the base of the much extended glossae. These are joined together along their mesial surfaces to form a single unit called by beekeepers the "tongue" and by entomologists the glossa, with full recognition of the fact that it is not a single structure but one constructed by the union of two separate units. This tongue is covered with many hairs and expanded distally into a small concave-convex labellum. A longitudinal groove runs along its posterior or ventral surface, whose margins, fringed with stiff hairs, are bent over the depression. It is flexible and contractile and though longer than the labial palps can be drawn back between them. There is no hypopharynx, and the ducts of the salivary glands open between the bases of the paraglossae and so at the proximal end of the tongue. When at rest, the proboscis is folded back under the head but in feeding it is brought forward and its parts pressed together lengthwise in such a way that the blade-like galeae of the maxillules make a roof over the tongue, while the long labial palps fit in along its sides. In this way a tube is constructed and brought into a position level with the mouth opening. The mandibles, surrounding the bases of the maxillules, serve to steady the proboscis and to hold it in position, their only contribution to the feeding process.

Feeding of bees can easily be watched under a dissecting microscope by shutting a bee in a glass tube with a drop of honey. If a little powdered carmine or other harmless dry dye is mixed with the honey, the flow of the liquid through the sucking tube can be followed. When the insect finds the honey, it plunges its proboscis into the drop until the ends of the galeae are submerged. The tip of the tongue is bent upwards and moved rapidly back and forth. Through capillarity the liquid moves into the end of the sucking tube and is continually drawn up it by the pumping action of the pharynx. Bees can also feed on sugar by first dissolving it in a secretion from the salivary glands which is spread over the crystals by the labellum. The saliva is discharged under pressure from the so-

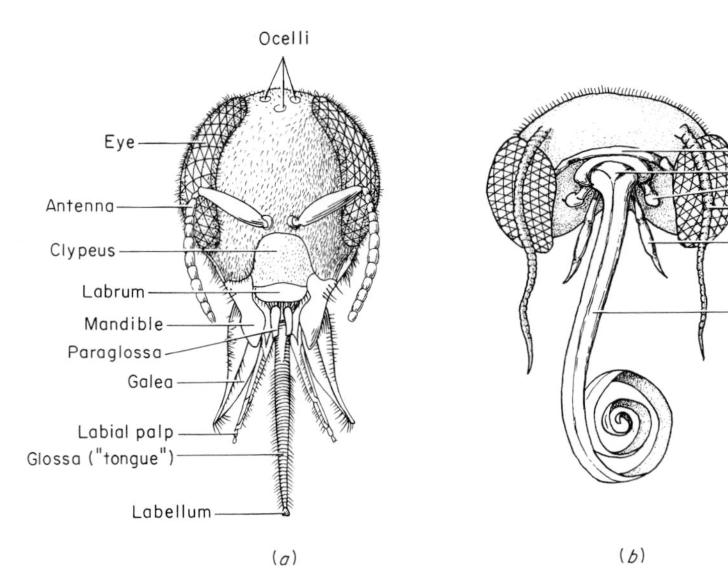

FIG. VII–26 *Insect mouthparts.* **(a)** *Mouthparts of a worker honeybee, adapted for chewing and lapping fluids.* (*After Ross.*) **(b)** *Mouthparts of a butterfly, adapted for sucking up fluids.* (*From Gardiner.*)

called "salivary pump" through the opening of ducts between the bases of the small paraglossae. Flowing through channels beneath the paraglossae, it reaches the groove on the ventral surface of the tongue and runs down it to the labellum, which is used to spread the secretion and also to draw up the dissolved sugar in the same way that honey is drawn up. Spines on the upper surface of the labellum are used to rasp the surface of the sugar crystals and so to expose them more rapidly to the salivary secretions. Imbibed nectar may also be ejected from the mouth, pass to the underside of the proboscis through the paraglossal channels, and then be sucked back again, after mixture with saliva flowing along these same channels. Liquid can, therefore, pass at different times in both directions along the paraglossal channels.

In Lepidoptera, it is the maxillules that have been modified to make a sucking proboscis, and in most species the mandibles are entirely absent (Fig. VII–26*b*). The labrum is a short plate and the labium is equally reduced in size and represented primarily by its palps, between which the proboscis, when at rest, is coiled in a flat spiral like a watch spring. The organ is formed by the greatly elongated galeae of the maxillules, whose laciniae are either very small or wholly absent. Each galea is grooved along its mesial surface and a tube is formed when the two are brought together and held firmly by interlocking hooks and spines. The tube is supported by chitinous rings, made by the apposition of half hoops in each galea. The length of the proboscis varies in different species; it is longest in certain members of Sphingidae, a family that includes hawk and hummingbird moths. In feeding, the proboscis is extended and the open tip of its tube brought into contact with an exposed fluid, which rises in the tube partly by capillarity but primarily by the suction exerted by contraction of the cibarial and pharyngeal muscles, which have combined to make an effective pump.

2. Mechanisms for ingesting internal (concealed) fluids

Blood, sap, or cell juices provide the principal source of food for many nematodes, some gastropods, and many arthropods—and for some species, the only source. To reach these fluids the animal must have means of penetrating through the protecting tissues and devices for abrading the integument. Boring or piercing through it are common among feeders of this kind. Mechanical devices may be supplemented by chemical ones, and enzymatic softening of tissues to facilitate penetration through them or to reduce them to a fluid state before they are sucked into the digestive tract is common practice among the invertebrates that have adopted this feeding habit.

a. IN NEMATODES

There are many species of phytoparasitic nematodes, which feed on the contents of cells either in roots or leaves. They use the hollow stylet to puncture the epidermis of the root in the same way that predaceous species use it to puncture the integument of an animal, and the juices are similarly sucked up by a pharyngeal pump. Some root-infesting species insert only the stylet into the tissue while the body remains outside in the soil; others insert the head and part of the anterior end of the body; and others move their entire bodies through the hole made by the stylet and do their sucking while embedded in a gall whose formation by the plant has been induced by their presence. The leaf-ingesting species seem to insert the stylets through the stomata and so reach the inner tissues without injury to the epidermis. Few plants can survive attack by nematodes, especially in numbers as large as have been reported. As many as 50,000 to 60,000 individuals have been counted on a single leaf of a Chinese primrose, and infestations have been almost as severe on leaves and roots of other plants.

Much of the destruction of crop and ornamental plants is attributable to these rapacious and prolific animals.

The actual food and the feeding mechanisms of zooparasitic nematodes are both open questions. Intestinal parasites, not attached to the mucosa, must feed directly on the gut contents, presumably by sucking it into their mouths, although the possibility of absorption through the body wall cannot be excluded. Identifiable substances, such as barium, charcoal, and starch grains, have been fed to animals infected with ascarids and later recovered from the parasites, providing evidence for their ingestion.

b. IN GASTROPOD MOLLUSCS Some opisthobranch gastropods feed on the contents of the cells of algae and may be very specific in the source of their food. *Hemaea bifida*, for example, selects the red alga *Griffithsia*, while *H. dendritica* prefers the green algae *Codium* and *Eryopsis*, which are also attacked by *Elysia viridis* and *Caliphylla mediterranea*. *Vaucheria* and *Rhizoclonium* are the preferred food of *Alderia modesta* and *Limapontia depressa*, but *L. capitata* feeds on two other genera, *Enteromorpha* and *Cladophora arcta*. A different species of *Cladophora, C. cocksi*, however, is the chosen food of *Acteonia*. In feeding, the alga is passed across the mouth and the radula is used to slit open individual cells in the filament or the thallus. In these snails, the radula has but a single tooth in each row. These are used one at a time to make the cuts through the cellulose walls, cell by cell. The contents of the vacuole is thus made accessible and drawn into the mouth by pumping action of the pharyngeal muscles.

Other fluid-feeding gastropods derive their nutriment from animals and feed on the blood and body fluids of sedentary worms, bivalve molluscs, crustaceans, and echinoderms. They seem also to be selective in the source of their food, since certain species are regularly found in association with particular hosts and virtually ectoparasitic upon them. Two species of *Odostomia* (*O. rissordes* and *O. scalaris*) feed most frequently upon *Mytilus edulis*, but three other species (*O. lukisii, O. unidentata,* and *O. plicata*) select the polychaete *Pomatoceros*. *Odostomia eulimoides* chooses scallops and *O. trifida*, the clam *Mya arenaria*. Members of the family Eulimidae feed on echinoderms only. None of these gastropods has a radula, and in eulimids there is no mechanism for cutting or tearing the integument. They have, however, glands in the proboscis that elaborate and discharge digestive enzymes upon any exposed soft tissues of the echinoderm, and the snail draws the resulting nutrient fluid into its mouth by means of its pharyngeal pump. The pyramidellids, which feed on bivalves, crustaceans, and worms, are provided with long, hollow chitinous stylets. These are developments of the jaws, or chitinous thickenings of the side walls of the buccal cavity common to all gastropods. In the fluid-feeding pyramidellids, the stylets are elongated and apposed to each other and lie in a tube formed from the soft margins of the buccal cavity. The cavity of the stylets is divided to make an upper tube, through which the juices are sucked, and a lower one, through which the salivary secretions flow from the mouth into the wound made by the stylets.

c. IN ANNELIDS Among annelids, certain leeches are fluid feeders; and there are some ectoparasitic polychaetes that are bloodsuckers; and a few species of oligochaetes, regularly found on earthworms' bodies, are believed to feed on their epidermal secretions. Leeches inhabit freshwaters, where they live in ponds, lakes, quiet streams, and marshes, and in the oceans, usually as ectoparasites upon fishes, sea turtles, crustaceans, and even dolphins. Some of them are quite specific in their hosts, others more general. Some, like the ectoparasitic oligochaetes, feed on surface mucus, blood, and tissue fluids from open wounds, which they draw up through their proboscides by means of a pharyngeal pump. Others (rhynchobdellids) that feed on vertebrate blood can reach it presumably by virtue of enzymes, discharged from a pore at the tip of the proboscis, that digest the integument and blood vessel walls. Only certain genera of the Hirudidae, or true bloodsuckers, are equipped with means for cutting through the tissues. In the oral cavity of these leeches are three cutting plates, or "jaws." These are oval in shape with sharp protruding serrated edges and are arranged in a triangular pattern, one dorsal in position and two lateral. In feeding, the leech attaches itself to the surface of the animal to be its food source by a sucker at the posterior end of its body and moves its anterior end about until a suitably vulnerable place on the integument is found. It then attaches itself firmly there by its anterior sucker, which surrounds the mouth, and brings its jaws into action. These are moved backward and forward by muscles and make three fine slits; a leech bite is characteristically triradiate. The blood, as it flows from the incisions, is prevented from clotting by an anticoagulant called hirudin that is contained in the secretions from the leech's salivary glands and discharged from the oral cavity when the incisions are made. The blood is then sucked up by the action of muscles attached to the pharyngeal and body walls. Bloodsucking leeches are intermittent feeders and after a full meal, which may increase their body weight fivefold, need not eat again for months. Well-fed leeches have been kept alive in the laboratory for as long as 2 years without having an additional meal.

The habits of the fluid-feeding polychaete *Ichthyotomus sanguinarius*, a syllid about 7 to 10 mm long, have been most carefully studied. It has been found attached to the fins, and most frequently the dorsal fin, of the eel *Myrus vulgaris*, taken from the Bay of Naples. This does not seem to be a specific association but one attributable to the habits of the eel, for any slow-moving fish, or one that remains stationary for any length of time, may be attacked. *Ichthyotomus* penetrates the skin and attaches itself, after penetration, by means of its jaws which have become highly modified as stylets. These, instead of being the crescentic structures found, for example, in *Neanthes*, are expanded distally into a "spoon," with toothed margins. These spoons are held together when the puncture is made but, once through the integument, are rotated so that the teeth are directed outward and moved apart like the blades of a pair of scissors. The skin of the host is torn as the stylets separate, and their final position in the tissues makes a very firm attachment for the worm, operating on the same principle as an expansion bolt. Clotting of the blood is prevented by secretions from two hemophilous glands in the anterior, or head, segments. These are thought to be modifications, at least as far as their secretions go, of cephalic glands found in worms that feed in other ways. In predatory species, they produce a poisonous, or toxic, substance.

d. IN ARTHROPODS In mandibulate arthropods, the mouthparts may be modified as piercing organs as well as suctorial tubes. There are a few ectoparasitic crustaceans that draw their nutriment from the blood or body fluids of fish or crinoids and also from other crustaceans. In these species, one or another pair, or pairs, of mouthparts are modified as long blades which puncture the integument, while the labrum and labium make the sucking tube. The copepod *Lepeophtheirus pectoralis* uses its antennae, maxillae, and maxillipeds to scrape and tear at the scaly surface of the fish from which it obtains its food. It then inserts its sharp mandibles and sucks up the blood. There are many fluid-feeding isopods that pierce the body wall of their hosts with mandibles and maxillules developed as stylets and suck the blood or tissue fluids through a labrum-labial tube. The cephalogaster, an enlargement on the dorsal surface of the head region, is believed to contribute to the suction.

Many insects have mouthparts adapted for piercing, as well as for sucking, and with them penetrate to the vascular tissues of plants as well as of animals. Aphids, coccids, and whiteflies aim their stylets with great precision at the phloem, which provides them with a fluid especially rich in sugars, but containing amino acids as well as other products of the plant's syntheses. Leafhoppers and cicadas strike either the phloem or the xylem vessels, from which they obtain a mixture of mineral salts and amino acids in more dilute solution than in the phloem vessels. In aphids, as in other hemipterans, the labium is extended as a rostrum (Fig. VII–27*a*). This may be of considerable length and in some species is longer than the main body of the insect. It is grooved dorsally and so makes a sheath for the two pairs of long, pointed stylets derived from the mandibles and maxillules. Each of the maxillary stylets is concave on its inner face, with the concavity divided by a longitudinal ridge, so that when they are apposed two very fine tubes are formed. The dorsal one of these is the food channel; the ventral one is an ejection duct through which salivary secretions pass outward. These secretions, when injected into the plant tissues, solidify and make a sheath around the inserted stylets. There are different physical problems involved in feeding from phloem or from xylem vessels. In the phloem, fluid moves under positive pressure, but in the xylem under negative pressure, because the principal force causing the ascent of sap is the transpiration pull. Phloem feeders, therefore, do not need to exert any suction, for the sap literally pours into their feeding tubes, and all they need to do is to swallow it. This has been demonstrated by cutting off the head of a feeding aphid and observing the flow of sap from the

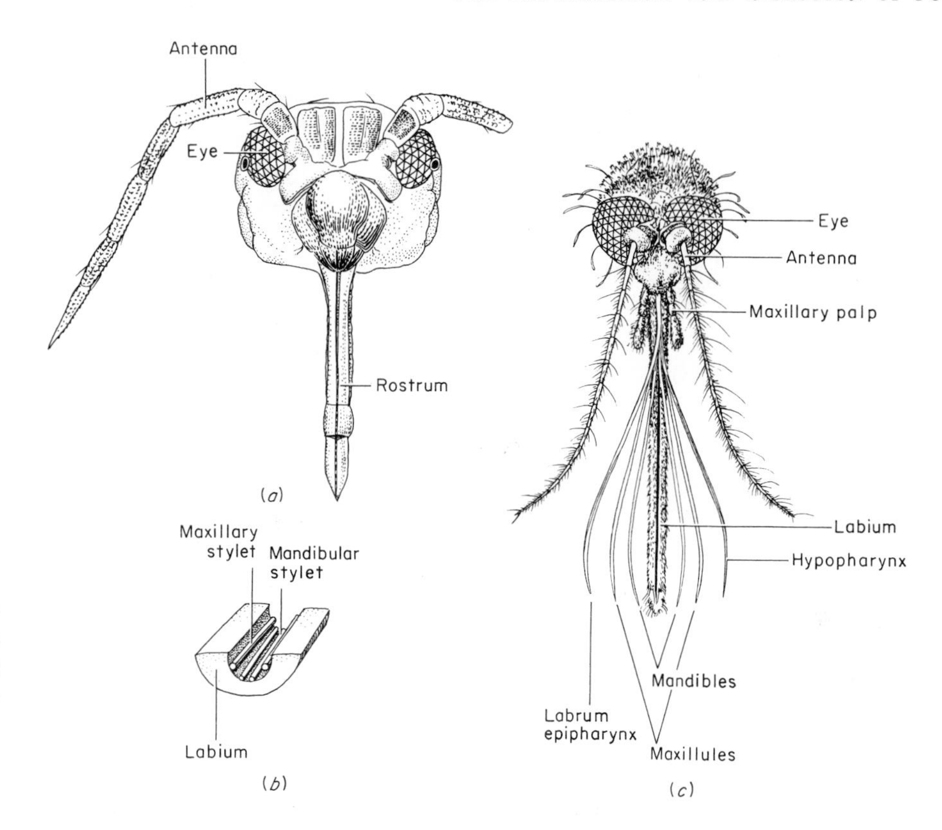

FIG. VII–27 *Insect mouthparts.* ***(a)*** *Head of an aphid, showing the rostrum. (After Esau.)* ***(b)*** *Section through the rostrum, showing the dorsal groove of the labium which encloses the mandibular and maxillary stylets.* ***(c)*** *Mouthparts of a female mosquito. (After Ross.)*

cut ends of the stylets left behind in the plant. Reasonably large collections of sap can be made from them, a situation that has been used to advantage in obtaining material for investigations of the composition of fluids in the phloem vessels. Species that feed on xylem must, on the other hand, pump up the fluid, and do so by action of their cibarial muscles.

The piercing-sucking mouthparts of Diptera are illustrated at their greatest development by those of the adult female mosquito (Fig. VII–27*c*). In *Culex*, for example, the labrum and epipharynx are combined into a long, sharp spear with a central groove; the mandibles and the galeae of the maxillules are similarly extended. The hypopharynx is long, slender and somewhat flattened. All of these are held together in the groove of the equally long labium which forms a sheath for them. The food channel is made by the apposition of labrum-epipharynx and hypopharynx. In feeding on animal blood, the tip of the labium, which is covered with fine sensory hairs, is first brought into contact with the skin. The mandibles and maxillules then make the puncture, and the labrum-epipharynx is inserted into the hole. Blood is then drawn into the mouth by the cibarial pump.

Chelicerate arthropods are all fluid feeders, for unlike the others in the phylum, they have not diverted any of their appendages to the function of mastication nor developed any new ones for this purpose. Their food is caught and held by the chelicerae, considered homologous with the antennae of crustaceans and insects (Fig. VII–28*a* and *b*). In scorpions (Scorpionida), harvestmen and daddy-longlegs (Phalangida), and many species of ticks and mites (Acarina), each chelicera has three segments; in solpugids, spiders (Araneida), and a few ticks and mites, there are two. The terminal segment is usually a pincer with one movable and one immovable process (Fig. VII–28*c* and *f*). The chelicera of spiders have only one process, which is movable and can be drawn back against the first or basal segment of the limb. In all but one family, this appendage contains a poison gland, whose duct opens near the tip of the fang, on its

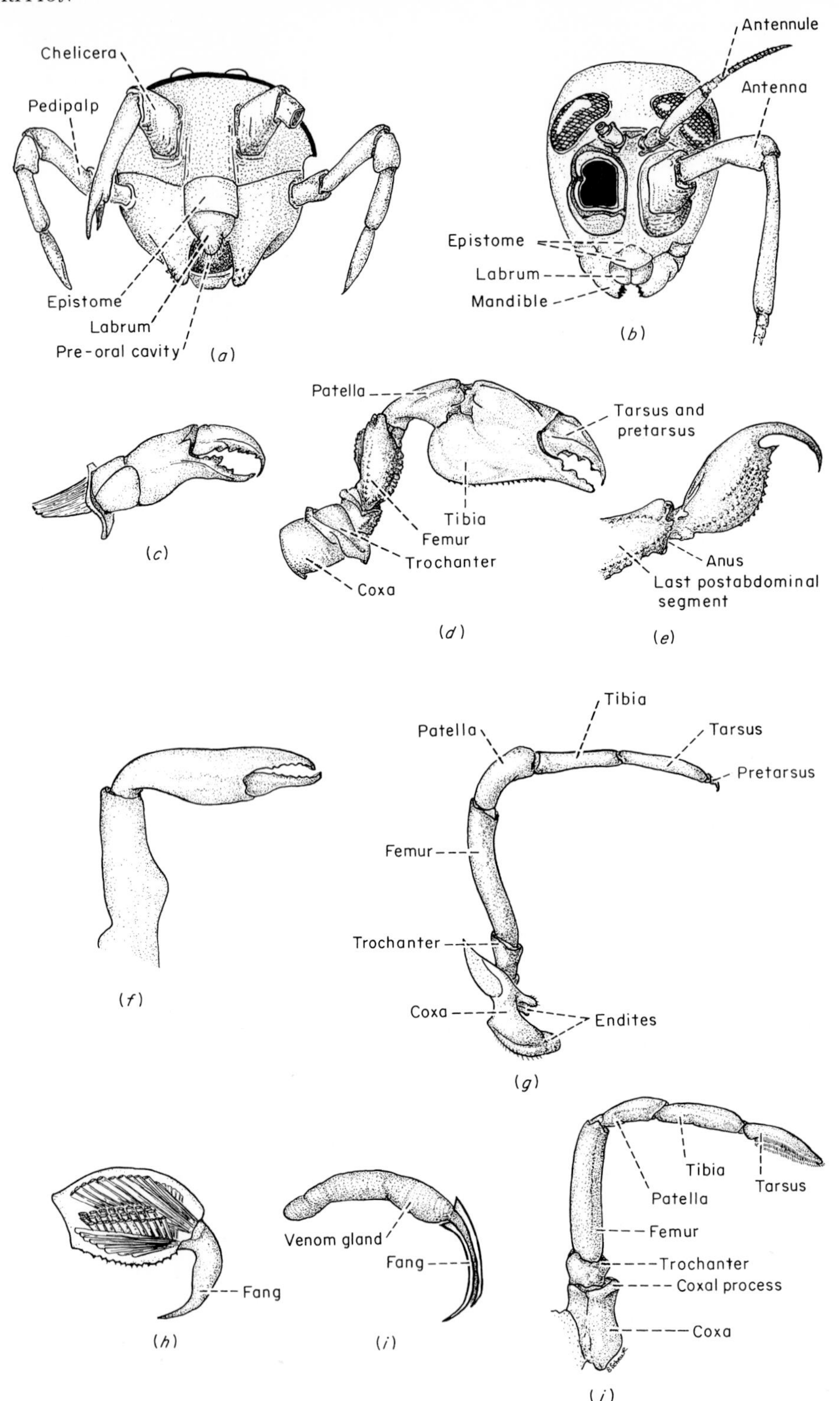

FIG. VII–28 *Feeding organs of arachnids. (After Snodgrass.)* ***(a)*** *Head of an arachnid, anterior view.* ***(b)*** *Head of an amphipod crustacean,* Talorchestia longicornis, *showing antennae in position of arachnid chelicerae.* ***(c)*** *Chelicera of scorpion* Pandinus. ***(d)*** *Pedipalp of* Pandinus. ***(e)*** *Sting of the scorpion* Centroides. ***(f)*** *Chelicera of the phalangid* Holosiro, *showing fixed and movable dactyl.* ***(g)*** *Pedipalp of the phalangid* Leiobunum. ***(h)*** *Chelicera of the tarantula* Eurypelma, *with no fixed dactyl.* ***(i)*** *Cheliceral fang and venom gland of* Eurypelma. ***(j)*** *Pedipalp of* Eurypelma.

convex side (Fig. VII–28*h* and *i*). In ticks and mites the termination of the chelicera is extremely variable; it may be a claw, a hook, a chisel, a saw, or, in blood-sucking species, a stylet that can penetrate through the integument of the host. The second pair of arachnid appendages, the pedipalps, are homologous with crustacean and insect mandibles but quite different from them in construction and in use. They have six segments, like the walking legs, and in many respects are similar to them (Fig. VII–28*g* and *j*). In scorpions they are chelate, the claw being formed from the fifth and sixth segments (tibia and tarsus, respectively) (Fig. VII–28*d*). The coxae of the pedipalps form the sides of the preoral cavity, which is overhung by the labrum; the sternum of the second somite, or some structure that has replaced it, makes the floor. There is no counterpart of the insect labium.

In feeding, predaceous arachnids grasp the quarry in their chelicerae, or in their pedipalps if these are chelate, and crush and tear it so that the soft parts are exposed (Fig. VII–29). Scorpions, like spiders, have poison glands and narcotize or kill their prey by injecting the venom into them. But the sting of a scorpion lies at the tip of the abdomen, presumably an appendage of the terminal segment to which it is articulated in such a way that it can be moved both vertically and laterally (Fig. VII–28*e*). The two poison glands lie in the swollen base of the sting and open individually near its tip. All arachnids suck up the exposed fluids of their prey by action of the pharyngeal muscles, which in most cases is augmented by that of the "stomach pump," actually an enlargement of the posterior part of the oesophagus. Phalangids alone ingest solid material as well as fluids, for pieces recognizable as parts of their prey have been found in their digestive tracts. The consumption of a spider, enclosed in a test tube with the phalangid *Platybunus corniger*, was observed under a microscope. The phalangid held the spider with its chelicerae and slit open the abdomen with one of them. Then, while one pincer held the cut open, the other dipped into the body repeatedly and pulled out pieces that it conveyed to the mouth, into which they were passed by the coxal endites of the pedipalps. When all of the soft parts had been eaten, the phalangid drew its chelicerae and pedipalps through its mouth and cleaned off any material remaining on them.

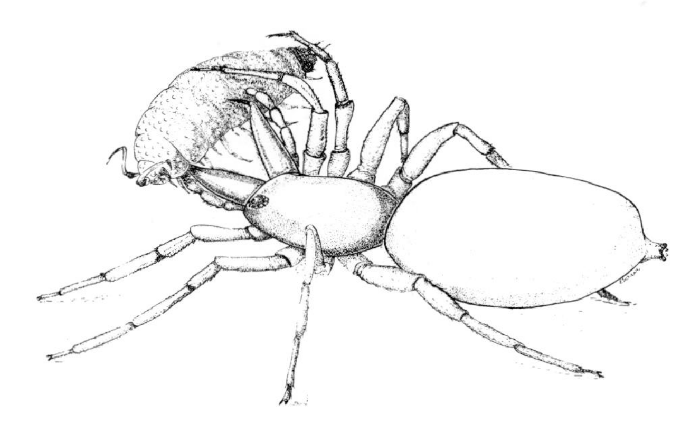

FIG. VII–29 *Predaceous spider. (After Bristowe.)* Dysdera *grasping a woodlouse (isopod crustacean) with its fang-like chelicerae, which not only hold the spider's victim but also pierce its chitinous covering.*

Some spiders, particularly those two species that feed on hard-shelled insects, like beetles, pierce the covering with their chelicerae and suck out the fluids directly. But most of them rely heavily on external digestion, reducing all the tissues of their prey that they can to a fluid state and sucking up the resultant nutrient liquid. The digestion is accomplished mainly by enzymes secreted by glands in the stomach and expelled through the preoral cavity onto the prey. These enzymes have been shown to be primarily proteinases and lipases and can complete the disintegration of the exposed parts of an animal, even one of considerable size, in a few hours. There are also glands in the coxae of the pedipalps that discharge their secretions into the preoral cavity, but tests of the amount and potency of their secretions, in comparison to those of the stomach, make it evident that they play a small and slow role in external digestion. In feeding in this way, a spider plunges its chelicerae into the body of its victim, narcotizes or poisons it, and then uses the chelicerae to tear the tissues apart so that the large drops of the spider's stomach secretions reach them all. The digested material is then sucked back by the spider and more enzymes expelled. The spider also uses its chelicerae to knead and squeeze the body of the prey so that no digestible tissue is lost. It has been reported that in $2\frac{1}{2}$ hr the body of a lizard $1\frac{3}{4}$ in. long was reduced by a South African spider to a small dry mass about $\frac{1}{4}$ in. in diameter. This the spider allowed to drop from its chelicerae.

Some spiders are highly selective in their choice of food and reject any food that is distasteful to them by spitting it out and then wiping their mouths against a leaf or a blade of grass. A patch of sensory epithelium in the gullet seems to act as the organ of taste, for substances that come in contact with this area are either accepted or rejected. A tarsal organ on the last segment of the legs or the palps is also chemosensitive and is used to test edible prey or potable water. It has been shown experimentally that a spider will drink only when water is in contact with the tarsal segment and not with any other surface of the body, and that it will not drink if this tarsal segment is coated over or sealed before the drop of water is applied. Further, tests have shown that this organ enables a spider to discriminate between water, a 0.7 percent solution of saccharine, a saturated solution of sodium chloride, and a 0.5 percent solution of quinine sulfate.

Mites and ticks (Acarina) are a very heterogeneous group, and one of wide distribution. They inhabit the littoral zones of the seas, freshwaters, and the land, where they live in soil, in humus and litter, and on plants of all kinds. They feed on plants and other arthropods and small animals. Ticks especially feed on the blood of vertebrates. Often they are cannibalistic. Their bodies are small and, unlike those of other arachnids, the prosoma is divisible into a more or less clearly defined gnathosoma, or capitulum, bearing the mouthparts (labrum, chelicerae, and pedipalps) and a podosoma bearing the legs. The prosoma is extended dorsally, above the chelicerae, as the epistome and ventrally as a buccal cone, within which lie the preoral cavity and the mouth. The coxae enclose this chamber, as they do in the other arachnids. The palps consist of three to five segments and vary greatly in form in different species. In some, they are small and slender and function as tactoreceptors. In others, they are large and are used to capture prey and to hold it after capture. The chelicerae also show many different modifications of the basic pattern. These can, however, be resolved into two main types, for either they are pincers or they are claws (hooks). The jaws of the pincers may be stout or in the form of curved needles, but they are always toothed. They are used to seize the food, as well as to tear it, or to lacerate the integument of the animal from which blood is drawn. Solid food is digested externally, or in the preoral chamber, by enzymes discharged from the mouth. Suction of the nutrient fluids, whatever their source, is accomplished as in other arachnids by muscular movements of the pharynx.

SELECTED BIBLIOGRAPHY

Microphagy

Allen, J. D.: Preliminary Experiments of the Feeding and Excretion of Bivalves, Using *Phaeodactylum* labelled with ^{32}P, *J. Mar. Biol. Assoc. U.K.*, vol. 42, pp. 609–623, 1962.

Atkins, D.: The Ciliary Feeding Mechanisms of the Entoproct Polyzoa and a Comparison with That of the Ectoproct Polyzoa, *Q. J. Microsc. Sci.*, vol. 75, pp. 393–424, 1932.

Atkins, D.: On the Ciliary Mechanisms and Inter-relationships of Lamellibranchs. I. Sorting Mechanisms, *Q. J. Microsc. Sci.*, vol. 79, pp. 181–308, 1936; II. Sorting Devices on Gills, pp. 339–374; III. Types of Lamellibranch Gills and Their Food Currents, pp. 375–422; IV. Cuticular Fusion, pp. 423–446.

Atkins, D.: On the Ciliary Mechanisms and Inter-relationships of Lamellibranchs. VI. The Pattern of the Lateral Ciliated Cells of the Gill Filaments of the Lamellibranchia, *Q. J. Microsc. Sci.*, vol. 80, pp. 321–435, 1938.

Atkins, D.: Ciliary Feeding Mechanisms of Branchiopods, *Nature*, vol. 177, pp. 706–707, 1956.

Cannon, H. B., and S. M. Manton: On the Feeding Mechanism of a Mysid Crustacean, *Hemimysis lamornae, Trans. R. Soc. Edinburgh*, vol. 55, pp. 219–255, 1927.

Dales, R. Phillips: Feeding and Digestion in Terebellid Polychaetes, *J. Mar. Biol. Assoc. U.K.*, vol. 34, pp. 55–79, 1955.

Duerden, J. E.: The Role of Mucus in Corals, *Q. J. Microsc. Sci.*, vol. 49, p. 591–614, 1906.

Fontaine, A. R., and Fu-Shiang Chia: Echinoderms: An Autoradiographic Study of Assimilation of Dissolved Organic Molecules, *Science*, vol. 161, pp. 1153–1155, 1968.

Fryer, G.: The Feeding Mechanisms of Some Fresh Water Cyclopoid Copepods, *Proc. Zool. Soc. London,* vol. 129, pp. 1–25, 1957.

Jørgensen, C. B.: The Rate of Feeding by *Mytilus* in Different Kinds of Suspension, *J. Mar. Biol. Assoc. U.K.,* vol. 28, pp. 333–344, 1949.

Jørgensen, C. B.: Quantitative Aspects of Filter Feeding in Invertebrates, *Biol. Rev.,* vol. 30, pp. 391–455, 1955.

MacGinitie, G. E.: The Method of Feeding of *Chaetopterus, Biol. Bull.,* vol. 77, pp. 115–118, 1939.

MacGinitie, G. E.: On the Method of Feeding of Four Pelecypods, *Biol. Bull.,* vol. 80, pp. 18–25, 1941.

MacGinitie, G. E.: The Size of the Mesh Openings in Mucous Feeding Nets of Marine Animals, *Biol. Bull.,* vol. 88, pp. 107–111, 1945.

Marshall, S. M., and A. P. Orr: On the Biology of *Calanus finmarchicus.* VIII. Food Uptake, Assimilation and Excretion in Adult and Stage V *Calanus, J. Mar. Biol. Assoc. U.K.,* vol. 34, pp. 495–529, 1955.

Orton, J. H.: An Account of the Natural History of the Slipper Limpet (*Crepidula fornicata*), *J. Mar. Biol. Assoc. U.K.,* vol. 9, pp. 437–443, 1912.

Orton, J. H.: The Mode of Feeding of *Crepidula,* with an Account of the Current-producing Mechanisms in the Mantle Cavity and Some Remarks on the Mode of Feeding in Gastropods and Lamellibranchs, *J. Mar. Biol. Assoc. U.K.,* vol. 9, pp. 444–478, 1912.

Orton, J. H.: On the Ciliary Mechanisms in Brachiopods and Some Polychaetes, *J. Mar. Biol. Assoc. U.K.,* vol. 10 (N.S.), pp. 283–311, 1914.

Orton, J. H.: The Mode of Feeding of the Jellyfish *Aurelia aurita* on the Smaller Organisms of the Plankton, *Nature,* vol. 110, pp. 178–179, 1922.

Paine, R. T.: Filter Feeding Pattern and Local Distribution of the Brachiopod *Discinisca strigata, Biol. Bull.,* vol. 123, pp. 597–604, 1962.

Pilson, M. E. Q., and P. B. Taylor: Hole Drilling by *Octopus, Science,* vol. 134, pp. 1366–1368, 1961.

Snodgrass, R. E.: The Feeding Organs of Arachnids, Including Mites and Ticks, *Smithsonian Inst. Misc. Collect.,* vol. 110, no. 10, pp. 1–87, 1948.

Southward, A. J.: Observations on the Ciliary Currents of the Jellyfish *Aurelia aurita, J. Mar. Biol. Assoc. U.K.,* vol. 34, pp. 201–216, 1955.

Stephens, G. C.: Uptake of Organic Material by Aquatic Invertebrates. I. Uptake of Glucose by the Solitary Coral *Fungia scutaria, Biol. Bull.,* vol. 123, pp. 648–659, 1962.

Stephens, G. C., and R. A. Virkar: Uptake of Organic Material by Aquatic Invertebrates. IV. The Influence of Salinity on the Uptake of Amino Acids by the Brittle Star *Ophiactis arenosa, Biol. Bull.,* vol. 131, pp. 172–185, 1966.

Taylor, A. G.: The Direct Uptake of Amino Acids and Other Small Molecules from Sea Water by *Nereis virens* Sars, *Comp. Biochem. Physiol.,* vol. 29, pp. 243–250, 1969.

Macrophagy

Barnard, J. L., and D. J. Reish: First Discovery of Marine Wood Boring Copepods, *Science,* vol. 125, p. 236, 1957.

Bovée, E. C.: Studies of Feeding Behavior of Amebas. I. Ingestion of Thecate Rhizopods and Flagellates by Verrucosid Amebas, Particularly *Thecameba sphaeronucleolus, J. Protozool.,* vol. 7, pp. 55–60, 1960.

Brown, H. P., and M. M. Jenkins: A Protozoan (*Dileptus*: Ciliata) Predatory upon Metazoa, *Science,* vol. 136, p. 710, 1962.

Burnett, A. L., R. Davidson, and P. Wiernik: On the Presence of a Feeding Hormone in the Nematocyst of *Hydra pirardi, Biol. Bull.,* vol. 125, pp. 226–234, 1963.

Carriker, M. R.: Comparative Functional Morphology of Boring Mechanisms in Gastropods, *Am. Zool.,* vol. 1, pp. 263–266, 1961.

Carriker, M. R., and D. van Zandt: Gastropod *Urosalpinx*: pH of Accessory Boring Organ while Boring, *Science,* vol. 158, pp. 920–922, 1967.

Christianson, R. G., and J. M. Marshall: A Study of Phagocytosis in the Ameba *Chaos chaos, J. Cell Biol.,* vol. 25, pp. 443–457, 1965.

Fontaine, A. R.: The Feeding Mechanisms of the Ophiuroid *Ophiocomma nigra, J. Mar. Biol. Assoc. U.K.,* vol. 45, pp. 373–385, 1965.

Fox, H. M.: Organic Matter in Natural Waters in Suspension or Adsorbed on Solid Surfaces, *Sci. Mon.,* vol. 80, pp. 256–259, 1955.

Hanks, J. E.: The Rate of Feeding of the Common Oyster Drill *Urosalpinx cinerea* Say at Controlled Water Temperatures, *Biol. Bull.,* vol. 112, pp. 330–335, 1957.

Hunt, O. D.: The Food of the Bottom Fauna of the Plymouth Fishing Grounds, *J. Mar, Biol. Assoc. U.K.,* vol. 13 (N.S.), pp. 560–599, 1925.

Jones, E. C.: *Tremoctopus violaceus* Uses *Physalia* Tentacles as Weapons, *Science,* vol. 139, pp. 764–766, 1963.

Kitching, J. A.: On Suction in Suctoria, *Proc. Symph. Colston Res. Soc.,* 7, pp, 197–203, 1954.

Knowlton, R. E., and J. M. Moulton: Sound Production in the Snapping Shrimps *Alpheus* (*Crangon*) and *Synalpheus, Biol. Bull.,* vol. 125, pp. 311–331, 1963.

Kohn, A. J.: Chemoreception in Gastropod Molluscs, *Am. Zool.*, vol. 1, pp. 291–308, 1961.

Miller, R. C.: The Boring Mechanism of *Teredo, Univ. Calif. (Berkeley) Publ. Zool.*, vol. 26, p. 41, 1924.

Myers, S. P., and E. S. Reynolds: Incidence of Marine Fungi in Relation to Wood Borer Attack, *Science*, vol. 126, p. 69, 1957.

Naylor, E.: The Diet and Feeding Mechanism of *Idotea*, *J. Mar. Biol. Assoc. U.K.*, vol. 34, pp. 347–355, 1955.

Ray, D. L., and D. E. Stuntz: Possible Relation between Marine Fungi and Limnoria Attack on Submerged Wood, *Science*, vol. 129, pp. 93–94, 1959.

Rudzinski, M.: The Fine Structure and Function of the Tentacle in *Tokophrya infusionum*, *J. Cell Biol.*, vol. 25, pp. 459–477, 1965.

Springer, V. G., and E. R. Beeman: Penetration of Lead by the Wood Piddock *Martesia striata, Science*, vol. 131, pp. 1378–1379, 1960.

Stephens, W. M.: The Snail That Floats on a Raft, *Sea Front.*, vol. 10, pp. 224–229, 1964.

Wyman, R.: Notes on the Behavior of the Hydroid *Corymorpha palma, Am. Zool.*, vol. 5, pp. 494–497, 1965.

Yagin, R., and Y. Shegenaka: Electron Microscopy of the Ectoplasm and the Proboscis in *Didinium nasutum*, *J. Protozool.*, vol. 12, pp. 363–380, 1965.

Yonge, C. M.: Feeding Mechanisms in Invertebrates, *Biol. Revs.*, vol. 3, pp. 21–76, 1928.

chapter VIII

DIGESTION

A. SITES OF DIGESTION: EXTRACELLULAR (LUMINAR) AND INTRACELLULAR

B. FUNCTIONAL AREAS OF COMPLETE DIGESTIVE TRACTS

C. EXTRACELLULAR DIGESTIVE ENZYMES

D. UNUSUAL ENZYMES AND DIGESTIVE PROCESSES

- **1. Digestion of structural polysaccharides**
    - a. CELLULOSE
    - b. ALGAL POLYSACCHARIDES
    - c. CHITIN
- **2. Digestion of scleroproteins**
    - a. KERATIN
    - b. COLLAGEN
- **3. Digestion of wax**

E. INTRACELLULAR DIGESTION

- **1. Endocytosis**
- **2. Lysosome concept and lysosomal enzymes**

F. INVERTEBRATES WHOSE DIGESTION IS EXCLUSIVELY INTRACELLULAR: PROTOZOA AND PORIFERA

G. INVERTEBRATES WHOSE DIGESTION IS BOTH INTRACELLULAR AND LUMINAR

- **1. Evolution of luminar digestion**
- **2. Cnidaria and Ctenophora**
    - a. ANATOMY OF DIGESTIVE SYSTEM
    - b. DIGESTION IN *HYDRA*
    - c. DIGESTION IN ANTHOZOANS AND SCYPHOZOANS
- **3. Platyhelminthes**
    - a. GENERAL ANATOMY OF DIGESTIVE SYSTEM
    - b. DIGESTION IN TURBELLARIANS
    - c. DIGESTION IN TREMATODES
- **4. Nemertinea: anatomy of digestive system**
- **5. Endoprocta**
- **6. Brachiopoda**

7. **Mollusca**
   a. GENERAL ANATOMY OF DIGESTIVE SYSTEM
   b. PROTOSTYLE AND EVOLUTION OF LAMELLIBRANCH CRYSTALLINE STYLE
   c. MIDGUT (DIGESTIVE) GLANDS
   d. AMEBOCYTES AS AGENTS OF DIGESTION
   e. COURSE OF DIGESTION
8. **Annelida**
9. **Arthropoda: Arachnida and Acarina**

H. INVERTEBRATES WHOSE DIGESTION IS EXCLUSIVELY LUMINAR

1. **Advantages of luminar digestion**
2. **"Aschelminthes": Anatomy of digestive system in rotifers and nematodes**
3. **Ectoprocta**
4. **Mollusca: Cephalopods and gastropods**
   a. DIGESTION IN GASTROPODS: *POMACEA; APLYSIA*
   b. DIGESTION IN CEPHALOPODS: SQUID
5. **Echiuroidea**
6. **Annelida**
   a. ANATOMY OF THE DIGESTIVE SYSTEM IN POLYCHAETES AND OLIGOCHAETES
   b. DIGESTION IN *SABELLA; AMPHITRITE*
   c. DIGESTION IN *LUMBRICUS:* CALCIFEROUS GLANDS
   d. DIGESTION IN *EUTYPHOEUS*: "HEPATO-PANCREATIC GLANDS"
   e. DIGESTION IN LEECHES
7. **Onychophora**
8. **Arthropoda**
   a. GENERAL ANATOMY OF DIGESTIVE SYSTEM
   b. DIGESTION IN ARACHNIDS
   c. DIGESTION IN DECAPOD CRUSTACEANS
   d. DIGESTION IN ISOPODS: *LIMNORIA AND LIGIA*
   e. DIGESTION IN AMPHIPODS: *COROPHIUM*
   f. DIGESTION IN INSECTS
      (1) Diversity of food and eating habits
      (2) Modifications of digestive systems as adaptations to food and eating habits
      (3) Digestion in certain homopterans, dipterans, lepidopterans, coleopterans

9. **Echinodermata**
   a. GENERAL ANATOMY OF DIGESTIVE SYSTEM
   b. COURSE OF DIGESTION IN ECHINOIDS; ASTEROIDS

A. SITES OF DIGESTION

Digestion, or the chemical disintegration of food, is accomplished through the agency of enzymes. These may be released from the cells in which they are synthesized and so act extracellularly, or they may be parts of, or attached to, cell structures and so act intracellularly. Extracellular enzymes may be discharged into the animal's environment and operate either at a distance from it or directly upon food in the immediate vicinity of the mouth, so that it is partially or wholly predigested before it is actually eaten (extracorporeal digestion); or they may be released into a special internal cavity or enteron and act upon food that has been ingested and, in many cases, macerated or triturated (luminar digestion). Intracellular digestion may occur in cellular cavities or vacuoles, or in association with cytoplasmic granules or membranes. The plasma membrane appears to be one with which digestive enzymes are associated and, since this is in direct contact with the external environment, its enzymes might be considered to be acting extracellularly. Yet, since they represent a part of the structure of the cell wall, it seems equally justifiable to consider them intracellular in operation.

Intracellular digestion is characteristic of protozoans and sponges, but it is also common among metazoan invertebrates in which it may be effected in either fixed or mobile phagocytic cells (amebocytes). In this respect more than in any other, the digestive processes of invertebrates differ from those of vertebrates, in which digestion, except for membrane digestion which may prove to be a universal phenomenon, is luminar only. The term intracellular digestion is reserved, in the following discussion, for the biochemical disintegration of particulate food that has been phagocytosed. The engulfment of material by other mechanisms of endocytosis is here considered absorption. Although such a distinction is not clear cut and is therefore artificial as many biological distinctions are, it is made for clarification of the nature of the two processes as they occur in a single cell or in different cells of the same tissue. In digestion of material that has been phagocytosed, the vacuole serves not only as a site for the decomposition of the food but also for absorption of the end products. These two events may go on concurrently, and the absorbed products, if not reduced to the compounds usually regarded as terminal ones of the action of digestive enzymes, may be further degraded in the cytoplasm, possibly by membrane digestion. In such circumstances, a distinction between digestion and catabolism would seem to have little meaning, nor would it be meaningful in cells or tissues of any kind that absorbed compounds of such complexity. Moreover phagocytic cells, whether fixed or migratory, may be the seat of the synthesis of substances that have been digested and absorbed into other compounds, of general or specific nature, that serve as stored reserves to be used as need requires. Such cells, or the tissues in which they predominate, may equally well be categorized as storage, absorptive, or digestive, since they subserve each of these three functions either concomitantly or sequentially. They have likewise an egestive function, in that they may eliminate undigested residue or metabolic waste by exocytosis.

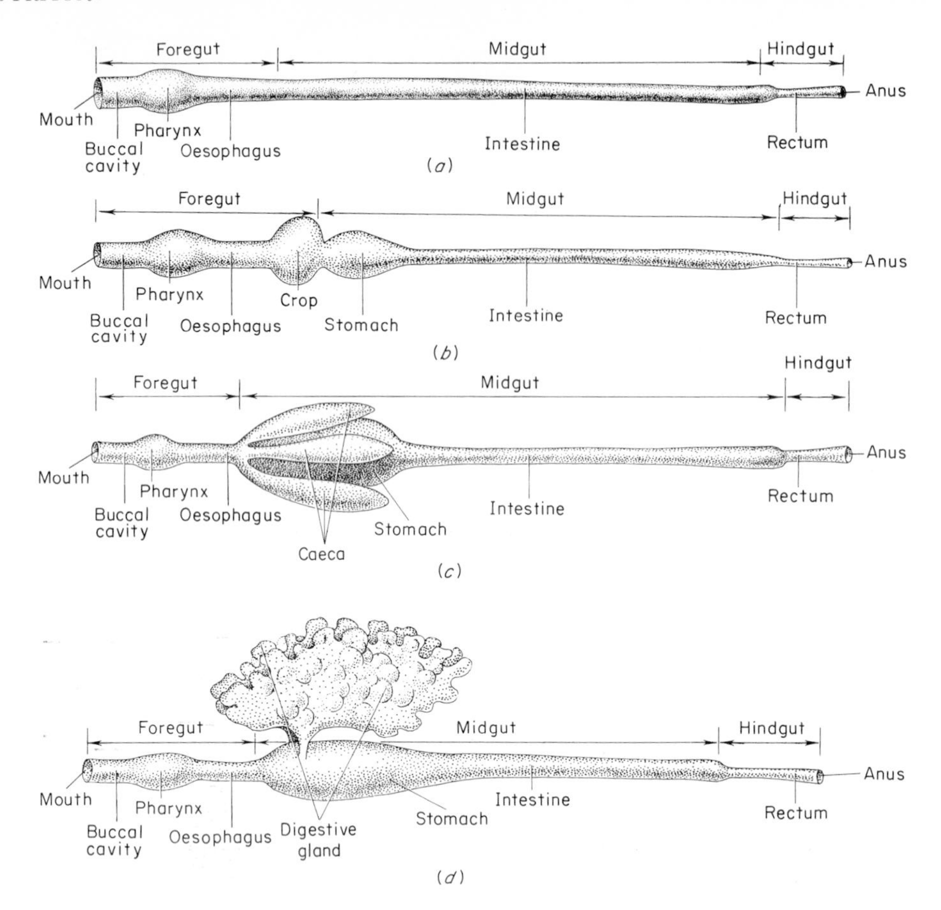

FIG. VIII–1 *The principal divisions of a complete digestive tract and their modifications.* ***(a)*** *The simplest plan, as it is found in nematodes, some worms, and the larvae of many insects.* ***(b)*** *A slightly modified plan, with well-defined crop and stomach.* ***(c)*** *A more modified plan, with an extensive crop consisting of several blind caeca extending some distance posterior to their origin from the foregut.* ***(d)*** *The plan as found in molluscs and arthropods (except insects), with a well-developed midgut gland.*

B. FUNCTIONAL AREAS OF COMPLETE DIGESTIVE TRACTS

In metazoan invertebrates constructed on a simple sac plan, like cnidarians and most platyhelminthes, there is little opportunity for the separation of functional areas within the digestive cavity. But in those constructed on the tube-within-a-tube plan, at least five such areas can be distinguished (Fig. VIII–1). Morphologically, a complete digestive tract is divided into foregut, midgut, and hindgut. The foregut and hindgut, as derivatives of the stomodaeum and proctodaeum, respectively, are lined with ectoderm, continuous with the epidermis. The midgut is lined with endoderm and is the principal region for the digestion of food and absorption of the terminal products of the digestive processes. The functional areas of the foregut include the buccal, or oral, cavity, into which the mouth opens; the pharynx; and the oesophagus.* The buccal cavity and pharynx are concerned with the intake of food, the oesophagus with its passage to the midgut. The preoral cavity of insects and arachnids is not homologous with the buccal cavity, for it is external to the digestive tract proper. The oesophagus may be short or long, relative to the size of the animal and the extent of its digestive tract, but it is usually

* Anatomical terms used in invertebrate morphology are often the same as those used in vertebrate morphology. This does not necessarily imply any structural homology nor, indeed, even a functional one. In order to avoid multiplicity in nomenclature, various areas of invertebrate digestive tracts are frequently (and here) designated by names familiar in vertebrate anatomy, without implication of homology in embryonic development or physiological function. Similarly, the term "lung" is frequently used to designate the respiratory areas of certain invertebrates, but no invertebrate has a "lung" either derived or designed like a vertebrate lung

tubular. The cells lining the foregut are frequently ciliated or mucus-secreting, but any digestion that takes place in it is by virtue of enzymes discharged from salivary glands whose ducts open there or by enzymes that are produced by cells situated more posteriorly in the tract and regurgitated into this region. The midgut usually includes the third and fourth functional areas, the stomach and intestine; the food mass is often subjected to stirring and churning while in the stomach and, in some macrophagous species, to grinding or trituration. The major part of digestion occurs in the stomach and intestine, and both may be sites for the absorption of the final products. The residue then enters the hindgut, which may include colon and rectum as the final functional areas. Here the feces are formed, primarily through the absorption of water. This is a process of particular importance to terrestrial species, with the exception of fluid feeders, since conservation of water is one of the principal demands their environment imposes; likewise it is an important process for animals of sessile or sedentary habit or who live enclosed in shells, to which fouling of the water by fluid or particulate feces could be disastrous. The feces are then eliminated through the anus, the posterior opening of the digestive tract.

In many species, and especially in eucoelomates where the secondary body cavity provides sufficient space, one or several of these areas may be enlarged as temporary storage, grinding, or macerating areas, or as areas in which digestion and absorption may occur. For example, the posterior region of the oesophagus may be expanded into a single pouch, generally known as a crop, or into a number of finger-like projections or caeca, in which the bulk of a meal may be stored. In these conditions the opening of the oesophagus into the stomach is usually guarded by a valve of some kind, which permits only intermittent passage of small amounts of food. The posterior region of the oesophagus may also be modified as a gizzard, or grinding "stomach." Often there are large diverticula from the midgut, which may be the sites of intracellular digestion, of the secretion of enzymes that are carried to other regions of the gut for operation or that operate in the lumen of the gland itself, and of the absorption of end products of such localized digestion. Diverticula of this kind have frequently been homologized with the vertebrate liver and referred to as "livers," or "hepato-pancreases," although the terms "digestive," "gastric," or "midgut" gland seem more appropriate since these organs are not structurally nor, in most cases, functionally homologous with either the liver or the pancreas of vertebrates. In many invertebrates, the aspects of liver function that deal with intermediary metabolism and storage of reserves are per-

TABLE VIII-1 Extracellular Digestive Enzymes Common to Animals

Category	*Action*	*Products*
Carbohydrases	Hydrolysis of carbohydrates	
Amylases	Hydrolysis of starch, or amylose, a storage polysaccharide of plants	Maltose (disaccharide) and glucose (monosaccharide)
	Hydrolysis of glycogen, a storage polysaccharide of animals	Glucose
	Hydrolysis of dextrins, or glucosans $[(C_6H_{10}O_5)_x]$	Glucose
Glucosidases	Hydrolysis of disaccharides and glucosides (compounds of glucose with other organic radicals)	
Invertase	Hydrolysis of sucrose	Glucose and fructose
Maltase	Hydrolysis of maltose	Glucose
Lactose	Hydrolysis of lactose	Glucose and galactose
Alpha glucosidase	Hydrolysis of glucosides containing the alpha form of glucose	Glucose

(*Continues on pages 284–285.*)

TABLE VIII-1 Continued

Category	Action	Products
Proteases	Hydrolysis of proteins by breaking peptide (C—N) linkages between amino acids in the chain	
Endopeptidases (proteinases)	Act on C=N linkages anywhere within the chain	
Pepsins[1]	Act specifically on bonds linking tyrosine or phenylalanine to other amino acids	Smaller chains of amino acids (proteoses, peptones, etc.)
Trypsins	Act specifically on bonds linking arginine or lysine (at carboxyl ends) to other amino acids	Polypeptides and amino acids
Chymotrypsins	Act on bonds linking tryptophane or phenylalanine to other amino acids, but differ from pepsins in that the cleavage is at the carboxyl end of the specific amino acid (tyrosine or phenylalanine) and not at the amino end, as in peptic digestion, e.g., 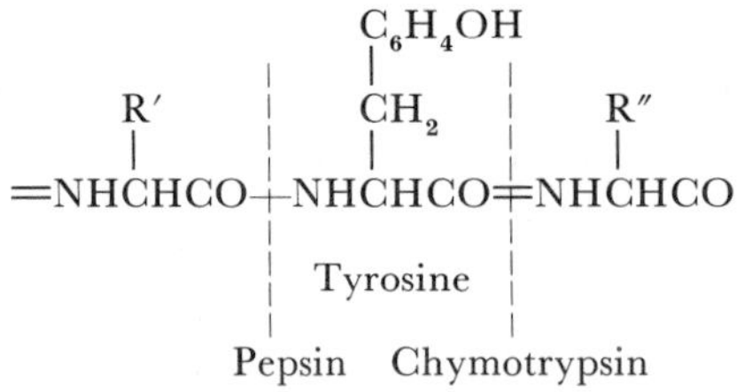	Polypeptides and amino acids

```
                       C6H4OH
                   |     |       |
        R'         |    CH2      |     R"
        |          |     |       |     |
   =NHCHCO--|--NHCHCO--|--NHCHCO
                   |             |
                   |   Tyrosine  |
                   |             |
                Pepsin    Chymotrypsin
```

Category	Action	Products
Exopeptidases	Act only on bonds joining terminal amino acid to main chain	
Aminopeptidases	Act on peptides and polypeptides to break peptide linkage at end of chain where there is a free amino group, e.g., 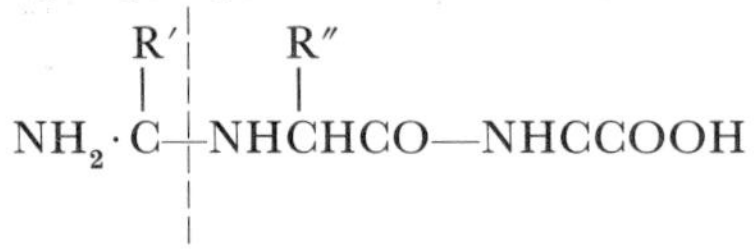	Amino acids

```
          R'|   R"
          | |   |
   NH2·C--|--NHCHCO—NHCCOOH
            |
```

(Continues on page 285.)

formed by trephocytes or other cells of mesodermal origin. Midgut glands are found in all classes of molluscs and in crustaceans, scorpions, and arachnids, but they are lacking in most annelids and in insects.

In a number of species in different phyla, the epithelium of the midgut proper secretes material that forms a membrane between it and the contents of the gut. This is the peritrophic membrane, which is typical of insects except Hemiptera but is also completely or partially formed in some annelids and molluscs, and in some copepods and caridean shrimps. Electron microscopy shows that the peritrophic membrane of insects representative of several different orders is a network of fibrils, made up of three systems of parallel strands oriented at angles of 60° to each other, with amorphous material in the interstices. Chemical analysis shows that the fibers are of chitin, whose secretion by endodermal cells is unusual; the interstitial material is largely, if not entirely, protein and can be removed by proteinases. The membrane may be secreted by all the cells of the midgut or by a group of them in a restricted area. Its function has

TABLE VIII-1 Continued

Category	*Action*	*Products*
Carboxypeptidases	Act on peptides and polypeptides to break peptide linkage at end of chain where there is a free carboxyl group, e.g., $NH_2 \cdot \overset{R'}{C}\text{—NH}\overset{R''}{C}\text{HCO} \mid \text{NHCCOOH}$	Amino acids
Dipeptidases	Act only on dipeptides in which the amino group of one acid and the carboxyl group of the other are unsubstituted	Amino acids
Esterases	Hydrolysis of fats and other esters, e.g., catalysis of reaction $R \cdot COOR' + H_2O \leftrightharpoons R \cdot COOH + HO \cdot R'$	
Lipase	Hydrolysis of neutral fats and oils with long-chain (eight or more carbon atoms) fatty acids	Glycerol and long-chain fatty acids
Nonspecific esterases	Hydrolysis of esters of short-chain (less than eight carbon atoms) fatty acids, and glycerol and other organic acids, or an inorganic acid with alcohol or carbohydrate	Organic or inorganic acid and alcohol or carbohydrate

1. Pepsin is so rare among invertebrates as to be regarded as virtually absent. This may be because of its low pH optimum (*c.* 1.5 to 2.0) and the fact that the pH in the gut of invertebrates is consistently higher than this.

been considered primarily protective, guarding the delicate epithelium of the gut from abrasion or other injury by rough or sharp particles that may be present in the food; it may also act as an ultrafilter. Experimental studies on peritrophic membranes of insects have shown that albumin and chlorophyll will pass through them as well as particles of colloidal gold 2 to 4 μ in diameter, but there is no specific information about its permeability to other substances. It must, however, permit exchange of materials between the epithelial cells and the lumen of the gut, for enzymes enter the gut cavity and the end products of digestion are absorbed by the epithelial cells. It may be an effective device to promote absorption, since the fluid in the space between it and the absorptive cells would be comparatively stagnant. The peritrophic membrane is formed either continuously or whenever a meal is taken and moves along the digestive tract as the food mass moves to be eliminated with the feces at the anus.

C. EXTRACELLULAR DIGESTIVE ENZYMES

Some broad generalizations about the complement of digestive enzymes of any animal can be drawn from observation of the food that it eats. If this provides it with the nutrients necessary for its maintenance, its growth, and its reproduction, it must be able to digest them and to assimilate the products. More specific and detailed information, comparable to that which is known about digestive processes in mammals, has been obtained for only a small fraction of the known invertebrate species. Several methods have been used in investigations of this kind upon species representative of the major phyla. One of these methods is to test for digestive enzymes present in the contents of the gut as a whole or of selected regions of it. The contents of the gut are sucked or carefully washed out and then mixed with compounds representing the various categories of food stuffs, proteins, fats, or carbohydrates.

If, after a suitable period of incubation under appropriate conditions, such substrates can be shown to have undergone partial degradation or complete decomposition into their constituent compounds, evidence is provided for enzymatic activity in the test material. Thus, the recovery of amino acids, or of peptides, composed of a few amino acids linked together, after exposure of the gut contents to a protein substrate such as gelatin or casein is diagnostic of the presence of proteolytic enzymes, or proteases. The degradation of starches to their component sugar molecules or of neutral fats and oils to fatty acids and glycerol is indicative of the activity of carbohydrases and lipases, respectively. After the presence of enzymes in these very general categories has been established, specific ones operative at specific points in the sequence of disintegrative steps can be identified by incubation with their specific substrates. Degradation of substrate can be detected by reduction in the turbidity or the viscosity of the solution and the end products of these enzymatic reactions can be identified by appropriate chemical tests, by spectrophotometry, and by chromatography.

Such analyses of the contents of the gut furnish evidence only of the digestive capacities of animals practicing extracellular digestion and none as to the origin of the enzymes involved nor of intracellular digestive activity. Extracellular enzymes may be exogenous as well as endogenous. They may be contributed by the food itself, since in nature this is eaten raw and may bring with it its own digestive enzymes or disintegrative bacteria. They may also be provided exogenously by symbionts living in the gut, in special chambers connected with it, or in mycetomes. Evidence for endogenous enzymes can be obtained only from a demonstration of their presence, or of that of their precursors, in the tissues or cells of the animal in question, when all possibility of contamination with microorganisms is excluded. Localization of enzymes within tissues can be determined by preparing extracts from them and testing these extracts against appropriate substrates. Such extracts can be prepared from the entire alimentary canal or from selected regions of it. After the contents have been thoroughly washed out, the gut wall is ground, or homogenized in a blender or similar apparatus, so that the cells are broken up. The homogenate or brei is then centrifuged to remove the grosser cell debris and any bacteria that may have remained even after careful, repeated washings. The enzymatic activity of the resulting extract is then tested, and its ability to decompose protein, carbohydrate, or fat is taken as an indication of its content of the appropriate enzymes. However, even with the most meticulous technique, there always remains some possibility that the extract may contain exogenous digestive enzymes that have been released by bacteria in the course of its preparation or even before it was started. There are, of course, also in the extract cellular enzymes that function in intracellular metabolic processes, and these must be distinguished from those that are generally regarded as digestive.

Such investigations as these have shown that the extracellular proteases, carbohydrases, and lipases of invertebrates do not differ in any significant ways from those of vertebrates. The course of digestion of proteins, carbohydrates, and fats is essentially the same in both, although the digestive enzymes of invertebrates operate in general more effectively at pH's somewhat higher than the optima for their counterparts in vertebrates. A summary of the extracellular enzymes common to animals is presented in Table VIII–1.

D. UNUSUAL ENZYMES AND DIGESTIVE PROCESSES

The selectivity and specificity in diet exhibited by some invertebrate species have suggested that they may synthesize and efficiently utilize enzymes other than those characteristic of animals in general. This question has been raised particularly in regard to species that feed exclusively upon plants and plant products; upon fur, wool, and feathers; and upon honeycomb. Attempts to answer it have been made by a number of experimental methods, but the answers in many cases have been equivocal or contradictory. In some instances, however, it has been established that there are special enzymes, or special conditions in the gut, that make possible the digestion of such generally resistant compounds as structural polysaccharides, structural proteins, and the esters of alcohols and fatty acids of high molecular weight.

1. Digestion of structural polysaccharides

Polysaccharides are stable macromolecules, made up of large numbers of monosaccharides, or their derivatives, linked together in linear sequence. A large proportion of the material ingested by phytophagous animals is the structural polysaccharide cellulose composed of beta-glucose units, for this is the principal component of the walls of plant cells. A lesser, but universal, constituent of these walls is pectin, another stable polysaccharide which on hydrolysis in vitro yields the hexose galactose and the pentose arabinose. The walls of woody plants also contain lignin, a mixture of substances not susceptible to the action of the usual animal enzymes. Red algae, on which many browsing invertebrates feed, contain relatively large amounts of agar, a polysaccharide made up of galactose units; brown algae, also an important food source for a number of invertebrate species, contain algin, made up of units of mannose. The walls of fungi contain chitin, a linear polymer of glucoseamine units. Invertebrates that feed on fungi, or upon animals with chitinous cuticles or cases, ingest considerable amounts of this material. All these substances are resistant to disintegration by ordinary means, but they are energy-rich substances and therefore valuable foods if they can be depolymerized and their constituents digested, absorbed, and metabolized.

a. CELLULOSE Although most animals can digest starch, a polysaccharide made up of alpha-glucose units, their enzymes as a rule cannot attack beta-glucose chains. (Alpha and beta glucose are isomers, whose structural differences are reflected in the difference in their optical rotations.) Many bacteria, on the other hand, have an enzyme, or enzymes, that can break beta linkages. It is well known that ruminant mammals depend upon symbiotic microorganisms in their digestive tracts to digest the cellulose of the food they eat and make the end products of this process available to them. A similar dependence upon enteric symbionts has been suspected of phytophagous invertebrates and has indeed been established for some species. This has been shown to be the case, for example, in termites and shipworms, both of which seem to derive some nutrition from the wood that they swallow as well as from the other plant products ingested along with it. Flagellate protozoans perform the digestion of cellulose for termites; bacterial symbionts do it for *Teredo*.

Proof of an endogenous cellulase depends upon the demonstration of cellulose digestion by sterile and uncontaminated animals, or extracts of their tissues. Failure to show such digestion is not, however, absolutely conclusive evidence for absence of the enzyme, for it may be present in such minute amounts that present procedures and instruments are not sensitive enough to detect the end products of its activity. The tissues of a number of species suspected of synthesizing their own cellulase have been tested for cellulolytic activity by incubating extracts from them with such substrates as finely shredded filter paper or Kleenex, cotton fibers, or powdered wood. Depolymerization or digestion of these materials has been determined by measuring the difference in viscosity of the mixture after a suitable period of incubation or by identifying by chemical or physical means the anticipated end products of the enzymatic reaction. It has been pointed out that substrates such as these are attacked only slowly and to a slight degree by cell-free enzymes, and that positive rather than negative results might be obtained if the substrate were presented in a form that would permit more direct and rapid access of the enzyme to it. Cellulose, swollen by treatment with phosphoric acid, and sodium carboxymethylcellulose, which is soluble in water, do permit such access and have been used in some experiments. Their use is, however, open to the objection that they are not natural products and so not comparable to the material that the animal ordinarily eats.

In addition to the Vampyrellidae, two genera of soil amebae, *Hartmanella* and *Schizopyrenus* have been shown to have an endogenous cellulase by the recovery of reducing sugars after incubation of extracts of the animals with sodium carboxymethylcellulose. Although these amebae feed on soil bacteria, it was shown that the microorganisms isolated from them and grown in pure culture had no effect upon this substrate. It therefore seems unlikely that they or their products contribute to the digestive processes, although the possibility is not excluded that in nature they may do so.

Some species of phytophagous nematodes produce cellulase by virtue of which they can disintegrate the walls of plant cells, into which the stylet can then be easily inserted. Cellulolytic activity has been demonstrated in homogenates of whole animals, so that the site of synthesis of the enzyme has not been localized. It is, however, presumed to be produced by cells in the pharyngeal glands. *Ditylenchus dipsaci*, which penetrates into onion bulbs, secretes a pectinase which macerates the tissues without injuring the cells.

A number of molluscs have been investigated in respect to their ability to digest cellulose. Although the evidence for endogenous cellulase is conflicting in several cases, and especially in that of the herbivorous garden snail *Helix*, it seems quite definite in that of the black turban snail, *Tegula funebralis*, which feeds on algae. Cellulolytic activity, measured by decrease in the turbidity of suspensions of finely divided filter paper, has also been reported for the oyster *Ostrea*, the mussel *Mytilus*, and the clams *Mactra* and *Mya*. It has also been demonstrated in the nudibranch *Aplysia*.

It seems probable that earthworms can also digest cellulose by virtue of their own enzymes and so to derive some value from the large amounts of it that they obtain in the leaf mold and humus they ingest. Seventeen different species of British earthworms have been tested for this capacity. After incubation of aqueous extracts of different regions of the gut wall with either sodium carboxymethylcellulose or finely divided cellulose, the viscosity of the mixture was lowered and reducing sugars were recovered from it. The site of production of the enzyme has been localized to a certain extent in *Lumbricus terrestris*, for extracts from the posterior half of the intestine showed only one-tenth the activity of those from the anterior half, and extracts from other regions of the gut wall showed no activity at all.

Extracts of different regions of the gut of the wood-boring isopod *Limnoria* have also been tested for cellulolytic activity, with very finely divided filter paper and carboxymethylcellulose used as substrates and viscosity measurements as indications of their degradation. Those from the midgut diverticula gave the only positive results, with all tests for contamination proving negative. Indeed, *Limnoria* is exceptional in being without any microorganisms in any region of its gut.

The most convincing demonstration of an endogenous cellulase has been obtained from experiments performed in vivo on the silverfish *Ctenolepisma longicaudata*, an apterous insect that feeds upon the bark of eucalyptus trees, in which it lives. It was found possible to rear completely sterile insects from sterilized eggs and to feed them cellulose labeled with C^{14}. The air these insects expired contained $C^{14}O_2$, giving uncontrovertible evidence that not only had the labeled cellulose been digested but its end products had been metabolized as well. Analyses of tissue extracts from various regions of the digestive tract showed that the midgut was the site of cellulase production. Although in nature the insects do harbor intestinal microorganisms, none of these, when isolated, was found to digest cellulose.

Various species of cockroach also synthesize cellulase in their salivary glands. This has been shown by experiments in which glucose has been recovered after incubation of extracts of these glands with 1 percent solutions of carboxymethylcellulose. It is true for the omnivorous *Periplaneta* as well as for the wood-eating *Cryptocercus*, which, in spite of its endogenous cellulase, depends largely upon the microfauna in its hindgut for the digestion of its food. It would seem that ability to synthesize this enzyme is a very old attribute of this very old group of insects that has survived even when the immediate need for it no longer exists or has been bypassed.

Positive tests for endogenous cellulase have been reported for tissues of a number of Japanese species of invertebrates. Extracts of selected tissues, carefully washed to rid them of bacteria, were incubated with sodium methylcarboxycellulose whose degradation was determined by reduction in the viscosity of the solution. Cellulase activity was demonstrated in extracts from the gut wall of seven of the eight species of annelids examined, from the digestive glands of all 18 species of crustaceans and 16 of the 19 species of molluscs, and from the pyloric caeca of 3 of the 14 species of echinoderms. None was shown by tissue extracts from the gut wall of 11 representative species of chordates, including fish and mammal.

b. ALGAL POLYSACCHARIDES The snail *Tegula* has not only an endogenous cellulase but also enzymes that can depolymerize algin and two other structural polysaccharides found in the littoral algae on which they feed—iridophycin from a red alga, *Iridophycus flaccidum*, and fucoidin from the brown alga *Fucus*. *Tegula* supports large populations of bacteria and yeasts in its intestine, but none of these when isolated and tested produced any alterations in the algal polysaccharides. Extracts from the tissues of the foregut and midgut of the snails did, on the other hand, effect their digestion. Also, extracts from the stomach of the giant chiton, *Cryptochiton stelleri*, digested fucoidin and laminarin, a polysaccharide of the sea lettuce *Laminaria*, but they did not digest agar or iridophycin. The browsing sea urchin *Strongylocentrotus purpuratus* has also been shown to be capable of degrading the structural polysaccharides of brown algae.

c. CHITIN A somewhat similar distribution of endogenous chitinase has also been reported. Acetylglucoseamine has been recovered from mixtures of finely divided chitin and extracts of the amebae *Hartmanella* and *Schizopyrenus* after over 2 weeks incubation. Extracts of the intestinal walls of 12 species of earthworms have also shown chitinase activity, indicated both by changes in the viscosity of chitosan and by the release of acetylglucoseamine from finely divided chitin after exposure to them. Chitinase is regularly present in the molting fluid of arthropods, where it depolymerizes the chitin of the old cuticle. Although this does not occur in the gut, it is a digestive process and also a nutritive one, for the end products of the enzymatic reaction are available for anabolic and catabolic processes.

2. Digestion of scleroproteins

Keratin and collagen are both very stable proteins which, like structural polysaccharides, are not susceptible to the usual digestive processes of animals and can be degraded in vitro only with quite drastic means. Yet both seem to serve as a source of food for certain invertebrates.

a. KERATIN Keratin is the principal component of fur, wool, and feathers. It has a high content of the sulfur-containing amino acid cystine and owes much of its stability to linkages between the sulfur atoms in adjoining polypeptide chains. When these disulfide bridges are broken, the keratin molecule becomes more soluble and more easily available to enzymatic action. Larvae of a number of genera of moths in the family Tineidae and of a few beetles in the family Dermestidae, and adults of the biting lice (Mallophaga) subsist on diets of wool, fur, and feathers. The course of digestion of keratin has been studied extensively in larvae of the clothes moth *Tineola*, where it was anticipated that a specific keratinase might be found. None could be shown by any means, and it has also been demonstrated that the bacteria in the gut have a negligible effect, if any, upon this protein. The ability of *Tineola* larvae to digest wool depends upon somewhat unusual conditions in the midgut, where a very low oxidation-reduction potential is maintained by means that are still undetermined. The oxygen supply is clearly deficient there, for this region of the digestive tract is very poorly supplied with tracheae. In addition, the pH in the midgut is 9.8, an alkalinity that is unusually high even for invertebrates. The conditions are such that the disulfide bonds of keratin are broken, reducing them to sulfhydryl groups. Once these disulfide bonds are broken, the polypeptide chains are vulnerable to attack by the usual proteases, and experiments in vitro have shown that those of the *Tineola* midgut operate most effectively at the high pH found there. Direct evidence that digestion of keratin does take place in the midgut has been provided by observations of living larvae in polarized light, for they are sufficiently transparent to permit the passage of a beam through their bodies. These observations have shown that there is a sudden and very marked decrease in birefringence of the wool fibers soon after they enter the midgut, indicating the disorganization of the linear arrangement of their constituent molecules and so of their fibrous structure.

Dermestid beetles seem to digest wool in a similar way. The midguts of their larvae are virtually without tracheae, and a low oxidation-reduction potential is maintained, with the same effects upon keratin as in the midgut of *Tineola* larvae. Mallophaga that infest birds can digest the keratin of their feathers, also because of the reducing conditions maintained in the gut. But those that infest mammals are unable to digest that of wool or hair and subsist upon the secretions and debris of the epithelial cells of the skin of their hosts.

b. **COLLAGEN** Collagen fibers are abundant in the connective tissue of vertebrates and are as resistant as keratin to the action of ordinary proteolytic enzymes. Larvae of the nematode *Strongyloides* are believed to have a collagenase, or collagenase-like enzyme, whose activity facilitates their migrations through the tissues of the animals in which the adults live as intestinal parasites. Extracts presumably containing the enzyme, while active on cartilage, have, however, little effect upon native collagen in vitro. Maggots of the blowfly *Lucilia sericata* live on meat or carrion and consume it completely, leaving no trace of connective tissue. Their digestion is external for they are incapable of ingesting solid particles. When the digestive fluid from bacteria-free larvae, raised under aseptic conditions from sterilized eggs, was incubated with collagen taken from the tendons of an ox, the collagen was completely dissolved. *Lucilia* larvae also live in the wool of sheep and even penetrate into the tissues, causing serious damage to infected flocks. These must, however, subsist either on the debris of epithelial cells or other material on the skin, or among the hairs of the sheep, for in vitro studies have shown that the digestive fluid they express has no effect upon keratin. Since this operates outside the body, the necessary reducing conditions could obviously not be established as they are in the midgut of *Tineola* larvae. The endogenous production of collagenase in the midgut of these insects during their early life seems quite definite.

3. Digestion of wax

Larvae of at least two species of moths, *Galleria mellonella* and *Achrois grisella*, live in beehives and eat the comb. About 40 percent of the substance of the comb is beeswax, a mixture of hydrocarbons, long-chain fatty acids, and their esters. The most abundant of these esters is myricyl palmitate, the ester of myricyl alcohol ($C_{31}H_{62}OH$), and palmitic acid ($C_{15}H_{31}COOH$). Since only about 50 percent of the wax ingested by *Galleria* appears in the larval excreta, it seems likely that the remainder is digested and absorbed, but there is no conclusive evidence of an endogenous cerase, or wax-splitting enzyme. Aseptically reared larvae of *Galleria* can digest stearic acid ($C_{17}H_{35}COOH$) and its ester with octadecanol [$CH_3(CH_2)_{16}CH_2OH$], but there is no evidence that they can hydrolyze myricyl palmitate. Indeed, it is more likely that the digestion of beeswax is performed by symbiotic microorganisms in the gut of the larva, for these on isolation are known to hydrolyze beeswax in vitro. Any nourishment the insects obtain from the ingested wax seems therefore to be derived from the activity of their symbionts, and even this is not a necessity, for larvae have been successfully reared on diets entirely without wax.

E. INTRACELLULAR DIGESTION

1. Endocytosis

Intracellular digestion takes place when nutrient material is engulfed by a cell and chemically disintegrated within it. The term "endocytosis" has been proposed as an inclusive one for the various ways that material may be taken in by cells, as opposed to "exocytosis," the various ways that material may be eliminated from them. Endocytosis therefore includes the engulfment of particulate matter by phagocytosis and of macromolecules by pinocytosis. Perhaps the most valid distinction that can be made between these two aspects of endocytosis is on the basis of the size of the particles engulfed. If they are larger than colloidal, with diameters greater than 100 nanometers (nm), phagocytosis is involved; if they are of colloidal dimensions, with diameters less than 100 nm, they are taken in by pinocytosis. Endocytosis is a necessary preliminary to intracellular digestion, although the materials that may be engulfed are not necessarily nor invariably nutritive. Most phagocytic cells, for example, will take up powdered carmine and graphite or small particles of polystyrene latex as readily as they will take up food. The engulfment of these substances, because they are not digested, has positive advantages in the study of intracellular digestion, for the course of vacuoles marked with them can be followed with the aid of a microscope and the grosser physical and chemical changes of foods during the progress of digestion can be observed.

The source and the nature of the enzymes that bring about changes in phagocytosed or pinocytosed nutrient material present many problems, which are still to a great extent unsolved. Two possibilities as to the source of the enzymes immediately present themselves: (1) that they may be continuously present in the cytoplasm and diffuse through the membrane of the vacuole soon after its formation, and (2) that they are synthesized only in response to the presence of food within the vacuole. In respect to the first possibility, it is difficult to reconcile the presence of free digestive enzymes in the cytoplasm and the integrity of the cell, unless these are in some way bound, inhibited, or partitioned off from the constituents of the cell that would seem to provide for them substrates as adequate as the cellular constituents they take in as food. If there is not some barrier to their activity, why is a cell not continually digesting itself? In respect to the second possibility, it is difficult to visualize how a cell can respond all at once with the battery of hydrolytic enzymes necessary to digest the variety of complex chemical compounds they are known to engulf.

2. Lysosome concept and lysosomal enzymes

Present evidence suggests that the first possibility is the more acceptable one and that cells do, indeed, have a store of enzymes capable of catalyzing hydrolytic reactions that effect the disintegration of molecules of many different kinds. These enzymes are contained within a membrane and constitute a category of submicroscopic particles that have been given the name of lysosome. These particulates can be isolated from cells by homogenization and differential centrifugation and, when ruptured, tested for their complement of enzymes. Such experiments have shown that they contain a variety of hydrolases, operating most effectively in an acid medium, which degrade carbohydrates, proteins, esters of phosphoric and sulfuric acids, and nucleic acids. The enzymes at present known to be associated with lysosomes are summarized in Table VIII–2.

Some few of these enzymes can be demonstrated by cytochemical means and so localized within cells. This demonstration involves the fixation of the cells or tissues by appropriate means, their embedding and sectioning, and exposure of the slides to an appropriate substrate for an appropriate period of incubation. It is necessary that the end product of the enzymatic reaction, if one has taken place, be made insoluble and treated so that it can be distinguished microscopically from other substances in the cell. There are various methods by which one or the other product of the enzymatic reaction can be so visualized. One of the most precise, and diagnostic, of these cytochemical tests is that for acid phosphatase. Since acid phosphatase seems to be universally associated with lysosomes, its demonstration within a cell particulate offers persuasive, if not compelling, evidence that other lysosomal enzymes are located at the same site. There are various problems remaining as to the nature and source of the lysosomal membrane, and of the enzymes contained within it, but there is some evidence that they originate in the region of the Golgi substance. Other problems concern the factors that contribute to the stability of the membrane during the life of the cell so that its contents of destructive enzymes are not freely liberated but retained until occasion requires them.

Studies of phagocytotic and pinocytotic vacuoles in cells by means of electron microscopy suggest very strongly that lysosomes contribute their hydrolytic enzymes to them and so make digestion of the contained material possible. Lysosomes seem to collect around a newly formed vacuole; the two membranes presumably fuse, and the vacuole is converted into a phagosome, in which digestion of particulate material takes place, or into a pinosome, in which digestion of macromolecules proceeds. There are many points that still need clarification in the "lysosome concept," but there is little reason to think that the mechanisms of intra-

TABLE VIII-2 Lysosomal Enzymes

Category	Action
Carbohydrases	
Alpha glucosidase	Hydrolysis of alpha glucosides (glycogen) to glucose
Beta glucuronidase	Hydrolysis of beta glucuronides } Components of mucopolysaccharides
Beta galactosidase	Hydrolysis of beta galactosides } Components of mucopolysaccharides
Alpha mannosidase	Hydrolysis of alpha mannosides } Components of mucopolysaccharides
Beta-*n*-acetylglucose-amidase	Hydrolysis of beta-*n*-acetylglucoseamides
Proteinases	
Cathepsins	
I	An endopeptidase, with specificity similar to that of pepsin but with optimum pH at 5.6 instead of 1.5 to 2.5, characteristic of the pepsin in the gastric juice of vertebrates
II	An endopeptidase, with specificity and activity like that of trypsin
III	An aminopeptidase, splitting peptide bonds where there is a free amino group
IV	An exopeptidase, splitting peptide bonds where there is a free terminal carboxyl group (carboxypeptidase)
V	An endopeptidase, with activities similar to those of chymotrypsin
Collagenase	Hydrolysis of collagen
Esterases	
Acid phosphatase	Splits inorganic phosphate from phosphate esters (including mononucleotides) at a pH of 5.0
Phosphoprotein-phosphatase	Splits inorganic phosphate from phosphoproteins
Aryl sulfatase	Splits sulfuric acid from sulfate esters
Nucleases	
Deoxyribonuclease	Hydrolysis of DNA to nucleotides
Ribonuclease	Hydrolysis of RNA to nucleotides

cellular digestion differ in any important respects in the various kinds of cells in which it is known to occur. Protozoans, amebocytes of invertebrate and vertebrate tissues, and fixed cells of the enteric epithelium of some invertebrates are all known to phagocytose particulate material and digest it in phagosomes. The extent to which they and other cells practice pinocytosis and the fate of the pinocytosed material remain to be elucidated.

F. INVERTEBRATES WHOSE DIGESTION IS EXCLUSIVELY INTRACELLULAR: PROTOZOA AND PORIFERA

In protozoans and sponges, internal digestion is of necessity intracellular. The grosser changes in the phagosomes of protozoans such as amebae, paramecia and similar ciliates can be followed with the light microscope, and changes in the chemical nature of their contents, at least as reflected in changes in pH, can be observed if the food is stained with a nontoxic dye that acts as an indicator. A paramecium, for example, that has been engulfed by an ameba can be seen to be folded up within a spherical vesicle large enough to accommodate it and to rotate within this confined area while the cilia continue to beat. After a short time its movements stop and it gradually loses shape and form, and the vacuole decreases in size and becomes filled with amorphous lumps of material. Much of the contents of the vacuole becomes liquefied; the vacuole shrinks and ultimately approaches the surface of the cell through which it erupts, discharging its contents, which include any parts of the prey that have not been digested, to the external environment. The term exocytosis has been applied to this event, the opposite of endocytosis, or phagocytosis in reverse. Changes in pH can be similarly detected by providing test animals, such as *Paramecium*, with a paste of heat-killed yeast cells stained with Congo Red. This is a dye that changes color in a pH range of 3.0 to 5.2, being blue in the lower pH's, red in the higher. Within the first 15 min after ingestion, the contents of the vacuole pass through three stages, or phases, in respect to the hydrogen ion concentration. The color is at first orange, indicating a pH of 5.0 or more, and then changes to deep blue, indicating a very acid condition and a pH that has been estimated to be as low as 1.4. In the last phase, it becomes orange again, with a pH of neutrality or slightly above (7.0 to 7.8). Histochemical tests show that this final period of alkalinity coincides with the highest activity of acid phosphatase, an anomalous situation that has yet to be understood and explained.

Detailed studies of the morphology of the vacuoles during the course of digestion have been made in a few species by electron microscopy. In amebae, ingestion may take place at any point on the surface and the food cup, when it is formed, is bounded not only by the plasma membrane but also, internally, by the additional surface coats characteristic of amebae. In *Pelomyxa*, the dense material of the surface coat disappears soon after the vacuole is formed. In *Paramecium*, the base of the cytostome is a single unit membrane; apparently the impact of any particles swept down the gullet is sufficient to initiate its invagination and the formation of a vacuole. Food vacuoles, have been designated in relation to the sequence of events within them as ingestion vacuoles (phagosomes), which are later converted to digestion vacuoles and finally to egestion vacuoles (Fig. VIII–2). These stages can be distinguished on the basis of the morphology of the vacuole, the degree of acid phosphatase activity, taken as an index of general hydrolytic activity, that can be demonstrated in association with it, and the condition of the food within it. They are of different duration in different animals and in individuals of the same species under different conditions. Ingestion vacuoles are usually spherical, with smooth contours, and contain a quantity of water that has been brought in with the food. In *Paramecium multimicronucleatum*, grown in a medium rich with bacteria, with India ink added as a marker, this stage lasts 5 min, on an average, and during it no acid phosphatase activity is discernible. About $1\frac{1}{2}$ min later, faint activity can be detected around the vacuole and frequently granules reacting intensely to tests for acid phosphatase are found near its periphery, sometimes connected to it by faint lines. The conversion of an ingestion to a digestion vacuole is accompanied by reduction in its diameter, through loss of water, and by intense acid phosphatase activity at its periphery, presumably when the lysosomes become applied to its wall and the release of their enzymes is in progress. But the ingested food may also bring in with it hydrolases as well as bacteria that contribute to the decomposition of dead material. There is therefore no justification for assuming that, except for protozoans maintained in sterile, axenic cultures, lysosomes provide the entire complement of digestive enzymes. Changes in the condition of the food have been followed by electron microscopy in *Tetrahymena* engulfed by *Pelomyxa*. The first effects of digestion are apparent in the outermost surface membranes of the ciliate; later its mitochondria, cytoplasmic matrix, and nuclei undergo disruption, and finally the cilia, kinetodesma, and the inner surface membranes. Trichocysts are not digested. As digestion of the other parts proceeds, a zone of amorphous material appears at the periphery of the vacuole just within its membrane. This membrane also shows changes; in all species studied in detail, it develops folds, indentations, and numerous

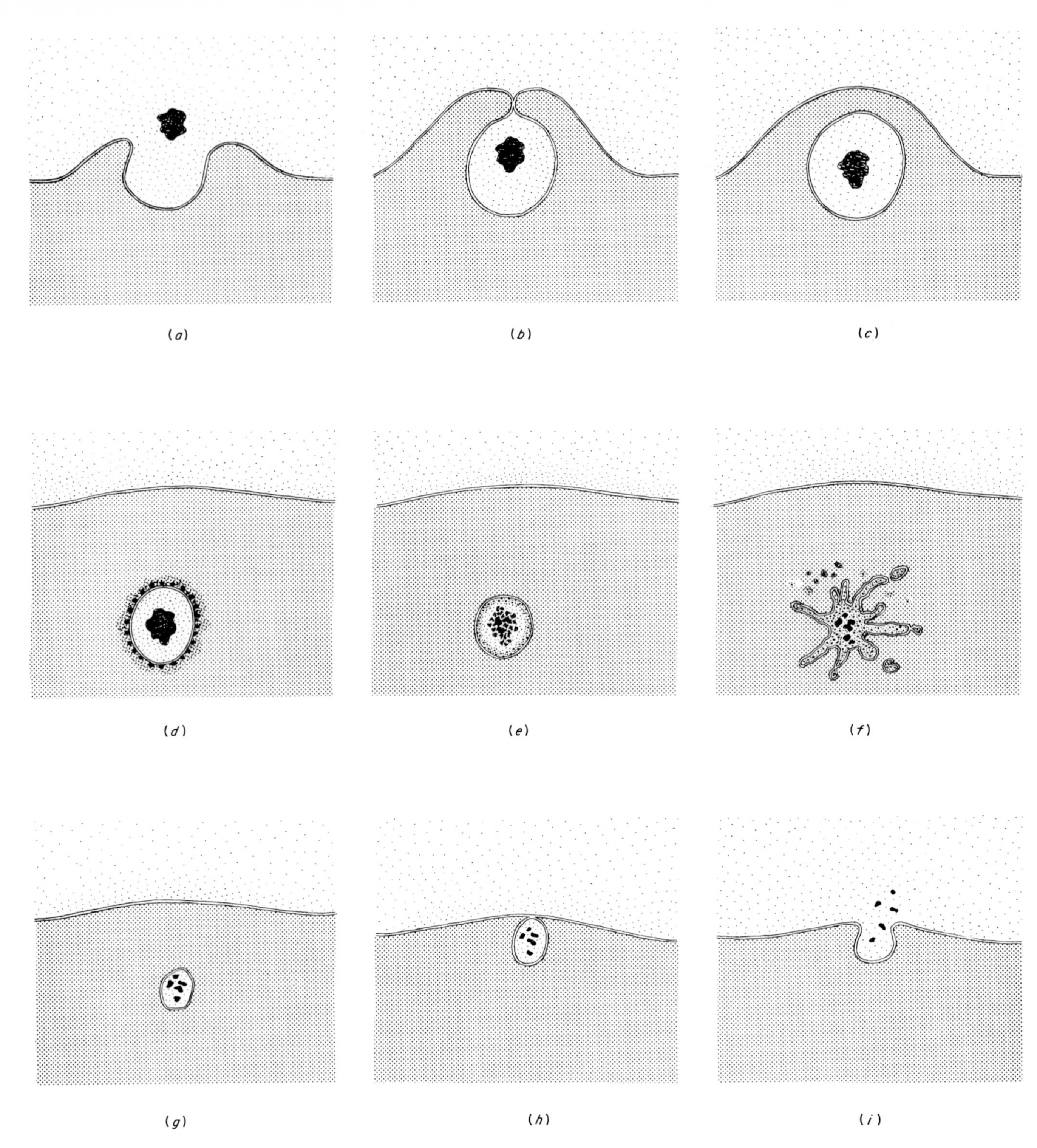
(a)
(b)
(c)
(d)
(e)
(f)
(g)
(h)
(i)

finger-like processes which extend at varying lengths into the surrounding cytoplasm and into the vacuole itself. The area of the previously smooth surface is thus vastly increased. The period of active digestion, lasting longer than any other in the life of a food vacuole, is also one in which absorption of the end products is in progress. In *Pelomyxa*, the appearance in the cytoplasm adjacent to a food vacuole of material in all respects identical, as revealed by electron microscopy, to that between the vacuolar wall and the disintegrating animal inside it, as well as in numerous small vesicles 0.2 to 9.5 μ in diameter, is interpreted as transference of these products directly through the membrane and also by pinocytosis or, on this scale, micropinocytosis. After digestion is completed, acid phosphatase can no longer be demonstrated in relation to the food vacuole, which then becomes an egestion vacuole and is eliminated from the body of the protozoan either, as in *Paramecium*, in a localized area, or cytopyge, or, as in amebae, at any point on the body surface.

Studies of digestion in sponges have been limited to a few species, and in these there seems to be some difference in functional capacities between the different kinds of cells. In the freshwater species *Spongilla proliferans*, for example, the choanocytes phagocytose food particles but digest them only partially, if at all. Ingested material passes to the base of the choanocyte and from it to an amebocyte, where digestion may be completed, or the final stages may take place in a cell to which the amebocyte has transferred the material

FIG. VIII–2 *The probable course of events in the transformation of an ingestion to a digestion and finally to an egestion vacuole.* ***(a)*** *Pseudopodial extensions enclosing food particle.* ***(b)*** *Apposition of pseudopodial processes.* ***(c)*** *Fusion of membranes of pseudopodial processes and formation of ingestion vacuole.* ***(d)*** *Early stage of digestion vacuole, with decreased diameter and lysosomes (?) applied to its surface.* ***(e)*** *Digestion of food particle and appearance of zone of amorphous material around periphery of vacuole.* ***(f)*** *Invagination and evagination of vacuolar membrane and appearance of amorphous material in cytoplasm.* ***(g)*** *Vacuole with undigested residue moving toward surface of the cell.* ***(h)*** *Egestion vacuole at surface of cell.* ***(i)*** *Fusion of membranes and elimination of material from egestion vacuole.*

it received. The morphological and chemical changes in the food vacuoles of the amebocytes are similar to those described for *Paramecium*. The amebocytes are also responsible for transferring the products of their digestion to other cells in the sponge body but apparently never do so to the choanocytes, which must have some means of providing for their own nutrition. It has been observed that they do accomplish the preliminary digestion of relatively large particles, such as blood cells and sperm, before passing them on to the amebocytes. Further evidence that they have a complement of hydrolytic enzymes has been provided by experiments in which the cells of the marine sponge *Halichondria panicea* were separated by centrifugation and the isolated choanocytes and amebocytes tested for enzymatic activity. It was found that the proteinase of the choanocytes had 8 times the activity of that of the amebocytes at a similar pH, while their amylase was 8½ times, and their lipase 15 times more active than their counterparts in the amebocytes.

G. INVERTEBRATES WHOSE DIGESTION IS BOTH INTRACELLULAR AND LUMINAR

1. Evolution of luminar digestion

A combination of intracellular digestion and of extracellular digestion is employed by cnidarians, ctenophores, most free-living and some parasitic flatworms, nemerteans, endoprocts, some brachiopods, most lamellibranch and some gastropod molluscs, a few annelids, some arachnids, and possibly some echinoderms. It seems likely that extracellular digestion, as a more sophisticated means of processing food, evolved from intracellular and through conditions of exocytosis. Enzymes that have functioned in the digestive processes must be represented in the contents of an egestion vacuole, as well as those portions of the food on which they had no effect. The discharge of the vacuole results, therefore, not only in the expulsion of waste that cannot be incorporated into the cell but also of hydrolases that have been synthesized by it.

When these are eliminated into the external environment, as they are by protozoans, they are lost to the organism, but were they expelled into an internal cavity they could operate on their specific substrates if present. Acoelous flatworms offer an illustration of a possibly transitional stage between the conditions of exocytosis in this sense and the discharges of enzymes unaccompanied by waste. The food captured by *Convoluta paradoxa*, for example, is immediately enmeshed in the syncytial network of the parenchyma into whose interstices zymogen granules, precursors of the digestive enzymes, are discharged. Digestion thus occurs in vacuoles between the cells of the syncytium, formed by the union of their pseudopodial processes, rather than in vacuoles within them. It is not difficult to envision a further step in the evolutionary process that would eliminate the pseudopodial meshwork as a closed syncytial gut and replace it with an open cavity into which zymogen granules could be directly expelled. The discharge of these granules might involve the disintegration of the cell that produced them (holocrine secretion) or the loss, by constriction, of greater or lesser amounts of its distal portion where the granules have accumulated (apocrine secretion). The products might also be periodically discharged without damage to the cell (merocrine secretion).

2. Cnidaria and Ctenophora

a. ANATOMY OF DIGESTIVE SYSTEM The digestive system in cnidarians is a hollow cavity lined by secretory phagocytic and absorptive cells. In Hydrozoa, this is a simple sac into which the mouth opens directly; in Scyphozoa, the mouth may open into a short gullet which leads to a central expanded area, the stomach, from which slender tubes, the radial canals, extend to the margin of the body where they join the circular, or circumferential, canal; in Anthozoa, the mouth opens into a stomodaeum, or pharynx, leading to the gastral cavity which is partially divided by septa (mesenteries) or into a number of chambers in free communication with the central cavity (Figs. II–7 and II–8). Each septum is an internal extension of the inner layers of the body wall and consists, therefore, of apposed layers of gastrodermis separated by mesoglea. The free edge of each septum is known as a septal or mesenterial filament; above the pharynx each filament is trilobed, with the central lobe consisting of nematocytes and zymogen cells, and the lateral lobes of ciliated, phagocytic cells (Fig. VIII–3). Movement of the flagella contributes to the circulation of fluids in the gastral cavity. Below the pharynx there are no lateral lobes. In the common sea anemone, *Metridium*, the middle lobe is greatly extended toward the base of the column as a long, slender thread, or acontium, that can, on occasion, be protruded through the mouth or through pores in the body wall.

b. DIGESTION IN *HYDRA* The histology of the gastrodermis and the nature of the digestive processes have been studied to the greatest extent in *Hydra* and in some of the Anthozoa, forms that can be satisfactorily maintained in the laboratory. The digestive region of *Hydra* occupies about the middle third of the column, in which techniques of differential staining reveal three major cell types, in addition to interstitial cells. In well-fed animals, about 72 percent of the cells in this region are columnar epithelial cells, usually with two to four flagella, each with its own basal body and with many vacuoles of heterogeneous content. The apical ends of these cells are broader than the basal and have many fine processes, identifiable by electron microscopy as branching and anastomosing microvilli. Their free surfaces are covered with a feltwork of fibrillar material, like the mucopolysaccharide coating of amebae, which provides in a similar way for the attachment of particulate material. Particles of ferritin, an iron-protein complex experimentally introduced into the gastrovascular cavity, adhere to this layer and become incorporated into pinocytotic vacuoles within the cells. The covering of these endodermal cells is quite different, in this respect, from that of the epidermal cells which serves to block the entrance of material rather than to facilitate it. The remaining 28 percent of the cell population is glandular, the majority (25 percent) cytologically like acinar cells of the vertebrate pancreas, while the remainder (3 percent) are mucus producing and cytologically like the mucous cells of vertebrates. After 12 days without food, the number of zymogen cells of a hydra increases by about 50 percent, while there is no significant change in the number of mucous cells.

Daphnia fed to starved hydras are broken down within 4 hr in the gastral cavity into small particles, which are primarily of protein although some may have a relatively high content of lipid or polysaccharide. This breakdown is accomplished by extracellular enzymes, which is followed by endocytosis of the particles and the completion of their digestion intracellularly. Proof of extracellular proteases has been derived from tests of the contents of the gastral cavity, obtained most effectively by electrically stimulating (60 to 80 volts current) the animals to regurgitate at an interval after feeding (about 5 hr) when the particulate material has been engulfed. The expelled material may therefore be expected to contain little but undigested food and extracellular enzymes. Protease activity, as measured by degradation of hemoglobin, was found to a considerable degree over a wide range of pH, with the maxima at pH 2.0 to 3.0 and at 7.0. This would suggest that at least two enzymes are present. Yet, although these extracellular enzymes may bring about degradation of some of the protein in the food of *Hydra*, there is also evidence that much of it is taken up in unhydrolyzed form and that its entire digestion is intracellular, a process that is of primary importance to *Hydra*.

Experiments in which hydras were induced to eat mouse liver labeled with S^{35} also show that protein is only partially decomposed while in the gastrovascular cavity, in whose contents there was no accumulation of peptides and free amino acids. By radioautography it was shown that the label was in the gastrodermis a few hours after feeding and by fractionation procedures that this was associated with large peptide molecules. Twenty-four hours after feeding, much of the radioactivity was demonstrable in the ectoderm, with proportionate decrease in the endoderm.

Events in the digestive process can be followed to some extent in animals that have been starved 12 to 18 days and then fed natural food, or test meals. Twenty-four hours after feeding with nauplii of *Artemia* many protein droplets can be seen in the basal two-thirds or three-quarters of the cytoplasm of the phagocytic cells. Variation in the reactions of these droplets to the same stains indicates that they represent different stages in the proteolytic process. They disappear completely after 10 days of fasting and, therefore, furnish a protein reserve upon which the animal can draw under stress.

There is no evidence of an extracellular lipase, and test substances, such as cod liver oil or oxyethylene oxypropylene polymer, fed to fasting animals as 7.5 percent emulsions are taken up by the phagocytic cells within 40 min. The smaller droplets are pinocytosed, the larger ones are phagocytosed. These events, like phagocytosis of the protein particles, take place only along the distal border of the cell and not along its lateral margins, which interdigitate with those of adjacent cells and are also bound to them by desmosomes. Twenty-four hours after the ingestion of lipid, 95 percent of the animals observed had great accumulations of droplets, almost exclusively within the gastrodermis, and for the most part concentrated in the basal two-thirds or three-quarters of the phagocytic and pinocytotic cells, although some of these might be literally filled with engulfed material. As with protein, there is marked depletion of this material in most animals that have fasted 12 to 18 days, although there is great individual variation in the extent to which the lipid reserves are utilized.

Cytochemical tests show that 24 hr after a meal of *Artemia* nauplii glycogen is present as dense masses of granules 200 to 400 Å in diameter throughout the cytoplasm of the flagellated cells, between the zymogen granules of the secretory cells, in the mucous cells, and in the mesoglea. Its appearance among the lipid droplets after starved animals have been fed only emulsified fat suggests that it may be formed locally in relation to lipid ingestion and that lipid may be converted to carbohydrate in the phagocytic cells. There is at present no direct or positive evidence for either event. Glycogen is likewise demonstrable in the intracellular spaces of both epidermis and endodermis of well-fed animals, giving some support to earlier ideas that the mesoglea serves as a medium of transport, since these intercellular spaces are in direct communication with the mesoglea.

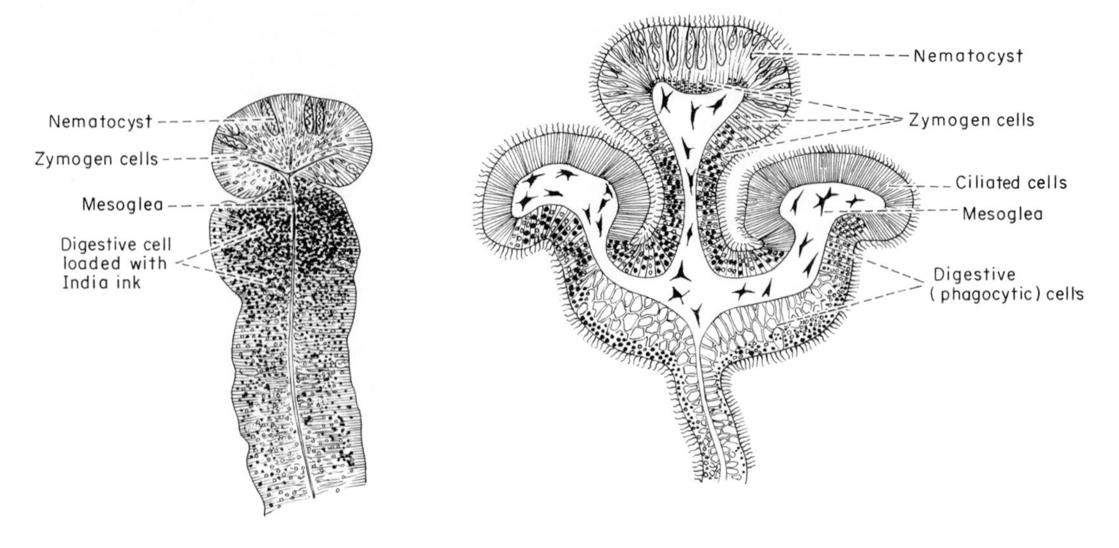

FIG. VIII–3 *Transverse sections of mesenteric filaments of Anthozoans.* ***(a)*** *Transverse section through the free edge of a mesentery of* Astrangia danae, *below the pharynx, after the polyp had been fed a mixture of pounded crab meat and India ink. (After Boschma.)* ***(b)*** *Transverse section through the free edge of a mesentery of* Sagartia, *above the pharynx.*

c. DIGESTION IN ANTHOZOANS AND SCYPHOZOANS Studies of digestion in anthozoans show that the septal filaments are the primary areas of both extracellular and intracellular digestion, and that extracellular digestion is of considerable importance to them (Fig. VIII–3). Extracellular digestion of proteins was conclusively shown 50 years ago when fibrin, enclosed in filter paper sacs, was put into the gastral cavity of *Anemonia sulcata*, for the fibrin was digested in spite of the fact that it was never in direct contact with the septal filaments. Strong protease activity has been found in extracts of the septal filaments of *Calliactis parasitica*, demonstrated by reduction in turbidity of suspensions of homogenized boiled egg white after incubation with them. In vivo, this anemone can digest 1 gram (wet weight) of gelatin pellets within 24 hr, at a temperature of 12°C. When polyps of the coral *Astrangia* are fed living copepods stained with neutral red, it can be seen that the crustaceans are caught in the coils of the filaments and quickly immobilized. Fifteen to thirty minutes after its ingestion the surface of a copepod, pressed against the filament, loses its color, and red vacuoles are evident in cells along the face of the septum. Two hours later, almost all the soft parts have been reduced to a state in which they can be taken up by the digestive cells, while the skeleton remains intact and is ultimately egested through the single opening of the gastral cavity. When the extracellular digestion is complete, the gastrodermal cells in general are loaded with red vacuoles. Small particles from test meals such as pounded crab meat mixed with India ink, carmine, or litmus as tracer substances are taken up directly by the phagocytic cells, which, within 1 hr after the ingestion of such a meal, are crowded with vacuoles. The particles mixed with litmus are blue (alkaline) while in the gastral cavity but turn red (acid) after being taken up by the digestive cells and remain so for about 48 hr, after which the vacuole becomes bluish. Conditions, as indicated by this means, seem therefore to parallel those in the gastric vacuoles of *Paramecium*. In *Astrangia*, the second alkaline phase is probably the period of absorption of the final products of digestion and certainly that of egestion of undigested material.

Digestion in scyphozoans follows a similar course. Extracellular enzymes are secreted by zymogen cells of the gastric filaments that extend from the septa between the gastric pouches, and the gastrodermis as a whole takes up the decomposed material and completes its digestion intracellularly. Similarly, in ctenophores, digestion is initiated extracellularly in the pharynx and completed intracellularly in the cells lining the stomach and canals.

3. Platyhelminthes

a. GENERAL ANATOMY OF DIGESTIVE SYSTEM In some free-living flatworms digestion is entirely intracellular, in some it is entirely extracellular, and in others both intracellular and extracellular. The means used by any given species seem more clearly related to the type of food that is eaten and to the efficiency with which it is mechanically disintegrated by muscular movements of the pharynx or other parts of the body than to any evolutionary trend. Probably the most primitive mechanisms are exhibited by acoelous turbellarians such as *Convoluta paradoxa* and *C. roscoffensis*. In rhabdocoeles, the short pharynx leads into a single sac-like cavity. In triclads, the digestive region of the gut has three primary branches, one extending anteriorly and two postero-laterally, with each branch having many lateral extensions. In polyclads, the enteron is a single central tube with many lateral branches, each of which branches again and again; the terminal ones sometimes anastomose. In all, the lining of the gut is a single layer of epithelial cells resting on a delicate basement membrane. Most of them are relatively tall, columnar ones, ciliated in some species, that contain numerous vacuoles with food in different stages of digestion and are clearly phagocytic. The uptake of macromolecules by pinocytosis has been shown in *Dugesia trigrina* and *D. dorotocephala*. After a period of starvation, the animals were fed raw kidneys from mice that had previously received intracardiac injections of horseradish peroxidase. Three hours after feeding, peroxidase could be demonstrated in vacuoles in the cells of the gut; the vacuoles reached their maximum size in 24 hr and disappeared completely 8 days after feeding. Peroxidase could also be found in vacuoles in cells of animals whose pharynges had been removed before exposure to solutions of peroxidase, suggesting that the protein must have been taken in first by the epidermal cells and that this is a possible means by which regenerating pieces may acquire their nutrients until the pharynx is formed and normal feeding resumed. The other cells of the gastrodermis are smaller and contain granular inclusions giving positive reactions to tests for proteins; these have been designated as "sphere cells" or "club cells" and are secretory in function, discharging an endopeptidase into the lumen of the gut.

b. DIGESTION IN TURBELLARIANS Extensive studies of the digestive mechanisms of some turbellarians have been made by starving freshly captured animals until their guts were empty and then following the fate of natural or test food ingested by them. Histochemical tests for hydrolytic enzymes have also been carried out. The rhabdocoele *Macrostomum* makes use of both luminar and intracellular digestion. *Daphnia* which are captured by the ciliated pharynx can be seen intact in the gut shortly after ingestion, but after 10 hr they have undergone disintegration and the pieces of suitable size have been phagocytosed. Yet another genus of rhabdocoeles, *Stenostomum*, which lives also in ponds and feeds on similar kinds of animals, breaks its food up mechanically by violent contractions of the gut and digests the pieces entirely intracellularly. A large ciliate may be broken up in 15 min into pieces that can be phagocytosed; a longer time is required for the disintegration of rotifers, whose soft parts are phagocytosed while the cuticle is eliminated through the mouth. The triclad *Polycelis*, living in fast running streams, processes its food in the same way and relies largely upon intracellular digestion, although the demonstration of proteases, of the type that initiate proteolysis, in the gland cells of the pharynx shows that chemical disintegration is at least adjuvant to mechanical. Some luminal digestion of protein may also occur, but the terminal stages, as well as the entire digestion of carbohydrate and fat, take place within cells. Histochemical tests show that boiled starch paste remains unchanged in the lumen of the gut but is quickly phagocytosed and digested in food vacuoles in the tall columnar cells, and in mesenchyme cells as well. Suspensions of cod liver oil stained red with Sudan III were introduced into the gut using gammarids as carriers; the oil droplets, after disintegration of the crustacean, were taken up unchanged by the phagocytic cells and digested in them, but they were not taken up by cells of the mesenchyme.

Prey of the marine polyclad *Leptoplana* is broken into small pieces by muscular movements of the entire body and of the pharynx before it is ingested. Digestion is begun in the lumen of the gut but is completed intracellularly. Digestion is wholly luminar in another member of this order, *Cycloporus*, that seems to feed exclusively on the tunicates *Botryllus* and *Botrylloides*. The zooids, sucked out of their cases, reach the gut almost undamaged but rapidly disappear from it with no indication of the engulfment of any particulate material. Luminar digestion of starch paste has also been observed in *Leptoplana*.

Histochemical tests for hydrolytic enzymes, specifically acid phosphatase, amino peptidase, and beta glucuronidase, have been carried out on the triclads *Dugesia tigrina* and *D. dorotocephala*. The animals were starved at least 3 weeks and then fed raw and boiled liver. No differences were found in enzyme activity in either case, making it clear that the enzymes demonstrated were endogenous. No activity of any of these enzymes was demonstrable in the cells of the gastrodermis of starved animals, although there was some amino peptidase activity in their mesenchyme cells, which increased as the fasting period was extended, suggesting that it was involved with utilization of the animal's reserves. Some acid phosphatase activity in the vacuoles of the phagocytic cells of the gut could be seen as early as 5 min after feeding, becoming more intense with time, so that 4 to 6 hr after the meal the vacuoles were very strongly reactive to the test. Twenty-four hours after the meal, there was intense acid phosphatase activity throughout the gastrodermis, evident in the cytoplasm of the phagocytic cells as well as in the vacuoles, but none in the mesenchyme. This intensity persisted for 48 hr, then diminished. Aminopeptidase activity could be demonstrated in the phagocytic cells 30 min after feeding and became most intense 24 to 48 hr later. No reaction could be obtained in the gastrodermal cells 7 days after feeding, although there was some in the mesenchymal cells. Tests for beta glucuronidase were positive in the gastrodermis 24 to 48 hr after feeding. The early appearance of acid phosphatase would seem to indicate the first step in the transition of an ingestion to a digestion vacuole, which, on the basis of the synchrony of maximum activity of the three enzymes investigated, is in full operation 24 to 48 hr afterward. These facts lend support to the lysosome concept of the source of intracellular hydrolases.

In turbellarian flatworms practicing intracellular digestion, either exclusively or in conjunction with extracellular, the events take place in phagosomes of the columnar cells of the gut. These may also act as storage cells, accumulating reserves, or they may pass the products of their digestive activity on to mesenchyme cells for resynthesis into reserves or for further digestion. Starch in the course of digestion was, for example, shown in mesenchyme cells of *Polycelis*.

c. DIGESTION IN TREMATODES

Less is known about digestive processes in trematodes, but present evidence indicates that they are both intra- and extracellular. Light and electron microscopy of the gut of *Fasciola hepatica* shows that there is little differentiation of cells and that most of them pass through absorptive and secretory cycles. Fine processes extend into the gut lumen from the free ends of these cells when in their absorptive phase which become less numerous in the secretory phase. *Fasciola*, like *Schistosoma* and *Polystoma*, feeds on the blood of its host and, like them also, begins the digestion of its food in the lumen of the gut. When reduced to soluble form it is engulfed, and further digestion proceeds within the cells of the gut wall. An indigestible residue, the iron-containing pigment hematin, accumulates in these cells and is eventually egested, either in vacuoles or with the complete disintegration of the cell. The egested material is the probable source of the extracellular enzymes, and, if it actually is, trematodes as well as acoelous turbellarians would furnish illustration of the transition from primitive to more advanced conditions of digestion.

In some other blood-feeding species and in some species that eat tissues other than blood, as well as mucus and other secretions and even fragments of the host's food, digestion is largely or entirely extracellular. In these blood-feeding species, hematin is not a by-product of hemoglobin digestion and the excess iron is eliminated in another form.

4. Nemertinea: anatomy of digestive system

Nemerteans employ both extra- and intracellular mechanisms. The events have been most closely followed in *Lineus ruber*, captured from its habitat under rocks at the midtide level off Plymouth, England. The digestive tract in all nemerteans is complete; in *Lineus* it consists of a buccal cavity opening into a foregut that occupies about one-tenth of the animal's body length and leads to the intestine that runs throughout the remainder to the terminal anus. The area of the intestine is greatly increased by paired, serially repeated pouches. The foregut is lined with a single layer of low cuboidal epithelium; the cells are ciliated and some of them are glandular, secreting mucus that serves as a lubricant for the ingested food. The wall of the intestine is a single layer of cells resting on a delicate basement membrane. The majority of these are columnar and ciliated, but between their bases are smaller, nonciliated cells containing zymogen granules. The zymogen cells are most numerous in the anterior region of the intestine, the proportion of them in relation to the ciliated cells gradually decreasing posteriorly; there are none in the unpouched region near the anus. The food, mainly small annelids and crustaceans with varying amounts of organic detritus, is swallowed whole. Living animals are killed almost immediately in the foregut, presumably by acid discharged into its lumen. Ten to fifteen percent of the cells in the foregut give a positive reaction to cytochemical tests for carbonic anhydrase, an enzyme known to be associated with the production of hydrochloric acid in the mammalian stomach. This similarity suggests that hydrochloric acid may be the lethal agent in the secretions of the foregut. Within minutes after its arrival in the intestine the food is reduced to a fluid mass, whose fine particles are engulfed by the ciliated cells. These cells would seem especially fitted for the propulsion of food along the intestine, but their phagocytic capacities have been quite definitely proved by the presence of starch grains in them after a test meal. It has been reported that when phagocytosis is in progress the cilia fuse to form pseudopods, a change that would lend support to the idea of close structural and functional relationships between these two types of organelles. It suggests also that the cells are not highly differentiated and are still labile enough to assume two quite different roles in the digestive process.

From cytochemical evidence it seems apparent that luminar digestion concerns only the initial degradation of protein and that its final breakdown, as well as the complete digestion of carbohydrate and fat, is intracellular. An endopeptidase has been demonstrated in the intestinal gland cells of starved animals. Thirty minutes after feeding, the majority of these cells no longer give a positive reaction to tests for this enzyme, indicating its discharge into the lumen, but 6 hr later, they are loaded with it. Aminopeptidase, lipase, and a presumed carbohydrase could be found only in the phagocytic cells several hours after starved animals had been fed. No acid phosphatase was found in any region of the gut, but a strong alkaline phosphatase reaction (pH optimum 8.0) was given by the gland cells immediately after feeding, when discharge of secretion was in progress. The ciliated cells of the intestine, but not those of the foregut, showed constant alkaline phosphatase activity along their free borders; after feeding and phagocytosis, the reaction was general throughout these cells.

5. Endoprocta

Although there is little detailed information about digestive processes in endoprocts, the histology of the stomach in *Berentsia* and *Pedicellina* suggests that in these two genera at least both luminar and intracellular mechanisms are used. In the gastric epithelium of these animals, there are cells that are typically glandular in appearance as well as tall columnar ones with microvilli on their free borders. In the pyloric region of the stomach of all endoprocts there are ciliated cells. Beating of the cilia rotates a firm strand, made primarily of mucus and fecal pellets, and draws it through the stomach. This strand acts as a food string, for particles adhere to it as they are swept into the mouth by the cilia on the lophophore. They become detached from it as they reach the relatively acid region of the stomach, for the viscosity of mucus is reduced with increased acidity and it is almost fluid at pH's below 5.5. Some digestion of these detached particles very likely takes place in the lumen of the stomach and the remainder within cells. Some particles may be directly engulfed and digested entirely intracellularly.

The principle of a rotating strand or a rod is a sound one mechanically, for by it power, generated in a localized area, can be economically transmitted to operate some distance from its source. It is a mechanism common among microphagous metazoans, especially the ciliary feeders, and has evolved in quite divergent groups of invertebrates as well as in protochordates. It has developed in all the lophophorate phyla (endoprocts, ectoprocts, brachiopods, and phoronids), in molluscs (gastropods, pteropods, and bivalves), in urochordates, hemichordates, and the cephalochordate *Branchiostoma*. This is clearly evidence of convergent evolution, correlated with the food and feeding habits of the animals. In addition to being a mechanism that provides mixing and movement of the food along the gut, independent of peristalsis, it serves as a sorting mechanism and a means of selecting and rejecting what is in the gut on the basis of size and nutritive value.

6. Brachiopoda

Brachiopods have a pharynx, an oesophagus, a large stomach with a digestive gland that opens into it by one or more pairs of ducts, and an intestine that in the Articulata ends blindly but in the Inarticulata opens through an anus. The passage of food within the digestive tract has been carefully followed in specimens of the inarticulate *Lingula unguis* young enough to have transparent shells, and the nature of the digestive enzymes and their sites of operation have been determined by analyses of the gut contents and of extracts of selected regions of its walls. The stomach is divisible into two regions—a large, thin-walled highly distensible anterior chamber, into which the ducts of the four digestive glands open, and a posterior (pyloric) chamber, whose diameter is small and whose thick walls do not permit any marked dilation. Three of the four digestive diverticula lie dorsal to the stomach and one ventral. These open into the stomach by wide ducts, not guarded by valves, which then divide and subdivide, each final subdivision terminating in a tubular or globular sac, or acinus. The entire canal is lined with simple columnar epithelium, ciliated everywhere except in the tubules or acini of the digestive diverticula. In the most posterior of the dorsal diverticula, and in this one only, there is a deep groove lined with low columnar, ciliated cells that courses through the duct system and continues into the posterior chamber of the stomach. Isolated mucous cells are scattered among the ciliated cells of the epithelium in general and groups of them are found in the oesophagus and posterior stomach. Migratory amebocytes are found in both the epithelia and the lumina of the stomach, in the digestive diverticula, and in the intestine. These become more numerous after a meal.

When specimens with transparent shells are fed yeast mixed with Congo Red, it can be seen that the particles are trapped in a mucous strand that is moved through the mouth into the pharynx by cilia on the cells of the lophophore and carried along to the stomach by the beating of the oesophageal cilia. The effective beat of the cilia along the ventral wall of the anterior region of the anterior chamber of the stomach is also posterior, but along the right wall it is dorsally directed and along the dorsal and left walls, ventrally, so that the contents are rotated. In the posterior part of this region, the ciliary currents on the right wall move directly posteriorly, while along the left and dorsal walls they are directed obliquely posteriorly, passing the openings of the more posterior of the digestive diverticula. Although the cilia in the ducts of the digestive glands create a peripheral inward current and an axial outward one, food is drawn into them and pumped out of them by contractions of muscles in their walls. These pumping actions are very vigorous after a full meal and may continue for some hours, during which rotation of the food in the anterior part of the stomach stops. Ciliary currents in the intestine are also directed posteriorly but the main propulsive force is given the fecal mass by peristaltic contractions of the walls. In the stomach and digestive glands and in the intestine, the ciliary currents serve more to agitate the contents than to move it along or in and out of the glands. Particles dropped into the ciliated groove of the posterior dorsal diverticulum of a dissected specimen became entangled in a mucous cord that was moved along the groove into the intestine where it became included in the fecal mass. The function of this groove in *Lingula* at least seems, therefore, to be concerned with the removal of excess or undesirable material rather than with the movement of a food string, although the requisites for such a mechanism are there. The stomach of *Lingula* has much in common with the midgut of ectoprocts and with the stomach of *Phoronis*, both of which move their food mass by rotation of a mucous cord.

Tests for enzymes in *Lingula* showed that protease and lipase were present only in extracts of cells of the digestive diverticula; neither was found in the contents of their lumina. On the other hand, amylase and other carbohydrases, such as lactase, maltase, and invertase, were found in the contents of their ducts and tubules as well as in the contents of the intestine. Digestion of proteins and fats would therefore seem to be entirely intracellular and that of carbohydrate extracellular. Phagocytosis by cells of the stomach, digestive glands, and intestine has been observed, with the greatest phagocytotic activity in cells of the diverticular tubules and acini, which appear to be the major sites of intracellular digestion. Egestion vacuoles have also been seen to be pinched off the free borders of these nonciliated cells and to be set free in the lumen of an acinus.

7. Mollusca

a. GENERAL ANATOMY OF DIGESTIVE SYSTEM The molluscan digestive tract consists typically of a buccal cavity, into which salivary glands open; a stomach with paired diverticula, the digestive glands; an intestine, often of considerable length and coiled around within the body; and a rectum, opening at the anus. This is the pattern found in *Neopilina* and in modern polyplacophorans and aplacophorans, but it has undergone profound modifications in members of the other classes. In lamellibranchs there is neither buccal cavity nor salivary glands, while in gastropods and cephalopods the buccal cavity is large enough to accommodate the odontophore, and the salivary glands are of considerable size and important in the ingestion of food and preliminary stages of its digestion. The stomach of the lamellibranch is capacious and has become a most efficient area for sorting the particulate material swept into the mouth from the mantle cavity. In less highly specialized molluscs, it is a slender sac tapering gradually off into the wide anterior region of the intestine; in the limpet *Patella* it is so reduced in size that it is only large enough for the openings of the two digestive glands (Fig. VIII–4).

b. PROTOSTYLE AND EVOLUTION OF LAMELLIBRANCH CRYSTALLINE STYLE Primitively, the epithelium of the digestive tract is composed of columnar ciliated cells. In the intestine the cilia beat transversely across the lumen and their motion rotates a rod of mucus, secreted by the epithelial cells, which projects into the stomach. Food moves along the oesophagus entangled in a mucous string which is propelled into the stomach by cilia of the oesophageal cells. As the food string meets the head of the intestinal mucous rod, it is wound around it in a tight spiral and so continuously drawn through the stomach as a rope can be drawn along by a windlass or a capstan (Fig. VIII–5). This rod is regarded as a protostyle, a forerunner of the more elaborate crystalline style that has evolved in lamellibranchs, and the area of the intestine in which it is formed and rotated is known as the style sac. A style sac and rotating rod are found in *Neopilina*.

The role of a protostyle is apparently entirely mechanical, effecting stirring and mixing of the food and entanglement of waste. Primitively, the gastric epithelium of molluscs is elevated in ridges or depressed in furrows. Food particles, detached from the mucous string, which is comparatively fluid as it passes through the stomach where the pH is on the acid side of neutrality, fall on these ciliated cells. The smaller, lighter ones are swirled about by the cilia on the ridges and are ultimately swept into the openings of the digestive

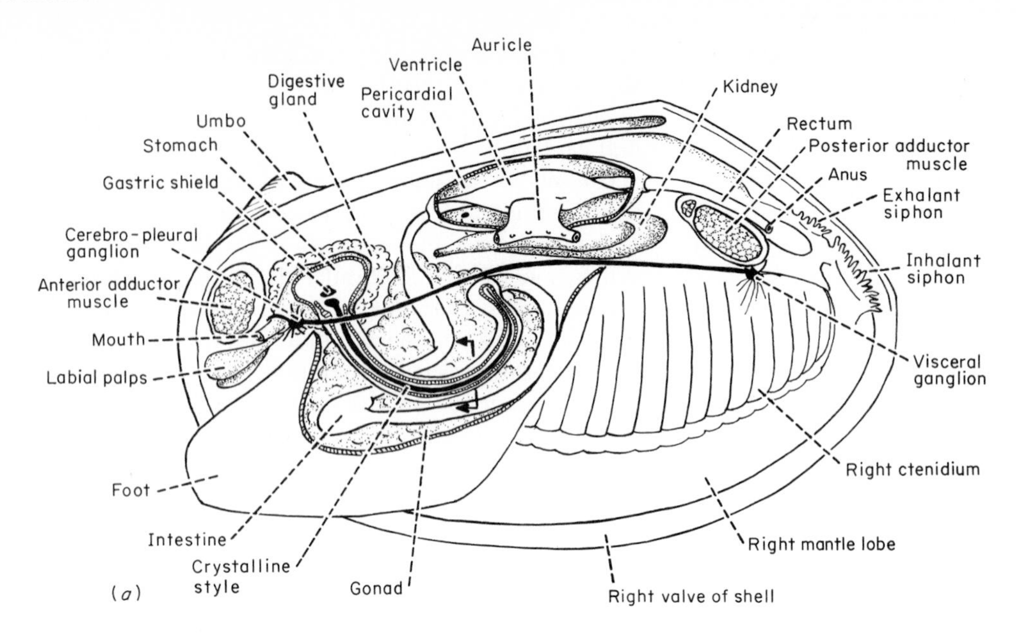

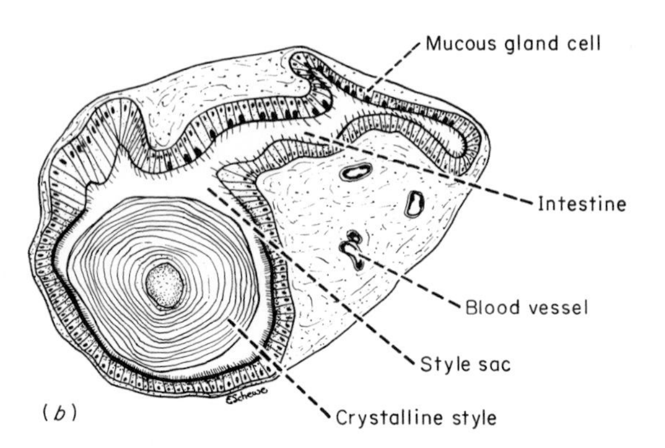

FIG. VIII–4 Anodonta grandis, *a freshwater mussel. (After Nelson.)* ***(a)*** *Internal anatomy.* ***(b)*** *Cross section through intestine at area indicated by arrows in (a), showing style sac and crystalline style.*

glands. The larger, heavier particles drop into the furrows and are swept into the intestine where they become firmly bound to the mucous rod, here very viscous because of the alkalinity of the intestinal contents. Bits of the rod, with this rejected material as well as waste adhering to it, are cut off from time to time as fecal pellets. These are moved along to the anus by cilia on cells in the more posterior region of the intestine. Rotation of the style has been seen in living gastropod larvae, young enough to be transparent, and clocked at 40 turns per minute in veliger larvae of *Struthiolaria* and 60 to 70 turns per minute in oyster spat.

In the evolution of molluscs using the style mechanism there has been a progressive separation of the style sac from the intestine proper. This has taken place through elevation of the wall of the intestine around the rod, so that it lies in a groove but is in free communication with the lumen. With further development the sac has become completely cut off from the intestinal lumen by apposition of the ridges (typhlosoles), and only the head of the style, which

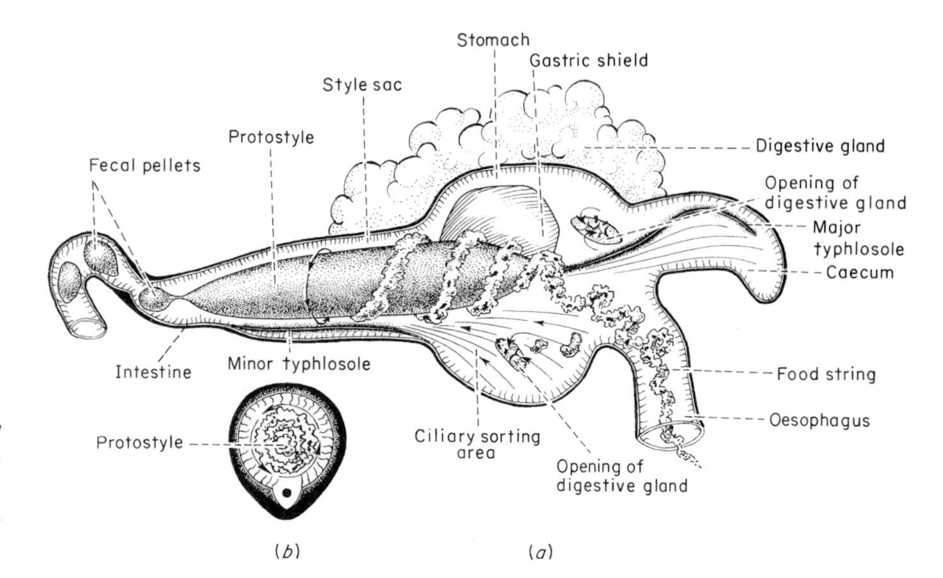

FIG. VIII–5 ***(a)*** *The generalized form of the digestive tract in an early prosobranch mollusc.* (*After Morton.*) ***(b)*** *Transverse section across the style sac.*

protrudes into the stomach, is in contact with the lumen of the gut. In this condition, the style no longer directly contributes to formation of the feces. Another adaptation, especially evident in species with a well-developed style mechanism, is a translucent plate covering an area of the gastric epithelium near the head of the style. This is called the gastric shield. It is essentially a localized peritrophic membrane, which is secreted in droplets by the underlying epithelial cells. Analyses of the composition of the shield in *Crassostrea virginica* and in the gaper clam *Schizothaerus* show that chitin is its major, if not its sole, component. Its function is to prevent abrasion of the delicate epithelium by hard particles that might be tossed or rubbed against the stomach walls by the head of the rotating style.

In lamellibranchs, the protostyle has evolved into the crystalline style, whose function is both mechanical and chemical. It has long been established that crystalline styles are a source of extracellular amylase and that the preliminary digestion of starch takes place in the lumen of the lamellibranch stomach. More recently it has been shown that styles isolated from oysters have a very rapid effect upon the red alga *Cryptomonas*, a natural source of food for them. Living cells of *Cryptomonas* disintegrated immediately on coming into contact with an undissolved style in a hanging drop preparation, yet those of another species, *Isochrysis*, showed no ill effects from it. It therefore seems likely that the styles of oysters contain, in addition to the more general amylase, an enzyme that can digest the particular polysaccharide of *Cryptomonas* but that has no effect upon the polysaccharide of *Isochrysis*. Solutions of styles from oysters and mussels reduce the turbidity of suspensions of finely divided filter paper, suggesting that they have a cellulase, and those from the clams *Mactra* and *Mya* erode the regenerated cellulose of which Visking casing is made. There is also experimental evidence to show that the styles of *Crassostrea virginica* and of *Modiolus demissus*, the ribbed mussel, provide extracellular lipase. In these experiments, emulsions of olive or peanut oil, stained with Sudan Black or Sudan III, were fed to the animals, and the stomach contents, withdrawn at intervals by means of a pipette, were tested for the conversion of neutral fat into fatty acids. When the withdrawn fluid was exposed to Nile Blue Sulfate, a dye that stains neutral fat pinkish orange and fatty acids blue, blue droplets were found in it. An extract from minced styles also hydrolyzed these fats, showing that the style was a source, if not the exclusive one, of lipase. There is no convincing evidence that, in any mollusc, the style provides a proteolytic enzyme. In a number of species the style has been found to harbor microorganisms, which may be responsible for some of the enzymes that have been attributed to it. A large spirochaete *Christospira*, for example, has been obtained from styles of at least 15 species of lamellibranchs, with which their symbiosis seems so complete that they have never been successfully cultured in vitro.

Styles of oysters have a fluid core enclosed in concentric layers of gelatinous mucoprotein. There are no critical data as to the origin of the style; presumably it is continuously secreted by cells of the typhlosoles because it does not diminish in size, although its surface is continually dissolving away in the acid contents of the stomach. In some species, cessation of feeding results in dissolution of the style, which is replaced when active feeding begins again. This is a cyclic event in animals living at high tidal levels. Mucous glands are, however, very generally distributed throughout the gut epithelium of molluscs, and there is no proof that those of the typhlosole are exclusively responsible for the construction of the style.

c. MIDGUT (DIGESTIVE) GLANDS The midgut (digestive) glands of molluscs are paired diverticula of the stomach opening by ducts into its posterior region (Fig. VIII–6). In lamellibranchs, these ducts are grooved and only the cells within the groove are ciliated. The effective stroke of the cilia is outward, creating an exhalant current that produces an inhalant counterpart in the nonciliated region so that a continuous circulation is maintained (Fig. VIII–7). These main ducts branch into nonciliated secondary ducts, each of which terminates in a blind tubule or, in some species, a globular acinus. Ducts and tubules are bound together by connective tissue containing isolated muscle fibers, and the whole gland is an irregular lobulated mass, often of considerable size and occupying a large part of the hemocoele. A few of the cells in the tubules are ciliated but the majority are not and contain secretory granules and phagosomes (Fig.

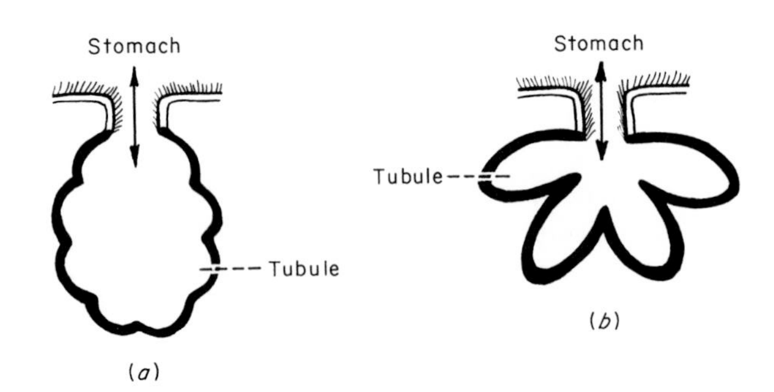

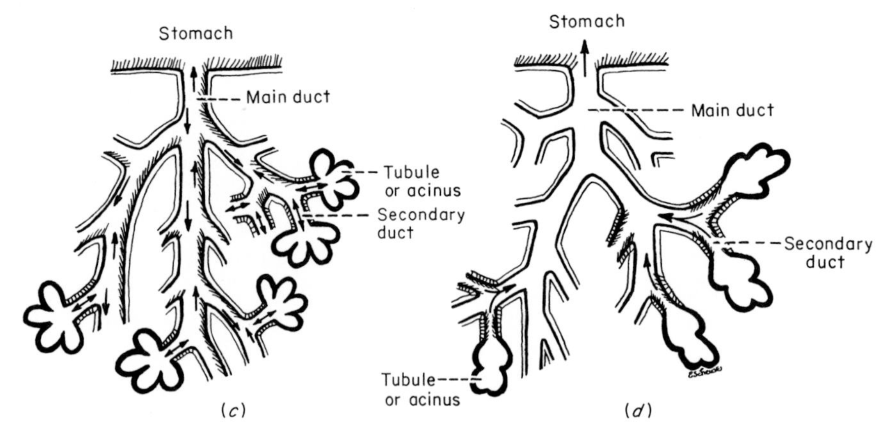

FIG. VIII–6 *The types of digestive glands in lamellibranch molluscs. (After Owen.) The double-headed arrows indicate movement resulting from muscular activity; the single-headed ones, ciliary currents.* ***(a)*** *Hypothetical primitive condition, found in many larval lamellibranchs.* ***(b)*** *Type of gland found in septibranchs and some eulamellibranchs.* ***(c)*** *Type of gland found in most eulamellibranchs.* ***(d)*** *Type of gland found in Nuculidae.*

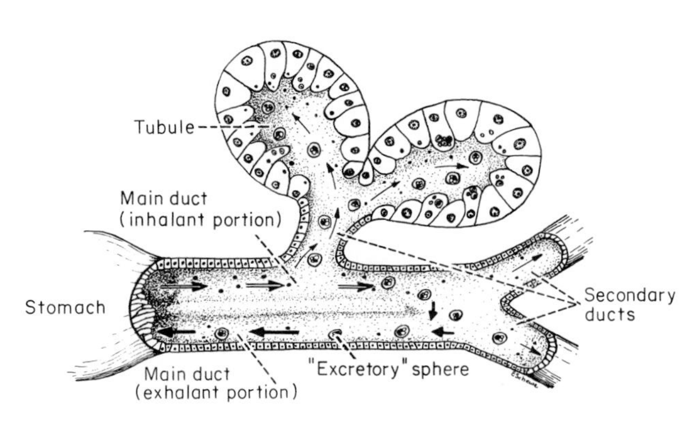

FIG. VIII–7 *The probable paths of circulation of fluids and particles in the digestive glands of Anisomyaria and Eulamellibranchia. (After Owen.) The heavy arrows represent ciliary currents; the arrows with double line, a nonciliary inhalant counterpart current; the fine arrows, movement due to the absorption of fluid by the tubule cells.*

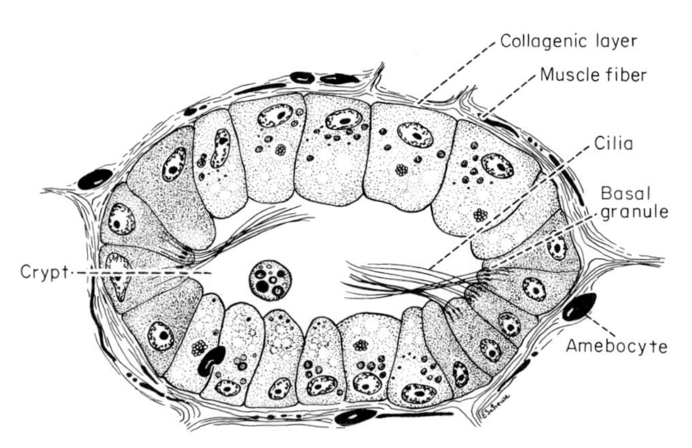

FIG. VII–8 *Transverse section through the tubule of the digestive gland of* Venerupis pellastra, *showing the arrangement of cilia and their relation to the cells of the crypts. (After Owen.)*

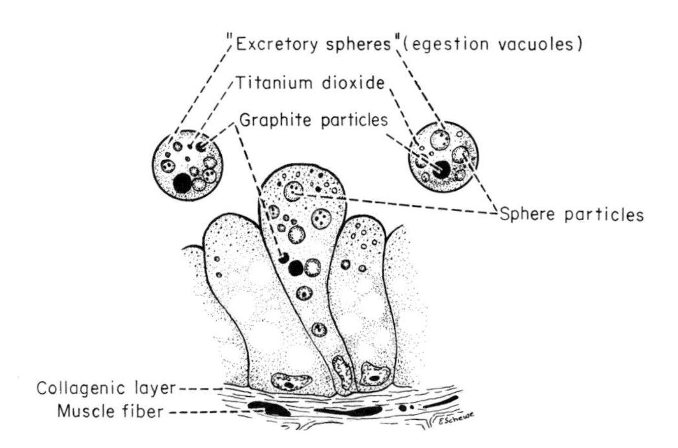

FIG. VIII–9 *Cells of a tubule of* Cardium edule *6 hr after feeding with a mixture of titanium dioxide and India ink. (After Owen.)*

VIII–8). They produce carbohydrases, proteases, and lipase, which are found in the stomach and intestinal juices as well as in extracts of the gland tissues. The relative proportions of the different types of enzymes vary to some extent with the habitual diet of the animal, proteases being more abundant in carnivorous species and carbohydrases in herbivorous ones. Cyclical and seasonal variations in activity are also known to occur in the digestive glands of many species.

There has been much conflicting evidence over the role of the digestive gland in intracellular digestion, but there is a considerable mass of experimental data to show that in lamellibranchs and some gastropods the cells of the tubules take up particulate material. Titanium dioxide, for example, introduced into seawater as a suspension of particles with diameters ranging from 0.5 to 2 μ was evident 3 hr later in the digestive glands of the four species of lamellibranchs tested (Fig. VIII–9). Movement of the particles into the digestive glands and of waste materials out of them is effected apparently entirely by ciliary currents and not by muscular movements such as those observed in *Lingula*, for although there are muscle fibers in the walls of the glands, they are neither sufficiently numerous nor sufficiently organized to function adequately in such a pumping mechanism. When the marine gastropod *Hermaea dendritica* was fed algae with chloroplasts of different size, these could later be recognized in the cells of the digestive glands by which they must have been phagocytosed. Similarly, phagocytosis by these cells has been demonstrated in the carnivorous gastropod *Pleurobrachaea* by the appearance of gold fibrin and carmine in them after these had been introduced into the digestive tracts of experimental animals. Moreover, India ink and iron saccharate, as well as carmine, have been shown to be taken up by the tubular cells of various lamellibranchs and gastropods, indicating their absorptive capabilities if not their phagocytic ones.

The digestive glands of pulmonate gastropods contain lime cells as well as digestive ones. In the Roman snail, *Helix pomatia*, these cells are smaller and less numerous than the digestive cells and are readily distinguishable from them optically because of their triangular shape and content of colorless calcium spherules (Fig. VIII–10). They give negative reactions to tests for such intracellular enzymes as beta glucuronidase, esterases, and aminopeptidase, to which digestive cells of starved and fed animals react positively in varying degrees. The calcium cells serve as depositories for the mineral which is used in construction of the shell. Experiments with species of *Helix, Arion,* and *Agriolimax,* fed lettuce with either P^{32} or I^{131} evaporated upon it, showed that the phosphorus was taken up by both lime and digestive cells. It was stored in the calcium spherules in the lime cells but transferred from the digestive cells to the hemolymph and dispersed to the tissues, where it was particularly concentrated in cells of the nerve ring, odontophore, salivary glands, mucous glands, especially those of the foot, and in the special calcium cells of the mantle. Iodine, on the other hand, was taken up to far less extent by the lime than by the digestive cells and was not accumulated by any tissue cells other than those of the kidney from which it was excreted. It was also eliminated from the digestive cells to become included in the fecal material. These facts indicate that the digestive cells may take up any substance but that its retention and distribution throughout the body depends upon its value to the organism and the demand for it.

d. AMEBOCYTES AS AGENTS OF DIGESTION

Migratory amebocytes are also important factors in molluscan digestion. These may wander from the blood or the connective tissue into the lumen of the stomach and phagocytose particles too large to enter the digestive glands or pinocytose macromolecules such as bovine hemoglobin, human gamma globulin, and albumin, as has been observed in the amebocytes of oysters. They are found afterwards among the cells of the ducts and tubules of the digestive glands, as well as in the blood and connective tissue, and in most molluscs play a significant role in the digestion and transport of food, and the transport of waste. In the quahog *Mercenaria* (*Venus*) *mercenaria,* small yellow granules have been seen in the amebocytes which are

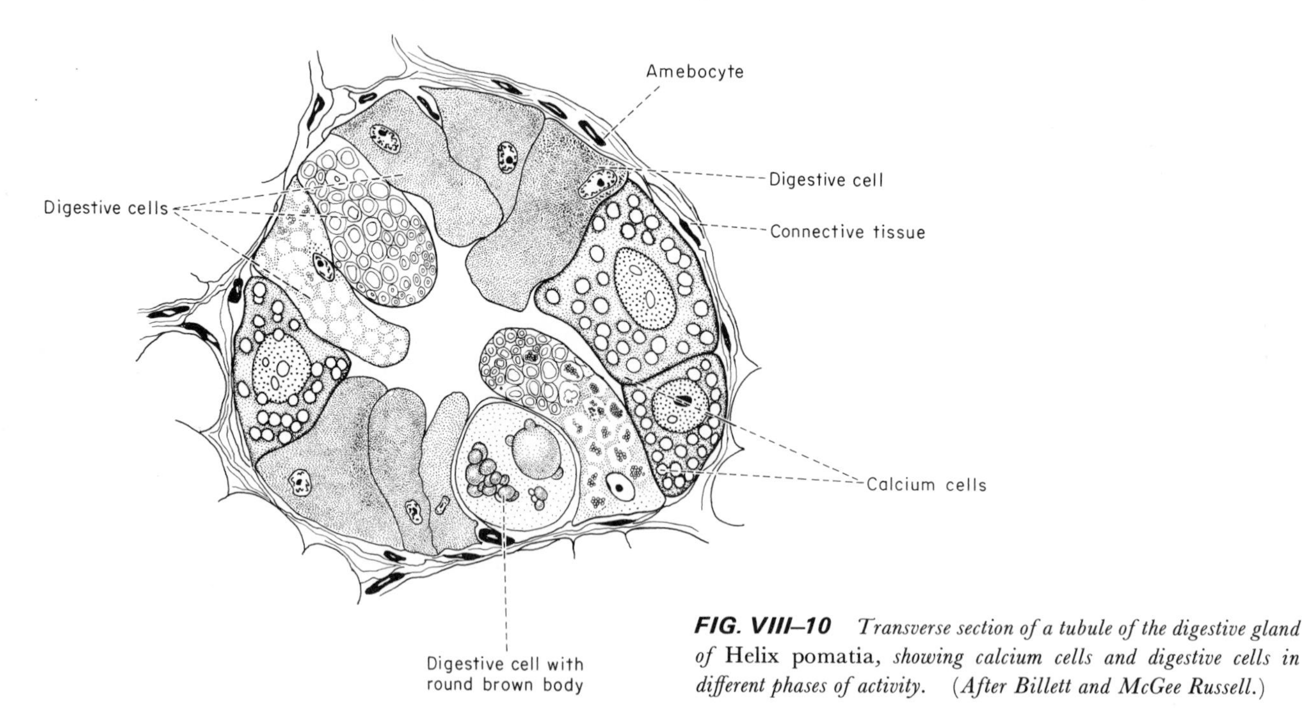

FIG. VIII–10 *Transverse section of a tubule of the digestive gland of* Helix pomatia*, showing calcium cells and digestive cells in different phases of activity. (After Billett and McGee Russell.)*

similar in all respects to the masses of yellow and brownish material packed between the epithelial cells of the gut. This pigment contains phospholipid and has been considered most likely to be a by-product of intracellular digestion in the amebocytes, for it is not found in animals that have gone without food for any length of time. Amebocytes laden with this material may migrate to the intestinal epithelium and deposit it there.

e. COURSE OF DIGESTION Protobranch and septibranch molluscs are exceptional among lamellibranchs in that they do not make use of intracellular mechanisms of digestion. In protobranchs that feed on bottom deposits, like *Nucula*, digestion is wholly luminar, for experiments with tracer particles introduced into the stomach give no evidence for phagocytosis by cells of the digestive gland or of the gut itself, and these molluscs have no amebocytes. They have a protostyle which is concerned only with the elimination of feces; the stomach is small but its wall is muscular and food is ground up in it. Amylase is secreted by cells of the style sac and other enzymes by those of the digestive glands (Fig. VIII–11). In eulamellibranchs, the stomach is less a digestive area than a sifting and sorting one, where ingested particles are graded according to size and mass and not at all, apparently, according to food value. Except for the initial step in the degradation of starch, all digestion takes place in cells of the digestive gland and in amebocytes. Egestion vacuoles, eliminated from cells of the digestive glands, are carried back into the stomach in the exhalant stream and then to the intestine where they are formed into feces along with any ingested material that has been sifted out by the elaborate sorting mechanism of the stomach. It is quite likely that they release some enzymes into the stomach or the intestine in their passage through them. Amebocytes that have engulfed material and are loaded with egestion vacuoles may migrate back into the stomach and disintegrate there. They may, therefore, also contribute some free enzymes before their debris is carried off to the intestine and included in the fecal material.

Gastropods that are ciliary feeders use means of digestion similar to those of eulamellibranchs. The stomachs of such deposit and suspension feeders as *Crepidula, Viviparus, Struthiolaria,* and *Vermetus* are likewise sorting areas, and the crystalline styles in them provide the only extracellular enzymes. All digestion, except the preliminary stages of the breakdown of starches, is intracellular. Some other gastropods that are not ciliary feeders also have a crystalline style, although their digestion is, on the whole, luminar as it is in a few lamellibranchs, the macrophagous gastropods, and all cephalopods.

8. Annelida

The digestive tract of annelids is a continuous tube running from mouth to anus, usually with a quite clearly defined oesophagus, stomach, and intestine, and in most terrestrial species oesophageal and

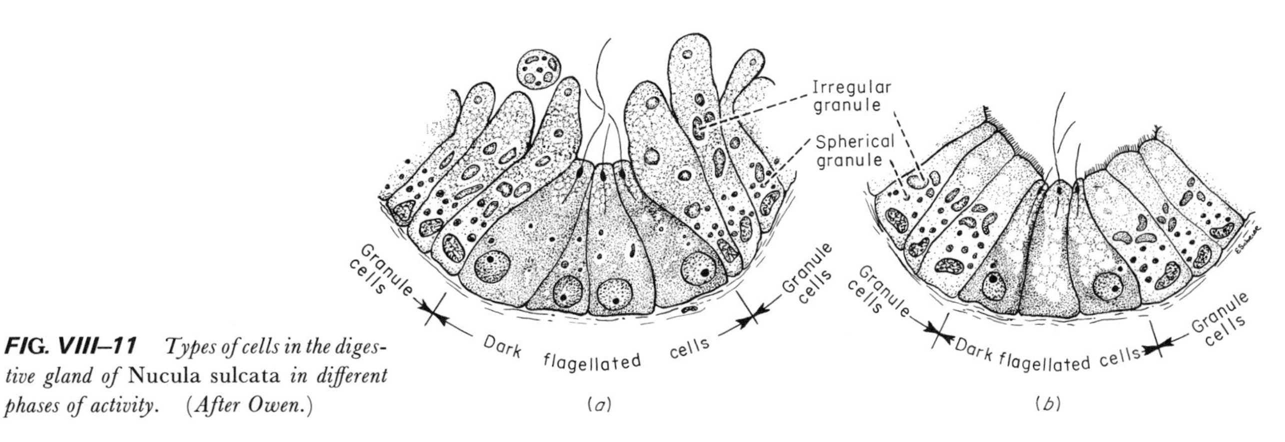

FIG. VIII–11 *Types of cells in the digestive gland of* Nucula sulcata *in different phases of activity.* (*After Owen.*)

intestinal gland cells. In the majority of worms, digestion is entirely luminar and occurs in the intestine, but it is possible that those that eat sand and soil digest intracellularly some of the material they take in. Experiments with *Arenicola marina* have shown that after forcible feeding of a suspension of India ink in seawater to narcotized animals, graphite particles were visible in the epithelial cells of the oesophagus and stomach but not in those of the intestine. The particles were eventually taken up by amebocytes, which were found clumped in the coelom, and ultimately returned by them to the lumen of the intestine and removed with the feces. Presumably, the same mechanism is used for the finer particles of detritus that are drawn in by the proboscis along with the mud and sand; this seems to occur chiefly in the postcardiac region of the stomach, where the mucus, with which the ingested material is lubricated, is less viscous and the action of the cilia of the epithelial cells keeps the mass in motion and brings it into contact with the phagocytic epithelial cells. As this region of the stomach is wider than other areas, movement along the gut is slowed down, which also facilitates phagocytosis. Amebocytes probably play an important role in the actual digestive processes of these annelids, for food particles engulfed by cells of the gastric epithelium move to the basement membrane and are taken up by migratory cells which pass into the bloodstream or among the peritoneal cells enclosing the blood vessels. Material that they cannot digest is brought to the epidermis, the intravascular tissue, or the coelom or gut lumen and egested there.

9. Arthropoda: Arachnida and Acarina

Although in most arthropods digestion is entirely luminar, in some of the arachnids the final stages of protein digestion are probably intracellular, for cells of the stomach diverticula become filled with darkly staining granules after a meal. These gradually disappear so that the cells are empty of them before another meal. Mites are exceptional among arthropods in their apparently exclusive use of an intracellular mechanism. After a meal, the endodermal cells of the midgut put out pseudopodia and engulf the ingested blood, which is then digested in vacuoles within the cells.

H. INVERTEBRATES WHOSE DIGESTION IS EXCLUSIVELY LUMINAR

1. Advantages of luminar digestion

Luminar digestion is characteristic of rotifers, gastrotrichs, kinorhynchs, nematodes, nematomorphs, ectoprocts, echiuroids, annelids, onychophorans, and echinoderms and is practiced also by some flatworms and molluscs; it is apparently universal among vertebrates. It is, on the whole, a more rapid process than intracellular digestion, and it has the added advantage of providing opportunity for division of labor among cells of the alimentary canal and of separating the functions of secretion, absorption, and storage. Moreover, special types of secretory cells may be localized in particular regions of the gut or in glands opening into it, so that digestion may proceed in a series of steps, the food reaching each specialized region in a condition most suitable for the action of the particular enzymes discharged there. It can therefore be a very efficient way of chemically processing food and many of the invertebrates employing it have highly specialized alimentary canals, well adapted to the rapid digestion of the kinds of food they eat.

2. "Aschelminthes"

There is little definitive information about digestion in the animals of the five phyla included under this heading. In most species of rotifers two conspicuous salivary glands open into the pharynx and two gastric glands into the stomach. All of these contain secretory cells whose products are discharged into the lumen of the gut where digestion seems to be entirely extracellular. Absorption of the end products takes place in the stomach, and the undigested residue is carried to the anus through the intestine.

The alimentary canal in nematodes is a straight tube running from mouth to anus and, typically, with buccal cavity, pharynx, intestine, and rectum. Secretions from the pharyngeal glands are involved in external, or extracorporeal, digestion rather than internal, especially in predaceous species. These secretions contain proteases, carbohydrases, and, in some species, lipase and other esterases, all of which operate in the preliminary or the complete digestion of the food before its actual ingestion. In *Ascaris lumbricoides*, secretory cells are more abundant in the anterior part of the intestine than in its midregion or hindregion, and secretion is to a great extent holocrine, although merocrine secretion has also been observed. Most of the hydrolysis of the ingested material, consisting of semidigested food, bacteria, and other contents of the host's digestive tract, takes place in the anterior part of the intestine, where amylase, maltase, protease, peptidase, lipase, and other esterases have been reported to be present. A somewhat wider range of enzymes has been found in some of the tissue-feeding parasites. *Trichuris*, for example, which burrows into the intestinal mucosa of its host, has been found to have amylase, invertase, lactase, and trehalase (an enzyme that splits the disaccharide trehalose into two molecules of glucose), as well as other glucosidases, a "mucopolysaccharidase," protease, and esterases other than lipase. In *Ascaris*, and very likely in other nematodes as well, absorption takes place in the midregion and hindregion of the intestine. Experiments in which ascarids removed from the host have been placed in solutions containing P^{32} incorporated in sodium phosphate show, on autoradioautography, that little of the isotope is present in the cuticle but a considerable amount in the cells of the intestine. Also, when isolated animals have been "fed" selected substances, it has been evident that both amino acids and simple sugars have been absorbed and converted in the tissues to glycogen and proteins characteristic of the body fluid. After being fed with horse hemoglobin, for example, worms showed an amount of hemoglobin in the body fluid greater than that of those on an ordinary diet, but since it had all the characteristics of ascaris hemoglobin it was evident that the horse hemoglobin had not been directly absorbed in the body fluid, but ingested, digested, and its components recombined into the animal's own kind of pigment.

3. Ectoprocta

The digestive tract of ectoprocts consists of a pharynx, a nonciliated oesophagus, a stomach, an intestine, and a rectum. The cells lining the pharynx and posterior part of the stomach are ciliated, and those of the stomach and intestine are glandular. There are muscle fibers in the outer wall of the gut, whose contraction is primarily responsible for the propulsion of food along it. Marine species make use of a rotating mucous cord to draw the food along. As far as is known, digestion is extracellular in all ectoprocts, although there is some evidence that in the marine species fat is digested intracellularly. The intestine is believed to be the site of absorption.

4. Mollusca: cephalopods and gastropods

Digestion is luminar in all cephalopods and some gastropods as well as in protobranch and septibranch bivalves. It has reached its greatest efficiency in the cephalopods, where there is separation not only of digestive and absorptive areas but also of those for different aspects of digestion.

a. DIGESTION IN GASTROPODS In the gastropods, digestion may take place in the lumina of the digestive glands, of the stomach, or of other parts of the gut. In any one of these areas the digestive enzymes are mixed together and their efficacy depends primarily upon the degree to which the food has been broken up before it reaches them and how thoroughly the finely divided particles are mixed with them during their stay in the digestive area. In some gastropods, the radula is the only macerating mechanism; in others, its action may be supplemented or even replaced by an internal triturating device. The phytophagous snails, *Pomacea canaliculata*, a prosobranch, and *Aplysia punctata*, an opisthobranch, have, for example, gizzards in which the plants that they eat are finely ground before encountering the enzymes or while the enzymes are operating upon them.

In *Pomacea*, a snail that feeds largely upon aquatic angiosperms, the gizzard is developed from the gastric shield area of the stomach, and, therefore, from the midgut. Food enters this region from the crop, is macerated there, and is digested in the posterior region of the stomach by enzymes synthesized in the zymogen cells of the digestive gland and discharged through ducts that enter a special region of the stomach called the vestibule. The products of digestion are carried in solution to the digestive glands where they are absorbed by the nonzymogenic cells. These cells may also function as excretory ones, eliminating waste into the lumen of the gland. Amebocytes, migrating into the lumen of the intestine and of the style sac, also absorb some of the soluble food, but no absorption by epithelial cells of the stomach or intestine has been observed. Compaction of the feces begins in the style sac and is completed in the intestine.

In *Aplysia*, which feeds on a variety of seaweeds, the food, lubricated by mucus discharged from cells in the buccal cavity, enters the crop in large pieces and is passed by peristalsis to the gizzard, which in these snails is developed from the posterior part of the foregut. It is lined with chitin, which is thickened locally to make four or five rows of alternating teeth. Posterior to this is a filter chamber, also chitin-lined, with about 24 scattered acicular teeth. The chunks of algae are processed in these two chambers until they are fine enough to enter the stomach. The stomach proper is small, essentially only the combined dilatations of the main ducts of the digestive glands; a blind pouch, or caecum, whose cavity is unequally divided by a large and a small typhlosole, opens from it. Histologically, four types of cells can be distinguished in the digestive glands: absorptive, secretory, excretory, and storage. The absorptive cells are not considered phagocytic, since they did not take up erythrocytes from dogfish blood that had been introduced into the digestive tract. The walls of the digestive diverticula are muscular, and their contractions and those of the stomach drive the digestive juices forward into the anterior region of the gut with a force strong enough to dislodge fairly large, partly macerated pieces of algae from the teeth in the gizzard and push them forward into the crop. These juices operate in the lumina of the oesophagus, crop, gizzard, filter chamber, stomach, tubules of the digestive glands, and anterior region of the intestine, and although they are all operating simultaneously, the triturating mechanism is so effective and the area in which digestion takes place so extensive that the exposure of substrate to enzyme is adequate to ensure almost complete digestion of all the food eaten.

Tests for enzymes of different kinds in *Aplysia* have shown that they are localized in the salivary and digestive glands. All molluscs, except lamellibranchs, have salivary glands, opening by ducts into the oral cavity or the oesophagus, but the secretions are not universally digestive. Amylase and protease, active on gelatin, are demonstrable in extracts from these glands in *Aplysia*; protease, amylase, invertase, lactase, and maltase have been found in extracts of the digestive glands and in fluid withdrawn from the digestive tract.

A compact rod of mucus is constructed in the caecum of *Aplysia*, in which fecal matter is entangled. This is moved into the intestine, where the feces are finally consolidated and propelled by cilia and peristaltic contractions to the anus for elimination.

b. DIGESTION IN CEPHALOPODS: SQUID Of all molluscs, squids show the greatest modification of the basic pattern of the digestive tract and the highest degree of specialization of digestive processes. Yet even within this comparatively small group, there are differences in the structural and functional picture of the digestive organs, so that no general account can apply strictly to any particular species. The major modifications, common to them all, are the division of the midgut gland into a large distal region, designated the hepatic portion because of certain functional similarities to the vertebrate liver, and a smaller proximal region, whose volume represents about one-tenth of that of the entire organ, designated the pancreatic portion; and the development of a large thin-walled caecum, a blind sac originating from the proximal region of the stomach, into which the main duct from the midgut gland opens and which connects also with the intestine. These modifications permit the separation of digestive enzymes and their sequential operation, and of regions for digestion, absorption, and collection of particulate waste. As adaptations which facilitate rapid processing of food, they are particularly advantageous to predatory animals that, like squids, are almost constantly in motion and depend for their food on chance encounters with schools of fish which must be eaten at once and in quantity to satisfy their nutritional needs (Fig. VIII–12).

The passage of food, marked with a mixture of

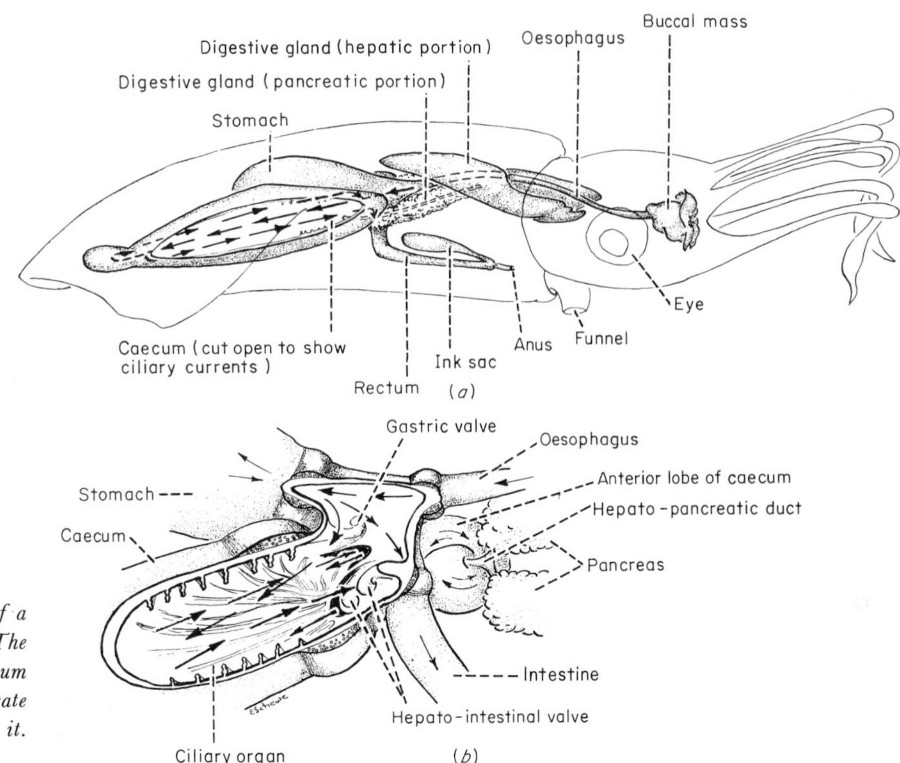

FIG. VIII–12 *The digestive tract of a squid. (After Williams.)* ***(a)*** *The entire alimentary canal, with the caecum shown as if opened. The arrows indicate the course of the main currents inside it.* ***(b)*** *The five-way caecal valve.*

carmine and iron saccharate and with Nile Blue Sulphate, along the digestive tract and the gross changes in its composition have been followed in young specimens of several species of squid whose transparency makes the gut visible. In this way it can be seen that a piece of food bitten off by the powerful jaws enters the stomach almost unchanged. Within half an hour, it is decomposed into fine particulate and liquid form and in this state passes into the caecum, where further digestion and absorption takes place. This rapid preliminary processing in the stomach, and transference to the caecum for final digestion and the major part of absorption of the terminal products, makes it possible for a squid to dispose of a medium-sized fish in 15 to 20 min by a series of rapid bites and swallows. As soon as food enters it, the stomach undergoes violent muscular movements and expands until it extends about one-third of the length of the mantle. At the end of the meal, when it is empty but caecal digestion is in progress, it becomes again a small, flaccid bag ready to receive the next consignment of food when the squid, in its ceaseless stalking, comes upon another feeding ground.

There is little specific information about the digestive enzymes, which seem to be produced exclusively in the digestive gland. Their synthesis and discharge by cells in the special region of the caecum known as the ciliated organ is questionable. There is no decisive evidence that the salivary glands, opening into the buccal cavity, contribute anything but mucus, as a lubricant, and tyramine, toxic to the prey; the oesophagus, lined with a protein-chitin complex, is a passageway only, moving the bites of food along by peristaltic contractions; the stomach, with a similar lining, receives digestive enzymes from the midgut gland, as does the caecum. The intestine is lined with ciliated cells, which keep a current directed toward the anus, and with mucus-secreting and absorptive cells. The cells in the rectum are not ciliated and in the distal region, near the opening of the ink sac, have long, retractile processes. Their function is enigmatic, but it has been suggested that they may phagocytose, secrete a coagulant of the ink that prevents its entry into the intestine, or be involved in reabsorption of water from the intestinal contents.

The efficiency of the digestive processes is referable to the construction of the midgut gland and of the caecum, the connections between them and other regions of the gut, and the high degree of coordination in their operation. The midgut gland originates as paired tubular evaginations of the caecal region. The distal portions of this evagination fuse to form a long, slender racemose gland, extending from the "neck" of the animal to the stomach and enclosing the oesophagus. This is the "hepatic" region of the midgut gland. The proximal regions fuse to form a lobulated organ, the "pancreatic" portion. Proteases and lipase have been localized in the hepatic portion, amylases as well as protease and lipase in the pancreatic portion. None of these enzymes has, however, been characterized, and further information about their specificities would be important and particularly significant in justifying the terminology of "hepato-pancreas" for the midgut gland of cephalopods. The secretions from the hepatic portion pass through the lumen of the pancreatic portion to the common duct that opens into the caecum, but the opening of the hepatic duct into the pancreas is guarded by a sphincter muscle and can therefore be opened or closed. Stimulus of food in the stomach apparently causes its opening, for the hepatic secretion is discharged into the stomach only when food is ingested. The pancreatic secretion is, however, discharged whenever the cells have accumulated it, and is stored in the caecum between meals. There is no evidence for absorption in any part of the gland nor for the direct entrance of digested material into it. The hepatic portion, besides being secretory, is also a storage area, receiving from the blood materials that can be synthesized there into glycogen and other reserves. This aspect of its functioning is one of the strongest arguments for its homology with the vertebrate liver, although there is no evidence that it secretes a surface active agent like bile. The rhythmic secretory activity of the pancreatic region has led to the speculation that it may have an endocrine as well as an exocrine function, discharging biologically active compounds into the blood with which it is in intimate contact through an extensive ramification of vessels among and between its lobules. Such similarity in function would likewise argue for its homology with the vertebrate pancreas.

The caecum is also a bipartite organ, with a spiral proximal region, lined with ciliated and mucus-secreting cells, and a long, straight distal portion, or sac, lined with ciliated and absorptive cells but without mucous ones. The proximal region of the caecum is complex, containing both the ciliary organ and valves that regulate communications between different parts of the midgut and between mid- and hindgut (Fig. VIII–13). This part of the caecum is twisted into a spiral of about $1\frac{1}{2}$ turns. The inner epithelium is folded into a number of shelves, or leaflets, that extend into the spiral and converge at the intestine; each leaflet is itself ridged with folds that run parallel to its free edge. Particulate material, both nutritive and waste, is collected between the ridges in mucous strings, which are gathered together into a main groove that runs around the columella, or central axis of the spiral, and is moved on into the intestine. The ciliary organ has been considered the homologue of the ciliary mechanisms of the lamellibranch midgut, although apparently functioning entirely as a collecting device and not at all as a sorting one. Usually only particles with diameters of 10 μ and less are found in the mucous strings, which apparently entrap any solid material in their vicinity, regardless of size or mass. The function of the organ seems to be the removal of all particles from the material that enters the caecum, so that only material in solution is passed to the distal sac, where absorption takes place. Homology with the style sac of other molluscs may be the appropriate one for this mechanism.

The opening of the caecum into the intestine is guarded by a curiously constructed valve. The inner margin of the pancreatic portion of the digestive gland runs along the columella of the spiral and opens at its apex; the opening between the caecum and the intestine is at the other end of the spiral, corresponding to the mouth of a snail's shell. A groove, bordered by two folds—the columellar ridge and the hepato-pancreatic fold—runs around the columella between these two openings. The groove can be converted to a tube by apposition of the hepato-pancreatic fold on the columellar ridge, so that direct connection can be

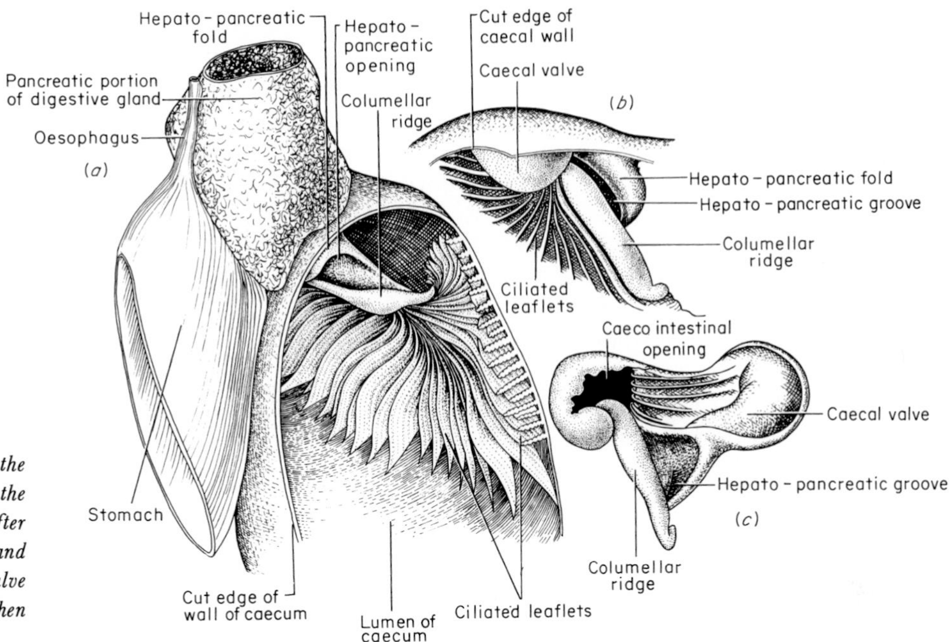

FIG. VIII–13 *Anterior end of the caecum of a squid, opened to show the ciliated leaflets and caecal valve.* (*After Bidder.*) ***(a)*** *The caecal valve and ciliated leaflets.* ***(b)*** *The caecal valve when shut.* ***(c)*** *The caecal valve when open.*

established between the hepato-pancreatic duct and the caecum. The opening of the caecum into the intestine has a thickened rim, and here the hepato-pancreatic fold is expanded in such a way that it can fit down over the rim, occluding the hepato-pancreatic groove from the caecum. The connection between the caecum and intestine remains open under these conditions so that the ciliary organ can continue to function in its removal of particulate material, while digestive juices cannot enter the caecum, but can pass freely into the stomach and intestine. Alternatively, muscles in the expanded region of the hepato-pancreatic fold, or caecal valve, can cause it to fit down tightly over the rim and clamp it there, so that the caecum is entirely cut off from other parts of the digestive tract. When the valve is open, and the caecal muscles relaxed, material can pass freely in and out of the caecum; when it is tightly closed, the caecum is isolated, so that any large masses that have not been reduced in the stomach and that have escaped from it are excluded; its partial closure permits the exit of mucous strings from the ciliary organ, but admits nothing to the caecum. The timing of these events is of the greatest importance to the squid's successful digestion and seems to be accomplished with great precision, regulated most probably by the splanchnic ganglion situated close to the stomach.

In the course of digestion the food first meets the secretion from the pancreatic region of the digestive gland, which has accumulated in the caecum and is driven into the stomach by strong contractions of the caecal sac. At this time the caecal valve is open, but closes afterwards until the contents of the stomach has been decomposed by the action of the pancreatic enzymes and the churning and mixing action of the stomach muscles until it is almost entirely in liquid form. When this state has been reached, the caecal valve opens and the fluid, with any small particles it may contain, is passed into the caecum. The undigested residue remains in the stomach and, if completely indigestible, is ultimately eliminated. In the caecum, any particulate material that has entered is assembled by the ciliary organ while the solubilized material enters the long caecal sac. Here it comes in contact with enzymes from the hepatic portion of the digestive gland, for the sphincter guarding the hepatic duct has relaxed and allowed its contents to pour through the common hepato-pancreatic duct into the caecum. After entrance of fluid from the stomach the caecal valve closes, so that caecum and stomach work independently without mixture of their contents or interference with each other's muscular movements. The combination of ciliary activity of its inner epithelium and muscular activity of its walls keep the contents of the caecum in constant motion, exposing it to the enzymes for digestion and to the epithelial cells for absorption of the end products. While absorption is still proceeding actively in the caecum, the caecal valve opens permitting some of the digested material to pass from caecum to intestine, where it is also absorbed. Absorption in the intestine continues for some time after absorption in the caecum has ended. At intervals, during the course of a meal, the stomach delivers more of its contents to the caecum for the continuation and completion of the digestive processes; and some time after a meal is over, or even after the end of a second meal, the undigested contents of the stomach, in a mucous mass secreted by the intestinal cells, is egested through the anus. When digestion is over, the hepatic sphincter closes and the secretions of that part of the gland are held in its ducts; the caecal valve opens and the secretions of the pancreatic portion of the gland enter the caecum.

The midgut gland of squids is different from that of other molluscs both in its morphological and physiological duality and in its lack of absorptive capacity. There is no evidence from experiments in which marked foods have been fed to squids that absorption occurs anywhere but in the caecum and intestine. Absorptive capacity seems to have been retained however in the midgut glands of octopods and cuttlefishes, which live quite different kinds of lives. Octopods are bottom dwellers, lurking under rocks and stretching out their tentacles to grasp prey that they can digest at leisure, their digestion of a meal occupying 12 to 14 hr, three times longer than that in the squid *Loligo*. Both octopods and *Nautilus* have a distensible crop which, after a meal, is packed with food, consignments of which are passed into the stomach from time to time. Digestion and absorption of a meal by the cuttlefish *Sepia*, partially a bottom dweller and much less active than the pelagic squid, may take 18 to 24 hr. Uptake of particulate material by the digestive gland of *Sepia* and of *Octopus vulgaris*, fed meat from crabs that had been injected with carmine, has been reported.

5. Echiuroidea

Digestion in echiuroids, from evidence obtained from the limited number of species investigated, is believed to be entirely luminar and to be effected chiefly in the midgut. The foregut, throughout which there runs a ciliated groove, is divisible into a pharynx, with thick muscular walls, a rather long oesophagus, and a crop. The midgut is in general long in relation to the body and is coiled within it. A small tube, called the siphon, runs along it just outside the muscle coats but beneath the investing coelomic epithelium. It opens into the anterior and posterior ends of the midgut, but its function is unknown. The hindgut has likewise a ciliated groove and is expanded posteriorly into a bulbous and muscular cloaca. Extracts of the wall of the midgut and fluid withdrawn from its cavity from specimens of *Ochetostoma erythrogrammon*, collected from muddy sand at intertidal level off the west coast of Singapore island, contain a protease, an esterase, and a lipase. Similar enzymes have been identified in *Echiurus echiurus*. The midgut fluid from individuals of *Ochetostoma* that had been starved at least 4 weeks had a higher content of enzymes than did extracts of the walls of their midguts, indicating that secretion is a continuous process and not dependent upon the presence of food in the alimentary canal.

6. Annelida

a. ANATOMY OF THE DIGESTIVE SYSTEM IN POLYCHAETES AND OLIGOCHAETES

The digestive tracts of most annelids are straight tubes running the length of the body, with buccal cavity, pharynx, oesophagus, stomach, intestine, and rectum as defined areas. In some few, however, there is no marked distinction between stomach and intestine and the tube is one of nearly uniform diameter from oesophagus to rectum. In others, the posterior region of the oesophagus is enlarged as a crop; this may be a thin-walled expansion of the alimentary canal, as it is in most earthworms, or a number of lateral, finger-like extensions, as it is in the blood-sucking leeches. In the common earthworm, *Lumbricus terrestris*, posterior to the crop is a gizzard or grinding stomach, with muscular walls and horny or chitinous lining. The midgut may be enlarged anteriorly as a stomach, followed by a tubular intestine, usually straight but in some species coiled, connected to the anus by a short hindgut. In *Lumbricus*, the area of the midgut is increased by a deep median dorsal groove, the typhlosole, which hangs down in the lumen and extends about two-thirds of its length. There is no typhlosole in the posterior part of the intestine in *Lumbricus* and none anywhere in the gut in *Megascolides australis*, the largest earthworm yet known. In *Megascolex* and *Eisenella* the typhlosole is only a small indentation of the wall, hardly extending at all into the lumen of the gut. In many species of annelids, glands open into the anterior region of the gut; midgut glands are typically absent, although diverticula opening into the midgut have been described in Indian earthworms of the genera *Eutyphoeus* and *Pheretima* and are considered homologous to the midgut glands of molluscs and arthropods (Fig. VIII–14). External to the lining epithelium of the midgut, in which there are both mucous and zymogenic gland cells, the wall is composed of muscle fibers, usually arranged in inner circular and outer longitudinal layers, connective tissue, and a covering of highly modified coelomic epithelium. These cells are tall and pear-shaped, and the free borders of their expanded ends, extending into the coelomic cavity, are often irregular. Because of the inclusions in them, many of which are yellowish-green in color, these cells have long been known as chloragocytes and study of their functions has attracted the attention of many investigators.

Annelids show great diversity in their feeding habits and in the types of food they ordinarily seek and ingest. Yet present evidence points to the fact that in all of them digestion is luminar, with the possible exception of *Arenicola* whose intestinal cells have been shown to be phagocytic. Movement of the food mass along the gut is accomplished primarily, even in most of the ciliary feeders, by peristaltic contractions of its muscular

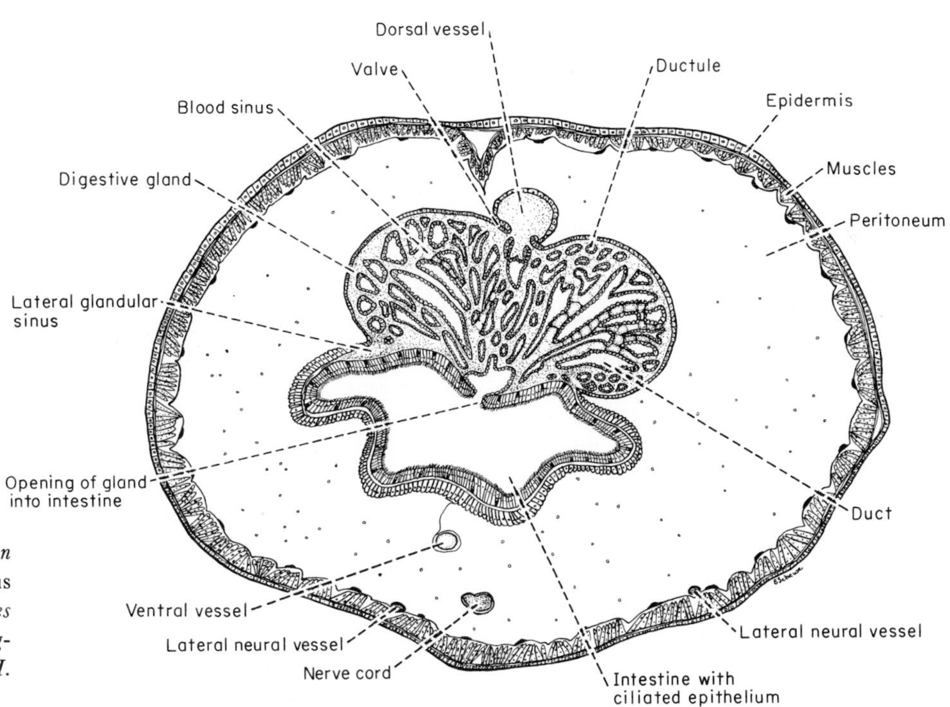

FIG. VIII–14 *Transverse section through the body of* Eutyphoeus waltoni, *showing the gland that lies dorsal to the digestive tract in segments LXXIX to LXXXIII. (After Bahl.)*

walls, although in some cases cilia are ancillary, especially in initiating passage from oesophagus to midgut.

b. DIGESTION IN *SABELLA*; *AMPHITRITE*

Feeding and digestion have been followed to some extent in the microphagous polychaete *Sabella pavonina*, a tube-dweller common in estuarine mud all around the British coast, and in the burrower *Amphitrite johnstoni*, one of the largest of British marine worms. Both of these feed by ciliary mechanisms, and in both the food particles are trapped in a mucous string derived to a great extent from cells in the epithelium of the oesophagus. In *Sabella*, the alimentary canal is a simple tube in which four regions—oesophagus, stomach, intestine, and rectum—are distinguishable. Its wall has a muscle coat in which circular and longitudinal fibers are intermingled, an extensive vascular sinus, and a lining of ciliated and glandular epithelium. Mucous cells are abundant in both oesophagus and rectum; in the intestine, the ciliated epithelium is raised in two large ventral folds, and in the stomach, it is either ciliated or secretory. It has been suggested that ciliation and secretion represent two phases of a single cell type, rather than two distinct types. Secretion in the stomach would seem to be merocrine, for large greenish-brown droplets have been described as protruding from the cells and being ultimately liberated from them into the lumen. Aqueous extracts of the walls of the gut, and of its contents, are found to contain amylase, protease, and, in lesser amounts, lipase. Reserve fat is stored in cells at the bases of the parapodia and particularly in cells in the coelomic fluid and connective tissue; glycogen as well as fat has been demonstrated in the eggs, but not in other tissues. Propulsion of the food string in *Sabella*, unlike that in other annelids, seems to be effected more by ciliary than by muscular activity, although weak peristaltic contractions of the gut wall have been observed.

In *Amphitrite* the oesophagus opens into a wide, thin-walled forestomach, posterior to which is a muscular hindstomach, lined with a peritrophic membrane, which opens into the intestine. Digestion is presumably entirely luminar, although most of the gland cells in the midgut look more like mucus-producing than zymogenic ones. Mucous cells are found in all regions of the gut except the hindstomach. Progress of the food mass along the gut is effected by a combination of ciliary activity and peristalsis, and absorption of the end products of digestion is believed to take place in the intestine.

c. DIGESTION IN *LUMBRICUS*

The post-pharyngeal gut of *Lumbricus terrestris* has been studied in detail and found to have more regional differentiations than are apparent in gross dissection. The oesophagus, which extends from segments VI to XIII, is divisible into three parts. In the anterior and posterior parts the epithelium is thrown into broad longitudinal folds, while in the middle part, which is shorter than the one preceding it, the wall is unfolded and the epithelium is ciliated as it is in the anterior but not in the posterior region. In segments X, XI, and XII are paired, pouch-like diverticula. Within the two posterior pairs, the epithelium is radially folded in leaf-like plates or lamellae and is in close association with blood sinuses derived from the main dorsal vessel, in which the blood flows from posterior to anterior. The epithelium of these lamellae secrete calcium carbonate, and the diverticula have therefore long been known as calciferous glands. The secretions are carried forward to the anterior pair of pouches which communicate directly with the oesophagus through ducts that are guarded by valves with mucous cells near their openings. The valves open to permit exit of the contents of the ducts, which then pass along the gut with the rest of the food mass and are finally included in the excreta.

CALCIFEROUS GLANDS

The secretion of calcium carbonate by these glands has been recognized for a century, but the manner in which they do it and the significance to digestive or metabolic processes have been matters of doubt and dispute. Cytologically, two types of cells can be identified in them, whose distribution divides each gland into two regions. Cells in the posterior region are distinguished by their mitochondria, which are of different types, and by deep infoldings of the plasma membrane on their proximal borders adjacent to the blood sinuses. Those in the anterior region are loaded with concretions of calcium carbonate which obscure or displace other cytoplasmic elements. The distal borders of the cells in both regions are ciliated. Alkaline phosphatase can be visualized in all the cells and carbonic anhydrase in far greater amounts than in other parts of the gut. It has been postulated that the two regions of the gland function somewhat differently in the secretion of calcium carbonate, a difference that is related to their position. The cells in the posterior region, the first to receive the blood as it comes from the posterior part of the intestine with its load of absorbed material, are thought to take up from the sinuses a fluid, almost identical in composition to the blood, which is then filtered into the lumen of the gland, proteins, and other macromolecules being retained in the cell. By mechanisms of active resorption, involving the alkaline phosphatase system, water and necessary salts are returned to the cell; the contents of the lumen thus becomes more concentrated and the calcium carbonate precipitates out in solid masses, by mechanisms probably similar to those operative in the deposition of a coral's corallum or a mollusc's shell. Because of the activity of the cells in the posterior region, those in the anterior region receive blood of slightly different composition, so that the calcium carbonate precipitates in the cytoplasm and the resulting crystals are discharged into the lumen through disintegration of the cell. The glands, although associated with the digestive system, seem to function in the excretion of excess calcium and in the regulation of salt and water balance, and of internal pH, rather than in the digestion per se. It has been believed, however, that the calcium carbonate liberated in the gut may have a neutralizing effect upon the food, which is often very acid, although the pH of the gut contents consistently remains within a pH range of 6.5 to 6.7. The fact that the precipitated calcium moves along the gut in crystalline form casts doubt upon its efficacy as a neutralizing agent of any real importance, while the size of the glands in all species inhabiting soils that are rich in calcium, which is greater than in those living where it is less abundant, argues for their role in its excretion as an excess or unwanted constituent of the food. Calciferous glands similar to those of *Lumbricus* are found in *Allolobophora caliginosa* and *Eisenia foetida*, earthworms of similar feeding habits and favoring areas with many limestone and other calcareous rocks. Feeding experiments, in which filter paper moistened with various common calcium salts was ingested by worms, show that the calcium may be derived equally well from carbonate, sulfate, phosphate, oxalate, chloride, and nitrate, as well as from a more natural source, pear leaves, used in these experiments for convenience only and not for any particular experimental rationale.

The epithelium in the crop of *Lumbricus* is likewise folded. The muscles of its walls undergo contractions stronger than those of the gizzard, which result in dilating and twisting movements that regulate the rate at which food is passed into this next chamber. The gizzard has a comparatively thick, and hard, lining and is the principal area for maceration of the larger pieces of organic material in the food. This enters the mouth intact but is lubricated by secretions from pharyngeal gland cells. The intestine extends from about the nineteenth segment to the rectum in the terminal segments, with a typhlosole along about two-thirds of its length. In the anterior region, the typhlosole is greatly folded, and there are many goblet cells, secreting mucus, on its lateral surfaces; elsewhere, the epithelium is composed of ciliated and zymogenic cells. When food is taken, cells in the anterior region of the typhlosole secrete a thin, translucent peritrophic membrane. This lacks the adhesiveness of mucus but gives negative results in tests for chitin, for which similar membranes in terebellid polychaetes give positive ones. The typhlosole of the more posterior segments has smooth walls and is entirely absent in the final third of the intestine.

d. DIGESTION IN *EUTYPHOEUS*: "HEPATOPANCREATIC GLANDS" Extracts of tissues from different regions of the gut of *Lumbricus*, and of other species of earthworms, incubated with starch, glycogen, cellulose, and inulin; with casein, gelatin, and chitin; and with methyl and ethyl butyrate have provided evidence of a regional separation of cells secreting lipase and the ordinary carbohydrases and proteases as well as some unusual ones. In *Allolobophora* and in *Lumbricus*, a protease is secreted by cells of a gland closely applied to the pharyngeal wall, although enzymatic secretion is in most cases confined to the cells of the midgut epithelium. There is also some evidence that in *Eutyphoeus waltonii* protease is secreted by cells in a pair of prominent, lobulated diverticula, comparable in construction to the midgut glands of molluscs and arthropods. There is no typhlosole in this region of the gut nor in that immediately posterior to it, and the epithelial cells of the diverticula are nonciliated and apparently secretory in character. There is no confusion of these glands with calciferous glands, for their secretion is not milky in appearance from calcium carbonate suspended in it, and there are true calciferous glands situated more anteriorly in the body. These so-called "hepato-pancreatic glands" of *Eutyphoeus* contract rhythmically and pour their secretions into the lumen of the gut where digestion takes place. Cells in caeca originating from the gut in segments XXII to XXVI of *Pheretima* secrete amylase, but in most worms digestive enzymes are synthesized in the epithelial cells of the gut proper and discharged from them directly into its lumen. In other earthworms that have been investigated, amylase is secreted by cells in the anterior region of the intestine, lipase by cells somewhat posterior to these, and proteases by those in the final half. Chitinase and cellulase as products of cells in the anterior region of the intestine have been found in a number of different species of earthworms with, in *Lumbricus*, the cellulase-secreting cells being particularly abundant in the anterior half of the intestine. Indirect evidence of the secretion of invertase has been derived from measurements of the optical rotation of light in a 20 percent solution of sucrose after incubation with worm castings; this was about twice that of solutions incubated with ordinary soil, but there is no proof that the effect was due to an endogenous enzyme.

The organic material in the food an earthworm swallows, which may contain fresh leaves as well as decayed or partially decomposed ones and also various products of the excretion or decomposition of animals, is digested by these enzymes at a pH that seems remarkably constant in the specimens in which it has been measured. Absorption of the end products takes place through the intestinal epithelium, and the undigested and unabsorbed residue, together with the soil or sand that was ingested, are eliminated as castings, so conspicuous in rich soil where earthworms flourish.

e. DIGESTION IN LEECHES The digestive tract in leeches consists of a muscular pharynx, whose epithelium contains mucus-secreting cells; a short oesophagus, into which a pair of salivary glands open; a midgut, represented by a stomach, often with a number of diverticula; and an intestine, with paired diverticula and a rectum. In blood-sucking leeches, the salivary gland secretes an anticoagulant and the food is drawn into the stomach and its diverticula by pumping action of the pharynx. Digestion is slow; blood may stay in the diverticula for several months during which it is gradually passed into the stomach and intestine for digestion and absorption. There is little definite evidence about the enzymes involved or the sites of their production and operation.

7. Onychophora

Digestion has been studied in four species of *Peripatopsis* captured and kept in the laboratory. These animals are carnivorous, taking a meal of small arthropods at intervals of about 2 weeks and digesting it in 18 hr or more. The digestive tract is straight with no midgut diverticula nor any distinct histological differentiations of cells in the midgut area. The foregut is chitin-lined and in the buccal cavity are the openings of the paired salivary glands, which are large structures extending two-thirds of the length of the animal's body. Amylases, active on starch and glycogen; proteases, active on casein, fibrin, and gelatin; and a peptidase that splits leucyl-glycyl-glycine have been demonstrated in extracts from these salivary glands, whose secretions are poured from the mouth on to the captured prey. Invertase and maltase; lipase, active on olive oil; esterases that split ethyl acetate and ethyl butyrate; amino- and carboxypeptidases and dipeptidase have been found in extracts of the midgut epithelium. This also secretes a delicate tubular peritrophic membrane which surrounds the undigested material and is egested with it through the anus. There is no indication of phagocytosis or intracellular digestion anywhere within the gut. The character of the enzymes and the behavior of the animals indicate that the first steps in digestion are extracorporeal, effected by the enzymes of the salivary secretions, and that the final stages, and absorption of the terminal products, take place in the midgut, along which the food mass is moved by peristalsis.

8. Arthropoda

a. GENERAL ANATOMY OF DIGESTIVE SYSTEM In most of the arthropods the digestive tract is straight, with little coiling. The outer layers of its walls are muscular, and food is propelled along it by peristalsis. The foregut and hindgut, continuous with the external epidermis, have, like it, a chitinous covering. Digestive processes have been studied to some extent in arachnids and to the greatest extent in crustaceans and insects, but even in these classes information is still scanty and there remains much to be obtained.

b. DIGESTION IN ARACHNIDS The pharyngeal region of the foregut of arachnids is developed as a pumping organ, in accordance with their habit of feeding on fluids that must be drawn into the digestive tract. These pumps vary in shape and size in different orders, but all have a thick muscular wall, composed of dilator, and usually of constrictor, muscles. Posterior to the pump is a narrow oesophagus, which leads into the midgut, or mesenteron. In many species the oesophagus is expanded, just before its connection with the mesenteron, into a second sucking organ, the postcerebral or "stomach" pump. The mesenteron has a small central cavity but many diverticula, which are regarded as expansions of the stomach and not as digestive glands comparable to those of crustaceans and other invertebrates. These diverticula are very extensive, usually occupying most of the abdomen and some of the cephalothorax; in some species, they even extend into the appendages and into the more anterior regions of the cephalothorax, or the head. In spiders the posterior part of the midgut is expanded as a cloaca (stercoral pocket) in which waste is stored. The hindgut is a short sclerotized tube connecting midgut and anus.

The inner epithelium of the stomach diverticula contains secretory (zymogen) and absorptive cells. Amylase, lipase, and proteinases are discharged from the secretory cells into the lumina of the diverticula, where digestion of food that has been drawn into the mesenteron takes place. In spiders that practice external, or extracorporeal, digestion, the digestive juice is expelled from the stomach onto the prey. It contains an endopeptidase similar to trypsin, an aminopeptidase, and a carboxypeptidase, lipase, and very probably amylase. Studies of enzymes synthesized in the maxillary glands of tarantulas indicate that they include proteinase, peptidase, amylase, and lipase. It is likely therefore that the secretions from the maxillary glands may at least initiate events of predigestion, which are completed by those from the midgut glands that are regurgitated into the stomach. The absorptive cells accumulate the end products of digestion by means other than phagocytosis and presumably act also as storage cells.

c. DIGESTION IN DECAPOD CRUSTACEANS In crustaceans, the oesophagus opens into an enlarged area, which in entomostracans, the smaller members of the phylum, is part of the midgut and so a stomach. In malacostracans, on the other hand, this extended region is the posterior part of the foregut and is divided into a large, anterior cardiac "stomach" and a smaller, posterior pyloric "stomach" that opens into the midgut, or mesenteron (Fig. VIII–15). The two regions are, therefore, not homologous in the two groups, into which the class is conveniently divided without any taxonomic significance. In most species of malacostracans, the cardiac stomach contains the gastric mill, a triturating device by which the food, torn off by the mouthparts and kneaded by them into a form in which it can enter the oesophagus, is ground into pieces fine enough to pass through the setose filter of the pyloric stomach. The wall of the cardiac stomach is usually quite heavily chitinized, and often calcified

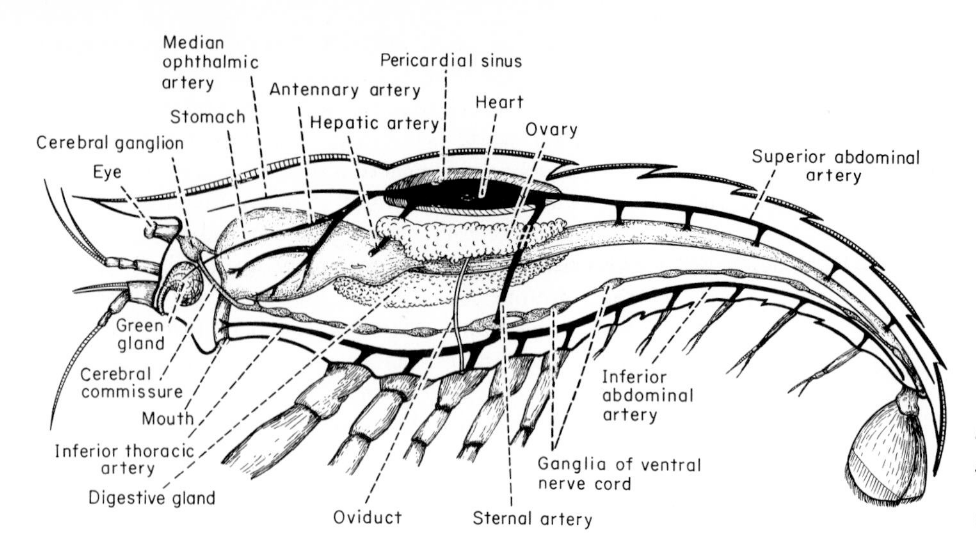

FIG. VIII–15 *Section in the sagittal plane of a female lobster* (Homarus americanus). (*From Gardiner.*)

in places; its gastric mill consists of three strong, calcareous teeth, of which two are lateral in position and one is dorsal. In lobsters, the lateral teeth are heavy bands, whose long axes are parallel to that of the stomach and whose upper margins are serrated. The shaft-like median tooth projects downward from the dorsal wall of the stomach between the posterior ends of the lateral teeth. Moved by strong muscles of the stomach wall, these teeth form an effective mill in which the food is ground again and again as it is moved around in the stomach. Pieces sufficiently small then pass through the aperture, guarded by coarse hairs, between the cardiac and pyloric stomachs; in the pyloric stomach, a complex arrangement of setae filters off the coarser particles among them, to be returned to the mill, but permits the finest ones to pass through into the midgut where the ducts of the digestive diverticula open. As a means by which food can be processed over an extended period of time, the gastric mill seems particularly adapted to the life and eating habits of sedentary crustaceans that can leave their burrows or hiding places under rocks to capture food and then retreat to them for its mastication and digestion. The mechanism is greatly reduced, or absent entirely, in some of the swimming crabs, whose mandibles are so constructed that they chew the food as soon as it is caught. It is also lacking in the isopod *Ligia oceanica*, whose eating habits demand rapid processing of its food.

According to information at present available, all the digestive enzymes of crustaceans are produced by cells of the digestive diverticula. These glands have the same architectural pattern as those of the molluscs, consisting of a system of branching ducts which terminate in blind tubules or alveoli. The lining of the tubules is a simple columnar epithelium containing secretory and storage cells. Histologically, the secretory cells are typically zymogenic; in the crayfish *Astacus* and the lobster *Homarus* they discharge their products by holocrine secretion and, therefore, must continually be replaced. In the other crustaceans studied, secretion is merocrine. The nature of the enzymes discharged into the ducts and the midgut has been explored principally in decapods. Digestive juice, essentially free from any food or digestive products, has been obtained from specimens of *Astacus*, the fiddler crab *Uca*, and the spider crab *Maja* by inserting a cannula into the stomach of an animal that has been starved for some time. The juice that flows out contains a mixture of hydrolytic enzymes, all of which must come into contact with the food at the same time. In the mixture there are proteases (proteinase, similar in its activity to trypsin, carboxypeptidase, and aminopeptidase), carbohydrases (amylase, maltase, and invertase), and an esterase. In *Nephrops*, the activity of this enzyme on olive oil as a substrate indicates that it is a lipase, which may also be the case in *Homarus* and *Palinurus*. The esterase of

Astacus, on the other hand, is more active on esters of lower fatty acids and lower alcohols than it is on fats and so is probably not lipase.

The midgut is the longest part of the digestive tract in *Nephrops*; it has dorsal and posterior diverticula, as well as paired digestive glands with muscular walls. The opening of each gland is guarded by two chitinous plates, which meet in the midline; and each plate has transverse chitinous rods with rows of setae on each side, forming an effective filter. The glands contract and expand during the course of digestion, expelling their contents into the midgut and drawing back material from the midgut fine enough to pass the filter. Secretions from the glands have been tested on starch, glycogen, sucrose, maltose, lactose, inulin, and raffinose; on olive oil and on fibrin, casein, and peptone. They have been found to contain amylases, sucrase, maltase, and lactase, but no inulase or raffinase; and lipase, proteinase, and peptidase.

Tests of extracts of different parts of the gut and of the digestive diverticula of the amphipod *Corophium* show also that the digestive diverticula are localized sites for the production of hydrolytic enzymes. Extracts of the two ventral diverticula gave positive results in tests for amylases, with starch and glycogen used as substrates, for invertase and for maltase; for a protease that could liquefy gelatin and a lipase active upon olive oil and condensed milk.

The storage cells of the digestive diverticula of decapod crustaceans contain glycogen granules and fat droplets. Their absorptive capabilities have been shown experimentally by the demonstration of iron in animals that had been fed iron saccharate or iron lactate; the iron accumulated in these cells only and could not be demonstrated in the secretory cells. It has also been shown that P^{32} is present in these cells after animals have ingested compounds containing it. The epithelium of the midgut proper likewise contains cells of this type, but no zymogen cells. Absorption by cells of the midgut has been shown in *Nephrops norvegicus*. Fine globules of fat were evident in the cells of the midgut, but not in those of the fore- or hindgut, 1 to 3 days after a suspension of olive oil had been injected into the cardiac stomach. There is no evidence that any of these absorptive and storage cells are phagocytic.

The question of absorption of fats in invertebrates is a puzzling one, since there is little evidence that any of them discharge bile salts into the lumen of the gut to reduce the surface tension of the fats and keep them in a finely divided state as vertebrates are known to do. Some of the decapod crustaceans seem, however, to have solved the problem in a similar way, for surface active substances have been found in the digestive juices of *Astacus*, *Homarus gammarus*, *Cancer pagurus*, and *Eriocheir sinensis*. These substances have the properties of bile acids, with the greatest resemblance to taurodeoxycholic acid, and are products also of the digestive glands. Secretion of such products, and of digestive enzymes, combined with the function of storage of glycogen and fats would seem to justify the designation of these glands in crustaceans as a hepatopancreas, in which are represented the physiological activities of the vertebrate liver and pancreas. The absence of phagocytic cells in them, which play so important a role in the digestive processes of microphagous molluscs, makes the analogy of the crustacean and vertebrate organs even closer.

d. DIGESTION IN ISOPODS The digestive tract of the boring and wood-eating isopod *Limnoria* is a straight tube. The posterior part of the foregut is developed as a gastric mill which serves the dual purpose of trituration and sorting, for it sifts out particles so that they do not enter the midgut glands. These are two pairs of long diverticula which represent the entire area of the midgut. Posterior to their openings the hindgut extends as a straight tube to the anus.

The midgut glands, like the entire alimentary canal, are enclosed in a delicate peritoneum beneath which are muscle cells and the lining epithelium. When extirpated, the glands undergo contractions and distortions of shape, but there is no evidence of rhythmic contractions in the living animal. The epithelial cells surrounding the lumen, which is filled with a clear, viscous fluid free of any visible particles, are histologically of two types, distinguished as alpha and beta cells. The alpha cells are large, binucleate, and extend deeply into the lumen of the gland. They are separated from each other by the smaller beta cells, which form a network between them. Both are in all probability secretory cells. The cytoplasm of the alpha cells contains numerous oil drops but, most conspicuously, large yellow tetragonal or bipyramidal crystals, whose long axes range from 1 to 28 μ. These contain protein, most probably a porphyrin, with iron bound so tightly to it that it cannot be released by chelating agents. As many as 27 such crystals have been counted in a single cell, usually in a perinuclear or basal location, although there is a considerable variation in number among individuals even of the same population. Observations with the electron microscope suggest that the large crystals are formed by aggregations of smaller ones that are produced in association with certain cytomembranes; these tiny crystals may have their precursors in dense bodies arising in cisternae of the Golgi complex. The smaller beta cells are also binucleate and are filled with oil drops and densely packed, electron opaque granules, quite distinct in size, shape, and color from the crystals of the alpha cells. Histochemically, the beta cells react positively to tests for sulfhydryl groups, both the reaction and its intensity being directly proportional to the amount of iron in the alpha cells. Various hypotheses have been proposed to explain the presence of the crystals in the alpha cells and their possible role in the isopod's economy. One of these hypotheses relates them to the secretion of cellulase but is weakened by the fact that their frequency varies markedly in different populations inhabiting different locations and that they are absent in all young animals, which would hardly be expected since individuals a few hours after their release from the brood pouch of their mothers begin to bore side channels from the parental burrows. More acceptable is the idea that the crystals represent accumulations of excess iron, which the animals cannot excrete and so deposit tightly bound with protein in the cells of the midgut gland, which thus serves as a repository for "excretory material" as well as the site of digestive enzyme synthesis and discharge.

The midgut gland is the only site of secretion of digestive enzymes, and the very small size of the animals presents great difficulties to the isolation and characterization of them. Dissected glands from more than 600 animals are needed to provide a bare milliliter of crude homogenate. Such homogenates, obtained from animals living in the cold waters of the Pacific Ocean off the coasts of Washington and Oregon and from the subtropical waters of the Mediterranean, have, however, been tested on a variety of carbohydrate substrates. The extract is active upon wood cellulose, filter paper, cellophane, carboxymethyl cellulose, methocel, starch, and glycogen, indicating its content of cellulase and amylase. It is also active on laminarin but not on algin, agar, pectin, or inulin. There is evidence also that it contains maltase, beta glucosidase, and galactosidase. These seem to be the principal digestive enzymes, for wood is the chief nutriment of *Limnoria*. Analysis of the wood of Douglas firs and of the fecal pellets of a number of animals boring through pieces of it show that about half the cellulose, and virtually all the polyuronide hemicelluloses and the noncellulosic polysaccharides, can be removed from the wood fibers during their passage through the gut. A source of protein, and proteolytic enzymes, would seem also essential to an adequate diet. The source may be the fungi with which wood exposed to seawater for any length of time is always infected. This possibility is substantiated by the fact that, while animals will bore into sterile wood, they cannot survive in it and sterile seawater any longer than they can survive in the complete absence of wood, which has been shown to be a period of 2 to 3 months. Moreover, the amino acid complement of *Limnoria*, as it has been identified by paper chromatography of hydrolysates of whole animals, differs significantly from that of southern pine heartwood into which the isopods frequently bore. Five of the thirteen amino acids found in *Limnoria* were missing in the wood—cystine, cysteine, lysine, serine, and threonine. These were, with the exception of cysteine, constituents of the mycelium of the marine fungus *Lulworthia floridana* on which experimental animals fed avidly in the laboratory. This would suggest that in nature *Limnoria* derives its protein from marine fungi, digests it, and constructs its own protein from the amino acids thus made available.

The isopod *Ligia oceanica*, although an omnivore in the laboratory, naturally feeds upon *Fucus*, emerging at night from the crevices in rocks well above high-tide level, where it lives by day, to browse upon the seaweed at half tide. Its feeding time is very limited and it must eat and process its food quickly in order to get a full meal every 24 hr, for it has no caeca or crop for the storage of undigested food. The posterior part of its foregut is a single chamber with a complex filtering device instead of a gastric mill, by means of which the fine particles of food, bitten off by the mandibles, are strained from the water taken in with them. The midgut is extremely short, represented almost entirely by three pairs of digestive diverticula. The zymogenic cells of these diverticula secrete amylases, active on starch and glycogen; glucosidases, active on sucrose, maltose, and raffinose; a protease, active on fibrin; and lipase, active on olive oil. Tests for cellulase have been unconvincing; those for fucoidin, negative. The enzymes are discharged into the lumen of the digestive glands, which have muscular walls and undergo rhythmical contractions that can be watched in specimens that have recently molted and are transparent. The contents of the gland is thus expelled into the foregut and hindgut, where digestion takes place. The products of this digestion, whether partial or complete, may then be sucked back into the digestive glands at each relaxation after contraction. Absorption of terminal products occurs here, as well as in the intestine, for iron can be demonstrated in the cells after the administration of either iron lactate or iron saccharate. The intestine, as part of the hindgut, is lined with chitin, which in permitting the passage of molecules of this size must function here as a peritrophic membrane.

e. **DIGESTION IN AMPHIPODS** The amphipod *Corophium volutator*, feeding also on algae, has a more typical crustacean digestive tract, with a cardiac stomach and gastric mill, and a pyloric stomach and filtering device. There are two pairs of digestive diverticula—a dorsal pair that runs anteriorly and a ventral pair that runs posteriorly. Tests for digestive enzymes in different parts of the gut show that the zymogenic cells are localized in these glands. Extracts of the walls of the ventral pair depolymerized starch and glycogen, hydrolyzed sucrose and maltose, and were also active upon gelatin, olive oil, and condensed milk, showing the presence of amylases, glucosidases, protease, and lipase in them. Tests for cellulase, using 1 percent sodium carboxymethyl cellulose as a substrate, were negative, as were those for algal polysaccharidases, when 1 percent solutions of fucoidin and laminarin were used as substrates. Both *Ligia* and *Corophium* must therefore derive their nutrition from components of the algae other than their structural carbohydrates.

f. **DIGESTION IN INSECTS** (1) **Diversity of food and eating habits** The insect diet is extremely varied, ranging from such dry foods as seeds, grains, and milled flour to fluids such as blood, sap, and nectar, and including such unusual materials as horn, fur, feathers, cork, pepper, beeswax, and cured tobacco, which are the exclusive choices of certain species. Termites have been known to eat rubber insulation; and some beetles, lead pipes. Since insects evolved together with plants, one might expect that primitively they were phytophagous and that polyphagy and zoophagy, along with the oligophagy and monophagy exhibited by those with highly restricted diets, were secondarily developed. However, the most generalized digestive mechanisms seem to be associated with omnivorous species and the most specialized with those that feed on plant products. The diversity of feeding habits in modern species, as well as ignorance of the mechanisms of digestion in the majority of them, make generalizations about these meaningless, beyond the fact that they occur through the agency of enzymes like those of other animals and in localized regions of the alimentary canal. The picture is further complicated by the fact that many insects change their eating habits, and therefore the food that they must digest, at least once during a life cycle, so that information gathered from larval or nymphal stages may not apply to the same individual as an adult. Moreover, in the lives of all insects there are periods when they do not eat at all, not because of satiety or scarcity of food but because they have arrived at an aphagous phase, either as adults or as juveniles. In some species, too, there are differences between the sexes in feeding habits and so in digestive procedures.

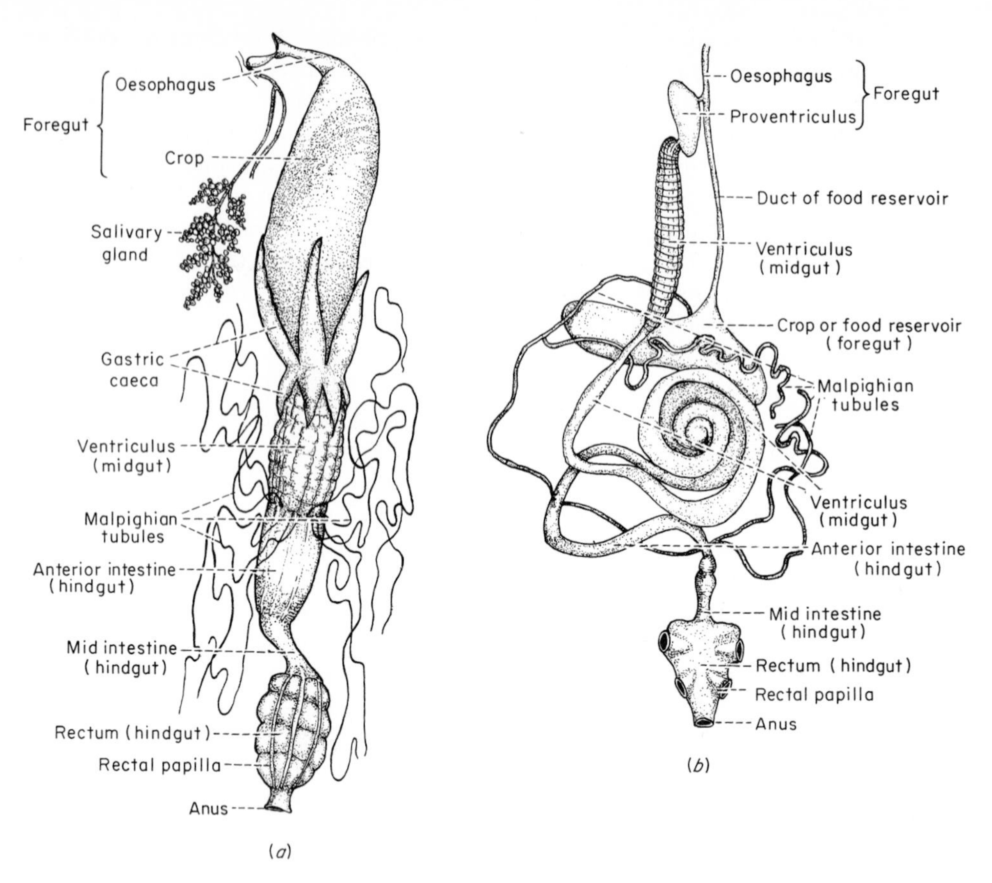

FIG. VIII–16 *Insect digestive tracts.* ***(a)*** *Dissection of the digestive tract of an adult grasshopper* (Acridium)*, which feeds on vegetation.* ***(b)*** *Dissection of the digestive tract of the adult blowfly* Calliphora, *which feeds upon fluids.* *(From Gardiner.)*

(2) Modifications of digestive systems as adaptations to food and eating habits The alimentary canal shows modifications that can be correlated to some extent with feeding habit (Fig. VIII–16). The most common modification is in length, for it is longer, in one region or another, in fluid feeders than in those that ingest solid food. Several other modifications of the basic plan have been recognized. In the basic, or primitive plan, as it is found, for example, in larvae of Lepidoptera and in Siphonaptera, the three primary divisions of foregut, midgut (mesenteron or ventriculus), and hindgut are evident, with paired salivary glands opening into the preoral cavity and Malpighian tubules arising at the junction of mid- and hindgut. These are long, slender blind tubules that ramify extensively throughout the hemocoele and are credited with excretion as a primary function. In some species they arise from a special region of the intestine called the ileum. Characteristic of insects also is the prolongation of the wall of the foregut into the lumen of the midgut and that of the midgut into the hindgut. These short extensions act as valves, permitting the passage of material posteriorly but, in general, preventing its forward movement from one region to another.

The simplest modification of this basic plan, as it is found for example in Orthoptera, is expansion of the foregut into a crop where food can be stored prior to its digestion (Fig. VIII–17*a*). In adult Lepidoptera and in some Diptera, such as mosquitoes, midges, and crane flies, this storage region is a good-sized diverti-

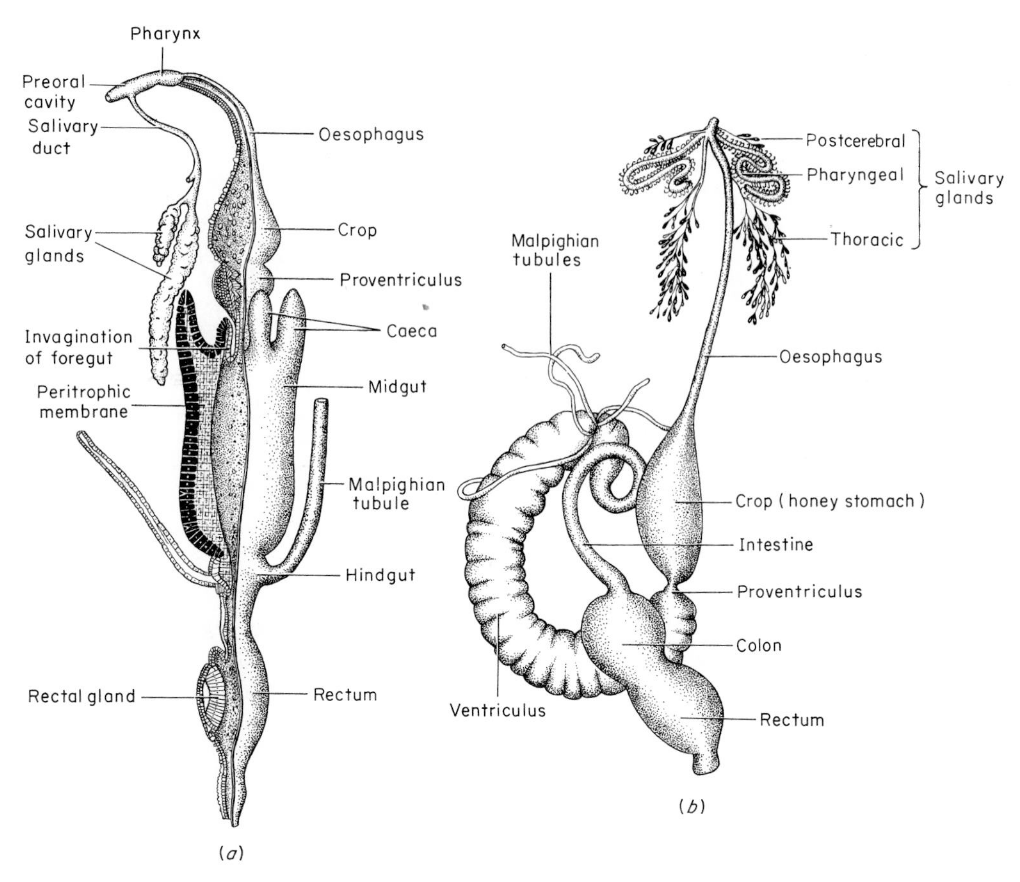

FIG. VIII–17 *Insect digestive tracts.* ***(a)*** *The generalized plan as it is found in some Orthoptera. On the right of the drawing the digestive tract is shown in surface view; on the left as if opened.* (*After Roeder.*) ***(b)*** *Surface view of the digestive tract of a honeybee.* (*After Borradaile et al.*)

culum of the foregut. In many insects, a proventricular region is inserted between the crop and the midgut. This is part of the foregut and is lined by cuticle, which may be thin and delicate or thick and heavy. In cockroaches, the proventriculus functions as a gizzard or triturating organ. In some species, its cuticle is prolonged into fine spines or hairs, so that it acts as a strainer or sieve. This is particularly characteristic of Hymenoptera, in which the proventriculus functions both as a valve and a sieve (Fig. VIII–17*b*). In bees, the mechanism is particularly complicated and is called the "honey stopper." Four pouches whose lips are covered with spines project from the anterior end of the proventriculus into the crop. The walls of the crop, when it is full of food, contract and expand rhythmically, keeping its contents in constant motion so that the pollen grains imbibed with the nectar are evenly distributed in the fluid medium. The pouches open one by one and the contents of the crop flows into each as it opens; when filled, the pouch collapses and the nectar is returned to the crop, or "honey stomach," while the pollen is caught in the meshwork of spines that can effectively filter off particles with diameters of 3 to 50 μ. At intervals, a ball of pollen is passed into the midgut and packed within the peritrophic membrane, through which any remaining fluid diffuses into the surrounding space. By this device, the proteolytic enzymes of the midgut can operate without the extreme dilution they would experience if the ingested food entered it directly.

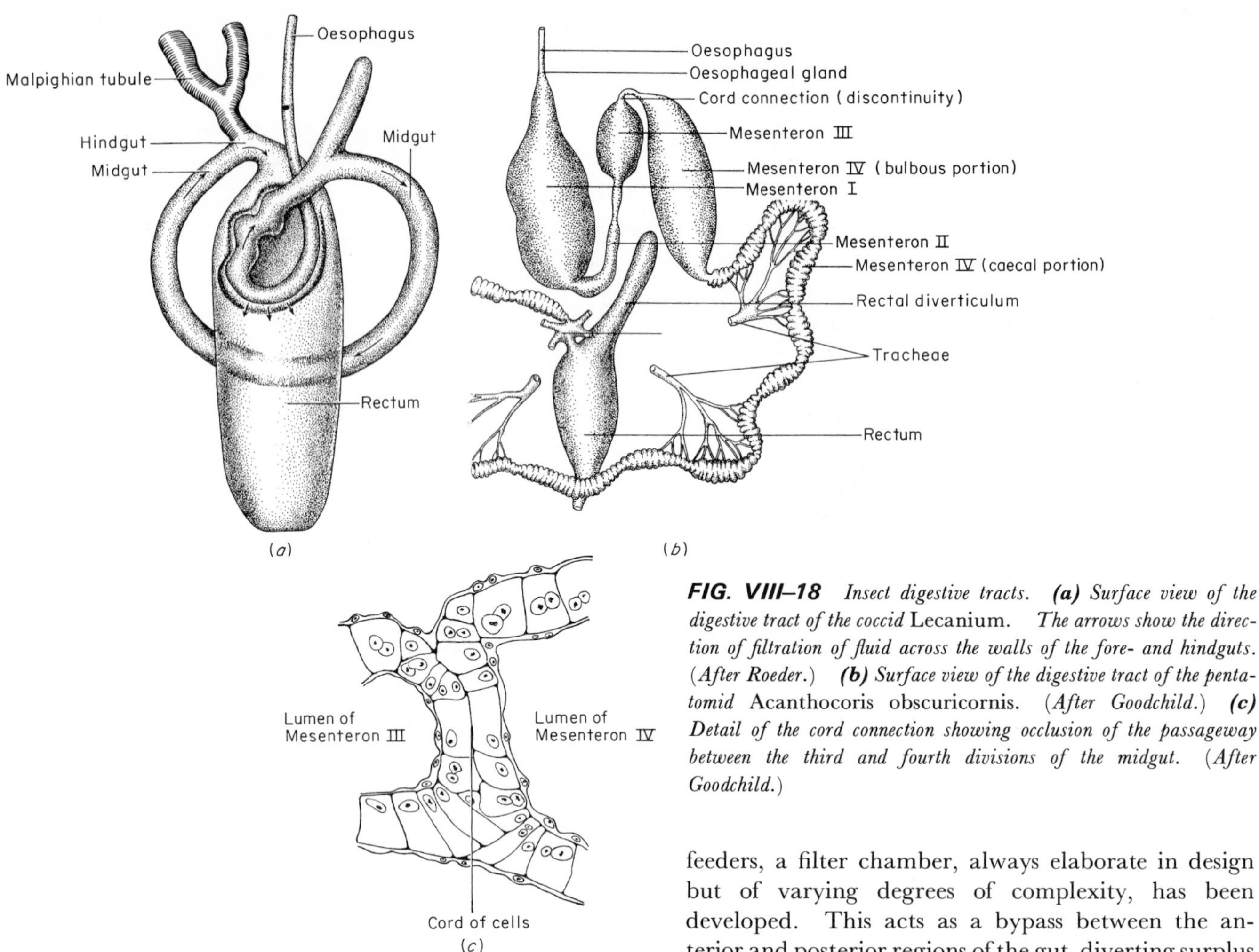

FIG. VIII–18 *Insect digestive tracts.* ***(a)*** *Surface view of the digestive tract of the coccid* Lecanium. *The arrows show the direction of filtration of fluid across the walls of the fore- and hindguts. (After Roeder.)* ***(b)*** *Surface view of the digestive tract of the pentatomid* Acanthocoris obscuricornis. *(After Goodchild.)* ***(c)*** *Detail of the cord connection showing occlusion of the passageway between the third and fourth divisions of the midgut. (After Goodchild.)*

In blood-sucking Diptera, in addition to the modification of the foregut as a crop, the midgut is greatly extended, giving a longer passage and a greater exposure of the food in the digestive and absorptive areas. There is also a well-defined rectal area in the hindgut.

The most pronounced modifications and greatest specializations are evident in Hemiptera, Homoptera, and the larvae of some Hymenoptera and Neuroptera. The midgut of Hemiptera, which for the most part feed on the juices of plants, is even more greatly extended than that of the fluid-feeding Diptera and is coiled within the body. In Homoptera, also fluid feeders, a filter chamber, always elaborate in design but of varying degrees of complexity, has been developed. This acts as a bypass between the anterior and posterior regions of the gut, diverting surplus liquid from the midgut so that the material that enters it is more concentrated than that which was ingested. In its simplest form, the filter chamber is a tubular connection between the posterior end of the foregut and that of the midgut, through which water from which most of the solutes have been filtered off flows toward the intestine without passing the digestive and absorptive areas. In complicated form, as it is found in the coccid *Lecanium*, it consists of a plexus, derived from the posterior region of the oesophagus and the anterior regions of the mid- and hindguts, that projects into a dilated rectal area of the hindgut (Fig. VIII–18). The importance of the filter chamber in water regulation in insects of this kind, whose problem is not its conservation but its elimination, has been clearly established.

In the larvae of some Hymenoptera and Neuroptera and in adults as well as nymphs of the pentatomid Homoptera, the lumen of the gut may be completely occluded in some place, or places, by a cord, or wall, of cells. This block is situated, in larval Hymenoptera and Neuroptera, at the junction of mid- and hindgut; in the pentatomids, or shield bugs, it may be between the anterior and posterior mesenteron. Four distinct regions can be recognized in this part of the gut. The first of these, whose opening into the oesophagus is guarded by a valve, is a capacious sac; the second is a slender tube, often looped; the third is, as the first, saclike, but considerably smaller; and the fourth, if it is well developed, is a tube whose length may exceed that of all the other regions combined and whose walls are extended into numerous small pockets or caeca. The discontinuity of the gut in these insects lies between the third and fourth regions; in some species there is a second block between the fourth region and the ileum. Occlusions like this are associated with a diet of concentrated fluids. The larvae of Hymenoptera and Neuroptera with discontinuous guts are fed material that has already been processed by the adults and from which there is little, if any, residue to be passed into the intestine. Homoptera with this modification are all sapsuckers, and it is not found in other apparently closely related species that are zoophagous.

Almost all species of insects have at least one pair of salivary glands at some stage of their lives. These may be tubular or acinar in their construction and are usually of considerable size, extending back into the abdomen. Their highest degree of structural complexity is found in hemipterans, where each gland may consist of several well-defined lobes. Their ducts open into the preoral cavity near the points of attachment of the mandibles, the maxillae, or the labium, and because of this they may be designated as mandibular, maxillary, or labial glands. The labial seem to be the most generally distributed and therefore the most important. Only exceptionally are all three found at the same time in one insect, although they may appear and become functional successively. The secretions collect in the salivarium, a pocket with muscular walls at the base of the hypopharynx, from which they are ejected to the exterior or onto food that has been ingested. The secretions are very diversified and by no means exclusively digestive. Also, their components may change during the life of an individual insect; the labial glands of the moth *Bombyx mori*, for example, produce silk during its larval life (when the salivarium acts as a silk press), and amylase and glucosidases during its adult life. Queens and drones of the honeybee *Apis mellifera* lack the oral glands that in workers secrete invertase, by means of which the sucrose of the imbibed nectar is hydrolyzed in the crop. Although drones eat large amounts of honey from that stored in the cells, neither they nor the queens participate in its manufacture, and the invertase active in the drones' digestion is produced in their midguts. In the fruit fly *Drosophila* and in some other Diptera, the salivary glands are functional only during larval life and are lost during metamorphosis of pupa to adult.

The midgut is, as in other tube-within-a-tube animals, the principal site of digestion and absorption, although enzymes regurgitated from it may cooperate with those from the salivary glands in hydrolyzing food in the crop. The area of the midgut may be greatly extended by caeca arising from any part of it, though most frequently the caeca emanate from the ends, in single or multiple pairs. In some insects these caeca are crammed with symbiotic microorganisms that contribute in one way or another to the nutrition of their host. The single layer of epithelial cells that lines the midgut may all be columnar in type; in Lepidoptera there are, in addition to these, cells with large vacuoles, resembling goblet cells, quite regularly interspersed between them. The columnar cells in these insects are believed to be absorptive in their function and the goblet-like cells to be secretory, discharging products other than mucus, but evidence for this is not definitive. A third type of cell is found in the epithelium of some insects; this is a cuboidal cell that is capable of division and multiplication. The epithelium is frequently elevated in folds or depressed in crypts of considerable depth. Secretion has been described as holocrine, merocrine, and apocrine, with the cuboidal regenerative cells replacing those that have been destroyed or exhausted as a result of their activities. In most insects there is a tubular peritrophic membrane; its presence in some fluid feeders and its absence in some species that eat solid food throws doubt upon the interpretation of it as a purely protective device.

(3) Digestion in certain homopterans, dipterans, lepidopterans, coleopterans Insects have progressed beyond most other invertebrates in their digestive mechanisms, in that areas of digestion and absorption, and in most cases those for storage, are separated. No epithelial cell, therefore, need function either simultaneously or successively as secretory, absorptive, and storage, as it must, for example, in even the most highly evolved mollusc. Although activity of any particular enzyme in any particular part of the gut is not evidence that it has been secreted in that place, there are anatomical, histological, and physiological grounds for believing that the ventriculus of many insects is divided into a number of different functional areas. Grossly, this division is most apparent in Homoptera where, as in *Lecanium*, there are striking regional differences. In *Dysdercus koenigii*, a coccid called the "cotton stainer" because of its practice of injecting a microorganism upon the cotton boll on which it feeds that turns the fibers red, the first area of the midgut has been shown to be the region for digestion of oligosaccharides; the second, that for the absorption of fructose; and the third, that for the absorption of glucose. In another coccid, *Eurygaster integriceps*, the absorptive cells are concentrated in the midregion. In other homopterans, such as the blood-sucking *Rhodnius*, the first area may be the site of water absorption as well as a reservoir for ingested food. Zonal differences are especially apparent histologically in Diptera and Lepidoptera. In adults of the fly *Tabanus*, for example, there are two distinct areas, and in *Glossina*, three. In the anterior part of the midgut of *Glossina*, the cells are small and stain only faintly with ordinary histological procedures; in the middle region, they are tall and stain deeply; in the posterior, they are intermediate in size between those of the anterior and midregions and uniform in appearance. Three zones are also recognizable in the midguts of larvae of the moth *Tineola*, where, in the anterior and posterior zones, cells of goblet type are very regularly interspersed between granular, columnar ones, while in the midregion the goblet cells are fewer in number and different in shape than those in other regions, and the columnar cells are nongranular. The goblet-like cells seem to function in the accumulation and storage of excretory waste, for they become filled with characteristically colored material in larvae that have been fed wool impregnated with salts of elements that form insoluble colored sulfides. These are produced from the hydrogen sulfide, liberated from cysteine in the breakdown of keratin, and the minerals introduced into the diet. Iron and copper are likewise accumulated in specific cells in the midgut of *Drosophila* and *Lucilia* larvae. In blowfly maggots, five distinct zones are recognizable in the midregion of the midgut. The second of these contains cells of two different types, one of which is packed with spheres of lipid and granules of glycogen while the other accumulates iron and copper from a diet rich in these elements. Histochemical tests reveal that there is acid phosphatase in the cytoplasm of cells of the first type, and esterases, dehydrogenases, and cytochrome oxidase in those of the second, where only weak acid phosphatase activity can be demonstrated. All these facts point to regional differences in functional activity in the midgut at least, which may be much more widespread than the present meager information indicates.

Physiologically, measurements of the hydrogen ion concentration in different regions of the gut of a number of different species show that while in many of them there is no great variation from one part to another when the diet is unchanged, in others there may be differences between different parts as great as two pH units. These differences could mean that different areas provided optimal pH's for the operation of different enzymes, thus effectively distributing their activities along the gut, regardless of their source. In adults of *Dysdercus*, the pH of the midgut ranged from 4.6 to 6.8, with the two last regions more acid than the first two. Secretions from the anterior part of the midgut of larvae of *Lucilia sericata* had a pH of 7.4 to 7.6, but those from adults had a range of 4.8 to 5.3, while those from the posterior region of both maggot and adult had one of 7.6 to 8.0. Recent investigations on the stable fly *Stomoxys calcitrans* and the housefly *Musca domestica* have shown that conditions in the anterior and middle regions of the mesenteron in adult third-instar larvae were such as to promote the activity of a protease comparable to vertebrate pepsin. This enzyme is active at pH's of 1.0 to 5.0 with an optimum range of 1.5 to 2.5 and is inactivated in a neutral or alkaline medium. Pepsin has been presumed to be totally absent in invertebrates in whose digestive tracts the pH is, on the whole, too high for its activity. Determinations of the pH in the midgut made from larvae of *Stomoxys* that were dissected after 2 days feeding upon an artificial diet mixed with indicator dyes, showed that it was 2.0 to 3.0 in the anterior midgut, 2.4 to 4.8 in the midregion, and 6.8 to 7.9 in the posterior. Extracts of isolated midguts hydrolyzed albumin most effectively at a pH of 2.4, an acidity about equal to that of mammalian gastric juice. The enzyme so far can only be considered pepsin-like, a proteinase with a low pH optimum, for the properties that would establish its identity with vertebrate pepsin have yet to be demonstrated.

In *Stomoxys* and some other Diptera that will ingest sugar solutions as well as blood, the two different fluids go at first to different regions of the gut, the blood going to the ventriculus and the sugar to the crop. In some species of mosquitoes, in which the crop consists of two dorsal and one ventral diverticula, sugar solutions go directly to the ventral diverticulum from which they are passed bit by bit into the ventriculus, where all the blood from a mixed meal goes directly. In the tsetse fly, *Glossina*, on the other hand, blood passes at once into both crop and ventriculus. The diversion of the two kinds of food in the insects in which it has been observed and experimentally tested seems to depend upon sense organs in the buccal cavity that can discriminate between the two types of food and equally upon nervous connections that effect the appropriate distribution of the stimuli received.

Typically, the hindgut of insects is divided into anterior and posterior regions, the ileum and rectum, respectively. The rectal epithelium is basically cuboidal, but in the majority it is modified as rectal glands, regions between groups of cuboidal cells in which the cells gradually become taller and taller and then taper off again to cuboidal dimensions, so that a small cushion or hillock is formed. In some, especially among Diptera and Hymenoptera, the pads are enlarged as papillae that project well into the lumen of the rectum. The hindgut is concerned primarily with the moulding of feces and their expulsion through the anus. Compaction of the feces is accomplished through the absorption of water from the material that enters the ileum from the midgut and of any solutes remaining in it; this is performed by the rectal glands. In this way, the hindgut functions in water regulation and in the preservation of osmotic balance in the insect body, as well as in the conservation of minerals and other valuable substances that would otherwise be lost.

The movement of material along the alimentary canal is through peristalsis, for the wall throughout its length contains muscle fibers, often arranged in clearly defined circular and longitudinal bands. In many parts of the gut, especially in the crop and ventriculus, rhythmic pulsations keep the food in motion, in some cases macerating it as well, but in all exposing it more fully to digestive enzymes and delaying its passage along the gut. Successive peristaltic and antiperistaltic contractions may also pass the food mass back and forth between two regions, keeping it moving between them for considerable periods of time.

The nature of the digestive enzymes has been studied in a fairly wide variety of insects, almost entirely in vitro, by the use of regurgitated contents of the alimentary canal, or of homogenates of the gut, or of entire animals. The small size of most insects has complicated studies of this kind, for only minute amounts of material can usually be obtained for them. The identification of the enzymes has been for the most part in general terms, based on their hydrolysis of substrates that may or may not be components of the natural diet, and few of them have been specifically characterized. Salivary secretions of most insects contain carbohydrases, almost universally amylases, but glucosidases have also been found, which include invertase, maltase, lactase, and, in the silkworm *Bombyx mori* at least, trehalase. In some insects the salivary glands secrete amylase only, but in others there may be various combinations of amylase and other carbohydrases with proteases. A protease, for example, has been found in the saliva of the milkweed bug *Oncopeltus fasciatus*, together with amylase, invertase, and lipase, and one with properties similar to those of trypsin in that of *Eurygaster integriceps*. *Dysdercus fasciatus* secretes peptidases in its saliva, as well as alpha and beta glucosidases and lipase, but apparently no amylase. The salivary secretions serve also to moisten and possibly to lubricate the food, especially when it is hard or dry, although the secretion of mucus is rare in insects. In some species they contain unusual enzymes such as cellulase, pectinase, and hyaluronidase. Those of blood-sucking insects often include an anticoagulant, an irritant, or an anaesthetic. Tsetse flies, for example, from which the salivary glands have been removed, will feed readily upon the blood of their mammalian victims, but the blood clots in the proboscis, proventriculus, or crop and so may never be wholly digested. The bite of the mosquito *Aedes stimulans* produces a red and itching wheal on mammalian skin, an antigenic reaction that is especially pronounced in sensitive persons. When survivors of the delicate operation of transection of the salivary duct sting human beings, there is no resultant wheal, although blood is drawn and forms a clot in the midgut as it does in normal mosquitoes. Yet in spite of the absence of wheals, the bites of experimental mosquitoes were more painful than those of controls, which are naturally accompanied by saliva, suggesting that one of its components might have an anaesthetizing or pain-reducing effect.

Secretion of amylases is not limited to the cells of the salivary glands, for in a number of species they are secreted by those of the midgut, along with proteases, lipase, and other esterases. Amylase has been found in midguts of the meal-feeding larvae of the beetles *Tenebrio molitor* and *Tribolium castaneum*, and in those of adult *Tribolium*; of adults of the flies *Lucilia cuprina, Calliphora vicinia,* and *Drosophila melanogaster*, and of the larvae of *Calliphora*. Most of the midgut proteinases of insects are trypsin-like; carboxypeptidases, aminopeptidases, and dipeptidases are also secreted by the zymogenic cells of this region. Lipase has been found in the contents of the midgut, and in extracts of midgut tissue, of a number of insects, but is apparently lacking in some others. An esterase that splits ethyl acetate has been demonstrated in the midgut of *Tribolium castaneum* and one that hydrolyzes ethyl butyrate in the blood-sucking fly, *Apaulina avium*. Lecithinase and cholinesterase have been found in that of the wax moth *Galleria mellonella*; cholinesterase has also been found in the midgut of the cockroach *Periplaneta americana*. There is no evidence that the digestion or absorption of fats is aided by the secretion of an emulsifying agent like the bile salts of vertebrates, although the presence of fatty acids seems to promote absorption of undigested fats through the chitinous lining of the foregut of *Periplaneta*.* From the studies that have been made, it seems likely that digestive processes in insects follow a pattern similar to those of vertebrates and that polysaccharides are ultimately reduced to monosaccharides, proteins to amino acids, and lipids to fatty acids, and that in these states they are absorbed. The possibility cannot be excluded, however, that digestion in vivo does not follow precisely the same course as that demonstrated in vitro, although the end products may be the same. Indeed, such a difference has been shown in the dusky cotton bug *Oxycarenus hyalinipennis*, a hemipteran that feeds on cotton plants. Invertase has been found in the salivary glands and in all regions of the midgut, whose action in vitro results in the immediate release of free glucose and free sucrose. Experimentally, insects were fed on small balls of cotton soaked in sugar solutions and the indicator dye Brom Thymol Blue so that the position of the ingested food in the gut could be seen at any time in dissected specimens. In these, it was shown by chromatography of the gut contents that while some free glucose and fructose resulted from the hydrolysis of sucrose in the ventriculus, a trisaccharide composed of two units of glucose and one of fructose was simultaneously formed. This oligosaccharide was gradually broken down into glucose and fructose, but its temporary synthesis in the gut revealed a difference in the operation of the enzyme under natural and artificial conditions, although the end products of sucrose degradation were the same.

Most insects seem to be provided with a wide range of digestive enzymes, which may or may not seem directly useful to them. Hyaluronidase, for example, can be demonstrated in extracts of the salivary glands of *Periplaneta americana*. This may have a digestive function, but its applicability to the feeding habits of an omnivorous scavenger such as the cock-

* The presence of emulsifying agents can be detected, but their identification has not been made. However, mixtures of fatty acyl-sarcosyl-taurines, similar to the acyl-taurines of industrial detergents, have been demonstrated in the gut of *Cancer pagurus*.

roach seems less direct than to those of insects and other invertebrates that penetrate the skin to obtain their food and to which a catalyst of the decomposition of hyaluronic acid, a component of the matrix of connective tissue, would be distinctly advantageous. There is little reason to suppose that the potential complement of digestive enzymes of any insect is limited to those most effective on the materials that predominate in its diet, although phytophagous species seem to make greater use of carbohydrases than zoophagous ones, in which there is greater proteolytic activity. In fact, no proteolytic enzymes have been demonstrated in some phytophagous insects such as the moth *Agrotis orthogonia*, whose larvae are the destructive cutworms, which seem to depend on the enzymes of their food for digestion of its protein. The fact that insects can obtain adequate nutrition from diets of different kinds, to which they may turn naturally or which may be imposed upon them in the laboratory, implies that genetically they are endowed with the capacity to synthesize a wide range of extracellular digestive enzymes, a capacity which may have been reduced or even lost in the most highly specialized feeders, which for either anatomical or physiological reasons are limited to specific kinds of food from specific plants or animals.

The midgut and its caeca or, in some insects, the caeca alone are the principal sites for absorption of the end products of digestion. Different areas of the midgut may be involved in the absorption of similar or of different molecules, but wherever absorption occurs it is believed to be done by cells that are also secretory, so that absorption and secretion are regarded as simultaneous, not alternate, events. The hindgut is also an important absorptive area, less for the absorption of sugars, fats, and amino acids than for water and other solutes. But, since digestive enzymes are rarely found in the excreta, it has been argued that those that are produced in excess are degraded and reabsorbed in the hindgut.

Absorption of monosaccharides by cells of the midgut epithelium has been shown experimentally in insects of several kinds. The rates at which glucose and fructose are absorbed differ in some insects; in many, the sugar in the hemolymph is in the form of trehalose, which suggests an immediate conversion of the absorbed glucose into this disaccharide and the consequent maintenance of a concentration gradient across the gut wall. Evidence has been produced that sugars are absorbed in the hindgut of larvae of the mosquito *Aedes* and fats in the hindgut of some Hymenoptera, but these seem to be exceptional cases. There is some evidence that fat may be absorbed, unchanged, by cells in the foregut. This was shown experimentally in cockroaches (*Periplaneta*) in which fat could be demonstrated in the epithelium of the crop, although ligation of the alimentary canal between fore- and midgut had prevented enzymes synthesized in the midgut from reaching the crop. Therefore, the ingested fat could not have been degraded before its absorption. As a rule, however, fats are hydrolyzed and their components absorbed in the midgut, where different regions may show different degrees of activity in this respect. For example, in blowfly larvae there is no absorption of fatty acids in a short region of the midgut that lies between longer anterior and posterior regions where absorption is active. Fat is absorbed only in the anterior half of the midguts of larvae of *Aedes*, but in the middle region in larvae of *Culex*; while in adults of this genus it is absorbed principally in the posterior region. In some roaches, the midgut caeca are the only sites of fat absorption. Amino acids resulting from protein digestion are also absorbed primarily in the midgut; the concentration of free amino acids in insect hemolymph, often as much as 50 times that in human serum, is one of its exceptional characteristics.

Materials absorbed in excess of an insect's immediate needs are resynthesized and stored, mainly in cells that are usually massed together in the so-called "fat" bodies. This term is a misnomer, for the cells are rich in carbohydrate as well as fat reserves and also contain some protein. These masses represent accumulations of trephocytes, which are also found free in the hemolymph. Increase in the numbers of trephocytes, in their size, and in their contents can be followed during the course of an insect's life. In insects with complete metamorphosis these increases are apparent during the successive larval stages, but during pupation the content of the cells is depleted as it is in adults during reproductive activity and as a result of starvation. Carbohydrate is stored as glycogen, for which the gut epithelium may also serve as a repository. Stored fats vary greatly in their composition in insects of different kinds, both in the fatty acids represented and in the proportions in which they are included in the fat molecules, with consequent differences in chemical and physical properties of the neutral fats. There is little storage of protein, and it is thought that the amino acids taken into the trephocytes are deaminated there and the carbohydrate moieties of their molecules converted into glycogen or fat.

9. Echinodermata

a. GENERAL ANATOMY OF DIGESTIVE SYSTEM The alimentary canal of all echinoderms, except ophiuroids, is a complete one. In ophiuroids it is a blind sac, for there is no anus. In crinoids and holothurians, it consists essentially of oesophagus and intestine, and the tube is relatively long and coiled within the body. It is also long and coiled in echinoids, with oesophagus, stomach, intestine, and rectum. A peculiarity of these echinoderms is the siphon, comparable to that of echiuroids, a tube that runs parallel to the stomach and connects the oesophagus with the intestine, bypassing the stomach. Its function is in doubt, but it may be concerned with the elimination of water from the gut contents. In asteroids, the digestive tract is short and runs directly oral-aborally, but there is a well-defined bipartite stomach. The cardiac stomach, into which the oesophagus opens, is thin-walled, with pouches extending into the arms. Aboral to the cardiac stomach is the much smaller pyloric stomach, into which open ducts from the digestive glands, or pyloric caeca, that extend into each arm. The intestine is short with, in some species, lateral outgrowths, the rectal caeca, just below the anal opening.

The histology of the digestive tract and the events of digestion are best known in a few of the echinoids and asteroids, but even here information is as yet scanty. Practically nothing is known about digestion in crinoids and ophiuroids, and very little about that in holothurians. Amylase, protease, and glucosidases have been demonstrated in extracts of the wall of the intestine of certain holothurians and of the rete mirabile, a mass of vascular tissue that surrounds part of the intestine. Since there is no direct histological evidence of zymogen cells in the wall of the gut, the assumption is made that the luminar enzymes are transported to the gut by amebocytes and released there. Whether the amebocytes are themselves the synthesizers of the enzymes or whether they incorporate those made in other cells and carry them to the digestive organs is problematical.

In the echinoids that have been studied histologically the wall of the gut has been found to be covered by the delicate coelomic epithelium, beneath which is a layer of connective tissue, nerves, and muscle fibers surrounding the innermost layer of columnar cells. In the oesophagus, this inner layer contains numerous secretory cells that discharge mucus and an acid secretion. In the stomach there are tall, secretory cells, most probably zymogenic. There are no recognizable secretory cells in the intestine or rectum, but amebocytes are plentiful in the gut wall in all regions. Extracts of the wall from a number of species have been found to contain amylases, proteases, glucosidases, and an alginase. Lipase is conspicuously absent. Hydrolytic activity is greatest in extracts of cells from the oesophagus and stomach, an indication that the major part of digestion takes place there, which is substantiated by data from experiments in which purple sea urchins (*Strongylocentrotus purpuratus*) were fed disks of the red alga *Iridaea flaccidum* labeled with C^{14}. These experiments showed that most of the radioactivity was lost from the lumen of the gut in the oesophagus and stomach before the material had ever reached the intestine. These regions seem, therefore, to be the major, if not the exclusive, sites of digestion and also of absorption. The extent to which amebocytes are involved in absorption and transport is not clear, although they have been shown to phagocytose particulate material. The coelomic (perivisceral) fluid is known to be a vehicle, and probably one of great importance, in the transference of the end products of digestion, for radioactivity is apparent there 6 hr after a meal carrying the C^{14} label. The gut epithelium is one site for the storage of absorbed material, although the main reserves may be localized in the gonads or in other tissues.

b. COURSE OF DIGESTION IN ECHINOIDS; ASTEROIDS The digestive enzymes of *Strongylocentrotus* are effective on a wide variety of foods. Starved animals will, in the laboratory, eat and digest almost anything from boiled eggs to cooked and raw vegetables of different kinds. Their preferred natural food is brown algae, and the alginase found in extracts of the gut is active in digestion of the polysaccharide of these algae. Ciliate protozoans and bacteria live in the urchins' gut but neither of them probably in sufficient numbers to contribute significantly to their host's digestion. Since it is impossible to rid the animals of their microfauna and flora, there is no critical evidence to exclude the possibility that they do help with the digestion, at least of complex algal polysaccharides.

The short gut of asteroids has the same general histological features as the gut of echinoids. It is covered by coelomic epithelium, has a layer of connective tissue, muscles and nerves, and a lining of columnar epithelium. These cells are flagellated, with mucous cells interspersed among them; in most species, zymogen cells are restricted to the digestive glands and are not found in the oesophagus, stomach, or intestine. The cardiac stomach is anatomically a complex organ. Narrow at the oral end where the oesophagus opens, it expands into five good-sized pouches, whose walls are thin and marked by a regular pattern of furrows and ridges. In predaceous species like *Asterias* and *Patiria* it is eversible and can be wrapped around the soft parts of prey, or a piece of food, while they are undergoing digestion through the agency of enzymes discharged from the zymogen cells in the digestive glands.

The histology of the digestive glands has been studied in greatest detail in *Asterias forbesi*, the common starfish of the Atlantic coast. Five ducts originate from the pyloric stomach and almost immediately branch, each branch running the length of the caecum, which extends almost to the tip of the arm, and giving off lateral branches that subdivide and ultimately terminate in pouches. In each arm there are therefore two caeca, each with a central tubular cavity known as Tiedemann's diverticulum. Like the other parts of the digestive tract, the glands are enclosed in coelomic epithelium, below which is a layer of connective tissue, muscles, and nerves; the inner layer is a sheet of tall, slender cells in which four types can be distinguished. These are current producers, storage cells, zymogen cells, and mucous cells. The current producers are the most numerous; they line Tiedemann's diverticula and the oral and aboral walls of their primary branches. Each cell bears a single long flagellum and the concerted lashing of the flagella on all of them moves a current throughout the gland. The storage cells are found in the lateral walls of the primary branches of Tiedemann's diverticulum and in the secondary and ultimate branches. They, too, bear a single flagellum, but in their cytoplasm a small amount of glycogen can be demonstrated and, in well-fed animals, many droplets of lipid. These are predominately triglycerides, for they stain pink with Nile Blue Sulphate. In these cells there are also globules of material that seems to be a polysaccharide-protein complex. Alkaline phosphatase activity can be shown at their free borders. Zymogen cells are found among the storage cells, but not in regions occupied by current-producing cells. Extracts of regions where there are zymogen cells show proteolytic activity when tested with such substrates as gelatin, fibrin, egg white, casein, peptone, and peptides. Amylases and lipase have also been reported to be produced by the secretory cells. So far as is yet known, there are no enzyme-secreting cells in the stomach or intestine of starfishes, although in some species, such as *Henricia*, they are more generally distributed than they are in *Asterias*.

That these glands are the seat of enzyme production has been proved conclusively by experiments on *Patiria miniata*, the omnivorous cushion or bat star of the Pacific coast of North America. Animals from which the glands had been removed ingested food but did not digest it, for it remained in the cardiac stomach in the same condition as that in which it had been eaten. It is generally accepted that, in intact animals, digestion takes place in the cardiac stomach, through the agency of the enzymes synthesized in the digestive glands and conveyed there by the flagellary currents.

The maintenance of such currents is of the greatest importance in digestion in starfishes, both to convey the enzymes to the site of their activity and to convey the products of this activity back to the digestive glands for resynthesis into storage compounds. In all species that have been investigated, the pattern of the currents seems to follow the same general course. They flow in an oral-aboral direction from cardiac to pyloric stomachs, then radially across the floor of the pyloric stomach into the pyloric ducts, along whose oral surface they flow outward. They flow outward also along the smaller ducts, in a circular pattern in their terminations, and aborally back into the central ducts and the pyloric stomach. In such species as *Asterias* and *Patiria* the outgoing current carries the enzymes discharged into the lumen of the digestive glands eventually to the cardiac stomach, where they come in contact with food that has been ingested or enveloped, and the ingoing current carries the end products of digestion back into the digestive glands where they are absorbed and resynthesized by the storage cells. Glycogen and lipid can be demonstrated in these cells, and C^{14} is concentrated in the digestive glands after glucose, glycine, and palmitic acid labeled with it have been ingested with the food.

In genera that are presumably particulate feeders, such as *Henricia*, there are very effective flagellary pumping organs known as Tiedemann's pouches, which originate almost at the junction of cardiac and pyloric stomachs. Extensions of their side walls divide these pouches into a number of parallel flagellated channels, each directed upward into the main duct of the digestive gland. Their construction is such that they can generate a current of sufficient strength to draw a stream of water containing material in suspension or solution from the cardiac stomach and drive it to the tips of the digestive glands, from which it returns in an axial stream. There is, in *Henricia*, a more complete separation of glandular and flagellate cells than there is in *Asterias*, a fact possibly to be interpreted as relative to its more specialized feeding habits. The cardiac stomach is small in relation to that of *Asterias*, with five radial and five interradial pouches, but no indication that it can be everted. The lumen of the stomach is almost occluded by five or six bulges of its wall that project inward above the interradial pouches. The epithelium of these vesicles contains secretory cells, both zymogen and mucous, as does that of the radial pouches of the stomach, of the pyloric stomach, and of the median and lateral ducts of the digestive glands. The radial pouches of the cardiac stomach are lined with flagellate cells, which contribute to drive a current toward Tiedemann's pouches, where the main propulsive force outward is generated. There are no Tiedemann's pouches in *Asterias*, but they are present in *Patiria*, although histologically different from those of *Henricia*. In *Patiria* they are not divided into flagellate channels, and the only differentiation of their walls is evident in parallel bands of mucous cells alternating with the flagellate ones. They do not function nearly as effectively as those of *Henricia* in producing a strong current, although there is some evidence that *Patiria* at times supplements its macrophagous diet with a microphagous one and so could make use of a mucoid-flagellary mechanism to bring its food to the sites of its digestion.

SELECTED BIBLIOGRAPHY

Digestive processes

Barrington, E. J. N.: Digestive Enzymes, in O. Lowenstein (ed.), "Advances in Comparative Physiology and Biochemistry," vol. 1, pp. 1–67, Academic, New York, 1962.

Boschma, H.: On the Feeding Reactions and Digestion of the Coral Polyp *Astrangia danae* with Notes on Its Symbiosis with Zooxanthellae, *Biol. Bull.*, vol. 49, pp. 407–439, 1925.

de Reuck, A. V. S., and M. Cameron (eds.): "Lysosomes," Little, Brown, Boston, 1963.

Eppley, R. W., and R. Lasker: Alginase in the Sea Urchin *Strongylocentrotus purpuratus*, *Science*, vol. 129, pp. 214–215, 1959.

Florkin, M.: Comparative Biochemistry, *Annu. Rev. Biochem.*, vol. 21, pp. 459–472, 1952.

Galli, D. R., and A. C. Giese: Carbohydrate Digestion in a Herbivorous Snail *Tegula funebralis*, *J. Exp. Zool.*, vol. 140, pp. 415–440, 1959.

Hinton, H. E.: Digestion of Keratin, *Sci. Prog.* (*London*), no. 164, pp. 674–682, 1953.

Hobson, R. P.: On an Enzyme from a Blow-fly Larva, *Lucilia sericata*, which Digests Collagen in Alkaline Solution, *Biochem. J.*, vol. 25, p. 1458, 1931.

Huang, H., and A. Giese: Tests for Digestion of Algal Polysaccharides by Some Marine Herbivores, *Science*, vol. 127, p. 475, 1958.

Lasker, R., and A. C. Giese: Nutrition in the Sea Urchin *Strongylocentrotus purpuratus*, *Biol. Bull.*, vol. 106, pp. 328–340, 1954.

Lasker, R., and A. C. Giese: Cellulose Digestion by the Silverfish *Ctenolepisma lineata*, *J. Exp. Biol.*, vol. 33, p. 542, 1956.

Marshall, J. M., and V. Nachmias: Cell Surface and Pinocytosis, *J. Histochem. Cytochem.*, vol. 13, pp. 92–104, 1965.

Mast, S. O.: The Food Vacuole of *Paramecium*, *Biol. Bull.*, vol. 92, pp. 31–72, 1947.

Muller, M., and I. Toro: Studies on Feeding and Digestion in Protozoa. III. Acid Phosphatase Activity in Food Vacuoles of *Paramecium multimicronucleatum*, *J. Protozool.*, vol. 9, pp. 98–102, 1962.

Ray, D. L., and J. R. Julian: Occurrence of Cellulase in *Limnoria*, *Nature*, vol. 169, p. 32, 1952.

Roth, L. E.: Electron Microscopy of Pinocytosis and Food Vacuoles in *Pelomyxa*, *J. Protozool.*, vol. 7, pp. 176–185, 1960.

Strunk, S. W.: The Formation of Intracellular Crystals in Midgut Glands of *Limnoria lignorum*, *J. Biophys. Biochem. Cytol.*, vol. 5, pp. 385–391, 1959.

Tracy, M. N.: Cellulase and Chitinase of Earthworms, *Nature*, vol. 167, p. 776, 1951.

Tracy, M. V.: Cellulase and Chitinase in Soil Amebae, *Nature*, vol. 175, p. 815, 1955.

Ugolev, A. M.: Membrane (Contact) Digestion, *Physiol. Rev.*, vol. 45, pp. 555–595, 1965.

Vieira, E. C., and B. A. Ledeira: Amylase of the Snail *Australorbis glabratus* (Mollusca, Planorbidae), *Comp. Biochem. Physiol.*, vol. 14, pp. 281–288, 1965.

Vonk, H. J.: The Specificity and Collaboration of Digestive Enzymes in Metazoa, *Biol. Rev.*, vol. 12, pp. 245–287, 1937.

Vonk, H. J.: Comparative Physiology (Nutrition, Feeding and Digestion), *Annu. Rev. Physiol.*, vol. 17, pp. 483–498, 1955.

Waterhouse, D. F.: Digestion in Insects, *Annu. Rev. Entomol.*, vol. 2, pp. 1–18, 1957.

Wharton, D. R. A., and M. L. Wharton: The Cellulase Content of Various Species of Cockroaches, *J. Insect Physiol.*, vol. 11, pp. 1401–1405, 1965.

Yokoe, Y., and T. Yasumasa: The Distribution of Cellulase in Invertebrates, *Comp. Biochem. Physiol.*, vol. 13, pp. 323–338, 1964.

Yonge, C. M.: Evolution and Adaptation in the Digestive System of Metazoa, *Biol. Rev.*, vol. 12, pp. 87–115, 1937.

Alimentary canal and digestion

Chuang, S. H.: Structure and Function of the Alimentary Canal in *Lingula unguis*, *Proc. Zool. Soc.* (*London*), vol. 132, pp. 293–311, 1959.

Dean, D.: A New Property of the Crystalline Style of *Crassostrea virginica*, *Science*, vol. 128, p. 837, 1958.

Feng, S. Y.: Pinocytosis of Proteins by Oyster Leucocytes, *Biol. Bull.*, vol. 129, pp. 95–105, 1965.

Fretter, V.: Experiments with P^{32} and I^{131} on Species of *Helix*, *Arion* and *Agriolimax*, *Q. J. Microsc. Sci.*, vol. 93, pp. 133–146, 1952.

George, W. C.: The Digestion and Absorption of Fat in Lamellibranchs, *Biol. Bull.*, vol. 102, pp. 118–127, 1952.

Goodchild, A. J. P.: Studies on the Functional Anatomy of the Intestines of Heteroptera, *Proc. Zool. Soc.* (*London*), vol. 141, pp. 851–907, 1963.

Graham, A.: The Molluscan Stomach, *Trans. R. Soc.* (*Edinburgh*), vol. 61, pp. 737–778, 1949.

Gressor, R. A. R., and L. T. Threadgold: A Light and Electron Microscope Study of the Epithelial Cells of the Gut of *Fasciola hepatica*, *J. Biophys. Biochem. Cytol.*, vol. 6, pp. 157–162, 1959.

Jennings, J. B.: Studies on Feeding, Digestion and Food Storage in Free-living Flatworms (Platyhelminthes, Turbellaria), *Biol. Bull.*, vol. 112, pp. 63–80, 1957.

Jennings, J. B.: Observations on the Nutrition of the Rhynchocoelan *Lineus ruber* (O. F. Muller), *Biol. Bull.*, vol. 119, pp. 189–196, 1960.

Jennings, J. B.: A Histochemical Study of Digestion and Digestive Enzymes in the Rhynchocoelan *Lineus ruber*, *Biol. Bull.*, vol. 122, pp. 63–72, 1962.

Jennings, J. B.: Further Studies on Feeding and Digestion in Triclad Turbellaria, *Biol. Bull.*, vol. 123, pp. 571–581, 1962.

Lentz, T. L., and R. J. Barrnett: Enzyme Histochemistry of *Hydra*, *J. Exp. Zool.*, vol. 147, pp. 125–150, 1961.

Lentz, T. L., and R. J. Barrnett: Surface Specializations of *Hydra* Cells: The Effect of Enzyme Inhibitors on Ferritin Uptake, *J. Ultrastruc. Res.*, vol. 13, pp. 192–211, 1965.

Mariscal, R. N.: The Adult and Larval Morphology and Lfe History of the Entoproct *Barentsia gracilis* (M. Sars 1835), *J. Morphol.*, vol. 116, pp. 311–338, 1965.

Morton, J. F.: The Functions of the Gut in Ciliary Feeders, *Biol. Rev.*, vol. 35, pp. 92–140, 1960.

Nelson, T. C.: On the Origin, Nature and Function of the Crystalline Style of Lamellibranchs, *J. Morphol.*, vol. 31, pp. 53–112, 1918.

Nelson, T. C.: Recent Contributions to the Knowledge of the Crystalline Style of Lamellibranchs, *Biol. Bull.*, vol. 49, pp. 86–99, 1925.

Newell, B. S.: Cellulolytic Activity in the Lamellibranch Crystalline Style, *J. Mar. Biol. Assoc. U.K.*, vol. 32, pp. 491–495, 1953.

Nicol, J. A. C.: Digestion in Sea Anemones, *J. Mar. Biol. Assoc. U.K.*, vol. 38, pp. 469–476, 1959.

Osborne, P. J., and A. T. Miller, Jr.: Uptake and Intracellular Digestion of Protein (Peroxidase) in Planarians, *Biol. Bull.*, vol. 123, pp. 589–596, 1962.

Owen, G.: Observations on the Stomach and Digestive Diverticula of the Lamellibranchs. Part I. The Anisomyaria and Eulamellibranchs, *Q. J. Microsc. Sci.*, vol. 96, pp. 517–538, 1955.

Owen, G.: Observations on the Stomach and Digestive Diverticula of the Lamellibranchs. Part II. The Nuculidae, *Q. J. Microsc. Sci.*, vol. 97, pp. 541–569, 1956.

Purchon, R. D.: The Stomach in the Eulamellibranchs, *Proc. Zool. Soc. (London)*, vol. 135, pp. 431–489, 1960.

Reid, R. G. B.: The Structure and the Function of the Stomach in Bivalve Molluscs, *J. Zool.*, vol. 147, pp. 156–184, 1965.

Rosenbaum, R., and B. Ditzion: Enzymic Histochemistry of Granular Components in Digestive Gland Cells of the Roman Snail, *Helix pomatia, Biol. Bull.*, vol. 124, pp. 211–224, 1963.

Rosenbaum, R., and C. I. Rolon: Intracellular Digestion and Hydrolytic Enzymes in the Phagocytes of Planarians, *Biol. Bull.*, vol. 118, pp. 315–323, 1960.

Shaw, B. L., and H. I. Battle: The Gross and Microscopic Anatomy of the Digestive Tract of the Oyster *Crassostrea virginica* (Gmelin), *Can. J. Zool.*, vol. 35, pp. 325–347, 1951.

Shaw, B. L., and H. I. Battle: The Chemical Composition of the Gastric Shield of the Oyster *Crassostrea virginica* (Gmelin), *Can. J. Zool.*, vol. 37, pp. 214–215, 1959.

van Weel, P. B.: The Comparative Physiology of Digestion in Molluscs, *Am. Zool.*, vol. 1, pp. 245–252, 1961.

Yonge, C. M.: On some Aspects of Digestion in Ciliary Feeding Animals, *J. Mar. Biol. Assoc. U.K.*, vol. 20, pp. 341–345, 1935.

Zacks, S. I.: The Cytochemistry of the Amebocytes and Intestinal Epithelium of *Venus mercenaria* (Lamellibranchiata), with Remarks on a Pigment Resembling Ceroid, *Q. J. Microsc. Sci.*, vol. 96, pp. 57–72, 1955.

Digestion and absorption

Agrawal, V. P.: Studies on the Physiology of Digestion in *Corophium volutator, J. Mar. Biol. Assoc. U.K.*, vol. 43, pp. 125–128, 1963.

Anderson, J. M.: Structure and Function of the Pyloric Caeca of *Asterias forbesi, Biol. Bull.*, 105, pp. 47–61, 1953.

Anderson, J. M.: Studies on the Cardiac Stomach of the Starfish, *Asterias forbesi, Biol. Bull.*, vol. 107, pp. 157–173, 1954.

Anderson, J. M.: Studies on the Cardiac Stomach of a Starfish *Patiria miniata* (Brandt), *Biol. Bull.*, vol. 117, pp. 185–201, 1959.

Anderson, J. M.: Histological Studies on the Digestive System of a Starfish, *Henricia*, with Notes on Tiedemann's Pouches in Starfishes, *Biol. Bull.*, vol. 119, pp. 371–398, 1960.

Anderson, J. M.: Aspects of Digestive Physiology among Echinoderms, *Proc. XVI Int. Cong. Zool.*, vol. 3, pp. 124–129, 1963.

Andrews, E. B.: The Functional Anatomy of the Gut of the Prosobranch Gastropod *Pomacea caniculata* and of Some Other Pilids, *Proc. Zool. Soc. (London)*, vol. 145, pp. 19–36, 1965.

Arthur, D. R.: The Postpharyngeal Gut of the Earthworm *Lumbricus terrestris, Proc. Zool. Soc. (London)*, vol. 141, pp. 663–675, 1963.

Bahl, K. N., and M. B. Lal: On the Occurrence of "Hepatopancreatic" Glands in the Indian Earthworms of the Genus *Eutyphoeus, Q. J. Microsc. Sci.*, vol. 76, pp. 107–127, 1933.

Bidder, A. M.: Digestive Mechanism of the European squids *Loligo vulgaris, L. forbesii, Alloteuthis media* and *A. subulata, Q. J. Microsc. Sci.*, vol. 91, pp. 1–44, 1950.

Brues, C. T.: "Insect Dietary," Harvard, Cambridge, Mass., 1946.

Chuang, S. H.: Digestive Enzymes of the Echiuroid *Ochetostoma erythrogrammon, Biol. Bull.*, vol. 125, pp. 464–469, 1963.

Eisner, T.: The Digestion and Absorption of Fats in the Foregut of the Cockroach *Periplaneta americana, J. Exp. Zool.*, vol. 130, pp. 159–182, 1955.

Farmanfarmaian, A., and J. H. Phillips: Digestion, Storage

and Translocation of Nutrients in the Purple Sea Urchin *Strongylocentrotus purpuratus, Biol. Bull.*, vol. 123, pp. 105–120, 1962.

Fraenkel, G.: Utilization and Digestion of Carbohydrates by the Adult Blowfly, *J. Exp. Biol.*, vol. 17, pp. 18–29, 1940.

Goodchild, A. J. P.: Studies on the Functional Anatomy of the Intestines of Heteroptera, *Proc. Zool. Soc.* (*London*), vol. 141, pp. 851–907, 1963.

Heatley, N. G.: The Digestive Enzymes of the Onychophora, *J. Exp. Biol.*, vol. 13, pp. 329–343, 1936.

Kermack, D. M.: The Anatomy and Physiology of the Gut of the Polychaete *Arenicola marina*, L., *Proc. Zool. Soc.* (*London*), vol. 125, pp. 347–382, 1955.

Lambremont, E. N., F. W. Fisk, and S. Ashrafi: Pepsin-like Enzyme in Larvae of Stable Flies, *Science*, vol. 129, pp. 1484–1485, 1959.

Mansour-Bek, K.: Extracellular Proteolytic and Lipolytic Enzymes of Some Lamellibranchs, *Nature*, vol. 158, pp. 378–379, 1946.

Nicholls, A. G.: Studies on *Ligea oceanica*. II. The Processes of Feeding, Digestion and Absorption, with a Description of the Structure of the Foregut, *J. Mar. Biol. Assoc. U.K.*, vol. 17, pp. 675–707, 1931.

Nicol, E. A. T.: The Feeding Mechanism, Formation of the Tube and Physiology of Digestion in *Sabella pavonina*, *Trans. R. Soc.* (*Edinburgh*), vol. 56, pp. 537–548, 1930.

Pavlovsky, E. N., and E. J. Zarin: On the Structure and Function of the Digestive Organs of Scorpions, *Q. J. Microsc. Sci.*, vol. 70, pp. 221–261, 1926.

Pickford, G. S.: Studies on the Digestive Enzymes of Spiders, *Trans. Conn. Acad. Arts Sci.*, vol. 35, 1942.

Riedl, I. B., and J. Simpson: Absence of Invertase in Queen and Drone Honey Bees, *Experientia*, vol. 17, p. 365, 1961.

Robertson, J. D.: The Function of the Calciferous Glands of the Earthworm, *J. Exp. Biol.*, vol. 13, pp. 279–297, 1936.

Saxena, K. N.: Digestion and Absorption of Carbohydrates in the Alimentary Canal of the Red Cotton Bug *Dysdercus koenigii* Fabr. (Heteroptera: Pyrrhocoridae), *Physiol. Zool.*, vol. 31, pp. 129–137, 1958.

Saxena, K. N., and P. Bhatnagar: Nature and Characteristics of Invertase in Relation to Utilization of Sucrose in the Gut of *Oxycarenus hyalipennis* (Costa) (Heteroptera: Lygaeidae), *J. Insect Physiol.*, vol. 7, pp. 109–126, 1961.

Vonk, H. J.: The Properties of Some Emulsifiers in the Digestive Fluids of Invertebrates, *Comp. Biochem. Physiol.*, vol. 29, pp. 361–372, 1969.

Yonge, C. M.: The Mechanism of Feeding, Digestion and Assimilation in *Nephrops norvegicus*, *Br. J. Exp. Biol.*, vol. 1, pp. 343–389, 1924.

chapter IX

FOOD HABITS AND DIETARY REQUIREMENTS

A. THE NEED FOR FOOD AND ITS USES

B. EVOLUTION OF DIETARY HABITS

1. Attraction to foods
2. Phagostimulation
3. Metabolic defects; nutritional mutants
4. Toxic compounds and detoxication

C. NUTRITIONAL ADEQUACY

1. Experimental conditions and types of diets
2. General dietary requirements

D. NUTRITION OF PROTOZOA

E. NUTRITION OF INSECTS

1. Special needs and special nutrients
2. Caste determination
3. The fat body
   a. CARBOHYDRATE METABOLISM
   b. LIPID METABOLISM
   c. METABOLISM OF NITROGENOUS COMPOUNDS

F. NUTRITION OF OTHER INVERTEBRATES

A. THE NEED FOR FOOD AND ITS USES

Capture and ingestion of food, digestion, and absorption are of little value to an animal unless the products absorbed contribute to its bodily needs. These contributions may be provision of compounds with stored energy that in the cells can be converted to chemical energy; they may be provision of compounds that are requisite for the construction or reconstitution of cellular components or products, or of those that contain atomic groupings that the animal cannot make itself and that are necessary to its biosyntheses; or they may be provision of compounds essential to the operation of its chemical machinery. An animal may be wholly dependent upon its diet for the fulfillment of each or all of these needs, or it may be able to make some of the compounds for itself if given appropriate precursors. The chemical changes, whether destructive (catabolic) or constructive (anabolic), constitute metabolism, the sum total of the energy transformations that go on in living things. In general terms, the environment is the ultimate source of all the reactants, gaseous, aqueous, or organic; but for animals the organic ones must come from the food that is ingested, since heterotrophs are incapable of synthesizing them from gases and minerals.

The dietary requirements of any animal are quite different, both qualitatively and quantitatively, at various stages of its life and in various phases of its activity. Energy sources are in greatest demand during periods of muscular exertion and of active growth; they are reduced when an organism is at rest and mature. Structural compounds are in greatest demand during periods of active growth and reproduction; the requirements for them are reduced when an animal is fully grown and reproductively inactive. Dietary requirements on the whole are minimal during periods of dormancy, when the level of metabolism is low and when the animal may be using its stored reserves.

Ability to get full value from food depends on an organism's equipment of enzymes to catalyze essential reactions. This, in turn, depends upon its genetic endowment, which determines whether or not it can synthesize those enzymes. But even with the appropriate enzymes the rates at which reactions take place within the body of a given animal may be too slow to produce required metabolites in amounts adequate for its needs, and the deficiency must therefore be made up from exogenous sources. Since its needs vary with the conditions of its life, what may be an adequate amount at one time may be markedly insufficient at another, when, therefore, reliance upon an exogenous source becomes imperative.

B. EVOLUTION OF DIETARY HABITS

The loss or, less probably, the acquisition of enzyme systems through mutation must have been an important factor in the evolution of dietary habits. Limitation in the range of foods eaten implies limitation in the means of processing them, and, similarly, exploitation of new foods implies either new developments of such means or induced activity of those inherently present but not previously used. It has already been pointed out that in the laboratory some animals will eat and thrive on foods they would never encounter in nature, but which they can obviously digest, absorb, and metabolize. This is perhaps not surprising in view of the fundamental similarities in biological composition and biological processes of all living things, but it is perhaps surprising that unusual foods should be recognized as potential nutriment, eaten, and often, apparently, relished. At least three factors are involved in selection of food: attraction, phagostimulation, and nutritional adequacy. An animal may be attracted to a food but fail, when in reach of it, to bring its feeding mechanism into action. It may be both attracted to a food and stimulated by it to the feeding reaction, but find it unpalatable, toxic, or otherwise nutritionally inadequate. Failure to attract a particular species, failure to arouse its feeding reaction, or failure to provide it with the materials needed for its growth and reproduction result in the elimination of that substance from its diet. Potential foods will also of necessity be eliminated from the diets of animals that cannot adequately metabolize certain compounds in them, for if the compound itself or products of its partial metabolism prove toxic, only the organisms which avoid that food will survive. Conversely, a substance that fulfills the conditions of attraction, phagostimulation, nontoxicity, and nutritional adequacy may become the one preferentially sought and eaten by a particular species.

The interplay of these factors in the establishment of a dietary habit is illustrated by the silkworm *Bombyx mori*, often cited as an example of a strictly monophagous species, since its larvae feed preferentially on young leaves of the mulberry *Morus alba*. They will feed, nevertheless, on other plants of the family Moraceae, such as osage orange (*Maclura aurantica*) and the paper mulberry (*Broussonetia papyrifera*) and even on lettuce and dandelion. The insects are attracted to the plants by products that have been isolated and identified as beta-gamma hexanol and alpha-beta hexenal, which are found in leaves of many other plants. They are also found in whale oil, giving it its characteristic odor. These substances can be recognized at some distance by *Bombyx* and probably by a number of other insects to which they also act as attractants. But some of the plants to which silkworms are drawn and some of those on which their feeding responses have been tested, whether or not they have attractants, are not touched by the caterpillars. Soy-beans and tea, for example, which have the specific attractants are not eaten at all, nor are hops, elms, and some species of fig. The biting reaction seems to be stimulated to activity by beta sitosterol, which is a nutrient indispensable to many insects. The cultivated fig (*Ficus carica*) is eaten avidly for a time, but larvae that eat it die even sooner than those that have not been given any food at all, probably because the leaves contain a substance that is toxic to them. Even with osage orange, which provides the next best nutriment to mulberry, only a small proportion of the larvae complete their development, and the silk of their cocoons is of poor quality.

It may be questioned whether an animal that will eat such a variety of plants is truly monophagous, yet it would seem that monophagy is really imposed upon *Bombyx*, since the leaves of *Morus alba* alone give the caterpillars all the materials they need for growth, spinning the silken cocoon, and the profound changes that occur during pupation and metamorphosis. Moreover, the age of the mulberry leaves, the time of day at which they are cut for cultured caterpillars, and the conditions of culture and fertilization of the trees are all factors affecting their nutritional value for the insects. Indeed, meteorological conditions, as they influence the climate, and soil conditions, as they influence the mineral composition and general luxuriance of a plant's growth, are directly significant to the nutrition of all phytophagous animals and so indirectly to zoophagous ones. Seasonal as well as diurnal variations in the food supply, and in the quality of that supply, are factors of no little importance to the adequacy of any diet at any given time and must have influenced the feeding habits of modern invertebrates.

1. Attraction to foods

Other invertebrates may be attracted to foods by volatile substances in them, or released by them, that in themselves have no nutritive value. For example, *Urosalpinx* has been shown to be led to prey by products of oxidative metabolism diffusing from it. Plant substances that may attract phytophagous insects other than *Bombyx* are the so-called "secondary" substances such as essential oils, alkaloids, tannins, and glucosides. These have probably been of considerable importance in establishing food habits and food preference, which must, however, ultimately depend upon the quality of the food and the value to be derived from it. The Colorado potato beetle, *Leptinotarsa decemlineata*, would be an insect of limited range and limited diet and not the agricultural pest that it is, if it had not been attracted to the leaves of *Solanum tuberosum*, when the plant was introduced into its locality, and had not found nourishment in them fully as adequate as in those of its original food, *S. rostratum*. Some other species of *Solanum* and other members of the family Solanaceae are avoided by the beetles, presumably because of their content of alkaloids that may be attractive but are undesirable nutritionally. Demissin, an alkaloid that can be extracted from the leaves of *S. demissum*, prevents development of ovaries in females, and the plant is therefore inadequate as a complete food source and is one that is not chosen. *Nicotiana tabacum*, another member of the Solanaceae, is avoided probably because of the toxicity of nicotine, an alkaloid that is synthesized in the roots and translocated to the leaves. Yet, *Nicotiana* is still attractive to the beetles, for they will feed freely on the leaves of plants grafted on potato roots but will avoid those of potato plants grafted on tobacco roots. The related petunia is also shunned by the beetles, for something in their leaves is highly toxic to the insects.

Recent tests of the attractiveness of the cucurbitacins, bitter substances that are almost universally produced by members of the squash and melon family of plants, have shown them to have positive attraction for *Diabrotica undecempunctata*, the spotted cucumber beetle. Fourteen different cucurbitacins are known, designated by the letters A through N and identified as belonging to the tetratriterpene group of chemical compounds. Cucurbitacins are toxic to livestock and are avoided by some insects. But they are specifically attractive to *Diabrotica*, which when given a choice between the fruit of a mutant strain of watermelon that was particularly bitter and seedlings of other melons and squash with lower cucurbitacin content, ate the watermelon pieces most voraciously. Bees (*Apis mellifera*) and yellowjacket wasps (*Vespula* spp.) avoided the very bitter melon but visited the other plants. The beetles also chewed vigorously on filter paper moistened with extracted cucurbitacins, showing preference for cucurbitacin B and interest in E and D, but avoiding I. These experiments with extracted principles give strong evidence for the fact that it is the triterpenoids of particular configuration, and not other features of the plant, that are the attractants. There is as yet no information as to their usefulness or value to the plant that produces them, except that because of the distastefulness of their flavor and their potential toxicity, they can protect it from predators other than the beetle, which is its particular pest.

Larvae of the moth, *Lymantria dispar*, will readily eat the leaves of a number of trees of different kinds, but only certain ones provide them with a suitable diet. When given apple, spruce, or larch, only 2 percent of a group of experimental insects died, but 18 percent died when fed beech leaves, 61 percent when fed pine needles, and 89 percent when fed alder.

In this connection the accidental discovery of the effects of different kinds of paper upon sexual maturation of the hemipteran *Pyrrhochoris apterus* is of some interest. A strain of these bugs which had been thriving in a laboratory in Prague were shipped to the United States, where with apparently the same care they failed to produce viable gametes. The difficulty was traced to the kind of scrap paper used to cover the bottoms of the cages. When the torn-up pieces came from paper toweling, the *New York Times*, the *Wall Street Journal*, the *Boston Globe*, and even *Science* and the *Scientific American*, the insects failed to reproduce; when they came from the *London Times* or from *Nature*, reproduction was normal. This effect could be attributed to the different sources of wood used in the manufacture of the paper and the probable presence in some woods of a substance that was deleterious to the insect's gonad and gamete development. A heat-stable substance, soluble only in organic solvents such as methanol, acetone, and ether, has been extracted from the wood of balsam firs, hemlocks, and yews but is not found in red spruce, European larch, or southern pine. This substance, not yet characterized, may be the active principle affecting the fertility of *Pyrrhochoris*. It may have some relation to the biologically active glucoside, quercitrin, which is found primarily in oak bark. Quercitrin is well known to the American wood pulp industry, for it is yellow and tends to discolor paper made from woods containing it. Its biological activity has been shown in its effect upon the phytoflagellate *Chlamydomonas*, for it has been found to be the precursor of two different substances, one of which is required by females and the other by males for formation of their respective gametes. Compounds with activities like those of the hormone secreted by the corpora allata of insects (Chapter XVII) have been found in extracts of various plants as well as of the tissues of many invertebrates. These compounds, like the natural hormone, promote larval development but suppress metamorphosis, thus keeping the insect in a juvenile state. Extracts of American newspapers and other paper products have been shown to contain an analogue of this hormone which, when administered to insects, effects, in Pyrrochoridae only, destruction of the developing egg cells. The "paper factor" has also been found to affect the metamorphosis of the red cotton bug *Dysdercus keonigii*. If the factor, extracted from Canadian balsams was applied to the surfaces of fifth-stage larvae in doses even as low as 0.05 or 0.01 micrograms (μg), most of them were unable to emerge from their larval cuticles, and the few that were able to emerge could not develop wings as normal adults. The possibilities of the factor being used as an insecticide are being tested in India, where the bug is of a good deal of economic importance in its effect upon the quality of the cotton marketed.

2. Phagostimulation

Evocation of feeding mechanisms depends upon appreciation of material as suitable food and initiation of the movements involved in biting, chewing, and sucking, or whatever device the organism employs, and swallowing. It is essentially translation of a chemical stimulus, received by a receptor organ that may be considered gustatory, to a nervous excitation, which is in some way so directed that the appropriate muscles are activated. Experimental investigations of the nature of the chemical stimulus have been conducted on invertebrates that feed on blood, such as leeches, ticks, and blood-sucking insects, and on some phytophagous insects, but the evidence from them is by no means uniform or definitive. In experiments with bloodsuckers, the animals are usually confined individually in tubes and offered test substances, in a series of concentrations, heated to the blood temperature of the hosts on which they customarily feed. They are separated from the test substance by an artificial membrane of some kind through which their mouthparts must penetrate in order to reach the food.

Feeding responses to different substances have been tested in the leech *Hirudo medicinalis*, which will readily suck sheep blood, plasma, and washed erythrocytes, from which they are separated by a "Silverlight" membrane made from the caecum of an ox, until they have increased their weight six- to eightfold. When this point has been reached they are satiated and detach themselves from the membrane. They reject water and aqueous solutions of glucose, but will suck glucose dissolved in 0.15 *M* NaCl in a concentration of 1 mg/ml. They will also accept *d*-galactose, *l*-arabinose, *l*-sorbose, and *dl*-glyceraldehyde in similar concentrations in salt solution, as well as *d*-xylose, but this only in a concentration 10 times that of the other sugars. But although they will suck these sugar solutions readily, they do not imbibe as much of them as they do of whole blood. The feeding response is not evoked by similar solutions of *l*-xylose, mannose, *d*-fructose, *d*-arabinose, *d*-ribose, sucrose, lactose, or turanose, or by those of any of the sugars dissolved in a solution of potassium chloride. This has led to the hypothesis that the Na^+ ion is essential to stimulation of the taste organs, which are most likely among the neuroepidermal cells that have been described as distributed generally over the surface of a leech's body but most numerous in the anterior region. It is suggested that sugar facilitates the entrance of the Na^+ ion into these cells and that its penetration triggers the nervous excitation that initiates the muscular activities involved in sucking.

Ticks will feed readily on whole blood at 38°C from which they are separated by a membrane of parafilm. *Ornithodorus tholozani*, the pajatuella of Southern California and Mexico, feeding in nature on the blood of man and other mammals, both wild and domestic, will feed also on a hemolysate of washed sheep erythrocytes. It does not respond to water alone, to sucrose, or to a saline solution isotonic with blood, but it does respond to one in which glutathione is dissolved in concentrations of 1×10^{-2} *M* and 1×10^{-3} *M*, approximately its concentration in mammalian blood. Indeed, the response to this tripeptide is so immediate and effective that a nymph 5 mg in body weight has imbibed as much as 30 mg of the solution in 10 min. Glutathione is also a phagostimulant to *Ornithodorus moubata*, the eyeless tampan of South Africa, and to *Argas persica*, the common fowl tick. It is also the substance first found to evoke the feeding reaction in *Hydra* (see page 231).

Tested on a variety of substances, each over a range of concentrations, the hemipteran *Rhodnius prolixus* responded positively to those that could act as phosphate donors, with the implication that translation of chemical stimulus to nervous excitation here involves phosphorylation. Observation of the bugs in feeding experiments in which they were offered a wide variety of substances, showed that they quickly approached any warm fluid with their proboscides extended and immediately began to probe through the rubber membrane separating them from it. A minute amount was apparently sucked up and thus brought into contact with the epipharyngeal sensillae, a group of highly innervated cells lying dorsal and anterior to the pharyngeal pump and responsive to chemical stimuli. If these were not appropriately stimulated, the proboscis was withdrawn in 3 to 20 sec, and the insect moved onto another spot and probed again until it finally gave up. If the sensillae were stimulated, the insect at once settled down and began to gorge itself, ingesting five to seven times its body weight. The substances most effective in inducing the sucking reaction were the di- and triphosphates of adenosine, guanine, inosine, cytosine, and uridine, in concentrations of 1×10^{-3} *M*. Riboflavin-5-phosphate and 5′-adenylic acid also induced positive responses. Mammalian red blood cells contain adenine nucleotides, mostly as ATP (adenosine triphosphate), in a concentration of 1.5×10^{-3} *M*, and it is likely that these are the substances that under natural conditions trigger the feeding response in *Rhodnius*.

The feeding reaction is induced also in mosquitoes by ATP, and by ADP (adenosine diphosphate) and AMP (adenosine monophosphate) as well. Riboflavin-5-phosphate is not effective with *Culex pipiens*, to which the nucleotide configuration and adenine moiety seem essential. Mosquitoes will also respond to sugar solutions and to washed erythrocytes of ox blood, but not to the plasma. What may be the activating substance in sugar solutions is not clear, but the feeding reaction is definitely evoked by it, and the fact that sugar is diverted to the crop while blood goes directly to the midgut can hardly account for the similarity of response with such disparate fluids.

Among other substances tested, the chelating agent EDTA (ethylene-diamine tetraacetic acid) induced feeding reactions in 122 out of 612 specimens of *Aedes aegypti* offered solutions of it. In consequence, it has been suggested that removal of Ca^{++} ions from the dendritic membranes of neurons conducting the stimulus from the gustatory cells to the appropriate centers depolymerizes the membrane and effects the conversion of the chemical stimulus to the nervous excitation.

Feeding responses of nymphs of the locust *Schistocerca* and the grasshoppers *Melanoplus bivittatus* and *Camnula pellucida* have been tested by offering them filter paper impregnated with various substances. Sucrose and glucose are powerful phagostimulants for locusts, which will also respond to wheat germ oil and an ethereal extract of bran. Grasshopper nymphs, especially in their fourth and fifth instars, are strongly stimulated to feeding activity by wheat germ oil and by phospholipids extracted from plants, especially lecithins and phosphatidyl inositol. These compounds, however, when included in synthetic diets of nymphs do not effect any significant improvement in their growth and development, except insofar as they promote ingestion of large amounts of food by fourth- and fifth-instar nymphs.

It would seem that specificity of feeding habit is closely related to specificity of phagostimulant, which would, in turn, depend upon specific sensitivities of the gustatory cells, wherever they were located, and upon the mechanism by which their response is conveyed to the nervous system. This is a complex and provocative problem.

3. Metabolic defects; nutritional mutants

Inability to metabolize particular substances must also have been an important factor in the evolution of dietary habits. Such an inability has been demonstrated in honeybees to which the sugar mannose was, more or less by chance, found to be lethal. In experiments designed to test the taste sensitivities of bees, it was found that bees that sucked from the dishes containing mannose solutions died, while those that sucked from dishes containing other hexoses showed no such ill effects. In further experiments designed to find an explanation of the effect of mannose, groups of bees were isolated and given either a 1 *M* solution of glucose, a 1 *M* solution of mannose, or water only. Within 3 hr, over 90 percent of those that had imbibed mannose were dead, while almost all of those taking glucose or water survived far longer. Investigations of the enzyme systems in homogenates of whole bees that had been fasted for 1 hr showed a deficiency in phosphomannoseisomerase, the enzyme that converts the phosphorylated form of mannose to its isomer, phosphorylated fructose, a conversion that is necessary to its further utilization. Phosphorylation is the initial step in the degradation of a hexose and is accomplished through the agency of hexokinase, an enzyme of wide, if not universal, distribution. The two reactions involved may be represented as:

$$\text{Hexose} + \text{ATP} \xrightarrow{\text{hexokinase} + \text{Mg}^{++}} \text{hexose-6-phosphate} + \text{ADP} \quad (1)$$

$$\text{Hexose-6-phosphate} \xrightarrow{\text{phosphohexoseisomerase}} \text{fructose-6-phosphate} \quad (2)$$

The bees proved to have fully adequate amounts of hexokinase and phosphoglucoseisomerase, but because of the deficiency of phosphomannoseisomerase in their cells, mannose-6-phosphate accumulated in them and interfered in a number of ways with essential energy-releasing reactions. The lethal effects of mannose could be directly attributed to this interference. Mannose has been found to be toxic also to species of bees other than *Apis mellifera* and to the wasp *Vespa vulgaris*, presumably for the same reason. This is but one of many known cases of possible "inborn error of metabolism," or inability to synthesize an enzyme essential to a key metabolic reaction. Attention to such metabolic diseases was drawn over 50 years ago by an English physician, A. E. Garrod. His recognition of these inborn errors gave impetus to the genetic studies of nutritional mutants in fungi and bacteria that have proved so rewarding. The destructive effects of this particular "deficiency disease" in bees can be avoided by elimination of mannose from their diets, just as those of diabetes, phenylketonuria, and similar disorders in man and mammals can be controlled and prevented by appropriate diets.

4. Toxic compounds and detoxication

The ability to detoxify, or to convert harmful substances into harmless ones that can be stored, further degraded, or excreted, must also have had significance in determining food habits, either by broadening the range of potential foods or by permitting an organism to make use of a food source denied to others. Detoxication occurs through oxidations, reductions, hydrolyses, and syntheses catalyzed by enzymes, each of which is specific for the particular type of reaction upon a particular substrate. The Pacific shipworm, *Bankia setacea*, for example, ingests a number of free aromatic aldehydes that are consistently present in timber. Among these, in the wood of both coniferous and deciduous trees, is vanillin (4-hydroxy-3-methoxybenzaldehyde), which is quite toxic to most animals. It has been shown that *Bankia* oxidizes the vanillin absorbed to innocuous vanillic acid, through the agency of a specific enzyme, vanillin oxidase, demonstrable in extracts of the caecum and the combined tissues of the stomach, digestive gland, and foot. Difficulty in dissecting the digestive tract away from other tissues has prevented more precise localization of the enzyme but the low activity exhibited by extracts of oyster tissue on vanillin suggests that it is not a digestive but a metabolic enzyme of bivalves, and one whose activity is sufficiently great in *Bankia* to permit that bivalve to exploit wood as a dietary source.

Detoxication mechanisms have been studied intensively in insects, partly for the practical purposes of evaluating insecticides and partly for the light such information may throw upon insect biochemistry and so, indirectly, upon nutrition. Some of the mechanisms insects have been found to use in detoxication of ingested compounds or their metabolites are the same as those of vertebrates, some are different, and some that vertebrates use have not been found in insects. Insects, for example, rarely, if ever, make use of reductions in detoxifying compounds, while oxidations and conjugations are very frequent means. Adult beetles of the species *Aromia moschata* and larvae of *Melasoma* and *Plagioderma* oxidize saligenin, a beta glucoside particularly abundant in the bark and leaves of willows, to salicylic acid, which can be excreted. Larvae of *Melasoma, Plagioderma*, and the brassy willow beetle, *Phyllodecta vitellinae*, but not the adults, convert the salicylic acid to salicylaldehyde. This, in larvae of *Phyllodecta*, is secreted in special glands on the dorsal surface of the body and acts as a repellant to other animals. Such by-products of detoxication, and, indeed, of other metabolic reactions, may account for many of the diverse repellant, attractant, and informative secretions of insects and other animals.

Oxidations may also be a first step in the processes of detoxication by conjugations, in which a molecule contributed by the organism is combined with a suitable functional group on an oxidative product or directly with a foreign molecule that has been absorbed.

The native component of the conjugate is, in insects, most frequently either glycine or glucose. *Locusta migratoria*, for example, conjugates glycine with its derived salicylic acid and can make the same combination with benzoic acid. The recovery of hippuric acid, the product of benzoic-acid-glycine conjugation, in the excreta of locusts after experimental administration of benzoic acid to them, gives evidence for this reaction and for the elimination of the product from the body. Vertebrates, and larvae of the mosquito *Aedes aegypti*, are known to treat benzoic acid in the same way. Such conjugation seems to be a common method of disposing of many other potentially harmful aromatic acids and probably some aliphatic ones as well. Phenols, which are in general more toxic than these acids, are detoxified through the formation either of ethereal sulfates or of beta glucosides. In the contribution of glucose to this reaction, insect biochemistry resembles that of plants more than of vertebrates, in which the native component of the conjugate is usually glucuronic acid and the product a glucuronide.

The ability to detoxify insecticides, whether eaten or absorbed through the surface of the body, has, in recent years at least, been of great significance in the natural selection and establishment of resistant strains, and much attention has been given to analysis of the mechanisms by which an insecticide enters the body and is handled after absorption. In some cases, a metabolite is more toxic than the original substance so that, because of a reaction within its body, an insect may poison itself in a way from which other animals, lacking the capacity to perform that particular reaction, are protected. A number of chlorinated hydrocarbons, the earlier ones of which are DDT (di-chlor-diphenyl-trichlorethane) and lindane (benzene hexachloride), have been most successful as insecticides, at least from the point of view of their first effects and their continuing ones upon persisting susceptible strains, but most deleterious in their overall effects. Animals with chitinous cuticles are particularly sensitive to DDT, which is both absorbed and concentrated by chitin. The colonial cnidarian *Obelia*, with a chitinous perisarc, is, for example, very sensitive to it, as are strains of insects without means of detoxifying it. The milkweed bug, *Oncopeltus*, dies very quickly after injection with DDT, but resistant strains of houseflies (*Musca domestica*), selected in nature and cultured for generations in the laboratory, can dechlorinate it, after absorption, to a dichlorethylene derivative, DDE. This is accomplished by a specific dehydrochlorinase, which has been isolated and purified and found, in vitro, to require as a cofactor either glutathione or cysteinylglycine. The rate at which the reaction proceeds increases with temperature, so that flies are less resistant at 16°C and thereabouts than at 20°C and above. The enzyme is present in nonresistant strains but has very low activity in vitro. Its function in ordinary metabolism is unknown, for the isolated enzyme shows no activity upon any natural substrate. The gene controlling DDT resistance in *Drosophila melanogaster* has been located on the second chromosome; this gene directs the formation of an enzyme that catalyzes the conversion of DDT to a compound at least six times less toxic to the flies than DDT and one that is readily degraded further in their bodies. Yet this metabolite, 1,1-bis-*p*-chlorophenyl-trichlorethanol, is very toxic to mites, and is used as an acaricide under the name of Kelthane. The heteropteran *Triatoma* can also metabolize DDT to a compound similar to Kelthane and to another derivative of the nature of a phenol. This capacity has been shown to be localized in the cellular microsomes and most probably to depend upon two different enzymes.

It is possible that DDT may exert its primary effect in susceptible insects by disturbing their amino acid metabolism. The hemolymph of insects is notably high in its content of free amino acids, which make an important contribution to its osmotic pressure. Their concentration, individually and collectively, is of great significance in maintaining osmotic stability. A drop in the total amino acid level has been shown in houseflies of susceptible strains after exposure to DDT and one in the single amino acid, proline, in the cockroach *Periplaneta*, an insect incapable of dechlorinating DDT. In larvae of *Aedes aegypti*, exposed 4 to 8 hr to DDT, the total amino acid level in the hemolymph increased in a resistant strain, but remained virtually constant in a susceptible one. This increase was shown by chromatography to be due almost entirely to an increase in the concentration of alanine, suggesting either an increased synthesis of this particular acid or a decreased rate in its metabolism.

Houseflies seem also able to dechlorinate lindane, for pentachlorocyclohexane can be demonstrated in their tissues in the course of the disappearance of the hexachlorocyclohexane. DDT dehydrochlorinase is not the enzyme involved, for in vitro it is inactive upon lindane. It may be that the detoxication is achieved through condensation of the benzene hexachloride with glutathione, or some other sulfhydryl compounds, since on alkaline hydrolysis, dichlorothiophenols are recovered.

C. NUTRITIONAL ADEQUACY

Even a cursory survey of the eating habits of invertebrates shows that the natural diets that are wholly satisfactory to different species vary very widely. Many problems are presented to the investigator who wants to find out why a particular diet is selected by a particular species, or what particular compounds are required to satisfy its nutritional needs. These problems arise not only from the great diversity of feeding mechanisms exhibited by invertebrates and the great variety of foods they consume but also from technical difficulties involved in conducting adequately controlled experiments to test the necessity and even the utilization of a given substance. Such tests can be made only by presenting the organism with a diet whose composition is both quantitatively and qualitatively known and which can be manipulated in such a way that, one by one, each compound in it can be increased, reduced, or withheld so that its relative importance can be estimated. Such defined diets are often not ones that experimental animals will accept, even when very hungry, and adjustments to make them attractive and palatable may well result in defeat of the primary purpose of the experiment.

Success in provision of a fully adequate synthetic diet is measured by survival and growth of the test animals. In protozoans, this is determined by increase in the number of individuals commensurate with that of a wild population and the maintenance of this growth rate in successive subcultures under similar conditions. In metazoans, it is determined by growth of individuals, evidenced by increase either in weight or in bodily dimensions, and by their juvenile development at a rate equivalent to the average rate of the same species in nature, or in the laboratory when given their natural food. It is measured also by the ability of test animals to pass through critical stages in their life cycles, such as molting and metamorphosis, and by their capacity to reproduce and give rise to as many healthy and vigorous offspring as they would in nature. Ideally, an experiment should be carried on for several generations to ensure that the diet is fully adequate for survival of the species, not only of the individual, and to obviate the possibility that a nutrient, essential but needed only in minute amounts, may be transmitted through the eggs. Ideally, too, the experiment should be conducted with a genetically homogeneous population and under strictly aseptic conditions. These requirements have been met in but a few cases and, for invertebrates, absolute dietary demands are known for a limited number of species of protozoans and insects, which have lent themselves particularly well to studies of this kind. Some of the results of such studies are presented later in this chapter.

1. Experimental conditions and types of diets

Conditions for controlled diets may be catalogued as follows:

I. Gnotobiotic (Greek *gnosis*, "knowledge," and *bios*, "life"): in which all the organisms (test animals, food, and associated fauna and flora) are identified and known.
   A. Axenic (Greek *a*, "without," and *xenos*, "guest"): in which the only organism is the test animal ("pure cultures").
   B. Synxenic (Greek *sun*, "with," and *xenos*, "guest"): in which there are one or more associated species, as monoxenic, dixenic, trixenic, etc.

II. Agnotobiotic: in which some or all of the associated organisms are unknown.

III. Synthetic: in which the food is made up of known amounts of chemical compounds.
   A. Holidic (Greek *holos*, "whole"): in which the chemical structure of each compound is known (defined medium).

B. Meridic (Greek *meros*, "part"): in which the chemical structure of all but a few components is known.
C. Oligidic (Greek *oligos*, "little"): in which the precise chemical structure of the components is not known.

Asepsis is obligatory for the most precise nutritional studies, since microorganisms are known to be capable of a broad spectrum of biosyntheses and may well be sources of nutrients to animals with which they are associated, either as contaminants of the medium or of the food, or as symbionts in or on the animal itself. It is, of course, debatable whether a symbiotic microorganism, so adapted to its host that neither can live without the other, is to be regarded as a contaminant or as much a part of the organism as its own cells and tissues. Where such associations are firmly established in nature, it seems questionable whether insistence on working with a completely sterile organism has any real validity.

It has been clearly shown in nutritional studies that the proportionate amounts of the ingredients included in a synthetic diet are critical factors in its success. Excess amounts of substances, in themselves essential as nutrients, may be actually harmful in inhibiting the utilization of others, and the ratio between antagonists must be adjusted to optimum levels. Conversely, one substance may have a "sparing" effect upon another, in that its inclusion in the diet supplants or supplements that of the other, which therefore can, and often must, be included in reduced amounts. Mixtures of all the amino acids known to be regular constituents of protoplasmic systems are usually more successful than those of only the 10 essential or indispensable ones, and peptide-bound amino acids and proteins more so than free ones. And in addition to chemical factors, certain physical conditions must also be met. Important considerations in provision of an adequate and acceptable experimental diet are the consistency and texture of the food, the time of day at which it is offered, and the temperature and humidity at that time; the method of dispensing it, and its accessibility, in terms at least of the animals being able to reach it and to consume it in as natural a way as possible.

Presumably in nature the food selected by a given species provides the necessary balance and proportionality even for monophagous ones. Possibly, however, polyphagy is a response to the need to obtain the requisite range and variety of nutrients or to get enough of those that are limited in distribution. Experimental restriction of the diet of polyphagous species is known to lead, in some cases at least, to atypical growth and behavior. Giants and monsters appear among suctorians that are fed only *Tetrahymena* for any length of time, and gigantism occurs also in populations of the ciliate *Stylonichia*, in which cannibalism has broken out; this condition is not unusual in laboratory cultures. When it happens in a culture of *Tetrahymena*, individuals several times the standard size appear in it. Neoplasms that occasionally develop in test organisms may also be related to the diet. Tumors have, for example, been frequent among individuals in cultures of the freshwater annelid *Enchytraeus* kept under axenic conditions in an oligidic medium of nutrient agar and heated liver extract. In other cultures, under the same conditions, entire populations have been destroyed by the sudden onset of lysis. Observations of the incidence of neoplasms in *Drosophila melanogaster* indicate relationship between genetic constitution and diet, for it was higher in flies of a tumorous stock given a rich diet than in those kept on a minimal ration. It has been shown by these studies that the relative proportions and amounts of amino acids and of vitamins in the diet can be manipulated in ways that lead to increases or decreases in the percentages of specimens developing tumors.

2. General dietary requirements

All animals have certain common nutritional requirements. These are listed in Table IX–1 and include some elements and some so-called "accessory food substances," or vitamins. Specific requirements, as they are known for certain groups of invertebrates, are listed in Table IX–2. The necessary elements, aside from carbon, nitrogen, hydrogen, and oxygen, are to a great extent obtained as mineral salts, as noted in Chapter I. Autotrophs can derive carbon and nitrogen from inorganic compounds (carbon dioxide, nitrates, and ammonium salts), but heterotrophs must depend upon organic ones. These vary in complexity, depending upon an animal's ability to synthesize complex compounds from simple ones. Some can use the salts of short-chain acids, such as acetic, for carbon sources and starting points for the construction of complex carbohydrates, but most require compounds of higher molecular weight (long-chain acids, fats, sugars, and starches). Amino acids, peptides, polypeptides, and proteins are the usual sources of nitrogen. Ten of the nineteen amino acids recognized as common constituents of the proteins most generally found in biological systems are listed in Table IX–1. These are the "essential" or indispensable amino acids, since they have been found to be so for the rat and several other animals whose absolute dietary requirements have been ascertained and which cannot synthesize them for themselves. The other nine [alanine, aspartic acid, cystine (and cysteine), glycine, glutamic acid, hydroxyproline, proline, serine, and valine] are considered "non-essential" (dispensable), since they can be synthesized by some animals at least. Tests in synthetic diets have shown that the isomeric form of the acid is highly

TABLE IX-1 Nutrients Known, or Presumed, to Be Essential to All Invertebrates

Nutrient	*Source*	*Use*
Carbon	Simple and complex carbon compounds	Construction of cells, tissues, and their products; sources of energy and reserves
Nitrogen	Simple and complex nitrogen compounds such as amino acids, peptides, polypeptides and proteins. Essential (indispensable) amino acids (10): Arginine, Histidine, Isoleucine, Leucine, Lysine, Methionine, Phenylalanine, Threonine, Tryptophan, Valine	
Other elements		
Oxygen	Bound in carbon and nitrogen compounds; free (as O_2) in air and water	Energy-releasing reactions
Phosphorus, Magnesium, Sulfur, Chlorine, Iron, Potassium, Sodium	Food, water, and aqueous solutions	Components of cell and body fluids

TABLE IX-1. Continued.

Nutrient	*Source*	*Use*
Vitamins	Food	Cooperate in metabolic reactions
Thiamin (B_1)		Essential to carbohydrate metabolism [can be synthesized by some invertebrates from one or both precursors (pyridine and thiozole)]
Cobalamine (B_{12})		Participates in methylations, probably in synthesis of proteins and methyl compounds
Biotin		Involved in carboxylations and decarboxylations of organic acids; in deamination of certain amino acids (aspartic acid, serine, threonine); in synthesis of citrulline and unsaturated fatty acids
Riboflavin		Component of several flavoprotein enzymes operating in cellular oxidations
Niacin (nicotinic acid)		Component of di- and triphosphonucleotides, which act as hydrogen acceptors in more than 50 known metabolic reactions
Pyridoxine (B_6)		Contributes to pyridoxal phosphate (coenzyme for transaminase functioning in protein synthesis); involved in fatty acid metabolism
Pantothenic acid		Component of coenzyme A; involved in methylations and so in metabolism of carbohydrates, fats, and proteins
Pteroylglutamic acid (folic acid, vitamin M, vitamin B_c, *Lactobacillus casei* factor, norite eluate factor)		Operates in methylations; required for synthesis of purines and pyrimidines
Thioctic acid (alpha lipoic acid)		Condenses with thiamin to form lipothiamin; operates in first step in decarboxylation of pyruvate

relevant to their adequacy. Only the levo form of most of them is accepted by animals, and the dextro form, which is not found in nature, is in most cases useless. There are, however, some exceptions. Larvae of *Tribolium confusum*, the flour beetle, can use *d*-lysine, *d*-methionine, and *d*-phenylalanine, which other beetles cannot, but *Tribolium* cannot use the *d*-forms of the other amino acids. *d*-Serine is extremely toxic to larvae of *Drosophila melanogaster*, and *l*-serine, although required by them, is slightly toxic and can be included in their diets in minute amounts only.

Certain substances, known as micronutrients because they are needed in such small quantities, are also common nutritional requirements. Though often complex in structure, they are not utilized as major sources of energy or of constructive materials, but provide specific molecular configurations that are essential to the operation of cellular enzymes or that contribute in other ways to intermediary metabolism. These include some of the vitamins, especially those of the B complex, as well as other compounds. A vitamin is broadly defined as a compound other than a carbohydrate, amino, or fatty acid, that has been shown to be a dietary requirement for at least one vertebrate animal. No such requirement has been shown by invertebrates for some of the compounds in this category, and some not included in it have been shown to be "vitamins" for invertebrates, for choline, inositol, and carnitine can fulfill vitamin-like functions (Table IX–2). Choline can be synthesized by some invertebrates from methionine and ethanol amine. It is a precursor of acetylcholine, mediating the passage of nerve impulses, and a component of the lecithin group of phospholipids; and, as a carrier and donor of methyl groups, it is also indirectly involved in methylations. Inositol, a cyclic polyhydric alcohol, is a constituent of some other phospholipids and is essential to the metabolism of fat and cholesterol. The biochemical role of carnitine (beta hydroxybutyrobutaine), a dietary requirement for some insects, has not yet been elucidated, although its structure suggests that it may, like choline, act as a carrier and donor of methyl groups.

Ascorbic acid (vitamin C) and the fat-soluble series designated in very general terms as A, D, E, and K, seem to have specialized functions for specialized cells and tissues and are not, like the water soluble vitamins, of almost universal need. Ascorbic acid can be synthesized from glucose in a series of steps and seems to be made in adequate amounts by most of those invertebrates whose precise nutritional needs are known. It is, however, a dietary requirement for hemoflagellates and phytophagous insects and must be included in the holidic diet on which the grasshoppers *Melanoplus bivittatus* and *Camnula pellucida* can be raised from egg to adult.

Although vitamin A has been found in the tissues of many invertebrates as a product of the metabolism of ingested carotene, it seems to have no essential role in them. No animal can synthesize it *de novo* and must obtain it either as such or as a readily convertible precursor. Beta carotene, astaxanthin, and echinone, a pigment found in the gonads of certain echinoderms, molluscs, and sponges, are all precursors for both vertebrates and invertebrates that feed on these animals. Vitamin A_1 has two more hydrogen atoms than vitamin A_2; its aldehyde is retinol, the key pigment in the visual cycle of vertebrates. It has been demonstrated by biological assay on rats in two species of nematodes (*Dictyocaulus viviparus*, the cattle lung worm, and *Anisakis physeteris*, a parasite in the stomach of sperm whales); in the red arrowworm *Eukrohnia fowleri*, (but not in Sagitta); in the annelid *Enchytraeus*; in two species of starfish; and in numerous molluscs and arthropods. In molluscs, the principal site of its accumulation is the digestive gland, but it is also found in other tissues of the visceral mass and, in cephalopods, in the eggs and the eyes, where it may have a visual function. It has also been found in the eyes of crustaceans, as well as in their digestive glands and exoskeletons, and in the eyes of some insects. Retinol has been extracted from the eyes of *Orconectes virilis*, a freshwater crayfish, and from those of honeybees. A visual function is indicated in the mosquito *Aedes aegypti*, for the eyes of specimens raised aseptically on a synthetic diet lacking both vitamin A and beta carotene were markedly deficient both functionally, as measured by electrical response to photostimulation, and structurally, as visualized in electron micrographs and compared to those of mosquitoes raised on a diet containing vitamin A. The other form of vitamin A, vitamin A_2, has never been obtained from an invertebrate.

A need for vitamin E (alpha tocopherol), required by rats for normal breeding, has been shown in five different arthropods. The cladoceran *Daphnia magna*, kept in a medium from which the vitamin had been eliminated, failed to grow and to reproduce at the optimum rate. These were resumed when the vitamin was added to the medium. Feeding experiments with the lepidopteran *Ostrinia nubilalis* show that it requires either corn oil or a mixture of linoleic acid and alpha tocopherol for normal growth and emergence. Unless wheat germ to the extent of 2 percent of the total diet is included in that of the orthopteran *Acheta domesticus*, eggs laid by the female fail to hatch, although in point of egg production these females may be as fertile as females on a complete diet. This failure is due to male sterility, for the spermathecae of females raised in the same cages as males, both on a deficient diet, are empty. Histological examination of the males shows that all stages of spermatogenesis proceed normally until spermiogenesis, but that in the absence of alpha tocopherol the transition from spermatid to spermatozoan is not accomplished. Reproduction is also affected in the coleopteran *Cryptolaemus mountrouzieri* and in the dipteran *Agria affinis*. Females of this fly have been raised successfully from egg to adult on a

TABLE IX-2 Nutrients Known to Be Required by Special Groups of Invertebrates

Nutrient	*Source*	*Use*	*Known dietary requirement for:*
Elements	Food, water, and aqueous solutions		
Iron		Component of cytochromes and hemoglobin	All protozoans and insects tested; all species with hemoglobin
Copper		Component of hemocyanin	All species with hemocyanin
Calcium		Shell and skeleton construction	Some protozoans, sponges, corals, and molluscs
Silicon		Shell and skeleton construction	Some protozoans and some sponges; can replace calcium in some cases
Manganese		Necessary for certain enzymatic reactions	Some protozoans and some insects
Amino acids	Food	Protein construction	
Cystine			*Aedes aegypti* (mosquito)
Glycine			*Tetrahymena pyriformis* (ciliate); *Rhodnius prolixus* (assassin bug); larvae of some dipteran insects [*Drosophila melanogaster* (fruit fly), *Aedes aegypti* (mosquito), *Calliphora erythrocephala* (blowfly), *Pseudosarcophaga affinis* (flesh fly)]
Glutamic acid			*Tetrahymena pyriformis* (ciliate)
Proline			*Phormia regina* (blowfly); males of *Blatella germanica* (cockroach)
Serine			*Tetrahymena pyriformis* (Strain E); males of *Blatella germanica* (cockroach)
Vitamins	Food		
Ascorbic acid (vitamin C)		Operates in oxidation-reduction reactions	Hemoflagellate protozoans; phytophagous insects [*Melanoplus bivittatus, Camnula pellucida* (grasshoppers), *Schistocerca gregaria* (locust) for growth beyond fourth nymphal instar]
Vitamin A	Carotene	Growth and pigmentation	Some insects [*Schistocerca gregaria* (locust), *Melanoplus bivittatus* (grasshopper) for normal pigmentation; *Aedes aegypti* (mosquito) for eye development and vision(?)]
Vitamin E (alpha tocopherol)	Leaves of many vegetables; wheat germ oil	Intracellular antioxidant	*Daphnia magna* (water flea), *Ostrinia nubilalis* (European corn borer), *Acheta domesticus* (cricket), males for reproduction; *Agria affinis* (fly), females for reproduction; *Cryptolaemus mountrouzieri* (ladybug)
Inositol		Growth	*Ephestia kuehniella* (flour moth)
Choline		Growth to maturity	*Blatella germanica* (cockroach), *Lasioderma serricorne* (cigarette beetle), adults; *Tribolium* (flour beetle), larvae; *Ephestia kuehniella* (flour moth) and *Ephestia elutella* (tobacco moth), larvae; *Phormia regina* (blowfly)

TABLE IX-2. Continued.

Nutrient	*Source*	*Use*	*Known dietary requirement for:*
Carnitine		Growth to maturity	*Tenebrio molitor* (darkling beetle), larvae; *Palorus ratzburgi* (small-eyed flour beetle), larvae; *Tribolium confusum* and *T. castaneum* (flour beetles)
Lipids	Food		
Cholesterol		Structural (component of cell membranes) lipid metabolism and synthesis of fatty acids	Some protozoans and most insects [*Anthonomus grandis* (boll weevil adults for egg production); *Dermestes maculatus* (carpet beetle), *Tineola pellionella* (clothes moth), *Bombyx* (silk moth), *Blatella* (cockroach), *Locusta* and *Schistocerca* (locusts)]
Linoleic acid		Energy metabolism	*Trichomonas* ("symbiotic" flagellate); some insects (*Ephestia* spp.); locusts; grasshoppers
Purines and pyrimidines			
Thymine, uracil, guanine		Construction of nucleic acids	Some protozoans and insects

complete diet, but their eggs, when alpha tocopherol was omitted, did not develop beyond the blastula stage.

No essential role in invertebrate metabolism has been attributed to vitamin K, although echinochrome, also a naphthoquinone, is structurally very similar to vitamin K_2. The D vitamins seem equally unnecessary to invertebrates; on the other hand, some species show definite requirements for steroids and some other lipids.

Steroids include a large number of compounds with fused carbon rings (Fig. IX–1). Cholesterol, a complex monohydroxy alcohol, is the most generally distributed animal steroid. Structurally, it is a component of cell membranes, and functionally, it operates in lipid metabolism. It is a dietary requirement for some protozoans and some insects, although several compounds related to it can be used effectively by others. Among these are ergosterol, obtained from the fungus ergot and from yeast, and stigmasterol, obtained from soybeans.

Certain long-chain fatty acids that are known to be essential to vertebrate nutrition have been found, so far, to be equally necessary to some invertebrates. Among these is primarily linoleic acid ($C_{18}H_{32}O_2$). Arachidonic acid ($C_{20}H_{32}O_2$) and linolenic acid ($C_{18}H_{30}O_2$) can be used effectively by insects, but are not required by them. Indeed, some insects can grow and develop, and some protozoans can live and

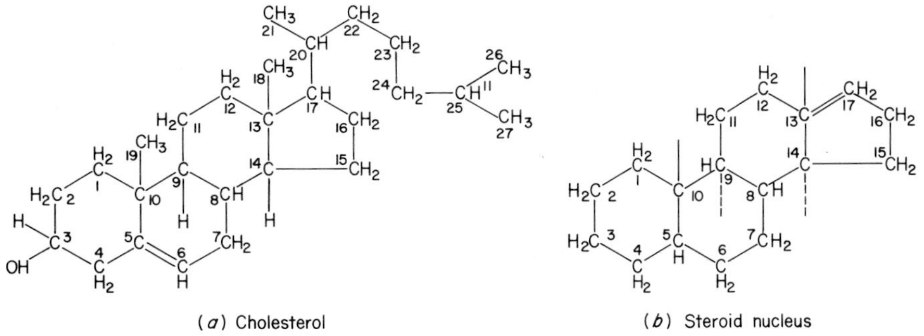

FIG. IX–1 *Structural formulae.* ***(a)*** *Of cholesterol.* ***(b)*** *Of the steroid nucleus: solid lines from carbons 10 and 13 denote substituents lying above the plane of the paper, in which the rings lie; dotted lines from carbons 9 and 14 denote substituents lying below the plane of the paper.*

(*a*) Cholesterol

(*b*) Steroid nucleus

multiply, in the complete absence of any dietary fat. Even larvae of the wax moth *Galleria mellonia*, that ordinarily feed on beeswax, have been reared successfully on diets completely devoid of fat in any form but containing the requisite cholesterol.

Some invertebrates have absolute requirements for purine and pyrimidine compounds, or at least those with a pyrimidine ring:

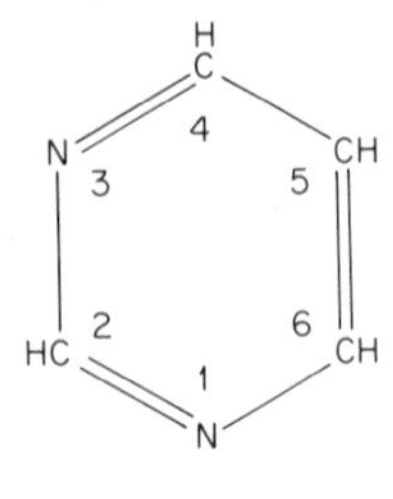

Pyrimidine

These configurations are essential to construction of amino acids, and invertebrates seem less capable of synthesizing them than vertebrates. Some insects have specific requirements for the pyrimidines, thymine or uracil, and some protozoans also require these or the purine, guanine.

D. NUTRITION OF PROTOZOA

Protozoa exhibit a very wide range of nutritional potentialities and dietary freedoms and restrictions. The chlorophyll-containing flagellates (chlorophytes) are autotrophic and, like plants, synthesize their carbon compounds from carbon dioxide and nitrate or ammonium ions as carbon and nitrogen sources, respectively. The colorless flagellates (leucophytes) require more complex carbon sources and use acids with carbon chains of various lengths, ranging from 2-carbon acetic acid (CH_3COOH) to 14-carbon myristic ($C_{13}H_{27}COOH$). They can also use a variety of alcohols, but no sugars. They differ, however, in which of these acids or alcohols they are able to use. *Chilomonas paramecium* can use acetic, butyric (C_3H_7COOH), valeric (C_4H_9COOH), caproic ($C_5H_{11}COOH$), and octylic ($C_8H_{17}COOH$) acids, but not propionic (C_2H_5COOH), isobutyric, or heptylic ($C_7H_{15}COOH$) acids; it can use ethyl (C_2H_5OH) and butyl (C_4H_9OH) alcohols, but not methyl (CH_3OH) or isobutyl alcohols. It can also use lactic (CH_3CHOH $COOH$) and pyruvic (CH_3CO $COOH$) acids. It can use ammonium as a nitrogen source, with acetate as its carbon source, only if the pyrimidine and thiazole of thiamin are supplied; otherwise it must have a complex peptone from which these moieties of the vitamin are presumably derived. *Polytoma caudatum* also requires thiamin, but can make it if thiazole alone is supplied; and of the fatty acids, uses only acetic and butyric acids as carbon sources. It cannot use any alcohol as a carbon source, nor nitrate as a nitrogen source, but it does use ammonium and peptone.

The fact that the hemoflagellates live in a sterile medium, the blood, makes them potentially desirable objects for exact nutritional studies, but the fact that they are specialized parasites makes data obtained from them of less general value than those from free-living species. The various trypanosomes that have been studied require glucose as a carbon source and some have been successfully cultured in media containing only glucose and peptone from meat. *Trypanosoma cruzi* has, however, no specific glucose requirement, but does require ascorbic acid and other growth factors. It, like some other species of *Trypanosoma* and of *Leishmania*, needs hemoglobin or its hydrolysate, probably for the porphyrin ring supplied from the hematin. Some other trypanosomes, whose principal phase is not in the blood of warm-blooded animals, also require hematin, and some do not. *Stegomonas* (*Leptomonas*) *fasciculata*, for example, a parasite in the digestive tract of a mosquito, has a definite requirement for hematin, while *S. oncopelti*, a parasite of hemipterans, and *S. parva*, a parasite of flies, have not and can be cultured on a glucose-peptone diet. *Stegomonas oncopelti* does, however, require thiamin and cannot satisfy this need from either the pyrimidine or the thiazole moieties alone, or together. *Stegomonas fasciculata* has also a specific thiamin requirement.

Although their natural diets seem far more complex, the nutritional needs of certain free-living ciliates are better known than those of any other protozoans. This applies especially to the common ciliate *Tetrahymena pyriformis*, which in nature feeds on bacteria and unicellular algae but has been axenically cultured for generations on a holidic medium. The organisms which can be maintained as a clone, or a population that has descended from a single individual by successive mitotic divisions, are therefore genetically homogeneous. The medium also fulfills the criteria for an ideal experiment in being sterile and chemically defined, and in supporting them indefinitely. It contains a variety of salts supplying the necessary chemical elements in appropriate proportions, together with some, as traces, more for insurance than for any presently known need. These include iodine and selenium, which are known to be requisites for higher animals but for which no specific need has yet been demonstrated in protozoans, as well as boron, molybdenum, and vanadium. Phosphorus is supplied as glycerophosphate, a compound that is more soluble than inorganic phosphates and that also has good buffering action. The medium also contains 13 amino acids, selected after their identification by fractionation procedures as constituents of the peptone on which the animals were known to flourish in autoclaved meridic media. The three additional to the ten "essentials" are glutamic acid, glycine, and serine. The pyrimidine, uracil, and the purine, guanine, are also included, guanine as its monophosphate ester, guanylic acid, which also contributes to the pool of available phosphate. Vitamins in the medium are niacin, pantothenic acid, thioctic acid, and biotin. Glucose is an indispensable component not only as an energy source but as a starting point for a variety of syntheses. Both lactate and acetate can "spare" glucose in reducing the amount of it required. No lipid need be added to this medium, although there are some strains of *Tetrahymena* that have shown a lipid requirement.

There are differences between strains, too, in their demands for specific sugars or amino acids. Some can make carbohydrate and fatty acids from acetate, and most can use fructose, mannose, and maltose, but not sucrose or other sugars, as sources of energy and of valuable intermediates. The 10 essential amino acids satisfy the minimal needs of Strain W of *T. pyriformis*, but growth is improved by the addition of serine or glycine. Strain E has, however, an absolute requirement for serine. Methionine, providing the major source of sulfur, is an absolute requirement for all species and strains but can be "spared" by cysteine or homocysteine. Likewise, arginine is "spared" by citrulline and ornithine, which the ciliates can combine into the amino acid.

Media refined to this extent have not yet been devised for carnivorous rhizopods and ciliates such as *Paramecium*, yet a good deal is known about the nutritive requirements of this particular genus. Besides the combination of salts and 10 amino acids needed by *Tetrahymena*, *Paramecium* requires a fatty acid, such as stearic or oleic, and a sterol, which may be of plant origin, such as stigmasterol or sitosterol. These ciliates do not utilize purines and pyrimidines well in the form of free bases, but more effectively as their nucleotides, that is, combinations of the base with phosphate and sugar that are the repeating units of the nucleic acid helices. The known vitamins that *Paramecium* requires are thiamin, riboflavin, niacin, pyridoxine, pteroylglutamic acid, and pantothenic acid. Yeast extract, or certain fractions of it, is used to supply nutrients that are necessary but not yet identified.

E. NUTRITION OF INSECTS

Insects most suitable for nutritional studies are those whose larvae live naturally in aquatic or moist habitats and whose culture media can therefore be sterile fluids, as well as those insects that habitually live on very dry foods which do not support the growth of microorganisms. Culture of larvae hatched from surface-sterilized eggs has been successful, in terms of adequate growth, pupation, and metamorphosis, in the orthopterans *Melanoplus bivittatus* and *Camnula pellucida*, the coleopterans *Tenebrio molitor* and *Palorus ratzburgi*, the lepidopterans *Ostrinia nubilalis, Ephestia kuehniella* and *E. elutella* and the dipterans *Drosophila melanogaster*, *Aedes aegypti*, and *Agria affinis*.

The synthetic diets on which insects have been successfully reared usually contain vitamin-free casein as a protein or amino acid source and glucose or sucrose as one of carbohydrate. The amino acid content of casein is known; it provides a number of amino acids in addition to the 10 "essentials," so that with it in a medium that supplies the other necessities satisfactory growth and development may be expected. Requirements for specific amino acids have been established for only a limited number of insect species. Nymphs of the cockroach *Blatella germanica* do not require phenylalanine, for they are apparently capable either of synthesizing its phenyl moiety or of deriving it from tryptophan. They have, however, an absolute requirement for the alanine moiety which must be supplied from an external source. The endogenous synthesis of phenylalanine may be due to activity of microorganisms in their mycetocytes. These symbionts may also account for incorporation of inorganic sulfur into proteins for, while sterilized individuals require methionine, nonsterile ones do not. With the exception of alanine and leucine, whose omission produces the most drastic effects, deficiency in any one of the other amino acids results only in a slower rate of growth, but not in mortality, suggesting that some amino acid synthesis is continually going on in the bodies of the nymphs. Males require serine and proline as well as the other amino acids for their optimal growth, but the females do not. Glycine is an absolute dietary requirement for larvae of *Drosophila melanogaster* and *Aedes aegypti*. The mosquito larvae cannot convert serine to glycine, a transformation that is effected by higher animals, and they also require cystine for full development. Those deprived of cystine cannot accomplish the final molt, from pupa to imago, either because of general weakness or because of some inadequacy in the molting fluid. Omission of glycine from the diets of larvae of the blowfly *Calliphora erythrocephala* and of *Pseudosarcophaga affinis*, a parasite of the spruce budworm, retards their growth and usually results in mortality during some juvenile stage. Some individuals have, however, been known to reach maturity without it. The black blowfly, *Phormia regina*, is the only dipteran known that does not seem to require glycine, but it, like the males of *Blatella*, does require exogenous proline. Other insects tested have absolute requirements for only the 10 indispensable amino acids, but their growth rate is improved when the medium contains the nine dispensable ones as well. Larvae of *Tribolium confusum*, for example, will survive and grow slowly in a mixture of the 10 indispensables only, but will grow and reproduce at a rate comparable to those on a casein diet if the other nine are also included. On the other hand, larvae of the hide beetle *Dermestes maculatus* and of *Tenebrio molitor* do not grow on mixtures of pure amino acids and require casein.

Experiments in which fifth-instar nymphs of the hemipteran *Rhodnius prolixus* were fed uridine-glucose, with the glucose labeled with C^{14}, show that these insects can synthesize from it the carbon chains of alanine, asparagine, cystine, glutamic acid, and serine. Since no radioactivity was detected in the arginine, glycine, histidine, isoleucine, leucine, lysine, phenylalanine, threonine, tryptophan, or valine recovered from the nymphs, these amino acids may be considered essential in the sense that they cannot be synthesized in the body, at least through the pathway common to the others.

The source of protein in the diet of homopterans that live entirely on the juices of plants has long been in question. Aphids are the most prolific of all insects, some species producing 40 to 50 generations a year, each consisting of 10 or more individuals. Because their food appears to be largely a solution of sucrose alone, it has been thought that they have some means of nitrogen fixation, very possibly through symbionts, to meet their obvious need for protein. Analyses of pure plant sap have been few, because of the difficulties of obtaining it in sufficient quantities; but correlations between the nitrogen and carbohydrate of that which has been analyzed and the honeydew secreted by aphids have been most revealing. The sap of the red currant bush, *Ribes rubrum*, has been found to have a higher nitrogen content than the honeydew of *Cryptomyzus ribis*, the aphid that feeds on it. And the phloem sap of willows has also a higher content of nitrogen than the honeydew of *Aphis saliceti*, but both have the same amino acids. Comparisons of the carbohydrate content of plant saps and aphid honeydews show that in this respect the two are almost identical. One additional, and different, sugar, alpha maltosylfructofuranose, is found in honeydew and not in sap, together with sucrose, glucose, fructose, trehalose, and melizitose, which they have in common. Alpha maltosylfructofuranose has also been found in the honeydew of the citrus mealybug, *Pseudococcus citri*, when it was fed on potato, and in that of *Aphis pomi* and *Eriosoma lanigerum*, which feed on apple trees. From these facts, it seems evident that plant sap does provide all the nutrients necessary to aphids, if sufficient quantities are ingested, and also that the carbohydrates, at least, undergo some conversion before they are channelled from the midgut and excreted as honeydew.

Carbohydrate usually forms a large part of the natural diet of insects, yet it is not essential to most of them. Many can convert ingested protein into fat and carbohydrate reserves to be called upon as sources of energy in times of stress. This has been most convincingly shown by larvae of *Aedes aegypti*, which, fed on casein after starvation, built up stores of glycogen and fat, as well as of protein; and stored glycogen, but not fat, when under the same conditions they were fed only alanine or glutamic acid. Yet some insects do require carbohydrate at some period in their life cycle and so are more or less restricted to diets rich in it. Among these are pests of dried foods, such as the larvae of *Tenebrio molitor*, *Ephestia kuehniella*, and *Oryzaephilus surinamensis*, the sawtoothed grain beetle, which feed on stored products with high carbohydrate content such as grains, cereals, and dried fruits. *Tribolium confusum*, a pest in grain elevators, mills, and groceries, will feed not only on all kinds of grains and dried fruits but also on beans, nuts, chocolate, and pepper; and *Lasioderma serricorne*, the cigarette beetle, finds pepper, raisins, ginger, and upholstered furniture, as well as dried tobacco, acceptable food.

Whether or not they actually need them, most insects do utilize a good number of different carbohydrates when offered them in test diets. The extent to which any nutrient in the diet of any animal is utilized can be calculated as the percentage of the amount digested or absorbed according to the formula:

$$\frac{\text{Dry weight of nutrient consumed} - \text{dry weight of nutrient excreted}}{\text{Dry weight of nutrient consumed}} \times 100 = \%\ \text{utilization}$$

The efficiency with which ingested material is converted to the animal's own living substance can also be expressed as a percentage by using the formula:

$$\frac{\text{Dry weight of animal}}{\text{Dry weight of nutrient consumed}} \times 100 = \%\ \text{gain in dry weight}$$

A number of insects have been tested for their capacities to make use of hexoses, pentoses, and alcohols that could presumably be directly absorbed, as well as of di-, tri-, and polysaccharides that would require digestion prior to absorption. Of all those tested, glucose seems the most generally acceptable, although a number of insects, including larvae of *Oryzaephilus* and *Ephestia* and adults of *Drosophila* will use fructose. Adult honeybees will also use fructose, but their larvae do so only poorly. *Oryzaephilus* will use lactose, but larvae of *Apis* and of *Ephestia* cannot. Adults of *Drosophila*, *Phormia regina*, and *Calliphora erythrocephala* can use mannose, but the larvae of *Tribolium* and *Tenebrio* cannot, nor can honeybees. Larvae of *Tribolium* and *Tenebrio* cannot use galactose either, nor can those of *Oryzaephilus*, *Ephestia*, or *Apis*, but adult blowflies can and adult fruit flies do, but only to a certain extent. The pentoses arabinose and fucose were not utilized by any of the insects tested; rhamnose was only slightly useful to adults of *Phormia* and *Apis*, and not all to those of *Calliphora* and *Drosophila*. Ribose was not utilized by *Tenebrio* larvae and only to a slight degree by adults of *Drosophila* and *Calliphora*; xylose was not utilized by larvae of *Oryzaephilus*, *Tribolium*, or *Tenebrio*, but was to some extent by adult blowflies and fruit flies.

Of the alcohols, sorbitol was not only utilized by adults of *Calliphora* but proved to be the carbohydrate they used most effectively. It was also utilized by larvae of *Ephestia*, *Tribolium*, and *Oryzaephilus*. Dulcitol was not used by larvae of *Tenebrio*, but mannitol was, as well as by larvae of *Tribolium* and *Oryzaephilus* and adults of *Calliphora*, *Phormia*, and *Drosophila*.

Sucrose, though not used by larvae of *Tribolium*, *Oryzaephilus*, or *Ephestia*, is used by adult blowflies, by *Drosophila*, and by larval and adult bees. Starch can be used by larvae of *Tribolium*, *Tenebrio*, and *Oryzaephilus*; to some extent by larvae of *Ephestia kuehniella*, but not at all by those of *E. elutella*; nor by larvae of *Apis*, although it is, but only poorly, by the adults.

Many adult insects, especially dipterans, hymenopterans, and lepidopterans, can live on aqueous solutions of acceptable and usable sugars alone. These will support their adult activities, but for all the important phases of their juvenile lives other nutrients are required. There is a range of acceptability and of usefulness in the sugars that support adults and in the concentrations of them in the solutions provided. Adult honeybees have lived well on solutions of glucose, fructose, sucrose, maltose, and trehalose; less well on melizitose; and cannot survive on lactose, galactose, raffinose, rhamnose, xylose, or arabinose. The hymenopteran *Macrocentrus ancylivorus*, a parasite of the oriental fruit moth, *Grapholitha molesta*, and the strawberry leaf roller, *Ancylis comptana*, make good use of fructose, glucose, maltose, and sucrose; little use of galactose; and still less of lactose. Concentrations up to 5 percent are usually best for any insect; too dilute a solution requires the insect to imbibe relatively enormous amounts of the fluid to get enough of the dissolved sugar, but too great a concentration has proved to have deleterious effects upon survival.

Vitamin requirements of insects have been tested in the conventional way by reducing the amount of each individually in test diets, or by omitting it entirely, and also by the administration of antivitamins, or vitamin antagonists. This latter method is of particular value with insects that harbor microorganisms as symbionts and from which, therefore, certain vitamins at least cannot easily be withheld. The unnatural conditions induced by sterilization of such insects may confuse interpretation of experimental results to a greater extent than those arising from the introduction into the body of a compound of molecular configuration closely resembling that of a given vitamin, which interferes with its operation in biochemical reactions by competing with the vitamin for available sites of activity. Defects arising from such interference, or inhibition of its action, are indications that the vitamin is synthesized endogenously.

At least five of the B vitamins—niacin, pantothenic acid, pyridoxine, riboflavin, and thiamin—are very general requirements of insects. Nymphs of the cockroach *Blatella germanica* require choline or else betaine, even when they have their bacterial symbionts, in addition to these five. Adults of the cigarette beetle, *Lasioderma serricorne*, require also choline and pteroylglutamic acid, but their larvae are apparently entirely independent of exogenous vitamins, with the possible exceptions of biotin and choline. The others are supplied to them in sufficient quantities by their intracellular symbionts. Larvae of *Tribolium* also require exogenous biotin, choline, and pteroylglutamic acid as well as the usual five, but those of *Oryzaephilus* need neither biotin, choline, nor pyridoxine. Larvae of *Tenebrio* and *Palorus ratzeburgi* are unusual in their requirements for carnitine, which seems also a necessity for *Tribolium confusum* and *T. castaneum*, if they are to reach maturity successfully. Carnitine can replace choline in the diet of larvae of *Drosophila melanogaster*, which require also biotin as well as the usual five others of the B complex. Larvae of *Ephestia kuehniella* and *E. elutella* have similar requirements, both of them needing exogenous biotin and choline as well as the "major five." Choline also is necessary for satisfactory postlarval development of *Phormia*. The postlarval stages are more demanding in their requirements for structural proteins than the earlier ones. Larval growth is adequate when analogues for choline are substituted in the diet, provided that these analogues have two methyl groups on the nitrogen atom and a hydroxyl group on the second carbon atom from it. There is some evidence that xanthopterin, a fluorescent yellow pteridine that is most likely a degradation product of pteroylglutamic acid, is an adequate substitute in the diets of these moths, if given in amounts 1,000 times that of the effective amount of folic acid. This is not true for larvae of *Aedes* and *Tenebrio*. *Ephestia kuehniella* is unique among insects in its requirements for inositol, at least for a satisfactory growth rate and survival time. No such need has been discovered in other insects tested, nor has any general requirement for cobalamine. Growth of *Tenebrio* larvae and pupal development of *Aedes* are reported to be better on diets containing small amounts of cobalamine than on those without it, and females of *Blatella* seem to require it for the development of viable eggs, although its presence or absence in the diet has no significant effect upon growth.

The only insect yet known to have an absolute dietary requirement for ascorbic acid is the desert locust *Schistocerca gregaria*, which does not survive beyond the fourth nymphal instar without it in the semisynthetic diet on which they are reared. Yet 46 generations of *Blatella germanica* have been raised on sterile culture media lacking it entirely, and the tissues of members of the last generation were found to contain as much of it as those of freshly captured specimens, nourished by their customary omnivorous diet. The possibility of intracellular synthesis by symbionts is not excluded in this case, yet the ubiquity of ascorbic acid in the tissues and hemolymph of insects not recognized as hosts of microorganisms suggests that synthesis of the vitamin is common practice in this group of invertebrates at least. Its function in them is not readily apparent, but recent experiments on *B. conjuncta*, a New Zealand species of cockroach, indicate that it may be concerned in the oxidation of tyrosine, an important phase of all animal metabolism.

Although *Blatella* has also been grown successfully for generations without either vitamin A or its carotene precursor, *Schistocerca* does seem to require beta carotene for growth and survival as well as for development of its characteristic coloration. Without the provitamin growth is slower than in individuals fed their natural diet of grass, and mortality is high in the fourth instar and usually total in the fifth. Moreover, as early as the second instar absence of natural pigmentation is apparent and is strikingly evident in insects that survive to the fourth or fifth nymphal stages. Addition of vitamin A to the diet improved the growth rate and decreased mortality but had no effect on coloration, suggesting that beta carotene alone can provide the requisite for pigmentation and that the growth factor may be a derivative common both to it and to vitamin A. Possibly the primary effect of beta carotene deficiency is disturbance in pigment metabolism, which has a secondary effect upon growth. Pigmentation in the grasshopper *Melanoplus bivittatus* is likewise abnormal when the nymphs are reared on a synthetic diet without β-carotene, although its absence does not interfere with their growth and development in other ways.

In contrast to their relative independence of exogenous sources of vitamins, insects have far greater dependence than other animals upon exogenous sources of lipids, especially sterols. A good many species are known to be able to synthesize a fairly wide range of fatty acids from carbohydrate, protein, or amino acid sources, but with the possible exception of *Ctenolepisma* sp., none is yet known that can synthesize a sterol for itself. They fail to do so when given such precursors as acetic or mevalonic acids, starting points of the synthesis in vertebrates, or any of the intermediates that are known to be in the biosynthetic pathway. Cholesterol is the most general insect requirement, although some species do not need it as such but can use closely related compounds. Adults of the cotton boll weevil, *Anthomus grandis*, require, for example, about 20 mg cholesterol/100 gram diet for sustained egg production and normal longevity. This may be the only insect known to require exogenous sterols as an adult, for most of them accumulate enough during their juvenile lives to provide for themselves as adults. The life span of *Anthomus* when reared in the laboratory is 58 to 89 days, but if the diet is deficient in cholesterol, it is cut to 13 to 15 days. Beta sitosterol, a sterol of plant origin which differs from cholesterol only by the addition of another CH_3 group on the side chain attached to carbon 17, can substitute for cholesterol in the diets of *Lasioderma serricorne, Oryzaephilus surinamensis, Tenebrio molitor, Tribolium confusum, Aedes aegypti, Drosophila melanogaster, Phormia regina, Ephestia kuehniella, Bombyx mori, Tineola bisselliella, Blatella germanica, Locusta migratoria, Schistocerca gregaria*, and a number of other species. Ergosterol can be equally well used by all these insects, with the exception of *Blatella,* which makes less good use of it than it does of beta sitosterol or of stigmasterol. Stigmasterol, differing from cholesterol both in a longer side chain and in a second double bond, between carbons 22 and 23, is also a satisfactory substitute for colesterol for *Drosophila, Ephestia,* and *Aedes*. But none of these is adequate for *Dermestes maculatus* or *Tineola pellionella*, whose larvae are therefore restricted to foods of animal origin. 7-Dehydrocholesterol is adequate for *Dermestes* and for some other insects but is not for *Bombyx, Blatella, Locusta,* or *Schistocerca*. On the other hand, *Dermestes* cannot use cholestanol, which differs from cholesterol only in the absence of a double bond between carbons 5 and 6, although *Blatella* and other insects can. Derivatives of cholesterol that have not a hydroxyl radical at position 3, such as cholestanone and cholestene, or which have an additional one, like 7-hydroxycholesterol, are not known to be used by any insect. It has been postulated, primarily from study of sterol metabolism in *Blatella*, that the single hydroxyl group at position 3 is the critical factor in the configuration of the molecule. This configuration is essential to the activity of cholesterol esterase, and esterification of the molecule is considered important to its absorption from the lumen of the gut, obviously a necessary preliminary to its metabolic utilization. In insects, in addition to their more general roles, sterols may be involved in sclerotization of the cuticle and, perhaps even more importantly, in the production of hormones that regulate growth and development. It has, for example, been shown that cholesterol labeled with tritium (H^3) is converted by fly larvae to ecdysone, the molting hormone (pages 721–722). Perhaps the primary use of cholesterol in insects is in the construction of cytomembranes. In homogenates of the roach *Eurycates floridana,* for example, it is found in greatest quantity in the particulate fraction.

Acute defects resulting from dietary deficiencies in fatty acids have been conclusively shown in only one genus of insects. On a fat-free diet, larvae of three species of *Ephestia* (*kuehniella, elutella, cantrella*) grow slowly and, though they ultimately pupate, do not eclose. This defect can be completely remedied by the addition of 4 mg linoleic acid/gram dry food, and partially so by doses from 0.3 mg/gram upwards. The visible effect of deficiency is malformation of the wings, from which some, to all, scales are missing. When the level of linoleic acid is just suboptimal, only a few scales are gone, but at lower levels, nearly all of them. Their loss is attributed to the fact that they stick to the pupal skin, which in turn implies a defect of some sort in the scales or in the molting fluid. This is ordinarily secreted between the pupal skin and the adult cuticle and so separates the imago from the pupal case and allows it to slip out easily. Linolenic acid is also effective in promoting growth and emergence, but arachidonic acid in amounts from 1.7 to 5 mg/gram dry food promotes growth but not eclosion. Linoleic acid deficiency can also be detected in wing formation and the final nymphal molt of locusts and grasshoppers, but the effects are by no means so drastic as they are in *Ephestia*. Cockroaches (*Blatella germanica*) show no defects when raised on a fat-free diet, but their first-generation offspring do show those of linoleic acid deficiency when their diet is also free of it. *Ostrinia nubilalis* and *Pectinophora gossypiella*, the pink boll weevil, have also been shown to have fatty acid requirements. *Tenebrio molitor* is known to synthesize its own linoleic acid, as probably many other insects also do, because they are known to synthesize other fatty acids, particularly during their larval or nymphal stages. These, stored largely in the form of neutral fats, in the fat body or other tissues, usually constitute a supply adequate to the demands of adult life. Some insects probably continue to synthesize fatty acids or to modify those that absorbed, as adults. Experiments with the cotton boll weevil (*Anthonomus grandis*), given acetate labeled with C^{14}, show that both larvae and adults can synthesize long-chain fatty acids such as palmitic, stearic, and oleic acids from it. Both adults and larvae, maintained in aseptic conditions, synthesized the same kinds of acids but in different amounts, for the larvae incorporated 60 percent of the absorbed C^{14} into oleic acid, as against 25 percent incorporated by the adults, but made only one-half as much stearic acid and one-third as much palmitic acid as the adults. The larvae were, however, unable either to synthesize linoleic acid from acetate or to convert even closely related long-chain acids to it. There is no evidence that it is a dietary requirement for them. They, like other insects, can make some modifications of absorbed fatty acids and so convert them to forms suitable for their own special needs, either metabolic or storage.

A few insects have been tested for their abilities to utilize fats and fatty acids given them in test diets. Larvae of *Ephestia kuehniella, elutella,* and *cantella* will use cod liver and wheat germ oils, as will those of a closely related species, *Plodia interpunctella*, and those of *Tineola bisselliella*. *Ephestia* and *Plodia* make poor use of butterfat, which larvae of the green bottle fly, *Phoenicia sericata*, use effectively. Larvae of *Aedes aegypti* do not use linoleic or linolenic acids when offered them in the diet, nor the triglycerides, triolein, tripalmitin, or tristearin. Oleic acid is not used by any insect tested and indeed has been shown to have detrimental effects upon the growth of some.

1. Special needs and special nutrients

Adequate nutrition during their development stages is particularly critical for insects that change their feeding habits after metamorphosis or that do not feed at all as adults. The young of such species must, therefore, eat enough not only for their immediate needs but to provide for their future ones. In many of these insects the reproductive organs and gametes are completely developed during juvenile life, and the adult is only an agent for the operation of the former and the deposition of the latter. Each larva or nymph must therefore provide for the next generation, at least for as long as the embryo subsists upon the yolk within the egg. All the nutrients needed for reproduction are accumulated during larval life by the fruit moth, *Laspeyresia molesta*, because the adult females lay as many eggs when given only water as they do when fed sucrose solutions. Adult mayflies, whose lives are limited to a few hours, do not eat at all and consume the foodstuffs they have gathered and stored as nymphs, while flying about, mating, and laying their fertilized eggs.

Many insects that do feed as adults require a special nutrient of one kind or another before oviposition and exhibit special hungers which are often correlated with specific behavior patterns. The best known of these are exhibited by mosquitoes and flies. Males of mosquitoes emerge from pupation fully mature and can live out their life spans on a diet of glucose alone. But females, at emergence, have only partly developed ovaries and require a source of nitrogen to complete ovarian development and reach maturity. Some species find this in mammalian blood and make use of their mouthparts, adapted for piercing and sucking, to obtain it. Some strains, on the other hand, are autogenous, but the source of the nitrogen, equally necessary to them at this time, is not definitively established. It has been suggested that *Aedes communis*, an autogenous species, derives it from autolysis of its flight muscles, and there is histochemical evidence that autogenous strains of *Culex pipiens* utilize the protein stored in the fat body. In this case, the nutrition of the larvae would be of paramount importance to the fecundity of the adult.

Many flies also require protein before oviposition. Females of *Lucilia sericata* must have at least two meals of meat if their eggs are to mature, and those of *Stomoxys calcitrans* do not oviposit when fed on washed red blood cells or on serum, but do if given blood from which only fibrin has been removed. Sex differences in the feeding behavior of *Phormia regina* were observed in flies kept in nylon net cages and offered either a 0.1 *M* solution of sucrose or protein (Difco brain heart infusion), or a choice between the two. Males, whether mated or not, ate protein in gradually increasing amounts for 4 to 8 days after hatching and took little or none thereafter. Virgin females behaved in the same way, although they ate rather more than the males. Mated females increased their protein intake after each bout of oviposition, but maintained a level of protein feeding like that of virgin females if there was any impediment to gametogenesis.

Such specific hungers doubtless reflect specific metabolic needs. Other insects also exhibit periods of "indispensable" nutrition, reflecting perhaps more general needs. Many caterpillars must feed a certain time after each molt for the next one to follow in due course, and the blood-sucking hemipteran, *Rhodnius prolixus,* also requires a meal before each of its nymphal molts. A fully fed and engorged insect may ingest blood up to seven times its body weight in a single meal, and fourth-instar nymphs will not molt until they have ingested at least 2.9 times their body weight of blood. A fifth-instar nymph can, in 15 min, take up as much as 300 mg of rabbit blood or more than six times its own weight. This meal fulfills two needs. One is a mechanical one that can be met by artificial diets that will, like blood, distend the insect's abdomen and so trigger off the hormone cycle that regulates molting and morphogenesis. The other is nutritional and can be met only by the specific food. Adult males also require a blood meal to produce viable sperm, and fertilized females must ingest at least 56.6 mg of blood (about one-fifth of a maximum adult meal) for production of eggs. Theoretically, the blood of all mammals should be perfect food since it has, in addition to its own salts and proteins, amino acids, sugars, fatty acids, vitamins, and other compounds that have been absorbed by the gut and passed into it. Rabbit blood is, however, inadequate nutritionally for *Rhodnius*, possibly because it does not provide all of the necessary vitamins or enough of them. It may be that this, too, is the reason that lice are not found on rats whose diet is deficient in pyridoxine, pteroylglutamic acid, or choline.

2. Caste determination

The effects of nutrition are quite dramatically evident in the polymorphic colonies of social insects and have been studied most fully in bees. There are three principal castes in honeybee societies: queens and workers which develop from fertilized eggs and are genetically female, and drones, which develop from unfertilized eggs and are genetically male. The drones' only function in the colony is to fertilize the queen; for this they are raised by the workers, but once it is accomplished they are left to starve to death and their bodies are swept from the hive. The queen, in turn, functions only as a producer of eggs and is generously fed by the workers. The sex organs of the workers are incompletely developed, for oogenesis is arrested at the stage when, in the queens, yolk is deposited. The failure of the workers' ovaries to mature seems to be influenced by a substance exuded from the mandibular glands of an ovipositing queen, as well as by their nutrition. This "queen substance," which has been chemically identified as 9-oxdec-2-enoic acid, is licked off the queen's body by the workers and distributed in the food. Workers do not mate although they occasionally do produce and lay eggs. According to their ages, they divide the work of the colony between them as nurses and caretakers of the hive, foragers and conveyors of food. Potentially, every fertilized egg could develop into a queen bee, but in the ordinary course of events in a hive only one does. This egg, though laid in an ordinary brood cell, is transferred to a larger one and the larva hatching from it is fed, after the first 2 to 3 days of its life, a food that differs both qualitatively and quantitatively from that given the other larvae. These, after the first 2 to 3 days, are fed bee milk, secreted by the pharyngeal glands of workers and nurses. These glands are not developed in drones and are vestigial in queens, but they are large in workers and especially so in the nurses. They consist of coiled tubules, longer even than the bee's body, which open at the base of the "tongue" so that the secretion from them can be passed directly to another bee or to a larva. Bee milk is low in fat; but rich in sugars, derived from nectar; and contains also protein, derived from pollen; and some, if not all, of the pollen's vitamins. The mineral content of all bee milk seems relatively constant, but the proportions of carbohydrate, proteins, and vitamins in it show a good deal of variation. The amount of carbohydrate, which varies with that of protein, never exceeds 50 percent, and the protein content is especially high in the milk that is fed larvae destined to become workers. The larva to be queen is fed a special brand of bee milk known as royal jelly and fed it in such abundance that there is always some left over in her cell when she begins to pupate. Royal jelly is fed for 2 to 3 days to all newly hatched larvae, but only to the queen throughout larval life. It is one of the richest sources of pantothenic acid and contains also thiamin, riboflavin, pyridoxin, niacin, biotin, pteroylglutamic acid, and inositol but no vitamin A and little ascorbic acid. Its content of pantothenic acid is 12 to 16 times that of the pollen from which the vitamins were obtained, and three times that of an equivalent weight of yeast or beef liver. The bee milk that is fed the worker larvae also contains pantothenic acid and other vitamins, but far less pantothenic acid than royal jelly, and this amount is decreased as the larvae grow older.

Experiments in which eggs have been artificially transferred to larger cells or substituted for the one selected by the bees to become the queen show quite clearly that caste differentiation is a nutritional and not a genetic one, for the larvae hatching from these are fed royal jelly by the workers and develop into queens. If the change is made during the course of larval life, the larva develops into a kind of intermediate between worker and queen, the degree of difference between a perfect queen and a true worker depending upon the time the change was made. Moreover if, in natural conditions, a queen dies, the workers at once set about raising another by moving an egg from a smaller, worker cell to the large queen cell and gorging the larva with royal jelly. But worker bees will not continue their brood rearing activities on a diet of sucrose or honey alone; they must be fed bee bread, which contains pollen as well as honey.

The wax with which bees build their hives is an exudate and represents a side product of their high carbohydrate diet. Wax is secreted as fine scales or plates from wax glands on the abdomen, but none is formed if the bees are fed as long as 8 days on pollen only. On the other hand, the composition of the wax of bees fed only sucrose is identical with that of bees fed honey, showing not only its origin from ingested carbohydrate but the conversion of different sugars to the long-chain fatty acids characteristic of beeswax.

Nutrition also affects caste determination and polymorphism in ants, although the relationship is not as clear and direct as it is in honeybees. There are two female castes in ant colonies, queens and workers, whose differentiation seems to have more of a genetic basis than a nutritional one (Figs. IX–2 and IX–3). It is possible, however, that the amount of yolk in the eggs from which they develop has some influence on their future form and role in the colony. Larvae of *Camponotus* or *Myrmica*, kept on half-starvation rations, develop into undersized queens or undersized workers. There are no intermediates such as those that can be produced in honeybee colonies by switching the diet of a worker larva from bee milk to royal jelly. The ant larva is hatched as a future queen or a future worker.

Yet there is great variability in the size and the form of worker ants which may express differences in their nutrition. This is evident in the development of soldiers in colonies of *Pheidole pallidula*, the Italian house ant. Soldiers are workers with overdeveloped heads and correspondingly large antennae and mouthparts. It has been shown that food is a decisive factor in the modification of a worker as a soldier, for in one group of larvae fed 10 days on honeydew or sugar solutions, none emerged as soldiers, while in another, fed dead termites for the same length of time, many emerged as such. The critical period is the second larval instar, for if at this stage sugar was substituted for the flesh of termites, no soldiers developed, and when meat replaced sugar at this time, they did. This result has been attributed to a special substance, extractable from the flesh of termites and initially called termitin, which has a vitamin-like effect on the ants in promoting mobilization of their reserves. Later, a substance with the same effect was extracted from the annelid *Tubifex* and from yeasts (torulae), and given the more general name of T-factor. It, like other known vitamins, is not specific and can influence the growth of animals other than ants.

Ants of the families Camponotidae and Dolichoderinae feed on honeydew and nectar and often collect

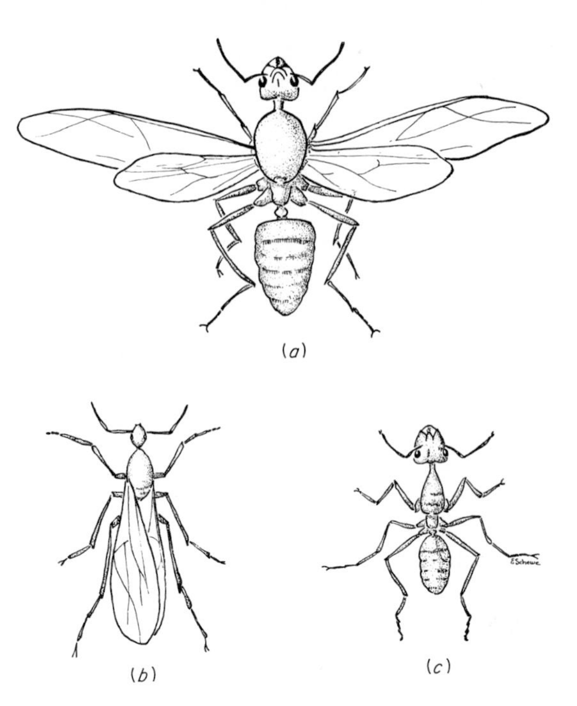

FIG. IX–2 *Castes of the carpenter ant* Camponotus herculaneus. *(After Goetsch.)* ***(a)*** *Winged female, which, after fertilization, sheds her wings and becomes queen of the colony.* ***(b)*** *Male, which lives only a short time and dies after fertilizing a female.* ***(c)*** *Worker, a sterile female.*

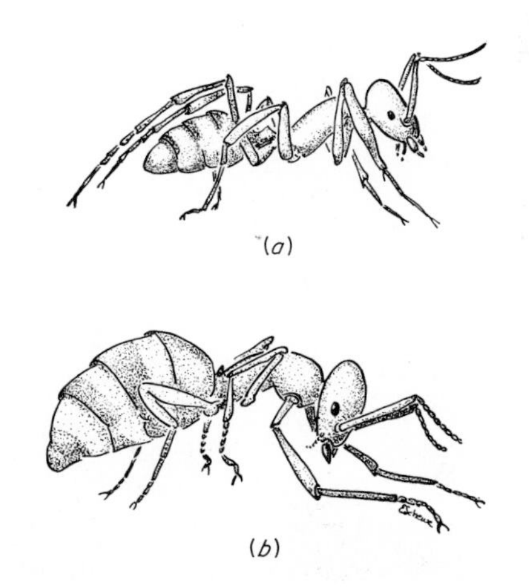

FIG. IX–3 ***(a)*** *Worker and* ***(b)*** *water carrier of the Chilean desert ant* Tapinoma antarctica. *(After Goetsch.)*

such enormous amounts that disposition of it becomes a problem. In these circumstances, when food is so plentiful, certain members of the colony—usually newly emerged workers—are selected to act as storage tanks. The workers, collecting as much honeydew as their bodies can hold and carry, lumber back to the nest and regurgitate the excess food into the mouths of the young females, whose crops become so filled that their abdomens are tremendously extended. This "caste" is known as "repletes" or "honey pots" and is readily distinguishable from other members of the colony by their grotesque shape and their inactivity (Fig. IX–4). Repletes of the American honey ant, *Myrmecocystus Mexicanus hortideorum*, found in areas extending from Mexico City to Denver, may have abdomens as large as ½ in. in diameter. These engorged individuals are incapable of movement and hang from the tops of the extra large chambers made for them. There may be as many as 300 repletes in a large colony of *Myrmecocystus*, from which the workers draw the stored nectar and honeydew as they need it. When a replete has been relieved of all its stores, it reassumes its ordinary slender form and becomes indistinguishable from other workers. Differentiation of the repletes is thus not nutritional *sensu stricto*, for they make no use themselves of the food in their crops, and the difference in their form and habits from those of the active workers are temporary, imposed upon them for the general benefit of the colony to which they belong.

FIG. IX–4 *Ant honey pots. (American Museum of Natural History.)*

3. The fat body

The importance of the fat body as the main site of the conversion of absorbed nutrients, the storage of nutritional reserves, and the release of them on demand has been established in a number of different species of insects. This information has been obtained by histological examination of the tissue and identification of the inclusions of its cells; by analysis of the composition of some of these inclusions by histochemical and chemical means; and by studies in vitro of the capacity of the tissue as a whole to effect synthesis, alteration, or degradation of specific substances, as well as those of the enzymatic activities of homogenates or extracts of it. Data from these investigations indicate that the role of the fat body in the metabolism of nutrients is similar to that of the vertebrate liver, with which it is considered analogous.

Observation of the fat body of the cockroach *Blaberus discoides* by means of electron microscopy reveals that the mass of cells is completely surrounded by a membrane about 3 μ in thickness. This membrane has a rather complicated structure, with an outer layer of closely packed granules and fibrils, a middle layer of more loosely packed striated fibrils arranged at approximately right angles to those of the outer layer, and an inner layer whose appearance is like that of the outer one. The plasma membranes of the cells are deeply infolded into the cytoplasm, where there are mitochondria but few cytomembranes, either of the endoplasmic reticulum or the Golgi element. This is surprising because the cells are known to be involved in protein synthesis as well as in that of lipid synthesis. Similar observations of the fat body of *Drosophila* larvae have revealed extensive systems of cytomembranes in its cells.

In newly hatched and young larvae or nymphs of all insects studied, the cells of the fat body contain few inclusions, but with age these increase in number and the cells increase in size and possibly also in number. The fat body of a well-nourished insect during the later stages of its juvenile life fills most of the space in the body not occupied by other organs. As maturity approaches it is usually reduced in size, especially in females because of its contributions to the yolk content of the eggs and, in both sexes, it is considerably crowded by the enlarging gonads. The inclusions in the cells represent materials that have been transferred from the hemolymph; these materials are products not only of digestion but also of histolysis of larval tissues, a process that is particularly active just before a molt. The inclusions are of carbohydrate, lipid, and nitrogenous compounds of various kinds, which are often associated with lipid in vacuoles or globules. In well-developed fat bodies, the cells may be so distended that it is impossible to make out their boundaries and the tissue has the appearance of a syncytium although it is not really one, for boundaries may again become evident when the cells are depleted of their reserves. This is the case in Orthoptera, Lepidoptera, Hymenoptera, and some Coleoptera. In Ephemeroptera, Trichoptera, some Diptera, and some Coleoptera the tissue is a true syncytium. In some insects also, some of the cells may be loaded with microorganisms that are consistently part of their organization.

In many species the fat body contains pigments which show through the transparent epidermis and so give color to the entire body. This is very apparent in some caterpillars whose coloration is due less to integumentary pigments than to those of the fat body. The pigments may be carotenes or their derivatives or, in leaf-eating insects, derivatives of chlorophyll. In the squash bug, *Tricopoda pennipes*, for example, the porphyrin nucleus of chlorophyll is degraded to a green pigment chemically similar to biliverdin.

a. CARBOHYDRATE METABOLISM Glycogen is the predominant polysaccharide in fat bodies, as it is in the vertebrate liver. This can be visualized as granules or amorphous masses by means of standard staining procedures and has been so demonstrated in the fat bodies of a variety of insects. In well-fed specimens of *Blaberus*, the cells are packed with granules of glycogen, which disappear almost completely after 3 days of starvation. Degenerative changes in the cells themselves can be followed during prolonged starvation. In the course of 30 days without food, amebocytes penetrate through the containing membrane of the fat body and can be seen just below it, in contact with the cells. The mitochondria degenerate and the cytoplasm becomes filled with concentrically lamellated and large, complex crystalline inclusions. Lysosome-like bodies can also be recognized in the cytoplasm. Degeneration is complete after 30 days starvation, but up till then the cells can be restored if the insects are fed.

The glycogen demonstrable in the cells of insects' fat bodies represents only a fraction of the carbohydrate taken into them. At least 90 percent of the monosaccharides in its blood are taken into the cells of the fat body of *Schistocerca gregaria* when the insect is at rest; lesser, but still appreciable, amounts are taken in when it is in flight. The greater part of these are made into the disaccharide trehalose and, as such, are returned to the blood and made available to the muscles and other tissues as a source of energy. In *Schistocerca*, as in *Phormia* and *Periplaneta*, the formation of trehalose takes place very rapidly. When glucose labeled with C^{14} was injected into the hemolymph of blowflies that had been starved for 24 hr, the blood withdrawn 30 sec later contained measurable amounts of labeled trehalose, and after 10 min about 90 percent of the injected sugar was recovered as trehalose. The enzymatic reactions involved in the course of this synthesis have been worked out with *Schistocerca* as a model. Mannose and fructose are also readily convertible to trehalose, but galactose is converted much more slowly and so probably by a different pathway. The glycogen that is made in the fat body can also be depolymerized to glucose in its cells and the glucose converted to trehalose before it leaves them. The conversion of glycogen to blood sugar is one function of the vertebrate liver, but in vertebrates the transportable sugar is glucose not, as it is in insects, trehalose.

In silkworms, conversion of monosaccharides to trehalose seems to occur elsewhere than in the fat body. It may take place in the hemolymph itself, for labeled trehalose was detectable there when fifth-instar larvae were injected with labeled glucose-6-phosphate. Very little of the labeled carbon was incorporated in the trehalose and glycogen of the fat body.

Another distinctive treatment of carbohydrate by insects is their conversion of glycogen to glycerol and at least two other alcohols, sorbitol and mannitol. This has been observed especially in species that at some developmental stage pass through a period of arrested metabolism known as diapause (pages 734ff), but it is not limited to them. In insects that winter over in diapause, glycerol begins to appear with the onset of cold weather, reaches a maximum and approximately constant level during the cold months, and is reconverted to glycogen in the spring, or at the termination of diapause. Yet the glycogen ⇆ glycerol reaction can be induced in nondiapausing insects, such as the carpenter ant *Camponotus herculeanus* and the mountain pine beetle *Dendroctonus monticolae*, anytime in the year by exposing them to low and high temperatures. The temperature dependence of the reactions would seem to indicate their survival value, with glycerol acting as a biological antifreeze and so protecting the insects from the effects of extreme cold that are fatal to the majority of them. But a survey of insects that winter over in the zero and subzero temperatures of western Canada, either in exposed or relatively protected places, has shown that there is no simple correlation between glycerol content and winter hardiness. No glycerol, for example, was detectable in the hemolymph of diapausing larvae of the gallfly *Euura nodus*, collected in February, and only 3 percent in that of larvae of the goldenrod gallfly, *Eurosta solidaganus*, though both of them survived through the bitter weather. There were also relatively large amounts of sorbitol in the hemolymph of *Eurosta* larvae, but there is no evidence that this alcohol does act as an antifreeze. Yet levels of glycerol as high as any found in winter-resistant insects were found in a number of species that are highly susceptible to cold. The hemolymph of first- to third-instar larvae of *Dendroctonus*, collected outdoors in December, contained 23 percent glycerol, as much as that of larvae of *Eurytoma gigantea*, a wasp parasitic on *Eurosta*. These larvae can survive not only long exposure to subzero temperatures but even immersion in liquid nitrogen. And concentrations as high as 32 percent were found in the hemolymph of larvae of *Rhabdophaga globosa*, a gall midge of willows, which is susceptible to cold. It is possible that some other change occurs in the cells of freeze-resistant species to make glycerol effective as an antifreeze for them and that this change has allowed them to take advantage of the glycogen ⇆ glycerol reaction as a protection against prolonged and extreme cold.

b. LIPID METABOLISM Globules and droplets of lipid can easily be recognized when a fat body, either fresh or fixed, is examined under the light microscope. The increase in number and in size of the fat deposits, as a larva and its fat body grow, shows that this tissue is probably a primary site of fat synthesis and accumulation. However, fat droplets are demonstrable in most of the other tissues but not in such abundance. They are found, for example, quite conspicuously between the fibrils of the flight muscles of aphids, and there are considerable amounts of fat in the eggs of most insects. In the fat body of the carpet beetle, *Anthrenus vorax*, the fat deposits consist of a central globule surrounded by smaller peripheral globules. Histochemical tests show that the central droplet contains neutral fat, for it becomes deep pink when exposed to Nile Blue Sulphate and also gives positive reactions for neutral fat when treated with other diagnostic stains. The peripheral globules are positive to tests for protein and for phospholipids (phosphatids), indicating that their contents may be a complex of the two substances. Such an association of lipid with protein seems to be common in cells of the fat body.

Phospholipids, or phosphatids, are considered the metabolically active forms of lipid, in contrast to triglycerides, or neutral fats, which are storage reserves and as such comparatively metabolically inactive. A phosphatid is a combination of a phosphorylated triglyceride, or phosphatidic acid, with the formula:

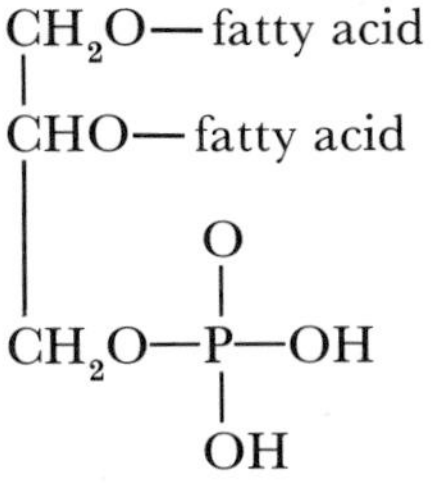

$$
\begin{array}{l}
CH_2O\text{—fatty acid} \\
| \\
CHO\text{—fatty acid} \\
|\quad\quad O \\
|\quad\quad | \\
CH_2O\text{—}P\text{—}OH \\
\quad\quad\quad | \\
\quad\quad\quad OH
\end{array}
$$

and a base that may be either a nitrogenous one, such as choline or serine, or a carbohydrate, such as inositol. In lecithins, which are the predominant phosphatids of mammalian cells, the nitrogenous base is choline. In cephalins, the base is either serine or ethanolamine, which can be derived from serine by the removal of the carboxyl group and converted to choline by the addition of methyl groups:

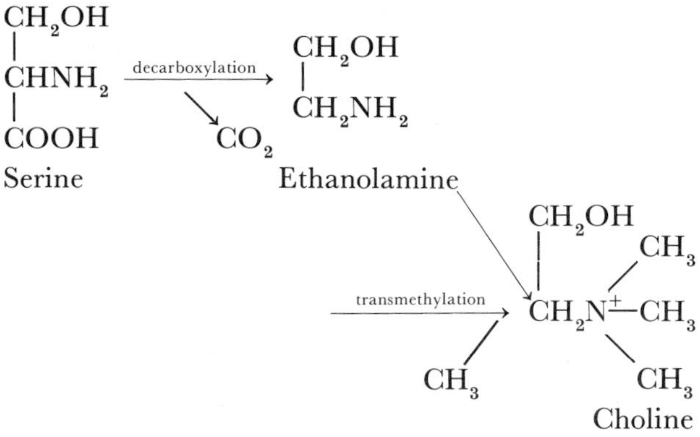

$$
\underset{\text{Serine}}{\begin{array}{l} CH_2OH \\ | \\ CHNH_2 \\ | \\ COOH \end{array}} \xrightarrow{\text{decarboxylation}} CO_2 + \underset{\text{Ethanolamine}}{\begin{array}{l} CH_2OH \\ | \\ CH_2NH_2 \end{array}} \xrightarrow[+\,3\,CH_3]{\text{transmethylation}} \underset{\text{Choline}}{\begin{array}{l} CH_2OH \\ | \\ CH_2N^{+}(CH_3)_3 \end{array}}
$$

Removal of the fatty acid from the second carbon in the glycerol moiety of a phosphatid converts it to the lyso- form. Lysolecithin, which is a potent hemolytic agent causing the destruction of red blood cells, is produced by the removal of oleic acid from the lecithin component of the erythrocyte. The reaction is catalyzed by a specific enzyme, phosphatidase A, known also as lecithinase. This enzyme is found in the venom of bees, as well as in that of scorpions and snakes, and is held to account, in part at least, for the destructive effects of their stings or bites.

Lecithins represent a smaller proportion of the total phospholipid in the cells of insect fat bodies, and of their other tissues, than they do in mammalian cells. The fat body of the flesh fly, *Sarcophaga bullata*, contains, for example, 51 percent phosphatidylethanolamine, 14 percent lysophosphatidylethanolamine, 21 percent phosphatidylcholine, and 13 percent lysophosphatidylcholine. A direct effect of diet upon the nature of the fat body phosphatids has been demonstrated in *Phormia regina*, which has a requirement for choline but can synthesize carnitine in amounts sufficient for its needs if there is adequate choline in its diet. In these circumstances, phosphatidylcholine is found in the fat body, but if choline is omitted from the food and carnitine substituted for it, a new phosphatide, phosphatidylmethylcholine, appears. Chromatographic analyses of homogenates of whole larvae of *Tenebrio molitor*, which has a dietary requirement for carnitine as well as for choline, show that phosphatidylcholine comprises 43 percent of the total phospholipid and phosphatidylethanolamine, 41 percent. In some other insects, phosphatidylcholine represents less than 20 percent of the total amount of lipid.

The conversion of protein and of carbohydrate to fat by insects is well known and is established on indirect evidence at least. It seems also clear that the cells of the fat body are the active agents in these transformations. Proteins must be the source of fats in *Anthrenus*, because its diet like that of other dermestids consists primarily of fur, feathers, wool, and silk. Moreover, it has been shown that extracts of the fat body of *Schistocerca* and the armyworm *Prodenia* can incorporate acetate labeled with C^{14} into fatty acids by a pathway that seems to be similar to that known to be followed in mammalian tissues, since the synthesis by extracts from insects requires the same cofactors. Mosquitoes and flies can synthesize fat from glucose, and female mosquitoes of the species *Aedes sollicitans* and *A. taeniorhynchus* do so in far greater amounts than do the males or either the females or males of *Musca domestica*. At emergence both male and female mosquitoes have about the same amount of lipid, of which phospholipids and triglycerides represent more than 90 percent. But during the first 6 days of adult life, the triglyceride level drops steadily in the males and increases steadily and rapidly in the females, while the level of phospholipid remains about the same in both. The amount of fat in the fat body of a female may reach an amount about 50 times that of the male, even when both sexes are feeding freely on the same glucose solution. Analysis of the fatty acids synthesized solely from glucose showed that palmitic, palmitoleic, and oleic acids made up at least 92 percent. The fact that only trace amounts of linoleic acid were found suggests that the mosquito does not make polyunsaturated fatty acids, on a glucose diet at least.

That degradation as well as synthesis of fat occurs in the fat body is shown by its disappearance during pupation, starvation, or periods of food scarcity, and, in females, during yolk formation. Also, lipase has been found in extracts of the fat body of a number of different insects. The hydrolyzed fat is released into the hemolymph as diglycerides and free fatty acids, which are transported complexed to protein. The form in which lipid is carried in the circulating fluid in insects is, therefore, as the form in which sugar is carried, different from that in mammals, in which it is transported as triglycerides, free fatty acids, or both.

c. METABOLISM OF NITROGENOUS COMPOUNDS The role of the fat body in nitrogen metabolism has been experimentally established in a number of insects. These events involve processing of amino acids in preparation for carbohydrate or fat synthesis as well as their linkage to form protein. There is little storage of protein as such in the fat body, as compared with that of glycogen or lipid, although some can usually be demonstrated histochemically in that of juveniles of all stages. The fat body is also known to be involved in the production of uric acid.

Uric acid is the usual end product of nitrogenous metabolism in insects as it is in birds. In many insects, cells loaded with crystals of various salts of uric acid are conspicuous elements in the histological picture of the fat body. These cells are not the only repositories for urates, however, for the crystals may also accumulate in the walls of the alimentary canal, especially of the hindgut; in the Malpighian tubules; and, in some insects, in the wings or in special organs, such as the lanterns of fireflies. In the silkworm, the amount of urate in the fat body is less than that in the Malpighian tubules, and in larvae of the rice moth, *Corcyra*, there is none at all in the fat body but much in the Malpighian tubules.

Synthesis of protein from labeled amino acids injected into the hemolymph has been shown to take place in the fat bodies of a number of insects. C^{14}-labeled glycine is incorporated into protein by the fat body of larvae of *Bombyx mori*, although not in amounts as great as that in their silk glands. The fat body of larvae and pupae of the moth *Hyalophora* (*Platysamia*) *cecropia* incorporated labeled leucine into protein; in larvae, the greatest amount was incorporated while the cocoon was being spun and in pupae, on the second day of adult development. The most rapid rate of incorporation of C^{14}-glycine into protein in the fat body of *Schistocerca* coincides with the period of yolk deposition and diminishes at the termination of oocyte development. Since reproduction in these females is a cyclical event and one under hormonal regulation, it would seem that this activity of the fat body must also be under endocrine control, indirectly if not directly.

The two fundamental reactions involved in amino acid metabolism are deamination and transamination (pages 478f). One type of deamination is an oxidative reaction catalyzed by amino oxidases and resulting in the removal of the amino group from the acid and the formation of a corresponding keto acid, with the release of ammonia:

$$\underset{\text{Amino acid}}{\mathrm{H{-}C(R)(COOH){-}NH_2}} + \tfrac{1}{2}O_2 \xrightarrow[\text{oxidase}]{\text{amino}} \underset{\text{Keto acid}}{\mathrm{R{-}C(=O){-}COOH}} + NH_3$$

A dehydrogenase, originally thought to be specific for glutamic acid and therefore called glutamic dehydrogenase, also catalyzes deamination of amino acids; this enzyme requires as a cofactor either NAD (nicotine-adenine dinucleotide, or coenzyme I) or NADP (nicotine-adenine dinucleotide phosphate, or coenzyme II). *d*-Isomers of the various amino acids are deaminated by amino oxidases more readily than the *l*- forms, an anomalous situation since *d*-isomers are of such rare occurrence in nature. Glutamic dehydrogenase, on the other hand, is active on *l*-isomers and its action, unlike that of the amino oxidases, is reversible, and conversion of an amino acid to its corresponding keto acid, and vice versa, is effected through the agency of this enzyme.

Transamination results in the transfer of the amino group from an amino acid to a keto acid, with formation of the corresponding amino acid. The reaction is reversible, is catalyzed by transaminases, and requires as a cofactor pyridoxal phosphate, the active form of vitamin B_6.

$$\underset{\text{Amino acid}'}{\text{H}-\overset{\overset{\text{R}'}{|}}{\underset{\underset{\text{COOH}}{|}}{\text{C}}}-\text{NH}_2} + \underset{\text{Keto acid}''}{\text{H}-\overset{\overset{\text{R}''}{|}}{\underset{\underset{\text{COOH}}{|}}{\text{C}}}=\text{O}} \xrightleftharpoons[\text{pyridoxal phosphate}]{\text{transaminase}} \underset{\text{Keto acid}'}{\text{H}-\overset{\overset{\text{R}'}{|}}{\underset{\underset{\text{COOH}}{|}}{\text{C}}}=\text{O}} + \underset{\text{Amino acid}''}{\text{H}-\overset{\overset{\text{R}''}{|}}{\underset{\underset{\text{COOH}}{|}}{\text{C}}}-\text{NH}_2}$$

In insects, alanine, aspartic and glutamic acids, or glutamine, a combination of glutamic acid and ammonia, appear to be the most active donors of the amino group, for the rate of transamination is markedly less with the other amino acids. Glutamic acid, which is readily synthesized by most animals, plays a key role in amino acid metabolism, acting as a central depot for the transference of amino groups.

Activity of these enzymes is shown in the fat bodies of a number of different insects, and most completely in *Schistocerca* and *Calliphora*. Homogenates of the fat body of *Schistocerca* convert leucine and other amino acids to keto acids and glutamate to glutamine, showing that they contain amino acid oxidases, glutamic dehydrogenase, and glutamine synthetase. They also show transaminase activity, as do those of *Calliphora*. The nature of these reactions as they take place in vitro seems in all essentials like that of similar reactions in mammalian tissues. It seems likely, therefore, that the pattern of amino acid metabolism in insects in general will prove to be much the same as that in mammals.

The metabolic pathways by which uric acid is formed in the insect fat body are almost completely unknown. Radioisotopic studies on pigeons have shown that glycine, serine, formate, carbon dioxide, aspartic acid, and glutamine amide are all precursors to its synthesis in birds, in that each contributes a specific atom or atomic grouping to the finished molecule of uric acid. When C^{14}-labeled formate was injected into the hemolymph of cockroaches (*Periplaneta*), uric acid was recovered from the fat body with the labeled carbon in the 2 and 8 positions, where they are found when similar experiments are performed on pigeons. Uric acid labeled in the same way was likewise recovered from intact, isolated fat bodies incubated in solutions containing C^{14} formate, but not from homogenates of the cells. Production of uric acid by isolated fat bodies of *Tenebrio* and *Prodenia* has been increased by the addition to the incubation medium of the purines xanthine and hypoxanthine, known to be intermediates in the conversion of adenine and guanine to uric acid in vertebrates. And a riboside, known also to be in the pathway of uric acid synthesis in birds, promoted its production in the fat body of *Calliphora*. A high concentration of guanase, the enzyme operating in the conversion of guanine, one of the purine components of nucleic acids, to xanthine has been found in the fat bodies of *Periplaneta*, the cockroach *Leucophaea*, *Tenebrio*, and *Prodenia*. Extracts of fat bodies of *Prodenia* larvae in their final instar have been found to contain adenase and xanthine oxidase as well as guanase, and those of *Galleria mellonella* all three of these enzymes. It is evident that purine metabolism does occur in the fat body and that its course, like that of amino acid metabolism, is very likely similar to that in vertebrate tissues, primarily the liver. The overall picture of metabolic events in the insect fat body does suggest very strongly that its cells and those of the vertebrate liver contribute in much the same way to the general operation of the body of which they are a part, although the two organs are wholly different in origin and in structure.

F. NUTRITION OF OTHER INVERTEBRATES

Very little is known about the nutrition and dietary needs of other groups of invertebrates. Attempts at maintaining cnidarians, flatworms, "aschelminthes," and other "lower" invertebrates for long periods under controlled dietary conditions have on the whole been unsuccessful, and little has been learned of the specific nutrients they require. Nor is there any exact information about the needs of annelids, molluscs, arthropods other than insects, or echinoderms; although in some cases analyses of their entire bodies or of isolated organs in respect to their protein, carbohydrate, and lipid content, coupled with knowledge of their eating habits, give some clue to the nature of their metabolic processes.

Hydras have been the objects of the most precise studies on cnidarian nutrition. These can be maintained indefinitely in the laboratory in an active state of growth and asexual reproduction on a diet of living *Artemia* nauplii, but their life span and reproductive activity are significantly reduced when they are fed frozen nauplii, and they do not survive on a diet of heat-killed ones. It is possible also to keep them in a flourishing state in a balanced salt solution supplemented with either fresh chick or beef liver extracts. On this diet the animals bud as prolifically as those fed living nauplii. Attempts to identify the active nutrient in bovine liver extract have been successful only to the extent of associating it with the crude protein fraction. That it is not a purified protein, like albumin or hemoglobin, nor an amino acid, carbohydrate, nucleic acid, nor vitamin has been shown by microinjecting salt solutions, semisolidified with agar, containing each of these nutrients into the gastral cavity of experimental animals. None of them promoted budding to the degree that it was exhibited in control animals fed either living nauplii or beef liver extract.

Similar oligidic diets have been used in keeping the rotifer *Lecane inermis* and several species of nematodes alive. Axenic culture of these animals requires isolation of eggs, their surface sterilization, usually by means of antibiotics, and, in most experiments, the separation of individuals from each other and from any possible source of contamination. Cultures of the free-living, hermaphroditic and self-fertilizing soil nematode *Coenorhabditis briggsae* have been kept for more than 3 years in a medium consisting of salts, amino acids, glucose, and certain vitamins of the B group, supplemented with chick embryo extract and a liver protein fraction. Manipulation of that medium has shown that this nematode requires the 10 essential amino acids, niacin, pantothenic acid, pteroylglutamic acid, riboflavin, and thiamin. Needs for biotin, cobalamine, and lipoic acid were not detected, but these may well have been adequately supplied in the supplements. In this medium, individuals reach maturity in 4 to 5 days at 20°C, but they do so in 3 days if bacteria are present in it. Larvae of the parasitic nematode *Trichinella spiralis* have been isolated from rat muscles by peptic digestion and cultured in a basic medium consisting of salt solution supplemented with inactivated rabbit serum and chick embryo extract. With the addition of a mixture of vitamins, including cobalamine, the larvae molted successfully and reached sexual maturity, conditions that were not attained in the basic medium alone. The addition of amino acids or of glucose did not promote their development and that of bile salts was deleterious to it.

Relationships between the diet of the host and nutrition of its parasites have been shown for ascarids and for cestodes. Polysaccharides and lipids are the principal reserves of nematodes and cestodes alike, both of which have also a relatively low content of protein. When a fowl parasitized by *Ascaridia galli* is starved, the glycogen reserves in the intestinal cells, hypodermis, and gonads of the nematode are rapidly depleted, and within 48 hr its total glycogen is less than 90 percent of the initial amount. Cystercercoid larvae of the rat tapeworm *Hymenolepis diminuta* grew into stunted adults in rats kept on a carbohydrate-free diet, and egg production in adults stops if the host's diet is deficient in vitamins of the B complex. It is, of course, possible that these effects are not direct ones upon the parasite but, rather, indirect ones, caused by general derangement of the host's metabolism because of its malnutrition.

It is evident from study of their natural life cycles that the nutritive requirements of larval cestodes may be different from those of adults. Two closely related species of freshwater fish parasites, *Triaenophorus lucii* and *T. crassus*, are, for example, often found together as adults in the digestive tracts of pikes, but the pleurocercoid larvae of *T. lucii* encyst on the surface of the livers of burbots (eelpouts) and those of *T. crassus* in the dorsal muscles of whitefish. Larval stages of these and other cestodes can be kept alive for several days in salt solutions containing glucose; in salt solution alone, their glycogen content drops, and unless glucose is restored, the larvae die. Parasitic trematodes will also survive in balanced salt solutions supplemented with glucose, and liver flukes have been kept alive as long as 60 hr in such solutions adjusted to a pH of 8.4.

The necessity of glucose to the survival of flukes and larval tapeworms and its relationship to the glycogen content of their tissues implies their ability to synthesize glycogen from monosaccharides, but there is no direct proof of this.

The question of lipid metabolism in these animals is an open one. It seems likely that the lipid in the tissues of *Ascaris* is nutritionally inert, for while its glycogen reserves were depleted during an experimental period of starvation, its lipid reserves were not. They might, however, have been drawn upon if the period had been extended. Lipid is stored in greater amounts in the Acanthocephala that have been investigated than in any other parasites; there are stores of neutral fat in nearly every tissue of the body and phosphatids and glycogen are found as well in the subcuticular cells, where lipase has also been demonstrated. Fat is likewise the principal food reserve in the nemertean *Lineus ruber,* for globules up to 5 μ in diameter are visible in the cells of the parenchyma, and there are also accumulations in special cells of the intestinal wall.

Most of the work on molluscan nutrition has been done with the practical aim of ascertaining the most suitable foods for rearing larvae on a large scale. Here again the nutritional requirements during juvenile stages may be quite different from those in adult life. Various algal diets have been tried on larvae of various bivalves, particularly those of *Ostrea edulis, Crassostrea virginica*, and *Mercenaria* (*Venus*) *mercenaria*. The most valuable algae, from the point of view of larval survival and growth, are members of the family Chrysophyceae, and of these *Isochrysis, Monochrysis*, and *Chromulina* seem to provide the requisite nutrients in assimilable form. Mixed algal cultures seem on the whole better than unialgal ones, supporting the idea that varied diets are on the whole better than restricted ones.

The value of plants as food for these animals, as well as for all others, depends upon the efficiency with which they are digested as well as upon the nutrients they supply. Crystalline styles isolated from adult oysters have no effect upon *Isochrysis,* although the alga supplies good nutrition to its larvae; the natural food of adult oysters is the red alga *Cryptomonas* (see page 305). Different plant foods have also been found to have different values for adult abalones (*Haliotis*) and for *Turbo cornutus*, as evidenced by differences in their rates of growth on diets of the different algae. The composition of the plants tested, and therefore any critical dissimilarities between those that are nutritionally good or nutritionally poor, are not known. *Helix*, when fed diets consisting of a complete salt mixture with casein, starch, glucose, lactose, dried yeast, and other additives, did not grow unless also given cholesterol and either leaf extract or cod liver oil, but while pteroylglutamic acid might be suspected to be the essential nutrient in these, it has not been identified as such.

A Puerto Rican strain of *Australorbis glabratus,* the molluscan host of the fluke *Schistosoma mansoni*, has been maintained axenically for a number of years on a diet that includes both autoclaved brewers yeast and formalin-killed bacteria (*Escherichia coli*) in a balanced salt solution. Neither bacteria nor yeast alone are adequate, nor are heat-killed bacteria in conjunction with yeast. Even with the optimum diet, growth of the snails is subnormal, but they do survive and reproduce.

What role the midgut glands have in storage and general metabolism is but little known. In some molluscs they may be the principal sites of carbohydrate and lipid reserves; in others, as in the chiton *Katherina tunicata*, the main storage areas may be special cells in the foot, mantle, or germinal epithelium. In *Katherina* these cells in the foot and mantle ridge are full of neutral fat, phosphatids, and glycogen, or a glycogen-like material, and the cells of the germinal epithelium are rich also in this material and in neutral fat. The stores in the germinal epithelium diminish during gametogenesis but the fat in the foot and mantle cells is retained during this period and lost only after prolonged starvation. The digestive gland in *Katherina* was at all times devoid of glycogen. But digestive glands may store carotenoid and chlorophyllic pigments. Large amounts of carotenoids have been found in those of *Loligo opalescens* and *Octopus bimaculatus*, and in that of the marine snail *Cerithidea californica*, where there were also large amounts of fat.

There is some evidence that the molluscan midgut gland may be, like the insect fat body, active in the metabolism of nitrogenous compounds. Amino acid oxidases have been demonstrated in extracts of the digestive glands of the gastropods *Lymnaea stagnalis* and *Helix pomatia*; the bivalves *Mytilus edulis, Cardium tuberculatus*, and *Anodonta* sp.; and the cephalopods *Sepia officinalis* and *Octopus vulgaris*. High concentrations of uric acid have been reported in the tissues of *Helix pomatia* and other land snails. *Otala lactea* synthesizes uric acid from the same precursors used by mammals, birds, and insects, and in intact snails the distribution of the carbon atoms is the same. Teased preparations of the digestive gland incubated in vitro incorporate formate labeled with C^{14} into uric acid, but with an aberrant distribution of the labeled carbon.

Much of the interest in annelid nutrition has centered on the chloragocytes, so conspicuous in earthworms, and the possibility of their functional homology with the molluscan midgut gland, the insect fat body, and the vertebrate liver. They are the principal storage areas in earthworms, but in polychaetes various regions of the body serve this purpose. The gut wall is the main repository for lipid in *Amphitrite*, *Terebella*, and *Nereis*, but some is also found in the body wall and in the coelomocytes. The greatest store of glycogen is in the body wall and coelomocytes of *Amphitrite*, the body wall of *Arenicola*, and the peritoneum of *Nereis*. In earthworms, the greatest store is in the chloragocytes, where in *Allolobophora caliginosa*, feeding on a plentiful natural diet, it represents on an average 62.3 μg/mg dry weight of the intestine. After a month's starvation this amount is reduced approximately one-half. Glycogen and lipid can be visualized histochemically in these cells. In *Lumbricus terrestris* the lipid components have been resolved chromatographically into cholesterol, phosphatidyl choline, phosphatidylethanolamine, and a complex one not yet identified. The characteristic deep gold color of the cells is attributed to carotenoids complexed with lipid and to flavones and flavenols derived from the plant material ingested by the worms. Their equally characteristic "mouldy hay" odor is attributed to dicoumarol, derived also from the plant material that is eaten. Electron microscopy has further revealed that some chloragocytes contain mica and muscovite, although the presence of these minerals may be entirely fortuitous, resulting from an unusual migration of them from the lumen of the midgut.

The role of chloragogue tissue has been in question ever since the recognition of these highly modified peritoneal cells. It has been held that their function is primarily excretory and that the chloragosomes in them represent insoluble products of metabolic waste. It is also believed that their function is primarily nutritive and that they receive end products of digestion directly from the intestinal epithelium, which they are in a particularly advantageous position to do, and make them available to other cells as occasion requires. This view is supported by the demonstration of large concentrations of alkaline phosphatase at the distal ends of the intestinal cells, suggesting that a mechanism of active transport exists between them and the adjacent chloragocytes.

The true function of chloragogue tissue may prove to be a compromise between these two concepts. Chromatographic analyses of fresh chloragocytes of *Allolobophora* show that they contain certain products of intermediary metabolism, such as glutamic acid, arginine, and ornithine, as well as relatively large amounts of ammonia; kynurenine, a degradation product of tryptophan, has also been found in them. The nature of all the inclusions identified in them suggests that they are cells active in the transformation of assimilated materials, storage, and eventual release to the system, but there is no definitive proof of this. They are known to detach themselves from the intestinal wall and to migrate into the coelomic cavity to become part of the population of coelomic corpuscles. Proponents of the excretory theory of their function believe that in this way they convey waste to the nephridia for elimination or to other tissues for deposition in insoluble form. The strongest proponents of the nutritional theory believe that the freed chloragocytes transform into trephocytes and as such carry and transfer their stores of reserves to tissues requiring them. The trephocytes may thus constitute a storehouse of materials derived from a primary store in the absorptive areas of the gut and a transport system that can convey them about the body. The mobilization of trephocytes containing typical chloragosomes at sites of rapid growth, such as gonads, wounds, and regenerating regions, gives added support to the nutritional concept of chloragocyte function.

The botryoidal tissue of leeches may have a similar use, but this has not been established. Fat and glycogen are stored in the coelomocytes of leeches, which may have the same origin as those of earthworms.

Virtually nothing is known of the nutrition of arachnids, crustaceans, millipedes, or diplopods. Crustaceans have been investigated chiefly in relation to their content of vitamins, especially of vitamin A. The greatest concentrations of this are found in free-swimming pelagic species, with euphausiid shrimps the richest in it. The other vitamins seem very generally distributed in the tissues of the representative species examined, with every indication that they, like vitamin A, are largely or entirely derived from the food. An especially high concentration of ascorbic acid, in respect to that in other tissues, has been found in the digestive glands of *Cancer* sp., *Eriphia spinifrons*, *Nephrops norvegicus*, and *Homarus*. The fact that it had entirely disappeared from the digestive gland of *Astacus* after 3 days starvation substantiates the idea that its origin is exogenous, but does not preclude the possibility of its synthesis somewhere in the body.

Effects of vitamin deficiencies have been shown in the branchiopod *Artemia salina*. Grown under aseptic conditions in a meridic medium, *Artemia* nauplii were found to require, in order to reach sexual maturity, thiamin, pyridoxine, riboflavin, niacin, pantothenic acid, biotin, pteroylglutamic acid, and putrescine (a product of the decarboxylation of arginine). In the absence of any one of these, growth was arrested at various stages in development. The most drastic effect was produced by the omission of thiamin, when the nauplii did not go beyond the early metanauplius stage, and the least drastic by the omission of pteroylglutamic acid, when the nauplii developed to an early juvenile stage. In all cases the animals remained alive 10 to 20 days after the arrest in their growth, but all ultimately died. Omission of either carnitine, inositol, or choline from the medium did not retard development, but with these three vitamins added to the others, females produced more eggs than they did without them. Experiments with antivitamins suggest that niacin, as well as alpha tocopherol, may be a dietary requirement for *Daphnia*, and also for the isopod *Oniscus*.

In starfishes the digestive glands are the principal storage organs between reproductive cycles. Prior to reproduction, their reserves are transferred to the gonads, which just before spawning season may increase 20 to 30 times their previous size. Most of the reserves in the digestive glands are in the form of lipid, which may constitute, on an average, 30 to 50 percent of their dry weight and even reach a maximum of 75 percent in some individuals. Protein is stored in lesser amounts, representing on an average 20 to 30 percent of the dry weight, and glycogen in still smaller quantities, representing only 1 to 2 percent of the dry weight. In mature gonads, protein contributes a larger share of the dry weight than lipid, yet the amount of glycogen stored in them is very small.

Comparisons, both quantitative and qualitative, of the lipid stores in the Pacific sea star *Pisaster ochraceus* and that of its principal food source, *Mytilus californianus*, indicate that both digestive glands and gonads are sites of lipid synthesis as well as of storage. The total lipid content of *Mytilus* is less than that of the digestive glands alone of *Pisaster*, while the glycogen content of the mollusc is much higher. Moreover, polyunsaturated fatty acids constitute a large proportion of the lipid in the mollusc but only a small fraction of that in the starfish. It has been shown that isolated pieces of digestive glands and of gonads taken from *Pisaster* at different times of the year and incubated with 1-C^{14} acetate incorporate it into lipid. The highest rate of incorporation by the digestive glands occurs in the fall and that of the gonads in the period between January and May. Although this evidence does not preclude the possibility that some of the dietary fat is directly stored, it does present a cogent argument for the seasonal activity of digestive glands and gonads in the construction of lipid different in character from that which was ingested.

In sea urchins, without digestive glands, the gonads are at all times the principal storage organs, although some reserves are found in the cells of the intestinal epithelium. Protein and lipid are stored in about equal amounts in the gonads of *Strongylocentrotus fransiscanus*, with glycogen representing, on an average, about 4 percent of their dry weight, an amount about equivalent to that in the human liver.

Invertebrate nutrition remains largely an unexplored field of biology but one whose further pursuit can hardly fail to be rewarding in the insight it would give to comparative physiology and biochemistry, invertebrate ecology, and speciation. Many invertebrates may well provide models of anabolic and catabolic pathways as yet uncharted or add confirmatory evidence to those that have so far been traced only in vertebrates, even only in mammals. Knowledge of their nutritive needs and how they are met would contribute greatly to understanding the distribution of invertebrates and their relations to their natural habitats, and could provide new ways of increasing or reducing specific populations. The information that has been gathered, however, serves to strengthen the concept of fundamental unity in biological processes, a concept of biochemical evolution that is an essential corollary to the all-embracing concept of organic evolution.

SELECTED BIBLIOGRAPHY

The need for food and its uses

Liebman, E.: On Trephocytes and Trephocytosis: A Study of the Role of Leucocytes in Nutrition and Growth, *Growth*, vol. 10, pp. 291–330, 1946.

Traeger, W.: The Nutrition of Invertebrates, *Physiol. Revs.*, vol. 21, pp. 1–35, 1941.

Evolution of food habits

Agosin, M., N. Scaramelli, L. Gil, and N. E. Letelier: Some Properties of the Microsomal System Metabolizing DDT in *Triatoma infestans, Comp. Biochem. Physiol.*, vol. 29, pp. 785–793, 1969.

Babers, F. H.: Development of Insect Resistance to Insecticides I, *U.S. Bur. Entomol. Plant Quar., Ser. E 776*, pp. 1–31, 1949.

Babers, F. H., and J. J. Pratt, Jr.: Development of Insect Resistance to Insecticides II, *U.S. Bur. Entomol. Plant Quar., Ser. E 818*, pp. 1–45, 1951.

Bramner, J. D., and R. H. White: Vitamin A Deficiency: Effect on Mosquito Eye Ultra Structure, *Science*, vol. 163, pp. 821–822, 1969.

Chambliss, O. L., and C. M. Jones: Cucurbitacins: Specific Insect Attractants in Cucurbitaceae, *Science*, vol. 153, pp. 1392–1393, 1966.

Clementson, C. A. B.: Effect of Certain Types of Paper on Sexual Maturation of the Insect *Pyrrhochoris apterus*, *Nature*, vol. 208, p. 510, 1965.

Dethier, V. G.: Behavioral Aspects of Protein Ingestion by the Blowfly, *Phormia regina* Meigs, *Biol. Bull.*, vol. 120, pp. 456–476, 1961.

Fraenkel, G.: The Raison d'être of Secondary Plant Substances, *Science*, vol. 129, pp. 1466–1470, 1959.

Galun, R., and S. H. Kindler: Glutathione as an Inducer of Feeding in Ticks, *Science*, vol. 147, pp. 166–167, 1965.

Galun, R., and S. H. Kindler: Chemical Specificity of the Feeding Response in *Hirudo medicinalis, Comp. Biochem. Physiol.*, vol. 17, pp. 69–73, 1966.

Hosoi, T.: Adenosine 5′ Phosphates as the Stimulating Agents in Blood for Inducing Gorging of the Mosquito, *Nature*, vol. 181, pp. 1664–1665, 1958.

Micks, D. W., M. J. Ferguson, and K. R. P. Singh: Effect of DDT on Free Amino Acids of Susceptible and DDT Resistant *Aedes aegypti* Larvae, *Science*, vol. 131, p. 1615, 1960.

Mittler, S.: Influence of Amino Acids upon Incidence of Tumors in the Tu^{50j} Stock of *Drosophila melanogaster*, *Science*, vol. 116, pp. 657–658, 1952.

Mittler, S.: Influence of Vitamins upon Incidence of Tumors in the Tu^{50j} Stock of *Drosophila melanogaster, Science*, vol. 120, p. 314, 1954.

Richards, A. G., and L. K. Cutkomp: Correlation between the Possession of a Chitinous Cuticle and Sensitivity to DDT, *Biol. Bull.*, vol. 90, pp. 97–108, 1946.

Saunders, S. A., R. W. Gracy, K. Schnackerz, and E. A. Noltmann: Are Honey Bees Deficient in Phosphomannose Isomerase?, *Science*, vol. 164, pp. 898–899, 1969.

Smith, J. N.: Detoxication Mechanisms in Insects, *Biol. Revs.*, vol. 30, pp. 455–475, 1955.

Sols, A., E. Cardenas, and F. Alvarado: Enzymatic Bases of Mannose Toxicity in Honey Bees, *Science*, vol. 131, pp. 297–298, 1960.

Nutritional adequacy

Chernin, S.: Cultivation of the Snail, *Australorbis glabratus*, under Axenic Conditions, in E. C. Dougherty (ed.), Axenic Culture of Invertebrates: A Goal, *Ann. N.Y. Acad. Sci.*, vol. 77, pp. 237–245, 1959.

Dougherty, E. C., E. L. Hansen et al: Axenic Culture of *Coenorhabditis briggsae* (Nematoda, Rhabditidae) with Unsupplemented and Supplemented Chemically Defined Media, *Ann. N.Y. Acad. Sci.*, vol. 77, pp. 176–217, 1959.

Fisher, L. R., and S. K. Kon: Vitamin A in the Invertebrates, *Biol. Revs.*, vol. 34, pp. 1–36, 1959.

Fisher, L. R., S. K. Kon, and S. Y. Thompson: Vitamin A and Carotenoids in Certain Invertebrates. I. Marine Crustacea, *J. Mar. Biol. Assoc. U.K.*, vol. 31, pp. 229–258, 1952.

Fraenkel, G., and M. Blewett: Linoleic Acid, Vitamin E and Other Fat Soluble Substances in the Nutrition of Certain Insects—*Ephestia kuehniella, E. elutella, E. cantrella* and *Ploidia interpunctella, J. Exp. Biol.*, vol. 22, pp. 172–190, 1951.

Meirovitch, E.: Studies on the *in vitro* Axenic Development of *Trichnella spiralis, Can. J. Zool.*, vol. 43, pp. 69–85, 1965.

Nicholas, W. L., E. C. Dougherty, and E. L. Hansen: Axenic Culture of *Caenorhabditis briggsae* with Chemically Undefined Supplements: Comparative Studies with Related Nematodes, *Ann. N.Y. Acad. Sci.*, vol. 77, pp. 218–236, 1959.

Pickett, C., and W. G. Friend: The Nutritionally Essential Amino Acids of *Rhodnius prolixus* (Ståhl) Determined with Glucose-U-^{14}C, *J. Insect. Physiol.*, vol. 11, pp. 1617–1623, 1965.

Provasoli, L., and A. D'Agostina: Vitamin Requirements of *Artemia salina* in Aseptic Culture, *Am. Zool.*, vol. 2, p. 439, 1962.

Sang, J. H.: The Quantitative Nutritional Requirements of *Drosophila melanogaster, J. Exp. Biol.*, vol. 33, pp. 45–72, 1956.

Sang, J. H., and R. C. King: Nutritional Requirements of Axenically Cultured *Drosophila melanogaster* Adults, *J. Exp. Biol.*, vol. 38, pp. 793–809, 1961.

Sedee, P. D. J. W.: Qualitative B Vitamin Requirements of the Larvae of a Blowfly, *Calliphora erythrocephala, Physiol. Zool.*, vol. 31, pp. 310–315, 1958.

Townsley, P. M., and R. A. Richey: Marine Borer Aldehyde Oxidase, *Can. J. Zool.*, vol. 43, pp. 1011–1019, 1965.

Waterhouse, D. F.: Axenic Culture of Wax Moths for Digestion Studies, *Ann. N.Y. Acad. Sci.*, vol. 77, pp. 283–289, 1959.

Weinstein, P. P.: Development *in vitro* of Some Parasitic Nematodes of Vertebrates, *Ann. N.Y. Acad. Sci.*, vol. 77, pp. 137–162, 1959.

Nutrition of Protozoa

Hitz, G. C.: Nutritional Requirements of Ciliates: Influences on Population Growth and on Morphology, *Am. Zool.*, vol. 2, p. 416, 1962.

Hutner, S. H.: Nutrition of Protists, in W. H. Johnson and W. C. Steere (eds.), "This is Life: Essays in Modern Biology," pp. 109–138, Holt, New York, 1962.

Nutrition of insects

Allen, R. R., and R. W. Newburgh: Phospholipid Composition of Fat Bodies of *Sarcophaga bullata, J. Insect Physiol.*, vol. 11, pp. 1601–1603, 1965.

Anderson, A. D., and R. L. Patton: *In vitro* Studies of Uric Acid Synthesis in Insects, *J. Exp. Zool.*, vol. 128, pp. 443–457, 1955.

Bricteux-Gregoire, S., Ch. Jeuniaux, and M. Florkin: Contributions à la biochemie du ver-à-soie. XXX. Biosynthèse de Tréhalose et de Glycogène a Partir de Glucose-l-phosphate, *Comp. Biochem. Physiol.*, vol. 16, pp. 333–340, 1965.

Briggs, M. H.: A Function of Ascorbic Acid in the Metabolism of an Insect, *Science*, vol. 132, p. 92, 1960.

Briggs, M. H.: Some Aspects of the Metabolism of Ascorbic Acid in Insects, *Comp. Biochem. Physiol.*, vol. 5, pp. 241–252, 1962.

Clayton, R. B.: The Utilization of Sterols by Insects, *J. Lipid Res.*, vol. 5, pp. 3–19, 1964.

Clements, A. N.: Studies on the Metabolism of Locust Fat Body, *J. Exp. Biol.*, vol. 36, pp. 632–640, 1959.

Dadd, R. H.: Ascorbic Acid and Carotene in the Nutrition of the Desert Locust *Schistocerca gregaria, Nature,* vol. 179, pp. 427–428, 1957.

Dadd, R. H.: The Nutritional Requirements of Locusts. I. Development of Synthetic Diets and Lipid Requirements, *J. Insect Physiol.*, vol. 4, pp. 319–347, 1960.

Dadd, R. H.: The Nutritional Requirements of Locusts. V. Observations on Essential Fatty Acids, Chlorophyll, Nutritional Salt Mixtures and the Protein and Amino Acid Components of Synthetic Diets, *J. Insect Physiol.*, vol. 6, pp. 126–145,1961.

Davis, G. F. R.: Quantitative Dietary Requirements for Histidine in the Sawtoothed Grain Beetle *Oryzaephilus surinamensis, Comp. Biochem. Physiol.*, vol. 28, pp. 741–746, 1969.

Earle, N. W., A. B. Walker, and M. L. Burks: Storage and Excretion of Steroids in the Adult Boll Weevil, *Comp. Biochem. Physiol.*, vol. 16, pp. 277–288, 1965.

Fraenkel, G.: A Historical and Comparative Survey of the Dietary Requirements of Insects, *Ann. N.Y. Acad. Sci.*, vol. 77, pp. 267–274, 1959.

Fraenkel, G., and M. Blewett: The Dietetics of the Clothes Moth, *Tineola bisselliella* Hum, *J. Exp. Biol.*, vol. 22, pp. 157–161, 1951.

Fraenkel, G., and M. Blewett: The Dietetics of the Caterpillars of Three *Ephestia* Species—*E. kuhniella, E. elutella* and *E. cantella*, and of a Closely Related Species, *Plodia interpunctella*, *J. Exp. Biol.*, vol. 22, pp. 162–171, 1951.

Friend, W. G.: The Gorging Response in *Rhodnius prolixus* Ståhl, *Can. J. Zool.*, vol. 43, pp. 125–132, 1965.

Friend, W. G., T. H. Choy, and E. Cartwright: The Effect of Nutrient Intake on the Development and the Egg Production of *Rhodnius prolixus* Ståhl, *Can. J. Zool.*, vol. 43, pp. 891–904, 1965.

Hill, L.: The Incorporation of C^{14} Glycine into the Proteins of the Fat Body of the Desert Locust during Ovarian Development, *J. Insect Physiol.*, vol. 11, pp. 1605–1615, 1965.

Hodgson, E., W. C. Dauterman, H. M. Mehendales, E. Smith, and M. A. B. Khan: Dietary Choline Requirements, Phospholipids and Development in *Phormia regina*, *Comp. Biochem. Physiol.*, vol. 29, pp. 343–360, 1969.

House, H. L.: Effects of Vitamin E on Growth and Development and the Necessity of Vitamin E for Reproduction in the Parasitoid *Agria affinis* (Fallen) (Diptera, Sarcophagidae), *J. Insect Physiol.*, vol. 12, pp. 409–417, 1966.

Kamienski, F. X., R. W. Newburgh, and V. J. Brookes: The Phospholipid Pattern of *Tenebrio molitor* Larvae, *J. Insect Physiol.*, vol. 11, pp. 1533–1540, 1965.

Lambremont, E. N.: Biosynthesis of Fatty Acids in Aseptically Reared Insects, *Comp. Biochem. Physiol.*, vol. 14, pp. 419–424, 1965.

Lambremont, E. N., C. I. Stein, and A. F. Bennett: Synthesis and Metabolic Conversions of Fatty Acids by the Larval Boll Weevil, *Comp. Biochem. Physiol.*, vol. 16, pp. 289–302, 1965.

Levenbrook, L.: Insect Biochemistry: Report of Seminar on Insect Biochemistry held in Chiba, Japan, *Science*, vol. 150, pp. 643–644, 1965.

Levinson, E. H.: Dietary Sterols in Insects: Evolution of Insect Sterol Requirements, *Proc. XI Int. Congr. Entomol., Vienna*, vol. 3, pp. 145–155, 1960.

Lipke, H., and G. Fraenkel: Insect Nutrition, *Annu. Rev. Entomol.*, vol. 1, pp. 17–44, 1950.

Meikle, J. E. S., and J. E. McFarlane: The Role of Lipid in the Nutrition of the House Cricket *Acheta domesticus*, *Can. J. Zool.*, vol. 43, pp. 87–98, 1965.

Nair, K. S., and J. C. George: A Histological and Histochemical Study of the Larval Fat Body of *Anthrenus vorax Waterhouse* (Dermestidae, Coleoptera), *J. Insect Physiol.*, vol. 10, pp. 509–517, 1964.

Nayar, J. K.: The Nutritional Requirements of Grasshoppers. I. Rearing of the Grasshopper *Melanoplus bivittatus* (Say) on a Completely Defined Synthetic Diet and Some Effects of Different Concentrations of B Vitamin Mixture, Linoleic Acid and β-carotene, *Can. J. Zool.*, vol. 42, pp. 11–22, 1964.

Nayar, J. K.: The Nutritional Requirements of Grasshoppers. II. Effect of Plant Phospholipids and Extracts of Bran on Growth, Development and Survival of the Grasshoppers *Melanoplus bivittatus* (Say) and *Camnula pellucida* (Scudder), *Can. J. Zool.*, vol. 43, pp. 23–38, 1964.

Naylor, A. J.: Possible Value of Casein, Gluten, Egg Albumen or Fibrin as Whole Proteins in the Diet of the Flour Beetle *Tribolium confusum* (Tenebrionidae), *Can. J. Zool.*, vol. 42, pp. 1–10, 1964.

Pierre, L. L.: Guanase Activity of the Symbionts and Fat Bodies of the Cockroach *Leucophaea maderae*, *Nature*, vol. 208, pp. 666–667, 1965.

Rockstein, M.: Some Aspects of the Intermediary Metabolism of Carbohydrates in Insects, *Annu. Rev. Entomol.*, vol. 2, pp. 19–36, 1957.

Schneiderman, H. A., and L. I. Gilbert: Control of Growth and Development in Insects, *Science*, vol. 143, pp. 325–333, 1964.

Sømme, L.: Effects of Glycerol on Cold Hardiness in Insects, *Can. J. Zool.*, vol. 42, pp. 87–101, 1964.

Stride, G. O.: On the Nutrition of *Carpophilus hemipterus* (Coleoptera: Nitulidae), *Trans. R. Entomol. Soc., London*, vol. 104, pp. 171–194, 1953.

Traeger, W.: Insect Nutrition, *Biol. Revs.*, vol. 22, pp. 148–177, 1947.

Van Handel, E., and P. T. M. Lum: Sex as a Regulator of Triglyceride Metabolism in Mosquito, *Science*, vol. 134, pp. 1979–1980, 1961.

Walker, P. A.: The Structure of the Fat Body in Normal and Starved Cockroaches as Seen with the Electron Microscope, *J. Insect Physiol.*, vol. 11, pp. 1625–1631, 1965.

Williams, C.: Selective Control of Insects by Juvenile Hormone Analogs, *Science*, vol. 152, p. 677, 1966.

Nutrition of other invertebrates

Allen, W. V., and A. C. Giese: An *in vitro* Study of Lipogenesis in the Sea Star *Pisaster ochraceus*, *Comp. Biochem. Physiol.*, vol. 17, pp. 23–38, 1966.

Greenfield, L., A. C. Giese, A. Farmanfarmaian, and R. A. Boolootian: Cyclic Biochemical Changes in Several Echinoderms, *J. Exp. Zool.*, vol. 139, pp. 507–524, 1958.

Jennings, J. B.: Observations on the Nutrition of the Rhynchocoelan *Lineus ruber*, *Biol. Bull.*, vol. 119, pp. 189–196, 1960.

Lane, C. E.: The Nutrition of *Teredo*, *Ann. N.Y. Acad. Sci.*, vol. 77, pp. 246–248, 1959.

Lee, T. W., and J. W. Campbell: Uric Acid Synthesis in the Terrestrial Snail *Otala lactea*, *Comp. Biochem. Physiol.*, vol. 15, pp. 457–468, 1965.

McWhinnie, M. A., and A. J. Corkill: Further Studies on Carbohydrate Metabolism in Decapod Crustacea, *Am. Zool.*, vol. 2, p. 431, 1962.

Nadarkal, A. M.: Carotenoid and Chlorophyllic Pigments in the Marine Snail *Cerithidea californica* Haldeman, Intermediate Host for Several Avian Trematodes, *Biol. Bull.*, vol. 119, pp. 98–108, 1960.

Nimitz, M. Aquinas, Sr., and A. C. Giese: Histochemical Changes Correlated with Reproductive Activity and Nutrition in the Chiton *Katherina tunicata*, *Q. J. Microsc. Sci.*, vol. 105, pp. 481–495, 1964.

Roots, B. I., and P. Johnston: The Lipids and Pigments of the Chloragosomes of the Earthworm *Lumbricus terrestris*, *Comp. Biochem. Physiol.*, vol. 17, pp. 285–288, 1966.

SUGGESTED READING FOR SECTION V

Cloudsley-Thompson, J. L.: "Spiders, Scorpions, Centipedes and Mites," Pergamon, New York, 1958.

Darwin, C.: "The Formation of Vegetable Mould through the Action of Worms, with Observations on their Habits," Murray, London, 1881.

Laverack, M. S.: "The Physiology of Earthworms," Pergamon, New York, 1963.

Lee, D. L.: "The Physiology of Nematodes," Oliver and Boyd, Edinburgh and London, 1965.

Ramsay, J. A.: "Physiological Approach to the Lower Animals," Cambridge University Press, Cambridge, 1952.

Schmitt, W.: "Crustaceans," The University of Michigan Press, Ann Arbor, 1965.

Smythe, J. D.: "The Physiology of Trematodes," Freeman, San Francisco, 1966.

Snodgrass, R. E.: "Anatomy and Physiology of the Honey Bee," McGraw-Hill, New York, 1925.

von Brand, T.: "Chemical Physiology of Endoparasitic Animals," Academic, New York, 1952.

section six

UTILIZATION OF FOOD

chapter X

TRANSPORT MECHANISMS

A. NECESSITY FOR TRANSPORT IN METAZOANS

B. MEDIUMS OF TRANSPORT: BODY FLUIDS

- **1. Inorganic (mineral) components**
- **2. Organic components**
  - a. PROTEINS
    - (1) Undefined
    - (2) Chromoproteins
      - *(a) HEMOGLOBINS AND CHLOROCRUORINS*
      - *(b) HEMERYTHRINS*
      - *(c) HEMOCYANINS*
      - *(d) OTHER CHROMOPROTEINS*
  - b. FREE AMINO ACIDS
  - c. CARBOHYDRATES
  - d. PIGMENTS OTHER THAN CHROMOPROTEINS
- **3. Cellular components**
  - a. TYPES
  - b. FUNCTIONS
  - c. NUMBERS AND ORIGINS
  - d. UNUSUAL CELLS

C. MEANS OF TRANSPORT

- **1. Relations of tissue and body fluids**
- **2. Movement of pseudocoelomic and coelomic fluids**
- **3. Movement of vascular fluids**
  - a. VASCULAR SYSTEMS
  - b. PUMPING CENTERS (HEARTS)

A. NECESSITY FOR TRANSPORT IN METAZOANS

Metabolic processes, like all chemical reactions, depend upon proximity of the reactants. Substrate and enzyme must come into contact with each other, and the end product of this relationship must be removed in order not to interfere with subsequent reactions. Removal may be brought about either by immediate conversion of one product to another or by its physical transport from the site of formation to another locale. In protozoans and in individual cells of multicellular systems the fluid cytoplasm and its cytomembranes provide a medium for this transport. In them, as in loosely assembled cell aggregrates, the cell surface provides an area adequate for the outward diffusion of unwanted metabolic products, whose accumulation could be destructive, as well as for inward diffusion of materials needed for metabolic processes. But for metazoans of any size and complexity of organization there must be some transport system to convey materials around the body from the sites of their entrance or production to the sites of their utilization or elimination. Tissue fluids moving through open spaces in the body function in this way in acoelomate and some pseudocoelomate invertebrates; coelomic fluid provides the primary transport system for echinoderms and acts as an accessory, or secondary, system in many of the invertebrates with a vascular system, in which the fluid is enclosed within tubes or other restricted spaces.

Any transport device is effective only if it can move throughout the body and carry metabolites, in solution, suspension, or physical or chemical combination, from area to area. Motility may be provided by ciliary activity, by general bodily movements, or by special propulsive muscular mechanisms and always imposes limitations upon the viscosity of the fluid. As conveyor of the end products of digestion, the system must make contact with the cells of the digestive area in such a manner and in such a condition that these can diffuse into it; as a conveyor of oxygen it must make contact with the surface of the body, or with special respiratory areas, in such a manner and such a condition that the gas may diffuse into it; as conveyor of storage and secretion products it must make contact with the sites of their deposition or production and also with the sites of their utilization or activity; and as a conveyor of metabolic waste it must make contact with the surface of the body, or with some special area of it, in such a manner and such a condition that these wastes, gaseous or otherwise, can diffuse from it into the environment. It must at all times remain essentially the same in volume, in osmotic pressure, and in hydrogen ion concentration to preserve homeostasis or what the distinguished French physiologist Claude Bernard (1813–1878) called constancy of the internal environment.* Fulfillment of these conditions is one measure of success in the organization of a metazoan body, and the regulatory mechanisms by which it is achieved are of particular interest in relation to the evolution of animals and their adaptations to external environments of different kinds. Transport systems represent one of the great coordinating and integrating mechanisms of the metazoan body and rapidly reflect any shifts or changes of metabolic patterns. Samples of such fluids are, therefore, revealing of metabolic states and metabolic activities. Similarly, introduction of test substances into them is a reasonably sure way of ascertaining whether these are taken up by cells and, if so, the effects they have upon them. Experimental alteration of their composition provides a means of determining the extent to which an organism can tolerate such changes and can compensate for them by one mechanism or another.

B. MEDIUMS OF TRANSPORT: BODY FLUIDS

Many invertebrates are of small size and the quantities of their body fluids are minute, measurable often only in microliters (1 μl = 0.000001 liter). This has been a major obstacle to analyses of these fluids, for it is often impossible to obtain enough from a single individual, or even from the pooled material of several individuals, to derive data that have any meaning. Modern developments in methods of microchemistry

* Many invertebrates do not preserve such constancy. Their body fluids increase and decrease in volume, and hence in concentration, depending upon conditions of their external environment. This question is discussed under osmoconformity and osmoregulation in Chapter XIV.

by electrophoresis, chromatography, and spectrophotometry are now providing means for separating and identifying organic compounds from samples of this order of magnitude. Some of these methods have already been applied to the study of the circulating fluids of some molluscs, annelids, and arthropods, especially insects, and may well be to those of other phyla about which information is much more scanty or even totally lacking.

It has, however, long been known that the body fluids of animals are complex solutions of both inorganic and organic compounds, whose composition has a considerable degree of uniformity in respect to some of the solutes but often a high degree of variability in respect to others.

1. Inorganic (mineral) components

The body fluids of animals are basically aqueous mineral solutions, containing all the principal salts of seawater. Those of representative invertebrates, as those of vertebrates, contain the inorganic ions Na^+, K^+, Ca^{++}, Mg^{++}, Cl^-, and SO_4^{--}. The average concentrations of these in seawater, expressed in millimolar concentrations (m*M*) are Na^+, 459.00; K^+, 9.78; Ca^{++}, 10.05; Mg^{++}, 52.50; Cl^-, 538; and SO_4^{--}, 26.50. But only in the body fluids of echinoderms are the concentrations of all these essentially the same as in seawater, for in those of other invertebrates one or more of them are present in greater or lesser amount. These differences imply that the animals have some means of regulating their concentrations, either by accumulation or by exclusion. The origin of these means and their subsequent employment have been of signal importance in evolution, especially in the exploitation and colonization of nonmarine environments.

Fluctuations in the ionic concentration characteristic of any species are known to occur with fluctuations in temperature, with season, with nutrition, and with physiological states, and, often, there are variations with age and sex. Absolute values have meaning, therefore, only with reference to the conditions under which they were obtained. In any circumstance, the concentrations are different from those of the cells with which the fluids are in contact. In general, cells contain more potassium and less sodium and chloride than the fluid surrounding them. This regulation of cytoplasmic concentration is a property of cell membranes, whose permeability to substances of all kinds has been subject to intensive investigation for many years, yet still remains to a great extent an unsolved problem. The regulation of ionic concentration in body fluids must depend finally upon the permeability of cell membranes, but it poses even broader problems of general bodily regulation, in which special structures or special areas may be concerned. These problems and the ways they have been met are discussed in a later section.

2. Organic components

Body fluids of invertebrates also contain organic components that are part of their constitution and qualitatively and quantitatively characteristic of them. Most significant among these constituents are proteins, amino acids, and sugars. All of these contribute to the osmotic pressure of the fluid, and the proteins, especially, with their capacities to act as buffers, contribute to regulation of its hydrogen ion concentration. In a number of species some of the proteins are colored and may function in oxygen transport. The fluids of some species contain nonprotein pigments, which may be derived directly from the food, and nitrogenous compounds, other than proteins and amino acids, which are products of cellular metabolism. These, like the end products of digestion, are more or less transitory components, or at least fluctuating ones, whose presence and quantity depends entirely upon the nutritional state of the organism.

a. PROTEINS Very likely there are proteins in the body fluids of most invertebrates but there is little specific information about them, apart from the chromoproteins that are of especial interest because of their relations with oxygen. Demonstrations, by incubation with appropriate substrates, of carbonic anhydrase in the coelomic fluids of *Sipunculus* and *Arenicola*, of amylase and phosphatase in the plasma of crustacean bloods, and of trehalase and tyrosinase (phenolase, phenol oxidase) in the hemolymph of insects provide evidence that enzymes may be represented among the proteins regularly present. There is also evidence that fibrinogen, or a protein like the fibrinogen of vertebrate blood, is a component of the blood of crustaceans and that in it there are also proteins that act as antigens or heteroagglutinins. Also, some of the bands in electrophorograms of the bloods of amphipod and isopod crustaceans and in the hemolymphs of insects react to specific stains in such a way as to identify them as glyco- and lipoproteins.

(1) Undefined Electrophorograms that have been made of blood taken from different species of arachnids and other arthropods show general similarities of pattern in the major taxonomic groups, but show also variations, often wide, in different species within a genus and even in different strains of a single species. Blood taken from two species of ticks, *Hyalomma excavatum* and *H. impeltatum*, was, for example, in both cases separable into 11 different bands, but the distribution of the bands was different and characteristic of each species. Striking differences have been found in the blood patterns of ticks of different genera in the family Argasidae, the number of bands ranging from three in *Argas reflexus* to twelve in *Ornithodorus moubata*. That the nature of the proteins expresses differences in the genetic endowments of the two genera is confirmed by the fact that there is no observable difference in the patterns of infected and uninfected specimens of *Ornithodorus*, which is a vector for the spirochaete *Borellia hispanicum*, the agent of relapsing fever in man. The presence of the spirochaete has apparently no effect upon the basic protein metabolism of the tick.

The hemolymph of all crustaceans examined shows a conspicuous fast-moving band identifiable as the chromoprotein hemocyanin, which accounts for 60 to 90 percent of the total hemolymph protein of the various species. In most of them there is a much smaller band of an almost immobile protein considered to be fibrinogen and between these two are bands representing other proteins with different velocities. These bands have different positions and are of different intensities in different species. In the amphipods examined and in the isopod *Idotea*, some of these intermediate bands have been identified as chromoproteins, probably carotenoid-protein complexes.

The uniformity of electrophoretic patterns shown by fluids obtained from different individuals of the same species and strain, and the overall similarity of those in the same genera, suggest that blood proteins may give some very reliable clues to phylogenetic relationships. There can be, however, a good deal of variation in specific patterns depending upon conditions in the organism at the time the blood was taken, so that comparisons are valid only when these conditions are known or standardized. The proteins of insect hemolymph, for example, vary quantitatively with developmental stage. Those in the hemolymph of *Periplaneta* are separable into 5 to 15 bands, the number increasing as the nymphs mature. In larvae of *Bombyx*, the hemolymph shows 2 to 3 bands, but their distribution changes with larval stage, with sex, with molting, and, in these insects, with infection-causing disease, such as nuclear polyhedrosis and flacherie. It is possible that the proteins of the body fluids represent an important food reserve and, therefore, resolve themselves into different patterns in accordance with different nutritional states. The hemocyanin in amphipods and isopods is considerably reduced in amount in animals that have been starved for some time, and reduction in hemolymph proteins has also been observed in insects after starvation and during the adult development of metamorphosing species. This loss has been attributed, in starving animals, to breakdown of protein in order to maintain the high concentration of amino acids characteristic of insect hemolymph and, in metamorphosis, to the contribution of materials for synthesis of protein in the adult. There is no knowledge of the sites of origin of hemolymph proteins in insects in general. However, isolated fat bodies of silkworms have synthesized, in vitro, proteins electrophoretically resembling those in the hemolymph of intact larvae, providing some evidence that this tissue may be at least one site.

On the other hand, consistent patterns are shown by different species of blood-sucking insects of the family Triatomidae, regardless of age, sex, food, or infection with *Trypanosoma cruzi*, for which they are vectors. Hemolymph taken from these bugs may show as many as 11 different bands, with numbers and patterns characteristic of each species. Differences in hemolymph proteins directly referable to the genotype have been shown in *Drosophila*, whose larval hemolymph has two bands that are different in position in standard individuals and in lethal mutants. Similarly, in *Culex pipiens* there is a difference in position of the single band in autogenous and anautogenous strains.

The gastrovascular fluid of the hydrozoan *Physalia physalis* has a protein content that shows little variation between individuals and tends to remain constant regardless of the size or nutritive state of an animal or the season of the year. This has been shown by digestion of fluid aspirated from the bulbs at the proximal ends of the fishing tentacles and chromatography of the resulting amino acids. At least 18 of these have been identified as constituents of the proteins, with concentrations ranging from 3.71 mg/ml of fluid, for glutamic acid, to 0.12 mg/ml for cystine. Since this fluid is moved around in a circumscribed space in the animal's body by continuous rhythmic contractions of the tentacles, it bears some resemblance to the transportive fluids of more highly organized animals and may serve similar functions. Conditions are quite different in the scyphozoan *Aurelia aurita* in which the extracellular fluid permeates the mesoglea and moves through it in a random way as the bell contracts. Its protein content, determined after obtaining the fluid by puncturing the ectoderm of the exumbrella and aspirating that which drains into the holes, is insignificant, but it is made up of the same amino acids as those in the proteins of *Physalia*, although two of these are present in only trace amounts. The total amino acid content of the extracellular fluid of *Aurelia* is about one two-hundredth to one five-hundredth that of the gastrovascular fluid of *Physalia*. The fluids of both these cnidarians contain also some ammonia and taurine ($H_2N \cdot CH_2 \cdot CH_2 \cdot SO_2H$), found also in the tissues of the alcyonarian *Renilla kollikeri*, the sea pansy. Experiments with radioactive tracers show that taurine, identifiable also in the tissues of many other invertebrates and in especially high concentrations in those of marine species, is derived from both cystine and methionine. It may, therefore, be endogenous and not necessarily of dietary origin.

(2) Chromoproteins The proteins that have been most fully investigated are those in which an atom of a metal, either iron or copper, is attached in some way to the polypeptide chain. Because of this they are colored, under certain conditions at least, and so are known as chromoproteins. They have been of special interest because of their resemblances, either structural or functional, to the hemoglobins of vertebrates. The chromoproteins of invertebrate tissue fluids fall into four general categories: hemoglobins (erythrocruorins), chlorocruorins, hemerythrins, and hemocyanins. Two of these, at least—the hemoglobins and chlorocruorins—are conjugated proteins with a nonprotein prosthetic group, which carries the metal attached to a globulin. In the hemerythrins and hemocyanins the metal is combined directly with the protein, but there are difficulties in defining the particular polypeptide to which it is attached as the prosthetic group. While, within each category, the metallic portion of the molecule seems to have similar configurations, the protein components can vary so widely that it is more accurate to refer to each kind of chromoprotein as groups of related compounds—hemoglobins, hemocyanins, etc.—than as molecules of a single type. Differences between those in a single category, as well as between those in different categories, can be determined by examination of their physicochemical properties, as well as of their chemical natures. They can be distinguished from each other by their sedimentation rates, isoelectric points, absorption spectra, and mobilities in an electric field, and by their molecular configurations and molecular weights. The molecular weight of the hemoglobin in the sea cucumber *Sclerodactyla briareus* is, for example, 23,600 and that of the lugworm *Arenicola marina* is 3,000,000. The hemocyanin of the crustacean *Pandalus borealis* has a molecular weight of 397,000, while that of the snail *Helix pomatia* is 6,680,000.

In general, the pigments of low molecular weight are contained within cells or corpuscles, while those of high molecular weight are dispersed in the fluid phase. This localization of smaller molecules within cells seems a good means of protection against their loss by elimination from the body along with waste molecules of comparable size that filter across excretory surfaces. It also serves to keep the total osmotic pressure of the fluid lower than it would be were the smaller molecules dissolved or suspended in it. There is a notable exception to this generalization—as is usual with biological generalizations—for in larvae of the midge *Tendipes* (*Chironomus*), the hemoglobin has a molecular weight of 31,400 yet is dissolved in the hemolymph. Excretion in insects, however, is not conducted by mechanisms similar to those in most other invertebrates.

(a) HEMOGLOBINS AND CHLOROCRUORINS

The hemoglobins (erythrocruorins) and chlorocruorins are of particular interest, because the prosthetic group of each is a cyclic tetrapyrrole to whose four nitrogen atoms is bound one atom of iron. The fifth of the iron's six coordination bonds is linked with a definitive group on the globulin (globin) and the sixth, under ordinary circumstances, either to a molecule of water or to one of oxygen. Similar ferroporphyrins constitute the prosthetic group of vertebrate hemoglobins, of myoglobins (muscle hemoglobins), and of certain cellular enzymes, the cytochromes, which are involved in the terminal events of cellular oxidations and are found in cells of oxygen-requiring (aerobic) organisms, of the peroxidases, operating also in cellular oxidations, and of catalase, the enzyme effecting decomposition of hydrogen peroxide. The ferroporphyrin of the hemoglobins is called heme and gives a reddish-yellow color to the fluid in which it is dissolved or to the cells in which it is contained; structurally it is very similar to, or even identical with, the prosthetic group of cytochrome b. The ferroporphyrin of chlorocruorin, called chlorocruoroheme, is unlike that of hemoglobin in that a formyl group is substituted for the vinyl chain in one pyrrole ring. The difference in the two configurations is shown by their structural formulae (Fig. X–1). The prosthetic group of chlorocruorin is identical with that of cytochrome a, the final enzyme in the oxidative chain, and in dilute solution gives fluids a greenish color, but a red one when concentrated. Chlorocruorins are very limited in their distribution among invertebrates and have so far been discovered in only four families of polychaetes—Sabellidae, Serpulidae, Chlorhemidae, and Ampharetidae—but in at least 21 species in these families. Yet even in these there is no uniformity. In all Chlorhemidae, the pigment is chlorocruorin; this is true also of the Sabellidae, with the single known exception of *Fabricia sabella* in which it is hemoglobin. Of the Serpulidae, *Spirorbis corrugatus* has hemoglobin in its blood, *S. miliaris* has no pigment at all, and all members of the genus *Serpula* have both hemoglobin and chlorocruorin. All of the Ampharetidae have chlorocruorin, with the exception of the genus *Melinna*, in which the pigment is hemoglobin. Chlorocruoroheme has been found free in the starfishes *Luidia* and *Astropecten*, which have not effected its conjugation with a protein to make a respiratory pigment.

The property of hemoglobin that is of particular biological interest is its ability to combine loosely and reversibly with oxygen. In this reaction a molecule of water attached to the iron of heme is replaced by a molecule of oxygen, and vice versa. The iron remains in the ferrous state so that the reactions are not oxidations and reductions, but oxygenations and deoxygenations. This property of reversible combination with oxygen is acquired by heme after its union with globin, for it does not occur with heme alone. The union between heme and globin is a firm one, firmer that that with the proteins of the heme enzymes. The reaction of these enzymes with oxygen is not reversible, a difference between them and hemoglobin that results from the nature of their proteins rather than of their prosthetic groups. Hemoglobin, therefore, in body fluids or in tissues, is potentially an oxygen carrier and oxygen donor and is so constituted as to be able to fill a vital role in aerobic respiration at both organismal and cellular levels.

In 1868 the English biologist Edwin Ray Lankester (1847–1929) proposed the term erythrocruorin for invertebrate hemoglobin, on the grounds that, while the prosthetic group was the same as that of vertebrate hemoglobin, the proteins were sufficiently different from vertebrate globin to make a distinction desirable and even necessary. The name erythrocruorin emphasizes the analogy of invertebrate hemoglobins with chlorocruorins, for the proteins of each contain proportionately more arginine and less histidine than is characteristic of vertebrate globin and have consistently lower isoelectric points than vertebrate hemoglobins. More recent investigations on vertebrate hemoglobins have shown, however, that there is a considerable degree of variation in the composition of their globins,

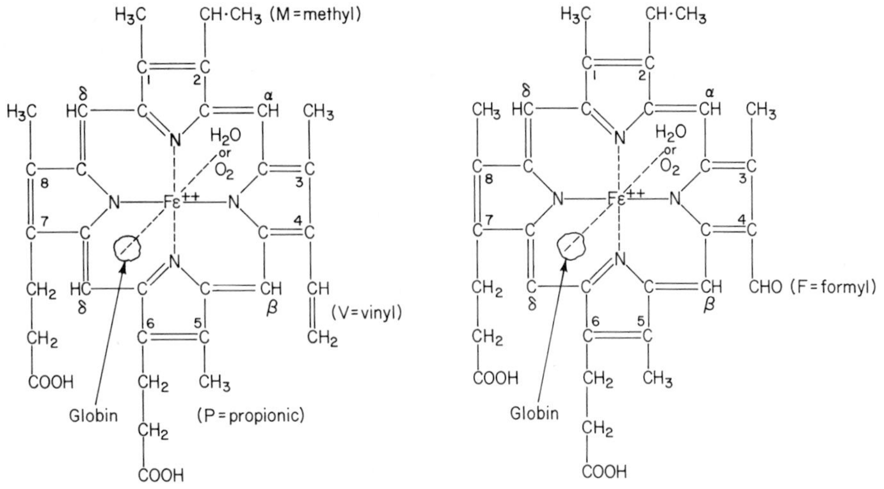

(a) Protoheme

(b) Chlorocruoroheme

FIG. X–1 *Structural formulae of **(a)** protoheme and **(b)** chlorocruoroheme. Theoretically there can be as many kinds of porphyrins as there are possible linkages to the eight carbon atoms at the outer corners of the four pyrrole rings. When each of these sites is occupied by a hydrogen atom, the molecule is known as porphin (see Fig. III-1). This configuration does not occur in nature but has been synthesized in the laboratory. Protoporphyrin is the configuration that becomes converted to heme when iron is joined to the four nitrogen atoms; chlorocruoroporphyrin is the configuration which by the same additions becomes chlorocruoroheme. Other naturally occurring porphyrins are uroporphyrin and coproporphyrin, so called because they were first isolated from human urine and feces, respectively. Uroporphyrin has four acetic and four propionic groups replacing the eight hydrogen atoms of porphin. Four isomers of this molecule are possible, but only two of them, uroporphyrin I and uroporphyrin III, are known to occur in nature. In uroporphyrin I the acetic groups are attached to the 2, 4, 6, and 8 carbon atoms and the propionic to 1, 3, 5, and 7. In uroporphyrin III the acetic side chains are located at sites 1, 3, 5, and 8, and the propionic at 2, 4, 6, and 7. Coproporphyrin has four methyl and four propionic groups, and again four isomers are possible, but only I and III exist in nature. Protoporphyrin, with four methyl, two vinyl, and two propionic side chains, has 15 possible isomers, but only one, protoporphyrin IX, has been found in animals. In ultraviolet light all porphyrins show intense red fluorescence, a property that makes it possible to detect amounts as small as 1 part in 1 × 10^9 parts of solvent. This property is lost when a metallic atom is inserted into the ring.*

not only in those of animals of different classes and species but even, as in man, of individuals in the same species and between that of the fetus and the adult. At least 22 different kinds of adult hemoglobin have been identified in man, some differing from each other only by a single amino acid in the polypeptide chain. Considering these differences as great as those between invertebrate and vertebrate hemoglobins, many biologists prefer to include both in the single category of hemoglobin.

Hemoglobin, as a pigment of body fluids, is more widespread among invertebrates than is chlorocruorin, but still is limited and highly sporadic in its distribution among the phyla and even the genera and species of different classes. It has not, as yet, been found in any member of the Porifera, Cnidaria, Sipunculoidea, Rotifera, Endoprocta, Ectoprocta, Brachiopoda, Chaetognatha, Onychophora, or Tardigrada. It is present, however, in the pseudocoelomic fluid of some nematodes; in the vascular fluid of all nemerteans examined, of a few bivalve and at least one genus of gastropod mollusc, of a number of polychaete and hirudine annelids and of most oligochaetes, of at least two echiuroids, of phoronids, and of a number of entomostracan crustaceans, where it is in solution in the blood of all notostracans, anostracans, and conchostracans; in that of many cladocerans, a few ostracods and copepods, and at least one branchiuran. It is present in the hemolymph of the larvae of one dipteran insect and, among echinoderms, in the coelomic corpuscles of some holothurians. In the annelids *Arenicola* and *Lumbricus* it is the only protein in the blood, but in others, as in the serpulids, it may be present with others, including chlorocruorin. Yet, in some species in which it is absent from the body fluids, it has been found in muscles, as myoglobin, or in cells of other tissues, and it is the only one of these chromoproteins that has been found there. It has been identified spectrophotometrically in the parenchyma cells of the freshwater rhabdocoele *Phenocora* and of the trematode *Fasciola*, and in the ciliates *Paramecium* and *Tetrahymena*. As "neuroglobin" it is present in the nervous tissue of many species, either in the glia cells or in the neurons. It is the pigment that gives the red color to the nerves and ganglia of the marine nematode *Amphiporus* and to the nervous tissue of *Busycon*, the whelk, and *Aplysia*, the sea hare, both gastropods whose blood contains only hemocyanin, as well as to the nerve cord of the polychaete *Aphrodite aculeata*, which has no chromoprotein of any kind in its body fluids. Hemoglobin may also be very generally distributed in the tissues of those invertebrates whose fluids also contain it. The echiuroid *Thalassema*, living in the tests of sand dollars, represents one extreme of this distribution, for in it hemoglobin is found in more different kinds of tissues than in any other animal. In *Thalassema* there is hemoglobin in the coelomic corpuscles and in the cells of the coelomic epithelium, ventral nerve cord, connective tissue, gonads, and eggs. *Scolopos*, a burrowing polychaete, illustrates the other extreme, for only its eggs contain measurable hemoglobin.

The conspicuous example of an insect with hemoglobin is the midge *Tendipes* (*Chironomus*), which, as a larva, has it in its nerve cells as well as in its blood and, as an adult, has it in the ovaries and eggs of the female. Other insects in which it has been identified are another midge, *Tanytarsus*, the botfly *Gastrophilus*, and some of the notanectid hemipterans, or water boatmen, where it is limited to the giant tracheal cells. Young larvae of *Gastrophilus*, living in the stomachs of horses, have hemoglobin in muscles, epidermis, and fat body; in older larvae, it is found only in the giant tracheal cells. Conditions that evoke the synthesis of tissue hemoglobin in *Gastrophilus* larvae are not readily evident. The pigment is found only in second and third larval instars, not in adults or first-stage larvae. It differs in molecular weight from the hemoglobin of the horse that is its host, so it is clearly a product of the larva's biosynthesis at certain times in its life. Why it appears at specific periods in the life cycle and not in others remains an open question.

Electrophoretic investigations of chironomid larvae have shown that even within a single species the hemoglobin may be of different molecular forms. Differences in mobility clearly separate four kinds in homogenates of whole larvae of *Chironomus thummi*, which, on further analysis, have been found to consist of two heme moieties and two dissimilar polypeptide chains. These chains are composed of 124 to 127 amino acids and are very different from those of mammalian hemoglobins. Most other chironomid species seem to have six or more distinct types of hemoglobin, with the exception of *C. atrella* which has only two. These findings suggest that several different genetic loci are simultaneously involved in the synthesis of the protein moieties.

Electrophoretic studies of the hemoglobins of the holothurians *Thyonella gemmata* and *Sclerodactyla* (*Thyone*) *briareus*, common in the intertidal zone of the Atlantic Ocean, have yielded information of similar kind. The species *T. gemmata* is separable, on certain behavioral, anatomical, and biochemical differences, into two populations, designated as "thins" and "stouts." Both have erythrocytes in the fluid of the perivisceral coelom and the water vascular system, but while there is no difference in the electrophoretic properties of the hemoglobin in these two compartments in members of the "stout" population, there is a difference in those of the "thin." In this sibling species there are eight distinct hemoglobins and three electrophoretically distinct polypeptide chains: A, B, and C. The erythrocytes of the water vascular system have a higher concentration of Hb II, with B chains only, than of other combinations, and those in the perivisceral coelom have a higher concentration of

Hb III, containing the A and C polypeptide chains. Erythrocytes of the "stout" sibling species have only five different hemoglobins, and two different polypeptide chains, the predominant one corresponding to the A pattern of the "thins." There is, however, no significant difference in the oxygen-binding capacities of the hemoglobins in members of these two populations. The erythrocytes of *Sclerodactyla* (*Thyone*) *briareus*, in the fluid of the water vascular system only, have also multiple hemoglobins and three different polypeptide chains, with electrophoretic mobilities corresponding to those of the A, B, and C chains of the "thin" population of *T. gemmata*. These three, then, would seem to be characteristic of the ancestral holothurian from which these two species have arisen. The reduced number of polypeptide chains and, correspondingly, of possible multiple hemoglobins in the "stout" sibling species of *T. gemmata* could, therefore, have resulted from the loss or inactivation of an ancestral gene that directed the construction of either the B or the C polypeptide chains.

This erratic distribution of hemoglobin, which has been described as defiant of all systematization, and the ubiquity of related chromoproteins, particularly the heme enzymes, raise many questions. Some of these are concerned primarily with the utilization of hemoglobin as a respiratory pigment by those invertebrates whose body fluids contain it, either in colloidal solution or within cells or corpuscles. Such utilization is common practice among vertebrates whose blood almost universally contains intracorpuscular hemoglobin. Notable exceptions are some fishes of Antarctic waters and the leptocephalous larvae of eels, whose bloods are colorless. Whether or not the hemoglobins of invertebrate body fluids act in the same way, or indeed function in respiration at all, are questions that have for years been of great interest to physiologists. Consideration of them, and of similar questions in respect to chlorocruorins, hemerythrins and hemocyanis, and the experimental evidence that has provided answers to some of them, are reserved for a later section dealing with respiration.

Other questions of paramount interest are those concerning the biochemical evolution of hemoglobins. The ability to synthesize tetracyclic pyrroles is widespread throughout the living world. Studies with radioactive compounds, both in vitro and in vivo, have shown that, although they are complex molecules, they can be synthesized from two relatively simple organic acids, glycine ($H_2N \cdot CH_2 \cdot COOH$), an amino acid that is readily synthesized by most animals; and succinic acid ($HOOCH_2 \cdot CH_2 \cdot COOH$), derived from acetic acid. They also arise from the oxidation of fats and carbohydrates and so are in almost constant production in the animal body. Similarly, the abilities to insert a metal, such as iron, magnesium, or copper, into the ring and so convert a porphyrin to a metalloporphyrin and to conjugate it with protein or a non-nitrogeneous compound are very general properties of living organisms. Chlorophyll, for example, is a metalloporphyrin with an atom of magnesium bound to the nitrogen of the pyrrole nuclei, conjugated with a long-chain unsaturated alcohol, phytol, attached to one propionic acid side chain. These abilities must have arisen very early in the evolution of living systems, for porphyrins derived from heme and from chlorophyll have been extracted from mineral oils, bituminous shales, and coal, deposited 300 to 400 million years ago. Porphyrins have also been detected in fossils of Eocene molluscs and are found in shells of modern ones. Shells of *Trochus*, a genus of tropical gastropods, contain uroporphyrin I, as do those of the bivalve *Pteria*, the pearl oyster. Porphyrins are also found in the integuments of a number of molluscs, even in those like the nudibranch *Tritonia* and the slug *Arion ater*, which have no shells. One with the characteristics of protoporphyrin IX prepared from sheep's blood is localized in the integument of the siphons of the wood-burrowing bivalve *Bankia setacea*. The siphons, which are the only parts of the animal that project from the burrow into the open water, are marked with patches of reddish cells, which are particularly prominent in the exhalant siphons of males at certain times of the year. Free porphyrin has also been found in the integument of some starfishes; in *Asterias rubens*, it has been identified as protoporphyrin, which is found also in the integuments of *Astropecten irregularis* and *Luidia ciliaris*, in which, quite surprisingly, there is chlorocruoroporphyrin as well. These are the only known instances of the occurrence of this configuration of porphyrin outside of the polychaetes, and in them it is not free but conjugated with protein. Free porphyrins of other kinds are, however, quite general in the tissues of annelids, especially of oligochaetes, which appear to be able to tolerate concentrations of them that in man and other vertebrates result in pathological conditions.

The sites of synthesis of the extracellular hemoglobin of body fluids are almost unknown. In polychaetes, the heart body, an intravasal plug or strand of loose, spongy tissue usually within the anterior part of the dorsal vessel, has been suspected as the region most active in synthesis of hemoglobin and of chlorocruorin, since various porphyrin intermediates have been identified in it. Electron-microscopic studies of crystalline hemoglobin prepared from the blood of *Arenicola marina* reveal a very distinctive crystalline pattern of paired hexagonal disks, each of which is composed of six more or less spherical subunits. This pattern is characteristic also of the chlorocruorin of *Sabella pavonina, Spirographis spallanzanii,* and *Stylaroides monolifer*; and of the hemoglobins of *Nereis diversicolor* and *Lanice conchilega*; of the oligochaete *Lumbricus* sp. and the leech *Hirudo medicinalis*; but not of the vertebrate hemoglobins similarly prepared and examined. This dissimilarity provides argument for the retention of the term "erythrocruorin" for invertebrate hemoglobins and the reservation of "hemoglobin" for the vertebrate chromoprotein. Recognition of the crystalline pattern of the hemoglobin of *Arenicola* has made it possible to identify the tissue in which it is formed. Electron micrographs of cells surrounding the vascular caeca, small blind vessels connected to branches of the dorsal vessel, show that at times they are packed with the hexagonal crystals. Adjacent to the vesicles containing the crystals are masses of ferritin, an iron-protein complex that is very generally believed to act physiologically as an iron storage mechanism. Similar deposits of ferritin are found in mammalian erythroblasts and in liver cells after the breakdown of the hemoglobin of senescent erythroplastids. The cells clustered around the vascular caeca of *Arenicola* are considered homologues of the chlorogocytes of oligochaetes, and their identification as hemopoietic tissue implicates chlorogocytes in hemoglobin synthesis and adds one more to the variety of metabolic processes attributed to them. In *Arenicola*, the synthesis of the chromoprotein appears to take place in the Golgi system of the extravascular cells. When complete, the swollen vesicles are packed with the pigment, which escapes in some as yet unknown way from the cells into the vascular caeca and so into the main channels of the vascular system.

Biochemical investigations of a number of polychaete species indicate that the course of heme biosynthesis in them is different, both quantitatively and qualitatively, from that in vertebrates, where the pathway has been known for some years. Essentially, these differences lie in the production of porphyrins in excess of those incorporated into heme and the conversion of these surplus by-products into metalloporphyrins that are distributed to the tissues and accumulate there. In *Arenicola*, coproporphyrin III has been found not only in the extravasal tissue but also in the nephridia and the integument. Uroporphyrin III has also been found in the integument as well as in the extravasal tissue. The metalloporphyrin coproporphyrin I and III and urohematin III, with iron in the ferric rather than the ferrous state, protoheme, and coproheme III have been found in these same tissues, in conjunction with hemoglobin in the intravasal tissues and the blood. Such metalloporphyrins, other than heme, do not arise in the biosynthetic pathway of higher vertebrates, nor is there any evidence that either uroporphyrin or coprophyrin are intermediates. No explanation has yet been forthcoming for the production by these annelids of porphyrins in excess of those in the established pathway, but one for the worms' tolerance of them is suggested by their habits of life. Porphyrins have photodynamic action and in light affect the metabolism of cells containing them in deleterious ways. The annelids in whose tissues they accumulate are all burrowers or tube-dwellers that spend the major parts of their lives, if not all of them, in darkness and are, therefore, protected from these damaging effects. That they are potentially damaging to them is shown by the difficulty of maintaining the animals in the laboratory when in light and is implied by the absence of porphyrins in the crowns of sabellids, the parts of the worms that are in nature exposed to light, although they may be distributed throughout other tissues. In the sabellid *Myxicola infundibulum*, for example, coproporphyrin III has been found in the nephridia and in the gut and body walls, as well as in the extravasal tissue. Protection from light is also afforded molluscs whose integuments contain porphyrins, for either their shells provide a shield or else, as is the case with the black slug *Arion ater*, which has the greatest accumulation of integumentary free porphyrins of any mollusc, a dense layer of dark pigment lies between the surface and the porphyrin deposits. It has been suggested that the porphyrin-containing cells of the siphons of *Bankia setacea* sensitize the animals to chemical conditions in the water, an adaptation that might be of particular advantage to them, especially to the males, in the spawning season.

It is, of course, possible that these tissue porphyrins are not endogenous but are derived from the animal's food. Although some of them may be, the demonstration of the synthesis of coproporphyrin III, in vitro, by homogenates of the body wall, gut wall, and chlorogocytes of *Lumbricus terrestris* gives evidence that an endogenous source is possible. Nor has any connection been established between diet and the so-called "gut hemes" that are regularly found in the digestive cavities of a number of polychaetes, various gastropods, and some crustaceans. These compounds are hemochromogens, or hemochromes, and consist of protoheme—even in animals whose bloods contain chlorocruorin or hemocyanin—joined with a nitrogenous compound, which is often, but not invariably, a protein. One of these, helicorubin, in which the nitrogenous part is a protein, has been found in considerable amounts in the guts of snails, even after 2 months on a diet of filter paper. It has also been found in the digestive gland of *Helix pomatia*, along with another somewhat similar hemochromogen.

All these facets of porphyrin metabolism suggest that organisms were endowed, very early in their existence on this planet, with a capacity to synthesize the tetrapyrrole super ring and that, in the course of evolution, mutation has effected numerous modifications of what may have been its primitive form. The genes concerned may also have been so profoundly modified in some species that the capacity has been lost, so that the animals lack porphyrins entirely. Ability to synthesize heme and to conjugate it with protein to make hemoglobin may have arisen independently in different species and be another illustration of convergence in evolution, or the ability may have been exploited particularly by those animals which could put hemoglobin to some physiological use. There are, for example, several species of invertebrates in which the synthesis of hemoglobin is initiated, or increased, by oxygen deficiency in their environment, permitting them to compensate for this lack. Increased synthesis is a recognized phenomenon in mammals, which respond to transportation from low to high altitudes by increased hematopoiesis. Man, for example, can increase his hemoglobin by 20 percent when he goes from sea level to a mountain top. The cladoceran *Daphnia hyalina*, a species that is colorless in its natural habitat of well-aerated lake water, synthesizes hemoglobin and turns pink when kept in the laboratory in water with low oxygen content. The hemoglobin content of individuals in the laboratory is inversely proportional to the amount of dissolved oxygen in the water in which they are kept. In nature, the hemoglobin content of *Daphnia* follows the same pattern, for ponds are different in their oxygen content and there are even variations in the same pond from month to month. Those that are oxygen deficient may become bright red because of the swarms of *Daphnia pulex* or other common species that have increased their hemoglobin synthesis, in response to decrease in available oxygen, by as much as 10 percent per individual. It has been shown histologically that the iron for this additional synthesis is absorbed from the midgut and accumulates in the fat bodies and ovaries, suggesting that these tissues are the principal sites of hemopoiesis. When, on return to oxygenated water, the animals lose their color as the hemoglobin is broken down, the iron is not conserved as it is in the vertebrate body and used again as need requires, but is eliminated. More must, therefore, be ingested and absorbed for future hemoglobin synthesis. The midgut caeca, fat body, and maxillary glands are all involved in hemoglobin breakdown and excretion of its products, but there is no clear information as to how this result is achieved.

The anostracan *Artemia* also synthesizes hemoglobin when in waters of high salinity, which are necessarily deficient in oxygen because of the low solubility of the gas in salt solutions. Synthesis of additional hemoglobin has also been observed in the snail *Planorbis* and the larvae of *Tendipes* (*Chironomus*) when the waters in which they are living become lower than usual in oxygen.

In these cases, hemoglobin synthesis is clearly regulated by environmental conditions, specifically those of oxygen availability to organisms that require it. It is perhaps not unreasonable to suppose, especially in the light of recent advances in knowledge of gene operation and gene control and regulation, that invertebrates other than these cited may still be able to synthesize hemoglobin but do not do so because they have other means of meeting whatever physiological role it may fill. Its respiratory functions are not nearly so well defined in invertebrates as they are in vertebrates, and it may well be that some invertebrate species, either through physiological mechanisms or through habits of life, have successfully bypassed it as a means of oxygen transport or storage, and delivery. Such events rather than the sudden acquisition or loss of the requisite genes, could account for its sporadic, rather than universal, distribution among them.

(b) HEMERYTHRINS Hemerythrins are also iron-containing chromoproteins, but in them the iron is joined directly to certain amino acids, probably attached as side chains to the main axis of the protein molecule. Like the hemoglobins and chlorocruorins, hemerythrins can combine reversibly with oxygen, but in the combination one molecule of oxygen is united with three atoms of iron, not to one as it is in the porphyrin-containing pigments. Hemerythrins are even more limited in their distribution than chlorocruorins, but in their small range are quite as sporadic as hemoglobin. They have never been found in solution, but always within cells or corpuscles; the molecular weight of the hemerythrin of *Sipunculus* is 66,000 and that of *Golfingia*, 119,400. Hemerythrin has been found in corpuscles in the pseudocoelomic fluid of *Priapulus*, in coelomocytes of the brachiopod *Lingula*, in coelomocytes and blood cells (hemerythrocytes) of sipunculids, and in the blood cells of a single genus of annelids, the polychaete *Magelona*. It has also been found in cells of the gut epithelium of *Sipunculus*. These are the only animals in which it is known to be synthesized, at least in detectable quantities. The color which it gives the cells containing it has been variously described as reddish-brown or violet. This is due to oxygenation of its iron, for it has no color when deoxygenated. In oxygenated hemerythrin of *Golfingia*, all the iron is in the ferric state; when deoxygenated, one-third of the iron is ferric and two-thirds ferrous. Studies of the hemerythrins of sipunculids have shown that each molecule consists of eight subunits, into which it can be dissociated by various means, and there is some evidence, although not definitive, that these subunits are of two kinds, differing from each other in amino acid content.

Electrophoretic and fingerprinting analyses of the hemerythrins in two species of sipunculids, *Golfingia gouldii* and *Dendrostomum cymodoceae*, have shown that there are differences between that in the coelomocytes and in the hemerythrocytes of the same individual, as well as differences between those of different individuals. While the vascular hemerythrin of all the individuals of *Golfingia* examined showed the same mobility in an electric field, pigment from the coelomocytes moved at a faster rate. Differences in the mobilities of hemerythrins taken from coelomocytes of different individuals of the same population distinguished them as "fast" or "slow" and these differences could be correlated with differences in the chromatographic fingerprints of their tryptic and chymotryptic digests, which showed one dissimilar peptide. More extensive studies of the distribution of fast and slow hemerythrins in a population of 181 specimens showed that in 73 individuals it was of the slow type, in 16 of the fast type, and in 92 its migration was intermediate between the two. In the light of current knowledge of the genetic control of protein synthesis, these results indicate that at least two different structural genes, or cistrons, direct the construction of the protein chains of hemerythrin and that in this species the gene directing the synthesis of coelomic hemerythrin has undergone mutation. The smaller number of individuals with fast hemerythrin indicates that its sequence of amino acids is determined by mutation of a "standard" to a recessive gene. Those individuals whose hemerythrin is of intermediate mobility are, then, heterozygotes resulting from the union of gametes from individuals with fast and slow hemerythrins, respectively. Fingerprints substantiate this, for those made from coelomic hemerythrin of these suspected heterozygotes show both peptides, the one characteristic of fast hemerythrin and the one characteristic of slow. Comparison of the fingerprints of vascular and coelomic hemerythrin of the Australian species *Dendrostomum cymodoceae* show that these have relatively few peptides in common and so differ quite radically from each other. Fingerprints of the hemerythrin of *Lingula reevei*, which is found only in the coelomocytes, as the animal has no vascular system, show certain resemblances to the hemerythrins of sipunculids, resemblances that are expressed in the general patterns of the chromatograms and in the positions of some peptides.

It has been postulated that human hemoglobins and myoglobins have evolved through mutations of an original common gene. This may also be the case with hemerythrins, and, if so, brachiopods and sipunculids may be of closer phylogenetic relationship than is generally supposed. It is difficult to account for the appearance of hemerythrin in *Magelona*, the only annelid known to have it, and the only one whose vascular chromoprotein is endocorpuscular. Of interest, too, is the physiological effect of such mutations, especially in relation to the oxygen-binding properties of the molecules resulting from them.

(c) HEMOCYANINS In hemocyanins, as in hemerythrins, a metallic atom, in this case copper, is directly bound to sites on amino acids of the chromoprotein. The copper, like the iron of the other three pigments, combines reversibly with oxygen, one molecule of oxygen being bonded to two copper atoms. According to present information, both atoms of copper are in the cuprous state in deoxygenated hemocyanin, but on its oxygenation, one is oxidized and becomes cupric, while the other remains cuprous. It is because of this that oxygenated hemocyanin is a deep blue color; with both atoms in the cuprous state, it is virtually colorless, or milk white. The proteins of hemocyanin are globulins, like those of the other chromoproteins and, like them, differ in their content and sequences of amino acids. There is, however, on the whole more uniformity between them than between hemoglobins. The differences in hemocyanins are reflected in differences in their molecular weights, as well as in their spectrophotometric and electrophoretic properties, their isoelectric points, and their solubilities. Single units, or molecules, of hemocyanin, each containing two atoms of copper, range in molecular weight from 50,000 to 74,000, but these units aggregate into macromolecules of various numbers with molecular weights ranging from 397,000 in the caridean crustacean *Pandalus borealis* to 6,680,000 in the gastropod mollusc *Helix pomatia*.

Hemocyanin is synthesized by many invertebrate genera and species and is the chromoprotein characteristic of the bloods of gastropod and cephalopod molluscs, of xiphosurans, arachnids, and malacostracan crustaceans, and is always in solution in their body fluids. Although the molecules must be formed in cells, they have never been found there as such. Hemocyanin is universal among cephalopod molluscs and almost so among gastropods—the notable exception being the pulmonate *Planorbis*, which has hemoglobin in its blood—and it is not found in lamellibranchs, whose pigments, if any, are hemoglobins. Yet in gastropods, like *Busycon*, whose blood contains hemocyanin, there may be myoglobins in the muscles and gut hemes in the digestive tract. It is curious, too, that hemoglobin, so widely distributed among the smaller crustaceans, is absent both in fluids and cells of the larger and presumably more highly evolved decapods. There is no chromoprotein in insect hemolymph, although proteins of high molecular weight have been demonstrated in them by microelectrophoresis.

Biosynthesis of hemocyanin requires the absorption and concentration of considerable quantities of copper. The copper content of the hemocyanin in the blood of marine animals is some 10,000 times that of the same volume of seawater, probably the primary source of the element for them. Undoubtedly, they derive some from their food, which must be the only source of it for terrestrial animals and the primary one for freshwater species. This ability to absorb, concentrate, and bind a metal, which even in minute amounts is toxic when free, must be common to all animals in whose fluids hemocyanin is found. Presumably, its union with amino acids of the protein is effected in the same way in all of them. Differences in hemocyanins are the result, like differences in hemoglobins, of differences in the genetically controlled sequential assemblage of amino acids into long-chain proteins. Divergence from a common pattern must then either expresss mutations of a common gene, as is believed to be the case in human hemoglobins, or the operation of different genes, as in the hemoglobins of chironomids and the coelomic hemerythrins of *Golfingia*.

(d) OTHER CHROMOPROTEINS Blue and orange chromoproteins, which are most probably conjugates of proteins and carotenoids, have been detected in the blood of several species of amphipod crustaceans and in that of the isopod *Idotea*. Electrophoretically, these make bands that lie between the fast moving hemocyanin and the slow moving fibrinogen. The blood of *Orchestia gammarella* gives two such bands, one of which (M_1) is always associated with a bright blue proteid, and the other (M_2) with an orange or pink one. *Marinogammarus marinus* has a third (M_3). The proportionate distributions of these pigments determine the color of the blood and so that of the animals. For example:

	M_1, %	M_2, %	M_3, %	*Color*
Orchestia gammarella	2.6	14.9	...	Orange
	7.8	10.3	...	Bright blue
Marinogammarus marinus	1.7	5.6	7.5	Bluish
	1.5	6.5	4.6	Green
	...	6.5	1.3	Pink

It is possible that these pigments participate in the mechanisms of color change, which these crustaceans undergo in accordance with color changes in the seaweeds among which they live off the English coast near Plymouth.

b. FREE AMINO ACIDS A good deal of attention has been given to the distribution of amino acids as such in the tissues of invertebrates, but not very much is known about it in their body fluids, with the exception of insect hemolymph. From the investigations that have been conducted on tissue amino acids, it is apparent that, in general, they are more abundant in the cells of invertebrates than in those of vertebrates, that their distribution there follows patterns characteristic of different species, and that the levels of concentration are, in general, higher in marine than in terrestrial or freshwater forms and higher also in generalized, or primitive, species than in the specialized ones of any class or order. There are, for example, more amino acids in the coelomic fluids of asteroids than of echinoids and of orthopteran than of dipteran insects. Although the patterns vary between species, each is fairly consistent, both qualitatively and quantitatively, within each group and is not altered to any significant extent by diet. This suggests that the acids are endogenous and have some physiological role which has not yet been satisfactorily assessed; osmoregulation is an obvious possibility. The free amino acids that are of most frequent occurrence and have been found in the greatest quantity in cells are glycine and the three that are involved in the events of the citric acid (Krebs, tricarboxylic acid) cycle of aerobic respiration. These three—alanine and aspartic and glutamic acids—can by transamination give rise, respectively, to pyruvate, oxalo-acetate, and alpha-keto-glutarate, all of which are intermediates in the oxidative reactions by which energy is released from carbohydrates and fats. Proline is another amino acid frequently found in invertebrate tissues and in concentrations considerably higher than in those of mammals, at least. Taurine is also often present in considerable quantities, especially so in the tissues of marine molluscs.

The distribution of free amino acids in the body fluids of invertebrates, so far as it has been determined, seems to follow a similar species characteristic pattern. Among the annelids, for example, the blood of the polychaete *Nephthys hombergi* has a total concentration of free amino acids amounting to 350 to 450 mg/100 ml (mmoles/kg), of which glycine represents about 50 percent, with alanine and proline the most abundant in the remainder. Starvation produces no consistent effect upon their concentration, so that it seems that maintenance of these levels is important to blood-tissue relationships and is preserved even at the expense of reserve and tissue proteins. Free amino acids in the blood of the echiuroid *Urechis caupo*, on the other hand, amount to only 80 mg/100 ml (mmoles/kg) of blood, a concentration which corresponds to that in the blood of the crayfish *Astacus astacus*. Fifteen of the common amino acids have been identified in the fluid phase of crayfish blood, in quantities ranging from 1.1 mg/100 ml (mmoles/kg) (histidine) to 29.8 mg/100 ml (mmoles/kg) (glutamic acid). Alanine [10.2 mg/100 ml (mmoles/kg)] is the next most abundant, and neither cystine nor methionine have been detected. The total free amino acid concentration in the blood of the European lobster *Homarus gammarus* is 64.1 mg/100 ml (mmoles/kg), and here glycine is the predominant one, with a concentration of 24 mg/100 ml (mmoles/kg), or four times that of its concentration in the blood of *Astacus*. The concentration of valine, on the other hand, is in *Homarus* about one-tenth that of its concentration in *Astacus*. No cystine, methionine, nor threonine has been found in *Homarus* blood, although there is both methionine and threonine in its muscles. There is, in both *Astacus* and *Homarus*, a steep gradient between muscle and blood in respect to free amino acids, the concentrations of each acid being always higher in the tissue than in the fluid. The total concentration of amino acids in crayfish muscle is 2,202.4 mg/100 ml (mmoles/kg) of tissue extract and in lobster muscle, 3,043.6 mg/100 ml (mmoles/kg) of tissue extract, but the proportionate distribution of the acids may not be the same in tissue and in blood. In *Astacus* muscle, for example, arginine is predominant, with glutamic acid next; in *Homarus* muscle, as in blood, glycine is the most abundant acid, although there is about 700 times as much arginine in the muscle as in the blood.

High concentrations of free amino acids, ranging from 293 to 2,430 mg/100 ml (mmoles/kg) are particular and distinctive characteristics of insect hemolymphs. Although in this respect the blood of *Nephthys* falls in the lower ranges of amino acid concentrations in insect hemolymphs, in general, these are far higher than those known for other invertebrates or for vertebrates, in which the total concentration averages about 50 mg/100 ml (mmoles/kg). In general, too, the amino acid level is lower in those insects whose post-embryonic development is a gradual progression through successive nymphal stages (hemimetabolous) than in those with a pupal period and an abrupt metamorphosis to adult form (holometabolous). The pattern of amino acids in the hemolymph is quite consistent for any species at any given period of life, but varies considerably between species and within them at different stages of the life history. Almost all the insects examined have high concentrations of glutamic acid and of proline and of either arginine, lysine, or histidine. Glycine and leucine are also very general in distribution. Altogether, more than 20 free amino acids and amino compounds have been identified in insect hemolymphs.

Although the same amino acids may be represented in the hemolymph of the larvae and pupae of holometabolous insects, their proportions are often very different in the different stages. There is, for example, more alanine and tyrosine than there is of other amino acids in the hemolymph of larvae of the fly *Calliphora*, but in the pupae glycine, glutamic acid, and valine increase in amount and tyrosine decreases. Fifteen of the common twenty amino acids have been identified in larvae and pupae of the moth *Sphinx ligustri*, but the concentration of alanine in the pupa is considerably greater than in the larva and that of glycine and of tyrosine considerably less. In fact, tyrosine and proline are the two acids for which the greatest ranges and variations within a species have been reported. Both are involved in the formation of new cuticle at the time of molting, proline with its actual construction and tyrosine with its sclerotization. Such variability, which in the circumstances is to be expected, emphasizes the importance of the selection of similar stages for comparative studies.

The total concentration and proportionate distribution of the amino acids, however variable, seems largely independent of food, although, of course, their maintenance at any level depends upon elimination or marked increase of their sources from the diet and upon its content of insecticides such as DDT (see page 348). The pattern characteristic of the larval stages of *Sphinx ligusti* did not, for example, vary significantly when the caterpillars were fed on different plants. The pattern in other insects has, on the other hand, been shown to change qualitatively, but not to any extent quantitatively, with a change in diet. In these cases, while the level of concentration of one amino acid or another might fall, its reduction was compensated by an increase in one or more of the others, so that the total concentration remained relatively the same. Of particular interest, in relation to diet and to biochemistry in general, is the fact that alanine in the hemolymph of *Oncopeltus fasciatus* is not the usual levo isomer, but the dextro form. This is not found in the milkweed seeds on which the bug feeds, so that it must be considered a product of its own metabolism.

c. CARBOHYDRATES The carbohydrates consistently found in body fluids are soluble sugars, most frequently the monosaccharides glucose and fructose. Their concentrations generally vary considerably with the metabolic state of the animal, although some organisms seem to preserve more or less constant levels whether starved or well-fed. This is apparently the case with the Indian scorpion *Heterometrus fulvipes* and the crabs *Cancer pagurus* and *Libinia marginata*, in which starvation does not produce a significant decrease in blood sugar. In *Cancer*, the values range from 50 to 300 mg blood sugar/100 ml blood, but in most of the other crustaceans investigated, great fluctuations in values have been reported. High or low levels depend not only upon the nutritional condition of the animal but upon the stage of its growth as well. Blood-sugar levels rise at the time of molting and are, in general, lower after a molt and during the intermolt period. It has been suggested that glucose is not a substrate for oxidation in the spiny lobster *Panulirus japonicus*, for it did not promote oxygen consumption by isolated tissues, and that it is present in the blood only in transport from the digestive gland, where glycogen has been depolymerized, to sites of chitin deposition. There is some evidence, although not entirely conclusive, that *Panulirus* oxidizes proteins rather than carbohydrates to build up its energy stores.

Trehalose is the predominant sugar in insect hemolymph, also in concentrations that vary within wide limits according to the condition of the organism. Levels as high as 1,000 mg/100 ml (mmoles/kg) of fluid are not uncommon, and the very high value of 6,554 mg/100 ml (mmoles/kg) has been recorded for the larva of a species of solitary bee. Marked changes coincide with changes in developmental stage. In mature larvae of the moth *Hyalophora cecropia*, for example, the level of trehalose is about 2,000 mg/100 ml (mmoles/kg) of hemolymph, but falls to approximately 600 mg/100 ml (mmoles/kg) in early pupae and to 150 to 300 mg/100 ml (mmoles/kg) during diapause, and rises again when adult development begins. Caterpillars of the hawkmoth *Celerio euphorbiae*, on the other hand, have no detectable trehalose in their hemolymph, and the sugar appears in it only when pupation begins, reaching a level of about 1,500 mg/100 ml (mmoles/kg), and of about 1,000 mg/100 ml (mmoles/kg) throughout diapause, but falling off again with adult development.

Larvae of the blowfly *Phormia regina* are also apparently without trehalose, but it is present in the hemolymph of adults, with levels that rise and fall with feeding and with flight activity. Marked decreases in the amount of hemolymph trehalose have been recorded also for *Bombyx* after several days of starvation.

d. PIGMENTS OTHER THAN CHROMOPROTEINS Pigments other than chromoproteins may give distinctive colors to the body fluids. In many cases these are not carried in the fluid phase but are contained within cells suspended in it. Dissolved pigments have been described in insects that are, in several species, carotenoids which give the hemolymph a yellow or orange color or mixtures of carotenoids with bilins which are blue. The hemolymph of *Leptinotarsa*, for example, contains as much as 14 mg carotene/100 ml hemolymph, an amount equivalent to its concentration in the green leaves that the insect eats. The green color of the hemolymph of some species of grasshoppers and caterpillars results from a combination of carotenoids, derived from the food, and of bilins, whose source is as yet unknown. It may be that in such phytophagous insects bilins are derived from ingested chlorophyll through separation of the phytol group and opening out of the porphyrin ring, but it is also possible that they are wholly endogenous. Half a century ago sex differences in hemolymph colors were reported, when it was observed that in larvae and pupae of males of *Bombyx mori* the hemolymph was colorless, while in females it was distinctly golden; and that in males of the noctuid moth *Xanthia flavago*, it was a very pale yellow and in females distinctly greenish. The genetic basis for this has recently been shown in the jack pine and spruce budworms, *Choristoneura pinus* and *C. fumiferana*, in which interaction of autosomal and sex-linked genes affects accumulation of the blue pigment. There is less bilin in the males, whose hemolymph is, therefore, yellow rather than green.

Within the past decade pigments known as aphins have been obtained from the hemolymph of 20 or more species of the family Aphididae. These pigments, found exclusively in the hemolymph in concentrations of about 0.8 percent, are salts of a compound which has been extracted and given the name of protoaphin. The deep reddish-purple color of the compound in solution is due to the anion of the ionized salt. Protoaphin has been found to be a glucoside, a combination of glucose with a polycyclic quinone; such a pigment is unusual, even unique, among animals. When the insects die, or are crushed, protoaphin goes through a series of changes giving rise to a number of derivative pigments of different colors, which fluoresce, although protoaphin does not. The first of these changes is an enzymatic reaction in which the linkage with glucose is broken leaving an unstable bright yellow pigment called xanthoaphin, which becomes converted to an orange one and finally to a stable vermilion pigment called erythroaphin. Erythroaphin, besides being strongly fluorescent, has a photodynamic effect, for *Paramecium* are killed by light when the pigment is added to their culture medium. Its properties resemble those of hypericin, a pigment obtained from the plant *Hypericum* (St. John's wort), which is also photodynamic and has a similar polycyclic quinone moiety. These derivative pigments are postmortem changes in protoaphin, and there is no evidence that they occur in the living animal.

Aphins that have been isolated from different genera and species of aphids and coccids have been found to be of two isomeric forms, one corresponding to the protoaphin originally obtained from the black bean aphid *Aphis fabae* and so called protoaphin fb, and the other corresponding to that obtained from the willow aphid *Tuberolachnus salignus* and called protoaphin sl. Lanigerin, the pigment in the hemolymph of the woolly aphid *Eriosoma lanigerum*, and strobinin, in that of the scale insect *Adelges strobi*, or kermes bug, are both aphins of the fb series. The origin of the aphins is not known, but their glucosidic nature suggests that they may be detoxication products, formed by conjugation of glucose with some potentially destructive plant product that has been ingested. It has been demonstrated in *Aphis fabae* that neither mycetomes nor symbiotic fungi are responsible for the synthesis of protoaphin, establishing it as a product of the insect's own metabolism and not of symbiotic microorganisms as has been suspected.

3. Cellular components

a. TYPES Various types of cells are suspended in the fluid phase of the body fluids of most invertebrates. These are designated as coelomocytes or hemocytes depending upon whether the fluid occupies a pseudocoelomic or coelomic space or is confined within a vascular system, either open or closed. Morphologically, the cells fall into two very general categories of agranulocytes (hyaline cells) or granulocytes, with inclusions of various kinds. Some of these inclusions may be pigments that may or may not be chromoproteins, but that give the fluid a distinctive color and, if the epidermis is sufficiently transparent, similar coloration to the body as a whole.

Studies with the light microscope have revealed that within these two general categories there are many different types of cells that are distinguishable from each other by differences in size, shape, nature and density of the inclusions, degree of independent motility, and phagocytic capacity. Fourteen different kinds of coelomocytes, for example, have been recognized in echinoderms. And the hemocytes of larvae of the southern armyworm *Prodenia eridania*, which presents probably the most complicated blood picture of any insect or any invertebrate, have been grouped into 10 main classes, together containing 32 different types. Comparative studies of the coelomocytes in different classes of echinoderms and of hemocytes in the different families of insects have led to the inference that they are more diverse in the more highly specialized members of a group than in the less specialized ones. There are, for example, fewer types in asteroids than in echinoids and in orthopterans than in lepidopterans.

Because of their pleomorphism, an extensive

terminology for coelomocytes and hemocytes has grown up, which is not always descriptive and is often confusing, for different names have been given to what may well be the same type of cell in different, or sometimes even in the same, species. Some of the difficulties in nomenclature have arisen from differences in the methods used for preparing the fluids or their cells for microscopic examination. Classical methods of fixation and staining used in vertebrate hematology, when applied to invertebrate fluids, have doubtless led to artifacts which may have been interpreted as standard cell types. Other difficulties have arisen from the nature of the tissue itself and its sensitivity to changes in the body as a whole, which may be reflected in its cell populations as well as in the composition of the plasma. In times of stress, such as spawning, molting, or injury, cells of one type may be mobilized and become dominant members of the population while those of another may at the same time virtually disappear, or an entirely new type may arise. In animals like arthropods, which pass successively through critical stages as they molt and, like some insects, pupate, and metamorphose, cell counts show that different types reach their peaks at different periods. Therefore, unless the stage in the life cycle is clearly determined and designated, data on cell populations may be statistically erroneous and misleading. Also, there is scant information about the origin of these cells in most invertebrate species and of their progressive development, if any, during their life in the fluid. The multiplicity of cell types described in many cases could represent transitional stages in differentiation and senescence of one or two basic types. There is no proof of this for any invertebrate animal, since techniques have not yet been developed that would permit continuous observation of presumed stem or mother cells through progressive differentiation into mature types. The use of phase microscopy on living cells in freshly drawn fluids has clarified matters to some extent for some species and emphasized the dynamic nature of the cytological picture. Extended use of electron microscopy and of cytochemical procedures should prove of great value in comparative morphological and physiological studies and in the further clarification of the entire picture, which offers a field wide open to productive investigation.

Agranulocytes are usually cells with large nuclei enclosed in a relatively small volume of clear cytoplasm in which there may be one or more vacuoles of different sizes. They are generally ameboid and often phagocytic. Their shapes vary with the number and kind of pseudopod extended at any time, and their general appearance alters with the number and size of the vacuoles within them. The prevalence of amebocytes in invertebrate tissues is well known, but their functions in body fluids are by no means fully defined. They are found in the mesenchyme of sponges and cnidarians, in lacunae in the parenchyma of flatworms, and, as giant cells, in the pseudocoele of nematodes. They have been found in the blood of at least three genera of brachiopods, in that of lamellibranch molluscs, and in the hemolymph of crustaceans and insects. It is probable that they are conveyors of enzymes throughout the body and that some of them contribute to the clotting of the fluid seems certain.

Granulocytes, in contrast, are cells with a larger volume of cytoplasm and relatively small nuclei. They also may be ameboid and phagocytic, but they are regarded primarily as trephocytes and conveyors of nutrients from tissue to tissue. They may also engage in the synthesis of specific products some of which may be pigments, like the hemerythrins in the hemocytes of sipunculids and the hemoglobins of the polychaetes *Glycera* and *Magelona*, the echiuroids *Urechis* and *Thalassema*, and the lamellibranch *Arca*. In *Magelona* and *Thalassema* the hemocytes are essentially corpuscles rather than cells, for, like erythroplastids of mammalian blood, they are enucleate and presumably terminal stages of originally active, nucleated cells. Coelomocytes in the pseudocoele of *Priapulus* contain hemerythrin as do those in the coeloms of *Lingula*, sipunculids, *Thalassema*, and the polychaete *Terebella lapidaria*. The hemoglobin-containing coelomocytes of *Thalassema*, unlike its hemocytes, are nucleated cells. The hemoglobin of the coelomocytes of *Terebella* is different spectrophotometrically from that in the blood plasma and is probably of different molecular weight; its total quantity in the coelomocytes is about half that in the blood. Coelomocytes of *Amphitrite johnstoni* contain hemoglobin, as well as carotene dissolved in oil drops and a dark brown pigment collected in granular masses near the nucleus. This combination of pigments makes the cells a dark reddish, bright orange, or brown color, depending upon their relative proportions. Cells in the integument contain similar pigments, but the general color of the worm is derived from the coelomic fluid as well as from the skin, for that is sufficiently transparent to allow the underlying color to show through. The juxtanuclear brown pigment has been characterized as partly hematin—that is, heme with iron in the ferric state—and partly an iron-proteinate that, however, gives negative reactions to tests for ferritin. Hemoglobin and hematin are not always together in the same cells, and in animals collected off the southeast coast of England, the hematin content was highest in late autumn and early winter (October–December), at the end of the spawning season.

The pigments of echinoderm coelomocytes are of considerable interest. Hemoglobin has been identified in those of many species of holothurians; and echinochrome, the generic term for the red, purplish, and brownish naphthoquinone pigments characteristic of echinoids, in the eleocytes (Greek *elaio*, "oil") of sea urchins and sand dollars. This pigment is deposited in the cells of many of their tissues and in their exoskeletons, along with spinochromes, which are not found in eleocytes or other cells. In sea urchins of the widely distributed genus *Arbacia*, coelomocytes of this category may be so crowded with spherical masses of bright red pigment that the coelomic fluid, as it is drawn or exuded from a wound, looks like blood. Chemical analyses have shown that in *Arbacia* the pigment is in its oxidized state, and in *Echinus*, in its reduced state. In consequence the coelomocytes of *Echinus* are colorless while in the sea urchin's body, but turn red when exposed to air.

It is problematical whether any of these pigments have any essential role in the general physiology of the animals in which they have been found. They may be involved to some extent in respiration, a function that has been suspected of echinochrome but never demonstrated for it. Because of its chemical similarities with vitamin K, an antihemorrhagic factor for vertebrates, it has also been thought to be involved in the clotting of echinoid coelomic fluids, but tests to demonstrate its role in clot formation have yielded negative results.

The ameboid coelomocytes of the sea urchin *Diadema antillarum* Phillippi become dark when exposed to air because of the development of melanin in them. Both enzyme system and substrate for this biochemical event are present in the coelomocytes, but it does not take place in the perivisceral coelom because of an inhibitor in the coelomic fluid. It is postulated, however, that the dark pigment found in the epidermal chromatophores is derived from these cells when they have migrated from the coelom to the integument and so out of range of the inhibitor.

Synthesis of products other than enzymes or pigments may be characteristic of coelomocytes in general and has been quite clearly shown to occur in certain arthropods. Injection of valine labeled with C^{14} into pupae of the moth *Platysamia cecropia* has shown that there is a marked increase in the level of blood protein carrying the label at the termination of diapause, a time during which the synthesis of blood proteins is negligible. Protein synthesis is active in all tissues throughout postdiapause development, and the appearance of the labeled compounds in the hemolymph somewhat later than they can be localized in other tissues by radioautography suggests that they are synthesized in the hemocytes and released from them into the fluid. Certain hemocytes in other insects have been found to contain glycogen, mucopolysaccharides, phospholipids, and ascorbic acid, as well as enzymes such as acid phosphatase, ATP-ase, lipase, nonspecific esterase, and alkaline DNA-ase. In other arthropods, products may be synthesized specifically at the time of molting. Granulocytes in the hemolymph of spiders, which reach a peak in number and activity just before the new cuticle is laid down, are involved in the secretion of the exocuticle to which they contribute nonchitinous materials. Histochemical tests have shown that they contain proteins and phenols, which are probably operative in the hardening of the new cuticle by a process of tanning similar to that which takes place in insects. At the time of molting a new type of cell emerges, which has been given the name of leberidocyte (Greek *leberis*, "exuvia") and is believed to be differentiated at this period of the life cycle from the ever-present amebocytes. Leberidocytes contain glycogen but neither protein nor phenols in demonstrable quantities. In the Haitian tarantula, *Phormictopus cancerides*, they first appear about 1½ weeks before a molt, and at the time of molting constitute about 50 percent of the hemocyte population. They penetrate into the epithelium and there secrete chitin, or at least its polysaccharide component, and they disappear after the molt is completed.

Cells with similar activities correlated with molting have been observed in the blood of some crustaceans in which the number of granulocytes regularly decreases before a molt. In *Carcinus meanas*, some of these enlarge, multiply by mitosis, and differentiate into cells containing vacuoles, in which both lipid and protein can be histochemically visualized. These lipoprotein cells reach their maximum number toward the end of an instar, when the uncalcified layer of the new cuticle is being laid down. Although they are evident throughout the hemocoele, they are especially numerous beneath the epidermis and the tubules of the digestive glands. Their congregation in these regions and the nature of their vacuolar inclusions suggest that they are synthesizing materials, which they contribute to the construction of the nonchitinous part of the cuticle. Like the leberidocytes of spiders, they disappear from the hemolymph at the conclusion of a molt, possible reverting to amebocytes, which increase in number at this time, or possibly degenerating, so that, in the next cycle, new ones must be formed from a new generation of amebocytes.

In crustaceans and in insects a specialized type of hemocyte produces substances of unknown chemical nature known as coagulins, since their liberation from the cells causes gelation, or coagulation, of the fluid medium around them. These coagulocytes are particularly fragile and have been called explosive or eruptive cells because of their property of immediate cytolysis when exposed to air or to a solid surface at the site of a wound. They are regularly present in the blood and hemolymph and must be replaced when they have liberated their product and in so doing destroyed themselves.

b. FUNCTIONS Transport of nutrients, assimilated from sites of digestion or storage, is a very general and important function of coelomocytes and hemocytes alike. They provide a means by which considerable amounts of material can be carried in macromolecular or in solid form without appreciable alteration of the physical character of the fluid, which would inevitably occur if the compounds were in solution. The trephocytic nature of these cells has been deduced from histochemical analyses of their contents in representative species of at least the major invertebrate phyla, for they have been shown to contain glycogen and oils as well as proteids. Changes, associated with different phases of metabolic activity, in the numbers of cells with such inclusions provide evidence of their role in nutrient transport. In *Amphitrite johnstoni* and other terebellid polychaetes, for example, the number of coelomocytes increases in spring and early summer during the first stage of gametogenesis and decreases just before spawning, when the gametes are fully grown. Their principal function may, therefore, be assumed to be conveyance of nutrients to the sites of gamete formation. Eleocytes of the oligochaetes *Lumbricus terrestris* and *Eisenia foetida* also migrate to the tissues and give up material to them, as well as to developing gametes and to the fluid in which they lie.

Similar relationships between these trephocytic cells and periods of natural growth and of wound healing and regeneration have been observed in individuals of many other species in other phyla. Echinoderms may be exceptional in this respect for there is no conclusive evidence, in starfish at least, that the coelomocytes convey nutrients around the body, although they might well be presumed to do so. Transport in these animals, on present evidence, seems to be provided primarily by the fluid medium, in which there is at all times a low level of organic solutes. It is postulated that there is a continual flux of these solutes between the coelomic fluid and the various tissues of the body, each giving up and receiving equivalent amounts to and from the other.

Some coelomocytes and some hemocytes are phagocytic. The giant cells in *Ascaris* engulf bacteria injected into the pseudocoele and ingestion of invading bacteria by phagocytic coelomocytes and hemocytes has been observed in many other species. One important aspect of the activity of these phagocytic cells is scavenging, thus removing from the fluid cell debris originating from the cytolysis of senescent coelomocytes or hemocytes, disintegrating juvenile or wounded tissues, and gametocytes that for some reason have failed to grow and mature. The coelomocytes of the bamboo worms *Clymenella torquata* and *Euclymene oerstedi*, for example, become particularly active phagocytotically in the late autumn, when the breeding season for these polychaetes is over, and ingest and digest large numbers of disintegrating gametocytes, unshed gametes, and other gonadal debris. The demonstration of a strong acid phosphatase reaction and of lysosome-like bodies in the phagocytic hemocytes of prepupae of *Calliphora erythrocephala* give evidence of the digestive capacities of these cells. Hemocytes of insects collect around foreign bodies that get into the hemocoele by one means or another. Hemocytes of larvae of *Ephestia kuehniella* will, for example, encapsulate most naturally invading parasites and the living tissues of other insects, as well as filaments of polyethylene that are experimentally implanted in their hemocoeles. They do not, however encapsulate eggs of the ichneumon fly *Nemeritis canescens*, their habitual parasite, whose females thrust the ovipositor through the integument of the caterpillar and deposit their eggs in its hemocoele. This protection is given the eggs by a layer of material that overlies the chorion which prevents adhesion of the hemocytes; when eggs from which this layer has been removed are inserted experimentally into the hemocoele, they quickly become encapsulated.

Phagocytosis and encapsulation are essentially protective activities of tissue fluid cells, as is, likewise, their participation in coagulation or clotting of the fluid. All organisms are in constant danger of fluid loss, especially through injuries that expose their coelomic or vascular spaces. Plugging the opening by rapid solidification of the fluid is critical to survival, most particularly in animals where it is under some pressure and tends to gush rather than to seep out, and there are few species without some such protective mechanism. Cooperation of tissue fluid cells in coagulation has been established for those invertebrates in which the phenomenon has been studied. Response to injuries leading to the exposure of body fluids is exhibited usually only by certain cells, and the response is different in different organisms. There appear to be three basic mechanisms: (1) formation of the clot by cells alone; (2) formation of the clot by cells and fluid phase together; and (3) formation of the clot in the fluid phase alone.

In the first case, cells may form a solid mass by agglutination or a meshwork by extending pseudopodia, which make contact with those of adjacent cells and may even fuse to form a syncytium. The fluid and, in many instances, nonpseudopodial cells are trapped in the interstices of the mesh. These cellular masses or meshworks provide a barrier to further outward flow of the internal fluid. This type of clotting has been observed in *Limulus* and in some crustaceans. In the second case, the cells agglutinate or form meshworks, and the plasma coagulates upon release of substances from the coagulocytes. In crustaceans the coagulin or coagulins released have a direct and immediate effect upon the fibrinogen in the plasma, which results in its gelation. This type of clotting has been observed in spiders and in insects. In the third case, the clot is formed by gelation of the plasma alone, which occurs when coagulin-containing cells disintegrate at the site of injury and so release their contents. Islands of gelated plasma are thus formed, which coalesce into a solid mass. Although cells of other types may be embedded in this, there is no evidence that they contribute in any way to its formation. This kind of clotting has been observed in crustaceans and insects and in echinoids and holothurians. The explosive or eruptive cells of arthropods, which exhibit their extreme fragility when exposed to the external environment, are the ones involved in this kind of clotting; but in echinoids, it is hyaline amebocytes that undergo cytolysis and release the coagulating substance, and in holothurians, the coagulin-producing cells are identified as filiform amebocytes, cells which are also phagocytic.

Very little is known of the chemistry of the clotting of invertebrate fluids. The complexity of protein patterns of many of them, as revealed by electrophoresis, suggests that some of these proteins may be involved in the coagulation process which may prove to be as complex, in terms of factors involved, as that of mammalian blood. Yet, in spite of this complexity and of the presence of hemocytes, the hemolymph of certain insects does not clot, and they are, therefore, vulnerable to mechanical injuries. Except for water scorpions (Nepidae) and giant water bugs (Belostomatidae), none of the cryptocerate Hemiptera have hemolymph that clots, nor have rove beetles (Staphylinidae), water scavengers (Hydrophilidae), and some cantharid coleopterans. The hemolymph of *Apis* and *Bombus* does not clot, nor does that of the larvae of *Tendipes* and of several adult dipterans. In none of these do the hyaline hemocytes undergo any alterations when exposed to the external environment.

C. NUMBERS AND ORIGINS There is little information about the origins of the body fluid cells, but the fluctuations in numbers during the life of individuals of the species in which counts have been made or numbers estimated show that they are cells that age and disappear and are replaced or increased by new ones. In the wax moth *Galleria mellonella*, for example, the number of cells in the hemolymph ranges from 4,000 per mm^3 to 1,500 per mm^3 as a function of age, and the number of freely circulating cells is very nearly doubled in 6 to 24 hr after an insect has been bled. In crustaceans, too, where counts have been made, the numbers vary with the molting cycle; in *Cancer magister*, just before a molt the cell count ranges from 25,000 to 30,000 per mm^3, and just after a molt, from 6,000 to 7,000 per mm^3. The blood of *Carcinus maenas*, just before a molt, may have as many as 40,000 to 50,000 cells per mm^3, but counts approaching the 5 million erythrocytes per mm^3, typical of the blood of human males, have not yet been recorded for any invertebrate. In any such fluid, the number of cells, like the quantity of colloidal particles suspended in it, has direct relation to its viscosity, and ideally a balance must be maintained between the effective number of cells and the consistency of a circulating fluid, however it may be put, or kept, in motion.

Coelomocytes of polychaete annelids are budded off the peritoneum lining the coelomic cavity. In *Amphitrite johnstoni*, these may at times be so numerous that the coelomic fluid has the consistency of cream. It is probable that the turnover in these coelomocytes is annual; as the spent ones disintegrate they are replaced by new ones, provided by the coelomic epithelium in general, and undergo differentiation and transformation, while free in the coelomic fluid. In *Clymenella* and *Euclymene*, the sites of coelomocyte production are more localized, and there are especial regions in the somatic peritoneum overlying large blood capillaries in the anterior segments, and even more extensive ones at the posterior end of each trunk segment, which are active centers of mitosis and from which the newly formed cells are set free. In the oligochaetes *Lumbricus* and *Eisenia*, the ameboid coelomocytes originate from the septal and parietal peritoneum and the nonameboid eleocytes from the visceral peritoneum, which is modified as chlorogogue tissue. The first-stage eleocytes are, therefore, newly freed chlorogocytes, which develop into typical trephocytes as they float in the coelomic fluid. In echinoderms, the parietal peritoneum and the axial organ, a strand of loose tissue that runs alongside the stone canal, have been suspected of being sites of coelomocyte production, but there is little definitive evidence of their source.

Centers of hemocyte production have been described in crustaceans, as well as, in some species, multiplication of hemocytes by mitosis while they are in the circulating fluid. In malacostracans these centers are usually small irregular lobules of tissue located somewhere in the thorax. In the entomostracans *Artemia* and *Branchippus* they are located at the bases of each pair of trunk limbs, near the leg muscles and clusters of large phagocytic storage cells. Mitotic figures are abundant in the hemocytopoietic centers, but none has been observed in the circulating cells. It seems likely that in Crustacea all hemocytes have a common origin and that morphological differences in the free cells represent stages in the differentiation of a single, original type.

In insects, hemocytes arise during embryogenesis from a median, ventrally placed strand of mesoderm and are set free into the epineural sinus. In postembryonic stages they may multiply by division while in the hemocoele, and they may also originate by proliferation and differentiation of cells in hemocytopoietic centers. These may be located near, or even attached to, the fat bodies in the abdominal region. In some instances, they are more distinctly localized. In *Calliphora*, the centers are regions of the hypodermis near the posterior spiracles; in *Drosophila*, they are thin epithelial sacs hanging down into the hemocoele in the anterior region of the larva. In both Lepidoptera and Diptera, the time of principal hemocyte production is just before pupation or in its very early stages; in Coleoptera, the cells are produced during larval life, but not at pupation.

d. UNUSUAL CELLS Among the varied types of cells found in body fluids there are sometimes unusual ones that are distinctive in appearance and limited in distribution among species. Occasionally giant cells, two to three times the diameter of even the largest of the others, appear in the hemolymph of crustaceans. These are normal in the connective tissue, but in abnormal circumstances get into the hemolymph and circulate there.

The giant cells, given the name of teratocytes, sometimes found in the hemolymph of insects parasitized by braconid wasps, are actually cells of the embryonic membrane of the parasite and not endogenous ones of the host. When the wasp embryo hatches from an egg laid in the body of the host insect, it breaks through the embryonic membrane that enclosed it, and afterwards the cells of the membrane dissociate and are released into the hemolymph of the host, either singly or in clusters of a few cells each. Free in the hemolymph, the cells round up and grow as the larva of the parasite grows. Their nuclei enlarge and their cell bodies become filled with lipid and, in some species, glycogen. Their diameters increase as much as 12 times the original dimension, and cells as large as 200 to 250 μ in diameter have been reported in the hemolymph of ladybugs parasitized by *Periletus coccinellae*. These cells are gradually consumed by the parasitic larva, for which they seem the principal, perhaps the only, food.

A very remarkable cell, or cell association, has been reported in a Madagascan cockroach, *Gromphadorhina portentosa*. In over 300 smears of hemolymph, large, crescentic anucleate bodies, averaging 25 μ in diameter, were seen close to, or even enclosing in pseudopodial-like extensions, smaller nucleate cells whose diameters averaged 8 μ. These were usually found in pairs, either very close together, but separate, or in a position that suggested that the larger body was actually engulfing the smaller one. Histochemical tests showed that the cytoplasm of the large body was rich in polysaccharides but apparently without lipids or nucleic acids. When engulfed, the nucleus of the small cell enlarged and persisted, although there was no evidence of integration between it and the foreign cytoplasm that had incorporated it. However, crescentic masses that failed to acquire nuclei disintegrated, while those that became nucleated continued as organized systems. The origin of the anucleate crescentic and apparently phagocytic bodies is unknown, but their relation to the small hemocytes and presumed acquisition of cell status by incorporating their nuclei is a most unusual phenomenon, and perhaps a unique one.

A peculiar type of coelomic corpuscle has been found in sea urchins and in at least one species of sea cucumber, *Stichopus californicus*. This is a more or less spherical body, with a large nucleus and a single flagellum-like extension, originating from a basal granule. These cells have been characterized as vibratile corpuscles, and it has been suggested that they are not true coelomocytes but invading parasites of some kind. Autoradiographs of the coelomic fluid of *Strongylocentrotus purpuratus* injected with tritriated thymidine have shown, however, that labeled vibratile cells begin to appear 24 to 29 days after injection, although the amebocytes and eleocytes show activity much earlier. Some cells in the peritoneum, axial organ, and other tissues are also labeled and were, presumably, like the amebocytes and eleocytes, synthesizing DNA and dividing. The tardy appearance of labeled vibratile cells is interpreted as evidence of their development from dividing cells, most probably in the peritoneum. This interpretation would negate the idea that they are parasites and establish them as endogenous and true, though unusual, members of the coelomocyte population. It has also been suggested that they function in keeping the coelomic fluid in motion by lashing of their vibratile processes, and so are of value to an animal like a sea urchin whose rigid test prevents bodily movements that would bring about displacement of internal fluids.

C. MEANS OF TRANSPORT

1. Relations of tissue and body fluids

In acoelomate animals no distinction can be made between body fluids and tissue fluids that are in direct contact with the cells and so constitute their immediate microenvironment. But in other metazoans the body fluids are enclosed in compartments—pseudocoeles, coeloms, or vascular systems—and thus are separated from tissue fluids by at least one layer of epithelial cells. The extent to which the gastral canals of hydromedusae and of most scyphozoan cnidarians and of ctenophores can be considered analogous to a vascular system is certainly debatable, but since in cnidarians the fluid does circulate in a more or less definite course and carries nutrients to outlying cells and metabolic products away from them, analogy is perhaps justifiable. The fluid is propelled by cilia on the gastrodermal cells lining the stomach and canals and must also be displaced to some extent by pulsations of the bell. In the moon jelly *Aurelia aurita* its course is from the gastral cavity through the unbranched adradial canals to the circular canal, from which it returns directly to the gastral cavity through the branched perradial canals and indirectly, via the gastric pouches, through the interradial canals (Fig. X–2). The oral surface, or floor, of each of the four arms of the gastral cavity that expand into the gastric pouches, has two ridges along its length, delimiting a central inhalant channel from two lateral exhalant ones. The gastrovascular fluid enters the adradial canals through the lateral channels and returns from the interradial canals through the central one. The entire course of circulation, which can be followed and measured by

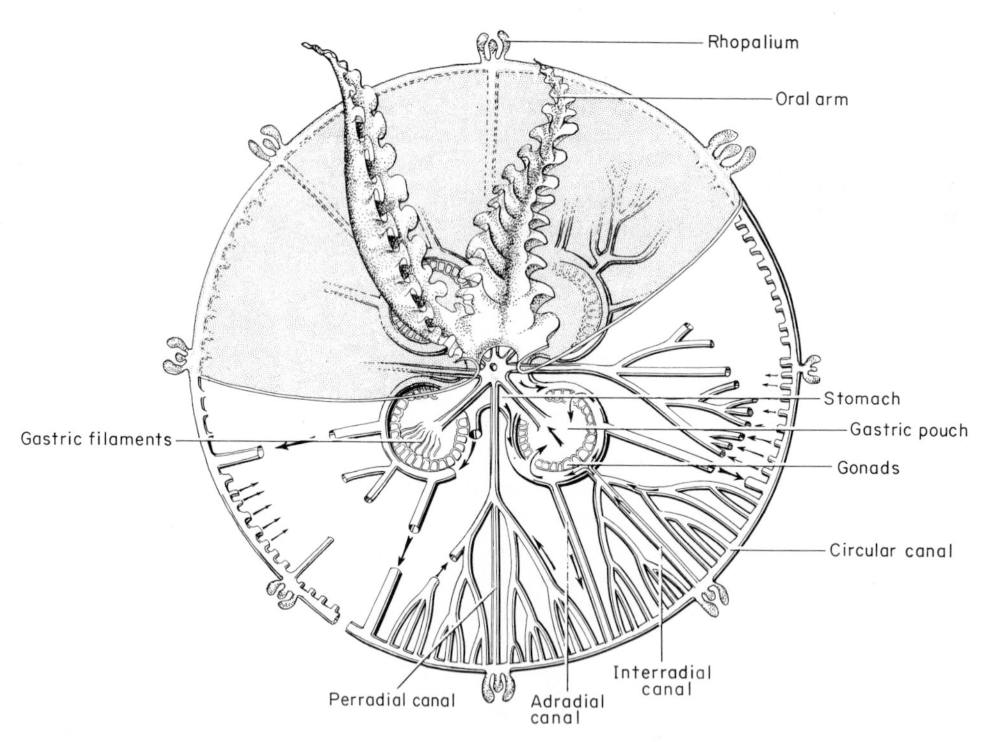

FIG X–2 *The canalicular system of* Aurelia aurita, *from the oral surface. The upper half of the diagram shows the exterior of the animal with the canals beneath the epidermis; the lower half shows the interior as if the epidermis were cut away. The complete system of canals is represented in only one quadrant. Arrows indicate the direction of flow of the fluid. In one quadrant, a portion of the gastric pouch is shown in cross section to reveal the inhalant and exhalant channels.*

injection of a dye or suitably tagged molecule, is accomplished in about 20 min, and the fluid that has made the passage and returned to the gastral cavity is swept out of the mouth and away from the body by the beating of cilia along median grooves on the four oral lobes that surround the mouth. In ctenophores, the canals leading from the gastral cavity end blindly so that a continuous circulation is not possible, but the fluid is kept in motion by ciliated cells on the distal (outer) sides of the canals.

In flatworms, extracellular fluid bathes the parenchyma cells and the organs embedded in them and is moved about by the bodily movements of the animal and, to some extent, by physicochemical events within it, such as diffusion and active transport across membranes.

2. Movement of pseudocoelomic and coelomic fluids

In soft-bodied animals such as nematodes, nemerteans, and annelids the fluids in the pseudocoele or coelom are kept in motion, although nondirected, by movements of the body as a whole and by those of the internal organs, primarily the digestive tract. These fluids serve also as a hydrostatic skeleton and their gross displacement is instrumental in locomotion. Minor displacement may be caused by the activity of excretory mechanisms, particularly by nephridia like those of annelids that have coelomostomes through which the coelomic contents are swept by ciliary action through tubules to the exterior.

Echinoderms are unique among coelomate animals in having a coelom that is divided into distinct cavities, typically five in number, each serving a different purpose. This separation, which is quite different from the septate partitioning of the coelomic cavity in annelids, occurs early in echinoderm embryology and in consequence of it the original single enterocoele is divided into a perivisceral coelom surrounding the digestive and reproductive organs; a hydrocoele that becomes the water vascular coelom with circumoral and radial canals and tube feet; an axocoele, enclosing the so-called "axial gland"; a perihaemal system that lies parallel to the water vascular system; and five pairs of genital sinuses. All of these are lined with a delicate peritoneum. The coelomic spaces are much reduced in size in ophiuroids, and the perihaemal system is best developed in holothurians and least so in asteroids. In this class, it is represented by an ill-defined space beneath the canals of the water vascular system, traversed by a strand of tissue called the hemal strand. The coelomic fluids are kept in motion by ciliary activity of the peritoneal cells and by movements of the tube feet. In soft-bodied holothurians, movements of the body wall and of the gut also bring about displacements of the fluid in the perivisceral coelom. In echinoids, activity of the muscles of Aristotle's lantern moves the fluid in the peripharyngeal or lantern coelom, a special division of the perivisceral coelom that surrounds the lantern. Recent evidence indicates that propulsive force is given to the coelomic fluids in general, in sea urchins at least, by contractions of the axial gland. The nature of this gland, or organ, has puzzled invertebrate physiologists for many years and a variety of functions has been attributed to it, in accordance with which it has been given a variety of more or less descriptive names, such as heart, kidney, brown gland, ovoid body, dorsal organ, and septal gland. Anatomically, it is a soft, elongated structure, usually brownish or purplish in color, that lies near the stone canal connecting the madreporite with the circular, or ring, canal of the water vascular system. In *Strongylocentrotus purpuratus*, the gland is hollow; in cross section, one half of its wall appears thin and membranous, and the other half much thicker and composed of fairly dense tissue interspersed with muscle cells. Its lumen communicates with that of the stone canal, and so with the water vascular system, and with that of the axocoele in which it lies, and with the perivisceral coelom through a narrow slit 2 to 4 mm long. Within the lumen of the gland, and extending along its length, is a tubular strand of tissue that sends numerous fine branches into the dense tissue of the wall. Aborally, this tubule, or vessel, ends in an ampulla that lies directly under the madreporite and is divided into two chambers. In living animals both the chambers and the vessel can be seen to pulsate at a rate of 4 to 8 beats/min. The beat begins in the distal of the two chambers of the ampulla; contraction of the next chamber follows and spreads in a peristaltic wave along the vessel. The surface of the gland is covered by a network of vascular tissue which comes together in a single strand that runs to the perioesophageal ring, from which another vessel arises that runs along the surface of the oesophagus and eventually becomes the inner hemal sinus, passing along the surface of the digestive tract and terminating, at one side of the rectum, in another pulsating chamber.

The axial gland thus represents, anatomically, a meeting point of the perivisceral coelom, the water vascular system, and the hemal system, and, functionally, a central contractile organ whose pulsations move coelomic fluid from the perivisceral coelom throughout the hemal system. Whether this is its sole function and whether it functions in the same way in other echinoderms are questions still to be answered. Since some echinoids can survive its removal, at least for several months, it may not be essential to the displacement of the coelomic fluids, for which the ciliary mechanism alone may suffice. The axial gland has also been found to be a site for disintegration of old amebocytes and for phagocytosis of cell debris and of foreign material injected into the perivisceral cavity or invading it by natural means. However it may come about, much importance is attached to the movement of coelomic fluids in echinoderms, since these provide the only means of transport throughout their complex and often large bodies. The intercommunications of the coelomic spaces demonstrated in echinoids provide for access of these fluids to the tissues with which the perivisceral fluid does not come directly in contact, and a means by which it and the solutes it gives and receives can be carried throughout the body.

3. Movement of vascular fluids

a. VASCULAR SYSTEMS Vascular fluids may be defined as those fluids that are contained in, and move through, a system of tubes and tubules derived from the primary body cavity and so distinct from the secondary body cavity or coelom. In general, the system of tubes consists of one or more longitudinal trunks, running parallel to the long axis of the body and giving off branches along their course. These branches may connect directly with other trunks or may divide and subdivide into vessels with thinner and thinner walls and smaller and smaller bores until they reach capillary dimensions. These smaller vessels may end blindly, as they do in serpulid worms and in phoronids, or they may reunite into larger vessels so that the system is a continuous one. Or the peripheral vessels branching off the central trunks may lead directly, or after branching a few or several times, into blood spaces or sinuses that replace the capillary beds. These spaces may be so large, as they are in arthropods and molluscs, that they occlude most of the true body cavity or coelom. The perivisceral space in these animals is thus a hemocoele, filled with blood or hemolymph, and not a coelom, filled with coelomic fluid and bounded by peritoneum. Whatever the pattern, propulsive force is given to the fluid contained within a vascular system by contraction of the walls of its vessels, a contraction that may spread as a peristaltic wave along their length or be localized in one or several areas which, therefore, perform the same function as a vertebrate heart.

Although nemerteans are acoelomate animals, with parenchyma that fills the space between the body wall and internal organs as it does in flatworms, they have a true vascular system and the fluid within it never comes in direct contact with the cells. In its simplest form, as it is found in the littoral paleonemertean *Cephalothrix*, it consists of two longitudinal vessels, one along each side of the digestive tract, that meet anteriorly and posteriorly in lacunar spaces in the parenchyma that are lined, like the sinuses of an open system, by a delicate membrane (Fig. X–3). Branches from these central vessels ramify among the tissues. The walls of the larger vessels are contractile, but the flow is not directed and its course may be first in one direction, then in another, chiefly in accordance with the bodily movements of the animal There are various elaborations of this basic plan among other genera and species in the phylum; in the heteronemertean *Lineus*, for example, there is a central median vessel on the dorsal surface of the digestive tract as well as the two lateral ones, which are connected with the dorsal by transverse vessels that run as half hoops around the digestive tract between its diverticula.

Annelids probably best exemplify an extensive and closed circulatory system with, again, various modifica-

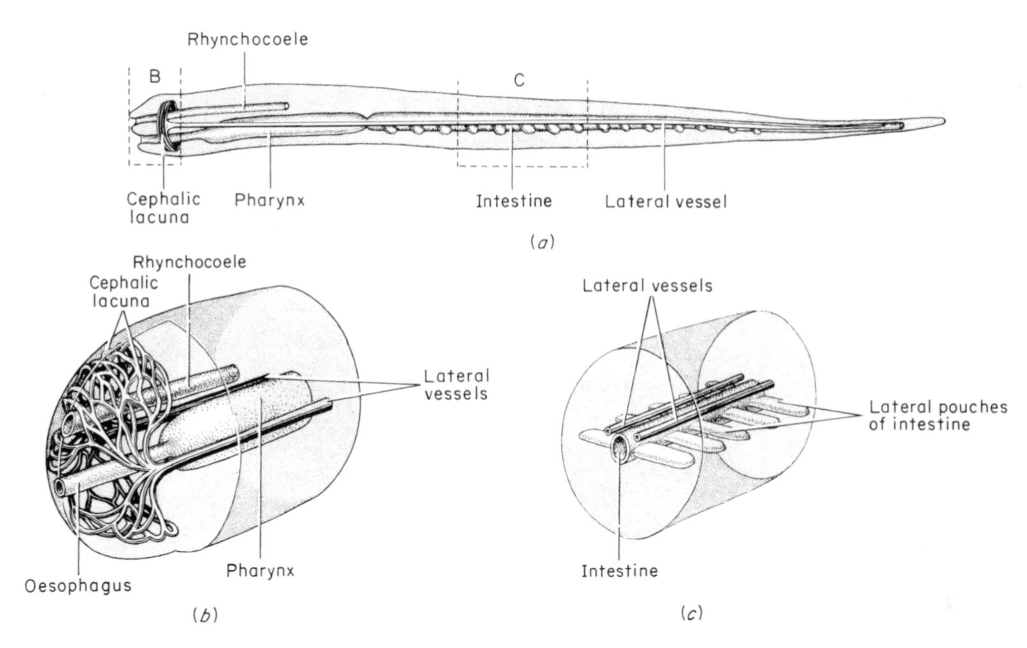

FIG. X–3 *The circulatory system typical of a nemertean.* ***(a)*** *Lateral view.* ***(b)*** *Enlargement of section B in a.* ***(c)*** *Enlargement of section C in a.*

tions of a general pattern (Figs. X–4 and X–5). This pattern consists of medianly placed longitudinal vessels, one dorsal to the digestive tract and the other ventral, connected at the anterior and posterior ends of the body by capillary networks that spread through the integument and wall of the gut. There may also be segmentally distributed transverse vessels between the dorsal and ventral aortae all along the length of the body or in certain areas of it only. There are no transverse vessels in sabellid and serpulid polychaetes, which are unique among annelids in that their blindly ending peripheral vessels are alternately full and empty as they receive blood from the central vessels and return it to them through the same channels. In other polychaetes and in oligochaetes capillary systems in the tissues provide connections between the larger vessels so that the blood circulates completely around the body. In *Arenicola* and in some other genera an extensive sinus partially or completely surrounds the gut instead of a network of small vessels and capillaries. This sinus usually connects with the ventral aorta by a single pair of vessels. Such a pattern is possibly a survival from that in primitive annelids, where there was probably a general gut sinus from which, in the course of annelid evolution, dorsal and ventral vessels became separated. In sabellids and serpulids there is no dorsal vessel in regions where the gut sinus is well developed.

The vascular system in rhynchobdellid leeches follows the pattern of that in oligochaetes, but in gnathobdellids and pharyngobdellids there is no true blood system. Its functions are performed by the intercommunicating coelomic sinuses characteristic of all Hirudinea. These have been created by the development of the botryoidal tissue and its invasion into the coelomic cavity so that all that remains of the typical annelidan coelom is a regular system of spaces and channels filled with coelomic fluid in which coelomocytes are suspended. This system is auxiliary to the blood system in rhynchobdellids, in which the coelomic fluid, containing hemoglobin in solution, is the equivalent in them of the blood of other annelids. It does not circulate in a directed course, but is moved irregularly by general body movements.

The walls of the larger vessels contain muscle fibers and the blood is pushed along as they contract

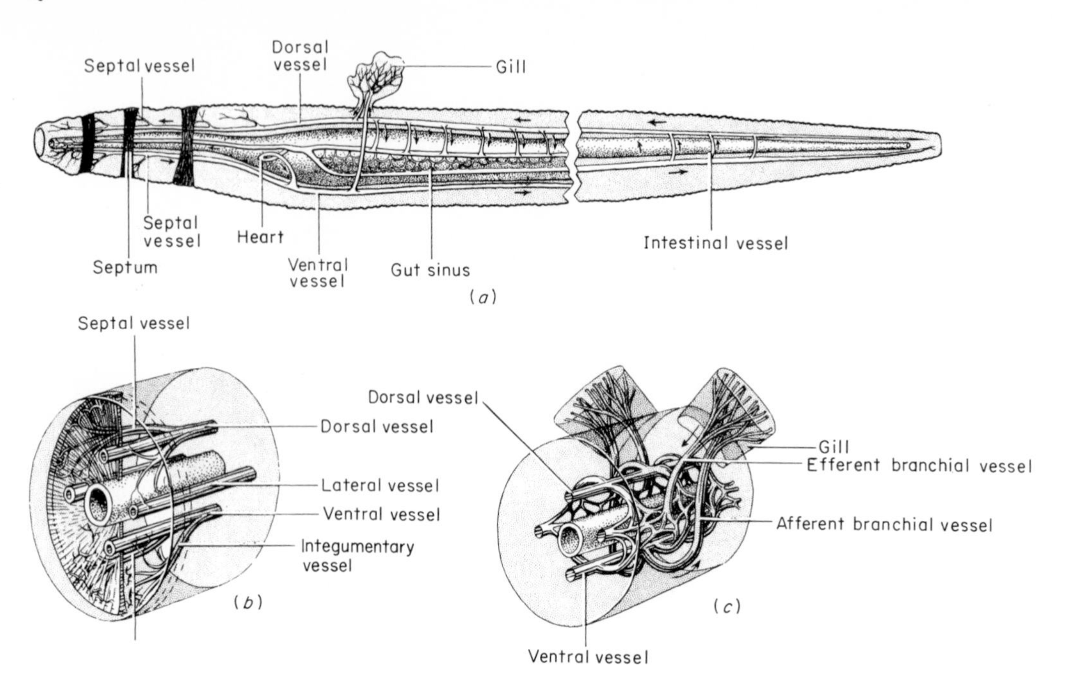

FIG. X–4 *The circulatory system of* Arenicola. ***(a)*** *Lateral view; only one gill is shown.* ***(b)*** *Section through anterior region showing a septum. The septum and septal circulation are shown in the left half of the diagram only and as if cut away in the right half.* ***(c)*** *Section through a region with a gill, showing the gill circulation.*

peristaltically. In most worms there are additional pumping stations, or hearts, with walls that are distensible and particularly muscular so that they expand as blood flows into them and contract to force it out. In all worms, except sabellids and serpulids, the course of flow is directed, the blood passing from posterior to anterior in the dorsal vessel and from anterior to posterior in the ventral. Branches from these vessels carry blood to, or receive it from, them and there is no separation of that which is especially rich in nutrients or metabolites and that from which these substances have been reduced or increased in amount as it passes through the tissues. With a few exceptions, no distinction can be made between arteries and veins in invertebrate circulatory systems, nor between arterial and venous blood. This has been a major obstacle to the study of blood physiology in this group of animals.

b. PUMPING CENTERS (HEARTS) The situation is improved, from the point of view of the physiologists at least, in molluscs, in which there is a central heart, dorsally placed and usually clearly divided into a pumping ventricle and one or more receiving auricles (Fig. X–6). The heart lies in one of the persisting regions of the coelom, the pericardial cavity, and, in primitive molluscs, is presumed to have had one ventricle, anterior in position, and two auricles posterior to it. A single main vessel, or aorta, led off from the ventricle and passed anteriorly out of the pericardial cavity to divide into branches that opened into large blood sinuses encompassing the various organs. The blood drained from these sinuses into channels which ran along the axes of the paired gills. Blood from these afferent vessels diffused into the gill filaments, where it came very close to the water in the mantle cavity and was collected in an efferent channel in each gill, which ran parallel to the afferent channel, and from it returned to the auricle of that side. A similar pattern exists in polyplacophorans and aplacophorans, in which the heart has but two auricles although receiving blood from numerous pairs of gills.

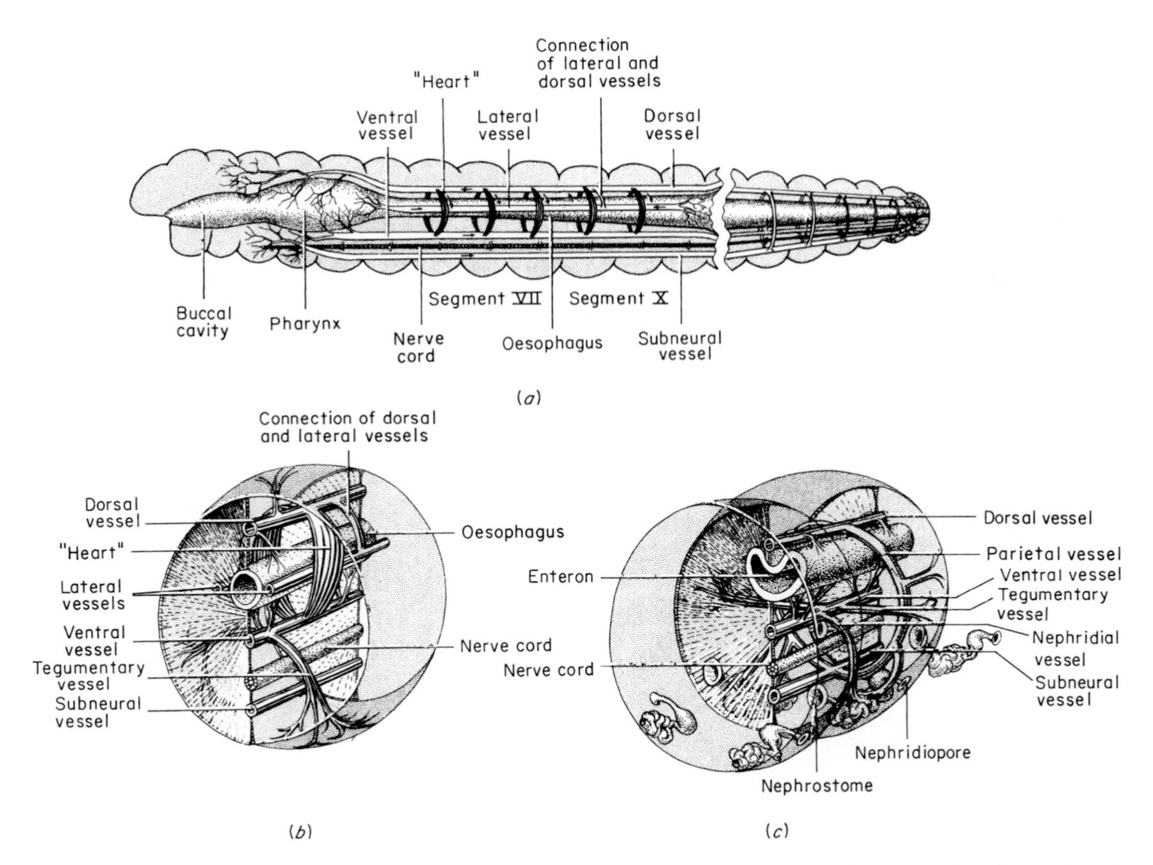

FIG. X–5 *The circulatory system of* Lumbricus. ***(a)*** *Lateral view.* ***(b)*** *Section through the region of the hearts (segments VII to XI). The two septa shown are intact on the left side only and represented as if cut away on the right side.* ***(c)*** *Section through a posterior segment, with the septa represented as they are in (b). Three pairs of nephridia are shown.*

Torsion in gastropods has altered the position of the heart and of the auricles and ventricles in respect to each other and, in some, has resulted in the suppression and loss of the right auricle as well as of the right gill. In pelecypods, the ventricle has become folded around the intestine in such a way as to make it appear that the gut ran through that part of the heart as well as through the pericardial cavity, and, in eulammellibranchs, a posterior aorta as well as an anterior one arises from the ventricle. In these molluscs, in which the mantle has almost entirely replaced the gills as an area of O_2—CO_2 exchange, branches of the anterior aorta carry blood to it, and the blood is returned from each lobe directly to the auricle of the same side. As a further modification of the primitive pattern, in all modern molluscs blood passes through the glandular portion of the excretory organs before going to the gills, so that the blood that returns to the auricles is different in composition from that anywhere else in the body. But the blood, even though moving slowly through the expanded areas of the sinuses, follows a definite course and makes a complete circuit through the body. Its main propulsive force is imparted by contraction of the ventricle, but in many gastropods and pelecypods there are accessory or "booster" pumping stations located along the main vessels.

The pattern is even more modified in cephalopods, in which there is a more extensive system of branching

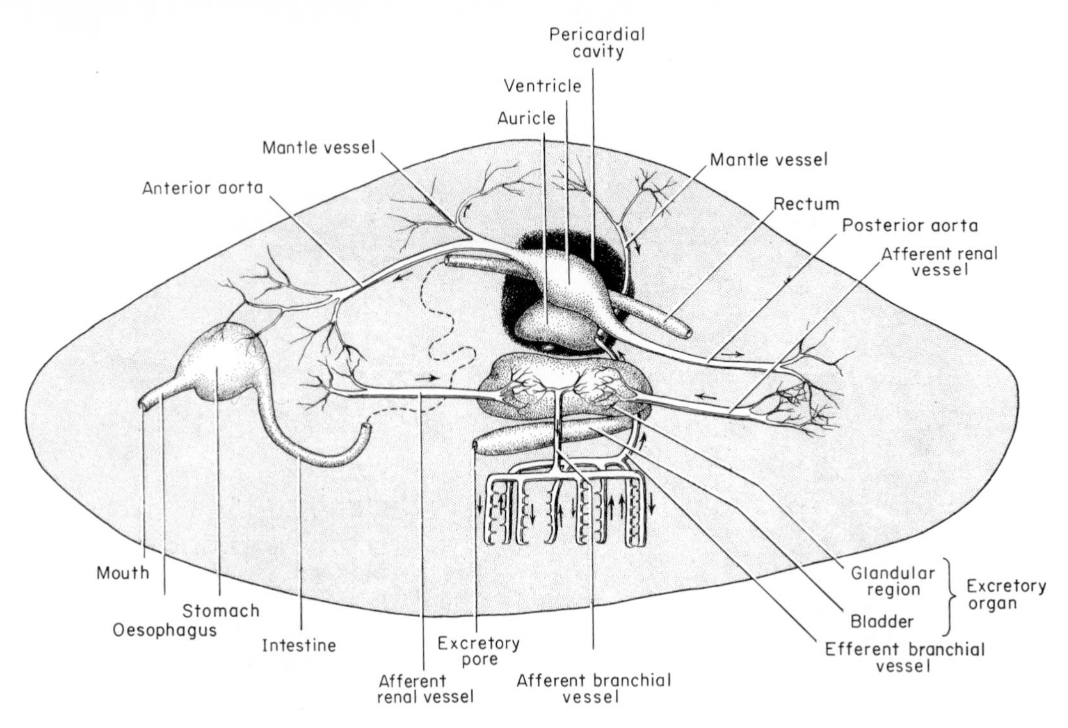

FIG. X–6 *The circulatory system of a lamellibranch mollusc* (Anodonta). *Arrows indicate the direction of flow.*

vessels than in any other of the molluscs (Fig. X–7). There is also a more complete and effective separation of oxygenated from nonoxygenated blood than there is in members of the other classes. This has been effected through the development of paired branchial hearts, lying at the bases of the gills lateral to the median systemic heart. Each branchial heart receives blood from all parts of the body; that from the anterior region passes through the excretory organs before entering one, but blood from the mantle and posterior viscera flow directly into each through mantle and abdominal veins on each side. On contraction of each branchial heart, blood is pushed into the afferent vessel of the gill on the same side and is distributed to capillaries in the filaments. It is collected from these capillaries into an efferent vessel, which empties into the corresponding auricle of the systemic heart and from the auricles passes into the ventricle. Contraction of the ventricle pumps blood into an anterior and a posterior aorta and so on its course throughout the body. The branchial hearts act not only as receptors of what may properly be called venous blood but also as auxiliary pumps to drive blood into the gills against the resistance of the capillary circulation in them. Microaspiration of blood from a branchial heart or from an afferent gill vessel will, therefore, provide a sample of blood different in quality from that taken from an afferent vessel or from the systemic heart and give the physiologist material with which to study changes in its composition that may occur in its passages through the viscera and the gills.

The vascular system of xiphosurans, arachnids, and arthropods is also an open one (Figs. X–8 and X–9). The main vessel is dorsal in position and may pump the blood by peristaltic contractions along its length or in a localized area or heart. The heart lies in a pericardial cavity which is not, as it is in molluscs, a coelomic space, but a part of the sinus system or hemocoele and receives blood from this cavity through one or several pairs of ostia. Thus, no distinction can be made between ventricle and auricles. The system is comparatively well developed in xiphosurans,

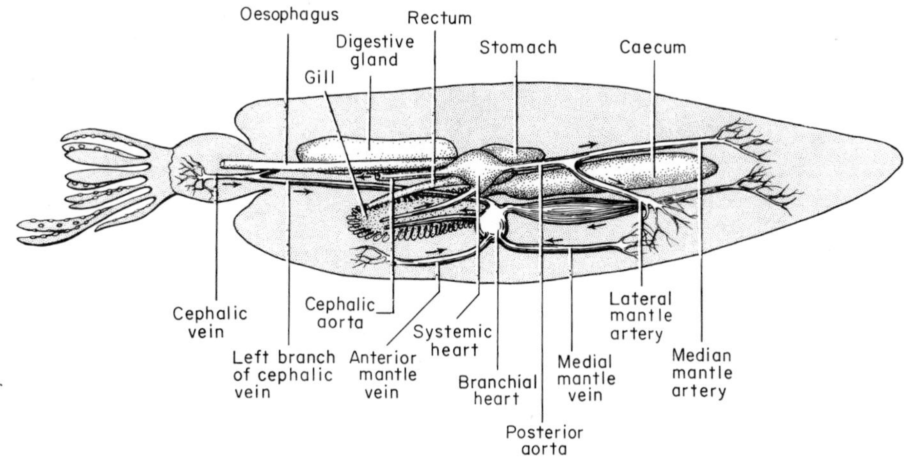

FIG. X–7 *The circulatory system of a cephalopod mollusc* (Loligo). *Arrows indicate the direction of flow.*

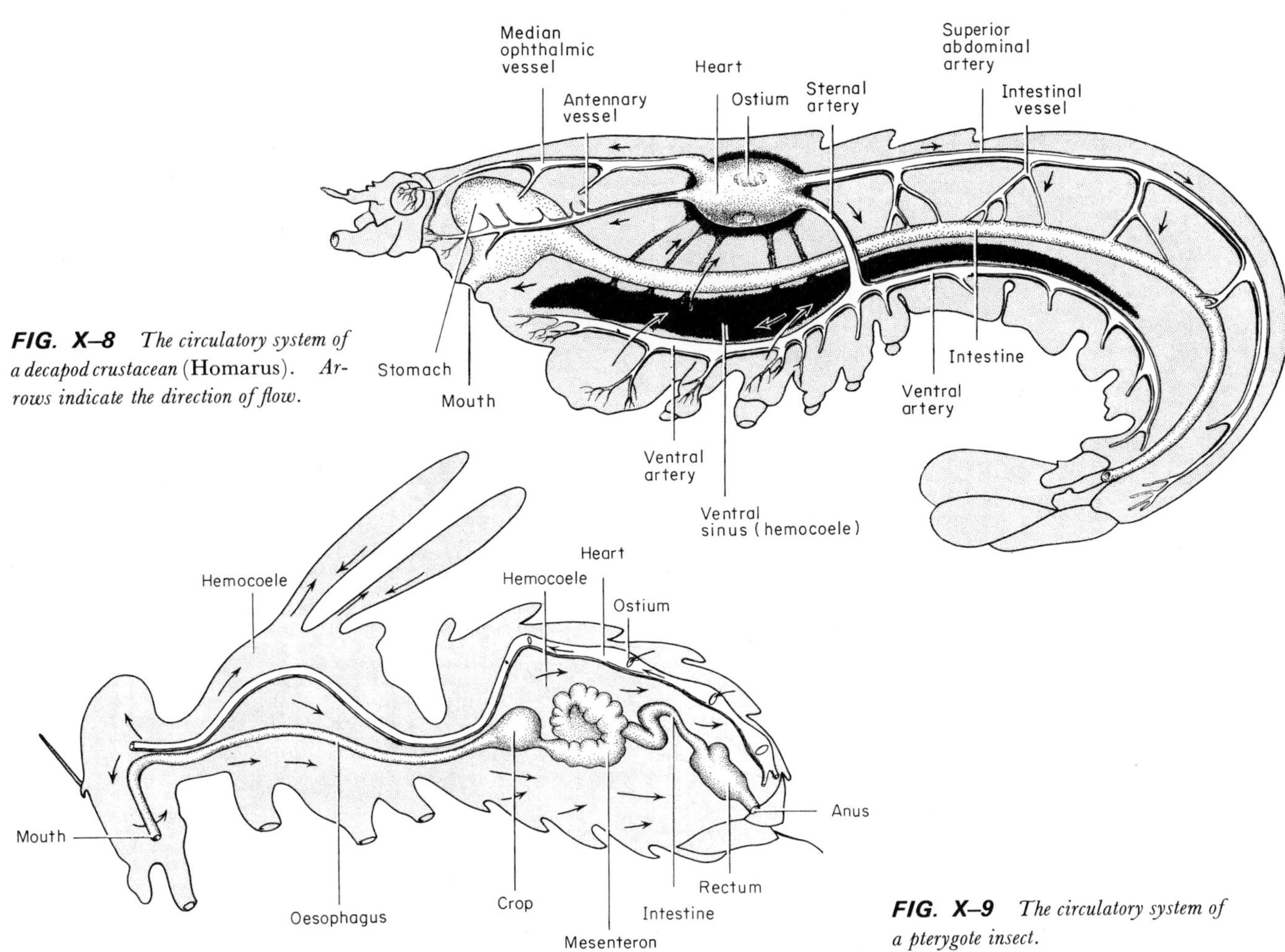

FIG. X–8 *The circulatory system of a decapod crustacean* (Homarus). *Arrows indicate the direction of flow.*

FIG. X–9 *The circulatory system of a pterygote insect.*

arachnids, and decapod crustaceans; poorly so in insects; and not at all in some entomostracan crustaceans like most ostracods, some copepods, and all cirripedes. In insects, it is represented only by a slender tube with delicate walls, lying along the inner surface of the dorsal body wall and opening into the spacious hemocoele by paired ostia, usually segmentally distributed in the abdominal region.

The heart of *Xiphosura* is a long tubular organ extending most of the length of the gut. Three anterior and four pairs of lateral vessels originate from it, all of which divide, with their branches subdividing, to make an extensive system of tubules throughout the tissues. The smallest subdivisions open into sinuses within the tissues which communicate with two large ventrolateral blood spaces. From these the blood passes into the gills and from them into the pericardial cavity and back into the heart through eight pairs of ostia. The ostia are guarded by valves so that flow into the pericardial cavity is prevented when the heart contracts.

The heart in arachnids is likewise a dorsal tube lying in a pericardial cavity but connected with it by only a single pair of ostia. From the heart a large vessel conducts the blood anteriorly through the prosoma and a somewhat smaller one carries it backward to the posterior abdominal segments. Small vessels lead directly from the heart to the anterior abdominal segments in which the heart itself lies. These main vessels subdivide and their smaller branches terminate, as in *Xiphosura*, in tissue spaces from which the blood collects in a ventral sinus that encloses the book lungs, the area of principal gaseous exchange. From this sinus and from the lungs the blood returns to the pericardial cavity to be recirculated when the heart muscle contracts.

The patterns of crustacean circulatory systems range from none at all to fairly complicated tubular systems with capillary beds as well as sinus spaces. In cladocerans the heart is either sac-like or tubular, and the only defined vessel is a short one leading anteriorly from it. This does not branch, and though the blood seems to follow a prescribed course it does so through sinuses and not through vessels. This is essentially the condition characteristic of insects. In isopods, there is a well-developed system of vessels originating from an anterior and several lateral vessels that arise from the heart to which the blood returns, as it does in other arthropods, from the pericardial cavity which receives it from the large ventral sinus. In decapods, the basic pattern is modified by the addition of several other main vessels and a true capillary network around the cephalic ganglion and the ventral nerve cord. The large vessels into which the capillaries reunite after ramification through the nervous tissue empty into the ventral sinus as do the terminations of the vessels that originate from the main trunks, and from this sinus the blood flows into the pericardial cavity and back to the heart.

These various patterns of circulation impose different conditions upon the propulsive organ or areas. Less resistance to flow is encountered in an open system than in a closed one, for the small size of capillaries and the extent of the capillary beds are principal factors in the obstruction to flow. Tissue and body sinuses offer less resistance and put less demand upon the muscles that pump the blood along its course. Very little is known about the nature and operation of pumping mechanisms in invertebrates other than molluscs, xiphosurans, and crustaceans in which the pump is centralized. Even in these groups information is far from complete and in some cases what there is is contradictory.

The rhythmicity and regularity of contraction in heart muscle has long been recognized and long presented problems to physiologists. Invertebrate hearts, like those of vertebrates, beat at rates characteristic of different species, but the rates of any can, in the intact and living animal, be altered, within limits, by conditions affecting the animal as a whole, such as temperature, activity, and other forms of stress. The rate of heartbeat in clams, for example, ranges from 0.2 to 22 strokes/min, and of squids and octopuses from 30 to 60 strokes/min. At 20°C, the rate in *Daphnia* is 240 to 450 strokes/min, and in the crayfish, 30 to 60 strokes/min. In Sphinx moths, the heart beats at a rate of 40 to 50 strokes/min when the insect is at rest, but at a rate of 110 to 140 strokes/min when it is in flight.

That the property of rhythmic contraction is inherent in heart muscle has also been known for a long time, but the means by which the timing of the contractions is regulated present many as yet unresolved problems. Current evidence indicates that in molluscs, insects, and echinoderms, as in vertebrates, the controlling mechanism, or pacemaker, is myogenic, originating in the muscles themselves; while in xiphosurans, arachnids, and crustaceans, it is neurogenic, originating in ganglion cells. Distinctions between the nature and the sources of the controlling mechanism can be made by a number of different methods whose combined data can provide reliable bases for judgment. Histological examination of the heart itself will tell whether there are neurons in its walls and, if there are, their extirpation, if this is possible, should affect the heart's action if the pacemaker is neurogenic. Records of the electrical waves of excitation that accompany the passage of a nervous impulse or a muscular contraction show differences between neurogenic and myogenic pacemakers, for the electrocardiogram of a neurogenic heart shows waves that are oscillatory, while that of a myogenic one shows waves that are large and slow. Application of certain drugs has also been used to give evidence of differences in the mechanisms. Acetylcholine is known to stimulate ganglion cells and, therefore, should have no effect upon a myogenic pacemaker, either in isolated hearts or in those *in situ* whose extrinsic nerve supply has been cut, and it would be expected to inhibit those whose nerve supply is intact. Ether in low concentrations has an opposite effect, inhibiting the responses of ganglion cells; and adrenalin and its analogues, which act as transmitters of the nervous impulse to smooth muscle fibers particularly, should affect the action of hearts whose pacemakers are myogenic. Pharmacological evidence, derived from experiments with such drugs, is even at best somewhat dubious, since there are so many variables in methods of their introduction into the body and in conditions inside and outside the cardiac system.

Histological examination of molluscan hearts has shown that in some gastropods, and in *Pecten* and a few other bivalves, neurons are distributed among the smooth and striated muscles of the heart wall. They have not been seen in the other species of gastropods and pelecypods studied and, indeed, not by some investigators even in those species in which others have reported them. In nudibranchs and the prosobranch *Fulgur* there are distinctly recognizable ganglia in the walls of the main vessels, and in cephalopods there is a ganglion in the tissue of the heart itself. The heart in all molluscs is innervated by fibers that originate in the visceral ganglion or in the cerebro-visceral commissures. Yet all these extrinsic nerves can be cut, and the cardiac ganglion in cephalopods removed, without significant alteration of the characteristic heartbeat. Moreover, isolated fragments of heart muscle continue to beat, when kept under optimal experimental conditions, in time with intact hearts.

All experimental evidence indicates that the pacemaker of the molluscan heart is myogenic, with the possible exception of *Fulgur*, in which it may be neurogenic. Gastropod, pelecypod, and cephalopod hearts, with distinct auricles and ventricle, probably work in the following way. As the auricles fill with blood they are distended and, at a certain threshold of extension, all the muscle fibers in them contract simultaneously, forcing blood into the ventricle. The muscles of the ventricle, when they are stretched by its distention, contract in a peristaltic wave. Blood is thus pushed out into the main vessels and does not reflux into the auricles, even though, except in cephalopods, there are no valves to prevent this. The hydrostatic pressure set up by contraction of the ventricle varies in different species. Measurements have shown that in *Anodonta* it is equivalent to that of 20 mm H_2O, but as great as 40 mm Hg in *Aplysia* and 80 mm Hg in cephalopods. Filling of the heart in diastole is believed to depend upon a pressure mechanism within the pericardial cavity. As the heart is fixed at both ends because the walls of its vessels and the pericardium grow together, its length is determined by the length of the pericardial cavity and its volume must remain constant. When the ventricle shortens in systole, the auricles must be stretched to a compensating length, and, conversely, as the auricles shorten, the ventricle must lengthen. The amount of fluid in the pericardial cavity is also fixed, for the only external connections of this chamber are small tubules communicating with the excretory organs, also coelomic spaces. Cilia in these channels beat in a direction that would prevent the entrance of any appreciable amount of pericardial fluid, and accordingly its escape from the pericardial cavity. When the ventricle contracts there is more area in the pericardial cavity to be filled by a constant volume of fluid, and, in consequence, suction is exerted on the walls of the auricles causing them to expand and to draw in blood from the efferent channels that open into them. Contractions of the auricles and ventricle must, therefore, be so timed that they alternate with a precision such that the total volume of blood in the heart is always the same. The heart is never entirely empty nor wholly full, and there is neither diastolic nor systolic arrest. Present evidence points to the fact that the controlling mechanism is a diffuse, not centralized, one and that all the muscle fibers in the heart possess pacemaker properties and are automatically capable of successive contractions that are not initiated by nerve impulses. The evocative agent is presumed to be a metabolite produced by the cells themselves to which the muscle fibers are sensitive when relaxed but to which they have an even lower threshold when stretched.

This self-regulated myogenic rhythm is, however, influenced by extrinsic nerves. Stimulation of certain fibers running to the heart slows down or even stops its beat, while that of other fibers accelerates it. Results of pharmacological experiments have led to the conclusion that action of the inhibitory fibers is mediated by acetylcholine, while that of the accelerating fibers may be mediated by adrenalin or adrenalin-like substances.

Conditions in the hearts of xiphosurans, arachnids, and crustaceans appear to be just the opposite. In these, as in the molluscan and vertebrate heart, the pacemaker is built in, for transection of nerves running to the cardiac tissue from ganglia in the body does not interfere with the heart's continued contraction. Experimental evidence shows, however, that in these animals the basic myogenic mechanism is dominated by a neurogenic pacemaker. Histological examination of their hearts has shown that in *Limulus* there is a well-defined and localized cardiac ganglion extending throughout the length of the heart and that in arachnids and some other arthropods there are neurons among the muscular elements of the heart wall. The heart of *Limulus* is particularly well suited to study of the pacemaker mechanism because of the localization of its nervous elements, and it has become more or less of a model for the mechanism in other chelicerate and mandibulate arthropods. The dorsal cardiac ganglion in *Xiphosura* develops after the heart is formed and is rhythmically beating in the embryo. When heart and ganglion are fully organized, the ganglion contains large unipolar nerve cells, which are most numerous in segments 4 to 6, and small bipolar and unipolar cells which are equally distributed along its length. Hearts of full-grown specimens of *Limulus* continue beating after removal of the dorsal ganglion, but the rate is slower than in the intact heart and the beat is peristaltic, indicating that under natural conditions it is regulated by events in the ganglion. Electrograms of isolated ganglia show 12 to 16 impulse volleys per minute, a frequency which is in close correspondence with that of the heartbeat in the living animal. It is postulated that the large unipolar neurons in the ganglion are the pacemakers and that impulses arise spontaneously in them and are distributed to the muscle fibers through an intermediary motor cell. The pacemaker neurons transmit their impulses to the motor neurons by means of acetylcholine, but the impulses arising in the motor neurons in response to this stimulation are transmitted to the muscle cells by means of adrenalin.

Although nervous and muscular elements are not so readily separable in the hearts of arachnids and crustaceans as they are in *Limulus*, electrical and pharmacological evidence indicates that a similar mechanism operates in them. The neurogenic pacemaker is influenced by extrinsic nerves, as is the myogenic one of molluscs. The extrinsic nerves that increase the frequency of impulses in the pacemaker neurons act through the release of acetylcholine, but the nature of the transmitter substance from the nerves that slow them down, or inhibit them, is still to be elucidated. There are probably also sensory fibers running from the heart to the pacemaker cells that relate the sequence of spontaneous impulses arising in them to stretching of the heart muscle. These impulses result in rhythmic contractions of the heart as a whole. In compact, saccular hearts, the contraction may be simultaneous throughout the organ; in elongated ones, it is generally peristaltic in character. Blood is usually driven only in a forward direction, and each systole is followed by a diastole that is effected by contraction of radiating strands of ligament-like tissue attached to the heart and to the wall of the pericardium. These may contract of their own elasticity or through the action of muscles in them. In some insects there is definitive evidence of muscular contraction, but this has not been established for other groups. However shortening may occur, its effect is to expand the heart and so to suck blood or hemolymph into it from the pericardial cavity with which it is in free communication through one or more pairs of ostia. When filled, the heart is again in a state to enter upon the next systole and send the fluid on its way through the vessels and sinuses.

The pulsating dorsal vessel of an insect is called its heart, but it is not the only organ for propulsion of the hemolymph or, indeed in some insects, even the main one. Its effect is augmented by that of other pulsating organs, or accessory hearts, located in the dorsal regions of the meso- and metathoracic segments, in the legs, and at the bases of the antennae. The thoracic hearts propel hemolymph into the wings; there is but one pair in Diptera. Leg hearts are characteristic of Hemiptera especially, and in these insects they lie in the tibia just distal to its articulation with the femur. Antennal organs have been particularly studied in nematocerid Diptera and are present in most, but not in all, families of this suborder. In *Culex* there are two such hearts, each giving off a vessel that enters the antenna of its side; each is connected by a muscle to the aorta arising from the heart. When this muscle contracts, hemolymph enters the vesicle through an opened valve and, when stretched, the vesicle contracts of its own elasticity as the muscle relaxes, and expels hemolymph into the antennal vessel. In these, as in the accessory hearts of other insects, the beat seems wholly independent of that of the dorsal vessel and of the other pulsating organs all of which may beat simultaneously though not synchronously.

The heartbeat in insects characteristically begins at the posterior end of the tubular heart and progresses anteriorly in a peristaltic wave, which moves at different rates in different insects and even in the same insect at different ages and, at the same age, under different conditions. Moreover, the direction of propagation may be partly or wholly reversed at certain periods of growth and development. This occurs, for example, in *Bombyx mori* when, for a definite period during pupation, a contraction wave begins at the third abdominal segment and proceeds anteriorly, while at the same time another begins at the fourth abdominal segment and proceeds posteriorly. Later, this double action stops and a complete reverse wave begins, progressing from the anterior end of the heart to the posterior. During later stages of pupation and even in the imago, reversal of beat may occur intermittently.

Pacemakers in insect hearts may be neurogenic or myogenic and at present there is little conclusive evidence upon which to base a distinction between the mechanisms operative in any insect. Failure to detect neurons in the hearts of a number of insects, or neural connections with the heart, has led to the assumption that the control of their beat is myogenic. The hearts of other representative insects have been shown to be well innervated, and their pacemakers may well be neurogenic. Yet in *Bombyx*, whose heart is innervated by a pair of lateral nerves originating in the ventral nerve cord and in the paired cardiac ganglia of the stomato-gastric system, the pacemaker may shift, when the beat is reversed, from one end of the heart to the other, or to an intermediate point, without any demonstrable connection with its innervation. It is, of course, not only possible but probable that the pacemaker mechanism may be different in different insects.

Evidence derived largely from pharmacological experimentation, indicates that contraction in the axial gland of *Strongylocentrotus purpuratus* is myogenic and attributable to the muscle cells in its walls, rather than neurogenic. Like the muscle of vertebrate hearts its contraction is inhibited by acetylcholine and accelerated by adrenalin.

SELECTED BIBLIOGRAPHY

Mediums of transport

Awapara, J.: Free Amino Acids in Invertebrates: A Comparative Study of Their Distribution and Metabolism, in J. T. Holden (ed.), "Symposium on Free Amino Acids (Amino Acid Pools)," pp. 158–175, City of Hope Medical Center, Elsevier, Amsterdam, 1962.

Boolootian, R. A., and A. C. Giese: Coelomic Corpuscles of Echinoderms, *Biol. Bull.*, vol. 115, pp. 53–63, 1958.

Chandler, A.: Causes of Variation in the Haemoglobin Content of *Daphnia* (Crustacea: Cladocera) in Nature, *Proc. Zool. Soc. London*, vol. 124, pp. 625–630, 1954.

Clark, M. E.: Biochemical Studies on the Coelomic Fluid of *Nephtys hombergi* (Polychaeta: Nephtydae) with Observations on Changes during Different Physiological States, *Biol. Bull.*, vol. 127, pp. 63–84, 1964.

Cosgrove, W. B., and J. B. Schwartz: The Properties and Function of the Blood Pigments of the Earthworm, *Lumbricus terrestris, Physiol. Zool.*, vol. 38, pp. 206–212, 1965.

Crossley, A. C. S.: An Experimental Analysis of the Origins and Physiology of Hemocytes in the Blue Blowfly *Calliphora erythrocephala, J. Exp. Zool.*, vol. 157, pp. 375–398, 1964.

Dales, R. Phillips: The Coelomocytes of the Terebellid Polychaete *Amphitrite johnstoni, Q. J. Microsc. Sci.*, vol. 105, pp. 263–279, 1964.

Deevey, G. B.: The Blood Cells of the Haitian Tarantula and Their Relation to the Moulting Cycle, *J. Morphol.*, vol. 68, pp. 457–492, 1941.

Ferguson, J. C.: Nutrient Transport in Starfish. I. Properties of the Coelomic Fluid. II. Uptake of Nutrients by Isolated Organs, *Biol. Bull.*, vol. 126, pp. 33–53 (I); 391–406 (II), 1964.

Fox, D. L.: "Animal Biochromes and Structural Colors," Cambridge University Press, New York, 1953.

Fox, H. M.: Chlorocruorin and Haemoglobin, *Nature*, vol. 160, p. 825, 1947.

Fox, H. M., B. M. Gilchrist, and E. A. Phear: Functions of Haemoglobin in *Daphnia, Proc. R. Soc. (London), B*, vol. 138, pp. 514–528, 1951.

Fox, H. M., S. M. Hardcastle, and E. I. B. Dresel: Fluctuations in the Haemoglobin Content of *Daphnia, Proc. R. Soc. (London), B*, vol. 136, pp. 388–399, 1949.

Fox, H. M., and G. Vevers: "The Nature of Animal Colors," Macmillan, New York, 1960.

George, W. C.: Comparative Hematology and the Functions of Leucocytes, *Q. Rev. Biol.*, vol. 16, pp. 426–439, 1941.

Gregoire, C.: Hemolymph Coagulation in Sixty-one Species of Insects, *J. Physiol.*, vol. 114, p. 43, 1951.

Gregoire, C.: Blood Coagulation in Arthropods. II. Phase Contrast Microscopic Observations on Hemolymph Coagulation in Sixty-one Species of Insects, *Blood*, vol. 6, pp. 1173–1198, 1951.

Gregoire, C., and M. Florkin: Étude au Microscope à Contraste de Phase du Coagulocyte, du Nuage Granulaire et de la Coagulation Plasmatique dans le Sang des Insectes, *Experientia*, vol. 6, p. 297–298, 1950.

Groskopf, W. R., J. W. Holleman, and I. M. Klotz: Amino Acid Composition of Hemerythrin in Relation to Subunit Structure, *Science*, vol. 141, pp. 166–167, 1963.

Hetzel, H.: Studies on Holothurian Coelomocytes. I. A Survey of Coelomocyte Types, *Biol. Bull.*, vol. 125, pp. 289–300, 1963.

Hetzel, H.: II. The Origin of Coelomocytes and the Formation of Brown Bodies, *Biol. Bull.*, vol. 128, pp. 102–111, 1965.

Hinton, H. E.: Insect Blood: The Giant Cells Sometimes Present in the Blood, *Sci. Prog. (London)*, vol. 168, pp. 684–696, 1954.

Holland, N. D., J. H. Phillips, Jr., and A. C. Giese: An Autoradiographic Investigation of Coelomocyte Production in the Purple Sea Urchin (*Strongylocentrotus purpuratus*), *Biol. Bull.*, vol. 128, pp. 259–270, 1965.

Jacobson, F. W., and N. Millott: Phenolases and Melanogenesis in the Coelomic Fluid of the Echinoid *Diadema antillarum* Phillippi, *Proc. R. Soc. (London), B*, vol. 141, p. 231–247, 1953.

Jones, J. C.: Current Concepts Concerning Insect Hemocytes, *Am. Zool.*, vol. 2, pp. 209–246, 1962.

Jones J. D.: The Functions of the Respiratory Pigments in Invertebrates, in G. A. Kerkut (ed.), "Problems in Modern Biology," vol. I, pp. 9–90, Pergamon, New York, 1963.

Kennedy, G. Y., and R. P. Dales: The Function of the Heart Body in Polychaetes, *J. Mar. Biol. Assoc. U.K.*, vol. 37, pp. 15–31, 1958.

Klotz, I. M., and T. A. Klotz: Oxygen-carrying Proteins: A Comparison of the Oxygenation Reaction in Hemocyanin with that in Hemoglobin, *Science*, vol. 121, pp. 477–480, 1955.

Lane, C. E., E. Pringle, and A. Bergere: Amino Acids in Extracellular Fluids of *Physalia physalis* and *Aurelia aurita, Comp. Biochem. Physiol.*, vol. 15, pp. 259–262, 1965.

Liebmann, E.: The Coelomocytes of Lumbricidae, *J. Morphol.*, vol. 71, pp. 221–249, 1942.

Lochhead, J. H., and M. S. Lochhead: Studies on the Blood and Related Tissues in *Artemia salina* (Crustacea, Anostraca), *J. Morphol.*, vol. 68, pp. 593–632, 1941.

Mangum, C. P., and R. P. Dales: Products of Haem Synthesis in Polychaetes, *Comp. Biochem. Physiol.*, vol. 15, pp. 237–257, 1965.

Manwell, C.: On the Evolution of Hemoglobin, *Biol. Bull.*, vol. 115, pp. 227–238, 1958.

Manwell, C.: Comparative Physiology: Blood Pigments, *Annu. Rev. Physiol.*, vol. 22, pp. 191–244, 1960.

Manwell, C.: Genetic Control of Hemerythrin Specificity in a Marine Worm, *Science*, vol. 139, pp. 755–758, 1963.

Morrison, P. R., and K. C. Morrison: Bleeding and Coagulation in some Bermudan Crustacea, *Biol. Bull.*, vol. 103, pp. 395–406, 1952.

Padmanabhanaidu, B.: Ionic Composition of the Blood and the Blood Volume of the Scorpion *Heterometrus fulvipes*, *Comp. Biochem. Physiol.*, vol. 17, pp. 157–166, 1966.

Padmanabhanaidu, B.: Physiological Properties of the Blood and Hemocyanin of the Scorpion *Heterometrus fulvipes*, *Comp. Biochem. Physiol.*, vol. 17, pp. 167–181, 1966.

Pilgrim, M.: The Coelomocytes of the Maldanid Polychaetes *Clymenella torquata* and *Euclymene oerstedi*, *J. Zool.*, vol. 147, pp. 30–37, 1965.

Read, K. R. H.: The Characterization of Radula Muscle Myoglobins from the Gastropod Mollusc *Busycon canaliculatum* L., *Comp. Biochem. Physiol.*, vol. 17, pp. 375–390, 1966.

Ritter, H., Jr.: Blood of a Cockroach: Unusual Cellular Behavior, *Science*, vol. 147, pp. 518–519, 1965.

Roche, J.: Electron Microscope Studies in High Molecular Weight Erythrocruorins (Invertebrate Hemoglobins) and Chlorocruorins of Annelids, in K. A. Munday (ed.), "Studies in Comparative Biochemistry," Pergamon, New York, 1965.

Salt, G.: Experimental Studies in Insect Parasitism. XIII. The Hemocytic Reaction of a Caterpillar to Eggs of Its Habitual Parasite, *Proc. R. Soc. (London), B*, vol. 162, pp. 303–318, 1965.

Sewell, M. T.: Lipoprotein Cells in the Blood of *Carcinus maenas* and the Cycle of Activity Correlated with the Moult, *Q. J. Microsc. Sci.*, vol. 96, pp. 73–83, 1955.

Skinner, D. M.: Incorporation of Labeled Valine into the Proteins of the Cecropia Silkworm, *Biol. Bull.*, vol. 125, pp. 165–176, 1963.

Thompson, P. E., and D. S. English: Multiplicity of Hemoglobins in the Genus *Chironomus* (*Tendipes*), *Science*, vol. 152, pp. 75–76, 1966.

Towe, K. M., H. A. Lowenstam, and M. H. Nesson: Invertebrate Ferritin: Occurrence in Mollusca, *Science*, vol. 142, pp. 63–64, 1963.

Townsley, P. M., R. A. Richey, and P. C. Trussell: The Occurrence of Protophoryrin and Myoglobin in the Marine Borer *Bankia setacea*, *Can. J. Zool.*, vol. 43, pp. 167–172, 1965.

Van Sande, M., and D. Karcher: Species Differentiation of Insects by Hemolymph Electrophoresis, *Science*, vol. 131, pp. 1103–1104, 1960.

Wieser, W.: Electrophoretic Studies on Blood Proteins in an Ecological Series of Isopod and Amphipod Species, *J. Mar. Biol. Assoc. U.K.*, vol. 45, pp. 507–523, 1965.

Wittenberg, B. A., R. W. Briehl, and J. B. Wittenberg: Hemoglobins of Invertebrate Tissues, *Biochem. J.*, vol. 96, pp. 363–371, 1965.

Wyatt, G. R.: The Biochemistry of Insect Hemolymph, *Annu. Rev. Entomol.*, vol. 6, pp. 75–102, 1961.

Wyatt, G. R., and W. L. Meyer: The Chemistry of Insect Hemolymph, *J. Gen. Physiol.*, vol. 42, pp. 1005–1011, 1959.

Yeager, J. F., and O. E. Tauber: On the Hemolymph Cell Counts of Some Marine Invertebrates, *Biol. Bull.*, vol. 69, pp. 66–70, 1935.

Means of transport

Boolootian, R. A., and J. L. Campbell: A Primitive Heart in the Echinoid *Strongylocentrotus purpuratus*, *Science*, vol. 145, pp. 173–175, 1964.

Clark, R. B.: The Blood Vascular Supply of *Nephtys* (Annelida, Polychaeta), *Q. J. Microsc. Sci.*, vol. 97, pp. 235–251, 1956.

Clements, A. N.: The Antennal Pulsating Organs of Mosquitoes and Other Diptera, *Q. J. Microsc. Sci.*, vol. 97, pp. 429–435, 1956.

Federighi, H.: The Blood Vessels of Annelids, *J. Exp. Zool.*, vol. 50, pp. 257–294, 1928.

Ferguson, J. C.: An Autoradiographic Study of the Distribution of Ingested Nutrients in the Starfish *Asterias forbesi*, *Am. Zool.*, vol. 3, p. 524, 1963.

Ferguson, J. C.: Nutrient Transport in Starfish. I. Properties of the Coelomic Fluid, *Biol. Bull.*, vol. 126, pp. 33–53, 1964.

Hanson, J.: Histology of the Blood System in Polychaeta and Oligochaeta, *Biol. Revs.*, vol. 24, pp. 127–173, 1949.

Hanson, J.: The Blood System in the Serpulimorpha (Annelida, Polychaeta). I. The Anatomy of the Blood System in the Serpulidae. II. The Anatomy of the

Blood System in the Sabellidae and a Comparison of Sabellidae and Serpulidae, *Q. J. Microsc. Sci.*, vol. 91, pp. 111–136 (I); pp. 369–378 (II), 1950.

Jones, J. O.: The Heart and Associated Tissues of *Anopheles quadrimaculatus, J. Morphol.*, vol. 94, pp. 71–123, 1954.

Krijgsman, B. J.: Contractile and Pacemaker Mechanisms in the Heart of Arthropods, *Biol. Revs.*, vol. 27, pp. 320–346, 1952.

Krijgsman, B. J., and G. A. Divaris: Contractile and Pacemaker Mechanisms of the Heart and Mollusks, *Biol. Revs.*, vol. 30, pp. 1–39, 1955.

Millott, N.: Axial Organ and Fluid Circulation in Echinoids, *Nature,* vol. 204, pp. 1216–1217, 1964.

Millott, N.: A Possible Function for the Axial Organ of Echinoids, *Nature,* vol. 209, pp. 594–596, 1966.

Phillis, J. W.: Innervation and Control of a Molluscan (*Tapes*) Heart, *Comp. Biochem. Physiol.,* vol. 17, pp. 719–740, 1966.

chapter XI

RESPIRATORY GASES AND THEIR TRANSPORT

A. OXYGEN AND LIFE

Transport of oxygen is a highly important function for body fluids. In living systems, energy is made available for biophysical and biochemical work by the combustion of organic compounds and, in the vast majority of them, oxygen is required for the completion of combustion and procurement of the compound's full energy value. Respiration is the term used for the oxidative degradation of organic compounds that takes place in cells through a long series of enzymatic reactions that result in a step-by-step transmutation of potential energy bound in the molecule to kinetic energy, manifested and measurable as work, heat, or sometimes as light. Respiration is often used in a broader sense to include the means by which an organism obtains oxygen from its external environment and conveys it to its cells for their oxidative reactions, but, *sensu stricto*, it refers to their oxidative metabolism only.

Development of respiration is generally considered as one of the major evolutionary events and as one that provided many opportunities for evolutionary change. It is now generally accepted that the first assembling of organic molecules took place in an atmosphere devoid of free oxygen, and that when Earth was condensed from the debris and dust that constituted it and the other objects in our solar system its surrounding atmosphere was composed primarily of water and water vapor, with some nitrogen, chlorine, carbon dioxide (probably 3.3×10^{-4} atm, or close to its present value), carbon monoxide, methane, ammonia, hydrogen, and hydrogen sulfide. This atmosphere afforded little protection from the radiant energy emitted by the sun, with wavelengths and intensity such that no biological system, as we know them, could have endured. But certain events could have occurred, and in all probability did, to change these conditions. Dissociation of water vapor into its constituent hydrogen and oxygen atoms through solar energy must have taken place, but this could not have gone on indefinitely and probably not even long enough to add more than an infinitesimal amount of oxygen to the atmosphere of the forming Earth, an amount that has been calculated as less than one one-thousandth of the amount in the air today, or about 0.02 percent of the gaseous mixture. The limiting factor to the continuing increase of oxygen by photolysis of water vapor is the tendency of the oxygen atoms, when released, to move away from the mass of the earth toward the outer regions of its atmosphere. As this happened, they formed a layer that absorbed some of the ultraviolet rays, thus screening the water vapor from them and continually reducing the rate of photolysis. Another event was also in progress that would change the atmosphere. This was the formation of ozone, a combination of three oxygen atoms (O_3) and a reaction for which the atmospheric conditions were favorable. Ozone made another gaseous layer surrounding the earth and shielding it from the full force of the sun's energy and the most destructive of its rays. Biological systems, as we understand them, could still not have survived the intensity of this radiant energy had they been exposed to it without further protection. But water provides another light screen and it is conceivable that as atoms were brought together into organic molecules at the surfaces of bodies of water, they drifted down into the lower depths where conditions were at least possible for, if not actually conducive to, their assemblage into cellular systems. Since no oxygen was available, these systems must have used other means than respiration to liberate energy or, more accurately, to redistribute it. Possibly they employed some form of fermentation (anaerobiosis), a process that does not require oxygen and, though less efficient than respiration in terms of energy made available, is one practiced by many modern microorganisms and by some metazoans for brief periods of time. Such organisms are known as anaerobes, either obligatory or facultative. The former are unable to utilize oxygen, and in some cases even unable to exist in its presence. The latter can survive for varying periods of time without oxygen. In anaerobiosis, the molecule of an organic substrate (most frequently glucose) is rearranged through a series of enzymatically catalyzed reactions (glycolysis) to yield two smaller molecules, as, for instance, two of lactic acid or one of ethyl alcohol and one of carbon dioxide. Other organic acids or alcohols may be end products of anaerobiosis. In any event, the substrate molecule is decomposed only partly, not wholly as it is when the end products of glycolysis are oxidized to carbon dioxide and water in respiration (aerobiosis). Only a fraction of the energy held in a glucose molecule becomes available under anaerobic conditions. Organisms that can perform the subsequent steps in its degradation are known as aerobes and by their utilization of oxygen in respiration can obtain nearly all the remaining potential energy of glucose (93 percent). Some invertebrates, such as intestinal parasites, can meet their energy needs by glycolysis alone, and excrete the end products. Others, as facultative anaerobes, can exist for a time by this means, if during this time their energy demands are not great. They can build up an "oxygen debt" which is paid off when oxygen again becomes available and the end products accumulated during their period of anaerobiosis are oxidized.

Respiration could not have arisen until there was at least some free oxygen in the atmosphere immediately surrounding the earth's crust. This was contributed by photosynthesizing organisms which, according to the fossil record, appeared in the pre-Cambrian period when conditions in pools and small quiet lakes, and even in shallow seas, were compatible, providing at depths of 30 to 40 ft enough light but still effective screening from the deadly rays of the sun. The contribution of these early photosynthesizing systems could hardly have been great enough to compensate for the loss to the upper atmosphere of the oxygen continually formed by the photolysis of water. But, as the ozone layer grew thicker and more dense, more of the sun's lethal rays were absorbed before they reached the water's surface, and photosynthesizing organisms could now survive in the upper levels. Therefore, they could spread into deeper lakes, and even into the seas, from which they had previously been excluded, for in large bodies of water there is enough turbulence at depths of 30 to 40 ft to have carried the first photosynthesizing systems out of the zone of their activities and survival.

Nearly a century ago Louis Pasteur (1822–1895), in his studies on wines and their diseases, made the observation that the microorganisms causing souring and spoilage could change from fermentation to respiration when the level of oxygen in their environment was about 0.2 percent of the total gaseous mixture, that is, no less than one-hundredth of that in the atmosphere today. Two energy-releasing metabolic pathways are thus open to them, between which they can shift according to the oxygen supply of their environments. Hence, they are known as facultative anaerobes, able to exist anaerobically in the absence of oxygen but using it for respiration when it is available. A level of about 0.2 percent oxygen is thus a critical one for them, determining which pathway shall be employed. When this was reached in the atmosphere of the primitive Earth, due almost entirely to the activities of aquatic photosynthesizing organisms, respiration could have supplanted fermentation as a means of supplying the energy necessary for life processes, at least in those systems with the genetic potential of developing the requisite biochemical pathways. Calculations indicate that this critical level could have been reached about the beginning of the Paleozoic era, and the Cambrian rocks, with their abundance of fossils of diverse types, testify that it was followed by an evolutionary explosion in which most of the invertebrate phyla became established. A second critical level was probably reached near the beginning of the Silurian period, when the concentration of atmospheric oxygen reached 2 percent, or one-tenth of its present level, and the protective screen of ozone had become sufficiently thick to permit plants and animals to live near the surface of the waters and even to migrate onto dry land. It is in the Silurian that the first air-breathing invertebrates made their appearance, so that it must be assumed that they had not only the mechanisms for cellular respiration but also for the admission of oxygen to their bodies and its movement to their tissues.

The spread of photosynthesizing systems in the periods that followed has added more and more oxygen to the atmosphere until now, in the Quaternary, it constitutes about 20 to 21 percent of the gas mixture. The level during the Carboniferous may have exceeded this, for that was a time when great areas of the land mass were covered with forests and other vegetation, and photosynthesis must have been at its maximum. That in the future the oxygen content may drop below its present level is not inconceivable, for massive deforestation, the building of great cities almost devoid of green plants, and the steady contribution to the atmosphere of carbon dioxide, carbon monoxide, and various noxious gases from furnaces, factories, and motor vehicles may well, in time, alter its composition significantly. However, since animals and plants at present on this planet live in an atmosphere in which there is more than enough oxygen and as the majority have evolved in such a way as to take advantage of it, the immediate question for the biologist concerns the means by which they do this and the adaptations that have arisen to implement respiration in different kinds of animals, in different kinds of environments, and in different kinds of circumstances. Yet, important as this metabolic process has become, there still are some organisms that cannot tolerate free oxygen at any concentration, some that habitually live in environments where its concentration is no greater than it was at the first critical level, and some that withstand extended periods of oxygen deprivation although they are ultimately dependent upon it for their survival.

Present evidence indicates that respiration is alike in all organisms in that similar enzymes are involved and similar pathways followed in the oxidative degradation of any given kind of nutrient. The entrance of oxygen into cells is also the same, for this is a diffusion phenomenon that depends upon a fluid medium and the existence of a gradient between an extracellular concentration of dissolved oxygen higher than the intracellular one. Carbon dioxide as the gaseous end product of respiration must diffuse out of a cell under conditions similar to those in which oxygen diffuses in—that is, according to a concentration gradient which in this case has its highest point inside the cell and its lowest outside. The velocity of diffusion depends upon the solubility of a gas in a given fluid at a given temperature, and the rate of diffusion, upon the steepness of the concentration gradient. The medium through which the diffusing gas must pass is also an important factor, for diffusion in biological systems is a passive, not an active, process. The diffusion constant—that is, the number of milliliters of a gas, at zero degrees Celsius and one atmosphere of pressure, which penetrate per minute through a thickness of one micron, when the pressure difference is one atmosphere—can be calculated for gases diffusing through different media at different temperatures. For oxygen, the diffusion constants at 20°C are 11.0 in air, 0.00034 in water, and 0.000013 in chitin. Carbon dioxide is 28 times more soluble than oxygen at 20°C, moves with a velocity 20 to 30 times that of oxygen, and has a diffusion constant, in water, about 25 times greater than oxygen. Also, it combines with water to form carbonic acid (H_2CO_3), which dissociates as $H^+ + HCO_3^-$, so that ionic transport is also a factor in its movements.

The means by which environmental oxygen is brought into diffusible range of the cells of multicellular systems are, on the other hand, various, and special areas or special mechanisms have evolved in most of those at the tissue level of organization. For diffusion alone is adequate only if the tissue is not more than 1 mm in diameter and no matter how great the external oxygen supply may be, the internal supply will be insufficient if the tissues are more than 0.5 mm away from an oxygen source. Diffusion from the external medium is, therefore, a limiting factor to the respiration of organisms of macroscopic dimensions and complex organization. It suffices for protozoans, for which the entire cell surface or, for those with shells, the exposed parts are in direct contact with the environment. It is adequate also for cell aggregates such as sponges and for such metazoans as cnidarians, ctenophores, platyhelminths, "aschelminths," ectoprocts, and endoprocts and for embryos of all species. But in most invertebrates, diffusion through the surface is supplemented by various means for conveyance of oxygen and carbon dioxide to and from the tissues, and, often, particular regions of the body have become most effective in gaseous exchange and function as primary respiratory areas.

B. RESPIRATORY AREAS

In some metazoan invertebrates with transport mechanisms, external gaseous exchange occurs across the entire surface of their bodies and is, therefore, known as integumentary or cutaneous exchange. In others, it takes place primarily across the surface of a special area, structure, or organ derived from the body wall and essentially an extension of it, either as an outgrowth or as an ingrowth. Outgrowths are designated by the very general term of "gills" ("branchiae") and ingrowths by that of "lung." No invertebrate lung is, however, developed or constructed in the same way as the alveolar lung of a vertebrate animal, although it serves the same purpose.

Whether an organ is truly respiratory in function can be proved by the effect of its removal, if such an operation is possible, or by preventing access of oxygen to it. If it is essential to respiration the effects of oxygen deficiency become apparent in the organism sooner or later. Uptake of oxygen in animals deprived of suspected respiratory organs can also be measured and compared to that in intact animals under the same environmental conditions. Decrease of oxygen consumption in experimental animals in relation to that of controls can be correlated with the importance to their respiration of the extirpated or blocked organ. But oxygen and carbon dioxide are not the only things that pass across such exposed areas, and other elements may diffuse inward and outward. Respiratory areas may, therefore, also be ones for the exchange of ions with the external environment, and for the uptake and loss of salts.

It is essential to the efficient operation of any respiratory surface that the medium with which it is in contact be frequently changed, so that the oxygen supply will not be reduced nor the carbon dioxide content increased beyond the limits necessary to maintain adequate concentration gradients. This renewal of water or of air is accomplished in a number of different ways. Water may be kept in motion by ciliary or flagellar activity, as it is in sponges, cnidarians, and ctenophores; in turbellarian flatworms and nemerteans with a ciliated epidermis; in rotifers, ectoprocts, and endoprocts where the food current is also a respiratory one; and in aplacophoran, polyplacophoran, lamellibranch, and some gastropod molluscs. Or it may be provided by movements of the body as a whole, as it is in the whip-like lashing of free-living nematodes; the writhing movements of some annelids; or the rhythmic or sporadic pulsations of most tubiculous annelids, of holothurians, and of some insects. Irrigation or ventilation may also be accomplished by movements of certain parts of the body, as those of the parapodia of nereid polychaetes and the appendages of crustaceans. Some very special adaptations are evident in insects that live aquatic lives either as juveniles or as adults, yet are unable to obtain oxygen from water and so are dependent upon continual contact with an oxygen source.

1. Integumentary (cutaneous) exchange

In nudibranch molluscs, most annelids, and some sipunculids and small arthropods, gaseous exchange occurs over the whole body surface. In earthworms, this surface is ordinarily kept moist by secretion of mucus from epidermal gland cells and by excretions from the nephridiopores, but under the stress of dehydrating conditions coelomic fluid may also be discharged through the dorsal pores adding to the surface film of moisture. Atmospheric oxygen in the air pockets in the loose soil in which earthworms customarily live is dissolved in this moist covering and diffuses from it into the integumentary capillaries of the vascular system. These form an extensive network and a system of looped tubules among and between the epidermal cells and so bring the blood very close to the surface (Fig. XI-1). Conditions are, therefore, suitable for diffusion of oxygen from the soil atmosphere into the transport system and of carbon dioxide from it into the air. Movements of the worm and air currents are sufficient to keep the air renewed in dry or damp soil, but the oxygen supply is much reduced after a rain when the air pockets are filled with water, which holds, in solution, much less oxygen than air. In these circumstances earthworms tend to come to the

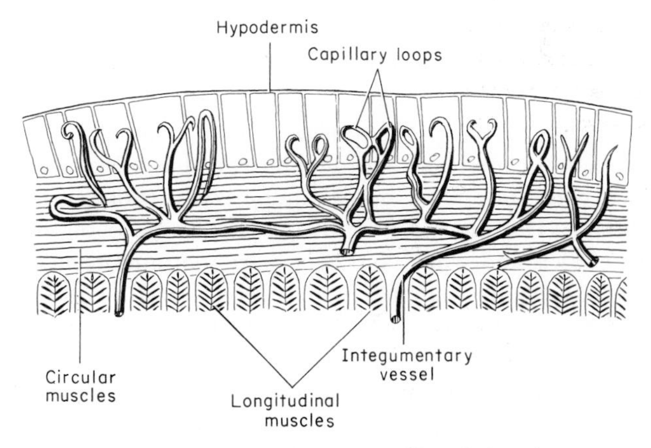

FIG. XI–1 *Section of the integument of* Lumbricus *in transverse section, showing the terminations of the integumentary vessels in the hypodermis. (After Laverack.)*

surface to expose their bodies to the upper air. Gaseous interchange in leeches is accomplished in the same way, the gases diffusing through the surface to and from the blood in rhynchobdellids and the coelomic fluid in gnathobdellids and pharyngobdellids.

In certain polychaetes, such as bamboo and scale worms (Maldanidae and Aphroditidae, respectively), the dorsal surface of the integument is the primary respiratory area. Certain modifications of the parapodia, perhaps primarily protective devices, are involved in the respiratory process, at least to the extent of directing a water current over the respiratory area. In scale worms the dorsal cirri of the parapodia on all, on most, or on alternate segments are developed into plate-like elytra (Greek *elytron*, "scale") that extend over the back of the worm like a roof. A current of water is kept moving through the space between this roof and the body by action of cilia on the cells of the dorsal epidermis and on the parapodia. In the sea mouse, *Aphrodite*, whose elytral roof is covered by a matted mass of long, slender notopodial setae through which the water is filtered, there are no cilia but a similarly directed current is maintained by movement of the elytra, which swing upward and then rapidly return to a horizontal position.

In sipunculids, diffusion to and from the coelomic fluid takes place across the entire body wall, composed of an external cuticle, epidermis, a delicate layer of connective tissue, circular and longitudinal muscles, and an inner layer of peritoneum. This wall, though structurally like that of earthworms, is much thinner and diffusion across it is adequate for the respiratory needs of these sedentary animals. This is true also for *Orchetostoma erythrogrammon*, a small echiuroid that lives in the intertidal zone in a burrow that it irrigates, when the tide is in, by peristaltic contractions of its body. Diffusion is general over the surface of the proboscis and trunk, and each part continues to take up oxygen after one has been severed from the other. Such experiments have shown that the uptake by the proboscis alone is 17 percent of that of the intact worm, but is apparently enough for its needs during the periods of low tide when the burrow cannot be irrigated. Experiments, in which the possibility of gaseous interchange across an internal surface, such as that practiced by *Urechis caupo*, has been eliminated by the insertion of an anal plug, have resulted in no diminution of oxygen uptake, providing cogent evidence that the integument is the primary respiratory area. Crinoids, and some of the burrowing and apodous holothurians, also employ cutaneous respiration, although in most holothurians the gaseous exchange takes place, as it does in *Urechis*, across an expanded region of the termination of the gut that acts as a lung.

Small adult crustaceans, such as free-living copepods, most ostracods, and many cirripedes, as well as larvae of all species prior to development of their adult respiratory organs, also obtain oxygen by diffusion across the general integument. The only decapod known to do so is the shrimp *Lucifer*. And even in crustaceans with developed respiratory organs, the integument may be responsible for a large share of the oxygen uptake. The pleopods of isopods are their principal respiratory areas, but it has been shown that in those living semiterrestrial lives 50 percent of the oxygen they use is taken in through the body surface, provided this can be kept moist. In the land hermit crab, *Coenobita*, there is an especially well-vascularized area on the antero-dorsal surface of the abdomen through which enough oxygen can diffuse to take care of the animal's respiration needs for a long time, for crabs deprived of their gills live for months after the operation.

Cutaneous diffusion probably accounts for a part of the oxygen intake of all insects and, probably, for a large part of their carbon dioxide output. It is particularly important to the respiration of larvae and nymphs of most species, whether they live in aquatic or terrestrial habitats. It accounts for 100 percent of the oxygen taken in by dragonfly nymphs and by some of those of the hemipterans and coleopterans whose entire life is spent under water. These are the plastron-breathing bugs and beetles that never come to the surface and derive their oxygen through a permanent layer of gas, the plastron, that envelops their bodies, held there by one means or another.

2. Gills (branchiae)

By definition a gill is an organ for gaseous exchange underwater; in practice, it is a term used for a number of structures of different kinds through which oxygen may pass from air as well as from water, provided that the surface of the gill is moist. Structurally, all gills are outgrowths of the body surface, which either project freely into the surrounding medium, as they do in annelids and the larvae of some insects, or are enclosed in chambers, as they are in molluscs and crustaceans. A current of water or air, in the case of aerial gills, is kept moving over them by ciliary action, their own motion, or the motion of the animal itself. In animals with vascular systems, they are plentifully supplied with blood vessels.

Probably the simplest form of invertebrate gill is found in starfishes and some sea urchins. These gills are also called dermal branchiae and papulae (Fig. XI–2*a*). They are projections of the body wall and perivisceral coelom that in asteroids extend between the skeletal plates of the disk and arms. Cilia of the epidermal cells keep water moving over them while those of the peritoneal cells keep the enclosed coelomic fluid in motion. These are, however, not the only sites of gaseous exchange, for this takes place also across the walls of the tube feet between the sea-water and the fluid in the water vascular system. Experiments in which the ambulacral grooves have been covered show that in *Asterias rubens* the tube feet account for about 60 percent of the oxygen taken up. The tube feet provide the sole areas for gaseous interchange in heart urchins and in sand dollars, neither of which have gills. In regular echinoids, the gills are branched sacs, larger and more distensible than they are in asteroids and very likely are their principal respiratory organs. They are located only on the peristome, from which five pairs extend into the water. Coelomic fluid from the lantern (peripharyngeal) coelom, a subdivision of the perivisceral coelom, is pumped in and out of them by movements of the compass muscles of Aristotle's lantern, and the water around them is renewed by their own movements and by ciliary activity of their epithelial cells.

Annelid gills may also be simple in construction. In *Branchiodrilus*, an aquatic oligochaete of tropical regions, each true segment has a long, slender filament extending from each side, and in *Branchiura* there are finger-like projections on the dorsal and ventral surfaces of the posterior segments only. *Dero* and *Aulophorus*, other common aquatic oligochaetes, have a circle of blunt projections around the anus. All of these structures are presumed to be gills and to have a true respiratory function, but there is no critical evidence of their importance to respiration relative to that of the rest of the body's surface.

In polychaetes with gills, except for terebellids, sabellids, and serpulids where they are developed from the peristomium, the gills are derived from the parapodia. In *Neanthes*, the dorsal lobe of each parapodium is particularly well vascularized and oxygen diffuses through its surface to the underlying capillaries from the water that is kept in motion by movement of the parapodium as a whole (Fig. VI–9). In many genera, only the dorsal cirrus of the parapodia on

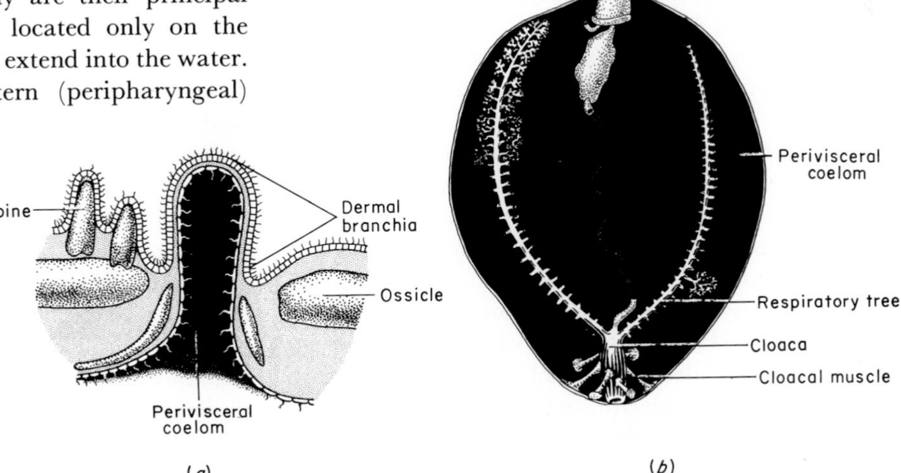

FIG. XI–2 *Respiratory mechanisms in echinoderms.* ***(a)*** *Section of the integument of a starfish, showing one dermal branchia.* ***(b)*** *A holothurian, showing the respiratory tree. The extensive branchings of the main tubes are shown only in part of the tree.*

certain segments has developed as a gill. This cirrus may be modified as a long, slender thread, a fleshy cone, or a flattened lobe over which water is kept moving by ciliary action. In *Eunice* and related genera the cirrus as such is not modified, but gills of different form, and characteristic of the genus, originate from its base. All are thin-walled and well supplied with capillaries.

The importance of their crowns to the respiration of fan worms is different in different species. Removal of the crown from *Sabella pavonina* does not result in significant reduction in oxygen uptake as long as the tube in which the posterior segments are contained is well ventilated. The fact that *S. pavonina* can autotomize its crown and live quite normally during the period of regeneration confirms the experimental evidence that it is not essential to respiration as long as the rest of the body has access to oxygen and cutaneous exchange can take place. The same operation on *Sabella spallanzani*, *Bispira voluticornis*, and *Myxicola infundibulum* results in a reduction of oxygen uptake as great as 60 percent or more. *Myxicola* lives in a tightly fitting gelatinous tube, which it does not irrigate, and must keep its crown exposed in order to obtain the necessary oxygen for its total respiration. This creates a hazardous situation for it, for the crown presents an attraction to predators. But *Myxicola* has an especially well developed giant nerve fiber system and can respond almost instantaneously to danger signals by retraction of its crown and withdrawal into its tube. The worm can survive for many weeks in a tube sealed at the hinder end if the crown can be exposed to seawater, indicating that its uptake of oxygen is sufficient to provide for the respiration of the entire body and indeed is essential to it. The Pacific sabellid, *Schizobranchia insignis*, on the other hand, can maintain its respiratory requirements when withdrawn into its tube, although it uses then about one-quarter as much oxygen as it does when fully expanded. Measurements and calculations have shown that the uptake of oxygen by the crown is mainly for its own respiratory needs, for in all fan worms the crown is active in food collection and filtration and operates at a high rate of oxidative metabolism. This condition in *Schizobranchia* has some parallelism with that in the echiuroid *Ochetostoma erythrogrammon*, in which the independent respiration of the proboscis is advantageous to the animal since this part of the body, extended by the coelomic fluid, can at low tide reach out into the damp sand while the remainder of the body is confined to the nonirrigated burrow.

The typical molluscan gill is a ctenidium, constructed as a central shaft, through which blood vessels run and from which there arise lateral branches, probably originally triangular blades that have in many modern species developed into slender filaments (Fig. VII–20). The epithelium of the filaments is ciliated, and in most molluscs they are supported by internal rods and have tributaries of the shaft vessels running through them. Typically, too, the gills are bipectinate, with filaments arising from both sides of the central axis, but in some gastropods they are monopectinate, with filaments on only one side. The gills extend from the body wall, to which they are usually attached along the length of their axes, into the mantle cavity so that they have some protection. This is particularly important in species that crawl along rough surfaces that might do damage to the delicate gills or that live in mud or sand that would foul them. Ventilation of the mantle cavity, which was the first distinctive structural feature to evolve in molluscs and which has been considered both their greatest asset and their greatest liability, at once their strength and their weakness, is effected in gastropods with gills and in lamellibranchs by action of the cilia on the surfaces of the gill filaments and by those on the epithelium of the hypobranchial gland. This is a mass of mucus-secreting cells on the roof of the mantle cavity of gastropods and some lamellibranchs. Clearing the water that passes through the gills of sediment, detritus, and fecal matter is of vital importance, especially to species that live in sand or mud, and has been achieved by the development of such mucous glands and the entanglement of particles in their secretions. In gastropods, too, ciliated cells of the osphradium contribute to the current. In the plicate gills of shelled opisthobranch gastropods, thick, fleshy folds have replaced the ciliated filaments, and the hypobranchial gland and osphradial cilia are the main current producers. In cephalopods, in which the gills are not ciliated and there is neither hypobranchial gland nor osphradium, the mantle cavity is ventilated by action of the mantle itself. When the radial muscles in its wall contract, the wall becomes thinner and the capacity of the mantle cavity becomes greater, so that water flows in around the collar and is later expelled through the funnel when the mantle muscles contract. In scaphopods, which have no ctenidia, the mantle wall is folded in certain regions; the folds are lined with ciliated cells that create an incurrent of water which is expelled by muscular contraction. In some limpets, the edges of the mantle have developed into pallial gills, which in some species are accessory to their single ctenidium but in others are the sole respiratory areas.

Probably in all molluscs the mantle is an area of some gaseous interchange; its importance in this respect has become increasingly great in bivalves where the gills have become increasingly adapted to food getting and food sorting. The mantle has been adapted to aerial interchange in terrestrial gastropods and its cavity has developed into a lung in the pulmonates. Unquestionably, in all molluscs it provides a sizable area for gaseous interchange, as does the body wall itself, but the relative importance of these areas and of the gills in total oxygen uptake is not certain.

In *Limulus* the abdominal appendages serve as gills. These, unlike the jointed, uniramous thoracic appendages are flat and broad and are fused together in the midline of the body. The first pair forms a membranous flap, or operculum, over the other five pairs, the so-called "book gills." The undersurfaces of these appendages are folded into deep pleats, resembling the leaves of a book, across which gaseous exchange takes place between the blood and the water that is constantly moving over them as the appendages wave back and forth. This motion also causes movement of blood in and out of the gill, for as the appendage moves forward, blood from the ventral sinus flows into it, and as it moves backward, blood flows from it into the pericardial cavity.

The prototype of the crustacean gill was probably also a globular or flattened lobe, an epipodite of the thoracic appendages. It is retained in somewhat this form in the syncarid *Anaspides*, where the two epipodites originating from the coxa of each pereiopod are sac-like. Blood vessels run along the margins of the sacs, with many transverse connections between them, so that the structure, fully exposed to the water of the lakes in which *Anaspides* lives, is well vascularized and well suited to gaseous interchange. Branchiopods and amphipods have similar vesicular gills; in the branchiopods they are developed on all the thoracic phyllopodia, but in the amphipods they may be limited to one or more of the thoracic segments.

In other aquatic crustaceans the gills are more complex in construction and, like those of molluscs, consist of a central axis from which side branches extend. These branches are of different types in different genera; they may be leaf-like, tree-like, lamellar, or filamentous in appearance. Afferent and efferent blood vessels run through the axis, and there is a network of vessels or blood spaces in their branches. The chitinous covering of the gill, as well as its epithelium, is usually thin and delicate, so that in their passage the gases diffuse a distance of only some 3 to 5 μ. Additional thoracic gills have arisen in many species, especially among the decapods. While the typical decapod gill is a podobranch, growing out from the coxa of each pereiopod, there may be in each segment also an arthrobranch, growing from the articulation of the coxa and the body, and a pleurobranch, growing from the side of the body wall, or pleuron. In decapods, also, the gills characteristically are covered by a flap of the carapace, the branchiostegite, that grows down on each side and forms a gill chamber (Fig. VI–27*h*). In other groups the gills are freely exposed to the water.

Four thoracic gills per segment is typical of modern decapods (one podobranch, two arthrobranchs, one pleurobranch), but there is no species with the theoretical maximum of 32. The tendency seems to have been toward reduction in gill number, even in those species that have remained strictly aquatic. One genus, *Callianidea*, has, however, added four pairs of abdominal gills to its thoracic complement. Isopods and stomatopods are atypical in that their gills are associated with their pleopods rather than their pereiopods. In stomatopods, the abdominal gills arise as tufts from the exopodites of the pleopods and the thoracic gills, from the coxae of the third, fourth, and fifth maxillipeds. In isopods, both rami of all pleopods may bear gills, whose surfaces are increased by folds, villi, or filaments, but in most species the number of appendages, as well as the rami bearing gills, are limited.

Ventilation of the gills, whether free or enclosed, is accomplished primarily by rhythmic beating of some or all of the appendages. In branchiopods, for example, all phyllopodia beat as do all the pleopods of isopods and stomatopods. In ostracods, the epipodites of the maxillules keep a current moving across the gills as do the scaphognathites, or gill bailers, of decapods. The importance of the action of the scaphognathite has been demonstrated in *Potamonautes perlatus*, a crab common in ponds and streams of the Cape Peninsula of Africa. These crabs can live on land, where diffusion of oxygen from the air across the gill surfaces is adequate for their respiration, but in water they asphyxiate if the scaphognathite is removed and ventilation of the gill chamber thus prevented. In other crustaceans the food current that is created by the appendages is also a respiratory one. The direction of flow through a gill chamber is regularly from posterior to anterior. In many species the margins of the branchiostegites are some little distance from the body, so that water enters the chamber from the sides as well as from the end. In macrurans the edges are closely apposed to the bases of the legs and the water enters only through small gaps between them. Reversal of flow is known to occur at intervals, possibly as a device to clear the opening, or the gills themselves, of accumulated silt. This reverse flow may be maintained for long periods by crustaceans that spend much time buried in the sand. In these species, too, there are often special channels for the incoming stream, analogous to the inhalant siphons of molluscs. In *Emerita analoga*, for example, the apposed antennules make such a channel, and the antennae, in the crab *Corystes cassivalanus*. Apposition of other appendages projecting from the sand form similar channels for other species of similar habit. In most of these, the water current follows its typical postero-antero course when the animal is not buried.

The number and size of the gills can be correlated with the habits and habitats of different species. A survey of the comparative gill area in 16 species of brachyurans, including those that live in the deep sea or offshore beyond the tide level and amphibious ones living in the intertidal zone or most of the time on shore, shows that there has been a tendency toward reduction in gill area per unit of body weight as the crabs have moved from a wholly aquatic existence to one in the intertidal zone or on land. Moreover, in the aquatic species, the gill area proves to be greater in the actively swimming pelagic portunids than in the sluggish spider crabs that live on the bottom of the sea. It is apparent, too, that the gills become increasingly less important to respiration in semiterrestrial and terrestrial crabs, their function being adopted by the walls of the gill chamber in the true land crabs, much as the function of the gills has been adopted by the mantle and mantle cavity in pulmonate molluscs. The robber, or coconut, crab, *Birgus*, and the hermit crab, *Coenobita*, have, for example, lived for months after removal of their gills. For the littoral species that have retained gills as the principal areas of gaseous exchange, there have been two major problems in adaptation to a semiterrestrial existence: (1) to keep their soft gills, without internal support, from collapsing when out of water, and (2) to keep them moist and ventilated. The first has been met by various devices to stiffen the gill filaments. In some species, they are heavily sclerotized, a limitation to the diffusion of gas across them, and in others, as in *Geograpsus*, there are "turgor cells" that keep the filaments extended at right angles to the central axis. The second problem has been met by devices that keep water in the gill chamber, so that after it is filled when the animal visits the sea, it remains wholly or partly so while the crab is on land. This adaptation has been possible because of the tightly fitting branchiostegites characteristic of crabs. The oxygen supply of the water in the gill chamber is renewed while the animal is on land by circulation of air through it. This circulation is maintained, as is the circulation of water in aquatic species, by the waving scaphognathites, which by their motion draw a current of air in through small spaces between the appendages and out through the aperture at the anterior end. Fiddler crabs, which burrow into sand at low tide, carry with them a bubble of air from which they draw oxygen during their time underground.

Isopods have also become well adapted to a terrestrial existence, and their gills show appropriate modifications. The exopodites and endopodites of their pleopods are flat, soft lobes; the exopodites overlap each other like shingles on a roof and so make a cover over the endopodites and a small cavity, or gill chamber, beneath them. In some species of land isopods, the first pair of exopodites is greatly enlarged and covers the gills of all the posterior appendages, much as the first pair of abdominal appendages of *Xiphosura* makes an operculum over the gill books. In semiterrestrial isopods such as *Trichoniscus* and *Ligia*, which live along the shore line, the gills are in no significant ways different from those of strictly marine species, but there are special adaptations in those Oniscoidea that have become as fully adjusted to terrestrial life and to aerial respiration as any crustacean, even the tree climbing coconut crab. The surfaces of the exopodites of *Oniscus*, the wood louse that lives in damp places on land, are depressed into an air-filled cavity, or lung. In the sow bugs *Porcellio* and *Armadillidium*, which can tolerate greater degrees of dryness than *Oniscus* can, the exopodites of the first two pairs of pleopods are traversed by fine, branching tubules filled with air, the pseudotracheae. In appearance, these lobes are quite different from the others; they are the "white bodies" that are known to function as lungs. Removal of the gills from these crustaceans has shown, however, that in *Porcellio* as much as 34 percent of the total oxygen taken up comes through the general integument, and 24 percent in *Armadillidium*. The white bodies in *Porcellio* may be of greater importance in carbon dioxide elimination than in oxygen intake, for their removal results in the animal's death.

The cuticle of land isopods is also adapted in other ways to their respiratory needs, for it is grooved in such a fashion as to form gutters in which water collects and from which it drips down over the pleopods. The main gutters run laterally along the body to the terminal abdominal appendages, or uropods. When the animal is in a damp place, it presses its uropods down causing water to move by capillarity into the lateral channels and so to the pleopods. In conditions of extreme aridity, water expelled from the mouth and flowing backward along the gutters serves the same purpose.

The young of many insects lead an aquatic existence, and a number of them have gills. Nymphs of stone flies, mayflies, and caddis flies, and the larvae of a few aquatic beetles, have leaf-like extensions of their abdominal or thoracic segments called tracheal gills, because branches of the tracheal system extend into them and oxygen from the water can pass across them into the tracheal system. Larvae of the lepidopteran *Cataclysta fulicalis* are unique in being, for one thing, aquatic in habit, and for another, in having gills without tracheae and with the hemocoele extending directly into them. Removal of the gills from nymphs of the caddis fly *Macronema* does not lead to any significant drop in their oxygen uptake, but the gills may have a true respiratory function in other species. One hypothesis of the origin of insect wings was based on the assumption that ancestral insects were aquatic and equipped with such tracheal gills for respiration (see pages 205–207).

3. Lungs

"Lung," like "gill," has become a very general term that is applied to a variety of respiratory structures. It can denote, in invertebrates, any inturned, infolded, or ingrown area of the integument, including the tubular tracheae characteristic especially of insects, where gaseous exchange takes place. The exchange may be, in water lungs, with water, or in aerial lungs, with air.

a. WATER LUNGS

Water lungs are respiratory cavities, usually associated with the digestive tract, into which water is periodically drawn and from which it is periodically expelled. *Urechis*, an echiuroid that lives in a U-shaped burrow in the sand or mud of the seashore, has, for example, a long, distensible hindgut whose walls are smooth and thin. Water is pumped in and out of this by muscular contractions of the cloaca, a derivative also of the proctodaeum. Determinations of the oxygen content of the water drawn in and that expelled have shown that there is 40 percent less oxygen, and more carbon dioxide, in the outgoing water than in the ingoing, making it apparent that this is really a respiratory device. It is controlled in such a way that there are about 30 "inhalations" preceding a single "exhalation."

Many holothurians also use their hindguts as respiratory organs. The respiratory tree of most sea cucumbers is developed from the cloaca as two main tubes that divide and subdivide to form an extensively branching system of blind tubules that extend all through the perivisceral coelom (Fig. XI–2*b*). Seawater is drawn in and out of the cloaca by action of its muscles and of those that attach it to the body wall and is pumped in and out of the branching tubules by muscles in the walls of the larger trunks. The pumping mechanism is a regulated one, for several dilations and contractions of the cloaca are required before the respiratory tree is filled. In the course of one "respiration," the cloacal muscles relax and the cloaca expands, drawing water in. Then its muscles, including the circular sphincter that closes the anus, contract, forcing water into the main trunks of the tree. Contraction of their walls moves the water on, and the cloaca relaxes again, draws in more water, and again contracts. Six or eight of these intake contractions may occur before the tree is filled, after which all the water in it is expelled in one general contraction, and the intake mechanism starts over again. Gaseous exchange between the circulating water and the coelomic fluid takes place in the finer branches of the tree, where the walls are thin. That this exchange is of major importance to the holothurian's respiration is shown by the fact that the oxygen consumption is reduced 50 to 60 percent in *Holothuria tubulosa* when the anus is blocked. The movements are related to the amount of oxygen in the water and are more rapid in stagnant waters than in those where there is movement and the oxygen supply is continually renewed. When the water in which it is living becomes very low in oxygen, *Holothuria* will lift its posterior end to the surface and draw air into the cloaca to supplement the oxygen supply.

The rectal gills of dragonfly nymphs have been believed to have a respiratory function, but the weight of experimental evidence is against this. These gills are developed as lateral outgrowths of the rectal portion of the hindgut into which water is pumped in and out in a manner resembling the cloacal pumping of *Urechis* and *Holothuria*. Fine terminations of the tracheal tubes are closely applied to the walls of these gills, but gaseous exchange across them becomes increasingly less important with increasing age of the nymphs. The integument and rectal glands seem the primary area for this, with the rectal gills functioning as rudders in locomotion.

In ophiuroids, in which the coelom is much reduced in comparison to its size in other echinoderms, the genital bursae serve as respiratory organs. These are invaginations of the body wall above the gonads, each of which communicates with the outside by a narrow slit through which water is drawn by action of flagella on the epithelial cells. The direction of their beat is such as to cause water to enter the bursa from the distal end of each slit, flow through the sac, and leave it at the oral end of the opening. The exchange of gases between the seawater and the perivisceral fluid takes place across the wall that forms the base of the bursa. The bursae also communicate internally with the gonads and at spawning receive the gametes from them. In some species they also act as brood pouches in which the embryos develop up to, and well past, metamorphosis.

b. AERIAL LUNGS These, like other respiratory organs, vary in their complexity and in their location in different kinds of invertebrates. In the Ugandan swamp worm *Alma emini*, there is a localized area of the dorsal body wall that acts as a lung. This worm habitually lives in a burrow in an environment of water-logged and decomposing plants where the oxygen tension must ordinarily be too low to support aerobic respiration. Often the posterior centimeter, or slightly more, of the worm, representing about one-thirtieth of its entire length, projects from the burrow and lies on the surface of the mud, where it is exposed to the air. The margins of these segments can curl over the dorsal surface to make a tube enclosing their greatly vascularized epidermis. Bubbles of air trapped in this tube are carried down into the burrow when the worm withdraws and provide it with the necessary oxygen for its stay underground, the oxygen diffusing from the trapped air to the capillary network in the integument.

Mantle and mantle cavity have proved assets to molluscs in their colonization of land habitats, for they are structures that could be adapted to aerial respiration and develop into lungs. This asset has been exploited particularly by the gastropods, some of which can live an entirely terrestrial existence even in very arid regions. The mantle cavity of apple snails (*Ampullaria insularum*) of Africa and South America is incompletely divided by a partition into two compartments, one of which is a gill and the other a lung. In pulmonates, the entire cavity acts as a lung, gaseous interchange occurring between the blood in the thin-walled vessels or sinuses that form a close-meshed network in the roof of the mantle cavity and the air with which it is filled; there are no ctenidia. The mantle cavity communicates with the exterior by a small opening, the pneumostome, that can be opened or closed and in land snails operates in this way more or less rhythmically. These lungs are diffusion lungs, for there is no mechanism for movement or circulation of air in them as there is in the ventilation lungs of terrestrial vertebrates and in the tracheae of some insects. The mantle cavity in the aboreal New Zealand slug *Athoracophorus* is reduced to a small chamber whose walls are produced into tubules that ramify through the adjacent blood spaces. The nature of these tubules has not been fully investigated and though their function may be respiratory, it has yet to be proved. If it were, these gastropods would be unique among molluscs in their use of a tracheal respiratory system.

Some pulmonates are amphibious and some, like the common pond snails *Lymnaea* and *Planorbis*, have returned from land to an aquatic existence. Lymnaeidae are not as fully adapted to this as Planorbidae and at intervals rise to the surface of the water to take air into the mantle cavity, which functions as a lung. In Planorbidae the mantle has little significance to respiration and its function has been taken over by an external gill developed as an enlargement of the lobe of the mantle. This enlargement is much folded and pleated, is external to the mantle cavity, and is never ciliated. All these features distinguish it from a ctenidium and give evidence that it is a secondary development as these snails have become readapted to life in water.

The gill chamber of anomuran crustaceans has served, for them, a purpose similar to that of the mantle cavity for pulmonate molluscs, by endowing them with an area that could be converted into a lung. In *Coenobita*, the floor of this chamber is much more extensively vascularized than it is in aquatic genera and is a respiratory area accessory to that on the ventral surface of the abdomen. In *Birgus*, the roof of the gill chamber is prolonged into deep folds that hang down into the cavity as well-vascularized respiratory tufts. In isopods, the gills with pseudotracheae and the white bodies of *Porcellio* are also diffusion lungs. These adaptations of organs, designed primarily for aquatic respiration, to organs of aerial respiration have not required any especial modifications of the typical pattern of the isopod vascular system, as the development of respiratory surfaces in the mantle cavities of molluscs and the gill chambers of crustaceans has done. In both these cases, blood from the gill circulation has been diverted to lacunar or sinus spaces in the walls of the derived lung.

Book lungs are characteristic of scorpions and are found in some other arachnids as well (Fig. XI–3).

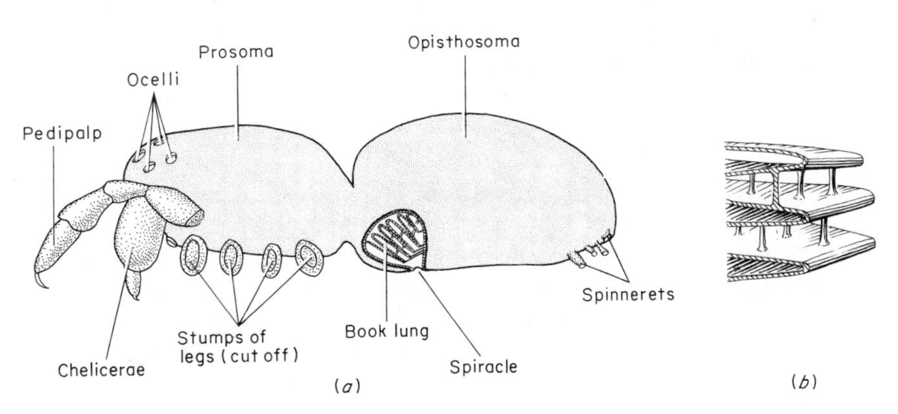

FIG. XI–3 *Respiratory mechanism of an arachnid.* ***(a)*** *Body of a spider, showing the location of one of the two book lungs.* ***(b)*** *Detail of a few lamellae and the bars that hold them apart.*

Very likely they originated as book gills, like those of *Xiphosura*, for they are fashioned in the same way by foldings of the wall of the respiratory surfaces. But in book lungs the lamellae lie in pockets with sclerotized walls that are turned in from the ventral surface of the abdomen. The inner walls of these pockets are folded into long, thin plates that are held apart by bars so that there are always air spaces between them. The outer walls are smooth and open to the exterior through small, usually oval, slits, the spiracles, through which air enters and leaves. Movement of the air may be expedited by dilation of the chamber brought about by contraction of a muscle on its dorsal wall; with subsequent relaxation of this muscle, the chamber returns to its original size and air is expelled from it. Gaseous exchange between the air and the blood in the lamellae occurs, as in other lungs, by diffusion across their broad surfaces. Removal of the book lungs from an African scorpion, *Opisthophthalmus capensis*, results in the complete cessation of oxygen uptake and consequent asphyxiation; if half of them are blocked by being painted over with rubber solution, the oxygen uptake is reduced to 70 percent of that of intact animals.

Tarantulas and trap-door spiders have book lungs in the second and third abdominal segments but in other Aranea there is but one pair, the posterior having developed into tubes or tracheae. In some species, the tracheal tubes are short and extend only into the abdomen, but in others they are long enough to reach into the head and legs, and they branch repeatedly forming a complex system and an extensive respiratory surface. They usually open to the outside by a single spiracle, representing the fusion of the originally paired openings and, in the species in which they are most greatly developed, may have a larger share in gaseous interchange than the single pair of book lungs.

C. **TRACHEAE** Tracheae are found in onychophorans, pseudoscorpions, opiliones, some mites and some spiders in which both pairs of book lungs have developed into tubular systems, in solpugids, in diplopods and chilopods, and in insects. They are air-filled tubes, open to the exterior through spiracles and developed in part at least as ingrowths from the surface of the body. They, therefore, fall into the general category of lungs, as these have been defined for invertebrates, and though a respiratory function is in every case presumed for them, it is in insects especially that tracheal respiration has been most closely studied and is assured. For them, the tracheal system is also a transport one, for the hemolymph has no part in the conveyance of gases.

The tracheal system of insects has a dual origin, the main tubes originating as cords of cells, about 20 μ in diameter, that grow in from the integument of the thoracic and abdominal segments, and their terminations, or tracheoles, originating from cells in the internal tissues (Fig. XI–4). Canals arising within the cords convert them to tubes whose walls consist of an epithelium continuous with the hypodermis and an inner lining continuous with the cuticle and, like it, coated with wax. It gives negative reactions to tests for chitin. This inner lining, the intima, is interrupted at intervals by taenidia (Latin dim. of *tainia*, "ribbon"). These are supporting structures that appear either band-like, helical, or beaded and that keep the wall from collapsing (Fig. XI–5), a matter of great importance to continual airflow. Typically, in each tracheate segment there develops a dorsal tube that carries air to the dorsal musculature and heart, a ventral tube that supplies the ventral musculature and nerve cord, and a median, visceral tube that supplies the digestive tract, fat body, and gonads. Latero-dorsal and latero-ventral trunks are formed by union of the segmental tubes, and branches from these trunks extend to all parts of the body and terminate in the tracheoles. These are fine tubules with diameters of less than 1 μ that are intracellular in origin, derived from processes extending from cells, the tracheoblasts, at the terminations of the tracheae. Channels arising within these processes become continuous with the finer terminations of the tracheal branches, so that the completely formed system looks as if these twigs had penetrated into the tracheal end cell. Careful study has shown, however, that this is not the case and that the tissues themselves contribute the tracheoles to the air transport system. Tracheoles can change their position and have been observed to be drawn to areas experimentally deprived of their normal tracheal supply by severance of the tracheal trunks or destruc-

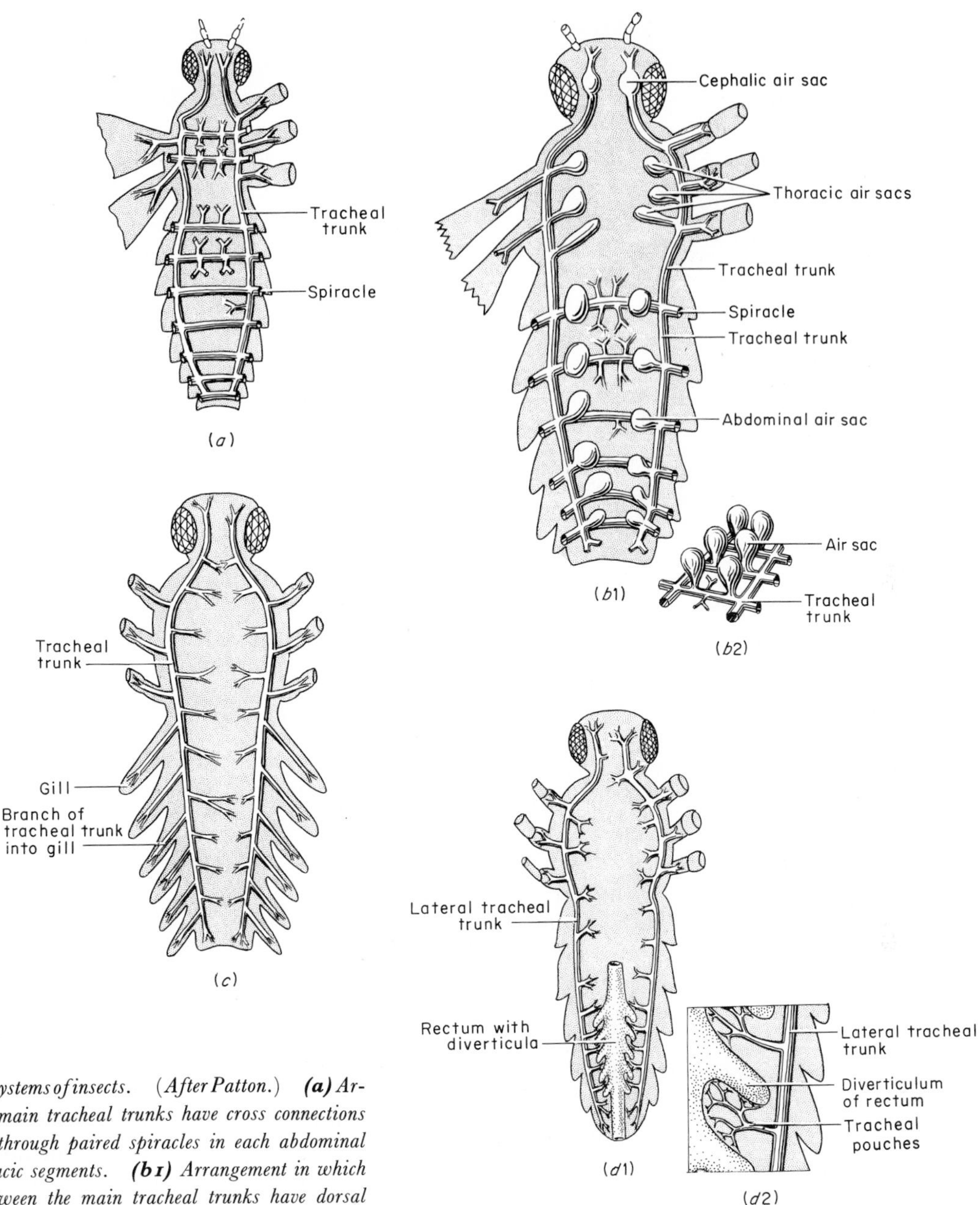

FIG. XI–4 *Tracheal systems of insects. (After Patton.)* ***(a)*** *Arrangement in which the main tracheal trunks have cross connections and open to the outside through paired spiracles in each abdominal segment and in two thoracic segments.* ***(b1)*** *Arrangement in which the cross connections between the main tracheal trunks have dorsal expansions, or air sacs.* ***(b2)*** *Detail showing the origin of the air sacs.* ***(c)*** *Arrangement of a closed tracheal system. There are no spiracles and in the abdominal segments branches of the trachea extend into lateral expansions of the body wall, or gills. This type of tracheal system is found in many aquatic insects.* ***(d1)*** *Tracheal system with rectal gills.* ***(d2)*** *Detail of a portion of the rectum, showing the extensions of the tracheae between the rectal gills.*

tion of the finer branches. Epidermal cells in the deprived area send out contractile filaments, sometimes as long as 100 μ, that become attached to the tracheoles and pull them into the new location.

The tracheal system develops during embryogenesis and during this time is filled with fluid, which,

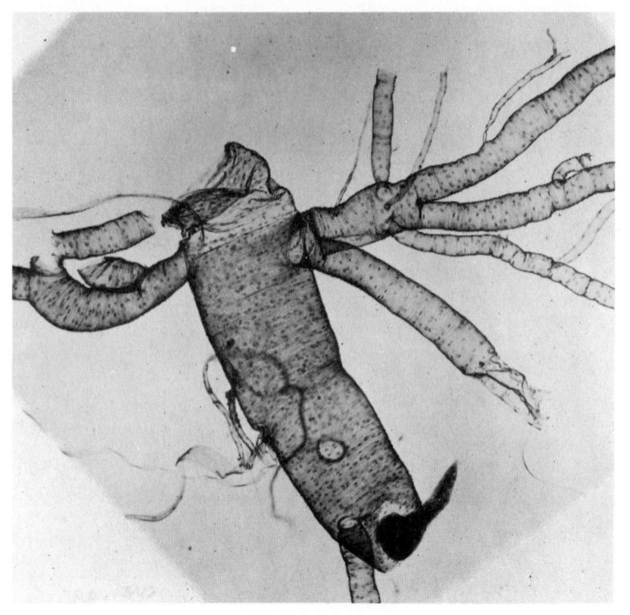

FIG. XI–5 *A tracheal trunk and its branches, showing the taenidia.* (*General Biological Supply House.*)

after hatching, is withdrawn into the tissues and replaced by gas. This replacement, which occurs at different times in different species, is independent of direct contact with air, for it may be accomplished while a larva is still bathed with the fluid contents of the egg. Gas usually appears first in the main tracheal trunks and later in their branches. In *Aedes* the postabdominal trunks are the first to be filled, then the main trunks from posterior to anterior, and, finally, the tinier branches. The whole system is gas-filled in about 12 min. But in *Sciara* the process may not be complete for several instars. Fluid remains in every case, and at all times, in the distal end of the tracheoles, and at each molt the whole tracheal system is filled with molting fluid as its cuticular covering, like that of the body as a whole, is shed and renewed. The source of the gas that replaces the molting embryonic fluids is not yet known, but the rapidity with which it can appear and the fact that it does so in submerged larvae suggest that it comes from some oxygen-releasing metabolic process, such as, possibly, the decomposition of hydrogen peroxide through the agency of the heme enzyme catalase:

$$2H_2O_2 \xrightarrow{\text{(catalase)}} 2H_2O + O_2$$

This is one of the most rapid enzymatic reactions known, for it has been shown that at 0°C 1 mg of pure catalase can produce 2,740 liters O_2/hr from hydrogen peroxide. This corresponds, in molecular units, to the decomposition of 4.2×10^4 mol H_2O_2/sec, at 0°C, by 1 mol of catalase.

Once gas filled, the tracheal system becomes a functioning respiratory organ. In some insects it operates as a diffusion lung, the air entering the open spiracles and moving through the tubes by diffusion only. In others, especially in large or active insects, it operates as a ventilation lung, in which air is moved in and out by various devices. These may be direct, involving the tracheae only, or indirect, involving bodily movements. In dipteran larvae the tracheae expand and collapse rhythmically, drawing air in through the spiracles as they expand and expelling it as they collapse. In many orthopterans, abdominal movements bring about inspiration and expiration, and in flying insects, movements of the thoracic segments coincident with flight displace the air within the tracheal tubes. For the most effective ventilation, there must also be some control of the spiracles to insure correlation of their opening and closing with the respiratory movements. This control is exhibited in a number of insect species.

Spiracles are slits that may be flush with the surface of the body, sunk below it, or elevated above it. They may lead directly into a main tracheal trunk or into a small atrium, which in some insects is provided with valves and devices for filtering the air. In insects they are usually surrounded by gland cells, which secrete an oily, water-repellent substance, and sometimes by hairs or bristles. The maximum number of spiracles in an insect is 20—two thoracic and eight abdominal pairs—but in different species, there is variation in their number, their position, and their use. They have the dual function of admitting air to the tracheal system and of preventing undue water loss from it, for the extensively ramifying tubes present a large area for possible evaporation.

In a number of insect species the longitudinal tracheal trunks are expanded into sacs whose walls are thin and not supported by taenidia (Fig. XI–4*b*). In *Periplaneta* such air sacs lie near the spiracles in each segment; in other insects, such as *Apis*, the segmental sacs have coalesced into several large ones. These increase the air capacity of the tracheal system (vital capacity of the insect) and must also give some buoyancy to a flying insect. Increase in vital capacity is perhaps the only advantage to be gained from air sacs that lie near the spiracles, but it has been postulated that the distal ones contribute to the circulation of air and its distribution to the tracheoles.

Respiratory movements have been studied in a number of insects, especially orthopterans, insects with ventilation lungs that lend themselves especially well to experimental studies. The course of air flow has been followed by introduction of fine graphite particles into the atmosphere and observation of the spiracles through which it enters and leaves the insect. The relation of particular spiracles to respiratory movements has been determined by blocking pairs of them successively and by correlation of the respiratory movements with changes in the volume of the thorax and abdomen of insects confined in chambers divided into two parts by a rubber diaphragm encircling the insect between its thorax and abdomen. Such observations have shown that, in grasshoppers, dorso-ventral flattening of the abdomen is largely responsible for ventilation of the tracheal tubes, while antero-postero telescoping of the abdominal segments effects it in hymenopterans and dipterans. In all cases, expiration is an active process of muscular contraction and inspiration a passive one, as return of the body to its original dimensions after a contraction is due to the elasticity of chitin. In some insects, air enters and leaves through all the tracheae indiscriminately; in others, there is a directed flow. The thoracic spiracles are open, for example, in the cockroaches *Periplaneta* and *Blaberus* during inspiration and the abdominal ones, during expiration.

The respiratory cycle has been most carefully followed in grasshoppers. It begins with expansion of the abdomen, which, under the conditions in which the observations were made, lasted about 0.25 sec; during the final 0.2 sec of this phase, the two thoracic pairs of spiracles and the first two abdominal pairs were open, while the last six pairs were closed. This is the period of inspiration, when air flows into the anterior region because of reduced abdominal pressure. Just before it is over, the spiracles close, sealing off the tracheal system. Then the compression phase begins, lasting about 1 sec. During this phase, all spiracles are closed and intraabdominal pressure rises as the tergo-sternal and intersegmental longitudinal muscles contract, reducing the volume of the abdomen. Maximum contraction is reached in the final, or expiratory, phase, when the abdomen returns to its preinspirational dimensions. This phase also lasts about 1 sec, in the first part of which all the spiracles are closed. In the last 0.3 sec, the posterior six pairs of abdominal spiracles open, while the others stay closed, and the air is expired through them. They continue to stay open during about the first 0.1 sec of the inspiratory phase.

Conditions in the air sacs have also been followed during the respiratory cycle and it has been observed that those near the spiracles fill during inspiration and empty and collapse during expiration. The distal sacs, on the other hand, collapse during inspiration and remain expanded during expiration. On the basis of a model illustrating air flow in such a system, it has been postulated that pressure built up in the distal sacs augments the intratracheal pressure of the compression phase in overcoming the resistance of the finer tracheae and helps to drive air onward into the tracheoles. The intratracheal pressure is transmitted from air sac to air sac because, as the air is pushed along the tracheae and into a sac, its volume becomes greater than the trachea leading to the next one can handle. The sac, therefore, expands until the pressure in it is enough to expand the tubule and let the air pass into, and fill, the sac immediately distal to it. The sacs, therefore, become sites of local pressure, helping to force air deeper and deeper into the tracheal system.

Gaseous exchange probably occurs throughout the tracheal system. The particular role of the fluid-filled tracheoles and their end cells is not yet clear, but it is generally believed that in all actively metabolizing tissues changes in osmotic relations between the hemolymph and the tracheolar fluid results in some loss of water from the tracheolar fluid. As the amount of fluid diminishes, the column of air moves a corresponding distance down the tiny tubule, bringing oxygen closer to the cells (Fig. XI–6).

All terrestrial arthropods, especially those with

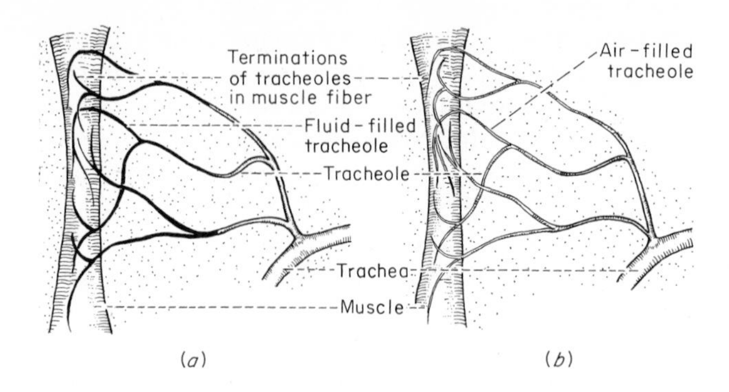

FIG. XI–6 *The presumed mechanism of tracheal respiration. (From Gardiner.)* ***(a)*** *Resting muscle. The tips of the tracheoles are filled with fluid, from which O_2 diffuses into the muscle fiber and to which CO_2 diffuses from the muscle fiber.* ***(b)*** *Muscle after contraction—changes in the osmotic relations between tracheolae and tissue fluids has led to the movement of water from the tracheoles and the passage of the air column further along them, facilitating gaseous interchange between muscle fiber and air tube.*

well-developed tracheal systems, are confronted with danger of death by dehydration. The perils are less to those whose tracheae are short and unbranched, like onychophorans and diplopods, even though they have no means of closing their spiracles, than to those whose tracheae are long and branching, like scorpions, chilopods, and insects. Insects with an open tracheal system, which lead an aquatic existence for all or part of their lives, are confronted with danger of death by drowning, if water enters their spiracles, or by asphyxiation, if gaseous oxygen is not available to them. Various adaptations have arisen that minimize these dangers. Many terrestrial species are nocturnal and so escape the full heat of the sun during the time of their activity, and many inhabit damp places and live in the bark of trees, under rocks, stones, fallen logs, and leaves, or under cover of forest litter or moist soil. Some protection is also afforded those that have sunken spiracles or ones guarded by tufts of hair, for air currents are deflected from them and the danger of evaporation accordingly reduced. Many devices have developed among insects to decrease the hazards of life in water. There is evident a general tendency toward reduction in the number of functional spiracles and fusion or displacement to favorable positions of those through which air can be admitted. Some insects have entirely closed tracheal systems and their external gaseous exchange is wholly cutaneous, their tracheae acting solely as conducting vessels. Those with open systems depend, for the most part, on some contact with the atmosphere or upon submerged water plants for their oxygen supply. Those that make contact with the atmosphere rise to the surface of the water, break its surface film, and stay there long enough to fill their tracheae with air, or else have some way of carrying air down with them to meet their respiratory needs while they are underwater. The development of hairs or bristles that are coated on one side with wax that repels water and, consequently, is not wetted, while the other, noncoated side, is, has been of great advantage to them. By virtue of circlets of these hydrofuge hairs that surround their posterior spiracles, mosquito larvae, for example, can break the surface film and expose them to the air, the hairs bending out so that their nonwaxy, hydrophile surfaces are in contact with the water and their waxy, hydrofuge surfaces spread out around the open spiracles. These larvae are thus dependent upon the capacity of their tracheal systems, and the completeness with which the air in them is changed by diffusion only, to support their respiration while they are underwater. They cannot stay submerged for more than a few minutes and habitually spend most of their lives suspended head down from the surface film. Some other dipterous larvae have spiracular devices for puncturing or slitting the tissues of aquatic plants and allowing the oxygen from photosynthesis to diffuse directly into the open spiracles. Others live near plants and open their spiracles, guarded by hydrofuge hairs, to the bubbles of escaping oxygen.

Pupae of insects that develop in water also have various devices that enable them to obtain oxygen during this period of active tissue resorption and replacement. Some beetles pupate in cases that contain gas; others have tubular extensions of the pupal case that make and retain contact with an oxygen source. Dipteran pupae make use of similar devices. Perhaps the most unusual of these are the spiracular gills or horns of crane flies (Tipulidae). These are structures developed from the spiracles, or from the integument immediately around them, as organs of either aquatic or aerial respiration. They thus permit the pupa and the adult just before its emergence (pharate adult) to obtain oxygen from the water, when it is high, and from the air when the water level drops below the position of the pupa. The respiratory horns of *Pseudolimnophila*, which as larvae live in rich mud and submerged decaying organic matter, are developed from the pronotum and may be as long as one-quarter the length of the pupal body. The distal ends of these horns are expanded as two flaps that spread apart as the pupa rises to the surface. These flaps, like the spiracular tufts of hydrofuge hairs, open out and guard a passage from the air to the pupal trachea, which extends beyond the forming, or formed, cuticle of the adult and connects the tube of the respiratory horn with the spiracle. By this device, air can diffuse in and out of the tracheal system of the pupa and pharate adult.

Adults of a number of species of Hemiptera and Coleoptera spend a great part of their lives underwater, surfacing periodically to ventilate their tracheal systems and collect stores of air to carry back with them when they submerge. When the back swimmer, *Notonecta*, for example, comes to the surface its abdominal spiracles, surrounded by hydrofuge hairs, are open and the tracheal system is flushed out and fresh air introduced by active ventilating movements. When the bug dives, the abdominal spiracles are closed and extra air is carried down trapped in a bubble under each fore wing. Underwater, the thoracic spiracles are open to this air pocket and oxygen can be drawn from it until it, as well as that in the tracheal system, is exhausted, when the insect must come again to the surface for replenishment. The water beetle *Hydrophilus* comes up head first and inspires air through its mesothoracic spiracles. These are open to the subelytral space, which is also filled with air before the beetle dives again into the water, where it can stay continuously for as long as 36 hr. The oxygen supply of these insects is increased not only directly by the reserve in the subelytral air pocket but indirectly by diffusion of oxygen into it. Ultimately the size of the bubble is reduced, and when its surface area becomes too small for adequate diffusion the insect must replace it. While this mechanism does not provide an indefinite supply of oxygen, it does enable an insect to stay underwater for long periods of time. *Notonecta* can stay for as long as 7 hr in fully aerated water. Yet the mechanism imposes certain conditions of underwater life, for the air bubbles increase an insect's buoyancy and unless it keeps actively moving about or clings fast to underwater vegetation or submerged rocks and stones, it will rise to the surface whether its respiratory needs require it or not.

The plastron of certain hemipterans and coleopterans that live perpetually underwater is not an air store like this, but a surface across which oxygen in the water diffuses continuously into the tracheal system. It, therefore, operates essentially as a gill and as its volume is very small and practically constant it does not increase the buoyancy of the insects to any great extent. This permits them to go deeper into the waters and to crawl about, or even stay quietly on the stones and pebbles at the bottom of streams and lakes with fast-moving currents. The plastron consists of a thin film of gas, which may cover all of an insect's body or be localized in particular areas, and which is held in place by hydrofuge hairs. The device has been most fully investigated in *Aphelocheirus*, a fairly large and very active bug found in rapidly flowing, clear streams. All of its ventral surface and much of the dorsal is covered by a thick pile of minute hairs, about 0.2 μ in diameter and 506 μ long. These hairs are regularly spaced, about 0.4 μ apart, and number some 2,500,000 per mm^2 of surface. Each is bent sharply at the tip, the entire system making a kind of slatted lattice over the cuticle supported by columns. The plastron appears first in fifth-instar nymphs, gaseous exchange in earlier stadia being entirely cutaneous. But when fifth-instar nymphs molt, the plastron is formed as the tracheae are filled after resorption of the molting fluid, so that it is not atmospheric air but gas derived from the same source as that which fills the tracheal system. This system is well developed in *Aphelocheirus* with the rosette-shaped abdominal spiracles opening into the plastron. They are protected by fine hairs continuous with the hair pile enclosing the plastron. The importance of the plastron to respiration has been shown by "painting out" the hairs that hold it in place with an agent that wets them, such as butyl alcohol. When the plastron of *Aphelocheirus* is so removed from the ventral surface the bug becomes quiescent but does not die, as diffusion of oxygen through the integument provides enough to maintain it at rest. Measurements have shown that this is about 16 percent the oxygen consumption of an active intact animal, the remaining 84 percent being accounted for by the plastron, across which there is continual inward diffusion of oxygen from well-aerated water and no imbalance in the outward diffusion of nitrogen.

The density of the hydrofuge hairs is an important factor in the preservation of the plastron. A pile with a density of 1×10^8 per cm^2 can withstand hydrostatic pressures up to 2 atm and permit insects like *Aphelocheirus* and some species of beetles and even weevils to live continuously underwater, and fill ecological niches inaccessible to other aquatic insects. Insects with hair piles of densities of 1×10^6 to 1×10^8 per cm^2, like the chrysomelid beetle *Haemonia* and some members of the beetle family Elmidae, have thicker plastrons, which probably provide them with a small reserve of oxygen and so serve as air stores as well as diffusion surfaces. The resistance of these hair piles is less than that of the thicker ones, but they will withstand pressures from 0.5 to 2 atm. Insects with them are not, however, so "perfectly" adapted to underwater life as are those with thicker hair piles and thinner plastrons. Possibly these have evolved from ancestors that lived along the banks of streams and lakes but were protected by hydrofuge hairs from the dangers of falling in, or that needed such protection when they entered the water to lay their eggs. There are modern riparian species that have such hairs but lead a terrestrial rather than an aquatic existence. Indeed, there are some species even within the same genus of weevils (*Phytobius*), that are fully adapted, by virtue of a plastron mechanism, to live underwater and others that, although with hydrofuge hairs, live entirely out of water and derive no evident benefit from them, unless it be protection from the rain.

C. OXYGEN AND CARBON DIOXIDE CONTENT OF NATURAL ENVIRONMENTS

By means of respiratory organs and coelomic and vascular systems, body fluids are brought close to the surface where they are separated from the environment, in most cases, only by thin layers of tissue. Diffusion occurs across these tissues to the extent that they permit it and, for the greatest efficiency of exchange, not only must the external medium be continuously renewed by some ventilating device but the body fluid must also be kept moving across the respiratory area. If this area is relatively large, the fluid may move slowly and still pick up and discharge its full quota of gas, but if the area is small, more rapid movement of the fluid is obviously desirable. To be useful in respiration, the fluid must be able to absorb oxygen at the pressure to which an organism is usually exposed in its natural habitat and to deliver it to the tissues under the conditions of oxygen pressure prevailing there. There is little specific information about the oxygen pressures in many natural habitats, other than aerial ones, and even less about those existing in the tissues of invertebrates. Yet from observations and experiments of various kinds, information has been gained from which deductions can be made about the relative values of body fluids to respiration, especially of those that contain the chromoproteins hemoglobin, chlorocruorin, hemerythrin, and hemocyanin.

Oxygen represents about 21 percent of the mixture of gases that constitute the Earth's present atmosphere, and carbon dioxide about 0.03 percent. The partial pressure of oxygen at sea level and thereabouts is, therefore, about 160 mm Hg (21 percent of 760 mm, or 1 atm) and of carbon dioxide about 0.23 mm Hg. The partial pressure of a gas in solution, usually referred to as its tension and expressed in terms of the amount, in milliliters, dissolved in 100 ml of the fluid, is determined primarily by its solubility, the composition of the fluid, and the temperature. Freshwater will dissolve more oxygen than salt water, and both will hold more in solution at low temperatures than at high. The greater solubility of carbon dioxide means that in the same conditions more carbon dioxide would be dissolved in any given fluid than oxygen. These properties of respiratory gases affect their distribution in natural waters, as well as their transport in body fluids, and both together impose limitations to the occupancy of certain habitats and so to the distribution of animals, both vertebrate and invertebrate.

The amount of dissolved oxygen in representative natural waters, when they are in equilibrium with air, are given in Table XI–1. These figures do not, however, necessarily represent the actual oxygen tension in any locality, since this varies with the degree to which the water is aerated by turbulence, photosynthesizing organisms, and other natural events. Indeed, even diurnal variations of considerable magnitude may occur in any body of water. A daily range in oxygen tension from 10 to 490 mm Hg has been reported from a drainage ditch near Leiden, Holland, in which there was an abundance of plant life and a considerable number of animal species. Although it is unlikely that these extremes are reached in many bodies of water, there is little doubt that aquatic animals are exposed to relatively large differences in the amounts of oxygen available to them at different times of the day, the month, and the year.

TABLE XI-1 Oxygen Content of Air and Waters of Different Kinds in Equilibrium with Air (ml O_2/100 ml fluid)

Medium	*Temperature, °C*	*ml O_2/ml fluid in equilibrium with air*	*Tension O_2, mm Hg*
Air	—	————	160
Water			
Fresh	0	1.03	7.8
	30	0.56	4.28
Salt			
0.2%	0	0.8	6.28
	30	0.45	3.42
3.5% (seawater)	20	0.54 (average) Range 0–0.85	4.10

Clearly, air is the most favorable medium for the absorption of oxygen by an animal, at least as far as the steepness of the concentration gradient is concerned. But the greatest number of invertebrate classes live in less favorable mediums, in waters of varying degrees of salinity and in ones that may be periodically or permanently stagnant. Habitats that are particularly low in oxygen content are deep waters; the muddy bottoms of ponds, lakes, and oceans; swamps, bogs, and the digestive tracts of animals, in which the oxygen content of the gaseous mixture may well be, at all times, considerably less than 1 percent (7.6 mm Hg).

D. THE TRANSPORT OF RESPIRATORY GASES

The quantity of oxygen that could be carried only in solution by a body fluid could hardly be adequate for the respiratory needs of any but inactive or sluggish animals, or ones in which the volume of fluid was large in comparison to the mass of tissue. The oxygen-carrying capacity can be increased if the oxygen is bound to a compound that is one of the fluid's regular constituents. This is the case in vertebrate blood, where the hemoglobins are known to be oxygen carriers, some more effectively than others; in man, for example, with "standard" hemoglobin, 98 percent of the oxygen transported from lungs to tissues is combined with hemoglobin. Recognition of hemoglobins and other pigments in the blood and coelomic fluids of invertebrates has raised the question of whether they, too, function in the same way, whether they have some other function, or, possibly, whether they have none at all. There are no grounds for assuming, a priori, that invertebrate hemoglobins perform in the same way that vertebrate hemoglobins do or that chlorocruorin, in spite of its chemical similarity to hemoglobin, acts as a respiratory pigment. The same reservations hold for the other pigments of invertebrate coelomic and vascular fluids. Investigation has shown, however, that invertebrate hemoglobins, chlorocruorins, hemerythrins, and hemocyanins do function as respiratory pigments, although in different ways and under different conditions, while the other chromoproteins of invertebrate fluids and the carotenoids, aphins, and echinochromes do not. Tracheate arthropods, most of which lead very active lives, have bypassed the necessity for a respiratory pigment by development of an internal system of tubes, which brings air almost into direct contact with their tissues.

1. The functions of respiratory pigments

There are three ways in which a pigment might conceivably be useful to respiration. It might, like vertebrate hemoglobin, continually take up, convey, and release oxygen to the tissues and thus act as an oxygen carrier; it might take up oxygen only at low tensions, functioning then at times when the level in the environment becomes for one reason or another so low that it is essential to the organism that the amount carried in solution be supplemented, although the dissolved oxygen might be fully adequate for its needs when the external oxygen tension is higher; or it might act as an oxygen store, holding the oxygen under ordinary conditions of an animal's life and releasing it only in times of extreme oxygen deprivation and acute respiratory stress. These possibilities have been tested with the pigments of a number of invertebrate species.

The potentialities of a pigment as an oxygen carrier can be ascertained by exposing it to different partial pressures of oxygen and measuring the amount combined when it is in equilibrium with the gas at each one of them. For invertebrate pigments, this must usually be done in vitro, either with samples of fluid drawn from a living animal, often after dilution with seawater, saline, or a buffer solution in order to secure a workable amount, or with solutions of the extracted and purified pigment. The properties of invertebrate fluids have been studied in vivo in some entomostracan crustaceans and insect larvae, which are sufficiently transparent to permit direct spectroscopic observation of the fluids in their bodies and so measurement of the amount of the oxygenated pigment at a given time and under given conditions. It has also been possible to make studies in vivo of the bloods of some decapod crustaceans and cephalopod molluscs, whose circulatory systems are such that pre- and postbranchial blood can be directly withdrawn. Indeed, the earliest observations of the oxygen-carrying properties of hemocyanin were made in this way a century ago, when it was seen that the blood of living specimens of *Sepia officinalis* changed from colorless to deep blue as it passed through the ctenidia and became oxygenated there.

The results of such measurements can be expressed in any one of three ways: (1) as the volume percent of oxygen held; (2) as the loading and unloading tension of the pigment; or (3) as the equilibrium (absorption or dissociation) curve of the pigment alone or of the whole blood. The first way, giving the volume of oxygen held by a 100 ml of fluid at any given external pressure, is useful in comparing the efficacy of different pigments, or of the same pigment in different circumstances. The second way is of particular value in interpreting the range of usefulness of a given pigment. By loading tension (t_l) is meant the pressure of oxygen with which the fluid or pigment is in equilibrium when 95 percent of the pigment is oxygenated, and by unloading tension (t_u), the pressure of oxygen when 50 percent of the pigment is oxygenated. These values have been found to vary greatly for different pigments, some of which become loaded and unloaded at much lower oxygen tensions than others, a difference which makes evident why some animals can live in conditions that would be virtually impossible for others. In the third way, the amount of oxygenated pigment, if the measurements are made spectroscopically, or the amount of oxygen bound by the pigment—that is, the percentage saturation—are determined when the fluid is in equilibrium with oxygen at different tensions, but at constant temperature and pH, and the values plotted. When the points are joined, a curve results from which can be read the amount of oxygen held by the fluid, or the pigment, at any pressure within the range tested. Such a curve reveals the oxygen affinity of a pigment, showing how tenaciously it will hold the gas as the external tension is reduced. A pigment of low oxygen affinity will release the gas at higher tensions than will one of high oxygen affinity and it will become fully oxygenated, or loaded, only at comparatively high external tensions. Comparisons of such curves made at different temperatures and different pH's show also the effect of these variables upon the properties of the pigment and permit predictions of how it might operate under different natural conditions.

If similar measurements were made and similarly plotted for a fluid without a respiratory pigment, where the oxygen could only be in solution, the result would be a straight, ascending line, because the amount of dissolved oxygen increases directly with the external pressure. The slope would also continue to rise indefinitely, while that of the equilibrium curve of a fluid with a respiratory pigment must flatten out when all the pigment becomes oxygenated. This flattening begins at different points—that is, at different pressures of oxygen—with different respiratory pigments and is a function not only of their combining power but of their concentration as well, since each molecule of pigment combines with a definite amount of oxygen. The volume percent and the total oxygen content of a fluid are also directly related to the concentration of the pigment in it. Where this is contained in corpuscles, the concentration is determined by the number of corpuscles as well as by the concentration of pigment in them. An equilibrium curve also rises more steeply than a solution curve, for respiratory pigments in every case increase the capacity of the fluid for oxygen. Like other proteins, they may act as buffers in regulating the pH, but as reactants with oxygen they act also as oxygen buffers in reducing the differences in oxygen tension due to absorption and liberation of the gas. This buffering effect, essentially the removal of oxygen from solution by its combination with the pigment, provides also for the preservation of a steeper gradient between the external medium and the fluid, and the fluid and the tissues, than would be the case were the oxygen in solution only. An invertebrate body fluid containing a respiratory pigment may contain 10 times as much oxygen as the fluid alone.

The shape of an equilibrium curve is often, but not always, influenced by the prevailing tension of carbon dioxide. The oxygen affinity of vertebrate hemoglobin is known to be reduced as carbon dioxide tension increases. This relationship, known as the Bohr effect, is an advantageous one to the vertebrate, for because of it the blood releases its oxygen more readily in the tissues, where it picks up carbon dioxide, than it otherwise would. This effect can be produced in vitro, when the equilibrium curve is shifted to the right as the carbon dioxide tension is increased or the pH reduced, meaning that a higher pressure of oxygen is necessary to achieve the same volume percent than is the case when no carbon dioxide is present, or if it had no effect upon the oxygen affinity of the hemoglobin. Some invertebrate pigments exhibit a Bohr effect of greater or lesser magnitude than that of vertebrate hemoglobin, some do not show it at all, and some show a reverse one, the oxygen affinity of the pigment becoming greater as the carbon dioxide tension of the medium rises. But, in general, carbon dioxide seems to influence the equilibrium curves of invertebrate pigments less than temperature does. Both the oxygen affinity of a pigment and the effect of temperature upon it must to a great extent determine the distribution of the animals possessing it, for those having a pigment of low oxygen affinity cannot live in oxygen-deficient environments, if they rely upon it at all for respiration, nor can any live outside the temperature range in which their pigment functions effectively.

It is clear that if a pigment is to play any part in respiration it must reach its maximum saturation at oxygen tensions that are at least of the same order of magnitude as those in the animal's environment. Moreover, the oxygen tension of the fluid must be sufficiently higher than that in the tissues to establish an effective diffusion gradient. Although the unloading tension of a pigment can be read from its equilibrium curve, conditions in vivo may be quite different from those in vitro, and there is no certainty as to what the oxygen tensions in the tissues really are. Yet the fairly wide distribution of tissue hemoglobins, which act as oxygen acceptors and storage centers, would suggest that in many cases a steep concentration gradient is maintained between fluid and tissue. Measurements of the oxygen capacity of an animal's body fluid and of its oxygen consumption under different conditions of activity combined with information from the equilibrium curve of its respiratory pigment provide indirect evidence for the delivery of oxygen from the fluid to the tissues. The laboratory studies that have been made of invertebrate hemoglobins, chlorocruorins, hemerythrins, and hemocyanins have shown that there is as much versatility in their biological properties as there is diversity in their chemical and physical ones. Field studies of animals in their natural habitats and laboratory studies of them where these habitats could be simulated have indicated a correlation between these biological properties and the conditions of an animal's life and revealed adaptations that make life possible in ecological niches of many different kinds.

a. THE BIOLOGICAL PROPERTIES OF INVERTEBRATE HEMOGLOBINS The hemoglobins of invertebrates differ in their biological properties from those of vertebrates particularly in their loading and unloading tensions and in their sensitivities to changes in carbon dioxide tension and pH. In general, their loading and unloading tensions are much lower than those of mammals and more in the range of those of aquatic vertebrates. Sample values for these are presented in Table XI–2. Reference to this and to the equilibrium curves of Figs. XI–7 to XI–10 will show that the hemoglobin of the lugworm *Arenicola*, for example, is fully oxygenated at pressures at which human hemoglobin is almost completely deoxygenated and that there is, likewise, much variation in the oxygen affinities of hemoglobins among the invertebrates tested.

The preferential affinity of hemoglobin, and the still higher affinity of chlorocruorin, for carbon monoxide, even in the presence of abundant oxygen, has been a valuable tool to study of the respiratory function of these pigments. The affinity for carbon monoxide of the chlorocruorin of the sabellid *Branchiomma* is, for example, 570 times that for oxygen. All hemoglobins and chlorocruorins will, however, become saturated with carbon monoxide in animals exposed to the gas at suitable pressures for suitable lengths of time. In no case has a pressure of carbon monoxide that will combine with all the hemoglobin or chlorocruorin in a body fluid been found to affect the operation of the heme enzymes, so that oxidative metabolism in the tissues is unimpaired even when the respiratory pigment is effectively blocked from combination with oxygen. After inactivation of its respiratory pigment by this means, an animal can be exposed to a wide range of external oxygen tensions and its oxygen consumption measured at each. The degree of its dependence upon the oxygen taken up and delivered by the pigment can thus be estimated and compared to the amount taken up and used in similar circumstances by control animals. This method is analogous to that by which dependence upon a respiratory organ can be estimated by removing it, or blocking access of oxygen to it. It cannot be employed in studies of hemerythrins or hemocyanins, for hemerythrin does not combine with carbon monoxide at all and hemocyanin does so only to a negligible extent.

A transport role, similar to that of vertebrate

TABLE XI-2 Loading and Unloading Tensions of the Hemoglobins of Representative Animals

	Animal	*Temperature, °C*	*CO_2 tension or pH*	*t_l, mm Hg*	*t_u, mm Hg*
Vertebrates	Man	38	40 mm Hg	90	27
	Codfish	14	0.3 mm Hg	70	15
	Carp	15	1–2 mm Hg	10	2–3
	Trout	15	1–2 mm Hg	43.5	17
Invertebrates					
Gastropod	*Planorbis*	12	————	10	1–2
Lamellibranchs	*Cardita floridana*	22–24	pH 7.5	45	17
	Noetia ponderosa	22–24	pH 7.4	40	12
Echiuroid	*Urechis caupo*	19	7 mm Hg	90–100	12
Polychaetes	*Arenicola marina*	17	pH 7.3	7–13	1.8
	Nephthys hombergi				
	Coelomic	15	pH 7.4	30	7.5
	Vascular	15	pH 7.4	12	5.5
Oligochaetes	*Allolobophora longa*	20	pH 7.3	17.5	6
	Lumbricus terrestris	20	pH 7.3	22.5	8
Holothurian	*Cucumaria miniata*	26	pH 7.5	4	8

hemoglobins, can be definitely attributed to the hemoglobins of those species in which blockage of the pigment with carbon monoxide has been found to interfere with ordinary oxygen utilization and metabolic activity that is dependent upon a supply of oxygen. For example, when specimens of the earthworm *Lumbricus terrestris* are exposed to carbon monoxide, their oxygen consumption drops significantly at all pressures higher than 8 mm Hg, indicating that the hemoglobin in ordinary circumstances of the worm's life carries and delivers oxygen continuously. The importance of the pigment in oxygen transport varies with the external oxygen pressure and with the temperature, but under no circumstances has it been found to carry more than half the total oxygen utilized, the remainder being supplied by that in solution (Fig. XI–7). Oxygen consumption is likewise reduced in leeches whose blood contains hemoglobin, after its saturation with carbon monoxide, at all concentrations of oxygen above 3 percent, while the respiration of leeches without hemoglobin is not altered after their exposure to carbon monoxide.

In specimens of the marine polychaete *Nereis diversicolor*, whose blood is saturated with carbon monoxide, oxygen consumption is reduced about 50 percent when the external oxygen concentration is 7.5 ml/liter and stops entirely when it is reduced to 3.3 ml/liter. The pigment thus seems to be of particular importance to them when the external oxygen tension is low, a condition they must frequently encounter as they lie buried in the sand, the oxygen bound with it providing half of that needed for respiration in these circumstances. Oxygen consumption is also reduced in the freshwater oligochaete *Tubifex tubifex*, after exposure to carbon monoxide. In these worms, whose hemoglobin has a high oxygen affinity, the reduction is about one-third at all external concentrations above 1 ml/liter. *Tubifex* lives in the muddy bottoms of ponds and lakes, where the oxygen tension is always low, and its hemoglobin probably at all times carries this fraction of the oxygen needed for respiration, the remaining two-thirds being supplied by that in solution.

In similar experiments with the pulmonate snail, *Planorbis*, blocking of the hemoglobin was found to affect its behavior. While fully active in water with an oxygen tension equivalent to 23 mm Hg, the snails became lethargic at tensions below this, down to 8 mm Hg, below which they became completely inert. This behavior could be interpreted as meaning that their hemoglobin acts as a carrier at all tensions or that, at tensions above 23 mm Hg, their oxygen requirements are satisfied by the gas that is in solution alone. *Planorbis* can tolerate foul water and can even endure extended periods of time out of water when, for one reason or another, it becomes stranded on shore. The oxygen consumption of the snails, in natural con-

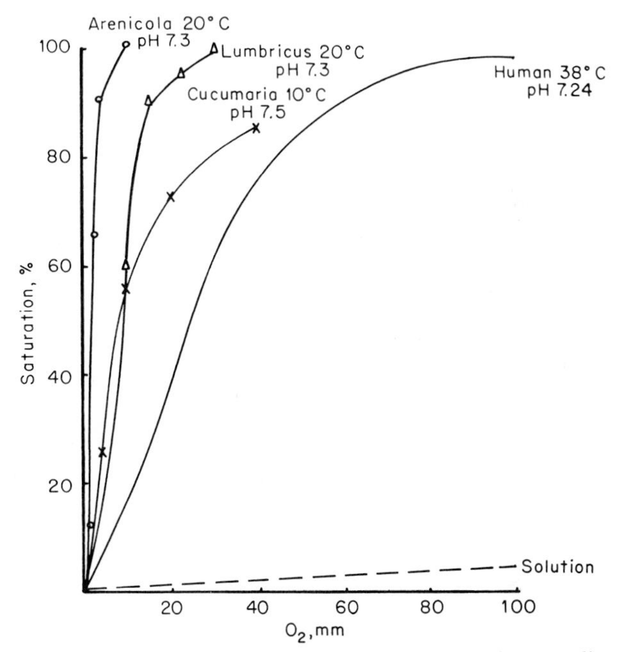

FIG. XI–7 *Equilibrium curves of human blood, of mammalian hemoglobin in solution, and of the hemoglobins of* Arenicola, Lumbricus, *and* Cucumaria. *The temperatures at which the determinations were made are in the range of the animals under natural conditions of life. (Data for curve of human blood from F. R. Winton and Sir William Bayliss, "Human Physiology," 6th ed., Churchill, London, 1968. Data for curve of hemoglobin in solution from W. H. Howell, "Textbook of Physiology," 15th ed., J. R. Fulton, ed., Saunders, Philadelphia, 1946. Data for curve for* Arenicola *blood from J. Barcroft and H. Barcroft, Proc. R. Soc. (London), B, vol. 96, 1924. Data for curve for* Lumbricus *blood from T. M. Haughton, G. A. Kerkut, and K. A. Munday, J. Exp. Biol., vol. 35, 1958. Data for curve of* Cucumaria *hemoglobin from C. Manwell, J. Comp. Cell Physiol., vol. 53, 1959.)*

ditions, has been measured as about 0.026 ml/(gm body weight) (hr) at 15°C, and the total oxygen capacity of their blood, on an average, as about 0.0081 ml. The oxygen in the blood alone, both in combination and in solution, could, therefore, meet a stranded snail's respiratory needs for only 18 to 19 min. Observations of snails in an atmosphere of pure nitrogen have shown, however, that their hemoglobin is not entirely deoxygenated for as long as 25 min. It has also been calculated that there would be sufficient air in the lung, even when closed off, to support respiration another 40 min. The snails should, therefore, be able to get along quite well for as long as 1 hr in the complete absence of external oxygen without resorting to anaerobiosis to meet their energy needs. The high oxygen affinity of their hemoglobin (t_l = 10 mm Hg; t_u = 1 to 2 mm Hg) means that it can take up oxygen from the air in the lung even when the pressure there falls as low as 21 mm Hg (2.8 percent of total atmospheric pressure). Underwater, *Planorbis* makes use of its gill as a respiratory organ and probably, under conditions of extreme oxygen deficiency, resorts to anaerobiosis and builds up an oxygen debt. Under ordinary conditions, the snails rise periodically to the surface to oxygenate their blood and refill their lungs, stimulated to do so presumably not by any accumulation of carbon dioxide in the pulmonary air, for that gas is eliminated elsewhere, but by decrease in their buoyancy resulting from the removal of oxygen from it. Under experimental conditions, when the external concentration of oxygen in the water was less than 23 mm Hg, they remained continually at the surface. It seems likely, therefore, that their hemoglobin functions both in storage and in transport.

The carbon monoxide method has been used also in assaying the value of hemoglobin to small crustaceans and to chironomid larvae. The fact that both *Daphnia* and *Artemia* respond to oxygen deficiency by synthesis of hemoglobin provides circumstantial evidence of their utilization of it in respiration. Its function has been deduced from experiments in which survival times and certain aspects of behavior of those with low hemoglobin content could be correlated with the same parameters of those with high hemoglobin content, under similar conditions of oxygen availability. The hemoglobin of *Daphnia magna* has a fairly high oxygen affinity with an unloading tension, at 17°C, of 3.1 mm Hg in the absence of carbon dioxide and of 4.9 mm Hg at a carbon dioxide tension of 7.6 mm Hg (1 percent carbon dioxide). Corresponding values for *Ciriodaphnia laticaudata*, which lives in fouler waters than *D. magna*, are 0.8 mm Hg and 1.0 mm Hg, respectively. They both, therefore, show a moderate Bohr effect. These hemoglobins take up oxygen at low tensions and could function as carriers under the conditions in which the cladocerans live. Tests of this possibility have been made by observing the differences between pink individuals and pale or colorless ones, in respect to their longevity in waters with different oxygen tensions; their feeding activity as measured by their rates of clearance of graphite particles or of the alga *Chlorella* from the medium; and by the fecundity and production of parthenogenetic eggs by the females. In every case, pink specimens survived conditions of oxygen depletion longer than pale ones, they moved their thoracic limbs faster and, therefore, collected more food, and they produced more eggs. Food collection and egg production decreased, in individuals of equivalent hemoglobin content, as the oxygen tension in their environment was decreased; at an oxygen tension of 20 mm Hg, their rate of clearance of *Chlorella* was no greater than that of pale individuals, but at 30 mm or above it was 20 percent greater. Food collection in pink animals exposed to carbon monoxide was also reduced, even at high oxygen tensions, as was their rate of locomotion, a function of the rapidity of movement of their antennae. Similarly, pink or red specimens of *Artemia salina* survived longer in poorly aerated seawater than colorless ones did, and those exposed to carbon monoxide showed a reduced rate of oxygen consumption as compared with unexposed controls.

The function of the hemoglobin in chironomid larvae is of particular interest since the pigment is exceptional in the hemolymph of insects. Extensive investigations have been conducted, especially on members of the plumosus group of this genus, which is worldwide in distribution and in which there are many subspecies. This diversity has led to some confusion in the interpretation of data from laboratory experiments because it is not always clear just which variety or subspecies has been used, or how their hemoglobins may differ. In nature the larvae of these midges live in tubes, usually constructed in the muddy bottoms or in other oxygen-poor regions of ponds and lakes. In the laboratory they will live in U-shaped glass tubes of suitable dimensions suspended in water from a platform covered with the mud or silt of their natural habitats. In these conditions, their habits can be watched and, because they are small enough and sufficiently transparent, direct spectroscopic records of their hemoglobin at different times in the cycle of their activities can be obtained. Such observations have shown that in well-aerated water a larva spends about half its time in irrigation of the tube, which it accomplishes by peristaltic contractions of its body. This is essentially a respiratory process, for oxygen is taken from the water as it passes through the tube. About 35 percent of the time is spent in the much more energy-demanding process of filter feeding, which requires secretion of the material of the mucous net, its construction, more vigorous irrigation of the tube than that required for respiration alone, and, finally, movement of the larva and consumption of the filter and its trapped particles. The remainder of the time is spent in complete inactivity, or rest. Filter feeding stops and the level of activity is reduced in larvae whose hemoglobin is saturated with carbon monoxide, showing that the hemoglobin is acting as a transporter of oxygen to maintain these activities. Filter feeding stops also in larvae with unblocked hemoglobin when the oxygen tension in the water is reduced to 45 mm Hg or below, but the periods of irrigation are extended until, at an oxygen tension of 7 mm Hg, they are almost incessant. Spectroscopic observations show that the hemoglobin is always partially oxygenated so that, at this tension, it is functioning continuously in oxygen transport. In complete absence of external oxygen, the larvae become immobile but can survive anaerobically for days and resume activity when oxygen is introduced into the medium and reaches a tension of 11 mm Hg and above. After a period of respiratory activity, filter feeding is resumed.

It, therefore, seems evident that the hemoglobin of chironomid larvae does not function in oxygen transport in well-aerated water but that at lower oxygen tensions it is an effective carrier. Moreover, spectroscopic examination has shown that it holds a store that would last a larva for 9 min and so tide it over its intermittent rest periods and even reduce the time that it would have to resort to an anaerobic mechanism during extended periods of oxygen lack. The value of hemoglobin to these larvae would, therefore, seem to be threefold. Its affinity for oxygen is higher than that of the hemoglobin of *Arenicola*, for unloading tensions of 0.177 mm Hg at 20°C and of 0.6 mm Hg at 17°C have been reported. This means that the pigment can transport oxygen at very low tensions and can, therefore, support respiration to the extent of making continual irrigation of the tube possible even when the external oxygen tension is as little as 7 mm Hg. Tensions well below that of water saturated with air will provide enough oxygen for filter feeding. The hemoglobin also provides a store for periods when the tube is not being irrigated and increases the rate of recovery from an extended period of anaerobiosis. Its presence in the hemolymph is, therefore, distinctly an asset to the larvae, for while it is of no use to them in well-aerated water, it does not exclude them from it and allows them to live and maintain their essential activities in waters where the oxygen concentration is no more than 2 ml/liter, and from which other insect larvae, less well endowed, would be excluded.

Hemoglobin can also be an asset to animals living at tidal levels, for the availability of oxygen changes drastically for them with high and low tides. Its utilization has been particularly thoroughly investigated in the polychaetes *Arenicola* and *Nephthys* and in the echiuroid *Urechis*. *Arenicola* makes U-shaped burrows in the sand, usually in such locations on the beach that they are covered by water only during high tide, and so are in open and exposed sand for hours at a time. Sometimes the burrows are made so far up shore that they are covered only when the tides are exceptionally high; there are reports of burrows made so much above ordinary tide levels that they are immersed but once a year, during the spring tides. Possibly in these cases, if the water level in the burrow falls low enough, the worms turn to aerial respiration, for those under observation in glass tubes in the laboratory in conditions simulating those of a burrow high on shore have been seen to draw bubbles of air down over their gills. However, in ordinary conditions when the tide is in, the worms irrigate their burrows intermittently by peristalsis of the body wall when a supply of water passes through the tubes. Rest periods and feeding periods of as long as 30 min interrupt irrigation, for these two activities cannot proceed simultaneously. Oxygen tension in the burrows is probably always low and undoubtedly always variable. Measurements of the interstitial water in the sand around the burrows at low tide gave an oxygen pressure of 6.17 mm Hg and in freshly filled burrows, one of 13 mm Hg. A hemoglobin with a high oxygen affinity is, therefore, the only one that could be of any use to an animal living in these conditions. The hemoglobin of *Arenicola*, according to its equilibrium curve, should be almost fully oxygenated at all times and has been shown to be so even after 5 hr of low tide (Fig. XI–7). Calculations of the oxygen consumption [0.031 ml/(gm)/(hr) at 10°C], of blood volume (approximately 0.382 ml/gm), and of the total oxygen capacity of the blood (0.037 ml/gm) show that the worm could continue its usual activities in the absence of external oxygen for about 21 min if it drew upon this store. There is no direct evidence that it does so, nor indeed that the hemoglobin functions in oxygen transport, although it is possible that it may be used during rest and feeding periods, when irrigation of the burrow is not in progress and contact of the gills with aerated water is impossible. It may, therefore, serve *Arenicola* in a way comparable to that in which hemoglobin serves chironomid larvae, by making it possible for them to continue a vital activity without recourse to anaerobiosis, even when the oxygen tension in the burrow is at a very low level.

Nephthys is another burrowing polychaete, but with hemoglobin in solution in its coelomic as well as in its vascular fluid. It does not make a U-shaped burrow but probes its way straight down into the sand, often in the same locality as *Arenicola*. *Nephthys*' burrow, open only at the upper end, is constantly irrigated by ciliary action, when the tide is in, but is sealed off from air and water, when the tide is out, by collapse of the walls at the top. The oxygen tension in the interstitial water is at best only about 7 mm Hg and must be even less if the sand contains much decaying organic matter. The equilibrium curves for *Nephthys*' hemoglobin, plotted for both the vascular and the coelomic hemoglobin, show that it has a relatively low oxygen affinity and that at an external tension of 7 mm Hg, the vascular hemoglobin would be less than 60 percent saturated and the coelomic, little more than 40 percent (Fig. XI–8). In a storage capacity, it could supply a worm, at its usual metabolic rate, for not much longer than 10 min, and although there is no direct evidence, it seems likely that the worms turn to anaerobic respiration during most of the time that the tide is out. At high tide, with well-aerated water coming into the burrow, the external oxygen tension would be high enough to make hemoglobin with these properties useful in oxygen transport, and it is probable that it does function in this way, even though at low tide it would be of no apparent use.

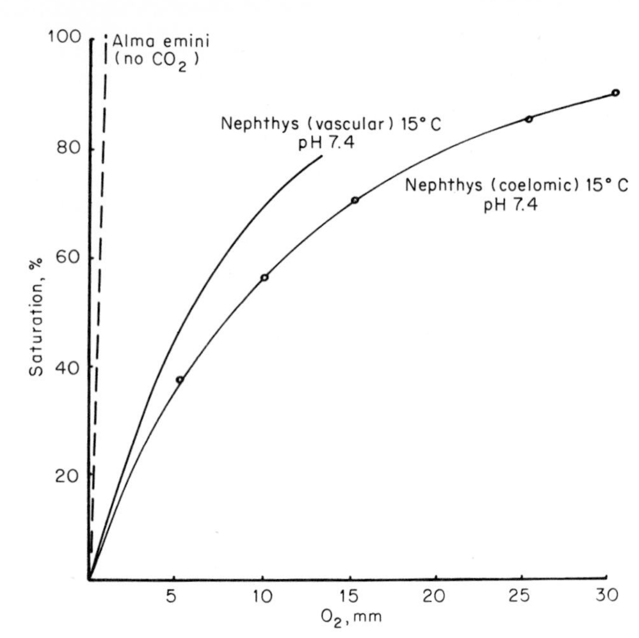

FIG. XI–8 *Equilibrium curves of the hemoglobins of* Alma emini *and the vascular and coelomic fluids of* Nephthys hombergii. *(Data for curve of* Alma *hemoglobin from L. C. Beadle, J. Exper. Biol., vol. 34, 1957. Data for curves for* Nephthys *hemoglobin from J. D. Jones, J. Exper. Biol., vol. 32, 1955.)*

The echiuroid *Urechis caupo*, living in estuarine mud flats along the coast of California, makes a U-shaped tube, like that of *Arenicola*, and ventilates it intermittently, when the tide is in, by peristaltic contractions of its body. These contractions provide a food as well as a respiratory current, with the greatest activity during the 10 to 20 min feeding periods.

Feeding and respiratory movements are followed by rest periods, lasting from 8 to 10 min to over 1 hr, during which the animals make no perceptible movements. In these periods and in the much longer ones when the tide is out, the water in the burrow is stagnant. The intervals when the tide is out and the flats and burrows are uncovered by water are 6 hr long, at least once every day, and as long as 18 hr during the high spring tides. *Urechis* has no vascular system, and its hemoglobin is contained within coelomic corpuscles. The oxygen equilibrium curve has been determined in undiluted coelomic fluid and shows that the oxygen affinity of the hemoglobin is comparatively low; its loading tension at 19°C, at a carbon dioxide pressure of 7 mm Hg, is 90 to 100 mm Hg, and its unloading tension, 12 mm Hg (Fig. XI–9). The minimal oxygen tension in the hindgut, when the animal is pumping water in and out of it, has been calculated at 100 mm Hg, so that at these times the hemoglobin could be fully

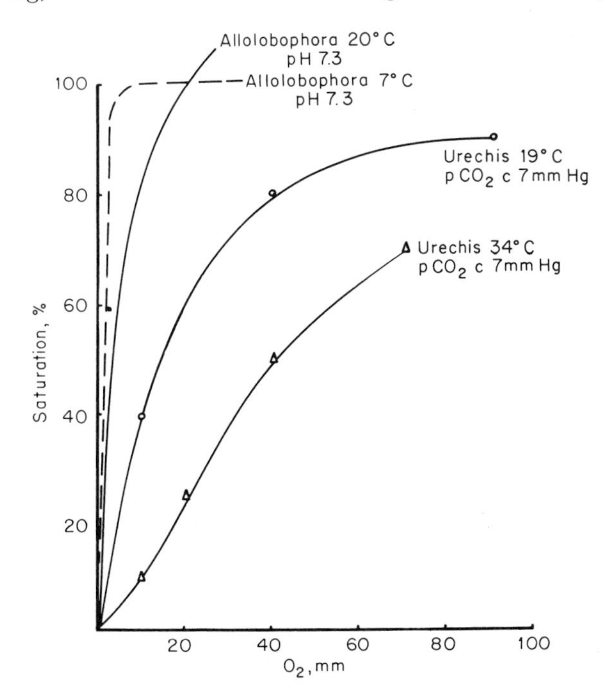

FIG. XI–9 *Equilibrium curves of the hemoglobins of* Allolobophora foetida *and* Urechis caupo *at different temperatures. (Data for curves for* Allolobophora *from T. M. Haughton, G. A. Kerkut, and K. A. Munday, J. Exp. Biol., vol. 35, 1958. Data for curves for* Urechis *from A. C. Redfield and M. Florkin, Biol. Bull., vol. 61, 1931.)*

saturated with oxygen. The mean oxygen consumption of specimens in well-aerated water is estimated as 0.013 ml/min, the volume of the coelomic fluid as 20 ml, and its total oxygen capacity as 4 vol percent (i.e., ml O_2/100 ml blood). Only about one-sixtieth of the total oxygen content would be needed for respiration at the usual rate of metabolism, and that in solution in the coelomic fluid would be more than adequate for this. There is, therefore, no reason to suppose that the hemoglobin of *Urechis* functions at all in oxygen transport as long as there is sufficient ventilation of the burrow and the hindgut. But the survival time of animals during the intervals when there is no irrigation seems inevitably to depend upon the hemoglobin. Calculations show that were there none, the available oxygen supply would not last a worm longer than $\frac{1}{4}$ hr, for the oxygen content of the coelomic fluid, which could not be greater than that of the water in the hindgut, would be about 0.074 ml and would supply the worm, at its ordinary metabolic rate, for about 5 to 6 min. This could be supplemented by the oxygen in the water in the hindgut and these combined resources could provide oxygen for about 8 min. But the total oxygen content of the coelomic fluid, when the hemoglobin is fully saturated, is about 0.8 ml, nearly 10 times that which could be carried in solution alone, and would supply an animal over a period of external oxygen deprivation for about 1 hr. If oxygen in the water already in the burrow were also available, there should be enough for an additional 2 hr supply, so that an animal should be able to meet the needs of its ordinary respiratory metabolism for about 3 hr, without any supplementation of its oxygen supply. It might be able to meet them even longer at the reduced metabolic rate of the periods when it does not pump water into its burrow and its hindgut, even when the flat is under water, and during the longer ones when it cannot do so because the flat is exposed. This amount of oxygen would be sufficient for the usual rest periods, but not for the extended ones of low tide. Measurements of the oxygen content of burrow water, during low tide, have shown that there is a rapid decrease during the first hour and that a minimum level of 0.06 ml/100 ml, corresponding to an oxygen tension of 14 mm Hg, is reached after an exposure of 4 hr. At this tension, the coelomic fluid would be about 60 percent saturated. During the fifth hour of exposure, there was a noticeable increase of oxygen in the burrow water, which continued throughout the duration of low tide. The probable cause of this is seepage of water into the burrow from the surrounding sand, due to physical conditions in the flat related to the outflow and subsequent inflow of water along it. Since the water in the burrow is thus replenished during the final hours the tide is out, *Urechis* is provided with an additional supply of oxygen, which is sufficient to keep it, at its reduced rate of metabolism, until its burrow is again under water and irrigation and feeding movements can be resumed. Its hemoglobin, therefore, while of no obvious value to it during periods when it is in well-aerated water and the coelomic fluid is saturated with oxygen, is of immense importance to it at other times, serving both as a store to be drawn upon in times when it is deprived of external oxygen and as an acceptor and transporter of the gas at low external tensions.

The function of hemoglobin in other invertebrates in which it has been identified is still problematical. The swamp worm, *Alma emini*, for example, lives in an environment with an oxygen pressure too low to support aerobic respiration and is presumed to meet its energy requirements by anaerobiosis. Yet it has hemoglobin in its blood and a means of external gaseous exchange in the vascularized regions of its posterior segments which it seems to make use of. The oxygen equilibrium curve of its blood shows that this is fully saturated at an oxygen pressure of 2 mm Hg; it is unaffected by biological levels of carbon dioxide tension and is shifted slightly to the right only when the tension is experimentally raised as high as 200 mm Hg (Fig. XI–8). *Alma* in the laboratory will live in well-aerated media, so that it is not wholly conditioned to life in the reducing atmospheres in which it is ordinarily found. Possibly it cannot sustain itself anaerobically indefinitely, because of accumulation of toxic by-products not easily disposed of in its stagnant environment. It may, therefore, need oxygen not so much for current energy requirements as for detoxication of these metabolites and may use its respiratory device and its hemoglobin to this end. Possibly the hemoglobins of nematodes, in solution in the fluids of their pseudocoeles, are useful in a similar way, for the properties of those that have been investigated are such as to make it unlikely that they contribute in any way to respiration. Some free-living nematodes inhabit water-logged soils, as *Alma emini* does, or environments that are equally low in oxygen or possibly devoid of it entirely, like the deep mud at the bottom of lakes and seas; these must practice anaerobic respiration. Intestinal parasites such as *Ascaris* and *Nippostrongylus* live in atmospheres of very low oxygen tension (the oxygen tension in the large intestine of the horse, for example, is 0.30 mm Hg) yet the hemoglobin of *Ascaris* has been found to have such a high oxygen affinity that it remains fully oxygenated even when the environmental oxygen tension is nil. It can be deoxygenated in vitro only slowly by chemical reducing agents such as sodium dithionite ($Na_2S_2O_4$). It, therefore, seems most unlikely that the pigment can contribute in any way to the parasite's respiration, and this is probably the case also with the hemoglobins of other nematodes. If not used in detoxication, they may well be metabolic products that are of no value to the nematode at all.

While the hemoglobins of some marine clams and of some holothurians have been identified and characterized, their biological functions have not been established, and investigation of their roles would be interesting and rewarding comparative physiological studies. Equilibrium curves have been plotted from spectrophotometric data for two Florida clams, *Noetia ponderosa*, an arcid clam with hemoglobin in erythrocytes, and *Cardita floridana*, with hemoglobin in solution in its blood (Fig. XI–10). These show that the hemoglobin has, in each case, a moderate affinity for oxygen, with unloading tensions in the vicinity of 17 mm Hg, which are not affected by pH. Neither of the clams is a deep burrower and except at low tide each has access to what is presumably an adequate supply of oxygen. It is possible, although not proven, that the oxygenated hemoglobin may serve as a store of oxygen to be used when the tide is out.

Although hemoglobins have been identified in representatives of at least three of the five orders of holothurians and an equilibrium curve plotted for that of *Cucumaria miniata* (Fig. XI–7), an intertidal species, there is no definitive evidence of their function in the respiration of any sea cucumber.

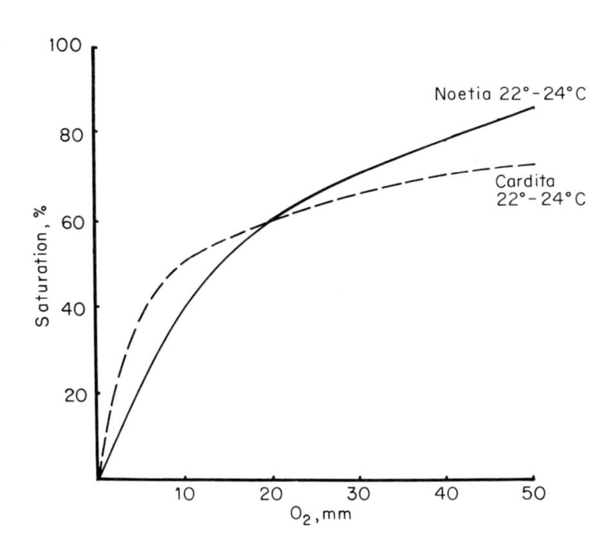

FIG. XI–10 *Equilibrium curves for the hemoglobin (2 percent) of the clams* Cardita *and* Noetia. *The curves are similar for* Cardita *at pH's 7.5 and 7.03, and for* Noetia, *at 8.24, 7.78, 7.40, and 6.68. (Data from C. Manwell, Comp. Biochem. Physiol., vol. 8, 1963.)*

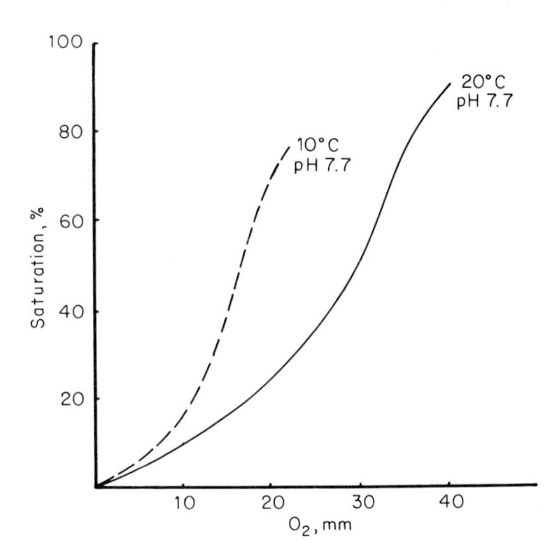

FIG. XI–11 *Equilibrium curve for chlorocruorin from* Spirographis *at different temperatures. (Data from H. M. Fox, Proc. R. Soc., vol. 111, 1932.)*

b. THE BIOLOGICAL PROPERTIES OF CHLOROCRUORINS Studies of the chlorocruorins of sabellid worms show that they, too, act as oxygen carriers and are especially suited to operate at comparatively high oxygen tensions. Equilibrium curves have been plotted for the blood of *Spirographis spallanzani* taken from vessels at the base of the crown and from the peri-intestinal sinus (Fig. XI–11). The pigment is not fully saturated even when in equilibrium with air, and its loading and unloading tensions are influenced by temperature, the curve being shifted to the right as the temperature rises. Thus, at 20°C, the loading tension is 40 mm Hg and the unloading, 27 mm Hg, while at 10°C the corresponding values are 26 mm Hg and 17 mm Hg, respectively. Its properties would permit the pigment to function as an oxygen carrier when the worm is in well-aerated water at about the extremes of summer and winter temperatures of its natural habitat. Experiments in blocking the chlorocruorin of *Sabella pavonina* show clearly its transport function in these worms, for oxygen consumption in those exposed to carbon monoxide fell off as the external oxygen tensions were reduced, and it could be deduced from the results that about one-third of the oxygen used was carried, at least at pressures down to 30 mm Hg, in combination with the pigment. The oxygen affinity of *Sabella* chlorocruorin is slightly lower than that of the chlorocruorin of *Spirographis*, for at 10°C its loading tension is about 26 mm Hg and its unloading, 9 mm Hg; it is likewise better suited to combine with oxygen at high than at low tensions. This might seem disadvantageous to a tubiculous worm, but under ordinary conditions, exposure of the crown and almost constant irrigation of the tube keep the respiratory areas of *Sabella* well covered with water.

c. THE BIOLOGICAL PROPERTIES OF HEMERYTHRINS The properties of hemerythrins have been investigated in a number of genera of sipunculids, and equilibrium curves have been obtained for that of *Sipunculus*, for the coelomic and vascular hemerythrins of *Dendrostomum* and *Siphonosoma*, and for the coelomic hemerythrin of *Phascolosoma* (*Golfingia*) *agassizi*. A curve has also been plotted for the hemerythrin of the brachiopod *Lingula* (Figs. XI–12 and XI–13). All of these pigments have relatively high oxygen affinities and are, therefore, suited to oxygen transport at low tensions. Sabellids live in mud or sand, from high

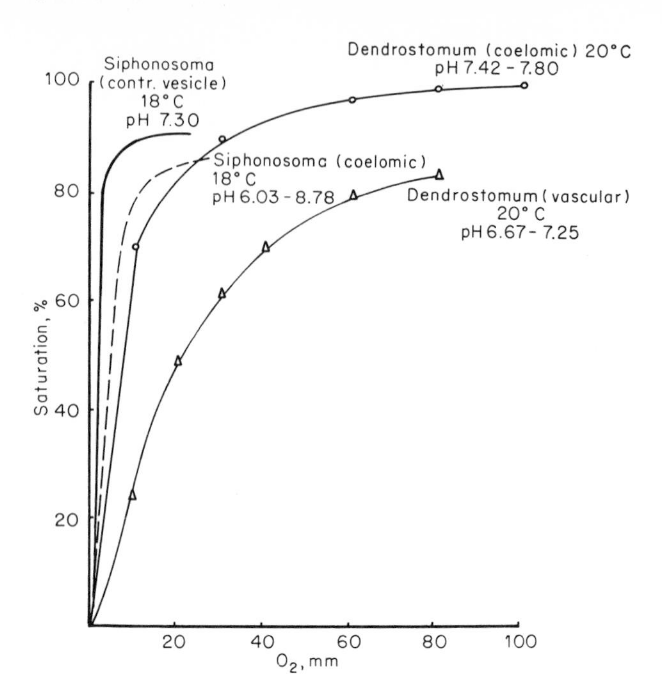

FIG. XI–12 *Equilibrium curves for the hemerythrins of* Dendrostomum *and* Siphonosoma. *There are no significant differences between curves plotted at different pH's.* (*Data from C. Manwell, Comp. Biochem. Physiol., vol. 1, 1960.*)

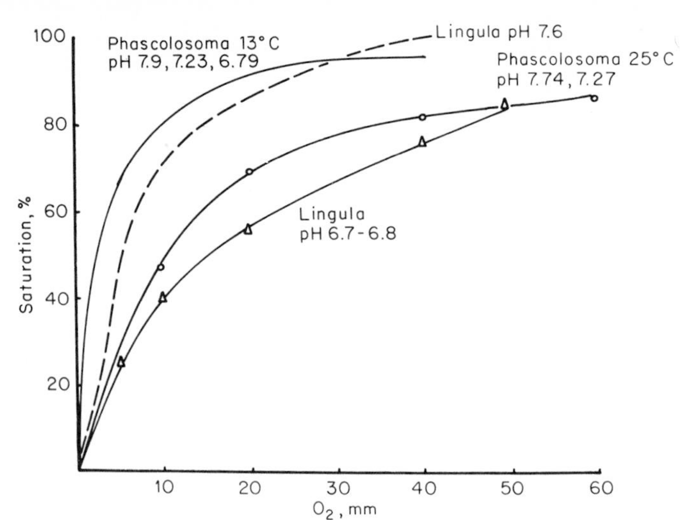

FIG. XI–13 *Equilibrium curves for the hemerythrin of* Lingula (*2 to 3 percent in* KPO_4 *buffer*) *and of* Phascolosoma (Golfingia) agassizi (*2 percent in K buffer*) *at different temperatures and different pH's.* (*Data for* Lingula *from C. Manwell, Science, vol. 132, 1960. Data for* Phascolosoma *from C. Manwell, Science, vol. 127, 1958.*)

tide level to depths of 15,000 ft, some in burrows that they excavate themselves, some in burrows abandoned by polychaetes, and some in crevices of rock or of coral formations. *Dendrostomum zostericolum* extends its tentacles from the burrow, and oxygen from the seawater diffuses across their walls into the superficial blood vessels. The hemerythrin in these vessels reaches its maximum saturation at an oxygen tension of 80 mm Hg and has an unloading tension of 15 mm Hg. The blood from the tentacular vessels flows to the main contractile vessel from which it is pumped into thin-walled diverticula that project into the coelom and are bathed by coelomic fluid. The coelomic hemerythrin has a higher oxygen affinity than the vascular (t_l = 25 mm Hg; t_u = 4 mm Hg), and it would be possible for it to take up oxygen from the vascular pigment and presumably transfer it to the tissues. The conditions are reversed in *Siphonosoma*, in which gaseous exchange occurs over the entire integument and the oxygen diffuses into the coelomic fluid. The vascular hemerythrin of *Siphonosoma* has a higher oxygen affinity than the coelomic, which has loading and unloading tensions of 10 mm Hg and 1 mm Hg, respectively, while those of the coelomic are 18 mm Hg and 2.0 mm Hg. Presumably the oxygen taken up from the water is combined with the coelomic hemerythrin and transferred from it to the vascular for delivery to the tissues. In well-aerated water, the oxygen in solution would probably be sufficient for ordinary metabolism, for the hemerythrin would be saturated, but it is possible also that the oxygenated hemerythrin acts as a store and transports oxygen when external tensions are low. In *Dendrostomum cymodoceae*, for example, in an oxygen-free experimental environment, the hemerythrin was completely deoxygenated in 6 hr, although this sipunculid can live anaerobically for as long as 5 or 6 days. The rapid depletion of its oxygen store makes it seem more likely that it uses its oxygen for aerobic respiration before turning to anaerobic than for detoxication of anaerobic by-products as has been postulated for the annelid *Alma emini*.

Observations on living specimens of *Lingula* indicate that its hemerythrin may also serve in storage as well as in transport of oxygen, because circulation of the coelomic fluid, containing hemerythrocytes, could be watched in the long, transparent peduncles. It appeared colorless for 20 to 30 percent of the time it was under observation, but for the remainder had the pink color characteristic of the oxygenated pigment. The absence of color may be taken as an indication of its complete deoxygenation and of the probability that it gives up all its combined gas during pauses in the respiratory and feeding activities of the animal.

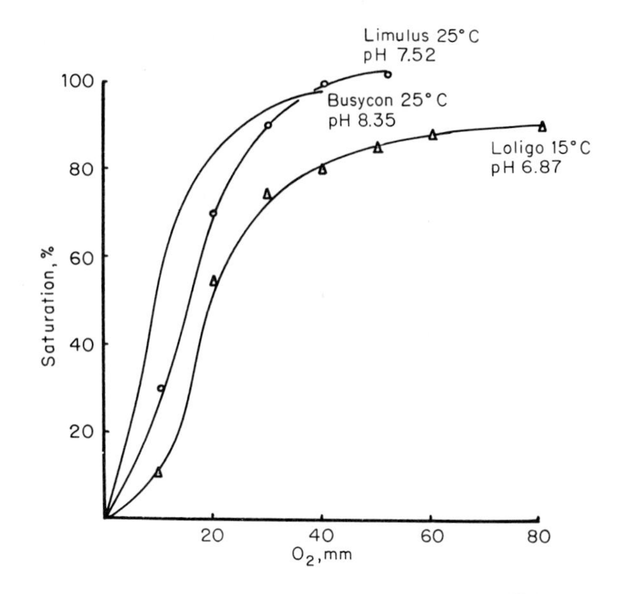

FIG. XI–14 *Equilibrium curves for the hemocyanins of* Busycon canaliculatum, Loligo pealii, *and* Xiphosura (Limulus) polyphemus. (*Data from A. C. Redfield and E. N. Ingalls, J. Comp. Cell. Physiol., vol. 3, 1933.*)

d. THE BIOLOGICAL PROPERTIES OF HEMOCYANINS Investigations of the role of hemocyanins in the respiration of the larger molluscs and crustaceans are simplified by the patterns of their vascular systems, which provide for the separation of freshly oxygenated (arterial) blood from that which has passed through the tissues and delivered oxygen to them (venous blood). It is possible, therefore, directly to determine the oxygen content of arterial and venous blood in a living animal, by methods that have been successfully applied to vertebrates, and to assess, from the differences between them, the measure of effectiveness of the pigment in respiration. Oxygen equilibrium curves have been plotted for the blood of representative molluscs and decapod crustaceans (Figs. XI–14, XI–15, XI–16, and XI–17), and from the combined data it has been established that hemocyanin does function in oxygen transport in the gastropod *Busycon*, the cephalopods *Loligo*, *Sepia*, and *Octopus* and in the American lobster, *Homarus americanus*, the spiny lobster *Panulirus interruptus*, the sheep crab *Loxorhynchus grandis*, as well as in the horseshoe "crab" *Xiphosura polyphemus*, and in *Pachygrapsus crassipes*, one of the smaller true crabs. The hemocyanins of these animals differ in their oxygen affinities, for in some it is high, operating most effectively at high tensions. They differ also in respect to the Bohr effect, for in some it is evident and even of great magnitude and in others it is reversed, the pigment holding its oxygen even more tenaciously as the carbon dioxide tension increases.

Busycon, the whelk, can be found from low tide mark to depths of about 100 fathoms and lives, therefore, in waters that are comparatively well aerated. At an environmental oxygen tension of 150 mm Hg, the oxygen tension of blood returning from the gills has been measured as 36 mm Hg, and that of the blood

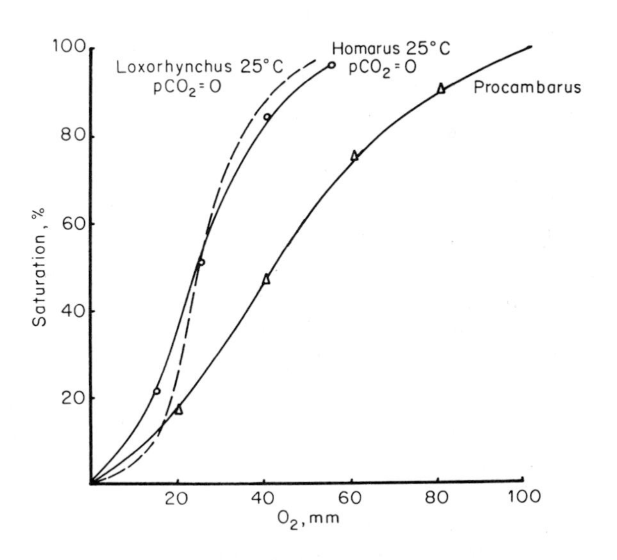

FIG. XI–15 *Equilibrium curves for the hemocyanins of the crayfish* Procambarus simulans, *the lobster* Homarus americanus, *and the crab* Loxorhynchus grandis. (*Data for* Procambarus *from J. L. Larimer and A. H. Gold, Physiol. Zool., vol. 34, 1961. Data for* Homarus *and* Loxorhynchus *from J. R. Redmond, J. Comp. Cell Physiol., vol. 46, 1955.*)

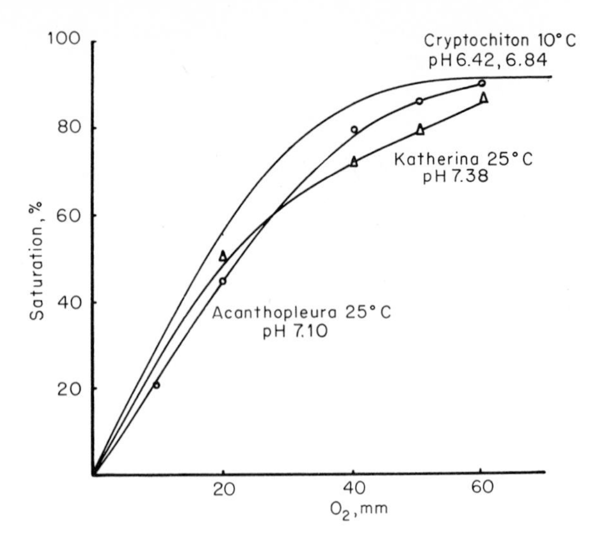

FIG. XI–16 *Equilibrium curves for the hemocyanins of three chitons:* Cryptochiton stelleri (*hemocoelic fluid diluted 3:1 with K buffer*), Acanthopleura granulata, *and* Katherina tunicata. (*Data for* Cryptochiton *from C. Manwell, J. Comp. Cell. Physiol., vol. 52, 1958. Data for* Acanthocephala *and* Katherina *from J. R. Redmond, Physiol. Zool., vol. 35, 1962.*)

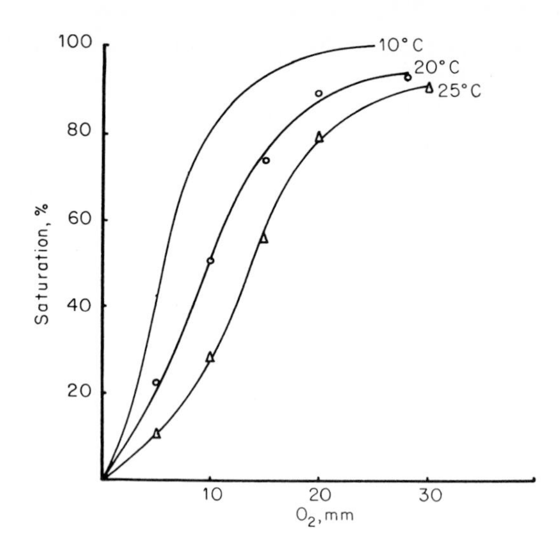

FIG. XI–17 *Equilibrium curves for the hemocyanin of* Panulirus interruptus *at pH 7.53 and three different temperatures.* (*Data from J. R. Redmond, 1955.*)

entering the gills as 6 mm Hg. These values correspond to 90 to 95 percent saturation of the hemocyanin, so that it is evident that more than 80 percent of the oxygen carried in the blood is delivered to the tissues. The volume of oxygen transported has been measured as 1.7 vol %, of which 0.14 percent represents that in solution. The remainder, 1.56 vol %, is bound to the hemocyanin, and the whole blood, therefore, carries and delivers nearly 17 times more oxygen than it could without the pigment, which is here obviously contributing to respiration. The hemocyanin of *Busycon* shows a reverse Bohr effect, a characteristic that could be of value to an animal living in such an environment as *Busycon's*, for it would enable the hemocyanin to hold its oxygen even when the external carbon dioxide tension rose, as it might well do at times in a whelk's immediate vicinity.

The hemocyanins of the cephalopods *Loligo* and *Octopus* differ markedly from each other in their oxygen affinities. That of *Loligo's* hemocyanin is low, with a loading tension of 80 to 100 mm Hg, while that of *Octopus* is high, with a loading tension of 7 mm Hg, under comparable conditions of temperature and carbon dioxide tension. Nevertheless, they both seem to function in similar ways in the respiration of the two animals, the differences between them showing adaptation, on the one hand, to the active pelagic life of squids and, on the other, to the more lethargic life of octopuses, which live on the sea floor and spend a great part of their time quiescent in crevices or under rocks. Squid hemocyanin has the highest Bohr effect recorded for any pigment in any animal. At 23°C, in the absence of carbon dioxide, its unloading tension is 36 mm Hg, but at a carbon dioxide tension of 6.5 mm Hg and the same temperature, it is 96 mm Hg. This must profoundly influence a squid's conditions of life, for any appreciable accumulation of carbon dioxide in its environment must result in an increase in the loading tension of its hemocyanin and a consequent dependence upon a great increase in the oxygen tension of the medium, or its inactivation as a carrier. Samples of blood drawn from the efferent gill vessels of living animals, whose mantle cavities were artificially irrigated to ensure maximum aeration of the water, showed an oxygen tension equivalent to 120 mm Hg and a carbon dioxide tension of 2 mm Hg. Samples of blood taken from the posterior venous sinuses showed an oxygen tension of 48 mm Hg and a carbon dioxide tension of 6 mm Hg. The figures for oxygen tension correspond to values of 4.7 vol % and 0.37 vol %, respectively, so that it is evident that the blood has lost 90 percent of its oxygen in passing through the tissues. Experiments in which the carbon dioxide tension in the mantle cavity was raised along with that of the oxygen tension have shown that animals die from asphyxiation even before the oxygen content of their blood has been reduced to the level of that in solution, which has been measured as 0.38 vol %. These experiments, taking advantage of the enormous Bohr effect of squid blood, are analogous to those that take advantage of the affinities of hemoglobin and chlorocruorin for carbon monoxide, because in both cases the effectiveness of the pigment as a carrier is blocked and so the value of it to respiration is established.

Arterial blood taken from specimens of *Octopus*, when immobile but actively irrigating their mantle cavities in seawater with an oxygen tension of 145 mm Hg, had oxygen tensions of 55 to 85 mm Hg, corresponding to an oxygen content of 4.6 vol %. Samples of venous blood showed an oxygen content of 0.4 percent and a carbon dioxide content of 6.6 percent. These figures are very close to those for the oxygen content of arterial and venous blood of *Loligo*, so that it seems likely that the two hemocyanins function in the same way in the two species, although one of them is oxygenated at high external tensions and the other at low.

Oxygen equilibrium curves have also been plotted for the hemocyanins of several polyplacophorans. (Fig. XI–16). All of them show relatively low oxygen affinities and low oxygen capacities. These do not exceed 2 vol % in two Japanese species (*Chiton tuberculatus* and *Acanthopleura granulata*) and in two species taken from Puget Sound (*Katherina tunicata* and *Mopalia muscosa*). The unloading tensions of the bloods of these four species are very close, ranging from 20 to 26 mm Hg when measured at 25°C; yet *C. tuberculatus* and *A. granulata* live in waters where the temperature averages 28°C, and the Puget Sound species live in waters with a temperature around 10°C. They must, therefore, in nature operate quite differently. The low oxygen affinities of these hemocyanins may reflect the modest demands for oxygen of these molluscs of sedentary habit, which have also large gill areas that are easily ventilated in their natural environments of shallow and well-aerated waters.

The hemocyanins of *Limulus* and the decapod crustaceans tested have comparatively high oxygen affinities. The oxygen equilibrium curve of the hemocyanin of *Limulus* resembles that of *Busycon*. Calculations show that it carries two to six times the amount of oxygen that could be transported in solution, and experiments show that, like the hemocyanin of *Busycon*, it has a reverse Bohr effect. There seems every likelihood, therefore, that it functions in the same way, providing the horseshoe crab with oxygen sufficient for its rather inactive life and protecting it from asphyxiation when the carbon dioxide tension of the environment becomes unduly high.

The crustaceans whose bloods have been studied also live rather sluggish existences. The oxygen affinity of their hemocyanins is comparatively high, yet the oxygen capacity of the blood is low, for the hemocyanin is never fully saturated with oxygen. Samples of prebranchial blood taken from the cephalic end of the abdominal sinus of *Homarus* and *Panulirus* and of postbranchial blood taken from the pericardial cavity show that in vivo their hemocyanin is never more than 50 to 70 percent saturated. Experimentally, the oxygen content of whole blood of *Panulirus* equilibrated with air was found to be 1.0 to 2.8 vol %, of which something less than 0.5 vol % was in solution. Yet blood taken from the pericardial sinus of living animals kept in well-aerated water had an oxygen content of only 0.82 vol % (54 percent saturated), and blood from the abdominal sinus one of 0.35 vol % (22 percent saturated). Thus, about 0.5 ml of oxygen is delivered to the tissues by each 100 ml of blood. This could be accounted for by that in solution alone were the blood in equilibrium with air. But the oxygen tension in the pericardial cavity, at 54 percent saturation, is only 7 mm Hg, even though, in the conditions in which the measurements were made, that in the water outside the animals was more than 100 mm Hg. At the oxygen tension in the pericardium, some 0.03 ml of oxygen could be in solution, an amount that would be negligible to the lobster's respiration. It, therefore, follows that of the 0.82 vol % of oxygen in the pericardial blood, 0.79 vol % must be bound to the hemocyanin, a value that is 26 times greater than that in solution. For each 0.47 ml of oxygen delivered to the tissues, 0.44 ml must be released from the hemocyanin and 0.03 ml derived from that in solution. The hemocyanin is, therefore, acting as a carrier, but one that is particularly adapted to conditions of low internal oxygen concentrations. Diffusion of oxygen through the gills of crustaceans and of *Xiphosura* is known to be slow, and in vivo the blood is at best carrying only a fraction of the oxygen that it might were conditions more suitable to its passage across the gill surfaces. Environmental tensions are, therefore, of less importance to these animals than the nature of their own respiratory organs, which appear to be the limiting factor to the efficiency of their respiratory pigments in the absorption of oxygen and may account, to a great extent, for their slow movements and sluggish habits of life. Data from *Pachygrapsus*, a small intertidal crab, show that its hemocyanin is as important to its respiration as the hemocyanins of the larger crustaceans are to them, and in much the same way.

The crustacean hemocyanins that have been tested show a slight Bohr effect, which is probably of little significance in their operation. Temperature seems of far greater importance in influencing their oxygen equilibrium curves than carbon dioxide or hydrogen ion concentration. The effect of temperature has been conclusively demonstrated in *Panulirus*, whose ordinary temperature range is from 10°C in the deep waters it inhabits in winter to 20°C in shallow water in summer. Reference to Fig. XI–17 will show that its unloading tension shifts from 5 mm Hg at winter temperatures to 10 mm Hg at summer ones. That of *Loxorhynchus* shifts from 5 to 15 mm Hg with a rise in temperature from 15 to 20°C. Similar sensitivity to temperature may well be typical of hemocyanins in general.

In summary it seems evident that these four types of pigments, so far as they have been investigated, are of value to the animals whose body fluids contain them. Exceptions to this generalization are the hemoglobins of nematodes and the annelid *Alma emini*. Each pigment may serve a different purpose, or the same purpose in a different way, but each increases the oxygen-carrying capacity of the fluid containing it and so, in one way or another, contributes an essential element to the aerobic metabolism of an organism.

2. The transport of carbon dioxide

Far less is known about the transport of carbon dioxide in invertebrates than is known about the transport of oxygen. In vertebrates, the plasma proteins are known to act as buffers in taking up the hydrogen ions released by dissociation of carbonic acid. They are also known to combine directly, through their free amino groups, with carbon dioxide, forming carbamates. The hemoglobin also contributes to the transport of carbon dioxide, acting as a buffer for the hydrogen ions liberated from the carbonic acid produced in the erythrocytes. Carbon dioxide diffuses readily into them and there rapidly combines with water to form carbonic acid under the agency of the enzyme carbonic anhydrase which is confined to the erythrocytes and is not found in the plasma. By these means a volume of carbon dioxide can be carried, bound, in the blood without a significant increase in its pH. Since, in most invertebrate bloods, the only proteins in solution are the respiratory proteins, they must provide the entire buffering or combining system of this kind. It has indeed been shown that the hemocyanin of *Xiphosura* acts both as a buffer and as a reactant with carbon dioxide. The protein-free blood of *Urechis* has no buffering capacity whatsoever, but in the coelomic fluid, at the concentrations of carbon dioxide normally occurring there, the carbon dioxide content is about evenly distributed between the plasma and the corpuscles. The distribution of carbonic anhydrase in the body fluids of invertebrates seems to be very limited. The enzyme has, in fact, been found only in the coelomic fluids of *Arenicola* and *Sipunculus*. It is, however, very generally distributed in their tissues which may be the principal sites for the formation and breakdown of carbonic acid.

SELECTED BIBLIOGRAPHY

Oxygen and life

Fox, H. M., and A. E. R. Taylor: The Tolerance of Oxygen by Aquatic Invertebrates, *Proc. R. Soc. (London) B*, vol. 143, pp. 214–225, 1955.

Urey, H. C.: Some General Problems Relative to the Origin of Life on Earth or Elsewhere, *Am. Nat.*, vol. 100, pp. 285–289, 1966.

Respiratory areas

Beadle, L. C.: Respiration in the African Swampworm *Alma emini* Mich., *J. Exp. Biol.*, vol. 34, pp. 1–10, 1957.

Chuang, S. H.: Sites of Oxygen Uptake in *Ochetostoma erythrogrammon* Leuckart and Peupell (Echiuroidea), *Biol. Bull.*, vol. 123, pp. 86–93, 1962.

Cosgrove, W. B., and J. B. Schwartz: The Properties and Function of the Blood Pigment of the Earthworm *Lumbricus terrestris*, *Physiol. Zool.*, vol. 38, pp. 206–212, 1965.

Dales, R. P.: Observations on the Respiration of the Sabellid Polychaete *Schizobranchia insignis*, *Biol. Bull.*, vol. 120, pp. 82–91, 1961.

Gray, I. E.: A Comparative Study of the Gill Area of Crabs, *Biol. Bull.*, vol. 112, pp. 34–42, 1957.

Hinton, H. E.: On the Structure and Function of the Respiratory Horns of the Pupae of the Genus *Pseudolimnomorpha* (Diptera, Tipulidae), *Proc. R. Entomol. Soc., London*, vol. 29, pp. 135–140, 1954.

Locke, M.: The Formation of Tracheae and Tracheoles in *Rhodnius prolixus*, *Q. J. Microsc. Sci.*, vol. 99, pp. 29–47, 1958.

Wells, G. P.: The Respiratory Significance of the Crown in the Polychaete Worms *Sabella* and *Myxicola*, *Proc. R. Soc. (London).*, *B*, vol. 140, pp. 70–82, 1952.

Wigglesworth, V. B.: The Role of Epidermal Cells in the "Migration" of Tracheoles in *Rhodnius prolixus* (Hemiptera), *J. Exp. Biol.*, vol. 36, pp. 632–640, 1959.

Yonge, C. M.: The Pallial Organs in the Aspidobranch Gastropoda and Their Evolution throughout the Mollusca, *Philos. Trans. R. Soc. (London)*, *B*, vol. 232, pp. 443–518, 1947.

Zoond, A., and E. Charles: Studies on the Localization of Respiratory Exchange in Invertebrates, *J. Exp. Zool.*, vol. 8, pp. 250–266, 1931.

Transport of respiratory gases

Borden, M. A.: A Study of the Respiration and of the Function of Hemoglobin in *Planorbis corneus* and *Arenicola marina*, *J. Mar. Biol. Assoc. U.K.*, vol. 17, pp. 709–738, 1931.

Ewer, R. F.: On the Function of Hemoglobin in *Chironomus*, *J. Exp. Biol.*, vol. 18, pp. 197–205, 1941.

Ewer, R. F., and H. M. Fox: On the Function of Chlorocruorin, *Proc. R. Soc. (London)*, *B*, vol. 129, pp. 137–153, 1940.

Florkin, M.: Comparative Biochemistry, *Annu. Rev. Biochem.*, vol. 21, pp. 459–472, 1952.

Fox, H. M.: The Oxygen Affinity of Chlorocruorin, *Proc. R. Soc. (London)*, *B*, vol. 111, pp. 356–363, 1932.

Fox, H. M.: The Function of Chlorocruorin in *Sabella* and of Hemoglobin in *Lumbricus*, *Nature*, vol. 145, pp. 781–782, 1940.

Fox, H. M.: The O_2 Affinities of Certain Invertebrate Hemoglobins, *J. Exp. Biol.*, vol. 22, pp. 161–165, 1945.

Fox, H. M.: The Hemoglobin of *Daphnia*, *Proc. R. Soc. (London)*, *B*, vol. 135, pp. 195–212, 1948.

Fox, H. M.: The Effect of O_2 on the Concentration of Haem in Invertebrates, *Proc. R. Soc. (London)*, *B*, vol. 143, pp. 203–213, 1955.

Fox, H. M., and E. Phear: Factors Influencing Hemoglobin Synthesis by *Daphnia*, *Proc. R. Soc. (London)*, *B*, vol. 141, pp. 179–189, 1953.

Graskopf, W. R., J. W. Holleman, and I. M. Klotz: Amino Acid Composition of Hemerythrin in Relation to Subunit Structure, *Science*, vol. 141, pp. 166–167, 1963.

Johnson, M. L.: The Respiratory Function of the Hemoglobin of the Earthworm, *J. Exp. Biol.*, vol. 18, pp. 266–277, 1942.

Jones, J. D.: Observations on the Respiratory Physiology and on the Hemoglobin of the Polychaete Genus *Nephthys*, with Especial Reference to *N. hombergi*, *J. Exp. Biol.*, vol. 32, pp. 110–125, 1955.

Jones, J. D.: The Functions of Respiratory Pigments in Invertebrates, in G. A. Kerkuk (ed.), "Problems in Biology," vol. I, pp. 11–89, Pergamon, New York, 1963.

Leitch, I.: The Function of Hemoglobin in Invertebrates with Special Reference to *Planorbis* and *Chironomus* Larvae, *J. Physiol*, vol. 50, pp. 370–379, 1916.

Manwell, C.: Oxygen Equilibrium of *Phascolosoma agassizi* Hemerythrin, *Science*, vol. 127, pp. 592–593, 1958.

Manwell, C.: Oxygen Equilibrium of *Cucumaria* Hemoglobin and the Absence of the Bohr Effect, *J. Comp. Cell. Physiol.*, vol. 53, pp. 75–83, 1959.

Manwell, C.: Oxygen Equilibrium of Brachiopod *Lingula* Hemerythrin, *Science*, vol. 132, pp. 550–551, 1960.

Manwell, C.: The Chemistry and Biology of Hemoglobin in Some Marine Clams. I. On the Distribution and Properties of the Oxygen Equilibrium, *Comp. Biochem. Physiol.*, vol. 8, pp. 209–218, 1963.

Manwell, C.: Genetic Control of Hemerythrin Specificity in a Marine Worm, *Science*, vol. 141, pp. 755–758, 1963.

Read, K. R. H.: The Hemoglobin of the Bivalved Mollusc *Phacoides pectinatus* (Gmelin), *Biol. Bull.*, vol. 123, pp. 605–617, 1962.

Redfield, A. C., and M. Florkin: Respiratory Function of the Blood of *Urechis caupo*, *Biol. Bull.*, vol. 61, pp. 185–210, 1931.

Redmond, J. R.: The Respiratory Function of Hemocyanin in Crustacea, *J. Comp. Cell. Physiol.*, vol. 46, pp. 209–246, 1955.

Redmond, J. R.: The Respiratory Characteristics of *Chiton* Hemocyanins, *Physiol. Zool.*, vol. 35, pp. 304–313, 1962.

Thorpe, W. H., and D. J. Crisp: Studies on Plastron Respiration. I. The Biology of *Aphelocheirus* (Hemiptera) and the Mechanism of Plastron Retention, *J. Exp. Biol.*, vol. 24, pp. 227–329, 1947.

Thorpe, W. H., and D. J. Crisp: Plastron Respiration in Aquatic Insects, *Biol. Rev.*, vol. 25, pp. 344–390, 1950.

Walshe, B.: The Function of Hemoglobin in *Chironomus plumosus* under Natural Conditions, *J. Exp. Biol.*, vol. 27, pp. 73–95, 1950.

chapter XII

OXIDATIVE METABOLISM

A. METABOLIC WORK
  1. Basal metabolism
  2. Measurements of metabolic work: Respirometry, Q_{O_2}, RQ
B. OXIDATION-REDUCTION REACTIONS
  1. Definition
  2. Energy-rich bonds: ATP
C. OXIDATIVE METABOLISM OF CARBOHYDRATE
  1. Glycolysis
  2. Oxidative decarboxylation (Krebs cycle, citric acid cycle, tricarboxylic acid cycle)
  3. Oxidative phosphorylation
  4. Energetics
  5. Roles of mitochondria and of cytoplasm
  6. Alternative pathways
    a. PENTOSE SHUNT
    b. ANAEROBIC METABOLISM AND OXYGEN DEBT
    c. ALTERNATIVE PATHWAYS IN INVERTEBRATES
      (1) Means of verifying "classic" pattern
      (2) Anaerobic metabolism
      (3) Carbon dioxide fixation and succinate formation
      (4) Utilization of pentose shunt
      (5) Glyoxalate cycle
      (6) Alpha glycerophosphate dehydrogenase in insect flight muscle
    d. OXIDATIVE METABOLISM AND MITOCHONDRIAL DEFICIENCIES
    e. ELECTRON TRANSPORT IN INSECT FLIGHT MUSCLE
D. OXIDATIVE METABOLISM OF FATS
E. OXIDATIVE METABOLISM OF PROTEINS
  1. Deamination and utilization of carbon residues
  2. Urea synthesis: ornithine cycle
F. OXIDATIVE METABOLISM AND THE SYNTHESIS OF CARBOHYDRATES AND LIPIDS

A. METABOLIC WORK

1. Basal metabolism

The amount of oxygen utilized by an aerobic organism is an index of its metabolic work; the greater the metabolic activity of an individual, the greater its use and demand for oxygen. This activity is not necessarily overt, but may be that of cells and tissues in performance of the biochemical processes essential to their maintenance as living systems, involving the preservation of their membrane systems and other structural constituents; their osmotic relationships with their environments; and, in the case of growing cells, the construction of new material and, often, cell division. All these operations require the expenditure of energy. The minimal energy requirements, and so the minimal oxygen demands, are those of an animal as completely at rest as any living system can be—quiescent and with its biochemical operations reduced to the lowest level consistent with maintaining the integrity of its organization. This minimal level represents its basal metabolism, and the amount of oxygen used under these conditions is a measure of its least energy requirements. Many invertebrates protect themselves, when oxygen in their environments is deficient or when conditions are otherwise unfavorable to their absorption of oxygen, by reduction of their activity to, or close to, this level by withdrawing into their shells, cases, or tubes, or taking shelter in various places where they can remain until the external oxygen supply again becomes adequate for greater activity. The most extreme reduction of this kind occurs during diapause in insects, a time of almost completely arrested metabolism arising, in some species of insects, at one time or another in their life cycles (see pages 734–740).

In respect to their responses to changes in external oxygen tension, animals are regarded as oxygen regulators or oxygen conformers. An oxygen-regulating organism is independent of external oxygen pressures and maintains an even level of metabolism with falling oxygen tension down to a critical level, below which its oxygen consumption falls off rapidly. The critical pressures are different for different organisms, ranging, for example, from 4 mm Hg for fiddler crabs to 100 mm Hg for oysters. The oxygen consumption of oxygen conformers, on the other hand, increases steadily with increasing external pressure and falls steadily with its decrease. There is no hard and fast distinction between these categories and there are many invertebrates intermediate between the extremes of regulation and conformity.

2. Measurements of metabolic work: Respirometry, Q_{O_2}, RQ

The oxygen consumed by an animal, a group of animals, a piece or pieces of tissue, under given conditions, can be measured in a respirometer. The oxygen consumed is customarily expressed in milliliters or microliters (1/1,000 ml) per unit time per unit of weight of tissue and is designated as Q_{O_2}. In measuring the respiration of protozoans or of metazoans so small that there is a significant source of error in weighing them individually, known numbers of specimens are used and the aggregate values obtained reduced to those per capita. The development of highly sensitive balances and of microrespirometers is eliminating some of these difficulties and should make individual measurements possible, with a satisfactory degree of accuracy.

The amount of oxygen consumed and of carbon dioxide eliminated in a given period of time provides an index not only of metabolic activity during that time but also of the substrate that is being oxidized to provide the energy and of the means by which the energy is released. The substrates are derived from the food that is eaten and digested and so may be end products of any one of the principal categories of foodstuffs—carbohydrate, fat, or protein. These may be the immediate products of digestion or may be ones that are mobilized from the storage tissues. Initial steps in the oxidation of sugar, fatty acid, and amino acid are in each case different, requiring different amounts of oxygen and yielding different amounts of energy per unit of substrate oxidized. The ratio of carbon dioxide evolved to oxygen consumed is known as the respiratory quotient ($RQ = CO_2/O_2$) and gives an indication of the nature of the substrate oxidized and the extent to which it is reduced. When glucose, for example, is burned or completely oxidized in a furnace or calorimeter, 6 mol of oxygen is required for each mol of sugar and 6 mol of carbon dioxide is evolved, plus 686 Cal of heat. The respiratory quotient is, therefore, 1 ($6CO_2/6O_2 = 1$) and values close to this are usually taken as indications that carbohydrate is being oxidized in an organism. The RQ for fat oxidation is 0.71. Protein, or amino acids, must be deaminated before oxidation of their carbon framework; this leaves a nitrogenous by-product to be metabolized and, usually, eliminated. The RQ of protein is 0.93 if the nitrogen is converted to ammonia (NH_3), but it is 0.83 if the nitrogen is converted to urea [$(NH_2)_2CO$], in which carbon and hydrogen are bound. Strict application of these values to living systems must be qualified, since it would be based upon the assumptions that only carbohydrate, fat, or protein are being oxidized at any one time, that gaseous exchange results only from the complete oxidation of a substrate, and that carbohydrate, fat, and protein are the only substrates used. But the foods of an animal are usually mixtures of carbohydrate, fat, and protein, and interconversions of carbohydrate into fat and of fat and protein into carbohydrate, and of carbohydrate into protein go on in organisms, and these factors consequently result in elevation or lowering of the RQ. Moreover, organic acids often serve as substrates, and the resulting RQ varies with the oxygen content of the molecule. If, for example, acetic acid (CH_3COOH) is oxidized, the RQ is 1, but if the substrate is propionic acid (CH_3CH_2COOH), the RQ is 0.85, and if butyric acid, ($CH_3CH_2CH_2COOH$), 0.8. Organic acids are products of anaerobic respiration, so that RQ values may vary after an animal has resorted to this means of energy production in times of oxygen deprivation, and the organic acids that have accumulated in its tissues are oxidized, and the oxygen debt paid off. With due recognition of these qualifications, the respiratory quotient remains a useful indication of what is going on in the overall oxidative metabolism of an animal.

B. OXIDATION-REDUCTION REACTIONS

1. Definition

The oxidation of a compound is essentially a flow of energy, involving a sequence of oxidation-reduction reactions. Oxidation is any process in which hydrogen atoms (H) or electrons (e^-) are removed from a compound. It follows that they must be received or accepted by another compound, which thereby becomes reduced. Under aerobic conditions, oxygen is the final recipient of the hydrogen and the electrons and becomes reduced to water, but there are many transfers from compound to compound before oxygen is reached in the long series of reactions that are known to take place in the oxidative metabolism of cells. In the course of these reactions, there is no liberation of "free energy"—that is, of energy not manifested as heat or work—but, instead, a generation of energy-rich bonds, primarily the phosphate bonds of adenosine triphosphate (ATP) that provide the activation energy that directly or indirectly drives most of the energy-requiring processes of cells.

2. Energy-rich bonds: ATP

Adenosine triphosphate is a complex compound that is of wide, probably universal, distribution in animal and plant cells and in bacteria (Fig. XII–1). Its concentration in animal cells, from which it has been extracted and purified, is approximately 0.5 to 2.5 mg/ml of cell fluid. It has been synthesized in vitro, and the product has shown all the activity and characteristics of the native compound. A molecule of ATP contains the purine base adenine, a 5-carbon sugar, or ribose, and three molecules of phosphoric acid.

The symbol ~ denotes a so-called "high-energy bond," a term that is somewhat confusing in its implication that the energy is *in* the bond and can be set free when the bond is broken. This is not the case. Rather, bond energy is, in terms of physical chemistry, the energy required to break a bond. In the sense in common usage among biochemists, the energy value of a bond denotes the standard free energy change of a reaction; a high-energy bond means that the difference in energy content between the reactants and products of a reaction is relatively great.

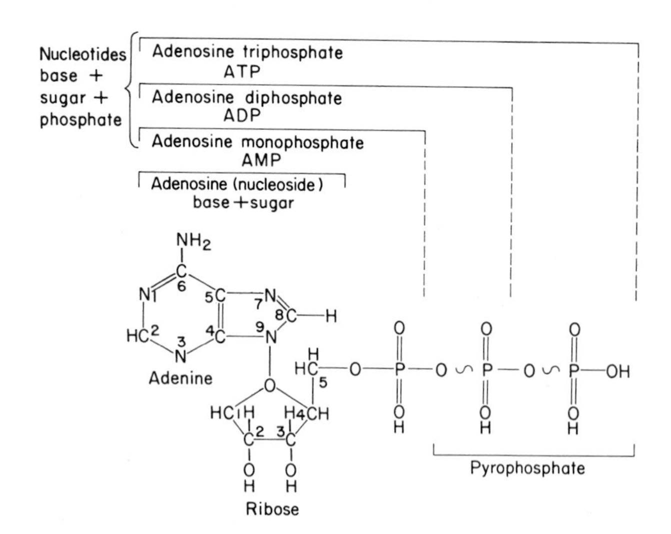

FIG. XII–1 *Structural formulae of adenosine phosphates.*

Adenosine diphosphate (ADP) is formed when the terminal phosphate group is split off from ATP at the pyrophosphate bond, the link between two energy-rich phosphates; ATP can be reconstituted by the addition of a phosphate group to ADP. Similarly, when ADP loses its terminal phosphate, the resulting product is AMP, the nucleotide adenosine monophosphate, a component of nucleic acids, in which the bond between the phosphate and the sugar is one of low energy. Each molecule of ATP thus can supply two, but not three, energy-rich units, and serves as a depot for the energy generated in oxidative reactions.

C. OXIDATIVE METABOLISM OF CARBOHYDRATE

The oxidative metabolism of carbohydrate has been most intensively studied in vertebrate muscle and liver, but there is good reason to believe that the course of events is, by and large, similar in aerobic cells of all kinds. It consists of a series of sequential reactions, the majority of which are catalyzed by enzymes and in which the product of one reaction serves as substrate for the next. More than 20 enzymes are known to be involved in the complete oxidative process, most of which have been extracted and purified and their activity tested in vitro. A corresponding number of intermediates are known. It is convenient to think of oxidative metabolism as comprised of three major events: glycolysis, oxidative decarboxylation, and oxidative phosphorylation. The main steps in these events are outlined in Figs. XII–2 to XII–4 and summarized below in an abbreviated and oversimplified resume of an exceedingly complex process, whose unravelling is the result of many experiments logically conceived by brilliant minds and conducted during the past half-century.

1. Glycolysis

In glycolysis, glucose, either derived directly as a product of digestion or from the depolymerization of stored glycogen, is converted to two molecules of a 3-carbon compound, pyruvic acid, which exists in the cell in the form of a salt, pyruvate. According to the scheme known, from the pioneer experimenters, as the

Embden-Meyerhof pathway, this conversion is effected by the initial phosphorylation of glucose, one molecule of ATP contributing a phosphate group to each molecule of glucose (Fig. XII–2, equation 1). Mg^{++}

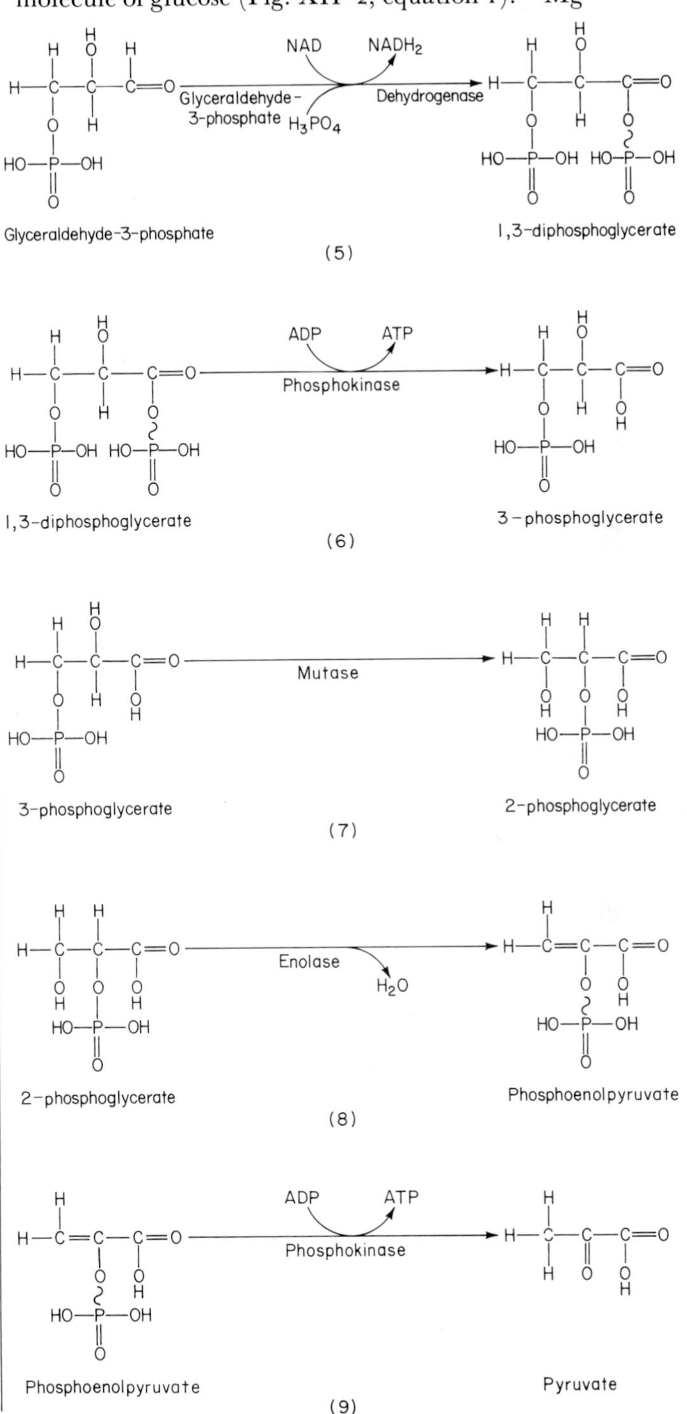

FIG. XII–2 *The principal events in glycolysis.*

is required in this reaction, as is probably also the case in all reactions involving the transfer of phosphate from or between nucleotides. By molecular rearrangement, the glucose-6-phosphate formed in this way becomes fructose-6-phosphate (Fig. XII–2, equation 2) to which another phosphate group is contributed by ATP and attached in the 1 position (Fig. XII–2, equation 3). This so alters the stability of the molecule that it breaks apart into two molecules of triose phosphate, the 3-carbon compounds glyceraldehyde-3-phosphate and dihydroxyacetone phosphate (Fig. XII–2, equation 4). The latter is readily convertible to glyceraldehyde-3-phosphate. An enzymatic reaction then occurs in which another phosphate radical is added to the molecule of glyceraldehyde-3-phosphate, the donor this time being inorganic phosphate (P_i) and not ATP, and two atoms of hydrogen are removed from it. The resulting compound is 1,3-diphosphoglycerate (Fig. XII–2, equation 5). The hydrogens are accepted by a cellular constituent originally known as coenzyme I, later as diphosphopyridine nucleotide (DPN) but more recently identified as nicotine-adenine dinucleotide (NAD). This nomenclature has been officially adopted by an international commission. The compound contains the purine nucleotide adenylic acid, phosphate, and the vitamin nicotinic acid as its amide. The triphosphate of this compound, originally known as TPN but now as NADP, is also a hydrogen acceptor and carrier. In both, the pyridine ring is the significant part of the molecule, because it can accept two electrons and a proton. This reaction is coupled with an energy-conserving one, in which the phosphate group attached to the 1-carbon atom is transferred to ADP and a molecule of ATP is thus produced. The 3-carbon compound remaining is 3-phosphoglycerate (Fig. XII–2, equation 6).

In another molecular rearrangement, the phosphate group shifts from the 3 to the 2 position (Fig. XII–2, equation 7) and the resulting 2-phosphoglycerate then loses a molecule of water (Fig. XII–2, equation 8). This leads to a redistribution of energy in the molecule and the formation of a high-energy phosphate bond in the compound resulting from it, phosphoenolpyruvate. The phosphate is then transferred to ADP with the formation of one molecule of ATP and one of pyruvic acid (Fig. XII–2, equation 9).

2. Oxidative decarboxylation (Krebs cycle, citric acid cycle, tricarboxylic acid cycle)

The next major event, oxidative decarboxylation, is cyclic, for it depends on a key compound which is reformed after a series of enzymatic reactions in the course of which a succession of intermediate products is produced (Fig. XII–3, equations 9a to 18). This organized series of reactions is known as the citric acid cycle, the tricarboxylic acid cycle (TCA), or the Krebs cycle, after the Nobel laureate Sir Hans Adolf Krebs (1900–), who, with his colleagues, did much of the pioneer work in its analysis. It begins after the removal of the carboxyl group from pyruvate and the consequent formation of a 2-carbon compound, acetaldehyde (Fig. XII–3, equation 9a), which combines with an enzyme, coenzyme A (CoA) to form acetyl-coenzyme A (Fig. XII–3, equation 10). Coenzyme A is composed of adenosine and a derivative of the B vitamin pantothenic acid, which has a terminal thiol (SH) group. This is the group concerned in the activation of acetate to acetyl-CoA. The combination of coenzyme A with acetaldehyde results from a number of rather complicated reactions involving several cooperating substances. The bond between acetate and the sulfur of CoA is a high-energy bond, and the acetyl group is thus in an activated or energized form. This is the only form in which carbon compounds can enter the Krebs cycle and their oxidation progress. The cycle is initiated by the condensation of the activated acetate with oxalo-acetic acid, a 4-carbon dicarboxylic acid, resulting in the construction of citric acid, a 6-carbon acid with three carboxyl groups (Fig. XII–3, equation 11). In this reaction coenzyme A is removed, the energy of its bond being used in the condensation. Oxalo-acetic acid can be formed by the addition of carbon dioxide to the methyl group of pyruvic acid, as has been shown by the use of carbon dioxide labeled with C^{14} or C^{13}. This may be one of the several ways in which it is made in cells and becomes available for this critical step in oxidative metabolism.

The removal of a molecule of water from citrate converts it to *cis*-aconitate (Fig. XII–3, equation 12), and the addition of a molecule of water to the intermediate results in isocitrate (Fig. XII–3, equation 13). In the next step in the cycle, which is an oxidation,

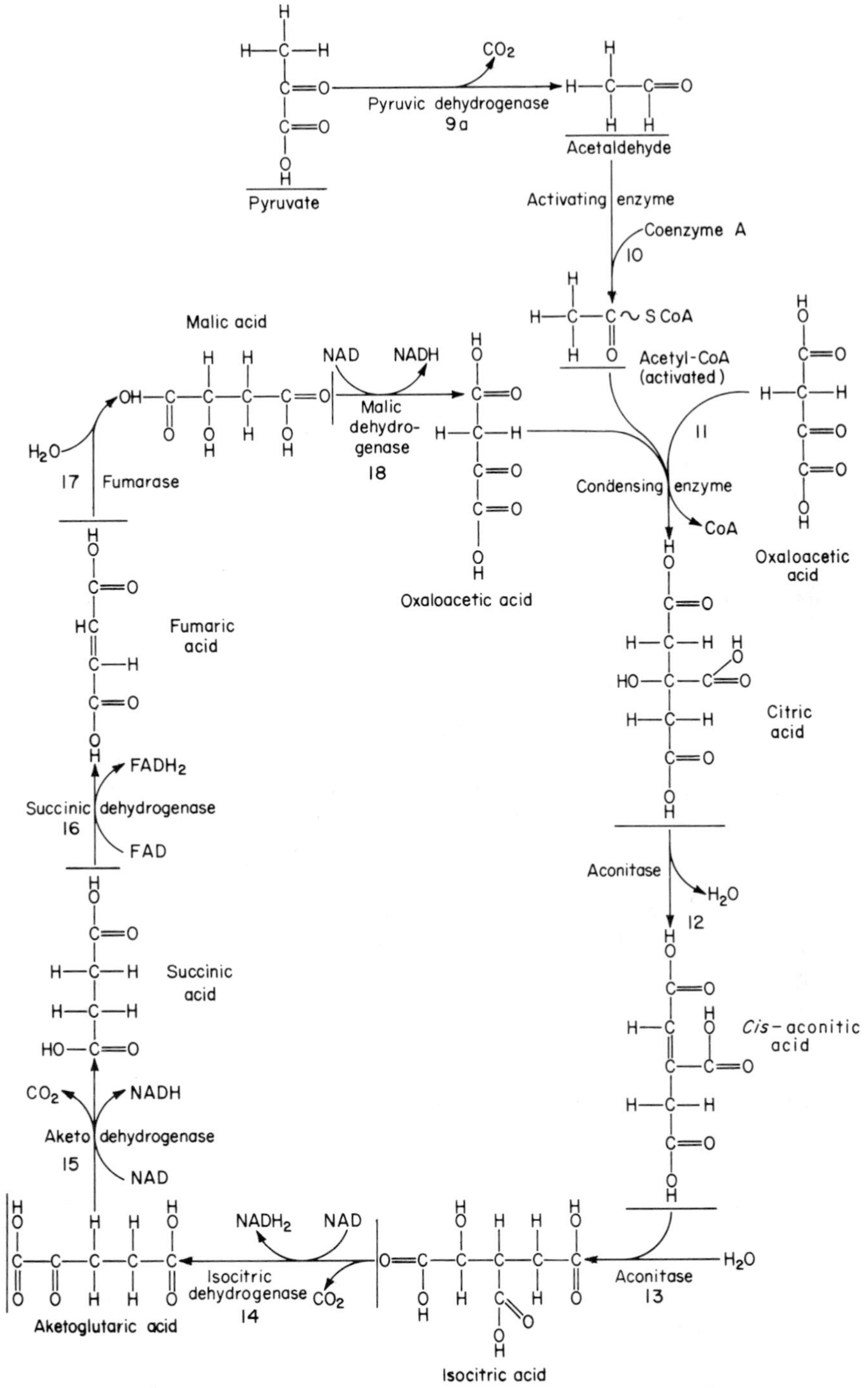

FIG. XII–3 *The principal events in the Krebs cycle.*

two atoms of hydrogen are removed from isocitrate and accepted by NAD, and a molecule of carbon dioxide is set free, the first that is produced in the course of oxidative metabolism (Fig. XII–3, equation 14). This leaves a 5-carbon compound, alpha-keto-glutaric acid, which then loses hydrogen to NAD and sets free another molecule of carbon dioxide (Fig. XII–3, equation 15). This reaction results in the formation of a 4-carbon compound, succinic acid. The loss of hydrogen from succinic acid, which, in this case is accepted by the catalyzing enzyme, succinic dehydrogenase, produces a molecule of fumaric acid (Fig. XII–3, equation 16). Succinic dehydrogenase belongs to the category of flavoproteins (FP) and consists, in addition to the protein, of flavin adenine dinucleotide (FAD), which contains the vitamin riboflavin. FAD acts as a hydrogen acceptor in much the same way that NAD does. Fumaric acid then adds a molecule of water to become malic acid (Fig. XII–3, equation 17). This then becomes dehydrogenated, with NAD acting as the acceptor, and oxalo-acetic acid is thus reformed (Fig. XII–3, equation 18), to condense again with new pyruvate entering the cycle as the acetyl-coenzyme A complex.

Thus, at the beginning of the cycle, a 2-carbon compound derived from a 6-carbon sugar was introduced to combine with a 4-carbon compound to make another 6-carbon compound. This underwent various transformations, in the course of which two molecules of carbon dioxide were produced, three molecules of NAD and one of FAD were reduced, and the key carbon compound, oxalo-acetic acid, was reconstituted. The cycle, therefore, can continue as long as acetyl-coenzyme A is fed into it and as long as there are hydrogen acceptors for the hydrogen atoms removed in the dehydrogenation (oxidation) reactions. The first condition can be met while there are, in the organism, carbon compounds to give rise to acetaldehyde, and the second can be met by mechanisms that remove hydrogen from the acceptors and restore them to their oxidized condition and readiness to receive hydrogen again. These mechanisms are provided by the third major event of oxidative metabolism, oxidative phosphorylation.

3. Oxidative phosphorylation

In this final major event, there is a transference of electrons from the three reduced NAD molecules and the reduced FAD molecule to the enzymes that constitute the cytochrome system, and finally to oxygen. Five of the cytochromes have been identified and, in the order in which they accept the electrons, are known as cytochromes b, c_1, c, a, and a_3. The iron in the heme of these enzymes can exist either in its oxidized (Fe^{+++}) or its reduced (Fe^{++}) state. Each in its oxidized state can accept an electron, thereby becoming reduced, and each in its reduced state can donate an electron, thereby becoming oxidized ($Fe^{+++} + e \rightleftharpoons Fe^{++}$). Each cytochrome, like other reducing agents, has a characteristic electron pressure and, like other oxidizing agents, a characteristic electron affinity. The tendency is for electrons to flow from donors of high electron pressure, or high negativity, to those of lower electron pressure, or high positivity.

For each molecule of glucose that has passed through the previous events of oxidative metabolism there are 12 pairs of hydrogen atoms bound to hydrogen acceptors. Two pairs, set free in the early stages of glycolysis when glyceraldehyde-3-phosphate was converted to 1,3-diphosphoglycerate, were accepted by NAD. Two other pairs, bound also to NAD, originated at the union of acetaldehyde with coenzyme A to form acetyl-coenzyme A. The remaining eight pairs originated within the Krebs cycle, two when isocitrate was converted to alpha-keto-glutarate, with NAD the receptor, two when alpha-keto-glutarate was converted to succinate, with NAD again the receptor, two when succinate was converted to fumarate, with FAD the acceptor, and two when malate was oxidized to oxalo-acetate, NAD being the receptor in this reaction.

Electrons cannot enter the cytochrome system directly from NAD, but only from flavoprotein (FP); some flavoproteins can transfer hydrogen directly to oxygen, bypassing the cytochrome system entirely. In passing through this system, however, the hydrogen atoms bound to NAD are transferred to FP, then lose their electrons and become hydrogen ions (H^+). The electrons from FP are accepted first by oxidized cytochrome b, and then transferred to oxidized cytochrome c_1. Electrons from c_1 then pass to cytochrome c, then to a, and so on to a_3. In the presence of this enzyme, known also as cytochrome oxidase, the electrons reunite with H^+ and with oxygen to form water. Oxygen thus forms the final acceptor of electrons, and the reduction of oxygen constitutes the final step in the entire oxidative sequence.

At each step, the enzyme losing the electron becomes oxidized and so returns to the state in which it can again accept an electron from a donor with higher electron pressure. Each cytochrome can accept only one electron at a time and, therefore, each must react twice to accomplish the oxidation of a previous member of the chain. Since, also, two electrons are required to reduce one atom of oxygen and since oxygen, in air or in solution, is in molecular form, as O_2, only one-half molecule of oxygen is reduced each time a pair of electrons passes along the chain, and four pairs are necessary to reduce one molecule of oxygen and make two molecules of water (Fig. XII–4). The 12 pairs contributed by the reduced nucleotides and flavoprotein, plus the 12 hydrogen ions, can account for the formation of six molecules of water and, as six molecules of carbon dioxide were produced in the two revolutions of the Krebs cycle necessary to oxidative decarboxylation of the two molecules of pyruvate derived from each molecule of glucose, the overall equation for the oxidation of glucose may be written $C_6H_{12}O_6 \rightarrow 6CO_2 + 6H_2O$.

All the enzymes cooperating in these reactions and all those of the Krebs cycle are known to be localized in the mitochondria, and measurements of the rates of electron transport in preparations of intact, isolated mitochondria have shown that the maxima are attained only when phosphate and ADP are included in the medium and that both these substances are used up in the process. The formation of ATP is now known to be concomitant with electron transport in the cytochrome system; hence, the name oxidative phosphorylation for the coupled reactions of this event. From such quantitative studies it has been calculated that in the passage of a single pair of electrons from $NADH_2$ to oxygen, three molecules of ATP are formed from ADP and inorganic phosphate. At each transfer of electrons there is a decline in free energy, related directly to the drop in electron pressure. The largest drops occur when hydrogen is transferred from $NADH_2$ to FP, and when electrons move from cytochrome b to cytochrome c_1, and from cytochrome a_3 to oxygen. At these three steps, high-energy intermediates are formed which donate high-energy phosphate groups to ATP. In this way the energy of the oxidation of reduced FP is parceled out in a series of small packets and conserved in the high-energy bonds of ATP.

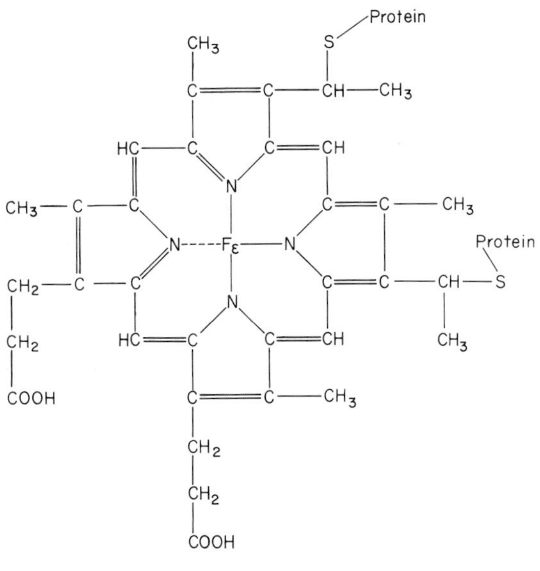

Cytochrome C

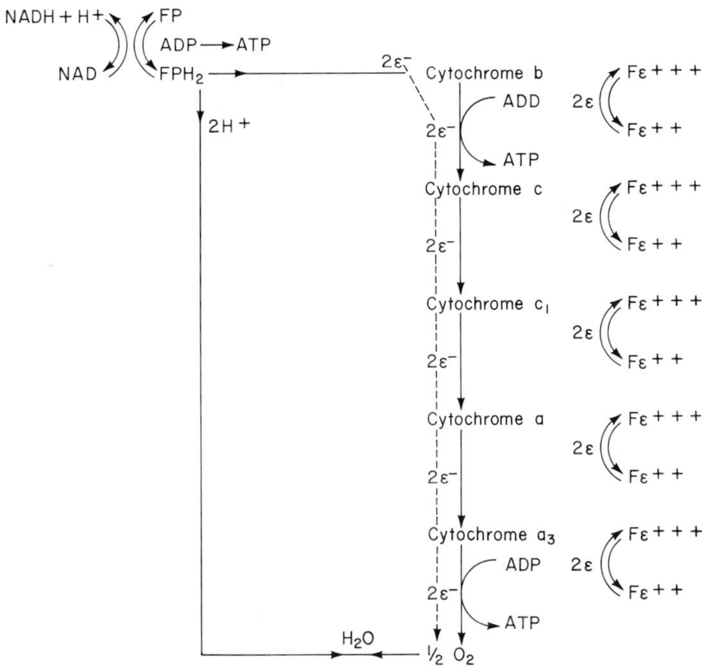

FIG. XII–4 *The principal events in oxidative phosphorylation.*

4. Energetics

Oxidative phosphorylation thus accomplishes the restoration of the principal hydrogen acceptors of the entire sequence and so ensures its continued operation as long as the initial substrate, the enzymes, ADP, and inorganic phosphate are available. It also accomplishes the generation of ATP, which is of signal importance to the energetics of the cell. Since two molecules of glyceraldehyde-3-phosphate, and ultimately two molecules of pyruvate, are derived from each molecule of glucose, four molecules of ATP are produced for each molecule of glucose that passes through the event of glycolysis, which is concerned primarily with splitting the glucose molecule and preparing the resulting fragments for entrance into the Krebs cycle. But two molecules of ATP were expended in glycolysis, one in the initial phosphorylation of glucose and one in the diphosphorylation of fructose-6-phosphate. The net gain is, therefore, two molecules of ATP. There is no gain in ATP from the Krebs cycle, in which the carbon framework is dealt with and through carboxylations and dehydrogenations oxidized to carbon dioxide, with corresponding reduction of hydrogen acceptors or carriers. There is a net gain of 36 molecules of ATP from the transfer of the 10 pairs of electrons from NADH to FP and their passage along the cytochrome chain, together with the pair from $FADH_2$ formed when succinate is reduced to fumarate.

These 38 molecules of ATP represent energy equivalent to 266 Cal, for the formation of each requires an input of energy equivalent to 7 Cal. This represents about 40 percent of the 686 Cal that are released as heat when 1 mol of glucose is burned in a calorimeter. Actually the efficiency of oxidative metabolism in vivo may be greater than 40 percent, for the energetics of the process have been worked out from studies, in vitro, of muscle and liver homogenates or slices, and the values in vivo may well be higher.

5. Roles of mitochondria and of cytoplasm

Within the past 30 years the magnitude and the nature of the contribution of the mitochondria to cellular operations, suspected since the beginning of the century, have been elucidated and appreciated. Techniques for their isolation, purification, and disruption have made possible the study of their enzymatic capacities as wholes or in fragments and have shown that their enzymes are very precisely localized as structural and functional units of the membranes, in spatial arrangements analogous to assembly lines in a factory. Electron microscopy and microspectrophotometry are providing evidence for the distribution of the cytochromes on the cristae, the membranes that project from the inner membrane of a mitochondrion into its lumen (Fig. IV–1). These enzymes seem to be collected in groups or clusters that have been given the name of oxysomes. The enzymes of the Krebs cycle seem equally precisely arranged in other parts of the organelle. These arrangements provide for maximum efficiency in the operation of the enzymes, because the product of each reaction, as substrate for the next, is in the immediate vicinity of the appropriate enzyme. The universality of mitochondria in cells of animals at all levels of structural organization and evolutionary development suggests that they must have arisen as discrete units of cytomembranes early in the history of cellular systems and been retained throughout its later course because of their tremendous advantage to the economy of the cell. Not only are they sites for the formation of the primary store of energy, ATP, but they are also common meeting grounds for the metabolism, both catabolic and anabolic, of carbohydrate, fat, and protein.

The enzymes concerned with glycolysis are, on the other hand, not bound to any cell particulate but are distributed throughout the cytoplasm, for in homogenates of tissues their activity is exhibited in the fluid, nonparticulate fraction. The intermediates of glycolysis are, however, all phosphorylated compounds and as such cannot pass across the cell membrane. Therefore, while they may diffuse freely in the fluid phase of the cytoplasm, they do not diffuse out of it. This condition contributes also to the efficiency and rapidity of cellular metabolism. In the anabolic events of synthesis of polysaccharides and lipids, larger molecules are being compounded from smaller ones and tend to stay in the area in which they are formed. But the assembly of amino acids into polypeptides takes place also in association with definite cell particulates, the ribosomes, in accordance with the blueprints provided by the code inherent in the DNA of the organism. It would seem, therefore, that as cells became the highly organized systems that are recognized today, the biochemical events requiring the greatest specificity and precision were assigned to particular organelles, while the more generalized ones were left to the fluid phase of the cytoplasm.

6. Alternative pathways

a. PENTOSE SHUNT Alternative pathways of oxidation-reduction are, of course, possible, although from the present evidence passage through the Krebs cycle and oxidative phosphorylation seems the most common, in vertebrates at least. The "pentose shunt" is one alternative glycolytic pathway that also seems in common use, often in conjunction with the Embden-Meyerhof pathway. In this, there are produced from three molecules of glucose-6-phosphate two molecules of fructose-6-phosphate and one of glyceraldehyde-3-phosphate (Fig. XII–5). NADP is used as the hydrogen acceptor in the series of reactions by which the six carbon sugars are oxidized to six carbon acids, which are then decarboxylated to form five carbon sugars. These pentoses can then interact with each other to form the end products, fructose and glyceraldehyde, which can then enter the glycolytic sequence, or they can be used in the construction of other molecules. Ribose, for example, is of especial importance in the synthesis of nucleotides. This shunt has the advantage of having two oxidative steps rather than the one of the Embden-Meyerhof pathway, as well as of having pentoses as intermediates. Intermediates other than ribose are known to be utilized by photosynthesizing organisms in the fixation of carbon dioxide, a means of returning carbon to biological systems and one that is now known to be quite widespread and practiced by other than photosynthesizing species.

b. ANAEROBIC METABOLISM AND OXYGEN DEBT Glycolysis is essentially an event that can proceed in the absence of oxygen. Under aerobic conditions its product, pyruvate, is oxidized to carbon dioxide and water through the events of the Krebs cycle and oxidative phosphorylation; under anaerobic conditions, other products result. These may be organic acids of various kinds. Pyruvate may, for example, be reduced to lactate ($CH_3CHOHCOO^-$) through the agency of lactic dehydrogenase, the hydrogen for this reduction coming from NADH formed in the glycolytic sequence and thus restored for future operations without the agency of the cytochrome system (Fig. XII–5, box 2). Lactic acid is produced in this way in the muscle cells of vertebrates, which experimentally have been shown to contract for some time in the complete absence of oxygen. The gain in ATP for muscle cells is slightly higher than that for other cells that carry glycolysis through to lactate, for muscles start the event from glycogen rather than glucose, and glycogen is depolymerized by phosphorylation with inorganic phosphate. The molecule of ATP used in the phosphorylation of glucose is, therefore, not expended and the net gain in ATP is three molecules per each 6-carbon unit of glycogen ($\equiv$21C). This yield of ATP is very small compared with the 38 molecules gained from the aerobic oxidation of glucose, or the 39 that muscle cells would gain under aerobic conditions, and an equivalent energy yield would require the degradation of far greater amounts of carbohydrate. It has been shown, in fact, that to accomplish the same amount of work an anaerobic cell must use 20 times as much glucose as an aerobic one.

Muscle will not contract indefinitely in the absence of oxygen but, after stopping, will begin contractions again when oxygen becomes available to it. Lactate is retained in the cell and can, like pyruvate, be prepared for entrance into the Krebs cycle and so can be completely oxidized with a gain of 39 molecules of ATP per 6-carbon unit. The oxygen required for this represents the debt the muscle has incurred during its period of anaerobiosis, or when it is producing pyruvate, during intense activity, faster than the oxidative mechanisms can handle. A number of aerobic invertebrates can also build up oxygen debts in their muscles and other tissues, during activity or periods of oxygen deprivation, and pay them off in longer or shorter order. Their ability to survive anaerobically for extended periods of time depends upon their tolerance of the end products of glycolysis or their capacity to inactivate or eliminate them. These end products may be a variety of organic acids

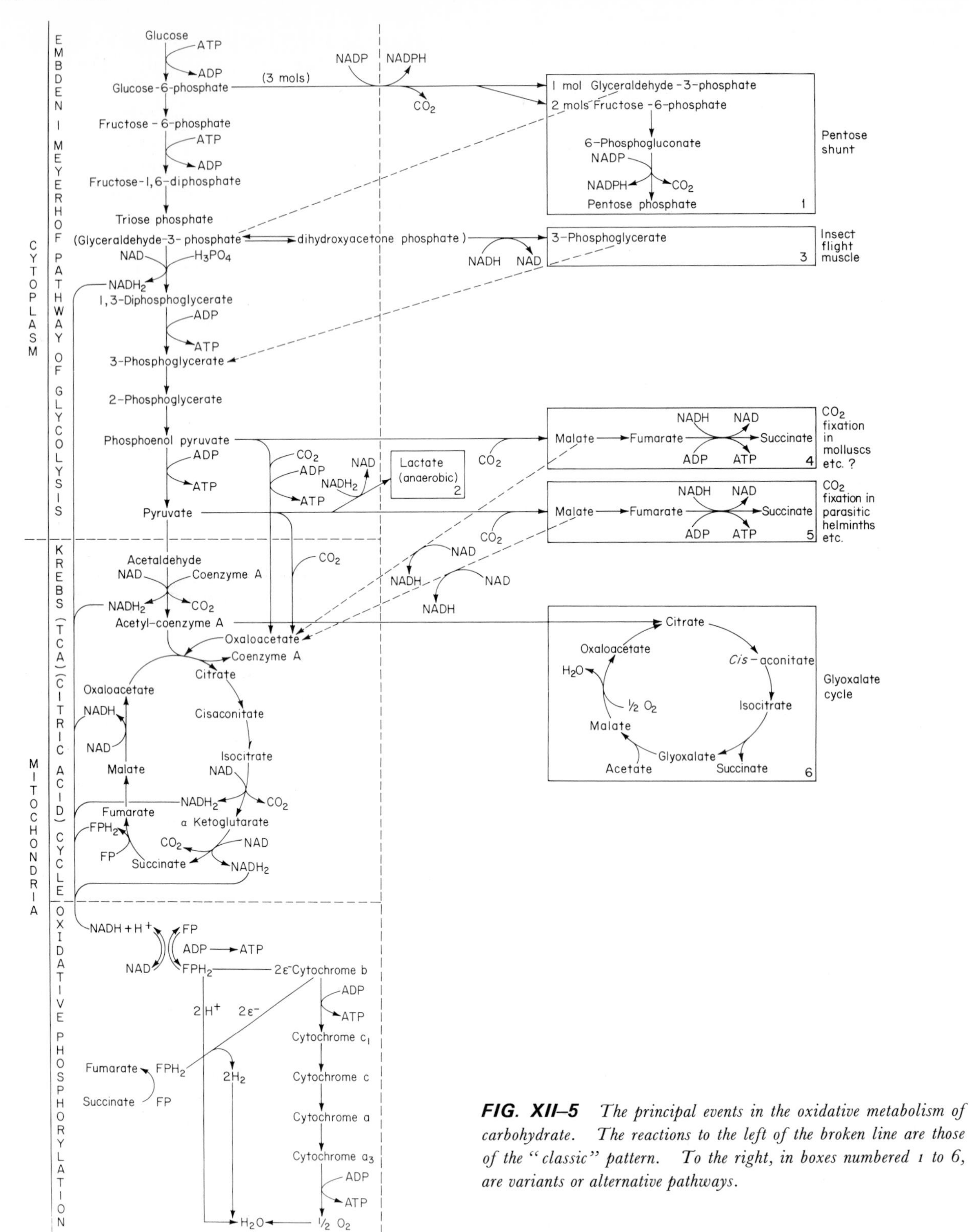

FIG. XII–5 *The principal events in the oxidative metabolism of carbohydrate. The reactions to the left of the broken line are those of the "classic" pattern. To the right, in boxes numbered 1 to 6, are variants or alternative pathways.*

other than lactic, and the oxygen debt can be paid off only if their derivatives can enter the Krebs cycle. In a number of animals, these are not oxidized but eliminated from the body by some means, with a consequent loss of the energy that might have been derived from them.

c. ALTERNATIVE PATHWAYS IN INVERTEBRATES (1) Means of verifying "classic" pattern The course of oxidative metabolism outlined above has been worked out primarily in mammalian tissues and yet is generally accepted as representative of vertebrates of all kinds and the "classic" pattern of events. It should be recognized, however, that while the data from these studies show possible pathways, they do not necessarily prove that these are the only ones or even, indeed, that they are the pathways followed in vivo. Nor do they prove that these are the pathways of oxidative metabolism in invertebrates, evidence for which can only come from exact repetition of the experimental procedures employed for vertebrates using analogous invertebrate tissues. But the ability of pieces of muscle, digestive gland, gut epithelium, or other organs taken from invertebrates, or of homogenates and fractions of homogenates from them, to carry out oxidations with the production of the same intermediates and the same end products shows that they, too, might follow the same course in vivo. There are other methods of verification as well. One of these is the use of compounds that are known to interfere with, or wholly inhibit, the action of specific enzymes operating in the oxidative processes in vertebrates. These are listed in Table XII–1, and sensitivity of an invertebrate system to any one of them would provide circumstantial evidence that the course of oxidative metabolism, at least up to the inhibited step, is similar to that established for vertebrate systems; conversely, insensitivity would give evidence that an alternative pathway is, or could be, used. Increased respiration and utilization of oxygen following the addition to a tissue preparation of any of the recognized intermediates in the events of oxidative metabolism would indicate that the same substrates were being used and, therefore, a similarity in the procedures. This would also be evident in the identification of cellular enzymes extracted from invertebrate tissues with corresponding ones from a vertebrate source.

TABLE XII-1 Inhibitors of Oxidative Metabolism

Inhibitor	*Enzyme Inhibited*	*Reaction Blocked*
Arsenite	Pyruvic dehydrogenase	Pyruvate → acetaldehyde
	Alpha-keto-glutaric dehydrogenase	Alpha-keto-glutaric acid → succinic acid
Fluoride	Enolase	2-Phosphoglycerate → phosphoenolpyruvate
	ATP-ase	ATP ⇌ ADP + H_3PO_4
Fluoroacetate	Aconitase	Citrate → *cis*-aconitate → isocitrate
Iodoacetate	Glyceraldehyde-3-phosphate dehydrogenase	Glyceraldehyde-3-phosphate → 1,3-diphosphoglycerate
Malonate	Succinic dehydrogenase	Succinate → fumarate
p-Chloromercuribenzoic acid	Succinic dehydrogenase	Succinate → fumarate
	Glyceraldehyde-3-phosphate dehydrogenase	Glyceraldehyde-3-phosphate → 1,3-diphosphoglycerate
Actinomycin A	DPNH cytochrome reductase	Reoxidation of cytochrome b
Cyanide	Cytochrome oxidase	Transference of electrons from cytochrome a_3 to O_2
Carbon monoxide	Cytochrome oxidase	Transference of electrons from cytochrome a_3 to O_2
Amytal		Transference of electrons from NADH to FP

The use of radioactive isotopes, especially those of carbon, have also been used in tracing the steps and determining consistency with, or deviations from, the established pathways. Chromatography has also proved useful in the identification of substrates and products as has spectrophotometry, particularly in the analysis of the operation of the cytochrome system, for these enzymes have different absorption maxima in their reduced and oxidized states.

Although such studies have been made only on a limited number of invertebrates, there is evidence that the "classic" pattern is modified to a greater or lesser extent in some invertebrates and that it is followed, apparently without modification, in others. The investigations to date have centered particularly upon glycolysis in animals living in environments poor in oxygen and the entire course of oxidative metabolism in insects, for the movements of their legs and wings involve the expenditure of relatively large amounts of energy and a high rate of metabolism in the muscles responsible. These two conditions represent extremes in the mechanisms of oxidative metabolism.

(2) Anaerobic metabolism The Embden-Meyerhof pathway of glycolysis leading to pyruvate seems to be followed in the crustaceans using carbohydrate as a main source of energy and in the tissues of some insects, although all the enzymes concerned have not yet been isolated. Under anaerobic conditions, lactic acid may be produced in invertebrate tissues as it is in those of vertebrates, but other organic acids are known also to be end products. Valeric, propionic, caproic, and butyric acids are excreted by enteric parasitic helminths, although the steps of the metabolism of the carbohydrate from which they are derived are not known. From the point of view of bioenergetics, this is a very wasteful procedure, for much potential energy is still held in molecules of this kind. But these animals can afford to follow it, for they live in an abundance of immediately available nutrient and expend little energy in muscular activity.

(3) Carbon dioxide fixation and succinate formation Another departure from the pattern of anaerobic respiration as exhibited by vertebrate muscle is recognized in a number of intestinal parasites, including trematodes, cestodes, nematodes, and the acanthocephalan *Monoliniformis dubius*. In all of these, succinate, rather than lactate, is the end product of glucose degradation (Fig. XII–5, box 5). These animals live in environments of very low oxygen tension, a condition that would virtually impose upon them an anaerobic mechanism of energy storage and release.

In *Ascaris*, pyruvate produced by glycolysis along the Embden-Meyerhof pathway is combined with carbon dioxide to form malate, which is then converted to fumarate, which, in turn, is reduced to succinate, with NADH acting as hydrogen donor. The transfer of electrons from NADH takes place through the agency of FP and results in the generation of ATP; thus, the reduction of fumarate to succinate accomplishes both the regeneration of NAD from the NADH formed during the oxidation-reduction step of glycolysis and the generation of ATP. This reaction can be accomplished in vitro by mitochondria isolated from muscles of the body wall. The major products of anaerobic respiration in *Ascaris* are succinate, alpha methyl butyrate, and alpha methyl valerate. Depletion of the glycogen stores in specimens removed from the guts of their hosts and kept in a suitable non-nutrient medium indicates that these compounds are derived from carbohydrate.

The canine whipworm *Trichuris vulpis*, in anaerobic respiration, produces succinate by a similar route, in addition to lactate, propionate, and a number of other fatty acids. Incubation of whole individuals of *Hymenolepis diminuta*, a tapeworm living in the small intestine of rats, with C^{14}-glucose and with $NaHC^{14}O_3$ has provided evidence that they metabolize glucose in the same way and that the conversion of glucose to succinate through pyruvate is the major route of their carbohydrate degradation. In spite of the fact that they will take up oxygen in an environment in which it is plentiful, they seem not to make use of the Krebs cycle as an energy-yielding pathway and their metabolism remains predominantly anaerobic. There is also evidence that a similar aberrant pathway is followed by the fluke *Schistosoma* and by *Monoliniformis*, an acanthocephalan living its adult life in the intestine of rats and its larval stages in the hemocoele of cockroaches and other insects. Incubation with C^{14}-glucose, 1:4C^{14}-succinate, and 2C^{14}-acetate of pieces and homogenates of its body wall indicate that the main route of glucose metabolism is through the carboxylation of pyruvate and the ultimate formation of succinate.

The importance of carbon dioxide to animals employing such respiratory mechanisms is obvious. An atmosphere with at least 5 percent carbon dioxide has been found to be essential to the development of *Ascaris suum* and to the prolonged survival of *Trichinella spiralis* in axenic culture. The intestinal habitats of these animals provide them with such environments; carbon dioxide in the intestinal gas of rats, for example, ranges from 17 to 88 percent, depending to a great extent upon the kind of diet they are fed and the time after feeding that the gas analyses are made. The attainment of activation energy by a metabolic pathway involving carbon dioxide fixation and the production of succinate distinguishes the metabolism of the tissues of the parasite from that of its host and may represent an adaptation to a parasitic mode of life. But carbon dioxide fixation is by no means limited to intestinal parasites and has indeed been found to be a common phenomenon among free-living marine invertebrates of many kinds. Incubation, either of whole animals or of pieces of their tissues, in seawater with $NaHC^{14}O_3$ has shown significant uptake of labeled carbon by the sun sponge *Hymeniacidon heliophila*, the sea anemone *Aiptasia pallida*, the flatworms *Bdelloura candida* and *Stylochus zebra*, the nemertean *Cerebratulus lacteus*, the polychaete *Chaetopterus variopedatus*, the horseshoe crab *Limulus* and the true crab *Callinectes sapidus*, the ectoproct *Bugula neritina*, the brachiopod *Lingula unguis*, the holothurian *Leptosynapta inhaerens*, and the oyster *Crassostrea virginica*. After homogenization of the tissues at the conclusion of the incubation period, extraction of the organic acids, and their separation by paper chromatography, it has been found that in the majority of animals tested the labeled carbon was incorporated primarily in succinate, fumarate, and malate. *Leptosynapta* was an exception for in it the label was fairly uniformly distributed among all the intermediates. Labeled lactate was also found in some species, as well as other acids of the Krebs cycle. This suggests that the pathway of glucose degradation in these free-living species may be similar to that in parasitic helminths, but does not preclude the possibility of the operation of a complete Krebs cycle. Yet incubation both aerobically and anaerobically of mantle tissue taken from three bivalve molluscs—*C. virginica*, *Rangia cuneata*, and *Modiolus* (*Volsella*) *demissus*—in a medium containing labeled glucose has shown that the carbon atoms of succinate are derived from glucose and that their incorporation is accomplished in the absence of oxygen. These facts would seem to rule out the operation of a complete Krebs cycle and to indicate that it is modified in these animals in a way similar to that in intestinal parasites.

With molluscs, the question arises of the step in glycolysis at which carboxylation occurs and precisely which intermediate is the starting-off point and combines with carbon dioxide to form the dicarboxylic acid that is finally converted to succinate. Incubation of tissue taken from the mantle of *C. virginica* with propionate labeled with C^{14} has resulted in recovery of the label in succinate, which, in turn, is converted to other intermediates of the Krebs cycle. When the tissue is incubated with labeled bicarbonate, the C^{14} is recovered first in malate, suggesting that the course followed is similar to that in *Ascaris* and related parasites. Evidence that carboxylation occurs at a step in glycolysis prior to the formation of pyruvate has been obtained from similar incubation experiments in which the mantle tissue of oysters was used as well as that from *Rangia* and *Modiolus*. Maximum amounts of carbon dioxide were incorporated from a medium containing $NaHC^{14}O_3$ when phosphoenolpyruvate was added to the mixture, indicating that carboxylation occurs at this level in the glycolytic sequence (Fig. XII–2, reaction 9). Supporting evidence for this is provided by the fact that about 10 times more labeled succinate has been recovered from pieces of the mantle of *Rangia* incubated with labeled glucose than from those incubated under similar conditions with labeled pyruvate. Moreover, high values for phosphoenolpyruvate carboxylase, the enzyme catalyzing the reaction in which carbon dioxide is combined with this 3-carbon derivative of glucose, have been found in marine molluscs and some other invertebrates, including *Hymenolepis diminuta*, and the possibility that carboxylation occurs at the phosphoenolpyruvate level is not excluded by existing evidence. Carboxylation of phosphoenolpyruvate through the mediation of a kinase results in the formation of oxalo-acetic acid and the liberation of a phosphate radical to combine with ADP, and the consequent formation of ATP.

There thus have been demonstrated two possible mechanisms for the production of succinate from glucose avoiding the Krebs cycle: (1) interruption of the Embden-Meyerhof pathway at the level of phosphoenolpyruvate and the conversion of this compound either to oxalo-acetic acid through the activity of phosphoenolpyruvic carboxykinase or, through the activity of phosphoenolpyruvate carboxylase, ultimately to succinate; and (2) continuation of the pathway to pyruvate with carboxylation of this, either through the agency of pyruvic carboxylase to give oxalo-acetic acid or of that of malic dehydrogenase to give malic acid. The dicarboxylic acids resulting from either of these mechanisms can subsequently pass through the necessary steps to yield succinate. Each represents a modification or curtailment of the Krebs cycle. Both lead to the formation of ATP and are anaerobic processes, and the animals employing them, so far as is yet known, are those that live in environments with an adequate supply of carbon dioxide, either as gas or as carbonate, and a limited supply of oxygen, for the lamellibranch molluscs used in these experiments close their shells tightly and so experience periods of several hours when they have no access to external oxygen.

(4) Utilization of pentose shunt Utilization of the pentose shunt has been demonstrated also in parasitic helminths such as *Ascaris lumbricoides*, the larvae of *Trichinella spiralis* and *Echinococcus granulosus*, and in insects. Incubation of homogenates of cysts of *Echinococcus*, taken from livers of newly slaughtered sheep, with various intermediates of the Krebs cycle has resulted in an oxygen uptake greater than that of homogenates in nonsupplemented media, indicating that they can carry through a complete Krebs cycle. This has also been proved for insect tissues, but the relative importance of the two pathways—pentose shunt and Krebs cycle—has not been established.

Tracer experiments with sugar labeled with C^{14} show that the larvae of both queen and worker honeybees (*Apis mellifera*) use the pentose shunt to a high degree. The weight of the larvae of both castes increases by several hundred percent between 72 and 96 hr of larval life, and utilization of the pathway can provide them with carbon fragments for synthesis not only of ribose sugars but also of lipids and some amino acids. Possibly a high degree of participation of this pathway is characteristic of many, or all, organisms in the most rapid stages of their development.

It seems apparent that in the cockroach *Periplaneta*, at least, the pentose shunt is of greater value in the syntheses of nucleic acids than it is in energy production. This may be true of insects in general. The overall operation of the Krebs cycle has been unequivocally demonstrated in several different species of insects, as evidenced by increase in oxygen uptake when Krebs cycle intermediates are added to the incubation medium and by its decrease when standard inhibitors are introduced. Both in insects and in *Echinococcus*, the enzymes of the Krebs cycle are located in the mitochondrial fraction of a homogenate and those of the pentose shunt, like those of the Embden-Meyerhof pathway, in the supernatant. The brachiopod *Lingula*, although known to fix carbon dioxide (*L. unguis*), seems also to have a complete Krebs cycle, because extraction of enzymes from *L. reevei* has shown that it has all those requisite for it. *Lingula* is particularly high in succinic dehydrogenase, in comparison to the oyster (*Crassostrea virginica*) and the mussel (*Modiolus demissus*); the activity of the enzyme in *Lingula* being 185 times that in *Crassostrea* and about four times that in *Modiolus*.

(5) Glyoxalate cycle Studies on the biochemistry of nematodes have demonstrated their utilization of another cycle, the glyoxalate cycle, a recognized pathway in many types of microorganisms (Fig. XII–5, box 6). Two molecules of acetate are required for each revolution of the cycle. One enters in the form of acetyl-coenzyme A which, as in the orthodox Krebs cycle, is converted to citrate, then to *cis*-aconitate and to isocitrate. In the glyoxalate cycle, the isocitrate becomes converted to glyoxalate (CHO · COO—), with succinate as a by-product, and the glyoxalate combines with a second molecule of acetate to form malate, which is then converted to oxalo-acetate by the addition of one atom of oxygen and the loss of one molecule of water. The key enzymes in this cycle are malate synthetase and isocitrate lyase. These have been demonstrated in four free-living species of nematodes: *Coenorhabditis briggsae, Panagrellus reduvivus, Rhabditis anomala,* and *Turbatrix aceti*, but they have not been detected in *Ascaris*. Their presence in animals at the nematode level of organization is of some interest, since hitherto they have been considered limited to plants, but the significance and metabolic function of the glyoxalate cycle in these animals is still a matter of speculation.

(6) Alpha glycerophosphate dehydrogenase in insect flight muscle Although the complete Krebs cycle has been demonstrated in the muscles of a number of insects and the accumulation of lactate found in those of *Gastrophilus, Tenebrio,* and *Periplaneta*, the activity of lactic dehydrogenase is low compared to that in mammalian or avian muscle. An alternative means of reoxidizing reduced $NADH_2$ seems more common in insects. This is through another NAD-coupled reaction catalyzed by the enzyme alpha glycerophosphate dehydrogenase, in which alpha phosphoglycerate is derived from dihydroxyacetone phosphate and reduced NAD is reoxidized. The activity of this enzyme in the flight muscles of insects such as *Phormia, Apis, Bombus,* and *Melanoplus* is far greater than that of lactic dehydrogenase, but far less than that of lactic dehydrogenase in the abdominal muscles of the lobster and the skeletal muscles of the crayfish (*Cambarus*) or of the rat. Multiple forms of alpha glycerophosphate dehydrogenase (isozymes) have been found in two moths (*Hyalophora cecropia* and *Samia cecropia*), as multiple forms of lactic dehydrogenase have been found in mammalian and avian tissues.

The activity of lactic dehydrogenase is greater than that of alpha glycerophosphate dehydrogenase in larvae of *Tendipes* (*Chironomus*), which are reported to excrete lactic acid. The leg muscles of *Locusta* and of *Belostoma*, and the walls of the midguts of *Periplaneta*, *Bombyx*, and *Orthodera* have all been found to be rich in lactic dehydrogenase. This may be related to the fact that the midgut tissue of most insects and the leg musculature of jumping and aquatic ones probably experience periods of oxygen deficiency and, therefore, depend to a considerable extent upon a glycolytic source of energy. This dependence is also evidenced by the paucity of mitochondria in these muscles and their deficiency in cytochromes. The thoracic muscles of flying insects, on the other hand, with a high rate of metabolism and large energy demands, are abundantly supplied with tracheae and with mitochondria, larger than those of the other tissues and distinguished as sarcosomes. These organelles have provided much of the material for study of the events of the Krebs cycle and of oxidative phosphorylation, because of their large size (1 to 4 μ in diameter) and the ease with which they can be teased out of fibrillar muscle, thus avoiding the drastic treatment of homogenization. Possibly the decrease in lactic dehydrogenase activity and the enhancement of that of alpha glycerophosphate dehydrogenase in these muscles is correlated with their need for the rapid and immediate production of energy. It has been shown, indeed, that alpha glycerophosphate is oxidized by flight muscle sarcosomes more rapidly than any other substrate, 10 times more rapidly, for example, than is succinate.

Comparisons of the activity of alpha glycerophosphate dehydrogenase in the thoracic and leg muscles of bumble bees (*Bombus terrestris*), insects whose locomotion is almost exclusively flight; of praying mantises (*Orthodera ministralis*), which can fly but more often walk; of katydids (*Caedicia simplex*), which usually walk and rarely fly; and of a long-horned locust of New Zealand, the wingless tree weta *Hemidenina thoracica*, have shown that, in the first three, the thoracic muscles contain much more of the enzyme than do the leg muscles. The amount in the flight muscles of the bee, expressed as units of activity per gram of fresh weight (i.e., the amount producing an initial rate of oxidation of 0.01 μmol $NADH_2$/min), has been found to be four times that in the flight muscle of mantises, slightly more than five times that in those of katydids, and almost 10 times that in the thoracic muscles of the weta. These insects have almost twice as much of the enzyme in their leg as in their thoracic muscles, while in bees the amount in the leg muscles is about half that in the flight muscles. Such a correlation between the amount of enzyme and the locomotor habits of an insect furnishes another illustration of biochemical evolution.

The course of the Embden-Meyerhof pathway in insects may thus result in production of alpha glycerophosphate, pyruvate, or lactate. In each case, NAD is reconstituted from NADH at some step in the sequence of reactions, but those coupled with oxidation have, in each case, different enzymes and different substrates.

d. OXIDATIVE METABOLISM AND MITOCHONDRIAL DEFICIENCIES Deficiency in mitochondria or structural defects in them might be expected in organisms showing modifications of the Krebs cycle and of oxidative phosphorylation. Electron microscopy has proved this supposition to be correct. In *Monoliniformis*, for example, the mitochondria have relatively few cristae, although most of the succinic dehydrogenase activity is found in the particulate fraction of homogenates. This fraction is also insensitive to actinomycin A, suggesting that either cytochrome b, whose reoxidation it blocks, is not included in the electron transport system or the established system is not used at all. There is also deficiency in cristae in the mitochondria of *Ascaris*, which have been found, by microspectroscopy, to be lacking in cytochromes c_1 and a_3, although small amounts of cytochromes a and b can be detected in them. Cytochrome a_3 has been found in *Hymenolepis*, yet larvae can be grown to normal adulthood under wholly anaerobic conditions. A cytochrome system has been detected in *Trichuris*, although the mitochondria have fewer cristae and the anaerobic method of respiration would imply that the mechanism for oxidative phosphorylation is not of great importance to these animals.

e. ELECTRON TRANSPORT IN INSECT FLIGHT MUSCLE The presumed pathways of electron transport in insects have been followed by microspectrophotometry, particularly in the mitochondria obtained from the flight muscles of adult houseflies, and from the somatic muscles of the cecropia moth at all stages of its development and from its nonmuscular tissue in larval, diapausing pupal, and adult stages. The electron transport pathways in the flight muscles both of flies and moths seem to follow the pattern established for vertebrates. Although no flavoproteins have been isolated from insect tissues, their presence has been detected spectrophotometrically. It is, therefore, probable that transfer of electrons from NADH is mediated through FB as it is in vertebrates, although there may be other enzymes with similar properties acting between NADH and the cytochromes. It has been suggested that vitamins K and E, as well as certain quinone derivatives, may operate in electron transport. Other possible enzyme systems are the diaphorases, which have been found in association with the sarcosomes of flight muscles and the mitochondria of the tissues of the midgut. Diaphorases dehydrogenate NADH but, in vitro, do not transfer electrons to cytochrome c. It is, of course, possible that they may do so in vivo, when they have not been subjected to extraction procedures, or that the electrons may be accepted by other, as yet undetermined, cell constituents. The chemical similarities between vitamin K and the echinochrome of sea urchins is of interest in this connection.

The particulate fraction of homogenates of midgut tissues from the larvae of cecropia moths has a cytochrome with an absorption band different from that of the others. The name cytochrome b_5 has been assigned to this heme compound, which is associated with the ribosomes and not, like the other cytochromes, with the mitochondria. A similar absorption band appears in the spectrum of the particulate fraction from homogenates of mammalian livers. Transfer of electrons from cytochrome b_5 to cytochrome c has been demonstrated in vitro, but there is no critical evidence for its occurrence in vivo. This might, indeed, be impossible in a cell because of the spatial separation of the ribosomes from the mitochondria in which the efficiency of electron transport rests upon the close proximity of the cytochromes in the sequence. It is possible, also, that cytochrome b_5 mediates the transfer of electrons directly from FP to oxygen, bypassing the cytochrome chain entirely. This possibility is supported by the fact that reduced cytochrome

b_5 can be oxidized by molecular oxygen under conditions in which the cytochrome pathway b to a_3 is blocked. The possibility that electrons may also be transferred from cytochrome b_5 to cytochrome c is supported by the fact that the addition of cytochrome c to homogenates of midgut tissue increases the oxidation of NADH forty-fold and that this increase is maintained in the presence of actinomycin A, but markedly reduced by that of cyanide. The established and presumed pathways of electron transfer in insect tissues is represented in Fig. XII–6.

D. OXIDATIVE METABOLISM OF FATS

The Krebs cycle operates also in the oxidation of fats and proteins and in the conservation of their energy in ATP. The course of events in the oxidation of fatty acids is not as firmly established as that of the oxidation of glucose, in vertebrates at least, but the initial steps are generally accepted as a combination of the fatty acid with coenzyme A, a reaction that requires the utilization of one molecule of ATP, followed by a series of dehydrogenations of the fatty acid at its beta carbon atom, that is, the next but one to the carboxyl group. In these beta oxidations, FAD or NAD acts as hydrogen acceptor. All of the reactions can be carried out by mitochondria isolated from vertebrate livers, so that in all probability the entire process takes place in vivo in the mitochondria. The reactions may be summarized by the following equation, where R represents a chain of carbon atoms with hydrogen or hydroxyl groups attached to them.

$$R\text{—}CH_2CH_2\overset{\beta}{C}H_2\overset{\alpha}{C}H_2COOH + \text{coenzyme A} + ATP \rightarrow R\text{—}CH_2CH_2CH_2CH_2CO \sim SCoA + AMP + \text{pyrophosphate} \quad (1)$$

$$R\text{—}CH_2CH_2CH_2CH_2CO \sim SCoA + FAD \rightarrow R\text{—}CH_2CH_2CH{:}CHCO \sim SCoA + FADH_2 \quad (2)$$

The product of reaction 2 is then hydrated, forming a hydroxyl group on the beta carbon atom, and this reaction is followed by an oxidation, with NAD acting as hydrogen acceptor, with the formation of a double bond with oxygen at the beta carbon atom.

$$R\text{—}CH_2CH_2\overset{\beta}{C}H{:}\overset{\alpha}{C}HCO \sim SCoA + H_2O \rightarrow R\text{—}CH_2CH_2\overset{\beta}{C}HOH\overset{\alpha}{C}H_2CO \sim SCoA \quad (3)$$

$$R\text{—}CH_2CH_2\overset{\beta}{C}HOH\overset{\alpha}{C}H_2CO \sim CoA + NAD \rightarrow R\text{—}CH_2CH_2\overset{\beta}{C}{:}O\overset{\alpha}{C}H_2CO \sim SCoA + NADH_2 + H^+ \quad (4)$$

In the presence of additional CoA the molecule is split at the oxygen double bond and a 2-carbon frag-

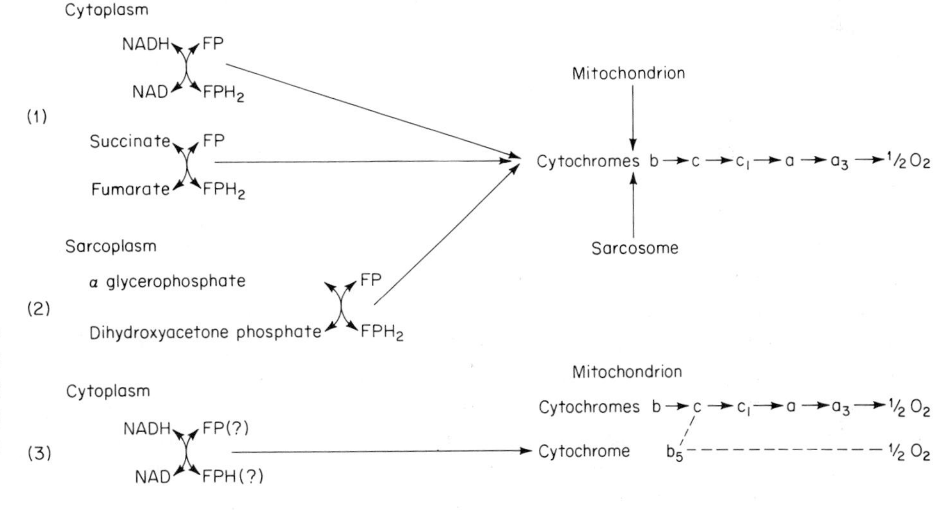

FIG. XII–6 *Pathway of electron transfer in insect tissues: 2 represents the predominant pathway in flight muscles; 3 represents the presumed pathways through cytochrome b_5 in midgut tissues.*

ment is released as acetyl-CoA, while the remainder unites with another molecule of coenzyme A and repeats the process of beta oxidation until another 2-carbon fragment is split off. A 4-carbon acid, butyric acid (C_3H_7COOH), after the series of reactions 1 to 4, is directly converted to acetyl-CoA, which in the presence of CoA splits into two molecules of acetyl-CoA. The reaction may be represented as

$$CH_3C{:}OCH_2CO \sim SCoA + CoA \rightarrow 2CH_3C{:}O \sim SCoA$$

Most of the naturally occurring fatty acids are ones with even numbers of carbon atoms, so that by successive splitting off of 2-carbon fragments in this way, half as many acetyl-CoA molecules will be formed as there are carbon atoms in the fatty acid chain. These run into high figures; from palmitic acid, for example, with 16 carbon atoms, eight molecules of acetyl-CoA can be obtained and nine from either stearic or oleic. Since fats are composed of three molecules of fatty acid, usually different and most often long-chain ones, it is clear that fat has a higher metabolic energy value than carbohydrate. The oxidative phosphorylation of palmitic acid, for example, yields 130 molecules of ATP. Beta oxidation results also in the production of large numbers of acetyl-CoA molecules, which must be dealt with if they do not directly enter the Krebs cycle. They may be used as starting points in the synthesis of lipid or of acetyl derivatives such as acetylcholine. If not so utilized, they may give rise to the so-called "ketone bodies," acetoacetic acid, beta hydroxybutyric acid, and acetone, which are known to accumulate in the blood of mammals when large amounts of fat are being oxidized and which, at concentrations above a certain level, have pathological effects.

While most of the naturally occurring fatty acids have even numbers of carbon atoms, there are some, such as propionic acid, C_2H_5COOH, with odd numbers. Propionic acid will combine, as other fatty acids do, with CoA to make propionyl-CoA, but in this form it cannot apparently be metabolized further unless it takes up carbon dioxide and is converted to succinic acid. Fatty acids with higher uneven numbers seem, like those with even numbers, to split off, by beta oxidation, 2-carbon acetyl-CoA groups until they are down to the 3-carbon propionyl-CoA stage, when this remaining fragment combines with carbon dioxide and succinate is formed.

E. OXIDATIVE METABOLISM OF PROTEINS

1. Deamination and utilization of carbon residues

Protein also provides an energy source, for after its digestion and deamination of the resulting amino acids the carbon residues may be oxidized. Most of the essential amino acids are glucogenic—that is, they give rise to glucose which can be treated in the same way as glucose from any other source. Some are ketogenic, giving rise to ketone bodies, though it is doubtful if any one of them except leucine is exclusively so.

When deamination takes place as an oxidative process, either NAD or NADP can act as hydrogen acceptor and the amino acid is enzymatically converted to its corresponding keto acid with the formation of one molecule of the reduced nucleotide and ammonia, in accordance with the equation

$$R\text{—}CHNH_2COOH + NAD \rightarrow R\text{—}C{:}OCOOH + NADH + NH_3$$

The resulting keto acid is available for conversions that will lead it into the Krebs cycle or into biosynthetic rather than degradative pathways. The ammonia, as a potentially toxic substance, may be directly excreted, or detoxified by conversion to urea or to uric acid, or retained in the cell by combination with other compounds. From these, it may later be released and utilized in the synthesis of amino acids or other important nitrogen-containing compounds. The amino acids that are synthesized are, obviously, from the nutritional point of view, nonessential, for, as endogenous ones, they need not be supplied by the diet. The Krebs cycle is involved in amino acid synthesis as it is in that of urea and uric acid, for both alpha-keto-glutaric and oxalo-acetic acids are utilized in these events.

Alpha-keto-glutaric acid can be converted to glutamic by the addition of a hydrogen atom provided by NADPH and an amino group provided by ammonia, in accordance with the equation

$$HOOC \cdot C{:}O \cdot CH_2CH_2COOH + NADPH + NH_3 \rightarrow HOOC \cdot CHNH_2CH_2CH_2COOH + NADP + H_2O$$

Once the nitrogen of ammonia has been incorporated into amino nitrogen it becomes available to other carbon skeletons and by transamination can be attached to them. The union of glutamic and oxalo-acetic acids, for example, and transference of the amino group from the first to the second results in the restoration of alpha-keto-glutaric acid and the formation of aspartic acid, as follows:

$$HOOC \cdot CHNH_2CH_2CH_2COOH + HOOC \cdot C{:}OCH_2COOH \rightarrow HOOC \cdot C{:}OCH_2COOH + \underset{\text{Aspartic acid}}{HOOC \cdot CHNH_2CH_2COOH}$$

In this transamination, as in all others, the vitamin pyridoxine is an essential cofactor. It is probable that other Krebs cycle intermediates also provide carbon frameworks for the construction of amino acids by transamination. From those that are synthesized in this way and those that are derived directly from the diet and not deaminated, specific proteins can be constructed.

2. Urea synthesis: ornithine cycle

Urea is formed in a cyclic process, the ornithine cycle, whose events are directly coupled with the Krebs cycle also, through alpha-keto-glutaric and oxalo-acetic acids, in their relationships to the formation of glutamic and aspartic acids, and through fumaric acid, which is a by-product of the ornithine cycle. To start this cycle, aspartic acid condenses with citrulline

$$\begin{array}{l} O \\ \| \\ C \cdot NHCH_2CH_2CH_2CHNH_2COOH \\ | \\ NH_2 \end{array}$$

to form argino-succinate, which is then split into a molecule of arginine

$$\begin{array}{l} HN{:}CNHCH_2CH_2CH_2CHNH_2COOH \\ \quad\;\; | \\ \quad\;\; NH_2 \end{array}$$

and one of fumaric acid, HOOCCHCHCOOH. In these two reactions, one molecule of ATP is utilized, leaving one molecule of AMP and one of pyrophosphate. The fumarate enters the Krebs cycle and the arginine is hydrolyzed in a reaction catalyzed by the enzyme urease. The resulting product is ornithine

$$\begin{array}{l} CH_2CH_2CH_2CHNH_2COOH \\ | \\ NH_2 \end{array}$$

Citrulline is reformed and the cycle is kept going by the reaction of ornithine with carbamyl phosphate. This is made from ammonia and carbon dioxide in a single, but complicated, reaction, requiring the expenditure of two molecules of ATP. In all, three molecules of ATP are used in each revolution of the ornithine cycle. Glutamic acid and carbamyl phosphate are also used in the synthesis of uric acid. Enzymes of the ornithine cycle, as well as carbamyl phosphate synthetase, mediating the biosynthesis of carbamyl phosphate, have been identified in the soluble fraction of homogenates of the gut tissue of *Lumbricus terrestris*, and utilization of carbamyl phosphate for the synthesis of citrulline by tissue extracts from a number of invertebrate species has been demonstrated in vitro. The results of these experiments make it seem likely that detoxication of ammonia by conversion to urea or to uric acid proceeds in invertebrates in the same way that it has been shown to do in vertebrates. Studies of the enzyme complement of the mantle of *Lingula reevei* indicate that this brachiopod may also have a complete ornithine cycle.

F. OXIDATIVE METABOLISM AND THE SYNTHESIS OF CARBOHYDRATES AND LIPIDS

Synthesis of carbohydrates and lipids is also keyed into the events of oxidative metabolism. While most of the reactions in these are reversible ones, in vitro, anabolism is not catabolism in reverse. Glycogen, the principal form in which carbohydrate is stored in animals, and other polysaccharides may have as precursors hexoses, most frequently glucose, but also lactate, glycerol, or amino acids after their deamination and a series of reactions of various lengths. In lipid synthesis, acetylcholine is the chief precursor, and since this is derived from carbohydrate and from protein, as well as from fat, an animal may synthesize and store fat on a diet rich in carbohydrate or protein and low or entirely deficient in fat. The interconvertibility of all these substances has long been recognized, but it is only within the past 30 years that it has been given a sound biochemical basis.

All the reactions that go on in cellular metabolism are essentially transfers or transductions of energy. Through oxidative metabolism, a large fraction of the energy bound in the macromolecules of carbohydrates, lipids, and proteins is transferred to ATP; some is transduced to heat or to work. In biosyntheses, macromolecules are again built up and energy bound in them, but at the expenditure of chemical energy. This may be directly derived from ATP or from other nucleoside triphosphates containing the purine base guanine, or either of the pyrimidine bases uracil or cytosine. These nucleosides, like those containing adenine, can exist in the cell as diphosphates and as such can accept a phosphate group from ATP and become triphosphates and, as such, provide the chemical energy for some reactions. Thus GDP → GTP; UDP → UTP; and CDP → CTP. UTP is the energy donor in the union of glucose molecules to form glycogen and some of the other polysaccharides and oligosides, such as trehalose. CTP provides the activation energy for certain steps in the biosynthesis of phosphatids and GTP for the step in protein synthesis in which an amino acid becomes incorporated into a peptide chain. These nucleosides may yet be found to contribute energy to other reactions, but, from present evidence, the high-energy bond of ATP seems the one most generally utilized.

G. PHOSPHAGENS

Another aspect of the universality of cellular metabolism is the utilization of phosphate. Ultimately, the source of phosphate for phosphorylations must be mineral salts taken in as food or drink. In the cell, inorganic phosphate furnishes some of that available, but there are also other donors, or phosphagens. These are defined as labile compounds that can contribute phosphate groups to ATP or other nucleotides. All of them are derivatives of guanidine (Fig. XII–7). At present seven are known of which two, phosphocreatine (PC) and phosphoarginine (PA), are most widely distributed. The other five, phosphoglycocyamine (PG), phosphotaurocyamine (PT), phosphohypotaurocyamine (PH), phospholumbricine (PL),

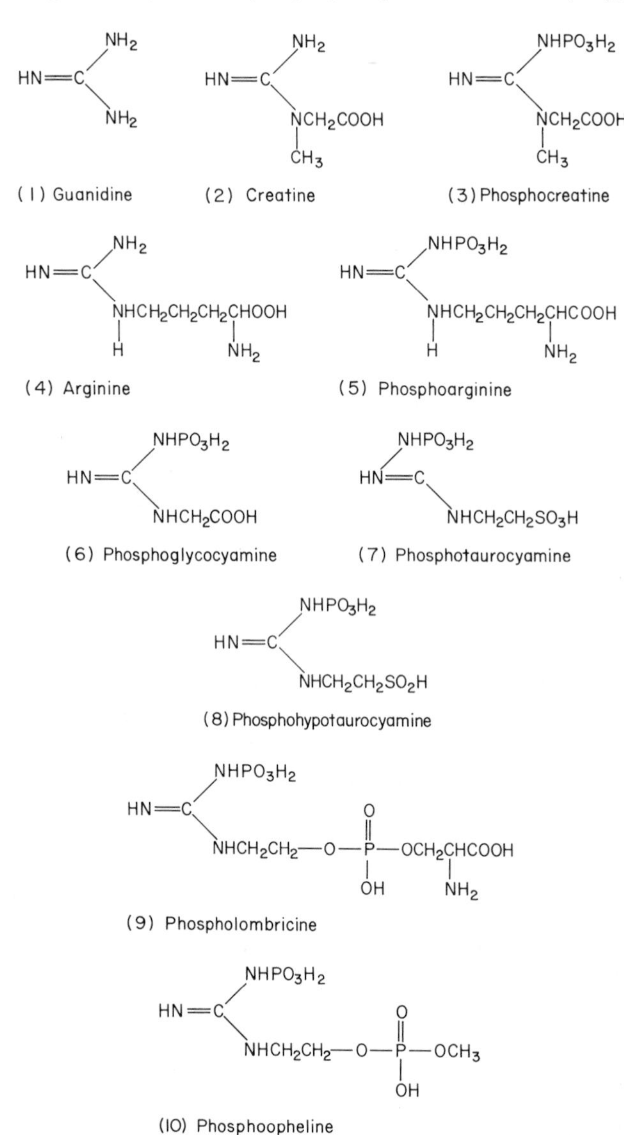

FIG. XII–7 *Structural formulae for guanidine and its phosphagen derivatives.*

and phosphoopheline (PO) have been found so far only in annelids and sipunculids. This does not mean that they, or similar compounds, may not also be in the tissues of other animals not yet investigated in this respect.

The first phosphagen to be demonstrated was PC, found in 1927 in extracts of frog muscle and, a little later, in the muscles of representative vertebrates of all kinds. But it could not be identified in similar preparations of the muscles of the invertebrates studied; instead, phosphoarginine was obtained from the crayfish and all other crustaceans and insects tested, and from the muscles of representative molluscs and of a number of echinoderms. This biochemical difference between vertebrates and invertebrates seemed to be so clear cut that the terms "creatinate" and "acreatinate" were suggested as substitutes for "vertebrate" and "invertebrate," or "chordate" and "nonchordate." It was presumed that a biochemical mutation had occurred in the ancestral stock, enabling individuals with it to synthesize creatine, phosphorylate it, and use it as a phosphagen, thus starting the vertebrate line. But the demonstration of phosphocreatine, through more extensive comparative studies, in the muscles of several annelids, of ophiuroids, and of some holothurians and echinoids made this premise a doubtful one. Phosphocreatine has also been found in at least one species of sponge and creatine in the tissue fluids of some cnidarians. PC is the exclusive phosphagen in several species of the polychaetes *Glycera*, in *Hermione hystrix*, and in *Onuphis teres*, and is found in conjunction with PA in the sea urchins *Echinus esculentus, Heliocidaris erythrogramma*, and *Strongylocentrotus lividus*, and in the holothurian *Cucumaria frondosa*. Phosphoarginine, on the other hand, has not been demonstrated in vertebrates, in which PC still seems to be the phosphagen they have in common, and the presence of PC in echinoderms has been considered supporting evidence for the echinoderm-chordate line of evolutionary descent.

But the source of the creatine in invertebrates is an open question and the probability that it is exogenous rather than endogenous is strong. In mammals, creatine is known to be synthesized by transfer of the amidine group of arginine, or some other amino acid, to glycine and by transmethylation, through methionine, of the resulting glycocyamine to creatine. The key enzyme in this reaction is transamidase, through which the initial carbon atom of arginine, with its attached NH_2 and NH radicals, is split from the rest of the molecule and then attached to glycine. This enzyme might be expected to be found in any organism synthesizing creatine and making use of its phosphorylated form as a phosphagen. Yet it has not been demonstrated in the invertebrates with PC in their tissues, although it has been found in some in which there is no creatine at all. Creatine is known to be excreted by some marine fishes, and it has also been found in soils, so that an exogenous source is available both to marine and to terrestrial invertebrates. Experiments have also shown that it is taken up from dilute solutions and accumulates in the tissues of a number of invertebrates, even those that do not phosphorylate it. It is possible, then, that its distribution among invertebrates is entirely fortuitous and that there is no special biochemical or evolutionary significance to be attached to it. It is perhaps surprising that with arginine so readily available to all animals which synthesize urea the vertebrates should have found it useful to convert this to creatine and to employ PC rather than PA as a phosphagen.

Extensive studies of phosphagens in annelids during the past decade have shown that they make use of compounds other than PC and PA. Although among polychaetes PC is the most widely distributed phosphagen, PG is the only one in several species of the "errant" group and either PA or PT in different species of the "sedentary" group. PG is the phosphagen, for example, in *Nereis diversicolor*, and PT in *Arenicola marina* and *A. assimilis*. PL has been isolated from the earthworms *Lumbricus terrestris* and *Allolobophora* sp, and also from the polychaete *Ophelia bicornis*. Some species have double phosphagens, as some of the echinoderms have. PC and PT have been found in the tissues of three serpulids—*Pomatoceros triqueter, Protula intestinum*, and *Serpula vermicularis*—and PA and PC in those of *Travisia forbesii*. It is quite evident that there is no phosphagen "characteristic" of the annelids, and probably not of invertebrates in general, whatever may be the origin of the phosphocreatine in them. It is clear, too, that the distribution of the phosphagens bears no relation to phylogeny, nor to habitat. Two species of *Ophelia*, for example, may live side by side, but there is PL in *O. bicornis* and PC in *O. neglecta.*

H. BIOLUMINESCENCE

Bioluminescence is also a manifestation of the biological transduction of energy and an offshoot from the main pathway of respiration. It results from the conversion of chemical into radiant energy of wavelengths of 400 to 700 nm, and so within the part of the spectrum to which the human eye responds. It is a property exhibited by some bacteria and fungi, as well as by a number of species of animals living in marine or terrestrial habitats. The only freshwater animal yet known to luminesce is a pulmonate gastropod native to New Zealand, the limpet *Latia neritoides*. There are luminescent protozoans, sponges, cnidarians, ctenophores, at least one species of nemertean, molluscs, annelids, arthropods, echinoderms, and fishes. The property of luminescence is so erratically distributed among the species in these phyla that it might be explained either as the result of convergent evolution or the survival of a process that was once common to all biological systems but that has been preserved only in those on which it conferred an advantage. It must be recognized, too, that not all animals from which light is emitted are self-luminous; some are secondarily so, for their light originates from bacteria or fungi associated with them.

Although Spallanzani in the eighteenth century had observed the remarkable fact that luminous medusae, when dried, would glow again after they were moistened, little progress was made in understanding bioluminescence until more than a century later. Then, in the latter part of the nineteenth century, a French physiologist, Raphael Dubois, made some studies upon the photogenic tissues of an elaterid beetle, *Pyrophorus*, and the boring clam, or piddock, *Pholas dactylus*. He found that a cold water extract of the luminous tissues taken from the inhalant siphon and the edge of the mantle of the clam continued to emit light for several minutes after the operation, but that the same tissues, after immersion in hot water, failed to luminesce at all. Pursuing the matter further, he discovered that the addition of a hot water extract to a cold water extract that after a time had ceased to glow started it luminescing again. He concluded from this that two substances were concerned in production of the light, one of them heat labile and the other heat stable, and assuming that the first was enzymatic in nature, called it luciferase and the other, its substrate, luciferin.

Subsequent investigations have shown that there are different luciferins in different animals. All of them are associated with the energy-liberating mechanisms of the cell and all of them are fluorescent molecules that become activated by the shift of an electron from a lower to a higher energy level, when there is energy available for this. As the excited molecule reverts to its unexcited state by return of the displaced electron to its former orbit, a quantum of light, or photon, is emitted. These events may be represented schematically as

$$
\begin{aligned}
R \text{ (reactants)} &\rightarrow P^* \text{ (excited product)} \\
&\rightarrow P \text{ (ground state of product)} \\
&\qquad + h\nu \text{ (quantum of light energy)}
\end{aligned}
$$

The intensity of the light emitted depends upon the number of quanta liberated per second and its color upon the nature and chemical configuration of the excited molecule, the conformation state of the protein enzyme, and the location of the photogenic material. In metazoans, the photogenic cells, or photocytes, are usually grouped together, sometimes in specific organs, which are so placed that the light must pass through tissues that act as screens or filters. The light-emitting process is an extremely efficient one; quantitative determinations of the amount of the excited product and the amount of light produced in the flash of a firefly have shown that one photon is emitted per molecule of excited product. Thus all the energy of the reaction is manifested as light and none is dissipated as heat nor "captured" in ATP.

The luminescence of animals may be either intracellular or extracellular. In intracellular luminescence, all the reactants are contained within the cell, in extracellular, they are brought together in secretions discharged from different cells or groups of cells. The reactants within the cells are sometimes distributed as visible granules in the cytoplasm, often quite definitely localized. In the dinoflagellate *Noctiluca*, for example, which is responsible for much of the luminescence of the open sea, the granules are scattered around the periphery of the globular cell and along the stout tentacle. Particles, called scintillons, have been isolated from *Gonyaulax polyedra*, an armored dinoflagellate of the Pacific Ocean; these can be caused to flash simply by lowering the pH of their medium. Electron micrographs show that there are many such particles in an individual *Gonyaulax*, for the most part aggregated around the periphery of the cell.

The photocytes of the scyphozoan *Aequora* lie in groups just below the endoderm of the ring canal, and the emitted light appears in myriads of points around the margin of the bell. In ctenophores, the photocytes lie along the meridional canals. In ophiuroids, the only class of echinoderms known to be luminous, the photocytes are located in the spines, in close proximity to mucous glands, but the light is intracellular so that this juxtaposition seems wholly fortuitous. In some insects, in both larval and adult instars, they are collected into rather complex organs. The photocytes of fireflies are derived from the fat body and are collected on the ventral surfaces of the last few segments of the abdomen. Several layers of cells loaded with white, opaque crystals of uric acid enclose the mass of photogenic cells in a cup, whose opening is toward the surface of the insect's body. These urate cells act as a reflecting layer for the light, so that it is all directed outward. The photogenic organ is well supplied with tracheoles, which traverse the layers of urate cells and terminate near, or in contact with, the photocytes (Fig. IV–10). In larvae of the midge *Arachnocampa luminosa*, the glowworm of New Zealand, the photogenic organs are modifications of the ends of the Malpighian tubules, the only known instance of the adaptation of these organs to light production. The photogenic organs, or photophores, of pelagic shrimps have reached what is perhaps the highest degree of specialization. These are distributed over the appendages and the thoracic and abdominal sterna, and the light from them is directed downward. In *Meganyctiphanes norvegica*, the photocytes are enclosed in a cup of reflecting material surrounded by pigment. At the opening of the cup, there is a crystalline mass that probably functions as a lens, and the outer surface of the organ is covered by a modified cuticle that forms a cornea. The back of the cup is well innervated so that the organ, in its construction, resembles a vertebrate eye, although it is a structure specialized for the production, rather than the reception, of light.

Luminescence is extracellular in *Pholas*, in the pulmonate gastropod *Latia*, in the polychaete *Chaetopterus*, and in the ostracod *Cypridina*. The luminous secretion of *Pholas* is produced by two layers of gland cells. The distal layer of these secretes mucus, and in the inner layer, some 500 μ removed from the surface, there are two types of photocytes. The more numerous of these contain large granules, the other photocytes contain smaller granules. Both the mucous and the photogenic cells discharge their secretions through necks that open on the surface of the siphon and on the edge of the mantle. The mucus acts as a carrier for the photogenic substances, and the whole mass glows with a bluish-green light. The luminous organs of *Latia* lie in the furrow separating the foot from the mantle and, like those of *Pholas*, contain two types of cells. The luminous slime emitted from them is of a yellowish-green color. The photocytes of *Chaetopterus* are particularly abundant on the aliform notopodia of segment XII, but they are distributed also on the more posterior notopodia and on the peristomial tentacles. They lie close to mucous glands, whose pores open directly on the surface. When the products of the cells of both types are discharged, the worm is covered with luminous slime. In *Cypridina*, the photogenic cells are localized on the upper lip near the mouth and contain yellow granules of luciferin and colorless ones of luciferase. Both dissolve on contact with water, when they are squeezed out through openings of the glands by contraction of muscle fibers that run from the body wall to the oesophagus, passing between the gland cells. Probably, mucus is also secreted to contribute to the mass that shines with a vivid blue light.

The difficulties of isolating photogenic organs or photocytes from many of the luminescent animals have so far limited investigations of the biochemical events involved in the phenomenon to only a few organisms. The most complete studies have been made upon *Cypridina* and the American fireflies *Photinus* and *Photurus*, and upon bacteria, for from all of these it is possible to obtain enough material to study the reaction in vitro and to analyze the reactants. But in no case is the information as yet complete. The *Cypridina* system seems to be the simplest one, requiring as reactants only luciferin, luciferase, water, and oxygen, with the light reaction following the general course

$$\underset{\text{Luciferin}}{2LH_2} + O_2 \xrightarrow[\text{luciferase}]{} 2L^* + 2H_2O$$

$$L^* \rightarrow L + \text{light}$$

Chemical analysis of the luciferin is now complete and its synthesis in vitro recently achieved. It is an indole derivative. Just a small amount of this substrate mixed with the enzyme will produce quite a bright color; it has been reported that one part of dried Cypridina in 400 million parts of water will emit a perceptible light. During World War II, Japanese soldiers carried dried *Cypridina hilgendorfi*, a species abundant in Japanese waters, about with them, because by moistening a few of the little animals in the palms of their hands they could make enough light to read dispatches when the risk of using a flashlight was too great.

Firefly luciferin has been extracted, purified, and crystallized. Fifteen thousand lanterns of *Photinus pyralis* have yielded some 9 mg of luciferin, which has been chemically identified and also synthesized in the laboratory. The luciferase has also been extracted, crystallized, and identified as a euglobulin with a molecular weight of about 100,000. In vitro, the light-producing reaction requires ATP, Mg^{++}, and oxygen, as well as the luciferin and luciferase, and has been shown to follow the general course

$$\text{Luciferin } (LH_2) + \text{ATP} + Mg^{++} \xrightarrow[\text{luciferase}]{} \text{luciferyl-adenylate } (LH_2\text{-AMP}) + \underset{\text{Pyrophosphate}}{\text{POP}} \quad (1)$$

$$LH_2 \cdot \text{AMP} + \text{POP} \rightarrow L^* \cdot \text{AMP} + 2H_2O \quad (2)$$

$$L^* \cdot \text{AMP} \rightarrow L \cdot \text{AMP} + \text{light} \quad (3)$$

Furthermore, the $L \cdot$AMP appears to become bound to the enzyme in a stable configuration that precludes further light production until it is broken. This may be done by the intermediation of coenzyme A, forming a luciferin-coenzyme A complex ($L \cdot$CoA), which then undergoes reduction to form $LH_2 \cdot$CoA, restoring LH_2 to the system when this complex is disrupted. The total reaction can be used to test for the presence of ATP in cells or tissues, or in homogenates of them, for the introduction of the luciferin and luciferase will evoke a flash if even minute amounts of ATP are available. Firefly lanterns also contain kinases that can effect the transfer of phosphate from CTP, GTP, and UTP to ADP.

The luciferin of bacteria is probably an aldehyde complex of dihydroflavin mononucleotide, and the luciferase is a flavin enzyme. The aldehyde of the luciferin complex is a long-chain compound with not less than six carbon atoms. The light reaction, as it has been followed in vitro with dried powders of luminous bacteria, seems to follow the general course

$$\text{FMN-H}_2 + \underset{\text{Aldehyde}}{R\text{CHO}} \rightarrow \underset{\text{Luciferin}}{\text{FMN-H}_2 \cdot R\text{CHO}} \quad (1)$$

$$\text{FMN-H}_2 \cdot R\text{CHO} + O_2 \xrightarrow{\text{luciferase}} \text{FMN}^* + R\text{COOH} + H_2O \quad (2)$$

$$\text{FMN}^* \rightarrow \text{FMN} + \text{light} \quad (3)$$

FMN may be reduced again by DPNH with the agency of the enzyme diphosphopyridine nucleotide oxidase, so that the light is continuous rather than intermittent. In animals, the light is usually given off in discrete flashes and a steady glow from an animal suggests that it is bacterial in origin. The luminescence of some cephalopods and, among vertebrates, of some teleost fishes is attributed to symbiotic bacteria. In at least four genera of cephalopods (*Loligo*, *Sepiola*, *Rondeletia*, and *Eypyrmna*) there are special glands, some of whose cells are packed with photogenic microorganisms. These glands may be invaginations of the mantle epithelium, or they may, in females, be associated with the accessory nidamental glands. They usually lie in close association with the ink sac. In *Sepiola*, each gland, which has a reflecting layer and a lens-like body, opens into the mantle cavity by a small papilla, and bacteria from its central cells have been isolated and cultured. While bacteria may be responsible for the light emitted by such cephalopods as these, that emitted by *Watasemia scintillans*, a squid that appears seasonally in certain areas along the coast of the Japan Sea, is intracellular and produced by the cells themselves. Electron-microscopic studies of the large photophores at the tips of the paired ventral arms have produced clear-cut evidence that the rod-like bodies in the photogenic cells are not microorganisms.

In these three systems (*Cypridina*, fireflies, and bacteria), which are as yet the best understood of all luminescent ones, the light reaction depends upon the presence of oxygen. This can be strikingly demonstrated in cultures of luminous bacteria, which cease to produce light when the oxygen supply is diminished, but which glow brilliantly again when oxygen is supplied. Some photogenic reactions, on the other hand, appear to be independent of oxygen. Luminous extracts of the ctenophores *Beroe, Eucharis,* and *Mnemiopsis*, for example, as well as those of the medusae *Aequora* and *Pelagia* and of the radiolarians *Thalassicola* and *Colozoon,* will continue their light production in the total absence of molecular oxygen. This relationship has been studied particularly in *Mnemiopsis*, which continues to be brightly luminescent for hours even under conditions of strict anaerobiosis, but it is still not clear whether the oxygen is bound for long periods in some form that is not available to ordinary respiratory catalysts but that can participate in the photogenic reaction.

There is a high degree of specificity in luciferins and luciferases. In general, luciferin obtained from one species will not react with luciferase obtained from another, unless the phylogenetic relationship between them is close. This is not surprising, since the chemical configurations of the luciferins are different, as is known from chemical analysis and shown by differences in the wavelengths of the light produced. Nevertheless, a cross reaction has been achieved between the luciferin and luciferase of *Cypridina* and the corresponding substrates and enzymes of the marine fishes *Apogon ellioti* and *Parapriacanthus beryciformes*. Although the enzymes are different, there is structural similarity in the substrates. The discovery of dead, but still luminous, *Cypridina* in the stomachs of some specimens of *Parapriacanthus* suggests that in these fishes the luciferin may not be endogenous, but derived from the ostracods on which they feed. If, however, it is endogenous, the use of closely related and similar compounds for light production by animals widely separated phylogenetically is open to interpretation as an instance of convergence in evolutionary development.

It has been postulated that bioluminescence is a concomitant of the chemical reactions that were most useful in removing oxygen from living systems. This removal would be of great importance, and, indeed, even essential, to those forms of life which are presumed to have originated in the reducing atmosphere surrounding the earth early in its history and to which even small amounts of oxygen might well have been toxic. The most efficient way to remove oxygen is to reduce it, with the consequent formation of water. This is an exergonic reaction, coupled, in oxidative phosphorylation, with the endergonic one of ATP synthesis. The most probable reducing agents in early anaerobic living systems are those compounds that were already part of their hydrogen transport mechanism. When these compounds mediate the reduction of oxygen to water, enough energy can be liberated to excite a molecule of a fluorescent substance and cause it to emit light. The energy of the reaction is thus not preserved in high-energy phosphate bonds, but is lost as light. All the luciferins so far identified are associated with the flow of energy in the cell, and it is a possible hypothesis that when organisms evolved that could tolerate free oxygen and even make use of it, some retained their oxygen-removing, light-producing mechanisms. According to such an hypothesis, luminescence in modern species represents a vestigial biochemical process that has persisted sporadically among animals because it had certain survival values for them.

These values may lie in the means of protection or of communication that light production offers. The emission of light may startle, and so frighten, a predator or it may serve to attract predators of a predator, thus protecting the luminescent animal itself. In some animals, where luminescence is limited to certain parts of the body, it may draw a predator's attention away from more vulnerable areas. In some worms, for example, the anterior region is the most luminous and is also the most readily regenerated part of the body. A predator may, therefore, be attracted to this part, while the more important posterior region where the reproductive organs lie, is autotomized and so escapes. Or, possibly, the light illuminates objects for the animal producing it and so enables it to see those, especially in the depths of the ocean, that are useful or dangerous to it. It may also serve as a lure, attracting prey or members of the opposite sex. It seems to act as an attractant of prey for *Arachnocampa*, living in dark caves and mine shafts in New Zealand, Tasmania, and New South Wales. All instars of *Arachnocampa* are luminous, with the probable exception of adult males. The larvae emit a light more intense than that of the imagos, and many of them glowing together make the interior of a deep cave as bright as day. They construct platforms of silk, from which hang down silk threads as long as 2 ft, with viscous droplets at intervals along them. The larvae lie on the platforms and glow, their light attracting other midges and mosquitoes that habitually live and breed in the damp caves. These become entangled in the pendant threads; the glowworms then wind up the threads and eat the ensnared prey.

The role of luminescence as a communication or signalling device seems quite evident in some of the polychaete worms and in fireflies. Both males and females of the Bermudan fireworm, *Odontosyllis,* are extracellularly luminescent during the spawning period, which occurs with great regularity in the summer months about an hour after sunset on the second, third, and fourth days after a full moon. At this time, individuals of both sexes come toward the surface. The females swim around in circles, giving off a bright glow which lasts 10 to 20 sec and which is emitted in such rapid sequence that it appears continuous. The males, flashing intermittently, come up from the deeper waters into the center of this luminous circle and approach the females. If the glow has faded by the time a male reaches a female, he waits until it has begun again. Both then shed their gametes and the proximity of the two animals at this time ensures a high percentage of fertilization. Males of the firefly *Photinus* fly about, flashing regularly with a flash of 0.2 sec duration at intervals of about 5 sec, while the females remain quiescent on the ground. Each female responds to the signal of the male by giving a flash lasting about 0.4 sec about 2 sec after his. The male's attention is thus drawn to the female, and repetition of these signals will ultimately bring him to her for copulation. The timing of the signals has been shown to be of distinct importance in the recognition of individuals of the opposite sex.

Even with such advantages as these, it is an open question why the property of luminescence should have been retained only in certain cells of the animals exhibiting it and why lights of different wavelengths should be emitted by the same animal, implying the utilization of different substrates at least. The two photophores of *Pyrophorus*, for example, that are situated on the prothorax give a greenish light and the single one on the ventral surface of the first abdominal segment, an orange-yellow one. South American railway worms, larvae of the beetle *Phrixothrix*, have an extraordinary appearance when glowing, for their heads emit a deep red light and the spherical, segmental photophores along the sides of the body, a greenish one. If these different colors are indicative of the use of different luciferins, the biochemical differentiation of cells must be different in each kind of photophore as well as different from that of the other, nonluminous cells of the body.

In protozoan and metazoan animals, light production is a response to stimulation, and in the majority can be regulated in one way or another. Some millipedes are continuously luminous and glow both by day and by night, although their light may fluctuate in intensity by as much as 40 percent and become very weak, but not entirely absent, during molting. It is intensified above the usual range if the animal is mechanically stimulated or chemically irritated, but how or why this happens is still unknown. In protozoans, intermittent flashing follows mechanical or electrical stimulation in the laboratory. The flash of a dinoflagellate usually lasts not more than 0.1 sec and can be induced by jarring or shaking the container or by passing an electric current through the water at intervals to produce successive electric shocks. The response can be watched under the microscope in individual *Noctiluca* and shows an increase in intensity with increase in intensity of the stimulation. If touched very gently with a needle, the cell glows only at the point of contact; with more vigorous prodding, luminescence spreads all around the periphery of the cell and to the tentacle. Luminescent radiolarians will also respond to mechanical and electrical stimulation, but their glow is of longer duration and lasts for

1 to 2 sec. In luminescent sponges, the response can be evoked by mechanical stimulation, but not by electrical. Presumably in nature, most of the luminous aquatic animals are mechanically stimulated by movements of the waters set up by waves, winds, and surface currents or by the passing of ships and other moving objects. An exception to this generalization, however, is the photosynthetic dinoflagellate *Gonyaulax,* which exhibits an inherent diurnal rhythm of luminescence which it follows regardless of external conditions.

In luminous metazoans, appropriate stimulation and conduction of nerve cells leads to the response. This stimulation may be directly upon the photogenic cells or indirectly upon them through the responses of cells adjacent to them. Experimentally, either contact or electrical stimulation initiates the response in cnidarians and ctenophores at the point where the stimulus is applied. With sufficiently strong stimulation, the response spreads throughout the photogenic areas, in all directions from the starting point. The conduction of the impulse is thus not polarized. In hydroid colonies, such as *Obelia* and *Campanularia*, the photocytes are localized in the polyps and the light spreads along the whole colony as successive zooids are directly stimulated to produce their light. In sea pens, luminescence may be limited to the polyps only or may be produced over the entire surface of the colony, spreading uniformly throughout it even after incisions have been made in the rachis, as long as a bridge of tissue is intact. Cutting a meridional canal in a ctenophore, on the other hand, prevents the occurrence of luminescence distal to the cut. Contact stimulation at any point in an intact animal results in an even spread of light all along the meridional canals, which would indicate that the nervous impulse passes over the apical pole where it has connections with all eight meridians and that its passage is checked along any given meridian if that canal is transected.

There is experimental evidence that the luminescense of polychaete worms also is under direct nervous control. Rhythmic flashing can be induced by electrical stimulation of the nerve cord. In polynoid worms the photocytes are localized on the scales or elytra, and an isolated elytrum can be similarly stimulated to rhythmic flashing. This has been done in *Acholoe astericola*, where the flashes first occur at a rate of about 5 to 10 per sec, but quickly become established at a steady rate of 1 to 2 per sec, as long as the ganglion in the elytrum itself remains intact. If this is removed, rhythmic flashing stops, and the response is obtained only after each individual electric shock. Mechanical stimulation also evokes the emission of light from glowworms and fireflies, as does electrical stimulation of the nerve cord. The eggs and pupae of fireflies are continuously luminous, but the flashing of larvae and adults is intermittent. Larval and adult photophores are quite different structures, for the larval light organ is reconstituted from migratory luminous cells. This presents some further questions with regard to the specific differentiation of these cells.

At least two general theories have been proposed for the mechanism of nervous stimulation of the photophore. According to one, the production or nonproduction of light depends upon the availability of oxygen, and the control is exercised indirectly through the tracheole system. Such a theory presupposes that oxygen can reach the photocytes only by way of the tracheal system, yet it has been shown that the light organ of the beetle *Phengodes* has no tracheal supply. It is reasonable, also, to suppose that even in those insects where tracheoles can be demonstrated in the photophore, oxygen cannot be entirely cut off even if the tracheoles are shut down. There could hardly fail to be some diffusion from the hemolymph or through the cuticle to the photogenic cells. These theories also imply that during the day, when light is not emitted by fireflies, no oxygen is available to these cells, nor possibly to neighboring ones. According to other theories, all the components necessary for light production are present at all times, but the reaction is prevented by the complexing of an enzyme with some compound that inhibits its action, as $L \cdot$AMP is believed to be complexed with luciferinase. Such a complex could be broken, and the inhibition released, by a nervous impulse that would set another series of reactions in train, such as those that may bring about the activity of coenzyme A and its binding with luciferin, and the ultimate reduction of luciferin in preparation for the next photogenic reaction. Possibly, acetylcholine, released through the passage of the nerve impulse, is hydrolyzed to acetic acid, which, in the presence of coenzyme A, leads to the release of pyrophosphate which breaks the inhibitory $L \cdot$AMP-enzyme complex.

Indirectly, the emission of light, whether intracellular or extracellular, may be controlled by neuromuscular mechanisms. This is thought to be the case in *Arachnocampa*, but the evidence is somewhat ambiguous. The larvae are easily alarmed, and when they are startled, light emission immediately stops. It may also stop, but more slowly, when they are gradually brought into light. The abrupt cessation has been thought to be the consequence of a change in position of the terminations of the Malpighian tubules, where the photogenic cells are located, by contraction of longitudinal muscles extending the length of the abdomen. Their contraction would draw the terminal segment of the abdomen forward and under a strip of pigment that lies along the dorsal surface of the abdomen, and, therefore, could act as a screen. Other observations indicate that the light is also simultaneously extinguished, which would suggest that nervous control was directly upon the photocytes. Control of light emission by screening devices or by rotation of the photophore has been described in luminous fishes, but no mechanisms of this kind have yet been satisfactorily established for invertebrates.

The nervous control of extracellular luminescence may be either direct or indirect. The histology of the photophore of *Cypridina* is such as to indicate that contraction of the muscles between the gland cells causes extrusion of their contents and so mixing of them in the seawater and the consequent emission of light. In *Pholas*, on the other hand, nerve fibers enter the photophore and ramify among the photogenic cells, so that stimulation here is direct.

It has been shown experimentally that the source of the stimulus to the nerves controlling the photophores or photocytes may be electrical or mechanical. In nature, it may be mechanical and photic, depending upon the presence of tacto- and photoreceptors. The fact that decapitated fireflies are incapable of continued flashing and that intact ones begin their photogenic reactions only after daylight has reached a certain level of low intensity is strong evidence for the effect of light upon their overall pattern of photogenesis. Similarly, the flashing of the female seems to be a visual response to the flashing of the male.

I. FACTORS INFLUENCING RESPIRATORY METABOLISM

Although it is impossible to detach one aspect of metabolism from another, so intricately interwoven are the pathways in every organism, it is convenient to consider the factors influencing basal metabolism as intrinsic and extrinsic ones. Intrinsic are size and weight, sex, age, and stage in the life cycle, responsiveness, and, in some animals, hormonal regulation. Extrinsic factors include such environmental conditions as temperature; the availability of oxygen as determining the need for ventilation of the respiratory areas and so the activity of flagella, cilia, or muscles; and, for aquatic animals, the conditions in the waters in which they live, especially in respect to motion and salinity. These are all variables that collectively, but rarely singly, are reflected in the metabolic rate and the minimal requirements for oxygen. Very few precise and quantitative investigations have been made of the influence of these factors on the basal metabolic rates of invertebrates; indeed, these rates are not even known for the majority of species. There is a great deal of disparity in the data from the investigations that have been made, as might well be expected from the number of variables that are involved and the difficulty of reducing them to levels desirable for adequately controlled experiments.

1. Intrinsic factors: size and weight, stage in life cycle

SIZE In general, the oxygen consumption of smaller animals is less than that of larger ones, although the metabolic rates of the smaller may be greater. Such comparisons are valid only when made on the basis of the amount of living tissue in the animals studied, for size may vary greatly with the water content of the body and weight with the amount of inert material deposited as shells or other external, as well as internal, structures. Difficulties inherent in investigations of this kind are illustrated by some of those made upon crustaceans. In the case of the isopod *Ligia oceanica*, for example, the oxygen consumption per unit of length was found to be constant for individuals over a size range of 0.95 to 3.1 cm, but showed a decrease per gram with increasing weight in individuals with a weight range of 0.04 to 1.03 g. The rate of oxygen consumption in the lighter individuals was thus about three times greater than that of the heavier ones. A possible explanation for these results may be in the relations of dimension and frequency of movement of the pleopods, because the size of these appendages bears a constant relation to the length of the body and their beat is more rapid in small animals than in large ones. In general, smaller crustaceans, like copepods, cladocerans, and euphausiid shrimps, have a higher metabolic rate than the larger ones, like lobsters and some of the bigger crabs, but the small ones are usually members of the plankton and are almost constantly swimming, while the larger ones are benthic in habitat and live sedentary lives.

STAGE OF LIFE CYCLE The stage in the life cycle at which the measurements are made is also of considerable importance in the interpretation and comparison of results, for the metabolic needs of an animal in the active phases of its growth are different from those at other times. Those of animals that periodically undergo molts are quite different just before and after a molt than they are in the intermolt periods, and in crustaceans, oxygen uptake has been found to increase markedly immediately before exuviation or emergence from the old shell. In *Maja squinando* the concentration of hemocyanin in the blood reaches its maximum just before molting and rapidly drops to trace amounts immediately afterwards. This provides at least circumstantial evidence for an increased need and utilization of oxygen at this critical stage of the crab's life. Since molting in crustaceans is known to be under hormonal control, the level of the molting hormones in the blood is also a factor influencing oxygen consumption.

The most marked differences in oxygen consumption at different ages and different stages are exhibited by insects, especially by holometabolous ones, in which the period of pupation is always one of reduced metabolic activity and that of diapause one in which it is minimal. The oxygen consumption of worker bees from the time of emergence to an age of 22 to 23 days has been calculated on a unit weight basis. The individual weights of the specimens ranged from 80 to 17 mg, with a good deal of variability among the individuals of any one age group, due largely to differences in the volume of the contents of the honey sac and rectum. Oxygen consumption was found, on the whole, to increase with age, beginning with the first week after emergence. Microspectroscopic examinations of mitochondria isolated from the thoracic muscles of emergent bees have shown that they are deficient in cytochrome c, because, relative to the amount of cytochrome a, they have only 43 percent of that in the mitochondria of bees 1 week old, or older. Diapausing pupae of *Platysamia cecropia* consume about one-fiftieth the oxygen that larvae do just before their pupation, and the resistance of the pupae to cyanide and carbon monoxide suggests that the cytochrome system is deficient in them also. The utilization of oxygen would, in both these insects, necessarily be restricted at these stages of their lives.

2. Extrinsic factors: temperature, oxygen tension, and need for ventilation, conditions in aquatic environments (water movements, salinity)

TEMPERATURE The relation of temperature to oxygen consumption in different species is of particular interest, for it is of great importance in determining their geographical distribution. Although invertebrates are poikilotherms and do not maintain a high body temperature as birds and mammals do, they make some adjustment of their temperatures to that of their surroundings and undoubtedly compensate for different external temperatures, within limits, by homeostatic mechanisms of some kind. Many species are known to become acclimated after exposure for a time to a temperature above or below that in which they ordinarily live and in the laboratory can adjust to and tolerate fairly wide extremes of heat and cold. The general conclusion has been reached that the basal metabolic rate is higher in aquatic species living in tropical waters than in those living in arctic or boreal ones. But this is not a simple relationship between temperature and oxygen consumption, for the oxygen content of warm waters is less than that of cold and conditions of oxygen deficiency as well as of temperature are involved. Moreover, in species with respiratory pigments there is an additional variable, for the oxygen affinity of the pigment is, in many cases, different at different temperatures.

Relationships between temperature and oxygen consumption have been established for a number of molluscs. The rate of oxygen uptake in *Brachiodontus demissus plicatulus*, for example, a bivalve that lives in salt marshes with a geographical range from Prince Edward Island to South America, has been found to increase with elevation of the ambient temperature up to 35.2°C, beyond which the animals did not survive. The rate in *Mytilus edulis*, which lives in the intertidal zone with a winter range from Greenland to Cape Hatteras, reaches a maximum at 20°C. Since the waters off Cape Hatteras may in summer be as warm as 26°C, *Mytilus* is virtually excluded from them during the hot months of the year, and these molluscs are far more abundant in the consistently cooler waters of the New England and Canadian coasts. In cold water, *Mytilus* pumps water in and out of the mantle cavity faster than it does in warmer waters, possibly compensating for the temperature difference by this increased ciliary activity.

Similarly, a marked increase in oxygen consumption with increase in temperature has been found in two species of the crayfish *Orconectes*. That of *O. immunis*, whose habitats are temporary roadside ditches, showed nearly a fivefold increase at 30°C over that at 16°C—from 0.044 ml/(g)(hr) at 16°C to 0.212 ml/(g)(hr) at 30°C. The oxygen consumption of *O. nais*, living in permanent ponds and running streams, increased from 0.054 ml/(g)(hr) at 16°C to 0.184 ml/(g)(hr) at 30°C. In both cases, the measurements were made when the medium in which the animals were kept was fully saturated with oxygen and in both cases the rate of oxygen consumption fell off when the ambient temperature rose from 30 to 35°C. Reduction, at the higher temperature, of the amount of dissolved oxygen and increase in the oxygen affinity of the hemocyanin are conditions also to be taken into account.

OXYGEN TENSION AND NEED FOR VENTILATION OF RESPIRATORY AREAS The prevailing tensions of oxygen and carbon dioxide also influence the rate of oxygen uptake in imposing the necessity for increased ventilation of the respiratory areas and so for increased activity of the organs or organelles concerned in the process. Illustrative of these conditions are the comparative figures for oxygen uptake and ventilation activity obtained from three species of terebellid polychaetes taken from the waters of the Pacific Ocean off the northern coast of the United States and studied in glass tubes in the laboratory. Specimens of *Eupolymnia heterobranchia*, a worm that habitually lives in the mud and under rocks, with a fresh weight of about 3 g apiece, took up oxygen, at 13°C, at a rate of 0.09 ml/(g)(hr) and pumped water through their tubes at a rate of 19.7 ml/(g)(hr). *Thelepus crispus* and *Neoamphitrite robusta*, species that live in clearer and better aerated water, of similar weights, consumed oxygen at a rate of 0.06 ml/(g)(hr). The pumping rate of *Thelepus* was 12.6 ml/(g)(hr) and that of *Neoamphitrite*, 9.1 ml/(g)(hr). The higher respiratory rate of *Eupolymnia* may, therefore, be related to its need for more vigorous ventilation of its tubes and respiratory surfaces, a need that is continuously imposed upon it by the oxygen tension in its immediate environment.

Low oxygen tensions lead also to increased ciliary activity on the part of bivalve molluscs, producing a flow of water through the mantle cavity that may be several times greater in foul waters than in well-aerated ones. Direct measurements of respiratory and pumping rates in *Mercenaria mercenaria* show that oxygen is consistently removed as water flows through the mantle cavity. The requirements for oxygen may be of greater importance in regulating the flow than the requirements for food. It is estimated that 3 liters of water must be pumped through the mantle cavity to provide an animal with 1.0 ml of oxygen. Ventilation of the mantle cavity of cephalopod molluscs may also be increased as much as tenfold by a reduction in oxygen content of the water around them. Increase in its carbon dioxide content results also in increase in the frequency and amplitude of their respiratory movements. Carbon dioxide tension also regulates the opening and closing of the pneumostomes of pulmonate gastropods and of the spiracles of insects, as well as the frequency and depth of their respiratory movements. In an atmosphere of 3 to 5 percent CO_2, the pneumostomes of pulmonates remain open permanently. The concentrations of carbon dioxide resulting in increased ventilation vary with different species of insects. The response at low concentrations is opening of the spiracle; if the concentration increases, abdominal movements become more frequent and greater in extent. Most insects are anaesthetized at concentrations of carbon dioxide above 25 percent, but there is at least one mutant strain of *Drosophila melanogaster* that is highly sensitive to it and is killed at concentrations far below this. The control of these respiratory movements is mediated by the central nervous system, whose sensitivity to external stimuli of this nature determines the responsiveness of the animal to such environmental stresses and its ability to protect itself from them.

CONDITIONS IN AQUATIC ENVIRONMENTS—WATER MOVEMENTS Aquatic invertebrates inhabiting swiftly flowing water characteristically have rates of metabolism higher than those of the same or closely related species living in ponds, pools, or other quiet waters. For example, the Q_{O_2} of isopods (*Asellus aquaticus*) taken from a fast-moving stream has been found to be about twice that of individuals taken from a slowly moving one. The rate for those in a swift stream has been found to be 1,085 $mm^3/(g)(hr)$ [108.5 ml/(g)(hr)], and for those in the slow one, 489 $mm^3/(g)(hr)$ [48.9 ml/(g)(hr)]. *Asellus leptodactylus*, a lake dweller, has a Q_{O_2} of 7.0 $mm^3/(g)(hr)$ [7.0 ml/(g)(hr)]. There is a similar difference between the larvae of two different species of caddis flies with different habitats. Larvae of *Hydropsyche* sp., living in swift waters, have a Q_{O_2} one and one-half times greater than those of *Melanna* sp, living in ponds. An even greater difference is exhibited by nymphs of mayflies. The nymphs of *Baetis rhodani* live in fast water and take up oxygen at a rate of 2,571 $mm^3/(g)(hr)$ [257.1 ml/(g)(hr)] in winter, at a temperature of 10°C, and in late summer, at a temperature of 16°C, at a rate of 5,619 $mm^3/(g)(hr)$ [561.9 ml/(g)(hr)]. This is, respectively, four times and three times the rates for nymphs of *Chloeon dipterum*, which live in ponds and take up oxygen at 10°C, at a rate of 600 $mm^3/(g)(hr)$ [60.0 ml/(g)(hr)] and at 16°C, at a rate of 1,740 $mm^3/(g)(hr)$ [174.0 ml/(g)(hr)]. The differences represent the differences in the amount of work these animals must perform in order to maintain their positions in the environments in which they live and to escape being carried away from their preferred locality by the water currents to which they are exposed.

SALINITY Performance of work of a different kind is expressed in the differences in oxygen consumption of animals living in the sea and in brackish and freshwater. This is osmotic work, an expenditure of energy necessary for the preservation of an internal environment whose ionic composition and osmotic pressure may be quite different from that of the external environment. The differences are also due probably in large part to increased utilization of oxygen by the tissues, because of their increased water content through diffusion of water into the cells from the more dilute external medium. The differences in oxidative metabolism under these conditions may again be illustrated by measurements made on isopods and other crustaceans. The Q_{O_2} of *Idotea*, a marine genus of isopods, is 204 $mm^3/(g)(h)$ [20.4 ml/(g)(hr)], about one-half that of *Asellus aquaticus* from freshwater streams with slow currents, and about one-quarter that of *A. aquaticus* from freshwater streams with strong currents. Comparable differences have been found for amphipods of the genus *Gammarus*. Measurements made on the males of three species, in which all the individuals were about 13 mm long, have shown that the Q_{O_2} of *G. marinus*, a species living along the seashore, is 340 $mm^3/(g)(hr)$ [34.0 ml/(g)(hr)] and that of *G. locusta*, of similar habitat, is 207 $mm^3/(g)(hr)$ [20.7 ml/(g)(hr)], while that of *G. pulex*, an inhabitant of freshwater, is 501 $mm^3/(g)(hr)$ [50.1 ml/(g)(hr)]. Oxygen consumption has been found to increase also in animals habitually living in salt or brackish water that are experimentally exposed to different dilutions of their natural mediums. Specimens of *Metapenaeus monoceros*, a prawn common in the sea and estuaries along the coast of Madras, all with weights of about 3.5 g, showed increased oxygen consumption with decrease in the salinity of the medium. For individuals taken from the sea the respiratory rate increased from a basal level of 0.67 ml/(g)(hr) to one of 0.80 ml/(g)(hr) at a dilution of 50 percent and to 0.914 ml/(g)(hr) at a dilution of 25 percent, and to 0.90 ml/(g)(hr) in tap water. For those taken from the brackish water of an estuary, whose concentration was about 50 percent that of seawater, the basal respiratory rate of 0.6 ml/(g)(hr) increased to 0.8 ml/(g)(hr) after their transference to seawater, to 0.71 ml/(g)(hr) in 25 percent seawater, and to 1 ml/(g)(hr) in tap water. For both groups of individuals the increase in oxygen consumption would seem to reflect the energy expended in making adjustments to an environment more dilute or, in the case of the estuarine species, more concentrated than their natural environment.

Experiments with other species have yielded, on the whole, similar results. The freshwater shrimp *Palaemonetes* increases its oxygen consumption when in seawater, and the barnacle *Lepas* does so when immersed in seawater of various dilutions. The swimming crab *Carcinus mediterraneus* showed an increase in oxygen consumption from 93.6 μl/(g)(hr) [0.936 ml/(g)(hr)] in seawater to 127.5 μl/(g)(hr) [1.275 ml/(g)(hr)] in 50 percent seawater, and a related species, *Callinectes sapidus*, under the same conditions, an increase from a basal level of 97.3 μl/(g)(hr) [0.973 ml/(g)(hr)] to one of 150.4 μl(g)(hr) [1.504 ml/(g)(hr)] in 20 percent seawater. The highly adaptable wool-handed crab, *Eriocheir sinensis*, on the other hand, shows no difference in respiratory rate when exposed to seawater of increasing dilutions, and the marine spider crabs, *Maja verrusa* and *Libinia emarginata*, reduce their oxygen uptake in 50 percent seawater from rates of 54.2 μl/(g)(hr) [0.542 ml/(g)(hr)] and 37.1 μl/(g)(hr), respectively, to ones of 28.2 μl/(g)(hr) [0.282 ml/(g)(hr)] and 26.0 μl/(g)(hr) [0.260 ml/(g)(hr)]. These are animals that do not ordinarily invade brackish waters, for they are incapable of osmoregulation and they die within a few hours after immersion in seawater at a dilution of 50 percent.

Exposure to seawater of different concentrations of isolated crustacean tissues, such as gills, muscles, digestive glands, and excretory organs, results in most cases in an increase in their respiration over a basal rate established in the water of the natural habitats of the animals. Gill tissue has been of particular interest in this connection, because the gills, as the exposed and permeable areas of a crustacean's integument, are believed to be organs greatly involved in osmoregulation. There is, however, no increase in oxygen consumption by excised gills of *Eriocheir* in freshwater over that in seawater, nor is there any difference in water content in intact animals taken from the sea or from freshwater. This fact provides support for the hypothesis that the increased oxidative metabolism of other crustaceans in dilute media reflects the state of hydration of their tissues more than their need for energy to be expended in osmotic work.

3. Locomotor activity

The factor causing the greatest increase in oxidative metabolism is locomotor activity, for this demands expenditure of energy in the alteration of the state of the contractile elements that result in movement of locomotor organelles or organs. The extent of this activity varies greatly in animals of different habits, but to greater or lesser extent is common to all in the mechanisms for the ingestion of food, whether they actively search for it or whether, as sessile or sedentary individuals, they merely take in that which is in their immediate vicinity. But almost all animals can be roused to brief periods of active locomotion by stimuli, either internal or external, that make a change of location desirable or even imperative to them. Such stimuli may be hunger, thirst, need for a mate, changes in the temperature, intensity of light or the humidity of the environment, or the proximity of a predator from which escape is necessary. Some invertebrates are actively moving the greater part of their lives in their quests for food, mates, nesting places, or new areas to colonize; and migratory species periodically make long trips from one locality to another, frequently covering great distances with only brief periods for rest and refreshment. In these animals, as in those that are only sporadically active, measurements of the oxygen utilized and of the food and storage reserves consumed while locomotion is in progress give a clue as to the amount of energy expended; determinations of the RQ show the nature of the principal foodstuffs that provide the substrates for the oxidative reactions. These measurements are even more difficult to make, in the case of many invertebrate species, than those of their basal metabolism, and it is hardly surprising that exact quantitative data are available for so few of them. The most revealing have been obtained from flying insects, for these are best suited to studies of this kind. They can be made to fly in respirometers, under controlled conditions of nutrition and environment, for specified periods of time or until they are exhausted. These flights may take place in cages, where the insect can fly freely within the confined area, or in flight mills, with the insect tethered to a support as it is when measurements are made of the frequency of the wing beat. The insect may be allowed to fly in quiet or in turbulent air or in a wind tunnel in an air pressure simulating either a head or a following wind of known velocity. Such measurements have been made for certain species of Diptera, Hymenoptera, Lepidoptera, and Orthoptera, leading to the generalization that their respiration in flight may be 50 to 100 times greater than it is at rest.

According to the nature of the food or of the reserves utilized, representative insects of these orders are divided into those that derive their energy sources primarily from carbohydrate and those that derive them from fat. *Drosophila, Musca,* and *Apis,* whose RQs are close to 1, are examples of carbohydrate users; and the migratory locust *Schistocerca* and the monarch butterfly *Danaus,* whose RQs are in the neighborhood of 0.75, at least during sustained flight, are examples of users of fat. Histological study of the tissues of resting *Drosophila* reveal dense deposits of glycogen in the fat body, haltere knobs, indirect flight muscles, and proventriculus, with small accumulations of fat in the midgut and fat body. These carbohydrate and fat reservoirs disappear concurrently when the insects are starved, but when they are flown to exhaustion the fat reserves are untouched though there is no glycogen left in the flight muscles or proventriculus and it is greatly depleted in the fat body and the haltere knobs. After a period of rest, an exhausted *Drosophila* can fly again for a short time without feeding, until its glycogen reserves are completely used up. If fed glucose, such a fly will begin flight in $\frac{1}{2}$ to $\frac{3}{4}$ min, and it can be calculated that 1 μg of glucose is sufficient to sustain flight for about 6 min. A meal of fructose, maltose, or sucrose will also enable an exhausted insect to fly again, but only after a somewhat longer interval. Galactose and xylose make brief flights possible, but an insect cannot sustain activity on these sugars, and lactose and sorbose are apparently ineffective in providing the necessary substrates. Of the various sugars tested on *Drosophila,* only glucose, fructose, mannose, maltose, sucrose, and trehalose were oxidized by the flight muscles at rates comparable to the oxidation of glycogen. Measurements of oxygen uptake by *D. repleta* give an average Q_{O_2} of 27.8 μl/(g)(min) [1.67 ml/(g)(hr)] for individuals at rest, and one of 351 μl/(g)(min) [27.06 ml/(g)(hr)] for those in flight, with a wing beat frequency of 10,920 strokes/min. With greater frequency of beat, the Q_{O_2} is greater, reaching as much as 576 μl/(g)(min) [34.56 ml/(g)(hr)], about a twentyfold increase. The wing beat cycle has been found to be a reliable index of the rate of energy expenditure; it has been calculated that in *Drosophila* each stroke represents the oxidation of 2.5×10^{-7} g of glycogen, with a liberation of energy equivalent to 10×10^{-5} Cal/g of muscle. At this rate of glycogen consumption, an individual weighing about 3.5 g would use approximately 0.1 mg of glycogen per hour of flight, representing about 3 percent of its total body weight and about 15 percent of the weight of its wing muscles. Duration of flight is, therefore, limited by the amount of the glycogen reserves, and after these are depleted, the fly must be fed an acceptable sugar and, for sustained flight, allowed time to build up its reserves again. The feeding experiments have shown that certain sugars are available immediately as energy sources and that some are more suitable as energy sources than others. It seems evident that under ordinary conditions glycogen is the principal substrate for oxidative metabolism in *Drosophila* and that this can be directly metabolized in the muscle itself, with the fat body contributing its reserves in the form of a transportable sugar after those of the muscles are reduced. These reserves from the fat body and other tissues can be readily mobilized, and the flies, even in vigorous and sustained flight, never accumulate more than a small oxygen debt which is paid off in a minute or two.

Relationship between the level of trehalose in the hemolymph and the frequency of wing beat has been established in the blowfly *Phormia regina,* indicating that in these flies, although the glycogen reserves in the flight muscles may be directly oxidized, the principal site for the depolymerization of glycogen is elsewhere and that the muscles make use of the resulting product, trehalose. In these experiments, male flies were fed on 1 *M* solutions of glucose for 4 to 5 days after eclosion and then made to fly in tethered flight; the frequency of wing beat was measured stroboscopically and the carbohydrates in the hemolymph assayed chromatographically and colorimetrically at different frequencies of beat. It was found that wing beat frequency fell during prolonged flight, dropping from a mean rate of 11,200 cycles/min at the start of the flight to 7,460 cycles/min when the flies were exhausted. Concurrently, while the consistently small concentration of glucose in the hemolymph did not change significantly, that of trehalose dropped from one of 20 g/liter to one of 6 to 7 g/liter. After injection of trehalose and a 3 min period of rest, exhausted individuals were able to fly again, with a frequency of wing beat that could be correlated with the amount of trehalose injected. Injections of 210 μg restored the wing beat of individuals to a frequency of 10,950 cycles/min, or about 99 percent of that of fresh, fully fed individuals. Half this amount (105 μg) and one quarter of it (52 μg) sustained wing beat frequencies of 9,750 and 8,800 cycles/min, respectively. It has been calculated that *Phormia* consumes carbohydrate at the rate of about 15 μg/min, and at this rate even the maximum amount of trehalose recorded for the hemolymph (about 200 μg) could provide substrate for only a few minutes of flight. The sugar, then, must be continually replenished during prolonged flight, and all evidence indicates that it is derived from the fat body, where glycogen is depolymerized and trehalose discharged into the hemolymph. The speed and efficiency with which this takes place are in this case limiting factors not only to the duration of the fly's flight but to its speed as well, for the concentration of this sugar in the hemolymph determines, in great measure, the rate of energy expenditure by the flight muscles as exhibited by the frequency of their contraction.

The flight muscles of bees also metabolize sugars directly, deriving them from the nectar on which the adult worker feeds. The oxygen consumption of a bee in flight may be more than 50 times greater than it is at rest. At 21°C and at rest, the Q_{O_2} of *Apis mellifica*, for example, has been recorded at 132 μl/(g)(min) and in flight, as 1,679 μl/(g)(min). At lower temperatures, there is an even greater expenditure of energy, and Q_{O_2}s of 4,400 μl/(g)(min) and of 5,200 μl/(g)(min) have been recorded for bees flying at temperatures of 11 and 18°C, respectively. With the exception of mannose, proved toxic to them because of their deficiency in phosphomannoseisomerase, the sugars usable as substrates by dipterans are equally valuable to bees, and the limitations to their flight are set by the quantity of nectar that they can imbibe and hold rather than by its quality. Bees have a cruising range of 29 miles when flying at a speed of 5.6 miles/hr, but they can fly in short bursts as fast as 9.4 miles/hr. Even short periods of flight at this speed would reduce their range, because of the greater depletion of their energy sources in rapid flight than in slower flight.

Some doubt is thrown upon the generalization that dipterans make use of carbohydrate exclusively as an energy source by experiments that have been made on mosquitoes. There are three possible sources of this kind for adult female mosquitoes—the nectar and blood on which they feed and the reserves accumulated during their larval stages. These reserves can be visualized in the fat bodies of fully grown larvae as granules of glycogen and droplets of fat and protein. The reserves of glycogen are calculated to be equivalent to about 0.12 mg of glucose. Female mosquitoes of *Culex pipiens*, form *berbericus*, that have not been fed, have been observed to fly a mean total distance of 5,339 m over a period of 5 days, covering the greatest distance on the first day and progressively less and less distance until on the final day they accomplished only one-fifth of that of the first day. Females allowed to feed to repletion on blood before being put on the flight mill flew more than twice the mean total distance that unfed mosquitoes flew, or those fed on solutions of glucose. This suggests that products of the digestion of the blood were used as sources of energy, and since the amount of sugar in the 3 mg of blood ingested would not have been sufficient to account for the additional substrate, it would seem that this was provided by the lipids or proteins.

A great deal of interest is attached to the oxidative metabolism of locusts and butterflies, which are capable of such long and sustained flights in their migrations. The Q_{O_2} of mature males of the migratory locust

Schistocerca at rest is 0.63 liter/(kg)(hr) and in tethered flight rises to 10 to 30 liters/(kg)(hr). The speeds and duration of their tethered flight are of the same order of magnitude as those which have been observed in natural swarms in the field. The RQ of resting insects is 0.82, but it drops after the first ½ hr of flight to 0.75, indicating that at this time fat is the primary substrate oxidized. Fat represents about 10 percent of the wet weight of a fully developed insect, and the fuel value of fat is three times that of carbohydrate. An insect metabolizing fat would, therefore, obtain about three times the energy that an insect would from metabolizing the same weight of nectar.

In nature, a swarm of locusts preparing to start flight in the morning spend some time milling about in groups before taking off en masse in the direction of the migration. From analyses of the glycogen and fat content of tethered insects before and during flight, it has been found that glycogen is expended in the preliminary period, after which the fat reserves are used. The undirected milling and surging of a swarm in nature seems then to represent a warming-up period during which the glycogen stores are quickly mobilized and utilized as they are in dipterans, but after which the metabolic pattern switches to that of fat degradation. A locust of average size and weight, with fat representing 10 percent of this, should be able to fly continuously for at least 10 hr, with 20 hr the upper limit for an exceptionally fat one. Flights from dawn till dusk are usual for a migrating swarm, but sufficient time for feeding and for rest is essential to them, and their rapacious appetites are matters of fact as well as of fable. It has been calculated that a swarm must eat as much as one to one and one-half times its own weight every day, and probably three times as much as this is actually eaten. Large swarms, whose weights have been estimated as 13,000 to 19,000 tons, could be responsible for the combustion of some 500 to 800 tons of fat, the equivalent, in terms of calories, of that which could be accomplished by about 1,500,000 human beings. Even medium-sized swarms, about one-tenth the size of the large ones, are responsible for the consumption of huge amounts of vegetation and stored grains, and the transformation of their potential energy into forms that can be used in the biological processes of muscular contraction and the mechanical work of flight.

Observations of locusts in tethered flight have shown that they accumulate an oxygen debt, which is paid off in an hour or two. This is far longer than the time taken by *Drosophila* to pay off any debt it may incur, and it has been suggested that a surplus of ketone bodies resulting from the intensive breakdown of fats may account for the longer recovery period in *Schistocerca*. There is a considerable amount of evidence that the fat used in flight is metabolized directly in the muscles of the locust, as glycogen is metabolized directly in those of *Drosophila*. The oxidation of fat is, in any event, a somewhat slower process than that of carbohydrate, and it has been generally assumed that the rate of its oxidation is not fast enough to provide the energy necessary for the rapid wing movements of a dipteran in flight. However, the thoracic muscles of *Drosophila* do not work at a higher rate than those of *Schistocerca*, and fat is used by *Drosophila* as an energy source for its leg muscles when walking or crawling. Possibly, the differences between the patterns of migratory locusts and lepidopterans—for the wing muscles of migrating butterflies also utilize fat—and those of flies and bees may lie in the sarcosomes. It may be that those of the users of carbohydrate have, in the course of dipteran and hymenopteran evolution, lost the ability to initiate the breakdown of the higher fatty acids, a step that would be essential to their introduction into the Krebs cycle and subsequent oxidation.

SELECTED BIBLIOGRAPHY

Oxidation-reduction reactions

Chance, B., and D. E. Parsons: Cytochrome Function in Relation to Inner Membrane Structure of Mitochondria, *Science*, vol. 142, pp. 1176–1180, 1963.

Lehninger, A. L.: "Bioenergetics: The Molecular Basis of Biological Energy Transformations," W. A. Benjamin Inc., New York, 1965.

Lehninger, A. L.: The Transfer of Energy within Cells, Ideas in Modern Biology, in J. A. Moore (ed.), "Proceedings XVI International Congress of Zoology," vol. 6, pp. 173–204, Nat. History Press, Garden City, N.Y., 1965.

McElroy, W. D.: "Cell Physiology and Biochemistry," Prentice-Hall, Englewood Cliffs, N.J., 1964.

Simpson, J. W., and J. Awapara: The Pathway of Glucose Degradation in Some Invertebrates, *Comp. Biochem. Physiol.*, vol. 18, pp. 537–548, 1966.

Oxidative metabolism of carbohydrate

Agrosin, M., and Y. Repetto: Studies on the Metabolism of *Echinococcus granulosus*. VII. Reactions of the Tricarboxylic Cycle in *E. granulosus* scolices, *Comp. Biochem. Physiol.*, vol. 8, pp. 245–261, 1963.

Bishop, S. H., and J. W. Campbell: Carbamyl Phosphate Synthesis in the Earthworm *Lumbricus terrestris, Science*, vol. 142, pp. 1583–1585, 1963.

Bryant, C., and W. L. Nicholas: Intermediary Metabolism in *Monoliniformis dubius* (Acanthocephala), *Comp. Biochem. Physiol.*, vol. 15, pp. 103–112, 1965.

Bryant, C., and W. L. Nicholas: Studies on the Oxidative Metabolism of *Monoliniformis dubius* (Acanthocephala), *Comp. Biochem. Physiol.*, vol. 17, pp. 825–840, 1966.

Chefurka, W.: Intermediary Metabolism of Carbohydrates in Insects, in M. Rockstein (ed.), "The Physiology of Insects," vol. II, pp. 582–660, Academic, New York, 1965.

Dixon, S. E., and R. W. Shuel: Respiration of Queen and Worker Honey Bee Larvae on Differentially Labelled Glucose C^{14}, *Comp. Biochem. Physiol.*, vol. 30, pp. 105–110, 1969.

Hammen, C. S.: Carbon Dioxide Fixation in Marine Invertebrates. V. Rate and Pathway in the Oyster, *Comp. Biochem. Physiol.*, vol. 17, pp. 289–296, 1966.

Hammen, C. S.: Metabolism of the Oyster *Crassostrea virginica, Am. Zool.*, vol. 9, pp. 309–318, 1969.

Hammen, C. S., D. P. Hanlon, and S. C. Lum: Oxidative Metabolism of *Lingula, Comp. Biochem. Physiol.*, vol. 5, pp. 185–191, 1962.

Hammen, C. S., and P. J. Osborne: Carbon Dioxide Fixation in Marine Invertebrates: A Survey of Major Phyla, *Science*, vol. 130, pp. 1409–1410, 1959.

Jahn, T. L.: Respiratory Metabolism, in G. N. Calkins and F. M. Summer (eds.), "Protozoa in Biological Research," pp. 352–403, Columbia, New York, 1941.

Kitto, G. B., and M. H. Briggs: Relationship Between Locomotory Habits and Enzyme Concentration in Insects, *Science*, vol. 135, p. 918, 1962.

Kmetec, E., and E. Bueding: Production of Succinate by the Canine Whipworm *Trichuris vulpes, Comp. Biochem. Physiol.*, vol. 15, pp. 271–274, 1965.

Rockstein, M.: Some Aspects of Intermediary Metabolism of Carbohydrate in Insects, *Bull. Brooklyn Entomol. Soc.*, vol. 45, pp. 74–81, 1950.

Rothstein, M., and H. Mayoh: Nematode Biochemistry. VIII. Malate Synthetase, *Comp. Biochem. Physiol.*, vol. 17, pp. 1181–1188, 1966.

Saz, H. J., and O. L. Lescure: Interrelationships between the Carbohydrate and Lipid Metabolism of *Ascaris lumbricoides* Egg and Adult Stage, *Comp. Biochem. Physiol.*, vol. 18, pp. 845–857, 1966.

Scheibel, L. W., and H. J. Saz: The Pathway for Anaerobic Carbohydrate Dissimilation in *Hymenolepis diminuta, Comp. Biochem. Physiol.*, vol. 18, pp. 151–162, 1966.

Wilmoth, J. H.: Studies on Metabolism of *Taenia taenaeformis, Physiol. Zool.*, vol. 18, pp. 60–80, 1945.

Phosphagens

Baldwin, E., and W. H. Yudkin: The Annelid Phosphagen: With a Note on Phosphagen in Echinodermata and Protochordata, *Proc. R. Soc. (London)*, vol. 136, pp. 614–631, 1950.

Ennor, A. H., and J. F. Morrison: Biochemistry of the Phosphagens and Related Guanidines, *Physiol. Rev.*, vol. 38, pp. 631–574, 1958.

Hobson, G. E., and K. R. Rees: The Annelid Phosphagens, *Biochem. J.*, vol. 61, pp. 549–552, 1956.

Hobson, G. E., and K. R. Rees: The Annelid Phosphokinases, *Biochem. J.*, vol. 65, pp. 305–306, 1957.

Stephens, G. C., J. F. van Pilsum, and D. Taylor: Phylogeny of the Distribution of Creatine in Invertebrates, *Biol. Bull.*, vol. 129, pp. 573–581, 1965.

Thoai, N., and Y. Robin: Distribution of Phosphagens in Errant and Sedentary Polychaeta, in K. A. Munday (ed.), "Studies in Comparative Biochemistry," pp. 152–161, Pergamon, New York, 1965.

Thoai, N., and J. Roche: Sur la biochemie comparée des phosphogènes et leur repartition chez les animaux, *Biol. Rev.*, vol. 39, pp. 214–231, 1964.

van Pilsum, J. F., and G. C. Stephens: The Phylogenetic Significance of Distribution of Creatine in the Invertebrates, *Am. Zool.*, vol. 4, p. 286, 1964.

Bioluminescence

Bassot, J. M.: On the Comparative Morphology of Some Luminous Organs, in F. H. Johnson and Y. Haneda (eds.), "Bioluminescence in Progress," pp. 557–610, Princeton, Princeton, N. J., 1966.

Buck, J. B.: Studies on the Firefly. I. The Effects of Light and Other Agents on Flashing in *Photina pyralis* with Special Reference to Periodicity and Diurnal Rhythm, *Physiol. Zool.*, vol. 10, pp. 45–58, 1937.

Buck, J. B.: The Anatomy and Physiology of the Light Organ in Fireflies, *Ann. NY Acad. Sci.*, vol. 49, pp. 397–482, 1948.

DeSa, R., J. W. Hastings, and A. E. Vatter: Luminescent "Crystalline" Particles: An Organized Subcellular Bioluminescent System, *Science*, vol. 141, pp. 1269–1270, 1963.

Eckert, R. Subcellular Sources of Luminescence in *Noctiluca*, *Science*, vol. 151, pp. 349–352, 1966.

Harvey, E. N.: "Living Light," Princeton, Princeton, N.J., 1940.

Harvey, E. N.: Bioluminescence, in M. Florkin and H. S. Mason (eds.), "Comparative Biochemistry," pp. 545–591, Academic, New York, 1960.

Hastings, J., and D. Davenport: The Luminescence of the Millipede *Lumiodesmus signorae*, *Biol. Bull.*, vol. 113, pp. 120–128, 1957.

Johnson, F. H., Y. Haneda, and E. H. C. Sie: An Interphylum Luciferin-Luciferase Reaction, *Science*, vol. 132, pp. 422–423, 1960.

Johnson, F. H., O. Shimomura, and Y. Saiga: Luminescence Potency of the *Cypridina* System, *Science*, vol. 134, pp. 1755–1756, 1961.

Kanda, S.: The Luminescence of a Nemertean, *Emplectonema kandai* Kato, *Biol. Bull.*, vol. 77, pp. 166–173, 1939.

Lloyd, J. E.: Aggressive Mimicry in *Photinus*: Firefly Femmes Fatales, *Science*, vol. 149, pp. 653–654, 1965.

McElroy, W. D.: Chemistry and Physiology of Bioluminescence, *Harvey Lect.*, pp. 240–265, 1956.

McElroy, W. D., and H. H. Seliger: Mechanisms of Bioluminescent Reactions, in W. D. McElroy and B. Glass (eds.), "Light and Life," pp. 219–257, McCollum Pratt Symposium, Johns Hopkins, Baltimore, 1961.

McElroy, W. D., and H. H. Seliger: Origin and Evolution of Bioluminescence, in M. Kasah and B. Pullman (eds.), "Horizons in Biochemistry," pp. 91–101, Academic, New York, 1962.

Nicol, J. A. C.: Luminescence in Polynoid Worms, *J. Mar. Biol. Assoc. U.K.*, vol. 33, pp. 65–84, 1953.

Nicol, J. A. C.: Observations on Luminescence in *Renilla* (Pennatulacea), *J. Exp. Biol.*, vol. 32, pp. 299–320, 1955.

Nicol, J. A. C.: Histology of the Light Organs of *Pholas dactylus*, *J. Mar. Biol. Assoc. U.K.*, vol. 39, pp. 109–114, 1960.

Nicol, J. A. C.: Animal Luminescence, in O. Lowenstein (ed.), "Advances in Comparative Physiology and Biochemistry," vol. I, pp. 217–274, Academic, New York, 1962.

Okada, Yô K.: Observations on Rod-like Contents in the Photogenic Tissue of *Watasenia scintillans* through the Electron Microscope, in F. H. Johnson and Y. Haneda (eds.), "Bioluminescence in Progress," pp. 611–625, Princeton, Princeton, N.J., 1966.

Rhodes, W. C., and W. D. McElroy: Enzymatic Synthesis of Adenyl-oxyluciferin, *Science*, vol. 128, pp. 253–254, 1958.

Seliger, H. H., and W. D. McElroy: Chemiluminescence of Firefly Luciferin without Enzyme, *Science*, vol. 138, pp. 683–685, 1962.

Factors influencing respiratory metabolism

Allen, M. D.: Respiration Rates of Worker Honey Bees of Different Ages and at Different Temperatures, *J. Exp. Biol.*, vol. 36, pp. 92–101, 1959.

Chadwick, L. E., and D. Gilmour: Respiration during Flight in *Drosophila repleta* Wollaston: The O_2 Consumption Considered in Relation to the Wing Rate, *Physiol. Zool.*, vol. 13, pp. 398–410, 1940.

Clegg, J. S., and D. R. Evans: Blood Trehalose and Flight Metabolism in the Blowfly, *Science,* vol. 134, pp. 54–55, 1961.

Clements, A. N.: The Sources of Energy for Flight in Mosquitoes, *J. Exp. Biol.,* vol. 32, pp. 547–554, 1955.

Dales, R. P.: Oxygen Uptake and Irrigation of the Burrow by Three Terebellid Polychaetes: *Eupolymnia, Thelepus* and *Neoamphitrite, Physiol. Zool.,* vol. 34, pp. 306–311, 1961.

Ellenby, C.: Body Size in Relation to Oxygen Consumption and Pleopod Beat in *Ligia oceanica, J. Exp. Biol.,* vol. 28, pp. 492–507, 1951.

Fox, H. M., and B. G. Simmonds: Metabolic Rates of Aquatic Arthropods from Different Habitats, *J. Exp. Biol.,* vol. 10, pp. 67–74, 1933.

Hamur, A., and H. H. Haskin: Oxygen Consumption and Pumping Rates in the Hard Clam *Mercenaria mercenaria, Science,* vol. 163, pp. 823–824, 1969.

King, E. N.: The Oxygen Consumption of Intact Crabs and Excised Gills as a Function of Decreased Salinity, *Comp. Biochem. Physiol.,* vol. 15, pp. 93–102, 1965.

Krogh, A., and T. Weis-Fogh: The Respiratory Exchange of the Desert Locust (*Schistocerca gregaria*) before, during and after Flight, *J. Exp. Biol.,* vol. 28, pp. 344–357, 1951.

Loveland, R. E., and D. S. K. Chu: Oxygen Consumption and Water Movement in *Mercenaria mercenaria, Comp. Biochem. Physiol.,* vol. 29, pp. 173–184, 1969.

Rao, K. R.: Oxygen Consumption as a Function of Size and Salinity in *Metapenaeus monoceros* Farb. from Marine and Brackish Water Habitats, *J. Exp. Biol.,* vol. 35, pp. 307–313, 1958.

Read, K. R. H.: Respiration of the Bivalved Molluscs *Mytilus edulis* L. and *Brachidontes demissus plicatulus* Lamarck as a Function of Size and Temperature, *Comp. Biochem. Physiol.,* vol. 7, pp. 89–102, 1962.

Weiss-Fogh, T.: Fat Combustion and Metabolic Rate of Flying Locusts (*Schistocerca gregaria* Forskål), *Philos. Trans. R. Soc.* (*London*), *B,* vol. 237, pp. 1–36, 1952.

Wiens, A. W., and K. B. Armitage: The Oxygen Consumption of the Crayfish *Orconectes immunis* and *Orconectes nais* in Response to Temperature and to Oxygen Saturation, *Physiol. Zool.,* vol. 34, pp. 39–54, 1961.

Wigglesworth, V. B.: The Utilization of Reserve Substances in *Drosophila* during Flight, *J. Exp. Biol.,* vol. 26, pp. 150–163, 1949.

Williams, C. M., L. Barnes, and W. H. Sawyer: The Utilization of Glycogen by Flies during Flight and Some Aspects of the Physiological Ageing of *Drosophila, Biol. Bull.,* vol. 84, pp. 263–272, 1943.

Yurkiewicz, W. J., and T. Smyth, Jr.: Effects of Temperature on Oxygen Consumption and Fuel Utilization by the Sheep Blowfly, *J. Insect Physiol.,* vol. 12, pp. 403–408, 1966.

chapter XIII

EXCRETION; IONIC AND OSMOTIC REGULATION

A. **EXCRETION**
   1. **Excretory areas**
      a. CONTRACTILE VACUOLES
      b. NEPHRIDIA
      c. UROCOELES
      d. MALPIGHIAN TUBULES
      e. OTHER MECHANISMS
   2. **Excretory products**
   3. **Tests for function of excretory areas**
   4. **Factors in the operation of excretory systems**

B. **WATER; IONIC AND OSMOTIC REGULATION**
   1. **Water**
   2. **Ionic regulation**
   3. **Osmoregulation**

C. **MECHANISMS OF EXCRETION; IONIC AND OSMOTIC REGULATION**
   1. **Tests for osmoconformity or osmoregulation**
   2. **Presumptive mechanisms of osmoregulation and conditions requiring it**
   3. **Nonregulators and nonconformers: stenohaline species**
   4. **Osmoconformers**
      a. MARINE PROTOZOANS
      b. SIPUNCULIDS
      c. MARINE GASTROPOD AND BIVALVE MOLLUSCS
   5. **Osmoregulators**
      a. BY WATER STORAGE
      b. BY WATER ELIMINATION
         (1) By contractile vacuoles
         (2) By Nephridia: flatworms, rotifers, annelids

(3) By urocoeles

(a) MOLLUSCS: CEPHALOPODS, FRESHWATER BIVALVES, FRESHWATER AND TERRESTRIAL GASTROPODS

(b) CRUSTACEANS

(i) Branchiopods

(ii) Isopods

(iii) Decapods

(4) By Malpighian tubules

(a) INSECTS

(b) MYRIAPODS AND ARACHNIDS

A. EXCRETION

By definition, excretion means the separation of waste matter from tissues and body fluids and its elimination from the organism. The waste to be eliminated is the result of cellular metabolism. It is, primarily, the end products of catabolic processes but may also include unusable or undesirable by-products of anabolism as well as of catabolism. The nature of the end products is related to the nature of the material that is metabolically degraded or assembled. Carbon dioxide and water as end products of oxidative metabolism are among them, as well as nitrogenous compounds resulting from the deamination of amino acids prior to oxidation of the carbohydrate residues. Chief among these nitrogenous compounds are ammonia, urea, and uric acid, and according to which of them is proportionately greatest in the mixture of excreted products, animals are conveniently classified as ammonotelic, ureotelic, or uricotelic. Included among excretory products are organic acids, as partial products or by-products of oxidative metabolism, and the products of nucleic acid degradation, for these complex compounds are broken down as well as synthesized by cells. While the nitrogen from their pyrimidine bases is usually released as ammonia or incorporated into uric acid, the purines often are unchanged and, if not utilized in further syntheses, excreted as such.

All these compounds must be transported in solution or in suspension from the areas in which they are produced to the areas from which they are eliminated, and excretory mechanisms are, therefore, closely related to the availability of water. In most animals the excretory products are flushed out of the body in a volume of water whose loss may be advantageous in some circumstances, deleterious in others. Along with these products and the water in which they are transported there may be salts and nutrients that are of value and whose loss, if it occurs, must be compensated for in one way or another. Excretion is, therefore, closely bound up with nutrition, and even more with the preservation of stability (homeostasis) in the internal medium of multicellular animals, and with their ionic and osmoregulation.

1. Excretory areas

Any exposed surface of an animal's body may serve as an excretory area, as well as one through which water and its solutes can enter or leave the body. The entire integument of animals without exoskeletons probably takes care of the final elimination of much of their nitrogenous waste as well as of that of carbon dioxide. Gills and the alimentary canal may be excretory as well as respiratory in their function, but in many invertebrates there are tubular systems, diffuse or compact, to which the function of excretion has been attributed. These systems include contractile vacuoles; nephridia; the urocoeles that are the "kidneys" of molluscs; the antennal and maxillary glands of crustaceans and the coxal glands of arachnids; and the Malpighian tubules of insects and myriapods. It has been generally assumed, partly because of their tubular nature and liquid content, that these are all functionally analogous to the vertebrate nephron, but in only a limited number of cases has experimental evidence fully justified this assumption. In some instances, it is apparent that they are more concerned with regulation of the internal environment than with excretion per se; in some others, they appear to be primarily excretory and in others to combine both functions.

a. CONTRACTILE VACUOLES These are intracellular vacuoles filled with a clear fluid which is periodically or rhythmically discharged by contraction or collapse of the vacuole's walls. Typically they are found in freshwater protozoans and in freshwater sponges, but there are some marine ciliates in which they are regularly present. Electron microscopy of contractile vacuoles in *Ameba proteus*, and in *Paramecium aurelia* and *P. caudatum*, has revealed details of their fine structure. The vacuole is bounded by a layer of electron-dense material and lies in an area of cytoplasm in which mitochondria are particularly abundant. As a vacuole is filling, many tiny droplets, bounded by membranes similar to that of the vacuole, can be seen close to it. The contents of a vacuole is derived in part from these feeder vacuoles, which in *Ameba* are incorporated into it, and in part from substances diffusing inward or transported across its membrane. Contractile vacuoles are not fixed in position in *Ameba* but may be seen anywhere in the constantly shifting endoplasm. In *Paramecium*, their locations are more or less fixed. In the species studied there are usually two, one anterior in position and the other near the posterior end of the cell. They are also more elaborate in design than those in *Ameba*, for fluid collects and enters the vacuole through a system of canals. Short injection canals, 0.5 to 0.7 μ in length, lead into the vacuole from ampullae, into each of which there opens a radial canal. As these radial canals fill with fluid, they can be seen at the magnification of an ordinary light microscope, for their diameters increase from about 100 to 300 Å to as much as 4000 Å. During the period of filling, or diastole, the radial canals are in direct connection with finer tubules, about 200 Å in diameter, that form a network in the adjacent cytoplasm. These fine tubules are also in direct, and apparently continual, connection with the extensive endoplasmic reticulum, but when the radial canals are in systole, these connections are broken. Possibly, these fine tubes correspond to the feeder droplets of *Ameba*. While the radial canals are in diastole, the contractile vacuole is in systole, its collapse effected by fibrillar elements which in cross section resemble ciliary threads. These contractile fibrils extend from the distal end of each ampulla over the injection canal and vacuole to the ejection canal which they spiral around. This single canal leads from the vacuole to an excretory pore on the surface; when the vacuole is filling, it is closed off from the excretory canal by a membrane which is ruptured when the vacuole contracts.

b. NEPHRIDIA The term "nephridium," a diminutive of *nephros*, the Greek word for "kidney," is used for a variety of organs considered excretory in function. A nephridium, in a more limited and indeed more accurate meaning, is the primitive excretory organ of triploblastic animals, draining metabolic waste from the mesenchyme or mesoderm. In the course of the evolution of the animals that still retain them as primary excretory organs, they have undergone modifications and changes of one kind or another, and in some cases osmoregulation has been added to their primary function. They are absent in some invertebrates; in others, they have been replaced by a different kind of organ with similar functions. In those genera in which they are retained they may be divided into two major categories: protonephridia and metanephridia.

Basically a protonephridium consists of a duct with an external opening (nephridiopore); the duct is a passage through hollowed-out cells that resemble the porocytes of sponges but lie end to end. Internally, the terminal cell of a duct is a cyrtocyte, either a solenocyte or a flame bulb, embedded in the mesenchyme or mesoderm. In solenocytes a single long flagellum arises at the base of the tubular region of the cell and extends into the nephridial canal. In flame bulbs there is, instead, a tuft of cilia that do not reach the nephridial ducts. Protonephridia in their simplest form are found in the larval stages of modern annelids and molluscs, in which they consist of a single cell, serving both as flame bulb and duct, or at most of three or four cells and a nephridiopore. There is usually a pair of such organs in a larva, which is replaced in adult worms by more elaborate ones and in molluscs by organs of different origin. Protonephridia in one form or another are characteristic of adult flatworms, rotifers, gastrotrichs, endoprocts, nemerteans, some acanthocephalans, kinorhynchs, polychaete annelids, and cephalochordates (Fig. XIII–1). The elaborations are usually elongation, coiling, or often branching of the primary duct, with each of the finer ductules ending in a solenocyte or flame bulb, which thus appear to lie in clusters. The ducts may also be histologically differentiated along their length and are frequently enlarged, usually near the nephridiopore, as a bladder in which products to be eliminated may be temporarily stored.

Metanephridia are characteristic of coelomate animals only and are distinguished from protone-

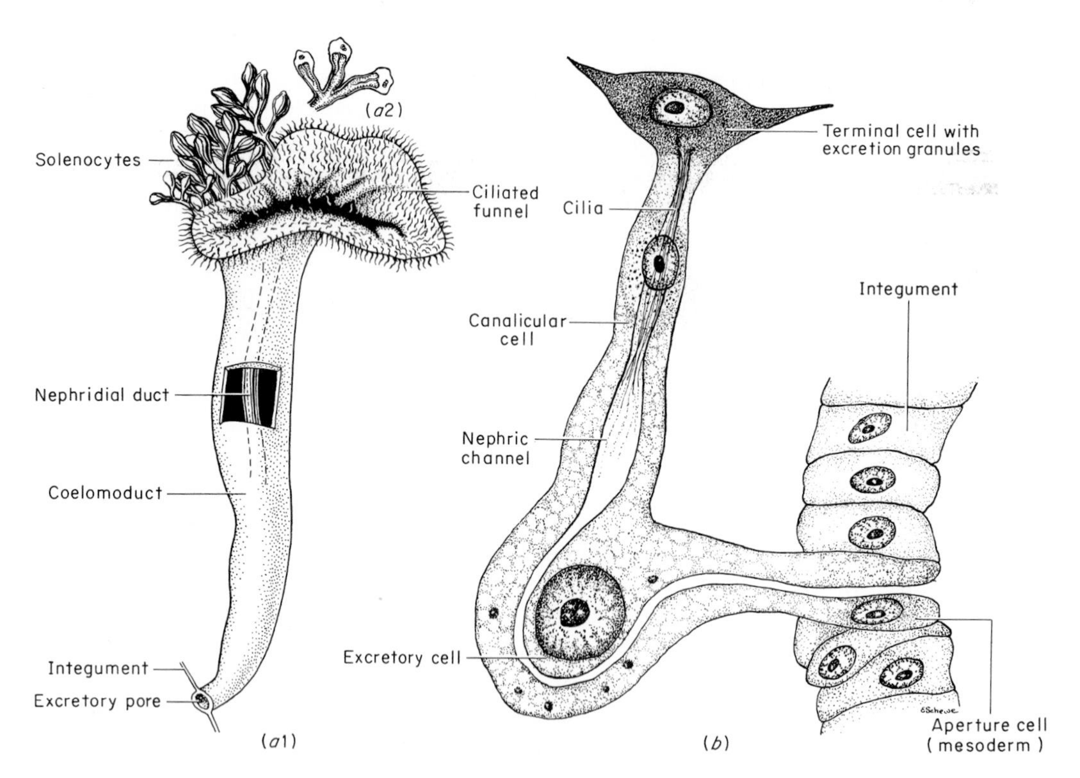

FIG. XIII–1 *Protonephridia.* ***(a1)*** *Protonephromixium of the marine annelid* Phyllodoce paretti. *(After Goodrich.) A small area of the coelomoduct is represented as if the wall were cut away to show the nephridial duct.* ***(a2)*** *Three solenocytes represented as if opened to show the flagellum inside each.* ***(b)*** *Protonephridium of the snail* Lymnaea stagnalis, *composed of four cells. (After Meisenheimer.)*

phridia in that their inner terminations lie in the coelom. Where the coelom is divided into compartments, these terminations lie in the segment anterior to the one containing the major part of the tubule and the nephridiopore; they may be open (nephrostomes) or closed. The ducts, like those of protonephridia, may be modified by elongation; by structural and functional differentiations along their length; by their union into a common duct or a common vesicle; or by changes in the position of the nephridiopore. Typically, and probably primitively, there was a single ventral external opening for each nephridium, but in many modern species excretory products from several or many nephridia are discharged through a single pore that may be variously situated. Also, the nephridiopore may open into some part of the alimentary canal (enteronephric system) rather than directly to the exterior (exonephric system). There seem, indeed, to be endless modifications of structural detail in an organ that still unmistakably conforms to a basic plan.

One important modification that has occurred in some coelomate invertebrates with nephridia is the combination of nephridium and coelomoduct to form a nephromixium. Coelomoducts were primitively genital ducts, conveying the gametes from the coelom in which they were formed or shed to the outside. Each had an internal opening (coelomostome) and an external opening (genital pore) with a duct connecting the two. In primitive segmented animals, a pair of coelomoducts may be supposed, like nephridia, to lie in every trunk segment. In the course of evolution, associations are believed to have arisen between these two systems to produce a combined organ with both excretory and genital functions. Such associations are best exemplified in annelids in which there may be protonephromixia and metanephromixia as well as mixonephridia. These will be defined and described in the discussion of annelid excretory organs.

Protonephridia are characteristic of flatworms. In free-living genera, with the exception of acoeles that have no such organs, the tubules of the protonephridia form a net, from which fine capillaries branch, each terminating in a flame bulb or a solenocyte. Electron microscopy of the terminal cells of *Stenostomum* sp. has shown that two flagella, moving together, extend through the tubular portion of each cell. The walls of the tubule have the typical cyrtocyte structure, with a fine membrane covering the interstices between the supporting rods. The number and position of nephridiopores are different in different species. The entire nephridial system is most extensive in freshwater species and is less well developed, or indeed may be wholly absent, in marine ones [Fig. XIII–2*a* (1) and *a* (2)].

In trematodes, as exemplified by *Fasciola hepatica*, there is a single nephridiopore and one main canal with side branches, each terminating in a cell with fine, arborizing processes extending from its free borders. In miricidia, the wall of the tubule of each terminal cell is supported by a double row of rods, so placed that those of the inner row are between those of the outer row. The tubule is thus effectively delimited from the surrounding parenchyma. This concentric arrangement is comparable to the two concentric collars observed in some species of Choanoflagellata, and adds confirmatory evidence to the concept of homology between the collars of choanocytes and the tubules of terminal protonephridial cells. Both represent morphological variations of a single cell type.

In cestodes, there are two main lateral canals extending from the first to the last proglottis, with cross connections in each proglottis. Smaller canals extend from the main ones into the parenchyma, each of the smallest ones terminating in a flame bulb.

Rotifers have typically one pair of protonephridia. In *Asplanchna priodonta*, each member of the pair consists of four flame bulbs, 10 to 18 μ long, which open into a greatly convoluted tubule that leads into a contractile vesicle into which the tube from the other side also opens. In *A. brightwelli*, there are 15 to 20 flame bulbs in each such protonephridium, and also a single contractile vesicle into which the tubules of each side open and which, in turn, opens into the cloaca. The flame bulbs lie in the pseudocoele, anchored to the wall by delicate filaments. This general pattern is probably typical of the system in rotifers, representing specialization of the more primitive condition in triclad and polyclad flatworms (Fig. XIII–2*b*).

Gastrotrichs have also one pair of protonephridia. Their internal terminations, as they have been revealed by electron microscopy in *Chaetonotus* sp., are composed of two closely apposed cyrtocytes, each with its own

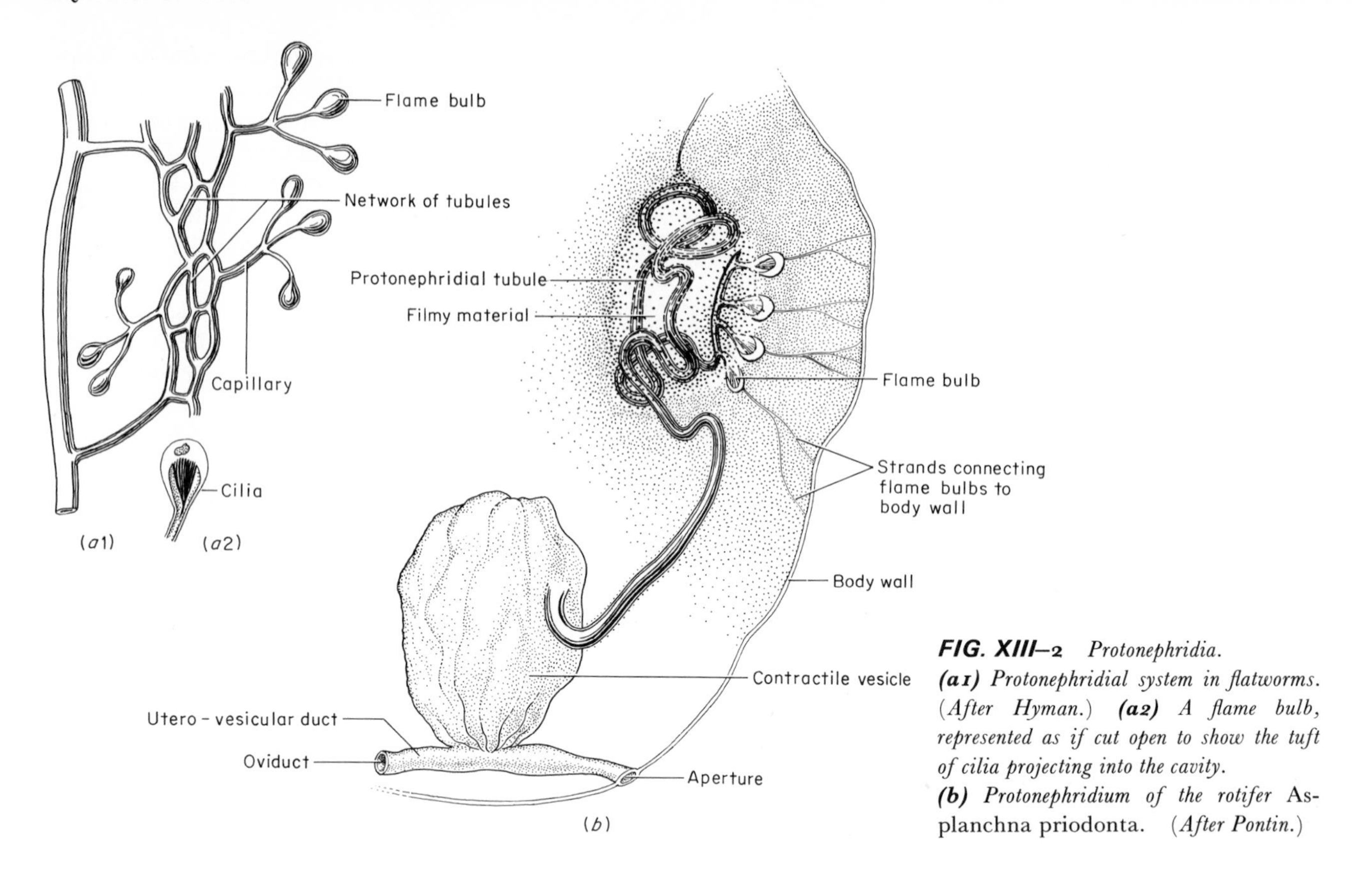

FIG. XIII–2 *Protonephridia.* ***(a1)*** *Protonephridial system in flatworms.* (*After Hyman.*) ***(a2)*** *A flame bulb, represented as if cut open to show the tuft of cilia projecting into the cavity.* ***(b)*** *Protonephridium of the rotifer* Asplanchna priodonta. (*After Pontin.*)

flagellum and both enclosed in a common membrane. This is unusual, but the double structure operates as one unit and serves the same filtration and transportive functions as single terminal cells. The protonephridial canal is coiled, and is differentiated into an anterior capillary region and a posterior glandular one.

In endoprocts, the pattern is that of two ducts, each ending internally in a single flame bulb, which join to form a common tubule that opens at a single nephridiopore. Some acanthocephalans (family Oligacanthorhynchidae) have also a pair of protonephridia, each terminating internally in 30 or more flame bulbs; the nephridiopore opens into the genital duct. In kinorhynchs, there is also a single pair of protonephridia; the tubules are short and each terminates in a solenocyte with one flagellum.

In nemerteans, there may be many, even hundreds, of individual protonephridia, each with a cluster of flame bulbs embedded in the parenchyma, a duct, and a nephridiopore. In some species, the smaller tubules from the flame bulbs on either side of the digestive tract unite into lateral tubes that run alongside the gut and open to the outside through a pair of nephridiopores. In *Cephalothrix major*, a littoral nemertean of the California coast, there are 300 or more individual nephridia in the anterior part of the body, each opening by a nephridiopore on the dorso-lateral surface of the body wall. Each terminates internally in a flame bulb, which lies close to a blood space and whose cilia lie in a slender stalk that opens into a larger tubule. This tubule coils through the parenchyma and leads into a narrow efferent duct that runs straight through the muscles and epidermis of the body wall (Fig. XIII–3). In *C. spiralis*, there is a single pair of nephridia situated anteriorly, each with a cluster of about 50 flame bulbs that lie in the lumen of a blood space attached to the inner surface of its wall. In the terrestrial genus *Geonemertes*, protonephridia number-

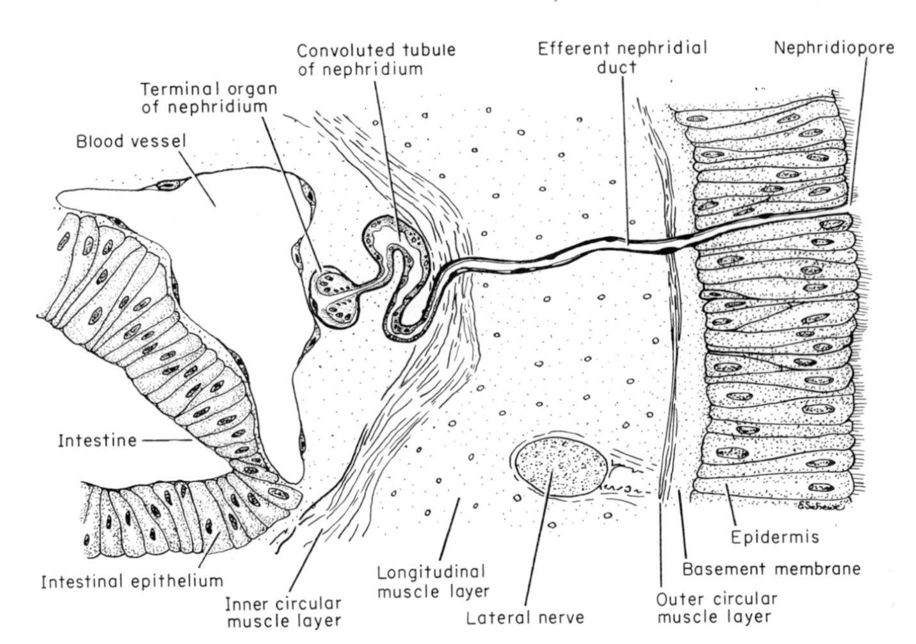

FIG. XIII–3 *Protonephridium of the nemertean* Cephalothrix major. (*After Coe.*) *This type of nephridium is uncommon among nemerteans in which the tubules usually terminate in many flame bulbs.*

ing in the thousands are distributed throughout the entire length of the body, each having its own efferent duct and nephridiopore. Internally, each tubule ends in a cluster of 6 to 10 solenocytes; the smaller tubules from these unite in a much convoluted duct, which becomes thinner-walled and straightens out as an efferent duct that opens through a nephridiopore. In some species of *Baseodiscus*, the system is enteronephric, with the ducts opening into the oesophagus.

Metanephridia are characteristic of phoronids, sipunculids, echiuroids, brachiopods, and annelids, although some polychaetes have protonephridia. In *Phoronis*, a pair of these organs opens into the coelom by one or two nephrostomes and to the exterior by a pair of posteriorly placed nephridiopores. Both excretory and genital products are discharged through them. In sipunculids, there is typically a single pair and in echiuroids, one, two, or up to hundreds of pairs. There are also in echiuroids a pair of hollow, sometimes branched, diverticula arising from the rectum and known as anal sacs. These have many ciliated funnels that open into the coelom and lead into the cavities of the sacs. Material that passes into them, as it does into the nephrostomes, is eliminated through the anus, not the nephridiopores. In brachiopods, there may be one or two pairs of metanephridia, with ciliated funnels opening into the coelom and nephridiopores discharging into the mantle cavity.

Polychaete annelids may have protonephridia, metanephridia, or nephromixia. The only family in all of whose members metanephridia and coelomoducts are entirely distinct is that of the Capitellidae, a group of sedentary worms in which the primitive pattern of a series of metanephridia and coelomoducts, each independent of the other, is retained in some trunk segments at least. In some of the genera of all the other families, association between nephridium and coelomoduct has resulted in protonephromixia, metanephromixia, and mixonephridia. In a protonephromixium, the coelomostome is grafted onto the canal of the protonephridium, and the protonephridial canal serves for genital, excretory, and, possibly, osmoregulatory functions. Such organs of dual origin are found in Phyllodocidae, Nephthydidae, Alciopidae, and Glyceridae. In a metanephromixium, the coelomostome has united with a metanephridium, and the combined organ can serve genital and excretory functions either concurrently or at different times. Such organs are found in syllids. In a mixonephridium, the nephrostome has presumably become occluded and lost, and the coelomostome has become completely fused with the internal end of the nephridial tubule. Such replacement of nephrostome by coelomostome has resulted in a urogenital organ usually capable of performing both genital and excretory functions. Such organs are general among polychaetes and are found in almost all families of sedentary worms as well as in the errant Aphroditidae and Eunicidae.

Phyllodoce and *Glycera* may be taken as illustrative of worms with protonephridial systems. In *Phyllodoce paretti*, there are typical protonephridia in the anterior segments of the body and protonephromixia in the posterior [Fig. XIII–1*a* (1)]. The solenocytes of the protonephridia project into the coelom and have been described as sticking out of the nephridial wall like pins from a pin cushion. This arrangement is similar to that in the cephalochordate *Branchiostoma lanceolatum*, although in the lancelet the cell bodies are not free in the coelom but rest on the walls of the glomerular capillaries. It is also comparable to conditions in the nemertean *Cephalothrix spiralis* in the attachment of flame bulbs to the walls of the blood vessel into which they project. In *Glycera* there are protonephromixia in every segment except the 15 to 20 anterior ones and the most posterior. The solenocytes of *Glycera* are distinctive in that the tubular portion is bent back upon the cell body and opens into the duct of the nephridium lying beneath it. These solenocytes have been seen to detach themselves from excised nephridia, straighten out and swim about in a suitable medium. Under these conditions, their resemblance to choanoflagellates is striking.

In nereids there are metanephridia in all but the most anterior and posterior segments. They have no connection with the coelomoducts, which have been converted into "dorsal ciliated organs" (ciliophagocytal organs). These have no external openings, but the coelomostomes function in collecting coelomic corpuscles loaded with waste which is phagocytosed in the part of the duct remaining. Dorsal ciliated organs are characteristic also of some glycerids and nephthyids with protonephridia, as well as of hesionids with metanephridia like nereids. The modified coelomoducts in these worms are excretory rather than genital in function. The loss of the latter function may be correlated with the reproductive habits of the worms, which shed their gametes by dehiscence of the body wall.

Syllids typically have metanephridia in every nephridial-bearing segment. In most species, however, in the fertile segments and at sexual maturity of the worm, the coelomostomes have become so intimately associated with the open ends of the metanephridia that the composite organs (metanephromixia) cannot well be distinguished from mixonephridia. This condition is another illustration of the difficulties in making hard and fast distinctions in biological categorizations.

Most tubiculous polychaetes have mixonephridia as adults. A general tendency for reduction in number of these urogenital organs is evident. In *Arenicola* there are only six pairs, in the fifth to tenth chaetigerous segments. Each organ has a large, richly vascularized preseptal funnel and a short duct that after passing through the septum expands into a bladder with muscle fibers in its walls before opening at the nephridiopore. During the breeding season, the bladders may be tremendously distended with reproductive cells. In some other genera there is only a single pair of nephridia in the anterior region, with the tubules united into one excretory duct that opens on the head segment. This is undoubtedly an adaptation to a tubiculous existence, for material eliminated through the nephridiopore is discharged outside the tube. Excretory products are thus prevented from fouling it, and gametes have a chance to escape into the open water. Other adaptations to life in a tube are also apparent. In *Owenia*, for instance, the mixonephridia are segmental and open externally, but into lateral canals. Each of these is formed by junction of the walls of an epidermal groove, which opens to the outside on the sixth segment, discharging there the combined contents of all the mixonephridia of one side.

All oligochaetes have metanephridia, but the modifications of their structure and patterns are many and various (Fig. XIII–4). The organs may be exo- or enteronephric, and the nephrostomes may be open or closed and single or multiple. If enteronephric, each metanephridium may open into a common excretory duct. In the genus *Lampito* this runs along the dorsal wall of the digestive tract and opens into it by a ductule in each segment in which there are nephridia (segment XIX and posteriorly) (Fig. XIII–5). In *Allolobophora antipae* and *Hoplochaetella khandalensis*, the

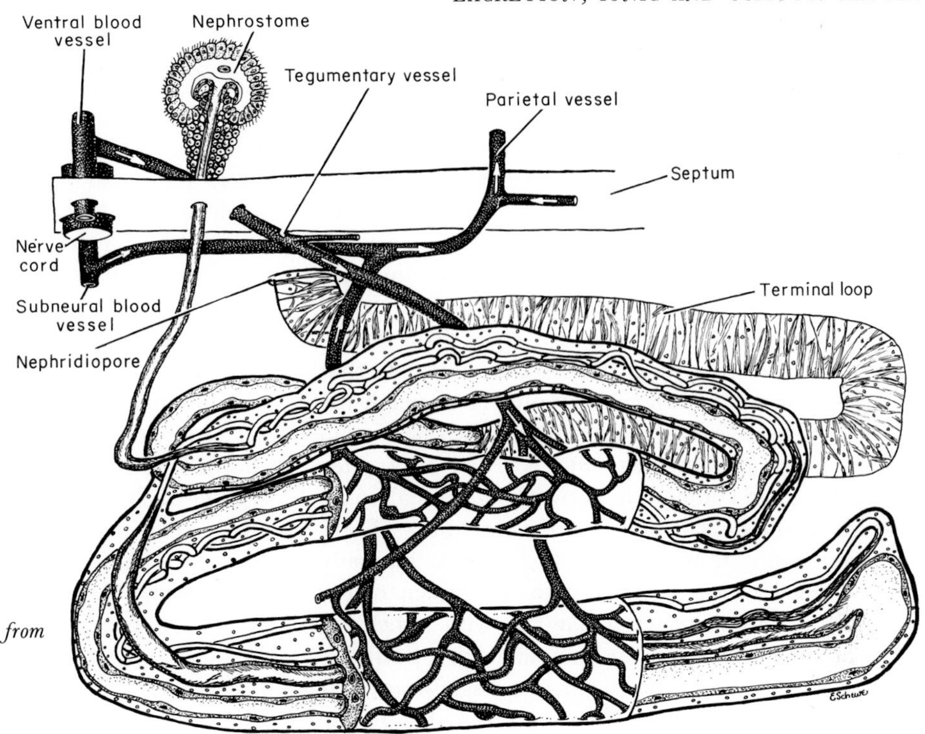

FIG. XIII–4 *A single nephridium from the earthworm* Lumbricus.

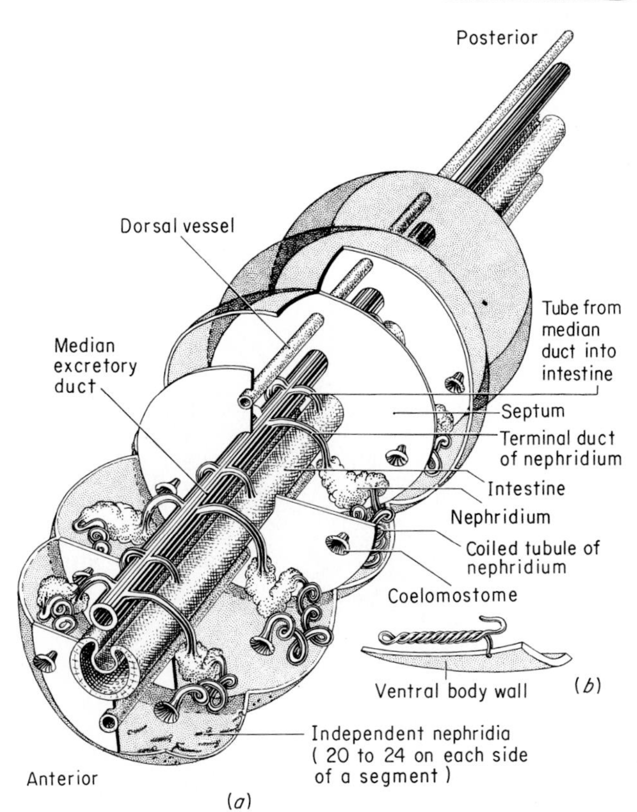

FIG. XIII–5 *Excretory system in* Lampito mauretii. (*After Bahl.*) ***(a)*** *Each segment has a pair of large nephridia whose terminal ducts lead to a common median duct lying dorsal to the intestine. Paired tubules lead from this duct into the intestine. There are also 20 to 24 small independent nephridia on each side of a segment, extending along the underside of the septum and opening through minute pores on the ventral body wall.* ***(b)*** *Detail of an independent nephridium.*

segmental nephridia in the posterior region open into longitudinal canals which run along the inner surface of the body wall and discharge their collective waste at the inner end of the protodaeum, where gut and body wall meet (Fig. XIII–6). Excretory waste in these worms is discharged through the anus. In *Pheretima*, the anterior nephridia lie in segments IV, V, and VI; the canals leading from their multiple terminations join a common duct in each of these segments. The three pairs of ducts so formed run anteriorly and open into the buccal cavity, so that excretory waste from them is eliminated through the mouth. The posterior nephridia of *Pheretima* begin at the septum between segments XV and XVI; they discharge their contents into two longitudinal tubes that lie on the dorsal surface of the intestine. There are cross connections between these tubes, but only one connection per segment with the lumen of the intestine. This connection may be with either the right or the left duct and usually alternates from one to the other along the segments, suggesting that possibly primitively there were two and that in the course of evolution one member of the pair has been lost from each segment. The waste from these posterior nephridia is discharged through the anus. The obvious advantage of an enteronephric system to worms living in arid environments is the opportunity for conservation of water, for the gut provides an extensive area for absorption of water from the excretory fluid.

Open and closed nephrostomes may be found in one individual, and even in one segment of an individual. In *Euthyphoeus foveatus*, for example, in each of the segments posterior to the intestinal diverticulum there is a pair of large open exonephridia lateral to the ventral vessel; and extending along the posterior face of the septum on each side, there is a row of much smaller, closed ones.

Multiple nephridia—that is, those in which the tubule terminates internally in several or many nephrostomes—have presumably arisen through divisions in the cord of embryonic cells from which the coelomic termination was derived (meronephridia). These are extremely diverse in size and in arrangement; they may be exo- or enteronephric and open or closed. There are both open and closed meronephridia in *Euthyphoeus foveatus*, and open meronephridia in *Lampito*. Another type of multiple metanephridia are "tufted nephridia." These may have arisen through branching of the inner end of a tubule or by union of a number of meronephridia. They may also be open or closed, exo- or enteronephric. In *Pontoscolex corethrurus* they are open and exonephric, and in *Pheretima*, closed and enteronephric in segments IV, V, and VI. Each nephridium in these segments has hundreds of terminations discharging into the common segmental canal.

In oligochaetes, the coelomoducts retain their identity and function as gonoducts. Correlative with the tendency in oligochaete evolution to limitation of the number of gonads, and their restriction to a few segments, is restriction of the coelomoducts to the fertile segments. There are usually not more than four pairs of testes and three pairs of ovaries in any individual. Each organ may be served by its own gonoduct (coelomoduct), or those on one side may

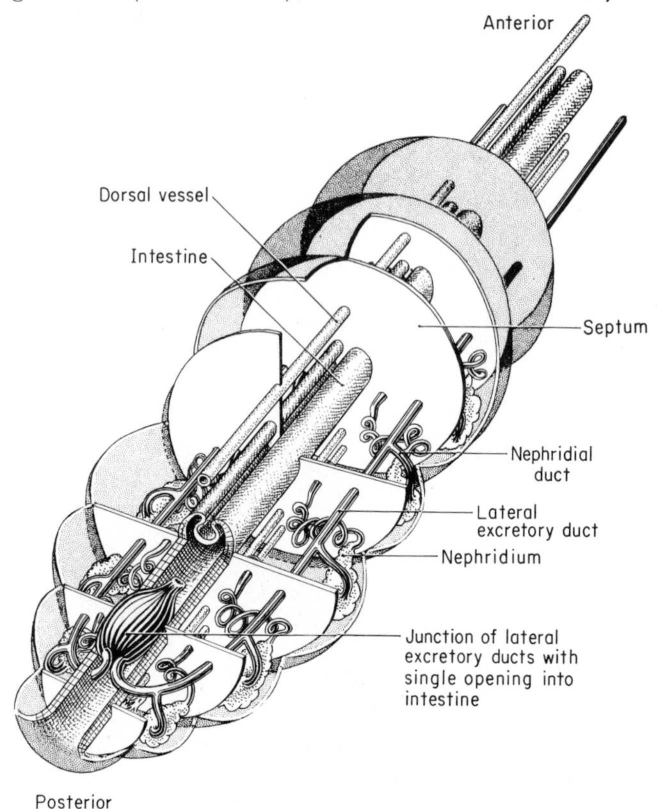

FIG. XIII–6 *Excretory system of* Allolobophora antipae. (*After Bahl.*)

transmit their products through a common duct and a single gonopore. There is thus a separation of the two functions that are combined in the nephromixia of polychaetes.

In leeches there are 10 to 17 pairs of nephridia, confined to the midregion of the body. Each nephridium has a nephrostome that lies in one of the sinuses of the much reduced coelom and leads into an enlarged capsule. The tubular portion is constructed, like the tubules of a protonephridium, of cells placed end to end with a continuous intracellular canal. Fine tubules, also intracellular and embedded in the mesoderm, lead into these canals, which may or may not be connected with the nephrostome and its associated capsule. Each opens to the exterior through a nephridiopore. In members of the family Hirudidae there is neither nephrostome nor capsule, but the blind ends of the tubes lie close to sinus spaces whose coelomic fluid contains hemoglobin in solution.

c. UROCOELES In adult molluscs and crustaceans portions of the coelom, greatly reduced by development of the hemocoele, persist. Two of them have undergone changes that have led to their identification as kidneys or renal organs. In molluscs other than gastropods, the coelomic spaces and their coelomoducts are paired, but in snails torsion has resulted in reduction or total loss of one member of the pair, usually that on the right side. In all the paired organs, connections remain between those of the two sides, and in all of them, paired or unpaired, there is a connection with the pericardial cavity, another of the surviving coelomic spaces. The general form of the molluscan excretory organ is that of a large-bored tube, much of whose inner wall is folded or pleated and whose duct opens exteriorly either into the mantle cavity or outside it. In bivalves the tube is folded upon itself in the form of a U; the walls of the proximal region, which opens into the pericardial cavity and forms the lower limb of the U, are glandular, while those of the distal region, forming the upper limb, are ciliated (Fig. XIII–7*a*). The glandular portion in particular is often referred to as the organ of Bojanus, although this term is also used for the entire organ. In cephalo-

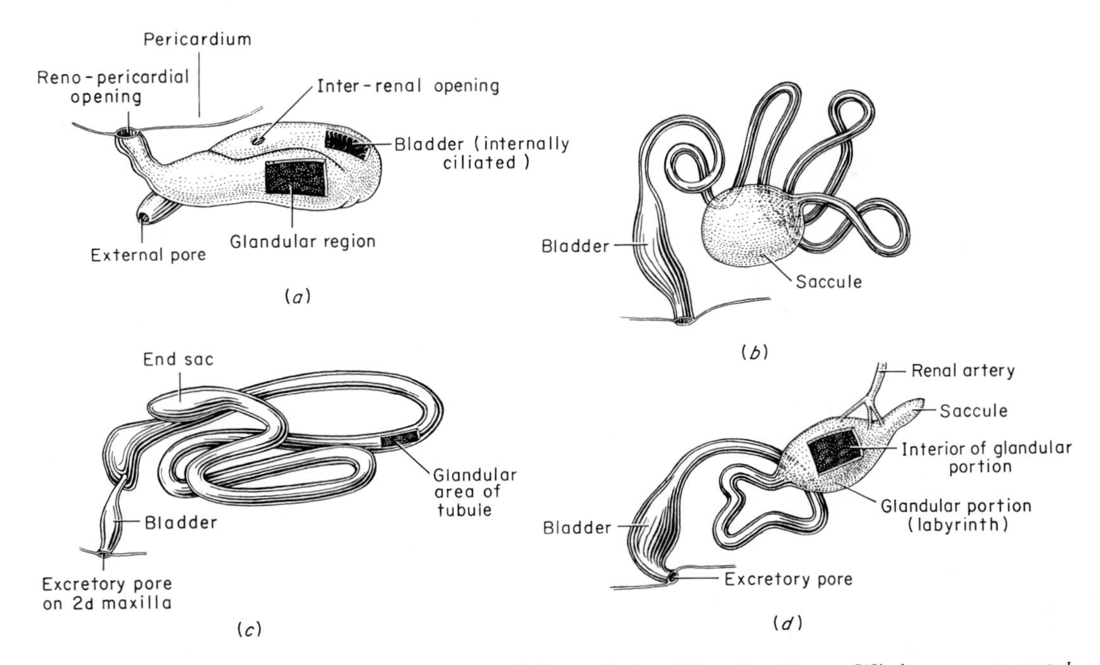

FIG. XIII–7 *Urocoeles.* ***(a)*** *Organ of Bojanus of a lamellibranch mollusc. Windows are represented in the glandular and bladder portions to show the nature of their interiors.* ***(b)*** *Coxal gland of an arachnid.* ***(c)*** *Maxillary gland of the tadpole shrimp* Triops (Apus), *a nostracan crustacean. A portion of the tube is represented as if cut open to show its glandular nature.* ***(d)*** *Antennary gland of a decapod crustacean.*

pods the organs are sac-like in appearance, with the reno-pericardial aperture enclosed within the sac. The glandular portion, which can be seen through the thin walls of the sac, contains numerous branched diverticula of the veins that run through it. These diverticula hang down into the sac and are known as the renal appendages.

The antennal and maxillary glands of crustaceans and the coxal glands of *Xiphosura* and arachnids are also urocoeles, which have not, however, retained connections with the pericardial cavity. The organs in crustaceans are alike in design, the distinction between them being based on the location of the nephridiopore and whether it lies in the basal segment of the antenna or the maxilla (Fig. XIII–7*c* and *d*). Typically, they consist of an end sac and a tubule of varying length, usually with glandular walls. In some species the tubule expands into a bladder before its opening at the nephridiopore. The cephalocarid *Hutchinsoniella* has a pair both of antennal and of maxillary glands, and in eight of its thoracic segments there are structures that resemble them and are open to interpretation as organs of similar function. If this interpretation proves to be correct there would be good evidence that primitively the urocoeles in crustaceans were segmental in their arrangement. Freshwater ostracods have both antennal and maxillary glands in their early stages. The antennal gland, consisting of an end sac and an intracellular duct of only three cells, reaches its maximum development in the fourth larval stage and then degenerates, while the maxillary gland, with an intracellular duct of four cells, persists. In other crustaceans there may be a reversal between juvenile and adult stages. Larvae of many branchiopods and copepods, for instance, have antennal glands, but the adults have maxillary glands. The situation in decapods is the reverse, for a larval decapod has maxillary glands and an adult, antennal glands.

In decapods, the walls of the end sac are composed of a single layer of epithelial cells and are richly supplied with blood from branches of the antennal artery. The walls of the proximal region of the tubule, the labyrinth or green body, consist also of a single layer of cells whose basal membranes are greatly folded and whose apical surfaces have many microvilli. Mitochondria are abundant in the cytoplasm between the basal infoldings, and the cytological picture of this area resembles very closely that of the proximal tubule of the vertebrate kidney. In freshwater species a nephridial canal of varying length leads from the labyrinth to the bladder or directly to the excretory pore. The cytology of this canal, as revealed by electron microscopy, is much like that of the distal tubule of the vertebrate kidney.

The urocoeles of *Limulus* and arachnids are known as coxal glands, for their openings are not located on the head appendages as they are in crustaceans but on the coxae of different segments in different species. They are not homologous with the coxal glands of some crustaceans, which are not urocoeles but aggregations of glandular cells in the bases of the legs. There are four pairs of coxal glands in *Xiphosura*. The four on each side open into a common lateral chamber, which leads into a coiled tubule that expands distally into a bladder and opens at a nephridiopore on the coxa of the posterior pereiopod of that side. The proximal portion of each organ is an expanded sac in close association with a blood space or sinus. In embryo scorpions there are five pairs of segmental coelomic cavities lying in segments 3, 4, 5, 6, and 8. All of them have coelomoducts, but only those in segment 5 reach the surface and open to the exterior (Fig. XIII–7*b*). This pair alone persists in the adult as excretory organs for the other coelomoducts disappear in the course of development and the coelomic sacs are resolved into the mesenchyme. Spiders have two pairs of glands, one pair opening on the coxae of the first pair of legs, the other on the coxae of the third pair. In web-building and hunting spiders, considered more highly evolved than others in the class, the coxal glands have become reduced to greater or lesser extent. Their function has been taken over, presumably, by Malpighian tubules, which are present also in scorpions that have well-developed coxal glands. Opiliones have one pair of such glands, opening between the coxae of the third and fourth pair of legs, and acarinae have one to four pairs. In some mites, there is also a pair of Malpighian tubules, but in others, there are only Malpighian tubules and no coxal glands.

Electron microscopy of the coxal glands of scorpions has revealed an ultrastructure similar to that of the vertebrate nephron. The gland consists of a central, thin-walled capsule, homologous with the end sac of the crustacean excretory organ, which constitutes the medulla of the scorpion's excretory organ. Its cortex consists of a long, thin tubule that is twisted and coiled around the medulla. The wall of the tubule is made of a single layer of flat, prismatic cells with an extensive endoplasmic reticulum and with mitochondria aligned between the membranes. The free borders of the cells are extended into many microvilli, a feature associated with actively secreting or absorbing cells.

d. MALPIGHIAN TUBULES Malpighian tubules are typical of insects and are found in all hexapods, except collembolans, aphids, and some thysanurans. They are found also in myriapods and some arachnids (Fig. XIII–8). They are slender, blind tubules extending in pairs from the junction of mid- and hindgut into the hemocoele within which they are often greatly coiled and twisted. Whether they originate from the proctodaeum, into which they open, or from the mesenteron, or from the undifferentiated region between proctodaeum and mesenteron is not yet certain. While the primitive number in insects seems to have been six, in modern species there may be only two, or as many as 250, the number in *Schistocerca*. In Orthoptera and a number of other orders, the distal ends of the tubules are free and the junction of the proximal ends with the gut is simple, with or without ampullae. In Coleoptera and Lepidoptera, the distal ends are reflected back along the gut and connected to its surface in the region of the rectal pads by a membrane. Histologically, the tubules are composed of large epithelial cells, appearing in cross section as a ring of six or eight, with microvilli on their luminal borders extending well into the tubules. These cells are enclosed in an elastic tubular membrane and, in some species, in thin layers of longitudinal and circular

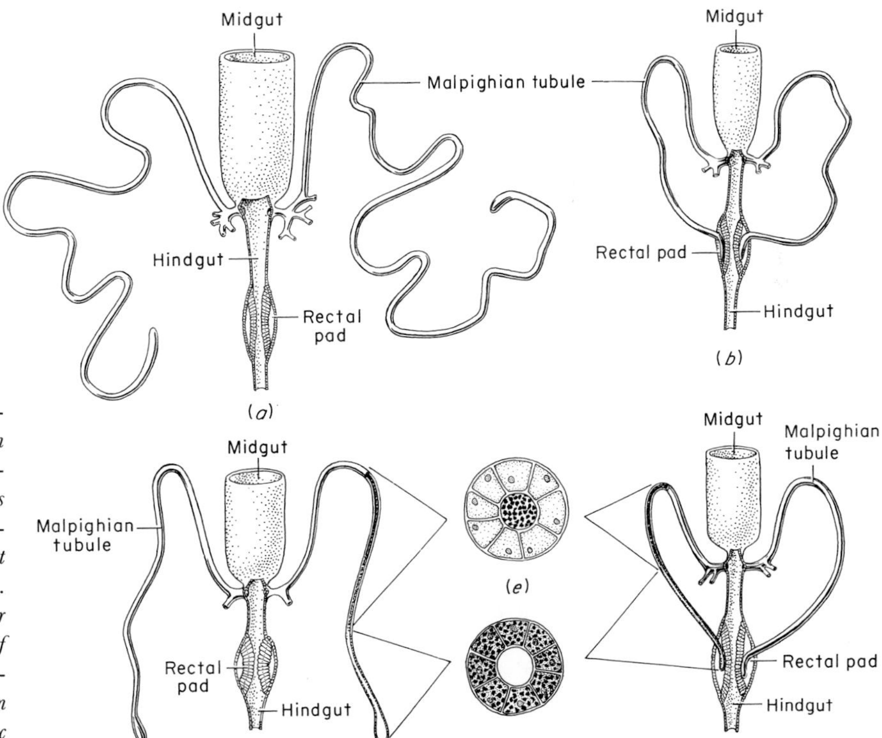

FIG. XIII–8 *Malpighian tubules in insects. (After Patton.)* ***(a)*** *Orthopteran type, with tubules ending freely.* ***(b)*** *Coleopteran type, with distal ends of the tubules inserted on the rectal glands.* ***(c)*** *Hemipteran type, with tubules ending freely but showing differentiation along their length.* ***(d)*** *Lepidopteran type, showing linear differentiation of tubules and insertion of their free ends on rectal glands.* ***(e)*** *Diagrammatic cross section of proximal region of tubules* c *and* d. ***(f)*** *Diagrammatic cross section of distal tubules in* (*c*) *and* (*d*).

muscle fibers whose contraction causes movement of the walls. The epithelial cells, in some species, seem differentiated along the length of the tubule. In *Rhodnius*, and probably in all other hemipterans, the cells of the upper two-thirds of each tubule, which is free distally, are full of granules and the fluid in the lumen is clear. The cells of the proximal one-third, on the other hand, are almost free of granules and the fluid in the lumen is full of them. A similar distinction is evident in the tubules of many Lepidoptera.

e. OTHER MECHANISMS Nematodes are peculiar among "aschelminths" in being without typical protonephridia. In some free-living genera, such as the phytophagous *Monohystera* and the marine *Enoplus*, there is a single large cell lying in the pseudocoele below the pharynx. This cell, known as a rennette cell, opens to the exterior through a pore, is glandular in appearance, and is believed to be excretory in function (Fig. XIII–9). Its role in excretion and osmoregulation has been established in some freshwater species. In other species the rennette cell, or a pair of them, is associated with a system of tubules or canals. In the free-living species of *Rhabditis* and in *Rhabdias*, a parasite in the lungs of amphibia and reptiles, two lateral, longitudinal canals, connected with a pair of rennette cells by a cross bar, run anteriorly and posteriorly, so

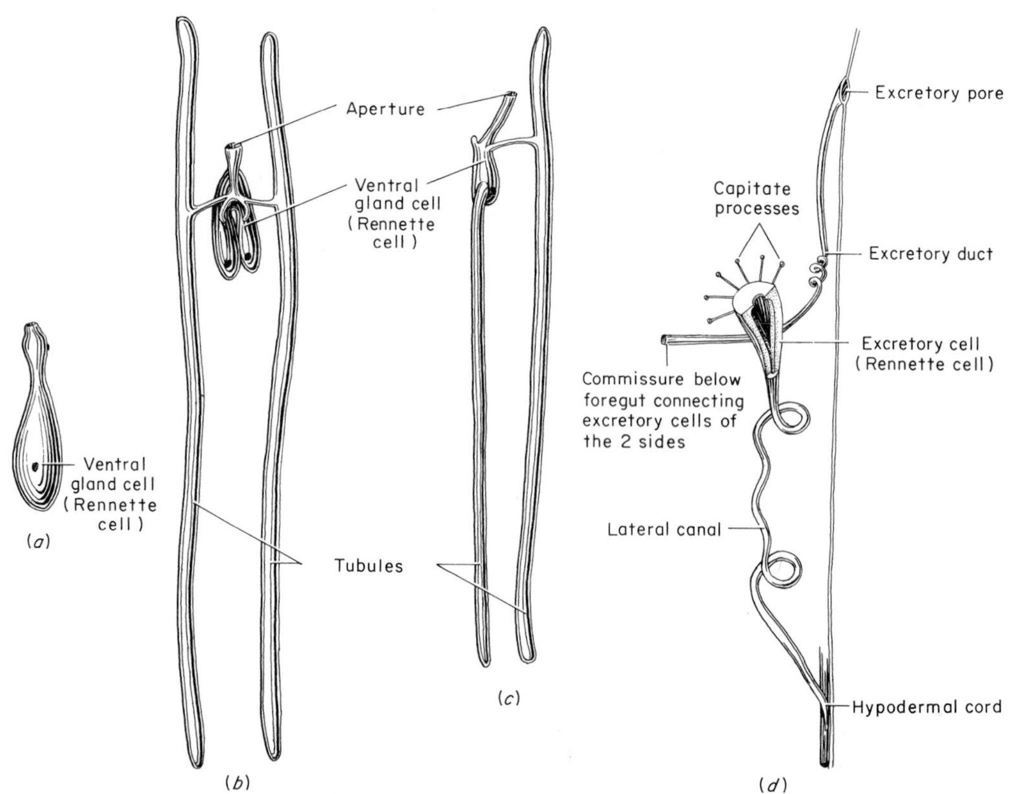

FIG. XIII–9 *Excretory systems in nematodes. (a, b, and c after C. Lee; d after L. Smith.)* ***(a)*** *Glandular type, consisting of a single ventral gland cell (rennette cell), opening to the exterior through a pore. Found in* Chromadorina, Monhysterina, *and* Enoploidea. ***(b)*** *Tubular type, as found in* Rhabditis, Rhabdias, *and* Strongylina *in general, with longitudinal tubes connected by a cross bar in the form of the letter* H. ***(c)*** *Tubular type characteristic of ascarids, with tubules reduced anteriorly.* ***(d)*** *Left half of the excretory system of* Panagrellus redivivus. *The excretory cell has been represented as if cut open to show the cilia inside.*

that the system has the form of the letter H. The canals open to the exterior by a single pore at the end of a short tube extending anteriorly from the cross bar. In *Ascaris* there is but a single rennette cell, and the part of the tubular system anterior to the cross bar of the H is much reduced so that the system looks like an inverted U. In other genera, such as the parasitic oxyuroids, there are no rennette cells, but only canals, and in others, such as *Anisakis*, parasitic in the stomachs of marine mammals and of birds, there is only one rennette cell and one longitudinal canal. In the free-living species *Panagrellus redivivus*, which has been successfully cultured in fermenting whole wheat flour, capitate processes, of unknown function, radiate from the anterior borders of the two rennette cells, and the tubes leading from them are much coiled and applied to the wall of the gut until they diverge from the mid-line and become embedded in the lateral hypodermal cords. A single duct, originating from the left rennette cell, extends forward and opens at a pore anterior to the pharynx (Fig. XIII–9*d*). It is probable that the nematode excretory system had its origin in the rennette cell and not in a protonephridium.

In many invertebrates, localized or migratory cells store insoluble wastes and have, therefore, been given the name of nephrocytes, functioning presumably as storage kidneys, or kidneys of accumulation. In molluscs the pericardial epithelium is in places thick and glandular in appearance. In the more primitive gastropods and in many lamellibranchs, these glandular areas lie in the pericardial wall above the auricles; in the more highly evolved gastropods, they lie on the side walls of the pericardium and in some lamellibranchs, as exemplified by members of the freshwater family Unionidae, they form extensions from the pericardium to the tissues of the mantle and are known as Keber's organs. In cephalopods, they constitute the glandular appendages of the branchial hearts.

The chlorogocytes of oligochaete annelids and the botryoidal tissue of leeches are believed to function as nephrocytes. A similar function is attributed to the fixed clusters of peritoneal cells, called urn cells, in sipunculids. Each of these clusters, which enclose a central vacuole, is capped by a single ciliated cell. The urns may become detached and move about in the coelomic fluid, and similar free urns have been seen in echiuroids. In arachnids, large cells clustered in parts of the prosoma and the opithosoma accumulate injected dyes, and so are thought to be nephrocytes. Similar cells are found in the axes of the gills and the bases of the legs of crustaceans; the latter are known as coxal glands. Zenker's organ in isopods contains concretions of urates and uric acid and acts, presumably, as a kidney of accumulation. In insects the pericardial glands are considered nephrocytes.

No tubular system to which an excretory or an osmoregulatory function can be attributed has been demonstrated in cnidarians, ctenophores, ectoprocts, or echinoderms. In apodous holothurians such as *Synapta* there are ciliated funnels or urns attached to the bases of the mesenteries. Coelomocytes containing particulate and probably nonparticulate waste pass through the funnels into the body wall. Coelomocytes very likely play some role in the transport and elimination of soluble waste in all coelomate invertebrates, as do fixed cells that act as repositories for products unusable for further metabolic events.

2. Excretory products

The fluid in these tubular systems is known as urine and is a mixture of substances. Analyses of these mixtures and determinations of the components in them that are actually excretory products present a number of difficulties. Chief among these is the means of collecting the urine in sufficient quantity and without contamination from other sources. This, as with the collection of blood and hemolymph, has been achieved in only a limited number of invertebrates whose excretory organs are especially suited to it. In many cases the nature of the material excreted has been deduced from analyses of the medium in which an animal or group of animals has been kept for a sufficient length of time to have its composition altered by the accumulation of excreted waste. This does not ensure that all the substances are excretory products, for some may have their origin in surface secretions or in digestive waste. This is perhaps a matter of greater importance in consideration of the function of presumed excretory organs than of the phenomenon of excretion as such, for any metabolic product that is discharged from the body by any route is essentially excreted from it. Secretion of integumentary coverings or coats, deposition of pigments in cells or outside them, of guanine in chromatophores and in the cells of the reflecting layers of photogenic organs, and of pterines in the wings of butterflies are all illustrations of the removal from a living system of waste products produced by it yet retained in association with it either because they are neutral in effect or because they have proved to be of positive survival value.

Ammonia is the product of nitrogenous metabolism that is most generally excreted by invertebrates. Although they are on the whole more tolerant of ammonia than vertebrates are, it is potentially toxic to them and they must get rid in one way or another of much of that which results from deamination and other reactions. A certain concentration of ammonia in body and tissue fluids is valuable to any animal in preserving the acid-base balance and in maintaining an alkali reserve. This concentration is, in general, higher in the body fluids of invertebrates than in those of vertebrates. Concentrations of 0.7 to 2.0 mg/100 ml (0.07 to 0.2 mM) have been found in the hemolymph of snails and of 1.6 to 1.8 mg/100 ml (0.16 to 0.18 mM) in that of the lobster *Homarus*. In *Sepia*, concentrations of 2.8 to 4.8 mg/100 ml (0.28 to 0.48 mM) have been reported. Yet levels above 0.1 mg/100 ml (0.01 mM) of fluid are rarely found in cold-blooded vertebrates, and even lower levels are characteristic of mammals. A concentration of 5 mg/100 ml (0.5 mM) of blood, only slightly above that customary in *Sepia*, is, for example, lethal to a rabbit.

Ammonia is very soluble in water and any amount above the level of an animal's tolerance is either flushed out of its body in solution or detoxified by conversion to urea or uric acid. Urea is also very soluble in water but is a less toxic compound than ammonia. Uric acid is only slightly soluble and essentially nontoxic at the concentrations at which it can be held in solution. So definite are the relations that have been established between the nature of the primary excretory product and the conditions of an animal's environment that it is apparent that nitrogen metabolism is an important adaptive character. The mechanisms possible to a species, at least insofar as production of its waste products are concerned, have undoubtedly been significant in determining the range of habitats possible to it and the extent to which it might explore and take advantage of new ones.

Ammonotelism is associated with aquatic environments, since the elimination of ammonia requires a relatively large volume of water for its dilution when being flushed out of the body, as well as for its dissipation afterwards. Ammonotelism is characteristic particularly of invertebrates that live in salt, brackish, or freshwater, such as protozoans, cnidarians, sipunculids, and polychaete worms; and freshwater leeches; bivalve, nudibranch, and septibranch molluscs; and crustaceans and echinoderms. Ureotelism is characteristic of animals that live where water is neither particularly abundant nor unduly restricted. Few invertebrates are primarily ureotelic, although most of them excrete small quantities of urea and some oligochaetes may excrete it as a primary product under certain conditions. Uricotelism is associated with an environment in which water is scarce and its conservation in the body important. Uric acid is the primary excretory product of terrestrial gastropods and of insects, although certain mutant strains of *Drosophila* are deficient in the enzyme xanthine dehydrogenase, and hypoxanthine is their principal excretory product. That of arachnids is guanine, a derivative of nucleic acids that in most animals undergoes further degradation. Chromatographic and spectrophotometric analyses of the excreta of 34 species of spiders have established that 78 to 87 percent of the nitrogen they excrete is bound in guanine. This is probably an economical way for animals with naturally a high protein diet to eliminate nitrogen, for four N atoms are bound in the purine ring of guanine and another is held in the amino group. Guanine is even less soluble in water than uric acid, a property that would also be of importance to terrestrial animals as sensitive to water loss as are arachnids.

Other compounds may also be present in varying amounts. Crustaceans and echinoderms, for example, regularly excrete appreciable amounts of amino acids. This procedure, like the excretion of other organic acids and molecules of large size, is from the point of view of energetics a wasteful one, for such compounds represent a source of energy that is lost unless they are fully oxidized.

Like other aspects of metabolism the excretory picture and excretory products may alter with age and with other changes in the conditions of an animal's life. Diet, naturally, is an important factor in influencing the nature of the products, for an animal whose food is largely carbohydrate has less nitrogen and fewer nitrogenous products to excrete than one whose diet is rich in protein. The gribble *Limnoria*, which feeds on wood, has no excretory organ at all nor have any products of the catabolism of nitrogenous compounds been detected in it. *Paramecium* excrete ammonia when on their natural diet of microorganisms, but do not excrete nitrogen in any form when on experimental diets consisting exclusively of starch. Earthworms excrete ammonia primarily when on a diet of decaying leaves, but when they are starved the principal product is urea. A starved *Eisenia* will excrete almost twice as much urea a day as a fed one, a starved *Allolobophora* about four times as much, and a starved *Lumbricus* 30 to 40 times as much. An increased rate of urea production in response to starvation is also exhibited by rats, but earthworms are unique in their ability to shift from ammonotelism to ureotelism at all stages of feeding and starvation. All the enzymes of the urea cycle have been demonstrated in cell-free extracts of tissue from whole guts of *Lumbricus*, four of them in the extramitochondrial fraction. These apparently come into action when the level of nutrition falls and are inactivated again when it rises. Shifts from ammonotelism to ureotelism may occur also during the embryogenesis of vertebrates, but these are permanent, not reversible, ones, and once the urea-forming pattern has been established, urea remains the primary excretory product of most fishes, amphibia, and mammals.

There are changes also in the excretory products during the life cycle of some invertebrates. It has been shown, for example, that the excretion of nitrogen is greatly reduced in the crab *Carcinus maenas* just before a molt, and in females, just before oviposition. Seasonal changes may also occur and are most marked in aestivating, hibernating, and diapausing animals. The Indian apple snail, *Pila globosa*, is most active during the rains and aestivates in the dry season by burrowing into the mud and remaining quiescent there. Measurements of the uric acid in the excretory organs of specimens collected during the rainy season showed a content of 7.68 mg uric acid/g dry weight of the animal. Similar measurements of specimens after 2 months of aestivation in damp jars in the laboratory showed a maximum accumulation of 144 mg uric acid/g dry weight. When these snails were given access to water again, the uric acid content of their excretory organs dropped to the level of those originally collected during the period of activity.

3. Tests for function of excretory areas

The function of tubular systems and other presumed excretory areas can be tested in a number of ways. Extirpation, except for organs that are compact and localized, is not feasible, but the effects of blocking the external orifices of the tubules or, in some instances, of ligaturing the tubules can be observed. Another method is introduction into an organism, by injection or other means, of soluble, nontoxic substances, which it does not metabolize to an appreciable extent, if it does so at all. Dilute solutions of "vital" dyes, such as neutral red and Janus green, and of fluorescent dyes, have been used. If, after sacrifice of the animal, the dye has been found to be taken up or concentrated by particular organs or cells, it has, by definition, been excreted by them. Recovery of an injected substance in the fluid in the lumen of a tubular system, or in that eliminated from its external orifice, also provides evidence of excretory function. Substances that have been successfully used in tests of this kind on vertebrates have also been used on invertebrates. Among these are the polysaccharide inulin, the pentose sugar xylose, and the nitrogenous compound creatinene. Inulin, a polysaccharide of plant origin and composed of some 30 or more fructose units, is known to filter through the glomerulus of the vertebrate kidney, but it is not resorbed during passage of the filtrate along the tubules, probably because of the large size of the molecule (mol. wt *c.* 5,000). Its injection into the blood, until a known concentration is reached, and measurement of the proportionate amount collected in the urine provide evidence of filtration by the system and data for calculation of the amount of fluid filtered in a given time. Compounds with radioactive atoms have also been used, particularly for study of the movement of materials across cell membranes.

More direct and precise methods are through collection and analysis of the tubular fluid. These can only be applied to animals whose presumed excretory organs are so constructed that a fine cannula or catheter can be inserted in them and enough fluid withdrawn to be adequately analyzed. Micromethods that are being developed both for obtaining samples of the fluid and for analyses of minute quantities are, however, enlarging the area of exploration. These methods, whether macro or micro, make possible determinations of the difference, if any, in composition of the fluid as it passes through the tubular system, a matter of great importance to full understanding of its operation and contribution to the preservation of internal stability.

4. Factors in the operation of excretory systems

Separation of material from body or tissue fluids necessitates its passage across cell membranes. Water, salts, and other solutes of appropriate physicochemical constitution also pass across them, so that the fluid in the excretory tubules may contain, initially at least, much that is of value to the organism as well as waste to be removed. This passage of material may be effected by means of diffusion, active transport, secretion, or filtration, provided that there is a head of pressure adequate for this. The tendency to assume that invertebrate mechanisms operate in the same way as vertebrate ones has led to the implication that the urine of invertebrates is produced in the same way as it is in the vertebrate kidney, by filtration, secretion, and selective reabsorption. Yet in only a limited number of invertebrate species are conditions known to be possible for ultrafiltration to take place between body fluid and tubular system, like that known to occur between glomerulus and Bowman's capsule of the vertebrate nephron. Such conditions do prevail between the vascular system and the urocoeles of molluscs and some crustaceans, such as *Astacus*, and probably also in annelids with open nephridia. But it is a question whether filtration is a factor in urine formation in invertebrates with protonephridia or Malpighian tubules.

In addition to filtration, material is known to enter the tubules of the vertebrate nephron by secretion. This, in the terminology of renal physiology, means transport of substances in solution across cells from the tissue or body fluid to the lumen of the tubule. In this sense, it does not imply any transformation of the substance such as that associated with secretion in its more general meaning: the elaboration and discharge of specific products by a cell or a group of cells. Another recognized factor in the operation of the vertebrate kidney is resorption of certain substances as the fluid passes along the tubular system, with consequent alteration of the composition of the initial filtrate. This filtrate must have essentially the same composition as the blood, without its proteins and other unfilterable molecules. Conservation of material of value to the animal is accomplished by resorption, the passage of substances across cells in a direction opposite to that of secretion. It is of the utmost importance in the restoration of nutrients that have filtered into the tubules, as well as in the restoration of water and salts which would otherwise be lost and lead to disruption of the ionic and osmotic stability of the body.

These problems are also common to invertebrates, but are met by them in a variety of ways. In some, the means are essentially like those characteristic of vertebrates, but in others, separation of wastes and restoration of useful and essential materials are accomplished by tissues other than those of the tubular organs. And in some there is no evident means of such regulation.

B. WATER; IONIC AND OSMOTIC REGULATION

1. Water

The Greek poet Pindar (522?–443 B.C.) wrote "the best of all things is water." It is indeed the best and most essential of things to living systems on earth. It is the universal biological solvent and the medium in which most of the reactions of cellular metabolism take place. Other physical properties of water, especially its thermal properties, enhance its value to living systems. Animals as well as plants contain a large proportion of water both in their cells and in their extracellular fluids. The water content of animal tissues may be as great as 80 to 95 percent by weight, varying with the type of tissue, the age and metabolic activity of the individual, and similar factors. In general, the water content of the more simply organized aquatic species is greater than that of terrestrial ones. The body of a scyphozoan, for example, may be 95 to 98 percent water, while that of an earthworm is about 84 percent. Similarly, water accounts for 60 to 90 percent of the weight of an insect larva and only 45 to 65 percent of that of its imago. Moreover, a current of water is virtually continually flowing through the body of every animal. Some is taken in with the food; some is formed in its metabolic reactions; and some is lost through its excreta and, in terrestrial species, through evaporation from exposed body surfaces. It is vitally important to an organism that its water content, both intracellular and extracellular, be nearly constant, or at least that it should vary only within a moderately small range, so that neither its tissues nor its body fluids are diluted or concentrated beyond the limits of its tolerance. The limiting factor to increase in water content is the dilution at which enzyme systems can effectively operate and carry on sufficient metabolism for active life. Different organisms have different limits of tolerance both for water gain and for water loss. Some of the smaller invertebrates can withstand long periods of dehydration and emerge from a dehydrated or encysted state with full vitality when water is again available. Well-known illustrations of this are found among protozoans, rotifers, copepods, and cladocerans. Moss-dwelling tardigrades, or water bears, have been known to survive desiccation for as long as 7 years, and larvae of the chironomid *Polypedelium vanderplanki* for 18 months in the laboratory. These midges breed in rock pools in Northern Nigeria that are alternately flooded and dried up. When the pool is dry the larvae lose all their moisture and reduce their respiration markedly, although some air is retained in the tracheae. In experimental desiccation in the laboratory, 90 percent of the individuals became fully reactivated after 18 months of apparent total dehydration. But, in general, there are ecologically few factors that limit the distribution of animals as much as the availability of water, and most of them have means of controlling the water content of their bodies, as well as the distribution and concentration of salts and ions in their tissues and their body fluids. The precise habitats of animals result, however, from the combination of many factors. Although an animal may be physiologically able to inhabit a number of different environments, one may be selected over another. The availability of food is one factor, correlated with the development of special feeding habits; the presence or absence of predators and a variety of other conditions may make one area more desirable than another. The mole crab, *Emerita*, for example, although a nonregulator in respect to its water and ionic content, can tolerate salinities of 75 to 125 percent seawater for at least 24 hr, so that it probably could live in estuaries and in tide pools. But its feeding habits cause it to live on beaches at the tide level where the outgoing waves wash over it.

2. Ionic regulation

Ionic regulation is a very important aspect of the relations of an organism with its environment and its adjustment to it. It expresses the control that an organism has over the entrance or exit of ions from its cells and body fluids, so that it can retain some at higher or lower concentrations than they are in the external environment and can, indeed, preserve a higher or lower concentration of certain ones in its cells than in the extracellular fluids, and vice versa. Most invertebrates have some degree of ionic regulation, although echinoderms as a group seem to have little and are restricted to marine environments and even to those where the salinity is about the same as that in their coelomic fluids. The composition of these coelomic fluids is very nearly that of "average" seawater. Most sponges and cnidarians are similarly restricted, although a few species of each have been able to meet the conditions of life in freshwater and keep the ionic concentration of their cells higher than that of their surroundings and different from it. Some degree of ionic regulation is shown by the marine scyphozoan *Aurelia aurita*, whose tissue fluid has slightly higher potassium and considerably lower sulfate concentrations than the water in which it lives. Regulation of the concentration of particular ions is more marked in other marine invertebrates from which data have been obtained. The concentrations of K^+ and Ca^{++} are, for example, higher in the hemolymph of the mussel *Mytilus* than in the surrounding seawater, and in the cephalopod *Sepia*, the concentration of K^+ is considerably higher and that of sulfate very much lower. The coelomic fluids of polychaete worms also have a higher concentration of K^+ than has seawater. Marine arthropods show a wide range of regulatory capacity in respect to different ions. Differences in the ionic composition of their hemolymphs and seawater are most apparent in respect to Na^+, K^+, and Ca^{++}, of which the level is usually higher in the hemolymph than in the water, and to Mg^{++} and SO_4^{--}, of which the level is usually lower in the hemolymph than in the water. The concentration of Cl^- is usually about the same in both fluids. There is little variation in the distribution of ions among different individuals of the same species, but a good deal between those of different species (Table XIII–1). The concentration of Mg^{++} is, for example, greater in the hemolymph of the spider crab *Maja squinando* and of the sponge-bearing crab *Dromio vulgaris* than it is in that of the lobsters *Homarus gammarus* and *Palinurus elephas*. Possibly this reflects differences in their ways of life, for spider crabs are notably sluggish in activity and in response, while lobsters are somewhat more lively. High concentrations of Mg^{++} are known to have an inhibitory effect upon neuromuscular transmission, and the concentration of Mg^{++} in the hemolymph of *Maja* is six times greater than it is in the hemolymph of *Homarus*.

Ionic concentrations in the body fluids of freshwater species are lower than in those of their marine counterparts, but invariably higher and often different from the concentration in their environments. Accumulation and concentration of ions is obligatory for animals living in freshwater, where the content of dissolved minerals is always less than in the ocean and, indeed, may be almost negligible. In Table XIII–2 are shown the concentrations of the major cations in the hemolymph of two representative freshwater invertebrates, the bivalve *Anodonta* and the arthropod *Cambarus*, and the concentrations in the waters in which they customarily live.

TABLE XIII-1 Concentrations of the Major Inorganic Ions in the Body Fluids of Representative Molluscs, Annelids, and Crustaceans, and the Ratios between Them

	Na^+, m*M*	K^+, m*M*	Ca^{++}, m*M*	Mg^{++}, m*M*
Mollusca				
Mytilus	502.00	12.50	12.50	55.60
	41 :	1 :	1 :	5
Sepia	460.00	23.70	10.40	56.90
	48 :	2 :	1 :	5
Annelida				
Arenicola	459.00	10.10	10.00	52.40
	46 :	1 :	1 :	5
Pheretima	41.00	19.00	4.00	3.00
	14 :	6 :	1 :	1
Arthropoda				
Cancer	459.00	10.00	12.00	22.00
	46 :	1 :	1 :	2
Homarus	455.00	9.00	17.00	9.00
	50 :	1 :	2 :	1

TABLE XIII-2 Concentrations of the Major Cations in the Hemolymph of the Mussel *Anodonta* and the Crayfish *Cambarus*, and in the Mediums in Which They Live

	Na^+, m*M*	K^+, m*M*	Ca^{++}, m*M*	Mg^{++}, m*M*
Anodonta sp.	13.90	0.28	11.00	0.31
Freshwater medium	0.48	0.059	2.70	0.375
Cambarus	146.00	3.90	8.10	4.30
Freshwater medium	0.65	0.01	2.00	0.21

Terrestrial invertebrates have the double problem of conservation of water as well as of ions, for they are in almost constant danger of dehydration and the consequent concentration of their body fluids beyond functional limits. They do, however, maintain ionic concentrations of the same order of magnitude as fresh-water species, obtaining the salts largely from the food that they eat and the water that they drink. The concentration of Na^+ in the blood of the earthworm *Pheretima* is, for example, 41 m*M*; of K^+, 19 m*M*; of Ca^{++}, 4 m*M*; of Mg^{++}, 3 m*M*; and of Cl^-, 14 m*M*. There are no significant differences in the content of K^+, Ca^{++}, and Mg^{++} in the blood of an Indian scorpion *Heterometrus* from that of other terrestrial animals, but its content of Cl^- is higher than that of Na^+, a reversal of the conditions found in most insects and in birds and mammals. The Cl^- content, moreover, is higher in females than in males, which seems also to be the case in the shore crab *Carcinus maenas* and the field crab *Paratelphusa* sp.

The composition of insect hemolymph has received especial attention from comparative physiologists. Analyses of it obtained from a number of species in several orders show that there is great variation in its ionic composition from species to species. The sample figures presented in Table XIII–3 are indicative of this diversity and of the extremes in values that have been obtained. Perhaps these are not of great significance from the comparative point of view, since the insects were in different stages of development and their nutritive and physiological states unspecified, but they are illustrative of some of the problems that are raised by this diversity. Some of these are concerned with the nature of the tissues bathed by the hemolymph and with their accommodation to such a wide range of environments.

Potassium levels seem, in general, characteristically higher in the hemolymph of phytophagous insects than in that of carnivorous or omnivorous ones. Comparison of the K^+ concentration in the hemolymph of the blood-sucking bug *Triatoma* and the leaf-eating grasshopper *Stenobothrus* show this difference (Table XIII–3). It is a predictable one since leaves and plant juices contain relatively more potassium than animal tissues and fluids, but it is an interesting and provocative one, especially in relation to the generally accepted ideas of neuromuscular functioning. In vertebrates and in other invertebrates, as far as is at present known, neural mechanisms depend upon

TABLE XIII-3 Concentrations of Major Inorganic Ions in the Hemolymph of Various Insects and the Ratios between Them

	Na^+, m*M*	K^+, m*M*	Ca^{++}, m*M*	Mg^{++}, m*M*
Orthoptera				
Locusta migratoria	60	12	17	25
(N)	5 :	1 :	1 :	2
Stenobothrus	61	62		
stigmatious (A)	1 :	1		
Periplaneta	161	7.9	4	5.6
americana (A)	40 :	2	1 :	1
Hemiptera				
Cimex lectularius (A)	139	9		
	15 :	1		
Triatoma magister	133	5		
(A)	26 :	1		
Diptera				
Stomoxys calcitrans	128	11		
(A)	11 :	1		
Aedes aegypti (L)	100	4.2		
	25 :	1		
Lepidoptera				
Bombyx mori (L)	14	40	18	50
	1 :	3 :	1 :	3
Ephestia kuehniella	17	60		
(L)	1 :	4		
Coleoptera				
Melolontha vulgaris	6	49		
(A)	1 :	8		
Tenebrio molitor (L)	86	45		
	2 :	1		
Hymenoptera				
Apis mellifera (L)	6	24	4	8
	1 :	6 :	1 :	2

A, adult; L, larva; and N, nymph.

an extracellular sodium-potassium ratio (Na/K) that is greater than 1, and an intracellular ratio that is less than 1. In human serum, for example, Na/K = 29 and in muscle, 0.33. When neuromuscular events are initiated, K^+ moves out of the tissue into the surrounding fluid and moves back in again in the inactive phase, restoring the gradient between cell and medium that puts the tissue in a state of excitability. Conduction in isolated nerves of most animals can be blocked by the addition of potassium to the standard saline solution in which they are kept. But those of *Locusta migratoria* are exceptional in that they tolerate concentrations of K^+ as great as 140 m*M* for as long as 2 to 3 hr before conduction stops. In vivo, concentrations that would block it in most other animals are natural ones to them, for their usual daily intake of potassium is 4.5 mg/day. This has been calculated from the area of grass (*c.* 25 cm^2) that a fifth-instar nymph eats in the laboratory in a 12 hr period. Since these insects feed intermittently and irregularly, there must be successive rises and falls in the level of K^+ in their hemolymph; indeed, it has been shown that during a short period of starvation the K^+ concentration can decrease as much as 50 percent. These fluctuations may have a direct bearing upon the excitation of their muscles and so of their overt behavior. It has been postulated that high levels of K^+ in the hemolymph during feeding keeps them in a state of locomotor inactivity, but that when this level falls the neuromuscular mechanisms of the extensor tibialis of the legs come into operation and the insects begin to hop. This is the regular means of linear progression for nymphs on the march, so that all those that reach the same physiological state in respect to depletion of K^+ in the hemolymph move in concert from an exhausted feeding ground. Thus mass migration to a new one begins.

It has been postulated, further, that all insects whose hemolymph has a high K^+ content lead lives of comparative inactivity and are sluggish when they do move. This may be of selective advantage, enabling them by quiescence to escape the notice of possible predators and by activity to move when they do not ingest enough potassium to keep a hemolymph-tissue ratio conducive to inactivity, so that they are saved from starvation by migration.

The peculiarities of insect hemolymph, in respect to ionic concentration, are by no means limited to those of Na^+ and K^+. The blood of all vertebrates and the body fluids of many aquatic invertebrates show molar ratios of $Na^+/K^+/Ca^{++}/Mg^{++}$ that are approximately 50:1:1:1, and the relative concentrations of the bivalent cations in the tissue-fluid relationships are quite as important as those of the monovalent ones. Reference to Table XIII–3 will show that while the cation ratios in *Periplaneta*, for example, are more or less orthodox, those in *Bombyx* and *Apis* are distinctly unorthodox, even bizarre, and invite further investigation and interpretation.

3. Osmoregulation

Ionic regulation is an important part of osmoregulation, although not all animals with ionic regulatory abilities have osmoregulatory ones as well. Osmosis, derived from the Greek word *osmos*, meaning "impulse," is a special case of diffusion, or diffusion under special circumstances. As defined in *Webster's New International Dictionary*, it is "The flow or diffusion that takes place through a semi-permeable membrane (as of a living cell) typically separating a solvent (as water) and a solution or dilute solution and a concentrated solution and thus bringing about conditions for equalizing the concentration of the components on the two sides of the membrane because of the unequal passage in the two directions until equilibrium is reached."* A semipermeable membrane is, strictly speaking, one through which only water can pass and which acts as a barrier to all other molecules. If, then, an aqueous solution of any kind is separated by such a membrane from pure water, the rate of diffusion of water molecules from the compartment containing the pure water will be greater than the rate of their diffusion from the compartment containing the solution. Thus, there will be essentially a flow of water across the membrane into the solution. A hydrostatic pressure that just prevents this flow defines the osmotic pressure of the solution. This is a function of its concentration, which is usually expressed in terms of the quantity of solute per unit weight (kilogram) of water. The most useful and generally accepted way of expressing the quantity of a solute is in terms of gram molecules or moles (M). This gives the number of particles of any given solute in a solution, for according to Avogadro's number, each mole of a substance contains 6.025×10^{23} particles. For convenience, the concentration of ions in solutions containing electrolytes are also expressed as moles, mole in these conditions being accepted as synonymous with gram ion.

The osmotic pressure of a solution can be measured in a number of ways and, when it is known, its osmotic concentration can be calculated. This concentration is usually expressed as osmoles* and is not the same as, or identical with, its chemical concentration, but, from biological considerations, it is of greater importance. The physical properties of a solution, such as its vapor pressure, boiling point, and freezing point are so intimately bound together or, in terms of the physical chemist, colligated, that they are collectively and proportionately altered by changes in its concentration. Measurement of any one of these properties becomes, therefore, a measure of its osmotic pressure. The method most commonly used, because it is the simplest and probably the most accurate, is measurement of the freezing point. This can be made with great precision in a volume as small as 1×10^{-5} ml. Pure water freezes at 0°C, and aqueous solutions at temperatures below this, depending upon the concentration of their solutes. Depression of the freezing point below zero, symbolized as Δ, is, therefore, a measure of the concentration of solute particles in a solution and thus of its osmotic pressure. The Δ of seawater is 1.85°C, that of brackish water ranges from 0.2 to 0.5°C, and that of "average" freshwater is 0.01°C. Since Δ depends upon the number of solute particles, rather than upon their kind, two solutions of different chemical composition may have the same osmotic pressure. When this is the case they are said to be isosmotic. If one has a greater osmotic pressure than another, it is considered hyperosmotic to it, and hypo-osmotic if its osmotic pressure is less.

Osmotic relations in biological systems are extremely complicated. The complications arise from the nature of biological solutions and the properties of biological membranes. Ideal conditions such as those of a simple solution separated from pure water by a semipermeable membrane never prevail in living systems. Biological solutions are complex ones containing many different kinds of solutes, and natural waters, although not nearly so complex, are mixtures of different solutes. Biological membranes are not truly semipermeable ones, but permit the passage of molecules other than water. Their permeability to molecules of different kinds varies with the nature of the membrane and even in what might be considered the same membrane under different conditions.

If there is an inequality in concentration of solvent or solute on two sides of a biological membrane, the effective movement or flux of the solvent will be from the more dilute solution to the more concentrated one, and that of the solute will be from the more concentrated to the more dilute solution. Effective flow across the membrane may then be in two directions in respect to water and solutes, rather than in one as in an ideal situation. The solutes may well include organic molecules as well as mineral salts and their ions.

The rate of diffusion of ions depends upon their mobility and in respect to this the most common inorganic ions can be arranged in such a series as $K^+ > Na^+ > Ca^{++} > Mg^{++}$, and $Cl^- > SO_4^{--}$. Their diffusion is regularly from areas of higher to those of lower electrochemical potential, and equilibrium is reached when this potential is zero. But there are conditions in biological systems when this is not the case, for biological solutions contain electrolytes other than mineral salts and in many cases a membrane is impermeable to one ion or another, or less permeable to one than the other, of a dissociated molecule. If, for example, K^+ is dissociated from protein on one side of a membrane and from KCl on the other, there can never be equal distribution of ions on the two sides if the protein anion cannot cross the membrane. As Cl^-, moving along the concentration gradient, accumulates in the compartment with the protein, the solution becomes negatively charged, and hence K^+ is attracted back into it and so moves against its concentration gradient. This "uphill" movement of K^+ will continue until the tendency for Cl^- to diffuse from the KCl solution into the K-proteinate solution

* An osmole is the total number of moles of particles per liter of solvent.

balances the tendency for K^+ to move back into the proteinate solution.

Another factor influencing the movement of ions across biological membranes is in the nature of the membranes themselves. All of them seem able to move cations against an electrochemical potential. This phenomenon, known as active transport in distinction to passive transport along a diffusion gradient, is of the greatest importance to the preservation of the distinction and organization of biological fluids. Both of these characteristics would be lost if diffusion and osmosis continued until equilibrium was reached. The thermodynamically unstable conditions within the living organism can only be maintained through metabolic work. The mechanism, or mechanisms, of active transport have not yet been resolved, although various hypotheses have been proposed. Most of them invoke the concept of a carrier molecule, or carrier molecules, built into the membrane, each one of which can combine with an ion on one side of the membrane and release it on the other. The ion thus transported must either be carried in association with one of opposite charge or exchanged for one of similar charge on the carrier. The molecules that act as carriers are not known, but that they are different for different ions is evident. There are substances that will inhibit the transfer of ions of one kind but not those of another, and ions are selectively moved from one side of a membrane to the other. Carrier molecules are also believed to move ions along, not against, their electrochemical potential, a phenomenon that is known as facilitated transport and increases the rate of transfer across a membrane permeable to them.

It requires energy to move water and solutes against a concentration gradient and to move ions against a gradient of electrochemical potential. The energy so expended is reflected in the oxygen consumption of an animal, for this increases in accordance with the work that is done. The biochemical events that link oxidative metabolism to the mechanisms that maintain osmotic constancy in the internal aqueous environment are not known, but these mechanisms represent part of the basal metabolism of an organism and, for aerobes, part of their constant oxygen requirements. They are essential to the preservation of an environment in which cells can properly function and with which they are essentially isosmotic. Much of the osmotic pressure of a cell's contents is probably attributable to small organic molecules, such as amino acids, hexoses, and the like, and less to inorganic ions, which are in the main responsible for the osmotic pressure of body fluids. Some of this is attributable, however, to organic molecules of larger dimensions that are carried in suspension in the medium and account for its colloidal osmotic pressure.

In general, the concentration of free amino acids in the tissues of invertebrates is higher than it is in those of vertebrates. There are, however, marked differences between major groups, and even between species of the same genus, both in the total amount of intracellular amino acids and in the proportionate amounts of different ones. The total content is higher in marine species than in freshwater or terrestrial ones. Taurine and glycine predominate in most of the species so far studied, which include representatives of Porifera, Cnidaria, Mollusca, Annelida, Arthropoda, and Echinodermata. The concentration of taurine is especially high in marine molluscs, in some polychaetes and in the cnidarian *Renilla*, the sea pansy. In other species the concentration of glycine may equal that of taurine; or glycine, proline, or another acid may predominate. The individuality of invertebrate species and the diversity between them is well illustrated by findings in two species of *Arenicola* in respect to taurine. There are 3.28 μmol/g fresh weight in *A. cristata*, but none has been detected in *A. marina*.

Taurine is a derivative of methionine via cysteine, involving biochemical reactions probably common to invertebrates and vertebrates. Cysteine can be oxidized to products that can give rise to taurine. It has been shown that these steps are accomplished in different ways by two species of bivalve molluscs, again demonstrating the diversity of invertebrate procedures. In *Rangia cuneata*, a clam of brackish waters, the pathway from cysteine to taurine is through cysteic acid; in the marine mussel, *Mytilus edulis*, it is through hypotaurine, an intermediate not yet found in any vertebrate studied, although taurine is a constituent of bile salts.

However it is produced, taurine constitutes the major part of the osmotically active molecular pool in most marine molluscs. Its low concentration, or entire absence, in the tissues of terrestrial and freshwater species may be due not to inability to make it but to failure to conserve it. The low intracellular content of amino acids characteristic of freshwater invertebrates correlates with the relatively low osmotic concentration of their tissue fluids, in comparison with conditions in marine species, and with the nature of their freshwater environments. Besides conserving such small organic molecules, marine organisms have the ability to remove them from the external medium considerably more rapidly than can freshwater ones. That the rate of removal is related to the salinity of the medium has been demonstrated experimentally. Euryhaline nereid polychaetes take up glycine only at moderate to high salinities of the water in which they are immersed, and fail to do so to any appreciable extent in water that is diluted to low salinities. Similar conditions have been observed in the brittle star *Ophiactis arenosa*.

Whether acquired by internal metabolic reactions or from the external environment, the intracellular amino acid pool provides marine invertebrates with a mechanism that can regulate the osmotic pressure in their cells according to changes in the osmotic concentrations of their body fluids. The postulate that it is the small organic compounds rather than inorganic ones that are in the main responsible for the preservation of osmotic equilibrium between cells and body fluids stems from comparative studies on the crustaceans *Eriocheir* and *Carcinus*, when the former was exposed to freshwater and the latter to salt and brackish waters. The role of amino acids as osmoregulators in the axoplasm of the nerves of several invertebrates has been investigated and quite firmly established in squids and cuttlefish. From these and similar experiments it seems apparent that successful osmoregulation is achieved in a wide range of invertebrate species by means of fluctuating pools of small organic molecules, a device that seems not to have been exploited by vertebrates.

C. MECHANISMS OF EXCRETION; IONIC AND OSMOREGULATION

The extracellular fluids of most marine invertebrates are isosmotic with seawater and with the cells with which they are in contact. This, of course, does not mean that they are of identical composition with either in respect to particular solutes or ions. Invertebrates that live in brackish or freshwater have body fluids that are in varying degrees hyperosmotic to their external environments. The differences are related to the salinity of the brackish water and the hardness or softness of the fresh, and are also specific ones, for the ionic concentration in some species is regularly higher than that in some others. In general, the body fluids of these animals are more dilute than those of marine species and are, therefore, hypoosmotic to seawater. The concentrations of solutes in the fluids of terrestrial animals vary even more widely from species to species and may give some clue as to the habitats of their immediate ancestors that succeeded in migrating to land. The blood of earthworms has a lower osmotic pressure than the hemolymphs of the terrestrial isopods *Oniscus* and *Porcellio*, suggesting a freshwater ancestry for the worms and a marine one for the crustaceans. Similarly, there are some decapod and brachyuran crustaceans that live in the sea and are exceptional among marine animals in that their hemolymphs are hypoosmotic to seawater. Concentration of the hemolymph of the crab *Pachygrapsus* is, for example, about 90 percent that of seawater, suggesting that their ancestors may have been freshwater inhabitants that made their way into the sea.

When aquatic animals are transferred from their natural environments to different waters, they are usually put under conditions of osmotic stress. Some cannot survive such a change and, therefore, in nature are prevented from migrating to areas where the salinity of the water is greater or less than that of their natural habitats. Others have proved more able to endure and survive considerable degrees of osmotic stress, either as osmoconformers or osmoregulators. Osmoconformers are those whose body fluids change in concentration with that of their aqueous environment and that, having a high tissue tolerance, can survive such changes as long as their basic metabolic functions can proceed effectively at the dilutions or concentrations to which they are thus subjected. Osmoregulators are those that have means of keeping the concentration of their body fluids much the same, in spite of changes in that of their external environments. Gradations between these two extremes of lability and stability in respect to internal concentration are frequent, and an organism may conform to external conditions in one situation and regulate in another. But, in general, osmoconformers can tolerate greater variations in their internal environments than can osmoregulators, and osmoregulators can tolerate greater variations in their external environments than can osmoconformers.

1. Tests for osmoconformity or osmoregulation

Such tests can be made by exposing aquatic animals to media more dilute or more concentrated than their natural environments. The experimental media most generally used are dilutions or concentrations of standard seawater, or "Normal Copenhagen Sea Water" with a total salt concentration of about 3.43 percent. The concentrations of the experimental media may be expressed as percentages of the standard (25 percent, 50 percent, etc.) or in terms of total salinity or chlorinity. This is usually given as the amount of solute in a liter of water; so, as salinity, is expressed as S ‰ and as chlorinity as Cl ‰. A possible weakness in this experimental procedure lies in the fact that sudden immersion in a medium of different salinity may be a shock to which an animal cannot adjust, as it might were the transition more gradual and a closer approximation to natural conditions. Loss or gain of water under experimental conditions can be determined by differences in weight of an animal before and after the experiment, and loss or gain of solutes by measurements of differences in concentration of test solutions of known concentration after immersion of an animal for a suitable period of time. Particular sites for water and solute loss or uptake can be tested by blocking suspected areas off by one means or another or by exposing isolated tissues, and when possible organs, to various media of known composition and concentration. Loss or gain of materials can be determined by analysis of the tissue after exposure to the test medium for a suitable length of time and by comparison of its content with that of similar tissue made at the beginning of the experiment.

Studies of this kind are ways of discovering how invertebrates meet the problems of osmotic stress in the variety of environments they now inhabit. These problems arise because all living systems are subject to loss or gain of water and its solutes by virtue of the permeability of their exposed surfaces. They apply to all stages in the life cycle, from egg to adult. Among aquatic species the stress is minor for those that live in the sea and are isosmotic, or very nearly so, with the water around them, but it is major for those that live in freshwater and are in danger of flooding through an influx of water from their hypoosmotic environments. It is probably maximal for invertebrates living at tidal levels and in estuaries where they must frequently adjust to conditions that are alternately essentially marine or freshwater. For terrestrial invertebrates the problem is primarily one of water loss by evaporation of the fluid that exudes from their exposed surfaces and the damage consequent to dehydration.

2. Presumptive mechanisms of osmoregulation and conditions requiring it

These problems conceivably could be met in a number of ways. One of these would be reduction in the area of permeability by investing as large a part of the body as possible with some kind of impermeable covering, such as a skeleton or a cuticle. Another would be reduction in the permeability of the exposed areas, a physiological protection against loss or gain of water and loss of valuable solutes that might be effected by a reduction in the number of carrier molecules or their inactivation under certain conditions. Secretion of water or salts, either inward or outward, against a concentration gradient is also a means of their acquisition or disposal. A system that would remove excess fluid would be of signal value to animals living in fresh or brackish waters, even though this might result in the loss of valuable solutes. Loss of salts in this way could be compensated for by their absorption through some part of the excretory system, through the gut, or from the external medium through some specialized surface area of the body. Food and drink is the primary, but not the only, source of salt and water replacement for terrestrial animals which can only survive if their elimination of water is reduced to a minimum. This could be effected in at least two ways: (1) by resorption of water from the tubules of the excretory system, if this is present; and (2) by biochemical adaptations leading to the production of end products of nitrogenous metabolism of low solubility, whose removal, therefore, does not require any considerable volume of water. Observations of animals under natural conditions and evidence from those that have been subjected to experimental tests show that invertebrates have exploited all these possibilities and employ one or several of these means to preserve homeostasis.

The permeability of exposed areas of the body is an important factor in osmoregulation that concerns animals with exoskeletons as well as those without them. The penetration of organic compounds of small molecular dimensions has been demonstrated in a variety of soft-bodied invertebrates (see Chapter VII), but even the cuticular coverings of arthropods permit passage of water and some solutes. Different degrees of permeability of the exoskeletons of selected decapod crustaceans have been demonstrated experimentally by removing disks from the carapaces and sealing them in tubes in such a way that their internal surfaces were in contact with 100 percent seawater and their external, with 50 percent seawater. The flux of salt and water across them could thus be measured. While change in the amount of water in the two compartments was negligible in all cases, salts did move across according to the concentration gradient which simulated conditions between the crustacean's hemolymph and a more dilute medium. It was evident that the exoskeletons of some species were far more readily permeable than those of others. Most permeable were those of the marine crabs *Pugettia producta* and *Cancer antennarius*, and least so were the exoskeletons of the freshwater crayfish *Cambarus* and the crabs *Pachygrapsus* and *Hemigrapsus* spp. that live in brackish waters. A somewhat similar series of gradations has been found in isopods, in which the degree of permeability has been found to be greatest in *Asellus*, a genus that never leaves the water; less in *Ligia*, which is amphibious but always lives close to water; still less in *Oniscus*, a terrestrial genus that must, however, inhabit damp places; and least in *Armadillidium*, which can live for some time in dry air, rolling itself into a ball to protect the moist surfaces of its gills from evaporation. Water loss through the cuticle is known to occur in other terrestrial arthropods, as well as absorption of water from moist air. Some protection from transpiration is afforded those insects whose cuticles are covered by wax, but if this is abraded accidentally or experimentally or its physical properties changed by elevation of temperature, dehydration can follow rapidly.

3. Nonregulators and nonconformers

Some marine invertebrates are unable to adjust to environmental fluctuations of any magnitude. Their distribution is therefore limited to waters whose salinities have osmotic pressures close to those of their body fluids. Such animals are stenohaline (Greek *stenos*, "narrow") and, though having some degree of ionic and volume regulation, are unable to survive if the concentration of their medium is appreciably raised or lowered. Among these are cnidarians, brachiopods, and echinoderms. Tolerance to dilution or concentration of intracellular fluid is low in marine cnidarians, and though they have means of some ionic regulation, they have little control of their water content and soon die in waters that are much diluted or concentrated. Little is known of the osmotic relations of brachiopods, but their restriction to a wholly marine existence implies that they are very limited in control of their internal environment and must remain in waters with salinities not far from 35‰.

The almost rigid tests of echinoderms, with the exception of holothurians, prevents swelling of their bodies to any great extent, although increase in weight of starfishes put into 55 percent seawater shows that some has entered their coelomic fluids and tissues. One species of *Asterias*, *A. vulgaris*, is strictly stenohaline and is not found in areas where at low tide the salinity falls below 27.4‰. Experimentally, it will live for a short time in water with a salinity of 22‰. *Asterias forbesi* can be acclimated, by gradual exposure to increasing dilutions, to seawater with a salinity of 17‰ and will live in this for some time but cannot survive long in lower salinities. The European species, *A. rubens*, is far less restricted in its habitats, for it is found both in the North Sea and in the Baltic. Its distribution in the Baltic extends as far east as Riegen Island, beyond the range of other echinoderms. Specimens taken from the North Sea, at the mouth of the Thames River, can only tolerate dilutions down to a salinity of 23‰. They are incapable of osmoregulation, for measurements of the osmotic pressures of the fluids in the perivisceral coelom and the water vascular system show that these are always in equilibrium with the external medium. Those in the Baltic live and reproduce in waters with a salinity as low as 8‰, although their "spawning intensity" is not as great as those that live in the North Sea. Their larvae, however, must be able to survive dilution to this extent, beyond the limits of those of most starfishes and other echinoderms, although bipinnaria of a Japanese species, *A. amurensis*, are known to tolerate waters with a salinity of 13‰. It seems likely that two races or strains of *A. rubens* have become not only geographically but physiologically separated and that the Baltic race has become acclimated to waters of low salinity from which the North Sea race, and other echinoderms, are still excluded, because of their inability either to regulate the concentration of their coelomic fluids or to tolerate such dilutions of them.

An analogous distinction into physiological races genetically determined seems to exist in *Ferrissia rivularis*, a limpet that lives in freshwater creeks and streams. This is exhibited by differences in the calcium content of individuals from four different populations living in soft-water streams whose calcium concentration was 10 to 20 mg/liter. The calcium content of the soft parts and shells of individuals living in one of these streams averaged 82 mg/g wet weight, and that of individuals living in another, 121 mg/g wet weight. Samples from populations living in other soft-water streams had values intermediate between these. In one, the average calcium content was 109 mg/g wet weight and equivalent to that of individuals taken from a creek with hard water whose calcium concentration was 60 to 70 mg/liter. The race from the soft-water stream must have expended more than six times as much energy to take up the same amount of calcium from the softer water. These data suggest that there are basic differences between the different races in their abilities to remove calcium from their environments, and that the distribution and growth rate in waters of different degrees of hardness reflect the efficiency with which this is accomplished, which, in turn, determines the areas in which the different strains can live.

4. Osmoconformers

Living things are never perfect osmometers, because none is simply a bag of salts within a semipermeable membrane, yet many marine invertebrates, as osmoconformers, behave like osmometers, swelling or shrinking in hypoosmotic or hyperosmotic solutions. The eggs of echinoderms and marine annelids come closest to being osmometers, but even these show some capacity for ionic regulation, indicating that their membranes are not truly semipermeable, but differentially so. The surfaces of all marine invertebrates are permeable to salts or their ions, as well as to water, and in hypoosmotic media loss of salts prevents their uptake of water to its theoretical maximum, so that some volume regulation is in effect.

a. MARINE PROTOZOANS Marine protozoans are isosmotic with their media and are osmoconformers within the limits of their tolerance to internal dilution or concentration. The luminescent dinoflagellates *Noctiluca*, which have diameters up to 500 to 1,000 μ, are of particular interest in respect to the composition of the fluid in the large intracellular spaces between the protoplasmic strands connecting the central endoplasmic mass with the peripheral ectoplasm. *Noctiluca* floats on the surface of the sea and must, therefore, have a specific gravity below 1.024, the average for seawater. Experimentally, it has been found that the protozoans will just float in seawater diluted to a specific gravity of 1.014 and will rise through successive layers of increasingly dilute seawater, increasing in volume as they do so, until at dilutions with specific gravities of 1.012 to 1.007 they will burst and sink. Only a few substances are known that have lower specific gravities than NaCl and yet could be isosmotic with seawater. Among these is ammonium chloride. *Noctiluca* will sink in an isotonic solution of NaCl—that is, in one which does not alter its volume—but will just float in a solution containing 0.8% NH_4Cl and 2.9% NaCl. It seems very likely, therefore, that the buoyancy of *Noctiluca* results from combination with Cl^- of the ammonia released from its nitrogenous metabolism, yielding a salt which, in solution in the intracellular spaces, keeps the animal in osmotic equilibrium with seawater but permits it to float.

b. SIPUNCULIDS Many multicellular marine invertebrates are osmoconformers, some with slight capacity for volume regulation, others with more. These animals are euryhaline (Greek *eurys*, "wide"), able, because of the tolerance of their tissues, to live actively, to reproduce, and to mature in a broad spectrum of salinities. Among those with little ability to regulate volume is the sipunculid worm *Golfingia*, whose body weight decreases or increases when transferred to seawater more concentrated or more dilute than that of its natural environments. The change in volume is rapid, and equilibrium is reached in 2 to 8 hr, after which the new volume remains constant for 2 days or more. The worms can survive indefinitely in 160 percent seawater. On return to standard (100 percent) seawater, after exposure to various dilutions or concentrations, the volume of a swollen or shrunken worm approaches its original dimensions, but does not precisely reach them. Analyses of the coelomic fluids of worms taken from the sea reveal more Na^+, K^+, and Ca^{++}, and less Mg^{++} and SO_4^{--} in them than in seawater, showing ionic regulation. With gut and nephridia ligatured, the weight of worms immersed in diluted seawater increases more rapidly than it decreases in those exposed to concentrated seawater, indicating that inward permeability of the body wall to water is greater than outward. Salt exchange through gut and nephridia has been demonstrated in another osmoconforming sipunculid, *Dendrostomum zostericolum*. Such exchange probably accounts in large part for the volume regulation exhibited by these worms. *Dendrostomum* can tolerate loss of fluid up to 36 percent of its ordinary body weight when exposed to concentrated seawater and recover when returned to 100 percent seawater.

c. MARINE GASTROPOD AND BIVALVE MOLLUSCS Marine gastropod and bivalve molluscs are also osmoconformers and show some volume regulation, principally also through leakage of salts to the external medium. Soft-bodied gastropods such as *Aplysia*, the sea hare, and *Onchidium*, a pulmonate, increase in weight when exposed to diluted seawater, but never to the theoretical limit that would be possible, nor does their reduced weight in concentrated water reach the theoretical minimum. *Aplysia* attains equilibrium with 75 percent seawater in 2 to 3 hr, but after this and its initial gain, loses weight until it may become lighter than it was before transference to the dilute medium. That this decrease in weight is due to loss of salts has been shown by analyses of the chloride content of the hemolymph and the external medium at the beginning and the end of the experiment. Possibly the final drop in weight is the result of muscular contraction and consequent expulsion of water from the tissues.

The limpet *Acmaea limatula*, whose natural habitat is the intertidal zone, is an osmoconformer, for its hemolymph becomes isosmotic with seawater over a range of salinities of 17 to 50‰. Other snails and bivalves have some protection from the immediate effects of changes in the concentrations of their environments, for they can close themselves in their shells. This is for them an important adaptation to life in fluctuating environments. Determinations of the freezing points of the hemolymph and urine of the marine snail *Turritella communis* and of a littoral species, *Littorina saxitalis*, show that both are isosmotic with the external medium over their ranges of tolerance. In diluted seawater, active specimens of *Littorina* reached equilibrium with the medium in about 1 hr, but the adjustment was slower and equilibrium was not reached for several days if the snails were withdrawn into their shells. The survival time of snails in seawater diluted to the limit of their tolerance may thus be prolonged, but not indefinitely, for they cannot stay continually retracted because of their needs for oxygen and the elimination of waste.

Hydrobia ulvae, a small snail living in mud flats and salt marshes along estuaries, is also an osmoconformer and tolerates a salinity range as great as 6 to 94‰. When active animals were transferred from 100 to 60 percent seawater or from 60 to 100 percent seawater, equilibrium of the internal with the external medium was achieved in about 1 hr, but not for several days if the animals were retracted. The tolerance of *Hydrobia* is so great that in concentrated seawater it can stand 100 percent increase in the concentration of its hemolymph for some time before death ensues.

The range of habitat of the bivalves *Mya arenaria*, *Cardium edule*, and *Mytilus edulis* show that each species can inhabit waters of very different salinities. *Mytilus*, for example, is found in the North Sea, where the salinity is about 30‰; in the brackish waters of the Baltic, with a salinity of about 15‰; and in the Gulf of Finland, with a salinity of about 4 to 5‰. In each of these areas the osmotic pressure of the hemolymph conforms to that of the surrounding water. The growth rate of *Mytilus* in dilute waters is reduced, for adults taken from the Baltic are smaller than those that have developed in waters of greater salinity. This suggests that they are under some osmotic stress when living in the more dilute water. Studies of their respiratory rates under different conditions of external salinity are also indicative of this. Mussels taken from the North Sea maintained an approximately constant respiratory rate when exposed to more dilute seawater down to a concentration of 25‰ salinity, below which the respiratory rate declined rapidly. Mussels taken from the Baltic showed a maximum respiratory rate at a salinity of 5‰ and reduced rates in both more concentrated and more dilute media. Bivalves, like operculate snails, can protect themselves for a time at least from such stress when the change in salinity of the medium is a rapid one. *Mytilus* can preserve an undiluted hemolymph for several days after transference from 100 percent seawater to lower concentrations, if the valves of the shell are tightly closed. If the hemolymph becomes at all diluted under these conditions it does so very slowly. But if the valves are propped apart, the hemolymph of specimens transferred from water of $\Delta 2.08°C$ to water of $\Delta 1.36°C$ becomes isosmotic with the medium in about 4 hr. Isolated tissues of *Mytilus* have been found to function well in a wide range of dilutions and concentrations of the hemolymph. Muscle from the heart and cilia on the gills beat regularly in concentrations of seawater between about 40 and 60 percent.

5. Osmoregulators

a. BY WATER STORAGE Migration from the seas to freshwater or to land, or vice versa, could not have been accomplished by invertebrates unless mechanisms for osmoregulation had evolved along with the locomotor, respiratory, and other devices that made exploration of these areas possible. Transition from the sea to freshwater and from freshwater to the sea, and from salt and freshwaters to land, and back again, have been successfully made by species in several different phyla. Studies of marine, estuarine, freshwater, amphibious, and terrestrial species, especially among molluscs, annelids, and crustaceans, show a general progression in osmoregulatory ability toward greater and greater effectiveness in those most completely adapted to life in freshwater and on land. They have shown, too, that there is no precise uniformity in the mechanisms by which these adaptations have been achieved. They are of extreme interest as being indicative of the different evolutionary steps that may have occurred to make these transitions possible. Anatomically, the progression of osmoregulatory ability can often be correlated with the development of more extensive, elaborate, and differentiated excretory organs; physiologically, with the specialization of certain tissues for active transport, secretion, or absorption of solutes; and biochemically, with changes in the metabolism of proteins and nucleic acids and the production of different end products.

Conditions in brackish and freshwaters are met in either of two ways: by storage of the excess water that enters the body through exposed areas, or by pumping it out. The latter is the more effective way, and adaptations of what may have been primitive excretory or drainage devices as osmoregulatory ones are evident in many nonmarine species. Water storage has been demonstrated in *Procerodes* (*Gunda*) *ulvae*, a triclad flatworm living in the tidal zone of estuaries of small streams where it is successively exposed to the water of the Atlantic Ocean, to the freshwater of streams which are known to have a high content of $CaCO_3$, and to conditions of salinity intermediate between these. When put into soft tap water the animals swell to about twice their initial volume in 1 hr, after which their volume drops slightly but they die in 48 hr. They swell, but much more slowly, in hard tap water and in dilute seawater, in which the degree of expansion is proportionate to dilution. Microscopic examination of specimens after a time in dilute seawater reveals large clear vacuoles in the gastrodermal cells, representing the collection of water that has flowed through the swollen parenchyma. The addition of calcium to soft water with a Ca^{++} content <3 mg/liter prevents swelling that would otherwise occur, but ions other than Ca^{++} do not. An animal in dilute seawater, with storage vacuoles, reaches equilibrium in some way, possibly by decrease in permeability. This may be a reflection of a change in its calcium metabolism, since the addition of calcium to soft water will protect it from a fatal influx. Possibly an increase in water excretion through the protonephridial system is the equilibrating factor. When in equilibrium with dilute seawater, the oxygen consumption of animals is greater than it is when they are in 100 percent seawater. Deprivation of oxygen or introduction of cyanide increases the degree of swelling, indicating that the mechanism in operation to prevent dilution of the tissues is an energy-expending one. They are, however, vulnerable to salt loss, the rate of which during periods of exposure to dilute media has been determined by measurements of the electric conductivity of the surrounding solution. Salts begin to be lost immediately after immersion in dilute water; 25 percent of the total initial salt content is gone from an animal after it has taken up a volume of water equal to its initial volume. Reduction of salt content to about 6 to 10 percent of that in seawater is the limit of viability for worms surviving in waters containing calcium.

Similar storage vacuoles are seen in the gastrodermal cells of freshwater *Hydra*. These vacuoles enlarge when individuals are transferred from pond to distilled water; after a few hours' exposure to distilled water, all the cells fall apart and the animal disintegrates entirely. *Pelmatohydra* and *Chlorohydra* lose weight, but survive, in very dilute seawater. Study of the gastrodermis and epidermis show that both are highly permeable to water. By use of radioactive tracers it has been shown that Na^+, K^+, and Br^- are actively absorbed from media in which they are present in low concentrations. Ionic regulation has thus been demonstrated, and may be the means by which these freshwater cnidarians succeed in remaining hyperosmotic to the waters in which they live.

b. BY WATER ELIMINATION **(1) By contractile vacuoles** The majority of nonmarine invertebrates make use of pumping devices. In freshwater protozoans, the contractile vacuole serves this purpose. Indirect evidence for the role of contractile vacuoles in osmoregulation is the fact that marine species do not as a rule have them; nor do marine sponges, although contractile vacuoles appear and operate in the few freshwater species. Marine protozoans acquire contractile vacuoles when they are put into diluted seawater; conversely, when freshwater protozoans with contractile vacuoles are transferred to salty waters, the vacuoles disappear. *Vahlkampfia calkinsi*, for example, an ameba parasitic in the digestive tract of oysters, has been successfully cultured on agar made up in seawater. In these conditions the ameba, presumably isosmotic with the medium, has no contractile vacuoles, but in specimens transferred to agar made up in tap or distilled water, two or more vacuoles developed and pulsated regularly while the animals continued what was an apparently normal existence. *Ameba verrucosa*, a freshwater species in which contractile vacuoles regularly fill and discharge, can be cultured in 10 percent seawater. When the concentration was gradually increased, pulsation of the contractile vacuoles became slower, and in 50 percent seawater, none were formed. They reappeared when the animals were put back into freshwater. Also, the rate of pumping of the contractile vacuoles in the suctorian *Podophrya* is some 10 times greater when the animals are feeding than when they are not, and the volume of their output is equivalent to the decrease in volume of the ciliates, principally *Paramecium* and *Colpidium*, that they have ingested. Concentration of the food, by expulsion of water from it, would result in increased protoplasmic water in the suctorians were it not eliminated in some way. The rate of output of the vacuoles is less at the end of a feeding period than during it, but is still higher than just before feeding begins.

Direct evidence of the osmoregulatory function of these organelles has been obtained by micromethods applied to *Ameba proteus*. Freezing point determinations of the minute amounts of fluid, about 1×10^{-4} µl, that can be withdrawn from a contractile vacuole by means of a micropipette, give an average concentration of 32 milliosmoles for the vacuolar fluid. Similar determinations show the average osmotic concentration of the cytoplasm to be 101 milliosmoles, and that of the surrounding medium, 6 milliosmoles. The vacuolar fluid is thus distinctly hyperosmotic to the medium and hypoosmotic to the cytoplasm. The mechanisms by which these relations are produced and maintained are not yet elucidated. It is possible that the fluid in the vacuole is formed in two successive and interrelated steps, the first being secretion of solute, accompanied by the passive movement of water, and the second, resorption of solute, leaving a less concentrated solution behind in the vacuole. If the vacuolar membrane has the properties of the plasma membrane, through which K^+ passes inward against the concentration gradient and Na^+ outward, K^+ may be actively transported from the initial vacuolar fluid into the cytoplasm, and Na^+ in lesser amounts from the cytoplasm into the fluid. In this way the cytoplasm could be kept hyperosmotic to the contents of the vacuole. Such secretion and absorption may be separated from each other both in space and in time, for possibly the smaller feeder vacuoles, filling before the large one, are the sites of secretion, and the large one the site of absorption. Both these processes require the expenditure of energy. In freshwater peritrichs, demonstration that the volume of water increases and the operation of their vacuolar system stops in the presence of cyanide implies that they are dependent upon oxidative metabolism. This may be typical of conditions in other protozoans.

(2) By Nephridia: flatworms, rotifers, annelids

Nephridia serve as the pumping system in metazoans with tubular organs of this kind. In acoelomate and pseudocoelomate animals with protonephridia and without circulatory systems, such as flatworms, rotifers, and endoprocts, the urine in the tubules can hardly be an ultrafiltrate, for the difference in pressure would not be great enough to permit ultrafiltration. It is possible that the movement of flagella and cilia in solenocytes and flame bulbs creates a negative pressure in the tubule that would tend to draw fluid into it. The observation that the tubules in the miracidia of the liver fluke *Fasciola hepatica* collapse when flame cell activity stops gives some support to such an hypothesis. There is evidence that secretion plays an important role in the excretion of adult flukes, for material is taken up into cells along the entire length of the tubules by pinocytosis and discharged into the lumen in droplets containing a mixture of granular material and of lipid.

FLATWORMS AND ROTIFERS Because of their minute size, quantitative studies of these systems have so far proved impossible, but there is indirect evidence, from their extent and their histology, that they do operate as excretory and osmoregulatory organs. There is, for example, a euryhaline species of flatworm, *Gyratrix hermaphroditus*, with strains that inhabit fresh, brackish, and salt waters. There is no protonephridial system at all in the marine strain and that in the freshwater strain is more elaborate than it is in the brackish. In the freshwater strain the system consists of many tubules, which terminate proximally in flame bulbs and on each side lead into a collecting tubule that is expanded distally into an ampulla with thick walls that opens into a bladder and thence to the outside through the excretory pore. In the brackish water strain, the ampulla and bladder are absent, and the collecting tubule opens directly at the excretory pore. These differences suggest that the tubular system is concerned, in the freshwater forms, not only with elimination of water and its solutes but with reabsorption of some of them and that in the brackish water strain, where salt conservation is not so essential and the resorptive area is missing, the function of the system is the excretion of a more or less dilute urine, in accordance with the concentration of the external medium. In both cases the intercellular fluids would be kept hyperosmotic to the external environment. In the marine strain they are probably isosmotic. In *Rhynchodemus terrestris*, a turbellarian that lives in humid places on land, the branched terminal ducts of the protonephridia terminate in large flame bulbs about 10 μ long and 4 μ wide. The cells in the terminal bulbs are not ciliated, but those in the greatly convoluted canals into which they open and that extend all through the parenchyma are. Histochemical tests show that these cells are rich in alkaline phosphatase, which is not demonstrable in the flame bulbs and the nonciliated cells of the terminal canals. This is strong circumstantial evidence for the activity of these cells either in secretion or in absorption and for their role in modification of the filtrate as it passes along this part of the canalicular system. There is a similar histochemical differentiation in the many thousands of protonephridia in the terrestrial nemertean *Geonemertes dendyi*, in which the distribution of alkaline phosphatase is similar to that in the tubules of the vertebrate kidney, an implication of analogy in their operation. The cilia in the flame bulbs beat more rapidly when the animal is immersed in water and fluid enters the body, a fact that provides additional support for the idea that the protonephridial system is involved in osmoregulation.

Very little is known of the osmotic behavior of trematodes. Experiments with *Fasciola hepatica* cultivated in vitro show that it is viable in media whose osmotic pressure varies from Δ0.40 to 0.81°C. Spermatogenesis is adversely affected in media whose freezing point depression is greater than 0.72°C, so that they are not wholly tolerant of external concentrations as high as this. The osmotic relations of the parasites in vivo is not known, but the fact that they show little or no water regulation under experimental conditions in vitro would suggest that they are essentially isosmotic with the intercellular fluids in the livers of their hosts and with their bile. Depression of the freezing point of bile from uninfected sheep is 0.57°C and from infected sheep is variable, reaching a maximum of 0.94°C in heavily parasitized animals.

Studies of the protonephridial system in various species of *Asplanchna*, a cosmopolitan genus of freshwater rotifers, have shown that the number of flame bulbs is greater in species with relatively large surface areas than it is in the smaller ones. Detailed observations have been made of the activity of the cilia in the flame bulbs and of the frequency of pulsation of the contractile vesicles in *A. priodonta* and *A. brightwelli*, collected from ponds in London and in Denmark. These little animals are so transparent that such observations can be made with a light microscope on specimens that are slightly flattened in a depression slide or chamber, which can be filled either with water from their native ponds or with a saline solution of any desired concentration. The frequency of the beat of the tufts of cilia in the flame bulbs can be very accurately measured with a stroboscopic device, a method similar in principle to that used for measuring the frequency of insect wing beats. Ciliary beat and vesicular pulsation have approximately the same frequencies in specimens of *A. priodonta* taken from Regent's Park Lake in London and in those from ponds in Denmark, when they are in their natural waters, although these are quite different in their "hardness." The water from Regent's Park Lake has a total ionic concentration of 38.7, in terms of milliosmoles NaCl/liter and has a calcium content of 200 to 260 mg/liter. The Danish water, whose calcium content is 30 mg/liter, has a total ionic concentration of 8.7 milliosmoles NaCl/liter. There is a difference between the two species of rotifer in their frequencies of ciliary beat and pulsations of the contractile vesicle. These are less in *A. priodonta*, each of whose protonephridia has four flame bulbs, than it is in *A. brightwelli*, each of whose protonephridia has 15 to 20 flame bulbs. There are also significant differences in these rates when the ionic concentration of the water is changed. With increased concentration, the activity of the flame bulbs and that of the contractile vesicle are less than they are in natural waters, but return to their usual level when the concentration is decreased to that of pond water. Increase of temperature within a range of 10 to 25°C results in increased activity of the flame bulbs and contractile vesicle. In *A. priodonta*, the rate of ciliary beat rose from about 488 strokes/min at 11°C to about 2,644 at 25°C, and the frequency of pulsation of the contractile vesicle trebled between 10 and 22°C. All these facts would indicate that the protonephridial system is concerned with osmoregulation, but give no clue as to its possible role in the excretion of nitrogenous waste, most of which, if not all, may be eliminated through the body surface. It has been estimated that the internal concentration of *A. brightwelli* is such that it is isosmotic with a medium whose concentration is equivalent to 100 to 160 milliosmoles NaCl/liter. These values are comparatively high in relation to others that have been obtained for the few invertebrates investigated in this regard. It is possible that a high internal osmotic pressure is a necessity for rotifers such as *Asplanchna*, whose hydrostatic skeleton depends upon the maintenance of substantial turgescence in their otherwise unsupported, though small, bodies.

ANNELIDS Annelids have become adapted to a wide range of environments and their nephridia show structural and functional differences accordingly. Among polychaetes these features have been particularly investigated in nereids, since some species, as osmoconformers, are restricted to life in waters of specific salinities while others, as osmoregulators, can live in the fluctuating environment of estuaries, in seas as dilute as the Baltic, and even in freshwaters. *Nereis pelagica* and *Perinereis cultrifera* are entirely marine and show no means of osmoregulation, while *N. diversicolor*, which lives in estuaries, has mechanisms that enable it to preserve an internal environment hypertonic to the external. The nephridia of *P. cultrifera* are simpler in construction than those of *N. diversicolor*, which are long, coiled tubules offering ample area for resorption of salts, presumptive evidence that such resorption takes place. Differences in permeability of the integuments of these two species have been demonstrated by the use of radioactive isotopes. It has been shown that the exchange of Na^{24} between seawater and body fluid is much more rapid in *Perinereis* than it is in *Nereis*. This would suggest that the surface of the former is more permeable than that of the latter and that the gain or loss of water through it would take place more quickly, and so probably fatally, in the one than in the other. Although not continuing to be isosmotic with 100 percent seawater, the blood and coelomic fluid of *N. diversicolor* remain hyperosmotic to dilute seawater in nature and in the laboratory, where the worms have lived in dilutions as extreme as 1.4 percent. They have been cultured and have completed full life cycles in seawater with a salinity of 7‰, and juvenile stages have been subjected to even greater dilutions than this. Prelarval and larval stages are critical ones, most sensitive to the effects of dilution. Young worms are the most resistant and some have even survived in freshwater for several months. The ability of mature worms to regulate is lost if cyanide is added to the medium, if oxygen is deficient, and if there is no calcium, conditions which suggest that such protection is attained only at the expense of work and reduction of surface permeability through calcium metabolism. *Neanthes virens* probably regulates in a similar fashion, but not so effectively. Its permeability to water is three times greater than that of *N. diversicolor*, and tracer experiments have shown that the exchange of Cl^- between environment and body fluid is more rapid in the former than in the latter. In diluted seawater, *N. diversicolor* absorbs Cl^- from the medium, and its permeability to ions is less than it is in 100 percent seawater. Both these protective devices enable it to live an estuarine existence. Experiments on chloride exchange in three other estuarine species, *Nereis limnicola*, *N. succinea*, and *N. vexillosa*, have shown that uptake of salt across the body surface is not a major factor in keeping their body fluids hyperosmotic to an external environment of low salinity. The source of salts in these worms may be food, or possibly there may be resorption from the urine.

Arenicola, living in similar habitats, is on the other hand osmotically labile and has no known means of controlling the concentration of its body fluids. It swells in diluted seawater to an extent according to the dilution, and its body fluids become isosmotic with the external medium. Adaptation of these worms to an estuarine existence rests on the ability of their tissues to function at higher levels of dilution than can those of most animals. At the other end of the scale, in respect to osmoregulation, is *Neanthes limnicola*, a species that can live in freshwater as well as in brackish and whose nephridia are long, coiled tubules providing, like those of *Nereis diversicolor*, an extensive area for possible resorption.

Oligochaetes are essentially freshwater animals and even earthworms, surrounded by the moist mucopolysaccharide coating on which their life depends, live in an aquatic environment of sorts. Nothing is known about the excretory and osmoregulatory mechanisms of strictly aquatic species such as *Tubifex*, but detailed studies have been made of a number of earthworms. The nephridia of all of these are long, coiled tubules, usually well vascularized. Filtration into them might occur in three ways. In those that are closed, it might take place directly from the blood, indirectly from the blood via the coelomic fluid, or from the blood and from the coelomic fluid independently. In open nephridia, filtration might be through the nephrostome, which, acting like a coarse sieve, would permit passage of proteins and even of particulate matter such as cell debris. Filtration from the blood directly into the tubule is also a possibility.

Although living in a moist environment, an earthworm is never fully hydrated and on transfer to tap water both *Lumbricus* and *Eisenia* have been found to gain about 15 percent in weight in some 5 hr and to recover their original weight in a shorter time when returned to soil. It has been shown that salts as well as water pass readily across the integument. Dilute salt solutions become depleted of Cl^- after worms have been immersed in them for a time. This fact has been taken as evidence of active transport of the ion into the body fluids. Other sites of salt and water loss are the gut, the dorsal pores, and the nephridia. Much of the nitrogenous waste is eliminated through the surface; some is excreted by the nephridia and is included in the urine; and some is incorporated in the chlorogocytes that act as kidneys of accumulation. There is no principal excretory product common to earthworms and varying amounts of ammonia, urea, uric acid, and protein have been found in the urine of individuals of different species and in that of those of the same species under different conditions of environment and quantity and quality of nutrition. This information, obtained from analyses of samples of urine collected en masse from a number of worms over a period of time, may be misleading because some of the contents may come from sources other than the nephridiopores.

The most precise study of nephridial functioning in annelids has been made on *Lumbricus terrestris*, a species to which it has been possible to apply methods used for the analysis of renal functioning in vertebrates. Determinations of the freezing points of samples of blood, coelomic fluid, and urine, collected by micropuncture at different levels of the tubule, have shown that the urine is considerably modified in its passage from nephrostome to nephridiopore. Such collections have been made from the nephridia *in situ* in living worms and in those that have been excised and kept in saline solutions. By these methods it has been shown that the blood is slightly hypoosmotic to the coelomic fluid and that the urine in the terminal part of the nephridium is distinctly hypoosmotic to both. It is isosmotic with the coelomic fluid in the part of the tubule just below the nephrostome; its osmotic pressure drops somewhat during its passage through the narrow tubule and very markedly in the wide tubule, where it is estimated that four-fifths of its solutes are returned to the blood. By the time it enters the bladder it has reached its minimal concentration and there is no further change before its final elimination through the nephridiopore. As a possible mode of functioning, it is suggested that in *Lumbricus*, and probably in other earthworms with open nephridia, protein, urea, ammonia, water, Cl^-, Na^+, and K^+ enter the tubule from the coelomic fluid through the nephrostome; that uric acid, urea, ammonia, and salts filter into the narrow, distal tubule from the blood; that salts may be added to it in the middle tube; and that the major site of resorption is the wide tube, where protein, K^+, Na^+, Cl^-, and some water are separated from it and returned to the blood against the osmotic gradient. This leaves a hypoosmotic urine, consisting of water, ammonia, urea, and uric acid, and possibly other nitrogenous compounds, to enter the bladder. The rate of urine production in *Lumbricus* has been found to be 25 ml/(kg)(hr), and such a copious secretion of a urine hypoosmotic to the body fluids could be an adequate means, for a freshwater animal, of preserving its internal environment at a relatively constant osmotic pressure. If the earthworm nephridium does perform in this way, its analogy with the excretory unit of the vertebrate kidney is apparent.

In earthworms with enteronephridia, the gut plays a prominent part in absorption of water. This role has been established for the Indian earthworm *Pheretima postuma*, in which there are three kinds of nephridia: (1) innumerable closed integumentary nephridia, opening externally through pores along the whole length of the body; (2) closed pharyngeal nephridia, opening into the pharynx; and (3) open nephridia, with septal funnels in the coelom and tubes that open into the intestine. There are thus three routes for water elimination: integument, mouth, and anus. All three are used when the worms are fully hydrated and immersed in water in which they can live indefinitely and which passes freely across the integument. Their use has been demonstrated by removing specimens of *Pheretima* from water, mopping them dry, and observing the emission of fluid from all these orifices at once. There are six known genera of Indian earthworms with enteronephridia, all of them inhabitants of regions where the monsoons soak the earth for 3 months of the year and the hot sun parches it for the remaining 9 months. During the rainy season, exonephric and enteronephric species live side by side in the upper levels of the soil, but when the dry season begins, the species with only exonephridia burrow deep into it, conserving their water and salts by reabsorption from the tubules. But enteronephric species like *Pheretima* can stay near the surface throughout the year, the gut functioning, as in other animals, in the absorption of water and salts from its contents, which has, in this case, an additional load in the material discharged into it from the open nephridia. When specimens of *Pheretima* are removed from dry soil and examined (and *Pheretima* can live in sandy soils where there is little or no water and from which other worms are virtually excluded), no emission of water from the anus can be detected and only minimal amounts from the mouth and integument. Moreover, in contrast to the feces of *Lumbricus* and other worms with only exonephridia, which are evident on the surface as soft, damp castings, the feces of *Pheretima* and other worms with enteronephridia are hard and dry. It is thus evident that the gut operates as an osmoregulatory organ, contributing far more than the nephridia to the conservation of water and the preservation of osmotic stability within the animal when it is in a dry and potentially dehydrating environment.

(3) By urocoeles Among molluscs and arthropods with urocoeles, there are species that are marine, estuarine, brackish, and freshwater as well as terrestrial in habitat. These phyla, therefore, offer an excellent opportunity for comparative studies of ionic and osmoregulation and of the adaptability to different conditions of an excretory organ constructed from a coelomic space. From the species that have been investigated, it seems evident that the common mechanism of urine formation is ultrafiltration from the vascular system into the pericardial cavity and modification of the filtrate after its passage into the urocoele.

(a) MOLLUSCS: CEPHALOPODS, FRESHWATER BIVALVES, FRESHWATER AND TERRESTRIAL GASTROPODS All cephalopod molluscs are marine and have no problem of water elimination or water conservation. Tests have shown that the hemolymph, pericardial fluid, and urine of *Sepia*, *Eledone*, and *Octopus*, representative genera of the class, are isosmotic with each other and with seawater, although differing from it and from each other in respect to nearly all the ions. Production of urine ceases when the reno-pericardial duct is ligatured, providing cogent evidence for the origin of urine from the pericardial fluid. The source of the pressure that would make ultrafiltration possible is problematical, but the fact that in *Octopus dofleini* the rate of formation of the pericardial fluid rises from a base level of 5 to 10 ml/hr to values as high as 48.6 ml/hr 1 hr after the injection of 100 ml of seawater or of hemolymph shows a definite relationship between blood volume and urine production. Although pressure in them would be adequate, filtration through the thick, muscular walls of the systemic or branchial hearts is unlikely, nor does it seem possible from the renal appendages, for the hydrostatic pressure in them is probably too low. The most acceptable current hypothesis is that it occurs in the branchial heart appendages, the homologues of the pericardial glands in bivalves. Although it has not been possible to measure directly the pressure in the appendages, it can be measured in the branchial hearts. Such measurements in *Octopus dofleini* have shown that the rate of filtration is related to the hydrostatic pressure within the heart and that the production of pericardial fluid stops when that pressure corresponds to one of about 4 ml of water. This is about equivalent to the colloidal osmotic pressure of the blood, attributable to its content of protein and other molecules of colloidal dimensions.

It is possible also to insert catheters into the renal organs and pericardial cavity of an octopus and to take samples, either intermittently or continuously, of the fluid in them. Analyses of the pericardial fluid and urine of *Eledone* have provided evidence for resorption of K^+, Ca^{++}, Mg^{++}, and Cl^- and for secretion of Na^+ and SO_4^{--} as the fluid passes through the urocoele. Water and ions are taken in through the gills and other permeable areas, including the gut. There is no evidence for resorption of water in the urocoele, and the fact that inulin injected into the bloodstream is recovered in the same concentration in the urine may be taken as evidence that no water is lost from the initial filtrate. Ammonia is the principal nitrogenous product excreted, and it is eliminated in comparatively large amounts. Analyses of urine collected continuously from *O. dofleini* showed an ammonia content that ranged from 5 to 65 ml/liter over the period that the collections were made. The urine showed also a lower content of Na^+ and Cl^- than the pericardial fluid and an increased content of K^+, Zn^{++}, Cu^{++}, and SO_4^{--}, all of which are secreted by cells in the walls of the urocoele, as is the ammonia.

Freshwater and, most likely, the majority of brackish water and estuarine bivalves and gastropods are able to keep their internal fluids hyperosmotic to their external environments. Renal functioning in nonmarine lamellibranchs has been most fully investigated in the freshwater mussel *Anodonta*. The urine is formed initially by ultrafiltration directly from the ventricle and auricles to the pericardium. The hydrostatic pressure in the ventricle corresponds to that of 6.0 ml of water, and is adequate for ultrafiltration, since the colloidal osmotic pressure of the hemolymph is equivalent to the hydrostatic pressure of 2.8 ml of water. The total osmotic pressure of the pericardial fluid is the same as that of the hemolymph, although, without its hemocyanin and other proteins, the colloidal osmotic pressure is lower. Analyses of samples of pericardial fluid and of urine, drawn directly from the pericardial cavity and the urocoele, show that the ultrafiltrate has the same Ca^{++}, Cl^-, and PO_4^{--} content as the hemolymph but that some salts are resorbed after it passes through the reno-pericardial aperture and into the urocoele. Cl^- is resorbed in greatest quantity, and the final chlorine content of the urine is about one-half that of the blood. As a consequence of this resorption, the urine becomes hypoosmotic to the hemolymph, having only about 60 percent of its ionic concentration. Protein and nonprotein nitrogenous waste products are added to the urine in its passage through the urocoele, and the nitrogen content of the urine becomes three to four times that of the hemolymph. This suggests that in *Anodonta* water is resorbed from the filtrate. A large volume of urine is filtered off each day, in spite of the fact that the osmotic concentration of the hemolymph (44 milliosmoles/kg H_2O) is the lowest yet recorded for any animal and corresponds to that of only 4 to 5 percent seawater. The high rate of water flow through the body is presumably related to the large surface-volume ratio of this mussel, for which the exceptionally large gills probably account in large measure. It is possible, too, that the exposed surfaces are highly permeable and that salts as well as water pass freely across them. This intake of salts would supplement renal resorption in compensating for those lost from the hemolymph in the ultrafiltration process. The total osmotic work of *Anodonta*, under ordinary conditions of life, has been calculated as 0.014 Cal/hr, representing about 1.2 percent of the total metabolic work of a mussel weighing 60 g. Of this osmotic work, 0.0131 Cal/hr is performed at the body surface and 0.0014 Cal/hr in the excretory organ. Were the ionic concentration of the hemolymph of *Anodonta* higher—as high, for example, as that of 100 percent seawater—the osmotic work would be far greater. One aspect of the evolutionary adaptation of lamellibranchs to life in brackish and freshwaters may, therefore, have been reduction in the ionic concentration of their body fluids and consequent reduction of the metabolic work required to preserve their physiological constancy.

Osmotic concentrations in the hemolymph of the small clams *Dreiseena* [a recent migrant from marine waters (Baltic, Black, and Caspian Seas) of low salinities to freshwaters] and *Pisidium* are 54 milliosmoles/kg H_2O, only slightly higher than that of *Anodonta*. In contrast, osmotic concentration in the hemolymph of the freshwater pulmonate *Lymnaea peregra* is 124 milliosmoles/kg H_2O, or about three times that of *Anodonta*. Yet this snail has made a successful adjustment to freshwaters. The most important phase of this adjustment may be a particularly high capacity, experimentally demonstrated, for salt absorption through the body surface, resulting in a continual replacement of those salts lost through filtration. The two mechanisms exemplified by *Anodonta* and *Lymnaea* represent different ways in which originally marine molluscs have adapted themselves to lives in freshwaters.

The adaptation of *Anodonta* is, however, not complete, for in its juvenile stages it is unable to osmoregulate. This may be true of other bivalves who live as adults in freshwater. *Anodonta* has met this problem by a highly specialized way of juvenile development.

The embryos, after hatching in the autumn, are brooded in the gills of the mother until the following spring. Then they are set free and begin a protected and parasitic existence for 3 to 12 weeks, embedded in the gills of a freshwater fish. During this time the internal organs complete their development and the "kidney" becomes capable of producing hypoosmotic urine. Were it not for this second adjustment to a freshwater existence, *Anodonta* could not have become a successful colonist of lakes and streams. Parallel conditions are found in crustaceans that as adults meet the demands of life in freshwater, but that must return to the sea to breed, because their eggs and larvae survive only in marine environments.

Colonization of the land, which gastropods have been the only molluscs to achieve, presents a primary problem of water conservation. A urocoele with greatly folded walls and a long, narrow and highly convoluted excretory duct provided with sphincter muscles is typical of the renal organs of land snails. Electron micrographs of the bladder region of *Helix pomatia* show ciliated cells and cells with deep microvilli and many cytomembranes between which mitochondria are abundant. This cytological picture is much like that of the distal tubule of the mammalian kidney, where resorption is known to occur, and the structural analogy makes the idea of a functional one attractive. A free flow of urine, such as that in freshwater molluscs, is probably rare in land gastropods under the ordinary conditions of their lives, and much of the fluid in the urocoele is resorbed. Correlative with this is a biochemical shift in nitrogenous metabolism from ammonotelism to uricotelism, and a progressive tendency toward increased production of uric acid can be traced in the adaptation of gastropods to a wholly terrestrial existence. There is a greater proportion of uric acid in the nitrogenous products excreted by land pulmonates than in those of aquatic species, and in the genus *Littorina*, at least, a greater proportion in the species habitually living high on shore than in those living nearer the water. Hibernating and aestivating snails void no urine at all; uric acid accumulates in the bladder and is excreted when activity is resumed.

In snails, as in other molluscs, the initial step in urine formation is ultrafiltration. In the giant land snail *Achatina fulica*, whose heart and renal organs can be studied *in situ*, filtration seems to occur directly into the cavity of the urocoele from branches of the renal artery with which the glandular part of the organ is liberally supplied. In *Lymnaea*, on the other hand, filtration takes place from the heart into the pericardial cavity as it does in bivalves. The hydrostatic pressure in the heart is equivalent to that of 8 ml of water, while the colloidal osmotic pressure of the blood is equivalent to 2.5 ml of water, a difference that would make ultrafiltration possible. The pericardial fluid of *L. peregra* has been found to be isosmotic with the hemolymph and the final urine hypoosmotic to it, with a concentration only 70 percent that of the hemolymph. The filtrate in *Achatina* also becomes hypoosmotic to the hemolymph in its passage through the urocoele, with a difference in osmotic concentrations of hemolymph and urine expressed by freezing point determinations of $\Delta 0.462$ for hemolymph and $\Delta 0.285$ for urine. Presumably salts as well as water are resorbed from the filtrate. Resorption of glucose has been demonstrated, for blood sugar values averaging 20.7 mg/100 ml have been obtained from 11 animals, whose urine sugar levels averaged 10.7 mg/100 ml. Concentration of injected dyes by cells in the urocoele gives indirect evidence that secretion as well as absorption is in operation. Resorption is probably promoted during dry periods, when conservation of water is more than usually important, by closure of the sphincter muscles of the excretory duct, thus preventing the escape of fluid and increasing the pressure in the lumen of the urocoele.

Survival during periods of drought depends upon behavioral as well as physiological adaptations. Some snails withdraw into their shells remaining, as *Otala lactea* does, above the ground and exposed to full sunlight. A thick shell and a tight-fitting operculum protect the tissues from desiccation. Species like *Helix aperta* (Born), with thinner shells into which they cannot wholly withdraw, go underground or retreat to some moist niche. During this time the hemolymph becomes concentrated to some extent, more through evaporation than flow of urine or secretion of mucus. When it rains, the snails emerge and take up water through the gut and the exposed integument, primarily that of the foot. Studies of a number of species of Helicidae have shown that after a rain the blood is diluted again to a level that remains more or less constant and characteristic of each species until another dry period occurs.

(b) CRUSTACEANS (i) Branchiopods Very little is known of the regulatory abilities of branchiopods, which live in brackish or freshwater ponds and appear sporadically in transient ones, where they are subjected to extreme variations in temperature and osmotic stress. Some species, such as *Branchionecta paludosa* and *B. schantzi*, are found in Alpine ponds where the water comes from melting ice and so is nearly pure. Other species may appear in desert ponds, where both temperature and evaporation rates are high and the water becomes increasingly concentrated. Comparison of the regulatory capacities of two species of tadpole shrimps, *Triops* (*Apus*) *longicaudatus*, an American species, and *T. cancriformis*, a Eurasian one, have shown interesting differences between them. These notostracans inhabit rice fields, where in California and in Spain they have done great damage to the crops in uprooting seedlings in their search for food. In the mud they find bacteria, protozoa, small worms, insect larvae, and other tiny creatures which they devour. The Cl^- content of the hemolymph of *T. longicaudatus* (56.2 mM) is almost 20 percent higher than that of *T. cancriformis* (42.7 mM), which can satisfy its salt requirements from its food. But specimens of *T. longicaudatus* lost both Na^+ and Cl^- from their hemolymph when put into deionized water with an adequate supply of their natural food, while the Mg^{++} and Ca^{++} values remained almost unchanged. The mean survival time of individuals in these conditions was 10.2 hr, and none lived longer than 15 hr. In natural pond water, an internal Na^+ and Cl^- concentration higher than that of the external environment was maintained against a high concentration gradient. These data indicate that there is active uptake of salts by this species of *Triops*, which must obtain some from its medium in addition to that from its food. The fairy shrimp *Chirocephalus* and the water flea *Daphnia* likewise depend more upon active uptake to maintain their salt balance than upon their diet. The osmotic concentration of the hemolymph of *Daphnia* is, however, low (average $\Delta = 0.27°C$), so that the salt demand is not very heavy. The freshwater isopod *Asellus aquaticus* and the amphipod *Gammarus pulex* also have hemolymphs of relatively low osmotic concentration (average $\Delta = 0.50$ to $0.55°C$), which they maintain by active uptake of salts from the environment. The branchiopod *Branchippus*, on the other hand, seems to get its salts mainly from its diet, for specimens will live in distilled water as long as they have a source of food.

The branchiopod *Artemia salina* is of special interest in connection with crustacean osmoregulation because it can live and breed over an exceptionally wide range of salinities. The survival time of individuals exposed to fresh or distilled water is short; but in waters of salinities ranging from 0.26% NaCl to about 30% NaCl, they maintain an osmotic concentration in the hemolymph that is virtually independent of the medium, and they function well within the entire range. The osmotic concentration of the salt lakes in which the shrimps are found may be so great that Δ ranges from 13.5 to 15°C, yet that of the shrimps' hemolymph is not more than 1.2 to 1.6°C. It has been shown that Na^+ and Cl^- account for most of the osmotic pressure in the hemolymph, which is hyperosmotic to media more dilute than 25 percent seawater and hypoosmotic to those more concentrated. The permeability of the body surface is, in general, very low, and water and salts pass through it only in negligible amounts. The epithelia of the gut and the first 10 of the 11 pairs of gills are, on the other hand, extremely permeable; these are the areas responsible for water and ionic regulation. *Artemia* takes in water continually both orally and anally. This has been demonstrated by keeping animals in water in which phenol red was dissolved, and the concentration of dye in the lumen of the gut, which reached seven times that of the medium, gave proof that water was absorbed from it. While the osmotic concentration of the fluid in the gut is greater than that of the hemolymph, the concentration of Na^+ and Cl^- is consistently less than in the hemolymph, implying passage of these ions across the gut epithelium and into the hemolymph. The importance of the gills in ionic regulation has been shown by experiments in which their epithelium has been damaged by exposure to potassium permanganate, after which adult animals are unable to osmoregulate and can live only in isosmotic media. The absorptive capacity of the gill epithelium has been demonstrated histologically by visualization of silver in the cells after immersion of the gills in solutions of silver nitrate. Confirmatory evidence has also been supplied by electron microscopy, which has revealed a cytological picture comparable to that of other established salt transference areas.

From these lines of evidence it is presumed that the gills excrete excess salt when the shrimps are in concentrated media and that they absorb it when in dilute waters, keeping the NaCl content of the hemolymph relatively stable. Salts, as well as water, enter the hemolymph through the gut epithelium; the excess water entering from dilute solutions is probably eliminated through the maxillary glands, which very likely also operate in the removal of excess divalent cations when the shrimps are in highly concentrated media. Beyond this they seem to have no part in osmoregulation, and the mechanism employed by *Artemia* is essentially that of the marine teleost fishes, which also swallow water, absorb water and salts through the gut epithelium, and excrete salt through their gills as a means of preserving osmotic stability in their blood. This may be another illustration of convergent evolution. *Artemia* can breed in dilute as well as in concentrated water, because its nauplius is equipped with a dorsal, or neck, organ which degenerates when the gills develop and which apparently functions in a similar way in salt absorption and excretion. How *Artemia* adjusts to the conditions of extreme alkalinity it encounters in some of its habitats is still an open question.

(ii) Isopods Isopods are also of particular interest because of the wide range of their habitats. There are marine, littoral, freshwater, and terrestrial species. Little is known of the means by which their adjustments to these different conditions have been made, but the terrestrial species are the most successful of all crustaceans in adaptation to life on land. *Ligia oceanica* is a littoral species, living under rocks at or above the high-tide level. Its hemolymph is hyperosmotic to ordinary seawater, and it can maintain this hyperosmoticity for as long as 13 hr in dilute seawater. It can even survive 8 to 36 hr in distilled water, although signs of oedema are apparent after a short time. These signs are also evident after exposure to 25 percent seawater. In concentrated seawater, the concentration of the hemolymph is equal to, or even greater than, that of the external environment. Its capacity for ionic regulation is apparent, for the concentrations of Na^+, K^+, and Cl^- are consistently higher in the hemolymph than in ordinary seawater, the Ca^{++} concentration is much higher, and that of Mg^{++} and SO_4^{--} are much lower. Although no excretory organs of any kind have been identified in the gribble *Limnoria*, it is capable of maintaining a hemolymph hyperosmotic to its external environment over a wide range of test salinities. *Limnoria tripunctata*, an English species found in Southampton Water, can tolerate estuarine conditions, although low salinities reduce its boring activities. Measurements of the osmotic concentration of the hemolymph of freshly collected specimens showed that it was hyperosmotic to full strength seawater. This concentration was maintained over a wide range of salinity in animals exposed to seawater from 5 to 39‰. In natural conditions, *Limnoria* probably never experiences such a range, because the openings of its burrows are almost always protected from sun and air when the tide is out by the masses of algae growing on the piles into which the gribbles bore. Yet, experiments have shown that they could endure it because of their powers of osmoregulation, but how this is accomplished is entirely unknown.

Asellus aquaticus is a freshwater isopod, whose hemolymph is far less concentrated than that of marine species and even than that of terrestrial ones, and corresponds in this respect to that of *Astacus*. Its maxillary gland has a complex structure and probably produces a dilute urine. The daily urine flow has been calculated as 20 percent of an animal's body weight and doubtless results in some loss of salts. Uptake of sodium from the medium has been shown in specimens previously depleted of salt, so that it is likely that there are compensatory mechanisms. The exoskeleton of *Asellus* is more permeable than that of any other species of isopod tested, and uptake through the body surface and against a concentration gradient could occur, and probably does.

Rather more is known about conditions in terrestrial species, at least as far as their water relations are concerned. *Oniscus, Porcellio*, and *Armadillidium* are all genera that habitually live on land, almost of necessity in moist places. Their cuticles lack the wax coating that protects many insects from dehydration, and for most of them the rate of water loss through transpiration is high. Some of this is replaced by their food, but there must be other sources as well. It has been shown that not only do they drink water but they can imbibe it through anus as well as mouth from any free water surface. This has been demonstrated both by comparative weight studies and by the use of colored water and subsequent dissection of animals to which it has been offered. Proof that water was taken in through the anus has been provided by experiments in which the mouth was blocked by electric cautery, yet colored water was found in the body. All isopods tested, except *Ligia*, could take in water through the mouth from a block of damp plaster of Paris, in quantities sufficient to offset water loss through transpiration in an atmosphere with a relative humidity of 85 percent. Since oral and anal uptake of water occurs only in living animals, it must be an active, not a passive, process and one by which they can obtain water from moist soil. They also protect themselves from water loss by behavioral procedures, moving into damp crevices, under rocks and stones, and even, because their bodies are so small, taking cover under moss and damp grass. All the isopods excrete ammonia as the principal product of their nitrogenous metabolism, but the terrestrial species in some way spare or suppress this so that the total amount of nitrogen excreted, calculated as 0.3 mg/(10 g body weight)(day), is less than that of marine, semiterrestrial, and freshwater species. That they excrete ammonia at all, rather than a less soluble compound, is unusual among terrestrial invertebrates and provides an exception to the generalization that in any group of them, adaptation to a life on land has been accompanied by a change in their protein metabolism.

(iii) Decapods Renal function and osmoregulation have been most extensively studied in the larger decapods, whose size facilitates sampling of the hemolymph and of the urine from different parts of the excretory organs. Presumably, the urine is formed by ultrafiltration directly from the blood vessels into the end sac, although no capillary system has yet been demonstrated there, and it is debatable whether there is enough hydrostatic pressure in the renal vessels for such a process. There is little direct evidence, either pro or con, from actual pressure measurements, of which only a very few have been made. Hydrostatic pressure of the hemolymph in the first pereiopod of the crayfish *Astacus* is comparable to that of about 37 ml of water, while its colloidal osmotic pressure compares with the pressure of 27 ml of water. The osmotic pressure of the urine is equivalent to the hydrostatic pressure of only 4 ml of water. In *Homarus americanus* the hydrostatic pressure in the heart is only slightly greater than that in the hemocoele, the difference being equivalent to the hydrostatic pressure of 9 ml of water, and an even smaller difference has been reported for *Cancer irroratus*. *Homarus* never produces more than a small volume of urine and in some conditions none at all. This is known to be the case in specimens kept in pounds, in which anuria lasting as long as 1 month has no apparent ill effects. Indirect evidence for ultrafiltration has been provided, also in a very limited number of cases, by comparisons of the ionic concentration of the hemolymph with that of the fluid in the end sac. The Cl^- content of the fluid in the end sac and labyrinth of *Astacus* is the same as that of the hemolymph. In *Palaemon serratus*, a shrimp that ordinarily lives in brackish water, the concentrations of ions in the end sac are essentially the same as those in the hemolymph in specimens kept in full strength seawater. The results of experimental injections of inulin into the hemolymph of *Homarus americanus* and the shore crab *Carcinus maenas* also provide indirect evidence for ultrafiltration, for the inulin can be quantitatively recovered in the urine of lobsters kept in seawater of various dilutions. The amount, after an initial injection, decreases equally and progressively in both fluids over a period of 24 hr. In *Carcinus*, during an observation period of 4 days, the urine-blood ratio (U/B) of inulin steadily increased until all that had been injected was excreted. Moreover, glycosuria can be induced in lobsters by injection of glucose into the hemolymph. Under ordinary conditions the blood-sugar level is 24 to 40 mg/100 ml, and there is no sugar in the urine, but when the blood-sugar level is raised to 100 mg/100 ml, sugar does appear in the urine.

While the case for ultrafiltration seems well established in *Astacus*, where there is adequate hydrostatic pressure, and satisfactory correspondence between the composition of the hemolymph and the primary urine has been shown, the means by which the urine is initially formed in other crustaceans is not so clear. In the majority of them, secretion may well be the process wholly or partly responsible. In *Procambarus clarkii*, for example, the U/B ratio of inulin, after its injection, shows a range of 2 to 5, which might be interpreted either as a result of water absorption or as the secretion of inulin, possibly along with some filtration. Histologically the cells of the labyrinth resemble the secretory cells of other renal organs. Their free ends have deep microvilli, and large clear vacuoles arise from them and project into the lumen, suggesting that water is transferred through them into the tubule. Urine production stops in *Procambarus* when the crayfish are put into waters of greater salinity than their usual freshwater habitats, and the osmotic concentration of the tubular fluid rises. The principal experimental evidence for secretion has been derived from injection of dyes, of which a wide variety have been used. The results indicate that some dyes, essentially those of the carmine group, are secreted into the cells of the end sac, and so excreted, and that those of the indigo group are secreted into the tubule cells. Some dyes are concentrated in both parts of the organ. Molecular size, as well as other physical characteristics, influences the behavior of dyes in tissues, and results of such experiments are by no means conclusive, either of secretion to the exclusion of filtration or of the sites at which secretion may take place.

There is a considerable body of evidence to show that the urine, however it may be initially formed, is, in some species, modified during its passage through the excretory organs. This modification is concerned primarily with ionic regulation, which seems to be the primary function of crustacean antennal glands, although not limited to them. These glands play little or no part in the elimination of nitrogenous waste. This waste is principally ammonia, although small amounts of urea and even smaller amounts of uric acid have been detected in some species. Ammonia is lost by diffusion, most probably to the greatest extent through the delicate epithelium of the gills, although the alimentary canal and the digestive gland in particular are not excluded as possible sites. After complete removal of the antennae from *Carcinus*, which includes removal of the basal segment in which the urocoele lies, the crab's total output of ammonia was found to be only slightly less than that of intact animals, proving that the contribution of the renal organ to ammonia elimination is at best minimal but not indicating where the elimination does occur. Evidence for the gills as the main site for this (which on the basis of their histology is a reasonable assumption) is deduced from experiments in which the excretory pores and the openings of the gut have been blocked in specimens, from which, nevertheless, both ammonia and urea were still eliminated. Accumulation of injected dyes in the cells of the digestive gland gives some evidence that this organ may function, at least temporarily, in excretion. The integument may also act as a storage excretory organ, waste products of nitrogenous metabolism, especially uric acid, accumulating there to be eliminated at each molt. Nephrocytes probably also dispose of some nitrogenous waste.

Aquatic crustaceans whose osmoregulation has been investigated can be arranged in a series, a sliding scale of adaptability to conditions of osmotic stress. In most marine species, the hemolymph is isosmotic, or nearly so, with seawater; and the urine is isosmotic with the hemolymph, showing that the renal organs have no part in osmoregulation. Indeed, there is little capacity for osmoregulation in these animals, although ionic regulation seems to be general Species living habitually in brackish waters, with salinities ranging from 30 to 0.5‰, produce urine isosmotic with their hemolymph, yet maintain a hemolymph hyperosmotic to the external medium. Estuarine species produce urine that is isosmotic with their hemolymph, whose ionic concentration remains relatively constant over the wide spectrum of salinities to which they are exposed. Freshwater crayfish are exceptional in producing urine that is hypoosmotic to their hemolymph, which is, in turn, hyperosmotic to the external medium.

Although not all marine crustaceans have hemolymph that is isosmotic, or nearly so, with seawater, they all maintain a balance of ions that is more or less constant and show ionic if not osmoregulation. The hemolymph of many prawns and of some isopods is consistently hypoosmotic to 100 percent seawater; the hemolymph of *Pachygrapsus marmoratus*, for example, is only as concentrated as 90 percent seawater. The distribution of ions in the urine of these animals is different also from that in their hemolymph, although both fluids are isosmotic. There is little difference in Na^+ and Cl^- content, but, in general, fluid taken from the antennal glands is higher in Mg^{++} and SO_4^{--}, and lower in K^+ and Ca^{++} than the hemolymph. In *Homarus*, the Mg^{++} concentration of the urine is almost 13 times greater than that of the hemolymph and in *Palinurus*, nearly five times greater. In *Maja* the difference is slight, as might be expected, but the concentration of SO_4^{--} in this crab's urine is three times greater than in its hemolymph. In *Homarus* there is five times as much sulfate in the urine as in the hemolymph, but in *Palinurus* there is not much difference. The differences that are evident, however, indicate that the antennal glands selectively excrete ions, removing some but not others from the hemolymph, and that the selection is different in different species.

There must be some mechanism to compensate for the loss of ions and water excreted by the antennal glands, if the hemolymph is to preserve its osmotic relations with seawater and with the urine. Various lines of evidence show that this mechanism is the active and selective uptake of ions through some part of the body surface, most probably to the greatest extent through the gill epithelium. Water and ions may be obtained in the food and absorbed from the gut and some water may also be derived from metabolic reactions. Yet the fact that ionic regulation is maintained for weeks by marine species deprived of food argues against the diet as an important source to them. Moreover, the production of urine does not stop when the mouth and anus are blocked, yet fluid accumulates in the antennal glands and animals gain weight when the renal apertures are plugged. Metabolic water would hardly be sufficient to account for the daily output of urine as measured in starved animals, which averages 5 to 10 percent of the body weight. These data provide strong presumptive evidence that the external surface is the area through which water and ions enter in sufficient quantities to preserve the ionic balance and water content of the hemolymph. The calcified integument of the general body surface of marine crustaceans is more permeable to water than that of crustaceans that live in other aquatic environments, but the most permeable areas of all are the gills, which offer a surface of lightly chitinized epithelium. Oxygen and carbon dioxide are known to pass across these cells, and it has been shown that Na^+ and Cl^- are taken up against a concentration gradient by isolated gills of the crab *Eriocheir*. This is an active process and requires the expenditure of energy, for it stops when the oxygen supply to the cells is cut off or when they are exposed to cyanide or excess carbon dioxide. Additional evidence for the activity of the gill epithelium in ionic regulation in crustaceans in general has been provided by biochemical studies. Homogenates of the gill epithelium of the tropical land crab *Cardiosoma guanhumi* have, for example, revealed an ATP-ase system in them similar to that to which the active transport of cations has been attributed in a variety of animal tissues. Also, it has been shown that the enzymatic activity of mitochondria isolated from the gill epithelium of four species of marine crabs increased with the decrease in osmotic concentration of the suspending medium, indicating an increased rate of oxidative metabolism.

Crustaceans such as these that are ionic but not osmoregulators are stenohaline, restricted to life in waters of average and fairly constant salinity. *Maja*, for example, when exposed to seawater of 20 percent salinity gains as much as 3 percent in weight in the first hour. Its hemolymph becomes isosmotic with the external medium over the entire viable range of dilutions, reaching the limits of tolerance at dilutions over 20 percent. Since it can neither osmoregulate nor endure much dilution of the fluids bathing its tissues, its range of habitat is limited.

Life in brackish waters demands either a high degree of tolerance of dilution of the body fluids or an effective means of osmoregulation. Both these factors make it possible for *Carcinus maenas* to live in the brackish regions of estuaries. *Carcinus*, when exposed to 15 percent seawater, can maintain an osmotic concentration in its hemolymph comparable to that of 60 percent seawater and survives well in these conditions. Its tissue operate with adequate efficiency at this dilution, and regulation is possible down to this level of external salinity. These crabs cannot, however, maintain an adequate internal concentration in freshwater and, therefore, cannot live far up estuaries where they might be exposed to river water. They are intermediate in their regulatory capacities between the stenohaline marine species and those that are truly euryhaline. In contrast to *Maja*, specimens of *Carcinus* do not gain weight in water of 20 percent salinity, suggesting either that their body surface is less permeable or that the excess water is removed at a rate comparable to its rate of entrance. Indirect evidence of lower permeability in *Carcinus* is provided by experiments in which the excretory openings were blocked in animals put into diluted waters. In 50 percent seawater, the weight increased less than 2 percent/hr, indicating a slow rate of water entrance. Comparison of the calcium content of the hemolymph in specimens of *Carcinus* and of *Maja* immersed in calcium-rich seawater showed that the Ca^{++} content of the hemolymph of the marine crab increased more rapidly than did that of the shore crab. The concentration of iodide in the hemolymph of *Carcinus*, after a 5 hr exposure to seawater to which iodine had been added, was considerably less than that of *Hyas*, a spider crab. Iodine was also lost more rapidly from *Hyas* than from *Carcinus* when both were put into iodine-free water. It seems likely, then, that the integument of *Carcinus*, as representative of brackish water crustaceans, is less permeable to water and to ions than that of strictly marine species and that such low permeability offers them some measure of protection from osmotic stress. The urine of *Carcinus* is isosmotic with its hemolymph, which, in turn, is kept hyperosmotic to the medium. Flow of urine increases as the salinity of the medium decreases, with as much as a tenfold increase in animals kept for a time in freshwater. There is evidence, also, of resorption of water and of ions from the urine. The U/B ratio of inulin, after its injection into the hemolymph, is in *Carcinus* and some other estuarine crustaceans, 2:3, presumably due to resorption of water that has entered the end sac. In these animals also the concentration of Na^+ is lower in the urine than it is in the hemolymph, so that this ion must be resorbed from the initial fluid. Since their antennal glands have no nephridial canals, such as those found in freshwater species, these crustaceans provide a natural experiment demonstrating the function of the labyrinth, for resorption must take place there. There is also evidence of an active uptake of ions from the external medium, for the concentrations of defined salt solutions are reduced after specimens of *Carcinus*, previously exposed to distilled water to reduce their salt content, have been kept in them for some time.

Thus, there seem to be three mechanisms operative in *Carcinus* that enable the crabs to keep the osmotic concentration of their hemolymph above that of their external medium. One of these is the comparatively low permeability of the general integument; a second is the increased fluid output, largely from the renal organs but probably also from the gut; and a third is absorption of ions across both renal and extrarenal surfaces. This is an active process, for the oxygen consumption of the crabs increases with the dilution of the water in which they live. Yet the evidence for this is not entirely conclusive, as animals subjected to such experimental conditions often struggle to get out of the unfavorable environment and so consume oxygen in their muscular exertions.

Greater powers of osmoregulation are shown by crustaceans that can live far up in estuaries and even in the freshwaters of the rivers that enter them. These capacities have been particularly investigated in *Eriocheir sinensis*, for the adults, coming to the sea to spawn, live in salt water as well as in fresh after their migration upstream at a juvenile stage. In both of these environments, as well as in the intermediate conditions along the estuary, *Eriocheir* produces urine that is isosmotic with its hemolymph, yet this is hyperosmotic to the external medium when the crabs are in brackish or freshwaters. Ammonia is the principal end product of nitrogenous metabolism and is eliminated through the general body surface or the gut, and not to any appreciable extent by the antennal glands. Nor have these organs any decisive role in osmoregulation, for the ionic concentration of the hemolymph is maintained by active absorption of ions through the gill epithelium, compensating for renal and extrarenal salt loss. The urine flow per hour in specimens averaging 60 g in weight has been found to be 1.5 ml/(kg)(hr), and the Cl^- lost in the urine, 0.42 mmol/

(kg)(hr). If the renal Na^+ loss is equal to the Cl^- loss, then 0.42 mmol/(kg)(hr) of Na^+ is lost via the antennal glands. But the mean rate of Na^+ loss from specimens exposed to deionized water is 2.1 mmol/(kg)(hr) so that the extrarenal loss of sodium, approximately 1.7 mmol/(kg)(hr) is even greater than the renal.

Measurements of salt depletion in test solutions in which specimens of *Eriocheir*, or their isolated gills, have been immersed have demonstrated that Na^+, K^+, Cl^-, bromide, cyanate, and thiocyanate are taken up by the gill epithelium against a concentration gradient and that Na^+ can be absorbed independently of Cl^-, and vice versa. This process is inhibited by poisons of the respiratory enzymes and by oxygen deprivation and is, therefore, an active one. Although the rate of ion absorption in *Eriocheir* is greater than it is in *Carcinus*, it provides sufficient uptake only in hard waters, and is not adequate in soft waters. This may be one reason why the crabs, which are so successful in populating the rivers of Germany and France, have not established themselves in those of Scandinavia.

It has been shown, too, that renal and extrarenal salt loss varies with different conditions, both internal and external. *Eriocheir* conserves magnesium, for example, when in freshwater but excretes it preferentially when in salt. The total salt loss is greater in molting than in intermolt individuals, whose integuments are less permeable. This may also be the case in mature and ovigerous females, which must return to the sea to spawn and which die in a few days when transferred from seawater to fresh. Neither their eggs nor early larvae can adjust to conditions of salinity less than 30‰, and it is not until the megalops (postlarva) stage is reached that the juvenile crabs can make their way up the estuary to the freshwater in which they spend the greater part of their adult lives.

Some decapods have become completely adapted to life in freshwater and not only live but breed there. Some of these are occasionally found in brackish water of low salinity to which they can adjust. *Pacifastacus leniusculus*, the crayfish of western North America, can, for example, regulate its internal environment over a wide range of external salinities and is found in bays as well as in streams. Possibly, one aspect of the adjustment of crayfishes as a group to freshwater environments is the absence of a larval stage in their life history, so that they need not return to the sea, as *Eriocheir* must, for spawning and the early postembryonic stages of their young. This need has also been circumvented, although in some different way, by freshwater branchiopods, for their eggs hatch out as nauplii that live and complete their development in the same waters the adults live in. However, the nauplius stage is also bypassed in the development of the American lobster, yet neither lobsterling nor adult can survive in diluted seawater.

It might be assumed that the determining structural and functional factor in adjustment to a freshwater existence is an extended tubular system with a nephridial canal, since this is the pattern of the excretory organ in most of the species examined. But there are some species of the prawn *Palaemonetes* that are fully adapted to life in freshwater and have no nephridial canals. These crustaceans excrete urine that is isosmotic with their hemolymph, indicating that the excretory organ has no part in osmoregulation. The urine excreted by crustaceans with nephridial canals is hypoosmotic to their hemolymph, indicating osmoregulatory activity by that part of the excretory system. The morphological and functional differences are strikingly illustrated by the amphipod *Gammarus*. *Gammarus locusta* lives in salt water, has no nephridial canal, and excretes isosmotic urine; *G. pulex* lives in freshwater, has a nephridial canal, and excretes hypoosmotic urine. The excretory organs of the crayfishes *Astacus* and *Procambarus* also have nephridial canals and the animals excrete hypoosmotic urine. It has been shown, by determinations of the chloride content in the end sac, labyrinth, nephridial canal, and bladder, that the Cl^- concentration of the fluid in the tubule is equivalent to that in the hemolymph until it reaches the nephridial canal, when it drops off sharply. Correlation between ability to excrete a fluid less concentrated than either the hemolymph or the initial filtrate and the presence of a nephridial canal as an additional segment to the more generalized excretory organ seems clear, but not that between the presence of a nephridial canal and adaptation to a freshwater habitat. Production of dilute urine is not essential to this, although it is undoubtedly advantageous, since from the point of view of energetics, absorption by the renal organ of a substance that has been filtered from the blood is more economical than its absorption from a medium as dilute as freshwater by some other area of the body.

Also, hemolymph of comparatively low osmotic concentration is not essential to life in freshwater, although this is in general a characteristic of freshwater animals. The hemolymph of the crab *Potamon* that habitually lives in freshwater has, for example, a Δ of 1.1 to 1.2°C. *Potamon* can also live quite successfully in seawater and is exceptional in being able to stand immediate transfer from freshwater to 100 percent seawater. Its hemolymph becomes isosmotic with a medium with a Δ of about 1.8°C, and the urine it excretes is isosmotic with its hemolymph under all conditions. The crayfishes, on the other hand, have far less concentrated hemolymph. The Δ of *Astacus* hemolymph is 0.6 to 0.8°C, and its urine has about 10 percent the concentration of its body fluids. This is true of *Procambarus* as well. It is not yet certain whether the change in concentration of the urine as it passes along the renal tubules of crayfishes results from resorption of salts or secretion of water. It seems evident that in *Astacus* there is selective excretion of ions and that the loss of Cl^- is greatest, at least in terms of absolute amounts. It is less relative to the original concentration in the hemolymph than the loss of Mg^{++}, which is less in absolute amounts, but the concentration of Mg^{++} is lower in the hemolymph than that of Cl^-. The exoskeleton of *Astacus* is far less permeable to water than are those of marine or brackish water decapods, but active uptake of ions has been demonstrated in both *Astacus* and *Procambarus*, presumably through the gill epithelium. These crayfishes will take up Cl^- from tap water and from solutions of NaCl as dilute as 1 m*M* or less, reducing their chloride content almost to zero. K^+ and SO_4^{--} are not absorbed and Ca^{++} only in the immediate postmolt stages. *Astacus*, unlike *Potamon*, cannot stand abrupt transition from its natural waters to seawater, but can be gradually acclimated to salt waters. It can live over a month in seawater with an osmotic concentration of Δ1.3°C and will even remain homoisosmotic in waters of greater concentration as long as it survives. Its osmoregulatory mechanisms break down, however, at osmotic concentrations with a Δ of 1.9°C or greater.

Invasion and successful habitation of freshwaters by crayfishes seems then to have resulted from a combination of relatively low permeability of their exoskeletons, relatively low osmotic concentrations of their body fluids, the ability of the excretory organ to produce a urine more dilute than the hemolymph, and ability of the gill epithelium to transport salts selectively. Because their body surfaces are comparatively impermeable, the danger of a great influx of water is minimized and because their body fluids have a comparatively low osmotic concentration, less work has to be done to compensate for any salt loss, renal or extrarenal. The role of the excretory organ of crayfishes in osmoregulation seems to be an active one, attributable mainly to the operation of the cells in the nephridial canal, where either secretion of water or absorption of ions, or both, may take place. There is evidence, from experiments on *Cambarus*, of secretion of inulin into the tubules by cells of the nephridial canal.

Some crabs are amphibious, living along the shore in the intertidal zone or above it and spending varying amounts of time entirely out of water. *Pachygrapsus crassipes* lives among rocks above the low-tide level along the Pacific coast and spends a good deal of time out of water, for the most part at night, moving over the exposed mud. *Grapsus grapsus* lives in similar habitats but is more diurnal in its activity than *Pachygrapsus* and spends more time on the mud by day. Ghost crabs of the genus *Ocypode* burrow high on sandy beaches, digging their burrows as much as 4 ft deep in order to reach moist sand. They emerge from their burrows and run over the sand by day, but their longer periods of activity are at night. Fiddler crabs of the genus *Uca* make burrows at the high-tide level that are not more than 1 ft in depth and spend much of their time by day scuttling over exposed mud flats and beaches. The crabs that are most terrestrial in habit are the tropical hermit crabs *Coenobita* and *Birgus*, and the brachyurans *Cardiosoma* and *Gecarcinus*. *Coenobita*, native to the East Indies, lives almost entirely out of water and uses half-coconut shells, joints of bamboo, and even broken lamp chimneys as substitutes for the molluscan shells commonly inhabited by hermit crabs, since the shells are not readily available to individuals living so far from the water's edge. The robber or coconut crab of the South Seas, *Birgus latro*, not only lives on land but is often very active in full sunlight, although primarily nocturnal in its habits. In some districts these crabs live close to the sea, but in others they have been found on stony plateaus 30 or 40 yd above the shore line. And in the Solomon Islands specimens have been found at elevations of 300 ft. The big blue-gray crab *Cardiosoma* lives in fields and woods. In Puerto Rico they have been damaging to the rice crops, for they have burrowed into the hard, dry clay beneath the dikes built to hold water for the rice seedlings and have turned the solid bottom into a sieve through which the water quickly runs away. They have also been destructive to agriculture in Florida by eating vegetable seedlings. *Cardiosoma guanhumi* can live several miles from the sea, in burrows or hardwood hammocks that they excavate among the roots of trees, especially fig trees. But they can also live totally submerged in water, either fresh or salt, that may fill their burrows. Sometimes, after spawning, they look for high places and have been known to climb to the top of mangrove trees and even of chimneys. *Gecarcinus* is the most terrestrial genus of all. These land crabs live in forests along the coasts of Africa and Central America, also at a considerable distance from the sea, which they visit only once a year for spawning. All of these amphibious and terrestrial crabs are similarly bound to the sea, but members of genera other than *Gecarcinus* visit it quite frequently, and not only in the breeding season.

The conditions which these crabs must meet are different from those of estuarine species. When in water they may be exposed to even greater extremes of salinity, extending for unpredictable periods of time, not just between tides. It has been recorded, for example, that fiddler crabs, after a torrential rainfall, were exposed to water with a salinity equivalent to that of 17 percent seawater and, at the other extreme, on a hot, sunny day, when evaporation was rapid, to water with a salinity equivalent to 170 percent seawater. That they can adjust to such extremes shows greater power of osmoregulation. They actually have no osmotic problems when they are on land, since they are not in an aqueous environment. The danger for them then, as for all terrestrial animals, is loss of water by evaporation and consequent concentration of salts in their body fluids. Most of this water loss is across the shelled areas of the body, although comparative studies have shown that the shells of terrestrial species are less permeable to water than those of aquatic species. Ability to absorb water against a concentration gradient and to excrete salts is, therefore, of especial importance to their ways of life.

Experimentation to determine their abilities and means of regulating their internal environments present certain difficulties. Most of the crabs cannot tolerate continuous or even prolonged immersion and must be allowed to come out of an aqueous medium at intervals, so that experiments cannot be conducted in just the same way as those on marine and estuarine species. *Birgus* drowns when kept in water for a day or two, and *Gecarcinus* even sooner than that. Nevertheless, tests for ionic and osmoregulation have been made on a number of species.

In natural conditions the hemolymph of *Pachygrapsus* is slightly less concentrated than seawater, although the concentration increases and nearly doubles just before an animal molts. In experimental conditions, adjustment to either hyperosmotic or hypoosmotic media is rapid, and animals preserve their internal stability over a wide range of salinities. In dilute media, the urine excreted is slightly hypoosmotic to the hemolymph; in concentrated media, it is isosmotic. The antennal glands seem, therefore, to have little part in osmoregulation. When the crabs are in air, it is possible that water may be secreted through the gill epithelium into the branchial chambers, and the fluid in these chambers may be a source of ions, replacing those that are lost in the urine or through the body surface and the gut. The branchial epithelium lining the gill chamber has been shown to be an additional site of ion entrance in some crabs, supplementing that of the epithelium of the gills themselves. The value of this source of salts is questionable, however, since the volume of fluid in the branchial chambers is so small in relation to that of the body fluids that its contribution would probably be negligible. Extravascular salt pools have also been postulated as possible sources of ions which could be drawn upon when the crabs are in extremely dilute water. Their replacement in the salt pool might be effected when the animals are in waters of high salinity.

Measurements of the concentration of Mg^{++} in the urine of *Pachygrapsus* and of *Uca* as well, compared with measurements of its concentration in the hemolymph and external medium, suggest that the antennal glands, while inoperative in osmoregulation and in all other aspects of ionic regulation, control the concentration of this particular ion. The concentration of Mg^{++} in the urine of specimens of *Pachygrapsus* exposed to 175 percent seawater is some four times greater than that in the external medium and 10 times greater than that in the hemolymph. There is evidence that the bladder is concerned with the regulation of Mg^{++}, as well as with absorption of glucose from the filtrate, indicating that this is a physiologically active part of the excretory system and not merely a reservoir in which urine may be temporarily contained. Similarly, the urine of *Uca* has been found to contain slightly more than twice the concentration of Mg^{++} in the hemolymph and contains more Mg^{++} than seawater or the fluid in the branchial chambers, while the fluid in the gut has a concentration of Mg^{++} about equal to that in the urine. The gut epithelium may thus function both in the excretion of salt as well as in its absorption. In *Uca*, the gut fluid is hyperosmotic to the hemolymph and has been found to have less Na^+ and Cl^- than average seawater, more Mg^{++} and K^+, more than twice as much Ca^{++}, and six times as much SO_4^{--}. It seems apparent, therefore, that the water taken in with the food and some of its salts are selectively absorbed by the gut epithelium, making this region important in osmoregulation and more important in ion regulation than the antennal glands.

Since crabs of more terrestrial habits are only occasionally exposed to water, the questions raised by their adjustments are concerned primarily with the extent of their tolerance of increased internal concentration resulting from loss of water by evaporation and with their means of replacing water so lost. Their body fluids, like those of freshwater and intertidal species, are regularly hypoosmotic to seawater. The ionic concentration of the hemolymph of *Birgus* in natural conditions is that of 85 to 90 percent seawater, but it can tolerate concentrations over a range equivalent to those of 70 to 119 percent seawater. These crabs drink water, and put it in their branchial chambers without being immersed in it. The salinity of their hemolymph varies with the salinity of the water available to them. If under experimental conditions robber crabs are allowed only freshwater, the concentration of their hemolymph may drop as low as that of 64 to 74 percent seawater, and if allowed only seawater, it may rise to values as high as 118 to 123 percent seawater. Although *Birgus* drowns in a day or two when totally immersed in water, it is capable of a high degree of regulation as long as it can survive the respiratory stress; and in dilute seawater can maintain an internal osmotic concentration that is greater than that of the medium; and in concentrated seawater, one that is lower, or hypoosmotic to it. Similarly, the salt content of the hemolymph of *Coenobita* varies with that of the available water, and the crabs can tolerate a range of Na^+ concentration in the body fluids from 82 to 220 percent of the usual level. Specimens of a Madagascan species, *C. cavipes*, were kept in the laboratory for 4 days without access to any water, and others were kept for a similar length of time with access either to fresh or to seawater. At the beginning of the experiment, the average ionic concentration of the hemolymph was equal to that of 90.8 percent seawater. At its conclusion, the hemolymph of the crabs without any water was hyperosmotic to seawater, with an average concentration equal to that of 116 percent seawater; the hemolymph of those with access only to seawater had an average ionic concentration equal to that of 105 percent seawater; and the hemolymph of those with access only to freshwater had an average concentration equal to that of 77 percent seawater, lower than that of animals in their natural environments. The periodic visits of *Coenobita* to the sea are not, therefore, primarily for the purpose of obtaining water for ionic regulatory purposes, for which they can obtain adequate amounts by drinking waters of lower salinity. The gut epithelium seems in these crabs to be the important tissue in preserving stability in concentration of their body fluids, and their food contributes significantly to the inorganic ions of the hemolymph. Experiments with *C. perlatus* deprived of food but with full access to freshwater have resulted in a reduction of Na^+, K^+, and Mg^{++} in their hemolymph.

Cardiosoma is more resistant to drowning than other land crabs and by standard experiments testing osmoregulation has been found to be a very effective regulator. The hemolymph of individuals captured on land is hypoosmotic to average seawater, and this concentration is maintained even when the animals are immersed in full strength seawater. The urine of specimens in natural conditions is isosmotic with the hemolymph, indicating that the antennal glands have no part in ionic regulation, but concentration of Mg^{++} in the urine has been demonstrated in individuals exposed to salt waters. In its native Madagascan habitats, *C. carnifex* has been observed to scoop water up from shallow puddles with its first pair of pereiopods and drop it into its mouth. It showed no preference for water of any especial salinity and drank equally readily from any pool in a range of salinity of 25 to 118 percent seawater. *Cardiosoma guanhumi*, a species native to Caribbean shores and those of southern Florida and living in a wide range of salinities, has survived for considerable periods of time in the laboratory in distilled water, in tap water, and in 100 percent seawater. The animals in distilled water showed regulation of Na^{+} and K^{+} in their hemolymph, urine, and gastric fluid for a period of 7 days, without measurable change in oxygen consumption. The Na^{+} and K^{+} concentrations in the hemolymph and urine increased 60 percent in those kept for the same period of time in 133 percent seawater and 300 percent in those exposed to 175 percent seawater. The mechanism of ionic regulation thus appears to break down when the crabs are in concentrated media, although a 60 percent increase in oxygen consumption over the usual rate in animals in 133 percent seawater suggests that more energy than usual is being expended toward one end or another.

Absorption through the surface seems to be a significant means of regulation in the species of crabs that are most terrestrial in habit. *Cardiosoma* has been shown to absorb water from a substrate dampened with either fresh or salt water. Animals that had been desiccated for 48 hr by exposure to circulating air, but out of the sun, and then put into sand dampened with either fresh or salt water were found to gain weight after a time and to reduce the osmotic concentration of their hemolymph and urine. These changes were greater in crabs in sand moistened with seawater than in sand moistened with freshwater, but the ionic concentration in the fluids of the former were lower than that of the seawater used to dampen the sand, showing that water had been taken up against the concentration gradient. *Gecarcinus*, which naturally visits the sea only once a year, can live indefinitely in sand dampened with freshwater, but dies from dehydration in sand dampened with salt.

It is evident that decapod crustaceans exhibit a wide range of regulatory ability and that those that are capable of osmoregulation employ a variety of methods to maintain homeostasis. The function of the antennal glands in this connection is negligible in marine species whose hemolymph is isosmotic with the medium; minimal in brackish and estuarine species, except in respect to regulation of certain ions, especially Mg^{++}; and maximal in freshwater crayfishes. Gut, gill, and branchial epithelium are of varying importance in different species in similar habitats as either absorptive or secretory areas. Elimination of water and salts is vital in the conditions of life to which some crustaceans may be exposed; their conservation or absorption is equally vital in others. There is still a great deal to be learned about the areas where these events occur and the mechanisms by which they are effected, but it is clear from information already gained that there is no uniformity in either. The gills are of proved importance to *Eriocheir* in the absorption of salts when the crabs are in freshwater, but *Birgus* and *Gecarcinus*, which have also problems of salt balance, can live for months after their gills have been removed.

(4) By Malpighian tubules The operation of Malpighian tubules has been most fully studied in insects, some of which are particularly favorable for experimentation. Of especial interest are those which, as fluid feeders, imbibe large quantities of water with their food, those that live in arid environments and subsist on dry food, and those that as juveniles or adults inhabit waters that are either fresh or salt. These different conditions impose different conditions of water balance, and ionic and osmoregulation and are met in different ways by different species.

(a) INSECTS Hydrostatic pressure in the hemolymph of insects is practically nil; the urine is formed by diffusion, given a sufficient concentration gradient of solute between the hemolymph and the tubular fluid, and by secretion. Histologically, the cells of the tubules are secretory in character, and there is also evidence that ions, especially K^+, move across them against both concentration and electrochemical gradients. Experiments in which Malpighian tubules have been drawn out of the body of an insect and observed in a drop of hemolymph to which phenol red or another suitable dye has been added show that the dye passes across the cells into the tubular fluid which becomes colored, indicating passage of the dye and water in one direction across the wall of the tubule.

The connection of the Malpighian tubules with the digestive tract and the consequent mixing of the tubular fluid with the contents of the hindgut make collection and analysis of the urine itself difficult. In a few species, however, it has been possible by micropuncture to obtain enough of the fluid from different regions of the slender tubule to make measurements of its osmotic concentration and compare them with that of the hemolymph. The osmotic pressure of the latter varies with different species, but it is, in general, higher in terrestrial than in aquatic insects and higher in those that live in dry environments than in those to which moist air is a necessity. The osmotic pressure of the hemolymph is largely due to its content of free amino acids and little to the salts dissolved in it. In general, the urine is hypoosmotic to the hemolymph, and its ionic content is quite different. The chief product of nitrogenous metabolism in insects is uric acid, which is thought to enter the tubules as one of its more soluble salts and then to be precipitated out in crystalline form when water and base are resorbed into the hemolymph.

The fluid-feeding aphids have no Malpighian tubules, but their highly specialized digestive systems provide a bypass for the water so that no appreciable amount is absorbed from the gut to dilute the hemolymph. In heteroptera, lacking the filter chambers characteristic of homoptera and feeding on plant juices, gastric caeca may serve as organs for the excretion of water. The blood-sucking hemipteran *Rhodnius* has proved to be exceptionally favorable material for study of the osmoregulatory system in a fluid-feeding insect. Taking as it does a large quantity of food at comparatively long intervals, the effect of an immediate influx of water into the hemolymph via the gut can be observed, as well as the long-term operation of the excretory system between the insect's meals. Some three-quarters of the water ingested with a blood meal is excreted by the Malpighian tubules within 3 to 4 hr. This rapid and continuous production of a dilute urine flushes out the tubules and clears them almost completely of their previous contents. It also reduces the persistent osmotic stress to which the insect would be subjected were the excess fluid not quickly eliminated. Some other fluid-feeders protect themselves in a similar way. The tsetse fly, for example, voids a volume of dilute urine equal to 70 percent of the volume of the total insect in 1 hr after filling its stomach with blood.

Observation of the tubules in living specimens of *Rhodnius*, dissected so as to expose them intact, shows that they are functionally differentiated. The upper (distal) two-thirds of each contains clear fluid, while the lower (proximal) one-third is filled with granular material and is opaque. If, some hours after a meal, a section of the proximal portion is blocked off from the rest of the tubule by ligatures of soft wax, the area between the ligatures is clear, while in the region distal to it, granular material accumulates and the tubule becomes distended. If a ligature is placed at the junction of the proximal and distal regions, the distal region remains clear but becomes appreciably distended, and when one is placed at the junction and another one a little distance below it, the region distal to the junction remains clear while that between the ligatures is clear but not swollen. These results are interpreted to mean that fluid enters the distal region of the tubule and leaves it in the proximal one-third, concentrating its contents and leading to precipitation of uric acid. The movement of neutral red introduced into the hemolymph has confirmed this interpretation, for it can be seen that the upper portions of the tubules soon become stained and that the fluid passes along them and loses its color after it reaches the proximal region. The fluid in the distal regions has been found to be slightly alkaline in reaction, while that in the proximal region is acid, confirmation also of the interpretation that uric acid enters the distal region as a soluble salt from which the base is removed in the proximal region and returned to the hemolymph, along with much of the water, while the precipitated uric acid is voided with the feces.

This resorption of water in the proximal part of the tubule may be peculiar to *Rhodnius* and an adaptation associated with its manner of feeding. For in other terrestrial insects investigated, which include representatives of Orthoptera, Lepidoptera, and Diptera as well as other Coleoptera, the rectum is the principal site of resorption, and tubules and hindgut together constitute the internal regulatory system. The tubules serve to collect materials from the hemolymph and the hindgut to restore some of them to it. In the stick insect *Carausius* (*Dixippus*) for example, phenol red injected into the hemocoele can be seen to enter the Malpighian tubules, which remain colored along their entire length, and the fluid does not lose its color until after it has entered the hindgut. Analyses of the hemolymph, and the tubular and rectal fluid show that while the osmotic concentration of the hemolymph is slightly higher than that of the urine, the osmotic concentration of the rectal fluid is two or three times greater. In respect to ionic concentrations, however, hemolymph and urine are strikingly different. Measured in millimoles per liter, there is about twice as much sodium in the hemolymph as in the urine, but the urine contains eight times as much potassium as the hemolymph and the rectal fluid more than twice as much as the urine. *Carausius*, as a herbivorous insect, consistently has a higher level of potassium in its hemolymph than is usual among carnivorous or omnivorous species, yet excretes a great deal of that which it ingests. That the transport of potassium takes place along the entire length of the tubule has been demonstrated in isolated pieces taken from all parts, which equally readily transport the ion from the medium to the fluid in the lumen.

The usual rate of urine production by *Carausius* is about 0.07×10^{-3} ml/(mm^2 surface area)(min), and further production ceases when the pressure in the lumen of the tubules reaches a level equivalent to that of about 20 mm H_2O. At the usual rate the tubules then secrete the equivalent of all the body water each day, most of which is returned to the hemolymph across the walls of the rectum. (This has a parallel in the absorption of water by the gut in earthworms with enteronephridia.) It has been calculated also that 95 percent of the sodium and 80 percent of the potassium secreted into the tubules is resorbed in the hindgut. Evidence for this part of the system providing the main site for resorption has been provided also by experiments with radioactive isotopes on larvae of *Tenebrio molitor*. Na^{24} introduced into the hemolymph as $Na^{24}Cl$ was recovered in the saline solution bathing the rectum, which, still attached to the midgut and tubules, had been dissected out of the body into a drop of salt solution made up with nonradioactive NaCl.

Water loss through transpiration from exposed and permeable surfaces is a major problem to all terrestrial insects. Most of them, unlike crustaceans, have cuticles covered with wax, or a waterproofing grease, which reduce the dangers of evaporation from the general body surface, but also minimize the possibility of absorption of water, a condition that might be of advantage to a terrestrial insect. It is possible, however, that passage of water may be accomplished more easily in one direction, from the outside to the inside, than in the other, from the inside to the outside. Studies of the ultrastructure of insect cuticles, more detailed even than those that have already been made, may provide an explanation for this functional asymmetry. For there is evidence that insects can absorb water from the air, especially from atmospheres of high humidity but still not saturated with water vapor. Starved mealworm larvae have been shown to regulate their water content in air with relative humidities below about 80 percent; in air in which the relative humidity was above 88 percent, their water content increased and they gained in weight. The rate at which water enters their bodies under these conditions may be as great as 0.4 mg/(cm^2)(hr), in contrast to the rate of water loss in dry air, which is estimated to be at a maximum 0.08 mg/(cm^2)(hr). Starved nymphs of the grasshopper *Chortophaga* were also found to gain weight when kept in air with a relative humidity of 95 percent and over; after 6 days of starvation the insects died, but all those that had been kept in atmospheres with relative humidities of 82 percent or above were found to have increased their water content. And prepupae of the rat flea *Xenophylla brasiliensis* can take up water from air with a relative humidity as low as 80 percent. Presumably, its entrance is through the integument in general, but there is no definitive proof of this such as has been provided from experiments on ticks. The castor bean tick *Ixodes ricinus* can take water vapor from air with relative humidities down to 92 percent, and this removal continues even when the spiracles are blocked. It stops, however, if the ticks are deprived of oxygen or exposed to cyanide, showing it to be an active process. *Ornithodorus moubata* also takes up water from the air under conditions of humidity or aridity that preclude diffusion as the process involved. Asphyxiation prevents uptake in *Ornithodorus*, but so, in this species, does blocking of the spiracles. It is presumed that in ticks the epidermal cells, in general, are concerned in the inward transport of water from air that is not fully saturated with water vapor, and that by reduced secretory activity they also operate to conserve body water when the relative humidity of the ambient air is low. The same phenomenon may be true for those insects that can absorb water vapor.

In addition to the general integument as a site of water loss through transpiration, respiratory areas are always surfaces through which water and salts may leave the body. Various means of reducing this danger have evolved in terrestrial insects, along with other adaptations to their life on land. These protective devices include depression of the spiracles below the surface of the body, tufts of hairs or bristles that deflect air currents from them, and devices for closing them. Most insects can obtain enough water and salts from their food and drink to replace those that may be lost by transpiration or excretion, with the exception of those that live in environments where water is not available and that eat very dry food. Mealworm larvae, for example, will live and grow on a diet of dry bran containing at most 1 percent water, when kept in closed containers with no access to liquid water. The feces of these insects are dry and powdery and, presumably, all the water that passes along their Malpighian tubules is conserved by resorption in the hindgut. Some of this water is derived from metabolic reactions, which is also the case in flour beetles, dermestid beetles, and other feeders on essentially dehydrated foods. The extent to which metabolic water is really useful in osmoregulation has been debated for many years, but that it is a source for these insects, which have no other evident means of obtaining fluid, seems unquestionable. The fact that it is a useful source to *Tenebrio* larvae does not weaken the evidence that they can absorb water from moist air, for the amount taken in by starved animals was far in excess of any that could have been provided by the metabolism of stored materials. More debatable are the questions of whether insects under stress of dehydration increase their metabolic rates in order to produce more water or shift from carbohydrate to fat degradation to the same end. This event has been reported in the blowfly *Glossina*, in which genus starving individuals in very dry air were found to metabolize fat rather than protein or carbohydrate. Such a shift could be an economical one for an insect needing to obtain water, because gram for gram the total oxidation of fat yields almost twice as much water as the total oxidation of carbohydrate.

Insects that live in aquatic environments have quite different problems and ones that they share with other aquatic invertebrates. Some insects, like dragonflies, mosquitoes, and midges, spend their juvenile lives in water; others, like diving beetles, live all their lives underwater with occasional visits to the surface. The larvae of some mosquitoes and midges live in freshwater, those of others in estuaries, in the sea, and sometimes in water of even higher salinity than this. Aquatic beetles inhabit freshwaters and like other animals in such environments have the problems of an influx of water through permeable areas and of an environment to a greater or lesser extent deficient in salts. Air-breathing insects, like beetles and dipteran larvae and pupae, are in less danger of water gain than species whose gaseous exchange takes place through gills, because their respiratory areas, presumably their most permeable ones, are not exposed to the water; but all of them have the problem of salt conservation or replenishment. Saltwater dwellers have the common problem of water loss and salt gain. And both fresh- and saltwater species that feed on other aquatic animals take in comparatively large volumes of water with their food, which if absorbed into the hemolymph must be eliminated in some way.

The means by which these conditions have been met have been investigated particularly in larvae of the alderfly *Sialis*, which have an extensive tracheal gill system; in larvae of the midge *Chironomus thummi*, which also make use of water respiration; in the air-breathing larvae of various species of the mosquito *Aedes*, and of the flies *Coelopa frigida* and *Ephydra* spp.; and in adult water boatmen, homopterans of the genus *Corixa*.

Larvae of *Sialis* are active predators on other aquatic insects and apparently depend entirely on their food for replacement of ions that leak out. Salt loss is greatest through the gills and almost negligible through the urine. There is very low permeability to Cl^-, for it was 6 weeks before the concentration of Cl^- in the hemolymph of larvae kept without food in distilled or tap water was reduced from the ordinary level, equivalent to 0.8 to 1.0% NaCl, to a level equivalent to that of 0.06% NaCl. Its restoration to the ordinary level after a few days' exposure to 1% NaCl suggests that the integument is somewhat permeable to ingoing Cl^-, although almost completely impermeable to outgoing. That this is not an active uptake of ions is shown by the preservation of the ordinary level of Cl^- in the hemolymph of specimens exposed even for several days to 100% CO_2.

The hemolymph of larvae of the midge *Chironomus thummi* has an osmotic pressure about equal to that of a solution of 0.9% NaCl. In spite of the fact that their integument is very permeable, they will remain homeo-osmotic in external environments that range from 0 to 0.8% NaCl, but have been found to lose weight in media whose salinity is greater than 1% NaCl. If ligatures are placed around the anterior and posterior ends of the body, blocking off the gut so that water cannot be eliminated by that route, the larvae gain weight in hypoosmotic solutions, suggesting that the gut is ordinarily the principal region for the elimination of excess water. These larvae, like those of mosquitoes, have anal papillae, elongated, leaf-like extensions of the integument surrounding the anus. Ligature of these organs, or their destruction by silver nitrate, prevents the uptake of Cl^- and Br^- which can be accomplished by intact *Chironomus* larvae from solutions as dilute as 0.001 *M*. Active uptake, as well as elimination, of water is, therefore, a factor in osmotic and ionic regulation in *Chironomus*.

Larvae of the mosquito *Aedes* are air breathers and are of especial interest in relation to insect osmoregulation, because some species inhabit freshwater and some, salt. Correlation between form and function in osmoregulatory organs is evident in the extent of the development of the anal papillae in the two different environments. In freshwater species, they are much larger than they are in saltwater ones. In species such as *A. argenteus* that, although naturally freshwater dwellers, can be gradually acclimated to life in salt water, the papillae become progressively smaller as adaptation proceeds. Proof of the function of the papillae has been derived in a variety of ways from a variety of insect species. For example, isolated papillae have been found to swell when immersed in tap water and to shrink in Ringer's solution. When the anal papillae are destroyed, larvae of *A. aegypti* become osmotically inert. It has been shown, too, by destruction of the papillae, that most of the exchange of Na^{24} between the larvae and the medium occurs through them, and only a small fraction through body wall and gut. Studies of their ultrastructure support the experimental demonstrations of their function and suggest that their intake of ions may be regulated by variations in the surface area of the plasma membrane on the cuticular side of the papilla. The cytological picture is much like that of the gill epithelium of *Artemia*. Electron micrographs of the cells in the anal papillae of larvae that had been raised in a NaCl solution with a salinity of 0.039 percent have shown that the plasma membrane on the cuticular side is greatly folded, with numerous mitochondria in close association with the folds. The plasma membrane on the opposite, basal surface of the cell is invaginated to form an anastomosing network. On the other hand, the corresponding plasma membranes in larvae that had been raised in 0.67% NaCl, which is approximately isosmotic with their hemolymph, or in solutions of 1% NaCl, which is hyperosmotic to it, are reduced in area and have comparatively few mitochondria associated with them. The distribution of mitochondria under these different conditions and the demonstration of abundant alkaline phosphatase in the cells of larvae raised in hypoosmotic solutions implies that ion absorption by these cells is an active process that requires the expenditure of energy. It is possible, too, that there are different mechanisms for the uptake of cations and of anions, and of different cations, for the movements of K^+ and Na^+ are independent of each other. The rate of secretion of

K^+ in *Sialis* is, for example, more than 10 times that of Na^+.

Unlike the integument of *Sialis* larvae, that of the larvae of *Aedes* is practically impermeable, with the exception of the anal papillae where ionic transport takes place. Ion uptake by these structures has been proved also in the larvae of *Culex* and, in nymphs of the dragonflies *Libellula* and *Aeschna*, by homologous structures, the rectal gill plates at the bases of the rectal gills. These nymphs feed on aquatic insects and small fish, taking in some water with each meal.

Larvae of *Aedes detritus* and *Culex fatigans* live in waters with salinities up to nearly three times that of seawater, and in the laboratory they can adapt to solutions of NaCl from 0 to 10 percent. The anal papillae are vestigial and relatively impermeable, or absent entirely. Since the larvae are air breathers, they would have little problem of osmoregulation were it not for the large amounts of water and salts that they take in with the suspended particles of food that they ingest. The minimal osmotic pressure of the hemolymph of *A. detritus* is equivalent to that of an 0.8 percent solution of NaCl, but when the environment is changed from 0 to 6% NaCl, the osmotic pressure of the hemolymph rises until it reaches a level equivalent to that of a 1.4 percent solution of NaCl. Water and salt transfer take place across the gut wall, and the excess is eliminated through the Malpighian tubules. When in seawater, the fluid in the rectum is hyperosmotic to the hemolymph and almost isosmotic with the external medium, but when the larvae are kept in freshwater their rectal fluid is hypoosmotic to the hemolymph. These facts show that the cells of the rectal glands absorb either salt or water, as the situation demands. The urine of *A. aegypti* is isosmotic with the hemolymph or very slightly hypoosmotic to it, but the rectal fluid is markedly hypoosmotic to it, indicating the absorption of solutes by the rectal glands.

Larvae of the flies *Coelopa frigida* and *Ephydra* spp. can also live in concentrated saline solutions. When they are in dilute media, the rectal fluid is hypoosmotic to the hemolymph, but it is hyperosmotic to both hemolymph and external environment when the medium is as concentrated as triple strength seawater or 10 *M* NaCl.

The Malpighian tubule–rectal gland system thus operates in ionic and osmoregulation in all of these insects, the main differences between them being in the source of the water and the ions. Whatever the source, water and salts are passed from the gut, the anal papillae, or corresponding organs into the hemolymph; and homeostasis is maintained, in part at least, by the active excretion of water by the Malpighian tubules and its resorption, or the resorption of salts, by the rectal glands. A compensatory mechanism for preserving constancy in the osmotic concentration of the hemolymph has been demonstrated in *Sialis*, *A. aegypti*, *Culex pipiens*, and *Helodes minuta* and *marginata*, two species of beetles whose larvae also have anal papillae. This is effected by changes in the nonprotein nitrogen content of the hemolymph, which shifts in such a way that the total osmotic pressure is not affected by considerable degrees of dilution or concentration of its salts. These shifts are primarily in the concentration of free amino acids at the sacrifice of the plasma proteins. It has been shown, for example, that under conditions of osmotic stress the content of the plasma proteins in *Sialis* may be reduced to one-twentieth of the usual level, while the osmotic concentration of the hemolymph remains approximately constant. By such adjustments, much of the burden of preserving stability in the osmotic concentration of the hemolymph would be removed from the absorptive and excretory areas.

Much less is known about osmotic regulation in adult aquatic insects. As air breathers, their respiratory surfaces are not exposed to water and so are not necessarily sites of water or ion exchange. There is no direct evidence that their integuments have a high degree of impermeability, but such an assumption is a natural one. Different degrees of regulation have been observed in three species of *Corixa*, one of which, *C. lugubris*, inhabits brackish water, while the other two, *C. distincta* and *C. fossarum*, live in freshwater. The freshwater species proved to be less effective osmoregulators than the brackish one and could preserve the homeoosmoticity of their hemolymphs only in pond waters. But *C. lugubris* was found to remain homeoosmotic over the salinities to which it was ordinarily exposed, which range from 0.1 to 1.5% NaCl, and to be hypoosmotic to its environment at greater salinities. There is no evidence that the gut is effective in this regulation, which is presumed to be the result of variations in permeability of the integument in different circumstances. Increased oxygen consumption when the concentration of the medium is above or below 0.6 percent salinity suggests that these changes in permeability result from some active, energy-demanding process.

In a number of insect species, the Malpighian tubules have muscle fibers in their walls and are, therefore, motile. This motility may be of value in changing the position of the tubules in the hemocoele, thus bringing them into more general contact with the hemolymph, or it may be most effective in mixing the contents of tubules and propelling it along them.

(b) MYRIAPODS AND ARACHNIDS The integument of myriapods, like that of isopods, has no waxy nor waterproof coating, so that the terrestrial species are vulnerable to water loss. The primary product of the nitrogenous metabolism of centipedes is, like that of insects, uric acid, which likewise is formed in the Malpighian tubules. Like isopods, centipedes are for the most part restricted to moist habitats, damp soils, and humid crevices or niches under rocks and stones. Only one family of chilopods, the Geophilomorpha, contains species that live in the littoral zone and are subjected to frequent immersion in seawater. *Hydroschendyla submarina* is found on the English seashore down to midtide level, so is often washed over by incoming waves; *Strigamia maritima* lives in the upper zone of the tide level. As its diet is almost exclusively small marine invertebrates, particularly *Orchestia gammarella*, whose body fluids are isosmotic with seawater, these chilopods ingest relatively large amounts of salt water. The osmotic pressure of the hemolymph of these two centipedes is about equivalent to that of 45 percent seawater, as is also that of a strictly terrestrial species, *Haplophilus subterraneus*. Exposure of specimens of these three species to seawater, and to two dilutions of it (10 and 60 percent), showed that there were different degrees in their tolerance and in their ability to regulate in them. *Haplophilus* could survive only 4 to 8 hr in 100 percent seawater; 12 to 21 hr in 60 percent seawater; and 30 to 46 hr in 10 percent seawater. *Strigamia* survived 12 to 24 hr in 100 percent seawater; 12 to 36 hr in 60 percent seawater; 36 to 72 hr in 10 percent seawater; while *Hydroschendyla* lived 12 to 36 hr in 100 percent seawater; 36 to 48 hr in 60 percent seawater; and 48 to 84 hr in 10 percent seawater. In each case there was evidence of salt and water loss, but no regions of increased permeability could be demonstrated by the use of the silver nitrate technique, although it was evident that the littoral species were less permeable than the terrestrial one. A marked difference in the size of the salivary glands in relation to the total body size was evident, for the glands in the littoral species were far larger than those of the terrestrial. This suggests that they may be the site of salt secretion and the areas active in osmoregulation in these centipedes.

There is no definitive information about the excretory products of millipedes, nor of their means for preserving stability in their internal environments. The excretion of guanine by spiders is undoubtedly an adaptation to their terrestrial mode of life, as is the protective layer of wax over their integuments. They, like insects, are better adapted to life on land than myriapods. Spiders, although apparently unable to absorb water through their integuments as ticks can, obtain the necessary water from the more or less fluid food that they eat and by drinking it. Indeed, they can even remove capillary water from soil. It has been shown that wolf spiders can suck up water against pressures as great as 600 mm Hg, so that like terrestrial isopods they can obtain water from regions where it is bound, not free. Conservation, as well as imbibition, of water and its dissolved salts are, therefore, important factors in osmoregulation in the spiders that have Malpighian tubules.

SELECTED BIBLIOGRAPHY

Anderson, J. F.: The Excreta of Spiders, *Comp. Biochem. Physiol.*, vol. 17, pp. 973–982, 1966.

Bahl, K. N.: Studies on the Structure, Development and Physiology of the Nephridia of Oligochaetae. VI. The Physiology of Excretion and the Significance of the Enteromorphic Type of Nephridial System in Indian Earthworms, *Q. J. Microsc. Sci.*, vol. 85, pp. 343–389, 1945.

Bahl, K. N.: Excretion in the Oligochaeta, *Biol. Revs.*, vol. 22, pp. 109–147, 1947.

Dresel, E. B., and C. J. Moyle: Nitrogenous Excretion of Amphipods and Isopods, *J. Exp. Biol.*, vol. 27, pp. 210–225, 1956.

Hyman, L. H.: Textbook Planarians and the Reality, *Am. Biol. Teach.*, vol. 18, pp. 124–127, 1956.

Kitching, J. A.: Osmoregulation and Ionic Regulation in Animals without Kidneys, *Symp. Soc. Exp. Biol.*, vol. 8, pp. 63–74, 1954.

Kümmel, G.: Terminalorgan der Protonephridien: Feinstruktur und Deutung der Funktion, *Z. Naturforsch.*, vol. 13, pp. 677–679, 1958.

Kümmel, G.: Zwei Neue Formen von Cyrtocyten: Vergleich der Bisher Bekannten Cyrtocyten und Erörterung des Begriffes "Zelltyp," *Z. Naturforsch.*, vol. 57, pp. 172–201, 1962.

Kümmel, G., and J. Brandenburg: Die Reusengeisselzellen (Cyrtocyten), *Z. Naturforsch.*, 16 B, pp. 692–697, 1961.

Lal, M. B., and B. B. Saxena: Uricotelism in a Snail, *Nature*, vol. 170, p. 1024, 1952.

Mitchell, H. K., E. Glassman, and E. Hadorn: Hypoxanthine in Rosy and Maroon-like Mutants of *Drosophila melanogaster*, *Science*, vol. 129, p. 268, 1959.

Pappas, G. D., and P. W. Brandt: The Fine Structure of the Contractile Vacuole in *Ameba*, *J. Biophys. Biochem. Cytol.*, vol. 4, pp. 485–488, 1958.

Pedersen, K. J.: Some Observations on the Fine Structure of Planarian Protonephridia and Gastrodermal Phagocytes, *Z. Zellforsch. Mikrosk. Anat.*, vol. 53, pp. 609–628, 1961.

Pontin, P. M.: The Osmoregulatory Function of the Vibratile Flames and the Contractile Vesicle of *Asplanchna* (Rotifera), *Comp. Biochem. Physiol.*, vol. 17, pp. 1111–1126, 1966.

Rasmont, R., G. Vandermeersch, and P. Castiaux: Ultrastructure of the Coxal Glands of the Scorpion, *Nature*, vol. 182, pp. 328–329, 1958.

Schmidt-Nielson, B.: Comparative Morphology and Physiology of Excretion, in J. A. Moore (ed.), "Ideas in Modern Biology," pp. 391–426, Natural History Press, Garden City, N.Y., 1965.

Schmidt-Nielson, B., and D. Lewis: Invertebrate Mechanisms for Diluting and Concentrating the Urine, *Ann. Rev. Physiol.*, vol. 25, pp. 631–658, 1963.

Schmidt-Nielson, B., and R. O'Dell: Studying the Contractile Vacuole by Micropuncture, *Science*, vol. 139, pp. 606–607, 1963.

Schneider, L.: Elektronmikroskopische Untersuchungen über das Nephridialsystem von *Paramecium*, *J. Protozool.*, vol. 7, pp. 75–90, 1960.

Smith, L.: The Excretory System of *Pangrellus redivivus* (T. Goodey, 1945), *Comp. Biochem. Physiol.*, vol. 15, pp. 89–93, 1965.

Tsube, I., and P. W. Brandt: An Electron Microscope Study of the Malpighian Tubules of the Grasshopper *Dissosteira carolina*, *J. Ultrastruct. Res.*, vol. 6, p. 28, 1962.

Ionic and osmoregulation

Allen, K.: Amino Acids in the Mollusca, *Am. Zool.*, vol. 1, pp. 253–262, 1961.

Avens, A. C.: Osmotic Balance in Gastropod Molluscs. II. The Brackish Water Gastropod *Hydrobia ulvae* Pennant, *Comp. Biochem. Physiol.*, vol. 16, pp. 143–153, 1965.

Avens, A. C., and M. A. Sleigh: Osmotic Balance in Gastropod Molluscs. I. Some Marine and Littoral Gastropods, *Comp. Biochem. Physiol.*, vol. 16, pp. 121–141, 1965.

Beadle, L. C.: Comparative Physiology: Osmotic and Ionic Regulation in Aquatic Animals, *Ann. Rev. Physiol.*, vol. 19, pp. 329–358, 1957.

Beadle, L. C., and J. Shaw: The Retention of Salt and the Regulation of Non-P Fraction in the Blood of the Aquatic Larva *Sialis lutaria*, *J. Exp. Biol.*, vol. 27, pp. 96–109, 1950.

Binyon, J.: Salinity Tolerance and Permeability to Water of the Starfish *Asterias rubens* L., *J. Mar. Biol. Assoc. U.K.*, vol. 41, pp. 161–174, 1961.

Binyon, J., and J. G. E. Lewis: Physiological Adaptations of Two Species of Centipede (Chilopoda, Geophilomorpha) to Life on the Shore, *J. Mar. Biol. Assoc. U.K.*, vol. 43, pp. 49–55, 1963.

Brandenburg, J.: Elektron Mikroskopische Untersuchung

des Terminal Apparates von *Chaetonotus* sp (Gastrotrichen) als Ersten Beispiels einer Cyrtocyten bei Askelminthes. *Z. Zellforsch.*, vol. 57, pp. 136–144, 1962.

Brandenburg, J., and G. Kümmel: Die Feinstruktur der Solenocyten, *J. Ultrastruct. Res.*, vol. 5, pp. 437–452, 1961.

Burton, R. F.: Aspects of Ionic Regulation in Certain Terrestrial Pulmonata, *Comp. Biochem. Physiol.*, vol. 17, pp. 1007–1018, 1966.

Camien, M. N., H. Sarlet, G. Duchâteau, and M. Florkin: Non ,Protein Amino Acids in Muscle and Blood of Marine and Freshwater Crustacea, *J. Biol. Chem.*, vol. 93, pp. 881–885, 1951.

Cannon, H. G.: On the Segmental Excretory Organs of Certain Fresh Water Ostracods, *Philos. Trans. R. Soc. (London)*, *B*, vol. 214, pp. 1–28, 1925.

Cannon, H. G., and S. M. Manton: Notes on Segmental Excretory Organs of Crustacea, *J. Linn. Soc. Zool.*, vol. 36, pp. 439–456, 1927.

Coe, W. R.: Unusual Types of Nephridia in Nemerteans, *Biol. Bull.*, vol. 58, pp. 203–206, 1930.

Copeland, E.: Salt Transport Organelle in *Artemia salina*, *Science*, vol. 151, pp. 470–471, 1966.

Croghan, P.: The Survival of *Artemia Salina* (L) in Various Media. The Osmotic and Ionic Regulation of *Artemia salina* (L), pp. 213–233. The Mechanism of Osmotic Regulation in *Artemia Salina* (L): The Physiology of the Branchiae, pp. 234–242; The Physiology of the Gut, pp. 242–249, *J. Exp. Biol.*, vol. 35, 1958.

Danielli, J. F., and C. F. A. Pantin: Alkaline Phosphatase in Protonephridia of Terrestrial Nemertines and Planarians, *Q. J. Microsc. Sci.*, vol. 91, pp. 209–213, 1950.

Edney, E. B.: Woodlice and the Land Habitat, *Biol. Revs.*, vol. 29, pp. 185–219, 1954.

Eltringham, S. K.: Blood Concentrations of *Limnoria* (Isopoda) in Relation to Salinity, *J. Mar. Biol. Assoc. U.K.*, vol 44, 675–683, 1964.

Goodrich, E. S.: The Study of Nephridia and Genital Ducts since 1895, *Q. J. Microsc. Sci.*, vol. 86, pp. 113–392, 1945.

Gross, W. J.: Osmotic Responses in the Sipunculid *Dendrostomum zostericolum*, *J. Exp. Biol.*, vol. 31, pp. 402–423, 1954.

Gross, W. J.: An Analysis of Responses to Osmotic Stress in Selected Decapod Crustacea, *Biol. Bull.*, vol. 112, pp. 43–62, 1957.

Gross, W. J.: Trends in Water and Salt Regulation among Aquatic and Amphibious Crabs, *Biol. Bull.*, vol. 127, pp. 447–466, 1964.

Gross, W. J.: Glucose Absorption from the Urinary Bladder of a Crab, *Comp. Biochem. Physiol.*, vol. 20, pp. 313–317, 1967.

Gross, W. J., and M. E. Conlee: Water Uptake by a Semiterrestrial Crab, *Am. Zool.*, vol. 2, p. 412, 1962.

Gross, W. J., and P. V. Holland: Water and Ionic Regulation in a Terrestrial Hermit Crab, *Physiol. Zool.*, vol. 33, pp. 21–28, 1960.

Gross, W. J., R. C. Lasiewski, M. Dennis, and P. Rudy: Salt and Water Balance in Selected Crabs of Madagascar, *Comp. Biochem. Physiol.*, vol. 17, pp. 641–660, 1966.

Herreid, C. F., II: Water Loss of Crabs from Different Habitats, *Comp. Biochem. Physiol.*, vol. 28, pp. 829–840, 1969.

Herreid, C. F., II: Integument Permeability of Crabs and Adaptation to Land, *Comp. Biochem. Physiol.*, vol. 29, pp. 423–430, 1969.

Hinton, H. E.: A New Chironomid from Africa, the Larvae of Which Can Be Dehydrated without Injury, *Proc. Zool. Soc. London*, vol. 121, pp. 371–380, 1951.

Horne, F. R.: Some Aspects of Ionic Regulation in the Tadpole Shrimp *Triops longicaudatus*, *Comp. Biochem. Physiol.*, vol. 19, pp. 313–316, 1966.

Hunter, W. R., M. L. Apley, A. J. Burky, and R. T. Meadows: Interpopulation Variations in Calcium Metabolism in the Stream Limpet *Ferrissia rivularis*, *Science*, vol. 155, pp. 338–340, 1967.

Kerley, D. E., and A. W. Pritchard: Osmotic Regulation in the Crayfish *Pacifastacus lenuiculus*, Stepwise Acclimated to Dilutions of Sea Water, *Comp. Biochem. Physiol.*, vol. 20, pp. 101–113, 1967.

King, E. N.: Oxidative Activity of Crab Gill Mitochondria as a Function of Osmotic Concentration, *Comp. Biochem. Physiol.*, vol. 17, pp. 245–258, 1966.

Kitching, J. A.: Contractile Vacuoles, *Biol. Revs.*, vol. 13, pp. 403–444, 1938.

Kitching, J. A.: Contractile Vacuoles, *Symp. Soc. Exp. Biol.*, vol. 6, pp. 145–146, 1952.

Knox, B. E., and E. M. Pantelouris: Osmotic Behavior of *Fasciola hepatica* L. in Modified Hedon-Fleig Media, *Comp. Biochem. Physiol.*, vol. 18, pp. 609–617, 1966.

Kromhout, G. A.: A Comparison of the Protonephridia of Fresh Water, Brackish Water and Marine Specimens of *Gyratrix hermaphroditus*, *J. Morphol.*, vol. 72, pp. 167–181, 1943.

Lane, C. E., E. Pringle, and A. Bergere: Amino Acids in Extracellular Fluids of *Physalia physalis* and *Aurelia aurita*, *Comp. Biochem. Physiol.*, vol. 15, pp. 259–262, 1965.

Locke, M.: Permeability of Insect Cuticle to Water and Lipids, *Science*, vol. 147, pp. 295–298, 1965.

Lockwood, A. P. M.: The Osmoregulation of Crustacea, *Biol. Revs.*, vol. 37, pp. 257–305, 1962.

Maluf, N. S. R.: Secretion of Inulin, Xylose and Dyes and Its Bearing on the Manner of Urine Formation by the Kidney of the Crayfish, *Biol. Bull.*, vol. 81, pp. 235–260, 1941.

Martin, A. W.: Recent Advances in Knowledge of Invertebrate Renal Function, in B. T. Scheer (ed.), "Recent Advances in Invertebrate Physiology," pp. 247–276, University of Oregon Publications, Eugene, Oregon, 1957.

Ogelsby, L. C.: Water and Chloride Fluxes in Estuarine Nereid Polychaetes, *Comp. Biochem. Physiol.*, vol. 16, pp. 437–466, 1965.

Pantelouris, E. M., and L. T. Threadgold: The Excretory System in the Adult *Fasciola hepatica*, *Cellule*, vol. 64, pp. 63–67, 1963.

Pantin, C. F. A.: The Adaptation of *Gunda ulvae* to Salinity. I. The Environment, *J. Exp. Biol.*, vol. 8, pp. 63–73, 1931.

Pantin, C. F. A.: The Adaptation of *Gunda ulvae* to Salinity. III. The Electrolyte Exchange, *J. Exp. Biol.*, vol. 8, pp. 82–94, 1931.

Pantin, C. F. A.: The Nephridia of *Geonemertes dendyi*, *Q. J. Microsc. Sci.*, vol. 88, pp. 15–25, 1947.

Parry, G.: Osmotic and Ionic Regulation in the Isopod Crustacean *Ligia oceanica*, *J. Exp. Biol.*, vol. 30, pp. 567–574, 1953.

Potts, W. T. W., and G. Parry: Osmotic and Ionic Regulation in Animals, Pergamon, New York, 1964.

Potts, W. T. W., and M. Todd: Kidney Function in the Octopus, *Comp. Biochem. Physiol.*, vol. 16, pp. 479–489, 1965.

Quinn, D. J., and C. E. Lane: Ionic Regulation and Na^+-K^+ Stimulated ATP-ase Activity in the Land Crab *Cardiosoma guanhumi*, *Comp. Biochem. Physiol.*, vol. 19, pp. 533–544, 1966.

Ramsay, J. A.: The Osmotic Relations of the Earthworm, *J. Exp. Biol.*, vol. 26, pp. 46–56, 1949.

Ramsay, J. A.: The Site of Formation of Hypotonic Urine in the Nephridium of *Lumbricus*, *J. Exp. Biol.*, vol. 26, pp. 65–75, 1949.

Ramsay, J. A.: Osmotic Regulation in Mosquito Larvae, *J. Exp. Biol.*, vol. 27, pp. 145–157, 1950.

Ramsay, J. A.: Osmotic Regulation in Mosquito Larvae: The Role of the Malpighian Tubules, *J. Exp. Biol.*, vol. 28, pp. 62–73, 1951.

Ramsay, J. A.: The Excretion of Sodium and Potassium by the Malpighian Tubules of *Rhodnius*, *J. Exp. Biol.*, vol. 29, pp. 110–126, 1952.

Ramsay, J. A.: Active Transport of Potassium by the Malpighian Tubules of Insects, *J. Ecp. Biol.*, vol. 30, pp. 358–369, 1953.

Ramsay, J. A.: Active Transport of Water by the Malpighian Tubules of the Stick Insect, *Dixippus morosus* (Orthoptera, Phasmidae), *J. Exp. Biol.*, vol. 31, pp. 104–113, 1954.

Ramsay, J. A.: The Excretory System of the Stick Insect *Dixippus morosus* (Orthoptera, Phasmidae), *J. Exp. Biol.*, vol. 32, pp. 183–199, 1955.

Ramsay, J. A.: The Excretion of Sodium, Potassium and Water by the Malpighian Tubules of the Stick Insect, *Dixippus morosus* (Orthoptera, Phasmidae), *J. Exp. Biol.*, vol. 32, pp. 200–216, 1955.

Robertson, J. D.: Ionic Regulation in Some Marine Invertebrates, *J. Exp. Biol.*, vol. 26, pp. 182–200, 1949.

Robertson, J. D.: Further Studies on Ionic Regulation in Marine Invertebrates, *J. Exp. Biol.*, vol. 30, pp. 277–296, 1953.

Shaw, J.: The Permeability and Structure of the Cuticle of the Aquatic Larva of *Sialis lutaria*, *J. Exp. Biol.*, vol. 32, pp. 330–352, 1955.

Shaw, J.: Ionic Regulation and Salt Balance in the Aquatic Larva of *Sialis lutaria*, *J. Exp. Biol.*, vol. 32, pp. 353–382, 1955.

Shaw, J.: The Control of Salt Balance in the Crustacea, *Symp. Soc. Exp. Biol.* (Homeostasis and Feedback Mechanisms), vol. 18, pp. 237–254, 1964.

Simpson, J. W., K. Allen, and J. Awapara: Free Amino Acids in Some Aquatic Invertebrates, *Biol. Bull.*, vol. 117, pp. 371–381, 1959.

Sohal, R. S., and E. Copeland: Ultra Structure Variations in the Anal Papillae of *Aedes aegypti* (L) at Different Environmental Salinities, *J. Insect Physiol.*, vol. 12, pp. 429–439, 1966.

Spencer, J. O., and E. B. Edney: The Absorption of Water by Woodlice, *J. Exp. Biol.*, vol. 31, pp. 491–496, 1954.

Stephens, G. C., and A. W. Raghunath: Uptake of Organic Material by Aquatic Invertebrates. IV. The Influence of Salinity on Uptake of Amino Acids by the Brittle Star *Ophiactis arenosa*, *Biol. Bull.*, vol. 131, pp. 172–186, 1966.

Topping, F. L., and J. L. Fuller: The Accommodation of Some Marine Invertebrates to Reduced Osmotic Pressure, *Biol. Bull.*, vol. 82, pp. 372–384, 1942.

Treherne, J. E.: The Exchange of Labelled Sodium in the Larva of *Aedes aegypti* L., *J. Exp. Biol.*, vol. 31, pp. 386–401, 1954.

Virkar, R. A.: The Role of Free Amino Acids in the Intracellular Isosmotic Regulation in the Sipunculid *Golfingia gouldi*, *Am. Zool.*, vol. 5, pp. 660–661, 1965.

Weil, E., and C. F. A. Pantin: The Adaptation of *Gunda ulvae* to Salinity. II. The Water Exchange, *J. Exp. Biol.*, vol. 8, pp. 73–81, 1931.

Wigglesworth, V. B.: The Physiology of Excretion in a Blood-sucking Insect, *Rhodnius prolixus*. I. Composition of the Urine, pp. 411–427. II. Anatomy and Histology of the Excretory System, pp. 428–442. III. The Mechanism of Uric Acid Excretion, pp. 443–451, *J. Exp. Biol.*, vol. 8, 1931.

Wigglesworth, V. B.: On the Function of the So-called "Rectal Glands" of Insects, *Q. J. Microsc. Sci.*, vol. 75, pp. 131–150, 1932.

Wigglesworth, V. B.: Function of Anal Gills in Mosquito Larvae, *J. Exp. Biol.*, vol. 10, pp. 16–26, 1933.

Wigglesworth, V. B.: The Adaptation of Mosquito Larvae to Salt Water, *J. Exp. Biol.*, vol. 10, pp. 27–37, 1933.

Wigglesworth, V. B.: Transpiration through the Cuticle of Insects, *J. Exp. Biol.*, vol. 21, pp. 97–114, 1945.

Wilson, R. A.: The Protonephridial System in the Miricidium of the Liver Fluke *Fasciola hepatica, Comp. Biochem. Physiol.*, vol. 20, pp. 337–342, 1967.

chapter XIV

SPECIAL METABOLIC PRODUCTS AND THEIR UTILIZATION

A. PIGMENTS (BIOCHROMES)
   1. Melanins
   2. Ommochromes
   3. Pterins
   4. Color patterns

B. TOXINS, VENOMS, AND REPELLENTS
   1. Oral introduction
   2. Parenteral introduction
      a. PHARMACOLOGICALLY ACTIVE COMPONENTS OF VENOMS
      b. MEANS OF INTRODUCTION
   3. Salivary venoms
   4. Contact venoms
   5. Repellents (defensive secretions)

C. PHEROMONES (PHERORMONES)

Not all metabolic products are excreted or stored as nutritive reserves. Some are deposited within or between cells as pigments or collected in glands as substances that may be harmful or offensive to other organisms and so ward them off, or that may serve as other means of communication between individuals. As essentially inert metabolic products, it is likely that these have been retained rather than eliminated because they have had some positive survival or adaptive value to a species. The precise localization of some of them suggests that their production is also precisely localized and is, in this way, analogous to the localization of bioluminescence, also serving as a means of communication, and raises similar question about the specialization of particular cells in the performance of what may once have been generalized biochemical reactions.

A. PIGMENTS (BIOCHROMES)

There is virtually no species without some kind of color somewhere in the body, whether exogenously or endogenously derived. Some of the pigments responsible for this coloration have established physiological functions, such as the respiratory and visual pigments, yet others seem functionally inert. Exogenous pigments have their direct sources in the food that an animal eats or in parasites or symbionts associated with it. The undigested chlorophyll and carotenoids, ingested first or second hand by herbivorous or carnivorous animals, and the hemoglobin, ingested by bloodsuckers, may all undergo some metabolic changes before deposition of the final product, but this product could not have been formed were the complex compound not obtained as such from an outside source. Endogenous pigments, on the other hand, represent the terminal products of some important inherent biochemical reactions, granting of course that the initial substrates are ultimately derived from digested food. Endogenous biochromes of some sort and in some location are universal among animals, except for the comparatively rare albinos. That their production is under genic control was established some 30 years ago in the classical work on eye color in *Drosophila melanogaster* and the demonstration that genes direct, step by step, the formation of pigments. This work initiated the area of biochemical genetics, leading to our present understanding of the fundamental nature and activity of the genetic material. Organisms deficient or defective in a gene responsible for any step in the sequence of reactions that leads to production of an endogenous pigment are deviates from the usual type, either with different colors or color patterns or else colorless albinos.

1. Melanins

Established endogenous pigments are the melanins, ommochromes, and pterins, all products of nitrogenous metabolism. Of these, the melanins are probably the most widely distributed and include a group of chemically ill-defined black, dark brown, tan, or sometimes yellowish or reddish nitrogenous compounds, found either in solution or in granular form within cells or in the matrix between them, and sometimes in their exoskeletons. Ommochromes are so called because they were first identified in the eyes (Greek *omma*, "eye") of *Drosophila* and some other arthropods, although they are not, as far as is now known, involved in photoreception. They are also found in chromatophores of molluscs and crustaceans and in the integuments of insects. They are yellow, red, or violet-brown in color and are chemically divisible into two main groups: the ommatins, acidic nitrogenous compounds of low molecular weight and alkali labile; and the ommins, nitrogenous compounds of high molecular weight and alkali stable. Pterins are, likewise, nitrogenous compounds and are white, yellow, orange, or red. They were first identified as the pigments giving color and color patterns to the wings of pierid butterflies, but are now recognized as of wide distribution among plants as well as animals.

Melanins arise from oxidation of tyrosine through the agency of the enzyme tyrosinase. The oxidative reactions in insects have been shown to follow the course outlined in Fig. XIV–1, which is probably a representative one for all invertebrates. The indole quinone, which is the final product in the chain of reactions, is polymerized and often associated with a protein as the chromogenic moiety of a complex

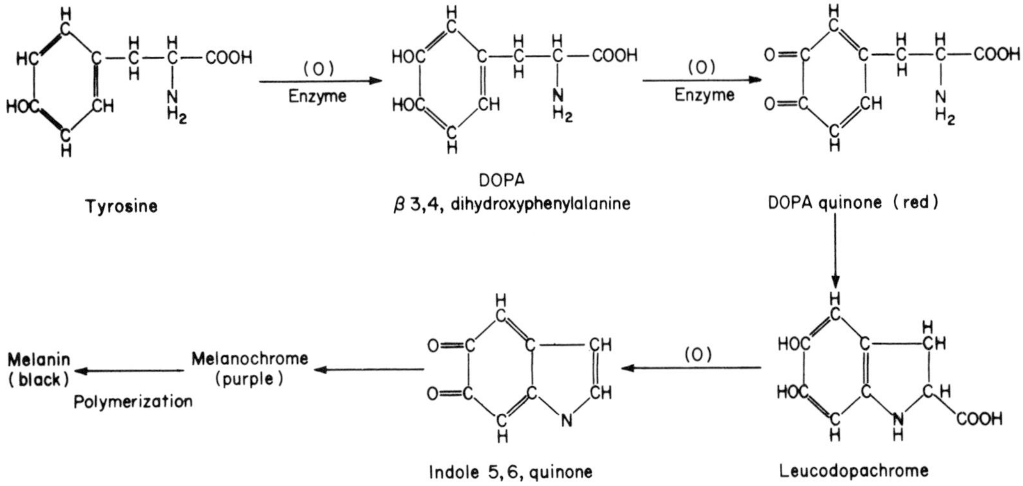

FIG. XIV–1 *Steps in the oxidation of tyrosine to melanin. An intermediate (not shown), which is red at pH 8.6 and below and green at higher pH's, is known as hallochrome and is found in the polychaete* Halla parthenopaea.

molecule. Since tyrosinase is an enzyme of quite general distribution in tissues, localization of melanin in particular areas of an animal's body suggests either a similar localization of the chromogenic substrate or of inhibitors that block the reactions by which the pigment is produced. Not all dark pigments are melanins, but these are known to be produced by most marine animals and by insects, as well as by plants and terrestrial vertebrates. They frequently occur in chromatophores along with other pigments but are often, also, the only pigment either in cells or around them. Melanins are most apt to be found in superficial tissues that are exposed to light, where they may well serve a protective function by absorbing the sun's rays and so screening underlying cells from incident light and heat. They may also be effective in heat exchange between organism and environment.

2. Ommochromes

Ommochromes also originate from the metabolism of an aromatic amino acid, in this case usually tryptophan, one of the 10 "essential" amino acids. Steps in the sequence of reactions from tryptophan to xanthommatin that are presumed to occur in vivo in invertebrates are represented in Fig. XIV–2. So far ommochromes have not been reported in vertebrates. The chemistry of the ommins is less well known that that of the ommatins, as they are extremely difficult compounds to work with. At least one of them, a dark violet-brown substance called ommin A, has been isolated from the heads of silkworms, in which there are at least four others. Upon hydrolysis, ommin A yields 3-hydroxykynurenine and a pigment with the characteristics of an ommatin. A possible relationship between ommochrome production and melanogenesis is indicated by the fact that tyrosine in the presence of DOPA will catalyze the oxidation of 3-hydroxykynurenine to xanthommatin. Like the melanins, ommochromes may be conjugated with proteins and in this form, or independently, are found in visual organs and chromatophores together with other pigments. Ommochromes are also present in the molting fluids of a number of insects.

3. Pterins

The nucleus of the pterin molecule is pteridine, formed by fusion of a pyrimidine and a pyrazine ring. Pteridines are of widespread occurrence among animals and

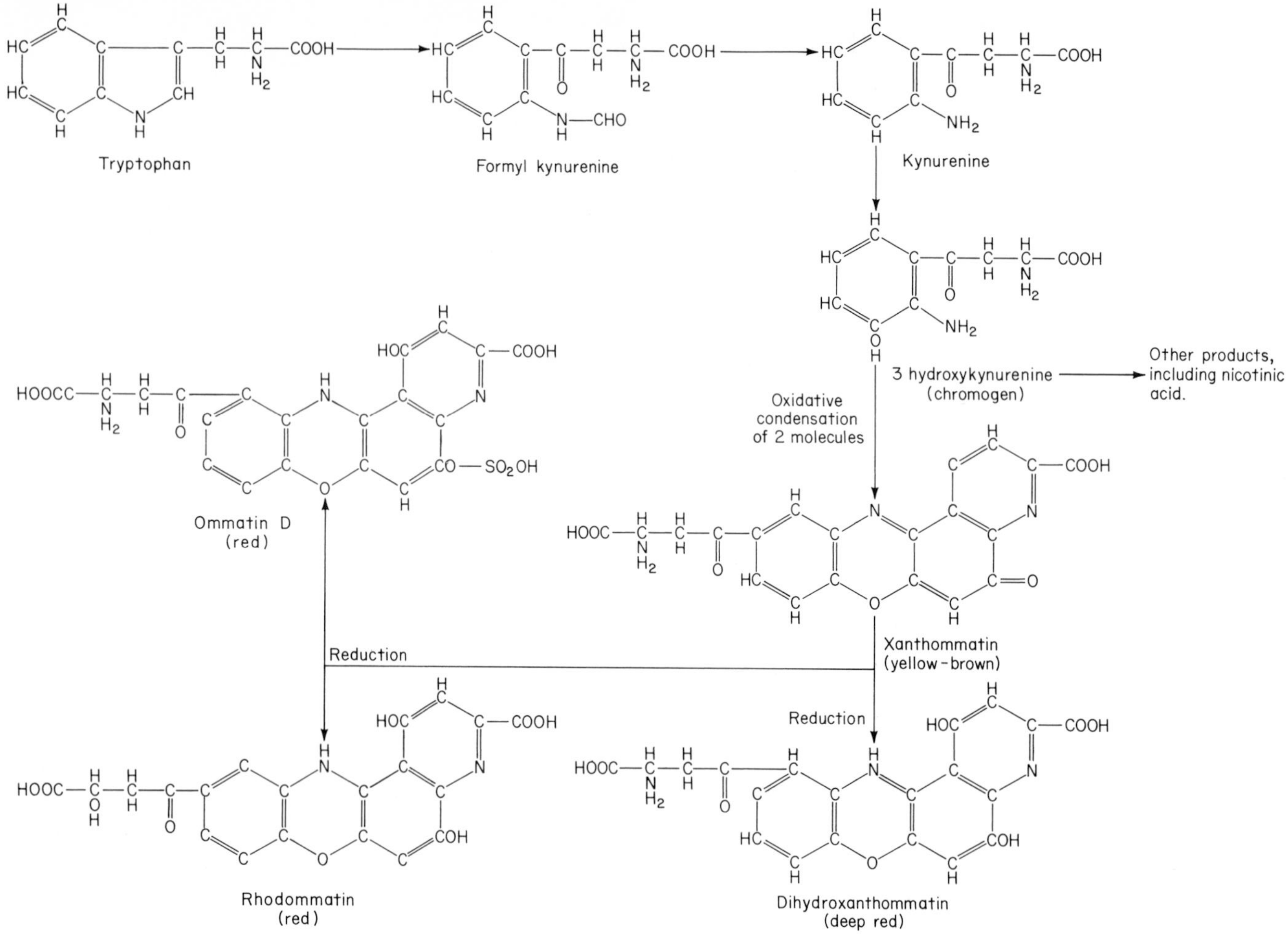

FIG. XIV–2 *Steps in the conversion of tryptophan to ommatins.*

are components of such functionally important compounds as the vitamins riboflavin and pteroylglutamic acid. Most of the pterins are derivatives of 2-amino-4-hydroxypteridine and differ from each other in the substituent groups at carbons 6 and 7. The structural formulae for several of these are represented in Fig. XIV–3. The majority of them are photolabile and fluorescent. In spite of their wide occurrence the steps in their biosynthesis are still unknown. Synthesis of the pteridine ring from purines may be accomplished in animals in vivo in the same way that it can be in vitro, but there is no conclusive evidence that it is. Certain moulds are known to be capable of synthesizing riboflavin from the purine guanine, with a pteridine as an intermediate and the direct precursor of this biologically active, soluble, yellow pigment with vivid green fluorescence in ultraviolet light. Experiments on feeding purines labeled with C^{14} to caterpillars have resulted in the recovery of two pterins bearing the label from wings of the adult butterflies. It is possible, also, that pterins are derived from flavins with which they are chemically closely related.

Pterins are very generally found in insects, not only in the epidermal scales of pierid butterflies, from which they were first isolated, but also in the integuments and eyes of other species. Kynurenine, rather

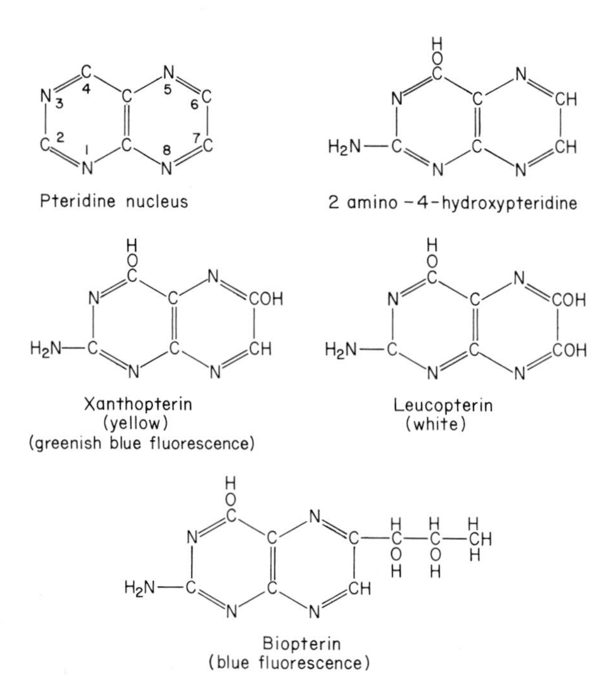

FIG. XIV–3 *Structural formulae for the pteridine nucleus and three of the commonly occurring pterins.*

than a pterin, is accumulated in the wings of papilionid butterflies. Xanthopterin is the pterin of widest distribution and has been identified in animals of several different phyla as well as in insects. In insects, but not in other animals, it is frequently associated with leucopterin, a product of its oxidation. Pterins, along with ommochromes, are found in the eyes of crustaceans as well as of insects where they may prove to be effective in vision. They are present in crystalline form in the hypodermis of many Hymenoptera, where they are responsible for the yellow color characteristic of this order of insects. Biopterin is a component of the royal jelly of honeybees and has been found to promote growth in some microorganisms. It has also been found as an excretory product in human urine.

4. Color patterns

When endogenous pigments such as these are formed or deposited in the superficial tissues of an animal they give it color and color patterns. The color is determined by the nature of the biochrome and its intensity by the amount that is formed and deposited. Both of these characteristics reflect the metabolism of the organism—the color, the reactions that take place and the intensity, the rate at which they occur. Patterning is the result of specific localization of the product or of different products. Both color and pattern vary with species differences and, within a species, may vary with age, sex, and sometimes with season, for temperature and humidity affect metabolism. Differences in coloration with age are shown, for example, by the moth *Cerura vinula*, whose larvae have dark brown markings, attributable to xanthommatin. These change abruptly to bright red just before pupation when the xanthommatin is reduced to dihydroxyxanthommatin. The gypsy moth *Porthetria dispar* offers a classic example of sex differences in color, for the females are almost white, with irregular dark markings on their wings, and the males are uniformly brown. There is a difference in the chromosomal complement of the two sexes, the females, as in other Lepidoptera, being heterogametic. The effect of temperature upon coloration is illustrated by the grain bug *Chlorochroa sayii*, in which a single individual may undergo complete color change in a few days. The bug is either olive drab or bright green, depending upon the environmental temperature. An extreme case of the effect of humidity is found in an African butterfly, *Precis octavia*. In the wet season the insects are bright orange with black markings, and in the dry season they are black with blue markings. While no change occurs in a single individual, as it does in *Cerura* and *Chlorochroa*, offspring of a black and blue female are all orange and black if the eggs hatch out in the rainy season.

Color differences of this kind are unlike the transient and often rhythmic shifts in coloration that are due to migration of pigments already formed rather than to changes in their chemistry or the rate of their formation. Such transient color changes occur in walkingstick insects, whose bodies are dark by night and light by day, owing to migration of pigments in specialized areas of the integument. They are also well known in molluscs and crustaceans, where they are under humoral control and result from alterations in the distribution of pigments deposited in the chromatophores (see Chapter XVII).

The type of pigment and the deposition of fixed ones into various patterns may be of distinct survival value to animals, providing them in many cases with protection from attack by predators. Concealing coloration or camouflage may be achieved in a number of ways. The pattern of pigmentation may be a disruptive one, so that the contours of the animal are difficult to see and tend to be confused with the background. The black and white striping of cuttlefish is illustrative of this. Other patterns may be ones so like an animal's habitual environment that it blends in with its background and is virtually invisible as long as it remains motionless (Fig. XIV–4). The bodies of walkingstick insects are, for example, long and slender and so like the color of the branch or twig to which they cling that they can hardly be distinguished from it. Many lepidopterans in larval, pupal, and adult stages resemble the bark or leaves of the trees they frequent and so are concealed when feeding on them, attached to them, or resting on them. Although in many butterflies the upper surfaces of the wings are brightly colored and patterned disruptively, the undersurfaces are like the object on which the insect habitually comes to rest, so that when the wings are raised over its back it becomes one with its background. In some harmless insects, too, the colors and patterns resemble those of noxious or poisonous species, by virtue of which they are avoided by predators that have learned to leave the harmful ones alone. These camouflaging and mimicry patterns have become established in the long course of evolution, selective advantage being given to those species that through fortuitous mutations have acquired them. How this selection may have come about is suggested by the spread of melanic forms of certain Lepidoptera that has been observed in the industrial areas of Great Britain. Before the darkening of vegetation and buildings by soot and smoke with the advent of the industrial revolution in the latter part of the eighteenth century, the light forms of these species were sufficiently well protected to survive,

(a)

(b)

(c)

FIG. XIV–4 *Concealing forms in insects.* ***(a)*** *Giant stick insect of the Trobriand Islands, barely detectable from the twig on which it is resting.* ***(b)*** *A tree-hopper,* Umbonia orizimba, *which looks like a thorn.* ***(c)*** *The long-horned grasshoppers* Chlorophylla, *whose fore wings (tegmina) closely resemble leaves. Upper figure: Female from British Guiana. Lower figure: Female from Peru. (American Museum of Natural History.)*

but they became conspicuous targets to predators when they were forced to alight on dark backgrounds. The occasional dark mutants were, on the other hand, well concealed and survived to propagate and become dominant in the population, replacing the lighter forms, which can, nevertheless, still flourish in less polluted areas.

There are many illustrations of successful camouflage in invertebrates. Some of them have achieved this by disruptive or mimetic coloration; some by structural devices that prevent their casting a shadow and so revealing themselves; and some by mimicking the bodies of noxious or inedible species and so, like those with warning, or aposematic, coloration, protecting themselves from attack (Fig. XIV–5). For instance, buprestid beetles of the genus *Acmaeodera*, found in Arizona and Florida, visit flowers and resemble hymenopterans of similar habit. The beetles use their membranous hind wings only in flight; their fore wings have a color pattern like that of the abdomens of the bees and wasps that visit similar flowering plants. These elytra are coupled together and held permanently over the beetles' backs. *Acmaeodera* is without organs of attack or defense, but derives some protection from predators through its mimicry of stinging insects.

B. TOXINS, VENOMS, AND REPELLENTS

Production of substances that may be poisonous or repulsive to other animals is widespread among invertebrates. Recognition of invertebrate biotoxins goes back to Babylonian and Assyrian cultures and, because of their effects upon man, have excited world-wide medical interest. Study of their precise nature and of their specific effects upon the individual is in its comparative infancy, although it has progressed to the extent that the chemistry and the pharmacology of some of the toxins are known, a matter of prime importance in the treatment of victims of poisoning of this kind. Other offensive or defensive products may be ones that do not cause severe injury to other animals but tend to ward them off or drive them away. These may be compounds that are malodorous, irritating on contact with the integument, startling, or frightening, or they may be ones with a bitter, acid, or otherwise unpleasant taste that make the animal repugnant to would-be predators.

Animal biotoxins, like animal biochromes, may be exogenous or endogenous. Exogenous toxins may be phytotoxins of plant origin or zootoxins of animal origin. Their biological significance is but little understood. In nature, zootoxins may be effective only after they have been taken into the body by way of the mouth or introduced parenterally by some special puncturing or stinging device, or they may be effective on contact.

1. Oral introduction

Although in nature parenteral is probably of greater significance than oral, introduction of biotoxins from

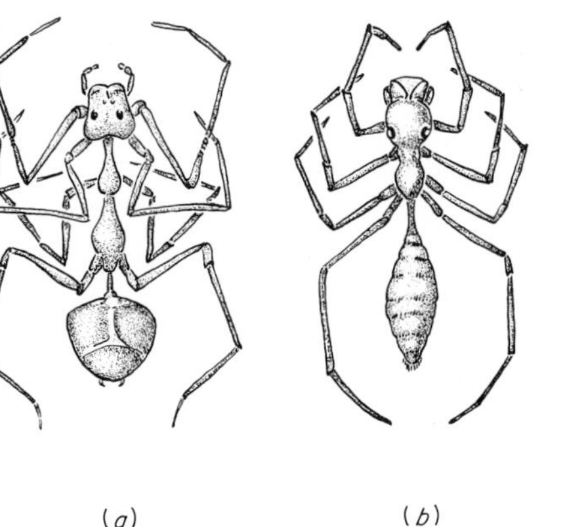

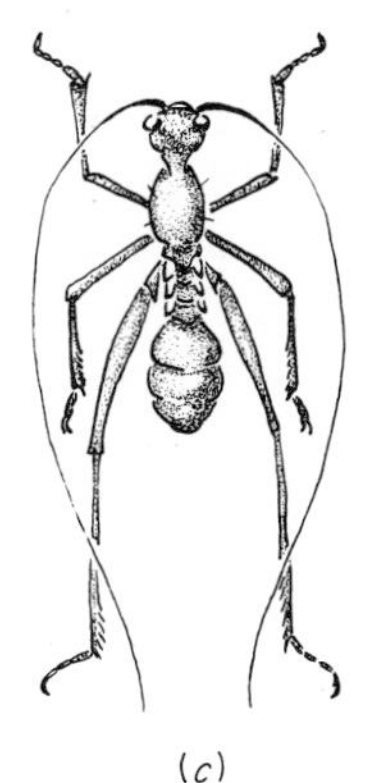

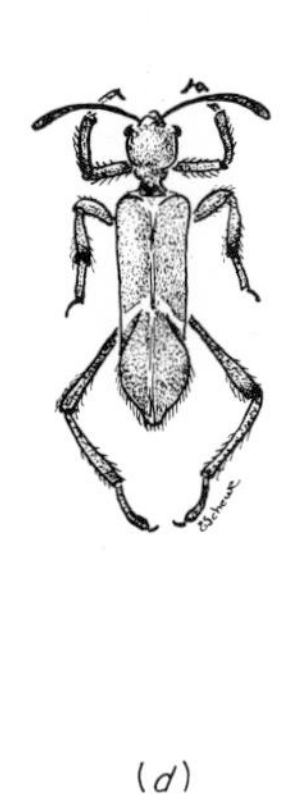

FIG. XIV–5 *Ant mimics.* (*After Portman.*) ***(a)*** *A spider,* Myrmicium, *with body resembling that of an ant.* ***(b)*** *A spider,* Myrmarachne, *which walks on only three of its four pairs of legs, holding the anterior pair close to its prosoma in semblance of antennae.* ***(c)*** *Larva of an African grasshopper,* Myrmecophana phallax, *which lives among ants in its early instars. The three pairs of bright patches on the dorsal surface of the anterior part of its abdomen give the appearance of a narrow waist like that of an ant.* ***(d)*** *A beetle,* Myrmecomaea*; the diagonal stripes on its abdomen simulate a waist.*

ingestion of poisonous invertebrates has directed a good deal of attention to this route. Outbreaks of poisoning in human beings after eating bivalve molluscs have been reported from Biblical times and a warning about eating "shellfish" is given in Leviticus (11:9–12). In the eighteenth, nineteenth, and twentieth centuries outbreaks were numerous in Europe and Great Britain; the British Medical Journal for 1857 contains a report of the poisoning of an entire ship's crew after eating mussels. The poison is exogenous in the molluscs, but endogenous in the dinoflagellates on which they feed. Two genera of Dinoflagellata, *Gymnodinium* and *Gonyaulax*, are known to synthesize toxins that are deadly to fish, crabs, shrimps, and other plankton feeders if the protozoans are eaten in sufficient numbers. Population outbursts occurring sporadically among them make them at times dominant members of the plankton, a condition that can be recognized by discoloration of the water or by exceptionally brilliant luminescence. "Red tides" caused by concentration of dinoflagellates in the plankton, and the effects of them, have been familiar phenomena for centuries. The Book of Exodus (7:20–21) describes how "all the waters that were in the rivers were turned to blood. And the fish that were in the rivers died; and the river stank, and the Egyptians could not drink of this water and there was blood through all the land of Egypt." Indians along the Pacific coast of the United States, where such population explosions are particularly severe, reportedly watched for periods of exceptional luminescence in the ocean and warned their fellows not to eat shellfish at that time. Concentrations of *Gonyaulax* in Monterey Bay have reached densities of 20 to 40 million individuals/ml seawater. In 1948, millions of fish and shellfish were killed off the coast of Florida by eating dinoflagellates, which were so numerous there that the water was thick and oily, and deep yellow in color.

The chemistry of the dinoflagellate toxin is not yet fully known, although it has been obtained in purified form from the mass maceration and fractionation of hundreds of mussels at a time. In mussels, and in clams, the digestive glands have the greatest accumulation of it, but in other molluscs it may be lodged principally in the gills or the siphons. Its effect on man is primarily upon the nervous system, affecting the respiratory and vasomotor centers of the central system, and the cutaneous tactile endings and the neuromuscular junctions of the peripheral, thus causing paralysis. It may also have gastrointestinal effects or erythematous ones.

There are also reports of the poisoning of humans from eating the viscera of abalones or entire whelks. *Neptunea* is a genus of whelks that are frequently eaten by native inhabitants of the Pacific islands and the Asian mainland. The poison, which is localized in the salivary glands and mixed with other secretions from them, has been identified as tetramine, tetramethyl ammonium hydroxide (Fig. XIV–8). Its effect is similar to that of the dinoflagellate poison, for by blocking the myoneural junctions it causes paralysis. In the natural conditions of the whelk's life the toxin probably serves it by paralyzing its prey. Callistin poisoning in man may occur from eating *Callista brevisiphonata*, a Japanese species of bivalve of the family Veneridae, which is harmful only during the spawning season when the toxin, whose pharmacologically active principal may be histamine or choline, accumulates in the ovaries. The ripe gonads of echinoderms, eaten quite generally in the Orient, may also contain substances that are toxic to man. A substance known as asterotoxin has been extracted from the starfish *Asterina pectinifera*. This is very toxic to fish and has been chemically identified as a saponin, a glucoside with a bitter taste. Asterotoxin causes hemolysis in individuals into which it is introduced.

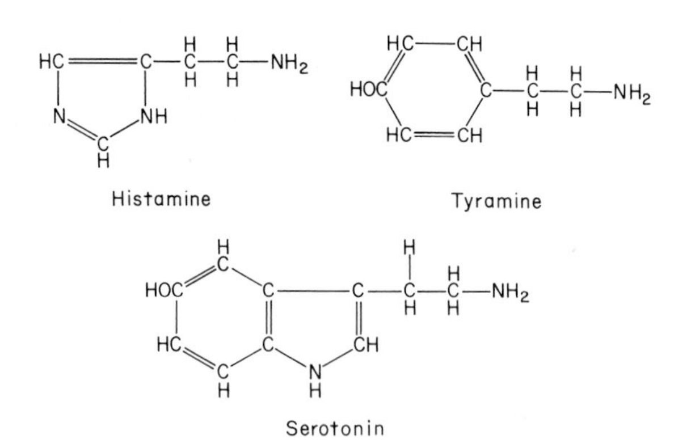

FIG. XIV–6 *Structural formulae of the tertiary amines histamine, serotonin, and tyramine.*

Many holothurians produce an oral poison, which has been given the name of holothurin and of which the active principal may be a triterpenoid sapogenin.

Insects, which provide a major source of food for many animals including birds and insectivorous mammals, may also carry in their tissues substances that make them dangerous, or distasteful, to potential predators. Histamine, for example, has been found in the hemolymph of a number of moths of different species. In *Spilosoma lubricepeda* the concentration is as great as 700 μg/g of body tissue, and 750 μg/g in *Hypocrita jacobaeae*. Acetylcholine, sometimes in association with histamine and other choline esters, has been demonstrated in the hemolymph of some moths. The abdominal tissues of females of the tiger garden moth *Arctia caja* contain a substance, not yet identified, whose potency is so great that injection of 5 to 10 μg of a saline extract will kill a guinea pig in 2 to 10 min. Their cervical glands also contain a choline ester with properties similar to those of acetylcholine. Birds, when offered these moths, have been seen to peck at them once but immediately expel them from their beaks which they then wipe clean. After one such encounter, a bird will invariably avoid *Arctia* as prey, recognizing it by its coloration.

The hemolymph or tissues of many phytophagous insects contain toxins of plant origin that make them dangerous to predators. The hemolymph of the locust *Poikilocerus bufonius*, for example, contains a digitalis-like compound, as does that of the butterfly *Danaus plexippus*. In the course of their evolution, insects carrying such substances must have developed mechanisms that enable them to withstand the pharmacological effects or else to detoxify the compounds, just as they must also have developed means for handling their own endogenous toxins.

2. Parenteral introduction

Venoms, the metabolic products of many animals both vertebrate and invertebrate, are mixtures of pharmacologically active substances that are communicated by biting, stinging, or other means of puncturings or penetrating the integument. They were most likely conserved primarily because of their use in tranquilizing, anaesthetizing, or killing prey, but in a number of species this offensive role seems to be less important than a defensive one. The pharmacological effects and the potencies of invertebrate venoms can to some extent be observed in nature, or in the laboratory by confining a known or suspected venomous species in a container with other animals upon which it may prey or from which it may have to defend itself. In many cases, these qualities have been tested upon animals with which the invertebrate would not come into contact in the ordinary course of events and upon vertebrates. Their effects upon the natural prey or enemies of the species under investigation have been deduced from such experiments. Indeed, in many cases the toxicity of an invertebrate substance has been discovered only in the course of routine tests upon laboratory animals.

The bite or sting of an invertebrate is often painful to man and evokes pain responses in experimental animals. In some cases, introduction of sufficient quantities of venom into human beings, like the oral introduction of poisons, causes severe pathological reactions, and even fatal ones. These effects upon man and upon his domestic animals have, for practical reasons, spurred research upon the nature and toxicity of invertebrate venoms, which are, of course, equally of interest for biological reasons alone. Some of the questions raised are concerned with the generality or specificity of their effects; the immunity of an individual to its own toxin or to that of other members of its species; and the immunity of some species to toxins that are potent poisons to others.

a. PHARMACOLOGICALLY ACTIVE COMPONENTS OF VENOMS The recognized pharmacologically active substances in venoms of both vertebrate and invertebrate origin are products of nitrogenous metabolism—tertiary and quartenary ammonium compounds in which there are substituent groups for three or four of the hydrogen atoms in a molecule of ammonia; various peptides; and proteins without recognized enzymatic properties. There may also be enzymes, such as hyaluronidase and phospholipase, which soften the tissues and so promote spread of the venom; in some, lecithinase has been identified, an enzyme that causes release of lysolecithin and hemolysis by destruction of the erythrocyte wall. The most generally recognized tertiary amines are histamine, serotonin, and tyramine (Fig. XIV–6). Some venoms are known to contain substances that act as histamine releasers when they reach the tissues, rather than histamine itself, but the end results are similar. The quarternary ammonium compounds that have been identified are tetramine and various choline esters, especially acetylcholine and, in at least one genus of molluscs, urocanylcholine or murexine, which, like histamine, contains an iminoazole ring. Another choline ester resembling beta, beta dimethyl acrylylcholine has been found in the cervical glands of imagos of *Arctia caja*.

Pharmacologically active peptides are known as kinins, a generic term which was introduced when a substance was discovered in the venom of the wasp *Vespa vulgaris* that closely resembled bradykinin, a peptide with nine amino acids that is released when human serum globulin is treated with snake venoms or with trypsin. Its name is derived from its effect upon isolated preparations of smooth muscle, for it causes them to contract slowly (Greek *bradys*, "slow"). Kinins are now known to be of wide occurrence in venoms, and the number, kind, and sequence of amino acids in several of them are known.

All these compounds are pain producing. They may also have effects upon the muscular and nervous systems, causing contraction of smooth muscle particularly, and so hypertension, and increased excitability of nerves. In some cases alterations in membrane permeability are indicated, which may be the common basis for their effects. All that have as yet been identified are compounds that are produced in the course of the ordinary metabolism of many animals and at least some plants. They are, or they belong to, categories of compounds not infrequently found in animal urines, although they are essentially intermediates in a series of metabolic reactions or byproducts of them. Correlative with their conservation are certain anatomical, structural, and behavioral adjustments of the organism that provide for precision in emission of the toxin and assurance that it will reach a vulnerable area of a vulnerable animal.

The nonenzymatic protein components of venoms are the truly toxic ones, causing paralysis which may or may not be followed by death. In sensitive or sensitized humans they often cause severe allergic reactions and anaphylactic shock. Implication of polypeptides or proteins of low molecular weight as the actual agents of toxicity has been derived from observations that some venoms no longer paralyze after they have been heated; after they have been exposed to trypsin or chymotrypsin; or after the proteins have been precipitated out by chemical means. It is probable that the various components of any venom act synergistically and that their multiple action accounts for the severity of the effects. These may be directly upon muscle membranes, not necessarily in the region of neuromuscular junctions; evidence for this has been offered by the response of isolated muscles to some cnidarian and wasp venoms. The toxin may also act directly upon the central nervous system, for nerve cords isolated from crayfish, on exposure to venoms, show first an increase in spontaneous electrical activity and then a complete cessation of it.

Some invertebrate venoms contain also toxic carbohydrates, the sapotoxins. These belong to the chemical category of saponins, glucosides that on hydrolysis yield sugars and sapogenins, an inclusive term used for the nonsugar component without defining its nature. Saponins, not all of which are toxic, are produced by a number of different kinds of plants and, like tannins and alkaloids, are considered "secondary plant substances" with as yet no proven role in the plant's economy. It is not without interest that the venoms of many arthropods contain some of these "secondary plant substances," whose efficacy in the attraction of insects to them as a food source has already been pointed out in Chapter IX.

b. MEANS OF INTRODUCTION Parenteral introduction of venoms is by means of stings, bites, or punctures by spines. Any of these may be structures or organelles especially designed for this purpose or ones that have other uses and have become adapted to this one through their connection with venom-producing cells or glands. Devices for introduction of venoms include cnidarian nematocysts; the detachable radula teeth of certain gastropod molluscs; the stinging apparatus of scorpions, wasps, and bees; and the spines of some echinoderms and caterpillars.

Cnidarians are notorious for their stinging capacities and venomous species are found among hydrozoans, scyphozoans, and anthozoans. Some of these are specifically mild stingers and some severe ones, and in some there is seasonal variation in the degree of severity of their stings. Their venom is localized in the nematocyst and is introduced into the victim by discharge of the penetrants and their lodgment in, or their penetration through the integument. The penetrants of *Physalia* are reported to be discharged with such force that they pierce through heavy-gauge surgical gloves. The effect of the toxin is to immobilize the prey, sometimes by killing it, so that it can be easily transported to the mouth by the tentacles. This effect is usually the result of parenteral injection, but it may also occur if the toxin is introduced orally. Yet certain animals feed on cnidarians without ill effects, possibly by digesting or in some way detoxifying the toxic proteins before they reach vulnerable areas or by leaving the nematocysts undigested and transferring them intact, as some flatworms and nudibranch molluscs do, to sites in their own bodies where they can be used in offense or defense. Recent techniques for the isolation of undischarged nematocysts, by maceration of whole animals or selected parts, and filtration and differential centrifugation of the nematocyst-containing filtrate, have made possible the study of the components of their venoms uncontaminated by products of other cells. This was a confusing factor in earlier investigations. Active principals in them have been identified as histamine, serotonin, and tetramine, in addition to polypeptides. The histamine content of the tentacles of the anthozoans *Actinia equina* and *Anemonia sulcata* have been found to be 20 to 150 μg/g dry tissue, which is more than in any other part of the body; and a histamine-releasing substance, called thallasine, has been recovered from the tentacles of *A. equina* and from those of the scyphozoan *Cyanea capillata*. Nematocysts of *Hydra* contain tetramine, which has a paralyzing effect upon crustaceans and speedily causes the death of *Daphnia* upon which *Hydra* ordinarily feeds. Extracts of nematocysts from other cnidarians, when injected into specimens of *Uca* and *Hemigrapsus* as test animals, cause them to autotomize their legs as do injections of pure tetramine.

The sting of the sea wasp *Chironex fleckeri*, a cubomedusan of north Australian waters, has neurotoxic effects similar to those of tetramine and also causes paralysis. This is probably the most venomous invertebrate alive and the marine animal that is most dangerous to humans at least, for it can kill a man in a few minutes and even in as short a time as 30 sec. These jellyfish, although solitary in habit, are particularly dangerous because their bodies are the same light blue color as the shallow waters they frequent and so translucent that they are barely visible to swimmers and divers. Moreover, although their bells are not especially large, with diameters of about 7.5 cm and a maximum height of 11 cm, their tentacles are unusually long and extend several meters down into the water. They may, therefore, make contact with a large area of a human body, and discharge of their many nematocysts could well introduce a lethal amount of venom. In addition to paralysis, stings of *Chironex* are accompanied by pain, wheals, redness, and a burning sensation at the site of nematocyst penetration. These symptoms would indicate that the venom also contains serotonin, a potent pain producer, and if not also histamine, something that induces the release of histamine in the tissues.

The venom of *Physalia*, the Portuguese Man-of-War, is also very potent, and multiple stings from the long tentacles of a single individual are fatal to the fish on which these siphonophores feed, as well as, occasionally, to human beings. It is reported that, like cobra venom, one thimbleful is enough to kill a thousand mice, and it has been found toxic to all multicellular animals tested, mammals as well as invertebrates. To the protozoans *Paramecium* and *Tetrahymena*, on the other hand, this venom seems to be not only nontoxic but an acceptable source of carbon. It contains tetramine, and nine other components of its toxin have been revealed by paper chromatography. All of these are peptides, both qualitatively and quantitatively different in their amino acid composition. The toxin must be relatively stable while in the nematocyst, for much of the potency remains in specimens that have been washed up on shore, even after they have become completely dry. And nematocysts isolated in the laboratory have been known to remain reactive for at least 2 weeks, even after contact for that length of time with tables or clothing on which they have been accidentally spilled.

Cnidarians are of greatest danger to human beings

in tropical or warm temperate waters, where venomous species are particularly abundant and where an individual in the water may be subjected to a barrage of nematocysts simultaneously discharged from many members of the population. The effect of the toxins on man ranges from mild dermatitis to almost instant death, depending upon the number of nematocysts reaching the target, for the severity of the reaction is, in general, proportional to the amount of toxin injected. The greatest number of human fatalities from the stings of cnidarians have been in the Indo-Pacific region between North Australia and the Asiatic mainland, but cases along the southern coasts of the United States are not uncommon. Warnings againt *Physalia* especially are quite often posted along the Florida bathing beaches when conditions are such as to bring the siphonophores inshore.

Another highly specialized device for the injection of venom is found in cone shells, gastropod molluscs of the family Conidae, in the suborder Toxiglossae of the order Neogastropoda, which includes the most advanced prosobranchs. Most species of Turridae and Tenebridae have poison glands located in the head and opening into the pharynx by long ducts, but the Conidae are the most deadly of all the Toxiglossae. Cones are nocturnal predators in shallow waters, some species feeding on worms, some on other gastropods, and some on fish. They paralyze their prey, which is often larger than themselves, before swallowing it whole. Their injection mechanism consists of a venom bulb, where the secretion is formed, lying dorsal and posterior to the rostrum; a duct that leads from it into the pharynx; the radular sac; and only the marginal rows of lateral teeth, for lateral and median rows are missing. Cones cannot sting unless their heads are out of their tall and conical shells, for in stinging, the proboscis is protruded with the radular teeth emerging from it. These teeth are specialized—long, spear-shaped, and detachable. In action, one of them, previously released from the radular sac, is propelled forward by the combined activity of cilia on the epithelial cells and peristalsis of the radular sac. The base of the tooth is held by the proboscis, while the free end hooks into the prey (Fig. XIV–7). The tooth is then drawn back into the buccal cavity, accompanied by the prey, which has been paralyzed by the venom previously squeezed into the radular sac by contraction of the venom bulb and its duct. Each tooth is swallowed with the prey, and if it is not, the tooth is lost. In either event a new one is released and readied for the next attack. People may be poisoned by careless handling of living cones which are frequently collected because of the beauty of their shells, yet the only species believed dangerous to humans are those that feed on fish. The components of the venom have not yet been identified, but pain, a burning sensation, and numbness at the site of the puncture have been described by those who have been stung. The toxic principle is not destroyed by heat and only loses some of its potency after incubation with trypsin. Amines other than serotonin, which has not been identified in cone venom, are in all probability present and its ultimate effect is paralysis, apparently by blockage of neuromuscular transmission.

Most scorpions, bees, and wasps, and some ants are venomous, injecting their poison by means of stings. The effect of this is occasionally fatal to humans, and in the United States about 25 deaths a year may be attributed to the stings of venomous arthropods. Scorpions are nocturnal predators on insects, spiders, and other scorpions, paralyzing their prey by injection of the venom before grasping it with their chelicerae and tearing it into bits with the pedipalps. The poison glands lie in the abdomen and their ducts open into the sharp terminal claw at the end of the telson (see Fig. VII–28). In stinging, the abdomen is arched forward over the back of the scorpion, which stabs its victim as it lies or passes in front of it. Venom emission can be induced by manual or electrical stimulation, so that with proper precautions scorpions can be "milked" and their pure venom obtained. Probably the most complete evidence for proteins as the agents of toxicity has come from work on two North African species, *Androctonus australis* and *Buthus occitans*. The venom of each of these contains two active basic proteins, which share almost equally in the total toxic activity of the scorpamines, a generic term for scorpion venoms. One of these proteins, isolated from the venom of *A. australis*, has a molecular weight of about 16,000 and contains 17 amino acids. The other, also with 17 amino acids, but different ones, has a molecular weight of 11,000. Both of these,

FIG. XIV–7 Conus striatus *with its proboscis extended and in contact with a fish (blenny). The radula tooth by which the fish is harpooned lies at the tip of the proboscis. (Courtesy of Dr. Alan Kohn; reproduced by permission of the National Academy of Sciences.)*

when tested on isolated tissues, induced contraction of smooth muscles, spastic paralysis, and increased excitability of the central nervous system. Two species of scorpion found in the southwestern part of North America, *Centruoides sculpturatus* and *C. gertsch*, are also venomous, their stings resulting in spastic paralysis with the action of their venom partly on the neuromuscular junctions and partly on the spinal cord. The venom injected by a single sting of either of these two species can be lethal to man, but that of other scorpions of the southwest usually produces local rather than systemic effects.

The venom of *Leiurus quinquestriatus* has been found by paper chromatography to contain serotonin in amounts equal to 2 to 4 μg/mg dry weight of venom, the largest concentration yet reported in any biological medium. There is much less than this in the venom of *Buthus minax*, and none has been found in the venoms of nine species from South America and the North American southwest. Venoms from Indian scorpions contain a variety of enzymes, proteases, 5′-nucleotidase and phosphodiesterase, but no phospholipase. Scorpamines have, in general, some of the characteristics of snake venoms and may be equally fatal to humans. It has been reported that in parts of North Africa deaths from scorpion stings are more frequent than those from poisonous snakes, but in the United States they are accountable for only 4 percent of those annually attributable to arthropods.

The sting in the females of most species of Hymenoptera is a modified ovipositor, an organ derived from the terminal segments of the abdomen and primarily one used in egg laying. Some male wasps have pseudostings, structures that resemble the true stings of the females and that can at least prick through the surface of another animal. But only the females have poison glands, located in the abdomen and consisting usually of a venom sac and a duct from it to the stinging mechanism; so-called "acid" and "alkali" glands open into the duct and contribute to the mixture of substances in the venom. One family of tropical bees, the Meliponidae, are without stings at all. Among the ants, members of the family Myrcinae have a stinging apparatus like that of honeybees and use it to inject venom from their abdominal glands, but in the most widely distributed family, the Formicinae, the stings are degenerate and the ants injure their prey by biting and then spraying their toxins upon the wound.

Primarily the function of the hymenopteran sting is tranquilization of prey and is used by solitary wasps for this purpose. In the social species this function has been subordinated to that of defense of the colony. Solitary wasps use their stings to paralyze other insects and spiders with which they stock their brood cells, thus providing their larvae, immediately after hatching, with a supply of fresh food. This practice is correlated with the development of patterns of behavior that are extremely precise and that lead to insertion of the sting in the most vulnerable part of the prey's anatomy. Many reports have been made by careful observers of wasp behavior in the field and in the laboratory describing the accuracy with which the attack is made, so that with one or two stings a larger insect or a spider can be almost instantly subdued. The digger wasp *Philanthus*, for example, feeds as an adult upon nectar; but when making its nest in summer, the female attacks pollen-gathering bees, because of which it has earned the popular name of bee-wolf. *Philanthus triangulum* has a special preference for honeybees and when confined in small containers in the laboratory with them can be seen to insert the sting, invariably and with unerring accuracy, through the unsclerotized membrane on the ventral side of the bee, behind the coxa of either member of the first pair of legs. The result is immediate paralysis of the proximal segments of the pro- and mesothoracic pairs of legs. The entire process of stinging lasts for about 30 sec, after which the wasp turns the bee over so that its ventral side is uppermost and, holding the bee beneath its body, squeezes its abdomen forward so that the honey sac is broken. The wasp then licks up the exuded contents and subjects the punctured area to malaxation, beating on it with its proboscis. This massage softens the now immobilized prey, which the wasp carries off to her nest in the ground, carefully putting one bee in each brood chamber and depositing an egg upon it. *Priocnemis*, another digger, performs a similar feat with spiders but usually makes several attempts at stinging before the final *coup de grace* on the ventral surface of the prosoma. This results in deep and permanent paralysis of the spider, which is carried off to the nest and deposited in a chamber along with an egg. The larva hatches out in 3 to 4 days and at once begins to feed upon the spider, which it can consume in 1 week, although spiders removed from a wasp immediately after capture and kept under observation in the laboratory have remained immobile but alive for as long as 33 days. The action of the toxin would appear to be upon the somatic rather than the visceral muscles of the prey, for their bodily functions continue although they are incapable of motion. Their hearts go on beating, although irregularly, for as long as 37 hr after a sting.

Braconid wasps lay their eggs in the bodies of other insects, either larval or adult, and the young of these wasps develop parasitically in their hosts, feeding off their tissues and destroying them after they have been immobilized by the female wasp's sting. Experimentally, it has been found that as little as 1 part of the venom of *Bracon hebetor* (*Habrobracon juglandis*) will produce permanent paralysis in late-instar larvae of wax moths (*Galleria mellonella*) but that even after complete immobilization, the hearts of the caterpillars kept on beating. Studies of the effect of this venom on larvae and adults of *Philosamia cynthia* indicate that the paralysis may result from a blockage of neuromuscular transmission at a presynaptic site, enabling the wasp to reduce its natural prey—*Galleria*, *Ephestia*, and *Ploida*—to a state of quiescence while it deposits its eggs.

Such precise localization in the insertion of the sting as that exhibited by these wasps would seem to require special sensory receptors that would enable them to detect the exact site. In *Philanthus*, and in other digger wasps that have been examined, the distal ends of the sheath lobes of the sting that guide the shaft in the course of stinging have some 70 sensory hairs, 3 to 4 μ in diameter and 45 to 90 μ in length, as well as two additional types of compression and stretch receptors (campaniform sensillae) in the dorsal wall of the sheath bulb (see page 620). These are all highly specialized touch receptors that respond to soft unsclerotized spots in the cuticle of the prey. The wasps can be induced to sting in places other than the usual ones if tiny windows in the chitin are cut in various spots of the body. This has been done not only as an experimental test of their sensitivity but also as a test of the hypothesis that the site of the sting is selected so that the venom may be inserted directly into one of the important central ganglia of the nervous system. Although after a sting in the customary place there is immediate paralysis of the front and middle pairs of legs, it is a full minute before the proximal segments of the hind legs are paralyzed, and half an hour before their pretarsi are. When, experimentally, a window is cut in the mesonotum and the wasp stings there, a bee can still move forward for 30 sec or more, but after 50 to 90 sec the coxae and femora of all the legs are paralyzed, while the tibia and basitarsi still kick for another 15 min. When the site of the sting is on one side of the thorax, the legs on that side become paralyzed before those on the other, and when it is in the abdomen, paralysis of the legs is delayed for as long as 3 to 4 min, and the bees can recover from it and walk again. Histological examination of the composite thoracic ganglion has shown that there are no significant changes in the areas innervating the second and third thoracic segments for 2 to 3 hr after insertion of the sting, but that the glia cells and neuropil (see page 663) in the area of the first segment show some minor cytoplasmic changes from the normal condition, although the cell bodies themselves show none. Twenty-four hours later, the glia cells and neuropil in the second and third areas are similarly affected. It would seem, therefore, that the sting does not penetrate to the ganglion nor the venom affect it directly, but rather that it has a progressive effect as it would if it were spread through the body by means of the hemolymph. Additional evidence for its progressive dissipation and gradual effect upon the somatic muscles comes from experiments in which newly hatched wasps, which have developed but little venom, have been induced to sting bees near their tegulae, small scale-like structures on the mesothorax covering the bases of the first pair of wings. Immediate paralysis of the wings follows a sting in this location and, after some minutes, paralysis of the coxae and the femora of the legs. If the poison were introduced directly into the mesothoracic regions of the ganglion, the wings and legs of that segment should be immobilized simultaneously. Yet the tibia and segments of the legs distal to them are never paralyzed, and after about 1 hr, the bee can recover full motility and fly again.

Although in nature *Philanthus*' attack is specifically upon bees, its venom is not specific to the natural prey. Tests have shown that it is effective upon spiders and representatives of all insect orders, including its own, but not upon *Pelacus variegatus*, another sphecid that uses *P. triangulum* as one of its hosts. Yet the wasp is immune to its own toxins, possibly because of a substance in the hemolymph that inhibits their action upon its muscles. If, however, a potential victim has a chance to turn upon the wasp and sting it, the bee venom can kill *Philanthus*.

A worker bee can sting only once, for most of the stinging apparatus is left behind in the victim when the bee flies away after inserting it. The loss damages the bee so, probably through depletion of its hemolymph, that it survives only a day or two after stinging. Queen bees, which have rather more slender stingers than workers, can sting several times and occasionally do so in self-defense when attacked by their natural enemies. Wasps can sting repeatedly, and the attack of a colony, when all the females are incited to defend the nest and sting as often as they can strike, can be a very serious affair to other animals, even those of considerable size.

Stinging behavior is so deeply ingrained in the behavior patterns of wasps that a solitary species, *Microbembex monodonata*, goes through all the motions of stinging its prey, although this prey consists of dead or disabled arthropods of all kinds that do not require immobilization.

The venom of social wasps and bees has been investigated more fully than that of other insects. It is possible to collect pure venom from several hundred individuals at a time by stimulating them electrically. When an insect is appropriately confined, the stinger can be made to empty over the well of a depression slide, after the insect's recovery from the carbon dioxide anaesthesia necessary for its preliminary handling. Serotonin is present in the venom of social wasps, with the largest amount yet reported being in the European hornet *Vespa crabro*, in which quantities as great as 19 mg/g of dry venom sac have been reported. Free histamine, in amounts of 14 to 30 mg/g dry weight of venom sac have also been found in it and a lesser amount, 4.3 mg/g dry weight of venom sac, in *V. vulgaris*, the commonest of yellow jackets in the northeastern United States. Histamine is also a component of the venom of honeybees. The venom of *V. crabro* contains acetylcholine in amounts equal to 18 to 50 mg/g dry weight of venom sac, but none has been found in the venom of *V. vulgaris*. "Wasp kinin," first detected in the venom of *V. vulgaris*, is distinguished from bradykinin by different chromatographic behavior and sensitivity to trypsin, as well as by greater pharmacological activity on the isolated rat uterus. It is different also from the kinin in hornet venom, which is not inactivated by trypsin and, when tested on guinea pig ileum, is less active. There are two kinins other than wasp kinin in the venom of *V. vulgaris*, but the composition of neither of them is as yet known. Their pharmacological effect, as exhibited by test animals, is lowering of blood pressure and contraction of smooth muscle. Wasp venoms, with the exception of *Philanthus*, are not known to contain paralytic polypeptides.

Mild electric shocks will cause honeybees to discharge their venom, so that collections of reasonably large quantities can be relatively easily obtained. The venom of *Apis mellifica* contains two toxic polypeptides, which have been isolated and their amino acid composition determined. One of them, apomine, is made up of ten amino acids and has a molecular weight of 2,100. There are twelve amino acids in the other, mellitin, which has a molecular weight of 5,700. Both have a wide spectrum of pharmacological activity. Apomine has been shown to increase the permeability of the walls of capillaries, and to induce hypermotility, hypersensitivity, and death with convulsions in rats and mice. Mellitin is even more potent in increasing capillary permeability and also evokes contraction of smooth muscle, impairs oxidative phosphorylation, and leads, ultimately, to flaccid paralysis. In vitro, it will release histamine, will lyse red blood corpuscles, and release serotonin from blood platelets. Bee venom also contains hyaluronidase and a phospholipase. It is, of all the invertebrate venoms, probably most like those of vipers and, drop for drop, is equally potent.

The venoms of stinging ants are difficult to obtain in any quantity, and little is yet known about their composition. The toxic agent in most of them seems to be a polypeptide or a protein of low molecular weight. The venom of the red bull ant of Australia, *Myrmecis gulosa*, which is one of the large species, contains free histamine and hyaluronidase. Although no kinins as such have been isolated from the venom, it does exhibit kinin-like activity, and a hemolytic, heat labile fraction has been isolated from it which has some of the properties of mellitin. The toxic and also hemolytic agent in the venom of the fire ant *Solenopsis saevissima* has been identified as an amine and called solenamine. Tests of its effect upon other insects have shown that it is highly toxic to fruit flies, houseflies, the termite *Kalotermes* sp., the boll weevil, and the rice weevil. The reaction of human beings to the stings of a fire ant is usually a local one, with edema and pain at the site of the wound, but occasionally there are systemic ones, with fever and allergic reactions. In a very few cases, the results have been fatal. All in all, venomous Hymenoptera account for some 65 percent of the total human fatalities annually attributed to the bites or stings of arthropods.

Formic acid ($H \cdot COOH$), the lowest in the series of fatty acids, is accumulated in the venom glands of ants of the subfamily Formicinae. This is stored in the venom sac and sprayed over prey or intruders in the nest after they have been bitten by the mandibles. The prey may be caterpillars or adult moths and butterflies, beetles, or other insects. These are often so much larger than the ant itself that mass attack is necessary to subdue them. Formic is quite a strong acid, with a distinctive smell and a very irritating effect, producing blisters on the skin of humans who come in contact with it. It is lethal to small invertebrates that receive enough of it and in lesser amounts repellent to them.

Other means of parenteral introduction of venoms are spines and stiff hairs or bristles. Some echinoderms are parenteral poisoners of this kind and local or generalized pathological effects result from contact with their spines or pedicellariae. *Acanthaster planci*, the crown-of-thorns sea star, lives in the Pacific Ocean from Polynesia to the Red Sea and is the only venomous asteroid yet known. It is a large starfish, more than 60 cm in diameter, with 13 to 16 arms on which the spines may be more than 6 cm long. The epidermal cells covering these are glandular; those that are basophil secrete mucus and those that are acidophil are believed to be the venom producers. Many echinoids have venom-secreting cells associated with their spines and pedicellariae. The venomous species are members of the families Echinotheridae and Diadematidae. In general, their spines are long, slender, hollow, and very brittle, so that they break off readily when they come in contact with an object and, if this is a living animal, are left behind in the flesh while their poison spreads through the tissues. Poison sacs or bags lie at the tips of the secondary oral spines of *Araesoma thetidis*, the tam-o'-shanter urchin of Australian waters and *Aesthenosoma varium*, the leather urchin found in the Indian Ocean and the Gulf of Suez. Their toxin is presumably secreted by the cells lining them and stored in them until used to subdue prey. Nothing is known of the chemistry of the venom of these spines and pedicellariae, but laboratory tests have shown that it induces muscular paralysis in many animals, which is particularly evident in their signs of respiratory distress.

Some caterpillars have stinging spines, stiff, hollow bristles that may be variously distributed on their bodies. Serotonin has been found in large amounts in those of the saturniid moth *Automeris* sp., but the nature of the actual toxins in these or in the urticant hairs of other caterpillars is not known.

3. Salivary venoms

Oral glands of a number of invertebrate species secrete toxins as well as digestive enzymes. The saliva of many predaceous nematodes has a paralytic effect upon their prey, which may result from a toxin in it. Nothing is known of the toxic principal, and immobilization of the protozoans, rotifers, tardigrades, small oligochaetes, and other nematodes, which provide the usual food for the predatory species, might be the consequence of their loss of turgor through leakage of fluid through the puncture wound made by insertion of the stylets.

Some gastropod and cephalopod molluscs also immobilize their prey with secretions from their salivary glands, in which, in some cases, the pharmacologically active agents have been identified. Those of two species of *Neptunea*, *N. antiqua* and *N. arthritica*, have a content of tetramine equivalent to 7 to 9 mg per gland, which accompanies the digestive enzymes that are discharged onto the prey after it has been wounded by the radula. There may be other active substances as well, but the tetramine that is known to be there may account for the poisoning suffered by humans after eating these whelks. The anterior and posterior salivary glands of certain octopods have been found to secrete a variety of substances, in addition to digestive enzymes, that include toxic polypeptides. The posterior salivary glands of *Eledone moschata* secrete serotonin in amounts of up to 300 to 500 μg/g of gland, as do those of *Octopus vulgaris* taken from the Mediterranean. No serotonin has been found in the secretions of these glands in *O. macropus*, in which, however, there is histamine as there is also in those of *E. moschata*. Tyramine has been detected in the post-salivary gland secretions of *O. vulgaris* and a number of other octopods, and there is acetylcholine as well in those of *O. vulgaris*.

A kinin called eledoisin has been isolated from the posterior salivary gland secretions of two species of *Eledone*, *E. aldrovandi* and *E. moschata*. This is a peptide with 11 amino acids which have been identified and their sequence determined. Eledoisin has a potent stimulating effect upon all preparations of vertebrate gastrointestinal muscle on which it has been tested, similar to that of another undeconal peptide isolated from a South American amphibian, *Physalaemus fuscumaculatus*. The amino acids in eledoisin are different from those of the amphibian kinin, but their pharmacological effects are the same. This suggests that the activity of such substances may be due to the positioning of particular amino acids or of particular atomic groupings in them rather than a general chemical similarity. A substance known as cephalotoxin has been extracted from secretions of the posterior salivary glands of *O. vulgaris*, which induces flaccid paralysis in animals that the octopus bites. Experimentally, it has been found to modify the electrical activity of the isolated central nervous system of crabs. Chemically, it may be a glycoprotein, a hexoseamine associated with a protein.

The poison glands of spiders are developed embryologically as invaginations of the chelicerae, growing in most species until they extend well back into the cephalothorax of the adult (see Fig. VII–28). The venom is injected into the prey when the fangs pierce its integument and the walls of the poison gland contract, forcing their contents into the wound. The effect of the toxin is almost immediate upon insects. Large spiders can kill birds and mice, although not very quickly, for the toxin is produced only in small amounts and what there is spreads slowly through the bodies of animals of this size. A number of species of spiders are poisonous to humans, occasionally with fatal results. The bite of black widow spiders, members of the genus *Lactrodectus* that is common in temperate as well as in tropical regions, is especially feared. The toxic properties of its venom have been located in the protein fraction, which can be separated by electrophoresis into two components, A and B. The A component has been found experimentally to cause instant paralysis of houseflies and the B component also to have a paralytic effect, but a slower one. They differ, too, in that the paralysis caused by A is reversible, if the dose is not a large one, but that caused by B is permanent, regardless of the quantity injected. In man, the effect of the bite of a black widow is pain and muscular rigidity, the abdominal muscles being especially affected if the bite is in the lower half of the body; the shoulder, back, and chest muscles, if it is in the upper half. Various concomitant symptoms suggest that other pharmacologically active agents accompany the paralyzants. The venom of the Brazilian wolf spider, *Lycosa raptoria*, has a hemolytic effect, and there is local necrosis at the site of the wound which gradually spreads to the adjoining tissues. Reports from Sao Paulo state that several hundred persons are bitten each year in this area of Brazil. The brown spider *Loxosceles reclusus*, native to Missouri and Arkansas, is also poisonous to humans, producing local necrosis in the region of the bite. Analyses of the venoms of several species of spiders have shown that they have a number of components, including gamma amino butyric acid, histamine, serotonin, a polyamine called spermine, proteases, hyaluronidase, phosphodiesterase, and polypeptides of various compositions. Not all of these are included in the venom of a single species, and the present evidence is that the toxic and lethal effects are due to the polypeptides primarily, the other components, whatever they may be, acting as adjuvants in the destruction of the tissues and so in the spread of the venom.

Ticks may introduce poisons when they bite, as well as the pathogenic organisms for which many of them are known to be vectors. The bite of *Dermacentor andersoni*, for example, causes a localized effect known as tick paralysis. Similar local reactions following the bites of blood-sucking insects such as mosquitoes, sandflies (*Culicoides furens*), and reduviid bugs like *Anilus cristatus* indicate the introduction of tissue active agents as well as of anticoagulants. Secretions from the salivary glands of the reduviid *Platymeris radamantus* contain at least six proteins, three of them with trypsin-like activity, active hyaluronidase, and a weaker phospholipase.

4. Contact venoms

Gastropods of the family Muricidae do not bite or rasp their prey, largely barnacles and bivalves, but drill through their shells to reach the soft parts. Their hypobranchial glands secrete a complex of substances that relax the animal inside. They have no means of injecting these secretions, which are unusual in apparently penetrating into the tissues and reaching the muscles in some way of their own. The snails, in capturing prey, wrap the foot around it so that it is held in a kind of pocket and in this enclosed space the soft parts may be exposed to an accumulation of glandular secretions for some time. The nature of the secretion has been most fully studied in the genus *Murex*, long known as the source of Tyrian purple. It is initially colorless, or a very pale yellow, but contains a photolabile chromogen that becomes brilliant violet-red when exposed to sunlight. The chromogen is presumably formed in cells in the medial region of the gland where there is a concentration of secretory cells containing colorless or yellowish granules. The gland also secretes mucus and the tranquillizing agent. This has been named murexine and is chemically identified as urocanylcholine, a combination of choline and urocanic acid, a derivative of histidine (Fig. XIV–8).

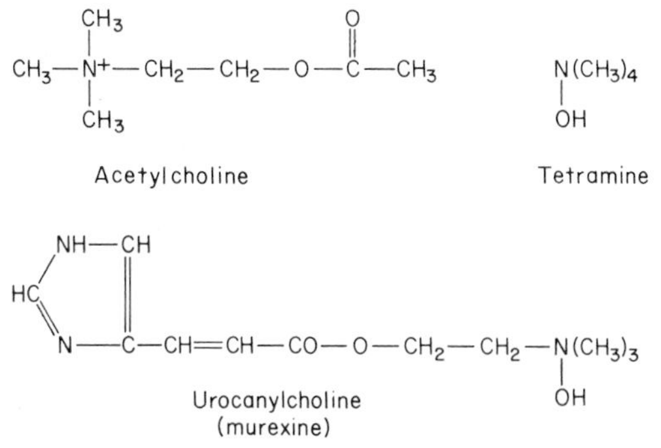

$$CH_3-\overset{CH_3}{\underset{CH_3}{N^+}}-CH_2-CH_2-O-\overset{O}{\overset{\|}{C}}-CH_3$$

Acetylcholine

$$\underset{OH}{N(CH_3)_4}$$

Tetramine

$$\text{(imidazole ring: } NH-CH,\ HC,\ N-C) -CH{=}CH-CO-O-CH_2-CH_2-\underset{OH}{N(CH_3)_3}$$

Urocanylcholine
(murexine)

FIG. XIV–8 *Structural formulae of tetramine, acetylcholine, and murexine.*

The secretion has an odor that is extremely unpleasant to humans and may be offensive also to other animals. The malodorous substances are mercaptans, probably released in the course of the production of the chromogen.

Some sponges also are malodorous and produce substances that are toxic to man on contact and may induce severe dermatitis. Most of these are species found in West Indian waters, but *Microciona prolifera*, which lives in the Atlantic Ocean, and *Suberites*, which lives also in the Gulf of California and in European seas, are also poisonous on contact. Nothing is known of the chemistry of the toxin or of its effect on other animals.

5. Repellents (defensive secretions)

Numerous invertebrates produce substances that act at a distance rather than after introduction into or direct contact with the body. These may serve as a means of communication between individuals of the same or of different species. The message to other species may be one of warning, causing retreat or inactivation of a predator or an aggressor, or it may be one of recognition between members of the same species, often between those of opposite sex. The same secretion may act in either way, for those that are odorous may be noxious or repellent when in high concentration, pleasant and attractive when in low.

Release of a noxious substance may so condition the environment of organisms producing it that their immediate vicinity becomes uninhabitable to individuals sensitive to them. A well-known illustration of this effect is the diffusion from killer strains of *Paramecium aurelia* of a substance, named paramecin, that is lethal to sensitive strains of the same species. Several kinds of molluscs and arthropods secrete and discharge compounds that, if not actually toxic to members of other species, are at least repugnant to them. These secretions are produced in special glandular organs, distinct from the venom-producing ones, and are discharged in response to traumatic stimuli. Under natural conditions the stimulus is usually the presence of a predator or an aggressor and may be a visual, an olfactory, or possibly a mechanical one. Experimentally, discharge may be prompted by contact, pinching, or pricking. The structure of the glandular organs, the nature of their secretions, and the means of their discharge are best known in some gastropod and cephalopod molluscs and in some myriapod and insect species, but further exploration may show a wider distribution of such defense mechanisms among other invertebrates.

Although diverse in their locations and in the chemical nature of their secretions, the organs show a general uniformity of pattern but a good deal of variation in detail. They develop as invaginations of the integument and may be single, double, or multiple organs located anywhere in the body. Typically, they consist of a sac, or reservoir, in which are collected and stored the products of cells in its walls or in close association with them. A duct leads from the reservoir to the external opening through which the product is expelled. This may ooze out onto the surface of the animal, making it repugnant to another, or onto that of a nearby attacker. More frequently it is discharged with some force so that it is shot or sprayed over a considerable distance. This may be effected by muscular contraction of the reservoir or of structures into which it opens; or in arthropods, by its compression through distortion of the body, increase in pressure of the hemolymph or of air in the tracheae, or, in some few, by chemical reactions in the organ itself and the generation of gas so that the secretion is discharged with explosive force. In some beetles and caterpillars the gland is contained within an eversible structure, whose opening may be trained on a particular object.

When irritated, and also when feeding on *Velella,* the pelagic gastropod *Ianthina* releases a purple secretion from its hypobranchial gland which seems to prevent discharge of the siphonophore's nematocysts. Two glands in the mantle cavity of the sea hare *Aplysia* combine to produce a secretion that is purple in color, fetid in odor, and toxic in action. The biochrome is produced in large unicellular glands in the mantle that open by numerous ducts into its cavity, where it is mixed with mucus and with aplysin, a toxic substance discharged from large unicellular glands in the floor of the mantle cavity. Nothing is yet known of the chemistry of aplysin, but when it is injected experimentally into frogs, they become paralyzed. The discharge is considered a defensive one, since it occurs only when the mollusc is irritated. The thick, dark secretion diffuses slowly through the water, screening the animal and driving others away by its odor, which is so offensive to man that it is assumed to be so to all animals. Laboratory observations have shown that the natural enemies of *Aplysia* are repelled by the secretion and move rapidly out of range of it.

Many dibranchiate cephalopods protect themselves in a similar way, by discharge of secretions from their ink glands. The reservoir, or ink sac, lies along the rectum into which its duct opens. The glandular cells are in the dorsal wall of the sac and their continuous secretion keeps it full of a colorless fluid in which brown or black granules of sepia are suspended. Sepia is a melanin, possibly in combination with protein. Mucus is also mixed with the secretion, causing its slow diffusion when it reaches the water. Upon appropriate stimulation, the ink is released from the ink sac into the rectum and so into the mantle cavity from which it is expelled through the funnel, often with considerable force. The dark cloud emerging may distract the attention of a predator and give the cephalopod a chance to change to a concealing color and to dart away. In some species, photophores or bacterial light organs are embedded in the ink sac, and a secretion that is luminous is similarly released into the mantle cavity and expelled through the funnel. This may also distract the attention of an enemy or alarm it so that it retreats or at least desists from attack. Possibly, too, the luminescent discharge may at other times serve as a signalling device between the sexes, as the flashing of fireflies does.

Although apparently absent in crustaceans, defensive and repellent organs have been identified in some scorpions and are widespread in myriapods and insects, in some of which they have reached a high degree of refinement, both in structural detail and in operational finesse. Their variability in morphological detail, as well as their sporadic distribution among species, suggest that they have arisen many times, independently, in the course of arthropod evolution, except in the crustacean line. They may be fairly simple in their organization and segmentally arranged, as they are in myriapods, or they may be localized in the head, the thorax, or the abdomen. They may be present and effectively used in one or several genera of an order, yet be absent in others. Their effectiveness depends upon the speed and accuracy with which the secretion can reach its target and the sensitiveness of the target animal to its components. These are almost always odorous, so that they may drive a chemosensitive animal away. They are volatile and evanescent and though they may temporarily incapacitate an animal with which they come in contact, the animal usually recovers in a matter of seconds or of minutes. Some have been found to be lethal to small insects or to other animals of minute size. It may take hours or even days before a reservoir, depleted of its contents, can be refilled; and massive or continuous discharge leaves animals with organs such as these defenseless during that time. Some myriapods and some beetles have very elegant devices for production and discharge of gases and powders, but for the most part the secretions are liquids, frequently emulsions, and are discharged as an aerosol spray. In some cases they harden or solidify when exposed to the air and their effect upon an aggressor is, therefore, mechanical rather than pharmacological, impeding its motion by covering it with a viscous or gelatinous film. The secretions are usually mixtures of compounds of low molecular weight (30 to 200), straight-chain fatty acids or their derivatives, and aliphatic and aromatic hydrocarbons. In all, 31 different substances have been positively identified as constituents, the most common of which are benzoquinones, formic, acetic, and caprylic acids, derivatives of butyric acid, *trans*-2-hexenal, and *trans*-2-decanal. The secretion of the pentatomid bug *Nezara viridula*, the southern green stinkbug, contains at least 18 aliphatic compounds, including hydrocarbons, aldehydes, ketones, and esters. At the other extreme is the cockroach *Eurycotis floridana*, whose secretion has a single component, *trans*-2-hexenal, which seems to be invariably present in the odorous substances produced by invertebrates. The mandibular glands of the harvester ant, *Veromessor pergandei*, produce principally benzaldehyde, a compound also produced by millipedes, but not hitherto found in insects. These harvester ants, native to the southwestern United States and Mexico, use this secretion most effectively as a defense against fire ants, although it is ineffective against workers of their own species.

Many of the compounds in these substances are unusual in animals but are common secondary plant substances. Hexenal, for example, is proven to be a food attractant to *Bombyx*, and salicylaldehyde is a component of the repellent secretion of *Phyllodecta*. They are found in carnivorous as well as in herbivorous species; in their defensive secretions some of them may be noxious, others may be adjuvant to these in promoting their dispersal or penetration, while others appear to be functionally inert. Their chemical nature has been accurately determined from analysis, by chemical and physical means, of the pure secretion where it has been possible to obtain it by milking the animal or by extirpating the glands; in other cases, it has been deduced from tissue analyses of areas in which the glands are located or even of entire animals too small to make other methods practicable. Their discharge has sometimes been observed in nature, but more often from watching animals confined in cages with known or suspected predators. It is apparent, from both means of observation, that the stimulus to discharge is a highly localized one and that the secretion is usually directed with great accuracy toward its source. In the case of paired glands, for example, only the gland on the side of the animal nearest the stimulus may be activated; and in the case of multiple glands, only the glands on one side, or even only an individual gland, may respond. Sometimes, moreover, the animal may assume a posture suitable for the most effective direction of the secretion.

The diversity of the nature of the secretions and of the means of their expulsion, characteristic of arthropods in general, is illustrated in even so small a group of closely related species as the millipedes. Their glands are paired and segmentally arranged, being absent only in the most anterior and posterior segments. Secretions from these glands in polydesmids are principally hydrogen cyanide; in chordeumids, mixtures containing phenols; and in julids, spirostrepsids, and most spirobolids, mixtures whose principal components are *p*-benzoquinones. The secretion of the spirobolid *Rhinocricus insulatus* from Panama, obtained by milking, has been found by spectrophotometric and chromatographic analysis to consist of the aldehyde *trans*-2-dodecanal and of quinone in the ratio of 2.5:1. These millipedes are practically invulnerable to attack by such natural predators as ants and grasshopper mice (*Onychomys torridus*), yet in the laboratory a number of them were eaten in rapid succession by the box turtle *Terrepene ornata*, which seemed to suffer no ill effects. Another spirobolid, *Orthocricus arboreus* from Puerto Rico, emits an odorous yellowish quinone-containing solution that tans human skin. This may ooze from the apertures of the glands, which have a mechanism controlling their opening and closing, or, on occasion, can be sprayed over a distance of 30 cm.

The cyanogenic apparatus of the polydesmid *Apheloria conjugata* is as highly developed as any secretory defense mechanism of invertebrates yet known. Each gland has two compartments, whose cuticular walls are overlaid with secretory epithelium differently modified in different regions. The inner compartment, the reservoir, is relatively large and membranous and is connected by a narrow duct to the outer compartment, whose walls are rigid. The duct between the two compartments is guarded by valves, which can be opened by a muscle inserted on the wall, but the external orifice of the outer compartment is permanently open. It is possible to excise the glands from frozen millipedes and to separate the two compartments. Spectrophotometric and chromatographic analysis show that the contents of the reservoir is a stable cyanogenic compound, mandelonitrile, the cyanohydrin of benzaldehyde. This is gradually dissociated into benzaldehyde and the gas HCN, when it comes into contact with the substance in the reservoir, a catalytic agent not yet identified. This contact is made when the reservoir is compressed and the valve between the two chambers is opened by contraction of the muscle. The secretion, as it is emitted, is an emulsion in which dissolved HCN forms the continuous phase and benzaldehyde and undissociated mandelonitrile, the discontinuous one. It oozes from the apertures of the glands and the lethal gas diffuses from it into the surrounding air. The glands of millipedes that do not discharge HCN are orthodox in design, with but a single chamber, so that it is most likely that the compartmentalized pattern is an adaptation to the nature of the secretion, providing an extracellular mechanism for the production of a respiratory poison. *Apheloria* seems also to have a higher tolerance to HCN than other arthropods, a physiological adaptation manifested by its ability to outlive those confined with it in a cyanide killing jar.

Similar protective adaptations have been found in certain ground beetles of the family Carabidae. Species of *Brachinus*, the bombardier beetles that are commonly found under rotten logs and decaying pieces of wood, have a pair of abdominal glands, each one of which has two compartments. The inner one, which is the larger of the two, contains an aqueous solution of hydroquinones (hydroquinone and methylhydroquinone) and hydrogen peroxide, and the outer chamber, a mixture of catalases and peroxidases. The two chambers are shut off from each other by a valve, whose opening is controlled by a muscle. When the beetle is aroused, the contents of the inner chamber is passed through the valve into the outer one, where the hydroquinones, under the agency of the peroxidases, are oxidized to their respective quinones, and the hydrogen peroxide, under the agency of the catalases, is decomposed. The oxygen released increases the pressure in the outer chamber so that its quinones are expelled with explosive force, accompanied by a small but audible detonation. These are the actual defensive agents for they are strongly repellent to many natural predators of *Brachinus* in any circumstances, but their effect is enhanced by the energy liberated as heat in the chemical reactions that take place in the outer chamber. Measurements have shown that the spray is ejected at a temperature of 100°C, giving it a thermal as well as an irritant effect. Indeed this can be appreciated independently as a burning sensation by a human who receives the spray on some region of the general integument, ordinarily impervious to the effects of quinones. Moreover, *Brachinus* can spray at least as many as 29 recorded times, in rapid succession and with a high degree of accuracy, for its abdomen can be rotated to direct the openings of the glands toward the attacker. The compartmentalized pattern of the defensive gland appears to be peculiar to bombardier beetles, as it is to cyanogenic millipedes, and is not characteristic of carabids in general, although at least two others are known to emit hot quinones—one a species of *Metrius* and the other a member of the tribe Ozaenini from Panama. In others, so far as is known, the defensive organs consist of a single chamber, the reservoir, with associated gland cells, whose secretions consist largely of methacrylic acid (87 to 94 percent), sometimes mixed with tiglic acid (6 to 13 percent), which are, respectively, unsaturated analogs of isobutyric and 2-methylbutyric acids.

Other refinements of defensive organs are devices for precision in aiming the secretions so that they reach the target with full force. Whip scorpions (Pedipalpi), for example, have no poison glands as the Scorpionidae have, but emit a spray which they can direct with great accuracy. The two glands of *Mastigoproctus giganteus*, a predaceous species that lives under logs, boards, and other objects in a range from Florida to Arizona, open at the tip of a short, revolvable knob like a gun turret. This is situated at the end of its long flagellum. In directing its spray, the animal usually brings its whole body, or its abdomen, into an appropriate position so that the knob can be turned to direct the opening of the glands toward the target. The spray is a mixture of acetic acid (84 percent), caprylic acid (5 percent), and water (11 percent), and smells so strongly of vinegar that the popular name for these arachnids is vinegaroon. The irritating agent is acetic acid, but caprylic acid, as a wetting agent and wax solvent, is an important adjuvant to its effectiveness. Caprylic acid, in the concentration in which it is found in the reservoir of the gland and in the spray when it is first emitted, would not be effective either in disrupting the cuticle of an arthropod attacker or in spreading the spray over its surface. But as the secretion travels through the air toward the target, acetic acid evaporates from it, and when it reaches its goal the relative concentration of caprylic acid has increased to 10 to 20 percent and is great enough to make it effective in dissolving the wax and wetting the surface. *Mastigoproctus* is in some way protected from the effects of its own spray, for that which falls on its body in the course of discharge has no visible results. Probably it does not penetrate the integument, as it does that of other animals, or possibly the species has a means of handling acetic acid in concentrations as high as those in which it may enter the body.

Some insects have eversible organs from which a defensive secretion can be emitted and directed. Among these are caterpillars of the swallowtail butterflies (Papilionidae), all of which have a cervical organ known as the osmeterium because of the unpleasant odor emitted from it (Greek *osme*, "smell"). This is a forked, two-pronged invagination of the integument just behind the head, which is everted when the larva is disturbed and projects outward like two bright orange horns. Its eversion is brought about by increase in pressure in the hemolymph and its retraction, after the disturbance is over, by retractor muscles attached to it. When experimentally stimulated by poking or pinching, a caterpillar rears up into a threat or attack position and, if the stimulus is strong enough, arches its body toward the site of stimulation and everts its horns. Control of their protrusion is shown by the fact that they may be everted together, either partially or fully but equally, depending upon the strength of a single stimulus or its repetition, and that when the stimulus is unilateral, the one on the stimulated side may be everted farther than the other. The secretion from the glands drips from the openings at the tips of the horns and may fall directly upon an attacker or be wiped off on it by movements of the caterpillar. The major components of the secretion have been identified as isobutyric and 2-methylbutyric acids, which are found in the ratio of 70:30 in larvae fed on the leaves of carrot or fennel and in a ratio of 52:48 on those fed parsnip leaves. The identity of the acids in the secretions of caterpillars raised on these three different diets is indicative of their endogenous origin and eliminates the food plants as their direct source. The secretion is demonstrably effective against ants, for in the laboratory, caterpillars which were put near the entrance to the nest of a colony of *Formica polyctena* were successful in repelling small numbers of attacking ants. When bitten by an ant, the caterpillar would instantly turn its anterior end with the osmeterium extruded to the site of the bite and rub its horns against the ant. The ant would equally promptly run off, stopping every now and then when out of range to clean off the irritating fluid by rubbing itself against the ground. When the supply of the defensive secretion was exhausted after repeated attacks, or an initial attack by large numbers of ants, the caterpillars became vulnerable to the ants' spray and were killed and dragged off by them to their nest.

Some tenebrionid beetles also have means of directing a defensive spray with great precision. The construction of the organs follows the same general pattern in all genera of the family that have been examined, but differs from that of the carabids. They lie in the posterior third of the abdomen and open to the exterior by narrow slits that can be dilated through the action of muscles. The reservoir is partly divided into two lobes. The distal one ends blindly and has a patch of glandular cells over its anterior surface. The proximal lobe opens into the duct that runs to the external aperture and opens also to numerous fine tubules that lead from a loose, irregular mass of gland cells extending all over its dorsal surface. The wall of the reservoir is not muscular and discharge of its secretion is presumably through a local increase in pressure of the hemolymph resulting from constriction of the abdomen. The defensive behavior of *Eleodes longicollis*, a large black beetle found in the deserts of southern Arizona, and the nature of its secretion have been fully studied. *Eleodes* is nocturnal in habit, is herbivorous, and, without hind wings, is flightless, and so doubtless relies heavily upon its defensive secretion for escape from its numerous predators: vertebrates and other arthropods. When alarmed, a beetle assumes an alert posture, stopping whatever activity it is engaged in and pointing its abdomen upward (Fig. XIV–9). It may remain motionless in this position for some time, but if further danger threatens, it discharges a yellowish-brown liquid from the openings of the defensive glands which lie on either side of the anus. The secretion, which is both malodorous and volatile, usually drips from them, but it can be shot out as a fine spray over a distance as great as 50 cm. It is an emulsion, with one phase containing a mixture of quinones (*p*-benzoquinone, 2-ethyl-1,4-quinone, and 2-methyl-1,4-quinone), hydrocarbons

FIG. XIV–9 *The beetle* Eleodes longicollis *in its defensive attitude.* (*Courtesy of Dr. Thomas Eisner.*)

(1-undecane, 1-tridecane, and probably 1-nonene), and caprylic acid, and the other phase containing water with free glucose. The quinones are the principal defensive components and serve to repel most of the natural predators of *Eleodes*. But grasshopper mice have learned to circumvent *Eleodes*' defenses. They do this by seizing a beetle as soon as it assumes its alert posture and holding it upright so that the tip of the abdomen is pressed against the ground, which, therefore, receives the secretion instead of the mouse. The momentary pause before spraying begins makes *Eleodes* also vulnerable to the toads that prey upon it, for by the quick motion of its tongue a toad can catch and swallow a beetle before the discharge begins. If this happens in the toad's stomach, no ill effects seem to follow. Toads, however, will avoid bombardier beetles for their response is so rapid that the spray strikes the toad's tongue as soon as it touches the beetle with the result that the anticipated meal is immediately rejected.

Similar constituents of the secretion, but with a different kind of spray mechanism, have been described in a cockroach, *Diplotera punctata*. Secretory cells surround paired dilatations of the tracheae leading from the spiracles of the second abdominal segment. These produce *p*-benzoquinone and two of its derivatives that are stored in the tracheal dilatation, which thus substitutes for the reservoir of most defensive organs. When experimentally stimulated, one organ, or more rarely both, is brought into action and the secretion is discharged by air pressure as a fine spray directed toward the source of the stimulation. The efficacy of the spray against potential predators has been tested by confining roaches individually in chambers with a single carabid beetle, praying mantis, or lycosid spider. They were similarly confined with a nest of the ant *Pogonomyrmex badius*. All of these predators attacked, and the roaches responded immediately by ejection of the spray from the spiracle nearest the attacker. Roaches that had just completed a molt were unable to do this and were eaten at once, for at the time of the molt the tracheal chamber and its contained secretion is shed along with the rest of the cuticle and the insect is defenseless until it is again filled. Ants and beetles were repelled by a roach capable of spraying and moved out of range as fast as they were able to, but their retreat was impeded by periodic disruption of normal leg movements, when they either could not move at all or did so only haltingly. Each of these attacks lasted not more than 2 min, after which the repelled predator resumed its flight. The spray of young nymphs was ineffectual against spiders, which ate them at once, but that of older nymphs and of imagos drove away these aggressors, though not the mantids, which seemed wholly unaffected and ate all the roaches with which they were confined whatever their age.

The southern walkingstick insect, *Anisomorpha buprestoides*, at all stages of its life produces a secretion in two elongated thoracic glands that open just behind the head. The active principle of the secretion has been identified as a terpene dialdehyde and is called anisomorphal; chemically, it is similar to the nepetalactone of catnip. The secretion is discharged when the glands are squeezed by contraction of the muscles around them and can be directed either unilaterally or bilaterally with great accuracy at a target 30 to 40 cm away. Natural targets are predators such as ants, beetles, mice, and birds, but discharge can be experimentally induced by pinching an insect's legs or tapping its body steadily. The glands of females contain enough secretion for five consecutive bilateral discharges, but those of the males only enough for one or two. The constant association of males with females may be an adaptation to this difference, both insects acting together in defense as long as the male's supply holds out, but after this the male relying on the female for protection. Depleted glands of females are refilled and ready for action again in a week or two, but in the interval the insects are without this means of defense.

When ants, carabid beetles, mice, and blue jays were confined in cages with *Anisomorpha*, the ants and beetles were sprayed as soon as they bit the walkingstick insect and raced away from it, trying to clean off their bodies as they ran. The mice, when they got close enough to sniff at the walkingstick, were similarly sprayed and similarly retreated from it. Insects whose glands had become depleted were captured and eaten. Of all the animals tested, the blue jays alone induced the discharge when they were about 20 cm away from the insect and so without actual contact with it. Experimental devices for inducing discharge without contact were ineffective, so that it seems as if there must be some particular means by which the insect is aroused by the proximity of a jay. The discharge was directed at the bird's head, and a jay that had been hit immediately flew off to its perch and rubbed its head against the feathers on its back, while it winked rapidly with its nictitating membrane to clear its eyes. Jays soon learned to avoid these insects, and even 2 to 3 weeks after an encounter with one would not leave their perches if a walkingstick was in the cage.

Pentatomid hemipterans are so well known for potency of their evil-smelling secretions that they go by the common name of stinkbugs. The odor is due to the volatile aldehydes in the secretion, stored in abdominal glands in nymphs and the ventral metathoracic ones of imagos. The secretions consist of saturated and unsaturated aldehydes, with chains of 6 to 10 carbon atoms and the lipophilic substance *n*-decane which acts as a spreading agent. They are contact poisons to ants, the principal predators on pentatomids, as well as to other insects, causing paralysis, sometimes followed by death. Ants and other insects that do not come into direct contact with the secretions are repelled by their odor and move out of range. Penetration of the aldehydes seems to be directly through the cuticle, and though the *n*-decane probably facilitates penetration by disrupting the chemical bonding of the cuticular lipids, it is not essential. Application of the secretion to adult *Calliphora* resulted in the death of the flies in 15 to 20 sec. These were chosen as test insects because they have only two stigmata, caudally placed and guarded by lipophobic valves, so that the possibility of the entrance of the secretion through the tracheal system is minimized. After application of each of the aldehydes separately, death was longer in coming, depending upon the length of the carbon chain. Those with the greater number of carbon atoms penetrated more slowly than those with less, but after injection of any one of them directly into the hemolymph, paralysis followed in the same length of time, regardless of the length of the chain. Pentatomids are vulnerable to their own poisons, for if the reservoir is punctured by a needle, the secretion enters the hemolymph and the bugs are paralyzed at once. But they are naturally protected from them, for if the secretion is applied to the surface of their bodies, in the same way as to the surface of *Calliphora*, they are not affected. Although the main route of its entrance in vulnerable insects is through the cuticle and epidermis, some of the secretion undoubtedly enters through the stigmata and disperses rapidly through the body from the tracheal system. Pentatomids are also protected from this, for the openings of their tracheal tubes are surrounded by mushroom-shaped elevations of the integument that prevent the spread of a secretion applied to their bodies.

Scent glands are highly developed in adults, both male and female, of the rice stinkbug, *Oebalus pugnax*, but only poorly so in nymphs. The secretion, which may be ejected unilaterally or bilaterally depending on the site and strength of the stimulus, emerges in large droplets that leave orange spots wherever they land. It contains the saturated hydrocarbon *n*-tridecane (60 percent) and *trans*-2-heptenal (40 percent). The identification of these substances in 1960 was the first instance of the isolation of them from an animal, although their widespread occurrence in plants was previously well known.

The coreid hemipteran *Acanthocephala femorata*, a large brown squash bug that feeds on citrus plants in the southeastern United States, secretes *trans*-2-hexenal in the large round gland situated in its metathoracic and first abdominal segments and opening to the outside through a pair of pores between the sternal and pleural plates of the metathorax. The secretion is ejected, in both nymphs and adults, as fine droplets which the adults can spray as far as 20 cm. There are muscles controlling the opening of the apertures of the ducts from the glands, but there is none in the walls of the reservoir. Since injection can be experimentally induced by flexion of the metathoracic legs, it is probable that the force effecting it is increased pressure in the hemolymph. This secretion, like others containing hexenal, repels fire ants when they are brought into the vicinity of *Acanthocephala*. The secretion of the cockroach *Eurycotis decipiens*, like that of *E. floridana*, contains *trans*-2-hexenal, but unlike that of *E. floridana*, it contains lactones and *d*-gluconic acid as well. These remain as a residue after the volatile hexenal and water have evaporated away.

C. PHEROMONES (PHERORMONES)

Communication between individuals of the same species may be made by means of substances that have been given the name of pheromone from the Greek word *pherein*, "to carry," and *hormon*, "exciting." They may be effective at a distance, on contact, or orally, in each case eliciting particular and appropriate responses on the part of the individual that receives the message. Their action may be as releasers, instigating some particular pattern of behavior, or as primers, initiating metabolic, morphogenetic, or when such possibilities exist, endocrine activities in the recipient. Like hormones, which act within the body of the organism that produces them, pheromones, which act on the bodies of other organisms, are effective in minute quantities. It seems probable that in the course of the evolution of such chemical signalling devices, adaptations for them have arisen both in the sender and the receiver. Communication by such means demands not only production of a volatile and active compound in one individual but also development of appropriate and often specific receptors in another. In some of the most completely analyzed cases, substance and receptor seem highly specific and most precisely adjusted to each other, but in others the adaptation seems to be one-sided: on the part of the recipient only, which can discriminate between a number of compounds emitted by senders, responding to only one of them at a time.

As releasers, pheromones may serve to attract individuals to each other, to enable them to distinguish members of their own species from those of another, to follow a trail toward a source of food or to the nest, or to mark out territorial rights. Although these pheromones, as well as the primers, have been most extensively studied in insects, their existence in other invertebrates has been demonstrated and must certainly be suspected to be of very general occurrence until proved absent.

Females of the slipper limpet *Crepidula fornicata*, for example, release in some way a substance of still unknown chemical composition that is species specific, for it attracts motile larvae or newly metamorphosed juveniles to them. There they settle on the female's shell. *Crepidula fornicata* habitually lives in linear associations of as many as 12 individuals and, like others of its genus and like many bivalves and other snails, is hermaphroditic, the individuals changing from maleness to femaleness with advancing age. In these chains, the first in the series are females and the last ones, males, while those intermediate are intersexes in the course of transition from males to females. The advantage of the attraction and the association may be related either to reproduction, ensuring the fertilization of sessile females, or to nutrition, the combined ciliary currents of a group being more effective in bringing enough food for all than a single ciliary current would be for a single individual.

Sexual dimorphism in the echiuroid *Bonellia* has been attributed to bonellein, a pigment produced by the females which inhibits larval development; larvae coming in contact with it become males, minute in comparison to the plum-sized females. There is, however, evidence that the basis for sex determination is inherent and resides in the cytoplasm of the eggs, indicated by differences in the number, distribution and staining capacity of the granules. In about half the eggs taken directly from the uterus these granules are more rapidly decolorized by treatment with NaOH or KOH after staining with methyl green than in the others. These eggs, on fertilization, develop into males.

In the case of *Crepidula*, the substance produced by the female acts both as a releaser and a primer, diffusing through the water and attracting the young limpets to her, and subsequently directing their morphogenesis. In the case of *Bonellia*, there is as yet no proof that the meeting of juveniles with mature females is anything but chance, although once their attachment has taken place, their morphogenesis is directed toward maleness.

The most fully investigated releaser pheromones at the attractants and trail markers of certain insects. That they have been particularly investigated is due partly to the fact that the phenomena with which they are associated are readily observable in nature, to their accessibility, and to the practical advantages of using them to detect the presence of a given species in a given locality and of controlling it if it is a desirable or an undesirable one. The existence of sex attractants was first established when it was observed that males would come from considerable distances to a caged or tethered female. A single caged female of the pine sawfly *Diprion similis*, for example, attracted over 11,000 males from the field in which the cage was placed.

The first attractant to be positively identified was that of *Bombyx mori*, an insect that has been an object of intense investigation from many aspects because of its commercial importance. This pheromone was extracted from the tips of the abdomens of virgin females, 500,000 of them yielding 12 mg of the pure substance. This has been chemically identified as hexadeca-10-*trans*,12-*cis*-dien-1-ol and given the name of Bombykol (Fig. XIV–10*a*). It, or more probably its immediate precursor, which may be a fatty acid of the same chain length, is produced in specialized epidermal cells, situated laterally at the tip of the abdomen, and can be visualized in them as large lipid droplets in the cytoplasm. The cuticular boundaries of these cells, from which both cuticle and Bombykol are secreted, are greatly folded with deep indentations extending far into the cytoplasm, which has all the characteristics of that of an actively secreting cell. Trace amounts of the Bombykol precursor, which differs from other fats of the moth's body in its absorption in the ultraviolet, are evident in these cells 4 days before a female emerges from pupation. They are filled with it 1 day before her emergence. It is discharged from the cells of the imago and, evaporating

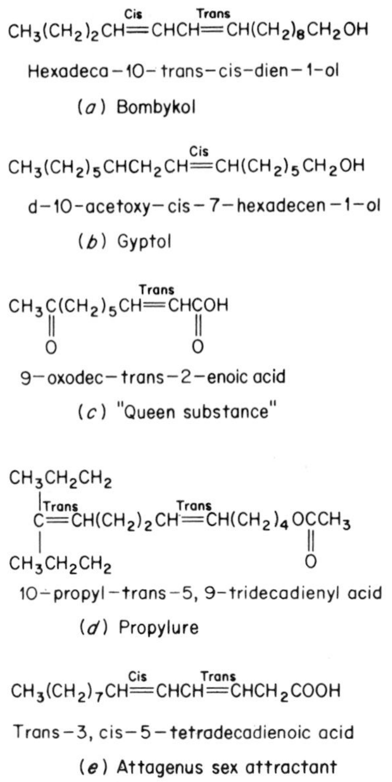

$$CH_3(CH_2)_2CH\overset{Cis}{=}CHCH\overset{Trans}{=}CH(CH_2)_8CH_2OH$$

Hexadeca-10-trans-cis-dien-1-ol

(*a*) Bombykol

$$CH_3(CH_2)_5CHCH_2CH\overset{Cis}{=}CH(CH_2)_5CH_2OH$$

d-10-acetoxy-cis-7-hexadecen-1-ol

(*b*) Gyptol

$$CH_3\underset{\underset{O}{\|}}{C}(CH_2)_5CH\overset{Trans}{=}CH\underset{\underset{O}{\|}}{C}OH$$

9-oxodec-trans-2-enoic acid

(*c*) "Queen substance"

$$\begin{array}{l} CH_3CH_2CH_2 \\ \;| \\ C\overset{Trans}{=}CH(CH_2)_2CH\overset{Trans}{=}CH(CH_2)_4O\underset{\underset{O}{\|}}{C}CH_3 \\ \;| \\ CH_3CH_2CH_2 \end{array}$$

10-propyl-trans-5, 9-tridecadienyl acid

(*d*) Propylure

$$CH_3(CH_2)_7CH\overset{Cis}{=}CHCH\overset{Trans}{=}CHCH_2COOH$$

Trans-3, cis-5-tetradecadienoic acid

(*e*) Attagenus sex attractant

FIG. XIV–10 *Structural formulae of insect sex attractants*

from the surface of her body, reaches special sense receptors on the antennae of the males. Female moths lack these receptors and so are insensitive to their own lure and after fertilization cease to produce it. It is a species-specific substance, for no males but those of *B. mori* are attracted by it.

The sex attractant of the gypsy moth, *Porthetria dispar*, likewise secreted by lateral abdominal glands, has also been extracted and chemically identified. Twenty milligrams of the pure substance were obtained from 500,000 virgin females and identified as *d*-10-acetoxy-*cis*-7-hexadecen-1-ol and given the name of Gyptol (Fig. XIV–10*b*). This is not species specific, for males of the nun moth *P. monarcha*, are also attracted by it, and it has some slight attraction for males of *Bombyx mori*. On the other hand, extracts from females of the nun moths attract only males of their own species.

Males of many butterflies and moths have special structures known as "hair pencils" by means of which the sex pheromones are transmitted by contact. These organs are usually concealed in infoldings of the integument at the tip of the abdomen, but when everted project like a pair of fine brushes with bristles outspread. In the queen butterfly *Danaus gelippus*, they are everted when a male overtakes a female in their courtship flight and are moved over her head and antennae. In this act, fine particles with which the hairs are covered are brushed onto the antennae and its sensory pegs stimulated, causing the female to alight and be receptive to copulation. Two components of the particles, secreted by gland cells at the bases of the hairs, have been chemically identified. One, the ketone component, is a crystalline pyrrolozidione which is the agent that acts as a releaser. The other, the diol component, is a viscous terpenoid alcohol that makes a sticky coating on the particle, so that it adheres to any surface with which it comes into contact, but has no excitatory effect. It is possible that both may be elaborated *de novo* by the butterflies or else that they may be transformed from closely related compounds obtained in the food. Synthesis of the diol from *trans-trans*-farnesol has been accomplished in vitro. The secretion from the glands of the hair pencils of *Lycorea ceres*, a butterfly from Trinidad, contains the ketone but, instead of the diol, two aliphatic acetate esters. Electroantennograms show that the sensory pegs of the queen butterfly are as responsive to this secretion as they are to that of their own species. On the other hand, they are totally unresponsive to the secretion of *D. plexippus*, the Monarch, which analysis has shown to be lacking in the ketone component. Yet Monarch antennae are sensitive to the ketone from *Lycorea* and from *D. gelippus*.

Queens of the honeybee *Apis mellifera* also attract males from considerable distances. Their attractant is produced in the mandibular glands, paired organs located at the bases of the mandibles and discharging their secretion through ducts. These glands are far larger in queens than they are in workers or drones. The compound 9-oxodec-*trans*-2-enoic acid (Fig. XIV–10*c*) has been extracted from them and has proved attractive to drones, whose receptors are sensillae placodea on the antennae. Only certain of the cells in these sense organs are specialized for response to this particular substance, the others serving general olfactory purposes. Unlike silk moth females, queen bees can appreciate their own attractant scent and, therefore, may possibly be able to control it. Also, they continue to produce it after fertilization.

A compound identified as 10-propyl-*trans*-5,9-tridecadienyl acetate (Fig. XIV–10*d*) has been extracted from the last two to three abdominal segments of virgin females of the pink bollworm moth *Pectinophora gossypiella*. This attracts males of the same species and elicits in them the excited flight, rapid wing vibrations, and upward curving of the abdomen, characteristic of their mating reactions. The name propylure has been proposed for it, which is the first, and so far the only one in the category of sex attractants with propyl branching. The sex attractant of females of the black carpet beetle *Attagenus megatoma* has also been extracted, purified, and identified as *trans*-3,*cis*-5-tetra-decadienoic acid (Fig. XIV–10*e*). Male beetles exposed to this exhibit the same pattern of behavior as that when they are exposed to paper disks with which females have been in contact, but somewhat less energetically, suggesting that some factor in the total emission enhances the stimulus of the sex attractant alone.

Valeric acid has recently been found to be the sex attractant for the click beetle *Limonius californicus*. The bodies of virgin females contain unusually large amounts of this (> 100 μg per individual), possibly in bound form from which the volatile acid is released at mating time. Concentrated solutions are repellent to males, but dilute ones (0.1 percent) lure them within 10 sec from a distance as great as 12 m and arouse great sexual excitement in them. After mating, the females deposit their eggs in the ground where they develop into larvae known as "wireworms" that are extremely destructive to vegetable crops. Still more recently (1968) identification of the principal component of the sex attractant of the western pine beetle *Dendrotonus brevicomis* has been reported. This is a complex cyclic compound, exo-7-ethyl-5-methyl-6,8-dioxabicyclo-[3.2.1]octane, for which the name exo-brevicomin is proposed. It was isolated from frass collected from logs of ponderosa pine infested with the beetles, and after its structure was known, it was synthesized in the laboratory. The synthetic product has proved to be attractive to male beetles in amounts as little as 1 μg.

Most of these attractants have been synthesized in the laboratory, making possible tests of the potency of the natural as well as of the synthetic products, and of their possible isomers and derivatives. These tests may be made in the field or in the laboratory and similar tests may be made of other compounds chemically related, or quite dissimilar, to the natural product. Field tests provide more natural conditions but are subject to a number of variables, which may alter the results profoundly. These variables are primarily temperature, light, humidity, and the number of other odors that may also be in the air. Atmospheric conditions affect the diffusion of a volatile substance, and what is an attractant at low concentration, may be a repellent at a higher one. The results of field tests are, therefore, often confusing or contradictory and are not easily replicated.

Laboratory tests, although providing better means of control and quantitation, are open to objection on the grounds that an animal's behavior in the laboratory and after the handling necessary to its capture is often quite different from that in its natural environment. And, either in the field or in the laboratory, the habits of the insect under investigation must be known and proper adjustments made for them. The synthetic attractant of *Apis* queens, for example, was found to lure drones only when it was suspended at least 15 ft above the ground, a condition that is consistent with the mating behavior of bees, for drones are attracted to a queen bee only when she is flying at a height that varies with wind velocity. These nuptial flights are almost always on warm, sunny afternoons. Gypsy moths rarely mate on cloudy days, and preparations that are attractive on sunny days prove ineffective on dark ones. Yet, in good weather, males may be drawn $\frac{1}{4}$ mile to the natural product or to synthetic Gyplure in an amount as minute as 1×10^{-12} μg. Some insects, too, are known to be receptive to attractants only at certain times of the day; adults of the southern armyworm, *Prodenia eridania*, for example, respond to their specific attractants only between the hours of 3 to 5 A.M. and are quite indifferent to them at other times. The times at which the females emit the scent also vary with the species. The moth *Clysiana ambiquella*, for example, discharges her secretion only between 2 and 6 A.M.; *Sparganothis pilleriana*, only between 11 A.M. and 4 P.M.; and *Lobesia botana*, the grapevine moth, in the evening.

In the laboratory, the potency of a female sex attractant may be estimated by the response of the male to graded concentrations of it and the critical range determined as that which draws him to the female and elicits characteristic courtship and mating behavior. Location of the receptor organs can be ascertained by tests, in which various parts of the body are touched by, or brought close to, a fine glass rod dipped in the attractant and their sensitivity to it determined. By these means the antennae have been discovered to be the chemosensitive areas in *Bombyx* and *Porthetria* and other members of the Lepidoptera, as well as in insects of various other orders, but not in all the species tested. Insertion of minute silver electrodes into a male's antenna, one at the tip and the other at the base, and amplification of the electrical potential set up when the sensory hairs are stimulated has provided a means of assaying the effectiveness of a compound as a sex attractant 1,000 times more sensitive than any method based on behavioral response. By such "electro-antennograms" it is possible to estimate the minimal and maximal amounts of the evaporated substance, in terms of number of molecules that stimulate the hairs of the male and evoke movement toward the female, and courtship and mating behavior. This has been done for Bombykol, and similar records of attractants for other insects will doubtless be forthcoming.

Although in most of the cases studied it has proved to be the female only that produces the sex attractant, both males and females of mealworms and cockroaches do so. The secretion from the female of *Tenebrio* is, however, more potent than that of the male. The secretions from the glands of the German cockroach *Blatta germanica* act more as a means of recognition between the sexes and an instigator of mating behavior than as an attractant. Meeting of the cockroaches seems to be by chance, but when they do meet, a female can identify a male by the pheromone produced in his tergal glands and released when he raises his wings. She eats this, and courting and copulation follow. The pheromone of the American cockroach, *Periplaneta americana*, is likewise produced by both sexes, and incites them to courtship and mating. There is no evidence of a ducted gland in these cockroaches, and the substance, which has not yet been completely identified chemically, appears to be produced by differentiated areas of the integument, primarily on the head but to some extent on other parts of the body. This has been deduced from experiments in which pieces of the integument taken from various parts of the bodies of females were held over the antennae of males. Those from the head elicited particularly strong responses, and that from the frons elicited the complete courtship display observed when a virgin female and a male come together. In concentrations even lower than those that act as attractants and stimuli to males, the pheromones of cockroaches may act as lures to members of their own or other species, collecting them in one area. A very potent pheromone is produced in the abdomen of males of *Nauphoeta cinerea*. This has been named seducin and identified as a polar neutral lipid of low volatility—its effect upon the female is to attract her to the male and induce in her the precise mating behavior necessary to copulation.

Release of a female's sex attractant is a controlled procedure, as is evident from behavioral studies of the males, as well as from assays of the amount obtained from the glands at different times, even at different hours of the day. The means of this control is as yet unknown, but it may, in some insects at least, be exercised through hormones. This probability is indicated in the results of experiments on the Cuban cockroach *Byrsotria fumigata*. After removal of the corpora allata, the organs known to produce a growth-regulating hormone (see page 722), female roaches did not produce the pheromone that leads to precopulatory behavior of the male. The stimulus to the corpora allata in this case is unknown, but it has been shown that an emanation from oak leaves identified as *trans*-2-hexenal acts upon specific sense cells in the antennae of females of the oak silkworm *Anthereae polyphemus*, attracting them to the tree and inducing the release of their sex attractant. At concentrations between 1×10^{-5} and 1 percent *trans*-2-hexenal extracted from oak leaves induces protrusion of the tip of the abdomen in females and exposure of the glands producing the sex attractant and presumably their discharge. This range is critical for the females, because their overt response, known as "calling," is not evoked by concentrations of the aldehyde outside it. It is, however, not critical for the males, because higher concentration do not affect their response to the pheromone released by the females. Once stimulated by the oak factor, a female will discharge her attractant even after her antennae have been cut off, but not after removal of her corpora cardiaca, another hormone-secreting organ of insects (see pages 719–720). It seems likely, then, that in this case at least the external stimulus imparted by the oak factor is mediated in some way by that organ before it exerts its final effect—discharge of the sex attractant. As this volatile substance diffuses through the air, males are lured from some distance to fertilize the females, whose eggs are laid on the trees and whose larvae feed upon the leaves. This delicately adjusted tripartite arrangement ensures that the female is in the vicinity of an oak tree when she is fertilized and ready to lay her eggs and that the larvae hatched from them have their particular food immediately accessible to them. A similar phenomenon is evident in certain bark and timber beetles, whose females are attracted by the volatile terpenes exuded by damaged or recently felled trees. These terpenes act as "primary" or "host" attractants, directing the flight of the female beetles, when they are ready to mate, toward the trees. The sex attractant that the females then release draws the males to them, and hence to the trees, under whose bark the eggs will be laid and the young will develop. Production and reception of lures such as these represent highly specialized systems by means of which insects are helped to realize their enormous reproductive potential, one of the greatest assets to species survival for animals whose lives are so full of hazards. The western pine beetle *Dendroctonus brevicomis*, for example, attacks ponderosa pines in masses, killing them in order to reproduce. The effect of exo-brevicomin, the pheromone produced by the females, is enhanced by myrcene, a terpene that is a natural component of the oleoresin of the pines. It is the combination of these two substances that attracts the beetles in such extraordinarily large numbers that they can destroy a full-grown tree. Fewer are attracted by exo-brevicomin alone, although this by itself is strongly attractive to another beetle, *Temnochila virescens chloridia*, a predator on *D. brevicomis*. *Temnochila* females must feed on the bark beetles in order to reproduce, and their larvae consume *Dendroctonus* larvae. In this case the sex pheromone of one species attracts not only the males of the same species but the females of a different one, and the host attractant, myrcene, is responsible for collections of beetles so numerous as to bring about the tree's destruction and also to provide an abundance of food for the destroyers' natural predators.

Pheromones may serve to bring individuals of the same sex together as well as those of opposite sexes. This, as already noted, is believed to be the case in cockroaches and has been demonstrated experimentally in honeybees. If a queen is removed from a swarm and caged in another location some distance away, workers from this swarm have been observed to find and settle around her within a few hours. Pieces of filter paper, on which the heads of several queens had been crushed, also proved a means for the orientation of swarms deprived of their queens, leading to the assumption that the substance that draws the workers to a "lost" queen emanates from glands in her head.

An identification hormone is secreted by Nassanoff's gland, a small mass of glandular tissue situated at the dorsal tip of the abdomen and discharged by worker honeybees when they return to the hive after an expedition in which they have found a new source of food. They report the location of this food in a precise series of movements, or dance, which has been carefully followed and its pattern analyzed. They discharge it also after prolonged absence from the hive and by means of it other members of the society recognize them as one of their own and tolerate their presence, although repelling any intruders that do not emit the distinguishing scent.

It has also been shown that certain habitual relationships are preserved by means of particular attractants. These are not pheromones, in their defined sense, because they act between members of different species. This distinction becomes a fine one if the association is such that survival of one member of the partnership depends upon its proximity to the other. This is the case in the sea anemone *Calliactis parasitica* that lives attached to the gastropod shells inhabited by the hermit crab *Dardanus arroso*. It is attracted by some substance in the shell and its periostracum, which, though not yet chemically identified, is not digested by trypsin and is insoluble in a wide range of organic solvents. Experimentally, it has been shown that pieces of shell, or of periostracum alone, will set in motion the series of reactions by which *Calliactis* transfers from shell to shell, when the crab moves from one to another, or by which the anemone moves from a surface on which it may have been arbitrarily placed to a shell. Thus, although under natural conditions the crab assists in settling its commensal on its shell, neither the crab nor its complex behavior pattern is necessary to bring the anemone to its habitual place of attachment. The "shell factor" acts at a distance, but the anemone's tentacles must make contact with the shell if further steps in the process of attachment are to take place. Although nothing is known of the nature of the receptors involved, it is apparent that information transmitted, probably also through the "shell factor," and received by the tentacles, causes the anemone to release its pedal disk from its previous site of attachment, bring it to the shell, or piece of shell, and then release its tentacles and erect its column.

Another anemone, *Stomphia coccinea*, also responds to a factor in molluscan shells. Preferential settling of this anemone upon the shells of the mussel *Modiolus modiolus* has been observed, although the association is not one as closely integrated as that of anemones with hermit crabs. *Stomphia* will transfer from another surface to the shell of a living mussel, or to an empty one. But it does not make this maneuver to a shell from which the organic compounds have been removed by boiling in alkali. There is, however, no information as to the properties of the *Modiolus* "shell factor," which may or may not be similar to that of the organic material in gastropod shells that attracts *Calliactis*.

The use of pheromones as trail markers is highly developed in ants and has been carefully studied in the leaf-cutting or town ant of the genus *Atta texana*. In this species the function of the poison gland has been diverted to production of a trail-marking substance. This has been experimentally removed from the poison sac as a clear, viscous liquid which solidifies on exposure to air. Worker ants, males, and virgin queens readily followed artificial trails made by a match stick moistened with the liquid and drawn across a piece of paper. In nature, the scouts that find a desirable supply of green leaves mark the trail to them as they return to the nest by touching their abdomens to the ground every 2 to 3 mm of their route and discharging minute amounts of the substance through the sting. Other workers will then follow the trail to the leaves and mark it again as they return. Trails may be several hundred feet long, and the same trail may be followed for several months. Experiments have shown that the substance retains its full potency as a trail marker for at least 5 months at room, or at subzero (−120°C), temperature, if kept in a tightly closed container.

The roach *Attaphila fungicola*, one of the few inquilines that live intimately with the ants, responds also to this pheromone, as well as to that of *Trachymyrmex septentrionalis*, a related species whose nests are often superimposed on the larger ones of *Atta*. In the laboratory, nymphs as well as adult roaches followed artificial trails, keeping their maxillary palps in almost constant contact with the trail. Though their antennae moved vigorously as they progressed along the trail, they never touched the marking substance. The roaches are less sensitive to this than are the ants and followed artificial trails made from crushed sacs of the minor workers less readily than those of the major. Investigation of the poison sacs at different stages of an ant's life has shown that they are not developed at all in larvae, are present but empty in pupae, are about half full in teneral workers, and are turgid in mature ones. Tests of the potency of their contents, made by exposing minor workers to artificial trails, showed that the contents of the turgid sacs from mature workers was most potent, that of the partially filled sacs of teneral workers moderately so, while the crushed sacs of pupae elicited no overt responses from the ants under observation.

Fire ants (*Solenopsis*) and house ants (*Pheidole*) also lay trails by means of pheromones. These are not secreted by the poison gland itself, but by Dufour's gland, which lies near the termination of the poison gland duct and discharges its contents likewise through the sting. Muscles at the base of the poison sac can close off the poison duct, so that it is anatomically possible for the contents of Dufour's gland to be discharged without mixture with venom. Fire ants lay their trails by dragging their abdomens, with sting extended, along the ground, and when a worker finds a new source of food or a better nest site, it returns to the colony, marking its homeward path in this way. Artificial trails made from the crushed poison glands, or other tissues, do not arouse other workers, but those made from extirpated Dufour's glands are immediately followed. The chemistry of the secretion from these glands is not yet known, but like other attractants, it is a volatile substance and, in still air, diffuses to form a semiellipsoid active space, a sort of canopy, under which the ants travel.

The minor workers of *Pheidole fallax* are the trail layers, and although the soldiers readily follow a trail that has been laid, they do not themselves participate in this activity of the colony. Dufour's gland is large in the minor workers and very small or entirely absent in the soldiers. The poison vesicle is, on the other hand, greatly hypertrophied in this caste, occupying about one-third of the entire abdominal cavity. A substance accumulates in it which has been identified as skatole, an indole compound with a distinctive and unpleasant smell. The poison glands of minor workers are not unduly large, and these individuals do not emit an odor appreciable to man.

Trails laid by some other ants, and some other insects, are marked by anal emissions, not true pheromones. That of the ant *Lasius fuliginosus*, for example, is an aqueous solution containing uric acid, proteins, and polysaccharides, essentially excretory and fecal material rather than a glandular secretion. In this species the secretion from Dufour's gland acts as an alarm substance, inciting other workers to intense but random activity. Likewise, fecal material in the hindguts and excreta of mature and fed males of the bark beetle *Ips confusus* provides the attractant responsible for the aggregation of individuals at a newly discovered source of food.

The operation of pheromones as primers is most evident in the social insects, where they have been demonstrated to influence metabolic and morphogenetic processes. Because of this, the term "ectohormone" is sometimes used for them to distinguish them from pheromones that elicit overt behavioral responses from another individual and from endohormones that influence metabolic and morphogenetic processes in the individual that produces them.

Caste differentiation in the most advanced social

insects—honeybees, termites, and some ants—is determined by primer pheromones, apparently acting through the endocrine system of the individual receiving the chemical message. The worker caste in honeybees is maintained in a colony with a queen through a complex of at least three pheromones, collectively called "queen substance." This complex inhibits development of the ovaries of workers, and one of its components is the same substance that acts as the queen's sex attractant. In addition to its effect upon ovarian development in workers, it keeps them from constructing queen cells, a task upon which they embark shortly after the loss of a queen from the hive, in order to rear another larva as her successor. It has been found that there are regularly about 100 μg of 9-oxodec-*trans*-2-enoic acid in a queen, and it has been calculated that she must discharge about 0.1 μg/(day)(worker) in order to prevent the rearing of other queens. A second component consists of volatile esters, two of which have been identified as methylphenylacetate and methylphenylpropionate. This inhibits ovarian development in workers, but does not affect their behavior toward rearing a larva to be a queen. Both of these together contribute to the "queen odor" by which other members of the hive recognize the presence of a queen and distinguish her from others in the society. It is possible, but not proven, that these pheromones operate through the corpora allata, for it has been observed that these organs increase in size in workers for several days after loss of their queen and the consequent disappearance of her scent and inhibiting pheromones. It is possible that a third substance is produced by the queen elsewhere than in her mandibular glands, which may act directly upon the gonads of the workers and prevent their development. This is licked from the body of the queen and, since it enters the body orally, may have a direct biochemical effect upon the reproductive system and not be mediated through the nervous system.

Little is known of the mechanisms of caste differentiation in other Hymenoptera, but they have been investigated and are better understood in some of the termites. All of these insects live in colonies and there is no known solitary species. They show a high degree of polymorphism and of social organization in the division of labor among members of a colony. It is usual for each colony to have two reproductive individuals, a king and a queen, and a more or less constant number of workers and soldiers. In the more primitive species these may be of either sex, but in those at higher levels of evolutionary development workers are of one sex and soldiers of the other. In *Kalotermes flavicollis*, one of the primitive species, the larvae develop from the fertilized eggs of the queen, which, after her nuptial flight with the king, loses her wings and settles down as one of the two reproductives of the colony. Each larva has the potential of developing as any one of the different members of the colony, but which kind it will become is not evident until the third molt, for first-, second-, and third-instar larvae show no morphological differences. Fourth instars may, however, differentiate as soldiers, becoming presoldiers at their fifth molt and soldiers at their sixth. They may also undergo another larval molt and, as fifth instars, molt into presoldiers, supplementary reproductives, or pseudergates (Greek *pseudo*, "false," and *ergates*, "worker"), which are the actual workers in the society. Pseudergates may continue to molt, but without any further growth or differentiation of parts, as long as the colony retains its reproductives. These molts have been called stationary molts, but if the reproductives are removed from the colony, the pattern of molting may change and pseudergates develop into supplementary, or replacement, reproductives. But some fifth-instar larvae may also continue on a path toward adult development, molting into first and second nymphal instars. In the course of these molts, wing pads develop. Both first- and second-instar nymphs may develop into supplementary reproductives or into presoldiers and soldiers. A first-instar nymph may also regress, through one molt, into a pseudergate, and a second-instar nymph progress to become a winged adult, one of the pair of reproductives that can fly off, mate, and found a new colony in another place.

The control of these different levels of differentiation, the preservation of castes, and the transformation of members of one caste into another has been shown to be directed by pheromones produced by the reproductives and possibly also by the pseudergates. Although the chemical nature of these substances and the precise location of the cells that produce them are unknown, there is evidence that both reproductives elaborate and discharge substances that inhibit the development of ovaries in the larvae, and so keep a large proportion of the colony at the pseudergate stage by inhibiting also further growth and morphogenesis. When a colony is deprived of its reproductives, either by natural death or experimental removal, absence of the inhibiting substance or substances results in the transformation of many of the pseudergates into supplementary reproductives, the number depending upon the number in an intermolt condition in which they are competent to respond to the chemical stimulus. Those that are not in the appropriate condition continue their stationary molts and remain pseudergates. A first- or second-instar nymph must be in a similarly receptive intermolt stage before it can transform into a supplementary reproductive. When a pair of reproductives produced by either caste begin to reproduce, the pseudergates somehow become aware of this and change their behavior pattern so that they kill all supernumerary supplementary reproductives, leaving a single pair to dominate the future development of castes.

Both of the functional reproductives produce pheromones that inhibit the development of pseudergates into supplementary reproductives. These are believed to be different substances: that of the queen being inhibitory only and that of the king without effect. Each complements the other, however, in such a way that together they are fully effective. Queens also produce a pheromone through which the presence of another queen can be detected and hostile behavior initiated, so that the one will kill the other. The same phenomenon is true for kings, which detect, attack, and kill any other king that may appear in the colony. A pheromone from the king also promotes production of the queen's inhibitory substance, or substances, and one from the queen has a similar effect upon the king.

Experimental evidence has shown that the inhibitory pheromones must be transferred directly from the reproductives to the pseudergates. If the royal pair is separated from the pseudergates by a barrier of double wire mesh, permitting the passage of vapors but not contact between the two castes, the result is essentially that of removal of the reproductives, and the pseudergates may molt into supplementary reproductives and a pair of them survive. If the barrier is only a single layer of mesh, permitting the antennae of one caste to reach the bodies of the other, the pseudergates at the appropriate stage develop into supplementary reproductives, all of which are killed. This suggests that the pheromone that primes the molt is different from the one that releases aggressive behavior.

Results of experiments with barriers that prevent the pseudergates from reaching different parts of the bodies of the reproductives and with extracts made from different parts of their bodies indicate that the active inhibitory substances are formed in the head or thorax and that they are eliminated with the feces, as materials from either the Malpighian tubules or the gut. They thus may pass directly to the pseudergate or indirectly from one pseudergate to another through its digestive tract. That they act through the endocrine system of these larvae is indicated by the fact that the corpora allata are small in those pseudergates that molt into supplementary reproductives and that this caste can be caused to molt into presoldiers and soldiers by implantation of the corpora allata of reproductives.

In what is considered the more highly evolved termite genera, the potentiality of developing into a particular caste is limited to one or the other sex, and castes are, therefore, determined at fertilization. Only in the phylogenetically lower genera is the plasticity of *Kalotermes* and the influence of pheromones evident.

Polymorphism in the social insects is associated with differences in metabolic as well as morphological potentialities. The different sexes and castes produce different odors and different pheromones. Thus, while the soldiers of *Pheidole fallax* emit the unpleasant odor of skatole when they are excited, there are no detectable amounts of any indole compound in the minor workers, and the trail-marking pheromone they emit is quite different in nature. The workers in a colony of *Acanthomyops claviger* produce monoterpene aldehydes with strong odors which act as defensive or alarm substances, while the males have quite a different odor emanating from a mixture of compounds. Investigation of two species of *Lasius* has shown that the workers do not smell but that both pleasant and unpleasant odors diffuse from the males. The source of these is the reservoir of the mandibular glands; when the heads of these ants are crushed, a pleasant fragrance can first be detected and, later, the unpleasant one of skatole. Chromatograms of material from the crushed heads of *L. neoniger* and *L. alienus*, and from those of *A. claviger* show that the different species produce different proportions of the same group of compounds, simple terpenes and terpene derivatives. These are, in general, the alarm substances of insects and are also produced by many plants, accounting for the fragrance of their flowers or leaves. Variation in the proportions of these compounds allows the males of each of these ant species to emit their own distinctive odor, which may serve to identify them or to attract others to them. This demonstrated diversity in the biochemistry of the three species opens the general possibility of a wide variety of mixtures, and a wide range of scents, that could be used for communication and the conveyance of information of many different kinds.

Another illustration of this biochemical diversity is offered by honeybees, for the queen alone produces the form of decanoic acid which acts as her sex attractant and inhibits the development of other queens. The workers do not, and when the synthetic compound carrying a radioactive label is fed to them, it is evident that they convert it, within 72 hr, to inactive substances, principally 9-ketodecanoic acid, 9-hydroxydecanoic acid, and 9-hydroxy-2-decanoic acid. They therefore do not emit the specific pheromone, but the possibility exists that these derivatives may be included in the food they give the queen and that she, and she alone, can perform the enzymatic processes that reconvert them to the active form. Were this the case, she would be spared the necessity of synthesizing the long fatty acid chain in order to keep her store of *trans*-9-keto-2-decanoic acid at the level necessary for its effective action in suppressing other queen rearing by the workers.

It is evident, therefore, that among invertebrates, especially from present evidence among insects, certain metabolic products may act as chemical signals and provide a valuable means of communication, supplementing or supplanting visual, auditory, tactile, or other devices. Since most insects produce a number of volatile substances, there exists the possibility of a good number of different combinations and of a variety of signals being given even by a single individual. More than one "bit" of information may, therefore, be contained in one emission. The usefulness of this means of communication rests upon the sensitivity, and primarily the olfactory sensitivity, of the recipient for which the signal has meaning; and in the course of evolution of the animals that rely upon this means, the production of pheromones, receptors for them and means of interpreting the signals, must have arisen within or among species. Further, mechanisms must have arisen that enable the recipient to distinguish between the variety of chemical stimuli to which it is inevitably exposed, for not only are secretions of this kind usually mixtures of volatile substances but in the atmosphere there are many from other sources. However these signalling mechanisms originated and developed, there can be little doubt that they represent a pattern of metabolism and a production and conservation of metabolic compounds that have tremendous survival value to the species possessing them.

SELECTED BIBLIOGRAPHY

Pigments and patterns

Kettlewell, H. B. D.: Insect Survival and Selection for Pattern, *Science*, vol. 148, pp. 1290–1296, 1965.

Lederer, E.: Les pigments des invertébrés (a l'exception des pigments respiratoires), *Biol. Revs.*, vol. 15, pp. 273–306, 1940.

Portman, A.: "Animal Camouflage," The University of Michigan Press, Ann Arbor, 1959.

Silberglied, R. E., and T. Eisner: Mimicry of Hymenoptera by Beetles with Unconventional Flight, *Science*, vol. 163, pp. 486–488, 1969.

Stephenson, E. M., and C. Steward: "Animal Camouflage," Adam and Charles Black, London, 1955.

Venoms

Adams, K. R., and C. Weiss: The Occurrence of 5-Hydroxytryptamine in Scorpion Venom, *J. Exp. Biol.*, vol. 35, pp. 39–42, 1958.

Bisset, G. W., J. F. D. Frazer, M. Rothschild, and M. Schacter: A Pharmacologically Active Choline Ester and Other Substances in the Garden Tiger Moth *Arctia caja* (L), *Proc. R. Soc. (London), B*, vol. 152, pp. 255–262, 1960.

Eisner, T., and Y. C. Meinwald: Defensive Secretion of a Caterpillar (*Papilio*), *Science*, vol. 150, pp. 1733–1735, 1965.

Halstead, B. W.: "Poisonous and Venomous Marine Animals of the World," vol. I, "Invertebrates," Government Printing Office, Washington, D.C., 1965.

Johnson, R. M., and H. L. Stahnke: Chromatographic Comparison of Scorpion Venoms, *Science*, vol. 132, pp. 895–896, 1960.

Lane, C. E., and E. Dodge: The Toxicity of *Physalia* Nematocysts, *Biol. Bull.*, vol. 115, pp. 219–226, 1958.

McCrone, J. D.: Spider Venoms, *Am. Zool.*, vol. 9, pp. 153–156, 1969.

Meldrum, B. S.: The Actions of Snake Venoms on Nerve and Muscle: The Pharmacology of Phospholipase A and of Polypeptide Toxins, *Pharmacol. Rev.*, vol. 17, pp. 393–443, 1965.

O'Connor, R., W. Rosenbrook, Jr., and R. Erickson: Hymenoptera: Pure Venom from Bees, Wasps and Hornets, *Science*, vol. 139, p. 420, 1963.

O'Connor, R., W. Rosenbrook, Jr., and R. Erickson: Disc Electrophoresis of Hymenoptera Venoms and Body Proteins, *Science*, vol. 145, pp. 1320–1321, 1964.

Pick, T., and E. Engels: Action of the Venom of *Microbracon hebetor* Say (*Habrobracon juglandis*) on Larvae and Adults of *Philosamia cynthia* Hübn, *Comp. Biochem. Physiol.*, vol. 28, pp. 603–618, 1969.

Pick, T., and R. T. S. Thomas: Paralysing Venoms of Solitary Wasps, *Comp. Biochem. Physiol.*, vol. 30, pp. 13–32, 1969.

Welsh, J. H.: Composition and Mode of Action of Some Invertebrate Venoms, *Annu. Rev. Pharmacol.*, vol. 4, pp. 293–304, 1964.

Repellents

Aneshansley, D. J., T. Eisner, J. M. Widom, and B. Widom: Biochemistry at 100°C: Explosive Secretory Discharge of Bombardier Beetles (*Brachinus*), *Science*, vol. 165, pp. 61–63, 1969.

Blum, M. S., R. D. Crain, and J. B. Chidester: *trans*-2-Hexenal in the Scent Gland of the Hemipteran *Acanthocephala femorata*, *Nature*, vol. 189, pp. 245–246, 1961.

Blum, M. S., F. Padovain, A. Curley, and R. E. Hawk: Benzaldehyde: Defensive Secretion of a Harvester Ant, *Comp. Biochem. Physiol.*, vol. 29, pp. 461–466, 1969.

Blum, M. S., J. G. Traynham, J. B. Chidester, and J. D. Boggs: *n*-Tridecane and *trans*-2-heptenal in Scent Gland of Rice Stink Bug *Oebalus pugnax*, *Science*, vol. 132, pp. 1480–1481, 1960.

Eisner, T.: Defensive Spray of a Phasmid Insect, *Science*, vol. 148, pp. 966–968, 1965.

Eisner, T., H. E. Eisner, J. J. Hurst, F. C. Kafatas, and J. Meinwald: Cyanogenic Glandular Apparatus of a Millipede, *Science*, vol. 139, pp. 1218–1220, 1963.

Eisner, T., F. McHenry, and M. M. Salpeter: Defense Mechanisms of Arthropods. XV. Morphology of the Quinone-producing Glands of a Tenebrionid Beetle (*Eleodes longicollis* Lec), *J. Morphol.*, vol. 115, pp. 355–400, 1964.

Eisner, T., and J. Meinwald: Defensive Secretions of Arthropods, *Science*, vol. 153, pp. 1341–1350, 1967.

Schildknecht, H., K. H. Weiss, and H. Vetter: Die Abwehrstoffe einiger Carabiden, inbesonderer von *Abax ater*, *Z. Naturforsch.* 17b, pp. 439–447, 1962.

Wheeler, J. W., J. Meinwald, J. Hurst, and T. Eisner: *trans*-2-Dodecenal and 2-Methyl-1,4 Quinone Produced by a Millipede, *Science*, vol. 144, pp. 540–541, 1964.

Woodring, J. P., and M. S. Blum: The Anatomy, Physiology and Comparative Aspects of the Repugnatorial Glands of *Orthocricus arboreus* (Diplopoda: Spirobolida), *J. Morphol.*, vol. 116, pp. 99–108, 1965.

Attractants

Barth, R. H., Jr.: Hormonal Control of Sex Attractant Production in the Cuban Cockroach, *Science*, vol 133, pp. 1598–1599, 1961.

Bedard, W. D., P. E. Tilden, D. L. Wood, R. M. Silverstein, R. G. Brownlee, and J. O. Rodin: Western Pine Beetle: Field Response to Its Sex Pheromone and a Synergistic Host Terpene Myrcene, *Science*, vol. 164, pp. 1284–1285, 1969.

Butler, G. C.: Insect Pheromones, *Biol. Revs.*, vol. 42, pp. 42–87, 1967.

Cavill, G. W. K., and P. L. Robertson: Ant Venoms, Attractants and Repellents, *Science*, vol. 149, pp. 1337–1345, 1965.

Eisner, T.: Spray Mechanism of the Cockroach *Diplotera punctata*, *Science*, vol. 128, pp. 148–149, 1958.

Erspamer, V., and O. Benati: Identification of Murexine, *Science*, vol. 117, p. 161, 1953.

Gilbey, A. R., and D. F. Waterhouse: The Composition of the Scent of the Green Vegetable Bug *Nezara viridula*, *Proc. R. Soc. (London), B*, vol. 162, pp. 105–120, 1965.

Jacobson, M., and M. Beroza: Chemical Insect Attractants, *Science, vol.* 140, pp. 1367–1373, 1963.

Jacobson, M., and M. Beroza: American Cockroach Sex Attractant, *Science*, vol. 147, pp. 748–749, 1965.

Jones, W. A., M. Jacobson, and D. Martin: Sex Attractant of the Pink Bollworm Moth: Isolation, Identification and Synthesis, *Science*, vol. 152, pp. 1516–1517, 1966.

Low, F. H., E. O. Wilson, and J. A. McCloskey: Biochemical Polymorphism in Ants, *Science*, vol. 149, pp. 544–549, 1965.

Meinwald, J., Y. C. Meinwald, and P. H. Mazzocchi: Sex Pheromone of the Queen Butterfly: Chemistry, *Science*, vol. 164, pp. 1174–1175, 1969.

Miranda, F., and S. Liseitsky: Scorpamines: The Toxic Proteins of Scorpion Venoms, *Nature*, vol. 190, pp. 443–444, 1961.

Moser, J. C.: Inquiline Roach Responds to Trail-marking Substance of Leaf-cutting Ants, *Science*, vol. 163, pp. 1048–1049, 1964.

Moser, J. C., and M. S. Blum: Trail Marking Substance of the Texas Leaf-cutting Ant: Source and Potency, *Science*, vol. 140, p. 1228, 1963.

Novák, V. J. A.: "Insect Hormones," Methuen, London, 1966.

Pliske, T. E., and T. Eisner: Sex Pheromone of the Queen Butterfly: Biology, *Science*, vol. 164, pp. 1170–1172, 1969.

Rathmayer, W.: Paralysis Caused by the Digger Wasp *Philanthus*, *Nature*, vol. 196, pp. 1148–1151, 1962.

Remold, H.: Scent Glands of Land Bugs, Their Physiology and Biological Function, *Nature*, vol. 198, pp. 764–768, 1963.

Riddiford, L. M.: *trans*-2-Hexenal: Mating Stimulant for *Polyphemus* Moths, *Science*, vol. 158, pp. 139–140, 1967.

Ross, D. M.: Complex and Modified Behavior Patterns in *Calliactis* and *Stomphia*, *Am. Zool.*, vol. 5, pp. 573–580, 1965.

Rudinsky, J. A.: Scolytid Beetles Associated with Douglas Fir: Response to Terpenes, *Science*, vol. 152, pp. 218–219, 1966.

Schacter, M.: Kinins of Different Origins, *Ann. N.Y. Acad. Sci.*, vol. 104, pp. 108–116, 1963.

Schacter, M.: Kinins. A Group of Active Peptides, *Annu. Rev. Pharmacol.*, vol. 4, pp. 281–292, 1964.

Schneider, D.: Chemical Sense Communication in Insects, *Symp. Soc. Exp. Biol.*, vol. 22, pp. 273–297, 1966.

Schneider, D., and U. Seibt: Sex Pheromone of the Queen Butterfly: Electroantennogram Responses, *Science*, vol. 164, pp. 1173–1174, 1969.

Silverstein, R. M., R. G. Brownlee, T. E. Bellas, D. L. Wood, and L. E. Browne: Brevicomin: Principal Sex Attractant in the Frass of the Female Western Pine Beetle, *Science*, vol. 159, pp. 889–891, 1968.

Silverstein, R. M., J. O. Rodin, W. E. Burkholder, and J. E. Gorman: Sex Attractant of the Black Carpet Beetle. *Science*, vol. 157, pp. 85–86, 1967.

Stürckow, B., and W. G. Bodenstein: Location of the Sex Pheromone in the American Cockroach, *Periplaneta americana*, *Symp. Soc. Exp. Biol.*, vol. 22, pp. 851–853, 1966.

Thomson, D. L.: The Pigments of Butterflies' Wings. I. *Melanargia galatea*, pp. 73–74; II. Occurrence of the Pigment of *Melanargia galatea* in *Dactylis glomerata*, pp. 1026–1027, *Biochem. J.*, vol. 20, 1926.

Thomson, D. L.: The Pigments of Butterflies' Wings. I. Gastropods, *J. Mar. Biol. Assoc. U.K.*, vol. 39, pp. 115–122, 1960.

Vité, J. P.: Sex Attractants in the Frass from Bark Beetles, *Science*, vol. 156, p. 105, 1967.

Wilczynski, J. Z.: On Egg Dimorphism and Sex Determination in *Bonellia*, J. Exp. Zool., vol. 143, pp. 61–75, 1960.

Wilson, E. O.: Source and Possible Nature of the Odor Trail of Fire Ants, *Science*, vol. 129, pp. 643–644, 1959.

Wilson, E. O.: Chemical Communication in Social Insects, *Science*, vol. 149, pp. 1064–1071, 1965.

SUGGESTED READING FOR SECTION VI

Baker, J. W. B., and G. E. Allen: "Matter, Energy and Life," Addison-Wesley, Palo Alto, 1965.

Baldwin, E.: "An Introduction to Comparative Biochemistry," 3d ed., Cambridge University Press, Cambridge, 1948.

Edney, E. B.: "The Water Relations of Terrestrial Arthropods," Cambridge University Press, Cambridge, 1957.

Ehrensvard, G.: "Life: Origin and Development," The University of Chicago Press, Chicago, 1962.

Evans, H. E.: "Wasp Farm," Natural History Press, Garden City, N.J., 1962.

Florkin, M.: "A Molecular Approach to Phylogeny," Elsevier, New York, 1966.

Green, D., and R. Goldberger: "Molecular Insights into the Living Process," Academic, New York, 1967.

Portman, A.: "Animal Camouflage," The University of Michigan Press, Ann Arbor, 1959.

section seven

MECHANISMS OF INTEGRATION

chapter XV

SENSE ORGANS: RECEPTION OF EXTERNAL AND INTERNAL STIMULI

A. SOURCES OF STIMULATION

All cellular systems respond to manifestations of energy in their environments. The energy may be mechanical, chemical, or radiant, but whatever its kind an organism responds to it in one way or another. The basic events in the excitation of any animal cell are (1) absorption of energy; (2) conversion (transduction) of that energy into electrical energy; and (3) production of a slow generator (receptor) potential. All present evidence links the first steps to events in the cell membrane, in which a change in permeability occurs permitting the inward passage of cations, probably Na^+ as the most important, with a consequent relative depolarization of the membrane. Opposite conditions lead to its hyperpolarization and inhibition rather than excitation. Depolarization results in the generator potential which spreads electronically throughout the cytoplasm. The mechanism by which mechanical, chemical, or radiant energy is transformed to electrical is still unknown and remains a central problem of neurophysiology. In receptor (sensory) cells, the generator potential initiates a series of repetitive discharges, or impulses, in the axon of the neuron, which may be transmitted to other cells in contact with the axonal terminations.

In unicellular systems and cell aggregates, all exposed surfaces may be equally receptive to the energy impinging on them, but these animals may also have localized areas or organelles that are particularly adapted to absorb energy of a certain kind. In multicellular systems, entire cells or certain areas of cell surfaces may be more receptive to one form of energy than another and so become specialized receptors (sense, or sensory, cells). It is possible that the basic mechanism of energy absorption is similar in all receptors, their apparent specificity depending upon the nature of particular sites in the membrane that are especially compatible with a certain form of energy. In multicellular systems, there are also neurons adapted for acceptance and conduction of the excitation aroused in a receptor and for its transmission from one conducting cell to another, or to an effector. It has been known for years that conduction along the axon of a neuron is a propagated disturbance, an electrochemical change that can be measured in terms of frequency and duration, and that an axon conducts either at its maximum or not at all ("all" or "none"). Methods that have been developed for inserting electrodes on or in a nerve and for amplifying and visualizing the disturbance have made such measurement possible. Some of the most important and pertinent information about neural conduction has been derived from electrophysiological studies of this kind on the giant axons of squids. It is possible, too, to insert electrodes and cannulae into the nerves of some of the large invertebrates and to hold them there permanently, so that recordings can be made and test substances introduced while the animal continues its ordinary activities. By microtechniques, electrical responses in individual sensory neurons can be detected, their frequencies measured, and the effects of different substances upon them observed. Such measurements have shown that while there are no recordable discharges from some sensory cells until they are specifically stimulated, from others there is a continuous low level of discharge.

The "resting potential" of certain sense cells has been measured, as well as the level of their depolarization, which has been found to be different in different kinds of cells. The critical depolarization level in the slow adapting cells of the stretch receptors of the lobster *Homarus* is, for instance, 10 millivolts (mv) and of the fast adapting cells 20 mv. Since the resting membrane potential is measured at 70 mv, this means that in the first case repetitive discharges (firing) are initiated at a membrane potential of 60 mv and in the second, at a membrane potential of 50 mv. This concept of a critical level of depolarization applies to sensory (receptor) cells of all kinds, and there are differences (gradations) in the level at which the firing of their impulses begins. Frequency of firing is related to the intensity of the stimulation; discharge cannot be a continuous process because some time must be allowed the membrane to return to its polarized (resting) condition in order to be excitable (depolarizable) again. During this period, the neuron is in a refractory state, and because of this, neurons are in general limited to frequencies of about 1,000 discharges per second. The more intense the stimulus, the more frequent within this limit is the firing of a receptor cell, providing the organism with a means of distinguishing certain aspects of the stimulation, such as distance, size (in terms of mass or of numbers), and the like.

The response of a cell to environmental energy is thus a series of graded events, each contingent upon the occurrence of a previous one at a level critical for its effectiveness. Consequently, messages from its environment are to a certain degree coded, a condition that is clearly of considerable value to an animal in making adjustments to those external conditions of which it receives information through neural mechanisms, and so to its survival.

In multicellular systems, the immediate environment of a receptor cell may be within the body and the source of the stimulating energy internal rather than external. Therefore, a distinction is made between exteroceptive and proprioceptive sensory cells—that is, between those that receive external stimuli that would be shared by other individuals in the same environment and those that receive stimuli from conditions within the animal's own body, and so are peculiar to it.

Under natural conditions, stimulation of any kind may result in behavioral responses by the organism. These may be simple random movements called kineses (Greek *kinesis*, "motion"), or they may be directed ones. These are called taxes (Greek *tassein*, "to arrange") in motile organisms, tropisms (Greek *trope*, "turning") in sessile or fixed ones. Taxes and tropisms may be positive or negative, depending on whether the direction of movement is toward or away from the source of the stimulus. In animals of simple organization, the excitation aroused by the stimulus may be transmitted directly from sensory to effector cell; this is the case in the induction of muscle activity in cnidarians. But in the majority of multicellular invertebrates the excitation set up in the peripheral sensory cells is conducted along afferent neurons to those more centrally situated, in most cases in chains or cords, from which it is relayed to efferent neurons and through them to effector organs. This relay may be a simple and immediate one from afferent to efferent neuron, or it may be indirect, the excitation being transmitted to other neurons in the central system before it reaches the effector neurons and the effector cells. These intermediary neurons provide a means whereby the excitation can be spread and directed. The central system can, therefore, act as a relay, and a sorting, center. Messages conveyed from a peripheral sensory area in one part of the body can be conducted via the central cords anteriorly and posteriorly along their length and so reach efferent neurons in quite a different part. Messages reaching the central system can also be selected and relayed to the appropriate efferent neurons to produce the most effective and suitable response. Integration and coordination are, therefore, major functions of the nervous system in a multicellular body and are as important to its success in this respect as are the contributions of its circulating fluids. The complex behavioral patterns of animals with complex nervous systems result from the combination of many stimuli simultaneously reaching them and the integration of responses to them into patterns of activity that are advantageous to the organism and contribute to its success in meeting the conditions of its environment and its life. One of the interesting and challenging aspects of invertebrate behavior is the degree to which these patterns may be fixed or modifiable. If the functional connections within the nervous system are inalterably set, the response must always be the same—reflex action. But if these connections are plastic and can be changed, alternative responses are possible and the animal has a choice of which response to make. This does not in any way imply a conscious choice. Yet studies of behavioral responses of a number of species of invertebrates have shown that they can select between several possible overt responses, can modify a pattern of behavior, can profit by experience, and can learn.

In addition to being a receiving, transmitting, and integrating mechanism, the nervous system is also secretory. Neurosecretion and the importance of neurohaemal organs in the regulation of certain activities of invertebrates are well established. Continued research along these lines may very probably expand this aspect of neurophysiology in significant ways. The present information about neurosecretion and its proven effects in invertebrates are presented in Chapter XVII.

B. SENSE (RECEPTOR) CELLS AND ORGANS

1. General characteristics

Since most nervous activity depends upon the initial reception of an appropriate and adequate stimulus, receptor cells and receptor organs in metazoans are first to be considered. These may be broadly classified as mechanoreceptors, chemoreceptors, and photoreceptors. Mechanoreceptors may be either exteroceptive or proprioceptive, the former responding to touch and contact, to gravity and pressure, either constant or intermittent, and the latter responding to stretching or compression of the tissues with which they are associated. Chemoreceptors respond to chemical compounds that reach them either as solutes or as gases. Photoreceptors respond to radiant energy both in the range of the spectrum visible to man and outside it. Receptors responding to radiant energy in the form of heat and to humidity have also been found in some invertebrate species.

Electron micrographs of a variety of different kinds of sense cells in both vertebrates and invertebrates have shown an ultrastructure common to many of them. This is particularly evident in the distal process, or dendrite, in which there is a segment with fine fibrils or tubules arranged, with some modifications, in the peripheral pattern typical of cilia and flagella. The central filaments are usually absent, but in several cases, basal bodies and rootlets have been demonstrated. Even in nematodes and arthropods, the two classic examples of animals believed to be totally devoid of any ciliation, the ciliary pattern can be seen in the processes of certain sense cells. It might be presumed that the prototype of all sense cells was a ciliate or flagellate cell, possibly with its sensory areas concentrated near the base of its locomotor organelle. Loss of motility in this organelle and development of special properties in its membrane, either in localized areas or over its entire surface, that gave it a special capacity for the absorption of different forms of energy would be steps leading to its differentiation as a sensory neuron. Implicit in this differentiation would also be the development of mechanisms for transduction of this energy into electrical energy and conduction of the current along the axon as a message to be transmitted to another cell. These messages differ quantitatively in accordance with the intensity of the stimulus, but qualitatively they are all alike, and it is only the surface of the sense cell that determines what the nature of a message may be. It is clear that some sense cells, and the organs of which they are a part, are highly specialized for absorption and transduction of one particular form of energy, but in other cases such specialization is not readily apparent, if it exists at all. No morphological distinction can be made, for example, between the individual sense cells that are variously distributed over the body of a sea anemone, except possibly on the basis of the length of their flagella, yet the animal gives particular responses to particular kinds of stimuli. It is quite possible that in animals at this level of organization the same sense cell may respond to stimuli of different kinds. Perhaps even those receptors that seem to have a high degree of specificity do so, too, although for them one form of energy may be the primary stimulant. Duality of function has been demonstrated in several instances in vertebrates, for example, where temperature sensitivity is combined with mechanoreception. It has also been shown in some invertebrates, notably in the crayfish *Procambarus clarkii* where cells in the abdominal ganglia are photoreceptive as well as tactoreceptive. A similar functional duality has been shown in ganglion cells of the mollusc *Aplysia*.

Specific sensitivities of particular cells or organs have been established by behavioral observations and electrophysiological methods in only a limited number of invertebrate species. In many of them, such specialization has been assumed because of morphological similarities between cells of presumed function and those whose major sensitivities have been proved, and because the responses of the animals concerned indicate that they are receptive to corresponding stimuli. Epidermal hairs, associated with sense cells, are, for example, of very common occurrence among invertebrates; in some, these have been proved to have a tactile or chemosensory role, but as yet their utilization in this way in other animals is only conjectural. The following discussion of receptors includes primarily those of proved function.

2. Mechanoreception

All cells are more or less sensitive to deformation of their surfaces by pressure, which may be either constant over a period of time or intermittent as pulses of varying frequency. There are, however, in many animals sense cells to which pressure is the primary stimulus. These are unipolar or bipolar neurons that may be generally distributed over the body or assembled in localized areas or definitive organs. The simplest mechanoreceptors responding to external stimuli are free nerve endings that lie between or beneath epidermal cells and are directly stimulated by contact of the epidermis with an object; somewhat more complex ones are those that lie at the base of a hair or bristle and are indirectly stimulated by its bending or displacement. Such tactile hairs are of very wide occurrence among invertebrates and have been shown to be involved in orientation to gravity, in postural orientations, in the reception of vibrations in water and in air, as well as in contacts with other objects (Fig. XV–1). Proprioceptive sense organs are also mechanoreceptors, responding to pressures and tensions originating within the body.

a. EXTEROCEPTORS: TACTORECEPTORS AND MECHANISMS OF ORIENTATION All organisms are characteristically oriented in respect to gravity and, when displaced from their natural positions, tend to readjust and reassume them. For the majority, this position is a horizontal one, with dorsal surface uppermost. Some, with sessile, burrowing, or pendant habit, spend most of their time in a vertical position and some, like heteropod molluscs, fairy shrimps, and back-swimming beetles, with ventral surface uppermost, are, by conventional standards, always upside down. In Eumetazoa, tactoreceptors are important in maintaining an animal's orientation, and contact with the substrate or medium provides the stimulation that keeps most invertebrates in their characteristic postures. Yet protozoans and sponges, without defined neural mechanisms, also show characteristic orientations and behavioral responses to touch and to contact. In some eumetazoans there are, in addition to the generally distributed tactoreceptors, specialized organs of equilibration, or statocysts. These have been found in representatives of all major groups of nonchordates, but not in all species of any one group.

(1) Statocysts A statocyst is basically an infolding of the epidermis, lined with receptor cells bearing stiff hairs or immobile cilia on their distal surfaces. It may be a superficial indentation, whose opening to the exterior has some sort of protection; a deep pit with a tubular connection to the exterior; or a closed vesicle filled with fluid. It contains a statolith, usually a ball of smaller concretions secreted by the cells of the invaginated epithelium or a ball made of foreign

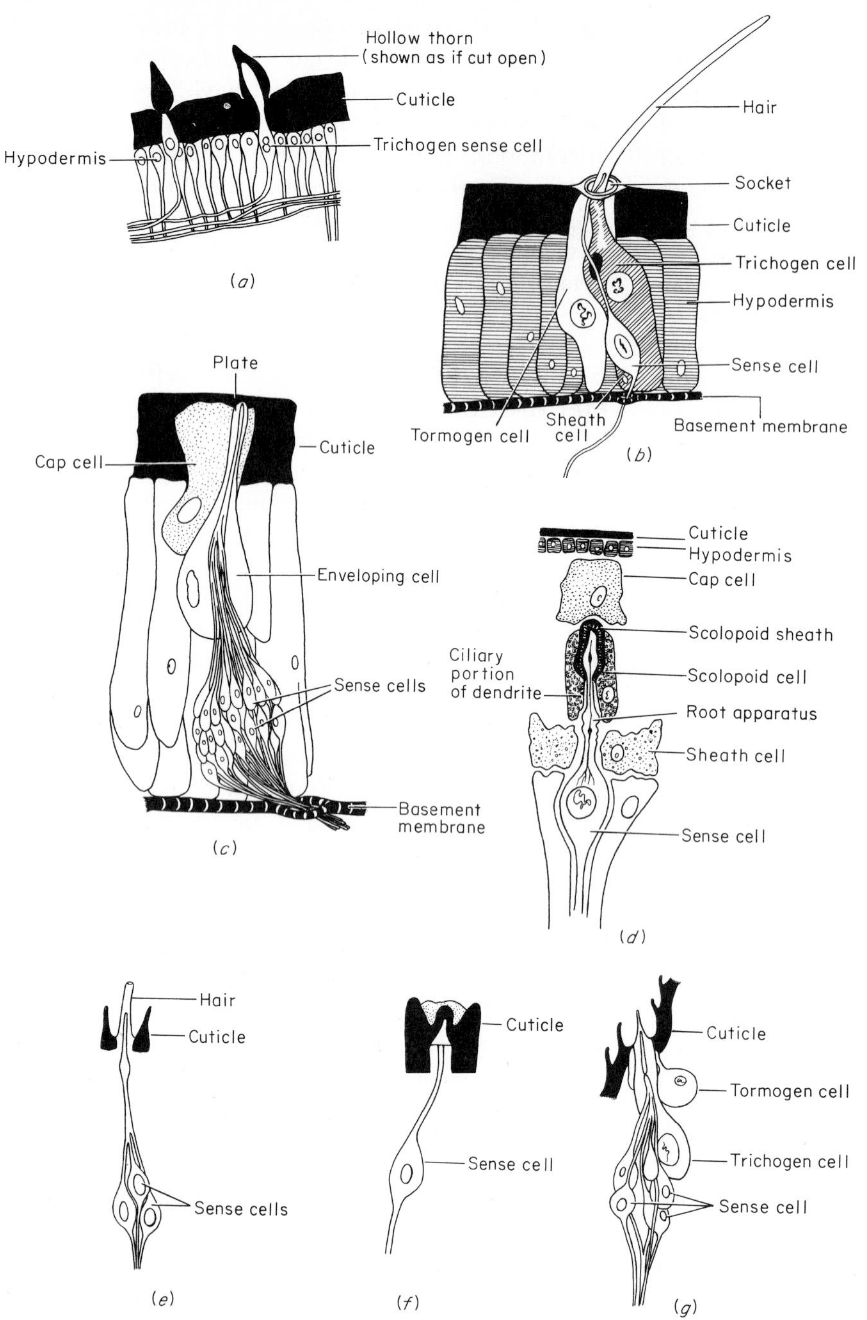

FIG. XV–1 *Sensory hairs.* ***(a)*** *Trichogen sense cell in the dorsal part of an elytrum of the scale worm* Sthenelais picta. *This is called a trichogen sense cell because it secretes the cuticular thorn, as well as receives and transmits stimulation. One thorn is shown as if cut open to reveal the distal process of the sense cell secreting and innervating it.* (*After Hanström.*) ***(b)*** *Sensillum trichodeum of an insect.* (*After Wigglesworth.*) ***(c)*** *Vertical section of a sensillum placodeum (hair plate) from the antenna of a honeybee.* (*After Snodgrass.*) ***(d)*** *Scolophophorous organs found in the chordotonal sensilla of insects.* ***(e, f, and g)*** *Sense organs from the antennae of* Bombyx. (*After Bullock and Horridge.*) ***(e)*** *Sensillum basiconicum;* ***(f)*** *sensillum campaniformis;* ***(g)*** *sensillum coeloconicum.*

material, even of loose sand grains. The discrete particles of such a ball are held together by gelatinous or adhesive material secreted by cells of the statocyst. A statolith may be balanced on the apices of the sensory hairs, or several small ones may lie free in the cavity and roll about in it when the position of the animal changes. The primary function of statocysts has been demonstrated principally by operative techniques, consisting of their destruction, their extirpation, or the severance of the nerves leaving them. In at least one classical experiment on crayfish, the function of the statocyst was shown by the substitution of iron filings for the sand grains ordinarily present in its cavity and the use of a magnet to move them about when the animal was in its natural horizontal position. When the position of the statolith, or statoliths, changes, the hairs of the sensory cells are bent, and this displacement leads to the events that bring about transduction of the mechanical energy into electrical energy. Connections of these cells with neurons of the second or succeeding orders determine the righting reactions that follow.

Statocysts are common among medusoid cnidarians, but absent in polypoid forms. They may be pits or closed vesicles (Fig. XV–2). In hydromedusae, they are usually situated at the base of the velum, between the tentacles. In scyphomedusae there is one at the tip of each rhopalium, the club-shaped structure at the end of each perradial and interradial canal in which the specialized sense organs are collected. The sense cells are long ones, with sensory hairs that are in contact with lithocytes, special cells peculiar to the cnidarian statocyst (Fig. XV–2*a* and *b*). A lithocyte contains a freely moving concretion made of calcium carbonate or calcium sulfate and some organic material. There may be a one-to-one arrangement of sense cells and lithocytes, or especially in organs with a large number of lithocytes, each one may be associated with several sense cells. If a jellyfish is tilted so that its body is oblique rather than parallel to the surface of the water or to the substrate, vigorous contractions of the musculature lead to reorientation in its natural plane. This results from pressure put upon the sensory hairs by displacement of the concretions in the lithocytes; the excitation so aroused is transmitted to the muscles and expressed by their contraction.

The apical organ of a ctenophore contains a single statolith, a ball of small calcareous concretions that are first formed within the endoplasmic reticulum of the cells at the base of the pit. The statolith is balanced on the tips of four groups of stiff, usually immobile cilia, each group standing at the apical termination of a ciliated groove, which divides, with each branch running out of the apical organ to one of the rows of swimming plates. Actively moving cilia cover the free surfaces of the cells in the depression, and cilia also form a protective covering over the statolith. At regular intervals one of the balancer cilia supporting the statolith gives a rapid beat toward the center of the group. The beat of the balancer is propagated as a wave along the ciliated groove and so out to one of the pair of comb plates with which it is in contact. Thus, waves of excitation, visible as ciliary waves, pass along the ciliated grooves, the frequency of the waves being controlled by the load on the balancer cilia. When an animal is tilted, it brings itself back into its position of equilibrium by adjustment, through this mechanism, of the frequency of waves on the uppermost and lowermost rows of comb plates.

Statocysts have not been reported in flatworms in general, although a closed vesicle containing a single concretion has been discovered in a primitive turbellarian *Hofstenia*, and they have been described in *Convoluta roscoffensis*. Similarly, a pair of them have been found in a Japanese species of *Lingula*, lying near the anterior adductor muscle, but they either are absent in other brachiopods or have not yet been described.

A pair of statocysts is a feature common to all groups of molluscs. These lie near the pedal ganglion in gastropods and bivalves and on either side of the composite ganglion, or brain, in cephalopods. They are rudimentary in chitons and absent in sedentary Ostreidae. Usually they have the form of closed vesicles, although in the Mytilidae connection with the exterior is retained through a narrow tube (Fig. XV–2*c*). They may contain a single statolith or numerous small ones. It has been shown, in the scallop *Pecten*, that there is a difference in the members of the pair, for the left one is larger than the right and contains a single statolith, and some of its sensory hair cells have long cilia, others short ones (Fig. XV–3). The cilia on the sensory cells of the right statocyst are all of the same length, like the shorter ones of the left statocyst, and the statoliths are small and numerous. *Pecten* usually swims with its dorso-ventral axis at an angle of about 45° to the vertical and with its left side and left statocyst uppermost. This posture may be related to the difference in the two organs, for removal

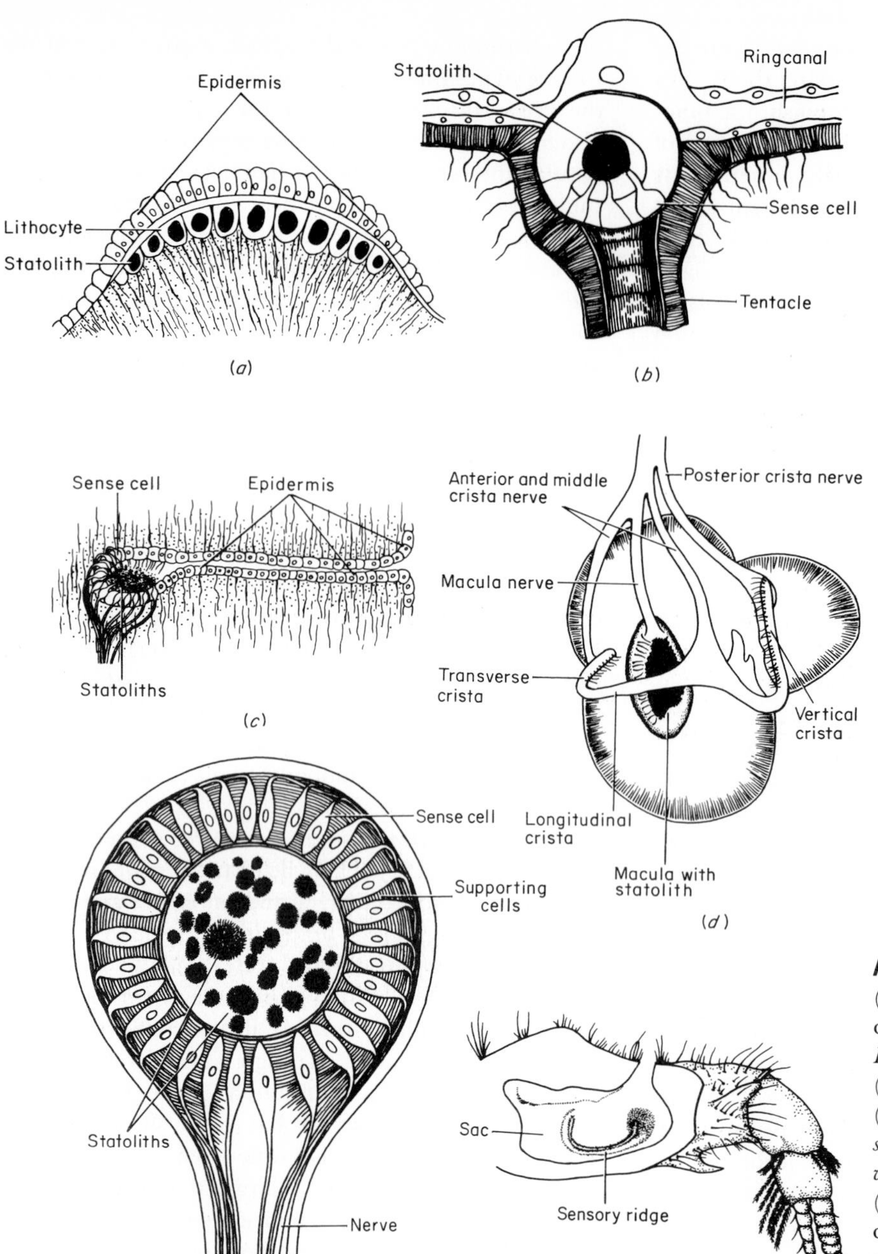

FIG. XV–2 *Statocysts.* ***(a)*** *Statocyst (open type) of the hydromedusa* Mitrocoma, *shown as in optical section.* (*After Bullock and Horridge.*) ***(b)*** *Statocyst (closed type) of the hydromedusa* Obelia. (*After Bullock and Horridge.*) ***(c)*** *Open statocyst of the mussel* Mytilus. (*After von Frisch.*) ***(d)*** *Statocyst of* Octopus. (*After Young.*) ***(e)*** *Statocyst of* Arenicola ecaudata. (*After Bullock and Horridge.*) ***(f)*** *Statocyst in the antennule of a decapod crustacean.* (*From Gardiner.*)

of the right statocyst makes little difference to its swimming posture, which is completely upset by removal of the left one.

The structure of the statocysts in cephalopods is considerably more elaborate than it is in the other classes of molluscs. Three differentiated areas, with three types of hair cell receptors, can be recognized in the statocysts of *Octopus vulgaris* (Fig. XV–2*d*). One of these areas, the macula, is oval in shape and contains ciliated cells; it is placed vertically on the wall of the

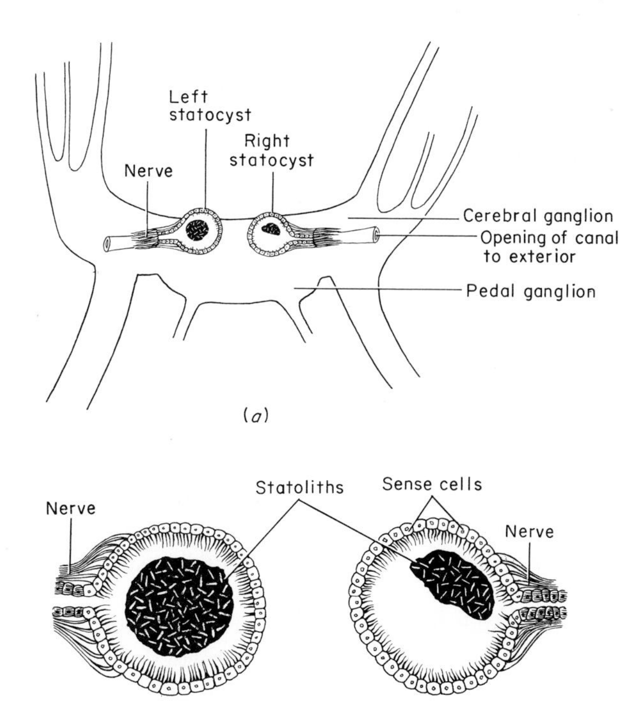

FIG. XV–3 *Statocysts of* Pecten inflexus. *(After von Buddenbrock.)* ***(a)*** *Diagrammatic representation from the dorsal aspect of the cephalic ganglia, showing the location of the two statocysts.* ***(b)*** *Diagrammatic representation of a median section through the statocysts.*

statocyst and the cilia support a statolith. Another, the crista, is a long ridge of ciliated cells, lying along the wall in three planes—longitudinal, transverse, and vertical. Each of these sections is subdivided into three subsections with cilia bearing statoliths. The rest of the wall of the sac, not occupied by macula and crista, contains scattered ciliated cells. Removal of one member of a pair of statocysts does not affect the behavior of the animal to any marked degree, but removal of both has drastic effects upon its swimming and its crawling. A statocystless animal seems unable to compensate for the jet that is its ordinary method of propulsion and somersaults rather than darts through the water. When attempting to crawl, it seems unable to control the posture of its head and stumbles and tumbles along.

There is experimental evidence to indicate that the statocysts of *Octopus vulgaris* contain rotation receptors similar to those of the labyrinth of the vertebrate ear and that centrifugal forces set up by rotation affect also the statocysts of molluscs that crawl or burrow. It is probable that the crista in the *Octopus* statocyst detects angular acceleration, as do the less highly organized statocysts of actively swimming gastropods and bivalves.

Information is scanty about statocysts in most of the annelids, but they have been found in sedentary and tubiculous polychaetes. In some families there may be five or six pairs, in others as many as twenty, located dorsally in the anterior segments usually near the gills or the notopodia. They may be superficial in position or deeply sunk beneath the epidermis, and they may be open or closed (Fig. XV–2*e*). In *Arenicola marina*, they are a pair of invaginations that penetrate deeply into the longitudinal muscles of the head and are connected to the exterior by narrow canals. Each contains a mixed assortment of foreign particles—sand grains, diatom shells, and spicules of various kinds—which enter the canal as the worm burrows through the sand. *Arenicola* starts to burrow head downwards, and worms with intact statocysts make compensating movements if the aquarium in which they are kept is tilted. If, for example, the aquarium is tilted through an angle of 90°, the worm turns through 90°, preserving its positive geotaxis, and continues its downward course. Removal of one statocyst or transection of the nerve from it does not affect this adjustment, but removal of both, or transection of both nerves, prevents it.

Almost all decapod crustaceans have statocysts, one in the basal segment of each antennule. They are found also in a few other crustacean groups, but are not universal among them. In myaids they lie in the bases of the uropods and are either closed or open sacs. Each contains one large statolith secreted by the cells of the sac. Statocysts have been found on the telson of burrowing isopods. In decapods, the opening of the sac is a narrow slit, fringed and guarded by hairs (Fig. XV–2*f*). In crabs, it is functionally closed; their statoliths are composed of fine sand grains held together by a sticky secretion and are either free in the cavity or attached to the cilia of the sensory cells. The statoliths are replaced after each molt as the lining of the cavity is shed at ecdysis along with the rest of the cuticle. In crayfish and lobsters, there are rows of sensory cells along the floor of the sac with which the statoliths are in contact either permanently or when moved by gravity. The distribution of these rows is characteristic of each species. In crabs there is some differentiation of the sensory cells, for some have free sensory cilia, not associated with the statolith, that are stimulated by movement of the fluid in the vesicle and so act as rotation receptors like those in the statocyst of *Octopus*.

Because of their position in the basal segment of the antennules, the floor of the decapod statocyst is inclined at an angle of about 40° when the animal is in its normal horizontal position, and there is an equal and constant gravitational pull upon the statoliths in each. If the angle is changed by tilting the body, the animal makes compensatory righting movements. Direct recordings from the branch of the antennular nerve that originates in the statocyst have shown that continuous impulses pass along it. Those from one side are balanced by those from the other, so that the animal remains on an even keel, but if one statocyst is injured or removed, it lists toward the uninjured side.

(2) Sensilla trichodea in insects Insects have no specialized organs such as these but derive their information about gravitational force through a variety of receptors. Photoreceptors have a large share but mechanoreceptors are most certainly also involved. All insect sense organs have fundamentally the same components and consist of a modified area of the cuticle, deposited by special epidermal cells, and of one to many neuroepithelial cells. The cuticle may be modified as a relatively long and slender hair, a short conical peg, a small plate, a thin oval or circular membrane of relatively large size, or, in the case of photoreceptors, a transparent cornea. The simplest insect sense organs are known as sensilla. Tactoreceptors, or sensilla trichodea, are found on all surfaces of the body. They are found in the greatest numbers on the legs, antennae, and mouthparts, and between the body segments and joints of the appendages. The simplest insect sensillum trichodeum consists of four cells: a trichogen or hair-forming cell, a tormogen or socket-forming cell, a bipolar neuron, and a neurilemma or sheath cell (Fig. XV–1*b*). The development of such a sensillum in the hemipteran *Rhodnius* has been described as follows. Two of the four cells derived from two divisions of a single epidermal cell enlarge markedly. One of these large cells is the trichogen which sends out a long process that projects above the surface of the epidermis and becomes the hair. The other is the tormogen which forms an articulating socket around the base of the hair. Two processes arise from the third cell, which becomes the sensory neuron; one of these processes extends distally to the base of the hair as the dendrite, while the other extends proximally as the axon and makes connections with a nerve from the central system. The remaining cell becomes the neurilemma cell sheathing the axon. All four cells lie on the basement membrane that is continuous beneath all the epidermal cells and is interrupted only where the axon passes through it to make internal connections. The basement membrane in some cases is depressed to form a capsule around the base of a sensillum. The dendrite of the neuron usually enters a cuticle-like tube of uncertain origin, called the scolopoid sheath (Greek *skolops*, "pointed instrument"), which may enclose it for varying distances before its termination. All the various kinds of sensilla found in insects are variations of this basic plan.

In a sensillum trichodeum the process of the trichogen cell becomes a long slender hair covered by cuticle. The hair is itself rigid but can be moved in its socket so that it can be bent at an angle with the body. The simplest sensilla trichodea, the mechanoreceptors, are innervated by a single neuron, but others may be innervated by several neurons and have sensory functions other than mechanoreception. Those on the labellum of the blowfly *Phormia*, for example, are known to have a single neuron responding to touch, while the others are primarily chemosensitive. In the mechanoreceptors, bending of the hair in its socket deforms the surface of the neuron resulting in its activation. Collections of individual sensilla trichodea, containing as many as 100 hairs with their associated neurons and known as hair plates, or sensilla placoidea, are common features among insects (Fig. XV–1*c*). In some insects, those on the antennae and other parts of the body have been shown to signal information about the gravitational force, while those between the body segments and between the joints of the appendages give information on the position of the body and so contribute to regulation of its posture.

There are in honeybees, for example, hair plates on the prothoracic segment, at the junction of the thorax and head, and also at the junction of the metathorax and abdomen. When the bee is in a horizontal position on a horizontal surface, the hairs of the prothoracic hair plate are in even contact with the head. When it climbs up a vertical surface, its head bends downward toward the sternum and the more ventral hairs of the hair plate are deflected; similarly, when it climbs downward, the head is elevated and the more dorsal hairs are deflected. Likewise, when the insect turns, the hairs on the right or the left side of the hair plate will be deflected, according to the direction of turning. Thus the bee has a delicate and sensitive means of appreciating its position in respect to gravity and its angle on a vertical plane. This is of tremendous importance to bees, for workers impart information about food sources by means of precise movements, or dances, oriented with respect to gravity. If the nerves to these hair plates are cut, or the head is prevented from moving when the body does, the responses to gravity and the communication dances are impaired. The posterior hair plates have been shown to have the same function, but this is apparently subordinate to that of the anterior set, because they are not able to substitute for them after they have been put out of action experimentally.

(3) Chordotonal sensilla A chordotonal sensillum, so called because it was first thought to be a resonant element like a stretched string and, therefore, comparable to the vocal chords of mammals, consists of a bipolar neuron and two or more companion cells stretched in linear order between two points in the body wall (Fig. XV–4*a*). The proximal portion of the dendrite of the neuron is enclosed by one of its companion cells, the accessory cell, and the rest of it by the other, the cap cell, that attaches it to the body wall. The tip of the dendrite is enclosed in a scolopoid sheath, at whose apex is a dark, peg-shaped body. The axonal end of the neuron is attached to the wall by a ligament-like strand of tissue, and the axon passes inward to terminate in connection with another neuron. Positioned as they are, chordotonal sensilla can respond to changes in extension or compression of the flexible cuticle and so are aroused when this is bent, stretched, or caused to vibrate. They may be exteroceptive or proprioceptive.

The simplest chordotonal organs contain a single sensillum, but there are those in which the numbers may run from ten to thousands (Fig. XV–4*b* and *c*). Johnston's organ, which was first described over a century ago in the antennae of mosquitoes, may contain from 10 to 100 sensilla, each of which shows some modification of the basic pattern, particularly in the absence of an apical cap and in the fraying out of the scolopoid sheath into numerous fine fibrils that are attached to the membrane of the joint. An insect's antenna is constructed of a basal segment, the scapus, articulated with the head; a more distal segment, the pediculus, articulated with the scapus; and a shaft, or flagellum, with varying length and number of immovable segments. In some insects the distal segment of the flagellum bears dorsally a bristle, or spine, called the arista (Latin *arista*, "beard of grain"). The flagellum can be moved as a whole by muscles that originate in the scapus and are inserted on the pediculus. The sensilla of Johnston's organ are stretched between the base of the pediculus and the membrane between this segment and the base of the flagellum (Fig. XV–5*b*) and are thus well placed to register movements of the flagellum in relation to the pediculus.

Johnston's organ is implicated in the control of flight in aerial insects and of swimming in aquatic ones. In both, initiation of movement depends upon the tarsal reflex, first described over 30 years ago when it was observed that flight in Diptera began as soon as the insects were caused to lose contact with their substrates. In most insects, too, contact with a substrate stops these movements. Some insects such as *Phormia*, *Calliphora*, and *Vespa* fly continuously after their legs have been amputated. Air currents striking the antennae cause *Calliphora* to bring its legs up into the characteristic flight position and also to maintain its flight and orient itself to the direction of the wind. Experiments have shown that the joint between the scapus and the pediculus is the most important one

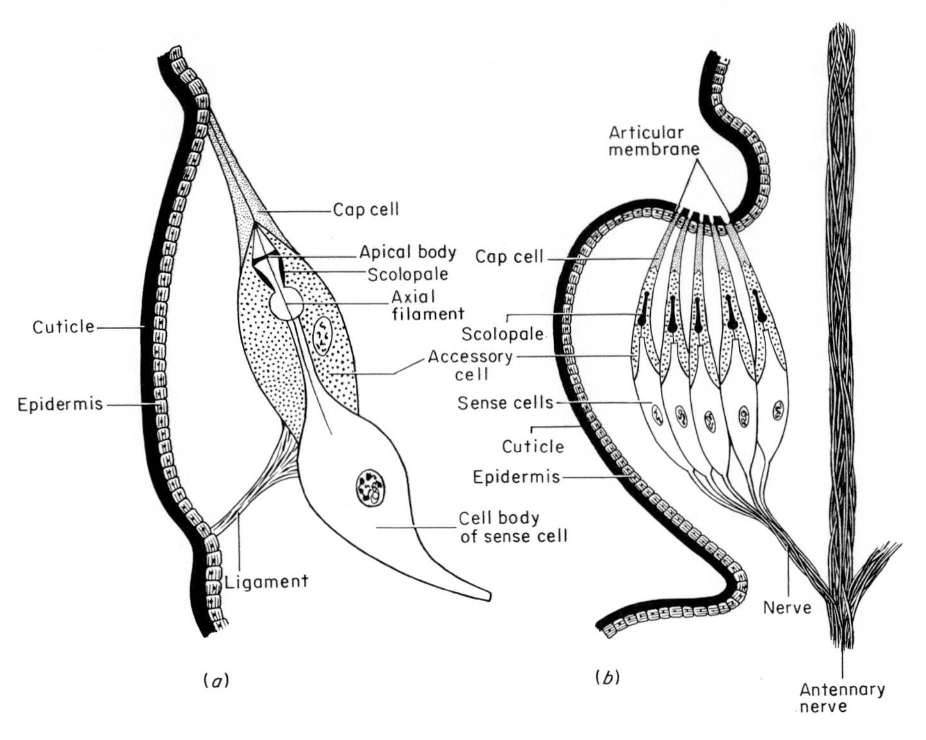

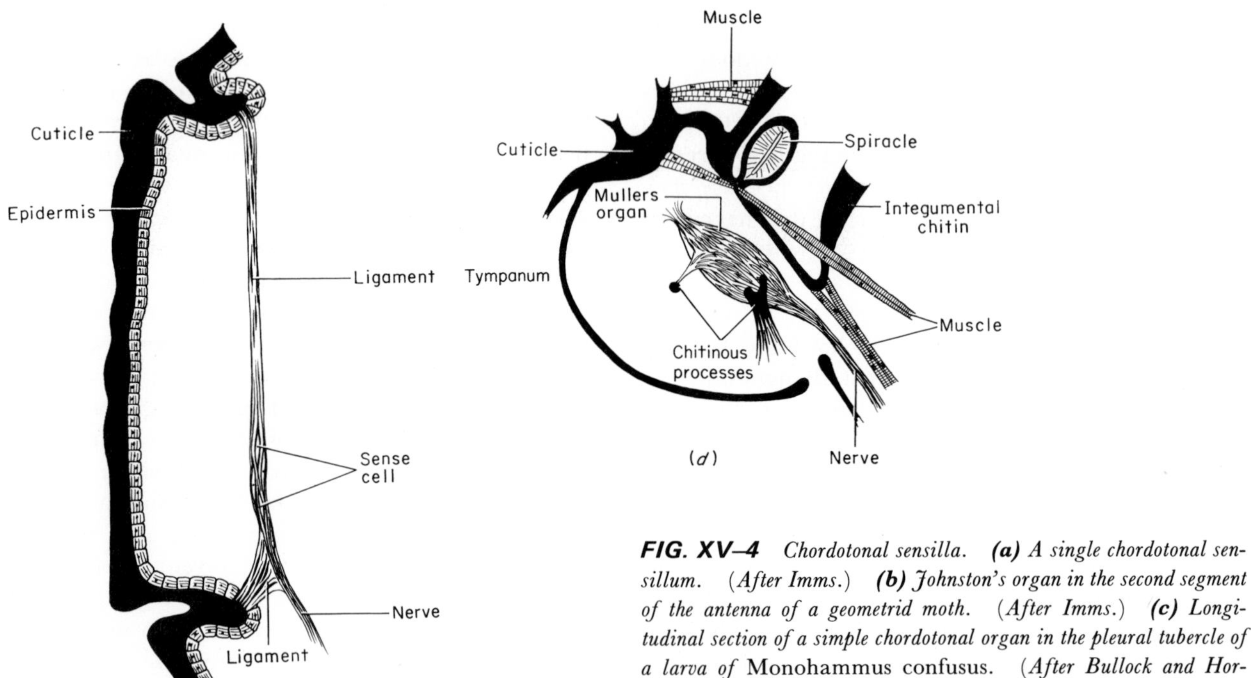

FIG. XV–4 *Chordotonal sensilla.* ***(a)*** *A single chordotonal sensillum.* (*After Imms.*) ***(b)*** *Johnston's organ in the second segment of the antenna of a geometrid moth.* (*After Imms.*) ***(c)*** *Longitudinal section of a simple chordotonal organ in the pleural tubercle of a larva of* Monohammus confusus. (*After Bullock and Horridge.*) ***(d)*** *The tympanic organ of* Schistocerca americana. (*After Imms.*)

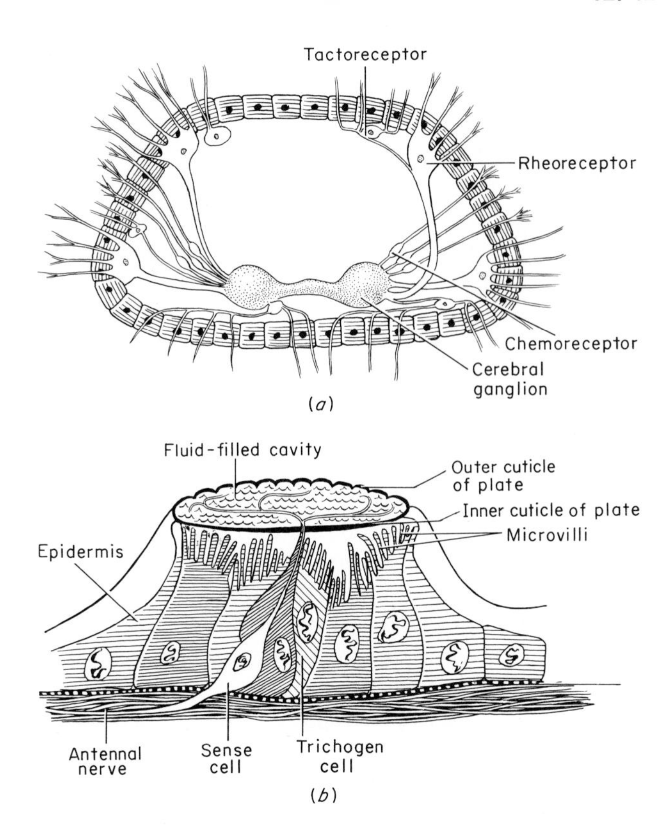

FIG. XV–5 ***(a)*** *Cross section through the head of the flatworm* Mesostoma, *showing four of the eight rheoreceptors, two groups of chemoreceptors, and four tactoreceptors.* (*After Bullock and Horridge.*) ***(b)*** *Section through the plate organ in an antennule flagellum of an aphid. Only one sensory neuron is shown.* (*After Slifer, Sekhon, and Lees.*)

for initiation and sustenance of the fly's active movements, which are registered by Johnston's organ there. The funiculus of the short antenna, with its attached arista, provides means for measuring air currents; mechanoreceptors at the joint between funiculus and pediculus signal the extent of rotation of the former. The effects of such movements are relayed to the appropriate muscles and result in changes in the path of the wing tip according to the air speed. Similarly, the locust *Schistocerca* can be made to fly by blowing a jet of air on its head. There are 10 groups of sensilla there which are receptive to the movement of the air. If the jet is directed at the front of the head, the locust will begin to fly; if it is directed at an angle to the head, the insect turns to one side or the other and corrects for yaw. Johnston's organ seems to supply information about speed and also to cooperate in the maintenance of flight, for movement of air against the antennae results in excitation of its sensory cells.

Tarsal reflexes also initiate swimming movements in aquatic insects. This has been definitely shown in the giant water bugs *Lethocerus americanua* and *Benacus griseus*, which begin to swim as soon as they are lifted off the substrate. Once in motion, and probably also at rest, they receive information about water movements through their antennae, and Johnston's organ is implicated as the rheoreceptor (Greek *rheos*, "current"). The diving beetle *Dytiscus* holds its antennae at a slight angle to its midline when swimming, and these are bent when it turns in its course or encounters a flow of water of different direction or different velocity. The whirligig beetle *Gyrinus* also receives information about water movements through this organ. Swarms of these beetles can be seen skimming rapidly over the surface of the water, each individual darting rapidly from place to place in apparently undirected movements. Yet they never collide with each other and seem to locate their fellows in the swarm, as well as their prey, through disturbances set up in the water by their movements. As each beetle skates along, it keeps its antennae in contact with the surface film through which these disturbances spread as waves.

(4) Vibration receptors Organisms of all kinds show that they are sensitive to periodic movements in their media. Protozoans exhibit taxes or tropisms when vibrations are set up in the water around them by shaking or jarring the container that holds them. The response of *Gonyaulax*—increase in intensity of luminescence—to such vibrations is now well known. Mechanoreceptors in eumetazoans at all levels of organization seem to be stimulated by vibrations of low or high frequencies. Cnidarians have been shown to be sensitive to small disturbances in the water around them, the disk of the sea anemone *Calliactis* and the hypostomal region of polyps of the colonial hydroid *Syncoryne* being the areas of their bodies most responsive to such vibrations. Detailed study, including electron micrography, of *Leucothea* (*Eucharis*) *multicornis*, a common ctenophore of Mediterranean waters, has shown that the surface of its body is covered with finger-like processes, about 1 cm long, at whose tips are sensory cells terminating in nonmotile cilia with a 9 + 2 pattern of filaments and an onion-like root structure. These act as very sensitive receptors of water displacement, in response to which a single finger can shoot out as an independent effector. Presumably vibration detection is a means of food location for the ctenophore, for a small crustacean hit by a projected finger is immobilized, possibly by secretion of some toxic substance.

In arthropods, whose responses to vibration have been most intensively studied, receptors may be located in the joints, as are their proprioceptors, and are excited by movement of the joint through vibration of the substratum, the water, or the air.

Arachnids and insects have localized vibration receptors; in both groups this particular kind of sensitivity seems to be of especial importance to their ways of life. In arachnids these receptors are known as lyriform or "slit" organs, so called because they seem to be made up of a number of slits of varying length arranged parallel, or nearly parallel, to each other, so that they resemble the strings of a lyre. Actually they are not open, for the "slit" is covered by a delicate cuticle. The sensory cells lie beneath this cuticle, to which their distal processes are attached, and are stimulated by alterations in its shape (Fig. XV–6*a*). Simple or compound organs of this type are found in nearly all arachnids, located in the joints of the legs and other appendages, on the sterna of the cephalothorax, and on the sting of scorpions. Recordings from electrodes placed on the sensory nerves of legs amputated from two Ceylonese arachnids—the scorpion *Heterometrus swammerdami* and the amblypygid *Phrynichus lunatus*—show responses to forced movement at the coxa-trochanteral, trochantero-femoral, femoro-tibial, and tibio-tarsal joints, indicating the function of these organs as mechanoreceptors. Detection of prey by web-building spiders, in which lyriform organs are situated at the joints of the legs but not in the tarsus, depends upon mechanoreceptors responsive to vibrations in the web caused by the struggles of insects caught in its viscid threads, which put the spider's legs under stress and cause them to bend. The victims are most frequently dipterans, coleopterans, and neuropterans that blunder into the web in the course of their flights. Lepidopterans are usually able to escape when they encounter a web because of their scale-covered wings. Small moths and butterflies can fly through the meshes of most orb webs, but if they do come in contact with the viscid threads their scales stick to them and are pulled off and left behind as the insect flies on unharmed. Large ones have greater difficulty because their whole bodies may come in contact with the viscid threads, but after a brief period of fluttering they, too, can usually escape, leaving their scales stuck on the threads. Insects not protected in this way, or by equally readily detachable hairs, or by waxy surfaces to which the adhesive secretion does not stick are less fortunate and are held in the web until the spider comes to eat them. Some spiders construct a web with a special signal thread of extra thickness and tensile strength on, or close to, which they rest while waiting for a signal from the captured prey. When a spider gets this, its behavioral response is to move rapidly toward it, protected from capture itself by the oily film on its legs, and either to eat the prey at once or to preserve it for later consumption by wrapping it in a case of silk.

The walking legs and pedipalps of spiders have movable hairs associated with two or more sensory cells. These organs have been called trichobothria because of the flask-shaped structure below the cuticle through which each hair passes (Fig. XV–6*b*). In the house spider *Tegenaria* there is a row of trichobothria on the tarsus and metatarsus of each appendage, and four rows on the dorsal and lateral aspects of each tibia. Measurements of the action potential in the sensory cells have shown that there is no activity when the hair is in its resting position and that only movements out of this position are registered. Electron microscopy has shown that each hair on the lateral rows is innervated by at least two receptor cells, while those of the dorsal row are innervated by three. In the lateral rows, bending of a hair outward activates one cell and bending of it inward activates the other, while there is no response to proximal or distal bending. In the dorsal rows, the nerve cells show the greatest activation with proximal, distal, and lateral displace-

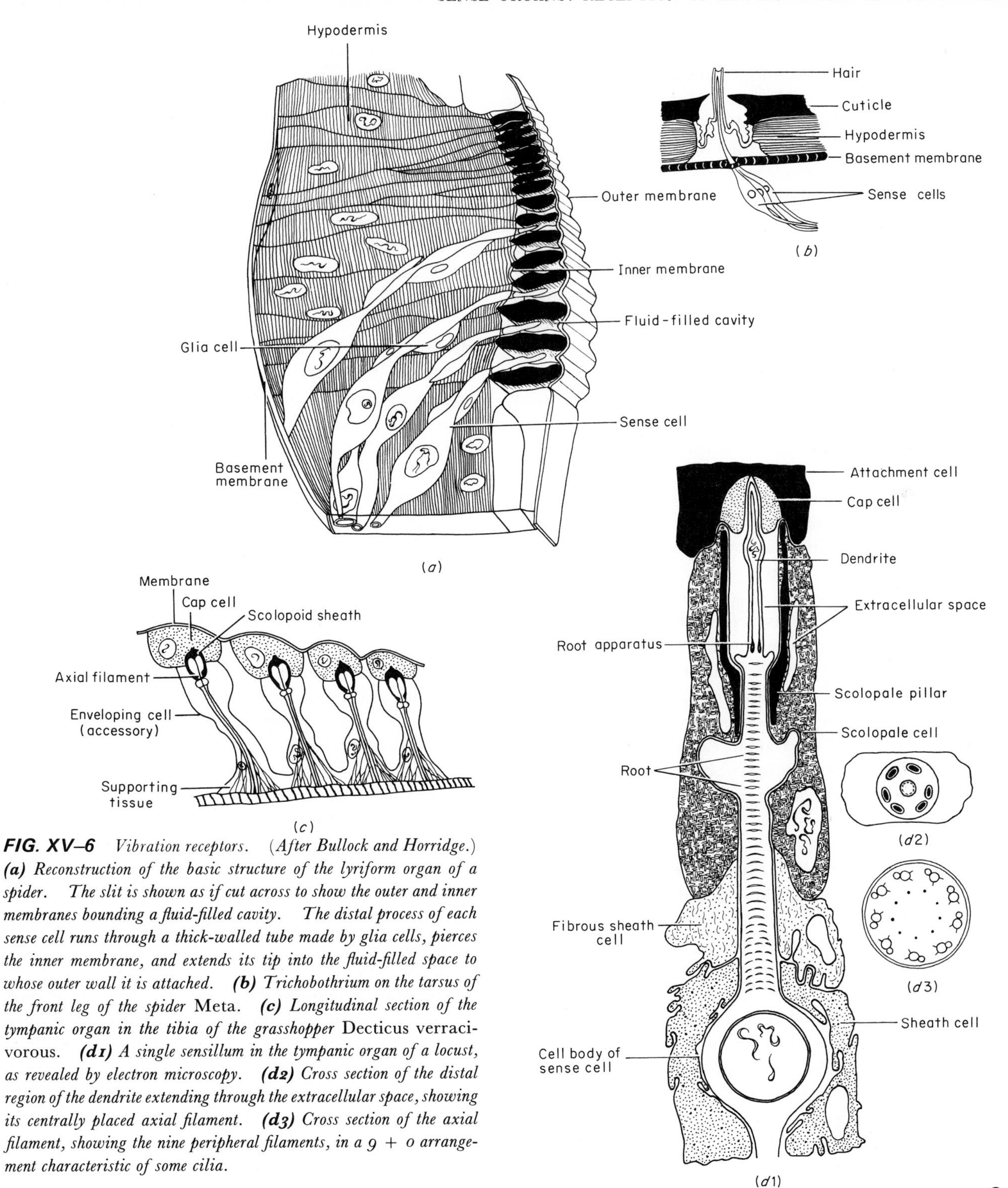

FIG. XV–6 *Vibration receptors. (After Bullock and Horridge.)* ***(a)*** *Reconstruction of the basic structure of the lyriform organ of a spider. The slit is shown as if cut across to show the outer and inner membranes bounding a fluid-filled cavity. The distal process of each sense cell runs through a thick-walled tube made by glia cells, pierces the inner membrane, and extends its tip into the fluid-filled space to whose outer wall it is attached.* ***(b)*** *Trichobothrium on the tarsus of the front leg of the spider* Meta. ***(c)*** *Longitudinal section of the tympanic organ in the tibia of the grasshopper* Decticus verracivorous. ***(d1)*** *A single sensillum in the tympanic organ of a locust, as revealed by electron microscopy.* ***(d2)*** *Cross section of the distal region of the dendrite extending through the extracellular space, showing its centrally placed axial filament.* ***(d3)*** *Cross section of the axial filament, showing the nine peripheral filaments, in a 9 + 0 arrangement characteristic of some cilia.*

ment of the hair. There is no critical evidence that these organs register sound, even though they oscillate when near a source of sound. Moreover, spiders can be attracted to the sound of a tuning fork, indicating that they are receptive to air-borne vibrations as well as to those of the threads in their webs.

(5) Sound production and sound reception

Only in insects is there positive evidence of the detection of sound and of receptors that are analogous to the auditory organs of vertebrates. The statocysts of some land crabs are slightly different structurally from those of closely related aquatic species, and it is possible, although there is no decisive evidence, that they may be organs of phonoreception as well as of equilibration. The fact that these crabs can make shrill, creaking noises by rubbing ridges on their large left claws against a ridge on the last joint of the left second pereiopod suggests that there may be communication between individuals by means of sound production and sound reception.

Indeed, production of sound is so general among crustaceans that it seems as if it must be a medium of communication between them. The mantis shrimp *Squilla empus*, for example, emits a loud rasping noise when it rubs its uropods against the undersurface of its telson. The ghost crab *Ocypode macrocera* can produce sounds audible to the human ear and distinguishable as three different tones by rubbing the surface of the larger jaw of its chela against a boss on the basal segment. The inner surface of the jaw has a row of five-toothed striae set transversely across it. The teeth are of different lengths in three different regions, which probably accounts for the three different tones, but there may be even more tones than these that are not appreciable to the human ear. Tape recordings of the sounds emitted by *O. ceratophthalamus* on the coast of Mozambique give three short rasping sounds followed by a rapid rattle. These sounds seem effective in establishing territorial rights for a crab, which makes and inhabits its personal burrow from which it will repel others by making noises that range from soft intermittent peeps to a long, high-pitched growl if an intruder is not put off by the gentler sounds. This behavior is probably important to survival of the crabs for they tend to congregate in great numbers on the same beaches where they are in constant danger from predatory birds and mammals. When danger threatens, each crab scuttles to a burrow, which is not large enough to hold more than one, and should others try to crowd into it, all might be in danger of asphyxiation. If the sounds that an occupant makes can be apprehended, they would give very clear warning that the burrow was full and that there was no room for another. These crabs only stridulate when they are in their burrows.

Scorpions, too, can produce sounds, although there is no critical evidence that they can hear them. In some species the sound is made by rubbing the chelicerae against the ventral side of the carapace and in some others, by scraping the tip of the tail across the first caudal and the last mesosomal tergites. One species, *Scorpio maurus*, and two of its subspecies, *S. maurus fuscus* and *S. maurus palmatus*, which are native to Israel, make thudding noises by striking the posterior half of the mesosoma against the ground. All of these sounds seem to be protective devices and are made in preparation for a fight between two individuals or when the presence of a predator is somehow detected.

Phonoproduction and phonoreception have been most intensively investigated in insects. From these studies it has been clearly established that in many species these mechanisms provide means of communication that are as meaningful to them as any limited form of vocal interchange. Some adult orthopterans, lepidopterans, hemipterans, and neuropterans have specialized paired tympanic organs sensitive to sonic or ultrasonic vibrations. Basically all such organs consist of a thin cuticular membrane, or tympanum, in close proximity to a tracheal sac and overlying a group of chordotonal sensilla called Muller's organ (Fig. XV–4*d*). They vary, in different species, in the number of sensilla, the position of these in relation to the membrane, and the position of the organs in the body of the insect. There are as few as two sensilla in each tympanic organ of certain moths and as many as 1,500 in cicadas. In cicadas, they are attached to the edge of the tympanic membrane, in short-horned grasshoppers and locusts to its center, and in long-horned grasshoppers and crickets to the tracheal sac rather than to the membrane. The structurally simplest tympanic organs are those of the Lepidoptera, where each lies in a cavity on the posterior part of the metathoracic segment, with two chordotonal sensilla attached to the inner surface of each tympanic membrane, their axons forming part of the tympanic nerve. In Orthoptera, the organs are situated in the tibia of the first pair of legs, either flush with the surface as in crickets or sunk below it as in long-horned grasshoppers. There are about 70 sensilla in each leg, arranged in a row on the dorsal side of the anterior trachea. Electron microscopy of these sensilla in *Locusta migratoria*, in which there are about 80 in each organ, has shown that the axial fiber of each sensory cell, which extends freely in the extracellular space between scolopoid sheath and cap cell, has a 9 + 0 or 9 + 1 arrangement, or the 9 + 2 fibrillar arrangement typical of cilia and flagella [Fig. XV–6*d* (1) to *d* (3)]. In these insects transformation of effector organelle into a receptor must have taken place early in their evolution, for it does not occur during their ontogeny and there is no stage in their embryological development where motile cilia are evident, yet the fine structure of the sense cell indicates that at some period insects or their ancestors did have them. In cicadas the tympanic organs lie on the ventral surface of the first and second abdominal segments. The proximal ends of the numerous sense cells are attached to a point on the periphery of the tympanic membrane and their distal ends to the cuticle of the body wall.

There is little definitive information about the way tympanic organs work. Their structure implies that they are displacement rather than pressure receptors, in the sense that they are activated by the periodic compressional waves resulting from the displacement of particles from positions of equilibrium by some local disturbance, rather than by the accompanying pressure changes. Such receptors register the direction of sound, which pressure receptors cannot, since there is no directional component in pressure changes. These waves cause similar periodic movements of the tympanic membrane which acts like a diaphragm or drumhead. Possibly the displacements are transmitted to the cap cells of the sensilla and through them to the sense cells, but it is by no means certain that this is so. It is clear, however, from behavioral responses as well as from direct recordings from the tympanic nerves that the disturbance results in nerve impulses that are transmitted to the central nervous system of the insect under investigation.

A number of ingenious experiments have been devised and conducted to test the function of tympanic organs. It is well known and indeed a matter of ordinary observation that insects can produce a variety of sounds in a variety of ways. Some do it by tapping their heads or other parts of their bodies against a solid substrate; some do it by vibrations of their bodies; and some do it by snapping their wings when in flight or by snapping particular parts or organs. Some do it by the forcible expulsion of air or other gases from apertures of various kinds or by stridulation (Latin *stridere*, "to rasp or whistle"), a noise made by rubbing one part of the body against another.

The value to an individual in making such sounds depends upon their apprehension by another; that is, the nature of the waves must be such that they are recorded by the phonoreceptor of another individual of the same, or different, species. As a means of communication, insect sounds are used mainly in mating, in defining territorial limits, and in defense. Similarly, reception of sounds made by predatory species may be of tremendous value in helping an insect to locate the predator and to escape from it. This is perhaps particularly important to night-flying insects, especially those that are preyed upon by bats. Insectivores in this family of flying mammals give short, sharp cries or chirps, with sonic and ultrasonic components when they are hunting and find their prey by echolocation. The predominant frequencies in these sounds are about 15,000 to 60,000 cycles per second (Hz), which are also in the range of the maximum sensitivity of the tympanic organs of noctuid moths, a common prey of bats. The behavior of a moth pursued by a bat shows that the bat cries are heard, for the insect's course of flight is changed when it is exposed to them. Moreover, direct recordings from the tympanic organs of captive moths show neural impulses arising after the chirp of a real bat or a sound that simulates it. Adults of *Prodenia eridania* respond to stimulation from a wholly ultrasonic source. Even moths without tympanic organs seem sensible to sound, for several species of Sphinx moths of the genus

Celerio were observed, when feeding from flowers, to fly away rapidly when a high-pitched tone was made near them. Electrical recordings from the connectives between the supra- and suboesophageal ganglia of captive specimens of *C. euphorbiae* showed the passage of impulses when they were exposed to ultrasonic vibrations. Operative procedures have led to location of the sound receptors at the articulation of the labial palps with the prementum. They are probably mechanoreceptors, stimulated by displacement of the palp. However, perception of sound is not universal among the sphingidae, for responses were not obtained from a number of other species put to similar experimental tests.

The sounds made and demonstrably heard by orthopterans seem mainly to be concerned with courtship and mating. Most sounds are made by stridulation, either by drawing a scraper on one wing across a file on another or by rubbing a toothed ridge on a leg against a wing. The first proof of hearing in insects was obtained from experiments on the cricket *Lyogryllus campestris* and the long-horned grasshopper *Thamnotrizon* (*Pholidoptera*) *apterus*. Females of these species were confined in a room out of sight and smell of the males, whose chirps were transmitted to them through a telephone. In spite of the distortion of sound inherent in this means of transmission, the females responded to it by moving toward the instrument, as in nature they are attracted to chirping males. Moreover, males of *Thamnotrizon*, whose song consists of an individually performed prelude and coda between which there is an alternating duet, sang their parts in concert when similarly separated and the chirps telephoned from one room to another. Young males sang their duets with artificially produced sounds having frequencies ranging from 400 to 28,000 Hz, but older ones would not. Nor did males sing in concert after removal of their tympanic organs, although they still made some responses to sound. These experiments showed conclusively that intact insects heard the sounds made by others and that, while the tympanic organs were the main centers of audition, they were not the only ones. Chordotonal organs not associated with tympanic membranes and sensilla trichodea are also sensitive to sound waves, giving insects some residual auditory powers even after loss of their tympanic organs.

Other experiments have shown that sounds are not only heard, but very accurately located. Unmated female crickets were observed to move in an almost straight line over a distance of about 10 m to singing males they could not see. This ability may be related to the position of their tympanic organs and to the fact that in walking the legs are moved forward through an angle of about 50°, through which the tympanic organ likewise swings. Its sensitivity varies with the angle at which the sound strikes it. Impulses from the organ on the side of the sound source diminish in frequency as the source passes through and out of the zone of greatest sensitivity and could lead to movements that would steer the insect toward the sound. The abdominal tympanic organs of locusts have been found to be similarly differentially sensitive and probably operate in ways similar to those of crickets in the location of the source of a sound audible to them.

Recording of impulses in the tympanic organs, or in the nerves leading from them, show that they have not the frequencies of the sound waves used to stimulate them. Although there is definitive evidence that insects can distinguish between different sounds, it must be concluded that this is not done according to their frequencies, as is the case in the human ear. It is possible that pulse frequency is the main factor in insect discrimination, for in the telephone experiments this would remain constant however other features of the sound waves might be altered and distorted.

The only insect known to hear in water is the water boatman, the hemipteran *Corixa*. Both males and females have tympanic organs and the males stridulate. The females respond to their chirps by swimming irregularly in circles, presumably in this way attracting the attention of the males, for copulation follows. As with crickets and grasshoppers, the chirping of one male incites others to follow suit, and they will sing in chorus, not only with each other but also with artificial sounds having frequencies of 2,000 to 4,000 Hz. The responses of females to males, and of males to each other, are eliminated when the tympanic organs are destroyed. The drumming of cicadas seems also to bring the two sexes together, but very little is known about their tympanic organs or, indeed, about those of any hemipteran.

In all these species the calling sound of the males serves to bring individuals of both sexes together.

When a lone male begins his stridulation, other males within hearing distance answer him and then each other and move closer together. The males, therefore, tend to assemble in groups to which females are attracted by the mating songs and so their fertilization is made more certain.

Experiments with orthopterans from which both tympanic organs had been removed showed that they were still somewhat responsive to sounds, indicating that they had sound receptors other than these specialized ones. There is evidence, too, of audition in species in which such organs have not evolved. Delicately balanced hairs could conceivably be deflected by sound waves, as well as by other forms of mechanical stimulation, and chordotonal sensilla could also respond to them even when not associated with a tympanic membrane. Experiments with a number of insects suggest that both these devices may operate in the detection of sound. Some of the early tests were made upon caterpillars, long known to respond to sounds with frequency ranges of 32 to 1,000 Hz by becoming motionless, by freezing, or by violent movements of their anterior segments. These responses were abolished when the hairs covering their bodies were immobilized by dusting them with flour or weighting them with drops of water. The fact that smooth-surfaced caterpillars give similar responses to similar sounds leaves open the question of whether the hairs of the fuzzy ones are really the actual detectors. Yet there is good evidence that the sensilla trichodea on the cerci of orthopterans have an auditory function. Those on the cerci of *Periplaneta americana*, for example, can be seen, when watched under a microscope, to be moved by currents of air too slight to be felt by a human. They are also moved by sound waves with frequencies between 50 to 3,000 Hz. Electrical recordings of nerves from the sensilla show that the frequency of the impulses synchronizes with the frequency of the sound in the range of 50 to 400 Hz, so that discrimination between sounds in this range can be made in the same way as in the human ear. At higher frequencies the responses are asynchronous with the stimulus, as they are in the tympanic organs of grasshoppers. The responses are much reduced, or no longer given, when the hairs are weighted with Vaseline. Electrical recordings from nerves from sensilla trichodea in other orthopterans suggest that they, too, may be sound-sensitive, but the data are equivocal in that there is no certainty that the recorded impulses originated in the sensory hairs alone.

Johnston's organ is highly developed in mosquitoes and gnats and contains large numbers of chordotonal sensilla. The humming and buzzing made by these insects in flight is well known, and there are many observations that the sounds serve as means of communication between individuals. Males of *Aedes aegypti* respond to the sound made by the wings of a flying female by taking off, if at rest, and flying toward her, and when reaching her, starting mating operations. This behavior can be initiated by artificial sounds of similar frequencies; the effective range of frequency varies with the age of the male. That for young, unmated males about 48 to 50 hr after emergence is, on the average, 400 to 700 Hz, while that for males 200 to 300 hr old, which presumably have previously mated, is 275 to 700 Hz. This mating response is abolished if the flagellum of the antenna is prevented, by various means including its removal, from moving on the pediculus. These experiments seem clearly to establish the importance of Johnston's organ to these insects in the reception of air-borne sound waves. The frequency of matings in *Drosophila melanogaster* are reduced unless the females' antennae are intact. The results of experiments involving removal of the antennae or immobilization of their parts have led to the conclusion that the arista acts as a sail, moving in response to the wing vibrations of the male and thereby twisting the flagellum with consequent stimulation of Johnston's organ.

Chordotonal sensilla in other locations may also be sensitive to sounds. Those in the legs may pick up vibrations in the substrate on which an insect is standing. Queen bees, for example, have been shown to respond, by characteristic piping, to similar artificial sounds transmitted through the substrate as well as to the piping of other queens. It is possible that all bees use sound, as well as smell and precise patterns of movement, in communicating with each other. A female *Drosophila* may also detect the sounds of a courting male through the chordotonal sensilla in the femurs of her legs. These sounds are made by vibration of one or both wings, by flicking the wings, and in some other way that does not entail detectable movement of any part of the male's body. The sounds produced by *D. persimilis* have a frequency of oscillation and a pulse repetition rate that is quite different from that of *D. pseudoobscura*. This difference may well be detected by females, for though the two species occupy the same habitats and share the same food, only very rarely, in nature, do they mate with each other. Production and detection of different sounds may in this case be a significant mechanism of reproductive isolation.

b. PROPRIOCEPTORS (1) Myochordotonal organs of crustaceans Proprioceptors are sense organs that respond continuously to tensions in the body and signal changes in length and compression of the tissues with which they are associated. The messages they convey in invertebrates are analogous to the kinesthetic sensations arising in the muscles and joints of the human body and give information of the relation of one part of the body to another, of position, and possibly of conditions in the internal organs such as the digestive tract, so that the animal will be aware of hunger or repletion. Definitive evidence of the existence of specialized organs of this kind was derived first from electrophysiological and histological studies of the dorsal thoracic and abdominal musculature of crustaceans. They have been found in decapods, stomatopods, amphipods, isopods, and mysids. Each organ consists of a thin muscle, in one area of which myofibrils are replaced by connective tissue fibers, and of a nerve cell, whose dendrites terminate in this intercalated tendon (Fig. XV–7*a*). The axon of each of these sensory cells runs to the central nerve cord where it divides, with one branch passing anteriorly to the cerebral ganglion and the other posteriorly to the last abdominal ganglion. These receptors are found in pairs at the dorsal surfaces of the extensor muscles in the abdominal segments and the free thoracic segments. They are absent, or retrograde, in the thoracic segments that are fused and lack motility. In *Homarus vulgaris* and *Palinurus vulgaris*, a small number of a somewhat different and perhaps more primitive type of receptor has been found in some of the thoracic muscles. These have multipolar nerve cells, designated as N-cells, whose dendrites spread out all over muscles with a full complement of myofibrils. These are the attractor muscles (M. attractor epimeralis) of the epimeron and the lateral thoracico-abdominal muscles. Recordings from electrodes in the dorsal proprioceptor organs have shown a continuous discharge of impulses that increases in frequency when there is any alteration in the dimensions of the muscle with which they are associated. One member of the pair in each segment acts as a slow, tonic receptor and the other one as a fast, phasic receptor. These mechanoreceptors are known as myochordotonal ones, since each has a specific muscle component.

(2) Stretch receptors Proprioceptors are also located in the joints of the appendages. Although constructed on the same principles, they vary in detail and in the information they primarily convey. These differences are illustrated by the following examples. A series of proprioceptors have been found in the walking legs of representative decapods and their fine structure has been particularly investigated in the shore crab *Carcinus maenas* [Fig. XV–7*b* (1) to *c*]. Each consists of a strand of elastic connective tissue in contact with bipolar cells organized as chordotonal sensilla. The cell bodies of the sensory cells lie in, or near, the connective tissue strands with their distal processes extending into a scolopoid tube. The tube is extracellular, but apparently secreted by one of the accessory cells that also secretes the scolopale, an intracellular agglomeration of fibrous material. The distal processes of the sensory cells, at the point where they enter the scolopale, contain an axial filament with transverse striations. In the chordotonal organ of the coxopodite-basipodite joint there is no difference in the fine structure of the filaments of the two cells, but in those at other joints the filament of one cell, in the part that passes through the scolopale, has a segment with the nine peripheral filaments of a typical cilium, but lacking the central pair. The filament of the other neuron has no such ciliary segment, but both terminate in a distal segment with microtubules. These two neurons probably respond to different conditions in the joint, one to extension of the connective tissue strand and the other to its shortening. It has also been suggested that one may be a position receptor and the other a movement receptor. There may be, in all,

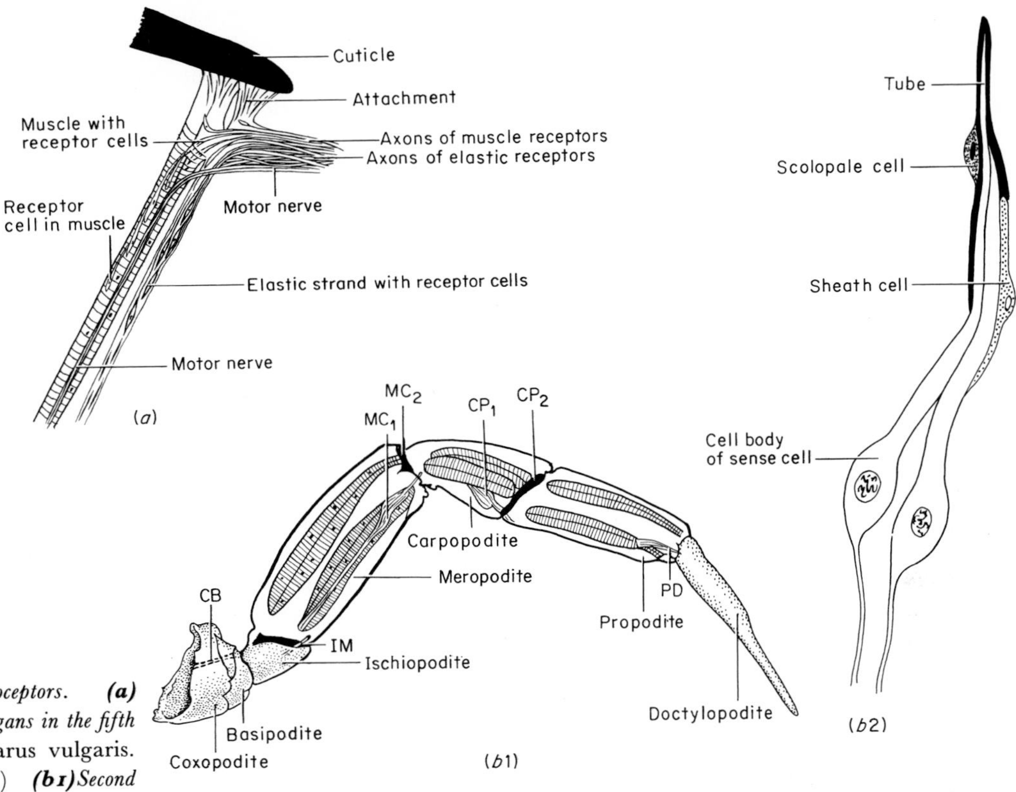

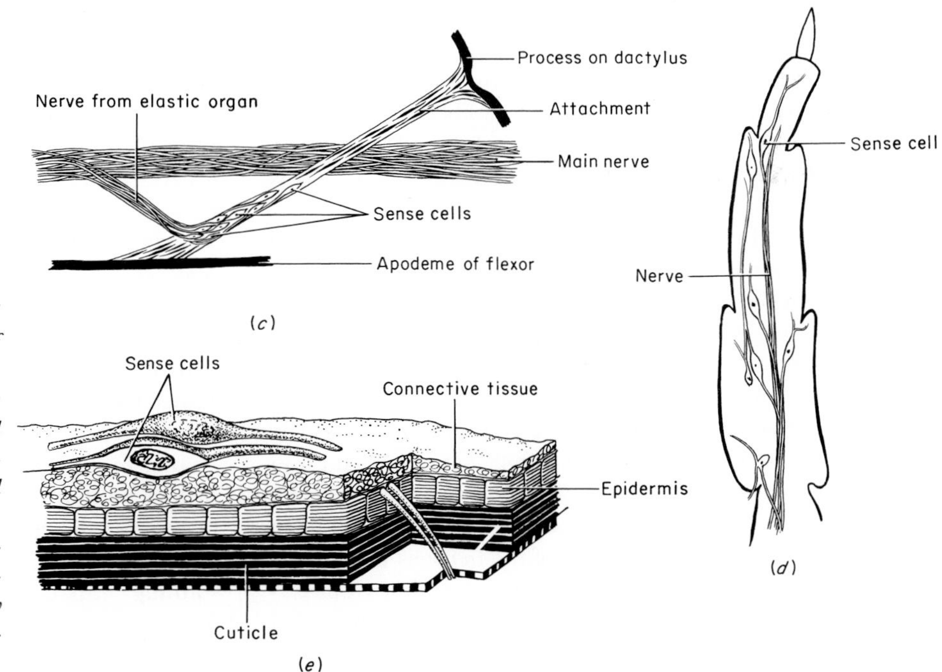

FIG. XV–7 *Proprioceptors.* ***(a)*** *Muscle and elastic sense organs in the fifth thoracic segment of* Homarus vulgaris. (*After Bullock and Horridge.*) ***(b1)*** *Second left pereiopod of* Carcinus maenas *as seen from the front and as if transparent, to indicate the positions of the larger tendons and chordotonal organs.* (*After Whitear.*) *CB, coxopodite-basipodite chordotonal organ; IM, ischiopodite-meropodite chordotonal organ; MC1 and MC2, meropodite-carpopodite chordotonal organs; CP1 and CP2, carpopodite-propodite chordotonal organs; PD, propodite-dactylopodite chordotonal organ.* ***(b2)*** *Detail of an isodynal scolopidium in the CB chordotonal organ of* Carcinus. (*After Whitear.*) ***(c)*** *Proprioceptor* (*elastic sense organ*) *in the PD joint of* Carcinus. (*After Bullock and Horridge.*) ***(d)*** *Positions of the chordotonal cells in the antennules of the amphipod* Caprella. (*After Bullock and Horridge.*) ***(e)*** *A pair of sensory cells in the IM chordotonal organ of* Petrochirus californiensis. (*After R. C. Taylor.*) *The sensory cells extend across the junction of two segments.*

some 30 or more neurons in one of these chordotonal organs, which, on evidence from electrical recordings of impulses in them, are under tension at all positions of the joint and react to unidirectional displacement of the distal segment with equal sensitivity over the total arc of possible movement. Chordotonal organs isolated from the joints of *Carcinus* respond to vibrations and to sudden changes in length by a burst of impulses.

In *Cancer magister* there is one myochordotonal organ in each walking leg, in addition to the chordotonal organs at the joints. The action of the single myochordotonal organ and the two chordotonal organs at the joint between meropodite and carpopodite has been particularly investigated. Ablation of the chordotonal organs has shown their necessity for coordinated movement in the crab's locomotion and has indicated that they provide specific information about action at the joint. It is postulated that the myochordotonal organ, on the other hand, functions primarily in setting the level of excitability of the neurons innervating the flexor and extensor muscles of the legs, and so controls the posture and attitudes of the animal. The sensory system at this joint would thus have the double function of detecting movement at the joint and determining excitability of the neurons concerned in it.

Similar organs have been found in the antennae of the amphipod *Caprella*, the antennules of the lobster, and the antennae of a hermit crab *Petrochirus californiensis* (Fig. XV–7*d*). In the decapods the organ lies in the basal segment of the flagellum. In *Petrochirus* a dorsal and a ventral set of bipolar neurons are embedded in the connective tissue at the distal end of this segment (Fig. XV–7*e*). There are about 50 of these in the dorsal set, about 30 in the ventral. Their processes extend across the joint between the basal segment and the one immediately distal to it and are fixed in the connective tissue beneath the exoskeleton. From silver-stained histological preparations, it appears that the dendrites of a pair of cells enter and end in a single scolopoid structure, as the two neurons of each unit of the chordotonal organs of *Carcinus* do. But in *Petrochirus* it is the processes of the neurons themselves that extend across the articulation, while in *Carcinus* it is the elastic tendon in which they are embedded that does. Flexion or compression arising from movement of the flagellum initiates impulses in the neurons, while stretching inhibits responses from them. Similarity in the operation of this chordotonal organ with the lateral line system of vertebrates suggests that it may detect not only body displacements but water movements, with a possible acoustic function.

The four structures believed to function as proprioceptors in the isopod *Ligia oceanica* are simpler in construction than those of decapods, consisting of only one or two cells suspended by two or three processes between the exoskeleton and either a nerve or a muscle. Each unit is either position-sensitive or movement-sensitive, with, therefore, functional similarity to the proprioceptors of *Carcinus*.

Stretch receptors are common among insects, where they have been found in thoracic and abdominal segments in Odonata, Orthoptera, Hymenoptera, Lepidoptera, and Diptera. Histologically, each is a multipolar neuron enclosed in a capsule of connective tissue and thus resembles both the muscle spindle and Pacinian corpuscle of vertebrates. The receptor may be associated with connective tissue or with a modified muscle cell. Experiments have shown that when the receptor cell is under no tension, no impulses can be recorded from it, but when it is under minimal tension, in a condition approximating that in the body, it discharges impulses at the rate of 5 to 10 per sec. This rate increases markedly when the receptor unit is stretched.

Insects have also other stretch receptors associated with the cuticle (Fig. XV–1*f*). These are campaniform sensilla, so called because of the bell-shaped elevation of the cuticle over the single bipolar neuron attached to its concave or undersurface (Latin *campana*, "bell"). They are found wherever the cuticle may be subjected to strain and are especially plentiful in joints where stresses arise from muscular contractions. Electrophysiological studies have shown that these sensilla respond primarily to pressure on the cuticle, from which it may be inferred that change in the shape of the bell-shaped cap and diminution of its diameter excites the neuron below it. In honeybees the cuticle of the cap, like that of the hair-plate sensilla, contains a protein that has special staining capacities; is digested by trypsin; and, physically, has a high degree of deformability, has perfect elasticity, and shows birefringence under stress. It has been given the name of resilin. In both these stretch receptors, electron microscopy has shown that the distal segment of the nerve process is connected to the proximal part of the fiber by a ciliary segment.

Campaniform sensilla in the halteres of Diptera have been proved to operate in the stabilizing functions of these organs. And in the locust *Schistocerca gregaria* two groups have been identified on the ventral surfaces of both the fore and the hind wings. In the fore wing there is a proximal group of 20 to 21 and a distal one of 62 to 65 located along the subcostal vein. There is a single group of 65 to 72 on the subcosta of the hind wing, spreading over an area of 10 mm. The frequency of impulses from these organs increases when the wing is stretched along its long axis. Tests of the flight capacity of locusts in a wind tunnel, after destruction of the sensilla by cautery, have shown that the sensilla are essential to the regulation of twisting of the fore wing during the constant lift reactions locusts must perform in flight.

Stretch receptors have been demonstrated histologically in the foregut of the black blowfly *Phormia regina*. These are bipolar neurons located in the branch of the recurrent nerve that innervates the foregut. In insects, this nerve runs posteriorly from the frontal ganglion, lying below and slightly anterior to the cerebral ganglion, and gives off branches to various parts of the digestive tract and to the aorta. Ganglion and nerves together comprise the stomodael (stomatogastric) nervous system, often considered the homologue of the sympathetic or autonomic system of vertebrates. Electrical recordings from the nerve show increased frequency of impulses when the foregut is expanded by oral introduction of food, the impulses coinciding with the peristaltic contractions of the gut. No such increase was evident when the midgut or the hindgut was distended by anal introduction of food. These foregut stretch receptors seem to be in large measure responsible for the feeding behavior of *Phormia*, which ingests a constant amount each day even when the supply is overabundant. The system probably works in such a way that the receptors signal the cerebral ganglion of the extent and duration of peristalsis in the foregut and so, indirectly, of the fullness of the crop, since its contents pass through the foregut on the way to the midgut for digestion.

Proprioceptors have also been demonstrated in xiphosurans and in chelicerate arthropods. The articular membrane of the coxa-trochanteral joint in the walking legs of *Limulus polyphemus* contains large, single multipolar cells embedded in connective tissue underlying the epidermis of the joint. The dendrites of these neurons end in the epidermal cells, and their axons are included in the distal gnathobase nerve that connects with the central nervous system. Electrical recordings from these nerves show that they give both tonic and phasic responses, the impulses increasing in frequency with increasing tension at the joint. In the house spider *Tegenaria atrica* a small nerve in the leg that branches from the main one in each coxa seems to serve the same function. This nerve contains some eight axons from nerve cells clustered below the femur-patellar hinge, as well as a number of axons from sensory neurons associated with hairs on the patella. That it is primarily proprioceptive, and not a mechanoreceptor attuned to vibrations, has been shown by transection experiments. After it has been cut, a spider still reacts to vibrations of insects dropped into the meshes of the web. In scorpions, the lyriform organs have been shown to respond to stresses in the cuticle, acting in this way like the campaniform sensilla of insects.

Specialized proprioceptors such as these have not yet been found in other invertebrates, whose behavior, nevertheless, is often such as to indicate transfer of information from one part of the body to another. This seems to be the case in sea anemones and in echinoderms, for example, in which posture and orientation to gravity are maintained most probably through mechanical effects of the position of various parts of the body. In earthworms, neurons of four kinds have been found in the muscle layers, two of which may be proprioceptive. One of them is a large cell that lies in the muscles between the dorsal and ventral chaetae of each segment, with a process running toward the chaetae; it is thus ideally situated to register changes in their position. The other is morphologically like the superficial sensory cells and lies in the circular muscles. Although there is no direct evidence that they are proprioceptive, worms do respond by contraction to stretching of their bodies, and the location of these neurons is very suggestive of their mediation in these responses. Even in animals with statocysts, it is probable that proprioceptors supplement the action of these specialized organs by helping the animal to maintain its position in respect to gravity and to give the appropriate responses when this is altered. Further information about the wider distribution and operation of mechanoreceptors of this kind, however, awaits further investigation.

3. Chemoreception

Behavioral studies have made it very evident that chemosensitivity is widespread among invertebrates, but the location of the sense cells concerned is known in relatively few of them and the way in which the response is invoked remains conjectural. Appropriate substances may excite chemosensitive cells when directly in contact with them or when, from a distance, the molecules of a soluble substance diffuse through water or those of a volatile one diffuse through air. Contact stimulation in every case involves solution at the site of the sensory cells, but in distance stimulation, the bulk of the substance may be remote from the animal with only its diffusing particles reaching it. Attempts to draw homologies between these aspects of invertebrate chemoreception and the human sensations of taste and smell are fruitless, since there is no means of determining if such sensory distinctions exist in invertebrates and since, in cases observed among some of them, the same sensory areas respond both to contact and to distance chemostimulation. There is, however, a difference between contact and distance chemoreception, in that much less of a substance is required for distance chemoreception than for contact, for distance chemoreceptors are sensitive to minute amounts of a volatile compound. Their high degree of sensitivity and the fact that they are composed of many neurons distinguish them from contact receptors. It is clear that invertebrates can discriminate between different chemical stimulants, showing indifference to some and either positive or negative reactions to others. One hypothesis of the mode of operation of a chemical stimulant relates the shape of the molecule to its effectiveness, assuming that there must be some conformity between it and a site on the membrane of the sensory cell in order for the changes to take place that are ultimately recordable as impulses. The nature of the behavioral responses to stimulation of an animal's chemoreceptor cells depends, like those from other sources, upon the connections that are made with other neurons.

Evidence from studies of invertebrates at all levels of organization shows that chemoreception is used in the location of food and in its acceptance or rejection when found; in the selection of habitats and of mates; in the recognition of partners in host-parasite and host-commensal relationships; and in the recognition of members of the same species as friends and often of those of other species as enemies, especially as predators. It is a primitive and highly important sense, and it is not surprising that the chemoreceptor cell should be one of the simplest of sensory cells and similar in many animal groups (see Fig. IV–33). It is a bipolar neuron with fine processes that project through a pore in the epidermis (and exoskeleton if the animal has one) and that are often associated with stationary cilia or with fine hairs. Such cells may be scattered individually over the surface of an invertebrate's body or collected in groups in one region or another.

Protozoans, without any defined surface areas of chemosensitivity; sponges, without recognized nerve cells; and cnidarians, in which no morphological or functional distinction can be made between their dispersed sensory cells, all respond in specific ways to contact and distance chemostimulation. This has been shown experimentally by the introduction of substances of various kinds, both organic and inorganic, into the medium and observation of the animals' reactions to them. To some of these substances, primarily to those organic compounds that might serve as food, their reactions are positive, and the animals move toward it if it is at a distance or begin feeding reactions if it is in contact with their bodies. Likewise, they exhibit avoidance or rejection reactions to some substances. The carnivorous sea anemone *Anemonia sulcata*, for example, moves its tentacles and opens its mouth in typical feeding behavior when substances associated with protein are presented to it. Mouth opening can also be induced by other compounds, such as skatole and glucose, which do not evoke the tentacular response, suggesting that the tentacular chemoreceptor cells are more selective than the oral. The direction of ciliary beat on the oral disk of

Metridium marginatum, also a carnivore, is changed from outward to inward when bits of filter paper soaked in glycogen, peptone, and certain amino acids are placed upon it. Sucrose and glucose have not this effect.

a. CHEMOSENSITIVITY IN PLANARIANS In most planarians there is a definite localization of chemoreceptors, although in some species they may be scattered over the body (Fig. XV–5*a*). In those in which they are localized, they lie in a pair of ciliated pits or grooves along the sides of the anterior region of the body. Turbellarians can be attracted from some distance to a piece of meat fastened upstream from them. They do not give this response after ablation of the ciliated organs, which are obviously of importance to them in location of their food. Because of the symmetrical position of these organs and the fact that water is constantly swept through them by the action of the cilia on the cells, planaria can proceed in an almost straight line toward the source of food; as they begin to move toward it, their heads wave from side to side, but as they get nearer and the stimulus becomes stronger and both chemoreceptors are, presumably, equally excited, these movements stop and the course is a direct one. *Dugesia dorotocephala* can discriminate between amino acids, being positively chemotactic to lysine and glutamine at threshold concentrations of 5×10^{-6} *M*, and negatively so to asparagine, aspartic, and alpha-keto-glutaric acids. This selective sensitivity may be an asset to them in distinguishing the organisms on which they prey.

b. CHEMOSENSITIVITY IN MOLLUSCS The behavior of molluscs gives clear indication of chemoreception. In the buccal cavity of monoplacophorans, aplacophorans, polyplacophorans, scaphopods, and some primitive gastropods, there is a subradular organ composed of very tall epithelial cells with many nerve endings. In more highly evolved gastropods, contact chemoreceptors are especially numerous on the cephalic tentacles, but are present also in the buccal region and on the foot. In bivalves with siphons, these extensions of the mantle edge are well supplied with chemoreceptors, and in those without them, the pallial tentacles on the middle fold of the mantle margin are chemoreceptive. In both cases they are concerned with contact reception.

The osphradium, an organ found in all molluscs except aplacophorans, scaphopods, nudibranchs, and terrestrial pulmonates, is still enigmatic as far as its type of sensitivity and its functions are concerned. Osphradia are round or oval patches of sensory epithelium situated in the mantle cavity near the gills. In bivalves the patches are small and lie usually on the undersurface of the posterior adductor muscle, but they may extend along the posterior ends of the gills. They are paired in chitons and in gastropods with two gills, but single in those gastropods in which one member of the pair is suppressed. The ciliated epithelium in the osphradium is usually folded in plates or leaves, thus increasing the area of a confined space, and the sensory cells are typical of those of chemoreceptors. The position of the organs is such that currents of water entering the mantle cavity sweep over them, and so they may well act as distance chemoreceptors. There is, however, no direct evidence for this, and it is also possible that they may be tactoreceptors, responding to the contact of solid particles and so giving information to the molluscs of the amount of sediment in the water. Their absence in aquatic pulmonates and in those that have adopted a strictly pelagic existence makes it seem more likely that this, rather than chemoreception, is their primary function and that they act as water samplers of all kinds.

In many cephalopods there is a pit, or a tubercle, on each side of the head, between the eye and the opening of the mantle cavity, richly supplied with sensory cells. Because of its location, in the path of the inhalant respiratory stream, and because of its histological similarity to the osphradium of gastropods, it is assumed to be homologous to the osphradium. There is, however, no critical evidence that this is the case or that the sensory cells are chemoreceptive.

There is, on the other hand, abundant evidence that other molluscs do react to chemical substances reaching them from a distance. This is of great importance to those that are mobile and search for food, and numerous observations have shown that both carnivorous and herbivorous gastropods can detect the food of their choice when some distance from it and direct their movements toward it. The carrion eater *Nassarius obsoletus* has been seen to move upstream in a current flowing over mutilated fish as much as 75 cm away from it, although exhibiting no rheotaxis in the absence of the fish. The herbivorous *Littorina obtusa* can locate *Fucus* from a distance of 100 cm, probably through sensitivity to metabolites emitted by this alga. Different species of *Fucus* and the alga *Ascophyllum* have different degrees of attractiveness for the snail, indicating that it can discriminate between the metabolites they produce. This seems also to be the case with commensal molluscs that find their hosts by chemoreception of specific products. *Modiolaria*, a genus of bivalves that is commensal with tunicates, can, for example, be attracted by very dilute solutions of tunicin extracted from the tests of their hosts. Secretions or emanations from members of the opposite sex act as attractants to members of the same species, and there is some evidence that distance chemoreception is operative in the homing of *Helix pomatia*, which, after transportation from the site of its capture to another some distance away, can return with a high degree of accuracy to it.

Chemoreception plays an important part also in recognition of predators and in escape from them. *Nassarius*, preyed upon by some starfishes and sea urchins, exhibits the same patterns of avoiding behavior when the predator is at a distance as it does when the posterior surface of its foot is touched by a predator. The behavior of the snail includes pushing the shell forward until it lies over the head, raising the anterior part of the foot, and extending the body so that the shell falls forward and to the right, so that the mollusc zigzags rapidly away from the echinoderm. This reaction is not elicited by nonpredatory asteroids, but it can be by the isolated tube feet of predators. The effectiveness of the tube feet in eliciting a reaction is diminished by boiling them, suggesting that the substance to which the snails are chemosensitive is a protein. Typical avoidance reactions on the part of other molluscs have been shown when they have been put into seawater in which their particular predators have lived. These reactions have also been shown by representatives of various species when brought into contact with a predator. The pulmonates *Physa fontinalis* and *Lymnaea* sp. will, for example, be indifferent to the presence of the predatory leech *Glossosiphonia* when it is crawling on their shells, but will give avoidance reactions if the foot or tentacles are touched by it. Other gastropods, too, exhibit escape reactions when certain areas are stimulated by contact with a predator. Chemoreception both on contact and at a distance seems an important protective device for these gastropods.

c. CHEMOSENSITIVITY IN ANNELIDS The tendency to cephalization, or concentration of exteroceptive sensory neurons and coordinating centers in the anterior region, is quite evident in the localization of sense organs in most annelids. In predatory, burrowing polychaetes, paired patches of sensory cells or a pair of ciliated pits, the nuchal organs, are situated in the head. The nuchal organs are lined with ciliated columnar epithelium and sensory cells. They are not found in ciliary feeders, and their ablation or destruction in macrophagous species results in failure to locate or capture food, indicating that they function as distance chemoreceptors and give information to the worms of the presence of prey. In earthworms, cells histologically similar to those in planarians, in which chemoreception has been demonstrated, are grouped in small tubercles that project from the surface of the body, their processes extending through the cuticle. In construction these resemble taste buds of vertebrates (Fig. XV–8*a*). They are found in all the segments, but are most numerous in the prostomium where, in *Lumbricus*, as many as 700 per mm^2 have been counted. There are similar sense organs also in the buccal wall and pharynx. Over a century ago Charles Darwin demonstrated the ability of earthworms to find leaves on which they feed and to discriminate between those of different kinds and different degrees of freshness and decay. This may be through contact or distance reception.

The condition of the soil is of great importance to earthworms and an ability to sample it in regard not only to its content of organic material and potential food but also to its texture, calcium content, and acidity is without doubt a great asset to them. Recordings of action potentials in the segmental and prostomial nerves of earthworms show a generalized

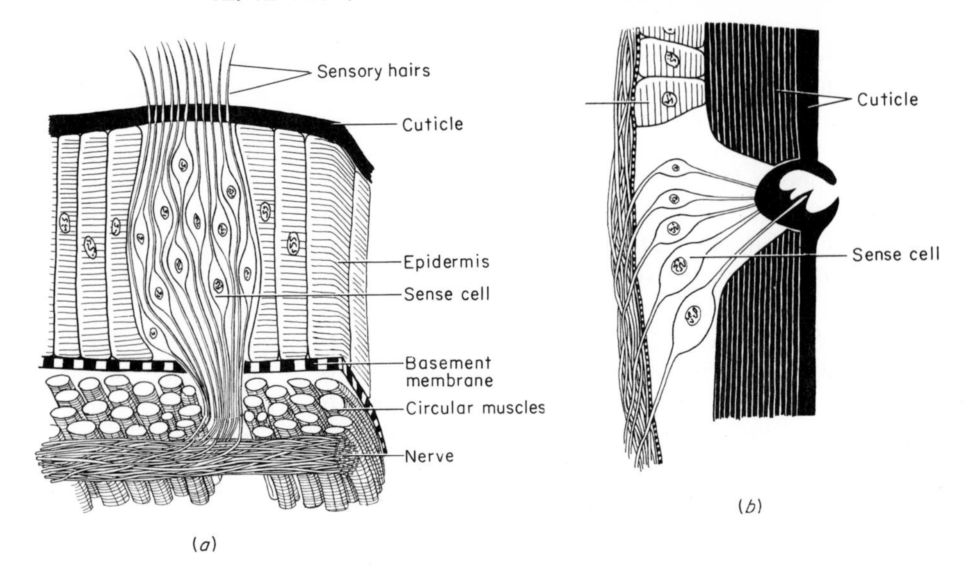

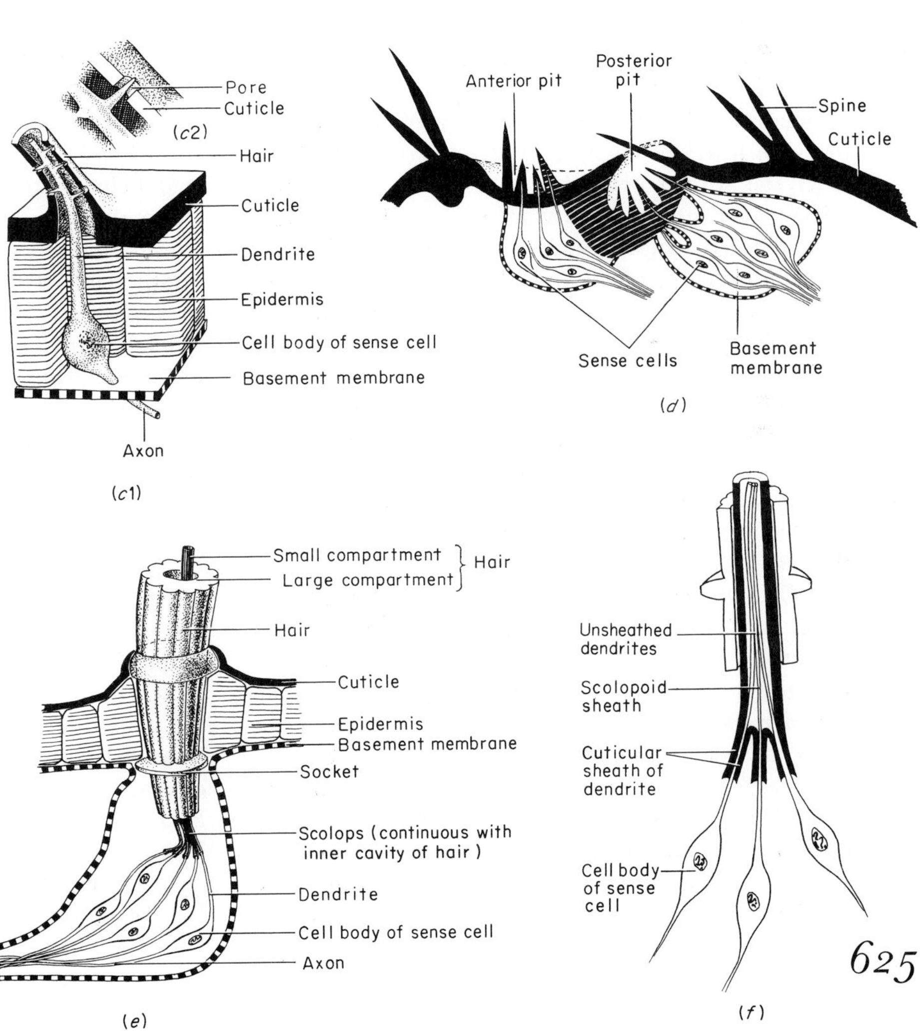

FIG. XV–8 *Chemoreceptors.* ***(a)*** *Sense bud in the epidermis of* Lumbricus. *(After Bullock and Horridge.)* ***(b)*** *Tarsal organ of a spider. (After Blumenthal.) One of the spikes in the depression has been represented as if opened to show how the dendrite extends up into it.* ***(c1)*** *Scheme of an insect chemosensory sensillum. (After Boeckh, Kaisslinger, and Schneider.)* ***(c2)*** *Detail of one process of a dendrite to show how it makes contact with the air through a pore in the cuticle.* ***(d)*** *Haller's organ in the tick* Ixodes reduvius. *(After Lees.)* ***(e)*** *Schematic representation of a single chemosensory hair from the labellum of* Phormia. *(After Wohlbarsht.)* ***(f)*** *Detail of the scolops, showing the individual sheaths of each dendrite which unite to form the scolopoid sheath common to all of them.*

sensitivity to tactile stimuli and a more localized one to chemical stimuli. Earthworms can discriminate between alkaloids, sugars, and acids, as well as between leaves of different trees and the state of their decomposition. They will not feed on pine needles coated with gelatin containing quinine or acids such as malic, oxalic, and phosphoric above certain concentrations, but they readily accept them if the gelatin contains sugars. In *Lumbricus terrestris* this chemosensitivity is greatest in the prostomium, while that to salt (NaCl) is distributed over the surface of the entire body, as is tactile sensitivity. Free nerve endings in the integument, as well as the special sense organs, probably act as chemoreceptors.

Earthworms can also detect degrees of soil acidity and differ in their tolerance to it. Acid sensitivity is distributed generally over the body surface, and action potentials recorded from fibers teased out of exposed segmental nerves show it to be related to individual ones that do not respond to salt, quinine, or sucrose. *Allolobophora longa* will not burrow into soil with a pH below 4.5; *Lumbricus terrestris* will not enter soil with a pH below 4.1, but *L. rubellus* will tolerate one with an acidity as high as pH 3.8. Acid soils are deficient in calcium, a factor which must enter into soil preferences and tolerances of earthworms, since they depend upon the availability of some calcium for their well-being and even their existence. These conditions vary from place to place in the ground through which an earthworm moves and are of great importance in determining the distribution of worms. Their tactile and chemosensitivities direct them away from hard, very acid soils, where there is little or no suitable food, and tend to keep them in those soils that are loose in texture and rich in nutriment.

Chemoreception in the choice of habitat is indicated in *Eisenia foetida*. Worms of this species congregate in compost heaps and will move from clean sand to that which is mixed with humus, the presence of which they must detect by the odors diffusing from it. Chemosensitivity is apparent also in leeches in which isolated cells, with bristles projecting through the cuticle, are scattered generally throughout the epidermis. Cells of similar type are collected in segmentally distributed papillae. The function of the papilla is not yet clear but the responses of blood-sucking leeches to freshly drawn blood with which they are not in direct contact indicates that they are in some way informed of its proximity to them. The fish leech *Hemiclepsis marginata*, even when hungry, makes no response to wet filter paper waved over its body, but starts at once on hunting maneuvers if the paper has been rubbed over the surface of a carp, and does so, too, if mucus scraped from the surface of a tench is introduced into the water around it. The duck leech *Theromyzon tessulatum* begins its searching behavior when the secretion of the bird's uropygial gland is put into the water with it. Selectivity in response to substances of the host is shown by *Hemiclepsis*, which are attracted and attach themselves to fish and to the salamander *Amblystoma*, but only rarely to the frog *Rana temporaria* and never to *R. esculenta*.

d. CHEMOSENSITIVITY IN *XIPHOSURA*

Chemosensitivity has been established in *Limulus* and is well known in arthropods. In *Limulus* electrical recordings from the proximal gnathobase nerve show that their fibers carry chemical messages, and tactile and thermal ones as well. Contact and distance chemical stimulation of the sensory spines on the surface of the gnathobases of the walking legs result in propagation of impulses along these nerves, while the distal ones, as has already been pointed out, respond to stimulation of proprioceptors situated in the articular membrane. The chemoreceptors respond strongly to amino acids and to extracts of marine bivalves, but not at all to solutions that to humans are salty, sweet, or sour.

e. CHEMOSENSITIVITY IN ARTHROPODS

Behavioral responses of arthropods to chemical stimuli are well known. The sites of the receptors have been determined largely by amputation experiments, in which pairs of appendages have been removed and the consequences, in respect to chemical stimulation, observed. In aquatic crustaceans there are many delicate sensory hairs on the appendages, usually arranged in rows among the stiffer tactile hairs. These are especially numerous on the antennules (where they occur in groups known as aesthetascs), the antennae, and the mouthparts, but are found also on all the other segmental appendages, the telson, and the gills. In the stomatopod *Squilla*, the gills have the greatest number of them. Histological examination shows that each of these sense organs consists of a group of bipolar neurons whose dendrites extend within the hair along its entire length. In terrestrial crustaceans, there are small plates on the surface instead of hairs.

Electrical measurements made of the receptors on the inner rami of the antennules, of those on the chelae, and on the first two pairs of pereiopods of the crayfish *Cambarus bartonii sciotensis* show a constant resting potential and a marked increase in frequency of discharges when solutions of compounds considered savory to humans replace the pond water used in control recordings. Even more precise measurements can be made from the tufts of sensory hairs on the first two pairs of walking legs, for a microelectrode can be put in contact with a single hair. Significant increase in rate of impulses are recorded when 0.25 *M* glycine is applied to the hair. Similar recordings from the chemoreceptors on the legs of brachyurans show response to stimulation by other amino acids as well. Among these hairs are certain ones specifically sensitive to trimethylene oxide, a compound present in fish muscle and released on its decay. This sensitivity would be of especial importance to scavengers in guiding them to a source of food. Experiments on two related species of crabs common on the west coast of the United States, *Cancer antennarius* Stimpson and *C. productus* Randall, have shown that the chemoreceptors on their pereiopods do not respond to trimethylamine below a concentration of 0.10 *M*; above this, their responses increase rapidly. The compounds most effective in exciting these receptor cells were DL-alpha-butyric acid, taurine, alpha-glutamic acid, and serine; with sugars (glucose, sucrose, trehalose), peptides, proteins, and certain salts (KCl, NH_4Cl, and Na acetate) virtually ineffective. A relationship between the configuration of the molecule and its effectiveness as a stimulant is indicated by these experiments.

Behavioral experiments provide indirect evidence of crustacean chemosensitivity. When filter paper soaked in meat juice is brought into contact with the thoracic legs or the mouthparts of shrimps (*Palaemon serratus*) that have been blinded, the animals begin feeding movements. A blinded shrimp will also find food in its vicinity, indicating distance as well as contact chemosensitivity. The crab *Carcinus maenas*, made to run in the air over wet blotting paper, starts searching, grasping, and feeding behavior if one of its pereiopods touches a place that has been soaked in meat juice. Evidence for chemoreceptors in the gills is derived from the grasping movements made by the posterior pereiopods when a piece of filter paper soaked in meat juice is held in the respiratory current. In other experiments involving successive amputation of the appendages, the outer rami of the antennules have been found to be the most sensitive to the presence of food at a distance, but a crab will still show typical feeding reactions when the antennules are removed. The sense organs on the antennae, mouthparts, and chelae carry on the chemosensory function adequately, but removal of these in order progressively reduces the animal's responsiveness, though it does not entirely eliminate it.

With aquatic animals, all stimulating compounds must be in solution, so that it is impossible to make a functional distinction between their contact and distance chemoreceptors. But with terrestrial animals some of the stimulating substances may be in the air as gases, and responses to these imply the possession of distance receptors.

In the majority of spiders, depressions in the tarsi of all the legs are covered with papillae innervated by sensory neurons (Fig. XV–8*b*). These tarsal organs have been proved, by behavioral observations, to be chemosensory (see Chapter VII, page 276). The spider *Araneus diadematus*, for example, will back away from a drop of clove oil or turpentine 1 to 5 mm away. Food at a distance will elicit feeding reactions. Male spiders can also detect through these organs whether a female has passed over a given area, and odors from young female spiders evoke courtship reactions in males. In ticks, Haller's organ lies on the dorsal side of the first pair of legs (Fig. XV–8*d*). This is a compound structure consisting of an anterior pit with peg-like sense organs, known to be humidity receptors, and a posterior capsule, also with peg-like sense organs that respond to chemical stimuli. *Ixodes reduvius* turns away from a drop of citronella, but moves right up to it after amputation of its forelegs and turns aside only after its palps have made contact with the oil.

Chemoreceptors in insects, in addition to sensilla trichodea, are sensilla placodea, or pore plates, sensilla basiconica, and sensilla coeloconica (Fig. XV–8*c1*, *c2*, *e*, and *f*), variously distributed over the antennae, the maxillary and labial palps, or their homologues, the legs and in females, the ovipositor. Each sensillum placodeum is composed of 12 to 18 bipolar neurons, whose dendrites have a ciliary segment from which a fibrous strand extends and terminates at a minute pore in a thin circular or oval plate lying flush with the cuticle to which it is attached by a delicate membrane. (See Fig. XV–1*c*.) The pores are arranged around the periphery of the plate. The number of these sensilla varies in different species of insects and in different areas of the same insect; almost 3,000 have been counted on each antenna of honeybees, where they are mingled with many other sense organs of different kinds. A single antenna may have a total of 20,056 sensilla upon it. Sensilla basiconica are variable in size, in shape, and in number of associated neurons. In general, they may be characterized as thick-walled or thin-walled; in each, the dendrites of the neurons extend into truncated cones, or pegs, that project from the surface of the cuticle. (See Fig. XV–1*e*.) There are usually about five bipolar neurons in a thick-walled sensillum basiconicum whose dendrites, enclosed in a scolopoid sheath, extend to a pore at the tip of the peg. In a thin-walled sensillum of this kind, there are about 40 to 60 neurons, whose dendrites are enclosed in a scolopoid sheath only until they reach the surface of the body; there the sheath bends and stops, but the dendrites emerge from it, through minute openings, and continue up into the tip of the peg, where they terminate at tiny pores. Sensilla coeloconica lie in pits below the general cuticular surface with the pores of the pegs opening into the concavity. (See Fig. XV–1*g*.)

Electron micrographs of sensilla placodea and basiconica have shown the fine structure of these organs in some insects and have demonstrated that the distal terminations of the sensory neurons are in contact with the air. They may be covered by a thin film of liquid but, even so, molecules diffusing from a volatile substance would have at most only this aqueous barrier to pass before making contact with the exposed part of the neuron's membrane.

Detailed studies of the sensilla placodea on the antennae of aphids suggest that they may be chemosensory organs, but there is as yet no positive proof of this. In the vetch aphid *Megoura viciae* there is, on the long terminal subsegment of the flagellum of each antenna, a circular flattened sense organ surrounded by a ring of small hairs with split tips. Close to it are six to seven small sensilla coeloconica. There is also a sensillum placodeum on the second subsegment. The sensory neurons below the plate are either single or grouped in twos or threes. The dendrite of each has a ciliary segment, beyond which it branches many times after passing through the inner layer of the cuticle. Groups of fine neurofilaments from these branches enter minute openings in the outer cuticular layer as, probably, the actual receptor surfaces of the neuron.

By very delicate and precise techniques it has been possible to make recordings from individual cells innervating the sensilla basiconica on the antennae of moths, beetles, grasshoppers, flies, and cockroaches; from those innervating the sensilla placodea of bees and wasps; and from those innervating the sensilla coeloconica of flies, grasshoppers, and bees. The recording electrode may be a glass capillary or fine tungsten wire, inserted through the epicuticle close to the base of the sensillum with the reference electrode introduced into the hemocoelomic space of the same segment or into another one. Or the reference electrode can be placed directly on the sensory neuron. Recordings show that all of the cells tested respond to a fixed spectrum of odor stimuli, on the basis of which they can be characterized as generalists or specialists. Examples of generalists are the sensilla basiconica of moths and the sensilla placodea of honeybees. Tested with a variety of volatile compounds blown over the antennae from pieces of fluted filter paper soaked in various dilutions of different substances, each of the many cells in these sensilla was found to have its own stable spectrum of odor sensitivity, with some substances eliciting impulses and others depressing the basic flow. Of three neighboring cells, one might be excited, one inhibited, and one not affected by the same substance. There was also some overlap between the spectra of different cells in a single sense organ. In sensilla basiconica of males of the moth *Antheraea pernyi*, for example, three cells consistently gave strong responses to geraniol to which some others gave only slight ones and still others none at all, while to yet others these aliphatic alcohols were strongly inhibitory. And only one cell was found to respond at all to eugenol, the oil of cloves, and then only feebly, while to others the compound was ineffective or inhibitory.

Examples of specialists are the chemosensitive cells of male moths and honeybee drones that respond only to the sex attractant of females of their own species. Tested with Bombykol and its isomers, with cycloheptanon and the alcohols terpineol and linalool, the chemosensitive sensilla trichodea of the males of Bombyx showed strong responses to Bombykol, slightly less strong ones to its isomers, weak responses to cycloheptanon and terpineol, and very weak and rarely occurring ones to linalool. Just the opposite results were obtained from females, whose neurons rarely gave even weak responses to Bombykol and its isomers, and to cycloheptanon, but gave moderately strong ones to terpineol and linalool.

The most detailed study of insect chemoreceptors has been made on the labellar hairs of *Phormia regina*. These are excited on contact with particular chemical substances. They have a limited but variable number of neurons, one of which has been proved to be a mechanoreceptor. One, two, three, or as many as four of them may be chemosensory. Of the four, one is known to be excited by inorganic salts, another by water, another by sugars, while the function of the fourth is still unknown.

The structure of the receptor, as revealed by light and electron microscopy, is quite complicated (Fig. XV–8*e* and *f*). The surface of the hair has longitudinal grooves and is almost completely covered with wax which extends out over the adjacent surface of the labellum. There are two cavities within the hair, an outer larger one surrounding an inner smaller one; both of these extend the length of the hair. The neurons lie beneath the epidermis, enclosed in a capsule of the basement membrane. The dendrites of each of these acquire a relatively thick, cuticular sheath, all of these sheaths uniting to form the scolopoid sheath that encloses all the dendrites and that becomes continuous with the inner cavity of the hair. In the scolopoid sheath and in the cavity of the hair the dendrites of the chemosensory neurons are naked and separated from each other by only a thin film of extracellular fluid. The dendrites of the mechanosensory neuron are unsheathed as they pass through the scolops but become sheathed again at its distal end and terminate in a chordotonal-like structure at the junction of the hair with its socket.

Electrical activity in the hair has been recorded. Installation of the recording apparatus seems to have little effect upon *Phormia*, for a fly will tolerate its presence for at least 1 week during which recordings can be taken and a variety of substances tested on the same hair. Behavioral responses can also be observed; protrusion of the proboscis is the one regularly following presentation of acceptable stimuli.

Evidence from experiments of this kind indicates that the receptor sites are mainly at the tip of the dendrite and that there are specific ones for specific substances. If the apparatus is so adjusted that the test substances are in contact with the side of the dendrite, there is a longer time interval between their application and the electrical response than there is when they are in contact with the surface of the tip, suggesting that they must travel an appreciable, although minute, distance before they reach their specific sites. Although these recordings are extracellular, differences between the impulses from different neurons can be determined on the basis of their magnitude and interaction with each other. In most of the hairs tested, impulses from the salt receptor neuron are the largest and most easily identified. This neuron responds preferentially to salts of all kinds and, although stimulated by some anaesthetics, is not excited by acids or bases. Results of experiments testing its sensitivity to solutions of NaCl and of the alkali halides show that both the anion and the cation mediate the effectiveness of a salt's stimulation. The sugar receptor neuron is excited only by carbohydrates, but differently by different ones. The configuration of the sugar molecule seems to be the significant factor, suggesting that conformity between it and its receptor site or sites on the membrane of the dendrite is a necessary condition to successful stimulation. One of these sites appears to accept molecules of the glucose type, and certain experiments with a series of *d*-glucose derivatives have indicated that the combination of membrane site and sugar molecule is with the C_3 and C_4 hydroxyl groups, for if these are blocked, the sugar becomes ineffective as a stimulant. It is of interest that, in other insects as well as in *Phormia*, there seems no correlation between the acceptability of a sugar and its nutritive value. Honeybees, for example, will accept mannose although it is essentially toxic to them because of their inability to metabolize it.

Evidence for chemoreceptors on the ovipositors of females has been obtained primarily from experiments on parasitic Hymenoptera that lay their eggs with great precision and specificity in the bodies of other insects on which the larvae feed. Isolated abdomens of these wasps make typical avoidance reactions when the valvulae, three pairs of elongated processes extending posteriorly from the terminal segments of the abdomen, are dipped into solutions of substances that have elicited rejection behavior from other insects when applied to their antennae, mouthparts, or tarsi.

Chemoreceptors on different parts of an insect's body may have different degrees of sensitivity to the same substance, and those on one part may be sensitive to substances that those on others are not. In general, in flies the labellar chemoreceptors are more sensitive than the tarsal, for responses from them can be induced with solutions of lower concentrations. And the proboscis of bees can detect lower concentrations of solutions than can their antennae, and these are in turn more sensitive than the legs. Chemoreceptors on the legs of the fly *Calliphora vomitoria* do not respond to lactose, while those on the oral lobes do but are much less sensitive to sucrose than those in the tarsi.

4. Photoreception

The primary event in photoreception is the absorption of a quantum of light energy by a molecule capable of being raised to a higher energy level when the photon is captured. This alteration in the molecule starts a train of events, which are in themselves independent of light and which lead to transduction of the radiant energy into electric energy. Light is that part of the electromagnetic spectrum with wavelengths of 200 to 1,000 nm, of which the region with wavelengths of 450 to 750 nm is "visible light" in that it is in the range of sensitivity of the human retina. That with wavelengths above or below this can, however, be measured by optical means. The range between 400 and 200 nm is that of ultraviolet, and that between 750 and 1,000 nm is that of infrared. The molecules absorbing radiant energy in the total light range are pigments and in all specific photoreceptor cells so far studied these are carotenoids conjugated with a protein. The most fully analyzed photosensitive pigments are rhodopsins in the eyes of vertebrates, cephalopod molluscs, and arthropods. These differ from each other primarily in their protein components, the opsins, and to some extent in their carotenoid components, the retinenes. Because of these differences they differ in their absorbing capacities, showing, with spectrophotometry, different absorption maxima in different ranges of the spectrum. Very little is known about other visual pigments or other photosensitive molecules in invertebrates, some of which respond to stimulation by light in the absence of any defined photoreceptor or any evident pigment.

In considering photoreception, distinction should be made between the general sensitivity of all protoplasmic systems to radiant energy, including light, and the localized sensitivity of organisms with photoreceptor organs. Distinction should also be made between photoreception as such and vision, or effective image formation. While lenses, with which many invertebrate visual organs are equipped, may constitute a dioptric system that forms images, their contribution to vision is negligible unless the image falls directly upon the photosensitive area. Whether they do form images, and the focal distance at which this occurs, can be determined indirectly by measurement and calculation, and directly by photography, using excised lenses for those of the camera. If the optical properties of the lens are such that the image is formed before or behind the photosensitive area, the organ can act only as a photoreceptor that registers different degrees of light intensity and so the passage or movement of objects in the visual field of an animal. They cannot act as eyes that register the shape and size of an object that may be stationary as well as moving.

Experiments with octopuses and with bees have shown that these invertebrates can discriminate between shapes, up to a point at least, and have led to the conclusion that their eyes register some kind of image, but perception of form in others is still an undemonstrated possibility.

a. DIFFUSE PHOTOSENSITIVITY Some protozoans of simple organization, with neither identified photosensitive pigments nor localized photoreceptors, give consistent and fixed responses to light. Amebae, for example, stop moving when the intensity of the light to which they are accustomed is changed, presumably because of alterations in the thickness and elasticity of the plasmagel and its rate of conversion into plasmasol. This may be a local effect if brighter light is directed at one particular region, or a general one if the entire body is illuminated. Some adjustment must follow, for if the light stimulus is continued an ameba will begin to put out pseudopodia again and to move about, the length of time needed for the adjustment depending upon the intensity of the light. Once adjusted to a certain level, locomotion can be stopped again by changing this.

Many metazoan invertebrates, even those with specialized photoreceptor organs, show general diffuse or dermal sensitivity, which is often greater in some areas than in others. In bivalve molluscs the edges of the mantle, the tips of the siphons, and the gills are usually the most sensitive regions. This has been demonstrated both by their behavior in response to illumination and by electrical recordings from their mantle nerves. Photoreceptor cells similar to those scattered through the epidermis of earthworms have been reported in the clam *Mya arenaria*, but in others it is the integument as a whole that seems responsive. Earthworms show general dermal sensitivity, although the greatest concentration of their photoreceptor cells is in the anterior region of their bodies. Echinoderms also exhibit general dermal sensitivity, and starfishes still react positively to light after removal of the eyespots at the tips of their arms. Most holothurians are generally photosensitive and in the genus *Holothuria* this may be related to the presence of a greenish-yellow pigment in the epidermis. This is one of the two pigments found in the integument of these sea cucumbers, whose sensitivity to light has been found to be related to the amount of it deposited there. These animals give generalized, and negative, responses to light, withdrawing their entire bodies from its source, as do most hydroids, sea anemones, and tubiculous worms; yet, colonies of *Eudendrium* bend toward the light, as does the anemone *Cerianthus*, in which it has been shown that the number of degrees through which the animal turns is proportional to the logarithm of the intensity of the incident light. Other animals with dermal sensitivity may give localized responses. In the clams *Mya* and *Pholas*, for example, only the siphon retracts when light is shone upon the animal, and in asteroids and echinoids, spines, pedicellariae, and podia may respond individually or in groups, depending upon the intensity and the direction of the incident light.

It has been shown by electrophysiological techniques that the visceral ganglia of the mollusc *Aplysia* and the abdominal ganglia of crayfishes are excited by light directly and not through the mediation of a photoreceptor of any kind. The giant nerve cells in the excitable ganglion of *Aplysia* contain granules of pigments that have been identified as carotene proteins and a heme protein. The heme protein shows three main absorption bands, one at 579 nm, one at 542 nm, and one at 418 nm, while the maximum absorption of the carotene proteins is at wavelengths of 490 nm and 463 nm. The fine structure of compound bodies in the innermost sheath cell of the giant fiber in the abdominal ganglion of the crayfish *Cambarus virilis*, consisting of lamellae, multivesicular bodies, and dense granules, so resembles that of known types of photoreceptors that it suggests that they may be responsible for the photosensitivity directly demonstrated in the ventral nerve cord. There is evidence also that the light-sensitive pigment in the sixth abdominal ganglion of *Procambarus clarkii* is different from that in the animal's compound eyes. Nerves in the sea urchin *Diadema antillarum* seem also to be directly stimulated by visible light. These may be protected by dermal chromatophores whose pigment disperses in light and concentrates in darkness, but it is the nerve cells themselves that are excited by the light.

b. LOCALIZED PHOTOSENSITIVITY AND PHOTORECEPTOR ORGANS Localized photoreceptors in invertebrates are either eyespots, or ocelli, relatively simple in their construction, or eyes with a high degree of structural complexity. Compound eyes, composed of varying, often large, numbers of separate visual units are characteristic of arthropods, while the pinhole camera type of eye of cephalopod molluscs is constructed like a vertebrate eye and lacks only some of its finer mechanisms of accommodation. All photoreceptors have in common a dioptric system—that is, a transparent covering and often a lens—a photosensitive substance, and light-absorbing pigment of some kind that shields the photosensitive substance from light coming toward it from certain directions. Those of metazoans have also a transducing and conducting neuron in which the photosensitive substance presumably is localized.

(1) Eyespots The simplest kind of photoreceptor organelle is found in phytoflagellates, of which *Euglena's* may be taken as a model (Fig. XV–9*a*). At the base of one of the strands that converge to form the flagellum there is a lateral swelling, the parabasal body, enclosed within the flagellar membrane. Opposite this is a crescentic body filled with droplets of a reddish pigment, chemically a lipid, that has been identified as astaxanthin. The parabasal body fits into the concavity of this pigment spot, which absorbs light and so shades it when the flagellate is oriented in such a way that the parabasal body is away from the source of light. The parabasal body has long been regarded as the light-sensitive element of the organelle, although there is neither direct physiological nor biochemical evidence for this assumption. The magnitude of the photochemical reaction believed to take place there in some way governs the direction of the beat of the flagellum, so that the protozoan's movement is always toward the light. This is of the greatest importance to the photosynthesizing strains and can be easily demonstrated by simple experiments. If, for example, a culture of one of them is put into a glass jar that has been covered, except for a small area, with some opaque material, in a very short time this translucent area will appear bright green because of the congregation of flagellates there. The green stripe, or pattern, will persist for some time even when the entire jar is uncovered and exposed uniformly to light. Euglenoids without the pigmented eyespot but with a paraflagellar body show similar positive phototaxis, but those without both do not. These behavioral differences provide indirect evidence for the role of the paraflagellar body in photoreception.

(2) Ocelli and eyes *(a) CHARACTERISTICS OF PHOTORECEPTOR CELLS* The ocelli and eyes of metazoan invertebrates are collections of photosensitive cells that are usually sunken below the rest of the epidermis in paired shallow depressions or in deeper cups. The photosensitive cells, which constitute the retina, may extend around the sides of the cup or be more or less completely localized at its base. Electron micrographs of the retinal cells of a considerable number of representative invertebrates have shown that the surface of the sensory cell is greatly increased by infoldings or outfoldings of the membrane as microvilli or lamellae that extend either inward into the cytoplasm or outward into the intercellular space. These may originate either from the cilium or the cell body of the sense cell. Ciliary processes are characteristic of the photoreceptors of cnidarians, chaetognaths, asteroids, and vertebrates. In cnidarians and asteroids, the microvilli extend outward in a tangled mass of fine processes; in cnidarians, from the surface of the shaft of the cilium; in asteroids only from its base. In ctenophores, there are four crescent-shaped groups of round bodies symmetrically placed around the floor of the apical organ. Each member of such a group in *Pleurobrachia pileus* is a lamellate body composed of the many outwardly extending microvilli of about 12 cilia in which there is, internally, the typical ciliary pattern of nine peripheral filaments. The anatomical resemblance between the cells that bear these lamellae and the photosensitive cells of other invertebrates with ciliary microvilli offers circumstantial evidence for photosensitivity in ctenophores. There is, however, as yet no proof of this and the fact that ctenophores migrate vertically upward and downward in the plankton according to diurnal light changes may be attributed to factors other than photosensitivity.

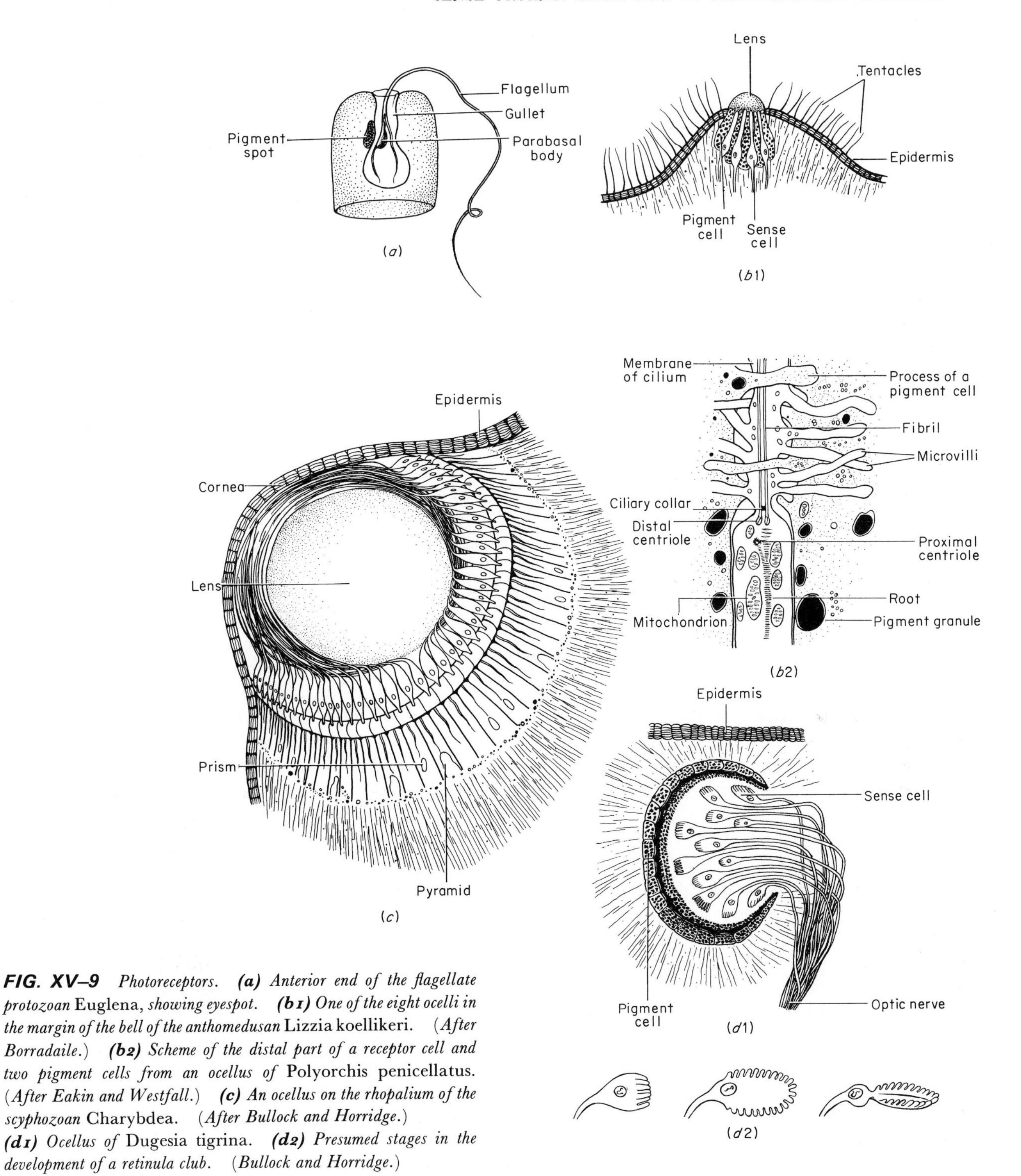

FIG. XV–9 *Photoreceptors.* ***(a)*** *Anterior end of the flagellate protozoan* Euglena, *showing eyespot.* ***(b1)*** *One of the eight ocelli in the margin of the bell of the anthomedusan* Lizzia koellikeri. *(After Borradaile.)* ***(b2)*** *Scheme of the distal part of a receptor cell and two pigment cells from an ocellus of* Polyorchis penicellatus. *(After Eakin and Westfall.)* ***(c)*** *An ocellus on the rhopalium of the scyphozoan* Charybdea. *(After Bullock and Horridge.)* ***(d1)*** *Ocellus of* Dugesia tigrina. ***(d2)*** *Presumed stages in the development of a retinula club. (Bullock and Horridge.)*

In chaetognaths and vertebrates the membrane is folded inward. In *Sagitta scrippsae* the microvilli form long slender tubules oriented parallel to the long axis of the shaft; these occupy the distal region below which the typical filamentar pattern is evident. There is, in this proximal region, also a unique body composed of granules and tiny rods whose function is undefined but which may be a light-trapping device. In vertebrates the infolded microvilli form closely packed disks in the distal segments of both rods and cones, at right angles to their long axes; this region is connected to the proximal region by a ciliary segment.

Extensions of the membrane originating from the cell body, and not from the cilium, are found in flatworms, rotifers, annelids, molluscs, and arthropods. In flatworms, cephalopod molluscs, and arthropods, these are straight, lateral projections that lie parallel to each other, but in polychaete annelids, they are irregular and twisted together. In the flatworm *Dugesia* and in pulmonate gastropods, they are straight and originate only from the distal surface of the sensory cell. In the rotifer *Asplanchna brightwelli* the evaginations likewise originate from the distal surface of the cell, but are flattened and curved lamellae that form a rounded structure that fits into a concavity of the pigmented cell. In every case, the evaginations originating this way form a stack of fine tubes or plates, often tightly packed together, which constitute a specialized part of these sensory cells called the rhabdomere. In the eyes of some invertebrates—notably cephalopod molluscs and arthropods—the rhabdomeres of adjacent sensory cells are closely apposed and form a composite body, the rhabdom (Greek *rhabdos*, "rod").

Whatever their origin, these membranous surfaces are believed to be the site of the critical photochemical reaction, providing as they do extensive plane surfaces for the distribution of the photosensitive molecules. The abundance of mitochondria near these tubules suggests that intensive oxidative metabolism is associated with photoreception and transmission of the excitation. They also provide the most efficient arrangement for the absorption of photons, an arrangement that has a parallel in the construction of plant chloroplasts in which the molecules of chlorophyll, the photon-capturing pigment, are packed along the membranes of lamellae, and of mitochondria, where the cytochromes are localized on the cristae. The evolution of lamellate bodies such as these and the ones in the photoreceptors of animals, and the evolution and genetic control of photosensitive pigments raise some of the most provocative questions in respect to the physiological relationships between living things.

Some evolutionary and phylogenetic significance may also be attached to the derivation of these surfaces. On this basis, a distinction can be made between ciliary and rhabdomeric types of photoreceptors, and a review of their distribution in invertebrate phyla makes it seem evident that those in the deuterostome line have developed the ciliary type and those in the protostome line, the rhabdomeric. As usual, there are exceptions to such a generalization. The sabellid polychaete *Branchiomma vesiculosum* has, for example, at the tip of each of its branchial filaments an eye composed of some 40 to 80 receptors, separated from each other by masses of pigment and each covered distally by a lens cell (Fig. XV–12*e*). In the proximal cytoplasm of each sensory cell is a cavity filled with about 400 closely packed tubules. The presence of a corresponding number of typical ciliary basal bodies from which nine fibrils extend to each tubule indicates their ciliary origin and so a deviation from the rhabdomeric type of photoreceptor generally considered typical of annelids and animals in the annelid line of evolution.

(b) PHOTORECEPTORS OF CNIDARIANS

Among cnidarians, only medusae have localized photoreceptors, the polypoid individuals showing diffuse sensitivity to light. Ocelli are of sporadic occurrence among hydromedusae and are even less frequent among scyphomedusae. In the species that do have them, they range in complexity from patches of sensory and pigmented epithelium flush with the surface of the body to cup-like depressions lined with chromatophores and sensory cells on which light is concentrated by a lens. There is no definitive evidence that the simplest of these are indeed photoreceptors, but the construction of the cup ocelli strongly suggests their function as such. The ocelli of hydromedusae are located either on each of the tentacle bulbs or on the margin of the bell; in scyphomedusae, they are situated either singly or in groups in the rhopalia. In the hydromedusan *Polyorchis penicellatus*, they lie on the margin of the bell at the bases of the tentacles. They are evident as comparatively large depressions, which are lined with epithelium differentiated as tall chromatophores containing dark pigment that is presumably melanin and as sensory cells. These are rather less numerous than the melanocytes, and each has a long process originating from its distal border and a slender axon from its proximal border. Nine peripheral and two central fibrillae extend the length of the distal process, which has also many microvilli, twisted and tangled together, extending from the surface of its shaft [Fig. XV–9*b*(2)]. The pigment cells have also numerous microvilli, devoid of pigment, that intermingle with those of the distal process of the sensory cells. The similarity between the ultrastructure of the sensory cells in these ocelli and those of other photoreceptors is striking and suggests very strongly a similarity in their function.

In *Tiaropsis*, another hydromedusan, the pigment cells are gastrodermal in origin, and the sensory cells lie beneath them with their axons emerging through the opening of the cup, making the retina an inverted one comparable to that of the vertebrate eye. Each of the eight ocelli on the margin of the bell of *Lizzia koellikeri*, also a hydromedusan, is a patch of sensory and pigment-containing cells, covered externally with an elevation of the cuticle presumably functioning as a lens [Fig. XV–9*b*(1)]. The ocellus of the scyphomedusan *Charybdea* is as complex in construction as any among cnidarians. Each rhopalium has four simple ocelli, and according to species, one or two cup ocelli. Each of the latter is lined with a stratified layer of pigment cells and two kinds of sensory cells, one prismatic and the other pyramidal (Fig. XV–9*c*). There is a cellular biconvex lens and a vitreous mass in the concavity of the cup; this is covered externally by modified epidermis that acts as a cornea. The ocelli of hydromedusae are without corneas and the cup is open to the exterior.

(c) PHOTORECEPTORS OF PLATYHELMINTHS Most flatworms, except cestodes, have ocelli at some stage of their lives. These vary in complexity from simple eyespots, such as are found in rhabdocoeles, to deeply sunk cup ocelli with varying numbers of sensory cells embedded in the mesenchyme or in the optic ganglia, such as are found in turbellarians. Two or three eyespots, situated anteriorly, are found in adults of some species of ectoparasitic trematodes and often in larvae of both endo- and ectoparasitic species. These consist of a single sensory cell embedded in a cup of pigment cells. The absence of ocelli in cestodes is correlated with their strictly endoparasitic existence.

In triclad turbellarians there is usually one pair of ocelli, but in polyclads there are always more than six pairs, and in some species many more than this. They are especially numerous in land planarians, where more than 1,000 have been counted in a single individual. The most simply constructed turbellarian ocelli have one to four sensory cells whose distal portions are sunk in a cup of pigment cells and whose proximal processes pass through the opening of the cup as the optic nerve that enters the cerebral ganglion [Fig. XV–9*d*(1)]. This is, therefore, an inverted retina, like that of *Tiaropsis*. With light microscopy the distal regions of the sensory cells appear as terminal knobs with striated borders; electron microscopy has shown the striae are in actuality lamellae running parallel to the long axis of the knob and exposing some 16 to 20 membrane surfaces per micron. Little is known of the light-sensitive substance presumably built into them. Spectrophotometric examination of extracts from the eyes reveal two absorption bands, one at 510 nm, which may be due to rhodopsin, and another in the near ultraviolet at 370 nm. The more complex terminations of the sensory cells (retinal clubs) of planarians may be derived from these terminal knobs. Retinal clubs are ovoid extensions of the sensory cells in which the striae run in two rows, oriented so that their long axes are at right angles to the long axis of the club, as if the knob of the more primitive cell had become extended and its tip invaginated [Fig. XV–9*d*(2)]. These retinal clubs of land planarians' ocelli extend through the pigment layer, so that the retina is an erect, not an inverted, one. No planarian ocellus has a lens, and the construction of the organ is such that light from only one direction can reach the photosensitive cells, passing through the transparent epidermis without concentration or focusing.

Experiments have shown that flatworms without ocelli will respond to light, but that animals with paired photoreceptors orient themselves so that both are equally stimulated. Planarians whose ocelli have been destroyed exhibit random movements when illuminated, which increase as the intensity of the light is increased. Recordings from microelectrodes inserted into the ocelli of *Dugesia* show that there are negative potential changes in response to illumination. It, therefore, seems apparent that flatworms with intact ocelli can distinguish both the intensity and the direction of a light source. *Dendrocoelum lacteum*, when exposed to uniform illumination from above, makes random, nondirectional movements until by chance it gets into the area of dimmest light, where its activity is reduced and so it stays there. When specimens are kept in darkness for as long as 18 hr, the terminal knobs of their sensory cells swell markedly. Examination of the fine structure of these swollen knobs reveals a tremendous increase in the number of large vacuoles in the axial cytoplasm, as well as an increase in the volume of the microvilli. This distention is evident all along their length but is often greatest at the tips which become expanded into tiny bulbs. When worms so dark-adapted are exposed to the diffuse sunlight of a summer day, the swelling goes down and the knobs become ovoid again as the vacuoles are reduced in number and the microvilli become slender and elongate. Ocelli of specimens kept in continual darkness deteriorate, and may even degenerate entirely. A proposed explanation for these changes in darkness and in light is that a photosensitive substance is continually produced in the axial cytoplasm of the sensory cells and stored in vacuoles, from which it is released to the membranes of the microvilli where it is broken down in light. Under ordinary conditions of the flatworm's life, production and disintegration are balanced, but in prolonged periods of darkness production exceeds disintegration and the substance accumulates in vacuoles. Degradation and destruction of the ocellus may be due to cytotoxic agents released when the photochemical reaction takes place. Again, under ordinary circumstances, the amount of these may be too small to produce other than an excitatory effect, but in an ocellus in which the photosensitive substance has accumulated in darkness and then undergone rapid disintegration on reexposure to light or in one in which it has continually accumulated beyond the possibility of storage, the end products or unused precursors may produce destructive effects.

(d) PHOTORECEPTORS IN NEMERTEANS, "ASCHELMINTHS," AND ECHINODERMS

Most haplonemerteans and heteronemerteans, and a few paleonemerteans, have two to several hundred inverted-pigment cup ocelli like those of planarians. In some species, each has a lens. The ocelli are located in the dermis, the musculature, or the cephalic ganglia.

"Aschelminths" show general dermal sensitivity, but some rotifers and nematodes have ocelli. Those in rotifers have been described as inverted pigment cups with sense cells, but there is some doubt if they are actually so structured. It is possible that there is no distinction between sensory and pigment cell or that the organ may be a syncytium. A few freshwater and free-living species of nematodes have a pair of ocelli sunk in pits on the sides of the pharynx. Each pit opens to the exterior, and at its base is a cuticular lens embedded in a mass of pigment. In the marine species *Leptostomatum* the lens is formed from the external covering of the pharynx and the ocellus is completely enclosed in its walls. Nothing is known of the innervation or the physiology of these structures, and their function as photoreceptors is still problematical.

Ocelli lie at the tips of the arms in many species of starfish, although they, like other echinoderms, show diffuse dermal sensitivity. Each ocellus is a cup-like depression lined with pigment and sensory cells, and without a lens. The sensory cells, as they have been revealed by electron microscopy in the ocelli of *Henricia leviuscula*, *Leptasterias pusilla*, and *Patiria miniata*, are columnar, with bulbous processes at their distal ends that extend into the lumen of the cup beyond the distal margins of the pigment cells. Numerous microvilli that are greatly interwoven originate from the proximal ends of the processes. Each of these has the filamentous structure of a cilium, with basal body, basal plate, and rootlets. In *Henricia* there are nine peripheral fibrils and no central one; in *Leptasterias* there are eight peripheral and one central. Starfishes in which the ocelli have been destroyed still move toward a light, but as they near it their orientation toward it is less accurate than that of intact animals. If lights from two sources shine upon an intact animal, its course is toward one or the other, either at once or after a short intermediate passage between the two, indicating the dominance of the arm nearest the light in directing the path of the animal.

(e) PHOTORECEPTORS IN MOLLUSCS

Photoreceptor organs in molluscs run a gamut of complexity, from simple shallow depressions as those in the limpet *Patella*, through deep open cups and closed vesicles, to the highly complex pinhole camera eye of cephalopods. They are as varied in their locations as they are in their construction. They are absent in monoplacophorans and aplacophorans, but in polyplacophorans they may be found on the dorsal surface of the shell plates. Presumably, they have been derived from bodies known as aesthetes, clusters of sensory cells below a thickening of the cuticle. These are arranged more or less regularly in rows and are most numerous near the growing points of the shell plates. In *Callochiton*, pigment cups have developed in the interior of some of the aesthetes, and the cuticular cap has formed a lens. These ocelli are small in comparison with the larger and more complex ones of genera such as *Acanthopleura* and *Tonicia*. In these the pigment is outside the aesthete and the lens is formed independently of the cuticular cap. Beneath it is a vitreous layer surrounded by the sensory cells. It is because of their gross structure that these organs are considered ocelli and believed to be photosensitive, although there is no confirmatory evidence for this from studies of either their fine structure or their electrophysiology. Behavioral observations of chitons show that they are light sensitive not only in the regions of the aesthetes but over the entire exposed dermal surface.

Most gastropods have paired cup-like ocelli or well-developed vesicular eyes. In prosobranchs these are elevated on stalks or ommatophores which originate near the bases of the cephalic tentacles. The ommatophores are retractile and capable of bending in one direction or another so that their photoreceptor organs can be either withdrawn or brought to converge upon an object. The burrowing prosobranch *Terebellum terebellum*, which has no tentacles, has been seen to protrude its ommatophores alternately while burrowing in soft sand so that the eye at the tip of each is just above the surface. The protruded ommatophore remains extended until the siphon, which the snail uses as a plough, breaks surface. At this point in the burrowing operations, the other ommatophore is extended and the first withdrawn, to emerge again when the next advance has been made and the siphon again breaks surface further along the shallow burrow. In opisthobranchs the ocelli, or eyes, usually lie beneath the epidermis and on the midventral surface of the cerebral ganglion; many species in this subclass have secondary photoreceptors in various regions of their bodies. In pulmonates the eyes lie either at the bases of the single pair of noninvaginable tentacles (Bassommatophora) or at the tips of the posterior of the two pairs of invaginable tentacles (Stylommatophora) into which they can be withdrawn when the tentacles are inverted (Fig. XV–11*a*).

In prosobranchs like *Patella*, *Haliotis*, and *Fisurella*, the ocellus is a shallow cup or pit, lined with pigment and sensory cells and open to the exterior. In *Patella* the cavity of the pit is filled with seawater, but in *Haliotis* it is filled with some secreted material that may act as a lens (Fig. XV–10*a* and *b*). In the top shell *Trochus*, the aperture of the cup is much smaller than it is in other species with open ocelli; and in other prosobranchs, the cup is completely closed over and sunk below the surface as an optic vesicle. In *Littorina littorea*, whose eyes may be taken as models of the closed vesicle type in the prosobranchs, the vesicle is ovoid and so placed on the outer side of each cephalic tentacle that its long axis is at a right angle to the long axis of the tentacle. The vesicle, perfectly visible through the transparent integument, is lined with pigment and sensory cells and the distal part of its concavity is filled with a spherical lens that is noncellular, homogeneous, and translucent, behind which is a vitreous body, probably more supportive than refractive in function (Fig. XV–10*c* and *d*). The vesicle is surrounded by a large sinus of the hemocoele, except for a strand of tissue through which the axons of the sensory cells pass to make contact with more proximally situated nerve cells and except for the anterior region where it is in contact with the integument. This covering of the vesicle has two layers, an outer conjunctiva made of transparent columnar epithelial cells and an inner cornea of transparent collagen fibers, whose inner ends

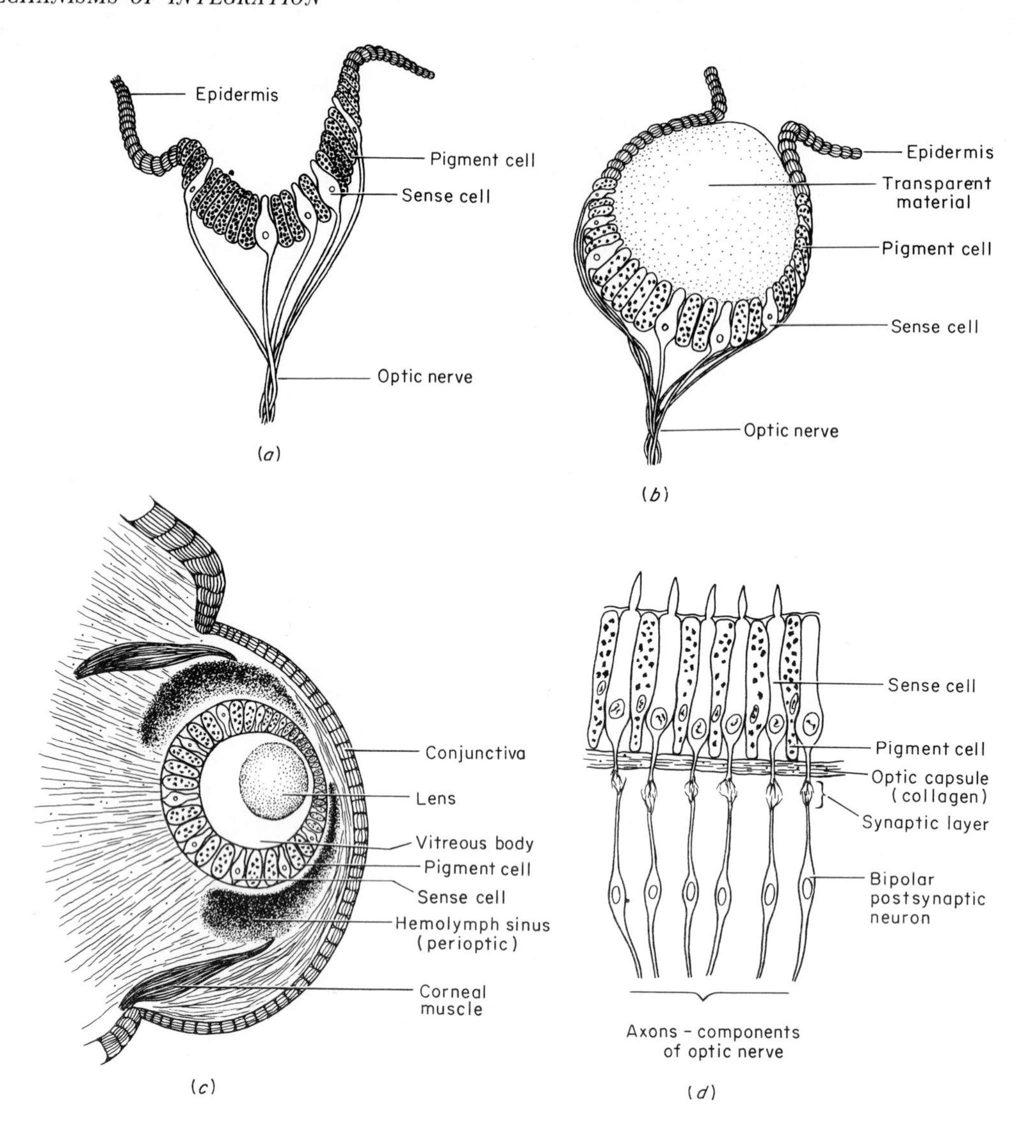

FIG. XV–10 *Molluscan photoreceptors.* ***(a)*** *Section through an ocellus of* Patella. (*After Parker and Haswell, vol. I.*) ***(b)*** *Section through the eye of* Haliotis. (*After Bullock and Horridge.*) ***(c)*** *Section through the eye of* Littorina littorea. (*After Newell.*) ***(d)*** *Vertical section of the retina of the eye of* Littorina. (*After Newell.*)

are attached to the cornealis muscles. These are a series of radial muscles around the periphery of the swelling on the tentacle made by the visual organ. The distal surface of the lens is separated from the cornea by a region of the wall of the vesicle in which the epithelial cells are transparent; this region constitutes the pupil, with a fixed diameter of 55 to 60 μ. The retina is an erect one, with the distal processes of the sensory cells facing the light rays, which enter, through the pupil, from above and in front of the head and are refracted by the lens. The visual cells, which are packed in among the supporting chromatophores along the sides and the base of the vesicle, are tall ones that become shorter as they approach the pupil, where there are only supporting cells without pigment. The distal end of a visual cell is elongated and cone-shaped and projects beyond the other cells into the vitreous body.

The most highly specialized gastropod eyes are those of heteropods, fast-moving pelagic prosobranchs

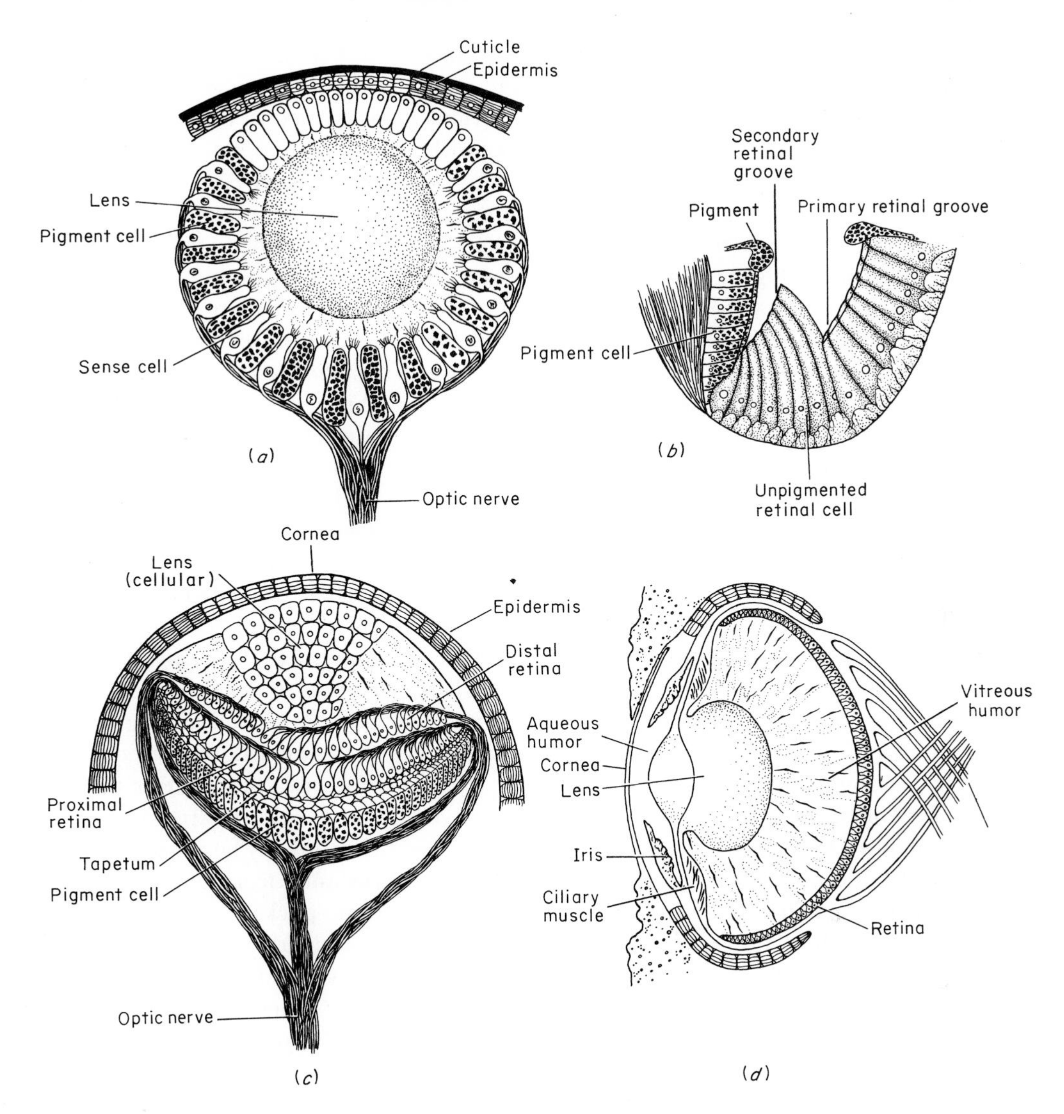

FIG. XV–11 *Molluscan photoreceptors.* ***(a)*** *Section through the eye of* Helix, *a stylommatophorous pulmonate with eyes on the tips of the tentacles. (After Bullock and Horridge.)* ***(b)*** *Section through the retina of the eye of the heteropod* Carinaria mediterranea. *(After Wilbur and Young, vol. II.)* ***(c)*** *Section through the eye of* Pecten. *(After Bullock and Horridge.)* ***(d)*** *Section through the eye of* Octopus. *(After Wilbur and Young, vol. II.)*

that swim upside down. The eyes of *Carinaria mediterranea* are very large and have the form of slightly projecting tapered cylinders that can be moved by muscles. Beneath the convex cornea is a spherical lens; the wall of the optic vesicle is composed mainly of chromatophores, except for a relatively large area of the dorsal surface and the band of photosensitive cells at its base. These lie in a groove, divided into two secondary grooves, one dorsal and the other ventral in position (Fig. XV–11*b*). The cells in the dorsal groove become progressively taller toward the center of the vesicle, while those in the ventral groove are of uniform height. All of them have their distal surfaces directed toward the chromatophores. This "ladder" arrangement makes it possible for light reflected from an object to fall upon two surfaces and, therefore, to give the mollusc some information about its distance away. Light can enter the eye through the forwardly directed cornea and be concentrated by the lens, and also through the dorsal "window," allowing the snail to detect movement in the water below it, an obvious advantage to predators such as these that swim with their ventral surfaces uppermost.

Some bivalves have eyes of varying degrees of complexity on their siphons or on the edges of the mantle. The most highly developed are those of scallops. In *Pecten*, they lie in the middle fold of the mantle and are clearly visible, when the valves of the shell are open, as about 100 bright blue spots. The unique features of these eyes are the double retina and the tapetum, which is responsible for their intense blue color (Fig. XV–11*c*). Beneath the multicellular lens is a thin connective tissue septum, separating it from the distal retina. This is an inverted one, for the axons of the sensory cells pass between the cell bodies and the lens to form the distal ramus of the optic nerve. Below the distal is the proximal retina, also an inverted one. Each of these is wholly independent of the other, and no nervous connections between them have ever been demonstrated. Beneath the proximal retina is the tapetum, or argentea, composed of layers of minute plates of guanine separated from each other by thin sheets of cytoplasm. Because of the high refractive index of guanine, the tapetum is a light-reflecting layer, doubling the efficiency of the eye in that it enables the visual cells to be stimulated twice by the same light source—once by light falling directly upon them and again by light reflected from the tapetum. The color of the eyes, like that of the blue sky, is due also to this reflected light. In some species of *Pecten*, the eyes are equally distributed along the edge of the mantle, but in others, specifically *P. jacobaeus* and *P. maximus*, the left mantle lobe has more eyes than the right, and the left lobe of the visceral ganglion is larger than the right one.

Electrical recordings from the nerves of the distal and proximal retinas, separately, have shown that the proximal retina responds to illumination with a discharge of impulses, while the distal retina responds in this way only when the intensity of illumination is reduced or illumination is stopped entirely. This "off" response is dependent upon the previous "on" response of the sensory cells in the proximal retina and stops again when the eye is reilluminated. The organ functions in detection of movement, but gives no response to a stationary object. Only a single image is formed in this eye and that one is reflected from the tapetum upon the sensory cells of the distal retina. The lens does not form an image and none falls on the proximal retina.

Although the paired eyes of the most primitive cephalopods are simple in construction, those of modern decapods and octopods have reached a high degree of evolutionary development. The eyes of *Nautilus* are cups without lenses, open to the exterior through a small, and to some extent, adjustable aperture and borne on short stalks. The retina, bathed by seawater, is supported by chromatophores and the entire organ is little different from the eye of *Patella*. The eyes of squids, cuttlefish, and octopuses are, on the other hand, closed vesicles, each with a cornea separated from a spherical and movable lens by a chamber containing aqueous humor, an iris diaphragm that can alter the size of the pupil and hence control the amount of light admitted, and at the base of the cup, a retina (Fig. XV–11*d*). The cup is filled with fluid like the vitreous humor of the vertebrate eye. The distal segments of the sensory cells vary in length, but in general are unusually long; those in the eyes of *Octopus* may be 200 μ in length, about one-fiftieth of the distance from lens to retina. The sensory cells are close-packed; the maximum number in *Octopus* is estimated as some 70,000 per mm^2 and in *Loligo*, 50,000 per mm^2. Electron micrographs show that many microvilli originate from them, oriented at right angles to the long axis and forming rhabdomeres of closely appressed microtubules. These, in the retinas of *Octopus vulgaris* and *Sepia officinalis*, form rhabdomes analagous to those in the compound eyes of arthropods. In the cephalopods, the rhabdomeres of four radially arranged cells form a rhabdome. Between the rhabdomeres are processes of the chromatophores, in which the dark pigment granules migrate distally when the eye is illuminated and proximally when the animal is in the dark. The axons of the sensory cells cross over before they enter the optic ganglion, those from the ventral half of the retina passing dorsally and those from the dorsal half passing ventrally. Bundles of these nerves go through the choroid coat, a layer of connective tissue, blood vessels, and muscles that encloses the optic vesicle, and through apertures in the scleroid coat, a thin layer of firm, cartilage-like material that separates this from the surrounding tissues. Nerves from the optic ganglion run outward to the eye, innervating its complex musculature.

The paired eyes of cephalopods are situated laterally, just behind the tentacles. The eyeball can be moved by a system of external muscles, and the lens can be moved inward by contraction of the ciliary muscles and outward by that of muscles in the choroid coat. This movement of the lens has been generally accepted as a means of accommodation by which the animal could bring objects at different distances into focus. It is possible, however, that given the sensitivity of the animal's tentacles to objects close at hand and the optical conditions in seawater, where the intensity of light is low and there is scattering because of turbidity, a focusing mechanism is of little advantage. The effective use of the intrinsic eye muscles may be to prevent distortion of the eyeball when an animal suddenly and rapidly accelerates, as it often does in the capture of prey or in escape maneuvers. The iris can also be moved by its own set of antagonistic muscles, innervated by nerves originating in the central nervous system, and the pupil reduced or enlarged according to external light conditions. Bright light will cause reduction of the pupil to a narrow slit. Reaction of the two eyes is independent, for if a light is directed at the left eye, its iris alone moves to close the pupil while that of the right eye is unaffected, and vice versa. Conditions other than light also induce reactions in the iris muscles. In *Sepia*, the pupil is opened wide in sexual display; and in *Octopus* and *Eledone*, when the animals are startled by sudden movements close to them. In each event the eyes appear larger, possibly as attractive or warning signals.

The photosensitive substance in the eyes of cephalopods, presumably distributed along the membranes of the rhabdomeres, has been identified as a rhodopsin with properties very similar, or identical, to those of vertebrate rhodopsins. In these it has been established that the action of light—and the only action of light—is to convert the 11-*cis* isomer of retinaldehyde (retinal), the chromophore attached to the opsin characteristic of the species, to its all-*trans* configuration. Subsequent reactions are "dark" ones, leading to the separation of the retinal from the opsin. The *cis* isomer is believed to fit closely into a section of its opsin partner, an interlocking that is disturbed when it becomes the all-*trans* form. In consequence, the protein molecule progressively opens out, exposing new groupings and ultimately releasing the retinal, again as the 11-*cis* form of retinaldehyde. Somewhere in the course of these events bleaching of the chromophore occurs, as well as excitation of the sensory neuron and transduction of this excitation into the electrical events that are propagated along the axon and the nerves with which it makes connections. The released 11-*cis* retinal recombines with the opsin to form the chromophore-protein complex ready for another photochemical change, so that the process is a cyclic one and could, theoretically, continue indefinitely. There is direct evidence from experiments on the extracted substance that such a cycle goes on in the eyes of squids, and presumably of other cephalopods, and also in the eyes of the arthropods in which rhodopsin has been identified.

The cephalopod eye is capable of forming a clear image underwater, and these molluscs, like the predators in other classes, depend primarily on vision for the capture of prey and recognition of the presence of enemies. The eyes of *Octopus* are so placed that their visual fields overlap slightly, both in front of the animal and behind it, and an animal usually turns its head sideways to look at its prey, using only one eye at a time. Decapods usually use both eyes in aligning themselves with a victim, taking advantage of their binocular vision before they strike it. It is clear from behavioral studies that cephalopods can judge distance with a high degree of accuracy and, once prey is sighted, can bring themselves into the proper position to reach it with their tentacles. *Sepia*, for example, will back away from a prawn dangled close to its head until it reaches the correct distance for an effective strike. Even newly hatched cuttlefish, never before confronted with food, will do this, indicating that it is an inherent rather than a learned reaction. Observations upon octopuses kept in aquaria show that they attack only moving objects and will emerge from their shelters at one end of the tank and glide toward live crabs that have been dropped into the water at the opposite end. When an octopus gets 15 to 20 cm away from the crab, it gives a rapid jump forward and traps the prey under the expanded interbrachial web. Sometimes it pauses before the final jump, undergoes a rapid color change, and jerks its body about as if to startle the crab before its final pounce. An octopus is not so aroused by a stationary object, even in the shape of a crab, nor by a motionless shrimp, from which the covering sand has been blown away by jets from the octopus' funnel, until the shrimp begins movements to bury itself again. Training experiments with octopuses, in which an animal is rewarded for a "correct" response and punished by mild electric shocks for an "incorrect" one, have shown that they can discriminate between the size of objects of similar shapes, even if the areas or the outlines are of the same size. There is also indication that these molluscs can be trained to discriminate between colors and that they can distinguish between disks painted green, blue, yellow, or black, and also between lights of these different colors. The intensity of the light is a complicating factor in the interpretation of the results of these experiments, and there is neither histological nor cytological evidence of any differences in the distal segments of the sensory cells, which might be expected if differences in the wavelengths of light were to be detected. Nor has any biochemical difference been found in the pigments extracted from the retinas. Only a single form of rhodopsin, with a maximal absorption band at about 490 nm and another smaller one near 350 nm, has been extracted from cephalopod retinas; this is similar to the rhodopsin of the rods of vertebrate eyes. Color vision in these molluscs is thus still not definitive, although behavioral studies make it seem highly probable.

(f) PHOTORECEPTORS IN ANNELIDS AND ARTHROPODS Development of photoreceptor organs in the annelid-arthropod line of invertebrate evolution has followed a somewhat different course. Although, like echinoderms, annelids in general exhibit diffuse dermal sensitivity, some species, especially of polychaetes, have localized photoreceptor organs. These vary from aggregations of photoreceptor cells into ocelli, as in leeches, or the seemingly compound eyes of the tubiculous polychaete *Branchiomma*, to the camera-type eye of the pelagic polychaete *Alciope* (Fig. XV–12). In nereids there are two pairs of photoreceptor organs on the first body segment. These are cup-like depressions, covered by a cornea of two layers, the outer a thickened translucent cuticle and the inner a layer of large epithelial cells. Beneath this is a lens, made from the greatly folded interdigitating processes of chromatophores that, with sensory cells, form the walls of the cup. The sensory cells in *Nereis vexillosa* are tall columnar ones whose distal processes, as much as 12 μ long, extend to the inner surface of the lens. Microvilli, in disorderly arrangement, extend from the sides of the sensory cell, in which there is a basal body and striated rootlets but, in the fully developed structure, no trace of ciliary filaments. In young larvae of three segments, however, each photoreceptor cell has a distinct cilium at its distal end. The microvilli do not arise from this, for the eye is of the rhabdomeric type. The fate of the cilium has not been traced but presumably it is lost, leaving behind its basal body and rootlets. The paired eyes of *Alciope* lie at the ends of short processes extending from the first segment and can be moved by muscles and turned in different directions. The lens, moreover, can be retracted, affording some means of accommodation for focusing on near and distant objects. The retina is an erect one, with the axons of the photoreceptor cells passing directly into a ganglion. The axons of the second-order neurons in this ganglion form the optic nerve that runs to the cephalic ganglion.

Each of the two eyes in the onychophoran *Peripatus* is a closed vesicle, sunk just below the epidermis which is here modified as a cornea. The cavity is filled with a lens and the wall is composed of chromatophores and sensory cells. The retina is an erect one, and the axons of the sensory cells come together to form the optic nerve, without making connections with second-order neurons in an optic ganglion.

In *Limulus* and in arthropods, with the exception of the median eye in those crustaceans in which there is such an organ, the sensory cells are monopolar, with axons but no dendrites, and are arranged in a sheaf called a retinula. An exception is the eccentric cell in the retinula of *Limulus*, which has a dendrite. The number of cells in a retinula vary in different species and even from retinula to retinula in the same eye of a single individual. The rhabdomeres of the retinular cells form a rhabdome, a shaft in the center

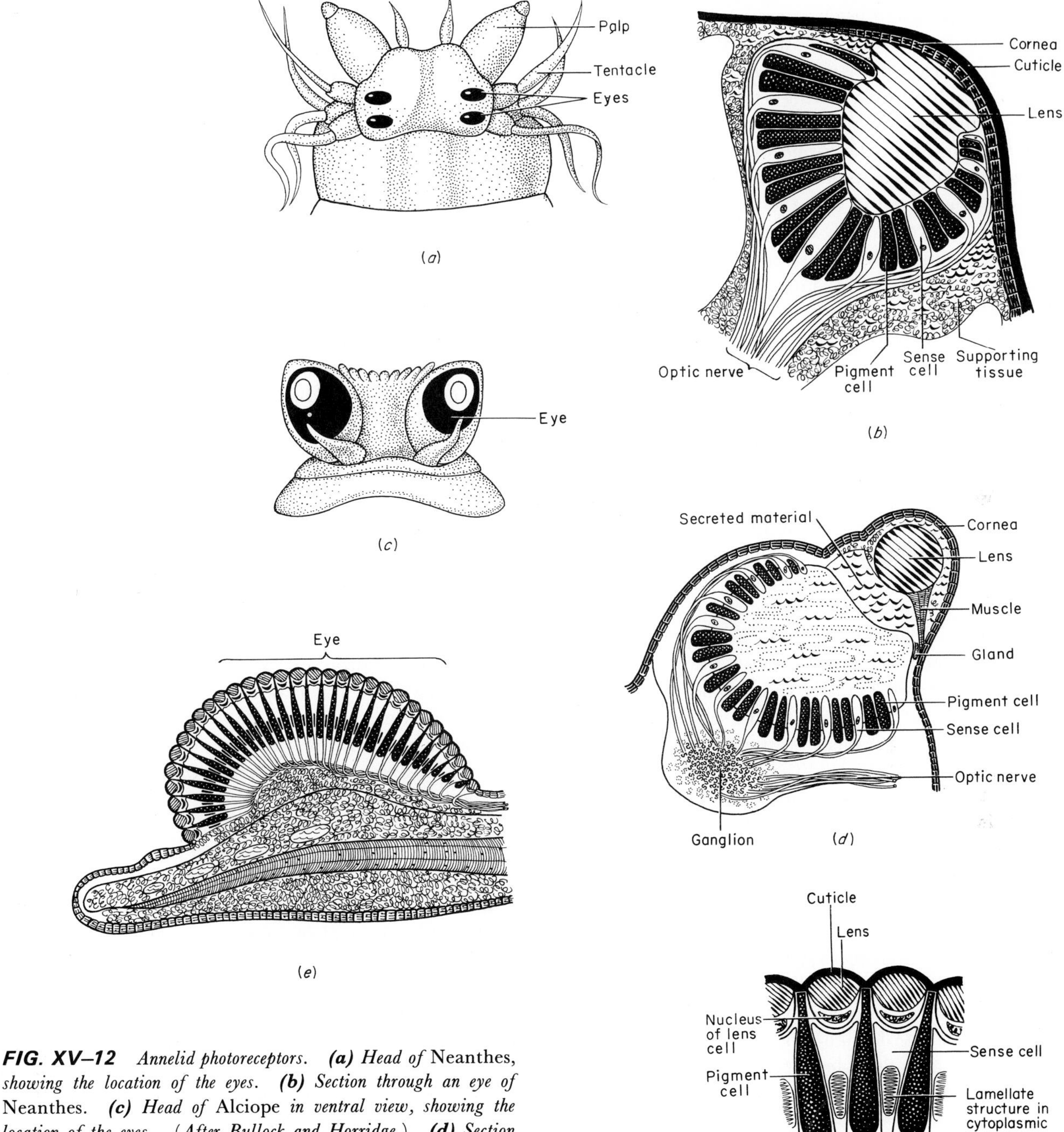

FIG. XV–12 *Annelid photoreceptors.* ***(a)*** *Head of* Neanthes, *showing the location of the eyes.* ***(b)*** *Section through an eye of* Neanthes. ***(c)*** *Head of* Alciope *in ventral view, showing the location of the eyes.* (*After Bullock and Horridge.*) ***(d)*** *Section through the eye of* Alciope. (*After Bullock and Horridge.*) ***(e)*** *Tip of a branchial filament of* Branchiomma vesiculosum, *showing the location of the eyes.* (*After Lawrence and Krasne.*) ***(f)*** *Detail of two visual units from the eye of* B. vesiculosum.

of the bundle. In *Limulus* there are 10 to 20 retinular cells, and one eccentric cell, in each bundle. Each of the two lateral eyes is made up of about 700 such bundles, each retinula having its own cornea and lens. Such a unit is known as an ommatidium (Fig. XV–13*a*). The ommatidia of the xiphosuran eye are distinctly separated from each other, with no pigment cells between them, and in this and other respects the eye differs so markedly from the compound eyes of crustaceans and insects that it seems most likely that they have evolved independently. The eccentric cell, unlike the other cells in the retinula, is bipolar, has a dendrite but no rhabdomere, and lies below the others and to one side of the central axis of the sheaf. The distal process (dendrite) of this cell runs along the rhabdome. The eccentric cell is not photosensitive, but excitation of the other cells in the retinula that are, but do not conduct, is transmitted to it and conveyed along its axon to cells in the optic nerve. Electrophysiological experiments show that the optic nerve cells are cross-connected and that when one cell is stimulated, another is proportionately inhibited. This serves to increase contrast in the image and is the principle upon which devices to produce sharper pictures on television screens and to improve radar reception are based. Horseshoe crabs have been using it for some millions of years.

Results of experiments designed to test the sensitivity of *Limulus's* eyes to different ranges of the spectrum show electrical responses to a wide range of wavelengths, including those in the ultraviolet and in the infrared. The eyes also respond electrically to polarized light, and since this is so, it may be that these animals, like some crustaceans and insects, take advantage of it in directing and guiding themselves. Although there is no direct evidence of this from behavioral studies, in theory all they would need to pursue a straight path from place to place would be a small patch of blue sky, for, although direct sunlight is not polarized, a considerable fraction of all the scattered and reflected light in the atmosphere is.

Limulus has also a pair of ocelli lying middorsally in the anterior region of its carapace. Each is a cup-shaped depression of monopolar sensory cells, above which the hypodermis is modified as a cuticular lens. These are photosensitive, for an animal will still respond to light when the more elaborate lateral eyes are completely covered.

The eyes of arthropods are either ocelli, small cups with a single cornea, or compound eyes, composed of many ommatidia set close together so that the surface of the eye looks like a fine mosaic. Compound eyes are found only in crustaceans and insects, often in conjunction with ocelli. In both crustaceans and insects the ocelli of larval or juvenile stages may be replaced by compound eyes in adults. There is a good deal of diversity in the details of the organization of the visual organs of chelicerate arthropods, but the median eye of a spider may be taken as illustrative of their general organization. Spiders typically have eight eyes, an anterior median pair on what is anatomically the first somite and three pairs on the second somite. In *Tegenaria* each median eye is a shallow cup whose cavity is filled with a lens separated from the underlying photosensitive cells by a layer of translucent epidermal cells. This has originated through apposition of the cells at the rim of the cup as invagination proceeded and their differentiation into "glassy" or vitreous cells. The retina is made up of sensory cells supported by glia cells and in these median eyes is an inverted one, although in those of other spiders and in the lateral eyes it may be an erect one (Fig. XV–13*b*). The rhabdomeres of two adjacent sensory cells form a rhabdome, with fine processes of chromatophores penetrating between the rhabdomeres. The rhabdomes of spiders, of *Xiphosura*, and of the centipede *Scutigera* are morphologically similar. Electron micrographs of the eyes of the spider *Lycosa* show deep invaginations of the membrane at the bases of the microvilli that form the rhabdomeres. These indentations end within the cell in many microvesicles or multivesicular bodies, giving a picture so reminiscent of conditions of a cell engaged in pinocytosis that it suggests the possibility of this phenomenon taking place in photoreceptors. In *Lycosa* also there is a tapetal layer from which the light is reflected back on the rhabdomes, which may number, in different species of this genus, from 100 to 4,500. Another refinement of the eyes of spiders of this genus are dorsal and ventral muscles originating in the wall of the head and inserting on the eye. These may be effective in regulating the focus by compression of the glassy cell

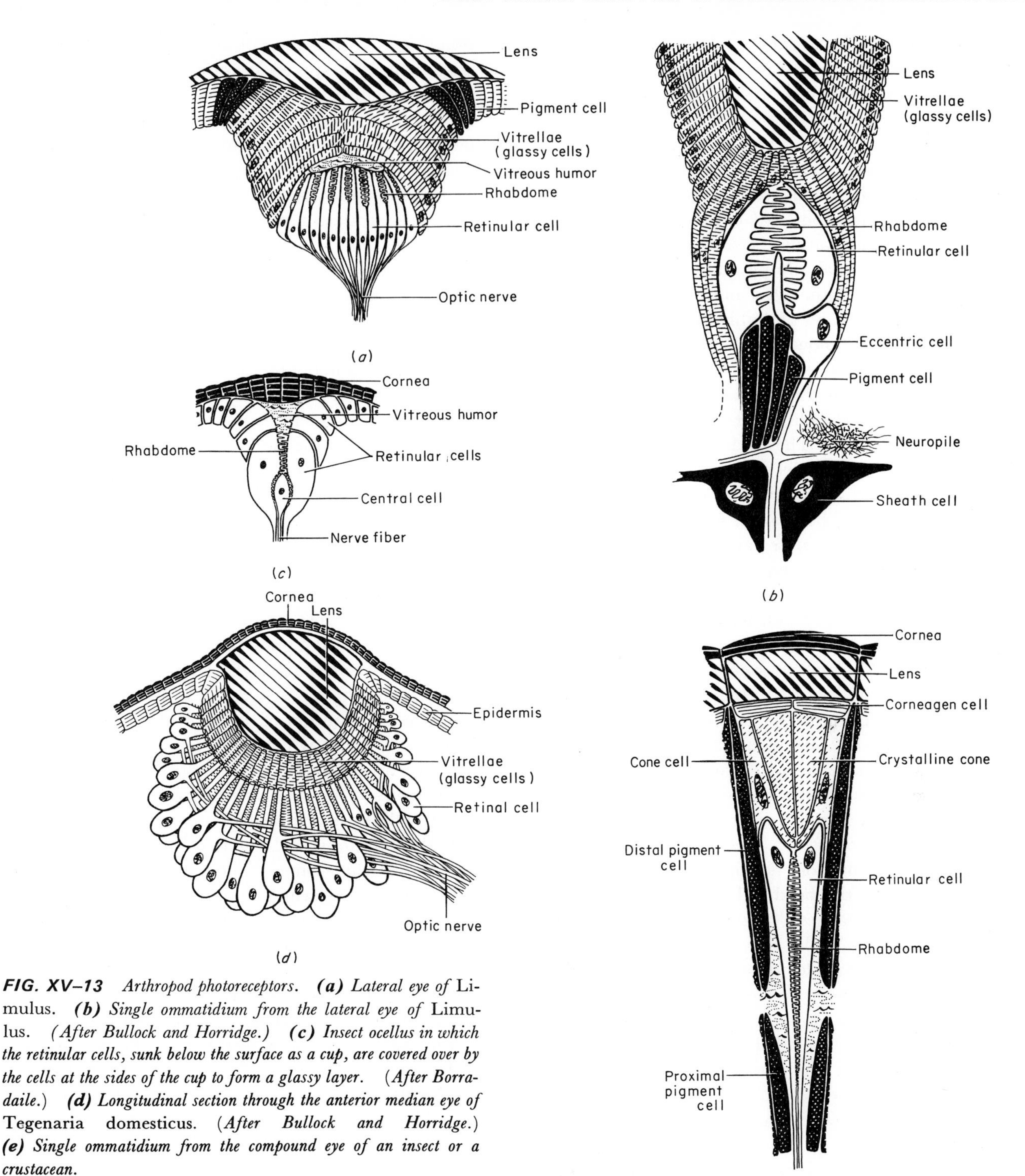

FIG. XV–13 *Arthropod photoreceptors.* ***(a)*** *Lateral eye of* Limulus. ***(b)*** *Single ommatidium from the lateral eye of* Limulus. *(After Bullock and Horridge.)* ***(c)*** *Insect ocellus in which the retinular cells, sunk below the surface as a cup, are covered over by the cells at the sides of the cup to form a glassy layer. (After Borradaile.)* ***(d)*** *Longitudinal section through the anterior median eye of* Tegenaria domesticus. *(After Bullock and Horridge.)* ***(e)*** *Single ommatidium from the compound eye of an insect or a crustacean.*

layer. In species of other genera, either the dorsal or the ventral muscles may be lacking and the remaining one be used to tilt the eye downward or upward, thus changing the direction of the lens in respect to an object. Such ocular musculature is unusual in arthropod eyes. The lateral eyes of spiders have no muscles, but are usually so positioned that they point in different directions and cover a wide visual field. The rapidity and accuracy with which spiders make responses to visual stimuli indicate that they have excellent vision, although if an image is formed it must be a mosaic one, and probably blurred.

Nothing is known of the spectral sensitivity of spiders' eyes, but the ocelli of females of the two-spotted spider mite *Tetranychus urticae* respond to wavelengths in the near ultraviolet and in the green of the visible spectrum. The independent variations in responses to these different wavelengths implies that there are separate receptor systems for them. Evidence from both behavioral and anatomical studies suggests that the anterior pair of eyes, which are sensitive both to near ultraviolet and to green light, are used as scanners of the visual field, while the posterior pair, sensitive only to ultraviolet, receive light from one direction alone.

In mandibulate arthropods the eyes may be single cups, each with several retinulae, as they are in the ocelli of insects, or groups of similar cups placed contiguously, as they are in myriapods, or they may be true compound eyes. In crustaceans like shrimps, lobsters, crayfish, and crabs, where the head is covered by the anterior edge of the carapace and the projecting rostrum, the compound eyes are borne on movable stalks originating from the sides of small lobes on the dorsal wall of the head. These eyestalks, although not true appendages, have two parts, or segments. The distal one is the movable portion, being well supplied with muscles that originate on the smaller proximal segment. By means of these, the eye at its tip can be turned in any direction and the eyestalk itself somewhat extended. In other mandibulates the eyes, whether compound or simple, are flush with the surface of the head and are directed toward an object only by movement of the head or the entire body of the animal.

There is a single median ocellus in the nauplius larva of crustaceans. This persists in the adults of many entomostracans but is vestigial in malacostracans. In most species it consists of three cups each with pigment cells and sensory neurons whose axons leave their distal surfaces, making the retina an inverted one. Each cup has its individual lens, but all are covered by a common cornea. Nauplius eyes are believed to be light-sensitive, but there is no definitive evidence of this, nor any information about their operation.

Insect ocelli are of two principal kinds, both in respect to position and to structure. Lateral ocelli, or stemmata, are typical of larvae; and dorsal ocelli, located on the frontal region of the head, are present in the adults of many species, together with the paired, sessile compound eyes. Lateral ocelli vary in complexity from pigment spots with refractive bodies to organs very like the dorsal ocelli. There may be a single ocellus on each side of a larva's head, or as many as six or seven of them may be grouped together. There is usually, but not invariably, a layer of glassy cells below the cornea and the lens; this is lacking in the ocelli of larvae of Coleoptera and of some Diptera. Beneath the glassy, or vitreous, layer is a retinula consisting of some 7 to 12 sensory cells and their rhabdomes. The axons leave the proximal ends of the sensory cells and pass inward, making the retina an erect one (Fig. XV–13*c*).

The distribution of dorsal ocelli is sporadic, but in the species in which they are present, there are usually three forming a triangle between the compound eyes. The basic structure is similar in all of them, with the sensory cells grouped together in units of two, three, or four, each cell in the unit contributing to the central rhabdome. The axons of these cells are very short and terminate just below the retina in association with dendrites of second-order neurons in the optic lobes of the brain. The axons of these second-order neurons run in the ventral nerve cord. In some species there is a layer of glassy cells between the retina and the cornea. The cornea itself may be flat on both surfaces, or thickened to be planoconvex or biconvex. There may be, as there is in the dragonfly *Sympetrum*, a tapetal layer below the retina.

The irregular distribution of dorsal ocelli among insect species raises a question as to their function. Although the dioptric system is such that an image could be formed, calculations show that it would fall some distance behind the retina and, therefore, be of no value to the insect's vision. The results of the electrophysiological studies that have been made are diverse and not easily resolved into a common meaning. They have, however, shown that the ocellus of the dragonfly is particularly sensitive to light, giving measurable responses when the illumination is as low as 1×10^{-5} footcandles (fc) and that the dorsal ocelli in the various insects tested are sensitive to the same range of wavelengths as are the compound eyes. In the cockroach and the dragonfly, continuous impulses along the ocellar nerve are recorded when the insects are in the dark; these stop when the ocelli are illuminated. This would suggest that the ocelli signal changes in the intensity of light and so may be of particular value in regulating the overt and metabolic activity of an insect through each 24-hr period. Their size and position on the head and the position and shape of the retina are such that they cover a wide visual field and so could be efficient and effective light gatherers. Some behavioral observations have shown that without its ocelli an insect's orientation toward light is disturbed. The blowfly *Calliphora*, for example, ordinarily moves toward a source of light; and first- and second-instar nymphs of *Locusta*, when adapted to darkness, move at right or acute angles to a light source, while those that are adapted to light walk straight toward it. In both insects phototactic reactions were reduced when the ocelli were painted over with an opaque lacquer. In *Calliphora* the disturbance was manifest only when the intensity of the light was high and in the nymphs of *Locusta* only when they were light-adapted. An explanation of the altered phototaxes could be that occlusion of the ocelli causes a decrease in the sensitivity to light of the anterior part of the compound eye, the ommatidia in this region being less readily excitable when the ocelli are no longer in operation.

The ocelli of myriapods are constructed on the same pattern as those of insect larvae and consist, like them, of a cornea, a cuticular lens, and a sunken retinula of the erect type. In some diplopods there are several such organs on each side of the head, well separated from each other, but in the Julidae, they are close together and covered by a common cornea. The inner surface of the cornea is thickened over each unit as a lens that projects into each cup, above the retinular cells, emphasizing the individuality of the parts of this unicorneal organ. Geophilomorph chilopods are usually eyeless, but the eyes of Lithobiomorpha, while unicorneal, have 1 to 20 units and those of Scutigeramorpha up to 200, each enclosed in a sheath of chromatophores. This gives the eye of a centipede like *Scutigera* the appearance of a compound one, although structurally it is far more like the ocelli of holometabolous insects than the compound eyes of adult crustaceans or insects. Virtually nothing is known of the functioning of these organs, although in general myriapods are negatively phototactic, and experimental observations have proved that *Lithobius* and *Julus* will walk directly toward a dark object, whereas they move in random fashion in diffuse light. Theoretically, the multiple eye of *Scutigera* should function like the apposition eyes of adult crustaceans or insects, and detect changes in light and shade and, therefore, objects moving in its visual field, without forming any image of them.

The surface of a truly compound eye appears faceted, each facet being the convex cornea of one of its pyramidal ommatidia. In crustaceans the facets are rectangular, in insects hexagonal, and they may vary in size even in a single eye. The number of facets, and hence the number of ommatidia, varies in different species from less than one hundred to many thousands. There are, for example, an estimated 20 to 25 in each eye of the isopod crustacean *Armadillidium*, about 3,000 in the stomatopod *Squilla*, some 14,000 in the lobster *Homarus*, and as many as 28,000 in the eye of a dragonfly. The same cells that secrete the cornea also secrete the thickened lens beneath it (Fig. XV–13*e*). These cells may remain in position below the lens after their secretory activities are over, as they do in crustaceans, or they may move out from beneath it, synthesize dark pigment, and become the corneal pigment cells, as they do in insects, except those of the most primitive orders. In most compound eyes a crystalline cone occupies the expanded distal region of the ommatidium, below the lens and above the retinula. This is the product of four "Semper" cells that form within their cytoplasm a body composed of ground substance and glycogen because of which their nuclei are displaced to the periphery. These intracellular bodies fuse to form a translucent rod, the homologue of the glassy layer of ocelli. The compound eye has, therefore, a duplex dioptric system, consisting of the corneal lens and this crystalline cone. Eyes with well-developed crystalline cones are designated as eucone, in distinction to acone and pseudocone types. In an acone eye no vitreous material is secreted and the cells, themselves transparent, occupy the position of the crystalline cone in the eucone eye. Such eyes are typical of certain insects, notably many Diptera, most Hemiptera and Dermaptera, and some Coleoptera. In pseudocone eyes, which are characteristic of many insects, a soft transparent mass or a liquid vesicle is actually secreted by the cells and deposited outside them.

This dioptric apparatus lies above the retinula, consisting most usually in crustaceans of 13 to 14 sensory neurons and in insects of 7 to 8. These are radially arranged around a central axis occupied by the rhabdome. In the majority of compound eyes that have been examined, one of these cells is shorter than the others, is possibly rudimentary, and is eccentrically placed. The number of retinular cells in the ommatidia of different species of crustaceans varies. In amphipods and branchiopods there are five, in the isopod *Oniscus*, seventeen, but in *Ligia* there are eight, seven of which constitute the retinula proper and the other an eccentric cell. In decapods there are seven retinular cells and one eccentric cell. This is the number and arrangement characteristic of most insects; but there are some species in which there are six tall retinular cells, a shorter one basal in position, and others in which the sensory cells are arranged in two courses, with four cells in the upper rank and three in the lower. In both classes, the retinular cells rest upon a basement membrane which has small openings through which their axons pass. These axons make connections with second-order neurons in the optic ganglia, some almost immediately upon entering it and others after passing a distance through it.

Each ommatidium contains chromatophores, usually a distal set surrounding the crystalline cone and known as iris cells and a proximal set surrounding the retinula and known as retinular pigment cells. Processes from these extend upward between the sensory cells. They are modified sheath cells containing a dark pigment, presumably melanin, with light-absorbing properties and, frequently, pale or colored pigments with reflecting properties. In most cases the pigment in the cells moves proximally or distally in response to different intensities of light. The retinular cells themselves may also contain similar pigments that exhibit similar movements. In some eyes there are cells containing reflecting pigment that form a tapetum below the basement membrane. Reflecting pigment is especially abundant in the eyes of crustaceans that habitually live in darkness or in dim light. The eyes of deep-water species of the pelagic shrimp *Sergestes* (*S. grandis* and *S. tenuiremis*) that live in the aphotic zone are, for example, especially rich in reflecting pigment. Those of *S. articus* that live in the photic zone have little or none, but large amounts of absorbing pigment is deposited in their retinular cells.

The retinular cells may lie immediately beneath the crystalline cone, and indeed in contact with it, or they may be separated from it by a space filled with crystalline material through which light may pass directly or be refracted at an angle to the longitudinal axis of the ommatidium. These relationships between cone and retinula make differences in the optical properties of the ommatidia in relation to each other and distinguish apposition from superposition eyes (Fig. XV–14). In an apposition eye, when each retinula is completely surrounded by light-absorbing pigment, the only rays of light from an object that can reach the photosensitive areas of the neurons are those that are parallel to its long axis; all others will be absorbed by the pigment. Photographic experiments have shown that the dioptric system is such that small erect images can be formed at the base of each crystalline cone. The number of these discrete images would be the same as the number of ommatidia through whose dioptric system light rays reflected from the object could pass, so that the composite would be a mosaic of bits and pieces of the object that would have no significant relation to its form. In a superposition eye, where the retinula is further removed from the crystalline cone and is not shielded by pigment, oblique as well as parallel rays of light can reach the sensory cells, so that a single retinula may receive light entering through as many as 30 adjacent corneas, lenses, and cones. Moreover, the image formed by the dioptric system of an ommatidium becomes larger the greater its distance from the base of the cone, so

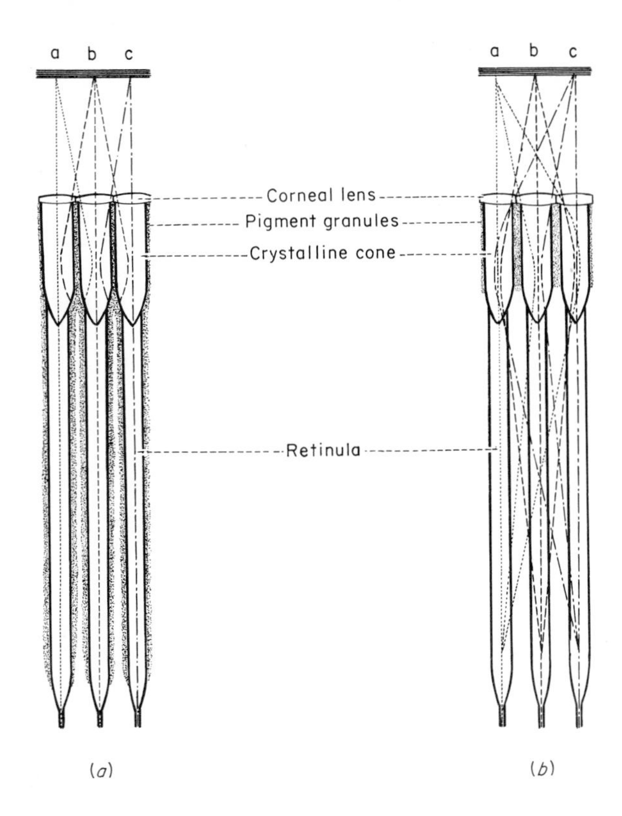

FIG. XV–14 *The course of light apposition and superposition images.* (*From Gardiner.*) ***(a)*** *Only rays of light from an object that can reach the retinula are those parallel to its long axis* (*broken lines*). *Oblique rays are absorbed by the pigment around each crystalline cone when that is fully dispersed in the chromatophores.* ***(b)*** *Oblique as well as parallel rays reach the retinula when the pigment is concentrated.*

that there may be overlapping of these images upon neighboring rhabdomes, with the image formed upon any one of them being blurred and confused, like pictures superimposed upon a single photographic plate.

Both types of eyes are found in crustaceans and insects. Superposition eyes are characteristic of most decapod crustaceans, but the hermit crab *Pagurus* has apposition eyes. Superposition eyes are also characteristic of insects that are active by night or in the dim light of dusk and dawn; apposition eyes are most frequent in diurnal insects. Yet the ordinarily superposition eye can become an apposition one by distal migration of pigment in the proximal chromatophores so that each retinula becomes as effectively isolated as in the typical apposition eye. This happens regularly in some nocturnal insects that are also active in the daytime and whose eyes can become light- or dark-adapted according to the intensity of the light to which they are exposed. In Lepidoptera, pigment movements are in direct response to changes in light intensity; in nocturnal species, these movements can be induced in 3 to 17 min by ultraviolet as well as by visible light. In the codling moth *Carpocapsa*, the movement begins $\frac{1}{2}$ to 1 hr before sunrise or sunset, regardless of the length of the day, and takes about 1 hr for completion. In many cases, migration of the pigment follows a rhythm that is independent of external conditions, as many other animal and plant processes are known to do.

In both crustaceans and insects there are many types of ommatidia intermediate between those just described, especially in relation to the distance of the crystalline cone from the rhabdome and, in crustaceans especially, in the nature of the rhabdome itself. Two types of rhabdome are recognized. In the open type the rhabdomeres are not in contact with each other, while in the closed type they are fused. In crustaceans, other variations in the rhabdome have been found. It may be atypical, either hypotrophied or hypertrophied, or it may be absent entirely. There seems to be little correlation between these differences and the type of vision of the animal, or with its habits of life. For example, deep-sea and burrowing crustaceans may have ommatidia in which the rhabdomes may be either hypo- or hypertrophied.

In general, compound eyes seem particularly suited to register changes in light and shade and so movements of objects rather than to form images. Only in the firefly *Lampyris* is there definite evidence of the formation of a superposition image. The visual range of a compound eye is related to its size, its position, and its movability. Each is approximately a hemisphere and each ommatidium, therefore, points in a direction different from that of the others. In crustaceans whose eyes are on movable stalks, the visual field can be increased and shifted by the turning of the stalk in one direction or another. In those whose eyes are sessile, and in insects, the visual range is determined by the size of the eye. An insect with large lateral eyes may see objects ahead of it, lateral to it, and behind it, covering two large, usually non-overlapping, visual fields. In insects like mantids, whose eyes are situated frontally, the visual areas of the median regions do overlap and to this extent the insect has binocular vision. Those of the lateral regions do not overlap and signal conditions independently of each other. The pigment in the median regions of the mantid eye is so distributed that they are of the apposition type, while the lateral parts are of superposition.

Only in insects is there any evidence of the ability of the compound eye to resolve patterns and discriminate between different forms. Bees that have been presented with bowls of water or a sugar solution placed on white cards with different black figures upon them will, after a time, associate a particular figure with food and almost unerringly fly to that card and light upon it, no matter what its position is in relation to the others, or whether there is a bowl upon it, or, if so, whether the bowl is empty or contains water or a sugar solution. It is not clear if the form has any meaning for the bee, as humans understand formal recognition, or if the distinction between the figures is based by the bee upon their dimensions and the extent of the areas where black and white are in sharp contrast.

The construction of ommatidia and, in some cases, the histologically demonstrable differences in the retinal cells, raises the question of division of labor among them. In the ommatidia of the isopod *Ligia*, for example, there is one eccentric cell and two of the other seven are considerably smaller than the other five. It may well be supposed that they serve different functions. The spectral sensitivity of the eyes of a number of crustacean and insect species has been tested by both behavioral and electrophysiological methods. Responses obtained by either means do not necessarily mean that the animal under investigation has color vision and recognizes colors as such, but rather that the visual unit is more readily excited by wavelengths in one range of the spectrum than by those in others, and so distinguishes between them.

Crustaceans are the only marine invertebrates in which color discrimination has been postulated, principally on the basis of behavioral observations. The hermit crab *Pagurus* will, for example, select specifically colored shells for its home, and amphipods and shrimps change color in response to backgrounds of different colors, even in illumination of constant intensity. The freshwater cladoceran *Daphnia* is negatively phototactic to short wavelengths and positively phototactic to long wavelengths, as has been shown by experiments using monochromatic light. This has been demonstrated in several species of copepods, as well as of cladocerans, whose habits of swimming downward in blue light and upward in yellow have been described as color dances. It is a habit that serves to keep them in the plankton, their food source, since aggregations of phytoplankton filter out the shorter light waves.

Behavioral observations of insects also give evidence of their color discrimination, for many of those that visit flowers for nectar select ones of particular hues. Insects, in general, are sensitive to the shorter waves in the spectrum. The visible spectrum of ants, for example, extends into the far ultraviolet, although the sensitivity of bees is in the near. Moths and some other insects will not fly to a piece of bright blue glass, but will fly to one that allows ultraviolet rays to pass through it. Most insects seem to be insensitive to red, at least to its deeper shades, but the moths *Pieris* and *Vanessa* visit deep red flowers in preference to those of other colors and will even fly to ones made of red paper. Bees have been trained to associate food with different colors, and from these experiments, it is apparent that they can only distinguish blue and yellow from others in the visible spectrum and that they also respond to wavelengths in the near ultraviolet which they recognize as color. The results of color-matching tests in which bees, trained to come to a selected color, were then exposed to mixed colors have led to the conclusion that in their eyes are receptors maximally sensitive to ultraviolet, other receptors maximally sensitive to blue-violet, and still other receptors maximally sensitive to green-violet.

More precise data on spectral sensitivity and the division of labor among retinular cells have been obtained by electrophysiological means. In these experiments, a recording electrode is inserted into the cornea or the retina of an eye and a reference electrode is inserted somewhere else in the body, from which the changes in potential produced in the retina when light strikes it can be measured and recorded as an electroretinogram. By this means it has been determined that the maximal sensitivity of the eye of the crayfish *Procambarus clarkii* is at wavelengths of 550 to 560 nm. In some insects it has been possible to obtain recordings from individual retinular cells. These, in worker bees, have confirmed the existence of at least two of the receptors postulated on the basis of color-matching tests, for maximal responses to green and ultraviolet were obtained from different cells in a single retinula. All the retinular cells in the eyes of drones have, on the other hand, the same spectral sensitivity with the peak in the blue region of the visible spectrum. Similar experiments have revealed that there are at least three different types of receptors in the single retinula of *Calliphora*, one maximally sensitive to green, one to blue, and one to yellow. The comparative magnitudes of the responses in each case are such as to suggest that five of the retinular cells are receptors for green, one for blue, and one for yellow. The eye of the cockroach *Periplaneta* has also more than one type of color receptor, for the dorsal area is more sensitive to ultraviolet than it is to blue-green, and the ventral area is less so. In dragonflies, likewise, the dorsal and ventral ommatidia are different, in structure as well as in spectral sensitivity. There are four retinular cells in each dorsal ommatidium, six in each of the ventral ommatidia. The different peaks of absorption evident in electroretinograms of the dorsal and ventral sets indicate that in *Libellula quadrimaculata* there is one type of photoreceptor in the dorsal ommatidia, which is sensitive to blue, and two in the ventral, one sensitive to green and the other to red. Similar studies with *L. luctuosa* suggest that in this species there are two types of photoreceptor in the dorsal ommatidia and three, possibly four, in the ventral.

Observations in the field, and under experimental conditions in the field and in the laboratory, have provided strong evidence that compound eyes are sensitive to polarized light and that the directional orientation of animals with them results from analysis of planes of polarized light. This has been shown in aquatic crustaceans, both entomostracans and malacostracans, and perhaps most dramatically in honeybees. Bees can distinguish different quadrants of the sky in which the plane of the polarization of sunlight differs, the polarization pattern being determined by the relationship of the direction of the patch of blue sky to the position of the sun. The orientation of a worker bee's communication dance on its return to the hive, by which it informs other workers of the location of a desirable foraging ground, depends upon such a patch of blue sky in its range of vision; the precision of the dance is upset if a piece of polaroid is held between the bee and the sky. The bees behave as if the sun were at right angles to the plane of polarization of the light; and in the laboratory under a polarizer, they orient at the same angle with respect to the plane of polarization as they do to corresponding quadrants of blue sky. There is also abundant evidence that other insects, crustaceans, and spiders make use of polarized light for their directional orientation, which raises the question of how they do this. Conceivably, the analysis could be made outside the eye by some external condition, or within it. If within it, the analyzer might be either in the dioptric system or in the retinula. The weight of present evidence is that it is in the retinula, and actually in the rhabdomeres, whose structure and presumed orientation of molecules could make this possible. This evidence has been derived from electrophysiological, polarization optical, and electron-microscopic studies. The physiological studies have shown that single retinular cells can discriminate the plane of polarized light; intracellular records from the eyes of the flies *Lucilia* and *Calliphora* have shown that the amplitude of the generator potential developed in them on exposure to polarized light depends upon its plane of polarization. Electron-microscopic studies have shown that in insect eyes the microvilli of the retinular cells have different orientations, and so the rhabdomere might be the analyzer. The answer to the question may not be so clear cut as these observations would seem to imply, for recordings made from the eyes of honeybees similar to those made from the eyes of flies have failed to show any differences in magnitude of the potentials developed in different planes of polarized light, or in nonpolarized light. Yet, whatever the experimental answer may prove to be, it is clearly evident that animals with compound eyes can detect differences in the plane of polarization of light, and act accordingly.

Photosensitive pigments have been isolated from a number of arthropod species with compound eyes. That in the eye of *Homarus americanus*, extracted from isolated rhabdomeres, has been identified as rhodopsin, with an absorption maximum at 515 nm, and with properties in all respects similar to those of cephalopod and vertebrate rhodopsins. Approximately nine-tenths of all the vitamin A in a lobster's body is concentrated in its eyes, and virtually all of this may be in the 11-*cis* form. There are also astaxanthin and small amounts of other colored carotenoids in their eyes, but there is no doubt that photoreception is mediated through rhodopsin and in the same way as in the vertebrate eye. Rhodopsins have also been extracted from the eyes of euphausiid shrimps, with absorption maxima at 462 nm, and from those of the crab *Callinectes hastatus*, with an absorption maximum at 480 nm. The situation in the insect eye is far less clear. Small amounts of retinal have been extracted from the heads of honeybees, but not specifically from their eyes. As these represent a very large proportion of the material in the head, it may be assumed that they are the source of the retinal. Some of this is bound to a protein and forms a photosensitive pigment with its absorption maximum at 440 nm. This is only one of the visual pigments presumably located in the compound eyes and ocelli of bees, but it is the only one of them yet known to contain retinal. The others have not been extracted and are recognized only through differences in spectral sensitivity as revealed by electroretinograms. Indirect evidence for a retinal-containing pigment in the eyes of *Musca domestica* comes from experiments in which flies were raised for at least 15 generations on a diet lacking carotenoids of any kind. Electroretinograms of adults of the later generations showed a significant decrease in the sensitivities of their eyes and ocelli to wavelengths both in the near ultraviolet and in the visible regions of the spectrum. This partial "blindness" could be prevented by the addition of beta carotene to the larva's food, and the fact that it did not become total, even in the fifteenth generation of flies reared on a carotenoid-deficient diet, can, perhaps, be attributed to the synthetic activities of microorganisms, which, in spite of aseptic precautions and procedures, might have survived the methods used for sterilization of the eggs or have reinfected the adults of each generation.

More specific data on the effects upon vision of a prolonged diet deficient in vitamin A have been obtained from histological studies of the eyes of adults of the tobacco hornworm *Maduca sexta*, a night-flying moth. Electrophysiological studies have shown that their maximum sensitivity is to wavelengths of 500 nm, suggesting that their photosensitive pigment is a rhodopsin. Observations of the vision of adults developed from larvae reared for several generations on a diet lacking vitamin A, and electroretinograms obtained from them, have revealed severe visual defects. Examination by light microscopy of excised retinas showed extensive histolysis of the epithelium as well as of the underlying nervous and connective tissues. Other aspects of the growth and development of the moths, and their reproductive capacities, were in no way impaired by absence of the vitamin, and the pathological conditions in the eyes of the parents could be corrected in the next generation if their larvae were fed a natural diet of tobacco leaves or a synthetic one supplemented with vitamin A. These results would seem to indicate a direct dependence upon an exogenous source of the vitamin for normal retinal development and vision in these insects, showing both their incapacity to synthesize it and its utilization as the carotenoid moiety of their photosensitive pigment.

Whatever their photosensitive pigments may prove to be and wherever they may be distributed in the body—whether diffusely in their tissues or localized in specific organs— the activities of invertebrate animals, like those of all other living systems, are profoundly influenced by light. In metazoans with other sense organs, the overt responses result from the combination of impulses received not only from photoreceptors but from one or more of the other sensory organs as well. Directional orientations, for example, are determined not only by visual stimuli but by mechanical ones from statocysts or other mechanoreceptors and certainly, in some cases, by chemoreceptors. Light is, however, an extremely important quality of the environment that affects all living systems and exacts from them numerous behavioral adjustments for their own protection and for their success in finding food, mates, and suitable dwelling places.

5. Thermoreception

Heat is another form of radiant energy to which all living things are exposed in varying degrees and to which they give definite behavioral responses. There is a great deal of indirect evidence for temperature sensitivity in invertebrates, but little is known about specific receptors even in the most highly evolved of them. In nature, species seek and find habitats in which the temperature is within a range optimal for them, for temperature affects all physiological processes and is one of the most variable of environmental conditions. If the temperature exceeds the upper limit of an optimal range, or falls below the lower one, and the animal has no means of escape or protection, it goes into a stupor and becomes immobile. It may recover if it is returned to its optimal range; otherwise, death ensues for a number of physiological reasons. Invertebrates vary widely in their temperature tolerances, for some live habitually in extreme cold or extreme heat, or can or cannot withstand sudden exposure to extremes for varying periods of time.

The temperature "sense" of an animal that leads it to seek the optimal range can be demonstrated by exposing it to a temperature gradient. This is most simply done by confining it, or several of its species at a time, in a tubular or circular container in which the temperature can be controlled so that it is either uniform throughout or warmer at one side or one end than at the other. The animals thus find themselves in a temperature gradient in which they are free to move until they find the temperature best suited to them, where they come to rest. In such experiments this is referred to as the "preferred" zone, without any anthropomorphic implications; it is usually in a fairly narrow range of temperature and not any fixed point on the scale.

Protozoans as well as metazoans at all levels of organization show behavioral responses to temperature differences. *Paramecium*, for example, in a tube in which the temperature is 19°C throughout distribute themselves uniformly along it, but if the temperature is different at the two ends, they congregate at the one where it is nearest the optimum for them, established as 24 to 28°C. Thus, if the temperature at one end is 38°C and it is 26°C at the other, they collect at the cooler end; but if at one end it is 25°C and at the other it is 10°C, they collect at the warmer end. This is accomplished through typical avoidance reactions, such as those they exhibit when in their random movements they encounter a solid object, a drop of oil or acid, or some similar obstacle or repellent. Their sensitivity is such that they will collect in their preferred zone in a tube 10 cm long with only a 3°C difference in temperature between the two ends.

The temperature senses of metazoan invertebrates have been tested in similar ways. Of particular interest are the reactions of ectoparasites that infest birds and mammals, since temperature must be an important factor in their selection of a host. The leech *Hirudo medicinalis* will move toward a tube of water warmed to 33 to 35°C and, in a gradient, will collect in a zone with a spread of less than 2°C. This assemblage is broken up when the temperature is made uniform throughout the container. Blood-sucking insects show similar responses to warm areas and have provided experimental material for at least the gross localization of their temperature receptors. This has been accomplished through amputation experiments. *Rhodnius prolixus* will move directly from a distance of 4 to 5 cm toward a warm tube, either 15°C above the ambient temperature or at 37°C. When the warm air from the tube first strikes it, the bug waves its antennae about and then points them toward the tube and, when it reaches it, crawls up on it, extends its proboscis, and explores the surface with its labium. Even with one antenna removed, the bug will react to the temperature gradient and respond by approaching the tube and crawling up it, but its directional orientation is disturbed when both antennae are amputated. From air temperature measurements it is evident that the temperature receptors on the antennae, which may be sensilla placodea, are sensitive to changes of less than 0.5°C. Similarly, intact *Aedes aegypti* will fly to a warm surface as readily as they will to human skin, but will not do so if their antennae have been amputated down to the last three segments. Their receptors seem sensitive to temperature differences of 0.5 to 1°C. The body louse *Pediculus humanus corporis* is less sensitive, but even after removal of its antennae will move toward a warm tube and follow it if it is moved about. It seems likely that their thermal receptors are distributed over the body as well as, probably, on the antennae. The sheep tick *Ixodes ricinus* will move toward an object 10 mm away that is heated 12°C above the temperature of the surrounding air. When first stimulated, the ticks hold their legs out in front of them in a "questing" attitude and keep them so as they move up the temperature gradient, sometimes becoming so aroused as they approach the warm object that they lose their footing and tumble over.

Experiments with other non-blood-sucking arthropods have shown similar responses of attraction to optimum temperatures and avoidance of those outside this range. Avoidance reactions are shown, for example, by the beetle *Dorcus parallelipipedus* to temperatures of 40 to 42°C. After removal of its maxillary palps, these reactions were not manifested until the upper limit of the temperature gradient was 42 to 44°C. After removal of the maxillary palps and the three terminal segments of the antennae, this limit was raised to 44 to 48°C, which was essentially that for specimens from which the maxillary palps and all segments of the antennae had been amputated. This has led to the conclusion that the specific receptors, if there are such, are located on the maxillary palps and the terminal segments of the antennae. In the beetle *Otiorhynchus ligustici*, the tarsi as well as the antennae seem to be the sites of the receptors; in the bug *Pyrrhocoris*, the antennae seem the principal sites but in *Lygaeus* they are elsewhere on the body as well as on the antennae. In the walkingstick insect *Carausius* they seem to be strictly localized on the dorsal surface of the fourteenth segment of the antennae of adults.

Since much of the behavior observed in these insects involves avoidance reactions away from unfavorable temperatures as well as positive orientations toward preferred ones, it is not possible to tell whether the receptors are sensitive to temperature as such or whether there is also an element of pain associated with a temperature above or below the preferred one. As there is no criterion by which pain, or degrees of pain, can be assessed in invertebrates there is no means of determining, except by subjective interpretation of the animal's behavior, whether the receptors concerned are strictly thermoreceptors or nociceptors as well (Latin *nocere*, "to hurt").

Difficulties in technique have limited electrophysiological experiments that might help to resolve this question, and also that of the strictly thermal specificity of the receptors as well as their precise localization. Recordings from the gnathobase nerves of *Xiphosura* have shown responses to a sudden rise in temperature of 3 to 5°C, and those from nerves in amputated legs of *Periplaneta americana* showed responses when the tarsi were warmed above 30°C or cooled below 10 to 13°C. The sensitive regions were found to be located in the first, second, and third tarsal segments and are probably cold receptors, since the activity of the nerves was greater at temperatures below 10°C than at higher ones and increased as the temperature was progressively reduced. The receptors proved sensitive to a drop of 1°C over the critical ranges below 13°C. A similar pattern of response has been reported for nerves from the first pereiopod of *Cambarus*. More precise recordings are those from the chemoreceptor hairs of blowflies, but these cannot be interpreted as thermal responses per se but rather as modulations of the responses of the salt receptors. In some of these cells the frequency of discharge was decreased when the temperature around the cell body was elevated 0.5°C or more and increased when it was reduced to the same extent. This did not occur when solutions applied to the tip of the chemosensory hair were correspondingly warmed or cooled, but only when the temperature change directly affected the region of the neuron where the impulses were initiated. In other experiments it has been found that the response increased with increase in the ambient temperature and that warming and cooling the tip of the sensillum, while recordings were being made from its side, raised and lowered, respectively, the frequency of the "spontaneous" action potentials along the axon. Moreover, the response on stimulation with appropriate salts increased with increase in humidity in the test chamber, while the temperature was held at a constant level. In no case yet has a specific thermoreceptor been identified in either an aquatic or a terrestrial invertebrate, in spite of abundant evidence of their thermal sensitivity.

The effects of temperature are so manifold that it is very likely that in a temperature gradient other receptors, especially those involved in the directional orientation of an animal, are made more or less sensitive to the specific stimuli ascribed to them. This might apply particularly to chemoreceptors, although the rate at which events in the excitation of any sensory cell and transduction of the energy absorbed take place at preferred temperatures may well be selectively altered at suboptimal and supraoptimal ones. It is likely, too, that in small animals, at least, the effect of temperature may be directly upon the central nervous system or that in any animal of any size it may be mediated through free nerve endings in the integument, rather than through any specific receptor or receptor organ.

6. Hygroreception

Humidity is another environmental condition to which all organisms are exposed and one that is of particular importance to terrestrial animals which give appropriate responses to the water content of the atmosphere around them. These responses may, however, be general ones to their degree of hydration or dehydration up to the limits of their tolerance before complete and lethal desiccation, rather than responses mediated by specific hygroreceptors. Movements of animals in humidity gradients have been tested and found, in some at least, to depend upon the physiological state of the animal in respect to its water content. According to this, they move in such a gradient, as they do in a temperature gradient, to a preferred zone. Earthworms, for example, sample the terrain with their prostomiums and turn away from dry soil. This avoidance probably results from stimulation of free nerve endings in the prostomium through loss of water. Some terrestrial arthropods prefer moist atmospheres and some, dry ones, and their aggregation in the preferred region seems to result more from the extent of their activity in the less favorable zones of a gradient than to any directed orientation toward the favorable one. The isopods *Porcellio scaber, Oniscus asellus*, and *Armadillidium vulgare*, which require moisture for their respiration among other needs, are very active in dry air and in their random movements tend to get out of it and, if they can, into a moister atmosphere where they are far less active and so tend to stay there. The centipede *Scutigerella* behaves in the same way. Conversely, adults of *Tenebrio molitor* are quieter in atmospheres of low relative humidity than they are in those of high and will collect in the drier zone of a humidity gradient. They are actually motionless four-fifths of the time in the drier zone of a gradient of 94 to 100 percent relative humidity. Humidity receptors are believed to be the sensilla basiconica on their antennae, as are those of the flour beetles *Tribolium confusum* and *T. castaneum*. The humidity reactions of these beetles can be correlated with the distribution of these sensilla. Removal of eight of the eleven segments from the antennae of *Tenebrio* adults results in loss of their responses to humidity, indicating the localization of the humidity receptors in these terminal segments. In *Tribolium* they seem to be localized on segments 7 to 11; with only one or two of these sensilla basiconica remaining, the beetles can still distinguish between 0 and 100 percent humidity. Intact flour beetles that have reached an appropriate degree of dehydration can locate a moistened piece of filter paper and move almost directly toward it, but they follow a circular path if one antenna has been removed. Sensilla basiconica may be hygroreceptors in other beetles as well, but there is less proof for this than in these two genera. Sensilla in other insects that have been tentatively designated as hygroreceptors are tuft sensilla on the ventral surface of the thoracic segments of housefly larvae and the guard hairs of spiracles of the tsetse fly. One of the neurons in the labellar chemoreceptors of *Phormia* has been shown by electrophysiological experiments to be sensitive to water.

The mechanism of the reception of humidity is not known, but an explanation has been offered for the operation, in respect to it, of the hairs recognized as humidity receptors in the anterior pit of Haller's organ in *Ixodes reduvius*. Possibly uptake or loss of water alters the curvature of these hairs, thereby stretching and stimulating the associated neurons. Although there is no observational or experimental evidence for this, it does remain an attractive possibility, and its validity could be put to observational test.

In scorpions there are numerous sensilla on the pectines, comb-like organs peculiar to them that lie on the sternal plate of the ninth body segment. On the ventral face of each tooth of a pectine, there are a series of small elevations, each a hollow cavity containing the terminal processes of many sensory cells. These structures on the pectines have been shown to be sensitive to the texture of the substrate on which a scorpion is put and may be humidity receptors.

SELECTED BIBLIOGRAPHY

General

Batham, E. J., C. F. A. Pantin, and E. A. Robson: The Nerve Net of the Sea Anemone *Metridium senile* (L): The Mesenteries and Column, *Q. J. Microsc. Sci.*, vol. 101, pp. 487–510, 1960.

Dethier, V. G.: "The Physiology of Insect Senses," Methuen and Co., London, 1963.

Kennedy, D.: The Initiation of Impulses in Receptors, *Am. Zool.*, vol. 2, pp. 27–44, 1962.

Vinnikov, J. A.: Principles of Structural, Chemical and Functional Organization of Sensory Receptors, *Cold Spring Harbor Symp. Quant. Biol.*, vol. 30, pp. 293–299, 1965.

Mechanoreception

Alexander, C. G.: Structure and Properties of the Mechanoreceptors in the Pereiopods of *Ligia oceanica*, *Comp. Biochem. Physiol.*, vol. 29, pp. 1197–1205, 1969.

Alexander, R. D.: Acoustical Communication in Arthropods, *Annu. Rev. Entomol.*, vol. 12, pp. 495–526, 1967.

Alexandrowicz, J. S.: Receptor Elements in the Thoracic Muscles of *Homarus vulgaris* and *Palinurus vulgaris*, *Q. J. Microsc. Sci.*, vol. 93, pp. 315–346, 1952.

Alexandrowicz, J. S.: Muscle Receptor Organs in the Paguridae, *J. Mar. Biol. Assoc. U.K.*, vol. 31, pp. 277–287, 1952.

Alexandrowicz, J. S.: Receptor Organs in Thoracic and Abdominal Muscles of Crustacea, *Biol. Revs.*, vol. 42, pp. 288–326, 1967.

Buddenbrock, W. von: Die Statocysten von *Pecten*; ihre Histologie und Physiologie, Zool. Jahresbericht, *Abt. Allg. Zool. Physiol Tierernaehr.*, vol. 35, pp. 301–356, 1915.

Burkhardt, D., and M. Gewecke: Mechanoreception in Arthropods: The Chain of Stimulus to Behavioral Pattern, *Cold Spring Harbor Symp. Quant. Biol.*, vol. 30, pp. 604–614, 1965.

Cohen, M. J.: The Crustacean Myochordotonal Organ as a Proprioceptive System, *Comp. Biochem. Physiol.*, vol. 8, pp. 223–244, 1963.

Esch, H., I. Esch, and W. Kerr: Sound: An Element Common to Communication of Stingless Bees and to Dances of the Honey Bee, *Science*, vol. 155, pp. 320–321, 1965.

Gelperin, A.: Stretch Receptors in the Foregut of the Blowfly, *Science*, vol. 157, pp. 208–210, 1967.

Gettrup, E.: Sensory Mechanisms in Locomotion: The Campaniform Sensilla of the Insect Wing and Their Function during Flight, *Cold Spring Harbor Symp. Quant. Biol.*, vol. 30, pp. 615–622, 1965.

Gorner, P.: A Proposed Transducing Mechanism for a Multiply-innervated Mechanoreceptor (Trichobothrium) in Spiders, *Cold Spring Harbor Symp. Quant. Biol.*, vol. 30, pp. 69–73, 1965.

Horridge, G. A.: Non-motile Sensory Cilia and Neuromuscular Functions in a Ctenophore Independent Effector Organ, *Proc. R. Soc.* (*London*), *B*, vol. 162, pp. 333–350, 1965.

Horridge, G. A.: Relations between Nerves and Cilia in Ctenophores, *Am. Zool.*, vol. 5, pp. 357–376, 1965.

Hughes, D. A.: Behavioural and Ecological Investigations of the Crab *Ocypode ceratophthalamus*, *J. Zool.*, vol. 150, pp. 129–143, 1966.

Laverack, M. S.: Responses of Cuticular Sense Organs of the Lobster *Homarus vulgaris* (Crustacea). I. Hair Peg Organs as Water Current Receptors, *Comp. Biochem. Physiol.*, vol. 5, pp. 319–326, 1962.

Manning, A.: Antennae and Sexual Receptivity in *Drosophila melanogaster* females, *Science*, vol. 158, pp. 136–137, 1961.

Parry, D. A.: The Small Leg Nerve of Spiders and a Possible Mechanoreceptor, *Q. J. Microsc. Sci.*, vol. 101, pp. 1–8, 1960.

Pringle, J. W. S.: The Function of the Lyriform Organs of Arachnids, *J. Exp. Biol.*, vol. 32, pp. 270–278, 1955.

Pumphrey, R. J.: Hearing in Insects, *Biol. Revs.*, vol. 15, pp. 107–132, 1940.

Roeder, K. D.: Auditory System of Noctuid Moths, *Science*, vol. 154, pp. 1515–1521, 1966.

Roeder, K. D., and A. E. Treat: Ultra Sonic Reception by the Tympanic Organ of Noctuid Moths, *J. Exp. Zool.*, vol. 134, pp. 127–157, 1957.

Roeder, K. D., A. E. Treat, and J. S. Vandenberg: Auditory Sense in Certain Sphingid Moths, *Science*, vol. 159, pp. 331–333, 1968.

Rosin, R., and A. Shulov: Sound Production in Scorpions, *Science*, vol. 133, pp. 1918–1919, 1961.

Sleigh, M. A.: The Co-ordination and Control of Cilia, *Symp. Soc. Exp. Biol.*, vol. 20, pp. 11–33, 1966.

Slifer, E.: The Fine Structure of Insect Sense Organs, *Int. Rev. Cytol.*, vol. 11, pp. 125–158, 1961.
Slifer, E. H., and L. H. Finlayson: Muscle Receptor Organs in Grasshoppers and Locusts (Orthoptera, Acrididae), *Q. J. Microsc. Sci.*, vol. 97, pp. 617–621, 1956.
Slifer, E., and S. S. Sekhon: Fine Structure of the Sense Organs on the Antennal Flagellum of the Honey Bee, *J. Morphol.*, vol. 109, pp. 351–381, 1961.
Stephens, W. M.: The Incredible Horseshoe Crab, *Sea Front.*, vol. 10, pp. 131–138, 1964.
Taylor, R. C.: The Anatomy and Adequate Stimulation of a Chordotonal Organ in the Antennae of a Hermit Crab, *Comp. Biochem. Physiol.*, vol. 20, pp. 709–717, 1967.
Thurm, U.: Mechanoreceptors in the Cuticle of the Honey Bee: Fine Structure and Stimulus Mechanism, *Science*, vol. 145, pp. 1063–1065, 1964.
Thurm, U.: Insect Mechanoreceptors. I. Fine Structure and Adequate Stimulus, pp. 75–82; II. Receptor Potentials, pp. 83–94, *Cold Spring Harbor Symp. Quant. Biol.*, vol. 30, 1965.
Waldron, I.: Courtship Sound Production in Two Sympatric Sibling *Drosophila* species, *Science*, vol. 144, pp. 191–193, 1964.
Wenner, A. M.: Communication with Queen Honey Bees by Substrate Sound, *Science*, vol. 138, pp. 446–447, 1962.
Whitear, M.: Chordotonal Organs in Crustacea, *Nature*, 187, pp. 522–523, 1960.
Whitear, M.: The Fine Structure of Crustacean Proprioceptors: The Chordotonal Organs in the Legs of the Shore Crab *Carcinus maenas*, *Phil. Trans. R. Soc. London, B*, pp. 245–325, 1962.
Wiersma, C. A. G.: Movement Receptors in Decapod Crustacea, *J. Mar. Biol. Assoc. U.K.*, vol. 38, pp. 143–152, 1959.

Chemoreception

Barber, S. B.: Chemoreception and Proprioceptors in *Limulus*, *J. Exp. Zool.*, vol. 131, pp. 51–69, 1956.
Boeckh, J., K. E. Kaissling, and D. Schneider: Insect Olfactory Receptors, *Cold Spring Harbor Symp. Quant. Biol.*, vol. 30, pp. 265–280, 1965.
Case, J.: Properties of the Dactyl Chemoreceptors of *Cancer antennarius* (Simpson) and *C. productus* Randall, *Biol. Bull.*, vol. 127, pp. 428–444, 1964.
Coward, S. J., and R. E. Johannes: Amino Acid Chemoreception by the Planarian *Dugesia dorotocephala*, *Comp. Biochem. Physiol.*, vol. 29, pp. 475–478, 1969.
Dethier, V. G.: The Physiology and Histology of the Contact Chemoreceptors of the Blowfly, *Q. Rev. Biol.*, vol. 30, pp. 348–371, 1955.
Dethier, V. G., and L. E. Chadwick: Chemoreception in Insects, *Physiol. Rev.*, vol. 28, pp. 220–254, 1948.
Gillary, H. L.: Stimulation of the Salt Receptor of the Blowfly *Phormia regina*. I. NaCl, pp. 337–350; II. Temperature, pp. 351–357; III. The Alkali Halides, pp. 359–368, *J. Gen. Physiol.*, vol. 50, 1966.
Laverack, M. S.: Tactile and Chemical Perception in Earthworms. I. Responses to Touch, Sodium Chloride, Quinine and Sugars, *Comp. Biochem. Physiol.*, vol. 1, pp. 155–163, 1960.
Laverack, M. S.: Tactile and Chemical Perception in Earthworms. II. Responses to Acid pH Solutions, *Comp. Biochem. Physiol.*, vol. 2, pp. 122–134, 1961.
McIndoo, N. E.: Chemoreceptors of Blowflies (Calliphora), *J. Morphol.*, vol. 56, pp. 445–476, 1934.
Schneider, D.: Chemical Sense Communication in Insects, *Symp. Soc. Exp. Biol.*, vol. 20, pp. 273–298, 1966.
Schneider, D.: Insect Olfaction: Deciphering System for Chemical Messages, *Science*, vol. 163, pp. 1031–1037, 1969.
Slifer, E., S. S. Sekhon, and A. D. Lees: The Sense Organs on the Antennal Flagellum of Aphids (Homoptera) with Special Reference to the Plate Organs, *Q. J. Microsc. Sci.*, vol. 105, pp. 21–29, 1964.
Wilson, E. O.: Chemical Communication in the Social Insects, *Science*, vol. 149, pp. 1064–1071, 1965.
Wohlbarsht, M. L.: Water Taste in *Phormia*, *Science*, vol. 125, p. 1248, 1957.
Wohlbarsht, M. L.: Receptor Sites in Insect Chemoreceptors, *Cold Spring Harbor Symp. Quant. Biol.*, vol. 30, pp. 281–288, 1965.

Photoreception

Behrens, M. E.: The Electrical Response of the Planarian Photoreceptor, *Comp. Biochem. Physiol.*, vol. 5, pp. 129–138, 1962.
Bruno, M. S., and D. Kennedy: Spectral Sensitivity of Photoreceptor Neurons in the 6th Ganglion of the Crayfish, *Comp. Biochem. Physiol.*, vol. 6, pp. 41–46, 1962.
Carlson, S. D., H. R. Steeves, III, J. Van de Berg, and W. E. Robbins: Vitamin A Deficiency: Effect on Retinal Structure of the Moth *Manduca sexta*, *Science*, vol. 158, pp. 268–270, 1967.
Cornwell, P. B.: The Functions of the Ocelli of *Calliphora* (Diptera) and *Locusta* (Orthoptera), *J. Exp. Biol.*, vol. 32, pp. 217–238, 1955.

Eakin, R. M.: Lines of Evolution of Photoreceptors, in (D. Mazia, and A. Tyler eds.), "General Physiology of Cell Specialization," pp. 393–425, McGraw-Hill, New York, 1963.

Eakin, R. M.: Evolution of Photoreceptors, *Cold Spring Harbor Symp. Quant. Biol.*, vol. 30, pp. 363–370, 1965.

Eakin, R. M., and J. A. Westfall: Fine Structure of Photoreceptors in the Hydromedusan *Polyorchis penicellatus*, *Proc. Nat. Acad. Sci. U.S.A.*, vol. 48, pp. 826–833, 1962.

Eakin, R. M., and J. A. Westfall: Further Observations on the Fine Structure of Some Invertebrate Eyes, *Z. Zellforsch.*, vol. 62, pp. 310–332, 1964.

Fernandez-Moran, H.: Fine Structure of the Light Receptors in the Compound Eyes of Insects, *Exp. Cell. Res., Suppl.* 5, pp. 586–644, 1958.

Goldsmith, T. H.: On the Visual System of the Bee (*Apis mellifera*), *Ann. N.Y. Acad. Sci.*, vol. 74, pp. 223–229, 1958.

Goldsmith, T. H.: Fine Structure of the Retinulae in the Compound Eye of the Honey Bee, *J. Cell Biol.*, vol. 14, pp. 489–494, 1962.

Goldsmith, T. H., R. J. Barker, and C. F. Cohen: Sensitivity of Visual Receptors of Carotenoid Depleted Flies: A Vitamin A Deficiency in an Invertebrate, *Science*, vol. 146, pp. 65–67, 1964.

Goldsmith, T. H., and D. E. Philpott: The Microstructure of the Compound Eyes of Insects, *J. Biophys. Biochem. Cytol.*, vol. 3, pp. 429–441, 1957.

Hama, K.: A Photoreceptor-like Structure in the Ventral Nerve Cord of the Crayfish, *Anat. Rec.*, vol. 140, pp. 329–336, 1961.

Horridge, G. A.: Presumed Photoreceptor Cilia in a Ctenophore, *Q. J. Microsc. Sci.*, vol. 105, pp. 311–317, 1964.

Hubbard, R., D. Bownds, and T. Yoshizawa: The Chemistry of Visual Photoreception, *Cold Spring Harbor Symp. Quant. Biol.*, vol. 30, pp. 301–315, 1965.

Jander, R., and T. H. Waterman: Sensory Discrimination between Polarized Light and Light Intensity Patterns of Arthropods, *J. Cell. Comp. Physiol.*, vol. 56, pp. 137–159, 1960.

Kennedy, D., and E. R. Baylor: Analysis of Polarised Light by the Bee's Eye, *Nature*, vol. 91, pp. 34–37, 1961.

Land, M. F.: Activity in the Optic Nerve of *Pecten maximus* in Response to Changes in Light Intensity and to Pattern and Movement in the Optical Environment, *J. Exp. Biol.*, vol. 45, pp. 83–100, 1966.

Land, M. F.: A Multilayer Interference Reflector in the Eye of the Scallop *Pecten maximus*, *J. Exp. Biol.*, vol. 45, pp. 433–448, 1966.

Lawrence, P. A., and F. B. Krasne: Annelid Ciliary Photoreceptors, *Science*, vol. 148, pp. 965–966, 1965.

McEnroe, W. D., and K. Dronka: Color Vision in the Adult Female Two-spotted Spider Mite, *Science*, vol. 154, pp. 782–784, 1966.

MacRae, E. K.: Fine Structure of Planarian Photoreceptor Cells, *Am. Zool.*, vol. 3, p. 548, 1963.

Miller, W. H.: Morphology of the Ommatidia of the Compound Eye of *Limulus*, *J. Biophys. Biochem. Cytol.*, vol. 3, pp. 421–428, 1957.

Miller, W. H.: Derivations of Cilia in the Distal Sense Cells of the Retina of *Pecten*, *J. Biophys. Biochem. Cytol.*, vol. 4, pp. 227–228, 1958.

Miller, W. H.: Visual Photoreceptor Structures, in J. Brachet and A. E. Mirsky (eds.), "The Cell," vol. IV, pp. 325–364, Academic, New York, 1960.

Millott, N.: Animal Photosensitivity with Special Reference to Eyeless Forms, *Endeavour*, vol. 16, pp. 19–28, 1957.

Press, N.: Electron Microscope Study of the Distal Portion of a Planarian Retinular Cell, *Biol. Bull.*, vol. 117, pp. 511–517, 1959.

Roggen, D. R., D. J. Raski, and N. O. Jones: Cilia in Nematode Sensory Organs, *Science*, vol. 152, pp. 515–516, 1966.

Röhlich, P., and L. J. Török: Elektronmikroskpische Untersuchungen des Auges von Planarien, *Z. Zellforsch.*, vol. 54, pp. 362–381, 1961.

Röhlich, P., and L. J. Török: The Effect of Light and Darkness on the Fine Structure of the Retinal Clubs in *Dendrocoelum lacteum* (Turbellaria), *Q. J. Microsc. Sci.*, vol. 103, pp. 543–548, 1962.

Ruck, P.: On Photoreceptor Mechanisms of Retinular Cells, *Biol. Bull.*, vol. 123, pp. 618–634, 1962.

Ruck, P.: The Diversified Visual System of the Dragon Fly, *Am. Zool.*, vol. 4, p. 277, 1964.

Trujillo-Cenoz, O.: Some Aspects of the Structural Organization of the Arthropod Eye, *Cold Spring Harbor Symp. Quant. Biol.*, vol. 30, pp. 371–382, 1965.

Wald, G.: Molecular Basis of Visual Excitation, *Science*, vol. 162, pp. 230–239, 1965.

Wald, G., and P. Brown: Human Color Vision and Color Blindness, *Cold Spring Harbor Symp. Quant. Biol.*, vol. 30, pp. 345–361, 1965.

Wald, G., and R. Hubbard: Visual Pigment of a Decapod Crustacean: The Lobster, *Nature*, vol. 180, pp. 278–280, 1957.

Wolken, J. J.: Studies of Photoreceptor Structures, *Ann. N.Y. Acad. Sci.*, vol. 74, pp. 164–181, 1958.

Wolken, J. J.: The Photoreceptor Structures, *Int. Rev. Cytol.*, vol. 11, pp. 195–216, 1961.
Wolken, J. J., and G. J. Gallik: The Compound Eye of the Crustacean *Leptodora kindtii*, *J. Cell Biol.*, vol. 26, pp. 968–972, 1965.
Yamamoto, T., K. Tasaki, Y. Sugawara, and A. Tonosaki: Fine Structure of the Octopus Retina, *J. Cell Biol.*, vol. 25, pp. 345–359, 1965.
Yoshida, M., and N. Millott: Light Sensitive Nerve in an Echinoid, *Experientia*, vol. 15, pp. 13–14, 1959.

Temperature reception

Kerkut, G. A., and B. J. R. Taylor: A Temperature Receptor in the Tarsus of the Cockroach *Periplaneta americana*, *J. Exp. Biol.*, vol. 34, pp. 486–493, 1957.
Murray, R. W.: Temperature Receptors, in O. Lowenstein (ed.), "Advances in Comparative Physiology and Biochemistry," pp. 117–170, Academic, New York, 1962.
Roth, L. M., and E. R. Willis: Hygroreceptors in *Tribolium*, *J. Exp. Zool.*, vol. 116, pp. 527–570, 1951.

chapter XVI

TRANSMISSION OF STIMULI AND NEURAL INTEGRATION

A. FUNCTIONS OF A NERVOUS SYSTEM

B. PATTERNS OF NERVOUS SYSTEMS

1. **Plexuses**
2. **Stomodael (stomatogastric) systems**
3. **Giant fibers and giant cells**
4. **Synapses and transmission across them**
   a. ELECTRIC AND CHEMICAL TRANSMISSION
   b. TRANSMITTER SUBSTANCES (NEUROHUMORS)
5. **Neuromuscular transmission**

C. NONNEURAL CONDUCTION

D. SPONTANEOUS ACTIVITY AND PACEMAKERS

E. NERVE NETS

F. CENTRALIZED NERVOUS SYSTEMS

G. FUNCTIONAL RELATIONSHIPS WITHIN CENTRALIZED SYSTEMS

A. FUNCTIONS OF A NERVOUS SYSTEM

The type and scope of an animal's sense organs permit it to scan and sample its environment, and the impulses initiated in them and transmitted to the effector organs account for its overt behavior. Although this may be dictated by one predominating source of stimulation, rarely if ever is the response the effect of a single one. For in nature, many stimuli of different kinds and from different sources are constantly impinging upon every organism, and even in the laboratory under the most precise conditions of experimentation, it is impossible to reduce stimulation to that of one specific source. The integration of the impulses aroused by these various stimuli and their transmission to the appropriate effector organs is the function of a successfully operating central and peripheral nervous system. It is in the central system that coordination between the sensory input of both external and internal origin and motor output takes place, usually through connections with interneurons. These connections may be fairly direct and simple, involving only one interneuron, or they may be elaborate, involving several or many interneurons and providing a variety of possible circuits. Or, more rarely, the impulse may be transmitted directly from afferent to efferent neuron, in a simple reflex arc. When such connections occur they may be within the central system or outside it, but in general, the central system supplies mechanisms for more complicated and sophisticated neuronal integration.

B. PATTERNS OF NERVOUS SYSTEMS

1. Plexuses

The principal patterns of invertebrate nervous systems may be categorized in very general terms as diffuse or centralized. In a diffuse pattern, neurons or their processes form a plexus or a net, while in a centralized one, they form chains or cords, usually running parallel to the long axis of the body. "Plexus" is a general term for a tangle of nerve processes oriented principally in one plane so that they constitute a layer of nervous tissue. There may, or may not, be functional connections between the processes. Plexuses are of wide occurrence among animals of all kinds and are frequently found in the integument, the gut wall, the muscles, and the pericardium of invertebrates with concentrated nervous systems. "Nerve net" is more limited in meaning and nerve nets are more limited in distribution. In a nerve net, which is characteristic particularly of cnidarians, the neurons are bipolar, tripolar, or multipolar, with no distinction between axon and dendrite. Their processes may, therefore, more appropriately be called neurites. In a nerve net, conduction between neurons can be in either direction, and any one neuron can substitute for another. Although the connections between adjacent neurons may be discrete and separate from each other, in some instances there is fusion between the neurites of neighboring cells so that the net is a continuous mesh. Also the neurons may be more numerous in some regions than in others and the mesh there more closely woven, but there are never the discrete nodular masses of nerve cells, or ganglia, that are characteristic of centralized systems.

In these systems each nerve cord is a longitudinal concentration of neurons. It may be little more than the thickened and straightened strands of a general plexus, or it may be an assemblage of many nerve cells, forming a conspicuous trunk with ganglionic enlargements along its length. A ganglion may be defined as a discrete collection of nerve cells, usually delimited from surrounding nonnervous tissues by a connective tissue sheath and located either in the course of a cord or in some part of the body peripheral to it. The terminations of axons in a ganglion may make connections with neurons in the ganglion itself, or the axons of neurons whose cell bodies lie in the ganglion may leave it and run longitudinally in the cord as connectives between successive ganglia or transversely as commissures between those laterally placed. Afferent nerves may have their terminations within ganglia and efferent ones, their cell bodies, with their axons leaving the ganglion to contribute to the bundles of nerve fibers that constitute the peripheral nerves. These bundles may contain only the axons of afferent nerves or only those of efferent nerves, or, as mixed nerves, both of these. These nerves, the superficial sense organs, the nerves innervating them, and the ganglia with which they may be directly associated constitute the peripheral nervous system, as opposed to the central nervous system, of an animal.

Invertebrate ganglia are usually so organized that the cell bodies of the neurons lie at the periphery of the nodular mass whose center is occupied by a tangle of fine nerve fibers, either dendrites or the terminal arborizations of axons, known as a neuropil (Greek *neuron*, "nerve," and *pilos*, "felt"). The neuropil forms the main mass of an invertebrate ganglion and is usually the only area where interneuronal synaptic connections occur. Analysis of conditions in the neuropil is difficult, but some order, both morphological and physiological, is beginning to be brought out of the intricate tangle of fibers. Microscopically, it can be distinguished as unstructured, either plexiform or diffuse, or structured, either glomerular or stratified, with in each case a defined pattern in the arrangement of the fibers. Electrical measurements have shown it to be the site of the most complex integrative activity, by means of mechanisms that are still problematical and evidently more inclusive than those that have been elucidated for other functional connections of neurons. Within a neuropil there exists the possibility of a multiplicity of circuits which a nervous impulse may follow.

Typically, the nerve cells in an invertebrate ganglion are unipolar, although their axons may divide into numbers of collaterals with a corresponding increase in the number of possible connections. Exceptions to this are the first-order giant cells of the squid and the cardiac ganglion cells of crustaceans, for in both these cases the cells are multipolar. The fibers in the neuropil are very close together, with less than 100 to 200 Å separating their membranes, and their junctions are axo-axonic (Fig. XVI–1*d*). Variations in the number of cells within the ganglia reflect

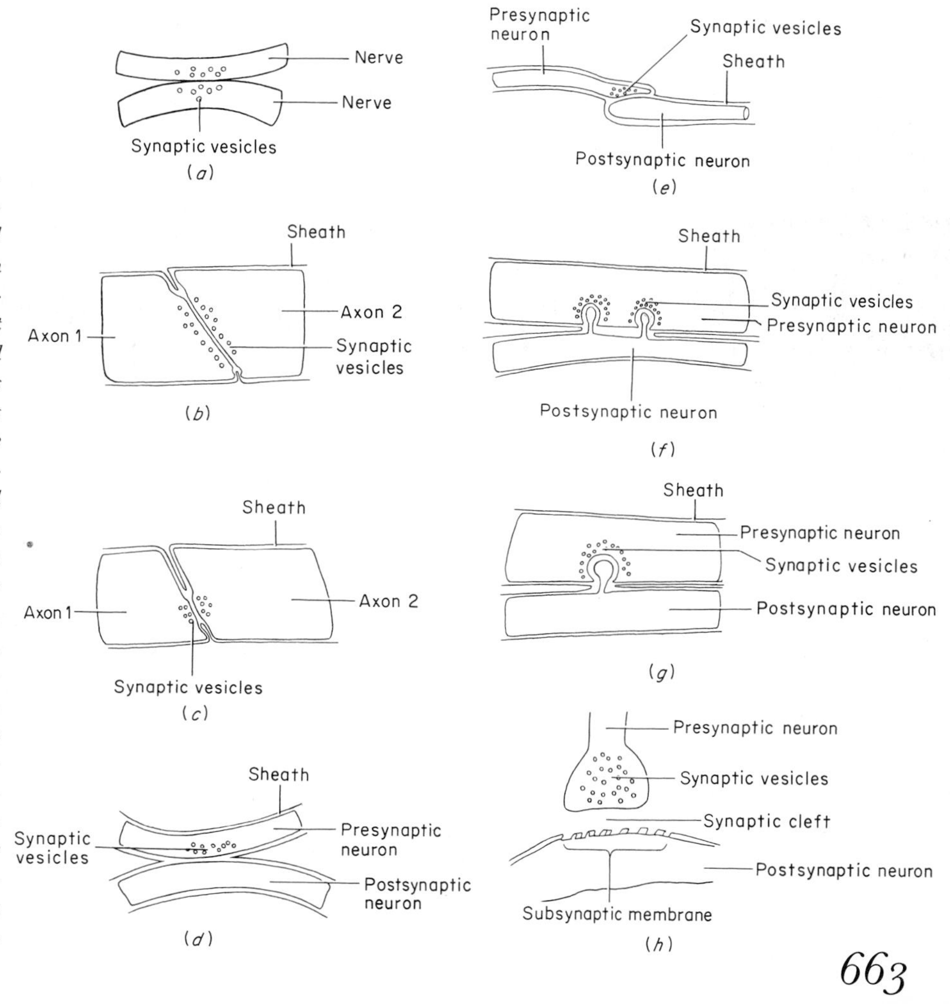

FIG. XVI–1 *Types of synapses in invertebrates as revealed by electron microscopy. (a, b, c, d, e, f, g, after Bullock and Horridge; h after Florey.)* ***(a)*** *Axon-axon synapse as found in the nerve net of the jellyfish* Cyanea. *The axons are without sheaths.* ***(b)*** *Axon-axon synapse as found in the septal synapses of the giant fibers of an earthworm. The axons have sheaths and contact is made across almost the entire expanse of the junction.* ***(c)*** *Axon-axon synapse as found in the septal synapses of giant fibers of crustaceans. The contact area is more limited than in b. In a, b, and c, transmission may be in either direction and is probably electrical.* ***(d)*** *Axon-axon synapse typical of a neuropil. Transmission is unidirectional.* ***(e)*** *Axon-dendrite synapse typical of a neuropil.* ***(f)*** *Synapse of a crustacean giant fiber to a motor neuron with postsynaptic neuron invaginated into presynaptic neuron.* ***(g)*** *Synapse between giant fibers in the stellate ganglion of a squid, with postsynaptic fiber invaginated into the presynaptic.* ***(h)*** *Diagrammatic representation of conditions at a synapse where transmission is unidirectional.*

differences in the complexity of neural organization between different genera, and their positions upon the phylogenetic tree. In the mollusc *Aplysia*, for example, the total number of cells in the central ganglia is estimated at 1×10^4, while in the crayfish *Procambarus clarkii* it is 6×10^6 and in *Octopus* 168×10^6.

2. Stomodael (stomatogastric) systems

In molluscs, annelids, and arthropods there is also a stomodael, often called stomatogastric, nervous system. It is essentially that part of the peripheral system which innervates the anterior region of the alimentary canal. These nerves connect with the most anterior ganglia of the central nervous system and may contain either afferent, efferent, or integrative fibers. Although invertebrate stomodael systems are often homologized with vertebrate autonomic systems, the composition of their nerves makes a distinction between them, for those in the vertebrate autonomic system are composed largely of efferent fibers from the visceral organs, with few, if any, afferent ones.

3. Giant fibers and giant cells

Giant fibers, formed by union of the axons of several different neurons, have already been mentioned in Chapter IV. They are found in some molluscs, annelids, and arthropods and, because of their size and the comparative ease with which, in most cases, they can be exposed or isolated, have been of signal advantage in neurophysiological research. In segmental animals like annelids and arthropods, each giant fiber is really a series of units, for the continuity of the axoplasm is interrupted between segments by membrane partitions that cross the axon completely. Transmission, however, seems virtually unimpeded. Giant cells are of quite common occurrence among invertebrates and have been reported in nemerteans, molluscs, arachnids, myriapods, and insects. The most intensively investigated of these are those in the visceral component of the complex abdominal ganglion of *Aplysia*, in which, in large specimens, the diameter may be nearly 1 mm. "Giant" is, of course, a relative term, used in respect to the size of other similar objects. The diameters of giant cells range from 25 to 800 μ, and the diameters of their axons may be proportionately large. The diameter of the axon of the giant cell of *Aplysia* is, for example, 178 μ. The length of the axon is not relevant, for in large vertebrates those of non-giant neurons may be measured in meters.

4. Synapses and transmission across them

Conduction of the impulse along a neuron and its transmission to another cell have proved to be problems as challenging and perplexing as the problems of its initiation. There is no reason to suppose that conduction in invertebrate neurons is in any way different from that in vertebrate neurons, but it is probable that transmission may be effected by a greater variety of mechanisms. The functional connection between a neuron and another cell, whether neuron or effector cell, is usually made through a synapse (Greek *synapsis*, "conjunction"). There may, however, be nonsynaptic connections in nerve nets where there is continuity between the nerve cells. In either case there is a transmitting and a receptor cell. When the impulse crosses a synapse the transmitting cell is designated as the presynaptic cell and the receiving as the postsynaptic. The membranes of the two cells are separated by a space, the synaptic cleft, which electron microscopy has shown to range, in various types of synapse, from 100 to 200 Å in width. While in vertebrates, synapses between neurons are between the terminal arborizations of the axon of one cell and the dendrites of another, and transmission is in one direction only, in invertebrates, they may also be between adjacent axons, and transmission may be in either direction. Most synaptic areas are, nevertheless, between the axon and dendrite arborizations in a neuropil, and transmission is unidirectional. In most cases, both in invertebrates and in vertebrates, there is an accumulation of vesicles, the synaptic vesicles, in the presynaptic neuron, although, particularly in the case of axon-axon synapses, there may be vesicles on both sides, unequal or approximately equal in number (Fig. XVI–1).

a. ELECTRIC AND CHEMICAL TRANSMISSION

Transmission of the impulse across a synapse may be electric or chemical. Both involve movement of ions across the membrane of the postsynaptic cell and so imply transitory changes in the permeability of its membrane. In the case of electric transmission, the change in ionic concentration in the synaptic area, consequent upon the arrival of the nervous impulse, is in itself adequate to cause the passage of ions across this membrane. These ionic movements may then alter the excitability of a postsynaptic neuron to the extent of inducing a nervous impulse in it. Electric transmission probably occurs at the axon-axon synapses of cnidarian nerve nets and at the septal segmental end unions in the giant fibers of earthworms and of crayfish (Fig. XVI–1*a* to *c*). Here the synaptic cleft is narrow and the delay in transmission is short.

Chemical transmission is probably a more general mechanism, involving the production of a transmitter with a specific effect upon the membrane of the postsynaptic cell. The concept of chemical transmission is based upon histological, cytochemical, biochemical, and pharmacological evidence. Histological evidence is derived from observations of the accumulation of specifically staining material in the synaptic vesicles of transmitting cells, their discharge into the synaptic area, and the absence of vesicular structures in most postsynaptic cells. Specific staining, or cytochemical, reactions of the vesicular material provide evidence for synthesis of different substances by different neurons, indicating that they are not only secretory cells but cells biochemically differentiated for the production of particular secretions. It has been possible to identify some of these compounds by biochemical analysis of isolated nervous tissue and by spectrophotometric analysis of extracts from them or of the fluid with which they are bathed or perfused during an experimental period. Where specific identification has not been possible, pharmacological experiments in the application or injection of known compounds and correlation of their effects with that of the passage of a nervous impulse under ordinary conditions of stimulation have provided presumptive evidence for similarity if not identity of the test compound and the natural neurohumor.

The concept of chemical transmission implies also that the neurohumor must in some way be removed from the synaptic area after it has accomplished its task. Conceivably, it might diffuse away sufficiently rapidly, but enzymatic degradation would certainly be a more efficient means of its disposal. Revelation in the synaptic area of an enzyme or enzyme system capable of splitting the molecule of a suspected transmitter would, therefore, also provide circumstantial evidence for its synthesis and operation in the tissues concerned. Moreover, where the operation of such an enzyme is suspected, specific antagonists or inhibitors of its action may be introduced. If these are effective in blocking the action of the enzyme so that the response in the effector cell continues beyond the usual limit of time, the nature of the transmitter, for which the enzyme is specific, may be deduced.

The concept also implies, besides specificity in the biosynthetic potentialities of the presynaptic cell, specificity in the receptivity of the postsynaptic cell. This specificity must reside in the membrane, since it is changes in the permeability of the membrane and in the passage of ions across it that lead to activation of the postsynaptic cell, whether nerve, muscle, gland, or other effector. It is presumed that patches of the membrane of the postsynaptic cell immediately adjacent to the terminals of the transmitting neurons have, in addition to structural molecules and the carrier molecules involved in active transport of ions, specific receptor molecules with groupings adapted to bind a particular transmitter substance. These patches have been designated subsynaptic membranes. They are different in their molecular composition from other areas of the membrane of a postsynaptic cell (Fig. XVI–1*h*). Subsynaptic membranes may be of two types—excitatory or inhibitory. The response of an excitatory subsynaptic membrane to the action of a transmitter agent is a general and marked increase in permeability to such ions as Na^+, K^+, and Cl^-, consequent depolarization of the membrane, and increase in excitability of the receptor cell. The response of an inhibitory subsynaptic membrane is increase in permeability to only one or two ions, essentially K^+ and Cl^-, either singly or together, with consequent hyperpolarization of the membrane and reduction in the excitability of the receptor cell. Some postsynaptic cells have either excitatory or inhibitory subsynaptic membranes, and some have both. Those of the stretch receptors of crustaceans are, for example, only inhibitory, while those of muscle fibers are both inhibitory and excitatory. Whether the passage of a nervous impulse across a synapse arouses a cell to activity or prevents activity in it depends, therefore, more upon the nature of the subsynaptic membrane than upon the transmitter substance, for the same neurohumor may have either an excitatory or an inhibitory effect.

b. TRANSMITTER SUBSTANCES (NEUROHUMORS) Two substances whose role as agents in synaptic transmission in vertebrates has been fully established are acetylcholine and adrenalin (epinephrin) or noradrenalin (norepinephrin). Neurons producing acetylcholine are designated as cholinergic and those producing adrenalin or one of its analogues as adrenergic. This distinction is a fair one, since a cholinergic neuron does not synthesize adrenalin nor any of the other suspected transmitter substances, nor does any neuron synthesizing one of these produce acetylcholine. In vertebrates, motor neurons innervating somatic muscles, preganglionic autonomic neurons, and postganglionic neurons of the parasympathetic system are cholinergic, while postganglionic neurons of the autonomic system and probably certain neurons of the brain stem are adrenergic. Similarly, subsynaptic membranes, whether excitatory or inhibitory, may be designated choliceptive or adreceptive as indicative of their adaptation to either the acetylcholine or the adrenalin molecule. Other less well authenticated transmitter substances in vertebrates are 5-hydroxytryptamine (serotonin) and the unidentified substances I and S. Tryptaminergic neurons have been reported from the vertebrate central nervous system, as well as I-neurons, with an inhibitory effect, and S-neurons, although these are primarily sensory ones.

As yet little is known about transmitter substances throughout the great range of invertebrate species. Cholinergic neurons have been demonstrated in molluscs, specifically in the cardioinhibitory fibers of the clam *Mercenaria mercenaria*, in the inhibitory fibers of *Aplysia*, and in the optic and cerebellar ganglia of cephalopods. Neither adrenalin nor noradrenalin has been found in the nervous tissue of any mollusc yet examined. Acetylcholine seems also to be the transmitter at the neuromuscular synapses of sipunculids and leeches, and of onychophorans and of holothurians, at least in the retractor muscles of *Stichopus*. The sensory neurons of decapod crustaceans have also been shown to be cholinergic, but the fibers innervating their muscles produce different neurohumors. The evidence for production and utilization of acetylcholine in these cases has been derived from its identification in the perfusion fluid of nerve-muscle preparations after stimulation of presynaptic neurons; from the initiation of nerve action in them upon application of exogenous acetylcholine to the system; from the potentiation of nerve action by action of anticholinesterases such as eserine and prostigmine, thus blocking the removal of the transmitter by the enzyme that naturally degrades it; and, in the case of the onychophoran *Opisthopatus costesi*, from observation of the effects of extracts of its nervous tissue upon the hearts of clams and the muscles of sea cucumbers. This in every respect paralleled that of acetylcholine and made possible quantitative bioassay of the amount in the nervous tissue of the onychophoran. This, in the motor nerve fibers, was found to be ± 20 percent when tested on both cardiac muscle of the clam and somatic muscle of the sea cucumber.

Acetylcholine and acetylcholinesterase have been found in insect ganglia, but there is no direct evidence of its action as a transmitter; indeed, any insect transmitter has yet to be identified. Confusion in the findings already reported may have arisen from differences in techniques in administering test substances to nerve muscle preparations. When, for example, acetylcholine at concentrations of 1×10^{6} M is locally applied to the exposed flight muscles of the fly *Sarcophaga bullata* there is no recordable difference in potential in them when the thoracic ganglion is stimulated electrically. This would indicate that acetylcholine is not the transmitter at the neuromuscular junction. Yet when the compound, at the same concentration, is injected in the abdomen of a fly, depolarization of the muscle cell without stimulation follows, suggesting that it is mimicking the effect of nerve stimulation and that, after injection, it reaches the muscle membrane in a way that it does not upon external application.

The presence of acetylcholine and acetylcholinesterase has also been reported in protozoans; in nonnervous and nervous tissue in species of flatworms and nemerteans; as well as in those species of molluscs, annelids, and crustaceans in which their functions in transmission have been established. Pharmacological experiments have shown that the muscles of the body wall of annelids are sensitive to acetylcholine, for the dorsal musculature of *Lumbricus terrestris, Arenicola marina, Brachiomma vesiculosum*, and *Hirudo medicinalis* contract upon application of the compound at concentrations of 1×10^{-4} and 1×10^{-5} *M*. The dorsal muscles of the onychophoran *Peripatopsis* respond similarly to acetylcholine at a concentration of 3×10^{-6} *M*, but give no response to adrenalin.

Adrenergic neurons have also been found in some invertebrates, although it seems likely that substances closely related to adrenalin and noradrenalin are in most cases the transmitters rather than either of these compounds themselves. Neither of them has been found in the nervous systems of the molluscs, crustaceans, or echinoderms examined. Adrenalin, together with small amounts of noradrenalin, has been found in the nervous system of earthworms, from which it may be inferred that some of their neurons are adrenergic, although their motor neurons are cholinergic. Adrenalin, in pharmacological experiments, has, however, been shown to be a potent activator of a number of organs in a variety of invertebrates, but it has not been established as a natural neurohumor.

There is, on the other hand, a considerable body of evidence that 5-hydroxytryptamine (serotonin), widely distributed in the nervous system of vertebrates as well as in invertebrates, is a transmitter substance. This has been derived from cytochemical studies, in which, by special staining procedures, catechol amines can be visualized and localized in particular regions and even in particular cells, and from pharmacological experiments in which serotonin has been added to the perfusing fluid of nerve-muscle preparations. The cytochemical procedure is to expose freeze-dried pieces of selected tissues to formaldehyde gas with which certain monoamines are known to condense to form fluorescent compounds. Serotonin gives off a yellow fluorescence, while adrenalin, noradrenalin, and some other catechol amines give off a green one. By means of a fluorescent microscope technique, the nature of the fluorescence can be recognized and the substance responsible accurately localized; and by means of spectrophotometry or chromatography, it can be chemically identified. Serotonin has, for example, been found in seven cells in the ganglia along the ventral nerve cord of *Hirudo medicinalis*, and in a number of cells both in the ganglia and in the connectives of the ventral nerve cord of *Lumbricus terrestris*. The amount in the ventral nerve cord of *Lumbricus* has been found to be 10.4 μg/g of cord. It has also been localized in the nerve cord of the polychaete *Nephthys*. Pharmacological experiments have shown that it imitates the action of the cardioaccelerator nerves in lamellibranch molluscs and the inhibitor nerves of their muscles. It also accelerates the frequency of heartbeat in various species of crustaceans, but reduces it in myriapods and chelicerates. It is, however, without effect, either by injection or application, upon the neuromuscular response of the flight muscles of *Sarcophaga*. The mechanism of its operation at the neuromuscular junctions where it is effective is unknown, but there is good reason to believe that tryptaminergic nerves are not unusual among invertebrates. In addition to the seven yellow fluorescing cells in the ganglia of *Hirudo* are two large fluorescing processes entering from the periphery. The nature of their fluorescence, which is bright green, suggests the presence of one or more catechol amines, as yet unspecified.

5-Hydroxytyramine, or dopamine, is another catechol amine that is implicated as a neurohumor. This has been found by fluorometric and chromatographic means to be present in the ganglia of at least three species of gastropod and seven species of lamellibranch molluscs. In *Mercenaria mercenaria* high levels of dopamine were found in the cerebropleural, visceral, and pedal ganglia, separately assayed, while no appreciable amount was found in any other tissue examined, such as gill, mantle, heart, and intestine. Neither adrenalin nor noradrenalin could be detected in the ganglia or other tissues. While dopamine is known to be a precursor in the synthesis of noradrenalin, this does not seem to be its role in these 10 molluscs, in which there would appear to be tyraminergic neurons. Dopamine may also function as a neurohumor in sea anemones. Cytochemical study of pieces taken from the tentacles, oral disks, column wall, and mesenterial filaments of *Metridium senile* and *Taelia felina* has shown that only those from the tentacles, when examined under the fluorescent microscope, fluoresce. The distal regions of these, especially, exhibit a bright green fluorescence, most marked in those of *Metridium*. Dopamine, like adrenalin and noradrenalin, gives off such a green fluorescence. Since, as in molluscs, neither adrenalin nor noradrenalin has ever been detected in cnidarians, the source of the green fluorescence in the tentacles may be tentatively identified as dopamine, at least until evidence to the contrary is forthcoming. The distribution of the fluorescence in the tentacles seems to indicate that the compound is localized in the nerve cells, and possibly in the sensory nerves which are sufficiently concentrated in the distal regions of the tentacles to make visualization of their products possible. Similar concentrations of neurons can be found in the rhopalia of scyphozoan medusae, which are here so dense and discrete that they might almost qualify as ganglia. A substance has been extracted from those in *Aurelia aurita* that, when applied to an intact specimen, accelerates the rhythm of the pulsations of its bell, as it does that of the hydromedusan *Phialidium*. It has, on the other hand, no effect upon such standard material for neurohumor assay as the hearts of clams and crustaceans, nor does it affect peristalsis in the crustacean hindgut. Similar extracts from nonnervous tissue and the mesoglea of *Aurelia* produced no change in the contractions of these test tissues nor in the rhythm of pulsation of medusan bells, results that indicate that if this substance is indeed a neurohumor it is not like any of those known or suspected for other organisms. Chromatography has not yet yielded any information about its chemical nature. Electron microscopy of these ganglionic concentrations has revealed their density and also the presence, in the sensory neurons, of numerous vesicles that are concentrated on both sides of the synaptic contacts, providing circumstantial evidence for the secretion of a neurohumor. Yet neither fluorescence microscopy, chemical extraction, nor pharmacological assay has yet produced conclusive evidence for the production of neurohumors or chemical transmission in hydrozoan polyps.

Another presumed transmitter substance is gamma-aminobutyric acid (GABA), a decarboxylated form of glutamic acid, which appears to operate only at inhibitory subsynaptic membranes and may be the I-substance extracted from mammalian brains. Its effect has been demonstrated upon the hearts and skeletal muscles of crustaceans, where it imitates stimulation of the inhibitory nerves. In two crustaceans, however, *Procambarus clarkii* and *Pacifastacus lenuisculus*, dopamine has been found to have a more powerful inhibitory effect than GABA, and dopamine can also act as an inhibitor for certain neurons in the snail *Helix aspersa*. The body of evidence for GABA as a neurohumor is nevertheless growing.

Particularly cogent evidence for the inhibiting action of GABA at neuromuscular junctions is provided by studies of its effect, compared with that of other compounds extracted from crustacean nerves, upon muscular activity in the pereiopods of *Homarus americanus*. In crustaceans, inhibition of neural activity takes place peripherally, not centrally as in vertebrates and, in some cases at least, in insects. In the lobster's pereiopods two efferent axons lie side by side, one conveying excitatory impulses, the other inhibitory ones. These can be separated from each other and individual recordings made of their action potentials. Furthermore, long segments of each can be dissected out and chemical analyses made of them. Of 10 substances extracted from crustacean nerves, GABA proved to be the most effective blocker of excitation, with taurine and betaine next, and beta-alanine, alanine, homarine, glutamine, and aspartic acid effective in lesser degrees. (Glutamic acid was the only excitatory substance found in the extracts.) Analyses of the inhibitor axons showed that they contained GABA in amounts equal to 0.5 percent of their wet weight, while none could be extracted from the excitatory axons.

Inhibitory impulses in mammals are known to be blocked by picrotoxin, a substance ($C_{30}H_{34}O_{13}$) found in the berry of an East Indian vine, *Anamirta cocculus*. When this is injected into cats, the animals go into convulsions. Perfusion of the claws of the crayfish *Orconectes immunis* with this drug results in contraction of the opener muscle when both the excitatory and inhibitory nerves are stimulated, indicating blockage of the inhibitory impulses. That the block is at the myoneural junctions rather than in the nerve itself was shown by the fact that the inhibitory effect of the stimulated nerve was not altered after it was soaked in a solution of picrotoxin, but was prevented when the drug was applied to the muscle. Assuming that the nerve releases a neurohumor that combines with the inhibitor sites at the myoneural junction, it is most probable that picrotoxin competes with it for these sites and at adequate concentrations (above $1 \times 10^{-6}\ M$) successfully blocks this combination, and so inhibition.

A similar effect of picrotoxin has been observed in the auditory synapses of the grasshopper *Gampsocleis burgeri*, where inhibition takes place in the central nervous system. In this insect there are two large auditory fibers in the central cord. One of these, the T-fiber, receives impulses from the tympanic nerves, and the other, the C-fiber, receives them from the sensilla trichodea on the cercus. These sensilla are sensitive to sound, but play only a minor role in the insect's response to stridulation, for which the tympanic organs are mainly responsible. Impulses from these organs are transmitted through the tympanic nerves to the prothoracic ganglion and from there to the T-fiber, which extends from the cerebral to the metathoracic ganglion; they are therefore conducted anteriorly and posteriorly. Inhibitory action between the tympanic nerves of the two sides has been demonstrated by transection experiments. When a T-fiber receives an excitation at the prothoracic ganglion from the tympanic nerve on one side, it also receives inhibitory impulses and a weaker excitation from the nerve on the opposite side. The grasshopper therefore receives not only information of calling individuals in its vicinity but some information about the direction from which the calls are coming. This discrimination must be of considerable value to it. Application of picrotoxin to the prothoracic ganglion blocks the inhibitor action, and impulses along the T-fiber increase in frequency. GABA acts reversibly as an inhibitor of excitatory impulses. From pharmacological studies such as these, it is concluded that the synapses of the T-fiber are of two kinds—excitatory ones with the tympanic nerves and inhibitory ones with interneurons that are activated by the tympanic neurons. Presumably, the neurohumors liberated at these synapses are different, or their combining sites are, so that in the one case the subsynaptic membrane is depolarized and in the other case it is hyperpolarized.

There is some evidence also that glutamate may function as a transmitter in invertebrates. Some of this has been derived from its identification in the perfusates of nerve-muscle preparations of *Helix*, *Carcinus*, and *Periplaneta* after stimulation of the nerve. The amount of glutamate in the perfusate (Ringer's solution) revealed by chromatography after electrical stimulation of the nerves in the preparation was proportional to the number of shocks given in any one experiment. Moreover, addition of glutamate to the perfusing solution, without stimulation of the nerves, resulted in contraction of the muscle. The effect of glutamate upon nerve-muscle preparations of the cockroach is antagonized by GABA. Glutamate, either injected into the abdominal cavity of *Sarcophaga* or applied locally, caused partial depolarization of the muscle membrane, with a similar but weaker effect than that of acetylcholine. Glutamate, therefore, meets the requirements of a transmitter substance, for it can initiate contraction in a muscle, it is produced in amounts proportional to the frequency of stimuli passing along a nerve, and it can be easily synthesized and destroyed. Its synthesis from glucose has been demonstrated in isolated snail nerves, and its rapid metabolism in tissues is an established fact.

Studies of the application of glutamate to neuromuscular junctions in the abductor muscle of the dactyl of the first or second pereiopod of the crayfish lead to the conclusion that the L form alone mimics the transmitter substance. Small doses excite, large doses inhibit, activity at these junctions. The fact that D-glutamate is ineffective suggests that the configuration of the molecule is significant and that the spatial arrangement of the atoms at the alpha carbon is critical for its binding. The L-glutamate-sensitive spots in the muscle are very circumscribed and most probably are identical with the normal neuroreceptors, implicating glutamate as the natural transmitter substance. Since the response to GABA in these muscles is different from that to glutamate, different receptors would seem to be involved, with those sensitive to glutamate being highly specific for the particular form of a particular molecule and becoming desensitized by large doses of it.

5. Neuromuscular transmission

Cells whose activation is most readily observed and in which the degree of response can be measured with a considerable degree of accuracy are contractile cells, which are indeed the most important effectors that an animal has. Muscle-nerve preparations have long been standard experimental material for study of neural conduction and transmission. The neurons conveying impulses to the muscles make up the motor nerves, in which the fibers may be either excitatory or inhibitory, and also either fast or slow, depending on whether they cause a quick contraction, or twitch, or a slow and usually sustained one. These are relative terms and are often applied to the nerve fibers themselves, especially in the case of muscles with multiple innervation where contrasts, similar to those that can be made between the effects of stimulation of an excitatory or an inhibitory nerve, can be made between the effects of a fast and a slow one. The responses of fast fibers are usually phasic, occupying a short period of time, while those of slow fibers are tonic, contributing to the "tone" or state of chronic contraction of a muscle.

A good deal of attention has been given to the terminations of motor neurons in relation to the muscles they supply. Most of the neuromuscular junctions that have been demonstrated in invertebrates seem to be simple terminals of the branches of axons, similar to those with other effector cells. Nematodes are unique in that their nerves do not send branches to the muscles; the muscles send processes out to reach the nerves, but the terminations of the nerves in them are believed to be like those found in most other invertebrates. The muscles of some invertebrates are doubly, or multiply, innervated, for a single muscle fiber may be supplied by more than one axon. Intracellular recordings from the proboscis retractor muscles of sipunculids indicate, for example, that the muscle fibers are innervated by fast and slow nerve fibers, and similar experiments on the sabellid worm *Myxicola* suggest that its locomotor fibers are also doubly, or multiply, innervated. Arthropod muscles, also polyneural, are markedly different from vertebrate muscles in that but few motor axons pass to them. Each neuron, whose cell body lies in the central nerve cord (except for the stomodael system), innervates many muscle fibers, sometimes even those in two functionally distinct muscle blocks. The limb muscles are, for example, supplied by not more than five axons; in crustaceans there are usually three, and in insects two, although some muscles of *Locusta* have three (Fig. XVI–2). The axons are multiterminal, dividing one (diplotomy) or several times (polytomy) after reaching the muscle, with their endings close together on the muscle fiber. Motor end plates resembling those found in vertebrate muscle have been described in coleopterans; often these receive more than one nerve fiber. End plates have also been seen in Homoptera and Orthoptera. These are syncytial structures, with small nuclei and granular cytoplasm, that splay out like a claw over the surface of a muscle, at intervals of 40 to 80 μ. They are frequently associated with tracheoles, which are close to them or seem actually within them (Fig. XVI–2*c*). Thickenings of the nerve fibers on the muscles of insects, known as Doyère cones, have been believed to be specialized end organs, but they are more probably points where the multiterminal axons branch before penetrating the sarcolemma, and the real terminations are probably processes too fine to be revealed by conventional means. This may also be the case for the end plates described in crustaceans and in other insects.

C. NONNEURAL CONDUCTION

It is an established fact that in multicellular systems of all kinds excitation can travel from cell to cell in the absence of any demonstrable nervous elements. This happens, for example, in sensitive plants of the genus *Mimosa*, in which the effect of a tactile or thermal stimulus applied to a single leaflet, or small group of leaflets, spreads to others, and those on the same branch, or even all those on the entire plant, show the typical response of curling and drooping. Moreover, the developing vertebrate heart begins to beat before any nerves have reached it, as a propagated excitation that begins at one point in the muscular sac and spreads over it. Among invertebrates the most

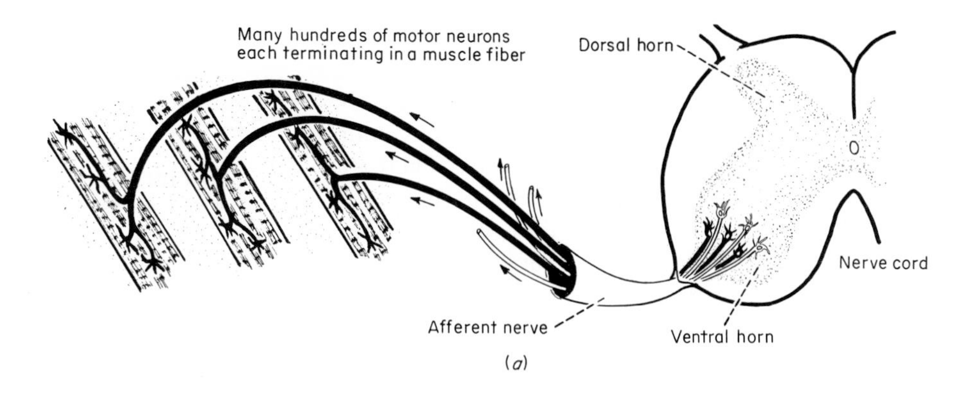

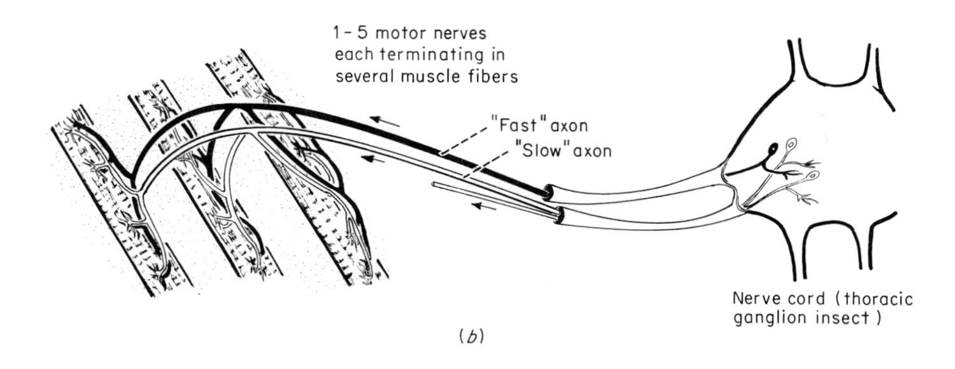

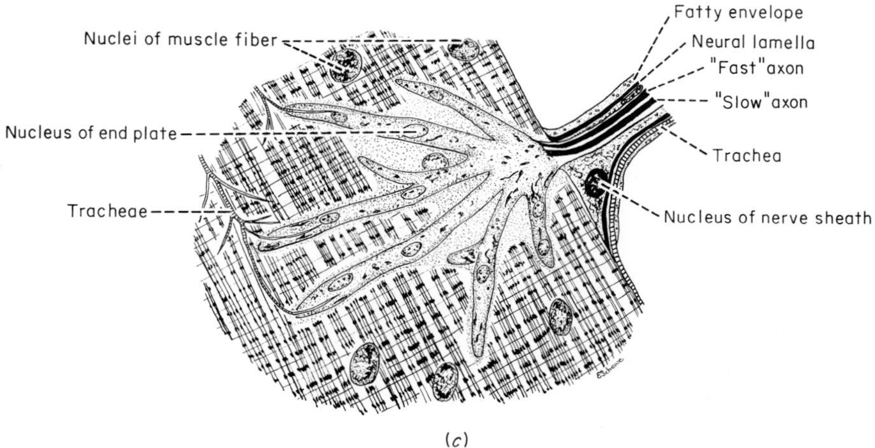

FIG. XVI–2 *Innervation in muscles. (From various sources.)* ***(a)*** *The innervation of vertebrate muscle. The motor neurons are shown in solid black, the inhibitor nerves in outline.* ***(b)*** *The double innervation of arthropod muscle, based on the extensor tibiae unit of the locust.* ***(c)*** *Detail of a motor end plate in locust muscle. The nerve terminals probably extend beyond the points shown in the drawing.*

clearly demonstrated cases of nonnervous conduction are in the simple epithelia of hydrozoans. This has been particularly carefully investigated in the epithelium covering the upper surfaces of the nectocalyces of siphonophores, and it has also been demonstrated in hydrozoan medusae. The epithelium of the exumbrella of a nectocalyx can quite readily be stripped off the underlying mesoglea for histological examination and physiological experimentation. In *Hippopodius hippopus*, for example, it is partly cellular and partly syncytial, with the cellular region lying near the stalk attaching the nectocalyx to the main body of the colony and the syncytial peripheral to the cellular. There is a sharp line of demarcation between these two regions, but the cellular merges into the contractile epithelium of the stalk through a transition zone of cells in which muscle processes are but poorly developed, while the syncytial makes contact with discrete epitheliomuscular cells at the margin of the bell. All the cells in this upper epithelium are alike—about equal in size, with a cytoplasm full of secretory granules, and without muscle processes. There are no interstitial cells in it, no cnidocytes, and no nerve cells, at least none that can be demonstrated by any of the standard neurohistological techniques. Yet when the upper surface of an isolated nectocalyx is touched, or stimulated electrically by the passage of a current through the water without contact with an electrode, the rim of the bell rolls in, partially or completely closing off the subumbril cavity, and the stalk may contract, drawing the nectocalyx upward and closer to the main mass of the colony. Such behavioral responses clearly indicate conduction of a local excitation through the homogeneous sheet of epithelium, an indication that is confirmed by electrical recordings made on isolated strips of tissue. These reveal nondirectional propagated depolarizations of its cells, conduction velocities of 20 to 50 cm/sec, and a refractory period of 2 to 3 msec, following external stimulation of the exumbrella at any point on its surface. It seems likely that, in nature, contact stimulation anywhere on the external surface of a nectocalyx results in responses protective of this particular member of the colony, such as its withdrawal and closure, and of the whole animal by the induction of swimming movements. These result from spread of the excitation to the epitheliomuscular cells of the subumbrellar epithelium, whose contractions and expansions cause the pulsations of the bell by which the colony is slowly propelled along, assisting the air and water currents that are mainly responsible for its passage through the water. Similar studies on other siphonophores have yielded evidence of similar nonnervous conduction. Potentials have also been recorded from the simple epithelia of the medusae of *Sarsia*, *Euphysa*, and *Phialidium*, after tactile or electrical stimulation, and there is evidence that the excitation initiated there may spread and be transmitted to nerve cells, arousing them to activity.

D. SPONTANEOUS ACTIVITY AND PACEMAKERS

There is no indication of endogenous or rhythmic activity in the bells of these hydrozoans, for contraction occurs only after external stimulation. In many tissues of other invertebrates, however, there is a rhythm of activity, although not always continuous, that can be attributed to the periodic transmission of a spontaneously arising nervous impulse. This may have its origin in a group of neurons, in a single neuron, or even in one particular region of a neuron, any one of which may serve as a pacemaker—that is, the ultimate source of the excitation which sets the tempo of activity for the cells that it controls. There may be several pacemakers operating for the same or for different tissues, and each may operate at its own pace, independent of the others. The impulses in them arise without any apparent stimulus, either external to the organism or internal to it, and must, at least in the light of present knowledge, be considered autogenic. Recordings from different areas of isolated neurons have shown that the membrane may have different properties in different regions, one area of the membrane being active while others are not. This has been most clearly demonstrated in the right giant nerve cell in the abdominal ganglion of *Aplysia depilans*. Branches of the axon of this cell extend, without synapses, through the right pleural, pedal, and cerebral ganglia. Recordings from microelectrodes placed at different points along its length have located the pacemaker in its proximal part, in a region where the intrinsic excitability of the membrane is especially high and where it is capable of showing periodic changes in potential. Data such as these have expanded the concept of the neuron as a simple on-off element to one that is capable of a wide range of activities, some of which may follow extended and intricate time courses. It is apparent, too, from investigations on other animals that the site of a pacemaker may shift from place to place and from time to time but that, whatever it may be and wherever it may be, it is a part of the nervous system capable of spontaneous, as yet unexplained, activity that can be communicated to other cells, coordinating their activity.

E. NERVE NETS

The nerve net as a pattern of neural organization is best known in cnidarians and in asteroids, whose dermal branchiae are innervated through a demonstrated nerve net (see Fig. IV–34*a*). The criterion of a nerve net is that conduction in any direction should be possible, the excitation spreading among the neurons adjacent to the point of stimulation in a nondirectional, nonpolarized fashion. Its existence depends upon experimental evidence that demonstrates that the impulse is conducted around cuts or other recognized neural blocks—that it can, so to speak, turn corners. Such evidence has greater validity than histological evidence, for even the most minute and scrupulous microscopic examination cannot make distinctions between a nerve net and a plexus of any other kind. Histology can, of course, provide evidence for a diffuse rather than a centralized system, but it is its functional rather than its morphological properties that characterize a nerve net.

Experimental evidence for nerve nets in cnidarians was provided nearly a century ago by the English biologist G. J. Romanes (1848–1894). In experiments on *Aurelia aurita* he found that deep incisions into the disk of the medusa, made either radially or circumferentially, did not interrupt propagation of the wave of contraction that spreads around the bell after a local stimulation, nor did removal of considerable portions of it. This was true as long as a strip of tissue not less than 1 mm in width remained intact. This would not be the case in a centralized system where transection of a nerve cord or trunk would effectively prevent passage of an impulse to the tissues innervated by it. These experiments of Romanes, which have been confirmed again and again on *Aurelia* and other jellyfish, showed that the impulse is not transmitted along a linear chain of neurons, but radiates from its point of origin. Such an unpolarized, diffuse mechanism, where alternative pathways are possible, clearly would be of advantage to a radially symmetrical animal sensitive to stimulation over all its perimeter.

Histological techniques have made it possible to visualize the types and distribution of neurons in cnidarian nerve nets. On the basis of their location, general structure, and the kinds and arrangements of vesicles in the cell body and its processes, three types have been demonstrated in *Hydra*, an animal generally considered to have one of the most primitive of nervous systems and one probably representing the basic pattern of cnidarian systems. These cells have been designated ganglion cells (i.e., interneurons), sensory cells, and neurosecretory cells in the more limited definition of the term. The ganglion cells lie at the bases of the epitheliomuscular cells, with their processes extending between the epithelial cells and terminating in bulbous enlargements. They are bipolar, tripolar, and occasionally multipolar, often with an especially long neurite that could qualify as an axon. In the smaller cells this process may be 15 μ+ in length and in large ones as much as 30 μ+. They are primarily the cells of the nerve net and are especially numerous in the proximal quarter of each tentacle, in the hypostome, and in the basal disk. Sensory cells lie between the epitheliomuscular cells and are thin, elongated, and often unipolar. They are most abundant on the tentacles, especially in the proximal quarter, and are absent, or not recognizable, in the basal disk. Neurosecretory cells are similar to the ganglion cells in location, in size, and in shape, but differ from them in their content of dense, membrane-bounded granules and in the apparent absence of any direct or synaptic connections between them and other neurons or effector cells, except possibly the nematocytes. These neurons seem to be especially concentrated in the hypostome.

Although morphologically there seems to be but one continuous nerve net in the various species of *Hydra* that have been examined, electrophysiological experiments give evidence of two conducting systems in the column. One of these is characterized by potentials of small amplitude, the other by those of large. There is, therefore, a functional distinction of pathways and through, rather than wholly diffuse, conduction even at this level of neural organization. Electrical recordings from the stalk of the colonial hydroid *Tubularia larynx* have also provided evidence for three parallel, but distinct, conducting systems there, each of which is nonpolarized. One of them controls depression and extension of the distal tentacles of the polyps, another triggers the potentials that appear spontaneously in the constricted neck region of each polyp, and the third gives rise to large, slowly propagated potentials in the stalk.

Two nerve nets are demonstrable histologically in scyphozoan medusae. One of these is a network of giant fibers, which is recognized as controlling symmetrical contractions of the bell of *Cyanea.* The other, a more diffuse one, can increase the frequency of the pulsations of the bell by acting in some way upon the pacemakers located near the rhopalia. The impulses that bring about rhythmic contractions of the bells of jellyfish are initiated in one of these and spread to others following the muscular contractions, so that they are brought into harmony with the instigating pacemaker. Both nerve nets in *Cyanea* are through conducting systems. A third one has been found by electrophysiological means. This is a slowly conducting one, with impulses passing at a velocity of about 2 cm/sec, and may well be a nonnervous one. Behavioral studies have shown that there is interaction between the nets and that locomotion, tentacle and manubrium movements, and righting reactions following stimulation of the statocysts are mediated by all three. Yet the mechanisms involved in the locomotion of scyphomedusae seem simpler than those in hydromedusae, where there appear to be multiple pacemakers, superimposed conducting systems, and other complexities of neuromuscular operation.

The motor neurons of anthozoans are organized as a network. Behavioral observations have produced evidence for at least two conducting systems in the column and oral disk, and electrophysiological experiments show fast and slow ones innervating the retractor muscles that run along the mesenteries, although there is no evidence of a double nerve supply to them. In *Metridium senile* and in a related New Zealand species, *Mimetridium crystum*, both of which may be considered typical of this class of cnidarians, the networks consist of bipolar, and occasionally tripolar, neurons in synaptic conjunction with each other. Their processes lie in the intercellular spaces between the epitheliomuscular cells, not in the mesoglea, and seem to terminate in bulbous enlargements on the muscle field. The neural meshwork over the retractor muscles in the mesenteries is far denser than it is over the parietal muscles; in *Mimetridium* there are many nerve processes, each about 2 μ in diameter, overlying the fast-contracting muscles, while over the slow-contracting muscles they are fewer in number and more widely scattered, with diameters of about 1 μ. Their parallel orientation over the retractor muscles is also evident, which, together with their size, would make for fast conduction of an impulse along them and rapid response of the animal. A mesh of giant nerves, with neurites 4 to 7 μ in diameter, lies in the oral disk; some of its neurons run around the disk, at right angles to the muscular processes of the ectodermal cells, while others diverge toward the bases of the tentacles, their processes running up into them. Each tentacle receives four to six such nerve processes, about 2 to 3 μ in diameter, which run along its length and are occasionally crossed by much finer neurites whose cell bodies may lie in the tentacles themselves. This constitutes the motor innervation of the tentacles, in which there are many sensory cells whose neurites may make contact with each other, with the neurites of the main network, or possibly directly with the muscle processes. In *Metridium*, nets of sensory cells are most conspicuous in the tentacles, the oral disk, and the parietal regions of the mesenteries nearest the body wall, but sense cells are also numerous on the oesophagus and the mesenterial filaments.

Two distinct networks can be microscopically distinguished in the external epithelium of the hydrozoan *Vellella.* One of these is an open one, whose neurons have fine processes that run independently of each other for 2 mm or less; the other is a closed one, whose neurons have large processes that unite in a syncytium in which the microfibrils of different neurons intermingle. The epitheliomuscular cells with which these nerves make contact are limited to the float of the colony and to its circular projection, or mantle, and to the tentacles, gastrozooids, and gonozooids. Both plexuses are found in these regions, but the open system is only poorly developed wherever the epithelium is invaginated. The continuity of the closed system implies that it operates as a nerve net, but there is as yet no experimental proof of this.

In the light of such histological and physiological evidence it seems unjustifiable to speak of the cnidarian nervous system as a nerve net. Rather, it is a system comprising several nerve nets, some of which are through conducting systems mediating quick responses, and some of which are diffuse ones, mediating slow responses. Although each net may represent a separate conducting system, there is both structural and functional integration between them, and while one may at times act as a separate unit in mediating a local response, they act together in mediating a general one.

The existence of a subepidermal nerve plexus, as well as of radial nerves and a circumoral nerve ring, has long been recognized as characteristic of the neural organization of echinoderms. But it is only recently that experimental evidence has been produced to show that in asteroids this plexus operates as a nerve net in mediating the responses of the dermal branchiae, or papulae, to tactile stimulation. In intact starfishes, these structures are retracted either when directly touched or when the epidermis near them is touched. They do so also after electrical stimulation of the epidermis in their vicinity, as well as when a shadow passes over them. This latter reaction depends upon the integrity of the radial nerves, but the withdrawal response to contact is independent. Experiments on several different species of starfish, from different parts of the Pacific Ocean, have shown this, for when the radial nerves were destroyed and cuts made through the epidermis and the underlying nerve plexus, papulae on both sides of the cut would retract after stimulation of the epidermis on one side of it. The responses were such as to show that the impulse radiated from the point of its origin and travelled around the cut to reach the retractor muscles of the papulae on the side opposite this point. Papulae some distance from it were not involved, for there is a limit to the area through which the excitation can spread, particularly if the corner it must turn is a sharp one such as that between two parallel, overlapping cuts a few millimeters apart. The maximum spread was through a radius of about 20 mm, exhibited by a species of *Culcita* from Eniwetok in the Marshall Islands, a starfish of which the total diameter, in large specimens, is about 22 cm. The papular response of these starfishes is the only instance in echinoderms where conduction through a nerve net is known to occur. Although there is a subepidermal plexus also in echinoids, it is not a nerve net, for the impulses that bring about convergence of the spines in response to tactile stimulation of the epidermis will not travel around a cut and do not "turn corners."

F. CENTRALIZED NERVOUS SYSTEMS

It seems likely that centralized nervous systems evolved from nerve nets, for the anatomical arrangement, if not the functional operation, of these persists in ctenophores, in the most primitive acoelous flatworms, in polyclad flatworms and in the peripheral system of others, and of nemerteans, endoprocts, phoronids, ectoprocts, and echinoderms. The pattern also persists in important regions of more highly organized invertebrates, for example, in the feet of gastropod and the arms of cephalopod molluscs, and in the gut plexuses of annelids, arthropods, and most molluscs. In subepithelial networks the axons of the neurons form a fibrous layer between the basement membrane and the epithelial cells, among which their cell bodies lie, and conduction in them is polarized, proceeding from neuron to neuron in one direction only. Thickening of the nerve fiber layer of a subepidermal plexus is a first step toward formation of a nerve cord; later steps may well have been the sinking of these thickened strands deeper into underlying tissues and the collection of nerve cells into ganglia. Subsequently, concentration of ganglia in the anterior region, associated with localization of important sensory areas there, and the organization of well-defined peripheral nerve trunks have resulted in the complex centralized systems of the most highly organized invertebrates.

In flatworms other than the most primitive acoeles and polyclads there are well-defined longitudinal nerve cords. In some acoelous species there are five pairs of these, situated dorsally, dorso-laterally, laterally, ventro-laterally, and ventrally. In others, some of them may be missing, indicating a tendency toward reduction in number and greater centralization with evolutionary development in the phylum. Even in primitive Acoela, where the entire system is in the pattern of a plexus, there is a concentration of nerve cells in the anterior region, forming a fairly compact mass constituting a cephalic ganglion or "brain" which is an important center of neural activity. In other turbellarians there is a reduction in the number of longitudinal nerve cords; marine triclads usually have three or four pairs, but in freshwater species and in most rhabdocoeles the ventral pair alone persists (see Fig. II–10). These may be greatly enlarged at their anterior ends to form the cephalic ganglia, or these ganglia may be separate concentrations of neurons adjacent to the anterior terminations of the longitudinal cords. There are connections between these cords and the peripheral subepidermal plexus, of which they are essentially a part.

In polyclads the plexus has moved in to lie below the muscle layer and, though in general more dense on the ventral side than on the dorsal and with some linear thickenings, is not compacted into the longitudinal cords characteristic of triclads. There is, however, a concentrated area in the anterior region which constitutes a cephalic ganglion. Transection experiments have shown that the system is not a nerve net. If a median longitudinal cut is made in an animal like *Polycelis*, extending from the posterior end of the body to the cephalic ganglion, stimulation at a point in the posterior region at one side of the cut leads to contraction of the muscles on the opposite side, indicating that the excitation has travelled anteriorly and been relayed through the cephalic ganglion to cells on the other side. But if, in addition, a transverse cut is made on one side, severing even one of the linear thickenings, there is no such contralateral response, indicating that the impulse travels anteriorly along a prescribed route and that its passage is blocked if this is interrupted. One pathway cannot substitute for another, as in a nerve net, and the impulse does not "turn corners."

The pattern of the nervous system of trematodes is basically that of turbellarians. Typically, there are three pairs of longitudinal cords connected with the cerebral ganglia—one dorsal pair, one lateral pair, and one ventral pair. These are connected by numerous commissures, of which there may be as many as 40 to 70, and are continuous with the submuscular nerve plexus. In cestodes there is a nerve ring in the scolex and three pairs of longitudinal cords extending throughout the chain of proglottids. Their nervous systems, therefore, remain as part of their flatworm organization that has been retained even in their wholly parasitic mode of existence.

In nemerteans there is a greater concentration of cells in the plexus into ventral nerve cords and cerebral ganglia than in flatworms. Paired dorsal and ventral ganglia, connected by dorsal and ventral commissures, form a nerve ring at the proximal end of the rhynchodaeum, the ectodermal invagination that leads from the proboscis pore to the proboscis. Two large lateroventral nerve cords run from the sides of this four-lobed ganglion along the whole length of the body. Branches from them, spaced at short intervals, pass to the body wall and to the cord of the opposite side and, as concentrated regions of it, both cords are intimately connected with the nerve plexus.

In endoprocts, as represented by *Loxosoma*, there is a single large median ganglion situated between the stomach and the vestibule. Six pairs of nerves are connected with it, three pairs running to the tentacles and three pairs to the stalk by which this solitary species attaches itself to the substrate. In phoronids the epidermal nerve plexus is thickened into a ring at the base of the lophophore; from this, nerves arise that innervate the tentacles as well as a large lateral nerve, actually a single giant nerve fiber, that runs posteriorly on the left side of the body. In some species there is a smaller nerve, of similar constitution, on the right side. The musculature of the body wall is innervated through the subepidermal nerve net. In ectoprocts, there is a ganglionic mass on the dorsal side of the pharynx from either side of which a nerve emerges to encircle the pharynx as a nerve ring. Nerves from the ganglion and the ring innervate the various organs and all are connected with the subepidermal nerve plexus. The ganglionic mass of marine species is somewhat smaller than that of freshwater ones; in these species two large nerves, unusual in organisms at this level of development because they have ganglionic enlargements along them, extend into the arms of the lophophore. In other respects the systems are alike in all members of the phylum.

Concentrations of the subepidermal plexus in echinoderms conform to their general pentameric pattern. There is a circumoral nerve ring from which conspicuous strands, the radial nerves, extend out into the arms of starfish and along corresponding ambulacral areas in members of the other classes. These constitute the superficial central system, beneath which there is a less conspicuous deep one, following the same pattern. Less conspicuous also than the radial cords and central ring of the superficial system is a similar aboral condensation of the plexus. There are also other local condensations, such as the nerve rings in the ends of the tube feet in starfishes and smaller nerves that run to various organs and structures in echinoderms of all classes. The role of the central nervous system in coordinating the activities of the tube feet in starfishes has already been noted (Chapter 6), as well as evidence for the control of their papulae through the diffuse subepidermal nerve net. In other echinoderms, as in echinoids, the peripheral system is represented mainly by the subepidermal plexus containing sensory, association, and motor neurons with polarized connections between them.

The nervous systems of invertebrates in other phyla are characterized anatomically by the presence of distinct ganglia and functionally by polarized conduction in the cords and the nerves originating from them. In rotifers there is a conspicuous ganglionic mass overlying the mastax from which varying numbers of paired nerves pass anteriorly and also two cords, with ganglionic enlargements along them, that run posteriorly along the length of the body in a more or less ventro-lateral position. Nerves emerging from these cords pass to the various organs and, in some species, to peripherally located ganglia. In nematodes a nerve ring with a single dorsal ganglion, and paired lateral and paired ventral, ganglionic enlargements encircles the anterior region of the digestive tract. A number of nerves pass anteriorly from it and, from its posterior surface, a pair of ventral, a pair of dorsal, and a pair of lateral cords. The ventral pair fuse shortly after they leave the ring and run posteriorly as a single large, median, ventral nerve. The dorsal pair have been shown to be composed of afferent fibers and the ventral pair of efferent ones. In some species there are also pairs of ventro-lateral and dorsolateral nerves, making a total of five pairs in all and a pattern similar to that of some of the turbellarian flatworms.

The basic pattern of the molluscan nervous system is exemplified by the chitons, in which the neural organization is similar to that of *Neopilina* and probably not very different from that of the ancestral type (see Fig. II–16). The chiton central nervous system consists of a nerve ring encircling the oesophagus with two pairs of cords running posteriorly from it and giving off nerves along their length (Fig. XVI–3). One pair emerging from the ventral surface of the circum-oesophageal ring lies ventrally and its nerves innervate the foot. The members of the other pair run along the sides of the body and unite posteriorly; the nerves from them innervate the mantle and the visceral organs. In addition to these two cords (pedal and pallioviseral, respectively), nerves from the circumoesophageal ring pass anteriorly to innervate the buccal cavity and the subradular sac.

In gastropods, pelecypods, and cephalopods there are well-defined ganglia, in some species quite distinct and well separated from each other and in others clustered together in a composite mass in which the separate components are often difficult to distinguish. The ganglia common to each of these classes are a

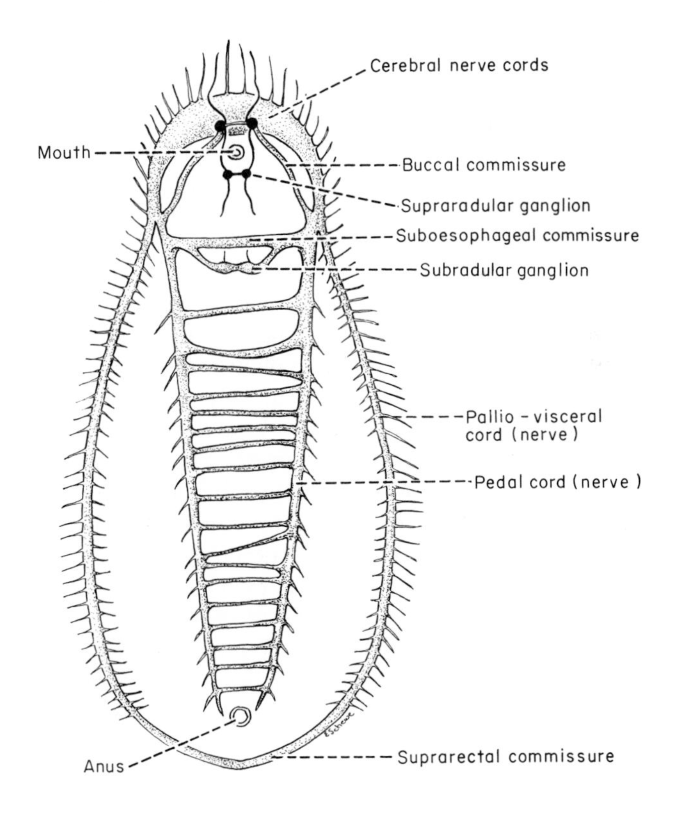

FIG. XVI–3 *Nervous system of* Ischnochiton winckworthi. (*After Grassé.*)

cerebral pair, lying dorsal to the oesophagus; a pedal pair, typically situated in the foot; and a visceral pair, situated among the visceral organs. The pedal ganglia are connected to each other by a commissure and each to the cerebral ganglion of its side by a cord. The visceral ganglia are likewise connected together and to the cerebral ganglia. In some species there are ganglionic enlargements along the course of each cerebrovisceral connective, or visceral nerve. These are a pleural ganglion which innervates the mantle and a parietal ganglion which innervates the gills as well as the mantle. There are usually, too, a pair of buccal ganglia anterior to the cerebral ganglia and connected to them by cords and to each other by a commissure.

In gastropods, torsion has greatly altered the fundamental basic bilateral symmetry of the molluscan pattern and displaced the parietal and visceral ganglia from one side to the other, besides twisting the two visceral nerves into a figure eight (see Fig. II–21). The left one passes across to the right side under the gut, the right one to the left side over the gut. The ganglia are also brought further forward in the body, and the parietal ganglion that was on the right before torsion, and after it is on the left, is placed higher in the visceral mass than its mate; they are, therefore, sometimes designated as supra- and infraparietal ganglia. This also brings them nearer the cerebral ganglia with which in the most highly evolved prosobranchs they are closely associated. In the genus *Busycon*, for example, cerebral, pleural, parietal, and pedal ganglia together form an eight-lobed mass around the oesophagus, with the two visceral ganglia removed but a little distance from it. Shortening and virtual disappearance of the connectives between the ganglia seem to have been features associated with torsion, and indeed with evolutionary development of molluscs in general. This is evident not only in the concentration of linearly placed ganglia that were separated some distance from each other as they are in the chitons but in the apparent fusion, or at least very close apposition, of the two members of a pair so that they appear as a unified mass rather than as discrete bodies. In *Patella*, for example, and in *Busycon*, the two visceral ganglia are compacted in this way, although remaining separate from the common mass encircling the oesophagus. But in pulmonates the visceral ganglia are included in this, and all traces of connectives between them and the parietal and pleural ganglia have disappeared.

Symmetry has been restored to some degree in opisthobranchs that have undergone partial or complete detorsion, for the visceral nerves are untwisted. In *Patella*, the circumoesophageal complex contains cerebral, pedal, and pleural ganglia, with the visceral removed some distance from it (Fig. XVI–4). Both visceral ganglia are fused into a common mass. In *Aplysia*, the left parietal ganglion is entirely wanting and the remaining right one lies close to the visceral, forming the abdominal ganglion in which the giant cell lies.

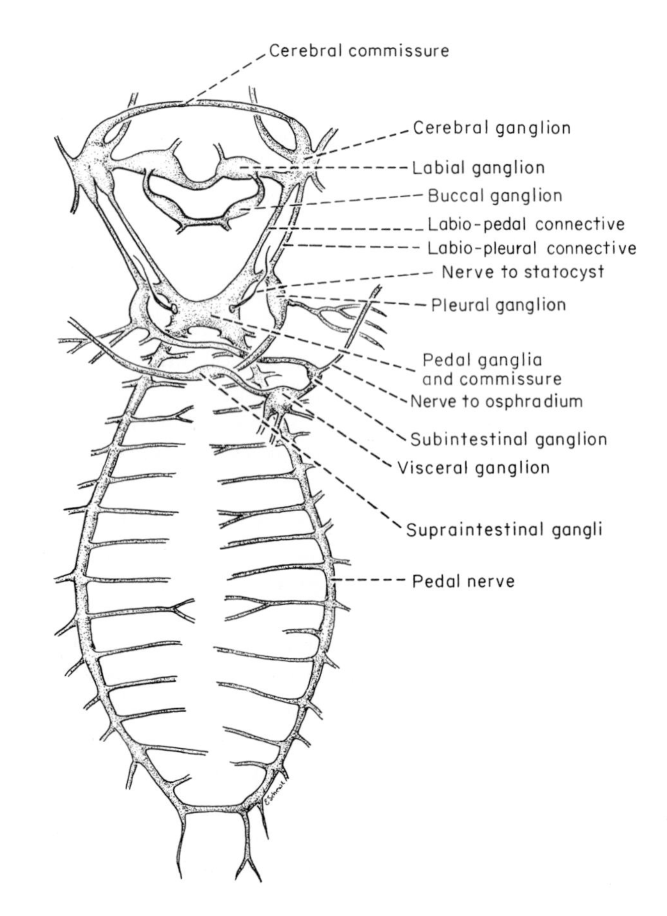

FIG. XVI–4 *Nervous system of* Patella. *(After Hänstrom.)*

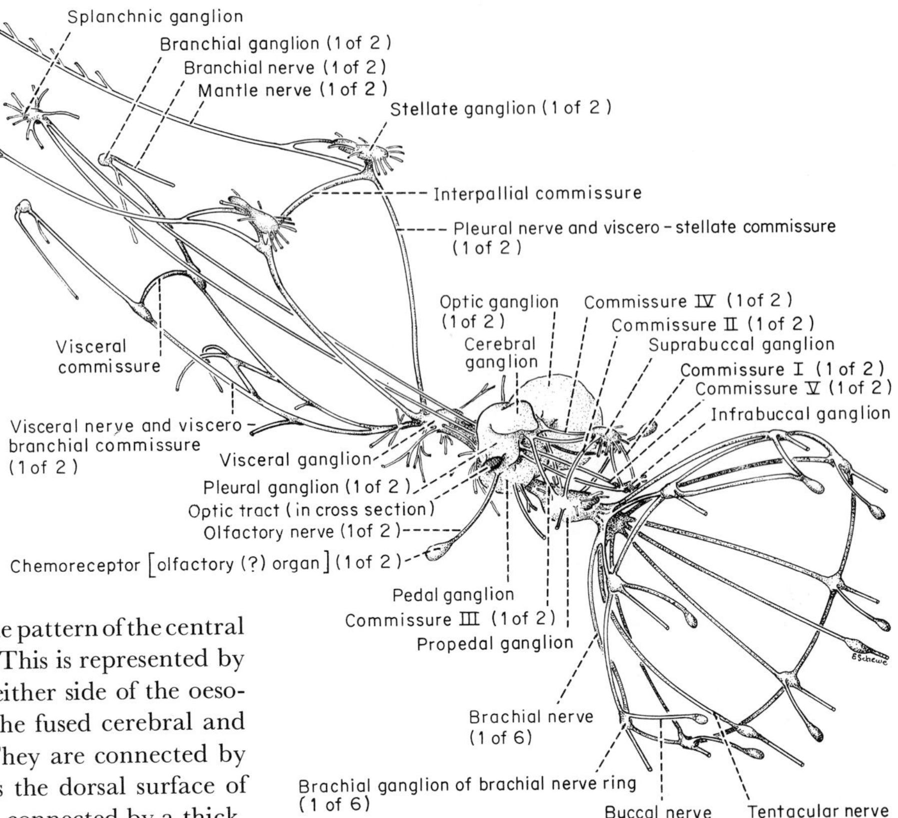

FIG. XVI–5 *Nervous system of a squid* (Loligo). (*After Williams.*) *The commissures are numbered I to V: I, suprabuccal-infrabuccal commissure; II, suprabuccal-propedal commissure; III, cerebro-propedal commissure; IV, cerebro-suprabuccal commissure; V, bucco-splanchnic commissure.*

Bilaterality is preserved in the pattern of the central nervous system of pelecypods. This is represented by a pair of ganglia, one lying on either side of the oesophagus and each consisting of the fused cerebral and pleural ganglia of that side. They are connected by a commissure that passes across the dorsal surface of the oesophagus, and each is also connected by a thick, long cord to a visceral ganglion, lying posteriorly, and similarly to a pedal ganglion situated more ventrally in the fleshy mass of the foot. Both of the visceral and both of the pedal ganglia lie quite close to each other, with only a short commissure connecting the members of each pair. The tendency to cephalization, or anterior concentration of the main ganglia, does not seem to have arisen in bivalves.

This tendency is, however, very marked in cephalopods in which the typical molluscan ganglia are collected together in a large mass that surrounds the oesophagus and may well be called a brain. The mass is, moreover, almost entirely enclosed in firm, cartilage-like tissue, forming a brain case and providing protection for this very vital part of the animal's anatomy, which is the seat for coordination of its sensory and motor responses, for retention of information, and for any learning ability it manifests.

The central nervous system of the squid is less compact than that of cuttlefish or octopuses. In *Loligo*, the nerve cells are collected in 31 ganglia, all but one of which, the splanchnic ganglion, are paired (Fig. XVI–5). In these lie all the cell bodies of the neurons in the central system. The circumoesophageal ring is formed by the two cerebral ganglia, which lie dorsal to the oesophagus, the two pleural ganglia at its sides, and the two pedal and visceral ganglia ventral to it. These eight ganglia are so completely fused together that it is impossible to distinguish their precise boundaries, in spite of the lobulated appearance they give to the complex. Slightly anterior to this is a buccal ring, composed of paired and fused propedal ganglia, which are joined to the pedal ganglia by a broad connective; two large optic ganglia attached to the pleural ganglia by the optic tracts; a pair of fused suprabuccal ganglia; and a pair of fused infrabuccal ganglia. These are interconnected by commissures so that they virtually encircle the anterior end of the oesophagus. A minute ganglion is situated in the anterior and dorsal side of each optic tract; this is the ganglion pedunculi, which is connected to the pleural ganglion alone, and to it by only a few fibers. Distal to the buccal ring is the brachial ring, made up of four pairs of ganglia lying in the bases of the arms and connected to each other by commissures. Another pair of ganglia lie in the bases of the tentacles, but these are not connected to those in the arms. All these ganglia, with the exception of the pleural, have the construction typical of invertebrates, with peripherally placed cell bodies and central neuropil. The pleural ganglia are exceptional in that they seem to be a mass of fibers only.

A pair of long connectives extend from the infrabuccal ganglion to the single splanchnic ganglion situated posteriorly among the viscera. Two pairs of connectives join the visceral ganglion with two other pairs of posteriorly situated ganglia, the branchial and the stellate. The branchial ganglia lie at the bases of the gills, each sending a nerve to the gill and the branchial heart of its side. The large stellate ganglia lie further anteriorly, embedded in the mantle wall anterior to the gills. Their size, and the size of the large nerves emerging from them, make them clearly visible through the thin covering of the wall. These ganglia are relay stations and regions for synapses of the giant fiber system. In *Loligo* this begins with a pair of giant cells in the posterior part of the pedal ganglia (the first-order giant cells), whose axons enter the visceral ganglia, fuse, and then divide into several branches that end in synaptic junction with the axons of other giant cells of the visceral ganglia (the second-order cells). Some of the fibers from these nerves form the commissures to the stellate ganglion of each side and synapse there with neurons of the third order whose postganglionic fibers innervate the muscles of the mantle. Other second-order fibers run to the retractor muscles of the funnel and to those between the head and the posterior part of the mantle. There are 8 to 10 pairs of mantle nerves, made up of third-order giant fibers, whose diameters may be as much as 600 μ. These are formed by the fusion of 300 to 1,500 axons of neurons of ordinary size. The largest of these innervate the more remote parts of the mantle, ensuring simultaneous contraction of all parts of its wall.

There is a somewhat greater degree of concentration of anterior ganglia in the central nervous system of cuttlefish. In *Sepia* the brachial ganglia lie closer to the central mass than they do in *Loligo*, and are barely distinguishable from the pedal complex. The optic lobes are considerably smaller than they are in *Loligo*, where they are one of the most conspicuous parts of the brain. There is, however, the same separation of stellate, branchial, and splanchnic ganglia as in squids, and these lie some distance posterior to the circumoesophageal mass. This mass is divisible into a number of fairly distinct lobes. The most dorsal is the vertical lobe and, directly beneath it, the subvertical. Anterior to this, proceeding dorsoventrally, are the superior frontal, the anterior basal, and the inferior frontal lobes. The posterior basal lobe lies beneath the vertical at the posterior end of the supraoesophageal mass, and the two optic lobes lie laterally. The most anterior lobe of the suboesophageal mass is the brachial, then the pedal, and slightly dorsally, between these two, the anterior chromatophore lobe. The posterior part of the pedal ganglion is the magnocellular lobe, where the cell bodies of the giant fibers lie, and posterior to this, the palliovisceral, with the posterior chromatophore lobe just above its midportion and the fin lobe above that and immediately below the oesophagus. These lobes of the suboesophageal mass have been so named because of the areas they are known to innervate. The anatomical divisions make the brain particularly suitable to experimental study of the localization of particular functions in particular areas and of the effects of damaging or removing any one of them upon the complex of responses that are the natural behavior of the animal.

In octopods, all the ganglia are collected in the circumoesophageal ring, which contains an estimated 168 million cell bodies, all of those in the central system. This ring is marked off externally into a number of lobes, with distinct differences in their microscopic anatomy. The most conspicuous of these in the supraoesophageal mass are the optic lobes that project laterally from the central portion and are connected to it by stalk-like commissures. These contain 75 percent of all the ganglion cells. Each stalk has two small lobes, a pedunculate and an olfactory. The central part of the supraoesophageal mass is the vertical lobe, with four longitudinal grooves along it marking off five ridges. Anterior to this are the superior frontal lobe, the inferior frontal lobe beneath it, and the subfrontal lobe below that. The buccal lobe marks the anterior limit of the brain; posteriorly, this lobe divides into two lateral inferior frontal lobes, posterior to which is the anterior basal lobe and posterior to this, the posterior basal lobe comprising a number of anatomically distinct regions.

Lateral to the oesophagus are the magnocellular lobes, areas that are homologous to the magnocellular lobes of the decapods, although octopods have no giant fibers. Beneath the magnocellular lobes, forming the suboesophageal part of the brain, are the brachial, pedal, palliovisceral, vasomotor, and chromatophore lobes. The whole mass, therefore, represents a higher degree of concentration of nervous elements than is evident in the decapod members of the class. These different anatomical arrangements are open to interpretation as expressions of a higher degree of evolutionary development in octopods than in decapods, but also as expressions of adaptations to body form and different ways of life. For the comparatively slender, elongated body of the rapidly moving pelagic squid an extended, elongated pattern is probably best suited, as is a more condensed one for the chunkier body of a cuttlefish that moves slowly along the bottom of the sea. Even greater concentration is in conformity with the globular body of an octopus that leads a comparatively sedentary life in the cracks and crevices in which it makes its home. The giant fibers of squids and cuttlefish may well compensate for the distances between control centers and effector organs, ensuring rapid transmission of impulses between them.

The brain of a tetrabranch is considerably simpler in its construction than that of a dibranchiate cephalopod. The brain of *Nautilus* consists of but three pairs of ganglia, connected by bands of nervous tissue containing cell bodies as well as fibers, and is not, as in *Octopus*, divided into lobes. The supraoesophageal pair of ganglia are homologous to the optic lobes of other cephalopods, but the homologies of the two suboesophageal pairs have not been established. These are known as the anterior and posterior suboesophageal lobes; from the anterior, nerves pass to the tentacles and buccal regions, and from the posterior, to the shell muscles, the gills, and the visceral ganglion.

Most molluscs have a stomodael nervous system, although this has not been described in pelecypods. In chitons, nerves from the buccal ganglia, lying along the buccal commissures, pass to the walls and muscles of the buccal cavity and pharynx; to the salivary glands; to the radular sheath; to the stomach; and perhaps to all the rest of the alimentary canal. In gastropods the stomodael system contains both sensory and motor nerves. It has its central connections with the buccal and visceral ganglia, the nerves from both anastomosing and contributing to the plexus that covers the entire wall of the gut. Most of the cells of this plexus are unipolar, but occasional bipolar and multipolar ones have been demonstrated. Innervation of the digestive and salivary glands through this system has been described. In cephalopods the nerves of the stomodael system connect with the infrabuccal and palliovisceral ganglia and, as in gastropods, enter into the plexus that covers the gut and its derivative organs.

The pattern of the central nervous system in the annelid-arthropod line is quite different from that in molluscs. It is fundamentally a segmental one with, however, the same tendency toward cephalization that is evident in molluscs. Most species in this line have a giant fiber system through which, as in molluscs, conduction is very rapid and responses to impulses travelling along it are almost immediate.

In annelids the common pattern is that of a pair of cerebral ganglia lying dorsal to the gut, connected to a pair of ventral ganglia by commissures that surround the gut, and, succeeding this pair, a sequence of ganglia situated ventrally, one per segment, and joined to each other by longitudinal connectives (Fig. XVI–6). The cerebral ganglia may lie in the first segment, or more posteriorly; in the earthworm *Lumbricus* they lie in the third segment and represent the fused ganglia of that segment and the two anterior to it. The connectives between this and the first ventral ganglion, or the suboesophageal ganglion, may themselves be ganglionated. The central system of annelids contains neuroglia, or supporting cells, as well as conducting neurons.

In the more primitive members of the phylum there is a distinct separation between the two ventral cords. This is the case in serpulids and Sabellidae among polychaetes and in the Aelosomatidae of the oligochaetes. These two chains are usually connected by transverse commissures between the ganglia of

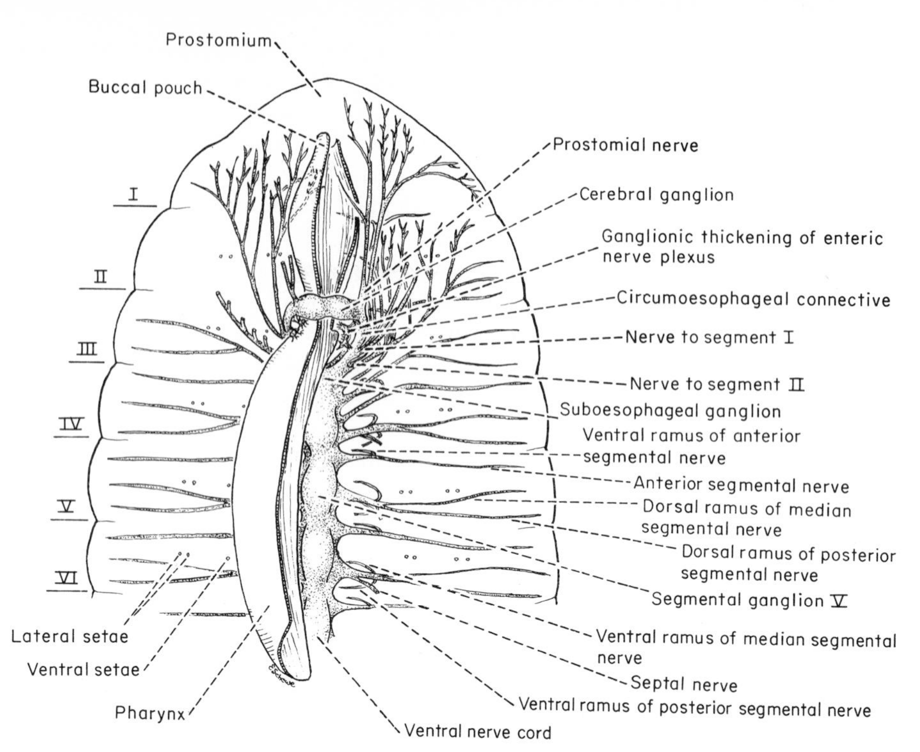

FIG. XVI–6 *The first five segments of* Lumbricus terrestris *showing the nervous system from the dorsal aspect. (After Hess.) The pharynx has been shown as if half cut away to show the ventral cord below it.*

opposite sides, giving the system a ladder-like appearance. In some polychaetes a separate pedal ganglion lies near each parapodium, connected to those in the next segment, on the same side, by longitudinal connectives, so that there are in effect four longitudinal and parallel nerve cords. In some species, there are no longitudinal connectives, but the pedal ganglia of the same segment are united by transverse commissures. There are various degrees of approximation and fusion of the paired ventral cords. In *Nereis*, they lie side by side in the midline, enclosed in a common connective tissue sheath, and in *Arenicola* they are fused into an apparently single cord. In *Arenicola* there are no ganglionic enlargements along this, and the cell bodies of the neurons are distributed along its entire length. In most oligochaetes, with the exception of Aelosomatidae, the ventral cords are fused into a single one. The pattern in leeches is, in general, typical, except for the fact that the supraoesophageal mass appears as a continuous ring rather than a bilobed structure, and the suboesophageal mass is large, formed by the fusion of four or five pairs of segmental ganglia. There is also fusion of ganglia at the posterior end of the body and a conspicuous neural mass there.

Peripheral nerves extend from the ganglia and, in most cases, from the cord as well. There may be as many as 16 pairs from the cerebral ganglia of polychaetes, depending on the number of sense organs localized in the head region. In *Lumbricus* there are three pairs, innervating the first, second, and third anterior segments and sending nerves to the prostomium as well. There are usually two to five pairs arising in each segment, but where ganglia are fused there are more. In *Lumbricus* there is no connection between the central system and the subepidermal nerve plexus, made up of sensory and motor neurons. These latter innervate the circular muscles of the body wall and with them constitute a system producing local responses that seem independent of any central direction or control.

The giant fiber system of annelids may consist of one or of many neurons. Among polychaetes it is best developed in tubiculous forms, to which rapid response to external stimuli by retraction into the tube of the tentacles or entire body is a necessary condition of survival. In *Myxicola* the range in diameter of a single giant axon is 500 to 700 μ, but diameters as great as 900 μ have been recorded. This fiber originates in the supraoesophageal ganglion from two especially large cells and lies on the dorsal side of the ventral cord, contributing to a large portion of its bulk. There are none in the Aelosomatidae, but they are regularly found in other oligochaetes. In *Lumbricus* the giant fibers, each containing a number of closely apposed axons, run all the way from the pharyngeal region to the terminal posterior segments. There are five of them—three lying in the middorsal region of the cord and two in the midventral. At 10 to 12°C, the speed of conduction in the median fiber is 17 to 25 m/sec; in the lateral ones, it is 7 to 12 m/sec.

A stomodael system is common in annelids and is even found fully developed in the archiannelids *Protodrilus* and *Saccocirrus*. It is highly developed in errant polychaetes with protrusible proboscides and is considerably simpler in sedentary species. It is well developed in oligochaetes and leeches. It consists of a plexus on the wall of the anterior part of the gut that is connected with the cerebral and first ventral ganglia or with connectives between them. There may also be connections with the more posterior part of the enteric plexus and with the segmental nerves of the posterior regions. Most probably there are sensory, motor, and integrative, or association, neurons in this system, the motor nerves in errant polychaetes providing a complicated mechanism for protrusion of the proboscis. Efferent nerves in other parts of the system may be concerned with maintenance of tone in the gut musculature and with secretion of digestive enzymes.

The tendency for anterior migration of the segmental ganglia and their concentration into a single mass, or brain, to which all the important sensory areas are connected is very evident in chelicerate arthropods and in *Limulus*. There are three regions, or areas, in the cephalic portion which have topographic, if not embryological, significance. The most anterior of these is known as the protocerebrum, which receives nerves from the eyes and the frontal organs; the next is the deutocerebrum, in which lie the sensory and motor centers of the antennae; and the third region, which is either ventral, inferior, or caudal to the other two, is the tritocerebrum with nerves to the labrum and stomodael system, and from which arise the postoral connectives. In some arthropods, one of these areas may be missing; there is, for example, no deutocerebrum in arachnids, antennaless arthropods.

In each of these areas quite well defined separate neuropils are histologically distinguishable, and connections within them may, therefore, be elaborate and intricate. All parts of the central nervous system of arthropods contain glia, or supporting, cells as well as ganglion cells or interneurons (see Fig. IV–36). In the protocerebrum of the arthropod brain there are characteristically five types of neuropil masses. Three on each side of the protocerebrum, arranged in linear order from the periphery toward the center, constitute the optic ganglia between the retina of the compound eyes and the brain proper and provide circuits for impulses originating in the eyes. In all arthropods these are similar in number and in position, but vary considerably in detail and the extent of their development. A neuropil that lies in the midventral region of the protocerebrum, directly adjacent to the oesophagus, is known as the central body. The protocerebral bridge is an elongated, median neuropil, crossing the anterior region of the brain. The corpora pedunculata, or "mushroom bodies," so called because of their shape, are two lateral groups of interneurons and neuropils, whose dorsal regions are cap-shaped, below which a stalk runs dorso-ventrally and expands into two small lobes. Bodies like these are found also in the supraoesophageal ganglia of annelids and of *Limulus*. Neuropils at the roots of the nerves of the simple eyes constitute the ocellar centers. In the deutocerebrum of mandibulate arthropods there is a conspicuous transverse commissure or bridge connecting the two sides and a large olfactory-globular tract of fibers running from the antennal lobes to the corpora pedunculata. In crustaceans, the olfactory lobes lie on each side at the level of the antennule nerves. The tritocerebrum contains the antennary neuropils and a transverse neuropil forming a bridge between the two halves. In crustaceans a tegumentary neuropil is also evident, between the antennary neuropils, representing the arborizations and connections of the small tegumentary nerves.

In *Limulus* the cephalic mass, which encircles the oesophagus like a collar, is composed of the ganglia of all the segments anterior to the abdomen. A ventral nerve cord extends posteriorly from this, without ganglia until it reaches the abdomen, in which there are five segmental ones. The lateral nerves extending from these innervate the appendages of the abdominal region, while all those of the cephalo-thoracic region are innervated by nerves emerging from the circum-oesophageal ring. Even greater concentration is evident in arachnids, in which the cerebral and the anterior ventral ganglia form a circumoral mass. There is a clear tendency in this group of chelicerates toward forward migration of all the ganglia in the ventral chain and their incorporation into a single ganglionic mass. Scorpions show transitional stages in this condensation and spiders, final ones. In scorpions the composite anterior ganglion innervates the organs of the cephalothorax, and from this a cord extends into the abdomen. In general, there are three ganglia in the preabdomen and four in the post-abdomen, or tail, the last one of which probably has resulted from the fusion of two ganglia. In *Thelyphonus*, one of the Pedipalpi, there is but a single abdominal ganglion, posteriorly placed at the end of the abdominal nerve (Fig. XVI–7). The posterior abdominal segments and the tail are innervated from this, but all other parts of the body are innervated from the composite anterior mass. In spiders, all of the ganglia are incorporated into a single mass lying around the oesophagus (see Fig. IV–7).

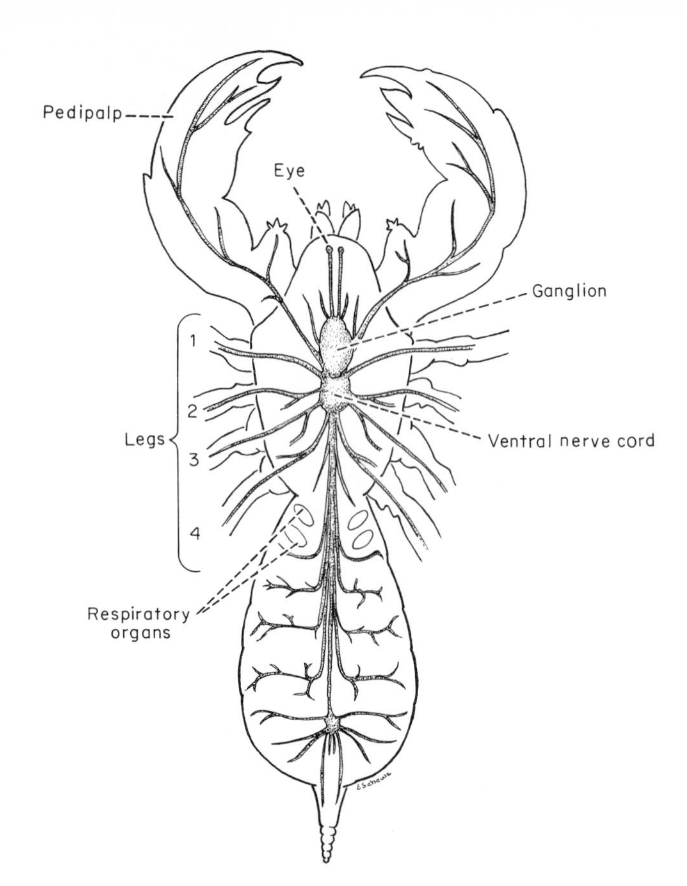

FIG. XVI–7 *Nervous system of the whip scorpion* Thelyphonus caudatus. (*After Hänstrom.*)

Among mandibulate arthropods the same tendency is manifest in crustaceans and insects, but not in myriapods in which the pattern is like that of annelids, with a supraoesophageal pair of ganglia and a ventral chain with one ganglion per segment. These and the connectives between them are closely apposed or fused together, so that it appears as a single system, although it is really a double one. In the simpler crustaceans, the duality of the system is very evident. In a branchiopod such as *Triops* (*Apus*), for example, a pair of ganglia, from which the optic nerves emerge, constitute the supraoesophageal mass, joined to the first ventral ganglia, which lie some distance from each other, by connectives that encircle the oesophagus. From these connectives emerge nerves that enter the antennules and the antennae, identifying them as deuto- and tritocerebral areas. The two cords of the chain of ventral ganglia, a pair in each segment, are also well separated from each other, with each ganglion connected to its mate in that segment by two commissures. Peripheral nerves emerge from these ganglia, those of the anterior segments innervating the mouthparts and those of the more posterior ones, the trunk appendages.

In copepods, on the other hand, there may be a very high degree of concentration of the ventral ganglia, with all of them uniting in a common sub-oesophageal mass. In some barnacles—the pedunculate species—there is a chain of four ventral ganglia posterior to the suboesophageal mass, but in sessile species all the ganglia have fused into a common mass.

Different degrees of cephalization are evident also among decapods. In crayfishes and lobsters, for example, the cephalic ganglia, as their nerve supply indicates, represent the fused ganglia of the segments bearing the antennules and antennae; the optic nerves are also connected to them, and each of these bears four ganglionic enlargements as it passes through the eyestalk (Fig. XVI–8). Each of the connectives between the cerebral ganglia and the suboesophageal mass has also a ganglionic enlargement from which emerge nerves of the stomodael system. These connectives are united by the tritocerebral commissure, crossing just behind the oesophagus, and join in the suboesophageal ganglion. This is also a fused one, composed of the ganglia of the segments that bear the mandibles, maxillules, maxillae, and the first two pairs of maxillipeds. So close to this that it seems almost a part of it lies that of the segment that bears the third pair of maxillipeds. Posterior to this are five segmental ganglia whose nerves innervate the five pairs of pereiopods and, posterior to these, a chain of six abdominal ganglia, the terminal one supplying

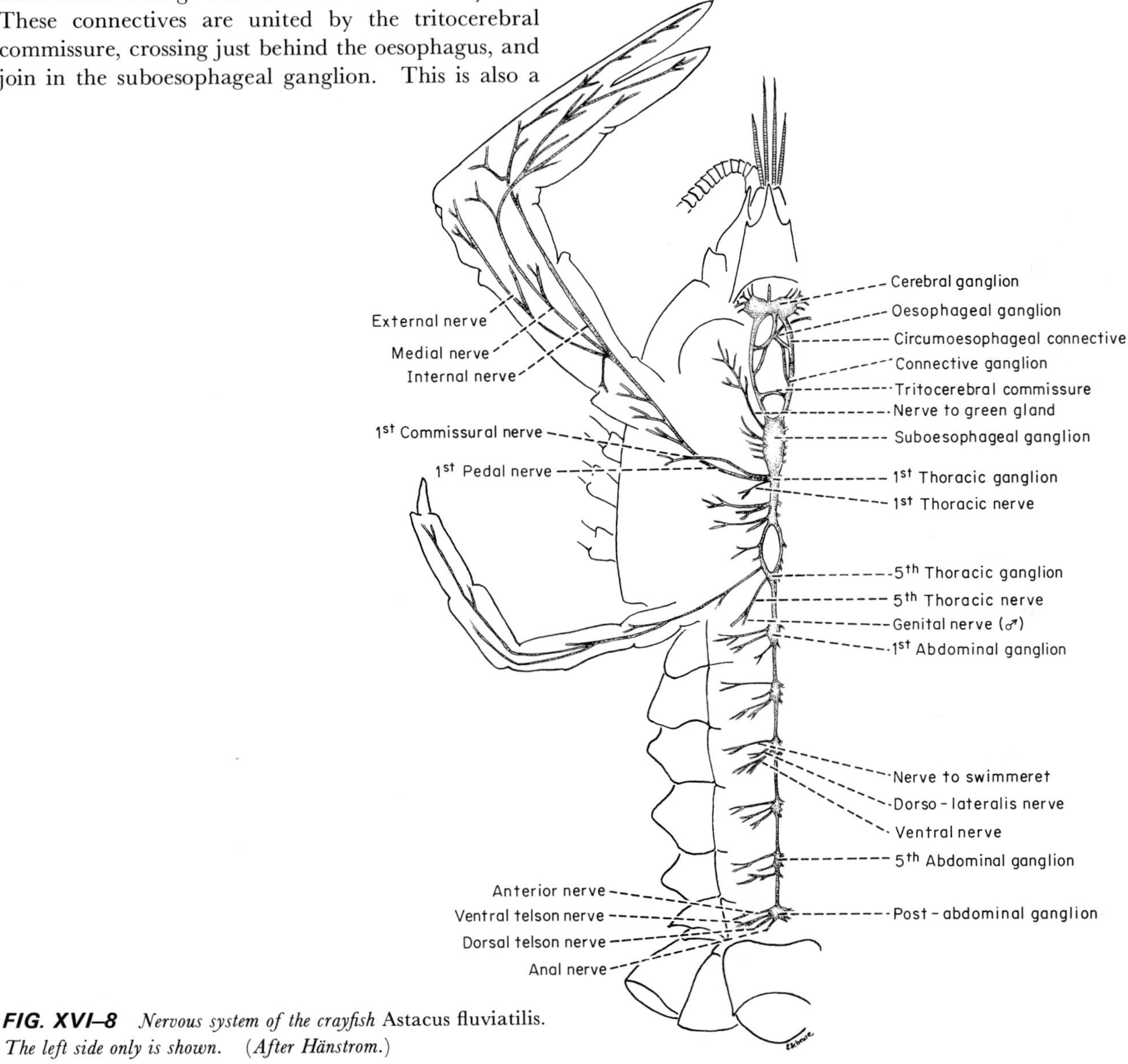

FIG. XVI–8 *Nervous system of the crayfish* Astacus fluviatilis. *The left side only is shown. (After Hänstrom.)*

nerves to the telson as well as to structures in its own segment. The two chains of ganglia are closely apposed through most of the length of the cord, but between the fourth and fifth thoracic ganglia the connectives diverge to permit passage of the sternal artery, a vessel that runs directly dorso-ventrally from the heart and divides below the nerve cord into a ventral abdominal and a ventral thoracic artery. In this region the duality of the nervous system is patent. The thoracic part of the ventral chain lies in a kind of vault made by the endophragmal system and so is protected in a way comparable to that in which the vertebrate nerve cord is protected by the vertebral column.

The system is considerably more concentrated in crabs, for in them all the thoracic ganglia are fused into a common mass which lies in the fourth or fifth thoracic segment, so that its connectives with the cerebral ganglia are considerably extended. From this ganglion emerge nerves that innervate the mouthparts as well as the thoracic appendages. A small nerve also leaves this ganglion and passes posteriorly into the reduced abdomen, but has no obvious ganglionic enlargements.

All crustaceans examined have well developed stomodael systems. In branchiopods, two nerves from the circumoesophageal commissures extend to the gut and ramify over its surface; in ostracods, a group of nerves originate from the posterior region of the oesophageal ring. A medial ganglion lies along the lateral loop of these, from which nerves go to the stomach, the aorta, and the gastric ganglion situated further posteriorly. In decapods, the nerves from the commissural ganglia unite in the midline in a single ganglion, which also receives a tiny nerve from the cerebral ganglion. Nerves from this median ganglion go to the oesophagus and the stomach.

Giant fibers have been described in some groups of decapods, but they are absent in brachyurans. Their course is best known in *Homarus* and in *Procambarus*. In the lobster there is a group situated in the middorsal region of the cord, whose cell bodies lie in the anterior part of the supraoesophageal mass. The axon of each crosses to the other side in the posterior part of the brain and then proceeds down the connectives and through the suboesophageal ganglion to the last ventral one. As they pass through the connectives the diameters of these fibers increase markedly and then gradually taper off. Diameters of 200 μ have been measured in those of the crayfish in the region of the connectives, which diminish to 150 μ or less in the last abdominal ganglion. Lateral giant fibers lie in the dorsal region of the cord, flanking the medial ones and often quite close to them. The cell bodies of these fibers lie in the ventral part of each segmental ganglion and their axons run anteriorly, after decussation in the ganglion and passage upward from ventral to dorsal side, and each terminates in conjunction with the axon of a neuron whose cell body lies in that ganglion. This association is so intimate that superficially there seems to be but one continuous axon. This lateral set of fibers thus forms a system rather more complex than the median one, and different from it in that the diameter of the individual fibers is greater in the cord (where it may be 200 μ) than in the commissures, where it is about 80 μ, and in that the impulses travel postero-anteriorly rather than antero-posteriorly.

A distinctive feature of the crustacean central nervous system is the paucity of neurons in it. The total number of neurons in all the ganglia of the central system of *Procambarus clarkii* is somewhat less than 30,000 in contrast to the 168 million in the circumoesophageal ring of *Octopus*. But the synaptic connections of interneurons are such that each receives impulses from many others and transmits them to many other cells. Crustacean neuromuscular relationships have been studied more intensively than those of any other group of invertebrates, and almost as intensively as those of vertebrates. Anatomically, they are distinguished by the fact that all the muscles in the body are innervated by only a few motor neurons, whose cell bodies lie in the ganglia of the central system and each of whose axons terminates on several muscle fibers. These axons may carry different messages as, for example, for a slow or a fast response or, in the case of decapods at least, for suppression or inhibition of response.

The general architectural plan of the insect nervous system closely resembles that of the crustacean and shows a similar tendency toward concentration of segmental ganglia, which is also apparent even within

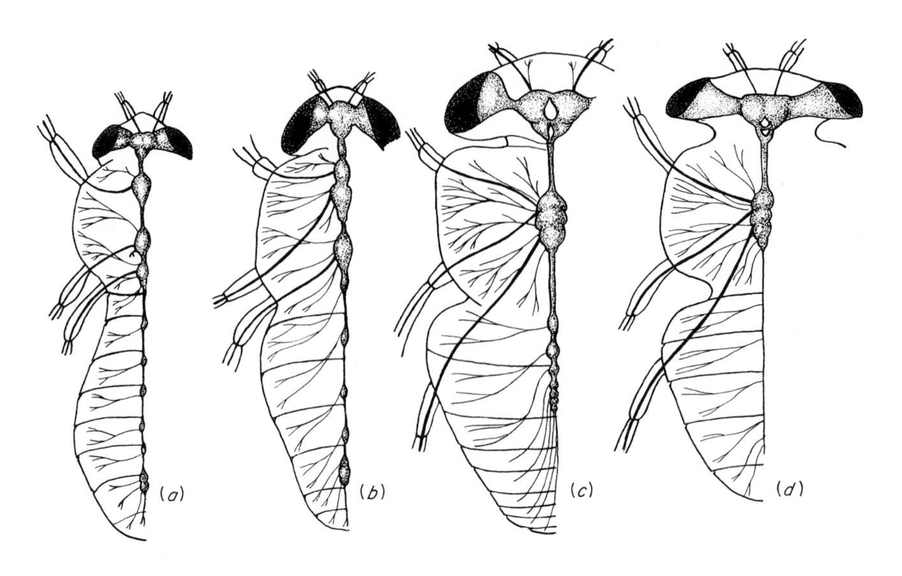

FIG. XVI–9 *Nervous systems in four members of Diptera, left side only. (From Gardiner.)* ***(a)*** *The midge* Tendipes (Chironomus), *in which the three thoracic ganglia and five of the abdominal ganglia are separate. The ganglia of the two terminal abdominal segments are fused into one.* ***(b)*** *The dance fly* Empis, *in which the ganglia of the first two thoracic segments are fused into one, as are those of the last three abdominal segments.* ***(c)*** *The horsefly* Tabanus, *in which all three thoracic ganglia are fused into one and all the abdominal ganglia concentrated in the first two abdominal segments.* ***(d)*** *The blowfly* Sarcophaga, *in which all segmental ganglia are fused into a single mass in the thorax.*

a single order (Fig. XVI–9). Insect muscles, like crustacean muscles, also have multiple and polyneural innervation, more than one axon terminating on a single muscle fiber on which each axon also has many terminations. There is also similarity between crustaceans and insects in the comparatively small number of neurons constituting their central systems and the small number of motor fibers to their muscles. In insects there are only one or two nerve fibers per muscle, in contrast to the tens or hundreds to even a small muscle in a vertebrate.

The brain of an insect shows the regional areas characteristic of arthropods in general, but in addition, in the extreme antero-dorsal part of the protocerebrum, just above the central body, there is a region known as the pars intercerebralis. This is a large cell mass, whose anterior median cells contribute to the ocellar tracts while the others are distinguished by their size and the number of vesicles in their cytoplasm. The brains of different species of insects vary from each other in size and in the relative sizes of their component parts; it has been said that there is as much divergence between the brains of locusts and wasps as there is between those of frogs and humans. The variations in insect brains are related chiefly to differences in development of different parts whose size reflects their functional importance to the insect's way of life. The mushroom and central bodies together, for example, represent half the total mass of a worker ant's brain, and one-fifth that of a bee's.

The suboesophageal ganglion is a complex of three pairs—those of the segments bearing the mandibles, maxillae, and labium. There are all degrees of fusion of ganglia and connectives in the ventral cord. In the thysanuran *Machilis*, for example, the cord is clearly double; there are three separate ganglia in the thorax and eight in the abdomen, the first seven being independent segmental ganglia and the last, a fusion of the ganglia of segments 8, 9, and 10. In the cockroach *Periplaneta americana* there are three large separate thoracic ganglia and six abdominal ones, some of these representing fusion of segmental ganglia. It is within the order Diptera that there is the greatest diversity in arrangements of the segmental ganglia of the ventral chain. In *Tendipes* (*Chironomus*) for example, all the segmental ganglia are separate, with the exception of the terminal abdominal one, which is made up of those of the last two segments. Others in the order show increasing degrees of concentration and fusion, up to the condition in houseflies and stable flies, in which all the ganglia of the ventral chain are united in one large mass in the thorax, through which the abdominal as well as the thoracic segments are innervated.

In some insects, a giant fiber system has been demonstrated, notably in *Periplaneta* and *Locusta*, in nymphs of the dragonflies *Anax* and *Aeschna*, and in *Drosophila*. There are six to eight giant fibers in *Periplaneta*, with diameters of 20 to 60 μ; their cell bodies lie in the last abdominal segment and their axons run forward without interruption to the thoracic ganglia, diminishing in diameter as they advance. There are four such fibers in *Locusta*, with diameters of 8 to 15 μ; they, too, originate in the last abdominal, and possibly also in the third thoracic, ganglion. There are six to seven fibers in *Anax* nymphs, with cell bodies lying in the last abdominal ganglion. These fibers differ from those of the cockroach and locust in that their course forward is not a continuous and uninterrupted one, for they make synaptic connections in both the abdominal and the thoracic ganglia. While the diameters of these fibers are small in comparison to those that have been measured in molluscs and annelids, they are considerably greater than those of other neurons in the insects, and given the size of an insect's body in relation to that of a worm's or a squid's, they are probably of proportionately the same order of magnitude. Their location in the posterior part of the abdomen, and the anterior course of the impulses set up in them in response to external stimuli, is consonant with the distribution of sense organs in the terminal segments, the cerci of cockroaches and the paraproct hairs of dragonfly nymphs being well supplied with receptors to whose stimulation an immediate response is desirable.

The stomodael system of insects typically consists of two nerves that emerge from the circumoesophageal connectives, pass forward, and unite in the frontal, or buccal, ganglion that lies ventral and slightly anterior to the supraoesophageal mass. From this ganglion, nerves run anteriorly to innervate the labrum, and a recurrent nerve runs posteriorly to the medianly situated hypocerebral ganglion, from which mixed nerves emerge and spread out over the foregut. This part of the peripheral system is concerned with movements involved in the ingestion and digestion of food and, therefore, with the coordination of many sensory inputs and motor outputs that may be extremely complicated, effecting rejection as well as acceptance movements of the mouthparts, regurgitation of food from the crop at appropriate intervals, and regulation of other delicately adjusted reactions of feeding and digestion. In some insects there is also a proctodael system, for which the nerves originate in the last, or the last two, abdominal ganglia and spread out over the hindgut.

G. FUNCTIONAL RELATIONSHIPS WITHIN CENTRALIZED SYSTEMS

Functional relationships within a centralized system can be determined by observations of the effects upon an animal's responses of transection, extirpation, or destruction of particular regions, or of electrical stimulation of them, in experiments similar to those that have been performed on vertebrates. Many invertebrates, however, because of their size or because of particular conditions in their neural organization, are not suitable for precise surgical or physiological techniques. For these it has been possible, in some cases, to determine the functional relations of nerve fibers by following the course of degeneration of those whose axons have been severed from their cell bodies by more or less gross operative procedures. From the comparatively few species from which significant data have been obtained, it is, nevertheless, apparent that while the central system acts as a relay and coordinating center for afferent and efferent impulses, there are wide differences in the degrees to which it controls or directs their pathways and so the animal's responses. There are also differences in the areas where this control, if any, is exercised. There are wide differences, too, in the extent to which invertebrates seem to be able to store information or experience to condition their future responses. The effects of habituation to a repeated situation or of training to respond to a predetermined experimental one, rather than to other possible alternatives, have been manifested by a number of invertebrate species subjected to training under experimental conditions. In the most extensive experiments that have been made upon this aspect of functional relationships within the central nervous system, it has been possible, primarily by surgical methods, to localize the seat of reminiscence, or "memory," with a considerable degree of accuracy. Detailed discussion of these relationships is beyond the scope of this book, but since they define the range of an animal's responses and behavior they have tremendous importance in their biology and must, therefore, have some mention, however brief. They represent, too, one of the most fascinating and challenging aspects of comparative neurophysiology.

The results of numerous experiments with turbellarian flatworms suggest that they may be able to store information and that past experiences may condition their future behavior. Flatworms naturally seek dark areas and will move from light to shade, provided their movements are unimpeded; contact of the anterior end with an object will, however, stop their forward progression. Polyclads, exposed alternately to 5-min periods of light and 30-min periods of darkness, were touched at the anterior end every time they moved during the period of light exposure, and so were effectively stopped. After hours of such treatment they made fewer attempts to get out of the light and thus were touched fewer times, a quantitative means of assessing the extent of their "reminiscence." This behavior continued some time without further training. Specimens from which the cerebral ganglia had been removed could not be trained in this way, indicating the significance of the ganglia as an associative and retention area. The triclad *Dugesia*, on the other hand, seems to "learn" even without cerebral ganglia; an explanation for this may reside in the more complicated and centralized "brain" of the polyclad. Since some experiments seem to have shown that "training" can be transferred from a trained to an untrained individual by cannibalization of the former by the latter, a hypothesis has been proposed that information is stored in the tissues on a molecular level and not necessarily in the associative circuits in the nervous system.

Molluscs, especially gastropods and cephalopods, have proved especially suitable objects for functional studies. In some of these it has been possible to trace the course of fiber tracts and neural pathways in the ganglia and lobes of the brain, as has been done for the brains of different vertebrates. In gastropods, the cerebral ganglia are coordinating centers for the mechanisms involved in locomotion, respiration, and cardiac activity. If these ganglia are removed, or their connectives cut, there is a general rise in excitability and, in some cases, increased locomotor activity, indicating that they exert an inhibitory influence upon other parts of the nervous system. The pedal ganglia are centers specifically for locomotion. The subepidermal plexus in the foot appears to be under their control and is also influenced by local, and perhaps remote, sensory stimuli. The pedal ganglia act as inhibitors of the effects of impulses traversing the plexus, which tend to keep the muscles of the foot in a state of contraction. If the pedal ganglia are removed from a specimen of *Helix*, the foot muscles shorten and remain fully contracted until the paralyzed animal finally dies. Relaxation, however, follows stimulation of the stumps of the pedal nerves distal to the ganglion, suggesting that in the intact animal impulses continually coming from the pedal ganglia counteract the effect of those induced in the plexus by peripheral stimuli, and so control the waves of contraction that normally spread along the foot. There is evidence also that in *Helix* the cerebral ganglia exert some control over both the inhibitory and excitatory impulses from the pedal ganglia, their own responses being influenced by impulses from the sense organs and other sensory areas. There is, as well, some evidence that the pedal ganglia are responsible for activation of the various glands, especially those producing mucus and odorous substances. Little is known about the precise function of the visceral ganglia in gastropods.

The cerebral ganglia of bivalves have a direct motor function as well as being a coordinating center for complex reflexes. The pedal ganglia control the activities of the foot in locomotion and, in *Mytilus*, spinning of the byssus threads. These activities are independent of the cerebral ganglia; an isolated foot of *Mytilus*, with its pedal ganglia intact, will still perform ordinary creeping movements. The visceral ganglia of these molluscs control shell opening and closing, which in *Mytilus* will continue in an animal from which the cerebral ganglia have been removed. The pedal ganglia are, however, necessary to this reflex and often rhythmic action, which requires retraction of the foot as well as contraction of the adductor muscles of the shell. In *Anodonta*, the visceral ganglia can control the closing of the shell valves, but not their opening; for this, the cerebral ganglia are necessary. The visceral ganglia also have an effect upon cardiac activity, for their electrical stimulation inhibits heartbeat.

Much more detailed work has been done upon functional relationships in cephalopods than in either gastropods or pelecypods, and extensive experiments have been conducted upon the motor effects of brain stimulation and brain lesions in *Sepia* and in *Octopus*. Electrical stimulation of the vertical, subvertical, superior or inferior frontal, dorsal part of the basal, and outer part of the optic lobes of *Sepia* provokes no motor responses, but extirpation of them, or transection of their connections with other parts of the brain, renders an animal incapable of learning and of training, either by repetition of experimentally produced situations or by repetition of those naturally occurring in the course of its life. Stimulation of the buccal lobe in *Sepia* results in movements of the buccal mass and radula; stimulation of the anterior basal lobe, in movements of the head, eyes, arms, and funnel; and of the posterior basal lobe, in extension of the tentacles, respiratory and swimming movements through contractions of the mantle muscles, and movements of the fins, as well as in changes in the distention and contraction of the chromatophores. Anatomical evidence has shown that neither of these basal lobes innervates the musculature directly, the impulses arising in them as a result of this stimulation being relayed through the suboesophageal parts of the brain. Stimulation of the inner regions of the optic lobes, or of the optic tracts, of *Sepia* lead to a wide variety of responses similar to those of the basal lobes, which is to be expected from the anatomical arrangements, since the fibers in the optic tracts run, for the most part, to the basal lobes. With both optic lobes removed, an octopus acts like a blinded animal, but there are no profound disturbances in its moving or feeding. If one lobe is removed or one optic tract cut, the natural posture of the animal is affected and it tends to bend toward the intact side. But lesions of the basal lobes cause profound differences in posture and in movement, as do those of the suboesophageal portion. If the entire supraoesophageal portion of the brain is removed, a cuttlefish moves continuously, always swimming backward, for the few days that it survives this operation, and an octopus does not move at all, and cannot eat, though it may survive for several weeks. During this period, its respiration continues, for much of the time, in the same way as in an intact animal, but with occasional periods of inactivity alternating with strong movements of the mantle muscles.

When the brachial lobe of a cuttlefish's brain is stimulated electrically, all the arms move; when the anterior part of the pedal lobe is, the two most dorsal pair of arms are raised upright in the "attention" position of an intact animal beginning to stalk its prey, and sometimes the fifth pair is extended laterally, another manifestation of the natural "attention" position. Stimulation of the posterior part of this lobe results in movements of the head, the funnel, and the fins; and stimulation of the lateral part, in eye movements on the stimulated side. If an anterior chromatophore lobe is stimulated, the chromatophores on that side of the head and arms expand and the skin papillae are erected. Stimulation of the posterior chromatophore lobes leads to expansion of the chromatophores on the funnel and the mantle. Contraction of the mantle, movements of the funnel, ejection of ink, and vasomotor responses follow stimulation of the palliovisceral lobe, and fin movements that of the fin lobe, the movements being limited to the fin on the side to which the stimulus is applied.

From observations and experiments such as these it may be concluded that in the suboesophageal lobes of a cephalopod's brain are the lower motor centers, concerned with innervation of particular organs or groups of organs vital to the animal. But it is in the supraoesophageal portion, and especially in the basal lobes, that coordination of these impulses takes place and converts what would otherwise be random responses into unified and meaningful patterns of behavior. That the pattern is influenced and can be modified by impulses from even higher centers has been shown by experiments on *Octopus*, in which the effects of training and learning have been demonstrated. The animal receives its information from the external world mainly through its visual and tactile sense organs and can learn both by sight and by contact to avoid or approach certain objects. An octopus can be taught, for example, by punishment through mild electric shocks not to attack a crab when it sees it along with a flat, white, square object, although it will continue to attack this natural prey when it is presented alone. When an octopus has had sufficient experience with the negative stimulus—that is, the stimulus to which it cannot respond with impunity—it may remember for 2 to 3 days not to attack a crab and a square, even without further experiences during this interval. If the training is reinforced by repetitions of the experience three times a day, the animal may refrain from attacks on the negative object for as long as 6 days. Hunger has apparently no effect upon this response, for a trained octopus that has been kept without food for several days still will not strike at a crab when it is accompanied by a square. Partial removal of the vertical lobe of the brain does not interfere with its memory; an octopus with as little as 10 percent of the vertical lobe still intact will still show the effects of its training. But if the entire lobe is removed, memory of its unpleasant associations with crab and square is lost, and octopuses so operated upon were found to attack the negative stimulus more often than intact animals before they had learned not to. The injured animals continued to do this as long as they remained alive, in spite of receiving electric shocks every time they did so.

The main connections of the vertical lobe system are shown diagramatically in Fig. XVI–10. In this it can be seen that afferent impulses from the eyes reach the vertical lobe only through a relay in the medial part of the superior frontal lobe. Transection of the superior frontal–vertical lobe tract of fibers (number 15 in the diagram) resulted in total loss of memory in most of the specimens so treated, while removal of the lobe resulted in failure, on the part of most of the specimens, to attack either crab alone or crab with square.

Memory is, however, to some extent retained in animals with lesions of the vertical lobe when they are trained by reward and punishment to discriminate between two different geometric figures, rather than between a figure and a living object such as a crab. They learn more slowly than intact animals, requiring a greater number of trials before they show the "cor-

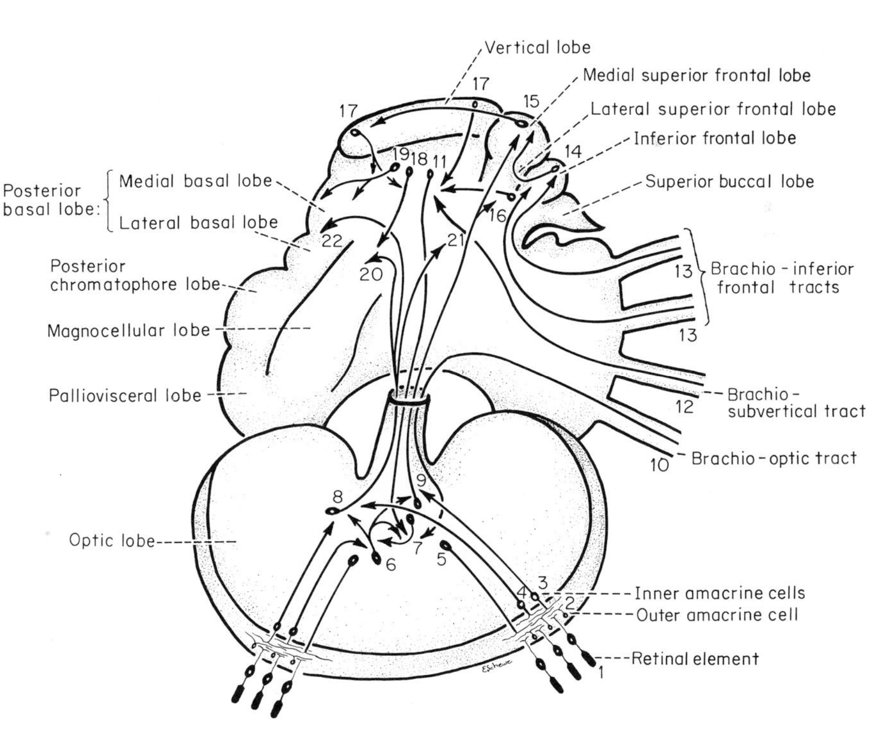

FIG. XVI–10 *Diagrammatic lateral view of the brain of an octopus, showing the main connections of the vertical lobe. (After Boycott and Young.) The optic lobe is displaced downwards. The sense cells of the retina of the right eye are represented as connecting with amacrine cells near the outer surface of the optic lobe. These amacrine cells are comparable to those without long processes or axons found in the vertebrate eye. The cells and tracts are numbered 1 to 22: 1, retinal element; 2, outer amacrine cell; 3 and 4, inner amacrine cells; 5, cell body in inner part of optic lobe sending process back to deep retina; 6 and 7, "association cells" ceep part of the optic lobe; 8, efferent tracts from optic lobe to anterior (21) and posterior (22) and magnocellular lobes (20); 9, optic–superior frontal tract; 10, brachio-optic tract; 11, subvertical-optic tract; 12, brachio-subvertical tract; 13, brachio–inferior frontal tract; 14, inferior frontal–superior frontal tract; 15, superior frontal–vertical tract; 16, lateral-superior frontal–subvertical tract; 17, vertical-subvertical tracts; 18, subvertical-magnocellular tract; 19, tracts from subvertical 1 to subvertical 2 and 3; 20, optico-magnocellular tract; 21, optico–anterior basal tract; 22, optico–posterior basal tract.*

rect" behavior, and the degree of their success in trial performances is directly proportional to the amount of vertical lobe tissue remaining after an operation. But with the entire vertical lobe removed they can be taught to discriminate even if the trials are made as long as 2 hr apart. This is not the case with injured animals in the crab-square situation, that require trials no longer than 5 min apart to show any effects of training.

The effects of tactile learning are somewhat different. Octopuses can be trained to discriminate between cylinders of the same material that differ from each other only in the distribution of longitudinal grooves along them—essentially in the proportion of grooves to flat surfaces. Blinded octopuses can be taught to do this, as well as those from which the optic lobes have been removed, but removal of the vertical lobes has much the same effect upon tactile as upon visual learning. The critical area for tactile learning seems to be the inferior frontal lobe, and lesions of this, or transections of its connections with other parts of the brain, result in failure of the animals to profit from favorable or unfavorable tactile experiences.

Dominance of the central over the peripheral system is very evident in all molluscs, but the superiority of the octopod brain exceeds that of the ganglionic systems of members of other classes. It is to this perhaps more than to any other feature of their organization that they owe their survival, for their cumbersome bodies make pursuit of prey or escape from predators difficult. Nor have they excretory or respiratory mechanisms that permit them to endure environments whose salt or oxygen concentrations vary to any marked degree, and the lack of a shell prevents them from even temporarily protecting themselves from such vicissitudes, for they cannot shut themselves off from the external world as gastropods and bivalves can when unfavorable conditions arise. The complexity of their neural organization may perhaps compensate for these other defects, which must have been the primary factors in keeping them from the adoption of new ways of life and the exploitation of new areas which they might eventually have colonized.

In annelids, the courses of certain pathways have been traced histologically in some species and their connections established. Operative experiments have shown that in earthworms the suboesophageal ganglion is the principal center of motor control, for all movement stops when this is removed or destroyed, although local stimulation will induce spontaneous movements. Movement is just as well coordinated in a worm without cerebral ganglia as in an intact one, but the animal is unable to correlate it with external conditions and moves without regard to environmental stimuli. There is some evidence that intact earthworms can be trained to follow a prescribed path in a T-shaped maze, after repeated encounters with rough sandpaper and electric shocks at the "negative" end of one of the short arms and with dark, moist soil, inviting them to burrow, at the other or "positive" end. Specimens from which the first five segments had been removed still showed the results of this training, but after their complete regeneration the animals required retraining. This was, however, accomplished more quickly than before the operation, a fact which is of some interest in relation to the results of training flatworms that had eaten their trained fellows.

Among arthropods, neural connections have been most fully studied in crustaceans and insects, but even in these classes information is far from complete. Although the complex and highly coordinated behavior patterns of arachnids are well known, functional relationships within their nervous systems are virtually unknown. Experiments like those that have been so informative for molluscs and vertebrates are not easily performed on crustaceans, for the size and structure of the brain are by no means so favorable for experimental surgery as is that of a cephalopod mollusc, and the absence of defined tracts and the complicated arrangements of the interneurons make transections unfeasible. They, too, exhibit very complex behavior in their relations to others of their kind, to enemies, to predators, and to prey, in their search for habitations and their construction or adoption of shelters, and in their responses to physical stimuli from their external environments. Behind such behavior must lie complex mechanisms of neural coordination and integration that have yet to be elucidated. Crustaceans also show some capacity for retention of the effects of past experiences and of behavior conditioned by them.

Lobsters and crabs (*Carcinus*) will not again enter a trap in which they have once been caught and from which they have been released. After being fed meat soaked in coumarin, an extract from the tonka bean that smells like vanilla, *Carcinus* and *Eriocheir* will approach anything with the smell of coumarin, even though the object may be totally inedible. Hermit crabs can be trained, by repeated rewards of food, to enter a dark box, whether food is there or not, against their natural tendency to seek light. Isopods and decapods have been taught to follow prescribed paths in mazes, where a wrong turn resulted in punishment and a correct one in a reward of food. The effects of this training are retained for several days without reinforcement with other trials. Nothing is known of the centers in the brain where the learned information is stored, but experiments have shown that the animals depend upon visual and tactile stimuli to guide them in the "right" course. Crustaceans from which both eyes have been removed make more errors when tested again after a test-free period than do intact animals that have had similar training and a similar length of time for retention; and animals without antennae as well as eyes tend to turn before they should in a maze which, as intact animals, they had learned to run with a high degree of accuracy. Yet a specimen of *Ocypode*, the ghost crab that makes periodic excursions from shore to sea to wet its gills, which was trained instead to climb upon a block of wood and immerse itself in a bowl of water, continued to climb the block and go through the motions of wetting itself even when there was no bowl there.

In spite of the high degree of coordination manifested by insects in their reactions and behavior, little is actually known about the mechanisms of this integration, less even than that known about them in crustaceans. In spite of the relative accessibility of the insect nervous system, operative techniques have proved difficult of performance with the degree of accuracy necessary to produce conclusive results. This is partly because of the small size of most insects and because of the multiple and polyneural innervation of their muscles. Stimulation experiments have so far also been somewhat unsatisfactory and other methods, such as that of following nerve degeneration after local lesions or blocking of suspected pathways, hardly less so. Experiments on the larger insects have, however, yielded some significant information, and the future development and application of micromethods may well be expected to yield more.

Decapitation of an insect or destruction of its supraoesophageal ganglia has little effect upon the coordination of its movements, but its activity is increased. Cleaning of forelegs and antennae go on indefinitely in an ant, for example, in which the brain exerts an inhibitory influence upon motor activity, as well as acting as a coordinating center for sensory and motor impulses. More precisely localized lesions indicate that the center for the inhibition is in the mushroom bodies, for locomotor activity continues for hours in crickets and grasshoppers after these parts of the brain have been injured or destroyed without damage to other parts. These, and the central bodies, have been found to be essential to stridulation in males. A cricket will sing in a normal manner if its brain is intact and only one connective to the second thoracic ganglion is left, for this ganglion controls the necessary movements of wings and legs. Removal of even part of the mushroom bodies and destruction of the central body neuropil leaves the insect unable to stridulate at all. But if only the dorsal part of the protocerebrum is injured, a male cricket will stridulate until it becomes completely exhausted. The mushroom bodies appear to act as centers for the integration of sensory impulses arising especially in the visual and auditory organs, which provide the insect with the information that determines whether it will sing or remain silent. Male crickets also begin stridulation when carrying a ripe spermatophore that they are ready to present to a female; the information for this onset of song seems to be carried via the abdominal cord, for its transection results in failure to sing the mating song or start on courtship maneuvers. The central body seems to be the center that determines the temporal sequence of wing movements involved in stridulation as well as in flight, for electrical stimulation of this region of the brain results in atypical wing movements. Stridulation and flight movements are, of course, closely related, and it may be that the production of sound is an adaptation of flight movements, a special case that has arisen through connections with inhibitory neurons in the brain that bring about more or less rhythmical interruptions of them. The thoracic ganglion is the center for organization of the basic pattern of wing and leg movements, but it cannot, by itself, provide the pattern that makes the song of a cricket. Many other behavioral mechanisms seem to be coordinated at the segmental level, as are reflex actions, for isolated segmental ganglia with their peripheral nerves and the muscles they innervate can give rise to spontaneous coordinated movements. The supraoesophageal ganglion, however, is the controlling center for all the more complex forms of behavior which, in insects, are manifold.

The capacity of insects to learn is well known and well documented. Cockroaches and beetles can be trained, by means of punishment and reward, to follow a prescribed course in a maze, and retain the effects of this training for days. The experiments of Karl von Frisch and others on training bees to return from flight almost unerringly to a selected color or geometric figure within the range of their discriminatory powers have already been mentioned. The bees depend upon visual information stored in their brains for such learned behavior, and experiments have shown that in them, as well as in ants also trained in the performance of certain acts, lesions of the mushroom bodies result in loss of memory. But it is possible that even segmental ganglia can store information. This would seem to be the case in the thoracic ganglia of cockroaches, for example, from experimental evidence derived from those that had been decapitated and confined in such a position that every time a leg was moved down below a certain point it came in contact with an electrode and received a mild shock. After 30 min of such experience, the insect no longer moved its legs down, but always up, thus avoiding the shock. It continued to show this avoiding reaction for days, even without further reinforcement, indicating the establishment of a conditioned response through repeated association of an action with unpleasant, if not painful, consequences.

In most cases, however, the responses of invertebrates seem constant and inevitable, even those that are elaborate and complicated as well as those at the reflex level. What have been defined as "fixed action patterns" are stereotyped behavioral responses that, nevertheless, depend upon reception of stimuli from a variety of sources and movements of various kinds, but that are repeated again and again in essentially the same form under appropriate circumstances. Courtship maneuvers, assumption of characteristic alert or threatening attitudes, cleaning of various parts of the body, chirping of crickets, and other actions of this kind are all illustrations of such fixed action patterns, in which the impulses follow genetically

FIG. XVI–11 *A circular column of workers of the army ant* Eciton burchelli, *aroused by emergence of a mature pupal worker brood. The ants move in concentric circles in their cage, with their antennae in contact with the substrate, scenting the trail that has been laid and reinforcing it by their own secretions as they move over it. (American Museum of Natural History.)*

determined physiological properties, such as timing, frequency, and threshold values. Although some invertebrates seem bound by their neural mechanisms always to respond in one way, and in one way only, to a given situation, others appear to have greater freedom. The trail-following ant, for example, has no alternative but to follow the path laid down by its leader, for the impulses from its chemosensitive sensilla, stimulated by the trail-marking odor, are so channeled that they dictate the complicated movements of the entire insect for as long as the olfactory stimulus persists (Fig. XVI–11). In those invertebrates where there is greater plasticity in the integrative mechanisms, there is the possibility of choice. This may be an entirely fortuitous choice, determined by the entire set of conditions to which the animal is exposed at that time and, from the point of view of its survival and well-being, may equally well be an unfortunate as a fortunate one. But it is also evident that in some, the choice may be made as the result of past experiences, when the animal has learned the consequences of an unfortunate choice and, also, in some instances, that of a fortunate one. How often such "learning" occurs, or has occurred, in nature remains problematical, but it may be assumed that fixed action patterns and the ability to modify them through experience have had selective value and that those that are exhibited by invertebrates today are ones that have, on the whole, contributed to the survival of their species.

SELECTED BIBLIOGRAPHY

Patterns of nervous systems

Batham, E. J.: The Neural Architecture of the Sea Anemone *Mimetridium crystum*, *Am. Zool.*, vol. 5, pp. 395–402, 1965.

Binstock, L., and L. Goldman: Giant Axon of *Myxicola*: Some Membrane Properties as Observed under Voltage Clamp, *Science*, vol. 158, pp. 1467–1469, 1967.

Bullock, T. H.: Functional Organization of the Giant Fiber System of *Lumbricus*, *J. Neurophysiol.*, vol. 8, pp. 55–71, 1943.

Bullock, T. H.: Comparative Aspects of the Superficial Conduction Systems in Echinoids and Asteroids, *Am. Zool.*, vol. 5, pp. 545–563, 1965.

Burnett, A., and N. Diehl: The Nervous System of *Hydra*, *J. Exp. Zool.*, vol. 157, pp. 217–250, 1964.

Child, C. M.: "Origin and Development of the Nervous System," The University of Chicago Press, Chicago, 1921.

Davey, K. G.: Neurosecretion and Molting in Some Parasitic Nematodes, *Am. Zool.*, vol. 6, pp. 243–250, 1966.

Ewer, D. W.: Networks and Spontaneous Activity in Echinoderms and Platyhelminths, *Am. Zool.*, vol. 5, pp. 563–572, 1965.

Hanström, B.: "Vergleichende Anatomie des Nervensystems der Wirbellosen Thiere unter Berüchsichtigung seiner Funktion," J. Springer, Berlin, 1928.

Horridge, G. A.: The Nerves and Muscles of Medusae. I. Conduction in the Nervous System of *Aurellia aurita* Lamarck, *J. Exp. Biol.*, vol. 31, pp. 594–600, 1954.

Horridge, G. A.: The Nerves and Muscles of Medusae. II. *Geryonia proboscidealis* Erscholtz, *J. Exp. Biol.*, vol. 32, pp. 555–568, 1955.

Lentz, T. L.: Fine Structure of the Nervous System of *Hydra*, *Anat. Rec.*, vol. 145, p. 334, 1963.

Lentz, T. L., and R. J. Barrnett: Fine Structure of the Nervous System of *Hydra*, *Am. Zool.*, vol. 5, pp. 341–356, 1965.

McCann, F. V., and R. W. Reece: Neuromuscular Transmission in Insects: Effect of Injected Chemical Agents, *Comp. Biochem. Physiol.*, vol. 21, pp. 115–124, 1967.

Mackie, G. O.: The Structure of the Nervous System in *Velella*, *Q. J. Microsc. Sci.*, vol. 101, pp. 119–131, 1960.

Maynard, D. M.: Organization of a Neuropil, *Am. Zool.*, vol. 2, pp. 79–96, 1962.

Nicol, J. A. C.: The Giant Axons of Annelids, *Q. Rev. Biol.*, vol. 23, pp. 291–323, 1948.

Nicol, J. A. C.: The Giant Nerve Fibres in the Central Nervous System of *Myxicola* (Polychaeta, Sabellidae), *Q. J. Microsc. Sci.*, vol. 89, pp. 1–46, 1948.

Nicol, J. A. C., and J. Z. Young: Giant Nerve Fiber of *Myxicola infundibulum*, *Nature*, vol. 158, p. 167, 1946.

Parker, G. H.: "The Elementary Nervous System," Lippincott, Philadelphia, 1919.

Passano, L. M.: Primitive Nervous Systems, *Proc. Nat. Acad. Sci. U.S.A.*, vol. 50, pp. 306–313, 1963.

Suga, N., and Y. Katsuki: Central Mechanism of Hearing in Insects, *J. Exp. Biol.*, pp. 545–558, 1961.

van Harreveld, A.: Doubly, Triply and Quintuply Innervated Crustacean Muscle, *J. Comp. Neurol.*, vol. 70, pp. 285–296, 1939.

Wiersma, C. A. G. (ed.): "Invertebrate Nervous Systems," The University of Chicago Press, Chicago, 1967.

Willey, R. B.: The Morphology of the Stomodael Nervous System in *Periplaneta americana, J. Morphol.*, vol. 108, pp. 219–262, 1961.

Wilson, D. M.: The Connections between the Lateral Giant Fibers of Earthworms, *Comp. Biochem. Physiol.*, vol. 3, pp. 274–284, 1961.

Young, J. Z.: The Giant Nerve Fibers and Epistellar Body of Cephalopods, *Q. J. Microsc. Sci.*, vol. 78, pp. 367–386, 1936.

Conduction

Aiella, E.: The Fate of Serotonin in the Cells of the Mussel *Mytilus edulis, Comp. Biochem. Physiol.*, vol. 14, pp. 71–82, 1965.

Bacq, Z. M.: L'acétyl choline et l'adrénaline chez les invertébrés, *Biol. Revs.*, vol. 22, pp. 73–91, 1947.

Barnes, W. J. P., and G. A. Horridge: A Neuropharmacologically Active Substance from Jelly Fish Ganglia, *J. Exp. Biol.*, vol. 42, pp. 257–267, 1965.

Bullock, T. H.: The Invertebrate Neuron Junction, *Cold Spring Harbor Symp. Quant. Biol.*, vol. 17, pp. 267–273, 1952.

Dahl, E., B. Falck, C. Mecklenberg, and H. Myhrberg: An Andregenic Nervous System in Sea Anemones, *Q. J. Microsc. Sci.*, vol. 104, pp. 531–534, 1962.

Eccles, J. C.: Excitatory and Inhibitory Synaptic Action, *Harvey Lect.*, pp. 1–24, 1955.

Ewer, D. W., and R. Van den Berg: A Note on the Pharmacology of the Dorsal Musculature of *Peripatopsis, J. Exp. Biol.*, vol. 31, pp. 497–500, 1954.

Florey, E.: Comparative Physiology: Transmitter Substances, *Annu. Rev. Physiol.*, vol. 23, pp. 501–528, 1961.

Florey, E.: Recent Studies on Synaptic Transmitters, *Am. Zool.*, vol. 2, pp. 45–54, 1962.

Florey, E.: Comparative Physiology: Neurotropic and Myotropic Compounds, *Annu. Rev. Pharmacol.*, vol. 5, pp. 357–382, 1965.

Florey, E., and E. Florey: Cholinergic Neurons in the Onychophora: A Comparative Study, *Comp. Biochem. Physiol.*, vol. 15, pp. 125–136, 1965.

Gerschenfeld, H. M.: Chemical Transmitters in Invertebrate Nervous Systems, *Symp. Soc. Exp. Biol.*, vol. 20, pp. 299–324, 1966.

Horridge, G. A., and B. Makey: Neurociliary Synapses in *Pleurobrachia* (Ctenophora), *Q. J. Microsc. Sci.*, vol. 105, pp. 163–174, 1965.

Josephson, R. K.: Mechanisms of Pacemaker and Effector Integration in Coelenterates, *Symp. Soc. Exp. Biol.*, vol. 20, pp. 33–48, 1966.

Katz, B.: Neuromuscular Transmission in Invertebrates, *Biol. Revs.*, vol. 24, pp. 1–20, 1949.

Kerkut, G. A., L. D. Leake, A. Shapira, S. Cowan, and R. J. Walker: The Presence of Glutamate in Nerve-Muscle Perfusates of *Helix, Carcinus* and *Periplaneta, Comp. Biochem. Physiol.*, vol. 15, pp. 485–502, 1965.

Kerkut, G. A., C. B. Seddon, and R. J. Walker: Cellular Localization of Monoamines by Fluorescence Microscopy in *Hirudo medicinalis* and *Lumbricus terrestris, Comp. Biochem. Physiol.*, vol. 21, pp. 687–690, 1967.

Kravitz, E. A., S. W. Kuffler, and D. D. Potter: Gamma Amino Butyric Acid and the Blocking Compounds in Crustacea. III. Their Relative Concentrations in Motor and Inhibitory Axons, *J. Neurophysiol.*, vol. 26, pp. 729–738, 1963.

Kravitz, E. A., S. W. Kuffler, D. D. Potter, and N. M. Gelder: Gamma Amino Butyric Acid and the Blocking Compounds in Crustacea, *J. Neurophysiol.*, vol. 26, pp. 739–751, 1963.

Krijgsman, B. J., and G. A. Divaris: Contractile and Pacemaker Mechanisms in the Heart of Molluscs, *Biol. Revs.*, vol. 30, pp. 1–40, 1955.

Mackie, G. O.: Conduction in the Nerve-free Epithelia of Siphonophores, *Am. Zool.*, vol. 5, pp. 439–454, 1965.

Mackie, G. O., and L. M. Passano: Non-nervous Conduction in the Epithelia and Myoepithelia of Hydromedusae, *Am. Zool.*, vol. 7, p. 727, 1967.

Passano, L. M.: Neurophysiological Study of the Coordinating Systems and Pacemakers of *Hydra, Am. Zool.*, vol. 2, pp. 435–436, 1962.

Rude, S.: Monoamines in Leech Ganglia, *Am. Zool.*, vol. 7, p. 738, 1967.

Sweeney, D.: Dopamine: Its Occurrence in Molluscan Ganglia, *Science*, vol. 139, p. 1051, 1963.

Takeuchi, A., and N. Takeuchi: The Effect on Crayfish Muscle of Ionophoretically Applied Glutamate, *J. Physiol.*, vol. 170, pp. 296–317, 1964.

Tanc, L.: Transmission in Invertebrate and Vertebrate Ganglia, *Physiol. Rev.*, vol. 47, pp. 521–578, 1967.

Treherne, J. E., and J. W. L. Beament (eds.): "The Physiology of the Insect Nervous System," Academic, New York, 1965.

Van der Kloot, W. G., J. Robbins, and I. Cooke: Blocking by Picrotoxin of Peripheral Inhibition in Crayfish, *Science*, vol. 127, pp. 521–522, 1958.

Welsh, J. H., and M. Moorhead: The Quantitative Distribution of 5-Hydroxytryptamine in the Invertebrates, Especially in Their Nervous Systems, *J. Neurochem.*, vol. 6, pp. 146–169, 1960.

Functional relationships

Boycott, B. B., and J. Z. Young: A Memory System in *Octopus vulgaris* Lamarck, *Proc. R. Soc.* (*London*), *B*, vol. 143, pp. 449–480, 1955.

Corning, W. C., and E. R. John: Effect of Ribonuclease on Retention of Conditioned Response in Regenerated Planarians, *Science*, vol. 134, pp. 1363–1365, 1961.

Hartry, A. L., and P. Keith-Lee: Planarian Memory Transfer through Cannibalism Re-examined, *Science*, vol. 146, pp. 274–275, 1964.

Maynard, D. M.: Integration in Crustacean Ganglia, *Symp. Soc. Exp. Biol.*, vol. 20, pp. 111–149, 1966.

Wells, M. J.: A Touch Learning Center in *Octopus*, *J. Exp. Biol.*, vol. 36, pp. 590–611, 1959.

Wells, M. J.: "Brain and Behavior in Cephalopods," Stanford University Press, California, 1962.

Wells, M. J.: Learning in the Octopus, *Symp. Soc. Exp. Biol.*, vol. 20, pp. 477–508, 1966.

Westerman, R. A.: Somatic Inheritance of Habituation of Responses to Light in Planarians, *Science*, vol. 140, pp. 676–677, 1963.

chapter XVII

NEUROSECRETION; ENDOCRINE REGULATION AND RHYTHMS

A. NEUROSECRETION

All in all, nervous systems are collections of secretory cells, differing biochemically from each other in their abilities to produce activating substances that may serve as transmitter agents (neurohumors) and also activating substances that may have other uses in the body. These latter products are distinguished as neurohormones, as opposed to neurohumors. Transported through a fluid tissue they exert their effects at a distance from their sites of origin, rather than in their immediate vicinity. Cells producing neurohormones are generally designated neurosecretory, or, more specifically, neuroendocrine, cells, and their products are sometimes called incretions, rather than secretions. In invertebrates as in vertebrates some nonneural cells synthesize and discharge hormones, but in invertebrates, neural origin, on present evidence, seems more general than nonneural.

1. Criteria for neurosecretory cells

The criteria for neurosecretory cells, in this somewhat limited use of the word, are derived both from cytological and physiological data. The cytological criteria are (1) that the cell have the morphological attributes of a neuron, with a cell body having characteristic inclusions and extended processes with neurofibrils, one of which is axon-like; (2) that their axons do not make synaptic contact with other cells but end, often in swellings, in close association with some area of a body fluid, frequently a blood vessel or space, the combination constituting a neurohaemal organ; and (3) that their cytoplasm contain membrane-bounded vesicles or granules that give characteristic staining reactions to particular stains, like those of the vesicles in established endocrine organs of vertebrates (Fig. XVII–1). The physiological criteria are (1) that extirpation or destruction of such cells, which usually occur in groups, or of the areas or organs in which they lie produce an alteration of existing physiological conditions within the organism, a change in its internal environment that can be rectified or counteracted by implantation of the ablated organ or injection of a macerate or of an extract from it; and (2) that implantation of an organ suspected of neurosecretion into a normally functioning animal brings about alteration of existing physiological conditions either through enhancement or suppression of certain events.

These criteria are essentially ones that have been applied to vertebrate endocrine organs, and the investigations of hormones in invertebrates have followed techniques employed in the study and analysis of those in vertebrates. These techniques include extirpation, destruction, and implantation of suspected organs, injection of extracts from them, and transfusion of body fluids. Parabiosis and telobiosis, operations by which two or more individuals are joined together side by side or end to end in such a way as to ensure inter-

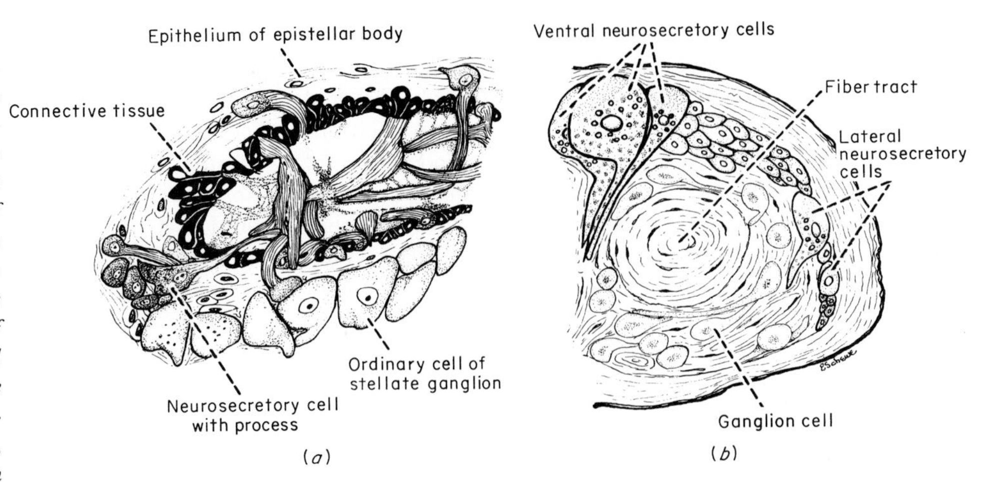

FIG. XVII–1 *Neurosecretory cells.* **(a)** *Sagittal section through the epistellar body of the octopus* Ocythoe tuberculata, *showing presumed neurosecretory cells.* (*After Young.*) *The epistellar body of the octopod cephalopods is the homologue of the giant fiber lobe of decapod cephalopods, associated with the stellate ganglion. The presumed neurosecretory cells of the epistellar body are innervated by a small nerve; removal of the epistellar bodies results in general muscular weakness.* **(b)** *Transverse section through the suboesophageal ganglion of an insect,* Imphita limbata, *showing ventral and lateral distribution of neurosecretory cells.* (*After Nayar.*)

mingling of their blood or body fluids, have also been used with each individual in a different state of neurosecretory activity or one of them intact and the other deprived of its suspected hormone-producing cells. Critical study of neurosecretion in invertebrates has been in progress only during the last two decades, although evidence was produced in the early 1920s that color changes in crustaceans resulted from the action of a substance originating in the eyestalks which is carried in the hemolymph and that pupation in caterpillars of *Lymantria dispar* was prevented when the brain was removed from a larva in its final instar. Yet the search for hormonal regulators in invertebrates, comparable to those demonstrated for vertebrates, for a long time seemed futile or, at best, unrewarding, since the early experiments on transplantation of gonads from insects of one sex to those of the other and injection of vertebrate hormones into invertebrates yielded negative results. It is in the last decade that the study of neurosecretion and neurohormones in invertebrates has made its most rapid strides, owing to refinements of microscopic, operative, and analytical techniques, both chemical and physical. Although arthropods have provided the most favorable material for this study, and more is known about the phenomenon in crustaceans and insects—the animals that offered the first evidence—than in any other groups, the range of investigation is expanding, and neurosecretion in invertebrates is not only accepted but recognized as widespread among them.

2. Nature of the secretory products

Cytochemical studies, while revealing the presence of granules presumed to be neurosecretory, give few clues as to their precise chemical nature. It seems most probable that the active molecule is in some way bound, either by chemical bonding or physical adsorption, to a carrier molecule that may well be the stainable substance. This appears to be a sulfur-rich protein, or polypeptide. In only a very few cases among invertebrates has the active substance been even identified as belonging to a particular category of chemical compounds; in some few the chemical structure is known and a compound with equal and similar activity synthesized. It has been established, however, that the product is formed primarily in the cell body, but possibly also along the axon, and that it may undergo some transformation as it travels down the axon to its terminal, or in the terminal itself. Discharge may occur at the tip of the axon, along its length, or from the cell body, and follows the propagation of an impulse along the axon. This has been demonstrated in the crab *Libinia marginata.* Stimulation of certain neurohaemal organs characteristic of decapod crustaceans (the pericardial organs) resulted in an accumulation of material in the medium in which the excised organs were maintained that had an excitatory effect upon the heart. The degree of this effect was proportional to the number of electrical stimuli applied to the organ, showing a clear relationship between the passage of an impulse and release of the neurohormone. The precise manner of discharge has not yet been clarified but there is no doubt that the product in some way gets into a fluid tissue which provides its transport. Although presumably it is in this way brought into contact with all the tissues, its effect is not indiscriminate but is limited to certain organs and tissues that are "targets" for it. Whether these tissues and organs alone take up the product or whether they are ones particularly responsive to it remain questions still to be answered. It is possible that its effect is to alter the permeability of the cell membrane, implying special receptor sites for special molecules. Neurosecretory cells are also characterized by their "action at a distance," for they are never in contact with, or even, often, in close proximity to, the responsive cells. This action may be a direct, or first-order, one, the product itself producing an effect upon the target cells; or it may be an indirect one, the product eliciting an effect upon a second, and possibly even a third, group of hormone-producing cells, so that the final physiological change may result from the production of secondary or tertiary hormones and represent a second- or third-order system. Even direct hormonal action is slow in comparison to that of short-range transmitter substances, and is also of longer duration, for it continues as long as the product remains in the circulating fluid at an active concentration—that is, at a level sufficiently high to influence the target organ. When the concentration falls below this threshold value, no effect is evident. The way, or ways, in which a hormone may influence the operation of a cell is not known, but it is probable that, like vitamins, hormones enter into biochemical reactions as cofactors of some kind and so control the direction the biochemical process may take. The fate of the hormone in the cells or in the body fluid is unknown, but as its effect is not apparent when it is no longer produced in sufficient quantities to maintain an active concentration, it must in some way be inactivated or disposed of.

The physiological processes that are affected are in general fundamental and long-term ones. These include such complex biological phenomena as growth, regeneration, reproduction, and development, and certain metabolic processes. In invertebrates, hormones are known to have myotropic effects, shown, for example, by alterations in the rate of heartbeat and of peristalsis in various organs; and chromatophorotropic ones, demonstrated especially in crustaceans where movements of pigment granules in the somatic chromatophores and in the retinal cells of the compound eyes have been definitively shown to be mediated by hormones. The fact that the secretions are, in many instances, produced periodically and that their effects are rhythmical also raises many problems in relation to the factors governing cyclic activities of all sorts.

3. Phylogenetic implications of neurosecretion

Among the problems posed by neurosecretion are those of its origin and possible phylogenetic implications. It may be a primary attribute of nerve cells, reflecting their differentiation from ectodermal cells and survival of the secretory activity common to this embryonic tissue, with evolving restriction or specialization of their secretory products. Or it may be a secondary development, with cells already differentiated as neurons becoming further specialized, biochemically, in their capacities to produce one or another neurohumor or neurohormone. Possibly, the carrier molecule represents a survival of the secretory products of the epidermal cell, to which, in the course of evolution, the special secretion of the neuron has been added. Although one of the morphological criteria for identification of a neurosecretory cell is absence of any demonstrable synaptic connections, conduction has been demonstrated in neurosecretory cells of some vertebrates and of some invertebrates (e.g., *Libinia*). It would seem, therefore, that they retain, or possess, one of the fundamental physiological properties of a neuron, although this is not exclusive to them as demonstrations of nonneural conduction have shown. It may also be argued that in the evolution of hormonal regulatory mechanisms, it is the target cells that have become specialized as receptors of neural products of a general nature or that the differentiation is reciprocal, neurons and target cells evolving together as mutually interacting systems. A similar problem is raised in relation to pheromones and receptor organs for them. However it may have come about, it would seem that in the long course of their evolution, the products of neurosecretory cells have been made use of in a wide range and variety of physiological systems in different animals, for which neuroendocrine and endocrine regulation have proved to be of especial value.

Possibly, a chemical mechanism is a more primitive mechanism of integration and communication within an organism than is a neural one, which can only operate in animals at the tissue level of organization, while cells below that, and even isolated cells, have shown themselves susceptible to chemical influences from others of their kind. This is evident, for example, in the reaggregation of dissociated cells and in the attraction of gametes of those species that shed their sex cells into the open water and in which fertilization is external but not wholly fortuitous. In the multicellular organism, chemical integration provides a means for correlation and regulation of events within it which are not immediate responses to stimuli but are long-term ones of far-reaching importance. These responses may bring about such synchronous or sequential occurrence of events as a process may require for its most successful operation.

B. EVIDENCE FOR NEUROSECRETION AND ENDOCRINE REGULATION AMONG INVERTEBRATES

Putative neurosecretory cells have been demonstrated by light microscopy and by cytochemical means in all the major invertebrate phyla investigated. These include Cnidaria, Platyhelminthes, Nemertinea, "Aschelminthes," Mollusca, Annelida, Arthropoda, and Echinodermata. It must be remembered, however, that a demonstration of membrane-bounded granules or vesicles accepting certain stains, although considered diagnostic, is not in itself a proof of hormone production, for these stains are indicative, rather than specific, and reveal a variety of products. Moreover, the stainable bodies might be synaptic vesicles containing neurohumors and not necessarily neurohormones. Only when cells conform to the physiological as well as the cytological criteria for neurosecretion can the evidence be considered definitive. The bulk of such evidence, to date, has been obtained from crustaceans and insects.

1. Cnidarians, platyhelminths, nemerteans, and nematodes

Cells that have been designated as neurosecretory in *Hydra* contain dense, membrane-bounded granules, and their neurites do not seem to terminate in relation to any other cell but to end in the intercellular spaces into which their products may presumably be discharged. The impossibility of isolating these cells and of treating them like a compact gland or tissue has precluded investigation of their function by standard procedures, but there is evidence that they are involved in the control of growth and regeneration. This relationship will be considered later in Chapter XIX. There are as yet no data to suggest that they regulate any other aspect of physiological activity in these cnidarians.

Neurosecretory cells have been demonstrated in a number of free-living flatworms. In *Dendrocoelum lacteum* they have been revealed by conventional methods in the cerebral ganglia, in the nerve cords and the nerves arising from them, and in the commissures between them. Their activity, as indicated by the rise and fall in number of stainable granules in them, seems to be greater by day than by night, showing a daily rhythm. In the polyclad *Polycelis*, these cells are more abundant in the cerebral ganglia than elsewhere in the nervous system. In the cestode *Hymenolepis* they lie in clusters in the rostellum and the cycle of their activity can be correlated with development of the adult tapeworm. They can be distinguished from other neurons even before the secretion is formed; they are bipolar cells, with one process extending anteriorly like those of sensory cells (which they may well be) and an axon that enters a nerve tract connected to nerves that lead to the lateral ganglia. The droplets first appear in the perikaryon of these cells about 3 days after the cysticercoid larva has entered the intestine of a rat, when the first proglottids are beginning to be formed, and are evident in the axon 16 to 18 days after infection, approximately coincident with the shedding of the first proglottid. This would seem to link the secretion in some way with reproductive activity, although in the light of present information no definite function can be assigned to it.

Neither can a definite function yet be ascribed to the neurosecretory cells that have been cytologically demonstrated in the cerebral ganglia of nemerteans, although removal of the anterior portion of *Lineus* results in premature maturation of the germ cells in the isolated posterior part, with those in the anterior following a normal course. This would suggest the involvement of the secretory product, or products, with reproductive activity. The cerebral organ, a mass of glandular cells in close proximity to a ganglion and the vascular system, a location suggestive of an endocrine function, shows a cycle of activity correlated with the times of spawning, indicative of a hormonal control of this event.

Neurosecretory cells have been described in the cephalic nerve ring of nematodes. In *Ascaris* these lie primarily in the lateral ganglia; in *Phocanema decupiens*, an intestinal parasite of seals, they lie in the dorsal and ventral ganglia. Most of the sense cells of the primary sense organs of *Ascaris* give staining reactions similar to those of the presumptive neurosecretory cells in the ganglia. Although it has not been possible to maintain specimens of *Ascaris* for any length of time outside the bodies of their hosts, last-stage larvae of *Phocanema* can be taken from the muscles of the cod, in which they have passed their previous phases of development, and kept in a culture medium in which their final molt to the adult stage will occur, an event which naturally takes place in the intestine of a host seal. It is thus possible to follow the course of this molt and to search for concurrent changes in the neurosecretory cells. They exhibit a cycle of secretion correlated with bursts of activity in the cells secreting the new, adult cuticle, yet the fact that the posterior half of the nematode (in which circulation of the pseudocoelomic fluid has been stopped by a ligature in the midregion) molts just as does the anterior half (from which the neurosecretory products have not been cut off) makes it impossible to accept the idea that the neurosecretory cells are directly involved in the deposition of the new cuticle. They may, however, be concerned in the shedding of the old, but critical evidence for this is still lacking.

2. Molluscs

Neurosecretory cells have been described in a wide variety of molluscs, including cephalopods, several lamellibranchs, many gastropods, and a scaphopod. Indeed, one of the earliest cytological reports of them in invertebrates was from the opisthobranchs *Aplysia* and *Pleurobranchaea*. None has yet been demonstrated in aplacophorans, monoplacophorans, nor polyplacophorans. In lamellibranchs and gastropods the presumptive neurosecretory cells lie in ganglia, usually in particular areas in most of them, but in some species, only in particular ganglia. Neurohaemal organs have been described in some gastropods and cyclical activity reported for the neurosecretory cells. There is little experimental evidence for the role of neurohormones in the organisms in which such cells or organs have been found, but what there is relates it principally to maintenance of water balance and reproduction. No active principal has yet been extracted.

Characteristic neurosecretory cells have been found in the cerebral, visceral, and pedal ganglia of the snail *Vivipara vivipara*, with numbers reaching a peak in June and July, the natural breeding season. In *Crepidula*, they have been shown in the cerebral ganglia only; the stainable granules in them are maximal in individuals at the intersex stage. They have been found in the cerebral, pleural, pedal, and buccal ganglia of the nudibranch *Haminaea*, in which a possible role in water regulation has been shown by operative removal of the different ganglia. A similar function has been attributed to the neurosecretory cells of *Patella*, for when the animals were immersed in hypotonic solutions these cells were empty of granules, but became full of them when the limpets were in hypertonic solutions. The cerebral, visceral, and parietal ganglia of pulmonates are especially rich in neurosecretory cells. In *Lymnaea stagnalis*, they are grouped in the cerebral ganglia in the lateral lobes and under the medio-dorsal and latero-dorsal bodies. These locations correspond to those also found in *Planorbarius*. A function in water balance and reproduction has been attributed to the neurosecretory cells in *Lymnaea*. Observation of the cyclic changes in them has led to the presumption that the medio-dorsal and latero-dorsal cells function in spermatogenesis, while another group, located in the caudo-dorsal part of the ganglion, function in oviposition. In extirpation experiments it was found that removal of the pleural ganglia resulted in an increase of water in the body. More specifically, removal of the left pleural ganglion did not lead to swelling, but removal of the right one did, suggesting that the right ganglion was the one principally responsible for secretion of the agent controlling water balance. Decrease in weight of a water-logged animal was evident very soon after injection of a fresh homogenate of the ganglion.

In *Helix* neurosecretory cells have been identified in all of the ganglia, including the buccal, and a neurosecretory-endocrine complex described in the optic tentacles. Amputation of the optic tentacles of the slug *Arion* is reported to result in a significant increase in the production of eggs, while injection of an extract from the tentacle reduced this to ordinary levels. In these experiments, also, injection of an extract of the brain into intact animals resulted in increased production of eggs, while injection of tentacular extract made no difference. The hypothesis has, therefore, been proposed that neurosecretory cells in the brain produce a hormone that stimulates egg production, while the glandular cells in the tentacle produce one suppressing it, the two acting together in an intact animal so that at an active concentration of the brain hormone eggs are produced, but when this level falls and the tentacular secretion reaches an active level, they are not.

Apparent neurohaemal organs have been observed in some of the pulmonates. In *Lymnaea* and *Planorbarius*, processes from the medio-dorsal and latero-dorsal neurosecretory cells end in tiny bulbs in the perineurium of a nerve next to a blood space. In *Helix* and *Arion* neurosecretory cells from the buccal ganglion have axons that end in relation to hemolymph sinuses.

Among lamellibranchs, neurosecretory cells have been found in the cerebropleural and visceral ganglia of *Mercenaria* (*Venus*) *mercenaria*. Neurosecretory cells have also been reported in the pedal ganglia of this clam, as well as in those of *Unio*. Altogether they have been found in representatives of four orders in this class of molluscs, although little is known of their function.

In cephalopods, cells with the histological characteristics of cells secreting hormones have been found in the optic glands, tissues of the vena cava, and, in *Octopus vulgaris*, the tissue around the infrabuccal ganglion and the buccal sinus. The optic glands, common to all cephalopods except *Nautilus*, are highly vascularized and innervated bodies lying in the optic stalks. Their secretion has been shown to function in sexual maturation of the animals, stimulating development of the reproductive organs and their products. The activity of the glands is suppressed by impulses transmitted to them by nerves originating in the brain, which are, in turn, activated by impulses from the eyes. In the octopuses *Eledone cirrosa* and *Octopus vulgaris*, nerves originating in the visceral lobe of the brain run to the vena cava. The axons of these nerves, which number some 2 million, unite in bundles and penetrate the muscle coat of the vein. Their terminations form a neuropil under the epithelial lining, which, because of unevenness in thickness of the neuropil, is ridged in the area where the nerves end. In *Sepia officinalis* the ridges extend into other veins adjacent to this area. Typical neurosecretory vesicles can be demonstrated in the neuropil, indicating the neurosecretory function of the nerves that constitute it. Electron microscopy of the juxtaganglionic tissue of *O. vulgaris* has revealed neurons filled with membrane-bounded, electron-dense granules similar to those in cells with established neuroendocrine functions. The axons of these neurons, also often filled with granules, end in direct apposition to the basement membrane of the buccal sinus. Although nothing is known of the nature of the secretion of their neurons, or of its effects, both their ultrastructure and their anatomical relationships are strongly suggestive of a neuroendocrine function.

3. Annelids

Neurosecretory cells have been demonstrated in the three major classes of annelids, but not yet in archiannelids. They are found in the ganglia, cords, and commissures of the central system, and in the stomodael system, and are of histologically different types. There are four different kinds in nereids and three in nephthyids; two of those in nereids may represent different stages in the activity of a primary cell type, but the three in nephthyids are considered in themselves primary cell types. There is also a neurohaemal organ at the base of the brain in polychaetes, which is particularly prominent in individuals in which the sex products are not yet ripe ("atokes," from the Greek *atokos*, "not having given birth"). Extirpation experiments on polychaetes indicate that neurohormones influence reproduction and regeneration, for removal of the anterior ganglia suppresses both these processes. The neurosecretory cells of the cerebral ganglia are believed to secrete a hormone that inhibits maturation of the gametes and, in species that pass through a heteronereid stage, the morphological changes characteristic of this transformation from atoky to epitoky (Greek *epitokos*, "fruitful," "having given birth"), or sexual maturity. Such a hormone would in effect be a juvenile hormone, keeping the individual in a permanently youthful state. Experiments on *Platynereis dumerilii* indicate that it is the titer of the hormone that determines attainment of sexual maturity. Extirpation of the cerebral ganglia (in which the neurosecretory cells are localized in one anterior and one posterior group) or of the prostomium from females with oocytes 80 to 140 μ in diameter slightly accelerates maturation, but their extirpation from females with smaller eggs impairs or prevents it. A worm so deprived will recover and attain sexual maturity if the prostomium from another worm is implanted in its coelom, but the time for recovery is longer if the donor is a juvenile rather than one approaching sexual maturity. Ultraviolet radiation of the cerebral ganglia of intact worms in which germ cells are already evident results in precocious maturation. These facts suggest that in juveniles the titer of the hormone is relatively high and suppresses occurrence of the two events by which sexual maturity is evidenced—maturation of the germ cells and metamorphosis to the epitokous heteronereid stage—but that when the titer is less, these synchronized events begin and continue to completion. Under natural conditions, then, sexual development would result from progressively decreasing activity of the neurosecretory cells in the cerebral ganglia, with, possibly, critical concentrations for each of the two events, which could account for their simultaneous occurrence under the influence of a single hormone. In nature, the factor leading to reduction of endocrine activity might be the photoperiod. Studies on regeneration have shown that mature specimens of *Nereis* will not reconstitute posterior segments that have been cut off if the cerebral ganglia are removed before their amputation or at any time in a 3-day period following it. These experiments and their results will be discussed in greater detail in Chapter XIX.

At least three types of neurosecretory cells have been found in oligochaetes. In several species, a-cells lie in the postero-dorsal region of the cerebral ganglia, b-cells at the junction of the cerebral ganglia and commissures, and u-cells in the suboesophageal ganglion, which contains also a fourth type, the c-cells, found in the ventral ganglia as well. The products are discharged into the vascular system, the nervous tissue being enclosed in a fine capillary network. Cyclic changes in the a-cells occur with the cyclic reproductive periods of *Lumbricus terrestris*, in which the gonads mature in the spring and early summer and afterwards regress till the following spring. There are no evident a-cells in juvenile individuals, but they differentiate as the animal grows older and become loaded with secretory granules in the spring reproductive period. They are discharged as the sex organs mature, and the number of a-cells is reduced.

If the cerebral ganglia are removed from a mature earthworm, it loses weight; the secondary sex characters, such as the clitellum, disappear; and egg laying ceases, but begins again after an interval of 8 weeks when the cerebral ganglia have regenerated. Removal of the suboesophageal ganglia leads also to loss of weight, but not until a week or more after the operation, and it is regained in the fourth postoperative week, while the capacity to form a cocoon is lost for 2 to 5 weeks. Egg laying stops for 7 to 17 weeks after removal of the supra- and suboesophageal ganglia and the circumoral commissures, and for 3 to 8 weeks after removal of the ventral nerve cord between segments IV and VI to VII, with the oesophageal complex remaining intact. There is histological evidence that two neurosecretory systems are involved in control of these events. Secretory material accumulates in the a-cells of the supraoesophageal ganglion when the suboesophageal ganglion is removed, suggesting that the hormones they produce do not act directly upon the reproductive system but upon the neurosecretory cells of the suboesophageal ganglion, exciting them to discharge their products. Removal of the suboesophageal ganglion alone leads to accumulation of secretory material in the a-cells of the supraoesophageal ganglion. This fact confirms the hypothesis that the neurohormone or neurohormones of the suboesophageal ganglion mediate the operation of the supraoesophageal ganglion and that the two together constitute a neurohormonal system of the second order.

Other extirpation experiments implicate neurohormones in the salt and water balance of *L. terrestris*. Specimens from which the supraoesophageal ganglia have been removed gain much more weight when placed in tap water than do intact animals, but those from which the suboesophageal ganglia alone have been removed, or in which the circumoesophageal commissures have been cut, gain no more weight than intact ones. Analyses of the coelomic fluid and blood of animals without supraoesophageal ganglia show a loss of Na^+, which is replaced after implantation of the ganglion into the coelom or after injection of a homogenate of one. The urine of animals without a supraoesophageal ganglion has a higher content of Cl^- than that of intact animals, sometimes 13 times as much. These results suggest that neurohormones affect the permeability of the body wall as well as the operation of the nephridia, possibly also through permeability changes.

The three types of neurosecretory cells that have been demonstrated in earthworms all seem to be operative in regeneration, and their apparent roles will be discussed in Chapter XIX. They are also implicated in the color changes that earthworms show, but what their target cells are or how their effects are exerted are matters not yet understood. Nor has an active principal from any of the neurosecretory cells been isolated or identified.

Two types of neurosecretory cells, designated as alpha and beta, have been shown in leeches, located in the supra- and suboesophageal ganglia and in the ventral nerve cord. In the duck leech *Theromyzon rude* these constitute about 5 percent of the total number of nerve cells in the two anterior ganglia. The alpha cells of *Theromyzon* and of *Hirudo* exhibit an annual cycle of activity, increasing in number and in their content of granules in the spring and early summer and diminishing again toward fall and winter. This correlates with the cycle of reproductive activity in these monoecious animals, in which there are both ovaries and testes in a single individual, with parallel and almost synchronous development of sperms and eggs, the events of spermatogenesis, however, slightly preceding those of oogenesis. If the supra- and suboesophageal ganglia of either of these leeches are removed before the beginning of testicular activity in the spring, sex cells in the testes not yet started on the process of gametogenesis do not mature, but those already undergoing gametogenesis complete it. These results imply that the ganglia have an effect upon gametogenesis, for which the alpha cells are held primarily responsible, but the nature of their hormone or its physiological action are still to be discovered.

4. Echinoderms

Neurosecretory cells have been found in the radial nerves and circumoral nerve ring of echinoderms, and their function in reproduction has been particularly investigated in starfish. A substance has been found in the radial nerves of eight species of asteroids, from American, Japanese, and European waters, that promotes the shedding of gametes from intact animals as well as from isolated fragments of their ripe gonads. This seems to be produced by cells in the ventral part of the radial nerves, where cells containing typical granules can be histologically visualized in animals in the reproductive phase. By isolation and fractionation techniques an active principal has been obtained, which, upon injection into a mature specimen or introduction into the medium in which pieces of gonads have been placed, induces maturation of immature germ cells and shedding of ripe ones. Similar extracts from other tissues do not produce these results. In isolated gonads, maturation of oogonia takes place only when they are in a fragment of the ovary; it never takes place in those that are free in the medium.

Chemically, the active principal has been found to be a polypeptide with direct effect upon its target cells. The fact that extracts of radial nerves are effective only at concentrations of 5 mg per 100 ml, and not at higher or lower ones, not only has established a threshold value for the active substance, but has led to the postulation of an inhibitor factor, which becomes effective at higher concentrations of the extract. This substance has been given the name of shedhibin and is obtained from mature starfish only, in greatest amounts just before spawning, and not from those whose gonads are not ripe, but nothing is known of its chemical nature. The shedding factor is neither sex nor species specific, and is found in the radial nerves of both males and females in the same amount throughout the year. It seems to have two physiological roles, one upon the contractile activity of the musculature of the gonad and the other upon the germ cells themselves, promoting their maturation. The concentration of the inhibitor, shedhibin, on the other hand fluctuates, reaching a maximum when the gonads are ripe and falling to an undetectable level after natural spawning has occurred. It is not unlikely, therefore, that natural spawning in starfish may be controlled by the level of shedhibin in the coelomic fluid and not by that of the shedding factor. Shedhibin may also account for the failure of different genera and species of starfish, living together in closely crowded assemblies and exposed to the same environmental conditions, to spawn simultaneously, since the level of shedhibin in the coelomic fluid of one species may be high, while low in that of another, in the one case counteracting the effect of the shedding factor and in the other having no such inhibitory action. Shedhibin has been extracted from the ovaries and testes of several starfish, and that from *Asterina pectinifera* has been chemically identified as *l*-glutamic acid.

5. Xiphosurans and arachnids

There are identifiable neurosecretory cells in the circumoesophageal nerve ring and ventral nerve cord of *Limulus*, and an extract of the nerve ring has been shown to have a chromatophorotropic effect upon the crab *Uca*, for injection of it causes darkening of specimens from which eyestalks have been removed. The nature of the active substance and its use, if any, in the horseshoe crab are unknown.

Arachnids have well-developed neurosecretory systems, those in spiders showing structural relationships as complex as those in insects. In general, the neurosecretory cells are grouped in bilateral masses, usually two groups in the protocerebrum and two in the tritocerebrum. Secretory activity in the protocerebral cells seems to coincide with molting and reproduction, implying their production of at least two different hormones operating on different groups of target cells. Neurohaemal organs have been found in all orders except pantopods.

The neurosecretory system of spiders has been especially carefully studied histologically, but little is known of its physiological effects. In addition to the neurosecretory cells found in all parts of the complex neural mass called "brain" there is, as there is in insects, a retrocerebral system (Fig. XVII–2). This consists of Schneider's organs, paired masses of axon terminations and glandular cells that lie posterior to the cerebral mass. The organs are innervated by a nerve that leaves the brain near the tritocerebral commissure and one that branches off the pharyngeal nerve, and, in some species, by an accessory nerve that may carry fibers from aboral neurosecretory cells. Secretion has been reported in all these nerves. The pattern in phalangids (opiliones) is similar, bodies known as Sokolow's organs being the homologues of Schneider's organs in araneids. The secretions of the nerves have been implicated in molting and in the attainment of sexual maturity. In pycnogonids, secretions from the nervous elements in Sokolow's organs are believed to influence cardiac function.

In myriapods, neurosecretory cells have been found in all parts of the central nervous system. Those in the protocerebrum send axons to a neurohaemal organ in the head. Their secretions are believed to influence molting and, in the chilopod *Lithobius* at least, to be active as well in directing the development of adult characters.

6. Crustaceans

a. PIGMENT MIGRATIONS Decapod crustaceans, especially brachyurans, have provided much of the experimental material for the study of neuro-

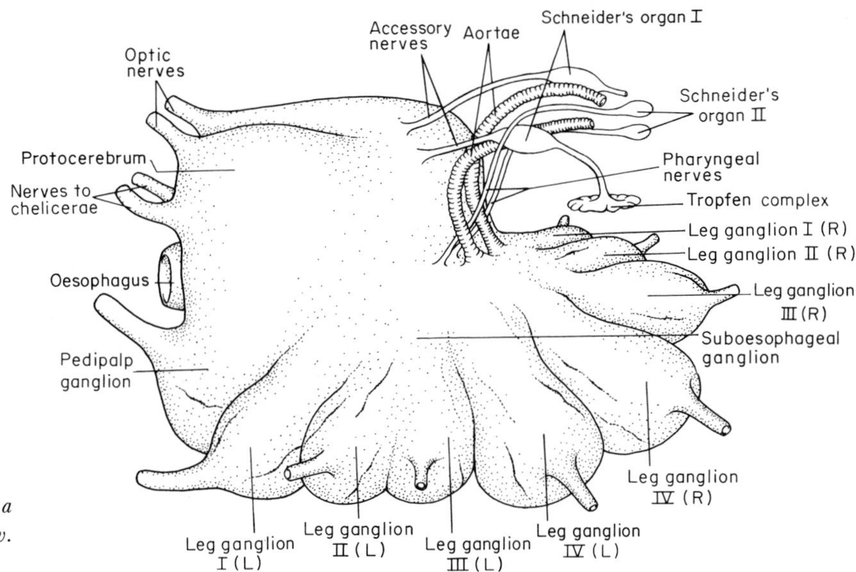

FIG. XVII–2 *Neurosecretory system of a spider, shown from the dorso-lateral view.* (*After Bullock and Horridge.*)

hormones in invertebrates and have provided definitive evidence of neuroendocrine cells in invertebrates and also of endocrine cells of nonnervous origin. Attention was first directed toward this whole area of invertebrate physiology by observations, made almost simultaneously by a German and an American investigator, that the color changes of shrimps in response to changes in their backgrounds were mediated through the hemolymph and not directly through the nervous system. This was demonstrated in two ways: (1) by cutting the nerves supplying the peripheral parts of the body, an operation that did not affect color changes in any area thus deprived of nerve supply; and (2) by blocking the hemolymph to corresponding parts through ligature of the vessels, a procedure that did alter the ordinary color responses. It is a matter of common observation that crustaceans respond to a light background by paling their bodies and to a dark background by darkening them, an adaptation that is of protective value in that, in each case, they become less readily discernible to potential predators. The changeover of a light-adapted animal to a dark-adapted one, and vice versa, takes place within an hour or two and is brought about by dispersion or concentration of the pigment granules in the chromatophores lying superficially in the hypodermis or more deeply in underlying tissues. The relative transparency of the exoskeleton and the tissues is a factor in the visible effect of these pigment migrations. Most somatic chromatophores contain an assortment of pigments—red, yellow, blue, brown, black, and white—which may be variously distributed in different chromatophores and in different combinations in individual chromatophores. Dispersion of white granules and concentration of colored ones results in the paling, or blanching, of an animal's body; and concentration of the white ones, with dispersion of the colored, in its darkening. Various degrees of coloration are brought about by differential distribution of the granules of different kinds, which may result from differences in the intensity of light falling upon an animal as well as by changes in the color and pattern of its background. This differential distribution varies among species, so that the response may be one of different character in different species as well as a graded one in individuals of the same species.

In addition to influencing somatic color changes, intensity of light is a major factor in migration of the pigment in the chromatophores associated with the eyes. Typically, in decapods, there are three sets of retinal pigments contained in different cells enclosing each ommatidium. The proximal pigment, which is black, is in the retinular cells (which contain all the pigment in more primitive crustaceans); the distal pigment, also black, is in chromatophores that surround the crystalline cone; and the light-reflecting pigment lies in another set around the retinular cells. The migrations of the retinal pigments, like the migrations of the somatic pigments, are different in different species. In *Palaemonetes*, for example, all three move outward or inward in response to differences in the intensity of the incident light; in *Astacus*, the reflecting pigment does not move, but remains always just above the basement membrane, while the other two move outward or inward; and in *Homarus* only the proximal pigment migrates, the distal and reflecting pigments remaining fixed in position. There is no nerve supply to either the somatic or the retinal chromatophores.

The search for an agent responsible for pigment migrations, and for its source, has been a long and often a confused one, due in part to the difference in reaction of different species and to different methods of investigation. It is not yet finished, although great strides have been made since the observation, in the 1920s, that injection of hemolymph from a dark-adapted specimen of the shrimp *Crangon* into a light-adapted one resulted in darkening of the latter even though it was on a light background. The first step was a classical kind of hunt for the tissue, or tissues, that produced the chromatophorotropin by extirpating various organs, making extracts of them, and injecting the extracts into test animals. None but extracts from the eyestalks produced any kind of color change. In *Palaemonetes*, a standard animal for such tests, injection of eyestalk extract into a dark-adapted animal resulted in its paling. Removal of the eyestalks led to darkening, through the centripetal migration of the granules of white pigment and centrifugal migration of red ones. Animals without eyestalks remained permanently dark unless extracts from eyestalks were injected into them. Removal of eyestalks is a rather drastic procedure, blinding the animal and depriving it of its principal sites of light reception, and more precise operative methods were indicated. Dissection and histological examination of the eyestalks revealed a thickened area on the dorso-lateral surface of the ganglia within them, the relay stations for impulses originating in the retinular cells. In decapods this area is flask-shaped, representing about one-hundredth of the total stalk tissue, and encloses a hemolymph sinus (Fig. XVII–3). In the early belief that it was a glandular structure it was given the name of "sinus gland," but later investigations showed that it is really a mass of swollen and intertwined axon terminations enclosed in a framework of supporting tissue. Secretory granules of different kinds can be cytologically demonstrated in these axons, whose cell bodies lie in various parts of the nervous system. The majority of them are clustered in a group called the X-organ in the medulla terminalis ganglion; some are in the optic peduncle and elsewhere within the eyestalk; and some in other parts of the central nervous system. There are five or six cytochemically distinct types of neuroendocrine cells in the X-organ of *Callinectes*; in the eyestalks and supraoesophageal ganglion of the amphibious grapsid crab *Sesarma* there are three, and four in the crayfish *Orconectes*. In species without eyestalks these organs lie in comparable regions of the cephalic

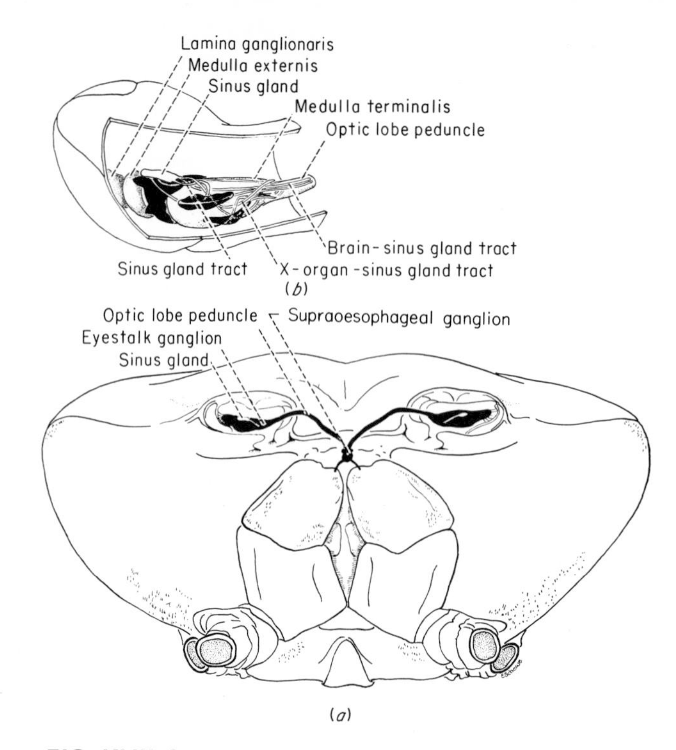

FIG. XVII–3 *Sinus glands and tracts in* Gecarcinus lateralis. *(After Bliss, Durand, and Welsh.)* ***(a)*** *Anterior view of an animal, shown as if dissected to expose the supraoesophageal ganglia and eyestalks.* ***(b)*** *Detail of right eye stalk. The blackened areas indicate the sites of neurosecretory cells.*

ganglia. "Sinus gland" is, therefore, actually a misnomer, for this structure is not intrinsically glandular but is a storage depot for products of the neuroendocrine cells; the name has, however, been retained for it, even though its true nature as a neurohaemal organ is now fully established.

This anatomical information made it possible to study effects of the ablation of the sinus glands and X-organs singly or in combination, without removal of the entire eyestalk, and also to study the effects of their implantation or the injection of extracts from them. Some species proved to be more favorable for precise operations of this kind than others. It soon became evident that the sinus gland stores and discharges into the circulatory system a complex of hormones which are involved in a wide range of physiological processes, including growth and molting, regeneration, reproduction, and various aspects of metabolism as well as chromatophore and retinal pigment migrations, and that the X-organ–sinus gland complex is a major part of a delicately balanced endocrine system comparable to that of the hypothalamic-neurohypophyseal system of vertebrates. Part of this system are the Y-organs that develop from ectodermal invaginations in the antennary segment in species with maxillary urocoeles and in the maxillary segment in those with antennary urocoeles. Postcommissural organs, pericardial organs, the gonads, and, in males, the androgenic, or vas deferens, gland are also part of the general endocrine system of crustaceans (Fig. XVII–4).

The individual behavior of chromatophores and of the pigments in them under similar conditions of light and background led to the postulation of a multiple hormone hypothesis, which presumed a number of distinct chromatophorotropins. This has since been substantiated by chromatographic and electrophoretic separation of eyestalk and sinus gland extracts, resolving them into a number of different active substances, including several different chromatophorotropins. These compounds are not species-specific, for injection of eyestalk or sinus gland extract from one species into an individual of another causes pigment movements in its chromatophores. There are probably at least five different chromatophoric substances, which can be distinguished biologically by their specific effects upon certain types of chromatophores and chemically by their different solubilities in alcohol. Little is known of their chemistry, but it is possible that they are polypeptides since those from *Palaemon* and from *Uca* are inactivated in vitro by trypsin, chymotrypsin, and other proteolytic enzymes. There is evidence, too, that there may be large precursor molecules, also with chromatophorotropic activity, that are broken down into smaller ones; this happens in vitro and may also do so in vivo. The fact, too, that extracts from the digestive gland, epidermis, and green glands degrade the active molecule into a state of ineffectiveness suggests that the hormones may be metabolized in the tissues after they have done their work.

Movements of the retinal pigments vary in different orders and genera, from the relatively simple inward and outward migration of dark pigment in the retinular cells of the amphipod *Gammarus* to the more complicated ones of the three different pigments in *Palaemonetes*. The hormones controlling these emanate from the sinus gland, but are in all probability different from the somatic chromatophorotropins, and may be different for each of the three pigments. In Fig. XVII–5 are shown the positions of the pigments in light-adapted and dark-adapted eyes of *Palaemonetes* (*a* and *b*) and in the eye of a specimen, preadapted to darkness, after injection of eyestalk extract (*c*). Ablation of the sinus gland in *Palaemonetes* results in migration of the distal pigment to its location in the dark-adapted eye; as long as 10 days after the operation, the pigment remains there even when the animals have been exposed to bright illumination for some hours. Injection of eyestalk extract causes the distal pigment to migrate to its position in the light-adapted eye. This is not the case with movements of the proximal pigment in *Palaemonetes*, nor does removal of the sinus gland affect them. It may be that this pigment is under hormonal control from a different source or is under direct nervous control, or that the melanophores are independent effectors. The reflecting pigment in the eye of *Palaemonetes* does respond to eyestalk extract, moving into the light-adapted position when the extract is injected into dark-adapted animals. Nothing is known of the chemical nature of these hormones nor of the specific factors that evoke their

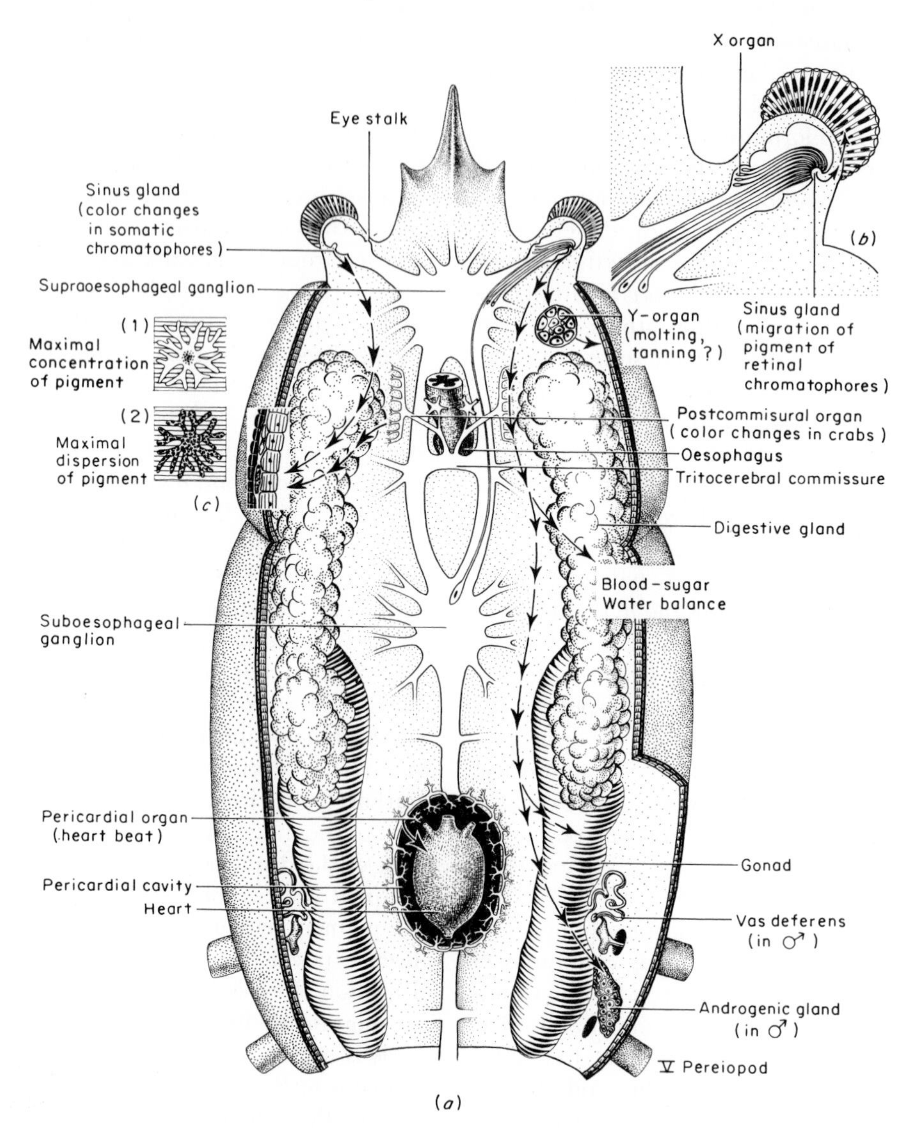

FIG. XVII–4 *Schematic representation of the hormonal system in decapod crustaceans. (From various sources.)* ***(a)*** *The cephalothorax of a generalized (composite) decapod with the carapace cut away to show sites of hormone production and the principal target organs. The arrows indicate the sources of the hormones and the organs they affect.* ***(b)*** *Detail of eyestalk and sinus gland, showing effect of its secretion on the retinal chromatophores.* ***(c)*** *Somatic chromatophores with pigment maximally concentrated (1) and maximally dispersed (2).*

release. Intensity of light is certainly a major factor, the location of the pigments in the eyes of an illuminated animal, by forming a screen around it, reducing the amount of light reaching a given rhabdome. Conversely, the location of the pigments in the eyes of an animal in the dark, or in dim light, permit the maximum amount of available light to reach each of the rhabdomes, for the rays passing through the crystalline cones of other ommatidia reach them and provide for their immediate reflection by the outward movement of the reflecting pigment.

There are many unresolved questions in respect to control of pigment migrations in crustaceans, both in the somatic chromatophores and in the retinal pigment cells. Among these is the problem of their rhythmicity, for in the chromatophores, especially, these movements may take place cyclically and independently of external light conditions. These responses may vary, too, under natural as well as under experimental conditions, with differences not only in the intensity of light but in its duration and in the relative length of the periods of light and darkness to which an animal is exposed, all of these apparently affecting the titer of the hormones in the hemolymph or, possibly, the sensitivity or receptivity of the target cells.

Neurohaemal organs other than sinus glands are also implicated in color changes. These are the postcommissural organs that have been found in a number of crustacean species and that have been particularly investigated, both structurally and functionally, in fiddler crabs. In *Uca* a pair of nerves originate from the anterior surface of the tritocerebral commissure and run antero-laterally, each member of the pair ending in a network of swollen axon terminations that lie lateral to the circumoesophageal commissures and in intimate relationship to a hemolymph vessel that runs dorso-ventrally (Fig. XVII–4). These are the postcommissural organs from which a hormone is liberated that influences migration of the pigment in these crabs' melanophores, the most abundant of their chromatophore types. Fiddler crabs exhibit a daily rhythm of color change and are darker by day than by night; this daily rhythm takes precedence over the background or other external conditions in determining the color of the animals. Analysis has shown that

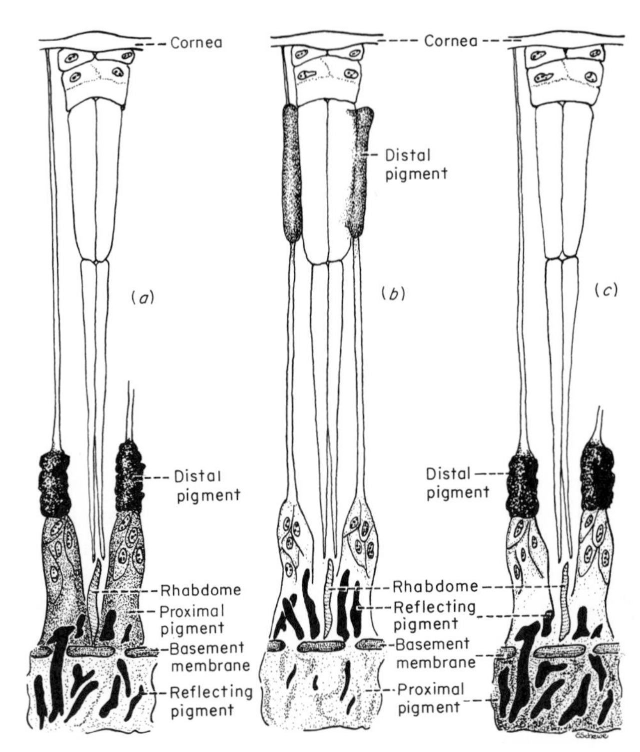

FIG. XVII–5 *Migration of pigment in the eyes of* Palaemonetes vulgaris. (*After Kleinholtz.*) ***(a)*** *Light-adapted eye.* ***(b)*** *Dark-adapted eye.* ***(c)*** *Eye from an experimental animal after preadaption to darkness and injection of extract from eyestalks of light-adapted animals.*

the dispersion of melanin is controlled by a hormone from the sinus gland, probably a peptide, but crabs deprived of their eyestalks may still show rhythmic movements of the pigment during certain months of the year. At other times they remain permanently pale. Electrical stimulation of the eyestalk stubs, however, results in dispersion of the dark pigment and to some extent also of the red pigment in the erythrophores. Injection of extracts from the postcommissural organs into animals in which the dark pigment was maximally contracted resulted also in its dispersion. Extracts from the postcommissural organs of the shrimp *Leander* have been separated electrophoretically into two major components, A and B. Injection of the A substance, obtainable also from sinus glands, results in concentration of the red and yellow pigments in the small and large chromatophores, while injection of substance B, obtainable only from the postcommissural organs, results in concentration of the red pigment in the large chromatophores, but dispersion of it in the small ones. Although the evidence is far from complete, it seems likely that the postcommissural organs have some effect upon somatic coloration and, in fiddler crabs at least, contribute to the maintenance of diurnal color changes.

Studies of color changes have been made at the cellular level in erythrophores of the prawn *Palaemonetes vulgaris*. Two hormones control the movement of red pigment in these cells. The one that controls concentration of the pigment is produced in the tritocerebral commissure, the other, which controls its dispersion, is produced in the abdominal nerve cord. Experiments show that sodium ions are necessary for a strong response of the chromatophore to the concentrating hormone, and calcium ions for a strong response to the dispersing hormone. Data from various tests have led to the postulate that the latter hormone causes an increase in the rate of flux of Ca^{++} into the chromatophore and that the increase in Ca^{++} concentration within the cell starts reactions through which cyclic AMP is produced from ATP; the AMP is the agent that triggers dispersion of the pigment. The action of the concentrating hormone, on the other hand, is to promote the operation of a physiological pump by means of which sodium ions from inside the chromatophore are exchanged with potassium ions from outside it; the high internal K^+/Na^+ ratio is responsible for concentration of the pigment. According to this postulate, the action of both hormones is upon the chromatophore membrane. The subsequent changes within the cell follow as consequences of changes in the membrane that determine the rate at which particular cations pass across it.

b. **MOLTING** Molting in decapod crustaceans is also influenced by hormones discharged from the sinus glands. Molting is not simply the act of emerging from the old exoskeleton, but a continuous series of events that go on during most of the life of an individual and lead to the actual act of ecdysis (Greek *ekdyein*, "to strip off"), or shedding of the old cuticular covering. Ecdysis occurs periodically in the life of a crustacean as a brief interruption of its ordinary activities, but preparation for it, in storage of reserves and changes in the integument, are continuous. It is a phenomenon associated with growth, which, in most arthropods, has the false appearance of being a discontinuous process. But anyone who has seen the soft wrinkled body of a crab or a lobster that has just crawled out of its old shell must recognize that multiplication of cells and differentiation of new tissues has been going on while it was still confined there, and that expansion of the body and hardening and strengthening of the new covering must occur before the very vulnerable "shedder" is ready to cope again with its environment.

The lives of crustaceans can, in relation to their molting cycles, which are usually characteristic of the species, be divided into at least four major stages. The first of these is premolt, or proecdysis, when the old cuticle becomes thinner and reserves, both organic and inorganic, are stored in the tissues with accompanying metabolic changes; the second is ecdysis, the splitting and shedding of the old covering and a rapid increase in size due to the immediate and rapid absorption of water; the third is postmolt, or postecdysis, when the exoskeleton is thickened by a new deposition of cuticle, toughened by tanning, and strengthened by the concentration of inorganic salts in it; and the fourth is intermolt, or interecdysis, a period of apparent quiescence as far as the visible events of molting are concerned but one in which those of the next molt are being anticipated and prepared for. The intermolt period varies in different species and at different ages in the same species. *Cambarus*, which has no larval stage but hatches as a minute crayfish, molts about every 12 to 13 days in the first year of its life and, later, usually twice a year, once in the spring and once in the autumn. In some crabs, such as *Maja* and *Carcinus*, molting takes place periodically until the full size of the animal is reached, after which molting does not occur again. *Homarus*, on the other hand, continues to molt throughout its entire life, although at longer and longer intervals as it grows older.

The first intimation of neuroendocrine control of molting was the observation that the intermolt period was significantly shortened in animals deprived of their eyestalks. Juvenile specimens of *Cambarus*, for example, molted at about 8-day intervals; mature specimens are not very viable after the operation, but some have lived long enough to molt two or three times at intervals of 15 to 20 days. Implantation of eyestalks from another crayfish counteracts these effects, so that the animal then molts with natural periodicity. Similar effects of eyestalk removal and reimplantation have been observed in a number of other decapods. Later experiments on crabs have shown that removal of the sinus gland alone does not induce precocious molting, but that removal of the X-organ does and that implantation of the X-organ in eyestalkless animals prolongs the intermolt period. It is, therefore, postulated that the X-organ produces a molt-inhibiting hormone that is stored in the sinus gland and released from it into the hemolymph in active concentration during postmolt and intermolt periods.

Molting is promoted by a hormone, or hormones, liberated from the Y-organs, small masses of glandular tissue that, in crabs, are situated ventrally and laterally to the eye socket, just above the junction of the branchiostegite with the cuticle of the lateral body wall at the anterior end of the branchial chamber. If these are removed during intermolt or in very early proecdysis, the animal will not molt again in any circumstances except after implantation of new ones, but if the Y-organs are excised when proecdysis is underway, the events of that molt proceed as usual. It seems apparent, then, that secretions of the nonneural endocrine Y-organs are necessary to trigger the molting process and that once this has begun, it will inevitably continue to its natural conclusion, the molt-inhibiting hormone from the X-organs having no effect upon it, even in the absence of the Y-organs. Clearly, the X-organ–sinus gland and Y-organ relationship is a complex one, reduction or cessation of secretion from the X-organs at the end of intermolt permitting the Y-organs to secrete and the titer of their secretions to rise to an active concentration in the hemolymph.

Release of the molting hormone is evident in a number of physiological changes. Early among these are accumulation of glycogen in the epidermis and in the subepidermal connective tissue, a rise in the level of Ca^{++} in the hemolymph, and the formation of gastroliths. These are modified portions of the anterolateral walls of the cardiac stomach—hence, ectodermal in origin—that have a protein-chitin matrix that becomes progressively mineralized as proecdysis advances. Later, new processes are developed by the epidermal cells to extend into the integumental spines, and the epidermis separates from the noncalcified proteinaceous membrane between it and the hard exoskeleton and begins to deposit chitin, thus diminishing its glycogen reserves. Concomitantly, reserves of organic material, primarily fat and glycogen, are accumulating in the digestive glands, for the animals are actively feeding. Magnesium and calcium phosphates are also stored in the tissues. There then follows a marked enlargement of the epidermal cells, which begin to secrete protein and chitin as well as a fluid, the molting (ecdysial) fluid, that contains chitinase and alkaline phosphatase; and the oxygen consumption of the animal increases. Feeding activities are reduced and stop entirely in the final proecdysial stages, when calcium is withdrawn from the old exoskeleton and from the reserves and deposited in the new exoskeleton, and ecdysial sutures open so that the animal, with a new soft exoskeleton conforming to the pattern of the old but also to the increased dimensions of the soft parts, can crawl out and begin on the events of postecdysis.

Many external factors influence the initiation of a molt and thus the cessation of production of the molt-inhibiting hormone by the X-organs and the production of the molting hormone by the Y-organs. Among these factors are light, temperature, nutrition, state of reproductive activity, and injury. Any one of them may be of especial importance to a particular species. The land crab, *Gecarcinus*, for example, which in nature molts when in its burrow, will not molt in its natural period when kept in full light, nor in the presence of other crabs. And the crayfish *Orconectes*, ordinarily molting in the spring, will not do so if kept in continual darkness. In general, molts are less frequent at temperatures between 14 and 17°C than they are at higher ones. Starved individuals do not molt, and it has been shown that females of *Crangon* "in berry," or bearing eggs, will not molt until the eggs have been hatched and the juveniles released from the pleopods. These egg-bearing females molt several weeks later in the spring than the males, yet removal of their sinus glands will induce molting. This suggests that the molt-inhibiting hormone from the X-organs is responsible for their failure to molt at the same time as the males and that there is some relationship between these organs and the reproductive state of the females.

The sinus gland seems also primarily responsible for the profound metabolic changes during proecdysis. Oxygen consumption is greater in eyestalkless animals than in intact ones, but returns to the level of the intact ones after implantation of eyestalks. Gastrolith formation during intermolt is induced in crayfishes by removal of the eyestalks, or of the X-organ complex in them. The increase of sugar in the hemolymph is believed also to be due to a hormone liberated from the sinus gland, called the diabetogenic hormone.

Another neurohaemal organ of decapods lies in the pericardium, as the pericardial organs. These are plexuses of nerve terminations attached to the pericardium and extending into the pericardial cavity (Fig. XVII–4). Their position is thus extremely advantageous for transfer of their secretions to the general circulation, and their neurosecretory nature has been established on both cytological and physiological grounds. Cytochemically several types of neurosecretory cells are recognized in them, the active principals in whose secretions are probably peptides. Their effect is directly upon the heart, increasing the amplitude of its beat and, in some species, the frequency. In other species, the frequency of beat is decreased by injection of extracts from the pericardial organs, although the amplitude is increased. There is evidence, too, for the presence of serotonin in these extracts, which has also an excitatory effect upon the heart. In *Homarus*, this seems to be mediated directly through the cardiac ganglion and to influence the pacemaker mechanism.

c. WATER BALANCE Water balance is also affected during proecdysis, and under natural conditions, relatively large amounts of water are taken in through the foregut or other areas. Much more is taken up if the eyestalks are removed, as has been shown in *Carcinus* and *Gecarcinus*. In the latter, with a terrestrial mode of life, the pericardial sacs are the principal areas of water storage and may function also in the actual uptake of water. These are membranous pouches that project from the pericardial sinus into the branchial chamber and just before ecdysis become so large that they can be easily seen in a living animal. These changes in water content and in the concentration of materials in the hemolymph impose upon the animal the necessity for osmoregulation, and so relate this vital physiological process to a hormonal control, at least during proecdysis. The location of the neuroendocrine cells involved is not yet known, aside from the fact that they lie in the central nervous system.

Evidence for neuroendocrine involvement in salt and water balance in intermolt periods has been derived from ligation and injection experiments in the crayfish *Procambarus clarkii* and the crab *Metapograpsis messor*. Ligation of the eyestalks leads to a rapid decrease in the osmotic concentration of the crab's hemolymph when the animal is in 25 percent seawater, the lower limit of seawater concentration to which it can ordinarily regulate. Since this fall in concentration can be to some extent prevented by injection of eyestalk homogenates, a contribution to osmoregulation by substances in the eyestalk is implied.

In postecdysis the external dimensions of a newly molted animal increase greatly, due principally to its uptake of water, which now becomes distributed through the tissues causing expansion of the soft exoskeleton. This is hardened by quinone tanning, through interaction of the proteins and quinones deposited in it. It is also thickened and strengthened by deposition of inorganic salts, primarily those of calcium. There is some evidence that the tanning process is influenced by a hormone from the eyestalks, which may possibly be the same as the molt-inhibiting hormone. The exoskeleton of the crayfish *Cambarellus shufeldti*, for example, from which the eyestalks have been removed early in proecdysis, does not become as dark as in intact animals, but the tanning process in it is hastened, and it becomes nearly as dark as controls after injection of eyestalk extract.

There is still much to be learned about the hormones in the eyestalks of decapod crustaceans and about those from still unknown sources that are involved in the complicated processes of molting. The specific target organs and the effect of the hormones upon them, their interaction with those from the reproductive organs, and various other aspects of their production and operation are matters still to be elucidated.

d. REPRODUCTION The reproductive organs of crustaceans are also sites of hormone production. The interaction between the molt-inhibiting hormone, or the molting hormone, and the reproductive state of female crayfishes has already been noted. It has been found, in general, among the species investigated that removal of the eyestalks from immature females results in rapid and precocious development of their ovaries, on the present evidence suspected of being through the activity of a hormone developed in the X-organ. In male crustaceans an androgenic gland has been identified as a solid strand of glandular cells located between the coxopodite of the last walking leg and the vas deferens, whose activity is controlled by centers in the eyestalk (Fig. XVII–4). If this gland is implanted in an immature female, her ovary develops as a testis and the secondary sex characters of the male, most evident in the modification of the first abdominal appendage as a passage for sperm, are acquired by her in the course of the next two or three molts. These altered females do not, however, become functional males, for although sperm ducts are formed, they remain solid cords and never develop a lumen. On the other hand, males from which the androgenic glands have been removed do not become females; if this operation is performed while they are still very young, their secondary sex characters do not develop further, but neither do characters of the other sex. It may be the destruction of this gland by invading parasites such as *Sacculina* that causes the virtual castration of infected males, for the parasite does not affect the females in this way, nor to this extent. If the eyestalks are removed from young males, the androgenic gland hypertrophies and testicular development is promoted. This effect is comparable to the promotion of ovarian development in eyestalkless females and points to a similar gonad-inhibiting factor released through the sinus gland.

7. Insects

Neuroendocrine controls in insects were suspected in the 1920s when it was noted that caterpillars did not pupate after their heads had been cut off or their cephalic ganglia removed. These controls, particularly those in relation to growth and development, have been investigated more thoroughly than they have in crustaceans, partly because of the comparative ease with which insects can be reared and manipulated in the laboratory but largely because of the practical value attached to comprehension of these mechanisms as a possible means of interfering with the natural course of an insect's development and so keeping in check populations of undesirable species. This has proved an incentive to research in insect endocrinology, as it has to insect nutrition. Another incentive, perhaps quite as strong, has been the expectation that insect endocrine systems might serve as models for those in other animals, particularly vertebrates, and that an understanding of them and their interactions would provide explanations of these phenomena that would be of general applicability. During the period in which this research has been intensively pursued, it has become evident that hormones, arising either from neurosecretory cells or from glandular tissue of nonneural origin, are involved in the postembryonic growth and development of insects, including molting and metamorphosis; in various aspects of morphogenesis and metabolism; and in reproduction and in overt activity.

a. MOLTING AND METAMORPHOSIS The postembryonic stages of an insect's life—that is, those stages between its emergence from the egg and the development of adult characters—are ones in which its soft parts increase in size, either by cell division or cell enlargement, necessitating shedding of its exocuticle at appropriate intervals and, except for apterygote species such as thysanurans and collembolans (if they are considered insects), gradual or abrupt metamorphosis, or morphological change. Apterygote insects are designated as ametabolous, as opposed to metabolous, pterygotes. The metabolous orders may be subdivided into those that are hemimetabolous and those that are holometabolous. Of the hemimetabolous orders, some, like mayflies and dragonflies, emerge from the egg as nymphs that do not bear close resemblance to adults and have an aquatic rather than an aerial mode of life, but their transition to the adult form, or imago, occurs in a series of steps during their juvenile life in which existing structures acquire adult form. Other hemimetabolous species, like cockroaches, grasshoppers, and bugs, hatch as nymphs that are essentially miniature adults, usually lacking only fully developed wings, gonads, and external genitalia, and gradually acquiring these particular adult features with successive molts. In holometabolous species, such as moths and butterflies, bees, wasps, ants, and flies, the insect hatches as a larva, quite unlike the adult in appearance and in habits, and grows in size without significant morphological changes during each successive instar, or intermolt stage, and at the end of the final larval instar becomes a pupa.

Pupation is a period of passivity and apparent dormancy, for the pupa does not move about, does not feed, and lives at a low rate of metabolism from the reserves it has stored. Radical internal changes go on, however, involving death and disintegration of larval tissues, and development of adult ones from imaginal disks, patches of embryonic cells that have remained in this condition during larval life and only begin their multiplication and morphogenesis during pupation. At the end of this the insect sheds its pupal covering and emerges as a winged and sexually mature adult. The metamorphosis is thus an abrupt rather than a gradual one.

Other periods of apparent rest, or interruption or reduction of activity, may occur at any time during an insect's life. Such a period is known as diapause (Greek *diapausis*, "interruption of work") and is a phenomenon that is especially frequent among arthropods, but by no means exclusive to them. It is common to many organisms that inhabit inconstant or fluctuating environments. The stage in an insect's life at which diapause occurs, whether embryonic, nymphal or larval, pupal or adult, is characteristic of the species and usually occurs at only one stage, but there are exceptions, and exceptions to the generality of diapause, for there are species and races within a species that do not regularly enter diapause at any time in their lives. There are, on the other hand, species in which diapause occurs regularly at two points in a life cycle. Illustrations of this, among lepidopterans, are the winter moth *Operaphtera brumata* and the leaf roller *Exapata congelatella*, which enter diapause in summer as pupae and in winter as eggs.

The first explicit evidence for hormonal control of growth and development in insects came from pioneer experiments on the blood-sucking bug *Rhodnius prolixus*. In the natural course of its development *Rhodnius* passes through five nymphal instars and emerges from the fifth ecdysis as an adult. The molts occur at a definite time after each meal; at 24°C there is an interval of 15 days between the time a fourth-instar nymph feeds and molts, and one of 24 days between its next meal, as a fifth-instar nymph, and the critical molt to adult form. If decapitated within 24 hr after feeding a nymph in any instar will not molt, although it may survive in this headless condition and in the same state of development for a year or more. If the operation is performed 3 to 4 days after a meal, however, some 80 percent of the headless nymphs molt, and if the operation is delayed as long as 5 days after a meal, all of them do, showing all the attributes of next stage nymphs. It seems evident, therefore, that something is produced in the head that promotes molting and that this product is produced, or transported throughout the body, slowly, so that there is a critical period during which the bugs are incapable of molting, but after which molting proceeds even in the absence of a head. That this is a substance carried by the hemolymph has been proved by parabiotic and telobiotic experiments, in which individuals at different stages were joined together either directly or by fine capillary tubes through which the hemolymph could flow without direct contact of the two bodies (Fig. XVII–6). It was found, for example, that when a fourth-stage nymph, decapitated shortly after a meal, was joined to another of the same stage that had passed the critical stage, both molted into fifth-stage nymphs. But if a fourth-stage nymph, before it had reached the critical stage, was joined to a fifth-stage one that had passed it, the fourth-stage nymph skipped its fifth nymphal stage and, like its partner, molted into an adult, albeit a small one. The same result was forthcoming when the brain of a fifth-stage nymph that had recently fed was transplanted into the abdomen of one of the fourth stage, decapitated before the end of the critical period. Premature or precocious molts of this kind could be induced even in first-stage nymphs, causing them to develop into dwarf adults. Even adults that do not ordinarily molt could be made to do so by attaching them to fifth-instar nymphs past the critical stage. Confirmatory evidence for the existence of a "brain

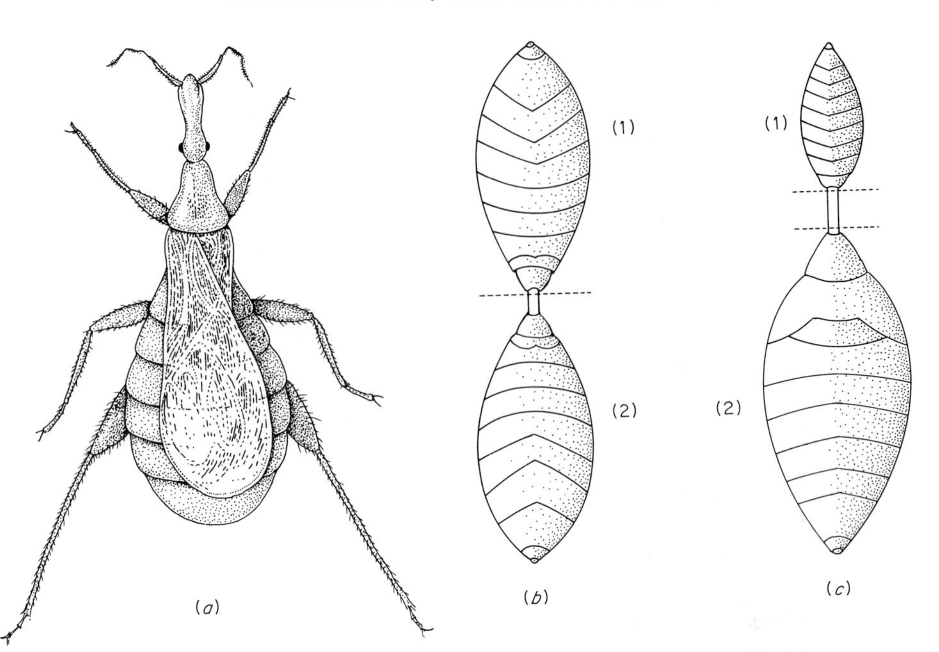

FIG. XVII–6 *Telobiotic experiments in* Rhodnius prolixus. (*From Gardiner.*) **(a)** *Adult* Rhodnius. **(b)** *Two fourth-instar nymphs united by a capillary glass tube. Nymph 1 was decapitated after the critical period and would not have molted unless attached to nymph 2.* **(c)** *Two nymphs of different ages united by a capillary tube after decapitation at the levels shown by the broken lines. Nymph 1 was in its first instar and decapitated before the critical phase; nymph 2 was a fifth-instar nymph and was decapitated after the critical period. After union, both nymphs molt, nymph 1 becoming a diminutive adult without passing through the remaining nymphal stages.*

substance" was provided by experiments on holometabolous insects, when it was shown that caterpillars of the moth *Lymantria dispar* do not pupate if they have been decapitated in their last larval instar, and that complete formation of the puparium by a final larval instar of the blowfly *Calliphora* is prevented when a ligature cuts off the circulation of the hemolymph from the head to the rest of the body.

Anatomical and physiological studies have shown that insects have well-defined hormonal systems, structurally and functionally paralleling those of crustaceans, but of even greater complexity. The pars intercerebralis of the insect protocerebrum contains neurosecretory cells, usually arranged in two groups of different numbers in different species. In *Bombyx*, there are 4 in each group; in *Calliphora* larvae there are 6; in cockroaches 15; in *Cecropia* 26; and over 200 in locusts. In addition to the median group there are, in some insects, lateral groups. The axons from the cells in each group run beside each other in two bundles that cross within the ganglion and then leave it to terminate in the two corpora cardiaca, neurohaemal organs that lie posterior to it (Fig. XVII–7). Transmission of material from the protocerebral neuro-

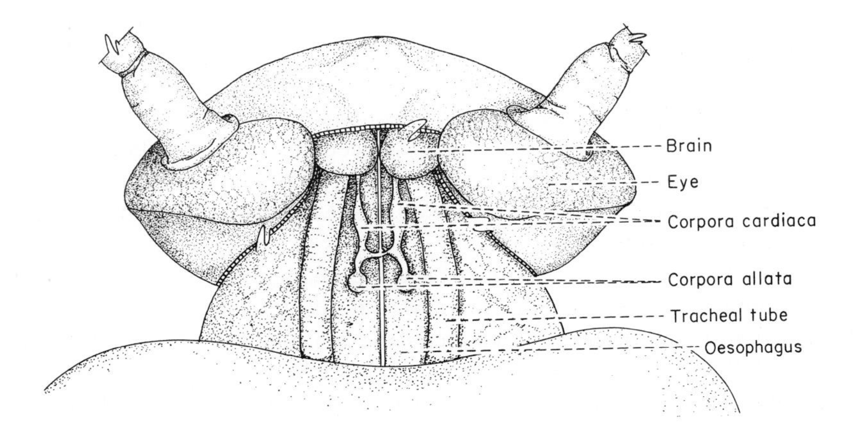

FIG. XVII–7 *Head of* Periplaneta americana, *showing the relation of the corpora cardiaca and corpora allata to other organs.* (*From Gardiner.*)

secretory cells to the corpora cardiaca has been proved by transections of the corpora cardiaca nerves and observation of the subsequent accumulation of secretory granules in the proximal ends of the stumps and of their absence from the corpus cardiacum to which the axons run. The corpora cardiaca contain secretory neurons and glia cells as well as the axon terminations of the cells in the pars intercerebralis. In the stick insect *Carausius* all these form a loosely knit tissue with ramifying extracellular spaces that are confluent with the hemocoele. Electron micrographs show that these bodies are sites of release of intrinsic as well as extrinsic secretions. This is accomplished by fusion of the membrane surrounding a neurosecretory droplet with the plasma membrane of the neuron and the formation of a vesicle which is later pinched off in a manner resembling exopinocytosis.

Posterior to the corpora cardiaca, to which they have nervous connection, are a pair of spherical or ovoid bodies, the corpora allata. These are of epidermal origin and are secretory. They, like the corpora cardiaca, may lie close together, and indeed in some species so close together that they seem a single body. They may also be so near the corpora cardiaca that they seem joined to it.

Other hormone-producing glands of epidermal origin are the prothoracic, or thoracic, glands that have been found in every pterygote insect examined. These arise early in embryogenesis as lateral invaginations of the ectoderm on each side of the maxillary segment, which grow down into the prothorax and, in some species, into the mesothorax. They are composed of two strips of glandular cells, histologically very similar to those in the corpora allata, and in many species, lie so close together in the midline that they seem a single structure. In some species, they are not compact or clearly demarcated glands, making location of the endocrine cells difficult. In cockroaches, for example, the cells are diffusely distributed among the thoracic tissues and in Hemiptera they lie in chains in the inner pair of lobes of the fat body.

In Diptera, corpora cardiaca, corpora allata, and prothoracic glands are united in a single compact mass called the ring gland that encircles the aorta (Fig. XVII–8). Its histology indicates that the cells in the dorso-median part of the ring represent the corpora allata, those in the lateral parts, the prothoracic glands, and those in the ventral part, the corpora cardiaca. The hypocerebral ganglion is also included in this complex.

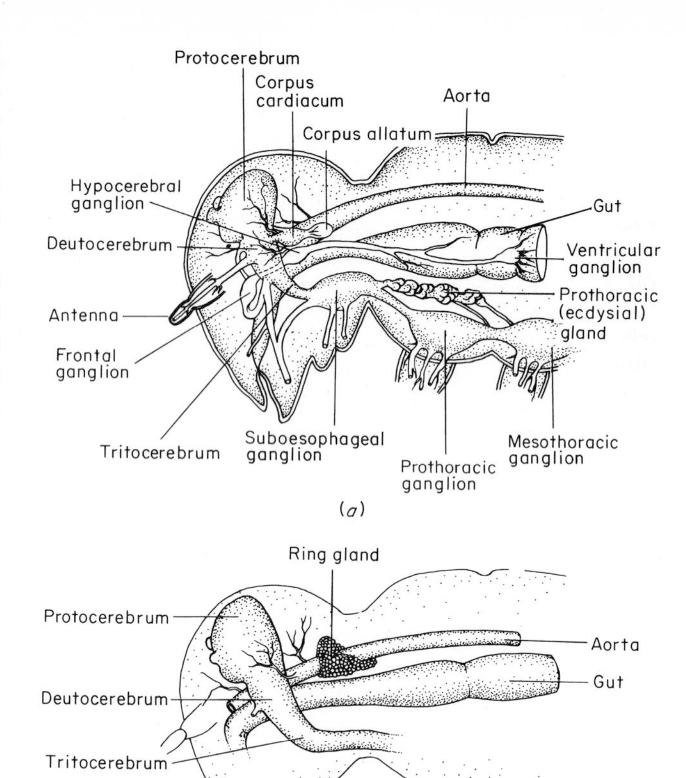

FIG. XVII–8 *Retrocerebral endocrine organs in insects. (After Bullock and Horridge.)* ***(a)*** *Head and prothoracic segment (in lateral view) of a generalized insect, showing the interrelations of the brain, stomatogastric nervous system, and retrocerebral organs.* ***(b)*** *Head and prothoracic segment (in lateral view) of a dipteran insect, showing the ring gland. The dorsally situated cells in this gland represent those of the corpora allata; the laterally situated, those of the prothoracic glands; the posterior group of cells lying ventral to the aorta represent those of the corpora cardiaca; and those more anterior, the hypocerebral ganglion, which is included in this complex.*

The neurosecretory cells of the cephalic ganglia and the retrocerebral organs—corpora cardiaca, corpora allata, and prothoracic glands—have been shown to be responsible for initiation and regulation of the complicated events of a molt. In some species, other glandular cells are known to be involved in molting. These are the ventral, or tentorial, glands, which are paired masses of cells that originate from epidermal invaginations in the ventrocaudad part of the head; the pericardial glands of phasmids; and the peritracheal glands, probably homologues of the pericardial glands, found in some families of Diptera. Histologically, their cells are similar to those of the prothoracic and tentorial glands.

The events of molting in juvenile insects are similar to those in crustaceans—that is, initially, detachment of the epidermis from the cuticle and its increase in area by a succession of mitotic divisions, followed by the secretion of a new and extensible cuticle and of enzymes that digest the inner layers of the old; waterproofing of the new cuticle; and emergence of a larva or nymph in its next instar or stadium. Some insects, like cockroaches and flies, immediately distend their larger bodies by taking in air, an inflation procedure comparable to distention of newly emerged crustaceans by intake of water, but one better suited to the needs of a terrestrial animal. The bodies of lepidopteran larvae remain undistended and wrinkled for some time, the contours smoothing out gradually with growth of the internal organs, especially of the fat bodies. After ecdysis, the cuticle hardens and darkens into its definitive color patterns, and the nymph or larva begins to feed and, in its intermolt period, to prepare for the next ecdysis. It is necessary for most insects to eat copiously before this in order to store up adequate reserves; starved individuals of most species will not molt, and if anything, such as decapitation or removal of the brain, cuts short this period of "indispensible nutrition" they do not molt in any circumstances (see Chapter IX). *Tenebrio* larvae, on the other hand, molt several times in the absence of any food, and unfed larvae of clothes moths have been known to molt as often as 40 times. In most insects, unlike crustaceans, the last juvenile molt is a critical one. In many hemimetabolous species, metamorphosis has taken place during the preceding intermolt and imaginal structures and characteristics have been laid down and established, so that the nymph emerges as a fully equipped adult. In holometabolous species the full-grown larva develops the pupal cuticle beneath the old and emerges from the last ecdysis as a pupa, and during its pupal life adult structures are organized and developed from the imaginal disks laid down in previous stages. These differences imply that there must be at least two hormones involved in the molting of insects, one to direct development of juvenile characters and one to direct development of adult, and that the balance in titer between the two determines which of these at any given time takes precedence over the other.

Searches for the source of these hormones in a wide variety of insect species, both hemimetabolous and holometabolous, have shown that there are, in fact, three sources and hormones of at least three types. The neurosecretory cells of the protocerebrum provide a hormone, originally called the growth and development hormone but perhaps better designated, in the light of recent information, as the activation hormone, for without it none of the events of molting take place. This has been demonstrated in numerous extirpation and implantation experiments. The substance may be released directly from the neurosecretory cells, or it may travel down their axons to the corpora cardiaca and be released there. The active principal may be a peptide, or possibly a steroid, possibly even cholesterol or a triterpenoid of related configuration. Both 7-dehydroxycholesterol and cholestanol have been shown to have "brain hormone" activity. The effect of this hormone is not directly upon the epidermis, but indirectly through the prothoracic glands. This was conclusively shown by numerous experiments.

Once activated, the prothoracic glands, or their homologues (tentorial, lateral cells in the ring gland, pericardial or peritracheal), discharge a hormone that conveys a message to the epidermal cells to start their proecdysial procedures. Originally known as the molting hormone, this is now generally called ecdysone. It was the first of the insect hormones to be isolated and chemically identified when, in 1954, an extract from the prothoracic glands of 500 kg of silkworm pupae yielded a sufficient quantity (25 mg) of the active substance to make its analysis possible. It proved to be a steroid with the structural formula represented in Fig. XVII–9. This was designated alpha-ecdysone and has been synthesized in a sequence of 22 steps from a derivative of stigmasterol. Subsequently, two other structurally related steroids have been isolated: 20-hydroxyecdysone with a hydroxyl radical rather than

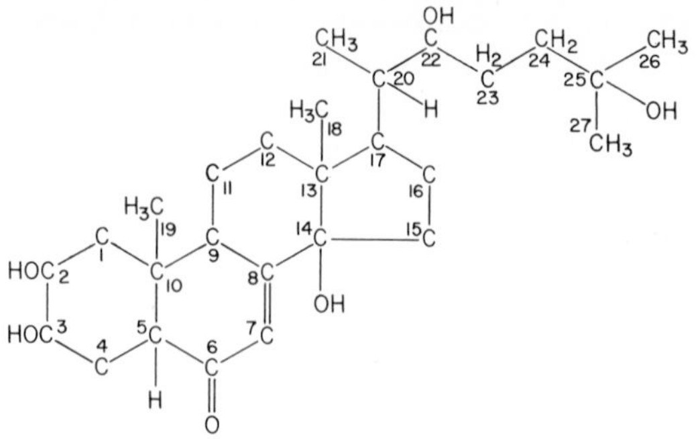

FIG. XVII–9 *Structural formula of alpha-ecdysone.*

a hydrogen on carbon atom 20, from the prothoracic glands of a number of different insects and from the Y-organs of crustaceans; and 20,26-dihydroxyecdysone with hydroxyl radicals on carbon 26 as well as carbon 20, from the prothoracic glands of pupae of the tobacco hornworm *Maduca sexta*. None of these is species specific, nor even class specific, for insect ecdysone injected into the hemolymph of a crustacean will cause it to molt and crustacean ecdysone will cause insects to molt. Nor are these steroid configurations peculiar to animals, for alpha- and 20-hydroxyecdysone have been obtained from the bracken fern *Pteridium aquilinum*, and similar if not identical steroids, also inducing molting in insects, from other ferns and from certain conifers. The suggestion has been made that the presence of these compounds in plants may have protective value for them, in that the growth processes of insects that feed on them would be interfered with, and so the number of predators reduced. Feeding tests in *Schistocerca*, however, reveal that there is no absorption of ecdysones from the gut and that it is only when the active compound is injected into the hemolymph that molting is induced. The adrenal glands and gonads of vertebrates also secrete steroid hormones, and the ubiquity and common physiological potency of this group of compounds make them of especial interest.

Histologically, the prothoracic glands have been found to undergo cyclical secretory activity that coincides with the events of proecdysis. They remain active during the juvenile life of pterygote insects; in *Pieris brassica* they reach their greatest development in the last larval instar, but degenerate and disappear 2 to 10 days after the imaginal molt. In other pterygote insects, they also disappear some time in adult life.

In apterygotes, the tentorial or ventral glands supply the molting hormone. Histological studies show that these undergo a cycle of activity that corresponds to that of the epidermal cells during proecdysis. These relationships have been particularly studied in the firebrat *Thermobia domestica*, a thysanuran that molts as an adult. Molts occur about every 5 days in an English strain that has been studied, about every 13 days in an American one. Activity is evident in the epidermis on the fourth or fifth day in firebrats that molt in the 5-day cycle, on the fifth or sixth day in those with a 13-day cycle, and in each case deposition of the new cuticle follows about 2 days later. Activity in the tentorial glands coincides with these periods of epidermal activity, reaching its maximum at the time of deposition of the new cuticle. Cyclical activity in the tentorial glands has been shown also in the walking-stick *Carausius*, and in *Locusta migratoria* it has been found that implantation of active glands at the beginning of the fourth or fifth instar accelerates the molting process. The hormone has not been identified, but the similarity in its effects to those of the prothoracic glands strongly suggests a similarity in their chemical nature. The tentorial glands degenerate after the imaginal molt in most pterygotes where molting does not continue into adult life. Exceptions to this are the workers and soldiers of termites in which the glands remain active throughout their lives.

The protocerebral neurosecretory cells, the corpora cardiaca, and the prothoracic glands, or their homologues, thus constitute a neurosecretory-endocrine system in insects essentially like that of the X-organ–sinus gland–Y-organ complex of crustaceans and similar, as the crustacean complex is, to the hypothalamic-neurohypophyseal system of vertebrates.

In insects, however, a third pair of glands is involved in metamorphosis and molting. Their counterpart has not yet been discovered in crustaceans. These are the corpora allata, which, during the juvenile stages of pterygote insects, secrete a product known as neotenin, the inhibitory or the juvenile hormone, since it prevents development of imaginal characters. The importance of this gland to the development of nymphs and larvae has been shown by ligation, extirpation, implantation, injection, and parabiotic experiments in a wide variety of insect species. If, for example, active corpora allata from third- or fourth-stage nymphs of *Rhodnius* are implanted in the abdominal cavity of a fifth-stage nymph, after the critical period of proecdysis, the insect molts as a giant, or sixth-stage nymph, not as an adult. Similarly, removal of the corpora allata from an earlier stage results in a precocious molt into a miniature adult. If, on the other hand, corpora allata are removed from fifth-instar nymphs, or during the final nymphal stage of other insects, there is no effect upon metamorphosis, and the molt proceeds with the insect emerging as an imago. Similar results have been obtained from similar experiments on holometabolous insects. In every case there is a critical period, as there is for ecdysone, before an active concentration of neotenin has been reached in the hemolymph; after this, extirpation of the corpora allata does not cause a precocious molt. This was made clear in pioneer experiments on silkworm larvae of a race that has five larval stages and pupates at the end of the fifth. If the corpora allata of a third-stage caterpillar are excised less than 34 hr after the previous molt, the larvae pupate, thus skipping two of their natural larval stages. But if this is done 44 hr or more after the previous molt, a caterpillar molts into the fourth larval stage, but pupates at the end of this, skipping one larval instar.

From numerous experiments such as these, it has become apparent that during the juvenile stages of an insect a balance exists between ecdysone and neotenin that determines the features of the next instar, whether juvenile or adult. A certain titer of ecdysone must be present for molting to occur at all, and when, by this means, a message reaches the epidermal cells, processes are set in motion for their multiplication, secretion of new cuticle, and other appropriate metamorphic events. But whether this cuticle is to be juvenile or adult is determined by the titer of neotenin in the hemolymph; if this is in active concentration, a message reaches the epidermal cells to lay down larval or nymphal cuticle and characters, but if the concentration is below its active level, the epidermal cells respond to ecdysone only and prepare for secretion of adult cuticle and the development of adult characters. Secretion of neotenin is reduced in the final juvenile stage, although the corpora allata do not degenerate (Fig. XVII–10).

Chemical identification of neotenin has only recently been made. A substance which, when injected into juvenile insects, has effects similar to those of neotenin has been synthesized in a two-step procedure beginning with the open-chain terpene alcohol, farnesol. A number of derivatives of farnesol have produced similar results; of these, farnesol methyl ether was found to be the most effective in preventing metamorphosis in *Rhodnius*. Studies of other related synthetic compounds have shown that the dihydrochloride of methyl farnesoate is some 10,000 times as potent as farnesol methyl ether and 100,000 times as potent as farnesol. Farnesol as such, and its aldehyde farnesal, have been found in the excreta of mealworms. This, in 1961, was the first discovery of a compound whose structure was known that acted as an insect hormone. Of added interest is the fact that the dimer of farnesol, squalene, is an intermediate in the synthesis of cholesterol and possibly of other steroids. Farnesol is not, however, neotenin, for experiments have shown that to produce the same results far more of it is needed (milligram or tenths of milligram amounts) than of neotenin, which is effective in micromilligram amounts. Various other derivatives or compounds chemically related to farnesol also mimic the effect of neotenin. Such compounds have been obtained from representatives of every animal phylum tested, with the exception of sponges and flatworms but including, interestingly enough, crustaceans, and from bacteria and a variety of multicellular plants. Dendrolasin, a product of the mandibular glands of the ant *Lasius fulginosus*, and geraniol, the active substance in the secretion of Nassanoff's glands in honeybees, both of which are chemically related to farnesol, have juvenilizing effects, as has a substance that has been extracted from the adrenal glands of cattle. The juvenilizing effect of the "paper factor" upon members of the Pyrrachoridae has already been mentioned (Chapter IX). It seems likely that compounds of this kind are, like steroids, widespread among living systems, in which their function, other than in insects, remains to be elucidated.

A substance has been extracted from male larvae of the moth *Hyalophora cecropia* that proved to be a *dl*-

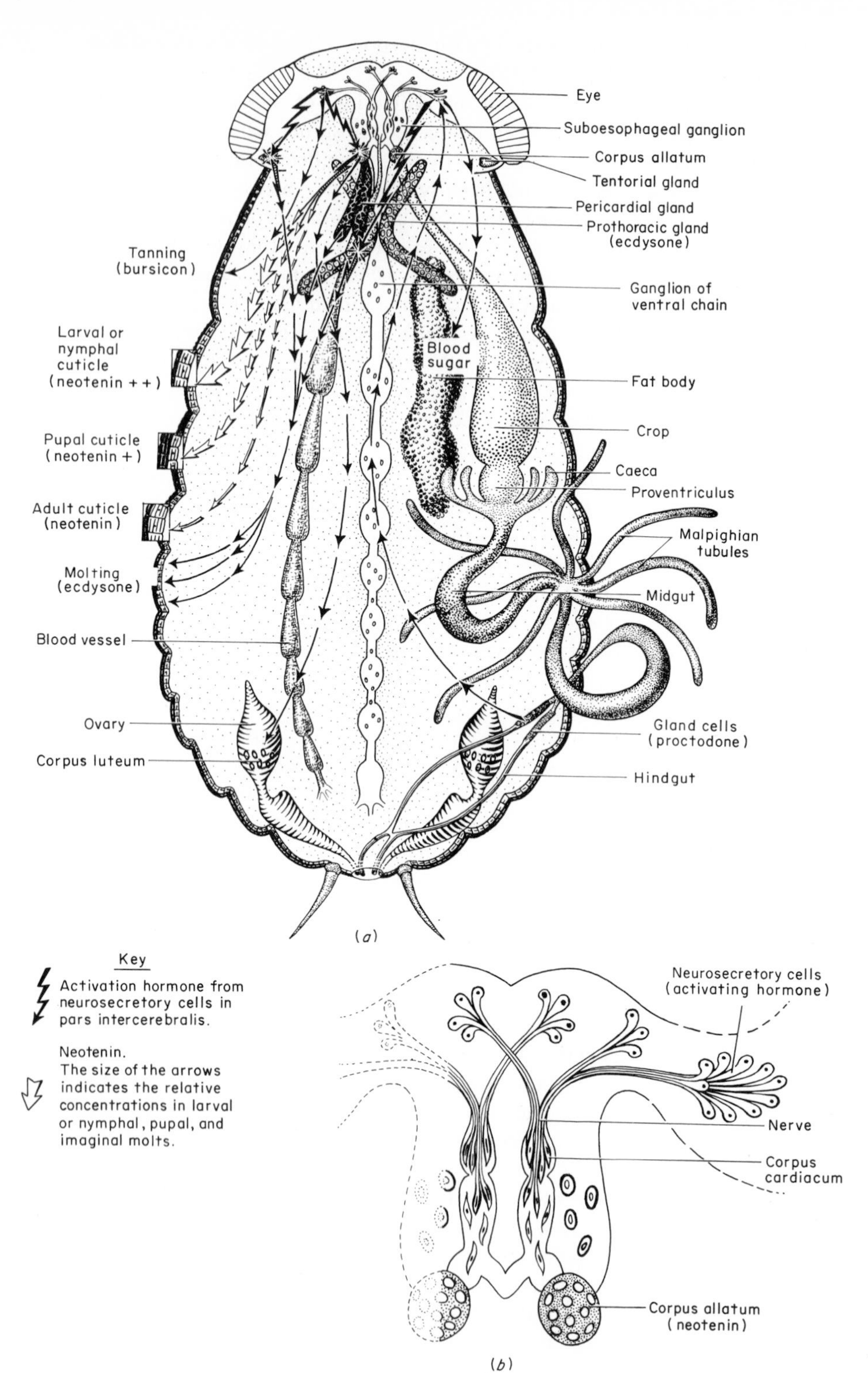

FIG. XVII–10 *Schematic representation of the hormonal system in insects. (From various sources.)* ***(a)*** *The body of a generalized (composite) insect, showing the sites of hormone production and the principal target organs. The arrows indicate the sources of the hormones and the tissues they affect.* ***(b)*** *Detail of pars intercerebralis, corpora cardiaca, and corpora allata complex.*

mixture of methyl-*trans,trans,cis*-10,11-epoxy-7-ethyl-3, 11-dimethyl-2,6-tridecadienoate, which has been identified as the juvenile hormone in this species. The compound has been synthesized and found to have the full potency of neotenin. Injection of the natural compound into third-instar larvae of the fly *Sarcophaga bullata* has been found to result in their failure either to form puparia or to continue development after about the third day of pupal-adult development. It seems likely then that the neotenin of lepidopterans is the same as, or very similar to, that of dipterans.

Major problems presented by hormonal controls in all organisms, in addition to identification of the active principals, are concerned with determination of the means by which an endocrine system is initially activated, how its products are transported to the target or responsive cells, and how they react to it. The information that has been gained about these aspects in insects may be equally applicable to other animals and contribute to the general solution of these problems. They therefore deserve consideration in some detail.

b. ACTIVATION, TRANSPORT, AND EFFECT ON TARGET ORGANS In molting and metamorphosis of insects, the key event is activation of the neurosecretory cells in the protocerebrum, either directly or via the corpora cardiaca, causing them to release the prothoracotropic hormone. In some insects the medial cells only are active in early instars, the lateral ones, if present, contributing their secretion late in postembryonic life, which suggests that they have a special role. In *Cecropia*, however, both secretions are needed for all molts, suggesting either that two hormones are involved or that one is complementary to the action of the other. Cells with different cytochemical properties have been demonstrated in these neuroendocrine groups. In the milkweed bug *Oncopeltus fasciatus*, for example, there are four types, designated A, B, C, and D cells, each differing from the others in certain morphological features, such as size and type of inclusions, and each showing different staining capacities. In most insects, these cells show cyclical activity, which reaches a maximum just before ecdysis, after which they regress.

Excitation of these cells probably has different origins in different insects. Nutritional state, sensory stimuli, humoral agents, and endogenous nervous rhythms have all been held accountable. In *Rhodnius*, it has been shown that the stimulus arises from distention of the gut and the consequent activation of the mechanoreceptors in the body wall. The excitation is conveyed to the protocerebrum through the ventral nerve cord, for cutting this prevents their secretion and, consequently, prevents molting, even when the gut is fully distended. That the stimulus derives from the quantity and not the quality of the fluid in the gut has been shown by experiments in which water was introduced into the gut; the distention resulting from this is just as effective a stimulus as a full meal of blood. Distention may also be the cause in some other insects, where impulses from stretch receptors in the gut, at the end of the period of indispensable nutrition, may be relayed to the protocerebrum via the frontal ganglia, for cutting the nerve from this to the cerebral ganglia prevents molting. It is not so for all, however, for some insects molt successfully when poorly fed or even starved. Other sensory stimuli, especially photic or thermal ones, may be responsible in these and other cases. There is evidence that in some insects a hormone secreted by glandular cells in the epithelium of the hindgut is the agent activating the protocerebral cells. This has been given the name of proctodone, and its effect has been demonstrated in larvae of *Galleria mellonella*, in which its transmission from the hindgut was prevented by ligatures tying off the posterior end of the abdomen. If these were applied to last instar larvae before they had begun to spin cocoons they did not pupate, but if they were applied after spinning had begun the caterpillars pupated in their regular fashion. Similar experiments on the corn borer *Ostrinia nubilalis* showed that diapause could not be terminated by the usually successful experimental means if the insects were ligated between the sixth and seventh abdominal segments and the supply of proctodone to the head cut off.

Once activated, however this may come about, the biochemistry of the neurosecretory cells is directed toward synthesis of the activation or prothoracotropic hormone. This is not the only hormone they produce. The identification of bursicon (Greek *bursicon*, "pertaining to tanning") as one of their products was first shown by ligation experiments on the flies *Calliphora erythrocephala*, *Phormia regina*, and *Sarcophaga bullata*. When, immediately after eclosion, a ligature was tied just behind the head, the cuticle of the entire body did not undergo the characteristic darkening and hardening known as tanning. Tying off other parts of the body had no such effect, but injection of hemolymph taken from another fly 30 to 60 min after its emergence from the puparium brought about tanning in the bodies of flies that had been ligated behind the head at an early stage. Actually the tanning agent reaches an active concentration in the hemolymph within 2 to 3 min after eclosion, and 15 min afterward is present in such amounts that it can be diluted 30 times and still induce tanning in an insect that has been decapitated or appropriately ligatured. Removal or destruction of the median neurosecretory cells of the protocerebrum has the same effect as decapitation or tying off the head. A substance with similar activity, and in even greater amounts, has been found in extracts of the compound thoracic ganglion, where it is especially abundant in the cells representing the abdominal ganglia. Neither it nor that from the neurosecretory cells of the protocerebrum is species specific, and both will promote tanning in other flies, as will hemolymph from newly molted nymphs and adults of *Periplaneta americana*, and from adults of *Tenebrio molitor*. This is not true of hemolymph taken from fully tanned flies. Bursicon has been isolated and chemically characterized to the extent of its identification as a protein or large polypeptide, with a molecular weight of about 40,000. It is thus distinct from the steroid activation hormone and differs from it also in that it is released only at the time of emergence of the adult. Its release is probably triggered by nervous stimuli from the periphery by way of the central nervous system, for if the nerves in the head of a newly emerged fly are cut, there is no tanning of its cuticle. The effect of bursicon seems to be directly upon the epithelial cells, directing them to complete the final step in molting and morphogenesis.

The targets for the activation hormone of the brain are the cells of the prothoracic glands and the corpora allata, which respond to its active concentration by secreting their own products, whose targets, in respect to molting and metamorphosis, are the epidermal cells. Evidence from injection and implantation experiments shows that neotenin also activates the prothoracic glands, and the hypothesis has been proposed that these organs, throughout the juvenile life of an insect, are kept at a low level of ecdysone production through the agency of neotenin and that they are roused into greater activity at molting by an active concentration of the prothoracotropic hormone. Moreover, ecdysone itself seems to activate the prothoracic glands, so that possibly they respond to three hormones: the neurohormone from the protocerebrum, neotenin, and their own secretion. The fact that they degenerate in adults when the concentration of neotenin is at its lowest supports the concept that their integrity during juvenile life depends upon the activity of the corpora allata.

It is generally accepted that the hemolymph is the principal means of transport of these secretory products, for injections or transfusions of hemolymph from a molting insect into a nonmolting one induces molting in the latter. In *Rhodnius* there is evidence that the hemocytes may be the actual transporters. Experiments in which the uptake of other substances has been blocked by injection of India ink, with which the hemocytes become engorged and afterwards clump together, have resulted in a delay in molting of 2 weeks past the usual time, implying interference with hormone action when these cells are put out of action and resumption of it only after a new generation has been produced. Electron microscopy has shown that in a number of insects there is neurosecretory material in the axons of neurons leaving the corpora cardiaca and running to various organs. In *Periplaneta americana*, for example, nerves from the corpora cardiaca to the aorta are so filled, and in the aphid *Myzus persica*, those to the muscles. This may be an accessory means of transport, especially adapted to insects whose fluid intake is great. In general, the hemolymph, either through its fluid or through its cellular components, seems to be the principal means, although it is possible, also, that the strands of connective tissue between the corpora allata, corpora cardiaca, and the prothoracic glands may serve as channels for the hormones each produces.

The effect of these hormones upon the target organs has been most fully studied in the epidermis. Observations that followed the distribution of labeled ecdysone injected into an insect have shown that it accumulated first in the epidermal cells and later in those of the fat body, with the nuclei of the epidermal cells showing the greatest radioactivity. The assumption followed that the principal sites of its action are the nuclei of the epidermal cells and that the cells of the fat body are engaged in elimination of excess hormone from the hemolymph. The discovery that ecdysone will induce puffing in the polytene chromosomes of the salivary glands of chironomid larvae supports the concept that its action may be intranuclear and even at the gene level. The natural phenomenon of puffing has been known for some time and recognized as local uncoilings and extensions of the chromatids that cause particular areas of the giant chromosomes, at different times, to look puffed or swollen. These puffs have been correlated with gene activity and the synthesis of messenger RNA, instrumental in the synthesis of proteins. Such puffing, in patterns characteristic of pupation, have been seen in the chromosomes of newly molted final-instar larvae of *Tendipes* (*Chironomus*) within 15 min after the injection of ecdysone. It has also been seen in other chironomids and, in vitro, in excised salivary glands bathed in hemolymph taken from insects at different developmental stages. The fact that the puffs arise in different areas of the chromosomes at different times makes a good case for ecdysone to be instrumental in turning on particular operons at particular times in the course of an insect's development and thus promoting the production of different kinds of RNA. However, it has also been shown that zinc and cadmium ions, as well as narcotics, will induce characteristic puffing patterns and that isolated pieces of chromosome also respond directly to these agents, which do not, however, induce molting in the intact insect. Further experiments along this line have shown that changes in the Na^+ and K^+ concentrations in the nucleoplasm will also bring about puffing and that the size and location of the puffs can be manipulated almost at the experimenter's will by manipulations in the amounts and relative proportions of these monovalent ions, for example, by increase in the concentration of K^+ in the nucleus or by decrease in that of Na^+. In the light of these findings, it seems more probable that the effect of ecdysone is upon the nuclear membrane, altering its permeability and so the intranuclear microenvironment, than directly upon the chromosomes. It is possible that neotenin works in a similar way. In either case, hormonal regulation in temporal control of the utilization and expression of information stored in the genome is implied, the bits* relevant to the condition of an insect at a particular time being brought out by the activation of some operons and the suppression of others. Thus, isogenic cells, like those of the epidermis, can be made to do different things at different times, depending upon the absolute concentrations and the relative concentrations of ecdysone and neotenin. In very general terms, the effect of the molting hormones upon the epidermal cells is to increase their synthesis of RNA, and of proteins, enzymatic and structural. In those species where there is multiplication of epidermal cells preceding a molt, synthesis of DNA and the materials involved in the machinery of mitosis must also be promoted.

c. MORPHOGENESIS Color changes are far more rare in insects than they are in crustaceans, but in some they are known to occur as morphogenetic adaptations, through the production of pigments of different kinds at different times in an insect's life. They also occur as physiological adaptations to external conditions, through the migration of pigment granules as in the chromatophores of crustaceans. These physiological color changes, short term and transitory, have been most thoroughly investigated in *Carausius morosus* and larvae of the mosquito *Chaoborus* (*Corethra*). There are four different pigments in the epidermal cells of *Carausius*, a green one and a yellow one that are uniformly distributed through the cytoplasm; an orange one that, when concentrated, forms a horizontal band at the level of the nucleus; and a dark one that moves proximally and distally in the cell. It is the migrations of these latter two pigments that cause the lightening and darkening of the integument, which, in these insects, occurs every 24 hr. Their bodies, therefore, are pale by day and dark by night, an adaptation of considerable protective value. By day,

* A "bit" may be defined as a unit of information resulting from a choice between two equally probable alternatives. It is a term introduced for the digital computer, as a unit of memory corresponding to the ability to store the result of choice between two alternatives.

the dark pigment is concentrated at the basal ends of the cells and the orange pigment in a band around the nucleus; by night, the dark pigment concentrates at the distal ends of the cells and the orange is uniformly distributed throughout them. If kept continually in light, the insects ultimately become permanently lighter and, conversely, if kept in the dark, become permanently darker, but it takes some time for the inherent rhythm of the pigment's migration to be altered. Hormonal control of these migrations and of the rhythm was suspected when histological examination showed that there were no nerve terminations in the epidermal cells. This suspicion was verified by ligature experiments on nymphs that prevented the hemolymph from reaching the posterior part of the body, which then showed no rhythmic color changes, and by removal of the cerebral ganglion, after which the entire body remained permanently light. This latter condition could be reversed if these ganglia from dark-adapted specimens were implanted, or if extracts from them were injected, into light-adapted ones. This substance proved, like bursicon, to be so potent, or to be produced in such excess, that the extract could be diluted 20 times and still be effective. Extracts from the suboesophageal ganglion were equally effective, and further experiments showed that neurosecretory cells producing compounds giving similar results were located in the deuto- and tritocerebral areas of the cerebral ganglia. It seems likely that at least two hormones are involved in the control of these color changes. One of these has been designated as D_1, derived from the pars intercerebralis and, in lesser amounts, from other parts of the central nervous system, and the other as C_1, from the deuto- and tritocerebrum.

The four large air sacs, two in the thorax and two in the seventh abdominal segment of larvae of *Corethra* are overlain with melanophores, which contract and expand. When the larvae are at the bottom of a pool with a dark background, these melanophores are maximally extended, so that the body of the transparent "phantom larva" is dark, but if they are transferred to a light substrate, the melanophores contract to little specks on the air sacs, and the body is generally light. Extirpation of different ganglia of the nerve cord, injection of their extracts, and chromatographic analyses of these have shown that they have two components with activity like that of C_1 and D_1 of *Carausius*, C_1 causing expansion of the melanophores and so darkening of the body and D_1 causing their contraction and consequent lightening of the body.

Hormonal control of morphogenetic color changes has been shown in the moth *Cerura vinula*, where the production of the red ommochrome just before pupation is prevented in those parts of the body which are cut off by a ligature that prevents the access of hemolymph to them. Injection of ecdysone into these regions leads to their reddening, which would indicate that this hormone in some way influences the biochemical conversion of brown xanthommatin to red dihydroxyxanthommatin.

Many morphogenetic effects are attributed to neotenin. Among these are phase polymorphism in locusts, seasonal polymorphism in aphids, and caste differentiation in *Kalotermes*. In *Schistocerca* and *Locusta*, individuals of the so-called "solitaria type" live sedentary, that is, nonmigratory, lives in geographical breeding zones. Periodically, these populations give rise to individuals of the migratoria type, which, as nymphs, begin to wander 8 days after emergence. The production of migratoria types can be correlated with climatic conditions, especially the seasonal occurrences of drought and high temperatures and, often, with reduction in food supply, either as a result of climatic conditions or of overpopulation. In their sedentary phase, these locusts are distinguished by their green color as well as by their habits. Implantation of corpora allata into young nymphs results in green coloration, even in conditions that would lead to production of drab brown migrants, so that the two phases become indistinguishable from each other.

In some species of aphids there is a regular alternation of winged with wingless forms, the winged generation that can migrate to other plants usually appearing after a succession of wingless generations that feed in a restricted area. It is reported that winged forms also appear when the food supply is reduced by weakening or sickening of the host plant or by overcrowding of the parasites when reproduction has been particularly successful. The hypothesis has been proposed that all aphids start development as potentially winged individuals, but that neotenin, whose production in the embryonic corpus allatum may be initiated through the hormonal system of the mother, prevents their development as such and that it is only when the supply is limited that the potentiality of wing development is expressed.

Other effects of neotenin on histogenesis have been observed in a number of species. In *Carausius*, removal of the corpora allata is followed by both degeneration and proliferation of tissues, sometimes occurring together in the same tissue. Atypical growth is occasionally evident, and tumors may be formed. Allatectomy also impairs the capacity for regeneration in some insects, but not in others. *Blatella germanica* is as capable of regenerating a missing part at any stage of its life without its corpora allata as it is with them. These histogenic and morphogenic effects are all extremely complex ones and are by no means as yet fully analyzed or understood, but they serve to illustrate the interactions and overall influences that hormones have upon all phases of insect development.

d. METABOLISM In addition to their involvement in the visible manifestations of growth and development, hormones are known to influence certain continuing and basic physiologic processes in insects. They may have their origin in the central nervous system, the corpora cardiaca, the corpora allata, or the prothoracic glands. One of the first metabolic effects to be recognized, in connection with molting and metamorphosis, was oxygen consumption. In *Rhodnius*, for example, this is at its lowest during intermolt, but increases when the prothoracic glands are activated, reaches its maximum just after ecdysis, and then falls off again. Similar changes have been observed in various lepidopterans during the different stages of their postembryonic development. These are in all likelihood due to the relative amounts of enzymes of the cytochrome system that are synthesized at these different stages and have been shown to be reflected in the number and size of the mitochondria. In *Rhodnius*, there are fewer mitochondria, and they are smaller, during intermolt, but they increase in number and in size when the prothoracic glands are activated and ecdysone is liberated into the hemolymph. A similar picture is presented by cecropia silkworms. The sarcosomes of the potato beetle in diapause, an event that occurs in their adult lives, and during which their respiratory rate is very low, show degenerative changes. The corpora allata also influence respiration, for allatectomy has been found to reduce oxygen consumption in a number of insects of different species; in *Calliphora erythrocephala* this amounts to a 19 percent decrease. The cycles of respiration correlated with cycles of reproduction and oviposition in *Pyrrochoris apterus* are eliminated with removal of the corpora allata, and the insects exhibit a constant level intermediate between the maxima and minima of intact insects. Removal of both corpora cardiaca and corpora allata results in an even lower level.

In a number of insects, extracts from the corpora cardiaca bring about significant increases in the rate of heartbeat and stimulate muscular contractions in the hindgut and Malpighian tubules. The active principal is believed to be a protein or peptide and, in *Periplaneta* through differential centrifugation of homogenates of excised corpora cardiaca, has been shown to be associated with a fraction containing granules that bear a close resemblance to the neurosecretory vesicles revealed by electron microscopy in neuroendocrine cells. Electrical stimulation of the cerebral ganglion of *Periplaneta* results in the release of a factor with high cardioaccelerator activity, whose derivation from the corpora cardiaca has been shown by removal of other parts of the retro-cerebral complex. In this cockroach, too, the corpora cardiaca release a diabetogenic hormone whose effects are comparable to those of the hormone released from the sinus glands of crustaceans. Injection of minute amounts of the extract from a single pair of corpora cardiaca into an intact cockroach results in a marked increase of trehalose in its hemolymph. Similar results can be obtained by injections of extracts from the cerebral ganglia, which make it clear that the diabetogenic factor is a product of neurosecretory cells situated there. Its effect is primarily mobilization of the glycogen stored in the fat body and its conversion to trehalose, possibly due to an increase in the concentration of active phosphorylase, which may be the direct effect of the hormone. Results from experiments on the incubation of excised fat bodies of the cockroach *Leucophaea maderae* with extracts of corpora cardiaca or with the intact glands indicate that it may act at two points in the pathway of carbohydrate metabolism. One of these is that leading to an increase in the rate of glycogen degradation, and the other is that leading to a diversion of the resulting hexoses from the glycolytic pathway at the phosphofructokinase stage (Figs. XII–2 and 3) and their combination into the disaccharide trehalose through activation of the enzyme trehalose-6-phosphate synthetase. Depletion of its carbohydrate reserves causes a shift in the endogenous metabolism of the fat body from a mechanism using carbohydrate as the main substrate to one using lipid, a secondary effect of the corpus cardiacum hormone. There is histological evidence of neurosecretory cells in the corpora cardiaca of *Leucophaea*, so that the hormone influencing carbohydrate metabolism in the fat body may be produced there.

There is also evidence for production of a diuretic factor by the neurosecretory cells in the cerebral ganglia of some insects. Destruction by cautery of the neurosecretory cells in the supraoesophageal ganglia of immature males of *Schistocerca gregaria* results in retention of water in the body with consequent dilution of the hemolymph. In mature males most of these cells, upon microscopic examination, appear filled with secretory granules, which are absent in specimens that have been injected with distilled water. There is, as yet, no evidence of any effect this hormone may have upon the cells of the Malpighian tubules. In no insect have these organs any nerve supply, so their activities must in some way be regulated through the hemolymph. In *Rhodnius prolixus* the large blood meals of fifth-stage nymphs are followed by a copious excretion of water, which has been correlated with the appearance in the hemolymph of a very active diuretic hormone. Ligation experiments and electron-microscopic observations have located the source of this in neurosecretory cells in the fused ganglionic mass in the mesothorax. The swollen axons of these cells run forward in the ganglion to a neuropil, where each divides. One branch runs deeply into the neuropil and ends there, possibly receiving stimuli from afferent nerves also terminating there. The other branch runs posteriorly along with axons of other neurons that form the abdominal nerve. The swollen endings of these branches from different neuroendocrine cells terminate in different abdominal segments, being much more numerous in the first three than in the more posterior ones. Direct effect of their action upon the Malpighian tubules has been demonstrated in experiments in which the tubules, still attached to the hindgut, have been dissected out of a nymph and immersed in a drop of hemolymph. Urine is immediately produced by a nymph that has just fed. The quantity, as measured by the volume of the drop that appears at the opening of the rectum, is at first large and then falls off, but if hemolymph from a freshly fed insect is added to that in which the tubules are bathed there is a new burst of urine production. This does not happen if hemolymph from an unfed specimen is added. Moreover, the tubules do not produce urine if the two abdominal nerves are ligated just after their emergence from the mesothoracic ganglion. In *Apis mellifica* injection of extracts from the corpora cardiaca promote retention of water, while those from the corpora allata lead to its elimination. Contradictory as some of this evidence may be, there is little doubt of the endocrine control of water balance in insects.

e. REPRODUCTION Full development of the reproductive organs and of the sex cells has been shown, in some insects at least, to be under hormonal regulation. The hormones concerned may be derived from neuroendocrine cells of the central nervous system or from the corpus allatum. In several species of insects, an increase in the number of neurosecretory granules has been noted in the cells of the pars intercerebralis of adult females in the period just before and during oviposition. In *Calliphora* a direct effect of an active principle secreted by them has been shown, for their removal or destruction results in incomplete development of the ovaries, the accessory glands that lead into the vagina and discharge into it products that provide the egg capsule as well as substances for attachment of the eggs, and also of the oenocytes, and the corpora allata. These conditions can be counteracted by implantation of actively secreting neurosecretory cells from another individual, but not by that of their corpora allata. In *Leucophaea maderae* three types of nerve cells, designated A, B, and C, can be histologically and cytochemically distinguished in the suboesophageal ganglia of females. Cells of the B type have also been found in other ganglia of the nerve cord in *Periplaneta americana* and *Blatta orientalis*. Ovariotomy of *Leucophaea* females between the third and seventh nymphal instars results in an accumulation of granules or vesicles in the B cells and a change in their staining properties. There are no such effects following castration of male nymphs. These results are suggestive of a specific relationship between the B cells and the ovaries, leading to discharge of their product at the appropriate time in the maturation of the females and response of the ovaries to it.

The influence of the corpora allata upon ovarian development was shown in *Rhodnius prolixus* early in the history of insect endocrinology. Extirpation experiments proved that they are indispensable to ripening of the eggs after the period when the follicle cells become active and contribute to the development of the eggs. After allatectomy the follicle cells degenerate, indicating that the action of the hormone is not directly upon the oocytes, but indirectly upon them through the follicle cells. Implantation of corpora allata restores activity in the follicle cells and promotes further oogenesis. Similar relationships have been shown in a number of other insects, including *Leucophaea*, *Melanoplus*, *Calliphora*, and the beetle *Dytiscus*. However, they do not seem to exist in *Carausius* or *Bombyx*. Full development of the accessory reproductive glands, in most insects tested, depends also upon the presence and activity of the corpora allata, although their secretory activity is not wholly suppressed after allatectomy. *Lucilia* and *Sarcophaga* are exceptions in that in these flies the corpora allata seem to have no influence in development of the accessory glands. Nor has any influence upon spermatogenesis or testicular development been demonstrated in the males of any species examined.

The connection between ovarian development and activity of the corpora allata raises the question of their production of a gonadotropic hormone as well as of neotenin. Existing evidence, however, contradicts such a possibility. There are, for example, no detectable cytological differences in the cells of a corpus allatum, suggesting that there is but a single product but not, of course, precluding the possibility that cells morphologically similar may produce different secretions at different times, or even simultaneously. More cogent evidence against dual or multiple hormone production comes from implantation experiments of corpora allata taken from individuals at different stages in their life cycles. Ovarian development arrested by allatectomy can be restored by implantation of corpora allata taken from any larval or nymphal instar of the same or another species; and, similarly, implantation into a juvenile of corpora allata taken from an imago has the same effects upon it as one taken from another insect at a similar stage of development. It seems most likely then that neotenin is the single hormone synthesized and discharged by these glands and that it is the target cells that change in the adult, the hormone's action being directed toward the follicular cells of the ovary rather than toward the epidermal cells of the integument. The comparatively low titer of neotenin in the hemolymph of adults compared to that of juveniles and the appearance of active follicle cells may be factors contributing to this shift.

There is, in no insect yet examined, any indication of a hormone produced by the gonads themselves. This was one of the facts contributing to delay in establishing hormonal regulation in insects, gonadal hormones being so general in vertebrates that it was confidently expected that they would be found in invertebrates as well, especially in those at so high a level of organization as insects. Failure to observe any effects upon sex differentiation from exchange of reproductive organs between males and females aroused skepticism about endocrine activity of any kind in insects and, by extension, in invertebrates in general.

Ovarian development may be promoted or inhibited by various external stimuli. In adults of *Schistocerca*, for example, release of neotenin and acceleration of the development of oocytes is induced in a number of ways—by the presence of males, by copulation, by enforced activity, and even by wounding. Messages about these conditions must, in some way, be conveyed to the central nervous system and the cerebral ganglia to bring the neurosecretory cells into action and effect release of the hormone that triggers activity in the corpora allata. Similarly, the secretion can be turned off by a neural message. Females of many species of cockroaches, for example, carry their eggs for several days in egg cases, or ootheca, protruding from their abdomens before depositing them in some spot suitable for further development and hatching of the egg inside (Fig. XVII–11). During this time no eggs are matured in their ovaries, which return essentially to the immature condition. In *Byrsotria fumigata*, it has been shown that the corpora allata are inactive in females carrying egg cases, but become active again after they have been shed. The inhibitory stimulus here would seem to be proprioceptive, initiated by distention of the abdomen by the relatively large egg case, for if the ootheca is removed or the ventral nerve cord severed, the corpora allata become active and egg cells are matured. Insertion of an artificial ootheca also inhibits activity of the corpora allata. Experiments on implantation of mature ovaries from other females into the abdomens of *Blatta orientalis* and *Blatella germanica* carrying egg cases have shown that the foreign ovary returns to the juvenile condition as the host's has done, clearly through some factor in the hemolymph. It has been suggested that this factor originated in the ovaries of the host, where the empty follicles form "corpora lutea" if its corpora allata are inactive at this time. Similar effects, however, result in these cockroaches from injection of various substances, including cholesterol, and there is as yet no information about the nature of the corpus luteum hormone, if such a one exists. At present neural hormones and neotenin are held responsible for ovarian development.

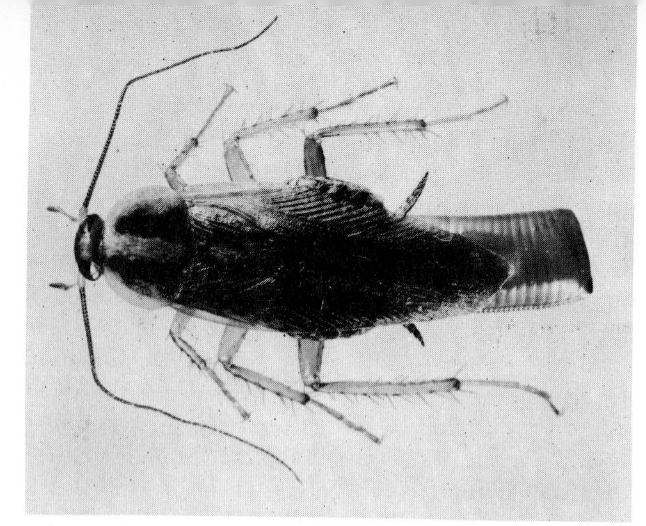

FIG. XVII–11 *Female cockroach* (Blatella germanica) *carrying ootheca (egg case). This species of roach carries the ootheca longer than any other, almost to the time of the hatching of the eggs.*

f. DIAPAUSE In addition to the regulation of active morphogenesis, hormones are involved in its arrest in insects, in the state known as diapause. The word diapause was first introduced into biology in 1893 to denote a stage in embryogenesis, specifically that of the grasshopper *Xiphidium ensiferum*, when the very young embryo is motionless after having moved through the yolk around the posterior pole of the egg and before it moves back to its original position. Later, in 1904, the word was used with different connotation to designate a period of arrested growth occurring at any period of an insect's life. Although during this period mitotic activity and differentiation is virtually at a standstill in most of the tissues, it is not a time of complete physiological inactivity. At this time certain changes take place, known as diapause development, which make it possible for the organism to resume its active growth and morphogenesis when diapause is broken or its usual activities if diapause takes place in the imaginal stage of an insect's life.

Diapause is not a universal phenomenon among insects, but in those in which it does occur regularly, it happens at a time characteristic for each species. This may be any time after formation of the blastoderm; it has never yet been recorded as occurring before this. It may, however, be interposed immediately afterwards, as it is in some lepidopterans and at least one cricket, *Homeogryllus japonica*. It may occur, as it does in silkworms and other lepidopterans and in the cricket *Gryllus mitratus*, in the course of early growth and mesoderm formation, but before the differentiation of tissues; or at the termination of this and just before active growth and establishment of the main organ systems begin. This is characteristic of a number of orthopterans, including *Gryllus commodus*, *Melanoplus differentialis*, and *Locusta migratoria*. Or it may occur just before hatching, as is the case in some coleopterans; some lepidopterans, as, for example, *Lymantria dispar*; and in *Melanoplus bivittatus*. It is evident that if it does intervene during embryogenesis, this is not at a time when mitotic activity is greatest or the differentiation of cells is in progress; once these events have started they cannot be stopped and must proceed to their completion. Or diapause may occur in larval, pupal, or adult life, manifesting itself in the first two instances as an arrest, of varying duration, in the ordinary developmental processes and, in the last, as delay in development of the reproductive organs with consequent hypertrophy of the fat bodies that supply the material for them. If it occurs during larval stages, it is most frequently in the last instars after the active feeding period has ended, but in some species it happens in earlier ones.

The duration of diapause varies also with the species and, indeed, in different individuals of the same species. The mean period in the grasshopper *Melanoplus* is, for instance, 40 days at 5°C, but in some eggs it takes only a few days for diapause development and in others it takes a hundred. Diapause may last 3 to 4 years in some sawflies and 6 to 7 years in the moth *Rothschildia jouella*. While a wide variety of stimuli, both natural and experimental, can bring about termination of diapause in some insects, the effective stimulus must be applied at some precise stage in diapause development when the organism is in a condition to respond to it. The coincidence of this receptive stage and the appropriate stimulus, under natural and experimental conditions, is the determining factor in breaking diapause. Grasshopper eggs, for example, collected just after they have been laid in late summer and brought into the laboratory where they can be kept at room temperature, never develop, but if they are refrigerated for several weeks, at 5°C or thereabouts, and then brought into room temperature, the majority will hatch in midwinter, long before their natural time and before their fellows left in the field. The period of cold is, however, essential to their diapause development and their ultimate hatching.

Although it is likely that all insects possess the potentiality of diapausing, as part of their genetic makeup, not all of them realize it. In some it is obligatory, with essentially all the individuals of each generation entering it at the time characteristic of their species, or of their strain within the species. In others, it is facultative, some generations manifesting it, others not. Distinction between facultative diapause, induced by external conditions such as extremes of temperature, and aestivation and hibernation is a subtle one, and possibly the periods of aestivation and hibernation, typical of some molluscs, annelids, and mammals, might legitimately be called diapause.

Whether this condition is obligatory or facultative in insects determines the number of broods produced annually, for if it occurs regularly in every generation there can be but one brood a season. Species or strains in which diapause is obligatory are perforce univoltine (Italian *volta*, "turn"), producing only one brood a year; others in which it is facultative may be bivoltine or polyvoltine, producing two or more broods in a season.

Environmental conditions are without doubt instrumental in setting in motion the internal events leading to diapause, which is a condition that allows the organism to survive adverse climatic conditions. Means for such resistance to environmental variations are evident among organisms of all kinds and are illustrated, among plants, by the formation of seeds and bulbs that can withstand cold and desiccation, and the dormant buds of woody plants that can be encased in ice in winter storms and burst into leaf or flower in the warm days of spring; and among animals, by the formation of gemmules in sponges, statoblasts in bryozoons, and winter eggs in cladocerans and rotifers. These responses may be evoked by the very conditions they enable the organism to survive, but preparations for them may anticipate the actual conditions, and the responses may continue after the conditions that evoked them have changed. In temperate regions, diapause usually occurs in the winter, but in tropical and subtropical ones, in the hot, dry season. Facultative diapause can be induced in diapausing species by artificial alterations of the environmental conditions.

One of the most important environmental conditions influencing diapause (for some insects the most important condition) is day length or photoperiod. There are, for example, "long-day" insects as there are long-day plants, and some short-day ones. The noctuid moth *Acronycta rumicis* is representative of a long-day species, for larvae that are exposed during the feeding period to light for 17 hr, or more, out of the 24 do not enter diapause, while nearly 100 percent of those that are kept in light for 15 hr, or less, do undergo it. Silkworms are representative of short-day insects; eggs of bivoltine or quadrivoltine strains that are kept in darkness, or up to 12 hr of light, develop into imagos that lay eggs that are pale in color and pass through embryogenesis, larval, and pupal stages without diapausing, while 70 percent of those exposed to light for as much as 17 hr, or even more, lay eggs that are dark in color and enter diapause at the characteristic time. Very precise relations between diapause and photoperiod have been observed, in the field and in the laboratory, in the pink cotton boll moth *Pectinophora gossypiella* that lays eggs on the outer surface of cotton bolls and is a serious pest to cotton growers in the southwestern United States and in Mexico. Larvae hatch from these eggs in about 5 days and burrow into the boll, staying a few days in the carpel and later moving further in until they reach the seeds, on which they feed. In summer, at the end of the feeding period, they eat their way out of the boll, drop to the ground, and metamorphose during a brief period of pupation, after which imagos emerge, mate, and start a new generation to repeat the cycle. But some of the eggs that are laid during the last two or three days of August hatch out as larvae that stay within the boll, spin a cocoon, and enter diapause. The number of these increases as the autumnal equinox approaches, and by mid-November practically 100 percent of the eggs develop into diapausing larvae, not to resume their growth and development again until after the vernal equinox the following spring. The appearance of the first diapausing larvae occurs with such precision every year that it is almost possible to set a calendar by it. The critical photoperiod for these insects is thus about $12\frac{1}{2}$ hr; when the days, from sunrise to sunset, are longer than this the larvae develop without arrest, but when they are shorter diapause intervenes in their larval life.

The emperor dragonfly *Anax imperator* is likewise an insect with facultative diapause. Final-instar nymphs may be found throughout the year, but they metamorphose to adults only between mid-May and the end of July. Nymphs that reach their final instar in early spring become imagos in 30 days, but those reaching this stage later enter diapause. Diapause can be experimentally induced in them at any time by exposure to constant photoperiods corresponding to the duration of daylight on May 12 or July 31, or to photoperiods that decrease each day by natural seasonal decrements.

Experiments have demonstrated that it is not the absolute time of exposure to light that is important in relation to diapause, but the duration of continuous light. For example, all larvae of *Acronycta* exposed to 9 hr of light and 15 hr of darkness enter diapause; 95 percent of those exposed successively to 9 hr of light, 9 hr of darkness, 3 hr of light, and 3 hr of darkness (represented conventionally as 9L 9D 3L 3D) enter diapause, but none of those exposed alternately to 6 hr of light and 6 hr of darkness (6L 6D 6L 6D) do, although the total number of hours of light (12) experienced in 24 is the same as that of those on the 9L 9D 3L 3D regime, virtually all of which went into diapause.

That the effect of light on diapause is not necessarily mediated through the visual organs has been shown in the moth *Antheraea* and the potato beetle *Leptinotarsa decemlineata*, for responses were obtained from individuals whose ocelli had been covered with enamel or destroyed by cauterization that were the same as those from intact individuals. It has been shown, too, that light of low intensity is effective. Embryos of *Bombyx*, for example, enter diapause when exposed to 0.01 fc, larvae of several other lepidopterans to an intensity of 1 fc, and those of the spider mite *Metratetronychus* to intensities of 1 to 2 fc. The pitcher plant midge *Metriocnemus*, a long-day insect, is sensitive to intensities as low as 0.0025 fc. *Metriocnemus* is a facultatively diapausing insect, with several broods in the spring, summer, and early autumn; the full-grown larvae enter diapause at the beginning of winter but delay their pupation until spring. The critical photoperiod for them is between 12 to 13 hr; larvae exposed experimentally to light for 12 hr, or less, remained in diapause and did not pupate, while those exposed for 13 hr, or more, did break diapause and pupate. Although their sensitivity to such low intensities of light indicates that in nature they respond to a part of the daily twilight period, their diapause is terminated in the spring when the days are actually shorter than when it was initiated in the autumn, suggesting that they must undergo some physiological conditioning to light during their diapause development. Presumably, there is, in all diapausing insects, a photolabile substance, not necessarily in their eyes, which is altered by exposure to light, white and also violet, blue, and blue-green; the latter color has been shown to be especially effective in *Bombyx*. A photochemical reaction of this substance then triggers the events that initiate or terminate diapause. The nature and location of this substance, in any insect, is as yet unknown.

Temperature is also a factor influencing diapause. In general, moderate temperatures tend to prevent it and low, to induce it, but this is by no means universal. There are, however, many species of insects within which there are geographical strains that differ in their diapause habits. Some are entirely free of it, some exhibit facultative diapause, and some, obligatory. This is the case, for example, with *Locusta*, for those living in the hottest areas of this insect's distribution do not diapause, while those living in the northern limits of its range regularly do. The spruce sawfly *Gilpinia polytoma* shows facultative or obligatory diapause according to the region it inhabits. The strain found in Connecticut, for example, is trivoltine, entering diapause in the autumn; the strain living farther north, in New Brunswick, Canada, is bivoltine, the last generation entering diapause at the onset of cold weather. The strain in the Gaspé is univoltine, with obligate diapause in each generation. In these insects the potentiality for diapause is realized in different ways in the different geographical strains and would seem to have undergone a certain amount of selective adaptation. Breeding experiments with *Bombyx* have shown that voltinism in them is determined by the genetic constitution of the female, involving three sex-linked and three autosomal genes, which are present in different combinations in univoltine and polyvoltine strains. Selection then would operate to establish in any strain the combination best suited to its survival.

Humidity is another environmental condition that has a bearing upon diapause, with, again, the generalization, also to be qualified, that arid conditions tend to bring it on and humid ones to break it. These relations have been carefully studied in *Melanoplus* where, during the first few weeks of development, water enters the egg through the hydropyle, a minute opening in the waterproof and otherwise protective membranes enclosing it. At about the end of the third week this uptake stops abruptly, and microscopic examination shows that the hydropyle has been sealed by a layer of wax. Diapause begins at this time and can be broken experimentally by dissolution of the wax. In nature, presumably the cells around the hydropyle destroy it, terminating diapause by permitting again entrance of water into the egg, which may take up as much as 2 to 3 mg during postdiapause development of the embryo.

The seasonal occurrence of diapause in phytophagous arthropods suggests that nutritional factors may be involved in its initiation, since the quality as well as the quantity of available food changes with the seasons. Females of *Metratetranychus*, for example, that have fed upon old and senescent leaves, or upon ones that are dying from injury by punctures of other mites, lay eggs in which diapause is obligatory, at the appropriate time in the life cycle, even when the photoperiod is long and the temperature high. Larvae of the apple codling moth *Carpocapsa* that in early summer feed on the green fruit do not enter diapause, but those that feed on ripened fruit do. Light is obviously a factor in this case. Yet very delicate adjustments between the availability of food, independent of photoperiod, and of other conditions essential to the survival of a species seem to have evolved. In many, the onset of diapause shortly precedes, or coincides with, the time when available food supplies become restricted, and its break, with the time they again become abundant.

Diapause may be synchronized with other aspects of an insect's life equally as important to survival of a species as its food supplies, yet indirectly related to them. The life history of the wheat blossom midge *Cantarinia tritici* is illustrative of this. The females, which live only 24 hr, lay their eggs in wheat ears and the larvae feed on them. The ovipositor of the female is so delicate that it can penetrate only the tissues of a newly emerged ear. The time of the female's release from diapause is, therefore, crucial to perpetuation of the species. If released too early, while the forming wheat ear is still within its sheath, or too late, after its tissues have hardened, she cannot lay her eggs there and must take her chances of finding an alternate, nutritionally less adequate host plant to feed her young.

Comparable adjustments have evolved between parasitic insects and their animal hosts. The hymenopteran *Trichogramma cacaeciae* lays its eggs in the eggs of a small moth *Cacaecia rosana*, which are deposited in masses of 30 to 60 toward the end of June and in July, on quince trees. This is a univoltine moth, and the eggs enter diapause early in their development and remain in this condition during the rest of the summer and throughout the winter, the young hatching out the following April. Its parasite *Trichogramma* is bivoltine and polymorphic, one generation of females having relatively large wings (macropterous) and the next having smaller ones (micropterous). During July the macropterous females lay their eggs in 8 to 10 of the centrally placed eggs of a *Cacaecia* egg mass, and these develop directly into maggots that consume the yolk and then enter diapause that lasts, like that of their hosts, until the following spring. Then the maggots pupate and emerge in March as micropterous imagos that lay their eggs in the remaining ones of the host's egg mass, which have now come out of their diapause. These *Trichogramma* eggs develop without arrest into macropterous females that are ready to lay their eggs in newly deposited *Cacaecia* egg masses. In the laboratory macropterous females of *Trichogramma* were induced—or forced—to deposit their eggs in those of a nondiapausing species of moth, *Mamestra brassica*. The maggots that hatched from these did not enter diapause, and five successive generations of macropterous females were produced with no interpolation of diapause or of a micropterous generation. It was winter when the sixth generation was hatched in the eggs of *Mamestra*, and essentially none survived. The development of a number of other parasitic hymenopterans has been observed to be as completely synchronized with the development of their hosts as is that of *Trichogramma* with *Cacaecia rosana*, an adjustment between morphogenetic events in two individuals that clearly has high survival value for the parasite at least.

In consideration of all these relationships, diapause can be regarded as a timing device by which the organism can apprehend change of seasons and make appropriate physiological adjustments. The mechanism of timing exhibited by diapausing insects has, in the many cases that have been investigated, proved to be a hormonal one, depending upon suppression or activation of the neuroendocrine, or endocrine, cells that provide the hormones regulating growth and development. In the case of diapause that occurs at an early stage of embryogenesis, as in *Bombyx*, conditions within the maternal endocrine system are the determining factors, for there is no evidence that at this stage the embryonic tissues produce hormones of the type under consideration; when diapause occurs in later embryonic, larval, pupal, or imaginal stages, the hormones concerned are the product of each individual itself.

In *Bombyx*, eggs that will enter diapause are visibly distinguished by their color. The type of egg that the female will lay is determined by a secretion from the neurosecretory cells in the suboesophageal ganglion that become active during her pupal phase. Extirpation of this ganglion at the beginning of the pupal instar causes a female to lay eggs that will not diapause in any circumstances, even in a univoltine strain. Conversely, implantation of a suboesophageal ganglion from either a male or a female of a univoltine strain into a female of a polyvoltine one, destined at that time to produce nondiapausing eggs, will induce them to diapause. This "diapause hormone" is not species specific, for implantation of suboesophageal ganglia from other species of moths with embryonic diapause, and even from the oak silkworm *Antheraea pernyi* with pupal diapause, will cause a female *Bombyx* to lay diapausing eggs. That activation of the neurosecretory cells in the suboesophageal ganglion is under control of the supraoesophageal ganglion has been shown by transection of the circumoesophageal commissures. When this is done in the pupal stage, a female moth will lay eggs that invariably enter diapause, indicating that this ganglion has an inhibitory effect upon the secretory activity of the neuroendocrine cells in the suboesophageal ganglion. Thus, although the "diapause hormone" is produced in the suboesophageal ganglion, conditions in the supraoesophageal ganglion determine whether it shall be released or withheld. The message from the supraoesophageal ganglion travels, in some way, through the circumoesophageal connectives and reaches the ovaries indirectly, for when the diapause hormone is not released, pale, nondiapausing eggs are matured and when it is, dark, diapausing ones.

Conditions within the mother are likewise instrumental in the induction of diapause in *Mormoniella* (*Nasonia*) *vitripennis*, a parasitic wasp with facultative diapause occurring in the fourth, or final, larval instar. *Mormoniella* is a chalcid that lays its eggs on the pupae of flies like *Sarcophaga bullata*, upon which they can be easily reared in the laboratory. The mature female lights upon a puparium, inserts her ovipositor through its tough coat, and deposits a generous number of eggs upon the pupa inside. These hatch in 2 to 3 days, the unfertilized eggs developing into haploid males and the fertilized into diploid females. At the end of 1 week, during which there have been four molts, the larvae under ordinary circumstances pupate, and remain in this condition for another week, during which there is a gradual development of adult structures and characters. Imagos emerge on the fourteenth day, search for fresh puparia into which they insert their proboscides, and feed for a short time, although they can live for as long as 6 days without food. About the second day after emergence the females deposit their eggs, and continue this the remainder of their short lives. If a larva enters diapause it does so at the end of the feeding period, at about the sixth day of its development, and does not pupate. Diapause may last for several months, or for as long as 2 years at room temperature, after which pupation ensues and adults emerge. The environmental factors responsible for diapause are photoperiod, temperature, and host deprivation, and are effective during the life of the mature female, not of the larva. Short-day lengths, low temperatures (10 to 15°C), and no access to fly pupae for a few days after emergence all induce females early in their reproductive life to lay eggs that will develop as diapausing larvae; while long-day lengths, higher temperatures (25 to 30°C), and immediate access to fly puparia result in the production of eggs that develop into adults in 2 weeks. Virgin females produce diapause larvae under the same conditions as fertilized ones, and a female may switch from the production of nondiapausing to diapausing larvae in the course of her life. Experimental evidence has made it quite clear that it is conditions to which a female is exposed during oogenesis, and not her offspring during their larval life, that set the pattern for diapause, and it is postulated that a chemical factor, in all probability a hormone, is transmitted from the mother through the egg to the larva. The actual mechanism might be through the endocrine system of the larva, since this is developed during larval life, but there is as yet no critical information of how the control is operated, although evidence from other insects in which diapause occurs at some period in postembryonic life indicates that it is due to deficiency of ecdysone.

Deficiency in ecdysone has been proved to contribute to the onset of larval and pupal diapause in insects in which arrest characteristically occurs at either of these stages and is, indeed, the critical factor in its induction and maintenance. Humoral control of diapause has been most thoroughly investigated in the giant silkworm moth *Hyalophora* (*Platysamia*) *cecropia*, in which, as in most other saturniids, it is usually obligatory. It begins in late summer at the time of the larval-pupal molt and may last for 6 to 10 months. Autoradiographic studies of the tissues of diapausing pupae injected at intervals with tritiated thymidine have shown that DNA synthesis and mitosis stop in all cells except the hemocytes, the reproductive cells, and certain cells of the cerebral ganglia; in these, mitosis may continue for about 1 week. Similar studies, using tritiated uridine to detect RNA synthesis, have shown that this stops at different times in different tissues. Ten days after pupation, epithelial, connective tissue, and muscle cells cease to synthesize RNA, and 3 weeks after, the cells in the fat body and wing buds follow suit; while the hemocytes, oenocytes, neurons, and cells of the Malpighian tubules continue its synthesis, and so that of protein, without any interruption. Physiological studies have shown that the oxygen consumption of a pupa is about one-fiftieth that of a larva just before pupation and that a pupa is insensitive to cyanide, carbon monoxide, and other respiratory poisons. This has been demonstrated also in grasshoppers in diapause. Investigations of the cytochrome content of the tissues of cecropia pupae have shown that in all but the somatic muscles cytochromes b and c are present in very low concentrations. Twenty-four hours after the onset of diapause, 95 percent of the cytochromes of the larval stages are gone from the tissues, and it is possible that at such low concentrations respiratory poisons have little effect. It has also been shown that the pupal tissues can shift from an aerobic mechanism of respiration to an anaerobic one, but that when their "anaerobic reserve" is used up they return to aerobiosis and again become sensitive to respiratory poisons. Another physiological manifestation of diapause in *Hyalophora* is cessation of spontaneous electrical activity in the neurons of the cerebral ganglia; these also fail to respond to electrical stimulation and most, if not all of the brain appears to be dormant, although the ganglia of the ventral chain maintain their natural spontaneous activity and respond normally to electrical stimulation. These conditions may be peculiar to *Hyalophora* and some other insects, for they are not universal. Brains of diapausing pupae of the geometrid moth *Bupalus piniarius*, for example, retain spontaneous electrical activity as do those of adult potato beetles. Such a difference is consistent with the observed facts of differences in internal conditions in insects that undergo diapause at the same stages in their life cycles and emphasizes again the individuality of organisms in their manifestations of biological events that are superficially alike.

The relationship between pupal diapause and the operation of the deutocerebral neurosecretory cells–prothoracic gland system has been proved by extirpation, implantation, parabiotic, and injection experiments. The direct cause seems to be failure of the neurosecretory cells of the pars intercerebralis to secrete the activation hormone, so that, consequently, the prothoracic glands do not secrete ecdysone, which is either absent from the hemolymph or present in such low titer as to be below an active concentration. Implantation of an active brain into a diapausing insect will cause it to break its diapause at any stage of diapause development, as will implantation of an active prothoracic gland or injection of suitable amounts of ecdysone.

Diapause in cecropia pupae kept at room temperature never terminates in less than 5 months, but if they are kept at a temperature of 3 to 5°C for 6 weeks or longer immediately after the onset of diapause, metamorphosis and emergence of imagos occurs in 1 month or 6 weeks after they have been brought into room temperature. Resumption of growth and development thus depends upon a period of chilling, which seems to affect particularly the neurosecretory cells of the brain. Implantation of cerebral ganglia from a chilled pupa into one in diapause, kept at room temperature, will bring about its imaginal development, and an unchilled pupa joined parabiotically to a chilled one will also break its diapause within a few weeks. Histological studies of the brains of diapausing pupae of the moth *Mimas tiliae* have shown that only after they have been put into the cold is there any indication of secretory activity in the cells of the pars intercerebralis. During the first 3 weeks of exposure to a temperature of 3°C, intracellular material accumulates in them and passes down their axons into the corpora cardiaca. Near the close of diapause development, these cells again become inactive. Thus, in these moths the deutocerebral neurosecretory cells require a period of chilling before they are able to secrete the hormone that is instrumental in bringing about the termination of diapause. That this is actually effected through secretion from the prothoracic glands has been proved conclusively by implantation experiments into isolated anterior and posterior halves of diapausing cecropia pupae. It is possible, by special techniques, to separate the head and thoracic regions of a pupa deprived of its brain, with both halves surviving in continual diapause for considerable periods of time. If a brain from a chilled pupa is implanted into the anterior half of such a bisected pupa, it undergoes the changes characteristic of metamorphosis, imaginal development, and emergence. If, on the other hand, a chilled brain is implanted in the posterior half, there are no such effects and its diapause continues indefinitely. Transection experiments at different levels along the pupal body showed that the critical factor was inclusion of the area with the prothoracic glands, which are essential to reinitiation of the temporarily arrested growth and development.

It, therefore, seems apparent that pupal diapause results from a shutting down of the operations of the neurosecretory cells of the pars intercerebralis and their consequent failure to liberate the activation hormone that promotes secretion of ecdysone by the prothoracic glands. Photoperiod, as well as temperature or other external conditions, is a factor contributing to this shutdown, which may very likely be brought about by stimuli reaching the secretory cells by nervous pathways. It is possible that the stimuli to these conducting neurons may be intrinsic as well as extrinsic. The shutdown may continue for varying periods of time, depending upon the species of insect, but in lepidopterans, at least, the most important factor for its termination and the revival of activity in the neurosecretory cells is exposure to low temperature for an appropriate period of time. This seems to have a direct effect upon the neuroendocrine centers.

It has been postulated that, in cecropia at least, the primary function of ecdysone is promotion of the synthesis of cytochrome c, whose continuous production may be a condition necessary to growth and development. Ecdysone is not, however, essential to all growth and metamorphosis, for adult insects, in which there is little or none, are capable of healing wounds and, within limits, restoring lost parts. Moreover, certain cells and tissues of diapausing juveniles, as has been mentioned, continue to synthesize DNA and to divide even when the supply of ecdysone is shut off. It is possible, of course, that there may be a certain amount of residual ecdysone circulating in the hemolymph adequate to promote growth activities in these particular tissues. Since the direct effect of ecdysone seems to be upon the cell nucleus, and directly or indirectly upon the chromosomes furthering the synthesis of different kinds of messenger RNA and so of different kinds of proteins and different kinds of enzymes, it may well be that termination of larval and pupal diapause is through a number of separate metabolic events controlled by enzymes. Of these events water balance may be as important to some insects as cellular respiration.

Conditions of adult or imaginal diapause are less well understood than those of pupal. In these cases diapause is primarily cessation of reproductive activity, with accompanying reduction in respiration and general overt activity of the insect. Secretion from the corpora allata is known to control development of the gonads in juveniles and maturation of the germ cells in adults; therefore, implication of the corpora allata in imaginal diapause is definitely indicated. Allatectomy in *Leptinotarsa*, in which diapause occurs in the adult stage, results in replication of all the physiological events and behavioral patterns typical of the natural event. These effects can be counteracted by implantation of the glands, although this has no effect upon beetles in natural diapause. This would suggest that there is some mechanism that inhibits activity of the glands in the intact insect when in diapause. That this can be broken by treatment with synthetic neotenin has been shown in the alfalfa weevil *Hypera postica*, which naturally enters a diapause lasting some 14 to 16 weeks. External application of the synthetic hormone, in doses of 50 to 100 μg, to the undersurface of weevils already 2 weeks in diapause resulted in resumed activity and feeding within 6 to 7 days, and in oviposition on the eighth day after treatment. This seems to have been due entirely to the exogenous hormone, for no activity of the insect's own corpora allata could be detected histologically, a condition consistent with the situation in the potato beetle described above. The controlling factor in the activity or nonactivity of the corpora allata seems to be, as with the prothoracic glands, derived from the neurosecretory cells of the protocerebrum. Diapause in *Leptinotarsa*, in *Pyrrhochoris*, and in aphids can be broken by implantation of active brains, or corpora cardiaca, or both together. The brain, therefore, seems to control the turning on and turning off of secretory activity in the corpora allata cells incident to the onset and termination of diapause, receiving its stimulus to do so from nervous impulses of proprioceptive or exteroceptive origins.

The neurosecretory cells of the brain are, therefore, the key factors in the initiation and termination of diapause occurring in postembryonic stages of an insect's life, as they are in its molting processes during its active growth. Their secretion is determined by external or internal conditions, or both, and messages about these are conveyed to the brain by afferent neurons. In the case of pupal, and probably of larval, diapause the message is transmitted by the activation hormone to the prothoracic glands, and the supply of ecdysone is reduced at the onset of diapause and restored at its conclusion. In the case of imaginal diapause, the message is conveyed by the activation hormone to the corpora allata, via the corpora cardiaca, with similar effects upon the supply of neotenin.

Ecdysone is also effective in breaking diapause in the winter tick *Dermacentor albipictus*. External application of alpha-ecdysone, or an analogue, to diapausing specimens terminates this state. This is the only instance yet reported of hormonal termination in arachnids, or in any organism other than an insect, but further investigation may well reveal others and establish it as a natural, not an experimental, event.

g. OVERT ACTIVITY Certain aspects of overt activity are under endocrine control, either directly or indirectly. In lepidopterans, pupation is regularly preceded by activity of the silk glands and the complex motions involved in spinning a cocoon; in some species, a cocoon of sorts is spun before each ecdysis. In dipterans, as well as in lepidopterans, the larva performs certain bodily movements that position it in the puparium or the cocoon. These are all incident upon the processes of molting and metamorphosis and are influenced by the same hormones that regulate them. Moreover, behavior in courtship and mating are related to the development of the reproductive system and the maturation of the gametes, both of which are events controlled by hormones. The role of ectohormones or pheromones in this behavior has already been mentioned (see Chapter XIV, pages 587–589), and it has been shown that in some insects the production of sex pheromones is controlled by their own endocrine system. This is the case in *Periplaneta americana* as well as in *Byrsotria fumigata* (see Chapter XIV, page 588). In both these cockroaches removal of the corpora allata shortly after the imaginal molt results in failure in production of the sex attractant and induction of courtship by the males consequent upon its release. The situation is slightly different in *Pycnoscelus surinamensis*, a species of cockroach in which there are two strains, one parthenogenetic in which the females produce eggs that do not require fertilization for development, and the other bisexual, in which the females require fertilization for successful reproduction. Allatectomy performed within 24 hr after the imaginal molt on females of the bisexual strain resulted in their failure to produce the sex pheromone, while it had no such effect upon females of the parthenogenetic race, which continued to produce it just as did females of the bisexual strain that had not been operated upon. The fact that the parthenogenetic females produce the pheromone at all—which induces courtship behavior in the males as successfully as that produced by females of the bisexual strain—and the fact that its production is free of endocrine control raise interesting questions about the relative values of sex pheromones to insects of different kinds and the values of hormonal regulation of them. Allatectomy is without effect upon production of sex attractants by the moths *Antheraea pernyi* and *Galleria mellonella*. Imagos of *Antheraea*, from which the corpora allata had been removed in the pupal stage after a period of chilling, tethered outdoors attracted males from a distance of 8 km and mated successfully. Similar results have been obtained from similar experiments with *Galleria*. Hormonal control of the production of sex pheromones is, therefore, not universal among insects. Those in which it has been demonstrated are those with comparatively long imaginal lives and a succession of mating periods, while those in which it has been proved lacking are ones with an adult life measured in hours, during which they do not feed and they mate only once. Hormonal control of communication within the individual, and between it and an individual of the opposite sex, may, therefore, possibly be correlated with the kind of life cycle characteristic of its species, with the possible implication that such control has been lost in those species to which it is no longer an asset to reproduction.

Neuroendocrine control of locomotor activity has been most convincingly demonstrated in *Periplaneta americana*. Like many other organisms, these insects exhibit a rhythm of activity, remaining quiet by day but by night running about and foraging for food. If they are kept in continuous light this rhythm disappears, and they may be equally active any time in the 24 hr of a solar day. The first evidence for endocrine control of their rhythmic activity was derived from parabiotic experiments, in which specimens made arhythmic by continuous exposure to light were joined back to back to rhythmic specimens from which the legs had been removed. The intact, arhythmic animals soon exhibited a natural pattern of rhythmicity, carrying the legless animals around on their back in the customary cycle of nocturnal activity and diurnal repose. The source of the active principal controlling this cycle of activity was traced, by cauterization of different parts of the nervous system, to four cells in the ventro-lateral surface of the suboesophageal ganglion. Implantation of this ganglion taken from a rhythmically active cockroach into the abdomen of one made arhythmic, or from which the head had been removed, is followed in a few days by rhythmic activity on the part of the host in a pattern conforming to that of the donor, for the implanted ganglion continues to secrete on its natural 24-hr cycle. When, however, the ganglion is chilled, secretion stops, or is reduced, during the period of chilling, and the natural rhythm is delayed for as many hours as the chilling was in progress. Natural activity begins soon after the beginning of darkness, registered by the ocelli. If a chilled ganglion is implanted in the abdomen of an intact roach that is operating on its natural rhythm, the secretions from the two are out of phase, and most of the cockroaches on which such an operation has been successfully performed have been found to develop tumors of the gut. These metastasize and also are transplantable to other roaches, producing tumors in them. Such results from imbalance or other disturbances in endocrine mechanisms are not unusual and have already been mentioned in connection with allatectomy and tumoral growth in *Carausius*.

It is apparent that the neurosecretory cells of the suboesophageal ganglion respond to changes in light intensity and operate on a 24-hr cycle adjusted for maximum activity at night. In *Periplaneta* their secretion is regulated by another hormone coming from the corpora cardiaca via small nerves that pass along either side of the suboesophageal ganglion. This second hormone does not seem to be secreted rhythmically as is that from the suboesophageal ganglion. Roaches are relatively inactive for a few days after each molt, and during this period no secretion travels along the nerves from the corpus cardiacum, and the hormone from the neuroendocrine cells in the suboesophageal ganglion is not active. If, however, during this period the ganglion is chilled, when activity is resumed after the molt, its cycle is retarded for as many hours as the ganglion was exposed to the cold. This is similar to the effects of chilling on the neuroendocrine cells of the cerebral ganglia, indicating that in both cases cold has a suppressive, but not a destructive, effect upon them.

C. RHYTHMS

1. Physical rhythmicity: solar and lunar

Periodic recurrences of natural phenomena on this planet are commonplace events. Solar rhythms result from Earth's rotation upon its axis, accomplished once every 24 hr, or a solar day, and from its revolution around the sun, now accomplished in $365\frac{1}{2}$ of these 24-hr periods. Because of its rotation, all parts of Earth are exposed to alternating periods of light and darkness, and because of its revolution in an elliptical orbit, to cyclical changes in photoperiods and so in seasons. Twice in the course of a complete passage around the sun there are times when the periods of light and darkness are everywhere of equal length. These are the vernal and autumnal equinoxes when the sun crosses the equator of earth. But after the vernal equinox, about March 21, in the northern hemisphere the periods of light, or daytime, become longer and the periods of darkness, or nighttime, become shorter by regular increments and decrements until, at the summer solstice about June 21 when the sun is farthest north from the equator, daytime is at its maximum length and nighttime at its minimum. In the arctic, daytime then is of almost 24 hr duration, but in latitudes between the pole and the equator it is less; the farther north, the longer the daytime. After this solstice, the periods of daytime shorten and those of nighttime lengthen, again by regular intervals until the winter solstice. This falls on about December 21, when the sun is farthest south from the equator, and nighttime reaches its maximum length and daytime its minimum. In the southern hemisphere the same conditions obtain, but in reverse, so that the longest day in the antarctic coincides with the shortest in the arctic. These changes in photoperiod, depending upon Earth's position relative to the sun, are fixed, measurable, and calculable ones, and they bring with them changes in the seasons—spring, summer, autumn, and winter—with changed conditions of temperature, rainfall, and humidity. These latter conditions are not nearly so predictable as are the changes in photoperiod, for they vary from place to place and from time to time with varying meteorological conditions. In general, however, in temperate zones, temperatures rise in the spring, reach their climax in the summer, decline in the autumn, and are lowest in winter. In tropical zones, the variation in temperature is less and seasonal differences are marked by alternating periods of drought and rainfall.

Lunar rhythms result from the moon's revolution around Earth, a passage accomplished every 29.5 days and exhibited in monthly cycles of its waxing, fullness, waning, and absence. The phases of the moon are, of course, relative to fixed positions on Earth; that is, the moon may be full in one region, but in its first or third quarter or absent in others. The moon reaches its zenith about 50 min later each day and its position relative to any area of Earth's surface affects not only the brightness or darkness of the night but also the tides of the oceans. These result from the rise and fall of the surface of the seas brought about by the attraction of the sun and the moon. Tides may occur once or twice in a lunar day of 24.8 hr, in alternating cycles of flood tide and ebb. The attraction, or gravitational pull, of the moon is greatest when it is at its zenith and lowest when it is at its nadir, so that once each month flood tides are especially high. When the moon and sun are in positions relative to each other and to Earth so that a straight line can be drawn between their centers and the center of Earth, their gravitational pulls are additive and spring tides occur. There is no seasonal significance of "spring" as applied to tides; the term has its derivation in the Anglo Saxon verb *springan*, meaning "to leap" or "bound" and denotes a tide that reaches the maximal level of a high tide. When the moon is in its first or last quarter, and therefore in such a position relative to Earth and the sun that a line drawn from the center of Earth to the center of the sun makes an angle of 90° with a line drawn from the moon's center to that of Earth, neap tides occur. The origin of this word is obscure, but "neap" is used to denote those tides whose level is minimal for a high tide; they are low high tides, not low tides. Obviously, both spring and neap tides occur twice each month in any given locality, as the moon passes along its orbit around Earth.

In addition to these cyclical events in the physical world there are cyclic variations in barometric pressure and in cosmic radiation. Barometric pressure shows a natural monthly rhythm as well as a daily one, for it usually falls after midnight, rises again in the late morning, falls in the afternoon, and rises again in the early evening. These rhythms are evident even in the pronounced fluctuations of barometric pressure that occur with different meteorological conditions and that are imposed upon the basic rhythm. Cosmic radiations, which also influence barometric pressure, show cycles of 24 hr and of 27 days, during which the showers increase and then decrease in frequency.

2. Biological rhythmicity

Biological systems also exhibit rhythmicity, often correlated with solar or lunar rhythms, or both. Thus, they may be expressed as annual or seasonal repetitions of biological events, or as daily ones, or as ones that coincide with certain phases of the moon or with the tides which particularly affect species living along the seashore and in estuaries. Precise observations of biological rhythms date from the early 1900s. One of the first reports, made in 1903, was that of the tidal rhythm in *Convoluta roscoffensis*, which was observed to emerge from the sand at low tide and to disappear at high tide. The notable feature of this behavior was that it persisted in specimens taken from the shore and kept in a laboratory aquarium where there were no daily changes in the level of the water. In 1906 it was reported that the sea anemone *Actinia aequora* expanded its oral disk and column at high tide and retracted them at low tide, and that this pattern of behavior continued for 3 to 8 days in animals in an aquarium, with a timing such that the movements were slightly in advance of the actual tidal changes in the nearby ocean. Locomotor activity in the snail *Littorina rudis* was found to occur every 15 days, coincident with the spring tides in its locality.

Such observations as these on the persistence of cyclical behavior in animals removed from their natural surroundings and in environments with a high degree of physical constancy suggested that they must have some means of estimating the passage of time independent of external events that might give them some clue as to the positions of the sun or the moon. The first controlled experimental evidence of animal chronometry was obtained in 1918 when it was found that rats and mice, mammals that are naturally quiescent by day and active by night, maintained the same pattern of behavior when kept in continual darkness, showing that they could measure the passage of time within a 24-hr period when free of the physical cycle of night and day. Since then, an abundance of observations and experiments with a variety of parameters have made it clear that persistent biological rhythms are not unusual, but very common, phenomena, and are exhibited in all groups of plants from bacteria to spermatophytes and in all groups of animals from protozoans to man. These rhythms may be expressed as cycles of growth; reproduction; locomotor and feeding activity; pigment migration and color change; oxygen consumption; production of light in luminescent species; and probably may occur in other ways yet to be investigated.

It has become apparent, too, that biological and physical rhythms may have different phase relationships to each other in different species of animals. Some animals, for example, are active by day, others by night; some are active at dawn, others at dusk. Some littoral species feed at the ebbing, some at the rising, tide. The fundamental question, raised early in relation to observed biological rhythms, is the extent to which they are independent of physical ones—whether organisms have their own self-sustaining clocks and calendars, or whether biological rhythms are induced by, or are dependent upon, the physical rhythms to which they often seem so closely adjusted. This is the question that has been put to experimental test by observations of animals removed from their natural environments and kept under "constant" conditions. "Constancy" in laboratory conditions implies exclusion of any differences in light; in temperature; and, in the case of terrestrial animals, in humidity; or, in the case of littoral and estuarine ones, in tides. Changes in barometric pressure or in cosmic radiation cannot be so readily controlled, and these must usually be considered variables in an otherwise "constant" environment. The results of such investigations seem to have made it clear that animals do possess their own timing devices and that their rhythms are endogenous rather than exogenous, although in the course of evolution they have in many instances become geared with physical rhythms, where such interlocking, or entrainment, has had survival value to the species. "Biological clock" is now a generally accepted term, and one that has been in use for nearly 20 years, during which manifestations of timing devices and of the mechanisms underlying them have been under active investigation in many laboratories all over the world.

Studies of animal chronometry follow three main courses: behavioral, in which the rhythms are followed under "controlled" conditions; physiological, in which experiments are devised to ascertain the way in which endogenous rhythmicity is translated into overall behavior; and biochemical and biophysical, aimed at resolving the cellular mechanisms involved in initiation and sustention of endogenous rhythms. Some of these mechanisms are known to be mediated, in multicellular animals, through endocrine or nervous systems, or both, but some are independent of such mediation. Rhythmic activities have been observed in bacteria, protozoans, and cell aggregates, as well as in cells in tissue culture taken from multicellular systems but free from all nervous or endocrine control.

a. BIOLOGICAL RHYTHMS AND SOLAR PERIODICITIES Many biological systems show annual or seasonal rhythms, as well as diurnal ones. The use of "diurnal" in this connection is ambiguous, since the word is also applied to events that occur by day as opposed to nocturnal events, or those that occur by night. The substitution of the term "24 hr" for diurnal is not entirely apt, for close timing of rhythms that follow such a general pattern has shown that recurrences do not take place at exactly 24-hr intervals, but only approximately so—at, perhaps, every 22, 23, 25, 26, 27, or even 28 hr. The term "circadian" (Latin *circa*, "about," and *diem*, "day") has been proposed as an alternative that is more exactly descriptive, and it is now in general use for those biological cycles in which there is a general coincidence with the rotational position of Earth in relation to the sun.

(1) Annual and seasonal rhythms Annual and seasonal rhythms are very evident in the overt activity, growth periods, and reproduction of some invertebrates. In temperate zones some species regularly aestivate in summer and hibernate in winter, becoming quiescent and metabolically less active during periods of heat and cold. And some species regularly change their habitats with the season. There are, for example, over 300 species of butterflies and many moths that are known to migrate regularly at definite times of the year. Migration, as opposed to chance redistribution, may be defined as a movement of many individuals, either en masse or separately, in a direction over which they have some control, which is continued long enough to take them away from their winter or summer habitat—in many instances a long way—to the one they customarily inhabit in the opposite season. The northern race of milkweed or Monarch butterflies, *Danaus plexippus*, for example, winters in Florida and other Gulf coast states, and summers in southern Canada and the northern United States. As autumn arrives they collect in bands, often numbering many thousands, and start their flight south on migration routes that are used year after year by different bands, settling on trees at night. So fixed is their course that they have been known to fly through railway tunnels and even through unfinished buildings with open window frames along their path. When they reach their destinations in Florida and southern California, they settle on certain trees, often using the same groups of them year after year, and pass the winter in a state of semihibernation. So precise is their timing that each year the date can be determined with a reasonable degree of accuracy by watching for the clusters to light on their habitual trees.

Annual cycles of growth are evident in the deposition of the hard parts of both animals and plants. Annual growth in the theca of a coral polyp, for example, is indicated by rings, like the growth rings in the wood of a tree, and daily growth is marked by striations in each ring. Counts of these rings in fossil corals have provided evidence corroborating the assumption of astronomers that while the length of the year, in terms of the time spent in one complete revolution of Earth around the sun, has been constant throughout geological time, the number of days included in this time span has decreased because of deceleration in the rate of Earth's rotation by 2 sec every 100,000 years. In the Cambrian period, according to calculations, the year would have had 424 days, each 21 hr long; during the Devonian, the number of days would have been reduced from 402 to 396, and in the Pennsylvanian from 393 to 390. Perfect correlation between the calculated number of days in a year and those recorded by annual rings of skeletal deposition in corals could, of course, only be attained if the corals grew continuously year after year with no breaks or appreciable delays in laying down their thecae, a condition of uniformity that is rarely achieved in any biological process. Yet examination of fossils of several extinct species from mid-Devonian strata of New York and Ontario have revealed a range in the number of striations within the annual rings of 385 to 410, and in specimens from the Pennsylvanian strata of Pennsylvania and of Texas, of 390 to 385.

The shells of lamellibranch and gastropod molluscs also show annual growth rings, often quite distinctly marked (Fig. III–12). Seasonal growth in arthropods has already been considered in relation to the cycles of molting in crustaceans and diapause in insects.

Seasonal dimorphism is manifested by certain insects. Spring and summer forms are recognizable in many butterflies by differences in pigmentation, generally accepted as reflecting differences in day length and the effect of this upon the biochemistry of pigment production. Females of the Great Southern White butterfly *Ascia monuste* are, for example, almost entirely white in the spring, with only a narrow margin of black on the wings, while in summer the body is uniformly dark. There may be intermediate types between these extremes, but in general, the difference between the short-day form and the long-day form is marked. Evidence from experiments in rearing these insects from egg to imago under different LD regimes indicates that the decisive factor in determination of the color pattern is the length of the period of light exposure, but does not entirely exclude effects of temperature; both of these factors may be operative in seasonal color changes in animals.

Seasonal periods of reproductive activity are evident in many invertebrates whose adult lives extend over a period of years. In temperate zones, late spring and early or midsummer are the usual spawning periods for molluscs, annelids, and crustaceans, with sometimes, in aquatic species, very precise timing of events in relation to tides, the phase of the moon, or other physical phenomena. Some molluscs, for example, though spawning irregularly throughout the year, always do so at precisely the same time of day. Great precision in spawning time is shown by polychaete worms of the family Eunicea. The eggs of *Diopatra*, for example, a worm native to the coast of North Carolina, are deposited on sandy beaches at night, or when the tide is high. *Eunice schemacephala*, in the waters off the coast of Florida, spawns only in the last quarter of the July moon. Perhaps the most dramatic spawning event is that of *Palolo siciliensis*, living off the coasts of Samoa and Fiji. All the worms reach sexual maturity at about the same day of the year and then cast off the epitokous portions of their bodies which swarm to the surface in countless numbers. Off Fiji this takes place when low tide occurs shortly before sunrise in the last quarter of the November-December moon. These worms are nocturnal in habit, and in their reproductive activity moonlight, although about one five-hundredth-thousandth as powerful, is, therefore, of more importance to them than sunlight. The quality and duration of this reflected light would seem to be the critical factor to the release of their reproductive cells. *Lysidice oele*, living in the waters off Amboina in the Malay Archipelago, spawns in March and April, two to three nights after the full moon. The evidence from neurosecretory involvement in sexual maturation in *Platynereis* supports the concept that lunar periodicity in spawning may be due to photoperiodic suppression of endocrine activity.

Terrestrial annelids also show seasonal reproductive activity. *Lumbricus terrestris* reproduces but once a year. The gametes mature in spring and early summer, after which copulation and external fertilization take place and the gonads regress and do not become active again until the following spring. Correlated with this seasonal reproductive activity is that of certain neurosecretory cells, the a-cells in the cerebral ganglia. The activity of the alpha cells in leeches and its relation to reproduction in these annelids has been mentioned previously. Reproductive activity in species in other phyla is usually greater at certain times of the year than at others, even in those that reproduce several times a year.

(2) Circadian rhythms Circadian rhythms are more easily studied in the laboratory and lend themselves more readily to experimental analysis than seasonal ones. One of the first and important facts to be established about them was their independence of temperature—the fact that they continued unchanged over a range of temperature in which other biological processes would be either speeded up or slowed down, for in general the rate of a biological reaction rather more than doubles for every 10°C rise in temperature, within the physiological range. Exposure to near-freezing temperatures will effectively stop, but neither damage nor destroy, the chronometric mechanism in animals ordinarily living at higher ones. Restoration to the customary temperature range will start it again on its regular cycle with a delay in phase equivalent to the duration of exposure to cold. The clock has thus been shown to be sensitive to temperature. As a mechanism for measuring photoperiod, it is also light-sensitive but light independent, in that it continues to run in constant photic conditions in animals accustomed to alternating cycles of light and darkness. The phase of the physiological clock can be changed by alterations in the natural photoperiod, as it can be changed by low temperatures, but its rhythm remains constant. The biological clock is a free-running system, inherent in the organism and clearly of survival value to it because of the adaptability of its phasing to external conditions and the possibility of its synchronization with them. The inherent rhythm of seashore invertebrates has, in the long course of their evolution, become harmonized with the tides in their immediate localities. Tidal rhythms, although physically expressions of lunar periodicities are in these animals essentially modified circadian ones.

Endogenous circadian rhythms have been demonstrated, among protozoans, in *Euglena*, *Strombidium*, and *Gonyaulax*, in which conditions have been particularly carefully investigated. The phototactic sensitivity of *Euglena gracilis* follows an approximately 24-hr rhythm, which persists even in specimens kept in continuous darkness at temperatures ranging from 17 to 33°C. The spirotrichous ciliate *Strombidium* swims freely in rock pools when the tide is low, but encysts at high tide. It excysts 18 hr later, regardless of tidal conditions, and remains free-swimming until the next high tide, when it encysts again to excyst 18 hr later.

Light production in the armored dinoflagellate *Gonyaulax*, abundant in the Pacific Ocean off the coast of California, follows a circadian rhythm. Their luminescence can be continuously measured by placing cultures, which are successfully maintained in the laboratory for years, before a sensitive photomultiplier tube. Undisturbed, the animals give off a faint luminescence, but when mechanically stimulated by bubbling air through the culture, the flashes are much brighter. In both cases the light emitted by night is brighter than that emitted by day. If the exposure to light is controlled so that the animals are exposed to 12 hr of light followed by 12 hr of darkness, the recordings show that the light emitted in the dark is 40 to 60 times brighter than that emitted in the light. This observation is substantiated by the fact that both luciferin and luciferase can be obtained in greater quantities from animals taken from the sea by night than from those collected by day.

Cultures kept in constant darkness preserve their rhythm for as long as 4 days, and very much longer if kept in light that is dim, yet strong enough to provide energy for photosynthesis. In both conditions the experimental cultures maintain a rhythm that is in phase with cultures kept in natural light. But continuous exposure to stronger light, such as bright daylight, upsets the rhythm, which can, however, be restored to its natural phase by giving the cells a period of darkness. The phase can be shifted if cultures grown in constant dim light are either exposed to bright light or kept for a time in darkness. After return to dim light, or to natural conditions, there is the same circadian rhythm of luminescence, but the treated cultures are now out of phase with the controls to the extent of the length of their exposure to light or to darkness. If, for example, this has been for a period of 6 hr, there is a 6-hr difference in the cycle of experimental and control cultures.

The circadian rhythm of *Gonyaulax* seems to depend on several factors. The influence of one of these is based on evidence that the degree of luminescence by night is related to the amount of light received by day. This would imply that the stores of energy drawn upon for the reaction in darkness must be replenished by the photosynthetic activity of the organism during the day. A second factor is the possible inactivation or inhibition of the luminescent system by light, so that it has only a fraction of its effectiveness during the day. A third factor is a possible rhythmic fluctuation in the sensitivity of the cell, which may be less by day than by night, for at night the cell responds to stimulation by increased frequency of flashes.

Gonyaulax exhibits two other circadian rhythms, which are expressed in the frequency of cell division and in the rates of photosynthesis. The frequency of cell division can be easily followed, because the daughter cells stay together for about 30 min after mitosis. Counts of the number of pairs in a culture at regular intervals throughout the light-dark periods show a sharp maximum just at the end of the dark period, but this does not mean that all the cells in the culture divide every 24 hr. While that may be the generation time for some individuals, for others it may be 48 hr or some higher integral multiple of 24. The peak of division, however, in cultures kept in alternating light and darkness, always coincides with emergence from the dark phase. In cultures kept in dim light and at constant temperature, the 24-hr rhythm of mitosis persists for long periods. A similar rhythm of photosynthesis has been established by measurements made for single cells in a microrespirometer.

Under constant conditions of light and temperature, the period of each of these four rhythms approximates 24 hr. The period may not bear any relation to solar time and is independent of temperature. The question naturally arises as to whether these four manifestations of periodicity are independent of each other or whether they are regulated by a single master control system within the cell, a question that is also posed by the rhythms of different events in other organisms that have the same or similar periodicities, but may run on different phases.

At least six rhythmic processes have, for example, been described in fiddler crabs, among the first animals to provide decisive evidence of biological chronometry. Fiddler crabs show an exact 24-hr, as well as a tidal, rhythm of dispersion of pigment in their somatic melanophores, the most abundant of their pigment-containing cells. Circadian rhythms are also manifested in their locomotor activity and their oxygen consumption; in their susceptibility to low temperatures; and, in animals deprived of their eyestalks, in their chromatophore responses. The cycle of blanching by night and darkening by day continues uninterruptedly for at least 2 months in intact fiddler crabs brought into a dark room and kept at temperatures ranging from 6 to 26°C. Yet the interlocking of this rhythm with the tidal rhythm and the imposition of the tidal rhythm upon it is shown by the fact that the maximum dispersion of the dark pigment coincides with the hour of low tide and its maximum concentration with that of high tide. Thus at Woods Hole, Massachusetts, there is an approximately 50-min difference each day in the time at which the pigment is maximally dispersed, and a curve representing pigment dispersal in relation to hours of the day is skewed toward the right as the time of low tide moves on into the afternoon and toward the left when it reaches the early morning hours. That crabs maintain the tidal rhythms of their natural habitats has been demonstrated by keeping specimens collected from different localities under identical conditions in the same laboratory. Crabs, for example, collected from a beach at Martha's Vineyard, Massachusetts, where the tide is later than at Woods Hole, and kept with Woods Hole crabs in the same dark room preserved their own rhythm of color change, as did those taken from the Woods Hole shores. As a test of the ability of crabs to measure time with no possible external indications of the period, crabs collected at Woods Hole were put into opaque wooden tubs, some of which were kept in a dark room at Woods Hole and others of which were flown to a dark room in Berkeley, California, about 51° of longitude and 3.3 hr of time away. Both tubs were opened simultaneously and the crabs in them kept under observation for 6 days, during which the clocks of both sets ran on Woods Hole time, the animals in Berkeley exhibiting consistently the Woods Hole tidal rhythm. Yet differences in temperature and in photoperiod alter the phase of this cycle, which remains, however, a 24-hr one. It is delayed for 6 hr in crabs that have been kept for that length of time in seawater at 2 to 3°C, and is similarly delayed in animals that have been kept in the dark and then, on three consecutive days, exposed to light from midnight until 6 A.M. If these are later exposed to light from 6 P.M. until midnight on three consecutive days, their clocks are set back another 6 hr and they exhibit a reversed rhythm, being dark by night and light by day.

At Woods Hole, high and low tides occur twice in a lunar day, but in the Gulf of Mexico only once. Yet *Uca pugilator* and *U. speciosa*, living on the Gulf shores, maintain a 12.4 cycle of color change, as does *Callinectes sapidus*. There is a phase difference of about 5 hr in the cycles of *U. pugilator* and *U. speciosa*, which is of interest in that they make their burrows at different levels on the beach, and the phase difference in their cycles corresponds to the difference in time that the water reaches their burrows.

Fiddler crabs are usually most active in the early morning hours and least so in the afternoon, but superimposed upon this circadian rhythm is likewise a tidal one, the crabs' maximum activity occurring about 2 to 4 hr before the hour of low tide on their native beaches. This is also their feeding time, and species of *Uca* from a region of semidiurnal tides—that is, ones in which there is a 12.4-hr interval between each high, or each low, tide—only come out to feed when low tide occurs in the daytime, but those from a region of diurnal tides—that is, ones in which there is a 24.8-hr interval between successive tides—come out at low tide whenever it occurs. This adjustment is a matter of survival to them, for if they had not made it they would go without food in the period that low tide occurs at night and so eat only every tenth day.

Continuous recordings of the metabolic rates of *Uca pugnax*, kept in constant temperature, illumination, and, in this case, barometric pressure, have shown their oxygen consumption to be maximal in the early morning, between about 6 and 9 A.M., and minimal in the afternoon, with a small increase between 8 and 11 P.M. Superimposed upon this is a tidal rhythm, the peak of oxygen consumption being reached 1 or 2 hr before the time of low tide and the lowest level about the time of high tide.

Although the rhythm of color change in *Uca* is temperature independent within a range of 13 to 30°C, it is temperature-sensitive in that it can be temporarily stopped by temporary exposure to a temperature of 5°C. Its relative sensitivity to chilling is a function of the time of day the exposure to low temperature is made, for it has been found that the rhythm is more strongly inhibited by a 2-hr exposure to cold between 8 and 10 P.M. than at any other time of the day, with the minimum inhibition occurring between 10 and 12 A.M.

The control of color changes, known to be exercised by secretions from the sinus glands, would indicate that their rhythm is an expression of periodicity in activity of the neurosecretory cells. Yet there is no corresponding cyclical variation in the hormonal content of the secretion in *Uca*, and, moreover, demonstration of an appropriate degree of pigment dispersion in the melanophores of a leg 60 min after its separation from the body of a crab that had been deprived of its eyestalks 48 hr earlier is extremely suggestive of some, if not total, independence of hormonal influence.

As with the different circadian rhythms manifested by *Gonyaulax*, the question arises as to whether these rhythms in crabs represent separate and independent timing mechanisms or whether there is a single master clock that controls them all. This is still an open question.

Other crustaceans have provided evidence of circadian rhythms. Several species of isopods show color changes like those of fiddler crabs, and these continue with the same periodicity for at least 2 months in animals kept in total darkness. Similar cyclic chromatophoric changes are exhibited by crayfishes and prawns, and a wide range of species show rhythmic migrations of their retinal pigments. In general, any one of the retinal pigments, or any combination of them, may undergo periodic movements, depending upon the species, and these may, again depending on the species, persist in constant light or in constant darkness as well as in light and dark. The distal retinal pigment in the eyes of the freshwater shrimps *Macrobrachium olfersii* and *M. acanthurus*, for example, concentrates at dusk and disperses at dawn under natural conditions, and these migrations continue in specimens kept in continuous light. Both the distal and the reflecting pigments in the eyes of the prawn *Anchistiodes antiguensis* maintain their natural cycles of migration in specimens kept for 2 weeks in continuous darkness, but in continuous light the cycle, though persisting, has a longer period and the range through which the pigments move is not so great.

Circadian cycles of activity are also exhibited by crayfish. *Astacus astacus* is ordinarily quiet by day and active by night, and follows the same pattern of activity during several weeks in constant darkness. The average activity of a group of *Orconectes virilis* recorded automatically during a 5-week period in total darkness was found to be greatest in the late afternoon and early evening, and least in the early morning hours. A similar rhythm has been observed in *Cambarus diogenes* and *Procambarus clarkii*. On the other hand, there is no demonstrable rhythmic activity on the part of the blind crayfish *Orconectes pellucidus* that has lived for generations in dark caves and has, presumably, become adapted to light that is perpetually dim.

Specimens of *Astacus* from which the eyestalks have been removed are completely quiescent, but become active again after injection of eyestalk extract. Eyestalk removal results, on the other hand, in continual activity in *Orconectes virilis*, in which it has been shown by transection of the optic nerve without injury to the other tissues of the eyestalk that the inhibitory influences to unrestrained activity are nervous in origin.

Crayfish also exhibit circadian cycles of oxygen consumption. It is at a maximum about 6 A.M. in the dwarf crayfish *Cambarellus shufeldtii*, both with and without eyestalks. Recordings from a group of *Orconectes clypeatus* show also a daily cycle, but with a difference in the times of the peaks. In about half the animals under observation, the time of maximum oxygen consumption was between noon and midnight, in the other half it was minimal at this time and maximal between midnight and noon. Both groups, therefore, operated on their own cycles, 12 hr out of phase with each other. Specimens from each group continued their characteristic patterns of oxygen consumption after removal of their eyestalks.

Another illustration of circadian rhythms in crustaceans is the vertical migration of zooplankton (see page 15 and Fig. I–6). This is a complex phenomenon, well recognized since 1817, in which extrinsic as well as intrinsic factors are involved. Observations of such mixed populations under controlled conditions of light have made it clear that the upward movement of some members is an expression of an internal timing mechanism, although in others it may be entirely a response to external conditions. Field studies have indicated that the upward movement of the plankters begins a little before dawn, as if the clocks, in those that have them, were slightly ahead of solar time.

Endogenous activity rhythms are manifested by a number of cnidarian species. Colonies of the sea pen *Cavernularia obesa* regularly expand just after sunset and contract again after a definite period of time, regardless of the position of the sun. This rhythm has been maintained for 5 months under constant laboratory conditions. And *Aurelia aurita* in the laboratory shows a persistent circadian rhythm of swimming activity. The cycle of the burrowing and emergence of the flatworm *Convoluta roscoffensis* has already been mentioned as one of the early recognized instances of rhythmicity in animals, as has the histologically demonstrable cycle of activity in the neurosecretory cells of *Dendrocoelum lacteum*.

A number of different rhythms are evident in molluscs. In the oyster *Crassostrea virginica*, the cycle of opening and closing of the valves of the shell can be resolved into circadian, tidal, and lunar rhythms, of which the tidal is the most conspicuous. This has been found to be the case in other littoral molluscan species. A circadian rhythm, expressed in locomotor waves and feeding activity, has been observed in the sea cucumber *Sclerodactyla* (*Thyone*) *briareus*, and similar feeding and burrowing rhythms have been found in holothurians of the Paloa Islands. The rhythm in *Sclerodactyla* continues without change in animals kept in continual darkness.

The rhythms of terrestrial invertebrates are not influenced by tides or other conditions in the water. Therefore, their patterns are rather simpler of resolution, if not of explanation, than those of aquatic animals. The earthworm *Lumbricus* shows clearly a circadian rhythm of activity which is maintained in both parts of an animal cut in two pieces, an observation that is consistent with an endogenous timing mechanism. Persistent circadian rhythms have also been observed in many arthropods other than crustaceans. The scorpion *Heterometrus fulvipes*, for example, is most active about 8 P.M. and least so between 8 and 10 A.M. A corresponding rhythm of spontaneous electrical activity has been recorded in its isolated nerve cord, and nerve cords tested at the time of minimal electrical response gave maximal responses when bathed with homogenates of the cephalothoracic nerve mass or with hemolymph taken from animals at the peak of their activity. A circadian rhythm of activity has been demonstrated in the tarantula *Sericoplema rubronitens* under constant laboratory conditions, and, similarly, the persistent activity rhythm of the millipede *Spirobolus* has been found to continue over a period of 15 days. Many insects show activity rhythms lasting for at least 10 days under constant laboratory conditions. The cricket *Gryllus domesticus* alters the time of its activity, and so sets the phase of its rhythm, according to the photoperiod, but the rhythm remains approximately a 24-hr one. The innate nature of these rhythms is shown by their continuance throughout the different stages of an insect's life. Young larvae of the beetle *Boletotheros cornutus*, kept in constant darkness and constant humidity for 3 months, developed to adults during this period while preserving their natural cycle of activity. The potato bug *Leptinotarsa decemlineata* has a circadian rhythm of feeding, evident in its larval stage, which is also expressed in the imago and so must have been carried through all larval as well as pupal stadia. As soon as a newly emerged adult has fed on a potato leaf or swallowed a drop of water the circadian cycle is revealed, for it next searches for food and ingests it with the same timing as in its larval stage. Possibly this rhythm is established in the adult only after its fluid content, which is low in the pupa, reaches a certain level and its tissues become bathed in hemolymph. This would suggest a hormonal regulation of its feeding activity, comparable to that which has been established for the rhythms of locomotor activity in cockroaches. And it has long been known that bees and ants return at the same times to the same feeding grounds; the "time sense" in these insects was an early observation of animal chronometry.

A very remarkable feature of insect timing is the synchrony of eclosion in certain species and the fact that it occurs always at the same time in the 24-hr period although it is an event that happens but once in an insect's life. This has been shown in chironomids, in the beetle *Ephestia* and in *Drosophila*. Eggs laid by the third generation of *Ephestia* females raised in constant conditions of light and temperature hatched together at the natural and characteristic time, and those of *Drosophila* females, descendants of 15 generations raised in constant light, hatched at their natural and characteristic time. On the other hand, there is no rhythm of eclosion in eggs from females reared in darkness, unless at a very early stage they had some exposure to light. This need be but a very brief one. A single unrepeated light signal of 1/2,000 sec is sufficient to set the cycle going in its natural rhythm for future generations. Egg laying is also precisely timed in some insects. Females of the mosquito *Aedes aegypti*, for example, always lay their eggs by day, usually between 2 and 3 P.M. under natural conditions, but if the light cycle of caged specimens is reversed so that they are illuminated during the solar night and are without light during the solar day, they lay their eggs at night. They become arhythmic if kept in total darkness, but exposure to light from 5 to 15 min restores the natural rhythm.

A particularly interesting aspect of rhythmicity in insects is the difference in their susceptibility to insecticides at different times of the day. This has been experimentally tested on the boll weevil *Anthomus grandis*, adults of which were exposed for 15 min, at different times of the day, to methylparathion, an inhibitor of the enzyme cholinesterase and an effective contact poison to the insects. Three sets of weevils were kept under different photoperiods: one set exposed to 10 hr, one set to 12 hr, and one set to 14 hr of continuous light out of the 24. Regardless of the photoperiod, all the weevils showed a cyclic sensitivity to the poison, the peaks of greatest sensitivity or of resistance occurring 6 hr apart. Thus, of those subjected to a regime of 10L 14D, the set in which the differences were most conspicuous, 10 percent were killed when exposed to methylparathion at the time of their maximum resistance and nearly 90 percent when exposed at the times of their maximum sensitivity, periods which occur four times in each 24 hr. The cause of the rhythm is unknown, but it might be related to a cycling of levels in the activity of cholinesterase. Rhythms of susceptibility to toxic agents have also been demonstrated in mammals, a matter of considerable therapeutic interest and probable importance in that a drug administered at one time of the day may be far more effective than a similar dose administered at another.

Rhythms of growth occurring within the solar day have been found in a number of invertebrate species, reflecting rhythms of metabolic processes. In addition to the annual growth rings of coral exoskeletons, daily depositions of material are also marked in them. Similarly, in the crayfishes *Cambarus virilis* and *C. immunis* daily deposition of calcium in the gastroliths can be traced, indicating a rhythm in the animals' control of the concentration of Ca^{++} in its tissues. The chitinous exoskeletons of *Schistocerca gregaria* and *Periplaneta americana* show lamellate and nonlamellate layers which can be distinguished not only morphologically but histochemically, mechanically, and morphogenetically. The chitin that is deposited at night is organized into plates or lamellae, but that which is deposited by day is not, so that each pair of lamellate and nonlamellate layers marks the formation of cuticle within each 24-hr period. In addition to cycles of secretory activity such as these, circadian rhythms of mitotic activity have been discovered in the growing periods and growth regions of both animals and plants.

b. BIOLOGICAL RHYTHMS AND LUNAR PERIODICITIES Reproductive activity in many animals seems to be correlated with lunar cycles, and the breeding times of many invertebrate species coincide with certain phases of the moon and certain relationships between the moon, sun, and Earth. The precise timing of the spawning of the palolo worm in relation to the phase of the moon at a particular time of the year has already been mentioned, and a similar relationship is shown by *Platynereis dumerlii*, native to the Mediterranean. This worm breeds between sunset and midnight during the first and third quarters of the moon in the months of May to September. Marine molluscs also show reproductive cycles coincident with lunar periodicities. *Littorina neritoides* is, for example, more prolific during spring tides, with more egg capsules deposited then than at other times; and oysters have distinct lunar rhythms of breeding.

Tidal rhythms of several kinds are evident in littoral species of molluscs. The mud snail *Nassa obsoleta* has, for example, a tidal rhythm of locomotor activity, and sedentary bivalves have rhythms of opening and closing the valves of the shell coincident with the tides. *Crassostrea virginica* opens its valves at high tide and closes them at low, and specimens taken from the waters of Long Island Sound and shipped to a laboratory in Evanston, Illinois, kept their accustomed rhythm for several days. Afterwards the phase shifted while the rhythm was maintained, the opening of the valves becoming coincident with the time that the moon reached its zenith in that latitude. A persistent tidal rhythm of opening and closing is also manifested by *Mercenaria mercenaria*, which shows besides a lunar rhythm of this activity, with the maximum degree of opening of the valves occurring every 29 days.

Mytilus edulis and *M. californianus* also show tidal rhythms in the rates in which they pump water across and through their gills, the maximum pumping rate occurring at high tide. The endogenous nature of this rhythm is shown by the fact that it remained constant for at least 1 month in specimens brought into the laboratory and exposed to a temperature range of 9 to 20°C. Yet its entrainment, or adaptation, to local tidal conditions has been shown by shipment of specimens of *M. edulis* from Woods Hole to Los Angeles, where, after 1 week in the California tidal zones, its phase shifted from the times of the Woods Hole tides to those of the new locality.

Rates of oxygen consumption in many littoral species have been shown to have frequencies corresponding to tidal, semilunar, or lunar periodicities. In general, the periods of maximum oxygen consumption in sedentary species occur shortly before high tide, and those of minimal consumption shortly before low. This has been found to be the case in the cnidarian *Actinia aequora*, in the annelid *Arenicola marina*, in the sea urchin *Paracentrotus lividus*, in the molluscs *Patella vulgata*, *Haliotis tuberculata*, and *Mytilus edulis*, and in the crustacean *Xantho floridus*. It was in this crab that the first systematic study of oxygen consumption in a crustacean was made, hourly determinations of oxygen consumption being taken on specimens kept in the laboratory for several weeks during which the tidal pattern of their respiration remained constant.

From data such as these it seems evident that invertebrates—at least those in the major phyla—have means of registering the passage of time. The intervals so registered may be of different lengths—within the 24 hr of the solar day, approximating it; within the lunar month or approximating it; and within, or approximating, the solar year. The fact that the intervals remain constant but the phase of their cycles may change is indicative primarily of two things: first, that the chronometric mechanism is endogenous and, second, that it may be entrained with conditions in the external environment arising from cyclical events in the physical world. Probably the rhythms of the innate, endogenous mechanism reflect some sort of oscillatory cellular process or processes. These may be of two kinds: autonomous systems whose oscillations decay if energy is not restored periodically from some source outside the system itself, and autonomous systems that are self-sustaining. But the precise nature of a biological clock is yet to be elucidated. Setting of the clock in accordance with solar or lunar cycles is a phenomenon that has occurred during the millions of years that biological systems have been adjusting themselves to conditions on earth. Such entrainment can also be accomplished experimentally, and for this photoperiodicity seems to be the critical factor, but the possibility that other physical conditions may be concerned is by no means excluded.

SELECTED BIBLIOGRAPHY

General

Bern, H. P.: The Secretory Neuron as a Doubly Specialized Cell, in D. Mazia and A. Tyler (eds.), "General Physiology of Cell Specialization," pp. 349–366, McGraw-Hill, New York, 1963.

Charniaux-Cotton, H., and L. H. Kleinholz: Hormones in Invertebrates Other Than Insects, in G. Pincus, K. V. Thimann, and E. B. Atwood (eds.), "The Hormones," vol. IV, pp. 135–198, Academic, New York, 1964.

Cooke, Ian: Electrical Activity and Release of Neurosecretory Material in Crab Pericardial Organs, *Comp. Biochem. Physiol.*, vol. 13, pp. 353–366, 1964.

Gabe, M., P. Karlson, and J. Roche: Hormones in Invertebrates, in M. Florkin and H. S. Mason (eds.), "Comparative Biochemistry," vol. VI, pp. 246–290, Academic, New York, 1964.

Hanstrom, B.: "Hormones in Invertebrates," Oxford University Press, New York, 1939.

Heller, H., and R. B. Clark (eds.): "Neurosecretion," 3d Int. Symp. Neurosecretion, University of Bristol, 1961.

Karlson, P., and A. Butenandt: Pheromones (Ectohormones) in Insects, *Annu. Rev. Entomol.*, vol. 4, pp. 39–58, 1959.

Karlson, P., and M. Lüscher: Pheromones: A New Term

for a Class of Biologically Active Substances, *Nature*, vol. 183, pp. 55–56, 1959.

Karlson, P., and C. E. Sekeris: Biochemistry of Insect Metamorphosis, in M. Florkin and H. S. Mason (eds.), "Comparative Biochemistry," vol. VI, pp. 221–241, Academic, New York, 1964.

Kleinholz, L.: Endocrinology of Invertebrates, Particularly Crustaceans, in B. T. Scheer (ed.), "Recent Advances in Invertebrate Physiology," pp. 173–196, University of Oregon Press, Eugene, Oreg., 1957.

Lane, N. J.: Elementary Neurosecretory Granules in the Neuron of the Snail *Helix aspersa*, *Q. J. Microsc. Sci.*, vol. 105, pp. 31–34, 1965.

Ralph, C. L.: Recent Developments in Invertebrate Endocrinology, *Am. Zool.*, vol. 7, pp. 145–160, 1967.

Scharrer, B. V.: Comparative Physiology of Invertebrate Endocrines, *Annu. Rev. Physiol.*, vol. 15, pp. 457–472, 1953.

Scharrer, E., and B. Scharrer: Neurosecretion, *Physiol. Rev.*, vol. 25, pp. 171–181, 1945.

van der Kloot, W. G.: Moulting and Growth, *Annu. Rev. Physiol.*, vol. 24, pp. 491–576, 1962.

Wigglesworth, V. B.: "The Control of Growth and Form," Cornell University Press, Ithaca, N.Y., 1959.

Wilson, E. O.: Pheromones, *Sci. Am.*, vol. 208, pp. 100–115, 1963.

Neurosecretion and endocrine regulation

Alexandrowicz, J. S.: The Neurosecretory System of the Vena Cava in Cephalopoda. I. *Eledone cirrosa*, *J. Mar. Biol. Assoc. U.K.*, vol. 44, pp. 111–132, 1964.

Alexandrowicz, J. S.: The Neurosecretory System of the Vena Cava in Cephalopoda. II. *Sepia officinalis* and *Octopus vulgaris*, *J. Mar. Biol. Assoc. U.K.*, vol. 45, pp. 209–228, 1965.

Andrewartha, H. G.: Diapause in Relation to the Ecology of Insects, *Biol. Revs.*, vol. 27, pp. 50–107, 1952.

Barber, V. C.: A Neurosecretory Tissue in *Octopus*, *Nature*, vol. 213, pp. 1042–1043, 1967.

Barth, R. H.: Insect Mating Behavior: Endocrine Control of a Chemical Communication System, *Science*, vol. 149, pp. 882–883, 1965.

Beck, S. D., and N. Alexander: Proctodone, an Insect Developmental Hormone, *Biol. Bull.*, vol. 126, pp. 185–198, 1964.

Belamarich, F. A., and R. Terwilliger: Isolation and Identification of Cardioexcitor Hormone from the Pericardial Organs of *Cancer borealis*, *Am. Zool.*, vol. 6, pp. 101–106, 1966.

Berry, S. J., A. Krishnakumaran, and H. A. Schneiderman: Control of Synthesis of RNA and Protein in Diapausing and Injured Cecropia Pupae, *Science*, vol. 146, pp. 938–940, 1964.

Bowers, W. S., and C. C. Blickenstaff: Hormonal Termination of Diapause in the Alfalfa Weevil, *Science*, vol. 154, pp. 1673–1674, 1966.

Bowers, W. S., H. M. Fales, M. J. Thompson, and E. C. Uebel: Juvenile Hormone: Identification of an Active Compound from Balsam Fir, *Science*, vol. 154, pp. 1020–1021, 1966.

Brown, F. A., Jr.: Hormones in Crustacea, *Q. Rev. Biol.*, vol. 19, pp. 32–118, 1944.

Carlisle, D. B., and P. E. Ellis: Bracken and Locust Ecdysones: Their Effects on Moulting in the Desert Locust, *Science*, vol. 159, pp. 1472–1473, 1968.

Carlisle, D. B., and F. G. W. Knowles: Neurohaemal Organs in Crustacea, *Nature*, vol. 172, pp. 404–405, 1953.

Carlisle, D. B., and F. G. W. Knowles: "Endocrine Control in Crustaceans," Cambridge University Press, New York, 1959.

Chaet, A. B.: The Gamete-shedding Substance of Starfishes: A Physiological-Biochemical Study, *Am. Zool.*, vol. 6, pp. 263–271, 1966.

Clementson, C. A. B.: Effect of Certain Types of Paper on Sexual Maturation of the Insect *Pyrrhochoris apterus*, *Nature*, vol. 208, p. 510, 1965.

Cooke, I.: The Sites of Action of Pericardial Organ Extract and 5-Hydroxytryptamine in the Decapod Crustacean Heart, *Am. Zool.*, vol. 6, pp. 107–122, 1966.

Corbet, P. S.: Environmental Factors Influencing the Induction and Termination of Diapause in the Emperor Dragon Fly *Anax imperator* Leach, *J. Exp. Biol.*, vol. 33, pp. 1–14, 1956.

Davey, K. G.: Neurosecretion and Molting in Some Parasitic Nematodes, *Am. Zool.*, vol. 6, pp. 243–249, 1966.

Davey, K. G., and W. R. Breckenridge: Neurosecretory Cells in a Cestode, *Hymenolepis diminuta*, *Science*, vol. 158, pp. 931–932, 1967.

Dearden, M.: Experiments on the Effect of Farnesol on the Development of Normal and Bar-eyed *Drosophila*, *J. Insect Physiol.*, vol. 10, pp. 195–209, 1964.

Evans, J. J. T.: Insect Neurosecretory Material Separated by Differential Centrifugation, *Science*, vol. 136, p. 314, 1962.

Fingerman, M.: Neurosecretory Control of Pigmentary Effectors in Crustaceans, *Am. Zool.*, vol. 6, pp. 169–180, 1966.

Fingerman, M.: Cellular Aspects of the Control of Physiological Color Changes in Crustaceans, *Am. Zool.*, vol. 9, pp. 443–452, 1969.

Fingerman, M., and Y. Yamamoto: Endocrine Control of Tanning in the Crayfish Exoskeleton, *Science*, vol. 144, p. 1462, 1964.

Fraenkel, G., and C. Hsiao: Hormonal and Nervous Control of Tanning in the Fly, *Science*, vol. 138, pp. 27–29, 1962.

Fraenkel, G., and C. Hsiao: Tanning in the Adult Fly: A New Function of Neurosecretion in the Brain, *Science*, vol. 141, pp. 1057–1058, 1963.

Fraenkel, G., C. Hsiao, and M. Seligman: Properties of Bursicon: An Insect Protein Hormone That Controls Cuticular Tanning, *Science*, vol. 151, pp. 91–93, 1966.

Hagadorn, I. R.: Neurosecretion and the Brain of the Rhynchobdellid Leech *Theromyzon rude*, *J. Morphol.*, vol. 102, pp. 55–90, 1958.

Hagadorn, I. R.: Histology and Histochemistry of Neurosecretion in the Medicinal Leech, *Hirudo medicinalis*, *Am. Zool.*, vol. 4, pp. 406–407, 1964.

Hagadorn, I. R., H. A. Bern, and R. S. Nishioka: Fine Structure of the Supraoesophageal Ganglion of the Leech *Theromyzon rude* with Special Reference to Neurosecretion, *Z. Zellforsch.*, vol. 58, pp. 714–758, 1963.

Hauenschild, C.: Influence of the Cerebral Hormone on the Post-embryonal Development of *Platynereis dumerlii*, *Ann. Endocrinol.*, vol. 25, pp. 49–56, 1963.

Hauenschild, C.: Die Hormonal Einfluss des Gehirns auf die Sexuelle Entwicklung bei dem Polychaeten *Platynereis dumerlii*, *Gen. Comp. Endocrinol.*, vol. 6, pp. 26–73, 1966.

Highnam, K. C., L. Hill, and D. J. Gingell: Neurosecretion and Water Balance in the Male Desert Locust *Schistocerca gregaria*, *J. Zool.*, vol. 147, pp. 201–215, 1965.

Ikegami, S., S. Tamura, and H. Kanatani: Starfish Gonad: Action and Chemical Identification of Spawning Inhibitor, *Science*, vol. 158, pp. 1052–1053, 1967.

Johnson, B., and B. Bowers: Transport of Neurohormones from the Corpora Cardiaca in Insects, *Science*, vol. 141, pp. 264–266, 1963.

Kamemoto, F. I., K. N. Kato, and L. E. Tucker: Neurosecretion and Salt and Water Balance in the Annelida and Crustacea, *Am. Zool.*, vol. 6, pp. 213–220, 1966.

Kamemoto, F. I., and J. K. Ono: Neuroendocrine Regulation of Salt and Water Balance in the Crayfish *Procambarus clarkii*, *Comp. Biochem. Physiol.*, vol. 29, pp. 393–402, 1969.

Kaplanis, J. N., M. J. Thompson, W. E. Robbins, and B. M. Bryce: Insect Hormones: Alpha Ecdysone and 20 Hydroxyecdysone in Bracken Fern, *Science*, vol. 157, pp. 1436–1437, 1967.

Kater, S. B.: Cardioaccelerator Release in *Periplaneta americana* (L), *Science*, vol. 160, pp. 765–766, 1968.

Kato, K. K., and F. I. Kamemoto: Neuroendocrine Involvement in Osmoregulation in the Grapsid Crab *Metopograpsus messor*, *Comp. Biochem. Physiol.*, vol. 28, pp. 665–674, 1969.

Knowles, Sir Francis: Crustacean Color Changes and Neurosecretion, *Endeavour*, vol. 14, pp. 95–104, 1955.

Knowles, Sir Francis, and D. B. Carlisle: Endocrine Control in the Crustacea, *Biol. Revs.*, vol. 31, pp. 396–474, 1956.

Krishnakumaran, A., and H. A. Schneiderman: Prothoracotropic Activity of Compounds That Mimic Juvenile Hormone, *J. Insect Physiol.*, vol. 11, pp. 1517–1532, 1965.

Lees, A. D.: "The Physiology of Diapause in Arthropods," Cambridge Monographs in Experimental Biology No. 4, Cambridge University Press, Cambridge, 1955.

Maddrell, S. H. P.: A Diuretic Hormone of *Rhodnius prolixus*, *Nature*, vol. 194, pp. 605–606, 1962.

Maddrell, S. H. P.: The Site of the Release of the Diuretic Hormone in *Rhodnius*—A New Neurohaemal System in Insects, *J. Exp. Biol.*, vol. 45, pp. 499–508, 1966.

Nishioka, R. S., I. R. Hagadorn, and H. A. Bern: The Ultra Structure of the Epistellar Body of the Octopus, *Z. Zellforsch.*, vol. 57, pp. 406–421, 1962.

Novák, V. J. A.: "Insect Hormones," Methuen, London, 1966.

Roth, L., and B. Stay: Control of Oocyte Development in Cockroaches, *Science*, vol. 130, pp. 271–272, 1959.

Schneiderman, H. A., and L. I. Gilbert: Substances with Juvenile Hormone Activity in Crustacea and Other Invertebrates, *Biol. Bull.*, vol. 115, pp. 530–535, 1958.

Schneiderman, H. A., and L. I. Gilbert: Control of Growth and Development in Insects, *Science*, vol. 143, pp. 325–333, 1964.

Schneiderman, H. A., and J. Horwitz: The Induction and Termination of Facultative Diapause in the Chalcid Wasps *Mormoniella vitripennis* (Walker) and *Trineptis klugii* (Ratzeburg), *J. Exp. Biol.*, vol. 35, pp. 520–551, 1958.

Schneiderman, H. A., A. Krishnakumaran, V. G. Kulkani, and L. Friedman: Juvenile Hormone Activity of Structurally Unrelated Compounds, *J. Insect Physiol.*, vol. 11, pp. 1641–1649, 1965.

Simpson, L. H., H. A. Bern, and R. S. Nishioka: Cytologic Observations on the Nervous System of the Pulmonate Gastropod *Heliosoma tenue*, with Special Reference to Possible Neurosecretion, *Am. Zool.*, vol. 4, pp. 407–408, 1964.

Simpson, L. H., H. A. Bern, and R. S. Nishioka: Survey of

Evidence for Neurosecretion in Gastropod Molluscs, *Am. Zool.*, vol. 6, pp. 123–138, 1966.

Sláma, K., and C. M. Williams: Juvenile Hormone Activity for the Bug *Pyrrochoris apterus*, *Proc. Nat. Acad. Sci. U.S.A.*, vol. 54, pp. 411–414, 1965.

Smith, U., and D. S. Smith: Observations on the Secretory Processes in the Corpus Cardiacum of the Stick Insect *Carausius morosus*, *J. Cell Sci.*, vol. 1, pp. 59–66, 1966.

Srivastava, U. S., and L. I. Gilbert: Juvenile Hormone: Effects on a Higher Dipteran, *Science*, vol. 161, pp. 61–62, 1968.

Ude, J.: Untersuchungen zur Neurosekretion bei *Dendrocoelum lacteum*, *Z. Wiss. Zool.*, vol. 170, pp. 223–255, 1964.

Watson, J. D. L.: Moulting and Reproduction in the Adult Firebrat *Thermobia domestica* (Packard) (Thysanura, Lepismatidae). I. The Moulting Cycle and Its Control, *J. Insect Physiol.*, vol. 10, pp. 305–317, 1964.

Welsh, J. H.: Neurohormones in Molluscs, *Anat. Rec.*, vol. 138, pp. 387–388, 1960.

Welsh, J. H.: Neurohormones of Mollusca, *Am. Zool.*, vol. 1, pp. 267–272, 1961.

Wiens, A. W., and L. I. Gilbert: Regulation of Cockroach Fat Body Metabolism by the Corpus Cardiacum *in vitro*, *Science*, vol. 150, pp. 614–616, 1965.

Wigglesworth, V. B.: The Breakdown of the Thoracic Gland in the Adult Insect, *Rhodnius prolixus*, *J. Exp. Biol.*, vol. 32, pp. 485–491, 1955.

Wigglesworth, V. B.: The Role of the Hemocytes in the Growth and Moulting of an Insect, *Rhodnius prolixus* (Hemiptera), *J. Exp. Biol.*, vol. 32, pp. 649–663, 1955.

Wigglesworth, V. B.: The Hormonal Regulation of Growth and Reproduction in Insects, *Adv. Insect Physiol.*, vol. 2, pp. 247–336, 1964.

Wright, J. E.: Hormonal Termination of Larval Diapause in *Dermacenter albipictus*, *Science*, vol. 163, pp. 390–391, 1969.

Rhythms

Adkisson, P. L.: Internal Clocks and Insect Diapause, *Science*, vol. 154, pp. 234–241, 1966.

Aiken, D. E.: Photoperiod, Endocrinology and the Crustacean Molt Cycle, *Science*, vol. 164, pp. 149–155, 1969.

Aschoff, J.: Comparative Physiology: Diurnal Rhythms, *Annu. Rev. Physiol.*, vol. 25, pp. 581–596, 1963.

Bennett, M. F., J. Schreiner, and R. A. Brown: Persistent Tidal Cycles of Spontaneous Motor Activity in the Fiddler Crab, *Uca pugnax*, *Biol. Bull.*, vol. 112, pp. 267–275, 1957.

Bode, V. C., R. DaSa, and J. W. Hastings: Daily Rhythm of Luciferin Activity in *Gonyaulax polyedra*, *Science*, vol. 141, pp. 913–915, 1963.

Brown, F. A., Jr.: Persistent Rhythmicity in Animals, *J. Nat. Cancer Inst.*, vol. 13, pp. 1384–1385, 1953.

Brown, F. A., Jr.: Biological Chronometry, *Am. Nat.*, vol. 91, pp. 129–134, 1957.

Brown, F. A., Jr.: Living Clocks, *Science*, vol. 130, pp. 1535–1544, 1959.

Bruce, V. G., and C. Pittendrigh: Endogenous Rhythms in Insects and Micro-organisms, *Am. Nat.*, vol. 91, pp. 179–195, 1957.

Bruce, V. G., and C. Pittendrigh: Resetting the *Euglena* Clock with a Single Light Stimulus, *Am. Nat.*, vol. 92, pp. 295–301, 1958.

Bünning, E.: Physiological Mechanisms and Biological Importance of the Endogenous Diurnal Periodicity in Plants and Animals, in R. B. Withrow (ed.), *A.A.A.S. Symp.* (*Pub. No. 55*), pp. 507–530, 1959.

Cole, C. L., and P. L. Adkisson: Daily Rhythm in the Susceptibility of an Insect to a Toxic Agent, *Science*, vol. 144, pp. 1148–1149, 1964.

Danilevskii, A. S.: "Photoperiodism and Seasonal Development of Insects" (J. Johnston, trans.), Oliver and Boyd, Edinburgh and London, 1965.

Enright, J. T., and W. M. Hamner: Vertical Diurnal Migration and Endogenous Rhythmicity, *Science*, vol. 157, pp. 937–941, 1967.

Fingerman, M.: Lunar Rhythmicity in Marine Organisms, *Am. Nat.*, vol. 91, pp. 167–178, 1957.

Fingerman, M.: Tidal Rhythmicity in Marine Organisms, *Cold Spring Harbor Symp. Quant. Biol. No. 25*, pp. 481–490, 1960.

Giese, A. C.: Reproductive Cycles of Some West Coast Invertebrates, *A.A.A.S. Symp.* (*Pub. No. 55*), pp. 625–640, 1959.

Harker, J.: Factors Controlling the Diurnal Rhythm of Activity of *Periplaneta americana*, *J. Exp. Biol.*, vol. 33, pp. 224–234, 1956.

Harker, J.: Diurnal Rhythms in the Animal Kingdom, *Biol. Revs.*, vol. 33, pp. 1–52, 1958.

Harker, J.: Endocrine and Nervous Factors in Insect Circadian Rhythms, *Cold Spring Harbor Symp. Quant. Biol.*, *No. 25*, pp. 279–287, 1960.

Hastings, J. W., and B. Sweeney: The *Gonyaulax* Clock, *A.A.A.S. Symp.* (*Pub. No. 55*), pp. 567–584, 1959.

Lees, A. D.: Photoperiodism in Insects and Mites, *A.A.A.S. Symp.* (*Pub. No. 55*), pp. 585–600, 1959.

Neville, A. C.: Daily Growth Layers in Animals and Plants, *Biol. Revs.*, vol. 42, pp. 421–441, 1967.

Pease, R. W., Jr.: Factors Causing Seasonal Forms in *Ascia minuste* (Lepidoptera), *Science*, vol. 137, pp. 987–988, 1962.

Rao, K. P., and T. Gropalakrishnareddy: Blood-borne Factors in Circadian Rhythms of Activity, *Nature*, vol. 213, pp. 1047–1048, 1967.

Richards, T. L.: Reproduction and Development of the Polychaete *Stauronereis rudolphi* Including a Summary of Development in the Superfamily Eunicea, *Mar. Biol.*, vol. 1, pp. 124–133, 1967.

Stephens, G. C.: Responses of the Diurnal Melanophore Rhythm of *Uca pugnax* to Changes in Temperature, *Biol. Bull.*, vol. 109, p. 352, 1955.

Stephens, G. C.: Twenty-four Hour Cycles in Marine Organisms, *Am. Nat.*, vol. 91, pp. 135–152, 1957.

Sweeney, B.: The Photosynthetic Rhythm in Single Cells of *Gonyaulax polyedra*, *Cold Spring Harbor Symp. Quant. Biol. No. 25*, pp. 145–148, 1960.

Sweeney, B., and J. W. Hastings: Effects of Temperature on Diurnal Rhythms, *Cold Spring Harbor Symp. Quant. Biol. No. 25*, pp. 87–104, 1960.

Williams, C. B.: Butterfly Migrations, *New Biol.*, vol. 9, pp. 58–76, 1950.

Withrow, R. B. (ed.): Photoperiodism and Related Phenomena in Plants and Animals, *A.A.A.S. Symp.* (*Pub. No. 55*), Washington, D.C., 1959.

SUGGESTED READING FOR SECTION VII

Buddenbrock, W. von: "The Senses," The University of Michigan Press, Ann Arbor, 1958.

Frisch, K. von: "Bees, Their Vision, Chemical Sense and Language," Cornell University Press, Ithaca, N.Y., 1950.

Frisch, K. von: "The Dance, Language and Orientation of Bees," Harvard University Press, Cambridge, Mass., 1967.

Lees, A. D.: "The Physiology of Diapause in Arthropods," Cambridge Monographs in Experimental Biology, Cambridge University Press, Cambridge, Mass., 1955.

Lentz, T. L.: "Primitive Nervous Systems," Yale University Press, New Haven, Conn., 1968.

Lindauer, M.: "Communication among Social Bees," Harvard University Press, Cambridge, Mass., 1961.

Parker, G. H.: "The Elementary Nervous System," Lippincott, Philadelphia, 1919.

Parker, G. H.: "Animal Color Changes and Their Neurohumours: A Survey of Investigations 1910–1943," Harvard University Press, Cambridge, Mass., 1948.

Romanes, C. J.: "Jellyfish, Starfish and Sea Urchins, Being a Research on Primitive Nervous Systems," Kegan Paul, Trench and Trubner, London, 1885.

Tinbergen, N.: "The Study of Instinct," Clarendon Press, Oxford, 1951.

Wells, M. J.: "Brain and Behavior in Cephalopods," Stanford University Press, Stanford, Calif., 1962.

section eight

REPRODUCTION, DEVELOPMENT, AND REGENERATION

REPRODUCTION

Reproduction, growth, and regeneration are different aspects of the same biological phenomenon, which is directed toward survival of life on Earth, perpetuation of existing species, and potentiation of new ones. Reproduction is the means by which an individual produces offspring and transmits to them a genetic constitution in which are encoded directions for development of the general characteristics, both morphological and physiological, of their kind and the particular characteristics that distinguish them as individuals. It provides for continuity in a species, or in a strain within a species, and, often, for variety and variation in it. From the variants that naturally arise there are selected, through their degrees of success in meeting the conditions of their immediate environments or of adapting themselves to new ones, those best fitted to survive. These will become the parents and ancestors of future generations, and possibly of new species. Success in reproduction, rather than fecundity per se, is the ultimate criterion for the continued perpetuation of a species. Growth is the means by which an individual attains the architectural pattern, the symmetry, and the size characteristic of its species and its immediate ancestry. Growth is a rapid and usually continuous process during the embryonic and juvenile life of a multicellular organism and may continue in its adult life at a reduced rate. It involves not only increase in the size and number of cells but also their assemblage and organization into specific parts and their differentiation as appropriately functional units. Regeneration is the aspect of growth that allows an organism to reconstitute and replace a part that has been lost, by accident or in the course of its natural life processes; regeneration is achieved through cell multiplication, histogenesis, and morphogenesis of the missing part or region.

Invertebrates show as much diversity in their methods of reproduction as they do in other ways of their lives. This diversity may be within different orders of the same class; elaborate genital systems and complicated means of production of sex cells and of bringing them together are by no means limited to those phyla generally considered to be in the upper levels of evolutionary development. Reproduction may be agametic, by fission, budding, or fragmentation; or it may be gametic, with production of special sex cells which generally must unite before a new individual can be produced. These cells are usually produced by different individuals of a species, designated as males and females, and the species then is bisexual, or gonochoristic (Greek *gonos*, "procreation," and *chorizein*, "to separate"). Some species, and in some instances individuals within a gonochoristic species, are unisexual, producing two types of gametes (macrogametes and microgametes) either concurrently or sequentially. Such individuals are known as hermaphrodites. Reproduction through gametic union is often referred to as "sexual," as opposed to "asexual," or propagation by other means. The terms "gametic" and "agametic" are considered preferable, however, and will be used in the following discussion.

Agametic reproduction is typical of a number of species, and gametic of others, but gametic reproduction occurs in some species of every invertebrate phylum. In some, there is an alternation of a generation, or of generations, of agametically reproducing individuals and those reproducing gametically; such a cycle is known as metagenesis. Development may be direct, without a larval stage, or indirect, through a larval stage or stages. Again, these differences in reproductive habit may be found in different orders within the same taxonomic class. Paedogenesis, polyembryony, and viviparity occur also in some species.

A. AGAMETIC REPRODUCTION

Agametic reproduction may be through division of a single individual into two (binary fission) or into several (multiple fission) parts. It may also occur through fragmentation, the separation of a small part of the body which will grow into a new whole or the breaking up of the body into several or many parts, from each of which a new individual arises. It may be by budding, the multiplication of somatic cells in particular parts of the organism, their morphogenesis into a new individual, and its detachment from the parent body. Or it may be through the production of special cells, or groups of cells, that are set apart from the parent body and develop outside, or inside, it into a new individual.

Parthenogenesis is a form of reproduction that does not involve gametic union (fertilization). It may therefore be included in the category of agametic reproduction in view of the distinction made above between gametic and agametic means. There is a certain ambiguity in this, however, since in some cases unfertilized macrogametes develop into new individuals; the term "asexual" is more applicable here. Such cases are illustrative of facultative parthenogenesis, in that the macrogamete ordinarily requires fertilization, but having in some way missed it is still capable of development. Parthenogenesis may also be obligatory; in this, the reproductive cells are of a special kind and essentially incapable of being fertilized. In metazoans, they are generally formed in the gonads and are known as parthenogenetic, or amictic, eggs, in distinction to macrogametes, or mictic (Greek *miktos*, "mixed"), eggs. In a number of species obligatory parthenogenesis is a natural event, occurring regularly, or at certain times of the year, or under certain conditions, and may be followed or preceded by gametic reproduction. Parthenogenesis may also be induced artificially in macrogametes, either by mechanical or by chemical means. Both methods have been widely used by experimental embryologists in investigations of the factors initiating development.

1. Fission and fragmentation

Fission is characteristic of many protozoans. Among flagellates, this is usually binary, the individual dividing lengthwise, although in dinoflagellates reproducing by binary fission the division is typically oblique. Some dinoflagellates reproduce by multiple fission. Most Sarcodina reproduce regularly by binary fission, as do ciliates, although in Sarcodina and Flagellata, especially, gametic reproduction may intervene in a population at certain times. With ciliates, the body divides transversely in binary fission. Multiple fission is characteristic of multinucleate rhizopods and of heliozoans.

In most cases reproduction of protozoans by fission involves mitotic division of the nucleus or nuclei and reconstitution of the cytoplasmic mass and organelles in the daughter cells. This represents growth and differentiation on the part of each cell, and also growth in the population of the species. In the case of flagellates with a single flagellum, or with few flagella, for example, each flagellum may duplicate before fission of the cell and the appropriate number be apportioned to each daughter cell, or it may be resorbed and each daughter cell develop its own. In multiflagellate species the flagella are usually distributed fairly evenly between the two cells, but in each case there must be an increase in cytoplasm to bring the half cell back to typical size and form, as well as replication of cytoplasmic particles. In ciliates with a high degree of organization, such as *Paramecium*, *Stentor*, and the like, the organelles of the anterior half must be replaced in the posterior half, and those of the posterior half, in the daughter cell derived from the anterior half (Fig. XVIII–1). In many cases this

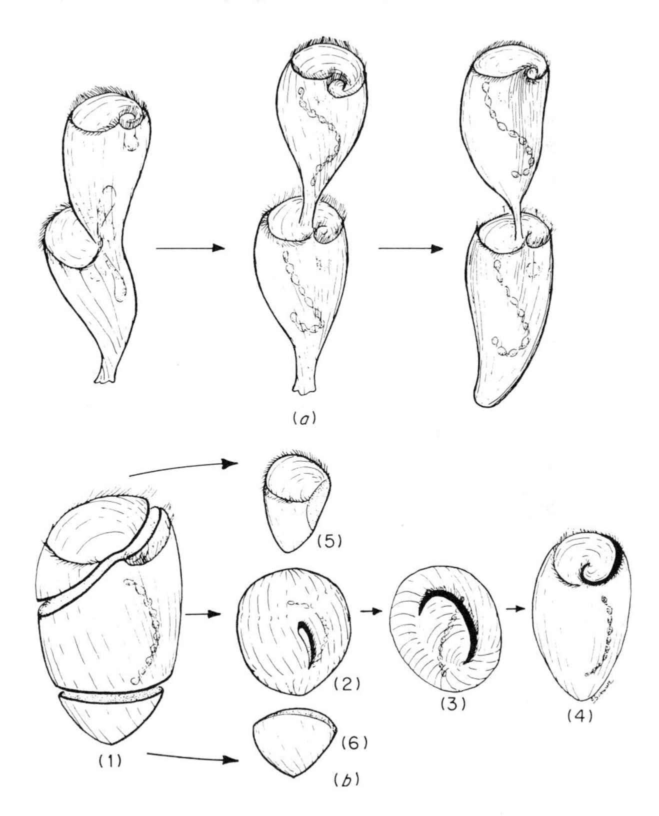

FIG. XVIII–1 *Division and regeneration in* Stentor coeruleus. *(After Korschelt.)* ***(a)*** *Binary fission. Organization of the vestibule is beginning in the posterior half well before its separation from the anterior.* ***(b)*** *Regeneration in nucleated and nonnucleated pieces. (After Tartar.) (1) The animal has been cut in three pieces, of which the central one only contains the nuclear apparatus. (2, 3, and 4) Regeneration of the nucleated piece. (5 and 6) Degeneration of the nonnucleated pieces.*

is accomplished before fission actually occurs, and its onset can be detected by the appearance of a second oral groove, gullet, and cytostome in the midregion of an actively swimming cell. Postfission events then involve construction of new cytoplasm, so that the half cell attains the dimensions of a typical individual of the species, replication of basal bodies and their organization or alignment in the pattern characteristic of the species, and the development of new cilia. In shelled rhizopods, at fission, one member usually retains the shell, while the other receives the materials for the formation of a new one. These may even be foreign particles, transferred with the separated cytoplasmic mass.

Fission raises the question of what constitutes a protozoan individual, for A, dividing into B and C, has lost its identity as A, although preserving it in its contribution to their mass and structure. All descendants of a single protozoan cell that have arisen through fission constitute a clone (Greek *klon*, "twig"), alike in their genetic constitution, since their nuclei are products of mitotic division, and in their cytoplasmic constitution, since this has been derived initially from the original cell and subsequently from replication of its materials. No clone, or no population of protozoans, can increase indefinitely, for limitations to the number are set by the space, food, and other physical necessities available to it. Is then a protozoan clone to be considered an individual, comparable to a cell aggregate or to a multicellular organism derived by successive divisions of a single cell? And can a multicellular organism be considered a clone of differentiated and specialized cells so morphologically and functionally integrated that they act as a single organismal unit?

Fission is a common means of reproduction in some groups of metazoan invertebrates, even though they may also reproduce gametically. Fission is, however, rare or unknown in other groups. In most of those in which it is frequent, there is a high degree of correlation between capacity for it and for regeneration, for both are phenomena requiring replacement of missing parts. Fission is a customary means of reproduction in some flatworms, nemerteans, phoronids, and annelids and is known to occur on occasion in cnidarians and echinoderms.

Fission in rhabdocoele flatworms has been most fully followed in *Stenostomum*. The plane of fission arises about the midregion of the body, where a second head appears and the two individuals then pull apart. It is suggested that in this case the anterior section represents the original individual whose integrity is preserved, and the posterior section essentially a new and younger one. The reason for this is that the anterior sections show progressive senescence, while the posterior sections have the attributes and characteristics of juveniles. Fission in *Stenostomum* may also be multiple, in that chains of two to seven individuals, called zooids, may arise through the development of new juveniles before the detachment of those earlier formed. In this case the first fission plane appears about two-thirds of the distance from the head to the tail. Both the anterior and the posterior sections increase in length as the new head is formed, and the second fission plane appears in the anterior section, but relatively more posterior in position than the first one. The third plane arises in the posterior section, but in this case relatively more anteriorly; the fourth arises just anterior to the second; the fifth anterior to the third; and the sixth between the first and second. Thus, there is a distinct tendency for new heads to arise closer to the ones recently formed than to the older ones. Such sequences of attached zooids may remain together for some time, but ultimately they separate and are set free as independent individuals.

The freshwater triclad *Dugesia* can reproduce by binary fission, although it commonly reproduces gametically. In fission, the plane of division usually lies just behind the pharynx and is foreshadowed, in *D. paramensis* at least, by the appearance of eyes and pharynx in this region. In the actual separation of the two parts, the posterior section stays firmly fastened to some fixed surface, while the anterior moves freely forward. Ultimately, the two parts are pulled apart and a larger anterior or head section is separated from a smaller posterior or tail section. Each then regenerates the parts it lacks, a complicated and highly organized procedure, for, among other things, cephalic structures must be reconstituted by the tail section and elaborate genitalia by the head section. Fission occurs less frequently in crowded cultures than in those in which there are comparatively small numbers of individuals in relation to the area available. This has been attributed to the effects of a thermostable substance, or substances, exuded from the flatworms. It has also been attributed to a feedback control from the cephalic ganglia. Higher rates of fission in decapitated specimens of *Dugesia dorotocephala* and *D. tigrina* than in intact ones or those with an equal amount of tissue removed from the tail region support this concept. Under such conditions, too, the heads regenerated by the posterior segments often have more than two eyes. Extra eyes are also formed in these heads when NH_4Cl is added to the water of an uncrowded culture, which has led to the hypothesis that accumulation of excretory products, principally ammonia, from many individuals may be the cause of this abnormality.

Fission is not uncommon among nemerteans, which also reproduce gametically. *Lineus vegetus*, for example, frequently reproduces by fragmentation and small sections from any part of the body back of the head develop equally well into minute but fully formed specimens. Among phoronids, *Phoronis ovalis* reproduces by transverse fission, and among cnidarians, the hydromedusan *Cladonema radiatum* reproduces by fission, dividing into two parts beginning with longitudinal division of the manubrium. Some anthozoans also divide longitudinally into two pieces. *Metridium* can reproduce by a kind of fragmentation. Pieces of the pedal disk are pinched off and become organized into minute polyps, which, in favorable circumstances, grow to full size.

Both polychaete and oligochaete annelids can reproduce by fission. This is a frequent phenomenon among syllids, serpulids, and sabellids. In the syllids *Autolytus* and *Procerastea*, the head of the new individual always appears on the anterior half of the fourteenth segment, but in other genera the location is not so definitely determined. As in *Stenostomum*, temporary chains of individuals may be formed, most frequently in the warmer months of the year. The tiny cerratulid *Ctenodrilus serratus*, which is found in debris off the coasts of California and the West Indies, reproduces for generation after generation by transverse fission, dividing into six or seven pieces, each of which becomes a new worm. In five generations, over 16,000 new individuals will have been produced from a single ancestor, provided all survive. At some stage in the life history of a population a generation appears that reproduces gametically and whose offspring start on a new series of fissions.

Fission is also characteristic of aquatic oligochaetes, especially in the families Aelosomatidae and Naididae (see Fig. VI–11*b*). Indeed, among some naidids this seems to be the sole means of reproduction, for sexual individuals have not been found or are so rare as to be negligible factors in propagation of the species. In true fission, new heads are formed before the two sections separate; there may be several planes of fission appearing in close succession resulting in chains of individuals. These worms may also undergo a process of fragmentation, in which the body breaks into a number of pieces, each of which regenerates a new head and tail while free in the water. A species of terrestrial and semiterrestrial oligochaete, identified as *Enchytraeus fragmentosus*, is reported to have no sex organs and to reproduce exclusively by fragmentation. Mature worms have been observed to break into 3 to 11 pieces, each containing about five segments, with each piece regenerating into a fully developed worm in about 10 days.

Some echinoderms are known to undergo spontaneous division into two parts. Although the life history of *Asterina burtoni* is not yet fully known, it is doubtful if it ever reproduces gametically and seems always to do so by fission. Specimens in all stages of division have been found in the Mediterranean, with the fission plane passing across the disk between the arms. Fission has also been reported in the asteroid genus *Linckia* and is common in the brittle star family Ophiactidae. Some holothurians are known to reproduce in this way. *Sclerodactyla* (*Thyone*), for example, can cut off the anterior tip of its body and the entire alimentary canal, giving rise to two individuals by replacement of the missing parts, and *Holothuria* frequently undergoes binary fission, the individual dividing transversely in the midregion.

Although fission would seem to be a simple way of propagation and a very effective way in regard to the numbers of new individuals produced, its interpretation poses many problems, some of which are the same as those raised by the phenomenon of regeneration. But the zone of fission is always one of active growth arising in an intact and apparently coordinated animal, not in one that has already lost a part of itself. What provokes this growth and morphogenetic activity and why it occurs in specified regions in some cases and more or less at random in others are still unanswered questions, which are raised also by reproduction through budding, another and very common means of agametic reproduction among invertebrates.

2. Budding

Budding as a means of reproduction is usual in sponges, cnidarians, endoprocts, and ectoprocts; and, among protozoans, in peritrichous ciliates and in suctorians. In many instances the buds remain attached to the original individual with the consequent formation of colonies in which all the members may be alike or differentiated as individuals fitted for performance of different functions. In other cases the buds are detached when fully developed, thus increasing the members of a population. Budding differs from fission in that the bud primordia are usually localized in particular regions of the parent organism, which does not necessarily lose its identity even when a colonial individual is formed. A bud, moreover, always develops from a small mass of cells and always under the influence of the parent body. Buds may be external, projecting from the surface of the parent, or internal, formed within depressions or cavities from which they are ultimately set free. Budding, for example, occurs along the stalk of a colonial peritrich such as *Carchesium* or *Epistylis*, the branches with their terminal zooids arising at definite intervals along the main stalks and usually at similar angles to them (Fig. XVIII–2).

In suctorians external buds appear as finger-like projections of the cytoplasmic mass; they may arise singly or in various numbers either concomitantly or successively. Each such projection contains an extension of the macronucleus and a mitotic derivative of the micronucleus. Internal buds, which are more common, may also be formed singly or in numbers. They appear in depressions in the cell, as similar extensions of cytoplasm with similar nuclear inclusions. In some species, particularly under unfavorable con-

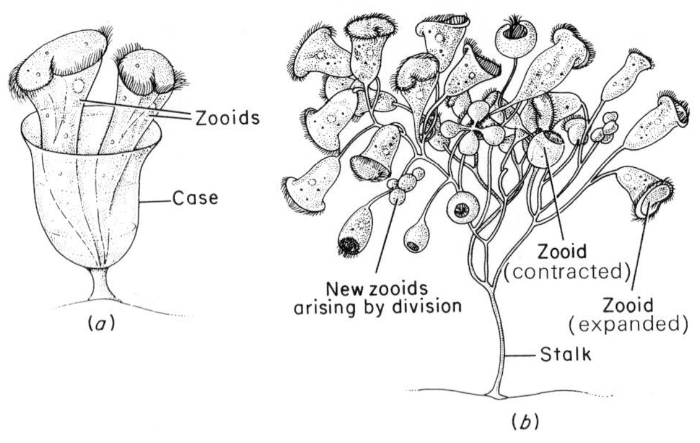

FIG. XVIII–2 *Colonial ciliates. (From Gardiner and Flemister.)* ***(a)*** Cothurnia patula, *with two zooids in a common case 50 to 80 μ tall.* ***(b)*** Epistylis, *with new zooids budding off the stalk.*

ditions, almost the entire cell body will be included in a single bud, so that nothing but the stalk and pellicle remain after the bud is released. Buds in most species are at first without tentacles and without cilia, but later they acquire a ring, or rings, of cilia by means of which they become mobile and swim away from the parent. The motile stage is a transitory one, for the ciliated individual soon settles down in another locale, loses its cilia, and develops the typical suctorian tentacles and sedentary mode of life. It is, however, during this brief period of existence, a means of distributing and dispersing the species.

External and internal budding is also a common way of reproduction in sponges. External buds appear first as small lumps on the sides of the parent body in which all the cell types characteristic of the species are represented. These lumps increase in size as the cells multiply until an individual with adult characters is formed; this may either break away from the parent or remain attached to it, forming the extensive colonies characteristic of most sponges, particularly the encrusting species.

Budding in cnidarians has been known ever since the pioneer observations of the Dutch microscopist Antony van Leeuwenhoek (1632–1723) and those of the Swiss biologist Abraham Trembley (1700–1784), whose studies on *Hydra* are classics in the history of biology (Fig. XVIII–3). *Hydra* has since provided much of the material for experimental investigations

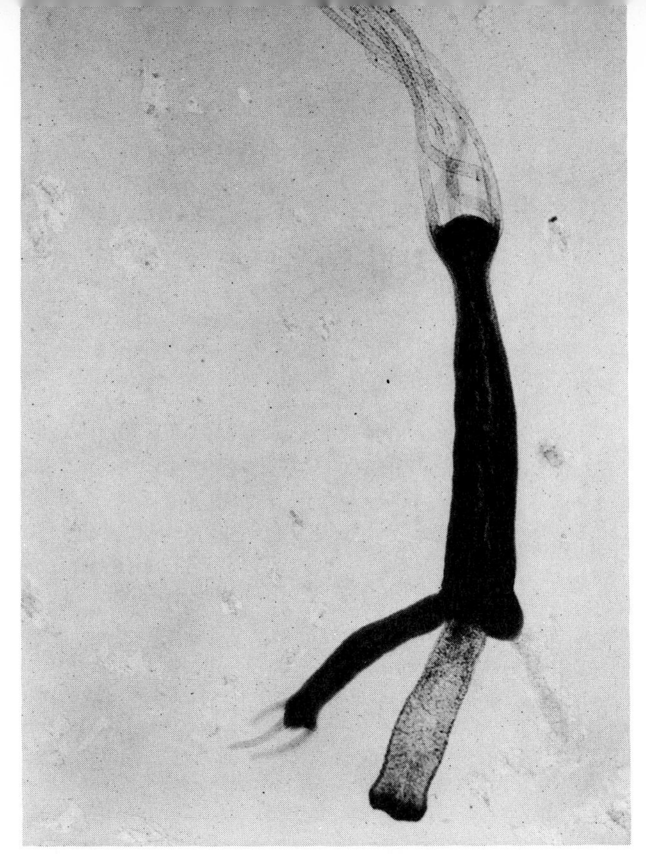

FIG.XVIII–3 Hydra *with bud.* (*General Biological Supply House, Inc., Chicago.*)

of growth and regeneration. Three regions of high metabolic activity, as measured by the relative capacity of the cells there to reduce methylene blue, are distinguishable in *Hydra*: the hypostome; the budding area, about midway between the hypostome and the basal disk; and the basal disk itself. These areas are also characterized by a greater abundance of interstitial and gland cells than are found elsewhere, as well as by a notable concentration of food reserves. Nutrition is important to agametic reproduction in *Hydra*, and well-fed individuals will develop buds rapidly and in considerable numbers. The first indication of a bud is an extension of the epidermis and the gastrodermis, making a slight swelling, on the column in the budding zone. This small bud stage is followed by the medium bud stage, when there is a conspicuous, hollow cylindrical extension of the column; and then by the short tentacled stage, when the primordia of the tentacles are evident as little knobs around the free end of the cylinder. During these stages there is continuity between the gastral cavities of bud and parent, but at the long-tentacled stage, when the tentacles have grown and the solid peduncle is formed, the connection is blocked off, and the bud, now able to detach, catches food and digests it for itself. Development of the bud involves cell multiplication, arrangement into the layered pattern typical of cnidarians, and differentiation of cell types, especially cnidoblasts, for the mature bud is equipped with nettle batteries similar to those of its parent. After a bud has reached the long-tentacled stage and differentiated its peduncle, another bud may arise on the opposite side of the column at the same level and a third above that. In very well fed hydras, two buds may develop simultaneously on opposite sides of the column, and a single individual may support several at a time and continue to bud rapidly, given optimal nutritional and environmental conditions, for weeks or even months. Indeed, a bud may bud while still attached to the parent and, as these individuals are set free, the population increases rapidly.

Colonial hydrozoans, both freshwater and marine, increase the numbers in a colony by similar processes of budding. In general, three kinds of buds may be formed: hydranth buds that remain attached to the parent colony; planuloid buds, or frustules (Latin *frustulum*, "little piece"), that separate from the parent and, like a planula larva, creep away, settle, and develop into a new budding and branching colony; and medusoid buds that develop into gametically reproducing medusae, in species such as *Pennaria* and *Obelia* being detached from the parent as free-swimming individuals and in those like *Tubularia* and *Sertularia* remaining attached and reproducing *in situ*. As the process of agametic reproduction has been followed in the freshwater hydrozoan *Craspedacusta sowerbi*, all three types of buds are formed laterally near the mid-region of the body of the polyp that lives either as a solitary individual or as a member of a small colony of two to seven individuals. The three types of buds, representing three types of morphogenesis, are formed successively, the hydranths being produced rapidly in the first 2 weeks that a young colony was under observation and again in the twelfth to fourteenth week; the frustules were produced most abundantly in the third to the sixth week, with an increase again in the fifteenth week; and the medusae were produced in the eighth to sixteenth week. The fact that both hydranths and frustules are produced in considerable numbers during the time that medusae production is at its height shows that the three different types of morphogenesis, although occurring sequentially, are not mutually exclusive, and that the development of one kind of individual does not necessarily inhibit or prevent that of another.

In hydroids, medusae buds may develop on the upright stem of a colony, on its stolon or creeping stem, or on the zooids. These can be either the feeding individuals (hydranths) or the nonfeeding, reproductive individuals (gonangia or blastostyles). The site of medusa bud production is characteristic of particular species. In some species of the genus *Syncoryne* and *Bougainvillia*, for example, they arise from the lateral walls of the stems; on other species of *Bougainvillia* and in *Campanulina*, from the upper surface of the stolons; in some species of *Syncoryne* and in *Tubularia*, *Pennaria*, and *Eudendrium*, from the walls of the hydranths; and in *Obelia*, from the gonangia. Medusae may also be budded off from the manubrium, the radial canals, the margin of the bell, or the bases of the tentacles of other medusae. They are, for example, budded off the manubrial wall in such species as *Sarsia prolifera* and *Podocoryne fulgans*; from the walls of the radial canals in *Eucheilota paradoxica*; and from the bases of the tentacles in *Niobe dendrotentaculata*.

Scyphozoan medusae reproduce gametically, but in some species such as *Aurelia aurita*, the planula that develops from the fertilized egg attaches itself to some substrate and becomes organized into a feeding polyp-like individual, the scyphistoma (Greek *skyphos*, "cup," and *stoma*, "mouth"). This, by a process of budding known as strobilation (Greek *strobilos*, "pine cone"), constricts off along its length a series of ephyrae, or young jellyfish, so that from a single polyp many new medusae are produced (Fig. XVIII–4). Once strobilation has begun, the scyphistoma, or now strobila, shows a series of annular constrictions and has the appearance of a stack of saucers, with the oldest of the potential jellyfish at the distal end of the series and the youngest at the proximal. After producing and setting free one lot of ephyrae, the scyphistoma may cease budding for a time, but begin again the following

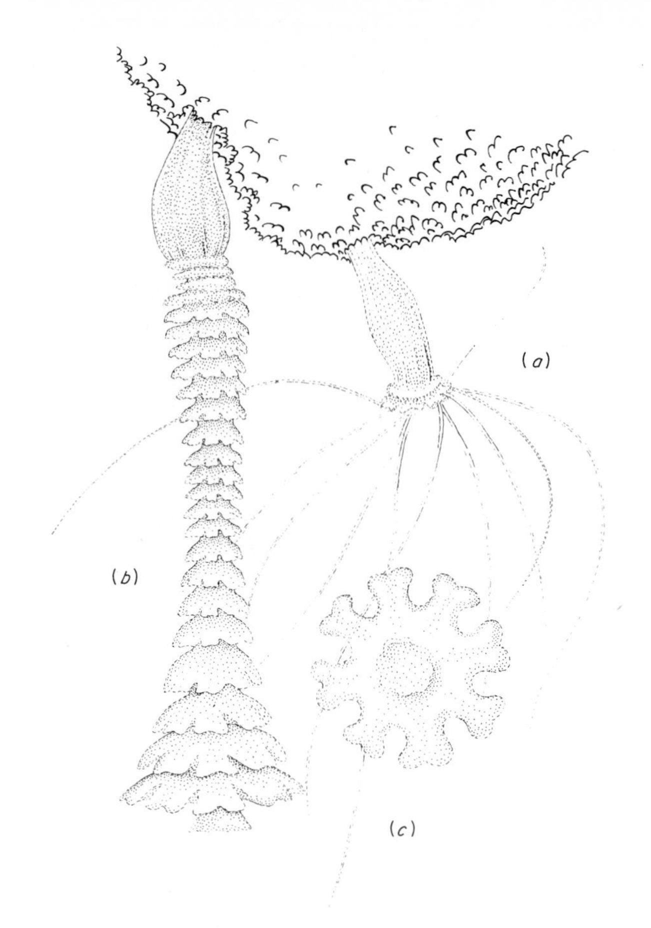

FIG. XVIII–4 *Strobilation in* Aurelia. (*After D. Wilson.*) ***(a)*** *Strobilation beginning in scyphistoma.* ***(b)*** *Strobila producing ephyrae* (*larval medusae*). ***(c)*** *Liberated ephyra.*

spring or under appropriate conditions. It has been shown experimentally that iodine in minute amounts ($1:10^8$ dilution) is necessary for a scyphistoma to strobilate. It has also been shown that this process can be induced by minute quantities of thyroxin, the first definitive demonstration of a morphological response on the part of an invertebrate to a vertebrate hormone.

Budding is common in all endoprocts, the buds originating on the stolon or the upright branches in colonial species and from the calyx in solitary ones. As the process has been observed in the colonial *Pedicellina cernea*, young zooids are successively formed from the tip of the stolon. This is divided into compartments by evenly spaced septa, with fertile, zooid-bearing segments separated from each other by infertile segments. There are two regions of growth in the terminal zooid. One of these is at the basal end and grows out as the stolon; the other is just below the calyx and this grows out as the new zooid.

Budding in ectoprocts is very similar to that in endoprocts. In *Bowerbankia* the buds arise some distance from the tip of the stolon, and the region of their origin is separated from the distal region by a septum, which develops after the buds have formed. The stolon tip continues to grow, and new buds are not formed until it has reached a definite length. The bud originates as an evagination of the ectoderm in which mesodermal cells are later included; several buds may develop together, in a spiral arrangement along the budding part of the stolon.

A *Bugula* (*B. avicularia*) colony originates by successive buddings from the primary zooid that remains attached to the substrate by its adhesive disk. Rhizoids grow out from this that creep along the substrate, as well as erect chains of zooids. Each chain is double, formed by zooids that lie side by side in pairs, but these double chains often branch into two double chains, separated from each other by definite and consistent angles. The entire colony is arborescent, with continuity, through the intervening zooids, between the zooid at the apex of each branch and the primary zooid at the base. Some biologists have found it convenient to think of each zooid as bipartite, composed of a cystide, or case, that includes the epidermis and the covering of the zooecium, and a polypide, essentially the visceral organs. The zooecium covering consists of protein (90 percent) and chitin (10 percent), impregnated, except for the apical growth regions of the buds and the pores where successive zooids communicate with each other, with calcite that represents 70 percent of the total dry weight of a colony. Colonies grow continuously from the tip of each apical zooid, which, like the others, is roughly cylindrical with a dome-shaped apex. Histologically, there are in this apex a group of spherical cells below the cuticle, separated by an almost cell-free space from a dense monolayer of spindle-shaped cells in the cylindrical region. These two groups of cells comprise the epidermal epithelium. The lumen in the center of the cylinder is traversed by a few cell strands that act as a transport system in supplying materials to the surface cells. A problem of growth is expansion of a body within a cuticle that is not shed, for a bud can only grow if the cuticle grows. But the cuticle is extended by the spherical cells at the apex, which move forward along the new wall as they make the cuticle. They do not divide, a condition that is unusual in a growing area, but their numbers may be supplemented from the spindle-shaped cells below them that do divide. There is a high level of metabolic activity in the cell strands in the lumen of the growing bud, as materials are passed along them to the apical cells. Under ordinary conditions of growth a bud elongates about 12 μ/hr, which represents a surface gain of about 3,600 μ^2/hr. The polypide, notorious for its powers of regeneration, increases in size by rapid mitotic divisions in its blastema region and, when the new cystide has reached its appointed dimensions, a transverse wall separates the new zooid from the old, leaving a small communicating pore between them. The new zooid then takes up its position as the apical one and adds to the height, or length, of its branch by a similar process of budding.

The effect of light upon growth has been clearly shown. Zooid buds are positively phototropic and rhizoid buds, negatively so. Under natural conditions, therefore, the zooid buds tend to grow toward the surface of the water, but when a beam of light is shone laterally upon a colony, it has been observed that the wall building cells on the apical zooid shift their position to the area of their maximal illumination and that the new zooid is, therefore, formed in the direction of the light. Determination of the action spectrum of the zooids shows that it corresponds to that of rhodopsin, although there is no visible pigment in the zooids.

3. Special reproductive cells, not gametes

Another means of agametic reproduction is through the segregation of special cells, or groups of cells, that are capable of developing into a complete individual. Some marine and freshwater sponges can reproduce by means of gemmules, which are formed throughout the year by marine species and grow at once into new individuals; they are formed only in the fall by some freshwater sponges and do not continue development until the following spring. Gemmules first appear in the mesenchyme as spherical or ovoid aggregations of amebocytes. In freshwater species the cells are binucleate and loaded with lipids and glycoproteins derived from other amebocytes acting as trephocytes. The mass is enclosed in a firm chitinous capsule, sometimes reinforced with spicules deposited by scleroblasts. Such gemmules can withstand protracted periods of drought and cold, and provide a means by which the sponge can survive the adverse conditions which a freshwater species living in temperate zones must inevitably encounter. During these periods the cells of the gemmule are dormant, usually still embedded in the framework of the dried or frozen body of the individual that produced them. They are also a means of dispersal of the species, for they may be carried considerable distances by one means or another. In the spring, or with water again available, the cells become active; emerge from an opening, the micropyle, in the capsule; and begin the processes of multiplication, histogenesis, and morphogenesis that result in a new sponge.

Sponges also reproduce agametically by formation of reduction bodies and by fragmentation. A reduction body is a mass of cells, representative of all the types in a mature sponge, left after the main mass has deteriorated and degenerated in consequence of unfavorable environmental conditions. Like gemmules, reduction bodies are a means of dispersion as well as of survival for a species, but unlike gemmules their development into a new sponge does not demand histogenesis, since all cell types are already present in the mass and their multiplication and morphogenetic distribution suffice to produce a new individual. And, in some species, tips regularly break off the branches of the sponge body and grow into a new sponge. Fragmentation may also occur through external causes such as strong waves or water currents or rough contact with a foreign body. Fragments so produced can grow into new individuals.

Although reproduction in ctenophores is typically gametic, a process of reproduction by segregation of pieces of tissue from the periphery of its body has been described in the creeping species *Vellicula multiformis*, and in some others. These pieces grow and differentiate into complete and perfect new individuals.

Statoblasts are bodies produced by freshwater ectoprocts, often in great numbers, that have the potentiality of developing into new individuals (see Fig. III–9). A single statoblast, or several, may develop on the funiculus of a zooid; each is composed primarily of coelomic epithelial and epidermal cells that have migrated to the site. The cell mass becomes enclosed in a case made of chitin and consisting of an upper and a lower valve. Statoblasts are usually formed continuously during the summer and fall, and can remain dormant for considerable periods of time. In some species there are hooks around the margin of the valves by which the statoblasts are fastened to some object; in some, the statoblast has air spaces so that it floats and, consequently, may be carried some distance from its parent. Statoblasts are resistant to cold and drought, and so serve to preserve as well as to disseminate the species. In the spring, or when conditions for germination are favorable, the valves of the case separate, and the contained mass of cells becomes organized as a zooid, which grows and again reproduces its kind by budding, by statoblasts, or by means of gametes.

Parthenogenesis is common among colonial flagellates. In *Volvox*, for example, a limited number of cells (gonidia) at the posterior pole of the sphere enlarge as they become filled with reserve nutriment and push into its hollow cavity. There they undergo cleavage and develop as individuals that escape from the parent colony, or parthenogonidium (see Fig. II–3). Since these cells are already haploid in chromosomal constitution, there is no question of meiosis.

Parthenogenesis is also common among rotifers, in which females predominate in any population and which are indeed the only representatives of some species. No males were ever recognized until 1848, and none has yet been found in the family Philodinidae. In the species in which there are males there is marked sex dimorphism, the males always being smaller than the females and often lacking complete digestive systems (see Fig. II–12). Except for the Philodinidae, which always produce amictic eggs that develop parthenogenetically into females, a female rotifer may produce amictic eggs, in which there is no reduction division and which develop into females, or mictic eggs that have undergone complete meiosis and are haploid. These, in a few days, may develop parthenogenetically into males or, if fertilized, store up large amounts of nutriment, become encased in a thick shell, and remain dormant for several months, developing subsequently into females. Some species are monocyclical, others dicyclical, in their reproduction. *Pedalion*, for example, reproduces sexually only in the autumn, and the species winters over as dormant eggs that in the spring hatch as females; *Asplanchna* has two sexual periods, one in the spring and one in the autumn. In laboratory studies on *A. brightwelli* it has been shown that females producing mictic eggs occur in a culture only when algae or other plant material are included in the diet. The active principal in the plants has been identified as alpha tocopherol (vitamin E), which in some way promotes the production by parthenogenesis of females whose eggs will undergo complete meiosis and hatch either as haploid males or, after fertilization, as diploid females. It is possible that the two periods of sexuality in *Asplanchna* may, therefore, be related to the type of food naturally occurring in their habitats and its seasonal content of alpha tocopherol. Biological conditioning of the medium through density of population has been held responsible for the increase in production of mictic females in *Brachinus calyciflorous*. In culture, more of them were produced in a population with 4.0 animals per milliliter than in one with 0.06 animals per milliliter. Possibly the concentration of a pheromone discharged from the animals determines the type of eggs the females will produce.

Natural parthenogenesis is recognized in several gastropod molluscs. In freshwater species such as *Potomopyrgus jenkensi* and *Campelona rufum* males are absent or extremely rare, and reproduction seems exclusively through females that reproduce parthenogenetically.

In some species of branchiopod and ostracod crustaceans, males are also extremely rare or entirely unknown and reproduction is through parthenogenetic females. As in rotifers, a series of generations of females that produce only amictic eggs may be followed by one that produces mictic eggs, which develop without fertilization into males and, after fertilization, into females that may in their turn produce amictic eggs. In *Daphnia pulex*, the common cladoceran of freshwater ponds and pools, females produce mictic eggs when the pool dries up or cold weather sets in. These so-called "winter eggs" contain relatively large amounts of nutritive material in the form of yolk, have thick shells, are resistant to dessication and to chilling, and can wait months before starting embryogenesis. The thin-shelled amictic summer eggs, on the other hand, contain relatively little yolk and develop rapidly into new individuals. A similar kind of life cycle is characteristic of many species of ostracods, although, as with cladocerans, bisexuality is unknown in some species. Even in those in which males may be produced in nature, successive generations of females forming exclusively amictic eggs have been kept in the laboratory for as long as 30 years.

Females of certain species of insects also produce amictic eggs that develop parthenogenetically. In aphids, sawflies, gall midges (cecidomyids) some chironomids, and a few orthopterans, parthenogenesis is usually apomictic, that is, without reduction, for there is only one maturation division, which is equational, or in a few species there may be two divisions, both of which are equational. Automictic parthenogenesis has been found in phasmids; in some homopterans, notably the whiteflies; in bagworms among the Lepidoptera; and in some few Diptera and Hymenoptera. In this type of amictic egg production, meiosis is complete, but the diploid number of chromosomes is restored to the egg capable of parthenogenetic development either by fusion of its reduced nucleus with that of the second polar body or by fusion of two cleavage nuclei.

The life cycles of a number of aphids have been carefully followed. These are often complicated, involving different hosts as well as sequences of parthenogenetic generations and the interpolation of sexual ones, yet aphids have the highest reproductive potential of any insect. Some can, in greenhouses, complete as many as 51 generations in a year, with each adult contributing about 85 young. In *Aphis rumicis*, fertilized eggs that have been laid in the autumn and have wintered over in this condition under the bark of the evergreen shrub *Euonymus* hatch out in the spring into wingless females, which produce apomictic eggs that develop within their bodies. To parthenogenesis is thus added the phenomenon of viviparity in this species. The females so produced ("fundatrices," sing. "fundatrix") may in the course of the summer give rise to a succession of generations of similarly apterous and parthenogenetic females that feed upon *Euonymus*. At some point in the series, winged females develop that fly to another host, often a bean plant, feed there, and in their turn produce amictic eggs that develop into females, and so on for a number of generations. Eventually, winged viviparous females appear in this parthenogenetic line, and also winged males. The females leave the bean plants and migrate to *Euonymus*, where they produce females that are wingless and develop mictic eggs, which are fertilized by sperm from the winged males, are laid in the autumn, and in the spring hatch as new fundatrices that start the cycle over again.

Both temperature and photoperiod influence the kind of individual produced. In aphids the proportion of males in a population can be increased above or reduced below the usual levels by exposure to high or to low temperatures and, in some species, by photoperiod. In spring and summer with long days and warm temperatures, viviparous, parthenogenetic females are produced, but with shorter and cooler days, sexuparae, or females whose offspring are gamete producers, appear. A biological timing system also seems to be involved, or possibly two of them, for it has been shown that the longer the time, irrespective of the number of generations, between a female and her fundatrix, the more susceptible she is to the influence of these external conditions. Since males, as offspring of sexuparae, appear a generation before the females and, therefore, earlier, there may be two chronometric systems involved.

There are some obvious advantages and disadvantages to agametic reproduction. The chief advantage lies in the minimization, or total avoidance, of chance, which plays a large part in gametic reproduction, in the meeting of gametes and of males and females. With agametic reproduction, perpetuation of the line is assured; and all the reproductive cells, or parts derived by fission, fragmentation, or budding, have equal opportunity of developing into new members of the species or the strain. Moreover, all the individuals are capable of reproducing new ones by themselves. In a genus, or species, where obligatory parthenogenesis is the rule and there are no males, or very rare ones, every member is independently reproductive and contributes individually to its propagation and perpetuation. The chief disadvantage lies in the loss of possibilities of variation in descent. Offspring produced agametically must inevitably have the same chromosomal composition and genetic endowment as the individual from which they sprang, with the only possibility of variation coming from mutation, in nature an event of rare occurrence. Thus, in a population reproducing agametically, exclusively or for many generations, the variants upon which natural selection can act are few and far between, as compared with those in a gametically reproducing population.

B. GAMETIC REPRODUCTION

Some species in every invertebrate phylum reproduce gametically, whatever may be their other means of propagation, and in some phyla and classes it seems to be the only means. It involves production of special cells, or gametes, and usually their union, or at least that of their nuclei, into a compound cell, or zygote. In gonochoristic species the gametes that unite in a zygote are produced by different individuals, often recognizably different in their anatomical features or their coloration and distinguished as males and females. In many cases among invertebrates, however, they are produced by the same individual, a hermaphrodite, either simultaneously or successively. Usually, there is some provision to prevent self-fertilization or to diminish the chance of its occurring, although in some sessile species, in which the individuals do not congregate, this very likely does take place. Cross-fertilization, or union of the gametes from two different individuals, is of most general occurrence. It may take place outside the bodies of the gamete-producing individuals, as external fertilization, or inside the female, as internal fertilization, a more precise means of ensuring gametic union.

1. Gametes and gametogenesis

Gametes, like the individuals that produce them, may be dimorphic with striking differences that distinguish a larger, usually immobile macrogamete, or egg, from a smaller usually active and mobile one, the microgamete, or spermatozoon. But gametes may also be alike in all visible features as are, for example, the isogametes of certain protozoans like the flagellate *Chlamydomonas*. Within the nucleus of each gamete is stored information for the progressive development of the characteristics, both morphological and physiological, of the organism at all stages of its life. This information is encoded in the helical molecules of DNA within the chromosomes and is decoded through cytoplasmic mechanisms whose elucidation has been one of the great triumphs of biological research in the past decade.

The majority of individuals arising from gametic reproduction have a double set of chromosomes, one set derived from each of the gamete nuclei that fused to form that of the zygote. Exceptions to this are the haploid individuals, with but a single set of chromosomes and that the maternal set, which arise from parthenogenetic development of unfertilized eggs, or among Protozoa, the individuals that arise from division of a potential gamete that has failed to undergo fusion. This is known to occur in *Actinophrys*, *Polytoma*, and *Chlamydomonas*. Exceptions also are the members of those species or strains in which the established number of chromosomes is some integral multiple, other than 2, of the basic or haploid set. Such polyploidy, as opposed to diploidy, although frequent in plants is uncommon among animals, yet it is known in some invertebrate genera as, for example, *Artemia*, where triploid and tetraploid strains occur in natural populations.

In order to preserve the double set (diploid number) of chromosomes throughout successive generations, there must be, at some point in the life cycle of every diploid individual reproducing gametically, a means of reducing the number of its chromosomes by half, so that its gametes contain but a single set, or the haploid (Greek *haploos*, "single") number. This is equally true of the flagellates, in which haploidy is characteristic of vegetative individuals and diploidy only of the zygotes. Reduction in chromosome number is, in either case, accomplished by meiosis, which occurs in diploid species during gametogenesis and in haploid species after gametic union (amphimixis) and before development of the new individual begins (Fig. XVIII–5).

Recombination in amphimixis, like segregation in meiosis, is in nature wholly a matter of chance, and every possible zygotic combination theoretically could be made. Thus, if both the macro- and microgametes in a hypothetical strain, whose diploid number is 8, represent all 16 possible different haploid combinations of chromosomes, in the absence of any crossing-over there are 256, or 2^8, possible zygotic combinations, none of which will be alike and only one of which will be the same as that of either parent.

Gametic reproduction thus provides a broad base for natural selection. Not all zygotic combinations are equally successful in meeting the conditions of a given environment, and the less successful tend to be eliminated from a population and the more successful to be established and to propagate. Yet certain ones, unsuccessful in one environment, may be eminently successful in another, flourish and so be perpetuated there. Length and breadth of wing, and capacity for sustained flight may, for example, be an asset to flying insects living inland, where they can safely cover long distances in search of food, mates, and habitation, but these attributes may be a liability to those living on a small island, where long flights might carry them out over waters from which they could not return. Hence, variants in wing size and flight capacity would have positive survival value in one situation, negative survival value in another. Gametic reproduction has in itself been selected; for in spite of the fact that it is a slower, more complicated and elaborate way of propagation than agametic, it is a far more common one. The fact that it seems to occur at times of environmental stress in animals whose usual means is agametic suggests that it has other advantages than provision for variation among descendants.

a. AMPHIMIXIS IN PROTOZOANS Among protozoans, amphimixis occurs, in nearly all species, at some stage in the life cycle of an individual or of a population, but in only a few of them are cells properly to be called "gametes" produced. There are various devices for accomplishing amphimixis. In some of the shelled rhizopods, like *Difflugia*, two individuals may come together and one leave its shell and move bodily into that of the other with which it fuses; the resulting zygote encysts. After a time the zygote becomes active and a new individual emerges from the cyst. Some of the foraminifera are dimorphic, in

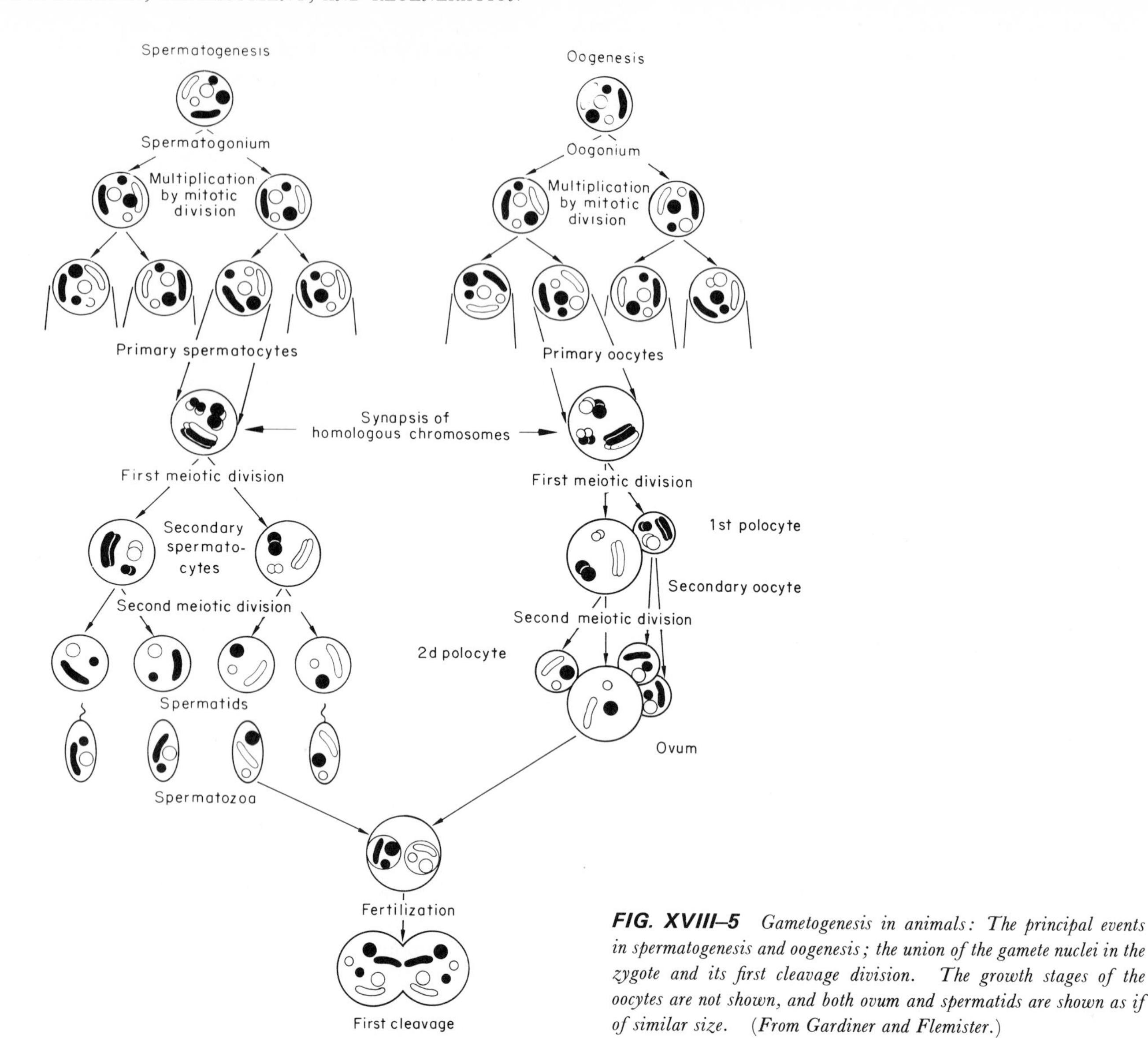

FIG. XVIII–5 *Gametogenesis in animals: The principal events in spermatogenesis and oogenesis; the union of the gamete nuclei in the zygote and its first cleavage division. The growth stages of the oocytes are not shown, and both ovum and spermatids are shown as if of similar size. (From Gardiner and Flemister.)*

that the first chamber, or proloculum, of the shell in some individuals is smaller than it is in others, making possible a distinction between microspheric and megalospheric forms. It has been shown that in some species the chromosomal content of the nuclei is diploid in the microspheric forms, haploid in the megalospheric. The microspheric forms are multinucleate at some stage of their lives and reproduce by multiple fission, the products developing as megalospheric individuals. These are uninucleate gamete producers, either by a rapid series of divisions, after which many small flagellate cells are set free that, as gametes, unite to form diploid zygotes and a new generation of microspheric forms, or by the formation of fertilization cysts. In this latter event, characteristic of some species, megalospheric individuals come together in a common cyst and form gametes that unite within the cyst, from which, after varying periods of time depending largely on external conditions, a new microspheric individual emerges. There is, thus, in such dimorphic species, a regular alternation of generations of microspheric forms, reproducing agametically through multiple fission, and of megalospheric forms, reproducing gametically either by means of free, motile gametes or by the formation of fertilization cysts.

Many protozoans undergo a process of endomixis, or autogamy, in which gametes, or gamete nuclei, arising in a single individual undergo fusion and, hence, through the antecedent meiosis, provide new combinations of genic material. This has perhaps a parallel in self-fertilization in hermaphroditic metazoans. Reproduction in the heliozoan *Actinophrys* illustrates both endomixis and amphimixis. In the case of the former, a single individual first retracts its axopodia and then forms a cyst, within which it divides. The nucleus of each daughter cell passes through two typical meiotic divisions; at each, one of the resulting nuclei is expelled from the cell, but the cell itself does not divide. This leaves one reduced nucleus in each cell; the cells unite and a diploid zygote results, which secretes a thick wall within which it may remain dormant for some time. Ultimately, it divides and the two cells escape from the wall as active individuals. In amphimixis, two different individuals encyst together and each divides, their nuclei undergoing the same series of divisions and expulsion of products as in endomixis. There are, thus, four haploid cells within the cyst. The cells derived from one individual may unite with a cell derived from the other, so that two zygotes are formed; or only two cells may be engaged in amphimixis and only one zygote formed. In this case the remaining cells develop parthenogenetically into new individuals.

The life cycles of sporozoans are, in general, complicated, with reproduction at some stages by multiple fission, through which numbers of motile cells are produced, some of which become gamonts and give rise either to micro- or to macrogametes (see Fig. III–5*b*). These unite and the encysted zygote undergoes a series of mitotic divisions resulting in numbers of new individuals.

In ciliates, amphimixis occurs through a process of conjugation, which has been most intensively studied in *Paramecium*. In this process, two individuals come together gullet to gullet in lateral association and an exchange of haploid nuclei is effected between them. While there is no visible distinction between the conjugating members in *Paramecium*, there is a hereditary physiological difference and a distinction can be made between mating types. Only members of compatible mating types will undergo conjugation, and this compatibility is first evidenced in cultures by agglutination of the free-swimming individuals that become clumped together in a common mass. Pairs of paramecia emerge from this mass and swim around, united by their apposed and sticky cilia and by a small region of fused cytoplasm. Nuclear changes involving both macro- and micronuclei occur in each of the conjugants simultaneously. The macronucleus breaks up into irregular masses and is ultimately resorbed; in multimicronucleate species, each micronucleus divides three times, two of these divisions, and probably the first two, being meiotic ones. In uninucleate species, the single micronucleus follows the same course. In each case, all but two of the micronuclei resulting from these divisions disintegrate. Those remaining are the gamete nuclei, with the reduced number of chromosomes; one of them remains in its cell as the stationary nucleus, homologous perhaps with the nucleus of a macrogamete, while the other, the migratory nucleus and perhaps homologous with that of the motile sperm, travels across the protoplasmic bridge between the two conjugants and fuses with the stationary nucleus of the other. The two migratory nuclei may pass each other as they travel across the bridge, but they do not fuse; fusion takes place only when each has reached the stationary nucleus of the other conjugant. The entire process of pairing, preparation of the nuclei for fusion, and exchange of micronuclei may occupy several hours, after which the two conjugants separate and, swimming freely around, reorganize their duplex nuclear apparatus from the zygote nucleus.

In sessile ciliates, such as *Vorticella*, there is a visible difference between the individuals that conjugate. One remains attached as the macroconjugant, while the other, as the microconjugant, breaks loose from its stalk, acquires a basal tuft of cilia, and swims around until it comes in contact, presumably by chance, with a macroconjugant to which it attaches. Nuclear exchange takes place, but the microconjugant does not detach itself and degenerates *in situ*. Only one member of the pair, therefore, survives, and the loss of identity by the microconjugant is, in a sense, similar to the loss of identity by a microgamete after its union with a macrogamete.

Ciliates also undergo endomixis, reorganizing their nuclear apparatus by processes similar to those preceding conjugation, but occurring in isolated individuals. The macronuclei break down and disintegrate; the micronuclei divide meiotically and two of the products fuse into a single diploid nucleus, from which are reorganized the macro- and micronuclei of the descendants arising from the immediately ensuing binary fissions.

b. ISOGAMY AND ANISOGAMY Gametogenesis in phytoflagellates results either in isogametes, morphologically indistinguishable from each other, or in anisogametes, with varying degrees of differences in size and mobility. Some species of the solitary genus *Chlamydomonas*, for example, produce isogametes by a series of mitotic divisions, for in their vegetative phase the individuals are haploid and meiosis occurs only at the zygote stage. Other species may produce anisogametes, a single cell dividing twice or as many as six times in rapid succession, but fusion is just as apt to occur between gametes of the same size as between those of different sizes. Gametes that fail to unite and form zygotes are capable of developing directly into new individuals. Colonial species usually produce anisogametes. In *Volvox*, for example, at times of gametic reproduction, macrogametes are formed by the separation of potential reproductive cells from the posterior pole of the colony; these enlarge, while floating in the central cavity, by accumulation of stored reserves. These cells are not as large as those that develop parthenogenetically, and are more numerous. Microgametes are formed by successive divisions of reproductive cells, also detached from the wall and free in the central cavity. As a result of these divisions, plates of elongated, biflagellate cells are produced. These cells break away from the plate, move out of the parent colony, and finding a colony with macrogametes in it, penetrate its wall and unite with them. Some species are monoecious (bisexual), both types of gamete being formed in the same colony, and others are dioecious (unisexual), macrogametes and microgametes being formed in different colonies. After union of the gametes, the diploid zygote secretes a thick protective wall about itself; and sometime later, its nucleus undergoes meiosis and then begins a series of divisions that result in construction of a new colony. These juvenile, miniature colonies are released into the medium upon disintegration of the parent.

In multicellular animals, anisogamy is the general rule, the macrogametes—eggs or ova—being derived from primordial germ cells, or gonocytes, through processes of oogenesis, and the microgametes—spermatozoa or sperm—through processes of spermatogenesis. Oogenesis and spermatogenesis are fundamentally similar in that each results in the production of haploid cells that, usually, have no future unless there is union between them. They are dissimilar in that the egg is engaged during much of oogenesis with growth in volume; with replication and organization of the cytoplasm and its particulate inclusions; and to varying degrees, with accumulation of nutrient reserves, collectively known as yolk. Moreover, at each meiotic division the cytoplasm of the oocyte is divided unequally, one small and abortive cell, or polocyte, being cut off at each; the first polocyte divides at the second meiotic division, so that four cells in all are derived from each oocyte. But as a result of these unequal divisions each gives rise to only one functional gamete, the ovum. This is usually spherical or ovoid and may be enclosed in membranes external to its plasma membrane and in a gelatinous capsule or shell whose characters vary according to species. These membranes are designated as primary, secondary, and tertiary according to their source. Primary membranes are formed by the egg itself, secondary membranes by the follicle cells in which, in some species, it is embedded, and tertiary membranes by the cells or glands of the genital tract along which it passes on its way to the genital opening. Meiosis may be completed before or after the egg has been shed and either before or after it has been penetrated by a sperm. The divisions may be arrested before formation of the first polocyte or before formation of the second and, in either case, are not completed until a sperm has entered the egg. Arrest at the beginning of meiosis is characteristic of nematodes, some annelids (*Neanthes* and *Thallesema*, in particular), the echiuroid *Urechis*, some molluscs (the lamellibranchs *Cumingea* and *Barnea*), and crustaceans. Arrest at metaphase of the first meiotic division occurs in the nemertean *Cerebratulus*, in the annelid *Chaetopterus*, and in insects. Meiosis is completed in the eggs of sea urchins and some cnidarians before they can be fertilized, but further divisions do not naturally take place until there has been union with a sperm.

c. DEUTOPLASMOGENESIS The conspicuous growth that distinguishes oogenesis from spermatogenesis occurs soon after the end of the multiplication period of the oogonia and after their differentiation as oocytes. Initially, there is little increase in cytoplasmic mass, but the nucleus undergoes premeiotic changes and increases in volume; at this stage the organelle is known as the germinal vesicle. Following this, there is, in the general course of events, synthesis of new cytoplasm and steady, though possibly slow, increase in volume of the cell as a whole. Later, the volume may increase abruptly in the active stage of yolk formation (deutoplasmogenesis or vitellogenesis) as nutrient material is deposited in the cytoplasm and the oocyte reaches its definitive size. This size varies greatly in different species, depending primarily upon the amount of yolk deposited; and the range in diameter of eggs of different species may be from millimeters to several centimeters. In general, the ova of species in which embryogenesis is rapid contain little yolk, while those of species in which embryogenesis is slow contain large amounts of yolk from which the developing embryos draw their nutriment. The eggs of starfish and sea urchins are, for example, small and, because of the paucity of yolk in them and the diffuse distribution of what there is, are so translucent that the events of sperm penetration, nuclear fusion, cleavage, and other processes can be watched in them. Because of this, they have provided much of the material for study by cell physiologists as well as by embryologists. The eggs of cephalopod molluscs, on the other hand, are large and heavily yolked; the eggs of *Sepia* reach some 2 cm in diameter.

Yolk may be synthesized entirely in the cytoplasm of the oocyte from raw materials provided by the fluid in which it is bathed, or it may be obtained in more or less final form from other sources or cells, with the finishing touches only added to it in the cytoplasm. In the first case, egg formation is distinguished as solitary, in the second, as alimentary. Chemically, yolk is broadly classified as carbohydrate, fatty or proteid, with eggs of different kinds containing more or less of one type than another. The eggs of the nematode *Ascaris*, the sipunculid *Phascolosoma*, the centipede *Geophilus*, and of many insects are, for example, exceptionally rich in glycogen, which is more or less uniformly distributed throughout the cytoplasm. In most cases, carbohydrate yolk is associated with protein in a mucopolysaccharide complex. This kind of yolk seems to be formed diffusely in the cytoplasm, without any specific relation to recognized organelles. Fatty yolk has, on the other hand, been variously described as arising, in different species, in association with the Golgi substance, with the mitochondria, with nucleolar extrusions, or with the yolk nucleus, a conspicuous and often transitory inclusion in the cytoplasm of oocytes of myriapods and spiders and some other invertebrates (Fig. XVIII–6*c*). In its final form, fatty yolk consists of globules of saturated fats and fatty acids, although phospholipids can often be cytochemically demonstrated in considerable quantities in early stages of fatty yolk formation, and to some extent in the finished product. Proteid yolk is usually laid down as a final phase of deutoplasmogenesis and, again, various cytoplasmic elements have been implicated in its formation. A detailed electron-microscopic study in the snail *Planorbis* has shown that it arises there in two ways. Proteid granules may originate from direct transformation of mitochondria, by detachment of the cristae from the outer membrane, their concentric arrangement within it, and the orientation of protein macromolecules in a crystalline pattern in the center of the mitochondrion. Or minute areas of the cytoplasm with their contained mitochondria may be isolated from the main mass by being enclosed in lamellae of the endoplasmic reticulum; protein molecules then accumulate within them

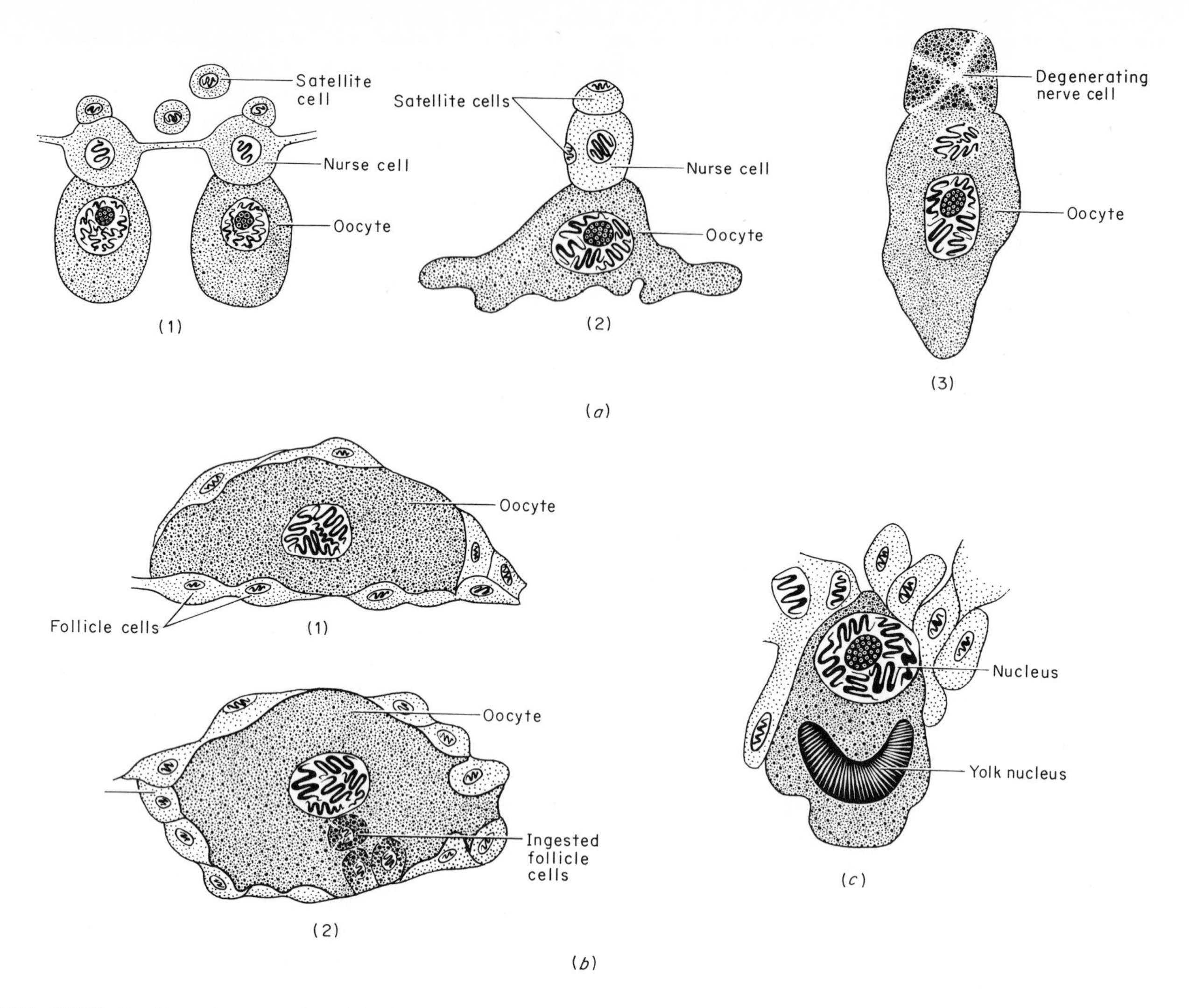

FIG. XVIII–6 *Deutoplasmogenesis.* ***(a)*** *Nutrimentary egg formation in the sponge* Sycon raphanus. (*After Dubosq and Tuzet.*) (*1*) *Two oocytes with nurse cells and satellites.* (*2*) *Larger oocyte with nurse cell and two satellite cells.* (*3*) *Oocyte with degenerating nurse cell, whose nuclear and cytoplasmic material has been taken in by the oocyte.* ***(b)*** *Egg formation in the snail* Helix. (*After E. B. Wilson.*) (*1*) *Young oocyte enclosed in follicle.* (*2*) *Older oocyte after ingestion of some follicle cells.* ***(c)*** *Oocyte of the spider* Pholcus phalangioides, *with yolk nucleus.* (*After Raven.*)

and likewise become oriented in crystalline patterns. There is also during the period of active deutoplasmogenesis a considerable increase in the content of cytoplasmic nucleic acid in the oocyte.

Deutoplasmogenesis is a time of intense synthetic activity on the part of the growing oocyte, and one in which great drains are made upon the nutritive resources and reserves of the parent. But in addition to the biochemical events and the meiotic divisions that are part of oogenesis, other changes take place. Among these are rearrangements of the cytoplasmic components of the oocyte. In eggs that have a large or a moderate amount of yolk, the nucleus is displaced toward one pole where it usually lies in a little pool of clear cytoplasm. This is designated as the animal pole and its antithesis as the vegetal pole. There may also be some redistribution of mitochondria and Golgi substance, and of various cytoplasmic granules exclusive of yolk. Of great importance both to fertilization and to morphogenesis of the future embryo is the organization of the cortex, or cortical layer. The boundaries of the cortical layer and its origin are ill-defined, but that it is highly organized is apparent from studies on its ultrastructure and its physiological importance. In lamellibranch molluscs and in sea urchins, special cortical granules have been demonstrated in the cortex. In lamellibranchs, these can be shown histochemically to consist of mucopolysaccharides; in sea urchins they contain an acid mucopolysaccharide with sulfuric acid residues as well as proteins, and are rich in sulfhydryl groups. Not only is the cortex of great significance throughout oogenesis as the area through which all substances that reach the interior of the oocyte must pass but also it must have properties that enable the egg to preserve its integrity in the new environments to which it is exposed after it leaves the ovary. The cortex seems also to possess a trigger mechanism which is set off in the natural events of fertilization by a sperm, and also by artificial activating agents, and which causes profound changes in some of its properties. Moreover, the cortex determines the polarity and symmetry of the embryo, and the early morphogenetic movements in it.

Most egg cells are surrounded by one or more membranes, although so-called "naked" eggs have been described in some sponges, cnidarians, and lamellibranch molluscs. The vitelline membrane is a primary egg membrane, formed essentially by gelation of the peripheral cytoplasm of the egg and some modification of its plasma membrane. The vitelline membrane may be thin and delicate or of relatively considerable thickness, and is sometimes composed of several layers. In some molluscs, worms, and echinoderms, it shows radial striations, possibly representing microvilli extending from the egg surface, and is thus distinguished as a zona radiata. External to this may be a jelly layer, also a primary membrane, as there is in the eggs of sea urchins. In many cases there is a small passage or micropyle through the primary membranes; in echinoderms, a protoplasmic projection extends through this reaching the surface of the jelly layer. The micropyle in other cases may represent the point of attachment of the egg stalk to the wall of the ovary and the scar left when this was broken and the egg set free.

Secondary egg membranes, formed by the follicle cells, are generally chitinous or horny in nature, and may be of complicated and intricate design. The chorion is such a secondary membrane; it is especially conspicuous in the eggs of cephalopods and insects. In cephalopods the follicle cells deposit tiny granules that enlarge and coalesce, at first into a rather coarse network, then into a finer one, and ultimately into a homogeneous sheet that makes a capsule around the egg. In insects the chorion is usually two-layered, often with an elaborately sculptured surface or with spines and other projections. The follicle cells degenerate after the chorion is formed, one of the last cytoplasmic events of oogenesis. Like the primary membranes, the chorion is usually pierced by a micropyle. Tertiary membranes are, in some species, superimposed upon the primary and secondary membranes and are acquired, sometimes after fertilization, as the egg passes along the genital ducts.

The egg is thus a highly organized system, a cell that is prepared for union with a sperm, whether its meiotic divisions have been completed or not and, after these have been, prepared for nuclear fusion and contribution of its quota of encoded information to the zygote and the embryo that will develop from it; prepared also for cleavage according to its inherent pattern, for control of the early events of embryogenesis, and for the provision of sustenance to the developing embryo as long as it needs it; and, also, protected from the stresses of environmental change by external coverings of various sorts. It is hardly surprising that, in spite of the high potential of gametocyte production natural to most species of invertebrates, only a relatively small proportion of invertebrate gonocytes should attain the degree of organization of a perfect egg, capable of development in an orderly fashion into a new individual with the characters and qualities of its species, yet with its own individuality. Many slips are possible and some may well occur between the initial differentiation of the oogonium and the completion of the finished egg.

d. SPERMIOGENESIS In most species there is little distinction between oogonia and spermatogonia in the early stages of gametogenesis, and in hermaphrodites both oogenesis and spermatogenesis may take place in the same individual, and even in the same gonad, concurrently. This would seem to show that the factors that determine whether a gonocyte shall become an egg or four sperms are subtle ones and that the balance between those that tip the scale toward oogenesis and those that tip the scale toward spermatogenesis is very delicate. It is with the appearance of the germinal vesicle and when the oocyte starts upon its pronounced growth and deutoplasmogenesis and the spermatocyte upon its divisions that morphological distinction can most readily be made. The role that the sperm plays in gametic union is quite different from that of the egg, and for this motility, either intrinsic or acquired, is essential, since it is the sperm that makes its way to the egg. There is a high probability of mischance while so doing, but the very large number of sperms produced, in proportion to eggs, by the individuals of most species gives some assurance that meetings will take place between a good many eggs and sperms. Spermatogenesis follows a course that will produce large numbers of gametes and includes numerous mitotic divisions of spermatogonia as well as the two meiotic divisions that convert each spermatocyte to four spermatids. Transformation of spermatid to spermatozoon (spermiogenesis) occurs after meiosis is completed and entails several events, involving migration of cytoplasmic organelles and of cytoplasm and, in some cases, sloughing off of some of the cytoplasm itself. Conspicuous in these events is development of the acrosome in association with the Golgi substance. The acrosome is a small vesicle that takes up a position at one pole of the nucleus, close to the nuclear membrane. At the opposite pole, a flagellum grows out and the bulk of the cytoplasm flows in this direction, leaving only a thin film around the nucleus and acrosome. The mitochondria become congregated in a mass at the pole of the nucleus opposite the acrosome; in some species they spiral around the proximal region of the axial strand of the flagellum. In consequence of these changes, four regions become clearly defined in the spermatozoa of most species: the head, including the nucleus and acrosome; the neck, containing two centrioles, one of which is the basal body of the flagellum; the middle piece, with the mitochondrial mass; and the tail, consisting essentially of the axial strand of the flagellum and its sheath, the plasma membrane (Fig. XVIII–7*a*). Typically, a spermatozoon is an active, motile cell with few reserves and its active life is usually short unless it meets an egg and fertilizes it. In this event it is perpetuated in the contribution of its chromosomes and their contained genetic information to the zygote and the subsequently developing individual. The sperm of some species are, however, viable for a long time; notable illustrations of this longevity are the sperm of bees and other Hymenoptera in which mating occurs in a single nuptial flight and the queens receive and store sperm to last them the rest of their lives. Ant queens have been reported to have laid fertile eggs for 15 years, and bee queens for 7 years, after a single mating.

Not all spermatogonia develop successfully into sperm. In insects there is always a fairly large proportion of abortive, immobile, or malformed sperm, and it is estimated that only one out of several million ever fulfills its destiny. Prosobranch gastropods regularly produce dimorphic sperm. One kind, designated as eupyrene, is a typical motile cell with an oval head and a long slender flagellum; the other, designated as apyrene, has no nucleus, an undulating membrane attached along its length instead of a flagellum, and a cytoplasm filled with large hexagonal bodies composed of albuminous material (Fig. XVIII–7*c*). Apyrene sperm, because of their large size in comparison with eupyrene sperm, are also known as giant sperm; in *Viviparus* they are over 100 μ long. They are motile, moving like spirochaetes with the undulations of their membranes and, though anucleate, have a function in fertilization in that, in a number of species, they act as carriers for the eupyrene sperm that attach to them by their heads and are transported in this way through the genital tract of the female. It is possible, too, that they act as nurse cells, or trephocytes, contributing the material in their hexagonal bodies to the eupyrene sperm, and that they may liberate some substance that facilitates penetration of these sperm into the eggs. In rotifers, there are also two types of sperm. One is the actual fertilizing agent and the other, small and rod-shaped, somehow assists in the introduction into

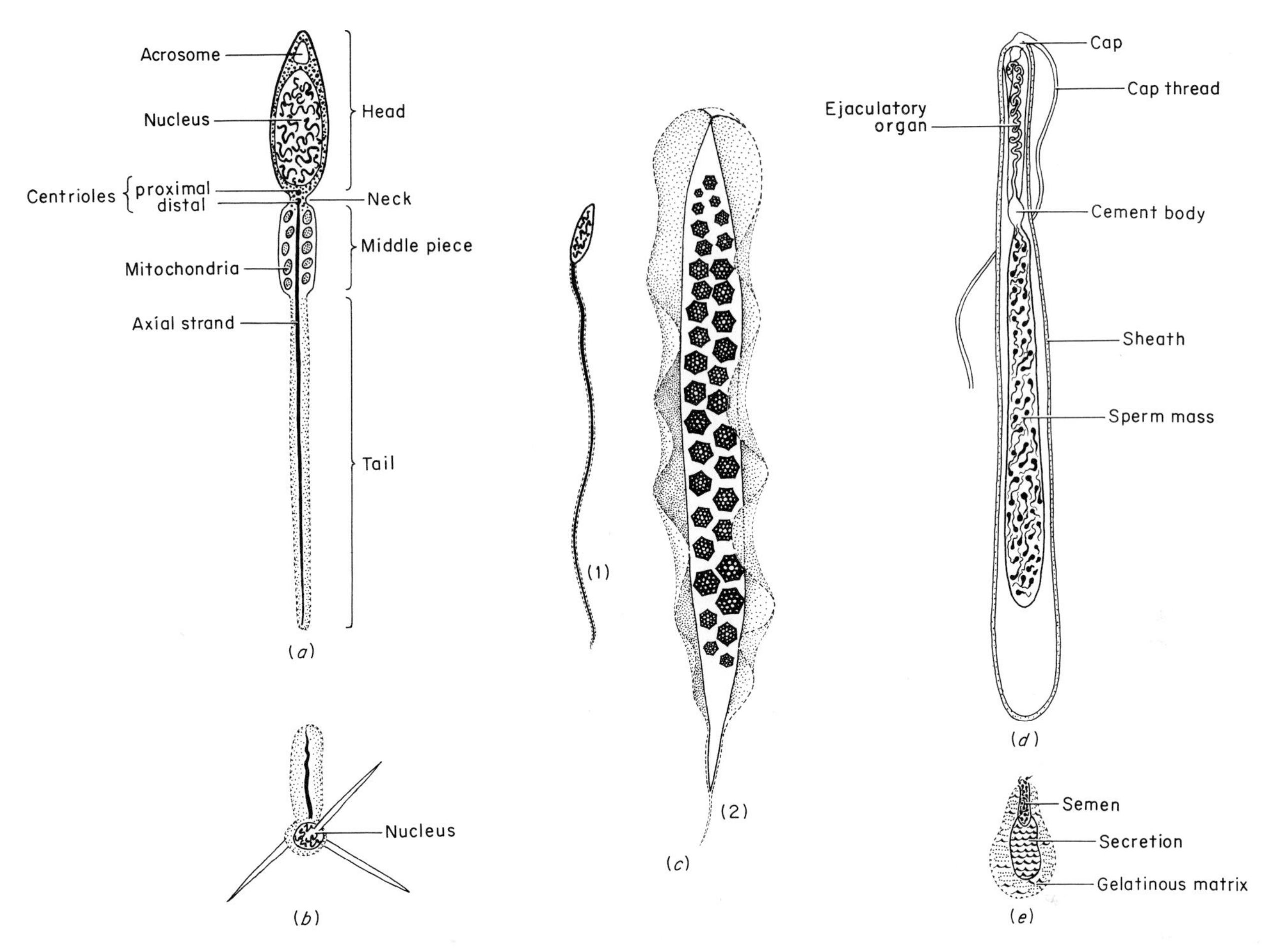

FIG. XVIII–7 *Spermatozoa and spermatophores.* ***(a)*** *"Typical" mature spermatozoon.* ***(b)*** *Nonflagellate spermatozoon of* Homarus. *(After E. B. Wilson.)* ***(c)*** *(1) Oligopyrene and (2) eupyrene sperm of the snail* Strombus bituberculatus. *(After Galstoff.)* ***(d)*** *Spermatophore of* Loligo. *(After Brown.)* ***(e)*** *Spermatophore of* Rhodnius. *(After Davey.)*

the pseudocoele of the female. The sperm of nematodes and of some crustaceans lack a flagellum (Fig. XVIII–7*b*); those of nematodes move through ameboid action while those of the crustaceans are immobile and depend upon a carrier for their transmission to an egg.

2. Diffuse and localized gametogenesis

a. DIFFUSE GAMETOGENESIS Gametes may be formed at various sites in an invertebrate's body, in which case their formation is said to be diffuse, or within specific organs, or gonads, in which case their formation is said to be localized. In sponges, the germ cells are derived from amebocytes whose derivation can, in turn, be traced to choanocytes. Indeed, in *Clathrina*, a primitive species, choanocytes have been observed to transform directly into oocytes, bypassing the amebocyte stage. In other species, however, an amebocyte that is to become a gametocyte usually moves through the choanocyte layer and comes to lie in the lumen of a canal or a flagellate chamber. If it embarks on oogenesis, it divides mitotically at least

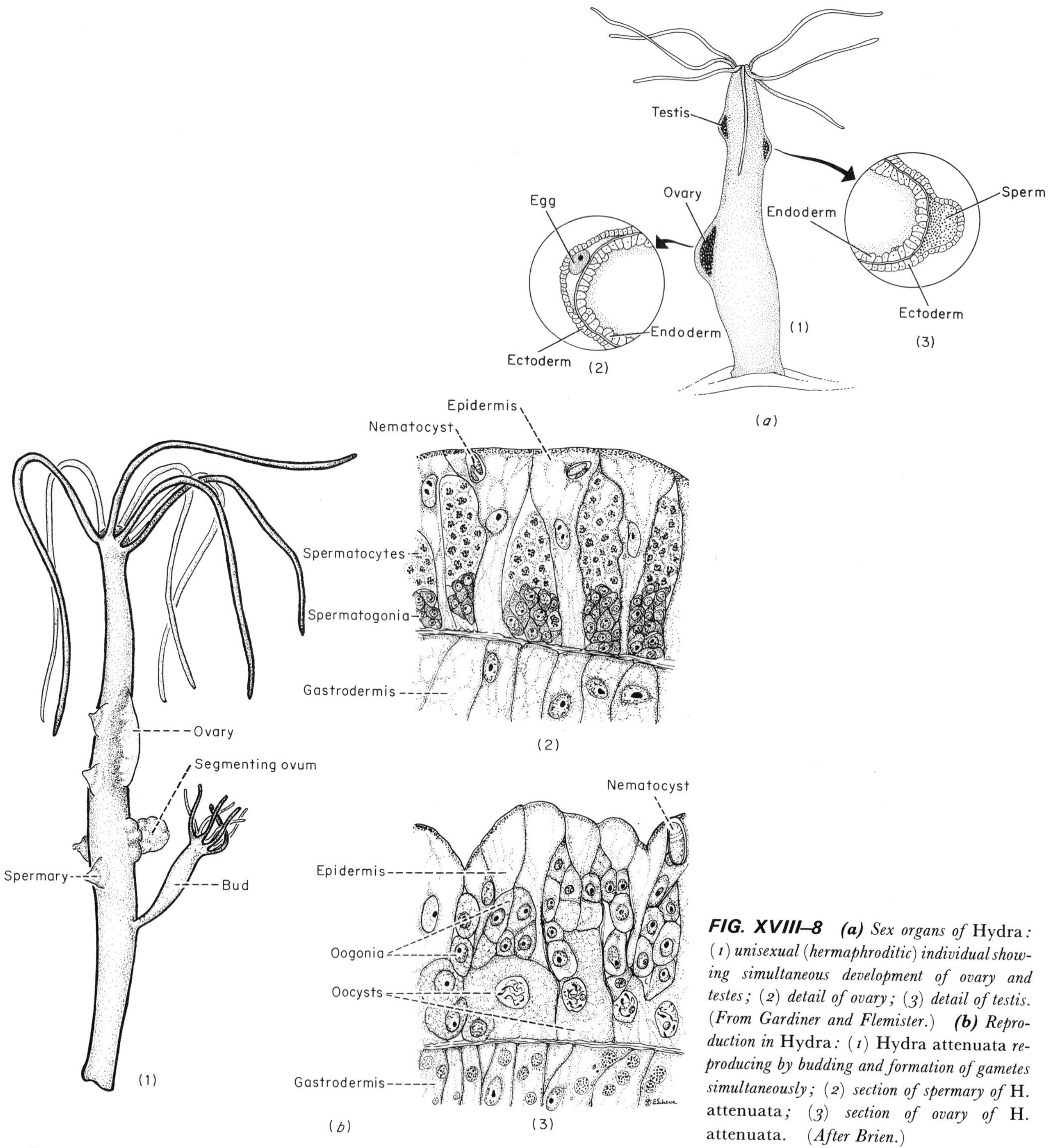

FIG. XVIII–8 ***(a)*** *Sex organs of* Hydra*: (1) unisexual (hermaphroditic) individual showing simultaneous development of ovary and testes; (2) detail of ovary; (3) detail of testis. (From Gardiner and Flemister.)* ***(b)*** *Reproduction in* Hydra*: (1)* Hydra attenuata *reproducing by budding and formation of gametes simultaneously; (2) section of spermary of* H. attenuata*; (3) section of ovary of* H. attenuata. *(After Brien.)*

twice to give rise to four or more secondary oogonia. These develop into oocytes, migrate back to the mesoglea between the choanocyte and the epithelial layers, and begin their growth. The material for this comes largely from other amebocytes or from choanocytes that the oocyte, still actively ameboid, ingests by phagocytosis. Gametocytes that will become sperm follow the same course of migration but divide rapidly many times to give rise to clusters of tiny flagellate cells.

The oocytes of hydrozoan cnidarians are also ameboid. They may, in different species, be derived from cells of the epidermis or the gastrodermis, or from undifferentiated interstitial cells. These migrate to specific positions among the epidermal cells and aggregate there, making a kind of transitory gonad (Fig. XVIII–8). In *Hydra*, usually only one such ovary is formed at a time, generally in the lower portion of the column, and only one ovum develops within it, although in some species, under optimal conditions, several ovaries, testes, and buds may be formed simultaneously. The other potential oocytes are engulfed by the successful oocyte and serve to provide it with material for deutoplasmogenesis. As the oocyte grows it extends from the side of the column, and the epidermis over it becomes greatly stretched and attenuated. It finally breaks, making an opening through which the sperm can pass. The source of the spermatocytes is similar; these aggregate to form temporary spermaries, or testes, somewhat higher on the column than the single ovary. Successive divisions of the cells result in masses of flagellate sperm that escape through rupture of the overlying epidermis (Fig. XVIII–9).

The general course of gametogenesis is similar in medusoid individuals although the position of the temporary gonad is more strictly localized in different places in different species, most commonly either below the radial canals or on the manubrium. In scyphozoans and anthozoans the gametocytes aggregate in the gastrodermis rather than the epidermis and, in scyphozoans, grow and mature either on the floor of the gastric pouches or in the walls of the septa between them (see Fig. X–2). In anthozoans the gastral septa are the sites of the temporary gonads and of gametogenesis; the oocytes at first lie between gastrodermal cells but as they enlarge they sink into the underlying mesoglea; in some species, however, they still retain their connection with the surface epithelium by a slender stalk.

In some acoelous flatworms the gametocytes, arising from ameboid cells of the mesenchyme, migrate through the parenchyma and come to lie in bands extending longitudinally along the body, with no demarcation between them and the surrounding tissue. Acoeles are hermaphroditic; both eggs and sperm are matured in the same individual and, indeed, in the same band of reproductive cells. Gametogenesis in

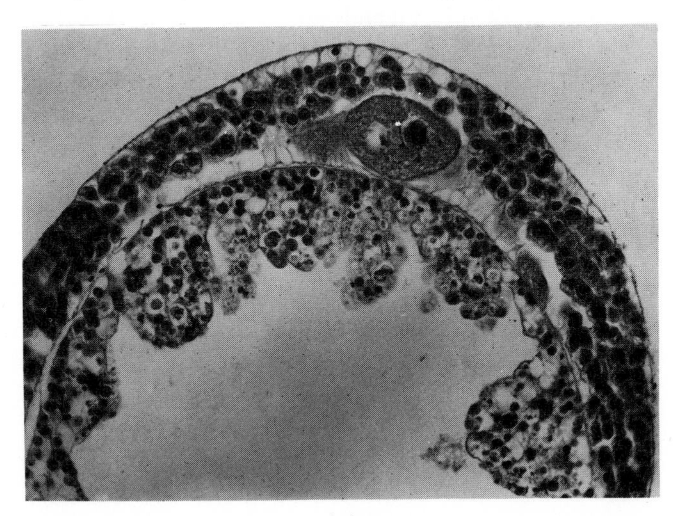

(a)

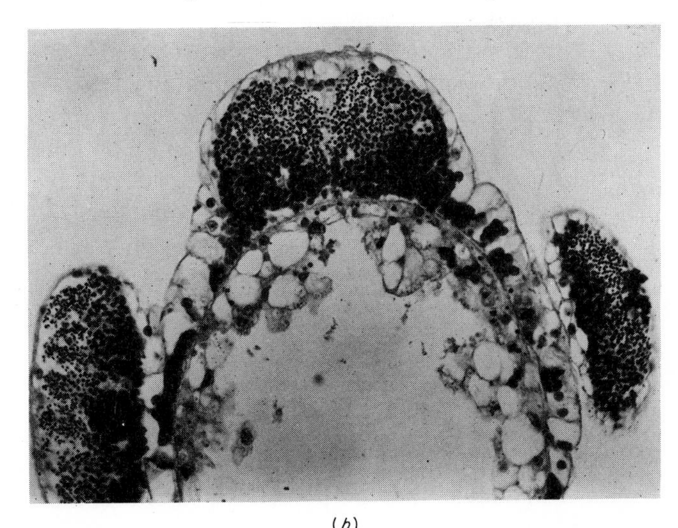

(b)

FIG. XVIII–9 **(a)** *Ovary and* **(b)** *testis of* Hydra. (*General Biological Supply House, Inc., Chicago.*)

some nemerteans is also diffuse, with the germ cells differentiating from parenchyma cells and aggregating in groups. These aggregations are enclosed in a thin membrane so that each is, in a way, a delimited temporary gonad. Numbers of these structures are arranged in linear order on either side of the body; in species with intestinal diverticula they lie between the pouches.

The gametocytes of annelids arise from coelomic epithelium, from different regions in different species (see Fig. IV–37). In some of the polychaetes and the Aelosomatidae of the oligochaetes, gametogenesis is diffuse, with the gametocytes proliferated in any segment of the body, either from the septa or from the epithelium surrounding the gut or blood vessels. In many nereids and scale worms of the family Aphroditidae and in eunicids, gametes may be produced in all but the head segments, but in others in certain segments only. In sabellids and serpulids, only the epithelium of the abdominal segments is gamete producing, and in *Arenicola* only the epithelium of segments V to X, inclusive. Whatever their origin, the gametocytes undergo their growth and maturation while floating freely in the coelomic fluid. The oocytes derive their nutriment from this, increasing in size, in *Arenicola*, for example, from cells with diameters of 15 to 20 μ to cells with diameters of 90 to 150 μ. The spermatocytes undergo their divisions, both mitotic and meiotic, in a similar environment in the males of the species, for the majority of polychaetes are unisexual. Some few sabellids and serpulids, however, are bisexual, or hermaphroditic, but their eggs and sperm are usually produced and developed in different segments. Among the oligochaetes of the family Aelosomatidae, which like other oligochaetes are hermaphroditic, gametes may be proliferated and developed in nearly all the segments.

b. LOCALIZED GAMETOGENESIS Localized gametogenesis is far more common than diffuse. Even in those species in which the gonads regress between reproductive periods to such an extent that they are hardly visible, they are permanent structures and the only sites of gamete development. The gonocytes may, however, originate in other parts of the body and migrate to the gonads before beginning their active multiplication and differentiation as eggs or sperm. This differentiation may take place in separate gonads—in ovaries for egg production and testes for sperm production—or within a single organ, or ovotestis. The gonads may be simple in their construction, indeed little more than centers for gametogenesis, or more or less complex. In ctenophores the construction is very simple, the gonads forming two bands, one an ovary, the other a testis, on the wall of each meridional canal. This is the case also in the earthworm *Lumbricus*, where there are two pairs of testes, or sperm-producing centers, on the posterior faces of the septa between segments IX and X, and X and XI, and a pair of ovaries, or egg-producing centers, on the posterior face of the septum between segments XII and XIII. In other cases, the gametocytes may be enclosed in a capsule whose wall may be a delicate one of peritoneum or a thicker one, reinforced by muscle fibers and supporting tissue. Such gonads may be of various shapes and sizes; in nematodes and crustaceans, they are long and cylindrical. In nematodes, they have a central cavity and may extend the whole length of the body, and even double back in it; in crustaceans, they are solid and at times of reproductive activity may extend far anteriorly and posteriorly. The reproductive organs of female insects consist of a number of such extended tubes, each an ovariole, slender at their distal ends but much distended at their proximal ends as the eggs within them grow (Fig. XVIII–10). Those of males are, likewise, multiple tubules, which like the ovarioles are enclosed in an investing sheath of tissue which gives them a compact external appearance. The gonad may originate, as it does in molluscs, arachnids, and echinoderms, as an evagination of the coelom, forming a hollow organ of various shapes in whose cavity the gametocytes undergo all or part of their development.

In solitary egg formation, the oocyte obtains its nutriment and materials for deutoplasmogenesis from the maternal fluids and tissues in general, while in alimentary, there are special cells, some of them derived from presumptive oocytes, which as nurse cells or follicle cells provide the nutriment directly. No hard and fast distinctions can be made between these types, for oocytes whose development seems to be of the soli-

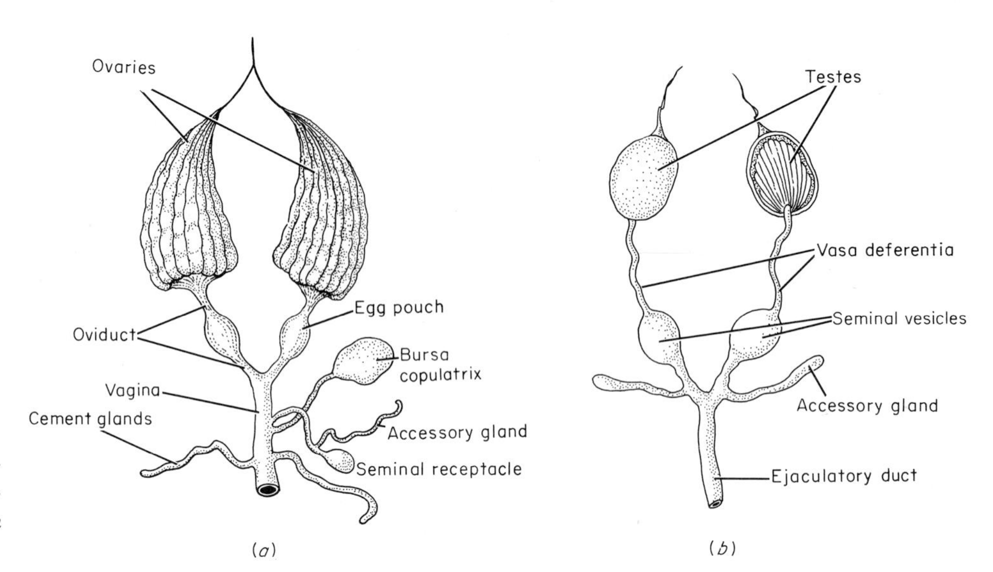

FIG. XVIII–10 *"Typical" genitalia in insects.* ***(a)*** *Genitalia of a female.* ***(b)*** *Genitalia of a male.* *(From Gardiner and Flemister.)*

tary type have often been found to be accompanied by, or closely associated with, one or more other cells which could quite well be designated nurse cells. Moreover, in certain species of gastropod molluscs, some of the presumptive spermatozoa never develop as functional, fertilizing cells but contribute instead to the sustenance of those that do develop as such. This has interesting implications in relation to the basic similarities of oogenesis and spermatogenesis, however pronounced the dimorphism of the final products may be.

Solitary egg formation is illustrated by conditions in lamellibranch molluscs and echinoderms. The gonads of lamellibranchs are hollow sacs with walls of columnar epithelium. In some species they are paired, in some, fused in the midline into a single organ. Cells of the epithelium enlarge to become gametocytes and as they do, project into the lumen of the gonad into which they are finally released. In oogenesis the enlarging gametocyte remains for a time connected with the basement membrane of the epithelium by a stalk along which adjacent cells of the epithelium may extend. These may contribute some nutriment to the growing oocyte, but the main source of it is the ovarial fluid—essentially coelomic fluid—with which the oocyte is bathed. It has been reported that the oocytes of small freshwater clams of the genus *Sphaerium* phagocytose some of the cells along their connecting stalks. The amount of yolk in the eggs of lamellibranchs varies greatly. It is relatively small in species that discharge their eggs directly into the water and in which embryogenesis is rapid and planktonic larval life long, and greater in those that incubate their eggs in brood chambers in which much or all of the larval life is spent.

Most lamellibranchs are unisexual, but some 4 percent of the species whose reproduction has been studied are hermaphroditic, a single individual producing eggs and sperm either simultaneously or consecutively. Among simultaneous hermaphrodites are such marine genera as the scallop *Pecten* and the cockle *Cardium*, and the burrowing clams *Thracia* and *Pandora* and also the freshwater clams *Anodonta*, *Sphaerium*, and *Pisidium*. In *Pecten*, *Cardium*, and the freshwater clams, eggs and sperm are formed in different parts of the same gonad, but in *Thracia* and *Pandora* there is a well-defined ovary and a well-defined testis. The wood-boring *Teredo diegenesis* is also a simultaneous hermaphrodite, although a small proportion of the individuals in a given population may be functional males all their lives. In consecutive hermaphrodites there is a change of sex, in terms of the type of gametes produced, with time. This change may occur once in the life of an individual, or successively once each year, or at even shorter intervals. In a population of *Mercenaria mercenaria*, 98 percent of the clams when they first reach maturity are males, but later in their lives their gonocytes develop into eggs and they become functional females. A small proportion of them, however, continue to produce only sperm and remain functional males. Other consecutive hermaphrodites of this type are the woodborers *Bankia setacea* and *Xylophaga dorsalis*. *Teredo navalis* is a consecutive hermaphrodite, with the potentiality of changing from male to female several times in the course of its life, but mortality in the female phase is so great that the change in most individuals occurs only once. Oysters are particularly interesting in respect to changes in sex. In *Ostrea edulis* the gonads first produce sperm but later eggs and then shift to the production of sperm again. In terms of sex differentiation the sequence of male-female phases is not fixed and may follow a pattern of male-female-female or male-male-female. The warmer the water in which the oysters are living, the shorter the interval between female phases. It is possible, too, that metabolic differences account for the shift in the developmental pathways the gonocytes will follow. All have the potentiality of becoming eggs or sperm, and it has been suggested that when protein metabolism predominates in the parent, oogenesis is promoted and when carbohydrate metabolism is dominant, spermatogenesis occurs. The accumulation of unusable products from one source could induce the changeover from one type of metabolism to the other, so that a rhythm would be established with which the sex change would be in accord.

In the American oyster *Crassostrea virginica* the gonads, on first maturity of the individuals, produce for the most part sperm, the next year eggs, and so on in regular annual alternation. Histological examination of the gonad of a young specimen shows that the outer or cortical layer is composed of oogonia surrounding the spermatogonia. It is probable that external conditions, such as temperature, promote or retard the development of one or the other kind of gametocyte, but in *C. virginica* they do not alter the inherent sequence of sperm and egg production. The fact that the two types of gametes may be produced in the same gonad either simultaneously or consecutively in bivalves such as these cited serves to emphasize how delicate is the balance between the factors that determine whether a gametocyte shall enlarge and grow as an oocyte or divide to give rise to a number of spermatocytes and eventual spermatids.

Echinoderms are typically dioecious, although externally there is little distinction between males and females. The ovaries are sacs with a large lumen bounded by a flat epithelium from which the gametes are differentiated (see Fig. IV–37*b*). As the oocytes enlarge they extend into the lumen and ultimately become detached and complete their growth there. In most species, little yolk is formed and the eggs are small, spherical, and clear; in some species, they contain distinctive pigments. The sperm, likewise, develop from spermatogonia that are set free in the lumen of the testis; and spermatocytes and oocytes alike are bathed in the fluids of their respective gonads, deriving their nutrients primarily from this source. There are very often other cells in the lumen of the gonad, accessory cells or disintegrating gametocytes, that probably contribute some nourishment.

Gonochorism is stable in some species, but labile in others. This latter condition may be manifested in individuals by synchronous or consecutive development of ovaries and testes with their characteristic gametes. Large numbers of hermaphroditic specimens of the gonochoristically labile sea urchin *Strongylocentrotus purpuratus* have, for example, been collected at Palos Verdes, California. In one of the specimens examined, two of the five gonads were testes, and three were ovotestes. In these, the ovarian region was in the aboral half of the gonad and the testicular region in the oral. When eggs and sperm were removed from such an ovotestis and brought together, fertilization followed and normal larvae developed from the zygotes.

Large numbers of hermaphrodites have been found in certain populations of the echinoids *Lytechinus variegatus* and *Tripneustes esculentus*. Labile gonochorism has also been demonstrated in *Strongylocentrotus drobachiensis*, *Arbacia luxula*, *Paracentrotus lividus*, and *Sphaerechinus granularis*, with varying degrees of aberrant gonad development. Gonochorism is also labile in some asteroids, in some ophiuroids, and in holothurians.

Seasonal conditions, geographic location, and temperature are all factors that seem to influence the direction of gametocyte differentiation in echinoderms. The greatest numbers of hermaphrodites have been found in populations of *Lytechinus variegatus* and *Tripneustes esculentus* that have spent the early stages of their lives in exceptionally cold waters. *Asterina gibbosa* is regularly hermaphroditic in the northern part of its range; individuals in the southern part are wholly differentiated as males or females. These observations suggest that low temperatures are the basic factors in causing hermaphrodites to arise in genetically labile gonochoristic species.

Alimentary egg formation is described as follicular, nutrimentary, or a combination of both. In follicular egg formation, the oocyte is completely, or almost completely, surrounded by one or two layers of cells, in many cases also derived from the germinal epithelium; in nutrimentary egg formation, the oocyte is associated with one or two cells, also similar in their origin, designated as nurse cells, which contribute their substance to it or are wholly ingested by it. In combined follicular and nutrimentary oogenesis, nurse cells may be included with the oocyte in the follicle or, as is the case in many insects, may lie outside it and contribute their material to the growing oocyte through cytoplasmic channels or the trophic or nutrient cords.

Even in species with diffuse egg formation, oogenesis may be follicular or alimentary. In some sponges, for example, mesenchyme cells are grouped around the oocyte in a kind of follicular arrangement, and in sea anemones and nemerteans the oocyte, as it grows, may push apart the more densely arranged mesenchyme cells and so come to lie in a kind of pocket of them. In sponges, too, egg formation may be nutrimentary. In *Sycon raphanus*, for example, the choanocytes in the region of the flagellate chambers where oocytes are growing either enlarge as nurse cells or degenerate as so-called "satellite" cells that are phagocytosed by the oocyte (Fig. XVIII–6*a*). In some other species of sponges, the choanocytes are ingested without any preliminary changes of this kind.

Follicular egg formation is typical of aplacophoran, polyplacophoran, gastropod, and cephalopod molluscs (Fig. XVIII–6*b*). In *Chaetopleura apiculata*, for example, the follicle cells enclosing each oocyte are initially flat and squamous in character, but as oogenesis proceeds they send out projections toward the oocyte, while from the free surface of each a cone develops that gives a jagged, spiny appearance to the chorion made by the follicle cells. The eggs of cephalopods contain large amounts of yolk, with the nucleus displaced toward one pole of the ovoid cell. They are developed within a follicle, which is more elaborate than that in other species. From a few cells that initially surround the oocyte there arises, by repeated divisions, a layer of columnar epithelial cells, the membrane granulosa. This is enclosed within the theca, a layer of connective tissue cells; and between the theca and the membrane granulosa is a network of blood vessels, passing from their origin in the genital vessels through a connecting stalk of the theca to the wall of the ovary. Deep folds of the membrane granulosa push into the oocyte, and material secreted from its cells passes into the oocyte and is converted into yolk. When deutoplasmogenesis is almost completed, the membrane secretes a secondary membrane, the chorion, which envelops the egg in a firm capsular coat, thicker at the animal pole than elsewhere. When deutoplasmogenesis is complete the folds of the granulosa disappear, the follicle bursts, and the egg leaves the ovary.

In annelids the follicle is less elaborate, but the developing oocyte is surrounded by a layer of cells derived also from the coelomic epithelium. Some of these cells, in some species, develop into nurse cells as well. In *Ophryotrocha*, where oogenesis is diffuse, the cells lie in pairs, one of which becomes the definitive oocyte and grows at the expense of the other (see Fig. IV–37*a*). In *Myzostoma* there are two such nurse cells, one placed at each pole of the egg, and in *Tomopteris* the potential gametocytes lie in groups of eight, one cell becoming the oocyte and the others being sacrificed to its growth.

Follicular development is characteristic of most arthropods. In spiders the oocyte is invested with a thin layer of epithelium, but in scorpions, mites, and myriapods, the follicle is considerably thicker. In myriapods some of the potential oocytes do not develop into eggs but degenerate after they are quite well along in oogenesis and are engulfed by others that continue their growth and development. Various means of nutrimentary egg formation have been described in branchiopod crustaceans. In *Daphnia*, cells are released from the germinal epithelium of the ovary in groups of four, in linear order. In some instances the third cell in the row becomes the definitive oocyte and the others, nurse cells; or else all four may degenerate and the group develop as a nutrient chamber whose fluid contents are used for the growth of another oocyte. Amictic eggs usually originate from a group of four, but the mictic, winter eggs, use from 2 to 12 such groups, or nutrient chambers, for their more extensive deutoplasmogenesis.

Each ovariole in an insect ovary consists of a terminal filament at its distal end, proximal to which is the germarium where the gametocytes arise by multiplication of cells, and proximal to this the vitellarium, containing oocytes surrounded by their follicles whose cells are of somatic origin (see Fig. IV–37*c*). Deutoplasmogenesis takes place in the vitellarium, the oocytes increasing greatly in size. After deutoplasmogenesis is completed, the oocyte enters the pedicel, the terminal region of the ovariole, from which it later passes into the oviduct.

The importance of the follicle to development of the insect oocyte has been demonstrated in females of cecropia silkworms, in which direct passage of proteins from the hemolymph through the follicle has been shown by both tracer and immunological techniques. In these insects there are intercellular spaces in the follicle surrounding the oocyte and its five nurse cells. When ovaries dissected from females were immersed in hemolymph from pupae that had been injected with leucine carrying H^3 and were later sectioned for autoradiography, it was found that the label could be visualized in the oocytes and in the intercellular spaces of the follicles after 30 to 60 min. The label was found to disappear in the same period of time when the ovaries were immersed in ordinary, unlabeled hemolymph, indicating that a two-way exchange between the hemolymph and ovaries was possible. Presumably, the proteins, having crossed the follicular wall through the intercellular spaces, are, on contact with the oocyte membrane, taken in by pinocytosis, the pinocytotic vesicles later fusing with each other and with yolk spheres in the cytoplasm to form large protein yolk spheres up to 20 μ in diameter that are characteristic of these eggs. Similarly, two antigens, known as antigen 3, a carotenoid protein, and antigen 7, the "female protein," have been identified in the hemolymph of female moths; the quantity of these in the hemolymph diminishes as deutoplasmogenesis begins in the oocytes, in which the antigens can later be identified in concentrations greater than in the hemolymph. The concentration of antigen 7 in an oocyte is, for example, 20 times that in the hemolymph. The follicle also provides a means of entry for microorganisms that are regularly transmitted through the eggs in those insects that depend on them for satisfaction of their nutritive needs. In *Blatta orientalis* it has been shown by electron microscopy that the bacterial symbionts in the mycetocytes (see Fig. IV–39*c*) migrate to the follicle cells and enter them. From these they in some way get into the egg, where they become enclosed in membranes in the peripheral cytoplasm and are transmitted through it to the developing embryo.

Triclad flatworms with localized egg formation are unique in that the yolk is not included in the oocyte or ovum but is contained within the shell surrounding the egg in the form of yolk cells derived from the usually numerous yolk (vitelline) glands whose ducts open into the oviduct; the products of these ducts are discharged into the oviduct and move along with the eggs that have left the ovary. This arrangement may represent a separation, in the course of their evolution, of the germ-cell-producing region of the ovary, or germarium, from the nurse-cell-producing region, or vitellarium. In most turbellarians, several egg cells with many yolk or nurse cells are enclosed within a common capsule; and in trematodes and cestodes, there is usually a single egg with several nurse cells. The nurse cells are consumed during the development of the embryo, as the deutoplasm of other eggs is consumed during embryogenesis.

In rotifers, germarium and vitellarium are contained within the same ovary, but are distinct parts of it. This condition may foreshadow the complete separation characteristic of flatworms. In most female rotifers there is but one of these compound reproductive organs, a small germarium containing some 10 to 20 gonocytes only, setting a limit to the number of eggs a female may produce. The vitellarium usually consists of eight fused cells with large nuclei, whose substance is contributed to the developing oocytes.

3. Gonoducts and discharge of reproductive products

As important to the propagation of a species as successful gametogenesis is the meeting of gametes of different kinds. This entails their liberation from the gonad and, in external insemination, from the body of the parent or, in internal insemination, some means whereby the sperm can reach the egg. In the latter case, there must be a way for the fertilized egg to reach the exterior or place within the body of the mother where it can remain for the appropriate period of embryogenesis. Provision for these needs have been met in various ways by different species of invertebrates; in some by simple means, in others, by means with different degrees of elaborateness.

The simplest way is clearly by direct release of the gametes into the external medium, which of necessity must be an aquatic one through which a flagellate cell can move. This is essentially the means in cnidarians, ctenophores, some acoelous flatworms, some polychaete annelids, and some echinoderms. In cnidarians the sperm are released through an opening in the epithelial covering of the testis and find their way to an egg. The fertilized eggs usually remain *in situ* for at least the early cleavages and then are set free by rupture of the temporary ovary wall. Such simple methods are also found in ctenophores, but as these are hermaphrodites and the two types of gonads not far apart, self-fertilization most likely occurs; cross-fertilization does also, the sperm from one individual reaching the eggs of another, probably by way of the mouth to the canalicular system. In acoelous flatworms, in which gametogenesis is diffuse, the eggs escape through the body wall, but in some species there are sperm ducts in which the sperm are collected and through which they pass. Some acoeles also have an external opening, a gonopore, that opens into a blind sac or tubule functioning as a vagina, for sperm from another individual are introduced into it during copulation. The sperm then travel through the parenchyma to the egg cells, but no acoelous turbellarian has an oviduct and the fertilized eggs always leave through rupture of some area of the external epithelium.

Methods such as these do not necessarily cause destruction of the parent body, but in those polychaetes—some nereids, syllids, and eunicids—in which the gametes escape by dehiscence, the parent is sacrificed and dies after its gametes are liberated for external fertilization. This calamity is avoided by epitokous species, where only the more posterior segments are fertile and are detached from the sterile anterior ones that survive to produce another generation of gametes and offspring. The palolo worm is an example of this procedure.

Otherwise, the gametes leave the body of the parent through ducts that may be developed in association with the gonads specifically as gonoducts or, in some coelomate animals, by coelomoducts that have taken over the function of gamete transmission exclusively or that combine this with other functions. In either case the pattern may be simple or complex, with the development of storage, muscular, and glandular areas along the course of a duct and often special devices for the transmission, reception, and storage of sperm; for the secretion of tertiary membranes around the egg after fertilization; and, in males of some molluscan and many arthropodan species, for the construction of spermatophores or packets in which many sperm are contained for transmission in one way or another to the females. In some species, too, the females brood their young and set their future offspring free only after they have attained various degrees of development; for this there must be special places in which the eggs are held during their development. Even within a phylum, or a single class within a phylum, the genitalia may be simple in some members, complex in others.

Such variation is illustrated by turbellarian flatworms. The simple conditions in acoeles has already been mentioned; in other orders, the genital system is far more complicated and shows, in different species, many modifications of a basic plan. This consists of a variable number (usually many) of testes, which are rounded, capsular bodies within which spermatogenesis takes place. They are scattered along the length of the body among the parenchyma cells. A small duct, the vas efferens, leads from each testis, the vasa efferentia on each side uniting in a common duct, the vas deferens, with the two vasa deferentia uniting in the midline in a muscular sac, the penis. Often the vasa deferentia are expanded, just before their union, into a seminal vesicle in which sperm are held prior to ejaculation. The female system—for almost all flatworms are hermaphrodites—consists of a single ovary, or of two, with an oviduct leading from it. The oviduct is expanded, just beyond the ovary, into a seminal receptacle and, distal to this, the ducts from the vitellaria open into it. The wall of the oviduct is thickened and glandular at its junctions with the vitelline ducts. These glands are the shell glands whose secretion as it is poured over the eggs makes, or helps to make, a firm covering over them. Both penis and oviduct terminate in the genital atrium, a chamber which communicates with the exterior through the genital pore, variously located on the ventral side of the body.

There are certain accessory structures associated with the genital atrium. In rhabdocoeles, two blind sacs, called uteri, extend from it, in which eggs are stored before they are extruded. The bursa copulatrix is a small muscular sac opening also into the genital atrium, in which sperm received at copulation are temporarily stored. These organs are not developed in triclads; instead, there are a stalked gland organ and a muscular gland organ. The invariable absence of eggs from the stalked gland organ makes its presumed homology with the uterus of other flatworms doubtful.

Cross-fertilization is the rule in the hermaphroditic species. In copulation two individuals come together, ventral side to ventral side, and with genital pores apposed, and the penis of each is inserted into the genital atrium of the other, with a mutual exchange of sperm. In rhabdocoeles the sperm pass into the bursa copulatrix, in triclads into the stalked gland organ, before moving up the oviduct into the seminal receptacles where fertilization takes place. The zygote then travels down the oviduct, receiving the contributions of the vitellaria and the secretion from the shell glands.

This basic plan is also found in trematodes, the principal modification being greater development of the shell glands into an ootype, where the eggs are encapsulated. The posterior region of the oviduct distal to the ootype acts as a uterus and has been given that name, but structurally it is the homologue of the uterus neither of rhabdocoeles nor of cestodes.

In cestodes there is, in every proglottid beyond the neck region, a complete reproductive system constructed along the lines of this basic plan. Progressive changes in its development can be seen in the successive proglottids of a single individual. Fertilization may take place between different proglottids of the same worm and the fertilized and finished eggs collect in the uterus, which, in cestodes, is a blind duct that leads out of the ootype. In a mature proglottid, the reproductive organs as such degenerate and the segment is filled with the much enlarged and lobulated uterine sac distended with eggs. Ripe proglottids are detached from the string of proglottids that constitutes the bulk of the body of a cestode and are expelled with the feces of the primary host, the eggs continuing their development when taken in by a secondary host.

In nemerteans, short ducts develop from the gonads only after the gametes are ripe and are absent in immature individuals, as well as between reproductive periods in adult ones. The males and females of *Lineus* and *Amphiporus* secrete common masses of slime in which both eggs and sperm are embedded, and in which they meet. In female rotifers a short oviduct leads from the ovary and vitellarium into the cloaca, through which the eggs are expelled to the outside. In males, in those species where there are such, a ciliated vas deferens runs from the testis to the gonopore, corresponding in position to the cloacal opening of the female, for males with a reduced or totally absent digestive system have no cloaca. Glandular cells, in masses along the vas deferens, discharge a secretion into it, and in most cases the termination of the vas deferens is modified as a penis or intromittant organ which can pierce the body wall of a female and introduce the sperm hypodermically into her pseudocoele.

Male nematodes usually have a long vas deferens leading directly from the testis and expanding distally into a seminal vesicle. Beyond this the tube is modified as a muscular ejaculatory duct into which a number of glands discharge adhesive secretions; it terminates in a cloaca into which also open two pouches containing stylet-like spicules which can be protruded into the lumen of the cloaca and out through its opening. In females an oviduct leads from each of the two ovaries and is expanded distally into a uterus whose upper end often serves as a seminal receptacle.

The two uteri open into the vagina, a muscular tube that leads to the exterior at the gonopore. Most nematodes are dioecious and fertilization is internal. In transmission of sperm, the cloacal spicules of a male are extruded through the cloacal aperture and inserted into the gonopore of a female to hold it open while the ameboid sperm pass through and into the vagina, from which they move to the seminal receptacle, or upper end of the uterus, and inseminate the eggs there. A secretion from the walls of the uterus makes a shell, often sculptured, around the eggs, which are retained in the uterus for some time before their deposition and even, in some species, during the early stages of embryogenesis. Hermaphroditic species, which are found among terrestrial nematodes, are protandric; that is, the male organs and sperm develop before the female organs and eggs. They are regularly self-fertilizing and the sperm are stored in the seminal vesicles until the eggs are ready for fertilization.

In dioecious endoprocts the gonoducts are simple and lead from the gonads to the gonopore, uniting before they reach it into a single tube. In monoecious species, where there are two ovaries and two testes, all four ducts unite before they reach the gonopore. The means of insemination in these invertebrates has not been followed, but it is possible that the eggs and sperm meet while the eggs are still in the ovary; when they are laid they are covered by a comparatively thick membrane whose substance is secreted by the walls of the gonoduct.

In these acoelous and pseudocoelous representatives of the invertebrates the gonoducts, whether transitory or permanent, have developed as independent structures in relation to the gonads, but in coelomate animals the possibility is open for coelomoducts to serve as gonoducts—indeed, this was perhaps their primary function—and even those that have been modified for excretion are in some species also used as gonoducts. In brachiopods the gametes are liberated from the germinal epithelium into the coelom and pass to the outside through the nephridia. In ectoprocts, as in some of the polychaete annelids, there are no gonoducts at all and the way in which a meeting is effected between the gametes that are set free in the coelom is not clearly known. Many species of ectoprocts are hermaphroditic, and possibly self-fertilizing, but nothing is known of the way the sperm reach the eggs in dioecious species, nor in monoecious ones if cross-fertilization occurs. The fertilized eggs escape from the coelom by a special opening near the lophophore and may be shed directly into the surrounding water or held during embryogenesis in special sacs evaginated from the coelom, called embryo sacs, or in special external chambers called ovicells.

Early molluscs had no separate gonoducts and the gonads opened into the pericardium from which the gametes entered the coelomoducts and were shed directly into the sea. This is the condition in *Neopilina* and in some aplacophorans; in scaphopods and in such gastropods as *Patella*, *Haliotis*, and *Fissurella*, the gametes reach the exterior by passing through one kidney, and in primitive lamellibranchs such as *Nucula*, through both. Yet in chitons there are new and separate ducts leading from the gonocoeles and opening to the exterior by special pores. There are two gonoducts, although the gonads themselves have fused in the midline to form a single organ. Eggs and sperm meet in the mantle cavity, into which the sperm are drawn in the inhalant current, and the fertilized eggs, usually held together in strings of mucus secreted by the walls of the gonoduct, leave the mantle with the exhalant current.

New ducts have also developed in other molluscs. In those female gastropods in which there is but a single gonad a duct has been developed along the right lobe of the mantle in close association with the nephridium on that side. This is anatomically the left nephridium, its position having been shifted from the left side to the right in consequence of torsion. The genital duct has thus three components: the gonoduct proper, the nephridium, and the new duct. Special glands have been differentiated along this new (mantle) duct (probably initially an open furrow), whose secretions enclose the eggs in an albuminous coat, often a tough capsule but, in some species, a jelly mass or ribbon. This is usually pale and translucent, but in nudibranchs, especially, the jelly is variously colored, corresponding to the color of the parent. The egg strings of *Coriopsis fulva*, for instance, are yellow; those of *Anisodoris nobilis*, orange; and those of *Rostanga pulchra*, red-orange; in each case not only like the color of the parent but also like that of the sponges upon which the parent habitually feeds. Furthermore, the duct has, in many cases, developed a seminal vesicle, a pouch between the albuminous and capsule glands, for storage of sperm; and a bursa copulatrix for reception of the penis of the male. In dioecious species, common among the prosobranchs, the vas deferens of the male connects with, or opens into, the penis, developed as a fold of the body wall in the head region, just behind the right tentacle. There are also glands associated with the vas deferens with whose secretions the sperm are mixed before reaching the penis. These seem to have the function of prostate glands (Greek *prostates*, "one who stands before") and are often referred to as such.

In the higher prosobranchs, fertilization is always internal and copulation between individuals of different sexes takes place. But in some species this is not possible, either because of their sessile habit, and so the virtual isolation of the sexes, or because of structural peculiarities, such as those found in the Turritelidae and Cerithiosidae. In *Turritella*, a marine genus of ciliary feeders, the entrance to the mantle cavity is guarded by a fringe of pinnate tentacles that prevents the entrance of sediment and also, of course, of a penis. In *Cerithiopsis*, whose visceral mass is long and tightly coiled within a tall spiral shell, the mantle cavity is very narrow and the gonopore of the female too far up the spiral for a penis to reach it. In snails such as these, the males have no penis and the sperm are shed into the water in the vicinity of a female and drawn into her mantle cavity with the inhalant stream. Another adaptation to this kind of situation is the development of apyrene sperm that act as carriers. In *Clathrus*, another snail with a tall and conical shell, many eupyrene sperm, attached to an apyrene one in an arrangement known as a spermatozeugma, are carried by its undulations into the oviduct of a female.

Some prosobranch and all opisthobranch and pulmonate gastropods are hermaphrodites, most of them simultaneous ones. The single gonad produces both eggs and sperm, and the genital duct is appropriately modified for passage of gametes of both kinds. In general, this is accomplished by the effective separation of the duct leading from the gonad into two channels, by means of grooves. From one of these the sperm are directed into the vas deferens, that part of the system derived from the mantle, which leads to the penis. From the other, the eggs are directed into the pallial region of the female tract, which is divided into an oviduct and a vagina. The vagina is connected to a blind pouch, the seminal receptacle, in which the sperm received in copulation are in all probability temporarily stored, for cross-fertilization seems to be the rule.

Gastropods with internal fertilization can deposit zygotes protected by various secondary or tertiary membranes. The eggs may be laid in jelly masses, each in its own capsule enclosed in a common mass or string, or the outer covering may be firm and tough, highly resistant to mechanical injury and impermeable to water and its solutes. Familiar along the sands of the Atlantic and Pacific coasts are the sand collars constructed by moon shells such as *Natica*, *Polinices*, and *Lunatia*. These are coiled or collar-shaped sheets of mucus in which multitudes of sand grains are entangled, which are shaped over the snail's rounded shell as the eggs are laid beneath the surface and pushed up above it as oviposition goes on. Collars have been found that are as much as $3\frac{1}{2}$ in. wide and 6 in. across the coil. Familiar also are the egg strings of the whelk *Busycon*, made up of a series of disk-shaped, parchment-like capsules regularly spaced along a central cord. Each capsule contains 20 or more eggs, and as many as 70 capsules have been counted on a single string. The capsules and their connecting cord are secreted by the nidamental gland, a conspicuous structure on the right mantle wall surrounding the oviduct. The strings are fastened to some object in the water, but may become detached by strong winds or currents and carried up on shore. Oviposition and construction of the fluted egg cases require intensive activity on the part of the snail. It has been observed that a whelk in an aquarium discharges a disk from the genital aperture about every 2 hr; the whole operation for a string as long as 70 disks would then occupy some 5 days. Yet the number of eggs produced by an individual snail, and the speed at which they are laid, seem often to be fantastically high. The sea hare, *Aplysia*, for example, lays eggs in capsules that appear at the rate of 230 per min. A count of the eggs inside the capsules shows that this represents a rate of oviposition of 41,000 eggs per minute. *Aplysia californica* spawns some

480,000,000 eggs soon after copulation, which may all develop into planktonic larvae.

The reproductive system in lamellibranchs is far simpler than in gastropods. The gonoducts are short and nonglandular and in eulamellibranchs have lost connection with the excretory organ; they open separately as individual ducts into the mantle cavity, always near the excretory pore. Copulation does not occur, and the genital products are shed into the mantle cavity, from which sperm are swept out in the exhalant current, and sperm from another individual are swept in in the inhalant current. Gametic union takes place in the mantle cavity, and the zygotes may be carried out at once or held there for periods of their embryogenesis. Most lamellibranchs are extremely prolific; *Mytilus edulis* may spawn as many as 12 million eggs, but the vicissitudes to which they and the larvae that could develop from them are exposed are so great that only a small proportion ever becomes adult mussels.

In cephalopods the sexes are always separate. In each, there is a single gonad which opens into the much reduced coelom. In octopods there are two gonoducts, but in *Nautilus* only the right one is functional; the left one is vestigial. In other members of the order only the left duct persists. The oviduct, except in octopods, is a straight tube into which the eggs that are set free in the coelom are swept through a ciliated funnel at its proximal end and which opens, at its distal end, into the mantle cavity near the anus. An oviducal gland—the only glandular area of the tube—discharges an albuminous secretion around each egg as it passes along the duct; in squids and cuttlefish one or two pairs of nidimentary glands, developed along the ventral wall of the body near the genital pore, release a proteinaceous secretion that envelops the eggs. In octopods, all the egg membranes are secreted by the oviduct. The genital duct in males is considerably more complicated and is highly specialized for the formation of spermatophores, cases in which many spermatozoa are packed and that are transferred to the female. These are constructed in the seminal vesicle, a dilatation of the long and much coiled vas deferens. Each spermatophore is a narrow tube of chitin, about two-thirds of which is filled with a dense mass of sperm (Fig. XVIII–7*d*). The spermatophores of *Loligo* are about 16 mm long, and a single male may produce 12 or more per day, and store up to 400 of them at a time in a large pouch, called Needham's sac, surrounding the seminal vesicle and opening into the mantle cavity. Two mucilaginous glands, situated just before the junction of the vas deferens and the seminal vesicle, secrete the axial material of the spermatophore, that is, the mucus in which the sperm are embedded, the cement body at the top of the sperm mass, and the coiled spiral filament that extends from this to the upper end of the spermatophore. Around the sperm mass are an inner tunic and an outer capsule of chitin separated from each other by a space filled with fluid under some pressure. These coverings are secreted by gland cells in the walls of the seminal vesicle. The partially completed spermatophore is then passed into an evagination of the seminal vesicle, sometimes called the prostate, where the outer coats are hardened, and then into the "finishing gland," where it is capped with chitin. As the spermatophore passes through the termination of the duct this cap is drawn out into a long, slender thread. The finished spermatophore then passes into Needham's sac from which, at the time of mating, it is withdrawn by a specialized arm given the name of hectocotylus. In *Loligo* the modification of an arm as a hectocotylus, which in this case is the fourth on the left side, involves only a change in a few of the suckers to accommodate attachment of the spermatophores, but in other cephalopods the modification is more extensive. In some squids all the suckers may be lost from the hectocotylized arm, which is covered instead with a sticky membrane. In octopods the third arm on the right becomes the hectocotylus with a tip that is spoon-shaped and a groove that runs from this to the base of the arm. The hectocotylus draws a number of spermatophores out of Needham's sac and, after some preliminary courtship maneuvers on the part of both the male and the female, inserts itself into the mantle cavity of the female or, sometimes, deposits its load of spermatophores on a glandular area near the buccal membrane. When the spermatophores are pulled out, their caps are loosened and the spiral filament dislodged. The sperm are released by elastic contraction and osmotic activity of the capsule wall, the cement body emerging first and serving to attach the sperm mass to the wall of the mantle near the opening of the oviduct, or to the buccal membrane. The sperm become active when the wall of their containing capsule is ruptured and they come in contact with the water in the mantle cavity. At this time the eggs are shed and the gametes meet. The male then, in most species, withdraws the inserting arm, but in *Argonauta* and two genera of octopods, *Ocythoe* and *Tremoctopus*, none of which make use of spermatophores, the hectocotylized arm is detachable and is left in the mantle cavity of the female, where it moves about for some time. It was because of this, an observation made early in studies of the mating behavior of cephalopods, that the name of hectocotylus was given to the modified arm, for it was thought to be a parasitic worm living in the mantle cavity and even given a taxonomic status and the name of *Hectocotylus octopodis* by the great French anatomist and paleontologist Georges Cuvier (1769–1832), who included it in an established genus of parasites.

The fertilized eggs are laid in gelatinous masses or clusters, usually attached to some object in the water, or in the case of *Argonauta* in a specially constructed bivalve shell of great delicacy and beauty, to which their popular name of paper nautilus is attributed. This encloses many eggs, for the female argonaut lays several thousand at a time. Other cephalopods are not so prolific, although a 5 lb octopus, caught just after she had laid her eggs, was found to have deposited strings containing some 45,000 of them; and in the egg piles of *Loligo* as many as 50,000 eggs have been counted. But these squids gather together at the mating period and the females lay their eggs in communal masses, so that this number does not represent the fecundity of a single individual.

Polychaete annelids, exclusive of those species in which the gametes are liberated by dehiscence and of the epitokous ones, use their nephridia for the passage of gametes to the exterior. The number of pairs of nephridia employed for this purpose depend upon the number of fertile segments. The gametes are moved through the nephridia by ciliary action, although in some worms bodily contractions and increase in hydrostatic pressure in the coelom are contributory factors. Polychaetes are almost exclusively dioecious, with external fertilization, but copulation has been observed in *Capitella*. In most species the eggs and sperm are shed freely into the water, but in some the eggs are laid in gelatinous masses that are attached to some object or to the substrate. The mucous secretion is believed to be produced by gland cells situated along the nephridial tubes. Some tubiculous species attach their egg masses to the inner walls of their tubes.

All oligochaetes are hermaphroditic and, with the exception of the Aelosomatidae, their gonads are limited to a few specific segments. New ducts have developed in relation to these, and the fertile segments have both gonoducts and nephridia. The vasa deferentia may be straight or coiled, and in some species their terminal portions are glandular. At their proximal ends they are expanded into large ciliated funnels, which are enclosed, as the testes are, in seminal vesicles that develop as pouches from the body wall. The sperm, originating in the testes, mature in the seminal vesicles and are swept into the funnels of the vasa deferentia and so transported down the ducts to the male genital aperture on the ventral side of the body, which usually lies several segments posterior to the testes. The oviducts are in general short, and each opens into the coelom by a ciliated funnel attached to the anterior face of the septum separating the segment in which the ovaries lies from the next posterior one. The duct passes through this septum to the female gonopore on this segment and is expanded, just after it penetrates the septum, into a small, sac-like diverticulum. Cross-fertilization occurs, with two worms coming together in copulation, ventral side to ventral side, and the anterior of one worm directed toward the posterior of the other. In most genera, except for *Lumbricus*, this brings the openings of the vasa deferentia opposite those of the seminal receptacles, individual pouches not connected with other parts of the reproductive system, which open to the exterior. The two worms are bound together by a viscid secretion from the cells of the clitellum, a glandular region of the epidermis including, according to species, a varying number of segments that makes a conspicuous saddle-shaped area on the dorsal, lateral, and, to varying extents, the ventral surfaces of the body. In *Lumbricus terrestris* and related species, there are two longitudinal grooves on the ventral surface of the body along which the sperm travel from the openings of the vasa deferentia to the openings of the seminal receptacles; the girdle made by the clitellum prevents their escape. After mutual exchanges of sperm, and its entrance into the seminal receptacles, the two worms separate, and a few days later each prepares to secrete a cocoon in which the eggs will be laid. This is made of mucus secreted by the epidermal cells of the anterior segments, which make a tube encircling them, and by the clitellum which secretes a chitinous material as well as a milky-looking albuminous substance. The ring thus formed moves forward toward the prostomium of the worm; as it passes the openings of the oviducts, eggs are discharged into it, and further anteriorly, as it passes the openings of the seminal receptacles, the sperm received in copulation. As the ring slips forward over the prostomium its open ends constrict and seal off the contents—the fertilized eggs and the albuminous secretion around them. Earthworms deposit their cocoons in the ground, and aquatic worms either attach them to some vegetation or leave them in the mud or debris on the bottom of their habitats. The embryos develop within the cocoon and emerge from it as minute, young individuals.

The reproductive systems of arthropods are usually single, paired, or multiple sets of tubules from which straight and simple, or more elaborate glandular and diverticulate, gonoducts and accessory passages conduct the gametes to their places of fertilization or to the exterior. The transfer of sperm is in many species effected by means of spermatophores and preceded by courtship maneuvers that often last hours or days. The spermatophore may be inserted in or near the genital aperture of the female or deposited in some place that she must pass over, as is the case, for example, in scorpions. The construction of spermatophores, which takes place in the genital tract of the male, entails a high degree of metabolic activity, especially of protein metabolism, on his part, since mucoproteins constitute a large part of most spermatophores. In some crickets, for example, it is estimated that the protein included in the spermatophores, and hence the protein loss to the individual insect, amounts to 40 percent of the body weight.

All arachnids are dioecious and in male scorpions the testes are long tubules, branched and interconnected, with a vas deferens leading from each main tubule into a genital atrium. Each vas deferens expands into a seminal vesicle before entering the atrium, and this is extended into two diverticula, directed posteriorly on each side, with accessory glands that are connected with the vas deferens. In females, the two ovaries are also long interconnected tubules, with a short oviduct leading from each into the genital atrium, and each is expanded, just before entering into this chamber, into a seminal receptacle. The spermatophore of the male, constructed presumably in the seminal vesicles and equipped with chitinous hooks, is deposited on the ground, after a considerable period of courtship perambulations of male and female together, which is indeed a kind of dance for the female is held by the pedipalps of the male and the two move together. The male guides the female over the spermatophore which becomes attached to her external genitalia by its hooks. The sperm are somehow released to enter her seminal receptacles and fertilize the eggs. In ovoviviparous species, the eggs are retained in the body of the female during the greater part of their development and the young hatch after they are laid; in viviparous species, complete development takes place within the body of the mother and the young emerge from it as tiny, juvenile scorpions.

The pattern of the reproductive system in spiders is similar. In females there are two ovaries, each a tubule lined with epithelial cells some of which give rise to oogonia and others to gland cells that secrete gelatinous material. The two oviducts unite in a tube, commonly called the uterus, which opens into a chitinous invagination of the body wall, the vagina, which, in turn, opens to the exterior at the genital aperture. Connected to the vagina, usually at its junction with the uterus, are two or more blind pouches, the seminal receptacles, and some accessory, sac-like glands. These, in most spiders, have their own openings to the exterior through which the sperm packets of the male are passed, the orifice of the vagina being used for the emission of eggs. The system in the male is simpler, with two tubular testes connected with vasa deferentia that open together at a single genital pore. The means of sperm transfer is peculiar to spiders, most probably related to the size and voracious habits of the females. The male first spins a minute platform, sometimes only a single strand of silk, and deposits a droplet of seminal fluid upon it. He then scoops this up in his pedipalps, whose tarsal segments are highly modified for this purpose and for the transference of sperm to a female. The tarsal segment of a male's pedipalp has been compared to a flexible medicine dropper, for within its chitinous covering is a bulb-like reservoir from which extends a long, tubular ejaculatory duct which terminates in a projection called the embolus. When not in use the apparatus is folded in a groove, the alveolus, on one side of the tarsus, but increase of hemolymph in the tarsus causes the embolus and bulb to be projected from it. With his pedipalps full of semen the male spider is ready to transfer it to the female. Transference usually involves elaborate patterns of behavior directed toward protecting the male from attack by the larger and more predaceous female. Various devices are employed by males of different species—signalling their presence and identity by waving of the palps or the legs in special ways; or in orb spiders, plucking at a strand of the web in a manner that will indicate his presence and not that of an insect to be used as food; or presenting a captured insect rolled up in a ball of silk to engage the female's attention while he is simultaneously slipping his droplet of semen to her. However recognition and the approach of the male to the female is accomplished, in successful mating the semen is inserted into the seminal receptacles of the female and there stored until she lays her eggs, which may be some time later. After inserting the sperm, the male hastily retreats, but not always fast enough to escape being caught and eaten by the female. In oviposition, the female spins a small sheet of silk and deposits her eggs upon it and after this is done, the semen is discharged upon them from her seminal receptacles. A covering sheet of silk is then spun over them and the edges of the two fastened together. This capsule is usually covered by more silk and made into a ball or cocoon that, again according to species, is attached to the web, hidden in the burrow or some suitable crevice, or retained on the spinnerets and carried around by the female until the spiderlings are hatched. The number of eggs laid by a female varies also according to species. The female black widow spider *Lactrodectus mactans*, for example, deposits 25 to 900 eggs in a single cocoon and makes 1 to 9 cocoons per season, a potential of 8,100 offspring at a maximum.

Almost all crustaceans are dioecious, with the exception of barnacles which are hermaphrodites practicing cross-fertilization. The gonads of both sexes are usually long, tubular structures, also usually paired, with gonoducts, basically simple in pattern, that lead from them to the genital openings on one or another pair of appendages or on one of the sternites. In some species the females have seminal receptacles, sometimes arising near the base of the oviduct but more frequently situated as ectodermal invaginations with their own external apertures either on the same segment as those of the gonoducts or on one close to them. In gametic reproduction copulation is the rule, and the sperm is often transferred by means of spermatophores. In some species the sperm is nonflagellate and immobile, requiring such a mechanism. There may be modification of certain appendages as clasping or intromittant organs. In branchiopods, for example, the antennae or the antennules may be modified in the males as clasping organs, by which a female is held close to the male while sperm is introduced, or trunk appendages may be adapted for this purpose. *Dolops ranarum*, an ectoparasite on fishes inhabiting inland waters in Africa, is unique among branchiurans in making use of spermatophores for sperm transference and unique also among crustaceans in developing them in accessory glands independent of the reproductive tract. The male uses his thoracic appendages to insert a spermatophore in the genital aperture of a female, where it is caught and held by hooks near the end of the duct leading to the spermatheca to which the sperm migrate, after which the empty spermatophore is shed.

In sessile barnacles like *Balanus balanoides* the ovaries lie in the basis, or undersurface of the body, or in the mantle, or carapace. In the stalked goose barnacles they lie in the stalk, or peduncle. The two oviducts open on the bases of the first pair of thoracic appendages, or cirri. The testes are situated in the "head," or cephalic region, the two vasa deferentia running from them uniting in a long penis that can be extended from one individual into the space between the carapace and the internal organs (mantle cavity) of another. The barnacles' habit of living close together in large associations makes such a procedure and consequent cross-fertilization possible. In some pedunculate species there are dwarf males that enter the mantle cavities of other individuals and attach there. These, like the males of rotifers, are not only much smaller than other members of the species but are also deficient in certain organs, such as all or part of the digestive tract and all or some of the thoracic appendages which are customarily devoted to feeding.

In female decapods the reproductive system consists usually of two tubular ovaries, connected by a cross band, each with an oviduct that opens to the exterior on the base of one of the thoracic appendages. In all but brachyurans these are simple tubes, but in crabs there is a glandular seminal receptacle and a vagina at the distal end of each. In most other female decapods there is a single seminal receptacle, a chamber derived from the most posterior of the thoracic sternites, with an independent opening to the exterior. The testes of the male are likewise paired, tubular structures with a cross connection between them, but the vasa deferentia, often looped and coiled, are always glandular before their termination at the genital pores on a segment other than that with the female aperture. The presence and presumed significance of the androgenic gland has already been mentioned. The vasa deferentia are concerned with the passage of sperm from the testes and the construction of spermatophores, for insemination in decapods is always effected by means of them. In *Cambarus affinis*, for example, a male turns a female over on her back and grasps all her pereiopods in the chela of one of his fifth pereiopods, making her incapable of movement. He then presses the appendages of his first abdominal segment, which are modified as grooved stylets, against the openings of her seminal receptacles, while sperm entangled in the mucus that makes the spermatophore passes into them from his gonopores. After this he releases her, and some days, or even weeks later, the female cleans her pleopods free of all debris or adherent plants or animals in preparation for oviposition. Some 200 to 400 eggs issue from her gonopores and are directed toward her immaculate pleopods, the sperm leaving the seminal receptacles as the eggs pass its opening and so becoming mixed with the egg masses. The eggs are entangled in mucus secreted by the pleopods and are carried on them during their embryogenesis. After laying her eggs, the female crayfish usually backs into a crevice and while "in berry" gently waves her pleopods back and forth so that the water around the developing eggs is constantly changed. The procedure of copulation and oviposition in lobsters and crabs is fundamentally similar to that in *Cambarus*. In those crayfish and other species that have no seminal receptacles, the spermatophores are deposited by the males on the sternites of the female between the last two pairs of pereiopods. There is greater risk of failure of insemination in this manner of sperm transfer than in the more precise way of introducing it into a seminal receptacle, but it is an effective one, for the spermatophore acts as a container for the sperm until the eggs are shed into the water and the sperm set free to swim to them.

Myriapods are also dioecious. In female millipedes and centipedes there is but one ovary. In millipedes the oviduct enlarges into a uterus, which divides into two vulvae, pouches that can be pushed out through the genital opening. A slender seminal receptacle arises from each vulva. There are no such comparable structures in centipedes, but on each side of the female genital aperture are two very small appendages called gonopods. Male millipedes have a pair of testes, centipedes anywhere from one to two dozen. In many millipedes, each vas deferens terminates in a penis; in others the two unite into a single penis. In centipedes with more than one testis there is a pair of vasa deferentia with which they are all connected, and a single penis, opening between a pair of gonopods. In millipedes, certain of the legs, the number and actual pair or pairs varying with the species, serve as gonopods. These the male fills with sperm and in copulation brings the female and himself into such relative positions that his gonopods are opposite her vulvae into which the sperm are transferred. The eggs are inseminated as they are laid; from 10 to 300 may be deposited at a time, depending on the species, directly in the ground or in capsules constructed from excrement or from regurgitated material. Some millipedes build a nest for their eggs. Little is known of how sperm is transferred in centipedes; the eggs may be laid singly or in masses of 15 to 35 in the soil or in cavities in decayed wood.

Insects show many variations of a basic plan of reproductive system (Fig. XVIII–10). The vasa deferentia of males are dilated into seminal vesicles, and associated with the vasa deferentia are numerous accessory glands of mesodermal origin, opening into them by means of ducts, as well as glands of ectodermal origin. The vasa deferentia terminate in an ejaculatory duct and an intromittant organ. In females the two ovaries connect with their respective oviducts that unite in a common duct which distally becomes the vagina and serves, in most insects, as a bursa copulatrix. The number of ovarioles in an ovary varies according to species; in some Diptera there is but a single one, in Lepidoptera there are typically four, but in some Isoptera there may be more than 2,000. In Lepidoptera the bursa copulatrix has become almost completely separated from the rest of the system; in some of them, completely. The seminal receptacle, or spermatheca, is a diverticulum of the common oviduct or of the bursa; the sperm received in copulation are stored in it. There are, in addition, various accessory glands contributing to construction of the egg membranes and opening into the vagina; one opens into the passage connecting the seminal receptacle with the vagina. In both males and females the posterior sclerites are modified as external genitalia, in the one sex for transmission of the sperm and in the other for its reception and for oviposition. It is the ovipositor that has in some species developed into a stinging apparatus.

In some insects—regularly in those of more primitive orders and sporadically in those in the more highly evolved ones—spermatophores formed from secretions of the accessory glands, in the intromittant organ, are used for the transference of sperm (Fig. XVIII–7*e*). In others, free spermatozoa are directly introduced into the reproductive tract of the female, either close to the spermatheca or actually into it. The males of these species have penes that are somewhat longer than those of spermatophore producers. The insemination of *Rhodnius* is illustrative of the first method and that of the milkweed bug *Oncopeltus* of the second. In *Rhodnius*, spermatozoa introduced into the bursa are drawn up into the spermatheca by active contractions of the oviducts, believed to be initiated by secretions contained within the spermatophore that stimulate the peripheral nervous system and the oviducal nerves. Males of *Oncopeltus*, by means of very long penes, insert their semen directly into the spermathecae of the females where the spermatozoa are immediately in position to fertilize the eggs as they pass along the oviducts.

The question of the way in which spermatozoa are released from a spermatophore is not entirely resolved. In Lepidopterans there are spines on the inner cuticular surface of the bursa which may well be responsible for tearing the walls of the spermatophores placed there. In other insects the spermatophore may be partially or wholly digested away by proteolytic enzymes secreted by the glandular walls of the bursa. In *Rhodnius* the spermatophore, after the spermatozoa have been set free, acts as a plug to close the distal parts of the genital ducts while the sperm is being propelled to the spermatheca. Among phasmids and mantids the bulbous part of the structurally rather elaborate spermatophore remains outside the genital opening of the female while the long narrow neck is inserted in her ducts. Female mantids have been observed to eat the spermatophore even before the sperm have been discharged from it and ants to carry off spermatophores from female walkingsticks. In both cases insemination and consequent fertilization of the eggs are prevented, but the predator obtains some valuable nutrients. Some protection against the loss of a spermatophore is shown by insects such as the tree cricket Oecanthus, for the females lick secretions from the metathoraces of the males for the first 15 min after a similar partial insertion of the spermatophore, occupying them and presumably satisfying their hunger for a time long enough for the sperm to be released and to travel to the spermathecae. The empty spermatophore is then expelled from her body, as seems also to be the case in a number of other species of insects. In some, however, the spermatophores seem to be completely digested in the bursa, thus providing the female with a source of valuable protein.

Movement of spermatozoa from bursa to spermatheca may be accomplished in other insects, as it has been shown to be in *Rhodnius*, by contractions of the oviducts induced in some way or other by the presence of the spermatophore. In other insects, especially in Lepidoptera, the spermatozoa seem to travel up the oviducts of their own accord, through active movements of their flagella. What directs them to the appropriate place in the female reproductive system remains an open question; chemotaxis and rheotaxis are both possibilities. There is a question also as to the source of energy which the minute spermatozoa, with few reserves, must expend in their passage up the oviducts and in the activity they display when stored in the spermathecae of the females. This is particularly cogent in the case of long-lived spermatozoa such as those of hymenopterans, but it is raised also by the activity of the spermatozoa of *Rhodnius*, which may continue for as long as 1 month after copulation. In some insects, an exogenous source may be supplied by the semen itself, glands of the male reproductive system discharging a fluid secretion in which the sperm move and from which they may well derive some nutriment. In most insects the walls of the spermathecae are glandular and may secrete materials that can be used by the spermatozoa as an energy substrate, but this is still subject to proof.

In some heteropterans insemination is hemocoelic, that is, the spermatozoa are introduced into the hemocoele of a female, from which some of them find their way to her gonads, while others may be phagocytosed and so provide her with some nutriment. In the damselbug *Alleorhynchus plebejus*, for example, sperm introduced into the bursa in an orthodox fashion migrate between the muscle fibers of the oviducts to the ovaries where fertilization takes place. Some are displaced during this migration into the hemocoele where they are phagocytosed. In some other nabids the sperm are introduced into the hemocoele from the bursa by the intromittant organ of the male which pierces its wall. Those that are not phagocytosed travel to the ovary. In some of the Cimicidae, or bedbugs, the spermatozoa may be introduced directly into the hemocoele through the body wall of the female by a kind of hypodermic impregnation such as that practiced by some rotifers. Part of the external genitalia of the male bedbug is modified as a stylet which is inserted through the body wall of the female at any point, but at different ones at each copulation. Or, as in *Cimex lectularius*, there may be a specialized region, or spermalege, known also as the Organ of Berlese, where the sperm is always introduced (Fig. XVIII–11). The spermalege is a delimited mass of cells in the ventral region of the fifth segment, beneath which the cuticle of the sternite is indented in a kind of pocket through which the stylet of the male is inserted. Some of the spermatozoa introduced at each copulation are phagocytosed by the cells of the spermalege; others collect near the periphery of the cell mass and the wall separating it from the hemocoele breaks down, allowing the sperm to pass into the hemolymph. They are carried through this to the reproductive tract and enter the sperm reservoirs, pouches of the oviducts that act as spermathecae but are not structurally homologous with the spermathecae of other insects, presumably by penetrating their walls. They remain there until the female has her next blood meal—which may represent a considerable period of waiting—after which they migrate through the walls of the oviducts to the ovaries and fertilize the eggs there. In

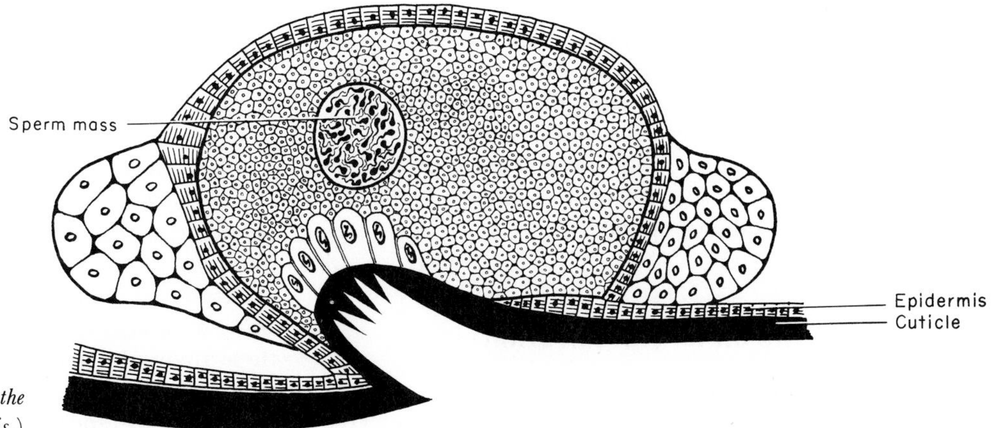

FIG. XVIII–11 *Section through the spermalege of* Cimex. (*After Davis.*)

some other species of bedbugs, the spermalege forms a tissue bridge between the main cell mass and the ovary, through which the spermatozoa travel. In spite of this complicated means of insemination, and the inevitable loss of many sperm through phagocytosis, a female bedbug can lay 200 to 300 eggs within 2 or 3 months, and when conditions are favorable for her and for the development of her offspring can give rise to three to four generations within a year. The fecundity of females of other insects is various; some may lay comparatively few eggs, others several hundred or several thousand. The extremely high reproductive potential of aphids has already been noted.

The reproductive systems of echinoderms are structurally simple. Fundamentally, they conform to the pentamerous plan characteristic of adult echinoderms, although some show loss of some of the five and so reduction in number. Holothurians are unique in the phylum in having but a single gonad. The gonoducts are short, often very small and inconspicuous, possibly transitory, passages through which the gametes pass to the exterior or to special areas where fertilization occurs and the embryonated eggs are brooded for shorter or longer periods of time. In ophiuroids the gametes are discharged into the genital bursae by rupture of the wall beneath them. External fertilization is general among echinoderms, the gametes being shed into the seawater and meeting there. The gonads arise from a circular strand of tissue called the genital rachis and consist of a germinal epithelium enclosed in peritoneum; between these layers are muscle cells. Between reproductive periods the gonads may regress to such a degree that they are almost undetectable, but become bigger as gametogenesis progresses until they occupy a large part of the ambulacral areas. The gonads of a ripe sea urchin, for example, may represent two-thirds of its total tissue weight. Most echinoderms are extremely prolific; a single asteroid can spawn as many as 2½ million eggs in a single reproductive period.

4. Gametic union (fertilization)

Gametic union involves meeting of the gametes, attachment of the microgamete to the macrogamete, the penetration into and activation of the macrogamete by the microgamete, and, finally, syngamy, or fusion of the two gamete nuclei. These events are generally subsumed under the general term "fertilization"; the crucial event is syngamy, conferring upon the zygote full capacity for development and inheritance according to its kind.

In biparental inheritance, propinquity of males and females is an important factor for fertilization. There are various devices that ensure this and make fertilization more probable. In aquatic species that shed their gametes into the water and where fertilization is external, males and females usually reach maturity at the same time, responding synchronously to external and internal factors that induce gametogenesis and simultaneous ripening of their reproductive cells. Combination of the external factors—primarily, temperature; photoperiod; and, sometimes, presence of members of the opposite sex—and of the internal factors—neural activity or hormonal secretion, or both—bring about locomotor movements directed toward the search for mates and, in many cases, instigation of signalling devices of one kind or another. For species that tend to live in close associations or in colonies in which the members are independent, such movement is not essential or even important, but it is for those that live more isolated lives. Many aquatic species not ordinarily living in shallow or quiet waters often seek them at times of reproductive activity and shed their gametes into relatively undisturbed surroundings. These migrations and those that bring deep water and burrowing animals to the surface, as in the swarming of marine worms; the emergence of earthworms from their burrows on damp, warm spring nights, when with their most posterior segments still anchored at the mouth of the burrow, they swing the rest of their bodies over the moist earth until they find another worm to mate with; and the chasing of females by males and courtship maneuvers on the part of both sexes that occur in many species are all responses to information from internal sources that the gametes are ready for union. Signalling devices that tend to bring the sexes together are also such responses; among these are production of light by luminescence, production of sound by a variety of devices; and emission of the chemical substances that act as sex attractants. In some species the gametes are not shed, nor preparations for reproduction completed, until both sexes are together. Males of the sea anemone *Sagartia*, the mollusc *Patella*, and the sea urchin *Lytechinus* shed their sperm naturally only in the presence of females. Spawning by one male sea urchin induces spawning in others and in the females around him, through some as yet undisclosed means of communication. Preliminaries to copulation and the actual act itself are essential to initiation of spermatophore development, and the early stages of its formation, in *Locusta migratoria migratoriodes*. This behavior has been shown experimentally to be controlled by the nervous system: initiation of courtship requires the cephalic ganglia and the receptors of antennae, palpi, and eyes; approach of male to female and his mounting of her require receptors and effectors of head and thorax, as well as the central nervous system; copulation is controlled by responses of the abdominal ganglia to stimuli received from anterior centers, as are the initiation and early stages of spermatophore formation, although later stages appear to be independent of the anterior centers and to depend upon the integrity of the abdominal ganglia alone.

Whether gametic union is preceded by precise patterns of behavior and whether insemination is internal or external, a central problem of the phenomenon is how sperm and eggs are finally brought together. Both chemotaxis and random movement of the sperm have been held accountable. Although chemotaxis is known to operate in bringing the macrogametes of various ferns, and of higher plants as well, to the macrogametes, there is little definitive evidence of its activity in this respect among animals. However, dark field cinephotomicrography has shown that sperm of two species of colonial hydroids (*Campanularia flexuosa* and *C. calceolifera*) become very active when in the vicinity of mature female gonangia of their own species and congregate around the apertures at the distal end of each as if attracted there. Tissue around the aperture produces a substance responsible for these directed movements of the sperm. Extracts of this tissue taken from gonangia of *C. calceolifera*, made in seawater or in alcohol, have the same effect upon sperm as the intact organs, and sperm will even aggregate around the end of a pipette that has injected the extract into a suspension of sperm. How widespread such chemotaxis may be among animals is not known and at present a kinetic theory, based on movement of the sperm and chance collisions with the eggs, seems more acceptable than a chemotactic one. Internal fertilization increases the chances of their meeting; indeed, in some species, especially of insects, special devices almost ensure, but do not guarantee, this. In many insects the egg is so precisely oriented in the oviduct that its micropyle is directly opposite the opening of the spermatheca, and in its passage along the duct, it pauses there long enough to receive a spermatozoon. What causes release of the spermatozoa at the precise time is still an unsolved problem. Even in species with external fertilization, the great numbers of sperm that are discharged into the water or over the eggs that have been laid make the chances of collision highly probable. The greater the number of sperm, within limits, in relation to the number of eggs, the greater the chances of collisions between them. There is, however, an optimum concentration at which these chances are maximal, and beyond this they become less, for the sperm may impede each other's movements and collide with each other instead of with an egg. Nevertheless, even with the most efficient means of bringing the gametes together, many sperm fail to meet or penetrate an egg, and many receptive eggs remain unfertilized.

For obvious reasons the actual course of fertilization has been most intensively studied in species that shed their gametes into the water, from which they can be collected and observed. Much of the material for these studies has been provided by echinoderms, polychaetes, and molluscs, for their gametes can quite easily be obtained in large numbers during the breeding season when the animals can be induced to spawn under laboratory conditions and the union of their gametes can be followed under suitable magnifications. Moreover, at any stage in the process samples can be removed for fixation and preparation for study by light and electron microscopy. Sea urchins and sand dollars have provided especially favorable material, for their eggs are relatively free of yolk and so, translucent; moreover, their meiotic divisions are completed before they are shed and, unlike the conditions in starfish and most other invertebrates, the events of fertilization are not delayed or complicated by further divisions of the egg nucleus. The fact that the bulk of present information about a most critical biological event has been obtained from such animals as these emphasizes the importance of invertebrate material in the study of biological problems common to all organisms.

Grossly, fertilization of an echinoderm egg can be followed as a series of happenings that begin with the increased activity of the sperm in the vicinity of the egg, their temporary aggregation into small clumps, contact of one of them with the surface of the egg, followed within seconds by a reaction initiated at the point of contact that spreads over the entire periphery of the egg and results in the elevation of a membrane around it. This is known as the fertilization membrane, separated from the surface of the enclosed egg by the perivitelline space, filled with fluid (Fig. XVIII–12*a* and *b*). In a matter of minutes after this, the sperm can be seen inside the egg with astral radiations emanating from its proximal centriole; and its rotation, so that the middle piece is directed inward, and its migration along a radial path toward the center of the egg can be followed. Pictures taken by time-lapse photography show that the egg nucleus also migrates toward the sperm nucleus. There are also conspicuous streaming movements in the cytoplasm, evidenced by the movement of pigment granules or other visible particles to what seems to be, because of the regularity of their final disposition in different eggs, predestined areas. Once the fertilization membrane has been raised, no other sperm seem able to enter the egg, although many may be in its vicinity and even immobilized on its periphery. When egg and sperm nuclei come close together they both swell and appear to be in contact and even to merge, after which they move together to the center of the egg. The sperm aster gives rise to the achromatic figure of the first cleavage division, the chromosomes of both nuclei align themselves on the spindle, and their replicates are separated in the first mitotic division of the developing embryo.

Such a description will serve as a model for fertilization in general, although there are variations of this pattern in those species in which the events have been followed, and but little is known of that in invertebrate species where fertilization is internal. There are, in all cases, many common problems, among which are:

1. The extreme specificity of the process, both in regard to tissues and to individuals. Sperm do not fertilize cells other than eggs, and ordinarily only eggs of their own species.
2. The mechanism underlying attachment of the sperm to the egg.
3. The mechanism underlying elevation of the fertilization membrane and the significance of this membrane.
4. The mechanism of sperm penetration.
5. The causes of the cytoplasmic movements that follow elevation of the fertilization membrane, and the relations of the consequent displacement of cytoplasmic materials to subsequent events in the egg.
6. The prevention of the entrance of more than one sperm, or the block to polyspermy, in eggs that are naturally monospermic, and the failure of this block in eggs in which dispermy occasionally occurs or that are naturally polyspermic.

Some of these problems have been at least partially resolved by close observation of the natural events and by experimental alterations of them, and by chemical and physical analyses of the components of egg and sperm cells that might be involved.

FIG. XVIII–12 *Fertilization, cleavage, and development of a sea urchin. (From Gardiner and Flemister.)* ***(a)*** *Ripe egg enclosed in its jelly coat. The two polocytes are visible below the jelly layer; the egg nucleus lies in the cytoplasm at the animal pole and the pigment granules are collected in a band at the vegetal pole.* ***(b)*** *Gametic union. (1) Penetration of spermatozoa through the jelly coat. The sperm first reaching the egg surface is shown in solid black. The arrows indicate spread of the fertilization impulse from the point of this contact around the periphery of the egg. (2) Formation of the reception (fertilization) cone. In this and subsequent drawings the jelly coat and unsuccessful spermatozoa are omitted. (3) Elevation of the fertilization membrane. The area between this and the egg cortex is the fluid-filled perivitelline space. The arrow indicates the path of the sperm toward the egg nucleus.* ***(c)*** *Cleavage stages,* *showing in 4 and 5 the localization of the pigment in the blastomeres in the vegetal hemisphere.* ***(d)*** *Blastula.* (1) *Exterior of a young blastula, showing localization of pigment in the cells between the equator and lower pole of the spherical cellular mass.* (2) *Section through a young blastula, showing the central cavity, or blastocoele. The pigment-containing cells are shown with heavy black borders.* ***(e)*** *Gastrulation.* (1) *Section through an early gastrula, showing beginning of migration of cells (primary mesenchyme) into the blastocoele.* (2) *Section through a later gastrula, showing invagination of pigmented cells, the blastopore, and first indication of an archenteron.* (3) *Section through a later gastrula, showing increase in length of the archenteron, formation of secondary mesenchyme at its blind end, and construction of skeletal rods by primary mesenchyme cells.* (4) *Exterior of a gastrula of the same age as that represented in 3, showing its uniform ciliation.* ***(f)*** *Larval stages.* (1) *Young larva from the right side. The second opening (mouth) of the archenteron can be seen. Because of more rapid growth on one side of the body than on the other the two openings are brought to lie on the same side of the body, distinguishing its ventral from its dorsal surface. Through the translucent body wall the right coelomic cavity can be seen, already constricted into anterior and posterior coeloms and hydrocoele. The strongly ciliated external cells are shown as forming a continuous band, extending down the sides of the larva and across its ventral surface above the mouth and above the anus.* (2) *Slightly older larva from the ventral side. In this and the next two drawings (3 and 4) internal structures other than the digestive tract are not shown.* (3) *Older larva (ventral view), showing growth of ciliated band and development of the larval arms.* (4) *Young pluteus (ventral view).* (5) *Pluteus at the beginning of metamorphosis seen from the left side and showing the "echinoderm rudiment," developed over the hydrocoele, in which are included those larval structures that are retained and developed in the adult. The others are discarded when metamorphosis is complete.*

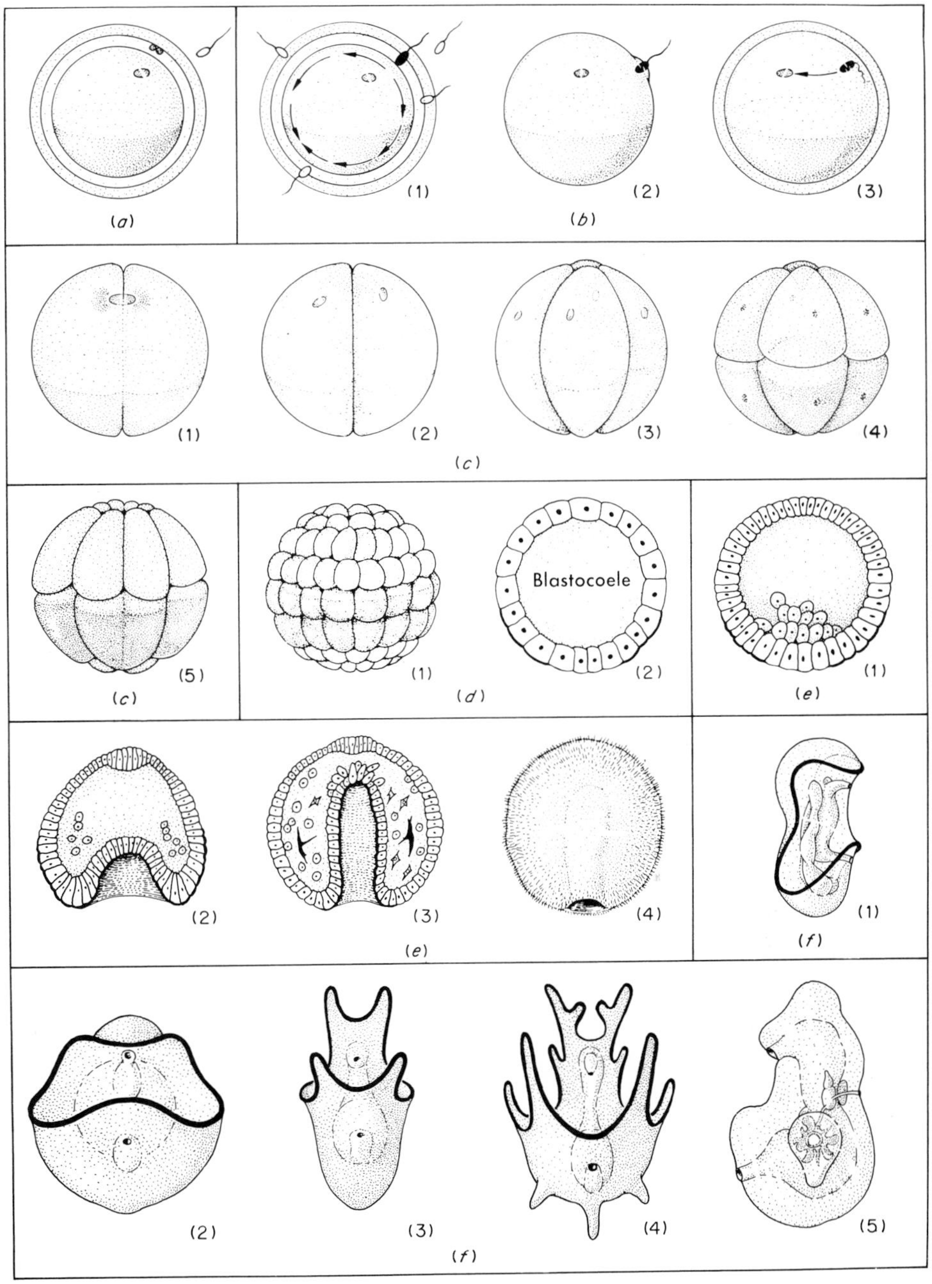

Fertilization is a process peculiar to gametes and does not occur between cells of other kinds, or between gametes of the same kind, except possibly under unusual and abnormal circumstances. Other tissue cells may aggregate and form pairs, multicellular associations, and even syncytia, but these procedures are not comparable to those of fertilization. In general, too, fertilization occurs only between gametes of the same species, though this is not universally true: interspecific fertilization can in some cases be induced under laboratory conditions and very likely occurs in nature as exceptions to the general rule. This may account in animals for the origins of some species as interspecific hybrids, as this is known to have happened in the evolution of a good many plants.

Observations made in the first quarter of this century on the behavior of the sperm of the sea urchin *Arbacia* and the polychaete *Neanthes* in the vicinity of their respective eggs revealed that they clumped together in small masses, in a fashion comparable to the agglutination of antigen by antibody, a reaction well known to immunologists. It was further observed that characteristic sperm agglutination occurred in water, conventionally known as egg water, in which eggs had been standing for some time, but from which they had been removed before introduction of the sperm of that species. These observations led the American biologist Frank R. Lillie (1870–1947) to the hypothesis that an antigenic substance contained in the eggs and exuded from them combined with an antibody-like substance in the sperm of their own species causing agglutination. He gave the name fertilizin to the isoagglutinin and the name antifertilizin to the sperm substance, which is also in the eggs. Assuming that the fertilizins of different species are different, that only egg cells produce fertilizins, and that those of different species combine only with the antifertilizin of the same species, a satisfactory explanation consistent with accepted immunological fact and theory could be offered for the restriction of the effect of fertilizin to sperm, and only to sperm of the same species. Such agglutination is believed to result from a chemical combination between fertilizin and antifertilizin, based on the structural complementarity of the molecules, or of certain sites on the molecules, which fit together according to the lock-and-key analogy of antigen-antibody and enzyme-substrate reactions. Both fertilizin and antifertilizin are believed to have several, indeed many, such reactive sites, so that one molecule of fertilizin can combine with several sperm at the same time, and several sperm can likewise combine simultaneously with one molecule of fertilizin.

Investigation shows, however, that in echinoderms at least the agglutination reaction deviates from orthodox antigen-antibody reactions. There is in echinoids a spontaneous reversal of agglutination, evidenced by the fact that clumps of sperm break apart a few minutes after their formation and that the cells will not subsequently agglutinate when transferred to new egg water. A possible explanation for this is that the fertilizin molecule is polyvalent and, after combination with antifertilizin, is split into monovalent fragments with only one reactive group, and that the sperm surface is changed when its reactive groups are saturated with these monovalent fragments. Then it can no longer combine with polyvalent fertilizin, and the sperm do not cluster together since they are no longer bound to common molecules. Such reversibility is not typical of immunological reactions. Nor is the fact that the agglutination reaction in starfishes requires a third factor, in addition to fertilizin and antifertilizin. A variety of chemical agents will serve for this if they have the common property of combining with metals and so of removing metal ions from the medium. Presumably these ions block or inactivate the combining sites on antifertilizin.

Egg water alone elicits reactions other than agglutination upon sperm of the same species, an implication that there may be several substances besides fertilizin that diffuse from the eggs. One experimental approach to the problems of fertilization has been extraction and tests of materials from both eggs and sperm. Immunological tests have revealed four different antigens in the eggs of the sea urchin *Paracentrotus lividus*, although not all of them may be concerned in fertilization. But such findings suggest that the cortex of most eggs is a very complicated structure in respect to its specific proteins. The most fully studied component of eggs or egg water is fertilizin, which has proved to be a glycoprotein and to be located in the jelly coat of sea urchin eggs—indeed, it may comprise the entire jelly coat—and in the jelly coat of the keyhole limpet *Megathura crenulata*. It may be, as well, in the more superficial layers of the egg as cytofertilizin, which is possibly also its site in eggs that lack a jelly coat. There is positive evidence for fertilizin in four of the major animal phyla—molluscs, annelids, echinoderms, and chordates—but as yet there has been no conclusive demonstration of it in others. Yet it may exist in a variety of forms, some insoluble, and a variety of methods may be needed for its extraction and for tests of its presence and activity. The mating type substances of *Paramecium* are comparable to insoluble fertilizin, for they cause specific agglutinations but do not diffuse freely into the water. Rather, they form coatings on the cilia and so bind the cells together when they become agglutinated in the preliminary phase of conjugation. Failure to demonstrate fertilizin in some of the eggs in which it has been looked for raises the question of its absolute necessity to fertilization and suggests that it may possibly facilitate the process rather than be indispensable to it. The fact, too, that sea urchin eggs deprived of their jelly coats are still capable of being fertilized, though requiring heavier insemination than usual, supports the idea of a facilitating role, although the possibility remains that the methods of jelly removal have not been 100 percent effective and that some fertilizin may remain, or that the cytofertilizin may be sufficient for the agglutination reaction.

Other effects of egg water upon sperm are, in some cases, increase in their motility and oxygen consumption, and most importantly, structural changes in them. Increased motility when in egg water or in the vicinity of eggs has been noted in the spermatozoons of sea urchins, of several molluscs, and of *Neanthes*, but not of starfish. Increased oxygen consumption has been shown in the sperm of some echinoderms and some molluscs when in egg water, but this is not true for all the species that have been examined and is not a universal phenomenon.

That agents such as these, acting at a distance upon other cells, can be obtained from eggs has led to the concept of gamete hormones, or gamones, which, according to their source in egg or sperm have been designated gynogamones or androgamones. Gynogamone I attracts and activates spermatozoa; gynogamone II and jelly coat substance are synonyms for fertilizin. Androgamone I inhibits sperm activity; and androgamone II, jelly-precipitating factor, and sperm receptor are synonyms for antifertilizin. Antifertilizin has been identified as a protein and, like fertilizin, may have a facilitating rather than an essential role in fertilization.

The conspicuous structural changes which sperm undergo in egg water, or in the presence of eggs of their own species, as they have been observed in molluscs, annelids, and echinoderms, center in the acrosome and the middle piece, which becomes displaced to a postero-lateral position in respect to the head from the posterior position it previously occupied. These changes are collectively called the acrosome reaction and have been most intensively studied by light and electron microscopy in various echinoderms and polychaete annelids. The importance of the acrosome reaction to fertilization in these animals is so great that its occurrence is implied in other groups, where, however, it has not yet been actually observed.

In electron micrographs, the acrosome in a mature spermatozoon appears bipartite, composed of a distal globule of electron-dense material and a stalk that extends into an invagination in the nucleus. The acrosome reaction includes release of some substance that immediately disperses and the formation of a fibril or filament that protrudes from the head of the spermatozoon. The filament is straight in the majority of species examined, but in some it is curved or hooked. It is relatively rigid and of various lengths, again according to species. Its length shows some degree of correlation with the thickness of the barrier to be crossed before its tip can reach the surface of the egg or the immediately underlying cytoplasm, for it is contact with the tip of the acrosome filament that initiates activity in the egg. The acrosome filament of a sea urchin spermatozoon is about 1 μ long, not nearly long enough to extend through the jelly coat, but this is not necessary for the sperm swim actively through this coat to the egg surface. This is not the case with starfishes and holothurians, and their acrosome filaments are long enough to reach through the jelly coat. Those of starfish spermatozoa measure 22 to 28 μ; and in *Sclerodactyla briareus*, whose eggs have jelly coats some 55 μ thick, the filament is about 75 μ long, although at least one measurement as great as 90 μ has been obtained. The extended filament in this case is some 15 to 30 μ longer than the sperm's flagellum. Detailed electron micrographs of the acrosome reaction in polychaete worms have revealed that the "filament" is in actuality a tubular structure, so that the term "acrosomal tubule" is more exact than "acrosome filament."

The acrosome reaction can be induced only in mature spermatozoa, but by a variety of agents. The spermatozoa of *Neanthes* and of echinoderms, in general, almost invariably react to egg water of their own species, but those of gastropod molluscs and of oysters are comparatively insensitive to egg water of their own species. Seawater with a pH of 9 or over induces the reaction in sperm of echinoderms and some molluscs, and contact with a glass or collodion surface does so in some species; oyster sperm are especially sensitive to contact stimulation. The reaction can be elicited in sperm of *Mytilus edulis* by the addition of Ca^{++} to natural seawater, and the presence of Ca^{++} seems to be requisite to occurrence of the reaction in echinoderm sperm; in the absence of Ca^{++} the sperm of sea urchins do not attach to the surface of the egg in the radial orientation that is characteristic of natural fertilization, and often attach tangentially. The natural inducer of the acrosome reaction in any sperm is contact with one of the surface layers of an egg of its own species. This may be the outer surface of the jelly coat, as it is in asteroids and holothurians; or it may be the vitellin membrane, as it is in various molluscs, in the polychaete *Hydroides*, and in echinoids, for the sperm of all these pass bodily through the jelly coats. The point of contact is the site of changes in the egg that are immediately evident. The first of these is dissolution or lysis of the membrane by a substance discharged from the acrosome. These egg membrane lysins are the gamones, whose role in fertilization, on data obtained particularly from molluscs, is best understood. The lysins can be extracted from sperm, and that obtained from the sperm of *Megathura* has been shown to differ from antifertilizin in its chemical and physical properties. Both the jelly coat and the closely fitting vitellin membrane of *Megathura* eggs are lysed by extracts of the limpet's sperm. Evidence for sperm lysins has been obtained from other molluscs, such as *Mytilus* and *Haliotis*, and from at least one annelid, the serpulid *Pomatoceros*. In *Mytilus* the sperm lysins are believed to be located on the acrosome filament. Refractile bodies in the sperm of the acoele flatworm *Childia groenlandica* may have a lysosome-like role in fertilization. The eggs, after internal insemination, go through their maturation divisions and the prophases of the first cleavage division in the body of the parent. Arrest of division occurs at metaphase, and development does not proceed until the eggs are laid and come in contact with seawater. The refractile bodies, evident in secondary spermatocytes, lie in a double longitudinal row in the sperm tail; in their ultrastructure and cytochemistry they are similar to the granules in basophil leucocytes and mast cells. Possibly, they contain an activating substance, which is released into the egg cytoplasm and either contributes to the events of fertilization or acts later in removing the block to development, or both of these.

A conspicuous reaction on the part of the egg to contact with the tip of the acrosome is formation of a reception, or fertilization, cone (Fig. XVIII–13). This is an elevation of the egg surface around the acrosome filament. In starfishes and sea urchins, the cone has been observed to creep up this, engulfing the sperm which then sinks into the egg through a funnel-shaped crater in the reception cone. In sea urchins this happens 50 to 60 sec after attachment of the sperm, in *Neanthes*, about 48 min afterwards. The mechanism effecting passage of the sperm into the egg has not yet been elucidated; it does not seem to be by active movement on the part of the sperm, although this is the way in which echinoid sperm make their way through the thick jelly coat of the egg. Writhing movements of the successful spermatozoon have been seen just after its entrance into these eggs, but these movements stop after the sperm head has rotated and the sperm aster is formed. In all species examined, the entire spermatozoon enters the egg and there is no evidence that the tail is left outside, although in starfishes and other species studied it remains motionless after the sperm has made contact with the egg surface. Close observation of the acrosome filament

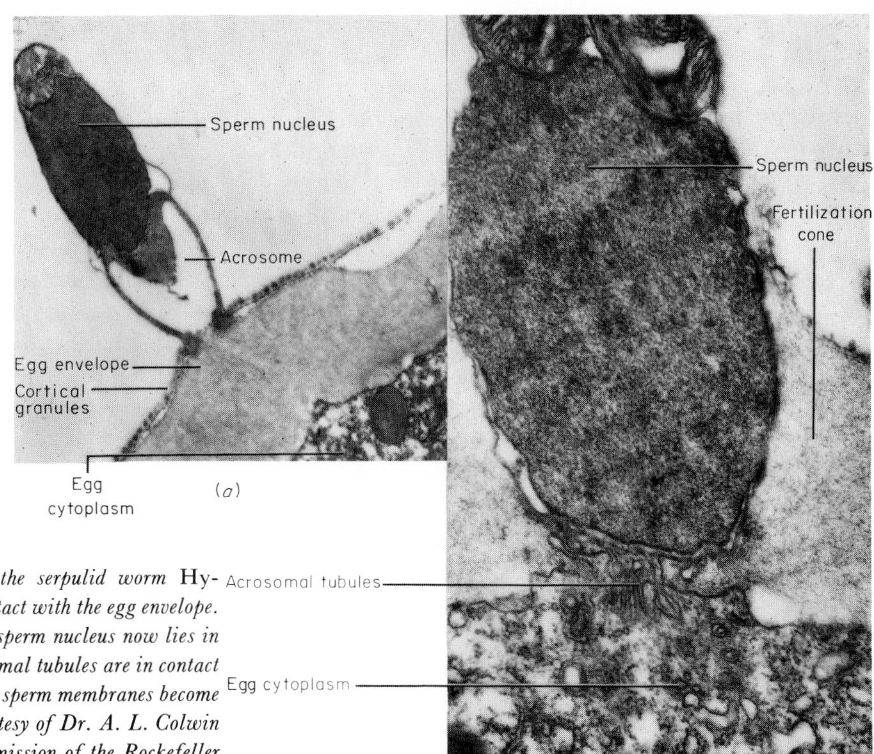

FIG. XVIII–13 *Acrosome reaction in the serpulid worm* Hydroides. ***(a)*** *Sperm immediately after contact with the egg envelope.* ***(b)*** *Formation of fertilization cone. The sperm nucleus now lies in the vitellin envelope of the egg and the acrosomal tubules are in contact with its plasma membrane. Later, egg and sperm membranes become continuous. (Electron micrographs by courtesy of Dr. A. L. Colwin and Dr. L. H. Colwin; reproduced by permission of the Rockefeller Institute Press from J. Biochem. Biophys. Cytol., vol. 10, 1961.)*

during passage of a spermatozoon into an egg shows that it is essentially unaltered in all dimensions until the sperm is well into the egg cytoplasm.

The site of sperm contact and formation of the reception cone is the focal point for profound changes that take place in the egg surface. Cortical cytoplasm seems immediately to flow in that direction and to be carried down into the interior of the egg with the entrance of the sperm, at which time the reception cone disappears. The "fertilization impulse" initiated at this site spreads rapidly over the surface of the egg, beginning when the tip of the acrosome filament is in contact with the egg surface proper. In echinoderm eggs, about 20 sec after this has happened, the cortical granules break down and the fertilization membrane is raised from the egg surface. Coincident with the passage of the fertilization impulse is the block to polyspermy; once the tip of the filament of one sperm has reached the egg surface and set in motion the events of reception, cone formation, and cortical reaction, other sperm do not or cannot enter.

In general, three types of cortex have been recognized in the eggs that have been studied: One is the type characteristic of echinoderms, where the cortical granules are extremely labile and the cortex undergoes drastic reorganization at fertilization or at maturation of the egg; the second is that characteristic of annelids and molluscs, where the cortical granules are more stable and do not undergo any visible change at this time, although their cooperation in the events is indicated; in the third, which may be typical of ascidians, there are no clearly differentiated cortical granules, nor indeed any evident cortical layer. Dissolution of the cortical granules is presumably effected by the sperm lysin, or lysins, known as androgamone III. It is not a protein. Breakdown of the cortical granules in echinoderm eggs can be brought about by a variety of chemical agents, such as urea, saponin, toluol, and butyric acid; and by some physical ones as well, such as ultraviolet light and pricking the surface. All these are activating agents, in that they bring about release of metabolic reactions previously blocked in the egg that are responsible for initiation of morphogenesis, but they are not fertilizing ones, although they may result in parthenogenetic development of an embryo. Such agents have been widely used by biologists in their analyses of the phenomena of activation and embryogenesis.

The cortical granules, after their disruption in echinoderm eggs, merge with the vitellin membrane and both are lifted off the surface of the egg as the fertilization membrane which subsequently undergoes some thickening and hardening. Beneath is a hyaline layer on the surface of the egg, separated from the fertilization membrane by the perivitelline space filled with a colloidal and, presumably, osmotically active solution whose components may be derived from the cortical granules. Cytofertilizin secreted at the time of fertilization is one of its components. There is no such conspicuous elevation of a fertilization membrane in the eggs of some molluscs and annelids.

Activation is also evident in the endoplasm. This is very clearly seen in the eggs of *Arbacia punctulata* in which, about 10 min after the acrosome and cortical reactions, all the chromatophores containing red echinochrome migrate to the surface and come to lie just beneath the cortex. It has been found possible to block spread of the fertilization impulse, so that the typical reactions following its initiation are restricted to a localized area. This can be done by raising the temperature, by application of nicotine or urethane, or by deformation of the egg by drawing it up into a capillary tube with a bore slightly less than the diaeter of the egg. In *Arbacia* eggs, in which the impulse has been blocked by one of these means, the chromatophores move only to the area along which it has passed, implying some sort of reaction between the activated cortical area and the underlying endoplasm, but how the one exerts its effect upon the other is still not clear. Streaming movements in the cytoplasm may account for movement of the nuclei as well as of cytoplasmic organelles. Electron micrographs have revealed displacement of the endoplasmic reticulum toward the surface, a fact that may be of great significance in starting the egg on a phase of active metabolism again. For, while the level of metabolism is very high in an oocyte during its period of growth and organization, this level is greatly reduced in the mature egg. Lillie related this to a block in the interchange between nucleus and cytoplasm, an interchange that is essential to metabolizing cells. Initiation of the cortical reaction in some way removes this block, although the metabolic processes in an activated egg follow different pathways than in an oocyte. There, they were directed primarily toward synthesis of yolk, while in the fertilized egg, they are directed toward its breakdown and the synthesis of new protein and of nuclear constituents essential for the cleavage divisions that shortly follow syngamy. The distribution, or redistribution, of endoplasmic reticulum may be of especial significance in this connection. Also, when the block is removed, oxygen consumption in the egg increases. This increase has been measured in sea urchin eggs, and it is seen to rise exponentially up to the early blastula stage, then to level off. A similar sequence has been shown in the eggs of *Rhodnius*.

There must also be some activation of the sperm after it has entered the egg. This is evident in the swelling of the sperm head, displacement of the middle piece, rotation of the sperm, and development of the aster. This activation does not occur until the egg has reached a certain stage of "ripeness." Sperm can, for example, enter the oocytes of sea urchins while the nucleus is still in the germinal vesicle stage, but their heads do not enlarge and no aster forms. These changes happen only after the germinal vesicle has broken down, implying delivery of something from it into the egg cytoplasm that has a releasing effect upon the spermatozoon. Evidence for this has also been derived from early experiments on starfish eggs separated into two pieces, with one piece containing the nucleus and the other without it. If the division was made while the germinal vesicle was still intact, only the piece with it could be entered by a sperm, but if the division was made after its breakdown, both pieces could.

Results such as this indicate not only that the egg must make some preparation for reception of the sperm, but that there is a period of unripeness during which more than one sperm can enter. Blocks to this, in eggs that are normally entered and fertilized by only one sperm, arise at different times in different species. Sea urchin eggs are protected against polyspermy only after they have completed meiosis, but those of starfish, molluscs, and annelids attain such protection while they are still in the oocyte stage. These differences seem to depend upon the time at which the cortical layer becomes organized and established as a specialized region in the egg. Eggs, however, appear to have several safeguards against polyspermy. One of these, in eggs that are surrounded by a jelly layer, is the filtering action of this layer in screening out the spermatozoa that swarm around an egg. Another is provided by the very exacting conditions of the interaction between the egg surface and the sperm. The question arises here as to whether the entire egg surface is equally receptive to sperm or whether the combining sites are localized in particular areas so that fertilization can be accomplished only by sperm that happen to strike there. There is some evidence that in sea urchin eggs certain equatorial or subequatorial zones are more responsive or receptive than others. And again as a safeguard against polyspermy are the events of the cortical reaction, which seem to provide a final and permanent barrier to the entrance of any other sperm than the one that has instigated them, a block that arises rapidly after contact has been made.

This block might be a physical one, provided by the elevated and hardened fertilization membrane, but in some eggs as, for example, those of certain molluscs and the annelid *Chaetopterus*, no fertilization membrane is raised and the cortical granules are not labile and remain intact. The barrier may lie in the hyaline layer or in the reorganized cortex, but wherever it resides it is, under ordinary conditions, an effective one against the admission of more than one spermatozoon. Yet in insects polyspermy is the rule, and many sperm may enter an egg, although only one unites with the product of the meiotic divisions of the egg nucleus that is the definitive gamete nucleus. The others may, however, undergo several divisions coincident with the divisions of the zygote nucleus, but they ultimately degenerate.

The problem of the block has been attacked in a number of ways, primarily by attempts to induce polyspermy by one means or another. It results sometimes, in starfish and sea urchin eggs, from very heavy insemination, especially in those that have been standing for some time and are consequently "stale." It can also be induced by dilute solutions of various chemical agents, such as nicotine, strychnine, or morphine, and by the action of trypsin. Additional spermatozoa will enter fertilized sea urchin eggs if they are put into calcium- or magnesium-free seawater. Increase in temperature, at least up to 31°C, is also conducive to polyspermy. These agents affect one or more of the natural barriers to the entrance of additional sperm, creating abnormal conditions in the egg and resulting in derangements of syngamy and subsequent cleavage. It has been observed in sea urchins that, in these conditions of pathological polyspermy, two sperm nuclei may unite with the egg nucleus and the triploid set of chromosomes be distributed at random in the ensuing mitotic divisions, with consequent abnormalities in the blastomeres.

The fact that activation of the eggs of sea urchins, starfish, annelids, and molluscs has also been accomplished by various artificial means, some chemical, some physical, indicates that all the elements necessary to one aspect of fertilization, activation, are contained within the eggs and that they are somehow held in check, or restrained from interaction, until liberated. In fertilization, release is effected through the cortical reaction incident to the acrosome reaction, but in the parthenogenetic development of mictic or amictic eggs, it must be by other means.

Eggs of different species respond differently to artificial activating agents. Starfish eggs and those of the annelid *Thalassema* are, for example, activated when the carbon dioxide content of their environmental seawater is slightly raised, but sea urchin eggs are not. Seawater made hypertonic by the addition of one or more of its natural salts is effective with most of the invertebrate eggs that have been so treated, and the lower fatty acids, in particular butyric acid, are as well. In some instances simply shaking the eggs, with consequent displacement of their contents, or the application of a weak electric current have proved sufficient to implement the cortical reaction and the events that naturally follow it. The addition of Ca^{++} to the seawater activates eggs of *Asterias glacialis*, the lamellibranch mollusc *Cumingea*, and the sipunculid *Phascolion strombi*, possibly effecting the release and promoting the operation of certain enzymes. These agents to some extent mimic the effect of the acrosome of a sperm, but it is highly problematical if they achieve it in the same way. The mitotic process is set in motion by such artificial means in eggs arrested at some stage in the course of meiosis, at least to the extent that these divisions are completed, although cleavage divisions may break down at some stage in embryogeny. Sea urchin and other larvae have, however, been successfully reared from artificially activated eggs, and some few sea urchins and starfish even carried through metamorphosis. But syngamy can only follow introduction of the sperm nucleus, and larvae resulting from artificial parthenogenesis, as from natural, are either haploid or have a diploid set of chromosomes entirely of maternal origin, resulting from replication without separation in an early division.

Observations of the events and results of fertilization and of artificial parthenogenesis indicate that successful fertilization involves activation of the egg, of the sperm that has entered it, and of mitotic processes in the egg. Mitosis results in the precise and equal distribution of replicates of the chromosomes contained in the nuclei of both egg and sperm to the daughter cells of which the embryo, the larva (if a larval stage intervenes), and the adult are constructed. Activation of the egg may also result in redistribution of cytoplasmic particles that thus are localized and come to be included in particular blastomeres. All these events are ones that are by no means fully understood as yet and are still open to experimental analysis. They might also be considered preliminary events, with the implication that fertilization is completed only when the gametes are formed in the adult developed from a fertilized egg and the maternal and paternal chromosomes in it sorted out and distributed to them at meiosis.

C. OVIPOSITION AND PROVISION FOR THE YOUNG

While the females of many invertebrate species shed their eggs apparently indiscriminately in the environment, others show varying degrees of maternal supervision and care. These range from selection of a more or less favorable site for their deposition and their development to incubation of them on or in the body of the parent. Free-living flatworms and many aquatic gastropods, for example, fasten their eggs, either in individual capsules or in common masses of jelly, to some object in the water, usually near the surface, so that they are protected from displacement from a favorable environment by water currents or other disturbances. This is, in most cases, all the care that is furnished them, but some gastropods protect them longer by crouching over the egg mass, and female octopods remain near theirs, guarding them from possible predators. Walkingstick insects drop their eggs anywhere in fallen leaves or litter on the ground, but mantises, closely related to them, enclose theirs, in the thousands, in characteristic capsules, which they attach to the twigs of low-growing bushes and then abandon them (Fig. XVIII–14*a*). Mosquitoes lay their eggs in flat masses, or rafts, that float on the surface of the water (Fig. XVIII–14*b*). Some terrestrial gastropods lay their eggs in pits in the ground, and earthworms deposit theirs in capsules in the soil, as do crickets and grasshoppers, whose sharp ovipositors penetrate to some depth so that the pods, each containing a number of fertile eggs, are placed well below the surface. Many other insects, although also abandoning their eggs after laying them, show extreme care in finding places for oviposition, selecting those that will provide immediate and suitable sources of food for their offspring when they hatch. As this is the same kind of food that they ate as juveniles or adults, selection of a site for oviposition is related to food preferences. Flesh flies, for instance, oviposit on raw meat, and fruit flies on decaying fruits. In phytophagous species, the choice may well depend upon the specific attraction between host plant and insect, the plant possibly being recognized at the time of reproduction by sensilla on the ovipositor. Some of these insects lay their eggs in or just beneath the bark of branches or stems, so that the larvae when they emerge have a food supply at hand. Those that oviposit in the autumn, whose eggs winter over, often orient them so that the head of the larva when it hatches in the spring is directed toward the fresh young leaves that have also just emerged.

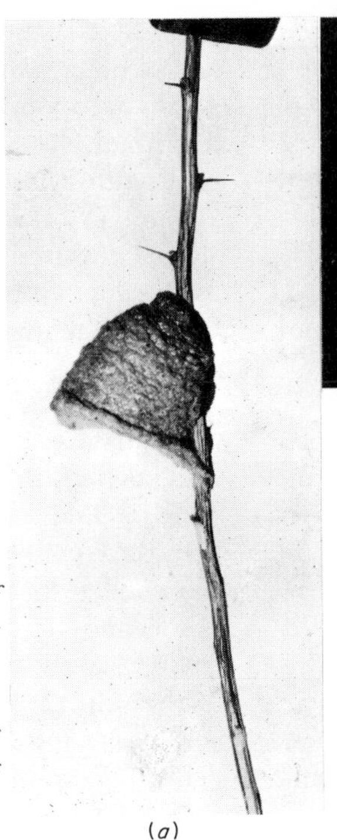
(a)

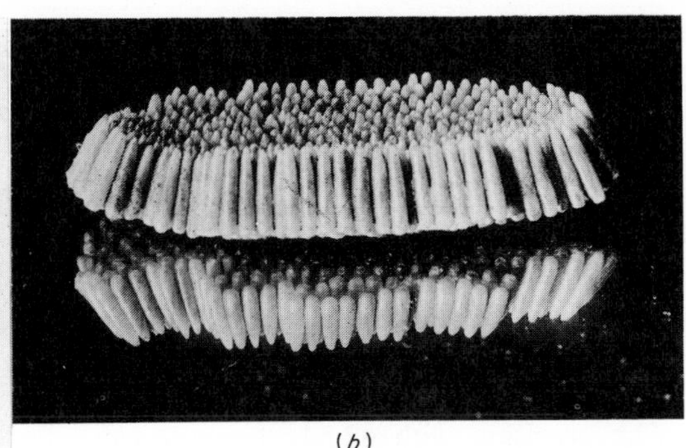
(b)

FIG. XVIII–14 *Egg cases of insects. (U.S. Department of Agriculture, Bureau of Entomology and Plant Quarantine.)* ***(a)*** *Egg case of the praying mantis* Tenodera sinensis. ***(b)*** *Egg raft of the mosquito* Culex pipiens. *Each cylindrical egg, in its protective coat, is cemented to its neighbors, and the whole mass is deposited on the surface of the water.*

Some lay their eggs in the ovaries of flowers of the host plant and the larvae feed on the ripened seeds or fruits. This is the reason why "worms" are found in cherries and apples and pears, and for the justification, given many years ago by a distinguished English naturalist, for the existence of earwigs; he contended that they are "most excellent mothers." So precise has the relationship become between ovipositing insect and host plant that in some cases the plant depends wholly upon this act for its pollination. The classical illustration of this is the relation between the moth *Pronuba yucasella* of Mexico and the southwestern United States and the yucca plant, whose sole means of pollination is the female moth that visits the flowers, and only these flowers, to lay her eggs, carrying a little ball of pollen between her maxillae from one flower to another and inserting it carefully in the pistil and then laying a few eggs there. In this way the moth ensures that the flower's egg cells will be fertilized, its seeds develop, and her larvae supplied with nourishing food as soon as they hatch. The larvae are limited in number and some seeds are left to mature and germinate, so that the species is perpetuated, another illustration of a delicately balanced adjustment that has evolved between the host as a source of food and the parasite consuming it.

Smyrna figs likewise are absolutely dependent upon insects—in this case the minute wasp *Blastophaga psenes*—for pollination. This wasp develops in galls on the flowers of male trees (caprifigs), and as the winged females emerge and crawl out of the flower they become covered with pollen. Some species of pollinating wasps have special concavities on their bodies in which the pollen is packed, but although there are two ventral longitudinal grooves on the head of *Blastophaga*, there is no evidence that these are special "pollen baskets." As the winged females visit the flowers of other trees in search of places to deposit their eggs, the pollen is dusted off on the pistils of the female flowers. This is vital to the Smyrna fig, which only produces pistillate flowers and would never be pollinated otherwise or produce mature fruit. The wasps do not lay their eggs in the female flowers, however, but only in those of caprifigs. Caprification has long been practiced in Asia Minor, and is now in commercial use in the United States, as a means of ensuring fertilization of Smyrna figs and the production of a valuable food crop. In caprification, caprifig blossoms with their galls are taken from the male trees when the wasps are about to emerge and tied among the branches of the Smyrna fig trees. The emerging females, well dusted with pollen, visit these trees first in their search for a place to oviposit. Although the flowers of the Smyrna fig tree are not suitable for this, and no galls are formed, the wasps accomplish their mission of pollination and so provide for a crop of figs.

The precision with which digger wasps capture prey and seal it in each cell in which an egg is laid has already been mentioned as has the precise timing relationships between insect oviposition and phases in the life cycle of particular plants.

Many invertebrates protect their eggs during incubation by brooding them. They may be held on the body in various ways or they may be covered over by it; or they may pass some or all of their developmental stages in chambers or brood pouches, either temporarily adapted to this purpose or specialized for it, or even in the reproductive tract itself. Among some species of cnidarians, the embryonated eggs are held on the tentacles; in others, there are brood chambers in which development takes place. In endoprocts, the internally fertilized eggs are enclosed in gelatinous capsules attached by short stalks to a region of the wall of the vestibule that is greatly thickened by the accumulation of food deposits there. In this region, known as an embryophore, the eggs develop, the embryos deriving their food from the stores in the wall. Most marine and all freshwater species of ectoprocts brood their young, either in embryo sacs, characteristic of the Phylactolaemata, or in ovicells, characteristic of the Gymnolaemata. Embryo sacs are developed on the body wall near the ovary; ovicells are highly modified zooids, which seem to provide very special conditions for incubation, for eggs removed from them do not develop normally, if at all. A few species of brachiopods brood their young, some using the mantle cavity for this, some the arms of the lophophore to which the eggs are attached, and some the nephridium which also serves as an oviduct.

Freshwater pelecypods typically hold their eggs in the water tubes of their gills for longer or shorter periods of their incubation. The young of *Anodonta* and of other members of the Unionidae are set free as glochidia, modified veliger larvae, with bivalve shells armed at the gape with sharp chitinous hooks and each with a long thread projecting through it. Development of a glochidium does not proceed further until it encounters a fish and attaches itself to its gill by its thread and the hooks of its shell. The fish responds to this attachment by building up a cyst wall around its visitor which thus becomes embedded in the tissues of the gill to complete its development there. Many glochidia may attach to a single fish, whose gills become distended with their load of developing molluscs, which remain as parasites upon it for 2 to 6 weeks before the young drop out of the cysts to the bottom of the pond to become mature only after several years. *Ostrea* and *Teredo* brood their young in the mantle cavity; young oysters are set free as "spat" and for further development must find a suitable site and conditions for attachment, the most propitious site being old or broken oyster shells. Some few polychaetes brood their eggs, either beneath the elytra, on the parapodia or, as in *Autolytus*, in a special sac on the ventral surface of the body.

Most decapod crustaceans, as has already been mentioned, carry their developing eggs on the hairs of their abdominal appendages, but in many branchiopods the eggs are held in brood pouches during their incubation. These pouches may be spaces beneath the carapace, as they are in cladocerans; ovisacs constructed at the time of oviposition by gland cells at the distal ends of the oviducts, as they are in anostracans; or overlapping exopodites of certain trunk appendages, as they are in notostracans and conchostracans. Females of the cladoceran *Daphnia* with eggs in the brood pouch are familiar sights in freshwater collections made in the warm days of spring and in summer. The summer eggs of *Daphnia* contain but little yolk and the embryos that remain there throughout their development are nourished by secretions from the wall of the pouch. Equally familiar in these collections are females of the copepod *Cyclops*, with loaded ovisacs extending laterally from the first abdominal segments. All pericarideans brood; the brood chambers are made by the apposition of oostegites, plates attached to the medial surfaces of the thoracic appendages, which meet in the midline and make an enclosure into which the eggs pass and from which they are released in a late stage of development.

Some 100 species of echinoderms are known to brood their eggs, sometimes only up to the larval stage, sometimes through metamorphosis. Some species of starfish simply hunch over the eggs they have shed and the embryos developing from those that have been fertilized. Forcipate species hold them on their paxillae, skeletal plates provided with several movable spines. *Leptasterias groenlandica* broods eggs in pouches of the cardiac stomach, and only eggs so brooded survive. Some sea urchins, the cidarids, hold their fertilized eggs around the peristome or periproct, making a kind of tent over them with crossed spines; others, the spatangoids, or heart urchins, have sunken pits or petaloids along the ambulacra in which the eggs incubate. A respiratory function has also been attributed to the petaloids. Many ophiuroids brood their young in the respiratory bursae that overlie the gonads, keeping them there through metamorphosis. It is not unusual to find a specimen with its bursae distended, or its aboral surface partly covered, with fully formed young of considerable size and with firm skeletal plates. In general, echinoderms with the habit of brooding their young live in the colder seas, although brooding ophiuroids are found in semitropical and tropical waters as well. More than 30 brooding species of holothurians have been found in south polar waters, some keeping their eggs on the ventral surface of their bodies, some on the dorsal, and some in special brood chambers that may be either shallow external depressions or deep internal sacs.

If valid distinction can be made between brooding species of invertebrates and ovoviviparous and viviparous ones, it must be based on the location in which the young develop—whether in brood pouches or in some part of the reproductive tract. Distinction between ovoviviparity and viviparity is also a tenuous one, for which the difference resides in the source of nutrient used by the developing embryo. In ovoviviparous species this is the egg itself, with its contained yolk; in viviparous ones, nutriment is derived directly from the tissues of the mother, by some special anatomical arrangement. There are both ovoviviparous and viviparous invertebrates.

Brachiopods that carry their eggs to full development in the nephridia serving as oviducts during the reproductive season are ovoviviparous, as are some littoral, freshwater, and terrestrial gastropods. All stages of development of the winkle *Littorina rudis*, the freshwater snail *Viviparus viviparus*, and the land snail *Helix* are completed within the terminal regions of the oviducts of their parents, and the young molluscs emerge fully formed and shelled. The only truly viviparous annelid is the polychaete *Ctenodrilus*, whose young are nourished throughout their development from the maternal blood vessels. The onychophoran *Peripatus* is both ovoviviparous and viviparous. Some species in this genus produce eggs that are large and full of yolk, which develop independently within the uterus, but some species produce eggs that are small and poor in yolk and these attach themselves to the wall of the uterus by a kind of placenta through which nutrients from the mother pass. All scorpions are also either ovoviviparous or viviparous, with similar species difference in the size of the eggs and the mode of development that obtains for *Peripatus*. The eggs of some develop in the maternal oviduct. Those that are heavily yolked complete this without other nutrient resources, but in those species in which the eggs are poor in yolk, processes extend from the oviduct to the midgut glands that absorb food for them and transfer it to the embryos. This is an arrangement, in principle, like that of a mammalian placenta, although the transference here is direct and not mediated by a vascular system. Some scorpions extend their care for the young even longer than the developmental stages, the females carrying them on their backs for varying periods of their juvenile lives, a habit that is also common in spiders. Some spiders also feed their young for a time on food they regurgitate for them.

There are also some viviparous insects. Included among these might be the cockroaches whose oothecae, after being molded into form by the ovipositor, are drawn back again into the oviducts; perhaps this is more a case of ovoviviparity than of viviparity. Another borderline instance is that of the tsetse fly *Glossina*. The single egg produced by a female is fertilized internally and then carried to the uterus where it hatches. Concurrently, there is enlargement and activity of the accessory glands, which open by a common duct into the uterus. The larva attaches itself to this opening and draws its sustenance from the secretions; it goes through its successive larval instars inside the body of the female, emerging in its final instar and immediately pupating. *Hippobosca*, an ectoparasite on birds and mammals, pupates in the uterus of its mother and emerges as an imago. In viviparous earwigs (Dermaptera) and aphids, in the cockroach *Diplotera*, and in a few other insects whose eggs are small and poor in yolk, a pseudoplacenta is formed, constructed from the external membranes of the egg and some region of maternal tissue—the follicle cells in the case of the earwig and the ootheca in the case of the cockroach. Presumably, nutrients pass through such a structure to the developing egg, but there is as yet no critical evidence for this. In some insects, the eggs are shed into the hemocoele and develop there, deriving their nutrients from this fluid, although not through placental or pseudoplacental structures.

D. PAEDOGENESIS AND POLYEMBRYONY

Hemocoelic development in insects is usually associated with two other reproductive anomalies—paedogenesis (reproduction at a larval or early developmental stage) and polyembryony (development of more than one individual from a single egg). These phenomena are not unusual among invertebrates, but are by no means exclusive to them, for there are paedogenetic and polyembryonic vertebrates as well. Trematode flatworms are both paedogenetic and polyembryonic, for both sporocysts and rediae are reproductive stages in their life cycles, the numerous germ balls budded off the inner surface of the walls of the former developing into rediae, and either another generation of rediae or a new one of cercariae arising similarly from the latter (see Fig. I–16). The gall midge *Miastor* and the beetle *Micromalthus* are illustrative of paedogenesis in insects as well as of hemocoelic viviparity. The midges, as adults, are oviparous and lay eggs that develop into larvae. If the larval environment is propitious, as it is for example in a rotting log, ovaries develop within the newly hatched maggots and oocytes are formed, each with a number of nurse cells. The mature ova break out of the ovaries into the hemocoele and begin parthenogenetic development, resulting in the production of 7 to 30 new larvae, which feed on the maternal tissues, especially the fat body, and may in turn produce more larvae. When fully mature the larvae penetrate through the maternal cuticle and pupate outside, finally emerging as males and females that mate and begin the cycle again. The paedogenetic larvae of *Micromalthus* are of two different kinds, reproductively. In the one, a larva produces a single large egg which develops, after a number of dissimilar larval molts, into a haploid male that seems, however, to be wholly nonfunctional. In the other, the paedogenetic larvae produce a number of larvae by hemocoelic viviparity. Some of these may become paedogenetic individuals, but others pupate and emerge as female imagos, which, however, have never been observed to mate. The unusual and curious features of such a life cycle bear further investigation.

Miastor and *Micromalthus* are also polyembryonic in that many individuals are developed from one that has arisen from a single egg, their production being delayed until a comparatively late developmental stage. Polyembryony at a later stage of development is characteristic also of cyclostomatous ectoprocts in which each egg that is attached to the special polypide known as a gonozoid—and only those that are attached to it develop at all—divides to form a ball of cells on whose surface lobes appear that become constricted off, each one continuing embryonic development and becoming a new individual. Among insects, many parasitic Hymenoptera and at least one species of Hemiptera are polyembryonic; in the latter, development of the embryos takes place in the hemocoele of the mother. In some parasitic hymenopterans that lay their eggs in the body of another animal, separation of the cells may occur in early cleavages, as it does in mammals; but in others, this may happen much later and as many as 2,000 embryos arise from a single egg.

E. METAGENESIS (ALTERNATION OF GENERATIONS)

Alternation of generations, as it occurs in plants for whose life cycles the term was originally used, implies a regular sequence of haploid generations with diploid generations. This is most clearly exemplified in mosses, where diploid sporophytes, whose sporocytes undergo meiosis, give rise to haploid gametophytes, sometimes readily distinguishable as "males" and "females." Union of the gametes of these individuals, produced without the usual events of gametogenesis, results in a diploid individual and the cycle sporophyte ($2n$)-gametophyte (n)-sporophyte ($2n$)-gametophyte (n)$\cdots$ continues. In some animals, there is also a regular succession of gametically and agametically reproducing individuals, but this does not involve haplo-diplo phases, except perhaps in the case of foraminiferans with megalospheric and microspheric generations, and dicyemids and orthonectids. The term "alternation of generations," in its original meaning, is therefore not properly applicable to them, although it is frequently so used. "Metagenesis" is a preferable designation for those animals in which a sequence of agametic-gametic reproduction is the rule. This is the case in some hydrozoans such as *Obelia*, in which a polypoid generation gives rise, by budding or other agametic means, to a medusa that produces gametes following the usual events of gametogenesis. The fertilized eggs develop into polyps and so the cycle is repeated as polyp-medusa-polyp-medusa. . . . But the question arises as to whether the polyp should be considered a persistent larval stage and only the medusa the fully evolved cnidarian, in which case the polyp, although a reproductive individual, could hardly be considered a "generation." This question is pertinent also in the case of digenetic trematodes, whose sporocysts and rediae, regularly appearing in the life cycle of such species as *Fasciola hepatica*, reproduce agametically. In other invertebrates, such as some rotifers and arthropods, gametic reproduction may intervene at irregular intervals in the life histories of those species that produce several generations agametically. This is not true metagenesis, but the interpolation of gametic or agametic reproduction into the life cycle when extrinsic or intrinsic conditions dictate. Such conditions are exemplified by rotifers; branchiopod, ostracod, and cladoceran crustaceans; and in those insects which reproduce parthenogenetically for an indefinite number of generations and then, sporadically or seasonally, reproduce gametically.

Many advantages accrue to life cycles with such lability, for the species derives both the genetic benefits of amphimixis and the reproductive benefits of rapid multiplication of its kind by a single individual during favorable conditions. Thus an organism may bud, or produce great numbers of rapidly developing parthenogenetic eggs, when the weather is warm and food is abundant. But advantage also comes to the species in which a gametic generation arises that is motile and can hunt, find mates, and move to new environments for their offspring to colonize. The great reproductive potential of the agametic generation is complemented by the evolutionary potential of the gametic, in dispersion of the species and the genic recombinations made possible to the zygotes, with consequent variation in the individuals developing from them, upon which natural selection may act and evolution progress.

SELECTED BIBLIOGRAPHY

Agametic reproduction

Bell, A. W.: *Enchytraeus fragmentosus*, a New Species of Naturally Fragmenting Oligochaete Worm, *Science*, vol. 129, p. 1278, 1958.

Berrill, N. J.: "Growth, Development and Pattern," Freeman, San Francisco, 1961.

Best, J. B., A. B. Goodman, and A. Pigon: Fissioning in Planarians: Control by Brain, *Science*, vol. 164, pp. 565–566, 1969.

Coe, W. R.: A New Species of Nemertean (*Lineus vegetus*) with Asexual Reproduction, *Zool. Anz.*, vol. 94, pp. 54–60, 1931.

Freeman, G.: Studies on Regeneration in the Creeping Ctenophore *Vellicula multiformis*, *J. Morphol.*, vol. 123, pp. 71–83, 1967.

Gilbert, J. J.: Mictic Female Production in the Rotifer *Brachionus calyciflorus*, *J. Exp. Zool.*, vol. 153, pp. 113–123, 1963.

Gilbert, J. J.: Rotifer Ecology and Embryological Induction, *Science*, vol. 151, pp. 1234–1237, 1966.

Gilbert, J. J., and G. A. Thompson, Jr.: Alpha Tocopherol Control of Sexuality and Polymorphism in the Rotifer *Asplanchna*, *Science*, vol. 159, pp. 734–736, 1968.

Kanatani, H.: Studies in Fission in the Planarian *Dugesia gonocephala*. I. Effects of Heparin on Occurrence of Fission, *J. Fac. Sci., Univ. Tokyo*, vol. 8, pp. 17–21, 1957.

Kanatani, H.: Further Studies on the Effect of Crowding on Supplementary Eye Formation and Fission in the Planarian *Dugesia gonocephala, J. Fac. Sci., Univ. Tokyo*, vol. 8, pp. 23–39, 1957.

Kille, F. K.: Regeneration of the Reproductive System Following Binary Fission in the Sea Cucumber *Holothuria parvula* (Selenka), *Biol. Bull.*, vol. 83, pp. 55–66, 1942.

Lenhoff, H. M., and W. F. Loomis (eds.): "The Biology of Hydra and of Some Other Coelenterates," University of Miami Press, Coral Gables, Fla., 1961.

Spangenberg, B. B.: Iodine Induction of Metamorphosis in *Aurelia, J. Exp. Zool.*, vol. 165, pp. 441–449, 1967.

Gametic reproduction

Boolootian, R. A., and A. R. Moore: A Case of Ovotestes in the Sea Urchin *Strongylocentrotus purpuratus, Science*, vol. 129, pp. 271–272, 1959.

Colwin, A., and L. Colwin: Sperm Entry and the Acrosome Filament, *J. Morphol.*, vol. 97, pp. 543–569, 1955.

Colwin, A., and L. Colwin: Morphology of Fertilization, Beginnings of Embryonic Development, *Pub. No. 48, A.A.A.S.*, pp. 135–168, 1957.

Dan, J.: The Acrosome Reaction, *Int. Rev. Cytol.*, vol. 5, pp. 365–393, 1956.

Davey, K. G.: "Reproduction in the Insects," Oliver and Boyd, Edinburgh and London, 1965.

Drew, G. A.: Sexual Activities of the Squid *Loligo pealii* (Les). I. Copulation, Egg Laying and Fertilization, *J. Morphol. Physiol.*, vol. 22, pp. 327–360, 1911.

Drew, G. A.: Sexual Activities of the Squid, *Loligo pealii* (Les). II. The Spermatophore: Its Structure, Ejaculation and Formation, *J. Morphol. Physiol.*, vol. 32, pp. 379–436, 1919.

Freyer, G.: The Spermatophores of *Dolops ranarum* (Crustacea, Branchiura): Their Structure, Function and Transfer, *Q. J. Microsc. Sci.*, vol. 101, pp. 407–432, 1960.

Galstoff, P. S.: Physiology of Reproduction in Molluscs, *Am. Zool.*, vol. 1, pp. 273–290, 1961.

Gregory, G. E.: On the Initiation of Spermatophore Formation in the African Migratory Locust *Locusta migratoria migratoroides, J. Exp. Biol.*, vol. 42, pp. 423–435, 1965.

Henley, C.: Refractile Bodies in the Developing and Mature Spermatozoa of *Childia groenlandica* (Turbellaria: Acoela) and Their Possible Significance, *Biol. Bull.*, vol. 134, pp. 382–397, 1968.

Horsfall, W. R., and J. F. Anderson: Suppression of Male Characteristics of Mosquitoes by Thermal Means, *Science*, vol. 133, pp. 1830–1836, 1961.

Kessel, R. G.: Micropinocytosis and Yolk Formation in Oocytes of the Small Milkweed Bug, *Exp. Cell Res.*, vol. 30, pp. 440–443, 1963.

Metz, C. B.: Specific Egg and Sperm Substances and Activation of the Egg, Beginnings of Embryonic Development, *Pub. No. 48, A.A.A.S.*, pp. 23–69, 1957.

Metz, C. B., and A. Monroy: "Fertilization," Academic, New York, 1967.

Miller, R.: Chemotaxis during Fertilization in the Hydroid *Campanularia, J. Exp. Zool.*, vol. 162, pp. 23–44, 1966.

Monroy, A.: "Chemistry and Physiology of Fertilization," Holt, New York, 1965.

Orton, J. H.: Observations and Experiments on Sex Change in the European Oyster (*O. edulis*), *J. Mar. Biol. Assoc. U.K.*, vol. 14, pp. 967–1043, 1926.

Raven, C. P.: "Oogenesis: The Storage of Developmental Information," International Monographs on Pure and Applied Biology, Macmillan, New York, 1961.

Rothschild, Lord: Sea Urchin Spermatozoa, *Biol. Revs.*, vol. 26, pp. 1–27, 1951.

Rothschild, Lord: The Behavior of Spermatozoa in the Neighborhood of Eggs, *Int. Rev. Cytol.*, vol. 1, pp. 257–263, 1952.

Rothschild, Lord: "Fertilization," Methuen, London, 1956.

Rothschild, Lord: Unorthodox Methods of Sperm Transfer, *Sci. Am.*, vol. 195, pp. 121–131, 1956.

Runnström, J., and T. Gustafson: Developmental Physiology, *Annu. Rev. Physiol.*, vol. 13, pp. 57–74, 1951.

Runnström, J., B. E. Hagström, and P. Perlmann: Fertilization, in J. Brachet and A. E. Mirsky (eds.), "The Cell," vol. I, pp. 327–398, Academic, New York, 1959.

Runnström, J., S. Lindvall, and A. Tiselius: Gamones from the Sperm of Sea Urchin and Salmon, *Nature*, vol. 153, pp. 285–286, 1944.

Schneider, D.: Normal and Phototropic Growth Reactions in *Bugula avicularia*, in E. C. Dougherty (ed.), "The Lower Metazoa," pp. 357–371, University of California Press, Berkeley, 1963.

Sotelo, J. R., and O. Trujillo-Cenoz: Electron Microscope Study of the Vitelline Body of Some Spider Oocytes, *J. Biophys. Biochem. Cytol*, vol. 3, pp. 301–311, 1957.

Stay, B.: Protein Uptake in the Oocytes of the Cecropia Moth, *J. Cell Biol.*, vol. 26, pp. 49–62, 1965.

Taber, S., III, and M. S. Blum: Preservation of Honey Bee Semen, *Science*, vol. 131, pp. 1734–1735, 1960.

Telfer, W. H.: The Route of Entry and Localization of

Blood Proteins in the Oocytes of Saturniid Moths, J. *Biophys. Cytol.*, vol. 9, pp. 747–759, 1961.

Telfer, W. H.: The Mechanism and Control of Yolk Formation, *Annu. Rev. Entomol.*, vol. 10, pp. 161–184, 1965.

Telfer, W. H., and M. E. Melius, Jr.: The Mechanism of Blood Protein Uptake by Insect Oocytes, *Am. Zool.*, vol. 3, pp. 185–192, 1963.

Thorson, G.: Reproductive and Larval Ecology of Some Marine Bottom Invertebrates, *Biol. Revs.*, vol. 25, pp. 1–45, 1950.

Tyler, A.: Role of Fertilizin with Fertilization of Eggs of the Sea Urchin and Other Animals, *Biol. Bull.*, vol. 81, pp. 190–203, 1941.

chapter XIX

DEVELOPMENT AND REGENERATION

A. CLEAVAGE
B. BLASTULAE
C. GASTRULAE AND GASTRULATION
D. EMBRYOGENESIS
E. POSTEMBRYONIC DEVELOPMENT
  1. Larval habitats
  2. Larval types
  3. Metamorphosis
F. ATYPICAL DEVELOPMENT
G. POLYMORPHISM
H. GROWTH AND REGENERATION
  1. Autotomy
  2. Problems of regeneration and approaches to their solution
  3. Protozoans: *Stentor*
  4. Sponges
  5. Cnidarians
  6. Planarians
  7. Nemerteans
  8. Ectoprocts
  9. Annelids
  10. Arthropods
  11. Echinoderms
  12. Other invertebrates
I. CONCLUSION

Successful fertilization is followed after longer or shorter intervals by cleavage of the zygote. This entails a series of mitotic divisions, sometimes in rapid succession, sometimes more slowly, by which the zygote is divided into cells or blastomeres and its cytoplasm partitioned among them. Cleavage is not a period of growth but one primarily of nuclear division and cytoplasmic segregation, and it terminates when a ball or mass of cells, the blastula, is formed. The number of cells constituting a blastula varies in different species, as does the relative size of the blastomeres. Morphogenetic movements then begin, certain cells moving into positions within the ball or cell mass from which the endodermis, the mesenchyme, and, in the case of animals with three primary cell layers, the mesoderm are differentiated. This is the process of gastrulation, after which embryogenesis begins: the organization of structures and organs and their development into the larval or the juvenile condition; this is a period of growth as well as of histogenesis and morphogenesis. When this state has been achieved, the new individual emerges as an entity with more or less independence, depending again upon the species and the extent to which protection is given the young. Postembryonic development involves active growth of the structures laid down and of the organism as a whole and, in the case of larvae or nymphs, changes of forms and acquisition of new structures. In every case the growth pattern conforms to that characteristic of the species and size, symmetry and the relation of parts, as well as the parts themselves, are determined by information encoded in the genes. Environmental conditions, in their broadest connotation, may affect the development of these characters, or the final expression of the genetic messages, but these alterations of the basic pattern are incidental to the individual, or to the individuals in a population equally exposed to such environmental influences, and unless these are so drastic as to cause changes in the genome itself are not passed on to succeeding generations. The adult state is attained when the gonads and genitalia develop and function in the production and transmission of gametes. This may be early or late in postembryonic development, and the life of the adult may be brief or of considerable duration. There is no comprehensive data on the longevity of invertebrates; as adults, mayflies live only a few hours, just long enough to mate and for the females to lay their eggs; the life spans of rotifers and small aquatic crustaceans may be at best only a few months and that of other invertebrates several or many years. There is, too, little information about senescence or aging in invertebrates, many of which seem to die through accident rather than from natural causes.

A. CLEAVAGE

Cleavage may be total (holoblastic) or partial (meroblastic), depending largely upon the amount and distribution of yolk within the egg. According to this, eggs may be classified as isolecithal, with little yolk and that more or less evenly distributed throughout the cytoplasm; centrolecithal, if their yolk is concentrated in the center of the egg; and telolecithal, if it is accumulated at the vegetal pole. Eggs of most echinoderms are isolecithal; those of most arthropods centrolecithal; and those of molluscs and annelids telolecithal, divisible also into those that are moderately telolecithal, with no great accumulation of yolk, and those that are strongly telolecithal, with large amounts of yolk and the cytoplasm restricted to a narrow peripheral rim and a small mass containing the nucleus at the animal pole.

In holoblastic cleavage, each furrow passes completely through the egg, separating the cells as units, although, depending upon the amount and position of the yolk, cytoplasmic division may lag behind nuclear division. In the centrolecithal eggs of Anthozoa, for example, cytoplasmic division begins only after the fourth nuclear division, when there are 16 nuclei in the peripheral cytoplasm. Cytoplasmic division is delayed also in the eggs of crustaceans, spiders, and insects, in which a multinucleate condition exists for some time before the cells are delimited. This is a process of cellularization similar to that postulated to have occurred in the origin of metazoa from multinucleate Protozoa.

Holoblastic cleavage may be equal or unequal, in that the cells formed as a result of division may be of similar or dissimilar size; early equal cleavages may be followed, as they are in sea urchins by unequal ones in which smaller cells or micromeres are cut off from large ones, or macromeres. Equal holoblastic cleavage is typical of sponges, cnidarians, ectoprocts, chaetognaths, and echinoderms; there are, as is usual in biological processes, some exceptions to the typical pattern. Unequal holoblastic cleavage is typical of ctenophores, platyhelminths, "aschelminths," nemerteans, brachiopods, molluscs, with the exception of cephalopods, onychophorans, and annelids. The differences between these two types in the deuterostomatous and protostomatous lines of invertebrate evolution have been mentioned in Chapter V in relation to phylogeny (see Fig. V–9). Cleavage of cephalopod and scorpion eggs is meroblastic, for they, like the eggs of birds and reptiles, are strongly telolecithal. The aberrant cleavage of other arthropods in its delayed cytokinesis eliminates them from any simple categorization.

Unequal cleavage is often spiral, resulting in early segregation of different areas of the zygote's cytoplasm with evident, or implied, differences, in the inclusions. That these are concerned, most likely as substrates or possibly cofactors in enzymatic reactions, in the differentiation of cells giving rise to different tissues has been shown by following their fates, when they can be identified, or by their destruction or removal. In nematodes, cleavage, though not spiral, is determinate, and from the first division on, the future germ cells are distinguishable from those that will form the other parts of the body; the germ line is thus early set apart from the somatic line. This has been most convincingly followed in *Ascaris megalocephala*, an intestinal parasite of horses with two different strains, *A. megalocephala bivalens* in which the primordial germ cells have only four chromosomes and the gametes, therefore, two, and *A. megalocephala monovalens*, in which the primordial germ cells have two chromosomes and the gametes one. The zygotes, therefore, have four or two chromosomes, respectively. These chromosomes are large and at the first cleavage are apportioned to each blastomere, one of which is slightly larger than the other. Before the next cleavage of this large blastomere a major part of each chromosome is eliminated from the nucleus, and the expelled masses can be recognized in the cytoplasm for some time, but gradually disappear from it. The remaining portion of each chromosome breaks up into a number of small bodies that, at the next cleavage of this blastomere, behave like orthodox chromosomes and are aligned on the spindle and distributed to the daughter cells in a perfectly regular fashion. This follows in all subsequent divisions of this cell. In the smaller of the two blastomeres, the large chromosomes remain intact and at division pass to its two daughter cells. In one of these blastomeres, however, elimination of chromosomal material takes place just as it did in the larger one of the first two blastomeres, and its daughter cells and their mitotic descendants have the same number of small chromosomes. When the other blastomere divides, the course of events is similar, with chromosomal elimination in one of the daughter cells and not in the other. There is thus preserved throughout cleavage one cell with large chromosomes like those of the zygote. This cell divides only a few times in the course of embryogenesis and becomes the parent cell of the genital organs and the germ cells of the adult; it is not included in the construction of any other organs.

Conditions within the egg determine the pattern of cleavage and, in some instances at least, characteristics evident in early embryogeny or even later in development. The egg is formed under the influence, and the information, of the maternal genotype and the effect of the paternal chromosomes introduced into the zygote may not become evident for some time. They have, for example, no influence whatsoever upon the development of females of nematodes of the genus *Rhabditis*, for here the sperm only activates the egg and then degenerates, and since syngamy does not take place the eggs develop essentially parthenogenetically. Maternal influence upon the cleavage pattern of the snail *Limnea peregra* is evident at the first two cleavage divisions, when the direction of coiling of the shell is determined by the direction in which the spindle swings and the blastomeres slide (see Fig. V–9*b*). In this species, as in other snails, there are strains in which the shell is coiled in a clockwise (dextral) or counterclockwise (sinistral) fashion. This difference is clearly evident in looking at a snail shell with its opening directed toward the observer to whom it is either on his right or his left, or an observer looking at its apex will see that the spiral follows the course of either a right-handed or a left-handed screw. *Limnea* is a self-fertilizing hermaphrodite but can also be cross-fertilized, and genetic tests have shown that dextrality is a characteristic that is dominant (+) to sinistrality (s). Homozygous dextral (+ +) or sinistral (ss) individuals produce only dextral or sinistral progeny, but in crosses made between dextral and sinistral individuals the F_1 generation all show the direction of coiling of the egg-producing parent. Thus, if the eggs (s) of a sinistrally coiled female are fertilized by sperm (+) from a dextrally coiled male, the cleavage and ultimate direction of coiling of the shell will be that of sinistral individuals, in spite of the presence of the dominant gene (Fig. XIX–1). In the reciprocal cross the spindle swings, and the shell coils, to the right, as would be expected in the presence of a dominant gene; but though the F_1 progeny in each cross is heterozygous (+s) and, so, genetically alike in respect to these two alleles, the genotype of the sinistral mother has imprinted the pattern for cleavage upon the cytoplasm of her eggs, and all her offspring of that immediate generation resemble her in respect to the direction of the coiling of their shells. The members of this generation from both crosses, however, form their eggs under the influence of the heterozygous genotype; all the members of the F_2 generation, after self-fertilization, are dextrally coiled, although some will be genetically heterozygous and some genetically homozygous for either the dominant or the recessive gene. In the F_3 generation, again after self-fertilization, there are three dextrally coiled individuals to every one that is sinistrally coiled, a mendelian ratio that would be expected in the F_2 generation but that fails to materialize because of the influence of the maternal genotype upon some as yet undisclosed feature of the cytoplasm of her eggs.

Another instance of maternal influence has been demonstrated in *Ephestia*. Some of these moths, carrying a dominant allele (+), are able to produce the pigment precursor kynurenin, while others, homozygous for its recessive allele (a) cannot. Kynurenin is diffusible and is included in the cytoplasmic components of eggs of females either homozygous or

Genotypes of parents								
	♀ ss				♂ ++			
Gametes	s				+			
Genotypes and phenotypes of F_1	s+ sinistral							
Self-fertilization								
Gametes	s				+			
Genotypes and phenotypes of F_2	ss dextral		s+ dextral		+s dextral		++ dextral	
Self-fertilization								
Gametes	s	s	s	+	+	s	+	+
Genotypes and phenotypes of F_3	ss sinistral		s+ dextral		+s dextral		++ dextral	
Ratio	1		: 3					

Genotypes of parents								
	♀ ++				♂ ss			
Gametes	+				s			
Genotypes and phenotypes of F_1	+s dextral							
Self-fertilization								
Gametes	+				s			
Genotypes and phenotypes of F_2	++ dextral		+s dextral		s+ dextral		ss dextral	
Self-fertilization								
Gametes	+	+	+	s	s	+	s	s
Genotypes and phenotypes of F_3	++ dextral		+s dextral		s+ dextral		ss sinistral	
Ratio	3						: 1	

FIG. XIX–1 *Results of reciprocal crosses of dextrally and sinistrally coiled snails* (Limnea peregra).

heterozygous for the dominant gene. If a heterozygous female is mated with a male homozygous for the recessive gene, the zygotes will be either heterozygous or homozygous for it, according to the scheme represented in Fig. XIX–2. Yet all the larvae show some pigmentation, though in half of them this fades after a few molts and adults with the genotype aa lack pigmentation. Enough kynurenin was, however, contained in the eggs and distributed to the cells in cleavage to permit further steps in the synthesis of a limited amount of pigment, but this stopped when the supply was exhausted and the doubly recessive individuals were, thereafter, colorless.

Another, and more abstruse, illustration of maternal influence upon early development has been shown in interspecific crosses of sea urchins in which the rates of cleavage and the characteristics of the skeletal rods supporting the larval arms differ in the species used. Under standard environmental conditions, the rate of cleavage has been found to be that characteristic of the maternal parent, whose influence extends in some cases up to the pluteus stage, as evidenced in the construction of the skeletal rods. These crosses have supplied interesting data on the times at which paternal genes make some of their influences evident.

Polarity of the embryo is, as a rule, related to polarity of the egg, anterior structures being developed from the animal pole and posterior from the vegetal. All animal eggs exhibit polarity, acquired during oogenesis and sometimes conveniently evident in the eccentric position of the nucleus or in the distribution of yolk or other visible inclusions. It may also be evident in the shape of the egg, as in the ovoid ones of some sponges, of cephalopods, and of insects. But, also, it may not be so readily apparent and exposed only when upset by accident or experimental design.

Genotypes of parents	♀+a		♂aa	
Gametes	+	a	a	a
Genotypes of F_1	+a	+a	aa	aa

FIG. XIX–2 *Results of matings in* Ephestia *of females heterozygous for gene a (incapacity for production of kynurenin) with males homozygous for the gene.*

Polarity is imposed on the oocyte, early or late in its development, primarily by its relations to the tissues surrounding it, whether in diffuse egg formation by adjacent cells or in localized, by those of the ovary or the follicle. In *Sycon,* for instance, the egg cell lies against the wall of a flagellate chamber, with its long axis parallel to the layer of choanocytes. The cleavage divisions cut across this axis; the short axis, at right angles to the choanocytes, is the main axis of the embryo. This presents an unusual condition, for the polocytes are extruded near one end of the long axis, and in most eggs the location of these abortive cells marks the animal pole. In cnidarians, the polarity of the eggs corresponds to that of the parental epithelium from which they were derived and which is, in its turn, related to their orientation with a basal end resting on the mesoglea and a free end directed away from it. In localized egg formation, polarity may be related to the point of attachment of the oocyte to the germinal epithelium of the ovary; this usually becomes the vegetative pole. In lamellibranchs this point is marked by the micropyle, but in echinoderms the micropyle develops at the opposite—the animal—pole by a pseudopodial extension of the cytoplasm that projects through the egg membrane and its outer layers. Conditions in the follicle may also have reference to the polarity of the egg, as is evident in cephalopods, where the granulosa is most strongly folded toward the vegetal pole of the oocyte, and in the beetle *Acanthoscelides,* where the closest contact between oocyte and follicle cells is made in the vegetal hemisphere. These external manifestations are merely landmarks that indicate an intrinsic polar axis which seems to lie, according to evidence from experimental embryology, in the highly organized cortex of the ripe egg. Displacement of visible inclusions by centrifugation or other physical means does not lead to abnormal development, either because there has been no fundamental disturbance of organization or because there has been some readjustment of any disturbances that may have been caused.

Bilateral symmetry which anticipates the symmetry of the embryo is sometimes apparent in the ripe egg. This is most obvious in the elongated eggs of cephalopods and insects; in those of other invertebrates it may become evident, after fertilization, in the rearrangement of certain inclusions, probably also reflecting an intrinsic symmetry that influences embryonic development.

Eggs, however, of different species differ markedly in the extent to which the cytoplasmic organization is definitive and localization so precise that cleavage results in blastomeres with qualitative differences in their cytoplasms. Separation, by mechanical or chemical means, of blastomeres at different cleavage divisions in different species and observation of their development as isolated cells have shown the extent to which their courses of development are specified. The first eight blastomeres of cnidarian eggs, for example, can develop independently into larvae with normal proportions; in some instances, these have resulted from blastomeres isolated at the 16-cell stage. The limit to normal development of these isolates seems to be set by the quantity, rather than the quality, of the material included in each blastomere, each apparently containing everything necessary for the complete development. Blastomeres of sea urchin eggs isolated at the two-cell and four-cell stage, and occasionally at the eight-cell stage, will develop into larvae that are smaller than usual, but perfect in their structure and proportions. Cleavage in such eggs as these is considered indeterminate, in that no blastomere up to these stages at least is inevitably destined to follow a fixed line of development, and each is totipotent as far as further differentiation goes. But conditions in eggs with determinate cleavage are not like this, and qualitative differences between blastomeres manifest themselves early. Though, for example, cleavage may continue in isolated blastomeres of mollusc or annelid eggs and they may even pass through blastula and gastrula stages, what develops from them are aberrant forms, far from perfect larvae. The isolated cells may give rise to the kind of structure they would have contributed to a perfect larva, thus expressing their own potentialities, but are incapable of developing in other ways. These differences reside largely in the cytoplasm rather than in the nuclei of the cells, for the nuclear divisions are usually equivalent in respect to chromosome distribution, whereas the cytoplasmic divisions segregate cytoplasmic components that provide the basis for evocation and action of different elements in the genome.

Oogenesis and the organization of oocyte and egg are thus critical factors to cleavage and embryogenesis and to the future individual. They are factors for which the egg-producing parent and her ancestry are primarily responsible, expressing in them not only her own genetic constitution but that of her progenitors and their evolution, and also the conditions to which she was exposed during her own development and maturity.

B. BLASTULAE

Cleavage results in the formation of a "blastula," a term derived from the Greek word *blastos*, meaning "germ" or "sprout." Blastulae are distinguished as "coeloblastulae," from the Greek word *koilos*, meaning "hollow," or as "stereoblastulae," from the Greek word *stereos*, meaning "solid," depending on whether they have a central, fluid-filled cavity, the blastocoele, or whether they are a solid mass of cells without such a cavity. The type of blastula is definitely referable to the organization of the egg from which it arises, particularly in respect to its polarity and distribution of the yolk. Coeloblastulae result from holoblastic and from meroblastic cleavage. In coeloblastulae such as those derived from holoblastic cleavage of the isolecithal eggs of some echinoderms, the blastocoele occupies the center of the hollow ball, whose wall consists of a single layer of ciliated cells, approximately equal in size [Fig. XVIII–12*d* (1) and (2)]. This little ball swims vigorously through the water. In the coeloblastulae of molluscs and annelids, arising from the holoblastic cleavage of moderately telolecithal eggs, the blastocoele is eccentric in position, displaced toward the animal pole. Its roof is formed of small cells and its floor of larger, yolk-filled ones. In the coeloblastulae of cephalopods arising from meroblastic cleavage of their strongly telolecithal eggs, the blastocoele is a narrow, slit-like cavity, roofed over by a disk of small cells; beneath it is the uncleaved yolk. Such a blastula, characteristic also of reptiles and birds, is known as a discoblastula. The blastulae of sponges are also coeloblastulae with a small blastocoele, but they represent an atypical condition, an amphi- or diploblastula in which the future choanocytes and pinacocytes are already delimited; it is, therefore, more a gastrula than a blastula.

Stereoblastulae may arise also after cleavage of isolecithal and moderately telolecithal eggs, if the inner ends of the blastomeres do not round off and draw apart from each other to make the blastocoele. Such a blastula is, for example, characteristic of alcyonarian cnidarians, although a coeloblastula is characteristic of the closely related zooantharians. Stereoblastulae are also products of cleavage in centrolecithal eggs; in those in which the cleavage furrows do not extend into the yolk, periblastulae may be formed, consisting of a rim of cells surrounding a central yolk mass. This is the antithesis of conditions in the ectolecithal eggs of some flatworms, where the yolk is not included in the eggs but surrounds them; their development is so aberrant that attempted homologies of their developmental stages with those of other invertebrates have little significance.

C. GASTRULAE AND GASTRULATION

Conversion of a blastula into a gastrula is marked by the onset of morphogenetic movements of the blastomeres. The term "gastrula" is derived from the Greek word *gaster*, meaning "stomach," and gastrulation, therefore, implies segregation of gastrodermal from ectodermal cells and establishment of two functionally different cell layers and of the future digestive area. Associated with it is the origin of another category of primary cell types, mesenchyme and mesoderm. Gastrulation can be performed in a number of different ways, of which the most common are invagination, epiboly, involution, delamination, and ingression. The method seems to be determined largely by mechanical conditions, related to the type of blastula and so, ultimately, to the distribution of yolk in the egg. While one of the methods listed above is usually typical of a given species, in some two of them may be combined to bring the cells of the blastula into their proper spatial relationships.

Some coeloblastulae gastrulate by invagination, the inturning or inpushing of cells at the vegetal pole [Fig. XVIII–12*e* (1) to (4)]. This may result in formation of a coelogastrula, with a hollow tube, blind at its inner end but open at the other through the blastopore, extending up into the blastocoele. The cavity of the tube is known as the gastrocoele (archenteron). Or it may result in a stereogastrula as, for example, in some nemerteans where the inturned cells form a solid mass in the blastocoele completely filling it. Gastrulation by epiboly (Greek *epiballein*, "to throw upon") usually follows unequal holoblastic cleavage, with the smaller cells at the animal pole moving over the larger ones at the vegetal pole and ultimately enclosing them, so that they come to lie within the blastocoele and form either a coelo- or a stereogastrula. In ctenophores, gastrulation is effected by both invagination and epiboly, for while the smaller apical cells are moving over the basal cells, these larger ones are concurrently moving into the blastocoele. Discoblastulae gastrulate by involution, cells in one area of the margin of the disk moving in beneath the others and thus constructing a two-layered disk. The location of the migrating cells is usually associated with the symmetry of the future organism and implies a difference in their potentialities established during cleavage. Involution results in a discogastrula, a special kind of coelogastrula with an incomplete gastrocoele, for this is not a tubular structure but one whose wall is a sheet of cells spread out over the yolk.

Delamination gives rise to stereogastrulae. It may occur either by tangential divisions of the blastula cells, resulting in an inner mass of cells that completely fills the blastocoele and is enclosed in an outer layer of cells, or by outward movement of some of the blastula cells in a sorting-out process of future gastrodermal from future ectodermal cells. Delamination in this way usually entails flattening of the cells that have moved outward, as they are, in appearance at least, stretched around the compacted inner core of those from which they have migrated. Gastrulation by ingression is the reverse of this, with cells migrating into the blastocoele and filling it. Ingression may be multipolar, with cells moving in from a number of places on the blastula wall, or it may be unipolar, with cells moving in from a single place; in this case, there is little distinction between ingression and invagination that results in a stereogastrula. Ingression, like invagination and involution, implies qualitative distinctions between the cells of the blastula, distinctions that may not be externally evident. Polar ingression, accompanied by delamination, is typical of hydrozoan gastrulation.

While gastrulation is an extremely critical event in the development of an organism, differences in the manner in which it is performed and the type of gastrula resulting are not of great significance, but rather represent adjustments to conditions arising from the presence of different amounts of deutoplasm and to the difficulties such inert material presents to free movement of the blastomeres. The critical factor is the creation of a two-layered system, with the positioning of cells so that they may realize their full potentialities of development and differentiation as ectodermis and its derivatives and gastrodermis and its derivatives. In stereogastrulae, as in coelogastrulae, the cells that have moved into the blastocoele will provide the walls of the archenteron, for the solid mass eventually hollows out or rearranges itself so as to make a tubular structure, which, even in the absence of a blastopore, will at some stage acquire an opening to the exterior.

In some cases, coincident with the events of gastrulation and always related to them, there is the appearance of a third primary tissue, the "middle" tissue, lying between the ectodermis and the gastrodermis. This may be in the form of mesenchyme with scattered cells and much intercellular substance, either fluid or more or less viscous, or of mesoderm with cells in contact or closely adjacent to each other, and hence epithelial in character. The middle layer may be derived from ectoderm cells or from endoderm; in species with determinate cleavage like that of polyclad flatworms, gastropod molluscs, and annelids, the middle layer stems from one particular blastomere, the mesendoblast, which is set apart early in cleavage (Cell 4D in Fig. V–9*b*), and is included in the group of cells occupying the blastocoele at the end of gastrulation. Cells of ectodermal origin are always mesenchymatous in character, and originate by ingression into the area between ectodermis and gastrodermis. This is the source of the cells in the mesoglea of sponges, cnidarians, and ctenophores. Cells that arise from endomesoderm may be mesenchymatous or epithelial in character. The time and sites of the origin of this middle layer differ according to species. It may appear early or late in the course of gastrulation, and its first appearance as primary mesenchyme may be followed by the later appearance of secondary mesenchyme or of mesoderm. The relation of this tissue to development of the coelom in eucoelomate species has been pointed out in Chapter II, pages 51–54, as has the relation of the blastocoele, as the primary body cavity, to that of the coelom, the secondary body cavity, and to the pseudocoele of Acanthoceophala, "Aschelminthes," and Endoprocta.

D. EMBRYOGENESIS

At the end of gastrulation, groups of cells with potentialities of developing in different ways are brought into positions where these potentialities can best be realized and where they are exposed to influences from overlying, or adjacent cells. The effect of certain cells and primary tissues upon others in directing or inducing their differentiation along particular lines has been brilliantly demonstrated by experimental embryologists, whose searches for the specific agents concerned in such induction are still in progress and vigorously pursued. Embryonic induction is another instance of communication between cells by chemical means, in this case transmission of information to differentiate in specific ways and to cooperate in the construction of specific organs. For at the end of gastrulation the main axes of the embryo are laid down, and histogenesis and organogenesis begin. Histogenesis of specialized tissues and subsequent organogenesis follow a general course that may be outlined as follows: From ectoderm are developed stomodaeum and proctodaeum; nerve cells and the epidermis with its manifold derivatives—glands, spines, hairs, etc.—and respiratory areas. From gastroderm (endoderm) are derived the inner wall of the digestive areas of the gut and the glandular structures that are associated with them. Mesenchyme and mesoderm give rise to supportive (connective), transportive, and muscular tissues; to the walls of the coelom in eucoelomate animals; and to at least parts of their reproductive and excretory systems.

The course of embryogenesis has been followed in detail in a number of invertebrate species. Each course is adapted to conditions within the species and to those under which development must proceed.

The variations in detail of the developmental processes, as would be anticipated among organisms so diverse as the invertebrates, are so great that adequate consideration of them can only be given in a book devoted exclusively to invertebrate, or to comparative, embryology. But, as in all organisms, embryogenesis in invertebrates is not accomplished by continuous growth and differentiation of all parts equally, but rather by periods of rapid growth and differentiation in certain parts and relative quiescence in others that in their turn become active at appropriate times. Development is thus a series of bursts and cessations of mitotic and metabolic activity in particular regions of the embryo and, normally, is a tightly controlled procedure resulting in the orderly arrangement of parts of the body with the symmetry and dimensions characteristic of the species. The nature of the factors arousing localized activity and of those suppressing it, and the mechanisms by which balance between the two is maintained are problems common to the developmental biology of all organisms. When the balance is in any way upset, and activity goes on when it should go off, or continues beyond the programmed time, or fails to begin when it should, development becomes abnormal; if the imbalance results from genetic changes, the altered pattern of development may be transmitted to the next and subsequent generations and thus subjected to selection pressures and either preserved or eliminated. Heterochrony, as a change in the timing of developmental processes, has been mentioned in Chapter V in respect to phylogenetic relationships.

Checks imposed during embryogenesis determine not only the symmetry of an organism in respect to its various parts but also their proportionate sizes and the dimensions of the organism resulting. The growth that takes place during this period of an individual's life results in most cases in an organism considerably larger than the egg from which it came and one composed of an indefinite number of cells. "Aschelminthes" are conspicuous exceptions to this latter statement, for the number of cells in the body of a juvenile or an adult is fixed within narrow limits. Based on counts of visible nuclei, there are, for example, 180 cells included in the epidermal syncytium of the rotifer *Epiphanes senta*, 19 in the pedal gland, 104 in all the musculature of the body, 28 in the protonephridia, 157 in the digestive tract, 8 in the yolk gland, 3 in the oviduct, and a total of 246 neurons. The number of reproductive cells in the ovary is variable, ranging from 10 to 20. The epidermis of the nematode *Rhabditis* contains 120 cells, the digestive tract 172, and the nervous system 200. In the parasitic Acanthocephala, the cells, with the exception of gonocytes, are also few in number and rigidly determined in position. This limit on cell number sets a limit to the size of juveniles and adults and precludes the possibility of further growth by cell division. Size limitations, but by no means so restrictive ones, are imposed upon other animals as well, for each species has an average size range at each stage of its development. The common earthworm of temperate regions *Lumbricus terrestris*, for example, hatches as a juvenile a fraction of an inch long and grows to an adult 8 to 12 in. long and some $\frac{3}{8}$ in. in diameter. Adults of the giant earthworms *Rhinodrilus fafneri* of Ecuador and *Megaloscolides australis* of Australia are some 7 ft long and 1 in. in diameter. Size limitations to parts are also imposed; most embryos are bilaterally symmetrical in all respects, any asymmetry characteristic of their species arising in postembryonic development. This is evident, for example, in gastropod molluscs, asymmetric after torsion; in decapod crustaceans, after the first few molts, when allometric growth of the first pair of pereiopods results in one that is much larger and heavier than the other; and in echinoderms, where the pentamerous pattern of the adult is imposed upon the bilateral symmetry of the larva, starting as the more rapid development of structures on one side of the body than on the other. Environmental conditions may also affect the relative sizes and relationship of parts. This can be readily demonstrated in the brine shrimp *Artemia* whose eggs, developing under different conditions of salinity, hatch out as individuals with remarkably different cephalic structures.

Retention of numbers of undifferentiated or embryonic cells throughout periods of active histogenesis, a common phenomenon in invertebrates, also presents challenging problems to students of developmental processes. These cells are remarkable in that their differentiation and to a considerable extent their multiplication are held in check throughout the entire course of embryogenesis. Precise localization of them in the imaginal disks of insect embryos and larvae, and their activation in the pupae to form specific adult structures dramatically illustrate the extent of the control that may be exercised during the development of an organism.

E. POSTEMBRYONIC DEVELOPMENT

Embryogenesis may be said to terminate with the emergence of the individual as a larva, a nymph, or a juvenile. In each case it is capable of feeding and, with the exception of those that are brooded, of living an independent existence. The larval stage may arise very early in the course of development, with much left to be done in construction of larval organs and characters before those of the adult are even begun, or the larva may be a well-finished product of embryogenesis. The dipleurula of an echinoderm is, for example, little more than a gastrula in an advanced stage, and it undergoes profound changes before it becomes a typical larva of its class and species. The trochophore of annelids and the veligers of molluscs are more advanced in their development. Insect larvae, when they hatch, although very different from the adults into which they will ultimately transform, have complete sets of larval organs and are highly organized and differentiated systems. Differences in habits of life as well as differences of form distinguish larvae from adults of their species. Nymphs, though less different in form, may also live in ways other than those of the adults of their respective species or in similar ways. Juveniles are for the most part miniature adults and at once exhibit the habits of adults.

In every major metazoan invertebrate phylum there are species with a larval phase in the course of their ontogenies. Cnidaria is an exception to this generalization for the planula, though often referred to as a larva, is really a late embryo. The significance of larval forms in relation to phylogeny has been mentioned in Chapter V, but whether such indirect development is a primitive characteristic of all metazoan invertebrates or an adaptive one that has arisen in the course of their evolutions remains a moot point. Larval stages have survival value to a species in permitting rapid embryogenesis and, therefore, providing possibilities for the production of large numbers of eggs with little yolk and so with minimal stress upon maternal resources and metabolism. They are also advantageous as a means of distribution of the species, for larvae may travel, either by their own means of locomotion or carried by air, by other animals, or by water currents, considerable distances from the site of their hatching before they become adults. Such dispersal is of especial importance to sedentary and sessile species, which as adults never move far, if at all, from one place. There are, on the other hand, many hazards in indirect development, for free larvae are exposed to environmental vicissitudes as well as to predation and the critical and potentially dangerous physiological conditions of their own metamorphosis. These dangers are averted with direct development.

1. Larval habitats

Larval life may be spent in the water, either salt or fresh, or in or on the ground. The habitat in most cases is that of the parents, but some terrestrial species in which the adults are primarily or entirely terrestrial lay their eggs in the water and the larvae live there; and some aquatic species in which the adults have successfully adapted to life in freshwater or in waters of low salinity return to the sea to spawn, and their larvae live there or near the mouth of an estuary. Illustrations of these adaptations already cited are the aquatic larvae of flying insects and the zoeae of land crabs and of *Eriocheir*. Aquatic larvae may become members of the teeming plankton, feeding on its minute organisms and becoming themselves food for surface-feeding animals of larger size; many perish in this way, as well as through the physical hazards of their environment—water turbulence, rapid and pronounced changes in temperature, etc. Larvae of other species may live attached to the substrate, or in tubes or cases that they make for themselves as the larvae of caddis flies do, or attached to the body of another animal in a temporary or permanent parasitic relationship as, for example, the glochidia of *Anodonta* and the cypris larvae of *Sacculina* (see Chapter V, pages 147–148). Terrestrial larvae may live and feed above the ground, as do the caterpillars of lepidopterans, the maggots of dipterans, and the nymphs of orthopterans; or they may live below the ground, like the grubs of beetles; or they, too, may live as parasites attached to some other organism. In some fully parasitic species there is no free-living stage in the entire life cycle, though the larvae may inhabit hosts different from those of the adults, or even a succession of different hosts, and metamorphose only when in the body of the final host.

2. Larval types

The amphiblastulae or diploblastulae of sponges and the planulae of cnidarians are considered larvae because of their external ciliation and mobility. Actually, the former, as their name implies, are coeloblastulae and the latter, stereogastrulae. The parenchymula, or parenchymella, larvae of some sponges are more planula-like in form. They are essentially stereoblastulae with the important difference that they show the peculiarities of sponge development, in that the future internal cells, or endoblasts, compose the ciliated outer covering; and the ectomesenchyme cells, which will produce the epidermal cells and the mesenchyme, compose the inner cell mass. In planulae these tissues are reversed and from the outset are in their definitive locations; their relations are attained in the parenchymula by migration of the outer cells inward and of the inner cells outward. Both amphiblastulae and parenchymulae after a day or two of free swimming settle down upon some smooth surface and gradually develop the features characteristic of their species. In a few kinds of calcareous sponges the larva transforms first into an olynthus, a hollow cylindrical object, attached at the base and with a single opening, the osculum, at the free end; the wall is composed of an outer layer of epithelial cells and of porocytes, and an inner layer of choanocytes, with mesenchyme cells in the gelatinous matrix between them, some of them already, as scleroblasts, secreting spicules. The olynthus, a transitory stage in the development of these sponges, is thought to represent an ancestral form from which all modern sponges have originated.

The planula is considered the "typical" larva of cnidarians and its position relative to the evolution of other metazoan types has been considered in Chapter V. In relation to cnidarian development, it, too, is but a brief stage in the life history of the individual, for it soon settles on some suitable or convenient substrate and develops mouth and tentacles characteristic of the polyp phase. A free planula stage is bypassed in the development of some hydrozoans and actinozoans; in the hydroid *Tubularia* it is passed in the gonophore or rudimentary medusa that remains a permanent part of the parent colony. Here the planula develops a mouth and tentacles and becomes an actinula (Greek *aktinos*, "ray"), after which it is released to creep along the bottom of the sea or join the plankton. It soon attaches to some substrate by its aboral end, grows in height, and develops buds from its sides as well as creeping stolons from its base, from which upright tubular structures arise. In this way the colonial structure typical of the genus is formed. In hydrozoans of the order Trachylinae, in which the polyp stage is suppressed, the larvae may be either free or parasitic. In *Gonionemus*, for example, so often used in teaching laboratories as an illustration of medusoid organization, the planula becomes a minute sessile individual; lateral buds develop which become constricted off when they are about four to five times as long as they are wide. They are not frustules, but stolon-like outgrowths. After a brief period of creeping about they reattach and develop medusae through formation of an entocodon (Greek *ento*, "within," and *kodon*, "bell"), an essential part of medusa production in hydrozoans. This occurs only at the tip; it is evident first as an evagination of ectodermis and gastrodermis in a strictly localized region and may be apparent in the bud even before it is set free from the planula. The ectodermis proliferates rapidly, forming a solid mass of cells, which is the entocodon, beneath which is a single layer of gastrodermis, in which cell division proceeds at a much less rapid rate. The entocodon grows proximally and acquires a cavity, the future subumbril space. The ectodermis continues to proliferate distally, making a mass of cells that also acquires a cavity; this is the tentacular cavity. The mass of tentacular epidermis separates from the apical epidermis, which takes no further part in construction of the medusa. Growth at an increased rate in the gastrodermis underlying the entocodon occurs at four points, the primordia of the future radial canals, which grow distally around the entocodon and at their ends fuse to form the circular canal. A fifth area of localized gastrodermal growth beneath the center of the subumbril cavity gives rise to the manubrium. In this way, through controlled and localized growth, the basic structures of the medusa are laid down, formation of the entocodon being, apparently, the critical event that instigates activity in the outer growth centers.

Some narcomedusae develop through actinulae that parasitize other organisms, sometimes other medusae. *Polypodium hydriforme* is parasitic in its actinula stage in the eggs of sturgeons. These fish are benthic in habit and the way in which they become infected is completely unknown, yet females caught on the spawning runs in the Volga River and opened before their eggs are laid usually contain a considerable number with actinulae in them. The stolon of the actinula is coiled in a spiral inside the egg and turned inside out so that the gastrodermis is in contact with the yolk; buds grow into the yolk as a series of evaginations from the stolon and while still inserted in it develop 12 tentacles. When the sturgeon spawns, the basal part of the stolon splits lengthwise, and the entire body, buds and all, everts so that the gastrodermis is again inside and the ectodermis, outside. Each bud, in everting, carries along some yolk in its gastral cavity and is thus provided with food until it acquires a mouth. The creature crawls along the river bottom on its tentacles; 12 more of these develop, after which it divides along the midline and two polyps with 12 tentacles each result from this fission. Each polyp divides into two individuals with six tentacles apiece; six more tentacles then grow out and several months after the sturgeon has laid her eggs four *Polypodium* polyps, with either 12 or 24 tentacles, have emerged from each egg that in some as yet obscure fashion became infected. The 12- or 24-tentacled individuals have been found on the river bottom, but never a six-tentacled stage. There are many points still to be investigated in this bizarre life history, which includes, among other things, polyembryony through budding and through longitudinal fission of a single polyp.

Although development in the majority of turbellarian flatworms is direct, some species of polyclads pass through a larval stage known as Müller's larva, named for the German biologist Johannes Peter Müller (1801–1858) whose biological interests were wide but particularly centered on marine and invertebrate biology. Müller's larva has a small ovoid body with eight ciliated lobes arising just below the equator [Fig. XIX–3*a* (1) and (2)]. One of these is median in position and overhangs the mouth, another is mid-dorsal, and the others are lateral. Their positions would indicate that they have arisen from cells $1a^2$ to $1d^2$ of the spirally cleaving egg (Fig. V–9*b*) and that they are, thus, homologous with the prototroch of a trochophore. Moreover, the cilia fringing each lobe are part of a continuous band that encircles the larva. Homologies have been drawn between this larva and ctenophores in arguments for phylogenetic relationship between them and flatworms, based to some extent upon the fact that the cilia on the lobes of the larva are laterally united into combs like those of the swimming plates of ctenophores and that the number of lobes, eight, corresponds to the number of ciliated bands in ctenophores. The larva of the polyclad *Stylochus* has, however, only four ciliated lobes and is known as Goette's larva, after Alexander Wilhelm Goette (1840–1922), a German professor and embryologist with a great interest in evolution. Müller's larva spends several days in the plankton and then sinks to the bottom. There, its transformation into the adult form occurs through elongation and flattening of the body, with reduction and, finally, disappearance of the ciliated lobes; development of a pharynx and adoption of the method of feeding of adult polyclads; and organization of the internal organs into the pattern typical of its species.

Trematodes develop indirectly, through a succession of larval stages often passed in different hosts. Such a life cycle, as it is known in the sheep liver fluke, is represented in Fig. I–16. The free-living miracidium is a short phase; and the longer ones of postembryonic development are passed in the more protected environments of another animal's body. Larval stages are also typical of most cestodes; each has distinctive ordinal or generic features. An early stage that they have in common is the onchosphere (Greek *onkos*, "hook"), an embryo with hooks at the posterior end (10 in Cestoidaria, 6 in Eucestoda), enclosed within two internal membranes of its own construction and the outer egg capsule. Encapsulated onchospheres, liberated from aquatic hosts, escape from the capsule, shed the outer of the two internal membranes, and enter the water as ciliated larvae known as coracidia (Greek *korax*, "raven," and *idon*, dim. suffix meaning "smaller"). The ciliated layer is shed after the coracidium is eaten by an appropriate intermediate host, usually an arthropod, and the larva attaches by its hooks to the wall of the gut and, in some species,

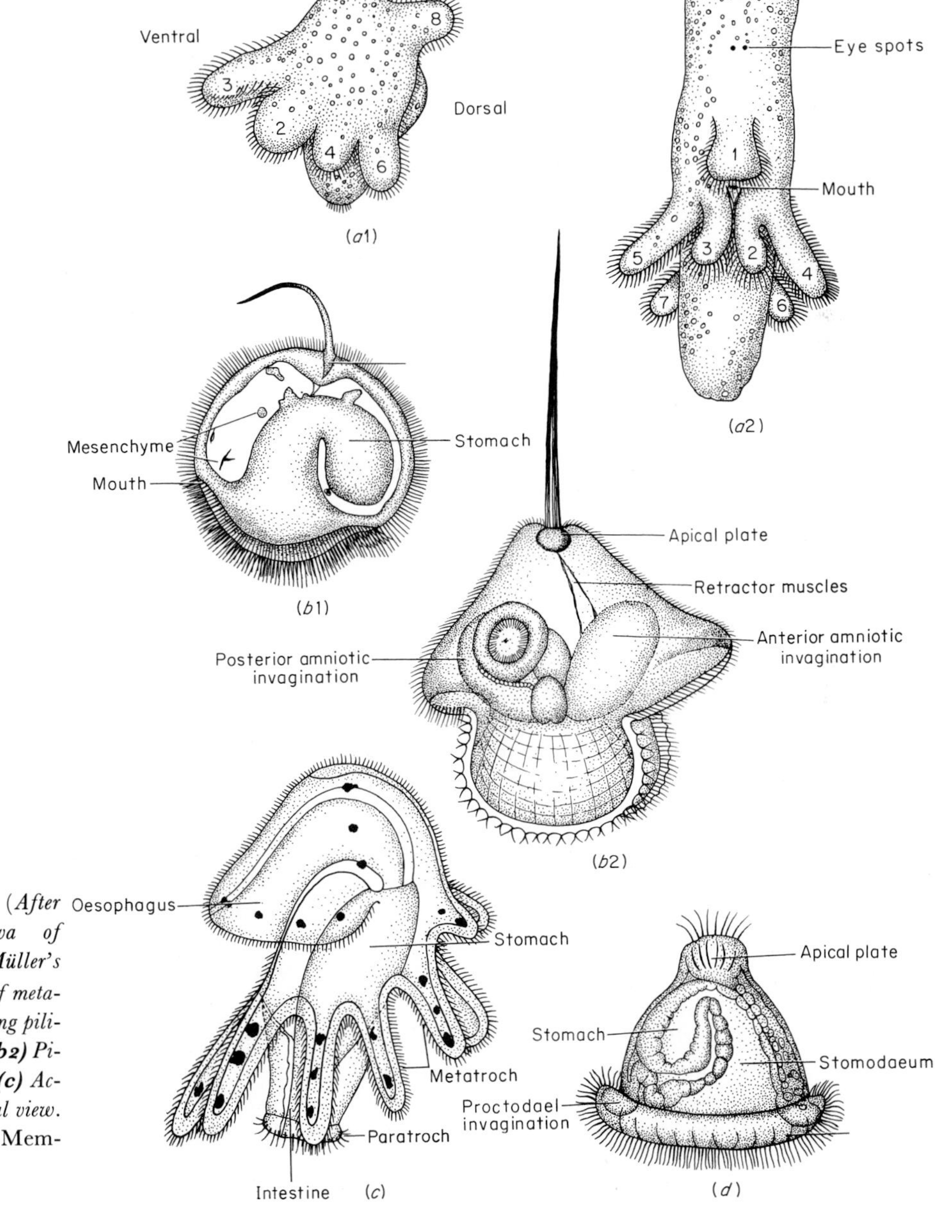

FIG. XIX–3 *Larval types. (After McBride.)* ***(a1)*** *Müller's larva of* Yungia, *lateral view.* ***(a2)*** *Müller's larva of* Yungia *at the beginning of metamorphosis, ventral view.* ***(b1)*** *Young pilidium of* Cerebratulus lacteus. ***(b2)*** *Pilidium just before metamorphosis.* ***(c)*** *Actinotrocha larva of* Phoronis, *lateral view.* ***(d)*** *Young cyphonautes larva of* Membranipora.

becomes a procercoid (Greek *kerkos*, "tail") by cell multiplication without much differentiation, enclosed in a thick cuticle. It remains in this state until the first host is eaten by a second one, usually a fish, in which the procercoid becomes a plerocercoid (Greek *pleros*, "full"), with an elongated body, a scolex with attachment organs, and primordia of the muscular, excretory, and nervous systems. The adult state is attained when the second host is eaten by the final one; this does not entail any drastic metamorphosis, but only the budding off of proglottids by strobilation and the development of reproductive organs in them. In species in which there is only one intermediate host, such as *Gyrocotyle*, which parasitizes elasmobranch fishes, the procercoid becomes an adult directly, but as such is essentially plerocercoid in habit and form, for it does not strobilate and remains a sac-like body in which reproductive organs develop. The onchospheres of other tapeworms develop either into cysticercoids or cysticerci. In each, the middle region of the onchosphere grows up around its anterior end, which later differentiates as a scolex, leaving the hooked tail free. In a cysticercus, the scolex is inverted into the fluid-filled cavity of the onchosphere or cyst; in a cysticercoid it is not. Cysticerci, known also as "bladder worms," may become very large through the accumulation of fluid in the cyst, amounting in some cases to a gallon or more. Beside the danger to the host in harboring a sac of such size with consequent compression of tissues and possible occlusion of passages, there is the added danger to another host of multiple infection from ingestion of a single onchosphere. Many heads are, for example, formed by a cysticercus of *Taenia coenurus*, which, lodged in the brain of a sheep, cause it to exhibit symptoms of "gid" or "staggers." The wild animal, wolf or fox, or the domesticated dog that kills and devours the sheep, made an easy prey because of its condition, acquires not one, but several, tapeworms when, in its gut, the scoleces are everted from the bladder and fasten to the mucosa. The parasites reach maturity by strobilation, and many thousands of ripe proglottids may be cast into the infected animal's intestine to be eliminated with its feces. Moreover, large bladder worms, known also as hydatid cysts (Greek *hydor*, "water") may bud off many more cysts, so increasing the infection of one host and providing multiple infection for the next.

Most heteronemerteans pass through a larval stage known, because of its external appearance, as a pilidium (Greek *pilidion*, "a little cap") [Fig. XIX–3*b* (1) and (2)]. At emergence from the egg, the embryo is ovoid, with a thickened band of cells having very long cilia, the prototroch, around its margin and a tuft of especially long and stiff cilia arising from a depression at the apical pole. Internally, there is a blind gut, with oesophagus and stomach, developed from the gastrocoele and mesenchyme cells in the blastocoele. The prototroch, which provides locomotion for the larva, soon grows downward on each side of the mouth as the marginal lappets, and the fully developed larva looks like a cap with ear flaps; hence, its name. As these changes are taking place externally, internally some of the mesenchyme cells differentiate as muscles, forming the retractor muscle of the apical plate, a sphincter around the mouth, and peritoneal muscles, which, in the lappets, effect their contraction and elevation so that they flap up and down. Nerve cells are also differentiated and form a band beneath the prototroch and its extensions into the lappets. Some of these are in close association with the tuft of stiff cilia at the aboral pole, composing an apical sense organ. At this stage the pilidium joins the plankton and swims and feeds there for about 2 weeks, after which changes begin that lead to its metamorphosis, a drastic and complicated procedure. Four ciliated invaginations of the ectoderm grow in on the flattened underside, two in front of the mouth on the right and left sides of the animal, respectively, and two in similar positions behind the mouth. These have been called amniotic invaginations and their inner portions, imaginal disks, without, however, any homology with the vertebrate amnion or the imaginal disks of insects. These invaginations grow until they extend upward over the surface of the stomach; then they fuse, enclosing the rudiment of the future nemertean in a cavity, whose inner wall will form its epidermis while the outer wall constitutes the amnion. One result of the growth of the amniotic invaginations is to divide the larva into an upper and a lower half. The latter contains the alimentary canal, and this part, cut off from the upper half, drops to the bottom of the sea and commences its life as a juvenile member of its species. The upper half, consisting of apical sense organ, larval ectodermis, prototroch, and lappets continues to swim about as long as its energy reserves can sustain it and then dies and disintegrates.

Some other heteronemerteans, including *Lineus*, develop through a stage known as the larva of Desor, named for a nineteenth-century European natural scientist, Eduard Desor. It has a spherical, ciliated body, without an apical sense organ or marginal lappets, and is regarded as a modified or simplified pilidium, adapted to a creeping existence or to remaining in the egg, as it does in *Lineus*.

The developmental history of Acanthocephala, as adults parasitic in the intestinal tracts of vertebrates, contains a larval phase, but never a free-living one. Fully developed embryos, known as acanthors, leave the host, in which the adults live, with its feces. The number of nuclei in the acanthor is about 200, most of which are compacted in a central mass, with a few larger ones scattered anteriorly and posteriorly to this. Cell walls are very indistinct, if they exist at all in what may well be a syncytium. The anterior end of an acanthor is marked by eight hooks or spines, and the whole body, only slightly larger than the zygote from which it came, is enclosed in a shell of complex structure. Acanthors represent the dispersive phase of the species and also a phase of developmental arrest, for there is no further growth, and the life cycle stops, unless the acanthor is eaten by the mollusc or arthropod that acts as its intermediate host. In the digestive tract of this host, the embryo hatches and penetrates the tissues to reach the hemocoele, moving by contractions of the very few muscles already differentiated but activated only at this time. Acanthors that reach the hemocoele begin development again as acanthellas. They grow, especially in length, and from the inner nuclear mass are differentiated primordia of the adult organs. This period of growth terminates with completion of the adult structures, and formation of a tough, leathery epidermis and an enclosing capsule, possibly a product of the host rather than of the parasite. This stage of development is known as the cystacanth and, like the acanthor phase, is one in which there is no further development unless and until the intermediate host is eaten by the final vertebrate host—fish, amphibian, reptile, bird, or mammal. Then the capsular covering of the cystacanth is digested, along with the tissues of the mollusc or arthropod; and the parasite, as a diminutive adult, attaches itself to the wall of the intestine, grows, matures its gametes, and, in time, produces a new generation of acanthors to take their chances of growing up.

Rotifers that are sessile as adults hatch as motile individuals that swim about briefly before attaching; these, too, are diminutive adults (juveniles) and not larvae. Nor are the juveniles of nematodes, although they molt four times before attaining adulthood. They have most of the characters of the adults and their molting is a consequence of growth of these structures within a nonexpansible cuticle, which no longer accommodates the enlarged body and consequently is shed.

Priapulids hatch as tiny larvae with certain rotifer-like features, such as an enveloping lorica and caudal appendages. The larva of *Halicryptus* has rings of spines at the anterior end, or future prosoma, which resemble the minute spines at the anterior end of a gastrotrich; and the larva of Priapulus has a terminal foot, like that of some rotifers, from which, as in rotifers, the caudal appendages originate. Larval life is comparatively long, but at its end the lorica is shed; and during a series of molts, the rotifer-like body grows into the elongated, unsegmented body of the adult priapulid that takes up its existence as a burrower in the littoral mud on the shores of cold seas.

Larvae of endoprocts are free-living, independent individuals. Those of *Pedicellina* look much like a trochophore, with an apical tuft of cilia above a ganglion, both constituting an apical sense organ; a prototroch, or ring, of large ciliated cells along the ventral margin of the body; and a preoral band of cilia below the mouth. The digestive tract is complete, with mouth and anus, and oesophageal, gastric, intestinal, and rectal areas. There is also an oral sense organ above the mouth, whose ganglion is connected to that of the apical sense organ, and some musculature. Larvae of *Loxosomella* look more like rotifers, with a corona-like expansion at the anterior end and a bifurcated tail. The larvae swim or creep about for a time, then attach by the prototroch and undergo drastic metamorphosis to the adult form, which includes rotation of the body through 180° as well as construction of new tissues and destruction of larval ones.

All ectoprocts pass through a larval stage, resembling in a number of respects a trochophore, with an apical sense organ, a ciliated prototroch, and a preoral band of large cells. They have also an adhesive sac at the posterior end of the body. In the larvae of ectoprocts that brood their young, the larval digestive tract is incomplete and nonfunctional and, though the larvae are set free in trochophore-like external form, their independent life is short with settling and metamorphosis occurring soon after their escape from the brood sac. In nonbrooding species, the larvae swim about for several months, feeding on small organisms. Larvae of some of these species are enclosed in a delicate, chitinous bivalve shell and are known as cyphonautes larvae (Greek *kyphos*, "bent," "humpbacked," and *nautes*, "sailor"), a name given to them when they were first found in the plankton and thought to be adults of a newly discovered genus (Fig. XIX–3*d*). Their true nature was revealed a century ago when they were seen in metamorphosis in a laboratory. These larvae have a well-developed musculature; nerve cells with muscular and sensory connections; and special glandular organs that are lacking in typical trochophores. They swim with the apical organ directed forward but when preparing for attachment after they have sunk to the bottom, the oral surface is directed downward and they appear to feel for a suitable place of attachment with the vibratile plume, a tuft of especially long and hook-shaped cilia that swings back and forth as they glide over the surface. When a suitable place is found, the adhesive disk everts and the larva becomes fastened to the substrate. Strong muscular contractions result in retraction of most of the larval organs, and histolysis of the larval tissues follows. What remains after their destruction is a thin-walled sac, containing the apical organ as a thick-walled invagination; and from this sac the future primary zooid is constructed.

The free-living larva of the Phoronidae is an actinotrocha, also originally thought to be an adult, independent organism. The actinotrocha likewise resembles a trochophore (Fig. XIX–3*c*). It has a complete digestive tract and, externally, a hood-shaped preoral lobe covered with fine cilia and with an apical plate, but without a ciliary tuft. Additional characteristics are an oblique ciliated band (metatroch), postoral in position and drawn out into a number of hollow ciliated tentacles that extend downward (posteriorly), and a ring of ciliated cells at the posterior end (paratroch). The actinotrocha grows in size during the several weeks of its larval life and, toward the end of it, acquires a deep invagination of the ventral ectodermis, about midway between mouth and anus, that extends inward and attaches to the intestine. This, as it grows, becomes thrown into loops, while the ectodermal pouch becomes comparably folded. Metamorphosis that follows is swift and drastic. Within 15 min the ectodermic sac is everted as the main part of the body of the adult, carrying with it the attached intestine bent into a U; the ciliated tentacles drop off leaving stumps from which the tentacles of the adult will grow out; and the preoral lobe disappears in some as yet undisclosed way. It has been reported that it is cut off, or autotomized, and, falling into the mouth beneath it, is digested. Possibly, its severance is not complete and the apical plate is left behind to become, as in the case of trochophores, the primordium of the anterior end of the adult. After metamorphosis the young phoronid sinks to the bottom, secretes its tube, and takes up existence as an adult.

Larvae of articulate brachiopods also bear some resemblance to trochophores and actinotrochs. The body of the larva of *Terebratulina*, for example, is constricted into three lobes, the anterior one of which is completely ciliated like the preoral hood of an actinotroch, but has an apical tuft of cilia. The second, or middle, lobe is the mantle lobe with two ciliated folds, one dorsal and one ventral, that are directed posteriorly; the third, or foot, lobe is smaller and is not ciliated at all. The digestive tract is incomplete, the larva cannot feed and its life span is measured in hours—at most, a day. Then it settles down, attaching by its foot lobe, which becomes the peduncle of the adult; the mantle lobes lose their ciliated covering and turn upward so that they enclose the anterior lobe and begin deposition of the shell substance. The tentacles of the lophophore are developed from the anterior lobe and the organs of the adult from larval primordia. Metamorphosis is, in this case, by no means as drastic as that of an actinotroch, and there is little destruction of larval structures but, rather, gradual transformation of them into those of the adult. Inarticulate brachiopods set free their young in a more advanced condition, with mantle lobes and shell valves already enclosing the anterior lobe; this lobe has a lophophore that acts as a locomotor organ for the juvenile, which, though free-swimming for a short time, undergoes no metamorphosis after its attachment as a sessile adult.

The most generalized trochophore is that of an archiannelid such as *Polygordius*, represented in Fig. V–10. This has, externally, an apical tuft of cilia overlying a sensory plate; prototroch, metatroch, and paratroch; a laterally placed mouth; and an anus on the flattened undersurface of the slightly dome-shaped body. Internally, it has a stomodaeum, stomach, and intestine; two pairs of protonephridia; and two defined, but delicate, muscle bands between the stomodaeum and the dorsal body wall. It is a free-swimming, feeding plankter. Its metamorphosis is a gradual one, accomplished while it is still swimming about. Metamorphosis is initiated by the activation of growth and division in three cells, derived from the third quartette of micromeres, which constitute the trunk blastema (Greek *blastema*, "bud" or "sprout"); and in two cells, derived from the fourth quartette of micromeres, which construct two long strings of cells, the mesodermal bands. Multiplication of these five cells, identifiable in the embryo and in the trochophore, results in an extension of the body at its flattened posttrochal surface, the mesodermal bands being covered by ectodermis derived from the cells of the trunk blastema and the intestine growing down between them. Thus, a tubular extension is inserted between prototroch and paratroch. A head blastema is derived from cells around the apical plate. The posttrochal extension soon takes on a segmented appearance as circular constrictions develop along it, and the apical tuft of cilia falls off and is replaced by a pit, which becomes almost enclosed by ectodermal outgrowths from which the future tentacles will develop. Concomitantly, histogenesis proceeds internally; and muscles and metanephridia, nerves and ganglia are differentiated. The larva now looks like a ball with a tail that is annularly constricted and continually growing in length, until the definitive number of segments is attained. Then its elongation is checked, and the most radical events of metamorphosis occur. These involve contraction of circular muscles underlying the prototroch so vigorous as to tear the body wall, and contraction of the longitudinal muscles anterior to it, pulling the head blastema downward until it makes contact with the trunk blastema. The torn prototroch is shed, and the internal larval structures undergo histolysis and phagocytosis. Later development is concerned with the construction of adult tissues and organs while the young worm burrows into the mud or soft sand that becomes its adult home.

Some polychaetes develop through a trochophore stage with the general characters of the trochophore of *Polygordius*, but with modifications characteristic of their kind. In some species, trochophores are free-swimming; in others, the comparable stage is passed within the egg. In most free-swimming trochophores the body is externally divisible into three regions: protrochal, containing the apical plate, the prototroch, and the mouth; pygidium (Greek *pygidion*, "little rump"), including the paratroch and the anal region; and a growth region between these two, which includes the trunk blastema. It is activation of cells in this region that leads to further growth and, finally, to metamorphosis. These regions are clearly evident in a 3-day-old larva of *Nereis limbata*, which has also the distinctive polychaete features of incipient parapodia supported by chaetae that project from them. *Stauronereis rudolphi*, a species with a wide range of habitats and known in coastal waters from Norway to the West Indies and from British Columbia to Chile, develops into a polytrochal pelagic larva 20 days after the egg is fertilized. The larva swims for about 1 week before settling on the bottom, by which time it has developed three segments bearing chaetae as well as functional jaws. As its later development has been observed in laboratory cultures, three more trunk segments and typical head appendages appear in the course of the next 2 weeks; and after about 57 days of larval life it has a total of 20 segments and a second pair of eyespots. Growth was more rapid for the following 50 days and gametes became visible through the translucent body wall. After this, the growth rate was slower, with the maximum number of 80 segments reached at 361 days, the worms by this time being fully adult and spawning.

Except for brooding species, all scaphopods and aplacophoran and polyplacophoran molluscs, some gastropods, and some pelecypods develop through a free-living trochophore, in all essentials like that of an annelid, but with molluscan features already showing in rudiments of shell gland and foot. In gastropods and most pelecypods, this larva becomes a veliger before metamorphosis (Fig. XIX–4). There is no such change in aplacophorans and polyplacophorans, and the trochophore metamorphoses directly into juvenile by growth and elongation of the posttrochal region; degeneration of the prototroch; and deposition of the shell plates by cells of the shell gland. As the juvenile takes up its location on the bottom, it orients itself so that the foot is pressed against the substrate and the elongated body extends parallel to the surface on which it crawls. The more advanced larval stage of gastropods, the veliger, is characterized by development of a dorsal elevation, the visceral hump, whose ectodermal cells are already secreting a delicate rudiment of the shell; development of a mantle fold at the margins of the area delimiting these cells as the shell gland; and marked enlargement of the velum, a distinctive feature of veligers and their principal locomotor organ. Internally, the digestive glands arise as two lobes from the embryonic stomach, the radula appears as a ventral evagination from the stomodaeum, the principal ganglia are organized, and some of the sense organs become evident. In some species, torsion occurs at the veliger stage, as has been described in Chapter II, pages 57–60. Metamorphosis is not drastic; it involves resorption of primary larval structures, such as the velum and telotroch, and growth of the foot, mantle, shell, head, and other structures and organs of the adult, whose primordia are already laid down in the veliger.

On the other hand, metamorphosis in a bivalve veliger, for example, like *Dreissenia*, the only freshwater bivalve to have a veliger stage and probably a recent (nineteenth-century) migrant from the Black Sea to the lakes and streams of Europe, is drastic and rapid. The cells of the velum degenerate and this part of the larva is sloughed off, while, concomitantly, the anterior region in which the mouth lies becomes progressively smaller, bringing the cerebral pit, from which the cerebral ganglia have become dissociated, closer to the mouth. Two lateral bands of ciliated epithelium originate from the rest of the cerebral pit and ultimately become the labial palps of the adult. The foot swings around so that it is directed forward rather than posteriorly, carrying with it the gill papillae developed on the lateral surfaces of the body. Other changes are definition and growth of the internal organs; growth and calcification of the shell; and secretion of the stout threads that constitute the byssus by cells of the byssus gland. It is by these threads that the adult attaches itself to some object in the water in its sedentary way of life. In other freshwater bivalves, the veliger stage is suppressed and the young emerge as glochidia (Fig. XIX–4*c*).

The larva of *Xiphosura* is called a trilobite (Fig. I–1*b*), because of its resemblance to the fossil trilobites whose position relative to the phylogeny of arthropods has been discussed in Chapter V. This little larva swims about and burrows in the sand and goes through a series of molts before it attains adult form. In the course of intermolt periods, the telson grows in length and in breadth and the final three pairs of gill books are added to the abdominal segments. It is some time before the diminutive horseshoe crab, into which the trilobite larva has been transformed, becomes fully adult, with mature and functioning reproductive organs, but during the interval between the last larval molt and full adulthood, it leads the swimming, crawling, burrowing, and scavenging life of its kind. Mites and pycnogonids also show a gradual transition from a so-called "larval" stage to the adult. The larva of a mite has only three pairs of legs and acquires the fourth pair, characteristic of arachnids, after a molt that transforms it into a protonymph. It goes through several nymphal stages, with molts at the end of each, before attaining all the characters of an adult. Pycnogonids hatch as protonymphs, with only three pairs of appendages: chelicerae, palps, and one pair of legs each consisting of but three segments. Growth of the trunk and addition of segments to the legs occur between a series of molts before the protonymph acquires the form and characters of an adult.

In every major group of crustaceans (except peracaridean), there is at least one genus with a free-living nauplius stage, the simplest form into which any crustacean egg hatches. The nauplii of ostracods

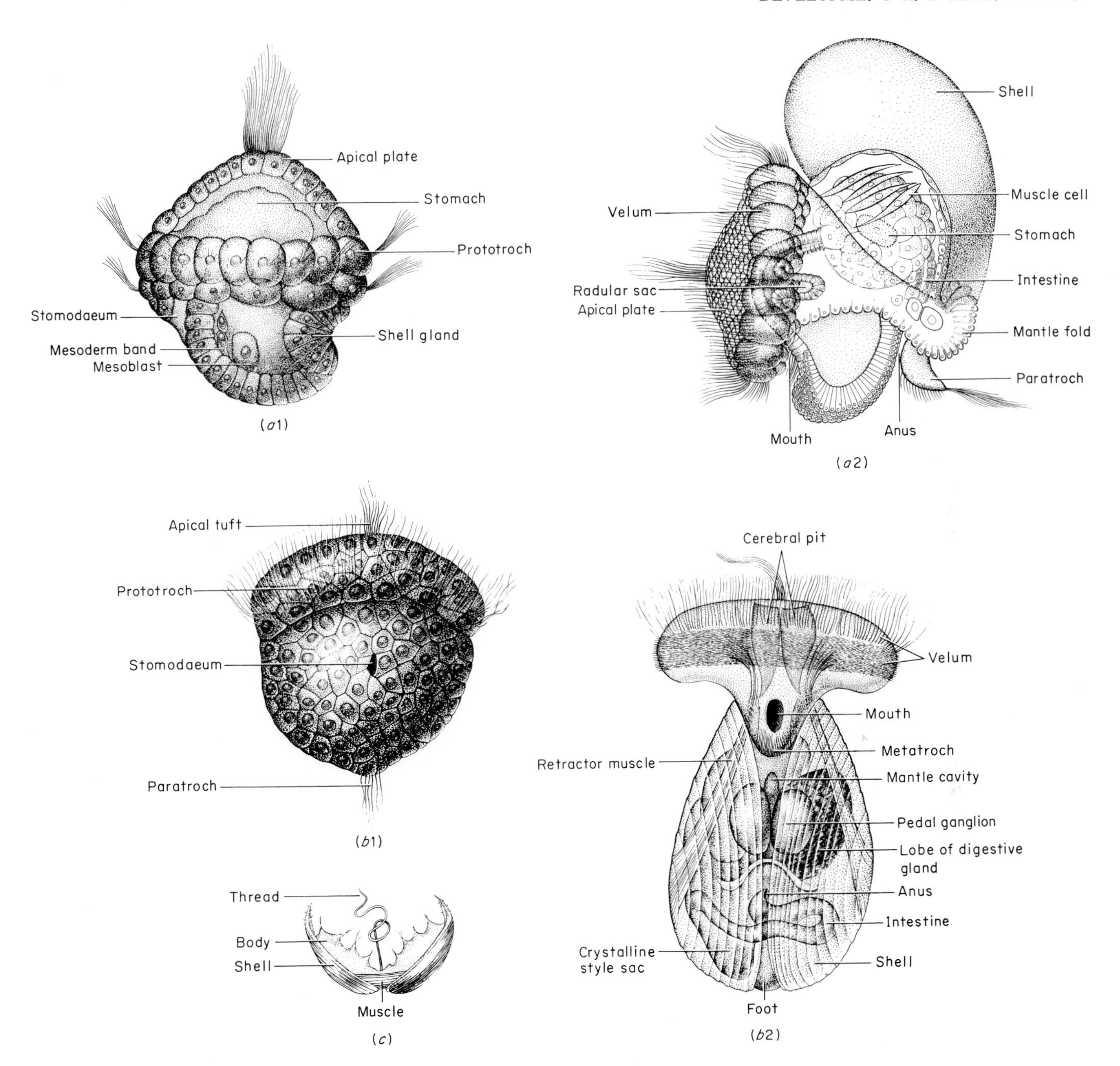

FIG. XIX–4 *Larval types.* ***(a1)*** *Trochophore of* Patella coerulea, *lateral view.* (*After McBride.*) ***(a2)*** *Veliger of* Patella coerulea *before torsion, lateral view.* ***(b1)*** *Trochophore of* Dreissenia polymorpha, *ventral view.* ***(b2)*** *Veliger of* Dreissenia polymorpha, *ventral view.* ***(c)*** *Glochidium of* Anodonta *before attachment.* (*After Raven.*)

show, upon hatching, a primary characteristic of their subclass in having a body enclosed in two flaps of ectodermis whose cells will later secrete the materials of its bivalve carapace. The various postembryonic stages of the decapod shrimp *Penaeus* have been outlined in Chapter V, pages 145–146, in relation to the evidence deduced from them for the principle of recapitulation. In some species that hatch as free-living nauplii transition to the adult form is gradual, the larvae emerging from each ecdysis having longer bodies and more pairs of appendages than before the molt. In a copepod like *Cyclops*, for example, several metanauplius stages follow its hatching as a nauplius, and the final metanauplius stage is succeeded by the first cyclops stage, in which the postlarva has the general form of an adult but no appendages posterior to the third pair of pleopods, nor any somites of the urosome, that part of the body which includes the last two thoracic and all the abdominal somites. These are added in five successive cyclops stages, along with their appendages; and at the final postlarval molt the animal has a body structure like that of the adult, with all its somites and appendages and their appropriate modifications. Growth in the nauplius occurs in the region between the third pair of appendages and the anus, and in the homologous area in metanauplius and succeeding cyclops stages, so that the posterior end of the body is continuously pushed away from the anterior. This is reminiscent of the addition of segments to the elongating body of an annelid trochophore, such as that described for *Polygordius*.

In some crustacean species the transition is abrupt, and there is a critical molt in the course of postlarval development from which the individual emerges with a different form and often different habits than it showed in the previous stage. This is characteristic particularly of parasitic and sessile species. The copepod *Lernaea variabilis*, for example, hatches as a nauplius and passes through the metanauplius stages as a free-swimming member of the plankton. In its first cyclops stage, however, it becomes a parasite on the gills of a flatfish and enters a "pupal" phase, during which certain regressive changes take place, followed by some progressive ones in which the body grows and the thoracic appendages, fewer in number than in free-living copepods, are reconstituted. At this stage the postlarva becomes motile and leaves its host. Males apparently die soon after copulation, but fertilized females seek another host, usually a member of the cod family, and attach to its gills. They then undergo drastic metamorphosis into individuals with long, worm-like tails extending from bodies that have modified head appendages and four pairs of segmented thoracic ones. In other copepods with parasitic habits, all postembryonic stages may be free-living and only the adults parasitic; or the nauplius and the adult may be free-living and the intermediate stages parasitic; or the early larval stages may be spent inside the egg, and the individual hatch in its third metanauplius stage, which is free-living, but later take up a parasitic existence, with another brief interval of independent life before attachment to the final host.

The parasitic cirripede *Sacculina* also undergoes drastic metamorphosis in its transformation from a cypris larva to the amorphous mass in the tissues of the crab to which it has attached. This life history has been briefly outlined in Chapter V, pages 147–148. Free-living barnacles also pass through a critical molt. They hatch as nauplii, in general, similar to those of *Cyclops* but with the dorsal integument extended as a triangular shield having two dorso-lateral spines, a caudal spine, and a ventrally pointing anal spine. With growth, the thoraco-abdominal portion of the body becomes divided into segments, each with a pair of bilobed appendages. The larva emerges from its critical molt in the cypris stage without a second pair of appendages, with the third pair much reduced, and with a bivalve carapace replacing the triangular dorsal shield. In this stage it swims about for a time but does not feed, and shortly settles down on a suitable object, attaching itself by a sticky secretion exuded from glands on the first antennae, retained during the critical molt. Another ecdysis then occurs, followed by a period of growth in which the epidermis of the ventral surface increases very rapidly, and the body, thus, is swung, or elevated, into a position in which its head is downward and its body upright in the water. The preoral region grows into the stalk of attachment; in barnacles like *Lepas* it elongates greatly into a long and stout peduncle. The bivalve shell is lost, the calcareous plates typical of the adult are deposited by epidermal cells of the carapace, and the thoracic appendages develop to assume their function as casting nets for food.

Peneid shrimps are the only decapods that hatch as nauplii; in all others, this stage is passed within the egg and the organism hatches in a more advanced larval form, none earlier than a zoaea, comparable to the fourth larval stage of *Penaeus*. Crabs hatch as zoaeae with distinctive features that identify them as crabs; the first free-larval stage of most lobsters is a mysis, although spiny lobsters hatch as phyllosomae, unlike any form in other ontogenies. These have long, spiny legs and slender bodies so transparent that they are commonly known as glass shrimps. Freshwater crayfish, as has already been mentioned, have no free-living larval stages and hatch as juveniles, diminutive adults whose size increases at each successive molt.

Zoaeae of crabs molt into megalops larvae in which the abdomen is large and carried extended, with paired appendages on every segment. With the final molt to adult form, the abdomen becomes much reduced in size and is flexed under the cephalothorax. In fiddler crabs, the first zoaea is a feeding, swimming plankter, and there are four zoaeal molts at weekly intervals. The fourth zoaea is twice as large as the first and the fifth is even larger and heavier; it does not swim but drifts and clumsily crawls along the bottom. The megalops that emerges after the critical molt of the fifth zoaea has a crab-like body, except for its extended abdomen, and rises to the surface where it becomes a powerful swimmer. This stage lasts 1 month without molts, but changes in form are in progress during it. The abdominal appendages, the pleopods, become much reduced in size and the megalops finally stops swimming and crawls into some crevice along the shore. In about 1 week there is an ecdysis from which the fiddler emerges in crab-like form but not as yet with the distinguishing characters of its genus. These are acquired, along with sex characters, in the course of four or more molts that follow the critical one. Six to seven months and some 15 molts are required to bring a fiddler from the zoaea stage, in which it was hatched, to maturity, during which time there is destruction of parts; growth; and organization of the bodily form and organs.

There are no abrupt changes in the postlarval life of lobsters such as *Homarus* and *Nephrops*. The mysis, at which they hatch, has most of the typical lobster features, except for rudimentary abdominal appendages. During its successive molts, these appendages grow and become typical biramous pleopods, modifications of the anterior pair distinguishing the two sexes. In the course of the molts, too, the biramous thoracic appendages of the mysis are transformed to the uniramous pattern of the adult and modified for their particular purposes. But there is no real metamorphosis, rather a gradual transition with each molt from larval to adult form. The phyllosoma larva of spiny lobsters, considered a highly modified mysis, does undergo a critical molt and metamorphosis. It has a very small abdomen without any trace of forthcoming pleopods, and there are only six thoracic appendages; the endopodites of the third maxilliped and the first three pairs of pereiopods are long and slender, while the first and second maxillipeds are very small. This larva is not a swimmer, but drifts along the surface with its long pereiopods stretched out to give it support. Both megalops and phyllosoma were given their names when they were found in the plankton and, like actinotrocha and cyphonautes, were considered to be adults of undiscovered species.

In millipedes and some centipedes, addition of segments and new appendages characterizes each postembryonic stadium. Most millipedes have, upon hatching, seven segments and add more with each molt until the adult number is reached. The spirobolid *Narceus annulans*, for example, starting life with seven segments, passes through nine segment-building stadia before its full complement of 51 to 59 segments is reached and usually two more molts before reaching maturity. The growth area for the new segments is a delimited region between the anal and the penultimate segment of each instar, and the segments are without legs when they first appear; their legs are developed during the next stadium and are apparent after the following ecdysis, along with the new, legless segments. This kind of growth seems to be common to all millipedes. Experimental studies on larvae of *Narceus* have shown that those poorly fed make fewer new segments during each stadium than those that are well fed, at least up to the seventh ecdysis, when poorly fed specimens add more segments than well-fed ones. They also added more at the eighth and ninth ecdyses, but not as many as at the seventh. In this way they were able to catch up with the better fed ones and end their larval stadia with a similar number of segments. Other millipedes compensate by adding additional segment-building stadia into their postembryonic life. These observations suggest that millipedes have some way of keeping track of the number of segments in their bodies at any time and of regulating their growth so that the typical number is finally attained. A similar kind of control is exercised by earthworms. Among centipedes, *Scutigera*, and *Lithobius*, and other members of their orders also hatch without their final complement of legs and add them in postembryonic stadia. A newly hatched *Scutigera* has four pairs of legs and, in the six molts through which it subsequently passes, acquires one more pair at the first molt and two more at the second, third, fourth, fifth, and sixth molts, making a total of 15. Four more molts are required before it reaches maturity, but no more segments and legs are added during these. These animals, too, are illustrative of the closely regulated growth that ensures conformity to the general bodily pattern characteristic of order, genus, and species.

Some consideration has already been given to the postembryonic development of insects and its regulation (Chapter XVII, pages 717–727). In hemimetabolous species, the nymph, upon hatching, may bear so close a resemblance to the imago that it can readily be recognized as belonging to a particular order. The nymph of a grasshopper, for example, is easily identifiable as an orthopteran. But in some, especially those whose nymphs lead aquatic lives and, therefore, are often called naiads, the resemblance is not so close and identification with the imago not easy. The naiad of a mayfly, for example, has a rather heavy, clumsy body, has small wing pads and tracheal gills on its abdominal segments, and may in its comparatively long life go through as many as 23 instars before its final nymphal molt. In its first instar, it is without gills, but has well-developed legs and biting mouthparts. By the third instar, slender gills have appeared and by the sixth, well-developed branched or lamellate ones and wing pads. It creeps about feeding on vegetation, its wings growing slowly during the successive stadia. At the end of its nymphal life it crawls up a plant on the water's edge and molts into a subimago, without gills and with two pairs of membranous wings. Metamorphic change has taken place before this molt, involving loss of the gills, reduction of the mouthparts to vestigial structures, loss of function of the digestive tract, and rapid development of the incipient wings. The subimago flies off and molts again, in some species almost immediately after the preceding molt, in others some hours later, and devotes the rest of its short life to mating and, in the females, finding water and ovipositing there. The change from naiad to imago in these insects is more abrupt than in orthopterans and other hemimetabolous orders, where the transition is accomplished gradually in a series of steps and not in a single leap, but not quite as abrupt as it is in holometabolous species.

Larvae of holometabolous insects bear little resemblance to their respective imagos and not very much to each other. They are, in general, grouped into three main types: campodeiform, eruciform, and vermiform. Campodeiform larvae, such as those of tiger, rove, and ground beetles, have well-developed legs, antennae, cerci, and mouthparts and derive their descriptive name from their likeness to adult thysanurans of the genus *Campodea*. Eruciform larvae derive their name from the Latin word *eruca*, meaning "caterpillar," and include such larval types as the grubs of Japanese beetles and the caterpillars of moths and butterflies (Figs. XIX–5*b* and XIX–6*a*). The eruciform larvae of lepidopterans often have prolegs on their abdominal segments; the true legs are short and stumpy and, as in all insects, are confined to the three thoracic segments. Vermiform larvae are without legs and are often referred to as "grubs" if they live in the ground or as "maggots" if they live on decaying flesh or vegetation. The larva of an ant would be considered a grub (Fig. XIX–7*f*) and that of a fly, a maggot. There are differences also in the types of pupae characteristic of different holometabolous orders and genera. In orders other than Lepidoptera and Diptera, they are typically exarate (Latin *exaratus*, "ploughed," i.e., grooved or furrowed), that is the legs and wing pads are free as they are in the pupae of ants and Japanese beetles (Figs. XIX–5*c* and XIX–7*b*). In Lepidoptera and most Diptera, they are obtect (Latin *obtectus*, "covered over"), with the legs, wing

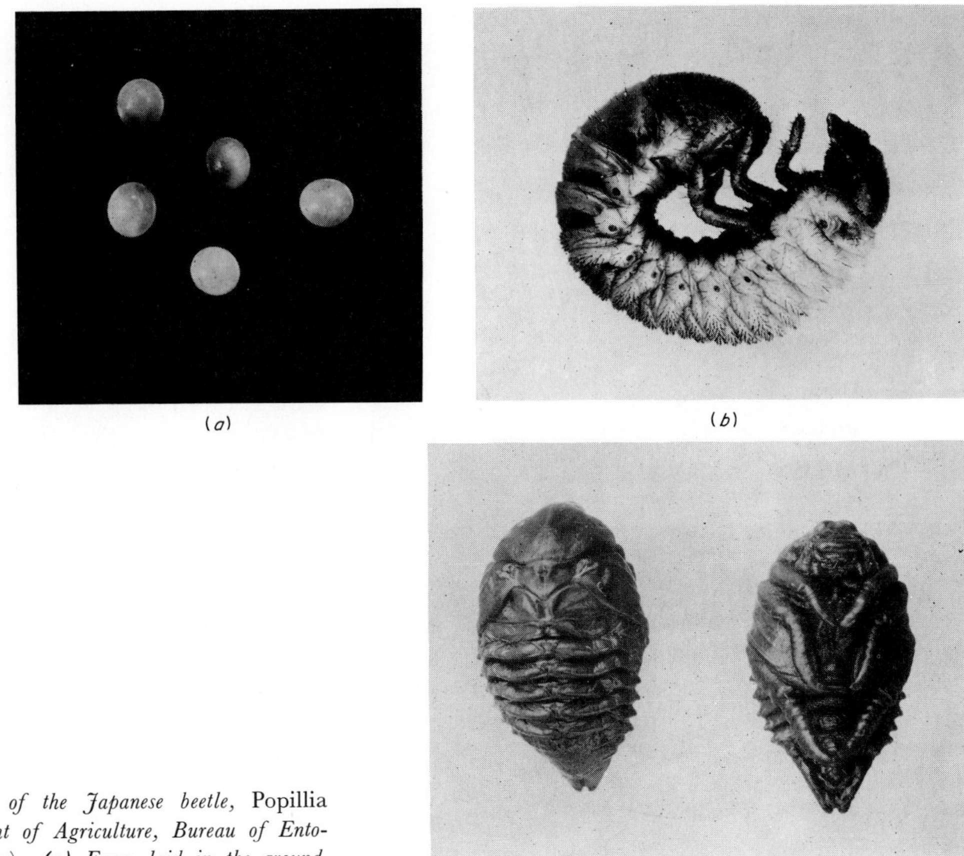

FIG. XIX–5 *Development of the Japanese beetle,* Popillia japonica. (*U.S. Department of Agriculture, Bureau of Entomology and Plant Quarantine.*) ***(a)*** *Eggs, laid in the ground.* ***(b)*** *Larva* (*grub*). ***(c)*** *Pupa: left, dorsal view; right, ventral view.*

pads, and antennae fused to the body wall (Fig. XIX–6*b*); but in some Diptera, in which the last larval integument is preserved as the puparium, they are coarctate (Latin *coarctatus,* "pressed together"). Metamorphosis in these insects is a cataclysmic event that involves destruction of larval tissues and some organs and construction of new ones as replacements from the imaginal disks. These disks are delimited areas of the ectodermis where the cells have remained undifferentiated and in a sense inactive, in that they have taken no part in the deposition of larval cuticle at the successive molts and have been insensitive to the effects of the hormone concerned.

The location of the imaginal disks is precise and mapped out during embryogenesis. Those of the mouthparts, adapted to feeding of a different kind from that of the larva, and of the legs arise in association with the larval appendages they will ultimately replace; the imaginal disks of the compound eyes, completely new structures that replace or supplement the ocelli of the larvae, originate in their appropriate positions on the head. In each case the destiny of the disk as a whole is determined, but the destinies of its individual cells or their descendants becomes fixed only in the course of construction of the part for which they are intended. Enlargement of parts already laid down, refinement of them into definitive forms and patterns, construction of new organs, and histolysis of the larval structures and organs to be discarded go on during the pupal stage, at the end of which an imago, different in architectural design, in coloration, and in habit, emerges in what seems a truly miraculous

(a)

(c)

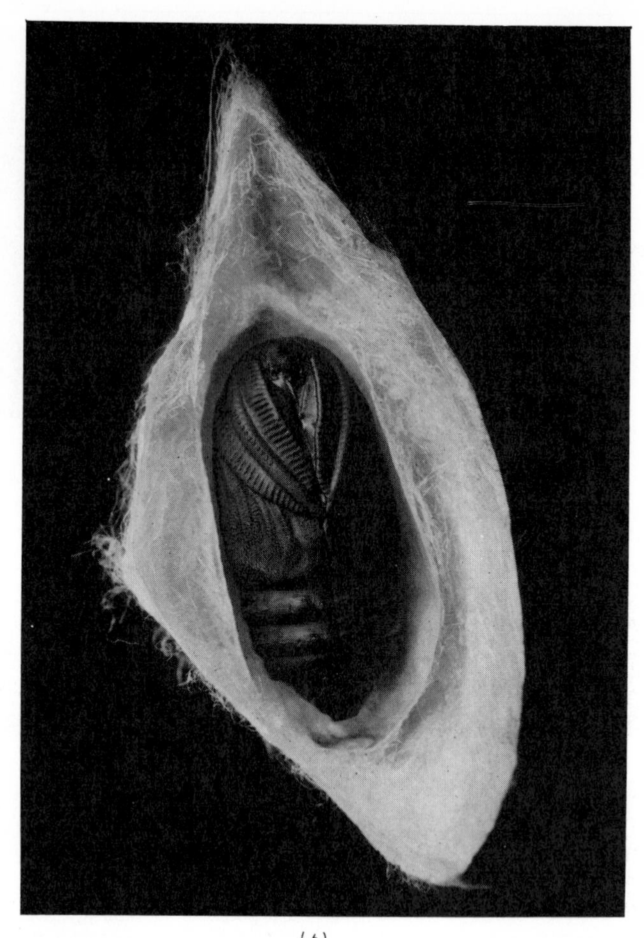

(b)

FIG. XIX–6 *Development of the moth* Antheraea polyphemus. (*Courtesy of H. Lou Gibson.*) ***(a)*** *Larva* (*caterpillar*). ***(b)*** *Cocoon opened to show the pupa inside, ventral view.* ***(c)*** *Imago* (*adult*), *male.*

change. Indeed, one definition of metamorphosis is "transformation especially by magic or sorcery." It seems especially mysterious and magical in insects because so much of it is hidden within the pupa and its case.

The swimming gastrulae of indirectly developing echinoderms are all very much alike—minute, cylindrical organisms, completely ciliated externally, with an embryonic gut marked off into oesophageal, gastric, and intestinal regions having stomodael and proctodael invaginations. The very early larvae (dipleurulae) are much alike, too, with ciliation limited at this time to a single continuous band of cells, a mouth, an anus, and coelomic vesicles visible through the body wall [Fig. XVIII–12*f* (1)]. They next begin to show the distinctive characters of their respective classes and become bipinnariae (asteroids), plutei (ophiuroids and echinoids), auriculariae (holothurians), and doliolariae (holothurians and crinoids). The bipinnaria has a pair of arms anterior to the mouth (preoral), a pair posterior to the mouth and anterior to the anus (postoral), and three pairs along the length of its growing and elongating body: an antero-dorsal pair, a postero-dorsal pair, and a postero-lateral pair; these are all ciliated. In some species there is a median dorsal arm as well as the two

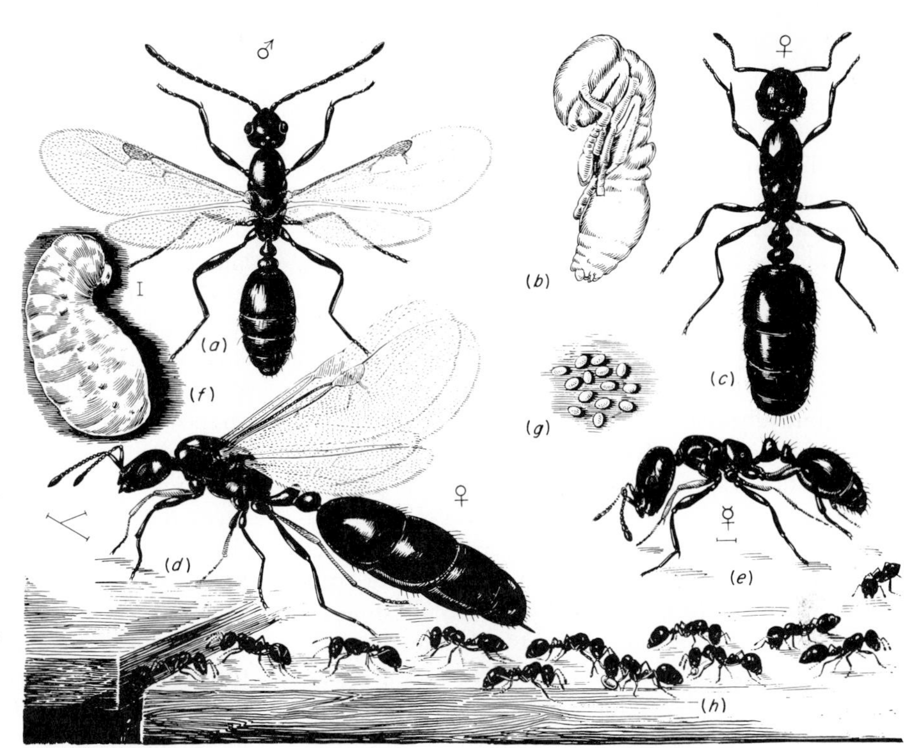

FIG. XIX–7 *Life history of the little black ant* Monomorium minimus. (*U.S. Department of Agriculture, Bureau of Entomology and Plant Quarantine.*) ***(a)*** *Adult male.* ***(b)*** *Pupa.* ***(c)*** *Adult female, wingless.* ***(d)*** *Adult female, winged.* ***(e)*** *Worker.* ***(f)*** *Larva (grub).* ***(g)*** *Eggs.* ***(h)*** *Group of workers on the march.*

preoral ones. A bipinnaria is a feeding, growing member of the plankton that after a time prepares for metamorphosis. In some species, three more short arms develop, one median and two lateral, at the anterior end in the region between the preoral and postoral part of the ciliated band. These arms have disks of gland cells that secrete an adhesive substance, and the larva, now known as a brachiolaria, attaches itself from time to time by these for brief periods of immobility. Internally, the enterocoeles enlarge and the rudiments of the hydrocoele appear, with growth on this side of the body, typically the left, preponderant over that on the other. Shortly after development of the brachiolarian arms, the larva sinks to the bottom and attaches itself firmly and permanently to the substrate by means of a sucker-like adhesive organ formed between the lateral brachiolarian arms. The anterior end of the larval body degenerates and becomes only a short stalk of attachment, for the starfish body develops from the rounded left side. The bipinnarian arms degenerate, and the mouth, the oesophagus, part of the intestine, and the anus undergo histolysis, to be reconstituted later; the adult stomach is an outgrowth of the larval stomach. The five radial arms first appear as thickened lobes on the part of the body overlying the hydrocoele, which grows out into them as the radial canals from which the first pairs of tube feet are soon developed. In many species the young starfish, now about 1 mm in diameter and with some skeletal plates, holds on to the substrate by these podia and gradually pulls itself away from the place of its original attachment, drawing out the short stalk into a slender thread that snaps in two. Metamorphosis is now complete, and the minute juvenile is prepared to take up the sedentary life of a starfish, to grow, and, in several years, to become a full-sized and mature individual.

The larvae of ophiuroids and echinoids are known as plutei, a name given them by Johannes Müller (1801–1858), a German scholar and teacher who devoted his last years to marine biology. When he found these little animals in the plankton he thought them adults of an as yet undescribed species and named them *Pluteus paradoxus*, taking the name pluteus from the word used for a painter's easel, for to him their form resembled that of an easel supported on outstretched legs. Later, their true position as echinoderm larvae was established and those of the two different classes were designated ophioplutei and echinoplutei. An ophiopluteus has a number of long, slender arms, each ciliated and supported internally by a calcareous rod that lengthens as the arm grows. The body is shorter and squatter than that of a bipinnaria and is roughly triangular in shape, with the postero-lateral arms extending far out from its sides. The antero-lateral and postoral arms are shorter than these, but all have skeletal rods. Metamorphosis of the ophiopluteus involves, like that of the bipinnaria and brachiolaria, extensive changes on the left side of the body, but these go on while the larva is still swimming and feeding, and it never settles down and attaches itself to anything. As metamorphosis proceeds, the larval arms and portions of the intestine are discarded as they are in asteroids. While this is in progress, the pluteus slowly sinks downward as it loses its organs of motility. The young ophiuroid reaches the bottom, having acquired the pentaradiate body plan and its tube feet, developed from the newly forming radial canals. At first the juvenile ophiuroid uses its podia to crawl about the substrate, but as it grows larger and older this method of locomotion is superseded by the writhing motion of the arms characteristic of this class of echinoderms, the so-called "sea serpents."

Echinoplutei, originating like bipinnariae and ophioplutei from an early larva with a continuous band of ciliated cells, also become members of the plankton and may spend some time there before metamorphosis begins. The general characters of an echinopluteus are represented in Fig. XVIII–12*f* (3) to (5). The arms, like those of ophioplutei, are long and slender and stiffened by skeletal rods of specific character, and all of them extend upward. Echinoplutei, too, metamorphose without attachment and also through growth of the left side of the body. In some species, metamorphosis takes place in the plankton; in others, as the arms grow longer and the skeletal rods heavier, the larva sinks to the bottom and undergoes a rapid and drastic metamorphosis, in some cases within 1 hr. Even before it sinks, the disk and underlying structures are evident on the left side of the body as the "echinoid rudiment," which is the part that is preserved and will grow into the adult sea urchin or sand dollar, the other larval parts and organs being discarded or resorbed as they are in brachiolariae and ophioplutei.

The auricularian larvae of holothurians are also plankters and look very much like asteroid bipinnarians; indeed, some asteroids pass through an auricularian stage before the bipinnarian. Later, however, the continuous ciliated band breaks up into three to five discrete bands which encircle the barrel-shaped larva like hoops. Because of its shape, it is now called a doliolaria, from the Latin word *doliolum*, meaning "a small cask," or sometimes, a banded larva. Still swimming, it undergoes a gradual transition to the adult sea cucumber of its genus and species, and when it has lost its ciliated bands by histolysis and, therefore, its means of locomotion, sinks to the bottom and assumes the adult way of life. Crinoids pass through a similar doliolaria stage, but omit the auricularian. Crinoid doliolaria have a short free-swimming existence; they soon sink downwards, attach themselves, and begin a rather slow metamorphosis which transforms them into small, stalked sea lilies. This is a transitory stage in the postlarval development of comatulids, the free-swimming feather stars that break loose from their stalks and become more or less mobile.

3. Metamorphosis

Metamorphosis, whether gradual or abrupt, is not merely a time of structural change, very evident externally and detectable internally, but also one of profound metabolic changes, in which the biochemistry of the organism may be radically altered. In many marine larvae, but not in all, attachment and metamorphosis is so synchronized that it would seem that they were both responses to the same set of conditions. Attachment is often very precise and dependent upon certain external conditions. Trochophores of the polychaete *Ophelia*, for example, settle only on sand grains on whose surfaces are living bacteria or other microorganisms in sufficient concentration and metamorphose normally only under these conditions; a film of dead or decaying organic material retards or actually inhibits their metamorphosis. A certain concentration of metallic ions seems necessary to the metamorphosis of some marine larvae; the importance of copper to oyster development has already been pointed out (Chapter I, page 23). In some instances, carbon dioxide tension seems to be a critical factor and temperature also has an effect. These latter factors may be the significant ones in the metamorphosis of larvae that transform without attachment; temperature, photoperiod, and activation or suppression of hormonal systems are of proved significance in the metamorphosis of insects (Chapter XVII, pages 717–725).

All metamorphosis involves destruction of some tissues and construction of others. In marine larvae, the tissues that are destroyed are for the most part those that have differentiated early, such as the ciliated bands of trochophores and veligers and the bands and arms of echinoderm larvae. This applies as well to holometabolous insects, where many of the tissues and structures functional in the first and later larval instars are destroyed in the pupa. These are, so to speak, aged tissues with cells apparently no longer capable of division and differentiation, whose senescence makes them vulnerable to phagocytosis by amebocytes, always numerous and active in tissues undergoing histolysis. In some cases, larval tissues are actually discarded rather than destroyed, as they are, for example, in actinotroch and many echinoderm larvae. In other cases what is spoken of as "disappearance" may really be autolysis, reduction to their ultimate components by hydrolytic enzymes released from their own lysosomes. Growth of new parts proceeds from localized centers as, for example, the growth region between the paratroch and prototroch in polychaete trochophores; that between the last pair of appendages and the anus in nauplii and succeeding larval stages of crustaceans; that between the penultimate and anal segments in millipedes; the imaginal disks of insects; the echinoid rudiment in sea urchins; and the homologous area in other echinoderms with complete metamorphosis.

Biochemical changes at metamorphosis are patent in ways of life and feeding habits that distinguish larvae from adults in many species. New enzymes and enzyme systems are brought into action concerned with the digestion of foods of different kinds and metabolism of the end products resulting. Synthesis of new enzymes implies alteration in the cellular mechanisms of protein production and the activation of genes whose information has not hitherto been decoded and utilized, a switchover from one kind of direction to another. This is dramatically illustrated in silkworms, where the salivary glands of the phytophagous larva are converted into silk-secreting ones when the cocoon is being spun, and then degenerate during pupation and metamorphosis. But it has been shown that in moths of the genus *Antheraea*, only the distal portions of the glands degenerate; the proximal, or anterior, portions remain and without further growth transform into specialized glands, whose cells secrete a fluid that acts as a solvent for the proteolytic enzyme cocoonase deposited in solid form on the surfaces of the maxillary galeae. The elaboration of this enzyme is also indicative of a radical change in the biochemistry of a localized group of cells. In diapausing pupae, the epidermal cells of the galeae are uniform in appearance, and for about the first 6 days after the beginning of adult development they all behave in the same way—retracting from the overlying cuticle and secreting molting fluid. Then some of the cells in the lateral half of each galea begin division and differentiate into three distinct types, two of which are involved in synthesis of cuticular structures while the third makes preparations for protein synthesis. Their nuclei enlarge greatly as a result of endomitosis and are, therefore, polyploid, and the machinery necessary for rapid protein synthesis is built up in their cytoplasms. During the eleventh to sixteenth day of adult development, they synthesize and store large amounts of cocoonase, an enzyme that can digest the material of the cocoon in which the pupa is completely encased, for moths of this kind make their cocoons without any external opening. The enzyme is discharged onto the surface of the galea through fine ductules, and forms white, solid, semicrystalline masses there. These are dissolved by the secretion from the persisting regions of the silk glands and so, when the moth is ready to leave its cocoon, a hole is digested in the cocoon at the head end of the imago through which it can push and pull its body. The functional change in the epidermal cells of the galeae and in the secretory cells of the silk glands represents a very complete alteration in their biochemistry as a specific adaptation to the conditions under which the moths pupate and emerge from the cocoon. These secretions are temporary ones, directed toward a specific purpose and a specific need of these particular moths, but the switchovers from larval digestive and metabolic enzymes to adult ones are permanent ones. Detailed study and analysis of such switchovers, either temporary or permanent, in invertebrates in transition from embryonic, larval, or juvenile stages to adult forms and functions would undoubtedly be rewarding. A parallel exists in the change from fetal to adult hemoglobin that is known to take place in humans at birth.

F. ATYPICAL DEVELOPMENT

Clearly, any biological procedure so complex and so carefully controlled as postembryonic development is subject to a variety of influences that may cause it to deviate from its normal pattern. Some of these influences are genetic and become permanent characteristics of a mutant strain. The significance of heterochrony as an expression of mutation has been pointed out in Chapter V, pages 145, 147. Other influences may be environmental, affecting only the individuals exposed to particular environmental conditions. Many examples of these could be cited, but only a few will serve to illustrate the fact. Rotifers of the species *Brachionus calyciflorus*, for instance, which are the natural prey of the carnivorous *Asplanchna brightwelli*, have a pair of long, movable postlateral spines, in addition to the normal complement of three pairs of short, rigid ones, if they have developed from eggs produced by individuals living in waters populated by *Asplanchna*. Investigation has shown that these waters are conditioned for *B. calyciflorus* by a substance released from *Asplanchna* that exerts its effect upon them during oogenesis. Eggs laid by *Brachionus* exposed to such waters develop into a generation with the additional long spines, but eggs laid by females that have not been so exposed and transferred to *Asplanchna*-conditioned water after their deposition do not. Chemical and physical studies on the *Asplanchna* factor indicate that it is a protein, probably unable to pass through the external membranes of newly laid eggs but able, when taken into the body of a rotifer in its reproductive phase, to enter the oocytes. The factor alters the usual pattern of development by inducing the formation of additional spines, but in no other discernible way. It differs from the inducers already known to embryologists in that it is exogenous, it is a product of a species different from that in which its effects are manifest, and it seems to act at cleavage. Its action represents another illustration of the biological conditioning of an environment, in this case with favorable effect upon the individuals responding, for specimens of *Brachionus* with long spines are not so readily eaten by their predators and so have greater chances of survival, although they only pass this protection on to their descendants if they themselves are continually exposed to the inducing substance.

The morphology of *Asplanchna sieboldi*, as well as its reproductive habit, is influenced by a dietary factor, transmitted to these carnivores through the algae-eating species on which they prey. The dietary factor has been identified as alpha tocopherol, and specimens given as food paramecia grown in cultures containing it produced offspring parthenogenetically that were characterized by three very distinct outgrowths, or humps, one posterior and two lateral. These were absent from specimens feeding on microorganisms low in alpha tocopherol content, a finding that suggests that the synthetic activity of the algae on which their prey ordinarily feeds is an environmental factor that influences their structural development, as well as their physiological, by directing their reproduction toward the formation of amictic eggs.

The influence of environmental conditions in metamorphosis of marine larvae, especially in relation to their attachment, has already been mentioned. Inability to settle in a laboratory aquarium is a possible explanation for the delayed and partial metamorphosis observed in specimens of the saddle oyster, *Anomia simplex*. Usually these molluscs when about 200 μ long drop to the bottom and begin to creep along it, using their strong and externally ciliated feet developed during their final veliger stages, until they find a rock or some other firm, clean surface. When they do, they attach by the byssus and complete their metamorphosis, of which the most obvious external events are disappearance of the velum and the foot, and appearance of the gills. Many specimens in the aquarium considerably longer than 200 μ were observed propelling themselves along the bottom by their still large and functional feet, although their gills had appeared and their vela disappeared. Moreover, while the larval shell was still visible, so also were the beginnings of the postlarval shell. It seems evident here that certain events of metamorphosis were delayed because of the inability of the larva to settle, due to either physical or chemical conditions in the aquarium environment. These conditions, whatever they may have been, did not affect all the specimens in the culture equally, for some went through a normal metamorphosis, beginning when they were about 200 μ long.

Delayed and precocious metamorphosis have also been reported in insects, resulting from a variety of unusual environmental conditions, such as extremes of temperature within the tolerance of the species in question, or biological conditioning of the environment. These reflect some disturbance of the hormonal balance to which particular target organs seem particularly sensitive at particular times. Delayed metamorphosis, expressed in continued growth of the larva, yet failure of the imaginal disks to differentiate and of the genital organs to develop, results in the condition known as metathetely (Greek *metathetos*, "changeable," and *telos*, "end"). Its converse, precocious development of certain adult structures by a larva, is known as prothetely. This has been reported in meal worms (*Tenebrio*), which sometimes develop wing pads and reach a condition intermediate between larva and adult. These prothetelous larvae may continue to grow and molt and to pupate but usually become imperfect imagos. Metathetely results from parasitic infections of larvae, as in those of the blackfly *Simulium* infected by nematodes, and has been reported also in larvae of the flour beetle *Ephestia* after exposure to an irritant gas emitted by the adults. Hysterotely (Greek *hysteros*, "later"), the relatively retarded differentiation and development of a structure or organ, is also an atypical pattern of growth.

The effect of temperature upon postembryonic development has been demonstrated experimentally in a number of cases. For example, larvae with the male genotype of *Aedes simulans*, a species of mosquito common in Canada and the northern United States, develop into functional males having all the usual secondary sex characters when they are reared, from the time of hatching, in water at a constant temperature of 24°C; but larvae from the same source reared in water at 29°C have all the external characters of females, except for a slight difference in appearance of the palpi, which in normal males have more bristles than in females. Internally, these genetic males, whose development is misdirected through exposure to high temperature, have female genital organs and may even be inseminated by normal males. This sex reversal occurs only if the male larvae are continuously exposed to the high temperature from the time of hatching onward, although an intersexual condition results from exposure to it during only the last 6 days of larval life, and minor defects in complete maleness and fertility from exposure for only the last 3 days. Development of the sex organs and genitalia is not affected in larvae of the female genotype by comparable changes in temperature; evidently, it is only those of the male that are sensitive to high temperatures and only the male characters which are suppressed.

Temperature shocks also affect the growth rate of parts in other insects, and their effects upon the development of eyes and wings have been studied especially in *Drosophila melanogaster*. Nutrition is another environmental factor profoundly influencing postembryonic development, either shortening larval life and promoting early metamorphosis or prolonging it and delaying metamorphosis. It is apparent also from experiments that certain organs and structures are more affected by nutritional deficiencies than others and at different times, additional evidence of differences in the growth rate of different parts at different periods of development. This has been quite clearly shown in studies of cyclomorphosis in cladocerans and rotifers, a term used to describe the seasonal changes in their bodily proportions occurring in nature. For example, various external protuberances, such as cephalic crests and peaks, spikes on the carapace, and elongated antennules, are characteristic of specimens of *Daphnia galatea mendotae* collected in warm seasons;

these are absent from those collected in cold weather. The heads of *Daphnia pulex* and *D. longispina* are round in midwinter but drawn out anteriorly as "helmets" in midsummer, with transitional stages between these two extremes of the relative proportions of head and body evident in specimens collected in spring and autumn. These differences have been correlated with the temperature of the water in the pond or lake in which the adults live and incubate their offspring. Temperature affects particularly the second half of embryogenesis; water turbulence also affects embryogenesis, and individuals developed parthenogenetically from females experimentally exposed to water turbulence had longer heads than the offspring of females kept in quiet water. Nutrition has been shown also to influence the pattern as well as the rate of growth. Well-fed *Daphnia* complete their fourth molting period in 100 hr, on the average; starved *Daphnia* in 167 hr; and the rate of head growth is considerably less in those ill fed than in those well fed, so that even at summer temperatures ($\pm 24°C$) no helmets are formed. These observations show not only the effect of environmental conditions upon structural development, and by implication upon physiological, but also the fact that certain regions—in the case of these *Daphnia*, the head—are more sensitive to them than are others and respond by an increase or a decrease in their natural growth rate.

G. POLYMORPHISM

One other aspect of postembryonic development in invertebrates to be considered is the polymorphism of colonial species, in which an individual, developing from a single fertilized egg, grows into a colonial adult with zooids of different morphology and function. This has, of course, its parallel in the polymorphism of social insects, discussed in Chapter IX, in relation to the nutrition of bees and ants, and in Chapter XVII, in connection with pheromones and caste differentiation in hymenopterans and isopterans. In both cases growth and differentiation of particular structures and organs in different individuals that represent the offspring of a single one are influenced by external factors. In the case of colonial polymorphism, as it is seen in hydrozoans and ectoprocts, different members of the colony that grows from a single zygote by budding develop along quite different lines under influences that are primarily internal, but may be tempered by external conditions.

Polymorphic colonies show an ascending scale of complexity in respect to the number of different types of zooids included in them and the degree of coordination between them. At the bottom of the scale are hydrozoan colonies that have only two types of zooids: gonozooids and gastrozooids. Medusae budded off represent a third type of zooid, but not one attached to the colony. Although in those species in which they are not set free or retained as degenerate members, they do represent a third type in the colonial organization. In such genera as *Obelia* and *Campanularia* the gonozooid is quite different in appearance from the gastrozooid, or hydranth. Gonozooids have no tentacles, no hypostome, and no mouth, while gastrozooids have one or more circlets of tentacles around the conical hypostome at whose tip the mouth lies. It is from the blunt, club-shaped gonozooids that the medusa buds are formed, while the gastrozooids are concerned with capture, ingestion, and digestion of the food that supports the entire colony. In genera such as *Syncoryne* there is less distinction between reproductive and nonreproductive zooids, for both are hydranths with mouths and tentacles. The nonreproductive hydranths are longer and narrower than the reproductive, with four separate rings of capitate tentacles; the reproductive hydranths, from whose bases medusae are budded off, are shorter and wider, with more tentacles crowded together. In some genera, such as *Hydractinia*, there is another type of zooid, the dactylozooid, or nematophore, which is of various forms in different genera, but, in general, is long and slender, without a mouth, and with clusters of nematocysts that are quite evident as little swellings along its length. In *Hydractinia* the gonozooids are dimorphic and can be distinguished as female, or egg producing, and male, or sperm producing, for the medusoid stage of this and related hydrozoans is vestigial and is represented only by its gonodal tissue in the gonozooid. Thus, four types of zooids can be recognized in such colonies.

Ectoproct colonies such as *Bugula* have feeding members, or autozooids, and two or three different kinds of nonfeeding members, or heterozooids. One of these is a very much reduced one that serves as an attachment disk to hold the colony to some fixed object; another is highly modified as an avicularium, so called because of its resemblance to a bird's head with a large beak. An avicularium may be either sessile or stalked and thus able to swing in various directions; the jaws of the "beak," a modification of the operculum, are moved by muscles and can open and snap together with considerable force. Avicularia are developed on the sides of the zooecia of autozooids and catch or ward off small animals, especially larvae of various kinds that might find the zooecium a suitable place for settlement and metamorphosis. A vibraculum is also a highly modified polypide, whose operculum has developed as a long movable bristle, or seta, that swings about and brushes away detritus and small organisms.

At the top of the scale of polymorphic complexity are the siphonophores, hydrozoan colonies that have reached the highest degree of specialization of zooids and integration and coordination of action between them. They are considered by some biologists to have reached the organ level of construction by converting whole individuals into organs, although they are at the cnidarian level of tissue differentiation. Siphonophores generally have six or more types of zooids, usually arranged in definite linear order either horizontally beneath the float, or pneumatophore, or hanging vertically from it. In pendant species the zooids are budded off a common stem that hangs down from the float. In addition to the float, which is itself a single modified zooid or several united together, there are nectocalyces, modified medusae whose pulsations help keep the colony afloat and move and steer it through the surface waters; gastrozooids, polyps without tentacles but with mouths; dactylozooids, mouthless polyps with a single long tentacle that is mobile and heavily equipped with nematocysts; and gonozooids. These different kinds of zooids are usually grouped in assemblages called cormidia (Greek *kormos*, "tree trunk"), each cormidium being covered, and separated from the next in line, by another type of zooid called the hydrophyllum, which is leaf-like or bract-like in appearance. In *Nanomia cara* and related species there is, in addition, a type of zooid known as a palpon, a small mouthless individual like a dactylozooid and always close to a gastrozooid.

In *Nanomia* there are certain independent effectors, but the zooids, although independent in some of their activities show a considerable degree of integration in others. At the cellular level the nematocysts, as in other cnidarians, are independent effectors as are the chromatophores whose pigment disperses in light and concentrates in darkness. In both cases the responses are to external stimuli and are independent of any other responses of the colony. At the zooid level, pulsation is the particular attribute of the nectocalyces and is performed by them only; ingestion and digestion are peculiar to the gastrozooids and, in *Nanomia*, their attendant palpons. Elongation and retraction is peculiar to the dactylozooids. Both nectocalyces and hydrophylla are capable of spontaneous detachment from the main body of the colony. Yet there is coordination and cooperation in the activities of the different zooids, mediated by the nervous system and probably by thorough conducting systems. Contraction of one dactylozooid may, for example, lead to elongation and writhing not only of the gastrozooid adjacent to it but of others further up or down the stem, for in *Nanomia* the zooids are pendant from the float. There is cooperation between the palpons and gastrozooids in digestion, and food from a gastrozooid may be pumped in and out of a palpon several times in the course of its enzymatic degradation. The gastrovascular cavities of both gastrozooids and palpons communicate with the gastrovascular cavity of the stem, their openings into it being guarded by valves that open and shut while the pumping action is in progress. The nectocalyces also show coordinated and integrated activity, acting in concert to propel the animal in one direction or to reverse its movement under instructions received through the conducting systems of the nerve net. Evidence for nonneural conduction in siphonophores has been presented in Chapter XVI, pages 671–672. In these most complex of all hydrozoans there is then not only differentiation and specialization of parts as represented by the different types of zooids in a single colony, which must be regarded as a single individual in light of its origin from a single fertilized egg, but also a fairly high level of coordination between them, so that the colony moves as an individual, feeds as an individual, and responds to generalized stimuli as an individual. Localized stimuli may lead to localized responses of single or small groups of adjacent zooids, just as localized stimuli may result in localized responses of individual parts of the bodies of noncolonial multicellular systems.

Polymorphism poses many problems of morphogenesis that are as yet unsolved, to whose answers studies of siphonophores might offer the most fruitful approach. But siphonophores are pelagic animals of semitropical and tropical waters, difficult to capture in good condition and impossible to keep for any length of time, or to rear, in the laboratory. Biologists have, therefore, turned to animals more amenable to observation and experimentation as material for the study of growth and morphogenetic processes.

H. GROWTH AND REGENERATION

Growth is an attribute of all living biological systems, manifested as an increase in size. This may come about through the addition of material to cells and, in multicellular systems, the addition of cells. Increase in the bulk of a cell through imbibition of fluid or through accumulation of storage material is not growth in the sense of its increase in structural and functional components that are permanent items of its cytoplasmic equipment. This is only accomplished through the active cooperation of nucleus and cytoplasm in the mechanisms that lead to replication of existing particulates or construction of new ones. Enucleate cells neither grow nor divide. This has been amply demonstrated by observations of end cells such as human erythrocytes, as well as of cells experimentally deprived of their nuclei. For example, an ameba from which the nucleus has been removed by microsurgery rounds up, puts out no pseudopodia to move or to feed, and does not divide, although it may preserve its integrity as a cytoplasmic system for some days or even a week or two. Irradiated cells in which chromosomal structure has been disrupted do not divide, a condition that is put to practical advantage in cancer and tumor therapy. Research in molecular biology continues to adduce evidence of nuclear-cytoplasmic interactions in active cells in synthesis of messenger, transfer, and ribosomal RNA and of the critical role of these nucleic acids in protein synthesis, upon which growth of cell substance, deposition of cell products, and cell division depend.

Growth as a biological process is a natural phenomenon whose interpretation has challenged scientists from classical to modern times. Biologists have hopefully turned to what seem to be the most simply organized animals for answers to its perplexing problems, which includes its regulation in the maintenance of body form, symmetry, and proportions. They have approached these problems in two general ways: (1) by observation of developmental processes and natural growth in intact animals and (2) by observation of growth induced by the experimental removal of parts. Much of the material for these investigations has been provided by invertebrates, particularly protozoans, sponges, hydrozoans, and turbellarian flatworms. Nemerteans, ectoprocts, annelids, arthropods, and echinoderms have also been subjected to this kind of study, as have vertebrates, especially urodele amphibians. All of these have shown some capacity for regeneration, or replacement of lost parts, a capacity that varies greatly in different phyla and genera. While the vast majority of animals are capable of protecting themselves by healing superficial wounds and so preventing loss of body fluids and exposure of delicate underlying tissues to immediate environmental stress, there are only some that can carry this repair to restitution of a part. In some cases, this is a replicate of the structure lost or one very close to it; in others, the new structure may be different from it. The former is a condition of homomorphosis, the latter of heteromorphosis, and both are equally provocative of analysis and potentially productive of clues to the interpretation of growth phenomena.

1. Autotomy

Regeneration has been so much studied experimentally that its occurrence as a natural event is apt to be overlooked or forgotten. In addition to its occurrence in fission, fragmentation, etc., it takes place after autotomy, a kind of self-amputation practiced by a number of invertebrates. Discarding of terminal hydranths is, for example, a functional part of the life cycle of some hydroids, of which *Tubularia* is a well-known example, that may have some positive selection value in dispersal of the gonophores with their contained gametes. Detachment of posterior, gamete-containing segments of epitokous polychaetes is a form of autotomy, as is also the detachment of posterior segments, not necessarily with ripe gametes, by some eunicids when they are disturbed, and of elytra by scale worms under similar conditions. The prosobranch mollusc *Gena varia*, living in the infratidal zone of the Red Sea, when touched autotomizes the part of its foot posterior to the shell margin. Regeneration begins several days later, and the missing part is restored. Decapod crustaceans autotomize injured appendages and, no matter which segment is damaged, break off the appendage at the breakage plane, just distal to the coxa, by flexion resulting from vigorous contraction of a special autotomizer muscle. This flexion serves both to stop the outward flow of hemolymph and to snap off the limb, and is a reflex following injury. It takes place in all five pairs of thoracic appendages in brachyuran crabs, in all but the last two pairs in hermit crabs, and in only the first pair in lobsters. Limbs that are not autotomized are also detached at the breakage plane, but must be pulled off by some external force. Some insects can also autotomize their legs after injury; the break here is usually at the level of the trochanter.

Some echinoderms are capable of autotomy and show very different degrees of capacity for it. It is a fairly common phenomenon among asteroids, ophiuroids, and crinoids, which readily cast off their arms after injury or even in response to unfavorable environmental conditions. Among starfishes, autotomy is most frequent among those species in which the arms are fairly distinct from the disk and is infrequent, if it occurs at all, in those in which the attachment is broad. *Asterias* and some other species autotomize an arm near the disk, usually at the level of the fourth or fifth pair of ambulacral ossicles regardless of the site of injury, but *Luidia* and *Brisinga* do so at any point along the length of the arm. Usually a starfish detaches an injured arm by fastening it firmly to the substratum with the intact tube feet and then moving the disk and the other arms away. Breakage takes place between ossicles, the stump heals over, and the missing portion is reconstructed. *Luidia* is a genus that drops its arms very readily, even while it is held in the hand after being picked up from the sand. Ophiuroids also do this so readily that they have earned the common name of brittle stars. Their arms are cast off at varying levels, and in some species the greater part of the disk may be cut off, too, leaving only the mouth frame. Crinoids cast off their arms when grasped and even when untouched if the temperature of the surrounding water exceeds the level of their tolerance or the concentration of oxygen drops below it. Echinoids are incapable of autotomizing or regenerating any of the major parts of their bodies, but some species respond to unfavorable environmental conditions by shedding their spines. *Echinus esculentus* and *Psammechinus miliaris*, for example, have been known to drop all their spines and then to make new ones when food was plentiful and environmental conditions favorable. A response of many holothurians to injury, rough handling, or unfavorable conditions, such as high temperatures or foul water, is evisceration, the expulsion of digestive tract, gonads, and one or both respiratory trees from the body. After a time the organs are regenerated, and the animal resumes its normal activities. Evisceration may be a seasonal event in some species, for in collections of *Stichopus regalis* in the Bay of Naples and of *Parastichopus californicus* from the Pacific Ocean off the coast of Washington, there were many more specimens without viscera taken in the autumn than at other times of the year.

Fission as a means of reproduction in holothurians, asteroids, and ophiuroids, and in representatives of other phyla as well, may perhaps be considered autotomy, in that it is a spontaneous division of the body into two or more parts and restitution of the missing portions. Both are responses of an individual to external and internal conditions that may be unfavorable, and both present similar problems in respect to the processes of growth involved in restoration of the parts that have been lost.

2. Problems of regeneration and approaches to their solution

In general, these problems are those of the cells involved in replacement and the controls that are exercised over the form, the size, and the architectural pattern of a regenerating part. An obvious question is the source of the cells: whether they come from a store of undifferentiated, formative cells with which the individual is inherently endowed or whether they are derived through dedifferentiation of already differentiated cells that can revert to a generalized type and later redifferentiate along a different course. In the one case, regeneration could be conceived of as limited, in multicellular systems, to those in which there is a demonstrable store of undifferentiated cells and, in the other case, to those in which some cells, at least, are not irrevocably specialized. Another related question is whether the cells arise by division of those that are either initially totipotent or that become so by dedifferentiation or whether previously differentiated cells migrate into the regenerating region. That is, is the regenerated part formed primarily through morphogenesis or through morphallaxis? Questions relating to controls of either morphogenesis or morphoallaxis are particularly relevant to all growth processes, and analyses of regenerative procedures are thus equally applicable to an understanding of embryogenesis and postembryonic development.

Solutions of these problems have been sought through dissociation of multicellular systems into their component cells to test for their capacity to regenerate an entire organism individually and independently, or after their association into groups; through surgical removal of parts and histological study of the course of events in their replacement; through transplantation and grafting of parts of regenerating and nonregenerating animals; through application of homogenates, or of extracts from them, of different parts of regenerating or nonregenerating animals to others, also regenerating or nonregenerating; and through application of compounds of known chemical composition suspected of affecting growth processes to regenerating specimens, in the search for inducing or inhibiting substances. These methods have been applied to vertebrates as well as to invertebrates, and results from them have yielded much information, although as yet not all that is needed for complete understanding and control of the basic and universal phenomenon of biological growth.

3. Protozoans: *Stentor*

Among protozoans, many of which have provided experimental material, the heterotrichous ciliate *Stentor* has been a most valuable subject for growth and regeneration studies. This is primarily because of its size, for it is 1 to 4 mm along its long axis, and because of the consistency of its cytoplasm, which is stiffer than that of most protozoans so that operations are technically simpler and less damaging through loss of outflowing endoplasm. There are also certain morphological features in *Stentor*, described most accurately by Trembley in 1744, that make it advantageous for this kind of experimentation. The anterior end, where the cytostome lies, is expanded into an adoral field surrounded by a band of cilia and membranelles, whose motion sweeps food particles into the gullet and cytostome (see Fig. VII–14). This region can be expanded and retracted so that a stentor's shape is sometimes like that of a trumpet and sometimes more like that of a barrel. There is a single contractile vacuole, usually to the right of the cytostome, which is considered ventral in position. The nuclear apparatus is dual, as it is in all ciliates; the micronuclei are minute, but the macronucleus is large and clearly evident in a living animal, even at low magnification, as a chain of small masses extending along the left side of the body. Experimental evidence from *Paramecium* as well as from *Stentor* has shown that the macronucleus contains the information for protein synthesis, the micronuclei being concerned primarily with storage of genetic material and the mechanics of cell division that ensure its transmission, and not at all with trophic functions. In *Stentor*, unlike *Paramecium*, the macronucleus is not reconstructed at division from daughter micronuclei, but is partitioned between the two daughter cells (Fig. XVIII–1*a*). The basal end of a stentor tapers into a disk-like foot by which the animal often attaches itself to some object in the water, but stentors are also free-swimming. Clearly visible also in a living stentor are some 100 dark stripes running the length of the body, separated by rows of cilia. Electron micrographs reveal that the stripes contain refringent granules, considered protrichocysts, which are colorless in *S. polymorpheus* and *S. roeseli*, black in *S. niger*, and blue in *S. coeruleus*, giving this species its beautiful color. In each species, the space between the dark bands contains a longitudinal row of cilia with their kinetosomes; a thread, the kinetodesma, running the length of the band just to the right of the kinetosomes; and, to the right of this, a strongly contractile fibril, the myoneme. These are all in the ectoplasm, but in the endoplasm are additional myofibrils in irregular arrangement. These stripes have a definite and consistent pattern, in that they are narrower and closer together in the region below and around the oral apparatus, which consists of buccal pouch, gullet, and cytostome, and become progressively broader, more widely separated, and longer, progressing from left to right around the body. The widest stripe, therefore, is adjacent to the finest stripe, and this region of juxtaposition is known as the ramifying zone, for it is the source of new stripes. This has been demonstrated particularly clearly in animals cut in half longitudinally; in each half, the usual number of 100 stripes is restored by division of the widest stripes that begins at their anterior ends and proceeds downward to the posterior.

It is in this zone, too, that the primordium of the new anterior region appears in fission or in regeneration. The first indication that fission is about to take place is the appearance of a small rift, or break, near the left boundary stripe of the ramifying zone, as if the bands were opening to let new structures through. Coincident with this is the condensation of the macronucleus into eight fused masses, still in linear order; as the rift extends and curves inward to form the boundaries of the adoral zone, the macronucleus condenses still more, into three lobes. A fission line then appears between the old and the new adoral zones. The macronucleus condenses into a single mass and the micronuclei divide mitotically. Then the macronucleus elongates, extending above and below the fission line, separates into two parts, and becomes lobulated again; a gullet develops in the new adoral zone and the fission line is completed, dividing the animal into two approximately equal parts, each with micronuclei, macronucleus, and oral apparatus. Under laboratory conditions, the whole process takes 3 to 4 hr, possibly triggered by mitotic preparations in the micronuclei but first externally detectable in the appearance of the rift in the longitudinal bands that marks the position of the adoral primordium.

Grafts made between specimens at different stages of their life cycles have provided evidence of induction in the processes of fission and organization of the new individual. If a small stentor, just after fission, is attached parabiotically to the body of one about to undergo fission, both divide, an adoral primordium appearing first in the larger, prefission member of the pair and some 10 to 90 min later in the smaller, postfission one, the time difference depending upon the extent of the area of contact between their fused bodies. The adoral primordium appears in the postfission member of the pair in the correct position, and a fission line in its correct position and at the same time as in the prefission member. If the two are joined along a considerable length of their bodies, the two fission lines, although arising independently, fuse into a single one, but whether they do or do not, both members divide simultaneously. Similarly, fusion between an individual that has just divided and one in the first hour of fission, when the adoral primordium and initial macronuclear contraction are evident, results in division of both in a perfectly regular fashion, but in similar grafts between postfission individuals and those in later stages of division the postfission member does not divide, while division is completed in the other. Whatever influence the partner just before fission or in its early stages has upon the one that has just completed fission seems, therefore, to be lost after the initial steps.

Regeneration experiments have also proved instructive in regard to growth control in *Stentor*. Early experiments, made at the turn of this century, on the regenerative capacities of fragments of animals literally shaken to pieces showed that those no greater than one sixty-fourth the size of a full grown specimen could reconstitute an entire animal and that this was an all or none procedure; either a stentor perfect in all respects was reconstructed from a fragment, or none at all. Later, by more precise methods of microsurgery, it has been shown that certain components are essential to regeneration of fragments and that pieces even as small as one-hundredth the size of the original individual, and from any part of it, will reconstitute new ones, provided these components are included in them. Notably, while the presence of a micronucleus is not essential for regeneration, although it is for future fission of a regenerate, some part of the macronucleus is (Fig. XVIII–1*b*). Even one lobe will suffice, and regeneration will be as rapid with a single one as with seven. At most stages, any lobe is adequate, but in animals cut into pieces just before they would have undergone fission, only the anterior lobes are competent to direct regenerative events. Indeed, it was *Stentor* that provided some of the early evidence for the necessity of a nucleus to the continued life of a cell, with recognition that the bipartite nature of the ciliate nuclear apparatus represents only morphological separation of two aspects of nuclear operation. Enucleate specimens of *S. coeruleus* survive and move for about 5 days, until their energy reserves are exhausted. In this time they retain certain morphogenetic capacities in that they can form new basal disks, after being free-swimming, form new contractile vacuoles, heal cut ectoplasm, and reorganize disrupted stripe patterns, but they cannot organize new oral primordia.

Transection experiments have also demonstrated polar differentiation in the stentor body for no matter where a cut is made a foot is always regenerated by an anterior piece and oral structures by the posterior. Polarity is also demonstrable in the left boundary stripe of the ramifying zone and, by implication, in the other stripes as well. In experiments in which various regions of this stripe were excised it has been found that the differentiation of adoral structures and of the adjacent contractile vacuole always occurs at its free anterior end and the differentiation of a foot always at its posterior end, and that any area along the stripe can mediate the appropriate differentiation, just as any area along the body can. This is presumably because of the presence of at least one lobe of the macronucleus and some part of the left boundary stripe, which makes the stripe capable of regenerating the other structures. Complete excision of this stripe does not lead to complete loss of regenerative capacity, but results in marked retardation of the regenerative processes; apparently another stripe takes over, after an interval, the properties and functions of the left boundary stripe.

Excision and reimplantation of the entire region where narrow and wide pigment stripes meet, known variously as the primordium site, the reorganization field, or the anarchic zone, have revealed its influence. If this region is transplanted to the dorsal region of another animal, new adoral structures and a new primordium are organized there. If such an animal is transected, it regenerates a double head and two primordia, and can divide and regenerate as such a doublet. If one of these primordia is removed, however, the animal reverts to the ordinary form with but one set of oral structures. But if both primordia are removed, double regeneration follows, for at each excision a primordium site originates in the area where pigment stripes of different widths are juxtaposed. Transplantation of pieces of the fine line sector from one *Stentor* into various areas of others results in development of new primordia wherever there is such a juxtaposition of wide and fine lines, but they do not appear when fine lines appose fine lines or wide lines appose wide lines. The proximity of fine line and wide lines thus seems to be the critical factor in setting up a ramifying zone of such influence.

The visible stripes of a stentor's body indicate very precise cortical topography and the high level of cortical organization inherent in these protozoans. The nature of the relationship between fine lines and wide lines that gives the primordium site its particular capacities, and the nature of the influence emanating from an oral primordium and inducing the formation of another, are matters still to be elucidated. What has been learned, however, about the cytoarchitecture of *Stentor*, especially in respect to its cortical organization, may well apply to other cells in which local differentiations of cytoplasmic or cortical functions have not such clear landmarks, but in which such specificity of cortical areas is implicit or has been indirectly demonstrated as, for example, in egg cells.

4. Sponges

The extent to which cells in an organism so simple in its organization and with so few cell types as a sponge are irreversibly differentiated has been investigated experimentally. These experiments have involved dissociation of cells from the aggregate and observations of their capacities to form a new sponge either as isolated individuals or as associations of them. The pioneer experiments of this kind were those of the

American biologist H. V. Wilson who, in 1907, dissociated the cells of the encrusting sponge *Microciona* by forcing pieces of it through fine-meshed bolting cloth, the heavy silk used in processing wheat flour. He saw that the separated cells, when the suspension was dropped on a clean glass surface, came together in clusters and that minute sponges were formed from such aggregates. Regeneration of this kind presents problems different from those in the regeneration of fragments or of restitution bodies, for in these, all types of cells are represented, and the questions asked in dissociation experiments are whether the descendants of a single cell have the capacities necessary for construction of a new sponge, and if not, how many different types of cells are required for this, and what ones are needed. In no case has a sponge been reconstituted from a single isolated cell. Individual cells spread out on a glass surface put out filopodia about 7 μ long, and when those from neighboring cells touch, they retract, drawing the cells together. Two cells in contact put out filopodia about twice as long, and three cells, filopodia 30 to 35 μ in length; and in aggregates of 100 to 200 cells the filopodia are 200 to 300 μ long. Thus, as the number of cells in a coalescing mass increases they reach out farther and farther and draw more and more cells into the aggregation. Cell contact is apparently essential to regeneration and depends upon specific particles in the cell surface, uniform in size and primarily polysaccharide in composition. From aggregates of about 2,000 cells a new sponge can be successfully reconstituted in 5 to 6 days. Such aggregates are about 1.0 to 1.5 mm in diameter; larger aggregates with diameters of 3 to 4 mm do not regenerate successfully. When the cells make contact with each other their plasma membranes fuse and the mass is surrounded by common ectoplasm while, internally, the endoplasms of the individual cells and the nuclei within them remain distinct. This membrane fusion is reminiscent of, and perhaps comparable to, the fusion of membranes that occurs during the entrance of a sperm into the cytoplasm of an egg. As the process of reconstitution of a new sponge has been followed in the fresh-water species *Ephydatia fluviatilis*, the central region of the aggregate undergoes histolysis and the debris is phagocytosed by amebocytes; it is only from the peripheral cells that the new sponge is organized. This may be because of the advantages in location these cells have, particularly in regard to oxygen exchange. As they become organized into the architectural pattern of a sponge, lacunae, forerunners of the canalicular system, appear in the necrotic central area.

There seems little discrimination in the initial contact of the cells, and those of all types may be drawn into an aggregation. Experiments on mixing cells dissociated from sponges of different species and different colors have shown, however, that after a time those of the different species do recognize each other and draw apart into groups of their own kinds. When, for example, cells dissociated from the red sponge *Microciona prolifera* were mixed with those from the yellow sponge *Cliona cellata*, the early aggregates contained islands of red and yellow cells which later segregated and, if conditions were favorable, regenerated sponges of their own kinds. Recognition of cells, not of their own species but of their own type, has been shown in experiments on *Ficulina ficus* in which several sponges were stained different colors with vital dyes and dissociated, and mixtures of their cells spread out on glass surfaces. Epithelial cells, choanocytes, and the various cells of the mesenchyme tended to sort themselves out and, at the beginning of aggregation, epithelial cells formed a boundary between the coalesced mass of other cells and the medium around them. Such mutual recognition of their own kind is shown also by vertebrate cells. If, for example, the limb and kidney primordia are removed from a chick embryo, and their cells dissociated and mixed, when those of the same kind make contact they stay together, but they move away after contact with cells of a different kind, so that there are formed small masses of limb or of kidney cells, but not of both. That such masses possess also the capability of organizing themselves into the general pattern of the structure to which they belong is demonstrated by the appearance of tubules in the kidney masses and the alignment of the cells in the limb masses as in the formation of cartilage. In reorganization of the sponge body from aggregated and coalesced cells, the cells divide and arrange themselves into layers, and, ultimately, into the patterns, of the sponge from which they came. It has been reported that in *Ephydatia* the cells dedifferentiate to a generalized state and then redifferentiate into the appropriate types. Epithelial cells, scleroblasts, and choanocytes are regarded as unipotent cells that after dedifferentiation can only redifferentiate along the lines of their original types. Amebocytes, on the other hand, are pluripotent in that, after loss of the characters that distinguish them as amebocytes, they can differentiate into any kind of sponge cell. Reconstruction of the sponge body, therefore, involves reassociation of isolated cells into groups in which all types of cells may be represented and which assume characteristic arrangements and multiply by division, probably using material from the autolyzed central cells as energy substrates, with amebocytes acting not only as scavengers but as sources for new cells as they are needed.

5. Cnidarians

Growth and regeneration have been extensively studied in *Hydra* and some colonial hydroids. *Hydra* presents particularly favorable material for it can be easily cultured in the laboratory and, if well fed, will bud prolifically for long periods; laboratory clones have been maintained for over 5 years. Also it is structurally comparatively simple, a more or less one-dimensional system uncomplicated by any specialized organs. The number of cell types in its body is limited and these are easily distinguished from each other; there are epithelio-muscular cells in the epidermis and, in the basal disk but no where else, gas-secreting cells and cells secreting an adhesive substance; there are cnidocytes, nerve cells, interstitial (I) cells, and gastrodermal gland cells of four kinds: peristomial mucous and gland cells, and gastric mucous and gland (zymogen) cells. The cells of the gastrodermis like those of the epidermis are basically epithelio-muscular in character, but are readily distinguished from them by their denser and more granular cytoplasm, and the presence in them of secretion droplets and, in most species, of symbiotic algae, either green or brown. Interstitial cells represent a permanent source of replacement cells, particularly of the cnidocytes whose life is over after discharge of their nematocysts, and of senescent gastrodermal gland cells. Their intermediate role in growth processes such as budding and regeneration is less clear. They are especially sensitive to radiant energy and can be destroyed by irradiation in doses that do not harm other cells; the two parts of a bisected hydra that has been deprived of its interstitial cells in this way will close the wounds and the basal piece will reconstitute a new hypostome and tentacles, but it will not live long afterwards since it cannot replace senescent cells. Such a regenerate can be revitalized, however, by insertion of a piece from a nonirradiated animal of the same species, for about 24 hr after such a graft has been made its interstitial cells begin to migrate in both directions along the mesoglea and bring a fresh supply to the regions deprived of them.

There is continuous replacement of cells within the body of an intact hydra, conferring, theoretically at least, potential immortality upon each individual. This is the case also in other cnidarians and in some flatworms, polychaetes, and echinoderms. In *Hydra* a zone of cells just below the hypostome has been held responsible for the continuous replacement and maintenance of constancy in body form and size, compensation for the continuous growth being due to disintegration and sloughing off of cells at the basal end (Fig. XIX–8). Such replacement was first shown by experiments in which specimens of the brown hydra, *Pelmatohydra oligactis*, were stained with Nile Blue

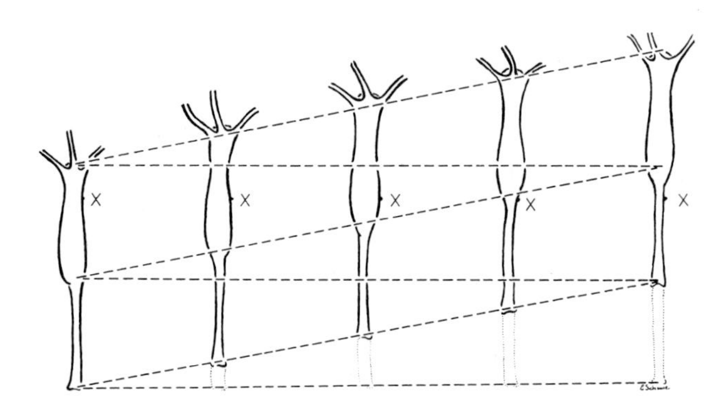

FIG. XIX–8 *Growth and maintenance of form in* Hydra. (*After Brien.*) *The progressive displacement of the point x shows growth at the oral end of the body, compensated for by disintegration and sloughing off of cells at the aboral end to preserve the size and size relations characteristic of a species.*

Sulphate and pieces from them grafted into the columns of unstained individuals. These pieces showed a progressive downward movement below the hypostome, and blue cells appeared at successively lower and lower levels of the column until, finally reaching the base in 4 to 5 weeks, they disappeared (Fig. XIX–9). At the budding zone, blue cells were incorporated in the forming bud. Cells from similar pieces grafted into the region of the hypostome moved upwards into the tentacles and disappeared at their tips. These results were confirmed by later experiments in which the possibilities of contributing effects of staining were eliminated. In these, the green hydra *Chlorohydra viridissima* was used, since it is possible to make these animals void their symbiotic algae and, therefore, become colorless in about 5 weeks by adding 0.5 percent glycerine to their culture water. Such animals were cut in two through the budding zone, and the anterior halves grafted onto the basal regions of untreated, green animals similarly transected. The buds formed on the grafted animals were sometimes half green and half white, but more often basically white with some green patches. These buds, when detached, budded in their turn, and some of their offspring had only one or two patches of green gastrodermal cells. The fates of these naturally colored cells could be directly followed without further operations. In 5 days green cells had moved from the growth region upward to the tips of the tentacles and in 2 to 3 weeks downward to the basal disk. At each site they were eliminated. Histological studies have revealed that in the basal migration of cells only the epithelio-muscular cells of the epidermis and the gland cells of the gastrodermis actually reach the basal disk. This and the peduncle are regularly devoid of interstitial cells, cnidocytes, and mucous cells, suggesting that cells of these types die and undergo histolysis before they reach these areas.

The results of these experiments would indicate that growth in the column of a hydra, exclusive of budding, is restricted to the subhypostomal growth zone, from which new cells are pushed upward into the tentacles and downward into the basal disk. But autoradiographic experiments using thymidine labeled with H^3 to mark nuclei in which synthesis of DNA is in progress and their mitotic derivatives, and using *Hydra littoralis* as the experimental animal, indicate that cell division takes place along the whole length of the column and is not restricted to one particular zone. Labeled nuclei were found at all levels along the axis of the column in both epidermis and gastrodermis, but there were few in the peduncle and virtually none in the tentacles or region distal to the hypostome. These observations are substantiated by counts of mitotic figures in histological preparations, which show that mitoses do occur in both epidermis and gastrodermis everywhere within the column, and with frequencies corresponding to the numbers of

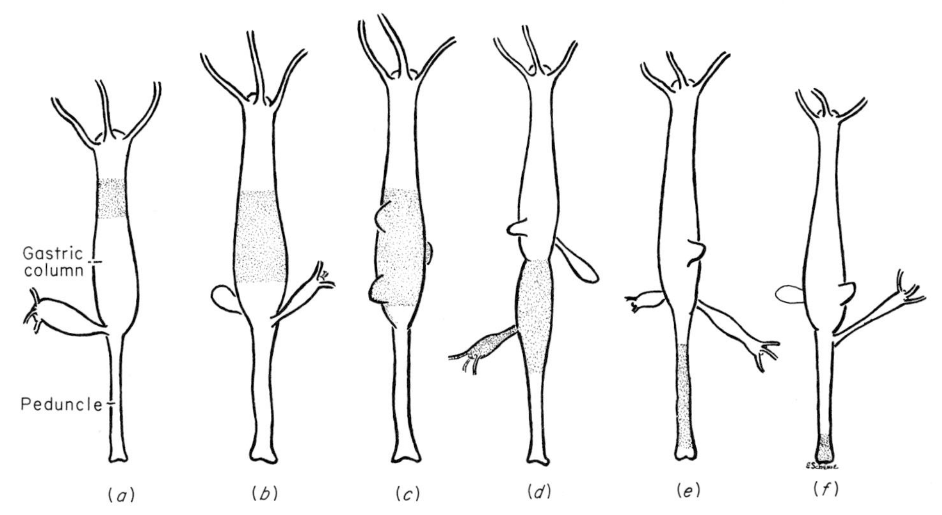

FIG. XIX–9 *Growth in* Hydra. *The progressive aboral displacement of a segment stained with Nile Blue Sulphate as growth in the subhypostomal region proceeds. (After Brien.)* ***(a)*** *Position of inserted blue strip at beginning of observations.* ***(b)*** *3 days later.* ***(c)*** *4 days later. The blue area is now in the budding zone and some of it is included in the buds.* ***(d)*** *7 days later.* ***(e)*** *11 days later.* ***(f)*** *17 days later. The blue zone has almost disappeared from the basal disk.*

thymidine-labeled nuclei. Mitoses are, however, rare in the distal region of the hypostome and in the tentacles and peduncle. Cell loss is greatest in the budding region, where epithelial cells move out into the buds, and next in magnitude in the outgrowth of tentacles; it is least in the oral and basal disks.

Both series of experiments show that maintenance of form in *Hydra* is a dynamic process, attained through constant loss of cells and their equally constant replacement. In question is the site of cell proliferation, not the types of cells dividing and contributing to the replacement—whether this is primarily a subhypostomal growth zone or whether growth is general along the length of the column. It is possible, considering the individuality of biological mechanisms, that one method or the other may be the predominant one in different species.

Hydra has tremendous capacity for regeneration, a fact that was recognized early by Trembley and has been capitalized upon by biologists ever since his experiments (Fig. XIX–10). It was shown at the beginning of this century that pieces as small as 1/432 mm^3, representing one two-hundredth the volume of an entire animal, could regenerate at least a mouth and one tentacle. Larger pieces, with diameters of 0.6 mm, can regenerate whole and perfect animals, although the capacity for regeneration varies with different regions of the body. Isolated pieces of the peduncle or of the tentacles do not regenerate readily, nor do those of the basal disk, although peduncles excised from extremely well-fed hydras, with cells stocked with nutrient material, can regenerate a hypostome and two or three tentacles. Also, it has been shown that the cells from peduncles of several animals, minced and mixed together, and also the cells from their basal disks, will aggregate within a few hours and after 24 hr construct a new individual with mouth and tentacles. Those reconstituted from minced peduncles grow to standard size, feed, and bud. Those reconstituted from basal disks are incapable of digesting food, for they lack gastrodermal glands and mucous cells, which in a normal animal are constantly being differentiated from interstitial cells which are absent from basal disks.

When a hydra is cut transversely into several pieces, the cut surfaces are closed over by contraction of the circular muscle processes of the cells on their margins, much as the lumen of a tube is closed by the action of a sphincter muscle. Cnidarians are different from other animals in that there is no immediate migration of cells to the site of the cut to form a blastema from which the new part will arise. The parts that are developed from a transected hydra show that there is polar differentiation in the column, for a basal region is regularly developed from the proximal surface (that nearest the foot) of each piece, and hypostome, mouth, and tentacles from the distal (that nearest the mouth). This polarity was originally attributed to a gradient of metabolic activity descending gradually from a high level at the anterior, or distal end, to a lower one at the posterior, or proximal end. Such a gradient was postulated from evidence obtained from experiments, primarily on hydras and flatworms, that showed progressive diminution in capacity to reduce methylene blue, a measure of oxidative activity, and of sensitivity to narcotics along the antero-postero axis. The theory

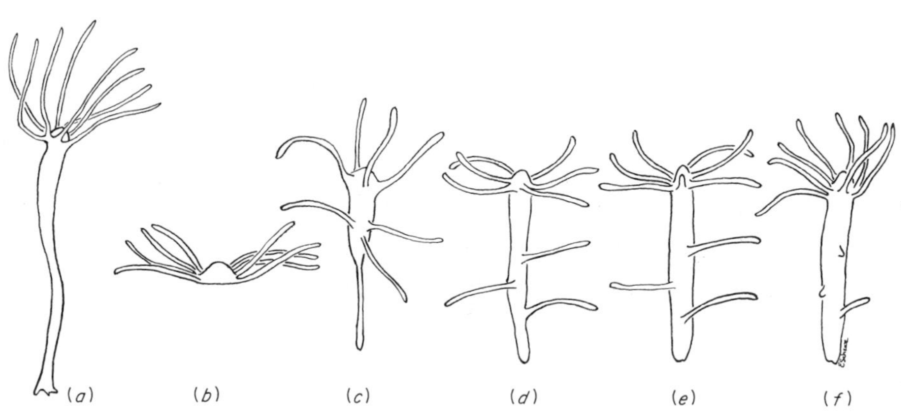

FIG. XIX–10 *Regeneration in* Chlorohydra viridissima. (*After Rand.*) **(a)** *Intact polyp with eight tentacles.* **(b)** *Head of polyp removed from column; 1 hr after the operation.* **(c)** *2 days after the operation.* **(d)** *4 days after the operation.* **(e)** *9 days after the operation.* **(f)** *13 days after the operation. The old tentacles that were misplaced by growth of the new parts are resorbed, and new ones have developed at the oral end.*

of metabolic gradients as controlling mechanisms in bodily organization was first proposed by the American biologist C. M. Child (1869–1954). Axial gradients of many kinds have been demonstrated in many organisms, although not always those of metabolic activity. In *Hydra* there are actually three regions of high metabolism—the hypostome, the budding region, and the basal disk—and the gradient is not a continuously descending one, but one with three peaks. Counts of interstitial cells taken along the columns of nonbudding hydras and the hydranths and stems of colonial hydroids show progressive diminution in number from distal to proximal regions, a histological gradient that is essentially an axial one. Polar differentiation in *Hydra* is strongly fixed, and most experiments to alter it have failed. Its inversion has been accomplished in grafts between *H. pirardi*, a large species native to Belgium, and the common brown hydra *Pelmatohydra oligactis*. If the subhypostomal region of *H. pirardi* is grafted to the peduncle of *P. oligactis*, cut clean away from the column and basal disk, the two types of tissue tolerate each other for almost 1 week, although there is no exchange of cells between them. During this period tentacles grow out from the distal (oral) region of the *pirardi* part of the graft, and a day after they appear, tentacles grow also from the proximal (basal) part of the *oligactis* graft. Similarly, tentacles grow also from the proximal end of a segment taken from the gastric region of *P. oligactis* grafted in reversed position to *H. pirardi*. There is, however, no reversal of polarity in pieces of *H. pirardi* grafted to *P. oligactis*, nor do tentacles grow out from the proximal cut surface of *pirardi* apposed to the distal cut surface of *oligactis*. Inverted pieces of *H. viridis* grafted to *P. oligactis*, on the other hand, show reversal of symmetry, suggesting that there are different degrees of potency in the effect of one species upon another. Polarity in *Tubularia* is much more labile than in *Hydra*, for hydranths are readily regenerated at both ends of a piece of stem, and if a piece of hydrocaulus is inserted in a stem in reversed position and a cut then made across it, a hydranth will regenerate at what is now the distal end but was originally the proximal. The polarity of a cut stem can also be reversed by raising the temperature of its proximal end above that of its distal, for following this a hydranth is regenerated at the proximal end.

The problem of the source of the cells involved in the construction of a regenerating part has been attacked in a variety of ways. Destruction of the interstitial cells in hydras by irradiation has shown that while they may contribute to a regenerating part they are not indispensable to its construction, however much they are to the continued renewal of cnidocytes and digestive cells. The potentialities of epiderm and gastroderm have been shown by isolation of these two layers and observation of their ability to reconstitute a whole organism. The layers can be separated by microsurgery, using very fine and very sharp knives, or by digesting away the mesoglea with proteolytic enzymes, such as pancreatin and trypsin, and then teasing the gastrodermis away from the epidermis with fine glass needles. It has been found that a new polyp can be produced in 5 to 6 days from isolated gastrodermis of *P. oligactis* and that after 7 days the regenerate is capable of feeding and pursuing the ordinary activities of its kind. Regeneration of complete polyps from isolated gastrodermis of *H. viridis* is accomplished in 5 days; this is a particularly favorable species for this kind of experiment, since its gastrodermal cells are so distinctively marked by their symbiotic green algae and also because, unlike most other species of *Hydra*, there are no mature cnidocytes in it. Isolated strips of its gastrodermis rounded up into balls and after about 5 hr the digestive cells on the periphery voided most of their algae and so became easily distinguishable from those in the center, which still retained their symbionts. After 8 hr, cells morphologically similar to interstitial cells were evident at the periphery, although histological examination of sample specimens had previously failed to reveal any. In 24 hr cnidoblasts were evident, some with fully formed nematocysts that were discharged on contact with the fixative when specimens were being prepared for histological study. These were more abundant in one half of the tissue mass than in the other, implying that polarity had been established and that the cnidoblast-rich distal end was already distinct from the cnidoblast-poor proximal end that would become the peduncle and basal disk. After 48 hr, the cells on the periphery had begun secretion of mucus typical of epidermal cells. These results are indicative of dedifferentiation of the gastrodermal cells and their redifferentiation into epidermal and interstitial cells, the latter then differentiating into typical cnidoblasts and cnidocytes with functional nematocysts.

Similar experiments on the freshwater colonial hydroid *Cordylophora* indicate that epidermal cells can reconstitute gastrodermal, for in experiments in which a terminal hydranth was cut off, the coenosarc pushed out of its surrounding perisarc and the epidermis and gastrodermis separated from each other, the epidermal pieces formed hollow spheres within 10 hr after their isolation, and in 5 to 9 days an entire hydranth was formed from them, with a typical gastrodermal layer. There is evidence also for the capacity of epidermis to form gastrodermis from experiments that have been performed on scyphistoma of *Aurelia aurita*. These have no cells corresponding to the interstitial cells of hydroids and have a comparatively thick mesoglea that makes separation of their orange-colored inner and colorless outer layers relatively easy. Pieces of epidermis about 1 × 0.5 mm regenerated into complete polyps after their cells had divided and given rise to a population of ameboid cells from which new gastrodermal cells were differentiated.

It seems likely then, that except for cnidocytes the cells in hydrozoans are not end cells, incapable of performing anything but specific functions, but that they have some lability and can return to, or give rise to, cells of generalized type and multiple capacities of differentiation, the final types to which they ultimately transform depending upon their location in the regenerating mass, and their organization in particular patterns and structures under influences from other cells and centers in the regenerate. This transformation is not one of cellular morphology alone, but of cellular biochemistry as well, and implies the repression, or even loss, of some enzymes and the activation or acquisition of others. An epidermal cell does not, for example, ordinarily form a nematocyst, but its dedifferentiated product can; nor does it elaborate and secrete digestive enzymes, but its dedifferentiated product can, instead of the mucus characteristic of epidermal secretion.

In colonial hydroids, growth and maintenance of form are not attained by the continual replacement of senescent cells, as in *Hydra*, but by loss and replacement of entire hydranths. In some species these are autolyzed and resorbed, but in *Tubularia* they are autotomized, a procedure probably imposed by its structure, for at the base of each hydranth the stem is almost wholly constricted by an annular invagination that leaves only a narrow connection between its tissues and those of the hydranth, through which there can be little migration of cells. This regression-replacement cycle is endogenous and occurs regularly but can be induced and influenced by exogenous factors. *Tubularia* colonies ordinarily discard only a few hydranths at a time, but when they are exposed to high temperature or low oxygen tension, they shed them all at once. *Pennaria* colonies undergo complete breakdown, resorbing all the hydranths, while the coenosarc retracts to the base of the hydrocaulus, leaving the empty tube of perisarc that breaks into pieces at definite points. When conditions again become favorable, an entire new colony is reconstituted from the clumped coenosarc. These colonies thus offer opportunity for study of natural regeneration as well as that following experimental amputations.

The question of the cells concerned in restitution of missing hydranths is not yet resolved. The first step in either natural or experimental regeneration is closure of the wound by contraction of the muscular processes of the epidermal cells. This is followed by migration of cells to the site. Time lapse microcinematography has shown that the epidermal cells become ameboid and, using the perisarc as a substratum, move along it dragging the gastrodermal cells with them. This is an active process on their part and dependent upon the availability of sufficient oxygen. The question is whether the new hydranth is reconstituted solely by morphallaxis of these already differentiated cells, their regression and redifferentiation, or whether interstitial cells are also involved. In either event it is apparent that a certain cell density is necessary to reconstitution procedures and that migration of cells to the construction site is essential. It has been shown by repeated amputations of hydranths that the capacity for their renewal is not indefinite and that, although isolated pieces of hydrocaulus will regenerate hydranths at both cut ends as often as 15 times in 74 days, with repeated amputations the process becomes slower and the regenerated hydranths smaller, due presumably to depletion of the stores of reserve cells. In *Tubularia*, the hydranth replacement cycle is an important element in the growth of each colony. As long as a hydranth is attached to its supporting hydrocaulus, growth in that part of the stem is suppressed, but when the regeneration process is underway, the stem below the regenerate elongates, sometimes as much as 3 mm/day, but this rapid growth stops as soon as the new hydranth has reached full development.

The basic problem of the control of growth has been approached in cnidarians by transplantation and grafting experiments. Transplantation of the hypostome, or of a newly forming bud, from one hydra to the column of another leads to the formation of a new polyp at the site of the transplant. But if the hypostomal region of one hydra—including tentacles and subhypostomal growth zone as well as the hypostome itself—is transplanted to the budding zone of another hydra, where, however, a bud has not yet begun to form, none will, until, in the usual process of growth, the transplant has become some distance removed from the budding area. If, on the other hand, the transplant is put near an already started bud, growth of the transplant will continue. By grafting experiments on *Tubularia* it has been shown that a newly regenerated, or a regenerating, hydranth will inhibit formation of another on the same piece of stem, although hydranths may be formed concomitantly on the same stem. The inhibitory effect of one hydranth upon another expresses itself as retardation in the rate of regeneration or as complete suppression of it. It can be prevented by blocking the customary circulation of fluid in the lumen of the hydrocaulus either by ligature or by insertion of a bubble of gas or oil. Combinations of host and graft in seven different stages of regeneration have shown that while there is no interaction between regenerates of the same age, older regenerates inhibit development of younger ones. Grafted pieces of stem without regenerating hydranths have no effect. It is apparent, therefore, that regenerating hydranths have an inhibitory effect upon the regeneration of others in their vicinity. This effect is evident about 30 hr after the initial operation, and the capacity to produce it increases during their development while their own sensitivity to inhibition decreases. In ordinary regeneration, there must be a balance between these two conditions—capacity to inhibit and sensitivity to inhibition—in order to avoid self-inhibition. There is a similar dilemma in the inhibition demonstrated by the hypostome of *Hydra*, for if this can prevent growth in the budding region, why does it not prevent growth in the subhypostomal region?

Evidence from experiments such as these points toward centers of active growth and metabolism as regions in which a substance, or substances, are produced that inhibit growth of similar structures nearby, and has led to the hypothesis of inhibition by diffusible substances that spread through the tissues adjacent to their origins and suppress growth wherever along the diffusion gradient the level is in the range of their active concentration. This hypothesis has been tested in a number of ways: by exposing cut stems of colonial hydroids to water that has contained numbers of excised hydranths, on the assumption that the inhibitor would diffuse from them and make "hydranth water," analogous to egg water; by extraction of homogenates of regions whose inhibitory effect has been established by operative and transplantation techniques and testing their fractions, obtained by chromatography or electrophoresis, for effect upon regenerating pieces; and by testing compounds that might be suspected of retarding or inhibiting growth.

Hydranth water has been found to inhibit regeneration of hydranths by excised stems of *Tubularia*, but although this effect is produced only by water that has contained hydranths and not isolated pieces of hydrocaulus, it is not entirely clear whether bacteria, almost inevitably present in such preparations, have not contributed some factor. Inhibitory substances have been extracted from homogenates of *Hydra* and colonial hydroids, but they have not yet been chemically identified, nor the site or mode of their action determined. Of the various known compounds tested, those most generally effective are nerve depressants such as chloretone, xylocaine, and sodium barbitol. Lipoic acid, the oxidized form of 6,8-dimercapto-octanoic acid (thioctic acid) has also proved inhibitory when cut animals are exposed to it up to 8 hr after the operation. Lipoic acid, or lipothiamine, an amine of thiamine and lipoic acid, is known to be a cofactor necessary to oxidative decarboxylation. Its most pronounced inhibitory effect is upon tissues in the most active phase of regeneration.

Increasing consideration is being given to the role of the nervous system in regeneration and growth processes in general. It has been shown in experiments on *Hydra littoralis* that a great increase in the number of nerve cells at the regeneration surface ordinarily is evident 12 hr after transection. This is not the case in specimens that have been exposed to such compounds as atropine, amphetamine, and xylocaine; nor, in histochemical preparations, do they show the same amount and distribution of such enzymes as acid phosphatase, adenosine-5-phosphate, and acetylcholinesterase as specimens that have not been so exposed, at least as long as the drug is present in the medium. It has been suggested that the nervous system in *Hydra* regulates the growth and differentiation of specialized cells through blocking delivery and discharge of neurosecretory materials to effector sites. This is substantiated by the observation that dense, membrane-bounded granules, presumably neurosecretory, which are demonstrable in some of the nerve cells of *H. littoralis*, accumulate in the terminations of the nerves after transection and are released into the intercellular spaces at the site of regeneration. It has been shown, too, that regeneration is abnormal in isolated midregions of hydras when these granules, after extraction and concentration by centrifugation, are introduced into the culture medium and allowed to remain there for 4 hr, before transference of the segments to a fresh medium without the granules. About 25 percent of the specimens so treated developed more than one hypostome and set of tentacles, sometimes at the distal end, the proximal end regenerating a base in the usual fashion; but sometimes at the proximal end, as well as the distal; or along the sides of the column, along with a distal head and a proximal base. These results suggest that the neurosecretory granules contain a growth-stimulating and form-regulating factor that may be operative in ordinary growth processes as well as in regenerative ones.

The stimulus to regeneration would seem to arise not only from exposure of the cut surface to the external environment but also from removal of inhibitions normally imposed on the cells there by the structures with which they are associated in intact organisms. This implies at least a dual set of controls in growth processes: one that promotes and one that inhibits migration, division, and differentiation of cells.

6. Planarians

Extensive studies of regeneration have been made on triclad flatworms, bilaterally symmetrical animals and morphologically more complex than hydroids. Many of them have high regenerative capabilities while others have little. Most of the species used in experimental studies are in the family Planariidae and, hence, are generally known as planarians, although few, according to present taxonomy, are in the genus *Planaria*. Many are, however, referred to in the older literature by their former names.

It has been known for many years that most planarians, when starved, live off their own tissues, which are autolyzed and resorbed in descending order of their importance to the life of the individual. First to be used are the reproductive organs and last the nerve cells, after which the animal dies. The loss is in the number of cells, not in the size of those that are not autolyzed. Before the nerve cells are gone, an animal can be revived by feeding, and in the course of resuscitation, the organs that have been sacrificed are reconstituted in ascending order of their importance to vital activities, with the reproductive organs the last of all.

An axial gradient of regenerative capacity is shown by the fact that complete new individuals are regenerated from pieces of planarians transected at any level of the body, with cephalic structures always arising from the anterior surface of the piece and caudal structures from the posterior (Fig. XIX–11*b*). If a specimen is cut into many pieces, those from the most posterior parts regenerate more slowly than those from the more anterior ones, sometimes abnormally and sometimes not at all.

Planarians differ from hydroids and in this respect more closely resemble vertebrates in that regeneration is preceded by formation of a blastema, easily recognized as a small bleb of clear cells at the cut surface, after this surface has been closed over by

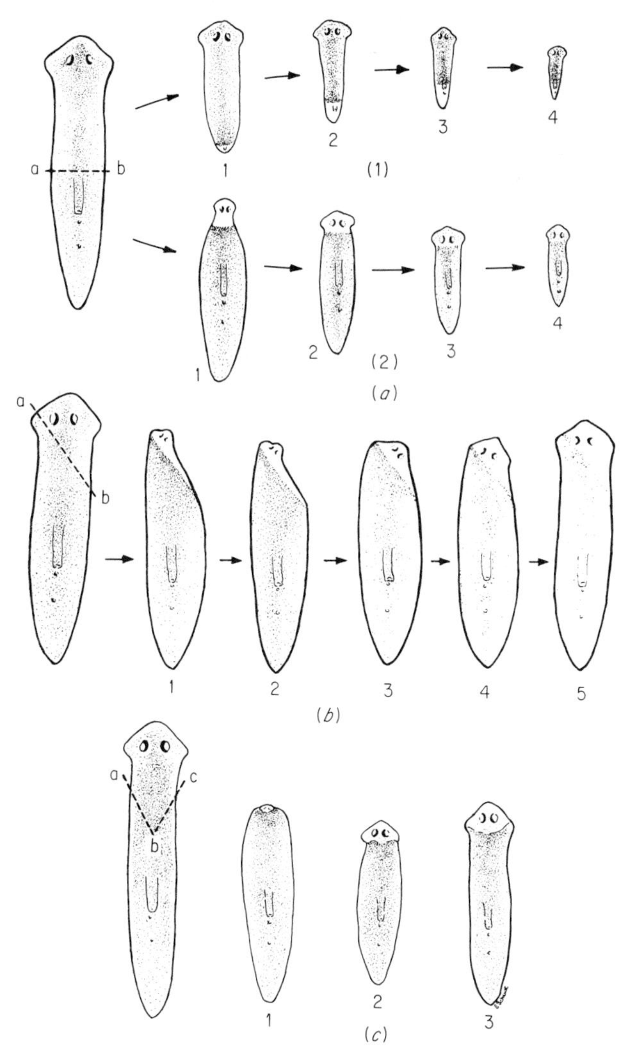

FIG. XIX–11 *Regeneration in planarians. (After T. H. Morgan.)* ***(a)*** *Regeneration of two parts of a flatworm transected across a–b. (1) Regeneration of anterior half; the last figure, 4, shows the regenerate 40 days after the operation. (2) Regeneration of the posterior half. 1: 7 days after the operation. 2: 18 days after the operation. 3: 24 days after the operation. 4: 40 days after the operation.* ***(b)*** *Regeneration of flatworm whose anterior region was removed by an oblique cut along the line a–b.* ***(c)*** *Regeneration of a flatworm whose anterior end has been removed by a wedge-shaped cut along the lines a–b–c.*

migration of epidermal cells. Blastemata are generally believed to be formed by migration of formative cells or neoblasts that are distributed throughout the body. They lie among the parenchyma cells and can be recognized by their oval or pear shape and histochemically distinguished from other cells by their high cytoplasmic content of RNA, characteristic also of cnidarian interstitial cells. Like interstitial cells, neoblasts are selectively destroyed by irradiation, disintegrating in 1 or 2 days after exposure to x-rays. If the entire population in a planarian is eliminated in this way, the animal is incapable of regeneration and dies in 3 to 7 weeks. But it can be revitalized by implantation of a piece from a nonirradiated animal, as implantation of new interstitial cells can revitalize a similarly deprived hydra. Moreover, if only a part of a specimen is exposed to x-rays and the remainder of its body shielded from them, the irradiated region becomes necrotic but the nonirradiated region survives. And if the anterior or posterior part alone is irradiated and the body then transected through the irradiated region, regeneration of the lost part follows. When these events are followed histologically, it is seen that the blastema is formed first of irradiated neoblasts that soon disintegrate and are then replaced by vital ones that migrate from the nonirradiated region across the irradiated zone, in which there remains only the debris of disintegrated cells and a few recognizable muscle and nerve fibers. The migration of healthy neoblasts into the necrotic zone, which they revitalize, and into the blastema, where they reconstitute the missing parts, has been followed in *Dugesia lugubris* stained with methyl green-pyronin, a recognized procedure for their visualization and identification. In animals whose anterior regions have been irradiated and that have then been transected behind the eyes, no stained cells are detectable in the necrotic zone for about 2 weeks after the operation; then they appear in its posterior region and gradually spread through it, finally reaching the blastema. In a number of experiments the time lapse between transection and the first signs of regeneration in the blastema could be correlated with the length of the irradiated area and the distance the healthy neoblasts had to travel to reach the blastema.

The selective staining of neoblasts has made it possible to determine the average number of them in any area of a planarian's body. Counts made in nonregenerating *Dugesia lugubris*, *Dendrocoelum lacteum*, and

Planaria vitta have shown that they decrease in number from anterior to posterior, corresponding to the postulated axial gradient and comparable to the distribution of interstitial cells in *Tubularia*. In *Dugesia* the gradient is not a uniform descent, but has two peaks: one in the region of the cephalic ganglion and one in the region immediately anterior to the pharynx. In specimens regenerating cephalic structures, the number of neoblasts in the prepharyngeal region is reduced, consequent upon their migration to the blastema, but is later restored by divisions of those that remain there.

Neoblasts do not represent a static reserve of embryonic cells that may become exhausted in a planarian's lifetime, but a dynamic one in which new members are constantly being added to the population by division and, possibly, also by dedifferentiation of more specialized cells. It has been shown, for example, that gland cells of the digestive tract can transform into neoblasts, a biochemical as well as a morphological dedifferentiation, for neoblasts not actually engaged in regeneration are low in acid phosphatase, an enzyme present at a high level of activity in gastrodermal cells. But neoblasts, whatever their source, can synthesize the enzyme for it is demonstrable in them, at high levels of activity, in regenerating regions. As far as their own capacities for differentiation go, they are at least pluripotent and most likely totipotent. It has been shown by in vitro cultivation of isolated blastemata that, while the destinies of the neoblasts are more or less fixed as early as 3 days after a blastema is formed, epidermis, rhabdoid cells, parenchyma, muscles, nerve cells, digestive tract, and gonads can arise from them. Moreover the blastemata are polarized, for one taken from the anterior region of a flatworm will develop only into a head, not an entire planarian, and one taken from the posterior region, only into a tail.

Induction and inhibition have been evoked as mechanisms underlying neoblast differentiation and structural organization in the blastema. Regulation of growth processes, evident in regenerating pieces, results presumably from a balance between these two mechanisms: the one directing differentiation along particular pathways and the other preventing it. Transplantation and grafting experiments have provided evidence for such interaction between tissues and organs, as has the application of tissue extracts. Pieces taken from one region of a flatworm's body and inserted into homologous regions of another do not provoke new tissue growth in the host, but those inserted in nonhomologous regions do. A piece taken from the cephalic region of a donor and inserted anywhere in the body of a host but its cephalic region provokes new growth. Unless the graft has been put in the postpharyngeal region, the new cells become organized as a head. If the graft is inserted in the postpharyngeal region, new tissue arises along the line of union of graft and host, and two new pharynges develop, either in the graft or the host tissue.

Additional evidence is derived from parabiotic experiments. If two individuals are divided longitudinally and the two halves of each are united side by side in perfect alignment, there is no growth of new tissue and a single complete individual results. When such an individual is transected at any level, a single symmetrical head is regenerated from the anterior surface of the posterior piece. But if lateral apposition of the two halves is not perfect and the anterior end of one is to a greater or lesser extent behind that of the other, head regeneration after transection is distorted. Such a result is shown in Fig. XIX–12*a*. If the experiment is performed in such a way that there is even greater disparity between the levels of the two partners than that figured, only one head regenerates, and that from the more anterior surface. A blastema may be formed along the more posterior surface, but if so, it becomes incorporated in the more rapidly developing anterior one.

These results indicate a difference in the time necessary to regenerate a head at different levels of the body. In most planarians the capacity for head regeneration, and therefore the time taken for it, diminishes along the length of the body from anterior to posterior. Excision and grafting experiments have shown that the time needed for head regeneration is specific for a given area, which has led to the concept of "time-graded regeneration fields." If a square or rectangular piece is cut out of a flatworm, a head regenerates on the lower surface of the window, even in the presence of an existing head, but the rate of its development and its ultimate size depend upon the level from which the piece was cut; the more posterior,

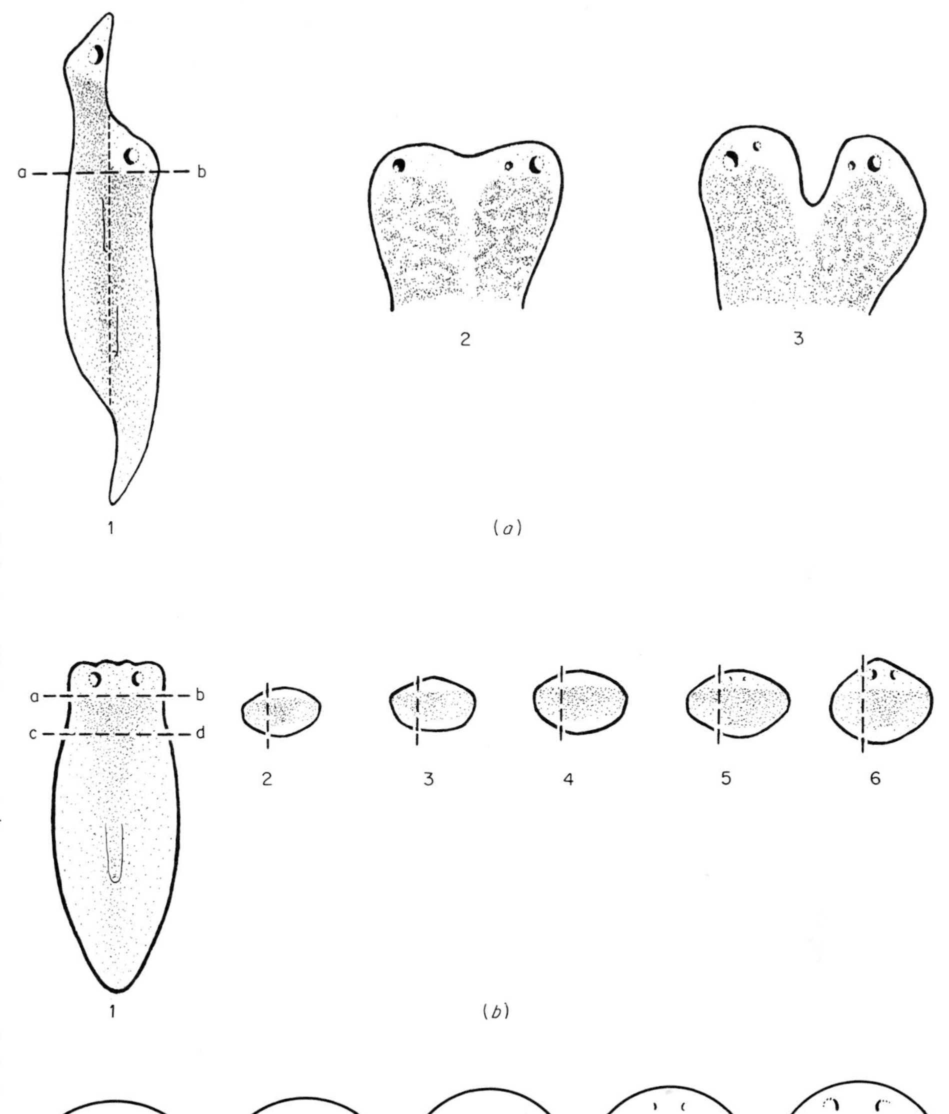

FIG. XIX–12 *Regeneration in planarians.* (*After Brønsted.*) ***(a)*** *1: Two individuals of* Dugesia lugubris *divided lengthwise and grafted together with their bodies transposed about one-quarter of the length along the animal. Four days after grafting the conjoint animal was transected along the line a–b. 2: 7 days after the operation. 3: 15 days after the operation.*
(b) *1:* Dendrocoelum lacteum, *showing location of the region, cut along lines a–b and c–d, removed for study of regeneration.*
2: Regenerating fragment 24 hr later.
3: Regenerating fragment 48 hr later.
4: Regenerating fragment 72 hr later.
5: Regenerating fragment 96 hr later.
6: Regenerating fragment 120 hr later.
The vertical line in 2 to 6 represents a cut, separating a lateral third of the regenerating portion from the remaining two-thirds in order to test for induction.
(c) *Lateral migration of inhibitor in the blastema.*

the slower the regeneration, and the smaller the resulting head. Similarly, pieces grafted from the body of one animal into that of another develop heads on their anterior surfaces, but they do this at the rate characteristic of the level at which they were in the donor's body, regardless of their position in the host's.

The accepted criterion for head regeneration is development of fully formed eyes. But head regeneration entails also organization of the cephalic ganglion and the ciliated organs. For all this to be accomplished there must first of all be some influence arising in the posterior region of a transected planarian to start the neoblasts on forward migration to make a head blastema. So far as is yet known, the cephalic ganglion arises in this blastema by autodifferentiation, but its inductive effect upon the development of the eyes has been clearly shown in *Polycelis niger*. In this flatworm, a row of eyes extends along the anterior margins of the body (the eye rim) almost to the base of the pharynx. If all the eyes and the cephalic ganglion are excised from a specimen of *Polycelis*, the ganglion regenerates from its blastema in 3 to 4 days, the eyes in 7 days, but if regeneration of the ganglion is prevented by repeated excision of its blastema every time it is formed, no eyes appear until 7 days after the last excision. And if the eye rim of *Polycelis* is transplanted near the ganglion of a specimen of *Dugesia lugubris* and the eyes subsequently excised, new eyes will regenerate in the *Polycelis* tissue although there is no nervous connection between them and the *Dugesia* ganglion. Eyes do not develop in similar strips implanted into the caudal region of the host, yet if a ganglion is implanted there as well, they do. Pieces of eye rim, from which the eyes have been removed, grafted anywhere in the body anterior to the base of the pharynx regenerate eyes, but those grafted more posteriorly do not. Moreover, it has been shown that in regenerating heads the eyes are formed before the optic nerves have reached them and that they may originate independently of neoblasts coming from the region of the ganglion, for irradiation of the cephalic region does not prevent their formation. This would imply that neoblasts from more posterior regions contribute to them and that their differentiation is directed by influences from the cephalic ganglion.

These results suggest that a substance diffusing from the ganglion induces formation of eyes and that it is effective only in the range of its diffusibility in active concentration. This hypothesis has been tested by the introduction of homogenates, or fractions of them obtained by centrifugation, of cephalic regions into cultures of specimens deprived of their eyes and cephalic ganglia. Eyes regenerate in the presence of the homogenate and in the absence of the ganglion, whose regeneration seems inhibited in these circumstances. It has been shown, too, that the active principle is not species-specific, for homogenates of *Dugesia lugubris*, *D. gonocephala*, and *Dendrocoelum lacteum* have the same effect upon *Polycelis* as homogenates of its own species. Homogenates from tail regions do not induce eye regeneration in specimens deprived of a cephalic ganglion, although after heating to 60°C for a few minutes, they do. This suggests that the inducing substance may be present in them in a bound or an inactive condition and manifests itself only when this is in some way altered.

Induction in areas other than the cephalic can also be demonstrated. When the head of a specimen of *Dugesia tigrina* or *D. dorotocephala* is grafted into the postpharyngeal region of the same animal, one or two pharynges may form there, although the animal still retains its original pharynx. More precise experiments, limiting the area of the grafted tissue, have shown that it is not the entire head but the region between the head and the base of the pharynx, or prepharyngeal zone, that provides the inducing agent for construction of the supernumerary pharynges. Regeneration of a pharynx, excised with its sheath from half of an animal transected at the posterior border of its prepharyngeal zone, takes place in normal position; moreover, a piece cut through the pharyngeal region from which the pharynx is then removed, regenerates a new one, even when regeneration of the anterior region is prevented by repeated amputations of the blastemata that would give rise to it. Reconstitution of the pharynx in an established pharyngeal region is, therefore, not dependent upon the presence of cephalic or prepharyngeal regions. A pharynx will also regenerate in the postpharyngeal region after transplantation of a head there and its subsequent extirpation and that of the supernumerary pharynx that has been formed. This suggests that during the time the transplanted head remained in the postpharyngeal region the tissues there were conditioned for pharynx formation—that they were "pharyngized," after which they could form a pharynx independently by autodifferentiation, just as the tissue of the natural pharyngeal zone was shown to be capable of doing. These results have led to a concept of zonal induction, on the assumption that a grafted head induces a prepharyngeal zone, and the prepharyngeal zone, a pharyngeal zone, which is then capable, even in a foreign position, of differentiating a pharynx and regenerating it after it has been experimentally removed. It is possible that zonal induction plays a large part in the regulation of regenerative growth in planarians.

There is some evidence also that gonads induce formation of the copulatory apparatus. This has been difficult to obtain because of the diffuse arrangement of the reproductive organs in planarians, and is not yet conclusive. But there are strains in the different species of *Dugesia* used for regeneration studies that reproduce entirely by fission and do not develop gonads or other genitalia. If the anterior region of a gametically reproducing individual containing sex organs, particularly testes, is grafted to the posterior region of an agametically reproducing individual without gonads or genitalia, typical copulatory organs develop there. The reverse graft has shown that the presence of genitalia in the posterior half of a graft between individuals of gametic and agametic strains leads to the development of testes in the anterior half when that is taken from an individual devoid of them.

The cephalic ganglion produces inhibitory, as well as inductive, agents. When, for example, a ganglion from one specimen of *Dugesia lugubris* was transplanted into the anterior region of another and the host was then transected anterior to the implant, in some of the individuals so treated a single head regenerated, the implanted ganglion assuming control of regenerative processes in the blastema. In others, however, the host regenerated a head, yet another was also regenerated around the implanted ganglion. Both had eyes, but there was no ganglion in the regenerate from the host, in its place there was only a mass of undifferentiated neoblasts. The implanted ganglion, which remained intact, induced formation of eyes in the host's blastema, but inhibited organization of its ganglion. Other regions of the body have been shown to have inhibiting influences by testing the effects of homogenates, or their fractions, made from different areas of the body upon homologous areas, and by grafting. A ganglion, for example, will not develop in the blastema of a headless planarian immersed in a medium containing an extract of cephalic tissues, although, as might be expected, eyes do. A complete head will, however, regenerate in such planarians immersed in extracts from caudal regions. Homogenates and extracts of pharyngeal regions inhibit regeneration of pharynges from specimens from which these have been removed. The hypothesis seems justified, therefore, that differentiated structures prevent differentiation of similar ones and that, in the course of regeneration, specific inhibitory substances are produced by a regenerating organ or region that prevent others like it from being organized at the same place at the same time. There would seem to be then in planarian regeneration a balanced succession of inductions and inhibitions by means of which the characteristic pattern of the regenerating part is laid down and provision made for replication of a missing part. Chemical communication between cells leads to the orderly allocation of available materials into a prescribed pattern.

The cephalic ganglion seems to be, in planarians, the master control center both for induction and inhibition, its influence decreasing in axial and lateral gradients. In Fig. XIX–12*c* is diagrammed the course of migration of the eye-inhibiting substance, showing progressive decrease in its effect as it moves from the central region to the peripheral regions. Diffusion of substances from centers where their concentration is maximum to regions where this is minimum may then underlie the axial gradient so clearly evident in these animals. Another more basic underlying factor may be the rate of protein synthesis, for it has been shown in experiments in which specimens of *Dugesia tigrina* have been incubated with carbon dioxide or glycine labeled with C^{14} that there is an axial gradient of incorporation of these compounds into proteins, although in the case of glycine the gradient is not a smoothly descending one from head to tail but has a slight peak in the caudal region over the level in the pharyngeal. The axial pattern is followed in regenerating pieces, although the head and tail blastemata show increased incorporation of labeled glycine. Polarity of these sections is upset if the worms before operation, or if the anterior surface of sections from untreated worms, are exposed to colcemide, a derivative of colchicine, at a concentration of 0.02 mg/ml, for heads develop on both surfaces. This does not happen if the section is from an anterior region and contains a head whose presence inhibits development of another on the posterior surface. Measurements of the rate of incorporation of labeled carbon dioxide in specimens treated with colcemide show that the gradient demonstrable in untreated animals is altered, or even obliterated. The polarity of cut sections can also be upset by exposing their anterior surfaces to chloramphenicol, at a concentration of 1 mg/ml, but, likewise, the presence of a head on a section prevents development of another at the opposite end. Both these compounds are recognized inhibitors of protein synthesis. The results of such experiments indicate that disturbance of the gradient of protein synthesis also disturbs the biological polarity of the animals and, conversely, that this is correlated in living animals and in regenerating pieces with differences in the rate of protein synthesis in different regions and the relative competition for nutrients that this entails. The appearance of a few two-headed individuals in cultures of an agametic strain of *Dugesia tigrina* shows that derangement of polarity may occur spontaneously, in the absence of any known environmental condition to which it might be attributed. In the case of these animals it may have been due to their unusually long length, about 30 mm, possibly too far for the head inhibitor to diffuse and reach the tail in effective concentration. One of the animals was seen to divide into a longer and a shorter piece, each with a pharynx and each with a head. The shorter piece regenerated a tail and underwent a series of ordinary fissions; the larger piece regenerated a head and pharynx at its posterior end. This bipolar individual behaved as two individuals, ingesting food and starting to move in different directions at the same time, the midregion being dragged along. The passivity of this region suggests that it was sufficiently inert to offer a barrier to diffusion of the head inhibitory substance, or perhaps the distance was too far; in either case, the appropriate message failed to get through.

7. Nemerteans

Nemerteans are like flatworms both in the ways in which regeneration is effected and in the differences in capacity for it in different genera and even in different species of the same genus. Most heteronemerteans have considerable powers of regeneration, distinctly an asset to animals so fragile that they break into many pieces even when gently handled. But in some, only the posterior part can be reconstituted from an anterior part, in some species following transection at any level and in others only if the cut passes through the midgut, so that some of this is included in the regenerating piece. *Lineus socialis*, a species common along the Atlantic coast, and *L. vegetus*, a Pacific species, have an almost infinite capacity for both anterior and posterior regeneration; and any fragment of the body, if of sufficient size, from either transverse or longitudinal cuts, will regenerate a complete new individual, provided that part of one of the longitudinal nerve cords is included in it. Thus, from a single specimen of *L. socialis*, which may be as long as 2 m, 100 new individuals may be reconstituted if it is cut into 100 pieces; and this process may be repeated again and again by partial regenerates, from which pieces are cut off until minute individuals are produced, about 0.15 mm long and 0.04 mm wide, less than one hundred-thousandth the size of the original. *Lineus vegetus* has a comparable capacity, but *L. ruber* is capable of only posterior regeneration, and only if the anterior ends of the nerve cords are in the section. In *L. pictifrons* and *L. rubescens*, the capacity extends further posteriorly, for in both, pieces of specimens cut through the foregut can regenerate, although when transections are made through successively more posterior regions of the foregut the probability of regeneration of anterior regions decreases. All pieces are, however, capable of reconstituting a missing posterior portion, and all pieces of all species, whether they contain a piece of the nerve cord or not, can heal a wound by migration of epidermal, or subepidermal, cells over the exposed tissues.

Replacement of posterior regions is not, strictly speaking, regeneration, for all the tissues and organs are represented at the cut surface, and replacement of the missing part comes about by division of tissues already differentiated and already located in organized structures. This growth is augmented by unspecialized cells that migrate to the growing points and differentiate along the lines of the cells with which they come in contact. But blastemata are not formed in the posterior regeneration of nemerteans, as they are in anterior regeneration, and the new structures are not primarily organized from undifferentiated cells.

As the process of regeneration has been followed in pieces of *L. socialis* taken from any part of the body, the first event after transection is violent contraction of the body, followed by migration of epidermal cells

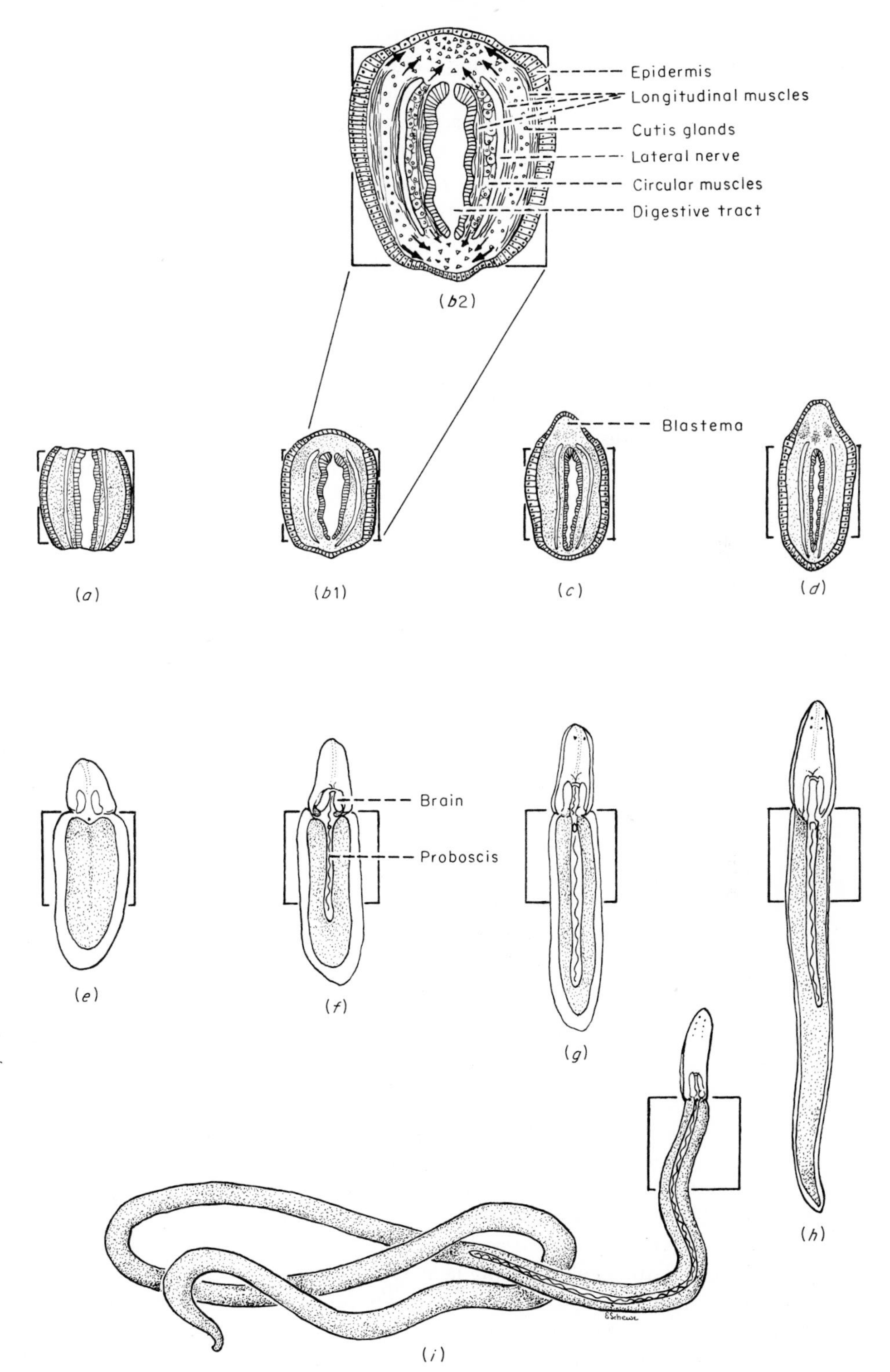

FIG. XIX–13 *Regeneration in nemerteans. (After Coe.) The rapid regeneration of a small fragment of* Lineus socialis, *which can be taken from any part of the body provided that a portion of one of the lateral nerve cords is included. The brackets indicate the size of the original fragment.* ***(a** to **d)*** *Healing of cut surfaces and formation of blastema at anterior end.* ***(e** to **h)*** *Differentiation of structures at anterior end and growth of body posteriorly.* ***(i)*** *Fully regenerated and reorganized individual.*

and neoblasts to both cut surfaces, where the epidermal cells form a thin surface covering (Fig. XIX–13). The neoblasts migrating to the posterior end become incorporated with the tissues there, but those that migrate to the anterior end form a blastema. Some regulative mechanism comes into operation, for the cells collect in three localized areas: two lateral and one median. The two lateral ones are situated close to the epidermis and soon become recognizable ganglia which then act as organizing centers and control the distribution and differentiation of the remaining, presumably totipotent, cells. After the ganglia are formed, the median mass of neoblasts begins its differentiation as proboscis and proboscis sheath, and two lateral invaginations of the epidermis, primordia of the canals of the cerebral sense organs, appear not far from the location of the new ganglia. Nerve cords grow out from the ganglia and the other cells of the blastema become organized as foregut, mouth (presumably induced by the foregut), blood vessels, and other cephalic tissues and organs. These, developed in the blastema without contact with their counterparts in the original piece, join up with them, the slender new nerve cords with the larger ones in the fragment, the blood vessels with the blood vessels, and the new foregut with the old digestive tract. Thus, by regeneration at the anterior end and growth at the posterior, a small new nemertean can be reconstituted from a piece no longer than 2 cm, which when it begins to feed, can grow into a nemertean of adult proportions and adult capacities.

Phagocytosis plays an important role in the nutrition of the reconstituting fragment, for cell debris is cleared up by phagocytes; digested; and its substance made available to the dividing and differentiating cells. The role of the nervous system is paramount in regenerative processes, and pieces of *L. socialis* lacking any part of the nerve cord heal but do not form blastemata and neither regenerate nor grow appreciably posteriorly. It seems likely that mobilization of the neoblasts depends upon messages received from the transected cords, which initiate the entire process of regeneration, the regulatory control being taken over by the cephalic ganglia as soon as they have arisen. In *L. socialis* and *L. vegetus*, the property of transmitting the message, or messages, is distributed all along the length of each lateral cord; in *L. pictifrons* and *L. rubescens*, it is more restricted, for only pieces containing sections of the nerve cord that lie anterior to the midgut can regenerate heads; and in *L. ruber* it is limited to the most anterior region of the nerve cords, essentially the ganglia, for only pieces that contain these are capable of anterior regeneration. Since neurosecretory cells have been histochemically demonstrated in the ganglia of nemerteans, it is possible, though as yet there is no definitive proof, that the transmission may be chemical.

There is also evidence for inhibition of regenerative processes comparable to that seen in planarians. This evidence has been derived particularly from experiments with homogenates or extracts of different parts of the body of *L. vegetus*. Blastemata do not develop on the anterior surfaces of sections taken from any region of the body when they are exposed to a supernatant obtained by centrifugation of the heads of other animals, but they do form and regenerate when exposed to supernatants from the tail region. Supernatants from the midbody inhibit formation of blastemata on sections posterior to the level of the midbody section used and also inhibit posterior replacement on sections anterior to this. Posterior replacement is also inhibited on head sections by homogenates made from any area of the body posterior to the mouth, but not by homogenates of heads, nor do these inhibit posterior replacement in sections taken from any other part. From these results it is postulated that regional differences in differentiation pattern exist along the length of the animal's body and that each region controls the differentiation of succeeding regions by preventing them from attaining the level of the preceding one. This hypothesis is consistent with the concepts of an axial gradient of differentiation, of zonal inhibition, and of communication from already highly differentiated areas to undifferentiated ones that instructs them not to replicate already formed structures.

8. Ectoprocts

Little has been done in the study of regeneration in ectoprocts, although their capacity for spontaneous regeneration is well known. It has been shown that after amputation of the lophophore, specimens of the freshwater phylactolaemate genus *Fredericella* will grow new lophophores from the residual organ systems, and that after amputation of the lophophore and detachment of the digestive tract, a new zooid develops from an adventitious bud originating at the site of attachment of the original oesophagus, representing construction of a complete individual from a small group of epidermal cells. Reconstitution of the entire anterior end follows transection of this part and, in the reconstituted portion, the very precisely determined distal budding area is included, making this as much a part of the general organization pattern, and presumably determined in the same way, as any of the structural organs of the zooid.

9. Annelids

Polychaete annelids have a high, but variable, capacity for regeneration, oligochaetes somewhat less, and leeches none at all. Segmented animals such as these have particular problems of growth in that there are local differentiations of some organs, such as gut and, in some species, reproductive organs, and serial repetition of others, such as parapodia, chaetae, and nephridia. In oligochaetes and some polychaetes, the number of segments is limited to an average range or, indeed, to a specific number for each species, and post-embryonic growth is expressed as increase in size of segments already formed and not by addition of new ones.

In some syllids an entire worm can be regenerated from a single segment from anywhere in the body, exclusive of the first and the last segment. Such regeneration follows naturally in spontaneously fragmenting species like *Ctenodrilus monostylus*, *Dodecaria cauleri*, and *D. fimbriatus*, which would imply that they have no axial gradient of regenerative capacity. Experimental isolation of single segments of *Chaetopterus variopedatus* have shown, on the other hand, that while any segment anterior to the fifteenth segment is capable of complete regeneration, segment XV and those posterior to it do not replace anterior structures, and only posterior regeneration is possible to them. Capacity for anterior regeneration is even more limited in the syllid *Autolytus pictus*, which by experimental transections has been found to be able to replace anterior segments successfully only if the cut is made anterior to the seventh setigerous one. Regeneration sometimes takes place after cuts made anterior to the eighth segment, but after cuts made posterior to this segment, replacement is incomplete and the regenerate atypical (Fig. XIX–14). There is, thus, in these

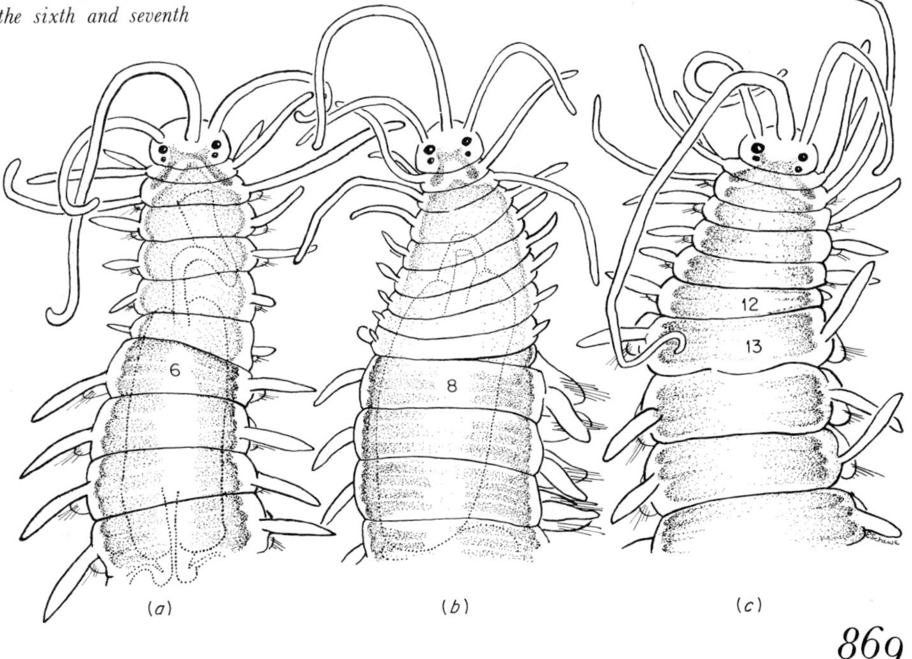

FIG. XIX–14 *Regeneration in annelids. (After Okada.)* ***(a)*** *Regeneration of anterior segments in the syllid* Autolytus pictus. *In this specimen the cut was made between the sixth and seventh setigerous segments, and complete restoration of the missing parts ensued, including a normal prostomium, a normal peristomium, and five body segments.* ***(b)*** *Regeneration in* Autolytus pictus *from which the first seven setigerous segments have been removed by a cut between the seventh and eighth segments. Maximum regeneration has taken place in this specimen, but this is not invariably the case when cuts are made at this level.* ***(c)*** *Regeneration in* Autolytus pictus *from which the anterior segments have been removed by an oblique cut across the posterior part of the twelfth segment. Here only the prostomium and fifth setigerous segments have been regenerated, and the mutilated twelfth segment has not replaced its missing portion.*

species an axial gradient of regenerative capacity, demonstrated by diminution of success in replacement of anterior segments by progressively more posterior ones.

In sabellids, anterior regeneration is never more than that of the three segments included in the head, but these may be regenerated from the anterior surface of any segment of the body. Intervening segments are, however, not reconstituted, but those of the posterior piece are transformed into the appropriate ones. In *Sabella pavonina*, for example, whose body consists of 3 head segments, 5 to 11 thoracic segments, and up to 300 abdominal ones, a cut through the abdominal region is followed by regeneration of a head on the cut surface and metamorphosis of the segments immediately behind it. Both thoracic and abdominal segments in an intact worm are setigerous, but in the thoracic region, the chaetae on the notopodia are bristle-like and those on the neuropodia are hooked, while in the abdominal, things are reversed, and the notopodial chaetae are hooked and the neuropodial, bristle-like. The transformation of abdominal to thoracic segments in a regenerating worm begins when the head blastema is formed, and it is recognizable in the degeneration of the hooked chaetae on the first two of the abdominal segments immediately behind it on the fifth postoperative day. On the sixth postoperative day, degeneration of those on the third, fourth, fifth, and sixth succeeding segments is evident; and on the tenth day, degeneration of those in the seventh and eighth; and finally, on the thirteenth day, degeneration of those on the ninth, tenth, and eleventh. New bristle-like chaetae appear in each segment on the third day after degeneration of the hooked chaetae. Posterior to the last segment to be transformed into a thoracic one, the abdominal segments remain unchanged and the regenerate, thus, has the external features of a typical worm. Internal metamorphic changes accompany these visible external ones, representing a progressive spread of morphogenetic influences from anterior to posterior, expressed here not as induction of new structures from undifferentiated cells or regulated growth from those already differentiated and organized, but of metamorphic change in them. This is quite a different procedure from that in *Chaetopterus variopedatus*, where an isolated segment XIV, or one anterior to it, remains as segment XIV, or the isolated one, and each newly regenerated anterior or posterior segment is typical of the segment in its position in an intact worm.

Most oligochaetes can regenerate anterior segments if the cut is made well in the anterior region, and posterior segments when the cut is made at nearly every level. *Eisenia foetida*, as thoroughly studied as any earthworm in respect to its regenerative abilities, usually reconstructs only three anterior segments when transected at the level of the fifth segment, four when the cut is made between the sixth and the tenth, and also fewer segments than the missing ones after cuts made between the tenth and twentieth; none is replaced after transections posterior to the twentieth segment. Posterior regeneration can be accomplished after cuts made nearly anywhere along the body, the most remarkable feature of which is that the number of segments added almost invariably corresponds to the number of those lost. *Eisenia* has typically 100 segments when it hatches and adds no more during its later life if uninjured. In posterior regeneration, segments are added in miniature only until they number 100; then their addition stops, almost as if the worm could count, and their increase in size begins. It has been postulated that information of the time to stop is transmitted electrically and derives from the difference in potential that can be measured along the length of the worm. The anterior end of *Eisenia* is electronegative to the posterior; after transection between the fiftieth and fifty-first segments, the difference in potential at the cut end decreased immediately from $+15$ to -11.8 mv, but as regeneration proceeded, the potential in the new segments steadily rose, reaching its original level in about 3 weeks, after which no more segments were added. This gives another facet to the theory of axial gradients, with a parallel in the differences of potential that have been measured in the roots and twigs of plants, where the growing apical regions are electropositive to more proximal ones. A similar mechanism may be behind the control that limits the addition of more than the requisite number of segments in other worms, for this is not related to the length or area of the regenerated part, but rather to the number of units in a linear series, as the difference in potential between a series of electric cells is related to their number.

According to observations made on a number of annelids, cells of various kinds form the blastemata. There is general agreement that after transection the wound is covered over by migrated epidermal cells. In some species, the basal cells of the epidermis seem to give rise to the blastema and the new tissues; in others, representing the majority of those investigated, neoblasts migrate to the site of injury, from their various locations in the body—in *Chaetopterus*, between the nerve cords; in *Nais* and *Lumbriculus*, along the septa. In *Eisenia foetida* it is the chlorogocytes that migrate, for those with large basophilic granules have been found in regenerating regions, reaching a maximum concentration there 9 hr after injury. They disintegrate, discharging their contents, largely phospholipids and glycogen, which may provide energy sources for the growing area. These cells, thus, act as trephocytes in storage, mobilization, and transference of nutrient material, as well as nephrocytes in storage and elimination of excretory waste.

The presence of the intestine is essential to the regeneration of posterior segments. This has been very clearly shown in experiments in which it has been removed from a number of segments anterior to the cut surface. New segments do not arise until the old intestine has grown through these and meets the cap of epidermal tissue at the cut surface. Then, as in regenerating worms in which the intestine has been left intact, the cap divides into a terminal pygidium, with little if any subsequent growth and a subterminal growing region. If the cut has been made in the midregion of the worm, the migrated epidermis covers over the open end of the intestine and a proctodaeum invaginates to meet it; this has been shown in *Lumbriculus*, *Tubifex*, and *Stylaria*. But if the cut has been made more posteriorly, the epidermis fuses with the wall of the intestine and the lumen remains open. In either case, complete regeneration depends upon the presence of the intestine. Such dependence has not been found in anterior regeneration, at least in the earlier stages. A stomodaeum first invaginates at the end of the blastema and from this the pharynx and anterior regions of the digestive tract are developed; these join up with the digestive tract in the remaining segments.

The absolute necessity of the nervous system to both anterior and posterior regeneration has also been very clearly shown. It was demonstrated 30 years ago in the sabellid *Myxicola aesthetica*, in which severance of the nerve cord alone at any level, without transection of the entire body, led to development of a new head at the site of the lesion. Furthermore, it was shown that heads regenerate from longitudinal pieces that contain the nerve cord but not necessarily the intestine. Similar dependence has been found in *Eisenia foetida*, in which deflection of the nerve cord to the dorsal body wall, or implantation of a piece of it there, provokes growth of new structures—a head if the operation is performed in the anterior segments and a tail if in the posterior. And removal of the central nerve cord from the posterior end of the earthworm *Rhyncelmis limosella* completely prevents regeneration in any part of the denervated region.

If the supraoesophageal ganglion of *Neanthes* is extirpated before amputation of the posterior segments, or at any time during the first 3 days following it, the wound heals but there is no regeneration. This follows, however, in the usual course if the ganglion is not removed until the fourth or a later day after the operation and, often but not always, if the extirpated ganglion is reimplanted in the animal within the first three postoperative days. Implantation of a cephalic ganglion from an intact, nonregenerating animal into a transected one from which the ganglion has been removed does not implement regeneration of posterior segments, but one taken from a worm in the course of regeneration of its own posterior segments leads to regeneration of posterior segments by the host. Histological study of the neurosecretory cells in the supraoesophageal ganglia of *Neanthes* and of *Nephthys* shows that they begin secretion 6 hr after amputation of the posterior segments, the signal for this coming to them through the nerves, not the blood. For if the latter were the case, an implanted ganglion from an intact worm would receive the same message and would secrete the substance called for, but regeneration does not follow such an implantation. Experimental evidence has been produced for an inhibitory message from the β cells of the ganglion, for if these are ablated shortly after amputation of the posterior end, regenerative processes are set in motion before they otherwise would be. It is generally believed that the secretions from the neurosecretory cells are discharged into the blood vessels with which the supraoesophageal ganglion is richly supplied and transported through the vascular system to the site of injury following chemical signals for repair and restitution.

The effects of various chemical agents known to be related to neural function and to other physiological processes have been tested on specimens of *E. foetida* transected between the seventieth and seventy-first segments. Acetylcholine and acetylcholinesterase have both proved to be inhibitors of posterior regeneration, suggesting that the concentration of acetylcholine is a factor in the limitation of growth and is regulated by some means. On the other hand, the neurohormones, epinephrine, serotonin, and tryptamine, and nerve depressants, such as chlorotone and xylocaine, are without effect. Lithium chloride, which is known to have pronounced effects upon both vertebrate and invertebrate embryos, is a strong inhibitor, but its effects can be reversed by treatment with sodium chloride, which might mean that Li^+ was acting as a nerve depressant by upsetting the normal Na^+ balance. Similar effects of Li^+ and Na^+ have been found in tentacle regeneration in *Hydra*. In *Eisenia*, the first observed effects of lithium are on migration, for cells fail to migrate as they do in untreated worms; possibly this migration is neurally controlled. Barbituric acid has also been found to be an inhibitor of regeneration, its effects first becoming apparent when the blastema is increasing in size and histogenesis is in progress. This may result from reduced nucleic acid synthesis. Iodoacetic acid, which prevents glycolysis through its inhibition of glyceraldehyde dehydrogenase, is also an inhibitor of regeneration, and it has been shown that glycolytic activity in the growing and differentiating blastema is about 80 percent higher than in the segment in intact worms corresponding to the position of the cut. Interference with this fundamental biochemical process might then be expected to interfere with regenerative, or any kind of, growth.

10. Arthropods

Regeneration in arthropods is pretty well limited to that of the appendages, the most vulnerable parts of their bodies. In crustaceans, an appendage injured in any segment is autotomized at the breaking plane, and a new one grows out from the stump. The first signs of it are evident at the molt following the injury and autotomy, when it usually appears complete, but in miniature, increasing in size at each succeeding molt. When legs are removed from the spiny lobster of the Pacific, *Panulirus interruptus*, at the next molt the regenerating limbs are about two-thirds the size of the original ones. These lobsters can survive simultaneous removal of all their pereiopods and antennae, and molt about 150 days later with a new but smaller set of appendages. It takes about the same time for restoration of smaller numbers of appendages.

Regeneration in crustaceans is often heteromorphic rather than homomorphic, and though an amputated or autotomized appendage is replaced, it is not always a replicate of the one lost. This is regularly the case with lobsters and crabs after removal of the large claw, when the new appendage in the course of a number of molts develops in the pattern of the other chela, while during the same period, the mate develops a heavy crusher claw in place of its usual pincers. The animal is, therefore, not left less able to defend itself as long a time as might be expected for full development of a large claw from a stump. Sometimes a completely different structure is regenerated, as is the case, for example, after transection of the eyestalk of *Palaemon*; the regenerate may not be an eye but an antenna (Fig. XIX–15).

After loss of an appendage the wound is quickly covered over by a clot of hemolymph and then by a cap of epidermal cells. Although other tissues are present below the cap, it is apparently from the epidermal cells alone that the new appendage is formed. This presents a problem similar to that in the regeneration of segments in annelids, for the appendages

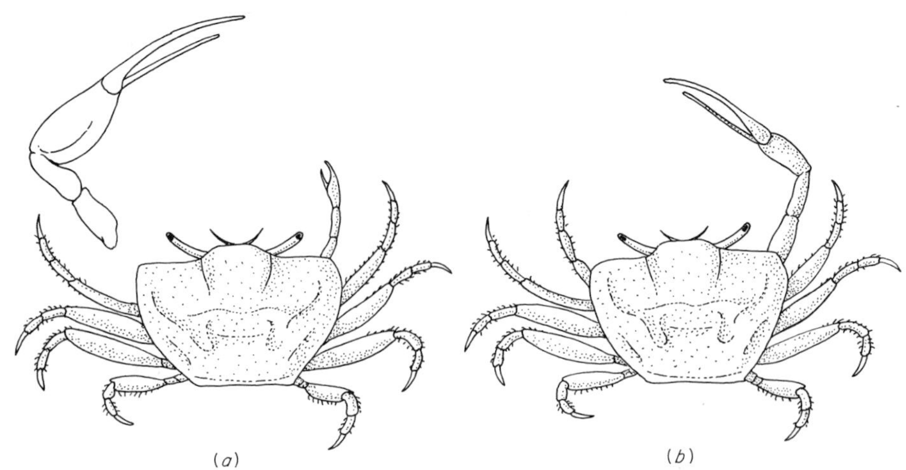

FIG. XIX–15 *Heteromorphic regeneration in* Uca. ***(a)*** *Crab in which the large "fiddling" claw of the left side has been lost.* ***(b)*** *The same crab after its next molt, showing replacement of the lost appendage by a small pereiopod and development of the first pereiopod on the right side toward a fiddling arm. (From Gardiner and Flemister.)*

of crustaceans have a characteristic number of segments as well as an architectural pattern typical of each. There is no definitive information about the mechanism by which the number of segments is controlled, nor about the mechanism by which instructions are transmitted from the stump of an autotomized crusher claw or fiddling arm to the appendage on the opposite side to begin to take its place. Severed nerves are present, however, in the stumps of all severed limbs, and neural messages may be conveyed from the injured to the uninjured member.

Crustacean fishermen take advantage of crustacean powers of regeneration, and indeed crustaceans do themselves. The large claw of stone crabs, *Lithodes*, is particularly prized as food; Florida fishermen break this off from the crabs they catch and then toss the rest back with the expectation—and usually the realization—that the next year they can catch the same crabs with regenerated claws and repeat the process as long as the crabs can continue to regenerate. And *Birgus*, the robber crab, makes a practice of pulling off the large chelae of other crabs to eat, but in so doing does not deplete its food supply, because its victims will grow others.

Many insects regenerate lost appendages during their larval and nymphal stages. Apterygotes, in which the tentorial glands do not degenerate and which molt throughout all their life span, can regenerate as adults, and some pterygotes have at least limited capacities to do so. Some mantids can regenerate lost appendages, at least in part; and parts of the genital tract—oviducts and vasa deferentia—can be reconstituted in some moths. Adults of *Dixippus*, however, cannot regenerate legs although they do so readily as nymphs, but as adults they can after the implantation of corpora allata. After allatectomy, nymphs cannot regenerate, which is hardly surprising in the light of the known effects of corpus allatum hormones upon insect growth and postembryonic development.

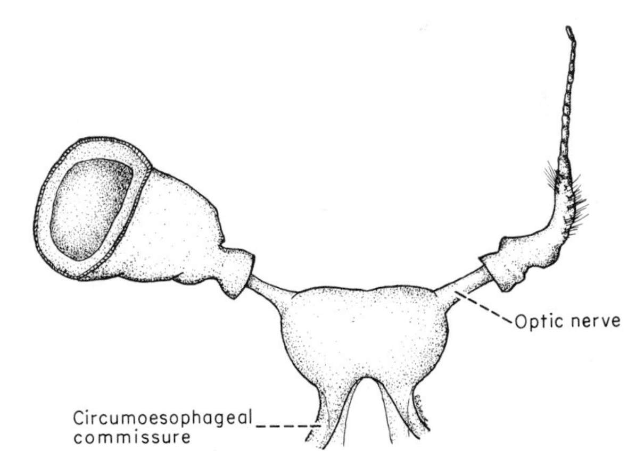

FIG. XIX–16 *Heteromorphic regeneration in* Palaemon. *(After Herbst.) The left eye has been removed by a cut close to the carapace and regeneration, taking place over a period of nearly 5 months in the dark, has produced an antenna-like structure instead of another eye.*

Immediately after injury, hemocytes collect at the site, phagocytose cell debris, and seal off the open area; then epidermal cells migrate to form a cap over the wound. It is from these cells that the new limb is organized, with the exception of the nerves, which grow into the blastema from their cut ends. However, experiments have shown that while nerves are necessary to differentiation and organization of muscles they are not for the construction of external structures and patterns. Legs and antennae, lacking nerves and muscles, have regenerated in the orthopteran *Sphodromantis* and the lepidopteran *Lymantria* after removal of nerves and ganglia from the stumps. But ordinarily the stump contains all the information, and the blastema the material, for construction of the missing part. This has been shown by experiments on caterpillars, in which a thoracic leg has been transplanted into the abdominal region and, after it has become established there, amputated so as to leave a bit of its proximal region as a stump. The regenerated limb is a thoracic leg, even in this foreign position, and a thoracic leg oriented as it would be in its natural place, although in transplantation it may have been differently positioned. Yet heteromorphosis is apparent also in insect regenerates, in which leg-like appendages have replaced amputated antennae; the frequency with which this happens seems to depend on the extent of the part removed, for heteromorphosis more often follows amputation of a large part, or all of an appendage, than that of a smaller part.

As is the case in crustaceans, recognition that regeneration has begun can usually be made at the next molt after the amputation, and its extent then and at succeeding molts depends on the time during the intermolt period that the operation was performed—that is, the time allowed for regeneration to proceed. Amputation in an early larval or nymphal instar usually results in regeneration of a limb of normal size, and amputation in later instars, in a smaller one. In some cases, after removal of a leg the entire structure may appear in miniature at the next molt and grow in succeeding ones, as is the case in caterpillars; but in others, as in nymphs of the damselfly *Agrion* and its relatives, the distal segments may appear first and the proximal ones at later molts.

11. Echinoderms

Starfishes, brittle stars, and sea cucumbers are now all well known for their capacities of regeneration. But this has not always been so, for oystermen, deploring the devastation of their oyster beds by predaceous starfish, used to dredge them up and after cutting them in two throw the pieces overboard in the anticipation that their debris and products of decomposition would provide rich food for the molluscs and that their oyster beds would become more and more luxuriant. On the contrary, what they did was to provide more predators, for after a time whole starfishes can be reconstituted from pieces of suitable size and composition. For most starfishes can regenerate an entire body from the disk and two or three arms, and some, like *Astrometris sertulifera*, a Pacific species, can regenerate an entire body from the disk alone. As far as is now known only one genus, *Linckia*, is capable of regenerating a new individual from only an arm, with no part of the disk. Specimens in the course of such regeneration are known as comets, in which there is extreme disparity between the surviving arm (the comet's tail) and the tiny new disk and arms (the comet's coma). Experimentally, it has been found that *Linckia columbiae*, a Pacific species that reaches a maximum diameter of 4 in., can reconstitute an entire animal from a piece of arm no longer than 1 cm, but that shorter pieces disintegrate. In some instances, when a new individual has been regenerated from a piece of arm longer than 2 cm, the regenerate after about 18 months casts off its five arms and regenerates a new set, while each autotomized arm also reconstitutes a new starfish. Larger pieces of arm may cast the new disk and arms, and begin regeneration of a new disk, and repeat this, decreasing in size each time until a better balance is reached between the old and the new arms. Unlike the condition in annelids in which the number of regenerated segments is rigidly controlled, a starfish may add extra arms in the course of regeneration or may omit some. Specimens of *Linckia* with four to eight arms instead of the usual five are often found in nature and have been observed in regenerates in the laboratory. The details of starfish regeneration have not been followed, but it seems that regenerating structures arise most frequently from outgrowth of the tissues in the stubs or from tissues comparable to those from which they developed in the course of embryogenesis. In many cases deposition of skeletal plates departs from the original pattern, both in construction and arrangement, which offers a means for recognition of a regenerated part.

Asterias forbesii, a species that everts its stomach when feeding, can regenerate this organ and may frequently do so in nature, for it must often suffer from injury or actual detachment. If surgically removed, it can be regenerated completely, with reestablished muscle connections, in 3 weeks. Chronological study of the course of its regeneration has shown that within 48 hr after its excision the cut edges have fused together and that within the next 24 hr the cells there are filled with debris, indicating that disintegrated cells have been phagocytosed by vital ones. Four days after the operation, cells in all layers at the sites of the cuts are in division, and many mitotic figures are evident in sections made at this time, although not in ones made earlier. Most of the new cells are added between the mouth and the level of the retractor muscles as primordia of the future stomach pouches. Histogenesis and organization of the reconstituted pouches continues in the following weeks. In the laboratory, starfishes that have accomplished this regeneration are able to ingest and digest amphipods and small snails. Similar experiments on extirpation of the pyloric caeca in such species as *Leptasterias pusilla*, *Patiria miniata*, *Henricia laeviscula*, and *Pisaster ochreus* have shown that complete morphological, and probably functional, regeneration takes place in 11 to 12 weeks in *Leptasterias* and *Pisaster*, in which the caeca lack Tiedemann's diverticula, and in a somewhat longer time in the other species, where the pattern is rather more complicated by these structures.

Brittle stars also have very complete powers of regeneration and seem often to utilize them, for it is unusual to collect specimens that are not obviously in the course of regenerating at least part of one or more arms. As far as present information goes, any part of an ophiuroid can regenerate an entire animal.

In spite of the fact that fission and evisceration are common phenomena in holothurians, little is known of the details of their regenerative processes or the controls exercised in them. Synaptids may divide into many pieces, but only the anterior ones can regenerate a new individual, indicating that there is some control of the process from this region. In regeneration of visceral organs after their spontaneous evacuation or after operation, new tissues grow from those still remaining in the body. In *Sclerodactyla* (*Thyone*) *briareus*, new gonads do not grow after extirpation of the entire organ—tubules and basis—but if only the tubules are removed, new ones may grow out from that part of the basis that contains gonocytes. Tubule formation seems dependent upon the presence of germ cells, for no sterile tubules have been found among those regenerated, and no tubule grows from a sterile proximal portion of a tubule that may have been left after the operation. In regeneration of the digestive tract, after spontaneous or induced evisceration, the torn surface of the gut heals over and the mesentery connecting this healed end with the cloaca thickens along its free margin. Later a lumen arises in it, continuous at its posterior end with the cloaca. This region acts as a center for growth of intestinal epithelium lining the lumen and for its extension anteriorly. At the anterior end, masses of cells in the tissue around the calcareous ring that is considered a rudimentary Aristotle's lantern act as a center for differentiation of intestinal epithelium there, and the solid cord is progressively invaded from these centers, until about 10 days after evisceration the two tubes join. The posterior half of a bisected specimen of *Sclerodactyla*, whether eviscerated or not, will regenerate all the missing parts and reconstitute a new animal, while anterior, eviscerated halves develop new stomachs and intestines, but no cloacas.

Regeneration of lost spines is not uncommon in sea urchins, which can also replace lost pedicellariae and tube feet. *Psammechinus miliaris* has been observed to drop all its spines, except those on its oral surface, when exposed to direct sunlight, and to regenerate them completely in 2 months, the first signs of reconstitution being evident 1 week after their loss. If a spine is broken but not entirely lost, something that happens quite often in the life of a sea urchin, it is repaired by distal growth of tissue and deposition of skeleton, with its characteristic grooves and ridges continuing along it in the ordinary pattern. But spines that have been totally lost and entirely regenerated may have different numbers of ridges and grooves, indicative of their new growth and responses to different conditions during it. A crack in the test can be healed, and if a small part is crushed it can be repaired, although it always remains as a deformation of the natural shape; a piece broken off a sand dollar test is not replaced, although the exposed tissues are healed over. Little is known of the capacities of echinoids to repair injuries to internal organs or to regenerate new ones, except in the case of the axial organ. This enigmatic structure has been operatively removed from specimens of *Arbacia punctulata*, which survive quite well without it, but do regenerate it. Following its excision, the stone canal and adjacent mesentery undergo dedifferentiation and histolysis. A blastema is formed from cells within the mesentery that divide and migrate to the site and from those in the surrounding coelomic epithelium, the lining of the stone canal, and the hemal vessel. From this blastema, a new axial organ and associated structures are differentiated and organized.

12. Other invertebrates

Regenerative capacities in other invertebrates are either unknown or very limited. "Aschelminthes," as Trembley showed in some of his experiments, are incapable of doing more than close a small wound and are very vulnerable to physical injury. Regeneration in molluscs is almost entirely limited to repair of the shell, when not too much of it has been destroyed; the ability of the gastropod *Gena varia* to regenerate its foot after autotomy has already been noted. Injury or removal of marginal regions of the shells of gastropods and pelecypods is usually followed by restoration of the damaged part, but holes drilled or cut in central regions are not so readily restored. Land snails are better at shell repair than aquatic ones, in which the hole must be covered before it can be restored. Exposed cells cannot take care of this.

The immediate response to shell injury is contraction of the muscles of the mantle, and repair cannot follow unless they relax enough to bring the mantle again under the hole. Formation of pearls by the pearl oyster *Pinctada martensii* and of similar concretions in other pelecypods follows introduction of a foreign object into the space between shell and mantle, which might indicate that the incentive to shell secretion is irritation of the mantle cells. The fact that so highly localized an injury as a minute hole drilled in a shell initiates shell secretion in a much more extensive region of the mantle raises the question of whether there is either a spread of the stimulus or some kind of communication between the irritated cells and their uninjured neighbors. This has something of a parallel in insects, for new and competent blastemata can be formed from a considerable area of epidermis around the base of a caterpillar's leg, and in order to prevent regeneration after removal of a leg not only the epidermis at its base, but the epidermis for some distance peripheral to the base, must be excised. This competence on the part of the cells of the molluscan mantle and the caterpillar epidermis is a kind of insurance of regeneration that provides valuable protection to the animals concerned, providing them with a pool of available cells larger than that actually needed for restoration of the lost or damaged part.

Shell regeneration involves both secretion of the organic matrix and its calcification. It frequently happens that the pattern of crystallization in the repaired area is not the same as that in the original shell. Aragonite is, for example, the usual crystalline form in the shell of *Helix*, but in a repaired area there may be calcite as well as vaterite, and in *Crossostrea virginica*, in which the usual form is calcite, both calcite and aragonite may be deposited in a mended region. Since the crystal pattern is an expression of the protein structure of the organic matrix, the biochemistry of the cells repairing an injury to the shell must be different to some degree from that of the cells that originally made the shell, if they deposit a different kind of protein. A more general shift in biochemical procedures is mobilization of the calcium reserves necessary to shell repair. Most of the calcium is drawn from the digestive gland, but some may come from other tissues or from the shell itself. Whatever the source, a change of calcium metabolism is indicated. Crustaceans also mobilize their calcium reserves when regenerating appendages, and even supplement them by eating the casts of others or the shells of those that they have, in their great need for calcium, attacked and killed.

Much remains to be learned about developmental and regenerative growth and maintenance of adult form and size in both Protozoa and Metazoa. Yet what is already known shows that there are features of developmental processes common to metazoans, and probably to plants as well. The phenomenon of induction, so brilliantly demonstrated by vertebrate embryologists, is one prominent also in invertebrate morphogenesis, to be regarded perhaps as a special case of intercellular communication. Neural mechanisms, whether hormonal or otherwise, seem of equal importance to growth in invertebrates and vertebrates, but not in plants, although hormones from other sources are known to control certain aspects of plant growth. Dominance of one part over another is characteristic of growth in all organisms, and the concept of organizers and progressive determination of fields in which certain areas of competency are delimited applies to regeneration as well as to embryogenesis in invertebrates and vertebrates. There are regenerative mechanisms that seem to be shared by all the organisms that have been investigated. Some species regenerate readily; yet to others, often closely related to them, regeneration is impossible. Common mechanisms are migration of cells to the wound, cleaning up of cell debris by phagocytes, mobilization of materials necessary as nutrients for the growing cells and as structural components of the reconstituting parts, and orderly arrangement of the cells in definite and prescribed patterns. In this latter phase, regenerative and embryological development are alike, and under the same directive forces and influences, for homomorphic regeneration of any part is a repetition of its development in ontogeny, as heteromorphic regeneration is a repetition of the development of a different part. But a regenerated structure develops, as a bud does, under influences from the parent body to which it is attached, and though it may originate from a blastema of pluripotent or totipotent cells, it never arises from a single cell as an embryo does, if a zygote is considered a single cell (see Chapter V, page 130). With repeated regenerations, the parent becomes depleted of its migratory cells and its reserves, and both it and the reconstituted part become smaller and smaller, though they are still perfectly proportioned. It is the perfection of proportion, consistent with the genetically prescribed pattern of the species, that expresses the regulation and control of growth and form, the riddle whose final solution is one of biology's major goals.

I. CONCLUSION

This book began with a statement of the diversity of invertebrate animals and went on to try to show some of the ways in which they have adapted, during their long evolutionary history, to the varied and varying conditions of earth. It concludes with evidence of this diversity, although only a sampling of that presently available; there is still much evidence hidden in the wealth of research subjects among invertebrates, a great deal of which is still untapped. It concludes also with evidence to be drawn of the fundamental unity of all biological systems in the basic mechanisms of life and living, for none transcends or infringes universal natural laws. Yet invertebrates have followed many paths within the confines of these laws in their manifold adaptations to many ecological niches. Ecologists say there are no slums in nature, and indeed there seems to be hardly a place on Earth which is not occupied as an acceptable, even a desired, habitat by some invertebrate species and in which it has found its place. Problems of overcrowding are met by migration and exploitation of new areas, by malnutrition with reduced virility and fertility, and by starvation, ways that seem hard and cruel to humanitarians but are natural ones. Balance of populations within a desirable habitat is adjusted to the available resources, unless upset by some external force. For there are calamities of such origin that invertebrates cannot meet and many of them perish. Species have become extinct in the past, and endangered ones may become so now and in the future, partly through their own limitations and partly through man's exploitation and destruction of them for his own purposes.

SELECTED BIBLIOGRAPHY

Postembryonic development

Berns, M. W., and W. T. Keeton: Regulation of Segment Building during the Postembryonic Development of a Common Millipede, *Science*, vol. 161, pp. 590–591, 1968.

Berrill, N. J.: Development and Medusa Bud Formation in the Hydromedusae, *Q. Rev. Biol.*, vol. 25, pp. 292–316, 1950.

Etkin, W., and L. Gilbert (eds.): "Metamorphosis," part I, "Invertebrates," Appleton-Century-Crofts, New York, 1968.

Hauenschild, C., and A. Fisher: Neurosecretory Control of Development in *Platynereis dumerlii*, in H. Heller and R. B. Clark (eds.), "Neurosecretion," p. 299, Proc. Int. Symp. Neurosecretion, Academic, New York, 1961.

Herlant-Meewis, H., and N. van Damme: Neurosecretion and Wound Healing in *Nereis diversicolor*, in H. Heller and R. B. Clark (eds.), "Neurosecretion," p. 287, Proc. Int. Symp. Neurosecretion, Academic, New York, 1961.

Kafatos, F., and C. Williams: Enzymatic Mechanism for the Escape of Certain Moths from Their Cocoons, *Science*, vol. 146, pp. 538–540, 1964.

Moment, G. B.: A Study of Growth Limitation in Earthworms, *J. Exp. Zool.*, vol. 103, pp. 487–506, 1945.

Moment, G. B.: On the Relation between Growth in Length, the Formation of New Segments and Electric Potential in an Earthworm, *J. Exp. Zool.*, vol. 112, pp. 1–12, 1949.

Moment, G. B.: On the Way a Common Earthworm, *Eisenia foetida* Grows in Length, *J. Morphol.*, vol. 93, pp. 489–507, 1953.

Richards, T. L.: Reproduction and Development of the Polychaete *Stauronereis rudolphi*, *Mar. Biol.*, vol. 1, pp. 124–133, 1967.

Regeneration

Anderson, J. M.: Observations on Autotomy in a Starfish *Asterias forbesi*, *Biol. Bull.*, vol. 111, p. 297, 1956.

Anderson, J. M.: Regeneration of Pyloric Caeca in Starfishes, *Anat. Rec.*, vol. 134, p. 526, 1959.

Anderson, J. M.: Regeneration of the Cardiac Stomach in *Asterias forbesi*, *Biol. Bull.*, vol. 119, p. 302, 1960.

Anderson, J. M.: Studies on Visceral Regeneration in Sea Stars. I. Regeneration of Pyloric Caeca in *Henricia leviuscula*, *Biol. Bull.*, vol. 122, pp. 321–342, 1962.

Anderson, J. M.: Studies on Visceral Regeneration in Sea Stars. III. Regeneration of the Cardiac Stomach of *Asterias forbesi*, *Biol. Bull.*, vol. 129, pp. 454–470, 1965.

Berrill, N. J.: Regeneration in *Sabella pavonina* (Sav.) and Other Sabellid Worms, *J. Exp. Zool.*, vol. 58, pp. 495–523, 1931.

Berrill, N. J.: Regeneration and Budding in Worms, *Biol. Revs.*, vol. 27, pp. 401–438, 1952.

Bodenstein, D.: Regeneration, in K. Roeder (ed.), "Insect Physiology," pp. 866–878, Wiley, New York, 1953.

Bodenstein, D.: Contributions to the Problem of Regeneration in Insects, *J. Exp. Zool.*, vol. 129, pp. 209–224, 1955.

Brien, P.: La Perennité Somatique, *Biol. Revs.*, vol. 28, pp. 308–349, 1953.

Brønsted, H. V.: Planarian Regeneration, *Biol. Revs.*, vol. 30, pp. 65–127, 1955.

Brønsted, H. V.: "Planarian Regeneration," Pergamon, New York, 1969.

Burnett, A. L.: Histophysiology of Growth in *Hydra*, *J. Exp. Zool.*, vol. 140, pp. 281–342, 1959.

Burnett, A. L.: The Maintenance of Form in *Hydra*, in D. Rudnick (ed.), "Regeneration," 20th Symp. Soc. Study Dev. Growth, pp. 27–52, Ronald, New York, 1962.

Burnett, A. L., and M. Garofalo: Growth Pattern in the Green Hydra *Chlorohydra viridissima*, *Science*, vol. 131, p. 160, 1960.

Campbell, R. D.: Cell Proliferation in *Hydra*: An Autoradiographic Approach, *Science*, vol. 148, pp. 1231–1232, 1965.

Clark, R. B., and D. G. Bonney: Influence of the Supraoesophageal Ganglion on Posterior Regeneration in *Nereis diversicolor*, *J. Embryol. Exp. Morphol.*, vol. 8, pp. 112–118, 1960.

Clark, R. B., M. E. Clark, and R. J. G. Ruston: The Endocrinology of Regeneration in Some Errant Polychaetes, in H. Heller and R. B. Clark (eds.), "Neurosecretion," Proc. Int. Symp. Neurosecretion, Academic, New York, 1961.

Coe, W. R.: Analysis of the Regenerative Processes in Nemerteans, *Biol. Bull.*, vol. 66, pp. 304–315, 1934.

Coe, W. R.: Regeneration in Nemerteans. IV. Cellular Changes Involved in Restitution and Reorganization, *J. Exp. Zool.*, vol. 67, pp. 283–314, 1934.

Costlow, J. D., Jr.: Metamorphosis in Crustaceans, in W. Etkin and L. I. Gilbert (eds.), "Metamorphosis: A Problem in Developmental Biology," pp. 3–42, Appleton-Century-Crofts, New York, 1968.

Fishelson, L., and G. Quidron-Lazar: Foot Autotomy in the Gastropod *Gena varia*, *Veliger*, vol. 9, p. 8, 1966.

Flickinger, R. A.: A Gradient of Protein Synthesis in *Planaria* and Reversal of Axial Polarity of Regenerates, *Growth*, vol. 23, pp. 251–271, 1959.

Galstoff, P. E.: The Ameboid Movement of Dissociated Sponge Cells, *Biol. Bull.*, vol. 45, pp. 153–161, 1923.

Galstoff, P. E.: Regeneration after Dissociation (an Experimental Study on Sponges), *J. Exp. Zool.*, vol. 42, pp. 183–221, 1925.

Gauguly, B.: The Differentiating Capacity of Dissociating Sponge Cells, *Arch. Entwick. Org.*, vol. 152, pp. 23–34, 1960.

Golding, D. W.: Endocrinology, Regeneration and Maturation in *Nereis*, *Biol. Bull.*, vol. 133, pp. 567–577, 1967.

Ham, R. G., and R. E. Eakin: Time Sequence of Certain Physiological Events during Regeneration in *Hydra*, *J. Exp. Zool.*, vol. 129, pp. 33–53, 1958.

Haynes, J., and A. L. Burnett: Dedifferentiation and Redifferentiation of Cells in *Hydra viridis*, *Science*, vol. 142, pp. 1481–1483, 1963.

Hobson, A. D.: Regeneration of the Spines in Sea Urchins, *Nature*, vol. 125, p. 168, 1930.

Humphreys, T.: Aggregation of Chemically Dissociated Sponge Cells in the Absence of Protein Synthesis, *J. Exp. Zool.*, vol. 160, pp. 235–246, 1965.

Humphreys, T.: Cell Surface Components Participating in Aggregation: Evidence for a New Cell Particulate (*Microciona prolifera*), *Exp. Cell Res.*, vol. 40, pp. 539–543, 1965.

Jacobs, J.: On the Regulation Mechanism of Environmentally Controlled Allometry (Heteroauxesis) in Cyclomorphic *Daphnia*, *Physiol. Zool.*, vol. 34, pp. 202–216, 1961.

Jenkins, M. M.: Bipolar Planarians in a Stock Culture, *Science*, vol. 142, p. 1187, 1963.

Kafatos, F., and N. Feder: Cytodifferentiation during Insect Metamorphosis: The Galea of Silkmoths, *Science*, vol. 161, pp. 470–472, 1968.

Kille, F. R.: Regeneration in *Thyone briareus* Lesueur Following Induced Autotomy, *Biol. Bull.*, vol. 69, pp. 82–108, 1935.

Kille, F. R.: Regeneration in the Genus *Holothuria*, *Annu. Rep. Tortugas Lab.*, pp. 93–94, Carnegie Inst., Washington, 1937.

Kille, F. R.: Regeneration of Gonad Tubules Following Extirpation in the Sea Cucumber *Thyone briareus* Lesueur, *Biol. Bull.*, vol. 76, pp. 70–79, 1939.

Kiortis, V., and H. A. L. Trampusch (eds.): "Regeneration in Animals and Related Problems," N. Holland Pub. Co., Amsterdam, 1965.

Lentz, T. L.: *Hydra*: Induction of Supernumerary Heads by Isolated Neurosecretory Granules, *Science*, vol. 150, pp. 633–635, 1965.

Lentz, T. L., and R. J. Barrnett: Changes in the Distribution of Enzyme Activity in the Regenerating *Hydra*, *J. Exp. Zool.*, vol. 150, pp. 103–118, 1962.

Lentz, T. L., and R. J. Barrnett: The Role of the Nervous System in Regenerating *Hydra*: The Effect of Neuropharmacological Agents, *J. Exp. Zool.*, vol. 154, pp. 305–327, 1963.

Liebman, E.: The Role of Chlorogogue in Regeneration of *Eisenia foetida* (Sav.), *J. Morphol.*, vol. 70, pp. 151–187, 1942.

Liebman, E.: New Light on Regeneration of *Eisenia foetida*, *J. Morphol.*, vol. 73, pp. 583–610, 1943.

Loosanoff, V. L.: Partial Metamorphosis in *Anomia simplex*, *Science*, vol. 133, pp. 2070–2071, 1961.

Massaro, E. J., and A. R. Shrank: Chemical Inhibition of Segment Regeneration in *Eisenia foetida*, *Physiol. Zool.*, vol. 32, pp. 185–196, 1959.

Millott, N., and A. Formanfarmaian: Regeneration of the Axial Organ of *Arbacia punctulata* and Its Implications, *Nature*, vol. 216, pp. 1136–1138, 1967.

Normandin, D.: Regeneration of *Hydra* from the Endoderm, *Science*, vol. 132, p. 678, 1960.

Osborne, P. J., and A. T. Miller, Jr.: Alkaline and Acid Phosphatase Changes Associated with Feeding, Starvation and Regeneration in Planarians, *Biol. Bull.*, vol. 124, pp. 285–293, 1963.

Pedersen, K. J.: Cytological Studies on the Planarian Neoblast, *Z. Zellforsch.*, vol. 50, pp. 799–817, 1959.

Spangenburg, D. B., and R. E. Eakin: Histological Studies of Mechanisms Involved in *Hydra* Regeneration, *J. Exp. Zool.*, vol. 151, pp. 85–94, 1962.

Spiegel, M.: The Role of Specific Surface Antigens in Cell Adhesion. I. The Reaggregation of Sponge Cells, *Biol. Bull.*, vol. 107, pp. 130–149, 1954.

Steinberg, M. S.: Cell Movement, Rate of Regeneration and the Axial Gradient in *Tubularia*, *Biol. Bull.*, vol. 108, pp. 219–234, 1955.

Steinberg, S. N.: The Regeneration of Whole Polyps from Ectodermal Fragments of Scyphistoma Larvae of *Aurelia aurita*, *Biol. Bull.*, vol. 124, pp. 337–344, 1963.

Swan, E. F.: Seasonal Evisceration in the Sea Cucumber *Parastichopus californicus*, *Science*, vol 133, pp. 1078–1079, 1961.

Tardent, P.: Axiale Verteilungsgradienten der Interstitiellen Zellen bei *Hydra* und *Tubularia* und ihre Bedeutung für die Regeneration, Arch. *Entwickl. Org.*, vol. 146, pp. 593–649, 1954.

Tardent, P.: Regeneration in the Hydrozoa, *Biol. Revs.*, vol. 38, pp. 293–334, 1963.

Tardent, P., and R. Tardent: Wiederholte Regeneration bei *Tubularia*, *Pubbl. Stn. Zool. Napoli*, vol. 28, pp. 367–396, 1956.

Tartar, V.: Pattern and Substance in *Stentor*, in D. Rudnick (ed.), "Cellular Mechanisms in Differentiation and Growth," pp. 73–100, 14th Symp. Soc. Dev. Growth, Princeton University Press, Princeton, 1956.

Tucker, M.: Inhibitory Control of Regeneration in Nemertean Worms, *J. Morphol.*, vol. 105, pp. 569–598, 1959.

Watanabe, Y.: Physiological Studies on a Fresh Water Triclad *Polycelis sapporo*, *J. Exp. Zool.*, vol. 109, pp. 291–329, 1948.

Whitten, J.: Metamorphic Changes in Insects, in W. Etkin and L. I. Gilbert (eds.), "Metamorphosis: A Problem in Developmental Biology," pp. 43–103, Appleton-Century-Crofts, New York, 1968.

Wilson, H. V.: On Some Phenomena of Coalescence in Sponges, *J. Exp. Zool.*, vol. 5, pp. 245–258, 1907.

Wilson, H. V.: Development of Sponges from Dissociated Tissue Cells, *Bull. Bur. Fish.*, vol. 30, pp. 1–30, 1911.

Wilson, H. V., and J. T. Penny: The Regeneration of Sponges from Dissociated Cells, *J. Exp. Zool.*, vol. 56, pp. 73–132, 1936.

Wolff, E.: Recent Researches on the Regeneration of *Planaria*, in D. Rudnick (ed.), "Regeneration," pp. 53–84, 20th Symp. Soc. Study Dev. Growth, Ronald, New York, 1962.

Wolff, E., T. Lander, and C. Ziller-Sengel: Le rôle de facteurs auto-inhibiteurs dans la regeneration des Planaires, *Rev. Suisse Zool.*, vol. 71, pp. 75–97, 1964.

Wyatt, G. R.: Biochemistry of Insect Metamorphosis, in W. Etkin and L. T. Gilbert (eds.), "Metamorphosis: A Problem in Developmental Biology," pp. 143–184, Appleton-Century-Crofts, New York, 1968.

Zwilling, E.: Formation of Endoderm from Ectoderm in *Cordylophora, Biol. Bull.*, vol. 124, pp. 368–378, 1963.

SUGGESTED READING FOR SECTION VIII

Austin, C. R.: "Fertilization," Prentice-Hall, Englewood Cliffs, N.J., 1965.

Baerg, W. J.: "The Tarantula," Lawrence, University of Kansas Press, 1958.

Berrill, N. J.: "Growth, Development and Pattern," Freeman, San Francisco, 1961.

Child, C. M.: "Individuality in Organisms," The University of Chicago Press, Chicago, 1915.

Goss, R. J.: "Principles of Regeneration," Academic, New York, 1968.

Hay, E. D.: "Regeneration," Holt, New York, 1966.

Huxley, Sir J. S.: The Individual in the Animal Kingdom, Cambridge, Cambridge University Press, 1912.

Huxley, Sir J. S.: "Problems of Relative Growth," Methuen, London, 1932.

Lillie, F. R.: "Problems of Fertilization," The University of Chicago Press, Chicago, 1919.

Raven, C. P.: "Morphogenesis: The Analysis of Molluscan Development," Pergamon, New York, 1958.

Tartar, V.: "The Biology of *Stentor*," Pergamon, New York, 1961.

Wigglesworth, Sir V.: "The Control of Growth and Form: A Study of an Epidermal Cell in an Insect," Cornell University Press, Ithaca, N.Y., 1959.

GENERAL AND SPECIAL REFERENCES

General

Barnes, R. D.: "Invertebrate Zoology," W. B. Saunders Company, Philadelphia, 1963.

Barrington, E. J. W.: "Invertebrate Structure and Function," Houghton Mifflin, New York, 1967.

Borradaile, L. A., F. A. Potts, L. Eastham, and J. T. Saunders: "The Invertebrate," 3d ed. (revised by G. A. Kerkut), Cambridge University Press, 1958.

Doflein, F. and Reichenow, E.: "Lehrbuch der Protozoenkunde," Gustav Fischer, Jena, 1927.

Dougherty, E. C. (ed.): "The Lower Metazoa," University of California Press, Berkeley, 1963.

Grassé, P. P. (ed.): "Traité de Zoologie," Masson et Cie, Paris, 1948–

Hyman, L.: "The Invertebrates," McGraw-Hill Book Company, New York, 1940–

Kaestner, A. (H. Levi, trans.): "Invertebrate Zoology," Interscience Publishers, New York, 1967.

MacGinitie, G. E., and N. MacGinitie: "Natural History of Marine Animals," McGraw-Hill Book Company, New York, 1949.

Meglitsch, P. A.: "Invertebrate Zoology," Oxford University Press, New York, 1967.

Pearse, A.: "Animal Ecology," McGraw-Hill Book Company, New York, 1939.

Pennak, R. W.: "Fresh Water Invertebrates of the United States," Ronald Press, New York, 1953.

Prosser, C. L., and F. A. Brown, Jr.: "Comparative Animal Physiology," 2d ed., W. B. Saunders Company, Philadelphia, 1961.

Ramsay, J. A.: "The Experimental Basis of Modern Biology," Cambridge University Press, 1965.

Russell-Hunter, W. D.: "A Biology of Lower Invertebrates," The Macmillan Company, New York, 1968.

Thompson, D'Arcy W.: "On Growth and Form," The Macmillan Company, New York, 1948.

Ward, H. B., and G. C. Whipple (2d ed., W. T. Edmonson, ed.): "Fresh Water Biology," John Wiley, New York, 1959.

Special

Albritton, E. C. (ed.): "Standard Values in Nutrition and Metabolism (Handbook of Biological Data, Fascicle 2)," W. B. Saunders Company, Philadelphia, 1954.

Baer, J. G.: "Ecology of Animal Parasites," University of Illinois Press, Urbana, 1951.

Boolootian, R. A. (ed.): "Physiology of Echinodermata," Interscience Publishers, New York, 1966.

Bullock, T. H., and G. A. Horridge: "Structure and Function in the Nervous Systems of Invertebrates," W. H. Freeman and Company, San Francisco, 1965.

Chandler, A. C.: "Introduction to Parasitology," 9th ed., John Wiley, New York, 1955.

Colt, H. B.: "Adaptive Coloration in Animals," Clarendon Press, Oxford, 1941.

Florkin, M., and B. T. Scheer (eds.): "Chemical Zoology," vols. I–IV, Academic Press, New York, 1967.

Fox, D. L.: "Animal Biochromes and Structural Colors," Cambridge University Press, 1953.

Fox, H. M., and G. Vevers: "The Nature of Animal Colors," The Macmillan Company, New York, 1960.

Fretter, V. (ed.): "Studies on the Structure, Physiology and Ecology of Mollusca" (22d Symposium of the Zoological Society of London and Malacological Society of London), Academic Press, New York, 1968.

Halstead, B. W.: "Poisonous and Venomous Marine Animals of the World, vol. I, Invertebrates," U.S. Government Printing Office, Washington, D.C., 1965.

Hanström, B.: "Vergleichende Anatomie des Nervensystems der Wirbellosen Tiere," Springer, Berlin, 1928.

Jahn, T.: "How to Know the Protozoa," Brown, Dubuque, 1949.

Korschelt, E., and K. Heider: "Textbook of the Embryology of Invertebrates," The Macmillan Company, New York, 1895–1900.

Kudo, R. R.: "Protozoology," 5th ed., Thomas, Springfield, Illinois, 1966.

Kume, M., and K. Dan: "Invertebrate Embryology," Nolit, Belgrade, 1968.

Lutz, F. E.: "Field Book of Insects," Putnam, New York, 1935.

Millott, N. (ed.): "Echinoderm Biology" (20th Symposium of the Zoological Society of London), Academic Press, New York, 1967.
Miner, R. W.: "Fieldbook of Seashore Life," Putnam, New York, 1950.
Morgan, A. H.: "Fieldbook of Ponds and Streams," Putnam, New York, 1930.
Nicol, J. A. C.: "The Biology of Marine Animals," Interscience Publishers, New York, 1966.
Ricketts, E. F., and J. Calvin (revised by J. W. Hedgpeth): "Between Pacific Tides," Stanford University Press, 1962.
Richards, A. G.: "The Integument of Arthropods," University of Minnesota Press, Minneapolis, 1951.
Roeder, K. D. (ed.): "Insect Physiology," John Wiley, New York, 1953.
Waterman, T. H. (ed.): "The Physiology of Crustacea," Academic Press, New York, 1961.
Wiersma, C. A. G. (ed.): "Invertebrate Nervous Systems," University of Chicago Press, 1967.
Wilbur, K. M., and C. M. Yonge: "Physiology of Mollusca," Academic Press, New York, 1964.

SOURCES FOR ILLUSTRATIONS

Agersborg, H. P. K.: *Quart. Journ. Micros. Sci.,* vol. 67, 1923.

Allen, R. D.: "The Cell," vol. II (J. Brachet, and A. E. Mirsky, eds.).

Atkins, D.: *Quart. Journ. Micros. Sci.,* vol. 79, 1936.

———: *Quart. Journ. Micros. Sci.,* vol. 80, 1938.

Bahl, K. M.: *Biol. Reviews,* vol. 22, 1947.

Bahl, K. N., and M. K. Lal: *Quart. Journ. Micros. Sci.,* vol. 76, 1933.

Barcroft, J., and H. Barcroft: *Proc. Roy. Soc. B.,* vol. 96, 1924.

Barnes, R. D.: "Invertebrate Zoology," W. B. Saunders Company, Philadelphia, 1963.

Beadle, L. C.: *Journ. Exper. Biol.,* vol. 34, 1957.

Berlese, A.: "Gli Insetti," Milan, 1909.

Bidder, A. M.: *Quart. Journ. Micros. Sci.,* vol. 67, 1923.

———: *Quart. Journ. Micros. Sci.,* vol. 91, 1950.

Billett, F., and S. M. McGee-Russell: *Quart. Journ. Micros. Sci.,* vol. 96, 1955.

Bishop, G. W.: *Journ. Morph.,* vol. 36, 1921–1922.

Bliss, D. E., J. B. Durand, and J. H. Welsh: *Zeit. Zellforsch.,* Bd. 39, 1954.

Blumenthal, H.: *Zeit. Morph. Okol. Tiere,* Bd. 29, 1935.

Bock, F.: *Zeit. wiss. Zool.,* Bd. 124, 1925.

Boeckh, J., K. E. Kaisslinger, and D. Schneider: *Cold Spring Harbour Sym.,* vol. 30, 1965.

Boettiger, E. G.: In "Recent Advances in Invertebrate Zoology" (B. T. Scheer, ed.), 1957.

Borradaile, L. A., F. A. Potts, L. Eastham, and J. T. Saunders: "The Invertebrata," Cambridge University Press, 1958.

Boschma, H.: *Biol. Bull.,* vol. 49, 1925.

Boycott, B. B., and J. Z. Young: *Proc. Roy. Soc. B.,* vol. 143, 1955.

Brien, P.: *Biol. Reviews,* vol. 28, 1935.

Bristowe, W. S.: *Endeavor,* vol. 13, 1954.

Brondsted, H. V.: *Biologeske Meddelser udgivet ag Det. Konlige Danske VidensKabernes Selskap,* Bd. 25, #3.

Bronn, H.: "Klassen und Orderung des Thierreichs," Friedlander u. Sohn, Leipzig, 1873.

Brown, F. A., Jr.: "Selected Invertebrate Types," John Wiley, New York, 1950.

Bullock, T. H., and G. A. Horridge: "Structure and Function in the Nervous System of Invertebrates," W. H. Freeman and Co., San Francisco, 1965.

Bullough, W. S.: "Practical Invertebrate Anatomy," Macmillan and Company, London, 1950.

Burch, P. R.: *Biol. Bull.,* vol. 54, 1928.

Calkins, G. N.: "Biology of the Protozoa," Lea and Febiger, Philadelphia, 1909.

"Cambridge Natural History" (S. F. Harmer, and A. E. Shipley, eds.), Macmillan and Company, London, 1906.

Carlgren, O.: *Biol. Centralbl.,* Bd. 25, 1905.

Carson, R.: "The Edge of the Sea," Houghton Mifflin, Boston, 1955.

Carter, G. S.: *Proc. Roy. Soc. B.,* vol. 96, 1924.

Caullery, M.: "Parasitism and Symbiosis," Sedgwick and Jackson, Ltd., London, 1952.

Chapman, G. B.: *Journ. Morph.,* vol. 95, 1954.

———: *Quart. Journ. Micros. Sci.,* vol. 94, 1953.

Coe, W. R.: *Biol. Bull.,* vol. 58, 1930.

———: *Biol. Bull.,* vol. 66, 1934.

Crofts, D. R.: *Proc. Zool. Soc. London,* vol. 125, 1955.

Cushman, J. A.: "Foraminifera: Their Classification and Economic Use," Harvard University Press, Cambridge, 1948.

Cushman, J. A.: *U.S. National Museum Bulletin,* no. 104, 1918.

Dahlgren, U., and W. A. Kepner: "A Textbook of the Principles of Animal Histology," The Macmillan Company, New York, 1908.

Davey, K. G.: "Reproduction in the Insects," Oliver Boyd, Edinburgh and London, 1965.

Deevey, E. S., Jr.: *Scientific American,* vol. 195, 1951.

Deevey, G. B.: *Journ. Morph.,* vol. 68, 1941.

Delage, Y., and E. Herouard: "Traité de Zoologie Concrete," Schleicher Frères, Paris, 1896.

Dendy, A.: *Quart. Journ. Micros. Sci.,* vol. 35, 1893.

Dennell, R., and S. R. A. Malek: *Proc. Roy. Soc. B.,* vol. 143, 1954.

Dethier, V. G.: "The Physiology of Insect Senses," John Wiley, New York, 1963.

Doflein, F., and E. Reichenow: "Lehrbuch der Protozoenkunde," Fischer, Jena, 1927.

Dubosq, O., and O. Tuzet: *Arch. de Zool. exper. et gen.:* T. 79, 1937–1939.

Duerden, J. E.: *Quart. Journ. Micros. Sci.,* vol. 49, 1906.

Eakin, R. M., and J. A. Westfall: *Proc. Nat. Acad. Sci.,* vol. 48, 1962.

Esau, K.: "Plants, Viruses and Insects," Harvard University Press, Cambridge, Mass., 1961.

Fawcett, D.: "The Cell," vol. II (J. Bracher, and A. E. Mirsky, eds.).

Florey, E.: *Amer. Zoologist,* vol. 2, 1962.
Fox, H. M.: *Proc. Roy. Soc. B.,* vol. III, 1932.
Galstoff, P. S.: *American Zoologist,* vol. 1, 1961.
Gardiner, M. S.: "The Principles of General Biology," The Macmillan Company, New York, 1952.
Gardiner, M. S., and S. C. Flemister: "The Principles of General Biology," 2d ed., The Macmillan Company, New York, 1967.
Gibbons, I. R., and A. V. Grimstone: *Journ. Biophys. Biochem. Cytol.,* vol. 7, 1960.
Gier, H. T.: *Biol. Bull.,* vol. 71, 1936.
Goetsch, W.: "The Ants," University of Michigan Press, Ann Arbor, 1957.
Goldschmidt, R.: *Zool. Anz.,* Bd 29.
Goodchild, A. J. P.: *Proc. Zool. Soc. London,* vol. 141, 1963.
Goodrich, E. S.: *Quart. Journ. Micros. Sci.,* vol. 83, 1942.
———: *Quart. Journ. Micros. Sci.,* vol. 86, 1945.
Grassé, P. P. (ed.): "Traité de Zoologie," Masson et Cie, Paris, 1948.
Gray, J.: "How Animals Move," Cambridge University Press, 1960.
Hanson, J.: *Journ. Biophys. Biochem. Cytol.,* vol. 3, 1957.
Hanstrom, B.: "Vergleichenden Anatomie der Nervensystem der Wirbellosed Tiere," Springer, Berlin, 1928.
Haughton, T. M., G. A. Kerkut, and K. A. Munday: *Journ. Exper. Biol.,* vol. 35, 1958.
Herbst, C.: *Arch. entwicklungsmech.,* Bd. 9, 1899.
Hermes, G.: *Zeit. wiss. Zool.,* Bd. 141, 1932.
Hess, W. N.: *Ann. Amer. Soc. Ent.,* vol. 10, 1917.
———: *Journ. Morph.,* vol. 40, 1935.
Howell, W. H.: "Textbook of Physiology," 15th ed., W. B. Saunders Company, Philadelphia, 1947.
Hungerford, H. B.: *Univ. Kansas Bull.,* vol. 32, 1948.
Hutchinson, G. E.: *Proc. Nat. Museum,* no. 78, 1930.
Huxley, S. E.: *Scientific American,* vol. 199, 1958.
Hyman, L.: "The Invertebrates," McGraw-Hill Book Company, New York, 1940.
Imms, A. D.: "A General Textbook of Entomology," E. P. Dutton and Company, New York, 1948.
Jahn, T. L.: "How to Know the Protozoa," Brown, Dubuque, 1949.
Jakus, M. A., and C. E. Hall: *Biol. Bull.,* vol. 91, 1946.
Janisch, E.: *Zeit. wiss. Zool.,* Bd. 121, 1924.
Jennings, H. S.: "Behavior of the Lower Organisms," Columbia University Press, New York, 1915.
Jones, J. D.: *Journ. Exper. Biol.,* vol. 32, 1955.
Jordan, H. E.: *Am. Journ. Anat.,* vol. 27, 1920.
Jordan, H. E.: *Biol. Centralblatt,* Bd. 31, 1911.
Kepner, W. A., and W. C. Whitlock: *Journ. Exper. Zool.,* vol. 32, 1921.
Kleinholz, L. H.: *Biol. Bull.,* vol. 70, 1936.
Korschelt, E.: "Regeneration," Borntraeger, Berlin, 1927.
Korschelt, E., and K. Heider: "Vergl. Entwicksges. der Tiere (nur bearbeitet von Korschelt)," The Macmillan Company, New York, 1936.
Krijgsman, B. J.: *Arch. für Protist,* Bd. 53, 1925.
Kudo, R. R.: "Protozoology," Thomas, Springfield, Illinois, 1939.
Lankester, E. R. (ed.): "A Treatise on Zoology," Adam and Charles Black, London, 1903.
Larimer, J. L., and A. H. Gold: *Phys. Zool.,* vol. 34, 1961.
Laverack, M. S.: "The Physiology of Earthworms," The Macmillan Company, New York, 1963.
Lawrence, P. A., and F. B. Krasne: *Science,* vol. 148, 1965.
Lee, D. L.: "The Physiology of Nematodes," Oliver and Boyd, Edinburgh and London, 1965.
Lees, A. D.: *Journ. Exper. Biol.,* vol. 25, 1948.
Lesperon, L.: *Arch. de Zool. exper.,* T. 79, 1937–1938.
Liebman, E.: *Experientia,* vol. 111, 1947.
Lochhead, J. H., and M. S. Lochhead: *Journ. Morph.,* vol. 68, 1941.
Lutz, F. E.: "Field Book of Insects," G. P. Putnam's Sons, New York, 1935.
MacGinitie, G. E.: *Biol. Bull.,* vol. 77, 1939.
MacGinitie, G. E., and N. MacGinitie: "Natural History of Marine Animals," McGraw-Hill Book Company, New York, 1949.
Mackie, T. T., G. W. Hunter, and C. B. Worth: "Manual of Tropical Medicine," W. B. Saunders Company, Philadelphia, 1945.
Manton, S. M.: *Soc. Exper. Biol.,* Symposium VII, 1953.
Manwell, C.: *Comp. Biochem. Physiol.,* vol. 1, 1960.
———: *Science,* vol. 127, 1958.
———: *Science,* vol. 132, 1960.
———: *Journ. Comp. Cell. Physiol.,* vol. 52, 1958.
———: *Comp. Biochem. Physiol.,* vol. 8, 1963.
———: *Journ. Comp. Cell. Physiol.,* vol. 53, 1959.
McBride, E. W.: "Textbook of Embryology," vol. I, The Macmillan Company, London, 1914.
McConnaughey, B. H.: *Univ. Cal. Pub. Zool.,* no. 55, 1949.
Meisenheimer, J.: *Zeit. wiss. Zool.,* Bd. 69, 1901.
Menzies, R. J., and A. H. Clarke, Jr.: *Science,* vol. 129, 1959.
Miall, L. C.: "The Natural History of Aquatic Insects," The Macmillan Company, New York, 1912.
Miller, R. C.: *Univ. Cal. Pub. Zool.,* vol. 26, 1924.
Miner, R. W.: "Field Book of Seashore Life," G. P. Putnam's Sons, New York, 1950.
Morgan, A. H.: "Field Book of Ponds and Streams," G. P. Putnam's Sons, New York, 1936.
Morgan, T. H.: *Arch. entwicklungsmech.,* Bd. 10, 1900.

Morton, J. E.: "Molluscs: An Introduction to Their Form and Functions," Harper and Brothers, New York, 1960.
Mueller, J. F.: *Trans. Amer. Micros. Soc.*, vol. 69, 1950.
Murray, M.: *Arch. exp. Zellf.*, Bd. 11, 1931.
Nayar, K. K.: *Biol. Bull.*, vol. 108, 1955.
Nelson, T. C.: *Journ. Morph.*, vol. 31, 1918.
Newby, W. W.: *Journ. Morph.*, vol. 69, 1911.
Newell, G. E.: *Proc. Zool. Soc. London*, vol. 144, 1965.
Nold, R.: *Zeit. wiss. Zool.*, Bd. 123, 1924.
Okada, Y. K.: *Arch. entwicklungsmech.*, Bd. 115, 1929.
Orlov, : *Zeit. Zellforsch. u. Mikr. Anat.*, Bd. 8, 1929.
Owen, C.: *Quart. Journ. Micros. Sci.*, vol. 96, 1955.
———: *Quart. Journ. Micros. Sci.*, vol. 97, 1956.
Parker, G. H.: "Animal Color Changes and Their Neurohumors," Cambridge University Press, 1948.
Parker, T. J., and W. A. Haswell: "A Textbook of Zoology," vol. I, The Macmillan Company, New York, 1921.
Parry, D. A.: *Endeavour*, vol. 19, 1960.
Pearse, A.: "Animal Ecology," 2d ed., McGraw-Hill Book Company, New York, 1939.
Pennak, R. W.: "Fresh Water Invertebrates of the United States," Ronald Press, New York, 1953.
Pitelka, D. R.: *Univ. Cal. Pub. Zool.*, no. 55, 1945–1949.
Pontin, P. M.: *Comp. Biochem. Physiol.*, vol. 17, 1966.
Portman, G. J.: "Animal Camouflage," University of Michigan Press, Ann Arbor, 1959.
Pringle, J. W. S.: "Insect Flight," Cambridge University Press, 1957.
Ramsay, J. A.: "Physiological Approach to the Lower Animals," Cambridge University Press, 1962.
Rand, H. W.: *Arch. entwicklungsmech.*, Bd. 819, 1899.
Raven, C. P.: "Oogenesis: The Storage of Developmental Information," The Macmillan Company, New York, 1961.
Redfield, A. C., and M. Florkin: *Biol. Bull.*, vol. 61, 1931.
Redfield, A. C., and E. N. Ingalls: *Journ. Comp. Cell. Physiol.*, vol. 3, 1933.
Redmond, J. R.: *Journ. Comp. Cell. Physiol.*, vol. 46, 1955.
———: *Physiol. Zool.*, vol. 35, 1962.
Roeder, K. D. (ed.): "Insect Physiology," John Wiley, New York, 1953.
Roskin, G.: *Zeit. Zellforsch. und Mikrosc. Anat.*, vol. 3, 1925.
Ross, H. H.: "A Textbook of Entomology," John Wiley, New York, 1956.
Russell, F. S., and C. M. Yonge: "The Seas," Frederick Warne and Company, New York, 1928.
Sanders, H. L.: *Proc. Nat. Acad. Sci.*, vol. 41, 1955.
Shaeffer, A. A.: *Journ. Exper. Zool.*, vol. 8, 1910.
Shearer, C.: *Quart. Journ. Micros. Sci.*, vol. 56, 1910.
Slifer, E., S. S. Sekhon, and A. D. Lees: *Quart. Journ. Micros. Sci.*, vol. 105, 1964.
Smith, D. E.: *Journ. Biophys. Biochem. Cytol.*, supp. to vol. 10, 1961.
Smith, J. E.: *Quart. Journ. Micros. Sci.*, vol. 88, 1947.
Smith, L.: *Comp. Biochem. Physiol.*, vol. 15, 1965.
Snodgrass, R. E.: "The Anatomy and Physiology of the Honey Bee," Comstock Publishing Company, Ithaca, N.Y., 1956.
———: "Principles of Insect Morphology," McGraw-Hill Book Company, New York, 1935.
———: "A Textbook of Arthropod Anatomy," Comstock Publishing Company, Ithaca, N.Y., 1952.
———: *Smithsonian Miscellaneous Coll.*, no. 80, 1927; no. 81, 1928; no. 82, 1929; no. 97, 1938; no. 110, 1948; no. 117, 1951–1952.
Stolte, H.: *Arch. für Protis*, vol. 48, 1924.
Storer, T. I., and R. L. Usinger: "Elements of Zoology," 2d ed., McGraw-Hill Book Company, New York, 1961.
Tartar, V.: In *Cell Mechanisms in Differentiation and Growth, 14th Sym. Soc. for Study of Dev. and Growth* (D. Rudnick, ed.), 1956.
Taylor, C. V.: *Univ. Cal. Pub. Zool.*, no. 19, 1920.
Taylor, R. C.: *Comp. Biochem. Physiol.*, vol. 20, 1967.
Tennent, D. H., and T. Ito: *Journ. Morph.*, vol. 69, 1941.
Thomas, A. P.: *Quart. Journ. Micros. Sci.*, vol. 23, 1883.
Thomas, H. J.: *Quart. Journ. Micros. Sci.*, vol. 84, 1944.
Tiegs, O. W., and S. M. Manton: *Biol. Revs.*, vol. 33, 1958.
Turner, C. D.: "General Endocrinology," W. B. Saunders Company, Philadelphia, 1966.
van Buddenbrock, W.: *Zool. Jahresber. Abt. Allgem. Zool. und Physiol. Tiere*, Bd. 35, 1915.
van Emden, M.: *Zeit. wiss. Zool.*, Bd. 134, 1929.
von Frisch, K.: "Biology," Harper and Row, New York, 1964.
Ward, H. B., and C. G. Whipple "Fresh Water Biology" (W. T. Edmonson, ed.), John Wiley, New York, 1959.
Wells, M. J.: *Soc. Exper. Biol. Symposium IV*, 1950.
Whitear, M.: *Phil. Trans. Roy. Soc. B.*, vol. 245, 1962.
Wigglesworth, V. B.: "The Control of Growth and Form," Cornell University Press, Ithaca, N.Y., 1959.
———: *Quart. Journ. Micros. Sci.*, vol. 76, 1933.
Wilbur, K. M., and C. M. Yonge (eds.): "Physiology of Mollusca," Academic Press, New York, 1966.
Williams, L.: "Anatomy of the Common Squid" (pub. under auspices of the American Museum of Natural History), E. J. Brill, London, 1959.
Wilson, D. P.: "They Live in the Sea," Collins, London, 1947.

Wilson, E. B.: "The Cell in Development and Heredity," The Macmillan Company, New York, 1925.
Winton, F. R., and L. E. Bayliss: "Human Physiology," Little Brown, Boston, 1962.
Wohlbarsht, M. L.: *Cold Spring Harbor Sym.*, vol. 30, 1965.
Woltereck, R.: *Arch. entwicklungsmech.*, Bd. 18, 1903.
Woodland, W.: *Quart. Journ. Micros. Sci.*, vol. 49, 1905–1906.
Young, J. Z.: *Quart. Journ. Micros. Sci.*, vol. 78, 1936.
———: *Proc. Roy. Soc. B.*, vol. 152, 1960.
Zawarzin, A.: *Zeit. wiss. Zool.*, Bd. 124, 1925.
Zenkewich, L. A.: *Journ. Morph.*, vol. 77, 1945.
Ziegler, H. E.: *Arch. entwicklungsmech.*, Bd. 7, 1898.
Zimmerman, W.: *Die Naturwiss.*, Bd. 13, 1925.

INDEX

Page numbers in *italics* indicate figures.

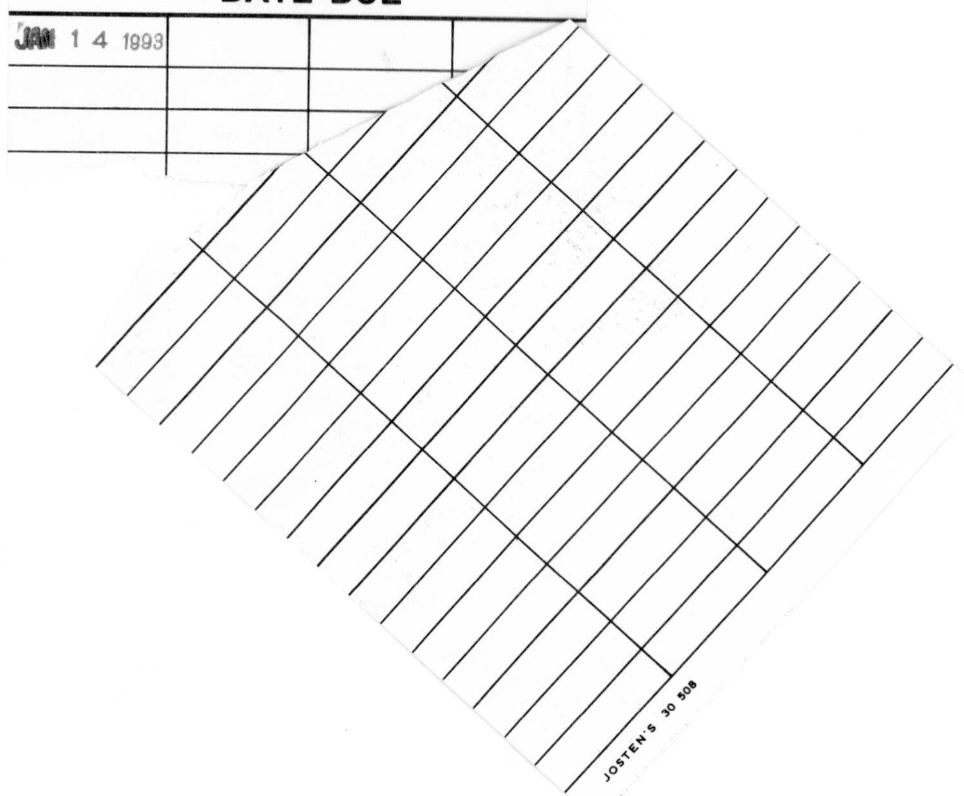
JOSTEN'S 30 508